PETROLEUM GEOLOGY
OF THE NORTH SEA

[illegible]CATED TO
THE LATE JOHN TAYLOR

Petroleum Geology of the North Sea

Basic Concepts and Recent Advances

EDITED BY K.W. GLENNIE
Honorary Professor,
Department of Geology & Petroleum Geology,
University of Aberdeen;
Formerly Shell UK Exploration and Production

FOURTH EDITION

b
Blackwell
Science

Editorial Offices:
Osney Mead, Oxford OX2 0EL
25 John Street, London WC1N 2BL
23 Ainslie Place, Edinburgh EH3 6AJ
350 Main Street, Malden
MA 02148 5018, USA
54 University Street, Carlton
Victoria 3053, Australia
10, rue Casimir DeLavigne
75006 Paris, France

Other Editorial Offices:
Blackwell Wissenschafts-Verlag GmbH
Kurfürstendamm 57
10707 Berlin, Germany

Blackwell Science KK
MG Kodenmacho Building
7–10 Kodenmacho Nihombashi
Chuo-ku, Tokyo 104, Japan

First published 1984
Second edition 1986
Third edition 1990
Reprinted 1992, 1994, 1996
Fourth edition 1998

Set by Semantic Graphics, Singapore
Printed and bound in Great Britain
at The Bath Press

A catalogue record for this title
is available from the British Library

ISBN 0-632-03845-4

Library of Congress
Cataloging-in-Publication Data

Petroleum geology of the North Sea:
basic concepts and recent advances /
edited by K.W. Glennie. — 4th ed.
p. cm.
Rev. ed. of: Introduction to the
petroleum geology of the North Sea.
3rd ed. 1990.
ISBN 0-632-03845-4
1. Petroleum—Geology—North Sea.
2. Geology—North Sea.
I. Glennie, K. W.
II. Title: Introduction to the petroleum
geology of the North Sea.
TN874.N78P48 1998
553.2′82′0916336—dc21 97–29316
CIP

DISTRIBUTORS

Marston Book Services Ltd
PO Box 269
Abingdon, Oxon OX14 4YN
(*Orders*: Tel: 01235 465500
Fax: 01235 465555)

USA
Blackwell Science, Inc.
Commerce Place
350 Main Street
Malden, MA 02148 5018
(*Orders*: Tel: 800 759 6102
781 388 8250
Fax: 781 388 8255)

Canada
Login Brothers Book Company
324 Saulteaux Crescent
Winnipeg, Manitoba R3J 3T2
(*Orders*: Tel: 204 224 4068)

Australia
Blackwell Science Pty Ltd
54 University Street
Carlton, Victoria 3053
(*Orders*: Tel: 3 9347 0300
Fax: 3 9347 5001)

Contents

M.J. FISHER, *Director, Nevis Associates, 34 West Argyle Street, Helensburgh, Scotland, G84 8DD*

D.C. MUDGE, *David Mudge Associates, 6 Kilmardinny Crescent, Glasgow, G61 3NR*

8 Jurassic, 245

J.R. UNDERHILL, *Professor of Earth Sciences, Department of Geology & Geophysics, The University of Edinburgh, West Mains Road, Edinburgh, EH9 3JW*

9 Cretaceous, 294

C.D. OAKMAN, *Colin Oakman Associates, 62 First Avenue, Netherlea, Glasgow, G44 3UB (Formerly Technical Director, Reservoir Research, Glasgow)*
M.A. PARTINGTON, *Petroleum Development Oman plc, P.O. Box 81, Muscat, Postal Code 113, Sultanate of Oman*

H.D. JOHNSON, *Enterprise Oil Professor of Petroleum Geology, Department of Geology, Imperial College of Science, Technology and Medicine, Prince Consort Road, London, SW7 2BP*

M.J. FISHER, *Director, Nevis Associates, 34 West Argyle Street, Helensburgh, Scotland, G84 8DD*

Foreword to the Fourth Edition

M.J. FISHER

It is 8 years since publication of the Third Edition of this book. In that time, our technical expertise and knowledge of the North Sea has continued to advance, accompanied by an ever-increasing volume of geological publications. The scope of this volume reflects these developments and the prefix 'Introduction' of earlier editions is no longer appropriate.

Changes in industry and academic practice have imposed even greater demands on authors' time than hitherto. Complete revision of the text has been underway for some years. The 8 year interval between editions was not intentional but represents the time required by some contributors to fulfil their commitments. Ken Glennie has worked tirelessly to ensure timely completion of this edition. It is a measure of the respect that we all have for him that, even after readings of the Riot Act and threat of 'ultimate sanctions', the most recalcitrant authors still hold him in such high regard. Even more than with any of the earlier editions, we all owe Ken an enormous debt of gratitude for seeing this volume through to publication.

Our appreciation must again be expressed to the parent companies and institutions of the various contributors to this volume and to the course on which it is based. We hope that this book is an adequate vindication of their support.

Sadly, the final stages of preparation of this edition were overshadowed by the sudden death on Saturday 4th October 1997 of John Taylor, the author of Chapter 6. John had provided the Zechstein chapter in all previous editions. He was an acknowledged expert in this field, as his long list of publications and even longer list of satisfied clients of V.C. Illing & Partners testify. Despite being confined to a wheelchair after contracting poliomyelitis on fieldwork in Ecuador as a young geologist, he was the most helpful and dependable of contributors. Even in his last, most difficult year, he was determined to complete the final revision of his chapter. Ironically this was only necessary because he had, predictably, met earlier deadlines that were superceded by the non-appearance of other chapters.

We would like to dedicate this book to the memory of Dr John Taylor; a gifted geologist, staunch colleague and inspiring man.

Foreword to the First Edition

R. STONELEY

This book is the outcome of a two-day short course held annually in London, and is based on the manual distributed to the course participants.

The course is arranged by the Joint Association for Petroleum Exploration Courses (UK) (JAPEC), an organization that come into existence in 1980 in response to demands from UK-based petroleum exploration companies for assistance with their training programmes. JAPEC is sponsored and supported jointly by the Geological Society of London, the Petroleum Exploration Society of Great Britain and the Department of Geology at the Imperial College of Science and Technology, University of London. It is run by an honorary committee drawn from the petroleum industry and from university geology departments, and *inter alia* now stages approximately eight short courses each year on a variety of topics.

The 'Introduction to the Petroleum Geology of the North Sea' was the first course to be arranged by JAPEC and was presented initially in the spring of 1981 under the energetic overall direction of K.W. Glennie. The contributors had all had considerable first-hand knowledge of the area. The course was designed as a rapid 'state of the art' overview for exploration geologists and geophysicists with little direct experience of the North Sea, but was also considered useful for those who wished to place detailed local knowledge in a basin-wide context. It thus provided the first systematic review of a basin which is undergoing rapid exploration and development. No attempt has been made to cover areas of the north-west European continental shelf outside the North Sea, and indeed the Norwegian, Danish, German and Dutch sectors are treated less fully than the British areas: this largely reflects the fact that none of the contributors has worked extensively in those sectors, but their descriptions are such that there are believed to be no significant gaps in the coverage.

With each annual repeat of the course, the content has been updated and modified in the light of constructive criticisms from the participants. This book may therefore be regarded as up to date to about the end of 1983, in respect of publicly released information and ideas.

The book falls into three broad sections. The first sets the scene with a summary of the history of exploration in the North Sea, which is followed by a review of the structural framework and pre-Permian development of the region.

The meat of the book is a series of descriptions, in stratigraphical order, of the depositional history and hydrocarbon-related rock units from the Permian to the Tertiary. The final section homes in on petroleum exploration with a review of that all-important factor, the oil and gas source rocks. The last chapter brings the preceding material together with a discussion of the various exploration 'plays': why are the oil and gas where they are, and how have they been found? What of the future?

Although both the JAPEC course and this book have been designed primarily for the benefit of those concerned with North Sea exploration, the book should have a wider appeal. As pointed out by Brennand in the opening chapter, our geological knowledge of the North Sea area was exceedingly sketchy when offshore exploration began in 1964. This then is a record of the geological knowledge gained almost entirely by the industry in a relatively short space of time, and it is a well-informed description of one of the world's major petroleum-bearing basins. It describes the geology of a large part of the north-west European region and therefore will be of significance to geologists in a broader range of disciplines: because the North Sea has been essentially an area of subsidence from the Late Palaeozoic to the present, its geological record has also proved to be important in unravelling some of the history of much of the surrounding, more positive, land areas.

With the passage of time, and with the modifications that have been made to the JAPEC course since its original presentation, there have inevitably been changes in the arrangement and authorship of certain sections. Our thanks, then, are due to those who participated in the past and who, whilst making no direct contribution to this book, have helped to make whatever success it may have: K.W. Barr (Consultant), C.E. Deegan (Hydrocarbons Unit, Institute of Geological Sciences), J.G.C.M. Fuller (Amoco Europe Inc.) and R.C. Selley (Imperial College of Science and Technology).

We sincerely thank the employing organizations of all contributors, past and present, for permitting their staff to give freely of their time and expertise. JAPEC was founded on the basis of industry self-help, and the industry has indeed been generous in its response.

Acknowledgements

The pace of exploration in the North Sea area has barely faltered since the First Edition of this book was published in 1984. By 1986, the amount of new information being released by the industry indicated that a Second Edition was needed, with the Third Edition following in 1990. Now, 8 years later, retirement or transfer elsewhere in the world has ensured that few of the contributors to the First Edition were available to bring us up to date for the Fourth Edition. All editions of the book have benefited from the contributors to earlier editions. It is apt, therefore, to acknowledge the efforts of these earlier writers: Tim Brennand, Berend Van Hoorn and Keith James, all then of Shell, who built up Chapter 1; Phil Richards of the Geological Survey, who wrote the Devonian chapter for the Third Edition; and Stewart Brown, also formerly of the Geological Survey but now the Director for Geological Research with PSTI in Edinburgh, who wrote all earlier versions of the Jurassic chapter; Jake Hancock, formerly Professor of Geology at Imperial College who wrote the earlier Cretaceous chapter; Bryan Lovell (British Petroleum) and Alan Parlsey, then Exploration Manager with Britoil and now with International Shell, who wrote earlier versions of the Cenozoic and Play chapters, respectively.

The amount of data now available on North Sea petroleum geology, together with a plethora of new hypotheses has resulted in the authorship of several chapters now being shared (Chapters 2, 7, 9 and 10). The new authors, Bob Downie, John Underhill, Colin Oakman with Mark Partington, Mike Bowman and Howard Johnson, have had a difficult task to rewrite chapters that make a major contribution in their own right—this, I think they have succeeded in doing. Chris Cornford, through the services of his wife and daughter, deserves special praise for the major task of compiling the References into one list; as a source of information, this is now valuable in its own right. My near-neighbour, geologist Jane Angus, is thanked for compiling the Index.

Finally, the book could not have been completed without the continued hard work put in at Blackwell Science by the Production Editor, Alice Nelson, and other staff.

Some Oil Industry Abbreviations and Conversion Factors

°API	Oil gravity, American Petroleum Institute degrees
bbl	Barrel
bcf	billion cubic feet ($\times 10^9$ ft^3)
GIIP	Gas initially in place
GOC	Gas–oil contact
GOR	Gas–oil ratio
GWC	Gas–water ratio
Ma	Millions of years (before present—BP)
mD	Milledarcies (permeability)
MMbbl	Million barrels
MMscf/d	Million standard cubic feet/day
ms	milliseconds
NGL	Natural gas liquids
OWC	Oil—water contact
scf	standard cubic feet
ss	below (sub-) sea level
STB	Stock tank barrel
STOIIP	Stock tank oil initially in place
tcf	trillion cubic feet ($\times 10^{12}$ ft^3)
TOC	total organic carbon
UKCS	United Kingdom Continental Shelf

Feet to metres	$\times$ 0.305
Cubic feet (ft^3) to cubic metres (m^3)	$\times$ 0.028
Cubic metres (m^3) to cubic feet (ft^3)	$\times$ 35.315
Cubic metres (m^3) to barrels (bbl)	$\times$ 6.290
Tonnes to cubic metres (m^3)	$\times$ 1.17
Tonnes to barrels (bbls)	$\times$ 7.34
Barrels to tonnes (t)	$\times$ 0.14

1 barrel of oil equivalent = 1 barrel of oil
1 barrel of oil equivalent = 1.446 bbls NGL
1 barrel of oil equivalent = 58 000 ft^3 natural gas

1 Historical Review of North Sea Exploration

T.P. BRENNAND, B. VAN HOORN, K.H. JAMES & K.W. GLENNIE

1.1 Introduction

Exploration in the North Sea was really initiated on 20 May 1964, when the German consortium spudded the first offshore well, Nordsee B-1. Even though Germany has no economic offshore development, the North Sea area in general has now become one of the most prolific hydrocarbon provinces of the world (Fig. 1.1). Total recoverable reserves found to date, including adjacent land areas, amount to some 100×10^9 barrels of oil, condensate, natural-gas liquids and oil-equivalent volumes of gas (Spencer *et al.*, 1996), of which almost 50% is found within UK waters. At the height of drilling activity, around 70 rigs were commonly active in the area and, although there has been a steady decline in average size of field discovered, particularly in the UK sector, new reserves are being added continuously.

The discovery of these large reserves of oil and gas would not have seemed possible at the end of the 1950s, when only a very limited geological knowledge existed of the north-west European continental shelf. Although the oil industry already had considerable experience in developing oil- and gas-fields offshore—notably in Venezuela, the Arabian Gulf, the South China Sea and the Gulf of Mexico—the rather small size of producing oilfields in east England, the Netherlands and northern Germany did not stimulate exploration in the more hostile and higher-cost areas offshore. Furthermore, the absence of an international agreement establishing sovereignty over the continental seas prevented offshore exploration.

By the late 1950s, those countries bordering the North Sea were negotiating their territorial rights to mineral resources below the seabed. Agreement was reached in 1958, with the Continental Shelf Convention in Geneva, which came into force in 1964 after sufficient countries had ratified it. By then, the industry had realized the very large size of the Groningen gasfield (Fig. 1.1), which, though discovered in 1959, was not established as a giant until 1963, when the well Ten Boer-1 was deepened to penetrate the gas-rich Rotliegend Sandstone near the field's gas–water contact (Stheeman and Thiadens, 1969). It is now known to have proved ultimate recoverable reserves of the order of 97×10^{12} ft^3 (2.7×10^{12} m^3) of gas, of which about half had been produced by the end of 1992 (Breunese and Rispens, 1996). The prospect of finding other gasfields of similar size provided the necessary economic and geological impetus for systematic offshore exploration.

The objective of this short introduction is not to summarize the chapters that follow, but to describe briefly the succession of major discoveries, technological advances and economic and licensing incentives that have contributed to today's view of the petroleum geology of the North Sea. The succeeding chapters may then be seen in the context of the overall effort of discovery, which began over 30 years before the publication of this edition. The emphasis in this chapter is on exploration in the UK sector of the North Sea; not only is half the North Sea under British jurisdiction, so far as petroleum regulations are concerned, but also well over half the exploration wells have been drilled in British waters. The exploration effort in other national sectors of the North Sea will not be neglected, however, because finds in one national sector stimulate exploration in similar geological settings of other sectors, thereby contributing to a more regional understanding of each of the hydrocarbon plays, the details of which are given in the stratigraphic chapters and summarized in Chapter 12.

After a brief summary of the licensing history and some of the general factors that have stimulated exploration since 1964, the development of the main plays is considered by emphasizing the successive contributions to knowledge made by drilling results, geological studies and seismic surveys. Following a historical summary of events preceding the beginning of offshore exploration, the review is treated in five shorter periods: 1964–70, 1971–76, 1977–85, 1986–92 and the period from 1993 onwards. Each period tended to be dominated by the pursuit of particular exploration objectives as new plays came to light, and each was roughly bounded by important UK and Norwegian events in the realm of acreage licensing (Figs 1.2 and 1.3) or changes to, especially, the UK tax regime.

1.2 Licensing of offshore acreage

The effective start of offshore exploration in different parts of the North Sea depended upon enactment of petroleum legislation by the countries claiming sovereignty over those waters.

For almost two centuries, 'territorial waters' had been limited by many Western countries to a distance of 3 miles (5 km) from their coasts and by up to 12 miles (20 km) by some others. Since 1945, however, many states have laid claim to the oil and other resources that lay between their coastline and the edge of the continental shelf, which is generally taken at a water depth of 200 m or 100 fathoms (600 ft). In the North Sea area, this precedent resulted in sovereignty over the mineral rights being extended in 1965–66 by bilateral agreements between the UK, Norway,

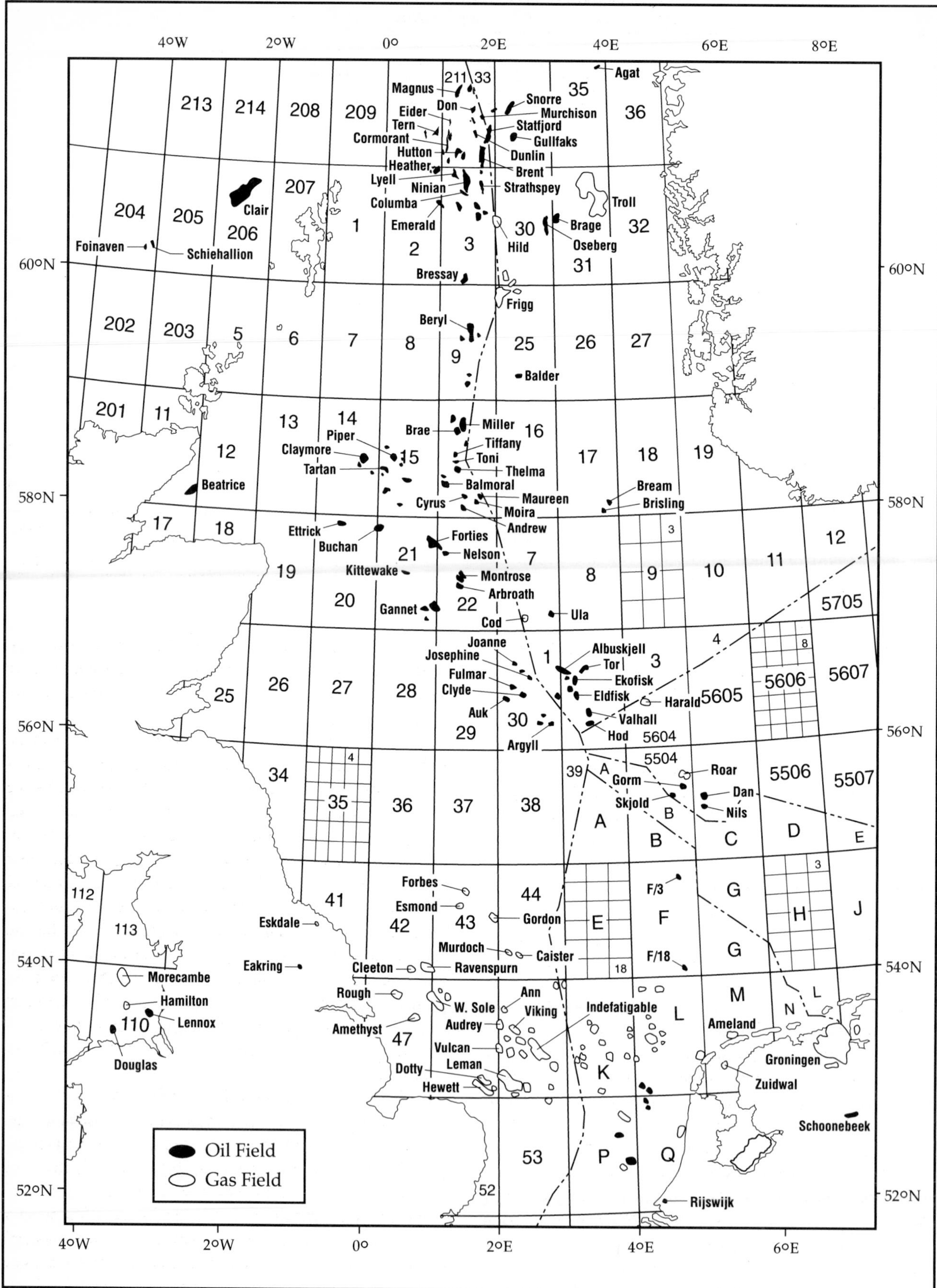

Fig. 1.1 Important named oil- and gasfields of the North Sea, with the number of blocks per quadrant highlighted for each of the national sectors.

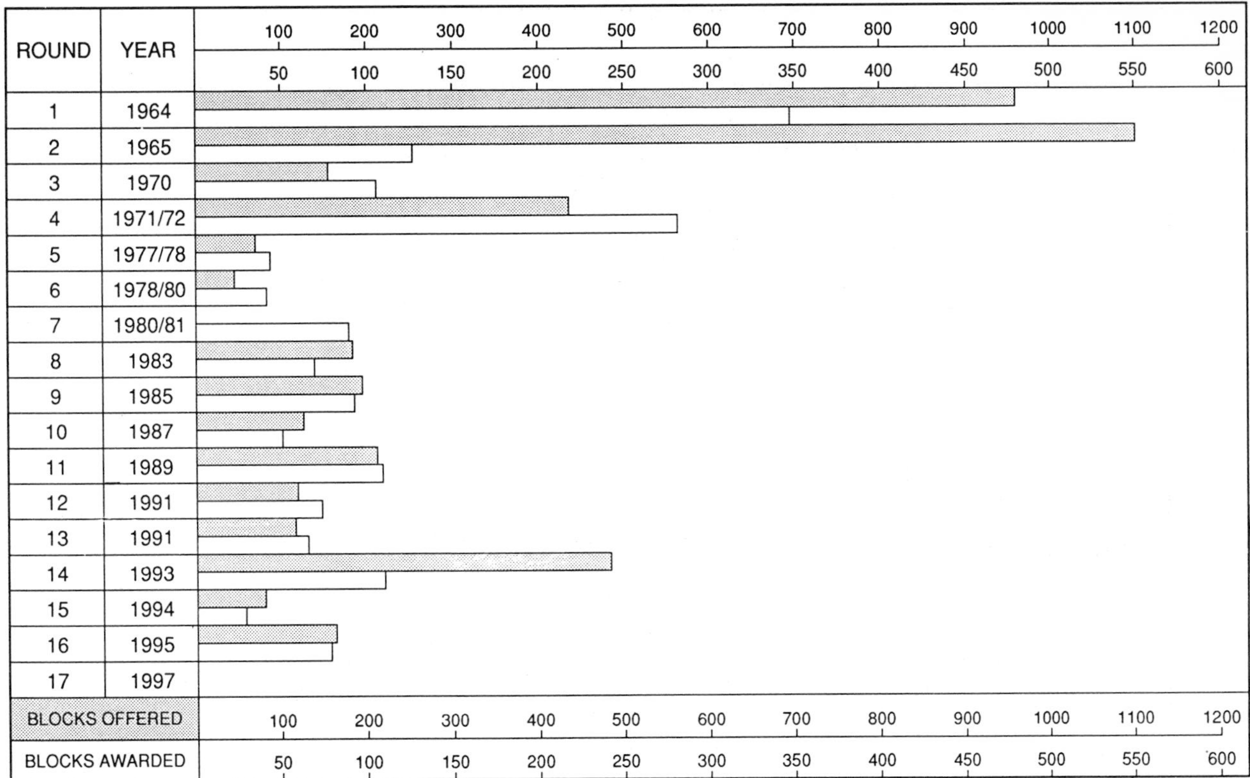

Fig. 1.2 Number of blocks offered and awarded in UK licence rounds. For most rounds there was a minimum statutory relinquishment of 50% after 6 years (67% for rounds 5 and 6 and 0–75% for 'frontier' rounds 13 and 14).

Denmark and the Netherlands to a median line halfway between them, following the rules established by the Continental Shelf Convention of 1958. The offshore boundaries between Denmark, Germany and the Netherlands, however, remained in dispute until 1970, when the International Court of Justice ruled in favour of the German claim for a larger part of the continental shelf.

Although the acquisition of geophysical data was able to proceed prior to the award of offshore exploration concession areas, the governments involved realized in 1963–64 that drilling could not proceed without the enactment of petroleum legislation. This legislation was to lay down the terms and conditions under which oil and gas could be searched for and exploited when found. These vary from country to country, and within countries have changed with the passage of time. Whereas the Danish government, in 1962, granted the exploration rights for the whole of their offshore area to one consortium, other countries offered individual concession areas of a relatively small size, ranging from about 200 km^2 (UK) through 400 km^2 (the Netherlands) to 550 km^2 (Norway).

In the UK, the Continental Shelf Act came into force on 15 May 1964 and paved the way for offshore licensing. Seventeen licensing rounds have been held to date, and much licensed acreage has subsequently been relinquished, both by statute and voluntarily (see Fig. 1.2). Licence awards are based upon the suitability of companies in the exploration field, their past record of activities and a negotiated level of work commitments (Walmsley, 1983).

In 1971, prior to the fourth round, the Petroleum Regulations, enacted in 1966, were amended to allow competitive cash tenders for blocks and to rationalize relinquishment rules.

An annual rent per square kilometre must be paid on each exploration licence awarded. For the first 6 years of the first and second rounds, this was £25, following which 50% of the licence had to be relinquished; the rental then rose yearly to a maximum of £290. Annual rentals increased with succeeding rounds, that for the sixteenth round being initially £410/km^2, rising to a maximum of £7050/km^2. In addition, for the early rounds, each licensee had a minimum work obligation (negotiated, depending on the known or expected geology) of wells to be drilled. For later rounds, the work obligation has varied; this is because the government has had to balance an apparently increasing ability to find hydrocarbons in mature exploration areas with the need to persuade companies to explore in high-risk 'frontier areas'.

In Norway, legislation was introduced in May 1965, and 15 rounds had been held up to the end of 1996, of which the fifth and seventh rounds did not take place in the Norwegian North Sea but were concentrated essentially north of 62°N (Fig. 1.3). Licences are granted for an initial period of 6 years, with a right of extension for 30 years following a 50% relinquishment of the original area. Several special awards (SA in Fig. 1.3) were limited to one or two blocks at a time. The Norwegian government has retained a large interest in the licences through the state oil company Statoil, and has issued only a limited number of blocks in each round, the greatest number being in the first and last.

Offshore Denmark was held entirely by the Danish Untergrunds Consortium (DUC) until January 1982, when 50% of the acreage was relinquished, thereby enabling the

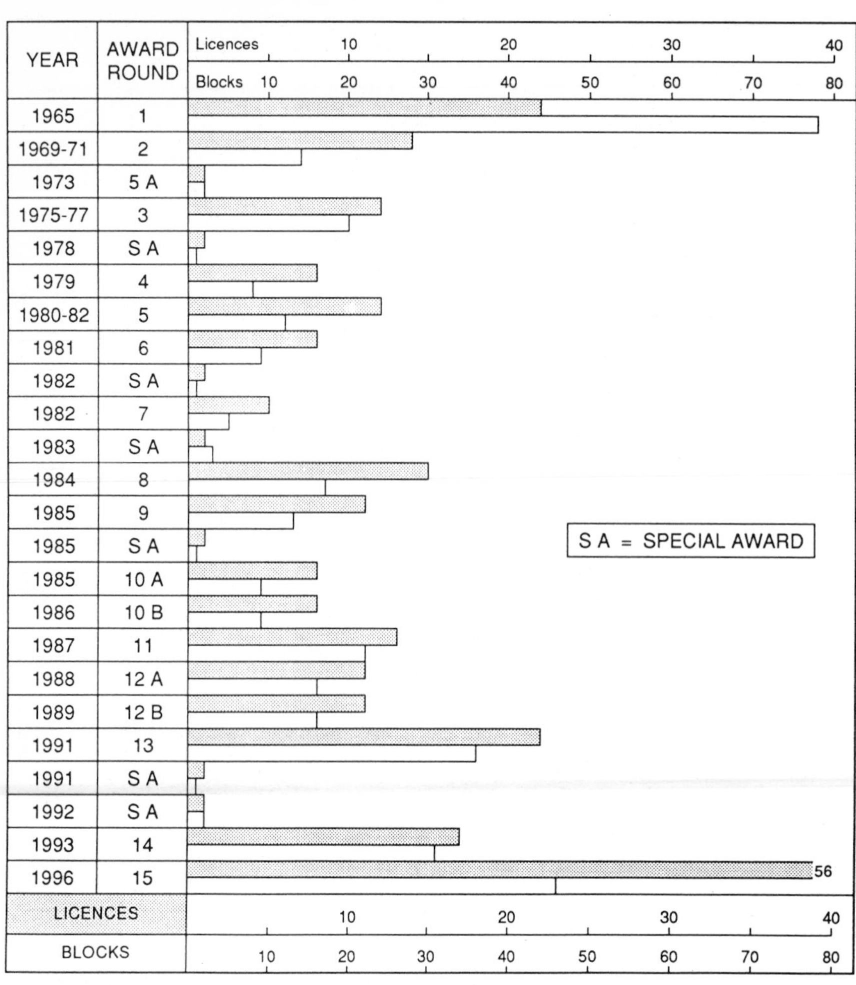

Fig. 1.3 Number of licences and blocks awarded in Norwegian licence rounds. Small unnumbered special awards were made in 1985, 1991 and 1992. There is a minimum statutory relinquishment of 50% after 6 years, with an additional 25% relinquishment 3 years later for rounds 1 and 2.

government to invite applications for surrendered acreage in 1983. A further 25% of the total acreage was relinquished in 1984. A second round closed in autumn 1985 and was followed by further reduction in DUC acreage on 31 December. The third round, which involved all open acreage, closed in March 1989.

As to offshore Germany, an initial permit granted the entire area to a consortium in 1964. Relinquished acreage has been on offer to other independent companies during rounds, but the perceived lack of prospectivity of the area has so far resulted in only modest industry response.

Exploration activities in the Netherlands offshore did not commence until 7 March 1968, when the government awarded a first round of licences. This delay was caused by prolonged negotiations on amendments to the proposed Continental Shelf Mining Act, introduced to Parliament in June 1964 but not implemented until August 1967. Since then, a large number of concessions have been licensed in successive rounds. In 1989, the seventh round included areas of quadrants L, M and N, which were previously closed for military purposes. There had been state participation of 40% in offshore gas production, but no participation in offshore oil production until 1976, when 50% state participation was introduced for both oil and gas; production licences last for 40 years (Petroleum Geological Circle, 1993).

Because of the sensitive environmental nature of coastlines, offshore acreage within 3 nautical miles of the coast of all countries is generally treated as 'landward areas' not forming a part of offshore licences; such acreage is subject to special conditions, which usually forbid active exploration. Exploration is commonly banned in other sensitive areas, such as the approaches to major ports.

1.3 General factors governing the growth of new knowledge

One of the most striking features of the discovery of the major oil and gas deposits of offshore north-west Europe has been the rapidity with which today's position was reached after the tentative beginnings in the mid-1960s. In only 32 years, nearly 4000 wildcats have been drilled, leading to nearly 800 hydrocarbon discoveries (both large and small, basic data on the more important being listed in the appendix to this chapter), with a success rate of about 29% (Spencer *et al.*, 1996). In the UK sector alone, there are over 200 named oil- and gasfields, which is more than the combined total in the Norwegian, Danish and Dutch sectors.

To these must be added many small hydrocarbon accumulations that are now starting to be developed because of a more favourable UK tax climate since 1993. While the most widely appreciated outcome of the enterprise was the discovery of large reserves of oil and gas, an additional, important product has been the unravelling of the hitherto unknown geology of a very large area of the north-west European continental shelf. Figure 1.1 illustrates the distribution of many of the oil- and gasfields that had been named by the end of 1995. Figure 1.4 contrasts the relative absence of geological knowledge in the North Sea in the

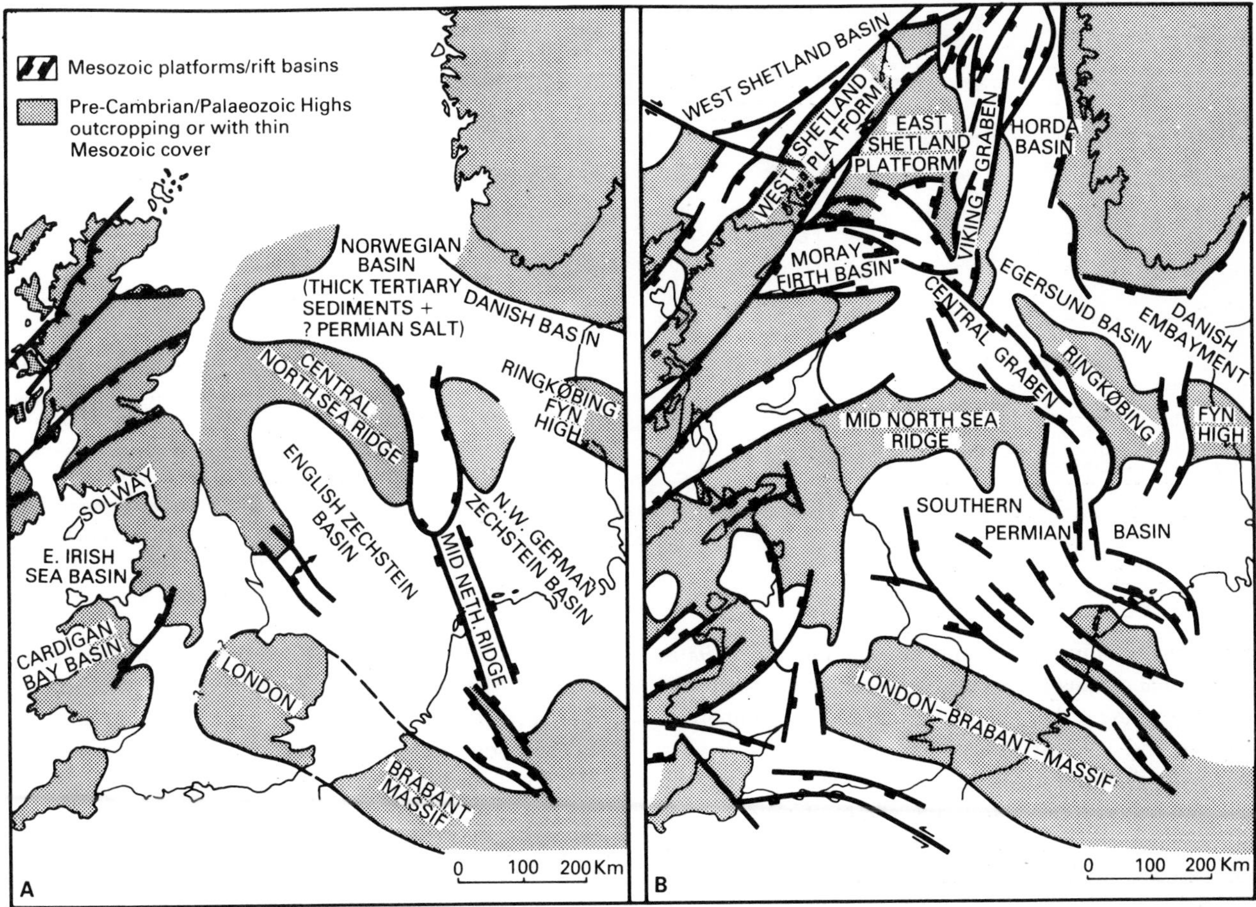

Fig. 1.4 The tentative tectonic framework of the North Sea. A. From 1964 to 1969: as understood immediately after the first phase of exploration in the southern North Sea. B. From 1970 to 1975 (adapted from Donovan, 1968; and Ziegler, 1982). Contrast these with the greatly improved position today, seen, for example, in greatly simplified form, in Fig. 1.17.

mid-1960s, immediately after the first phase of gas exploration in the southern North Sea, with the vastly improved position by the mid-1970s. Today, much more detail could be added.

To achieve this result required the shooting of well over 1 million kilometres of seismic lines across the North Sea area and the drilling of 3800 exploration and almost 1500 appraisal wells between 1965 and 1996 (excluding German waters, but including the Norwegian Atlantic margin), two-thirds of which were located in UK waters (Fig. 1.5A,B)—an effort that, in other major hydrocarbon provinces, has historically taken many decades. That it was achieved in this short time was due to a combination of several favourable factors.

Like most mineral 'rushes' in the past, that of the North Sea was fired by initial discoveries of a size that attracted the attention of the whole industry. The huge gas find at Slochteren in Groningen in 1959 and the rapid understanding of the distribution of its Carboniferous source and Permian reservoir and capping formations ensured that a major exploration campaign would ensue in the North Sea, and, significantly, in that part of the North Sea where it could most easily be undertaken—in the south. Ten years later, a billion-barrel (200×10^6 m^3) oil find at Ekofisk in the Norwegian sector provided what was needed to spur on the tentative northward forays into deeper and more exposed waters. The Forties discovery a year later provided a similar stimulus in UK waters.

The pace of exploration was certainly aided by the fact that the play was offshore from the start. This enabled a large coverage of relatively inexpensive seismic surveys to be obtained in the shortest possible time. Moreover, the drilling that followed, although it had to overcome severe physical and technical difficulties, was not hampered by lengthy planning applications and site preparations, as would have been the case onshore. The technology of drilling was fortunately developing at a rate which ensured that, when exploration interest moved northward, there were at least the beginnings of a capability to handle the problems there. While the physical conditions being met within the central North Sea were among the most difficult yet faced by the drilling industry, from both the weather and the supply points of view, the new generations of semi-submersible rigs eventually proved equal to the task. In the late 1960s, it would take an average of 3.7 string months to drill a northern exploration well. By 1976, the average time was reduced to 2.1 string months, a figure that has remained fairly constant since then. In the early phases of drilling in the north, winter activity was not normally attempted. By 1972, this seasonal barrier had been overcome by a number of operators, and since then, year-round mobile drilling has been the norm.

Advances in drilling techniques included the introduction of turbodrilling, advanced bits and single BOP stacks in the late 1970s, and computerized mud logging and

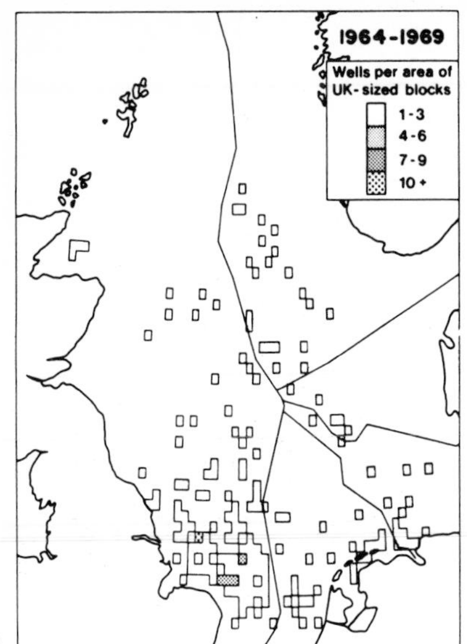

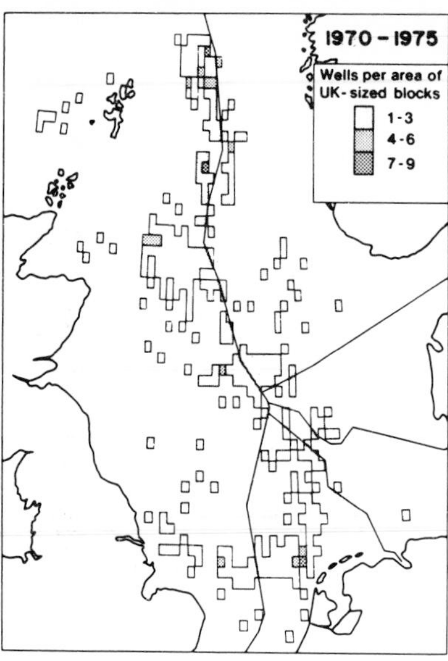

Fig. 1.5 A. North Sea drilling activity for the period 1964–69 (activity concentrated on the southern gas play) and 1970–75 (activity concentrated on the oil plays in the central and northern North Sea). B. (*Opposite*) Exploration and appraisal drilling activity in the various national sectors, 1965–96.

directional drilling and monitoring in the early 1980s. Directional drilling led inevitably in the late 1980s to the widespread use of horizontal drilling, with two important applications. The first application is within thin reservoirs where the greater formational surface area in contact with the production tubing limits the effects of water coning during production. The second is in reservoirs that have low permeability, the greater surface area permitting commercial rates of production in otherwise marginally commercial fields. The world's first successful horizontal drilling from a floating unit took place in 1988, in British Petroleum's (BP's) Cyrus Field. Logging procedures advanced with the introduction of the repeat formation test in 1977 and the use of tools run in combination in the 1980s, developments that reduced rig time and/or improved data gathering. Downhole measurement of resistivity and natural gamma radiation appeared in 1982. Together with advances in telecommunications, these developments now allow 'real-time' monitoring of remote drilling operations in head offices.

Production technology has also evolved. Subsea completions and the underwater manifold were introduced in the early 1980s, and several innovative surface facilities—the tension-leg platform, floating production systems and the first unmanned platform (Eider)—were developed during the mid- to late 1980s. The sharp drop in global oil prices in 1986 (Fig. 1.6) temporarily halted some developments, which later resurfaced after unit costs had been significantly trimmed.

Other developments affecting the economy of marginally commercial fields include the use of floating oil-storage tanks (exposed-location single-buoy mooring) for the efficient transfer to tankers of oil produced in volumes that were too small to warrant a pipeline, Shell's now defunct Brent Spar, which could store up to 300 000 barrels, and modified tankers for the temporary storing of rather larger volumes of oil; some other late technical advances will be discussed near the end of this chapter.

In the vanguard of exploration, geophysical surveying technology also made great strides during the late 1960s (Table 1.1A). The switch to non-dynamite sources of seismic energy in the late 1960s deprived the industry of an optimum signal, but vastly simplified its operations by eliminating the shooting boat—a major plus in the North Sea conditions. Very beneficial, too, in terms of suppression of 'noise' and in increased clarity of reflectors on seismic lines, was the greater ease in obtaining higher multiplicities. In 1966 fourfold coverage was usual, but by 1970 24-fold shooting was the rule. Coverage subsequently further increased to 48-fold, 96 channels by 1980, and to 120-fold, 480 channels by the late 1980s.

Also of key importance was the timely appearance of digital seismic recording techniques (Table 1.1A). At about the time of the discovery of Ekofisk (1969), this new technique was becoming widespread, and the following decade witnessed a remarkable improvement in the quality of seismic records. Digital recording came about as a result of the increase in the power and cost-effectiveness of digital computers. In the 1960s, deconvolution allowed much higher resolution to be achieved in time sections. This permitted seismic reflection sections to be tied with well-log data, thereby opening the way to detailed seismostratigraphy and the recognition of direct hydrocarbon indications. Lateral changes in lithology can now be mapped and, increasingly, complex fields are routinely resurveyed every 3–4 years to detect the position of fluid contacts and to enable more accurate targeting of infill wells (Dromgoole and Spears, 1997). In the 1970s, a corresponding improvement in spatial resolution was achieved by advances in seismic migration (Table 1.1B,C).

Late in the 1970s, a further significant advance was made by the introduction of three-dimensional (3D) survey techniques. Because of its density of coverage, and thus higher cost relative to 2D coverage, 3D surveys were used initially only for the development of fields that had already been discovered. With increasing confidence in their structural and stratigraphic definition, more regional 3D surveys were soon considered cost-effective and now spearhead most exploration programmes. Indeed, in the Netherlands, for example, after years of a declining rate of discovery, there were immediate successes following its first use for exploration around 1987 (Epting, 1996, Fig. 4).

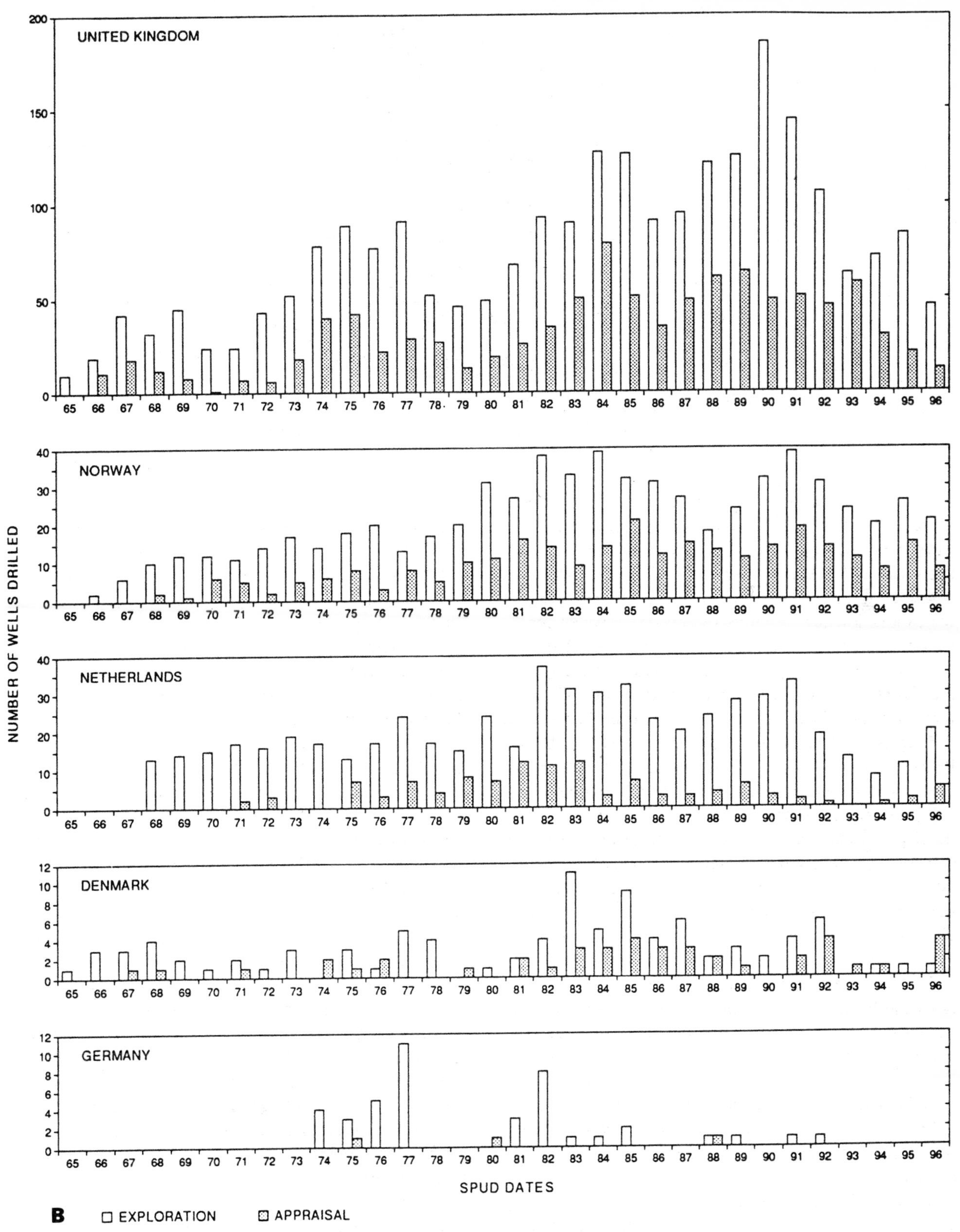

Figure 1.5 (*Continued*) B.

Rather than returning seismic lines for costly and time-consuming specialist reprocessing, the development of individual interactive workstations now enables experienced seismic interpreters, at little more than the touch of a button, to enhance and display in colour complex 3D data in the best way to highlight particular sedimentary or structural geometries; stages in basin development can be deduced by flattening particular horizons successively, and immediate hard-copy printouts make a permanent record (e.g. Eggink *et al.*, 1996, Fig. 17).

In contrast to many hydrocarbon provinces in other parts of the world, the intervals of interest in the North Sea area have an unusually wide stratigraphical range. Hydrocarbons have now been proved in reservoirs ranging from the Precambrian to the Tertiary (Figs 1.7 and 1.8), although

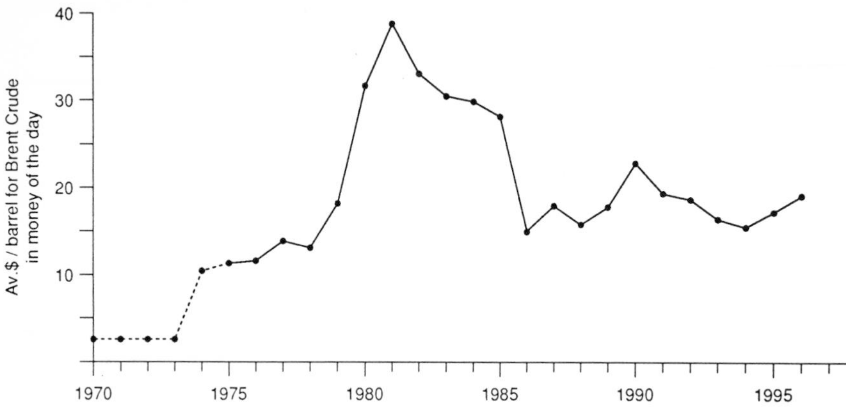

Fig. 1.6 Global prices of oil, 1970–96. Global prices about \$3.00 to \$3.50 from 1970 to 1973; 1974 price \$11.00, 1975 price \$12.

Table 1.1. Important developments in seismology, 1960s to 1996.

A *Seismic acquisition*	
Early 1960s	Airgun energy source (marine)
Late 1960s	Change from analog to digital recording
Mid-1960s	Acceleration-cancelling streamers; higher multiplicity; finer sampling; microelectronics
Late 1960s	Vibroseis (land)
Mid-1970s	Accurate radio positioning; satellite navigation
Early 1980s	3D acquisition over fields; water guns; shear-wave recording (land)
1990	Accurate streamer positioning; 3D multistreamer multisource recording
1993	3D walkaway VSP for enhancing seismic resolution of fields
1995	4D time-lapse seismic monitoring of reservoir with seabed arrays
B *Seismic processing*	
1960s–1990s	Continued rapid improvement in computer technology: ever-increasing cost-effectiveness
Mid-1960s	CMP stacking; digital recording; Wiener deconvolution; idealized mathematical models
Early 1970s	Section migration; autostatics
Early 1980s	Phase inversion; improved physical models; 3D
Mid-1980s	Depth domain processing
Early 1990s	Interactive processing
1991	3D three-step migration
1995?	Data-compression technology for transfer from ship to shore processing
C *Seismic interpretation*	
Early 1960s	Reflection-section phantoms
Mid-1960s	Stacked sections
Mid-1970s	Migrated sections; direct hydrocarbon indications; lithology recognition—well ties; improved 2D seismic increases ability to map ever smaller targets
About 1980	3D seismic for field development
1980	Seismostratigraphy, leading to stratigraphic plays
Mid-1980s	Interactive seismic interpretation on workstations
Late 1980s	3D seismic for regional exploration
1990	Amplitude v. offset
1992	3D visualization

VSP, vertical seismic profiling; CMP, common-midpoint.

very little has yet been found in the pre-Carboniferous. Figure 1.7 emphasizes the current minor role of reserves west of Shetland (but see Section 1.4.7) relative to the North Sea. Since the early 1970s, the improving seismic technology has continued to play a crucial role in clarifying what proved to be ever more complex structural and stratigraphic geology. As in the early stages of most petroleum plays, the larger, simpler fields were recognized early, but high-quality 3D seismology is now helping to define the smaller or more subtle structures.

Continuing improvements in seismic resolution are still as necessary as ever. Improvements in both seismic and drilling techniques were important factors that speeded up the gathering of new information. This brought about accelerated activity, because the faster data were obtained, the faster could new activity be planned and justified.

A new petroleum play without an economic push and without enough licence room within which to operate would not get far, however technically attractive. The rapid pace of North Sea development benefited from the start from mostly favourable conditions in both these areas.

As regards oil, the quadrupling of the price in 1974 (see Fig. 1.6), following the Yom Kippur war, provided a real incentive for non-Organization of Petroleum Exporting Countries (OPEC) countries to establish home-based sources of supply, and this boost for exploration came at a time when many large North Sea discoveries had already been made (see Section 1.4.4). Thus, although remaining prospects tended to be smaller and less straightforward than those that had been found previously and companies had become more clearly aware of the huge costs of North Sea oil development, exploration was able to continue without a downturn until the early 1990s.

As government priorities changed with the realization that enough oil was becoming available to satisfy their own national needs, and as oil prices rose, so did levels of taxation increase. In the case of the UK, by the end of 1982, North Sea oil taxation was a complicated system involving three tiers: royalty; petroleum revenue tax (PRT), which includes advance PRT (APRT); and corporation tax. Until 1975, when PRT was introduced, the government's marginal rate of tax was 58%. After the introduction of supplementary petroleum duty (SPD) in 1981, the marginal rate increased to over 90%. The replacement of SPD by APRT in 1982 reduced this to just below 90%.

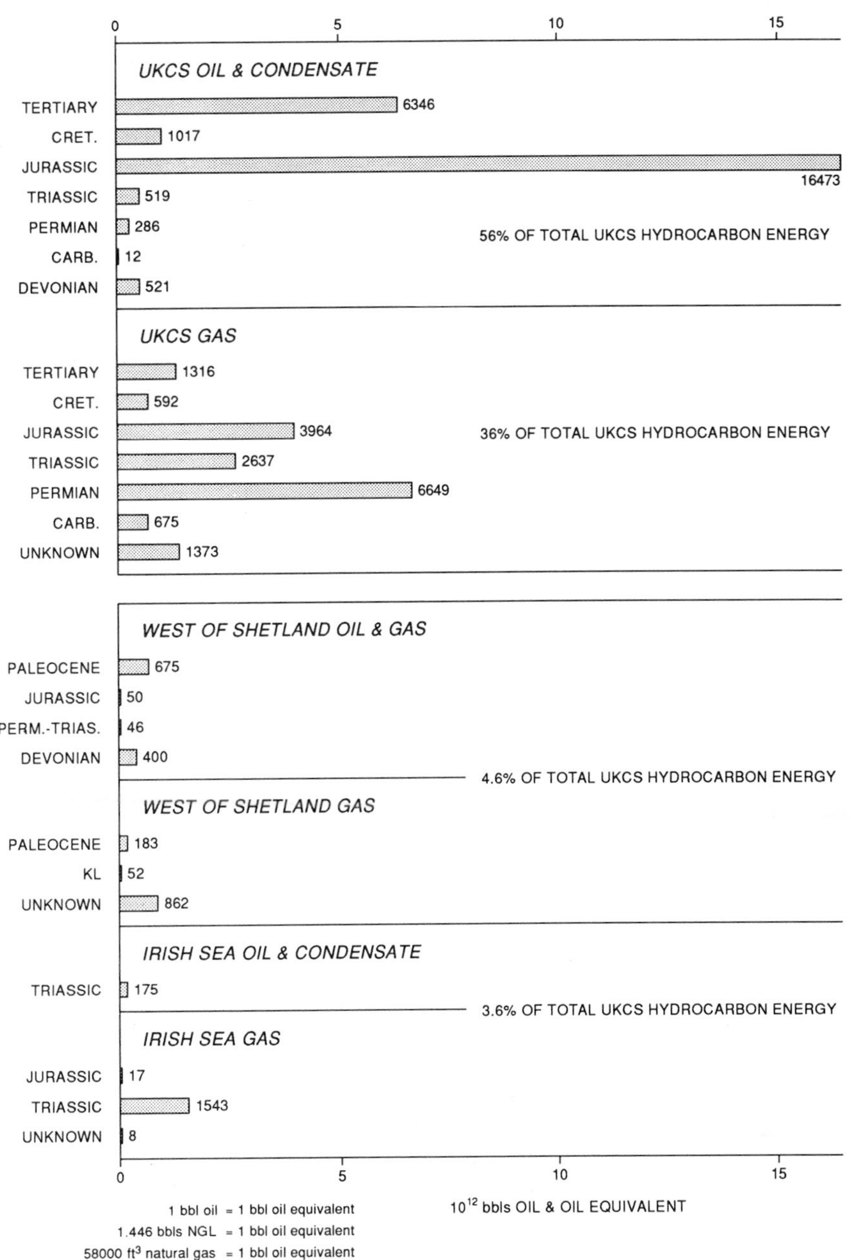

Fig. 1.7 Ultimate recoverable reserves of oil and gas in billions (10^9) of barrels of oil and oil-equivalent (gas) by broad geological age; UK continental shelf (UKCS), Irish Sea and west of Shetland.

From late 1980, the oil companies became increasingly concerned about the impact of heavy taxation, as well as the uncertainty caused by repeated amendments to the tax system. The industry's continuing interest in acquiring exploration licences was argued by some observers to be inconsistent with this concern. However, there is a significant difference between the decision to look for oil and the subsequent decision to develop what might be found.

New field developments in the UK fell to a very low level in the early 1980s. The oil industry and services expressed considerable concern at this. In the March 1983 Budget, the Chancellor responded by proposing changes that would reduce government revenues from oil and gas by more than £800 million over the 4-year period starting 1983–84, which would mean a substantial reduction in tax for future fields. The phasing out of APRT was a major feature of these changes, and in his autumn statement of 1987 the Chancellor announced the repayment of £300 million of this tax as a measure aimed partly at alleviating the negative effect of the dramatic decrease in oil prices in 1986 (see Fig. 1.6).

It must also be noted that the decline in exploration for gas in the southern sector of the UK North Sea in the 1970s was due to a lack of economic incentive, and not to an exhaustion of prospects. In the Netherlands offshore, where prospects are technically no better, exploration was maintained at a high level (see Fig. 1.5A,B), undoubtedly as a result of the higher value accorded to gas there (thermal equivalent of oil) than was obtainable in the UK during the 1970s. Indeed, since 1982, a higher price for new gas resulted in renewed and continuing activity in the UK southern North Sea. The 1988 Budget removed royalty and reduced PRT oil allowance for all fields given development consent since April 1982, bringing the southern North Sea into line with other areas and encouraging new field development. A further significant change on the gas scene

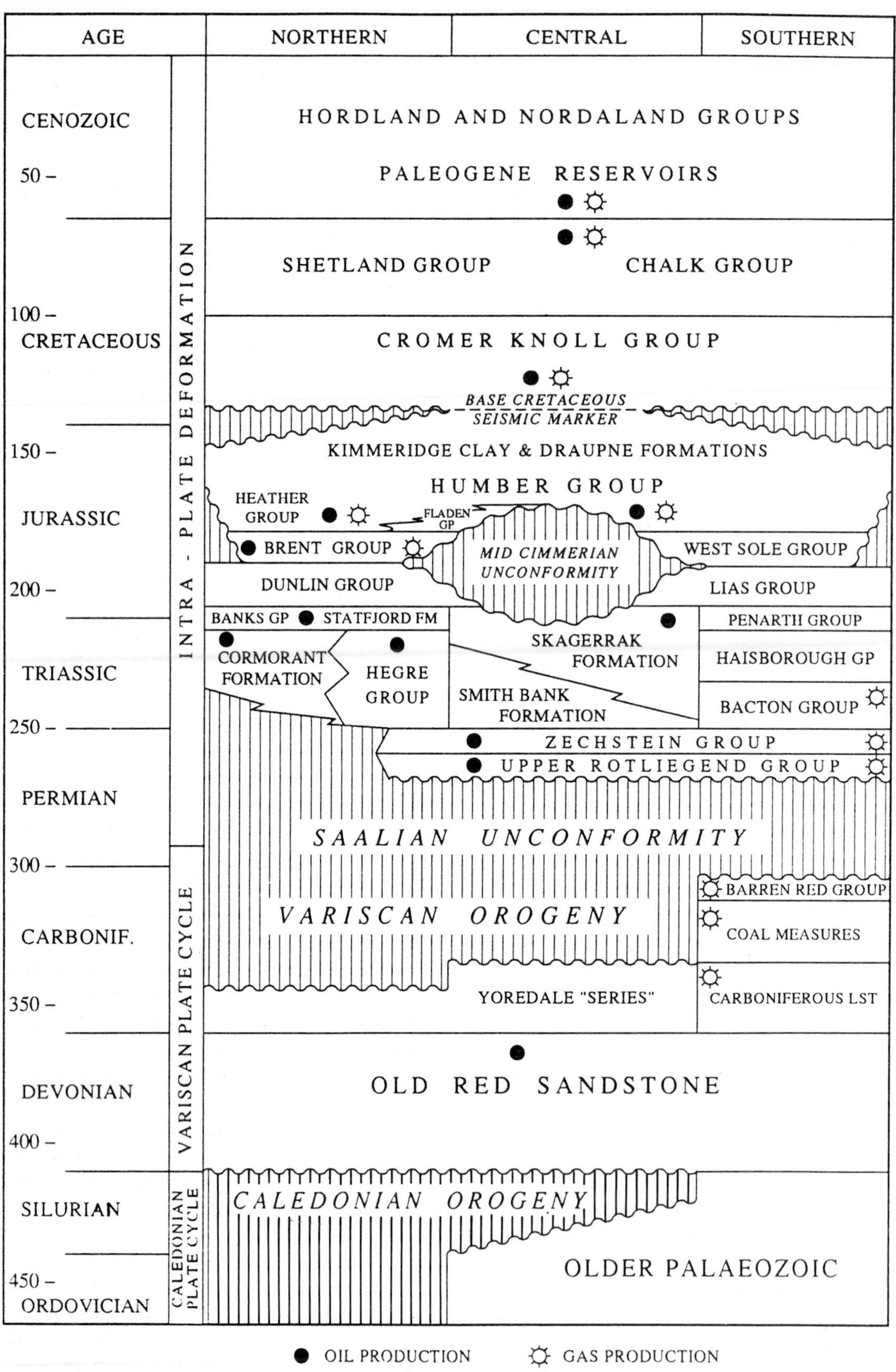

Fig. 1.8 Simple stratigraphic table, northern, central and southern North Sea, with geological ages of main oil- and gas-bearing reservoirs.

has been the removal of the British Gas purchasing monopoly. Gas from BP's Miller Field, in the central North Sea, and from Ranger's Anglia and Shell's Galleon fields, both in the southern North Sea, is now sold directly to power companies for the generation of electricity.

Presumably in order to maintain rates of oil production as far into the 1990s as possible, changes to the British PRT in 1993 favoured the development of otherwise marginally economic fields that had already been discovered. Exploration costs, however, could no longer be written off against oil production, which caused an immediate downturn in

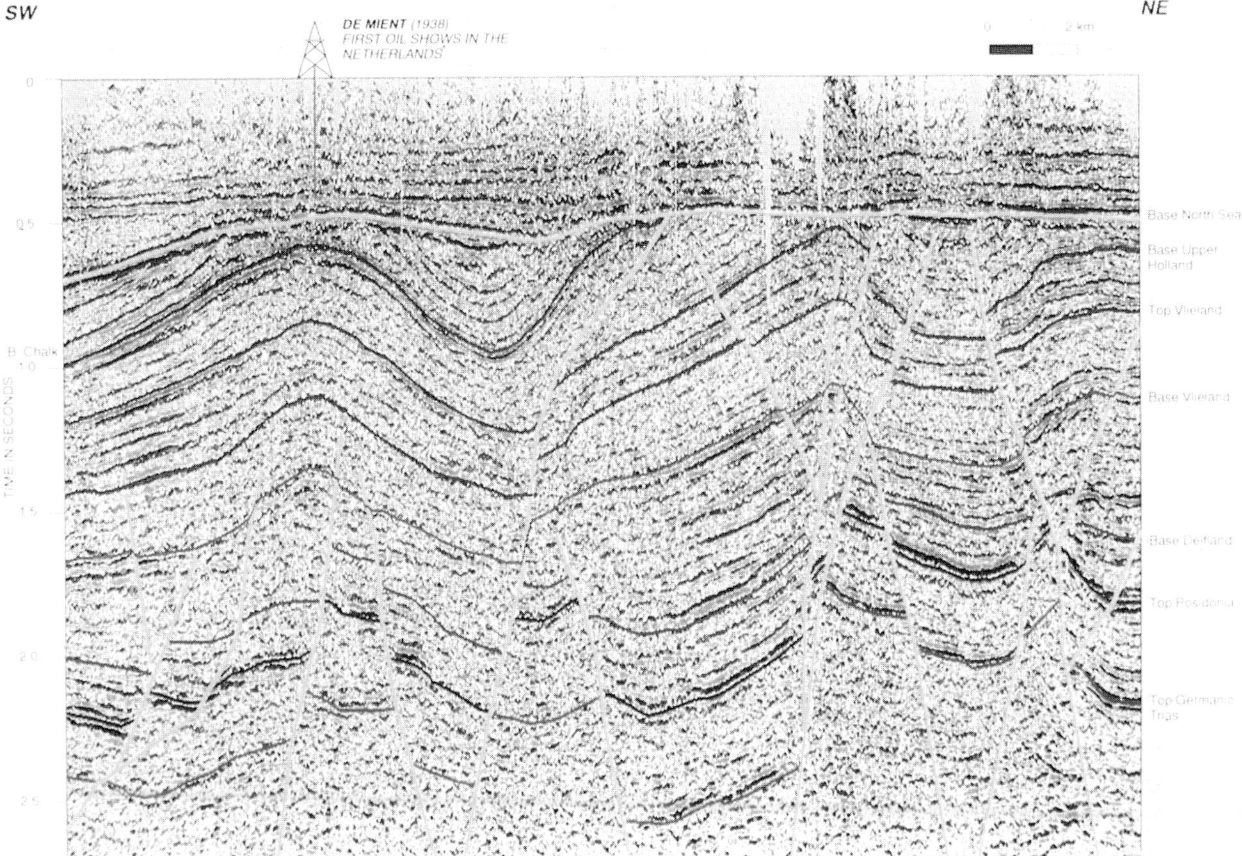

Fig. 1.9 Modern seismic line across the De Mient location, just north of The Hague, where oil shows were encountered in 1938 during a drilling demonstration (courtesy of Nederlandse Aardolie Maatschappij (NAM), the Netherlands).

exploration activity and the laying off or redeployment of exploration staff, although the drilling of already planned exploration wells continued.

The pace of North Sea exploration has been controlled to a considerable extent by the various rounds of licensing: for example, the general lag in Norwegian discoveries relative to those of the UK is a direct result of Norwegian licensing policy.

In the UK, the oil and gas industry is operated today entirely by the private sector, and it thus draws upon large pools of expertise and funds in a competitive environment. The discretionary award system, administered by the Department of Energy, considers applicants for acreage in the light of several criteria, such as financial and technical competence, track record and contribution to the industry's advancement. This system, with less stringent conditions applied to the higher-risk 'frontier areas' (see Fig. 1.2), has prompted a generally healthy response to individual rounds, and remains in very high regard after 17 rounds in over 30 years. The system has also contributed greatly to the rate at which the continental shelf has been evaluated.

Further factors that governed the development of exploration and production in the North Sea will be discussed in the appropriate historical phase of activity and, especially those of a technical, cost-cutting nature, in Section 1.4.7.

1.4 Main phases of North Sea exploration

1.4.1 The period before 1959

Exploration activities in the onshore areas surrounding the North Sea started in northern Germany, where, in 1859, oil was discovered in the Wietze Field near Hanover. Systematic exploration of similar salt domes and associated anticlines resulted in the discovery of some 70 fields, producing at an annual rate of 3×10^6 barrels (0.5×10^6 m^3), mostly from Lower Cretaceous and Jurassic reservoirs (Bentz, 1958, Schröder *et al.*, 1991; Glennie and Hurst, 1996). Gas was discovered by chance in 1910 in a water well near Hamburg, and some minor gas was subsequently found in Zechstein dolomites in a number of locations. The first indications of oil in the Netherlands were the result of serendipity. During the 1938 World Petroleum Congress in The Hague, oil shows were encountered during a demonstration of drilling at De Mient (Fig. 1.9) just north of The Hague. Exploration in the Netherlands had been initiated in 1935, leading to the discovery of the Lower Cretaceous Schoonebeek Field in 1943 (the largest onshore oilfield in western Europe) and a number of small gasfields (1948–53) in the north-eastern Netherlands, and in the establishment of a small production of oil in the western part of the country (Visser and Sung, 1958; Petroleum Geological Circle, 1993, 1996).

Exploration activities in east England go back to 1938, when BP drilled on the Eskdale anticline and found gas in Zechstein dolomites. Further exploration in 1939 of the Eakring anticline resulted in the discovery of low-sulphur

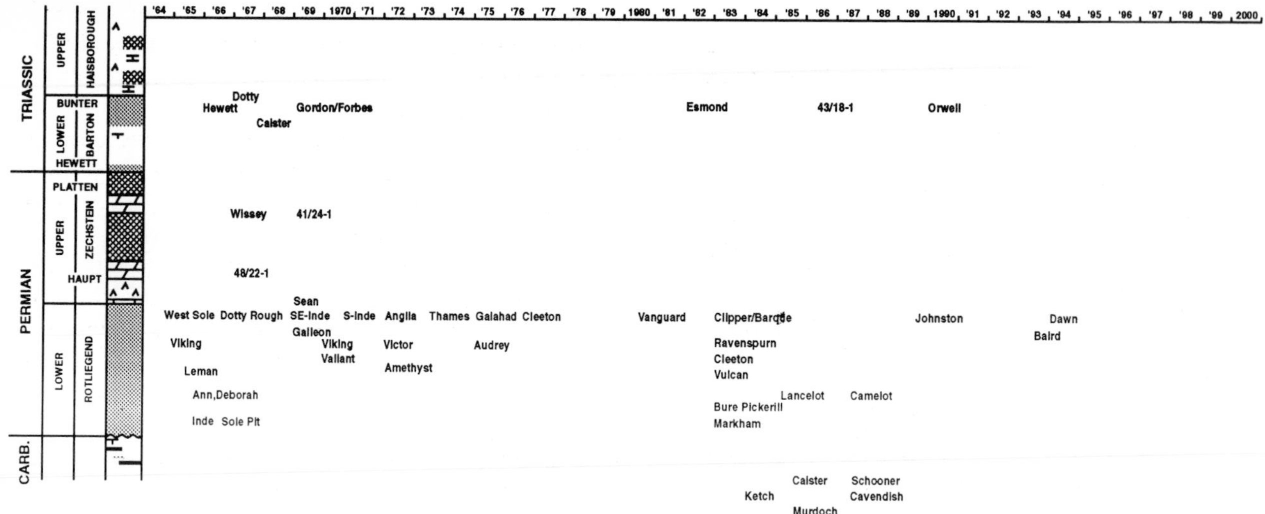

Fig. 1.10 Important gas discoveries, UK southern North Sea, 1965–96.

waxy crude in commercial quantities in Namurian and-Westphalian reservoirs (Kent, 1985). Seismic refraction work at Duke's Wood, Kelham Hill and Caunton defined other oil-bearing structures, which raised the total daily output in the East Midlands to a modest 2500 barrels (400 m^3). A second wave of exploration took place between 1953 and 1961, during which the Gainsborough oilfield, and 10 smaller fields, were found.

Scotland's now defunct oil-shale industry in the Midland Valley of Scotland was founded by James 'Paraffin' Young in 1851 by retorting oil from mined Boghead coal (torbanite); it functioned continuously for just over 100 years until put out of business by a combination of taxation and the low price of mineral oil. Oil seepages were known, and small quantities of hydrocarbons were encountered in wells drilled on the Cousland anticline near Edinburgh in 1937–38, but no commercial discovery was announced.

Oil seeps also occur in the south of England, in south Dorset, where drilling began as long ago as 1930. The small Kimmeridge oilfield was discovered in 1973. Geochemistry indicates that the Kimmeridge Clay, source of most North Sea oil, has not reached maturity in this area, and instead the Lower Lias is believed to source the oil.

1.4.2 The period 1959–64

The giant Groningen gasfield in the Netherlands was discovered by a Shell/Esso consortium in 1959, although its huge size was not realized until 1963. And, in 1962, another large field (2.7×10^{12} ft^3; 75.6×10^9 m^3 gas) had been discovered just to the south at Annerveen (Veenhof, 1996). With the technical ability available and an increasing tendency on the part of oil companies to explore offshore, the Groningen discovery provided the spur that finally jolted the industry into action in the North Sea. Regional geological correlation of the Groningen data suggested a distinct possibility that the gas-bearing Permian sandstone reservoirs could extend from the Netherlands, under the North Sea, to eastern England. Geophysical research work had already indicated that salt-induced structures similar to those present in onshore Germany and the Netherlands were also present in the offshore. From 1962 to 1964, various groups of companies carried out reconnaissance marine seismic campaigns, which, together with aeromagnetic surveys, gave a first and still rather hazy view of the geological configuration of the North Sea. At the same time, a few offshore wells were drilled as small outsteps into Dutch waters from fields onshore, beginning as early as 1961 (Kijkduin).

1.4.3 The period 1964–70

The first well drilled in UK waters, 38/29-1, located on the Mid North Sea High, failed to encounter hydrocarbons. However, commercial gas discoveries in the same Permian Rotliegend reservoir as Groningen were made in 1965 by BP at West Sole and by Shell/Esso, who found the giant Leman Field (Fig. 1.10; see also Glennie, 1997a). At this time, the disappointing well results in the German sector provided a better understanding of the geographical limits of the play, as these indicated that the Rotliegend sands were replaced by shale and evaporites towards the north.

Because of these early successes, and the relatively shallow water conditions enabling the employment of jack-up rigs, growth of new geological insight was initially greater in the south (see Fig. 1.5A) and the dimensions of the southern Permian basin and of the Rotliegend dune belt were soon established (Fig. 1.10; Glennie 1972; see also Chapter 5).

At that time, seismic resolution at depth was not very good (Fig. 1.11), and the Rotliegend target was commonly masked beneath the Zechstein salt. The larger structures at Hauptdolomite level, however, proved to be indicative also of large structures at the top Rotliegend. As time progressed, the remaining structures still to be tested were of progressively lower relief, and defining closure at Rotliegend level became more dependent upon accurate regional knowledge of the thickness to be expected of the Zechstein carbonates. Drilling had revealed that overlying the Rotliegend sands were thick shelf carbonates (Taylor and Colter, 1975; see also Chapter 6), which thinned northwards into a basinal facies, and the velocity gradient that

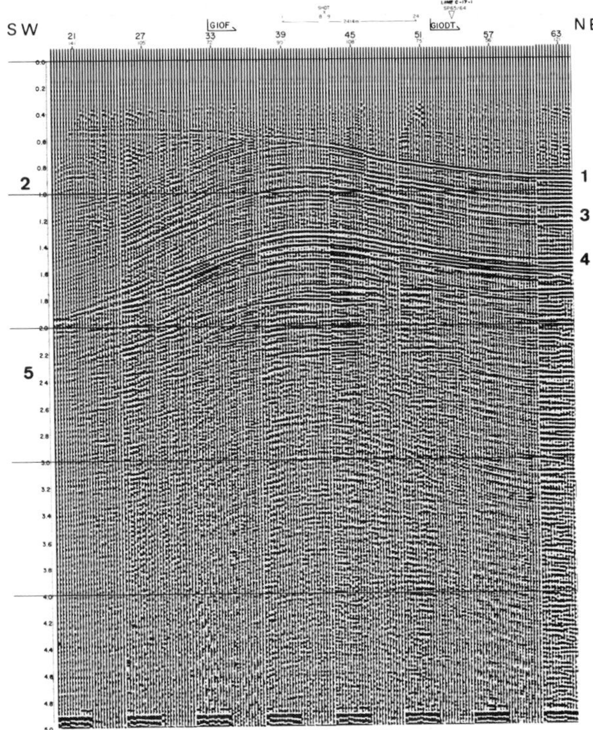

Fig. 1.11 Poor quality of the seismic used in 1965 to discover the southern North Sea gasfields, Viking area. 1: Base Chalk. 2: Jurassic. 3: Top Bunter. 4: Top Zechstein. 5: Top Carboniferous (courtesy of Conoco).

resulted imposed considerable effects on the depth conversion of seismic time sections.

By 1968 it was becoming clear that drilling a structure within the sand belt was no guarantee of success. In well 48/13-1, for instance, where the sands were the thickest so far found, they proved to be tightly cemented, even though gas-bearing over hundreds of feet. Further drilling and detailed petrological work showed that certain areas of the dune belt had suffered deep burial during the Mesozoic, with attendant porosity destruction by secondary silica and the growth of authigenic illite. Inversion of this basinal area occurred in the latest Cretaceous and mid-Tertiary, giving rise to broad structural highs, such as the Sole Pit. Other high areas, such as those to the east of the Indefatigable Field in Block 49/19, also proved disappointing: there, the sands were porous but unexpectedly devoid of gas. To explain this, it was necessary to invoke an unfavourable relationship between the timing of structure formation (probably Late Cretaceous), the onset of gas migration (Early Cretaceous) in the kitchen area further downdip and the absence of mature source rock within the high.

The slow enactment of petroleum legislation for the Netherlands offshore prevented drilling from taking place until 1968. There were immediate successes with the discovery of Rotliegend gas in Blocks K7, L2 and L12, which were followed by further discoveries in the K and L quadrants in the succeeding years (Fig. 1.12). Meanwhile, on land, development of the production potential of the Groningen field continued.

While the commercial limits of the Rotliegend sands appeared to be constrained to a relatively narrow east–west-trending zone in the southern half of the southern Permian Basin, the Triassic Bunter sands, of excellent reservoir quality, were found to be far more widely distributed and, furthermore, they were extensively structured by the underlying Permian salts (see Chapter 7). High hopes for a major Bunter play were short-lived, however, because the Permian evaporites proved in most places to be a barrier to gas migration from the Carboniferous source, and the traps were either completely water-bearing or only partiy filled with gas. Rare exceptions to this rule were where faulting or salt thinning had breached the seal. The only finds with Triassic reservoirs of immediate commercial interest were the Hewett Field, found by Arco in 1966 (Cumming and Wyndham, 1975), and the Dotty Field, discovered by Phillips in 1967. At Hewett, a basal Triassic sand, unknown elsewhere, was additionally gas-bearing. In the northern part of the southern Permian Basin, the small size of the Bunter gas discoveries made by Hamilton Brothers at Gordon (1969) and Forbes (1970) did not, at that time, lead to commercial development.

The carbonates of the Permian Zechstein were rated as a secondary objective for gas, following earlier experience in Germany and the Netherlands, and there were numerous instances where they were found to be gas-bearing during drilling for the deeper Rotliegend objective. So far, none of the offshore cases has proved to be of a size, or has indicated a sufficiently sustainable rate of production, to warrant commercial development. Some wells obtained respectable initial production rates from the Haupt- and Plattendolomite, sometimes with sizeable associated condensate/oil production (e.g. 48/22-1), but the high flow rates of both gas and liquids were typically short-lived.

While the Zechstein and Triassic intervals yielded little gas that could be developed, there was a rich yield of stratigraphic information. The lithostratigraphic correlation of individual carbonates, sands and evaporites matched surprisingly well far west of the German type areas. The new offshore evidence also provided the key to the integration of the Permian and Triassic basin-margin deposits of the UK with those of the continent. Regional work on these intervals provided the first published information from the offshore activities (Heybroek *et al.*, 1967; Geiger and Hopping, 1968). Our stratigraphic understanding of the Triassic and older intervals was further enlarged by the recognition of palynological correlations with known areas of continental Europe and England. A firm calibration has now been established with Alpine ammonite-controlled sections. Work on planktonic and arenaceous foraminifera was also undertaken on the new Mesozoic and Tertiary sections encountered in wells.

North of the Mid North Sea High, exploration interest focused on the Lower Tertiary, where seismic was interpreted to indicate 'growth-fault'-like features reminiscent of the prolific Gulf of Mexico and Niger delta hydrocarbon provinces. These 'growth faults' proved to be illusory in so far as they might have indicated deltaic rollover structural prospects. More significant, however, was the first discovery in Norwegian waters, when Phillips tested gas from Palaeocene deep-water sandstones in Block 7/I1 (Cod Field) in 1968 (D'Heur, 1987b). Further encouragement for this play was provided in 1969 by the result of Amoco/Gas Council's 22/18-1 well (Fowler, 1975). After

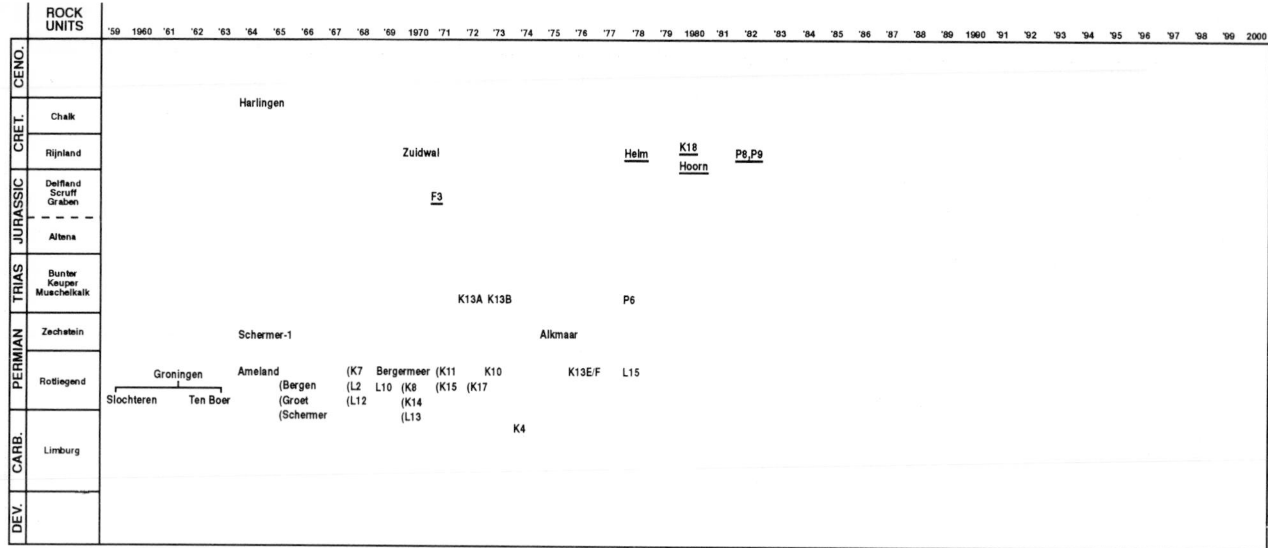

Fig. 1.12 Important oil and gas discoveries in Dutch onshore and offshore areas, from the discovery of the Groningen Field in 1959 to 1996. Offshore drilling began in 1968. Oilfields are underlined.

8 months with the Sea Quest to drill and test, encouraging oil flows were reported from Lower Tertiary sands. Although the sands were massive and porous, the oil column was rather small, therefore putting in doubt for some time the commercial viability of what was then called the Montrose Field. This structure is now known as the Arbroath Field, Montrose being an adjacent field separated by about 15 ft (5 m) of structural relief and a saddle 500 yards (460 m) wide (Crawford *et al.*, 1991).

In December 1969, Phillips, in Norwegian waters, established major flows of oil, with gas, from a 600 ft (183 m) oil column in the Danian Chalk in Block 2/4 (Byrd, 1975; Van den Bark and Thomas, 1981). This well, 2/4-A 1x, was in fact a redrill of an earlier one, which began to 'kick' when it encountered oil in Miocene carbonates. The Chalk's potential as a reservoir had already been indicated in Danish waters by the 1966 discovery of oil and gas at Anne (now Kraka; see the appendix to this chapter) and by the discovery of gas in Roar and Tyra (1968), and also in Norwegian waters by the Valhall oil discovery (1967; Fig. 1.13). The combined effects of salt diapirism and extensional tectonics causing extensive fracturation, together with high pressures and the presence of reworked chalk, are assumed to be the main reason for the unexpectedly favourable reservoir conditions.

The Norwegian Block 2/4 was quickly realized to contain a very large oilfield (Ekofisk: 1.49×10^9 barrels of oil; 4.5×10^{12} ft^3 of gas; Pekot and Gersib, 1987), and it proved to be a most important milestone with regard to oil exploration in the North Sea. It effectively closed a decade dominated by exploration for non-associated gas. From here on, the hunt for northern oil began in earnest, attracting most companies active in the international oil industry, and bringing back some American companies that had earlier left the area because of their poor success in finding gas.

During 1970, drilling in the UK northern area concentrated on fulfilling commitments in first- and second-round licences, which were due for 50% statutory reduction in 1970 and 1971 (see Fig. 1.2). The Lower Tertiary and Chalk now provided the main interest in the Central Graben area, while the Permian, Zechstein and Rotliegend were still sought for in the west and on or near the Mid North Sea High. By the end of the year, the atmosphere for exploration, which to some extent had become frustrated by the erratic sand development in the Lower Tertiary, and by tantalizing non- or poorly producible oil occurrences in the Cretaceous Chalk (no Ekofisk having appeared in UK waters), dramatically improved, with two clearly commercial discoveries late in the year.

British Petroleum's 21/10-1, the Forties-discovery well, found a gross interval of 118 m (386 ft) of oil in Lower Tertiary sands, which flowed at a rate of 750 m^3 (4720 barrels) per day with almost negligible gas (Walmsley, 1975; Carman and Young, 1981; Wills, 1991) and Shell/Esso's 30/16-1 well, the Auk Field discovery, produced light, low-sulphur crude at a rate of 940 m^3 (5920 barrels) per day from a thin zone of collapse-brecciated and vugular Permian dolomites (later wells found oil also in the underlying Rotliegend sands; van Veen, 1975; Trewin and Bramwell, 1991). The recoverable reserves at Forties (2.4×10^9 barrels) and at Auk (then thought about 50 but now 93×10^6 barrels) strengthened the hypothesis that generation of oil was likely to be optimum in the Central Graben area and from Mesozoic or Tertiary source rocks. Migration could be expected into older reservoirs, provided they were proximal to the Graben and were in a high structural position with respect to source rocks. The thick section of sands and the extensive core material obtained in the Forties Field provided valuable data for a sedimentological understanding of the nature and potential distribution of the unpredictable Palaeocene sands.

A third well in 1970 (Fig. 1.14) opened new and intriguing possibilities of deeper, pre-Cretaceous Mesozoic sands within the Central Graben domain. Phillips's Josephine discovery, 30/13-1, was the first in the North Sea to test oil from sands beneath the Chalk. A flow rate of 128 m^3 (800 barrels) per day was attained from a thin sand below

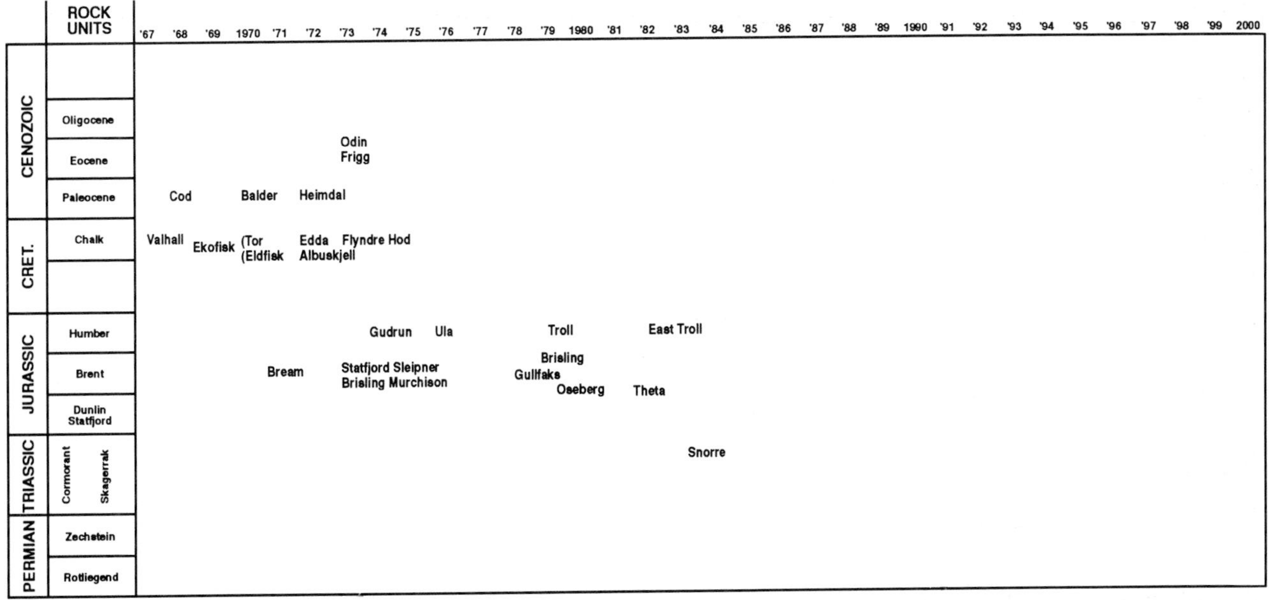

Fig. 1.13 Important oil and gas discoveries and approximate age of reservoir in the Norwegian sector of the North Sea, 1967–96.

3600 m (12 000 ft). This moderate flow rate and the thinness of the reservoir sands did not arouse great commercial interest at the time.

By the end of 1970, the centre of interest in UK waters had swung in a decided fashion away from gas in the Southern Basin, where, except perhaps for Conoco at Broken Bank, diminishing returns had set in for exploration. At that time, there was no sign of an improvement in the price for gas and, in the absence of an incentive to pursue what was becoming harder to find, exploration drilling in the south declined (see Fig. 1.5A). This trend was expressed in a relinquishment of first-round licences on 17 September 1970 that was greatly in excess of statutory requirements (75% rather than 50%).

1.4.4 The period 1971–76

The lack of exploration interest in the Rotliegend and Bunter gas sands in the UK was not matched in the Dutch sector, where oil companies had been allowed a major role in the marketing of gas, the price of which was competitively tied to the level of oil prices. This policy encouraged further gas exploration and development. Therefore, a whole string of gas discoveries continued to be made throughout this period in the K and L quadrants (see Fig. 1.12; see also the appendix to this chapter) and to the north-west of Amsterdam, both on land and straddling the coastline (van Lith, 1983). Gas was also found in Bunter sands in Block K/13 (Roos and Smits, 1983). In German waters, on the other hand, with the exception of a small gas discovery in Block A/6, the few Rotliegend and Bunter gas-bearing structures found since the mid-1960s were all too rich in nitrogen and carbon dioxide for their methane content to be marketed economically.

The Ekofisk discovery led to a series of oil finds in chalk reservoirs of the Central Graben in southern Norwegian (see Fig. 1.13) and Danish waters (see the appendix to this chapter, and also Chapter 9). While the UK third round of licensing came too early to catch the post-Ekofisk enthusiasm for northern oil exploration (the blocks being awarded in June 1970), the fourth round, in 1971, fully reflected the intense industry interest aroused by the discoveries. Following a period of very active seismic acquisition, which infilled the wide reconnaissance grids in the northern North Sea in the area south of the Shetland Platform and between the Forties Field and the Scottish coastline, a complex variety of structural elements began to emerge in increasing but never sufficient detail. The 'landscape' of buried topography, the surface of which appeared to be pre-Lower Cretaceous in age, created numerous objectives of potential interest in the fourth round of licensing. The shortcomings in seismic resolution beneath the ubiquitous unconformity, and the virtual absence of well penetrations below it, left the interpretation wide open as to what geological material composed the 'highs'. Interpretations ranged from the very favourable possibility of a Jurassic sequence, to those envisaging the less favourable Triassic, Devonian or older sequences.

The largest visible 'buried highs' were clearly in the northern North Sea. In the third round, Shell/Esso had acquired Block 211/29, covering what then had been a very loosely defined deep feature. The first well on Block 211/29, completed without testing before the fourth round closed in 1971, revealed that the pre-unconformity sequence contained good-quality oil-bearing Jurassic sands of deltaic origin and of good porosity. The Brent Field reservoirs were first tested in the second well on the structure the following year, and were proved capable of producing 6500 barrels (1040 m^3) per day of 38° API oil, with a gas/oil ratio (GOR) of 1550 standard cubic feet per barrel. An important gas cap was also present (Bowen, 1975; Struijk and Green, 1991). The discovery of the giant Brent Field (1.8×10^9 barrels (286×10^6 m^3) recoverable oil and 3.8×10^9 ft^3 (108×10^6 m^3) of gas) had a profound influence on exploration, and in succeeding years resulted in a spate of discoveries of a similar type in the northern North Sea.

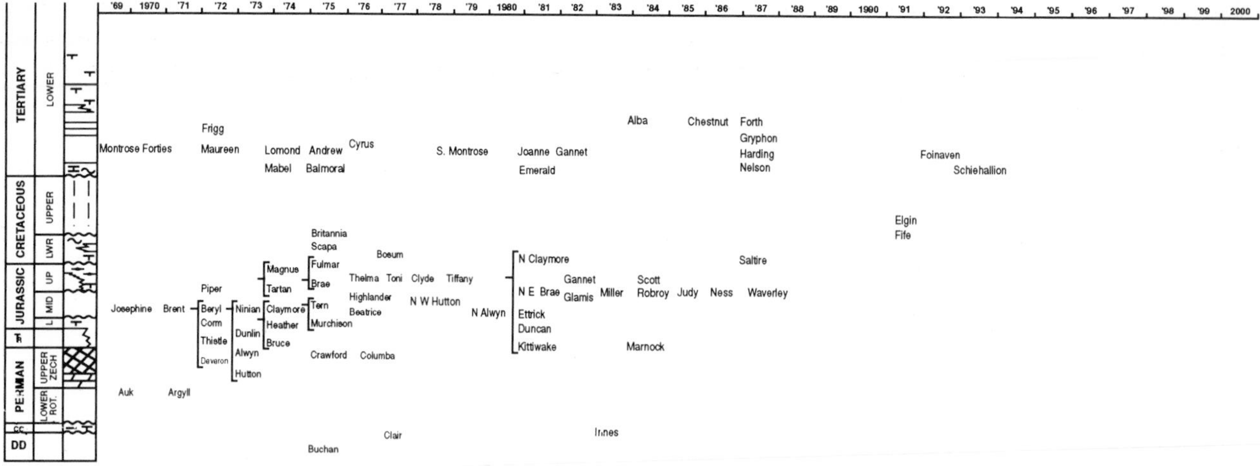

Fig. 1.14 Important oil and gas discoveries and approximate age of reservoir in the UK sector of the central and northern North Sea, 1969–96.

A second Auk-type field was also discovered in 1971 by Hamilton, the Argyll Field in Block 30/24, confirming the possibility of oil accumulations marginal to the Graben (Pennington, 1975). Here, both Permian Zechstein and Rotliegend reservoirs proved productive at attractive rates and, later, a part of the field was found to have oil in underlying Devonian sandstones as well (Robson, 1991).

The subsequent 4 years witnessed an exciting succession of discoveries in the East Shetland Basin in the northern North Sea, where the Brent-type 'buried highs' proved commercially oil-bearing in nine major fields (Fig. 1.14). These were resolved into tilted fault blocks, in which Upper Triassic to Lower Jurassic Statfjord sandstones are overlain by sealing shales of the Dunlin Group, followed by sandstones of the Brent Group overlain by the organic-rich Kimmeridge Clay, onlapped by Lower Cretaceous muds and marls. Additional drilling over the Median Line in the Norwegian sector resulted in the discovery of the Statfjord and Sleipner fields. Statfjord, discovered by Mobil in 1974, located mainly below Norwegian blocks 33/9 and 33/12 but extending into UK blocks 211/24 and 211/25, is the North Sea's largest producing field, with recoverable reserves of 3×10^9 barrels of 38–41° API oil (Kirk, 1980; Roberts *et al.*, 1987). The Sleipner gas field, located on the downfaulted eastern margin of the South Viking Graben in Norwegian blocks 15/6 and 15/9, is a salt-induced rollover at Middle Jurassic level. The Sleipner complex was proved to consist of several gas-bearing structures, of which one, to the east (Gamma, found in 1981; Pegrum and Ljones, 1984), involves Palaeocene sands. The total reserves of Sleipner are of the order of 6.7×10^{12} ft^3 (186×10^9 m^3) of gas and 280×10^6 barrels (45×10^6 m^3) of oil/condensate (Østvedt, 1987). Even further east, in the small Egersund subbasin, two marginal, Middle Jurassic oil accumulations in Bream and Brisling are interesting in view of the low maturity and limited kitchen area at Upper Jurassic level (D'Heur and de Walque, 1987).

The major drilling effort involved in these finds produced a wealth of new regional information on the Jurassic succession in particular. Knowledge of the sequences in east Yorkshire, Sutherland and east Greenland had for some time indicated scope for mid-Jurassic clastics in the northern North Sea. A prophetic illustration of this possibility is contained in a small inset map of the Middle Jurassic in Will's *Palaeogeographic Atlas* (1951), where, in the Bathonian, clastics were visualized as infilling a narrow north–south basin open to the Boreal Sea.

The generally shallow-water nature of the Lower and Middle Jurassic Statfjord and Brent sands, and the gentle transition upwards of the Statfjord sands from the underlying Triassic, itself of very widespread uniformity, encouraged the view that the regressive episodes they represented would be of regional extent; thus the resulting sands could be sought over most of the northern North Sea where Middle Jurassic was present. This model has been verified by extensive drilling, and the main limitation to the Middle to Lower Jurassic sand play is the depth to which these intervals have descended in the process of late Cimmerian (pre-Lower Cretaceous) faulting and Tertiary subsidence.

A new element in the Jurassic play appeared early in 1973, with the discovery (at the end of 1972) of Occidental's Piper Field (Williams *et al.*, 1975). Here the Upper Jurassic (Oxfordian) was in an attractive coastal sand facies, and oil-bearing (the potential of this play was further substantiated by the discovery in 1987 of the Waverley/Brunel Field). In 1974, oil was found in deep-water sands within the Kimmeridge Clay formation in Occidental's Claymore Field (west of Piper), and in BP's Magnus Field in the far north (De'Ath and Schuyleman, 1981). The Kimmeridge Clay, now widely recognized for its source-rock characteristics, had not been foreseen as a potential significant reservoir-bearing formation.

From the beginning of exploration in the northern North Sea, the coastal outcrops of Jurassic at Brora had received particular attention because of their possible relevance to the offshore area. However, the intervals of extremely coarse and ill-sorted debris of the Upper Jurassic sequence at these outcrops were thought to display neither the reservoir quality (porosity/permeability) needed for a productive reservoir, nor the persistence to suggest that such beds might be widespread. In 1975, further light was thrown on the commercial possibilities for the Upper Jurassic reservoirs by Pan Ocean's discovery in the Viking

Graben of the North Brae Field (Central Brae and South Brae were discovered in 1976 and 1977; the complex contains 572×10^6 barrels of oil and 830×10^9 ft^3 of gas). Here, a thick succession of conglomerates and sandstones was found oil-bearing to a maximum thickness of 450 m (1500 ft), in what was interpreted to be a suite of coalescing fans within which abrupt lateral lithological discontinuities occur (Harms *et al.*, 1981). Significantly, this development of Upper Jurassic clastics occurs at the level of the basal Cretaceous late Cimmerian unconformity at a point of high relative relief between the Graben and the platform area to the west.

In the same year, 1975, and far to the south, again in a location near to high pre-Lower Cretaceous relief, Shell/ Esso, at Fulmar (450×10^6 barrels), found Upper Jurassic oil-bearing sands in well 30/16-6. Additional impetus to this play in Norwegian waters was provided by BP's Ula discovery in Block 7/12, which found oil in shallow-marine Upper Jurassic sands deposited on the downthrown side of the main bounding faults on the east side of the Central Graben. Field reserves were established at 200×10^6 barrels (31×10^6 m^3) of oil with 2×10^9 m^3 (70×10^9 ft^3) of gas (1988 appraisal drilling raised these reserves to 330×10^6 barrels). As the main flush of discoveries related to the Middle Jurassic regressive sands dwindled in the mid-1970s, interest increased in the more difficult-to-predict Jurassic play.

Throughout this later period, a large amount of new data were accumulated on the Tertiary. Every well drilled to the Jurassic added new Tertiary information, and many wells had combined Jurassic–Tertiary objectives. A major gas discovery was made in 1971 in the Frigg Field (9.5×10^{12} ft^3 of gas) located in the South Viking Graben, straddling the Norwegian–UK Median Line (60 : 40). The reservoir rock consists of Eocene turbidite sands derived from the east Shetland platform to the west and deposited as deep-water fans and distal turbidites (Heritier *et al.*, 1980). Discoveries of oil in the Tertiary continued to be made at a moderate pace relative to that of the Jurassic. From 1972 to 1975, finds of oil in the Lower Tertiary were reported by Philips (Maureen and Andrew) and Amoco/Gas Council (Lomond).

Beyond the limits of the North Sea, gas of Carboniferous origin was discovered in the adjacent Morcambe North and South gasfields in the eastern Irish Sea. The first well, 110/ 8-2, drilled in 1969, was plugged and abandoned without its gas content being recognized by the operator (Gulf; see Dean, 1996). This acreage was acquired by Hydrocarbons GB (HGB, now Gas Council) in 1972. Hydrocarbons GB then confirmed the presence of gas in the Triassic Sherwood sandstone in 1974 with well 110/2-1, in what is now known as the South Morcambe Field (5.5×10^{12} m^3 (134×10^9 m^3) gas), and in the adjacent North Morcambe gasfield (Block 16/2b) in 1976 (Cowan, 1996). In Irish waters, in the Celtic Sea basin, about 1×10^{12} ft^3 of gas was discovered in Lower Cretaceous sediments in the Kinsale Head Field in 1971 (Colley *et al.*, 1981), the source rock was of Liassic age. West of Shetland, exploration began in 1972, and, although there were indications that 'there was oil in the system', no commercial accumulations were discovered during this period.

Throughout these years, the play concepts were dominated by the assumption that the only significant source rock for oil in the central and northern North Sea was the Upper Jurassic Kimmeridge Clay formation (see Chapter 11). On the basis of this assumption, a large part of the Inner Moray Firth was dismissed as being unprospective because the source rock was considered to be immature for oil generation. Mesa Petroleum's discovery of the Beatrice Field in 1976, proved, however, that explorationists have to keep an open mind for alternative possibilities.The field contains some 130×10^6 barrels (21×10^6 m^3) of recoverable oil in Middle Jurassic reservoirs, possibly derived from intercalated source rocks (Linsley *et al.*, 1980) with, perhaps, a contribution from lacustrine source rocks of the underlying Devonian (see Chapter 3, this volume). Further east, the potential of the Devonian Old Red Sandstone was illustrated by the discovery of the Buchan Field in 1974, which, although containing reservoirs with low porosity and permeability values, still had appreciable production rates (30 000 barrels/day) because of intense fracturing.

During this period also, notable developments in seismic technology took place, which enabled the Jurassic play to be pursued with no loss of momentum. An important growth in the practice of in-house post-stack processing enabled operators to concentrate upon the particular problems that applied in their areas. Advances in migration techniques and in velocity studies were aided by such in-house dedicated systems (Table 1.1). In biostratigraphy, the first half of the 1970s witnessed an enormous increase in the volume of new information (Table 1.2), and computer recording of data became mandatory. By the mid-1970s, calcareous nanoplankton provided the key to dating the Cretaceous and Danian chalks, while in the Jurassic the value of arenaceous foraminifera as environment indicators became recognized. In palynology, improvements in sample preparation enabled higher concentrations of microflora to be obtained, thus accelerating the establishment of a palynozonation based on dinoflagellates in the Jurassic and in parts of the Tertiary. Important, too, was the integration of palynological and palaeontological zonations in theTertiary sequence.

1.4.5 The period 1977–85

Activity remained generally high during this period (see Fig. 1.5B), with companies pursuing a variety of oil and gas/condensate plays in the central and northern North Sea and, towards the end of the period, returning to the southern North Sea gas province. Jurassic targets continued to provide considerable interest, but Lower Cretaceous, Palaeocene and Eocene plays became more prominent in later years. The seventh round of UK licensing, which closed early in 1981, put on offer most of the remaining prime and proved oil-exploration areas in the UK central and northern North Sea. This event, and the succeeding level of activities, underlined the fact that exploration had already reached a mature phase in much of the area. Consequently, by 1982–83, the acreage available for the Department of Energy to offer in the eighth round was largely undrilled and of a highly speculative nature with regard to its prospectivity. Exceptions were provided by

Table 1.2 Developments in geological applications to North Sea exploration, 1960 to 1996.

Sedimentology	
1960s	Heavy minerals—provenance studies; clay mineralogy (SEM, X-ray diffraction)
Mid-1960s	Aeolian/desert studies—Rotliegend palaeogeography
1970s	Rotliegend diagenesis; clay mineralogy: depth-related diagenesis models
Early 1970s	Turbidite fans (Forties Fm)
Late 1970s	Deltaic–paralic (Brent Gp) facies modelling—diagenesis; effects of salt diapirism on sediment deposition (Fulmar)
Early 1980s	Turbidite fans—Palaeocene–Eocene (Tay, Frigg Fms)
Mid-1980s	Fluvial deposits (Triassic, Carboniferous); palaeomagnetism; probabilistic image analysis; cathode luminescence
Late 1980s	Fission-track dating
Early 1990s	Sequence-stratigraphy analysis
Palynology	
1960s	Spores/pollen (Palaeozoic, Permo-Triassic, Middle Jurassic, Tertiary
Late 1960s	Spore coloration (organic maturity); chitinozoa (Palaeozoic); acritarchs (Palaeozoic, Mesozoic)
Early 1970s	Dinoflagellates (Mesozoic, Tertiary)
Late 1970s	Palynofacies
Palaeontology	
1960s	Benthonic/planktonic foraminifera (Jurassic, Cretaceous, Tertiary)
1970s	Diatoms (Tertiary)
Mid-1970s	Radiolaria (Jurassic, Cretaceous, Tertiary); calcareous nanoplankton (Jurassic, Cretaceous, Tertiary)
Late 1970s	Arenaceous foraminifera
Early 1990s	Correlation of microfossils to ammonite zones (Jurassic, Cretaceous)
Organic geochemistry	
1960s	Chemical oil and gas analyses
Late 1960s	Source-rock identification (humic, kerogenous) by macerals and maturity (spore colours, vitrinite reflectance)
Mid-1970s	Extract analysis (fingerprinting); oil typing
Late 1970s	Carbon isotopes
Mid 1980s	Rock-Eval
Non-biostratigraphic tools	
1990	Magnetostratigraphy
1992–93	Isotope stratigraphy (Sm/Nd); heavy-mineral correlations
1993	Chemostratigraphy

Sm/Nd, samarium/neodymium.

selected blocks for auction in those prime areas that had been left untaken in the 7th round, and some southern gas areas, the 1st significant offers in this area since the 4th round in 1972. The fiscal changes in the 1983 Budget induced a major investment response by the offshore industry, thereby stimulating exploration to high levels of drilling activity. As a consequence, the ninth round in 1984 attracted a record number of applications for both the 195 blocks offered on the usual discretionary terms (93 awarded) and the 15 offered for cash tender in mature areas (13 awarded).

Since 1977, exploration has pursued increasingly difficult objectives at Jurassic, Cretaceous and Tertiary levels. The seismic expression of traps and reservoir occurrences at Upper Jurassic levels is less straightforward than the Middle Jurassic horst and tilted fault blocks, or the large drape structures and salt-induced diapirs that constituted the main objectives in earlier periods. Growing experience in recognizing and drilling on the relatively modest seismic indications of small closures at the base Cretaceous level have led, however, to a number of discoveries, spread widely across the northern and central North Sea.

The Middle Jurassic play of the northern North Sea was followed by the discovery of small-scale satellites to larger developments, such as Mobil's Katrine (1977) in the Beryl Embayment.

Continuing the Brae-type play in the southern Viking Graben led to several discoveries in the 'T-Block' 16/17: Toni/Thelma (1976), Tiffany (1979) and SE Thelma (1980); effective porosity was low (10–11%) but permeability was in the 75–150 millidarcy (mD) range (Kerlogue *et al.*, 1995). Marathon, Conoco and BP extended this play into the deeper Graben in blocks 16/7 and 16/8, where the Miller Field (325×10^6 barrels) (Garland, 1993) and a number of gas/condensate discoveries have been made.

In the central North Sea, the Upper Jurassic continued to provide rewards, discoveries being made by Britoil at Clyde (1978), on the Western Platform by Shell/Esso (Gannet and Kittiwake fields) and around the fringes of the Forties and Jaeren Highs (e.g. 21/15a-2). Similar reservoirs, of shallow-marine origin, were also successfully tested by Hamilton in small fault traps adjacent to the Permian Argyll Field north of the Mid North Sea High (Duncan, East Duncan and Innes). The Piper play was further pursued in the Outer Moray Firth, leading to discoveries at Galley, Rob Roy, Ivanhoe and Glenn. A later trend was the increasing exploration for Upper Jurassic gas/condensate objectives in the deepest part of the Central Graben, where drilling units with specially equipped high-pressure stacks are needed to overcome formation pressures up to 7000 pounds per square inch (psi) above hydrostatic.

The Lower Cretaceous proved to be hydrocarbon-bearing and productive in the relatively limited area favouring turbidite sand deposition in the Witch Ground Graben and Fisher Bank Basin areas of quadrants 14 and 16. Discoveries included Gulf's Bosun (now called Kilda) find in Block 16/26 (1977) and Occidental's Claymore and Scapa fields in Block 14/19. Outside this region, with the exception of the Moray Firth area, the Lower Cretaceous has proved to be singularly poor in potential reservoir lithologies within UK waters. Oil-bearing Lower Cretaceous sands, however, have long been known onshore in the Netherlands (e.g. NAM's Schoonebeek Field near the German border, and the Rijswijk Field (Bodenhausen and Ott, 1981) in the vicinity of The Hague, which were discovered in 1943 and 1952 respectively). Offshore, Union discovered oil in their Helm (1976) and Hoorn (1980) fields, and Conoco in their K/18 Field (1980).

Exploration in the Tertiary increased in popularity dramatically as several large (200–400 × 10^6 barrel) accumulations were discovered. Reservoirs and traps at this level are often difficult to recognize and define on seismic, and the significance of shows at these levels was in some cases overlooked during earlier phases of exploration. Accumulations at shallow levels are prone to the loss of lighter fractions through leakage, biodegradation or washing by water or gas. It is also possible to have repeated charge from a kitchen of evolving maturity. Tertiary accumulations consequently range from gas/condensate to heavy oil, and some (e.g. Frigg) have gas above heavy oil. Heavy to light oil was discovered in the Outer Moray Firth—Balmoral (Sun, 1975), Glamis (Sun, 1982), Cyrus (BP, 1979), Mabel (Phillips, 1975), Drake (Superior, 1982)—and at the edge of the East Shetland Platform—Bressay (Chevron, 1978), Emerald (Sovereign, 1981). The Alba (Chevron, 1984) heavy-oil discovery in the Outer Moray Firth involved stratigraphic trapping in Eocene turbiditic sands. Horstad and Larter (1997) suggest that the gas charge to the Middle and Upper Jurassic reservoir of the Troll Field preceded that of the oil, which came from two different sources, in the west and from the north.

Activity in the UK sector of the southern gas basin remained depressed until 1982, when, in response to an economic improvement in the prospects for gas and in anticipation of the eighth-round offering, a significant jack-up drilling campaign resumed. One result of this renewed drilling activity was the discovery of Bunter gas in Block 43/13a (Esmond). Hamilton Brothers brought this field and two earlier Bunter discoveries, Forbes (43/8) and Gordon (43/15 and 43/20), into production in mid-1985.

Renewed activity brought the traditional Rotliegend play to a mature state. Many of the earlier discoveries in the Sole Pit area, such as Barque, Clipper and Galleon, where reservoir deterioration during former deep burial had imposed severe doubts on gas deliverability, were appraised during this, and brought into production during the succeeding two periods, using the economically enhancing benefits of horizontal drilling. Improved control on seismic velocity gradients in the complexly deformed overburden, coupled with enhanced resolution at the objective level, has resulted in a better depth conversion of Rotliegend structures located adjacent to the Sole Pit High (BP's Ravenspurn, Cleeton and Hyde; Conoco's Valiant, Vulcan and Vanguard; Arco's Yare, Thames and Bure) and most of the fields in this area are now in production. Further north, in the Silverpit Basin, where the Rotliegend is in a non-reservoir, shale/evaporite facies, attention focused on the underlying Carboniferous section (see Chapter 4). Several accumulations were appraised at this time; Ultramar's Markham Field came on-stream in 1992 and Conoco's Murdoch Field in 1993, but no development has been announced for the Shell/Esso Schooner and Ketch discoveries).The Carboniferous play attracted a similar level of interest in the adjacent J, K and L areas of the Netherlands.

Exploration activities in the Norwegian sector were determined by the relatively limited number of blocks awarded during a rapid succession of rounds (see Fig. 1.3). A major discovery was made at Gullfaks in Block 34/10, specially granted in 1978 to a group of Norwegian companies, with Statoil as operator. Gullfaks is a complex and highly faulted structure, situated on the Tampen Spur, with the Brent and Statfjord sandstones as main reservoirs. Recoverable reserves are of the order of 1.6 × 10^9 barrels (255 × 10^6 m^3) of oil and 3.5 × 10^{12} ft^3 (100 × 10^9 m^3) of gas, including the Gullfaks Alpha structure to the south of the main field in the same licence block.

The fourth round marked an important milestone in the exploration history of the Norwegian North Sea, with the discovery of the giant Troll gas- and oilfield (recoverable reserves of 44 × 10^{12} ft^3 of gas and 200 × 10^6 barrels of oil) made in 1979 in Block 31/2, operated by Shell. The Troll hydrocarbon accumulation, located in 350 m of water, is contained in Upper Jurassic shallow-marine sands in a series of fault blocks on the northern part of the Horda Platform. The gas occurrence, underlain by a thin, heavy-oil leg, is well marked by an extensive 'flat spot' recognizable on seismic (Brekke *et al.*, 1981; Rønnevik and Johnsen, 1984). The second important hydrocarbon accumulation in fourth round acreage was Norsk Hydro's Oseberg discovery (1979) in Block 30/6, a large easterly-dipping Middle Jurassic fault block on the eastern margin of the North Viking Graben, to the west of the Horda Platform (Badley *et al.*, 1984). Field reserves are estimated at around 1.5 × 10^9 barrels (250 × 10^6 m^3) of oil with 2.8 × 10^{12} ft^3 (79 × 10^9 m^3) of gas. In Norwegian waters, the Snorre Field, mostly located in eighth-round acreage awarded in 1984 (but discovered by well 34/4-1 in fourth-round acreage in 1979), has proved the oil potential of Triassic red beds near the northern end of the Tampen Spur (see also Chapter 7).

To the north of 62°N, the Norwegian fifth and seventh rounds (see Fig. 1.3) offered acreage along the Atlantic Margin, indicating a shift in emphasis as the northern North Sea began to reach a more mature state of exploration. Light oil and gas were discovered on Haltenbanken, while the Hammerfest Basin contains predominantly gas (Campbell and Ormaasen, 1987).

In Denmark, despite the breaking of its long-held monopoly, the DUC group, led by A.P. Moeller, is still the only company with established hydrocarbon production. The producing fields Dan, Gorm, Sjkold, Tyra, Rolf, Roar and Svend (the last two in production in 1996), all with Chalk reservoirs, contain about 63% of the total hydrocarbon reserves, which are estimated at some 1585 × 10^6 barrels (252 × 10^6 m^3) of oil and 5.1 × 10^{12} ft^3 (145 × 10^9 m^3) of gas (Energistyrelsen, 1997).

West of Shetland, the giant (straddling five blocks) but shallow (less than 2000 m) Clair Field was discovered in 1977. With several billion barrels of oil in place within the fractured Carboniferous, Devonian and Precambrian (Lewisian gneiss) reservoir, it represents the largest known oilfield on the continental shelf of north-west Europe. The oil is of intermediate gravity (25° API) and is possibly a mixture of biodegraded and late-flush lighter oil (see Chapter 11). The first well (206/8-1A) produced on test at a rate of 1500 barrels per day, but, because of a generally poor reservoir, this was never matched in the 10 appraisal wells drilled up to 1985. Recoverable reserves are currently estimated to be 400 × 10^6 barrels (a recovery factor of only 7%). Development had to be delayed until a method of achieving commercial rates of production could be found.

Increased cooperation between the holders of the five licences involved has been discussed by Coney *et al.* (1993), and may lead to eventual development; a prolonged flow test on appraisal well 206/8-10Z was announced for the autumn of 1996. Although early drilling results were disappointing, the field did indicate that oil had been generated right at the Atlantic Margin of the continental shelf, and that more productive reservoirs might be found elsewhere in that general area.

1.4.6 The period 1986–92

A worldwide glut of oil brought about a dramatic decrease in the price of oil in 1986 (see Fig. 1.6). This was reflected during the UK offshore tenth round by fewer companies seeking acreage, all of which was located in mature areas; only 51 of the 127 blocks on offer were awarded. However, the eleventh round, a larger round with 212 blocks on offer, saw an upturn of interest, with 125 applications being received, matching the interest generated by the seventh round. Moreover, the drilling commitments for this round were on average higher than in any previous round: 105 licences, involving 115 blocks or part blocks, were awarded. The government altered conditions slightly for this round; where no field development is announced during the first 12 years of the 30-year period following 50% relinquishment, the licence must be surrendered. Moreover, along with their applications, companies were asked to give details of activity in blocks awarded during the first to fifth rounds. Both of these measures were aimed at discouraging inactivity in licensed acreage. The twelfth and thirteenth rounds were both held in 1991, the former offering the usual acreage spread, while the latter (frontier) round concentrated on the Shetland area. For this round, the minimum relinquishment varied between 0 and 75%, depending on the number of wells drilled, and, in recognition of the fact that there may be no drillable prospects, the minimum work commitment was limited to seismic surveys.

The 1986 oil-price drop was followed in 1987 by a global stock-market collapse. The combined effect was a reduction of industry market values, which stimulated some companies to acquire proved oil assets and to strengthen their acreage positions by taking over other companies. Thus several well-known names were absorbed in 1988: Britoil (by BP), Acre Oil (British Gas), Blackfriars Oil & Gas (Ultramar), RTZ Oil & Gas (Elf), Texas Eastern (British Gas, Enterprise, Amerada Hess), Thomson North Sea (Lasmo), Tricentrol (Arco) and Whitehall Petroleum (Hess).

Declining revenue and uncertainty with respect to future oil prices led the industry to a careful review of development economics and the design of slimmed production costs. Oilfields that came into production in the UK sector were Petronella and Balmoral (1986), Clyde, Ness and North Alwyn (1987), and North Brae and Eider (1988). Oil prices hardened with the Gulf War of 1990–91, partly because of the threat to global supplies of Saudi Arabian oil, but also because Kuwait's production capacity was temporarily destroyed; the export of oil from Iraq was banned and was allowed to resume only late in 1996, and then only in limited volumes.

Exploration for the localized Tertiary deep-water sand fans that produce gas in the Frigg Field resulted in Ranger's 1986 discovery of heavy oil and gas in well 3/30a-3, on trend with and north of the field. This stimulated exploration to the south of Frigg, and in 1987 medium-gravity oil was discovered in the Gryphon (Kerr McGee) and Forth (BP) fields of Quadrant 9. Premier, in 1986, found Elgar, analogous to and south-east of Alba. Shell/Esso made a number of oil and gas discoveries in Eocene sands of the Gannet area along the western margin of the Central Graben. Early in 1988, Enterprise discovered light oil (36° API) in the Nelson Field, south-east of and in the same reservoir as the Forties Field.

At Jurassic levels, Mobil found Ness close to Katrine in the Beryl Embayment in 1986, while the 'T-Block' became the subject of renewed interest in 1988, when Agip's well 16/17-16 (Toni Field) tested unexpectedly high flow rates, totalling 26 000 barrels/day.

In the southern North Sea, Ultramar tested 60×10^6 ft^3 gas/day from their J/16-1 well in 1987 on the extension of the Markham Field into Dutch waters (Myres *et al.*, 1995). Production developments have included the installation of new platforms on the Leman Field, and the coming into production of several fields between 1986 and 1992: Sean, Arco's Thames, Yare and Bure, Audrey, Della, Conoco's North and South Valiant, Vanguard and Vulcan, and BP's Cleeton, Ravenspurn South and Amethyst, all of which are described in Abbotts (1991); other small fields include Anglia, Barque, Clipper and Markham. Since 1985, the Rough Field has been used for gas storage (Goodchild and Bryant, 1986; Stuart, 1991), ready for peak-demand periods.

Drilling results in Norwegian ninth- and tenth-round acreage have in general been disappointing, except for a few small Jurassic discoveries, such as Elf's Frøya Field (1987). More is expected from the 11 licences awarded in 1988 in the first part of the twelfth round, which followed the significant Phillips discovery in first-round Block 2/7, where Jurassic reservoirs beneath the Eldfisk Field produced 3560 barrels/day (566 m^3/day), confirming the extension into Norwegian waters of the high-pressure Central Graben Jurassic–Triassic play.

In Denmark, the combined Chalk and Zechstein reservoirs of the Rolf and Dagmar fields went on-stream in 1986 and 1991, respectively, while the purely Chalk Kraka field also came into production in 1991 (for further details, see Megson, 1992).

By the end of 1988 almost 9×10^9 barrels of oil and 24×10^{12} ft^3 of gas had been produced from UK waters; 33 oilfields and 23 gasfields were in production. Up to the end of the same year, Norway had produced around 3×10^9 barrels of oil and 10.5×10^{12} ft^3 of gas. Production continued to rise, although the rate was restrained in efforts to cooperate with OPEC. With the December 1988 start-up of Oseberg, the waterflood-enhanced Ekofisk production and the continuing development of Gullfaks, 1989 production was set to increase by about a third to nearly 1.5×10^6 barrels/day. The Netherlands had produced 105×10^6 barrels of oil and just under 6×10^{12} ft^3 of gas, while Denmark produced 186×10^6 barrels of oil. With regard to UK gas developments, increasing realism over the prices paid to producers and the end of British Gas's monopoly as a gas

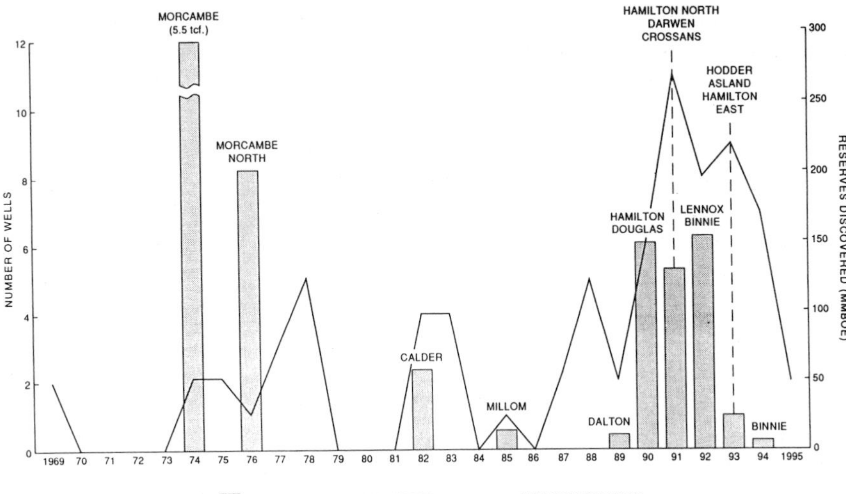

Fig. 1.15 Eastern Irish Sea basin, drilling activity and discoveries (courtesy of Steve Pickering, BHP Petroleum).

purchaser rekindled the industry's interest in finding and producing gas. In the eastern Irish Sea, for instance, renewed drilling activity in 1989 (Fig. 1.15) led to the discovery of several oil- and gasfields (e.g. Hamilton, Douglas and Lennox) over the succeeding 5 years, mostly to the south of the giant Morcambe gasfield.

Oil prices (see Fig. 1.6) are influenced mainly by developments in international supply and demand. Expectations about future price trends began to alter radically during 1981, when the world's oil demand dropped well below the level of 1979. By the end of 1981, oil was in such abundant supply internationally that its price was becoming increasingly slack. These developments undermined the assumption, which had dominated exploration strategy in the 1970s, that oil prices would continue to rise in real terms for the rest of the twentieth century. The dramatic fall in the price of oil in 1986 led much of the North Sea industry to examine closely its state of health. Some projects were put on hold and the costs of others were significantly reduced. During the late 1980s, the oil price was erratic and unpredictable, and in this uncertain climate the industry learned to plan ahead by simultaneously considering various price scenarios. Even so, interest in the North Sea remained high, as shown in the UK by the positive response to the eleventh round of licensing, the highest level of drilling, in 1988, since 1985, and the commissioning of 10 oilfields and three gasfields during 1990–92. Only four Norwegian fields were commissioned during this period, but they included the giant Snorre Oilfield, exceeding 700×10^6 barrels (116×10^6 m^3); see Hollander (1987).

1.4.7 The period since 1993

Perhaps the most significant event at the end of 1992, at least for the UK, Irish and southern Norwegian offshore areas of the Atlantic Margin, was the discovery of economically producible oil in Palaeocene reservoirs west of Shetland, with the well 204/24a-2 (Foinaven, 200×10^6 barrels; Cowper *et al.*, 1995). This discovery was followed in 1993 by that of Schiehallion (Block 204/25a, 340×10^6 barrels of oil and 500×10^9 ft^3 of gas); in 1994 Conival, with unconfirmed reports of around 5×10^{12} ft^3 of recoverable gas; and in 1995 Loyal (85×10^6 barrels of oil and 130×10^9 ft^3 of gas). In 1995, 14 exploration and nine appraisal wells were drilled west of Shetland. These discoveries followed almost 25 years of exploration activity, including the drilling of 90 wells. Well-developed structures on the Norwegian Atlantic Margin (Doré and Lundin, 1996) are considered likely exploration targets for Late Cretaceous or Early Cenozoic hydrocarbons.

Apart from the currently unexploited giant Clair Field (Carboniferous–Precambrian reservoir) and the small Victory gasfield (Lower Cretaceous reservoir), both drilled in 1977, almost a decade was to pass before Shell produced gas at 25×10^6 ft^3 per day in 1986 with well 206/1-2. In view of its gaseous nature, its location, the water depth (500 m) and the economic climate of the time, Shell relinquished the acreage; the small hydrocarbon accumulation is known by the present licence holders as Torridon (Texaco) or Laggan (Total). Three more small and currently undeveloped accumulations were found in 1990 at Strathmore (40×10^6 barrels of oil in a Triassic reservoir), and in 1991 Alligin (Palaeocene) and Solan (Upper Jurassic), each with recoverable reserves estimated at 50×10^6 barrels of oil. Although oil had clearly been generated and trapped in the area (various aspects of the area west of Shetland were discussed in several papers within Brooks and Glennie, 1987, e.g. Bailey *et al.*, 1987; Duindam and van Hoorn, 1987; Hitchin and Ritchie, 1987; Mudge and Rashid, 1987; Nelson and Lamy, 1987; and in Irish waters by Croker and Shannon, 1995; Shannon, 1996), the main drawback had been poor reservoir quality, considerable water depths (200–500 m) and, apart from Clair, the small size of the discoveries. Foinaven, Schiehallion and Conival have changed that relative gloom to a currently optimistic outlook.

Applications for acreage in the UK fourteenth round in 1993 were in three parts, and closed before the discovery of Foinaven was generally known. One part included acreage west of Shetland and had only a moderate response, while another, limited to the South-west Approaches, had a poor response. The fifteenth round, involving the central and southern North Sea, had only a moderate number of applications, but awards were made of about half the acreage offered in the wide-ranging sixteenth round in 1995, which again included the area west of Shetland.

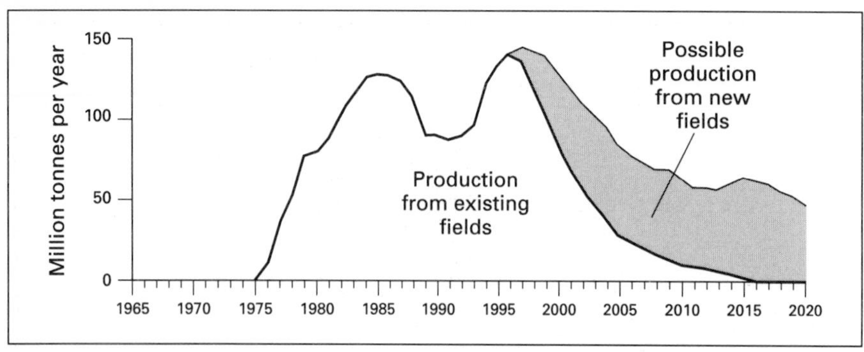

Fig. 1.16 UK oil production from 1975 and prediction to 2020 (courtesy of UK Offshore Operators Association).

Despite the less favourable tax conditions for exploration imposed in 1993, oil companies would go out of existence if they failed to acquire acreage to explore and, hopefully, in which to find new hydrocarbons.

Because of global oversupply, UK oil production peaked in 1985 at just over 127×10^6 t/year, and then suffered a sharp decline to two-thirds of that volume by 1991. Some of the drop in production in 1988 resulted from the closing in of fields that had produced through the destroyed Piper Alpha facilities (see p. 26), and the loss of production through the Fulmar system and, early in the next year, the loss of export through the Cormorant facilities. Following the Gulf War of 1990–91, production recovered to a volume of 139×10^6 t in 1996 (Fig. 1.16). After rising to a new peak in 1997, production is expected to decline steadily into the early part of the twenty-first century, unless it is replaced by oil from undeveloped and as yet undiscovered fields—see Daly *et al.* (1996) and, for a global outlook for the next 50 years, Kassler (1996). In part, these figures reflect the conflict between production from new fields coming on-stream and the diminishing contribution of giant fields, such as Forties and Brent, which are already in decline, although new recovery techniques have allowed estimates of their ultimately recoverable reserves to be revised upwards.

With the established oilfields reaching the end of their plateau production or even starting into their inevitable decline, the British PRT was altered in the 1993 Budget to help rectify the situation. These changes favoured the development of otherwise marginally economic fields that had already been discovered, to the detriment of exploration activity. The changes naturally resulted in a downturn in exploration activity and the laying off or redeployment of exploration staff. From 1993 to 1996, 17 mostly small UK oilfields and eight gasfields were commissioned. Once found, the economic need to develop hydrocarbon accumulations also applied to Norwegian waters, where nine oil- and an equal number of gasfields have been commissioned since 1992, with others planned for completion by the end of the century.

In Danish waters, Valdemar came on-stream in 1993; it will produce oil mainly from a Lower Cretaceous (Barremian) chalk, the first development at this level in the North Sea. Roar and Svend (which includes the former North Arne and Otto) came on-stream in 1996. The development of several other fields (Harald, Adda, Elly and Alma) is planned for the end of the century or early in the next. Gert, which has an Upper Jurassic reservoir, will be developed with the adjacent accumulation across the Median Line in Norway. A new departure for Denmark is the 1995 discovery (well 5604/20-1) of oil in a Palaeogene siliciclastic reservoir on the northern flank of the Ringkøbing High. The reservoir sands were sourced from the east and occupy an earlier scoured channel across the Chalk. This oil is in a combination stratigraphic trap with halokinetic structuring, and involves relatively long-distance migration (in excess of 20 km) from the graben margin.

During 1995, a total of 98 exploration and appraisal wells were drilled on the UK continental shelf, and these resulted in seven significant discoveries. And 244 development wells were drilled or commenced during the year, exceeding the previous highest during 1994 by 45 wells (Department of Energy Brown Book, 1996). The industry was becoming more active after several years in the doldrums. The Department of Energy estimated that UK's proved and probable reserves amounted to 3020 Mt (22×10^9 barrels) of oil and 1485 m^3 (52×10^{12} ft^3) of gas, with possible (undiscovered) additional reserves falling in the range of 990–4485 Mt ($7–33 \times 10^9$ barrels) of oil and $395–1412 \times 10^9$ m^3 ($14–50 \times 10^{12}$ ft^3) of gas (see also Daly *et al.*, 1996).

Some basic data on the oil- and gasfields of the North Sea and adjacent marine areas that were producing or under development in 1996 are presented in the appendix to this chapter. The locations of many of these fields are shown in Fig. 1.1 and the main basins that contain them in Fig. 1.17. The fields data provide a basis for the analysis of ultimately recoverable reserves, broken down by geological periods and by countries bordering the North Sea, given in Fig. 1.18. From this figure it is clear that the bulk of the hydrocarbons are found in Jurassic reservoirs, 54% of which occur in Norwegian waters and 46% in the UK offshore. On the other hand, the great bulk of Rotliegend gas occurs onshore in the Netherlands, while the UK has the greater preponderance of offshore Rotliegend gas. Note the differences in onshore and offshore field sizes implied by the relevant number of discoveries. Many of the factors affecting the discovery and development of the various hydrocarbon reservoirs are discussed in Chapter 12.

1.4.8 Future trends

With the possible exception of some of the large undrilled structures of the Atlantic Margin area, it is probable that in the adjacent North Sea (*sensu stricto*) most of the larger oilfields have already been discovered, and much of the area is now widely regarded as being at a mature stage of

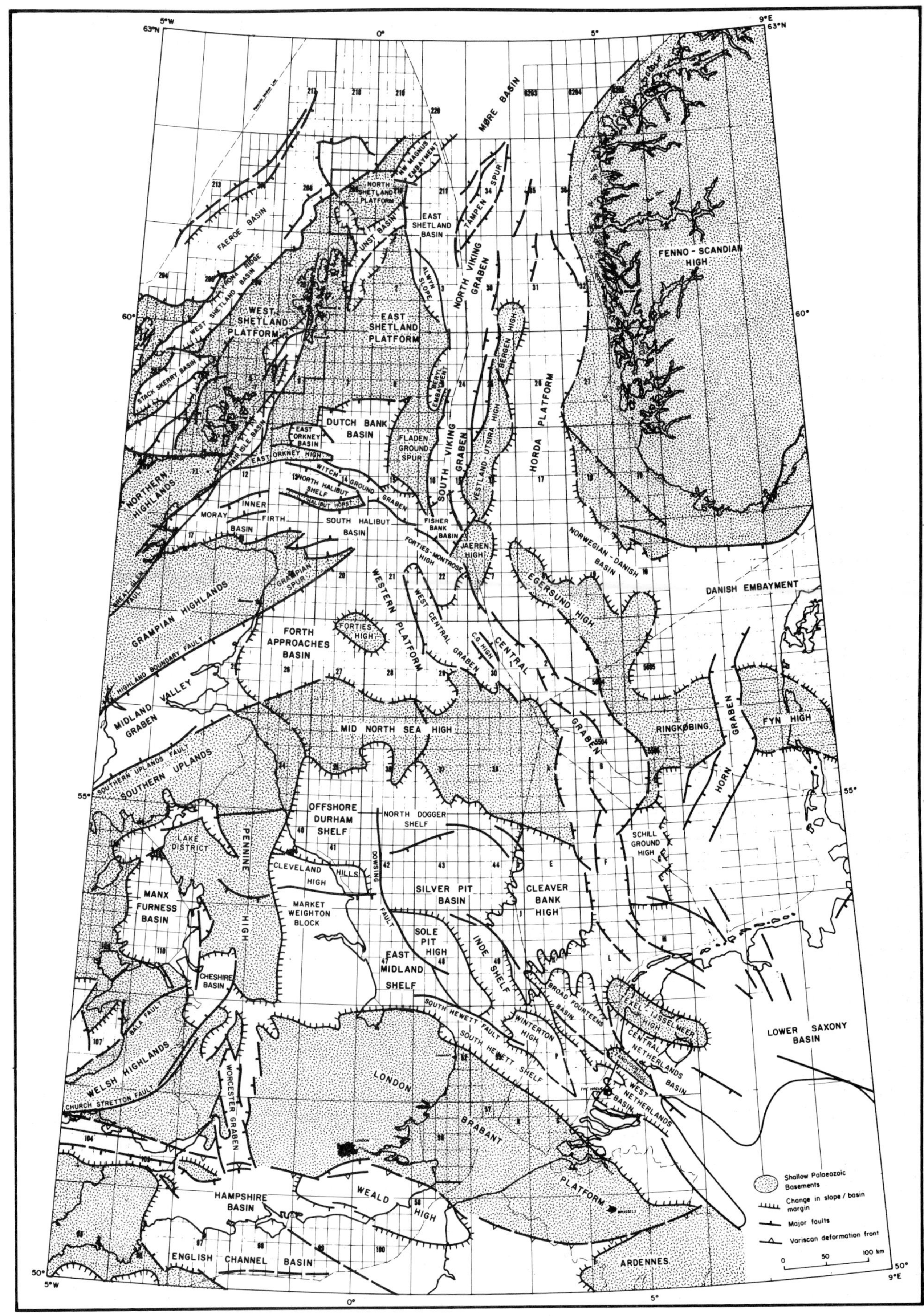

Fig. 1.17 The main prospective basins of the North Sea.

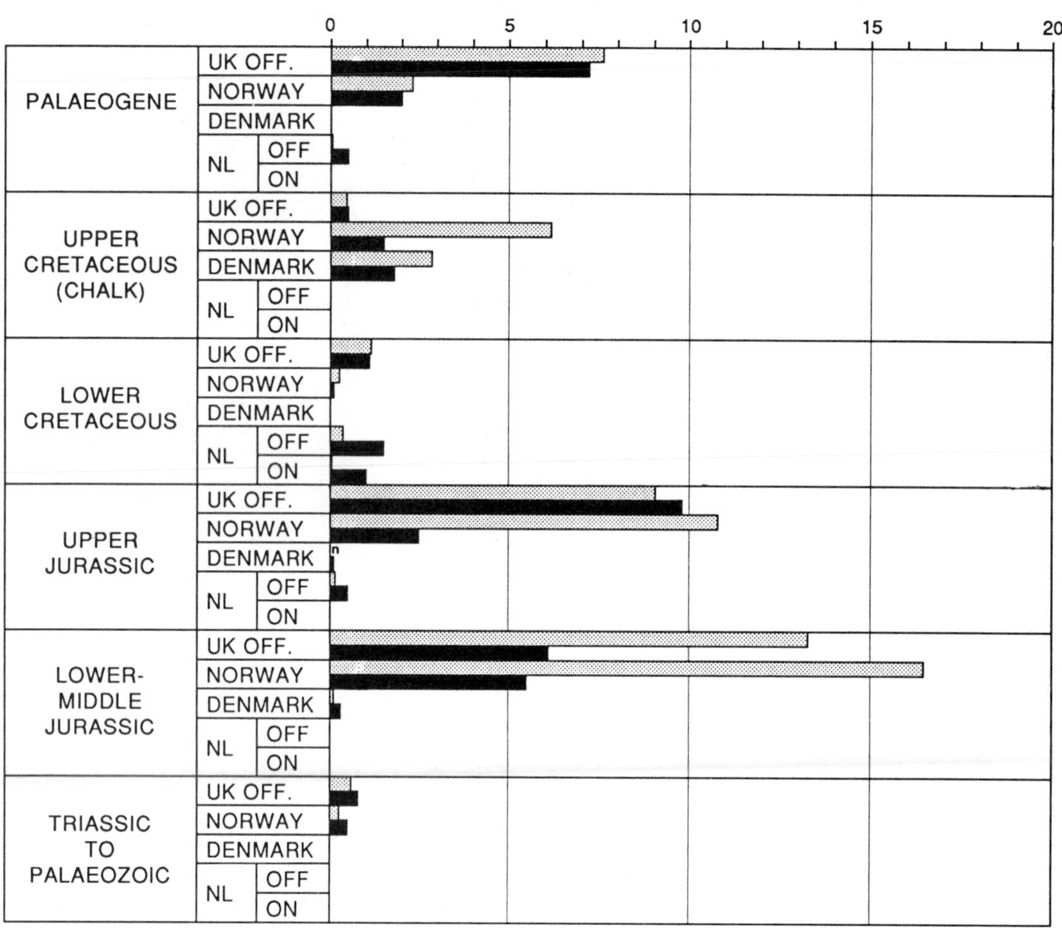

GAS OF CARBONIFEROUS ORIGIN IN BILLIONS OF BARRELS OF OIL EQUIVALENT

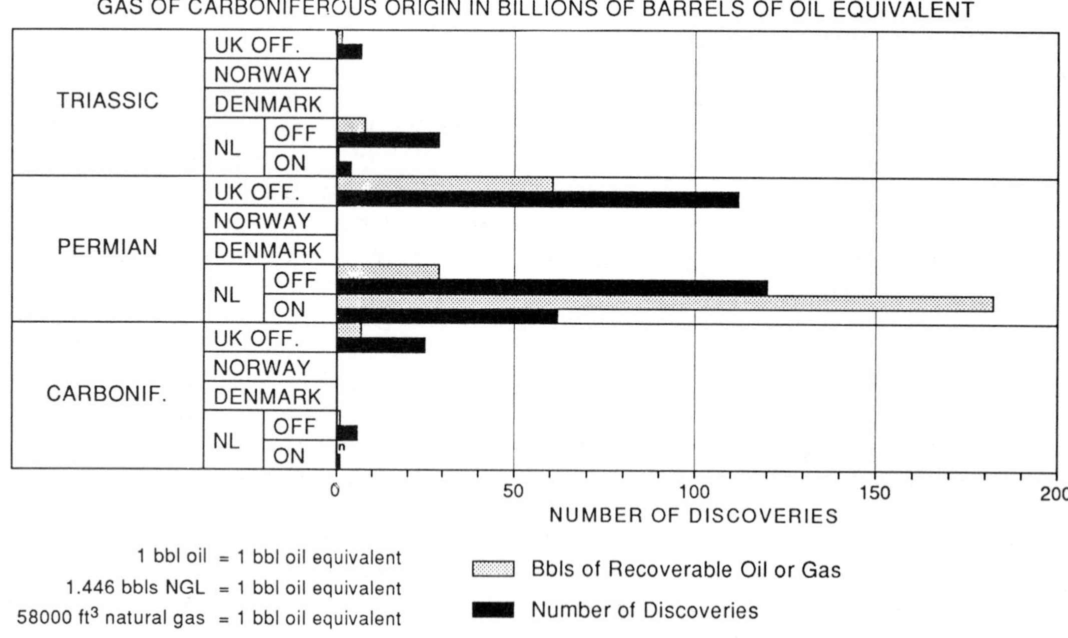

1 bbl oil = 1 bbl oil equivalent
1.446 bbls NGL = 1 bbl oil equivalent
58000 ft^3 natural gas = 1 bbl oil equivalent

Bbls of Recoverable Oil or Gas
Number of Discoveries
n Negligible

Fig. 1.18 Ultimate recoverable reserves in the North Sea by age and by country in billion (10^9) of barrels of oil derived from Jurassic source rocks and oil-equivalent gas of Carboniferous origin. Note how the Carboniferous gas is confined to the UK and Netherlands (NL) areas, with the bulk on the Dutch onshore (mostly Groningen). Based on Spencer *et al.* (1996).

exploration. Even so, several recent finds have been of the order of 100×10^6 barrels. Figure 1.19 illustrates the progress of exploration in the UK sector of the North Sea and indicates that the rate at which new reserves are found is quite healthy. Indeed, there is a growing confidence that much remains to be discovered in the North Sea.

It is noteworthy that, while discoveries continue to be

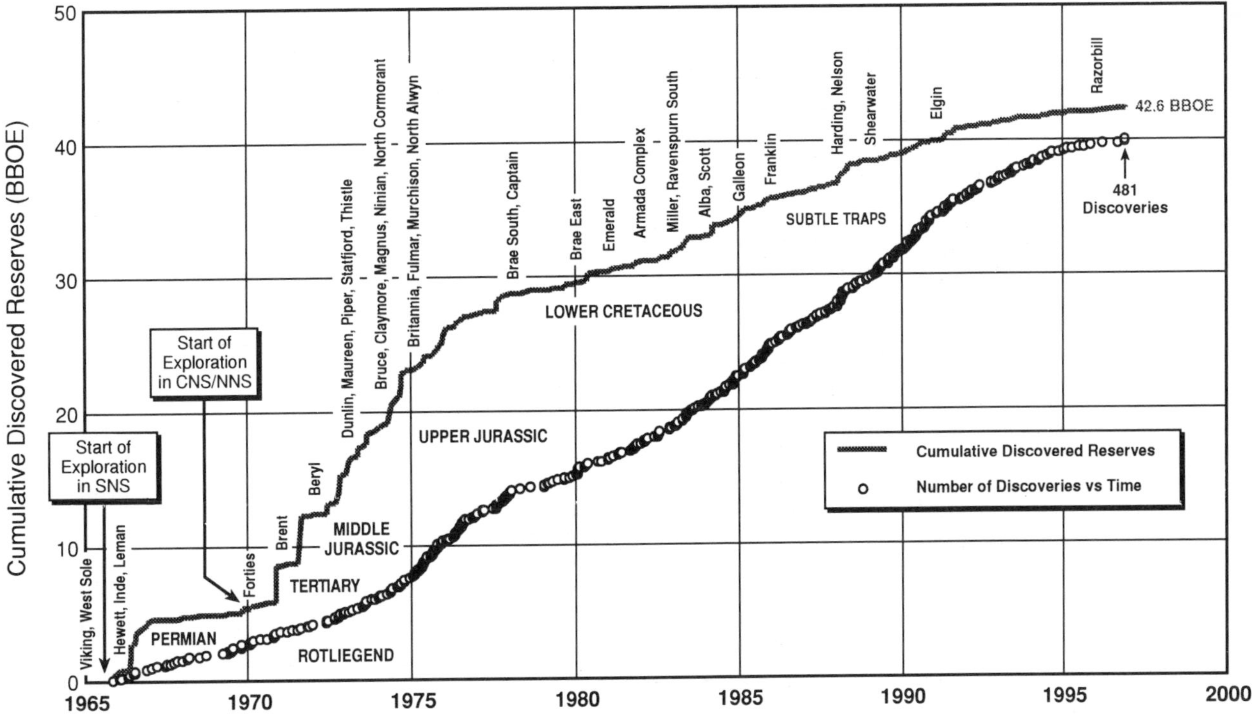

Fig. 1.19 Creaming curve of UK North Sea recoverable reserves, 1965–96 (modified from Shell UK data).

made in proved plays, new plays also surface, and the industry is not immune from surprises. In a historical review of North Sea exploration, Bowen (1989) emphasized that accepted exploration concepts and old data constantly need to be re-examined; such an exercise led to the discovery of the Nelson Field. As a corollary, Dean (1996) describes how, for a number of reasons, some fields (including Nelson) were not discovered until long after the original well on or close to the structural crest had been abandoned with oil shows.

Increasing attention is being focused upon so-called 'subtle' traps (e.g. Angus; Hall, 1992), usually involving stratigraphic closure and requiring very careful seismic definition, and upon satellite accumulations close to the infrastructure of developed fields.

The need to keep costs as low as possible and yet maintain production levels far into the future has led to two developments: cooperation between companies to make maximum use of existing infrastructures; and exploration for and development of smaller fields with relatively small recoverable reserves. The Shell/Esso Kingfisher Field, for example, with recoverable reserves of 56×10^6 barrels of oil and 368×10^9 ft^3 of gas, will be developed during 1997. Production will be treated on the Marathon-operated Brae B platform, with the export of oil and condensate through the Brae–Forties pipeline system. The gas from another Shell/Esso field, Schooner (Carboniferous reservoir), which came on-stream in 1996 and is sold on the deregulated market, is exported via the Conoco-operated Murdoch and Caister pipeline to the Theddlethorpe gas terminal, in Lincolnshire; there, its flow will be controlled by Conoco on Shell's behalf.

To realize the change in scale of development between many modern fields and their predecessors, compare the Kingfisher reserves with the $2.25–2.50 \times 10^9$ barrel ($359–398 \times 10^9$ m^3) Brent or Forties fields. The capital cost per barrel of oil recovered from these smaller fields may be higher than it is for the larger and established fields, but they have to be economically viable. Apart from the 1993 advantageous changes to the British PRT, many technological innovations help contain the costs associated with development of these new fields.

New recovery techniques are ensuring that a greater proportion of the oil or gas in place is recovered. The use of 4D time-lapse seismic analysis (Watts *et al.*, 1996) allows evaluation of the volume of oil or gas in place within reservoirs through lateral prediction (Hartung *et al.*, 1993). Also, the costs of developing new oilfields are being reduced by engineering advances, such as floating platforms, 'single-lift' facilities and subsea production systems (BP's SWOPS system, Shell/Esso's Underwater Manifold Centre and BP's diverless production system (DISPS)). Inherently low permeabilities have previously rendered many hydrocarbon accumulations uneconomic. This situation has been increasingly resolved since the late 1980s by the widespread use of horizontal drilling to achieve greater exposure of the well bore to the producing horizon.

New techniques can extend production in some fields far beyond their planned life. On Brent, depressurization, starting in 1997, will convert it from an oilfield to a gasfield by 2005. Before that happens, redevelopment will allow the production of an additional 350×10^6 barrels of oil before reduced reservoir pressure prevents further recovery. Careful study of the slumped crestal area of the Brent Field, based on 3D walkaway vertical seismic profiling (WVSP) and structural modelling, shows that 80×10^6 barrels of oil should be produced from this previously neglected area (Coutts *et al.*, 1996; van der Pal *et al.*, 1996); for a Central Graben example, see Arveschoug *et al.* (1995).

On a much smaller scale, and especially for single-well

and other small satellite fields, unmanned production facilities are becoming more widespread. Other new innovations include the use of coiled-tubing drilling, often in a sidetracked well. In a similar vein, a downhole pump in a sidetracked well, completed using coiled tubing, has boosted oil production from the small (originally 50, now 100×10^6 barrel) Auk Field, which first came into production in 1976. New horizontal wells are reaching previously untapped oil, thereby increasing recoverable reserves via existing platforms (see also Follows, 1997). Gannet will have two additional single-well satellites tied to it. Gannet E, with recoverable reserves of 23×10^6 barrels of relatively heavy crude, will have a downhole pump and a highly insulated pipeline system to keep the oil hot until it reaches the Gannet platform; the lighter crude of Gannet F (recoverable reserves of 19×10^6 barrels) will not need such pumps.

Given that ever smaller fields need to be explored and developed, advances are still being made in the application of seismic methods, in the field both of acquisition and of processing. On the acquisition side, 3D seismic coverage is now firmly established, allowing a much more detailed structural interpretation and understanding of the subsurface geology (Karnin *et al.*, 1991; Eggink *et al.*, 1996); indeed, Epting (1996) states that in the north-east Netherlands, the use of 3D seismic for exploration in the late 1980s reversed several years with declining volumes of newly discovered gas and resulted in the discovery of 5×10^{12} ft^3 of new Rotliegend gas within 5 years. This is of particular importance also for appraisal and development drilling. Lateral prediction and direct lithology determination from seismic data, together with interactive interpretation systems, provide a more cost-effective way of successfully predicting hydrocarbon fill and stratigraphic objectives (Bunche and Dromgoole, 1995; Barrett *et al.*, 1995). Three-dimensional surveys, repeated at intervals of 2–5 years, are now beginning to be used successfully to monitor the level of the oil–water contact in fields and to detect areas of bypassed oil production (Johnstad *et al.*, 1993). Improvements in offshore position fixing, critical for relating well site to seismic data, have made parallel progress.

Seismostratigraphic methods have an increasing part to play in the future. In the balance of remaining prospects for hydrocarbons, a stratigraphic component of trapping is more commonly recognized than used to be the case. In this respect, McGovney and Radovich (1985) underlined the importance of a rigorous examination of seismic data by providing a better understanding of the internal geometry of the Frigg fan complex. Seismostratigraphy, and its refined palaeontologically controlled basin-wide application to the more embracing sequence stratigraphy, began to dominate some aspects of geological interpretation in the late 1980s and early 1990s, first for understanding the relatively unfaulted Palaeogene sequences (Stewart, 1987; Vining *et al.*, 1993) and then for resolving the more complex Lower Cretaceous and older sequences (Partington *et al.*, 1993a; Steel, 1993; and others in the same volumes). Other non-biostratigraphic tools have also been developed: heavy minerals are now used for correlation within unfossiliferous continental sequences (Jeans *et al.*, 1993; Morton and Berge, 1995), as well as for deducing the provenance of sediments; and magnetostratigraphy (?1990), isotope (samarium/neodymium (Sm/Nd)) stratigraphy (Mearns, 1992) and chemostratigraphy (1995) are becoming increasingly valuable tools, especially in red-bed sequences.

Further developments in sedimentology, following up on the vast knowledge gained in the North Sea (see, for example, Johnson and Stewart, 1985), require even more to be integrated with seismostratigraphic interpretation to provide a better qualitative and quantitative reservoir prediction on both regional (exploration) and field (production) scales.

Deeper, older objectives in the Carboniferous and Devonian are relatively unexplored, and in several areas they are likely to claim increasing attention as the prime areas approach an even more mature stage of exploration. Although few exploration wells drilled in frontier regions, such as the Unst Basin (Johns and Andrews, 1985), the offshore continuation of the Midland Valley of Scotland (Forth Approaches), the Mid North Sea High and the Norwegian–Danish Basin, have been successful, the limited stratigraphic penetration does not allow a full condemnation of the potential of these areas (see Fig. 1.17).

Since the start of North Sea exploration, a great change has taken place in our understanding of its structural evolution. To account for the development of the North Sea graben systems, working hypotheses changed from the essentially static pre-plate-tectonic structural styles invoked during the 1960s and early 1970s to a debate involving the relative merits of crustal extension in keeping with the pure-shear model of McKenzie (1978) or of Wernike-style listric faulting (Gibbs, 1984a). With the improving quality of 2D seismic, followed by its 3D successor and the evolution of sequence stratigraphy, the earlier imaginative interpretations became much more constrained. Current hypotheses suggest that Mesozoic structural styles are controlled by a regional framework of basement rocks and inherited lines of weakness (Coward, 1990, 1993), and that the occurrence of inversion (Williams *et al.*, 1989) and pull-apart structures indicates that a degree of strike–slip movement must have been present (Bartholemew *et al.*, 1993), which was probably actuated by relative movement of Atlantic and Tethyan crustal plates and their effects on Alpine and Pyrenean orogenies and the evolution of the Atlantic Margin (Ziegler, 1988). These topics will be discussed in greater detail in Chapter 2.

The year 1988 was marred by incidents that resulted in loss of life and reduced production. The Piper Alpha tragedy, which resulted in the loss of 167 lives, mid-year, was later followed by the fire on the Ocean Odyssey and the loss of a further life. At the end of the year, production was lost from the Fulmar system and, early in 1989, production through the Cormorant facilities had to be shut down. These incidents refocused attention upon safety and the environment. The Department of Energy has stringent safety requirements and carries out rigorous inspection programmes involving all offshore activity. There is continued concern in some companies over the severe overpressures encountered by deep drilling in the Central Graben; for reasons, see Holm (1996). The successful development of such fields as Fulmar, however, has demonstrated that modern drilling procedures and equipment are capable of controlling such pressures and allowing the successful development of these fields.

Offshore installations and pipelines eventually outlive their purpose, and the Petroleum Act 1987 requires approved and funded abandonment programmes. Two installations, satellite structures in the West Sole and K13 fields, were removed from the North Sea with little comment. However, attempts in 1995 by Shell UK, with government approval, to dispose of the Brent Spar (a floating oil-storage and tanker-loading facility) by sinking it in the deep Atlantic, brought a somewhat hysterical (and dangerous) reaction from poorly informed 'Green' environmentalists, who boarded it at sea from a helicopter while it was being towed to the disposal site (an act of piracy). This resulted in the Spar being 'stored' in a Norwegian fjord until a new method of safe disposal could be devised. The oil industry as a whole does its utmost to prevent needless pollution. A balanced review of day-to-day pollution in the North Sea (fluvial effluents, as well as that caused by the industry) is given by Tromp (1996).

1.5 Concluding remarks

As a reminder, Tables 1.3 and 1.4 list some of the important events and engineering factors that helped make the North Sea area a major producer of hydrocarbons in the latter part of the twentieth century (for important seismic and geological data, see Tables 1.1 and 1.2).

A proper petroleum-geological understanding of the natural resources of the North Sea is crucial in judging what the policies should be regarding their future development. The debate concerning the eventual total recoverable hydrocarbons that took place in the mid-1970s between optimistic statisticians, on the one hand, and a perhaps overcautious industry, on the other, would have been more valuable had a better petroleum-geological understanding been available. On a field basis, Drumgoole and Speers (1997) go some way to resolving the problem. It is highly pertinent to note that many of the world's most prolific hydrocarbon provinces (California, Gulf of Mexico, Middle East, Nigeria, Venezuela), although discovered long before the North Sea and mostly of significantly greater hydrocarbon abundance, are still in production, are still being successfully explored and are still capable of producing new plays and additional reserves (Daly *et al.*, 1996). Although many fields, especially in UK waters, are already in decline, Kemp and Stephen (1996) predict that the UK will remain self-sufficient in oil beyond 2000, and that gas production will exceed gas demand well beyond that date.

Any interpretation is only as good as the available data and working hypotheses current at any one time. Working hypotheses should be revised, if only slightly, with each additional piece of evidence received, whereas, over the decades, important geological hypotheses have had a habit of being turned on their heads from time to time, leading to a complete new way of thinking (e.g. the recognition of turbidites in the late 1940s, or the now widely accepted hypothesis of plate tectonics after two centuries during which continents were thought to be immovable). Some of the working hypotheses used in this book in all good faith will certainly be replaced at some time in the future as new evidence stimulates or even demands a rethink. This is essential if new oil and gas are to be found in what today are unconventional or even unlikely reservoirs or trap locations.

Table 1.3 Significant events in the evolving hydrocarbon industry of the North Sea and adjacent areas.

1959	Slochteren-1: discovery of Groningen gasfield, the Netherlands
1963	Ten Boer-1: confirmation of the great size of the Groningen gasfield
1965	First offshore gas find: 1.8×10^{12} ft^3 West Sole
1969	Ekofisk (Norway): 1.5×10^9 barrel field (Chalk)
1970	Forties (UK): 2.5×10^9 barrel oilfield (Palaeocene)
1971	Brent (UK): 2.25×10^9 barrel oilfield (Mid Jurassic)
1971	Kinsale Head (Ireland): 2×10^{12} ft^3 gasfield (Lower Cretaceous)
1972	Piper (UK): 1×10^9 barrel oilfield (Jurassic)
1973	Frigg (Norway): 7.75×10^{12} ft^3 gasfield (Palaeocene–Eocene)
1973	OPEC price rises and Yom Kippur war made North Sea economic
1974	Morcambe gasfield (5.5×10^{12} ft^3) in eastern Irish Sea
1974	First oil west of Shetlands: 205/21-1A
1975	Devonian oil in small Buchan field (UK)
1979	Troll (Norway): biggest offshore gasfield (44×10^{12} ft^3 of gas and 200×10^6 barrels of oil) in Upper Jurassic sands)—long migration path
1981	Second phase of UK southern North Sea gasfield development began—price rise
1981	Midgard gas condensate (first hydrocarbons on Haltenbanken)
1981	Askeladd gas (first hydrocarbons in Hammerfest basin)
1984	Snorre (Norway): first giant Triassic oilfield (880×10^6 barrel field)
1986	Oil glut and worldwide drop in oil price—more economic field development
1987	Global stock-market collapse—many take-overs
1990	Saddam Hussein invaded Kuwait—oil prices improved
1992	Foinaven (UK): first commercial oil west of Shetland
1993	Changes to UK fiscal regime governing exploration and production

Table 1.4 Some important engineering factors aiding oil- and gasfield development: drilling, platforms and pipelines.

Early 1960s	Development of marine jack-up rigs
Late 1960s	Early semi-submersibles in North Sea
Mid-1970s	Semi-submersibles capable of year-round drilling in North Sea
Mid-1970s	Exposed location single-buoy mooring (oil storage) for tankers
Mid-1970s	Modified tankers for offshore oil storage
Early 1980s	Sea-bottom completion manifolds
Mid-1980s	Ability to drill safely through highly overpressured sequences
Late 1980s	Horizontal drilling revolutionized development of tight reservoirs and thin oil legs
Early 1990s	Unmanned production platforms for satellite fields
Mid-1990s	Coiled-tubing production wells; multilateral wells; logging while drilling

An encouraging feature of the discovery process we have reviewed has been the readiness of the industry to share its newly won knowledge. The conference on 'Petroleum and the Continental Shelf of Northwest Europe' (Woodland, 1975), held in London in 1974, set the standard for the amount and quality of information released by the industry, and was followed by conferences in London (1980, 1983, 1986, 1992, 1997), Stavanger (1984), Trondheim (1985, 1985), Oslo (1977), The Hague (1980, 1982) and Aberdeen (1995). The Geological Society Special Publications on thematic topics are also well received. Two volumes on the oil- and gasfields of Norway, published in 1987, were joined in 1991 by a similar volume on UK fields, compiled by the Petroleum Group of the Geological Society. Because these compilations and conference proceedings represent a concentrated wealth of geological information on the greater North Sea area, they are listed separately at the end of this chapter for the reader's convenience. The same desire to contribute to the earth sciences led to the publication of Ziegler's *Geological Atlas of Western and Central Europe* (1982a, 2nd edition 1990), which has helped considerably in the understanding of the North Sea's regional setting and history. The contents of this book are consistent with this practice of disseminating information.

1.6 Acknowledgements

Earlier versions of this chapter were published by permission of Shell Internationale Petroleum Maatschappij, The Hague, and Shell UK Exploration and Production, London. We are grateful to colleagues in London, Norway, Denmark and the Netherlands for their regional contributions. Steve Pickering (BHP) provided data on the eastern Irish Sea and Pat Shannon on the Irish offshore. Arthur Andersen (PSG) supplied considerable data for updating Figs 1.2, 1.3, 1.5B, and 1.7, as well as the basis for some of the text. The UK Offshore Operators Association (UKOOA) provided Fig. 1.16. Andy Hurst, Aberdeen University, corrected some errors. The drafting and updating of figures was undertaken at the Aberdeen office of Shell Expro and by Barry Fulton of Aberdeen University. The help of all these contributors is gratefully acknowledged.

1.7 Key references

Bowen, J.M. (1991) 25 years of UK North Sea exploration. In: Abbotts, I.L. (ed.) *United Kingdom Oil and Gas Fields, 25 Years Commemorative Volume.* Memoir 14, Geological Society, London, pp. 1–7.

Campbell, C.J. and Ormaasen, E. (1987) The discovery of oil and gas in Norway: an historical synopsis. In: Spencer, A.M. *et al.* (eds) *Geology of the Norwegian Oil and Gas Fields.* Norwegian Petroleum Society, Graham & Trotman, London, pp. 1–37.

Glennie, K.W. (1997) History of exploration in the southern North Sea. In: Ziegler, K., Turner, P. and Daines, S.R. (eds) *Petroleum Geology of the Southern North Sea: Future Potential.* Special Publication 123, Geological Society, London, pp. 5–16.

Glennie, K.W. and Hurst, A. (1996) Hydrocarbon exploration and production in NW Europe: an overview of some key factors. In: Glennie, K.W. and Hurst, A. (eds) *AD 1995: NW Europe's Hydrocarbon Industry.* Geological Society, London, pp. 5–14.

Petroleum Geological Circle (1993) Synopsis: petroleum geology of the Netherlands—1993. In: Rondeel, H.E., Batjes, D.A.J. and Nieuwenhuis, W.H. (1996) (eds) *Geology of Oil and Gas under the Netherlands.* KNGMG, Kluwer Academic Publishers, Dordrecht.

Ziegler, P.A. (1990) *Geological Atlas of Western and Central Europe.* Shell Internationale Petroleum Maatschappij BV, distributed by Geological Society, London, 239 pp.

1.8 Compilations and publications of North Sea conferences

Abbotts, I.L. (ed.) (1991) *United Kingdom Oil and Gas Fields, 25 Years Commemorative Volume.* Memoir 14, Geological Society, London, 573 pp.

Boldy, S.A.R (ed.) (1995) *Permian and Triassic Rifting in Northwest Europe.* Special Publication 91, Geological Society, London, 263 pp.

Brooks, J. (ed.) (1983) *Petroleum Geochemistry and Exploration of Europe.* Geological Society of London Special Publication 12, Blackwell Scientific Publications, Oxford, 379 pp.

Brooks, J. and Glennie, K. (eds) (1987) *Petroleum Geology of North West Europe* (2 vols). Graham & Trotman, London, 1219 pp.

Brooks, J. and Hardman, R.F.P. (eds) (1990) *Tectonic Movements Responsible for Britain's Oil and Gas Reserves.* Special Publication 55, Geological Society, London, 404 pp.

Brooks, J., Goff, J.C. and van Hoorn, B. (eds) (1986) *Habitat of Palaeozoic Gas in NW Europe.* Geological Society Special Publication 23, Scottish Academic Press, Edinburgh, 276 pp.

Croker, P.F. and Shannon, P.M. (eds) (1995) *The Petroleum Geology of Ireland's Offshore Basins.* Special Publication 93, Geological Society, London, 498 pp.

Donovan, D.T. (ed.) (1968) *Geology of Shelf Seas.* Oliver & Boyd, Edinburgh.

Finstad, K.G. and Selley, R.C. (eds) *Mesozoic Northern North Sea Symposium, Oslo, 1977.* Proceedings Norwegian Petroleum Society 6/1–6/26.

Glennie, K.W. and Hurst, A. (eds) (1996) *AD 1995: NW Europe's Hydrocarbon Industry.* Geological Society, London, 242 pp.

Hardman, R.F.P. (1992) *Exploration Britain: Geological Insights for the Next Decade.* Special Publication 67, Geological Society, London, 312 pp.

Hardman, R.F.P. and Brooks, J. (eds) (1990) *Tectonic Events Responsible for Britain's Oil and Gas Reserves.* Special Publication 55, Geological Society, London, 404 pp.

Hepple, P. (ed.) (1969) *The Exploration for Petroleum in Europe and North Africa.* Institute of Petroleum, London.

Hurst, A., Johnson, H., Burley, S.D., Cauhan, A.C. and Mackertich, D.S. (eds) (1992) *The Geology of the Humber Group, Central Graben and Moray Firth UKCS,* Special Publication 114, Geological Society, London, 350 pp.

Illing, L.V. and Hobson, G.D. (eds) (1981) *The Petroleum Geology of the Continental Shelf of NW Europe.* Hayden.

Kaasschieter, J.P.H. and Reijers, T.J.A. (eds) (1983) Petroleum geology of the southeastern North Sea and adjacent onshore areas. *Geol. Mijnbouw* **62**(1), Haarlem, 239 pp.

Kleppe, J., Berg, E.W., Buller, A.T., Hjelmeland, O. and Torsaeter, O. (eds) (1987) *North Sea Oil and Gas Reservoirs.* Norwegian Institute of Technology, Graham & Trotman, London, 352 pp.

Knox, W.O.B., Corfield, R.M. and Dunay, R.E. (eds) (1996) *Correlation of the Early Paleogene in N.W. Europe,* Special Publication 101, Geological Society, London, 480 pp.

Morton, A.C., Haszeldine, R.S., Giles, M.R. and Brown, S. (eds) (1992) *Geology of the Brent Group,* Special Publication 61, Geological Society, London.

Norwegian Petroleum Society. *The Sedimentation of the North Sea Reservoir Rocks: Proceedings Geilo Conference May 1980.*

Parker, J.R. (ed.) (1993) *Petroleum Geology of Northwest Europe: Proceedings of the 4th Conference* (2 vols). Geological Society, London, 1542 pp.

Rondeel, H.E., Batjes, D.A.J. and Nieuwenhuis, W.H. (eds) (1996) *Geology of Oil and Gas under the Netherlands*. KNGMG, Kluwer Academic Publishers, Dordrecht, 284 pp.

Scrutton, R.A., Stoker, M.S., Shimmield, G.B. and Tudhope, A.W. (eds) (1995) *The Tectonics, Sedimentation and Palaeogeography of the North Atlantic Region*, Special Publication 90, Geological Society, London, 309 pp.

Spencer, A.M. (ed.) (1991) *Generation, Accumulation and Production of Europe's Hydrocarbons*. European Association of Petroleum Geoscientists Special Publication No. 1, Oxford University Press, Oxford, 459 pp.

Spencer, A. *et al.* (eds) (1984) *Petroleum Geology of the North European Margin*. Proceedings Norwegian Petroleum Society, Graham & Trotman, London, 436 pp.

Spencer, A.M. *et al.* (eds) (1986) *Habitat of Hydrocarbons on the Norwegian Continental Shelf*. Norwegian Petroleum Society, Graham & Trotman, London, 354 pp.

Spencer, A.M. *et al.* (eds) (1987) *Geology of the Norwegian Oil and Gas Fields*. Norwegian Petroleum Society, Graham & Trotman, London, 493 pp.

Thomas, B.N. (ed.) (1985) *Organic Geochemistry in Exploration of the Norwegian Shelf: Proceedings Norwegian Petroleum Society Conference*. Graham & Trotman, London, 337 pp.

Woodland, A.W. (ed.) (1975) *Petroleum and the Continental Shelf of NE Europe*, Vol. 1. *Geology*. Elsevier Applied Science Publishers, Barking, 501 pp.

Ziegler, K, Turner, P. and Daines, S.R. (eds) (1997) *Petroleum Geology of the Southern North Sea: Future Potential*. Special Publication 123, Geological Society, London.

1.9 Appendix: North Sea fields: basic data

Field name	Block	Discov. date	Discov. well	Date on-stream	Reservoir	Trap type	Depth GWC OWC (mss) (T top res.)	Ultimate recov. reserves (10^6 m^3 O/C/N, 10^9 m^3 G)	References
Danish fields									
Adda	5504/8	1977	Adda-1	2000?	U & L Chalk	Inversion/Strat. trap			Megson, 1992
Alma		Mar. 77		2000?					
Dagmar	5504/15	1983	East Rosa-1	1991	Chalk/Ze	Salt pillar	1400 T	O 1 + G 0.24	
Dan	5505/17	1971	M-1x	1972	Pc Chalk 6	Dome over salt plug	1844 G	O 102	Childs and Reed, 1975
					KU Chalk 5		2001 O	G 23	Megson, 1992
Elly	5504/6	1984	Elly-1	2000?	Chalk/JU		3200 T		
							4000 T		
Gert	5603/27, 28	1984	Gert-1		JU		4900 T		
Gorm	5504/15, 16	1971	N-1	1981	Pc Chalk 6	Dome over salt plug	2100 T	O 46	Hurst, 1983
								G 23	
Harald	5604/21, 22	1980	Lulu-1	1992	Chalk/JM		2700 T	G 31	
		1983	W Lulu-1				3650 T	O 12	
Igor	5505/13	1968	G-1		Chalk		2000 T		
Kraka (Anne)	5505/17	1966	A-2	1991	Chalk	Salt swell	1800 T	O 4.7	
								G 1.3	
North Arne					JU/JM				
Regnar	5505/17	1979	Nils-1	1993	Chalk/Ze	Salt pillar	1700 T	O 1 + G 0.14	
Roar	5504/7	1968	H-1	1996	Chalk	Inversion	2070 T		Megson, 1992
Rolf	5504, 14, 15	1981	Middle Rosa-1	1986	Chalk/Ze	Salt pillar	1800 T	O 5	
Siri	5604/20	1995	Siri-1		Palaeocene	Strat. trap	2060 T		
Skjold	5504/16	1977		1982	KU Chalk	Dome over salt plug		O 0.29	Megson, 1992
South Arne	5604/25, 29, 30	1969	I-1		U & L Chalk		2710 T		
Svend	5604/25	75/82	T-1, Otto-1	1996	KU Chalk		2500 T		
Tyra	5504/11	1968	E-1	1984	KU Chalk	Dome over salt plug	2000 T	O 21	
	12							G 87	
Valdemar	5504/14, 11	1977	BO-1	1993 (N Jens)	U & L Chalk	Inversion/drape	2000 T	O 2.3	Ineson, 1993
		1985	N Jens				2600 T	G 1.2	
Dutch fields									
Ameland					P. Rotl			G	
F/3		1974						O	
F/18		1970						O	
Haven	Q1	1979		1982	KL Vlieland	Anticline thrust		O 10 + G	Roelofsen and de Boer, 1991
Helder									
Helm									Hastings *et al.*, 1991
Hoorn									
Kotter	K18a,b	1980	K18-1	1984	KL Vlieland			O 6	
Logger	L16a	1982		1985	KL Vlieland			O 5	Goh, 1996

Rijn	P15a	1982		1985	KL Vlieland			O 7	
K6	K6	1986		1972	P. Rotl			G	
K7	K7	1969	K7-1	1982	P. Rotl			G 7	
K8/K11	K8/K11	1970	K8-1	1978	P. Rotl			G 45	
K9	K9c	1985		1987	P. Rotl			G 1.5	
K9/L7	K9b/L7	1983		1987	P. Rotl			G 3	
K10	K10a	1979		1983	P. Rotl			G 12	
K12-A	K12	1975		1983	P. Rotl			G 9	
K12-B	K12	1982		1987	P. Rotl			G 11	
K13-A	K13	1972		1976	Tr Bunter			G 25	Roos and Smits, 1983
K13-B					Tr Bunter				
K13-E					P. Rotl				
K13-F					P. Rotl				
K14	K14	1970		1977	P. Rotl			G 16	
K15-A	K15, L13	1974		1979	P. Rotl			G 10	
K15-B	K15	1975		1984	P. Rotl			G 20	
K15-C	K15	1982		1987	P. Rotl			G 5	
L4	L4a	1974		1983	P. Rotl			G 11	
L7	L7	1971		1977	P. Rotl			G 13	
L8	L8a	1972		1988	P. Rotl			G 4	
L10	L10, L11	1970		1975	P. Rotl			G 42	
L11	L11b	1971		1986	P. Rotl			G 2	
L13	L13	1977		1986	P. Rotl			G 12	Frikken, 1996
L14	L14	1975		1987	P. Rotl			G 1.6	
P6	P6	1968		1983	P. Rotl			G 14	
P9	P9				KL			O 14	
P15	P15				KL			O 6	
Q8-A	Q8	1976		1986	P. Rotl			G 1.2	
German fields									
A6-B4	A6-B4		A/6-1		P. Zech			G	
Irish fields									
Admore	49/13, 14	1974	49/13-1		KL Sst				Naylor and Shannon, 1982; Naylor, 1996
Ballycotton	48/20	1989	48/20-1	1991	KL Sst	Inversion anticline		G 2.5 IIP	Murray, 1995
Connemara	26/28	1979	26/28-1		M-UJ Sst	Fault block	2330	O 32 IIP	MacDonald *et al.*, 1987
Helvick	49/9	1983	49/9-2		M-UJ Wexford	Faulted/dip closure	1865	O 0.5	Caston, 1995
Kinsale Head	48/20, 25 49/16, 17	1971	48/25-2	1978	KL Sst	Inversion anticline	Main 902 Lwr 966	G 45	Taber *et al.*, 1995 Colley *et al.*, 1981
Seven Heads	48/23, 24, 28	1974	48/24-1		KL Sst			O 0.3	Naylor and Shannon, 1982
	28							G 3	Naylor, 1996

Continued on p. 32

1.9 Appendix: North Sea fields: basic data (*Cont.*)

Field name	Block	Discov. date	Discov. well	Date on-stream	Reservoir	Trap type	Depth GWC OWC (mss) (T top res.)	Ultimate recov. reserves (10^6 m^3 O/C/N, 10^9 m^3 G)	References
Norwegian fields									
Agat	35/3	10/80	35/3-2		KL	Strat./struct.	3455	G 65P	Myreland *et al.*, 1981
							3561		Gulbrandsen, 1987
Albatross	7120/9	8/86	7120/9-1		JL-JM	Struct.		G 34P	
Albuskjell	1/6	10/72	1/6-1	1979	KU Tor	Struct./strat.	>3400	C 8	Watts *et al.*, 1980
					Ekofisk			G 18	D'Heur, 1987a
					Hod				
Askeladd	7120/8	8/81	7120/8-1		JL-JM	Struct.		G 52	Grung Olsen and Hanssen, 1987
Balder	25/11	5/74	25/11-1		Pc Heimdal	Strat.	1760	O 11–27	Frodesen *et al.*, 1981
					Balder				Hanslien, 1987
Brage	31/4	5/80	31/4-3	1993	JU-JL	Struct.	2030	O/G	Hage *et al.*, 1987
							2147	O	
							2381	O	
Byggve	25/5	3/91	25/5-4		Jim Brent				
Cod	7/11	6/68	7/11-1	1971	Pc Forties	Struct./strat.	>3070	O	Kessler *et al.*, 1980; D'Heur, 1987b
Draugen	6407/9	9/84	6407/9-1	1993	JU Frøya	Struct.	1638	O 175	Ellenor and Mozetic, 1986
Edda	2/7	9/72	2/7-4	1979	KU Tor	Struct., salt dome	3288	O 4	D'Heur and Michaud, 1987
					Ekofisk			G 2.2	D'Heur, 1986
					Hod				
Ekofisk	2/4	12/69	2/4-2	1971	KU Tor	Salt dome	3288	O 237	van den Bark and Thomas, 1981
					Ekofisk			N 13	Pekot and Gersib, 1987
					Hod			G 126	
Ekofisk West	2/4	12/70	2/4-6	1971	KU Tor	Struct./strat.	3000	O 48	D'Heur, 1987e
					Ekofisk			G 39	
					Hod				
Eldfisk	2/7	12/70	2/7-1	1979	KU Tor	Struct./strat.		G 48	Michaud, 1987
					Ekofisk				
					Hod				
Embla	2/7	6/88	2/7-20	1993	Red bed				
Frigg	25/1	7/71	25/1-1	1977	EO Frigg		1956	G 191	Mure, 1987a
Frigg O	25/2	9/73	25/2-1	1988	EO Frigg	Struct./strat.	1947	G 12.6	Heritier *et al.*, 1980
A + B							1956		Mure, 1987b
Frigg NO	25/1	5/74	25/1-4	1983	EO Frigg	Struct./strat.	1984	G 11	Mure, 1987c
Froya	25/5	8/87	25/5-1	1995				O	

Gullfaks	34/10	9/78	34/10-1	1986	JM Brent	Struct.	1947	O 220	Saeland and Simpson, 1982
					JL Statfjord		2043		Hazeu, 1981
							2090		Erichsen *et al.*, 1987
Gullfaks S	34/10	12/78	34/10-2	1994	JM Brent	Struct.		G	
					JL Statfjord				
Gullfaks V	34/10	6/91	34/10-34	1994		Struct.		O	
Gyda	2/1	3/80	2/1-3	1990	JU Ula	Struct./strat.	4160	O	
Heidrun	6507/7	6/85	6507/7-2	1995		Struct./strat.	2468	O 87	
								G 31	
Heimdal	25/4	12/72	25/4-1	1985	Pc Heimdal	Struct.	2150	G34	Mure, 1987d
								C 5	
Hild	30/7	7/77	30/7-6		JM Brent	Struct.	3700	G	Rønning *et al.*, 1987, 1987
Hod	2/11	12/74	2/11-2	1990	KU Tor	Salt dome		O	Hardman and Kennedy, 1980
					Ekofisk	Salt	2700	O 6	Norbury, 1987
Huldra	30/2	10/82	30/2-1	1998	JM Brent	Struct.		G	
Lille-Frigg	25/2	10/75	25/2-4	1994	JM Brent			G	
Loke	15/9	5/81	15/9-8	1993	Pc Heimdal	Struct./strat.		G	
Midgard	6507/11	12/81	6507/11-1	2000	JM Tomma	Horst	2488	G 87	Ekern, 1987
					JL Aldra		2499	C/N 17	
Mikkel	6407/6	2/87	6407/6-3					G	
Mime	7/11	6/82	7/11-5	1990	JU Ula			O	
Mjoelner	2/12	3/87	2/12-1	1995				O	
Murchison	9/6	8/75	211/19-2	1980	JM Brent			O	Engelstad, 1987; Warrender, 1991
Njord	6407/7	4/86	6407/7-1S	1994				O	
Odin	30/10	3/74	30/10-2	1983	EO Frigg	Struct./strat.	2025	G 32–36	Nordgård Bolås, 1987
Oseberg	30/6	9/79	30/6-1	1988	JM Brent	Fault	2700–2719	G 89	Larsen *et al.*, 1981
								O 203	Torvund and Nipen, 1987
Oseberg O	30/6	8/81	30/6-5	1995	JM Brent	Fault		O	
Peik	24/6	8/85	24/6-1					G	
Skirne	25/5	3/90	25/5-3	1997	JM Brent			G	
Sleipner O	15/9	7/81	15/9-9	1993	Pc Heimdal	Struct./strat.	2417	G 60	Pegrum and Ljones, 1984
					Maureen		2800		Østvedt, 1987
					JM Hugin				
Sleipner V	15/6	12/74	15/6-3	1996	JM Hugin	Dome over salt swell	3628	G 126	Ranaweera, 1987
								C 45	
Smørbukk	6506/12	2/85	6506/12-1	2000	JL–JM	Struct.	4450	O 60	Aasheim *et al.*, 1986
								G 83	
Smørbukk Sør	6506/12	7/85	6506/12-3	2000	JL–JM	Struct.	3990	O	
Snorre	34/4	12/79	34/4-1	1992	JL Statfjord	Fault	2595	O 116	Hollander, 1987
					TR Lunde				

Continued on p. 34

1.9 Appendix: North Sea fields: basic data (*Cont.*)

Field name	Block	Discov. date	Discov. well	Date on-stream	Reservoir	Trap type	Depth GWC OWC (mss) (T top res.)	Ultimate recov. reserves (10^6 m^3 O/C/N, 10^9 m^3 G)	References
Snøhvit	7121/4	10/84	7121/4-1					G 74	
Snøhvit N	7121/4	4/85	7121/4-2					G 3	
Statfjord	33/12	4/7	33/12-1	1979	JM Brent	Struct./strat.	2586	O 1036 P	Jones *et al.*, 1975
					JL Cook		2600		Kirk, 1980
					TR–JL Statfjord		2806		Roberts *et al.*, 1987
Statfjord N	33/9	2/77	33/9-8	1994	JM Brent	Fault	2680	O 36–66	Gradijan and Wiik, 1987
							2716	G 3.4	
Statfjord E	33/9	11/76	33/9-7	1994	JM Brent	Fault	2507	O 14.5–40	Nyberg, 1987
						Strat.		G 2.5–5.4	
Tommeliten *A*	1/9	2/77	1/9-1		KU Tor		3180	G 21	D'Heur and Pekot, 1987
B					Ekofisk			O/N 9	
					Hod				
Tor	2/5	11/70	2/5-1	1971	KU Tor		3292	O 18	D'Heur, 1987d
					Ekofisk			G 12–15	
					Hod				
Tordis	34/7	10/87	34/7-12	1994	JM Brent			O	
Trestakk	6406/3	11/86	6406/3-2					O	
Troll O	31/2	11/79	31/2-1	1996	JM Sognfjord	Struct.	1547	G	
							1551	O	
Troll V Gas	31/2	11/79	31/2-1		JM Heather	Struct.	1547	G 1252	Brekke *et al.*, 1981
							1559	N 44	Gray, 1987
								O 55	Horstad and Larter, 1997
Troll V Oil	31/2	11/79	31/2-1	1996	JM Fensfjord	Struct.	1543	G	
							1570	O	
Trym	3/7	1/90	3/7-4					G	
Tyrihans	6407/1	5/83	6407/1-2		JM Tomma	Horst	3658	G 35	Larsen *et al.*, 1987
N + S							3680	C 18	
								O 8	
Ula	7/12	9/76	7/12	1986	JU Ula	Fault/salt dome	3508	O 25	Bailey *et al.*, 1981; Home, 1987
Vale	25/4	8/91	25/4-6S					O	
Valhall	2/8	6/75	2/8-6	1982	KU Tor	Struct.		G 1	Munns, 1985
					Ekofisk	Strat.		N 0.3	Leonard and Munns, 1987
					Hod			O 6.3	
Veslefrikk	30/3	8/80	30/3-2	1989	JM Brent	Fault			
Vigdis	34/7	4/88	34/7-13	1998	JM Brent			O	
					JL Statfjord				
Visund	34/8	3/86	34/8-1	1998	JM Brent			G	
					JL Statfjord				
					TR Lunde				

UK fields									
Alba	16/26	Dec. 84	16/26-5	1944	Eo	Strat.	*c.* 1800	O 60	Newton and Flannagan, 1993
Alison	49/11	Feb. 87	49/11a-4		P Rotl	Struct.		G	
Alligin		1991			Pc			O 8	
Alwyn N	3/9a, 3/4a	Oct. 75	3/9a-1	1987	JM Brent	Tilted fault block	3231	O 28	Johnson and Eyssautier, 1987
					JL Statfjord		3580	O 22	Inglis and Gerard, 1991
Amethyst E	47/14a, 8a, 9a, 13a	Oct. 72	47/14a-1	1990	P. Rotl	Anticline	2670	G 23	Garland, 1991
Amethyst W	47/13a, 14a, 15a	Apr. 70	47/13-1		P. Rotl			G	
Andrew	16/28	June 74	16/28-1		Pc			O & G	
Anglia	48/19b	Dec. 85	48/19b-7	1991	P. Rotl	Struct.		G	
Angus	31/26	Mar. 83	31/26-3	1992	JU Fulmar			O 2	Hall, 1992
Ann	49/6	May 66	49/6-1	1993	P. Rotl	Struct.		G	
Arbroath	22/17, 18	Dec. 69	22/18-1	1990	Pc Forties	Struct.		O	
Argyll (abnd)	30/24, 25	Aug. 71	30/24-2	1975	P. Zech, Rotl, Dev	Tilted fault block	2600	O 15	Robson, 1991 Pennington, 1975
Audrey	49/11a 48/15	Mar. 76	49/11a-1	1988	P. Rotl	Complex anticline	2700	G 3	
Auk	30/16	Feb. 71	30/16-1	1976	P. Ze/Rotl	Tilted horst	2350	O 15	Brennand and van Veen, 1975; Buchanan and Hoogteyling, 1979; Heward, 1991; Trewin and Bramwell, 1991
Baird	49/23	Sept. 93	49/23-D5	1993	P. Rotl	Pop-up fault block	2473	G 2	
Balmoral	16/21	Aug. 75	16/21-1	1986	Pc Andrew	Diff. compact	2149	O 16	Tonkin and Fraser, 1991
Barque	48/13a, 14	May 66	48/13a-1	1990	P. Rotl	Salt-sealed horst	2727	G 38	Farmer and Hillier, 1991a
Beatrice	11/30a	Sept. 76	11/30-1	1981	JM Beatrice	Tilted fault block	1780	O 21	Linsley *et al.*, 1980; Stevens, 1991
Beinn	16/7a	Nov. 89	16/7a-30z	1994	JL-JM	Struct.		C 7	
Beryl 'A'	9/13	Sept. 72	9/13-1	1976	JU-TrL	Struct./Strat.	3292	O 127	Knutson and Munro, 1991
							3614		Robertson, 1993
Beryl 'B'	9/13	May 75	9/13-7	1984	JU Katrine	Struct./Strat.			Knutson and Munro, 1991
Big Dotty	48/29	Sept. 67	49/23-5	1976	P. Rotl	Struct.		G	Cooke-Yarborough, 1991
Birch	16/12a	Oct. 85	16/12a-8		J	Fault/Strat.		O & G	
Blair	16/21a	June 83	16/21a-8		Pc Andrew		2120	O	
Blenheim	16/21b	Nov. 90	16/21b-21		Tertiary			O	
Bosun	15/29				KL			O	
Brae C	16/7	Mar. 76	16/7-3	1989	JU Brae	Fault/Strat.	4092	O 10	Turner and Allen, 1991
Brae E	16/3a	Apr. 80	16/3a-1	1993	JU Brae	Dip closure	3580	C 45, G 42	
Brae N	16/71	May 75	16/71	1988	JU Brae	Fault/Strat.	3802	C 28, G 24	Stephenson, 1991

Continued on p. 36

1.9 Appendix: North Sea fields: basic data (*Cont.*)

Field name	Block	Discov. date	Discov. well	Date on-stream	Reservoir	Trap type	Depth GWC OWC (mss) (T top res.)	Ultimate recov. reserves (10^6 m^3 O/C/N, 10^9 m^3 G)	References
Brae S	16/7a	July 77	16/7a-8	1983	JU Brae	Fault/Strat.	4111	O 50	Harms *et al.*, 1981; Turner *et al.*, 1987; Roberts, 1991
Brent	211/29	July 71	211/29-1	1976	JM Brent	Tilted fault block	2610	O 359	Bowen, 1975
					JL Statfjord		2757	G 159	Bryant and Livera, 1991; Struijk and Green, 1991
Britannia	15/30	Sept. 75	15/30-1		KL Britannia	Strat.		C 24 (IIP)	
								G 121 IIP	
Bruce	9/8a, 9a, 9b	July 74	9/8-1	1993	JL-JM	Struct.	< 4000	C 19	Beckly *et al.*, 1993
Buchan (abnd)	21/1a, 2d, 5a	Aug. 74	21/1-1	1981	D ORS	Anticline	2680	O 19	Butler *et al.*, 1976
				ABN 1995					Edwards, 1991
Bure	49/28	May 83	49/28-8	1987	P. Rotl	Tilted fault block	2454	G	Werngren, 1991
Caister	44/23a	Jan. 68	44/23-1	1993	Tr Bunter	Faulted anticline	1400	G 11	Ritchie and Pratsides, 1993
	44/23	Feb. 85	44/23-4		C. Wphal CMeas		3703		
Camelot C	53/1a	June 87	53/1a-5	1989	P. Rotl	Tilted horst	1899		Holmes, 1991
Camelot NE	53/2	Mar. 88	53/2-7	1993	P. Rotl	Tilted horst	1956	G 6	
Camelot N	53/1a	Nov. 67	53/1-1		P. Rotl	Tilted horst	1934		
Captain	13/22	May 77	13/22-1	1996	KL Aptian	Horst block		O 48	
Carnoustie	22/17	Apr. 80	22/17-12		P. Zech	Struct./strat.		O	
Cavendish		89			C			G 19	
Chanter	15/17	Sept. 85	15/17-13	1993	JU Galley	Struct./strat.	3722	O 2 OIIP	Schmitt, 1991
						Piper	3987	G 27 GIIP	
Clair	206/7a	1977	206/8-1a		Carb	Partly exhumed fault blocks beneath Cret.		O 64	Coney *et al.*, 1993
	206/8				Dev				
	206/9a				PC Lewisian				
	206/12								
	206/13a								
Claymore	14/19	June 74	14/19-2	1977	JU Claymore	Tilted fault block	2638	O 54	Maher and Harker, 1987
Claymore C		June 72	14/19-1		C/P. Zech			O 13	Harker *et al.*, 1991
Claymore N		Nov. 74	14/19-6a		KL Sst			O 26	
Cleeton	42/29	Apr. 83	14/29-2	1988	P. Rotl	Faulted anticline	2270	G 20	Heinrich, 1991a
Clyde	30-17b	June 78	30/17b-2	1987	JU Fulmar	Rotated fault block	3831	O 24	Gibbs, 1984b; Smith, 1987; Stevens and Wallis, 1991; Turner, 1993
Conival								G 14?	
Cormorant N	211/21a	Aug. 74	211/21-2	1982	JM Brent	Tilted fault block	2790	O 100	Taylor and Dietvorst, 1991
	26a		211/26-1	1979			2600	G 6.5	
Cormorant S	211/26a	Sept. 72			JM Brent	Tilted fault block			

Crawford	9/28a	Apr. 75	9/28-2	1989	JM Brent;	Tilted fault block		O 2	Yaliz, 1991
					Tr Skagerrak	Strat./fault			
Crawford			9/28-2	1989				O	
Curlew					Pc K			O 11	
								G 7	
Cyrus	16/28	Oct. 79	16/28-4	1990	Pc Andrew	Compactional drape	2610	O 2	Mound *et al.*, 1991
Davy	53/5	Feb. 89	53/5a-2		P. Rotl	Struct.		G	
Dawn	48/29	Apr. 94	48/29-9		P. Rotl	Struct.		G	
Deborah	48/30	Aug. 68	48/30-7	1978	Tr Bunter	Anticline	1748	G	Cooke-Yarborough, 1991
Della	48/30	July 87	48/30-11z	1988	P. Rotl.	Anticline	1770	G 2	Cooke-Yarborough, 1991
Deveron	211/18a	Sept. 72	211/18-1	1984	JM Brent	Tilted fault block	2600	O 3	Williams, 1991
Don	211/18a	July 76	211/18-12	1989	JM Brent	Fault block		O 9	Morrison *et al.*, 1991
Donan	15/20a	May 87	15/20a-4	1992	Eo	Strat.		O 20	
Douglas	110/13	1990	11013-2	1995	Tr Ormskirk Sst			O 22	Trueblood *et al.*, 1995
Drake	22/5b	Sept. 82	22/5b-2		JM/JU Drake			G + C	
Dunbar	3/14a	Nov. 73	3/14a-1	1994	JM Brent JL Statfjord	Tilted fault block	3365	O 19	
Duncan (abnd)	30/24	Jan. 81	30/24-15	1983	JU Fulmar	Faulted anticline	2850	O	Robson, 1991
Dunlin	211/23a, 24a	Jul. 73	211/23-1	1977	JM Brent	Tilted fault block	2560	O 54	Baumann and O'Cathain, 1991
Eider	211/16	May 76	211/16-2	1988	JM Brent	Tilted fault block	2560	O 14	Wensrich *et al.*, 1991
Elgin		Oct. 91			JU			O 40 G 26	
Ellon	3/15	Aug. 73	3/15-1	1995	JM Brent	Faulted dip closure	3260	GC 6	
Emerald	2/15, 15a	Oct. 81	2/15-1	1990	JM Brent	Dip and fault	1599	O 38	Stewart and Faulkner, 1991
	10a				Pc			G 1.7	
Esmond	43/13a	June 82	43/13a-1	1985	Tr Bunter	Dome over salt	1454	G 11	Ketter, 1991a
Ettrick									
Everest	22/10a	Mar. 82	22/10a-2	1993	Pc Montrose	Strat./pinch-out	2628	G 18; C 5	Thompson and Butcher, 1991
Excalibur	48/17a	Feb. 88	48/17a-4	1989	P. Rotl	Fault block	2557	G 7	
Fife	31/26a	Apr. 91	31/26a-p		J			O	
Fleming	22/5b	Sept. 82	22/5b-2		T			G + C	
Foinaven	204/24a	Oct. 92	204/24a-2	1997	Pc	Struct.		O 37	
Forbes	43/8	Jan. 70	43/8-1	1985	Tr Bunter	Dome over salt	1754	G 3	Ketter, 1991a
Forties	21/10, 22/6a	Nov. 70	21/10-1	1975	Pc Forties	Over basement high	2217	O 398	Walmsley, 1975; Hill and Wood, 1980; Carman and Young, 1981; Wills, 1991
Frigg UK	10/1	May 72	10/1-1a	1977	Eo Frigg	Strat.	1955	G 176 (Nway)	Heritier *et al.*, 1981; Brewster, 1991
Fulmar	30/16, 30/11b	Dec. 75	30/16-6	1982	JU Fulmar	Dome over shale pod	3304	G 4.5	Johnson *et al.*, 1986
								O 68	Stockbridge and Gray, 1991

Continued on p. 38

1.9 Appendix: North Sea fields: basic data (*Cont.*)

Field name	Block	Discov. date	Discov. well	Date on-stream	Reservoir	Trap type	Depth GWC OWC (mss) (T top res.)	Ultimate recov. reserves (10^6 m^3 O/C/N, 10^9 m^3 G)	References
Galleon	48/20a	Dec. 85	48/20a-3a	1994	P. Rotl	Dip and fault	2314	G 40	
Galley									
Gannet A	22/21	Apr. 78	22/21-3	1993	Pc Tay	Salt diapir + drape	2220	O 10	Armstrong *et al.*, 1987
Gannet B	21/25	Sept. 79	21/25-1	1992	Pc Rogaland/ Forties		2060	O 25 + G	
Gannet C	21/30	Sept. 82	21/30-6a	1992	Pc Forties		1980	O 9	
Gannet D	22/21	Aug. 87	22/21-5z		Pc Andrew/Tay		2180	O 5	
							2500		
Ganymede	49/17	June 89	49/17-10		P. Rotl	Struct.		G	
Gawain	49/29a	Dec. 88	49/29a-7		P. Rotl	Struct.		G	
Glamis	16/21a	Nov. 82	16/21a-6	1989	JU Piper, Pc	Tilted fault block	3141	O 3	Fraser and Tonkin, 1991
Glenn									
Gordon	43/20	June 69	43/20-1	1985	Tr Bunter	Over salt swell	1661	G 5	Ketter, 1991a
Gryphon	9/18a	Jul. 87	9/18b-7	1993	Pc, Eo Balder	Dip closure	1648	O 17 + G	Newman *et al.*, 1993
Guillemot A	21/30				Eo Tay				
Guillemot B	21/24, 29				Eo Rogaland		2018		
Guillemot C	21/30			1988			2018		
Guillemot D	21/30		21/30-1	1969	Pc Forties		2024	O 7	Banner *et al.*, 1992
Guinevere	43/17b	Mar. 88	48/17b-5	1993	P. Rotl	Faulted anticline		G 2	
Hamilton	110/13	1990	110/13-1	1995	Tr Sherwood			O 80	
								G 7	
Hamish	15/21b	Jan. 88	15/21b-21	1990	JU Piper	Tilted fault block		O 1	
Harding	9/23b	Jan. 88	9/23b-7		Tertiary	Strat.		O + G	
Hawkins	22/5a	Oct. 80	22/5a-1a		Jurassic			G + C	
Heather	2/5	Dec. 73	2/5-1	1978	JM Brent	Tilted fault block	3307	O 16	Gray and Barnes, 1981; Penny, 1991
Hewett	48/28, 29	Oct. 66	48/29-1	1978	Tr Bunter	Flower structure	920	G 115	Cumming and Wyndham, 1975
	30				P. Zech		1280?		Cooke-Yarborough, 1991
Highlander	14/20	Apr. 76	14/20-5	1985	KL Sst	Struct./strat.	2909	O 12	Whitehead and Pinnock, 1991
					JU Piper				
Hudson	210/24a	July 87	21-/24a-3	1993	JM Brent	Tilted fault block		O 11	
Hutton	211/28, 27	Dec. 73	211/28-1a	1984	JM Brent	Tilted fault block	2960	O 28	Haig, 1991
Hutton NW	211/27	Apr. 75	211/27-3	1983	JM Brent	Tilted fault block	3941	O 17	Johnes and Gauer, 1991
Hyde	48/6, 47/10	May 82	48/6-25	1993	P. Rotl	Faulted anticline	2945	G .5	Steele *et al.*, 1993
Indefatigable	49/18, 19, 24, 25	June 66	49/18-1	1971	P. Rotl	Faulted horst	2705	G 78	France, 1975
									Pearson *et al.*, 1991
Indefatigable SW	49/23	Jun. 67	49/23-2	1989	P. Rotl	Faulted anticline		G 2	Pearson *et al.*, 1991
Innes	30/24	Apr. 83	30/24-24	1985	P. Rotl	Tilted fault block	4070	O 3	Robson, 1991

Ivanhoe	15/21a	Oct. 75	15/21-3	1989	JU Piper	Tilted fault block	2454	O 16	Parker, 1991
Joanne	30/7a	May 81	30/7a-1		Tertiary, Cret.	Struct./strat.		O	
Johnston	43/27	Apr. 90	43/27-1	1994	P. Rotl	Faulted anticline	3180	G 5	
Judy	30/7a	Aug. 85	30/7a-4a		Jurassic Tr	Struct./strat.		O + C + G	
Ketch					C West red beds				
Kilda (Bosun)	16/26	Feb. 75	16/27-1		KL				Guy, 1992
Kingfisher	16/8a, c	1972		1997	JU/JM			O/C 9	
								G 10	
Kittiwake	21/18	Sept. 81	21/18-2a	1990	JU Fulmar	Struct./strat.	3179	O 11	Glennie and Armstrong, 1991
					Tr Skag			G 1	
Lancelot	48/17a	Apr. 86	48/17a-2	1993	P. Rotl	Faulted anticline	2491	G 8	
Leman	49/26, 27, 53/1, 2a	Apr. 66	49/26-1	1968	P. Rotl	Anticline	2042	G 160	van Veen, 1975 Hillier and Williams, 1991
Lennox	110/15, 14	1992	110/14-3	1995	Tr Sherwood			O + G	
Leven	30/17b	Oct. 83	30/17b-9	1992	JU Fulmar	Struct.		G 1	
Linnhe	9/13c	Aug. 88	9/30c-40z	1989	JM Fladen	Tilted fault block		O 1.5	
Little Dotty	48/30	Aug. 69	48/30-1	1978	Tr Bunter	Faulted dip closure		G	
					P. Rotl				
Loyal								O 14	
								G 3.7	
Lyell	3/2	June 75	3/2-1	1993	JM Brent	Tilted fault block	3424	O 6	
Machar	23/26a	Apr. 76	23/26a-1		Pc, KU Chalk	Diapir		O 16 (O equiv.)	Foster and Rattey, 1993
Magnus	211/12, 7a	July 74	211/12-1	1984	JU Magnus	Tilted fault block	2800	O 114	De'Ath and Schuyleman, 1981; Shepherd, 1991a
Markham	49/5	July 84	49/5-2	1992	P. Rotl	Faulted dip closure	3219	G 20	
Maureen	16/29a	Feb. 73	16/29-1	1983	Pc Maureen	Dome over salt swell	2575	O 33	Cutts, 1991
Medwin	30/17b	May 79	30/17b-5	1994	JU Fulmar			O 0.5	
Miller	16/7b, 8b	Mar. 83	16/7-20z	1992	JU Kimm	Struct./strat.	3980	O 52	Rooksby, 1991; Garland, 1993
Moira	16/20a	May 88	16/29a-8	1990	Pc Andrew			O 1	
Montrose	22/17a, 18a	Nov. 71	22/18-2	1976	Pc Forties	Anticline	2514	O 16	Fowler, 1975; Crawford *et al.*, 1991; Shepherd, 1991b
Morcambe N	110/2, 3, 7a	1976			Tr Sherwood			G 34	Stuart, 1993
Morcambe S		1974			Tr Sherwood	Faulted anticline	1144	G 156	Bushell, 1986; Stuart and Cowan, 1991
Murchison	211/19a	Sept. 75	211/19-2	1980	JM Brent	Tilted fault block	2880	O 8	Warrender, 1991
Murdoch	44/22	Aug. 85	44/22-3	1993	C Westph. CM	Faulted anticline	3126	G 9	Ritchie and Pratsides, 1993
Nelson	22/11	Mar. 88	22/11-1	1944	Pc Forties	Over basement high		O 71	Whyatt *et al.*, 1992
Ness	9/13a, 13b	May 86	9/13b-28a	1987	JM Fladen	Tilted fault block	3035	O 6	

Continued on p. 40

1.9 Appendix: North Sea fields: basic data (*Cont.*)

Field name	Block	Discov. date	Discov. well	Date on-stream	Reservoir	Trap type	Depth GWC OWC (mss) (T top res.)	Ultimate recov. reserves (10^6 m^3 O/C/N, 10^9 m^3 G)	References
Ninian	3/3	Apr. 74	3/3-1	1978	JM Brent	Tilted fault block	3179	O 171	Albright *et al.*, 1980
								G 0.5	van Wessem and Gan, 1991
Orwell	50/26a	Feb. 90	50/26a-2	1993	Tr Bunter	Anticline		G 7	
Osprey	211/23a, 18a	Feb. 74	211/23-3	1991	JM Brent	Tilted fault block	2580	O 13	Ericksen and van Panhuys, 1991
Pelican	211/26	Aug. 75	211/264		JM Brent	Tilted fault block		O	
Petronella	14/20b	Feb. 75	14/20-1	1986	JU Piper	Tilted fault block	2230	O 3	Waddams and Clark, 1991
					JU Kimm		2330	G 0.3	
Pickerill	48/11b	Dec. 84	48/11b-4	1992	P. Rotl	Faulted anticline	2579	G 17	
Piper	15/17a	Jan. 73	15/17-1a	1976	JU Piper	Tilted fault block	2594	O 156	Williams *et al.*, 1975
					JU Sgiath			G 2	Maher, 1981; Schmitt and Gordon, 1991
Puffin									
Ravenspurn N	43/26	Oct. 84	43/26-1	1990	P. Rotl	Anticline/strat.	3020	G 35	Ketter, 1991b; Turner, P *et al.*, 1993
Ravenspurn S	42/30	Apr. 83	42/30-2	1990	P. Rotl	Faulted anticline	2760	G 18	Heinrich, 1991b
Rob Roy	15/21a	May 84	15/21a-11	1989	JU Piper	Tilted fault block	2419	O 11, G 1.8	Parker, 1991
Rough	47/8b	May 68	47/8-1	1975	P. Rotl	Faulted anticline	2910	G 11	Goodchild and Bryant, 1986; Stuart, 1991
Saltire	15/17	Jan. 88	15/17-16	1993	KL Valhall	Struct.	2066	O 22	Casey *et al.*, 1993
							3338		
Scapa	14/19	July 75	14/19-9	1989	KL Valhall	Struct./strat.	2575	O 15	McGann *et al.*, 1991; Harker and Chermak, 1992
Schiehallion	205/26a	93	205/26a-1		Pc	Struct.		O 68	Leach *et al.*, 1997
Schooner	44/26	Dec. 86	44/26-2		C Westph. R bed			G	
Scott	15/22	Jan. 84	15/22-4	1993	JU Piper	Complex fault blocks	3330	O 85	
Sean N	49/25a	Apr. 69	49/25-1	1986	P. Rotl	Faulted anticline	2607	G 6.6 + C	Ten Have and Hillier, 1986
Sean S	49/25a	Jan. 70	49/25-2	1986	P. Rotl	Faulted anticline	2603	G 5.4 + C	Hobson and Hillier, 1991
Sean E	49/25a	June 83	49/25a-5	1986	P. Rotl	Faulted anticline			
Shearwater		1989			J			O 33 G 31	
Solan		1991			JU			O 8	
Staffa	3/8b	July 85	3/8b-10	1992	JM Brent	Faulted dip closure	4024	O 8	
Statfjord UK	211/24b, 25b	Feb. 75	211/24-4	1979	JM Brent	Tilted fault block	2586	O 500	Roberts *et al.*, 1987
					JL Cook		2600	G 350	
					JL Statfjord		2806		
Strathmore		1990			Tr			O 6.4	
Strathspey	3/4	Mar. 75	3/4-4	1993	JM Brent	Tilted fault block		O 11 + G 10	

Tartan	15/16	Jan. 75	15/16-1	1981	JU Piper	Tilted fault block	3148	O 11	Coward *et al.*, 1991
							3709	O 8	
Tern	210/25a	May 75	210/25-1	1990	JM Brent	Tilted fault block	2360	O 40	van Panhuys-Siegler *et al.*, 1991
Thames	49/28	Dec. 73	49/28-4	1986	P. Rotl	Tilted fault block	2452	G 9	Werngren, 1991
Thistle	211/18	July 73	211/18-2	1978	JM Brent	Tilted fault block	2841	O 64	Hay, 1977
								G 0.2	Hallet, 1981; Williams and Milne, 1991
Tiffany	16/17	July 79	16/17-8a	1993	JU Brae	Fault and dip closure	3807	O 17	
Toni	16/17	Aug. 77	16/17-4	1993	JU Brae	Fault and dip closure	3627	O 7	
Trent					C Namur				
Tristan	49/29	June 76	49/29-2	1992	P. Rotl	Fault block	2403	G 1	
Tyne		1990			Carb			G 7	
Valiant N	49/16	Nov. 70	49/16-2	1988	P. Rotl	Struct.	2420	G 6	Pritchard, 1991
Valiant S	49/21	July 70	49/21-2	1988	P. Rotl	Struct.	2350	G 8	
Vanguard	49/16	Dec. 82	49/16-7z	1988	P. Rotl	Faulted anticline	2420	G 3	Pritchard, 1991
Victor	49/22, 17a	May 72	49/22-2	1984	P. Rotl	Faulted anticline	2675	G 27	Conway, 1986; Lambert, 1991
Victory					KL				
Viking A	49/12	Feb. 69	49/12-1	1972	P. Rotl	Horst	2761	G 80	Gray, 1975
							3091		
Viking B	49/17	Dec. 65	49/17-1	1972	P. Rotl	Horst		G	Gage, 1980
Viking C	49/16	Jan. 71	49/16-3	1972	P. Rotl	Horst		G 80, C 2	Morgan, 1991
Viking D	49/17	Feb. 73	49/17-9		P. Rotl	Horst		G	
Viking E	49/17	June 69	49/17-4	1972	P. Rotl	Horst		G	
Vulcan	49/21, 25	Apr. 83	49/21-6	1988	P. Rotl	Tilted fault block	2180	G 16	Pritchard, 1991
Waverley	15/21				JU				
Welland NW	53/4	Jan. 84	53/4a-5	1990	P. Rotl	Low anticline	4220	G 8	
Welland S	53/4	June 84	53/4a-6	1990	P. Rotl	Low anticline		G	
Wensun	49/28	Sept. 85	49/28-a4		P. Rotl	Struct.		G	
West Sole	48/6	Dec. 65	48/6-1	1967	P. Rotl	Struct.	2977	G 57	Butler, 1975
									Winter and King, 1991
Yare	49/28	May 69	49/28-3	1987	P. Rotl	Tilted fault block	2411	G 1	Werngren, 1991

G, gas; C, condensate; N, Natural gas fluids (NGL); O, oil; OIIP/GIIP, oil/gas initially in place; abnd, field now abandoned.
Sources of information: Norwegian Petroleum Directorate Annual Year Book (via A/S Norske Shell); Arthur Anderson Petroleum Services Group; Energistyrelsen, 1997; Department of Energy, 1996; Brooks and Glennie, 1987; Spencer *et al.*, 1987; Abbotts, 1991; Parker, 1993; and others.

2 Origin, Development and Evolution of Structural Styles

K.W. GLENNIE & J.R. UNDERHILL

2.1 Outline of the structural framework of the North Sea

2.1.1 Introduction and rationale

As a part of the north-west European continental shelf, the North Sea area has a long and complex geological history, with the later stages in its structural and stratigraphic development being largely controlled by its earlier history. The simple structural outline and associated historical development given here is intended to set the scene for the more detailed discussions given in later chapters. Emphasis is given to the pre-Devonian history, with the later Palaeozoic, Mesozoic and Cenozoic tectonics considered largely on the regional scale. This has been done in order to stress the effects of tectonic events that took place beyond the limits of the North Sea and to emphasize the main controls on the development and evolution of structural styles that have had a direct bearing on the basin's hydrocarbon prospectivity (Table 2.1).

2.1.2 Sources of data

Our understanding of the geology of the North Sea area is derived from two main sources:
1 The surface (and subsurface) geology of the surrounding land areas.
2 The subsurface geology of the North Sea itself.
By penetrating below the waters of the North Sea and its underlying cover of younger (Cenozoic and Late Cretaceous) sediments, we discover a relatively complex pattern of sedimentary basins and structural lineaments, which form a framework within which the hydrocarbons of the area are reservoired (Fig. 2.1). Most of the information that enables the structural and stratigraphic interpretation of the North Sea area to be deduced is based upon hundreds of thousands of kilometres of seismic data, which have been calibrated by the use of cores, side-wall samples, drill cuttings and wireline logs from an increasing number of exploration wells. Although giving unique insights into the evolution of this part of the north-west European continental shelf, it should be remembered that our knowledge of the component parts of the North Sea basin is still biased to areas covered by available detail and is thus dependent on the industry's assessment of any sub-basin's commercial prospectivity.

2.1.3 Evolutionary outline

A simple outline of the most important events in the geological evolution of the North Sea area can be given very briefly; they are listed in Table 2.1 and are highlighted in historical order as a series of cartoons in Fig. 2.2. The main controlling tectonic processes may be grouped under eight main headings, three of which relate mainly to original tectonic organization or to major plate-margin effects, such as ocean-creation and subduction–accretion processes; the other five relate to intraplate deformation:
1 *Precambrian events.*
2 *The Caledonian plate cycle*—the combined Late Cambrian to Late Silurian Athollian and Caledonian Orogenies (Fig. 2.2C). Prior to these events, the North Sea area comprised widely separated continental fragments in and marginal to different parts of the Early Palaeozoic Iapetus Ocean and Tornquist Sea (Fig. 2.2A,B).
3 *The Variscan plate cycle*, which lasted from Devonian to late Carboniferous times. Devono-Carboniferous rifting is possibly the result of adjustments between and along the margins of the formerly separate Laurentian and Scandinavian cratons. This is exemplified by Devonian reactivation of Caledonian lineaments (Fig. 2.2D) and by Early Carboniferous structural relief. The Late Carboniferous Variscan Orogeny marked the closure of the southern Proto-Tethys (or Rheic) Ocean and the creation of the supercontinent Pangaea (Fig. 2.2E; see also Fig. 5.23).

From the Permian to the present day, the British Isles and North Sea have largely lain in an intraplate setting. Despite this, the area has remained far from quiescent and at least five main regional structural events may be recognized.
4 *Permo-Triassic riftng and thermal subsidence.* Late Permian subsidence of the Moray Firth and the east–west-trending Northern and Southern Permian Basins (see Fig. 2.1) was possibly coeval with the initiation of subsidence in areas that were later to become the Viking and Central Graben systems (Fig. 2.2F). Subsequent Triassic to early Jurassic thermal subsidence was abruptly terminated by a phase of Middle Jurassic thermal doming.
5 *Middle Jurassic domal uplift*, in response to the development of a warm, diffuse and transient mantle-plume head, led to widespread erosion of the central North Sea area, volcanism and the subsequent development of a trilete rift system.
6 *Late Jurassic to earliest Cretaceous extensional tectonics*, which led to fault-block rotations and the formation of major structural traps within and adjacent to the Viking and Central Grabens. In contrast to areas west of Shetland, the phase of extensional basin development was followed by a phase of post-rift thermal subsidence in the North Sea during the later Cretaceous and Cenozoic.

Table 2.1 Simplified evolution of the Tethys and Atlantic Ocean tentatively related to North Sea post-Caledonian structural events. Note that the time-scale is not constant.

Ma*	Periods		Regional events: Tethys-related	Regional events: Atlantic-related		North Sea: Northern	North Sea: Moray Firth	North Sea: Central	North Sea: Southern	Ma*
0	Cenozoic	Miocene	ALPINE OROGENY	↑ SPREADING OF PRESENT MID ATLANTIC RIDGE		REGIONAL SUBSIDENCE CENTRED OVER GRABEN SYSTEM			↑	0
		Oligocene							Zechstein diapirism	
		Eocene		Plateau basalts		Doming of West Shetlands ↑	Uplift of Moray Firth & Scottish Highlands ↑	Renewed faulting in Central Graben / Inversion of Danish Embayment	Inversion of NW-SE trending sub-basins ↑	
		Palaeocene	PLATE COLLISION	SPREADING OF ROCKALL TROUGH EXTENDING NORTH TO NORWAY–GREENLAND SEA ↑		LATE RIFTING PHASE				
100	Cretaceous	Late	GRADUAL CLOSURE OF TETHYS ↑ / Rotation of Iberia					SUBSIDENCE in CENTRAL GRABEN / Extrusives in Danish Embayment		100
		Early		SEA-FLOOR SPREADING IBERIA–NEWFOUNDLAND	INTRAPLATE SETTING	Doming in N. Viking	Rapid subsidence of inner Moray Firth		Zechstein diapirism; Indefatigable erosion; Rapid Sole Pit subsidence	
	Jurassic	Late	SEA-FLOOR SPREADING IN TETHYS ↑	ONSET OF SEA-FLOOR SPREADING IN CENTRAL ATLANTIC ↑		ROTATIONAL FAULTING IN VIKING GRABEN ↑		DOMAL COLLAPSE & MAIN PHASE OF GRABEN FORMATION	RIFT & WRENCH TECTONICS	
		Mid				MAIN RIFTING PHASE	Volcanic activity in East	DOMING & LIMITED VOLCANIC ACTIVITY		
200		Early	RIFTING PHASE	RIFTING IN CENTRAL ATLANTIC ↑				Earliest Zechstein diapirism		200
	Triassic								4000 m Triassic in Polish Trough	
							DEVELOPMENT OF NW EUROPEAN BASIN/GRABEN SYSTEM	ZECHSTEIN FLOODING OF SUB-SEA-LEVEL BASINS		
	Permian	Late					SUBSIDENCE OF MORAY FIRTH, NORTHERN & SOUTHERN PERMIAN BASINS			
		Early	LATE HERCYNIAN WRENCH TECTONICS			E-W Scottish & Scanian dyke swarms: volcanics		Extrusion of L. Rotliegend volcanics		
300	Carboniferous	Stephanian	EARLY COLLAPSE OF VARISCAN FOLD BELT IN EUROPE ↑		VARISCAN PLATE CYCLE	↑			Inversion of Sole Pit Basin	300
		Westphalian D–A	VARISCAN OROGENY ↓	Rifting in Norway–Greenland Sea		RIFTING IN NORTH BRITISH ISLES ↓			2500 m U. Carboniferous VARISCAN FOREDEEP ↑	
		Namurian	PLATE COLLISION ↑	INITIATION OF NORTH ATLANTIC FRACTURE PATTERN		Renewed uplift of Scottish Highlands				
		Dinantian							BACK-ARC RIFTING ↓	
	Devonian	Late	STEPWISE CLOSURE OF PROTO-TETHYS	Possible strike–slip movement of Greal Glen Fault			Volcanics in Orcadian Basin	Marine Limestones in Auk & Argyll		
		Mid								
400		Early				Granites in Scottish Highlands	Subsidence in Orcadian Basin	Volcanics in S. Scotland	Granites in Lake District	400
						CALEDONIAN		OROGENY	Extrusion of L. Rotliegend volcanics	

*Ma, million years.

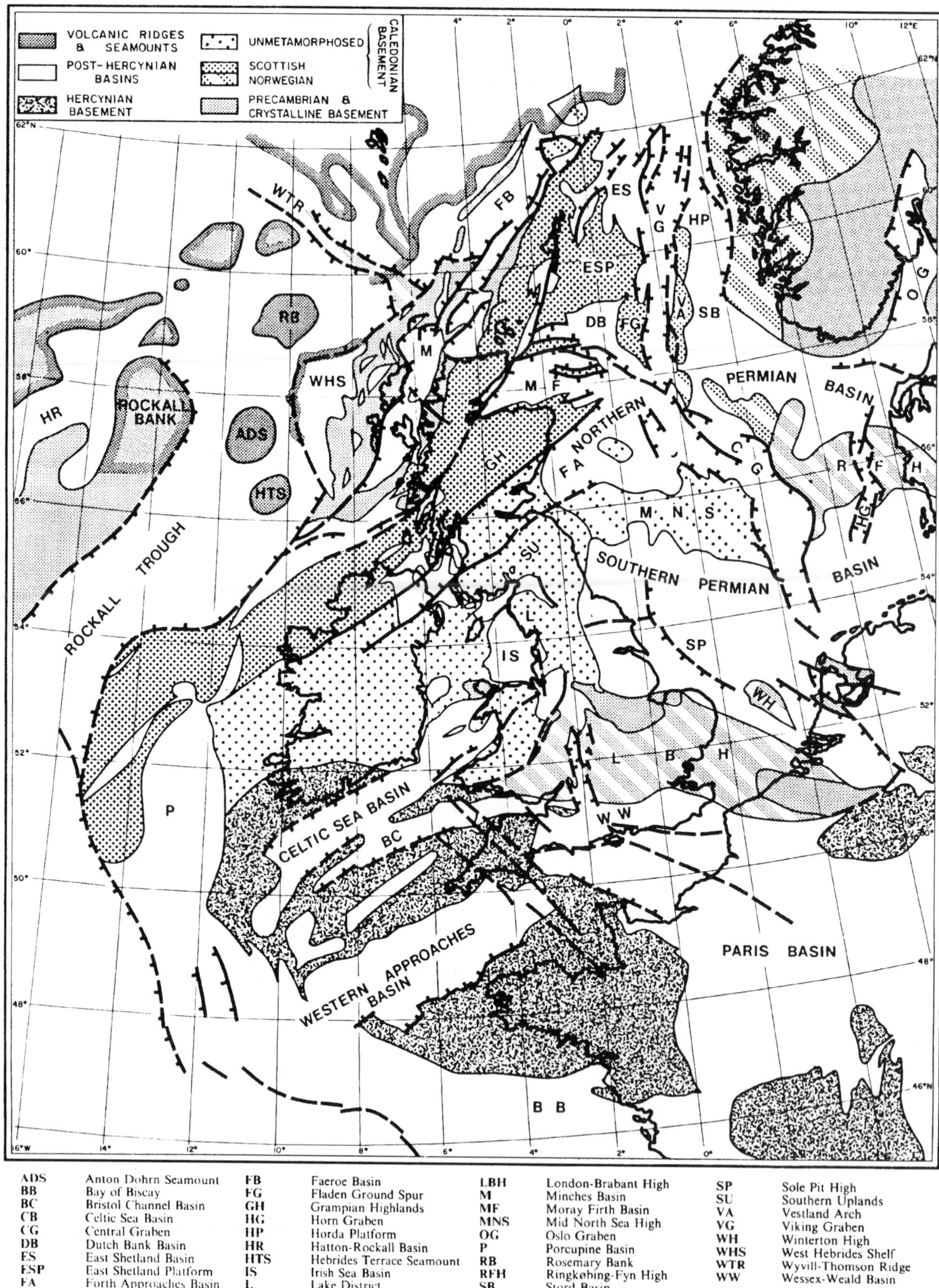

Fig. 2.1 Megatectonic map of the British Isles and surrounding areas depicting the main sedimentary basins and structural domains off the north-west European continental shelf.

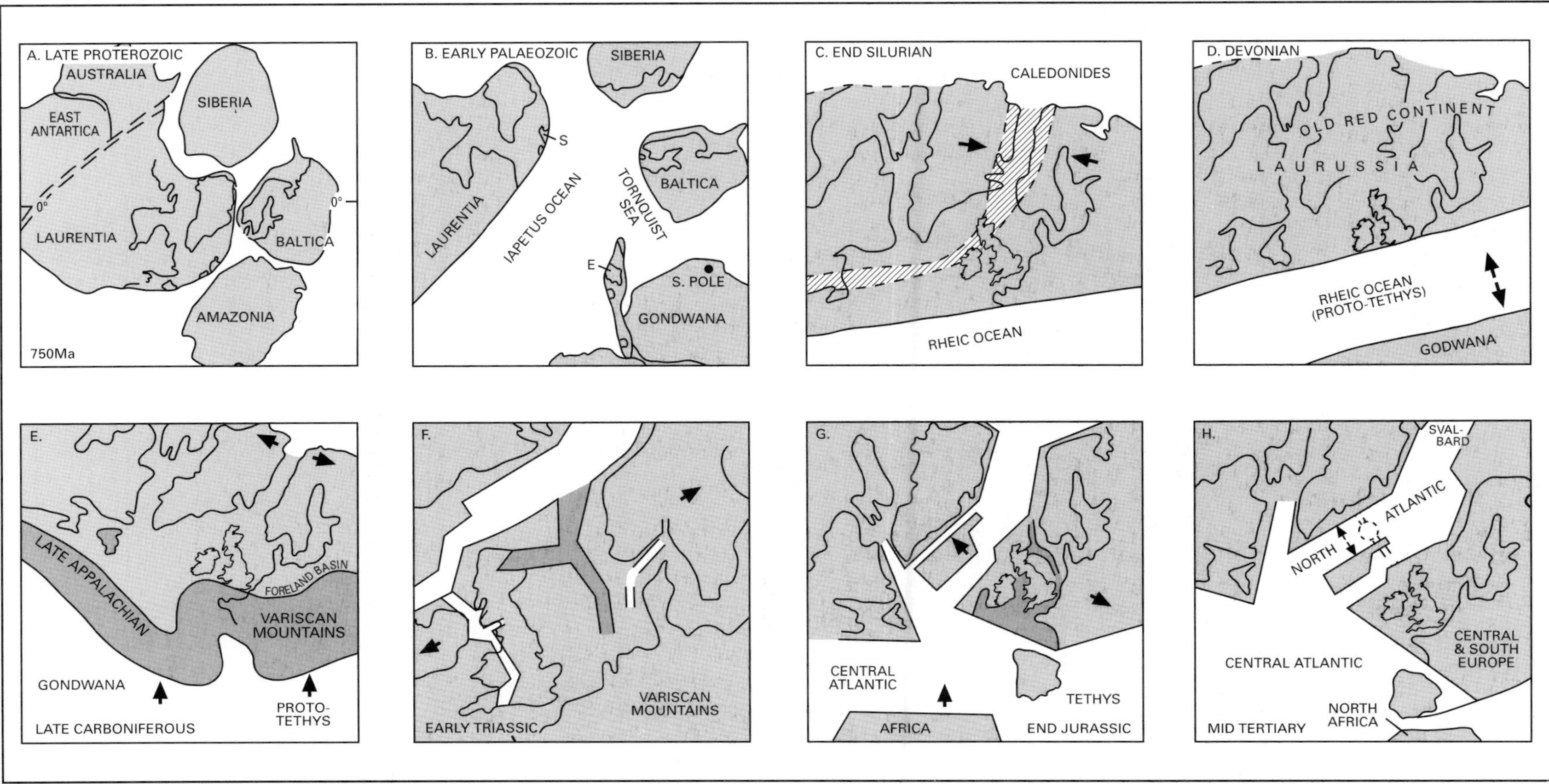

Fig. 2.2 Schematic diagrams depicting the main tectonic events that have shaped the North Sea basin during the late Proterozoic and Phanerozoic. In B, S = Scotland and E = England.

7 *Development of the Iceland hot spot and North Atlantic rifting.* In western parts of the UK, any North Sea tectonic influence was superseded during the Cretaceous by extension linked to the onset of sea-floor spreading in the North Atlantic Ocean (Table 2.1). Initially, this was along the line of the Rockall–Fareoes Trough (Fig. 2.2G) but it shifted to its present axis (Fig. 2.2H) by the mid-Tertiary. Opening of the Atlantic Ocean and the development of the Iceland hot spot were major factors in Cenozoic uplift and exhumation of the British Isles. The resultant regional tilt particularly affected the western rift arm in the North Sea, the Inner Moray Firth (Thomson and Underhill, 1993; Hillis *et al.*, 1994).

8 *Tectonic inversion of Mesozoic basin.* Creation of the Atlantic Ocean caused intraplate compression, leading to the tectonic inversion of former sedimentary basins across north-west Europe during the Late Cretaceous and Tertiary.

In addition to the main structural events listed above, regional tectonics had a further modifying effect on the tectonostratigraphic development of the area, mainly through the plate-tectonic control on climatic setting. Superimposed upon the changing pattern of crustal fragmentation and reunification was an overall slow northward passive drift of the continents. This drift took the southern North Sea area from south of the equator prior to the Carboniferous to its present location over halfway from the equator to the North Pole (Fig. 2.3; Habicht, 1979; Smith *et al.*, 1981). It had a latitudinal climatic effect on fauna and on sedimentation but cannot be considered of structural importance without relating this drift to other plate movements.

The combination of time- and latitude-related climatic changes and the pattern of structural deformation, erosion and sedimentation is responsible for oil and gas being in sedimentary reservoirs whose ages range from Devonian to Cenozoic. These changes also largely controlled the accumulation of prolific source-rock horizons during discrete time intervals. The bulk of the oil and gas in the central and northern North Sea basins is derived from just one source rock, the Late Jurassic Kimmeridge Clay, whereas most of the oil of the Netherlands and northern Germany is from another, Liassic source. In contrast, the southern North Sea gas province has been charged almost entirely from the Carboniferous Coal Measures.

2.1.4 Major structural features

Several structural elements are readily identified as a result of the tectonic events outlined in the previous section. As may be seen from the Top Chalk contour map (Fig. 2.4), the North Sea contains the site of an axis of considerable Tertiary subsidence, which is flanked by the positive areas of the British Isles to the west and Scandinavia (including the Danish peninsula) to the east. The southern margin is marked by the northern terrestrial limit of the Alpine foreland (the Netherlands and Germany), but, both in a physiographical and in a structural sense, the northern limit of the North Sea follows the edge of the north-north-east-trending Atlantic continental margin just beyond Shetland at about latitude 62°N.

The pre-Cretaceous structure of the North Sea area seems to be dominated by two east–west-trending basins, the larger Southern and the smaller Northern Permian Basin, which, as their names imply, came into existence during the Permian; they are separated by the Mid North Sea–Ringkøbing–Fyn system of highs and are surrounded by positive areas of older deformed rocks.

Cutting both basins and highs almost at right angles to the basins' axes is a system of grabens. The most important, in terms of both structural effect and its association with hydrocarbon accumulations, are the Viking and Central Graben systems. The Viking Graben lies north of the Northern Permian Basin and separates the Shetland Platform from the Fenno-Scandian High. The Central Graben, however, cuts both of the Permian basins and their intervening high.

Other areas of major subsidence include the Horn–Bamble–Oslo and North German systems of grabens, and the Moray Firth Basin, whose history of subsidence also began in the Permian. Beyond the North Sea area, other basins, grabens and half-grabens also developed at about this time (e.g. Manx–Furness, Solway and Ulster Basins in the British Isles); although following a pattern of subsidence differing from that seen in the North Sea area, their origins must be causally related.

Mining, drilling and seismic data reveal the presence of an even older sedimentary basin beneath the floor of the

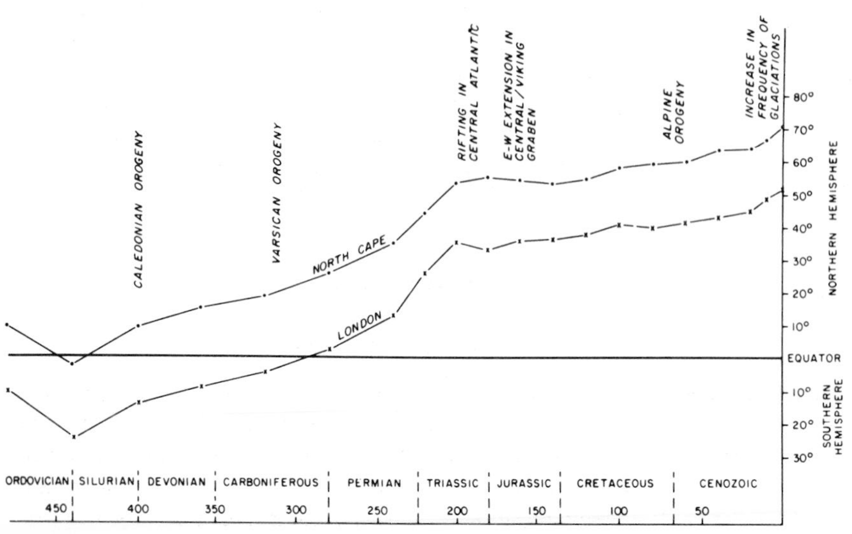

Fig. 2.3 A plot of the progressive northward drift of north-west Europe since the Ordovician. Major tectonic events seem to have had little effect on the steady rate of drift.

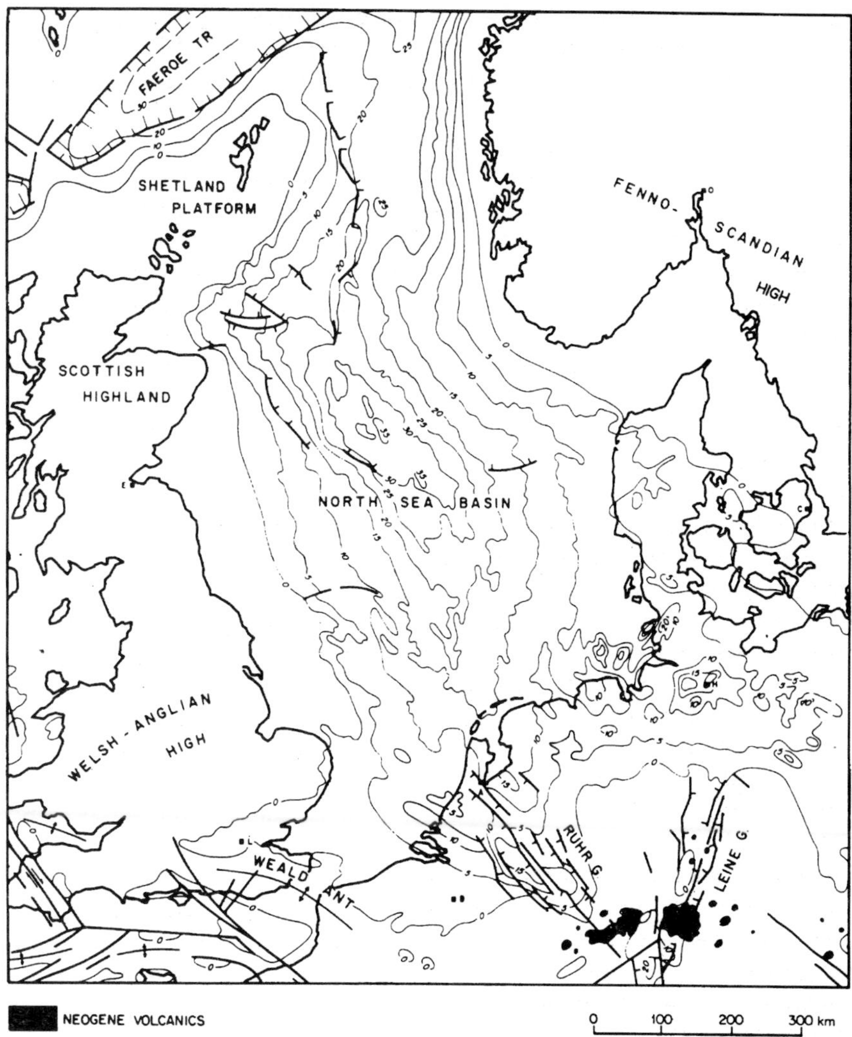

Fig. 2.4 Cenozoic Isopach map (contour interval in 100s of metres) (after Ziegler, 1990b).

Southern Permian Basin (Fig. 2.5). This deeper basin is important to the oil industry, because it contains abundant Carboniferous coal seams, which are the source rocks for most of the southern North Sea gas. With an ever-increasing cumulative thickness of coal, the surface of this basin extended as a broad plain southward from the Caledonian Highlands of Scotland and Norway; further south, the older Carboniferous sediments are of marine origin along the northern edge of the Variscan foredeep, which is now largely represented by eroded remnants of the Variscan Mountains (Fig. 2.5). Relative uplift and erosion following the Late Carboniferous Variscan Orogeny (e.g. Mid North Sea High) has reduced the area within which the Carboniferous Coal Measures are preserved.

Crossing southern Scotland obliquely and extending beneath the North Sea is the Midland Valley Graben, which virtually dies out in the Forth Approaches Basin before the Central Graben is reached; also the Southern Uplands, a positive area of strongly folded Lower Palaeozoic strata, seems to be continuous, as a structural unit, with the north-western part of the Mid North Sea High (Fig. 2.1).

Farther to the north, the folded and strongly metamorphosed Caledonian rocks of the Scottish Highlands and western Norway are wedged against the separate Precambrian cratons of the Hebrides Platform in the west and the Fenno-Scandian (Baltic) Shield to the east (Figs 2.1 and 2.5). During the Devonian, the synorogenic and post-orogenic Old Red Sandstone filled all the areas of relative subsidence within and beyond the Highlands with the erosional products of the Caledonian Mountains. The shape of the younger basins and highs, especially in the northern North Sea, has been strongly influenced by reactivation of Caledonide lines of weakness (see, for example, Johnson and Dingwall, 1981).

It is clear from the foregoing that the structural geometry of the North Sea area can be considered as the result of plate movements involving tension and compression in different directions at different times in its developmental history. The first of the cartoons outlining this history (Fig. 2.2A) indicates that it is important to return to the Precambrian to understand the development and evolution of structural styles more completely.

2.2 Pre-Devonian history

2.2.1 Precambrian beginnings

The history of the North Sea area prior to the Late Silurian–Early Devonian Caledonian Orogeny is known only in relatively simple terms. The results of studies in Britain, Scandinavia, Greenland and Canada, however, are beginning to give a logical, if still tentative, history of events

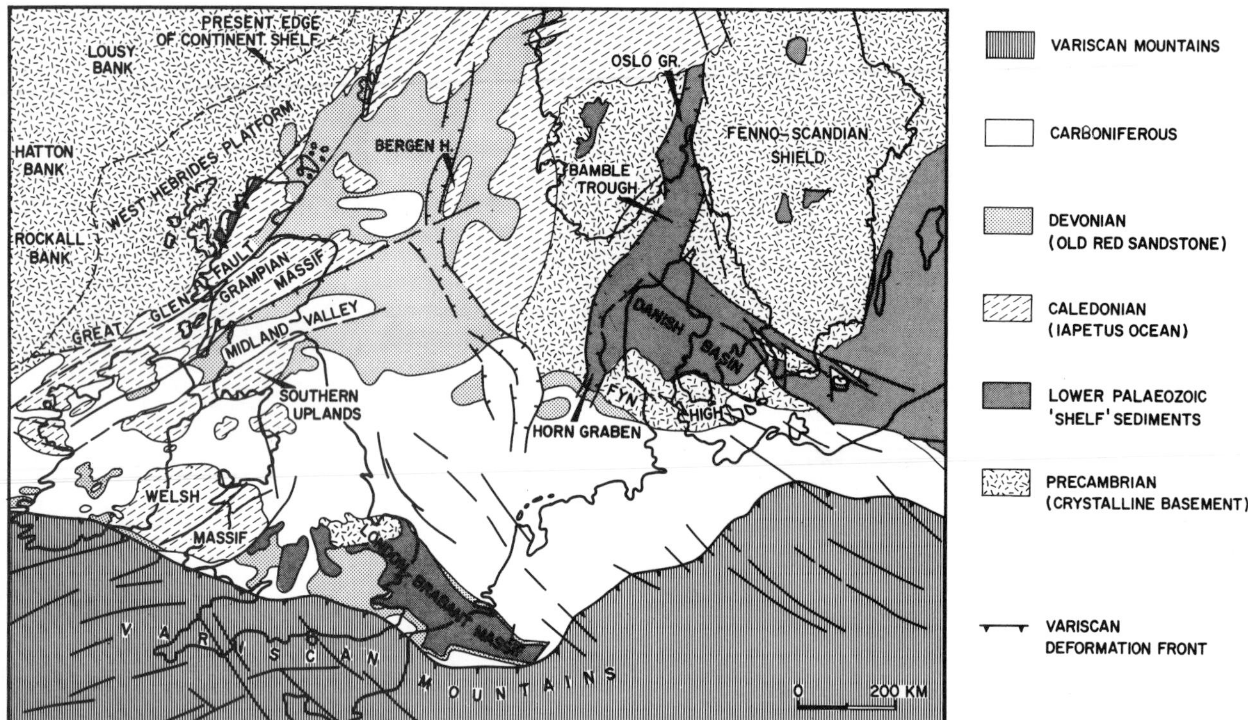

Fig. 2.5 Pre-Permian geological map showing the main Variscan, Caledonian and Precambrian basement elements of the British Isles and surrounding areas (modified from Ziegler, 1982a,b).

that have direct implications for the offshore area. An excellent starting-point for discerning this history is in the Scottish Highlands. The depositional and orogenic outline given here is an attempt to unify several somewhat diverging views and is based largely upon Johnson (1983a,b), Harris (1983, 1985), Barker and Gayer (1985), Kneller (1987), Barr *et al.* (1988), Rogers, G. *et al.* (1989) and Gibbons and Harris (1994).

The ages of Precambrian rocks in Britain are being carried further back as radiometric dating techniques improve and more rocks are dated (Gibbons and Harris, 1994; Harris *et al.*, 1994; Stephenson and Gould, 1995); the older the year of dating, the more suspect that age unless corroborated by other workers. The origin of the early Proterozoic ('Archaean') Lewisian Gneiss, so characteristic of the Outer Hebrides (Figs 2.1 and 2.6), extends back some 3000 million years (Ma) or more. Park *et al.* (1994) depict (from Winchester, 1988) how the Precambrian crystalline rocks of Scotland and northern parts of Ireland are related to similar rocks in Canada, Greenland and Scandinavia around 1500 Ma. Apart from the Lewisian, the metamorphic rocks of the Scottish Highlands can be divided into two major subgroups, the Moines and the Dalradians, which collectively cover a probable time span of deposition of some 400–600 Ma (Table 2.2); the precise time span and even the boundary between them are still in some doubt.

Deposition of these two metamorphic units was probably coeval with the still unmetamorphosed terrestrial Stoer and Torridon Groups of the north-west Highlands. Although Stewart (1982) dated deposition of the Stoer Group at less than 1000 Ma, it is now believed to extend back to about 1200 Ma (Rogers and Pankhurst, 1993). The younger Torridon Group, up to 6000 m thick, is dated by Stewart (1982) as around 780 Ma, but, as no lower limit is known, it is tempting to believe that its deposition continued into the later Proterozoic. Indeed, Johnson (1983a) implies that sedimentation of the Torridonian continued until around 700 Ma, when he considers deposition of the Dalradian to have begun.

It is now known that both the Moine and the Dalradian rocks are entirely Precambrian in age (already deformed Dalradian was intruded by the Ben Vuirich granite, with a crystallization age of 590 ± 2 Ma; Rogers, G. *et al.*, 1989). Dalradian sedimentation was formerly thought to extend to what used to be known as the Early Ordovician Grampian Orogeny, now the Athollian Orogeny (Table 2.2); for Caledonide orogenic events, see below.

According to Powell and Phillips (1985), the Moines were deposited during the time span 1200–1000 Ma, although Holdsworth *et al.* (1994) suggest a range of 1000–850 Ma as more probable. They were subjected to several phases of deformation and metamorphism (Table 2.2), so that they now consist mainly of psammitic metasediments, locally gneissic, and some calc schists. Their latest time of metamorphism was during the Early Silurian emplacement of the Moine Thrust towards the west-north-west, and the intrusion of many granites between then and the Early Devonian Period (Fig. 2.6).

East of the Great Glen Fault, the Moines (*sensu lato*) are subdivided by some workers into older and younger sequences, the Central Highland and Grampian Divisions, respectively (Barr *et al.*, 1988), which are separated by either an unconformity, a tectonic slide or both; for discussion, see Highton (1992). The Grampian Group is cut by muscovite pegmatites dated at *c.* 750 Ma. It seems increasingly likely that the only distinction between the Central Highland and Grampian Groups is in their degree of

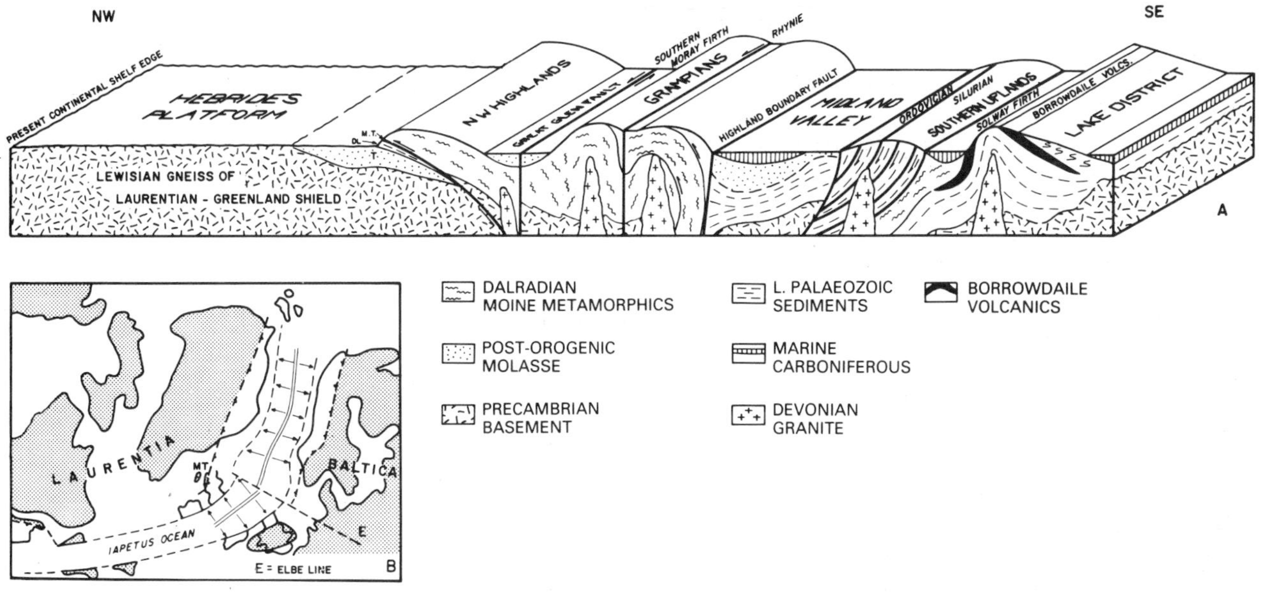

Fig. 2.6 A. Block diagram depicting the main Caledonian structures between the edge of the Hebrides Platform and the southern Lake District. B. Tentative reconstruction of the Iapetus Ocean. MT, Moine Thrust.

deformation and metamorphism (A. Crane, personal communication). Many workers, however, consider that there is no direct correlation between the Moines and the Central Highland Division across the Great Glen Fault (Kneller, 1987; Stephens and Gould, 1995; Table 2.2). Instead, the Grampian Group is now considered by many geologists to form the basal division of the Dalradians (Kneller, 1987; Table 2.2), separated from the overlying Appin Group by an unconformity. Carrying this reasoning to its ultimate conclusion, the Dalradians can be considered as including the Central Highland Division as its basal unit and the late Proterozoic Southern Highland Group at its top (the classification adopted here; Table 2.2), covering a time span of at least 200 Ma and possibly much more.

With this classification, Dalradian deposition possibly began at about the same time as the Moines of the Northern Highlands (± 1200 Ma) and continued until the Late Precambrian Orogeny (*c.* 600 Ma; see Rogers, G. *et al.*, 1989). Like the Moines, Dalradian deposition began in alluvial or shallow-marine environments, but the upper part of the supergroup includes deeper-marine turbidites, derived from the west and north-west, which were possibly deposited over oceanic crust. In partial conformity with the above line of reasoning, it is suggested that, despite some lithological similarities, the Central Highland Division of the Dalradians (Table 2.2) found east of the Great Glen Fault may not have been deposited in close proximity to the type area of the Moines west of that fault (Fig. 2.6). Indeed, as will be explained below, it might be simpler and more correct if the Moines are considered as being confined entirely to the area north-west of the Great Glen Fault and the Dalradians entirely to its south-east. Lindsay *et al.* (1989) point to the general similarity in lithostratigraphy between the Moines and Dalradians on opposing sides of the fault as possible evidence for its lack of importance as a terrane boundary, but admit that the contrast in structural styles would have to be explained.

At some time in the Late Precambrian (? 780 Ma age of Morarian pegmatites or older; Soper and Anderton, 1984), Laurentia and Fenno-Scandia (Baltica) separated to create the intervening Iapetus Ocean (Harland and Gayer, 1972; Fig. 2.2A,B). Supporting evidence from north Norway indicates that this could have happened in the range 800–750 Ma (see Drinkwater *et al.*, 1996). The Northern and Grampian Highlands of Scotland lay along the eastern edge of Laurentia, while southern Britain occupied a location within the pre-existing southern Iapetus (Proto-Tethys) Ocean off the northern edge of the megacontinent Gondwana. At this time, southern Britain and Gondwana were separated from Fenno-Scandia by the Tornquist Sea (Figs 2.2B, 2.7 and 2.8).

As pointed out by Coward (1990), the orogenic histories of the separate Northern and Grampian Highlands, the Midland Valley and Southern Uplands of Scotland, and their westward extensions into Ireland and the eastern Atlantic, are much easier to resolve if they are considered to have attained their present juxtaposition by left-lateral strike–slip movements spread over much of the pre-Devonian time span (Hutton and Alsop, 1996, Fig. 16; Snyder *et al.*, 1997, Fig. 1). All these units seem to have been associated with different parts of the Laurentian continental margin adjacent to the Iapetus/Laurentian–Cadomian Ocean. The progressive north-easterly accretion of these allochthonous terranes will have been limited in the east by whatever subduction mechanism controlled closure of the northern arm of the Iapetus Ocean between Greenland and Scandinavia. With final closure of the northern Iapetus Ocean, these different cratonic units may have had a slightly zigzag suture; crustal tension weakness of this suture probably helped to create the outline of the much younger Viking–Central Graben system, which seems to occupy the likely location of that suture in the northern North Sea area. This tentative model should be kept in mind when reading the rest of this chapter.

Table 2.2 Possible pre-Caledonian stratigraphic ranges of the different terranes of Scotland and the Iapetus Ocean. Very accurate modern radiometric dates are likely to alter some of these age ranges and their correlations.

Ma*	Caledonian Foreland	Northern Highlands	Grampian Highlands		Midland Valley	Southern Uplands	
400		'CALEDONIAN	OROGENY'				
415		Granites	Younger granites		Sediments		
425	Final movements of Moine Thrust						
440	Early movements of Moine Thrust				? ~~~~~~~~	Accretion Wedge	I
455		Metamorphism					A
	~~	~~~~~~~~	470 ATHOLLIAN				P
	Marine	?Grampian	Older granites	HBC†	?	? ~~~~~~~~	E
	?	Orogeny	Metamorphism				T
			510 OROGENY			?	U
				?			S
550	Sediments	Granites			?		
	?						O
600			Grampian Orogeny	~~~~	?Grampian Orogeny		C ?
			Southern Highland Group	D			E
			Argyll Group	A			A
650	?	?	Varangian glacial	L			N
			Appin Group	R			?
750	————	Morarian Event	~~ ?Unconformity ~~	A	Precambrian Basement?		–
				D			
	Torridonian Group sediments		Grampian Group	I			
				A			
1000	?			N			
	Stoer Group sediments	Deposition of Moines	Central Highland Division	?			
1200?	~~~~~~~~	~~~~~~~~	~~~~~~~~	~~~~			
1700	LAXFORDIAN OROGENY	Lewisian Inliers	?Lewisian Inliers				
2700	SCOURIAN OROGENY						
3300	OLDEST KNOWN LEWISIAN						

*Ma, million years.
†HBC, Highland Border Complex.

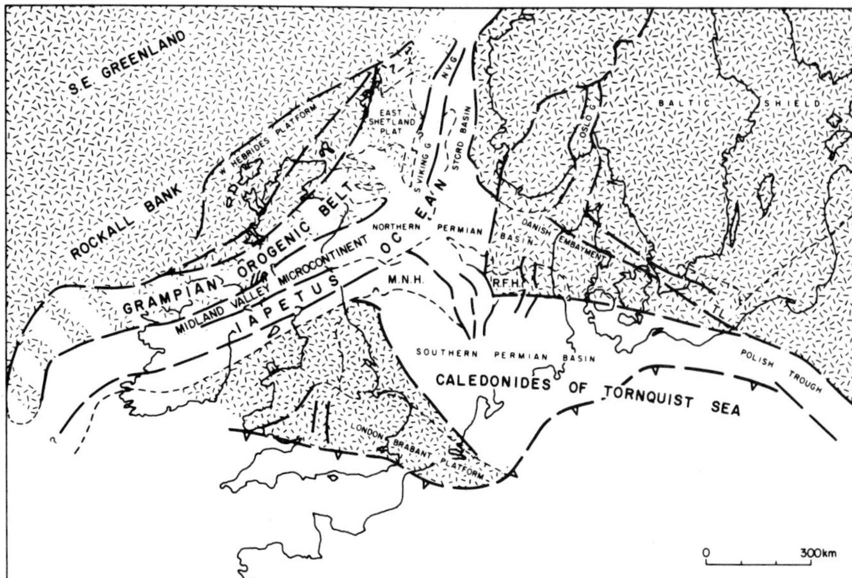

Fig. 2.7 Schematic diagram to illustrate the general setting for the Iapetus Ocean and the Tornquist Sea during Ordovician times relative to the younger Northern and Southern Permian Basins and Viking–Central system of grabens.

2.2.2 The Caledonian plate cycle

Early Palaeozoic development

The Early Palaeozoic history of north-western Europe is dominated by the Finnmarkian (northern Norway), Athollian (former Grampian; see Rogers *et al.*, 1989) and Caledonian orogenies, which are seen variously as the products of continent–ocean and continent–continent collision and major transpression.

Much of the marine sediments involved in the Athollian and Caledonian orogenies was deposited in the Iapetus Ocean. Anderton (1982) suggested that the opening of Iapetus was signalled by eruption of the Tayvallich Volcanics around 600 Ma (Late Precambrian) in an ensialic basin north of the Midland Valley. Whatever their true age, the Tayvallich Volcanics probably represent a phase of transtensional crustal separation within the Dalradian depositional domain, whereas the opening of the Iapetus Ocean could have been as far back as 850–810 Ma (Drinkwater and Alsop, 1996). Clearly, much has still to be learnt about the history of relative movement of the continental plates at this time (see Fig. 2.2A; Torsvik *et al.*, 1996).

A similar narrow basin is thought to have formed along the south-east margin of the Laurentian continent early in the Palaeozoic. Contrary to the interpretations of Henderson and Robertson (1982), this became the site of marine sediments of the Highland Border Complex (Table 2.2); associated serpentinites, gabbros and spilitic lavas suggest that the basin was floored by crust of oceanic type. If the concept illustrated in Fig. 2.7 is correct in principle, then the Highland Border Complex would represent sediments deposited in a narrow arm of the Laurentian–Cadomian part of Iapetus, following rift separation of the Midland Valley terrane from Laurentia. This event would have had no direct relationship with the northern leg of Iapetus between Laurentia and Fenno-Scandia, which could be much older, dating back, speculatively, to the Morarian Event or beyond (Table 2.2; Drinkwater and Alsop, 1996).

An increase in the volume of the world's midocean ridges is thought to have caused a Cambrian transgression over the older terrestrial equivalents of the Moine and Dalradian. This transgression is also recognized in the Baltic area, for instance, where the locally still-immature Alum oil-shale sequence was deposited (see Chapter 11). Unfortunately for Denmark, these rocks rapidly become postmature for oil further to the west (Thomsen *et al.*, 1983).

From the Mid-Cambrian onwards, the eastern margin of the Laurentian Plate occupied a near-equatorial location (Cocks and Fortey, 1982) and became the site of carbonate-platform deposition, with sediment starvation in more basinal areas. It has long been known that the faunas of the Scottish equivalent of this equatorial Cambro-Ordovician carbonate shelf (Durness Limestone, Figs 2.6 and 2.7) have a much greater affinity to those of the Beekmantown Limestone of Pennsylvania (see, for example, Phemister, 1960) than to the physically now much closer sequences of Wales. Cocks and Fortey (1982) used the climatic effects on assemblages of shallow-marine, planktonic and deep-water benthic faunas to reconstruct probable changes in the width of the Iapetus Ocean with time. In addition to the east coast of Laurentia, including northern Scotland, they found that the two highest nappes in the Trondheim area of Norway must also have lain close to the equator during the Early Ordovician. As the rest of Scandinavia and south Britain lay some 40–60°S of the equator, strike–slip movement between the site of origin of the high Trondheim nappes and the rest of Scandinavia seems likely. Using the provinciality of faunal assemblages as a guide, Cocks and Fortey (1982) recognized that Scandinavia and the Gondwana-derived microcontinents, including south Britain, must have been separated by a deeper-marine area, which they called the Tornquist Sea; this corresponds to the suture now marked by the north German–Polish Caledonides (Ziegler, 1982a). The Tornquist Sea had closed by the end of the Ordovician (Figs 2.7 and 2.9A).

Closure of the Tornquist Sea apparently did not give rise to a major fold belt. It is suggested here that the subduction geometry was such that the Tornquist Sea closed by oblique convergence between the Baltic and the Caledonides of central and eastern Europe. Support for this is found in a zone of low-grade metamorphism along the north-eastern

A

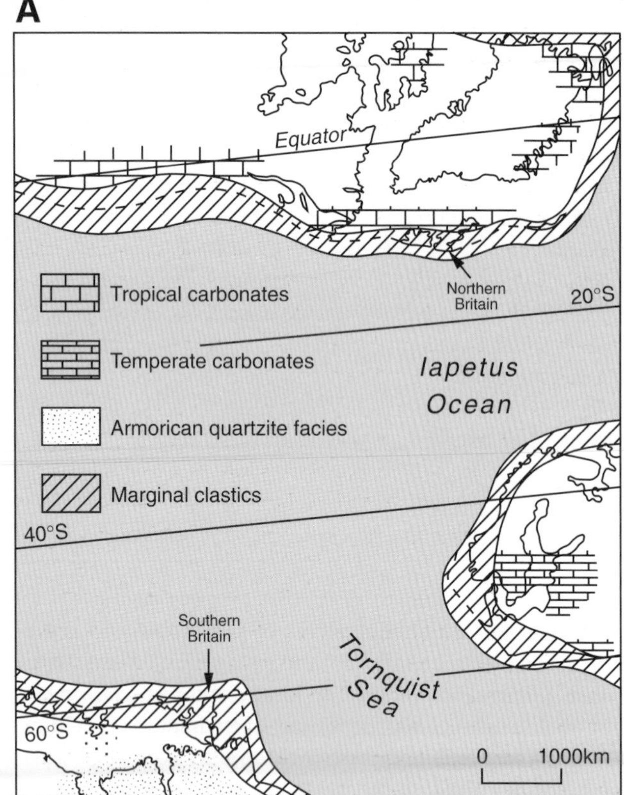

B

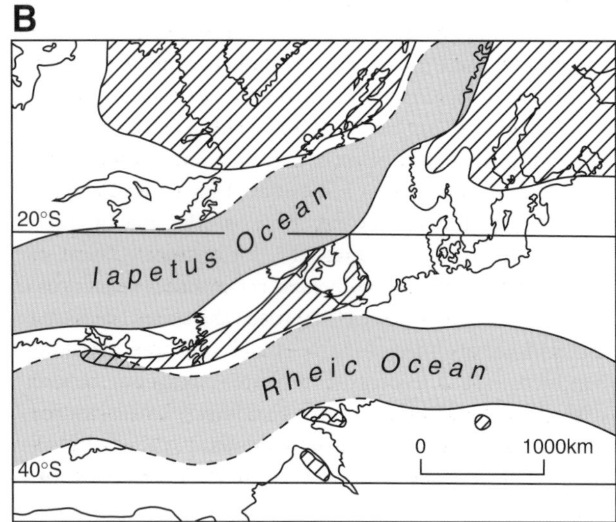

C

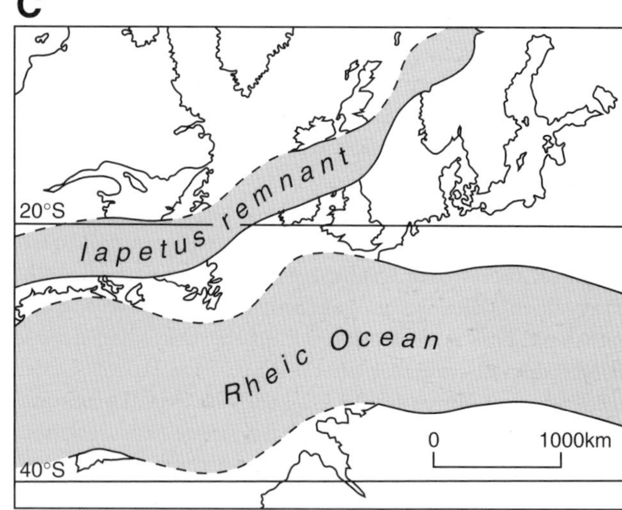

Fig. 2.8 Three stages in the oceanic separation of Britain during the Early Palaeozoic. A. Early Ordovician (Arenig). B. Early Silurian (Llandovery). C. Late Silurian (Ludlovian). Adapted from Cocks and Fortey (1982).

flank of the Polish Anticlinorium (site of the former Polish Trough), adjacent to the Tornquist–Teisseyre shear zone; this metamorphism resulted in the Alum Shale becoming overmature within the shear zone.

Radiometric ages (Ziegler, 1982a, 1990a) indicate that the southern limit of cratonic Scandinavia probably coincided with what, in the Permian, became known as the Ringkøbing–Fyn High. Other fractures associated with the Tornquist–Teisseyre shear probably created another intermediate zone of weakness along the northern edge of the Ringkøbing-Fyn High. This became activated only when a suitable transtensional stress pattern was generated in the Permian and Triassic, with subsidence of the Norwegian-Danish Trough.

Caledonide orogenic events

Some of the Late Proterozoic and Early Palaeozoic sedimentary and volcanic rocks of the Caledonides were deformed and metamorphosed more than once during the (?) 800 Ma of their developmental history, the older events being partly masked by the younger. We have already noted several metamorphic and deformation events within the Moine during its Precambrian history. Much of the Grampian region was subjected to another orogenic event, involving D1/D2 deformation just before the Cambro-Precambrian time boundary; because of the original definition by Lambert and McKerrow (1976), Rogers, G. *et al.*

(1989) suggest that this phase of deformation should now be known as the Grampian Orogeny. The main phase of Barrovian metamorphism in the Dalradian rocks of the north-eastern Highlands, which culminated in emplacement of the 'older' granites (see, for example, Johnstone, 1966; Bradbury *et al.*, 1976) was during the Early Ordovician; Rogers, G. *et al.* (1989) suggest that this event should now be known as the Athollian Orogeny. Most of these different times of deformation and metamorphism, and the ages of associated igneous activity in the Caledonides of the British Isles, are very clearly displayed on compilation maps in Harris (1985). It should be realized, however, that, with ever-increasing accuracy in radiometric dating, some of these ages, and the inferences made from them, may well be revised in the future.

Early in the Ordovician (about 500–460 Ma; Stephenson and Gould, 1995), the Dalradian of north-eastern Scotland was intruded by an almost rectilinear pattern of gabbro (the 'Older Basics' of the Huntly, Insch and Morven massifs; Ashcroft *et al.*, 1984), which suggests that either the sedimentary sequence in this area overlay oceanic crust or the fractures penetrated through continental crust to the upper mantle. Certainly, these rocks were subjected to late-emplacement shearing before 460 Ma. Gunn *et al.* (1996) show that the ultramafic rocks of the Portsoy Lineament contain elements that predate the 'Older Basics' and possibly represent subvolcanic portions of an island arc. If true, the lineament, which can be followed from the Moray Firth coast far to the south and south-west of the Grampian Highlands as a zone of imbricated metabasic rocks, would represent a former crustal separation of the Southern Highland Group and part of the Argyll Group from the essentially older rock units further to the west at some time between 595 and 653 Ma (Stephenson and Gould, 1995, Fig. 10).

The Grampian area was deformed and strongly sheared during the Athollian Orogeny, when the Midland Valley microcontinent collided with it transpressively, emplacing the Highland Border Complex at the same time. With deposition of the whole of the Dalradian now entirely within the Precambrian, former Dalradian sequences, such as the Lower Cambrian Leny Limestone, must now be considered as more closely allied to the post-Dalradian Highland Border Complex.

The Grampian Highlands must have been subjected to a period of considerable crustal thickening during the Athollian Orogeny. Constrained by grades of metamorphism and relics of presumed Devonian reddening, Watson (1985) suggests that 25–30 km of overburden was removed by erosion during the period between the Athollian and Caledonian orogenies, a figure not much greater than that proposed for parts of north-eastern Scotland by Hudson (1985). The whole Highland area was again deformed and suffered a further degree of metamorphism during the Late Silurian to Early Devonian Caledonian Orogeny.

As mentioned earlier, the Iapetus Ocean can be considered as having two different parts, separate closure of which resulted in orogenic events of different type.

1 Silurian closure of the northern Iapetus Ocean brought the Norwegian Caledonides into collision with those of Greenland and the Northern Highlands of Scotland. Products of this collision include the westward thrusting of the Moine over the Lewisian basement and its cover of pre-Silurian sediments (see Fig. 2.6). Eastward-dipping offshore reflectors, interpreted as thrust planes, have been identified on deep seismic reflection profiles just north of the Scottish mainland and Isle of Lewis (Brewer and Smythe, 1984; Snyder *et al.*, 1997). The thrusts extend to depths exceeding 30 km; extensional reactivation of these old lines of weakness later resulted in the creation of local half-grabens with a fill of Permo-Triassic sediments (see, for example, Fig. 12.72). The collision also involved obduction of the Unst ophiolites on to the Dalradian rocks of Shetland (Dewey and Shackleton, 1984; see also Mykura, 1976). The ophiolites have a fusion age of about 498 Ma, similar to that of the 'Older Basics' of the north-eastern Highlands. Even though cold when obducted, Flinn *et al.* (1991) and Flinn (1993) think that the obduction did not postdate fusion by long, thereby implying emplacement during the Athollian Orogeny.

2 On the east side of the North Sea, Brekke *et al.* (1984) indicate the emplacement of an ophiolite and overlying ensimatic arc sequence on to the margin of the Baltic Shield in the Sunnhørdland area of south-western Norway, probably during the Arenig; a younger sequence of back-arc volcaniclastics, cherts and turbidites were then superimposed on the older sequence while the Iapetus oceanic crust was subducted beneath the Baltic Shield. For outline discussion, see Nicholson (1979). The Tornquist–Teisseyre fault system had its origins in Ordovician closure of the Tornquist Sea against the Baltic Shield (see also Pegrum and Ljones, 1984); associated fault splays probably also created a zone of crustal weakness, extending across what was to become the Danish Embayment to the north of the Ringkøbing–Fyn High (see Fig. 2.7).

3 The southern Iapetus separated the microcontinents of south Britain and south-eastern Newfoundland (see Fig. 2.6B) from the Laurentian margin in north Britain and the Appalachians. The Athollian Orogeny was probably caused by a violent transpressive sinistral collision between the Scottish Highlands and a Midland Valley landmass that was underlain by rigid Precambrian basement (Longman *et al.*, 1979). A probable relic of this closure is the Highland Boundary Fault.

4 Late Silurian closure was also strongly affected by left-lateral, locally transpressive, strike–slip movements, which caused the juxtaposition of other allochthonous terranes of Laurentian origin (Dewey, 1982; Dewey and Shackleton, 1984). For example, close to the Highland Boundary Fault (see Fig. 2.6), slivers of older Palaeozoic rocks of the Highland Boundary Complex were thrust over the Dalradian during the Caledonian Orogeny; the Highland Border sequence lacks any of the large volume of material that must have been eroded from the Highlands since the Early Ordovician, and is therefore thought to have been brought adjacent to the Dalradian Highlands by strike–slip movements prior to obduction.

5 Early Ordovician uplift of an area in the vicinity of Newfoundland is likely to have provided a source for the conglomerates and turbidite fans that spread southward across the southern Iapetus Ocean to what is now the Southern Uplands (McKerrow and Elders, 1989). Northward partial subduction of these fans during the Late Silurian gave rise to the imbricate (McKerrow *et al.*, 1977; Webb, 1983) accretionary wedge of the Southern Uplands (Fig. 2.6). The Southern Uplands Fault may date from initiation of this accretionary wedge.

6 Closure of the Iapetus Ocean seems, at different times, to have been achieved by both north-west-directed and south-east-directed subduction (see, for example Phillips *et al.*, 1976, Fig. 4). Within the British Isles, the line of closure is marked by a suture that can be traced from the Shannon estuary into the north-east-trending Solway and Northumberland Troughs. Its straightness and lack of major disruption implies a strike–slip origin, with relatively slow convergence of the two sides of the ocean. This suture possibly meets those of the northern Iapetus and Tornquist Sea at a triple junction within the Central Graben, to the north of the Mid North Sea High.

7 South of the Iapetus suture in the British Isles, the Lake District of northern England and the Wicklow Mountains of south-eastern Ireland were the sites of extensive Mid-Ordovician andesitic volcanism (e.g. Borrowdale Volcanics, Fig. 2.6) typical of a subduction-related island arc. Late Silurian granites within the Lake District may be coeval with those postulated by Donato *et al.* (1983) to have intruded the Mid North Sea High. The current proximity of the Lake District portion of this arc to the line of suture can be accepted if the missing part of the accretionary wedge is assumed to have been removed by oblique strike–slip subduction; such a wedge is still preserved north of the Wicklow Mountains. The relatively straight nature of the Great Glen and Highland Boundary faults may be an indication of terrane boundaries formed during the accretion of discrete fragments of crust during the Late Silurian or Earliest Devonian.

With its assymetrical flanks, the broad structure of the Lake District (see Fig. 2.6) suggests that a zone of décollement exists between the pre-volcanic sedimentary sequence and the crystalline basement that is interpreted to underlie it. Such a zone would be in line with the partial subduction of this basement beneath the Southern Uplands accretionary wedge during the Late Silurian. This interpretation receives support from both Johnson (1984) and Bott *et al.* (1985), who suggest that the Iapetus suture dips beneath the Southern Uplands (see Figs 2.6 and 2.7; Leggett *et al.*, 1983, Fig. 3). The Lake District possibly represents a microplate distinct from the London–Brabant microcontinent, whose Lower Palaeozoic sediments are marked by a distinct lack of the strong Caledonian folding that characterizes the Welsh Basin, for instance. The limit of the London–Brabant microcontinent beneath the North Sea is uncertain, but might coincide roughly with the modern Dowsing Fault Zone (e.g. Fig. 2.7A).

Most of the major strike–slip faults that bound the different Caledonian terranes of the British Isles formed lines of weakness that were later reactivated whenever crustal stresses had to be relieved; such lines of weakness extend out to the graben system in the middle of the North Sea. For example, reactivation of the Tornquist trend has led to multiple phases of oblique-slip deformation in the southern North Sea.

2.3 Post-Caledonian structure and basin development

So far as the limited area of the North Sea is concerned, the details of its post-Caledonian sedimentary and structural development can be found in the succeeding chapters. This geological evolution was dependent to a considerable extent, however, on events that took place beyond the limits of the North Sea, and it is to the simple outline of their interrelationships that this section is devoted. Some of these correlations can be found in Table 2.1.

The eventual understanding of these events has depended to a very large extent on interpretation of the mass of seismic data of ever-improving quality that has been shot over most of the North Sea. In many areas, most post-Variscan seismic events can now be interpreted regionally with a remarkable degree of ease and accuracy, but, because of a lack of clear continuous reflectors, this ease of interpretation can rarely be extended to the seismic expression of older rock units in the area, except when dedicated to deeper crustal reflections (see Brewer and Smythe, 1984). The difficulty of interpreting deep seismic events is not helped by the general lack of sufficient drill penetration to provide ground truth for the interpretations.

2.3.1 The Variscan plate cycle

This cycle involved the development of syn-rift, post-rift and inversion-related stratigraphic megasequences that were controlled primarily by episodic rifting, periodic fault reactivation and eustatic sea-level changes.

Development of Devonian (Old Red Sandstone) intermontane basins

Closure of the Iapetus Ocean resulted in the creation of the megacontinent Laurussia (Figs 2.2, 2.6B and 2.9) and in uplift of a major mountain range that stretched from the southern United States to eastern Canada (Appalachians) through northern Britian to the northern end of the united Greenland–Scandinavian craton (Fig. 2.9A). Widespread Early Devonian granitic intrusions (e.g. the 'newer granites' of north-eastern Scotland; Johnstone, 1966) and associated volcanic activity testify to the anatectic remobilization of crustal material of continental type during orogenesis. Some of these intrusions were probably responsible for the local destruction of older Palaeozoic source rocks for oil and their conversion to graphite, as at Seathwaite in the Lake District (Parnell, 1982a).

Erosion in an almost vegetation-free continental climate that seems to have had a seasonal rainfall (Barrell, 1916) resulted in deposition of widespread fluvial sequences, which in some cases terminated in intramontane lacustrine basins (e.g. Lake Orcadie; Geikie, 1879). Fish beds in the lacustrine sediments of the Orcadian Basin are famed for the richness and diversity of the species they contain and have been considered as a potential source of oil in the Inner Moray Firth (Peters *et al.*, 1989). Reservoir rocks can be found in Mid and Upper Old Red fluvial and aeolian sandstones now exposed in Caithness, Orkney and Shet-

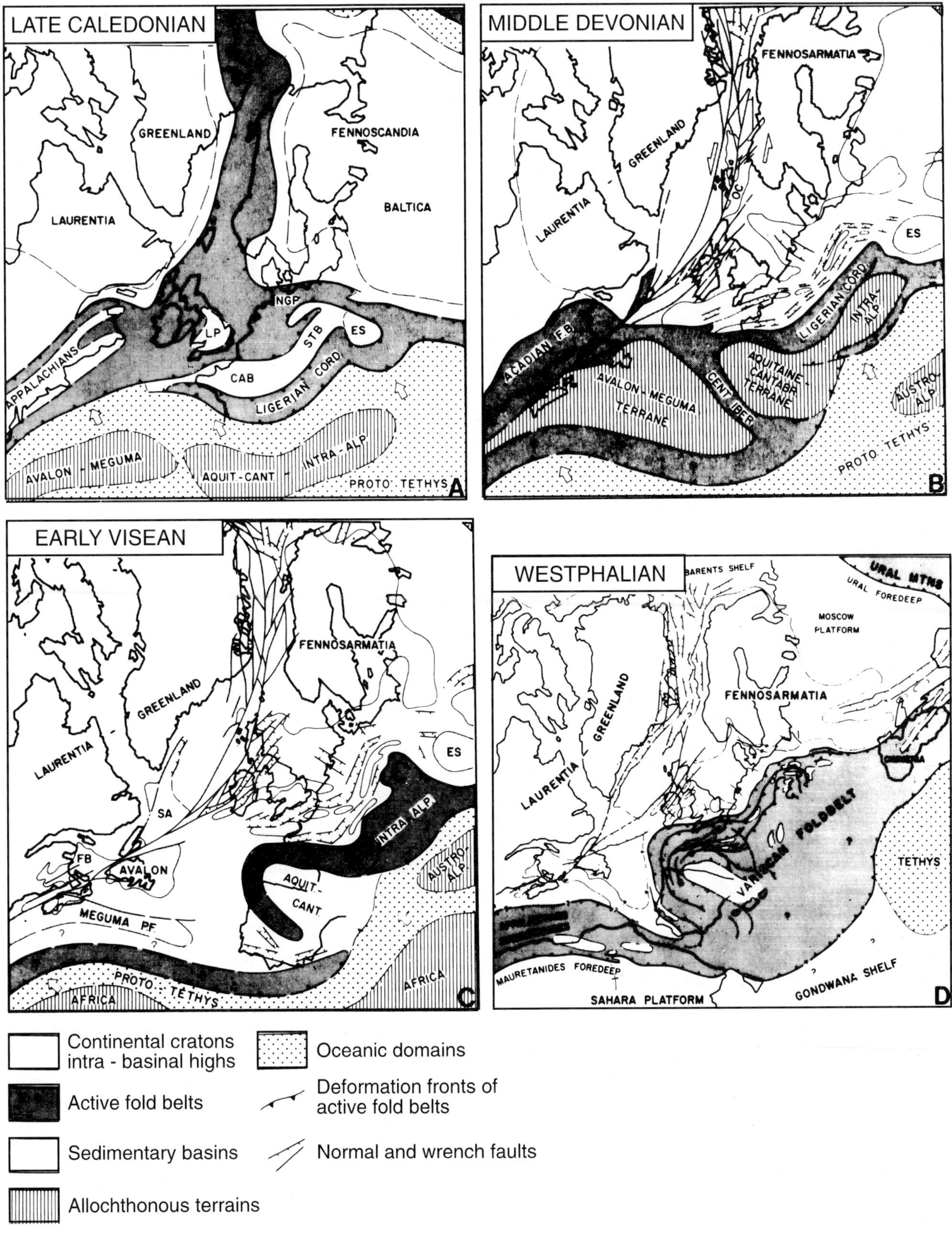

Fig. 2.9 Schematic palaeogeographic reconstructions depicting the tentative tectonic framework for Late Caledonian deformation, the Middle Devonian, Early Carboniferous (Early Visean) and Late Carboniferous (Westphalian). After Ziegler, 1990a.

land (Mykura, 1976, 1983a; Allen and Marshall, 1981) and which are locally impregnated with dead oil.

A major Old Red Sandstone basin (Fig. 2.10) seems to have occupied much of the northern North Sea area, a reflection, possibly, of crustal weakness inherited from the northern Iapetus suture between Laurentia and Fenno-Scandia. Along the southern margin of the newly formed Caledonian Range, early Old Red Sandstone drainage was parallel to the mountain front and, over a considerable distance, converged towards the present North Channel between Scotland and Ireland. A changing basin configura-

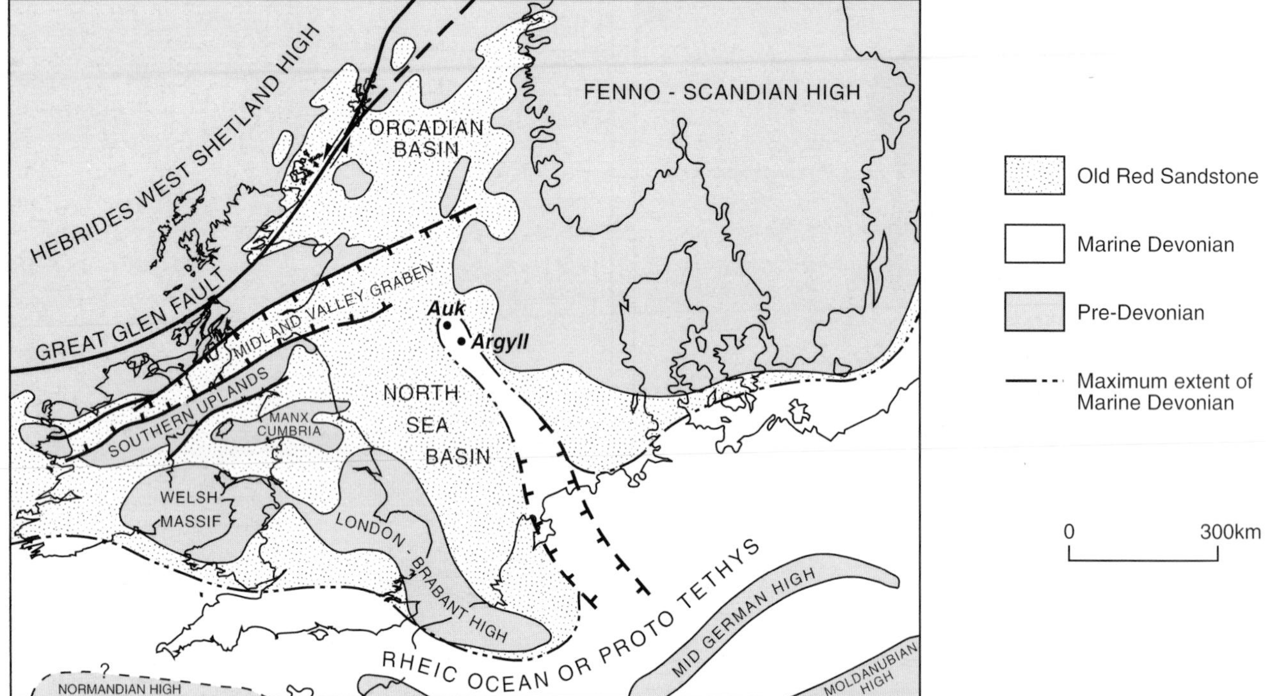

Fig. 2.10 Maximum extent of continental Old Red Sandstone and marine sediments of the Proto-Tethys (Rheic) Ocean continental margin (modified from Ziegler, 1982a).

tion resulted in late Old Red Sandstone sediment transport in the Midland Valley and Northumberland Trough being directed towards the North Sea area (Simon and Bluck, 1982). The Devonian (Proto-Tethys) sea lay across southern Britain (Fig. 2.10 and see also Fig. 3.3).

The Orcadian Basin (Fig. 2.10), within which the Middle Old Red Sandstone is up to 5 km thick, seems to have developed on both sides of the major Great Glen Fault (Figs 2.1 and 2.9B and see also Fig. 3.4), across which both the sense and amount of horizontal displacement is in dispute. Although originally given a post-Devonian sinistral offset of 110 km by Kennedy (1946), an offset of up to 2000 km in the same sense (see Fig. 2.9B) was proposed by van der Voo and Scotese (1981) on the basis of palaeomagnetic data. This proposal was hotly disputed by Donovan and Meyerhoff (1982), Parnell (1982b and Smith and Watson (1983), on the grounds that the unique facies patterns and biostratigraphy present on either side of the fault have not been recognized elsewhere. These latter authors therefore suggest that, if there has been any movement, it has been in a dextral sense and, especially since the onset of Middle Devonian time, of small amount (Rogers, D.A. *et al.*, 1989; Underhill, 1991a; Thomson and Underhill, 1993; Underhill and Brodie, 1993).

The deposition and deformation of the thick Old Red Sandstone sediments in the Orcadian Basin are probably the result of mostly closed drainage within an intramontane basin formed by tensional stresses associated with orogenic collapse of the Caledonian Mountains. Three horizons of marine fossils, however, have been reported from Orkney by Marshall *et al.* (1996), who suggest that the marine incursions entered Lake Orcadie from the Tornquist–Teisseyre Zone at times of maximum Devonian sea level.

To the south of the Old Red Sandstone continent, the Proto-Tethys or Rheic Ocean (Figs 2.9 and 2.10) had been in existence since the Early Palaeozoic, and would remain so until its closure led to the Late Carboniferous Variscan Orogeny (see Fig. 2.2E). Following a structural trend that possibly developed from the line of weakness represented by the northern Iapetus suture, an arm of the Mid-Devonian sea spread as far north as the Auk and Argyll oilfields, where shallow-marine carbonates were deposited. This embayment appears almost to coincide with the line later adopted by the Central Graben; again, repeated inheritance of an old line of weakness seems most likely.

Carboniferous basin development

With the slow northerly drift of Laurussia, Early Carboniferous sedimentation represents a transition from the relatively arid conditions of the southern hemisphere tropics that prevailed at the end of the Devonian to the more humid equatorial conditions of Coal Measure deposition (Habicht, 1979; see also Fig. 2.3). The patterns of sedimentation were strongly influenced by synsedimentary tectonics, with many of the strong north-west–south-east and north-east–south-west structural trends developed in the northern England Carboniferous being inherited from the mid-Palaeozoic Caledonian orogeny (Fig. 2.11). These fault trends were consistently reactivated throughout the Carboniferous in an extensional and, at the end of the Carboniferous, in a compressional sense (see Fig. 2.9C,D).

A number of transtensional basins and extensional grabens subsided during the Dinantian, including the Midland Valley of Scotland, the Shannon–Solway–Northumberland Trough (along the line of the old Caledonian suture) and the Bowland Basin (Figs 2.2, 2.11 and 2.12). Visean limestones locally reach 3000 m in the Solway–Northumberland Trough and yet are less than 500 m thick over the Alston

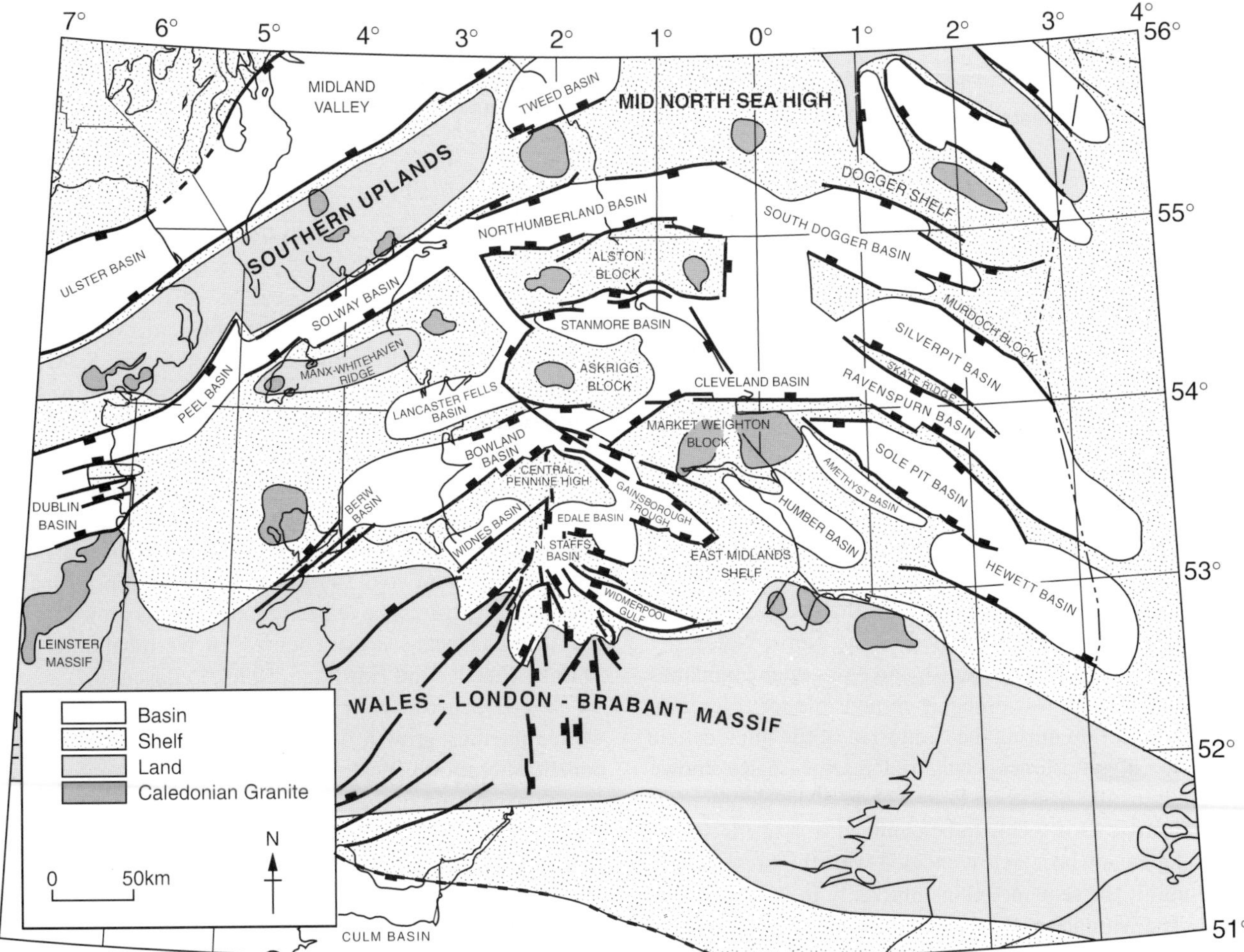

Fig. 2.11 Main tectonic elements that affected Early Carboniferous basin development. There is not only a strong Caledonian south-west–north-east and Tornquist north-west–south-east basement control, but also a control by granitic intrusions on the location of regional highs ('blocks'), between which major sedimentary depocentres ('troughs') developed (modified after Corfield *et al.*, 1996).

Block to the south (Taylor *et al.*, 1971; Fig. 17). Much of this relief was smoothed out by Visean sedimentation (Johnson, 1984; see also Fig. 2.9C). In some areas, such as the Midland Valley of Scotland, subsidence was accompanied by igneous activity under conditions of presumed crustal extension.

Erosion of the Caledonian Highlands and deposition of non-fossiliferous continental clastics in basinal areas continued to dominate the region north of the Highland Boundary Fault throughout the Early Carboniferous. South of the fault, and in continuation of the scene already set during the Late Devonian, a broad relatively flat plain occupied much of the southern half of the North Sea area and extended south to the axial parts of the Variscan foredeep basin. Marine conditions were established across the southern part of this plain, but, further north, suncracked shales and argillaceous dolomites of the basal Carboniferous Cementstone Group suggest deposition beneath inland sheets of water subjected to periodic desiccation (Cameron and Stephenson, 1985). In the Midland Valley of Scotland, the succeeding bituminous shales gave rise to Scotland's oil-shale industry, which began in 1851 and lasted just over 100 years (see, for example, Glennie *et al.*, 1987). Their deposition occurred within intrabasinal areas of differential subsidence related to oblique-slip transtensional movements on the Highland Boundary and Southern Uplands faults. Intervening highs, such as the Burntisland anticline, acted as barriers to sediment dispersal and occasional marine transgression from the south-east (Cameron and Stephenson, 1985).

Further east, the pre-Permian strata forming the floor of the Danish Embayment (see Figs 2.1 and 2.2) are mostly at depths that are too great (5000 m; Ziegler, 1982a) to form an economic target for the drill. It is still uncertain, therefore, whether or not they include a coal-bearing Carboniferous sequence; where calibrated by borehole data, only Silurian to Cambrian strata are present (i.e. Jutland, Zealand; Nielsen and Japsen, 1991). The reported presence of marine strata of Late Carboniferous age in the Oslo Graben (Olaussen *et al.*, 1982) poses the question of their former connection with the open ocean. These carbonates are of Westphalian age and relate to short-lived (glacio-eustatic) marine transgressions. The faunas indicate a possible connection with the Moscow Platform (Bergstrøm *et al.*, 1985), possibly via the Tornquist–Teisseyre shear zone.

Both graben development and igneous activity largely abated during the Namurian and ceased early in the Westphalian, presumably in response to north–south com-

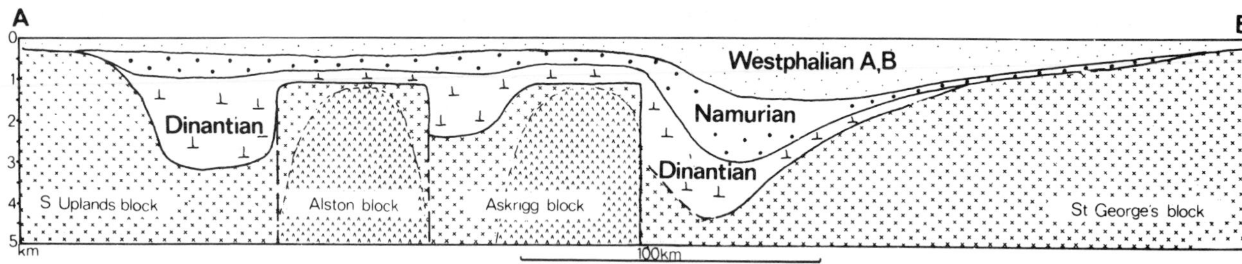

Fig. 2.12 Schematic north–south cross-section from the Southern Uplands, along the Pennine Axis to central Wales, showing the control that granite-cored blocks had on sedimentary thicknesses during Dinantian, Namurian and Westphalian deposition (after Leeder, 1982).

pression related to the Variscan Orogeny. The morphologically smooth and almost horizontal southern North Sea area of Westphalian time formed a marked contrast to the Early Carboniferous structural relief that must have been present between the Scottish border and North Wales.

Renewed uplift of the Scottish and Norwegian Highlands late in the Visean was probably responsible for the southerly progradation during the Namurian of the fluviodeltaic sequence of sandstones, coals and marine shales known generally as the Millstone Grit. The best development of the Millstone Grit seems to be confined to a north–south-trending zone whose axis coincided with the present Pennine uplift; the sequence thins markedly to both east and west. The Namurian is second only to the Westphalian in the economic importance of its coal seams (see Chapter 4).

Off the southern end of the Millstone Grit delta, a turbidite sequence was deposited in a deeper-marine environment, which also seems to have been the site of deposition of the source rocks from which the Eakring oil was derived. Separate development of local troughs has been recognized in the Grantham, Edale and Widmerpool areas. The first two follow a north-west–south-east Precambrian (Charnoid) structural trend, whereas the third aligns with the locally east–west orientation of the northern edge of the London–Brabant Platform (see Fig. 2.11).

The long history of the London–Brabant Platform as a positive area can be inferred from some of the southern North Sea wells, such as 47/29a-1, where Namurian strata overlie Ordovician rocks, and in Kent, where Westphalian Coal Measures onlap Silurian strata. The Platform was probably not transgressed by the sea until the Cretaceous, when it received a relatively thin cover of sediments (see Figs 9.2 and 9.8).

The Carboniferous of north-western Europe is very important to the hydrocarbon industry because it was during this period that the carbonaceous source rocks for the whole of the southern North Sea and Dutch–German–Polish gas belt were deposited. The main period of coal deposition was during the Late Carboniferous Westphalian stage, when coastal-plain paralic sediments prograded southward around (and only possibly across) the London–Brabant Platform. Up to 2500 m of strata were deposited locally, with a cumulative thickness of some 75–100 m of coal.

In the UK sector of the North Sea, the Westphalian sequence reaches a maximum thickness of some 1200 m in the vicinity of the Sole Pit axis of Late Westphalian inversion, whereas in the Netherlands (see Fig. 5.28) and Germany the total coal-bearing sequence (including Namurian) reaches over twice that thickness (Ziegler, 1977).

It was during the Westphalian to Stephanian time span that southern parts of the North Sea area and the Palaeozoic grabens of the British Isles were marginally affected by compressional deformations associated with the Variscan Orogeny. At least two of the anticlines, the Ravenspurn and Murdoch structures, have been interpreted as growth folds that acted as local sediment sources in Westphalian C and D times (Leeder and Hardman, 1990). Pedogenesis, caused by severe penetrative weathering associated both with synsedimentary growth of folds and postdepositional diagenetic alteration beneath the sub-Permian Unconformity, led to the creation of the 'Barren Red Bed' play of the Silver Pit area (Besly *et al.*, 1993; see also Chapter 4).

Variscan Orogeny

The Variscan Orogeny was the outcome of a collision between Gondwana and Laurussia. Both continents had been migrating passively northwards. The orogeny began in the Late Visean, when the more rapidly moving Gondwana collided with the southern margin of the slower-moving Laurussia (see Figs 2.2 and 2.9). It culminated in the creation of a mountain range that extended from north-west Spain, through Brittany and central Germany to southern Poland and beyond. By the Late Westphalian, the northern arm of Proto-Tethys (Rheic) Ocean had closed and some of the preserved Westphalian C terrestrial sediments of the southern North Sea area had a southern provenance.

The final closure of the Rheic Ocean occurred during the Late Carboniferous and led to development of the major west–east-trending mountain belt. Southern parts of Britain, which previously had formed part of the northern passive continental margin of the Rheic Ocean, began to experience rapid subsidence in a foreland-basin setting, ahead of a northerly migrating fold-and-thrust system. The resultant Culm Measures of south-west Britain subsequently experienced intense compressional deformation as it too was incorporated into the mountain chain. The northern limit to orogenic deformation is generally taken to be coincident with the 'Variscan Thrust Front', which extends from southern Eire, through South Wales to the Bristol area (see Fig. 2.5). The feature forms the effective southern limit to the Devono-Carboniferous play, because of the intensity of deformation and metamorphism experienced by rocks caught up in the mountain-building process (Smith, 1993). As a result of the Variscan deformation, the former northern passive continental margin of the Rheic Ocean was telescoped as the locus of thrusting migrated northwards. As a result, a series of east–west-trending thrusts formed in southern England, some of which are exposed in Devon and Cornwall while others form the basement to the Wessex Basin (Chadwick, 1993).

Regionally, in the late stages of the Variscan Orogeny, the rigid pre-Devonian margins of the Laurentian and Baltic components of Laurussia formed a re-entrant, into which the orogen was squeezed from the south (Figs 2.2 and 2.13). This stress pattern gave rise to closely spaced north-west–south-east-trending right-lateral wrench faults of mostly small offset, which transected much of the area between the north-east margin of the London–Brabant Platform and the Tornquist–Teisseyre line, including the eastern Variscan Mountains (see Fig. 5.5). These faults began the destruction of the Variscan Mountains east of the London–Brabant Platform almost as soon as they were formed, but the relatively rigid Platform partly protected the Variscan Range in western France from such shearing.

Foreland deformation

The effects of Late Carboniferous closure of the Rheic Ocean were not restricted to southern Britain, and deformation is now known to have affected large areas of the Variscan foreland (Fig. 2.14). Late Carboniferous contractional structures are well documented from the South Wales foreland basin, central and northern England (Underhill *et al.*, 1988; Fraser and Gawthorpe, 1990; Corfield *et al.*, 1996), the southern North Sea and as far north as the Midland Valley of Scotland, Easter Ross (Underhill and Brodie, 1993) and the Moray Firth (Thomson and Underhill, 1993). The structures display a range of orientations, which may be classified into three main basement trends: the Iapetus Domain, the Midlands Microcraton and the Tornquist Domain (Fig. 2.14; Corfield *et al.*, 1996). The

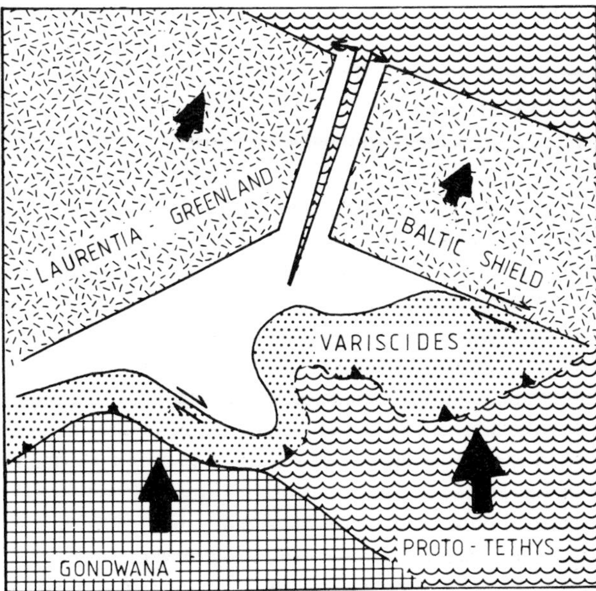

Fig. 2.13 Cartoon depicting the structural setting of the Variscan Orogeny resulting from the closure of the Proto-Tethys (Rheic) Ocean to the south.

relative orientation of the underlying basement grain and presence or absence of granitic pluton bodies seem to have predetermined the nature and severity of the compressional deformation. Those pre-existing structures that had a north-east–south-west Iapetus trend appear to have suffered the greatest amount of shortening, with many faults experiencing major reactivation and basin inversion.

Foreland deformation is seen to be the main trap-forming event for Carboniferous structures in the east Midlands and parts of the southern North Sea. All hydrocarbon discoveries made so far in the east Midlands display some element of Variscan inversion in their geometry. Their charge resulted from subsequent Mesozoic burial, ensuring that hydrocarbon generation after Variscan trap formation is the main control on the present-day distribution of hydrocarbons in the Carboniferous of northern England.

Subcrop patterns beneath the Permian illustrate the form that deformation took in the southern North Sea, with the formation of long-wavelength, north-west–south-east-trending folds (Fig. 2.15). Their existence controls the effective limits of several of the intra-Carboniferous plays, including the 'Barren Red Bed' play of the southern North Sea mentioned earlier (Besly *et al.*, 1993).

2.3.2 Post-Variscan basin development: the North Sea in an intraplate setting

With the possible exception of the Cenozoic, the North Sea area has affectively lain in an intraplate setting since the Early Permian. There were six main overriding events that affected the post-Variscan history of north-western Europe. Each of these has been influenced by the Pre-Permian geological and structural configuration inherited from the Precambrian, Caledonian and Variscan orogenies (see Fig. 2.5).

1 *Permo-Triassic postorogenic rifting*: following the Variscan Orogeny, intraplate extension led to the development of numerous Permo-Triassic basins and subsequent Triassic–Early Jurassic post-rift thermal subsidence.

2 Possibly the least influencial of the six on the structure of the North Sea was the *Late Triassic–Jurassic opening of western Tethys* (see Table 2.1) and the resulting resplitting of Pangaea into Laurasia and Gondwana.

3 *Middle Jurassic Doming of the North Sea* in response to the development of a transient mantle-plume head.

4 *Development of the trilete North Sea failed rift system*, consisting of the Moray Firth, Viking Graben and Central Graben rift arms. Subsequent Cretaceous and Cenozoic subsidence was largely, but not exclusively, controlled by a phase of post-rift thermal subsidence. Though an old line of weakness that intersected the trilete junction, the Tornquist–Teisseyre line seems not to have been active at the time of doming.

5 *The creation of the Atlantic Ocean* and the separation of Laurasia into North America and Eurasia (see Fig. 2.2G). A seaway between the Arctic and Central Atlantic had existed since the Early Jurassic (Hallam, 1977; Ziegler, 1988), but sea-floor spreading between the two continents was only partially achieved in the Middle Jurassic, at which time it was confined to the Central Atlantic. Spreading centres, established in the Rockall Trough and Labrador Sea–Bay of

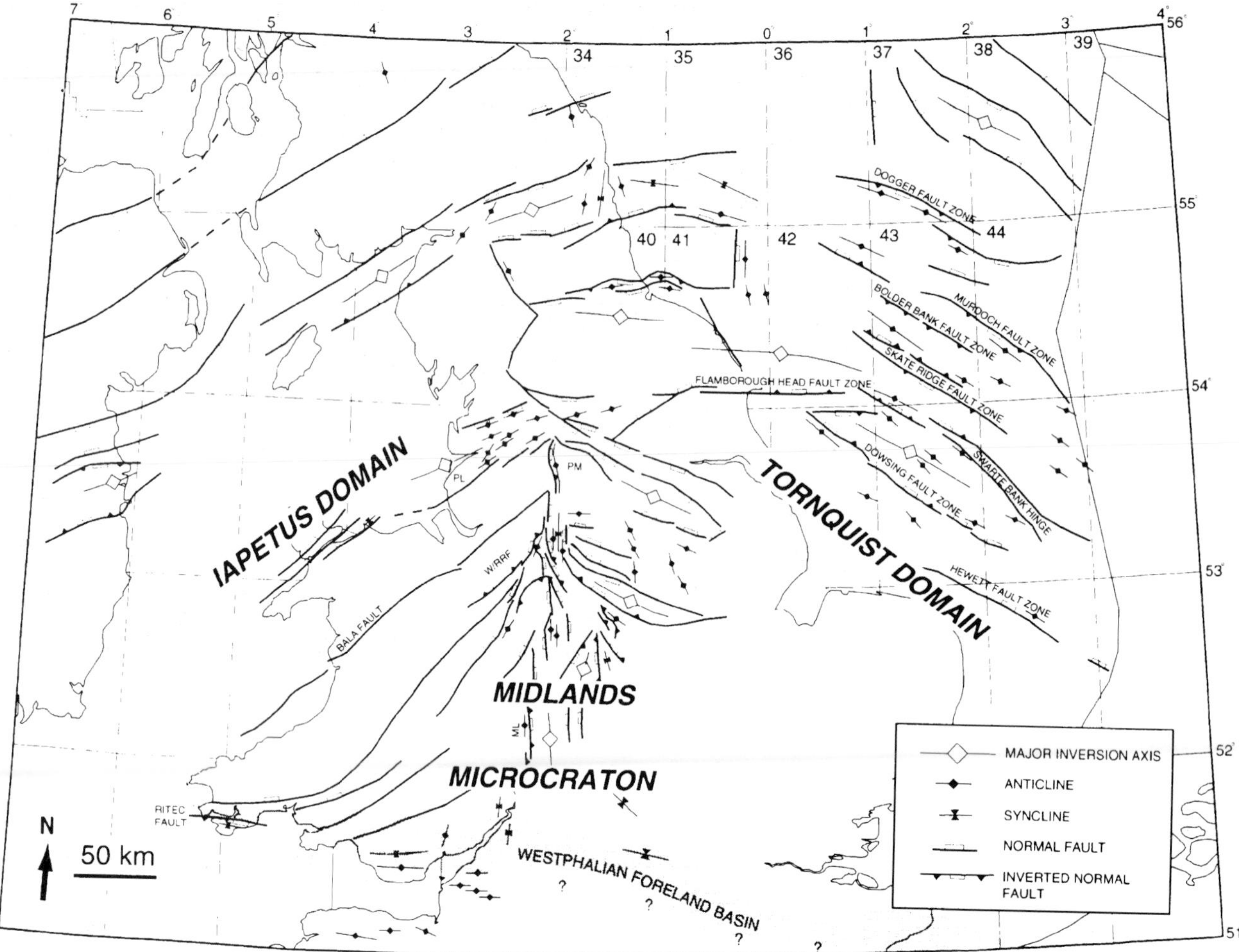

Fig. 2.14 Structural elements of the Variscan foreland. The main structures may be grouped into three main classes: those associated with Iapetus basement lineaments, those allied to the Tornquist trend and those within the Midlands Microcraton. The structures represent contractional (inversion) features caused by the effects of compression ahead of the Variscan Thrust Front (modified after Corfield *et al.*, 1996).

Biscay trends during the Aptain–Albian, had relatively short lives; the first ceased to spread at the end of the Cretaceous and the other early in the Oligocene (Ziegler, 1988). Complete crustal separation between North America and Europe was heralded by widespread Late Palaeocene (Thulean) volcanic activity in the flank areas of the Rockall–Faeroe Trough, southern Greenland and Norway (see also evidence in Bukovics *et al.*, 1984) and along the Reykjanes Ridge trend through Iceland.

6 The Cretaceous to earliest Tertiary *closure of the Tethys Ocean* separating Africa and Eurasia and the creation of the Alpine fold chain (see Fig. 2.2H). The combined effects of this and Atlantic opening placed north-west Europe into a compressional or transpressive stress regime, which led to important phases of basin inversion.

Early Permian extensional faulting and volcanism

Parts of the Variscan Mountains had begun to collapse during the late stages of the orogeny, with the development of several important intramontane basins. Strike–slip movements were already active during the earliest Permian, when the Variscan northern foreland was strongly affected. As might be expected from a stress pattern associated with complex block geometries, these movements were of both a transpressive and a transtensional nature. Horizontal displacement across the fault traces is thought to have been small, however, as major offsets cannot be demonstrated. With its northern edge anchored in the re-entrant made by the relatively rigid Laurentian, Scottish and Norwegian Caledonian basement (see Figs 2.1 and 2.13), strike–slip movement within the Variscan northern foreland is likely to have been limited.

Right-lateral transpressive movements resulted in inversion of former Carboniferous sub-basins, such as the north-west–south-east-trending Sole Pit axis (Figs 2.16 and 5.2–5.15; Glennie and Boegner, 1981). Transtensional movements on the other hand, gave rise to widespread mafic and intermediate volcanism (Lower Rotliegend; see Fig. 5.6), related to a pattern of both north-west–south-east- and conjugate north-east–south-west-trending grabens and half-grabens that developed across much of the foreland area. Accurate dating of Late Westphalian tuffs in Germany (Lippolt *et al.*, 1984) requires that the Carboniferous–Permian boundary be carried back to about 300 Ma (see review by Leeder, 1988) or perhaps 296 Ma (see Fig. 5.4; Menning, 1995). Thus, depending upon the age adopted for the Permo- Carboniferous boundary, the Lower Rotliegend volcanics are either entirely within the Permian or extend down into the Stephanian, as long believed.

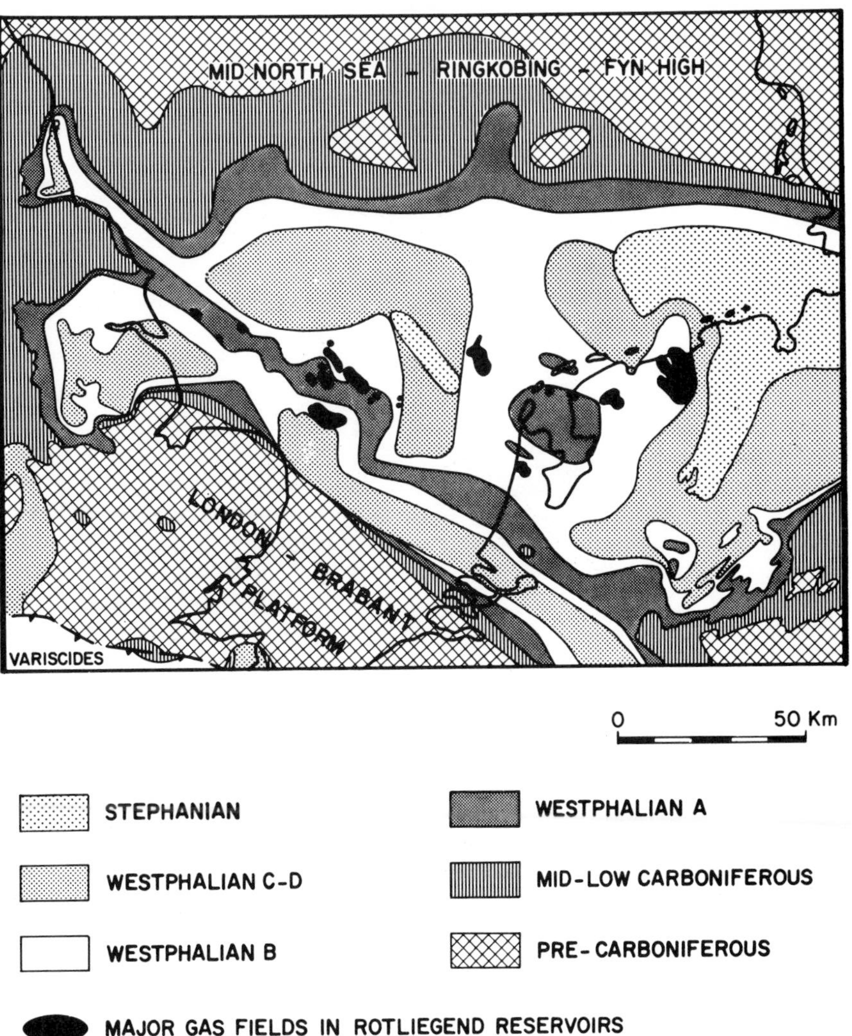

Fig. 2.15 Pre-Permian subcrop of the southern North Sea showing the importance of open folding in the Variscan foreland in preserving Late Carboniferous sequences.

Mafic volcanism was especially important in Germany, Poland and the eastern North Sea areas (see Fig. 5.6), where it was probably associated with right-lateral wrench movement along the old Tornquist–Teisseyre fault system (Ziegler, 1982b). The relationship between graben development and volcanism is exemplified by a cross-section through part of north Germany (see Fig. 5.7; Gast, 1988). Volcanics blanketed a faulted surface whose rocks varied in age between Namurian and Stephanian, indicating that a horst and graben relief involving erosion of the highs already existed before the start of igneous activity. A drilled thickness of over 3000 m of volcanics implies local caldera-related subsidence. To obtain such widespread volumes of mafic volcanics, the faults must have extended deep into the crust; it seems likely, therefore, that they were subvertical rather than, as depicted in Fig. 5.7, listric in nature. In a similar manner, an upper-mantle origin for the Lower Rotliegend basic volcanics and intrusions of the Midland Valley of Scotland is indicated by inclusions of magnesian peridotites and other ultramafic rocks (Cameron and Stephenson, 1985).

Volcanic activity at about the north-west limit of the Tornquist–Teisseyre fault system was involved in the development of the Oslo–Bamble–Horn Graben (see Fig. 5.6), which began to form in the earliest Permian (Oftedahl, 1976), possibly by exploiting an older Palaeozoic or even Proterozoic line of weakness (Russell and Smythe, 1983). This volcanic activity was preceded by deposition of shallow-marine, alluvial and aeolian sediments (Olaussen, *et al.*, 1982).

The orientation of the faults bounding the Moray Firth Basin, and especially the west-north-west–east-south-east-trending faults of the Witch Ground Graben, indicate a possible causal relationship with a northern splay of the Tornquist–Teisseyre shear zone via the Fjerritslev Fault (see Fig. 8.1; this splay fault is now effectively limited at its north-western end by the Bamble Trough). It is possible that the crustal fractures bounding the northern part of the Moray Firth Basin are artefacts of movements along the splay fault during Late Silurian closure of the northern Iapetus Ocean. Reactivation of the main Tornquist–Teisseyre shear zone may have been responsible for the west-north-west-trending transfer faults in the Danish sector of the Central Graben (see Fig. 2.2; Cartwright, 1987).

Volcanics also occur within the Central Graben, where the rift cuts the east-west-trending Mid North Sea–Ringkøbing–Fyn system of highs (Skjerven *et al.*, 1983), which have been dated at 265–267 Ma (Ineson, 1993; Glennie, 1997b, Fig. 3). The age of the Rotliegend volcanics on the flanks of the Central Graben probably heralds the initiation of structures that later defined the rift arm—in other words, the initial extensional movements of the

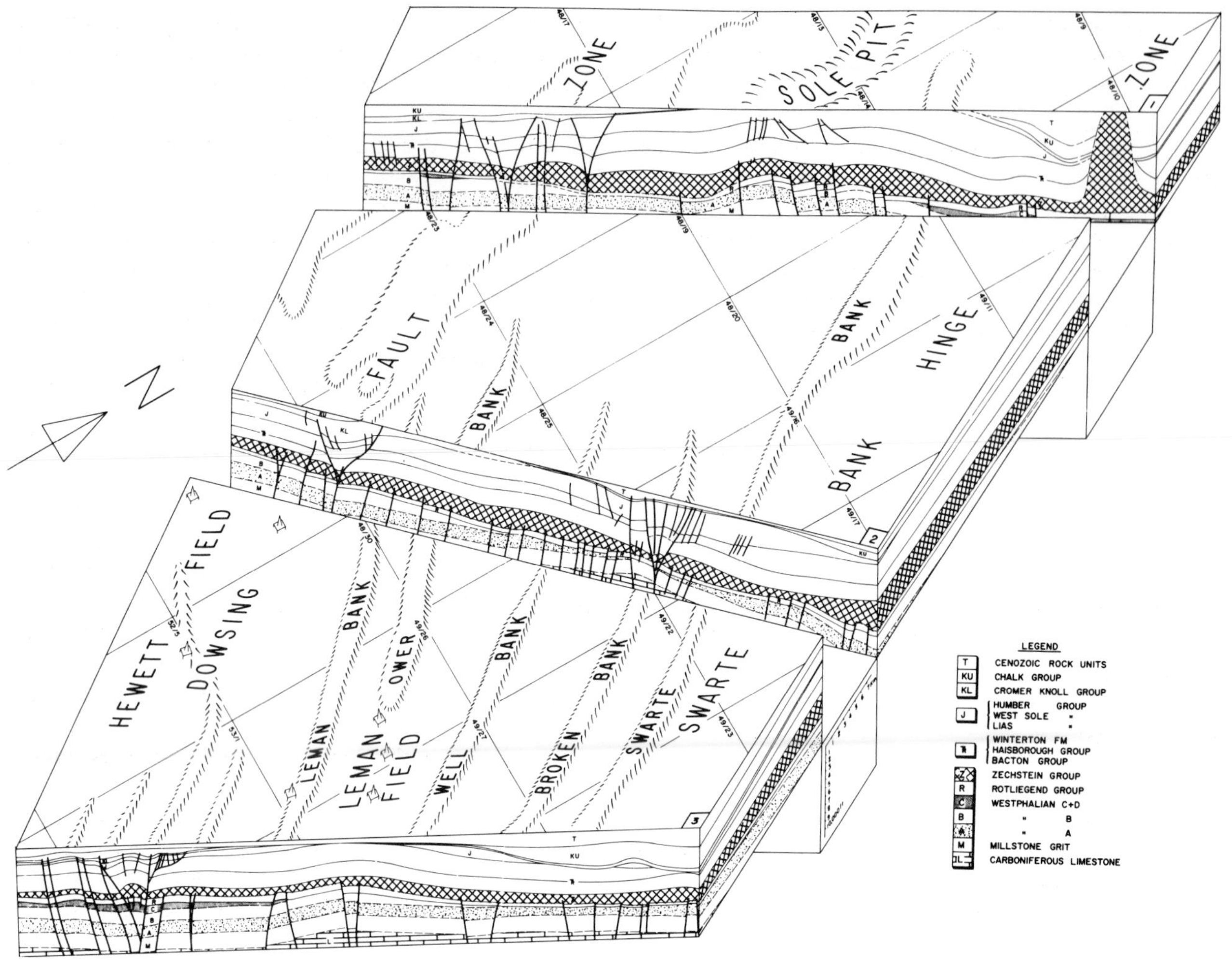

Fig. 2.16 Isometric block diagram of the Sole Pit area, UK southern North Sea. Interestingly, the area of Late Cretaceous inversion is underlain by an axis of Late Carboniferous inversion and is flanked by zones of probable strike–slip faults.

Central Graben. Their age also seems to match the time when sedimentation began in the Southern Permian Basin (see Glennie, Chapter 5). The Mid North Sea system of highs is underlain by granites (Fig. 2.11) and remained a positive area when the similarly orientated Southern and Northern Permian Basins began to subside either side of it.

Lower Rotliegend volcanics are not common in the western North Sea, a notable exception being the offshore extensions of the Great Whin Sill and associated east-north-east-trending dykes, which were intruded around 295 Ma (Robson, 1980; Glennie, 1997a,b). Apart from the Midland Valley of Scotland (Mauchline Volcanics; Cameron and Stephenson, 1985), they also occur in a few other isolated localities within the British Isles, such as Devon (Exeter Volcanic Series; Smith *et al.*, 1974; Edwards *et al.*, 1997) and north-eastern Ulster (Illing and Griffith, 1986). The granite plutons of Devon and Cornwall were emplaced diachronously between about 295 and 275 Ma (Chen *et al.*, 1993). The time span possibly reflects the geometry of Variscan plate collision at deep structural levels in this part of north-west Europe.

Confining themselves mainly to the Exeter Group sediments of south Devon, Edwards *et al.* (1997) show a clear affinity with the Rotliegend of the North Sea area. The Exeter Group can be divided into two parts: (i) a lower sequence containing volcanics that range in age between 291 and 282 Ma and are thus time-equivalent with the Lower Rotliegend volcanics of northern Germany; and (ii) an upper, entirely Late Permian sedimentary sequence, with a biostratigraphical span from about base Kazanian (260 Ma on their time-scale) to top Tatarian (Permo-Triassic boundary at 251 Ma). This evidence implies that deposition of the Dawlish Sandstone was probably coeval with that of the Zechstein, and the older formations with the Upper Rotliegend of the North Sea (see Chapter 5). The unconformity between these two sequences is compared to the Saalian Unconformity of Germany and the North Sea (see Fig. 5.4) and, like those areas, is estimated to have lasted about 20 Ma.

The volcanic-free south-western North Sea area is separated from the Variscan Mountains proper by the London–Brabant Platform, which is underlain by relatively rigid and undeformed Late Proterozoic Cadomian basement. The Variscan Deformation Front wraps around the southern edge of the London–Brabant Platform, whose rigidity was presumably sufficient to ensure that during the final stages of the Variscan Orogeny the area to its north was transpressional, thus causing inversion of the Sole Pit and other areas, instead of extrusion of volcanics. South of the Variscan Deformation Front, postorogenic intramontane, basins contain Stephanian and Autunian source rocks for both oil and gas (Kettel, 1989).

Although Rotliegend volcanics are associated with the Central Graben, they are not known in the vicinity of the Viking Graben. Even so, it seems that both grabens were initiated at about the same time, possibly in response to an attempt to create a Proto-Atlantic Ocean. The contrast in volcanic activity associated with the two grabens might be because the thicker continental Caledonian crust north of the Highland Boundary Fault was more rigid than the deep-marine turbidites of the North German–Polish Caledonides that underlie the central and southern North Sea. Thus the Early Permian volcanics of the Grampian Highlands Midland Valley of Scotland and Sunnhørdland area of south-western Norway (Dixon *et al.*, 1981) probably mark the northern limit of Lower Rotliegend volcanism.

The interpretation given here invoking an early Late Permian (267–265 Ma) initiation of the Viking and Central grabens, differs from that of Ziegler (1982a,b, 1990a), who believes that there is no evidence to support their opening any sooner than the early Triassic or perhaps latest Permian. The general parallelism of the Viking and Central Grabens with the half-grabens of Permian age in the west of Britain (see Fig. 5.1) is suggestive of a similar time of origin. Also, Glennie (Chapter 5) is impressed not only by the Late Permian age of the Inge volcanics (265–267 Ma) in the Central Graben but also by the presence of Upper Permian Zechstein salt within the southern 200 km of the Viking Graben (see, for example, Fig. 6.14; see also Fagerland, 1983), which is thick enough to act diapirically. Furthermore, he reasons that the rapid marine flooding of the Rotliegend desert basin by the Zechstein Sea must have involved the flow of water from the vicinity of the Arctic Circle via a pre-existing fracture system (Glennie and Buller, 1983). This fracture system was initiated in the far north during the Late Dinantian (see Table 2.1), but whether it had propagated southward to the North Sea area by the Early Permian is still open to question (see discussion in Haszeldine, 1984; Leeder, 1988). Ziegler (1990a, p. 77), however, argues that the thick Zechstein salt of the southern Viking Graben could also be explained as a downfaulted pre-rift sequence, whose lateral continuations were eroded following Mid Jurassic uplift of the graben flanks. Triassic strata now overlie Old Red Sandstone on these flanks. The origin of this and other associated graben systems will be discussed in the next section.

Permian basins and grabens

We have seen that around the Carboniferous–Permian time boundary, two different processes followed each other in fairly rapid succession. The first involved north–south compression and the creation of the Variscan Orogen, and probably ceased late in the Westphalian or early into the Permian. The other probably began very early in the Permian and involved east–west transtensional reactivation of through-going north-west–south-east-orientated lines of weakness inherited from the underlying Tornquist lineaments. The latter led to Lower Rotliegend volcanism and collapse of the Variscan Highlands, and later to subsidence of the Southern and Northern Permian Basins within the northern foreland. Figure 2.1 shows that the area between the Caledonian fold belts of Scotland and Scandinavia in the north and the Hercynian fold belt to the south is dominated by two east–west-trending basins, the Northern and the Southern Permian Basins. These basins are separated by the Mid North Sea–Ringkøbing–Fyn system of highs. Also, within the zone of strong Variscan folding in the south, another system of east–west- and north-east–south-west-trending horsts and grabens (e.g. English Channel, Southwestern Approaches, Bristol Channel and Celtic Sea Basins) probably all had their origins within the Permian. Whether this was in the Early or Late Permian is still uncertain.

Cutting the main basins and highs almost at right angles are the Central and Horn Grabens, which have northerly extensions in, respectively, the Viking and Oslo Grabens. East of the Horn Graben is a series of smaller grabens that also cut the Ringkøbing–Fyn High. In addition, to the west of the Pennine High, an arcuate series of half-grabens stretches from the Paris Basin in the south to the Minch Basins in the north, and probably had their continuation in east Greenland (see Surlyk *et al.*, 1984). It seems possible that, throughout the Variscan Orogeny, but especially during its later phases, the orocline formed by the Cantabrian Mountains acted as a northward-given wedge attempting to split the former Laurentian and Fenno-Scandian continents along the line of the northern Iapetus Suture and other zones of weakness (see Fig. 2.13). North–south-orientated grabens and half-grabens initiated by such a process would propagate from north to south, possibly beginning in the northern Proto-Atlantic in the latest Dinantian, and probably had a strong dip–slip component to their development.

Lorenz and Nicholls (1984) attribute both the collapse of the Variscan Highlands and basin subsidence within the northern foreland to thermal cooling; certainly, the maximum thickness of Rotliegend sediments occurs below the North German Plain (see Fig. 5.2) in an area where Lower Rotliegend volcanics are well developed. These authors calculate that emplacement of the Lower Rotliegend volcanics took about 5 Ma, a figure supported by Plein (1993). Thus, with volcanism beginning around 293 Ma, it would largely have ceased about 288 Ma. Sørensen (1986) provides evidence of similar Lower Rotliegend volcanism in the Danish Basin.

Permian basin fill

Evidence from North Germany indicates that a long (10–20 Ma) period of thermal uplift and erosion followed Lower Rotliegend volcanism (see Chapter 5). Sedimentation within the Southern Permian Basin probably began around 267 Ma, just prior to the Illawarra magnetic reversal (see Fig. 5.4). During the maximum of 16 Ma time span until the Permo-Triassic boundary at about 251 Ma, the deposition of over 2 km of Rotliegend desert-lake silts and salts was followed by about the same thickness of Zechstein evaporites, implying a rate of subsidence of around 250 m per million years. Thermal subsidence is unlikely to be the sole reason for creating such a high rate of subsidence. The geometry of the Permian basins indicates that regional transtension, between the north-east edge of the London–Brabant Massif and the Tornquist-Teisseyre fault system, was also a contributory factor. Thus subsidence of the Southern Permian Basin probably resulted from the combined effects of thermal cooling and the generation of an extensional or transtensional basin (Glennie, 1997b). In contrast, the Northern Permian Basin may have resulted entirely from the effects of extension or transtension along the southern, faulted margin of the combined Fenno-Scandian basement and Baltic Shield.

Following the Variscan Orogeny, much of Europe north of those mountains was an area of arid desert. With little rainfall and strong deflation, the high rate of subsidence in the Northern and Southern Permian Basins exceeded that of sedimentation throughout the Late Permian. By the time of the Permian Zechstein marine transgression, the deepest parts of these basins were occupied by desert lakes, whose surfaces were probably some 200–300 m below sea level.

After flooding, the Zechstein basins were still areas of relative sediment starvation. Especially during the early part of Zechstein deposition, the shallow-water basin margins were the sites of prolific manufacture of organic carbonate, but the floors of the basin centres were too deep, and therefore too dark, for the rapid organic growth of such carbonates; and, with low rainfall and little river supply, the rate of sediment influx was very low. However, with high temperatures, high rates of evaporation and limited access to supplies of both marine and fresh water, the enclosed Zechstein basins became giant evaporating pans, with deposition first of gypsum (now anhydrite), followed by thick sequences of halite and, finally, highly soluble potassic salts. Thus, although the Zechstein basins lacked a good supply of clastic sediment, they were eventually almost filled by largely chemical precipitates, of which halite had by far the greatest volume (see Chapter 6).

The Zechstein sequence is the product of five major depositional cycles, each of which shows the effects of increasing salinity with time; and some cycles contained similar subcycles. Each cycle began with a supply of normal sea water of relatively low salinity and ended with evaporation to probable dryness. Fossil evidence indicates that the water was most probably derived from the open ocean somewhere between Norway and Greenland, via a barred channel, the rate of supply varying not tectonically but with a global sea level that fluctuated in concert with the waxing and waning of the last of the Gondwana ice-caps (see also Chapter 5, Section 5.4.3).

The combination of structural and sedimentological controls on Carboniferous source and reservoir distribu-

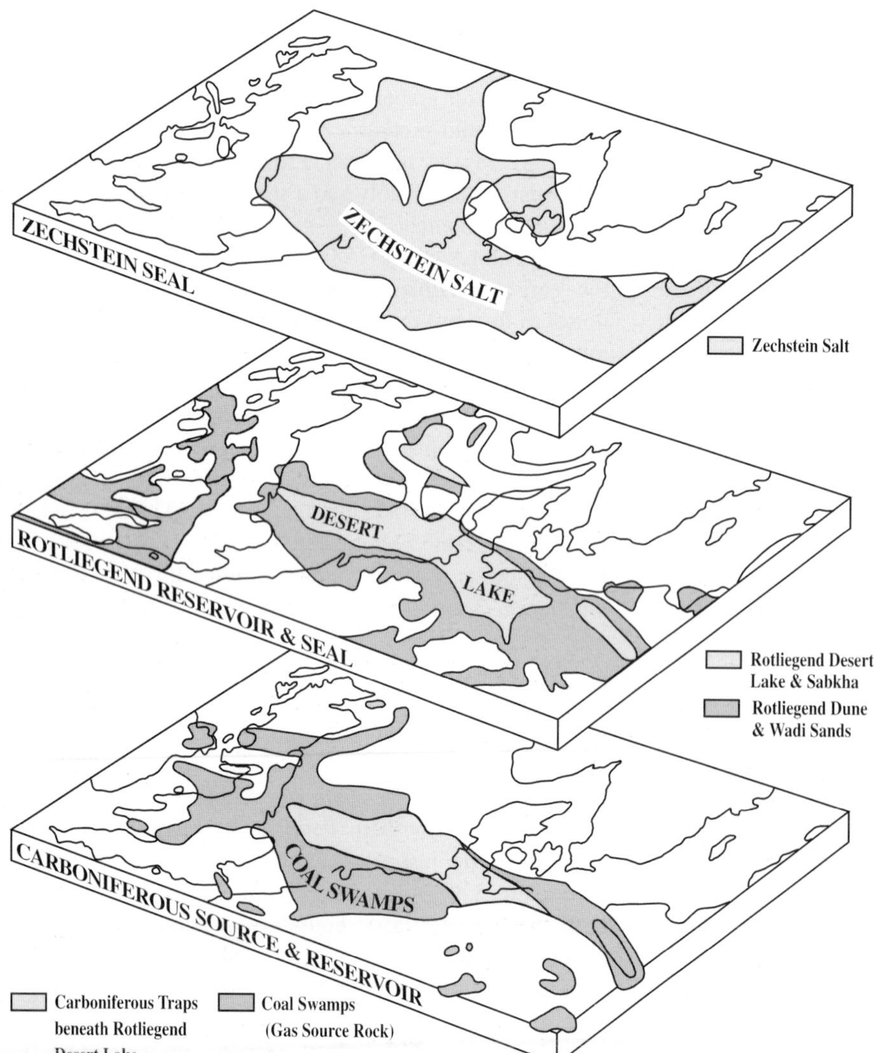

Fig. 2.17 Facies maps showing the close relationship between, and areal extent of, gas-prone Carboniferous source and reservoir rocks, Rotliegend reservoir and seal rocks and Zechstein evaporite seals.

tions, Variscan deformation, Rotliegend reservoir and seal deposition and, finally, the areal extent of the excellent Zechstein evaporite seal led to the southern North Sea having its prolific accumulations of gas (Fig. 2.17).

Controls on Triassic and Early Jurassic basin subsidence

During the Triassic, tensional stresses affected the crust and led to the extension of the earlier-formed network of half-grabens in and around the British Isles, which include the Cheshire, Wessex and Irish Sea basins (Fig. 2.18). The largest of the deep Triassic troughs was the East Irish Sea Basin, in which over 4 km of sediment were deposited (Jackson and Mulholland, 1993). This was not the only important Triassic depocentre, however, as over 5 km of sediments were deposited in the Danish Basin, along the Tornquist–Teisseyre trend, and a sedimentary column in excess of 6 km has been reported from areas west of Shetland (Ziegler, 1990a). Irrespective of their onshore or offshore location, the basins seem largely to have formed along the lines of inherited basement weaknesses. They were mostly filled by red-bed successions.

Later Triassic and Early Jurassic patterns of subsidence appear to be more uniform, implying that rifting gave way to a phase of thermal relaxation. Sedimentation in these post-rift sag basins became more and more marine-dominated, until flooding eventually occurred in the Early Jurassic (as recorded by the Blue Lias). Thermal subsidence appears to have continued until the Toarcian, when the North Sea region began to be influenced by a phase of uplift, which was ultimately to lead to the development of the 'Mid Cimmerian Unconformity'.

Development of the 'Mid-Cimmerian Unconformity'

The boundary between Middle and Lower Jurassic sediments is often marked by a widespread stratigraphic break, termed the 'Mid-Cimmerian Unconformity', which is apparent from the North Viking Graben to the southern North Sea and from the Moray Firth to the Danish Central Graben and its eastern flank area, where Upper Jurassic and even Upper Cretaceous strata overlie eroded Triassic sequences (Underhill and Partington, 1993, 1994; see also Fig. 2.36 in the appendix at the end of this chapter). Its development was associated with a pronounced depositional shallowing, marked by the onset of widespread non-marine fluviodeltaic deposition. The sediments laid down during the period following the unconformity form the reservoirs of the Brent Group in the North Viking Graben and the Fladen Group of the Moray Firth and Central Graben (see Chapter 8).

Subcrop pattern to the 'Mid-Cimmerian Unconformity'. Well correlations show that there is a systematic truncation of Lower Jurassic and older stratigraphic marker beds towards central areas of the North Sea (Fig. 2.19; see Chapter 8). Underhill and Partington (1993, 1994) have demonstrated that a series of concentric to elliptical sub-

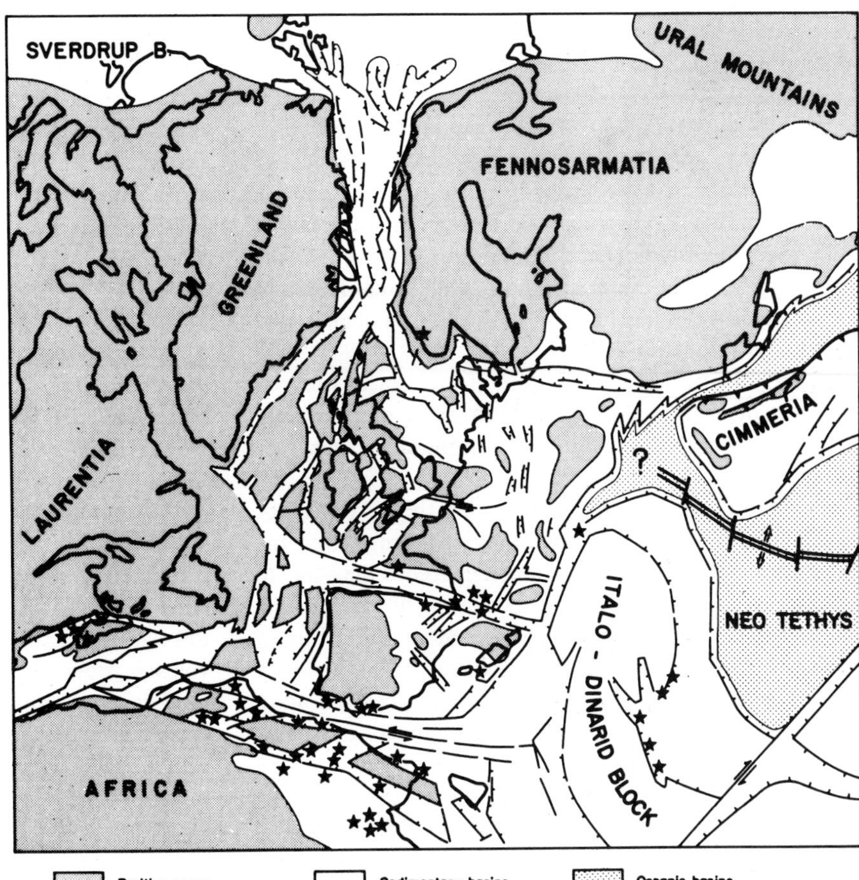

Fig. 2.18 Triassic rift systems of the North Atlantic area (after Ziegler, 1982a,b). Stars depict sites of volcanic activity.

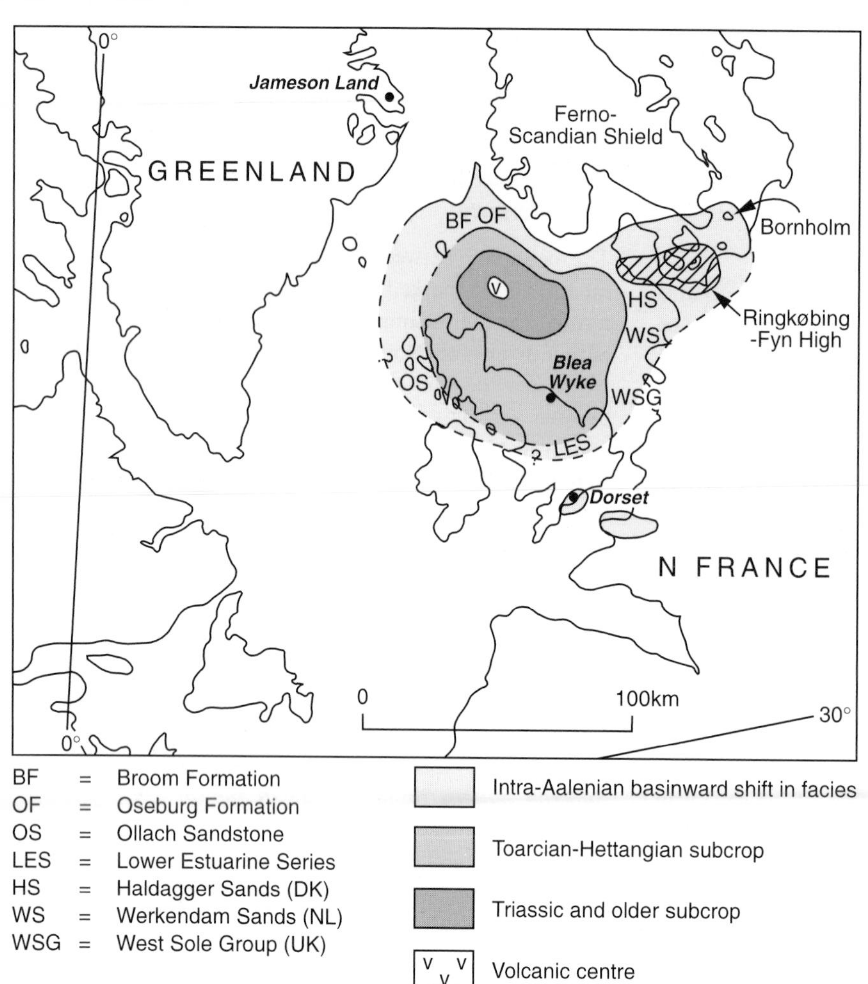

Fig. 2.19 Schematic diagram showing the maximum areal extent of erosion and sediment build-out associated with 'Mid-Cimmerian' thermal doming of the Central North Sea Dome (after Underhill and Partington, 1994).

crop patterns exists, with Triassic and older beds subcropping in the central region close to the triple junction.

Pattern of onlap on to the 'Mid-Cimmerian Unconformity'. Sedimentary units of Middle and Upper Jurassic age progressively onlap the unconformity, showing that the locus of marine flooding migrated towards the central area in a complex fashion (Fig. 2.20; see also Chapter 8). The patterns document initial drowning of the Brent Province in the North Viking Graben during the Aalenian and Bathonian (Underhill and Partington, 1993, 1994). They also document the separate nature of subsequent progressive marine incursions into the rift arms of the South Viking Graben, Moray Firth and Central Graben (see Chapter 8).

The most plausible explanation for the relationships highlighted by the subcrop and onlap patterns, given the lateral scale of the anomaly, is that they record the effect of widespread regional uplift and subsidence of much of the North Sea during the Jurassic. Furthermore, the nature of the stratigraphic relationships demonstrates that the uplift was of an approximately concentric nature, centred upon what was to become the North Sea rift junction. Consequently, the regional stratigraphic relationships support the existence of a 'Central North Sea rift dome' (Fig. 2.20), as originally proposed by Whiteman *et al.* (1975), and later by Hallam and Sellwood (1976), Eynon (1981), Ziegler (1982a,b, 1990a,b) and Leeder (1993).

The use of the more detailed sequence-stratigraphic methods outlined in Chapter 8 allows more accurate quantification of the affects of the dome throughout this important tectonic episode, and has direct application for a better understanding of its full implications for basin development in the North Sea.

Implications of 'Mid-Cimmerian events' for North Sea basin development

Thermal history of doming. Stratigraphical evidence suggests that the volcanic activity recorded by the Rattray Series and the Forties Igneous Province of the Central North Sea (Fig. 2.21) occurred during the Bathonian–Callovian after the development of the North Sea Dome. Radiometric dates, originally reported by Latin *et al.* (1990a,b), led Smith and Ritchie (1993) to invoke a complex history of igneous intrusion and local uplift. The latter workers used well data and seismic reflection profiles to locate at least three Jurassic volcanic centres within the volcanic province. Although the existence of more than one local igneous centre is not disputed, dependence upon radiometric data alone for the different times of their activity, without supporting evidence from volcanic outfall, is considered premature. Furthermore, it seems likely that individual volcanic centres occur on a quite separate (local) scale from their causal asthenospheric perturbation. As such, it is likely that individual intrusion events did occur,

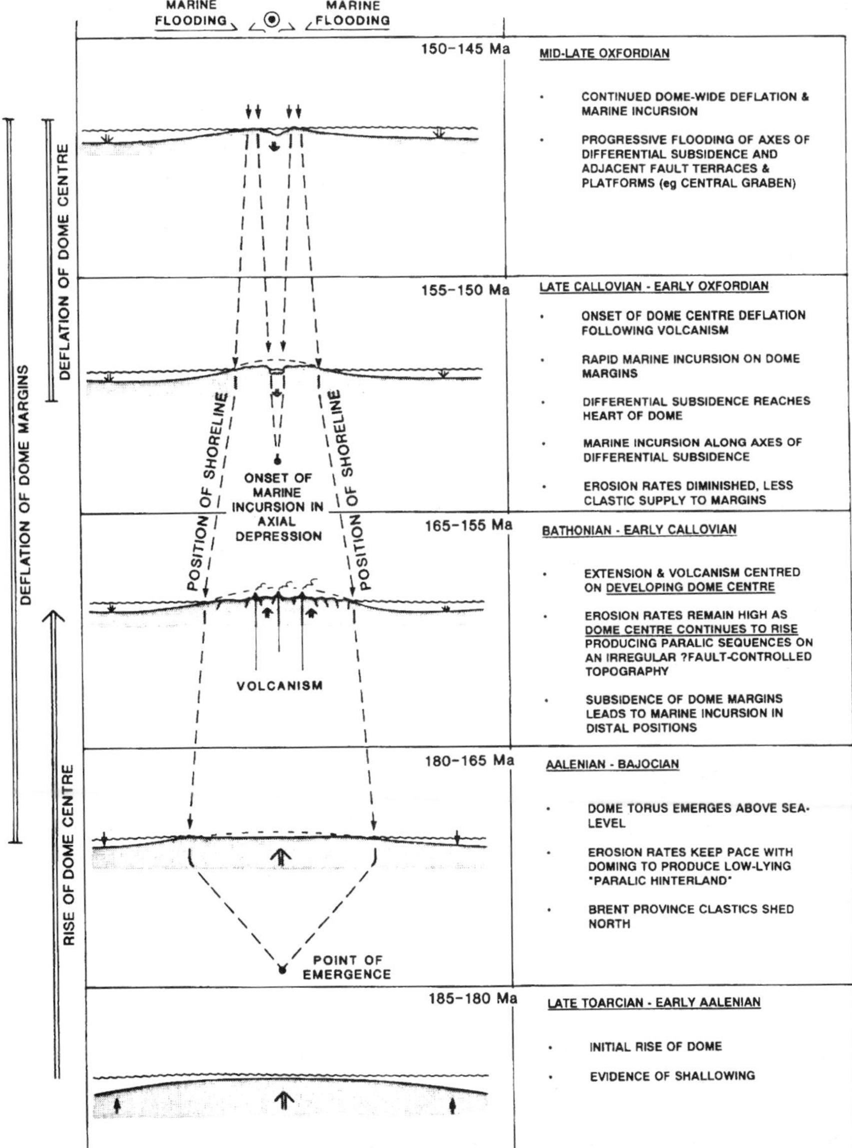
MARINE FLOODING
MARINE FLOODING
150–145 Ma
MID-LATE OXFORDIAN
CONTINUED DOME-WIDE DEFLATION & MARINE INCURSION
PROGRESSIVE FLOODING OF AXES OF DIFFERENTIAL SUBSIDENCE AND ADJACENT FAULT TERRACES & PLATFORMS (eg CENTRAL GRABEN)
155–150 Ma
LATE CALLOVIAN - EARLY OXFORDIAN
ONSET OF DOME CENTRE DEFLATION FOLLOWING VOLCANISM
RAPID MARINE INCURSION ON DOME MARGINS
DIFFERENTIAL SUBSIDENCE REACHES HEART OF DOME
MARINE INCURSION ALONG AXES OF DIFFERENTIAL SUBSIDENCE
EROSION RATES DIMINISHED, LESS CLASTIC SUPPLY TO MARGINS
165–155 Ma
BATHONIAN - EARLY CALLOVIAN
EXTENSION & VOLCANISM CENTRED ON DEVELOPING DOME CENTRE
EROSION RATES REMAIN HIGH AS DOME CENTRE CONTINUES TO RISE PRODUCING PARALIC SEQUENCES ON AN IRREGULAR ?FAULT-CONTROLLED TOPOGRAPHY
SUBSIDENCE OF DOME MARGINS LEADS TO MARINE INCURSION IN DISTAL POSITIONS
180–165 Ma
AALENIAN - BAJOCIAN
DOME TORUS EMERGES ABOVE SEA-LEVEL
EROSION RATES KEEP PACE WITH DOMING TO PRODUCE LOW-LYING 'PARALIC HINTERLAND'
BRENT PROVINCE CLASTICS SHED NORTH
185–180 Ma
LATE TOARCIAN - EARLY AALENIAN
INITIAL RISE OF DOME
EVIDENCE OF SHALLOWING
DEFLATION OF DOME MARGINS
DEFLATION OF DOME CENTRE
RISE OF DOME CENTRE
POSITION OF SHORELINE
POSITION OF SHORELINE
ONSET OF MARINE INCURSION IN AXIAL DEPRESSION
VOLCANISM
POINT OF EMERGENCE

Fig. 2.20 Schematic diagram showing the main events associated with the development and evolution of the Central North Sea Dome (after Underhill and Partington, 1993, 1994).

but the scale of the concentric subcrop pattern implies that a larger-scale mantle process provided the impetus for Jurassic volcanism.

All the stratigraphic relationships are consistent with volcanism extending across the time span between the development of the unconformity and onlap on to its central area (Middle Jurassic; Bathonian–Callovian; from approximately 160 to 150 Ma). If central uplift did indeed reach a maximum in the Bajocian/Bathonian and lithospheric thinning consequently also reached a maximum at this time, it would be consistent with the known generation of volcanics in the triple-junction area (Fig. 2.21) and the occurrence onshore of bentonites of the Fullers' Earth (Hallam and Sellwood, 1968; Jeans *et al.*, 1977).

The timing of uplift relative to volcanism and Late Jurassic rifting is consistent with an ideal Houseman and England-type (1986) model of active rifting, in which volcanism and stretching are consequent upon the impingement of a mantle-plume head at the base of the lithosphere (see Fig. 2.20). It cannot simply be a result of classic McKenzie-type (1978) passive rifting, in which volcanics postdate extension because of crustal thinning and adiabatic decompression. For the McKenzie (1978) model to be consistent with the data, volcanism should have occurred during the Late Jurassic or earliest Cretaceous, coeval with the stretching episode.

Although the stratigraphical data point to the presence of a plume head, Latin *et al.* (1990a,b) demonstrated that high mantle temperatures typical of plumes, such as Hawaii (i.e. with excess temperatures more than approximately 300°C), appear to be ruled out by the lack of extensive melting, which would have been a consequence of stretching the lithosphere above such a feature with stretching β factors of 2–2.5.

The largest amounts of melting (< 2%) resulted in the alkali basalts of the Forties province in an area where the β factor may have exceeded 2. Other off-axis areas of the rift saw smaller degrees of melting, with more undersaturated and extreme compositions, including nephelinites and ultrapotassic intrusive igneous rocks. Indeed, Latin *et al.* (1990a,b) have stressed that in the Forties province the observed melt compositions may only be reconciled with melting on the dry peridotite solidus at a normal (1280°C) temperature, assuming a relatively thin mechanical bound-

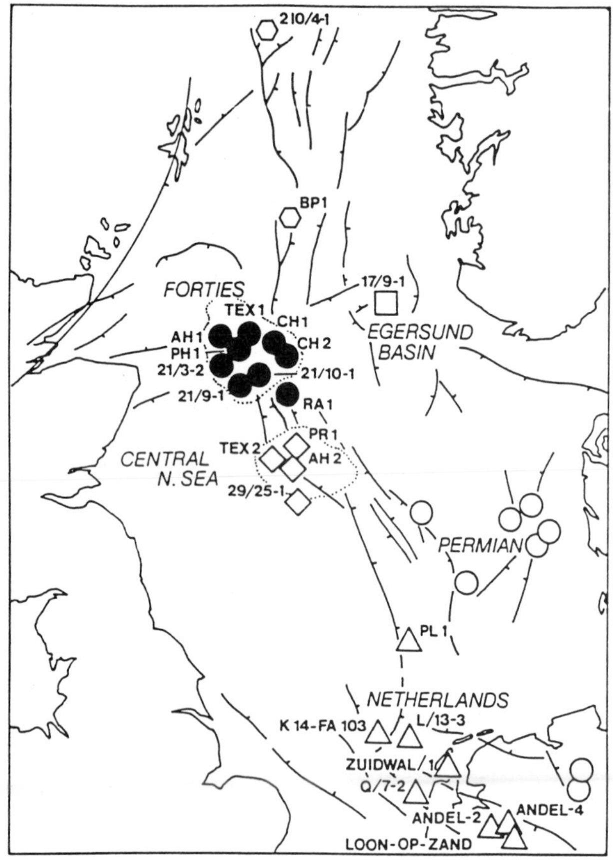

● Forties Province rocks; ◇ Central North Sea Mesozoic rocks; ⬡ Viking Graben occurrences; △ Netherlands on-shore and off-shore Mesozoic rocks; ○ Permian occurrences.

Fig. 2.21 Location of Mesozoic and Lower Permian volcanic rocks in the North Sea rift system (after Latin *et al.*, 1990b).

ary layer of 70 km. They have also shown that simple shear models of extension, using an initially planar detachment fault, cannot account for the existence or location of the magmatism.

That limited extensions characterized the Middle Jurassic is not disputed. Evidence exists to suggest that faulting controlled facies and thickness variations during deposition of the Brent Group and younger units. Although the amount of extension was enough to accommodate differential subsidence of the Viking Graben (as demonstrated by the onlap map; see Fig. 2.20), it was still considerably less significant than subsequent early Kimmeridgian rift events. Consequently, although McKenzie-type extension may have played a role in promoting adiabatic decompression and melting, it cannot be considered the main driving force behind the volcanism.

The transient nature of the uplift is striking, as is the occurrence of Bathonian–Callovian volcanics approximately 15–25 Ma after the onset of marginal onlap, and quite possibly after the onset of central subsidence. These features seem more consistent with the upward transport of a discrete plume head or 'blob' and its radial dissipation, rather than with the initiation of a permanent upward hot 'jet' (Griffiths and Campbell, 1991). Indeed, it is clear from the subsequent subsidence patterns and lack of any clear volcanic trail that no long-lasting 'jet' succeeded this uplift event. Griffiths and Campbell's (1991) experimental models, if reliable indicators of asthenospheric behaviour, also suggest that uplift, and even the initiation of subsidence, could precede the arrival of the plume head at the base of the lithosphere, as it would expand rapidly in area and thin as it rose (see Fig. 2.20). In this case, maximum uplift would occur when the thermal anomaly was still too deep to affect the presumed region of melt generation near the base of the lithosphere. In the experiments, the radial dissipation of a plume head culminates in unstable behaviour and its transformation into an expanding torus. This raises the possibility that no systematic temperature rise might ever be experienced in the central part of the asthenospheric column throughout a transient event, thus reconciling the melting arguments with the uplift history.

Dissipation of the excess temperature by lateral spreading of the head by the time extension actually occurs could have resulted in little or no excess melt production for a given β factor. However, the subsidence experienced by any part of the region will have been a complex combination of declining advective and cooling components in the asthenosphere, coupled with the isostatic and thermal components following lithosphere and crustal thinning. The implication is that subsequent Late Jurassic rifting occurred on the back of an earlier thermal event. The relatively rapid subsidence of the dome is itself indicative of an advective dissipation of the thermal anomaly, rather than its removal entirely by conductive cooling to the surface.

The conclusion is that the driving force for the regional uplift was a transient plume head or 'blob', which did not evolve to a focused 'hot' plume (see Fig. 2.20). Although the size of the elevation anomaly is large in plan view (approximately 1250×1500 km), it is of similar dimensions to those described by McKenzie *et al.* (1980) and McKenzie (1983), and is consistent with either a plume head or a steady-state jet (Parsons and McKenzie, 1978; Courtney and White, 1986; Griffiths and Campbell, 1991). The location of the dome's centre may indicate that the mantle-plume head was located in the vicinity of older crustal weaknesses related to the intersection of the Iapetus Suture and the Tornquist–Teisseyre line.

Structural development and evolution of the Late Jurassic extensional province

Structural styles. Seismic data reveal that the middle Oxfordian to early Kimmeridgian records the onset of a major phase of extensional activity in the North Sea Basin (Ziegler, 1982a, 1990a; Underhill, 1991a,b; Rattey and Hayward, 1993). The resultant accelerated subsidence during the Late Oxfordian and Kimmeridgian heralded basin deepening in all three rift arms (Fig. 2.22). Widespread deposition of sediments ascribed to the Humber Group records drowning as the former site of the Central North Sea Dome subsided. In addition to the obvious control on syn-rift sedimentation, this phase of extension caused the development of major faults, with consequent dissection and tilting of earlier sedimentary sequences (Fig. 2.23), which eventually led to creation of the major structural plays for which the North Sea is best known.

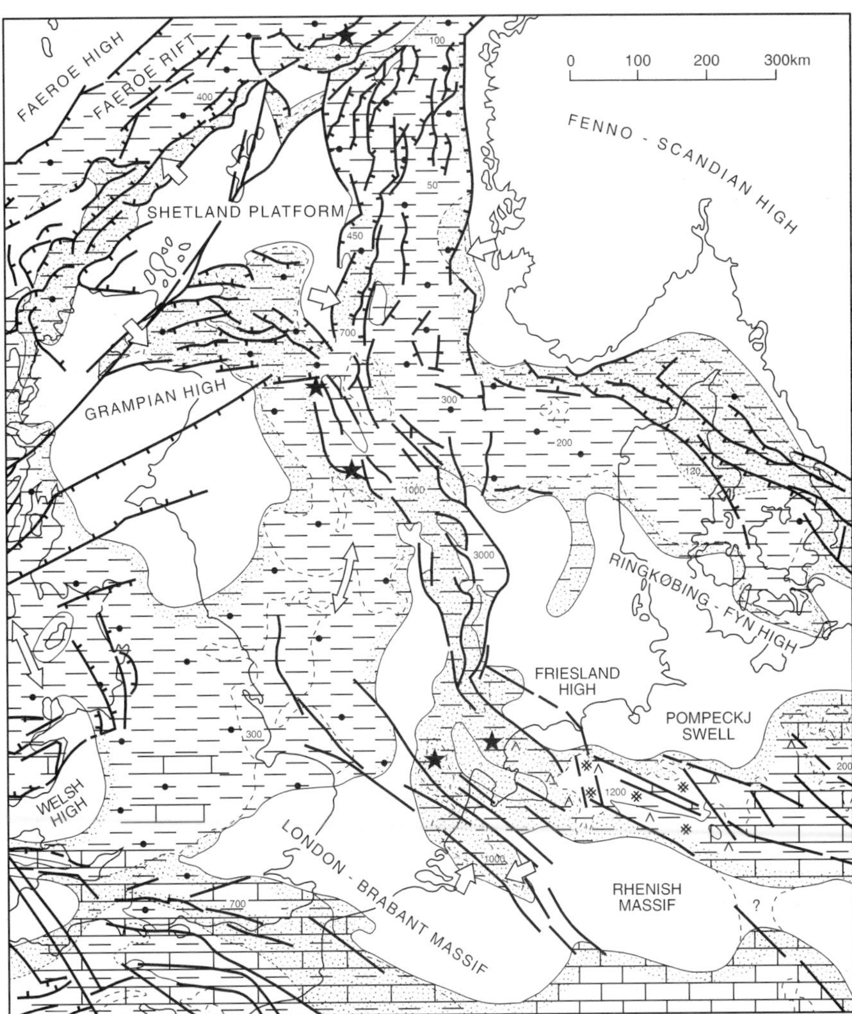

Fig. 2.22 Detailed palaeogeographic reconstruction of Kimmeridgian–Volgian of the North Sea (after Ziegler, 1990a,b).

Structures of the Viking Graben and Inner Moray Firth are consistent with an interpretation as extensional fault blocks, similar to those seen in other classic extensional provinces, which subsequently experienced thermally driven post-rift subsidence. The rift arms consist of numerous individual fault segments, which variously overlapped to form relay ramps or were linked by transfer faults (Fig. 2.24). In contrast to some early fault models (e.g. Gibbs, 1984a,b), it is now thought that the extensional faults had a planar geometry rather than a listric one (see Fig. 2.23). Like the Oligo-Miocene Suez rift and the modern-day East African rift, the North Viking Graben may have been characterized by major changes in fault-dip polarity along its length (e.g. Lee and Hwang, 1993).

Geometric, stratigraphic and sedimentary evidence suggests that each fault segment defined a fault block that was characterized by pronounced footwall uplift and hanging-wall subsidence (e.g. Jackson and McKenzie, 1983; Roberts *et al.*, 1990a,b; Yielding, 1990), which led directly to the development of numerous structural traps in areas like the Brent Province. Uplift of the footwalls of the extensional faults led in many cases to pronounced erosion and fault-scarp degradation (Underhill *et al.*, 1997), so that in some cases erosion has reached down to Triassic levels.

Although many of the Late Jurassic structures of the northern North Sea remain buried at depths in excess of 3 km, those of the Inner Moray Firth lie at much shallower levels, due to the effects of Tertiary regional uplift. As a consequence, the overall tectonic control and seismic definition of stratigraphic marker horizons is well imaged and better understood. Mapping of the middle Oxfordian to Early Cretaceous seismic interval highlights the role that major extensional faults (e.g. the Smith Bank Fault) had in controlling differential subsidence (e.g. Andrews and Brown, 1987; Underhill, 1991a). Cross-sections through the basin, restored for Late Jurassic times, further illustrate the similarities between the structures found in the Inner Moray Firth and those in the North Viking Graben (see Chapter 8 and Fig. 8.35; Underhill, 1991a,b).

The structural configuration of the Central Graben is made up of a number of intrabasinal highs (e.g. Forties–Montrose High), terraces (such as the Cod Terrace) and subbasins (such as the Søgne Basin). In contrast to the other two rift arms, extension in the east Central Graben was more complex, because of the influence of halokinesis. Movement on the basin-bounding faults led to rapid subsidence in the Eastern Trough of the Central Graben, which initiated differential flow of salt at depth (e.g. Erratt, 1993; Gowers *et al.*, 1993; Penge *et al.*, 1993). The presence of the underlying Zechstein evaporites allowed shallow detachments to develop. The main effect was to control the structural wavelength and hence trap size of Jurassic fault

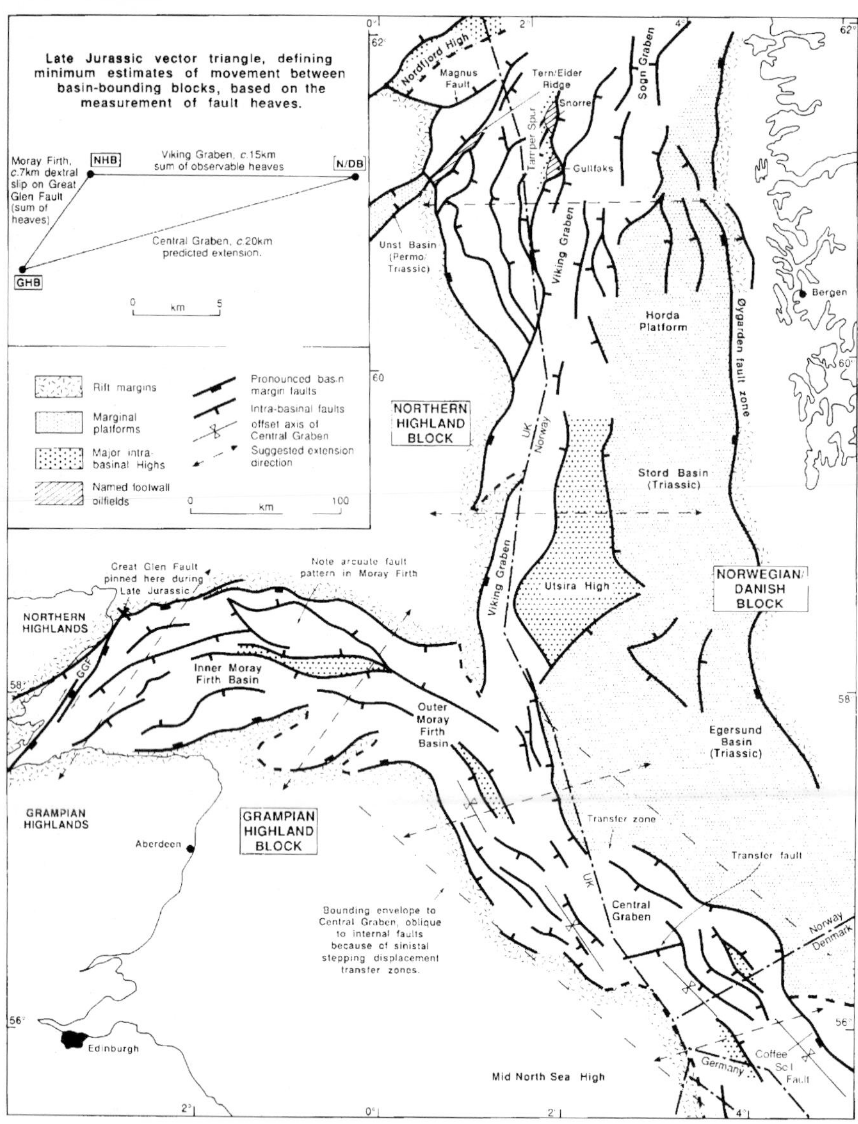

Fig. 2.23 Main extensional tectonic elements of the North Sea Domain. The arrows on the main diagram and in the insert show stress orientations as interpreted by Roberts *et al.* (1990c).

blocks, which were more limited in width (4–10 km) in the Central Graben than in the North Viking Graben, where no salt exists. The consequence for oilfield size has been considerable, with Central North Sea fields largely being one order of magnitude smaller (100–200 million barrels stock-tank oil initially in place (STOIIP) than their northern counterparts, such as Brent, Snorre, Ninian and Statfjord, which are all billion barrel STOIIP fields). An additional effect of the presence of salt was to induce regional evacuation from the graben centre, which eventually led to the development of salt pillows and localized diapiric intrusions during the Cretaceous and Tertiary.

In the southern North Sea, the Late Jurassic saw the deep burial of depocentres like the Sole Pit Basin, which ultimately led to the maturation of gas-prone Carboniferous source rocks. Synsedimentary extension may also be demonstrated in onshore areas, such as the Wessex, Weald and Cleveland basins, all of which were later to experience Cenozoic inversion.

The amount of stretching (β factor) experienced by the North Sea during the Late Jurassic rift episode has long been disputed. Investigations using forward modelling, post-rift flexural backstripping, the analysis of fault populations and cross-section reconstructions all suggest that the basin was stretched by between 10 and 20% during the Late Jurassic rift episode (β factors of 1.10–1.20; e.g. Roberts *et al.*, 1993). It is only within the graben axes that β factors locally rise to 1.4–2.0.

Orientation of Late Jurassic extensional stresses. Interpretation of regional three-dimensional seismic (3D) data has led some workers to suggest that the Late Jurassic structural evolution of the North Sea could be modelled under a uniform stress field. The goal of the structural analysis has been to determine and predict the nature and scale of structures likely to exist under a uniform regional stress in the main rift arms. However, the resultant interpretations have been extremely contentious and no model has received wide acceptance.

Initially, Roberts *et al.* (1990c) proposed that the dominant extensional-stress orientation was north-north-east–south-south-west, parallel to the Great Glen Fault, such that oblique-slip movements occurred in all three North Sea rift arms during their Late Jurassic evolution (see Fig. 2.23). These workers subsequently accepted that their conclusion was incorrect (see discussion in Thomson and Underhill, 1993), after it was shown that the Great Glen Fault had no role in Late Jurassic extension in the Inner

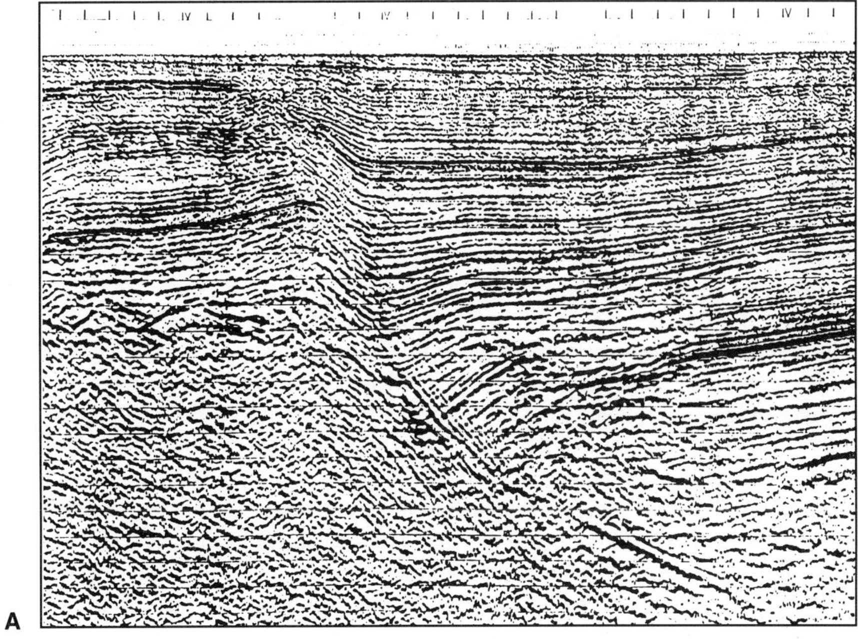

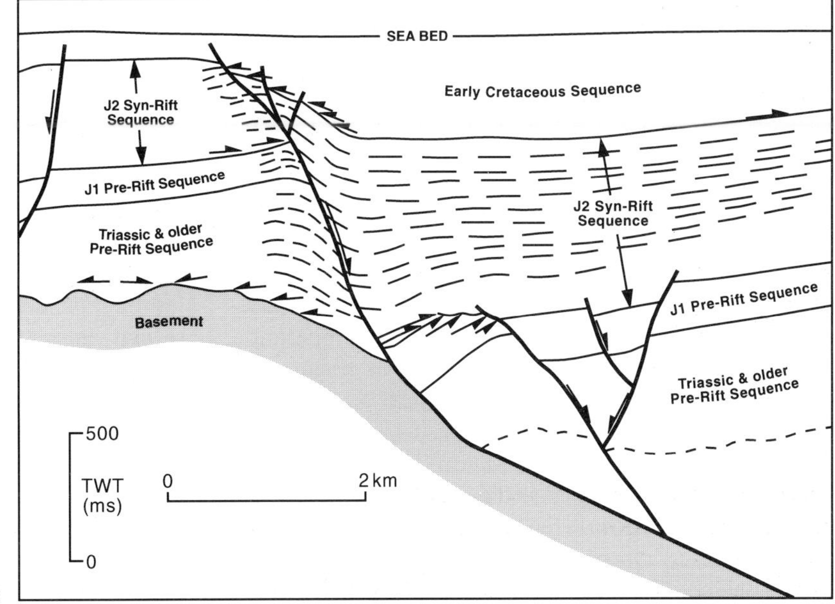

Fig. 2.24 Uninterpreted and interpreted seismic line from the Inner Moray Firth basin depicting the representative Late Jurassic syn-rift structural styles seen in the North Sea. The listric nature of the fault is more apparent than real, being caused by velocity pull-up deeper in the section. Depth conversion restored the fault to its proper planar shape.

Moray Firth (Underhill, 1991a,b; Thomson and Underhill, 1993).

While accepting Thomson and Underhill's (1993) Inner Moray Firth data, Bartholomew *et al.* (1993) interpreted the orientation, distribution and character of the structural styles found in the North Sea as being controlled ultimately by the reactivation of pre-existing shear zones, with repeated Mesozoic and Cenozoic movements being dominated by oblique slip. However, they considered the structural development of the elongated grabens to be linked to a regional stress pattern, which they thought was a dominant east–west extensional-stress system (Bartholomew *et al.*, 1993). More recently, Eggink *et al.* (1996) have built upon Bartholomew *et al.*'s (1993) work to suggest that tectonic movements in the Central North Sea were controlled by the relative orientation of the stress fields with respect to the fault strikes of older major Variscan fault trends. They suggested that the amount and direction of strike–slip movement was controlled by a gradual clockwise rotation of the minimum effective stress in the horizontal plane from approximately north-east–south-west to east–west in the first two phases. Within this framework, Eggink *et al.* (1996) suggest that halokinesis is only of local importance and serves to amplify the tectonically controlled structuration. They believe that their structural model explains the observed distribution of fields and structures in the Central Graben, as well as enabling prediction of structural development in its less-well-explored portions.

Although the regional models of Bartholomew *et al.* (1993) and Eggink *et al.* (1996) usefully explain some structures in the North Sea, synthesis of tectonic features from across the basin suggests to us that they cannot be placed into one all-encompassing structural model. That no single, unifying model readily explains structures seen in all three rift arms argues strongly *against* a simple regional stress orientation. Instead, it seems likely that each graben experienced its own unique stress regime, which in turn set up local intrabasinal stress fields as the trilete rift system developed.

Late Cimmerian 'inversion'. It has been widely believed that a short-lived phase of structural inversion occurred at the end of the Jurassic. These structures have been ascribed to 'Late Cimmerian compressional deformation' and are linked to the development of the 'Late Cimmerian Unconformity'. Leaving aside the fact that it has been demonstrated that the Base Cretaceous seismic marker (see Fig. 1.8) represents a highly condensed sedimentary succession rather than a conventional unconformity in basinal areas (Rawson and Riley, 1982; Rattey and Hayward, 1993), there are theoretical difficulties with a short-lived phase of structural inversion.

It now appears that some of the evidence for end-Jurassic contractional deformation may be explained by alternative tectonic mechanisms. In the South Viking Graben, for instance, salt appears to have acted as a key horizon of décollement, along which occurred mechanical decoupling of post-salt Triassic and younger strata from the pre-salt 'basement'. The onset of large-scale extension appears locally to have caused gravity gliding of post-salt sections to the west-north-west, with resultant low-angle extensional faulting upslope and salt-cored buckling downslope. Furthermore, it is now believed that many of the structural geometries ascribed to Late Cimmerian inversion may actually be consistent with synsedimentary extensional forced folding, related to the vertical and lateral propagation of normal faults (Withjack *et al.*, 1990) and local development of reverse faults in response to space-problems associated with creation of fault-tip propagation folds. Others may simply have formed in response to compaction, while yet others may be much younger compressional features, related to Cenozoic deformation rather than having formed during the Mesozoic. In conclusion, it remains unclear to us whether the popular notion that the end-Jurassic was marked by a short-lived phase of inversion is plausible. Given the weight of evidence for alternative explanations for the structural phenomena affecting Late Jurassic strata, it seems likely that the effects of Late Cimmerian structural inversion have been overstated.

Late Mesozoic and Cenozoic thermal subsidence

Cretaceous events. In general, the post-rift phase saw a time of tectonic quiescence in the North Sea Basin, with thermally driven subsidence following the earlier Late Jurassic rift events (Figs 2.4, 2.25 and 2.26). The main exception to this general picture appears to have occurred in the Witch Ground Graben, adjacent to the Halibut Horst, where syn-sedimentary hangingwall clastics accumulated (e.g. in the Claymore and Scapa fields of the Outer Moray Firth). Elsewhere, Neocomian and Barremian marls and shales blanketed the Upper Jurassic sediments and transgressed on to formerly exposed areas.

Aptian tectonic instability, resulting from the Austrian orogenic phase, resulted in renewed activity on faults, localized uplifts and fan-sand deposition, especially in the area of the former triple junction. Regional subsidence replaced the active tectonic events in the middle Albian, as a continuously rising sea level caused the inundation of emerged areas, such as the Fladen Ground Spur and Halibut Horst. Sea-level rise continued into the Late Cretaceous, with deposition of the Chalk.

Some of the deformation seen in the immediate North Sea area may reflect the greater tectonic activity that characterized areas west of Britain during the Cretaceous and Cenozoic, with the development of major extensional faults. The west Shetlands continental margin consists of a series of north-east–south-west-trending Mesozoic rift-related basins, formed along the north-west European continental margin in response to Atlantic rifting. The best-known and most representative basin in the area, the Faeroe–Shetland Basin, is bound to the south-east by the stable West Shetlands Platform and the Scottish Highlands Massif, and to the north-west by the loosely defined Faeroes Platform, which largely underlies the area of extensive Palaeocene lava flows. In practice, our geological understanding of the basin's sedimentary fill is limited by the areal extent of these flows to the north-west.

The Faeroe–Shetland Basin is characterized by the accumulation of up to 10 km of Cretaceous and Tertiary sediments and the widespread intrusion by sills and other igneous bodies (Mudge and Rashid, 1987). Structurally, the Faeroe–Shetland Basin is subdivided by a number of intrabasinal highs (such as the Rona, Flett (or Central) and Corona Ridges), which allow the definition of discrete depocentres within the broader basin (Ridd, 1983; Duindam and van Hoorn, 1987; Haszeldine *et al.*, 1987; Hitchin and Ritchie, 1987; Mudge and Rashid, 1987; Earle *et al.*, 1989). Gravity and magnetic data, together with structural information, demonstrate that the area is also dissected by several important, long-lived, but poorly understood, north-west–south-east-trending structural zones (e.g. the Judd Fault System; Kirton and Hitchen, 1987; Rumph *et al.*, 1993). These mostly affect the pre-Tertiary stratigraphic section and may be akin to the accommodation or transfer zones found in other rift basins (Mitchell *et al.*, 1993; Rumph *et al.*, 1993). Their occurrence effectively compartmentalizes the whole margin into discrete structural domains, each of which are believed to have a unique tectono-stratigraphic history (Ebdon *et al.*, 1995)

Early Cenozoic events west of Britain. Over 4500 m of Cenozoic strata are preserved in the depocentre of the Faeroe–Shetland Basin (Mitchell *et al.*, 1993). Broadly speaking, the underlying Late Cretaceous deep-marine sediments of the Faeroe–Shetland Basin are succeeded locally by early to late Palaeocene clastics in an overall upward-coarsening and shallowing unit (Mitchell *et al.*, 1993; Ebdon *et al.*, 1995). This regressive unit is over 2 km thick in places and is capped by a prograding late Palaeocene fluviodeltaic succession, which dominated the area until the early Eocene, when deep-marine conditions were re-established. The basin fill is remarkably similar to the lithostratigraphic relationships found east of the Scottish mainland (e.g. Outer Moray Firth), where overlapping submarine-fan clastics of the Montrose Group are disconformably overlain by fluviodeltaics of the Moray Group (Parker, 1975; Rochow, 1981; Mudge and Copestake, 1992a,b).

Regional seismic data show that Early Tertiary sedimen-

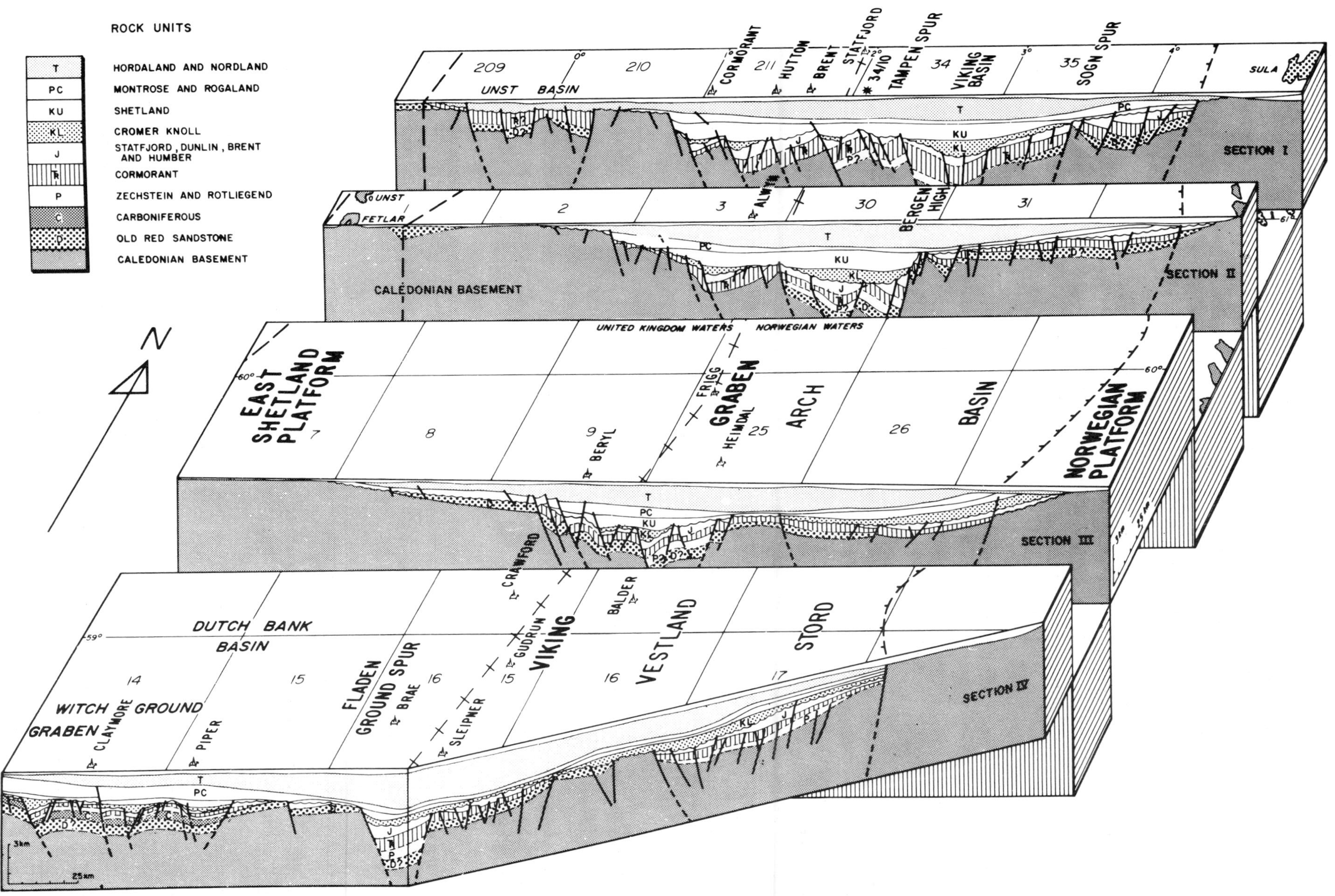

Fig. 2.25 Isometric block diagram of the northern North Sea serving to illustrate the structural development of the Viking and Witch Ground Graben areas. The pronounced post-rift steer-horn geometry is well illustrated.

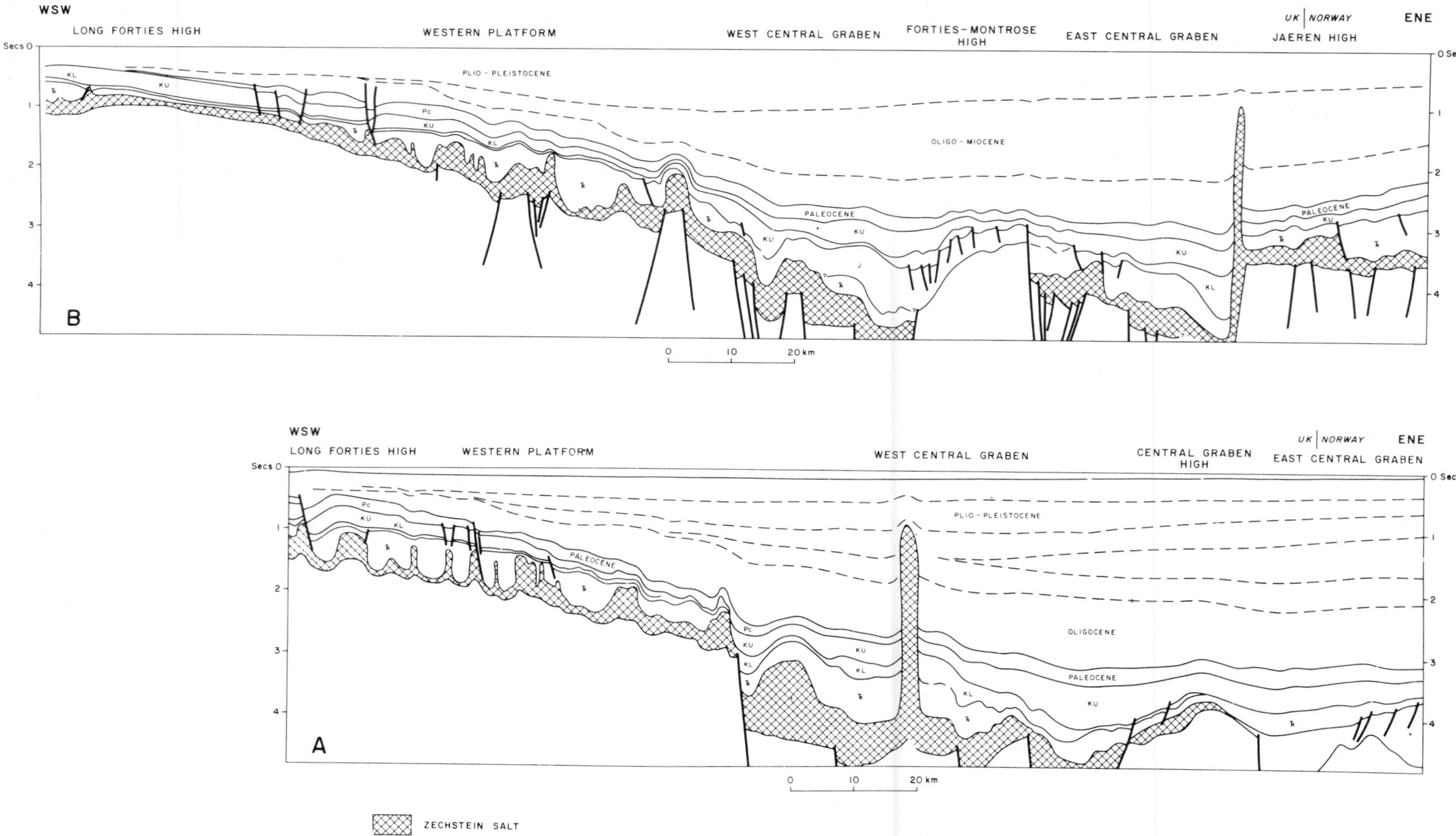

Fig. 2.26 Line drawings from representative seismic sections across the UK Central North Sea depicting the well-defined post-rift steers-horn geometry that characterizes the Cretaceous and Tertiary of the North Sea. Strong diapiric activity has occurred locally as a result of differential subsidence.

tation in the Faeroe–Shetland Basin has a marked 'steer's-head' geometry, centred upon an area previously termed the Flett Sub-basin, which lies between the Flett and Corona Ridges (Mitchell *et al.*, 1993) and formed largely in response to a period of post-rift thermal subsidence following Mesozoic extension (Duindam and van Hoorn, 1987; Turner and Scrutton, 1993). Some degree of irregular differential subsidence resulting from fault reactivation has been reported in the basin (e.g. Duindam and van Hoorn, 1987; Hitchin and Ritchie, 1987; Turner and Scrutton, 1993), possibly in response to initiation of the Iceland hot spot (Knott *et al.*, 1993; Clift *et al.*, 1995) and 'Thulean' volcanic activity (Mussett *et al.*, 1988; Ritchie and Hitchen, 1996) at the beginning of the Palaeocene, which has led to many of the Mesozoic structural elements imposing an important control on Early Tertiary sedimentary geometries and thicknesses.

Instigation of the Iceland hot spot and Tertiary uplift. Despite lying in an intraplate setting, the Cenozoic evolution of the British Isles was far from quiescent. Outcrop patterns of the British Isles, which show a general and progressive increase in the age of exposed rocks to the west, indicate that a regional tilt affected the region during the Cenozoic. The uplift caused many of the earlier sedimentary basins, such as the Permo-Triassic Irish Sea Basin, the Devono-Carboniferous Bowland, Pennine, Solway and Midland Valley basins and the Precambrian to Lower Palaeozoic Caledonian mountain belt, to be exhumed (see Fig. 2.1).

The evidence for the westerly increase in regional uplift and exhumation is best documented in the Moray Firth rift arm, where Cretaceous and Jurassic sediments progressively subcrop the seabed (Underhill, 1991a). Sonic velocities derived from the Cretaceous Chalk and the Late Jurassic Kimmeridge Clay successions indicate that the amount of erosion exceeded 1 km in western areas (Hillis *et al.*, 1994). Apatite fission-track analysis of sediments from the Northern Highlands and Western Isles suggests that between 1 and 2.5 km of erosion occurred across Scotland during the Cenozoic. It is generally considered that the regional tilt and consequent exhumation of Mesozoic and Palaeozoic basins was caused by igneous underplating related to the development of the Iceland hot spot during the Early Cenozoic (Brodie and White, 1994). Taken together with the stratigraphic evidence of shallowing, the melt is interpreted as resulting from the initiation of the mantle plume prior to continental breakup (White, 1989). Its effects extended across a region 2000 km in diameter (Fig. 2.27; White, 1989).

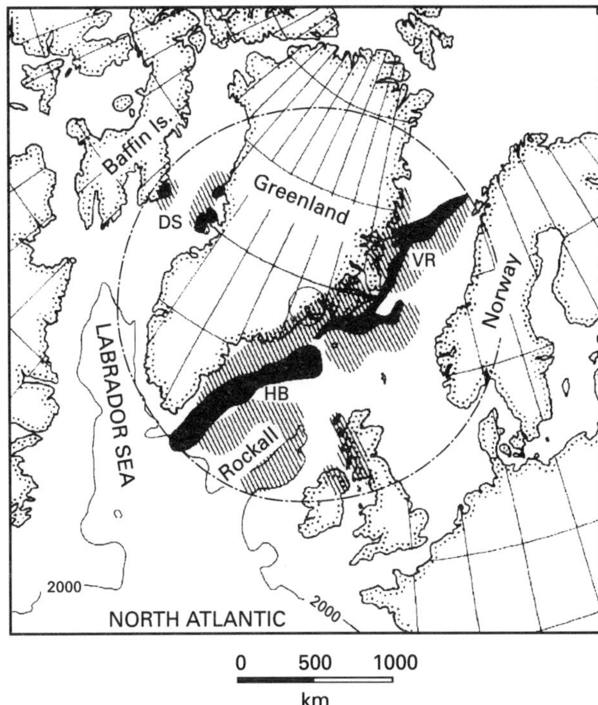

Fig. 2.27 Schematic diagram to illustrate the area affected by major volcanism and uplift associated with the development of the Iceland hot spot during Early Cenozoic times (after White, 1989). Volcanics, black; uplift, cross-hatched; HB, Hatton Bank; VR, Voring Basin. Outer circle, approximate limit to volcanic activity.

Structural inversion. Compressional deformation is known to have affected a wide area off north-west Europe during the Late Cretaceous and Early Cenozoic, when many former sedimentary basins experienced significant structural inversion (Fig. 2.28; Ziegler, 1990a). The effects of regional compression have persisted to the present day and borehole break-out data indicate that the North Sea area remains in north-west–south-east compression.

The structural effects of the initial Late Mesozoic and Early Cenozoic compression are well illustrated by the development of large antiformal structures above the former sites of Mesozoic basins (Fig. 2.29), such as the Sole Pit (Glennie and Boegner, 1981; Alberts and Underhill, 1991), the Broad Fourteens and West Netherlands Highs (van Wijhe, 1987b), the onshore Netherlands (Dronkers and Mrozek, 1991), Wessex (Fig. 2.30; Colter and Harvard, 1981; Stoneley, 1982; Butler, 1998; Harvey and Stewart, 1998), the Weald (Fig. 2.31; Butler and Pullan, 1990), and the Cleveland basins (Riddler and Hemingway, 1976), and the development and tightening of synforms affecting the Hampshire and London basins.

Many of the basin-bounding and intrabasinal faults in areas affected by compression display spectacular local structures, indicative of fault reactivation and tectonic inversion or the effects of oblique-slip and strike–slip faulting. The exact nature of displacement on the reactivated faults is largely a function of their orientation relative to the regional stress field. East–west-trending structures appear largely to have experienced structural inversion, while north-west–south-east-trending structures, like those of the southern North Sea, seem to have taken up the strain through strike–slip movement (see next section).

Examples of east–west-trending structures that have experienced tectonic inversion include the development of the Flamborough Head Disturbance on the southern margin of the Cleveland Basin, the Purbeck–Isle of Wight monocline on the northern margin of the Wessex Basin (Colter and Harvard, 1981; Stoneley, 1982; Underhill and Stoneley, 1998) and the Weald Basin (Butler and Pullan, 1990). Reversal of the sense of motion on the controlling former extensional faults has commonly led to the formation of spectacular parasitic folds in their immediate hangingwall (e.g. the Bempton Disturbance, Flamborough

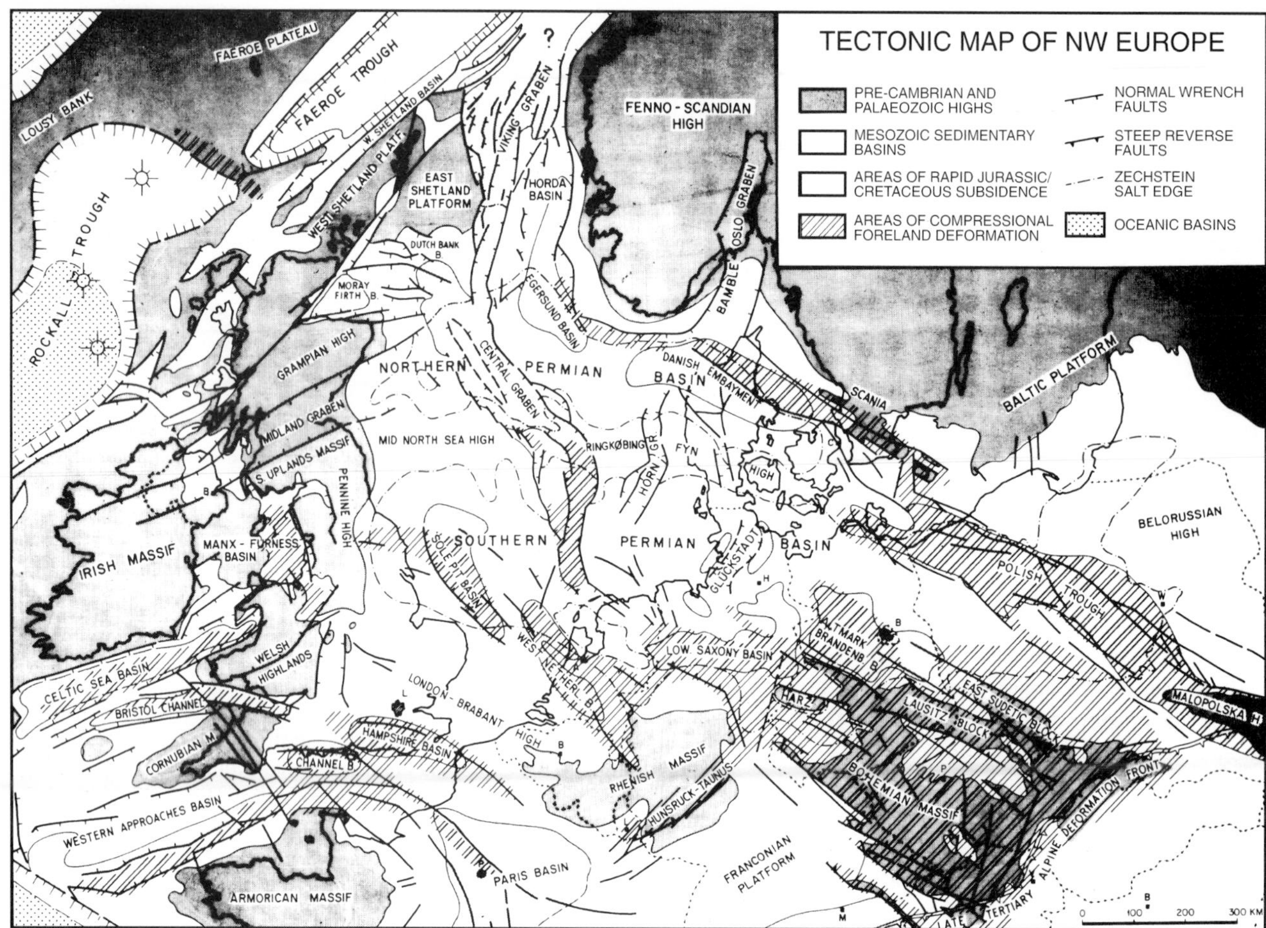

Fig. 2.28 Tectonic map of north-west Europe depicting the main sedimentary basins affected by Late Cretaceous and Cenozoic structural inversion related to the combined effects of Atlantic opening and Alpine collision (from Ziegler 1982a).

and the Lulworth Crumple, Stair Hole, Dorset; Underhill and Paterson, 1998).

Structural inversion of basins throughout southern Britain led to Late Palaeozoic and Mesozoic source rocks being lifted above their temperature window for hydrocarbon generation, leading directly to the effective switch-off of kitchen areas in the Cleveland, Wessex, Weald and southern North Sea basins. For their reservoirs to be hydrocarbon-bearing, many of the inversion structures have relied on remigration of hydrocarbons into them from breached palaeostructures.

Effect of oblique-slip movements on reactivated southern North Sea faults. As well as being the site of large-scale tectonic inversion, the southern North Sea Basin contains faults that show evidence for multiple phases of small-scale horizontal movement. Many of the predominant north-west–south-east-trending faults that now dissect the basin were probably initiated during collision across the Tornquist trend during the latest Carboniferous and earliest Permian, and were reactivated during later periods of tectonic activity.

The whole system of north-west–south-east-trending faults displayed in the southern North Sea area had a strike-slip component to them during their post-Variscan history. The presence of salt in the overburden is a complicating factor that enabled faults above the Zechstein to be decoupled from those affecting Late Palaeozoic strata (Fig. 2.32; van Hoorn, 1987a; Oudmayer and de Jager, 1993). On a smaller scale, the Triassic Röt Halite has had the same effect.

Despite the complicating effects of salt in the southern North Sea, modern seismic acquisition and interpretation techniques enable fault patterns to be accurately mapped and analysed. In cross-section, the seismic data often allow the identification of a single basement fault (often known as the principal zone of displacement), which branches upwards to form a fan of faults. These features are known as flower structures (e.g. Glennie and Boegner, 1981) and are characteristic of many strike–slip systems. Where the higher structure is dominated by extensional throws, the features are reminiscent of the fronds of a palm tree and are termed 'negative flower structures'. When they are dominated by contraction, the features often have a 'tulip-like geometry' and are termed 'positive flower structures'. It is very common for negative and positive flower structures to occur, and hence the sense of throw to alternate, along the same fault zone, as a result of irregularities in fault orientation (e.g. Dowsing Fault Zone).

In addition to the evidence from seismic sections, the expression of the Base Zechstein (Rotliegend equivalent) seismic marker in plan view commonly demonstrates the importance of en-echelon fault patterns, which are consistent with an interpretation as Riedel shears resulting from

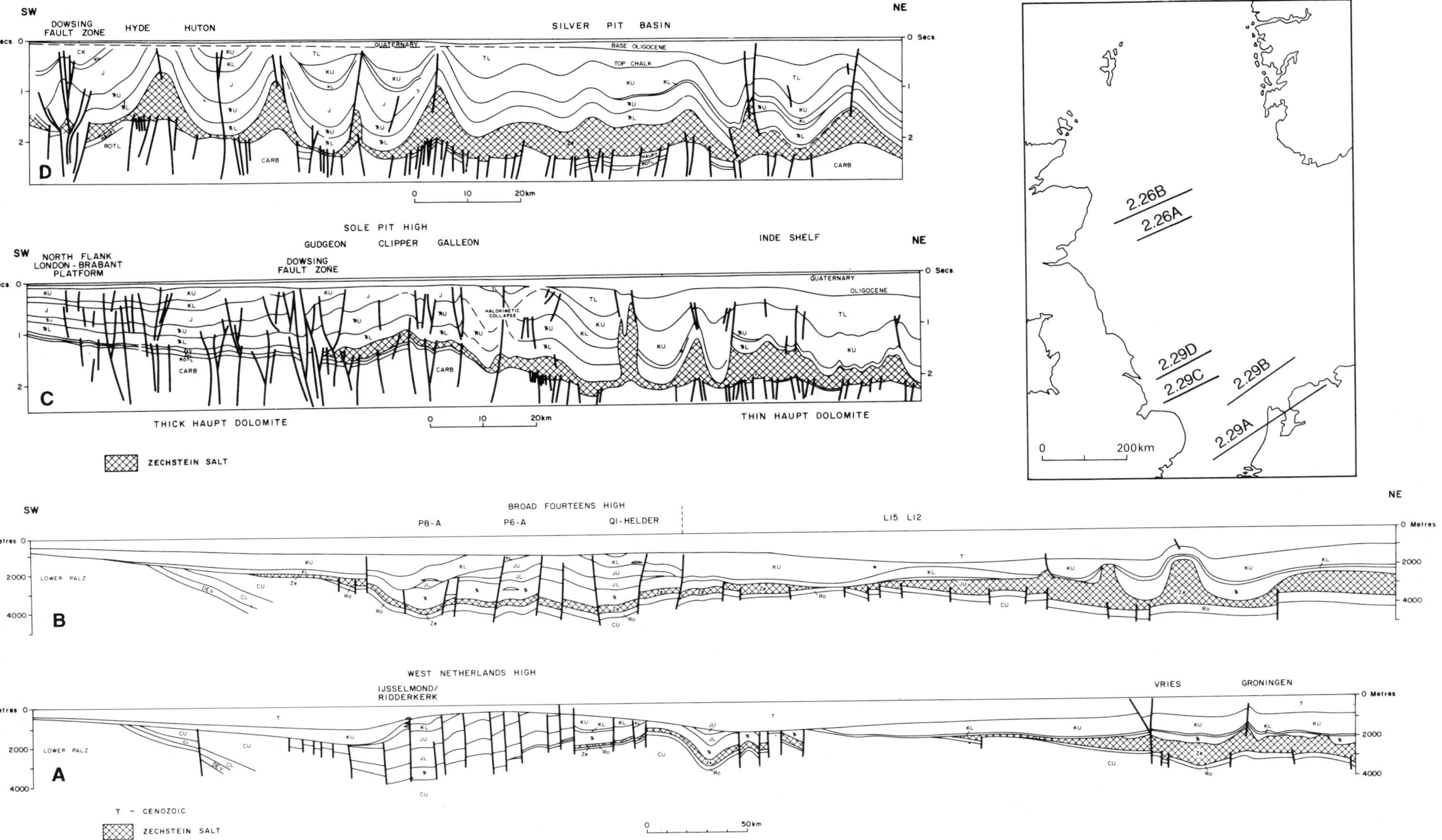

Fig. 2.29 (A & B) Cross sections of the Netherlands land and offshore areas. Note the inverted basinal structure of the Triassic, Jurassic and Lower Cretaceous strata of the Broad Fourteens High relative to the synclinal Permian sequence; also the marked Late Cimmerian erosion north-east of the Broad Fourteens and West Netherlands Basins. (Modified from IMNES and Netherlands Geological Survey wall chart, 1984.) Vertical scale in metres. (C & D) Line drawings from Shell-Expro seismic lines across the UK southern North Sea. Halokinesis has had a strong effect on especially the post-Jurassic sedimentation pattern. Many faults extend up into the Cretaceous where Zechstein salt is thin, or absent south-west of the Sole Pit High. Note the deep Late Cimmerian erosion in the north-east. Vertical scale in seconds 2-way time.

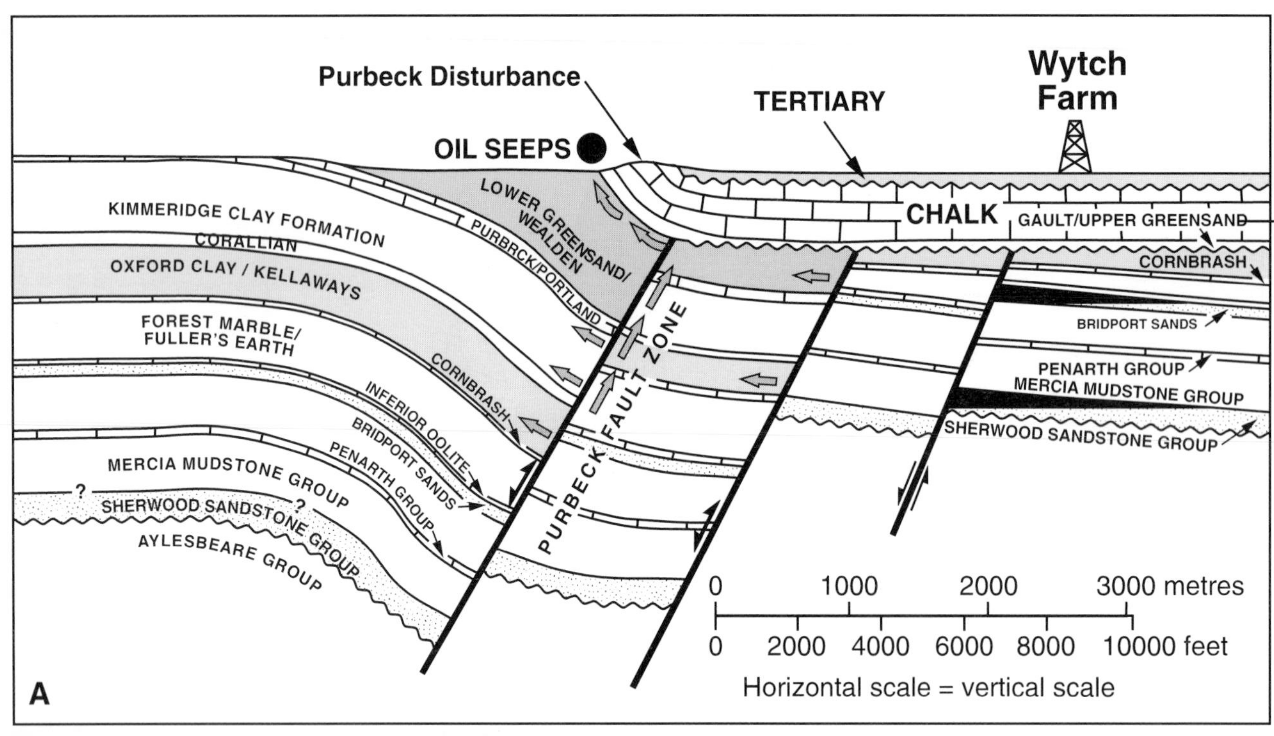

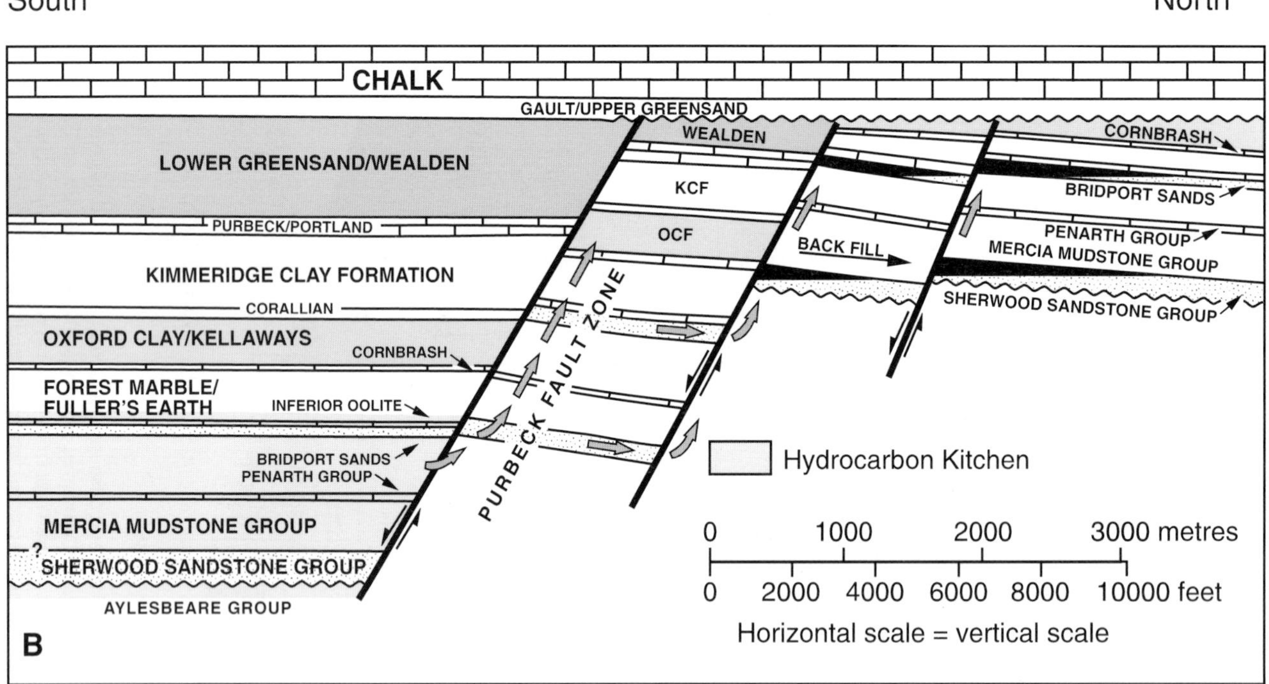

Fig. 2.30 Present-day and restored north–south-trending cross-sections from the Wessex Basin, southern England, showing the effects of basin inversion on the Purbeck Fault Zone. Cenozoic inversion has largely been detrimental to hydrocarbon prospectivity, since it switched off the main kitchen area and created structures that postdated charge. It is only in those footwall areas unaffected by the effects of Cenozoic deformation that hydrocarbon traps have lasted (e.g. Wytch Farm). (After Underhill and Stoneley, 1998; modified from Colter and Harvard, 1981.)

either transtensional or transpressional deformation (e.g. Oudmayer and de Jager, 1993).

Much of the transpressional deformation, and hence structural trap development, may indeed have occurred during the Late Cretaceous or Cenozoic in response to intraplate compression (Alberts and Underhill, 1991). Inversion from a basinal low to a gas-filled structural high in many structures probably resulted in the remigration of gas into them from adjacent structures. Were it not for the coherence of the Zechstein salt seal and its ability to re-anneal itself after fracture, much of the southern North Sea gas might have been lost as a result of Cenozoic

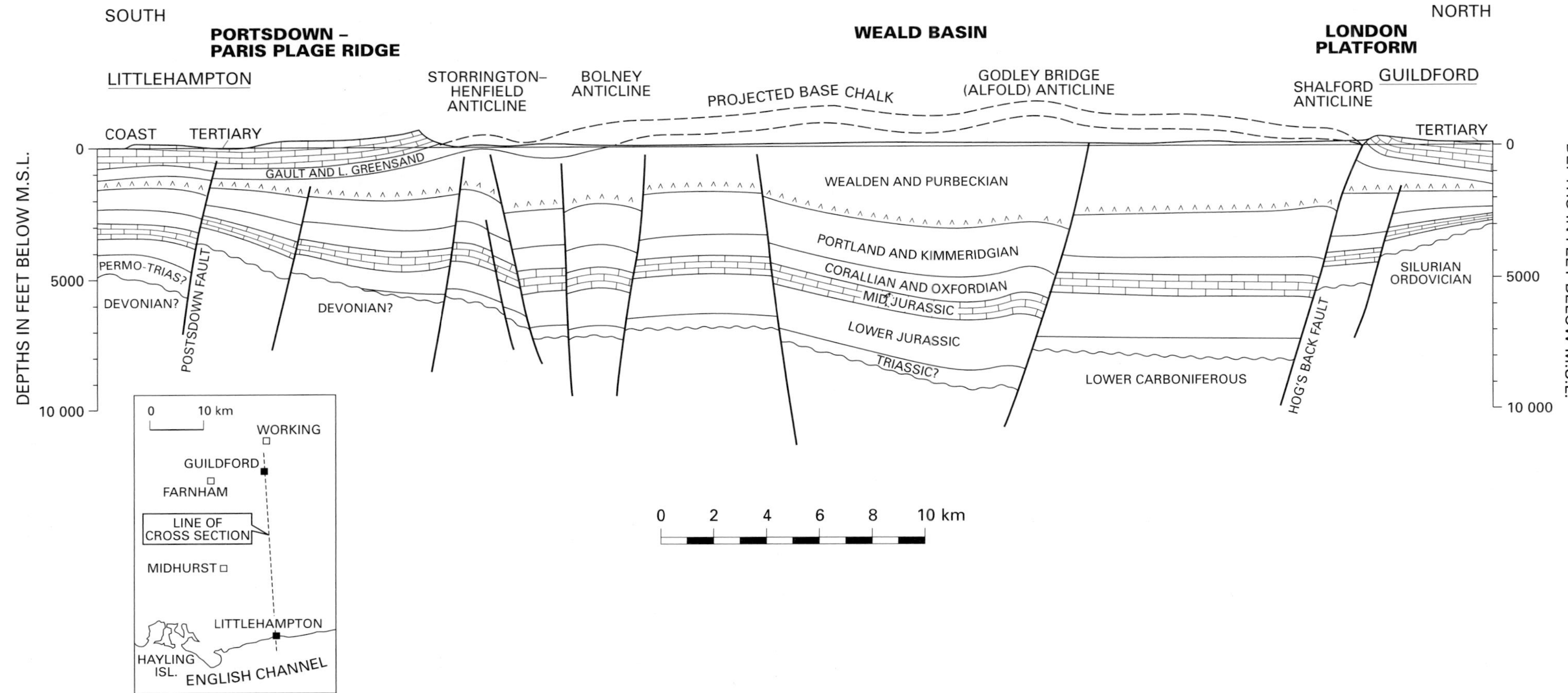

Fig. 2.31 Cross-section across the structurally inverted Weald Basin, southern England (after Butler and Pullan, 1990).

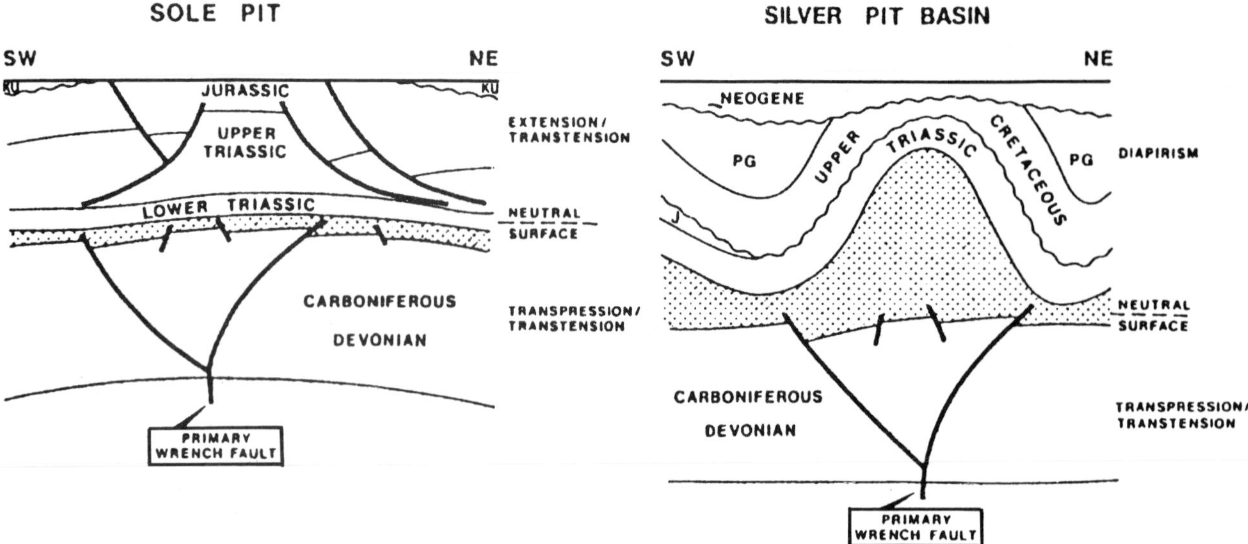

Fig. 2.32 Schematic diagrams depicting the main controls on structural styles in the Sole Pit and Silver Pit areas, southern North Sea. The diagrams emphasize the importance of the Zechstein evaporites in compartmentalizing strain (after van Hoorn, 1987a).

Fig. 2.33 North–south-trending seismic line in the Central North Sea showing the marked effect that diapirism has had on the Mesozoic and Cenozoic overburden (after Hodgson *et al.*, 1992).

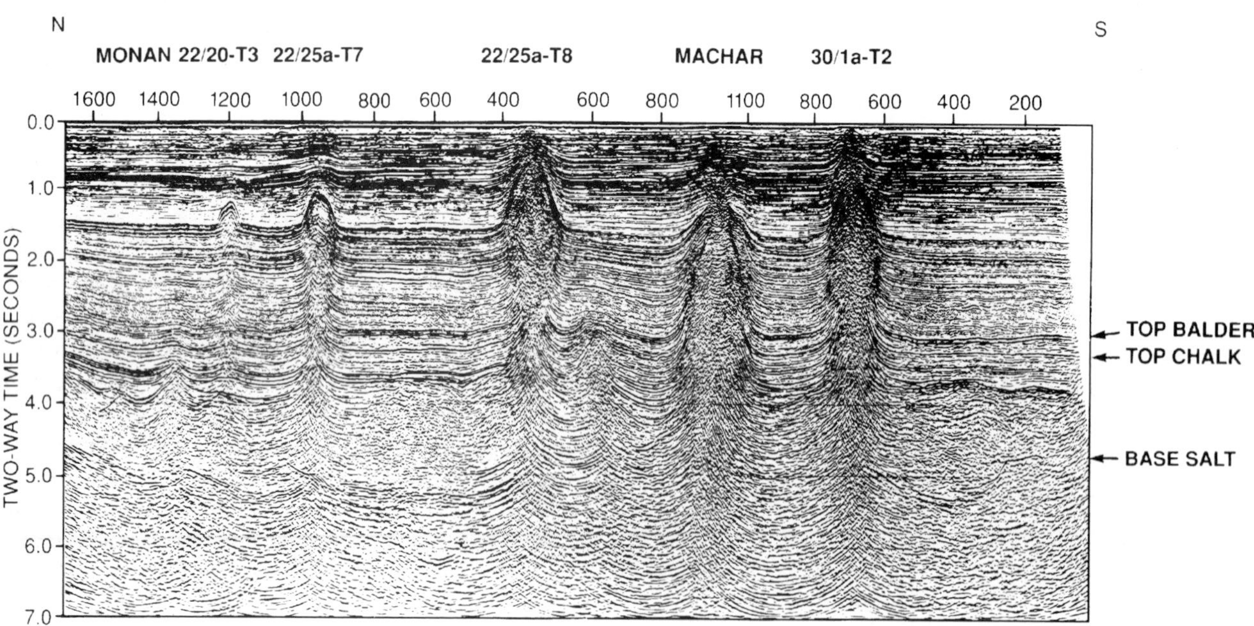

structural inversion and strike–slip faulting. As it is, it seems that only those areas of late uplift that lay relatively remote from gas-kitchen areas received only limited charge or no charge at all (e.g. the Cleaver Bank High; Alberts and Underhill, 1991).

Role of salt diapirism in North Sea post-rift structural modification. The presence of two extensive Late Permian salt basins has strongly influenced subsequent patterns of tectonics and sedimentation in the North Sea. The most obvious effects are seen in the southern North Sea, where Zechstein salt acts as the main regional seal for Rotliegend gas reservoirs (Fig. 5.31). Salt can flow where mobilization of the salt has occurred (see Chapter 6). Salt walls, pillows and diapirs developed and have affected sediment thicknesses within the Mesozoic and Cenozoic successions (see, for example, Fig. 2.26). Many of these salt walls have the same alignment as underlying fault patterns (see Fig. 6.12), leading to the assumption that diapirism was probably initiated by the destabilizing effects of fault movements. The timing of diapirism can often be deduced from the seismic stratigraphy within associated rim synclines (see Figs 6.20–6.22). Differences in the timing of diapiric activity in adjacent structures may be the outcome of differential fault activity within the substrate. When horizontal salt removal causes 'grounding' adjacent to a diapir (see, for example, Fig. 6.20), salt walls may ensue with very complex geometrical relationships within the rim synclines. The upward growth of some salt walls seems to have ceased only when the overburden of a Pleistocene ice sheet was replaced by Holocene sediment and sea water.

The grounding of Triassic strata above the Rotliegend in several areas has enabled the upward escape of gas from temporary accumulations. Where diapirs have created traps adjacent to these grounded areas, gas has migrated into the overlying Triassic Bunter Sandstone (e.g. the Forbes, Gordon and Esmond fields in UK waters and at least two of the K/13 structures of the Dutch sector; see Chapter 7).

Further north, in the central and northern North Sea, Mesozoic and Tertiary sediments accumulated above the Permian salt deposited in the northern of the two Zechstein basins. Salt withdrawal and dissolution over the flanks of the Central Graben during the Triassic and Jurassic led to the formation of a series of roughly north–south orientated salt ridges and intervening depocentres ('pods') filled with continental, mainly fluviolacustrine, Smith Bank shales (see Chapters 7 and 12). Further withdrawal of salt and the grounding of the pods created accommodation space for Upper Jurassic reservoirs, such as those that form the main reservoir in the Fulmar Field (see Chapter 8). The major Late Jurassic rift event in the North Sea utilized the inherent Triassic tectono-sedimentary grain, with the salt acting as an important detachment plane between the underlying rejuvenated Permian fault system and the incipient Late Jurassic faults.

Cretaceous thermal subsidence was locally enhanced by the effects of halokinesis in the region of the salt basin. Sediment accumulation led to differential loading of the Zechstein evaporites, to the initiation of prominent salt domes in the main areas of Late Jurassic rifting and, subsequently, to the formation of salt diapirs. These have continued to develop to the present day as a result of Late Cretaceous and Tertiary subsidence and sedimentation. The diapiric rise of evaporites has helped create structural traps for several major fields and discoveries in the Central North Sea Basin (Fig. 2.33), including Machar (Figs 2.34 and 2.35; Foster and Rattey, 1993). In Danish and Norwegian waters many of the Upper Cretaceous fields depended upon the diapiric rise of salt stocks and walls to fracture the Chalk reservoir, thereby permitting hydrocarbons to migrate into Danian carbonate reservoirs (Chapters 9 and 12).

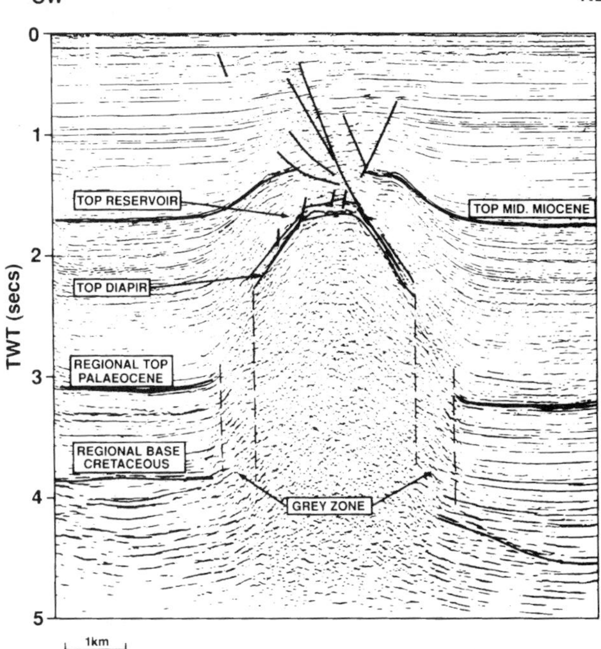

Fig. 2.34 South-west–north-east-trending seismic section across the Machar diapir, Central North Sea (after Foster and Rattey, 1993).

2.4 Conclusions

Structural inheritence has proved to be an important consideration in the development and evolution of structural styles in many hydrocarbon provinces. The prospective sedimentary basins of the North Sea and adjacent areas are no different and a knowledge of earlier tectonic events and structural trends helps in understanding subsequent structural effects more completely. Review of the main structures and knowledge of their regional plate-tectonic setting enables the tectonic development of the North Sea and surrounding areas to be considered under four main headings: Precambrian events, the Caledonian (Iapetus) plate cycle, the Variscan (Rheic) plate cycle and Post-Permian tectonic events related either to Atlantic Ocean opening or to structural events in an intraplate setting. Comparisons made between Precambrian and Palaeozoic structural lineaments and those tectonic elements that most notably influenced Late Palaeozoic, Mesozoic and Cenozoic basin development indicate that much of the younger history of the North Sea area, which was so crucial for the hydrocarbon habitat in the basin, was controlled by the

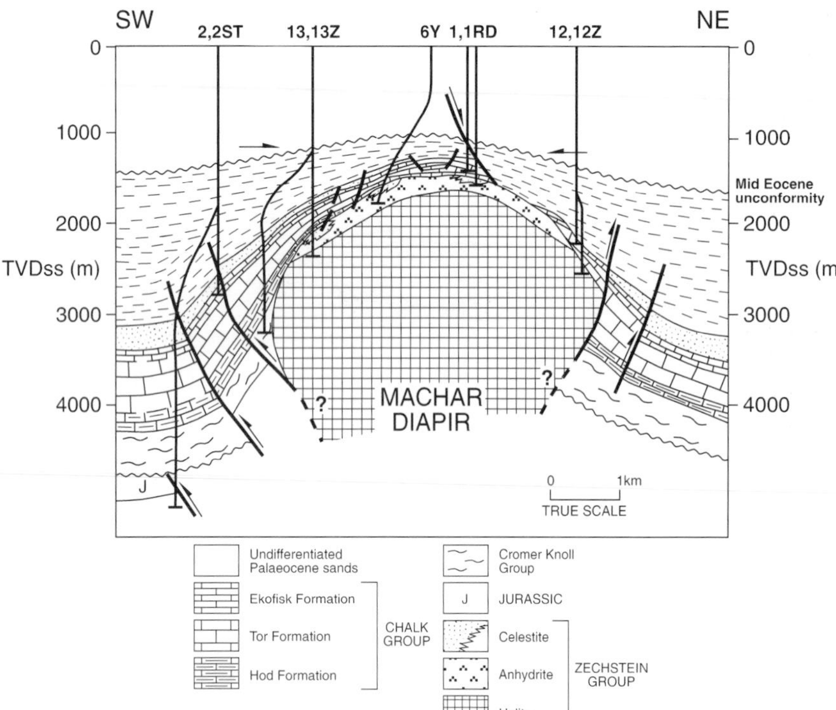

Fig. 2.35 Schematic cross-section across the Machar diapir, Central North Sea (after Foster and Rattey, 1993).

distribution of older, rigid, cratonic blocks and their intervening lines of weakness. Throughout the period that the North Sea has lain in an intraplate setting, the orientation of deeper structure relative to the prevailing stress orientation seems to have been fundamental in determining the nature and extent of deformation at any time, with non-orthogonal stress leading to a greater or lesser degree of oblique- or strike-slip movement on basement faults particularly in the Southern North Sea.

2.5 Acknowledgements

The earlier editions of this contribution were published by permission of Shell Internationale Petroleum Mij., The Hague. We are indebted to Peter Ziegler, not only for the use of several of his figures, but also for discussion and advice. Alan Crane contributed to our knowledge of the Scottish Precambrian geology. Ken Thomson, Richard Davies, Kevin Stephen, Dan Bishop, Nancye Dawers, John Dixon, Sarah Johnston, Dan McKenzie, Dave Latin, Neil McMahon, Jo Fleming, Susan Paterson, John Turner and Shona MacDonald are all acknowledged for furthering our knowledge of the Phanerozoic evolution of the North Sea and surrounding areas. Gerry White helped transpose or redraft the diagrams.

2.6 Key references

Bartholomew, I.D. (1993) Regional structural evolution of the North Sea: oblique slip and the reactivation of basement lineaments [and contained references]. In: Parker, J.R. (ed.) *Petroleum Geology of Northwest Europe: Proceedings of the 4th Conference.* Geological Society, London, pp. 1109–22.

Gibbons, W. and Harris, A.L. (eds) (1994) *A Revised Correlation of the Precambrian Rocks in the British Isles.* Special Report 22, Geological Society, London, 110 pp.

Hodgson, N.A., Farnsworth, J. and Fraser, A.J. (1992) Salt-related tectonics, sedimentation and hydrocarbon plays in the Central Graben, North Sea UKCS. In: Hardman, R.F.P. (ed.) *Exploration Britain: Geological Insights for the Next Decade.* Special Publication 67, Geological Society, London, pp. 31–63.

Knott, S.D., Burchell, M.T., Jolley, E.J. and Fraser, A.J. (1990) Mesozoic to Cenozoic plate reconstructions of the North Atlantic and hydrocarbon plays of the Atlantic margins. In: Parker, J.R. (ed.) *Petroleum Geology of Northwest Europe: Proceedings of the 4th Conference.* Geological Society, London, pp. 953–74.

Rattey, R.P. and Hayward, A.B. (1990) Sequence stratigraphy of a failed rift system: the Middle Jurassic to Early Cretaceous basin evolution of the Central and Northern North Sea. In: Parker, J.R. (ed.) *Petroleum Geology of Northwest Europe: Proceedings of the 4th Conference.* Geological Society , London, pp. 215–49.

Torsvik, T.H., Smethurst, M.A., Meert, J.G., van der Voo, R., McKerrow, W.S., Brasier, M.D., Sturt, B.A. and Walderhaug, H.J. (1996) Continental break-up and collision in the Neoproterozoic and Palaeozoic—a tale of Baltica and Laurentia. *Earth Sc. Rev.* **40**, 229–58.

Ziegler, P.A. (1990) *Geological Atlas of Western and Central Europe.* Shell Internationale Petroleum Maatschappij BV, distributed by Geological Society, London, 239 pp.

2.7 Appendix: seismic cross-section

(Fig. 2.36)

The composite line drawing of a seismic section in Fig. 2.36 can be divided into four major units, individual components of which have been dated regionally by the use of cores, cuttings and wireline-log correlations.

1 A Pre-Permian basement unit, the deepest parts of which contain only weak indications of internal structure. It has a surface relief of some 3 s two-way time, or around 3000–4000 m, depending on the acoustic velocity of the overlying strata. The strongest relief occurs on the flanks of the Central Graben. Regional evidence suggests that one basement reflector possibly represents the erosional contact between strongly folded Caledonian basement rocks and

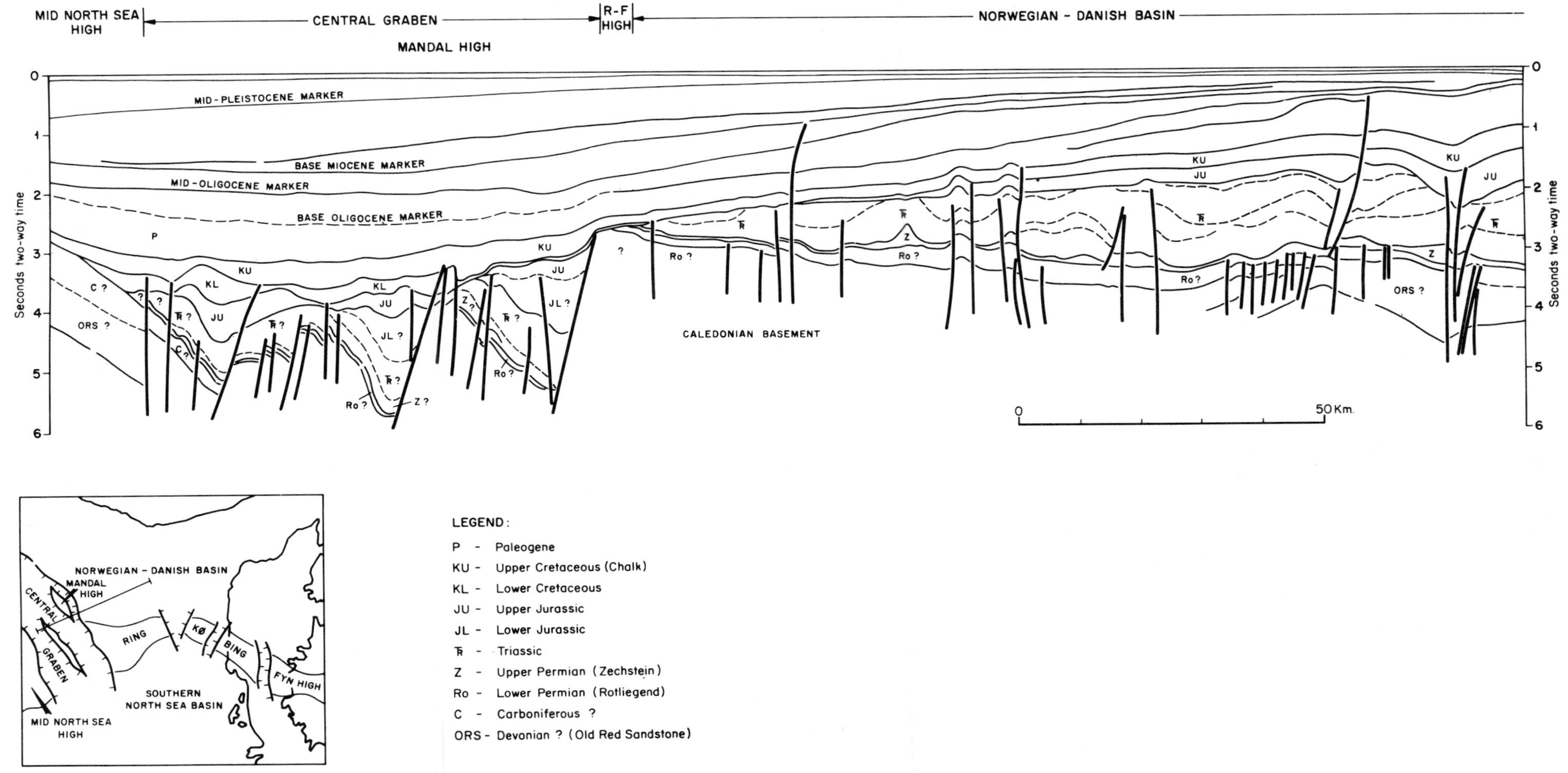

Fig. 2.36 Line drawing of a seismic line across the northern Central Graben and part of the Norwegian–Danish Basin. For fuller interpretation, see text of this appendix.

the overlying Old Red Sandstone. A character change within the basement sedimentary sequence of the Mid North Sea High indicates the possible presence of Carboniferous strata roughly conformable with the underlying Old Red Sandstone, an interpretation that is supported by well data.

2 Within the Norwegian–Danish Basin, the basement rocks are overlain by a truncated wedge of sediments, which, on the grounds of seismic and structural character, are believed to range in age from the Early Permian (Rotliegend) to the Late Jurassic (confirmed by well data). The Rotliegend has been penetrated by the drill along the northern edge of the Ringkøbing–Fyn High, but not in the basin centre. The Upper Permian Zechstein is recognized by its characteristic salt diapirs (many of which are now relics of collapse following salt solution at a later date), and its base forms a regionally correlatable marker. The pre-Zechstein strata presumably represent the Lower Permian Rotliegend, which here may be mostly in a desert-lake facies.

The post-Zechstein part of the sequence is largely Triassic in age, reaching an acoustic 'thickness' of about $1\frac{1}{2}$s two-way time. The geometry of internal reflectors indicates that Zechstein salt diapirism was already active locally during the Middle and possibly late Early Triassic; Zechstein salts still disrupt Triassic strata at a few localities.

The Triassic is strongly and uniformly truncated towards the Central Graben boundary fault and is overlain by a sedimentary sequence that has been dated by well correlation as Late Jurassic in age; and the Upper Jurassic itself wedges out before the margin of the Central Graben is reached, and is overlain by the Upper Cretaceous Chalk, which locally is separated from the basement by only a thin sliver of Lower Cretaceous.

3 Within the Central Graben, the basement is covered by a sedimentary sequence that ranges from Lower Permian (Rotliegend) to Lower Cretaceous. Here, however, the pre-Upper Jurassic sequences have been rotated into half-grabens, and both grabens and highs have been draped by Upper Jurassic and Lower Cretaceous sedimentary sequences of irregular thickness. There has been limited Zechstein diapirism.

4 Units two and three are overlain unconformably by the Upper Cretaceous Chalk, which attains its greatest thickness within the Central Graben. Most faults do not extend above the Lower Cretaceous surface, thus implying a much greater degree of tectonic calm after the event than before it. The few faults that show slight post-Cretaceous activity are all away from the flanks of the Central Graben. The overlying sequence records the relatively calm conditions of Cenozoic subsidence centred over the Central Graben. The prograding sequences of Early to Mid-Cenozoic strata indicate that a source of sediment must have been present to the east throughout that time span, with a major phase of outbuilding possibly coinciding with the Oligocene global low-stand of sea level.

It is clear from the cross-section that this part of the western flank of the Norwegian–Danish Basin underwent considerable uplift during the Mid-Jurassic, probably just before major fault-block rotation within the Central Graben. Together with the Central Graben, this same basin flank has been subsiding steadily since the later Late Cretaceous.

3 Devonian

R.A. DOWNIE

3.1 Introduction

The Devonian rocks of the North Sea Basin (Fig. 3.1) have generally been perceived to have little hydrocarbon potential, and the penetration of Devonian strata has commonly been taken as reason to terminate an exploration well. Despite this, a number of discoveries have been made, mainly in the UK sector (Table 3.1), where Jurassic-sourced hydrocarbons are trapped in nearby Devonian structural highs. In the Norwegian sector, the Embla Field (block N2/7) provides the sole discovery in sandstones of probable Devonian age; no Devonian discoveries are reported from the Danish, German or Netherlands sectors of the North Sea.

Without exception, all the fields and discoveries are relatively unusual in that the reservoir sandstones have low to, at best, moderate permeabilities and the flow of hydrocarbons is significantly assisted by the presence of open fractures. The presence of these dual porosity systems (fracture and matrix) has resulted in many of the discoveries being difficult to evaluate; for example, in the Buchan Field (block 21/1) resolution of these uncertainties increased the estimates of recoverable reserves from 50×10^6 barrels at the time of initial development, to 90×10^6 barrels by 1989 (Edwards, 1991), to recent estimates of 120×10^6 barrels (BP Exploration, 1995). Similarly, the giant Clair Field, to the west of Shetland, remains the largest undeveloped field on the UK continental shelf since its discovery in 1977, due to difficulties in proving whether economic production can be sustained (for discussion, see Coney *et al.*, 1993). The Stirling and West Brae accumulations on the Fladen Ground Spur have also been the focus of further drilling since their discovery, although proof of their commercial capabilities remains elusive. In comparison with younger stratigraphic intervals, the recoverable hydrocarbons in the Devonian reservoirs are volumetrically minor, although the discoveries are sufficiently numerous to indicate that Devonian sandstones provide an interesting, if high-risk target.

Table 3.1 Hydrocarbon accumulations in Devonian reservoirs of the UK continental shelf (UKCS) and Norwegian North Sea.

Discovery	Block	Hydrocarbons
Argyll	30/24	Composite Palaeozoic reservoir, up to 5500 barrels oil/day from Devonian interval (Robson, 1991). Field now abandoned
Buchan	21/1	120×10^6 barrels oil recoverable (BP Exploration, 1995)
Clair	208/6	Greater than 3×10^9 barrels oil in place (Johnston *et al.*, 1995), low recovery factor. Field under evaluation
Embla	N2/7	215×10^6 barrels minimum, possibly as high as 1028×10^6 barrels oil in place (Knight *et al.*, 1993)
Stirling	16/21	Maximum reported flow rate of 4334 barrels oil/day. Discovery under evaluation
West Brae	16/7a	Maximum reported flow rate of 3698 barrels oil/day. Discovery under evaluations

In addition to reservoir potential, the Devonian has some limited potential as a source rock; in the Inner Moray Firth, Devonian shales have been proved to have sourced at least part of the oil in the Jurassic Beatrice Field (Duncan and Hamilton, 1988; Peters *et al.*, 1989). Some limited gas potential may also be associated with Devonian coals, although these have been penetrated only in a single North Sea well, 38/3-1, located in the UK sector on the Mid North Sea High. Given a suitable thermal history, these would have been capable of generating gas; should these coals prove to be more widely distributed, they could provide a viable alternative source rock. It is thus clear that evaluating the factors which control Devonian source-rock distribution and maturity is important in the formulation of any exploration programme.

While Devonian rocks crop out widely in the UK and are present at subcrop over much of the North Sea Basin, only those in the Orcadian Basin and the Central North Sea (Fig. 3.1) have significant hydrocarbon potential, either as a reservoir or as a source. It is on these areas that this chapter focuses. The other areas of deposition are briefly summarized and comments are made regarding the factors that limit or exclude any hydrocarbon potential.

3.2 Regional setting

The tectonic controls on Devonian sedimentation are complex and a detailed review is beyond the scope of this chapter (see Chapter 2 for further discussion). In brief, Devonian sediments were deposited at the end of the Caledonian Orogeny and thus postdate the associated major compressional, deformational and regional metamorphism events. In some areas, however, deposition of Lower Devonian rocks predated or was coeval with em-

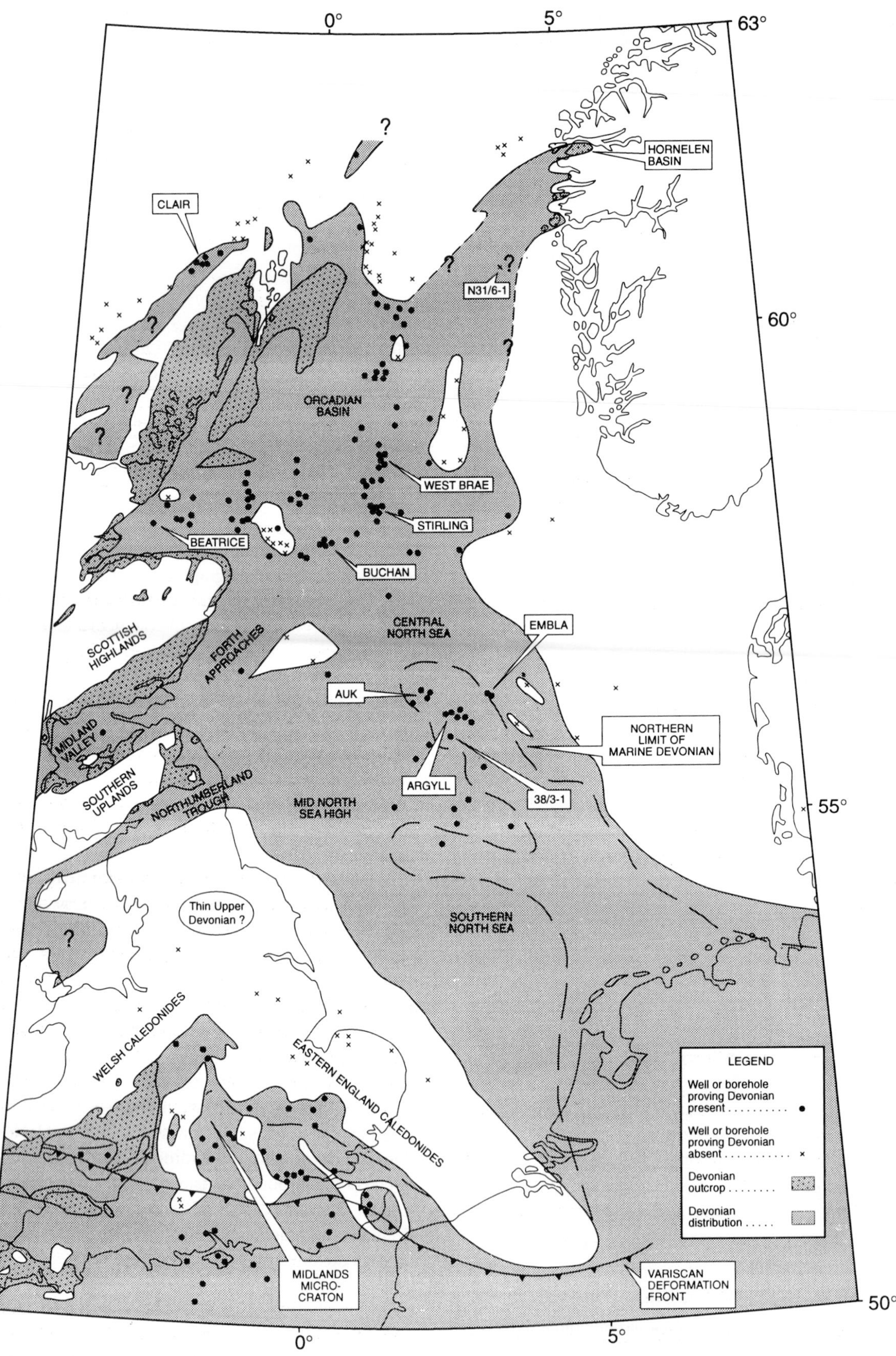

Fig. 3.1 Distribution of Devonian Old Red Sandstone (ORS) strata in the North Sea Basin and associated areas.

placement of post-tectonic Caledonian granites ('Last Granites': Reid, 1961). Examples of this are provided by the Glencoe and Northumberland areas, where Lower Devonian volcanics and volcaniclastics are intruded by granitic plutons (Fig. 3.2).

Overall, sedimentation was controlled by closure of the north-east–south-west-trending Iapetus Ocean in the Late Silurian to Early Devonian and the welding together of the major eastern and western continents, which in the British

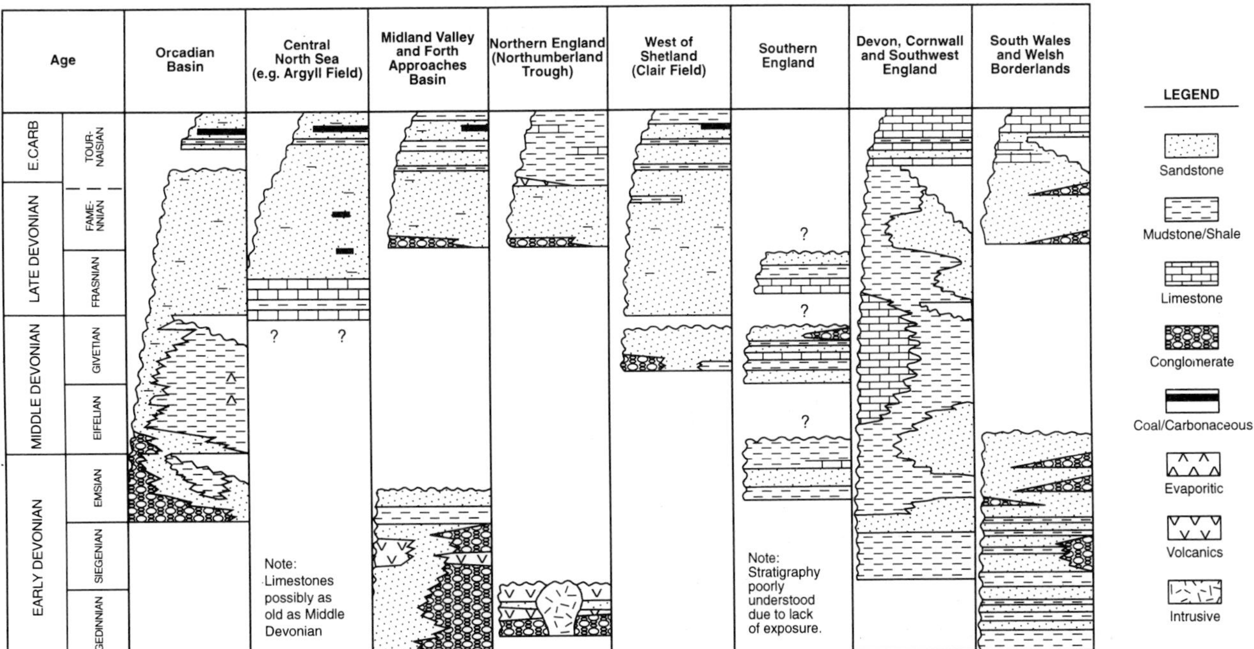

Fig. 3.2 Schematic Devonian chronostratigraphy of the UK and North Sea Basin.

Isles was associated with both north-west- and south-east-directed subduction. Sedimentation was concentrated in the areas between and around the adjacent Laurentia–Greenland and Fennosarmatia cratonic areas (Fig. 3.3), which straddled the Devonian equator. In the North Sea Basin, deposition largely occurred in a hot, arid continental setting, and alluvial fan, fluvial braid-plain and lacustrine environments predominated. The deposits of these environments are generally referred to as the Old Red Sandstone (ORS) and they form the bulk of the Devonian strata within the North Sea Basin. During this period the North Sea Basin drifted from approximately 20°S to 15°S (Tarling, 1985).

Coeval marine basins were developed to the south of the North Sea Basin, their deposits being present in the south and west of England and in continental Europe (Fig. 3.3). These marine facies comprise mostly shallow-water sandstones, shales and limestones, although, in south-west England, deposition in deeper water is evidenced by the presence of turbiditic sandstones. The shallow-marine facies periodically extended northward into the proto-Central Graben (Figs 3.1 and 3.4), where they have been penetrated in offshore oilwells.

3.2.1 Tectonostratigraphic summary

Evidence from the onshore UK shows that Devonian sediments range in age from the earliest (Gedinnian) to the latest (early Tournaisian) Devonian (see House *et al.*, 1977), although a complete succession is not present at any one locality (see Fig. 3.2). In general, Devonian rocks rest with unconformity on Silurian or older strata, with conformable Silurian–Devonian transitions recognized only locally (e.g. Midland Valley of Scotland, South Wales). The Devonian–Carboniferous boundary is commonly marked by a major unconformity, although again conformable successions are known (e.g. Midland Valley, Central North Sea).

The Devonian is divided into Lower, Middle and Upper divisions, which roughly equate to the Lower, Middle and Upper ORS, respectively; the latter spans the Late Devonian and part of the Early Carboniferous. These divisions are commonly separated by major erosional, commonly angular, unconformities (see Fig. 3.2), believed to have developed in response to north–south compressive events associated with the final stages of the Caledonian continental collision.

Offshore, the dating of Devonian sediments, in particular, and Palaeozoic red-bed sediments, in general, is hindered by the lack of definitive biostratigraphic evidence of age. For example, in the Argyll Field, while Devonian sandstones are undoubtedly present, they are overlain by lithologically similar sandstones of Permian age; distinction is difficult (cf. Robson, 1991, and Bifani *et al.*, 1987). Similarly, in the Embla Field in Norwegian block N2/7, Late Devonian ages are proved near the base of the sequence, in a marine-influenced, micaceous/argillaceous unit that overlies rhyolitic ?basement. The stratigraphically higher reservoir sandstones, however, are devoid of age-diagnostic microfossils. These are probably also Late Devonian in age, based on stratigraphic position and lithological character, although Knight *et al.* (1993) do not rule out the possibility of Early Permian or Carboniferous ages.

The distribution of Devonian strata in the UK and in the North Sea Basin, as shown in Fig. 3.1, provides information on the tectonic controls on sedimentation. The figured distribution was derived from evaluation of the many well and borehole penetrations (often reinterpreting the stratigraphy offered by the well operator) and seismic data, and by reference to published outcrop maps and documents, in particular those from the British Geological Survey (e.g. Cameron *et al.*, 1992). Equal note was taken of wells and boreholes that proved the Devonian to be absent and those

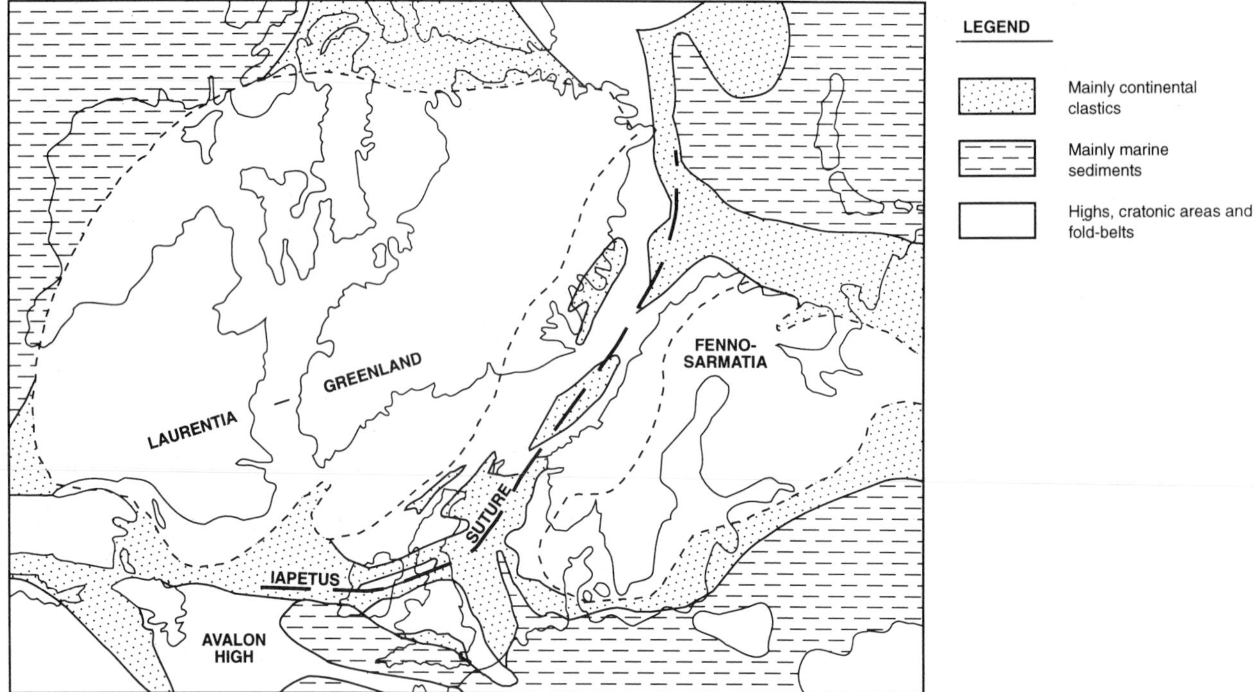

Fig. 3.3 Schematic distribution of Devonian sediments in the North Atlantic realm (modified and simplified from Ziegler, 1988).

that proved its presence. A significant feature of the map is the distribution of Devonian strata in the south of England and Wales, which appears to follow the outline of the Midlands Microcraton, as originally defined by Turner (1949) and later refined by Pharaoh *et al.* (1987). This microcraton is a northward-pointing, wedge-shaped area of Precambrian crust, with a relatively thin cover of Palaeozoic sediments, which were not strongly deformed during the Caledonian Orogeny. Soper *et al.* (1987) proposed that the Midlands Microcraton acted as a wedge-shaped 'rigid indenter', around which the Lower Palaeozoic rocks of the Caledonide belts were deformed during north–south compression. The observed distribution of Devonian strata is consistent with this model, and suggests that the northward collision deformed the north-east–south-west-trending Welsh Caledonides and north-west–south-east concealed Caledonides of eastern England into upland areas, which then shed sediment southward on to the stable Midlands Microcraton.

There is evidence of considerable strike–slip movement on some of the major fault systems that delineate the Devonian basins of the United Kingdom, although it is now generally considered that such movements were subordinate to extension during the Devonian (e.g. Roger D.A. *et al.*, 1989; Haughton *et al.*, 1990). It is now believed that the major Devonian basins of the North Sea Province were essentially extensional in origin, and were formed by the gravity-driven relaxation of northward-trending Caledonian thrust structures associated with the closure of the Iapetus Ocean (e.g. McClay *et al.*, 1986; Enfield and Coward, 1987; Norton *et al.*, 1987). Figure 3.1 clearly shows the north-east–south-west-trending Midland Valley, Northumberland Trough and Orcadian Basin, which parallel the north-west margin of the Midlands Microcraton.

In apparent variance with the above model is the Devonian of the Central and Southern North Sea. This forms a north-west–south-east-trending basin, approximately perpendicular to the northern Devonian basins described above. The axis of the North Sea Devonian Basin, however, is parallel to the north-east margin of the concealed Caledonides of eastern England (Fig. 3.1). It is possible that northward collision of the Midlands Microcraton resulted in thickening of the Caledonide crust, now buried below the central and southern North Sea, in a manner analogous to that proposed for the Scottish Highlands by McClay *et al.* (1986). With relaxation of compression, gravity-driven extension of the overthickened 'crustal welt', orthogonal to the principal Caledonide grain, may have started to operate by ?Mid–Late Devonian times. On the Mid North Sea High, seismic data clearly show evidence of extension, with the development of approximately north–south-trending half-graben structures, with a fill of Upper Devonian and Dinantian strata. Thickening of the Devonian strata on the downthrown sides (hanging walls) is evidence that these faults were active during the Devonian (Fig. 3.5).

The age of the North Sea Basin as a discrete structural entity has long been a topic of debate. The distribution of Devonian sediments discussed above (see Fig. 3.1) shows that a distinct basin (proto-North Sea) was clearly in existence by the Late Devonian.

3.3 Orcadian Basin

3.3.1 Basin overview

The Orcadian Basin was a large, complex area of deposition. Sediments crop out on the northern mainland of Scotland along the southern coast of the Moray Firth and in Caithness and Sutherland, as well as on the archipelagos of

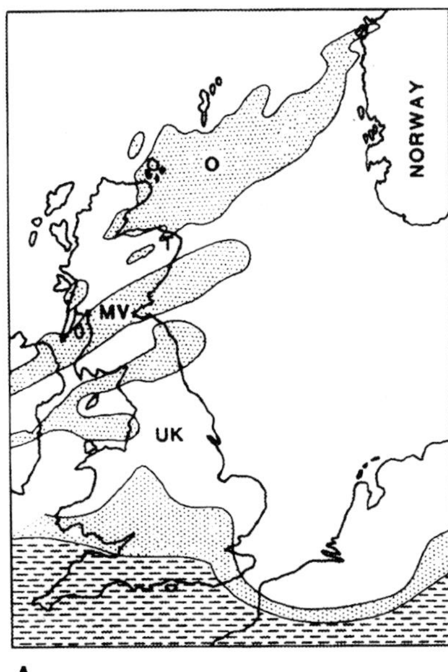

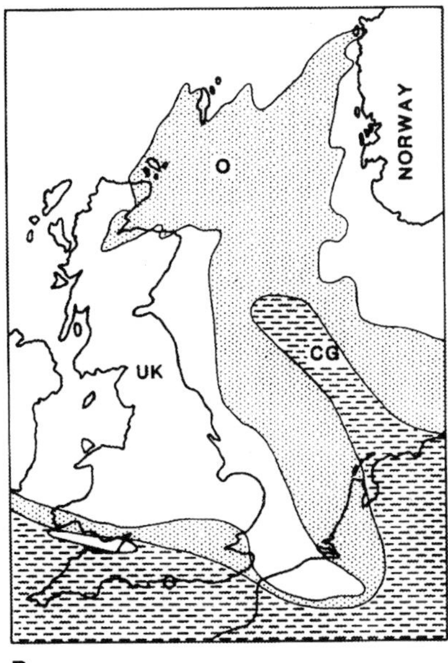

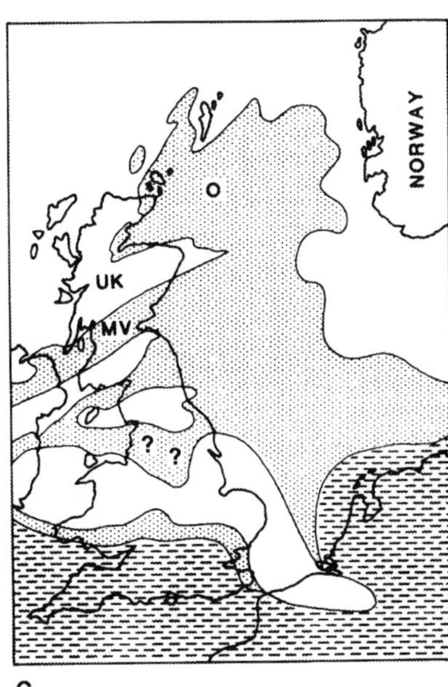

Fig. 3.4 Simplified Devonian palaeogeographic maps (modified from Ziegler 1982a). A. Early Devonian. B. Middle/Late Devonian. C. Late Devonian. T, Turriff Basin; O, Orcadian Basin; CG, Central Graben; MV, Midland Valley. Stippling shows terrestrial area and horizontal hatching shows marine area. From Richards (1990) with permission.

the Orkney and Shetland Islands (Fig. 3.6). Sediments are also present on the sea floor over much of the Shetland Platform, and have been penetrated beneath younger cover in numerous northern North Sea wells (see Fig. 3.1). Indeed, most of the records of ORS penetrations in commercial offshore wells come from the Orcadian Basin.

The eastward extent of the basin is unclear. It was suggested by Ziegler (1990a) that the Orcadian Basin sediments may have been contiguous with those of the Hornelen Basin of western Norway. Well data to support this suggestion are lacking, however, and the only released well penetrating to sufficient depth in the area of proposed connection (Norwegian well N31/6-1) contains Triassic red-bed sediments unconformably overlying basement rocks (see Fig. 3.1).

The western margin of the basin trends parallel to and to the west of the Great Glen Fault complex (Fig. 3.6) and its position may have been controlled by faults (Watson, 1985). The southern margin of the basin is onshore in the Grampian Region of Scotland and appears to cross-cut the Caledonide structural grain, but may have been defined by the position of a deep fracture (Watson, 1985), although the existence of any such fracture has yet to be proved. Many of the large-scale faults near the western margin of the basin (Fig. 3.6) have undergone strike–slip movement. Much attention in the literature has focused on the movement of the Great Glen Fault, with estimates of exceptional post-Devonian movements of 2000 km displacement, to much more reasonable estimates of 25–29 km dextral displacement (Rogers, 1987) but see Fig. 2.2A–D. Displacements on the other major faults in the vicinity are summarized in Fig. 3.6. Rogers D.A. *et al.* (1989) suggest a total post-ORS offset of 120 km along the strike–slip faults in Shetland.

In contrast to the other basins in Scotland, the Orcadian Basin received sediment throughout the Devonian, so that Lower, Middle and Upper ORS sequences are widely preserved, although there are local omissions (Fig. 3.2). Onshore, around the margins of the basin, the Lower ORS occurs in small, isolated, half-graben structures; the more extensive Middle ORS is unconformable above either Lower ORS or Caledonian basement, and the Upper ORS

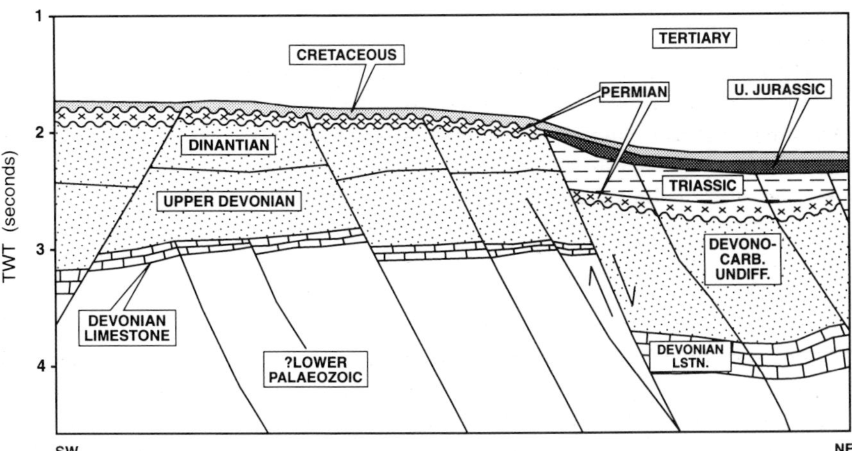

Fig. 3.5 Geoseismic cross-section through UK block 30/27, Mid North Sea High. Note thickening of Devonian limestones and Devono-Carboniferous fill into the downthrown hanging wall.

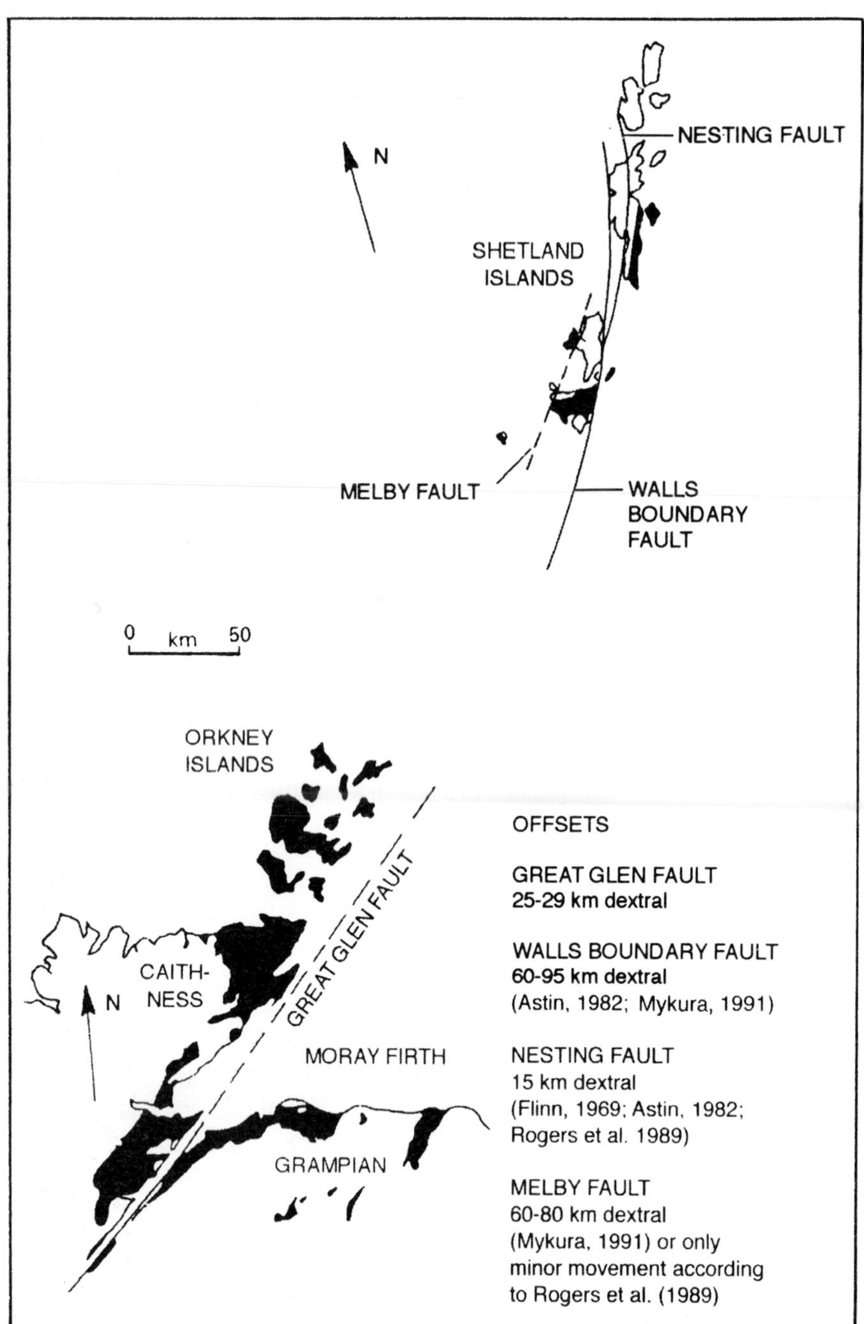

Fig. 3.6 Location of the major faults that complicate the western margin of the Orcadian Basin (modified from Rogers D.A. *et al.*, 1989). Shaded area denotes ORS.

is, at least locally, unconformable on folded Middle ORS (Mykura, 1991).

3.3.2 Stratigraphic summary

Detailed lithostratigraphic terminologies have been determined for the principal areas of outcrop in the Orcadian Basin (e.g. Fig. 3.7). These detailed subdivisions, however, are of little relevance with respect to offshore well penetrations because of difficulties in correlation and biostratigraphic dating of the commonly sandy well sections. An excellent review of the established onshore stratigraphy is given by Mykura (1991; see also references therein). Strata assignable to the Lower, Middle and Upper ORS are all present in the Orcadian Basin and, on the present well data set, this is the most detailed level of stratigraphic resolution practical for well-to-well correlation. Even this level of breakdown is often unachievable. The sole potentially recognizable horizon may comprise the Middle Devonian Achanarras/Sandwick 'Fish Bed', which is a prominent 'flood' horizon recognized from the Shetland Islands to the Moray coast (Trewin, 1986). This has been recognized offshore in well 9/16-3 (Duncan and Buxton, 1995); Marshall (1995) also suggests it can be recognized in wells 9/7-1, 12/29-2 and 13/22-1, using palynological assemblages, palynofacies and gamma-ray log character.

The Orcadian Basin continued to act as a depocentre into Carboniferous times; for example, in the Outer Moray Firth (e.g. Buchan Field), the uppermost ORS strata have locally been proved to be as young as Early Carboniferous (Hill and Smith, 1979). There is no apparent break in sedimentation, and a continuous transition from Devonian to Early Carboniferous red beds is implied. In parts of the basin, coal-bearing Dinantian strata, known as the Forth Formation (Leeder and Boldy, 1990), rest with presumed unconformity on ORS red-bed facies. Most Carboniferous

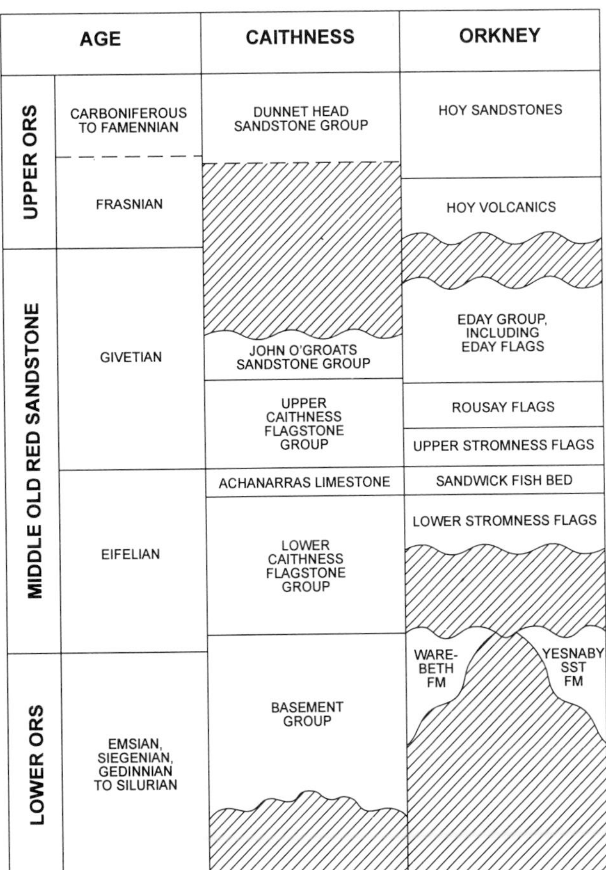

AGE
CAITHNESS
ORKNEY
UPPER ORS
CARBONIFEROUS TO FAMENNIAN
DUNNET HEAD SANDSTONE GROUP
HOY SANDSTONES
FRASNIAN
HOY VOLCANICS
MIDDLE OLD RED SANDSTONE
GIVETIAN
EDAY GROUP, INCLUDING EDAY FLAGS
JOHN O'GROATS SANDSTONE GROUP
UPPER CAITHNESS FLAGSTONE GROUP
ROUSAY FLAGS
UPPER STROMNESS FLAGS
ACHANARRAS LIMESTONE
SANDWICK FISH BED
EIFELIAN
LOWER STROMNESS FLAGS
LOWER CAITHNESS FLAGSTONE GROUP
LOWER ORS
WARE-BETH FM
YESNABY SST FM
BASEMENT GROUP
EMSIAN, SIEGENIAN, GEDINNIAN TO SILURIAN

Fig. 3.7 Simplified summary of the major ORS lithostratigraphic terms used onshore in the Orcadian Basin (after Mykura, 1991).

deposits, however, particularly onshore and in the Inner Moray Firth, are believed to have been removed prior to the Permian, by an erosive event(s) related to either the Variscan or Saalian Inversions recognized in the Central/Southern North Sea. The Devonian stratigraphy is briefly discussed below.

Lower Old Red Sandstone

Lower ORS sediments crop out around the margins of the Orcadian Basin, and rest with unconformity on eroded Caledonian basement. Deposition appears to have occurred in local basins, controlled by active faulting, into which alluvial-fan conglomerates and sandstones were shed (Mykura, 1983b; Mykura and Owens, 1983). In the Strathpeffer area, these passed laterally into a stratified, 'fetid' lake, which allowed the preservation of organic, predominantly algal material. Similarly aged lacustrine facies have also been proved in the Moray Firth wells 12/27-1, 12/27-2, 12/28-2 and 13/19-1 (Andrews *et al.*, 1990; Fig. 3.8), where they comprise mostly grey to reddish brown, laminated siltstones and claystones, with minor greyish brown, very fine-grained, silty, calcareous sandstones.

The lower member of the Yesnaby Sandstone Formation, Orkney Mainland, is of aeolian origin (Mykura, 1991), and Lower ORS aeolian sandstones possibly occur elsewhere in the basin. Many of the sandy ORS well penetrations in the Orcadian Basin cannot be dated, and it is possible that some of these may belong to the more proximal alluvial-fan facies of the Lower ORS. The thickness of the Lower ORS successions is believed to be very variable; an indication of maximum thicknesses is given by well 12/27-1, which proved 981 m of lacustrine sediments without reaching the base.

The offshore distribution of the Lower ORS sediments is difficult to ascertain from the sparse well data and because of the poor quality of seismic data, on which the ORS divisions generally cannot be distinguished. It is possible that the Lower ORS basins formed on the same sites as later Mesozoic basins, as postulated by Norton *et al.* (1987).

Middle Old Red Sandstone

The Middle ORS, where seen in outcrop, is more widely distributed than the Lower, and locally overlaps the Lower ORS at basin margins to rest directly on Caledonian basement. A regional unconformity is believed to separate the Lower and Middle ORS (e.g. Mykura, 1991), although some studies indicate that the observed unconformities are local phenomena and that the Lower–Middle ORS transition is generally conformable (e.g. Rogers D.A. *et al.*, 1989). Offshore, the distribution of Middle ORS strata is poorly understood because of the paucity of proved Middle Devonian dates in well penetrations, although wells 9/7-1, 9/16-3, 12/29-2, 13/22-1 and 13/24-11 allow a tentative palaeogeography to be drawn (Fig. 3.9).

The Middle ORS of Orkney and Caithness is notable for thick developments of lacustrine siltstones and shales, which were deposited under variably oxygenated to anoxic bottom-water conditions. Numerous sedimentological studies (discussed by Trewin, 1989) show the Middle ORS lacustrine succession to consist of stacked upward-shallowing cyclic deposits 5–10 m thick. The deeper-water deposits comprise laminated, organic-rich ('hot') calcareous siltstones 0.4–1.5 m thick (Rogers and Astin, 1991), in which fish fossils are commonly preserved. These are often referred to as 'fish-bed' laminites. Surrounding the main 'Orcadian Lake' depocentre, sequences of fluvial sandstones and alluvial-fan conglomerates were deposited by rivers and on alluvial fans draining from the Northern and Grampian Highlands (Mykura, 1991; Fig. 3.9). Minor aeolian deposits are locally recognized. Periodic expansions of the lake resulted in the deposition of mudstones on to the sandy lake margin, with lake expansion reaching a maximum at the level of the Achanarras/Sandwick Fish Bed.

In addition to lake expansion, there is also considerable evidence of the ephemeral nature of the 'Orcadian Lake'. Outcrop studies by Rogers and Astin (1991) suggested that the lake was ephemeral for most (*c.* 90%) of its history, although more persistent lacustrine conditions may have existed in the basin centre, believed to be located in the offshore areas. This ephemeral nature is shown by the presence of abundant desiccation features, such as 'mud cracks' and pseudomorphs after crystals of the evaporites gypsum and halite (Rogers and Astin, 1991). Thicker-bedded evaporites are also present in UK well 9/16-3, where a 31-m thick anhydrite unit is interbedded with a Middle ORS lacustrine sequence (Duncan and Buxton, 1995).

The thickness of the Middle ORS is currently the subject of some debate; outcrop studies have provided widely

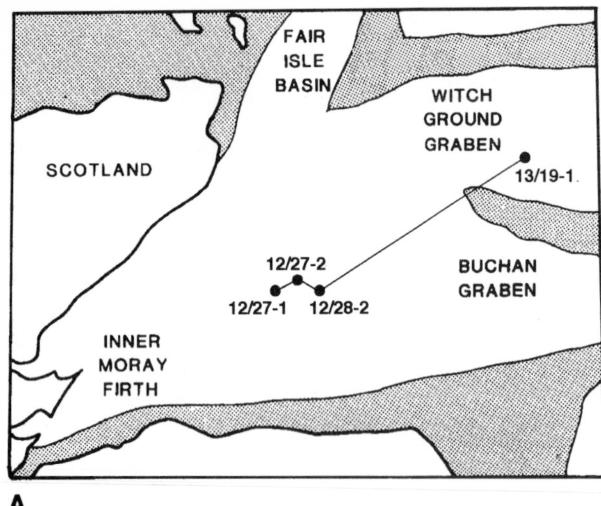

Fig. 3.8 A & B Correlation of Lower ORS and associated undifferentiated ORS sections from wells 12/27-1, 12/27-2, 12/28-2 and 13/19-1 in the Inner Moray Firth.

variable estimates, from 4750 m, based upon simple measurement of coastal sections (Donovan *et al.*, 1974), to a much more modest value of *c.* 890 m, based upon detailed field mapping (Astin, 1990). A value somewhere between these two extremes seems probable in the offshore basins. No released wells have penetrated deep into proved Middle ORS, although well 13/22-1 penetrated 268 m of sandstones and shales without reaching the base (Fig. 3.10).

Upper Old Red Sandstone

The Upper ORS occurs in a patchy coastal belt fringing the Moray Firth, specifically in the Elgin area, on the Tarbat Ness Peninsula (Ross and Cromerty), at Dunnet Head (Caithness) and on Hoy (Orkney Islands). The contact between the Middle and Upper ORS, where observed, comprises an unconformity; whether this is regionally developed is unclear (e.g. Rogers D.A. *et al.*, 1989). Offshore well evidence suggests that the Upper ORS is absent from the Inner Moray Firth, possibly due to erosion (Andrews *et al.*, 1990), although it is present in the outer Moray Firth, where it forms the reservoir for the Buchan Field (Hill and Smith, 1979; Fig. 3.11). Onshore, the Upper ORS is dominated by cross-bedded sandstones of fluvial origin, although aeolian intervals are recognized in the Hoy Sandstones and the Dunnet Head Sandstone Group (see Fig. 3.7). Offshore, the Upper ORS predominantly comprises fluvial braid-plain sequences, with non-reservoir mudstone interbeds generally more abundant than in the equivalent onshore strata. Onshore, the Upper ORS is locally thick: Mykura (1991) indicates a thickness in excess of 1000 m for the Upper ORS of Hoy. Seismic data commonly indicate great thicknesses of Devonian strata. For example, seismic data presented by Holloway *et al.* (1991) suggest that from 4700 to 6580 m of presumed Upper ORS are present in Quadrant 9.

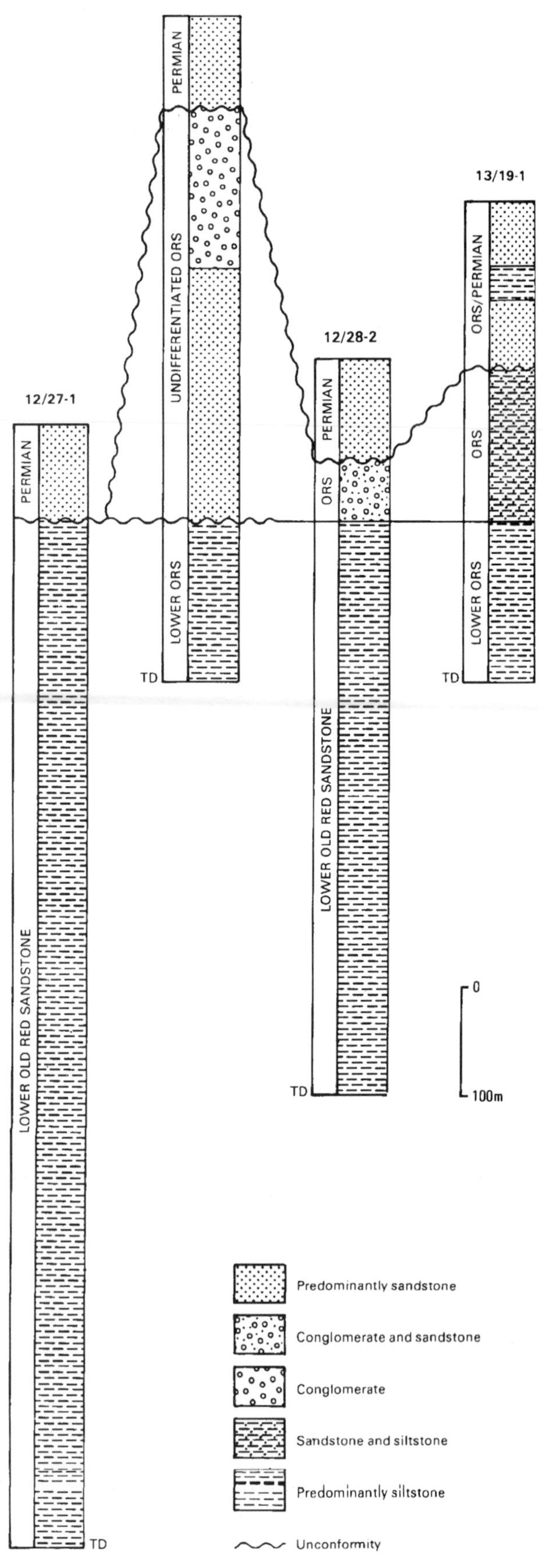

Fig. 3.8 *Continued.*

3.3.3 Reservoir potential

Rudaceous to arenaceous facies with potential as reservoirs are common in the Lower, Middle and Upper ORS. Lower ORS reservoir facies occur on the basin margin, but no reservoir facies of Lower ORS affinities have been proved

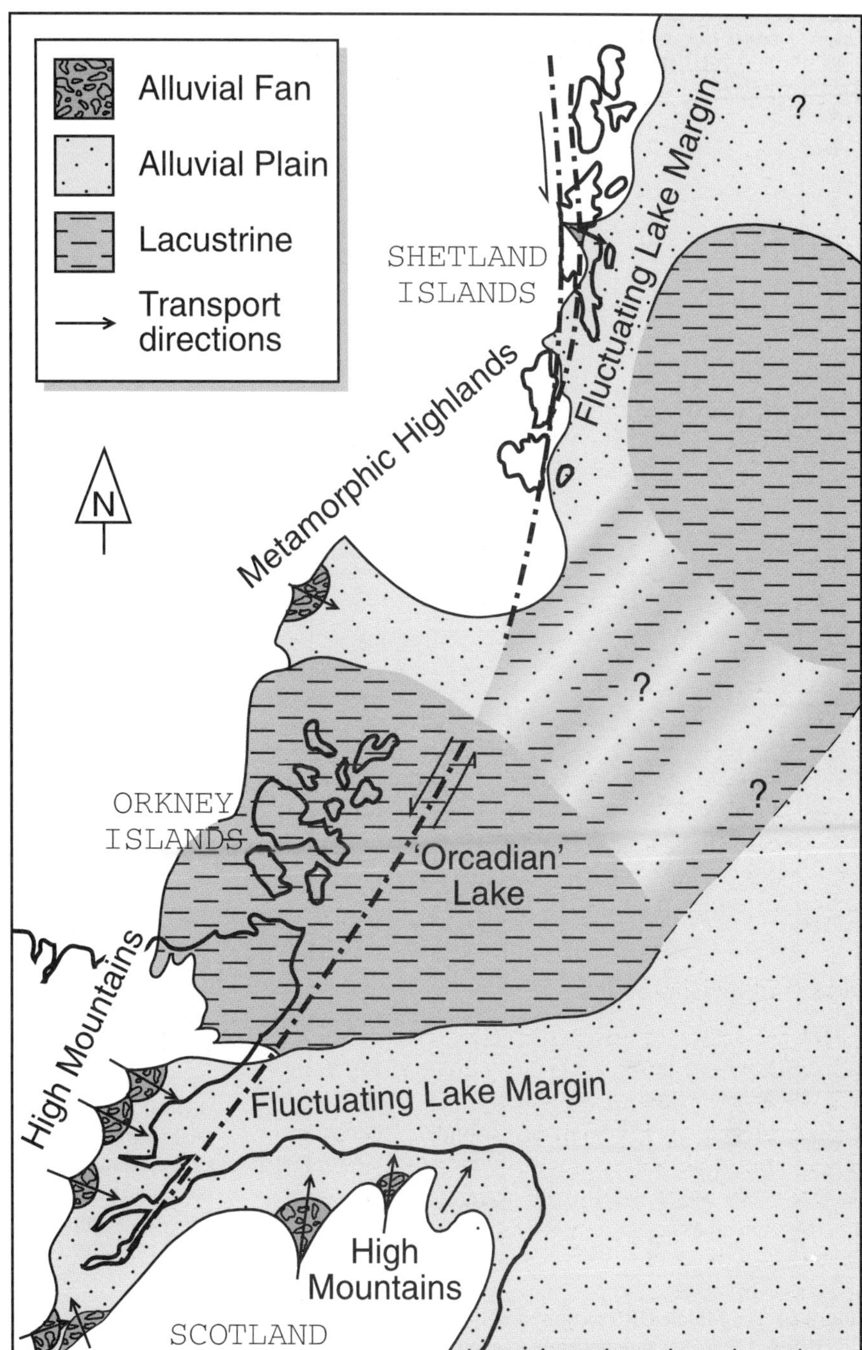

Fig. 3.9 A palaeogeographic sketch-map for the Middle ORS of the Orcadian Basin (after Mykura, 1991).

offshore, even though a number of undifferentiated sandstones have been drilled. By analogy with the onshore areas, Lower ORS reservoirs may be expected to comprise alluvial-fan conglomerates to fan-margin sandstones. In the Middle ORS, sandstones may be interbedded with and surround lacustrine shales (see Fig. 3.9). A good example of this is provided by well 13/22-1, where porous Middle ORS sandstones are interbedded with lacustrine shales on a 5–40 m scale (see Fig. 3.10). Most Middle ORS sandstones are of fluvial origin, although outcrop studies reveal some that are aeolian; the latter are less argillaceous than the fluvial, but are little more porous, since both are commonly carbonate-cemented (see below). The Upper ORS has the best intrinsic reservoir potential; offshore it is believed to be the most extensive of the three divisions (however, dating of these sandy divisions is difficult) and is dominated by a sandy braid-plain facies. These sandstones form the main reservoir interval of the Buchan Field (block 21/1) and are also believed to form the reservoir for the Sterling and West Brae accumulations on the Fladen Ground Spur.

Core-analysis poroperm data are limited, and the few suggest that reservoir quality is typically poor to, at best, moderate. For example, core-analysis measurements on outcrop sandstones collected from around the margins of the Orcadian Basin of mainland Scotland (Downie, 1989) showed a mean porosity of only 10.0%, based upon measurements on 59 sandstone samples. Similarly, measurements on 51 samples gave a mean permeability of only 7.3 millidarcies (mD) (Fig. 3.12). Indeed, the measured poroperm values are probably optimistic compared with

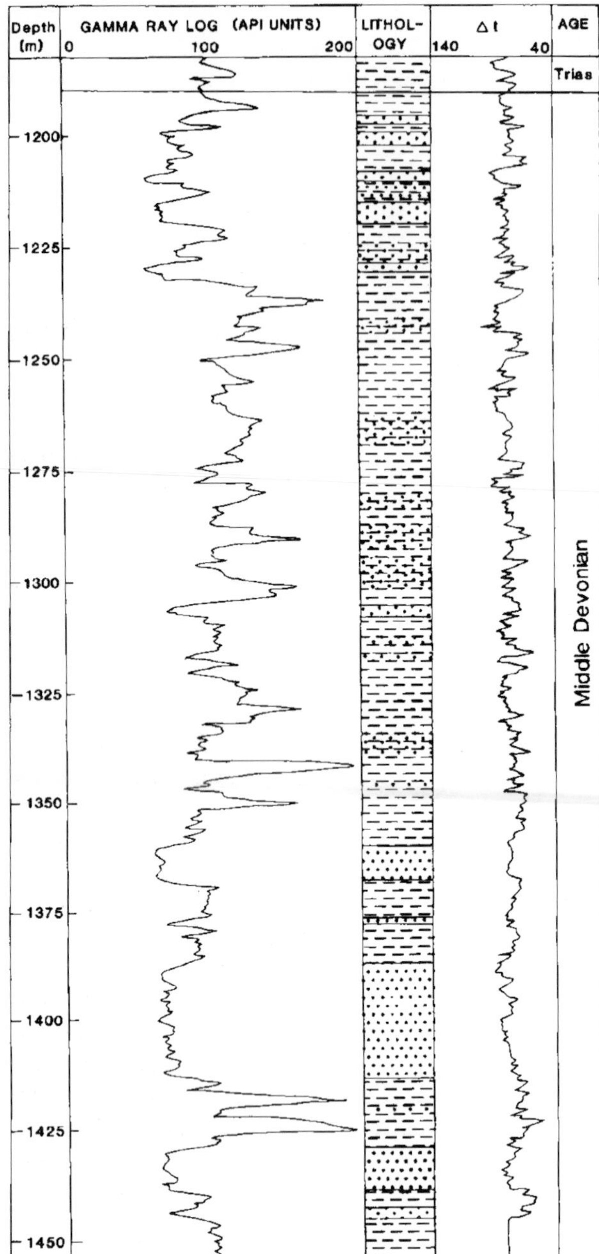

Fig. 3.10 The Middle ORS section in well 13/22-1. Note the interbedded lithology and moderate to good sandstone porosities, indicated by sonic velocities of 70–80 µs/ft.

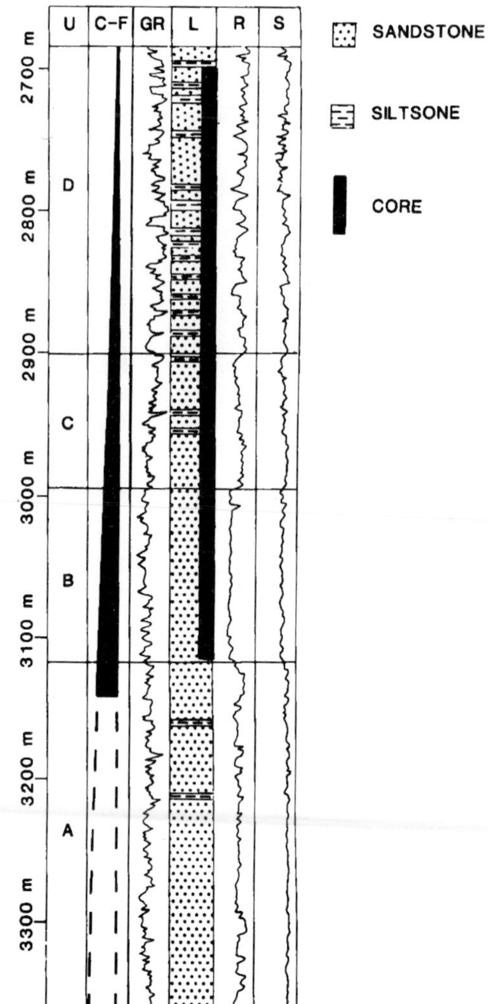

Fig. 3.11 The Upper ORS section in well 21/1-6 from the Buchan oilfield. The section shown is 675 m thick, and is divided into four units (A to D) on the basis of sand : shale ratio, facies variations and downhole log correlations. The section exhibits an upward increase in the number of siltstone and cornstone beds. U, unit; C–F, change of grain size (course to fine); Gr, gamma ray (0–200 API); L, lithology log; R, resistivity log; S, sonic log (140–40 µs/ft) (from Richards, 1985).

similar sandstones at reservoir conditions, because of the effects of surficial weathering and dissolution of carbonate cements. No significant differences in poroperm character were noted between similar sandstones from the Lower, Middle and Upper ORS divisions.

Few data concerning the reservoir quality of ORS sandstones in offshore well penetrations have been released. Edwards (1991), in a review of the Upper ORS Buchan Field, quotes an average porosity of 8.85% (range 7.1–10.9%) and generally low permeabilities of 0.1–2 mD. Limited core-analysis data have also been released for the undifferentiated ORS core in well 9/23-1, where the porosity ranges from 6.6 to 18.5% and permeability from 0.02 to 81 mD. Similarly, the presumed Upper ORS of the West Brae well 16/7a-5 has a helium porosity average of 4.6% and permeability average of 0.1 mD.

The most widely available data on reservoir quality are found in the suites of wireline logs released by the UK Department of Trade and Industry and the Norwegian Petroleum Directorate. Most penetrations of ORS sandstones in the Orcadian Basin show sonic velocities of less than 70 µs/ft, approximately equivalent to sandstone porosities of 12% or less, values comparable to the core-analysis measurements. There are a number of exceptions to this; some ORS well penetrations in the south of Quadrants 13 and 14 (e.g. 13/22-1; see Fig. 3.10) show sonic values in sandstone lithologies in the range of 70–80 µs/ft (or higher), suggestive of porosities of 12–22%. There is no clear pattern to the distribution of these wells, and a number of ORS penetrations between the above wells show more typical sandstone sonic velocities of less than 70 µs/ft.

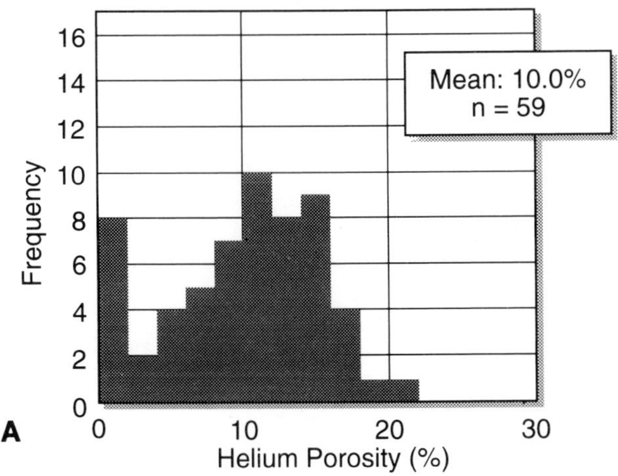

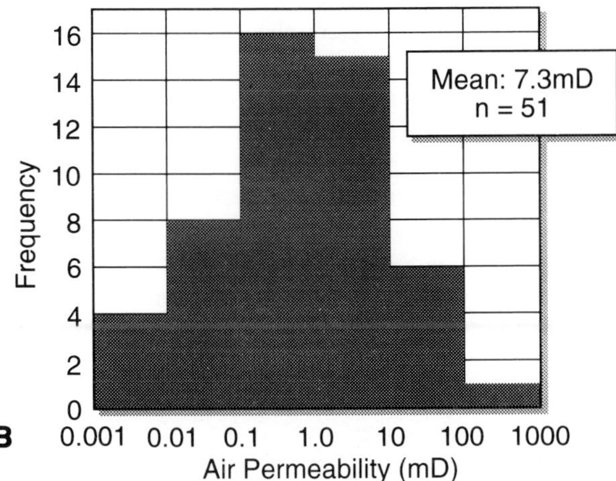

Fig. 3.12 Porosity (A) and permeability (B) measurements from Lower, Middle and Upper ORS outcrop samples from the Orcadian Basin (modified from Downie, 1989).

Controls on primary reservoir quality

It was suggested by Trewin (1989) that the poor reservoir quality, particularly in the Lower and Middle ORS, is largely because the sandstones are of first-cycle origin and contain common micas, feldspars and lithic fragments; by implication, their degradation and associated burial diagenesis impaired the reservoir quality. This is undoubtedly an important factor at the basin margins, where abundances of 30–80% feldspar are reported in thin section (e.g. Mykura and Owens, 1983). Regional evaluation of sandstones at outcrop and offshore (principally the Buchan Field), however, suggests that ORS sandstones generally have relatively mature quartzofeldspathic compositions (Fig. 3.13). Indeed, the compositions are not dissimilar to typical Jurassic reservoir sandstones of the North Sea, such as in the Bruce Field of blocks 9/8 and 9/9.

Petrographic studies suggest that the principal cause of poor reservoir quality is cementation. For example, in the Buchan Field, the principal cements—calcite, dolomite, quartz overgrowths and clays—commonly occupy over 16% of the bulk rock volume. Of these, syntaxial quartz overgrowths (up to 22%) are the most significant and, in the cleaner sandstones, often form an interlocking mosaic of quartz grains. Similar types and levels of cementation are present in the Stirling and West Brae accumulations. High levels of cementation are also observed in outcrop samples, although here calcite dominates over quartz overgrowths as the principal cement (Downie, 1989).

Fracturing

North Sea reservoirs of post-Devonian age in which open fractures are necessary to permit fluid flow are comparatively rare. Conversely, in the ORS reservoirs of the Orcadian Basin, open fractures in tight-matrix sandstones are present in all the discoveries: Buchan, Stirling and West Brae. In addition, open fractures are present in core from well 3/24-1, and cuttings evidence suggests fracture-fill cements to be present in ORS penetrations in Quadrants 3 and 9. Open fractures are also common in the Clair Field to the west of Shetland and the Argyll and Embla Fields of the Central North Sea.

Regionally, the well penetrations, in particular cored penetrations, are too few to show the overall distribution of fractures. The wireline log data from many of the Orcadian Basin wells suggest, however, that the sandstones have low porosities over most of their distribution and are therefore, by implication, tightly cemented, consistent with the petrographic observations in the cored wells. The intense cementation will not only have lowered the sandstone porosities, but also have caused the rocks to become more indurated and brittle. Open fractures are typically developed in brittle rocks with low intergranular porosity and can either be related to faults and folds or be part of a regional system (Stearns and Friedman, 1972). It is therefore reasonable to suggest that fracturing may be widespread in areas where the ORS sandstones have low porosities, notably the Outer Moray Firth, Fladen Ground Spur and Viking Graben. There is some evidence that fractures are, in part, related to a regional system caused by strain associated with the 'Variscan Inversion' event, although they could also be related to more local fault patterns. Significant open fracturing is not present in cores recovered from the Inner Moray Firth.

3.3.4 Source-rock quality

Most strata in the Orcadian Basin are clearly not of source-rock quality. Source potential is limited to the extensive lacustrine facies of the Middle ORS Caithness and Stromness Flagstones (see Fig. 3.7) and the more limited Lower ORS shales (e.g. Strathpeffer Group). Outcrop studies (e.g. Hamilton and Trewin, 1985; Marshall *et al.*, 1985) have shown that certain intervals of these sequences have good or even excellent source-rock quality, with total organic carbon (TOC) values of up to *c.* 4% (Fig. 3.14A). The kerogens in these rocks comprise predominantly amorphous organic matter of algal origin and are largely oil-prone. The best source-rock quality is associated with the deeper-water stages of the lacustrine depositional cycles, which resulted in the deposition of the thin 'fish-bed' laminites under anoxic conditions (Trewin, 1989). Rogers and Astin (1991) reviewed 18 measured sections of Middle ORS lacustrine strata in Orkney and Caithness. They found that 'fish-bed' laminites comprise from 1 to 16% (7% average) of the total lacustrine facies.

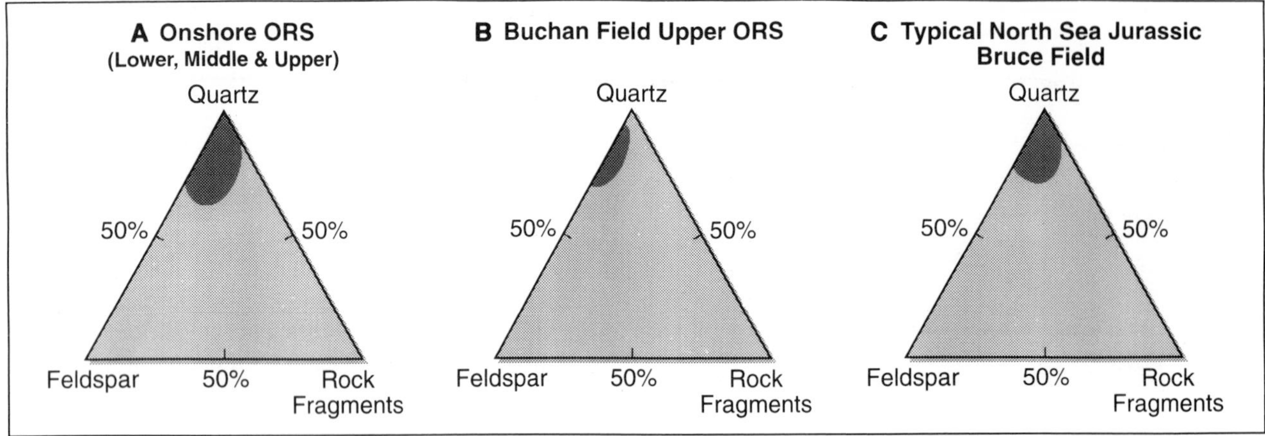

Fig. 3.13 Sandstone compositions for the ORS of the Orcadian Basin: (A) onshore outcrop samples; (B) Buchan Field samples; (C) Bruce Field Jurassic samples for comparison (ORS data from Downie, 1984. Bruce Field data from McBride, 1992).

Data from the offshore ORS are limited by the paucity of well penetrations, particularly those of argillaceous character. Wells 12/27-1 (Lower ORS) and 13/22-1 (Middle ORS) penetrated thick sections containing high proportions of argillaceous strata; Rock Eval pyrolysis of cuttings and core samples has show that their TOC values are similar, if slightly leaner, than the equivalent strata onshore, with a maximum TOC value of *c.* 2.5% (Fig. 3.14B). Wireline logs provide further indication of offshore ORS source potential. For example, a gamma-ray profile from well 12/13-1, presented by Kelly (1992), shows 210 m of a lacustrine sequence of presumed Middle Devonian age. The background gamma values typically range from 90 to 120 API, though 'hot-shale spikes' of 200–300 API are common (Fig. 3.15). These are believed to correspond to the 'fish-bed' laminites, and comprise about 14% of the illustrated section, comparable to the proportions described by Rogers and Astin (1991) for the onshore Middle ORS.

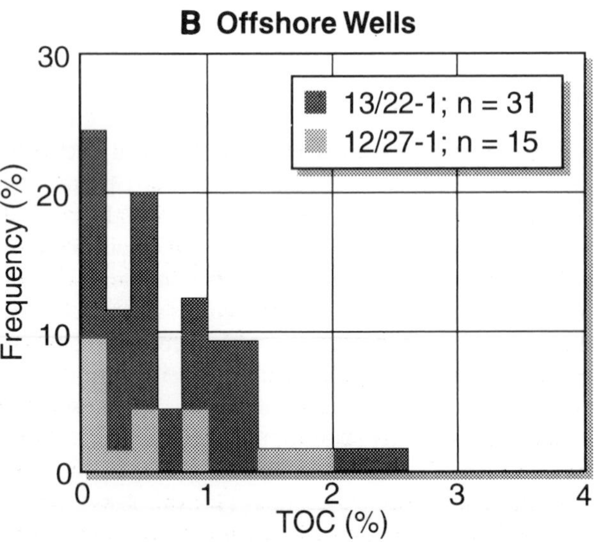

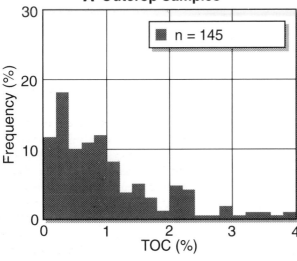

Fig. 3.14 Lower and Middle ORS total organic carbon (TOC) values from outcrop (A) and well (B) measurements (outcrop data from Marshall *et al.*, 1985).

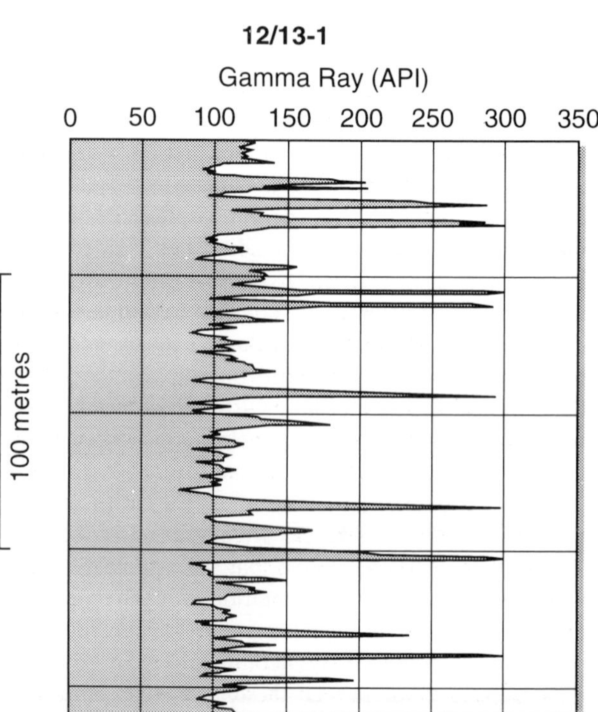

Fig. 3.15 Gamma-ray profile of the Middle(?) ORS section from well 12/13-1 (after Kelly, 1992). The gamma-ray spikes in excess of 200 API are believed to represent 'fish-bed' laminites.

3.3.5 Thermal history and hydrocarbon generation

Of particular relevance to evaluating the petroleum geology of the Orcadian Basin is understanding the thermal history experienced by the ORS sediments. There is now a growing body of evidence that much of the strata experienced elevated palaeotemperatures to levels considerably higher than might be expected during 'normal' heat-flow regimes. Evidence for this comes both from outcrop studies (reviewed by Hillier and Marshall, 1992) and from analysis of samples from offshore oil wells. Data from the latter are provided largely from unpublished regional studies undertaken by Halliburton.

Outcrop studies in the Shetland Islands and Fair Isle show maturities extending into the gas window and much higher, with vitrinite reflectance values of up to 7% R_0 (Hillier and Marshall, 1992; Marshall *et al.*, 1985). These exceptional values are attributed by the authors to thermal metamorphism caused by the Sandsting Plutonic Complex (and associated bodies), which intrudes into the Middle Devonian Walls Formation of southern Shetland. Similar studies in the Moray Firth/Caithness areas of mainland Scotland (Hillier, 1989) showed widely variable maturity parameters, from values of 0.6% R_o to values of 10.5% R_o adjacent to the Great Glen Fault. For a number of reasons, but primarily due to the lack of a relationship between stratigraphy and maturity, Hillier and Marshall (1992) similarly consider the maturity of the Caithness area to be a result of contact metamorphism adjacent to a large post-Devonian igneous intrusion(s), the presence of which is as yet unproved.

Many of the maturity data from the offshore wells also show evidence of considerable heating. In unpublished geochemical studies undertaken by Halliburton on ORS well penetrations in the Orcadian Basin, the quality of geochemical data was limited by the poor recovery of organic particles, although it was sufficient to show consistent jumps in the maturity at the top Devonian. Maturities were, in all cases, in excess of those predicted from present-day burial depths/temperatures (often considerably so), with maturities at top Devonian generally increasing rapidly from the Inner Moray Firth (immature/early mature for oil), eastward to the Fladen Ground Spur (postmature for gas). In addition to the maturity jump, the data showed steep maturity profiles in a number of wells, indicating that the palaeoheat flows were considerably in excess of those of the present day. Data from well 9/16-3 (Duncan and Buxton, 1995), while limited to three points in the ORS, suggest a similar gradient (Fig. 3.16). It must be stressed that in none of the wells are the geochemical data extensive, although confidence in their statistical significance is gained by the similar profiles found in each.

Modelling of these steep maturity gradients, utilizing commercial burial history and thermal modelling programmes, suggests heat flows of at least 3 heat-flow units (HFU) (120 mW/m^2), or higher, to be consistent with the data. Dependent on the thermal conductivities of the rocks, these heat flows are approximately equivalent to geothermal gradients of 100°C/km or greater, comparable to present-day geothermal fields (e.g. Cerro Prieto system of northern Mexico (Barker, 1991)). A geothermal gradient of this magnitude would require only limited depths of burial to produce the observed maturity profiles.

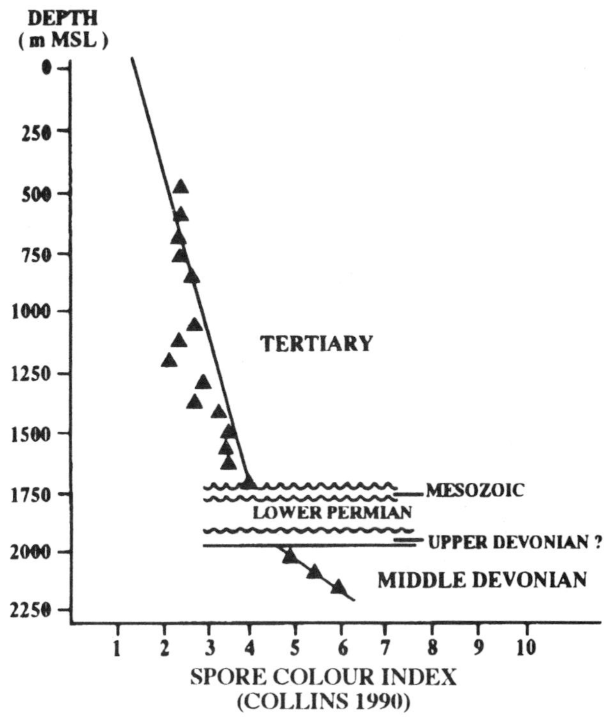

Fig. 3.16 Maturity profile of well 9/16-3 (after Duncan and Buxton, 1995). The Middle Devonian section shows an apparently higher maturity gradient than the Tertiary. Spore colour index of Collins (1990).

Some corroborative evidence of the high heat flow is provided by Glasmann and Wilkinson (1993), who undertook detailed clay-mineral analyses within Triassic, Permian and Devonian red-bed sandstones of wells 9/11-1, 9/12b-6 and 9/16-2. On the basis of relatively irregular and coarse-grained clay-mineral morphologies and oxygen-isotope composition, they argued that the ORS sandstones had been affected by a major low-grade metamorphic event, with maximum temperatures of 180–200°C. This temperature range is clearly incompatible with the present-day heat flow and burial depths of 1582–2615 m.

Timing

Of equal importance to the maximum temperature attained is the time of heating. This can be evaluated both from regional evidence and from direct measurement. With respect to regional evidence, the clay-mineral studies of Glasmann and Wilkinson (1993) show Permian sediments to be unaffected by high temperatures, implying that the heating had by then ceased. Additionally, in the Buchan Field, the uppermost ORS strata, of early Dinantian age, are shown to have been affected by the high heat flow, as evidenced by organic maturities considerably in excess of those predicted from their present-day burial depth. Taken together, these observations bracket the cessation of heating to a period between the early Dinantian and some time in the Permian.

Direct evidence for the time of cooling is provided in Caithness and the Orkneys by Apatite Fission Track Analysis (AFTA®—registered trade mark of Geotract International) from ORS and basement samples (results courtesy of Geotrack International). These reveal that cooling from palaeotemperatures in excess of 110°C began over most of the region between 300 million years (Ma) and 250 Ma (Late Carboniferous to Permian), while locally cooling began earlier (~350 Ma, Late Devonian to Early Carboniferous). In the 12/16-1 well in the Inner Moray Firth, AFTA and vitrinite reflectance data also reveal a Late Carboniferous cooling episode (Green *et al.*, 1995b). Further direct evidence is provided from well 9/16-3, where radiometric dating of authigenic illite from two separate Middle ORS sidewall core samples show the timing of clay authigenesis to be approximately Mid-Carboniferous (*c.* 320 ± 5 Ma; Duncan and Buxton, 1995) in both cases, which is taken to be the time of maximum heating.

In summary, both indirect regional evidence and the direct geochronometric measurements (AFTA, illite dating) indicate that the cessation of heating occurred possibly as early as the Late Devonian, but more probably during the Carboniferous, the exact timing probably varying from area to area.

Source of heat

As discussed above, the heating, at least locally, is probably the result of contact metamorphism related to the emplacement of large igneous intrusions. The emplacement of the 'Last' Caledonian Granites had essentially finished by the Early Devonian (*c.* 390 Ma). These are therefore unlikely to have been implicated in the maturation, particularly, of the Upper ORS strata, although they may be implicated for the Lower ORS. The Sandsting Plutonic Complex of the Shetland Islands, however, is worthy of some further consideration. This granite intrudes Middle ORS sediments of the Walls Formation and is clearly responsible for their thermal alteration (Marshall *et al.*, 1985). Its radiometric age of 360 ± 11 Ma (Mykura, 1991) indicates emplacement towards the end of the Devonian; hence it would have been capable of 'metamorphosing' Devonian sediments from any of the divisions, and possibly sediments as young as earliest Carboniferous. It is clearly too small, however (12 km in its longest dimension), to have caused the increase in heat flow observed elsewhere in the Orcadian Basin. Its presence, however, indicates the potential for high-level emplacement of Late Devonian granites elsewhere in the Orcadian Basin, although no others have, as yet, been discovered. Exceptionally high regional heat flows may also have been associated with this very late Caledonian plutonism.

Hydrocarbon generation

The elevated Late Devonian/Early Carboniferous heat flows will have caused much of the lacustrine source rocks to become mature and to have generated a significant proportion of their potential hydrocarbons. This maturation is the probable source of the abundant hydrocarbon staining/impregnations which are widely observed at outcrop throughout much of the Orcadian Basin (e.g. Parnell, 1983). Preservation of commercial quantities of these hydrocarbons to the present day, however, is highly unlikely, because seals to any accumulations were probably breached by numerous subsequent tectonic events that affected the basin, in particular Variscan, Kimmerian and Tertiary events.

While both the outcrop and well studies reveal widespread postmaturity as a result of Late Devonian/Early Carboniferous heat flows, they also show areas where ORS source rocks are likely to have remaining generative capacity (i.e. where maturities at the top ORS are within the early oil/oil window at the present day), and that further generation of hydrocarbons was possible if sufficient Mesozoic–Cenozoic burial occurred. There are, however, few areas in the Orcadian Basin where Mesozoic–Cenozoic burial has been sufficient to return any ORS source rocks back into the hydrocarbon window. A minimum value of approximately 3000 m burial is presumed to be necessary for this. Burial of the ORS to this depth has occurred in the main North Sea graben areas, such as the South Viking Graben, although any contribution from ORS source rocks (if present and assuming that they had residual generative capacity) will have been swamped by hydrocarbons from the prolific Kimmeridge Clay Formation. Over the East Shetland Platform, where Middle ORS source rocks are believed to occur, the depth to top ORS is typically less than 2000 m and hence there is little possibility for the 'recent' generation of hydrocarbons. Burial to > 3000 m occurs, however, in the Inner Moray Firth (Fig. 3.17), where maturity studies suggest the possibility of remaining residual post-Devonian generative capacity. This is the possible source of some or all of the hydrocarbons in the Jurassic Beatrice Field of block 11/30 (Peters *et al.*, 1989).

3.3.6 Summary of hydrocarbon plays

In the Orcadian Basin, sandstones are commonly tightly cemented, and the presence of open fractures is often necessary to form a viable reservoir. Fractures have been proved in a number of locations and are likely to be widespread across much of the basin, particularly on the Fladen Ground Spur, Viking/South Viking Graben and Outer Moray Firth. The accumulations discovered to date are all sourced from the conventional North Sea Kimmeridge Clay Formation. There is potential for further similar discoveries where ORS structural highs are located adjacent to mature Jurassic source rocks in lows, although any undrilled accumulations are likely to be small.

An alternative play comprises generation from ORS lacustrine source rocks, with hydrocarbon entrapment in either Devonian or younger strata, the latter being likely to have better reservoir quality. The sole discovery of this type comprises the Jurassic Beatrice Field (block 11/30) of the Inner Moray Firth (Fig. 3.1). This play is likely to be limited by two factors: firstly, the high Devono-Carboniferous heat flow, which will have limited the area and volume of potential source rocks able to generate hydrocarbons during Mesozoic–Cenozoic burial; and, secondly, there are few areas where Mesozoic–Cenozoic burial has been sufficient to return any ORS source rocks to the

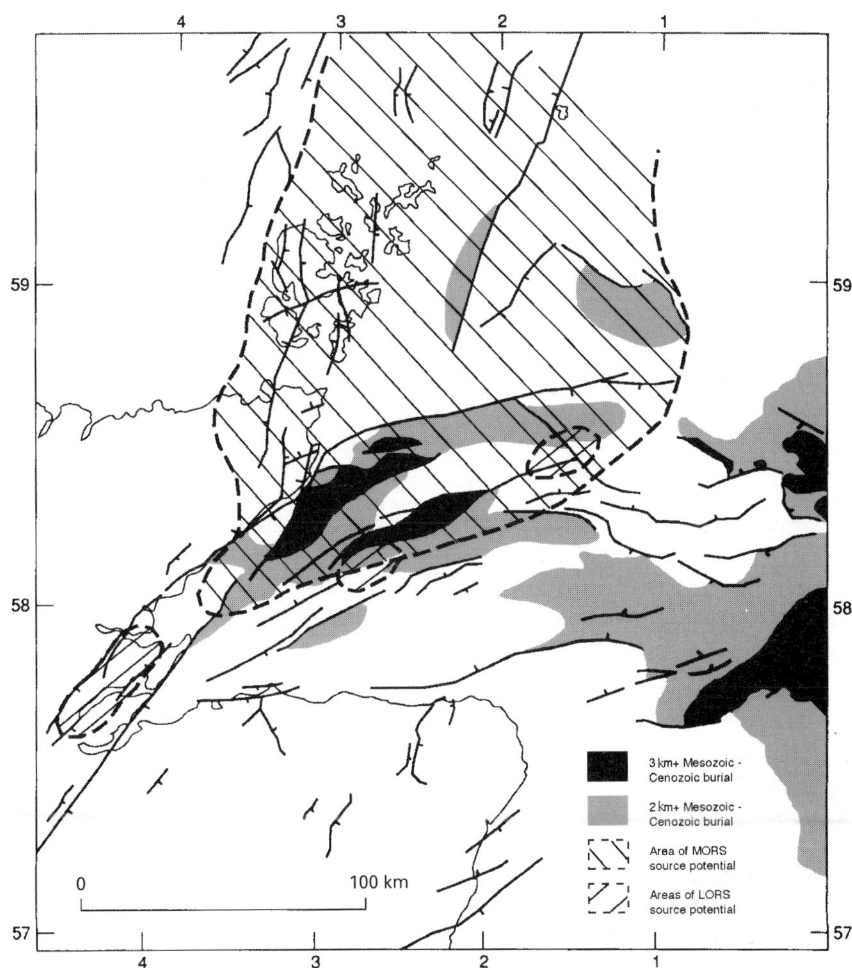

Fig. 3.17 Areas of the Orcadian Basin with 2 km+ and 3 km+ Mesozoic–Cenozoic burial. The hatched areas reflect regions with potential Lower and Middle ORS source rocks. (Modified from Trewin, 1989.)

hydrocarbon window (Trewin, 1989; Fig. 3.17). The sole area where both conditions are known to be satisfied is the Inner Moray Firth, where there is arguably potential for further discoveries sourced from ORS shales.

3.4 The Central North Sea

3.4.1 Basin overview

The term 'Central North Sea' is used here with reference to the Devonian of the North Sea, south of what is generally accepted to comprise the Orcadian Basin (south of approximately 57°30′N). This region includes the Mid North Sea High and the offshore extensions of the Midland Valley and the Northumberland Trough (see Fig. 3.1), although its relationship to these is unclear because of insufficient seismic cover and the absence of wells between the depocentres.

Little is known of the Palaeozoic history of this area. There is some indication that a proto-Central Graben existed in Devonian times, as evidenced by well and seismic data, which show that an arm of the proto-Tethys Ocean, which then lay to the south, spread northwards into the Central Graben (see Fig. 3.1). This seaway possibly followed a line of structural weakness, which was later exploited in the initiation of the Central Graben. Seismic evidence on the northern fringe of the Mid North Sea High shows Devonian strata to be present in major north–south-trending half-graben structures, with evidence of thickening into the hanging walls and thinning on to highs, implying that sedimentation was controlled by active rifting (see Fig. 3.5). It is these north–south structures that may have acted as lines of weakness, subsequently reactivated in Mesozoic times. Seismic evidence shows no significant break in sedimentation from the Upper Devonian to Lower Carboniferous fill of these half-grabens.

The Central Graben is a prolific hydrocarbon province, with a wide range of reservoir ages and types, although only two fields have Devonian reservoir horizons: Argyll (block 30/24) and Embla (Norwegian block N2/7). The Argyll Field (see Figs 3.1 and 3.18), located on the northern margin of the Mid North Sea High, is a complex, multi-reservoir Palaeozoic high, in which the principal reservoirs are Permian in age. These unconformably overlie Devonian (ORS) sandstones, which have produced up to 5500 barrels/day of oil (Robson, 1991). The nearby Permian Auk Field (block 30/16) is located in a similar Palaeozoic structural high, although no significant hydrocarbons are associated with the underlying Devonian sediments. The Embla Field was drilled on the Grensen Nose, which extends northward from the Mid North Sea High. The reservoir sandstones, while not definitely proved to be Devonian in age (see Section 3.2.1), comprise continental facies similar to the ORS strata in the Auk and Argyll Fields.

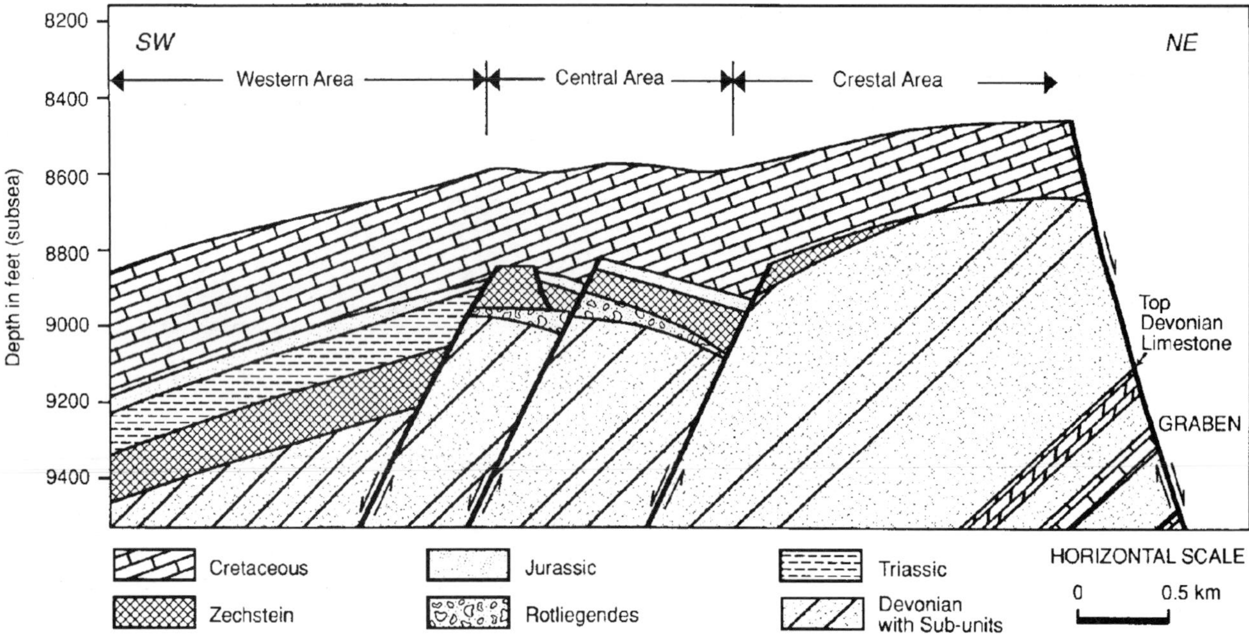

Fig. 3.18 Schematic cross-section of the Argyll Field (after Robson, 1991).

3.4.2 Stratigraphic summary

With the exception of the Auk and Argyll Fields, there are few well penetrations in the Central North Sea that can confidently be assigned to the Devonian; hence only general statements can be made regarding stratigraphy and depositional setting. The limited data suggest that rocks assignable to the Lower ORS do not occur (see Fig. 3.4A). Palaeogeographic reconstructions (e.g. Ziegler, 1990a) suggest that sedimentation commenced in the Middle Devonian with the deposition of marine limestones and shales in the basin centre, fringed by presumed coastal to fluvial facies (see Fig. 3.4B). The marine limestones have been penetrated in the Auk (Trewin and Bramwell, 1991) and Argyll (Robson, 1991) fields; in the Auk well 30/16-5, the limestones directly overlie schistose basement rocks, implying a position at or near the base of the Devonian sequence. In well 38/3-1 (Fig. 3.19), the limestones are cored and found to comprise fossiliferous lime mudstones, wackestones/packstones and rare grainstones, with a macrofauna including corals, bryzoans, brachiopods and crinoids. The sequence is interpreted as a forereef to reef transition. These limestones can be tied to two prominent seismic reflectors that are widely recognized across much of the Mid North Sea High, although they become difficult to recognize to the south of the high, where Permo-Carboniferous strata rapidly thicken into the Southern North Sea Basin. The limestone seismic reflectors are not recognized north of the Auk Field.

The exact age of this marine section is worthy of discussion. On biostratigraphical evidence, Pennington (1975) originally suggested a Mid-Devonian age for this unit in the Argyll Field. More recent palynological analyses on the limestone core from 38/3-1, however, suggest a Late Devonian (Frasnian) age to be more probable, based upon the recovery of rare specimens of *Archaeoperisaccus* spp. although these can range as old as latest Givetian. Some further evidence of a possible younger age is provided in the Embla Field. Here, well-sorted micaceous sandstones and silty mudstones are present towards the base of the sequence, overlying heavily altered rhyolitic rock (altered ?basement). These also contain palynological assemblages dating the strata to the Late Devonian (Frasnian). Of significance is the presence in the assemblages of marine microplankton (acritarchs and chitinozoa; Knight *et al.*, 1993), indicating that deposition occurred in a marine or marginal-marine setting. The correspondence in age between the micaceous sandstones and silty mudstones in Embla and the limestones in well 38/3-1 suggests that the former may represent a marginal-marine equivalent of the latter.

The sparse well penetrations, particularly those in the Auk and Argyll Fields, generally suggest that the marine section is overlain by a sandy, braid-plain to ephemeral lacustrine facies (e.g. Robson, 1991). These are typical Upper ORS-type continental deposits of presumed Late Devonian age. In well 38/3-1, however, the marine strata are overlain by a thick (*c.* 795 m) unit of very fine- to fine-grained micaceous sandstones, with thin interbeds of shales and coals (Fig. 3.19). The latter, clearly visible on logs and in cuttings, form beds about 0.5–1 m thick and appear to comprise from 1 to 2% of the total section. These coals have been ignored in palaeogeographic reconstructions by most workers, presumably because the age dating given on the released composite log was deemed to be erroneous, a Carboniferous age seeming more probable. Recent palynological re-evaluation of the main coal-bearing interval (*c.* 2755–3520 m) confirms the original operator age-dating, the highest occurrence of *Hystricosporites* spp. at 3048 m proving a Devonian age, and the frequent occurrence of this genus together with *Ancyrospora* spp. and common *Geminospora* spp. throughout the interval suggesting a Late Devonian age.

The coal-bearing unit was not cored, so the depositional setting can only be inferred. The lithologies are consistent with the development of a Late Devonian deltaic system,

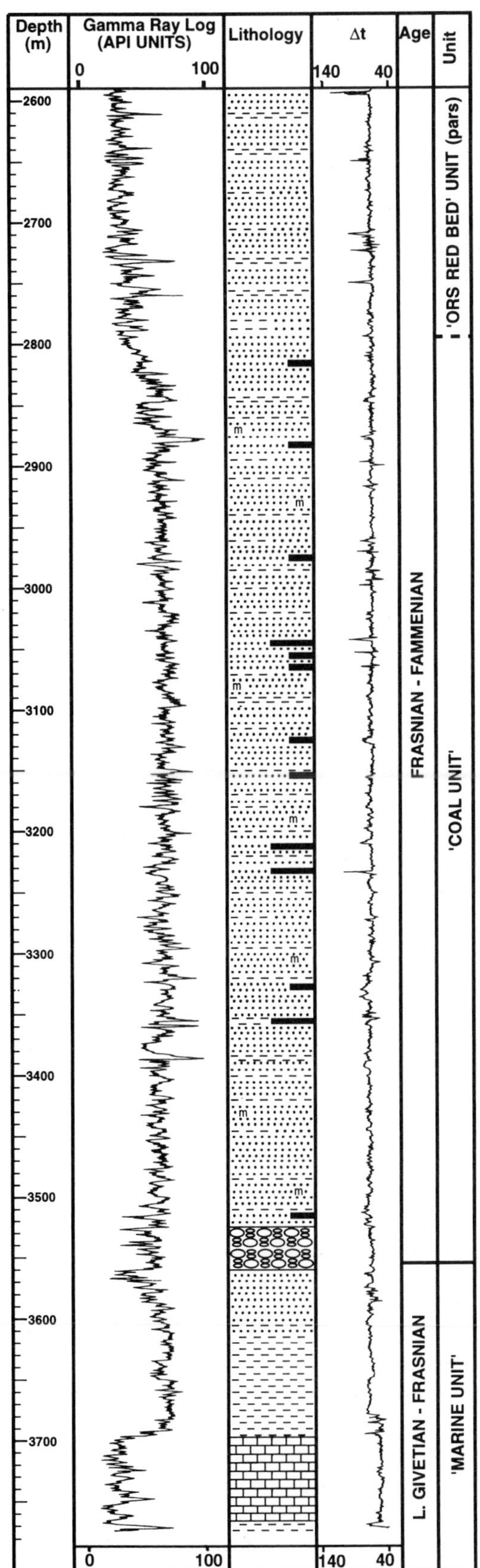

Fig. 3.19 The Devonian limestones and Upper Devonian coals in well 38/3-1.

which possibly prograded southward down the basin axis in response to a sea-level fall and retreat of the ?Late Devonian seaway. The considerable thickness of the unit suggests that it may be laterally extensive; alternatively, it may comprise the fill of a graben system of limited lateral extent.

3.4.3 Reservoir potential

Fluvial Upper ORS sandstones are believed to have been deposited across much of the Central Graben area (see Fig. 3.4C), Variscan erosion locally having removed the upper sequences. Despite this, it is anticipated that sandy, braid-plain facies are widespread in this region. Core analysis data from these sandstones are largely from the main hydrocarbon discoveries and suggest moderate to good poroperm character. In the Auk Field, Trewin and Bramwell (1991) quote an average porosity of 17% for the Devonian sandstones that underlie the Zechstein and Rotligend reservoirs. The Argyll Field (see Fig. 3.18) comprises a complex Zechstein, Rotliegend and Devonian reservoir, the latter two being locally difficult to distinguish between. Interpretation of the data from Bifani *et al.* (1987) and Robson (1991) suggests good poroperm properties from probable ORS strata, with porosities commonly over 15% and permeabilities ranging from tens of millidarcies to several darcies.

The underlying and laterally extensive marine limestones were cored in wells 30/16-5 and 38/3-1 and in both cases were tightly cemented, with no significant residual porosity. The possibility of more porous areas (reef knolls?) or areas with fracture porosity cannot, however, be discounted.

3.4.4 Potential source rocks

The Devonian of the Central North Sea and Mid North Sea High is believed to comprise mainly continental red-bed sandstones and associated shales. These have negligible source-rock potential. Conversely, the coals in well 38/3-1 have very good source potential, largely from type III gas-prone kerogens (see Fig. 11.5). Total organic carbons measured on three picked-cuttings samples of coal lithologies range from 22 to 35% (30.5% mean), and good potential yields are suggested by pyrolysis. The source potential of these coals is limited by their low abundance in the sequence (1–2%), although this is partly offset by the overall thickness of the coal-bearing interval (795 m). Analysis of the associated ?deltaic shales suggested minimal source potential (TOC 0.21–0.5%). Analysis of five samples of the marine shales associated with the limestones also suggested negligible source potential. Total organic carbons ranged from 0.04 to 0.13%, and the kerogens comprised mostly vitrinite and inertinite, with negligible generative capacity.

Maturity measurements in this well showed a maturity gradient higher than would be predicted from present-day burial depths and heat flow, indicating that, in common with the Orcadian Basin, the observed profile is a result of an elevated palaeoheat flow, though not to the same extreme level. The coals are mature to the main oil window near the top of the section and to the late oil/early gas window near the base (see, for example, Fig. 11.49).

3.4.5 Summary of hydrocarbon plays

The ORS sandstones of the Central North Sea are relatively porous and permeable, in comparison with those of the Orcadian Basin. They might therefore form attractive reservoir targets where structural configurations favourably juxtapose mature Jurassic source rocks against Devonian structural highs, such as in the Auk, Argyll and Embla Fields. Fracturing enhances the reservoir quality in these fields. The coals observed in well 38/3-1 would generate gas if buried to sufficient depth, and their presence should be considered in any exploration strategy in the area.

3.5 Other areas

3.5.1 Midland Valley of Scotland and Forth Approaches Basin

This north-east–south-west-trending basin, flanked to the north by the Highland Boundary Fault and to the south by the Southern Uplands Fault, contains Lower and Upper ORS, but no Middle ORS (see Fig. 3.4). The Lower ORS was deposited in alluvial fans and braided-river systems sourced from metamorphic and volcanic hinterlands undergoing rapid weathering and erosion. The general direction of flow of these Lower ORS rivers was along the Midland Valley towards the south-west (Bluck, 1978; Mykura, 1991), so that little of this sediment is believed to have entered basins situated in the area of the present-day North Sea. Drainage patterns appear to have changed by Late Devonian times, following uplift of the Midland Valley in the Mid-Devonian. Upper ORS fluvial systems here and in the nearby Scottish Border Basin flowed, at least locally, towards the east, in the direction of the present-day North Sea.

Little petroleum potential is believed to be associated with the Midland Valley ORS, primarily because of a lack of intraformational seals, the absence of known source rocks and a lack of significant burial subsequent to the Carboniferous. Additionally, the Lower ORS sandstones have little reservoir potential because of their high volcaniclastic content. The Upper ORS, conversely, is more quartz-rich, and measurements made by Browne *et al.* (1985) on Upper ORS outcrop and borehole samples show good porosities of 15–20% and permeabilities of 25–1500 mD. They predict such rocks to be widely distributed in the Midland Valley. These might form an effective reservoir for gas where they are suitably juxtaposed against mature Carboniferous source rocks, perhaps in the Forth Approaches Basin.

3.5.2 Northumberland Trough/Scottish Border Basin

This north-east–south-west-trending basin is broadly comparable to and parallel with the Midland Valley of Scotland. It too contains Lower and Upper ORS, and no Middle ORS, although the Lower ORS is believed to occur only in the east of the basin, where it comprises volcanics and volcaniclastics that are intruded by the Cheviot Granite (see Fig. 3.2). The Upper ORS crops out along the northern margin of the basin and is believed to underlie most of the Dinantian strata, which form the major basin fill. Little petroleum potential is believed to be associated with this basin, for reasons similar to those for the Midland Valley.

3.5.3 Southern North Sea

No Devonian sediments have been penetrated in this area, although seismic evidence shows Devonian strata on the Mid North Sea High dipping southward under the Permo-Carboniferous of the 'Southern Gas Basin'. The Devonian is mostly too deeply buried to play any role in the area's hydrocarbon prospectivity.

3.5.4 Southern Britain/Welsh Borders

The hydrocarbon potential of the Devonian of southern Britain south of the Variscan Deformation Front (see Fig. 3.1) is poor to zero, because of high-grade diagenesis and metamorphism. For example, well penetrations in UK Quadrants 86, 87, 98 and 99 commonly terminate in non-prospective Devonian pelites, quartzites, phyllites and hard shales. Despite this, Taylor (1986) suggests that unmetamorphosed Palaeozoic reservoir rocks may be present beneath thrust sheets, perhaps with the mylonitic thrust planes acting as seals. Such an eventuality, however, must be considered speculative.

North of the Variscan Deformation Front (see Fig. 3.1), the prospects are better but are still poor. No source rocks have been proved, and the predominance of non-marine and shallow-marine facies gives little comfort that large quantities of organic material are preserved anywhere. A number of well penetrations show the Devonian locally to have moderate porosities and hence some potential as a reservoir for hydrocarbons sourced from younger rocks.

3.5.5 West of Shetland

The Clair Field comprises the only proved occurrence of ORS strata to the west of Shetland, although similar strata are believed to occur in other half-graben basins west of the Shetland Islands (e.g. Duindam and van Hoorn, 1987; Ziegler, 1990a). The fill of the Clair Basin comprises typical ORS sediments of continental facies, alluvial fan to alluvial plain, with local and periodic development of aeolian dunes (Blackbourn, 1987; Allen and Mange-Rajetzky, 1992). Blackbourn originally suggested that the Clair Group was essentially one unit, ranging from Late Devonian to Early Carboniferous (Tournaisian) in age. Allen and Mange-Rajetzky (1992), however, suggest that the lower part of the sequence may be Middle Devonian in age (see Fig. 3.3).

The Clair Field sandstones are described as porous, and permeabilities of over 1000 mD are reported in aeolian facies (Blackbourn, 1981). Coney *et al.* (1993) described 10 reservoir units, and indicate that there is little variation in porosity between them; porosity averages range from 11 to 15%. Two of the reservoir units have significant proportions of aeolian interbeds, which have significantly better permeability than the others, with arithmetic means of 80 mD and 360 mD. These compare with averages of < 10 mD for the majority of the other units, largely as a result of the presence of clay minerals. Open fractures are common and these are important in contributing to the overall permeability (Coney *et al.*, 1993). The hydrocarbons reservoired in the field are derived from a Jurassic source, and no source potential is associated with the ORS sediments drilled to date. There is still a potential for further finds in Devonian strata on similar basement highs.

3.6 Acknowledgements

I thank Halliburton Reservoir Description for making information available for inclusion in this chapter. I particularly wish to acknowledge all my friends and former colleagues at Halliburton; they were responsible for generating much of these data and providing me with hours of invaluable discussion as to their significance. Without them, this compilation would not have been possible. I also thank GeoTrack for making available some results of an unpublished AFTA study on ORS outcrop samples from the Orcadian Basin. My thanks to Ken Glennie for useful comments on the style and content of the chapter, to Reservoir Research Ltd for assistance in its production, and to Phil Richards for allowing me to revise his original Devonian chapter, prepared for the previous edition of this book.

3.7 Key references

Astin, T.R. (1990) The Devonian lacustrine sediments of Orkney, Scotland: implications for climate cyclicity, basin structure and maturation history. *J. Geol. Soc. London* **147**, 141–51.

Duncan, W.I. and Buxton, N.W.K. (1995) New evidence for evaporitic Middle Devonian lacustrine sediments with hydrocarbon source potential on the East Shetland Platform, North Sea. *J. Geol. Soc. London* **152**, 251–8.

Hillier, S.J. and Marshall, J.E.A. (1992) Organic maturation, thermal history and hydrocarbon generation in the Orcadian Basin, Scotland. *J. Geol. Soc. London* **149**, 491–502.

Marshall, J.E.A., Brown, J.F. and Hindmarsh, S. (1985) Hydrocarbon source rock potential of the Devonian rocks of the Orcadian Basin. *Scot. J. Geol.* **21**, 301–20.

Mykura, W. (1991) Old Red Sandstone: In: Graig, G.Y. (ed.) *Geology of Scotland*, 2nd ed. Geological Society, London, pp. 297–344.

Trewin, N.H. (1989) The petroleum potential of the Old Red Sandstone of northern Scotland. *Scot. J. Geol.* **25**, 201–25.

4 Carboniferous

B.M. BESLY

4.1 Introduction

4.1.1 General background

Carboniferous rocks have an extensive subcrop beneath the Permian and post-Permian successions in the North Sea (Fig. 4.1), forming an extension of areas of outcrop and subcrop in the UK, Netherlands and German onshore areas. The Carboniferous sequence reaches aggregate thicknesses in excess of 9000 m. This great thickness, together with the presence of post-Carboniferous overburden of up to 3500 m, means that the Carboniferous succession in the offshore areas is still comparatively poorly known.

The Carboniferous is important in the petroleum geology of the North Sea for two main reasons:

1 It hosts very significant thicknesses of coal and carbonaceous shale, which act as the source rock for the gas accumulations in younger reservoirs in the Southern North Sea.

2 It contains important sandstone bodies, which are the reservoirs for Carboniferous-sourced gas in the Southern North Sea and may act as reservoirs for oil and gas derived from Mesozoic source rocks elsewhere in the North Sea.

4.1.2 Historical development

The early development of the petroleum industry in the UK was based on rocks of Carboniferous age. The production of pitch and heavy oil from surface seeps began in Shropshire in the seventeenth century. During the nineteenth century, oil was produced commercially from reservoirs discovered in coalmine workings in various parts of the English Midlands (Torrens, 1994), and production was also obtained by the destructive distillation of the boghead coals (torbanites) and oil shales that occurred in association with Carboniferous coal-bearing sequences (Hallett *et al.*, 1985).

Given the high content of organic matter in the Carboniferous, together with this long history of hydrocarbon seeps and oil-shale exploitation, it is not surprising that early exploration campaigns were largely targeted to Carboniferous objectives. The first commercial oil discovery in the UK, made in 1919 at Hardstoft in Derbyshire, produced from a horizon at the top of the Dinantian Carboniferous Limestone (Giffard, 1923). Subsequent exploration delineated an oil-bearing province in the east Midlands, most production coming from reservoirs in the upper part of the Namurian Millstone Grit and the basal part of Westphalian Coal Measures (Kent, 1985). Minor discoveries of oil and gas were also made in the Dinantian of the Midland Valley of Scotland, near Edinburgh (Hallett *et al.*, 1985; Scott and Colter, 1987). Accumulations in the east Midlands were generally small, the largest, at Welton, having recoverable reserves of about 20×10^6 barrels of oil (Rothwell and Quinn, 1987; Fraser *et al.*, 1990). In most accumulations, adverse reservoir conditions resulted in very low production rates.

In the Netherlands and Germany, exploration of the Carboniferous gathered momentum during the 1960s and resulted in the discovery of a number of gasfields with reservoirs in Westphalian C and D sandstones (Hedemann, 1980; Nederlandse Aaardolie Maatschapij and Rijks Geologische Dienst (NAM and RGD), 1980).

The history of hydrocarbon occurrence in the onshore Carboniferous influenced the first phase of offshore exploration. A number of the earliest offshore wells were drilled for Carboniferous objectives in the Southern North Sea Basin and on the Mid North Sea High. The results of these suggested that the Carboniferous was generally unprospective, having unpredictable development of very heavily cemented sandstone reservoirs, although possessing abundant source rocks, particularly for gas. Interest in exploring for Carboniferous accumulations was eclipsed by the rapid development of the Permian Rotliegend play, and the Carboniferous came to be widely regarded as economic basement in the North Sea. Many wells drilled at this time made Carboniferous penetrations, which in some cases were quite extensive, but these were not subjected to detailed study. The lack of interest in possible Carboniferous plays was compounded by the problems of seismic resolution and difficulties of mapping encountered in the North Sea and, in the UK, by the generally low gas prices that prevailed throughout the 1970s. In the Dutch sector, where gas prices were more favourable, a small number of wells tested gas in the Carboniferous in Quadrant K during this period, but results were generally disappointing, owing to the extent of diagenetic alteration of the sandstone reservoirs.

Since the early 1980s, a number of factors have contributed to renewed interest in the Carboniferous in the UK. The exploration of deeper, Palaeozoic objectives has been encouraged by the UK Department of Energy, and the Carboniferous has been perceived as a possible primary objective in a number of little-explored frontier areas. At the same time, improvements in the prospective prices for gas in the early 1980s led to a renewal of interest in the Southern North Sea at a time when few obvious closures remained undrilled in the sandstone-prone area of the Rotliegend. The latter factor, together with the indications of gas potential already found in the Carboniferous of Quadrant K in the Netherlands, has led to a high level of

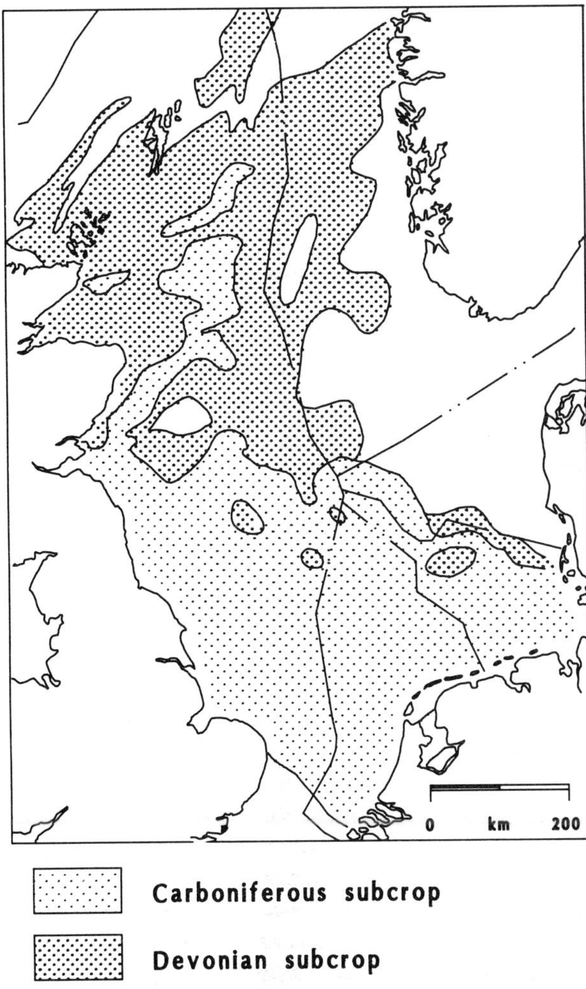

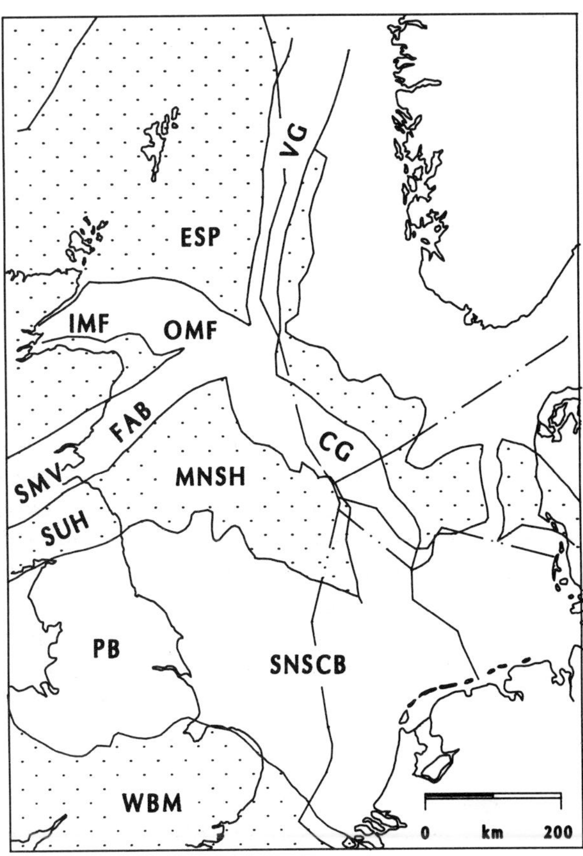

Fig. 4.1 Outline maps of North Sea area. A. Offshore subcrops of Devonian and Carboniferous rocks below the sub-Permian and sub-Mesozoic unconformities. B. Major present-day structural units referred to in the text. CG, Central Graben; ESP, East Shetland Platform; FAB, Forth Approaches Basin; IMF, Inner Moray Firth; MNSH, Mid North Sea High; OMF, Outer Moray Firth; PB, Pennine Basin Complex; SMV, Scottish Midland Valley; SNSCB, Southern North Sea Carboniferous Basin Complex; SUH, Southern Uplands High; VG, Viking Graben; WBM, Wales–Brabant Massif.

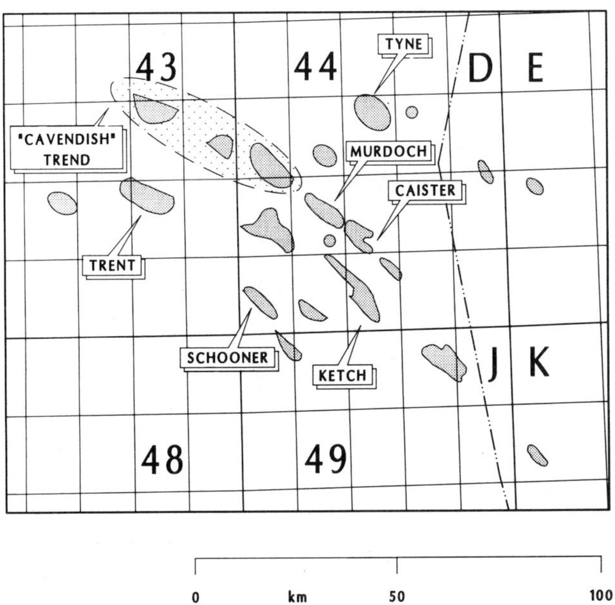

Fig. 4.2 Major Carboniferous gas discoveries of the Silverpit Basin and surrounding area of the UK and Netherlands Southern North Sea. For regional structural context, see Fig. 4.23.

exploration in the Southern North Sea, stimulated by early important discoveries made in eighth-round blocks 44/22 (Murdoch), 44/23 (Caister C) and 44/28 (Ketch).

In the period 1983–93, some 20 discoveries were made in the Carboniferous (Fig. 4.2), with inferred or published reserves of individual accumulations lying in the range $70–500 \times 10^9$ ft^3 of gas. Several significant gas-bearing trends have been established in the central and northern parts of Quadrant 44, and in the southern and central parts of Quadrant 43, containing fields having reservoirs in sandstones of Namurian and Westphalian age. An unexpected feature of these discoveries has been the high production rates that have been obtained from a sequence that had previously been expected to have poor reservoir properties. Wells in blocks 44/22 and 44/23 have been reported to have tested at rates of $16–65 \times 10^6$ ft^3 per day, rates that compare favourably with those obtained in the Rotliegend.

4.1.3 Nature of previous geological study

The Carboniferous rocks penetrated in many of the early wells drilled in the North Sea were not subjected to detailed stratigraphic or geological study at the time of drilling. The recent growth of interest in the Carboniferous has led to the re-evaluation of the stratigraphy and sedimentology of many of the early well penetrations, and many new wells have been drilled. Even so, there remain large gaps in the understanding of Carboniferous regional geology in the North Sea, and much still relies on the extrapolation into the offshore area of knowledge derived from sequences in the surrounding onshore areas.

Regional synthesis in the onshore areas is dominated by work carried out in parts of Britain in which the Carboniferous rocks are exposed. Elsewhere, especially on the European mainland, the Carboniferous is concealed by thick Mesozoic and Tertiary overburden. Until the 1980s, much work was concerned with the elucidation of local stratigraphies. This task was complicated by poor exposure, the great thicknesses involved and the inherent variability of a sequence containing both lithologically homogeneous and repetitive units and pronounced local changes of thickness and facies. Local studies were synthesized into a consistent regional stratigraphy in the late 1970s (George *et al.*, 1976; Ramsbottom *et al.*, 1978), on the basis of which regional syntheses of palaeogeography and basin subsidence behaviour have been produced (e.g. Miller *et al.*, 1987; Besly and Kelling, 1988; Leeder, 1988; Arthurton *et al.*, 1989). Since seismic reflection data have become accessible to academic researchers, structural and sequence stratigraphic studies have led to major improvements in the understanding of structural controls on sedimentation and the patterns of deformation of the sequence (Kimbell *et al.*, 1989; Ebdon *et al.*, 1990; Fraser *et al.*, 1990; Fraser and Gawthorpe, 1990; Chadwick and Holliday, 1991; Collier, 1991; Corfield *et al.*, 1996; Peace and Besly, 1997). In the same period, a stratigraphic synthesis of the Carboniferous in Germany was published by Jankowski (1991), and detailed modern descriptions of the upper parts of the Carboniferous sequence in the Netherlands have become available (van der Zwan *et al.*, 1993).

The Carboniferous, particularly in northern England, has been the subject of very thorough sedimentological studies. Both carbonate and clastic facies developments in the Dinantian have been widely studied (see reviews in Miller *et al.*, 1987; Arthurton *et al.*, 1989). The Namurian of the Pennine Basin has been one of the classic areas for the application of the facies-analysis technique in British sedimentology, forming the basis for a sequence of papers by Collinson and others, summarized in Collinson (1988), and more recently proving amenable to the application of sequence-stratigraphic concepts (Read, 1991; Maynard, 1992; Martinsen, 1993; Church and Gawthorpe, 1994; Wignall and Maynard, 1996). The sedimentology of the Westphalian coal-bearing and red-bed sequences remains only patchily studied in the UK (Besly, 1988; Guion and Fielding, 1988; Rippon, 1993), but has been extensively documented in Germany (David, 1990; Selter, 1990).

4.1.4 Stratigraphic methodology

The major biostratigraphic zonations of the Carboniferous of north-west Europe are summarized in Fig. 4.3. The methods employed in biostratigraphic subdivision depend to a large extent on the gross facies association present. The extent to which they are applicable in well penetrations is very variable. Three broad classes can be recognized: (i) marine-carbonate associations; (ii) basinal and deltaic associations; and (iii) red-bed associations.

Marine-carbonate associations

Traditionally, zonation in this association was achieved by the use of corals and brachiopods. More precise zonations have been achieved by the use of conodonts (Austin, 1973) and foraminifera (Fewtrell *et al.*, 1981; Strank, 1987). Both of these fossil groups are amenable to use on subsurface sample material, although there remains ambiguity in the definition of some stage boundaries (Ebdon *et al.*, 1990).

Basinal and deltaic associations

Throughout the Carboniferous, with the exception of the latest Westphalian and Stephanian, clastic sequences were deposited in predominantly freshwater or brackish conditions, even where deep-water sedimentary facies were developed. A very precise stratigraphy exists for such sequences, thanks to the repeated penetration of short-lived marine incursions, which deposited 'Marine Bands' (George *et al.*, 1976; Ramsbottom *et al.*, 1978). In the Upper Dinantian and Lower Namurian of Scotland and northern England, the marine intercalations gave rise to the interbedding of marine limestones and deltaic clastics—the so-called Yoredale facies. Owing to the rapid evolution of pelagic goniatites in the open ocean, which sourced these incursions, the Marine Bands have distinctive, often unique, goniatite faunas. In the Namurian and the Lower Westphalian, some 70 Marine Bands are recognized, each having a distinctive fauna. Further Marine Bands are recognized (particularly in the Westphalian) which lack a goniatite fauna but contain other distinctive organisms indicative of marine salinities. The highest widespread marine horizon recognized in Western Europe is the *Anthracoceras aegiranum* horizon, forming the Westphalian B/C boundary. In the UK, Marine Bands are present up to the boundary between the Lower and Upper Westphalian C, formed by the *Anthracoceras cambriense* horizon.

Although the Marine Bands offer the possibility of a refined subdivision of the Namurian and much of the Westphalian succession, they are very thin and are seldom cored in wells. Where they are either not recognized or not developed, various other organisms offer biostratigraphic possibilities that are of lesser and more variable precision. A conodont zonation exists for the Namurian (Higgins, 1976). Reasonably precise zonations exist for the Westphalian, based on non-marine bivalves (Trueman and Weir, 1946) and plant megafossils (Crookall, 1955). The former are unreliable after the mid-Westphalian C, owing to the development of facies-related forms. The most widely applicable zonation that is also effective in well penetrations is that based on palynomorphs (Clayton *et al.*, 1977; McLean, 1995). The accuracy of palynological zonation varies. In the Westphalian A to C, a precise zonation based on diagnostic palynomorph assemblages has been developed from the detailed work of Smith and Butterworth

SUB-SYSTEM	SERIES		STAGES	AGE Ma	GONIATITES ZONES		GONIATITES MARKER BANDS	SPORES	
SILESIAN		C	STEPHANIAN	300				NBM	P. novicus-bhardwajii Cheiledonites major
		B		303				ST	Angulisporites splendidus Latensina trileta
		A		305					
	WESTPHALIAN		WESTPHALIAN D	308				OT	Thymospora obscura Thymospora thiesseni
			WESTPHALIAN C	311			A. cambriense A. aegiranum	SL	Torispora securis Torispora laevigata
			WESTPHALIAN B		A	'Anthracoceras'	A. vanderbeckei	NJ	Microreticulatisporites nobilis Florinites junior
			WESTPHALIAN A	(315)	G2	G. listeri G. subcrenatum	G. subcrenatum	RA	Radiizonites aligerens
								SS	Triquitrites sinani Cirratriradites saturni
	NAMURIAN	C	YEADONIAN		G1	G. cumbriense G. cancellatum	G. cancellatum	FR	Raistrickia fulva Reticulatisporites reticulatus
		B	MARSDENIAN		R2	R. superbilingue R. bilingue R. gracile	R. gracile	KV	Crassispora kosankei Grumosporites varioreticulatus
			KINDERSCOUTIAN	319	R1	R. reticulatum R. nodosum R. circumplicatile	Hod. magistrorum		
		A	ALPORTIAN		H2	H. prereticulatus H. undulatum Hd. proteus	Hd. proteus	SO	Lycospora subtriquetra Kraeuselisporites ornatus
			CHOKIERIAN		H1	H. beyrichianum H. subglobosum	H. subglobosum		
			ARNSBERGIAN		E2	N. nuculum Cd. nitidus		TR	Stenozonotriletes triangulus Rotaspora knoxi
			PENDLEIAN	(326)	E1	E. bisulcatum C. malhamense E. pseudobilingue C. leion	C. cowlingense C. leion	NC	Bellispores nitidus Reticulatisporites carnosus

A

SUB-SYSTEM	SERIES	UK STAGES	BELGIAN STAGES	AGE Ma	CORAL/ BRACHIOPOD	GONIA-TITES	SPORES	
DINANTIAN	VISEAN	BRIGANTIAN	V3c	(326)	D2	P2	NC	(lower part)
						P1	VF	Tripartites vetustus Rotaspora fracta
		ASBIAN	V3b		D1	B	NM	Raistrickia nigra Triquitites marginatus
				(335)			TC	Perotriletes tesselliatus Schulzospora campyloptera
		HOLKERIAN	V2b–V3a		S2		PU	Lycospora pusilla
		ARUNDIAN	V1b–V2a		C2S1			
		CHADIAN	V1a	(350)	C1			
	TOURNAISIAN	COURCEYAN	Tn3		Z		CM	Schopfites claviger Auroraspora macra
							PC	Spelaeotriletes pretiosus Raistrickia clavata
			Tn2		K		NV	Verrucosisporites nitidus Vallatisporites vallatus
			Tn1	360			PL	Vallatisporites pusillites Spelaeotriletes lepidophytus

B

Fig. 4.3 A & B Major biostratigraphic subdivisions of the British Carboniferous (after George *et al.*, 1976; Clayton *et al.*, 1977; Ramsbottom *et al.* 1978). Absolute ages are from Lippolt *et al.* (1984); ages in brackets are interpolations of data of Lippolt *et al.* by Leeder and McMahon (1988).

(1967). Zonations in the Namurian (Owens *et al.*, 1977) and Dinantian (Neves *et al.*, 1972, 1973) are based on 'concurrent range zones' (defined by the overlap between the stratigraphic appearance of one species and the disappearance of another), which are difficult to apply to material derived from ditch cuttings. The palynological zonation loses precision in the late Westphalian and is not applicable in the Stephanian. The zonation based on palynology is very much less precise than that obtainable from the identification of marine markers, especially at key horizons, such as the Namurian/Westphalian boundary, and in the subdivision of the early Namurian and the mid-Dinantian. This may lead to difficulty in making detailed correlations in thick sequences of deltaic and non-marine strata, particularly in the offshore area, which remains sparsely drilled. This problem is discussed further in Section 4.8. There is, however, considerable potential for application of integration of palynology, lithostratigraphy and log-facies analysis in sequence-stratigraphic analysis (Davies and McClean, 1996).

It should be noted that almost all of the fossil groups used as markers for stratigraphic boundaries or stratigraphic intervals in the Carboniferous of north-west Europe have only a regional value (Bless *et al.*, 1987). Their use outside the fairly tightly constrained area of the Variscan foredeep basins is very limited.

Red-bed associations

Red beds are widespread in the Carboniferous of the North Sea and adjoining areas. Two types of red pigment development are recognized (Besly *et al.*, 1993), each presenting different stratigraphic problems: (i) primary red beds; and (ii) penetrative weathering.

Primary red beds, in which reddening took place at, or soon after, deposition, are widely developed in the early Dinantian and in the Westphalian and Stephanian. The early Carboniferous red beds are indistinguishable from those in the underlying Devonian, and yield few palynological ages. Correlation relies mainly on lithostratigraphic criteria. The late Westphalian and Stephanian red beds yield poorly developed, facies-related faunas of non-marine bivalves, annelids, branchiopods and gastropods. Reliable dates have been obtained only from plant macrofossil impressions (Wagner, 1983). A lack of preserved organic material generally precludes palynological zonation. Where palynomorphs have been recovered from the very high parts of the succession, their apparent ages differ from the more reliable plant macrofossils, suggesting the development of endemism (van der Zwan *et al.*, 1993).

The second group of red beds present in the Carboniferous are those formed by penetrative weathering of the denuded early Permian land surface under arid conditions. This type of reddening, described in detail by Trotter (1953, 1954) and Mykura (1960), affects all facies and stratigraphic horizons in the Carboniferous. Organic material was destroyed and siderite concretions and cements are replaced by haematite. Stratigraphic relationships are characteristically cross-cutting, reddening being preferentially associated with permeable units, and becoming less pervasive downwards. Red beds formed in this way may reach thicknesses of several hundred metres, depending on the depth of penetration of oxidizing conditions and the extent of denudation of the weathering mantle before the resumption of deposition.

It is difficult to discriminate between primary red-bed units of late Westphalian age and the secondary weathering mantle, especially when cores are not available. This problem is exacerbated by the local unconformable relationship between primary red-bed units and lower parts of the succession. The identification of both reversed and normal magnetization in red beds from the English Midlands (Besly and Turner, 1983) raises the possibility that a crude magnetic-reversal stratigraphy might be determined within the red-bed units, but the necessary research has yet to be carried out. Palaeomagnetism may, however, allow the distinction of Carboniferous from Permian red beds, since continuing northward drift caused a marked change in magnetic inclination during the Late Carboniferous to early Permian.

4.1.5 Stratigraphic nomenclature

The stratigraphy of the Carboniferous is complicated by the present state of its nomenclature. The very thick sequence, the complex basin geometries and stratigraphic relationships, and the long history of study in the different countries bordering the North Sea have resulted in an incomplete, often ambiguous, lithostratigraphic nomenclature, in which there is a multiplicity of formation and group names. Standard lithostratigraphic nomenclatures, recently proposed for the UK sector of the North Sea (Cameron, 1993a,b), are illustrated and discussed in more detail in subsequent sections of this chapter. A similar scheme is currently being prepared for the Netherlands sector by the Dutch Geological Survey. While the UK scheme has advantages of internal consistency, it is still not ideal. It proliferates ambiguity by allocating Formation names to units that have Group status in the onshore UK area, and some of the Formation names include units of obviously differing depositional environment. With further drilling and release of data, a more precise nomenclature will be possible, it already being evident that many of the lithostratigraphic units having informal Formation status in the UK onshore area can be recognized offshore.

4.2 Overview

4.2.1 Tectonic setting

The regional plate-tectonic context of the Carboniferous of north-west Europe has been summarized by Leeder (1988) and Coward (1993). The area lay to the north of the Variscan orogenic belt, an orogen that resulted from the collision of the African portion of Gondwanaland and the European portion of Laurussia. The initial deformation of this belt had started in the late Devonian, in the zone passing through Galicia, Armorica, the Massif Central, the Vosges and southern Germany. During the Dinantian, the Armorica and Massif Central areas were underlain by a northward-dipping subduction zone. Associated with this was a back-arc seaway, partly floored by oceanic crust, which stretched through south-west England and northern France and into central Germany. During the later evolution of the Variscan orogen, the deformation front migrated northwards and the back-arc seaway—the Rheno-Hercynian Zone—was closed and deformed into major thrust/nappe complexes. The loading imparted by these complexes led to the formation of a flexural foreland basin, which migrated northwards during the Namurian and Westphalian, reaching its maximum northward extent, in central England, by the latest Westphalian or Stephanian. Regional Variscan inversion reached these areas soon afterwards, probably from mid-Stephanian onwards, giving rise to widespread folding and block faulting. The later stages of inversion were probably influenced by west–east compression, related to collision tectonics in the Ural Mountain belt. Because no Early Permian sequence is preserved in most areas, the major unconformity that results from end-Carboniferous inversion is indistinguishable from the Saalian Unconformity at the base of the Upper Rotliegend.

The North Sea and adjoining onshore areas were largely peripheral to the orogen until its latest phases. However, they have a complex tectonic history, which was probably largely controlled by events within the orogen to the south.

During the Early Carboniferous, the stable 'Old Red Sandstone Continent', established in the Caledonian orogeny, was broken up by widespread crustal extension. Rapidly subsiding graben and half-graben ('basins') were established, which were separated by horsts and half-horsts ('blocks') on which condensed sequences accumulated. The largest of the stable 'block' areas consisted of stable basement massifs—the Wales–Brabant High and the Southern Uplands High (see Fig. 4.1B). The Wales–Brabant High separates the Carboniferous basin of northern England and its eastward extension into the UK North Sea from the Rheno-Hercynian zone and its northern-flanking flexural basins. It consisted of a rigid Precambrian block—the 'Midlands Microcraton'—flanked to the north by Lower Palaeozoic basinal sequences, which were deformed in the Caledonian orogeny (Soper *et al.*, 1987). The second major high, the Southern Uplands High, is also of Caledonian origin. Together with its offshore continuation—Mid North Sea High—it separates the depocentres of northern England and the Southern North Sea from those in the Midland Valley of Scotland and their offshore extensions. The location and orientation of the fault systems bounding the horsts in the extensional system appears to have been controlled to a large extent by basement heterogeneity, particularly the location of Devonian and Precambrian granite plutons (Bott, 1967), and by the tectonic grain inherited from earlier orogenic events (Fraser *et al.*, 1990, Figs 3 and 4).

The mechanism that drove the Lower Carboniferous extension and the orientation of the prevailing stress field remain subjects for debate. Leeder has argued extensively in favour of a north-west–south-east-orientated stress field, driven by back-arc extension to the north of the Variscan subduction system. Such an orientation cannot explain the reactivation of north-west–south-east-trending faults, widely observed in the east Midlands. Coward (1993) suggests that an additional tectonic component was provided by block rotation associated with the eastward lateral expulsion of a triangular basement block—named the North Sea–Baltic Block—lying between the Scottish highlands and the Wales—Brabant Massif and including most of the North Sea. Expulsion of this block was associated with progressive closure between the Acadian and Gondwana continents, and produced dextral shear along the northern margin of the Wales–Brabant Massif, giving rise to transtensional reactivation of north-west–south-east Caledonian lineaments. At the northern margin of the expelled block, in the Midland Valley of Scotland, sinistral shear produced the pattern of west–east extension observed by Stedman (1988) and Read (1988).

During the Late Carboniferous, differential subsidence associated with crustal extension gave way to a more uniform subsidence pattern. In northern England, this transition is regarded by Leeder and others to represent the onset of a phase of thermal subsidence. Minor inversion and syndepositional folding are widespread, causing local unconformities and onlap in the Namurian of the Scottish Midland Valley (Read, 1988) and the Northumberland–Solway Basin (Chadwick *et al.*, 1993) and in the Westphalian A and B in the English Midlands (Corfield *et al.*, 1996). These early precursors of the late Carboniferous Variscan inversion reflect the onset of compression related to the northward migration of the Variscan deformation front. From the mid-Westphalian B, compressional tectonics dominated the pattern of sedimentation: fault activity and folding in the depositional basin increased, with the development of regional unconformities in the late Westphalian C and Westphalian D. In the UK area, foreland flexural subsidence became established to the south of the Wales–Brabant Massif during the Namurian (Kelling, 1988), but spread to the north of this basement high only during the Westphalian D. In the Netherlands and Germany, where there was no basement high analogous to the Wales–Brabant Massif, the lack of a rigid basement buttress allowed the earlier establishment, during the Namurian, of a wide belt of flexural subsidence (Jankowski, 1991; Quirk, 1993).

The timing of the final Variscan inversion and deformation of the Carboniferous basin fills is poorly constrained, largely owing to the deep level of Variscan inversion. Corfield *et al.* (1996) and Leeder and Hardman (1990) demonstrate that some major folds were formed during the Westphalian C. Palaeobotanic dating demonstrates that sedimentation continued in parts of the UK onshore until at least the Cantabrian (early Stephanian: Cleal, 1978), and may have persisted locally into the Early Permian (Wagner, 1983). Coward (1993) suggests that inversion patterns were in part controlled by north-west–south-east compression associated with Variscan collision, and partly the result of reversal of the movement generated during the lateral expulsion of the North Sea–Baltic Block during the Devonian and earlier Carboniferous. Such reversal was associated with the closure of the Ural Ocean and associated tectonism in the Ural Mountains.

Throughout the Carboniferous, tectonic activity provided the prime control on sedimentary facies development. In the Dinantian, the initial positions of carbonate platforms and deep-water facies were dictated by the location of horst and half-horst structures (Grayson and Oldham, 1987; Collier, 1991). The flow patterns of systems depositing Lower Carboniferous clastics in Northern England were probably largely determined by the positions of active fault lines (Leeder, 1987; Leeder *et al.*, 1989; Turner B.R. *et al.*, 1993). In the Namurian, syndepositional tectonism may have been less important, although relic Dinantian highs, originally developed in tectonically controlled positions, were a significant control on patterns of basin fill (Collinson, 1988; Steele, 1988). By the Westphalian, deposition of sediment had brought about emergent or near-emergent conditions throughout the basin. Syndepositional fault activity and folding affected facies distribution in a subtle way, controlling the positions of splits in coal seams and the location and stacking patterns of channel sand bodies (Fielding, 1984a,b; Guion and Fielding, 1988; Corfield *et al.*, 1996). Finally, some of the facies patterns developed in the late Carboniferous red beds in the UK onshore area show a strong local tectonic control (Besly, 1988).

4.2.2 Palaeoclimate

Sedimentological and palaeobotanical studies provide abundant evidence for the nature of the Carboniferous palaeoclimate. The Tournaisian climate was markedly arid, witnessed by the occurrence of lacustrine sabkha deposits (Scott, 1986) and caliche (Andrews *et al.*, 1991) in the Cementstone Group of northern England and Scotland and of anhydritic marine sabkha deposits in the Tournaisian of central England (Llewellyn and Stabbins, 1970). Oscillation between semiarid and humid climates indicated by contrasting styles of karst development, is recorded in the early Visean by Wright (1990). The persistence of semiarid conditions with seasonal rainfall into the late Visean is indicated by the occurrence of calcretes in North Wales (Davies, 1991). Phytogeographical studies show a trend towards warmer and more humid conditions in the latest Dinantian and early Namurian (van der Zwan, 1981; Raymond, 1985; van der Zwan *et al.*, 1985), leading to a non-seasonal, humid equatorial climate in the Westphalian A to C (Parrish, 1982; Rowley *et al.*, 1985). From the Westphalian D onwards, a drying trend is apparent from the occurrence of caliche in red beds in the English Midlands (Besly, 1987) and in north Germany (Selter, 1989). By the Autunian, arid climatic conditions were established and were to continue throughout the Permian and Triassic.

The establishment and subsequent decline of the humid equatorial belt can be interpreted, as a first approximation, as charting the northward drift across the equatorial belt that accompanied the closure of the proto-Tethyan Ocean and the fusion of the Laurussian and Gondwanan land masses (Bless *et al.*, 1984). Climatic modelling carried out by Parrish (1982) and Rowley *et al.* (1985) demonstrates that this is an oversimplification, and that most of the observed changes in palaeoclimate can better be explained in terms of changes in atmospheric circulation, which accompanied the changing disposition of the continental land masses during continental collision and the subsequent growth of mountain belts. Monsoonal conditions during the Dinantian evolved into humid equatorial conditions partly as a result of latitudinal change, but also owing to continental amalgamation. The Late Carboniferous increase in aridity was caused initially by the growth of a rain-shadow to the north-west of the growing Variscan mountain chain, which sheltered the basins in the orogenic foreland from the prevailing trade winds, which blew from the proto-Tethyan Ocean to the east (Rowley *et al.*, 1985; Besly, 1987). The final assembly of the Pangaean supercontinent produced an elongate continental body covering a wide range of latitude, in which cross-equatorial air flow eliminated the equatorial humid belt.

4.2.3 Sedimentary cyclicity and sequence stratigraphy

Perhaps the most prominent aspect of the geology of the European Carboniferous is the very pronounced cyclicity present in many of the sedimentary sequences. This cyclicity has, at various times, been interpreted in terms of eustatic, tectonic and autocyclic mechanisms (for a review see Leeder, 1988). Eustatic explanations for cyclicity have largely been in terms of glacial eustatic controls, invoking a link between global sea level and the growth and decline of the contemporaneous Gondwana ice sheet.

In recent years, eustatic explanations for the sedimentary cyclicity have received wide publicity, in large part owing to the proposals that a number of major cycles—termed 'mesothems'—could be recognized (Ramsbottom, 1979). These cycles each comprised a number of smaller-scale cycles, termed 'cyclothems'. The proposed mesothems had a characteristic, asymmetric, form. The earlier cyclothems of the major cycle were claimed to be restricted to basinal areas having higher subsidence rates and, by implication, greater water depths, while the later cyclothems transgressed progressively larger areas of the more slowly subsiding, topographically elevated 'shelves'. Ramsbottom's hypothesis marked the first attempt to formulate a consistent sequence stratigraphy for any part of the north-west European Carboniferous, and has had wide influence as a framework for worldwide correlation of the Carboniferous (Ross and Ross, 1985), for regional stratigraphic subdivision (George *et al.*, 1976; Ramsbottom *et al.*, 1978) and as a framework for the erection of sedimentary facies models.

Recent studies of the sedimentology of the supposed mesothemic sequences have largely failed to confirm the hypothesis that such large-scale, eustatically controlled sequences exist (Barraclough, reported in Leeder, 1988; Holdsworth and Collinson, 1988). Their reality is also brought into question by the Carboniferous sequence stratigraphy of Ebdon *et al.* (1990), in which the seismically defined boundaries do not show a marked correlation with the mesothem boundaries. What is not in doubt is the reality of a strong eustatic control on sedimentary cyclicity at the level of the individual cyclothem or 'minor cycle'. Recent studies, both in the UK and in the USA, have suggested very convincingly that a glacio-eustatic control provides the best explanation for most Carboniferous minor cyclicity, calculated cycle periodicities possibly corresponding to known Milankovich periodicities (Collier *et al.*, 1990; Maynard and Leeder, 1992). At the same time a number of authors (Read, 1991; Maynard, 1992; Martinsen, 1993; Wignall and Maynard, 1996) have demonstrated that the sequence-stratigraphic concepts of van Wagoner *et al.* (1988) can be applied to the exposed deltaic successions of the Namurian and Westphalian in the onshore area. Although this work is at an early stage, it allows the recognition of the important control exercised on facies architecture by phases of eustatic lowstand, and offers the prospect of more rigorous prediction of the distribution of reservoir and source-rock facies (Church and Gawthorpe, 1994). A preliminary sequence stratigraphic analysis of the offshore Westphalian is presented by Quirk (1997).

4.3 Regional palaeogeographical development

4.3.1 Tournaisian

Following the terminal Caledonian deformation, probably in the Emsian (McKerrow, 1988), the topography of active sediment source areas was reduced through the late Devonian. By the time of the early Carboniferous transgression, the regional input of clastic material had become reduced. Shallow-marine facies became established very rapidly in the more distal parts of the basin area, in England and the southern part of the Southern North Sea. In the early phases of trangression, localized evaporitic facies were deposited (Llewellyn and Stabbins, 1970), but by Chadian times regionally extensive carbonate facies became established, giving rise to thick sequences of limestones. The onset of marine facies was diachronous, occurring just above the Devonian–Carboniferous boundary in southern Britain (George *et al.*, 1976), and somewhat later in northern England and southern Scotland, where significant thicknesses of Old Red Sandstone facies of earliest Carboniferous age mark a continuation of clastic sedimentation in areas proximal to the Caledonian source. The earliest Carboniferous rocks showing a marine influence in the northern area are in the Cementstone Group, a succession of interbedded shallow-marine limestones and alluvial deposits, with evaporite relics (Scott, 1986).

4.3.2 Visean

Two important palaeogeographical changes occurred progressively during the Visean. Firstly, clastic, mainly deltaic, sediments became dominant in the northern British basins and over much of the North Sea. Secondly, as Lower Carboniferous crustal extension proceeded, topographic differentiation between highs (horsts, tilt blocks) and corresponding areas of more rapid subsidence became pronounced. Evolution of the morphology of the rift system was accompanied by progressive onlap of high blocks that had not received sediment during the Tournaisian. Pre-Carboniferous basement in highs in North Wales and northern England is overlain by marine limestones of Arundian to Asbian age. This onlap was partly caused by a long-term eustatic sea-level rise.

The input of clastic sediments into basins developed during the Early Carboniferous was derived from several sources. Alluvial fans were sourced from the footwalls of the late Devonian and early Carboniferous rifts (Deegan, 1973; Fraser and Gawthorpe, 1990), and fairly small deltas were sourced from the remaining Caledonian Upland areas in Scotland (e.g. Leeder and Boldy, 1990). These sediment sources, however, cannot account for the very large volume of coarse clastic sediment, which spread gradually southwards throughout the Dinantian and early Silesian. The isotopic fingerprint of this sediment points to a source in the ?Grenville/Caledonian and Archaean terrains to the north and west of the British Isles (Cliff *et al.*, 1991). Coarse clastic sediment was consistently supplied to a point lying roughly in the centre of the present North Sea. The long duration of this sediment input point suggests that the Viking Graben was already in existence, and acted as a conduit of sediment into the basins of the North Sea and northern Britain (Coward, 1993). The cause of the initiation of the northern clastic source is enigmatic. There is currently little evidence of an orogenic event in these areas of sufficient magnitude to generate the very large volumes of sediment involved. Haszeldine (1984) regarded this material as derived from uplifts marginal to a contemporaneous rifting event in the proto-North Atlantic area, but Bristow (1988) has pointed out that the volume of sediment produced by such a mechanism would have been considerably less than that observed. It is possible that the necessary uplifts may have been related to a north-easterly extension of the major wrench system of the Maritime Provinces of eastern Canada (Leeder, 1988).

Areas lying to the north of latitude 54°N were dominated by clastic facies. Chadian to Asbian sequences are dominated by shallow-water deltaic deposits, in which major sand-rich fluvial units—the Fell Sandstone and correlatives—may record episodes of eustatic lowstand. Gradually rising sea level is recorded by an upward increase in the abundance of coal, in the Asbian, and by the onset of cyclic deltaic sequences containing marine-shelf limestone facies—the 'Yoredale facies'—in the Brigantian.

In areas of northern and central England and the Southern North Sea that were beyond the influence of the northerly-derived clastic input, topographic differentiation of the basin complex was prominent. Here, in the mid-Dinantian (Arundian to Holkerian), carbonate platforms became established at the crests of tilt blocks, separated by deep subbasins, in which turbiditic basinal sediments accumulated (Smith *et al.*, 1985; Grayson and Oldham, 1987). The basinal sediments deposited in the early phase of marine transgression into the extensional basins locally have kerogenous source-rock potential (Fraser *et al.*, 1990). To the north of 54°N, in areas affected by the clastic influx, syndepositional tectonism is recorded by differential sediment thicknesses in half-graben fills, and by the localization of channel facies and marginal alluvial-fan sediments (Leeder, 1987).

4.3.3 Namurian

At the beginning of the Silesian, there was a period of radical palaeogeographical change. In the early Namurian, the activity of most major subbasin-bounding faults ceased, and the tectonically controlled tilt-block subsidence pattern evolved into one of more uniform regional subsidence (Leeder, 1982, 1988). Locally, Dinantian carbonate platforms remained as relic highs, which were progressively onlapped (Collinson, 1988), but in some cases the change in subsidence pattern was accompanied by the collapse of these highs and very rapid deepening (e.g. Grayson and Oldham, 1987). At the same time, the influx of clastic sediment from the northern source area increased dramatically, probably owing to climatic change in the source areas. This put an end to carbonate deposition and initiated a sequence of major delta advances, which infilled the southern part of the Pennine Basin, eradicating the relic topographies inherited from the Dinantian and depositing the typical 'Millstone Grit' of the southern and central Pennines. To the south of the Askrigg Block, a well-defined succession of delta types can be recognized. The first phase of sedimentation involved the deposition of basinal mudstones, in places containing stratigraphically condensed sequences. These locally contain important oil source rocks. The early phases of fill of deep basins were effected by the development of turbidite fans and of major delta systems, which deposited thick (up to 500 m) upward-coarsening sequences (Collinson, 1988; Steele, 1988). These were succeeded by shoal-water deltas, which deposited thinner upward-coarsening deltaic units.

The infill of inherited basin topographies proceeded from north to south during the Namurian, marking the continuation of the advance of deltaic facies that had characterized the Dinantian further to the north. Episodes of turbidite deposition and slope progradation effected the infill of topographic lows in northern Lancashire and Yorkshire during the Pendleian and Arnsbergian. Similar major basin-filling episodes occurred in Yorkshire and northern Derbyshire during the Kinderscoutian, and in southern Derbyshire and Staffordshire during the Marsdenian. By the end of the Namurian, shallow-water to near-emergent conditions prevailed over the whole of the UK onshore area, and probably over much of the North Sea and onshore areas of mainland Europe.

In more northerly areas, to the north of the Askrigg Block, conditions of sedimentation during the Namurian were essentially similar to those that had already become established in the late Dinantian. Shallow-water deltas were sourced, ultimately, from the regional northern source, or from local sources in the relic Caledonian uplands. By the early Namurian (Pendleian–Arnsbergian), conditions were sufficiently emergent for the formation of major coal deposits in the Limestone Coal Group of the Midland Valley of Scotland. The deltas in Scotland and northern England had a greater marine influence than those further to the south, and the resulting sequence is a continuation of the Yoredale facies development of the late Visean, containing many interbedded marine limestones. The route of marine ingression is not known with certainty. In the mid- to late Namurian, localized intense deformation occurred in the Midland Valley of Scotland. Sequences of this age are in places condensed, consist mainly of fluvial facies and contain internal erosional unconformities. This deformation was related to a phase of dextral strike–slip deformation (Read, 1988). Elsewhere in northern England, the mid- and late Namurian sequences are generally condensed, and there is local seismic evidence for a phase of mild inversion (Chadwick *et al.*, 1993).

4.3.4 Early–mid-Westphalian

The establishment of widespread shallow-water deltaic sedimentation in the late Namurian heralded a major development of such conditions during the Westphalian A to early Westphalian C (van Wijhe and Bless, 1974; Guion and Fielding, 1988; Strehlau and David, 1989; David, 1990). Low-lying, near-emergent conditions existed over the whole of the Variscan foreland area. In the UK onshore area, a broad distinction can be made between periods of relative flooding, in the early Westphalian A and late Westphalian B–early Westphalian C, and periods of relative emergence, in the late Westphalian A–early Westphalian B and later Westphalian C. In the former, elongate deltas infilled shallow-water bodies on a basinwide scale: marine bands are common and regionally correlatable coals few. In the latter periods, much of the area was subaerially emergent. The dominant sedimentary motif was one of the infilling of freshwater lakes by deltas fed from a distributive channel network: marine bands are rare and there are major developments of coal in seams that can be correlated over very large areas. During this period, a significant change in sediment provenance occurred at some time during Westphalian A. The northern source, which had dominated sedimentation in northern England during the late Dinantian and Namurian, became less important, and the major input of sand into the basins of central and northern England came instead from the west (Glover *et al.*, 1996; Rippon, 1996). As a result, areas in the eastern part of northern Britain, which had formerly been located proximally in the depositional systems, became relatively sand-starved.

4.3.5 Mid–late Westphalian and Stephanian

The pattern of sedimentation established in the Westphalian A and B became modified during the Westphalian C, in response to the northward advance of the Variscan deformation front, to changes in sediment provenance in northern Britain and, possibly, to climatic changes. The onset of major deformation of the basin fill of the southern British portion of the Rheno-Hercynian zone can be dated accurately to have occurred in the early Westphalian C. The detritus shed from this uplift forms the Pennant Sandstones of South Wales and the coalfields of southern England (Besly, 1988; Glover *et al.*, 1996), but was prevented from penetrating into the Midlands and north of England by the barrier formed by the Wales–Brabant Massif. The deformation event is, however, recorded to the north of the Wales–Brabant Massif by the occurrence of local developments of red beds (Etruria Formation: Besly, 1988) and by a widespread unconformity at the base of the Westphalian D (Tubb *et al.*, 1986; Besly, 1988). To the east, in the Netherlands and Germany, no such barrier existed, and the deformation event is recorded by a marked increase in southerly-derived sandstones from the late Westphalian B onwards (David, 1987, 1990; Strehlau and David, 1989). The southerly-derived sandstones are best developed in Belgium and the Netherlands. In the latter area, they form the Tubbergen Sandstone Formation, of late Westphalian B to early Westphalian D age, which is locally an important gas reservoir in the eastern Netherlands (NAM and RGD, 1980).

In the north of the British area, a change of facies, from coal-bearing to red-bed alluvial sedimentation, started at about the Westphalian B/C boundary in the Midland Valley of Scotland. Little is known of the palaeogeography of these red beds, known informally as the 'Barren Red Group'. It appears that red-bed deposition spread southwards during the Westphalian C and D. In the Canonbie and Cumberland coalfields, red beds appear at the horizon of the *A. cambriense* Marine Band (mid-Westphalian C).

Later in the Westphalian, a major diachronous spread of red beds is recognized in the Midlands between the mid- and late Westphalian D, giving rise to two distinct, southerly-derived molasse sequences, whose lithostratigraphic nomenclature is currently poorly defined. The older of these, comprising the Halesowen Formation (Besly and Cleal, 1997), is of south-easterly derivation, composed mainly of recycled metasedimentary detritus derived from the deformation of the Rheno-Hercynian Zone. The younger unit, the Salop Formation (Besly and Cleal, 1997) consists of recycled Carboniferous and perhaps Devonian lithic material, probably derived from the erosion of the early phases of nappe emplacement in southern Britain. Both molasse units are characterized by the occurrence of caliche. The Halesowen Formation is of mid- to late Westphalian D age; the Salop Formation is probably not younger than Cantabrian (earliest Stephanian), although it might extend into the late Stephanian or Autunian (Wagner, 1983; Besly and Cleal, 1997).

Red beds are widespread from the mid-Westphalian D onwards in the Netherlands and Germany. In the Netherlands, they are known as the 'Barren Group' and are of Westphalian D to Stephanian age (NAM and RGD, 1980). A Stephanian age is widely accepted for the uppermost part of the red-bed sequence in north Germany (e.g. Hedemann *et al.*, 1984b). It is based, however, on a facies-dependent branchiopod and bivalve fauna, and a reinterpretation of the plant macrofossils suggests an age not younger than Cantabrian (Besly *et al.*, 1993). Although the base of the red-bed unit is diachronous (Schuster, 1971), broad patterns of diachronism and sediment provenance are not known. Such data as are available suggest that Westphalian D red beds in the Ibbenbüren area of West Germany were derived from the south. The sedimentology of the sequence is described by Selter (1990) and van der Zwan *et al.* (1992). It contains caliches and other features that allow a general comparison to be made with the Salop Formation of the UK onshore.

4.4 Carboniferous geology of the North Sea: introduction

Carboniferous rocks occur throughout the North Sea, forming extensions of known areas of occurrence in the adjoining land areas (see Fig. 4.1). The Carboniferous outcrop/subcrop pattern was established following the Saalian deformation event—the terminal Variscan event in the foreland area. Prior to this event, the whole area had constituted a single zone of deposition, albeit subdivided by more or less 'emergent' basement highs. Following the deformation, three broad areas of preserved Carboniferous rocks can be identified: the Forth Approaches and Moray Firth areas; the Central North Sea; and the Southern North Sea. These have distinctive patterns of facies development, post-Carboniferous evolution and general style of hydrocarbon prospectivity. Because of the differing prospectivity, the quantity and quality of data available differ markedly between the three areas. The Southern North Sea is the only area in which the Carboniferous has proved economic potential. For this reason, it is discussed in more detail than are the other areas.

4.5 Moray Firth and Forth Approaches areas

4.5.1 Stratigraphy and facies

Carboniferous sedimentary rocks are known from numerous well penetrations in the Western Platform of the Central North Sea and the Outer Moray Firth (Andrews *et al.*, 1990; Leeder and Boldy, 1990). The distribution of Carboniferous rocks (Fig. 4.4) reflects truncation of Variscan folds, and is related to basement fault trends of the Scottish Midland Valley in the Outer Moray Firth and to the Great Glen fault system in the Inner Moray Firth. At least in the Outer Moray Firth area, the Carboniferous sediments show a concentric pattern of subcrop, implying a

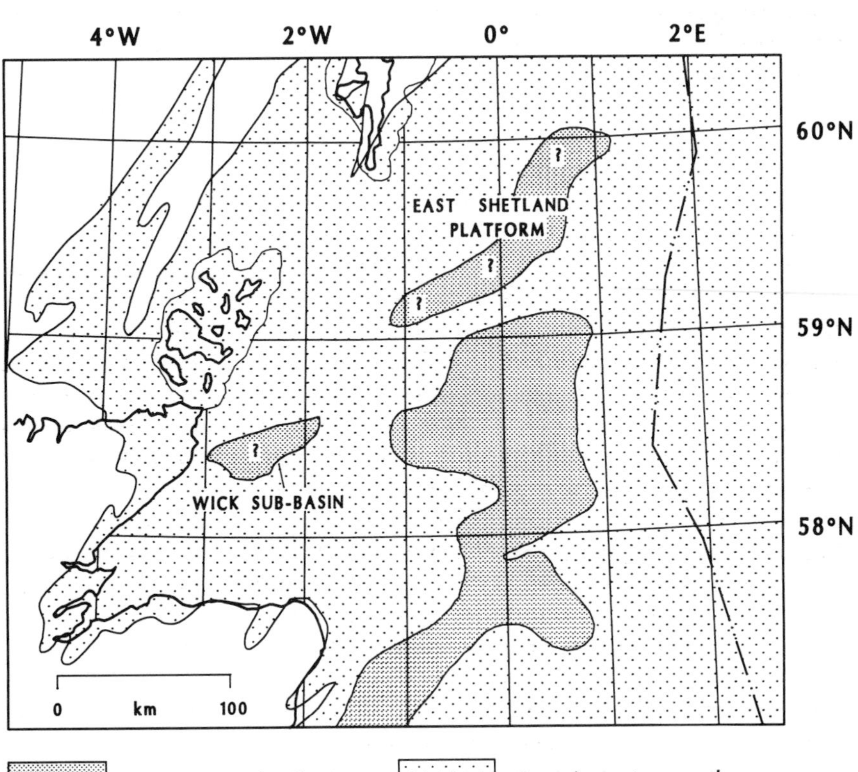

Fig. 4.4 Pattern of Carboniferous subcrops at the pre-Permian erosion surface derived from seismic interpretation: Moray Firth and adjacent areas. After Halliburton non-proprietary report.

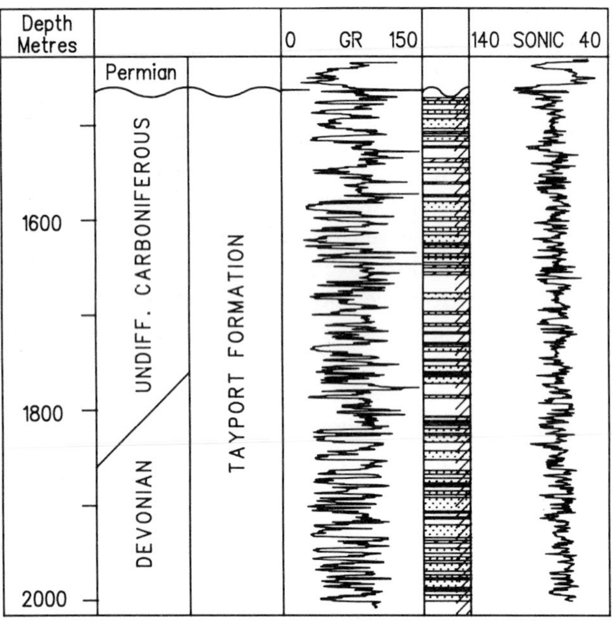

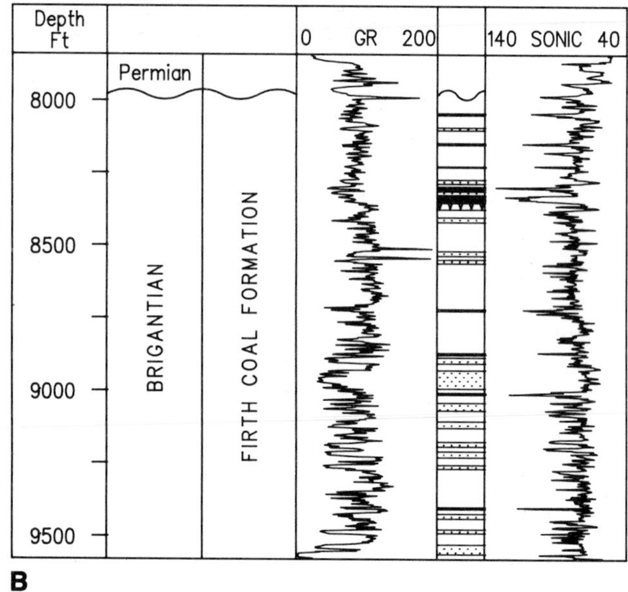

Fig. 4.5 Wireline logs of representative Carboniferous sequences in the Outer Moray Firth Basin. Diagonal ornament indicates red beds. Biostratigraphy after Halliburton (A); after Andrews *et al.* (1990) and Leeder and Boldy (1990) (B, C); and from completion log (D).

broadly synclinal distribution (Leeder and Boldy, 1990). The distribution of Carboniferous rocks in the Moray Firth area is partly affected by erosion of the crestal areas of later horst structures (Andrews *et al.*, 1990). Two isolated Carboniferous penetrations have been made in the South Viking Graben, in wells 16/12a-13 and 9/13a-22. Red beds of Westphalian age are reported from the latter (Cameron, 1993a). These occurrences support the speculation of Hedemann (1980) that Carboniferous rocks may have a wider than generally recognized distribution below the Viking and Central Graben systems. Carboniferous rocks are also present in the Forth Approaches Basin, where they form an offshore extension of the Scottish Midland Valley province, both in structural style and in facies. The extent of Carboniferous rocks further to the east and south is not known. They are generally absent from the southern part of the Western Platform area (Quadrants 27, 28 and southern part of 29).

In the Inner Moray Firth area, released well data are restricted to one penetration of olivine basalts and tuffs in well 12/23-1, which have been dated in the range 340–308 million years (Ma) (Visean to Westphalian). Seismic data suggest the preservation of a Carboniferous sedimentary sequence in the Wick Subbasin. No well penetration is yet released, but well 12/16-1 is known to have penetrated a section of Dinantian age (Owens *et al.*, 1995).

The sedimentary succession in the Outer Moray Firth consists of two distinct elements: a fluvial red-bed facies; and a widespread coal-bearing facies. These have recently been named as the Tayport and Firth Coal Formations, respectively (Cameron, 1993a), previously applied lithostratigraphic names (e.g. in Leeder and Boldy, 1990) being ambiguous. As in the onshore area, fluvial sediments of Old Red Sandstone facies extend up into the Lower Carboniferous, forming the Tayport Formation. Minor intercalations of carbonates in the red beds at the base of the Carboniferous may be local representatives of the 'cementstone' facies found in the Tournaisian of the Southern North Sea (well 14/8-1; Fig. 4.5A). Red-bed deposition may have persisted later than in the onshore area. The oldest ages obtained from the Firth Coal Formation are Arundian–Holkerian, and a Visean flora is recorded from the underlying Tayport Formation in the Buchan oilfield area (Hill and Smith, 1979). The Firth Coal Formation ranges in age from mid-Visean (Arundian–Holkerian) to mid-Namurian, the youngest rocks, of Namurian B age, having been found in well 20/4-2. The majority of penetrations are of Asbian to Brigantian age.

The detailed lithostratigraphy of the Carboniferous sequence remains poorly known, existing penetrations not allowing a detailed regional correlation. A minimum thickness of at least 1500 m is suggested (Andrews *et al.*, 1990). Data are insuffïcent to demonstrate the pattern of subsidence, although Thomson and Underhill (1993) suggest that extensional basin formation occurred in this area in a manner analogous to that documented in the onshore areas of northern England. Volcanic rocks are recognized in cuttings in many wells. It is possible that they may form a poorly defined regional-marker horizon in the upper part of the Dinantian sequence (Leeder and Boldy, 1990).

The sedimentary facies of the Firth Coal Formation are very similar to those of the alluvial/deltaic coal-bearing sequences in the Scottish Midland Valley, showing varying degrees of cyclic development of shale, siltstone, sandstone and coal (Fig. 4.5B–D). Log shapes show developments of both upward-coarsening progradational units and upward-fining, presumably channelized units. Coals locally form an

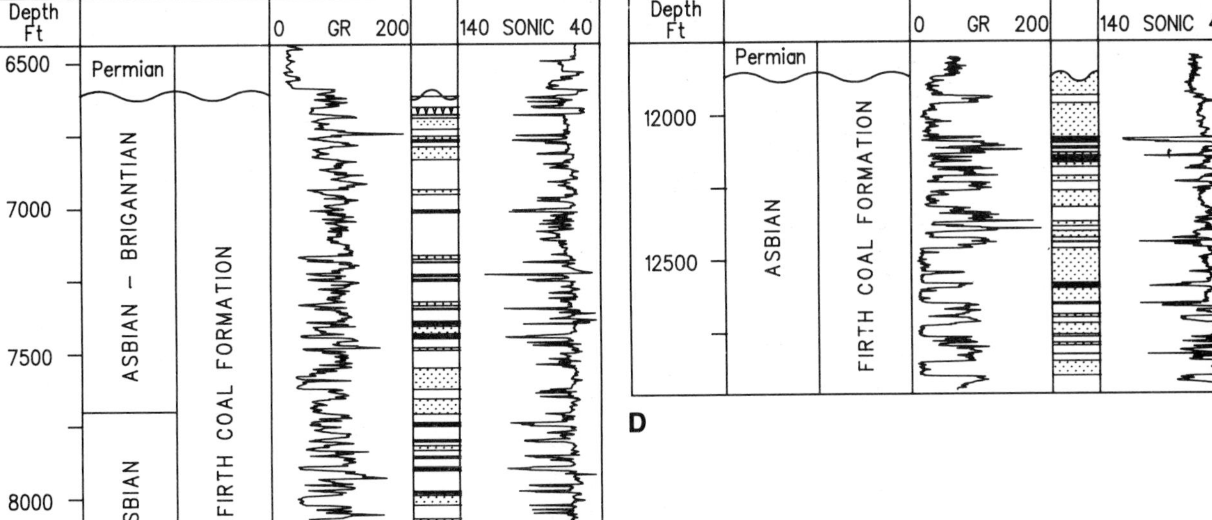

Fig. 4.5 (*Continued*)

important component of the sequence, being particularly thick and abundant in the Asbian. In well 15/19-2, seams of up to 4.3 m thickness are present. The pattern of sediment provenance is not completely documented. Leeder and Boldy (1990) infer local derivation of sediment from the Scottish highlands. The eastward increase in sand content demonstrated in their paper, and also observed in wells 20/10a-3 and 21/12-2b, suggests that the regional system fed by the Viking Graben (see Section 4.3.2 above) also contributed to the fill of these basins.

The extent of development of oil-source rocks remains unknown, but may be the subject for speculation on the basis of comparisons with the succession found in the onshore Midland Valley of Scotland. Rich lacustrine oil-source rocks are developed in sandstone-poor sequences in a small area of the Central Midland Valley, to the west of Edinburgh. The persistence of a lacustrine oil-source rock facies in this area was due to a combination of palaeogeographic constraints, in the form of volcanic edifices and structural highs, which led to a long-lived lacustrine system (Loftus and Greensmith, 1988). At the eastern end of the Midland Valley, in east Fife, coeval sequences contain high proportions of sandstone and a very poor development of lacustrine facies. A comparable contrast in facies has been recognized in the Moray Firth (Leeder and Boldy, 1990). Coals and sandstones are more abundant in the east of the Moray Firth area than in the west, suggesting the existence of a long-lived lake in the west of the area. Some of the shales in the western area have gamma-radiation levels in excess of 100 API and may have high organic contents.

In the Midland Valley, Upper Carboniferous rocks are preserved in major north–south-trending synclinal structures. A similar synclinal structure has been mapped in the offshore area in the Forth estuary (British Geological Survey (BGS) 1 : 250 000 Tay Forth sheet), and is inferred to extend north-eastwards into Quadrant 26 (Hedemann, 1980; see Fig. 4.4). As in the Scottish Midland Valley, the distribution and structure of the Carboniferous reflects the truncation of Variscan folds beneath the Saalian Unconformity, and is apparently controlled by the Midland Valley bounding faults. Well 26/7-1 (Fig. 4.6) penetrates 1090 m of Lower Carboniferous (Arundian–early Asbian) coal-bearing clastics, confirming the eastward continuation of the typical facies of the onshore area. The lower part of this succession is dominated by stacked fluvial-channel deposits with minor associated coal. These facies are identical to the onshore Calciferous Sandstone Measures (e.g. Cameron and Stephenson, 1985). Their allocation by Cameron (1993a) to the Tayport Formation seems misleading.

4.5.2 Petroleum geology

Although coals are abundant, they are generally of high-volatile bituminous rank and immature for gas generation. Higher maturities may be expected in the more deeply buried areas below the Mesozoic graben and in the south-eastern part of the subcrop area, in Quadrant 21.

Carboniferous sandstones form a minor reservoir unit for Jurassic-sourced oil in the Central Area of the Claymore Field (Maher and Harker, 1987; Harker *et al.*, 1991, Fig. 4). The sandstones are fine- to coarse-grained, and are extensively cemented by ferroan dolomite, which is in part replacive. Porosities and permeabilities are low to moderate (porosity 14–23%, permeability < 0.1–500 millidarcies (mD)). The low permeability, low sand-to-shale ratio and poor productivity have formerly rendered such sandstones

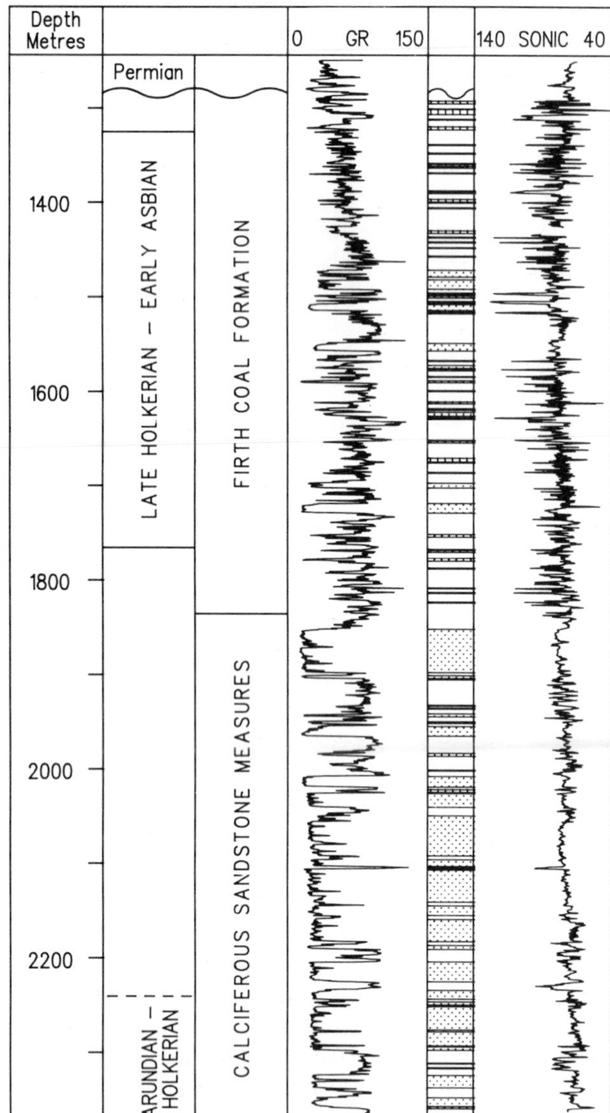

Fig. 4.6 Carboniferous succession found in the Forth Approaches Basin. Biostratigraphy from completion log.

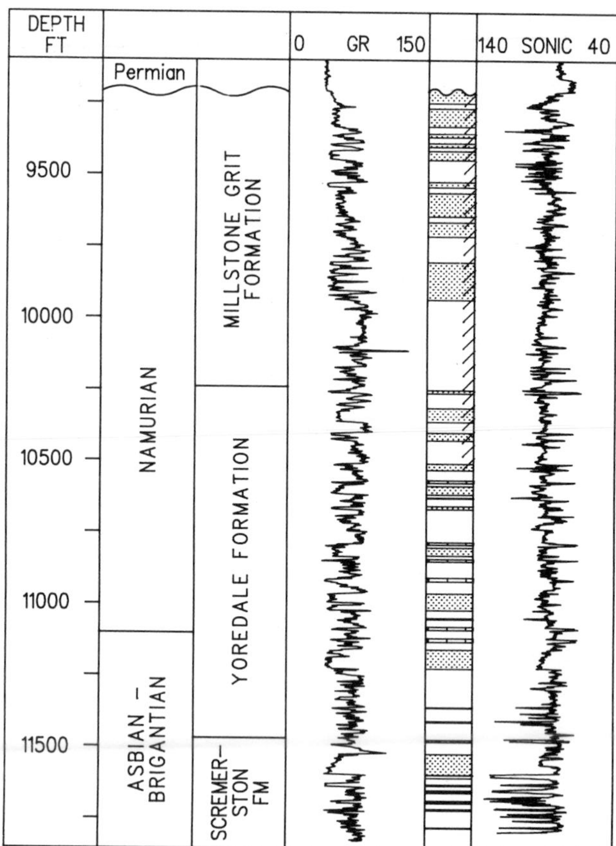

Fig. 4.7 Representative well penetration of Dinantian and Namurian, southern Central Graben. Diagonal ornament indicates red beds. Biostratigraphy by Halliburton Reservoir Description Services.

unattractive as potential reservoirs. There is a possibility that tectonic controls on sedimentation may result locally in concentration of fluvial-channel sands into zones of better reservoir potential: such a control has been described from the early Namurian in the Scottish onshore area by Read and Forsyth (1989).

The Carboniferous accumulation in Claymore is in a later structure, of Mesozoic age. Little is known of the potential for intra-Carboniferous accumulations in Variscan structures. The Variscan structure of the onshore area suggests that they may, in general, be rather small and complex (e.g. Hallett *et al.*, 1985; Scott and Colter, 1987).

The only well drilled to test Carboniferous prospectivity in the Forth Approaches Basin (26/7-1) was dry. Adjoining tests of Carboniferous-sourced plays relying on Devonian and Devonian/Rotliegend reservoirs (26/12-1 and 26/14-1, respectively) have also been dry. Analyses of the failure of these plays are not published, but problems of source-rock maturity and sealing may be inferred.

4.6 Mid North Sea High and south Central Graben

4.6.1 Stratigraphy and facies

Carboniferous rocks appear to have a much wider distribution in the central part of the North Sea than has hitherto been realized. Stratigraphic revision has allowed Carboniferous ages to be identified or inferred in a number of formerly undated pre-Permian well penetrations in the south Central Graben area (30/30-1, 30/30-2, 31/27-1, 39/2-1, 39/7-1; Norway 2/10-1; Denmark P-1, Gert-2) (Ofstad, 1983; Nielsen and Japsen, 1991). The pattern of subcrop in this area is uncertain, and it has been suggested (e.g. by Hedemann, 1980) that Carboniferous rocks may have a wide distribution beneath the deep parts of the Central Graben. Biostratigraphic data are sparse, and lithostratigraphic interpretation is partly by analogy with the succession proved in the Moray Firth area. A fourfold lithostratigraphy may be inferred, the upper 3 units of which share the names of units better defined in the Southern North Sea (Cameron, 1993b). These are: a sandstone-rich Tournaisian unit, a coal-bearing mid–late Dinantian unit (Scremerston Formation), a late Dinanatian to early Namurian clastic unit containing minor marine limestones (Yoredale Formation) and a sandstone-rich mid-Namurian unit (Millstone Grit Formation) developed in red-bed facies (Fig. 4.7). It is possible that the sandstone-rich red-bed unit forming the reservoir of the Embla Field may belong to the Carboniferous (Knight *et al.*, 1993), although a Devonian age seems more likely (see Chapter 3).

To the west of the Central Graben, in the Mid North Sea High, Carboniferous subcrops can be inferred from seismic data in Quadrants 26–28 and 34–38. Like the Southern Uplands High in the onshore area, of which it is a continuation, the Mid North Sea High was originally a Caledonian High, which was reactivated during the Variscan, and was probably a site of major condensation of the Carboniferous sequence. None the less, both areas also contain Carboniferous subbasins that have escaped denudation during subsequent Variscan reactivation of the high. The subbasins have fills of varying degrees of stratigraphic completeness, and are proved in the offshore by gravity studies (Lagios, 1983) and by a few well penetrations (36/13-1, 36/23-1, 37/10-1, 38/16-1). Palynological reappraisal of two well penetrations formerly regarded as Lower Carboniferous have shown that they may in part be of early Namurian age (36/23-1, 36/26-1). It is likely that some red-bed penetrations currently regarded as being of Devonian age can be reinterpreted to be of later Dinantian age, affected by pervasive penetrative oxidation. An example is provided by well 37/10-1. The mapping of Corfield *et al.* (1996) suggests that all these wells penetrate the shallow parts of fills of Dinantian half-grabens. Gravity studies (Donato *et al.*, 1983) also suggest the presence of at least three granites, of inferred Devonian age, which may, by analogy with the onshore sequence, have controlled the location of the bounding faults of these half-grabens.

4.6.2 Petroleum geology

The Dinantian (Asbian–Brigantian) coal-bearing unit is not mature for gas generation where penetrated in wells 31/26-1, 39/2-1 and 39/7-1. Maturities, however, do increase to the east. This unit is likely to act as a gas-source rock in the Danish and northern Netherlands Central Graben areas, where Carboniferous gas may charge younger reservoirs. Possibilities exist for reservoir development in Carboniferous sandstones, either charged from Carboniferous sources or, in structurally high locations, hosting hydrocarbons derived from deeply buried Jurassic source rocks in adjacent graben areas, in a manner analogous to the Devonian reservoir of the Argyll Field (Robson, 1991). Away from areas that have been affected by Mesozoic and Tertiary burial around the Central Graben, it is likely that the Carboniferous is immature for hydrocarbon generation, and prospectivity is low.

4.7 Southern North Sea

4.7.1 Tectonic framework

It is clear from gravity modelling (Collinson *et al.*, 1993) and seismic interpretation (Leeder and Hardman, 1990; Corfield *et al.*, 1996) that the Lower Carboniferous extensional terrain extends eastwards into the North Sea and the European mainland, zones of low subsidence being controlled by the strong north-west–south-east Caledonian basement grain and by the location of Caledonian plutons (Leeder and Hardman, 1990). Because of ambiguities in interpretation and modelling, there is not yet a consensus view of the disposition of horst and graben areas, and few data are available for the Netherlands sector. A compilation, based largely on Corfield *et al.* (1996), is shown in Fig. 4.8. There is agreement that 'shelf' areas, in which sequences are comparatively thin, include the South Hewett Shelf, lying to the south and west of the South Hewett Fault, and areas in the northern parts of Quadrants 42 and 44, forming the offshore extension of the Alston Block and the southern edge of the Mid North Sea High, respectively. Elsewhere, interpretations differ, although there is a measure of agreement as to the location of horst structures in Quadrants 47 and 48 and of basinal areas in Quadrant 42 (extension of the onshore Cleveland Basin), the northern part of Quadrant 43, the central part of Quadrant 48 and much of Quadrant 49. The Dowsing Fault, which forms a prominent structural zone in the Mesozoic overburden, is not recognizable in the pre-Zechstein sequence. Many of the other major fault lines affecting the Mesozoic and Tertiary succession do, however, appear to represent reactivation of faults that bounded elements in the Carboniferous rift province.

Ambiguities in interpretation of controls on the location of thick Carboniferous sequences result from the presence both of Dinantian graben fills and of extremely thick Silesian sequences. The two do not necessarily coincide. An example of the lateral variation that may be expected is seen in the Northumberland Trough in the onshore area (Fig. 4.8). This feature, which passes eastwards into the offshore area, is a markedly asymmetrical, southwards-dipping half-graben. It is bounded to the south by the Alston Block, a Dinantian horst cored by Devonian granite. The Dinantian sequence thickens northwards from *c.* 400 m of Asbian and Brigantian on the Alston Block to a complete Dinantian sequence of more than 4000 m in the hanging-wall area of the half-graben bounding faults (Kimbell *et al.*, 1989). Seismic data show clearly that pre-Asbian deposition was restricted to the half-graben area, with a uniform transgression of the horst area in the Asbian, after which continued fault movement led to a continuing thickness contrast. This area of very marked differential subsidence in the Dinantian is overlain by a comparatively thin Silesian sequence (Ramsbottom *et al.*, 1978; Leeder, 1982). Similar contrasts in Dinantian thickness and facies are known to strike out into the offshore area in the eastern extension of the Stainmore Trough (Fig. 4.8), proved by the exceptionally thick sequence (3443 m) of Arundian to Late Brigantian sediments penetrated in the Seal Sands-1 well near Middlesbrough (Dunham and Wilson, 1985, p. 14), and are inferred from onshore exploration data in the Cleveland Basin in North Yorkshire and the 'Humber Basin' in south Yorkshire and Lincolnshire (Fraser *et al.*, 1990; Fig. 3).

4.7.2 Carboniferous subcrop pattern

Difficulties in seismic imaging and the lack of well control in many areas cause problems in establishing an unambig-

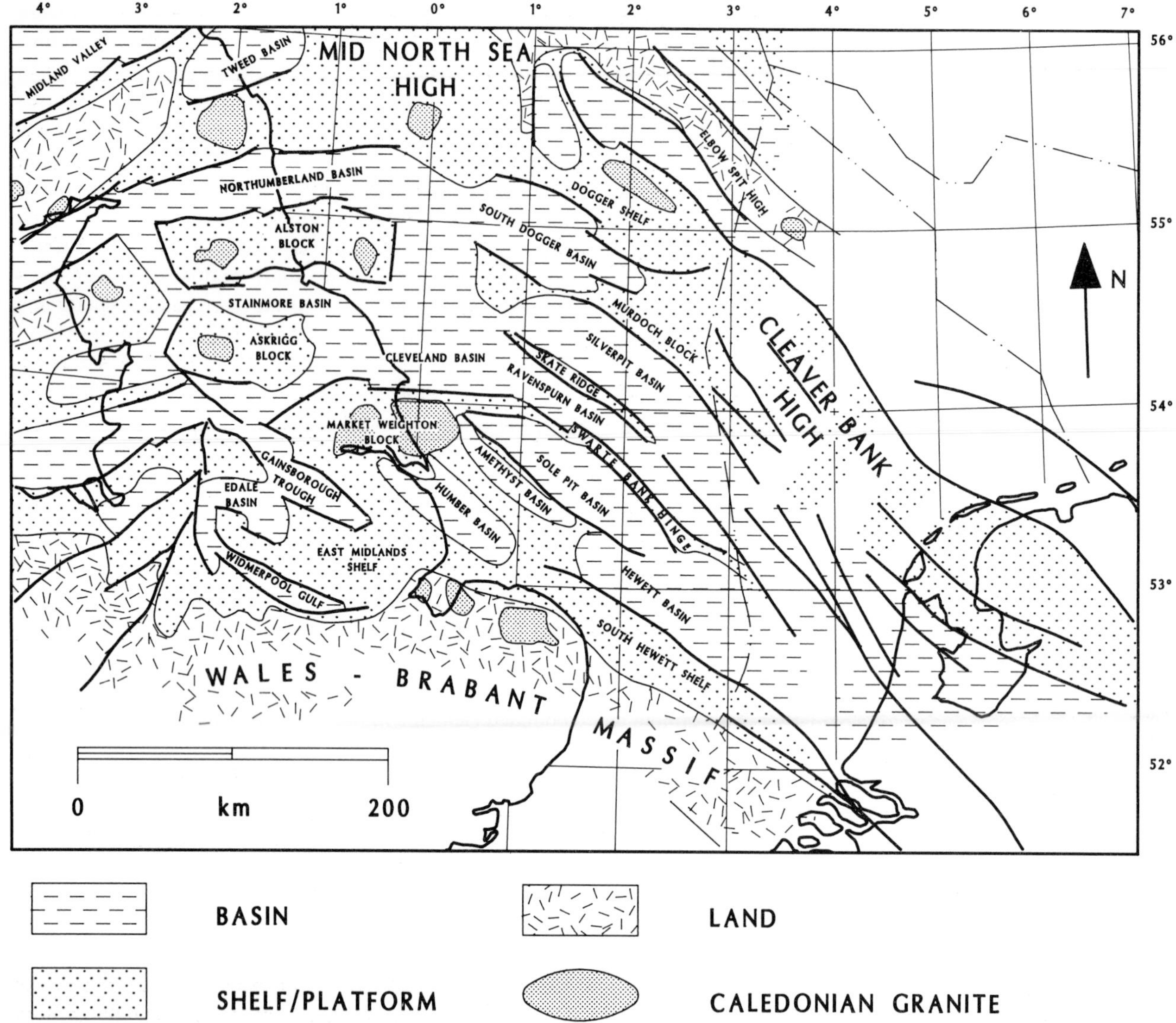

Fig. 4.8 Structural framework of lower Carboniferous graben and half-graben development. 'Highs' are located mainly over Devonian granite masses. UK area mainly after Corfield *et al.* (1996). Onshore Netherlands after Ziegler (1990a). Structural trends in Netherlands offshore area from Quirk (1993); positions of highs in this area are schematic.

uous map of the subcrop of the Carboniferous beneath the Saalian Unconformity. Maps published since the advent of modern seismic data (Leeder *et al.*, 1990a; Cameron *et al.*, 1992; Bailey *et al.*, 1993; Oudmayer and Jager, 1993) are mutually incompatible and do not cover the Netherlands sector, for which only the generalized compilation of Ziegler (1990a) is available. The subcrop map illustrated here (Fig. 4.9) represents a generalized compilation.

The subcrop pattern reflects the extent of inversion associated with the terminal Variscan events. In the extreme south of the area, in UK Quadrant 53, Carboniferous sediments onlapping the rigid Wales–Brabant Massif were little affected by Variscan folding, and the Westphalian D sequence subcrops the pre-Permian erosion surface. To the north, moderate to severe inversion of the Lower Carboniferous half-graben fills has resulted in a series of major anticlines and synclines. The subcrop consists mainly of Westphalian A and B Coal Measures, although locally Namurian and Dinantian inliers are present. Less severe inversion is apparent in UK Quadrants 42 to 44, where the Variscan deformation involved less uplift and large areas of Westphalian C to Stephanian subcrop the erosion surface. In general, areas affected by major Variscan inversion were also the sites of significant Mesozoic and Tertiary uplift. The extent of involvement in the post-Variscan inversions exercises an important control on the present levels of organic maturation of the Carboniferous.

4.7.3 Lithostratigraphy and facies

The lithostratigraphic nomenclature of the Carboniferous in the Southern North Sea is summarized in Fig. 4.10, and the facies architecture of the basin fill is summarized in Fig. 4.11.

Dinantian

The stratigraphy of the Dinantian in the Southern North Sea is poorly known from scattered well penetrations, supplemented by published seismic interpretations in the extreme south. Two broad areas of facies development can be recognized. In the northern and central parts of the area, the Dinantian is developed in a dominantly clastic facies, resembling the sequences seen onshore in the coastal sections of Northumberland. A section spanning the whole

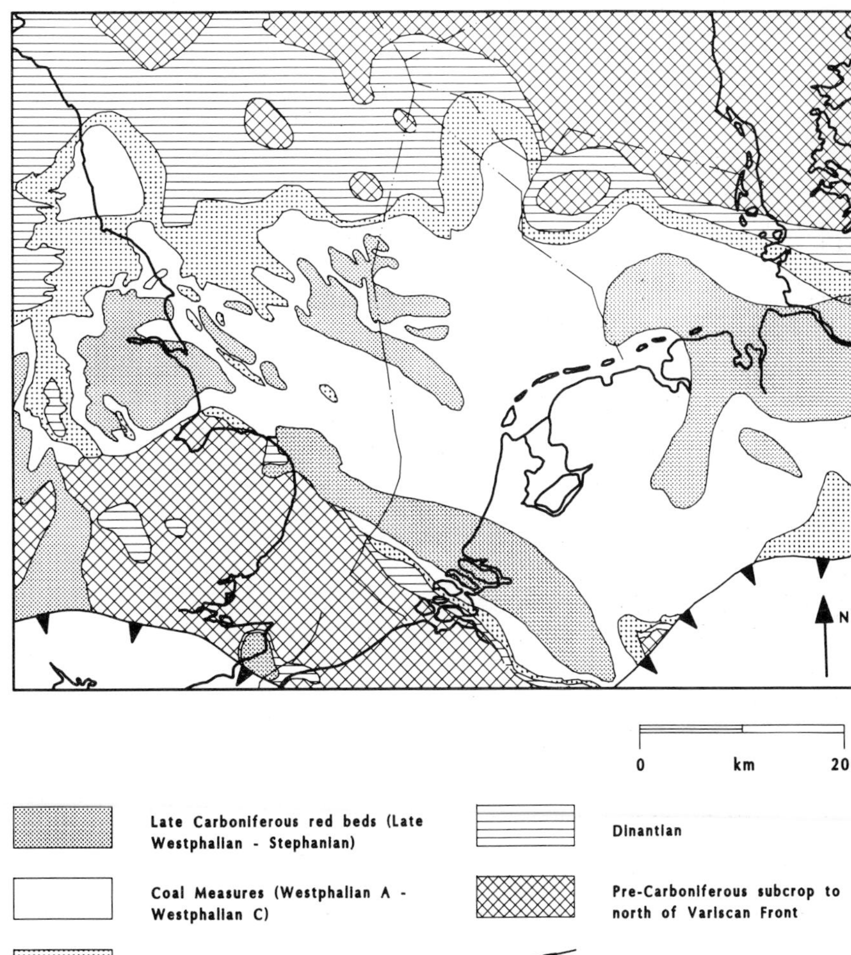

Fig. 4.9 Pattern of Carboniferous subcrops at the pre-Permian erosion surface, Southern North Sea and adjacent areas (compiled from Bailey *et al.*, 1993; Cameron *et al.*, 1992; Corfield *et al.*, 1996; Halliburton non-proprietary reports; Leeder and Hardman, 1990; van Wijhe and Bless, 1974; Ziegler, 1990a; maps and memoirs of British Geological Survey.) cf. Fig. 2.15.

of the Dinantian in this area is seen in wells A/16-1 in the Dutch sector (Fig. 4.12A) and 44/2-1 in the UK sector (Fig. 4.12B). The lithostratigraphic nomenclature of the onshore area of northern England is directly applicable. In 44/2-1, the base of the succession consists of continental red beds, of presumed late Devonian to Tournaisian age, which pass upwards into the Cementstone Formation, a late Tournaisian to earliest Visean marginal-marine sequence of fine clastics interbedded with limestones and dolomites. The overlying sequence in 44/2-1, of Arundian to early Brigantian age, comprises the Fell Sandstone and Scremerston Formations, consisting of thick, stacked, sandstone bodies, deposited by major braided fluvial or deltaic distributary channels, associated with coals in the upper parts. The highest part of the Dinantian sequence is penetrated in A/16-1. This penetration partly overlaps that of 44/22-1. The lower part of the sequence contains fluviodeltaic coal-bearing strata of the Scremerston Formation, developed here in a less sand-prone facies than in 44/2-1. This passes upwards into an interbedded clastic/carbonate sequence, comparable to that found in the Yoredale association of northern England and named the Yoredale Formation in the offshore area. Penetrations in 41/8-1, 41/20-1, 42/10a-1 and 44/7-1 suggest that Dinantian facies, at least in the Brigantian, are similar over most of the northern part of the Southern North Sea. To the south, in Quadrants 42–44, Brigantian sediments are developed in basinal mudstone facies (Fig. 4.13A), with minor turbiditic sandstones and limestones. The extent of these facies is as yet poorly defined, but they clearly represent a situation analogous to the basinal succession of the Bowland Basin in the northern onshore area in England.

In the extreme south of the Southern North Sea, on the South Hewett Shelf, the Dinantian is developed in a strongly contrasting marine-carbonate facies, best seen in UK wells 53/12-2, 53/12-3, and 53/16-1 and Dutch well S/5-1, and named the Zeeland Formation. The best penetration in the UK sector is in well 53/12-2, where 450 m of-dolomites and shallow-marine limestones are present (see Fig. 4.12C). The succession here is estimated, from seismic interpretation, to exceed 1 km in thickness (Cameron *et al.*, 1992), although it thins rapidly southwards on to the Wales–Brabant Massif through thinning and intra-Carboniferous erosion. These carbonate facies resemble those seen in the Midlands and North of England, where they are separated by basinal facies deposited in extensional half-grabens. It may be inferred that a similar pattern is developed under much of the Southern North Sea (see Fig. 4.11). There are, however, no released well penetrations over most of this area between latitudes 53° and 54°40′N, and the exact nature of the distribution of platform and basinal facies and of the lateral transition into the northern deltaic facies remains unknown.

Silesian—Namurian

Namurian facies and stratigraphy show a broad similarity to that of the UK onshore sequence. The northernmost

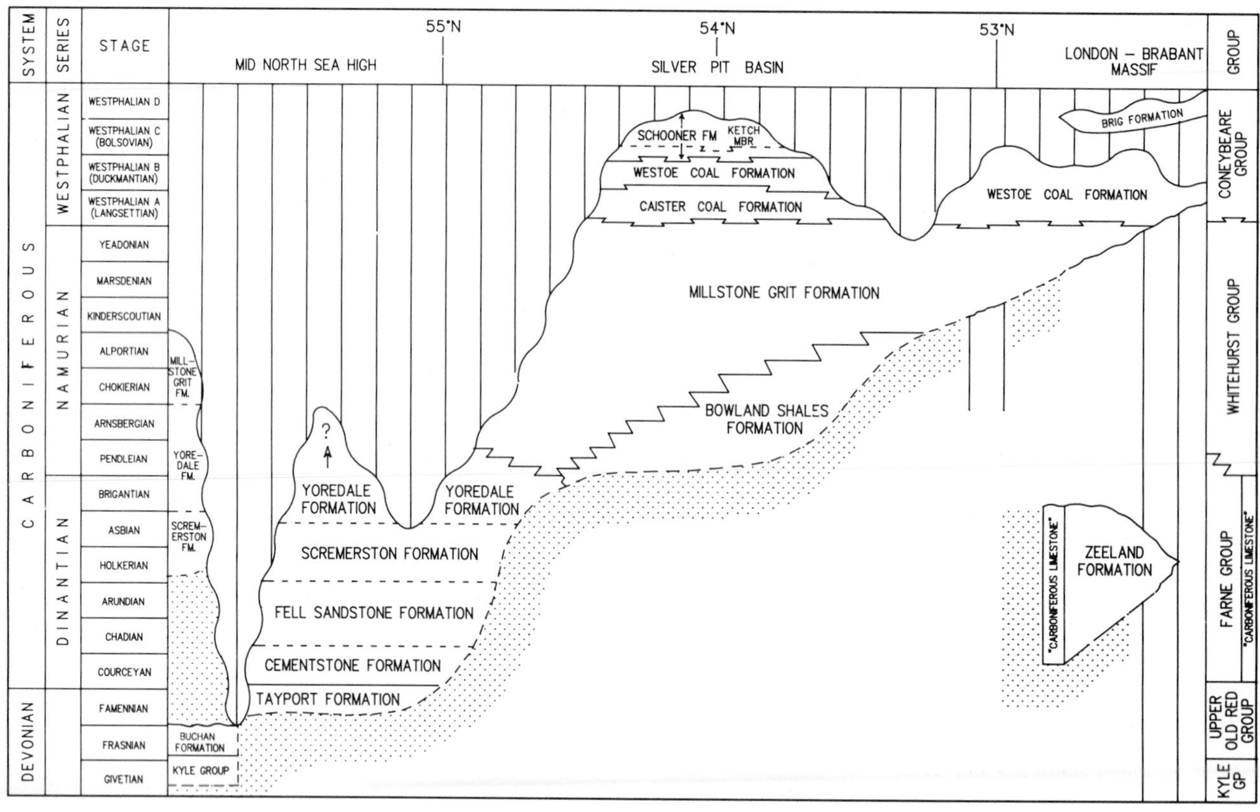

Fig. 4.10 General Carboniferous and Upper Devonian lithostratigraphy of the UK Southern North Sea (after Cameron *et al.*, 1993b).

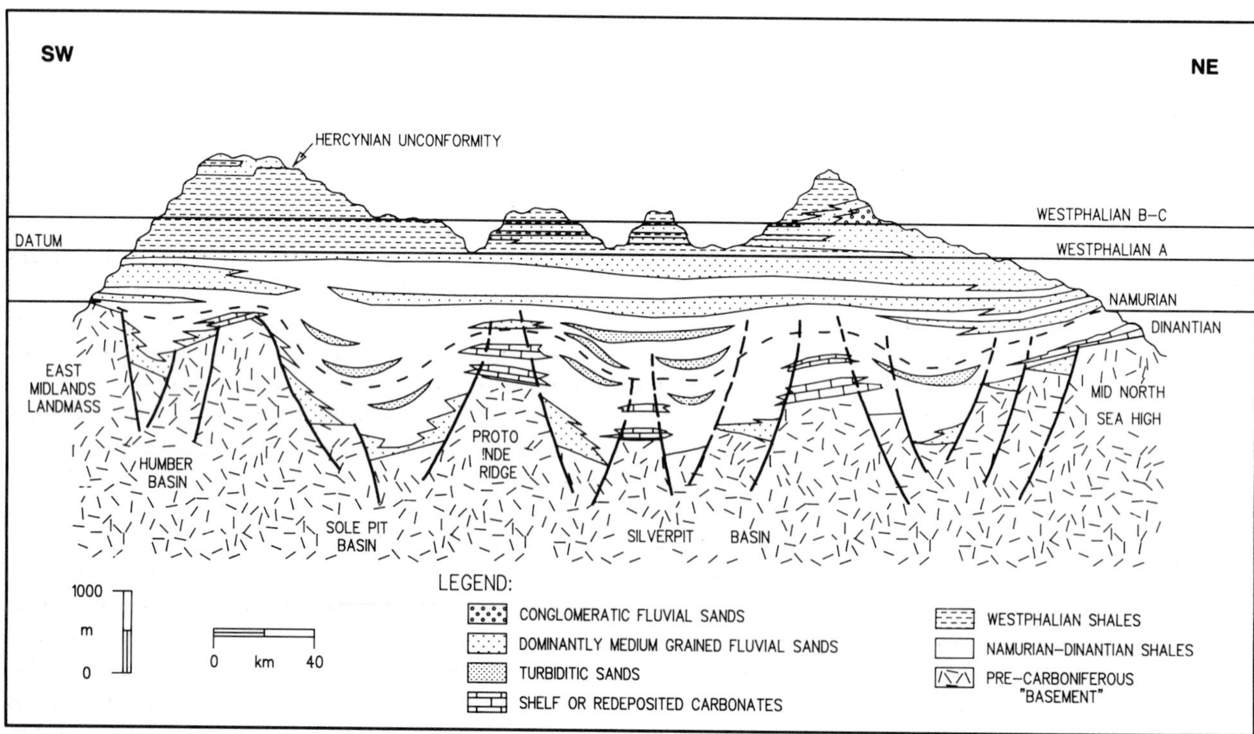

Fig. 4.11 Schematic stratigraphic cross-section of the Carboniferous, UK Southern North Sea, illustrating relationship of facies architecture to basement structure and highlighting reservoir development (after Bailey *et al.*, 1993).

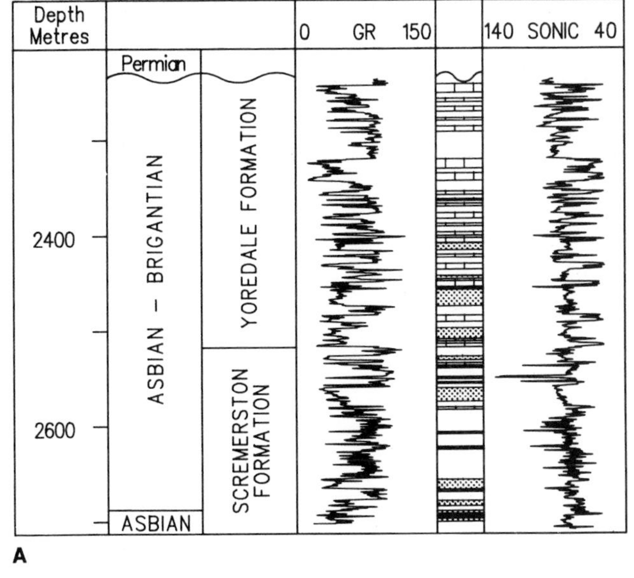

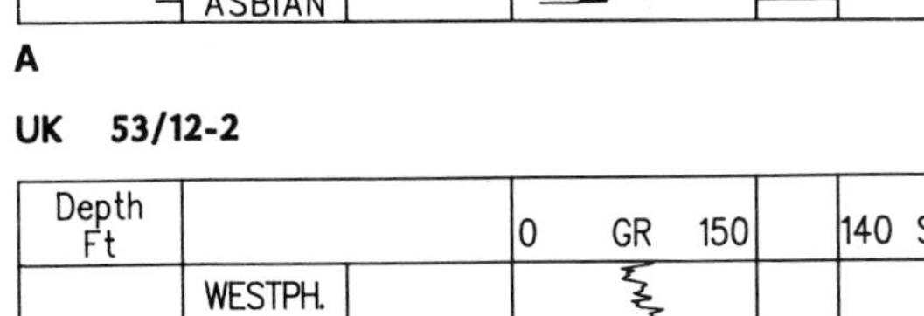

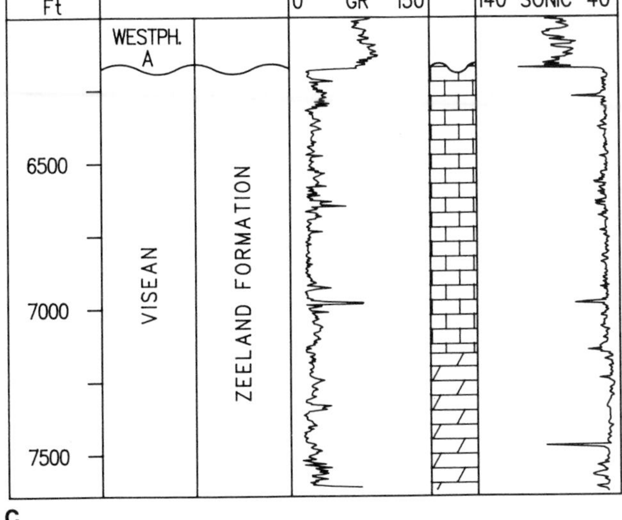

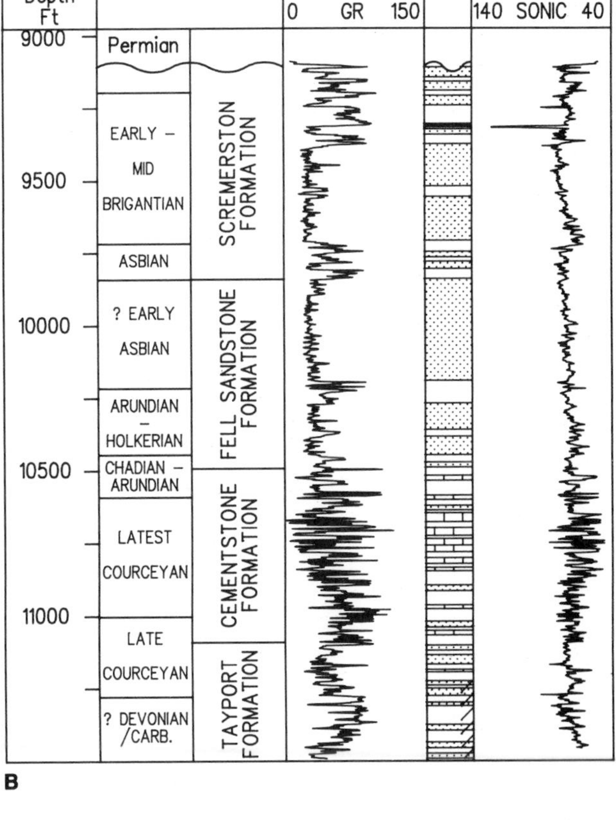

Fig. 4.12 Wireline logs of representative Dinantian (Lower Carboniferous) sequences in the Southern North Sea Basin. A and B. Mixed carbonate/clastic facies, Silverpit Basin. C. Carbonate platform facies, South Hewett Shelf. Diagonal ornament in B indicates red beds. Biostratigraphy by Halliburton Reservoir Description Services.

penetrations, in wells in Quadrant 36 and in wells 42/13-1 and 43/3-1, show early Namurian deltaic facies. The sequence in well 42/13-1 contains an Arnsbergian shallow-marine limestone, demonstrating an eastward extension of the 'Yoredale' facies. These rocks are similar to those in the underlying Brigantian, and are allocated to the same lithostratigraphic unit, the Yoredale Formation. In the remainder of the Southern North Sea area, the whole of the Namurian falls into one major lithostratigraphic unit—the Whitehurst Group—which encompasses two major, in part laterally equivalent, lithological units:

1 The Bowland Shale Formation, the older unit, consisting of basinal mudstones and turbidites, and representing the infill of residual topography inherited from the Dinantian rift system.

2 The Millstone Grit Formation, a sandstone-dominated succession deposited in a complex of shallow-water deltas, which succeeded the major basin-filling episode.

The relationship between the two formations implies a north-to-south pattern of progressive shallowing, similar to that described by Collinson (1988) in the UK onshore. Infill of deep water by turbidite and slope sequences occurred in the Pendleian in the Silverpit area (well 43/17-2: Fig. 4.13A; Collinson *et al.*, 1993) and the offshore Durham area (well 41/24a-2: Fig. 4.13B), while shallowing occurred slightly later in northern Quadrant 48 (well 48/3-3; Collinson *et al.*, 1993). A significant feature of the shallow-water deltaic sequences of Namurian B–C age is the occurrence of anomalously thick, stacked, channel-sandstone bodies (e.g. 11 575–11 775 ft in well 43/17-2: Fig. 4.13A). Recent fieldwork in the outcrop areas of northern England, coupled with the rigorous subsurface study of Church and Gawthorpe (1994), suggests that these bodies represent the deposits of braided fluvial channel

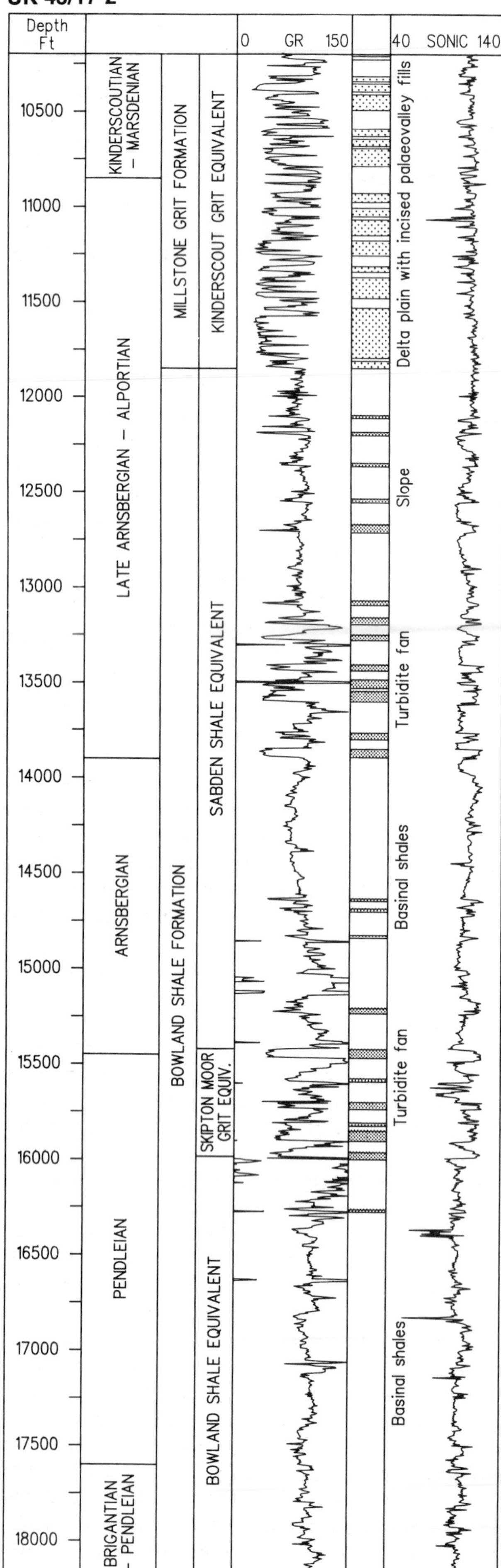

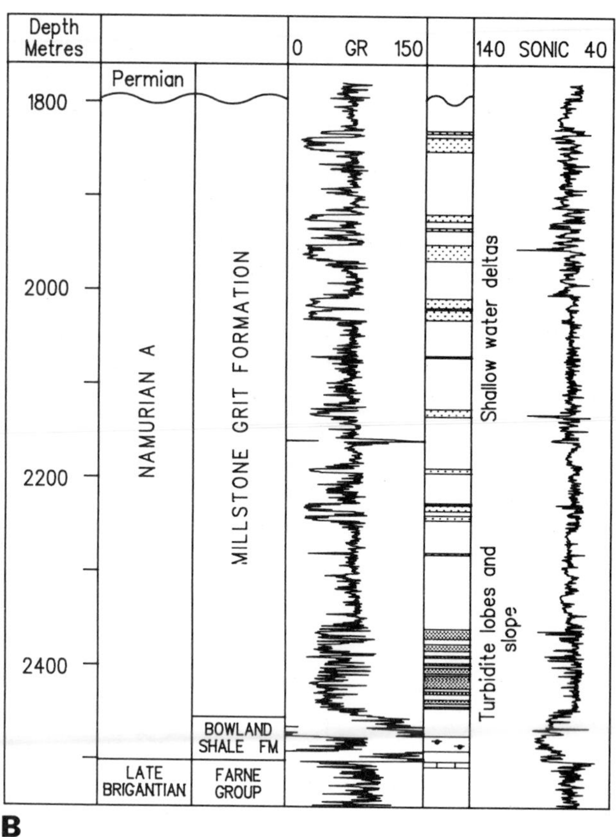

Fig. 4.13 Wireline logs of Namurian (Silesian) sequences in the Southern North Sea Basin. A. Representative sequence, comprising basinal shales and turbidites, a major shallowing-upward sequence, and shallow-water delta deposits containing one or more phases of palaeovalley incision. B. Early Namurian sequence, consisting of basinal sequence characterized by potential source-rock development, overlain by shallow-water delta deposits. Fine stipple indicates inferred turbidite units. Biostratigraphy by Collinson Jones Consulting (A) and Halliburton Reservoir Description Services (B).

systems that backfilled palaeovalleys incised during eustatic lowstands.

To the south, shale-dominated Namurian sequences are present in well 47/29a-1, where the Namurian onlaps onto Lower Palaeozoic at the northern edge of the Wales–Brabant High, and in Netherlands wells P/10–1 and Rijsbergen-1 (Fig. 4.14). These represent distal positions in the basin, never reached by a major deltaic system. This area had been characterized by comparatively deep water and a lack of sandstone deposition for much of the Namurian (Ziegler, 1990a).

In the South Hewett Shelf the Namurian is absent, the Dinantian being overlain by Westphalian A rocks. This situation is analogous to that found in the onshore East Midland area, where there was a phase of uplift during the Namurian (Strank, 1987). An interpretation published by Tubb *et al.* (1986), based on seismic data, suggests that the Namurian onlaps the northern edge of the Wales–Brabant High in the area of the South Hewett Fault.

Silesian—Westphalian

Westphalian coal-bearing rocks are known from a large number of well penetrations in both the UK and Dutch sectors. Currently released data do not, in most cases, allow reconstruction of total thicknesses for this association. The basal 100 m of the Westphalian generally comprises a sandstone-dominated sequence comparable to that in the topmost Namurian and is thus allocated to the Millstone Grit Formation. The remainder of the Westphalian is named the Conybeare Group, within which three broad lithostratigraphic units are recognized (Figs 4.11 and 4.14):

1 The Caister Coal Formation: a sandstone-rich, coal-bearing sequence, developed in the Westphalian A and earliest Westphalian B in the northern part of the Southern North Sea.

2 The Westoe Coal Formation: a coal-bearing sequence having a low sandstone content, forming the majority of the Westphalian B in the northern part of the Southern North Sea, and the whole of the Westphalian B and majority of the Westphalian A in the southern area.

3 The Schooner Formation: a sandstone-rich sequence, coal-bearing in its basal part, and passing upwards into primary red beds of the Ketch Member (see below).

The facies present in the coal-bearing associations are similar to those described from the Durham Coalfield by Fielding (1984a,b; 1986), consisting of lacustrine delta sequences showing upward-coarsening fills, capped by fluvial or distributary channel deposits and coals (Fig. 4.15). In some areas, the major channels were braided (Cowan, 1989). In the Silverpit Basin, these locally deposited, stacked, conglomeratic sand bodies can reach thicknesses of more than 30 m (Ritchie and Pratsides, 1993). Such thick sand bodies have few obvious analogues in the onshore sequence, with the possible exception of the basal Westphalian sandstone, which forms the reservoir in the Welton oilfield (Rothwell and Quinn, 1987).

In the early Westphalian A, sediment provenance was dominantly from the north, as in the Namurian. This resulted in a major southward decrease in the amount of sandstone in the sequence (see Fig. 4.14) and the lateral passage of the Caister Coal Formation into the Westoe Coal Formation. The relatively sand-poor nature of the Westoe Coal Formation also reflects the mid-Westphalian A change in dominant sand provenance in the basin (Rippon, 1996), which led to the Southern North Sea occupying a relatively more distal position in the regional sediment-transport path. Even in the more proximal areas in the north, some well sections show very low sandstone contents, suggesting that major distributary channels may have been localized by syndepositional faulting, in a manner similar to that inferred in the onshore sequence by Fielding (1984b) and Guion and Fielding (1988). Alternatively, the thick, stacked, channel-sandstone bodies in the proximal areas may be interpreted as a palaeovalley fill (Ritchie and Pratsides, 1993).

The red-bed sequence in the Southern North Sea has been reviewed by Besly *et al.* (1993). It is clear that the term 'Barren Red Group', generally applied to red beds occurring at the top of the Coal Measures sequence, encompasses three lithostratigraphically distinct units (Fig. 4.16):

1 Red beds developed by penetrative weathering beneath the sub-Permian unconformity.

2 A sequence of primary red beds of Westphalian C age, recognized in the northern part the UK sector and in the north-western part of the Netherlands sector.

3 A sequence of primary red beds with subordinate coal-bearing facies of Westphalian D age, recognized throughout the Netherlands sector and locally in the UK Silverpit area, and correlating with the red-bed sequence in the onshore Netherlands and western Germany.

Over most of the area, the uppermost 50 m or more of the Carboniferous sequence has been pervasively reddened by penetrative oxidation below the Permian desert land surface. Detailed petrographic studies by Cowan (1989) and Besly *et al.* (1993) show that permeable horizons up to 200 m below the unconformity have been reddened in this way.

The primary red beds in the Silverpit Basin are currently all named as the Ketch Member of the Schooner Formation (Cameron, 1993b), informally divided into lower and upper 'units', which represent the Westphalian C and Late Westphalian C/Westphalian D units, respectively (Fig. 4.17A). Relationships are complicated by the presence of a diachronous relationship between the Lower Ketch Unit red-bed facies and coal-bearing facies, demonstrated between Netherlands wells K/4-2 and K/5-1. In addition, an unconformity is locally developed between the red beds and the underlying Coal Measures (Leeder and Hardman, 1990; Corfield *et al.*, 1996). Folding contemporary with the deposition of the red beds led to onlap of the latter onto the partly eroded flanks of major fold structures, while conformable sequences occur in the adjoining synclines. The Westphalian C red beds of the Lower Ketch Unit seem comparable to the red beds of the same age found in the Midland Valley of Scotland and the Canonbie area of the Scottish borders. The later, Westphalian D, red-bed unit may be correlated with a red-bed unit named the Brig Formation (Cameron, 1993b), which occurs in the South Hewett Shelf area, overlying a dated Westphalian C coal-bearing sequence. This unit is in turn correlated with the onshore Westphalian D Keele Formation (Tubb *et al.*,

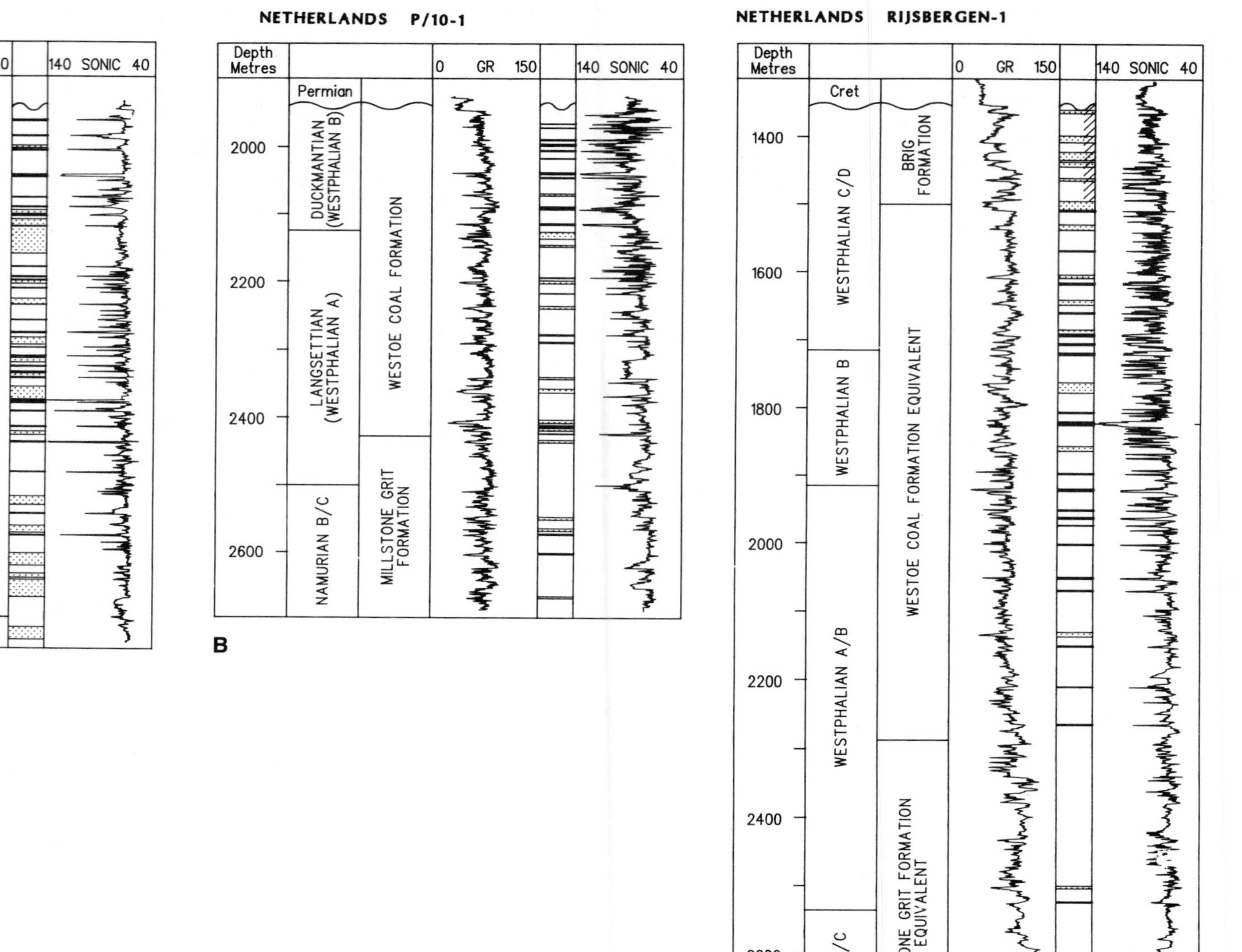

Fig. 4.14 Wireline logs of representative Westphalian A–B (Silesian) coal-bearing sequences in the Southern North Sea Basin, showing decrease in sand from north (left) to south (right). Logs of Rijsbergen-1 from NAM and RGD (1980). Biostratigraphy partly from NAM and RGD (1980) and partly by Collinson Jones Consulting.

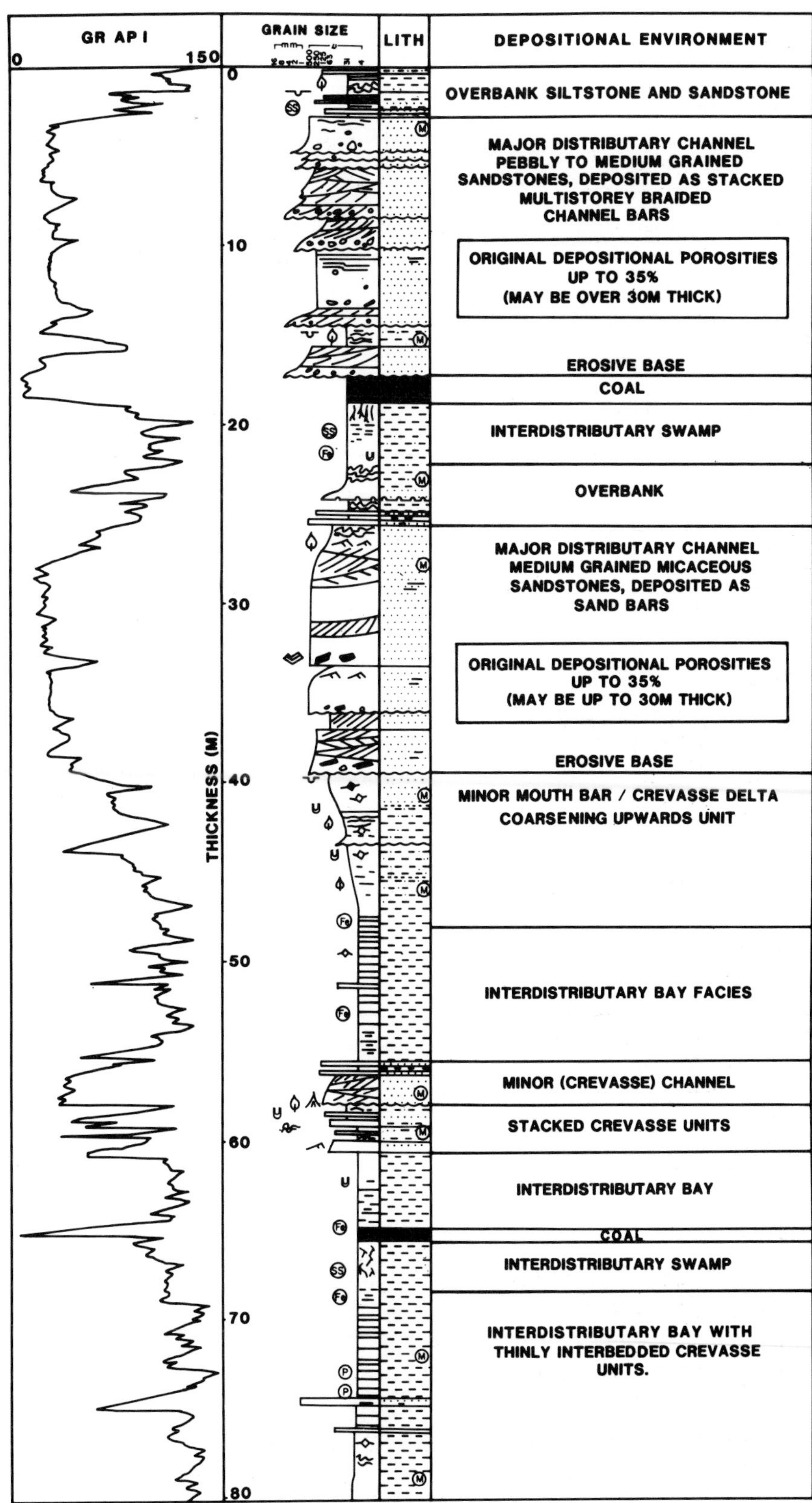

Fig. 4.15 Typical facies associations encountered in Westphalian A–C coal-bearing sequences, Southern North Sea. Generalized section based on penetrations in a number of unreleased wells in Silverpit Basin area. (From Cowan, 1989.)

1986)—now referred to as the Halesowen Formation (Besley, 1995; Besly and Cleal, 1997). More generally, the Westphalian D red-bed unit may be equated with the southerly- and easterly-derived Variscan molasse, which forms the SalopFormation of the UK onshore (Besly and Cleal, 1997). Thick successions of this younger unit are not present in the UK part of the Silverpit Basin, but have a widespread distribution in the Netherlands area. A typical sequence is illustrated in Fig. 4.17B.

All of the red-bed units are typical fluvial deposits, containing well-defined upward-fining channel bodies, overbank sheet sandstones and silty and muddy floodplain deposits. The sedimentology of the Lower Ketch Unit of the Silverpit area is described by Besly *et al.* (1993). The sequence contains single and multistorey braided-channel deposits and intensely rooted overbank deposits, which locally contain fairly mature palaeosols of ferruginous type (see Besly and Fielding, 1989). In some parts of the

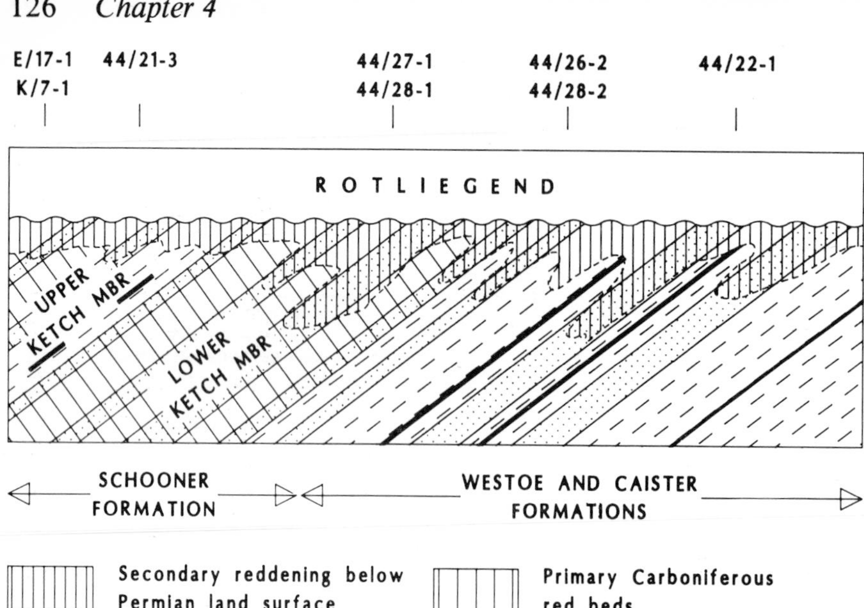

Fig. 4.16 Schematic relationship between contrasting types of red beds in the late Carboniferous, Southern North Sea (after Besly *et al.*, 1993). Well positions are schematic, illustrating nature of succession penetrated at different localities.

succession, the association of facies is strongly reminiscent of the waterlogged alluvial-plain facies encountered in the underlying Coal Measures, interpreted to have been deposited as a grey, Coal Measure association of facies, and to have subsequently undergone early diagenetic oxidation. The timing of this oxidation is constrained by the occurrence of interbedded mature primary red palaeosols. These suggest that deposition under waterlogged conditions was closely followed by falls in the water-table, during which well-drained soils formed and the pre-existing sediment was oxidized. Such base-level changes, almost certainly caused by tectonic activity, were responsible for valley incision, which is inferred to have imparted a shoestring geometry to the channel sandstone reservoir units. The Upper, Westphalian D, Ketch Unit contains finer-grained fluvial-channel deposits, and is characterized by the occurrence of caliche palaeosols, a typical feature of red beds of Westphalian D to Stephanian age in the UK onshore (Besly, 1987, 1988; Besly and Fielding, 1989) and in Germany (Selter, 1989).

The detrital composition of the sandstones in the red beds is still poorly documented. Sandstones in the Lower Ketch Unit (Westphalian C) are similar in composition to those in the earlier Westphalian and Namurian (Collinson *et al.*, 1993), reflecting continued input from the regional northerly source, perhaps intensified by the erosion of earlier Carboniferous deposits from early inversion structures within the basin (Leeder and Hardman, 1990) and by source-area uplift accompanying this inversion. Sandstones in the Westphalian D appear to be less feldspathic and richer in metamorphic lithic detritus than the typical Silesian arkoses of northern derivation. Comparison with the UK and European onshore successions indicates that these are derived from a Variscan source.

A final important aspect of the stratigraphy of the late Carboniferous red beds is the problem of recognizing the boundary between them and the overlying Permian, in areas where an obvious basal Rotliegend sandstone is not present. Both units are dominated by red mudstone and lack biostratigraphic markers that are easily recognizable in cuttings. This may have led to confusion between sand bodies in the Carboniferous and those in the Leman Sandstone Formation, with incorrect identification in some wells. Detrital and early authigenic clay-mineral assemblages may be of value in differentiating between the Carboniferous red beds, deposited in an inferred humid environment, and the desert-lake sediments of the Silverpit Formation. A distinctive sandstone composition may be another criterion for differentiating Carboniferous primary red beds from the overlying Permian.

Patterns of sandstone derivation further to the east remain largely unknown. It is, in particular, not clear whether the Westphalian B–D Tubbergen Sandstone extends into the Dutch offshore area.

4.7.4 Petroleum geology

Source rocks

The occurrence of source rocks in the Carboniferous is reviewed by Bailey *et al.* (1993) and by Cornford (Chapter 11). It is clear that the coal-bearing alluvial and deltaic sequences at all horizons in the Carboniferous contain large quantities of detrital plant material, either in coal seams or in disseminated form, and have the potential to form source rocks for gas. Coals are concentrated particularly in the Westphalian A to C, where they form between 5 and 8% of the sequence and have fair to excellent gas-source potential. Thinner developments of coal and carbonaceous shale in the upper part of the Namurian also have gas-source potential, but the middle and lower parts of the Namurian contain significant amounts of inertinite and have reduced potential for gas generation. Coals are also widespread in the Brigantian in the northern part of the Southern North Sea, although their source potential is little known.

Assumptions regarding the oil source-rock potential of the Carboniferous have, in the past, been based on the very rich oil shales developed in the Asbian–Brigantian of the Midland Valley of Scotland. As pointed out previously, this facies development is laterally restricted and may not be representative. The onshore sequence in Northumberland, in which there are only a few oil-shale horizons in a sequence up to 1400 m thick, is probably more typical for

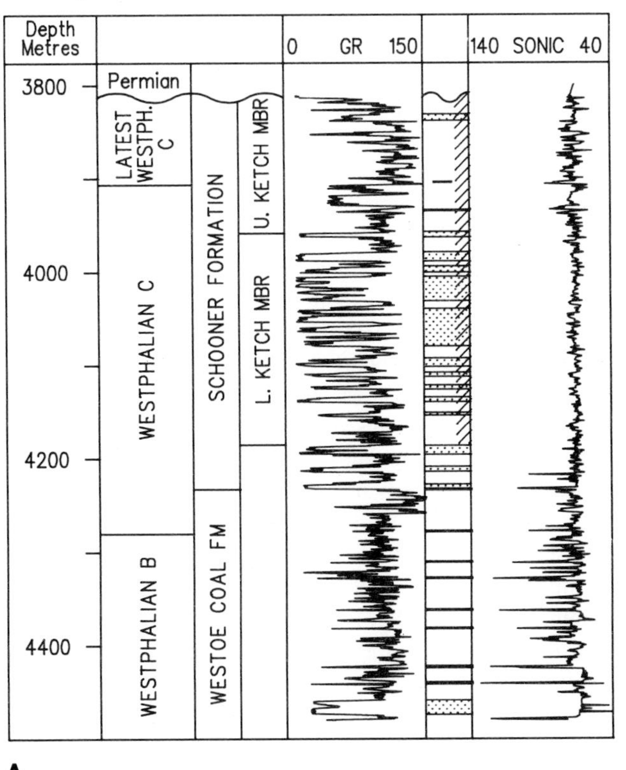

NETHERLANDS EMMEN-7

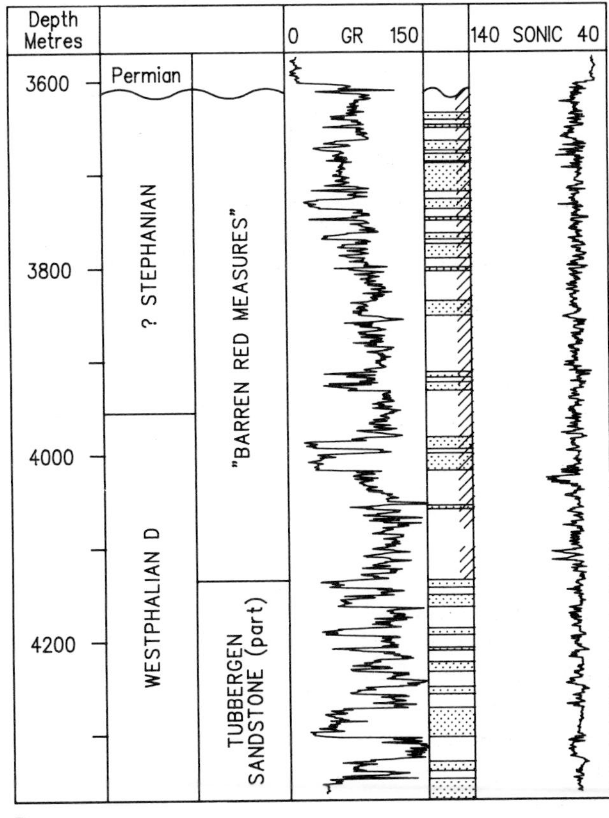

Fig. 4.17 Wireline logs of representative Westphalian C–Stephanian (Silesian) primary red-bed sequences in the Southern North Sea Basin. Diagonal ornament indicates red beds. Note interbedded nature of basal contact between red beds and coal-bearing sequence in Emmen-7. Logs and biostratigraphy of Emmen-7 from NAM and RGD (1980). Biostratigraphy of 44/21-3 by Conoco UK Ltd, cited by Besly *et al.* (1993).

the upper part of the Dinantian in the Southern North Sea. This is confirmed by the generally poor occurrence of oil-prone source material in the few available offshore Dinantian penetrations. Further to the south, where a deep-water carbonate association may be expected in Dinantian half-graben fills, oil-prone source material may, by analogy with the onshore sequence, be present in the Brigantian (see Fraser *et al.*, 1990). The Namurian sequence, in the onshore area, contains important oil-source horizons in the Pendleian and Arnsbergian in the southern part of the Pennine Basin (Fraser *et al.*, 1990). Source-rock deposition occurred where there were deep basins that were starved of clastic input. The Westphalian, in addition to the obvious source potential of its coals, contains variable but sometimes significant oil shales in the onshore sequence, both related to marine incursions and in freshwater, algal-dominated sections. In addition, many coals contain algal-rich cannel-coal horizons. This oil-prone material may be inferred to have been responsible for the condensate that is found in association with the Carboniferous-sourced gas. Little regional synthesis of Westphalian source type and potential has been carried out. On the basis of examination of carbon and nitrogen isotopic composition of Carboniferous-sourced gases, Kettel (1989) has suggested a general eastward decrease in oil-prone source material in the Westphalian of the Netherlands and Germany.

Maturation of the deeper parts of the thick Carboniferous sequences started before Variscan basin inversion (Hawking, 1978; Kirby *et al.*, 1987; Leeder and Hardman, 1990). Most of the hydrocarbons produced during this phase are likely to have escaped during the deformation and erosion that accompanied Variscan inversion. The maturation history of the Carboniferous source rocks is therefore largely dependent on their Mesozoic burial history (Kirby *et al.*, 1987; Leeder and Hardman 1990; see also Chapter 11). Several distinct zones of maturity can be recognized at the Carboniferous subcrop (Bailey *et al.*, 1993). In the marginal-shelf areas, the top Carboniferous is currently immature or marginally mature for oil generation (Vitrinite Reflectance (VR) = 0.5 – 0.8). Over wide areas in the centre of the Southern North Sea Basin, maturities are greater than VR = 1.00. Finally, much greater maturities are found in the Sole Pit/Cleveland High and in other areas that have undergone major Mesozoic or Tertiary inversion. It is clear that, outside these inverted zones, there is currently widespread generation of gas in the more deeply buried sections of the Carboniferous sequence. In most marginal areas of the basin, it is likely that lateral and vertical migration of large volumes of gas from the major active kitchen may have flushed out oil that accumulated at shallower depths.

Reservoirs

Sand bodies are developed throughout the Carboniferous sequence. The greatest potential for reservoir development is found in channel sandstones within the fluvial and deltaic sequences, particularly where these occur as stacked, multiple units (Bailey *et al.*, 1993). Major sandstone developments are identified by Bailey *et al.* (1993) and Collinson *et al.* (1993) in the Fell Sandstone Formation and Scremerston Formation (Arundian to Brigantian), the Millstone Grit Formation (mainly Namurian B–C), the Caister Coal Formation (Westphalian A–early Westphalian B) and the Ketch Member (Westphalian C):

1 *Dinantian.* Late Dinantian sandstones cored in A/16-1 have porosities of up to 14%, while log-derived porosities in this well and in 44/2-1 show maxima of over 20%.

2 *Namurian.* Reservoir quality in Namurian sandstones is strongly influenced by facies control on texture and diagenesis. In the Marsdenian Chatsworth Grit reservoir of the Trent Field strong contrasts in porosity and, particularly, permeability exist between two units of different facies. The lower unit, an incised palaeovalley fill, has average porosity of 22%, and permeabilities of 0.1–60 mD. The upper unit, a tidally reworked body, has lower average porosity (12%) but much better permeability (100–320 mD). Namurian reservoirs are also present in the Cavendish Field.

3 *Westphalian Coal Measures.* In the Westphalian Coal Measures, many of the sandstone bodies have very low porosities, particularly in the more distal, sand-poor southern areas, where they are finer-grained. Coarse-grained, multistorey, braided-channel facies, as described by Cowan (1989), are developed in the southern part of Quadrant 44 (Leeder and Hardman, 1990); where they act as reservoirs for the Caister C (Ritchie and Pratsides, 1993) and Murdoch Fields. Core poroperm data for the Murdoch Field show porosities in the major channel bodies ranging from 3 to 19%, with permeabilities of 0.1–11 mD (mean 1.87) (Besly *et al.*, 1993). Average porosity in the Caister C reservoir is 11% (Ritchie and Pratsides, 1993).

4 *Westphalian red beds.* Tubb *et al.* (1986) report average porosities of 14–20% in this group in the South Hewett Shelf area, with good permeability indicated by mudcake development. In the Silverpit area core porosities of 7–19% and permeabilities of 10–1000 mD are recorded for Lower Ketch Member sand bodies in K/4-1, and in the Ketch and Schooner Fields (Besly *et al.*, 1993; Mijnssen, 1997). The permeabilities are significantly better than those found in sandstones in the underlying Coal Measures (Besly *et al.*, 1993). Net : gross ratio is low and very variable in this unit. As a result gas recovery may be negatively affected by poor sand body connectivity (Mijnssen, 1997).

The reservoir characteristics of the Upper Ketch Unit (Westphalian D) are poorly known. Sandstones of this age in K/7-1 have porosities of less than 10%, probably reflecting their fine grain size and enhanced content of labile components. Throughout the succession, sandstone abundance decreases from north to south, and sand-rich fairways are inferred to be developed as a result of preferential location of major channel systems above the differentially compacted fills of Lower Carboniferous Graben structures (Collinson *et al.*, 1993; Fig. 4.18).

Carboniferous sandstone reservoirs in the Southern North Sea area have undergone a complex diagenetic history (Fig. 4.19; Hawkins, 1978; Cowan, 1989; Leeder and Hardman, 1990; Besly *et al.*, 1993), which has resulted in a major reduction in primary porosity, largely through compaction and the precipitation of large volumes of early quartz and dolomite cements. Virtually all of the observed porosity is secondary in origin, resulting from two phases of dissolution that affected carbonate cements and detrital feldspars. The first of these was caused by the flushing of the sequence by meteoric waters during Permian uplift. This process also caused the secondary reddening of the topmost part of the sequence. Further dissolution was brought about by organic acids produced by thermal decarboxylation. The

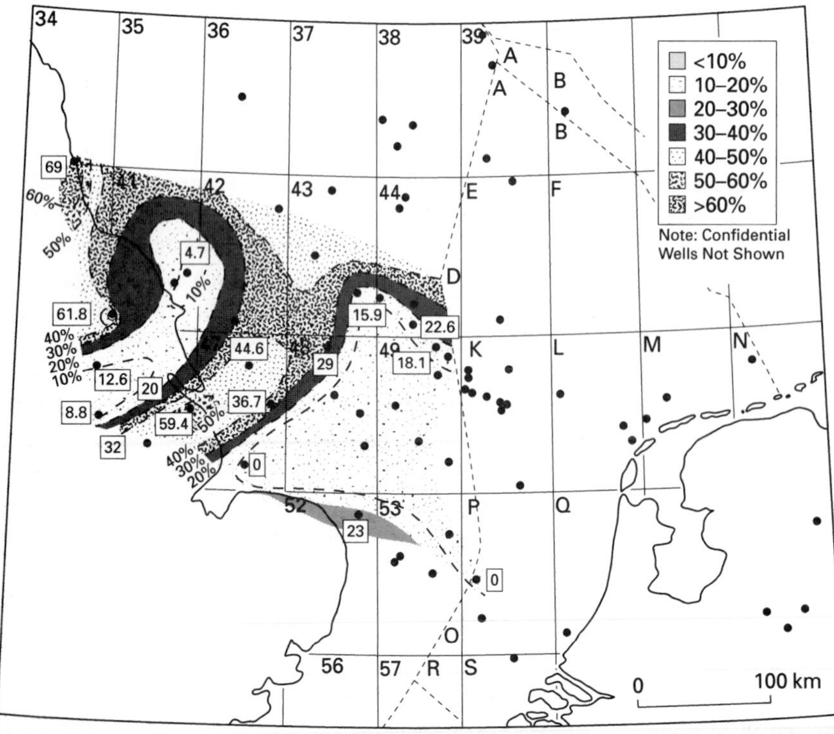

Fig. 4.18 Percentage of total interval thickness occupied by channel sandstones in Namurian B/C. Note the diminution of sandstone abundance over the Market Weighton Block. (From Collinson *et al.*, 1993.)

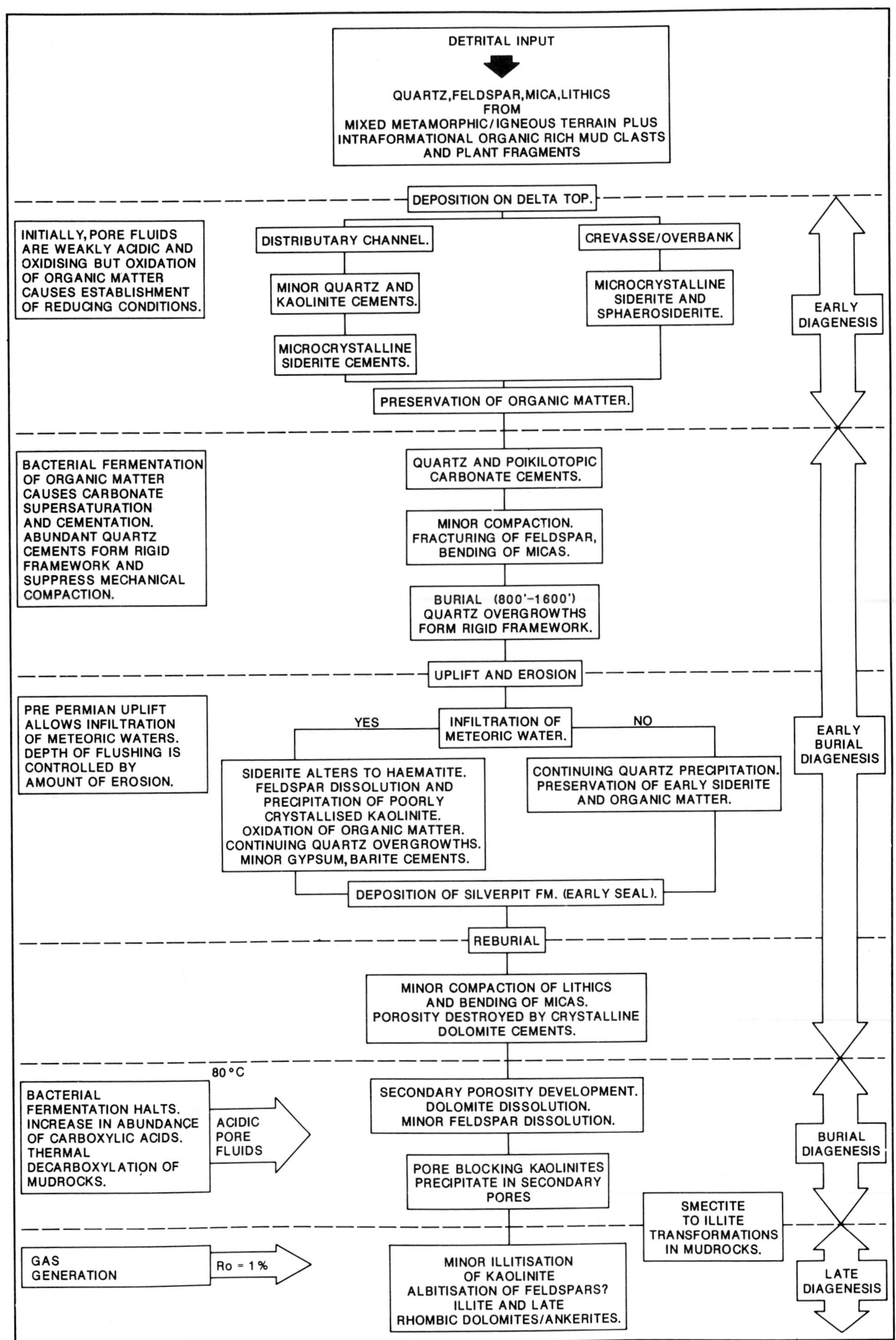

Fig. 4.19 Schematic diagram of the principal diagenetic reactions of the Upper Carboniferous sandstones of the Southern North Sea (from Cowan, 1989).

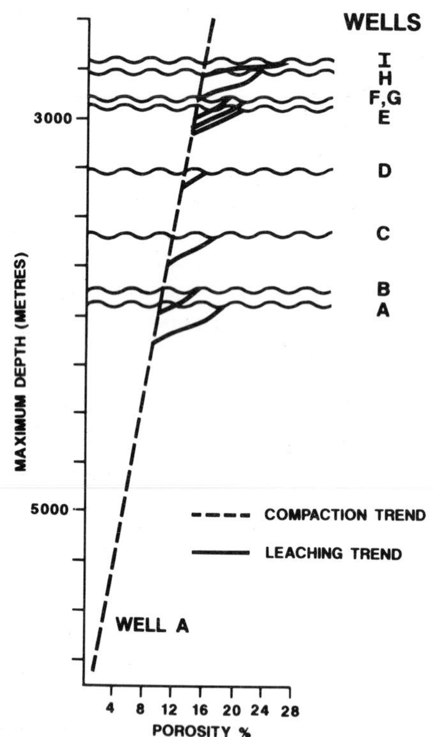

Fig. 4.20 Silverpit Basin Carboniferous sandstones. Plot of porosity versus maximum burial depth. (From Bailey *et al.*, 1993.)

secondary porosity is generally patchily distributed, often with poor pore interconnection. Permeability is reduced by pore-filling kaolinite, which was precipitated as a product of the second phase of dissolution. Porosity distribution is strongly related to grain size, potential reservoirs occurring only in medium- to coarse-grained sandstones. Bailey *et al.* (1993) note that porosity shows a linear decrease with increasing maximum burial depth (Fig. 4.20). Development of secondary porosity is enhanced in the weathered zone, extending for approximately 200 m below the base Permian unconformity.

Apart from the unpredictable distribution of porosity within sand bodies, the depositional facies architecture of sand bodies in the alluvial/deltaic facies associations has the potential to increase the complexity of Carboniferous reservoirs (Fig. 4.21). Coarse-grained sandstones with potentially higher porosities are restricted to the central high-energy zones of distributary channels, which themselves may have limited lateral extents and complicated patterns of channel stacking (Hawkins, 1978). Within individual reservoir-quality sand bodies, there is marked variation in permeability distribution, high-permeability layers being restricted to coarser-grained horizons. Reservoirs may thus show major variations in stratigraphy and reservoir properties over very short distances (Hawkins, 1978; Rothwell and Quinn, 1987; Storey and Nash, 1993). Initial development of the Murdoch Field will involve extensive use of inclined production wells to maximize the intersection between the well bore and the irregularly distributed high-permeability zones that contribute most production (Gunn *et al.*, 1993).

The reservoir potential of the marine limestones in the Dinantian is speculative. Sequences encountered in the South Hewett Shelf are extremely tight (Tubb *et al.*, 1986). These authors suggest that some reservoir potential might be found in bioherms. Exposed analogues in the UK onshore do not lend credibility to this suggestion. A further speculative possibility is that porosity may be present in the Dinantian carbonates in the extreme south of the South Hewett Shelf, where these are karstified at the Cretaceous weathering surface. Gas shows have been observed in wells drilled in this setting on the northern flanks of the London–Brabant Massif in Belgium.

Traps and seals

Two principal trap types are inferred: (i) Variscan anticlinal folds having an intra-Carboniferous seal; and (ii) structural traps formed during Mesozoic tectonic events, possibly modifying Variscan structures. In addition, it is possible that stratigraphic traps may occur beneath the sub-Permian unconformity. These trap types are illustrated in cartoon form in Fig. 4.22:

1 *Variscan folds* (A in Fig. 4.22). The morphology and distribution of Variscan folds in the Southern North Sea are described by Leeder and Hardman (1990). Major folds are largely absent in the South Hewitt Shelf area and in the

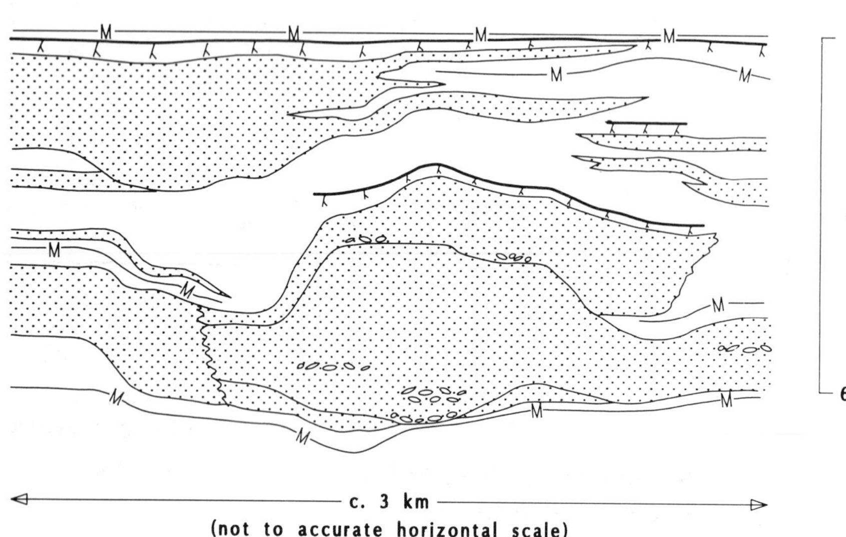

Fig. 4.21 Complex pattern of channel stacking leading to marked lateral variation in reservoir properties, base Westphalian A, Bothamsall oilfield, East Midlands (after Hawkins, 1978).

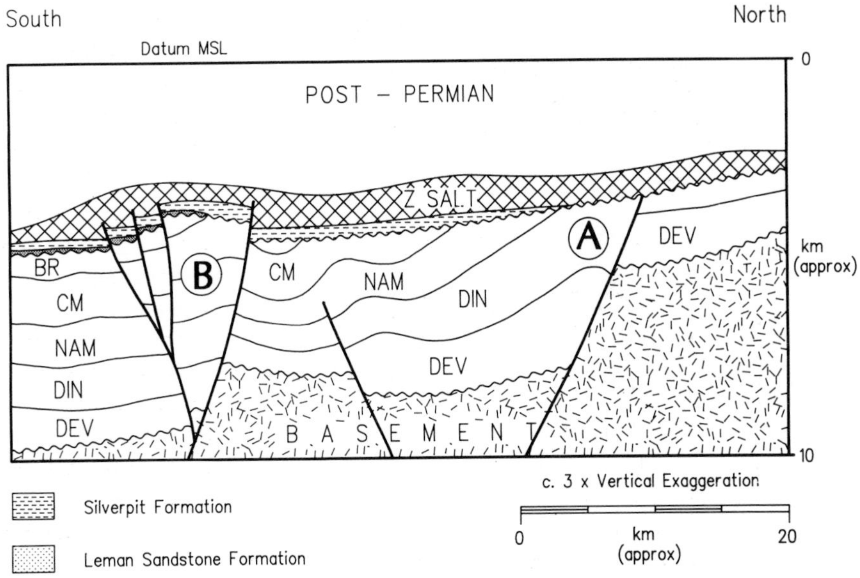

Fig. 4.22 Cartoon of style of structural development in a transect running from the Silverpit Basin to the southern edge of the Mid North Sea High, illustrating conceptual Carboniferous trapping styles. Note the regional seal formed by the Silverpit Formation. For discussion, see text. BR, red-bed facies of Schooner Formation; CM, coal measures of Schooner, Westoe and Caister Coal Formations; NAM, Namurian; DIN, Dinantian; DEV, Devonian.

North Dogger Shelf area of the Mid North Sea High, but are widely developed in the Silverpit Basin (Figs 4.23 and 4.24). Many of the major folds are related to west-north-west–east-south-east faults having reverse throw in the pre-Permian sequence, and are interpreted as having formed through Variscan inversion of the Dinantian half-grabens. Onlap of the Westphalian C red beds (Fig. 4.25) shows that the major folding occurred earlier than in adjacent onshore areas. Subsequent north-north-east–south-south-west-orientated late Carboniferous–early Permian extensional faults have the effect of compartmentalizing the fold structures (Oudmayer and de Jager, 1993; Fig. 4.26), producing a style of faulting very similar to that seen in the onshore coalfield areas. These extensional faults

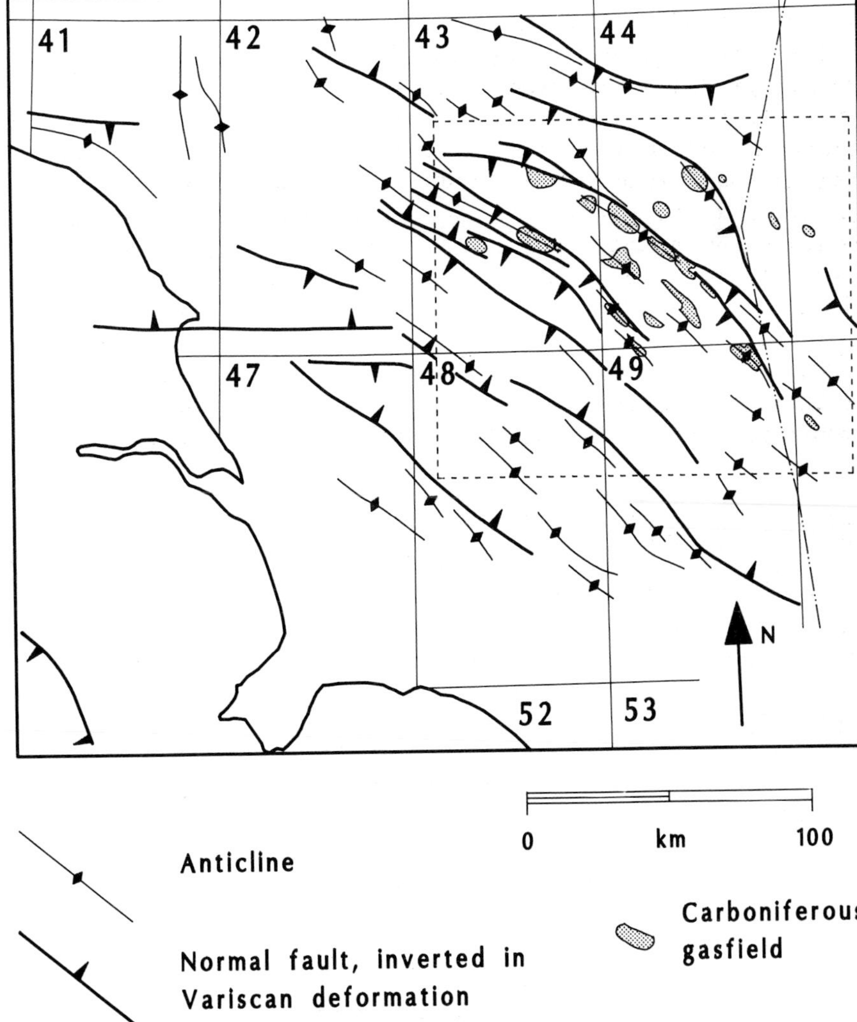

Fig. 4.23 Distribution of major Variscan folds and faults in the UK sector of Southern North Sea (compiled from Corfield *et al.*, 1996; Leeder and Hardman, 1990; Halliburton non-proprietary reports). Dashed outline indicates area covered by Fig. 4.2.

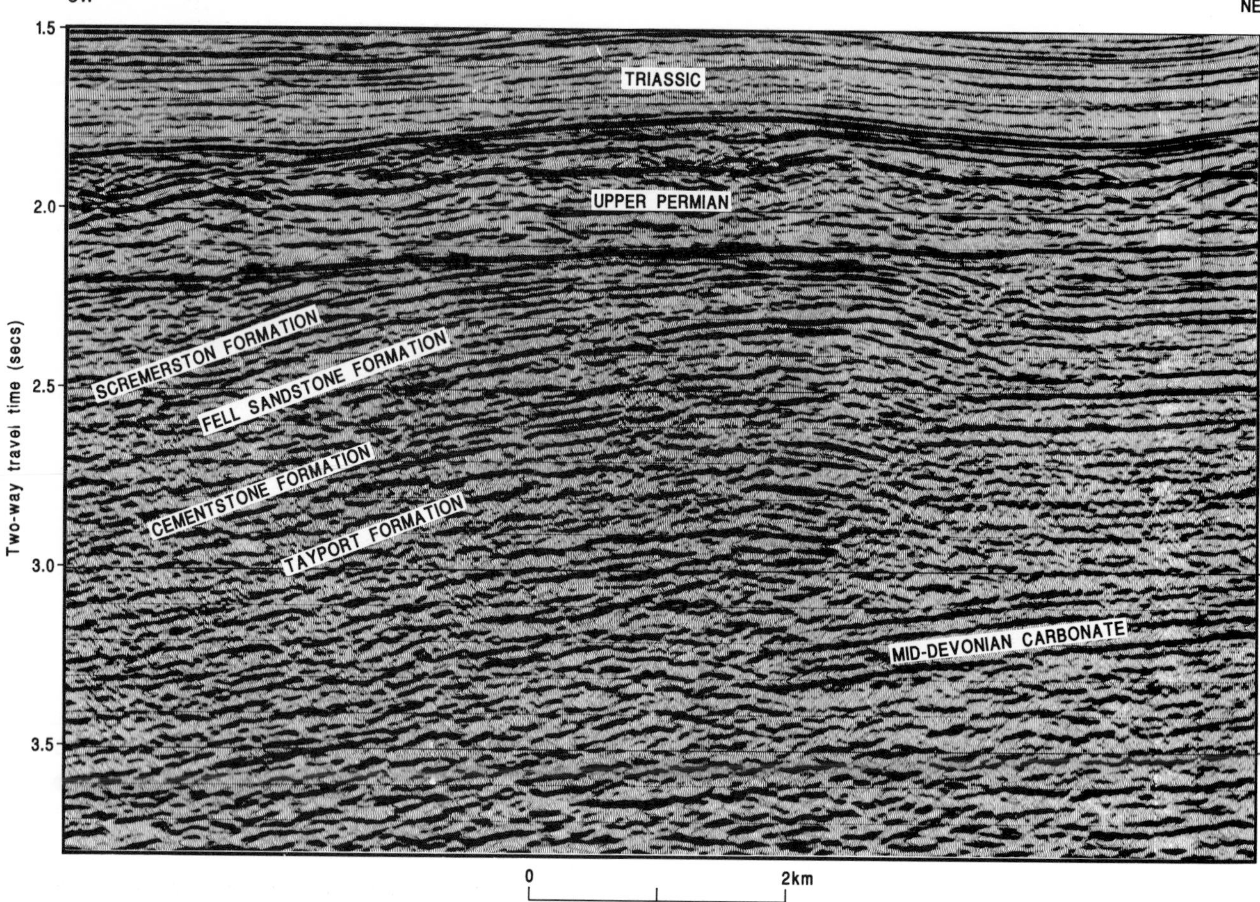

Fig. 4.24 Type A anticline, not modified by post-Variscan reactivation. Alternation of banded and transparent acoustic character enables recognition of Scremerston, Fell Sandstone and Cementstone formations. Thickness variations in these units demonstrate that the fold has formed by inversion of an early Carboniferous extensional structure. Seismic section provided by Intera Information Technologies Ltd.

have been incorrectly interpreted as syndepositional listric faults (see discussion of Hollywood and Whorlow, 1993, in Corfield *et al.*, 1996).

Traps within these folds rely on an effective intra-Carboniferous seal. Studies in the onshore area show that mudstone units intercalated in the late Carboniferous deltaic sediments may not always be efficient seals for gas (Fraser *et al.*, 1990). No convincing example of an effective intra-Carboniferous seal has yet been identified in the offshore area. The experience onshore suggests that such seals may have limited sealing capacity, which is likely to be further downgraded by fault juxtaposition of sandstone units.

2 *Post-Palaeozoic structures* (B in Fig. 4.22). The major basement structures that controlled the positions of Variscan inversion folds were transpressionally reactivated during the Cimmerian and Late Cretaceous/Early Tertiary. This modified the pattern of closure in pre-existing Variscan folds, creating pop-up structures with a number of possible trap configurations, sealed by the Permian Silverpit Formation. Where dip closure exists at base Rotliegend level, thin basal sand units below the Silverpit Claystone may form an additional reservoir unit. The Mesozoic inversion structures may also contain elements of Variscan dip closure and, where they involve large post-Variscan fault throws, may involve lateral fault seals where the Carboniferous is juxtaposed against the Silverpit Formation or the Zechstein salt. This is the typical trapping style developed in fields for which structural information has been published (Figs 4.25 and 4.27).

3 *Stratigraphic traps.* These may be developed where Carboniferous reservoir units subcrop the Silverpit Formation developed in a seal facies, or below the Zechstein. Such a trap configuration requires an effective intra-Carboniferous bottom seal.

All of the discoveries that have been announced or reported by scouting organizations can be inferred to fall into the first two types of trap described above. There is a strong relationship between discoveries and the occurrence of major Variscan inversion anticlines (see Figs 4.2 and 4.23), which control trends of gas accumulation in the Silverpit Basin.

4.8 Outstanding problems in Carboniferous exploration

It is evident from the preceding discussion that exploration of the North Sea Carboniferous is at little more than the frontier stage, with the exception of a small area in Quadrants 43 and 44 of the Southern North Sea. Continuing success is currently hampered by a number of difficulties.

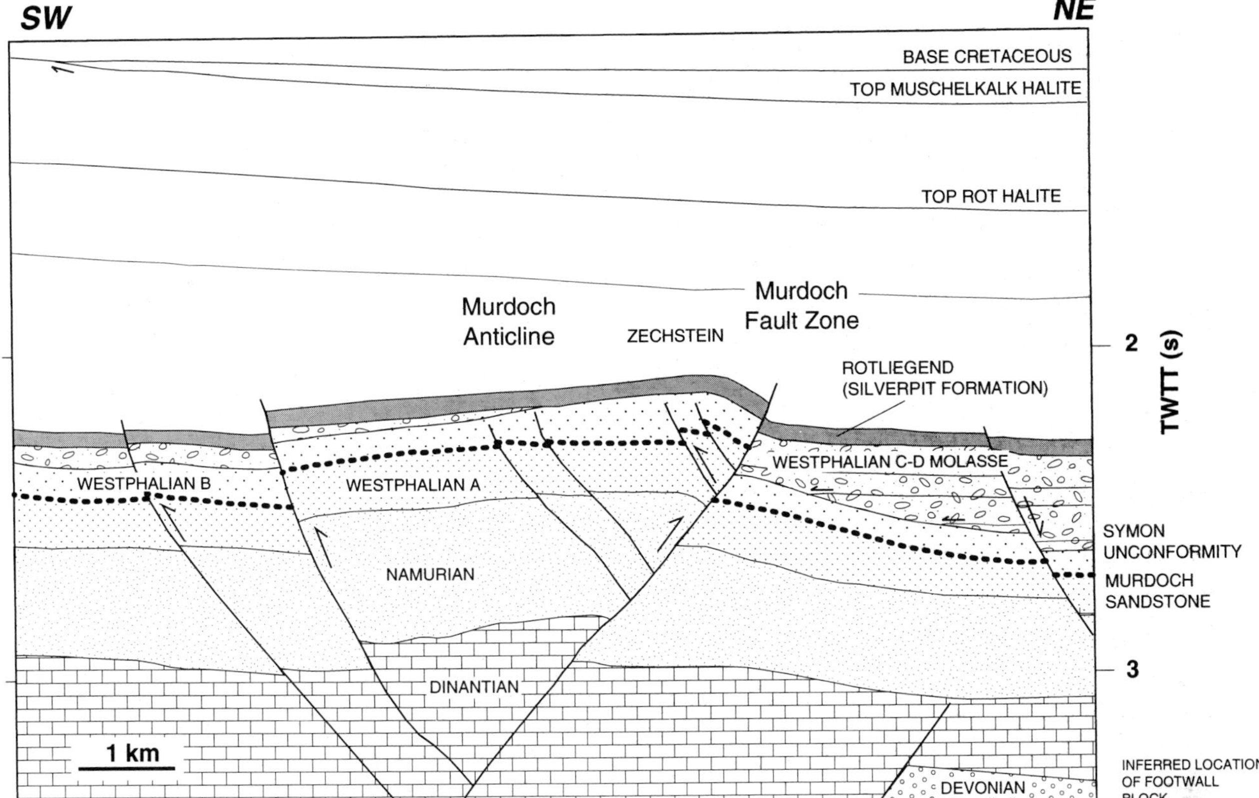

Fig. 4.25 Cross-section of the Murdoch Field (UK block 44/22)—a typical type B anticlinal structure, in which the Variscan closure has been modified by formation of a transpressional pop-up structure during the Tertiary. Note onlap of the Lower Ketch Member ('Westphalian C–D molasse') on to folded Westphalian A–B. (From Corfield *et al.*, 1996.)

4.8.1 Geophysical resolution and interpretation

Exploration of the Carboniferous is hindered by difficulties in seismic imaging. These are brought about by the complex overburden geology, which largely results from the behaviour of the Zechstein salt. Salt diapirism creates complex ray paths, which may be further complicated by rafting and brecciation of the Zechstein III Plattendolomite. Seismic resolution of the pre-Permian section is further impeded by the high reflectivity of the basal Zechstein Hauptdolomite and by the lack of acoustic impedance contrast between the Rotliegend sequence and the underlying Carboniferous. Where the Rotliegend rests upon a Carboniferous sequence lacking coals, particularly when the latter is formed by the late Carboniferous primary red-bed sequence, it is difficult or impossible to recognize the unconformable contact between the two (Figs 4.27 and 4.28). In these circumstances the mapping of closures at the Saalian Unconformity surface has to be undertaken by indirect means. The combination of the imaging difficulties enumerated above makes it difficult to map intra-Carboniferous horizons, although improvements in seismic-data quality now make it possible to derive more extensive lithostratigraphic data from seismic sections. In the Dinantian of the northern part of Quadrants 42–44, a marked contrast in acoustic response allows local differentiation and mapping of the Fell Sandstone Formation (Holkerian–Arundian) and the late Dinantian Scremerston Formation (see Fig. 4.24). Clearer images of the acoustically transparent zone at the top of the Coal Measure succession make recognition of the 'Barren Red Group' easier (Fig. 4.28). Recognition of these seismic facies becomes difficult in areas of pillowing and diapirism in the overlying Zechstein salt. At a smaller scale, Evans *et al.* (1992) claim that sandstone-prone zones in the Westphalian Coal Measures (and, by implication, in the underlying Namurian) can be recognized locally on the basis of phase and amplitude changes. Such features might represent individual channel complexes, but seem more likely to be formed by thicker, stacked sand bodies formed in incised valley fills.

Mapping of structures, both at top Carboniferous level and within the Carboniferous, is further complicated by difficulties of depth conversion (Clark-Lowes *et al.*, 1987). These again result from the presence of the salt. This, when present in diapiric form, creates a complex overburden velocity field, comprising velocity pull-ups due to the salt itself, and differential burial-related lithification in the post-salt sedimentary sequence. The latter effect is particularly pronounced in the Upper Cretaceous chalk (Davis, 1987).

The difficulties of mapping enumerated above are not the only problems encountered in economic evaluation of Carboniferous fields. Because the reservoir units are generally fairly thin, direct observation of hydrocarbon/water contacts is rare, which, in conjunction with uncertainties in the mapping of structures, leads to difficulties in volumetric evaluation of accumulations. Lateral variability in reservoir geology adds to this complexity.

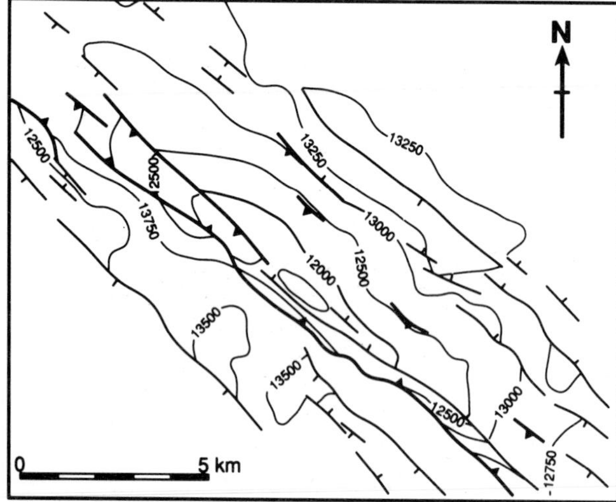

2D interpretation

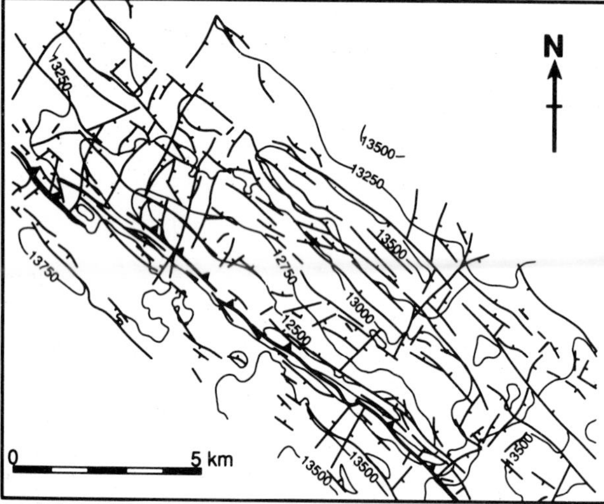

3D interpretation

Fig. 4.26 Maps of structure at top Carboniferous, based on two-dimensional (2D) and 3D seismic data. Intense development of small north-north-east–south-south-west cross-faults is identifiable on the 3D interpretation. (From Oudmeyer and de Jager, 1993.)

On a broader scale, it is still not possible to delineate areas in which differential subsidence occurred during the Dinantian, as the base of this interval is usually not identifiable on seismic. The basic structural framework of Carboniferous deposition is therefore still imperfectly known.

4.8.2 Stratigraphic and sedimentology problems

The penetration of very thick, monotonous sequences of deltaic sediments has created a set of subsurface correlation problems that differ from any previously encountered in the North Sea. In exposed successions, the stratigraphy of such sequences can be assessed by the identification of the succession of Marine Bands. At presents these can be identified in the subsurface only by extensive coring. Wireline-log correlation can be applied successfully in sequences of this type, but depends on a fairly dense well spacing and on the succession having been calibrated by identification of Marine Bands (e.g. Schuster, 1966). Even so, changes in facies may cause rapid lateral variations in the level of radioactivity in individual marine bands (e.g. Whittaker *et al.*, 1985, pp. 26–8). In the light of these factors, the application of simple peak-by-peak correlation seems of dubious value, especially in the red-bed sequences that lack marine bands (e.g. Schuster, in Hedemann *et al.*, 1984b). The experience in the UK onshore area suggests that additional correlation problems may be caused by extensive faulting in the Carboniferous succession.

One possible solution to these problems, adopted by Leeder *et al.* (1990a) and Hollywood and Whorlow (1993), lies in a multidisciplinary approach, combining geochemistry, spectral gamma-ray logging and conventional biostratigraphic techniques. Spectral gamma logging allows the discrimination between radiation peaks due to high concentrations of uranium, related to the presence of organic material, and those caused by radioactive potassium, concentrated in feldspar or mica. Studies of the onshore sequence show that uranium concentrations coincide with marine mudrocks, particularly where these have a high organic content, as a result of the development of anoxic conditions during marine highstands. Marine intervals may also be identified by the determination of the ratio of organic carbon to sulphur in mudrocks (Berner and Raiswell, 1984). With suitable correction to allow for the loss of organic carbon through hydrocarbon generation and expulsion, the combination of spectral gamma logging and geochemical analysis may allow the identification of the pattern of marine-band development in an otherwise thick and monotonous sequence.

The success of this approach depends critically on palaeontological calibration. A revised Namurian stratigraphy of well 48/3-3 is presented by Collinson *et al.* (1993), who demonstrate that the stratigraphic interpretation presented by Leeder *et al.* (1990a) relied on an incorrect biostratigraphic interpretation of a key palynomorph. Some progress is being made with the recognition in the subsurface of the Marine Band markers, on the basis of the recovery of diagnostic goniatites in cores or sidewall samples, but, as yet, no results have been published. Recent work (Maynard *et al.*, 1991) has also highlighted possible ambiguities in the identification and correlation of Marine Bands using wireline techniques. It is likely that the success of this approach will be increased by integration of gamma-ray interpretation with detailed palynofacies analysis (Davies and McClean, 1996).

The final area of uncertainty in the stratigraphy of the Southern North Sea Carboniferous results from the stratigraphic complexity of the relationship between primary red beds and the coal-bearing sequence, and from a lack of knowledge of the factors controlling the location of the major channel systems, which apparently form the best reservoirs. The recognition that many of the thicker sand bodies in the Namurian and Westphalian form the fills of incised palaeovalleys adds a further uncertainty to the correlation of sand bodies and prediction of reservoir quality in the succession. While further work on the onshore sequence may produce conceptual models that contribute to the understanding of these problems, solutions will not be found until data are available from many more wells.

A

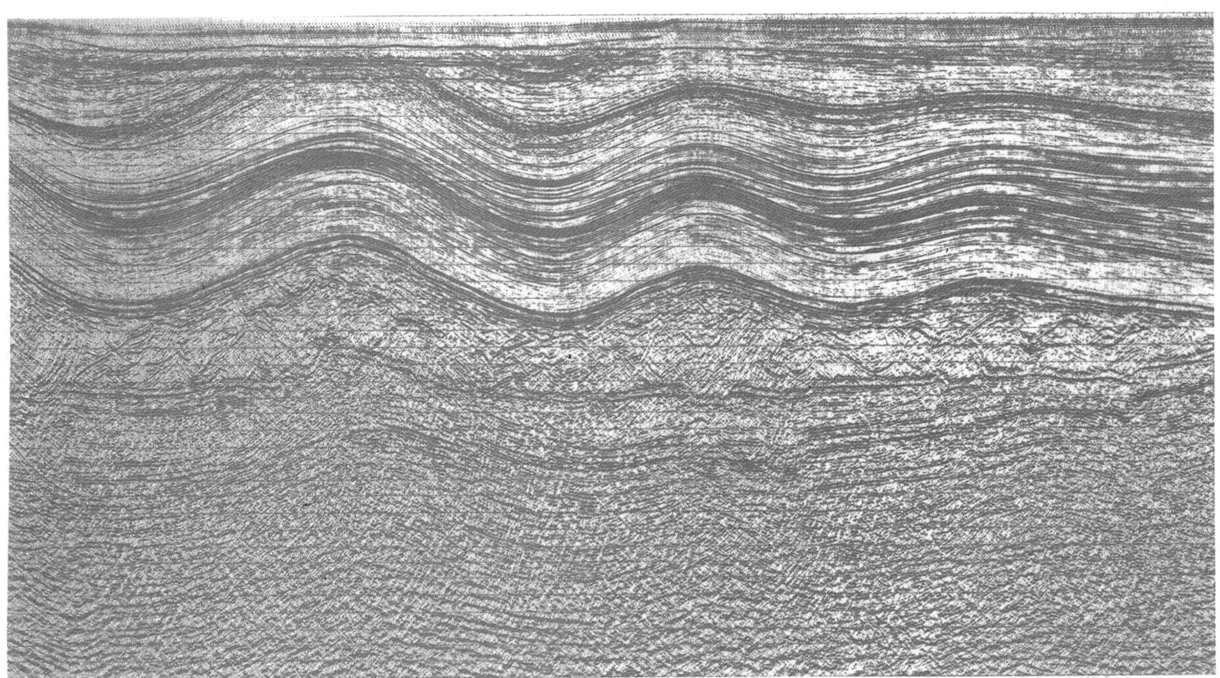

B

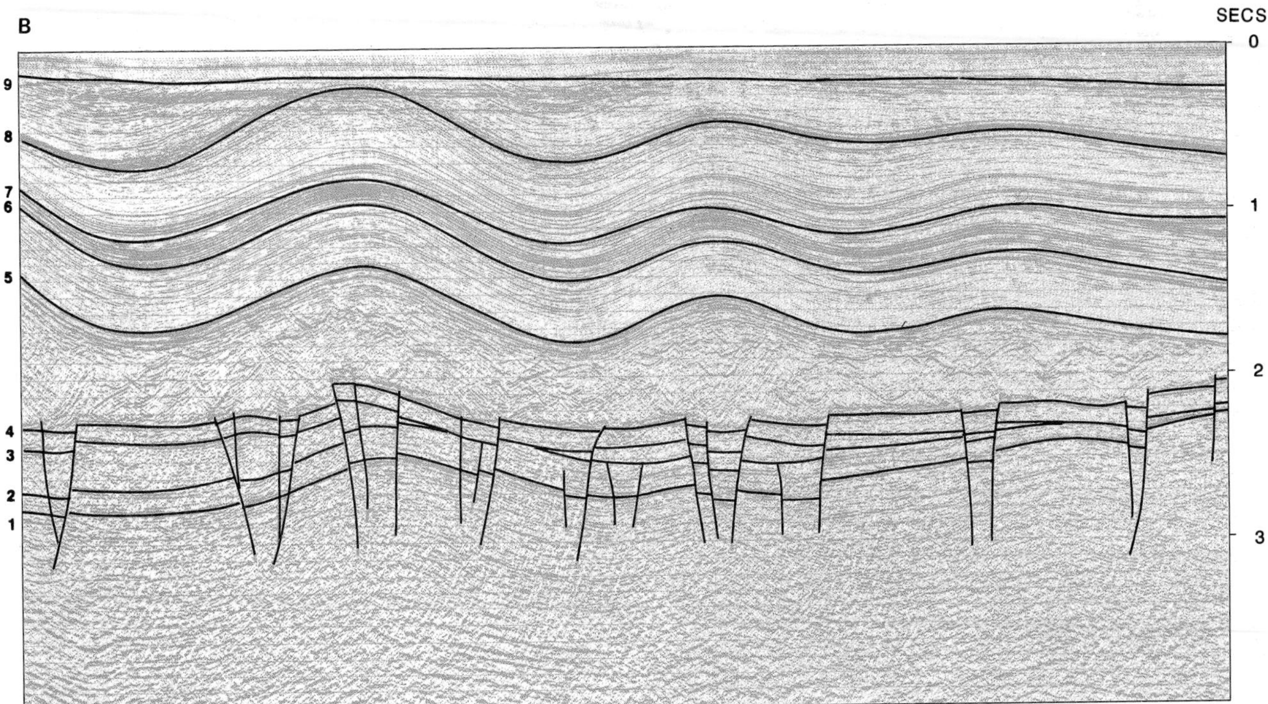

Fig. 4.27 Seismic reflection section across a Carboniferous prospect in the Silverpit area of the Southern North Sea. Note: (i) post-Permian modification of Variscan anticlinal structure; (ii) seismically transparent zone encompassing Rotliegend Silverpit Formation and underlying Upper Carboniferous primary red beds; (iii) lack of easily mapped seismic events within Westphalian coal-bearing sequences; and (iv) distortion of the image due to velocity pull-up beneath diapiric Zechstein salt. 1, top Caister Formation; 2, base Lower Ketch member; 3, top Carboniferous; 4, top Hauptdolomite; 5, top Bröckelschiefer; 6, top Röt Halite; 7, base Chalk; 8, base Tertiary; 9, base Oligocene.

4.9 Acknowledgements

This compilation could not have been written without the generous help of Halliburton Geoconsultants Ltd (formerly Gearhart Geo Consultants Ltd), who have kindly made available a number of non-proprietary reports on the Carboniferous of the North Sea, from which much of the unattributed factual information cited here is derived. Interpretations contained in this chapter are, however, the responsibility of the author.

BP Exploration, Conoco UK Ltd and Shell UK Exploration and Production are thanked for making unpublished

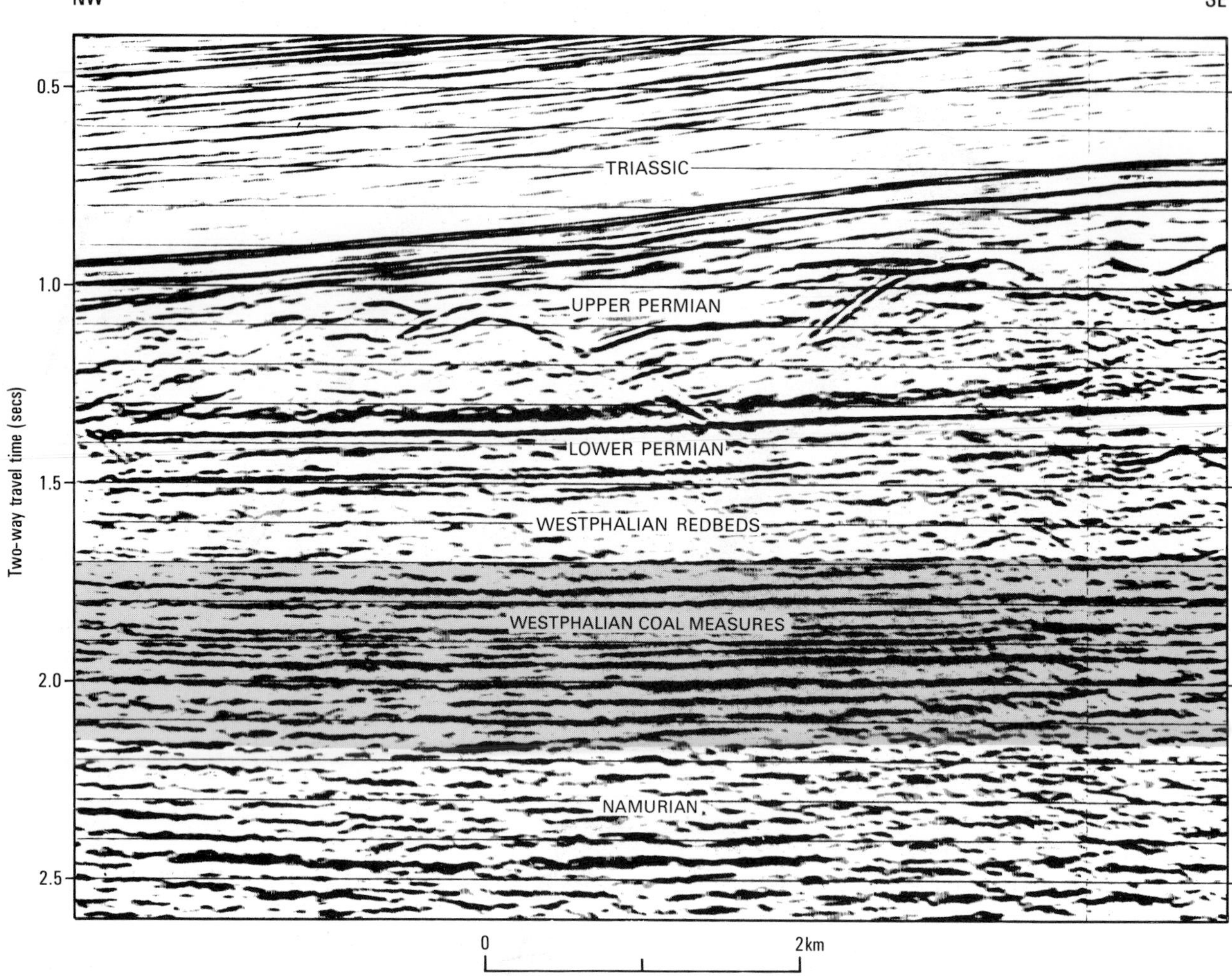

Fig. 4.28 Seismic section showing the character of the Westphalian coal-bearing and red-bed facies in the Silverpit Basin (from Cameron *et al.*,1992).

data available. Invaluable assistance has also been received from Don Cameron (British Geological Survey), John Collinson (Collinson Jones Consulting), Diana Cooper and Gillian Tester.

4.10 Key references

Bailey, J.B., Arbin, P., Daffinoti, O., Gibson, P. and Ritchie, J.S. (1993) Permo-Carboniferous plays of the Silverpit Basin. In: Parker, J.R. (ed.) *Petroleum Geology of Northwest Europe: Proceedings of the 4th Conference.* Geological Society, London, pp. 707–15.

Cameron, T.D.J., Crosby, A., Balson, P.S., Jeffery, D.H., Lott, G.K., Bulat, J. and Harrison, D.J. (1992) *United Kingdom Offshore Regional Report: the Geology of the Southern North Sea.* HMSO for the British Geological Survey, London.

Collinson, J.D., Jones, C.M., Blackbourn, G.A., Besly, B.M., Archard, G.M. and McMahon, A.H. (1993) Carboniferous depositional systems of the Southern North Sea. In: Parker, J.R. (ed.) *Petroleum Geology of Northwest Europe: Proceedings of 4th Conference.* Geological Society, London, pp. 677–87.

Fraser, A.J., Nash, D.F., Steele, R.P. and Ebdon, C.C. (1990) A regional assessment of the intra-Carboniferous play of Northern England. In: Brooks J. (ed.) *Classic Petroleum Provinces.* Special Publication 50, Geological Society, London, pp. 417–40.

Leeder, M.R. & Boldy, S.R. (1990) The Carboniferous of the Outer Moray Firth Basin, Quadrants 14, 15, Central North Sea. *Mar. Petrol. Geol.* 7, 29–37.

Leeder, M.R. & Hardman, M. (1990) Carboniferous geology of the Southern North Sea Basin and controls on hydrocarbon prospectivity. In: Hardman, R.F.P. and Brooks J. (eds) *Tectonic Events Responsible for Britain's Oil and Gas Reserves.* Special Publication 55, Geological Society, London, pp. 87–105.

5 Lower Permian—Rotliegend

K.W. GLENNIE

5.1 Introduction

5.1.1 Distribution and stratigraphy

The *Rotliegendes* (Rotliegend in English) is an old German term for the red beds that underlie the Zechstein. The classical Rotliegend sedimentary sequence was deposited in a post-Variscan basin that extended some 1500 km from eastern England to the Russo-Polish border, and has been referred to as the Southern Permian Basin (Figs 5.1 and 5.2). Seismic surveys and offshore drilling have shown that another, much smaller, Rotliegend basin occurs between the fragmented Mid North Sea–Ringkøbing–Fyn High and the Shetland and Egersund platforms; it is known as the Northern Permian Basin. A third area of Rotliegend deposition occurs in the Moray Firth Basin. Of the same approximate age of creation is a series of small half-grabens that stretch from south-west England to south-west and west Scotland; their fill of Permian sediment has been correlated with North Sea sequences by Smith *et al.* (1974) and Lovell (1983); see also Smith (1972a) and Smith and Taylor (1992). Wider-ranging correlations across northern Europe can be found in Stemmerik (1995) and Kiersnowski *et al.* (1995).

In the eastern half of the Southern Permian Basin, the Rotliegend can generally be divided into two distinct units, the Upper and the Lower Rotliegend (Fig. 5.3), which are separated by the Saalian Unconformity. The Lower Rotliegend is characterized by the presence of volcanic rocks of acid (rhyolites, ignimbrites) or intermediate type, whereas the rare volcanics of the Upper Rotliegend, again seen mostly in Germany, are basaltic. In Germany, the Upper Rotliegend has also been divided into two units (Figs 5.3 and 5.4); Upper Rotliegend 1 (UR1), which is confined largely to central Germany and is partly time-equivalent to the Saalian Unconformity; and Upper Rotliegend 2 (UR2), which occurs in the north German part of the Southern Permian Basin and, by implication, extends to the rest of it.

Because of current uncertainties in radiometric dating, the precise age and time span of Rotliegend deposition is in doubt. Most Rotliegend 'dates' depend on estimates rather than precise measurement, the control points occurring beyond the limits of Rotliegend deposition. This doubt creates minor problems of nomenclature but, more importantly, as discussed later in Section 5.4.2, also has implications for an inferred high rate of basin subsidence in the Southern Permian Basin.

According to Gast (1988), the total Rotliegend depositional time span in Germany extended from ~300 to 258 million years before present (Ma BP). As stressed by Leeder (1988), accurate radiometric dating of Late Westphalian

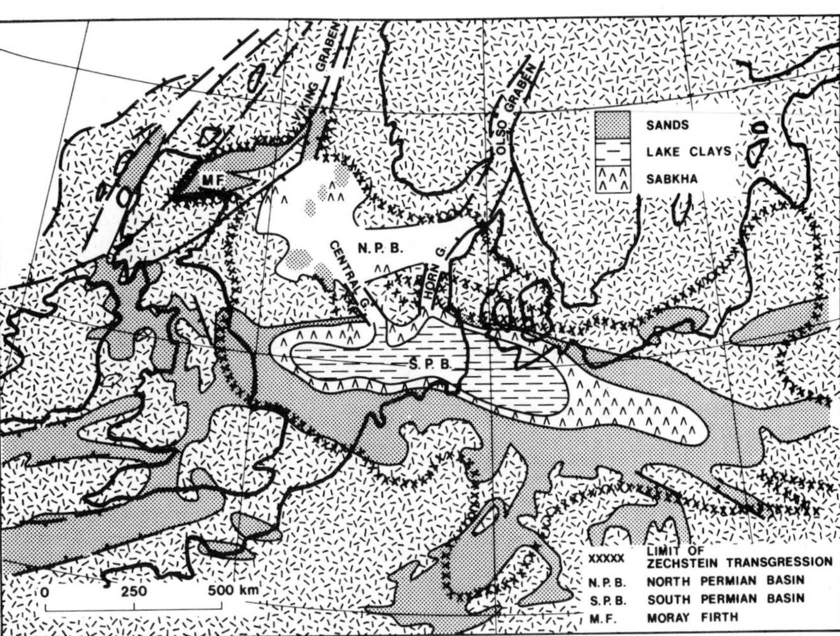

Fig. 5.1 Upper Rotliegend facies and palaeogeography, and limit of Zechstein transgression. Facies distribution poorly known in the Northern Permian Basin. (Modified from Glennie, 1972, and Ziegler, 1978.)

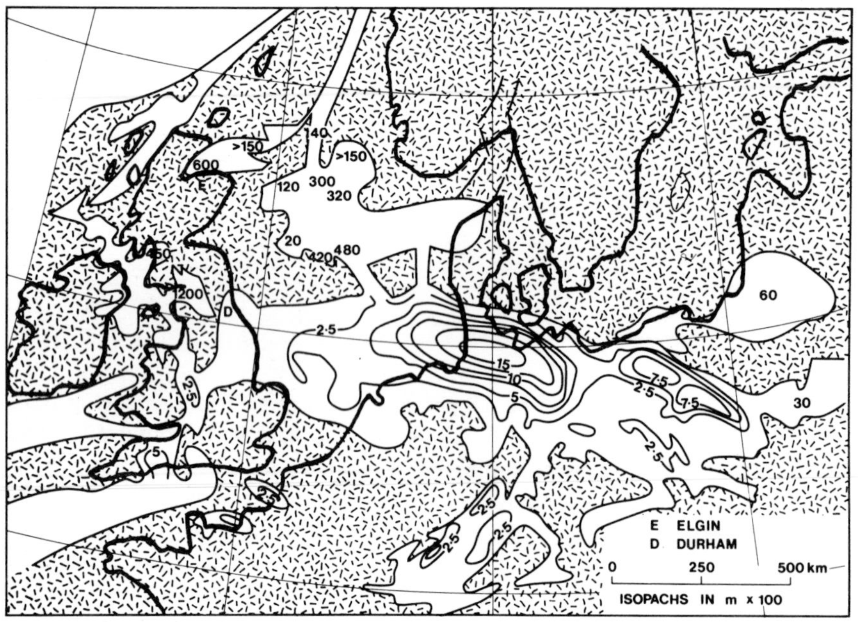

Fig. 5.2 Upper Rotliegend isopachs and isolated thicknesses (modified from Ziegler, 1980b).

tuffs in Germany (Lippolt *et al.*, 1984) carried the Permo-Carboniferous boundary back to a probable age of ~300 Ma (see also Menning, 1991). This latter date has been revised by Menning (1992, 1995) to 296 ± 2 Ma BP, although it must be stressed that this revision entails no new radiometric data. All existing dates for Lower Rotliegend sedimentation and volcanism range between about 288 and 300 Ma. Depending on the age adopted for the

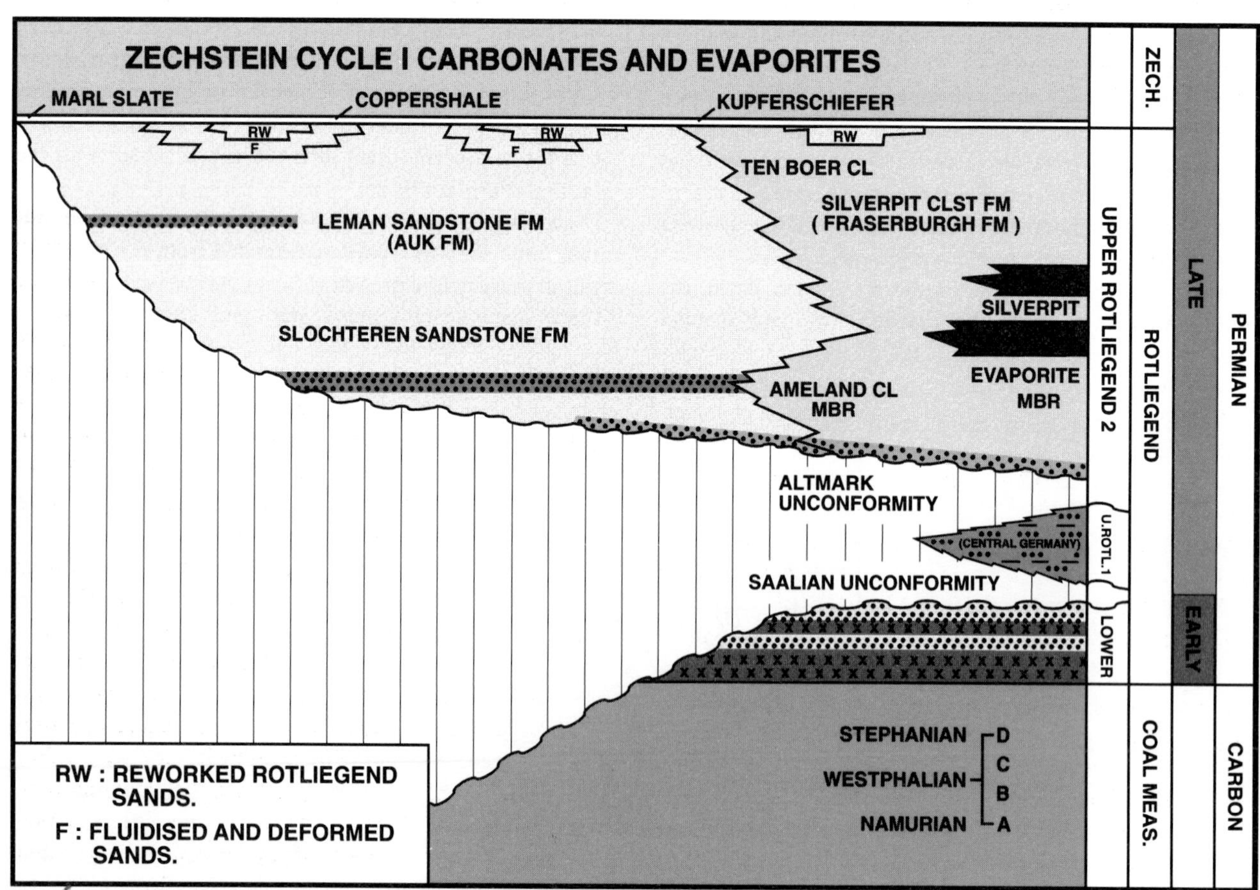

Fig. 5.3 Rock-stratigraphic diagram of the Permian Upper (UR2; see Fig. 5.4) and Lower Rotliegend groups of the North Sea area. Names in parentheses refer to formations in the Northern Permian Basin (cf. Fig. 5.1). Slochteren Sandstone and Ten Boer and Ameland claystones correspond to Dutch terminology. Dotted areas represent fluvial or mixed fluvial and aeolian sands. The Lower Rotliegend may not be entirely within the Permian (for discussion, see text).

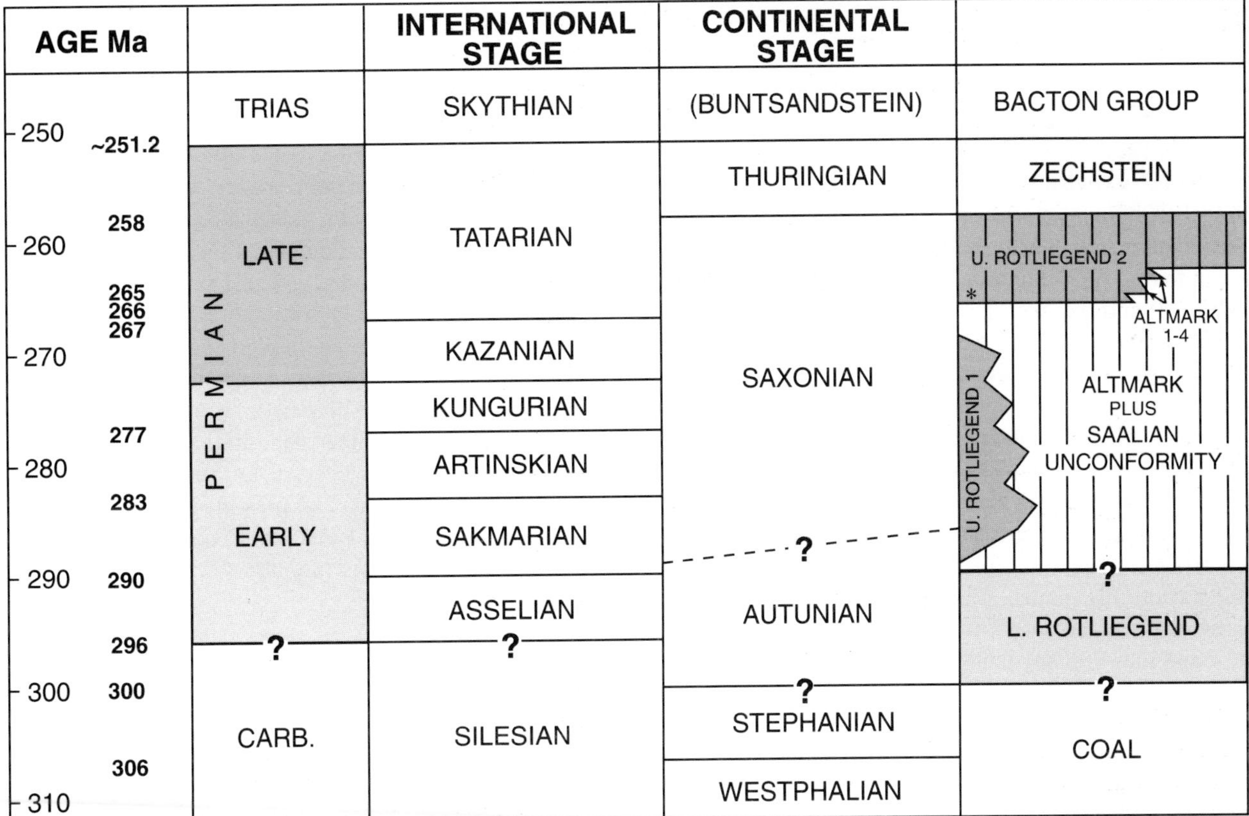

AGE Ma
INTERNATIONAL STAGE
CONTINENTAL STAGE
250
~251.2
260
258
265
266
267
270
277
280
283
290
290
296
300
300
306
310
TRIAS
LATE
PERMIAN
EARLY
?
CARB.
SKYTHIAN
TATARIAN
KAZANIAN
KUNGURIAN
ARTINSKIAN
SAKMARIAN
ASSELIAN
?
SILESIAN
(BUNTSANDSTEIN)
THURINGIAN
SAXONIAN
?
AUTUNIAN
?
STEPHANIAN
WESTPHALIAN
BACTON GROUP
ZECHSTEIN
U. ROTLIEGEND 2
*
ALTMARK
1-4
ALTMARK
PLUS
SAALIAN
UNCONFORMITY
U. ROTLIEGEND 1
?
L. ROTLIEGEND
?
COAL

* Possible age of Illawarra magnetic reversal (265 Ma BP)

Fig. 5.4 The Permian period seen in its possible time frame. Between the Lower Rotliegend and units of the Upper Rotliegend 2 is the long-lived Saalian Unconformity, which is partly coeval with Upper Rotliegend 1. In north-east Germany, Upper Rotliegend 2 was affected by four tectonically induced Altmark unconformities. (Based on Haq and van Eysinga, 1987; Gast, 1988; Leeder, 1988; Taylor, 1990; Menning, 1988, 1995; Gebhardt *et al.*, 1991; Schneider *et al.*, 1995.)

Permo-Carboniferous boundary (296 or 300 Ma), the Lower Rotliegend is either entirely within the Permian (as implied by Fig. 5.4) or extends down into the Stephanian.

At the other end of the Rotliegend time span, spores with Late Permian affinities have been found in Upper Rotliegend grey claystones in the Auk oilfield (Heward, 1991). Taylor (Chapter 6) suggests that Zechstein deposition took place entirely within the Late Tatarian, the youngest stage of the Permian. Based on a continuous sequence in China, the upper (Permo-Triassic) boundary is now considered to be at 251.2 ± 3 Ma (Claoue-Long *et al.*, 1991; Fig. 5.4), although there is no confirmation that the North Sea Zechstein–Bunter transition conforms to that date.

Recent work in Germany (Hoffmann *et al.*, 1989; Gebhardt *et al.*, 1991) indicates that the complete Rotliegend sequence of the Southern Permian Basin (UR2) was deposited within the early Tatarian, a time span that could be as short as 4 or 5 Ma (U. Gebhardt, personal communication, 1993) or as long as 10 Ma (Gast, 1993b). Indeed, Yang and Baumfalk (1994), working with deduced Milankovitch cycles, estimated that the much thinner Upper Rotliegend of the Dutch offshore took almost 11 Ma to be deposited. To judge from the range of Rotliegend dates published since, say, 1988 (Leeder, 1988; Menning *et al.*, 1988; Hoffmann *et al.*, 1989; Gebhardt, Schneider and Hoffmann, 1991; Menning, 1991, 1992, 1995; Schneider *et al.*, 1995), the above ages are likely to be refined still further, the general trend seemingly being towards a shorter time span for UR2 deposition. Menning (1995) prefers a time span for UR2 deposition of 8 Ma (266–258 Ma). The debate is not over, however, as Gradstein and Ogg (1996), on their chart of the Phanerozoic time-scale, give the Permian an age range from 248 ± 4.8 to 290 Ma BP, with the base of the Late Permian at 256 Ma. No doubt, other estimates will follow.

Apart from the Permo-Triassic boundary in China, which may not apply to the North Sea area, the other key marker is the Illawarra magnetic reversal, which has been recognized near the base of the German UR2 within the Southern Permian Basin and elsewhere in the world. Because it still lacks actual dating, its apparent age seems to hover between 256 and 266 Ma. Whichever of these short time spans for Upper Rotliegend deposition is the more correct, they must have important consequences for both basin development and sediment preservation, factors that will be discussed later. The stratigraphic framework now accepted by both industry and academia in Germany can be found in Plein (1995); it has far-reaching implications for adjacent areas of Europe (see, for example, Glennie, 1997b).

5.1.2 Rotliegend and hydrocarbons

Sandstones of the Upper Rotliegend form a most important reservoir rock for gas in the Southern Permian Basin (see Fig. 5.3). They contain some 4.46×10^{12} m^3 (157×10^{12} ft^3) of proved recoverable reserves, of which 1.2×10^{12} m^3 (42×10^{12} ft^3) are in offshore fields of the southern North Sea (Spencer *et al.*, 1986) and 2.74×10^{12} m^3 (97×10^{12} ft^3) are in the giant Groningen gasfield in the Netherlands (see Table 5.1; and Fig. 5.27, p. 163).

The source for all this gas is the Coal Measures of the underlying Carboniferous, which, depending on the temperature gradient, gave up its gas when buried at depths of between 4000 and 6000 m (Lutz *et al.*, 1975; van Wijhe *et al.*, 1980; see also Chapters 4 and 11, this volume), which could be any time from the latest Permian (Zechstein time) until the Present. The seal is formed by the overlying Zechstein sequence. The Zechstein cycle II (Stassfurt) halite (see Taylor, this volume, Chapter 6) is the most important individual seal, because it is regionally thick and is able to flow and thus heal any fault-induced fracture.

In the Northern Permian Basin, Rotliegend sandstones are oil-bearing in both the Auk and the Argyll fields (see Figs 5.37 and 5.38), the source here being the Upper Jurassic Kimmeridge Clay, which matured deep in the adjacent Central Graben.

5.1.3 Colour

As the name Rotliegend implies, these sedimentary rocks are mostly red-coloured. They acquired their colour post-depositionally when ferrous ions in the groundwater were oxidized to the ferric state. As shown by Walker (1967, 1976), beneath the surfaces of modern deserts, a diagenetic environment conducive to reddening is commonly present below the water-table. In that environment, Walker (1976) suggests that it probably takes thousands of years for a coating of red oxide to form on grains of terrestrial sand. Reddening, however, is probably mostly an early diagenetic event that is more closely related to periods of increased humidity and the formation of temporary soils than to aridity. Besly and Turner (1983), for example, show that the reddening of some Late Carboniferous soil horizons in central England occurred soon after deposition by the dehydration of detrital ferric hydroxides and the oxidation of ferrous iron. Illite age dating, however, indicates that in the Rotliegend of Germany, most clay-mineral growth took place at burial depths between 1.5 and 2.5 km (Platt, 1994). Thus, following their early reddening, the haematite-rich clays probably underwent further diagenetic changes with increased depth of burial.

In the Rotliegend of some North Sea wells, it is noticeable that the red colour is not continuous; it is replaced locally by green or white sediments, commonly with sub-vertical contacts, especially in the vicinity of fractures. These non-red colours are believed to have resulted from the passage of reducing groundwaters, which, in some cases, probably owe their origins to the compaction of coals and carbonaceous shales within the underlying Carboniferous Coal Measures.

The uppermost part of the Rotliegend sandstone sequence is commonly grey or white, which has given rise to the German names Grauliegendes and Weissliegendes. It is thought that Weissliegend sands were above the water-table and dry (and thus were incapable of reddening) until the rapid Late Permian Zechstein marine transgression. This event was soon followed by reducing conditions on the sea floor, which prevented reddening of the submerged Weissliegend (Glennie and Buller, 1983). This lack of reddening will be referred to again in Section 5.3.1.

Clearly, the colour of Rotliegend sediments cannot be taken as an indication of their desert origins. The palaeogeographic, climatic and diagenetic significance of red beds formed in desert and other depositional environments is discussed in considerable detail by Turner (1980). Sedimentological evidence will be presented below to support the desert origins of the Rotliegend sequences by comparison with modern deserts (see also Glennie, 1970, 1972, 1987).

5.2 The Lower Rotliegend

The Lower Rotliegend comprises an association of rocks that is predominantly volcanic in character but does include sedimentary sequences, especially in central Germany (Falke, 1971; Plein, 1978; Gast, 1988; Lützner, 1988; Gebhardt *et al.*, 1991; Stollhofen and Stanistreet, 1994; Schneider *et al.*, 1995; Figs 5.3, 5.5 and 5.6) and Poland (Depowski, 1978; Pokorski, 1989); these sequences were deposited mainly in fluvial and lacustrine environments under a climate that alternated between humid and semiarid, with only sparse evidence of aeolian activity.

The distribution of the Lower Rotliegend is relatively limited when compared with the Upper Rotliegend (cf. Figs 5.1 and 5.6). The Lower Rotliegend volcanics are best developed in northern Germany and Poland, where they were associated with collapse of the recently formed Variscan Mountains, and in the extensional Oslo Graben–Bamble Trough and Horn Graben areas (see, for example, Sørensen and Martinsen, 1987); the volcanics range from rhyolites and ignimbrites to basalts. In Germany, where their thickness ranges up to more than 2000 m, the times of volcanic activity cluster between 288 and 293 Ma BP (Plein, 1993). Michelsen and Nielsen (1993) show that in the north-west–south-east-trending Tornquist fault zone marginal to Fennoscandia, the Lower Rotliegend is confined to a half-graben and is almost entirely in a volcanic facies (minor red claystones); the Upper Rotliegend is absent and a thin Zechstein is in a siliciclastic facies.

Similar associations of Lower Rotliegend sediments and volcanic rocks of about the same age (loosely dated as Autunian), but covering much smaller areas, are known in France, south-west England, south-west Scotland and some of the flank areas of the Mid North Sea–Ringkøbing–Fyn highs. East–west-trending dyke swarms in the Midland Valley of Scotland and the Great Whin Sill (295 Ma; Randall, 1980) in north-east England, which should now probably be attributed to Early Permian rather than Late Carboniferous igneous activity.

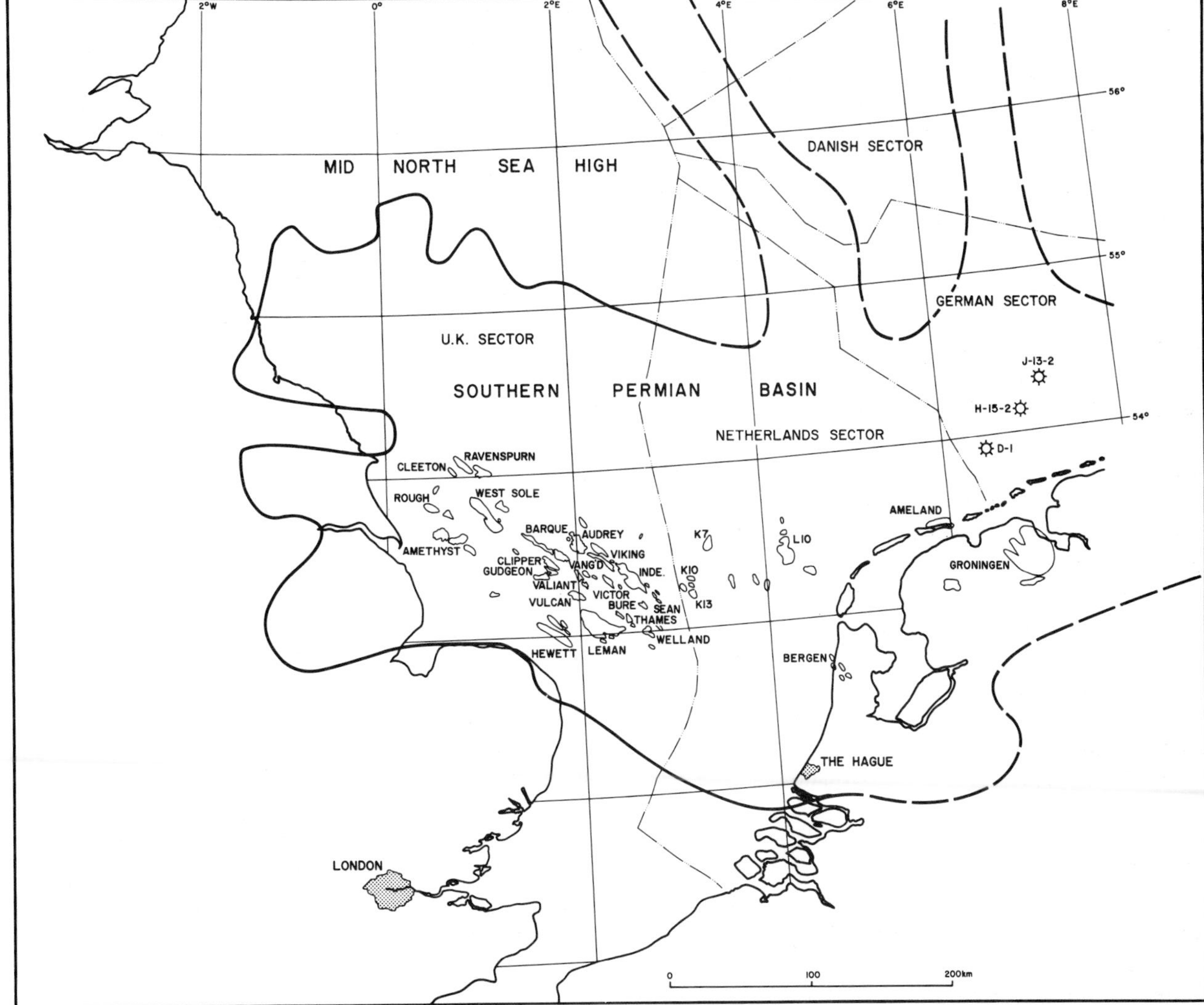

Fig. 5.5 Distribution of Rotliegend offshore gasfields within the Southern Permian Basin. The German offshore wells have a high content of nitrogen.

The Rotliegend volcanics within the UK part of the Central Graben (Quadrants 31 and 39) are now called the Inge Volcanics Formation (Cameron, 1993a). They comprise basalts and tuffs with interbedded mudstone and rare sandstone, reaching a thickness of almost 350 m in well 31/26-1. Similar rocks in the Danish part of the Central Graben have been dated at 265–267 Ma (Ineson, 1993), which may well indicate the time of opening of the Central Graben and the start there of UR2 sedimentation, rather than any direct connection with the Lower Rotliegend volcanics (Glennie, 1997b). It possibly also coincided with the start of UR2 sedimentation in northern Germany.

The rhyolites and ignimbrites of Germany indicate continental volcanism. The basic volcanic rocks, however, and their distribution adjacent to known or inferred faults suggest that their origin was related to the earliest transtensional movements, which would eventually cause the breakup of the recently formed Pangaea. By creating lines of weakness and associated thermal uplift and erosion, these movements were also the forerunners to both the Northern and Southern Permian basins, and possibly also to part of the graben system of the North Sea (Oslo and Horn grabens), Germany (see, for example, the grabens and half-grabens illustrated by Gast, 1988; Fig. 5.7) and Poland (Pokorski, 1989; Kiersnowski *et al.*, 1995); see also Ziegler (1990, encl. 17).

The timing of volcanism in the Central Graben (Ineson, 1993) indicates that initiation of the Viking–Central Graben system was later than the grabens in Germany, although not so late as suggested by Ziegler (1982a, 1990b), who believed that development of the Viking–Central graben system did not begin until the Triassic. Thus it seems likely that initiation of the graben systems of the Southern Permian Basin resulted from at least two phases of transtensional shearing, the first associated with extensive Lower Rotliegend volcanism during the final stages of the Variscan Orogeny or shortly thereafter, and the second possibly at the initiation of UR2 sedimentation.

During the short time span of some 5 Ma or so, early in the Permian, a series of roughly north–south- and north-west–south-east-trending grabens were developed across parts of north-west Europe. Most of these grabens are thought to have formed because of transtensional stresses

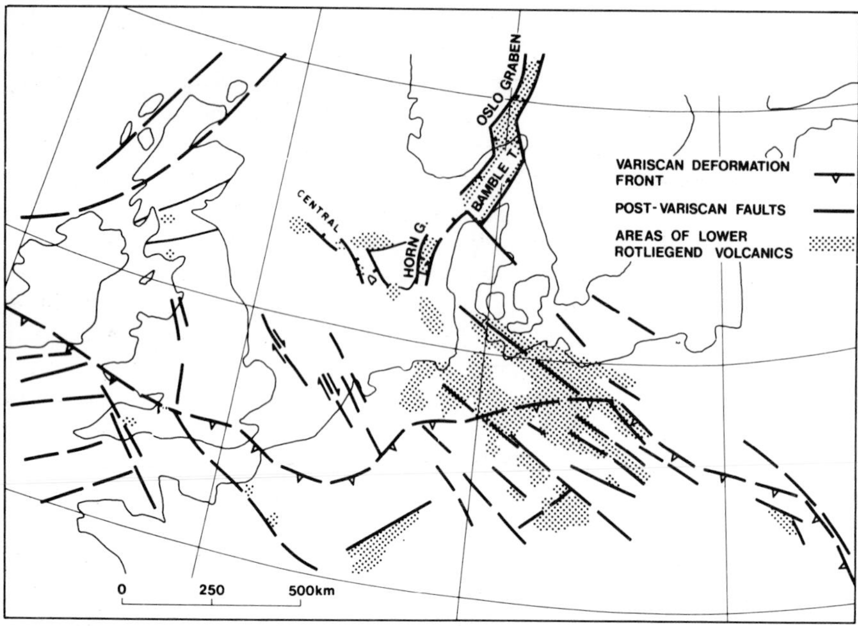

Fig. 5.6 Autunian fault patterns and distribution of Lower Rotliegend volcanics.

associated with the extrusion of deep-seated basic to acid volcanic rocks. The diversity of ages that are beginning to appear for this activity (290 Ma, Germany; 265 Ma, Central Graben) makes it likely that individual grabens of the north-west European system developed one at a time, although currently there are insufficient age data to discern any clear pattern.

Heating associated with Lower Rotliegend volcanism is probably responsible for the crustal uplift and erosion associated with the long-lived Saalian Unconformity (see Figs 5.3, 5.4 and Section 5.4). Some idea of the scale of the Permian differential vertical movements can be gained when one considers that, prior to the Zechstein transgression, erosion removed all the previously deposited Carboniferous strata over large parts of the Mid North Sea system of highs; and, adjacent to the Central Graben, pre-Zechstein erosion additionally cut deep into the Devonian Old Red Sandstone and, locally, older strata, thus implying even greater uplift (see Nielsen and Japsen, 1991: Novling-1, where the Zechstein overlies Silurian strata, and Rønde-1 and Slagelse-1, where the Silurian is overlain by the Upper Rotliegend). This erosion is thought to have resulted from transpression-related uplift initiated at the beginning of the Permian period.

The Late Westphalian inversion along the Sole Pit–Cleveland Hills axis (Figs 2.16 and 5.8) within the Southern North Sea Basin resulted locally in erosion of the total Westphalian (probably over 1000 m thick) prior to deposition of the Upper Rotliegend (see also Kent, 1981). This inversion was thought to have been caused by right-lateral transpression immediately following the Variscan Orogeny (Glennie and Boegner, 1981), but must now be considered in the context of the widespread Saalian Unconformity. Whether thermal heating was involved as well as transpression is less certain.

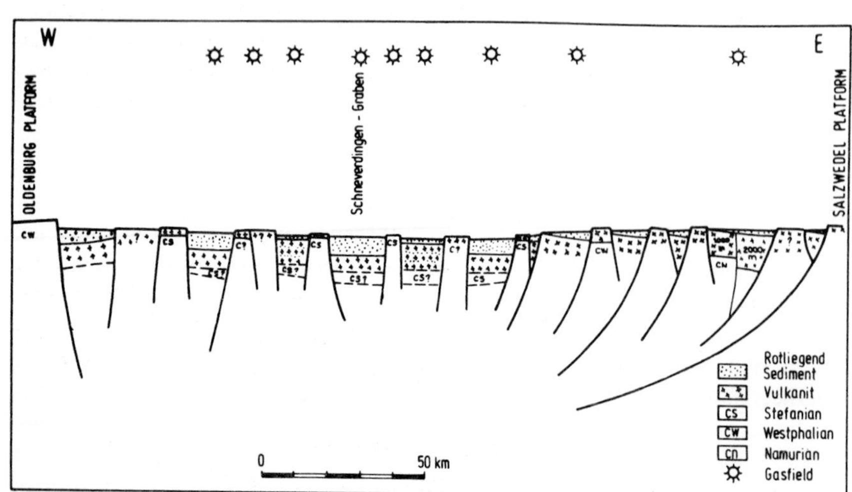

Fig. 5.7 Schematic cross-section through the system of small Permian grabens in the Lower Saxony Basin. Gas is present in the lower part of Rotliegend 2 sandstones in the Schneverdingen Graben and in the overlying Hauptsandstein. (Slightly modified from Gast, 1988.)

5.3 The Upper Rotliegend

5.3.1 Southern Permian Basin

The Upper Rotliegend of this basin is made up of four distinctive facies associations, which have been interpreted as the products of deposition in fluvial (wadi), aeolian, sabkha and lacustrine environments (Figs 5.9 and 5.10; Glennie, 1972; Marie, 1975; George and Berry, 1993, 1994), cores of three of which were illustrated by Glennie (1972; Figs 5.11 and 5.12). All four facies are widespread in the asymmetric Southern Permian Basin, whose floor sloped from south to north over much of its area (see Fig. 5.26). In the southern North Sea area, the sandy facies is referred to as the Leman Sandstone in UK waters, as the Slochteren Sandstone in the Netherlands (see Figs 5.3 and 5.9) and as the Haupt or Dethlingen Sandstone in northern Germany (Burri *et al.*, 1993); in the Northern Permian Basin it is the Auk Formation, and in the Moray Firth the Findhorn Formation (Cameron, 1993a). The clayey lacustrine facies in both the UK and Dutch areas is named the Silverpit Formation (Rhys, 1974; NAM and RGD, 1980) and the Hannover or Wechselfolge (Transition Beds) or Elbe Series in Germany (see Fig. 5.20). The characteristic lithologies that make up these formational units have been studied in cores and can be deduced from the wireline logs of, for instance, the type and reference sections depicted in Fig. 5.9 for different parts of the North Sea area.

Within both the Leman Sandstone and Silverpit Claystone formations, George and Berry (1993, 1994) recognize five distinct drying-upward cycles, which can be matched across the UK and Dutch sectors of the southern North Sea (see also Yang and Baumfalk, 1994, for the Dutch offshore). A simplified and slightly modified version of one of George and Berry's (1993) figures is given in Fig. 5.13. These cycles are considered to reflect long-term fluctuations in the Rotliegend climate.

Fluvial facies

The fluvial (wadi) sequences are characterized by the occurrence of curled clay flakes, indicating frequent subaerial exposure and desiccation typical of an arid environment, together with rip-up clasts of varying size. Some thicker clays seem to have had their cracks infilled with sand from above; others were injected with a slurry of sand and water from below to form sandstone dykes (see Fig. 5.11).

The sandstones directly below layers of clay flakes can display centimetre to decimetre foresets with low dip and commonly discontinuous laminae, which are interpreted as having been deposited by flowing water in shallow channels. These sandstones are locally conglomeratic (more proximal facies), with some of the contained pebbles consisting of red clay, similar in character to the bedded claystones; the rounded edges of these 'rip-up clasts' imply some distance of transport from the site of their original deposition. Other conglomerates seem to consist almost entirely of clasts of vein quartz (George and Berry, 1993). Unlike the associated aeolian sands, the fine-grained fluvial sandstones and silts are commonly micaceous, a factor that can give a clay-like 'kick' on a gamma-ray log (see, for example, Conway, 1986, Fig. 9).

Laminated sandstones locally grade up into apparently homogeneous sandstones that lack sedimentary structures. Some homogeneous sands contain large clay pebbles and are obviously of fluvial origin; others, however, grade down into sandstones with the well-defined laminae typical of aeolian sands, which are commonly interbedded with those of fluvial origin. Wadi sandstones tend to be noticeably more argillaceous than those of aeolian origin, and are generally well cemented with primary dolomite; they form good reservoirs for hydrocarbons only locally (e.g. in the Groningen Field; see below and Fig. 5.27).

Many of the above features indicate that these essentially fluvial sequences should be interpreted as the deposits of ephemeral streams in an arid or semiarid environment, and may thus be referred to as wadi deposits; some have aeolian interbeds. Wadi sandstones are common along the southern margin of the Rotliegend Basin (see Fig. 5.10) and especially in Holland, Germany and Poland (Ziegler, 1990a, encl. 18). They tend to be best developed low in the sequence (Stäuble and Milius, 1970; Arthur *et al.*, 1986). In the UK sector of the North Sea, the best development is in the west (e.g. the Rough Field: Goodchild and Bryant, 1986; Stuart, 1991; Amethyst: Garland, 1991; Ravenspurn South: Heinrich, 1991b; Ravenspurn North: Ketter, 1991b; Turner, P. *et al.*, 1993) where they have been transported to the north, east and north-east across the pediments of the East Midlands and South Hewett shelves (George and Berry, 1993) to become interbedded with sandstones of aeolian origin. A schematic diagram of how the fluvial sequences crossed the lake-margin sabkha in the Ravenspurn area is illustrated by Ketter (1991b). Wadi sandstones are well displayed in the subsurface of the east Netherlands (Bungener, 1969) and in outcrop in other

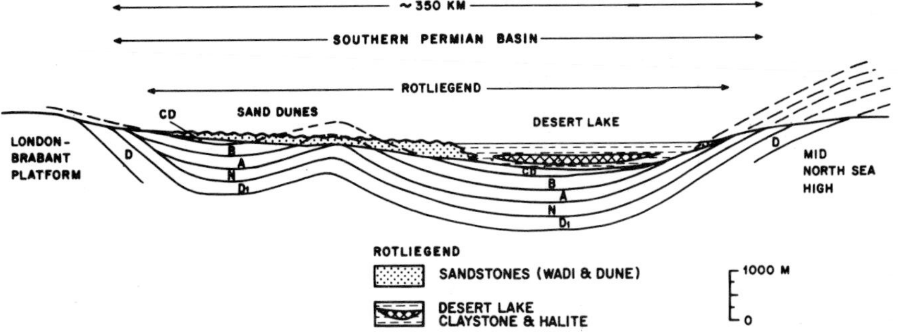

Fig. 5.8 Cross-section illustrating the relationship between the Rotliegend sedimentary basin and the eroded pre-Permian strata. Note the overall asymmetry of the basin and the axis of Late Carboniferous Sole Pit inversion. A, Westphalian A; B, Westphalian B; CD, Westphalian C–D; N, Namurian; D_1, Dinantian; D, Devonian. (From Glennie, 1983b.)

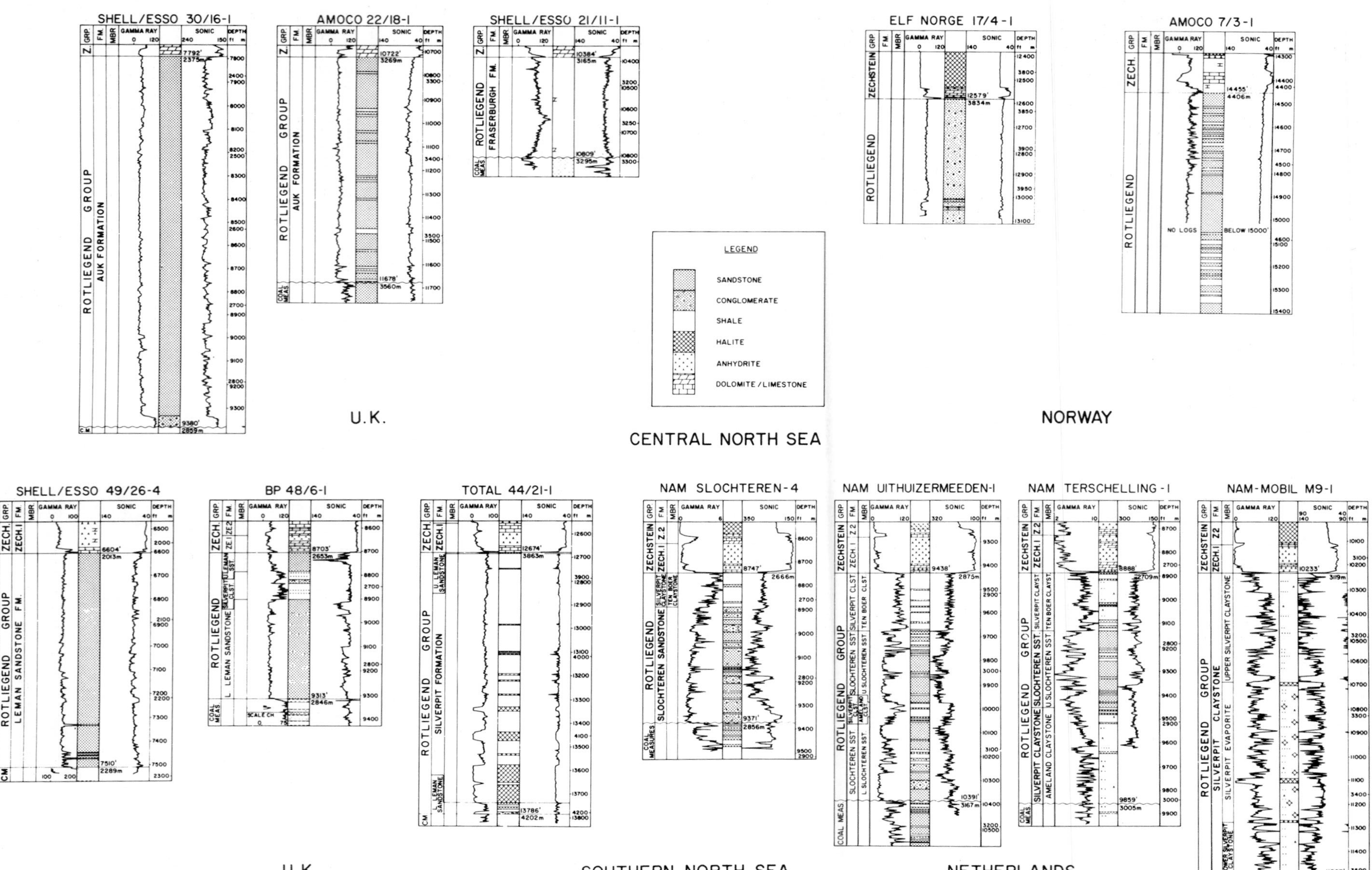

Fig. 5.9 Rotliegend type and reference-well sections.

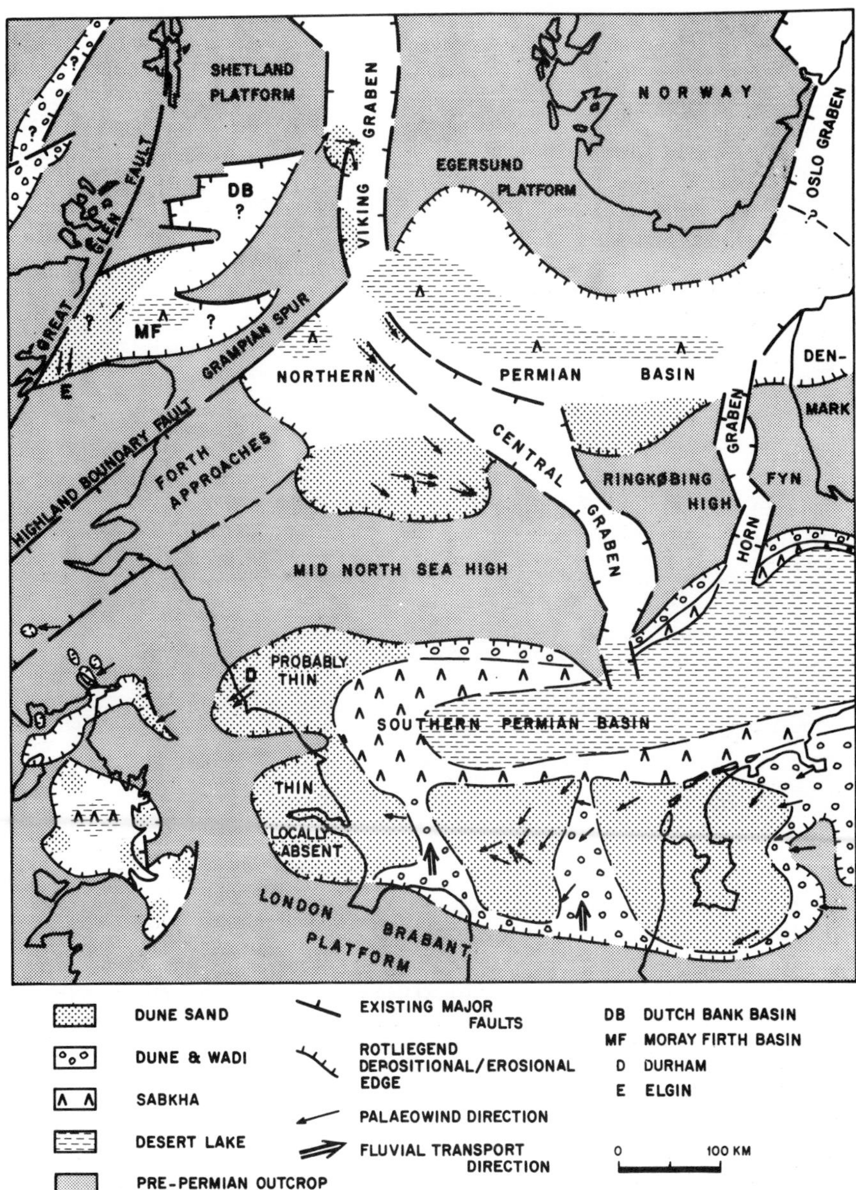

Fig. 5.10 Facies distribution and pattern of Early Permian (Rotliegend) winds in the North Sea area. Areas of dune sand are stippled.

basins with areas of Permian relief, as in Arran, Scotland (Clemmensen and Abrahamsen, 1983), south-west Scotland (Brookfield, 1980) and Devon, England (Laming, 1966).

In a series of palaeogeographic maps of the Dutch offshore, George and Berry (1994) show the interplay between tectonics and fluvial, aeolian and lacustrine sedimentation. They recognize both massflow- (dry type) and stream-flow- (wet type) dominated alluvial fans, as well as mass-flow-dominated fan deltas. Among other facies changes, they note that mass-flow sands grade into fan deltas when sediment progrades into a standing body of water. And fan-head entrenchment occurred during periods of tectonic stability, leading to reduced relief, which encouraged stream-flow processes and the deposition of multistorey channel-fill deposits.

Mention has already been made of the relatively good reservoir quality of the wadi sands and conglomerates found in the Groningen Field (porosity (ϕ) = 15–20%; k = 0.1–3 millidarcies (mD)); the same can be said of the Rough Field in the UK offshore (ϕ = 5–17%; k = 0.07–78 mD; Goodchild and Bryant, 1986; Ellis, 1993). This seems not to be the case with many other fields, where the wadi sands are more tightly cemented than aeolian sands (e.g. Leman Bank: Hillier and Williams, 1991; Barque: Farmer and Hillier, 1991a). The differences cannot be accounted for by variations, for instance, in their burial-related diagenetic history, as both fluvial and aeolian sands should be affected by the same deep burial. In areas such as Leman Bank or Victor (Conway, 1986), however, the wadi sands are much more tightly cemented than those of aeolian origin, suggesting a primary depositional origin for the cement. Thus it is likely that differences in the reservoir quality of the wadi sands stem from differences in their respective environments of deposition. The wadi sediments of Groningen were deposited in a basin-margin alluvial fan, whose surface is likely to have been well above the level of the permanent water-table. Wadi and aeolian sands alternated. In Rough, a similar situation occurred, although the fluvial sands are now thought to have been deposited by one major and two minor wadis that cut into the underlying and flanking dune sands (Ellis, 1993). Indeed, Goodchild and Bryant (1986) point to the rarity of adhesion ripple sands in the Rough Field as evidence of a deep water-table.

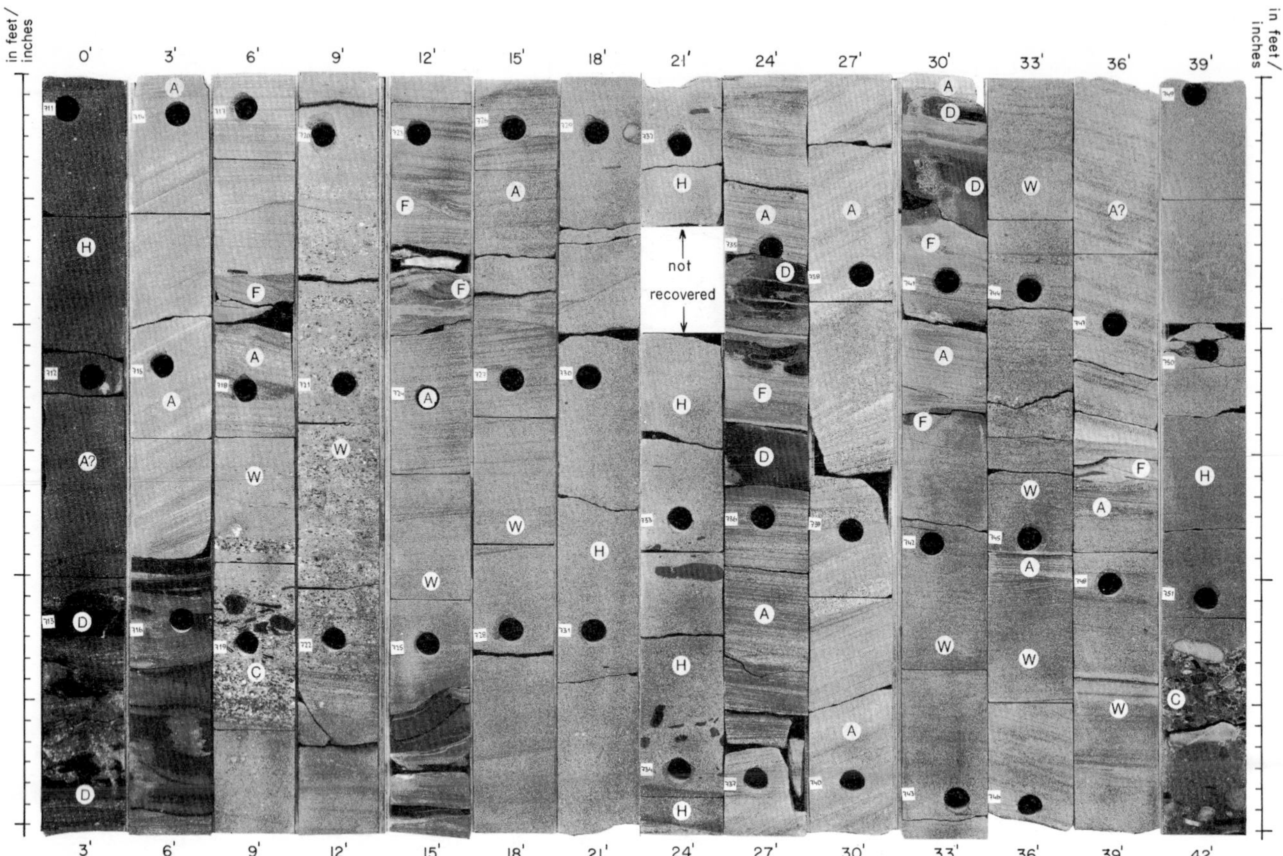

Fig. 5.11 A sequence of Rotliegend cores starting about 8 ft above Carboniferous (Westphalian) shales. The sequence includes alternations of conglomerates 'C' (including clay pebbles and flakes), curled and broken clay layers (black) that are probably still in their original site of deposition, sandstone dykes 'D', fluvial sands 'F', structureless (homogeneous) sands and well-laminated aeolian sand 'A'. Other weakly laminated sands 'W' could have been deposited by either wind or water. The numbered holes are of plugs cut for porosity/permeability determination. (From Glennie, 1972, reproduced by permission of the American Association of Petroleum Geologists.)

In both fields, by the time burial brought these beds to the level of the water-table, they were probably too far from the surface for evaporation, and thus cementation, to be effective.

In a more modern setting, James (1985) has shown that, in the outwash sands of the Quaternary Atherton Formation in northern Indiana, diagenetic alteration and cementation are limited to within 7 m of the present ground surface. Thus, in the better-drained Rotliegend localities, the sands and conglomerates were subjected to a minimum of cementation. In the low-lying desert plains between the Rough and Groningen areas, however, where associated adhesion ripples and anhydrite nodules testify that water was close to the surface during sedimentation, the wadi sands are well cemented with primary dolomite.

In both the Groningen and Rough Fields, the reservoir quality of the interbedded dune sands tends to be only slightly better than in those of wadi origin, whereas in Leman Bank and Indefatigable, for instance, only the dune sands form an effective reservoir. In contrast, Gage (1980) reports that, in the Viking Field, aeolian sands that are interbedded with those of wadi origin can also be strongly cemented, presumably because of near-surface saturation by carbonate-rich wadi waters. A similarly poor reservoir quality in fluvial sands is found in the Broad Fourteens Basin, where the porosity is in the range 4–15%, compared with the 15–23% of interbedded aeolian sands (Oele *et al.*, 1981).

Within graben settings in northern Germany, fluvial sands within the basal Upper Rotliegend Schneverdingen Group form reservoirs that are perhaps more important than those of aeolian origin (Burri *et al.*, 1993).

Wadi sediments are the source of virtually all aeolian sands, including those of lake-margin sabkha environments. With the absence of any sustained fluvial flow, there are no lake-margin deltas of any size, undoubted fluvial sequences in this environment commonly being only 10–50 cm thick and only locally exceeding a metre or more. In this respect, the apparent importance given to fluvial sands in the maps of Verdier (1996) is perhaps more an indication of where wadis supplied fluvial sediment rather than the volume of what can now be recognized in core or on wireline logs.

Notwithstanding the above, fluvial sands and conglomerates in important amounts do exist as the outcome of flash floods or seasonal rainfall. In the Annerveen Field, just south of the Groningen Field, for example, basal Rotliegend conglomerates make up a sequence, some 30–40 m thick, deposited in an alluvial-fan to braided-stream environment (Veenhof, 1996).

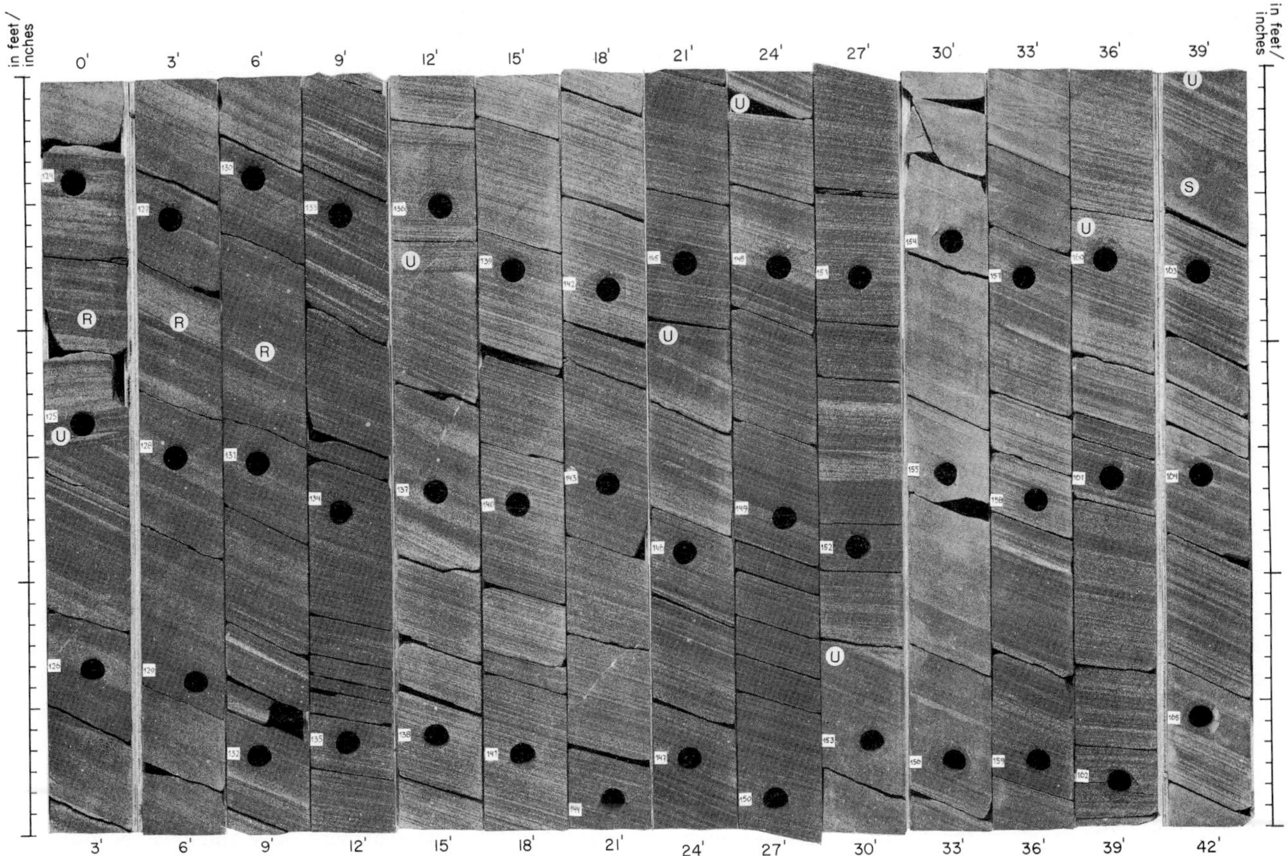

Fig. 5.12 Several sets of Upper Rotliegend dune sandstone separated by major bounding surfaces 'U'. Each set begins with subhorizontal laminae, the angle of inclination increasing upward until truncated by the next bounding surface. Ripple-drift sands occur at 'R', and a small slump at 'S' probably resulted from rainfall. Most of the inclined bedding probably represents grainflow sands near the foot of the avalanche slope of a transverse dune. (From Glennie, 1972, reproduced by permission of the American Association of Petroleum Geologists.)

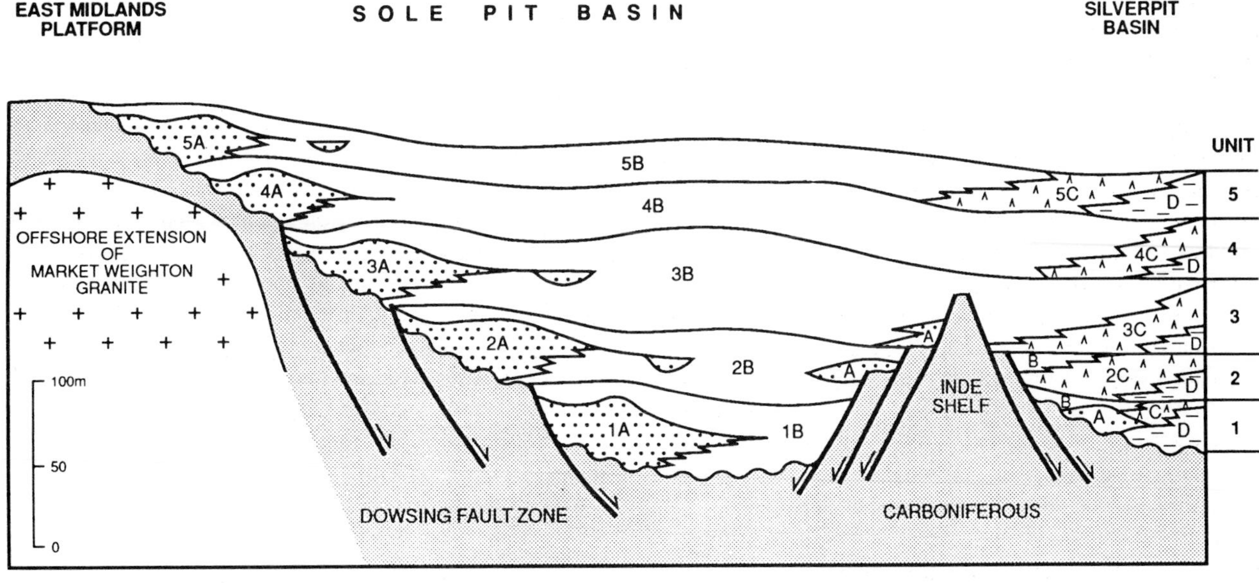

Fig. 5.13 A depositional model for the Upper Rotliegend 2 sedimentation in the UK Southern Permian Basin (simplified and slightly modified from George and Berry, 1993). Five drying-upward cycles are thought to represent climatic changes induced by high-latitude glaciations. The Sole Pit Basin is inferred to have extended by back-stepping of the Dowsing fault system. The faults are shown steeper than those of George and Berry (1993) and, unlike their version, the offshore extension of the Market Weighton Granite is not shown uncovered.

Aeolian facies

The reservoirs of most Rotliegend gasfields are dominated by sandstone of aeolian origin (Nagtegaal, 1979). The aeolian sands are most readily recognized when they conform to a series of criteria, including: (i) well-defined planar or trough-bedded strata, in which adjacent laminae commonly show sharp grain-size differences between that of very fine sand and some 1 or 2 mm; (ii) many intraformational unconformities, above which the laminae are commonly horizontal; there is an upward increase in the inclination of the laminae to an angle of some 20–25° before being terminated by the next low-angle truncation; and (iii) a lack of mica flakes. For other criteria, see Glennie (1987), Heward (1991) or George and Berry (1993). Sequences of almost continuous aeolian bedding that fit these criteria (see Fig. 5.12) locally reach thicknesses of 100–200 m or more.

Specific environments of aeolian deposition can be identified in core material on the basis of lamination type; these laminae are generally referred to as wind-ripple, grainfall, grainflow and horizontal dune-base or bottomset beds. As reservoir quality varies from one lamination type to another, their recognition can prove to be important in reservoir studies. They have been described in some detail by Heward (1991) for the Auk oilfield and by George and Berry (1993), who recognize the differences between dune-top and dune-base facies (commonly bimodal, more argillaceous and better cemented), which are separated by the more porous and permeable dune-core (foresetted) facies. The interdune sequences are commonly separated into dry (horizontal bimodal laminae), damp (more argillaceous with adhesion ripples) and even wet facies (some soft-sediment deformation). The latter two are well cemented and thus tight. The distribution and upward-drying nature of these facies is illustrated by George and Berry (1993; see also Fig. 5.13).

Wind-ripple structures form by the near-surface traction (saltation, limited rolling and surface creep) of sand grains across the desert surface under the influence of prevailing winds. As the size of the saltating sand grains varies with wind strength, these sands are usually finely laminated. Wind-ripple structures generally form on low-angle desert surfaces. They can also develop across the major leeward slipfaces of dunes. This happens under the influence of winds that are temporarily transverse to the prevailing wind. Wind ripples tend to form finely laminated sands typical of the low-angle bottomset sands in Fig. 5.12. During a study of the Middle Jurassic Page Sandstone in northern Arizona, Goggin *et al.* (1986) noted that wind-ripple permeabilities lay in the range <100–2600 mD; this compares with 250–3600 mD for grainflow sands and less than 400 mD for those of interdune areas.

Grainfall laminae result from the fallout of saltating sand grains on the leeward side of a dune's crest. The size of the contained grains varies with source as well as with the intensity of the sand-transporting wind. In large migrating dunes, grainfall laminae are seldom preserved, but they could be important in small dunes of an otherwise interdune area. In the crestal area of a dune, dry grainfall sands are not stable at angles greater than 34°; the angle of slope is reduced to about 30° as the sands slide down the avalanche slope, the larger and rounder grains rising to the surface, thereby creating inverse grading. At the time of deposition, grainflow sands probably had a porosity approaching 50%. A single grainflow can vary in thickness between < 1 and about 30 cm, and it toes out rapidly at the foot of the slope, where it grades into wind-ripple deposits. The thicker, higher-angle sands in Fig. 5.12 are interpreted as the product of grainflow.

Prosser and Maskall (1993) studied a core from the Auk Field that had been taken during deviated drilling. At one point, the core was parallel to the grainflow laminae of an avalanche slope for almost 5 m. Detailed studies, using a probe permeameter, showed that permeabilities could be correlated for distances of almost 1 m parallel to the sedimentary dip, which is about the length of individual pinstripe laminae. As expected from other studies (e.g. Goggin *et al.*, 1986), permeabilities in grainflow laminae (13–960 mD) were much higher than in wind-ripple sheet-sand laminae (0.75–489 mD).

Dune sands (see Fig. 5.12) form the main reservoir rock for gas in the Southern Permian Basin. At the time of deposition their porosities probably averaged around 42% (Hunter, 1977). Both in modern dunes and after diagenesis at considerable depths of burial, the actual reservoir quality (porosity and permeability) of the sand varies according to its original mode of deposition; higher porosity on the relatively high-angle avalanche slope and lower porosity in the low-angle dune base. In the Sean Field, for instance (see also Fig. 5.31), Ten Have and Hillier (1986) noted average porosities of 21% and average permeabilities of 650 mD in avalanche-slope sand; these reduce to averages of 17% and 166 mD, respectively, in dune-base sands. In marked contrast, associated wadi and interdune sabkha sands have averages of only 11% and 5 mD (Hobson and Hillier, 1991). Differences in reservoir quality of this type can be matched in most other Rotliegend gasfields.

The foresetted sandstones preserved between the horizontally laminated bottomset beds and the overlying plane of truncation generally represent preservation of only the lowest part, commonly only a few metres, of any one migrating dune, whose height may have been many times greater. Arthur *et al.* (1986), however, have recorded set thicknesses of up to 100 ft (30 m) in UK Block 49/28.

Analysis of the orientation of the aeolian bedding, both in outcrop and in wells (cores, dip-meter logs; Figs 5.12 and 5.14), indicates that probably only transverse dunes existed over most of the Southern Permian Basin (for a different interpretation, see George and Berry, 1993, Figs 6 and 16), with linear dunes limited mostly to basin margin areas, such as County Durham (see Figs 5.10 and 5.26); for a possible along-length transverse origin of the Durham dunes, see Steele (1983).

Within the Southern Permian Basin, the Late Permian winds blew from a general easterly direction (Glennie, 1972, 1983a; van Wijhe *et al.*, 1980; Steele, 1983; Luthi and Banavar, 1988), which, after correction for the rotation of north-west Europe since the Permian, suggests that the Rotliegend was deposited in a Trade Wind Desert of the northern hemisphere, similar to the Sahara of today (Figs 5.15 and 5.23). As can be seen from the facies distribution maps of Verdier (1996), aeolian sands seem to have been trapped preferentially to the west of the Gronin-

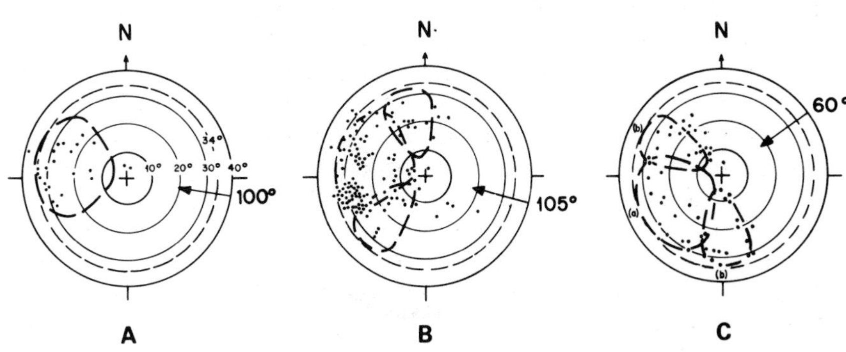

Fig. 5.14 Poles of bedding attitudes and deduced palaeowind directions derived from dipmeter logs and field outcrops of Rotliegend dune sand from the Netherlands, southern North Sea and north-eastern England. A. 24 dips (Lievelde-1). Wind from N 100°E. B. 127 dips (49/26-2). Wind from N 105°E. C. 54 dips (combined dip data from Field House and Crime Riggs sandpits. County Durham). Wind from N 60°E. Dip attitudes in area (A) are typical of transverse dunes and in (C) of longitudinal dunes. (From Glennie, 1983b.)

gen High and Friesland Platform, across which they had been transported (see van Wijhe *et al.*, 1980, Fig. 5; Petroleum Geological Circle, 1993, Fig. 10).

This interpretation is disputed by Sneh (1988), who believes that the Permian wind blew uniformly from the north across both the Northern and Southern Permian Basins, the dune axes of both areas being oblique to the wind, in a style described by Tsoar (1983) for a seif dune in Israel. There seem to be no large systems of modern dunes, however, that display such marked changes in axial orientation relative to the prevailing winds and thus provide a model for the interpretation that Sneh (1988) invokes for the Rotliegend. Furthermore, because the thick sequences of dune sands along the southern flanks of both the Southern and Northern Permian Basins had mud-rich desert lakes on their northern margins, it is difficult to imagine the derivation of clean aeolian sands from northern sources.

The sands of dry interdune areas are characterized by low-angle to horizontal wind-ripple laminae, commonly with a bimodality in grain size. In depositional areas that lie above the water-table, such laminae are referred to as bottomset beds where they grade up into the more steeply inclined grainflow sands left by the migrating dune (see Fig. 5.12).

Where deposition is within capillary reach of the water-table, wind blown sand and dust particles stick to the damp subhorizontal interdune surface to form the typical wavy lamination of adhesion ripples. Adhesion ripples in Rotliegend sands were first recognized in the Groningen Field. The local association of small blebs of anhydrite with the adhesion ripples testifies to the hot arid climate, which caused gypsum to crystallize within the sands. A crust of halite is present at least seasonally during deposition in modern desert adhesion-ripple environments but dissolves when the horizon passes below the water-table. The halite is hygroscopic and can trap saltating particles, which become deformed after burial solution. George and Berry (1993) differentiate between damp and wet interdune areas, the latter being recognized by the local occurrence of soft-sediment deformation and biogenic structures.

Where the water-table is close to the surface of some interdune areas (e.g. in Indefatigable; Fig. 5.16), adhesion ripples are common. Their absence (e.g. Leman) indicates a deeper water-table. Adhesion ripples become more common as the Silverpit desert lake is approached at the northern end of the Sole Pit Basin and in the Ravenspurn–Cleeton area (Heinrich, 1991a,b; Ketter, 1991a, b; Turner, P. *et al.*, 1993). In Victor and the V Fields, adhesion ripples are well developed in two field-wide zones (Conway, 1986; Pritchard, 1991), which can possibly be correlated with the Indefatigable Field.

Lacustrine facies

The lacustrine facies consists primarily of red-brown mudstone with minor siltstone. Several halite horizons occur in the lower half of the sequence in UK waters (Figs 5.3, 5.9, 5.17 and 5.26), but in north Germany and the adjacent offshore area, up to 16 beds of halite account locally for over 450 m out of a total of 1800–2000 m of the Upper Rotliegend (Figs 5.18 and 5.19). Indeed, the combined thickness of salts in northern Germany is sufficient for the sequences to have reacted diapirically (see, for example, Plein, 1978), the highly deformed lacustrine beds being known as the Haselgebirge Facies. As can be seen from cross-sections in Kockel (1995), the Rotliegend diapirs occur in the south-eastern German sector of the North Sea

Fig. 5.15 Permian latitudes of north-western Europe with superimposed wind directions.

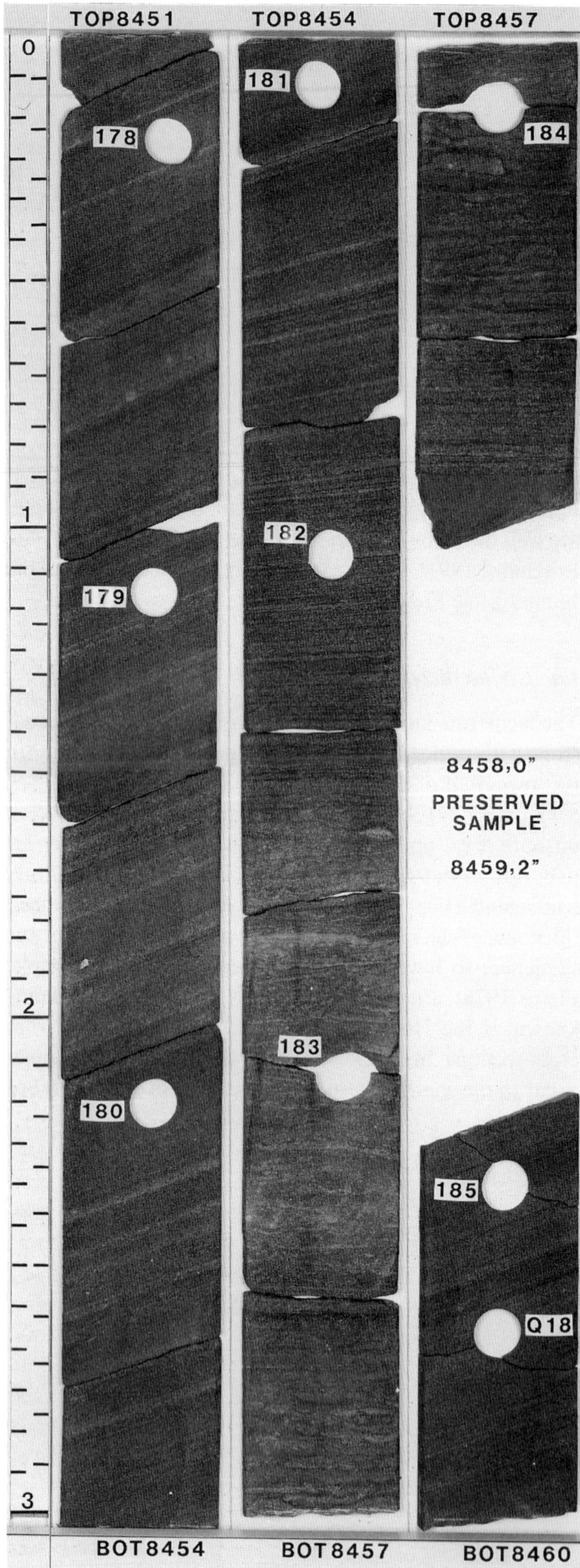

Fig. 5.16 Cores showing sands of wavy-laminated horizontal adhesion ripples overlain by ripple-drift sands, which grade up into those of grainfall origin on an avalanche slope. Rotliegend, Shell-Esso well 48/19a-4.

and extend eastward to the mainland. In the Glückstadt Graben, for example, at the southern end of the Danish peninsula, Rotliegend halite again occurs within the same diapiric structure as halite of Zechstein age (Best *et al.*, 1983, Fig. 4), mobilization probably being triggered by Triassic earth movements associated with the Hardegsen disconformity (see Chapter 7). The dominant north–south fault pattern of the area is intersected by some east–west-trending cross-faults. Diapiric uplift of the Rotliegend salt was probably assisted by upward drag at the time of maximum diapiric movement of the overlying Zechstein.

In parts of north Germany and the Netherlands, the desert lake facies extends much farther across the Hauptsandstein–Slochteren dune sands than over the equivalent facies in UK waters (Figs 5.3 and 5.20). The basin-centre areas of the lacustrine sequence contain no known sedimentary structures indicative of subaerial desiccation or erosion (see for example, Fig. 5.17), although this may, in part, reflect the lack of coring in commercially unproductive rocks. The desert lake is therefore believed to have been a constant rather than an ephemeral feature. At its fullest extent, including its sabkha margins, the lake must have covered an area of some 1200 km east–west by over 200 km north–south (see Fig. 5.1).

Although the lacustrine facies achieves a thickness of some 1800–2000 m in northern Germany, it was formerly thought to be almost devoid of fossils. Recent work in Germany, however, has shown that there are fossils associated with certain marker beds. Gebhardt (1994) believes that within at least the German part of the desert lake, salt precipitation was essentially a continuous process during deposition of the Havel Subgroup, resulting partly from the concentration of fluvial and near-surface groundwater. In the overlying Elbe Subgroup, salt precipitation was punctuated by several short-term incursions of less saline water of marine origin, which carried marine faunas into the desert lake. Beneath several halite horizons, the fossils are of lacustrine type. Above these halites, the fossils are of marine origin and are replaced upwards by fresh- to brackish-water faunas before conditions of high salinity again meant that aquatic faunas could no longer survive. Marine foraminifera (*Spirillina*) have also been described from the upper part of the Lower Sandstone member of the Leman Sandstone in two wells of the West Sole Field (Butler, 1975).

These marine faunas seem to indicate that limited marine flooding of the Southern Rotliegend Basin was possible on relatively rare occasions. The marine faunas of the Bänderschiefer, in the top metre or so of the German desert-lake sequence, have fossils with strong Zechstein affinities (see list in Plumholf, 1966), and can probably be attributed to the last, and perhaps biggest, of the temporary marine incursions just prior to the main Zechstein marine transgression.

The lacustrine facies lacks any reservoir development; to the contrary, over most of its depositional area, it forms an effective seal for hydrocarbon accumulations within the underlying Carboniferous.

Sabkha facies

Between the deposits of the desert lake and the more southerly depositional areas of the wadi and aeolian sands (see Figs 5.1 and 5.10) is a broad band of poorly bedded clays, silts and sands, up to 20 or 30 km wide, which display

Fig. 5.17 Core showing bed of halite that grades up into almost structureless red-brown argillaceous siltstone of the Rotliegend desert-lake facies. Shell-Esso well 44/28-2.

many of the features of a sabkha (e.g. mud cracks, sandstone dykes, adhesion ripples, anhydrite nodules). These features collectively indicate a humid depositional surface that was subjected to variable degrees of aeolian sedimentation and subaerial desiccation and deflation in an arid climate. The sabkha sediments represent the area that either was covered by water only during the maximum extensions of the desert lake (see Fig. 5.1; Wrigley *et al.*, 1993) or was within capillary reach of the groundwater for long periods of time.

Proximity of the surface to the water-table was the single factor controlling whether the sediment was to be wavy-laminated adhesion sands and silts trapped on a damp, salt-encrusted surface or horizontally laminated sheet sands and foresetted dunes that had migrated over a dry surface; that proximity will have resulted from climatic changes between hyperaridity and relative humidity, coupled with the accommodation space resulting from the interplay between subsidence and sedimentation (see also Section 5.4.2). A reduced rate of subsidence leads to a drier surface; an increased rate of subsidence could lead to the temporary development of local ponds or small lakes.

Sediment transport along the southern margin of the Rotliegend desert lake was under the general influence of east winds. At times of low water-table, a slightly deviant wind from the north-east would distribute deflated clays, silts and salts landward, whereas a wind from the south-east would spread sands, deflated from wadi channels, obliquely towards the lake margin, where water, either free or in the form of a damp surface, would eventually stop its distribution; we know that such variable winds existed (see, for example, wind rose, Fig. 5, in van Wijhe *et al.*, 1980). Gast (1991; see also Burri *et al.*, 1993) suggests that in Germany lake-shoreline sands extended westward under the influence of wind-driven currents and formed important reservoirs for gas. It seems more likely, however, that these sands are mainly of aeolian origin and were spread across the flat coastal sabkha as sheet sands and dunes in the drier areas and as sand-rich adhesion structures on the more humid surfaces see Fig. 5.20.

During lake and water-table lowstands, aeolian sands were spread across the sabkha surface, either as aeolian sand sheets and dunes or as sand-rich adhesion structures. If their porosity was not destroyed by primary cementation, such sands could retain reservoir quality when covered again by lacustrine claystones. The Dutch offshore Ameland Field is probably in the right palaeogeographic location to possess some reservoir sands of these types (cf. Figs 5.1, 5.5 and 5.10; see also George and Berry, 1994) and such facies associations have been described from the Ravenspurn and Cleeton areas (Ketter, 1991b; Heinrich, 1991; Turner, P. *et al.*, 1993). Martin and Evans (1988) describe the reservoir quality and the wireline-log response to the interplay between sandy and silty sediments deposited in the lake-margin environment of the North Viking and Venture gas accumulations.

The crust of halite that typifies modern sabkha surfaces will have been deflated by the Permian wind during periods of lake lowstand, or have dissolved; during the next high lake level or on passing below the water-table during burial, the halite will pass into solution. Gypsum crystals (later anhydrite nodules), on the other hand, which form some 20–50 cm below the sabkha surface because of ionic concentration of the groundwater during surface evaporation, would normally be preserved as nodules or even as horizons of cementation within interbedded sands. The absence of anhydrite nodules possibly indicates free water in the

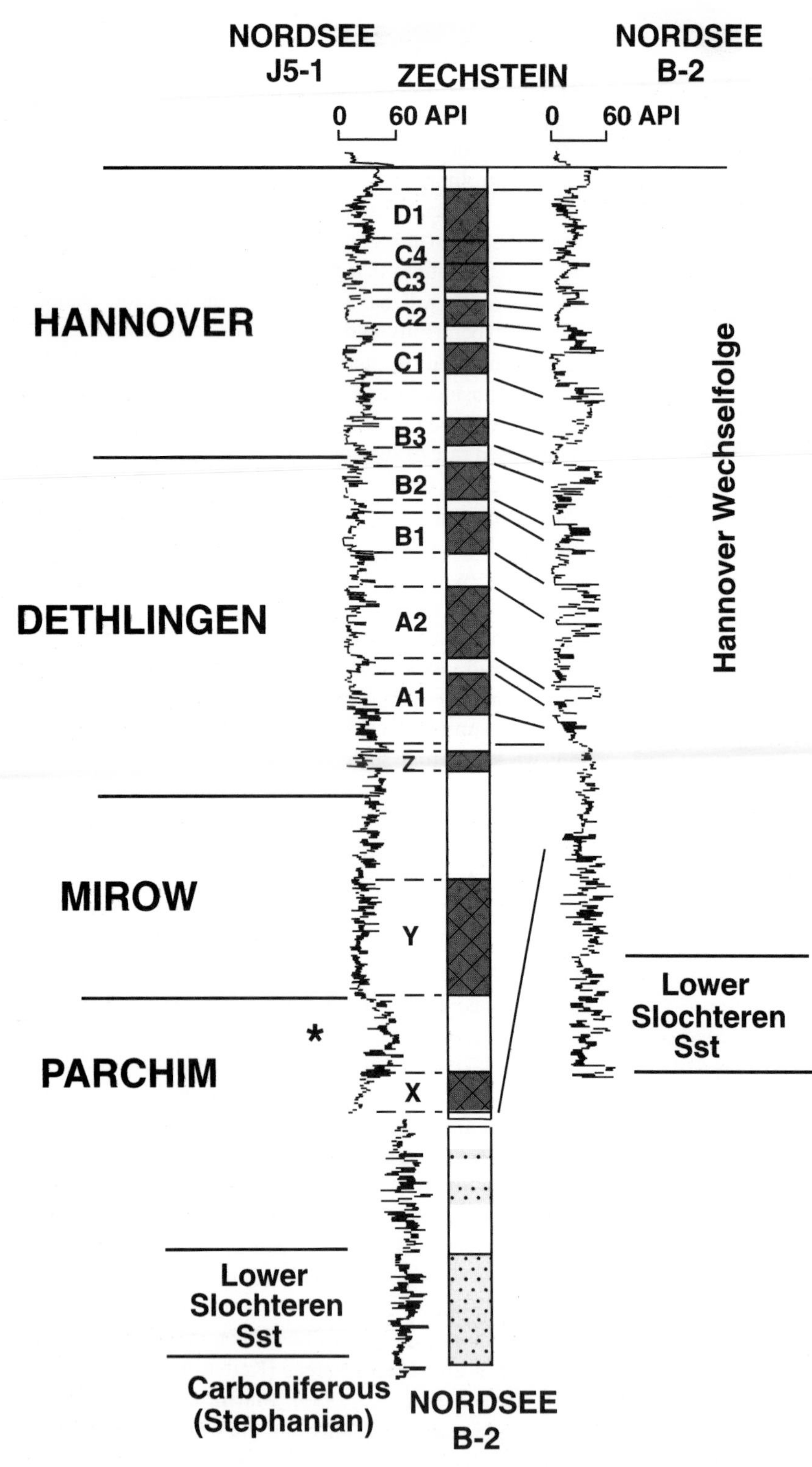

Fig. 5.18 The type log for the Upper Rotliegend desert-lake facies, Helgoland area, German North Sea; wells J5-1 and B-2, with identified halites X to D1. (From Hedeman *et al.*, 1984a, modified to show correlation with other UR2 formations of northern Germany and the approximate position of the Illawarra magnetic reversal (U. Gebhardt, personal communication, 1993).)

overlying sequence or subsidence and sedimentation that was too rapid for gypsum to be precipitated in recognizable nodules.

The Weissliegend

In the southern North Sea and the Netherlands, the Rotliegend sequence is developed, from south to north and from base to top, in the following generalized facies (see Fig. 5.10): (i) mixed wadi and aeolian; (ii) mainly aeolian; (iii) sabkha; and (iv) desert lake. The whole sequence is capped by the basal Zechstein Kupferschiefer, with, in most areas, only a few centimetres to at most a few metres of clearly marine-reworked sands between them (Fig. 5.3).

At many localities in the belt of dune sands, and for a thickness beneath the Kupferschiefer of up to 50 m (150 ft) or more, a sequence of uncoloured (Weissliegend), structureless sands alternate with and grade into both highly

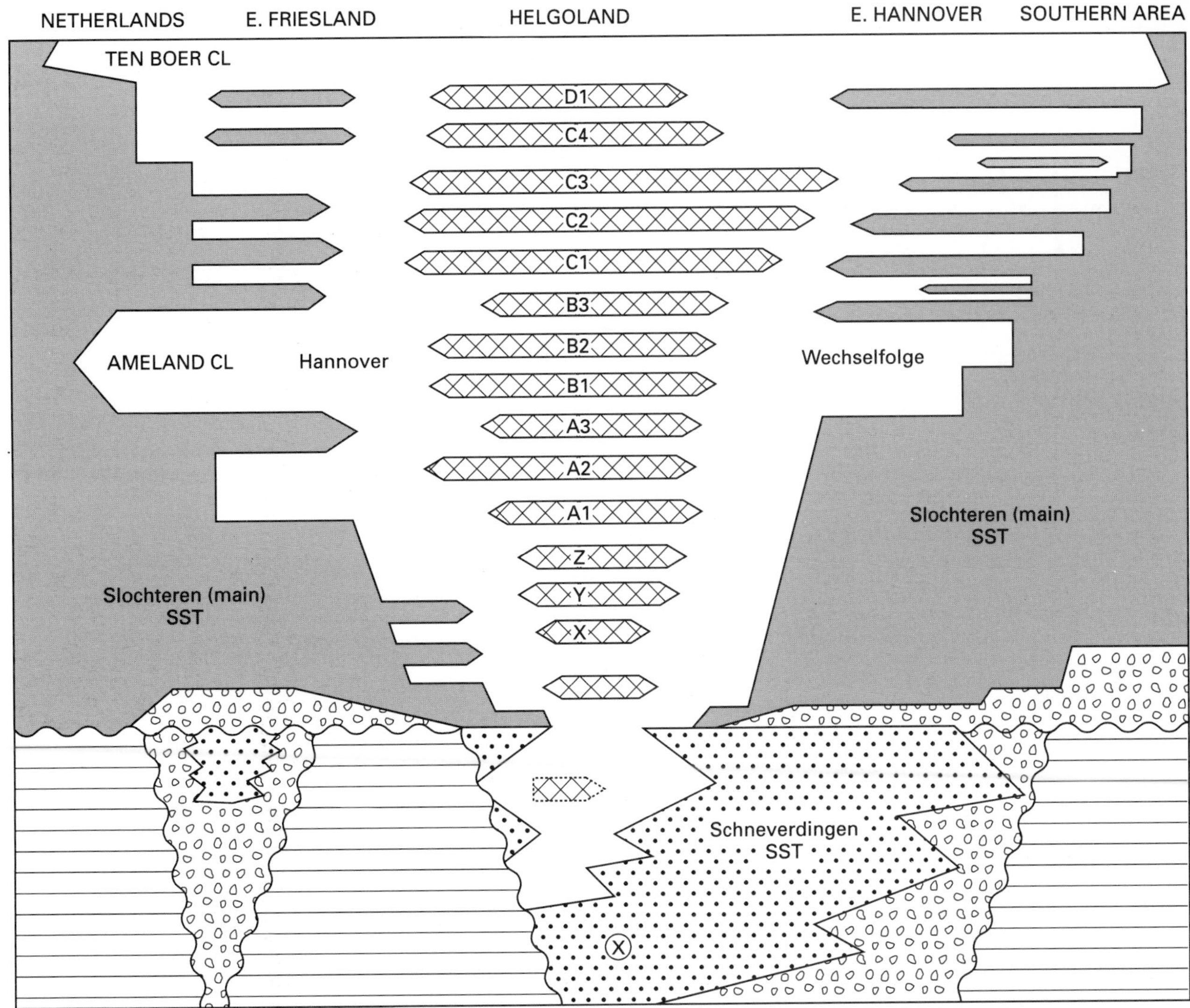

Fig. 5.19 The Upper Rotliegend 2 facies of north Germany and the German North Sea (modified from Hedemann *et al.*, 1984a), showing the approximate position of the Illawarra magnetic reversal in the lower part of the Schneverdingen Sandstone. The Rotliegend facies follow an arc from the Netherlands, through the southern offshore area, into northern Germany, so that the desert-lake shales and halites are flanked by the largely aeolian 'main' sandstones of the Slochteren Formation. The conglomerates in the lower part of the diagram are probably associated with early Altmark volcanicity (cf. Fig. 5.4).

deformed strata and beds in which the original aeolian laminae are only weakly preserved (Fig. 5.21). Yet, at other localities, such as the Indefatigable J Platform, less than a metre of marine-reworked basal Zechstein sand separates the Kupferschiefer from undeformed Rotliegend dune sand (see Glennie and Buller, 1983, Fig. 3). The extreme case is reached between the linear dunes (Yellow Sands) of County Durham, where the Kupferschiefer (Marl Slate) locally lies directly upon Carboniferous strata (Smith and Francis, 1967; Smith, 1980a), even though thin reworked sands occur on the crests and flanks of the adjacent high-standing dunes.

Some of the non-depositional features described above possibly resulted from a very fast Zechstein transgression and the escape of air trapped beneath the wetted surface of aeolian sands during the rapid rise of water level (Glennie and Buller, 1983). Such a rapid transgression was invoked by Smith (1970a, 1979) to explain the quality of preservation of the Yellow Sands in north-east England.

Alternatively, the deformation could result from slumping induced by flooding, so the structureless sands might represent the reworking of surface sands over distances that were too short for the development of sedimentary structures characteristic of that transporting medium (Glennie, 1972). Indeed, Arthur *et al.* (1986) report that, in some cases in Block 49/28, an irregular erosional boundary with the underlying dune sands can be recognized.

In the Northern Permian Basin in the vicinity of the Auk oilfield, Heward (1991) has shown that the 1–42-m-thick Weissliegend partially infills interdune areas in a fashion analogous to that of marine reworking, as described by Eschner and Kocurek (1988) for the marine transgression of a Jurassic desert in North America. Heward (1991) interpreted the combined massive, stratified and deformed sandstones, shales, dolomites and intraformational conglomerates as non-marine mass-flow deposits slumped and

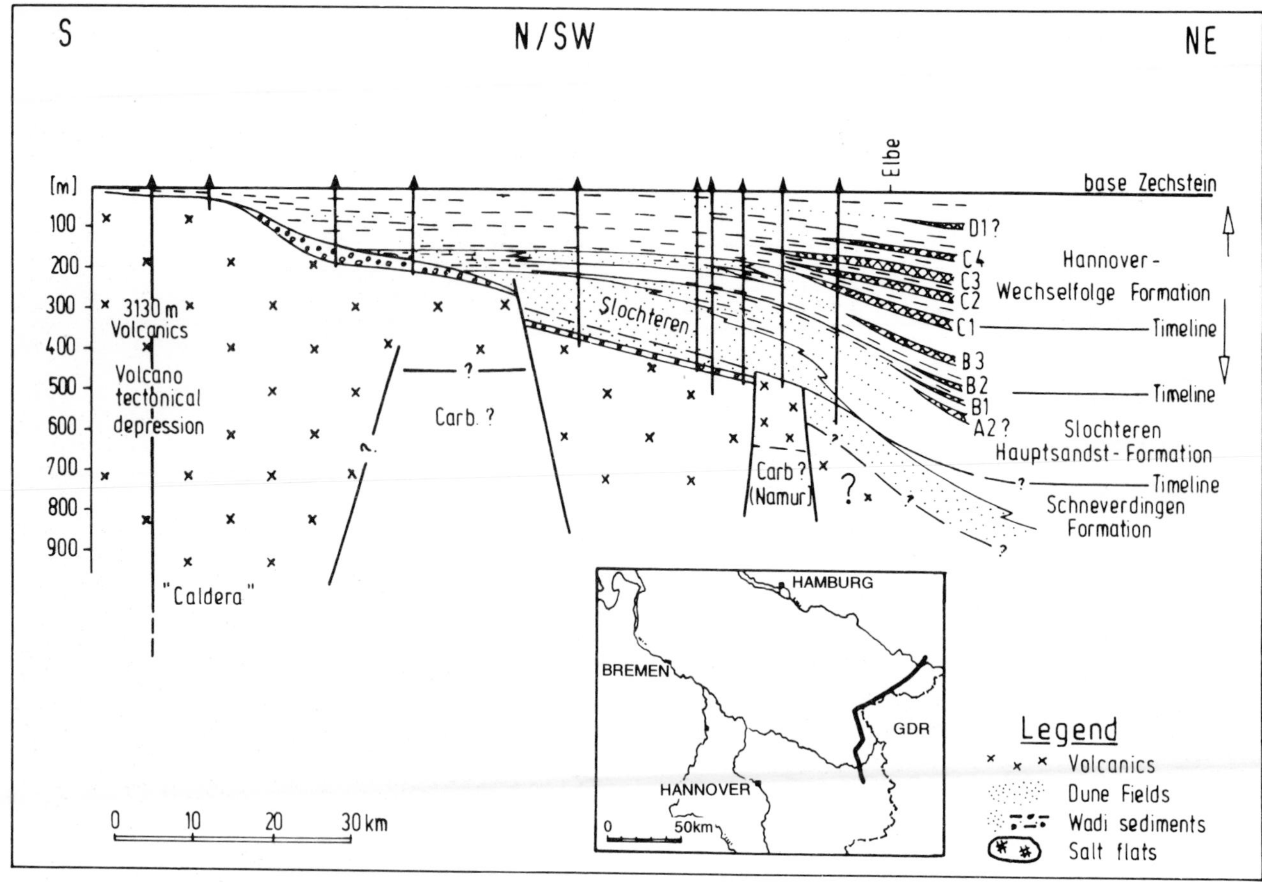

Fig. 5.20 Schematic cross-section showing correlation between basin-margin and basin-centre sequences of Rotliegend in northern Germany (slightly simplified from Gast, 1988).

washed into the interdune areas from upstanding dunes during intense rainstorms prior to the Zechstein transgression. In Auk, the grey, organic-rich laminated shales and limestones yield no evidence of marine faunas or floras, and are interpreted as interdunal lake deposits, the conglomerates being products of local reworking; these reworked sediments of the Auk area are thus truly Weissliegend and not basal Zechstein sandstones.

The considerable thickness (> 50 m) of some Weissliegend sequences with probable relics of former dune bedding would seem to preclude any large-scale marine reworking except locally. The local 'preservation' of relatively undisturbed dune bedding within a sequence of otherwise deformed or structureless Weissliegend sands (e.g. Fig. 5.21) may indicate the temporary return of aeolian deposition between Weissliegend rainstorms.

Recognition of the Weissliegend solely on the basis of colour can lead to difficulties. The downward change to red-coloured Rotliegend is commonly gradational. Also, in some cores, zones of green (partly reduced iron oxide) or white sandstones can occur in the middle of the Rotliegend sequence; many of these zones can be traced to uncoloured fractures. Discoloration was possibly caused in some cases by the upward migration of weak carbonic acid, associated with burial of the underlying Coal Measures. In the Auk Field, which is not underlain by Coal Measures, Heward (1991) reports that the Weissliegend sandstones are generally red below the oil–water contact and bleached and oil-stained above it. The reducing effect of early migrating oil can be inferred.

Glennie and Buller (1983) suggest that the general lack of a red colour in the Weissliegend sands is because, for the most part, they represent dunes that were above the water-table prior to the Zechstein transgression and thus were not in a diagenetic environment (Walker, 1976), where reddening could take place; after the transgression, the basin floor soon became a reducing environment, so that reddening of the sands could no longer occur (see Glennie, 1989, Fig. 4).

5.3.2 Moray Firth and Northern Permian Basins

The facies distribution of the Rotliegend in these two basins is still incompletely known. In the deeper parts of the Northern Permian Basin, the Rotliegend is rarely reached by the drill and it is not seen in outcrop. Using the limited available well data, an attempt has been made by Sørensen and Martinsen (1987) to map the facies distribution in the Danish Norwegian subbasin by use of a seismic-stratigraphy analysis. Although their map seems plausible, confirmation for most parts of the basin will have to await the drill.

On present evidence, both basins probably contain rocks of the same general sedimentary facies already recognized

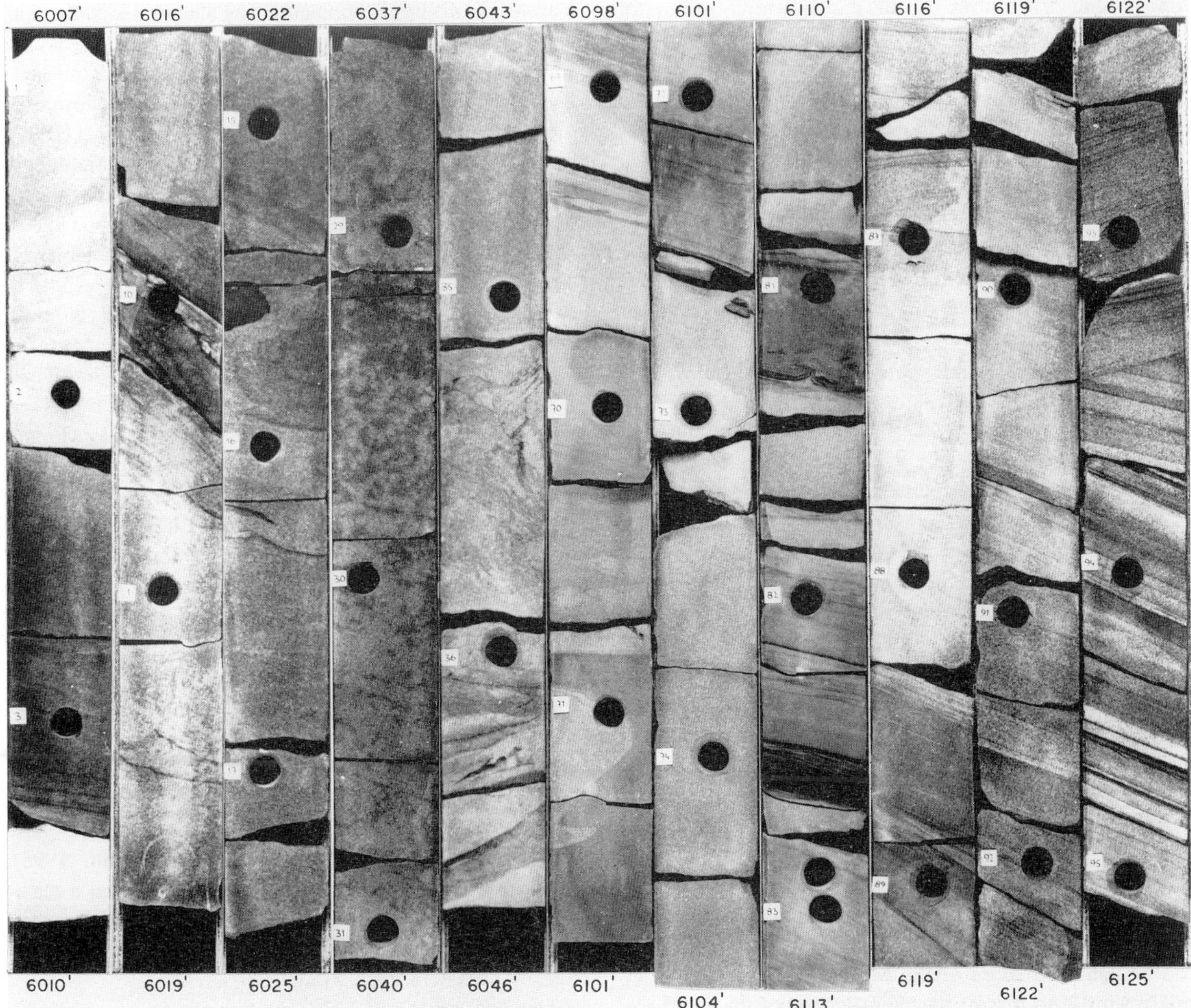

Fig. 5.21 Core photographs selected from the uppermost 125 ft (38 m) of sandstone immediately underlying the Kupferschiefer (at about 6000 ft) in Leman Bank discovery well 49/26-1. Typical aeolian bedding occurs below about 6120 ft. Most of the overlying sands are homogeneous but with deformation visible at 6023 and 6100 ft, fluid escape structures at 6044 and 6111 ft. A clay-rich bed separates weakly laminated sands at 6112 ft, with steeply inclined clay at 6017 ft. The mottled sandstone between porosity/permeability plugs 29 and 30 is enriched with pyrite.

in the Southern Basin (Weissliegend, fluvial, aeolian and sabkha, although the presence of bedded halite in the lacustrine facies has yet to be demonstrated). The sand-dominated sequence of the Northern Permian Basin in UK waters has been designated the Auk Formation and the shaly sequence the Fraserburgh Formation (Deegan and Scull, 1977; Cameron, 1993a; see Fig. 5.3), while the locally underlying Lower Rotliegend volcanics are known as the Inge Volcanics Formation (Cameron, 1993).

Much of the Auk sequence in the type locality is interpreted as aeolian dune sand. The shales of the Fraserburgh Formation in Shell/Esso well 21/11-1 contain adhesion ripples and also stringers of dolomitic and micaceous sandstone that are anhydritic. The depositional environment is interpreted as a dune-bordered sabkha (Deegan and Scull, 1977). Thus the Auk and Fraserburgh Formations are broadly similar to the Leman Sandstone and Silverpit Claystone formational sequences of the Southern Permian Basin (see Fig. 5.3). Similar sequences have been recognized in released Norwegian wells but, as yet, have no national names (see Fig. 5.9).

In the UK part of the Northern Permian Basin, the thickness of the Rotliegend changes rapidly from place to place, making correlation between wells very difficult. The differences may reflect deposition in small rotated half-grabens, such as those formed on the flank of the developing Central Graben (see, for example, Fig. 5.26). Such an interpretation could explain the derivation from adjacent fault scarps of the clasts of quartz and schist in the basal 14 m of the Rotliegend in the Shell/Esso well 30/16-1 (see Fig. 5.9) mentioned by Deegan and Scull (1977). In this respect, Heward (1991) describes the Rotliegend at Auk as a wedge-shaped accumulation that thins from 500 m in the west to 150 m in the east. Part of this thinning results from onlap on to gently tilted Old Red Sandstone, and more by truncation beneath the Chalk (see Fig. 5.38). Heward (1991) also notes that conglomerates in the Auk area are developed only locally, where they possibly infill early topographic depressions on the unconformity surface of Old Red Sandstone.

The Rotliegend of the Inner Moray Firth Basin is now named the Findhorn Formation (Cameron, 1993). Unlike the aeolian and sabkha sediments of the Auk and Fraserburgh Formations, the Findhorn Formation is largely in a fluvial facies. This formation seems to occupy a smaller area than the Rotliegend sediments illustrated by Andrews *et al.* (1990), who show its presence almost reaching the southern coast of the Firth. These authors show Rotliegend sandstone thicknesses varying between 500 and 700 m in the Inner Moray Firth, whereas a sandstone and mudstone facies occupies the Outer Moray Firth. Although not penetrated by the drill, the same facies is thought to reach a thickness in the vicinity of 1000 m in the East Orkney Basin (Andrews *et al.*, 1990). With intervening Rotliegend-free highs, such figures indicate considerable differential basin subsidence within the Moray Firth and East Orkney Basin and matching rates of Rotliegend sedimentation.

It seems likely that, in the Moray Firth Basin, Rotliegend sedimentation kept up with subsidence, as the overlying Zechstein is entirely in a shallow-marine facies. In the centre of the Northern Permian Basin, on the other hand, subsidence probably greatly exceeded sedimentation, and the succeeding Zechstein is dominantly in a basinal halite facies (see Chapter 6, Fig. 6.12; Taylor, 1981). Like the central parts of the Southern Permian Basin, subsidence permitted the halite to reach thicknesses over much of the Northern Basin that were great enough to allow later diapirism (see Fig. 6.12).

Glennie and Buller (1983a) interpreted the deformed bedding within the lower half of the Hopeman Sandstone on the southern shore of the Moray Firth as having an origin similar to that of the Weissliegend of the Southern Permian Basin; because most of these structures have a strong vertical component, they believed that deformation resulted from the vertical escape of air through wetted sands. This is disputed by several workers. Despite their vertical nature, Frostick *et al.* (1988), for example, prefer an interpretation involving fluvially induced slumping. Both they and Cameron (1993a) consider that these 'slumps' cannot be correlated with the Weissliegend, because fossil footprints above the deformation level have been dated as Latest Permian to Early Triassic (Benton and Walker, 1985), the Weissliegend (and underlying Rotliegend) being presumed to be much older (i.e. Early Permian). As described in the introduction, we now know that both the Zechstein and Upper Rotliegend of the North Sea basins were deposited within the latest Permian Tatarian stage, and thus the inferred age of deformation matches or was just prior to the Late Permian Zechstein transgression and could be very little older than the fossil footprints.

5.4 Historical development

Because age-diagnostic faunas and floras are largely absent from the Upper Rotliegend, especially in UK waters, it has only recently been realized that its deposition took place mostly, if not entirely, during the Late Permian (e.g. rare palynofloras with 'Late' Permian affinities in the Auk Field; Heward, 1991) and was separated from the underlying Carboniferous by the long-enduring Saalian Unconformity (see Table 1 in Brown, 1991). However, by using faunas found in rare grey beds within the different Rotliegend basins of Germany, Hoffmann *et al.* (1989), followed by Gebhardt *et al.* (1991) and Schneider and Gebhardt (1994), concluded that the Upper Rotliegend can be separated into two major stratigraphic units, the older UR1 and the younger UR2. The UR1, limited to some subbasins in central Germany, contains minor volcanics and is largely time-equivalent with the Saalian Unconformity (see Fig. 5.4).

Deposition of UR2 was probably confined to the lower half of the Tatarian stage. Some of the oldest sequences of UR2 have been described from subbasins of north-east Germany, where its deposition was punctuated by several tectonically induced unconformities, referred to as Altmark I–IV (Hoffmann *et al.*, 1989). Regional stratigraphic correlation (e.g. Gebhardt *et al.*, 1991; Schneider and Gebhardt, 1994) indicates that possibly the whole Upper Rotliegend sequence within the German part of the Southern Permian Basin belongs to UR2. The Schneverdingen Sandstone Formation (Mirow plus Parchim formations of the Havel Subgroup; Schröder *et al.*, 1995), which contains the Illawarra Magnetic Reversal near its base (Gast, 1991; see also Figs 5.19 and 5.20), is the lowest UR2 unit, and was first recognized within a graben setting. According to Menning (1991), the Illawarra reversal took place around 257 Ma BP, or 10 ± 4 Ma older than the then recognized Permo-Triassic boundary (alternatively, 9 or 8 ± 4 Ma before base Zechstein; R.E. Gast, personal communication, 1995); Menning (1995) now gives the Illawarra Reversal a maximum age of 265 Ma BP (approximately 14 Ma earlier than the presumed Permo-Triassic boundary or 7 Ma before the start of Zechstein deposition).

The Zechstein and UR2 are thought to be of similar duration (both contain a similar number of major glacially induced cycles: U. Gebhardt, personal communication, 1993; see also George and Berry, 1993, 1994), each seemingly being some 4–7 Ma long (Fig. 5.22); Menning (1995) prefers 7 (?5) Ma for Zechstein deposition and 8 Ma for the UR2. The Hannover and Dethlingen Formations represent only the upper two-thirds of the time span of UR2 sedimentation in Germany (see Fig. 5.20). The thickness of the UR2 lacustrine sequence in UK waters is much thinner than in Germany (300 versus > 1800 m). If both areas had similar rates of deposition, the UK Silverpit Claystone, and thus also the adjacent dune sands, could have been deposited during a time span of around 1–2 Ma (Glennie, 1997b, Fig. 4).

5.4.1 Climate

The change from the humid equatorial conditions under which the Carboniferous Coal Measures were deposited to the arid climate of Upper Rotliegend deposition was probably mostly an effect of the passive northerly drift of Laurasia (about 5 cm/year; see Fig. 2.3), much of it being coeval with the long-enduring (> 20 Ma?) Saalian Unconformity. The dominance of fluvial and lacustrine sediments of UR1 implies that for a large part of that time interval the climate within the Variscan Mountains was relatively humid. The Rotliegend Southern Permian Basin, however,

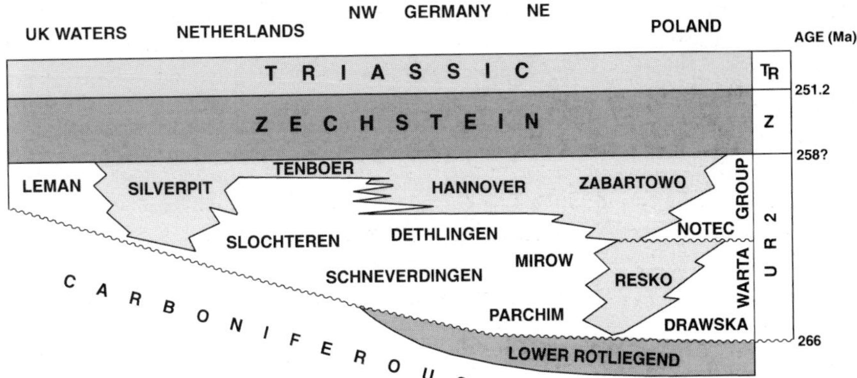

Fig. 5.22 Approximate correlation of Upper Rotliegend 2 sequences between the full sections seen in Germany and Poland and the more limited sections of the Netherlands and UK waters. These sequences are separated from the Lower Rotliegend by the Saalian Unconformity. The ages given are still very tentative.

eventually came to occupy a latitudinal position north of the equator, similar to that of the present North African and Arabian deserts (see Fig. 5.15). The Artinskian age of the aeolian Kreutznach Sandstone of central Germany and the possibility of strong deflation under arid conditions of the area that later became the Southern Permian Basin (Glennie, 1983b; see Fig. 5.8) imply that, as just suggested above, the Saalian Unconformity coincided with much of that climate change. By the time UR2 sedimentation began in the Southern Permian Basin, the area was already an arid desert. In this context, it is perhaps pertinent to note that halite was being deposited off northern Norway during the Mid-Carboniferous and Early Permian (Faleide *et al.*, 1984).

The recently created Variscan Highlands occupied a near-equatorial location, with the more northerly Rotliegend desert in the rain-shadow of humid Tethyan Trade Winds (Fig. 5.23). Throughout the earlier Permian, the mountains probably had regular tropical rainfall, to judge from the fluvial and lacustrine Lower Rotliegend and UR1 sediments of eastern Germany (Lützner, 1969; Gebhardt *et al.*, 1991) and Poland (Pokorski, 1989) and from the Stephanian to Autunian coals in Saarland (see, for example, Schäfer and Sneh, 1983) and Central France. A time-related reduction in the volume of fluvial sediments within the Southern Permian Basin may indicate that, during the earlier part of the UR2 depositional history, the local climate was becoming increasingly arid, although the sources of fluvial activity in the Variscan Mountains still had enough rainfall to keep the Silverpit desert lake in existence.

Well data, coupled with excellent quarry exposures of seif dunes in County Durham, indicate that many of the Rotliegend dunes attained a height of 50 m or more. Today's winds, however, seem to be incapable of constructing seif dunes with heights much greater than 10 or 20 m. The large seif dunes of Arabia and North Africa, possessing heights of some 100 m and wave lengths of 1 or 2 km, were constructed during the last Pleistocene glaciation (see, for example, Glennie, 1970, 1987, 1994). A likely reason for these size differences is that the large areas of high barometric pressure associated with major glaciations (Permo-Carboniferous, as well as Pleistocene) caused a concentration of the world's air-pressure belts towards the equator and thus created shorter distances between the zones of high and low pressure than is now the case (Fig. 5.24). The resulting wind systems probably had higher average velocities than now and were also colder than now, interpretations that are gaining support among many workers (e.g. Galloway, 1965; Bowler, 1976; Krinsley and Smith, 1981; Rea and Janacek, 1982; Petit-Maire, 1994).

With a very large ice-cap centred over Gondwana during much of the Carboniferous and probably all the Permian, meteorological conditions over the Rotliegend desert are likely to have been analogous to those of Pleistocene Trade Wind deserts: (i) strong persistent winds during periods of maximum glaciation; and (ii) a more humid climate during interglacials. Oxygen isotope data indicate that during the past 500 ka, glaciations peaked about every 100–120 ka (Boulton, 1993, Fig. 21.3). Two periods of dune-sand deposition in eastern Arabia, separated by major bounding surfaces, are dated tentatively as coeval with the last two glaciations, while fluvial gravels coincided with an earlier interglacial (Goodall T.M. and Pugh J.M., unpublished data, Aberdeen University, 1993). Whether the same cyclicity occurred with Permian glaciations and Rotliegend dune activity is not yet known. Hedemann *et al.* (1984a; see Fig. 5.18) and Gast (1991) illustrate a German North Sea gamma-ray log of halite-rich lake-centre equivalents of the Slochteren and Hannover Folge, with 11 halite horizons that are younger than the Illawarra Reversal (?10 ± 4 Ma older than the Permo-Triassic boundary). In Fig. 5.18, the stratigraphic terminology has been updated to show that the Illawarra Reversal lies just above the 'X' halite (U. Gebhardt, personal communication, 1993). If the time span of 4–7 Ma suggested earlier for the duration of UR2 is anywhere near the truth, then major climatic repetitions (Milankovitch cycles) would have occurred at intervals of the order of 300–500 ka (see Gast, in Plein, 1995).

The general uniformity of wind directions in both the Southern and Northern Permian Basins, as deduced from dip-meter data on dune sands in wells and in outcrop (Glennie, 1983a, 1994; see Figs 5.14 and 5.15), seems to support the idea of a strong, glacially induced, wind system. The regional wind pattern apparently has also been preserved with time (i.e. in the vertical sequence), although some effects of weaker and more variable interglacial winds can also be inferred locally (Heward, 1991). The glacially induced Pleistocene winds possibly persisted for the greater part of the year, instead of blowing for only a few hours or days at a time, as is now the case. With ice-caps the size of those found in Gondwana, the Permian dunes, like those of

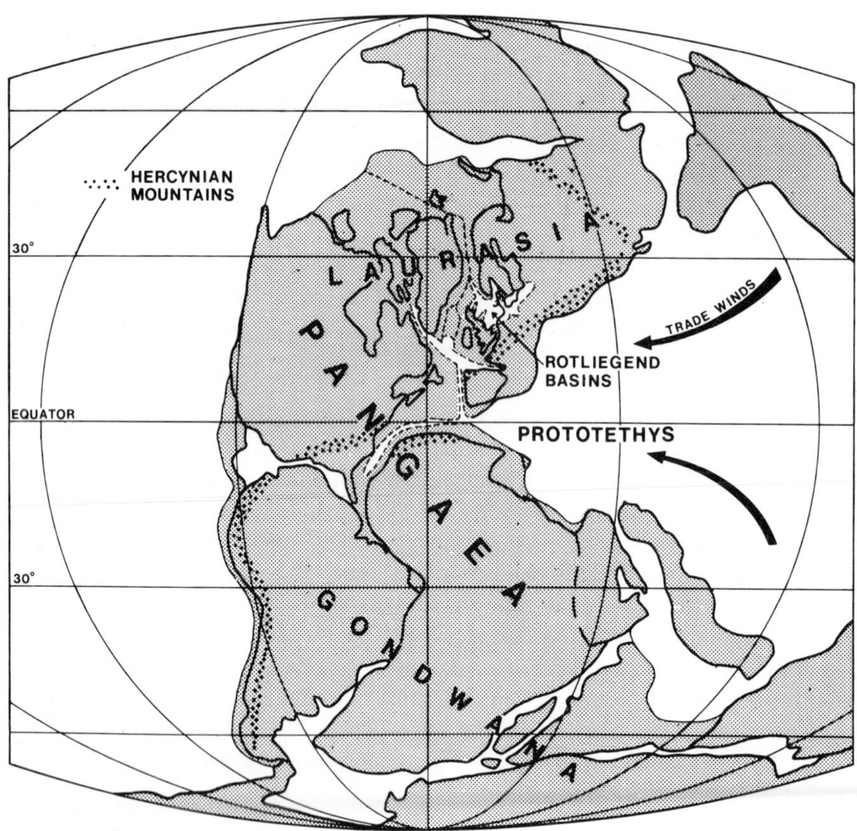

Fig. 5.23 Location of the Northern and Southern Permian Basins within the megacontinent Pangaea. The lines of crustal separation of the future Atlantic and Tethys oceans are shown. (From Glennie, 1986b.)

the Pleistocene glacials, are likely to have been built on a scale that is not seen today (Glennie, 1983a,b).

If these deductions are correct, the Rotliegend desert winds were probably strong and, because of their long continental route before reaching the western part of the Southern Permian Basin, also very dry. Thus the area of dune activity will have extended during Gondwanan glaciations, and strong evaporation will have reduced the size of

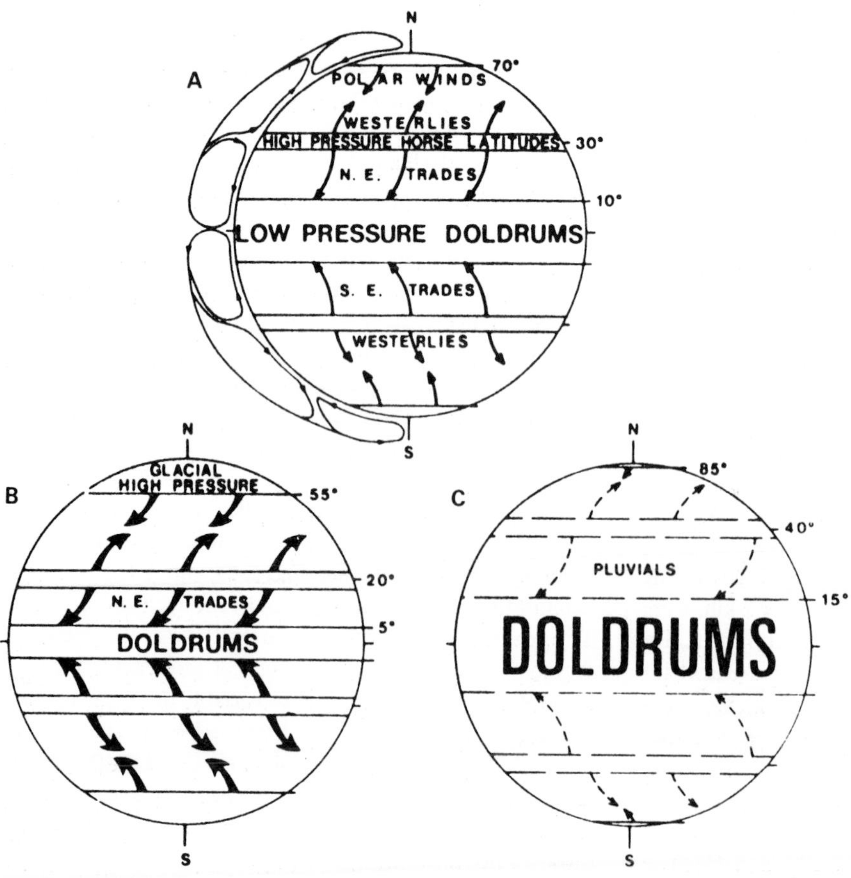

Fig. 5.24 Conceptual differences in width and location of the earth's air-pressure belts in relation to the size of Polar ice caps. A. Present. B. Glacials. Very strong constant winds built major dune systems. C. Interglacials. Very weak wind systems: strong convection influence near coasts and mountains; desert 'pluvials'.

the desert lake to a minimum and sabkhas would be subjected to strong deflation. The dying stages of the Permian Gondwana glaciations, on the other hand, should have coincided with higher global sea levels, the occasional marine incursion to the desert lake and a generally weaker wind system, with a resulting combination of higher convection-induced rainfall and less evaporation. This combination may explain the lateral extension of the Ten Boer lacustrine and sabkha facies of the Rotliegend, well displayed in the Groningen Field in the Netherlands (see Fig. 5.27) and in West Sole (Butler, 1975), for a period prior to the Zechstein transgression (but see Section 5.4.2). Furthermore, the Weissliegend slumps in the Auk area are interpreted by Heward (1991) as a product of desert rainfall. The Ameland Claystone in the Groningen Field possibly represents a similar but earlier glacial waning, although locally more rapid subsidence of the basin margin could also account for the apparent lateral extensions of the desert lake (compare Figs 5.3 and 5.20).

The general coincidence of aeolian activity and halite deposition implied above has been referred to elsewhere (e.g. Glennie, 1972, p. 1065); it is not supported by Gralla (1988), however, who correlates halite beds of the Rotliegend desert lake in north Germany with more southerly shale horizons (i.e. with increased fluvial input).

5.4.2 Basin formation

As described in Chapter 2 (Fig. 2.13), the North America–Europe plate began to collide with the microcontinents of southern Europe during the Visean. Closure of the intervening Rheic Ocean eventually led to the end of Coal Measure deposition on the coastal plains to its north. Along the southern edge of southern Europe, north-dipping subduction of Proto-Tethys was associated with folding and uplift of the Variscan Mountains. Construction of this mountain chain was completed during the final stages of collision between southern Europe and Gondwana and creation of the megacontinent Pangaea during the Westphalian (see Fig. 5.23).

The early collapse of the Variscan Highlands seems to have begun at a time when the Variscan orocline north of Iberia formed a wedge that pushed into the re-entrant between the rigid margins of the former Laurentian and Baltic cratons (see Fig. 2.13). This wedge was probably later responsible for creating, stage by stage, the whole system of roughly north–south-trending fractures that developed on either side of the line of weakness represented by the Caledonian northern Iapetus suture; these later developed as the Proto-Atlantic and Viking–Central system of grabens, together with an allied suite to both east (e.g. Horn and Oslo grabens) and west (e.g. Cheshire Basin, Vale of Eden half-graben; see Figs 5.1 and 5.2).

Widespread north-west–south-east transtensional movements utilized old lines of weakness along the Tornquist–Teisseyre Line and the north-east margin of the London Brabant Platform. A series of roughly north–south-trending grabens extended from Germany across the North Sea, where reaction with the Tornquist–Teisseyre zone of weakness resulted in the dog-legged north-north-east–south-south-west-trending Horn–Bamble–Oslo system of grabens. These movements resulted initially in eruption of Lower Rotliegend volcanics, not only in the eastern Variscan Mountains but also across its adjacent northern foreland. Further west, however, the relatively rigid London Brabant Platform seems to have protected the south-western North Sea area from major extensional movements and associated volcanism. Instead, the probable earlier transpressional conditions resulted in inversion of the Cleveland Hills and Sole Pit areas (Kent, 1980; Glennie and Boegner, 1981), where Permian erosion of Carboniferous strata locally extended down to the Namurian (see Figs 2.14, 2.16, 5.8 and 5.31).

During the long time span represented by the Saalian–Altmark Unconformity (up to ?30 Ma), the areas of thick Lower Rotliegend volcanism protected the underlying sequences of Stephanian rocks from erosion, whereas, further west in Dutch and UK waters, the Stephanian was partly preserved only within half-grabens or other areas of local down-warping (see, for example, Ziegler, 1990a, encl. 3). Elsewhere, there was considerable erosion of Carboniferous sequences over transpressional structures, such as the Sole Pit area, and especially over the Mid North Sea–Ringkøbing/Fyn High. The evidence from Germany indicates that deposition of the UR2 did not begin before about 266 Ma BP. If true, then what happened to the sediment eroded during the preceding 20–30 Ma or so? The Carboniferous shales were possibly removed by deflation and blown as dust clouds westward across North America (Glennie, 1983b), but coarser sediments would not be so readily disposed of. Even so, with a prevailing east wind, it seems likely that much of the sand fraction was also transported westward, first to the half-grabens of western England and then to what is now the continental shelf of western Europe.

If the Lower Rotliegend volcanic event was responsible for crustal uplift by thermal expansion, as seems likely, then the erosional products could have been deposited to both north and south of the uplift. Apart from subbasins within the Variscan fold belt of central Germany, which contain URl sediments, there is no known evidence to support such a hypothesis.

As the axes of the Northern and Southern Permian Basins and intervening Mid North Sea–Ringkøbing High are subparallel to the Variscan Orogen, their creation should be linked to that orogeny or, more probably, to the destruction of the orogen. Indeed, UR2 sedimentation in Poland began in the narrow west-north-west–east-south-east-trending Middle Polish Trough, parallel to the Tornquist–Teisseyre Line (Depowski, 1978; Pokorski, 1993); it is on structural trend with the Northern Permian Basin rather than with its western counterpart in Germany (Glennie, 1997b). Upper Rotliegend 1 sedimentation may have begun in northern Germany slightly earlier than in Poland (Gralla, 1993; Plein, 1993, Fig. 4); also, with time, it developed westward through the Netherlands (van Wijhe, 1987b) to the UK southern North Sea (van Hoorn, 1987a).

Upper Rotliegend 2 has a maximum thickness of up to 2500 m in north-east Germany (McCann, 1998). Because of a lack of firm ages, UR2 is estimated to have been deposited during a time span of between 4 and 7 or 8 Ma. In this area, therefore, postcompaction subsidence would have been within the range 0.6–0.3 mm/year (30–60 cm/10^3 years). For comparison, the long-term rate of late Pleistocene subsidence along the Texas coast south of Galveston Bay is 0.05 mm/year (Paine, 1993).

With the rates of Permian subsidence encountered in Germany, the apparent extension of the lacustrine sequences across the lake-margin sabkhas could have been as much subsidence-induced during an important period of transtensional movement subparallel to the Tornquist–Teisseyre line (together with a thermal component), as resulting from phases of wetter climate. This interpretation might explain why the equivalent of the Ameland Claystone (see Figs 5.3 and 5.27) in the Groningen Field is recognized in northern Germany but has no obvious equivalent in UK lake-margin fields (compare, for example, Figs 5.18 and 5.22 with Fig. 5 in Ketter, 1991b), and why the Hannover Wechselfolge of northern Germany contains lake-margin sabkhas (see Figs 5.19 and 5.20).

The alternative interpretation is that sedimentation in UK waters did not begin until after deposition of the Ameland Claystone, an interpretation supported by the relatively thin Rotliegend sequence in the UK area (cf. Figs 5.2 and 5.6). Apart from the Great Whin Sill, the western end of the Southern Permian Basin seems not only to have been protected from extensive Lower Rotliegend volcanism, but also to have suffered much less subsidence with increasing proximity to the north–south-trending Pennine Range; transtensional subsidence was concentrated within the Sole Pit Basin.

In stark contrast, many areas of modern desert dune sand located on stable, non-subsiding continents have only a low chance of future preservation. Indeed, dune and sabkha sequences in eastern Arabia that were deposited over 200 ka BP are still exposed or have recently become exposed by deflation, thus indicating a negligible rate of subsidence. A high rate of subsidence in the Southern Permian Basin implies a high initial preservation potential for the Rotliegend continental sediments; it also implies that, over much of the basin, the sedimentary sequence would be aggradational rather than progradational (see, for example, Shanley and McCabe, 1994).

With sporadic UR2 desert sedimentation controlled by intermittent glaciations over Gondwana, it is small wonder that the basin fill was unable to keep up with the rates of basin subsidence implied above, resulting in a continental basin whose floor was considerably below global sea level at the time of the Zechstein transgression.

5.4.3 Zechstein transgression

As shown earlier, the Weissliegend sediments of the uppermost Rotliegend also contain evidence concerning the probable rapidity of the Zechstein transgression.

It was along a combination of the Proto-Atlantic and North Sea fracture systems that the waters of the Zechstein transgression are presumed to have been transported from the Permian open ocean somewhere between Norway and Greenland (Fig. 5.25). A northern source for these waters is indicated by early Zechstein faunas that have boreal affinities. Although the Zechstein Sea eventually had a connection with Tethys south-east of Poland, that threshold was shallow, permitting only a limited exchange of faunas, and was not the source of flooding.

Surlyk *et al.* (1984) show that, in east Greenland, continental gravels were deposited in north–south-trending half-grabens. The mid-Permian development of eastward-hading faults was followed by a widespread marine transgression. These data suggest the possibility that, although the Zechstein transgression of the North Sea basins may have coincided with a high stand in sea level, the rapidity of its flooding could have been caused by fault-related subsidence somewhere between Greenland and Norway. Future evidence from the flank areas of the Atlantic Ocean may indicate whether the mid-Permian Proto-Atlantic rift was a narrow graben, as implied in Fig. 5.25, or was a broader arm of the ocean, as suggested by Callomon *et al.* (1972) and Taylor (Chapter 6, this volume, Fig. 6.1).

The Zechstein transgression probably started because a worldwide rise in sea level, coinciding with the end of a phase of Permian glaciation, permitted oceanic water to flow along a pre-existing fracture system. Because the surfaces of both the Northern and Southern Permian Basins were probably well below the level of the open ocean, once water began to flow south along the fracture the transgression continued until the level of the Zechstein Sea matched that of the open ocean (Figs 5.25 and 5.26).

The route by which the Zechstein waters entered the Southern Permian Basin is in dispute. Ziegler (1990a) suggests that flooding occurred around or across the Pennines from the Irish (Bakevellia) Sea, whereas Smith and Taylor (1992) suggest a northern route via the Fair Isle and Moray Firth basins, because of the sandy nature of the Zechstein in the latter area. Ziegler believes that the Central–Viking graben system did not develop until the Triassic. The presence of thick Rotliegend aeolian and sabkha sands in the southern Viking Graben (at least 380 m in the UK well 9/28-3), together with Zechstein halite thick enough to act diapirically, suggests, however, that an incipient graben already existed at the time of the transgression and is a more likely route for flooding the Permian basins from the north.

Because Zechstein halite occupies most of the Central Graben with a thickness sufficient to act diapirically (where insufficiently thick, this may have been reduced by later solution), Glennie and Buller (1983) suggested that the graben formed an active passageway between the Northern and Southern Basins and was probably the main route for flooding. Indeed, the Diamant-1 well, drilled in the Danish part of the Central Graben in 1986, contains almost 300 m of Rotliegend sediments (Nielsen and Japsen, 1991). The Horn and other grabens of the Ringkøbing–Fyn High provide additional potential routes.

Other workers have suggested that there are alternative (extra?) routes for Zechstein flooding of the Southern Permian Basin. Jenyon *et al.* (1984) seismically mapped a north–south-trending half-graben that crosses the Mid North Sea High and is sufficiently deep to contain Zechstein halite, rather than a shallow-water carbonate; and Smith and Taylor (1989, 1992) postulate the presence of a similar fault-bounded trough that crosses the high just east of the Northumberland coast. The similarity in trend

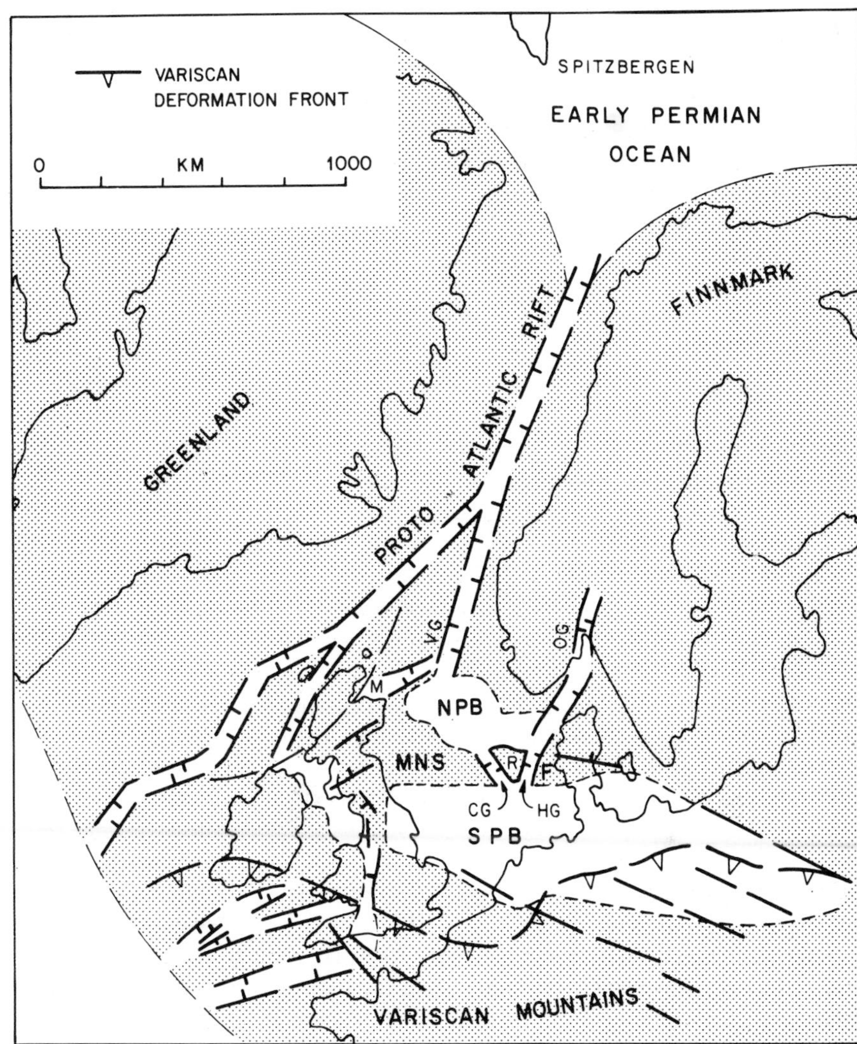

Fig. 5.25 Post-Variscan fault patterns. CG, Central Graben; HG, Horn Graben; M, Moray Firth; MNS, Mid North Sea High; NPB, Northern Permian Basin; OG, Oslo Graben; R, Ringkøbing–Fyn High; SPB, Southern Permian Basin; VG, Viking Graben. (Based on Russell, 1976, and Ziegler, 1978.)

of these proposed routes for the flooding of the southern Rotliegend basin strongly suggests that their origins were coeval. The greater floor depth of the present Central Graben (up to 10 km; Day *et al.*, 1981) and its content of diapirically deformed Zechstein halite suggests that it was also the deeper at the time of flooding and was therefore more likely to have been the route by which flooding of the Southern Basin was initiated.

Simple calculations show that, if the surface of the Rotliegend desert lay about 250–300 m below ocean level (a figure reached on other grounds by both Smith, 1970a, 1979, and Ziegler, 1982a), it would have required some 75 000 km^3 of water to fill the Southern Permian Basin and another 35 000 km^3 of water to fill the Northern Permian Basin (Glennie and Buller, 1983). Ignoring seepage and evaporation, these basins could have been filled in about 6 years if they were jointly flooded at the rate of some 50 km^3/day by utilizing a channel of water, say, 10 km wide and 20 m deep and with an average velocity of 3 m/s (Mississippi River in flood).

Such a flood of water into the Southern Permian Basin might have caused some initial scouring of the desert lake sediments where it debouched into the basin, but on the other side of the lake, 100 km or more away (Fig. 5.26), erosion will have been minimal. If the proposed flow rates are correct, the initial rise in water level will have been around 30 cm/day, and lake-side dunes, 50 m high, will have been covered with water in just over 150 days. Glennie and Buller (1983) thought it was this relatively rapid and continuous rise in water level that was responsible for air entrapment within, and deformation of, the Weissliegend upper part of the Rotliegend sedimentary sequence. Because the surfaces of these dunes received only limited reworking by wave action, the original shapes of the dunes were modified only slightly. As can be seen from exposures in north-east England, considerable relief is preserved, and the succeeding Kupferschiefer draped this dune relief. It would be unrealistic, therefore, to attempt a detailed correlation of Rotliegend reservoir sequences in the belief that the Kupferschiefer formed a horizontal datum plane.

The reworking of Rotliegend and older strata during the Zechstein transgression gave rise locally to a basal Zechstein conglomerate, which is commonly only a few centimetres thick. In the shallow-marine environment of the Durham area, the top few centimetres of the basal Zechstein breccia became colonized by Lingula immediately prior to deposition of the overlying Marl Slate (Bell *et al.*, 1979). In the environment of the former Rotliegend desert lake in northern Germany, the fossiliferous Bänderschiefer lies above the red-brown Rotliegend claystones and underlies about 2 m of red-brown marly clay, which Plumhoff (1966) places in the uppermost Rotliegend. These latter clays probably represent a reddish facies of the

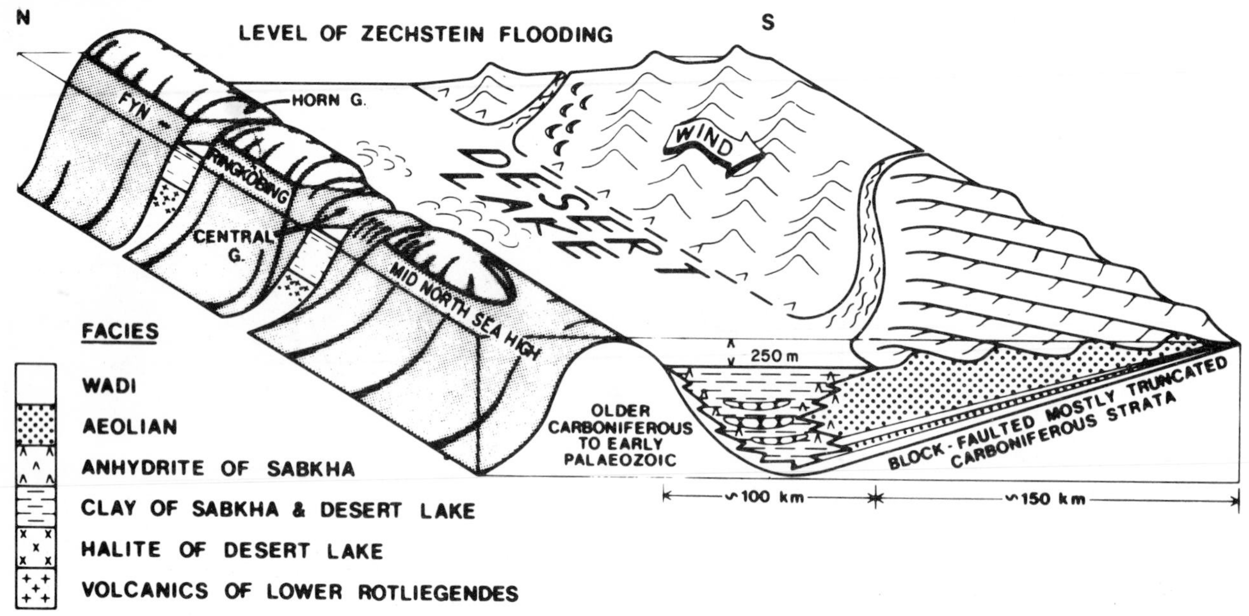

Fig. 5.26 Conceptual block diagram of the Southern Permian Basin and Central North Sea system of highs at the time of the Zechstein transgression. Zechstein Sea is presumed to have flowed into the basin via Central and/or Horn Graben. Note suggested change in dune style from transverse in centre of basin to longitudinal in western basin-margin location. (From Glennie and Buller, 1983.)

Kupferschiefer, known as the *Rote Faule*, the origin of which has been attributed to the local introduction of fresh water into the early Zechstein marine basin (Strakhov, 1962).

5.5 Hydrocarbon occurrences

5.5.1 Gasfields

Exploration in the hostile environment of the North Sea was triggered by the realization that there were sufficient reserves in the giant Dutch Groningen Field (Fig. 5.27) to alter the fuel economy of much of north-west Europe from a reliance on coal and oil to one based extensively on gas. Although discovered in 1959, it was then not seismically possible to see beneath the Zechstein salt, so it was not realized until several more wells had been drilled that the continued discoveries all belonged to the same field (te Groen and Steenken, 1968; see also introduction to Chapter 1). With a surface area of about 1000 km^2 and a porosity that ranged from 10 to 25% (permeability 0.1–1000 mD), the field was conservatively estimated in 1968 to have ultimate recoverable reserves of some 1650×10^9 m^3 (58×10^{12} ft^3) of gas (now believed to be nearer 2700×10^9 m^3 (97×10^{12} ft^3); Petroleum Geological Circle, 1993); the unexplored southern North Sea lay down the palaeowind, roughly due west of Groningen. The basic details of this giant are given in a series of articles by te Groen and Steenken (1968), van der Laan (1968), Bungener (1969) and Stäuble and Milius (1970). In 1962, before the great size of Groningen had been realized, one of the wells to its south discovered the Annerveen Field, which formed a separate trap with a mere 75.6×10^9 m^3 (2.5×10^{12} ft^3) of gas (Veenhof, 1996).

The search for Rotliegend gas beneath the North Sea began in UK waters in 1964. It soon resulted in the discovery of a series of important fields, West Sole, Viking, Leman and Indefatigable (see Fig. 5.5), of which the last two account for some 425×10^9 m^3 (15×10^{12} ft^3) of recoverable gas; these four fields have been described, respectively, by Butler, Gray, Van Veen and France in Woodland (1975), and again by Winter and King, Morgan, Hillier and Williams, and Pearson *et al.* in Abbotts (1991). The development of other relatively early discoveries, such as Sean North and South (see Fig. 5.34; see also Ten Have and Hillier, 1986; Hobson and Hillier, 1991), the Valiant, Vanguard and Vulcan group of fields (Pritchard, 1991), Victor (Lambert, 1991) and the Thames group of fields, was delayed until an upsurge in activity in the UK gas province in the early 1980s (see Chapter 1, this volume). Some basic data (block, discovery year and year on stream, depth to gas–water contact, recoverable reserves) involving the Rotliegend gasfields of the North Sea area are given in the Appendix to Chapter 1 (see also Table 5.1), while more detailed production data from UK fields can be found in the compilation tables at the back of Abbotts (1991).

All the producing Rotliegend gasfields of Poland, Germany, the Netherlands and the southern North Sea are underlain by Westphalian Coal Measures and overlain by Zechstein salt (see Fig. 2.17). The centre of the desert lake contains no reservoir rocks, but acts as a seal for gas in underlying Carboniferous reservoirs (Chapter 4, this volume; Bailey *et al.*, 1993). The lake-margin sabkhas in Germany are locally associated with aeolian sands, which form excellent reservoirs (Gast, 1991; Burri *et al.*, 1993), while similar sabkha sequences are associated with fluvial and aeolian sands in the reservoirs in the Ravenspurn–Cleeton area (Ketter, 1991b; Heinrich, 1991a, b).

Fluvial sands are second only to aeolian sands in the reservoirs of Groningen and Rough (Robinson, 1981, Fig. 6; Goodchild and Bryant, 1986; Stuart, 1991). In most fields, however (e.g. Amethyst; Garland, 1991), fluvial sands are

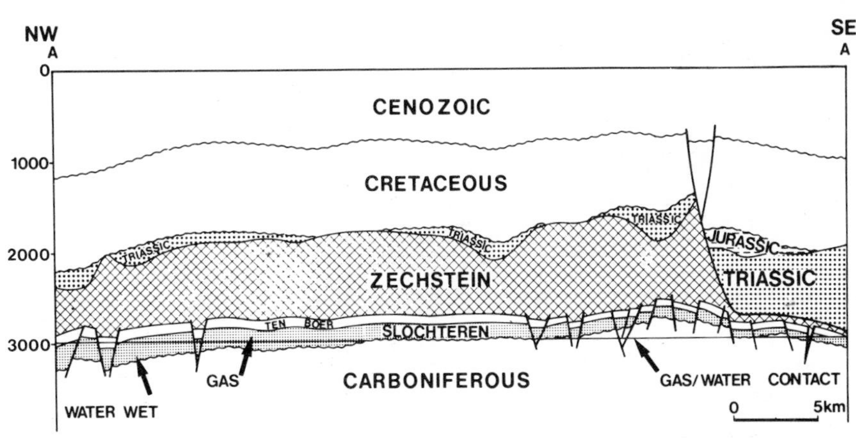

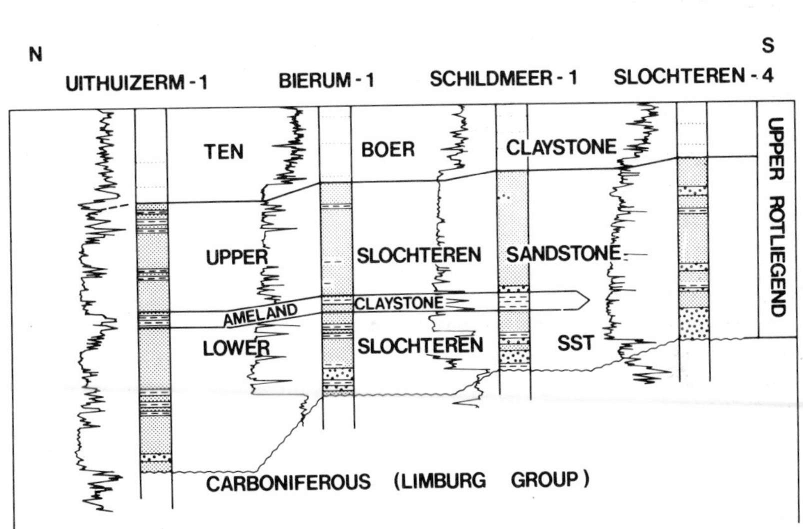

Fig. 5.27 The Groningen gasfield. The main reservoir comprises aeolian and fluvial sandstones and conglomerates, which fine to the north, with Carboniferous Coal Measures also above the gas–water contact locally. The Zechstein salt seal has been affected by varying degrees of diapirism. Pre-Cretaceous erosion removed virtually the total cover of Early Jurassic and Triassic strata, since when the area has subsided regionally about 2000 m and is now in its second phase of gas generation. (Well correlations modified from Stäuble and Milius, 1970.)

Table 5.1 Reservoir parameters, Rotliegend gas fields: Southern Permian Basin.

Field	Ultimate reserves ($m^3 \times 10^9$)	Reservoir depth (m)	Thickness (m)	Average porosity (%)	Air permeability (mD)	Major faces
UK						
Indefatigable	127	2400	35–130	15	0.1–2000	Dune
Leman Bank	298	1800	180–270	14	60	Dune
Rough	10	2800	30	15	1–1200	Wadi, dune
Sean	13	2600	35–70	20	5–650	Dune
Thames, Yare, Bure	13	2500	85–210	19	–	Dune
Victor	23	2600	115	16	130–300	Dune, wadi, sabkha
Viking	84	2800	150	16	60	Dune, wadi
West Sole	54	2700	125	15	<1	Wadi, dune
The Netherlands						
Bergen	–	2200	200–270	17	–	Dune, wadi
Groningen	2720	2750	70–240	17	0.1–1000	Dune, wadi
K8 FA	20	3300	200–270	10–15	–	Dune, wadi
K/13-E	–	2500	100–200	21	–	Dune
K/13-F	–	2500	100–200	21	–	Dune

generally too well cemented to form good reservoirs. Most commercially productive reservoirs are therefore confined to the aeolian facies. These factors result in the Rotliegend gasfields of the Southern Permian Basin being confined effectively to an east–west band some 50–100 km wide, stretching from the North German/Polish Plain, through the Groningen area to the east coast of England. Within this area, the ultimately recoverable reserves of gas approximate to 4.50×10^{12} m^3 (160×10^{12} ft^3). Including the contribution from Groningen and other Dutch land fields, this amounts to about 27% of the hydrocarbon energy of the North Sea area (Spencer *et al.*, 1996; see also Fig. 1.18).

Some gasfields of the Netherlands offshore that were strongly influenced by inversion movements have been

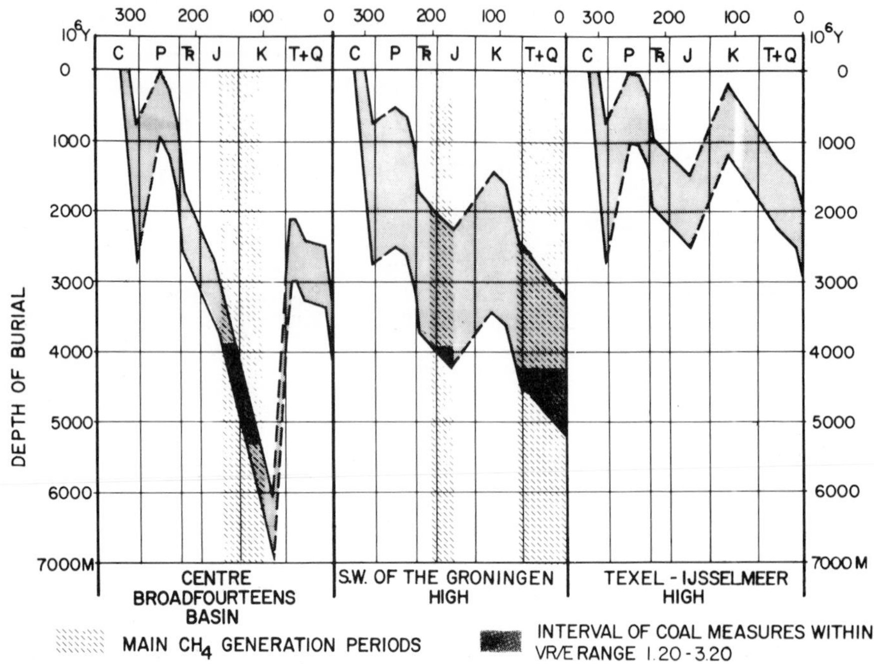

Fig. 5.28 Burial history of the Coal Measures in the Netherlands (from van Wijhe *et al.*, 1980).

described by Oele *et al.* (1981) for quadrants K and L (see Fig. 5.32) and by Roos and Smits (1983) for block K/13. The latter authors consider that Late Cretaceous inversion enabled the Triassic Bunter Sandstone in block K/13 to be charged from a gas-filled Rotliegend reservoir; the preservation of gas in two small nearby Rotliegend fields (structures E and F) supports this interpretation.

Similar fault movements probably caused the breakdown of the intervening Zechstein seal, and permitted gas to transfer from the Rotliegend sandstone to both Zechstein (Plattendolomite) and Triassic reservoirs in the Bergen area of the Netherlands (van Lith, 1983) and to Triassic reservoirs of the Hewett area on the margin of the Sole Pit Basin (see Figs 5.5, 5.31 and 7.6; Cumming and Wyndham, 1975; Cooke-Yarborough, 1991); gas was still retained in the Rotliegend reservoir of the adjacent small fields, Deborah, Big and Little Dotty, Della. Another Triassic beneficiary of Carboniferous gas via the Rotliegend is the Esmond, Forbes, Gordon group of small fields in Quadrant 43 (Bifani, 1986; Ketter, 1991a; see also Chapter 7). There, the Rotliegend is in the desert-lake facies and forms a much better seal than a reservoir; faulting initiated diapirism in the Zechstein halite and caused sufficient local salt withdrawal for the seal to be breached and permit the upward migration of gas from the Carboniferous to the Bunter reservoir.

With the North Sea's long history of vertical movements, it is not surprising that the Rotliegend reservoirs in the southern North Sea are highly faulted. Movement on many of these faults undoubtedly triggered diapirism in the overlying Zechstein salt; in turn, some idea of the time of fault activity can be deduced from the erosional and depositional history of rim synclines flanking the diapirs (see Figs 6.19 and 6.20). Because of the relatively high acoustic velocity of the halite, the rapid changes in thickness resulting from halokinesis considerably distort the shape of the underlying reflectors, and must be taken into account when converting time maps of the Rotliegend to depth (see, for example, Christian, 1969; Butler, 1975; Gage, 1980).

Gas generation and migration

Superimposed upon the three basic requirements for an oil- or gasfield—source, reservoir and cap rocks—is the general need for structural deformation to create a trap. It is axiomatic that a trap must be formed before migrating gas can be retained in a reservoir. In the Southern Permian Basin, some early traps apparently formed in areas of slower subsidence, whereas others resulted from early subsidence followed by inversion. In some of the latter cases, subsidence was continuous from the Late Permian (Tatarian) until the Late Cretaceous and mid-Tertiary, when a two-phase uplift of the order of 1–4 km took place (Figs 5.28–5.30). In other areas, such as the Cleaver Bank and Winterton highs, straddling the median line between British and Dutch waters, and also in the Groningen area, up to 3 km of erosion took place between the mid-Jurassic and fairly early in the Cretaceous (Hauterivian/Barremian). This was followed by a similar amount of subsidence between then and now (Figs 5.27, 5.29C and 5.31). All these movements were probably coincident with widespread but minor (Late Cimmerian) wrench faulting related to the final phases of graben development, which preceded the demise of the major North Sea graben system.

Gas generation has been shown by van Wijhe *et al.* (1980) to result from burial of the Coal Measures to depths of some 4000 m or more (for more detailed discussion, see Chapter 11). On the basis of tectonic reconstruction and Vitrinite Reflectance of the underlying coal, gas generation and migration south-west of the Groningen Field is considered to have begun during the latest Triassic (see Fig. 5.28). Following Jurassic–Cretaceous uplift, the Groningen gas kitchen is again within the generation window.

This general migration age receives support from the work of Lee *et al.* (1985, 1989). They used potassium/argon

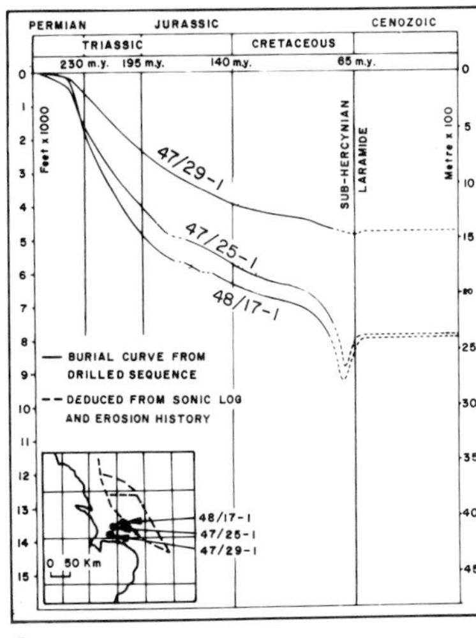

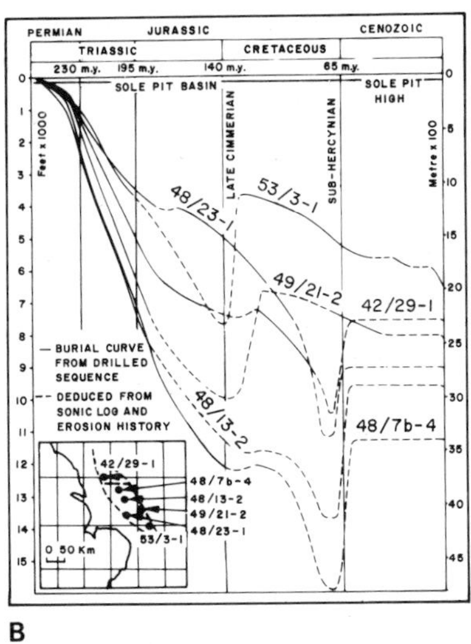

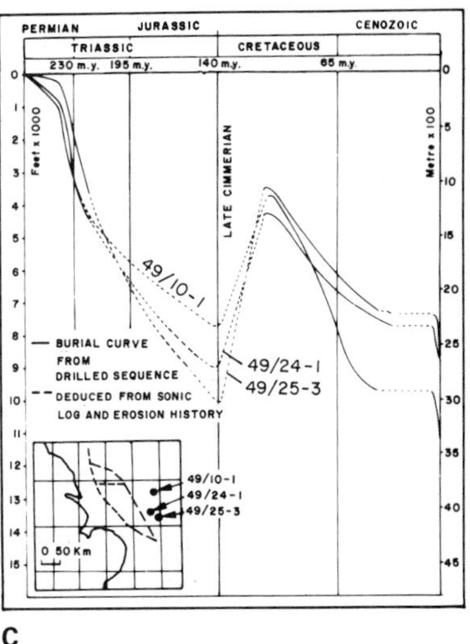

Fig. 5.29 Burial histories of the base Rotliegend at selected well locations, Sole Pit Basin, southern North Sea. A. West of Sole Pit Basin. B. Within Sole Pit Basin. C. East of Sole Pit Basin. (From Glennie and Boegner, 1981.)

(K/Ar) and oxygen isotope data to date the time of formation of depth-related diagenetic illite in the pore spaces of the Rotliegend sandstone reservoir (see 'Reservoir quality and diagenesis', below, and Fig. 5.30). They found that illite in the Groningen Field becomes more altered diagenetically and gets younger with increasing depth; they infer that this must indicate reservoir that was slowly filling with gas, illite ceasing to form as soon as the interstitial water was displaced downward by gas. The dating of diagenetic illite in water-wet reservoirs in the Broad Fourteens Basin indicates that its growth ceased around 140 Ma. Lee *et al.* (1985) interpret this as indicating that water in the reservoir was replaced by gas at that time, although end-Jurassic (Late Cimmerian) uplift could also be invoked to halt the process of diagenesis in flank areas of the basin (cf. Figs 5.28, 5.32 and 12.21A). If the first interpretation is accepted, the sandstone must again have become water-wet at a shallower depth when the gas escaped during inversion. Without knowing the burial history of the samples in greater detail, there must remain an element of doubt concerning a possible Late Jurassic gas-bearing phase at that locality.

The technique has also been applied to the UK southern North Sea. There, gas was formerly presumed to have been generated deep in the Sole Pit Basin during its long period of pre-inversion subsidence, and then to have migrated straight into adjacent highs of slower subsidence, such as in the Indefatigable area; during Late Cretaceous and Tertiary inversion, the gas was then thought to have remigrated back into a reservoir that was already damaged by diagenetic illite (e.g. Glennie, 1986b, and earlier editions of this book). How the gas managed to displace formation water in such a damaged reservoir at pressures around the same as found in the sandstones today was a puzzling feature of that interpretation, especially where permeability is so poor, because

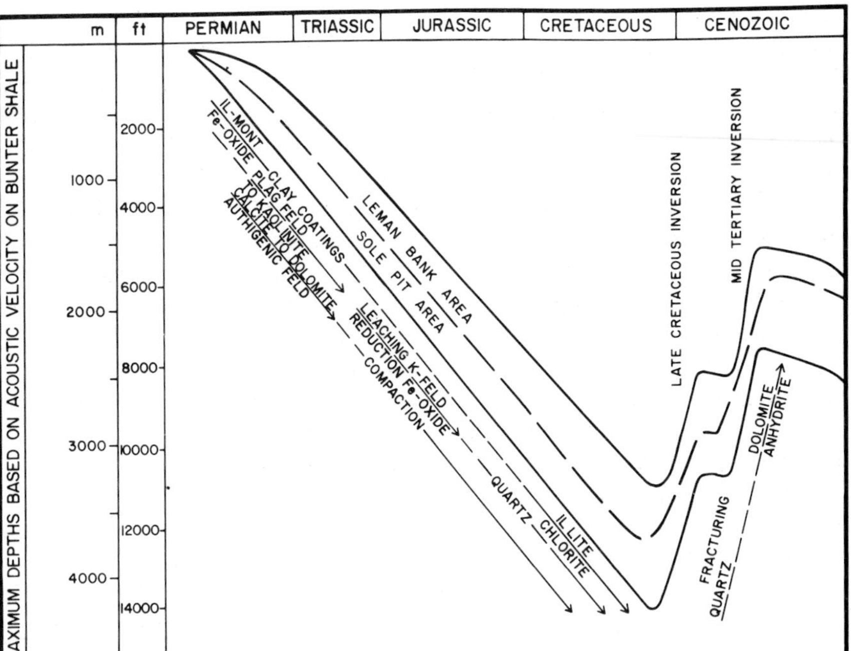

Fig. 5.30 Depth-related diagenesis in the Leman Bank and Sole Pit areas of the UK southern North Sea (after Glennie *et al.*, 1978).

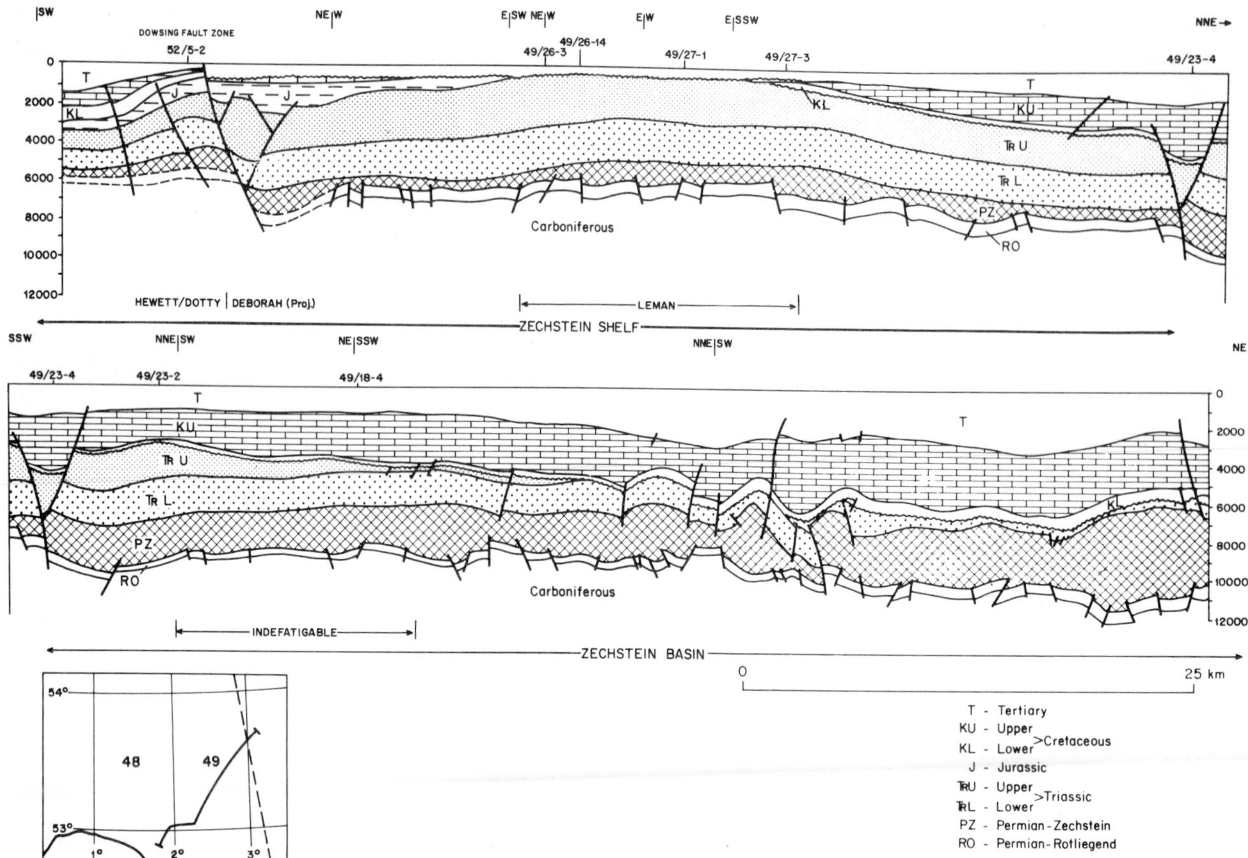

Fig. 5.31 Geological cross-section, UK southern North Sea. Note fault-bounded relief of Rotliegend sequence, thick cover of Zechstein salt over Indefatigable, and much thinner salt over Leman and adjacent to the Dowsing Fault. Note also the considerable mid-Jurassic erosion, which increased in amount towards the north-east (Cleaver Bank 'High'). Later subsidence resulted in deposition of an Upper Cretaceous and Tertiary sequence about equivalent in thickness to that lost during the mid-Jurassic phase of erosion.

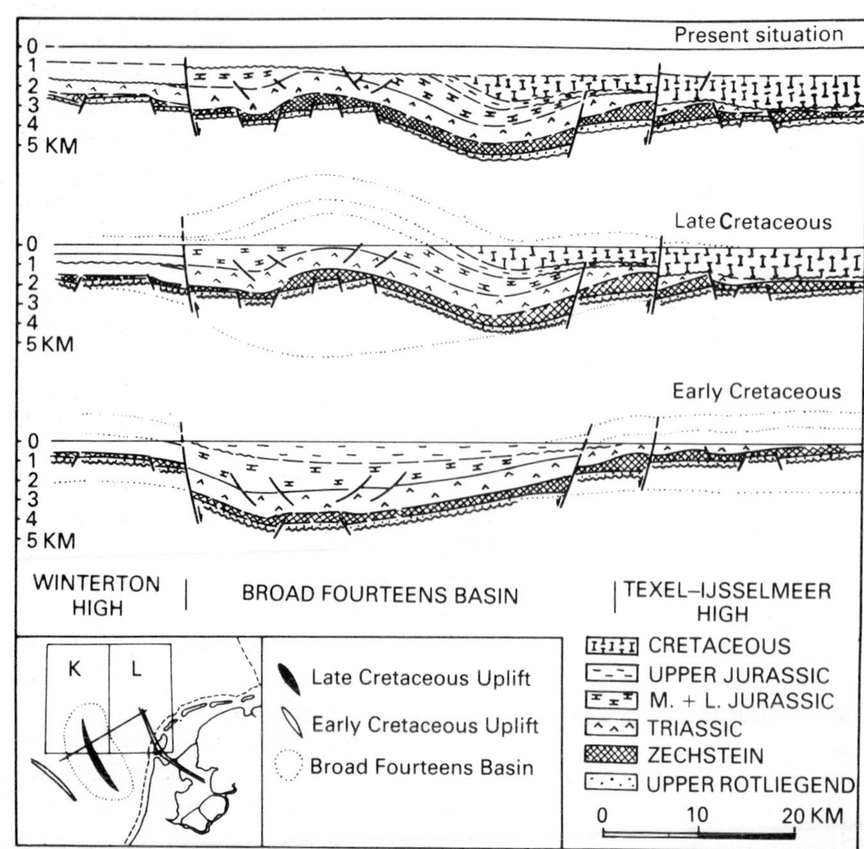

Fig. 5.32 Reconstruction of the burial history of the Rotliegend in the Netherlands offshore K and L Quadrants (from Oele *et al.*, 1981).

of strong diagenesis, that production rates from the matrix are subeconomic.

A more extensive analysis by Lee *et al.* (1989) has shown that illite ceased to form in the Leman Bank area around 170–175 Ma BP, long before it had reached its maximum depth of burial in the later Cretaceous (see Fig. 5.30). Because of an excellent cap rock of salt and very effective lateral fault-plane seals, the gas seems to have remained within the reservoir ever since. If correct, the possibility exists that gas may still be preserved in deep basinal locations.

This interpretation leads to further possibilities. Because of the considerable burial depth of the reservoir during the Jurassic (see Fig. 5.30), the migrating gas presumably had sufficient pressure to achieve entry through pore throats that were already partly blocked by illite filaments. As hydrostatic pressure was reduced during inversion, the gas will have expanded downwards until it reached a leak point in the flanking fault planes; all gasfields on the Sole Pit High seem to be filled to spill (leak) point.

The process of inversion has another aspect to the filling of reservoirs with gas. The steep fractures that compartmentalize the Leman Field have long been known to be filled with anhydrite. Oxygen isotope compositions of the anhydrite are interpreted by Sullivan *et al.* (1994) as indicating precipitation temperatures between 120 and 140°C, which coincides with the Rotliegend's maximum depth of burial of 3.5–4 km (see Fig. 5.30). These workers show that inversion uplift would have brought overpressured Carboniferous formation water into contact with Zechstein brines within the relatively underpressured Rotliegend reservoir. Not only was anhydrite precipitated along the fault planes to form effective lateral seals, but, as indicated above, methane may well have migrated into the already damaged Rotliegend reservoir under the influence of strong Carboniferous overpressures during the early phase of inversion. Pore-filling anhydrite in the upper part of the Rotliegend reservoir also dates from the same time as the fracture anhydrite.

Platt (1994) demonstrates that in Germany the Zechstein had little influence on Rotliegend cements except in the vicinity of horsts, and this probably applies to British and Dutch areas that were not affected by inversion. C. Cornford (personal communication) suggests the additional possibility that some of the gas generated during deep burial remained adsorbed within the coals themselves, and was released only after a reduction in pressure associated with uplift; migration in that case would be vertical into the damaged reservoir. In this context, Cornford (Chapter 11) has shown that, for the gas drainage area around the Sole Pit Basin, the proved reserves of some 693×10^9 m^3 (24.5×10^{12} ft^3) of gas could have been generated from a sheet of coal less than half a metre thick. With a cumulative thickness underlying the Southern Permian Basin of up to 75 or 100 m of coal, not to mention the much thicker, although less prolific, carbonaceous shale, there was clearly no shortage of gas to fill available traps.

In the area around Hamburg, Germany, Westphalian coals have reached the rank of anthracite (Bartenstein, 1979), and must by now have given up most, if not all, of their gas. This was an area of rapid subsidence in the latest Permian (e.g. 1800 m of Rotliegend desert-lake sediments, in contrast to about 200 m of sandstone in the Leman Bank Field) and Triassic (almost 9 km thick in the Glückstadt Graben; Best *et al.*, 1983, Fig. 4). Much of the succeeding Jurassic was removed during the 'Late Cimmerian' phase of erosion around the Jurassic–Cretaceous time boundary, preserved sequences being confined largely to the rim synclines of Zechstein diapirs, which were already active during the later Triassic (Best *et al.*, 1983, Fig. 4). Under these conditions of burial, gas generation in the Glückstadt Graben had probably begun already during the Triassic. With impervious desert-lake sediments immediately above, this gas must have migrated to the south to reach porous Rotliegend sandstones. To the north, the lack of a Rotliegend reservoir and the general absence of Zechstein seal over the Ringkøbing–Fyn High will have resulted in the escape of much gas to the surface (see also Fig. 2.17).

Further east, in the former East Germany, the production of nitrogen associated with the Lower Rotliegend volcanic rocks and associated metamorphism increasingly replaces methane, to the extent that methane represents no more than 40% of the giant Salzwedel field and that percentage reduces to less than 5% as the Polish border is approached (Müller *et al.*, 1993).

Reservoir quality and diagenesis

The original Rotliegend depositional facies quite naturally plays an important role in controlling the type and distribution of early diagenetic cements (Walzebuck, 1993); those sediments deposited in a dry environment usually retain better reservoir quality than those deposited in wet environments (Epting *et al.*, 1993). Some of these facies controls (e.g. nodular anhydrite of sabkhas and early quartz prisms; early carbonate cements and tangentially orientated illite platelets in alluvial deposits) are described by Gaupp *et al.* (1993) and Platt (1994) for the German Rotliegend and by Turner P. *et al.* (1993) for part of the UK North Sea.

With increasing depths of burial, porosity is progressively reduced, first by compaction and pressure solution and then by the growth of authigenic minerals, including chlorite and especially illite (see Fig. 5.30). At greater depths, especially as the reduced porosity approaches 10%, illite develops in a fibrous form; this has a very deleterious effect upon permeability, without any significant further reduction in porosity (Glennie *et al.*, 1978; Hancock, 1978; Nagtegaal, 1979; Glennie and Provan, 1990). Some fields, such as West Sole and Clipper, can still produce gas at commercial rates even when the reservoir has been badly damaged by fibrous illite. This is because the reservoir is crossed by a system of natural open fractures, formed during inversion, which provide very large areas of contact with the damaged pores (Winter and King, 1991; Franssen *et al.*, 1993). Where natural fractures are absent, the effects of diagenetic damage can now be overcome by directional drilling (Frikken, 1993). But, irrespective of the degree of authigenic mineralization, the better porosity and permeability will always be preserved in those sands that had the better depositional porosity (normally aeolian rather than wadi; avalanche rather than bottomset dune beds or adhesion ripple) and whose environment of deposition precluded porosity-destroying early cementation (e.g. away from any direct influence of the water-table, as in the Groningen and Rough fields).

In general, the best reservoir sands occur in the dominantly aeolian middle part of the Rotliegend sequence, with porosities in some fields ranging up to about 25% and air permeabilities with values in excess of 100 mD where not damaged diagenetically. This is largely an effect of environment of deposition, where the extensively developed foresetted aeolian sands of the middle sequences had their depositional porosities virtually unaltered by early diagenesis, in contrast to the lower mixed aeolian and wadi sequences, which suffered some early cementation. The reservoir in the Markham field straddles the median line near the northern limit of UK Quadrant 49 and the Dutch sliver of Quadrant J, close to the northern shale-out edge of the aeolian sands. Here, the dune sands have a porosity of 15–20% and a permeability of 10–1000 mD. In the adjacent sabkha sands, these figures are reduced to 5–17% and 0.1–10 mD, respectively (Myres *et al.*, 1995).

The porosity of the aeolian sands directly beneath the Kupferschiefer is commonly reduced for two different reasons: the original grain packing became tighter in the Weissliegend sands (the top 0–65 m), because of deformation and homogenization associated with the Zechstein transgression; but, whether deformed or not, the main porosity destroyer is dolomite cement that is believed to result from proximity to the directly overlying Zechstein carbonates. Locally, where the Rotliegend is juxtaposed with Zechstein evaporites, the reservoir quality is destroyed by an anhydrite pore fill. In the Leman Field, fault-related anhydrite possibly formed when brines of Zechstein origin invaded the lower-pressure Rotliegend reservoir at the start of inversion uplift in the Early Cretaceous (Sullivan *et al.*, 1994).

Robinson (1981) found that the ratio between horizontal and vertical air permeability in most aeolian and wadi sands is between 1 and 100. This distinction is not seen in the Weissliegend sands because of either a lack of distinct bedding or the presence of deformed bedding. Both these points are reflected by dipmeter logs, which generally show no dip data or, alternatively, only random dips (see, for example, Heward, 1991, Fig. 8). The thick Rotliegend sequence found in the Leman Bank gasfield comprises mostly foresetted aeolian sandstone similar to that seen in Fig. 5.12 (see also van Veen, 1975, Plate 1), which is interpreted as having been deposited largely on the avalanche slopes of a transverse dune. These inherently under-compacted sands had a naturally high primary porosity, in the range 42–47% (Hunter, 1977), which since deposition has been reduced to half or less by the effects of compaction and burial diagenesis (see Fig. 5.30; Glennie *et al.*, 1978).

By the use of cores and dipmeter logs, van Veen (1975) demonstrated a clear correlation between bedding type (e.g. bottomset and foreset beds) and the response to gamma-ray and FDC logs in the Leman field. Similar log responses have been noted in other North Sea fields, such as Sean (Ten Have and Hillier, 1986; see Fig. 5.33) and North Viking (Martin and Evans, 1988). They compare well with aeolian reservoirs of different age in North American fields (e.g. Lupe and Ahlbrandt, 1979; Ahlbrandt and Fryberger, 1982; Lindquist, 1988). In a new way of looking at boreholes, Luthi and Banavar (1988) related electrical borehole images of Rotliegend sandstones to aeolian bedding type, as well as to dipmeter logs.

Horizontally laminated bottomset beds have a higher argillaceous content, are generally more cemented and have much higher capillary pressures associated with the finest-grained laminae than with foreset beds. Such beds are more prone to later diagenesis and thus form semipermeable barriers to gas flow in producing wells and also limit water coning at the base of the reservoir to a minimum. Some of these statements were confirmed in a probe-permeameter study by Prosser and Maskall (1993) of a core in the Auk Field. The core was taken in a deviated well, and almost 5 m was cut parallel to the avalanche slope. They found that, at a microscopic scale, the main controls on permeability are grain sorting and porosity, the development of authigenic cements and the detrital and authigenic clay content. Clay minerals and cements are concentrated along pinstripe laminae, especially those of sand sheets.

To assist the prediction of bedding-related variations in production in the Leman Field, Weber (1987) made a study of the excellent exposures of the Permian De Chelly dune sandstones of Arizona (see also McKee, 1979), which, like those of the North Sea area, are of the transverse type. Weber (1987) established a rough relationship between the thickness of one set of dune bedding, its width and its downwind length of 1 : 50–100 : 200 (see Fig. 5.34A, B). Because of the very low vertical permeability of the bottomset beds, and the strong 1–100 difference in permeability parallel to and across the foreset bedding, Weber (1987) predicted that there would be a marked horizontal anisotropy in the foreset beds. The geometry of these relationships is depicted in Fig. 5.34C.

In contrast to the Leman Field, not only is the overall sequence much thinner in the Indefatigable area (about 80 m compared with over 200 m in Leman Bank; see also Fig. 9 in George and Berry, 1994) but some of the horizontally laminated dune-base sandstones are replaced by horizons of argillaceous and anhydrite-cemented adhesion-ripple sands deposited in interdune sabkhas; these are fairly effective barriers to vertical gas flow (see Fig. 5.35 for the poroperm quality of similar sands from the Sole Pit area). Their lateral extent, however, is probably more limited than that of the more permeable flanking dune sands.

The burial-related growth of authigenic minerals, and especially of illite, in areas of former deep burial, such as the Sole Pit area (see Fig. 5.30), has resulted in the porosity and permeability of the dune sandstones now being much lower on average than those of, say, the Indefatigable Field (compare Sole Pit well 48/13-2 with Indefatigable well 49/24-1 in Figs 5.29B, C; see also Table 5.1).

The capillary effects associated with smaller pore connections in the diagenetically more damaged reservoir of Leman Bank probably account for the overall differences in recovery factor between it and Indefatigable. The recovery factor is 80% of the gas in place in Indefatigable and 75% in Leman Bank. Similarly, Robinson (1981) attributes a relatively high water saturation in the Amoco well 47/15-1 (Amethyst area) to a reduction in the size of the pore throats related to the precipitation of authigenic minerals. It has been noted by Ten Have and Hillier (1986), however, that, in the Sean Field, pore-throat diameters at 50%

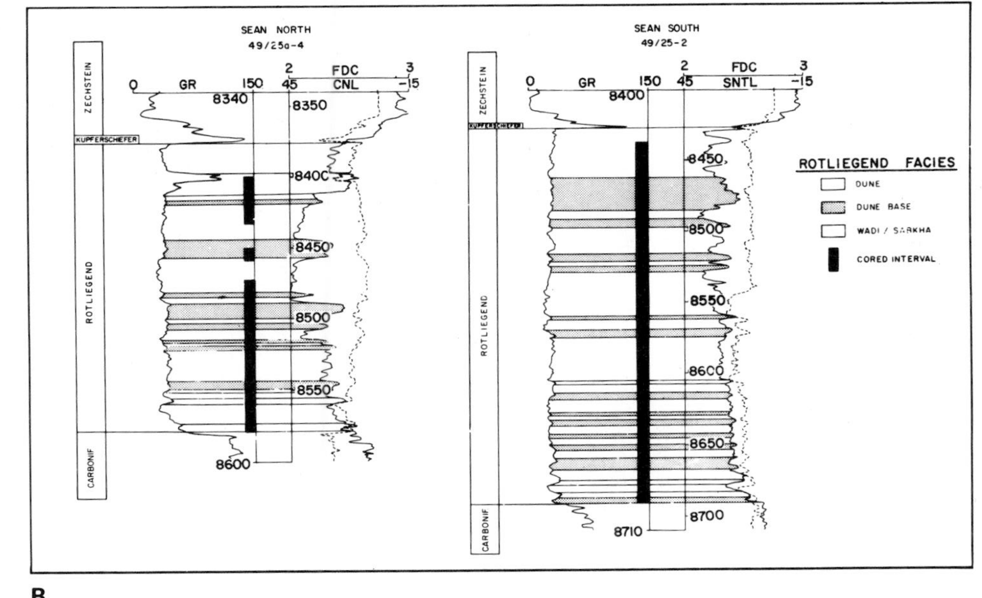

A

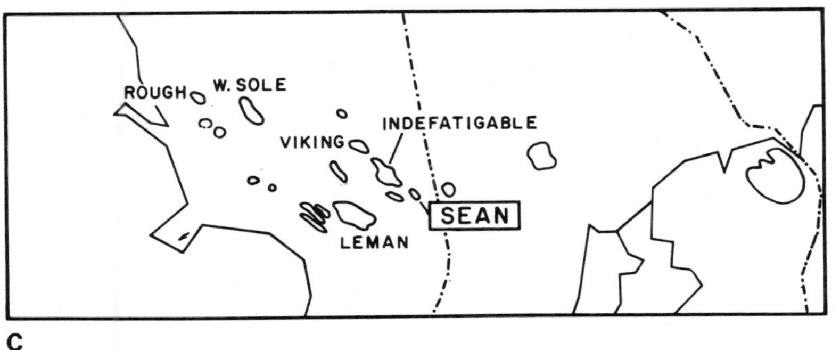

B

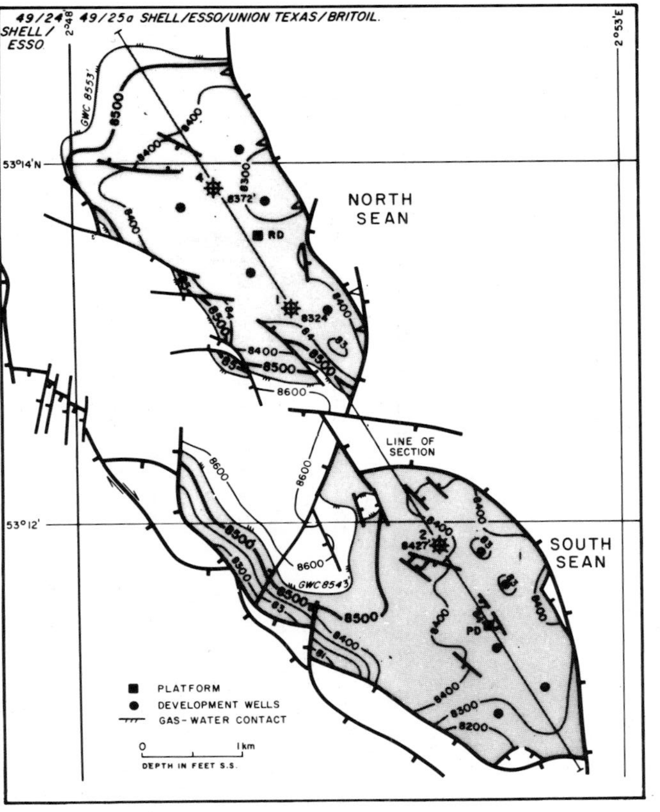

C

Fig. 5.33 The Sean gasfield. A, C. Two separate fault and dip-closed Rotliegend structures on trend with the larger Viking and Indefatigable Fields to the north-west. B. The character of the FDC log picks out the differences in porosity between avalanche-slope, dune-base and wadi/sabkha sands. Following Late Cimmerian uplift and erosion, the Rotliegend in Sean is now close to its maximum depth of burial. (From Ten Have and Hillier, 1986.)

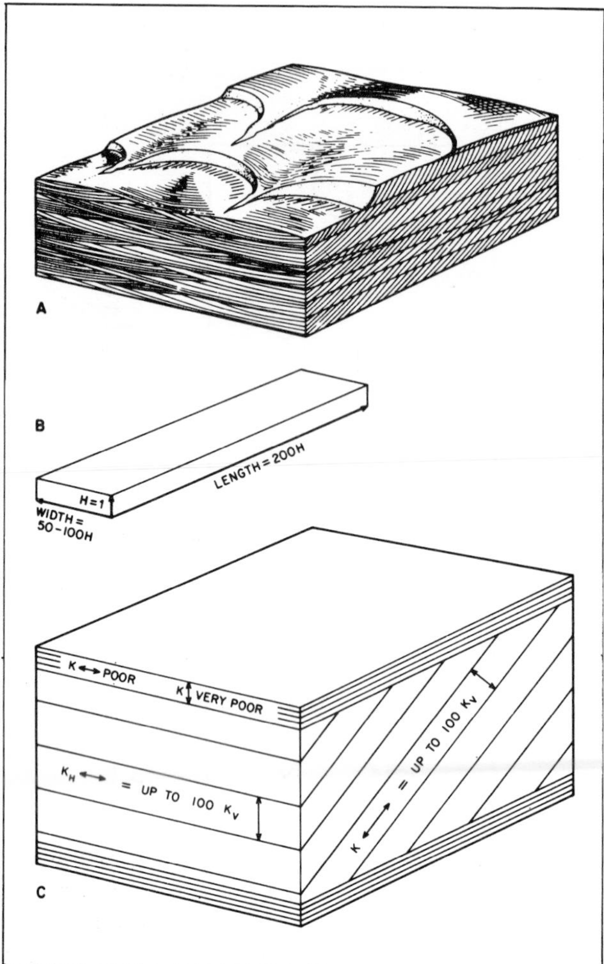

Fig. 5.34 A. Model of transverse dune sedimentation (from Weber, 1987), illustrating the basis for a height, width, length relationship for bedding sets of 1 : 50–100 : 200 seen in B. and C. Differences in permeability in bottomset beds and in foreset beds parallel to and perpendicular to lamination result in a marked horizontal anisotropy in the foreset beds (inspired by Robinson, 1981, and Weber, 1987).

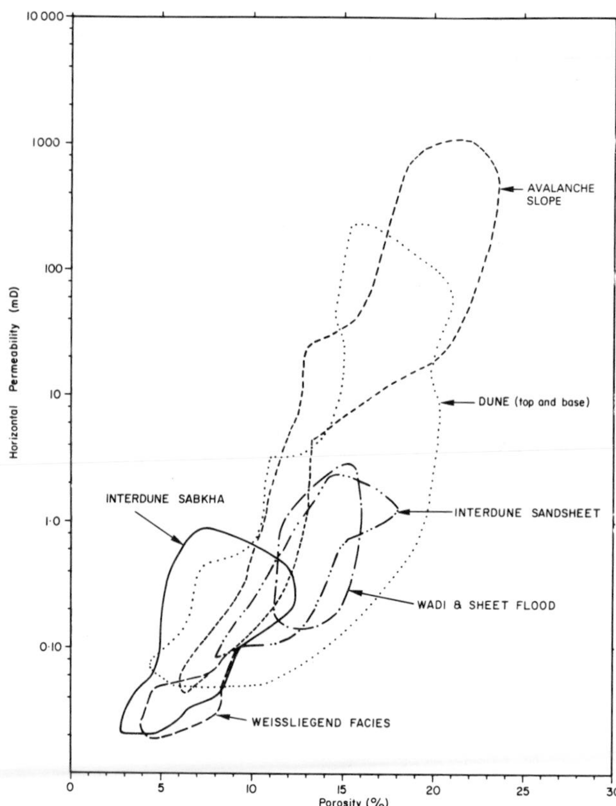

Fig. 5.35 Porosity–permeability relationships for different Rotliegend facies in the Sole Pit area. There is a progressive improvement in reservoir quality from interdune sabkhas to wadis, through dune bottom and top sets to sands of avalanche slopes, which have the best sorting, least clay content and most open primary packing; the Weissliegend facies has a carbonate cement, probably derived from the overlying Zechstein.

wetting-phase saturation vary from 11 μm in cross-bedded dune sand to 2 μm in dune-base sands and 0.3 μm in wadi and sabkha sands. Thus, apart from the diagenetic effects that are related directly to depth of burial, the amount of mineral growth around the pore throats also depends on the environment of deposition.

In this respect, Gaupp *et al.* (1993; see also Burri *et al.*, 1993, for summarized results) have noted that, in northern Germany, the lacustrine facies belts that parallel the edge of the Rotliegend desert lake controlled the distribution of early and shallow-burial cements. Although the larger dune fields form good reservoirs (David *et al.*, 1993), most of the north German Rotliegend gas and the best reservoirs occur in what are reported to be shoreline sands of the Wustrow Sandstone (Hannover Folge; Gast, 1991; Burri *et al.*, 1993). These sands, which are likely to be of sabkha type with a strong aeolian component, have pore-lining radial chlorite cement and good porosities, and form the best Rotliegend gas reservoirs in Germany (Gast, 1991). This is possibly because the chlorite, which probably originates from the alkaline waters of the sabkha groundwater, inhibits later cementation. The chlorite is especially abundant on the outer fringes of the sabkha's sandstone facies, where it interdigitates with red shales, and where the facies is downfaulted in grabens within the Hamweide High (see Burri *et al.*, 1993, Fig. 1), which are almost normal to the lake margin (Gaupp *et al.*, 1993).

The chlorite contrasts with the more notorious illite cement found in both aeolian and alluvial reservoirs. The generation of illite is associated with acidic-formation waters derived by compaction of Carboniferous sediments and the maturation of beds of coal. It is particularly abundant near major fault zones and, in UK waters, the permeability-destroying fibrous illite is developed at depths greater than about 3000 m (see Fig. 5.30). Gaupp *et al.* (1993) state that, in the German Rotliegend, permeabilities in illite-cemented sandstones do not exceed 1 mD, even where the porosity is above 10%. This contrasts with permeabilities of over 100 mD for sandstones with chlorite grain rims, and up to 400 mD for uncemented sandstones.

Early albite forms as overgrowths on detrital feldspars and as pore-filling crystals in sabkha sands, whereas kaolinite occurs in areas of shallow burial in the western part of north Germany and, together with dickite, where Rotliegend sands are juxtaposed with coal-bearing Westphalian strata. Farther from the fault zone, the kaolinite-rich rock commonly grades into an illite-rich zone (Gaupp *et al.*, 1993, Fig. 13).

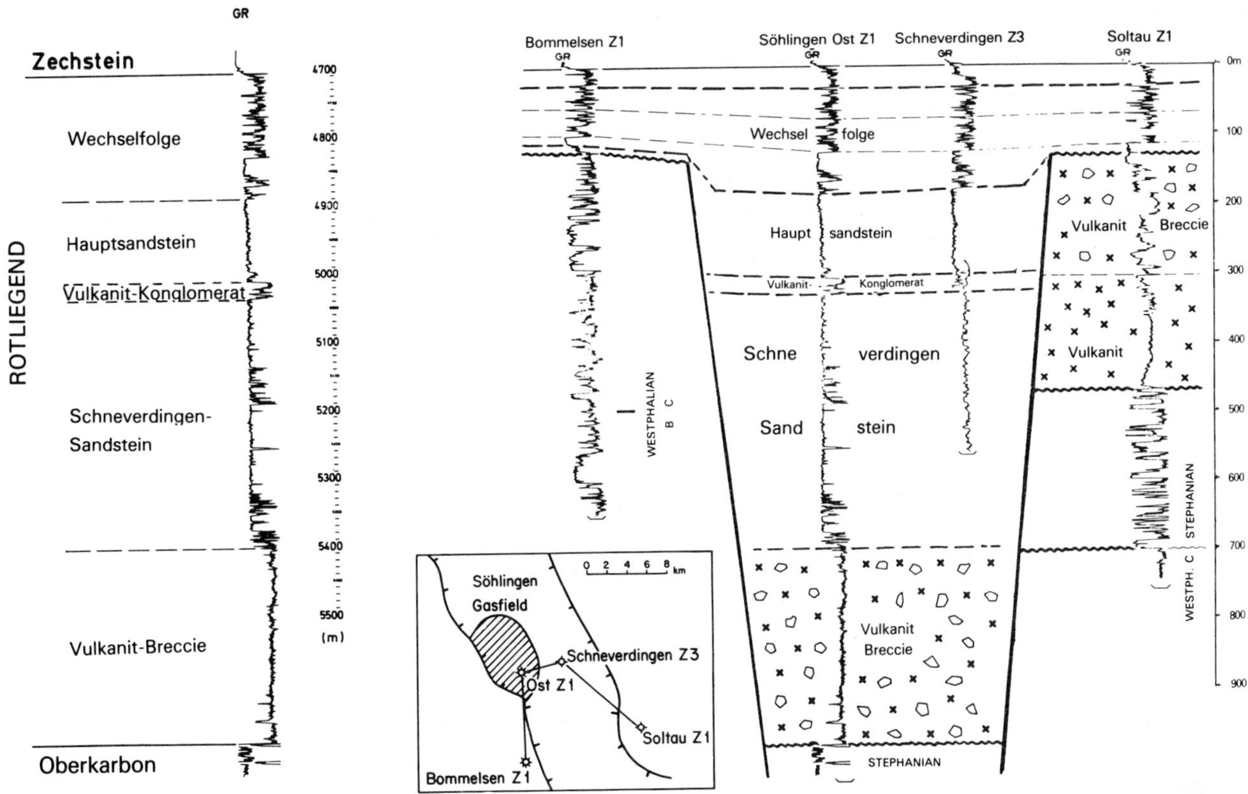

Fig. 5.36 Stratigraphic relationship and tectonic setting of the Upper Rotliegend Schneverdingen Sandstone, West Germany (slightly modified from Drong *et al.*, 1982).

A combination of fluvial, aeolian and sabkha sands form gas-filled reservoirs on a series of structural highs (e.g. Ravenspurn and Caister–Murdoch ridges) within the UK Silverpit Basin (Bailey *et al.*, 1993). Possibly because of the alkalinity of the groundwater and adjacent lake waters, permeability-destroying illite is not abundant, except in the south-west where fluvial clays enrich the basinal sediments (Heinrich, 1991a,b; Ketter, 1991b).

As we have seen, because of the effects of diagenesis, the quality of most gas reservoirs becomes poorer within particular facies belts or with increased depth of burial. It comes as a pleasant surprise, therefore, to find an example where early UR2 sandstones, both aeolian and fluvial, have better porosity and permeability at a depth of over 5000 m than the dune sandstones of the overlying Slochteren (Hauptsandstein) Formation (now Parchim Formation; see Fig. 5.22). This rather unusual situation is found in the Söhlingen gasfield within the Schneverdingen Graben in Germany, about 50 km east of Bremen (Drong *et al.*, 1982; Gast, 1988; Gralla, 1988; see Fig. 5.7). The almost 400-m-thick Schneverdingen Sandstone was deposited in an actively subsiding narrow graben (see Fig. 5.36) and has preserved porosities of up to 15% and permeabilities in the range of 1–10 mD, in contrast to the 1–5% porosity and <1 mD permeability of the overlying Hauptsandstein (now Mirow Formation).

Drong and his colleagues (1982) suggest that the difference in reservoir quality might reflect differences in climate between the times of deposition of these two Rotliegend sequences (less calcite and anhydrite cement in the Schneverdingen Sandstone). It seems just as likely, however, that the porosity differences resulted from graben-related facies control on early cements, with deposition well above the capillary reach of the water-table. It could also be related simply to differences in grain size, with stronger capillary retention of formation water (and hence stronger diagenesis on burial) in the finer-grained (0.1–0.25 mm) Hauptsandstein than in the coarser (0.25–0.5 mm) deeper sands of the Schneverdingen Sandstone; coarser sands always seem to retain a greater portion of their original porosity and permeability after burial diagenesis than do finer-grained sands.

Remote-sensing techniques are resulting in the discovery and development of Rotliegend gasfields. Kruis and Donzae (1993); for example, describe the 1990 discovery of the largest onshore gasfield (Grijpskerk) in the Netherlands since Groningen by the use of three-dimensional (3D) seismic for exploration; the lateral seal in this case is the Ten Boer Claystone. Seismic impedance modelling of 3D seismic can lead to an evaluation of the gas in place in both newly discovered and producing gasfields; this technique is removing much of the uncertainty of what is happening between wells and reducing the need for expensive appraisal wells (Hartung *et al.*, 1993; Krystofiak *et al.*, 1993; Veenhof, 1996).

5.5.2 Oilfields

There is a marked contrast between the hydrocarbon potential of Rotliegend reservoirs in the Northern and Southern Permian Basins; the Southern Basin is a gas province, while in the north only oil has been found. Although minor amounts of condensate and natural-gas

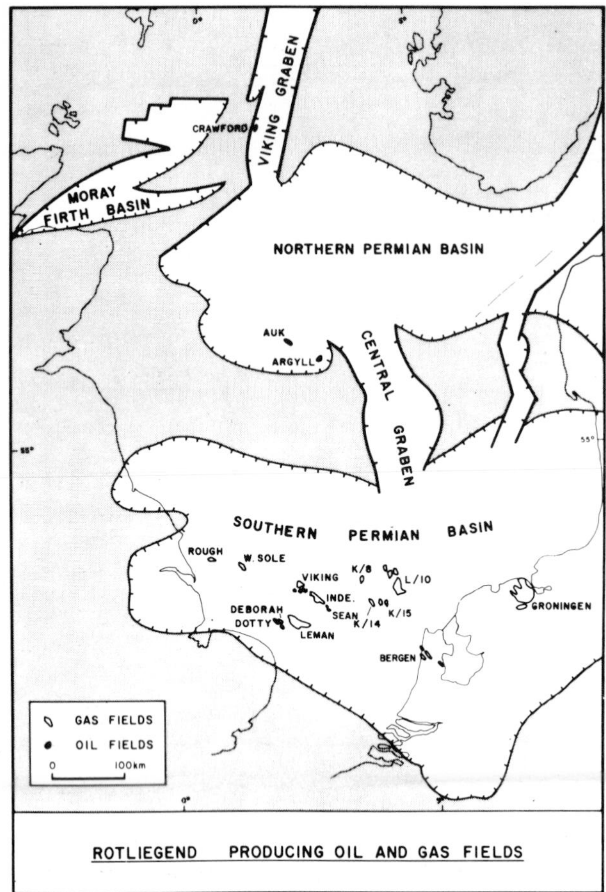

Fig. 5.37 Distribution of some Rotliegend producing oil- and gasfields.

liquids (NGL) are known from some Rotliegend structures in the Southern Permian Basin, they occur in the heavy fractions of gas of Carboniferous origin.

In the Northern Permian Basin, the Zechstein cap rock of salt is present over the greater part of the area (Taylor, 1986), but the all-important source rock for gas, the Carboniferous Coal Measures, are either completely absent (except, perhaps, in a limited offshore extension of the Midland Valley of Scotland) or are too poorly developed to form a viable source rock (see Chapter 4, Fig. 4.2). Thus, in this area, those Rotliegend reservoirs that have been penetrated by the drill are devoid of gas. Where the Rotliegend sandstones of the basin do form reservoirs for hydrocarbons, they have been brought into the correct geometric relationship with a much younger mature source rock for oil.

The Auk (Brennand and van Veen, 1975; Buchanan and Hoogteyling, 1979; Heward, 1991; Trewin and Bramwell, 1991) and Argyll (Pennington, 1975; Bifani *et al.*, 1987; Robson, 1991) oilfields in the central North Sea (Fig. 5.37, 5.38) produce much of their oil from the basal Zechstein carbonates. In the shallower parts of the fields, however, Rotliegend sandstones are also saturated with oil. Oil in the small Innes field is reservoired entirely within Rotliegend sands (Robson, 1991). These fields are situated close to the western flank of the Central Graben.

Pennington (1975) suggested that the oil found in the Argyll Field was derived from structurally adjacent Palaeocene shales. These shales, however, probably have not

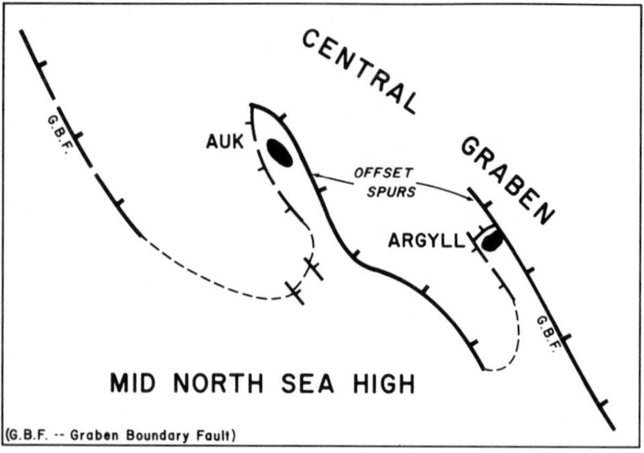

AUK

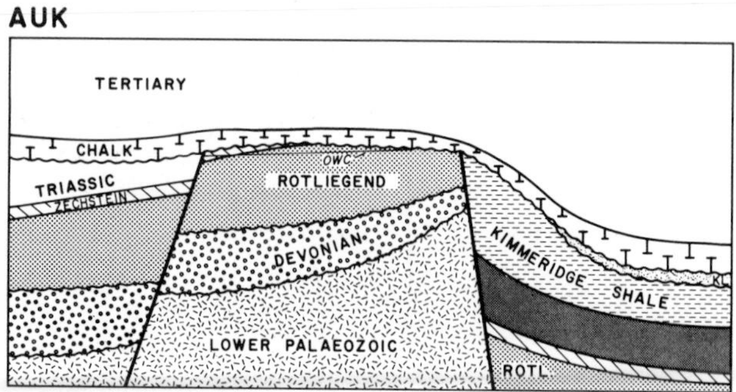

ARGYLL

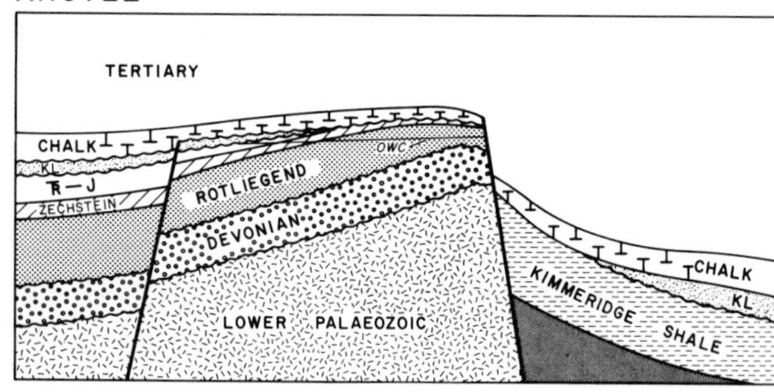

Fig. 5.38 Schematic structural setting and cross-sections of the Auk and Argyll oilfields. Source rock—probably Upper Jurassic Kimmeridge Shales. Reservoir rock—Zechstein dolomites and Rotliegend sandstones. (Modified from Brennand and van Veen, 1975, and Pennington, 1975.)

been buried sufficiently deeply to be mature. As with most of the fields in the central and northern North Sea, it is now certain that the oil-source rock for both the Auk and Argyll fields is the Jurassic Kimmeridge Shale (Robson, 1991; Trewin and Bramwell, 1991; see Fig. 5.38), which matured deep within the Central Graben sometime during the Mid- to Late Tertiary. The oil probably migrated up graben-flank faults (possibly even via the adjacent graben-edge Fulmar Field, which is close to spill point (Trewin and Bramwell, 1991), or around the noses of offset spurs, and became trapped beneath the general cover of relatively impervious Upper Cretaceous chalk (Fig. 5.38). This interpretation would seem to be supported by the occurrence of oil in the Crawford Field (block 9/29, west flank of the South Viking Graben) in Rotliegend (not producible), Triassic and Jurassic reservoirs (Yaliz, 1991, Fig. 2).

The Rotliegend in Auk has been divided by Heward (1991) into five distinct sedimentological units, the upper three being oil-bearing. The lowest unit comprises conglomerates, which are only locally developed. The overlying units are mostly of aeolian origin, with the Weissliegend represented by aquatically reworked dune sands resulting from pre-Zechstein rainstorms. Adhesion-ripple sands are not found in Auk, which indicates that both fluvial and aeolian deposition took place well above the water-table. As in the Southern Permian Basin, the best reservoir quality is found in slip-face sands, where the combination of medium grain size and good sorting results in high permeabilities.

In contrast to Auk, the Rotliegend sequence found in the Argyll Field seems to have been deposited in a much wetter environment, which includes shallow desert lake, alluvial fan and fluvial channel, as well as dune sediments. The sedimentary model deduced by Bifani *et al.* (1987) is that of a half-graben, filled with the deposits of an alluvial fan and terminal desert lake, which then became covered with the sands of transverse dunes; adhesion-ripple beds are interpreted as the deposits of a damp interdune area. The whole sequence is capped by homogeneous sandstones of the Weissliegend. As in Auk, the best oil productivity is from dune sands, which seem to be free of authigenic cements. Although the fluvial sands are also oil-bearing, many are likely to give up their oil efficiently only through a limited system of natural fractures or where they have a large areal contact with interbedded aeolian sands. The top and basal layers of the Innes reservoir consist mostly of fluvial and interdune sandstones, while dune sands form the middle layer and possess the best reservoir characteristics (Robson, 1991).

5.6 Acknowledgements

Permission to publish the first two editions of this chapter was given by Shell UK Ltd and Esso Petroleum Co. Ltd. These two companies are thanked for their continued support, and for releasing the core data shown in Figs 5.16 and 5.17. Shell UK Ltd also draughted the additional and updated figures found in the fourth edition.

5.7 Key references

Gast, R. (1993) Sequence stratigraphy of the north German Rotliegende. *AAPG Bull.* 77(9); 1624.

Gaupp, R., Matter, A., Platt, J., Ramseyer K. and Walzebuck, J. (1993) Diagenesis and fluid evolution of deeply buried Permian (Rotliegende) gas reservoirs, northwest Germany. *AAPG Bull.* 77(7); 1111–28.

George, G.T. and Berry, J.K. (1993) A new lithostratigraphy and depositional model for the Upper Rotliegend of the UK sector of the southern North Sea. In: North, C.P. and Prosser, D.J. (eds) *Characterization of Fluvial and Aeolian Reservoirs.* Special Publication 73, Geological Society, London, pp. 291–319.

Glennie, K.W. (1997) Recent advances in understanding the southern North Sea Basin: a summary. In: Ziegler, K., Turner, P. and Daines, S.R. (eds) *Petroleum Geology of the Southern North Sea: Future Potential.* Special Publication 123, Geological Society, London, pp. 17–29.

Goggin, D.J., Chandler, M.A., Kocurek, G.A. and Lake, L.W. (1986) Patterns of permeability in eolian deposits. In: Society of Petroleum Engineers and US Department of Energy (eds) *5th Symposium on EOR, April 1986, Tulsa.* SPE/DOE, Tulsa abstract 14893, pp. 181–8.

Heward, A.P. (1991) Inside Auk—the anatomy of an eolian oil reservoir. In: Miall, D. and Tyler, N. (eds) *The Three Dimensional Facies Architecture of Clastic Sediments and its Implications for Hydrocarbon Discovery and Recovery.* SEPM Series Concepts and Models in Sedimentology and Paleontology 3, Tulsa pp. 44–56.

Plein, E. (ed.) (1995) *Stratigraphie von Deutschland I. Norddeutsches Rotliegendbecken.* Rotliegend-Monographie Teil II, Courier Forschungsinstitut Senckenberg 183, Frankfurt, 193 pp.

6 Upper Permian—Zechstein

J.C.M. TAYLOR

6.1 Introduction

The Zechstein Group is a complex of evaporite and carbonate rocks of Late Permian (probably late Tatarian) age, which underlies a large part of the North Sea and north-west Europe (Fig. 6.1). Deposition was probably coeval with the Guadalupian to Ochoan sequences of the Delaware Basin, USA, with which there are interesting similarities and differences.

Exact correlation with the standard succession is difficult. It is possible that Zechstein deposition spanned the Permian/Triassic boundary, determined from a thin layer of bentonite in southern China, as 251.2 ± 3.4 million years (Ma) (Claoue-Long *et al.*, 1991). The duration of Zechstein deposition has been estimated at 5–7 Ma (Menning, 1995).

As one of the world's 'saline giants' the Zechstein Basin merits attention for the light it may cast on the origin of other thick evaporite sequences. So far as the oil industry is concerned, Zechstein rocks are significant on five counts: (i) generation of structure; (ii) cap rocks; (iii) structural information; (iv) reservoir rocks; and (v) source rocks.

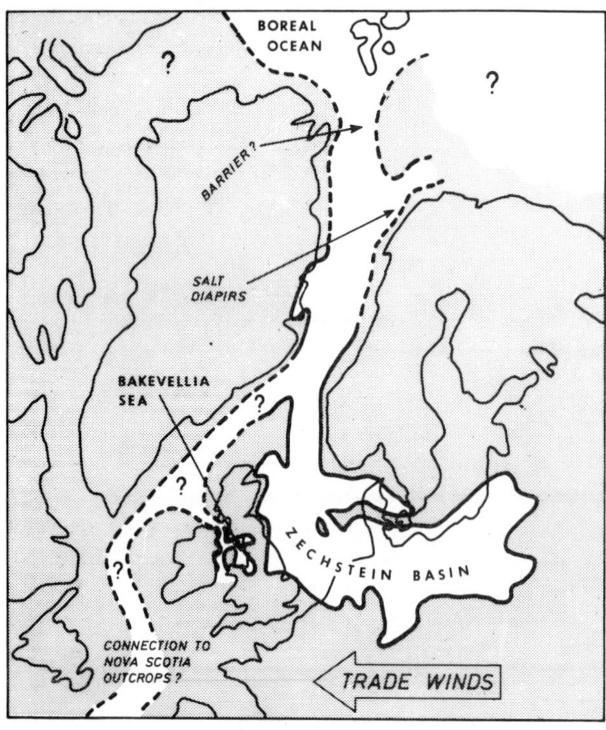

Fig. 6.1 Sketch map of Zechstein basins in relation to pre-drift North Atlantic area.

6.1.1 Generation of structure

Movement of Zechstein salt is responsible for closures in overlying strata. The giant Ekofisk field is only one of several in the Central North Sea which owe their existence to Zechstein salt diapirs. In recent years, closures resulting from salt migration, withdrawal and dissolution at depth have received wide interest as economic targets in and around the Central Graben. The ability of Zechstein salt to decouple younger from older rocks has been fundamental to the development of post-Zechstein structures and traps.

6.1.2 Cap rocks

The Rotliegend and some Carboniferous gasfields of the southern North Sea depend largely on the sealing efficiency of Zechstein salt.

6.1.3 Structural information

The top and base of the Zechstein commonly provide important seismic reflectors, which give clues to the structure of hydrocarbon-bearing strata above or below.

6.1.4 Reservoir rocks

Some Zechstein carbonates have good porosity and permeability or extensive fractures, providing commercial reservoirs on the Continent and in the North Sea.

6.1.5 Source rocks

Zechstein carbonates include potential source facies, although adequacy for the economics of the North Sea has yet to be demonstrated.

In addition, Zechstein evaporites constitute reserves of commercial potash salts, halite, anhydrite and gypsum in north-east England and on the Continent. The thick halite sections are used for leached-cavern storage of hydrocarbons and chemicals, and have been considered for the disposal of radioactive and other wastes.

6.1.6 General

The place of the Zechstein in North Sea petroleum geology was outlined by Kent (1967a,b) and, with special reference to its evaporites, by Brunstrom and Walmsley (1969). Correlation and nomenclature have been discussed by Pattison *et al.* (1973), Rhys (1974), Smith *et al.* (1974,

1986) and Deegan and Scull (1977). The lithostratigraphic nomenclature in the Central and Northern North Sea has since been revised by Cameron (1993a). The lithostratigraphic nomenclature has also been revised in the UK Southern North Sea and in the Netherlands (van Adrichem Boogaert and Kouwe, 1993). Smith (1980b) described the evolution of the English Zechstein Basin and provided a guide to the extensive earlier literature relating essentially to onshore areas and the Continent. The Late Permian palaeogeographical evolution of north-east England has been described by Smith (1989) and that of the UK onshore and offshore regions by Smith and Taylor (1992). The probable time equivalence of named units in different areas is given in Fig. 6.2.

An alternative approach to subdivision of the Zechstein, based on sequence stratigraphy, has been suggested by Tucker (1991), later discussed by Goodall *et al.* (1992). Strohmenger *et al.* (1996) have modified this model to accommodate recent German results.

For an introduction to evaporite/carbonate sedimentation and environments in general, see Schreiber (1987).

Well logs representing Zechstein successions typical of various regions of the UK North Sea are given in Figs 6.10 to 6.12, which also indicate the probable correlations. Figure 6.3 shows the locations of released wells used (together with other published data) in the construction of the maps in this chapter.

6.2 Distribution and general character

(Fig. 6.1)

Zechstein rocks occupy two east–west-trending subbasins, partly separated by the Mid North Sea–Ringkøbing-Fyn High (MNSH and RKFH) (Ziegler, 1981). These depressions have been attributed to thermal contraction following the end of Rotliegend volcanism, the subsidence continuing to the end of the Zechstein (Lorenz and Nicholls, 1984) or beyond. Thick salt occupies the basin centres, whereas carbonates and anhydrites are more abundant round the edges and over parts of the MNSH.

The Southern Salt Basin extends from eastern England through the Netherlands and Germany to Poland and Lithuania; it is well documented as a result of the search for hydrocarbons and other minerals. The Northern Salt Basin stretches from the Scottish Firth of Forth and Moray Firth across the North Sea and northern Denmark into the Baltic (see Fig. 6.23). It is fairly well delineated by seismic, which shows the group to be over 2000 m (6000 ft) thick in the Norwegian–Danish Basin and locally elsewhere (Taylor, 1981; Fig. 6.4), but less lithological detail is known than in the southern basin, since much of the group is below current target depths for the drill (Fig. 6.5).

From England to the Netherlands, the southern margin of the Zechstein Sea was formed by the London–Brabant Platform. In the north, arms of the sea occupied the Moray Firth, the southern end of the Viking Graben and probably other embayments in Norwegian waters. No halite has been reported north of 59°50′N, and perhaps the Zechstein Basin proper extended no further, although relatively few wells have passed through the equivalent interval in the North Viking Graben or East Shetland Basin.

Further north again, the Late Permian Foldvik Creek Group of the East Greenland Basin includes anhydrite and gypsum (Maync, 1961; Surlyk *et al.*, 1984; Stemmerik, 1987). These, however, appear to be of marginal rather than basin facies. The presence of thick black shales (of interest as potential oil sources), on the other hand, suggests restricted water circulation. Scholle *et al.* (1991) describe bioherms with > 100 m relief and a biota resembling that of the first Zechstein cycle in the Wegener Halvo sequence. Three subcycles, with rapid and extensive transgressions, are deduced. It is not yet clear whether these correlate with the later Zechstein cycles. If there was a final northern circulation barrier marking the limit of the Zechstein basin and controlling its cycles (Taylor, 1980 and in preparation), it seems to have lain south of Svalbard, where rocks, probably of equivalent age, consist of shales of unrestricted deep-marine shelf facies (for details, see Steel and Worsley, 1984). Under the Barents Sea, the Late Permian has been shown by drilling and seismic to consist of siliceous sponge-rich mudstones deposited over drowned shelf carbonates. The thick diapiric salt noted by King (1977) in the Norwegian Tromsø Basin (see Fig. 6.1) and other evaporites in that region appear to be of pre-Tatarian age, and therefore not Zechstein equivalents (Nilsen *et al.*, 1992; Spencer and Eldholm, 1993). The model resolves the puzzling absence of micro-taxa from the deep basin sediments of the Zechstein.

The basins of the Late Permian Bakevellia Sea occupy parts of the Irish Sea and adjacent Northern Ireland and north-west England (Pattison *et al.*, 1973; Colter and Barr, 1975; Illing and Griffith, 1986; Jackson *et al.*, 1987). They were probably fed from west of Scotland. Haszeldine and Russell (1987) predicted that evaporites of Zechstein age would be discovered in other basins from Svalbard south and west to the Porcupine Basin. Tate and Dobson (1989) have since confirmed that a thin Late Permian sequence, including dolomites and anhydrite and having marine microfossils, has been found in wells in the Donegal Basin off north-east Ireland.

The correlation of Late Permian formations in Cumbria with the Zechstein of the North Sea and the possibility of a cross-Pennine marine connection are still the subject of debate (Holliday, 1993a, 1994a; Jackson, 1994; Smith, 1994).

6.3 Outline of depositional pattern and typical facies

Figure 6.6 is a diagrammatic profile of the Zechstein in the UK Southern North Sea showing the principal lithofacies, together with the German-based nomenclature in common use. The traditional scheme of subdivision is adhered to here. Appraisal of alternative systems based on sequence stratigraphy is deferred until seismic interpretation has been discussed in Section 6.9.

Similar profiles described from the Netherlands (Brueren, 1959; Clark, 1980a; van Adrichem Boogaert and Burgers, 1983), Germany (Richter-Bernburg, 1959; Füchtbauer, 1968), Denmark (Sorgenfrei and Buch, 1964; Sorgenfrei, 1969; Clark and Tallbacka, 1980) and Poland (Depowski, 1978; Wagner *et al.*, 1981; Pokorski and Wagner, 1993)

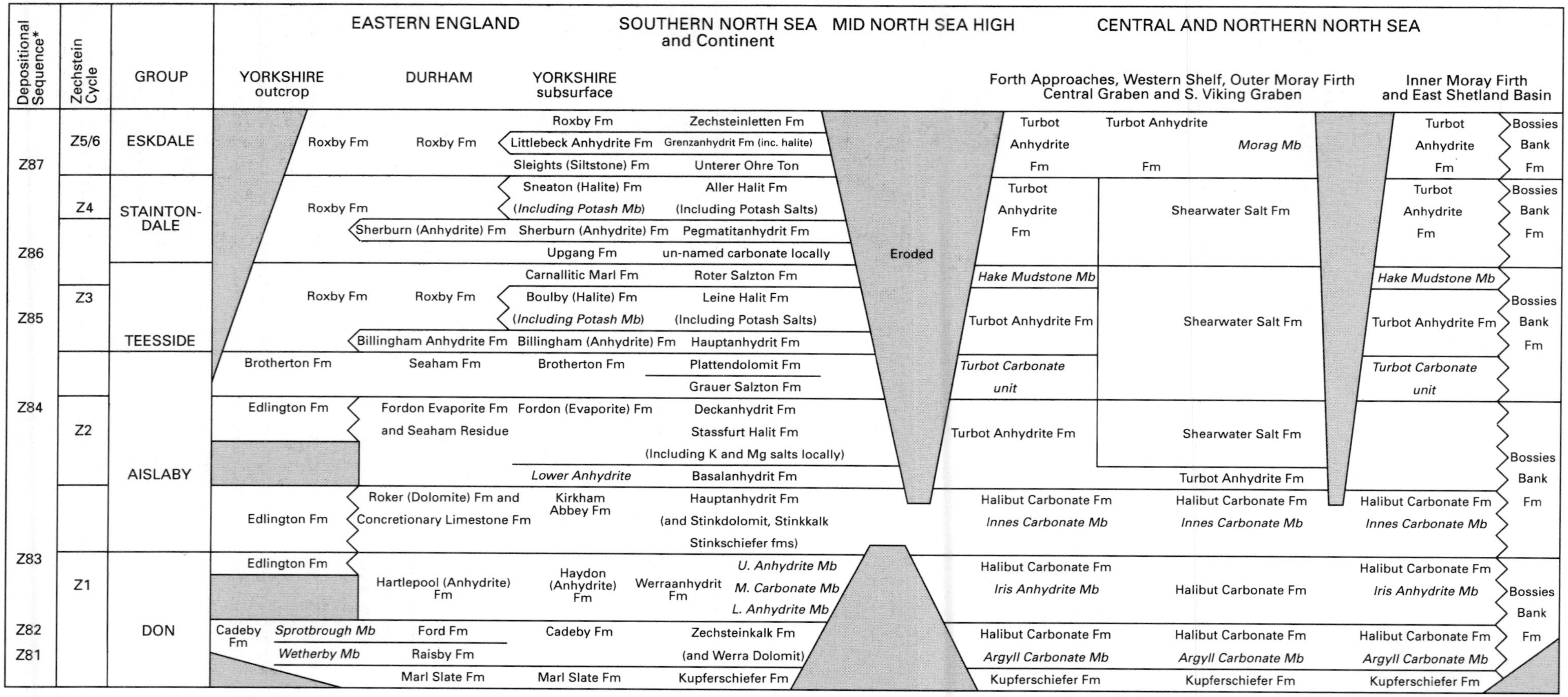

Depositional Sequence*	Zechstein Cycle	GROUP	EASTERN ENGLAND: YORKSHIRE outcrop	EASTERN ENGLAND: DURHAM	EASTERN ENGLAND: YORKSHIRE subsurface	SOUTHERN NORTH SEA and Continent	MID NORTH SEA HIGH	CENTRAL AND NORTHERN NORTH SEA: Forth Approaches, Western Shelf, Outer Moray Firth Central Graben and S. Viking Graben		CENTRAL AND NORTHERN NORTH SEA: Inner Moray Firth and East Shetland Basin	
Z87	Z5/6	ESKDALE	Roxby Fm	Roxby Fm	Roxby Fm Littlebeck Anhydrite Fm Sleights (Siltstone) Fm	Zechsteinletten Fm Grenzanhydrit Fm (inc. halite) Unterer Ohre Ton	Eroded	Turbot Anhydrite Fm	Turbot Anhydrite Fm *Morag Mb*	Turbot Anhydrite Fm	Bossies Bank Fm
Z86	Z4	STAINTON-DALE	Roxby Fm	Sherburn (Anhydrite) Fm	Sneaton (Halite) Fm (*Including Potash Mb*) Sherburn (Anhydrite) Fm Upgang Fm	Aller Halit Fm (Including Potash Salts) Pegmatitanhydrit Fm un-named carbonate locally		Turbot Anhydrite Fm	Shearwater Salt Fm	Turbot Anhydrite Fm	Bossies Bank Fm
Z85	Z3	TEESSIDE	Roxby Fm	Roxby Fm Billingham Anhydrite Fm	Carnallitic Marl Fm Boulby (Halite) Fm (*Including Potash Mb*) Billingham (Anhydrite) Fm	Roter Salzton Fm Leine Halit Fm (Including Potash Salts) Hauptanhydrit Fm		*Hake Mudstone Mb* Turbot Anhydrite Fm	Shearwater Salt Fm	*Hake Mudstone Mb* Turbot Anhydrite Fm	Bossies Bank Fm
			Brotherton Fm	Seaham Fm	Brotherton Fm	Plattendolomit Fm Grauer Salzton Fm		*Turbot Carbonate unit*		*Turbot Carbonate unit*	
Z84	Z2	AISLABY	Edlington Fm	Fordon Evaporite Fm and Seaham Residue	Fordon (Evaporite) Fm *Lower Anhydrite*	Deckanhydrit Fm Stassfurt Halit Fm (Including K and Mg salts locally) Basalanhydrit Fm		Turbot Anhydrite Fm	Shearwater Salt Fm Turbot Anhydrite Fm		Bossies Bank Fm
			Edlington Fm	Roker (Dolomite) Fm and Concretionary Limestone Fm	Kirkham Abbey Fm	Hauptanhydrit Fm (and Stinkdolomit, Stinkkalk Stinkschiefer fms)		Halibut Carbonate Fm *Innes Carbonate Mb*	Halibut Carbonate Fm *Innes Carbonate Mb*	Halibut Carbonate Fm *Innes Carbonate Mb*	
Z83	Z1	DON	Edlington Fm	Hartlepool (Anhydrite) Fm	Haydon (Anhydrite) Fm	Werraanhydrit Fm *U. Anhydrite Mb* *M. Carbonate Mb* *L. Anhydrite Mb*		Halibut Carbonate Fm *Iris Anhydrite Mb*	Halibut Carbonate Fm	Halibut Carbonate Fm *Iris Anhydrite Mb*	Bossies Bank Fm
Z82			Cadeby Fm: *Sprotbrough Mb*	Ford Fm	Cadeby Fm	Zechsteinkalk Fm		Halibut Carbonate Fm	Halibut Carbonate Fm	Halibut Carbonate Fm	
Z81			*Wetherby Mb*	Raisby Fm		(and Werra Dolomit)		*Argyll Carbonate Mb*	*Argyll Carbonate Mb*	*Argyll Carbonate Mb*	
				Marl Slate Fm	Marl Slate Fm	Kupferschiefer Fm		Kupferschiefer Fm	Kupferschiefer Fm	Kupferschiefer Fm	

Fig. 6.2 Probable time equivalents of stratigraphic units in North Sea and eastern England. Based on Rhys (1974), Taylor and Colter (1975), Clark (1980), Colter and Reed (1980), Taylor (1981), Smith *et al.* (1986), *Tucker (1991) and Cameron (1993a), and on the assumption that major sea-level and salinity changes were essentially synchronous across the whole region.

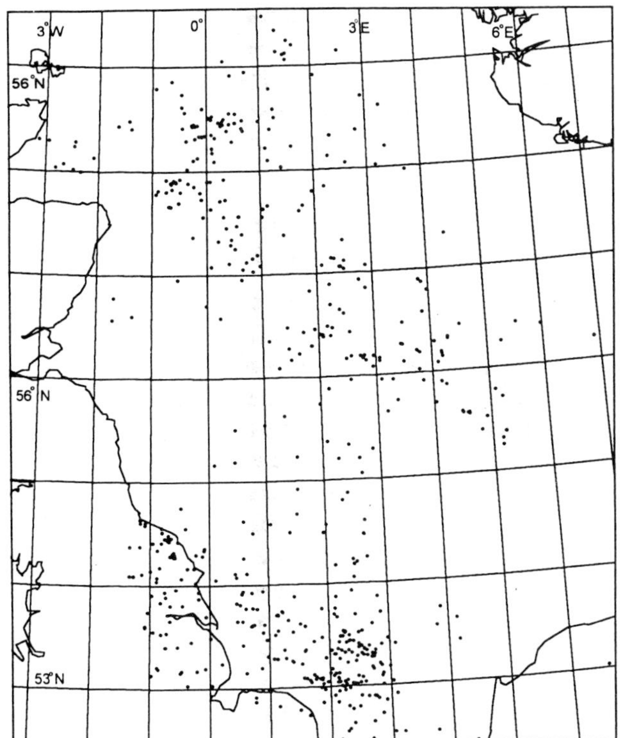

Fig. 6.3 Distribution of released wells used (together with other published material) in the construction of the maps in this chapter.

testify to the depositional unity of the Southern Salt Basinand lead to a model that is useful for exploration purposes. Examples from as far away as Poland, where more than 1000 Zechstein wells have been drilled since the 1960s, are therefore relevant to the Southern North Sea. It is believed, though not yet conclusively proved, that a similar depositional model can be applied to the Northern Salt Basin (Taylor, 1981).

Towards the close of the Permian, the whole region lay in the Trade Winds belt and the climate was predominantly hot and dry. By the end of Rotliegend time, aridity, combined with lowering of source areas, appears to have cut sediment supply until deposition lagged behind subsidence, resulting in inland depressions well below ocean level. Zechstein sedimentation began when these desert basins were flooded from the Boreal Ocean (see Fig. 6.1) by a combination of rifting and rise in sea level (Smith, 1970a, 1979, 1980b; Glennie and Buller, 1983; see Chapter 5). According to Doré and Gage (1987) rifting between northern Greenland and Svalbard first resulted in a restricted basin between east Greenland and Norway. Later propagation of the rift allowed catastrophic flooding of the Zechstein Basin proper.

The current stratigraphic divisions used in the Southern Salt Basin are based on the notion of cycles of increasing salinity. Each cycle is akin to the 'genetic stratigraphic

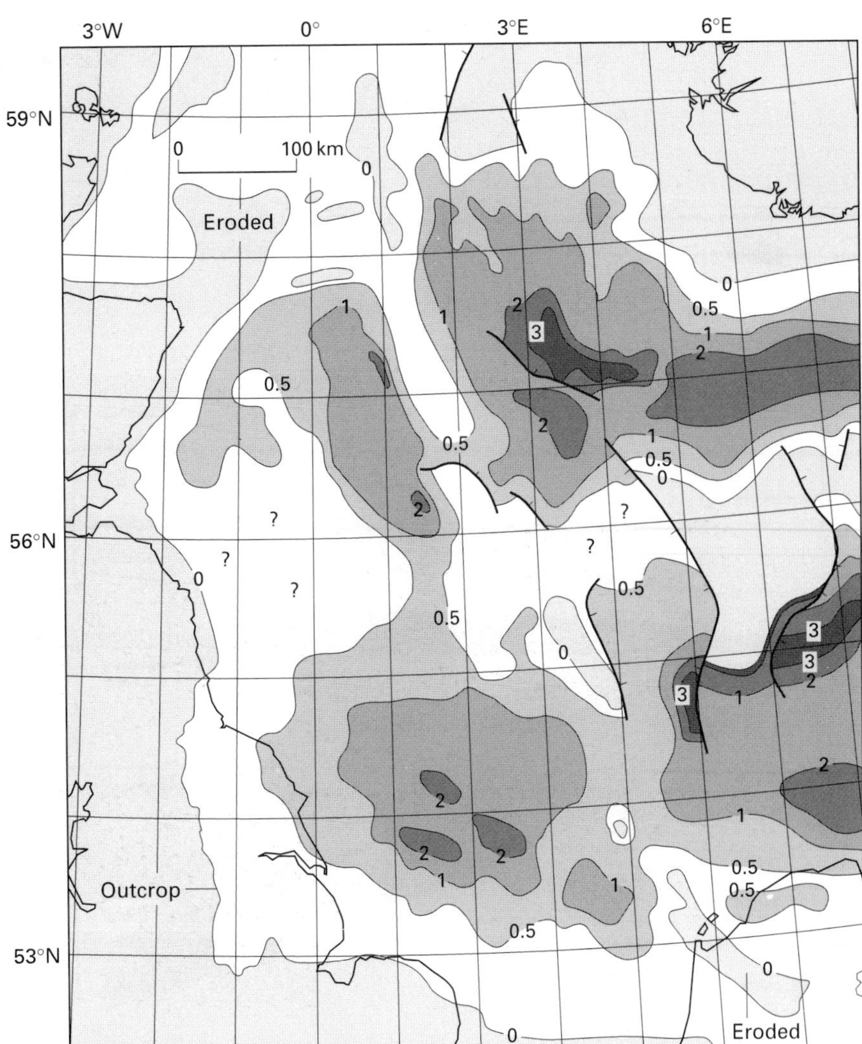

Fig. 6.4 Average thickness of Zechstein (km). In many areas, the 0.5 km line broadly marks the transition between the Z2 shelf facies, dominated by carbonates and anhydrite, and the basin facies, with thick mobile salt. Modified from Taylor (1981) with data from released wells (Fig. 6.2) and from Smith *et al.* (1993), van Adrichem Boogaert and Burgers (1983) and M.C. Geluk (in preparation).

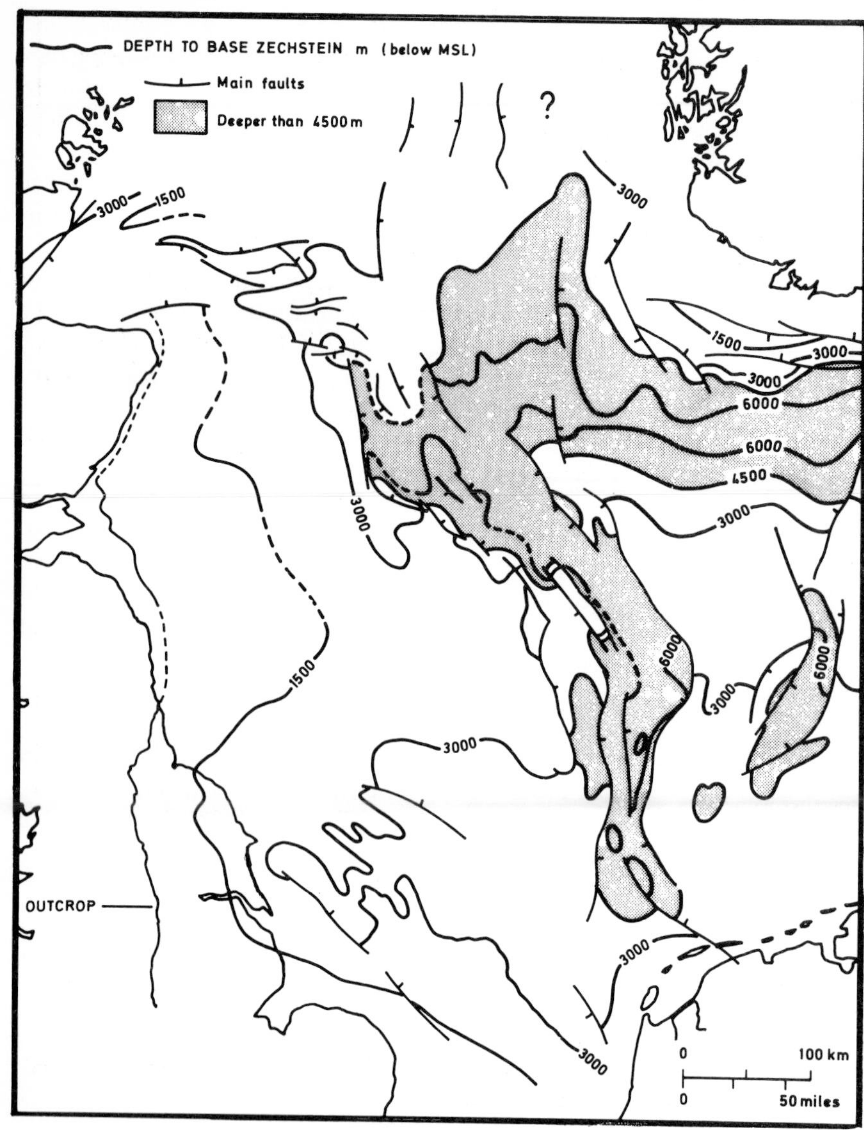

Fig. 6.5 Structure on base of Zechstein (generalized, mainly from Day *et al*., 1981).

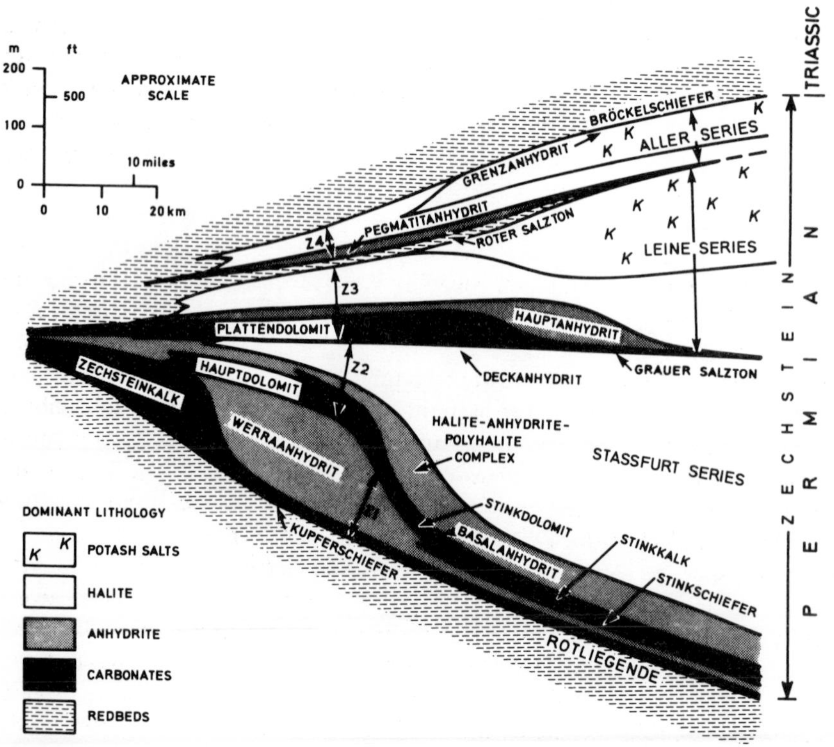

Fig. 6.6 Diagrammatic shelf-basin profile of the Zechstein in the Southern Salt Basin, showing the German-based nomenclature in common use.

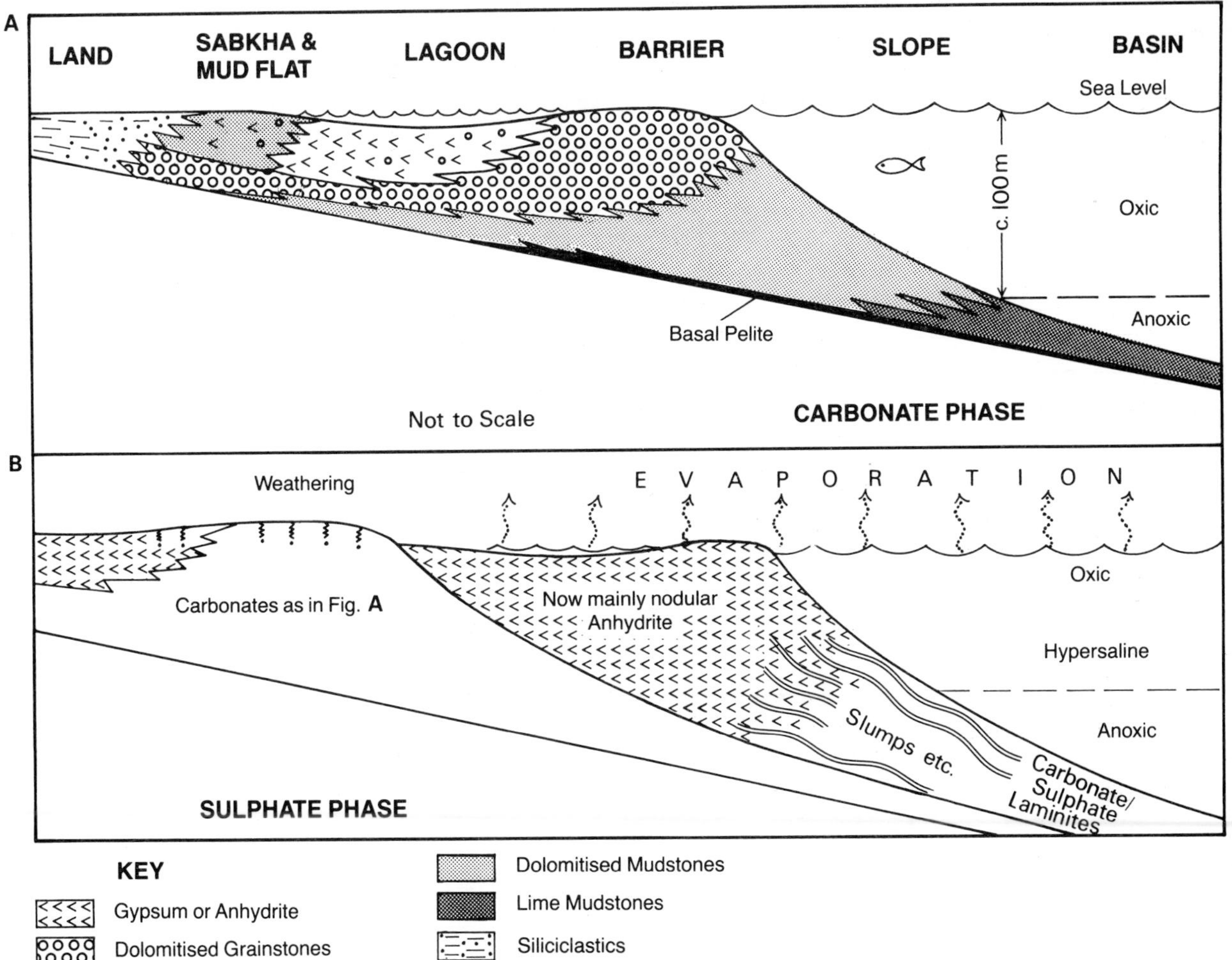

Fig. 6.7 Generalized model for a Zechstein carbonate–evaporite cycle. A. Carbonate phase, near-normal basin salinity; after Z1, the lagoon was probably hypersaline. B. Early part of evaporite phase, with deposition of thick calcium sulphate around edges of basin; after the Z1 evaporite phase, the deep waters of the basin were probably dense magnesium-rich brines. Basin halite and eventually potash salts were formed in the basin during Z2 and Z3. With continued subsidence, evaporites covered the shelf and in time filled the basin.

sequence' of Galloway (1989a). In the North Sea, four main evaporite cycles and a rudimentary fifth—Z1 to Z5—can be distinguished, as on the Continent. The 'ideal' cycle (Richter-Bernburg, 1955, 1959) reflects the influence of increasing salinity caused by evaporation due to restriction of sea-water inflow some time after an initial marine incursion (Fig. 6.7). It commences with a thin clastic member, passing upwards in turn through limestone, dolomite and anhydrite to halite, and finally to highly soluble salts of magnesium and potassium (the 'bitterns'). While this simple scheme is useful as a framework, there are many omissions and reversals of stages in the Zechstein cycles. It glosses over the distinction between basin-centre evaporites and marginal evaporites, which bear different time relationships to the carbonates. Furthermore, it may give the misleading impression that each cycle represents the intake of one basin fill of ocean water, followed by progressive evaporation towards dryness. As indicated briefly below, this cannot be the case. Some of the exceptions to the ideal cycle are attributable to the diagenetic origin of the present minerals, but most arise through the subtle interplay of changing climate, run-off, oceanic exchange, water depth, sedimentation rate and subsidence.

It has been suggested that the overriding cyclic control was eustatic variation in ocean level, amplified by the mediation of an ocean/basin barrier, which, whenever it was completely or almost completely exposed, allowed net evaporation and drawdown in the basin system behind it (Smith, 1970a, 1979, 1980b; Clark and Tallbacka, 1980). Volumetric considerations led Taylor (1980) to suggest the incorporation of an older idea—the assumption of permeability in the barrier—which allowed continuous but restricted reflux. This enabled brine concentrations to remain within the solubility limits of particular evaporite minerals for substantial periods. Under such conditions, the type of mineral precipitated is determined by climatic factors, chiefly relative humidity and temperature. The MacLeod evaporite basin in Western Australia appears to exhibit a present-day example of a seepage barrier at work and is being extensively researched (Logan, 1987). This model resolves the puzzling absence of marine micro-taxa from the Zechstein deep-basin sediments.

A deep-water, deep-basin model—long inferred from the Zechstein of Germany—seems to be required by the geometry of correlatable shelf and basin sequences in the North Sea, but reflux between the basin and the ocean is needed to explain the imbalance of anhydrite over halite in each cycle, together with the great thicknesses of these nearly

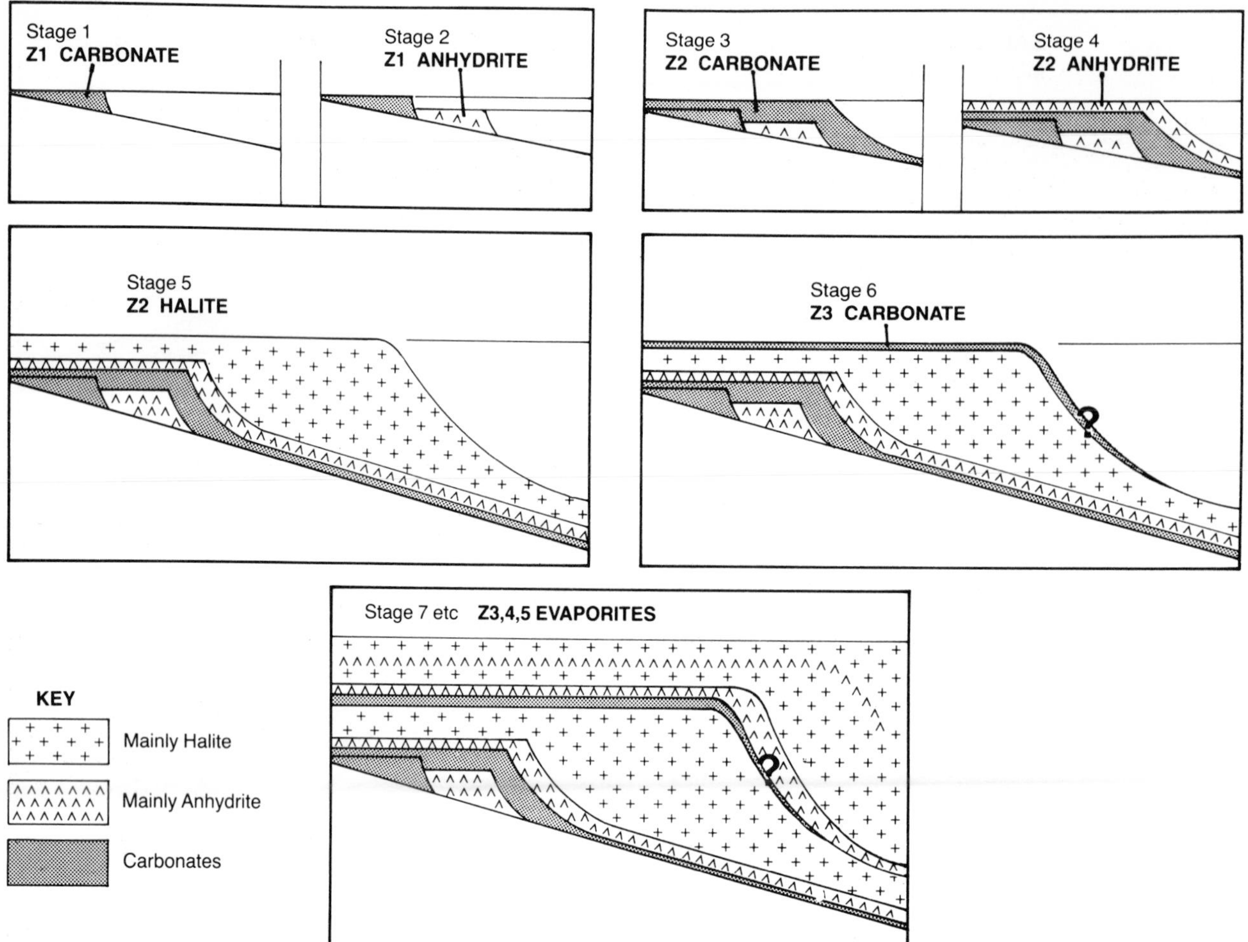

Fig. 6.8 Thumbnail sketches illustrating how the complete Zechstein sequence in the Southern Salt Basin was built up by repetition of the basic cycle.

monomineralic deposits. However, the shallow-water/deep-basin and shallow-water/shallow-basin models of evaporite deposition also have application. Intermittent partial or complete desiccation has been deduced for the first cycle (Taylor, 1980)—but also contested (Richter-Bernburg, 1986a). In the debate concerning the degree of drawdown in evaporite basins, it is important to remember that elementary hydrodynamic principles limit the difference between basin and ocean levels if reflux is to operate. In any event, the sedimentological evidence suggests that the bulk of the Werraanhydrit formed when the basin contained a substantial depth of brine (Taylor, 1980); similar processes of reflux and perhaps desiccation may have operated in subsequent cycles, for which we have fewer data. Marginal salina and sabkha deposits (early transgressive-system tract retrogradational evaporites and late highstand evaporites of Tucker, 1991) are recognizable in all cycles, and progressive shoaling of the whole basin is apparent in later cycles, as the rate of salt deposition periodically overtook subsidence.

The carbonates of the Southern North Sea overlap their early transgressive shaly beds towards the margin of the basin, where in many areas they pass into continental sandstones (see Fig. 6.6). The anhydrite formations do not generally extend as far from the basin centre as the carbonates. Halite formations are less extensive than anhydrite, passing landwards into clays, mudstones and siltstones, while potash zones make their appearance even further from the margin. The resulting 'bull's-eye pattern' (Hsü, 1972) in each cycle is not, however, simply the product of a sea shrinking under the influence of evaporation following a single flooding, for each phase—carbonate, anhydrite and halite in turn—persisted long enough for many basin volumes of sea water to yield their precipitates, during which significant basin subsidence took place.

The thumbnail sketches in Fig. 6.8 show how the 'standard' Southern North Sea Zechstein profile of Fig. 6.6 was built up in successive stages as the basin continued to subside. Variations on this profile occur where different amounts of subsidence occurred between or during cycles. Typical examples are given diagrammatically in Fig. 6.9. Figure 6.9A shows the relationship off the Norfolk coast and 6.9B that under North Yorkshire (Taylor and Colter, 1975), while 6.9C is found in inlets in the MNSH (e.g. Jenyon, 1988d). In parts of the Northern Salt Basin, it can be inferred that the Plattendolomit has dropped even further in the section (Fig. 6.9D).

Successive Zechstein cycles are increasingly 'evaporitic', significant carbonate deposition being confined to the first three cycles, of which only the first possesses a diverse marine fauna. In the UK sector, halite is abundant only in Z2 and later cycles, while potentially economic potash salts are concentrated in Z3 and Z4.

The marginal carbonates of the first three cycles in the

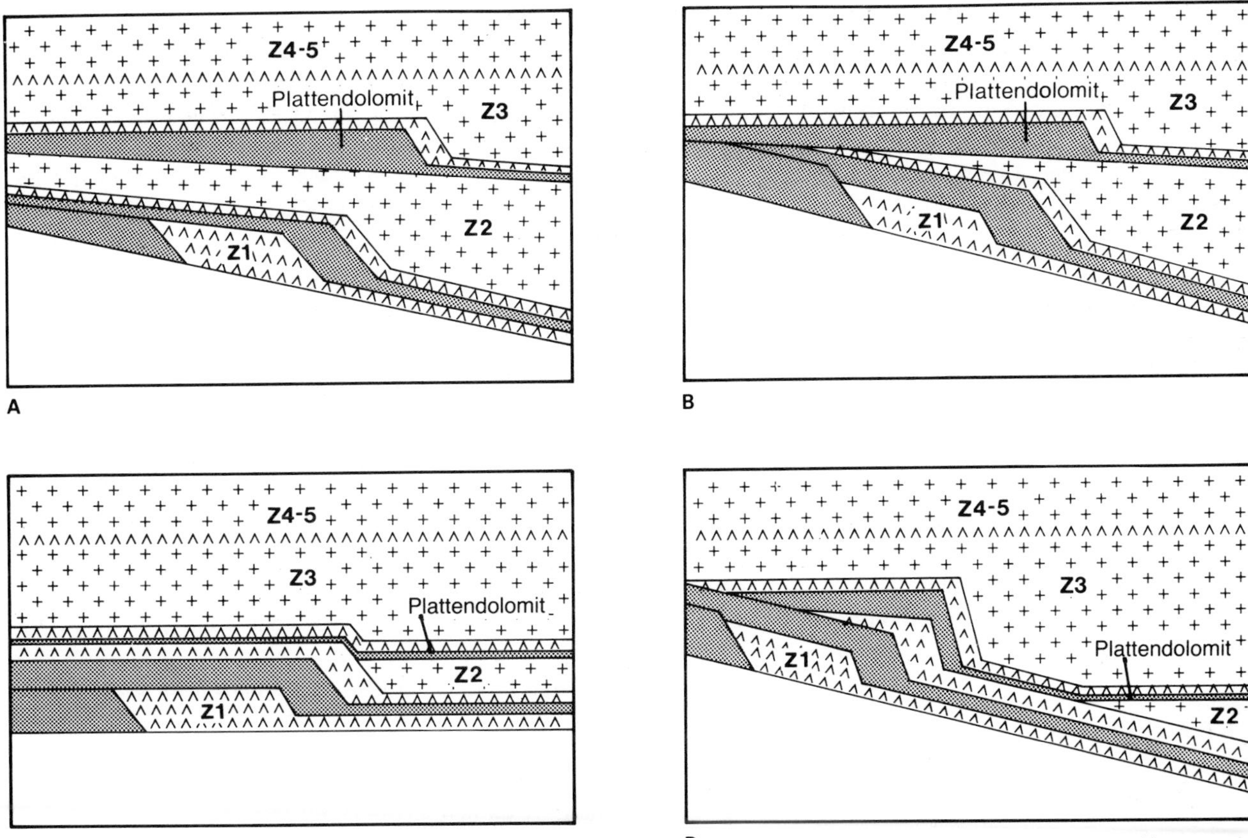

Fig. 6.9 Regional variations in Zechstein profile resulting from differing amounts of subsidence during or between cycles. A. Off Norfolk. B. North Yorkshire. C. Mid North Sea High. D. Parts of Northern Basin, e.g. Western Shelf.

Southern Salt Basin are complex wedges, each thickening for some distance into the basin and then thinning towards the centre (Figs 6.6–6.9). Following each transgressive phase, they formed by the progradation of a variety of near-shore facies across more uniform and finer-grained slope and basin facies. They represent highstand-system tracts in sequence stratigraphic terms (Tucker, 1991).

Carbonate production was most rapid in the well-aerated, warm, well-lit shallow water round the edges of the basin; consequently, sedimentation here easily exceeded subsidence, leading to the construction of broad shelves comprising barrier, lagoonal, intertidal and slightly emergent (including sabkha) environments. Reefs were common in Z1. Algae were particularly significant as sediment producers and binders, especially in the later stages of each carbonate phase, when salinities rose beyond the tolerance of many other taxa.

These shallow conditions favoured accumulation of light-coloured grainstones and boundstones with good primary porosity and permeability—properties tending sometimes to be preserved or enhanced by early conversion to microdolomite, and in places augmented by vadose and other diagenetic processes (Clark, 1986). Some Zechstein dolomites were generally thought to have originated in sabkhas, while the majority formed by refluxing magnesium-rich brines during the sulphate phases of later cycles. Magaritz and Peryt (1994) have studied the carbon-13 (^{13}C) and oxygen-18 (^{18}O) values of the Zechsteinkalk in Poland, and conclude that a significant proportion of the dolomites originated as a result of the mixing of meteoric waters and evaporative brines.

In contrast to the shelves, basin carbonates are thin, dark, compact, argillaceous or shaly micrites, mostly deposited well below wave base and commonly under anoxic conditions, normally offering no reservoir prospects. Their potential as source rocks is limited mainly by their thinness.

Slope carbonates—which represent a considerable proportion of the thickest part of each wedge—are intermediate in texture and composition, becoming progressively paler, less argillaceous and less shaly upwards. They are probably built mainly from the winnowings of the shelf, together with fine skeletal material, and appear to include carbonate turbidites and slumped material. Though commonly dolomitized, the predominant micritic facies tends to have poor intercrystalline porosity and hence mediocre permeability. However, better reservoir properties are possible in displaced coarse shelf material or as a result of diagenesis (Clark, 1986; Amiri-Garroussi and Taylor, 1992).

The geometry of the principal anhydrite phase (the lowstand marginal gypsum wedges of Tucker, 1991) of each cycle also testifies to centripetal accretion by progradation, as opposed to aggradation, as the main process of basin filling. A similar conclusion has been reached for the mixed halite/anhydrite/polyhalite/kieserite zone in the second cycle, shown in Fig. 6.6 (Colter and Reed, 1980). The same principle probably applies to later halite phases, but this is difficult to prove.

6.4 First Zechstein cycle (Fig. 6.13)

The initial flooding of the Zechstein Basin was accompanied by disturbance and minor reworking of uncemented Rotliegend sands (Smith and Francis, 1967; Glennie and Buller, 1983; Glennie, 1990b, see also Chapter 5).

6.4.1 Kupferschiefer Formation

The Kupferschiefer (copper shale), which marks the formal base of the Zechstein, is a dark grey 1 m (3 ft) sapropelic shale, distinguished on logs by a strong gamma-ray peak (Figs 6.10–6.12). It formed under anoxic conditions below the influence of waves (though not everywhere under particularly deep water) and it drapes minor elevations. It can be recognized across the floor of both Southern and Northern Salt Basins, in the Moray Firth and in some wells in the Viking Graben area, but is absent from parts of the MNSH and other highs, especially where the Rotliegend is missing. Although organic-rich and oil-prone, it is too thin to be a credible source rock.

6.4.2 Zechsteinkalk Formation and equivalents

The Kupferschiefer rapidly passes up into marine carbonates—the equivalents of the Zechsteinkalk and Werradolomit. The carbonates thicken from a mere 3 m (10 ft) or so of argillaceous and carbonaceous dolomites and limestones on the floor of the Southern Salt Basin (e.g. wells 48/22-2 to 41/8-1 and 44/14-1 in Fig. 6.10) to about 100 m (300 ft) of combined slope and shelf facies round the edge of the basin. Two transgressive–regressive subdivisions are recognizable, each comprising a transgressive-system tract followed by a highstand-system tract (Tucker, 1991). Onshore these are represented by the Raisby and Ford Formations in Durham and the Wetherby and Sprotbrough Members of the Cadeby Formation in Yorkshire. Both contain an abundant fauna of brachiopods, bivalves, bryozoans, crinoids and foraminifera in the marginal belt, but become impoverished basinwards and upwards. Diagenesis in the Cadeby Formation has been discussed by Harwood (1986) and Kaldi (1986b) and in the Raisby Formation by Lee (1993).

Giant sandwaves characterize the Z1 oolite-shoal facies, represented by the Sprotbrough Member in Yorkshire. These have been investigated by Kaldi (1986a), who suggests that the topographic highs on which they formed may have been provided in places by residual hills on the pre-Zechstein unconformity and perhaps elsewhere by differential subsidence over reefs in the Wetherby Member. An example of the grainstone facies in thin section is shown in Fig. 6.26A.

At the top of the poorly fossiliferous carbonates across the Southern Salt Basin floor, there is commonly a distinctive thin argillaceous packstone layer, containing oncoliths, accompanied by foraminifera and crinoids. It occurs from the UK (the Trow Point Bed; Smith, 1986) to Poland, though probably only on sea-floor elevations, and appears to reflect a temporary fall in water level of 90 m (300 ft) or more. Normal marine conditions were never re-established on the basin floor during the Zechstein after the period represented by this bed, although restricted marine faunas reappeared on the shelves in the early parts of Z2 and Z3.

Near the depositional margin in Nottinghamshire and off the Norfolk coast (e.g. in 53/12-2, Fig. 6.10), dolomite grainstones pass southwards into terrigenous sands. Further east, in Netherlands waters, the Z1 carbonates are seldom more than 45 m (150 ft) thick and are predominantly shaly. A high influx of clastics is noted in this area in later cycles also, and may be related to a major river system draining the London–Brabant Platform (van Adrichem Boogaert and Burgers, 1983; van der Baan, 1990; Geluk *et al.*, 1996).

Another area of clastic input is indicated in the northwestern part of the Moray Firth where 90–120 m (300–400 ft) of interbedded limestones, shales and dolomites, with oolites and terrigenous sands near the top, may correspond with the Z1 carbonates (Fig. 6.13). The lower part of this sequence has been renamed the Halibut Carbonate Formation by Cameron (1993a).

In the few released wells that have penetrated the Zechstein north of the Moray Firth (e.g. 8/27a-1, 9/16-1 and 210/15-1, Fig. 6.12), the proportion of clastics is so great that neither the standard cycles nor the Turbot Bank/Halibut Bank breakdown can confidently be recognized. The name Bossies Bank Formation has been introduced by Cameron (1993a) for these rocks.

Reefs

Bryozoan–algal reefs capped by stromatolites are characteristic of the shelf Z1 carbonates in the UK, Denmark, Germany and Poland, and may therefore be expected in the appropriate setting in areas not yet explored. Small patch reefs are a feature of the lower member at outcrop from Yorkshire to Nottinghamshire (Smith, 1981b), whereas a massive shelf-edge reef 100 m (300 ft) high and at least 35 km (20 miles) long was constructed in what became the Ford Formation of Durham (Smith, 1981a). Chance preservation of an undolomitized portion of the Z1 reef in north-east England has enabled Hollingworth (Hollingworth and Tucker, 1987) to make a detailed and beautifully illustrated reconstruction of the reef communities and to recognize the importance of seabed cementation by aragonite. This should provide clues to puzzling features in some more strongly altered examples.

Sheets of dolomitized and generally porous oolites spread shorewards from the Durham reef, and similar beds surround the patch reefs. Small pinnacle reefs occupying open shelf and fore-reef positions are also known on the Continent, where they are sited on local topographic highs, such as those commonly provided by the stumps of Lower Rotliegend volcanoes (Paul, 1980). Note that, if the model for Zechstein deposition adopted in this chapter is correct, reefs are unlikely to have grown far from the margins in what were originally the deepest parts of the basin, except where substantial local uplifts are present, because the initial transgression would have been too rapid for organisms to keep pace with the rise in sea level.

One North Sea area that appears to favour reef development is the MNSH, which hosted shoals through much of Z1 (and also Z2) The possible relationship between buried

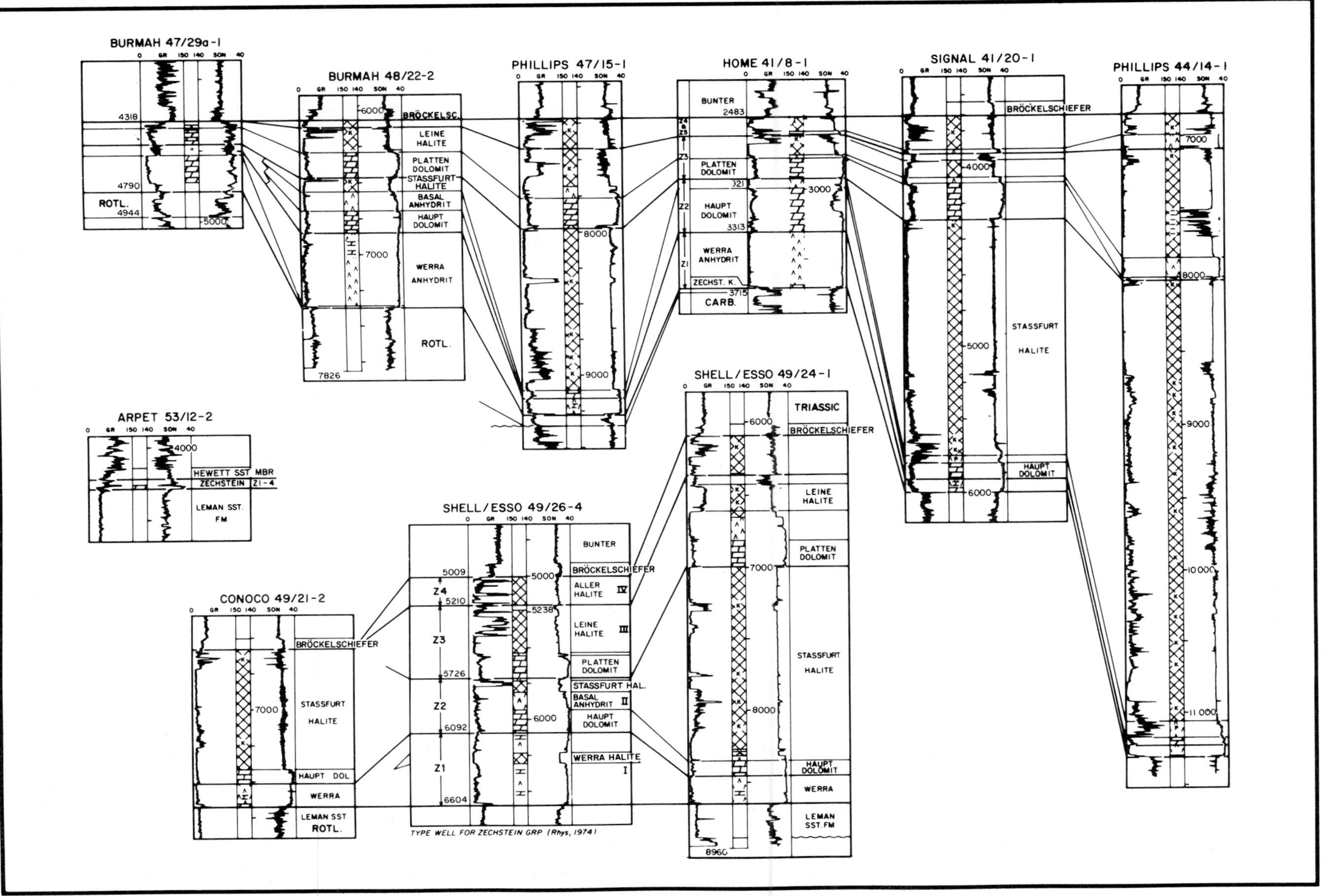

Fig. 6.10 Typical Zechstein logs from the Southern Salt Basin. Wells 47/29a-1, 48/22-2, 41/8-1, 49/26-4 and 53/12-2 represent 'shelf' sequences, typified by dolomite and anhydrite. Wells 47/15-1, 41/20-1, 44/14-1, 49/24-1 and 49/21-2 represent 'basin' sequences, dominated by thick halite. Note how, at the transition between these two types, the Z2 Stassfurt salts increase in thickness at the expense of the Werraanhydrit. Well 53/12-2 shows a thin marginal sequence, consisting only of dolomite and clastics, comparable with outcrops in Nottinghamshire. Well 49/21-2 has lost parts of the later cycles through salt movement. For legend see Fig. 6.12.

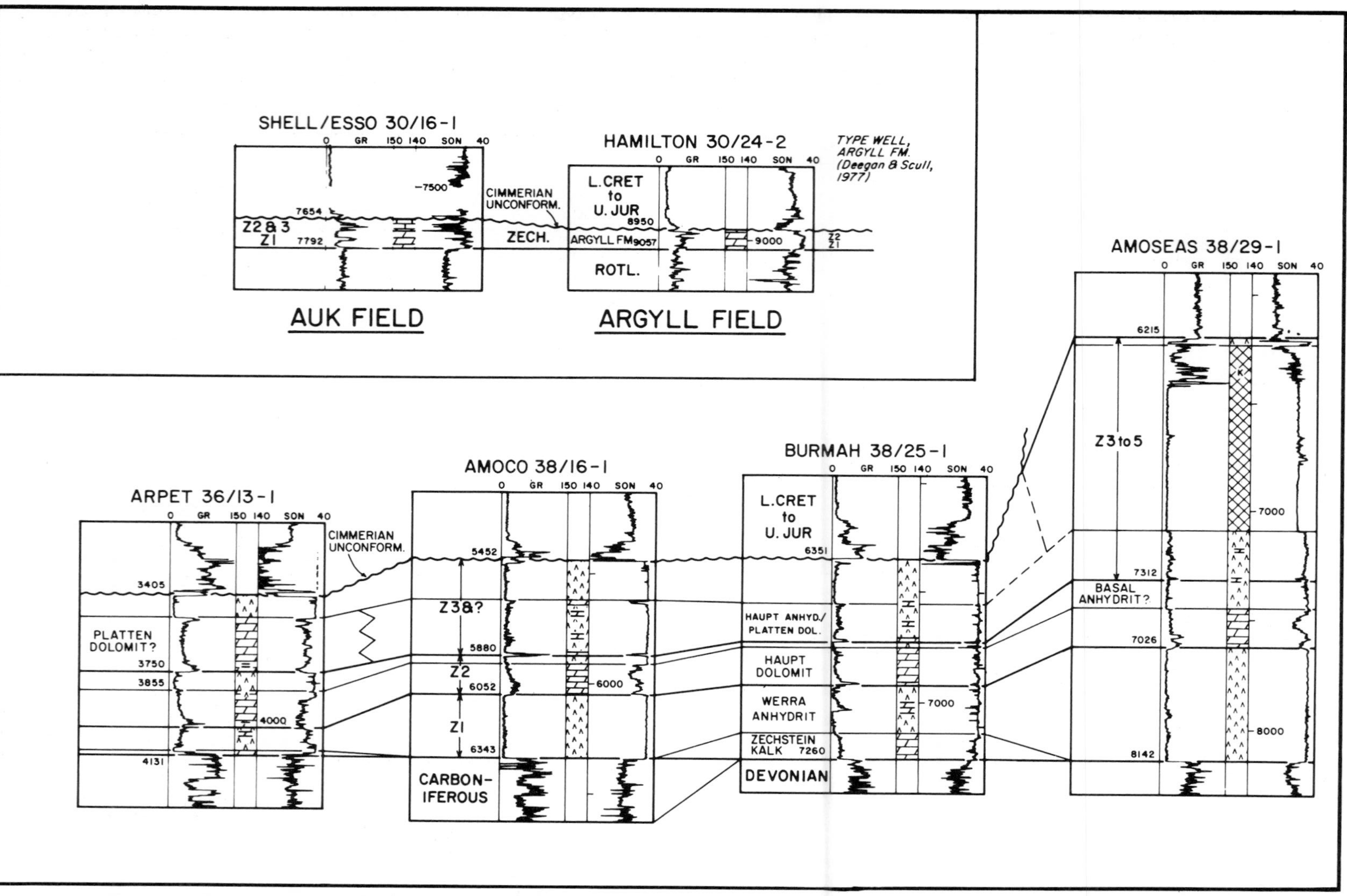

Fig. 6.11 Zechstein sequences from the Mid North Sea High often appear atypical. However, 'shelf' facies of the Hauptdolomit and upper parts of the Werraanhydrit can commonly be recognized. Some earlier units appear to be attenuated or absent as a result of late submergence (e.g. in 38/29-1, 38/16-2), while the more soluble salts may be missing as a result of dissolution under the Late Cimmerian unconformity (e.g. 36/13-1, 38/16-1, 38/25-1). The thin representative of the Zechstein in the Auk field is believed by the writer to have been a 'basin' sequence from which thick salt has been removed. The Argyll field, on the other hand, represents the eroded and leached remnants of a 'shelf' sequence (see Bifani, 1985); Z1, Z2 and, possibly, traces of Z3 carbonates are thought to be present. For legend see Fig. 6.12.

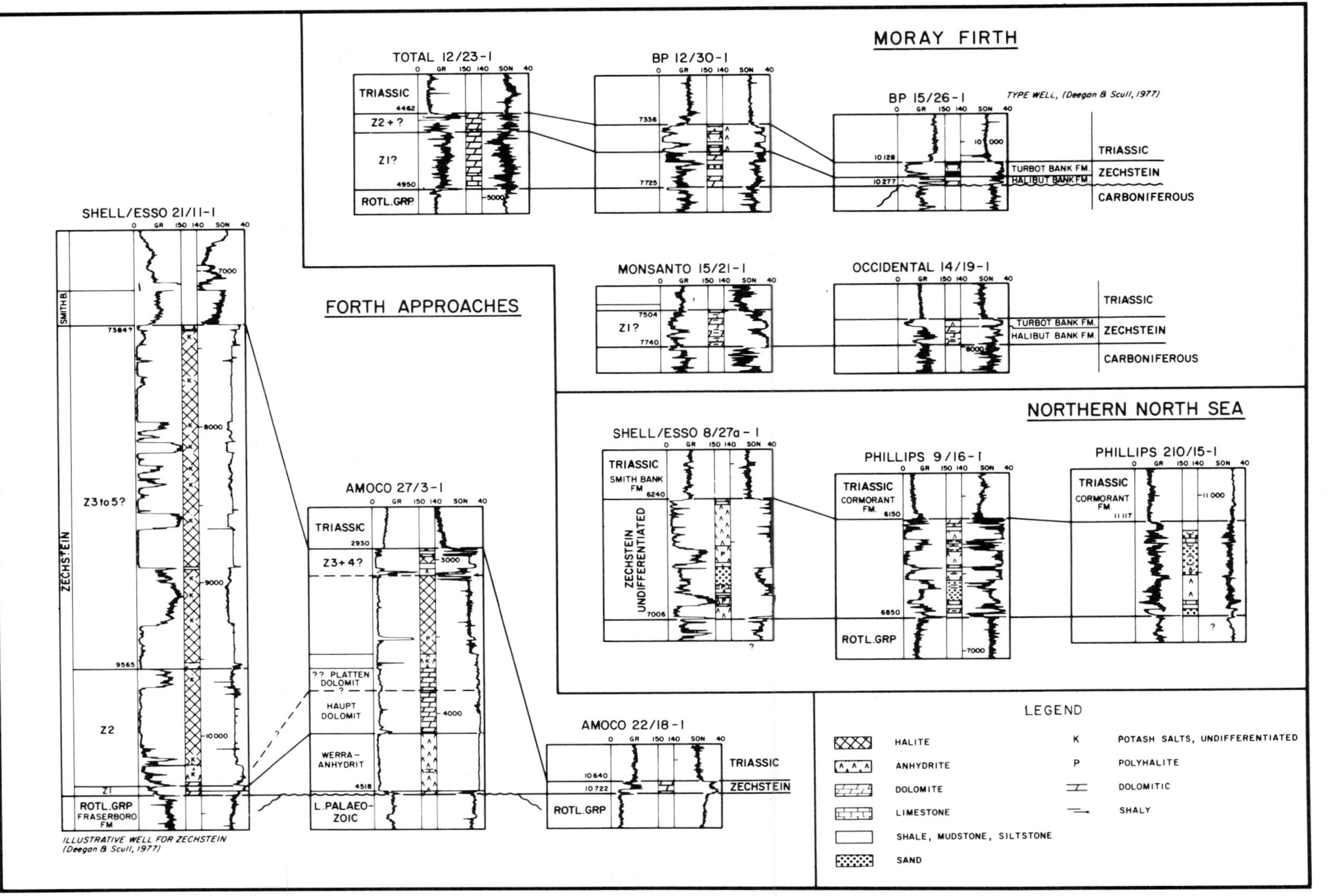

Fig. 6.12 Zechstein sequences from the Forth Approaches (21/11-1, 27/3-1, 22/18-1), Moray Firth (12/23-1, 12/30-1, 15/26-1, 15/21-1, 14/19-1) and further north (8/27a-1, 9/16-1, 210/15-1). Where thick salt is present (21/11-1), the sequence resembles that of 'basin' wells in the Southern Salt Basin, although the third and later cycles may be difficult to differentiate from the second. Note the similarity between 27/3-1 and 38/29-1 (Fig. 6.10) on opposite sides of the Mid North Sea High. Moray Firth sequences lack halite and have a high clastic content. They are divided into the Turbot Anhydrite (formerly Turbot Bank) Formation above and Halibut Carbonate (formerly Halibut Bank) Formation below (Cameron, 1993a). Clastic sequences of Zechstein age in the Moray Firth and other parts of the Northern North Sea (e.g. 210/15-1) are designated the Bossies Bank Formation by Cameron (1993a).

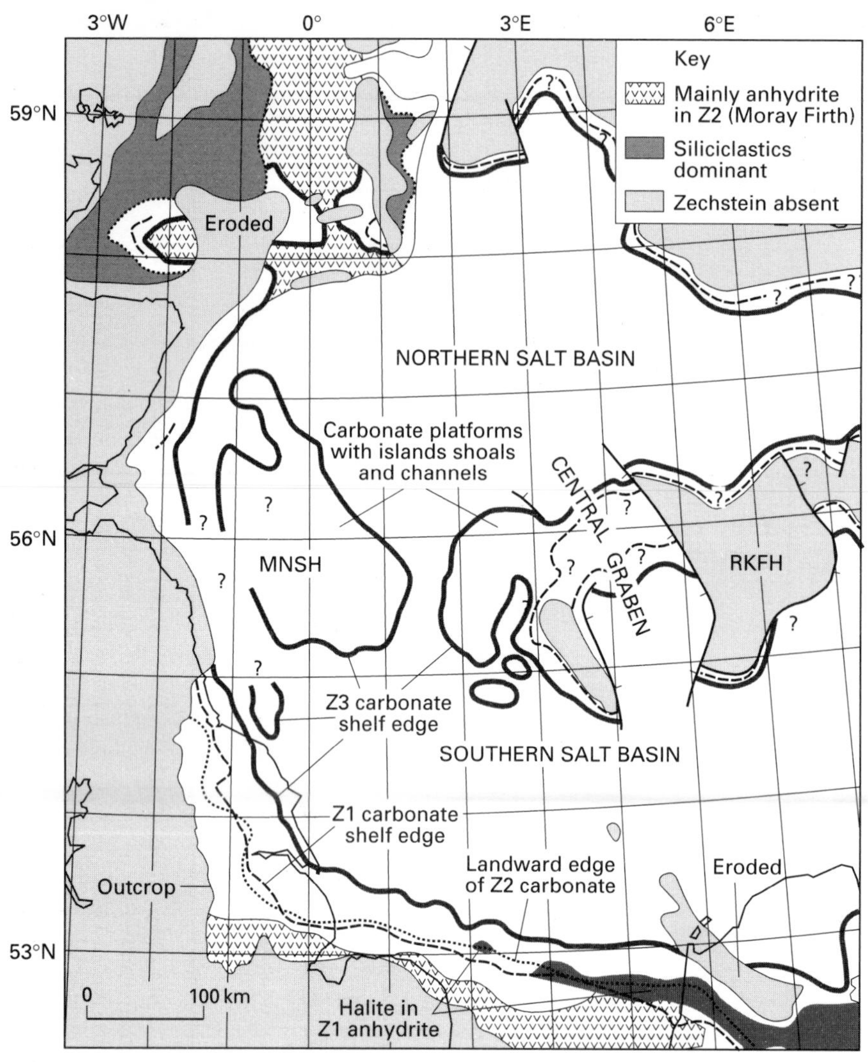

Fig. 6.13 Summary of significant facies distribution in Z1 and Z2. Compiled from released well data (Fig. 6.2), supplemented by information from Clark and Tallbacka (1980), Day *et al.* (1981), van Adrichem Boogaert and Burgers (1983), Jenyon *et al.* (1984), Amiri-Garroussi and Taylor (1987, 1992), Jenyon (1988d), Andrews (1990), Cameron (1993a), van Adrichem Boogaert and Kouwe (1993–94), and M.C. Geluk (in preparation). Uncertainties remain about the distribution of shelf facies in the Northern Salt Basin and the Central Graben and of the facies on Mid Graben Highs.

granites and shoaling in that area has been discussed by Donato *et al.* (1983). Jenyon and Taylor (1983, 1987) show seismic sections that may include carbonate build-ups, although many reef-like features have other origins (see later). Cores from the Argyll field on the northern flank of the MNSH show facies provisionally interpreted as fore-reef talus (Bifani, 1985).

Reservoir facies

Oolitic and oncolitic shelf dolomites of Z1 form commercial gas reservoirs in Poland (Depowski, 1981). Effective porosities of 6–13% and sometimes up to 30% are recorded, with permeabilities in the range 100–200 millidarcies (mD) and rarely 1000 mD. Similar facies should provide good reservoirs in the North Sea where traps and adequate mature hydrocarbon sources are available. Plugging by evaporite minerals tends to be less of a problem in the grainstones of Z1 than it is in most carbonates of Z2 and Z3, although in the Hewett gasfield anhydrite is the most troublesome cement (Southward *et al.*, 1993). In this field, the Z1 dolomite is relatively tight, with porosities of only 5–8% and permeabilities of < 1 mD (Cooke-Yarborough, 1991, 1994); productivity is closely related to fracturing.

Scholle *et al.* (1991) have described Upper Permian bioherms several 100 m across and high in east Greenland, constructed chiefly by bryozoans and algae. They seem likely to be of the same age as the Z1 reefs of the Zechstein Basin. Scholle *et al.* (1991) suggest that, if similar but larger build-ups exist in parts of the subsurface that have undergone an appropriate burial and thermal history, they could constitute hydrocarbon reservoirs north of the Zechstein Basin proper.

Patch reefs and associated dolomudstones in Denmark have a potential net pay of about 24 m, with porosities and permeabilities in the range 12–30% and 10–100 mD, respectively, according to Clark (1986).

At outcrop in England, the grainstones surrounding patch reefs generally have superior porosity and permeability to the reefs themselves, because of the reefs' micrite matrix.

6.4.3 Werraanhydrit Formation and equivalents

A sharp break, accompanied by features attributed to a sea-level fall and of desiccation, separates the Z1 carbonates from the Z1 evaporites—the Werraanhydrit and equivalents (Fig. 6.2). This surface and its correlatives constitute the first major (type 1) Zechstein sequence boundary of Tucker (1991). The Z1 evaporites form a peripheral lens up to 180 m (600 ft) thick around the Southern Salt Basin, contained mostly within the encircling wedge of Z1 shelf carbonates (Figs 6.6 and 6.13). They consist mainly of bluish-white displacive anhydrite (with 'nodular', 'chicken-wire' or 'mosaic' structure, and enterolithic bands) in a sparse brown microdolomite host, interspersed at intervals with thin anhydritic dolomite layers. The ratio of dolomite to anhydrite seems to decrease as the total thickness of the Werraanhydrit increases. On parts of the shelf, however, dolomite tends to predominate near the middle of the Werraanhydrit, and a distinct unit—the Middle Carbonate Member, about 40 m thick—is recognized in Denmark by Clark and Tallbacka (1980). It probably represents the transgressive-system tract of the Upper Werraannhydrit. The thick Werraanhydrit, illustrated by 41/22-2, 41/8-1 and 49/26-4 (the Type Well of Rhys, 1974) in Fig. 6.10 can be traced in wells across the MNSH to the Long Forties High (27/3-1, Fig. 6.12) and to the Moray Firth. From the MNSH northwards, it is now formally designated the Iris Anhydrite Member of the Halibut Carbonate Formation (Cameron, 1993a).

For brevity, the thick marginal Werraanhydrit and its equivalents described above are often referred to as 'shelf' or 'platform' type although they apparently represent a variety of predominantly shallow-water to possibly supratidal facies built out over slope and deeper-water facies. According to B.C. Schreiber (personal communication), anhydrite displacement structures and fabrics produced during frequent periods of emergence probably disguise earlier subaqueous features, during which gypsum may have formed. Pseudomorphs after gypsum are seen in wells in Denmark (Clark and Tallbacka, 1980), in quarries in Germany (Richter-Bernburg, 1985) and cores from Polish wells (Peryt *et al.*, 1993; Peryt, 1994). The Polish studies referred to emphasize the polygenetic origin of the Werraanhydrit. Minor bodies of massive halite, probably of salina origin, occur near the top of the Lower Werraanhydrit, and are found locally on the southern shelf off the Norfolk coast (e.g. in 49/26-4, Fig. 6.10). Further east these merge to form a continuous body in the southern Netherlands (van Adrichem Boogaert and Burgers, 1983; van Adrichem Boogaert and Kouwe, 1993–94; M.C. Geluk, in preparation). In southern Denmark, halite has been found near the top of the Z1 anhydrite (Clark and Tallbacka, 1980). In the Netherlands, Germany and Poland, further from the presumed connection to the ocean, the Werra halite is thick and eventually accompanied by potash salts (see Ziegler, 1989; van Adrichem Boogaert and Kouwe, 1993–94; Czapowski *et al.*, 1993).

Papers by Taylor (1980), Richter-Bernburg (1986a) and Langbein (1987) present differing conclusions about the origin of the Werraanhydrit. Its accumulation may have occupied a significant proportion of Zechstein time: Menning (1995) estimates 2 Ma for the Lower Werraanhydrit alone. Although not at first sight related to carbonate reservoirs, the sheer size and controversial origins of the shelf facies of the Werraanhydrit force themselves on one's attention. The unit is in fact extremely important to petroleum geology, its relief largely controlling the facies of overlying carbonates, while the contrast in velocity between it and the basin-centre halites is a major factor in depth-corrected mapping of the Rotliegend.

In both the Northern and Southern Salt Basins, the thickness of the peripheral anhydrite lens declines to only about 18 m (60 ft) across the floor, where four or five subcycles each commence with a layer of displacive anhydrite and grade up into dark brown to black bituminous, flat, submillimetre anhydrite/carbonate laminites. It has been deduced that these subcycles formed during successive periods of basin recharge, following episodes of evaporative drawdown (Taylor, 1980), although Richter-Bernburg (1986a) disputes appreciable changes in basin sea level because of lack of evidence of exposure at the basin margins.

The slope facies of the Werraanhydrit has been poorly sampled by wells in the UK sector, but, on the Continent, a variety of slump, mass-flow and turbidite features have been described (Schlager and Bolz, 1977; Meier, 1981; Peryt *et al.*, 1993; Peryt, 1994), which demonstrate the depth of water in the basin at the time.

Basic igneous rocks, once described as lavas and tuffs, in the lower part of the Zechstein of blocks 29/20 and 29/25 have since been reinterpreted as Middle Jurassic intrusives of the Puffin volcanic centre (Cameron, 1993a).

Reservoir facies

The carbonate layers within the basin laminites are normally tight apart from local vugs, but, on some highs bordering the Central Graben, Mesozoic uplift, erosion and weathering have dissolved the anhydrite layers, leading to creation of good secondary porosity and permeability in the Auk oilfield in block 30/16 (Brennand and van Veen, 1975). This lithofacies is included in the Halibut Carbonate Formation of Cameron (1993a).

According to Clark (1986), the Middle Carbonate Member of the shelf facies of the Werraanhydrit in southern Denmark has porosities of 5–20% and permeabilities of up to 11 mD, apparently as a result of late leaching.

6.5 Second Zechstein cycle (Fig. 6.13)

6.5.1 Hauptdolomit Formation and equivalents

On the floor of both the Southern and Northern Salt Basins the Werraanhydrit passes upwards with rapid transition into dark brown to black, bituminous, thinly laminated carbonates. The lower part—the Stinkschiefer—contains many thin shale layers and produces a distinctive gamma-ray marker. In the Southern Salt Basin, this unit merges into the overlying Stinkkalk, which is similar but less shaly. Limestone predominates over dolomite in the deeper parts of the basin, typified by large interlocking early diagenetic calcite crystals embracing several laminae. These crystals engulf much oil-prone sapropel. According to Leythauser *et al.* (1995) this type of association favours oil expulsion at high efficiencies. The thinness of the basin carbonates (only 9–18 m, 30–60 ft), raises doubts about their ability to release large volumes of oil. The deposits of the slope may hold more promise (D.B. Smith, personal communication).

Around the edges of the Southern Salt Basin and across the shelf formed by the Werraanhydrit, the Hauptdolomit or Main Dolomite (the shelf facies of the Kirkham Abbey

Formation of Yorkshire and the Roker Formation of Durham) consists of 30–90 m (100–300 ft) of shallow-water to intertidal dolomites (e.g. 48/22-2, 41/8-1, Fig. 6.10), typically consisting of very fine ooids and pelletoids with leached centres. Large pisoliths and algal (cyanobacterial) sheets characterize the barrier facies developed near the break in slope into the basin. A restricted fauna of bivalves, gastropods, foraminifera and ostracods occurs locally, but the reef frame-builders of Z1 have not been recognized. Nevertheless, basin-facing slopes of over 10° can be calculated from seismic sections in some areas in the North Sea (see Fig. 10 in Jenyon and Taylor, 1987; Fig. 4 in Amiri-Garroussi and Taylor, 1992) and are difficult to account for unless sediment-binding and/or very early cementation had imbued some measure of wave resistance. In onshore Poland, Z2 'pinnacle reefs' and 'atolls', with margin slopes of 11–14° can be recognized on seismic sections in fore-barrier positions (Antonowicz and Knieszner, 1984). Seven wells drilled on one such isolated platform have confirmed an atoll-like construction and facies distribution (Peryt and Dyjaczyn'ski, 1991).

The shelf facies of the Hauptdolomit is traceable across much of the MNSH and may be present in the Central Graben. It presumably occurs around the margin of the Northern Salt Basin, although it has yet to be widely recognized there. One occurrence, modified by uplift and evaporite dissolution, has been described by Bifani (1985) from the Argyll oilfield. More recently released wells demonstrate its presence in blocks 20/2 and 20/3 at the entrance to the Moray Firth and 26/14 in the Forth Approaches Basin. From the MNSH northwards, this dolomite is now formally called the Innes Carbonate Member of the Halibut Carbonate Formation (Cameron, 1993a).

The environments and lithofacies of the Hauptdolomit from wells in the Netherlands are discussed and well illustrated with cores and photomicrographs by Clark (1980a) and van der Baan (1990) and in Denmark by Clark and Tallbacka (1980). Sønderholm (1987) describes the Z2 carbonates and adjacent evaporites from wells in Denmark, deducing similarities between conditions in the Northern and Southern Salt Basins.

The slope facies of the Z2 carbonate is known mainly from outcrop in Durham and boreholes in north Yorkshire and on the Continent. It consists dominantly of grey-brown dolomitized lime-muds, commonly pelleted and in places burrowed, with ostracods, foraminifera, and occasional bivalves, the passage downwards and laterally into the basin facies being accomplished by coalescence of increasingly abundant calcite concretions (Taylor and Colter, 1975; Smith, 1980a). A detailed account and model of the slope facies and their diagenesis in south Oldenburg (northwest Germany) have also been provided by Huttel (1989) and in Poland by Peryt (1992).

Reservoir facies

The shelf facies of the Hauptdolomit provides commercial oil and gas reservoirs in Poland, Germany and the Netherlands, principally from oncolitic and oolitic beds in the barrier facies, from local highs on the fore-barrier and in back-barrier lagoons. Strohmenger *et al.* (1993a,b) recognize 27 subfacies and discuss their prediction and relationship to reservoir properties. Good primary porosity is commonly reduced by cementation, especially by anhydrite or halite (see Fig. 5.26B,C) but is locally enhanced by secondary dissolution. In Poland, porosities of 10–15% and permeabilities of a few to a few hundred millidarcies are considered normal, with porosities sometimes up to 25%, accompanied by permeabilities of 1–5 D (Depowski, 1981; Depowski *et al.*, 1981; see also Peryt and Dyjaczyn'ski, 1991). Clark (1986) discusses the diagenetic factors affecting petrophysical properties in Dutch and Danish wells; cores from Denmark have porosities of 15–30%, with permeabilities of up to 100 mD; net pays of 10–20 m can be expected in this area.

In the North Sea, the Hauptdolomit has been pierced by many wells along the southern margin of the basin off East Anglia. Despite a number of hydrocarbon shows, productivity has commonly been limited by anhydrite and salt plugging (e.g. in 48/22-2: Clark, 1986). This is probably due to penetration of an unfavourable depositional facies (see later).

The barrier zone has been encountered in a few wells onshore north of the Humber in eastern England, but so far the most prospective parts have generally been missed in wells drilled offshore to Rotliegend targets.

The slope facies commonly has poor porosity and permeability. Gas production at Lockton in Yorkshire is thought to have depended mainly on fractures, as in some German fields. However, gravity-displaced shelf material on the lower slope and basin floor could offer better prospects; D.B. Smith (personal communication) reports oolite lenses up to 60 ft thick and several hundred feet long in Durham. The reservoir potential of the slope facies has been championed by Clark (1980b, 1985, 1986), who has deduced that porosities of up to 30% and permeabilities of 1–100 mD were created by late diagenetic processes. Amiri-Garroussi and Taylor (1992) describe a variety of displaced facies with porosities of up to 26% and permeabilities of up to 100 mD.

6.5.2 Stassfurt evaporite equivalents

The bulk of the Z2 evaporites (Stassfurt Salze, Fordon Formation; see Fig. 6.6 and Fig. 6.2) consists of halite—the principal mobile component of Zechstein salt structures in the Southern UK Basin. Here, the evaporites were deposited across the Z2 carbonate shelves to varying extents in different areas, thickening markedly basinward of its edge (Fig. 6.12). This zone of thickening was essentially controlled by the thickness of the Werraanhydrit below, and coincides very roughly with the 0.5 km Zechstein isopach (see Fig. 6.4). The original thickness may have been in excess of 1400 m (4500 ft) in the basin centre (Christian, 1969).

From the landward margin of the shelf, red beds pass basinward into anhydrite above the Hauptdolomit, and are joined by foresetted units of halite, polyhalite and (beyond the break in slope) kieserite (Colter and Reed, 1980), forming a complex about 90 m (300 ft) thick, which spreads out across the floor beneath the main body of halite (Taylor and Colter, 1975; see also Figs 6.6. and 6.13. The Basalanhydrit proper is the lowest unit of this complex in basinal areas and has a gradational (interlaminated) boundary with the underlying Stinkkalk.

The presence of a widespread thin layer of potash minerals at the top of the Z2 evaporites suggests the overtaking of subsidence by salt precipitation and the construction of a broad shelf close to sea level, reaching far towards the basin centre. This levelling was responsible for considerable lateral uniformity during later sedimentation.

According to Day *et al.* (1981), Z2 salt probably filled the Central Graben and also occupied a breach in the MNSH further west (Fig. 6.13). Detailed seismic interpretation of the latter area (Jenyon *et al.*, 1984) confirms a salt-filled channel through Quadrant 37, controlled by a pre-Zechstein arcuate normal fault on its western side. It is thought also to have occupied a relatively deep channel through the western end of the MNSH, close to the Northumberland coast (Smith and Taylor, 1989), but it is doubtful whether it ever covered much of the intervening highs. Correlation of released wells shows that the Z2 evaporites begin to thin northwards just to the south of the MNSH and are only between 50 and 100 m (165 and 330 ft) thick over much of the Western Platform of the Northern Salt Basin (Taylor, 1993, Fig. 9).

The Z2 salt is believed to be thick in the deeper parts of the Northern Salt Basin, but cannot always be distinguished from evaporites of later cycles (Figs 6.9 and 6.14). Collectively these have been newly designated the Shearwater Salt Formation (Cameron, 1993a). They are absent from the Moray Firth and have not yet been demonstrated further north into the Viking Graben than UK block 9/10 (Cameron, 1993a).

6.6 Third Zechstein cycle (Fig. 6.14)

6.6.1 Plattendolomit Formation and equivalents

The Z3 carbonate (Plattendolomit or Platy Dolomite, equivalent to the Seaham Formation of Durham and the Brotherton Formation of Yorkshire, in the past widely referred to as the Upper Magnesian Limestone) is best known in the Southern Salt Basin. It overlaps a 1 m (3 ft) basal shale (Grauer Salzton or Grey Salt Clay) towards the basin margin. The shelf facies thickens inwards to a maximum of 75–90 m (250–300 ft) (see Fig. 6.6), before thinning to a few metres across parts of the basin floor, although it is not always recognizable there. The Grauer Salzton provides a sharp gamma-ray peak, though less strongly radioactive than the Kupferschiefer (e.g. well 49/24-1 in Fig. 6.10).

Much of the shelf facies of the Z3 carbonate consists of grey microcrystalline dolomite, with thin shaly layers. Sheets of the tubular calcareous alga *Calcinema permiana*

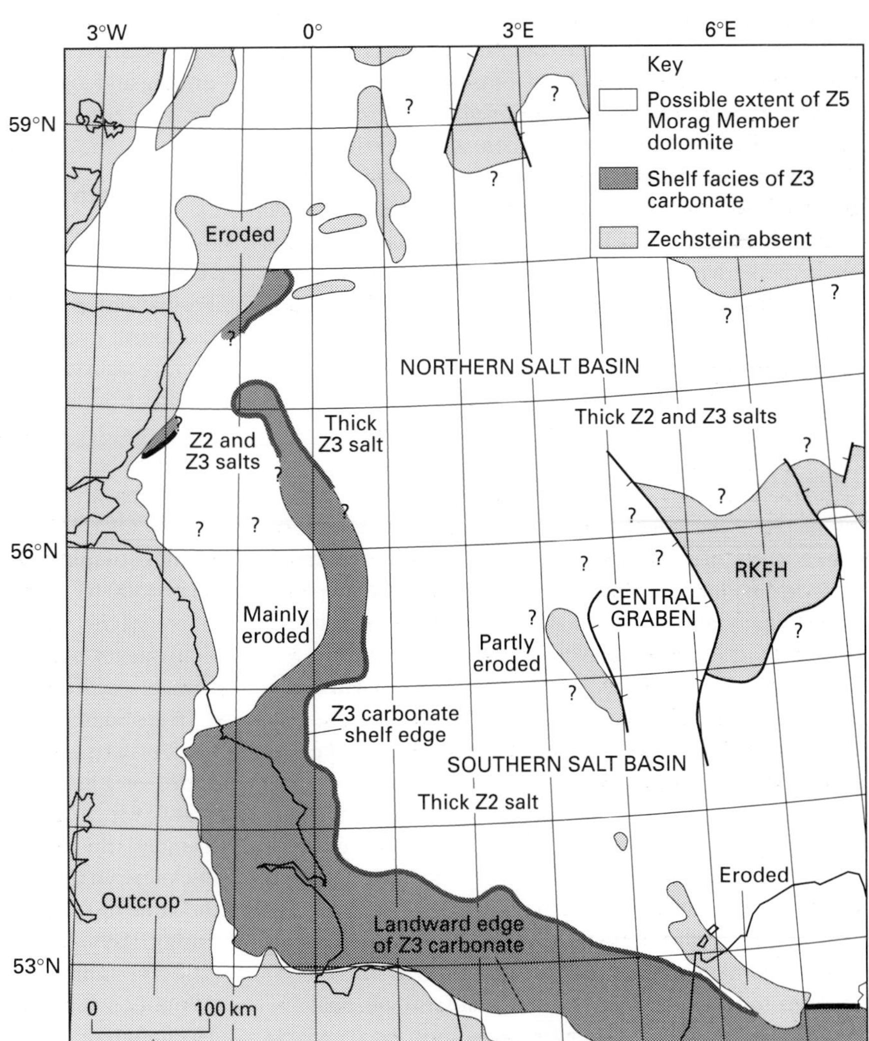

Fig. 6.14 Summary of significant known facies distribution in Z3 and Z5, with note on relative salt thickness of Z2 and Z3. Mainly compiled from released wells (Fig. 6.2).

are typical, and microfossils and stunted bivalves are locally abundant. This facies of the Plattendolomit appears to have formed in mainly shallow- but quiet-water (lagoonal or restricted shelf) conditions. An algal-rich shelf-edge barrier zone has been recognized in Poland (Depowski, 1981) but is possibly of lower relief than such features in earlier cycles. Calcite concretions similar to those of the Z2 slope carbonates, and probably representing a comparable environment, occur in the lower part of the Plattendolomit Formation. The *Calcinema* facies has not yet been positively identified on or north of the MNSH but may be present in the Moray Firth (Amiri-Garroussi and Taylor, 1987).

Towards the basin centres, the Grauer Salzton, Plattendolomit and Hauptanhydrit are much attenuated and difficult to recognize, particularly when displaced by movement of Z2 salt, as is commonly the case. An unfossiliferous but apparently correlative shale–dolomite–anhydrite unit has been found in several wells in the Northern Salt Basin, but so far a lack of cores has prevented verification of the correlation. In this region, the dolomite is referred to as the Turbot Carbonate Unit of the Shearwater Salt Formation (Cameron, 1993a).

Reservoir facies

Porous grainstones are common but thin at the extreme landward margins of the formation; they are also found elsewhere on the shelf, but porosity and permeability seldom approach those of the Hauptdolomit (Clark, 1986). Some gas production has been obtained from lagoonal facies at Lockton and Eskdale in Yorkshire. Plattendolomit production there and on the Continent is usually dependent on natural fracturing, but at Alkmaar, northwest of Amsterdam in the Netherlands, production is from leached algal and oolitic facies (van Lith, 1983). Studies offshore the Netherlands by Baird *et al.* (1993) have found the best reservoir conditions where shoal or shoreline facies coincide with faults, which have facilitated diagenesis by low-temperature hydrothermal fluids.

6.6.2 Third Zechstein cycle evaporites, Leine Halite Formation and equivalents

At the top of the Z3 shelf carbonates in the Southern North Sea, a series of sabkha cycles, with algal (cyanobacterial) mats and nodular anhydrite, lead upwards, in some places transitionally and elsewhere with a sharp break, into the Hauptanhydrit (Main Anhydrite). This thickens from 3 m (10 ft) or so over the shelf to some 45 m (150 ft) within the rim of the thickest carbonates.

The Z3 halite or Boulby Formation thickens from less than 30 m (100 ft) near its boundary with the encompassing red beds to around 120 m (400 ft) in the centre of the Southern Salt Basin; it becomes the dominant salt over the Western Platform of the Northern Salt Basin (Taylor, 1993).

Near the top of the unit, the thin Boulby Potash Member, consisting mainly of sylvite, is worked in Cleveland. It thickens towards the centre of the Southern Salt Basin by developing at successively lower levels, forming a variable halite/carnallite/polyhalite/mudstone complex (Smith and Crosby, 1979).

The evaporites show evidence of deposition in very shallow to emergent conditions and are capped by thin, salty and potassic red beds (the Carnallitic Marl Formation, Roter Salzton or Red Salt Clay); the terrigenous components die out basinward from the margins. An equivalent unit is apparent in some Northern Salt Basin wells, where thick zones of mixed potash salts in the underlying halite suggest broad similarity with the third cycle of the Southern Salt Basin.

Seismic sections suggest that salt was formerly far more extensive than now over the MNSH, having been lost by dissolution beneath a thick overburden from Triassic through to Tertiary time, particularly during the Early Cretaceous. Evidence from surrounding areas suggests that the lost salt had belonged to the third rather than the second Zechstein cycle.

Third Zechstein cycle salt may once have been more widespread in the Moray Firth area. According to Ziegler (1990a), seismic data suggest that the moderately thick Zechstein salts occupying the southern end of the Viking Graben form part of the downfaulted pre-rift sequences, which was eroded over the adjacent flanks of the graben as a consequence of their mid-Jurassic uplift. By analogy with the Long Forties and MNSH highs, the greater part of the eroded Zechstein section may have consisted of Z3 rather than Z2 halite.

Magnesium-bearing salts of Z3 are being exploited in the north-east Netherlands by solution-mining, using modern drilling and logging techniques developed by the oil industry (Coelewij *et al.*, 1978).

6.7 Fourth and fifth Zechstein cycles

(Fig. 6.14)

A thin, tight dolomite or magnesite unit (the Upgang Formation), perhaps representing the transgressive base of the fourth cycle in Cleveland and Yorkshire, has not yet been recognized offshore, but the overlying Sherburn Formation (formerly the Upper Anhydrite or Pegmatitanhydrit) can be followed widely around the edge of the Southern Salt Basin, and appears to be present as far north as Norwegian waters (well N17/4-1)

The Aller Halit (Sneaton Formation, formerly Upper Halite) occupies a slightly smaller area, thickening to about 90 m (300 ft) in the middle of the Southern Salt Basin, where sylvite, carnallite and red mudstone are developed. Equivalents north of the MNSH–RKFH are difficult to correlate with certainty.

In the Southern Salt Basin, the fifth cycle, separated from the fourth by a thin red mudstone, consists of the Grenzanhydrit or Littlebeck Formation (formerly the Top Anhydrite), which splits basinward to include a thin halite member, the whole totalling only about 6 m (20 ft) in thickness. In the Northern Salt Basin, what is thought to be the same cycle is represented by the Morag Member of the Turbot Anhydrite Formation. It consists of up to 60 m (200 ft) of anhydrite, followed in some wells by shale and dolomite, which may comprise a further cycle (Fig. 6.14). The dolomite, on average less than 16 m thick, is restricted to wells in the central and what was probably the deepest part of the Northern Salt Basin, mostly lying in the Norwegian sector, and at the southern entrance to the Viking Graben. It presumably occupies a depression incompletely filled by the Z4 halite, or caused by continued gentle subsidence. No diagnostic fossils have been reported and it is not clear whether it is of marine or limnic origin.

Minor sixth and seventh evaporite cycles have been reported in Germany and the Netherlands, but may be of continental rather than marine origin, and of Triassic age.

6.8 Regional variations

Besides the notable differences between typical sections of the 'shelf' and 'basin' areas, certain regions have their own characteristic Zechstein sequences. Because the Northern and Southern Salt Basins appear to have experienced similar cycles of flooding and evaporation, the distinctions are most easily explained by differences in uplift or subsidence before, during or after deposition, by local inflow of fresh water and accompanying clastics, in some instances by slumping, sliding or faulting, or in certain areas by post-Zechstein dissolution of evaporites.

The principal regions to be contrasted with the Southern Salt Basin are: (i) the MNSH–RKFH; (ii) the Western Shelf, Forth Approaches Basin and Long Forties High; (iii) the edges of the Central Graben and the intragraben highs; and (iv) the greater Moray Firth Basin.

6.8.1 Mid North Sea–Ringkøbing–Fyn High

Much of the MNSH appears to have been periodically submerged during the Zechstein, under water that was shallow relative to the Northern and Southern Salt Basins but in some channels and re-entrants up to about 200 m (700 ft) deep. Local emergent or very shallow areas formed the nuclei for wedges and sheets of carbonates and evaporites, which, during the first two cycles, aggraded to near the prevailing sea level and then spread rapidly (prograded) towards deeper water. Large slumped masses can be inferred on some seismic sections at the foot of oversteepened shelf margins bordering channels and inlets (Jenyon *et al.*, 1984); slumping can also be recognized in cores in similar situations (Amiri-Garroussi and Taylor, 1992). Some of the deep areas remained unfilled by carbonates and anhydrite, later filling with salts of the second, third and fourth cycles.

The maximum thickness of the Zechstein over the MNSH is about 600 m (2000 ft). It is not unusual for the upper and lower parts of the normal sequence to be missing, but the Z2 Hauptdolomit (Innes Carbonate Member), about 60 m (200 ft) thick, is generally identifiable in oolitic facies, sandwiched between thicker anhydrites (Turbot Anhydrite Formation and Iris Anhydrite Member). For instance, in 38/16-1 the Werraanhydrit rests directly on the Carboniferous and in 38/29-1 on the Devonian (see Fig. 6.11), suggesting that some sites were emergent or swept clear of sediment until the later part of the first cycle. According to Clark and Tallbacka (1980), the Zechstein shows an onlapping relationship to large portions of the RKFH too; in the Danish Arnum-1 well, the Werraanhydrit rests directly on metamorphic basement.

Lack of the upper parts of the Zechstein is generally the result of uplift and dissolution of the more soluble salts or erosion of the less soluble ones beneath the Cimmerian unconformities (e.g. 36/13-1, 38/16-1, 38/25-1, Fig. 6.11). However, in some channels and embayments the full Z3, Z4 and Z5 evaporite succession is preserved (e.g. 38/29-1, Fig. 6.11). There are other large areas of the MNSH where the presence of collapse structures in the Mesozoic, visible on seismic, indicates that salt was originally present but later dissolved (see also Section 6.10). It is believed that the same may be true of the RKFH.

The Plattendolomit is recognizable on seismic sections in channels crossing the MNSH but difficult to distinguish on the adjacent shelf areas, possibly because it rests directly on the Hauptdolomit or with only thin intervening anhydrite.

6.8.2 The Western Shelf, Long Forties High and Forth Approaches Basin

Correlation of the wells in this region suggests that in Z2 time it subsided less relative to sea level than the Southern Salt Basin, but subsided more during Z3 (Taylor, 1993). Over substantial areas of the Western Platform, the Plattendolomit equivalent rests directly on the anhydritic basal evaporite complex of the Z2 evaporites. The Z3 salt then dominates the section, forming pillows and diapirs. The only alternative explanation for the low position of the Plattendolomit equivalent would be withdrawal of underlying salt on a regional scale. While not impossible, this would be difficult to reconcile with the apparently unbroken and undisturbed nature of the dolomite, in contrast to its obvious deformation where Stassfurt salt withdrawal has occurred in the Southern North Sea (see ('X' in Fig. 6.19A).

The fault-controlled Long Forties High appears to have affinities with the MNSH, forming a Zechstein shelf or series of platforms. The few wells drilled on it suggest that it became a positive feature during Z1 time, accumulating thick anhydrite (the Iris Anhydrite Member or Werraanhydrit equivalent) As a result the Z3 carbonate rests directly on the Z2 carbonate, and the Z2 evaporites thin out against its flanks. During later subsidence, it became covered by thick Z3 halite.

To the west of the high, wells in the Forth Approaches Basin have found at least 600 m (2000 ft) of Zechstein, with the normal five cycles. Halite and potash salts are present in the second, third and fourth cycles. Intervals of quartz sand in the Stassfurt evaporite equivalent suggest active erosion of the neighbouring Scottish Highlands at that time.

6.8.3 Central Graben

Where the Central Graben intersects the northern flank of the MNSH, the Late Cimmerian Unconformity cuts down to the Z2 carbonates in shelf and slope facies in the Argyll oilfield, and probably in basin facies at Auk (see Figs 6.11 and 6.24). These abbreviated successions are designated the Halibut Carbonate Formation by Cameron (1993a). Generally thin and rather nondescript Zechstein successions are also found on the Jaeren High and horsts within the Central Graben, usually succeeded by Triassic beds (e.g. 22/18-1 on the Montrose structure, Fig. 6.12). Evidence from the Forties–Montrose High (Crawford *et al.*, 1991) indicates that its positive history can be traced back at least to the Carboniferous. The facies of the Zechstein is not yet clear. Salt is generally absent, and yet the carbonates are usually micritic, with little or no evidence of high-energy environments. Deep leaching may have removed all evaporites from what may formerly have been basinal sequences in at least some localities.

The nature of the Zechstein in the downfaulted section of the MNSH–RKFH (see Fig. 6.13) is uncertain. The narrow portion free from diapiric salt may include shelf carbonates and anhydrites of one or more cycles.

6.8.4 Moray Firth

The Zechstein of the greater Moray Firth region is generally less than 240 m (800 ft) thick and is characterized by abundant anhydrite, common clastics and absence of halite or more exotic evaporite minerals. Cameron (1993a) follows Deegan and Scull (1977), dividing the group into three formations: a dominantly anhydrite Turbot Anhydrite Formation above and a shaly carbonate, called the Halibut Carbonate Formation, below, generally resting on a thin, strongly radioactive Kupferschiefer Formation. Lower parts of the sequence are missing in some wells, where an onlapping relationship to older rocks is seen, for instance on the Grampian Arch and, as already noted, on the MNSH–RKFH.

Andrews (1990) provided an overview of the greater Moray Firth region, drawing on the extensive British Geological Survey (BGS) database. It appears that, as a result of post-Permian uplift and erosion, the Zechstein is absent from extensive parts of the southern half of Quadrants 13 and 14 (see Fig. 6.13). On the other hand, the Zechstein is thought to have originally extended across both the Halibut Horst and Fladen Ground Spur, but has since been removed.

The Kupferschiefer is commonly found beneath reddish siliciclastics of the Bossies Bank Formation in the west and north, demonstrating initial marine flooding of these areas. Zechstein siliciclastics are believed to occupy both the West Fair Isle and East Orkney basins according to Andrews (1990).

The Moray Firth region appears to have been a series of interlinked shallow basins during the Zechstein. It could be inferred that shelf-type sequences with dolomite grainstones, perhaps porous and permeable, might exist around the margins (see Fig. 6.13). This appears to be confirmed in the released wells 12/23-1 (see Fig. 6.12) and 20/2-1 and 2.

At the entrance to the Moray Firth, the Turbot Anhydrite Formation in well 20/8-1 includes in its lower part 22 m (72 ft) of halite accompanied by traces of potash salts, grading up into anhydrite. It provides a link between the halite-free anhydrite-dominated successions of parts of the Moray Firth and the typical halite-dominated successions of the Northern and Southern Salt Basins.

A further link between the standard Zechstein of the south and that of the Moray Firth is provided by 20/3-3, on the foundered eastward plunge of the Peterhead Ridge between the two areas. It is divisible into Turbot Bank and Halibut Bank Formations by comparison with adjacent Moray Firth wells, but the lower part shows strong affinities in thickness, lithology and log response with the Zechsteinkalk–Werraanhydrit–Hauptdolomit succession on parts of the MNSH and the western and southern margins of the Southern Salt Basin. The Innes Carbonate Member, the supposed equivalent of the Hauptdolomit is 38 m (128 ft) thick and is in part oolitic, though tight.

Beds of sand occur interspersed with Z2 halite in 20/10a-3, situated just to the west of the Buchan Horst and south of the extension of the Highland Boundary fault, suggesting elevation and erosion of the highs in this area during Z2 time.

The absence of halite from the Moray Firth may have a variety of explanations: (i) the diluting effect of the substantial inflow of fresh water, which is implied by the abundance of clastics—silt and sand as well as clay minerals; (ii) dilution by oceanic water, which may have bypassed the Viking Graben (the feeding channels for the Zechstein salt basins may have utilized depressed areas of the Shetland Shelf behind the tilted edges of the Viking Graben, as portrayed speculatively on maps in Smith and Taylor, 1992); (iii) the region may have been left high and dry by the evaporative drawdown accompanying the halite phases of precipitation; or (iv) halite may once have been present in places but later removed by circulating groundwater.

6.9 Interpreting Zechstein stratigraphy and lithology from seismic sections

Only certain formation boundaries within the Zechstein are capable of giving strong seismic reflections, and not all those that do so in basin areas necessarily do so over shelves or platforms, and vice versa. The reason can be understood by considering values of the reflection coefficient between different lithologies. As a first approximation the proportion of incident energy reflected at a boundary can be judged by using the simplified formula $R = (d_2v_2 - d_1v_1)/(d_2v_2 + d_1v_1)$, where R is the reflection coefficient and d_1 and d_2 are the densities of the upper and lower media, and v_1 and v_2 the velocities of the upper and lower media, respectively. Representative values are shown on the Zechstein profile in Fig. 6.15.

Neglecting formations that are too thin to have much effect on their own, the three main components of interest are halite, anhydrite and dolomite, which occur in that order downwards in a complete Zechstein cycle. For dolomites with 0–5% porosity, which are typical of basin areas and many slope and low-energy back-barrier environments, the reflection coefficient between anhydrite and the underlying dolomite (e.g. the Hauptdolomit and the Plattendolomit, or the Basalanhydrit and the Hauptdolomit) is in the range 0.05 to −0.05, almost an order of magnitude less than that between the overlying halite/anhydrite boundary, where $R = 0.3$. The coefficient for anhydrite overlying dolomite only becomes comparable with the halite/anhydrite value (though with negative sign, i.e. phase-reversed) where dolomite porosity rises to around 20–25%.

The foregoing is an oversimplification, since interference between the pulses returned from the tops and bases of the anhydrite and dolomite layers can add or cancel, depending on thicknesses and seismic frequency, while thin interven-

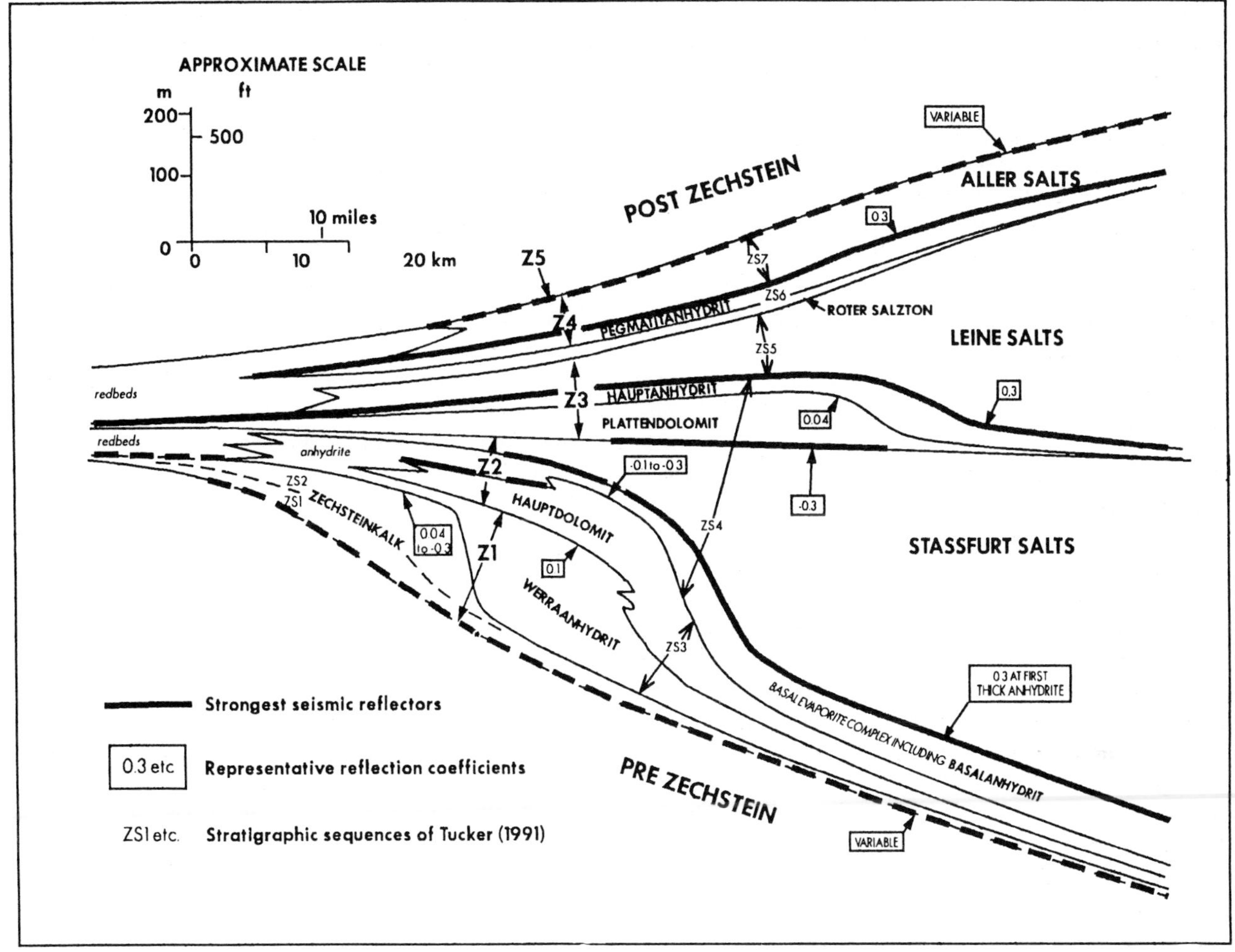

Fig. 6.15 Diagrammatic profile of Zechstein showing strongest seismic reflectors, with representative reflection coefficients, and also stratigraphic sequences of Tucker (1991). See Section 6.9 for discussion.

ing shale, halite or potassic beds can also influence the signal. Nevertheless, synthetic seismograms confirm the generalization. The strong reflections commonly attributed to the top of the 'Plattendolomit' and the Hauptdolomit' are in practice initiated at the tops of the Hauptanhydrit and Basalanhydrit, respectively. Characteristic compound signals from the Hauptanhydrit/Plattendolomit/Grauer Salzton and from the Basalanhydrit/Stinkkalk/Werraanhydrit/Zechsteinkalk/Kupferschiefer packages can usually be traced for long distances in basin areas. The Pegmatitanhydrit/Roter Salzton also returns a recognizable signal when thicknesses of the relevant units are favourable. Whether either the top or base of the Zechstein do so depends on the lithology and state of compaction of the strata above and below—in some areas in the Central North Sea, for instance, the velocity of the Trias is too close to that of halite for a strong reflection to occur at the top of the Zechstein.

It follows from the above that the Hauptdolomit proper is most likely to be visible when it has good reservoir properties. Maureau and van Wijhe (1979) and van Wijhe (1981) have described the use of synthetic seismograms to identify porous zones in the Hauptdolomit in the Netherlands. Unfortunately, in shelf areas where the Stassfurt evaporites are thin, interference from the 'Plattendolomit' and beds within the Stassfurt evaporites may completely suppress reflections from the top of the Hauptdolomit, even when it has significant porosity. This problem has been addressed by Mathisen and Budny (1990), Budny (1991) and Karnin *et al.* (1991) in relation to the Zechstein carbonate belt of north-west Germany and by Antonowicz and Knieszner (1984) in Poland.

Antonowicz and Knieszner (1984) suggest that the interval between the Plattendolomit and Hauptdolomit must be of the order of 60 m before it is possible to distinguish the top of the latter. Karnin *et al.* (1991) illustrate the effects of overlying beds on the seismic response of the Hauptdolomit. They also advocate the use of three-dimensional (3D) seismic for better structural definition and porosity prediction in the Zechstein, claiming that 3D seismic was able to discover almost twice as much gas as 2D seismic in the Harpstedt concession.

Matrix porosity and permeability are usually variable and difficult to predict in Zechstein reservoirs, and fractures can be essential for good well productivity. Improved seismic techniques are increasingly being used to overcome this problem. In the Hewett gasfield (Cooke-Yarborough, 1991, 1994), effective porosity has been mapped using acoustic impedance from surface seismic, and seismic and core data have been integrated into a tectonic model to predict zones of fracture porosity. Strohmenger *et al.*

(1993a,b) describe the prediction of facies and reservoir properties in Germany by combining data from cores, well logs and 3D seismic.

Recognition from seismic sections of the top of the Zechsteinkalk and of porous zones within it encounters difficulties similar to those of the Hauptdolomit.

Whether the top of the Hauptdolomit in its shelf facies can be identified on seismic sections or not, its presence can generally be inferred where thick tabular masses of Werraanhydrit are visible (van der Baan, 1990). The slope between basin and shelf is also generally identifiable. Antonowicz and Knieszner (1981, 1984) illustrate examples from Poland showing that it is possible to map Z2 carbonate barriers seismically from their morphology. Strohmenger *et al.* (1993a) describe how abrupt changes in Z1 anhydrite thickness, identified by seismic, can define facies boundaries in the Z2 carbonate.

Large isolated carbonate build-ups can be recognized on seismic sections. However, not all reef-like bodies on seismic sections are reefs. Many on the MNSH and perhaps other highs are residual masses of carbonates or anhydrite remaining after dissolution of more soluble evaporites (see Fig. 6.19C; Jenyon and Taylor, 1983, 1987). A different type of mound, up to 4 km across and 10 m high, shows up clearly below thick salt in Quadrant 29 on the Western Shelf (see Fig. 6.19D). They have been found to consist of swellings in otherwise thin Z2 evaporites beneath the Plattendolomit (Taylor, 1993). One tested by the Amerada well 29/27-1 contains a mineral identified from wireline logs as tachyhydrite, a highly soluble, hydrated calcium–magnesium chloride mineral rare in marine evaporites, which generally has a diagenetic origin, possibly associated with hydrothermal activity. These mounds have no reservoir potential.

6.9.1 Relevance to sequence stratigraphy

Tucker (1991) has reinterpreted the Zechstein in terms of 'Exxon-style' sequence stratigraphy, suggesting that the logical surfaces at which to divide the group are the breaks between the carbonates and the overlying anhydrites, representing times of major regression. The great practical advances made by sequence stratigraphy in siliciclastic rocks have been due to the fact that the Exxon-sequence boundaries tend to be prominent on seismic sections. It will be clear that this advantage does not hold in the case of the Zechstein, where the most consistent strong reflectors are the tops of the anhydrites, not the carbonates. It can be seen from Fig. 6.15 that, of the sequence boundaries chosen by Tucker (1991), only that between Zechstein sequence boundary 6 (ZS6) and ZS7 (the top of the Pegmatitanhydrit) is represented by a consistent strong seismic reflection.

Yet the bases of the classical evaporite cycles fare little better, for, although they generally display a strong contrast in acoustic impedance, an identifiable reflection is commonly masked by the train of waves from above. On the other hand, the bases of the classical cycles, being represented by thin gamma-ray markers or baseline shifts, are almost always the easiest points to correlate on wireline logs, as can probably be seen on Figs 6.10–6.12. The bases of the classical cycles generally represent marine flooding or transgressive surfaces. Despite these drawbacks, the detail now available from Germany (Strohmenger *et al.*, 1996) strongly supports the Tucker model. The best system for stratigraphic breakdown of the Zechstein remains the subject of debate, and in the writer's opinion awaits more concrete evidence about changes in relative sea level during evaporite phases. The ideas generated, however, have practical consequences for the prediction of reservoir facies (Tucker, 1991). M.C. Geluk, A. Plomp and van Doorn (1996) find, for instance, that, in the Southern Netherlands Basin, sandstones laterally equivalent to Z1 and Z4 evaporites were formed as lowstand deposits. They were shed further into the basin than sandstones in Z2 and Z3, which grade laterally into carbonates and represent highstand deposits.

6.10 Salt behaviour—halokinetics and dissolution (Figs 6.16–6.22)

The movement of Zechstein salt has become recognized as a major factor controlling sedimentation and hydrocarbon trapping in the Central North Sea. The resulting structural history is complex. Some knowledge of the rheological behaviour of salt is helpful in understanding why salt bodies developed where or when they did and to interpret correctly their shapes from seismic images. Some general points are therefore included here.

Zechstein salt structures are common where the evaporites are thick basinward of the carbonate/anhydrite shelves (compare Figs 6.4 and 6.13). Because the areas concerned are now relatively deeply buried (see Fig. 6.5), these facts might be taken to imply that a critical minimum overburden or a minimum salt thickness is necessary before rock salt will move, as was once commonly thought. Such is not the case. Work reviewed by Jackson and Talbot (1986) convincingly suggests that salt behaves as a non-Newtonian fluid (one in which the viscosity varies with stress). There is consequently no minimum stress below which it will not flow, although movement may be so slow as to be imperceptible, even on a geological time-scale, for its equivalent viscosity is extremely high—10^{17}–10^{20} Pa s.

When salt is buried, its bulk density remains constant at about 2.2 g/cm^3, whereas the densities of other sediments increase with depth. As a result, there is a depth below which the salt is less dense than the overburden and could rise by buoyancy. This depth lies between about 600 and 1000 m, depending on the sediment's original density. Such a system is metastable, but it still needs some extra factor to set the salt in motion. Thus, in parts of north-west Germany salt about 1000 m thick has been buried to 3000–4000 m without discernible movement (Sanneman, 1968) while in the Sole Pit Basin, Brunstrom and Walmsley (1969) estimated the overburden under which Zechstein salt started to move at only about 610 m. As Jackson and Talbot (1986) stated, salt movement can be initiated, retarded or accelerated by regional tangential forces that stretch, wrench or compress sedimentary basins and, while structures may evolve from low-amplitude concordant forms to high-amplitude intrusive and thence to extrusive types, they can stop growing at any stage.

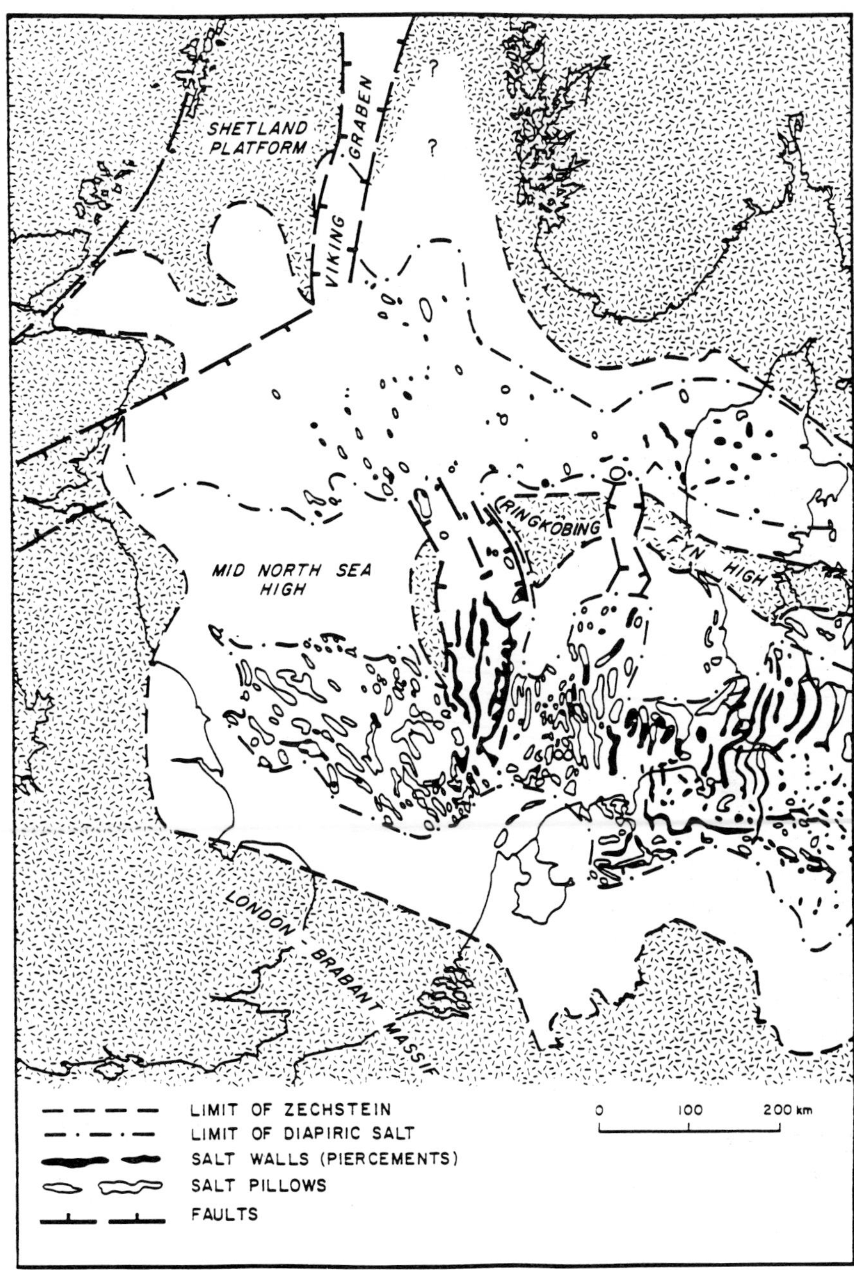

Fig. 6.16 Distribution of Zechstein salt structures (modified from Sorgenfrei, 1969, and Heybroek *et al.*, 1967).

6.10.1 Initiation of movement

The three principal factors that provide the stress necessary to trigger motion, separately or in combination, are differences in elevation at the salt/overburden interface, differential loading and crustal movements; the last two may operate at depths shallower than those at which buoyancy becomes effective. Of these, differential loading is probably the most nearly universal factor, and basement faults are thought to be common means by which the load differences have been brought about or localized. Tectonic control might be deduced from the alignment of most diapirs with known regional trends (compare Figs 6.16 and 5.6). In many workers' minds, the accumulating evidence from high-quality seismic puts beyond reasonable doubt the importance of basement faults as loci for many salt structures and Cimmerian (especially Late Cimmerian) episodes of disturbance as temporal controls. According to Graverson (1994), for instance, salt piercement in the Danish Central Graben develops at the intersection between graben-parallel faults and cross-cutting transverse faults in the basement (see also Remmelts, 1996).

The significance of basement faults was, however, early disputed by Christian (1969). More recently, Hospers *et al.* (1988) have claimed that in the Norwegian–Danish Basin only five out of 86 identified salt structures are genetically associated with basement faults. Many structures occur on trend lines paralleling the basin margin, and Hospers *et al.* (1988) consider that a modified form of Trusheim's (1960) halokinesis theory is likely, with salt ridges and diapirs developing as burial produces critical values of temperature and pressure along palaeodepth contours. Geil (1991, 1992) proposes an empirical model, also based on observations of salt structures in the Norwegian–Danish Basin. He believes that the differential overburden pressure needed to start movement is provided by the dip of the basin floor and the weight of local depocentres, rather than by subsalt faulting. Geil (1991, 1992) notes that substantial structures

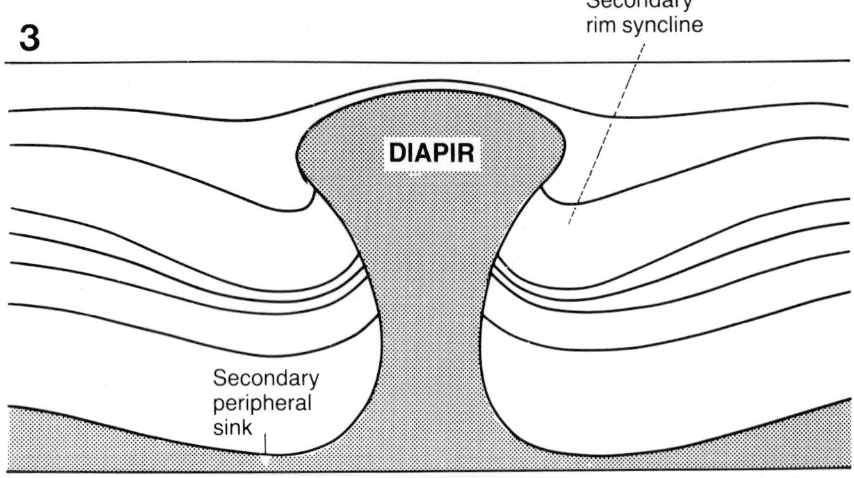

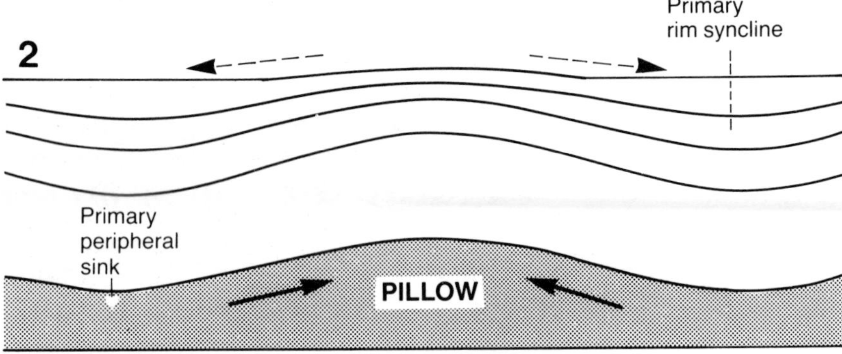

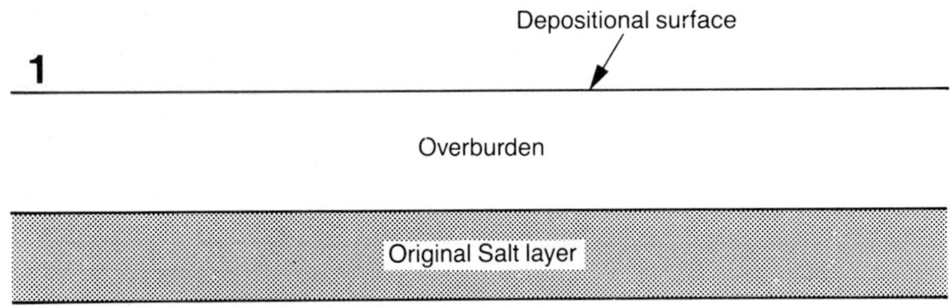

Fig. 6.17 Stages in development of salt pillow and diapir, modified from Trusheim (1960), Nettleton (1934) and others. This classic model appears to require that the overburden behaves essentially in a plastic manner. See Section 6.10.2 for discussion.

are not found unless present-day dips on the basin floor exceed 2–3°. The evidence for this is disputed by Madirazza (1992), who considers that movement of faults cutting the base of the Zechstein was responsible in the examples cited by Geil (1991, 1992).

It would be reasonable to suppose that the difficulty in recognizing faults (especially wrench faults) at depth beneath thick salt is responsible for the divergence of views. However, a different slant on the issue has been given by the model work of Vendeville and Jackson (1992a), which demonstrates that diapirs can form in response to extensional faulting in a brittle overburden, without the necessity for faults on the salt floor. Moreover, sand-box experiments by Oudmayer and de Jager (1993) show how faults may develop in the Mesozoic some distance from their causative sub-Zechstein faults.

A sensible conclusion from all the above would seem to be that subsalt faulting can and does activate salt structures, but may not in every case be necessary.

The complex relationships between salt structures and faults have been extensively discussed by Jenyon (1985a, 1986a,b and 1988b,c), using many seismic sections from the Southern North Sea as examples. Jenyon also deduced lateral flow of salt from the basin centre towards its margins, resulting mainly from the regional difference in loading.

Features that restricted the lateral movement, such as faults or the thickening wedge of underlying shelf carbonates and sulphates, provided another means of localizing salt structures, including a special class, basin-edge diapirism (Jenyon, 1985c, 1986a, 1987; Jenyon and Cresswell, 1987; see Fig. 6.19A). More recently, the movement of salt

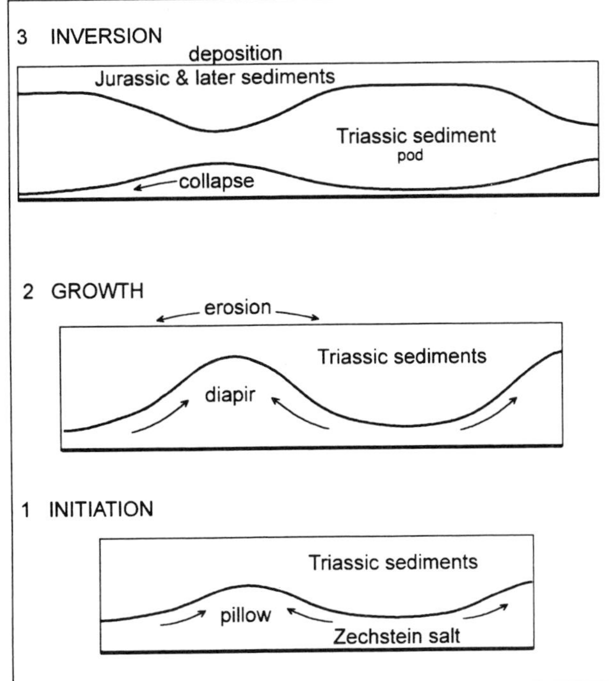

Fig. 6.18 Underlying notion explaining post-Triassic inversion (e.g. in the Central North Sea) due to the rise and collapse of Zechstein salt diapirs as a result of extension. No attempt is made here to indicate the faults in either the basement or the carapace, which would be required by brittle strata, or to differentiate between the effects of extension in the basement or of the carapace, or of both together. In practice, these and the tilt of the subsalt surface are essential variables, and differences in them are responsible for the complexity of actual examples and also for divergences of interpretation, noted in the text.

away from the margins towards the centre of the basin has been stressed.

Important information about the mechanism of lateral flow comes from cores through the Z3 halite in closely spaced brinefield boreholes in north-east England. Described by Smith *et al.* (1994b), it suggests that bulk movement occurred through differential lateral movement of large sheets and lenses of (mainly) undeformed halite that slid past each other at different rates, between glide planes consisting of intensely deformed and fragmented halite. This explains a long-standing puzzle: that, despite the bulk movement that occurred, distinctive markers on wireline logs through the evaporites can be correlated over large distances.

Diffusive mass transfer ('pressure solution') has been found to be an important internal mechanism by which salt deforms (Spiers *et al.*, 1984, 1986). This makes its strength enormously sensitive to variations in minute amounts of water, always present as inclusions and intergranular films, and suggests a speculative link between faulting and salt structures. If fault displacement occurs rapidly enough for the salt to exhibit brittle behaviour, one would expect the resulting fractures to introduce formation water into the adjacent salt mass, lowering its effective viscosity by several orders of magnitude.

6.10.2 Growth of salt structures

According to the 'classical' view (developed mainly in Germany), salt responds to any imbalance in the overburden by flowing sideways from the area of greater load to one that has a lighter burden, forming a salt 'swell' or 'pillow' and flanking 'peripheral sink' (see Fig. 6.17). Differential uplift and subsidence ensue, the products of erosion from the uplifted areas being deposited in the adjacent 'rim synclines'. Thus the imbalance is accentuated and the upward movement of salt is assisted. If it continues, piercement of the overburden may result in a 'diapiric' structure—a salt 'plug' or 'stock' or, if elongated, a salt 'wall'. Withdrawal of salt to form the diapir results in a secondary peripheral sink and secondary rim syncline. Theory, confirmed by model studies, shows that in plastic media the creation of a peripheral sink gives rise to new stresses, driving salt outwards as well as inwards and initiating a secondary ridge or group of domes; the process should then be repeated.

The steps described above are claimed to apply to the well-explored salt structures of north Germany. It is important to recognize that different factors may assume dominance at different stages in the evolution of salt structures, and may also differ from region to region. On the basis of recent experiments, Spiers (1994) suggests that buoyancy-driven pillow initiation can probably occur only by pressure-solution creep and that pressure-solution creep and dislocation creep probably alternate in importance during diapirism.

Trusheim (1960) noted a progression from pillows through stocks to walls, related to increasing depth to the base of the salt, while Sanneman (1968) found evidence for wave-like propagation of structures. However, the relationship between depth of burial and the type of structure has not been found to apply in the Southern North Sea, where tectonic influences seem to be more important (Brunstrom and Walmsley, 1969; Jenyon and Cresswell, 1987). In practice, most diapirs associated with major basement faults are strongly asymmetric (for example, see Fig. 6.19B).

Where the sediments flanking diapirs are not upturned, faulted or otherwise disturbed, the forcible extrusion of salt cannot be presumed. In such examples, it may be suspected that the diapir developed by 'down building' (Barton, 1933), in which the crest remained stationary at the surface while sediment built up in the subsiding flank areas. Alternatively, in the creation of elongated diapiric salt walls, tensional faulting of the overburden may be invoked, as in the model studies by Vendeville and Jackson (1992a,b). Their papers challenge many established views on salt structures and become of special significance in relation to the Central North Sea, and they have also been applied to the Southern North Sea (Bishop *et al.*, 1995).

The classical model of diapirism assumes a plastic overburden. In the alternative model envisaged by Vendeville and Jackson (1992a), assuming a brittle overburden, there may be three stages of diapir development: (i) a reactive stage, in which the salt rises by buoyancy in response to the local thinning caused by extension and

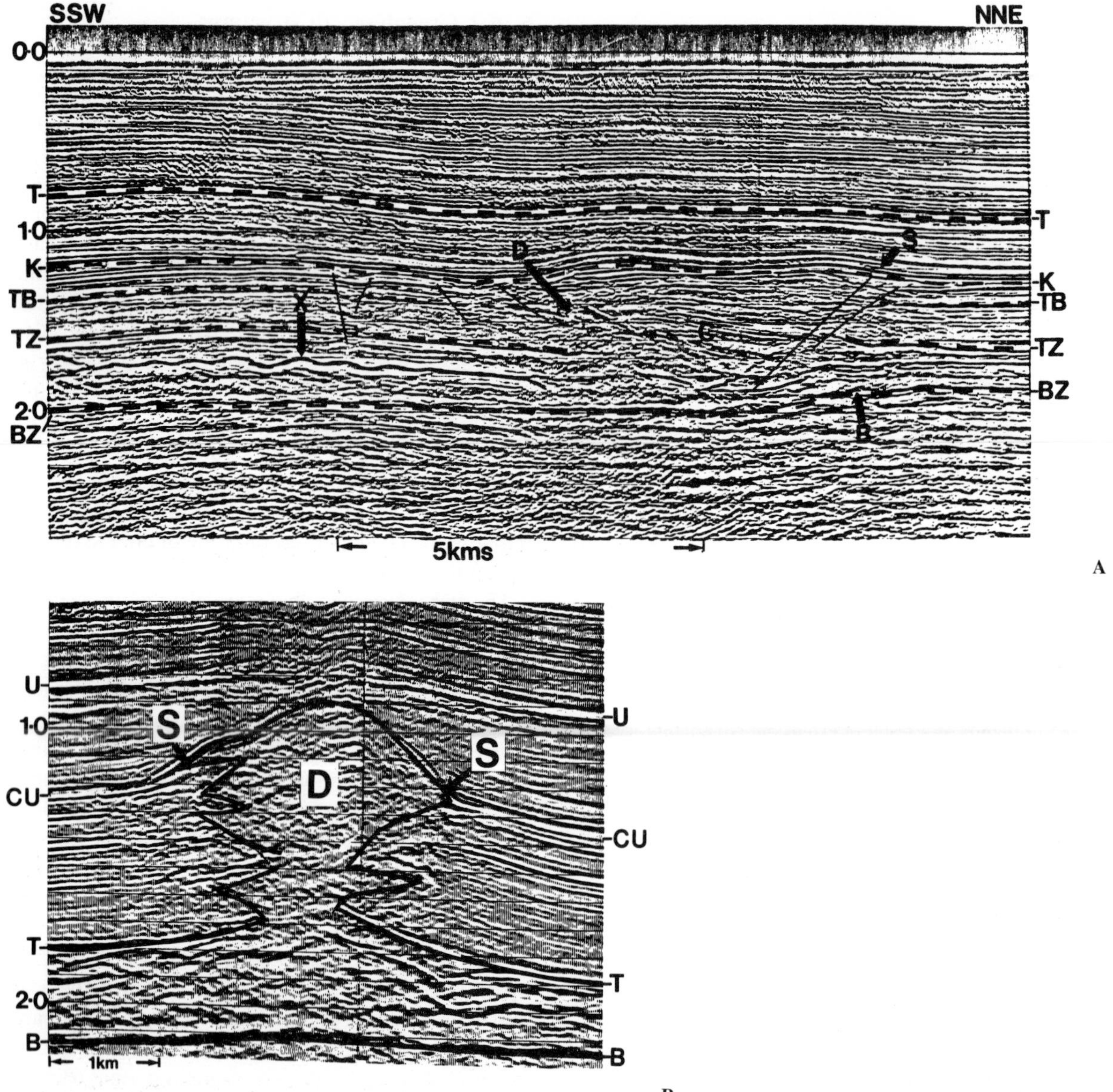

Fig. 6.19 Migrated seismic sections. A. Basin-edge diapirism. BZ, base Zechstein; TZ, top Zechstein; TB, top Bacton; K, Late Cimmerian Unconformity; T, base Tertiary. Arrowed B is essentially the Zechstein basin–slope transition; C is a rotated and subsided prism of post-Zechstein sediments; D is what is left of the primary basin-edge diapir; S is a slide zone produced by fracturing and salt injection. Notice the prominent wavy Plattendolomit reflection X between the TZ and BZ reflectors, which suggests lateral flow and shortening of the salt interval. (From Jenyon, 1985c.) B. Section across the short axis of an elliptical diapir. D, diapir; S, splay of bedding due to lateral injection of salt; B–T, salt-source interval; CU, Late Cimmerian Unconformity; U, Base Tertiary. (From Jenyon, 1986a.) C. (*Opposite*) Structures produced by retreat of salt due to edge dissolution, southern flank of MNSH. BZ, base Zechstein; TA, top basal carbonates; TZ, top Zechstein; CU, Late Cimmerian Unconformity; BT, base Tertiary. The salt interval TA–BZ has retreated to the left towards the centre of the Southern Basin; note drape of Mesozoic beds over residual hummocks. (From Jenyon and Taylor, 1987.) D. Reef-like subsalt feature R, at northern entrance to channel across the MNSH. B, base Tertiary; T, top Zechstein; U, Late Cimmerian Unconformity; TY, base Tertiary. Note absence of significant velocity contrast between R and adjacent salt, and remains of large overlying salt pillow truncated below U. (From Jenyon and Taylor, 1987.) Similar features in the area have been shown to be local swells in thin underlying Z2 evaporites; the thick overlying salt belongs to Z3 (Taylor, 1993).

normal faulting; (ii) an active stage, when the salt pierces through the thinned overburden, independently of further tectonic extension; and (iii) a passive stage, when the diapir reaches the surface; further growth keeps pace with sedimentation by downbuilding, as the base of the diapir sinks while the crest remains near the surface. Eventually, if regional extension continues and the supply of salt at the base of the diapir becomes exhausted or restricted, there may be a fourth stage, in which the widening diapir collapses (Vendeville and Jackson, 1992b).

Diapirs can rise at rates of up to 530 m/Ma, but more usually 200–300 m/Ma (Trusheim, 1960; Seni and Jackson, 1983). This has allowed ample time for Zechstein salt to rise from typical depths of 3000 m to the present seabed. While some structures have developed to this extent (see Fig. 6.22) many more have not, for a variety of reasons. Some salt pillows have remained quiescent since the Late Cretaceous.

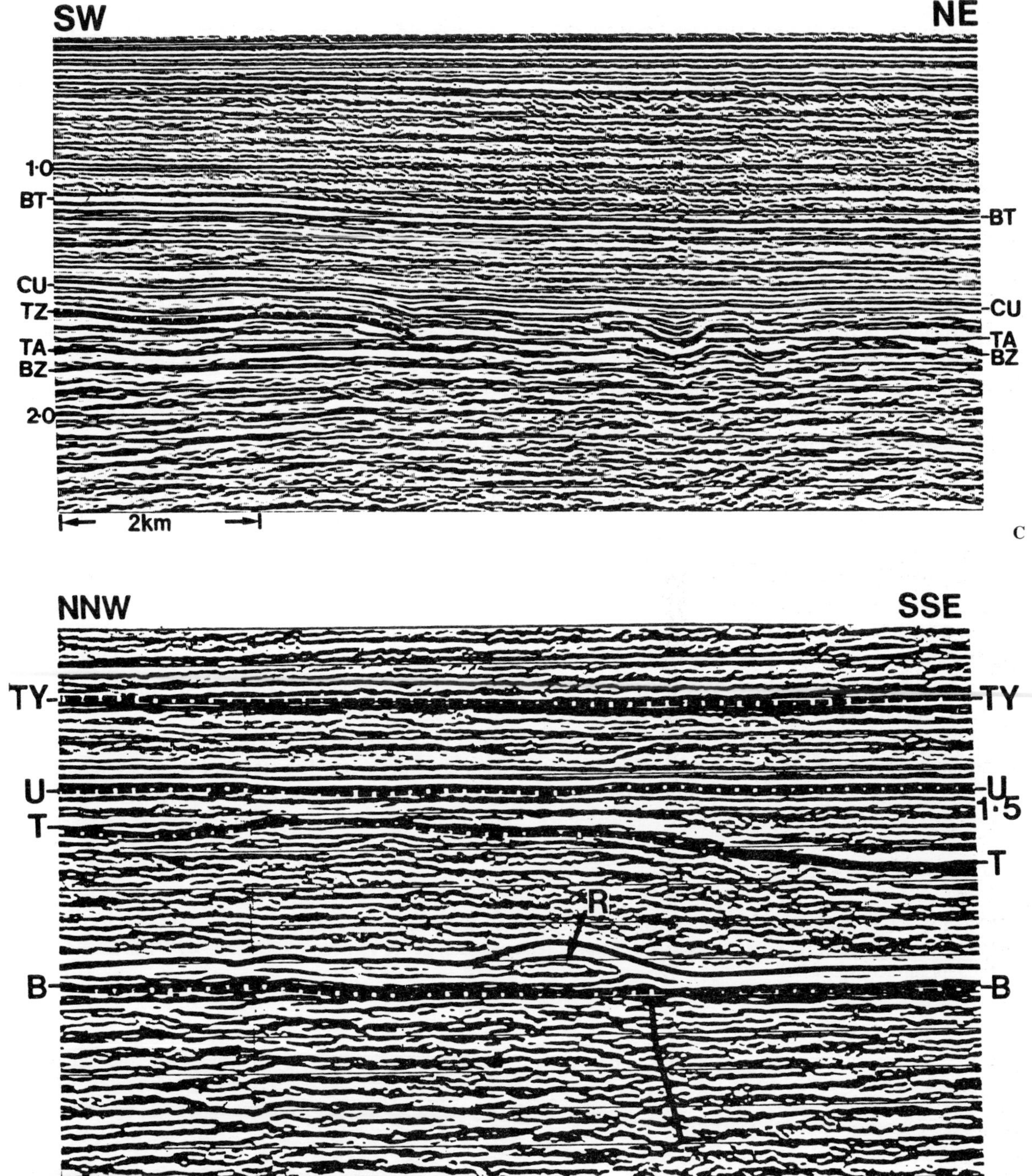

Fig. 6.19 *Continued*

6.10.3 Arrest of diapir growth

Salt swells may fail to grow perceptibly simply because they are too low; Jackson and Talbot (1986) suggest that, unless relief on the salt surface exceeds about 150 m, rates of flow due to buoyancy will be too small to be noticeable. If buoyancy is operative, growth will stop when the weight of the salt column balances that of the adjacent overburden—commonly when the top of the diapir is covered by 1 km or less of sediment, depending on its density.

The rise of a diapir may be halted by exhaustion of the source bed or the local closing together of the beds above and below. However, in his classic paper on the fluid mechanics of salt movement, Nettleton (1934) showed that the flow of material into a diapir can be cut off at an earlier stage simply by hydrostatic forces resulting from the formation of a peripheral sink, if the overburden is plastic and weak enough. But, if the overburden is brittle, the upward movement of a diapir will cease if it encounters layers of overburden that have sufficient strength to resist it. Salt may then intrude sideways into less competent strata (see Fig. 6.19B).

Finally, salt movement may be arrested or even reversed if the overburden on the original salt source is decreased by uplift and erosion. This is likely to have been an important factor around the Sole Pit inversion zone (see, for instance, Walker and Cooper, 1987) and has major consequences for the migration and trapping of hydrocarbons in the Central North Sea (see Section 6.10.7 and references therein)

6.10.4 Preferential movement of particular salts

Not all Zechstein salts flow equally. Whereas anhydrite and polyhalite are relatively immobile compared with halite, sylvite and especially carnallite flow more readily. Yet the thick and essentially pure halite of Z2 forms the bulk of most salt structures in the Southern North Sea (see examples on seismic sections in Jenyon, 1985b). It is not surprising that the underlying more rigid complex of anhydrite, halite and polyhalite is relatively undisturbed, but why Z2 halite moves in preference to Z3 and Z4 halite has yet to be discovered. Differences in crystal fabric and free energy of the crystal boundaries may be important. Viscosity falls dramatically with decrease in salt crystal size (Vendeville and Jackson, 1992a). D.B. Smith (personal communication) notes also that the Z3 and Z4 halites contain far more impurities than the Z2 halite in Cleveland. Water content is another factor that might be investigated, and others may be involved. Over the Western Shelf of the Central North Sea, where the Z2 evaporites are thin, it is the thick halite of Z3 which is diapiric (Taylor, 1993).

Unexpected juxtapositions of strata can arise because of the different rates at which salts move under the same stress. Highly complex examples, revealed by mining in the Harz–Thüringer Wald region of Germany, are illustrated by Nachsel and Franz (1983) and other examples are given by Richter-Bernburg (1980), and from Yorkshire by Talbot *et al.* (1982).

6.10.5 Examples of halokinesis in the Southern North Sea (Figs 6.19–6.22)

In the North Sea, timing of movement often differs in adjacent structures. This can be demonstrated most easily in the Southern North Sea. In Fig. 6.20, uplift of the overburden has resulted in removal of virtually the total Jurassic sequence from above the left-hand salt pillow, just prior to deposition of the Upper Cretaceous. Development of the right-hand pillow, on the other hand, did not start until during the Late Cretaceous. Salt movement continued after the earlier Tertiary. This example illustrates the importance of differing trigger events in causing movement of salt at one location rather than another, despite similar depth of burial and thickness. Note that the Plattendolomit does not extend over the greater length of the upper pillow surface. Also note the apparent uplift of the Hauptdolomit beneath the thickest salt section. This pseudostructure (due to 'velocity pull-up') is caused by the wedge of halite, which has a much higher interval velocity than the laterally adjacent sediments.

In Fig. 6.21 differences in the evolution of two adjacent diapirs are apparent. The stages of their development can be deduced from their flanking rim synclines. There is only

Fig. 6.20 Seismic section, North Sea, showing different times of initiation of two salt pillows (ZE). PL, Plattendolomit; TRL, Lower Triassic; TRU, Upper Triassic; JL, Lower Jurassic; KU, Upper Cretaceous; TL, Lower Tertiary; vertical scale = TWT in seconds. See Section 6.9.5 for further explanation.

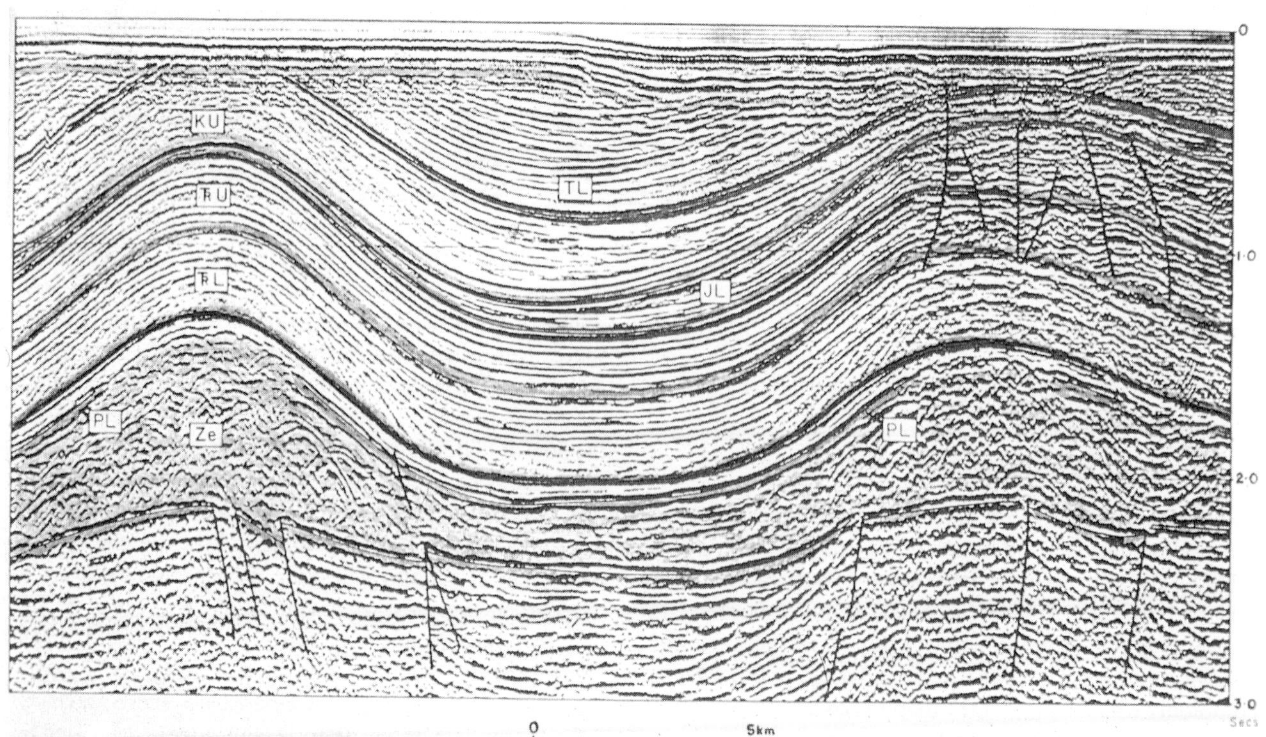

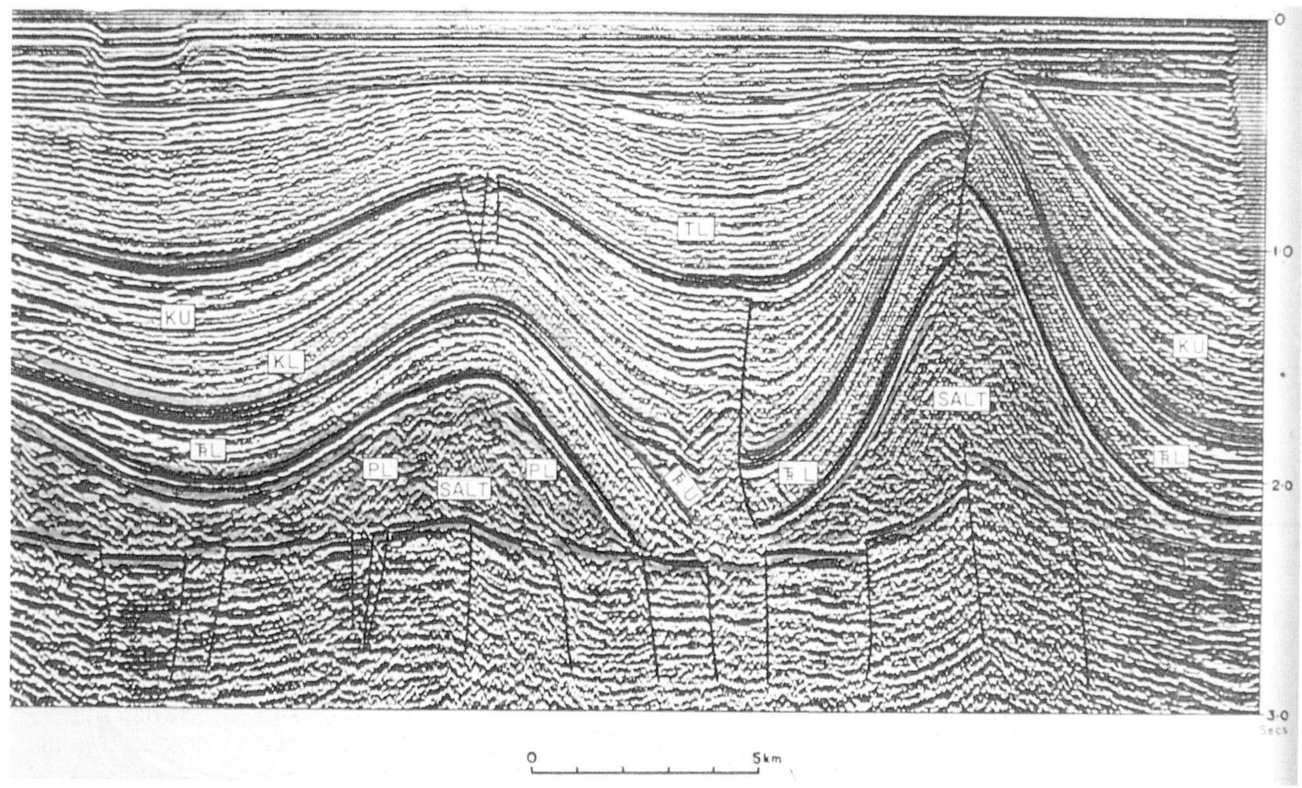

Fig. 6.21 Seismic section, North Sea, showing Mid-Tertiary end to diapiric growth caused by complete withdrawal of salt and rupture of intervening rim syncline. KL, Lower Cretaceous; other abbreviations and scale as in Fig. 6.20.

minor tensional faulting of the base Tertiary reflector over the left-hand pillow, whereas the crest of the right-hand diapir is more strongly faulted, with considerable relative vertical movement across the fault, accompanied by collapse of the overlying strata, possibly related to dissolution. By some time in the mid-Tertiary, salt movement had resulted in complete withdrawal from the area between the diapirs.

Fig. 6.22 Seismic section, North Sea, showing Cretaceous to Early Tertiary growth of salt wall dated by rim synclines. J, Jurassic; T, Tertiary. Other abbreviations and scale as in Figs 6.20 and 6.21.

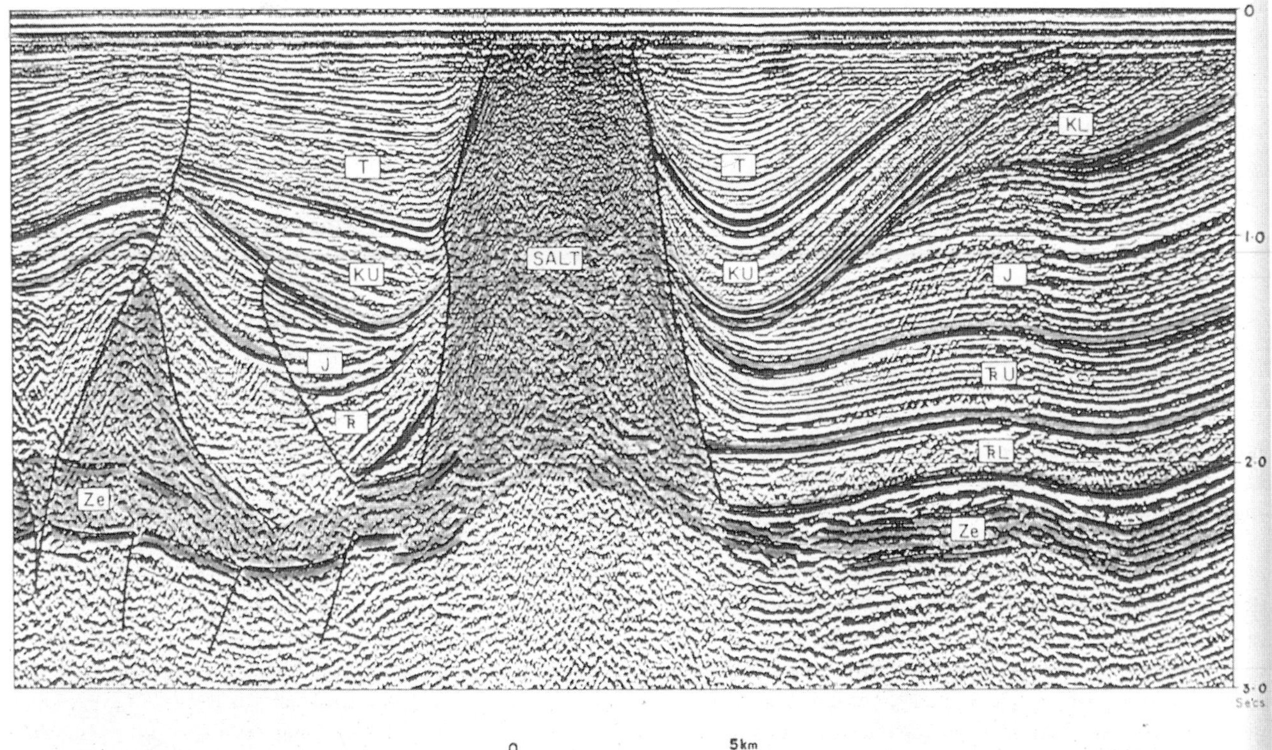

In Fig. 6.22, salt flow has resulted in the creation of a salt wall, which extends almost to the sea floor, where it has an apparent width of some 3 km. The structural relationships outlining the uplift and erosion of the flanking Jurassic strata indicate that the diapir went through a pillow stage during the Early Cretaceous. Salt movement continued well into the Tertiary but, to judge from the absence of salt in the section to the right of the salt wall, it has probably now ceased. To the left of the salt wall, diapirism has resulted in a pre-Cretaceous structural pattern that is almost too complex to decipher from seismic data alone.

Basin-edge diapirism (Jenyon, 1985c) is illustrated in Fig. 6.19A. A diapiric salt mass is thought to have formed in the region of D near the Zechstein shelf–basin transition in response to lateral flow. It rose in the late Triassic or Jurassic, and was lost by erosion and dissolution at the surface. Withdrawal of adjacent salt let down the thick prism of Mesozoic sediments, C.

Conventional seismic processing attenuates signals originating from steeply dipping structures. May and Covey (1983) describe inverse-modelling techniques, using very high stacking velocities, which give clear definition of diapir flanks, even where slightly overhanging; these techniques were not used in Figs 6.19–6.22. Figure 6.19B is an interpretation of the actual shape of one diapir, which shows lateral salt injection. Seismic-imaging problems and their solutions adjacent to and beneath salt diapirs have been reviewed by Reilly (1992).

6.10.6 Structural effects of evaporite dissolution

Dissolution of Zechstein evaporites beneath the Late Cimmerian (base Cretaceous). Unconformity has already been mentioned in connection with the creation of secondary porosity. Important structural effects can also be produced, especially where dissolution occurred beneath several hundred metres of overburden, as evidence suggests. Little attention was paid to this aspect after it was noted in the southern North Sea by Lohmann (1972), but with improving seismic quality it has become possible to distinguish dip reversal due to dissolution from that caused by salt flow (Jenyon, 1984a). This is easiest in areas where there is little salt disturbance, but the process has probably also been a contributing factor to structures in the Central Graben and Norwegian–Danish Basin.

Seismic evidence of the dissolution of formerly extensive halite over parts of the MNSH was illustrated by Jenyon and Taylor (1983, 1987; Jenyon, 1988a). The process leaves residual masses of evaporites and carbonates (see Fig. 6.19C), which may superficially resemble carbonate build-ups (and in some cases possibly are, although drilling has so far been discouraging). Substantial drape in overlying beds can result.

The effects of dissolution of evaporites on the fabric of overlying Zechstein carbonates have been described and illustrated by Smith (1972b, 1975). Some idea of the likely pattern of subsidence in the overlying Mesozoic may be gathered from an area of anhydrite dissolution in North Yorkshire, studied by Cooper (1986), although in this example leaching is taking place only at shallow depths.

6.10.7 Application to hydrocarbon exploration

The importance of structural traps in beds above Zechstein diapirs soon became obvious in the Central North Sea. In addition to Ekofisk, doming above diapirs has given rise to chalk fields at Eldfisk, Albuskjell, Tor, Hod and Valhall in the Norwegian North Sea, and to Tyra, Gorm, Skjold and Dan in the Danish sector, among others. Tertiary fields in crestal positions above diapirs include Maureen, Lomond, Cod and Scoter. The ETAP diapirs ring the Eastern Trough of the Central North Sea; their growth by active and passive phases during the Mesozoic, and possible late Tertiary piercement, has been described by Allan *et al.* (1994).

Less obvious than crestal traps, but increasingly important, are Mesozoic and Tertiary structures related to the lateral migration, withdrawal and dissolution of salt, and to the interaction between these processes and various styles of tectonics. A number of papers on these topics appeared at an EAEG/EAPG conference in Florence (1991) and in the conference on the Petroleum Geology of North-west Europe held at the Barbican, London (1992; see Parker, 1993). The theme was continued at the AAPG International Conference at The Hague, the Netherlands, in 1993 (see Remmelts, 1996; Alsop *et al.*, 1995), and at the meeting on Salt Tectonics at the Geological Society, London, in 1994. Many more examples are illustrated and discussed in the other chapters of this book despite the intense interest generated by the papers and earlier contributions by Vendevile and Jackson (1992a,b). A consensus view of the mechanisms at work has yet to emerge, and a concise summary is not feasible here. It is clear that the present stage of interpretation has only been possible since the availability of high-resolution 3D seismic, and much more of this, together with drilling, is needed to arbitrate between competing models. The following examples merely serve to illustrate some of the problems.

Hoiland *et al.* (1993) compare three models—thin-skinned extension, salt dissolution and a combination of the two—to explain similar structures on the Jaeren High, preferring the combination model. Errat (1993) describes the variety of structures found at the margins of the Central Graben and intragraben highs. Salt withdrawal related to basement extension from the Triassic to the Cretaceous is seen to be responsible for the network of pods and inter-pods. A similarity with the flower structures associated with wrench faulting is noted. Penge *et al.* (1993), writing about the East Central Graben, in contrast emphasize the role of post-Triassic extension, with Zechstein salt movement occurring as a passive rather than an active process. Sears *et al.* (1993) present a polyphase tectonic model, in which an element of oblique slip gives a better explanation of drilling results. Gowers *et al.* (1993) recognize seven tectonic phases in the structure of the Norwegian Central Trough, with major movement of Zechstein salt before the end of the Triassic. Sundsbø and Megson (1993) also find oblique-slip movement in the Danish Central Graben.

There are interesting differences between the structural styles developed above Zechstein salt on platform areas on the UK and Norwegian sides of the Central North Sea, despite their similar subsidence history. According to

Buchanan *et al.* (1996) the UK Western Platform is dominated by thin-skinned, gravity-driven detachment faulting, with reactive diapirs concentrated in the footwalls of large, listric, growth faults. In contrast, the Norwegian margin exhibits active diapirism, with crestal-collapse grabens that reflect symmetrical upwelling and down building. The difference may be related to a smaller regional tilt and greater overburden thickness on the Norwegian side.

Inversion structures (see Fig. 6.18) are a characteristic of the region and are responsible, for example, for fields along the south-western margin of the Central Graben. Smith *et al.* (1993) emphasize the importance of salt migration and withdrawal for changing sediment-distribution patterns. Thick Triassic sediment pods, deposited in the earliest salt sinks, later became highs, between which younger Mesozoic and Tertiary sediments accumulated, as salt withdrew or was dissolved from the original diapirs. Grounding of wedge-shaped pods on the Zechstein salt floor led to their rotation, causing dip reversals above.

A specific example of up-dip closure produced by the withdrawal of Zechstein salt and its extrusion up active fault planes beneath an Upper Jurassic sand reservoir is provided by Smith's (1987) model for the Clyde oilfield, on the margin of the Central Graben. In the same general area, the withdrawal of salt controlled the distribution of Triassic sedimentation in a rim syncline in the Fulmar field. Complete withdrawal of salt following deposition of the overlying Upper Jurassic sands caused the Triassic 'pods' to ground, with the creation of 'turtlebacks' and a trap structure in the Upper Jurassic Fulmar Sandstone (Johnson *et al.*, 1986). A similar history of salt movements controlled structural development of various reservoir rocks in parts of the Gannet cluster of fields (Armstrong *et al.*, 1987), and both sedimentation and trap formation of the Fulmar Sandstone in the Kittiwake field (Glennie and Armstrong, 1991).

Complete withdrawal of salt from areas close to large diapirs may provide the opportunity for hydrocarbons to migrate from the Carboniferous to post-Zechstein reservoirs. This is the case in the Triassic gasfields of Quadrant 43 (Bifani, 1986).

Rim synclines commonly preserve strata not found round about, and basin-edge diapirism appears to have preserved large elongated prisms of Mesozoic rocks, which are eroded elsewhere (Jenyon, 1985c). It is worth looking at individual cases to see whether potential source rocks might be present in volumes and at depths that could have enabled them to generate significant quantities of hydrocarbons. However, rim synclines are formed only when withdrawal of salt by rising diapirs is relatively fast. When withdrawal is slower, for instance during growth of diapirs by downbuilding, no rim synclines may be detectable.

Scaled analog modelling that incorporated basement faulting (Weston *et al.*, 1993) has succeeded in reproducing some of the features of Central Graben salt-generated structures. Alsop (1996) used glass beads to model deformation of overburden adjacent to diapirs.

The variety of structural, unconformity, stratigraphic and other subtle traps potentially associated with salt structures and salt dissolution deserve a chapter to themselves. For an expanded discussion of these and other fundamentals associated with salt behaviour, especially as seen on seismic sections, see Jenyon (1986a, 1990).

Fluids associated with salt masses influence the diagenesis of strata above or adjacent to the salt. Dronkert and Remmelts (1996) found halite cementation in Triassic sandstones to be limited to less than 1.5 km from a Zechstein diapir in Netherlands block L-2. For an account of the important structural, mechanical and chemical influences of Zechstein evaporites on chalk reservoirs in the Greater Ekofisk area, see Taylor and Lapré (1987).

In conclusion, the thermal conductivity of halite is more than twice that of most sedimentary rocks (Gussow, 1968) and hence modifies temperature distributions in adjacent strata. Petersen and Lerche (1994) present modelling that quantifies the timing and location of earlier onset of maturation in sediments near the top of salt structures, and the delayed conversion of trapped oil and gas in deeper sediments.

6.11 Practical problems

6.11.1 Identification of cycles

Because of the cyclic repetition of similar depositional facies in the Zechstein, it can be difficult to identify individual cycles (or even to tell how many are present) where the succession is incomplete—for instance, in marginal areas, such as parts of Yorkshire, where evaporites do not separate the carbonates, and in central-basin regions where shales and carbonates are unrecognizable between salts. The difficulty can be severe when only wireline logs and cuttings are available, as is commonly the case (see Section 6.11.2). It is possible that detailed microfossil, palynological and geochemical work may eventually resolve such problems.

The loss of thick salt by dissolution can lead to problems in assigning carbonates above and below to their respective cycles. Where the remaining carbonates are thin (as along the uplifted edge of the Central Graben), it can be very hard to tell from logs and indifferent samples whether one is dealing with a basinal or a marginal facies.

Cycles can sometimes be distinguished in carbonate sequences by the upward-diminishing radioactivity, which reflects their overall regressive character.

Where present, a diverse marine fauna, especially if including brachiopods, crinoids and bryozoans, is indicative of the Z1 carbonates, whereas beds consisting largely of *Calcinema* stems are diagnostic of Z3. Polyhalite is the commonest potassium-bearing mineral in Z2, but is often altered to gypsum in cuttings. Carnallite and red mudstones are more abundant in the basin evaporites of Z3 and Z4 than in earlier cycles. Wireline-log parameters for identifying the commonest evaporite minerals are given in Table 6.1. Quantitative results can be calculated by linear-programming methods (Ford *et al.*, 1974) Logs exploiting the 'photoelectric effect' now facilitate differentiation between limestone and anhydrite/dolomite mixtures. Computerized lithological interpretations have now reached a useful, though not infallible, level of reliability.

Table 6.1 Wireline-log parameters of common evaporite minerals (from various sources).

Mineral	Apparent density (g/cm^3)	Δt (μs/ft)	Apparent limestone neutron porosity (%)	Gamma ray (API units)
Halite (NaCl)	2.03	67	0	0
Anhydrite ($CaSO_4$)	2.98	50	0	0
Gypsum* ($CaSO_4.2H_2O$)	2.35	52.5	49	0
Polyhalite ($K_2SO_4.MGSO_4.2CaSO_4.2H_2O$)	2.79	57.5	15	180
Carnallite ($KCl.MgCl_2.6H_2O$)	1.57	78	65	200
Syvite (KCl)	1.86	74	0	500
Kieserite ($MgSO_4.H_2O$)	2.55	n/a†	n/a	0
Kainite ($4(KCl.MgSO_4).11H_2O$)	2.12	n/a	45	225
Langbeinite ($K_2SO_4.2MgSO_4$)	2.82	52	0	275

*The stability field of gypsum is such that it is rarely encountered at depths greater than 2000 ft, where its place is taken by anhydrite.
†n/a, no figures available.

6.11.2 Drilling problems

Unless saturated salt-inverse oil emulsion- or oil-based drilling muds are used, hole enlargement occurs in halite sections until the mud has become saturated at formation temperature, and even then potash salts continue to dissolve in any of the water-based muds. Undercompacted clays interbedded with the evaporites also wash out. Large cavities result. The cuttings returned consist only of hard, insoluble lithologies, and may remain unrepresentative and mixed for many hundreds of feet of further penetration. The cavities may later release their cuttings whenever hard formations are encountered; for example, red mudstones are often falsely recorded at the top of the Werraanhydrit.

Where basinal carbonates have been buried deeply enough for gas generation, local vuggy porosity and fractures can give rise to high-pressure/low-volume gas shows.

6.12 Hydrocarbon occurrences in Zechstein carbonates (Fig. 6.23)

A number of oil and gas discoveries have been made in Zechstein carbonates in and around the North Sea, and there is production from numerous onshore fields in the Netherlands, Germany and Poland, which provide useful analogues. Most finds on the Continent have been in oncolitic and oolitic rocks of barrier facies and tend to be concentrated near the basinward margin of the shelf in the Southern Salt Basin; the most significant are shown in Fig. 6.23. Gas finds predominate, and the Hauptdolomit is the most common reservoir. Gas has migrated from the Carboniferous to Z1 reservoirs, and to those in Z2 and possibly Z3 where the seat seals provided by the underlying evaporites have been breached. Oil and condensate discoveries have been confined to Z2 and Z3 carbonates, and are thought, from structural and geochemical evidence, to have been sourced by the finer-grained basin and slope facies of those cycles. Some details of oil/source-rock facies correlation in eastern Germany are given by Wehner *et al.* (1993), who also discuss the origin of the high percentages of nitrogen (N_2) and carbon dioxide (CO_2) that commonly occur in associated gases. While Zechstein source rocks have given rise to hydrocarbon accumulations that were worth exploiting onshore on the Continent, it has yet to be demonstrated that the same would apply in the North Sea.

6.12.1 Poland

By 1993, there were 90 gasfields and 12 small oilfields in the Zechstein of Poland. They are located in the Pomeranian areas, adjacent to the Baltic, and the Lubuska and Silesian regions, on the south side of the basin (Fig. 6.24). Total proved reserves are of the order of 300×10^6 m^3 of gas and 23.5×10^6 m^3 of oil. Oil production has ranged up to 730 barrels or, exceptionally, 2190 barrels/day from individual wells; gas flows of several thousand cubic metres a day to a few million cubic metres a day have been reported (Depowski *et al.*, 1981). Oils are naphthenic or paraffinic and sulphurous, with densities of 0.85–0.87 g/cm^3. Gases usually contain hydrogen sulphide and N_2; N_2 content in the west of the country locally reaches 90%. Some details of the petroleum geology and production of the Sulecia and Rybaki oilfields and the Tarchaly gasfield have been described by Depowski and Peryt (1985).

More recently, a promising new play has opened up, exploring isolated carbonate platforms basinward of the main shelf, surrounded and covered by salt (Peryt and Dyjaczyn'ski, 1991). The six drilled have all found gas and/or oil. The platforms have an atoll-like facies zonation; as usual, low-porosity lagoon sediments are surrounded by barrier zones having porosities of around 15%. High pressure gradients are reported (0.016–0.019 MPa/m). The Gorzyca field has a 40 m gas zone and 15–20 m oil zone. The gas has a high N_2 content and 0.5–2% H_2S.

6.12.2 Germany

According to Albertsen (1992), there were over 20 small Zechstein oilfields in production in Germany, mainly in the Pomeranian and south-east Brandenburg areas; there were older finds north of the Harz Mountains and in the Schleswig-Holstein region. Aggregate production, however, was by then only of the order of 1000 barrels of oil per day.

Gas is the principal Zechstein hydrocarbon, being produced near the Netherlands border and in Lower Saxony, where it is generally sour and rich in N_2. Reserves in Lower Saxony were estimated by Hauk *et al.* (1979) to total about

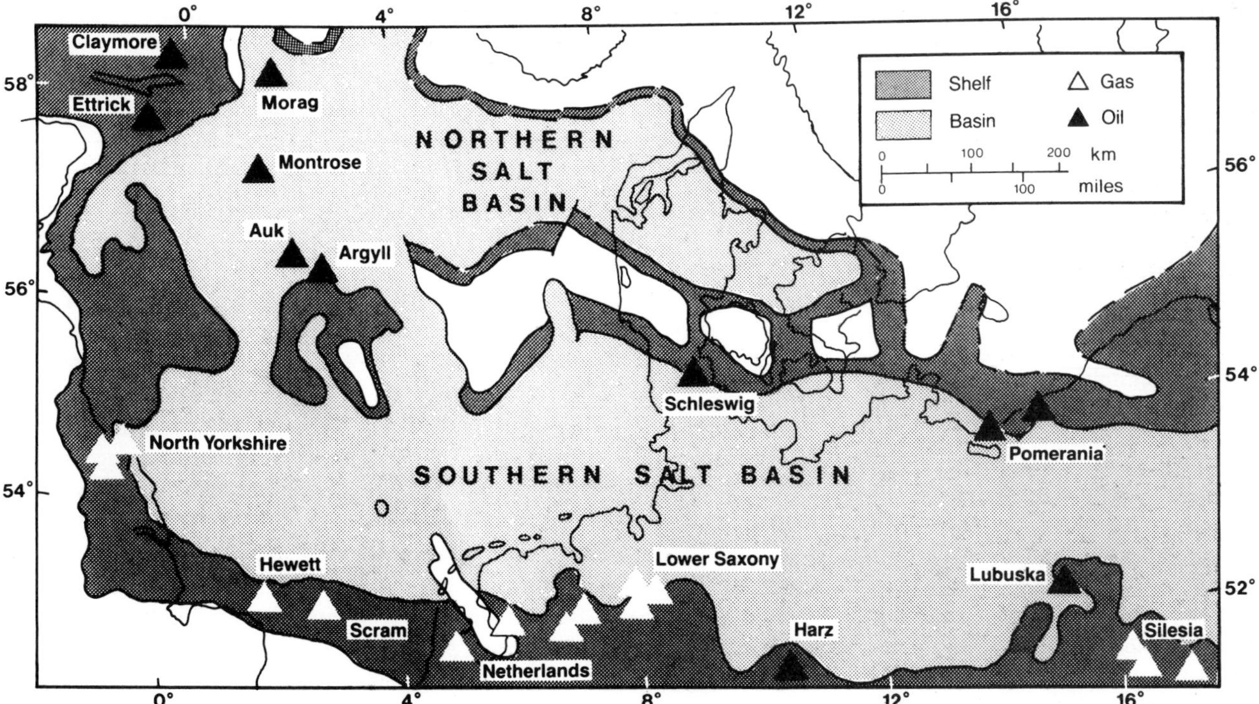

Fig. 6.23 Principal oil and gas occurrences in the Zechstein basin, generalized in mainland Europe.

200×10^9 m^3 (7000×10^9 ft^3); there were some 40 commercial fields, with reserves ranging up to about 20×10^9 m^3 (700×10^9 ft^3). The Hauptdolomit is the main reservoir. Horizontal drilling is being employed to overcome the disadvantage of poor Zechstein reservoir quality. Special-alloy steels are successfully used to combat high H_2S and CO_2 content. More details on the Z2 play are given by Strohmenger (1993b) and Strohmenger *et al.* (1995).

6.12.3 The Netherlands

Across the border in the east Netherlands, gas is produced from the Hauptdolomit in the Drenthe and Twente areas. Facies variations are commonly important in defining the productive areas (Maureau and van Wijhe, 1979; Clark, 1980a). The absence of Rotliegend reservoirs between the Coal Measures and the Zechstein is presumably also a significant factor. The gas contains H_2S, except in the Coevorden (east) and Schoonebeek fields. The barrier facies is again the most porous and permeable. The most productive wells (400 000 m^3/day) invariably drain from fractured reservoirs, whereas non-fractured reservoirs in similar facies may produce as little as 4000 m^3/day, according to van der Baan (1990), who states that dramatic improvements can be achieved through artificial fracturing.

Several gas discoveries have been made in the Z3 carbonates in the western Netherlands, with production in the Bergen field (van Lith, 1983), where the Z2 carbonates

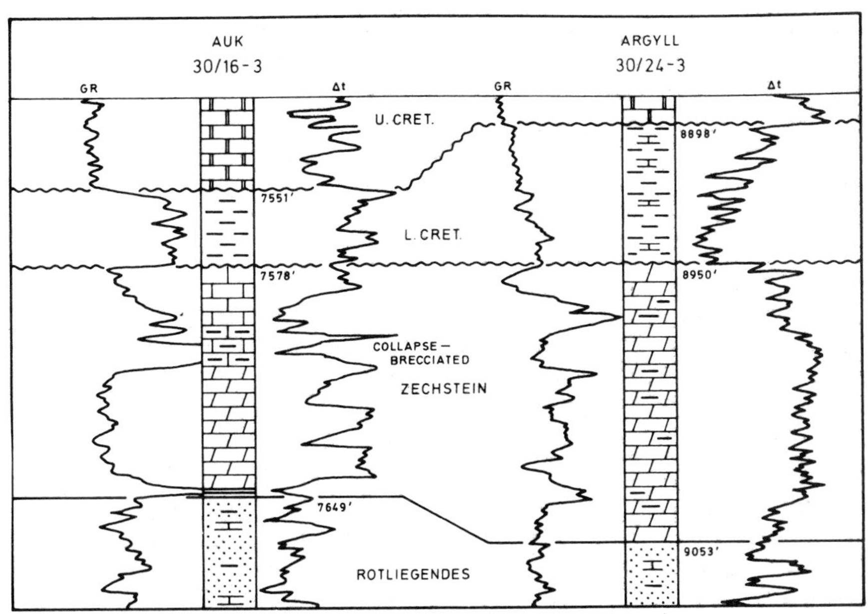

Fig. 6.24 Typical sections through Zechstein reservoirs in the Auk and Argyll oilfields.

are in non-reservoir facies. The gas is generally free from H_2S.

Despite being dwarfed by Rotliegend reserves, the Zechstein apparently remains an active exploration target and discoveries continue to be made (van der Baan, 1990). A combination of 3D seismic and detailed acoustic impedence modelling has led to further discoveries of gas in the Z2 carbonate (Van der Sande *et al.*, 1996). There are now some 20 gas discoveries, with cumulative reserves of more than 40×10^9 m^3 of gas (Casson *et al.*, 1993).

6.12.4 Denmark

Zechstein carbonates have been the main target for exploration in southern Denmark and on the islands of Lolland and Falster in the Baltic (Stemmerik *et al.*, 1987; Thomsen *et al.*, 1987). Average measured porosities range from 3 to 15%, accompanied by permeabilities of < 1–20 mD and locally as much as 700 mD. These variations result from complex diagenesis, especially leaching.

Eight wells drilled, plugged and abandoned in southern Jylland, with shows in four of them, were reported by Stemmerik *et al.* (1987). Gas (mainly N_2) was tested, generally in small amounts, but in the Løgumkloster-1 well at up to 15.4×10^6 ft^3/day, with some condensate. Source rather than reservoir rocks seems to be the problem with this play. Only seven wells had penetrated the full Zechstein sequence in the Northern Salt Basin (Thomsen *et al.*, 1987), and they were not in optimum positions; one had traces of oil. The most promising area for Zechstein exploration is stated to be north of the East North Sea High.

In the Danish Central Trough, according to Damtoft *et al.* (1987), seismic indicates that Zechstein carbonates favourably placed for leaching below a major unconformity may occur along the MNSH, in communication with Upper Jurassic source beds in the Outer Rough Basin.

Depth and some thickness data for the Zechstein in wells drilled up to 1990 in Danish onshore and offshore areas are given by Nielsen and Japsen (1991).

6.12.5 United Kingdom

In the North Sea, Auk and Argyll were the first Zechstein discoveries in the Northern Salt Basin, and are still the only North Sea fields to have had substantial sustained Zechstein oil production, although at Argyll this has now ceased. The structural setting of these fields on the edge of the Central Graben and the relationship of the Zechstein to the probable oil source (the Kimmeridge Clay) are outlined by Glennie (this volume, Chapter 5 and Fig. 5.38). Their favourable production characteristics owe much to uplift and the consequent proximity of the Late Cimmerian Unconformity, creation of which also removed the evaporites of Z2–Z5. This is particularly so in the case of the Auk field (Brennand and van Veen, 1975; see also the log of 30/16-3 in Fig. 6.24), where the reservoir is of unusual type. Extensive vugs and fractures sustain production, despite low matrix porosity, and appear to have originated mainly as a result of leaching of anhydrite laminae (from what Taylor, 1981, believed to be the tight basinal Werraanhydrit) with collapse, brecciation, and complex diagenesis of associated limestone and dolomite layers. Core analysis severely underestimates reservoir quality; measurements on core plugs range from 0.02 to 620 mD, averaging 53 mD, but formation permeabilities of tens of darcies have been calculated from production and build-up tests (Trewin and Bramwell, 1991).

At Argyll (Pennington, 1975; Bifani, 1985; Bifani and Smith, 1985; see also log of 30/24-3 in Fig. 6.24), the situation is slightly different. Whereas Auk may have occupied a basinal position during early Zechstein time, Argyll appears to have been sited on the Z1–Z2 shelf built out from the northern edge of the MNSH. Zechstein production was mainly from the fractured and collapse-brecciated Hauptdolomit, which here contains oolitic beds, which probably already possessed primary porosity, augmented later by dissolution of interbedded lagoonal or sabkha anhydrite layers. Some production was also obtained from reefal Z1 dolomites. The Z1–Z2 dolomites have flowed at rates in excess of 16 000 barrels of oil/day (Robson, 1991).

Reserves in such reservoirs are, clearly, difficult to determine. Indeed, parts of the Zechstein reservoir at Auk have produced more oil than was estimated to have been originally in place (Trewin and Bramwell, 1991) and communication with the Rotliegend reservoir is suspected. Recoverable reserves at Auk and Argyll were each estimated to have been about 90×10^6 barrels of 38° API oil.

Oil has been encountered in Zechstein carbonates in a number of wells in the Moray Firth area. Drillstem tests from Zechstein dolomites below the Jurassic Ettrick oilfield in the released well 20/2-2 produced at rates of 4414 barrels of oil/day through a $\frac{3}{8}''$ choke from a 40 ft interval, and 9198 barrels of oil/day through a $\frac{3}{4}''$ choke from a deeper, 130 ft interval. The complex diagenesis seen in cores has been described by Amiri-Garroussi and Taylor (1987); leaching and fracturing have contributed to the reservoir characteristics.

The setting for the minor Zechstein share of the oil production from the mainly Jurassic Claymore field in the Witch Ground Graben (Maher and Harker, 1987) is another example of a leached Zechstein reservoir. It is said that it can be mapped from seismic sections. The oil is found in the Halibut Bank Formation, which is 34–58 m thick. Freshwater leaching (?pre-Triassic) on the crest of the structure has contributed to porosities of 2–19% and permeabilities of 2–899 mD.

In the Morag field, oil has been found in an unusual type of Zechstein reservoir, capped by Triassic Smith Bank mudstones, below the Maureen oilfield in block 16/29. It is at the top of a salt plug in dolomite, believed by the writer to belong to the fifth or even a sixth Zechstein cycle, possibly of Triassic age. The dolomite is interpreted to have formed in conditions of very shallow water and been periodically exposed, and it has signs of synsedimentary deformation. It was probably fine-grained and tight originally, but fracturing, brecciation, recrystallization and dissolution have led to coarsening and to significant fracture and vug porosity. The changes appear to have resulted from the intermittent rise of the underlying Zechstein salt structure over a long period, which led to exposure during the Middle Jurassic. Although diagenesis played a significant part in the development of its reservoir properties, it is not a salt-dome cap rock of classic US Gulf Coast type. The latter are composed not of dolomite, but of calcite, created by reaction between anhydrite and hydrocarbons. Drill-stem tests in the 16/29a-A1 and A2 wells produced at rates ranging from 537 to 6250 barrels of oil/day.

In the Southern Salt Basin, gas, originally subcommercial, was found in the Hauptdolomit and Plattendolomit beneath the Hewett Triassic gasfield in blocks 48/29 and 48/30, with shows in some nearby blocks. The Main Hewett field (Cooke-Yarborough, 1991, 1994; Fig. 7.6) now produces from the Z1 Zechsteinkalk, which has a net/gross pay ratio of 0.66, average porosity of 5–8% and permeability of < 1 mD. The trap is structural, and was probably charged towards the end of the Jurassic or in the Early Cretaceous (Cooke-Yarborough, 1991, 1994). According to Southwood *et al.* (1993), who consider that the structure developed through Cretaceous–Early Tertiary compression, the dominant fracture trend across the Hewett field is north-north-east–south-south-west to north–south, and it probably formed during Late Jurassic to Early Cretaceous extension. Van Alstine and Butterworth (1993) claim that palaeomagnetic core orientation can accurately predict fracture trends in the Southern North Sea.

Gas and condensate have also been found in block 53/4, designated the 'Scram' field. Some excitement was once aroused by shows of oil tested at up to 2000 barrels of oil/day in 48/22-1, but an appraisal well found the equivalent interval to be plugged by anhydrite (Clark, 1986).

Onshore in England, gas was obtained for a limited period at Lockton and Eskdale in North Yorkshire from fractured Z2 and Z3 carbonates of relatively tight facies. According to Scott and Colter (1987), wells at Malton in Yorkshire and nearby Kirby Misperton were reported to have tested gas in 1985. These locations probably lie in the oolite shoal facies of the Z2 Kirkham Abbey Formation. The same authors also outline the problems encountered in the earlier ill-fated Lockton gasfield. In early 1989, Kelt UK announce a low of 9.5×10^6 ft^3/day through a 44/64″ choke from their Low Marishes well in the same area. Gas and condensate have been tested not far offshore in blocks 41/20, 41/24 and 41/25. Small discoveries continue to be made onshore.

6.13 Hydrocarbon potential of the Zechstein—conclusions

Considering the large number of wells that have been drilled through the Zechstein, particularly in the Southern North Sea and onshore UK, there have been few commercial discoveries in it. This has been responsible for a common perception that the Zechstein is a high-risk objective.

Yet it is easy to discover reasons for the lack of success. Most of the wells in the south have been drilled to deeper objectives and have not been ideally sited for Zechstein targets. Onshore, there are few structures significantly interrupting the regional basinward dip, so favouring fluid communication with outcrop. The best Zechstein reservoirs are in the Hauptdolomit and the Zechsteinkalk—but it can be difficult to recognize the tops of these formations on conventional seismic sections. Offshore, the principal belts of Rotliegend reservoirs lie either north or south of the Hauptdolomit shelf-edge barrier, so that wells pass through either tight lagoonal facies or tight basin facies in Z2. There are few penetrations far enough south to encounter the most porous parts of the Zechsteinkalk—but, significantly, the few include the main Hewett field.

Since whatever primary porosity there was in the lagoonal facies is generally tightly cemented by anhydrite and halite, and since it is quite challenging to imagine how carbonates surrounded by evaporite formations could fail to be cemented in this way, it is easy to fall into the trap of assuming that poor reservoir characteristics are the norm. This is a very partial view. The Hauptdolomit in wells on the MNSH commonly has excellent porosity and, while it might be suspected that this is due to Cimmerian weathering, high porosity exists in some wells where the formation is still covered by thick salt (38/24-1 for instance; Amiri-Garroussi and Taylor, 1992). Much more remains to be learnt about the preservation or re-creation of porosity in Zechstein carbonates. The example just cited and the overpressured hydrocarbon-yielding atolls in Poland described by Peryt and Dyjaczyn'ski (1991) suggest that formation fluids can circulate through what would, at first sight, appear to be closed systems.

Carbonate/evaporite depositional facies are particularly sensitive to small changes in relative sea level. The contemporaneous highs, which favoured high-energy facies, do not in general coincide with the much bolder structures produced by later tectonics, and so are difficult to detect, especially on vintage seismic.

Work carried out in the Hewett gasfield challenges the generally accepted logic of searching for high-porosity reservoirs. Since production there is mainly from fractures, efforts have been made to determine fracture patterns and to establish which fractures contribute most to gas production, by the integrated use of cores, dual laterolog, formation microscanner (FMS) gamma-ray spectrometry and spinner wireline tools (Cooke-Yarborough, 1994). Counter-intuitively, zones with high gamma-ray readings and high estimated water saturation turned out to have the highest productivity. Cooke-Yarborough (1994) suggests that this is because low porosity favours brittle fracture, clay laminae are susceptible to acid fracturing and uranyl ions are commonly precipitated in conduits. These observations could lead to reappraisal of many Zechstein wells formerly passed over as not worth testing.

Zechstein carbonates can be sourced externally from below or above, or from within. The generalized model for Zechstein deposition illustrated in Figs 6.7–6.9 and described in Section 6.3 carries the possibility that, on burial, hydrocarbons may be generated in the fine-grained euxinic basin and lower-slope sediments and migrate up-dip into the porous barrier zone, where they are sealed above and, in some cases, up-dip by evaporites. This seems to be confirmed by the known occurrences of Zechstein oil in Hauptdolomit reservoirs on the Continent, and there should be more discoveries to be made using this approach. However, there are two obvious snags with this model.

Firstly, the thinness of the Zechstein source beds sets a limit on the volume of hydrocarbons available, unless drainage from a very large area can be envisaged. Secondly, the same evaporites that could preserve reservoired hydrocarbons may have impeded their entrance from external sources. The evaporites also have the opportunity to oc-

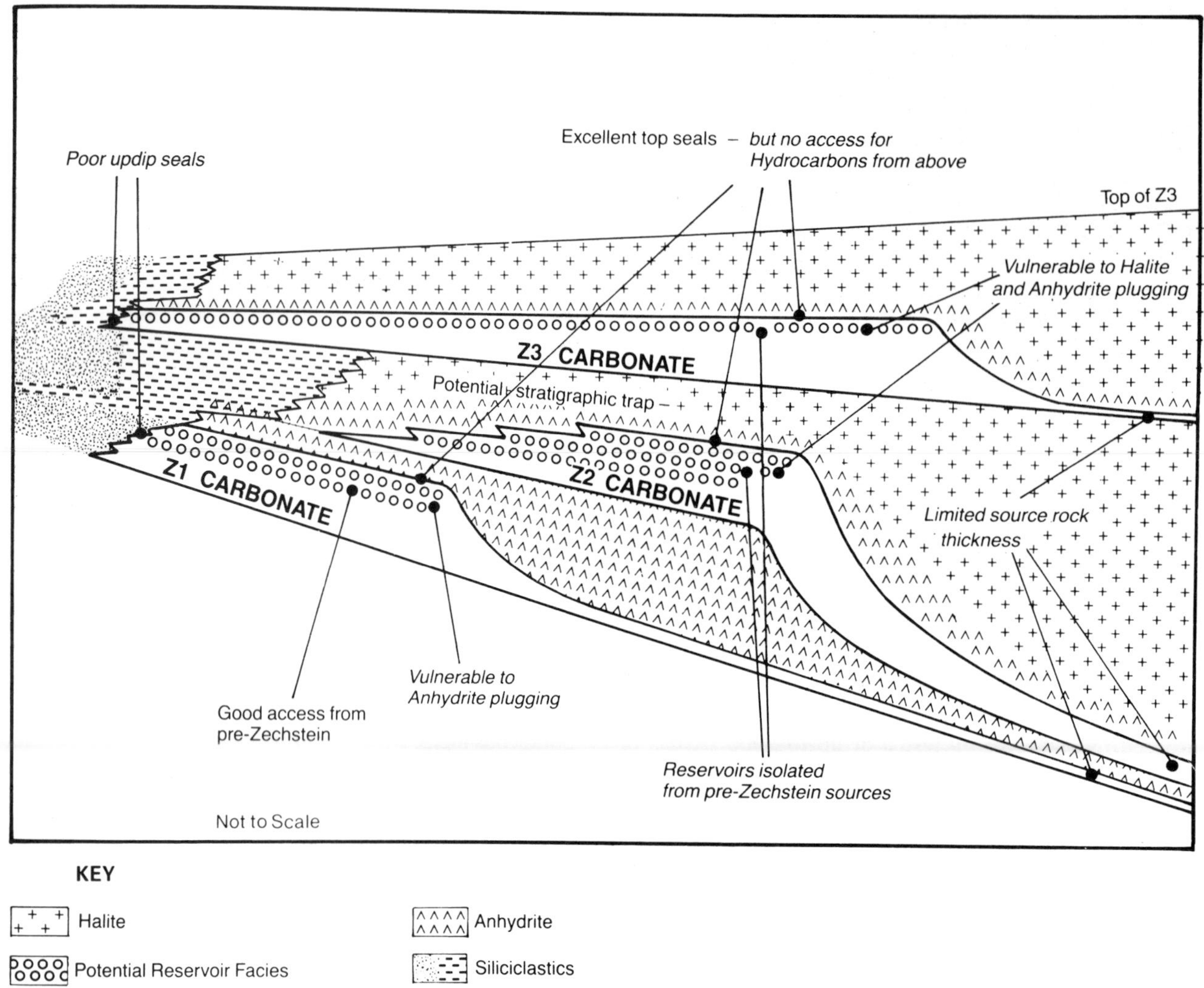

Fig. 6.25 Generalized diagram illustrating the advantages and disadvantages of each of the Zechstein carbonates as potential reservoirs. Not to scale.

clude the pores, notwithstanding the conclusions arrived at earlier. Porosity seems to have been extremely variable, even without cementation by evaporites; according to Peryt (1985), the main reason for the unpredictable reservoir properties of the Zechstein is the very irregular pattern of early meteoric-related diagenesis. All these factors operate to a different extent in the carbonates of different cycles; the chief positive and negative aspects for Z1, Z2 and Z3 are illustrated diagrammatically in Fig. 6.25. From this it may be deduced, for instance, that in the Southern Salt Basin, where the most likely mature source rocks lie below the Zechstein, the best primary reservoirs may be expected in the Z1 shelf carbonates, provided there is an adequate seal or structural reversal of dip in a direction away from the basin centre.

Future potential can therefore be suggested in the southern gas basin in the Zechsteinkalk along the depositional strike from the Hewett field and might be sought more determinedly in the shelf-edge barrier zone of the Hauptdolomit. It is not difficult to delineate these fairways in broad terms from existing wells (e.g. see Fig. 6.13; Clark, 1986) or with the help of seismic. Smith (1989) gives details of the thickness and facies of Zechstein units in the UK onshore area; the general position of potential stratigraphic traps along the extensive up-dip margin of the Hauptdolomit equivalent can readily be deduced from his maps (Smith, 1989, Fig. 10). A problem with this play is that there may be a rather broad transitional zone which is neither a good seal nor a good reservoir.

The search need not be confined to the shelves. There appear to be possibilities in carbonates on the slope (Clark, 1980b, 1986; van der Baan, 1990; Peryt, 1992) and in displaced carbonates at the foot of the slope (Amiri-Garroussi and Taylor, 1992). In all these plays modern high-resolution and preferably 3D seismic is likely to improve the chance of success. Intrabasin highs of atoll shape imaged on seismic on the northern margin of the Silver Pit Basin have resulted in local deposition of types of carbonates that may have reservoir potential. Analogues in Germany and the Netherlands suggest a narrow barrier zone with high reservoir potential.

On the MNSH and further north, pre-Zechstein sources become speculative and, over large areas, highly improbable. Northward-increasing proportions of land-derived material in the basin and slope carbonates may improve the viability of Zechstein source rocks, but sourcing from the Jurassic is the most hopeful option. All the most promising North Sea Zechstein discoveries have been in this category,

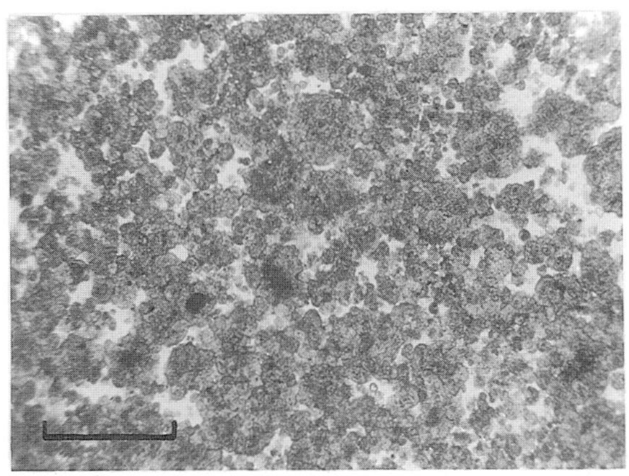

A

Fig. 6.26 Photomicrographs, all in plane-polarized light and same magnification. Bar scale = 0.47 mm. A. Excellent intergrain and intercrystalline porosity in dolomitized grainstone of the Z1 Cadeby Formation near outcrop, probably regained by dissolution of early evaporite cements; Askern water boring, 20.4 m (67 ft).

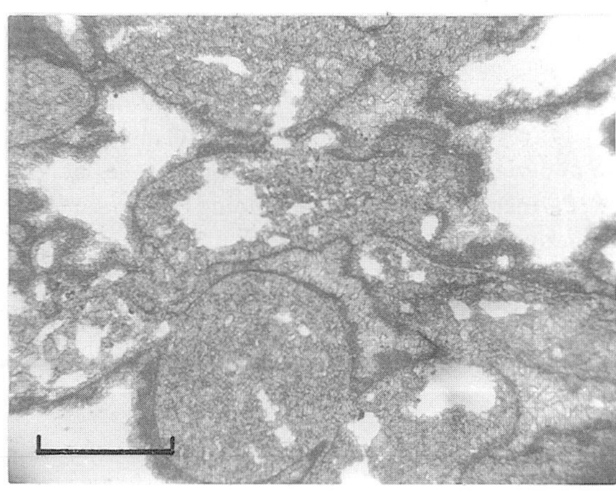

B

B. Dolomite oncoliths with partly leached centres and intergrain porosity entirely plugged by clear halite; Hauptdolomit, 47/18-1, 2125 m (6972 ft).

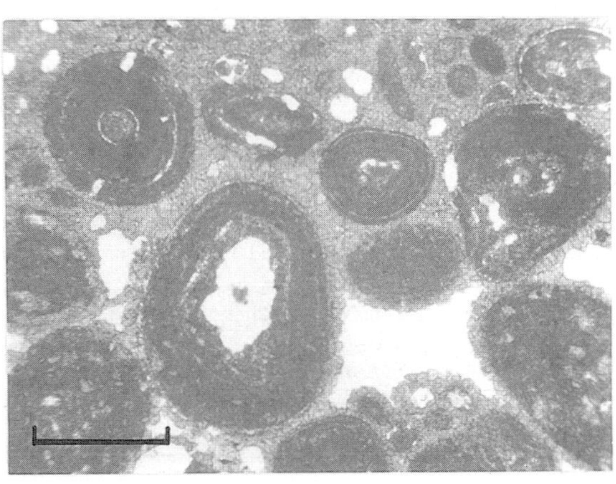

C

C. Dolomite pisoliths and ooliths with intergrain and leached-grain pores plugged by clear halite, same sample as B, illustrating varieties of grain type and size.

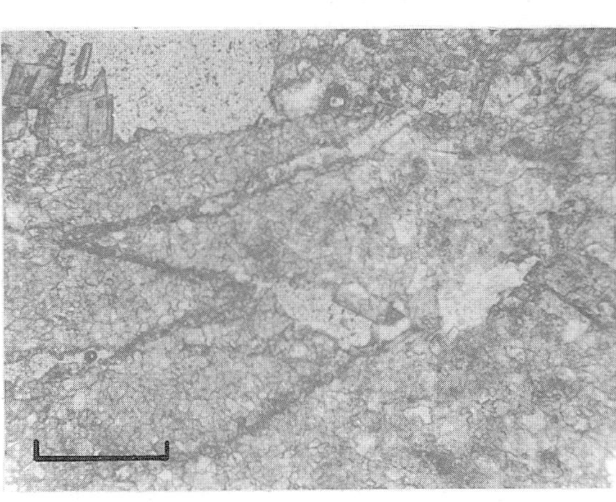

D

D. Fracture porosity in collapsed Zechstein dolomite, Ettrick field, 20/2-3, 3920.1 m (12 861 ft 4 in.).

with Auk and Argyll as the outstanding examples. It is noteworthy that the reservoirs of this high-productivity class were often in rather unpromising primary facies, and owe their high permeabilities and, in some instances, their porosities to leaching, collapse brecciation or fracturing. Their discovery has often owed much to serendipity.

Far from being a disadvantage, the original presence of evaporites becomes an asset in these secondary reservoirs. The association may be of three main kinds, depending on whether the evaporites were predominantly below the carbonates, interbedded or interlaminated with them or actually within the carbonate beds between or within grains, replacing grains or matrix or filling vugs and fractures. Leaching of each of these produces a different fabric, with different reservoir properties. Another important distinction is whether salt or anhydrite was present, for, whereas thick salt can readily be removed by circulating groundwater to depths of at least 900 m (3000 ft), massive anhydrite is, as a rule, dissolved with much greater difficulty, unless quite close to the surface, and its removal is strongly dependent on joints and fractures (Cooper, 1986).

Dissolution of evaporites from below Zechstein carbonates has been described by Smith (1970b, 1972b, 1975). It results in displacement and fragmentation on a spectacular scale as the dissolution slope retreats down-dip. Vertical breccia pipes may be created as residual caves collapse. Secondary voids may be blocked by washed-in sediment and by cements, particularly near the base of the carbonate formation, but joints and fractures remaining open can provide exceptionally high formation permeability.

Whether this is backed up by enough matrix porosity to sustain production depends on the porosity of the original carbonate. The original high porosities and permeabilities of grainstones can be restored by leaching of early intergranular evaporite cements. Figure 6.19A from the Cadeby Formation near outcrop is probably one such example. The possibilities where intragrain porosity is also involved, particularly where large pisoliths and oncoliths are present, as in Fig. 6.26 (B,C), may well be imagined.

The most promising secondary-reservoir conditions of all could result from leaching of intraformational, interbedded and underlying evaporites together. The potential for this combination is provided by the back-barrier facies of the Hauptdolomit and, to a lesser extent, the Plattendolomit.

First, second and third Zechstein cycle carbonate reservoirs charged with Jurassic oil are to be sought on the graben margins of the Central North Sea, on the intragraben highs and in parts of the Moray Firth. Locating areas with just the right amount of Cimmerian erosion is likely to be exacting, but may become easier as seismic quality improves. Dissolution slopes, which indicate the limits of salt removal, can be recognized on seismic (see Fig. 6.19C; Jenyon and Taylor, 1983, 1987). Leaching of Zechstein dolomites may be expected where salt has been dissolved from beneath grounded Triassic pods.

Finally, secondary reservoirs of the Morag type, perched above diapirs at the top of the Zechstein, are unlikely to be unique to that field. Others may occur over the deepest parts of the Northern Salt Basin (see Fig. 6.14). Beyond the North Sea there exists the possibility of a major new play in a reef barrier believed by Taylor (1980) to exist between Greenland and Norway. Further east, in Russian waters, this is already a prime producer. So far, it has been traced as far westward on seismic as 4°E along the edge of the Fennoscandian continental shelf (Ivanova, 1997) but is on trend with the zone predicated.

6.14 Acknowledgements

My thanks to Ken Glennie for seismic illustrations, reduced well logs, for suggesting improvements and additions to the text and much more. I am indebted to Phillips Petroleum Company UK Ltd for permission to include information on the Morag Zechstein dolomite, and to H.A. van Adrichem Boogaert and M.C. Geluk, of the Geological Survey of the Netherlands, for making available papers in preparation. I thank Malcolm Jenyon for permission to reproduce the seismic examples of Fig. 6.19 and for many helpful discussions. Denys Smith gains my thanks for suggesting improvements to the text and illustrations of an earlier edition, as well as for long-term exchange of information and views. I owe much to past colleagues at V.C. Illing & Partners for assistance, including core logging and data-gathering.

This chapter is dedicated inadequately to my late wife, Patricia Mary, without whose unstinting support from 1956 to 1987 it could never have existed.

6.15 Key references

Cameron, T.D.J. (1993) *Lithostratigraphic Nomenclature of the UK North Sea,* Vol. 4. *Triassic, Permian and Pre-Permian of the Central and Northern North Sea.* British Geological Survey.

Clark, D.N. (1980) The diagenesis of Zechstein carbonate sediments. In: Füchtbauer, H. and Peryt, T.M. (eds) *The Zechstein Basin with Emphasis on Carbonate Sequences.* Contr. Sedimentology No. 9, Schweitzerbart'sche Verlagsbuchhandlung, Stuttgart, pp. 167–203.

Ge, H., Jackson, M.P.A. and Vendeville, B.C. (1997) Kinematics and dynamics of salt tectonics driven by progradation. *AAPG Bulletin* **81**(3), 398–423.

Jenyon, M.K. and Cresswell, P.M. (1987) The Southern Zechstein Salt Basin of the British North Sea, as observed in regional seismic traverses. In: Brooks, J. and Glennie, K.W. (eds) *Petroleum Geology of North-West Europe,* Vol. 1. Graham and Trotman, London, pp. 277–92.

Kiersnowski, H., Paul, J., Peryt, T.M. and Smith D.B. (1995) Paleogeography and sedimentary history of the Southern Permian Basin in Europe. In: Scholle, P.A., Peryt, T.M. and Ulmer-Scholle, D.J. *The Permian of Northern Pangea,* Vol. 2. Springer-Verlag, Berlin, pp. 119–36.

Penge, J., Taylor, B. and Munns, J., in press. Rift-raft tectonics: examples of gravitational sliding structures from the Zechstein basins of northwest Europe. In: Boldy, S.A.R. and Fleet, A.J. (eds) *Petroleum Geology of Northwest Europe.* Proceedings of the Fifth conference.

Schreiber, B.C. (1987) Arid shorelines and evaporites. In: Reading, H.G. (ed.) *Sedimentary Environments and Facies.* Blackwell Scientific Publications, Oxford, pp. 189–228.

Schultz-Ela, D.D. and Jackson, M.P.A. (1996) Relation of subsalt structures to suprasalt structures during extension. *AAPG Bulletin* **80**(12), 1896–1924.

Smith D.B. (1980) The evolution of the English Zechstein basin. In: Füchtbauer, H. and Peryt, T.M. (eds) *The Zechstein Basin with Emphasis on Carbonate Sequences.* Contr. Sedimentology No. 9, Schweitzerbart'sche Verlagsbuchhandlung, Stuttgard, pp. 7–34.

Stewart, S.A. and Clark, J.A., in press. Impact of salt on the structure of Central North Sea hydrocarbon fairways. In: Boldy, S.A.R. and Fleet, A.J. (eds) *Petroleum Geology of Northwest Europe.* Proceedings of the Fifth conference.

Strohmenger, C., Antonini, M., Jäger, G., Rockenbauch, K. and Strauss, C. (1996) Zechstein 2 Carbonate Reservoir facies distribution in relation to Zechstein sequence stratigraphy (Upper Permian, Northwest Germany): an integrated approach. *Bull. Centres Rech. Explor-Prod. Elf Aquitaine.* **20**(1), 1–35.

Taylor, J.C.M. and Colter, V.S. (1975) Zechstein of the English sector of the Southern North Sea Basin. In: Woodland, A.W. (ed.) *Petroleum and the Continental Shelf of Northwest Europe*, Vol. 1. *Geology.* Applied Science Publishers, Barking, pp. 249–63.

Tucker, M.E. (1991) Sequence stratigraphy of carbonate–evaporite basins: models and application to the Upper Permian (Zechstein) of northeast England and adjoining North Sea. *J. Geol. Soc. Lond.* **148**, 1019–36.

Vendeville, B.C. and Jackson, M.P.A. (1992a) The rise of diapirs during thin-skinned extension. *Marine Petrol. Geol.* **9**, 331–53.

Vendeville, B.C. and Jackson, M.P.A. (1992b) The fall of diapirs during thin-skinned extension. *Marine Petrol. Geol.* **9**, 354–71.

7 Triassic

M.J. FISHER & D.C. MUDGE

7.1 Introduction

The Triassic spanned some 42 million years (Ma), ~248–206 Ma before present (BP) (Gradstein *et al.*, 1995), encompassing a significant period of Earth history. The breakup of Pangaea had begun with crustal thinning and rifting along the axis of the incipient Atlantic and the westward extension of Tethys. This was already establishing a new structural framework in north-west Europe, which would control deposition throughout the Mesozoic. In the North Sea, this extensional phase modified the structural pattern inherited from the Permian, and many Palaeozoic fault zones were reactivated as extensional features in conjunction with Permian–?early Triassic rifting (Fig. 7.1). In the Northern and Central North Sea, although the Triassic fault pattern has a dominantly north–south orientation, structural interaction with Palaeozoic structural trends generated a complex set of multidirectional basins. To the east of the North Sea basins, the fault-bounded Polish–Danish Trough was sited on the west-north-west–east-south-east orientated Tornquist–Teisseyre lineament.

Triassic rifting is most apparent in the basins in the west of Britain, and it is widely assumed that the Northern and Central North Sea Basins were similarly affected (Ziegler, 1978, 1982a, 1990; Badley *et al.*, 1988; Scott and Rosendahl, 1989; Thorne and Watts, 1989; Marsden *et al.*, 1990; Roberts *et al.*, 1990, 1993, 1995; Coward, 1995). Roberts *et al.* (1995) and Platt (1995) suggest that the Unst Basin and the contiguous Magnus half-graben represent the remnant of the 'Triassic Viking Graben', which was linked with the West Shetland basins. Largely because of overprinting from subsequent deformation, however, it has proved difficult to date or quantify precisely any Triassic element of late Permian–early Triassic lithospheric extension, block rotation and graben infilling in the North Sea. Roberts *et al.* (1990, 1993) calculate Triassic extension to be the equivalent of up to 70% of the total Mesozoic extension in the Viking and Central Grabens. In contrast, White (1990) and White and Latin (1993) maintain that the Triassic stretching episode is poorly constrained and is of minor importance in the evolution of the Viking and Central Grabens. The significance of Triassic extension in the East Shetland Basin remains unresolved, although Roberts *et al.* (1995) suggest that the scale of Jurassic and Triassic extension was comparable.

However, convincing evidence of Triassic rifting is available from some marginal areas (e.g. the Unst Basin/Magnus half-graben, the Egersund Basin, Sele High, the Horn Graben and the Norwegian–Danish Basin). In the Horda Platform (Lervik *et al.*, 1990; Marsden *et al.*, 1990; Steel and Ryseth, 1990; Coward, 1995; Roberts *et al.*, 1995), wedge-shaped packages of early Triassic sediments overlie a tilted basement topography and display sedimentary growth against the Horda/Øygarden Fault Zone. In the basins in the Danish Central Trough bordering the Mid North Sea High, Triassic sediments are generally conformable on the Zechstein but there is pronounced angular discordance between the Rotliegend and the Zechstein. Based on a series of investigations in the northern part of the Tail End Graben, Cartwright (1991) has proposed a non-rotational late Permian–early Triassic rifting phase.

In the Southern North Sea Basin, the pattern of subsidence continued from the late Permian (see Chapters 5 and 6, this volume), with the addition of some Triassic extensional faulting. The basal Triassic units are uniform in thickness throughout the basin, but there is evidence of increasing differential subsidence towards the end of the Triassic. Faulting of the Dutch Central Graben appears to have commenced in the late Triassic, and Kooi *et al.* (1989) record similar rift-related subsidence in the Broad Fourteens and West Netherlands basins.

To date, volcanic activity accompanying this restricted phase of rifting has been recorded only where the Central Graben breaches the Mid North Sea High and in south-west Norway (Ziegler, 1990b). Deeper sequences in postulated Triassic rift basins have rarely been penetrated. Away from areas of rift-related subsidence, the major depocentres reflect a phase of post-rift passive thermal subsidence modified by halokinesis in the Southern and Central North Sea Basins as Permian basin margins became progressively overstepped.

Any tectonic events should also be viewed in the context of global sea-level fluctuations, which, by modifying base levels, may have had a contributory effect in controlling sedimentation in the North Sea Basins. Ormaasen *et al.* (1980) and Haq *et al.* (1987) postulate a gradual rise of global sea level during the Triassic, with regressions in the Lower Triassic, the Middle Triassic and the early and late Upper Triassic. Schopf (1974) and Forney (1975) attribute this variability in sea level to tectono-eustatic factors, and Ziegler (1982a) further emphasizes the influence that contemporaneous volumetric changes in the midocean ridges would have exerted. Short-term variability in relative sea level is more difficult to ascribe. While acknowledging the absence of firm evidence of glacial deposits in the northern hemisphere, Ziegler (1990b) also stresses the role that glacio-eustatic sea-level changes may have played.

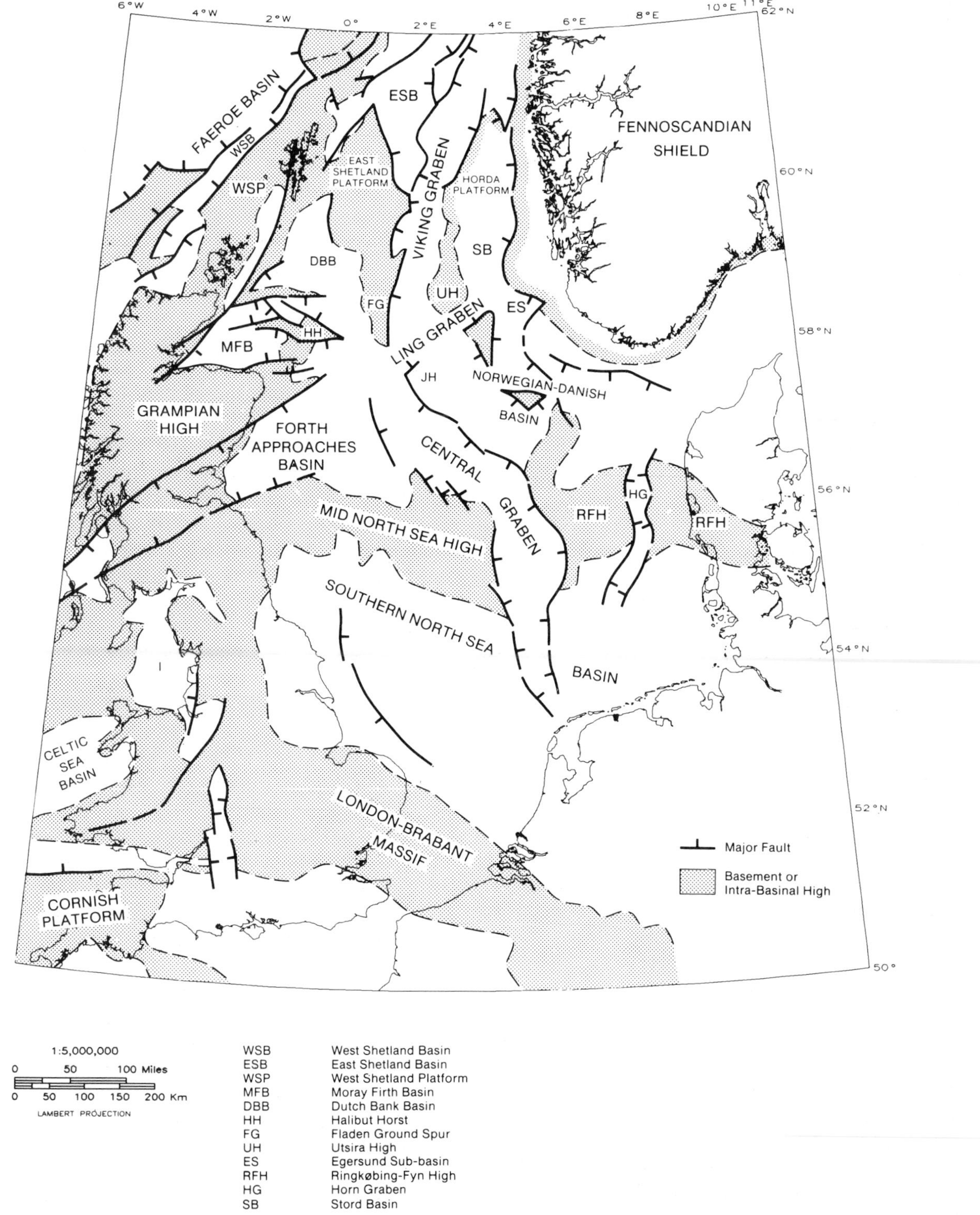

Fig. 7.1 Triassic structural elements.

The close of the Permian saw the end of widespread marine sedimentation and a return to dominantly non-marine depositional environments. With the withdrawal of the Zechstein seas in the North Sea basins, the already established continental and paralic environments were extended from the periphery into the centre of the basins. Wherever the basal Triassic overlies uppermost Permian, it does so with apparent conformity but often with an abrupt change of facies.

In the Southern North Sea Basin, the regional regression, accompanied by faulting and uplift at the Permian–Triassic transition, rejuvenated the physiographic profile but did not significantly modify the basin geometry. As a consequence, within the overall trend from coarse-grained Early Triassic sediments to fine-grained Late Triassic sediments, there is considerable lateral uniformity of facies. Halokinesis played a relatively minor role in the stratigraphic evolution of the basin. Towards the end of the Triassic, however, there is evidence that facies distribution was controlled locally by the development of halokinetically induced topography.

North of the Mid North Sea–Ringkøbing–Fyn High, regional tectonism throughout the Triassic may have played a much more important role in controlling sedimentation. In some subbasins, a number of major tectonic episodes, together with more restricted local events, are considered to have resulted in a succession of upward-coarsening cycles, with poor lateral facies continuity (Jakobsson *et al.*, 1980). Another significant factor in basin development in the Central North Sea was halokinesis. The Permian halites were initially mobilized during the Early Triassic by regional tectonism and later by overburden pressure (Ziegler, W.H., 1975; Hodgson *et al.*, 1992). Coward (1995) attributes the great variability in thickness of Triassic sediments on either side of the Central Graben to decoupling of the Triassic from the Zechstein salts. This led to reactive diapirism between thin-skinned rifts or to 'pillow-like' buckle folding above the salt. Whatever the mechanism, salt movement generated a pattern of major depocentres, separated by halokinetically induced highs parallel to the graben margins.

The Northern North Sea Basin lay to the north of the Permian salt basins, and Triassic sedimentation was not affected by halokinesis.

In all the North Sea basins, Triassic sediments are dominantly red beds, including representatives of alluvial-fan, fluvial, aeolian, sabkha, lacustrine and shallow-marine facies. The relative abundance and the interrelationship of the various facies are largely dependent on the tectonic setting. In tectonically active fault-bounded basins, alluvial fans merge with playa lakes or other basin-centre environments. In more stable basins, with lower physiographic profiles, the marginal alluvial fans may be separated from the basin-centre environments by broad flood plains (Hardie *et al.*, 1978). The Southern North Sea Basin is dominated by fine-grained clastics and evaporites, with coarse-grained clastics restricted to the Early Triassic. In the Central and Northern basins, coarse clastics predominate, especially in the Late Triassic, and evaporites are largely absent.

The mechanism responsible for the coloration of Triassic red beds has been the source of much speculation. Turner (1980) has reviewed the current hypotheses and favours a process similar to that described by Walker *et al.* (1978) in desert sandy alluvium. Briefly, this requires the post-depositional degradation of ferromagnesian minerals, supplemented by detrital ferric hydroxides to form the haematite pigment. Clay mica is the dominant component of Triassic fine-grained siliciclastic sediments and may sometimes be the sole component of the clay assemblage. Jeans *et al.* (1994) suggest that these micas originated in coeval desert soils, which were then eroded and redeposited as fine-grained detritus.

Triassic sediments undoubtedly accumulated in relatively arid environments. Clemmensen (1979) reviewed the available data and concluded that the North Sea basins lay in low northern palaeolatitudes (approximately 20°N) in a central trade-wind zone. Precipitation was probably seasonal and restricted to cloudbursts, which initiated short-lived sheet floods that drained into large, shallow, saline playa lakes or inland seas. Conditions varied in the three major basins, as fluvial sediments are much more common throughout the Triassic in the northern basins than in the Southern North Sea Basin. The dominant wind direction, deduced from thin intercalations of aeolian sands with fine-grained lacustrine sediments at Helgoland, was south-easterly. This is partially supported by Mader (1982), who records south-easterly to south-westerly wind directions in the more extensive Middle Buntsandstein aeolian sands of the Eifel area. Further refinement has been provided by van der Zwan and Spaak (1992), who have used climatic modelling to infer that monsoonal conditions were established in the Southern North Sea area during the Lower Triassic, reaching a peak towards the end of the Scythian.

Parrish *et al.* (1982), in a worldwide palaeoclimatic evaluation, provide evidence that precipitation in the North Sea Basin would have been considerably higher in the later Triassic than in the Early Triassic. As the dispersal of the Pangean plates moved the North Sea area further northwards towards 30°N (see Fig. 2.3), floral and faunal asemblages suggest a progressive amelioration of the climate. There is also widespread evidence of extensive vegetation at these palaeolatitudes throughout the Anisian to Carnian and Rhaetian. The associated higher rainfall may still have been subject to pronounced seasonality. However, the variations in palaeofloral assemblages ascribed to extreme climatic fluctuations by van der Zwan and Spaak (1992) may in part reflect the effects of base-level fluctuations on coexisting upland and lowland floras (Fisher, 1985).

Correlation of Triassic red-bed sequences has always presented problems. In north-west Europe, the diagnostic ammonite and bivalve assemblages of the Boreal and Tethyan marine provinces are absent. Similarly, the vertebrate remains used for correlation in continental facies are rare. Traditionally, lithostratigraphic correlations have been favoured and, although these can be applied with a high degree of confidence across the Southern North Sea Basin, interbasinal lithostratigraphic correlation is less satisfactory north of the Mid North Sea–Ringkøbing–Fyn High. Here, palynological zonations established in the type sections of the Tethyan and Boreal provinces are the most reliable means of correlation (Geiger and Hopping, 1968; Lervik *et al.*, 1989; van der Zwan and Spaak, 1992; Goldsmith *et al.*, 1995). One development that has had encouraging results in correlating monotonous and unfossiliferous mudstone sequences has been the use of diagnostic detrital and neoformed clay assemblages (Fisher and Jeans, 1982). This technique, in conjunction with heavy-mineral analysis, is now being successfully applied to correlations in the North Sea basins (Jeans *et al.*, 1993; Mange-Rajetsky, 1995).

Geophysical and petrophysical characterization of the Triassic is relatively straightforward in the Southern North Sea Basin but becomes more conjectural north of the Mid North Sea–Ringkøbing–Fyn High. In the Southern North Sea Basin, Rhys (1974) has illustrated the petrophysical characteristics of the lithostratigraphic units. Cameron (1993a) has amplified the petrophysical criteria for the recognition of all major lithostratigraphic units in the Central and Northern North Sea, originally documented by Deegan and Scull (1977) although here, definition of the base Triassic may be conjectural. Day *et al.* (1981) recognize a strong 'top Triassic' seismic event, which they correlate with the top of the 'Keuper shale' or Rhaetian sandstones, although, particularly on the Mid North Sea High, thin arenaceous Upper Jurassic sequences cannot be distinguished from the underlying eroded Triassic. Within the Triassic the 'top Bacton' reflector equates with the base of the Röt Halite and is of variable quality. This reflector is of considerable importance because it can be used, in conjunction with sonic log-derived velocities of the Bunter Shale Formation, to determine regional maximum palaeoburial depths for the Lower Triassic and Permian reservoirs (van Wijhe *et al.*, 1980; Hillis, 1995b). The base of the Triassic is equated with the 'top Zechstein' reflector in areas where the Zechstein facies is represented and has not been disrupted by halokinetics.

There have been a number of attempts to interpret the North Sea Triassic succession in terms of stratigraphic sequences (Gabrielsen *et al.*, 1990; van der Zwan and Spaak, 1992; Smith *et al.*, 1993; Steel, 1993). Much of this succession comprises non-marine facies that are not directly correlative with coeval marine facies. Shanley and McCabe (1994) have documented the problems of applying sequence-stratigraphic concepts to continental strata, but an additional limiting factor is the lack of a precise Triassic stratigraphic framework for boundary definition. Triassic palynofloral zones average 8.6 Ma, compared with 1.7 and 1.8 Ma for Jurassic and Palaeogene palynofloral zones, respectively, and this limits the confidence in interbasinal correlations. Detailed analysis of both Jurassic and Palaeogene sequences has also emphasized the relative significance of regional or local tectonic rather than global eustatic control of relative sea-level changes. As a consequence, it follows that attempts to correlate stratigraphic sequences based on global eustatic or glacio-eustatic events to the non-marine Triassic succession (van der Zwan and Spaak, 1992) may be severely constrained. Confident definition of stratigraphic sequences may be limited to local, intrabasinal events until a more precise and reliable stratigraphic framework is established.

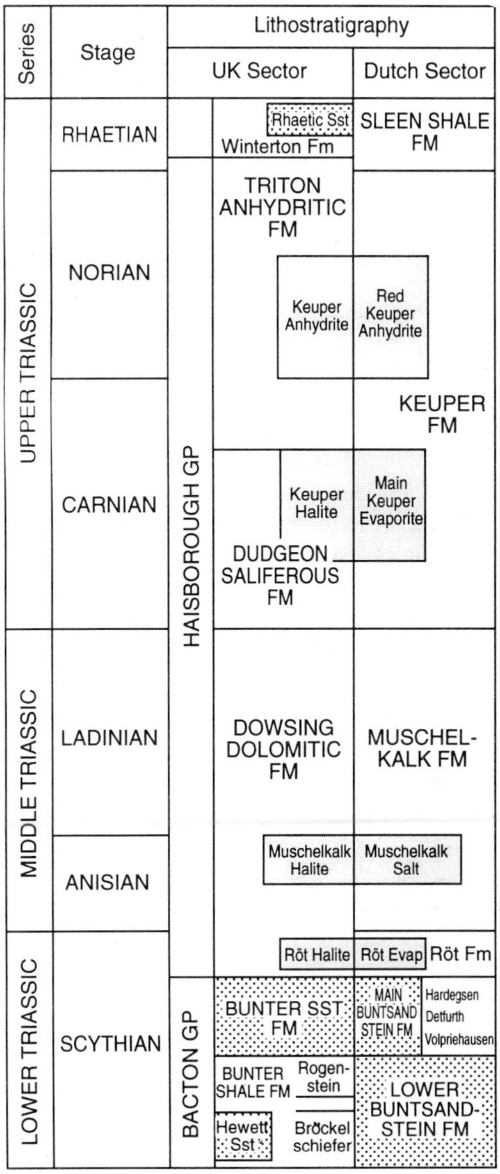

Series	Stage	Lithostratigraphy		
			UK Sector	Dutch Sector
UPPER TRIASSIC	RHAETIAN		Rhaetic Sst; Winterton Fm	SLEEN SHALE FM
	NORIAN	HAISBOROUGH GP	TRITON ANHYDRITIC FM; Keuper Anhydrite	Red Keuper Anhydrite
	CARNIAN		Keuper Halite; DUDGEON SALIFEROUS FM	KEUPER FM; Main Keuper Evaporite
MIDDLE TRIASSIC	LADINIAN		DOWSING DOLOMITIC FM	MUSCHEL-KALK FM
	ANISIAN		Muschelkalk Halite; Röt Halite	Muschelkalk Salt; Röt Evap; Röt Fm
LOWER TRIASSIC	SCYTHIAN	BACTON GP	BUNTER SST FM; BUNTER SHALE FM; Rogen-stein; Hewett Sst; Bröckel schiefer	MAIN BUNTSAND STEIN FM (Hardegsen, Detfurth, Volpriehausen); LOWER BUNTSAND-STEIN FM

Fig. 7.2 Triassic lithostratigraphy, Southern North Sea.

7.2 The Southern North Sea Basin

The Triassic Southern North Sea Basin occupied roughly the same geographical position as that of the Southern Permian Basin (Fig. 7.1). Residual positive structural features were the London–Brabant Massif to the south, the Mid North Sea–Ringkøbing–Fyn High to the north and the Pennine High to the west. The Central and Horn Grabens, which dissected the Mid North Sea–Ringkøbing–Fyn High, were initiated in the Permian. Rapid infill of these features continued throughout the Triassic, with up to 2000 m of sediments accumulating in the southern part of the Central Graben and up to 7000 m in the Horn Graben. Within the basin itself, the Sole Pit, Off Holland Low, Broad Fourteens and West Netherlands subbasins were major tectonically controlled Triassic depocentres.

Because of the economic importance of the underlying Permian and, to a lesser extent, the Triassic itself, this area has provided more information on the Triassic than any of the other North Sea basins. It was also the subject of two key studies: one, by Geiger and Hopping (1968), established a basin-wide stratigraphic framework and the other, by Brennand (1975), outlined the main facies distribution.

The sedimentary succession was originally described by Rhys (1974) and has been revised by Cameron *et al.* (1992) (Fig. 7.2). The Triassic succession falls naturally into three major groups:

1 The Bacton Group, representing a phase of largely coarse-grained clastic deposition, comprising red sandstones, shales and mudstones.

2 The Haisborough Group, a largely fine-grained clastic and evaporite sequence, with marked cyclicity.

3 The Penarth Group, which reflects the marine transgression that marked the passage from the Triassic to the Jurassic.

The typical log character of these rock units is illustrated in Fig. 7.3.

7.2.1 The Bacton Group

The contraction of the Zechstein Sea resulted in the progradation of already established marginal clastic sedimentary environments, typically represented by red,

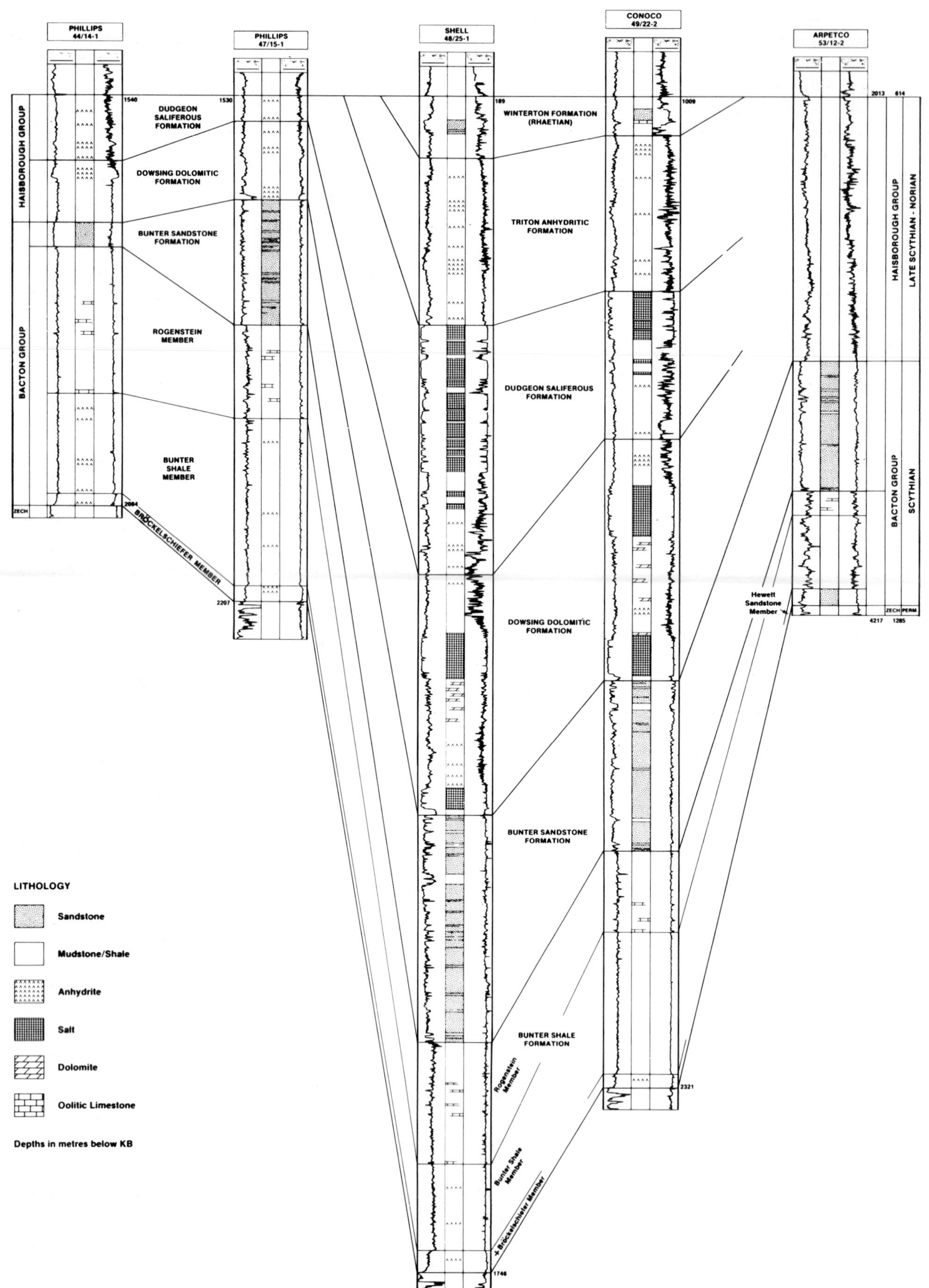

Fig. 7.3 Southern North Sea–Triassic well correlation.

argillaceous sandstone or siltstone, further into the basin. The Permian–Triassic boundary is traditionally associated with the distinct facies break where the basal Bunter Shale Formation overlies the basinal carbonates, anhydrites or halites of the Zechstein. However, at the margins of the basin, the basal member of the Bunter Shale Formation, the Bröckelshiefer Member, may be time-equivalent to Zechstein evaporites (Kiersnowski *et al.*, 1995). Fuglewicz (1987) estimates that this transition occurred during the mid to late Tartarian, predating the base Triassic by some 3 Ma. Local facies variations at the south-western basin margin include a poorly sorted, fine- to medium-grained quartzose sandstone recognized as the Hewett Sandstone Member (Cumming and Wyndham, 1975). This and smaller but similar sandy intercalations were probably conduited through a feeder system active in the Permian, which resulted from local faulting and uplift of the London- Brabant Massif to the south. The Triassic age of this basal sequence has also been questioned and Geluk *et al.* (1996) equate the Hewett Sandstone with the fourth Zechstein cycle (Z4) Sandstone of the Netherlands offshore.

The old Permian Basin was markedly featureless and correlation across the basin is facilitated by considerable lateral facies continuity (Fig. 7.4), which is reflected in the uniform petrophysical log characteristics (see Fig. 7.3). The Bunter Shale Formation displays considerable consistency in thickness and lithology. It is typically developed as an anhydritic red-brown mudstone, with some minor greenish shales. In the upper part of the formation, occasional calcareous intercalations, with distinctive ferruginous ooliths, characterize the Rogenstein Member. The Bunter Shale Formation reflects the maximum extent of an early Triassic playa lake or inland sea, which occupied the major part of the basin, where the Bunter Shale facies accumulated as lacustrine or flood-plain deposits. The 'lake' margin fluctuated considerably, as is evidenced by the often rapid alternations of lacustrine with sheet flood, aeolian or fluvial sediments approaching the basin periphery. Towards the UK basin margin, the Rogenstein Member disappears and the Bunter Shale Formation coarsens and thins rapidly to form the Amethyst Member (Cameron *et al.*, 1992). Onshore, in eastern England, the contemporaneous sediments were deposited largely within fluvial channels or as sheet floods; aeolian and alluvial-fan deposits are relatively rare. In the Netherlands, the Bunter Shale equivalent in the Roer Valley Graben is developed as massive sandstones and conglomerates.

With time, the marginal clastic sedimentary environments prograded into the centre of the basin so that, overall, the Bacton Group is represented by a gross upward-coarsening unit. This progradation was accomplished by a series of progressive encroachments into the basin and, as a consequence, the boundary between the Bunter Shale Formation and the succeeding Bunter Sandstone Formation is markedly diachronous.

The Bunter Sandstone Formation reflects further rejuvenation of the basin, probably brought about by a lowering of erosional base level, combined with increased rainfall in the sediment source areas. In western areas, the formation is represented by a relatively homogeneous upward-coarsening sand complex. Bifani (1986) described a depositional model to accommodate the facies variations displayed in the Esmond, Forbes and Gordon fields (Quadrant 43), reflecting the range of depositional environments common to the area. Within a coalescing fan complex, he recognizes braided-stream channel-fill, overbank or channel-margin, sheet-flood and lacustrine sediments. There is occasional evidence of cyclicity, but the individual cycles are poorly defined and may reflect restricted, local factors rather than widespread, regional events.

On the south-eastern margins, as in the west, the proximal facies of the Bunter Sandstone Formation is undifferentiated sandstone. In the basin centre and extending south-eastwards into the northern end of the Roer Valley Graben, however, the Bunter Sandstone Formation comprises up to three depositional cycles, each with a regressive basal sandstone unit. These have been correlated with the Volpriehausen, Detfurth and Hardegsen sequences of

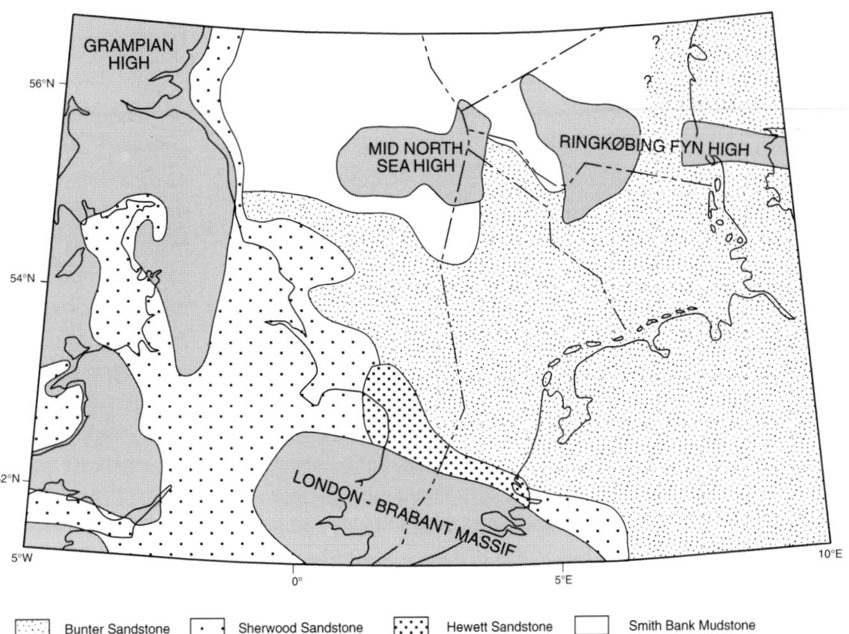

Fig. 7.4 Lower Triassic lithofacies.

Germany (see Fig. 7.2). Each cycle represents a succession of northward-prograding alluvial-fan/braided-fluvial complexes, separated by transgressive argillaceous flood-plain or lacustrine sediments. The Roer Valley Graben was the main fluvial conduit and the source areas have been identified as the Massif Central (van der Zwan and Spaak, 1992) or the Vosges (Geluk *et al.*, 1996). The generating mechanism of these cycles is not entirely clear, although erosional unconformities terminating each cycle suggest an element of tectonic rejuvenation. Van der Zwan and Spaak (1992), however, propose that climatic and sea-level fluctuations linked with Milankovitch cycles resulted in variations in the erosional effect of monsoonal activity on the source areas. A fourth cyclical unit, the Solling, is strictly part of the overlying Röt Formation and is largely restricted to the easternmost periphery of the basin.

The three major cycles vary in thickness, due to differential subsidence in individual basins. However, they all show evidence of contemporaneous thinning towards palaeohighs, and their areal distribution was further restricted by subsequent uplift and erosion. As a result, the Volpriehausen Sandstone, which was the most widely distributed and best developed of the basal sandstone units, remains the most attractive reservoir target in this formation. The dominant facies are of fluvial origin but aeolian sands, typically sheet sands, are more common north of the West Netherlands Basin (Ames and Farfan, 1996). In the Dutch F15-A gasfield, four lithofacies—dune, interdune, fluvial and lacustrine—have been recognized in the Volpriehausen Sandstone Member (Fontaine *et al.*, 1993). Here, the Volpriehausen is 37.5–41 m thick and comprises five lithological units. These exhibit upward-coarsening profiles, reflecting increasing aridity as the successive depositional environments grade from playa lake to dunes.

The northern margin of the basin, the Mid North Sea–Ringkøbing–Fyn High, acted as only a minor source of sediment. As a consequence, all the sandstone sequences thin significantly northwards from the southern margin, across the basin centre and towards the northern part of the basin. In the southern Horn Graben, however, the three classic upward-fining cycles of the Bacton Group are still well developed (Best *et al.*, 1983).

The major tectonic event that followed deposition of the Hardegsen cycle is of regional significance in continental European sections. Within the sheet-sand complex of the western and southern periphery and in the basinal areas, it is less discernible, although Geiger and Hopping (1968), following continental practice, identified the Hardegsen disconformity in a number of well sections. Perhaps more importantly, in terms of the evolution of the basin, fault activity associated with this event may have initiated the halokinesis that had a major effect in controlling later sedimentation in the North Sea Basins (Ziegler, W.H., 1975; Best *et al.*, 1983; Glennie, 1986). Zeigler, W.H. (1975) suggested that the Hardegsen disconformity could represent the Permian–Triassic boundary. Palynological evidence contradicts this view, as an assemblage from a presumed Volpriehausen equivalent has been demonstrated to be of Triassic, probably Nammalian (Dienerian), age (Fisher, 1979).

The effect of the Hardegsen event had been to generate topographic features that effectively subdivided the area into minor basins and highs. The period that followed saw the erosion of the highs and the accumulation of dominantly fine-grained clastics, whose facies and thickness were dependent both on proximity to the sediment source areas and on the local post-Hardegsen relief. It was in this environment that the youngest cycle of the Bacton Group, including the Solling sandstone equivalent, was deposited.

7.2.2 The Haisborough Group

The Haisborough Group (see Fig. 7.2.) represents a period of greater tectonic stability. Marine conditions were re-established in the basin and the lack of coarse clastic sediments and pronounced lateral facies continuity again reflect the overall low relief. Sedimentation was predominantly in distal flood-plain environments, alternating with coastal sabkha or shallow-marine environments.

Although there is evidence to suggest that erosion of the Mid North Sea–Ringkøbing–Fyn High had proceeded sufficiently to allow a partial sedimentary cover, it still provided enough relief to form the northern limit of the main basinal deposition. The subbasins on the northern margins of the basin, the Thirty Seven Basin, the southern Central Graben and the southern Horn Graben—all show evidence of having been in a distal location relative to any clastic source area during deposition of the Haisborough Group. It was only during the early Triassic history of the Horn Graben that there is evidence of the Mid North Sea–Ringkøbing–Fyn High having been a significant source of clastic material (Best *et al.*, 1983; Michelsen and Andersen, 1983; Cartwright, 1990).

The oldest member of the Dowsing Dolomitic Formation, the Röt Halite Member, has a thin basal transgressive unit, the Röt Clay, which is of considerable lateral extent. The overlying sediments include red and red-brown shales, with two variably developed halites. One halite, the Main Röt Evaporite, is extensively developed and, in the eastern part of the basin, a second, younger Röt Halite is also present. The thickness of the Röt Halite Member is somewhat variable within the basin. These variations result, in part, from thicker salt deposition in the residual depressions generated by the Hardegsen event. In this context, the origin of the salt is also of interest, because it has been postulated that it may be derived from leached and reprecipitated Zechstein halites exposed by the Hardegsen movements (Ziegler, W.H., 1975). Both the Röt transgression and the succeeding transgression, represented by the Muschelkalk Halite Member, entered the basin from the east via the Polish Trough. Holser and Wilgus (1981) cite the high bromide content of the Röt and Muschelkalk halites as an indication of marine origin. As these indicators suggest increased marine influence eastwards, towards the inferred incursion route, it is unlikely that Permian salts have made a significant contribution to either the Röt or Muschelkalk halites.

The typical shelly limestone facies of the classic German Muschelkalk stratotype sections is not represented in the North Sea. In the east of the basin, interbedded dolomites and silty shales sandwich a thick halite, but, towards the western part of the basin, the dolomites become discontin-

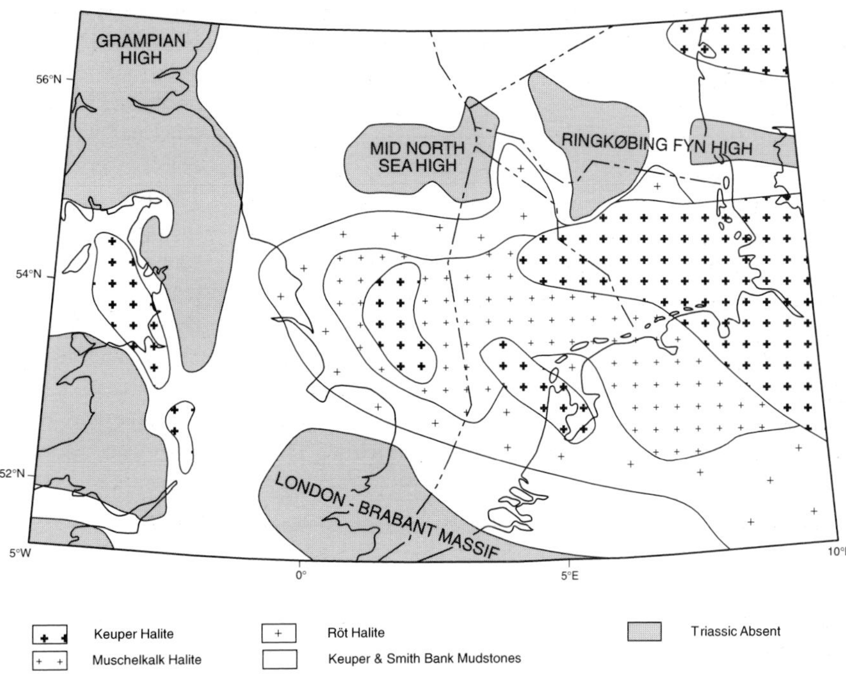

Fig. 7.5 Middle Triassic lithofacies.

uous and are eventually replaced by anhydritic mudstones and silty clays. On the northern margins of the basin, the Muschelkalk halite thins rapidly towards the Mid North Sea–Ringkøbing–Fyn High. It is represented in the Dansk Nordsø U-1 well but in the Nordsee R-1 and S-1 wells is largely replaced by a dolomitic–anhydritic unit with very little salt and in the Dansk Nordsø V-1 well by marlstone and calcareous grey claystone. Although the Röt and Muschelkalk halites are of comparable thickness, typically 70–150 m, the Röt halite basin was areally more extensive (Fig. 7.5). The Röt halite is present in both the Dansk Nordsø U-1 and Nordsee S-1 wells (Jacobsen, 1982; Best *et al.*, 1983).

The regional regression that followed the Muschelkalk transgression temporarily severed communication with Tethys. The initial result was a progressive reduction of salinity, with the establishment of brackish floras and faunas (Kozur and Reinhardt, 1969). With the eventual contraction of the Muschelkalk sea, clastic sedimentation resumed. Coarse-grained sediments are common only at the basin margins, and elsewhere uniformly fine-grained sequences of red shales and mudstones predominate. These sediments represent the Dudgeon Saliferous Formation, which, in the upper part, contains thick halite beds interbedded with red-brown mudstones. The Keuper Halite Member, the most widespread of these halites, is also the thickest, exceeding 300 m in some locations. It is very variable in thickness and this, together with the lateral impersistence of the thinner halites, reflects the degree to which the post-Hardegsen topography of swells and lows had been maintained or even accentuated by continued halokinesis. The Keuper halites therefore suggest numerous, scattered saline lakes, rather than the more continuous bodies of water envisaged for the Röt and Muschelkalk seas. The Keuper halites also differ from the Röt and Muschelkalk halites in that they may result from evaporation of incoming surface water or of near-surface groundwater and not from a major marine transgression. The route of the previous Triassic marine incursions via the Polish Trough was disrupted throughout the Carnian and Norian. Subsequent communications with Tethys were established intermittently only from the south.

The group terminates with the Triton Anhydrite Formation. This comprises variegated green and red mudstones with common anhydrite and rare dolomite beds. The anhydrite beds become extremely persistent in the upper part of the formation, where they constitute the Keuper Anhydrite Member.

Deposition of the Haisborough Group has been considered to have been largely continental. More recent investigations (e.g. Jeans, 1978) suggest that deposition occurred in subaqueous environments, commonly in shallow, hypersaline water. With additional evidence from the Triassic ichnofacies (Pollard, 1981), it is reasonable to postulate that the Southern North Sea Basin was never totally drained and that the Haisborough Group and, possibly, the younger part of the Bacton Group accumulated largely under marine or quasi-marine conditions. These studies also reveal considerable cyclicity in terms of flora, fauna, sedimentary structure, fabric and detailed mineralogy.

7.2.3 The Penarth Group

In the latest Triassic, the occurrence of diverse marine Rhaetian fossil assemblages and the disappearance of typically hypersaline clay-mineral asemblages in the basal grey shales of the Penarth Group indicate that more normal marine conditions had resumed. Hallam and El Shaarawy (1982) deduced from the faunal evidence that initially the Rhaetian Sea exhibited reduced salinities (25–30‰). It was not until early Liassic times that stenohaline faunas were successfully re-established in the North Sea Basin. The group represents a significant transgression, however, with sandstones and shales onlapping old structural highs. A sandy facies, which generally comprises interbedded silty sandstones and thin shales, is well developed offshore and is a distinctive correlation marker. It is generally overlain by a dark grey shale, with restricted marine microplankton assemblages, which passes conformably into the overlying Liassic sediments.

7.2.4 Economic geology

The Southern North Sea Basin is an established gas-producing area. The major discoveries are in Permian reservoirs, but 11—the Hewett, Dotty, F15-A, K13, P6, P15, P18, Caister B, Esmond, Forbes and Gordon fields—have Triassic reservoirs. The Hewett Gasfield (Figs 7.6 and 7.7) has been described by Cumming and Wyndham (1975) and Cooke-Yarborough (1991). It is located on a north-west–south-east trending, fault-bounded anticline. There are two Triassic reservoirs: the lower reservoir, the Hewett Sand Member, has 60 m gross pay with 21.4% porosity (ϕ) and 1310 millidarcies (mD) permeability (k); the upper reservoir, in the Bunter Sandstone Formation, has 98 m gross pay with $\phi = 25.7\%$ and $k = 474$ mD. Ultimate recoverable reserves are estimated at 105×10^9 m^3 (3.7×10^{12}ft^3). The two reservoirs contain gas of different compositions, the gas in the Bunter Sandstone Formation having a significantly higher hydrogen sulphide and nitrogen content. The adjacent field, Dotty, also contains gas in the Hewett Sand Member and Bunter Sandstone Formation.

The relative insignificance of the Triassic as an exploration objective in this prolific gas-producing area results from the lack of communication between the Carboniferous source rocks and the Triassic reservoirs. Although the Hewett Sand Member is only of local significance, the Bunter Sandstone Formation is an attractive target. Notwithstanding the effects of early diagenetic dolomite and of halite and anhydrite cements preferentially destroying the permeability of lithofacies with the best primary reservoir properties (Dronkert and Remmelts, 1996; Purvis and Okkerman 1996) it displays good reservoir characteristics (ϕ = 20–25%, k = 100–700 mD) over the major part of the basin. In the UK sector, it forms large structures—for

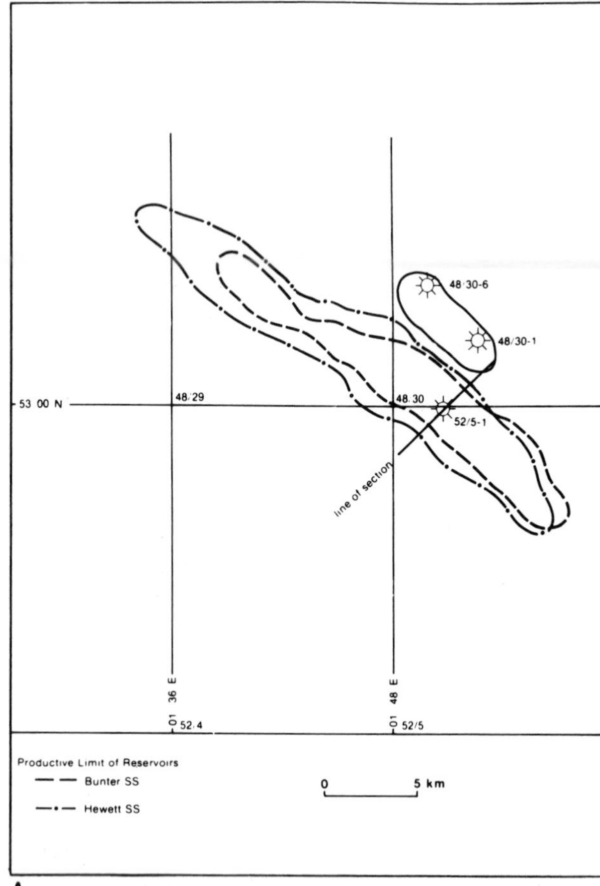

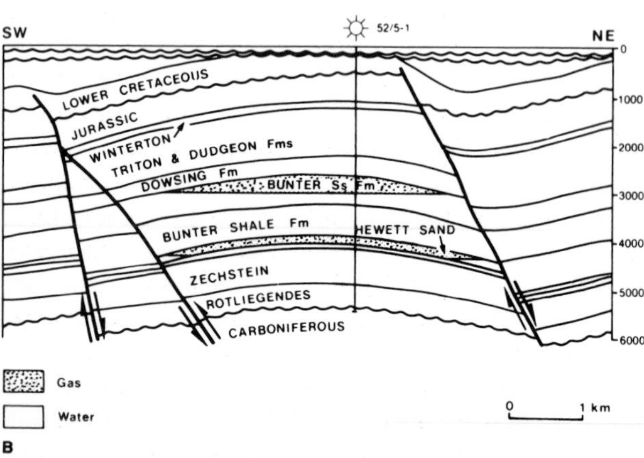

Fig. 7.6 Map (A) and cross-section (B) of the Hewett Field (after Cumming and Wyndham, 1975).

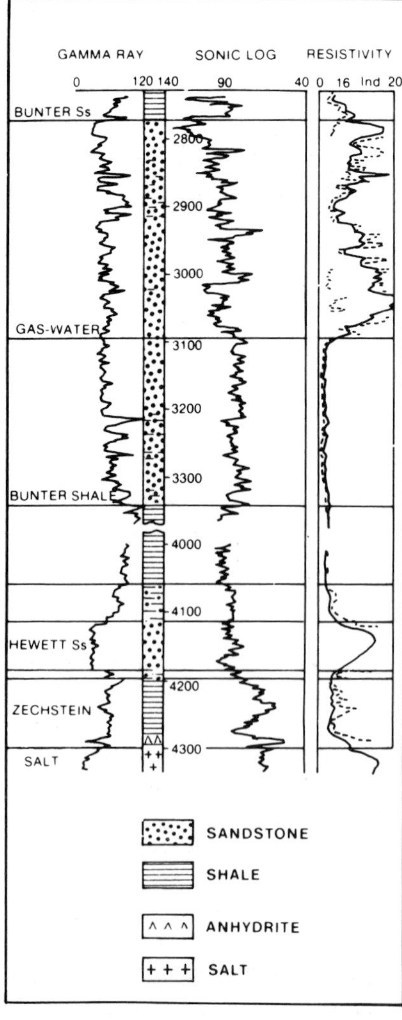

Fig. 7.7 Generalized well log of the Hewett Field (after Cumming and Wyndham, 1975).

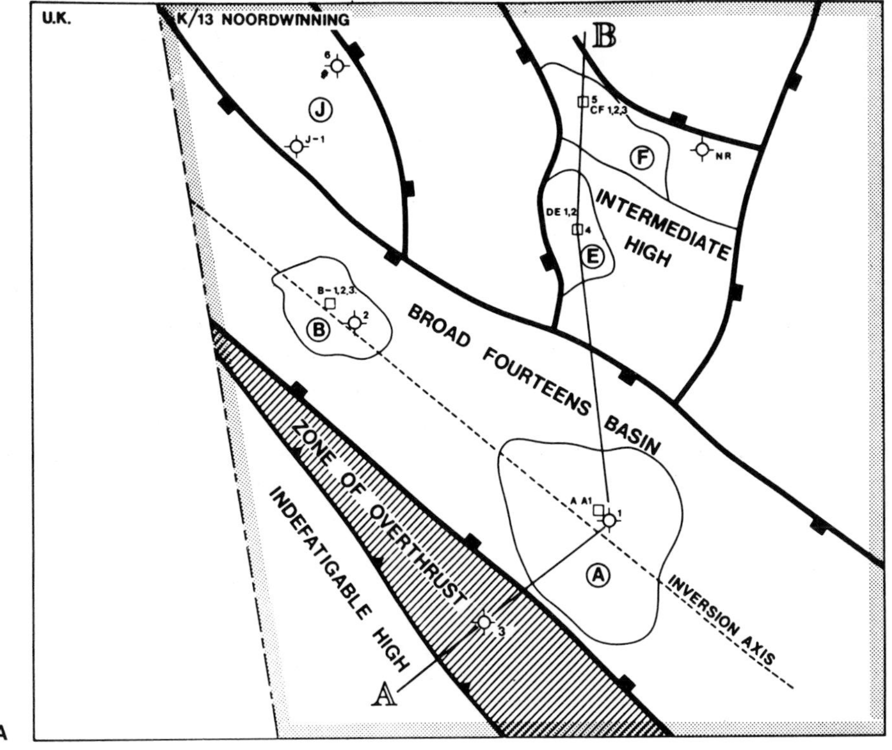

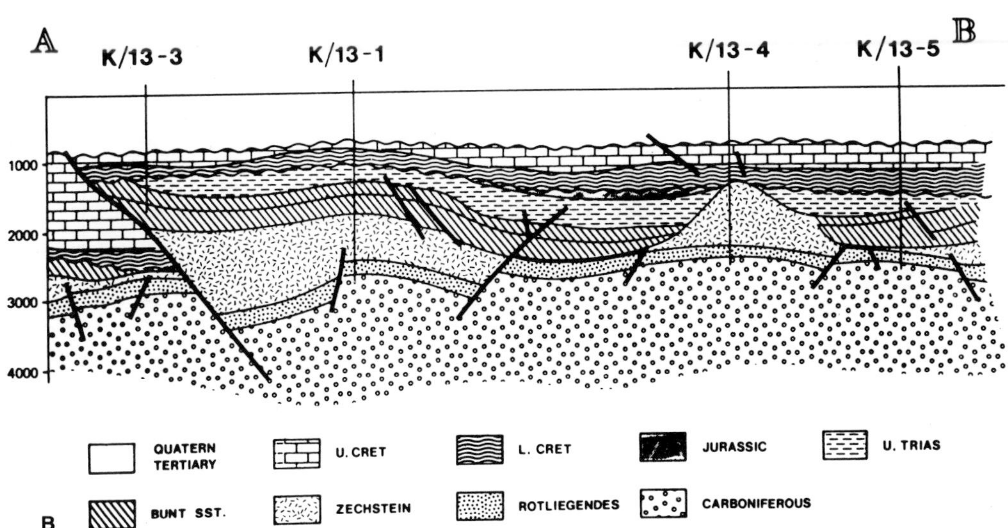

Fig. 7.8 Map (A) and cross-section (B) of K/13 (after Roos and Smits, 1983).

example, in Quadrants 43 and 44—relatively few of which contain significant volumes of gas. The migration barrier is the Zechstein evaporites, particularly the Z2 halite, which, even when the underlying Rotliegend reservoirs have been severely faulted, have flowed and sealed the fractures. Breaching of the seal was only accomplished by either considerable diapiric withdrawal of salt or, in the Dotty and Hewett fields on the periphery of the Zechstein Evaporite Basin, failure of thin evaporites to seal the fault conduits.

The F15-A gasfield is located on a terrace separated from the Dutch Central Graben and Schill Grund High by Zechstein salt walls. Gas reserves of 10×10^9 m^3 (0.37×10^{12} ft^3) are significantly overpressured and are reservoired in the Volpriehausen Sandstone Member in a southward-plunging turtle-back anticlinal trap (Fontaine *et al.*, 1993).

The Dutch offshore block K13 contains four gasfields: two are reservoired in the Rotliegend and two, K13-A and B, in the Bunter Sandstone Formation (Fig. 7.8A). Roos and Smits (1983) consider that gas generated from Westphalian source rocks in the Late Jurassic to Early Cretaceous was originally reservoired in the Rotliegend structures and sealed by Zechstein evaporites. Late Cretaceous inversion movements breached the seal, which had already been thinned by diapiric withdrawal, and gas remigrated from the Permian reservoirs along large reverse faults into the Bunter Sandstone Formation (Fig. 7.8B).

Salt withdrawal was also a significant factor in the formation of the Caister B (44/23a), Esmond (43/13a), Forbes (43/8a) and Gordon (43/15a: 43/20a) fields. The Bunter structures are located over salt 'swells', which have adjacent associated areas of thinned or absent Zechstein

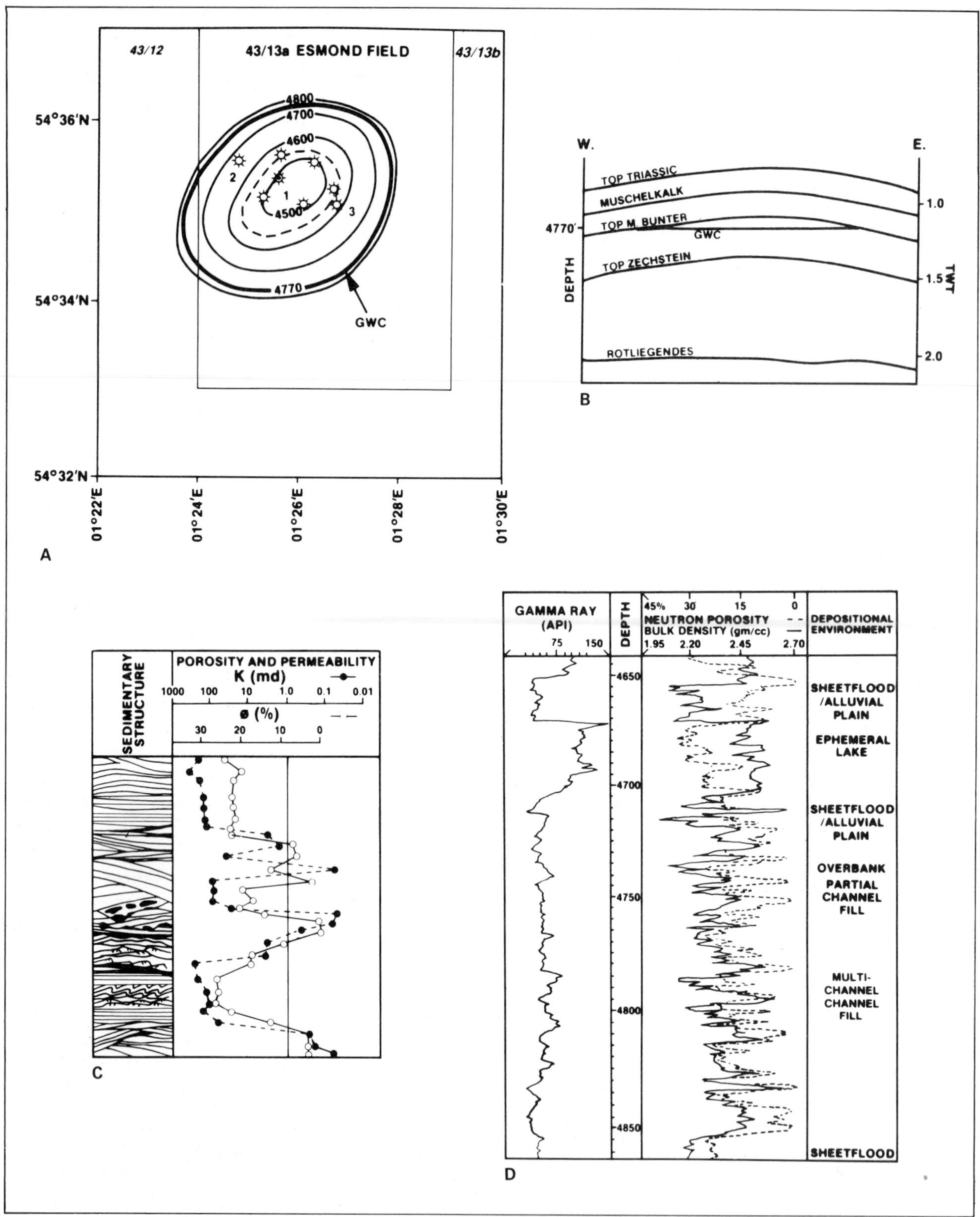

Fig. 7.9 Map (A), cross-section (B), generalized sedimentary structure (C) and well logs (D) in the Esmond Field (after Bifani, 1986).

halites, and none is filled to spill point. Where the thinned Zechstein seal has been breached by faulting, gas has migrated into the Bunter and has accumulated in the salt-controlled Triassic structures. Caister B contains reserves of 3.2×10^9 m^3 (0.12×10^{12} ft^3) in stacked sheet-flood sandstones, with fair reservoir characteristics ($\phi = 21\%$, $k = 100$ mD), and partially overlies Caister C, a faulted Carboniferous structure. In the Esmond, Forbes and Gordon fields, total combined reserves are estimated to be 15×10^9 m^3 (0.53×10^{12} ft^3). Esmond (Fig. 7.9), the largest of the three fields, is typical of these simple, unfaulted, dome structures. It contains a net pay of 80 m in a gross reservoir interval of 104 m. Porosity and permeability characteristics are typical of the formation, although Bifani (1986) and Ketter (1991a) have correlated significant variations in reservoir quality with facies (Fig. 7.9 C,D).

A characteristic feature of these four Bunter gas accumulations is that the nitrogen content is generally higher than in equivalent Carboniferous or Permian gas accumulations. In Caister B, the nitrogen concentration is 14.5%, compared with 6% in the Westphalian reservoir of the associated Caister C. In Esmond, Forbes and Gordon, the nitrogen concentrations are 8%, 12% and 16%, respectively. Nitrogen is a natural product of the Carboniferous source rocks and increases in concentration with increased maturity. The most likely explanation here, however, is that the higher mobility of nitrogen, compared with methane, results in differential concentration when long migration routes are involved. Hydrogen sulphide is also less common in Triassic reservoirs than in the Zechstein traps. Again, this may refect the low efficiency of long-range migration of this gas.

In addition to the fields already described, gas has also been tested from the Bunter Sandstone Formation in UK Block 43/18 and in the Dutch Blocks K17, L2, P2, P12, P14 and Q16 and from the Dutch onshore Bergen Concession and De Wijk and Waalwijk Fields. No other major reservoir units have been recognized offshore, although, in the De Wijk Field, Gdula (1983) and Bruijn (1996) also record gas production from the Rogenstein (average $\phi = 20\%$) and Lower Muschelkalk Members (maximum $\phi = 20\%$) in addition to marginally economic production from the Upper Röt Claystone Member. With the exception of the Volpriehausen Sandstone Member of the Bunter Sandstone Formation (average $\phi = 30\%$) and the Solling Sandstone Member of the Röt Formation (average $\phi = 15\%$), which have good primary reservoir properties, the reservoir potential of the other members resulted from leaching of syndepositional anhydrite during phases of Cimmerian erosion.

7.3 The Central North Sea Basin

The Central North Sea Basin lies to the north of the Mid North Sea–Ringkøbing–Fyn High and is located approximately on the site of the Northern Permian Basin (Figs 7.1 and 7.10). The Central Graben may have been initiated as an area of rifting in the early Permian (Hamar *et al.*, 1980), while basalts and tuffs of the Inge Volcanics Formation in UK Block 31, dated at 265–267 Ma in their Danish equivalents (Ineson, 1993), may identify an early Tartarian phase of opening. There is also a pronounced angular discordance between the Zechstein and the Rotliegend in the Danish Central Trough. Basins to the north of the Mid North Sea–Ringkøbing–Fyn High continued as active depocentres throughout the Triassic but, over the axis of the high, deposition was restricted. The combination of diapiric activity of Zechstein salts and mid-Jurassic truncation of Triassic sediments severely impairs reconstruction of extensional structures. Lervik *et al.* (1990) consider that the phase of early Triassic rifting that affected the North Sea region had only a minor influence in the Central Graben, while Roberts *et al.* (1993, 1995) emphasize the significance of pre-Jurassic extension.

Skjerven *et al.* (1983) record the north–south-trending Triassic rifting of the Central Graben in the southern Norwegian North Sea, possibly extending into the Danish Central Trough area, although their data quality does not enable precise dating of the rifting phase. Subsequently, Thorne and Watts (1989) recognized a major, albeit local, phase of extensional faulting in the Central Graben during the late Permian–early Triassic, followed by a period of regional subsidence, in which normal faulting was absent. Cartwright (1991) also recognizes a non-rotational rifting phase in the basins adjacent to the Coffee Soil Fault, bordering the Ringkøbing–Fyn High, commencing in the late Permian. Overall, the Triassic sediments in the Central Graben north of the Mid North Sea High do not appear to reflect major grabenal activity, and sedimentation could have resulted from passive thermal subsidence. The most convincing evidence of Triassic rifting is the development of a complex series of fault-bounded grabens in the Eger-

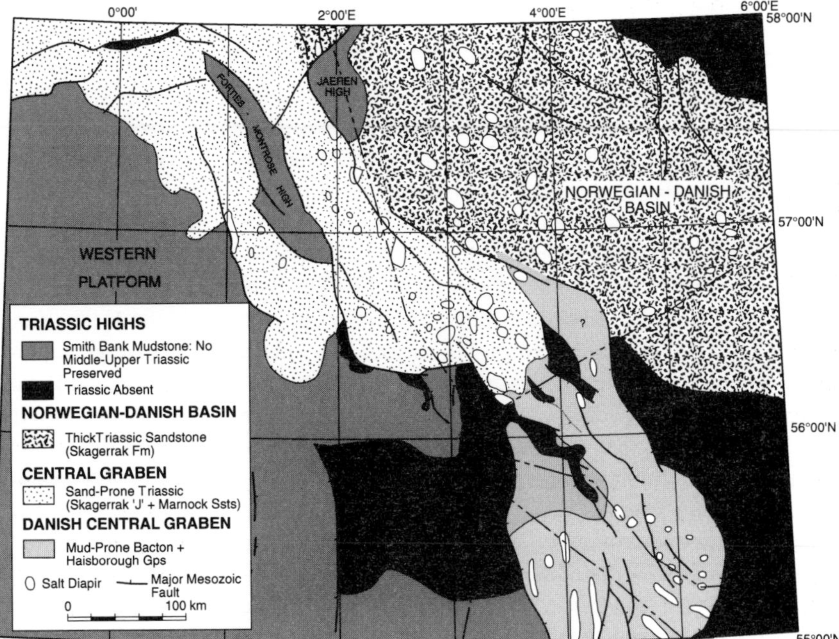

Fig. 7.10 Triassic tectonic elements, Central North Sea.

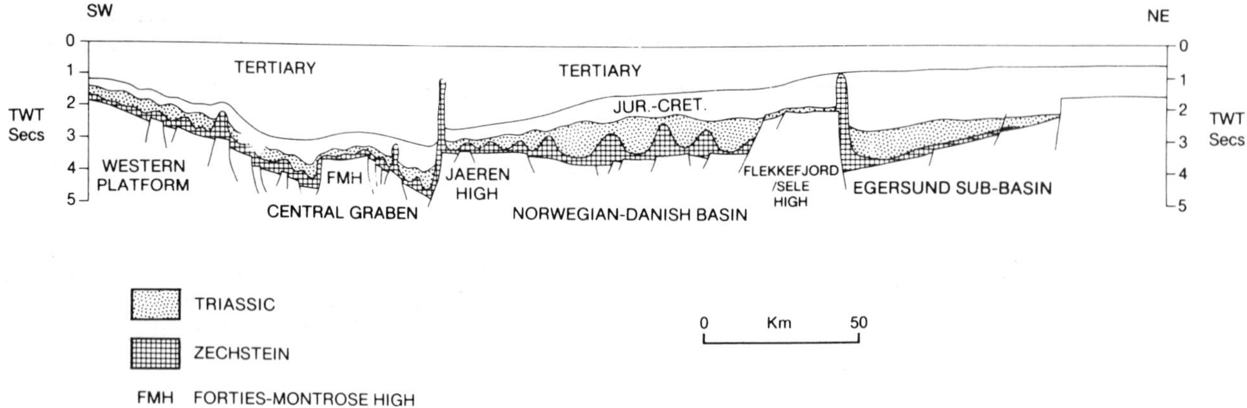

Fig. 7.11 Geological sketch section across the Central North Sea.

sund Basin and possible activation of the eastern bounding fault of the Danish Central Trough (Lervik *et al.*, 1990; Cartwright, 1991). The cross-section (Fig. 7.11) clearly illustrates the half-graben structure of the Egersund Basin; it is also apparent that the major Triassic depocentres were to the east of the Central Graben (see Fig. 2.1).

The Horn Graben is an extensional structure, transecting the Ringkøbing–Fyn High in a north-north-east–south-south-west orientation, which also underwent intense rifting and fault-controlled subsidence during the Triassic (Clausen and Korstgård, 1993). The graben is divided into northern and southern segments, which are underlain by Zechstein salt, although the transition zone between the two segments is devoid of salt. The northern segment is bounded by a westward-dipping main boundary fault and the southern segment has an eastward-dipping boundary fault.

North of the Mid North Sea High, where the Central Graben eventually adopted a north-west–south-east orientation, Sundsbø and Megson (1993) record a complex system of interconnected north–south-trending basins and half-grabens of Triassic age. There is no firm evidence either of Triassic rifting or to suggest that these structures fully breached the Mid North Sea High. Triassic sequences thin towards the High from both the Central and Southern North Sea basins by stratal condensing, with no indication of basal onlap or erosional truncation. With the exception of the Horn Graben, therefore, it would appear that the Mid North Sea–Ringkøbing–Fyn High remained an area of deposition largely restricted to marginal fracture zones throughout the Triassic (Cartwright, 1990). As a non-subsiding massif, although not necessarily of any significant relief, the high continued to be an effective barrier between the Central and Southern North Sea basins but only a relatively minor source of sediment.

Triassic sedimentation was controlled by the same deep structural elements that were important features during the Permian. In addition, there was a close relationship between Zechstein salt movements and Triassic thickness and lithofacies distribution. The pre-Triassic subcrop map (Fig. 7.12) indicates that Zechstein evaporites or carbonates formed the substrata for the Triassic basin. The residual areas of high relief on the western and southern margins provided only a relatively minor supply of coarse clastic material to the basin. Locally, sands accumulated adjacent to emergent Devonian and granitic highs in the Outer Moray Firth, at the southern end of the Fladen Ground Spur and where Rotliegendes sandstones subcrop the Triassic along the Auk–Argyll Ridge in the western part of Quadrant 30. Further south-east, Permian volcanics and volcaniclastics were exposed along the margin of the Mid North Sea High and on highs within the Danish Central Trough area. With these exceptions, sedimentation in the western and central parts of the basin was dominated by fine-grained clastics throughout the Early to mid-Triassic.

The major source of coarse clastic material was on the eastern margin of the basin, where the Fenno-Scandian Shield had an active history of uplift and erosion throughout the Triassic (Zeck *et al.*, 1988). Immediately to the west and south-west, the Norwegian–Danish Basin, which originated during the Permian, was an area of major Triassic fault-controlled subsidence (Skjerven *et al.*, 1983). Thick sequences of coarse clastic sediments accumulated in the Egersund Basin and the North Danish Basin, and towards the west and south these coarse clastics progressively onlap the basinal silts and muds. Adjacent to the Fenno-Scandian Shields, the coarse clastics may lie directly on pre-Triassic rocks.

Deegan and Scull (1977) recognized these two distinct facies in their lithostratigraphic subdivision for this area (Fig. 7.13). They referred the red, silty claystones to the basinal Smith Bank Formation, with the type section in BP 15/26-1, and the coarse, clastic sediments to the marginal Skagerrak Formation, with the type section in Petronord N10/8-1. Deegan and Scull (1977) were severely constrained in their lithostratigraphic analysis by a lack of well data, and Cameron (1993a) has subsequently revised the lithostratigraphy of the area and assigned the Triassic units to the Heron Group. The original formation names have been retained, with more clearly defined limits on their areal extent. In addition, the Skagerrak Formation has been subdivided into three sandstone and three mudstone members in the south Central Graben area, and the lithostratigraphic framework has been enhanced by the availability of accurate palynological dating (Goldsmith *et al.*, 1995). An interpretation of the complex facies relationships within the Central Graben is illustrated schematically in Fig. 7.14.

Thick accumulations of Triassic sediments occupy the area of the Northern Permian Basin and are therefore

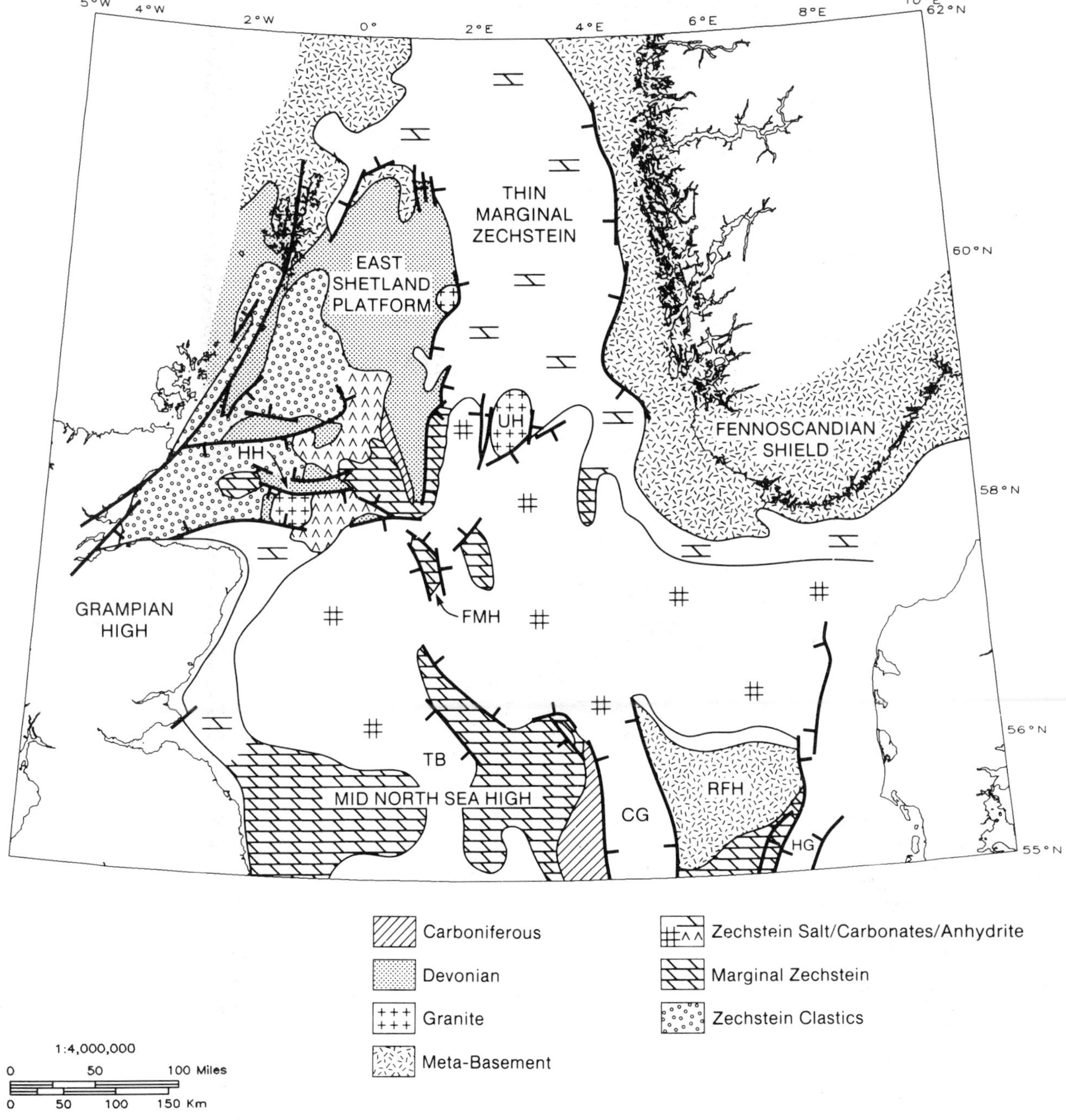

Fig. 7.12 North Sea pre-Triassic subcrop.

invariably underlain by extensive Zechstein evaporites. Johnson *et al.* (1986), Glennie and Armstrong (1991) and Hodgson *et al.* (1992) describe how the process of salt withdrawal created large areas of subsidence, in which standing bodies of water could persist. In this environment, thick pods of Smith Bank shales accumulated, until complete salt withdrawal grounded the pods on the underlying basal Zechstein dolomites and anhydrites or Rotliegend and subsidence ceased. With continued subsidence of the pod margins, many were converted to whalebacks.

A widespread angular discordance, which has been correlated with the Hardegsen event, marks the start of a period in the development of the basin that differs markedly from events in the Southern North Sea Basin. South of the Mid North Sea High, mid to late Triassic sediments are almost exclusively fine-grained clastics and evaporites; north of the High, sheet-flood and fluvial arenaceous sediments become increasingly dominant (Fig. 7.15). Throughout this period, salt tectonics continued to play a major role in controlling sedimentation, and Hodgson *et al.* (1992) link the evolution of the halokinetically controlled sediment pods with the pattern of distribution of sheet-flood and fluvial facies. They suggest that the sheet floods, which are relatively unconfined and show little channelling, were deposited by flood waters flowing over and between the sediment pods, with depocentres concentrated over subsiding pods. Channelling developed when the drainage became more restricted through the linking of pod synforms.

During the Early and mid-Triassic the basin appears to have been in communication with the Southern North Sea

Series	Stage	Inner Moray Firth		Central North Sea: Western Platform	Central North Sea: Central Graben		Norwegian-Danish Basin	Denmark	
UPPER TRIASSIC	RHAETIAN	HERON GROUP	STOTFIELD CALCRETE FM		*Marnock Sst*; Joshua Mdst Mbr; Josephine Sst Mbr	SKAGERRAK FM	GASSUM FM	GASSUM FM; VINDING FM	MORS GP
UPPER TRIASSIC	NORIAN	HERON GROUP	LOSSIEHEAD FM		*Skagerrak Mdst*; Jonathan Mdst Mbr	SKAGERRAK FM	SKAGERRAK FM	ODDESUND FM	JYLLAND GROUP
UPPER TRIASSIC	CARNIAN	HERON GROUP	LOSSIEHEAD FM		*Skagerrak Mdst*; *Skagerrak Sst*; Joanne Sst Mbr	SKAGERRAK FM	SKAGERRAK FM	TØNDER FM	JYLLAND GROUP
MIDDLE TRIASSIC	LADINIAN	HERON GROUP	LOSSIEHEAD FM		*Skagerrak Sst*; Julius Mdst Mbr	SKAGERRAK FM	SKAGERRAK FM	FALSTER FM	LOLLAND GROUP
MIDDLE TRIASSIC	ANISIAN	HERON GROUP	LOSSIEHEAD FM		*Skagerrak Sst*; Judy Sst Mbr	SKAGERRAK FM	SKAGERRAK FM	ØRSLEV FM	LOLLAND GROUP
LOWER TRIASSIC	SCYTHIAN	HERON GROUP	HOPEMAN SANDSTONE FM	SMITH BANK FM	SMITH BANK FM		SKAGERRAK FM; SMITH BANK FM	BUNTER SST FM; BUNTER SHALE FM	BACTON GP

Fig. 7.13 Triassic lithostratigraphy, Central North Sea.

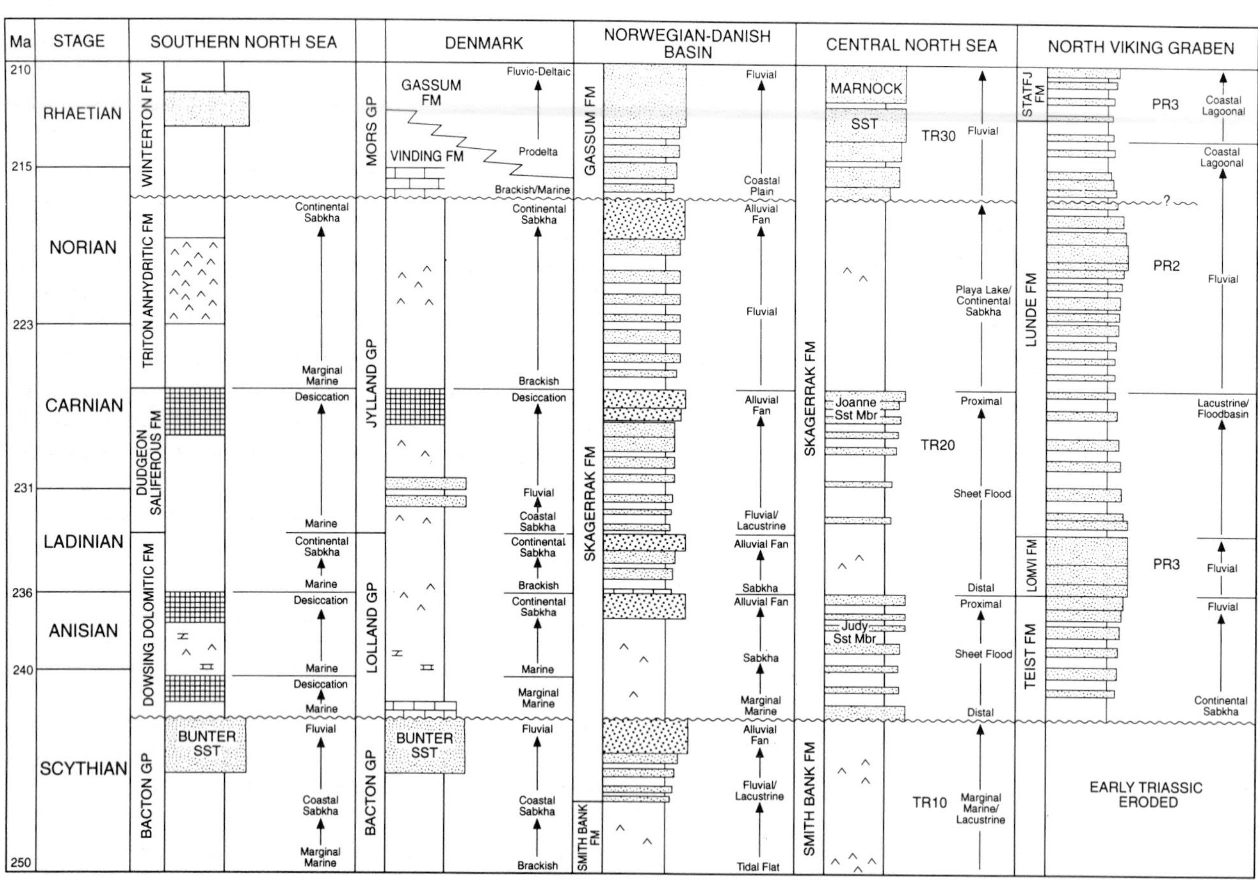

Fig. 7.14 North Sea Triassic stratigraphy and facies. Central North Sea sequences TR10–TR30 from Smith *et al.* (1993); North Viking Graben sequences PR1–PR3 from Steel (1993).

Basin. The similar lithofacies characteristics of the Smith Bank and Bunter Shale formations could be attributed to similarities in depositional environments rather than to contiguity of facies. However, although no major evaporite basin was developed north of the Mid North Sea High, there is firm evidence of the Röt and Muschelkalk incursions and this would support the case for restricted communication between the two basins. One further inference that may be drawn from this record of the limited expression of the Röt and Muschelkalk transgressions is that there was no extensive bathymetric low in the area of the Central Graben. Following the mid-Triassic, communications between the two basins were maintained in the Danish area and Goldsmith *et al.* (1995) correlate Carnian–Norian and Rhaetian mudstone intervals in Quadrant 30 with possible marine incursions from the south.

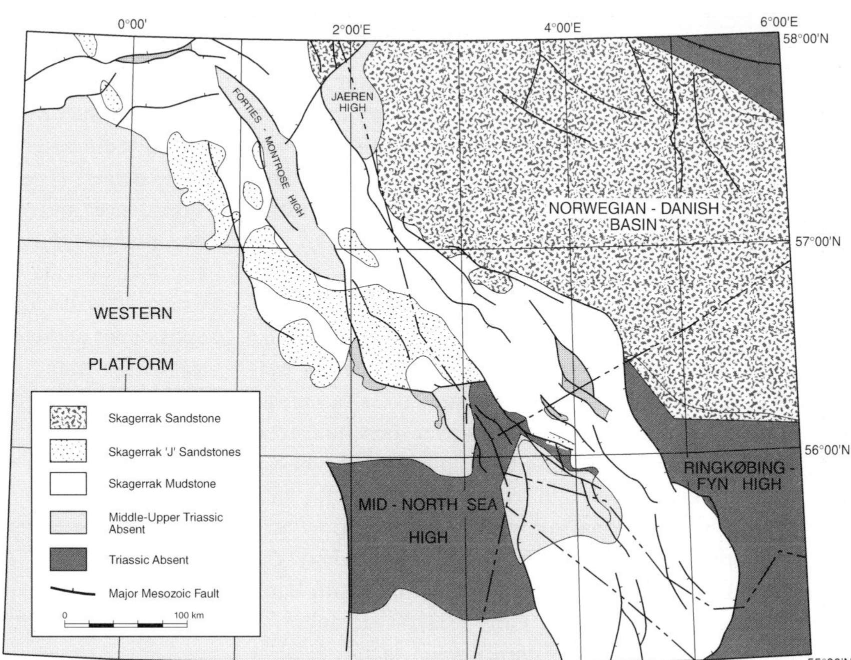

Fig. 7.15 Middle–Upper Triassic lithofacies distribution, Central North Sea.

Towards the end of the Triassic, the gross geometry of the basin displayed a southward tilt, with the regional high probably located close to the southern end of the South Viking Graben. The Central Basin was still topographically higher than the Southern Basin, however, because there is only a minor expression of the Rhaetian marine transgression north of the Mid North Sea High. The Late Triassic phase of sedimentation is characterized by evidence of increased run-off and sand transport, reflecting increased rainfall. The areas available for erosion, and hence clastic detritus, had not changed since the Early Triassic. It seems likely, therefore, that many of the Rhaetian sands may be reworked older Skagerrak sands (Fig. 7.16). The Late Triassic patterns of sedimentation also suggest that movement of the Zechstein evaporites was taking place on a large scale. Two mechanisms may have been responsible; dissolution or salt-wall 'sinking' associated with the generation of accommodation space by periodic extension. Surprisingly, if dissolution was the reason, Triassic halites are rare, and the dissolved Zechstein halites may have drained southwards into the Southern North Sea Basin. In this context it is interesting to speculate on the origin of the Keuper halites.

Invoking salt movements as the major control on Triassic deposition, rather than the previously accepted synsedimentary rifting models of Thorne and Watts (1989) and Roberts *et al.* (1991), also provides a useful working hypothesis for predicting sand distribution at both regional and local scales. In particular, the observed trend of increasing sand input into the region with time indicates that the Upper Triassic is likely to have the most prospective reservoirs. In areas of thin salt, such as over the

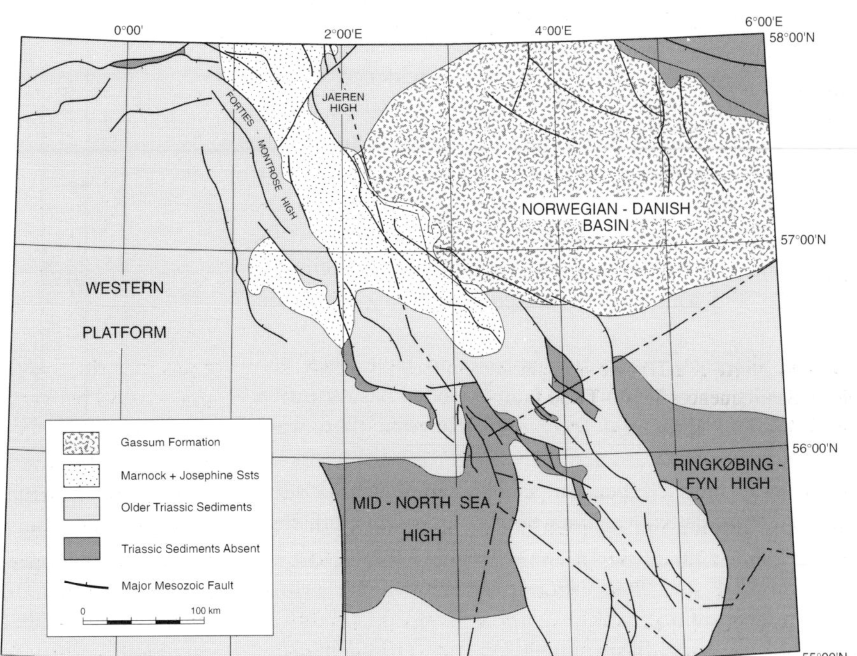

Fig. 7.16 Upper Triassic lithofacies distribution, Central North Sea.

Western Platform and the Forties-Montrose High, the sediment pods are likely to have grounded before the Middle Triassic. In these areas, only eroded Lower Triassic mudstones are preserved, with no sand potential. However, within the deeper parts of the basin, where the salt was thickest, sediment pods may not have grounded until post-Triassic times, allowing the accumulation of sand-prone Upper Triassic sediments.

7.3.1 The Norwegian–Danish Basin

The stratigraphy of the Norwegian–Danish Basin is outlined in Figs 7.13 and 7.14. The fine-grained clastics of the Smith Bank Formation are developed over the whole area, with the exception of the eastern periphery of the basin. The Formation is composed of red-brown, anhydritic, silty mudstones, with occasional sandstone stringers. No reliable biostratigraphic dating is available. However, where the Muschelkalk equivalent provides a datum, the Smith Bank Formation is assumed to be of Early Triassic age only.

In the area between the eastern margin of the Central Graben and the Norwegian–Danish Basin, a number of wells, including Norske Murphy N2/3-3, Esso N8/3-1 and Esso N9/8-1, have recorded Skagerrak sands directly overlying the Zechstein (Myhre, 1975, 1978; Riise, 1978). As the Smith Bank Formation is represented in adjacent well Norske Murphy N2/3-1, in A/S Norske Shell N17/11-1 and in Block N9/4 (Frodesen, 1979; Olsen, 1979), it is probable that these are examples of solution synclines, developed on salt-piercement structures separating pods of Smith Bank mudstones and acted as conduits for the distribution of Skagerrak sands.

The base of the Skagerrak Formation is seen as strongly diachronous and younging to the west. Regional evidence suggests that on the eastern margins of the basin, the Skagerrak Formation represents the lateral facies equivalent of the Smith Bank Formation. The interbedded conglomerates, sandstones, siltstones and shales accumulated primarily at the basin margins and adjacent to fault scarps as a westward-prograding, laterally extensive alluvial-fan sequence. In Scandinavian wells on the periphery of the basin, the Skagerrak Formation directly overlies the pre-Triassic rocks and there is no clear evidence for the existence of the Smith Bank Formation (Bertelsen, 1980). The Fenno-Scandian Shield was tectonically active in the Triassic and provided the major source of coarse clastic material for the basin from the crystalline basement and overlying Variscan thrust sheet (Zeck *et al.*, 1988). At the same time, the fault-bounded basins bordering the Shield to the west and south were repeatedly subsiding (Sørensen, 1986).

The Skagerrak Formation appears to have been geographically restricted to these fault-bounded basins until at least the end of the mid-Triassic. At that time, there was a progradation westward across the Fiskebank Subbasin and into the Central Graben. In general, the Skagerrak sandstones prograde over actively subsiding Smith Bank shale pods, but in some instances, as in Norske Murphy N2/3-3, Esso N8/3-1 and Esso N9/8-1, the sandstones directly overlie Zechstein.

In the Egersund Basin, Jakobsson *et al.* (1980) interpret the sedimentary succession in Phillips N17/12-1 as comprising six tectonically induced, coarsening-upward cycles. The depositional environments described are similar to those outlined by Glennie (1972) to account for the Rotliegend sedimentary facies in north-west Europe. The cyclical sequence (Fig. 7.17) commences with regional basinal subsidence and a marine transgression. Alluvium that accumulated during the Zechstein was redeposited in central lakes or marginal marine environments as uniform, brick-red shales of the Smith Bank Formation. Continued subsidence and erosion of the newly generated fault blocks stimulated rapid progradation of alluvial fans, with the deposition of poorly sorted conglomerates. The second

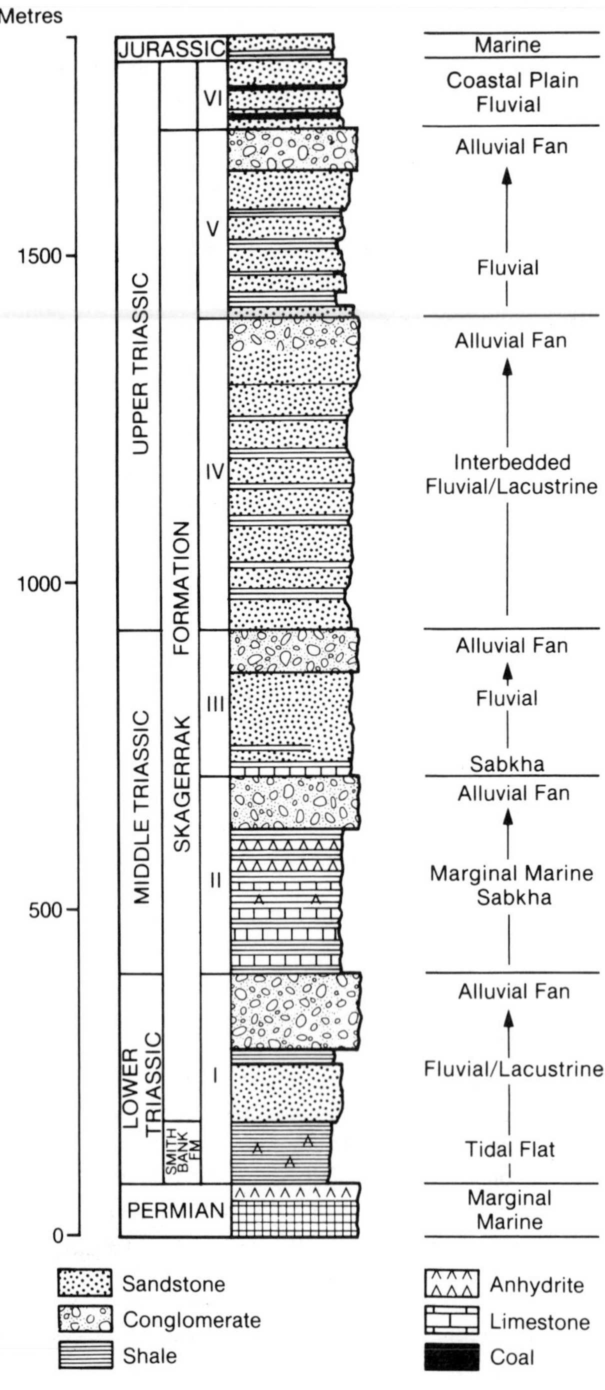

Fig. 7.17 Egersund Subbasin well N17/12-1 (after Jakobsson *et al.*, 1980).

cycle commences with a northerly extension of the Muschelkalk transgression through the North Danish Basin. Slow progradation of marginal alluvial fans is reflected in the transition from the interbedded limestone, evaporite, sabkha and fluvial sediments into a sandy conglomerate. The third cycle was dominated by aeolian sands and sabkha deposits, which were succeeded by a new fan outgrowth, represented by coarse-grained mass-flow conglomerates. The fourth and fifth cycles each represent braided-stream deposits, terminated by prograding alluvial-fan deposits. The sixth and final cycle, which should properly be assigned to the Gassum Formation, reflects the Rhaetian marine transgression. This entered the Central North Sea Basin via the North Danish Basin, and resulted in extensive deposition of fluvial sands and silts in the Egersund Basin. Jakobsson *et al.* (1980) record a sequence of grey, fine- to medium-grained sandstones with minor conglomerates interbedded with grey-brown silty shales and coals. This suggests that a system of large meandering rivers with extensive flood plains replaced the previously ephemeral braided streams as the marine transgression entered the basin from the Danish basins to the south.

Local facies variations are usually associated with tectonically active areas. Coarser clastic facies are developed where the Horn Graben breaches the Ringkøbing–Fyn High, but, in the Danish Central Trough, coarse clastics are rare. Michelsen and Andersen (1983) suggest that coarse clastic sediments may be restricted to the Tail End Graben, with a maximum development of up to 1000 m. In the northern part of the trough, in the Dansk Nordsø Q-1 well, 145 m of shales with minor interbedded sandstones have been assigned to the Smith Bank Formation. Overall, the virtual absence of coarse clastic deposits in wells suggests that the Danish Central Trough was relatively quiescent throughout the Triassic.

In contrast, the Horn Graben experienced a major phase of subsidence in the Triassic, when up to 4200 m of sediment accumulated in the centre of the graben. At the northern and southern ends of the graben, these events initiated halokinesis in the underlying Zechstein evaporites at the bounding faults and in the graben centre. Growth of these salt structures continued during deposition of the Muschelkalk, culminating in the development of diapirs in the late Triassic (Olsen, 1983) and of broad depositional pods that locally grounded on underlying Rotliegend. Geoseismic sections in Kockel (1995) show a saucer-shaped development of Lower and Middle Bunter (i.e. to top Solling Sandstone) within the German extension of the Horn Graben, indicating early salt withdrawal. Here, the total Triassic locally exceeds 7000 m in thickness and is unconformably overlain by probable Upper Jurassic. These sedimentary sequences are referred to the standard Danish succession by Clausen and Korstgård (1993) and to the Southern North Sea lithological units by Olsen (1983). This illustrates the key position of the Horn Graben as a link between the Central and Southern basins and its significance in the Triassic depositional history of the southern part of the Central Basin and especially of the Norwegian–Danish Basin. The Outer Rough Basin and the southern Tail End Graben would also appear to have been linked to the Southern North Sea Basin, while the Feda Graben and the northern Tail End Graben show more affinity with Central Basin sedimentary sequences.

7.3.2 Denmark

Bertelsen (1980) has described a comparable succession from the North Danish Basin (see Figs 7.13 and 7.14). In the north, the Skagerrak Formation may be up to 3000 m thick; it comprises interbedded red-brown, brown and grey, fine- to medium-grained sandstones, siltstones and claystones. The sandstones are arkosic, micaceous and, in the upperpart of the formation, glauconitic. These sediments accumulated initially as braided-stream complexes and grade southwards and south-westwards into distal facies of the Germanic province. The geographical location of the North Danish Basin at the eastern junction of the Southern and Central North Sea Basins is therefore critical in establishing the relationship of the Skagerrak Formation with the contemporaneous Germanic facies. Bertelsen (1980) has described the transitional facies and has proposed a new lithostratigraphic terminology to accommodate them.

In the southern part of the North Danish Basin, Bertelsen (1980) recognizes a Bunter Shale Formation contiguous with the Southern North Sea and North German Basin facies. North-westwards there appears to be a lateral transition into the Smith Bank Formation, while towards the basin margins the fine clastics are replaced by arenaceous Skagerrak facies. Overlying the Bunter Shale Formation, an undivided Bunter Sandstone Formation extends southwards into the North German Basin but to the north and north-east coalesces with the Skagerrak facies. Heavy-mineral analyses of the sandstones in the North Danish Basin and the northern part of the North German Basin indicate derivation from a common source to the north and north-east. These Bunter Sandstones must therefore be viewed as a distal development of the Skagerrak Formation.

The succeeding Lolland Group, comprising the Ørsler and Falster Formations, is equivalent to the Dowsing Dolomite Formation of the Southern North Sea Basin and represents the north-eastern marginal expression of the Röt and Muschelkalk transgressions. The red clastics of the Skagerrak Formation here grade southwards through distal flood-plain into sabkha, playa and shallow-marine/brackish-marine environments.

Increased subsidence in the Late Triassic (Carnian–Norian) created a hypersaline sea and, at the same time, the proximal facies appear to onlap the Fenno-Scandian Border Zone to the north and north-east. The Jylland Group is equivalent to the Keuper Formation (see Fig. 7.14) and is divided into the Tønder and Oddesund formations. The Tønder Formation comprises interbedded claystones and siltstones, with sandstones becoming more common in the upper part of the formation. With the retreat of the Skagerrak Formation facies to the north, these sediments are considered to be distal, coastal-plain deposits. The overlying Oddesund Formation consists of red-brown and variegated claystone, evaporites and siltstone, with two major halite members separated by a fluviodeltaic carbonaceous silty sandstone. The halites were deposited in substantial salt lakes that formed in the centre of the basin. Ephemeral lakes and sabkhas would also have occupied considerable areas. The fluviodeltaic sediments separating the halites may reflect a period of increased humidity and a consequent initiation of coarse clastic sedimentation in the salt lakes.

The sequence terminates with a pre-Rhaetian transgressive phase, represented by the Vinding Formation. This trangression entered the basin from the south and completely inundated the Ringkøbing–Fyn High, creating a large, shallow, brackish–marine sea, in which limestones, marlstones and dark grey shales accumulated. To the north and east, the contemporaneous Skagerrak facies comprise low-energy fluvial and marginal-marine red shales and arkosic sandstones. At the top of the Vinding Formation, the appearance of pro-delta siltstones and fine-grained sandstones marks the transition into the overlying Gassum Formation.

Sørensen (1986) has analysed the post-Palaeozoic subsidence history of the North Danish Basin and concluded that there were major subsidence episodes in the Early Triassic (Bunter/Ørsler) and Late Triassic (Tønder/Øddesund). The high subsidence rates during these episodes and the magnitude of overall Triassic subsidence are interpreted as rift-controlled, possibly a continuation of the pull-apart initiated in Rotliegend time and linked to widespread movement along the Tornquist–Teisseyre lineament.

The final, Rhaetian phase of Triassic sedimentation saw the progradation of fluviodeltaic environments from the eastern margins into the centre of the Central North Sea Basin. In the type sections of the North Danish Basin, Bertelsen (1978) describes the Gassum Formation as being characterized by three members. The lower member is an upward-coarsening unit with dark grey claystones, siltstones and fine-grained sandstones overlain by grey, fine, medium and coarse-grained, quartzose sandstones, with locally developed coal seams. This member represents delta-front deposits, grading upwards into distributary-channel sands.

The middle member reflects a major regional transgressive event, which drowned the delta plain and which correlates with the Contorta Beds of North Germany and the Westbury Beds of south-west England. The inferred transgression route, from the Germanic province to the south, implies that at this time the typically non-marine Central North Sea Rhaetian Basin was in communication with the predominantly marine Southern North Sea Rhaetian Basin. The most characteristic lithotype of the middle member is a dark carbonaceous claystone, which passes upwards into lighter-coloured grey and reddish brown, locally glauconitic siltstones and claystones of the upper member. These fine-grained clastic sediments represent a final regressive phase, during which limnic conditions were re-established in the basin.

7.3.3 United Kingdom Central North Sea

The stratigraphy of the UK Central North Sea area, which comprises the Central Graben, Forth Approaches Basin and Outer Moray Firth, is represented in Figs 7.13 and 7.14. The typical stratigraphy of the Central Graben is illustrated by selected wells in Figs 7.18 and 7.19. The Smith Bank Formation (Cameron, 1993a) is present throughout the area and is composed predominantly of the classic fine-grained clastic facies of the type section. In the Central Graben, occasional sandstone stringers may become thicker and more continuous locally, for example in the upper part of the section in block 30/7. In northern Quadrants 20 and 21, more extensive arenaceous intervals may occur at the base of the Formation.

Deposition of the Smith Bank Formation was largely under lacustrine or quasi-marine conditions, and cores frequently show evidence of bioturbation. Evidence of fluvial, sabkha and coastal-plain environments has also been recorded. The widespread and uniform nature of the dominant mudstone lithotype is comparable with the Bunter Shale Formation of the Southern North Sea. It is probable that communications were maintained between the two basins during the early Triassic through the Thirty

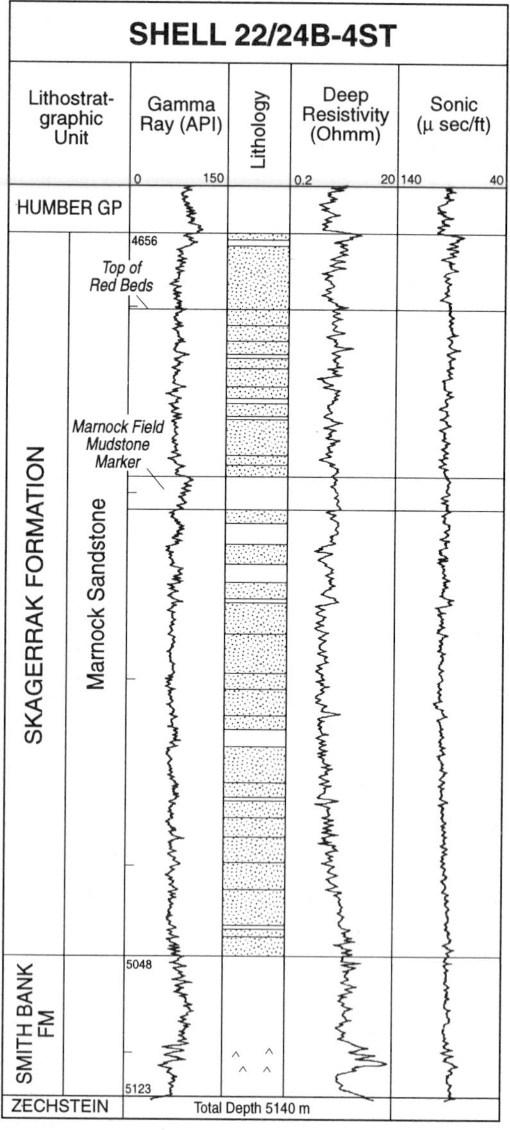

Fig. 7.18 Triassic (Heron Group) lithostratigraphy of well 22/24b-4ST, Central North Sea. Skagerrak Formation comprises sand-rich Marnock facies. The prominent mudstone horizon within the Marnock sandstone can be traced over much of Block 22/24. Depths in metres below KB; depth ticks at 100 m intervals. For key to lithology, see Fig. 7.3.

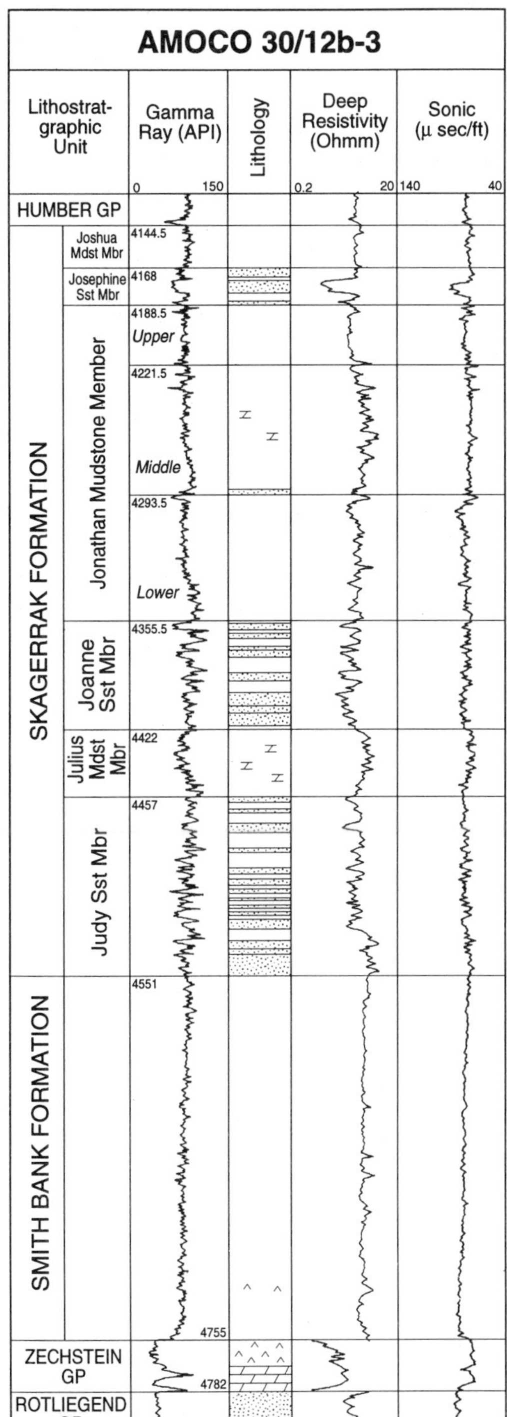

Fig. 7.19 Triassic (Heron Group) lithostratigraphy of well 30/12b-3, Central North Sea. Lithostratigraphic nomenclature after Cameron (1993a). Depths in metres below KB. For key, see Fig. 7.3.

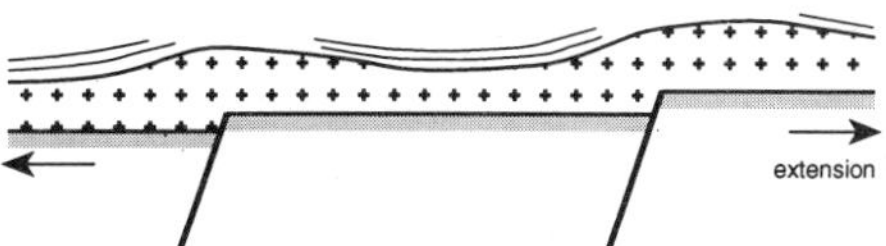

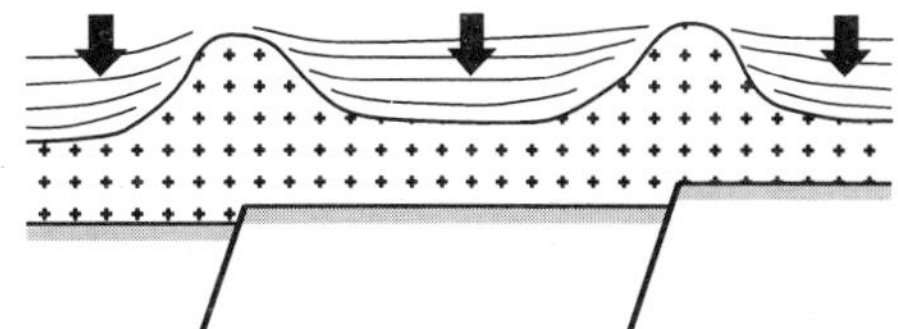

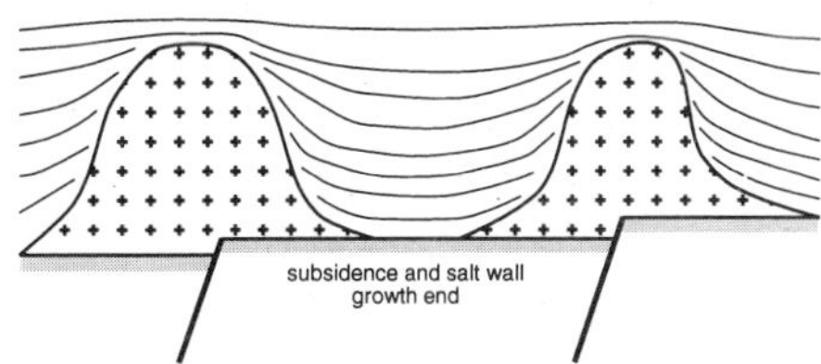

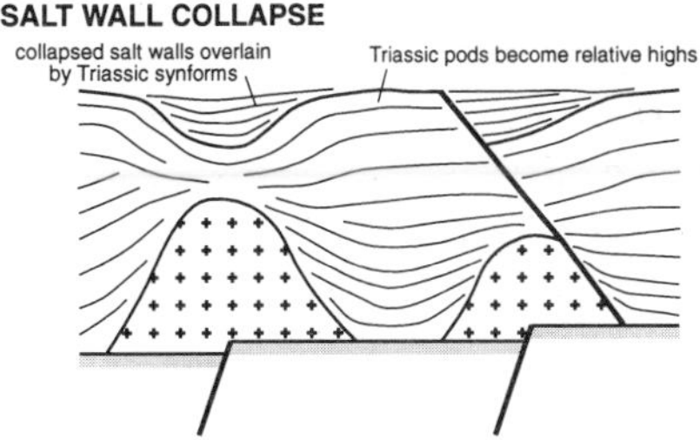

Fig. 7.20 Evolution of Triassic sediment pods and salt walls (after Hodgson *et al.*, 1992).

Seven Basin, the Central Graben and the Horn Graben. In the Central Graben, to the north of the Mid North Sea High and in the area of the Horn Graben, lateral equivalents of the Bunter Sandstone Formation may be inferred from lithofacies and stratigraphic relationships. No evidence is available to confirm that these are distal expressions of Bunter sands of the Southern North Sea Basin, and it may be that they were more locally derived but reflect a similar generative mechanism. There is also evidence of the Röt or Muschelkalk transgressions entering the Central Basin through both the Central Graben and the Horn Graben. There are no well data, however, that would support a major extension of the Röt or Muschelkalk Halite Basins beyond the southern ends of the Central and Horn Grabens. Olsen (1987) refers to minor Triassic evaporite halokinetics in the Mandal Complex, but the presence of Triassic halites has not been confirmed by well evidence.

Considerable thickness variation of the Smith Bank Formation is evident within the area. Up to 900 m has been encountered in well sections in the Forth Approaches Basin and over 1000 m is calculated from seismic data. The thickness variations often reflect post-Triassic uplift and erosion, although it is more likely that halokinesis in the Zechstein halites would have been the major factor in controlling patterns of sedimentation. Johnson *et al.* (1986) briefly outline this mechanism in their study of the Fulmar Field (30/16), and its regional significance has been elaborated by Hodgson *et al.* (1992) and Smith *et al.* (1993). Initiation of salt movements required only a few hundred metres of sediment loading and salt was progressively displaced from beneath the growing sediment pod to form salt pillows, walls or ridges (Fig. 7.20). The intervening topographic lows were progressively infilled with Triassic sediments and, once started, subsidence continued until all the salt was displaced from under the sediment pod. Pod subsidence and Triassic sedimentation ceased when the pod became grounded on the Base Zechstein Halite surface, as the pods could no longer provide accommodation space at the surface. The Triassic sediment pods are predominantly elongated in a north–south direction and are often flanked by salt walls (see Fig. 2.26A). The adjacent salt walls are often located over the crests of pre-Zechstein fault blocks. Growth of these salt walls also ceased, as salt supply was cut off by grounding of the pods. The thickness of the salt, which was controlled by end-Rotliegend palaeotopography, determined both the thickness of the sediment pods and the wavelength of the salt walls.

Gowers *et al.* (1993) consider that the mobilization of Zechstein salt during the Triassic had a major influence on the structural history of the Norwegian Central Trough. They document evidence of halokinetically induced rotation and differential subsidence of blocks of Triassic strata. In the East Central Graben, Penge *et al.* (1993) envisage similar, but more restricted, halokinetic control on the sedimentation of the Smith Bank Formation, with salt withdrawal being limited to local overdeepening of Triassic depocentres, with no significant formation of salt walls. They consider that the major salt-related movements were post-Triassic, largely as an accommodation process to regional extension. Høiland *et al.* (1993) also invoke a combination of halokinetics, salt dissolution and extensional tectonism in their analysis of the Jaeren High, immediately to the east of the East Central Graben. However, they imply a greater influence for the Triassic halokinetic component in the structural evolution of the Jaeren High.

The Skagerrak Formation, originally circumscribed by Deegan and Scull (1977), has been redefined in the UKCS by Cameron (1993a) and Goldsmith *et al.* (1995). The most significant revision is the recognition in the south Central Graben of three mudstone and three sandstone members. The oldest of these sandstone units, the Judy Sandstone Member (Anisian), is also the most widespread of the three sandstone members, occurring in northern Quadrants 29 and 30. It represents 'flash' sheet-flood deposits, 60–400 m thick and largely confined to depocentres on halokinetically controlled subsiding sediment pods. It has marine indicators at the top of the section, which may reflect evidence of the Muschelkalk incursion. The overlying anhydritic, grey, silty mudstone, the Julius Mudstone Member (Anisian–Ladinian), is 20–150 m thick and slightly more geographically restricted. It represents playa-lake and mudflat deposition during a temporary retreat of sheet-flood deposition into the north Central Graben. In general lithology and well-log character, it is similar to the age-equivalent upper part of the Dowsing Dolomite Formation of the Southern North Sea. The Joanne Sandstone Member (Ladinian–Carnian) is another 'flash' sheet-flood deposit, 70–490 m thick, which occupies the same areal extent as the underlying Julius Mudstone Member. The succeeding members, the Jonathan Mudstone Member (Carnian–Norian), which is tentatively correlated with the Triton Anhydrite Formation, the Josephine Sandstone Member (Norian–Rhaetian) and Joshua Mudstone Member (?Norian–Rhaetian), are all very limited in extent, being recognized at present mainly in blocks 30/1 and 30/6, 30/12 and 30/13 and 30/17. In addition, the predominantly non-red sandstones, previously correlated with the Gassum Formation or informally termed Marnock Sandstone Formation (Hodgson *et al.*, 1992), which overlie the main red-bed sequences to the east and north, respectively, are considered lateral equivalents of the Josephine Sandstone Member.

The inclusion of Anisian–Carnian sandstone and associated mudstone units hitherto assigned to the Smith Bank Formation (Deegan and Scull, 1977; Fisher and Mudge, 1990) has had the apparent effect of increasing the age of the first appearance of Skagerrak sandstones in the area. There are no other published records of pre-Carnian Skagerrak sandstones in the area beyond the limits of the Judy and Joanne Sandstone Members, and it could still be argued that these areally restricted units should be retained in the Smith Bank Formation. In support of this view, Jeans *et al.* (1993; C.V. Jeans, personal communication) suggest that the heavy-mineral fingerprint of these sandstones may be distinct from that of the Skagerrak Formation sandstones *sensu* Deegan and Scull (1977).

Elsewhere in the area, the Skagerrak Formation is represented by fluvial sandstones and subordinate, thinly bedded siltstones and mudstones. Poorly to moderately sorted sandstones predominate in the lower part of the section and are attributed to braided-stream, alluvial-plain and sheet-flood deposits. Sorting improves significantly where these grade upwards into channelized fluvial sands. Hodgson *et al.* (1992) link this development of fluvial facies to the subsidence history of halokinetically controlled sediment pods. Where the 'pods' containing thick Smith Bank shales were continuing to subside at this time, the sheet-flood systems would have preferentially followed these topographic lows. Towards the end of the Triassic, drainage became more restricted and channelized fluvial sands, reworking older Skagerrak fluvial and sheet-flood sands, became more abundant.

Because of the great thickness of underlying salt in the Eastern Trough, pod subsidence continued into the Late Triassic, creating a major depocentre. Drainage at this time was concentrated along the axis as an extensive north–south-trending braid plain. In the Marnock Field (22/24a), the Ula Field (N7/12) and blocks 22/12 and 22/13, the Upper Triassic reservoir is well sorted, but exceptional porosity has been preserved by a combination of disequilibrium overpressure and the formation of chlorite grain coatings, which have inhibited the formation of late quartz overgrowths. Authigenic illite can also impair reservoir quality and is especially prevalent in sections adjacent to salt walls and diapirs, where the critical temperature of 100°C has been exceeded. The interaction of chloritization with illitization is unclear, but early chlorite formation may again have an inhibitory effect (Smith *et al.*, 1993).

Glennie and Armstrong (1991), in a review of the Kittiwake Field (21/18), suggest that Skagerrak sands also may be preferentially conduited along solution synclines generated by dissolution at the crests of salt walls bordering grounded 'pods'. In these instances, Skagerrak sands would directly overlie Zechstein evaporite, rather than Smith Bank shales (e.g. Shell 21/18-2).

In the Outer Moray Firth Basin, exposed highs, including

the Peterhead Ridge and the Halibut Horst, were the source of sands deposited in the Witch Ground and Buchan Grabens. In the Claymore Field (14/19), these are of variable reservoir quality, being particularly micaceous, with detrital sericite and authigenic smectite and kaolinite locally constituting up to 40% of the rock (Spark and Trewin, 1986). Harker *et al.* (1987) record a sequence of red anhydritic siltstones and shales, assigned to the Smith Bank Formation, resting with apparent conformity on Zechstein evaporites. Locally, the Smith Bank Formation is unconformably overlain by interbedded micaceous sandstones, siltstones and shales, which are tentatively assigned to the Skagerrak Formation on the basis of stratigraphic position and lithological similarity. No stratigraphic contiguity with the Skagerrak Formation of the eastern part of the basin can be established, however. This coarse clastic deposition is interpreted as reflecting an initial phase in the opening of the Witch Ground Graben.

At its northern limit, the Skagerrak Formation is laterally transitional into the Cormorant Formation across the Crawford Spur. In the Crawford Field (9/28), the Skagerrak Formation is of late Carnian age, has a maximum thickness of 529 m and directly overlies marginal-marine Zechstein clastics (Yaliz, 1991). The sequence comprises a lower unit of alluvial-plain deposits, with associated fluvial-channel sandstones and extensive calcretes. The upper unit, which is recorded only in the eastern part of the field, is dominated by coarse-grained channel-fill sandstones, which accumulated on a west-north-west-flowing braid plain.

The major source of clastic material for the Skagerrak sands is considered to have been the Fenno-Scandian Shield, although more local sources in the Outer Moray Firth may also have contributed. Major structural features, such as the Ling Graben, may have provided an important east–west conduit, but there is also evidence of connections linked to salt-controlled structures across the platform areas bordering the eastern margins of the major axial depocentres. The mechanism that initiated the main phase of regional progradation is unclear. Beyond the area of the Judy and Joanne sandstones, no Skagerrak sandstones positively older than Carnian have been recorded from the Central Graben, so the timing of the progradation, approximately Ladinian–Carnian, is coincident with observed major changes in basin processes in the Norwegian–Danish Basin. It is also coincident with the uparching of the East European Craton, which effectively reshaped European Triassic palaeogeography. One further concurrent event was the likely increase in run-off as the area of the North Sea drifted into higher northern palaeolatitudes with increased annual rainfall (see Fig. 2.7).

7.3.4 Inner Moray Firth

The sedimentary succession of the Inner Moray Firth is represented onshore by the aeolian Hopeman and Lossiemouth Sandstones, which are separated by the fluvial sandstones of the Burghead Beds, with the 'Cherty Rock' completing the succession. Glennie and Buller (1983) and Glennie (Chapter 5, this volume) correlate the lower part of the Hopeman Sandstone with the Weissliegend, but this is disputed by Frostick *et al.* (1988) and Cameron (1993a). Cameron (1993a) has revised the lithostratigraphy of the offshore succession, retaining the Hopeman Sandstone Formation for a basal unit of aeolian and fluvial sandstones, and introducing two new units. The Lossiehead Formation, equivalent to the combined Burghead Beds and Lossiemouth Sandstones, is represented by sheet-flow, lacustrine and occasional incised fluvial–channel deposits; the Stotfield Calcrete Formation is equivalent to the 'Cherty Rock' and is interpreted as a palaeosol deposited during a period of low clastic input at the end of the Triassic. In Britoil 11/30-1 (Fig. 7.21), the typical offshore succession is of basal silty shales overlain by a thick sequence of sandstone, with interbedded shale, and capped by lacustrine sandy limestones, marls and mudstones of Rhaetian age. A similar succession is seen in Hamilton 12/26-1 (Jakobsson *et al.*, 1980).

Frostick *et al.* (1988) have interpreted the Inner Moray Firth Triassic succession by using a model derived from studies of East African continental-rift processes. They conclude that sedimentation occurred in a simple half-graben structure, which was controlled by dip-slope development on the Great Glen fault. The onshore sediments, on the unfaulted, southern margin of the half-graben are seen as proximal equivalents of finer-grained sediments, which accumulated under lacustrine conditions in the depocentre. Roberts *et al.* (1989, 1990a) propose a more complex history for the area, with a major Permian phase of fault-controlled subsidence. They suggest that subsequent Triassic synsedimentary faulting subdivided the area into smaller subbasins and that the wedge-shaped asymmetry central to the Frostick *et al.* (1988) thesis does not exist. In a third interpretation, Thomson and Underhill (1993), while conceding that the data are sparse, interpret the depositional history as having occurred in a basin, with broad-based subsidence following earlier extension. This implies a Triassic phase of thermal subsidence, following a minor phase of Permian rifting, and is consistent with observations of Upper Permian and Triassic seismic sequences dominated by concordant reflectors and basin-wide westerly thickening packages, with little variation adjacent to faults. In addition, they note the absence of any abrupt thickening across the Great Glen Fault, implying its negligible role in controlling basin development at this time.

7.3.5 Economic geology

No account of major hydrocarbon fields in the Triassic of the Central North Sea Basin has been published, although a number of significant discoveries, mostly in the upper part of the Skagerrak Formation, have been made. Oil has also been tested from probable Smith Bank sandstones in well 22/16-1. Unfortunately, this discovery has not been fully appraised and relevant reservoir data are lacking.

In the western part of the basin, fine-grained clastic sediments predominate and sands are largely developed only near the basin margins. In the Norwegian and Danish basins and in the Central Graben, thick sands are developed in the prograding Skagerrak Formation or adjacent to isolated 'highs' within the basins. The Josephine structure in Block 30/13 represents just such an intrabasinal high. It

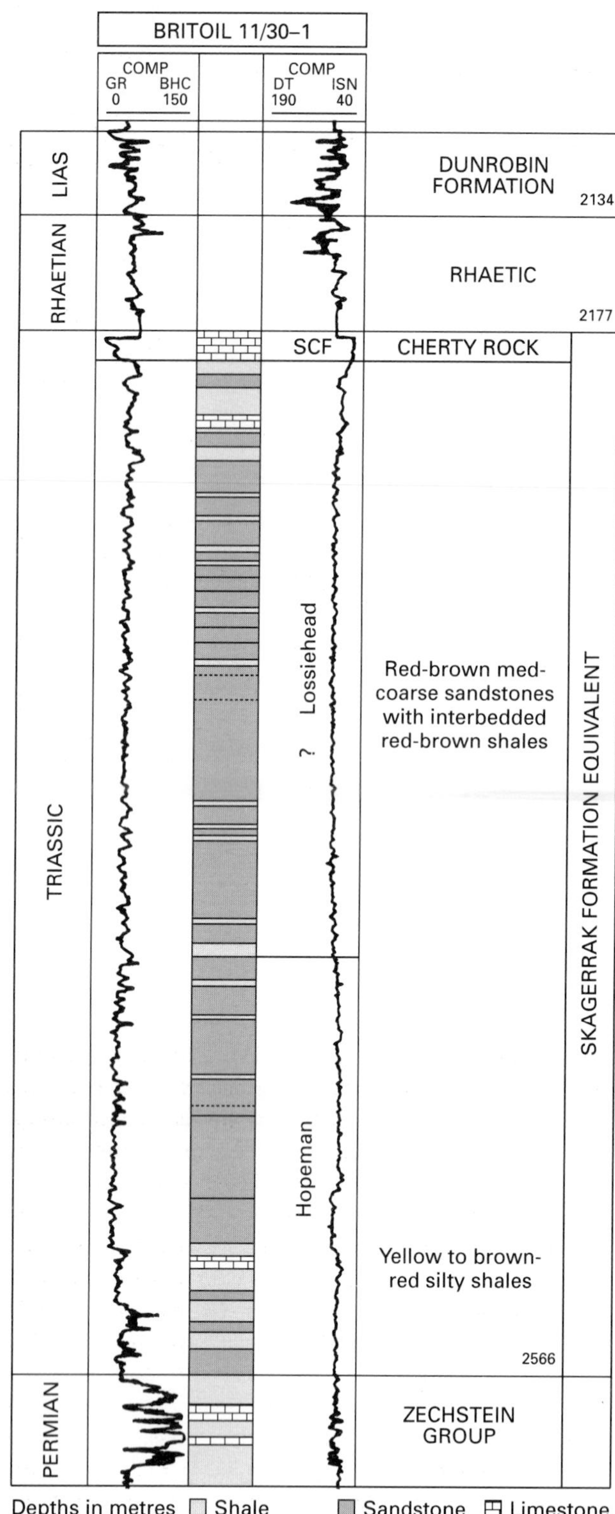

Fig. 7.21 Triassic stratigraphy of well 11/30-1, Inner Moray Firth.

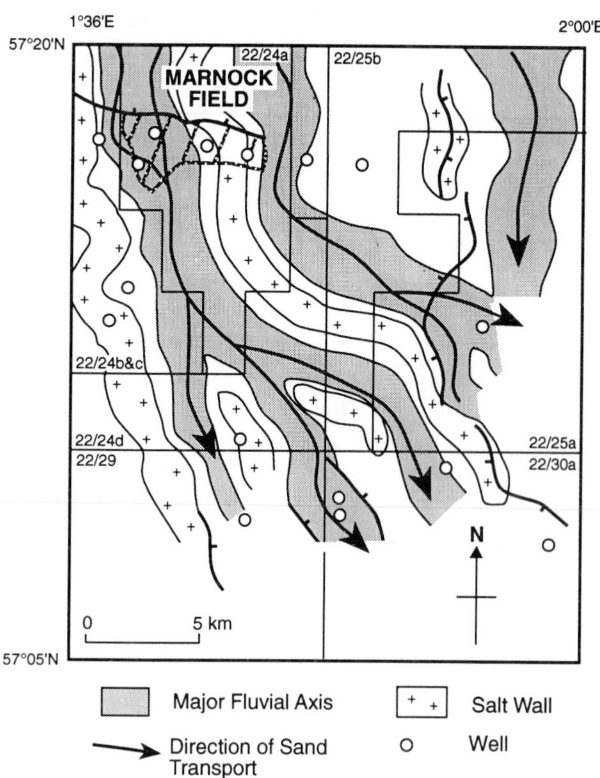

Fig. 7.22 Marnock Field area: distribution of Upper Triassic (Marnock) sandstones (after Smith *et al.*, 1993).

contains oil in the Josephine Sandstone Member, but this has not yet been confirmed as a commercial discovery. Wells in Blocks 22/19 (Fiddich Field), 29/8a (Acorn), 29/9b (Beechnut), 22/24a (Marnock) and 22/24b (Skua) have also tested oil or gas condensate from late Triassic sands of the Skagerrak Formation. The Marnock Field (Figs 7.22 and 7.23) has been announced as a major gas-condensate discovery, with reserves exceeding 30×10^9 m^3 (1.06×10^{12} ft^3). The Judy–Joanne complex in Blocks 30/7a and 30/12 has total hydrocarbons in place of 11.5×10^9 m^3 (0.40×10^{12} ft^3) of gas and 98×10^6 barrels of oil.

Glennie and Armstrong (1991) document a typical non-productive Skagerrak reservoir section in their study of the Kittiwake Field (21/18). Reservoir characteristics are not encouraging, with ϕ = 6–20%, k = 0.05–10 mD and net-to-gross ratios of 0.1 : 0.4. Compounding the problem, the sandstones contain numerous tightly cemented zones, generating poor vertical and lateral communication. The Crawford Field (Block 9/28) contains combined reserves of 9×10^6 barrels of oil in Triassic and Middle Jurassic sandstones in an area where the Skagerrak Formation is transitional with the Cormorant Formation at the southern end of the Viking Graben (Yaliz, 1991). Reservoir quality is fair in the upper of the two reservoir units, where stacked channel sands predominate (ϕ = 20–23%; k = > 100–2000 mD), but reservoir performance is limited by compartmentalization and horizontal stratification.

In all of these discoveries, the source of the hydrocarbons is the Kimmeridge Clay Formation, and the structural setting—Triassic sandstones in fault communication with downthrown Jurassic source rocks—is repeated in commercial fields in the Northern North Sea Basin (Fig. 7.24).

The limited success of exploration in the Triassic of the Central North Sea reflects lack of confidence in locating a suitable reservoir rather than adequate trapping and sourcing mechanisms. To the west, where the sheet-flood sands prograde into lacustrine shales, the sandstones tend to be ratty and heavily cemented; in the Central Graben, although the Skagerrak sandstones are better developed, they are fine-grained and again generally tight. The clastic

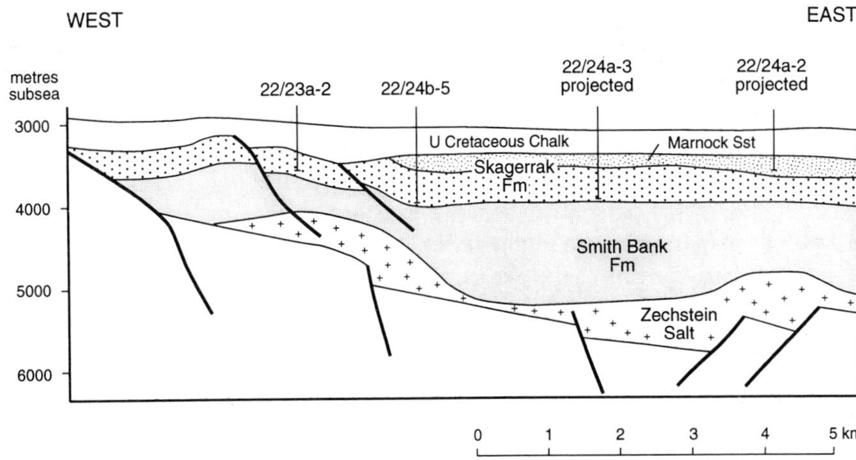

Fig. 7.23 Marnock Field geological cross-section (after Hodgson *et al.*, 1992).

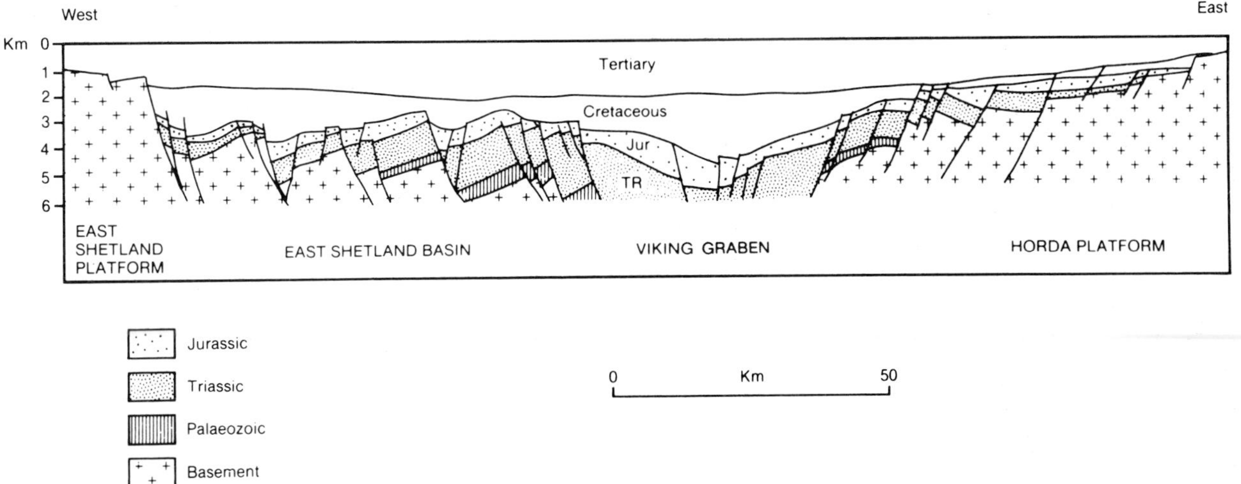

Fig. 7.24 Geological sketch section across the Northern North Sea.

alluvial-fan complexes of the Norwegian and Danish Basins are invariably tightly cemented.

Burley (1987), Humphreys *et al.* (1989) and Purvis (1990) have described various diagenetic processes that could contribute to the formation or preservation of porosity in sandstones of the type encountered in the Skagerrak Formation. Instances of secondary porosity have already been recorded, and it is possible that a more detailed correlation of lithofacies and burial histories with the application of diagenetic modelling may lead to the recognition of areas of enhanced reservoir development. In the Marnock and Fiddich fields, which lie on the flanks of the Forties–Montrose High, sediments buried at shallow depths during the Jurassic and Cretaceous had the potential for leaching by groundwater circulation during periods of uplift and erosion. These processes could also have taken place along the margins of the Central Graben system and over intragrabenal highs, such as the Forties–Montrose High, Josephine High and Mandal High.

7.4 The Northern North Sea Basins

The structure of the Northern North Sea Basins is dominated by north–south faulting, which caused the formation of deep and well-defined grabens (Fig. 7.24). The variations in thickness of the Triassic sediments in the grabens indicate differential subsidence, and the facies relationships are compatible with fault-controlled sedimentation. There is a general thickening of late Triassic sequences into the basin axes, which suggests deposition during thermal subsidence following an earlier rifting episode (Badley *et al.*, 1988; Marsden *et al.*, 1990). A late Permian–early Triassic rift event has been proposed by Ziegler, P.A. (1975, 1978, 1990b), Badley *et al.* (1988), Scott and Rosendahl (1989), Thorne and Watts (1989) and Roberts *et al.* (1995), but there is considerable uncertainty on the precise dating of this event and on the scale of the Triassic component (Gabrielsen *et al.*, 1990; Roberts *et al.*, 1995). Between the Horda Platform and the Unst Basin, there is no unequivocal evidence for major syn-Triassic faulting and block rotation, because of later tectonic overprinting. Roberts *et al.* (1995) estimate that average Triassic extension across the East Shetland Basin is comparable with later Jurassic extension (*c.* 15%) and suggest that, in the early Triassic, the Tern/Eider and Cormorant fault blocks comprised a horst uplifted in the footwall of major faults flanking the Magnus and Statfjord half-graben. On the Horda Platform and in the Stord Basin and Sogn Graben, however, convincing evidence for major rifting in the early Triassic is present (Lervik *et al.*, 1990; Marsden *et al.*, 1990; Steel and Ryseth, 1990; Roberts *et al.*, 1995). Here, wedge-shaped sediment packages with growth into north–south-orientated faults overlie a tilted basement topography, and Roberts *et al.* (1995) estimate that Triassic extension reached *c.* 40%.

The Northern North Sea Basins were separated from the Central North Sea Basin by a structural element in the area of the triple junction. This was most apparent during Rhaetian sedimentation, when the drainage pattern in the Viking Graben was from south to north while in the Central Graben it was from north to south. Earlier in the Triassic, the depositional system of the Skagerrak Formation had extended into the South Viking Graben as far as the Crawford Spur.

There are few published records of penetration of complete sequences of thick Triassic sediments, so the relationship of the Triassic with the underlying sediments and the effect of the Hardegsen movements on basin formation are therefore imperfectly understood. In thc Beryl Embayment, Ormaasen *et al.* (1980) indicate that in Quintana 9/17-1, deposition of continental arenaceous red beds with interbedded shales was continuous from the late Permian into the early Triassic. It can be assumed, however, from the evidence of basal Triassic conglomerates overlying Zechstein evaporites in the Beryl Embayment (9/13a-22), that in the basins bordering the Viking Graben—the East Shetland Basin to the north-west and the Stord or Horda Basin to the east (see Fig. 7.1)— Triassic deposition would have been dominated by alluvial fans at the fault-bounded margins, with finer-grained fluvial, flood-plain or lacustrine sediments in the 'lows'. In these areas, the transition from Permian sediments should be clearly defined on well logs. North of the Viking Graben, in the Møre Basin, the relationship between the Triassic and underlying Zechstein is again apparently conformable but with an abrupt transition from carbonates and argillaceous sediments to dominantly arenaceous sediments.

Series	Stage	Lithostratigraphy: Beryl Embayment	E Shetland Basin	Norway Viking Graben
L JURASSIC	SINEMURIAN	BANKS GP: NANSEN FM	NANSEN FM	NANSEN FM
L JURASSIC	HETTANGIAN	BANKS GP: STAT-FJORD FM	STAT-FJORD FM	STATFJORD FM: Eiriksson Mbr, Raude Mbr
UPPER TRIASSIC	RHAETIAN	HERON GROUP: Harris Mbr	HERON GROUP: CORMORANT FM	HEGRE GROUP: LUNDE FM
UPPER TRIASSIC	NORIAN	HERON GROUP: Harris Mbr / Lewis Mbr	HERON GROUP: CORMORANT FM	HEGRE GROUP: LUNDE FM
UPPER TRIASSIC	CARNIAN	HERON GROUP: Lewis Mbr	HERON GROUP: CORMORANT FM	HEGRE GROUP: LUNDE FM
MIDDLE TRIASSIC	LADINIAN	HERON GROUP: CORMORANT FM	HERON GROUP: CORMORANT FM	HEGRE GROUP: LOMVI FM
MIDDLE TRIASSIC	ANISIAN	HERON GROUP: CORMORANT FM	HERON GROUP: CORMORANT FM	HEGRE GROUP: TEIST FM
LOWER TRIASSIC	SCYTHIAN	?	?	?

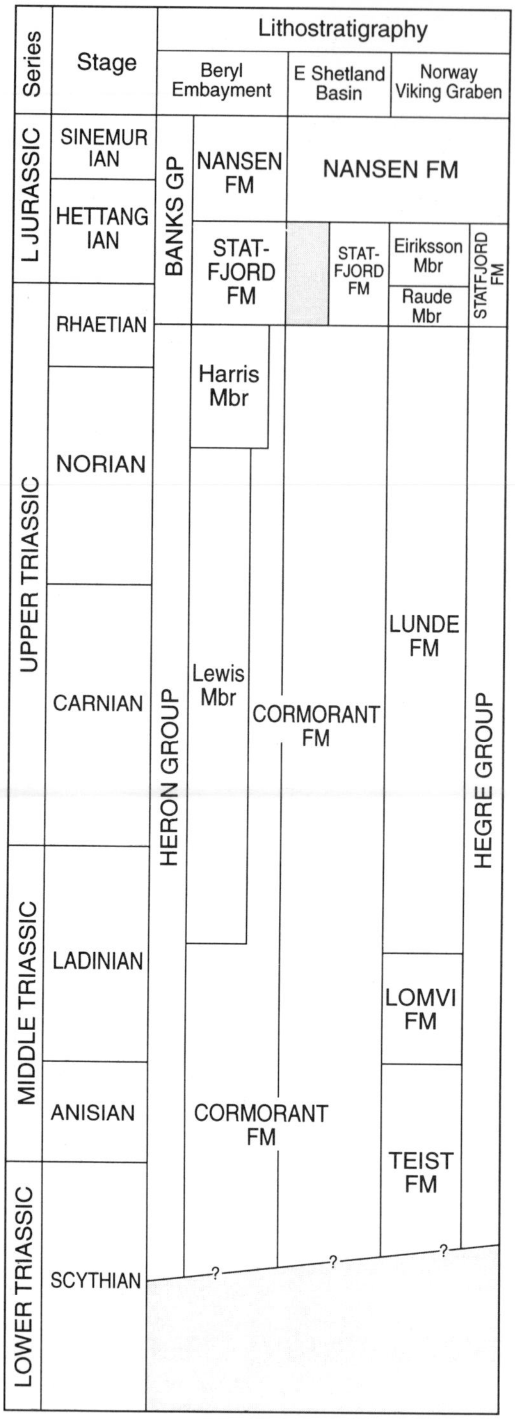

Fig. 7.25 Triassic lithostratigraphy, Northern North Sea.

7.4.1 The Cormorant Formation

With the data available to them, Deegan and Scull (1977) recognized only one Triassic formation in the Northern North Sea Basins, the Cormorant Formation. Subsequently, enough well data became available north of 60°N for Vollsett and Doré (1984) to propose a new unit, the Hegre Group (Fig. 7.25). In Deegan and Scull's (1977) original definition, the Cormorant Formation is typically composed of pinkish or white, fine- or medium-grained argillaceous sandstone with some red-brown siltstones and shales. Thicker sequences with coarser-grained sandstones and conglomerates overlain by more typical Cormorant Formation facies are developed towards the East Shetlands Boundary Fault. Because sedimentation was controlled by relatively locally generated relief, variations in thickness from one fault block to another are considerable and may be of the order of 2000 m. Similarly, both lithostratigraphic and biostratigraphic correlations are difficult in these sediments. Palynological evidence indicates a late Norian to early Rhaetian age for a relatively widespread argillaceous horizon near the top of the formation and a Rhaetian age for the higher beds (Brennand, 1975). The characteristics of the Cormorant Formation are illustrated in Fig. 7.26.

Near the top of the Cormorant Formation, in the area of the Beryl Field, Knutson and Munro (1991) record a sequence of massive sheet-flood sandstones, grading upwards into fluvial sandstones and lacustrine-delta mudstones. This is thought to represent a sheet-flood-dominated fan delta that prograded towards a lake in the north-west of the Beryl Embayment. This sequence is now termed the Lewis Member (Cameron, 1993a) and it is overlain by lacustrine mudstones of the Harris Member.

It is difficult to generalize about the depositional environments of the various facies in the Cormorant Formations because of the relatively complex tectonic setting that controlled patterns of sedimentation in the Viking Graben. Some of the more widespread and laterally persistent cyclical sandstone and shale facies, however, appear to have

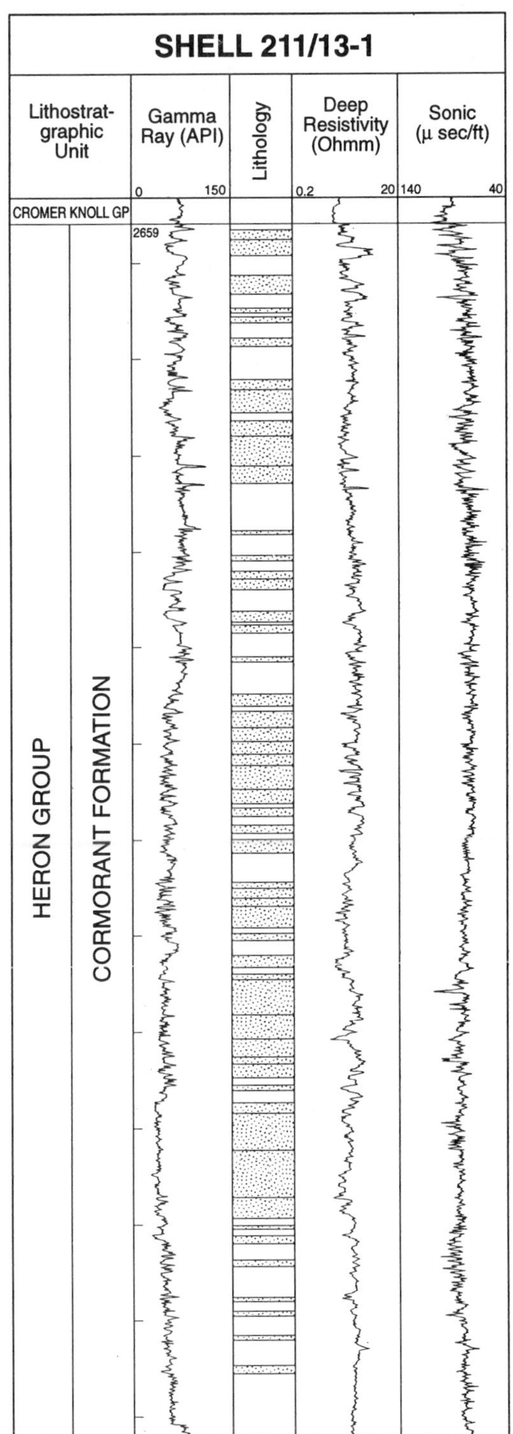

Fig. 7.26 Triassic (Heron Group) lithostratigraphy of well 211/13-1, Northern North Sea. Depths in metres below KB depth ticks at 100 m intervals. For key, see Fig. 7.3.

been deposited under shallow-water conditions, and the absence of marine organisms suggests a fluvio-lacustrine environment. The sandstones are usually poorly sorted, with an argillaceous matrix, and often show transitional boundaries with the interbedded shales. This poor sorting, together with the occurrence of oxidized plant debris in the shales and in the argillaceous matrix, is consistent with deposition in a low-energy environment. There is no evidence of winnowing or prolonged reworking, as would be found, for example, with flash-flood or alluvial-fan processes. It is therefore reasonable to assume that access to the main depocentres from the higher-energy basin-margin and fault-terrace areas was relatively restricted. A sedimentary model outlined by Clemmensen *et al.* (1980) is compatible with these observations. They suggest marginal alluvial fans, feeding through braided streams and stabilized distributary channels into a central, northwards-draining, elongate basin or coalescing series of basins, which would have included lacustrine and sabkha environments. The Beryl Embayment was the site of such a lacustrine depocentre. Dean (1993) describes a massive, bioturbated and cross-laminated sandstone facies, associated with finely interbedded sandstones and shales, conglomerates, siltstones and shales. These accumulated in fluvial and sand-flat environments adjacent to an extensive lake, whose fluctuating levels are linked to climatic cyclicity.

Towards the end of the Triassic, a transgression from the Boreal Sea resulted in the establishment of marine environments in the central part of the basins at the northernmost end of the Viking Graben, and of fluviodeltaic environments along the margins (Clemmensen *et al.*, 1980; Jakobsson *et al.*, 1980). In the central northern part of the Viking Graben there is, therefore, a conformable passage from the Cormorant Formation to the overlying Statfjord Formation.

7.4.2 The Hegre Group

Based primarily on evidence from wells in the East Shetlands Basin, Vollsett and Doré (1984) introduced a new unit, the Hegre Group, comprising the Teist, Lomvi and Lunde Formations, and proposed restricting the use of the term 'Cormorant Formation' to attenuated sequences on structural highs in the UK sector. Lervik *et al.* (1990) published additional well data from the Central and Northern North Sea, which further assisted in resolution of the lithostratigraphic subdivision of the area.

Cameron (1993a) considers the use of Vollsett and Doré's (1984) terminology in the UK sector potentially misleading, particularly because of problems in identifying the Lomvi Formation. As an example, he cites inconsistencies in correlation of wells in the Beryl Embayment by Frostick *et al.* (1992) and therefore proposes retaining the use of the Cormorant Formation. In the eastern parts of Quadrants 3 and 211, however, a sheet sandstone up to 150 m thick can be confidently assigned to the Lomvi Formation and in this area the tripartite subdivision of the Hegre Group could be applied.

A typical well that penetrates the group is illustrated in Fig. 7.27. The informal lithostratigraphic units and biostratigraphy are after Lervik *et al.* (1990). The group consists of interbedded white to red sandstones and red shales and claystones, associated with intervals dominated by coarse or fine clastic sediments. The base of the group has been identified by Lervik *et al.* (1990) in Norsk Hydro N31/6-1, and the upper boundary is clearly defined by the overlying Statfjord sands.

The Teist Formation is of Early to Late Triassic age and composed of interbedded sandstone, claystone and marl. In the type well (Mobil N33/12-5), the formation is represented by a gradual upward-coarsening succession. The formation has been encountered in all deep wells between

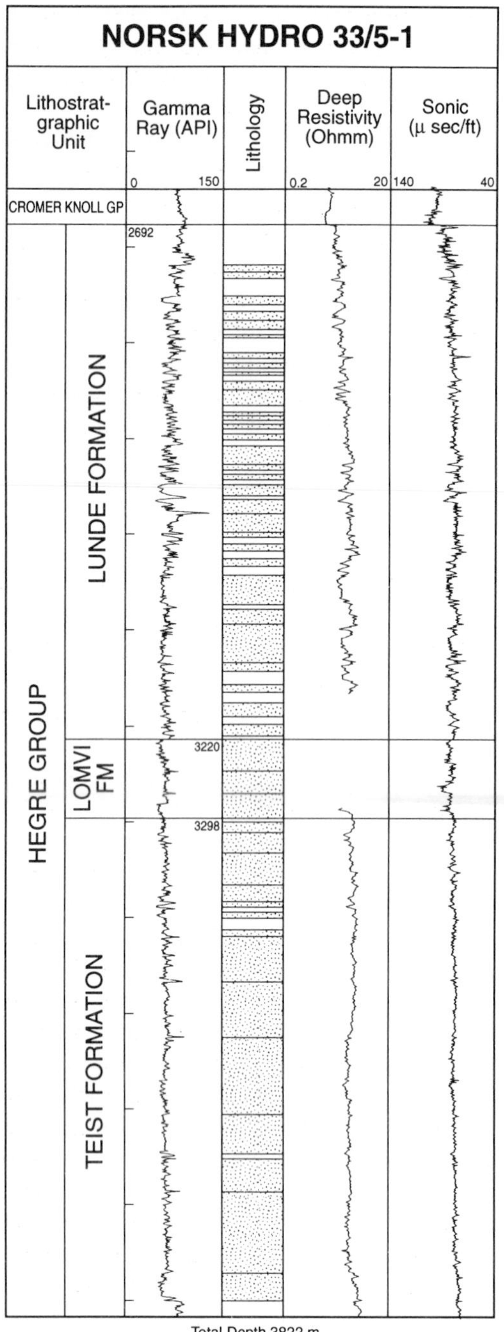

Fig. 7.27 Triassic (Hegre Group) stratigraphy of Norwegian well N33/5-1, Northern North Sea. Depths in metres below KB; depth ticks at 100 m intervals. For key, see Fig. 7.3.

the Brent Field and the Møre Basin and is presumed to be of continental origin. The sandstones are attributed to fluvial and aeolian processes, and the fine-grained sediments are assigned to overbank and lacustrine environments. At the base of this unit, in N31/6-1, Lervik *et al.* (1990) describe a 274-m-thick, red claystone unit overlying basement. They equate this claystone with the Smith Bank Formation of the Central North Sea and propose separate formation status for it.

The Lomvi Formation overlies the Teist Formation between Brent and the Møre Basin and is composed predominantly of fine- to coarse-grained kaolinitic sandstones. This formation is believed to be of fluvial origin and has been dated as Skythian to Anisian in Saga N34/4-4 and Ladinian in Norske Shell N31/2-4 by Lervik *et al.* (1990). The range of palynological dating may suggest that the Lomvi lithofacies is not restricted stratigraphically.

The boundary between the Lomvi Formation and the overlying Lunde Formation is marked by the appearance of a thick claystone unit. Although the Lunde Formation is defined as a sequence of interbedded fine- to coarse-grained sandstones, claystones, marls and shales, this basal unit, up to 300 m thick, is particularly distinctive and is consistent in thickness and lithology in the southern parts of Quadrants 211 and N33.

The Lunde Formation is assumed to have been fairly widespread. Its current distribution pattern, however, indicates its absence from structural highs, resulting from erosion or non-deposition. Scott and Rosendahl (1989) postulate that fluvial sedimentation in the southern part of the graben derived from a persistent high to the east, in the area of the Egersund High or the Horda Platform. Samarium–neodymium isotope analysis of sediments from the Snorre Field, however, suggests a south-westerly source (Mearns *et al.*, 1989). Deposition of this formation is considered to have taken place in lacustrine or fluvial environments. The fine-grained sandstones in Shell 211/13-1 (see Fig. 7.26), attributed by Lervik *et al.* (1990) to the upper part of the Lunde Formation, exhibit small-scale ripple cross-stratification and bioturbation and incorporate mud clasts and mud balls. The age of the formation is poorly defined but has a range of Ladinian to Rhaetian.

In the Snorre Field area, Nystuen *et al.* (1990) recognize 10 high-order, predominantly upward-coarsening log sequences in the Hegre Group and the Statfjord Formation. In the grossly upward-fining Upper Lunde Formation, they record five medium-order log sequences, which equate with reservoir zones. These medium- and high-order log sequences are believed to reflect responses to relative changes in base level and sediment input. The high-order log sequences exhibit considerable regional continuity and are identifiable in the Statfjord, Troll and Gullfaks field areas.

The relationship of the Hegre Group to the Cormorant Formation and to units in the Central North Sea Basin has been addressed by Lervik *et al.* (1990) and Cameron (1993a). Detailed correlation does not appear feasible on the currently published database.

7.4.3 The Statfjord Formation

The Statfjord Formation was originally recognized by Bowen (1975), although here it is used as defined by Deegan and Scull (1977). The transition from the Cormorant Formation to the Statfjord Formation is indicated in the type section by an upward-coarsening succession of variegated grey, green and red shales, interbedded with thin siltstones, sandstones and dolomites (Fig. 7.28). This unit, which is 60 m thick, constitutes the Raude Member (see Fig. 7.25) and represents a major transgressive sequence. The transgression was a relatively slow, progressive event, with the continental pattern of deposition established in the southern and more elevated parts of the basin persisting, while the Statfjord Formation slowly onlapped from the north with pronounced diachroneity. Chauvin and Valachi (1980) describe the depositional environment of the type

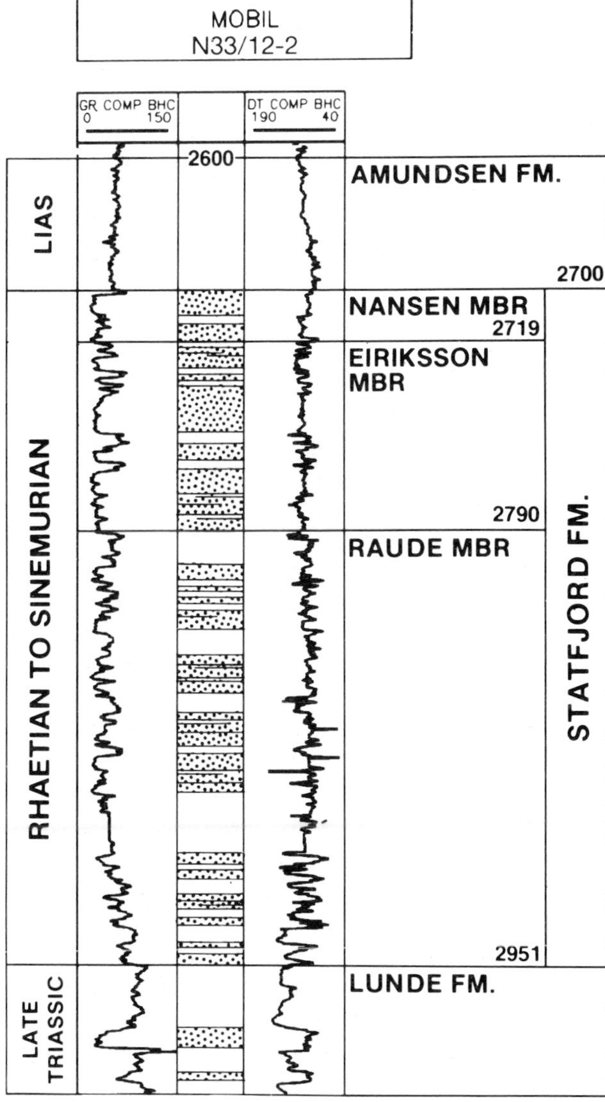

Fig. 7.28 Statfjord Formation type section Norwegian well N33/12-2 (after Vollsett and Doré, 1984).

Statfjord Formation as a flood plain with some meandering streams. This is closely comparable with the interpretation of the contemporaneous Gassum Formation in the Egersund Basin (Jakobsson *et al.*, 1980). In a later interpretation, however, Røe and Steel (1985), while acknowledging a fluvial source for most of the arenaceous units, suggest that the fine-grained units represent coastal-plain rather than overbank deposits. Their examination of core and log data from the Statfjord area indicates that the transition from the Cormorant Formation to the Raude Member was in response to marked fault activity and rapid basin subsidence. This resulted in the generation of coastal alluvial fans or fan deltas, which show evidence of repeated progradation and abandonment on to the coastal plain.

As already noted, Mearns *et al.* (1989) have used samarium–neodymium isotopic analysis to identify the source of the Statfjord Formation in the Snorre Field. Their evidence suggests that the Lewisian Shield, west of Shetland, is the most likely provenance, although this is difficult to reconcile with some of the sedimentological studies cited above.

The age of the Raude Member is generally accepted as Rhaetian, probably mostly late Rhaetian, but the paucity of diagnostic palynofloras and the problems inherent in defining the Triassic–Jurassic boundary on palynological criteria alone (Fisher and Dunay, 1981) do not allow an accurate dating of the upper boundary of the member. In the type area, there is a major shale break between the Rhaetian Raude and Jurassic Eiriksson members before the channel sandstones coarsen upwards into the thicker, more massive, low-sinuosity braided-stream facies of the Eiriksson Member.

7.4.4 Economic geology

The Northern North Sea Basins have proved to be a highly successful offshore exploration area, with major discoveries in the Cormorant Formation (Beryl and Tern fields), the Lunde Formation (Snorre Field) and the Statfjord Formation (Alwyn North, Brent, Gullfaks, Snorre and Statfjord fields).

The Beryl Field is located on a complex of horsts and tilted fault blocks in the Beryl Embayment, in the west-central part of the Viking Graben (Knutson and Munro, 1991). The Lewis Member of the Cormorant Formation is a secondary reservoir to the Jurassic accumulations and contains oil reserves of 85×10^6 barrels from total field reserves of 800×10^6 barrels and 45×10^9 m^3 (1.6×10^{12} ft^3) gas. The Cormorant Formation also contains significant reserves (20×10^6 barrels) as a secondary reservoir in the Tern Field, a tilted horst in block 210/25a in the East Shetland Basin (van Panhuys-Sigler *et al.*, 1991). The problem of modelling the complex distribution pattern of channel and sheet sands in a dominantly silty succession has been solved by high-resolution three-dimensional (3D) seismic technology, which visualizes the reservoir facies in 3D space. In the Penguin Field Complex, gas condensate has been recorded in the Cormorant Formation (Lunde Formation equivalent) of 211/13-1 (Brooks, 1977). The well, located on a north-east–south-west-trending horst, has 52 m of gross pay, with 21 m of effective reservoir.

The Lunde and Statfjord formations form significant reservoirs in the North Viking Graben, in particular in the area of the Tampen Spur, where the Alwyn North, Brent, Gullfaks, Snorre and Statfjord fields are located. The Lunde Formation is the major reservoir in the Snorre Field (Fig. 7.29). The fluvial sandstones have variable reservoir properties, which result from their energy of deposition. The higher-energy channel sandstones are usually 2.5–3 m thick, massive, fine- to medium-grained and well sorted, typically with a < 1-m-thick, calcareous, cemented, intraformational basal conglomerate. Reservoir quality is good, with $\phi = 25\%$, $k = 200$–3000 mD and clay content < 5%. These channel sandstones may occur singly or as stacked bodies up to 15 m thick. With decreasing energy of deposition, the sandstones become finer-grained, laminated and clay-rich, with a resulting loss of porosity (21%) and permeability (50–200 mD). These sandstones are interbed-

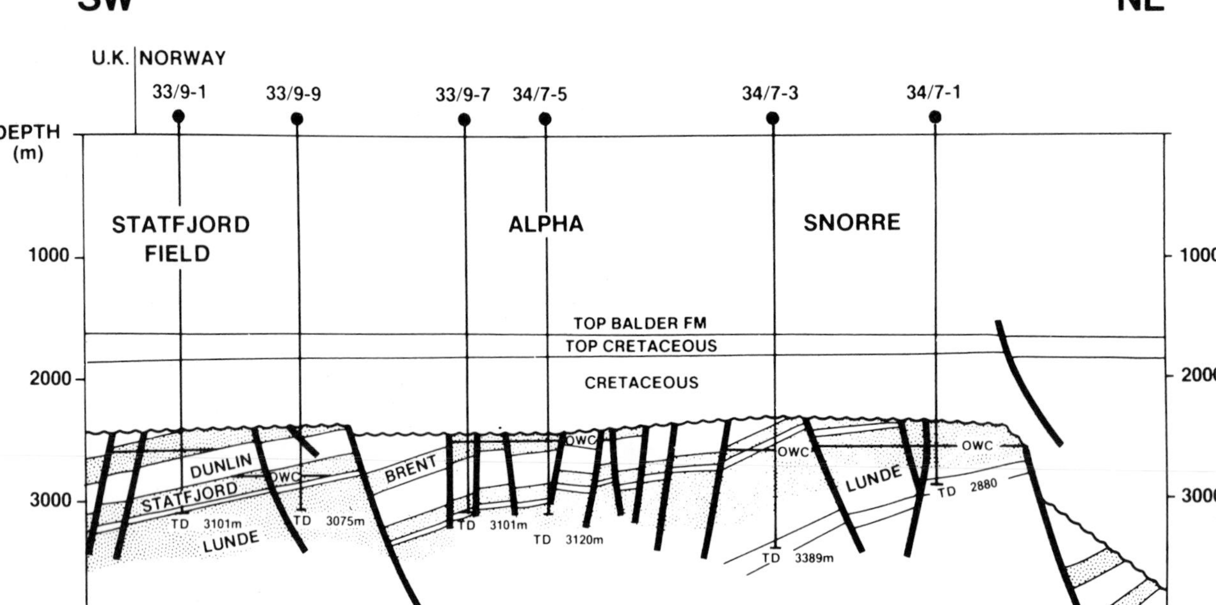

Fig. 7.29 Section across the Statfjord and Snorre fields, Northern North Sea (after Karlsson, 1986).

ded with non-channelized sands and overbank shales with thin (< 2 m) crevasse sands. Production tests have resulted in good flow rates, up to 1750 m^3/day on a 17.5 mm choke, and reserves of 485 × 10^6 barrels of oil are predicted to be reservoired in the Lunde Formation (Karlsson, 1986).

The Statfjord Formation is a major reservoir in the Alwyn North, Brent, Gullfaks, Snorre and Statfjord fields. In the Alwyn North Field, reserves of 46–48°API condensate in the Statfjord Formation reservoir a total of 27 × 10^6 barrels of oil, with 17.2 × 10^9 m^3 (0.6 × 10^{12} ft^3) of gas (Inglis and Gerard, 1991). In the Brent Field (Fig. 7.30), the Statfjord Formation has a gas cap of 140 m and an oil column of 130 m. Reservoir characteristics are good, with ϕ = 10–26% and k = up to 5500 mD (Bowen, 1975); oil reserves are estimated at 95 × 10^6 barrels with 33.9 × 10^9 m^3 (1.2 × 10^{12} ft^3) of gas (Struijk and Green, 1991). The Snorre Field production tests have resulted in good flow rates, up to 1750 m^3/day on a 14.3 mm choke being recorded from the Eiriksson Member; oil reserves in the Statfjord Formation are estimated at 245 × 10^6 barrels (Hollander, 1987). In the area of the type well (Statfjord Field), there is a conformable transition from the uppermost Lunde unit into the basal Raude Member. The lower unit of the Statfjord Formation is lithologically similar to the uppermost Lunde unit, consisting of shales and siltstones interbedded with generally thin (1–3 m), fine- to medium-grained channel sandstones. These sandstones increase in abundance upwards and may form stacked bodies up to 8 m thick. Reservoir quality is poor to moderate with net-to-gross of 18% and ϕ = 20.8%. A major shale break separates the Raude and Eiriksson members. The thicker, more massive, low-sinuosity braided-stream facies of the Eiriksson Member comprise fine- to very coarse-grained, poorly sorted, feldspathic to arkosic sandstones and contain both a kaolinitic matrix and calcite cement. Individual channels average 3 m thick but commonly occur as stacked sand bodies up to 10 m thick. They are interbedded with silty shales and sparse coals. Reservoir quality is generally good with ϕ = 20–30% and k = 1.3–2 D, and reserves are estimated at 288 × 10^6 barrels oil, with 10.8 × 10^9 m^3 (0.38 × 10^{12} ft^3) gas (Roberts *et al.*, 1987).

No significant discoveries have yet been documented in the Lomvi Formation. The reservoir characteristics are attractive, with porosities typically in the range 10–20%. With the knowledge that throws on the major graben faults commonly result in potential fault communication between mature Kimmeridgian source rocks and the Triassic, one can speculate on the possibility of undiscovered accumulations in the Lomvi Formation underlying the major Jurassic fields.

The source of the oil reservoired in the Triassic is the Kimmeridge Clay Formation. Oil generation, which is still in progress, probably commenced between the latest Cretaceous and Late Tertiary in the East Shetlands Basin and Viking Graben (Goff, 1983). The characteristic tilted fault blocks, with Triassic sediments faulted above Jurassic source rocks, therefore predate the main phase of oil generation and the faults themselves provide the migration pathways (see Fig. 7.24).

7.5 Adjacent areas

As drilling activity accelerates in offshore Norway north of 62°N, our knowledge of the Triassic is enhanced. Jacobsen and van Veen (1984) have established correlations not only with the North Sea but also with Greenland and Svalbard. In the areas of East Greenland, Svalbard and the Barents

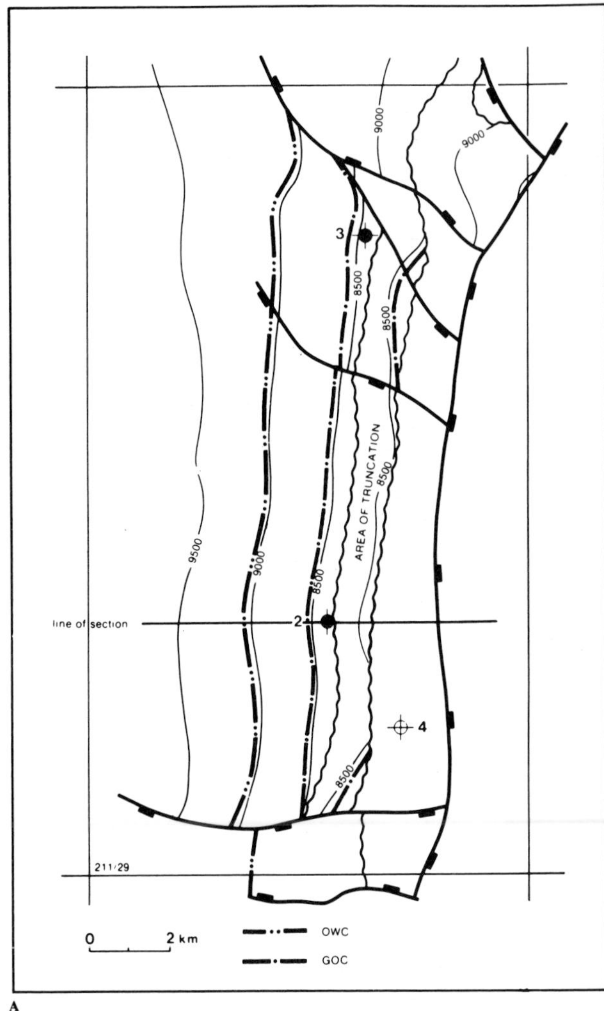

A

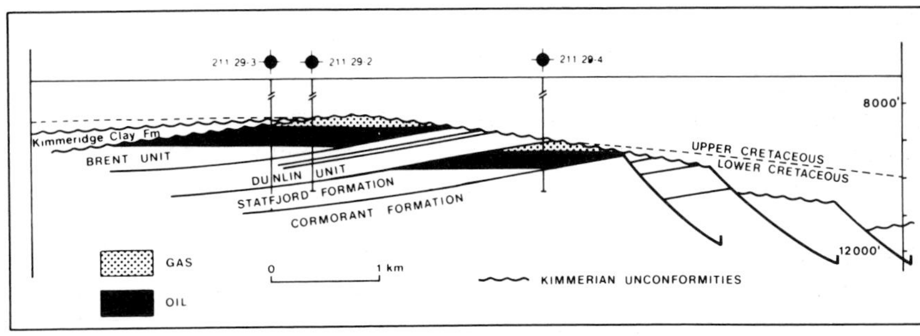

B

Fig. 7.30 Map (A) and cross-section (B) of Brent Field (after Bowen, 1975).

Sea, we know that marine sedimentation was more or less continuous in the Boreal Sea, and the rich source rocks in the Middle Triassic Bothneheia Formation of Svalbard (Mørk and Worsley, 1979) must be considered in any assessment of regional prospectivity in northern waters.

To the west of the Shetland Platform (see Fig. 7.1), Triassic sedimentation was active in the narrow, fault-bounded, south-west–north-east-trending West Shetland Basin complex (Booth *et al.*, 1993). Roberts *et al.* (1995) and Platt (1995) consider that these grabenal structures were originally contiguous with a 'Triassic Viking Graben', of which the Unst Basin and Magnus half-graben represent the remnant. Ridd (1981) has postulated the presence of over 1000 m of Permo-Triassic sediments adjacent to the Shetland Spine Fault System. The dominant lithologies are thin red-brown siltstones and shales, interbedded with white and grey, commonly calcareous sandstones. Coarser clastics, with sandstones and conglomerates, are encountered near the basin margins. If this is analogous with the Central and Northern North Sea basins, deposition probably took place in a rapidly subsiding asymmetric graben. Coarse clastic material accumulated in marginal alluvial fans, and finer clastic material, fed via fluvial channels across extensive flood plains with aeolian sands, accumulated in basinal lakes.

In the East Irish Sea Basin, significant oil and gas discoveries have been made in the Triassic Sherwood Sandstone Group (Fig. 7.31). Both oil and gas are sourced from the Namurian Hollywell Shale. The upper part of the Ormskirk Sandstone Formation is the major reservoir unit and comprises stacked sequences of aeolian dune facies, aeolian sand sheets and braided-fluvial facies (Fig. 7.32). In

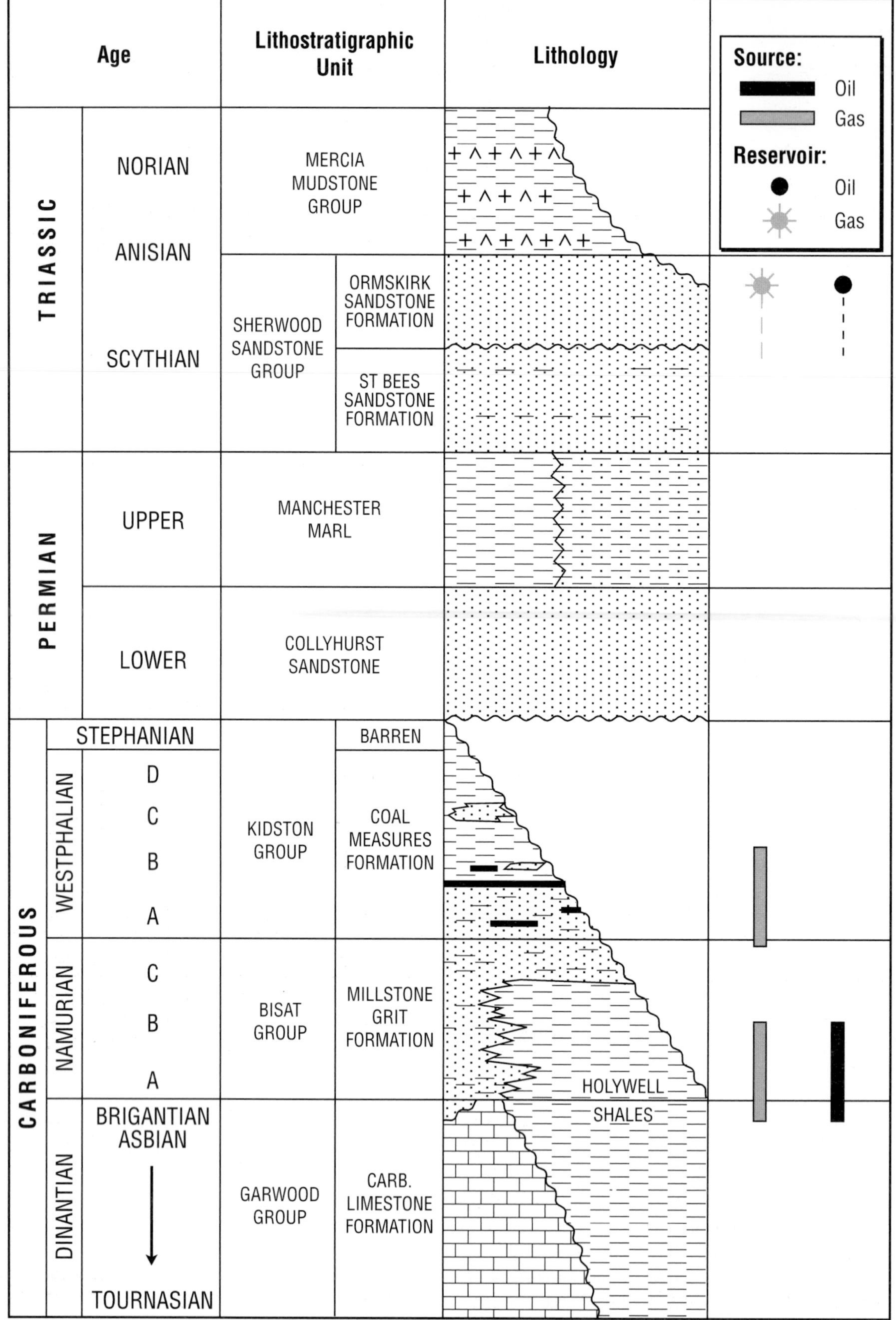

Fig. 7.31 Generalized stratigraphy, East Irish Sea Basin.

the south of the basin, reservoir quality is excellent, but, in the north, pervasive illitization has caused serious deterioration in reservoir quality. The South Morecambe Field is currently in production and has gas reserves of 130×10^9 m^3 (4.57×10^{12} ft^3) (Stuart and Cowan, 1991); the Douglas (oil), Hamilton (gas) and Lennox (oil and gas) fields (Trueblood *et al.*, 1995) came on stream in 1995 (Table 1.9).

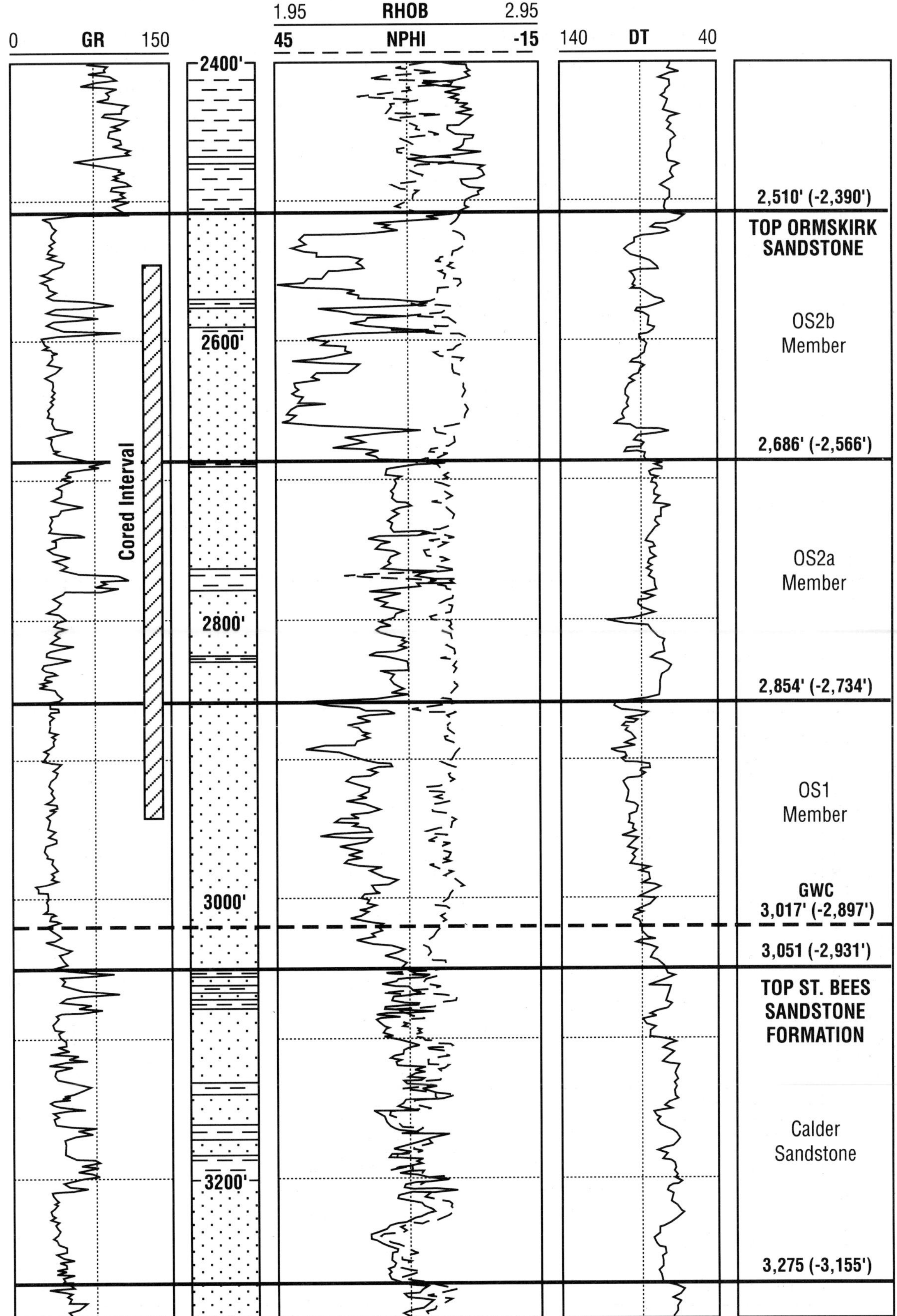

Fig. 7.32 Reservoir section and core-facies analysis, well 110/13-1.

On the south coast of England, the Triassic Sherwood Sandstone is the principal reservoir in the Wytch Farm oilfield. Total Triassic recoverable reserves are 250×10^6 barrels, of which 100×10^6 barrels are in an offshore extension of the field. An additional 50×10^6 barrels are reservoired in Lower Jurassic Bridport Sandstone and 'Frome Clay' limestone (Bowman *et al.*, 1993).

7.6 Summary and conclusions

Throughout the Triassic, sedimentation was predominantly of clastic red beds, accumulating in continental basins. South of the Mid North Sea–Ringkøbing–Fyn High, the sedimentary succession displays considerable lateral uniformity and includes thick halites. In the Central North Sea Basin, persistent halites are absent and there is marked diachroneity as the coarse, clastic Skagerrak Formation progrades westward over mudstones of the Smith Bank Formation. The Northern North Sea Basins also lack persistent halites, and the sediments display considerably less lithostratigraphic continuity than in the basins to the south.

Rarely in the North Sea has the Triassic realized its potential as a major exploration objective. In the Southern North Sea Basin, the Zechstein evaporites have effectively sealed off the Bunter Sandstone reservoirs from the Westphalian source rocks. In the Central North Sea Basin, it is only as the processes that control the distribution of good-quality reservoir are more fully understood that successful Triassic plays are being extensively tested. In the Northern North Sea basins, the structural relationship between source rocks and reservoirs and the timing of trap formation relative to oil generation are excellent. Unfortunately, the sandstones in the Cormorant Formation are commonly thin, impersistent and with poor reservoir characteristics. New 3D seismic technology may hold the key to the successful development of valuable incremental reserves in these Triassic reservoirs below established Jurassic fields. In the tilted fault blocks on the eastern margins of the Viking Graben, appreciable hydrocarbon reserves have been tested in the Lunde and Statfjord Formations and, again, additional potential may still exist for Triassic reserves below existing Jurassic fields.

Beyond the limits of the North Sea, the reservoir potential of the Triassic is clearly demonstrated in the giant Morecambe Gasfield in the eastern Irish Sea and in the Wytch Farm Field on the southern coast of England.

At the end of the Triassic, peneplanation was almost complete. The basins contained thick sediments that onlapped the eroded ‘highs’. The early Jurassic transgression rapidly invaded the vast continental flood plains and tidal flats of the North Sea Basins and re-established epicontinental stenohaline marine conditions in north-west Europe.

7.7 Acknowledgements

The advice and support of Dr Ken Glennie in the preparation of this and previous manuscripts is gratefully acknowledged. Stephen Pickering (BHP Petroleum) kindly provided Figs 7.31 and 7.32 and much of the information relating to the East Irish Sea Basin. In addition, MJF would like to thank Drs Robert Dunay (Mobil North Sea), Neil Hodgson (British Gas) and Christopher Jeans (University of Cambridge) for their valuable contributions during revision of the text.

7.8 Key references

Abbotts, I.L. (1991) *United Kingdom Oil and Gas Fields: 25 Years Commemorative Volume.* Memoir 14, Geological Society, London.

Bertelsen, F. (1980) Lithostratigraphy and depositional history of the Danish Triassic. *Danmarks Geol. Undersøgelse, Serie B* **4**, 1–59.

Cameron, T.D.J. (1993) Triassic, Permian and Pre-Permian of the Central and Northern North Sea. In: Knox, R.W.O'B. and Cordey, W.G. (eds) *Lithostratigraphic Nomenclature of the UK North Sea.* British Geological Survey, Nottingham.

Coward, M.P. (1995) Structural and tectonic setting of the Permo-Triassic basins of northwest Europe. In: Boldy, S.A.R. (ed.) *Permian and Triassic Rifting in North West Europe.* Special Publication 92, Geological Society, London, pp. 7–39.

Lervik, K.S., Spencer, A.M. and Warrington, G. (1990) Outline of Triassic stratigraphy and structure in the central and northern North Sea. In: Collinson, J.D. (ed.) *Correlation in Hydrocarbon Exploration.* Norwegian Petroleum Society, Graham and Trotman, London, pp. 173–89.

Rhys, G.H. (1974) *A Proposed Standard Lithostratigraphic Nomenclature for the Southern North Sea and an Outline Structural Nomenclature for the Whole of the (UK) North Sea.* Report No. 74/8, Institute of Geological Science, HMSO, London.

Smith, R.I., Hodgson, N. and Fulton, M. (1993) Salt control on Triassic reservoir distribution, UKCS Central North Sea. In: Parker, J.R. (ed.) *Petroleum Geology of Northwest Europe; Proceedings of the 4th Conference.* Geological Society, London, pp. 547–58.

Vollsett, J. and Doré, A.G. (1984) *A Revised Triassic and Jurassic Lithostratigraphic Nomenclature of the Norwegian North Sea.* Bulletin No. 59, Norwegian Petroleum Directorate, Stavanger. 53 pp.

8 Jurassic

J.R. UNDERHILL

8.1 Introduction

After over 30 years of exploration and production, the North Sea Basin has become a mature hydrocarbon province. Even though petroleum reserves have been found in reservoirs of various stratigraphic ages, most of the oil- and gasfields located outside the southern North Sea and Irish Sea basins are sourced by and/or reside in rocks of Jurassic age (Fig. 8.1). It is currently estimated that almost 50% of the North Sea's hydrocarbons (oil and oil equivalents) occur in Jurassic reservoirs (and over 70% if the Southern North Sea Gas Province is excluded). Of these reserves, 90% were found within the 5-year period that immediately followed the discovery of the Brent oilfield in 1971 (Spencer *et al.*, 1996).

The bulk of the Jurassic oil and gas is found in the northern North Sea. A sizeable proportion is also located within and flanking the central North Sea. In almost all these cases, the Late Jurassic Kimmeridge Clay Formation represents the main source rock. It is only in more peripheral areas that different Jurassic or other stratigraphic levels provide alternative source-rock horizons. For example, in the southern North Sea, neither the Kimmeridge Clay nor the oil-prone source rocks of Liassic age have much petroleum-generation potential, due to erosion or lack of maturity, and it is the gas-prone Carboniferous Coal Measures which give that area its hydrocarbon potential. The presence and burial of the Liassic source intervals in adjacent areas, such as the Netherlands, northern Germany, southern mainland Britain (e.g. the Wealden and Wessex basins) and northern France (e.g. Paris Basin), have, however, provided charge for numerous small discoveries, all of which make an important contribution to the local economy. More significantly, the Liassic also provides the main source for the Wytch Farm oilfield of Dorset, the largest onshore oilfield in north-west Europe.

The Jurassic of the North Sea area was deposited in an intraplate tectonic setting south of the weakly linked Laurentian, Greenland and Fennoscandian shield areas and north of the Tethyan Ocean, which was experiencing active extension (Ziegler, 1982a, 1990a). In Early Jurassic times, depending upon location, erosion of the shield areas and more permanent or transient land areas provided sediment to an epeiric sea, consisting of a complex array of shallow marine shelves separated by deeper troughs (Ziegler, 1982, 1990). In contrast, the dominating depocentres during Late Jurassic times, following a period of Middle Jurassic thermal doming (Underhill and Partington, 1993, 1994), were more focused and included the developing Viking, Central and Moray Firth rift systems, the Norwegian–Danish–Polish Trough and the Sole Pit and Cleveland Basins of the Southern North Sea (Ziegler, 1982a, 1990a). More stable regions, comprising the Scottish Highlands, the Irish–Welsh–London–Brabant Massif in the west and the Bohemian Massif in the east, continued to be important sources of sediment throughout.

The relatively complex structural and sedimentary history of the North Sea area, especially during the latter half of the Jurassic, has meant that correlation of geological events has often been difficult. However, given the hydrocarbon prospectivity of the Jurassic interval and the wealth of data obtained over the past three decades, it is perhaps not surprising that high levels of confidence now exist with Jurassic stratigraphic correlations, which enable the major reservoir sandstones to be placed in their correct chronostratigraphic positions. That a robust scheme has not existed until relatively recently is largely for historical and commercial reasons. The main motivation for unifying the stratigraphic framework for the North Sea Jurassic has been the ever-increasing need to understand and predict subtle subsurface facies relationships now that the basin has reached such a mature stage of exploration, with very few classic structural plays remaining.

8.2 Stratigraphic basis for Jurassic correlation

8.2.1 Lithostratigraphic-based schemes

Many of the stratigraphic schemes employed in the Jurassic of the North Sea have historically been based upon the principles of lithostratigraphy. Furthermore, most of these have been derived from the original and pioneering descriptions of Deegan and Scull (1977) for the UK North Sea and Vollset and Doré (1984) for the Norwegian sector. Updates of their schemes have been largely pragmatic and area-specific and this has led to a proliferation of local terms.

The lack of a simple and robust stratigraphic scheme has led to problems in communication and a lack of appreciation of vertical and lateral facies relationships between various clastic depositional systems in the North Sea. In some cases, the same sandstone units have been given different lithostratigraphic terms across license-block boundaries, while in others the same terms have been ascribed to units of very different ages. Recognition that a solution to this problem would immediately improve the understanding of temporal and spatial variations in subsur-

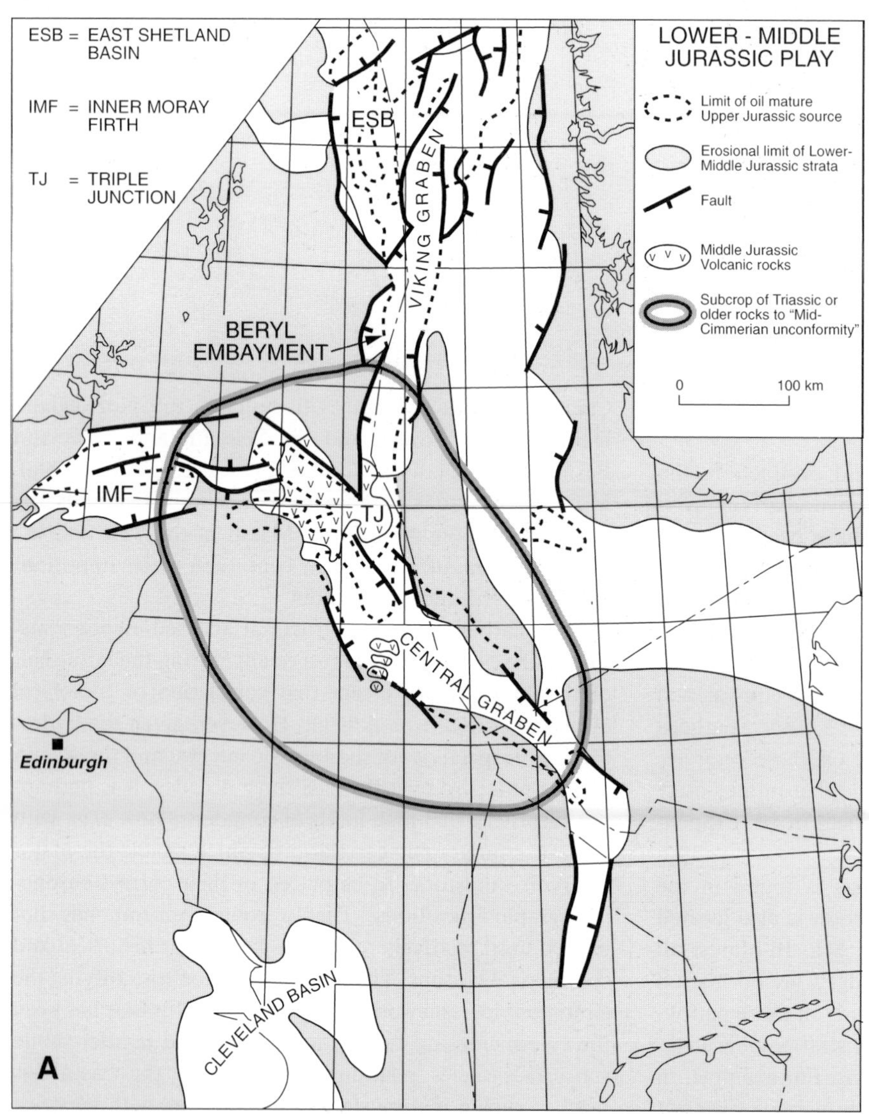

Fig. 8.1 Jurassic play maps depicting the spatial distribution of hydrocarbon discoveries (after Spencer *et al.*, 1996).
A. Lower–Middle Jurassic 'pre-rift' play map. The main plays exist outside the central North Sea Triple Junction area, which experienced Middle Jurassic uplift and erosion. Prospectivity is limited to the rift arms and adjacent platform areas lying close to the limit of oil mature Late Jurassic source rocks. Triassic–Lower Jurassic alluvial sandstones (Banks Group), shallow-marine sandstones (Cook and Johansen formations) and deltaic sandstones, derived from coeval erosion of the hinterland (Brent Group), form the main reservoir units in the East Shetland Basin (ESB) northern North Sea.
B. (*Opposite*) Upper Jurassic 'syn-rift' play map showing the distribution not only of shallow-marine reservoir sandstone play fairways (e.g. Piper and Fulmar formations) but also of 'deep-marine' basin-margin talus-fan and submarine-fan clastics (e.g. Brae Formation).

face lateral-facies relationships led to the introduction of a revised lithostratigraphic scheme (Fig. 8.2; Richards *et al.*, 1993) and to the recent development of new chronostratigraphic- and sequence-stratigraphic-based templates (Figs 8.3–8.6; Partington *et al.*, 1993a). Both schemes were introduced primarily with a view to unifying correlations and improving intercompany communication, in order to help reduce exploration risk and enhance ultimate recovery from existing fields.

8.2.2 Chronostratigraphic-based schemes

Introduction

Renewed realization of the value of integrating detailed biostratigraphical analysis in subsurface exploration and production studies has led to the development of a new, robust chronostratigraphic template for the North Sea Jurassic, which incorporates well-studied onshore exposures, together with the subsurface data. This in turn has led to an advance in the understanding of the temporal and spatial distribution of reservoirs.

All current chronostratigraphic schemes derived from sequence-stratigraphic analysis are underpinned by biostratigraphy and, as a result, are limited by its resolution. Although difficulties clearly exist with some aspects of the subsurface database, by virtue of sample spacing or poor recovery, a clear picture has emerged. Cross-reference has been made in all new schemes to the classic ammonite zonation, mainly through use of Woollam and Riding's (1983) macropalaeontological and palynological calibration of onshore sections of the UK.

Many workers have attempted to utilize the new concepts of sequence stratigraphy in conjunction with the biostratigraphic database in order to develop a usable and pragmatic stratigraphic scheme, which, where correctly applied, might have an important predictive capability. Varying degrees of success have been achieved by applying many of the original concepts of sequence stratigraphy. The biggest problems seem to emerge in those cases where interpretations are equivocal or through sticking rigidly to methods more appropriate to three-dimensional (3D) field exposures, for which they were originally developed. It has been demonstrated, however, that subsurface data need to be handled with care, since they are completely different from onshore exposures, being imperfectly sampled one- or two-dimensional data sets.

Sequence stratigraphic methods

The method adopted for the log-based sequence-stratigraphic analysis has consisted of correlating sedimen-

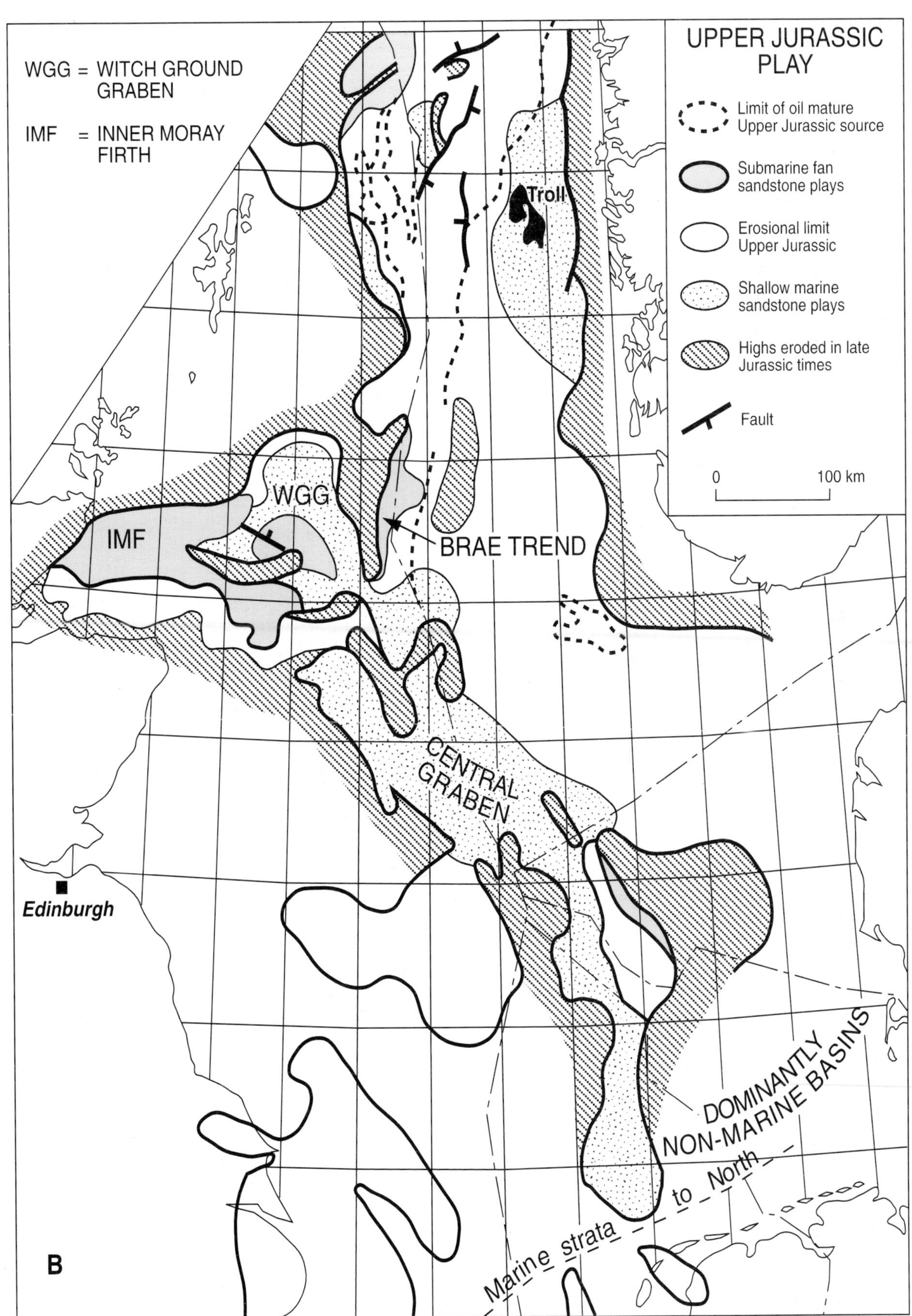

Fig. 8.1 *Continued*

tary facies deposited in basin-margin to basin-interior positions and placing them into a biostratigraphic framework. In marginal settings, the procedure has involved recognizing shallowing-upward and deepening-upward cycles of deposition from the electrical well logs, biostratigraphic studies and sedimentological descriptions of cores and outcrops (e.g. Stephen *et al.*, 1993; Underhill and Partington, 1994). An attempt is then made to separate the

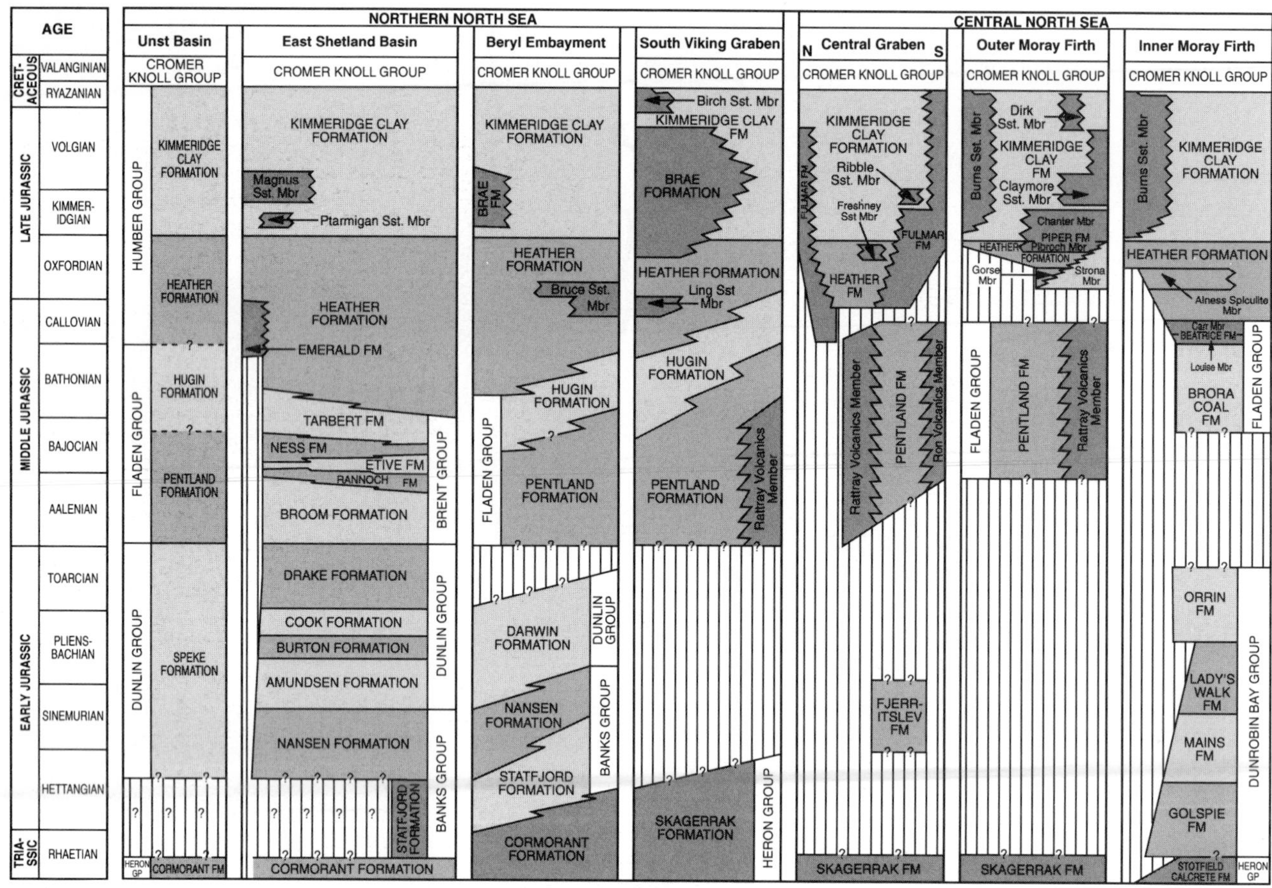

Fig. 8.2 Lithostratigraphic nomenclature employed in the northern and central North Sea Jurassic (after Richards *et al.*, 1993).

sedimentary cycles into their component stages of progradation, aggradation and retrogradation, by defining the points of onset of progradation, onset of aggradation and onset of retrogradation or time of maximum progradation (Fig. 8.7). The initiation of retrogradation equates with the marine-flooding surface of van Wagoner *et al.* (1990), and the period of retrogradation may be marked by the development of sharp ravinement surfaces (Swift, 1968; Nummedal and Swift, 1987), resulting from shoreface erosion (Fig. 8.7). The points of maximum retrogradation are commonly marked by marine shales (gamma-ray spike), which punctuate non-marine and shoreface successions and represent the maximum extent of marine flooding into non-marine environments (i.e. 'maximum flooding surfaces' (MFS) equivalent to surfaces marking the point of peak transgression or 'time of maximum flooding' (TMF) of Posamentier and Vail, 1988, and Posamentier *et al.*, 1988).

Although recognition of all surfaces with a correlative potential is important for subdividing a cycle of relative sea level and in producing the most accurate and unifying paleogeographic maps, the ease of recognition of marine shales developed during the period of maximum retrogradation or flooding in each cycle has been found to be the most practical for dating and confident correlation across the region (Fig. 8.7). Particular success has been possible where fossils and other criteria have been used to determine the maximum depth of water.

The detailed analysis of various offshore and onshore areas has demonstrated that marine shales were both numerous and distinctive throughout the Jurassic of the North Sea. The important shale-prone intervals also tend to record facies changes resulting from the starvation of terrigenous sediment in distal, basinal locations, where many are similarly marked by high radioactivity and can also be defined by high gamma-ray peaks and low sonic velocities on electrical well logs (Fig. 8.7). Recent studies have suggested that a greater accuracy in the identification of MFS may be achieved by the use of spectral gamma-ray cross-plots, rather than relying solely upon total gamma-ray counts.

The slow sedimentation rates associated with deposition of the shales seem to have commonly promoted authigenic mineral growth (e.g. phosphatic and pyritic nodules) and preserved high amounts of total organic carbon (TOC). Such surfaces are laterally extensive and evidently represent periods when sediment supply and current activity were minimal. These factors, together with associated distinctive and diagnostic faunal influxes and extinction 'bioevents' particularly of dinoflagellate cysts (dinocysts) and other micropaleontological species, including ostracods and radiolaria, make the candidate MFS even more readily identifiable and correlatable (Fig. 8.7). These characteristics suggest that they are equivalent to condensed sections (*sensu* Loutit *et al.*, 1988; Posamentier and Vail, 1988; Posamentier *et al.*, 1988; van Wagoner *et al.*, 1990), marine condensed horizons (MCH of Partington *et al.*, 1993a,b) or hiatal surfaces in the (bathymetrically) deepest marine sections. In the most distal basinward positions, the

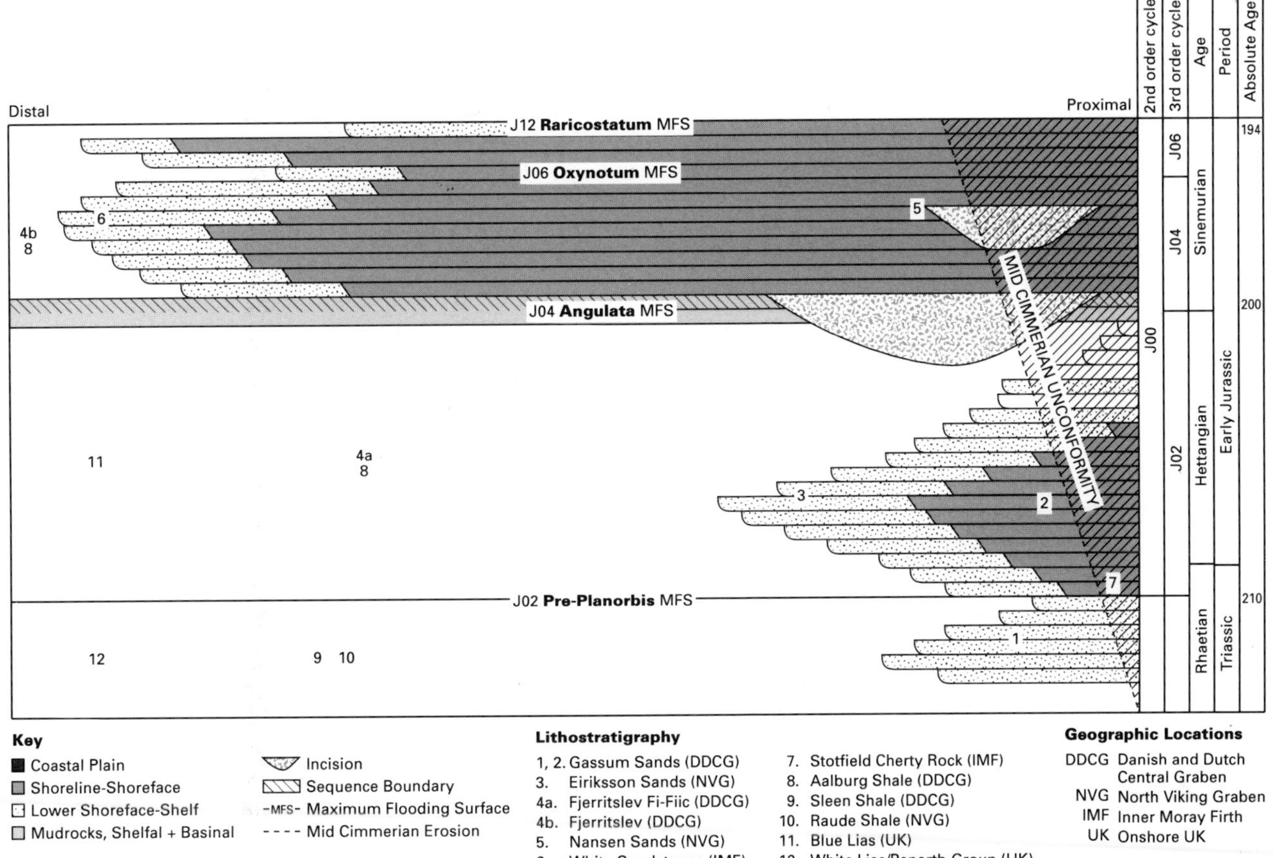

Fig. 8.3 Sequence-stratigraphic template for the Hettangian–Sinemurian succession in the North Sea Basin (after Partington *et al.*, 1993a). Recognition, biostratigraphic calibration and correlation of marine condensed horizons and maximum flooding events has enabled interregional definition of genetic stratigraphic sequences, which allow individual lithostratigraphic units, including important reservoir sandstones, to be placed in their proper stratigraphic context. The J numbers refer to terms used to identify individual maximum flooding events in British Petroleum's in-house stratigraphic scheme.

events often appear to merge into a stacked interval made up of several condensed sections (Fig. 8.7). Comparison of sections along basin-to-margin transects suggests that the condensed basinal marine horizons may be correlated with the times of maximum flooding on its margins (shown schematically on Fig. 8.7; see also Partington *et al.*, 1993a). As such, the horizons should be consistent with previous correlations of condensed sections at a third-order (0.5–3 million years (Ma)) scale, which have been used in part to document times when sea levels were at their highest in areas used to construct the current, popular, global sea-level charts (e.g. Haq *et al.*, 1987, 1988).

The exclusive use of the MFS approach as the primary boundary in stratigraphic subdivision differs from that most commonly employed in existing sequence-stratigraphic schemes (e.g. van Wagoner *et al.*, 1990), although these events occur on a similar frequency to the boundaries used by other workers. Van Wagoner *et al.* (1990) considered that the basic stratigraphic units are bounded by unconformities and their correlative conformities (sequence boundaries) and are subdivided by marine-flooding surfaces, which record shallowing and initial deepening events, respectively.

Despite the difficulties outlined above, attempts have also been made to correlate on the basis of the initial, confident recognition of sequence boundaries by some workers (e.g. Donovan *et al.*, 1993), but these have met with mostly limited success. The main reason for the disappointing correlations is that the sequence boundaries appear to be much more difficult to recognize, being more subtle features in an actively subsiding basin, where shale-on-shale contacts may dominate. Furthermore, although North Sea subsurface studies demonstrate that there is a potential to recognize sequence boundaries between any two maximum flooding events, they are extremely hard to identify in practice using electrical logs alone, and may also be difficult to recognize in core, particularly where there is a lack of deep incision or erosion or where basinward shifts cannot be easily ascertained. Consequently, their definition can often depend on very subtle variations and be highly interpretative, and there may be cases where they are indistinguishable from migrating channel bases or other autocyclic causal processes for the generation of sharp-based sands. Even where identified, sequence boundaries may also be very hard to date precisely, because by definition they have time duration. These problems seem unlikely to be restricted to the North Sea Jurassic and, rather than consider it unique in this respect, it seems probable that any area that experienced relatively rapid rates of basin subsidence will be marked by a relative suppression of sequence boundaries and the enhancement of MFS.

Further problems arise with existing methods because there is considerable difficulty in defining whether progra-

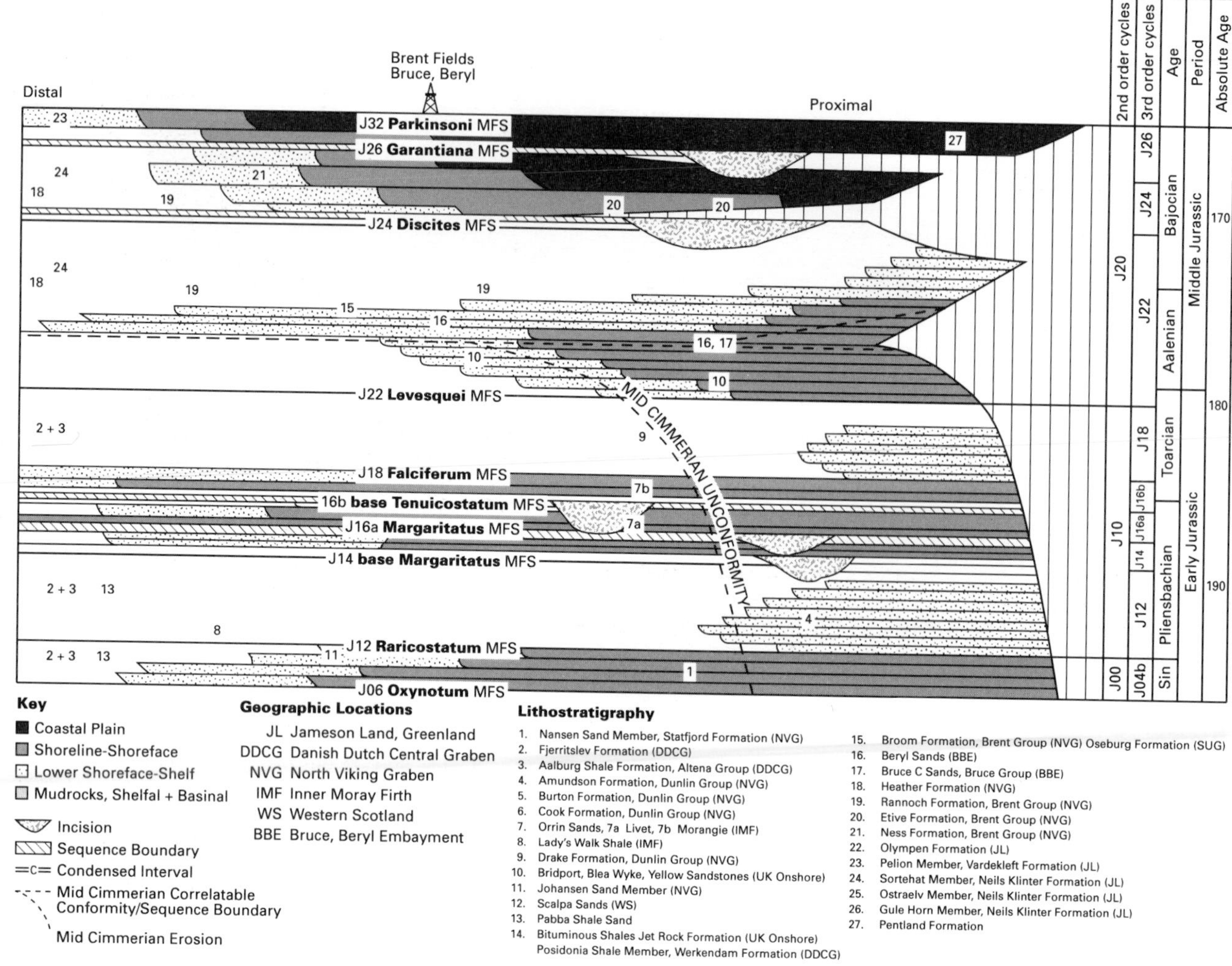

Fig. 8.4 Sequence-stratigraphic template for the Pliensbachian–Bajocian succession in the North Sea Basin (after Partington *et al.*, 1993a).

dation occurred during a relative rise in sea level or by a process of forced regression (*sensu* Posamentier *et al.*, 1992), which occurs during a relative sea-level fall (e.g. Hunt and Tucker, 1992). Similarly, the problems arising from less-precise placement of sequence boundaries in wells without core coverage make their interpretation non-unique and highly interpretative. Consequently, these difficulties often make subdivision of the sedimentary cycles (e.g. into their component system tracts) too tentative at present and this approach has not been widely adopted in subsurface studies to date. Because such problems of definition are likely to occur where similar data sets have been used to subdivide basin subsurface successions elsewhere, system-tract classifications in other basins also may need to be treated with caution.

In conclusion, as a result of the difficulties in correlating other surfaces regionally, the easily identifiable MCH and MFS have generally been used to subdivide the Jurassic section into depositional episodes or genetic stratigraphic sequences (see Figs 8.3–8.6), in a manner similar to that of Galloway (1989a,b). Although each MFS may also be thought to have the potential to be diachronous, particularly in areas with extremely thick sections controlled by high sediment supply and/or multiple sediment supply points (e.g. Pliocene–Pleistocene of the Gulf Coast), where biostratigraphic calibration of the events is irrefutable, only those surfaces that remained within tightly constrained dinocyst and ammonite biochronozones have been used in correlating the North Sea Jurassic (see Figs 8.3–8.6; Partington *et al.*, 1993a). In doing so, inaccuracies related to sample spacing, sample processing and detrimental drilling factors have been minimized, such that the data are generally constrained by sidewall and core samples and, where practicable, no reliance is generally placed on cuttings descriptions. Hence, the maximum flooding events can be considered to be effectively isochronous over the North Sea and adjacent areas within the bounds of the biostratigraphic resolution currently employed in Jurassic global stratigraphic correlations.

Genetic stratigraphic sequences of the North Sea Jurassic (see Figs 8.3–8.6)

The Jurassic succession may be subdivided and correlated with confidence using biostratigraphically constrained and regionally extensive MFS (see, for example, Figs 8.3–8.6; Partington *et al.*, 1993a; Underhill and Partington, 1993, 1994). Application of this approach has enabled a number of meaningful sequences to be traced across the North Sea area and allowed recent workers to produce a higher resolution of the temporal and spatial evolution of the

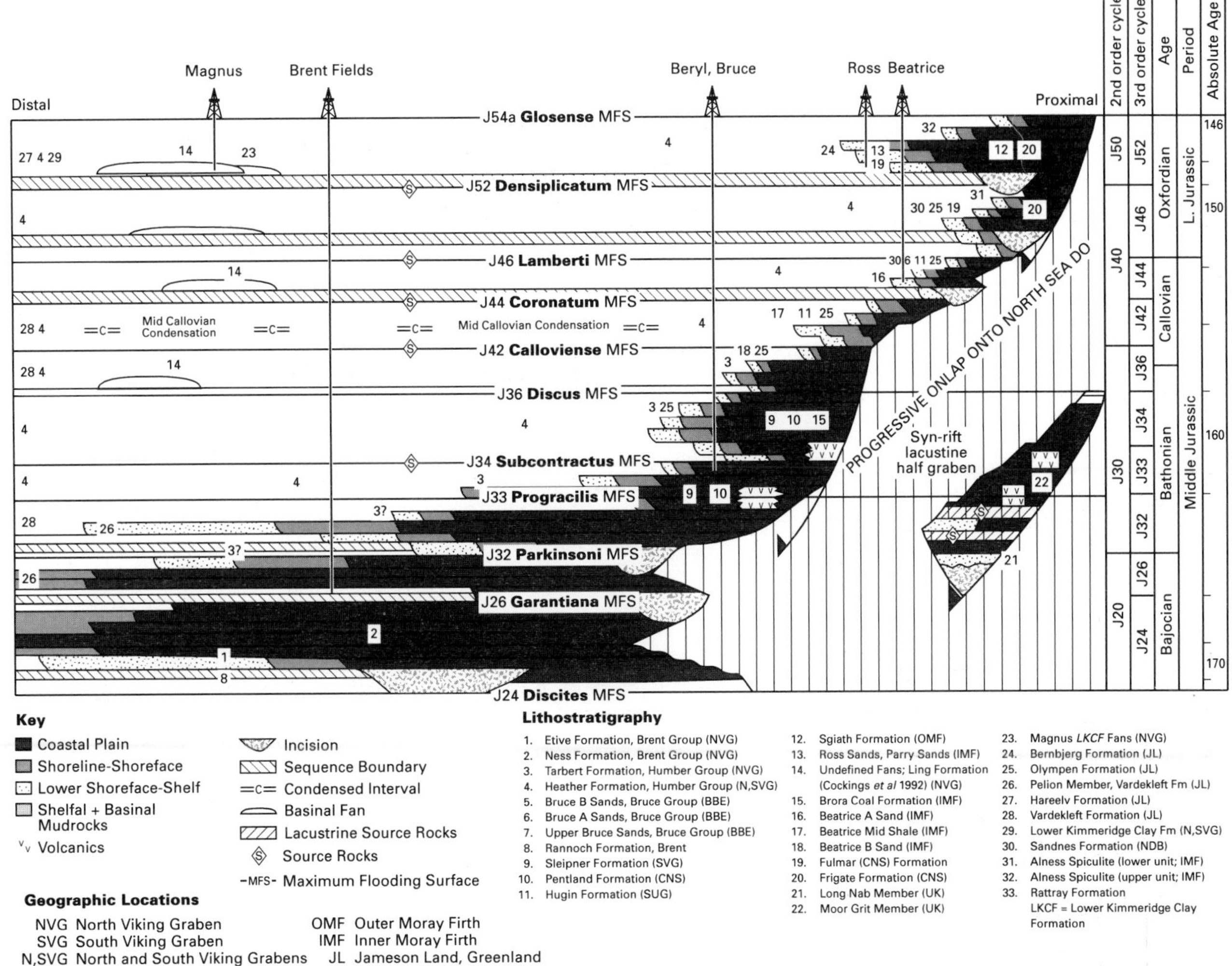

Fig. 8.5 Sequence-stratigraphic template for the Bajocian–Oxfordian succession in the North Sea Basin (after Partington *et al.*, 1993a).

basin, and hence develop an accurate understanding of the main controls on basin development and sediment dispersal throughout the Jurassic (e.g. Mitchener *et al.*, 1992; Partington *et al.*, 1993b; Rattey and Hayward, 1993; Stephen *et al.*, 1993; Underhill and Partington, 1993, 1994).

At least 33 regionally correlatable MFS are currently recognized in Jurassic sections from well logs and outcrop data, allowing the succession to be separated into 32 genetic stratigraphic sequences (*sensu* Galloway, 1989a,b) at present (see Figs 8.3–8.6) (Partington *et al.*, 1993a). More such surfaces are likely to be present and identified in the future. Each maximum flooding event has been named by reference to the standard ammonite biozonation scheme for the Jurassic (e.g. *eudoxus MFS*; *baylei MFS*; *glosense MFS*, etc.; Partington *et al.*, 1993a; Underhill and Partington, 1993), by reference to the seminal work of Woollam and Riding (1983), who undertook a comprehensive review of all UK ammonite-controlled Jurassic outcrops. Importantly, the maximum flooding events could also be defined by their diagnostic dinocyst extinction event (for details, see biostratigraphic charts in Partington *et al.* 1993a, and Veldkamp *et al.*, 1996). Indeed, the latter may be preferable, in view of the problems associated with exact dinocyst–ammonite correlations, for example in the offshore Late Jurassic (e.g. Boldy and Brealey, 1990), and, although the equivalent ammonite biozones have been used here, it is evident that some modification may be needed in the future.

Some improvement in correlation may also be necessary for the Middle Jurassic interval, where a lack of marine conditions precludes good ammonite control in many areas. It is unlikely, however, that Early Jurassic ammonite calibrations will change significantly, because of the spatially extensive, uniform nature of sedimentation at that time. Indeed, it is possible that some of the MFS identified in the Early Jurassic of the North Sea may be related to more widespread (e.g. platewide) or worldwide anoxic events. For example, the early Toarcian marine horizon within the *falciferum* ammonite zone equates with the Posidonia Shale and *schistes cartons* of Mediterranean Tethys (Fleet *et al.*, 1987). However, the extent and significance of others are less clear and correlation outside the immediate North Sea/north-west European domain remains to be proved. A broader understanding of their wider correlation (e.g. across plate boundaries and in marine basins formed under a variety of tectonic settings) will be invaluable for anyone wishing to attempt global correlations.

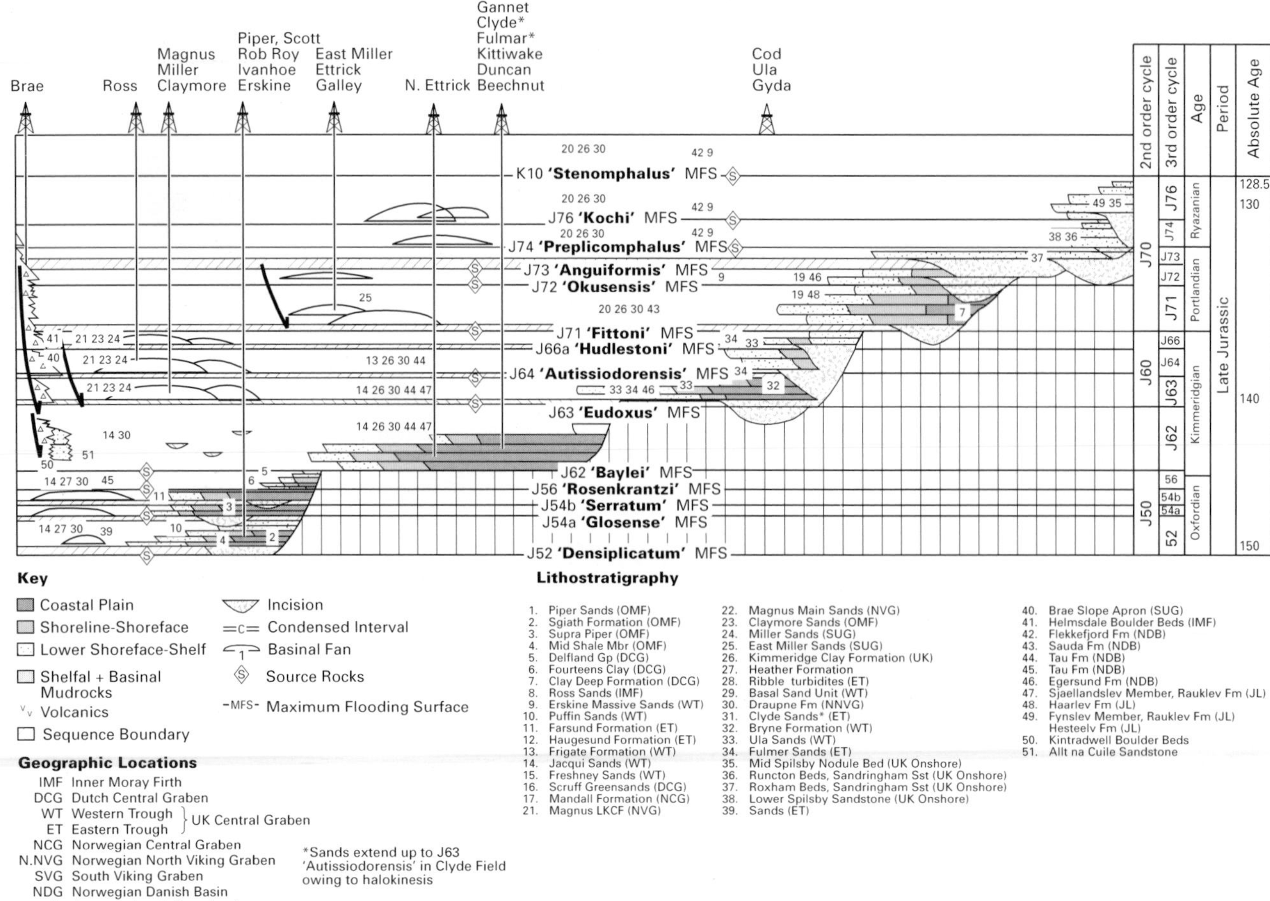

Fig. 8.6 Sequence-stratigraphic template for the Oxfordian–Ryazanian succession in the North Sea Basin (after Partington *et al.*, 1993a).

Importantly, the development of such a rigorous stratigraphic template potentially permits the identification and correlation of intervening unconformities and their correlative conformities (Exxon-type sequence boundaries). Their recognition allows for the evaluation of synchronous facies variations in the Jurassic succession and the construction of well-constrained basin palaeogeographies through time for component parts of a sequence. It is often advantageous, however, to produce them for the time of maximum progradation, since this is often well marked on logs and commonly reflects the limit to prospectivity, because it marks the point of maximum clastic build-out.

Given that MFS may be considered isochronous chronostratigraphic markers in the Jurassic (Partington *et al.*, 1993a), identification of these events on seismic reflection data gives additional control on basin architecture. Indeed, such potential has been realized in some areas where sufficient well data and good-quality seismic coverage show that condensed sections may be recognized by converging (primarily onlapping or downlapping seismic) reflectors. Although theoretically feasible for the whole Jurassic interval, the problems of seismic resolution in relatively thin stratigraphic successions often make it difficult for MFS to be confidently detected and correlated on most seismic reflection data with confidence.

Only the Inner Moray Firth currently allows accurate seismic imaging and correlation of such events, because Jurassic sequences are thick, shallow and covered by a dense grid of well-imaged seismic data (e.g. Vail *et al.*, 1984; Vail and Todd, 1981; Underhill, 1991a,b). Indeed, bounding surfaces defined by seismic-scale marine onlap have been recognized (Underhill, 1991b), which allows the subdivision of the Late Jurassic (mid-Oxfordian–Ryazanian) succession into five mappable seismic sequences (Underhill, 1991b). These surfaces appear to correlate with sequences recognized in other areas (e.g. South Viking Graben, the Outer Moray Firth and Magnus oilfield areas, Northern North Sea; Partington *et al.*, 1993a,b), where they have been shown to represent the most prominent maximum flooding events in the North Sea (Partington *et al.*, 1993b; Rattey and Hayward, 1993).

Controls on and effects of sediment dispersal

The stratigraphic results demonstrate the importance of sharp-based shallow-marine and shoreface clastics between the MFS. The shallow-marine aspect of these units suggests that the sand pulses represent extremely rapid basinward progradation of the shoreline into the basin over considerable distances. They are interpreted as representing third-order regressive intervals that punctuated the overall transgression, and they probably demonstrate the importance of forced regression (*sensu* Posamentier *et al.*, 1992)

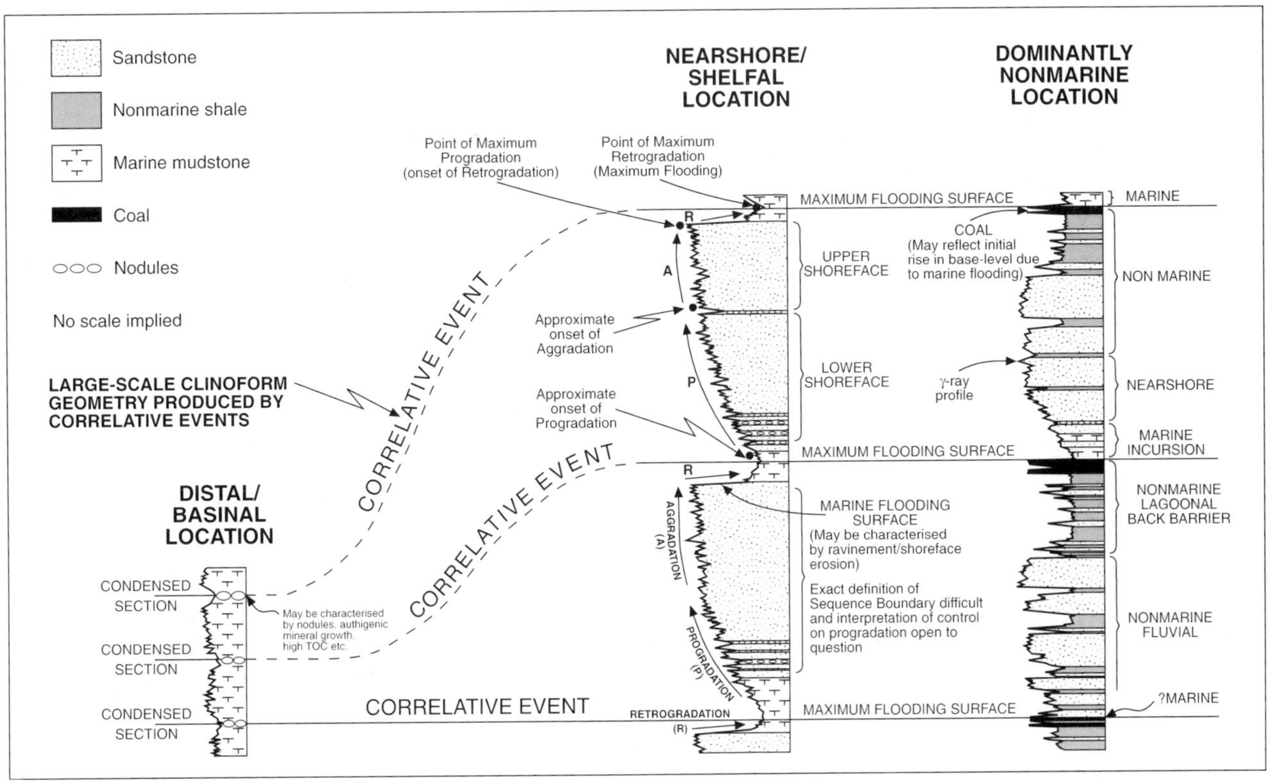

Fig. 8.7 Sequence-stratigraphic method of correlation between offshore, nearshore and dominantly non-marine depositional settings by the use of regionally correlatable maximum flooding surfaces and marine condensed horizons (after Underhill and Partington, 1994).

in controlling sediment dispersal and reservoir distribution in the North Sea Jurassic.

The controls on the timing of the development of third-order MFS and regressive pulses are poorly understood. Until studies of stratigraphic sequences are undertaken in areas outside the North Sea, it is not possible to discount a more regional or global control on sea-level fluctuations. It seems likely, however, that the events were influenced to a degree by tectonic events associated with Late Jurassic rifting and Middle Jurassic thermal doming.

8.3 Lower Jurassic stratigraphy and sedimentation

Lower Jurassic marine strata are widespread over the southern North Sea and adjacent areas. In the central North Sea, however, Lower Jurassic strata are known to be absent from the eastern parts of the Moray Firth, the South Viking Graben and the Central Graben. Their distribution away from core areas of the North Sea is believed to be in response to erosion resulting from post-depositional uplift of the central North Sea and South Viking Graben (see Chapter 2 and Section 4.4).

Despite the geographical limitations consequent upon erosion due to later tectonic events, when placed in a chronostratigraphic framework, deposition throughout the Early Jurassic was remarkably uniform. Taken together with the fully and open-marine nature of deposition and the lateral continuity of coarse clastic units, sedimentation is generally inferred to have taken place in a shallow, well-aerated epicontinental (epeiric) basin experiencing broad patterns of thermal-driven subsidence following the earlier Permo–Triassic rift events. It was only during latest Toarcian times that a pronounced shallowing was seen to affect parts of the North Sea area, which eventually led to a marked change in deposition from a shallow-marine to a dominantly non-marine fluviodeltaic setting.

Where preserved, units of Early Jurassic age have proved to be important hydrocarbon reservoirs. They have particular significance in the North Viking Graben, where they commonly form a level of prospectivity below Middle Jurassic Brent Group sequences, and in the Inner Moray Firth, where they comprise the lower reservoirs of the Beatrice Field. Despite the absence of Lower Jurassic sequences over much of the northern North Sea, comparison of the areas where they are present with onshore exposures of Britain and other subsurface sections in the southern North Sea allows the main controls on sediment dispersal to be determined. The spatial separation between the areas in which Lower Jurassic sediments are preserved, and their distances from classic onshore outcrops of mainland Britain, have meant that Hettangian–Toarcian sequences have often been treated independently of each other and have been ascribed different lithostratigraphic terms. In the following descriptions of Jurassic formations, sufficient lithological detail is presented to enable their recognition and, in some cases, criteria are given for defining their boundaries with other formations. More recently, and largely as a consequence of the regional extent of the epeiric seaway, considerable success has been achieved with sequence-stratigraphic-based evaluations, enabling excellent opportunities for local and regional subdivision and correlation of sedimentary units.

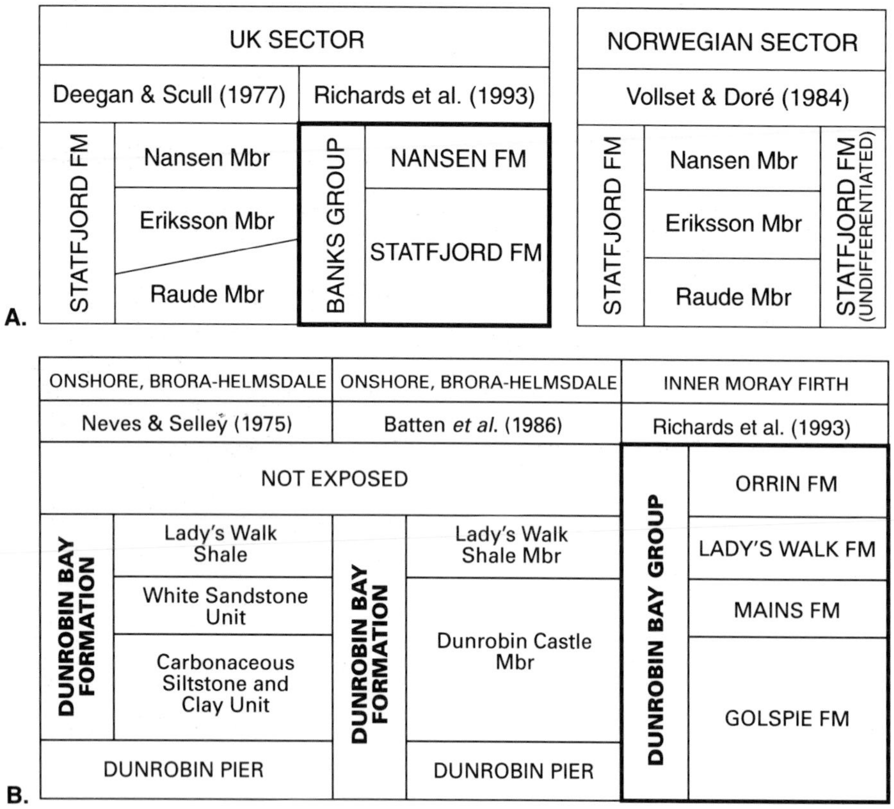

A.

UK SECTOR			
Deegan & Scull (1977)		Richards et al. (1993)	
STATFJORD FM	Nansen Mbr	BANKS GROUP	NANSEN FM
	Eriksson Mbr		STATFJORD FM
	Raude Mbr		

NORWEGIAN SECTOR		
Vollset & Doré (1984)		
STATFJORD FM	Nansen Mbr	STATFJORD FM (UNDIFFERENTIATED)
	Eriksson Mbr	
	Raude Mbr	

B.

ONSHORE, BRORA-HELMSDALE		ONSHORE, BRORA-HELMSDALE		INNER MORAY FIRTH	
Neves & Selley (1975)		Batten *et al.* (1986)		Richards et al. (1993)	
NOT EXPOSED				DUNROBIN BAY GROUP	ORRIN FM
DUNROBIN BAY FORMATION	Lady's Walk Shale	DUNROBIN BAY FORMATION	Lady's Walk Shale Mbr		LADY'S WALK FM
	White Sandstone Unit		Dunrobin Castle Mbr		MAINS FM
	Carbonaceous Siltstone and Clay Unit				GOLSPIE FM
DUNROBIN PIER			DUNROBIN PIER		

Fig. 8.8 Lithostratigraphic nomenclature for the Triassic–Lower Jurassic of the northern and central North Sea areas (after Richards *et al.*, 1993). A. Stratigraphic convention for the UK and Norwegian sectors. B. Stratigraphic convention for onshore exposures and offshore penetrations in the Inner Moray Firth Basin.

8.3.1 The Viking Graben

The Lower Jurassic of the UK Viking Graben has recently been subdivided into two units: the Banks Group and the Dunlin Group, while the Lower Jurassic of the Inner Moray Firth is referred to as the Dunrobin Bay Group (Richards *et al.*, 1993). The component parts of the stratigraphic succession are given in Figs 8.2, 8.3 and 8.8–8.10.

The Banks Group

The Banks Group is a newly introduced lithostratigraphic unit for the UK sector of the North Sea (Richards *et al.*, 1993), comprising a lower, Statfjord Formation containing the Triassic–Jurassic boundary and an upper, Nansen Formation. Its introduction has updated and replaced the previous lithostratigraphic divisions of Deegan and Scull (1977) and Vollset and Doré (1984), who referred to the same interval as the Statfjord Formation, which they subdivided into lower (Raude), middle (Eiriksson) and upper (Nansen) members (Fig. 8.11). Internally, the reservoir is difficult to correlate between wells, because of its highly heterogeneous nature.

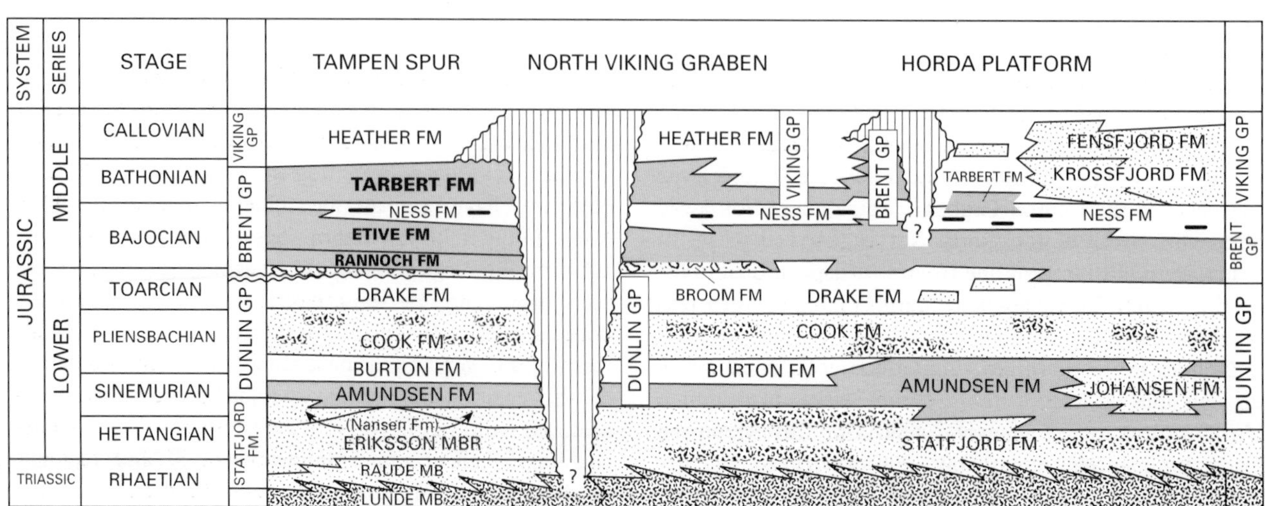

Fig. 8.9 Lithostratigraphic terminology of Lower and Middle Jurassic units of the Tampen Spur, Viking Graben and Horda Platform areas of the northern North Sea (after Marjanac, 1995).

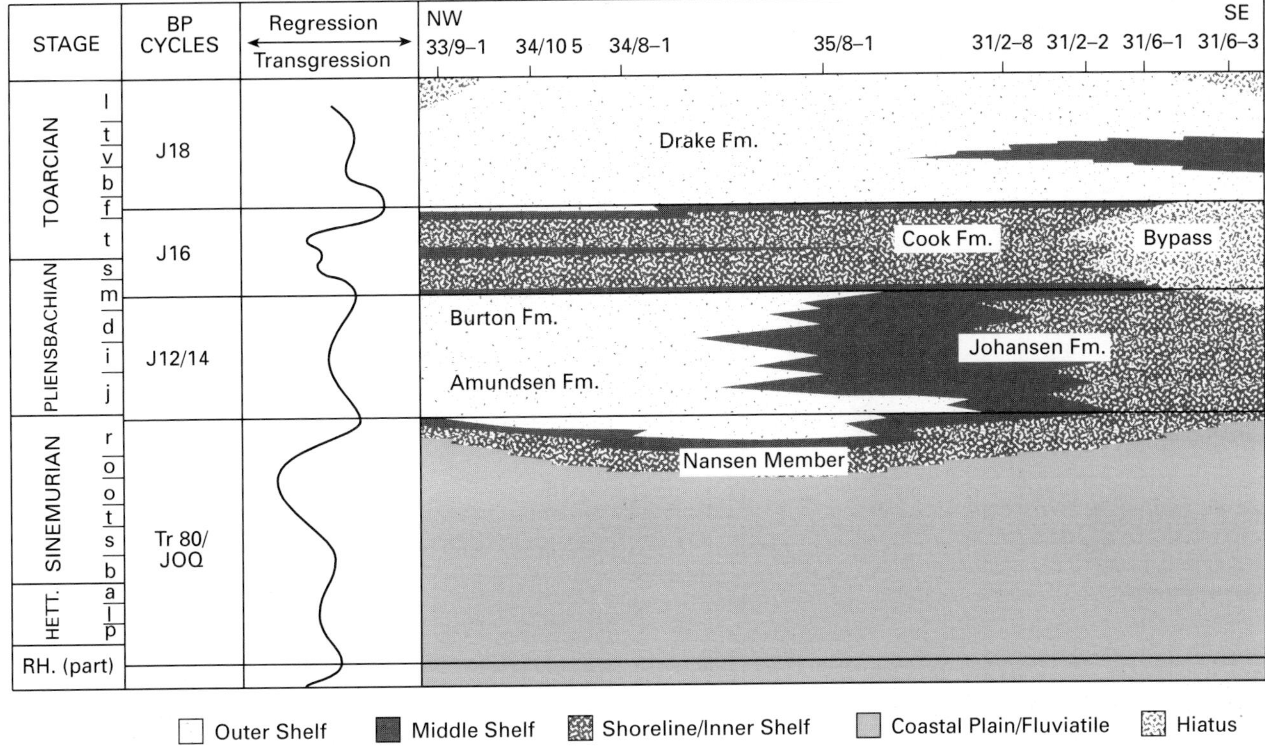

Fig. 8.10 Sequence-stratigraphic subdivision of the Dunlin Group in the northern North Sea (after Parkinson and Hines, 1995; based on Partington *et al.*, 1993a).

The Banks Group lacks a well-defined biostratigraphically based correlation, due mainly to its continental aspects, but is thought to have been deposited during Rhaetian–Sinemurian times in the East Shetland Basin. However, it is considered to be time-diachronous in the Beryl Embayment, where it has been ascribed a Hettangian–Pliensbachian age. Recent advances both in non-marine sequence stratigraphy and, more particularly, in heavy-mineral data have provided an independent basis for subdivision and correlation in the Statfjord and Nansen formations (Morton and Berge, 1995).

Statfjord Formation. Statfjord Formation sandstones form important reservoirs in the North Viking Graben (e.g. in the Alwyn, Brent, Statfjord, Gullfaks, Snorre, Oseberg, Veslefrikk and Brage fields) and they remain an important exploration target. Their reservoir potential is controlled by the distribution, density and stacking pattern of multistorey fluvial-channel sandstones, within successions of interbedded fluvial and interfluvial deposits (Ryseth and Ramm, 1996). Porosities in the Statfjord sandstones of the type locality average 22%. Permeabilities average 470 millidarcies (mD) at depths in excess of 2.5 km (Roberts *et al.*, 1987).

Deposition of the Statfjord Formation occurred as climatic conditions were changing from semi-arid to humid and as regional sea-levels were rising. Although its deposition immediately succeeded an important phase of Late Triassic (Carnian–Early Rhaetian) retrogradation within the Cormorant Formation (Steel, 1993), type wells from the Horda Platform and Tampen Spur in the Norwegian sector demonstrate that the Statfjord Formation records an episode of sediment progradation. Full development of the Statfjord Formation is restricted to axial regions of the Viking Graben to the east of the Hutton–Ninian trend (Figs 8.12, 8.13 and 8.14), where the sediments comprise fine- to coarse-grained cross-bedded sandstones containing evidence for channel incision. They were originally interpreted as having formed in a coastal-fan and fan-delta setting on a low-energy coastline (Røe and Steel, 1985). More recently, however, evidence from the finer-grained intervals, including mottled mudstones, rootlets and thin or scattered coals, together with the descriptions of the sandstones, has led to a fluvial-depositional setting being interpreted (Nystuen *et al.*, 1990; MacDonald and Halland, 1993), in which perennial braided streams cut across moderately to poorly drained low-lying interfluves (Ryseth and Ramm, 1996).

Nansen Formation. The term 'Nansen Member' was originally used to describe a unit of pale sandstones that lay above heterolithic units of the Raude and Eiriksson members of the Statfjord Formation and below shales belonging to the Amundsen Formation (Deegan and Scull, 1977). Studies in the Brent Province, beyond the Viking Graben axis indicate that the Nansen Formation has a wider distribution than the underlying Statfjord Formation, recording a progressive onlap onto the East Shetland Platform (see Fig. 8.13). The unit consists of light-coloured, fine- to coarse-grained, well-sorted, pebbly, quartzitic, calcareous sandstones. It is a time-transgressive shallow-marine sand that records the retreat and local ravinement of the Statfjord alluvial system. The diachroneity of the

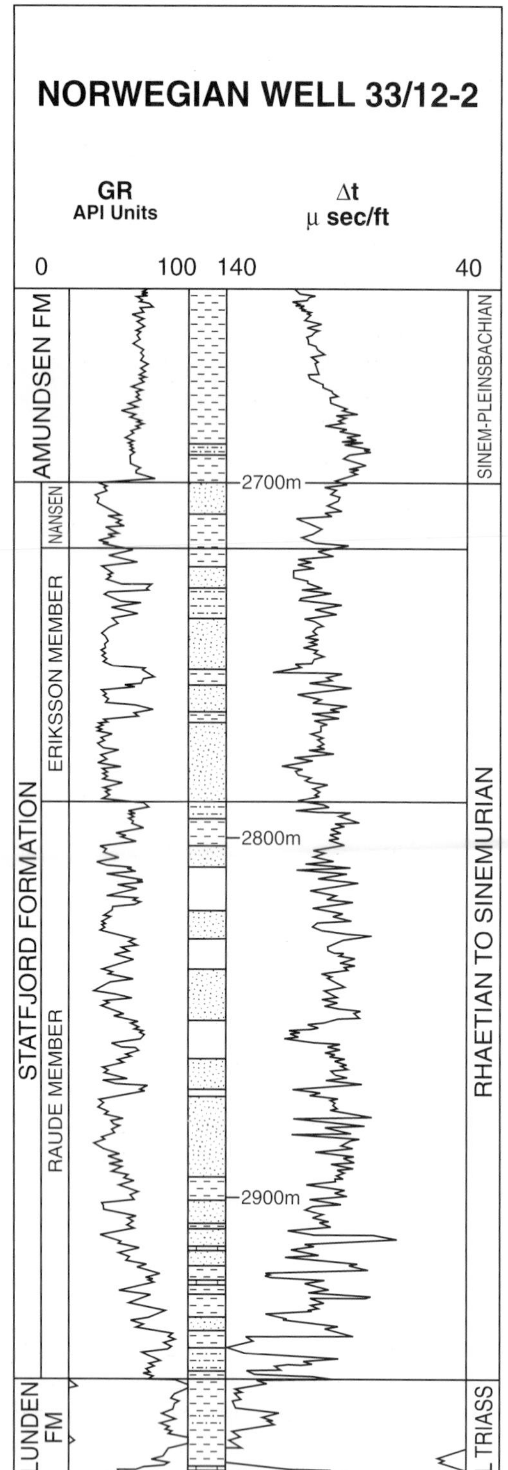

Fig. 8.11 Norwegian exploration well 33/12-2, depicting the previous breakdown of the Statfjord Formation into three main units: the Raude, Eiriksson and Nansen members (modified after Brown, 1984).

Nansen Formation suggests that, in part, it is laterally equivalent to the Statfjord Formation.

The Dunlin Group

Deposition of the Dunlin Group ranged from the Sinemurian to the Toarcian in most areas—for example, the North Viking Graben, where it has been separated into four main units: the Amundsen Formation, Burton Formation, Cook Formation and Drake Formation (Fig. 8.15). Some workers consider that the group is diachronous in the Beryl Embayment, where it is synonymous with the (undifferentiated) Darwin Formation (Fig. 8.15).

Subdivision of the shale-dominated Dunlin Group into its component parts has met with greatest success in the Viking Graben (Partington *et al.*, 1993a; Marjanac, 1995; Marjanac and Steel, 1997), where constituent sands form important reservoirs in some oilfields (e.g. Cook Formation in the Oseberg, Statfjord and Gullfaks fields). As with other sedimentary units, the succession may be subdivided using a genetic sequence-stratigraphic approach (Figs 8.3 and 8.16). Typical Dunlin reservoir sandstones, such as those in the Statfjord Field, have average porosities of 23% and permeabilities in the range 10–300 mD (Roberts *et al.*, 1987).

Amundsen Formation. The onset of Dunlin Group deposition is marked by a change from variably cemented sandstones of the Nansen Formation to bioturbated and carbonaceous mudstones and siltstones. The lowest unit, the Amundsen Formation, coarsens up from grey mudstones to siltstones and fine-grained sandstones, and is overlain by mudstones ascribed to the Burton Formation. The unit is found across the East Shetland Basin and exhibits overall easterly thickening to reach over 100 m in Total's Ellon Field (United Kingdom Continental Shelf (UKCS) Block 3/15).

Burton Formation. The Burton Formation comprises a dominantly mudstone succession, containing rare, millimetre-thick, lenticular, parallel and ripple-laminated sandstones. Although the level and diversity of bioturbation is low, the mudstones contain *Helminthoidea* traces throughout. The unit also contains numerous authigenic mineral-rich horizons (e.g. phosphate-, calcite-, chamosite-, glauconite- and siderite-rich grains) and at least one prominent and regionally-correlatable condensed section occurs at the top of the unit and consists of an early siderite-cemented bed, comprising scattered, rounded ooids, pellets and peloids. The Burton Formation evidently represents deposition in a low-energy shelf setting, which, due to its relatively low level of bioturbation, may have experienced partially restricted circulation and been prone to sediment starvation and condensation.

Johansen Formation. Evidence suggests that clastic progradation occurred during mid-Sinemurian to early Pliensbachian times off the Norwegian margin, which was coeval with deposition of the Amundsen and Burton formations. These sands form a clastic wedge, which is normally referred to as the Johansen Formation. Although they consist mainly of fining-up, sharp-based nearshore and inner-shelf deposits, there is a suggestion that more brackish water and alluvial environments exist in eastern areas.

Cook Formation. The Cook Formation is a laterally persistent sand-prone unit of Pliensbachian age, which is found encased in mudstones belonging to the Burton and Drake formations. At its thickest extent on the Bergen High, it

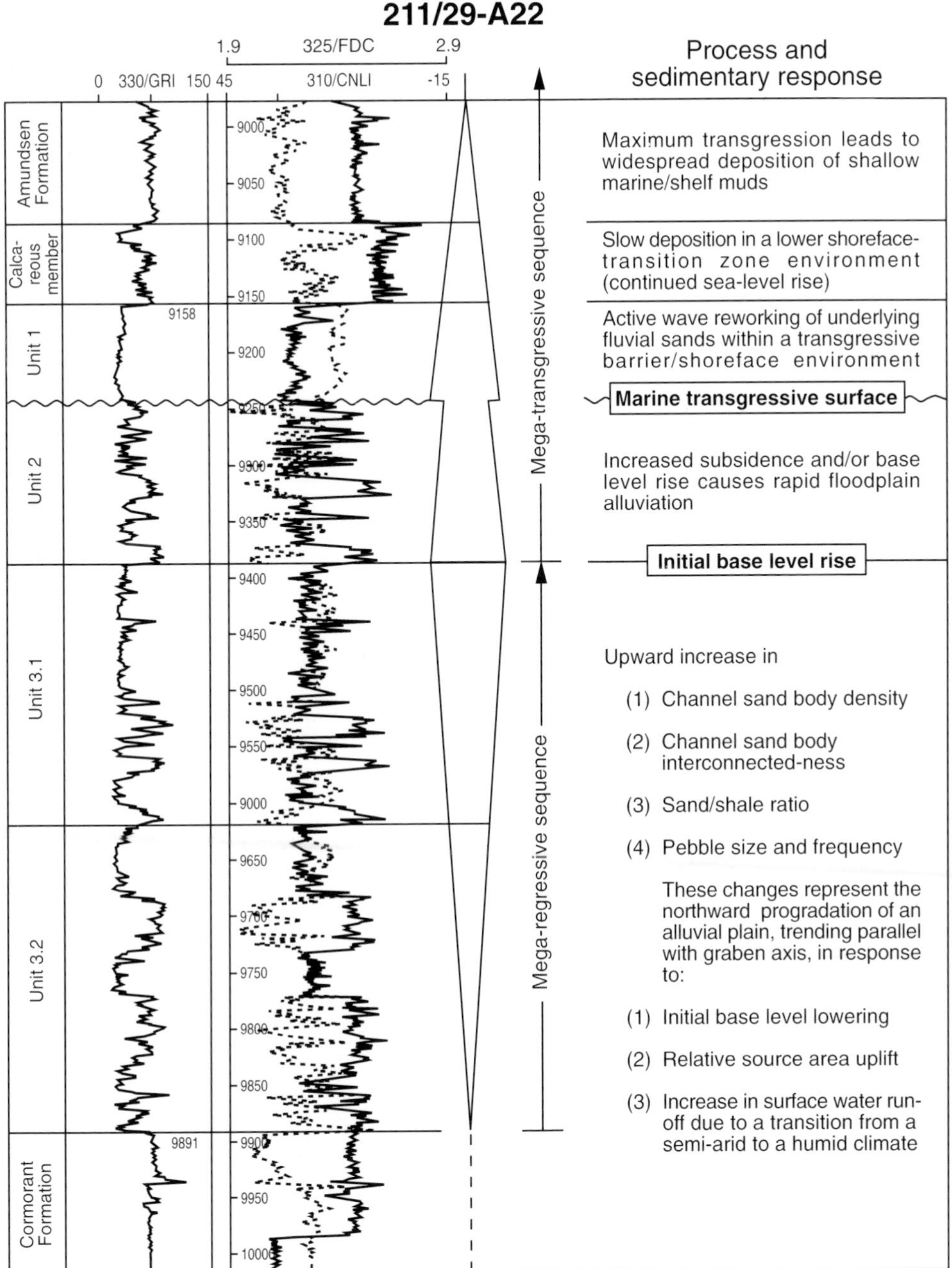

Fig. 8.12 Reservoir subdivision of the Banks Group, Brent Field, northern North Sea (modified after Johnson and Stewart, 1985).

consists of at least three component parts: a lower progradational (Cook A) unit, a thin intermediate mudstone (Cook B) and an upper progradational (Cook C) unit (e.g. Norwegian wells N30/6-3 and N30/6-6). A fourth unit is recognized in the Gullfaks Field (Erichsen *et al.*, 1987).

Most of the detailed sedimentological information has been derived from the Oseberg oilfield, where the Cook A unit consists of a distinctive upward-coarsening and cleaning cycle above the Burton Formation (Livbjerg and Mjøs, 1989). The coarsest units comprise medium-grained sandstones, which display flaser- and cross-bedding where locally unaffected by bioturbation. More commonly, however, reworking by trace-fossil assemblages, including common *Asterosoma*, *Siphonites*, *Planolites* and rarer *Monocraterion*, *Skolithos* and *Thalassinoides*, makes identification of sedimentary structures more difficult (Livbjerg and Mjøs, 1989). The upper part of the Cook A commonly contains reworked bioclastic material, glauconite and rolled-belemnite guards. The succeeding Cook B unit consists of a thin sequence of fine-grained clastics, containing parallel-laminated and *Helminthoidea*-burrowed mudstones and siltstones. The overlying erosive, massive and cross-bedded medium-grained sandstones, containing abundant *Asterosoma* traces, are ascribed to the Cook C unit. The Cook A and C units are interpreted as prograding shallow-marine sandbodies, separated by a transgressive mudstone in areas where it has not been eroded by incision of the Cook C unit.

Drake Formation. The terms 'Dunlin Shale Member' (Bowen, 1975) or 'Drake subunit' (Deegan and Scull, 1977) were initially introduced for the unit of sandy and calcareous grey marine mudstones lying between an underlying

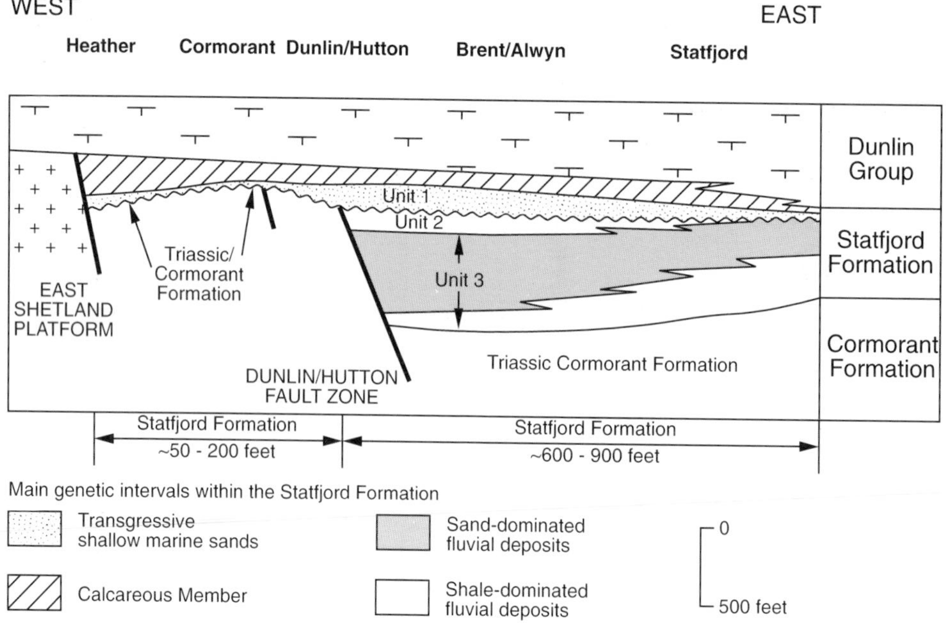

Fig. 8.13 Spatial distribution of reservoir units within the Banks Group, on the western margin of the North Viking Graben, northern North Sea, demonstrating the presumed structural control on component reservoir units of the Statfjord Formation (modified after Johnson and Stewart, 1985).

sand-prone package (Cook Formation) and units belonging to the Middle Jurassic Brent Group. This regionally correlatable package of sediment has since been ascribed to the Drake Formation. The formation commonly contains thin beds of oolitic sideritic ironstones. The upper parts of the unit may extend into the Aalenian but are commonly eroded by a surface defining the base of the Brent Group. The surface of erosion is often referred to as the 'Mid Cimmerian Unconformity' (Underhill and Partington, 1993,1994). Where preserved, the upper parts of the Drake Formation show evidence of progradation and shallowing above the acme of early Jurassic transgression (point of maximum retrogradation).

Darwin Formation. A new lithostratigraphic term has recently been introduced to describe the occurrence of bioturbated siltstones and fine sandstones in the Beryl and Bruce oilfields of the Beryl Embayment (Richards *et al.*, 1993). They lie directly above the Nansen Formation and are erosively overlain by non-marine sediments of the Pentland Formation. They are interpreted as Pliensbachian to Toarcian shallow-marine shelf deposits, equivalent to the Dunlin Group of the Brent Province.

Fig. 8.14 Sand-body architecture of the 'Statfjord Megasequence' in a north-easterly traverse from the Statfjord to Snorre fields (modified after Steel, 1993).

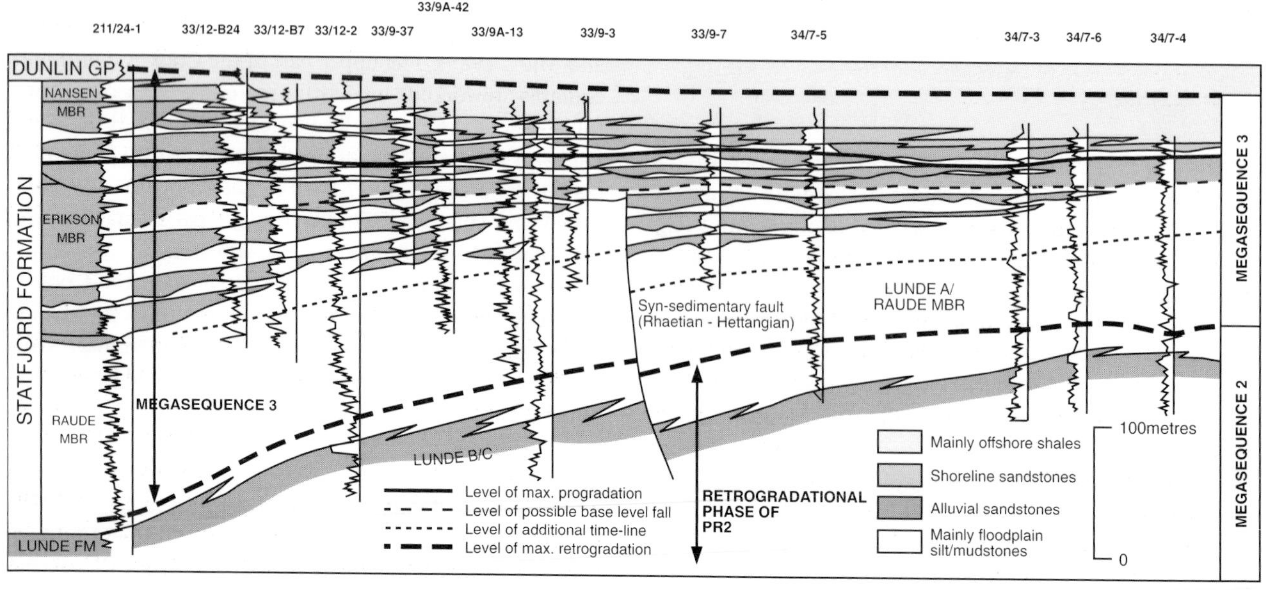

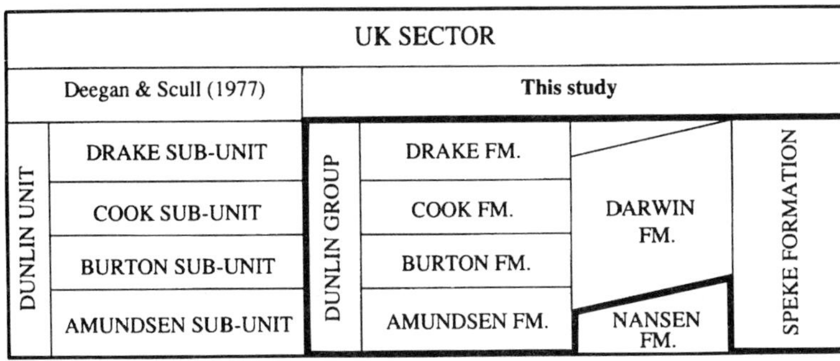

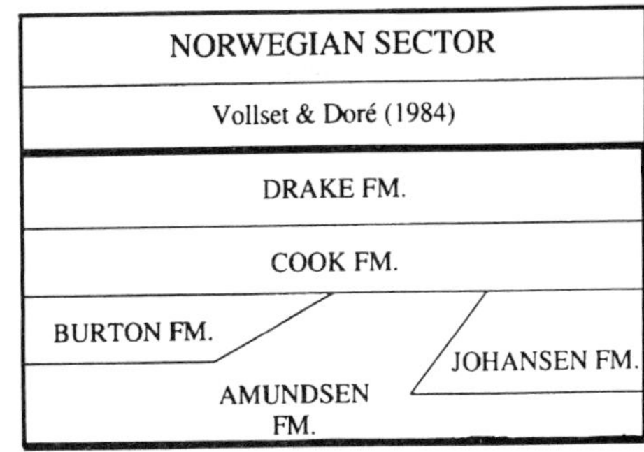

Fig. 8.15 Lithostratigraphic nomenclature for the Lower Jurassic Dunlin Group of the UK and Norwegian sectors of the North Sea (after Richards *et al.*, 1993).

8.3.2 The Inner Moray Firth

Dunrobin Bay Group

Lower Jurassic sediments have also been penetrated in exploration boreholes in western parts of the Moray Firth Rift Arm, where they form the prospective lower reservoir units in the Beatrice oilfield in UKCS license block, 11/30a. There, the siliciclastic sequence has recently been ascribed to the Dunrobin Bay Group, which comprises the Golspie, Mains, Lady's Walk and Orrin formations (see Fig. 8.8) and ranges in age from the Hettangian to the Early Toarcian (Richards *et al.*, 1993).

Onshore exposures along the neighbouring Sutherland coast also enable direct comparisons to be made with the subsurface data. The base of the succession is marked by an unconformity, across which there was a change in climatic conditions (Batten *et al.*, 1986). Underlying units belonging to the Triassic Lossiehead Formation are characterized by several pedogenic horizons, including the regionally extensive Stotfield Calcrete Formation silcrete (Peacock *et al.*, 1968; Naylor *et al.*, 1989; Cameron, 1993a). The variation in velocity resulting from the siliceous nature of the Stotfield Calcrete Formation causes an important acoustic-impedance contrast, which enables the base of the Dunrobin Bay Group to be traced regionally on seismic data.

Golspie Formation. Sediments previously ascribed to the informal varicoloured unit in offshore areas and to the Dunrobin Pier conglomerate and the carbonaceous siltstone and clay unit in onshore exposures are now assigned to the Golspie Formation (Richards *et al.*, 1993). The sequence fines up from conglomerates to red-brown, yellow, grey and green fissile and mottled calcareous interbedded siltstones and mudstones containing evidence of desiccation. The formation records the upward transition from braided-stream deposition on an alluvial fan to lacustrine sedimentation. A marine influence has been detected towards the top of the formation in onshore exposures at Golspie (Lam and Porter, 1977) and in the Lossiemouth Borehole (Berridge and Ivimey-Cook, 1967), above which there was a temporary return to freshwater conditions.

Mains Formation. Previously known as the White Sandstone unit, the term Mains Formation has been introduced to describe the sharp-based heterolithic unit of sandstones, muddy siltstones and black organic mudstones that lies between the lacustrine-dominated Golspie Formation and the marine Lady's Walk Shale Formation (Richards *et al.*, 1993). Internally, the unit comprises numerous upward-fining cycles of cross-bedded and current-rippled sandstones, capped by burrowed and rootletted mudstones. The formation is interpreted as comprising a series of fluvial and estuarine distributory channels. The channelized sandstones form the lowest reservoir units of the Beatrice Field, where they are known informally as the Grant Sand Member (Stevens, 1991) or the J sand reservoir unit (Linsley *et al.*, 1980).

Lady's Walk Shale Formation. The top of the Mains Formation is marked by a sharp change to deposition of the calcareous and micaceous mudstone of the Lady's Walk Shale Formation, which contains an abundant marine fauna, including ammonites, thin-shelled bivalves, crinoids and belemnites. The Lady's Walk Shale Formation displays a fining-upwards trend to a surface marked by prominent carbonate concretions and a high faunal diversity. Although early work suggested that this level was marked by an unconformity (the 'Mid Pliensbachian Event' of Andrews and Brown, 1987), more recent studies have shown that the stratigraphic break described was a function of sample spacing (Stephen *et al.*, 1993). Instead, the faunal diversity and abundance indicate that sedimentation was continuous but highly condensed. The point of maximum deepening may be correlated with the *Paltechiocerus taylori* subzone of the *U. jamesoni* ammonite zone. A shallowing in depositional facies occurs above this level and heralds the onset of delta progradation.

Orrin Formation. The base of the Orrin Formation is marked by a distinctive upward increase in sand content above the Lady's Walk Shale. The Orrin Formation is 45–60 m thick and records the progradation of a major deltaic system. The formation may be split into two units (termed the H and I reservoir sands in the Beatrice Field). The lower unit comprises an upward-coarsening subarkosic to quartz arenitic sandstone, which records initial shoreline progradation and the development of a major mouth-bar

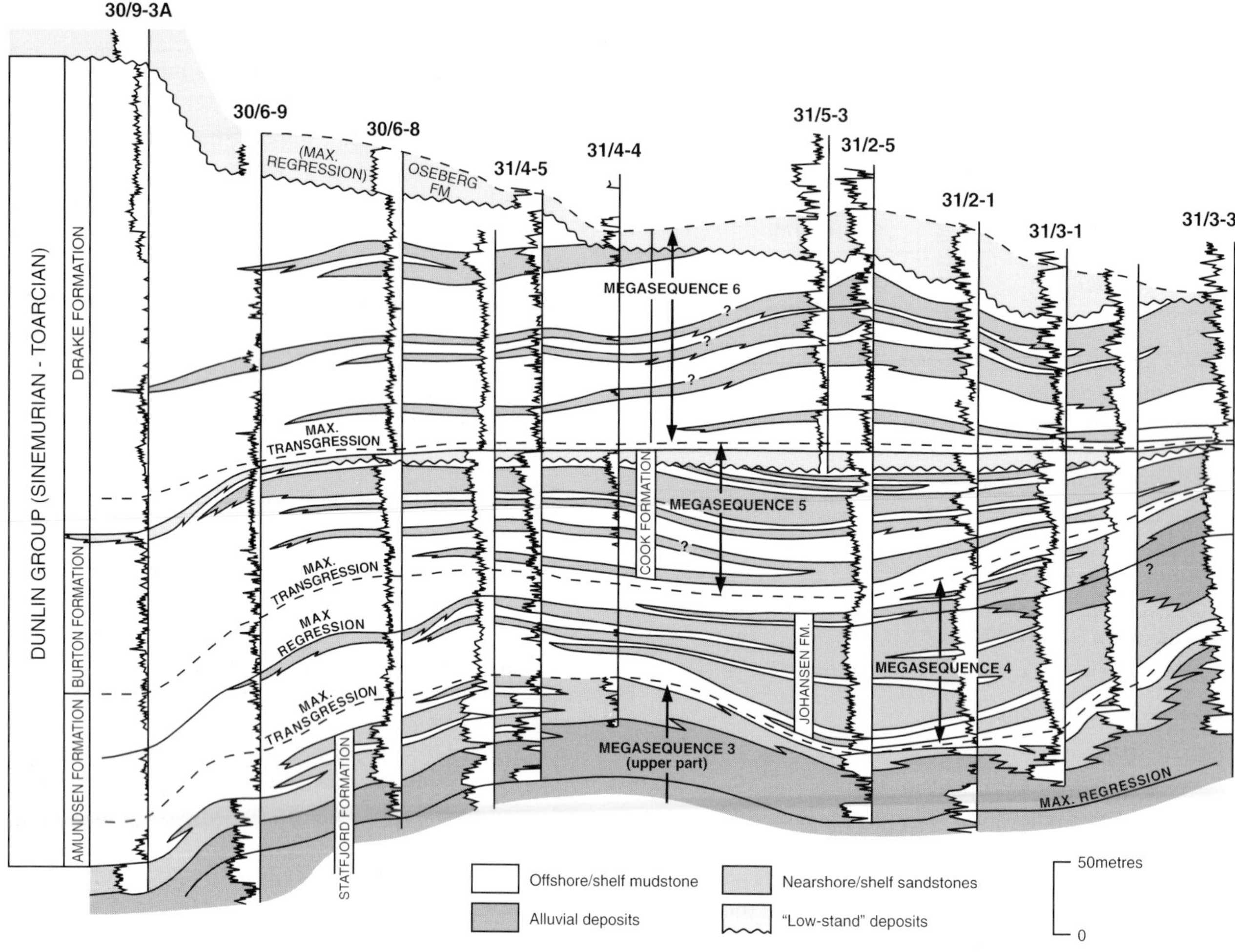

Fig. 8.16 Sand-body architecture of the component parts of the Dunlin Group, Horda Platform, northern North Sea (after Steel, 1993).

complex. The overlying unit is characterized by isolated fluvial channels, lagoonal sediments and coals. The top of the formation is marked by a regional unconformity termed the 'Mid Cimmerian Event' (Stephen *et al.*, 1993).

8.3.3 The Central Graben

As a result of subsequent erosion and incision, sediments with Lower Jurassic faunas and floras have been found in only a few wells in the Central Graben area, where they have been termed the Fjerritslev Formation (see Fig. 8.2), after the type-well section in the Danish subsurface. The sequence is laterally equivalent to the Dunlin Group of the northern North Sea, the Dunrobin Bay Group of the Moray Firth and the Lias Group in the southern North Sea and onshore Britain. It comprises dark grey calcareous mudstones, with occasional argillaceous limestone and fine-grained sandstones, interpreted as representing deposition in a low-energy, open-marine setting.

8.3.4 Southern North Sea and onshore exposures

Lower Jurassic deposition in southern Britain and adjacent areas was characterized by a distal marine, argillaceous sequence, ascribed to the Lias or Altena Group. The unit locally includes organic-rich source-rock horizons (e.g. the Toarcian Posidonia Shale or *schistes cartons*, the Jet Rock and Bituminous Shales of the Cleveland Basin and the Black Ven Marls of the Wessex Basin; Cornford *et al.*, 1988), all of which have generated hydrocarbons locally, such as in the Wessex, Paris, Broad Fourteens and West Netherlands basins.

Within UK waters of the southern North Sea, the Posidonia Shale was never buried sufficiently deeply to reach maturity. Depending on their subsequent burial, the Posidonia Shales of the west Netherlands basin have, however, locally sourced oil in Triassic (Bunter) sandstones (particularly where the units have been juxtaposed across faults), and Upper Jurassic to Lower Cretaceous reservoirs, such as in the Ijsselmonde–Ridderkerk field. Total recoverable reserves in the basin are estimated at just over 400×10^6 barrels of oil and nearly 200×10^9 ft^3 (5×10^9 m^3) of gas (de Jager *et al.*, 1996). In the Wessex Basin, the equivalent stratigraphic unit forms the source rock for the Lower Jurassic Bridport Sandstone and the Triassic Sherwood Sandstone reservoirs of the Wytch Farm and Kimmeridge oilfields (Colter and Harvard, 1981; Stoneley, 1982; Butler, 1998; McKie *et al.*, 1998; Evans *et al.*, 1998).

8.4 Middle Jurassic stratigraphy and sedimentation

8.4.1 Introduction

The boundary between Middle and Lower Jurassic sediments is often marked by a widespread stratigraphic break, termed the 'Mid Cimmerian Unconformity', which is apparent from the North Viking Graben to the southern North Sea and from the Moray Firth to the Ringkøbing–Fyn High and the Danish Embayment (Underhill and Partington, 1993, 1994). Its development was associated with a pronounced depositional shallowing, marked by the onset of widespread non-marine fluviodeltaic deposition. It is the sediments laid down during the period following the unconformity that form the widely distributed reservoirs of the Brent Group in the North Viking Graben and the Fladen Group of the Moray Firth and Central Graben. This section will firstly address the occurrence and distribution of the various Middle Jurassic lithostratigraphic units before returning to the evidence for the existence of and causal mechanisms for the 'Mid Cimmerian Unconformity'.

8.4.2 The Viking Graben

The Brent Group

The Brent Group makes up the most important hydrocarbon-reservoir sequence within the North Sea area. It comprises five constituent formations: the Broom, Rannoch, Etive, Ness and Tarbert formations, which form a broadly regressive–transgressive wedge of diachronous, coastal and shallow-marine sediment, which record the outbuilding and subsequent retreat of a major wave-dominated delta fed from the south (Fig. 8.17; Johnson and Stewart, 1985). In the Cormorant and Statfjord fields, the Etive and Lower Ness formations clearly have the best reservoir properties within the Brent Group, with porosities often in excess of 25% and permeabilities in the range 30–3000 mD (Taylor and Dietvorst, 1991).

The Brent Group clastic wedge locally exceeds 500 m in thickness and ranges from Aalenian to Early Bathonian in age. Chronostratigraphic relationships show that it is partly time-equivalent with the Hugin and Sleipner formations of the South Viking Graben (Underhill and Partington, 1993). Although the stratigraphic subdivision of the Brent Group into five formations has proved too simplistic for field-management purposes, recourse may still be made to the component Broom, Rannoch, Etive, Ness and Tarbert formations (Fig. 8.18) to illustrate the general sedimentological and palaeogeographical aspects of deposition.

The general palaeogeographical setting for a snapshot in time is illustrated in Fig. 8.19 (Budding and Inglin, 1981). Representative palaeogeographies have also been successfully produced using sequence stratigraphy (Mitchener *et al.*, 1992), which detail more exactly the progressive build-out and drowning of the delta front during Middle Jurassic times. The predictive capabilities of sequence stratigraphy were recently employed in the fifteenth Norwegian licensing round, in an attempt to locate and exploit deeper-water turbiditic-sandstone prospects, which might have been deposited ahead of the delta during the time of relative lowstand in sea level.

Broom Formation. The basal sequence of the Brent Group is termed the Broom Formation in the UK and the Oseberg Formation in Norwegian waters. It comprises medium- or coarse-grained cross-stratified marine sandstones and rare pebbly sandstones, which lie above an unconformity or correlative conformity. Their deposition is interpreted to be the result of single- or multistorey, progradational sandbodies, which formed part of an amalgamated series of fan deltas shed transversely into the East Shetland Basin (Graue *et al.*, 1987; Helland-Hansen *et al.*, 1992; Steel, 1993). Graue *et al.* (1987) have shown that the uppermost parts of the formation have a transgressive character and probably interfinger with the overlying Rannoch Formation (Steel, 1993).

Rannoch Formation. The progradational part of the Brent Group is marked by deposition of the Rannoch, Etive and Ness formations. The Rannoch Formation consists of coarsening-up, cleaning-up fine-grained micaceous sandstones, dominated by horizontal to low-angle and undulatory lamination (< 15° dip) characterized by mica-rich and mica-poor alternations and cut by steep-sided scours, filled by onlapping laminae (Scott, 1992). Its sedimentary characteristics lead to the interpretation of the Rannoch Formation as being deposited in a middle shoreface environment of the progradational Brent Group, which was affected by combined unidirectional and oscillatory flows in a high-energy, storm-dominated shoreface environment.

Etive Formation. Two sedimentary sequences commonly characterize the Etive Formation. One consists of an upward-coarsening profile from the underlying middle

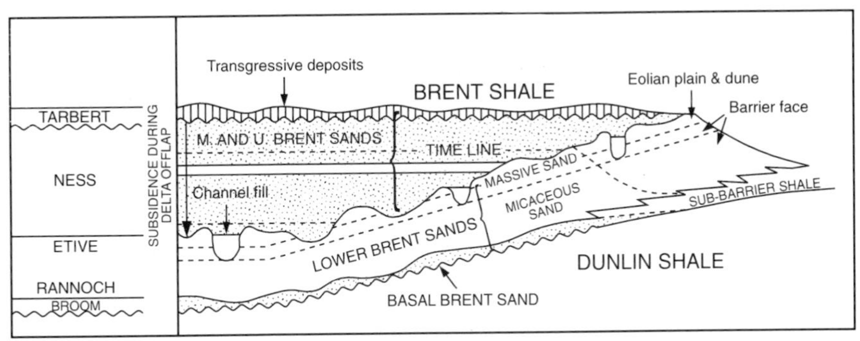

Fig. 8.17 Basic depositional framework of the Middle Jurassic Brent Group (after Johnson and Stewart, 1985).

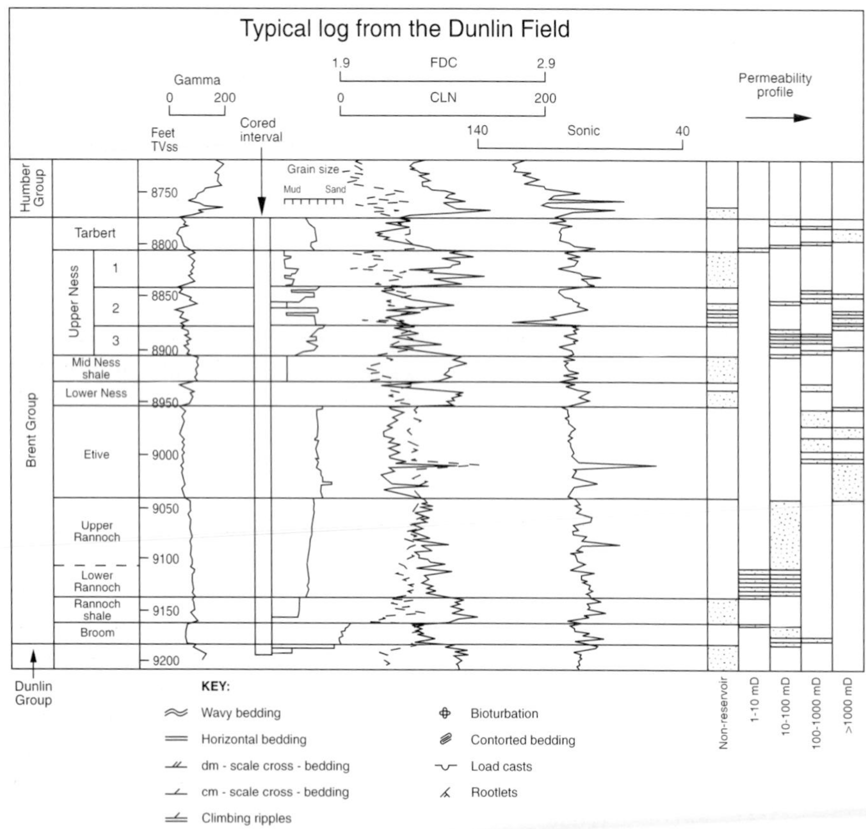

Fig. 8.18 Typical sedimentary sequences and corresponding wireline log and permeability profiles for reservoir sandstones of the Brent Group, Dunlin Field (after Johnson and Stewart, 1985).

shoreface and is dominated by parallel and low-angle laminations, interpreted as representing a prograding barrier beach. The second fines upward and is characterized by sharp-based, cross-bedded sandstones, marked locally by mud rip-up clasts and muddy laminations. The second facies association is interpreted as representing the deposits of a barrier-beach system, whose orientation was controlled by tidal-inlet channels. Overall, the Etive Formation is thought to represent complex 3D shoreline topography,

Fig. 8.19 Schematic three-dimensional palaeogeography depicting the main depositional environments and the resultant facies belts in the Brent Group of the Cormorant Field, northern North Sea (after Budding and Inglin, 1981).

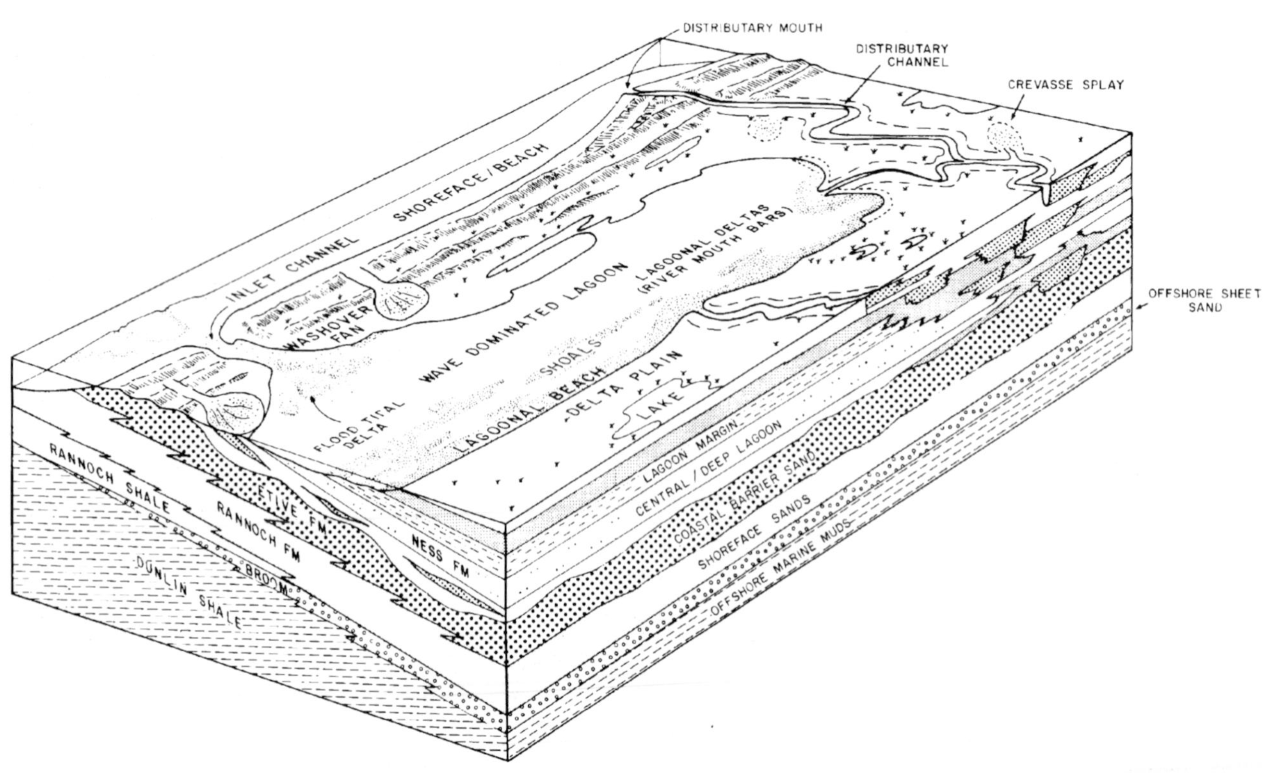

which records the development of a microtidal, wave-dominated barrier island, made up of nearshore bar and trough systems, cut by tidal-inlet channels, which pass up into a foreshore environment.

Ness Formation. The Ness Formation consists of a heterolithic sequence of interbedded sandstones, mudstones and coals, interpreted as having been deposited in a delta-top setting, containing a wide spectrum of subenvironments, including lagoonal muds, distributory channels, levees, mouth bars and lagoonal shoals. The Ness Formation is often subdivided into three component parts: a lower interbedded unit, the mid-Ness Shale and an upper sandstone-dominated unit. Overall, the Ness Formation is considered to record the last part of the main phase of northward progradation of the Brent delta and the onset of its retrogradation.

Tarbert Formation. The term 'Tarbert subunit' was originally introduced by Deegan and Scull (1977) to describe the strongly time-transgressive unit of fine- to medium-grained sandstones at the top of the Brent Group. The formation typically consists of grey to brown, relatively massive, fine- to medium-grained, locally calcite-cemented sandstones, with subordinate finer siliciclastic units and rare coals.

The units ascribed to the Tarbert Formation demonstrate a marked backstepping (retrogradational) profile, interpreted as resulting from a punctuated southerly-directed marine transgression of the former Brent delta (see Fig. 8.17). The transgressive pattern may also be followed eastward on to the Horda Platform and westwards on to the East Shetland Platform.

Evidence from the Brent Province suggests that the Tarbert Formation is a stratigraphic unit separate from the underlying Brent formations (e.g. Ninian Field; Underhill *et al.*, 1997). It is separated from them by an unconformity and shows a quite different spatial distribution, being restricted to downflank areas of the tilted fault blocks. It is interpreted as representing the early stages of footwall uplift and hangingwall subsidence, associated with extensional movements of the fault-block-bounding faults.

Fig. 8.20 Lithostratigraphic nomenclature for the Middle Jurassic Fladen Group of the UK and Vestland Group of the Norwegian sectors of the North Sea (after Richards *et al.*, 1993, Deegan and Scull, 1977, and Vollset and Doré, 1984).

UK SECTOR
Deegan & Scull (1977)
CENTRAL NORTH SEA
FLADEN GROUP
PENTLAND FM.
RATTRAY FM.
This study
NORTHERN NORTH SEA
HEATHER FM.
HUGIN FM.
PENTLAND FM.
CENTRAL GRABEN
HEATHER FM.
Rattray & Ron Volcanics Mbrs
PENTLAND FM.
OUTER MORAY FIRTH
Stroma Mbr
PENTLAND FM.
Rattray Volcanics Mbr
INNER MORAY FIRTH
HEATHER FM.
BEATRICE FM.
BRORA COAL FM.
NORWEGIAN SECTOR
Vollset & Doré (1984)
VIKING GRABEN
HEATHER FM.
HUGIN FM.
SLEIPNER FM.
CENTRAL GRABEN
HAUGESUND FM.
BRYNE FM.
VESTLAND GROUP

The Fladen Group (Fig. 8.20)

The definition of the Fladen Group has recently been expanded from that of Deegan and Scull (1977) to encompass four formations: the Pentland, Brora Coal, Beatrice and Hugin formations (Richards *et al.*, 1993). Use of MFS enables contemporaneous lithostratigraphic units to be directly correlated and compared with those of the northern North Sea (see Fig. 8.5).

Pentland Formation. Coal-bearing heterolithic clastic sediments of paralic aspect and broad Middle Jurassic age characterize large areas of the northern and central North Sea, including the Outer Moray Firth, Central Graben and South Viking Graben, and are generally known as the Pentland Formation. Coal-bearing units found in the Outer Moray Firth and previously referred to as the Skene Member of the Sgiath Formation (Harker *et al.*, 1993) are now described as the Stroma Member and are thought of as a subdivision of the Pentland Formation, rather than forming part of the overlying Heather Formation. Similarly, the earlier definition of the Sleipner Formation in the Beryl Embayment and South Viking Graben area has been superceded, with these paralic units now also being included in the Pentland Formation. However, the term Brora Coal Formation has been retained to describe a synchronous heterolithic unit of non-marine sandstones, mudstones and coals located in the Inner Moray Firth basin.

Volcanic rocks form an important component of the Pentland Formation. They have been ascribed to the Rattray Volcanics Member and the Ron Volcanics Member (Richards *et al.*, 1993). The Rattray Volcanics Member is found in and around the North Sea triple-junction area in Quadrants 15, 16, 21 and 22 (Fig. 8.21). It consists of lavas and volcaniclastic sediments. The lavas are undersaturated porphyritic, alkali olivine basalts (Dixon *et al.*, 1981; Fall *et al.*, 1982). A newly introduced term, the Ron Volcanics Member, describes a separate, thick volcanic succession centred on the Quadrant 29 area of the Central Graben and thought to form a discrete volcanic centre (Richards *et al.*, 1993; Smith and Ritchie, 1993). The Rattray Volcanics Member has been dated by Ritchie *et al.* (1988) as 153 ± 4 Ma, using argon/argon (Ar/Ar) methods.

Brora Coal Formation. The term Brora Coal Formation is applied to a heterolithic unit of non-marine sandstones,

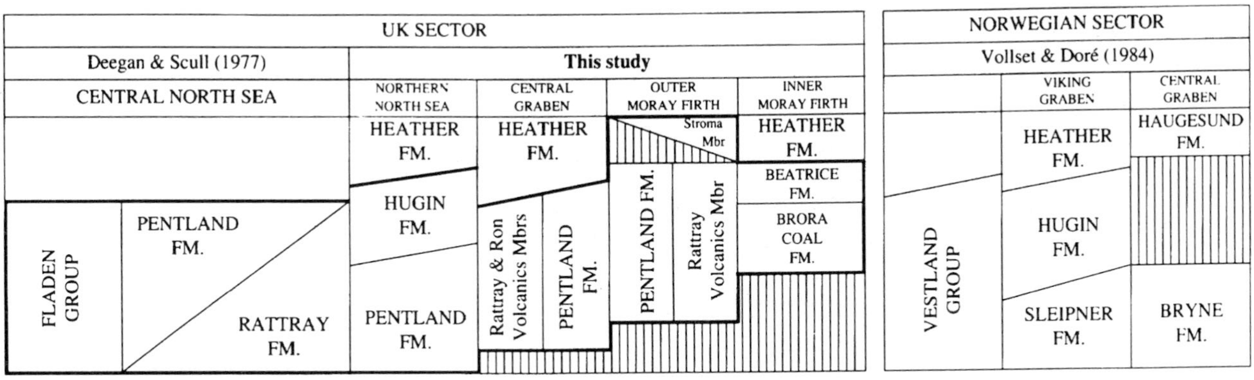

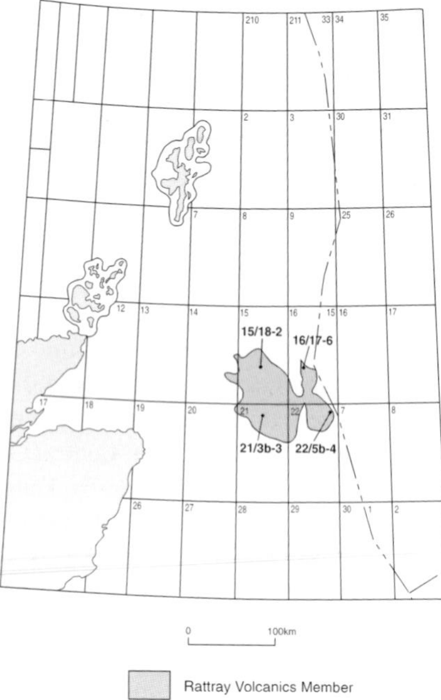

Fig. 8.21 Distribution of the Rattray Volcanics member, central North Sea (after Richards *et al.*, 1993). Rare radiometric ages support stratigraphical data to suggest that volcanic activity occurred during the Bathonian and Callovian (e.g. Ritchie *et al.*, 1988; Latin *et al.*, 1990a,b), thus postdating uplift of the central North Sea dome and predating the Late Jurassic rift episode. Although their timing and limited aerial extent and geochemistry suggest that the volcanics are not the products of a long-lived, hot, focused, mantle jet (Dixon *et al.*, 1981; Latin *et al.*, 1990a,b), they are consistent with a warm, transient, plume-head source (Underhill and Partington, 1993, 1994).

Fig. 8.22 Distribution of component parts of the Middle Jurassic in the Inner Moray Firth Basin (after Richards *et al.*, 1993). A. Brora Coal Formation. B. Beatrice Formation.

mudstones and coals, which lies between the Orrin Formation and the Beatrice Formation in the Inner Moray Firth (Fig.8.22A; Richards *et al.*, 1993) and in exposures of the Sutherland coast (Hurst, 1981; Stephen *et al.*, 1993). It is interpreted as being Bajocian to Early Callovian in age (MacLennan and Trewin, 1989). The formation contains reservoir sandstones in the Beatrice oilfield, where they are informally known as the C, D, E, F and G sands (Linsley *et al.*, 1980).

Beatrice Formation. The term Beatrice Formation has been introduced to describe the predominantly sandstone sequence that lies above the Brora Coal Formation and below the Heather Formation in the Inner Moray Firth (Fig. 8.22B). The name was taken from the Beatrice oilfield, which remains the only producing field in western areas of the basin. Lithostratigraphically, it has been subdivided into a lower, Louise Member and an upper, Carr Member. The Louise Member consists of a stacked succession of upward-coarsening units, commonly comprising a lower dark grey micaceous, carbonaceous and pyritic mudstone, with shell debris, through fine-grained flaser-bedded silty sandstones to fine–medium, cross-bedded and parallel-laminated sandstones, containing bivalves and belemnites. The Carr Member consists predominantly of sandstones in two or more well-defined upward-coarsening cycles. Stephen *et al.* (1993) have shown that units of the Brora Coal and Beatrice Formations show a progressive pinch-out towards the east, such that their spatial distribution does not extend to the Outer Moray Firth.

Hugin Formation. The Hugin Formation is the lithostratigraphic term originally introduced by Vollset and Doré (1984) to describe a diachronous set of marine sandstones and mudstones lying beneath marine mudstones of the Heather Formation and above the heterolithic continental

clastics of the Sleipner or Pentland formations of the South Viking Graben. As such, it is partly coeval with deposition of the Beatrice Formation of the Inner Moray Firth.

Middle and Upper Callovian open-marine turbidites have been described by Cockings *et al.* (1992), which are detached from the Hugin shelf and shoreline depositional systems and form a potential exploration target (Fig. 8.23). These sediments have been termed the Ling Formation, which consists largely of fine- to medium-grained massive and graded sandstones occurring in both the UK and Norwegian sectors of the South Viking Graben (e.g. wells 16/8a-4, N15/3-1 and N15/3-3.

Middle Jurassic sedimentation in the southern North Sea and adjacent areas

Middle Jurassic deposition in the southern North Sea is marked largely by the non-marine fluviodeltaic deposition of the West Sole Group. Its base is marked by a minor disconformity (correlative to the 'Mid Cimmerian Unconformity'). Onshore exposures at Blea Wyke, Ravenscar, in the Cleveland Basin, Yorkshire, demonstrate that the sequences immediately underlying the event mark pronounced basin shallowing prior to the onset of non-marine deposition, and give unique evidence for the rapidity with which shallowing occurred. Subsequent Middle Jurassic deposits of the Cleveland Basin belonging to the Scalby Formation have long been used as analogues for non-marine (Ness Formation) deposition in the Brent Province (e.g. Hancock and Fisher, 1981; Fisher and Hancock, 1985). Locally, continental (paralic) deposition extends up into the Callovian. For example, sandstones, ascribed to the Lower Graben Sand, that contain coal seams occur within the Dutch part of the Central Graben disconformably above the Aalenian, marine, Werkendam Shale Member (NAM and RGD, 1980; Underhill and Partington, 1994).

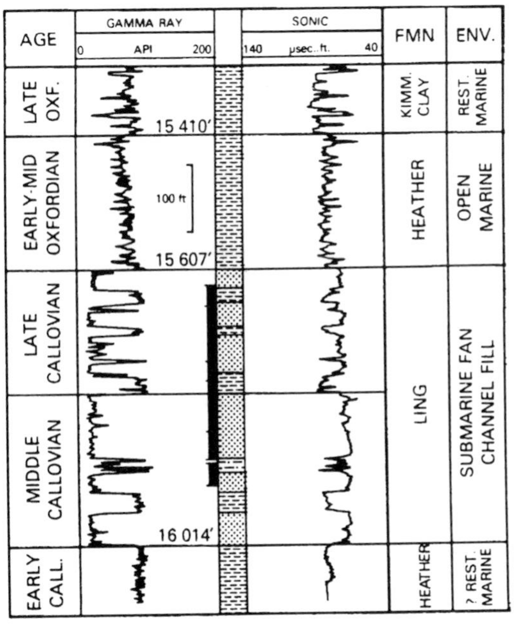

Further to the south, in the Weald and eastern parts of the Wessex Basin, Middle Jurassic deposition is marked by a facies change into a more carbonate-dominated sedimentary succession, which includes several potential reservoir intervals. Of these the most significant regionally is the Great Oolite, which forms the reservoir in the Humbly Grove, Horndean, Stockbridge, Storrington and Singleton discoveries. Other known Middle Jurassic reservoirs include the fractured Cornbrash at Kimmeridge Bay (Evans *et al.*, 1998) and the Frome Clay in Wytch Farm (Underhill and Stoneley, 1998).

Fig. 8.23 Gramma-ray and sonic-log profiles for well 16/8-4 showing geological ages and the main depositional environments of Lower Callovian to Middle Oxfordian sequences, South Viking Graben (after Cockings *et al.*, 1992).

8.4.3 Development and significance of the 'Mid Cimmerian Unconformity'

Subcrop pattern of the 'Mid-Cimmerian Unconformity'

Well correlations in the Inner Moray Firth (Fig. 8.24) and Viking Graben (Fig. 8.25) areas show that there is a systematic truncation of Lower Jurassic stratigraphic marker beds from west to east and from north to south, respectively. Triassic and older stratigraphic units immediately underlie the unconformity in central parts of the North Sea, adjacent to the area that was to become the site of the rift-arm triple junction during the Middle to Late Jurassic. A plot of the youngest stratigraphy preserved beneath the unconformity demonstrates that erosion caused progressive truncation of beds throughout the North Sea domain, such that a series of concentric to elliptical subcrop patterns exist, with Triassic and older beds subcropping in the central region close to the triple junction (Fig. 8.26). The scale and low-angle nature of the unconformity means that the Mid Cimmerian Unconformity is hard to detect on seismic data, and it is only in the Inner Moray Firth that it has been convincingly demonstrated using seismic (Ziolkowski *et al.*, 1998).

Mapping of the stratigraphical relationships above and below the 'Mid-Cimmerian event' has important implications for understanding the nature of the driving mechanism behind its formation. The deepest stratigraphic incision can be documented on the western margin of the South Viking Graben (i.e. Fladen Ground Spur and Utsira High) and on the Halibut Horst in the Moray Firth, where Devonian sequences subcrop the unconformity. Rather than interpreting these relationships as resulting from the doming episode, it is more likely that these subcrop patterns formed in response to the effects of a subsequent (Kimmeridgian) synsedimentary graben-margin (footwall) uplift. As such, they overprint and mask the effects of the earlier event in the area and it is interpreted that deposition of a complete Triassic–Early Jurassic succession probably occurred on the highs. Similar variations on the Horda Platform are also likely to be due to footwall incision.

Interestingly, the Middle Jurassic is marked by significant faunal provinciality in the north-west European domain (Callomon, 1979; Enay and Mangold, 1982; Ziegler,

LITHO. STRAT.
GEN. STRAT. SEQ.
12/21-3
12/22-2
12/29-1
12/30-1
13/27-1
GR
DT
0 150
40 140
HEATHER
J2
UPPAT
ALNESS SPICULITE
J1bv
J1biv
J1b
J1biii
J1bii
J1bi
BEATRICE
A SAND
M.S.
B SAND
BRORA COAL
J1a
MFS
TS
LOCATION MAP
50KM
LEGEND.
TS: TRANSGRESSIVE SURFACE
MFS: MAXIMUM FLOODING SURFACE
GR GAMMA RAY LOG (API UNITS)
DT SONIC LOG (MICROSEC/FT)
250
FT
0
SHOREFACE/FORESHORE SANDS
TIDAL/ESTUARINE SAND BARS
FLUVIAL DISTRIBUTARY CHANNELS ON A MARGINAL MARINE/NON MARINE DELTA PLAIN
UNCONFORMITY

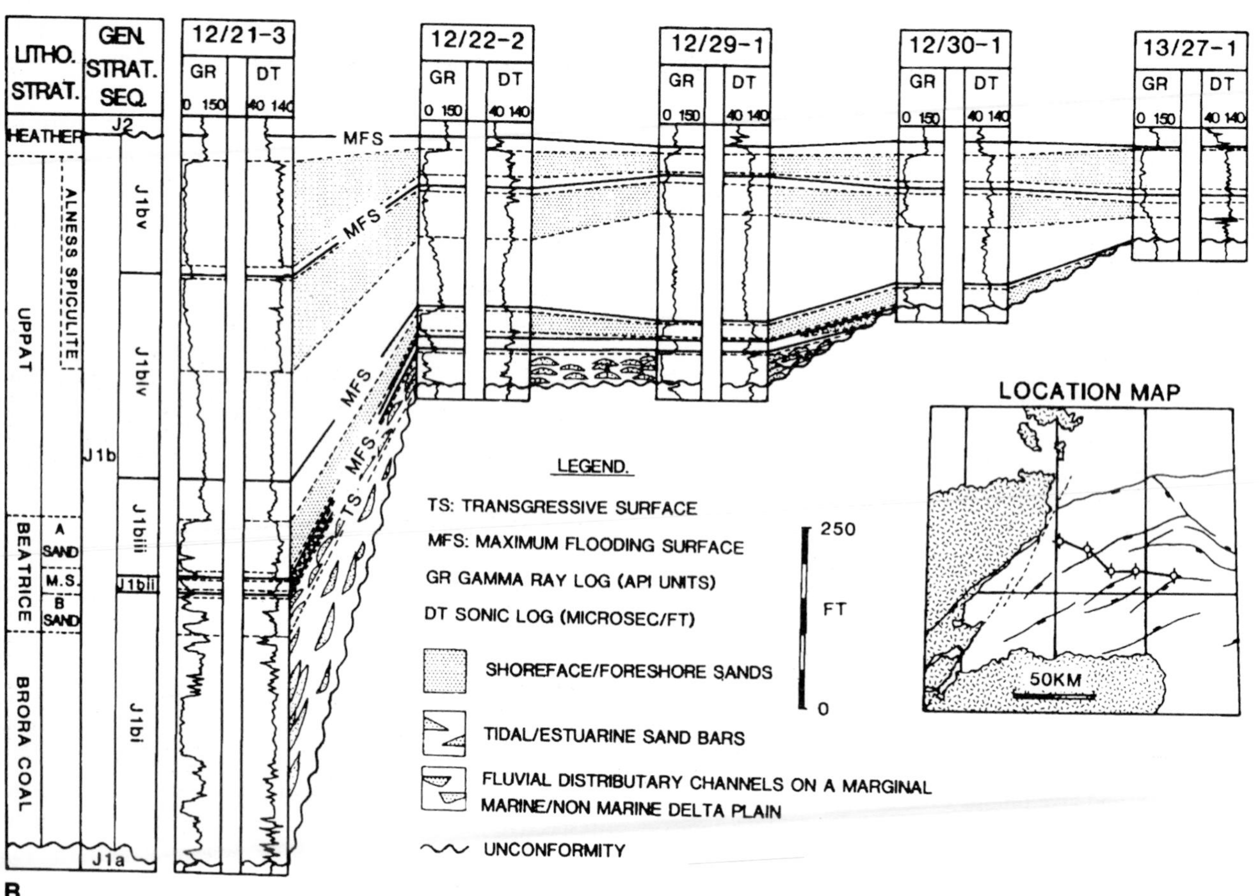

B

LITHO. STRAT.
GEN. STRAT. SEQ.
12/21-3
12/22-3
12/22-2
12/29-1
12/30-1
GR
DT
0 150
40 140
J1b
ORRIN
H SAND
I SAND
J1aiii
J1aii
J1ai
Depth in m
DUNROBIN BAY
LADY'S WALK SHALE
WHITE SANDSTONE
VARI-COLOURED
UNDIF. TRIAS.
TR
MFS
TS
100
ft
0
LOCATION MAP
50 km
TS TRANSGRESSIVE SURFACE
MFS MAXIMUM FLOODING SURFACE
GR GAMMA RAY (API UNITS)
DT SONIC LOG (MICROSEC/ft)
SHOREFACE/FORESHORE SANDS
OFFSHORE MARINE MUDS/SILTS
DISTRIBUTARY CHANNELS (FLUVIAL &/OR ESTUARINE)
ARGILLACEOUS BAY, LAGOON, ESTUARY, FLOODPLAIN DEPOSITS
UNCONFORMITY

A

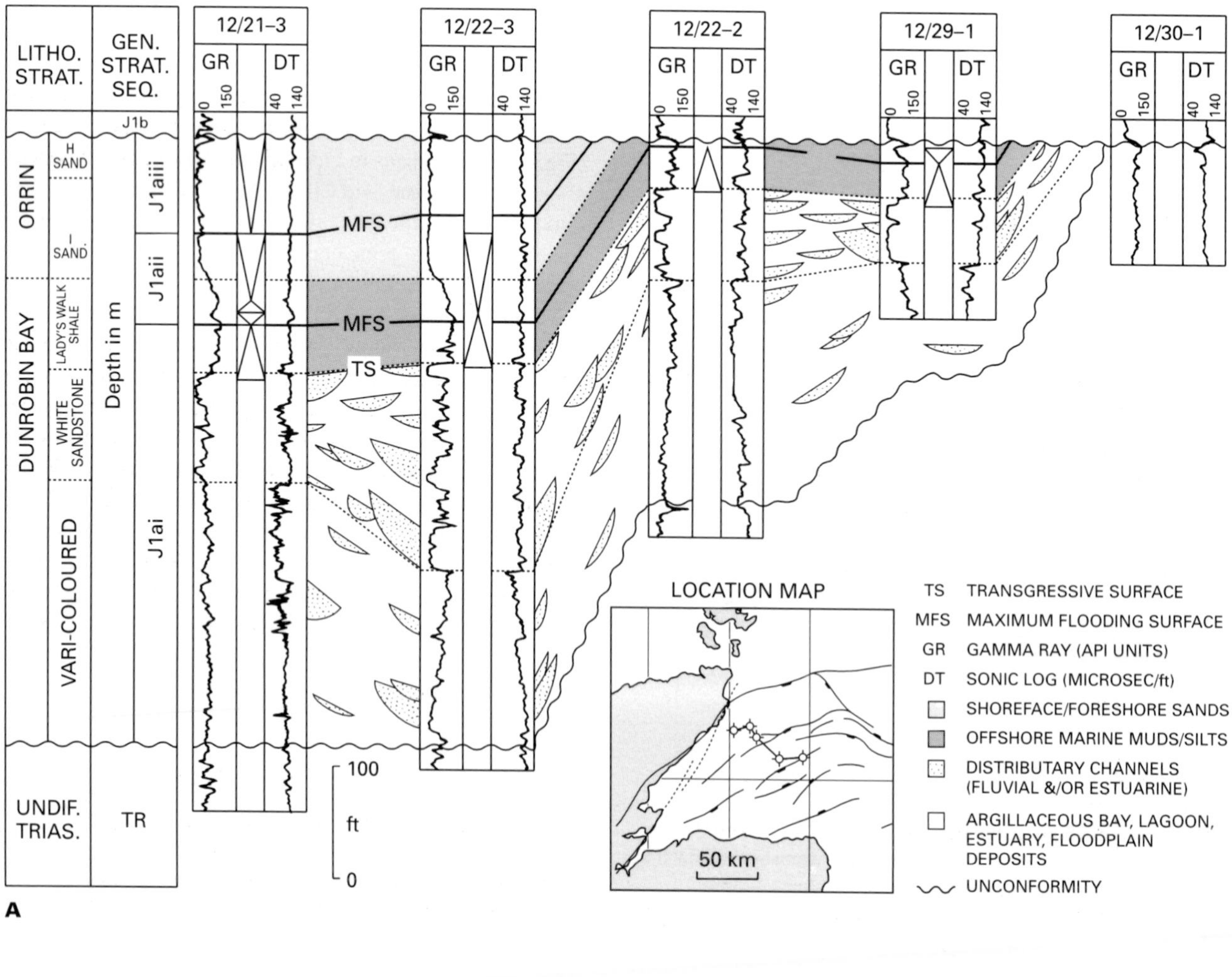

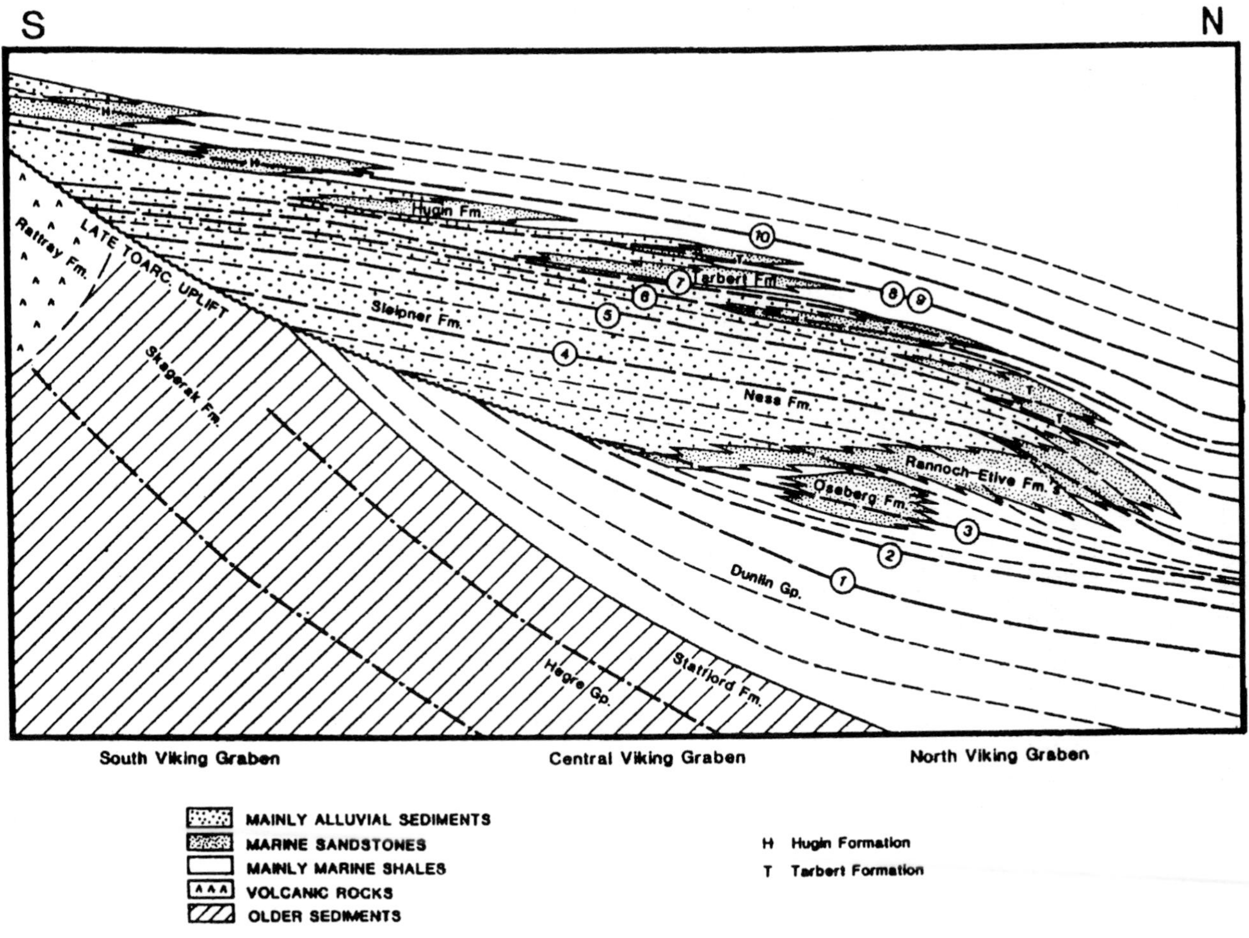

Fig. 8.25 Schematic north–south section through the Brent, Vestland and Dumlin Groups showing their constituent formations and timeline correlations (after Graue *et al.*, 1987, and Helland-Nansen *et al.*, 1992). The progressive truncation of the Lower Jurassic interval has been attributed to Late Toarcian–Aalenian uplift of the central North Sea (Underhill and Partington 1993, 1994).

1990a). The onset of this restriction in faunal exchange between southern, Tethyan and northern Boreal waters occurred during the Aalenian and is consistent with the development of a barrier coincident with the dome described here. Hence, it seems likely that this regional event was sufficient to have formed a significant barrier, separating the two areas until at least the late Bathonian or Callovian, when species mixing and diversification once more occurred between subbasins as the dome deflated. Unfortunately, the faunal provinciality, which persisted well into Late Jurassic times, has neither aided regional correlations nor helped comparisons to be made between different subbasins.

Onlap pattern on to the 'Mid Cimmerian Unconformity'

The extent to which progressively younger units may be traced continuously to apparently onlap the unconformity shows that the locus of flooding migrates toward the central area in a complex fashion (Fig. 8.27). The patterns document initial flooding of the North Viking Graben Brent Province during the Aalenian and Bathonian (Figs 8.27 and 8.28), and the separate nature of subsequent progressive marine incursions into the South Viking Graben and across the Moray Firth. Marine waters entered the Central Graben from the north during the late Bathonian–early Callovian

Fig. 8.24 (*Opposite*) Well correlation panels depicting the stratigraphic relations in the Inner Moray Firth Basin (after Stephen *et al.*, 1993). A. Lower Jurassic correlation panel depicting the progressive easterly erosional truncation of the Orrin and Dunrobin Bay formations of the Dunrobin Bay Group. B. Middle–Upper Jurassic correlation panel depicting the progressive easterly marine flooding and onlap of the Brora Coal and Beatrice formations of the Fladen group and the Heather formation (previously informally known as the 'Uppat Formation'). The unconformity defined by the erosional truncation and marine onlap has been termed the 'Mid Cimmerian Unconformity' (Underhill and Partington, 1993).

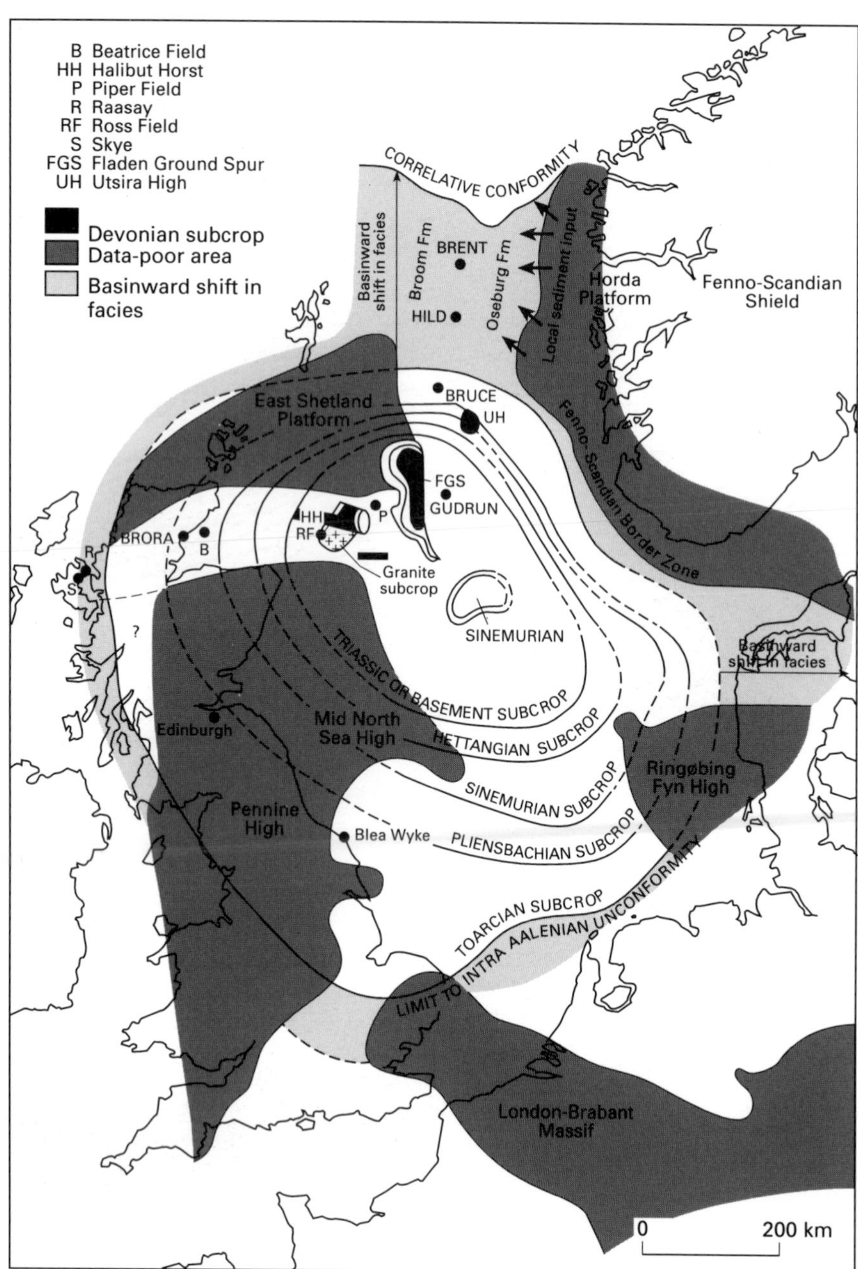

Fig. 8.26 Subcrop to the 'Mid Cimmerian Unconformity' (after Underhill and Partington, 1994). The circular to elliptical pattern of the subcrop has been attributed to the development of a central North Sea dome, which experienced uplift above a warm and transient mantle-plume head (Underhill and Partington, 1993, 1994). Erosion of the exhumed Triassic–Lower Jurassic sedimentary succession led to a coeval basinward shift in facies, marked by major progradation of fluvial–deltaic clastics (e.g. the Brent Group).

and progressed generally southward through the late Callovian and Oxfordian, reaching the Dutch Central Graben in the Late Oxfordian (Underhill and Partington, 1994). Although full marine conditions were restored over the North Sea during the early Kimmeridgian for the first time since the Liassic, a few isolated shelves and tectonic highs (e.g. Fladen Ground Spur) appear to have remained subaerially exposed.

The Central North Sea Dome

The most plausible explanation for the relationships highlighted by the subcrop and onlap patterns, given the lateral scale of the anomaly, is that they record the effect of widespread, but transient, regional uplift and subsidence of the North Sea during the Jurassic. Furthermore, the nature of the stratigraphic relationships demonstrates that the uplift was of an approximately concentric nature, centred upon what was to become the North Sea rift triple junction. Consequently, the regional stratigraphic relationships support the existence of a 'Central North Sea rift dome', as originally proposed by Whiteman *et al.* (1975), Hallam and Sellwood (1976), Eynon (1981), Ziegler (1982a, 1990a,b) and Leeder (1983). Stratigraphic evidence provided by the Pentland Formation suggests that the dome created by the thermal anomaly probably had a relatively low-lying but highly irregular regional relief, thereby allowing the accumulation of paralic sediments in areas that experienced some form of differential subsidence. It is now thought unlikely that the dome ever achieved elevations of 2 km (i.e. the presumed thickness of interval represented by the Mid Cimmerian Unconformity; Ziegler, 1982a).

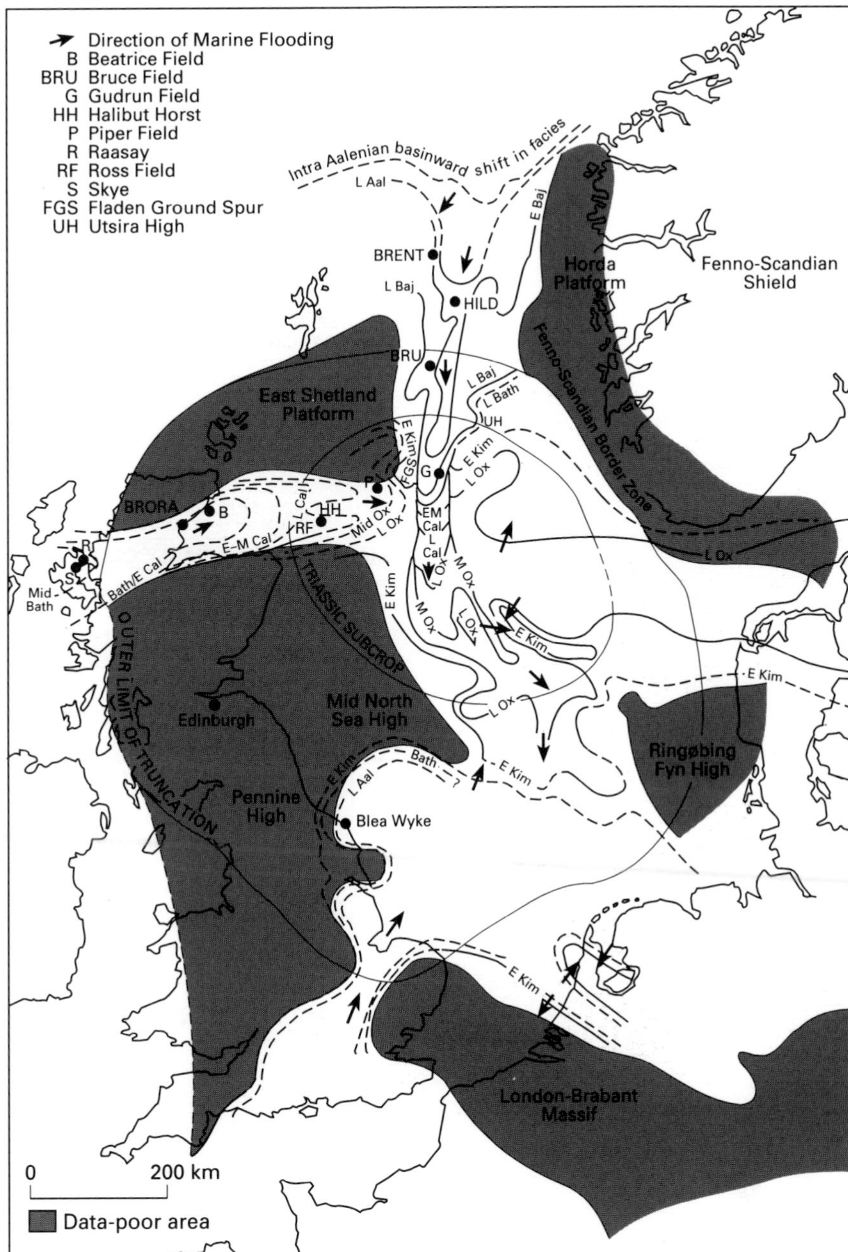

Fig. 8.27 Pattern of marine onlap on to the 'Mid Cimmerian Unconformity' (after Underhill and Partington, 1993, 1994). The pattern highlights the progressive nature of marine transgression down the incipient Viking Graben and across the Moray Firth before connection and subsequent south-east-directed flooding of the South Central Graben during the Early Kimmeridgian.

8.5 Late Jurassic stratigraphy and sedimentation

8.5.1 Structural development and evolution of the Late Jurassic extensional province

Seismic data reveal that Middle Oxfordian to Early Kimmeridgian times record the onset of a major phase of extensional activity in the North Sea basin (Ziegler, 1982a,b 1990; Underhill, 1991a,b; Rattey and Hayward, 1993; Thomson and Underhill, 1993). The resultant accelerated subsidence during the Late Oxfordian and Kimmeridgian heralded significant shoreline retreat and basin deepening in all three rift arms. Widespread deposition of sediments, ascribed to the Humber Group, records drowning of previous Callovian shoreline sites in the Viking Graben and Moray Firth and the introduction of marine sedimentation into the Central Graben for the first time as the former site of the Central North Sea Dome subsided. As well as the obvious control on synrift sedimentation, this phase of extension caused the development of major faults, with consequent dissection and tilting of early sedimentary sequences, which eventually led to the creation of the major structural plays for which the North Sea is best known.

Structures of the Viking Graben and Inner Moray Firth are consistent with an interpretation as extensional fault blocks, similar to those seen in other classic extensional provinces (e.g. modern-day Aegean Sea, Greece and the Oligo-Miocene of the Gulf of Suez), which subsequently experienced thermally driven post-rift subsidence (Fig. 8.29). Geometric, stratigraphic and sedimentary evidence suggests that each fault block was characterized by pro-

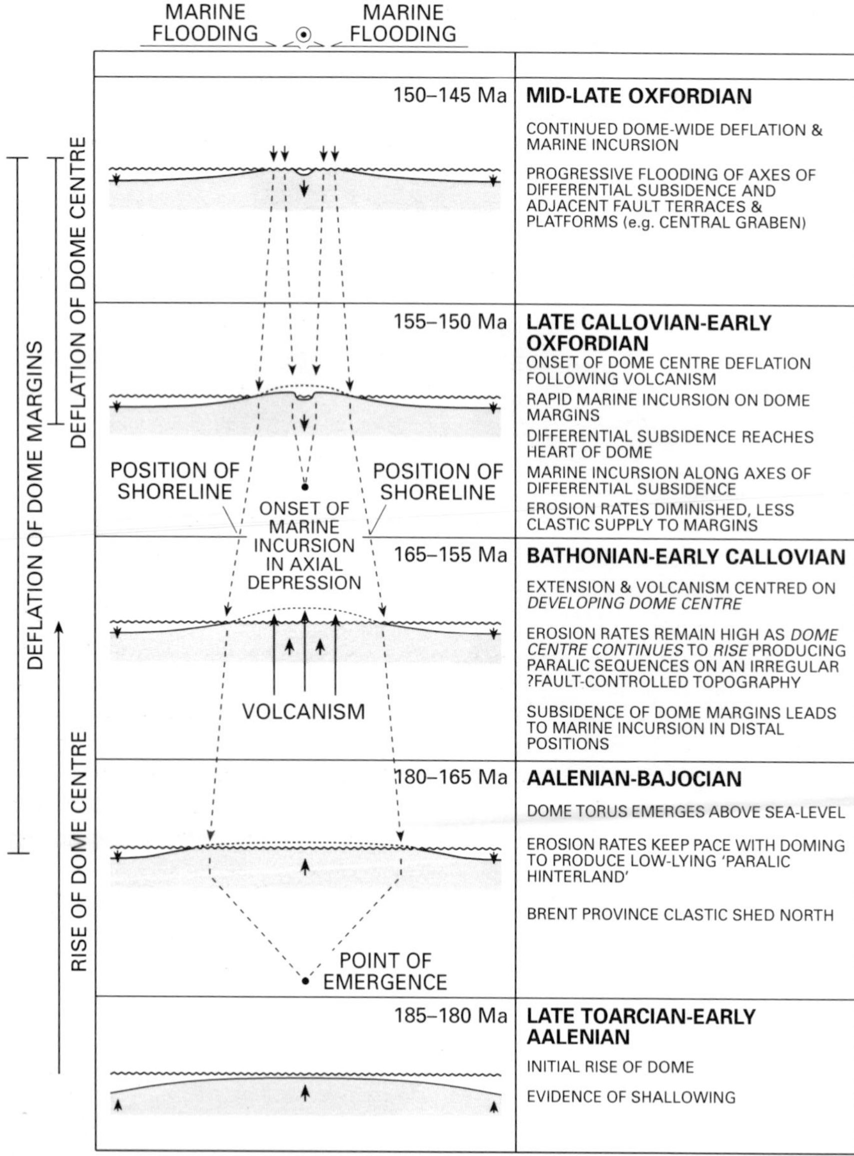

MARINE FLOODING
MARINE FLOODING
150–145 Ma
MID-LATE OXFORDIAN
CONTINUED DOME-WIDE DEFLATION & MARINE INCURSION
PROGRESSIVE FLOODING OF AXES OF DIFFERENTIAL SUBSIDENCE AND ADJACENT FAULT TERRACES & PLATFORMS (e.g. CENTRAL GRABEN)
155–150 Ma
LATE CALLOVIAN-EARLY OXFORDIAN
ONSET OF DOME CENTRE DEFLATION FOLLOWING VOLCANISM
RAPID MARINE INCURSION ON DOME MARGINS
DIFFERENTIAL SUBSIDENCE REACHES HEART OF DOME
MARINE INCURSION ALONG AXES OF DIFFERENTIAL SUBSIDENCE
EROSION RATES DIMINISHED, LESS CLASTIC SUPPLY TO MARGINS
POSITION OF SHORELINE
POSITION OF SHORELINE
ONSET OF MARINE INCURSION IN AXIAL DEPRESSION
165–155 Ma
BATHONIAN-EARLY CALLOVIAN
EXTENSION & VOLCANISM CENTRED ON DEVELOPING DOME CENTRE
EROSION RATES REMAIN HIGH AS DOME CENTRE CONTINUES TO RISE PRODUCING PARALIC SEQUENCES ON AN IRREGULAR ?FAULT-CONTROLLED TOPOGRAPHY
SUBSIDENCE OF DOME MARGINS LEADS TO MARINE INCURSION IN DISTAL POSITIONS
VOLCANISM
180–165 Ma
AALENIAN-BAJOCIAN
DOME TORUS EMERGES ABOVE SEA-LEVEL
EROSION RATES KEEP PACE WITH DOMING TO PRODUCE LOW-LYING 'PARALIC HINTERLAND'
BRENT PROVINCE CLASTIC SHED NORTH
POINT OF EMERGENCE
185–180 Ma
LATE TOARCIAN-EARLY AALENIAN
INITIAL RISE OF DOME
EVIDENCE OF SHALLOWING
DEFLATION OF DOME MARGINS
DEFLATION OF DOME CENTRE
RISE OF DOME CENTRE

Fig. 8.28 Schematic diagram depicting the development and evolution of the North Sea Dome from the Late Toarcian to the Late Oxfordian (i.e. prior to seismic-scale extensional rifting). The figure emphasizes the temporal and spatial changes in the feature through time. (After Underhill and Partington, 1993, 1994.) The volcanics of the North Sea are most compatible with an active rift model (Houseman and England, 1986) consequent upon the development and dissipation of a warm, transient plume head, rather than a McKenzie (1978)-type passive rift model.

Fig. 8.29 Structural cross-section extending from the Viking Graben across the Lomre Terrace and Horda Platform to the Troll area (after Stewart *et al.*, 1995). As well as demonstrating the effects of the main phase of Middle–Late Jurassic rifting, the section also illustrates the earlier Permo-Triassic rifting phase and its associated Lower Jurassic post-rift sag fill.

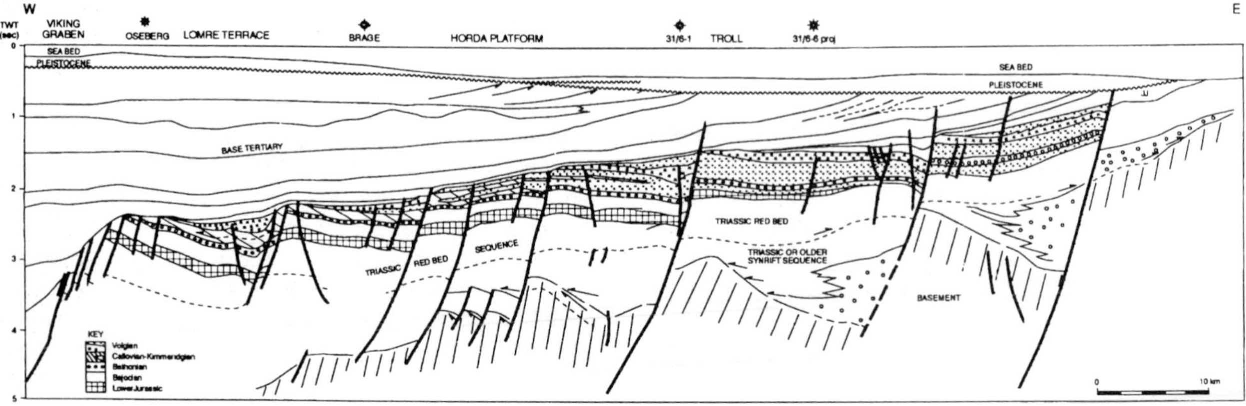

W
E
TWT (sec)
VIKING GRABEN
OSEBERG
LOMRE TERRACE
BRAGE
HORDA PLATFORM
31/6-1
TROLL
31/6-6 proj
SEA BED
PLEISTOCENE
BASE TERTIARY
TRIASSIC RED BED SEQUENCE
TRIASSIC RED BED
TRIASSIC OR OLDER SYNRIFT SEQUENCE
BASEMENT
KEY
Volgian
Callovian-Kimmeridgian
Bathonian
Bajocian
Lower Jurassic
0
10 km

nounced footwall uplift and hangingwall subsidence (e.g. Jackson and McKenzie, 1983; Roberts *et al.*, 1990a,b; Yielding, 1990) which led directly to the development of numerous structural traps in areas like the Brent Province (Figs 8.30 and 8.31).

Uplift of the footwalls of the extensional faults led, in many cases, to pronounced erosion and fault-scarp degradation. In some cases, erosion has reached down to Triassic levels (Figs 8.32 and 8.33). The amount and rate of uplift on any given fault block have been calculated using quantitative models (e.g. Yielding *et al.*, 1992; Roberts *et al.*, 1993). On all but the largest fault blocks, they argue that the rates of footwall uplift were low compared with erosion rates, thus implying that the crests of the structures would have been degraded faster than they would have done had they been subaerially exposed. Consequently, this might explain why meteoric flushing of the Brent Group in footwall areas, with its consequent impact on reservoir porosity and permeability characteristics, is restricted to only the largest tilted fault blocks (e.g. Snorre, Gullfaks and Murchison fields; Yielding *et al.*, 1992; Ashcroft and Ridgway, 1996).

Although many of the Late Jurassic structures of the northern North Sea remain buried at depths in excess of 3 km, those of the Inner Moray Firth lie at much shallower levels, due to the effects of Tertiary regional uplift. As a consequence, the overall tectonic control and seismic definition of stratigraphic marker horizons is well imaged and better understood. Mapping of the Middle Oxfordian to Early Cretaceous seismic interval highlights the role that major extensional faults (e.g. the Smith Bank Fault) had in controlling differential subsidence (Fig. 8.34; e.g. Andrews and Brown, 1987; Underhill, 1991a). Cross-sections through the basin and restored for Late Jurassic times further illustrate the similarities between the structures found in the Inner Moray Firth and those in the North Viking Graben (Fig. 8.35; Underhill, 1991a,b).

The structural configuration of the Central Graben is made up of a number of intrabasinal highs (e.g. Forties–Montrose High), terraces (such as the Cod Terrace) and subbasins (such as the Søgne Basin; e.g. Fig. 8.36). In contrast to the other two rift arms, extension in the East Central Graben was more complex, because of the influence of halokinesis. Movement on the basin-bounding faults led to rapid subsidence in the Eastern Trough of the Central Graben, which initiated differential flow of salt at depth. The effect was to induce regional salt evacuation from the graben centre, which eventually led to the development of salt pillows and localized diapiric intrusions during the Cretaceous and Tertiary. Such salt pillows created hydrocarbon-bearing structures in overlying Cretaceous Chalk and Tertiary reservoirs (see Chapters 9 and 10).

In the central North Sea, the presence of the underlying Zechstein evaporites allowed shallow detachments to develop. The main effect was to control the structural wave-

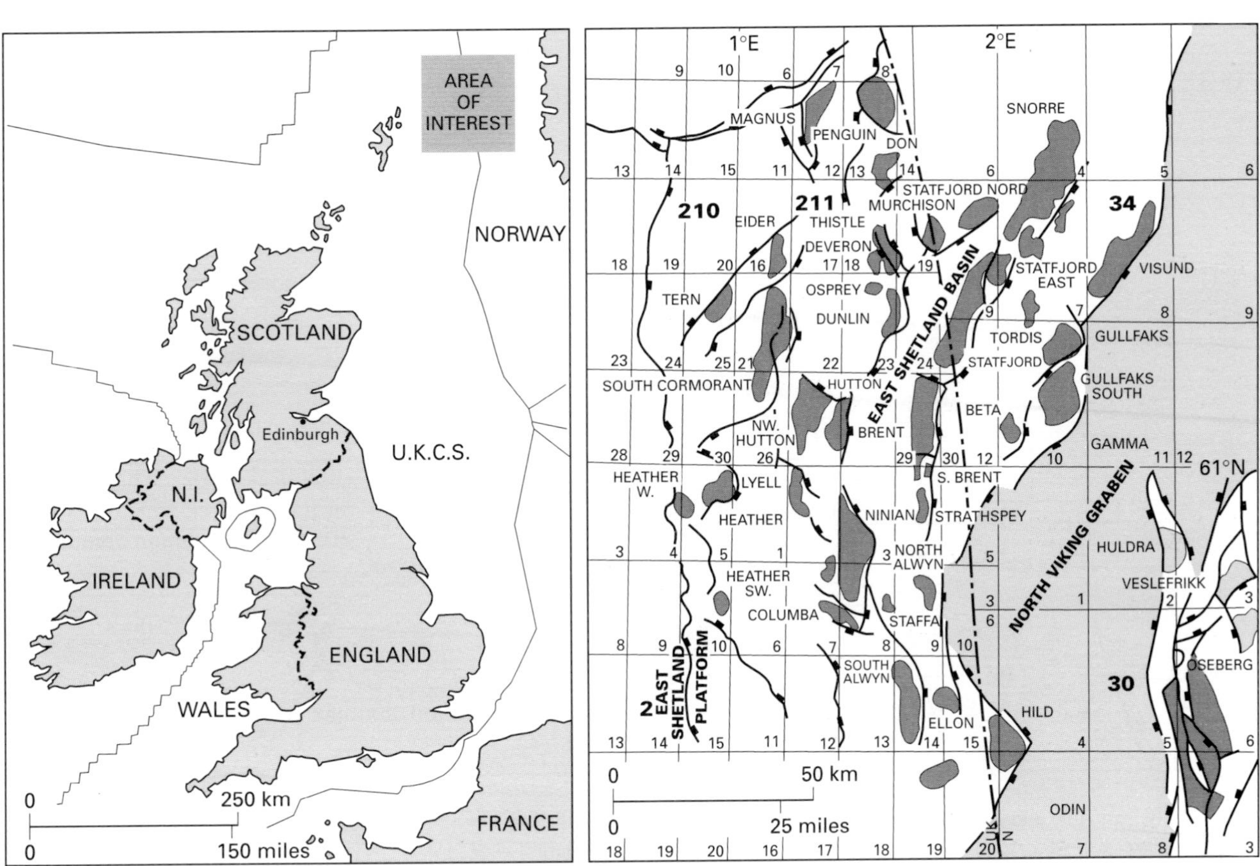

Fig. 8.30 Location maps depicting the site of oilfields in the Brent Province (after Underhill *et al.*, 1997). A. General location map. B. Distribution of main fields in the East Shetland Basin. Most are situated in tilted fault blocks and terraces to the west of the main North Viking Graben axis (tinted).

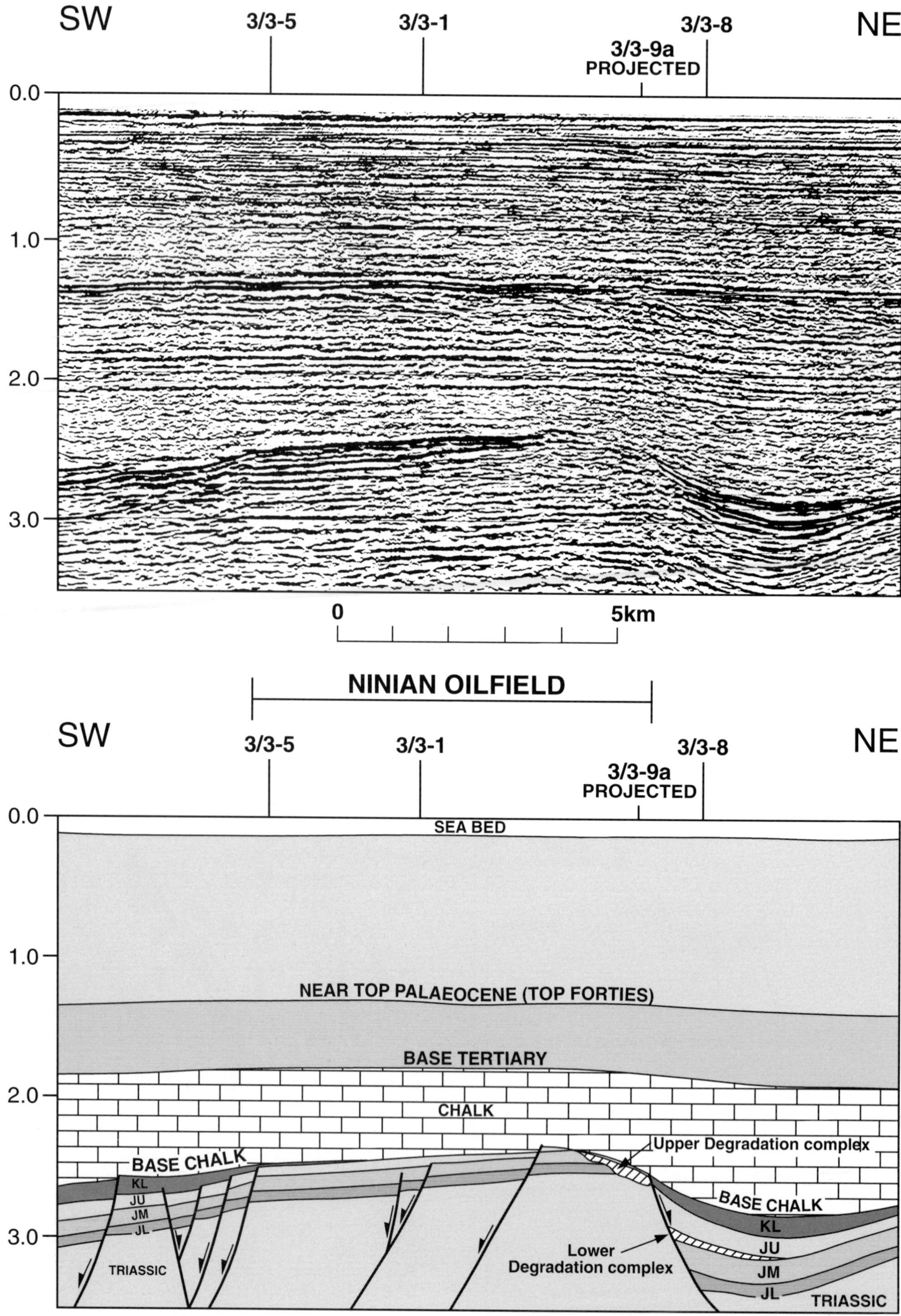

Fig. 8.31 Seismic line and geological interpretation of a representative east–west-trending section across the Brent Province depicting the nature of fault blocks in the East Shetland Basin. The section runs through the Ninian oilfield (after Underhill *et al*., 1997).

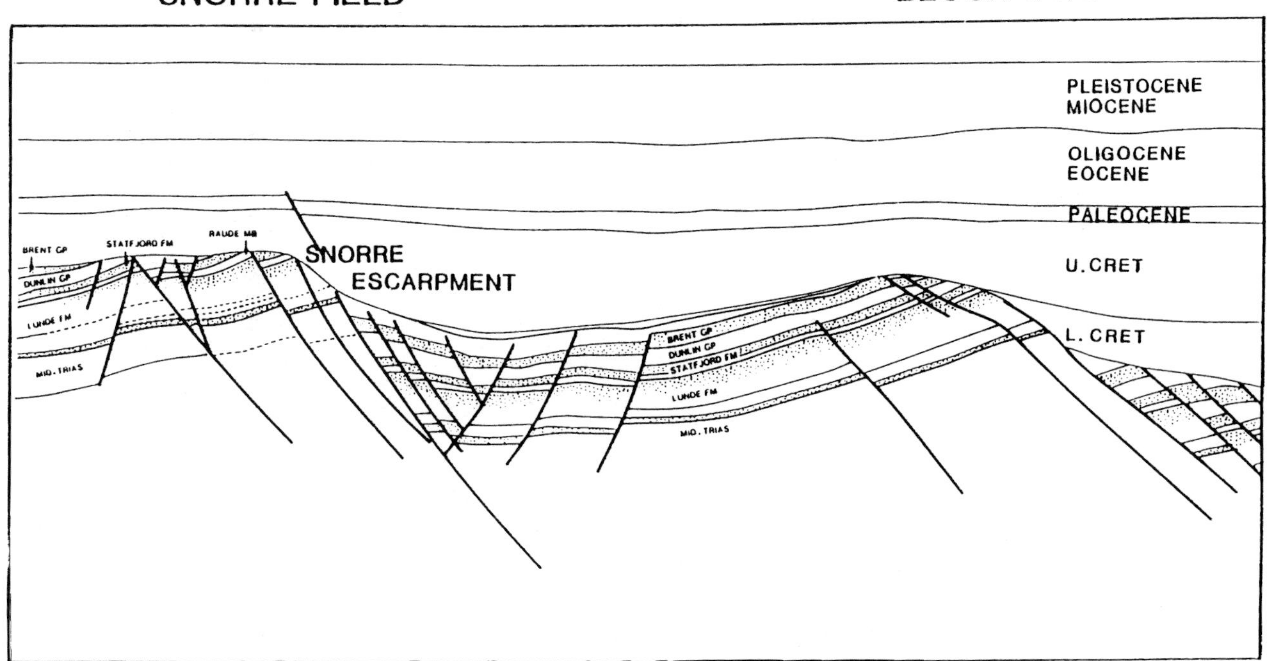

Fig. 8.32 Representative cross-section across the Snorre area, highlighting the scale, nature and typically deep erosional truncation of section seen in many of the major extensional fault blocks that characterize the Brent Province.

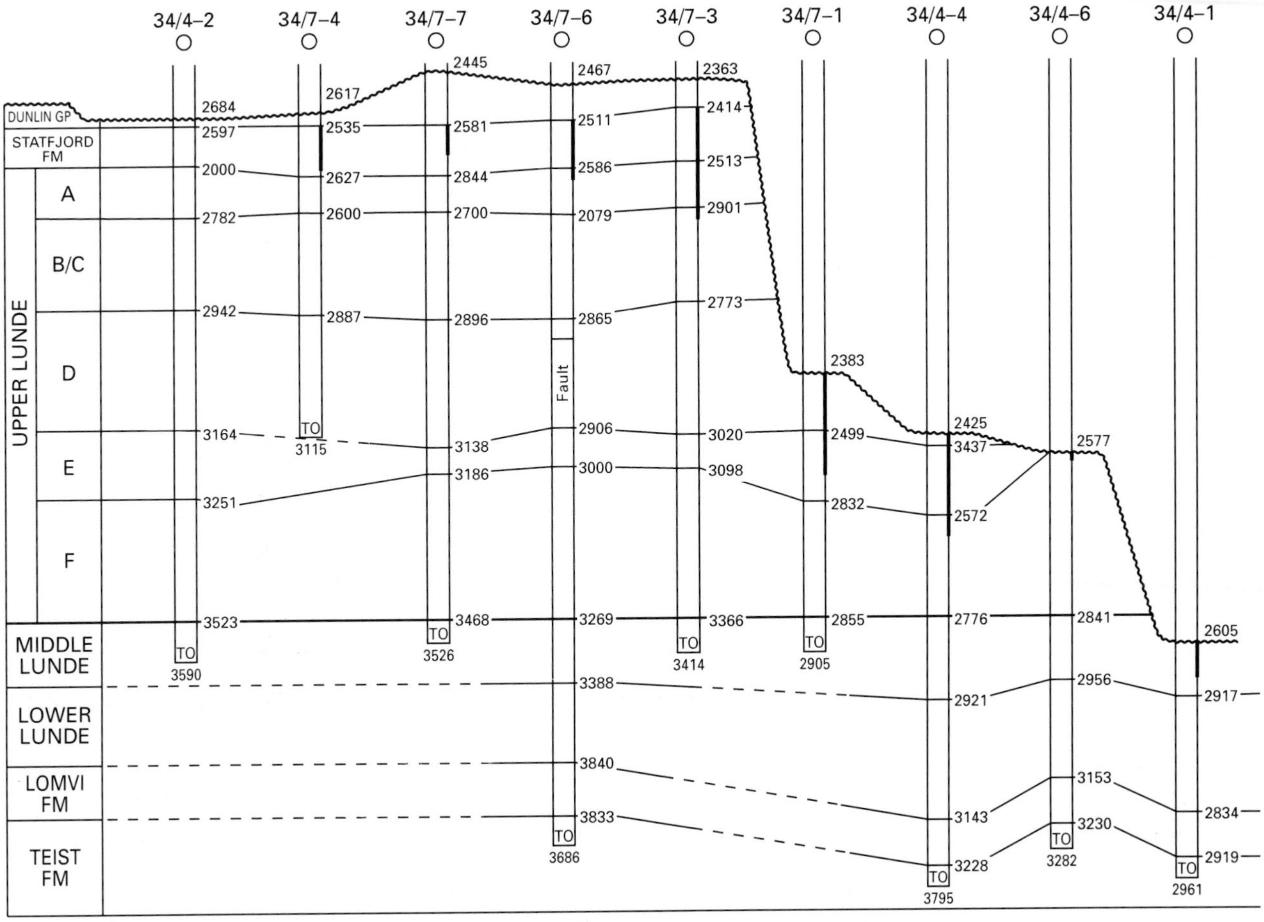

Fig. 8.33 Well correlation panel across the Snorre Field to show the progressive erosion resulting from the effects of footwall uplift on the main fault bounding the E margin of the field. Erosion reaches down into the Triassic middle Lunde Formation in the immediate footwall to the fault.

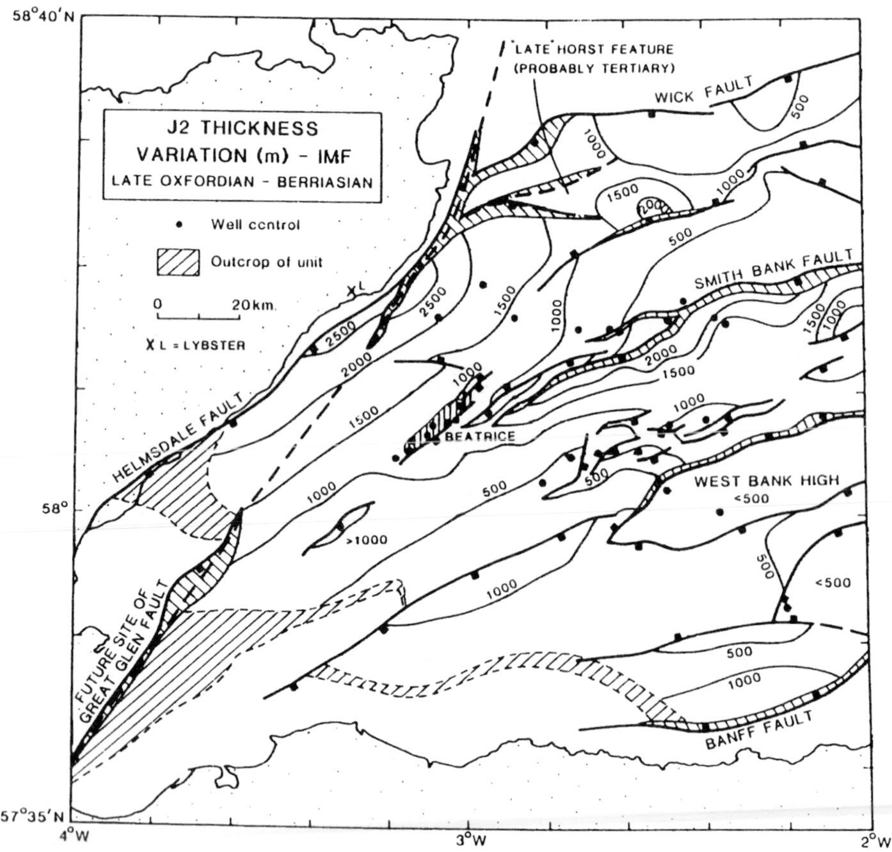

Fig. 8.34 Late Jurassic (Late Oxfordian–Ryazanian) thickness variations in the Inner Moray Firth Basin (after Andrews and Brown, 1987). The close spatial association between thickness variations and major extensional faults attests to the structural control on sediment accumulation. In the case of the Smith Bank fault, over 2 km of sediment accumulated in its immediate hangingwall, while its footwall was characterized by less than 500 m of sediment.

length and hence trap size of Jurassic fault blocks, which were more limited in width (4–10 km) in the Central Graben than in the North Viking Graben, where no salt exists (Fig. 8.37). The consequence for oilfield size has been considerable, with central North Sea fields being largely one order of magnitude smaller ($100–200 \times 10^6$ barrels stock-tank oil initially in place (STOIIP) than their northern counterparts, such as Brent, Snorre, Ninian and Statfjord, which are all 1×10^9 barrel STOIIP fields (Hodgson *et al.*, 1992).

Evidence for synsedimentary fault activity and its control on sediment dispersal and hydrocarbon prospectivity is neither restricted to the central and northern North Sea nor confined to offshore areas. In the southern North Sea, the Late Jurassic saw the deep burial of depocentres like the Sole Pit Basin, which ultimately led to the maturation of gas-prone Carboniferous source rocks. Synsedimentary extension may also be demonstrated onshore, where several 'swells' have been described across the stretch of Jurassic outcrop. The most pronounced of these is the Market Weighton High, the creation of which was associated with the renewed development and subsidence of the Cleveland Basin depocentre, beneath which, as in the case of basins of the southern North Sea, Carboniferous source rocks were buried to maturity in excess of 3 km.

8.5.2 Stratigraphic nomenclature

The Humber Group was defined by Rhys (1974) for the southern North Sea, where it consisted of a lower shale unit, a middle limestone unit and an upper shale unit, equivalent to sediments exposed on the Yorkshire coast. In the southern North Sea, the Humber Group is overlain by the Lower Cretaceous Cromer Knoll Group and is underlain by the West Sole Group. At that time, a tentative application to the whole UK North Sea was suggested.

In the central and northern North Sea, Deegan and Scull (1977) subsequently redefined the Humber Group as the mudstone-dominated unit lying above the coal-bearing, generally sandy Brent and Fladen Groups of the northern and central North Sea, respectively, and below the Cretaceous Cromer Knoll Group. The Humber Group is generally subdivided into two mudstone-dominated units (the Heather and the Kimmeridge Clay or Draupne Formation) and four sandstone-dominated units (the shelfal Emerald, Fulmar and Piper formations and the basinal Brae Formation). Additional sandstones, which lie as discrete units within the mudstone-dominated succession, have not been afforded formational status, despite being important reservoirs in several areas of the North Sea Basin. These include basinal sandstones, which occur within both the Heather and the Kimmeridge Clay formations. Those found within the Heather Formation are referred to as the Alness Spiculite, Bruce Sandstone, Freshney Sandstone, Gorse

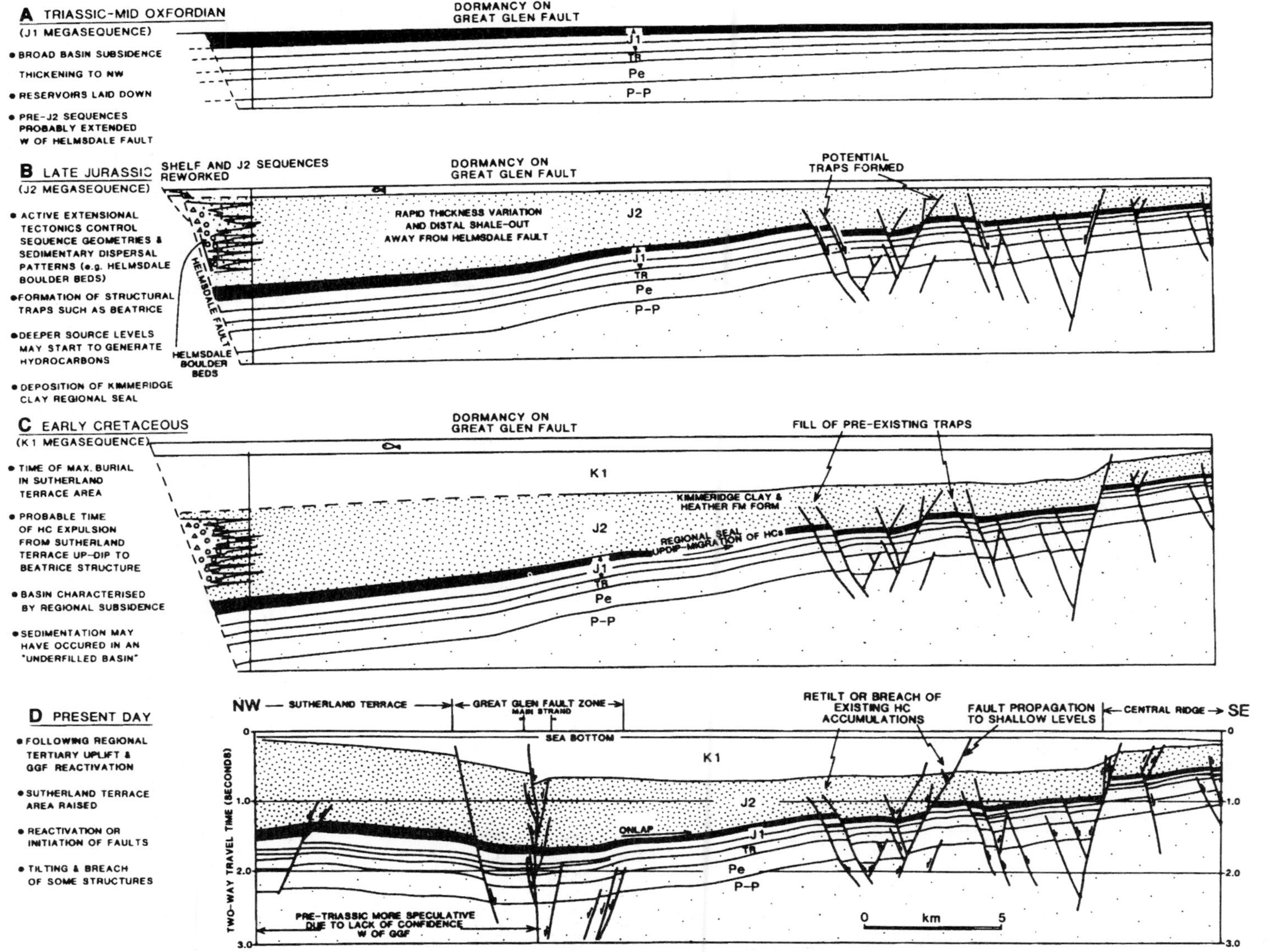

Fig. 8.35 Schematic series of cartoon sections to demonstrate the effects that the development and evolution of structural styles had on hydrocarbon prospectivity in the Inner Moray Firth Basin (after Underhill, 1991a). Exhumation of the Inner Moray Firth during the Cenozoic has led to the Jurassic subcropping the seabed and being exposed in a thin coastal strip along the East Sutherland coast. As such, the area provides a unique insight into the stratigraphy, structure and sedimentology of the North Sea Jurassic.

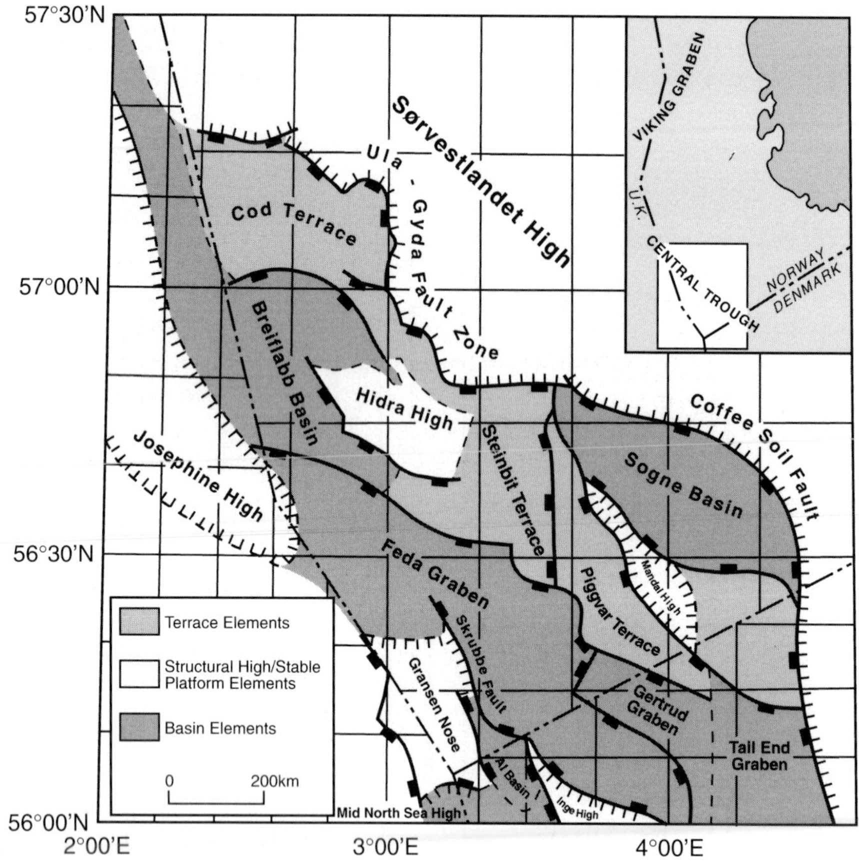

Fig. 8.36 Diagram depicting the main structural elements of the Central Graben in the UK, Norwegian and Danish sectors.

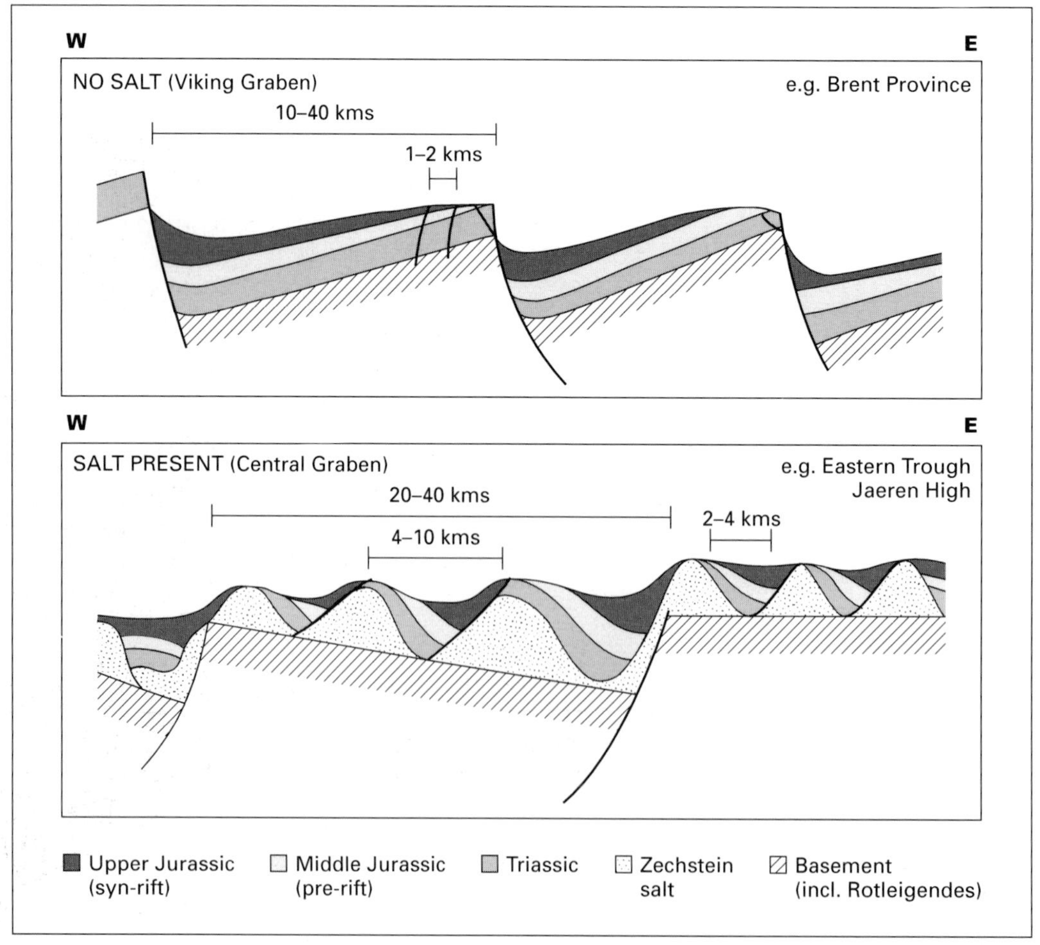

Fig. 8.37 Diagram depicting the influence of salt on fault trajectories, fault-block geometry and trap size. The effect of salt or its absence has been interpreted as the main reason why structures in the northern North Sea are up to an order of magnitude larger than those found in the central North Sea (Hodgson *et al.*, 1992).

and Ling Sandstone members. Those within the Kimmeridge Clay Formation have been termed the Birch Sandstone, Burns Sandstone, Claymore Sandstone, Dirk Sandstone, Magnus Sandstone, Ptarmigan Sandstone and Ribble Sandstone members, depending upon their geographical position in the basin (Richards *et al.*, 1993).

In the Norwegian northern North Sea, the Humber Group is referred to as the Viking Group, and consists of the Heather and Draupne formations in the North Viking Graben and the Krossfjord, Fensfjord and Sognefjord formations on the Horda Platform (Fig. 8.38; Stewart *et al.*, 1995). As in the UK, great success has recently been achieved in characterizing these units, using sequence stratigraphy, particularly in the Troll area on the Horda Platform (Fig. 8.39; Stewart *et al.*, 1995).

Revision of lithostratigraphy in the Norwegian sector by Vollset and Doré (1984) resulted in the introduction of new terms for the Upper Jurassic succession. The Humber Group was replaced by the terms Vestland and Tyne groups. In time, these two groups are partly equivalent, with the Vestland Group extending from Bajocian to Ryazanian, while the Tyne Group spans the Callovian to Ryazanian. The Tyne Group occurs throughout the Norwegian Central Graben, where it is subdivided into the Haugesund, Eldfisk, Farsund and Mandal formations. The Vestland Group is more restricted to the Vestland Arch area and southern parts of the basin, where it is commonly coeval with sandstones ascribed to the Ula Formation (Fig. 8.40).

Lithostratigraphic- and sequence-stratigraphic-based studies of the area have resulted in subdivision of the Haugesund Formation (Bergan *et al.*, 1989) and have clarified the temporal and spatial relationships between major sandstone units (Fig. 8.41). At formation level, the shallow-marine Ula Formation passes laterally into marine shales of the Haugesund (Callovian to Early Kimmeridgian) and Farsund (Kimmeridgian to Volgian) formations. In the west, the two shale units are separated by a sandy unit, the Kimmeridgian Eldfisk Formation. The radioactive shales of the Volgian–Ryazanian Mandal Formation represent the top of the Tyne Group.

Recent studies have also updated and clarified the Middle to Late Jurassic stratigraphy of the Danish Central Trough (e.g. Johannessen and Andsbjerg, 1993). The Jurassic succession is dominated by claystone, ascribed to the Farsund and Lola formations, which lie on the non-marine back-barrier and coastal-plain sediments of the Bryne Formation. Interbedded sandstones of Kimmeridgian and Volgian age are referred to the Heno and Poul formations (Johannessen and Andsbjerg, 1993).

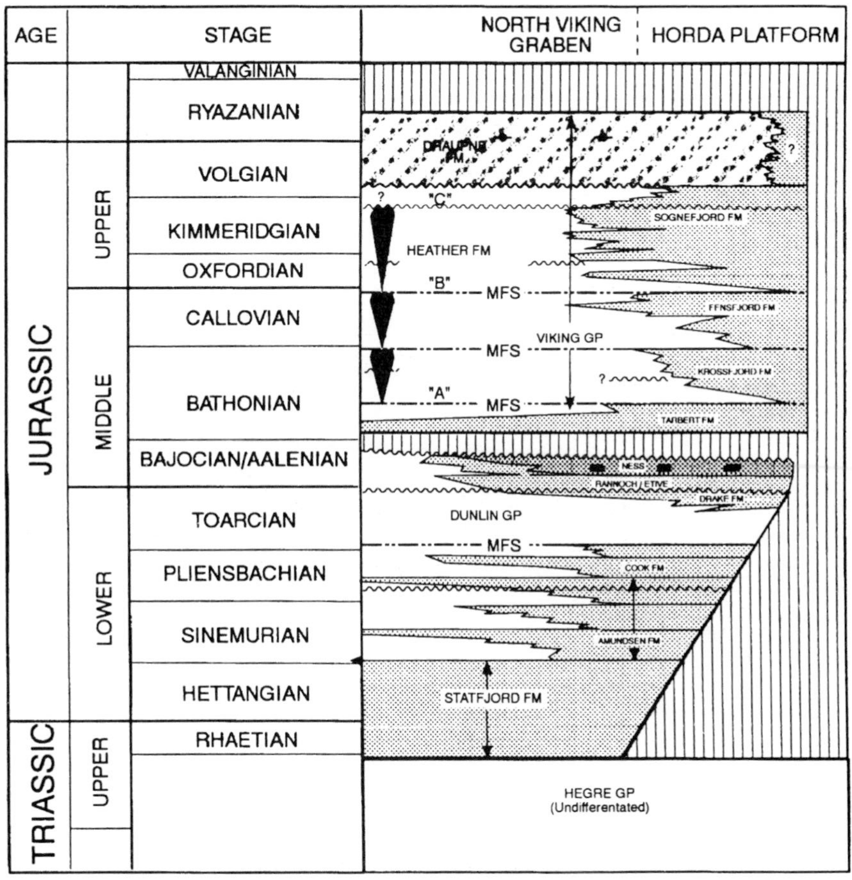

Fig. 8.38 Stratigraphic synopsis of the Jurassic of the Norwegian sector of the northern North Sea, emphasizing the effective use of maximum flooding surfaces (MFS) to define and correlate genetic stratigraphic sequences in the area (after Stewart *et al.*, 1995).

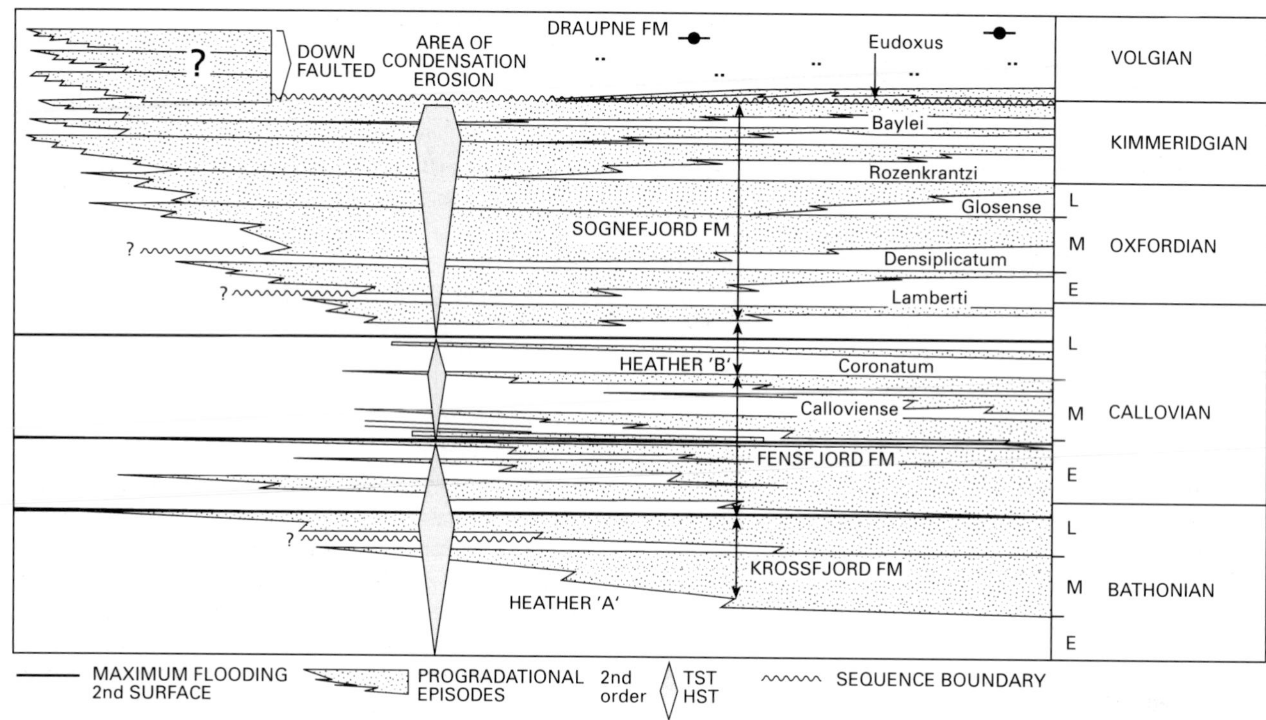

Fig. 8.39 Sequence-stratigraphic framework for the Bathonian–Volgian interval in the Horda Platform area, Norwegian North Sea, showing the sequence hierarchy and component stacking patterns displayed by the shallow-marine depositional system (after Stewart *et al*., 1995).

Fig. 8.40 Callovian to Ryazanian lithostratigraphy of the Vestland Group, Norwegian Central Graben, as established by Vollset and Doré (1984) for the Central Graben–southern Vestland Arch areas and subsequently modified by Bergan *et al.* (1989).

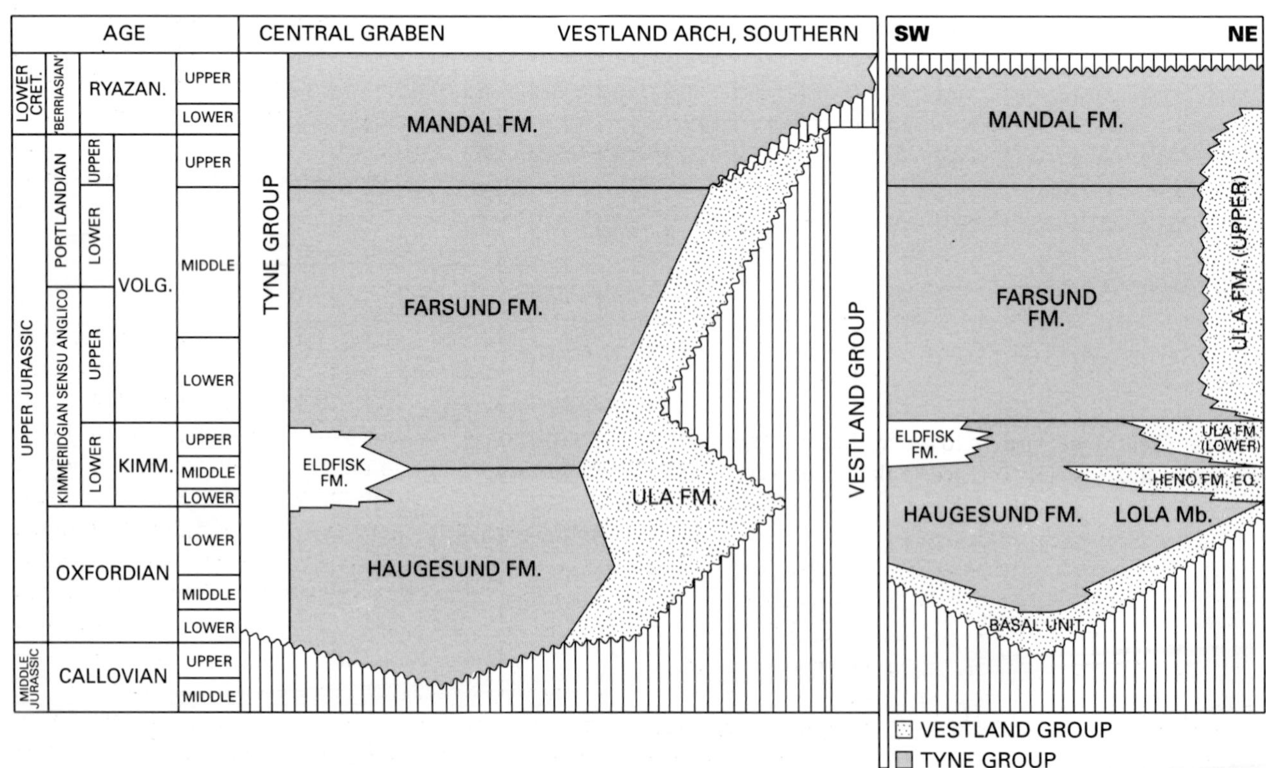

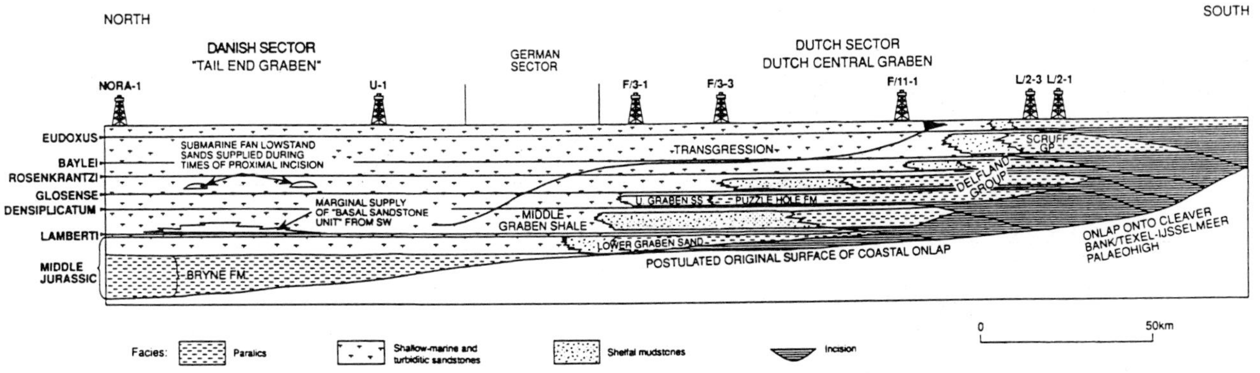

Fig. 8.41 Schematic north–south section along the length of the southern Central Graben depicting the main lithostratigraphic units in the Danish, German and Dutch sectors (after Underhill and Partington, 1994).

8.5.3 Deposition of the Humber Group

Heather Formation

The Heather Formation is the lithostratigraphic term used to describe grey, silty mudstones lying between the coarse clastics of the Brent and Fladen Group and the more organic-rich marine mudstones of the Kimmeridge Clay Formation (see Fig. 8.2). The boundary between the two units is commonly taken just below the first downhole ocurrence (FDO) of the dinocyst *S. crystallinum*. Several geographically distinct and prospective sandstone units occur interbedded with the Heather Formation. For example, at the Troll field (the largest offshore gasfield in the North Sea area), 63×10^{12} ft^3 (1788×10^9 m^3) gas and 6×10^9 barrels (970×10^6 m^3) of oil reside in Upper Jurassic Heather reservoirs of the Sognefjord and Fensfjord members. The main reservoir units consist of good-quality shoreface sandstones with porosities in the range of 19–34% and permeabilities that range between 1 mD and 10 D (Gray, 1987).

The most significant phase of synrift extensional activity affected the North Sea during Late Jurassic times, when units belonging to the Humber Group were deposited. The activity was concentrated on the three rift arms and led to a profound change in the style of deposition in these areas. The most notable result of the synrift activity was the development of significant fault scarps on the tilted fault blocks. Their occurrence led to the erosion and degradation of footwall areas and the accumulation of very coarse-grained clastic sediments in basin-margin positions.

The most pronounced examples of fault-scarp degradation occur on the flanks of several major tilted fault blocks, particularly oilfields within the Brent Province (e.g. Ninian, Brent and Statfjord fields; Lee and Hwang, 1993; Schulte *et al.*, 1994; Coutts *et al.*, 1996; van der Pal *et al.*, 1996; Underhill *et al.*, 1997), where it was coeval with deposition of the Heather Formation. Being in relatively elevated structural positions, the areas characterized by fault-scarp degradation are sites of remaining hydrocarbons in fields that have had long production histories. Indeed, interest in the hydrocarbon habitat of these areas is increasing, as production from the main field diminishes and the relative proportion of remaining potentially producible hydrocarbons lying in these structurally complex areas increases. By way of an example, it is currently estimated that at least one third of the oil remaining in the Brent field lies within its degradation complex of 'slumps' (Schulte *et al.*, 1994).

In the Ninian oilfield, the degradation affects the whole 25 km length of the field, extends over a width of 2 km from the main eastern-boundary fault (Figs 8.42 and 8.43) and attains a thickness in excess of 100 m (Underhill *et al.*, 1997). The complexes have recently been shown to consist of a series of semi-autochthonous to allochthonous rotated glide blocks and stacked submarine slides above a terraced, low-angled décollement surface (Fig. 8.44; Underhill *et al.*, 1997). There is some potential for improving understanding of the internal structural geometries of the degradation complexes using the latest walkaway vertical seismic profiling (VSP) (van der Pal *et al.*, 1996) or directional drilling techniques (Sawyer and Keegan, 1996). The blocks demonstrate an important down-dip decrease in sand content and reservoir quality, such that their lower parts often consist of disintegrated (and unprospective) sediment. Biostratigraphic information suggests that the degradation process was triggered by accelerated synrift extension during Callovian to Early Oxfordian deposition of the Heather Formation.

Fulmar Formation

The term 'Fulmar Sands' was informally introduced by Shell to describe shallow-marine sandstones of the western Central Graben in the late 1970s (Brown, 1984). Their distribution is now known to extend into areas of the Fisher Bank Basin and on to the Fladen Ground Spur. They are thought to be coeval with the Piper Formation of the Outer Moray Firth, from which they are geographically separated by a narrow high. Like the Piper Formation, the Fulmar Formation is diachronous and was deposited from the Callovian to the Volgian. As such, they interfinger with marine mudstones of both the Heather and the Kimmeridge Clay formations (see Fig. 8.2). The sandstones are interpreted as representing discrete progradational pulses within an overall transgressive (retrogradational) cycle,

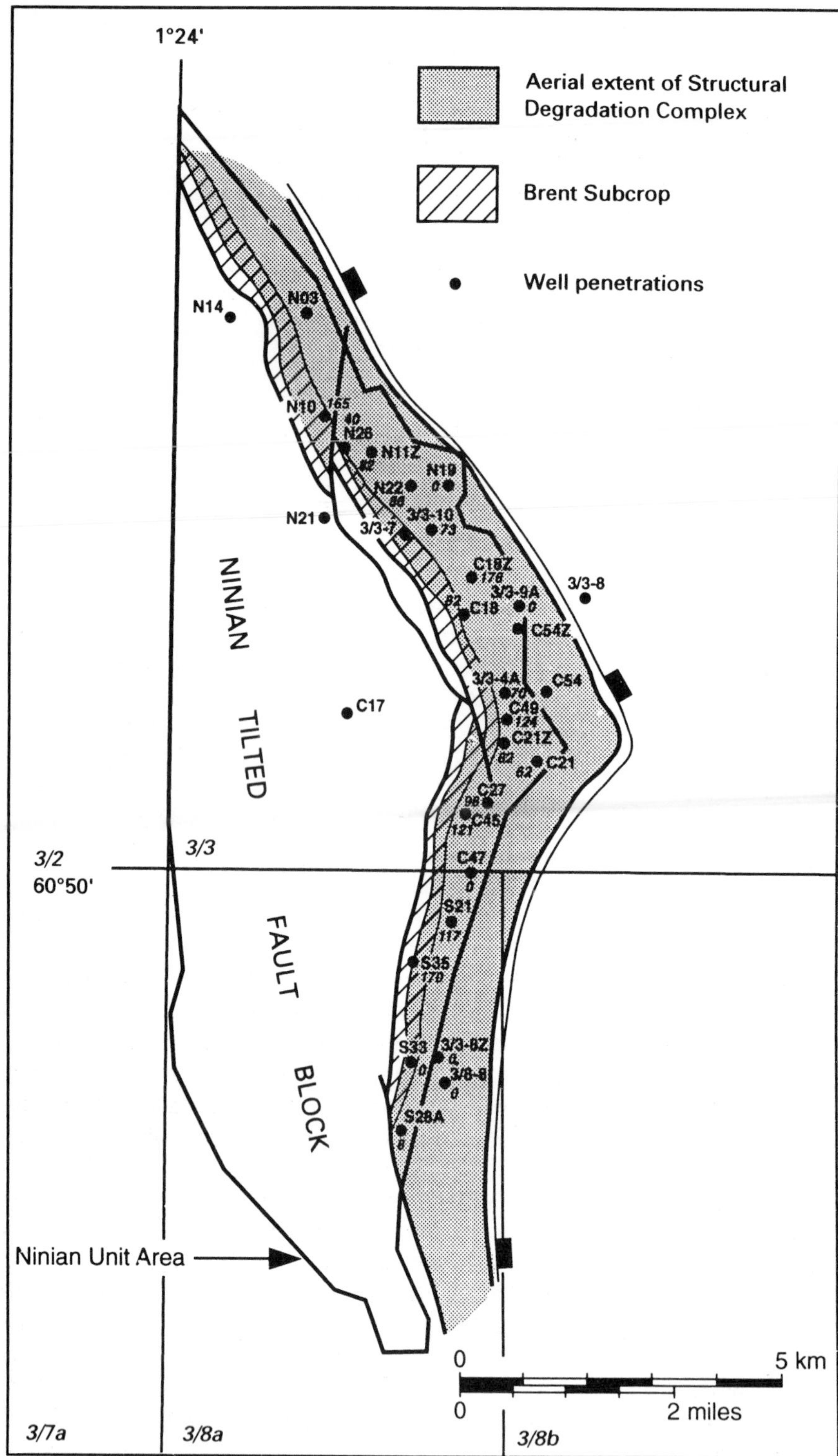

Fig. 8.42 Map of the Ninian oilfield showing the aerial extent of the footwall erosion and its associated structural degradation complex (after Underhill *et al.*, 1997).

which began soon after deposition of the Ness Formation of the Brent Group and lasted until at least the Late Volgian.

Although notoriously difficult to predict, the exact distribution of the Fulmar Formation sandstones along the edge of the Western Platform now appears to be closely related to halokinetic activity of Zechstein evaporites (Wakefield *et al.*, 1993) and erosion of basin-margin and intrabasinal highs, such as the Forties–Montrose High. Salt withdrawal during the Triassic led to the development of thick packages of claystones in rim synclines, termed 'Smith Bank pods'. Subsequent sedimentation above areas affected by salt diapirism was marked by coarse clastic deposition of the Triassic Skagerrak Formation, which is interpreted as recording the infill of topographic lows formed in response to groundwater dissolution of the salt pillows. This complex tectono-sedimentary behaviour left a legacy by also exercising a control on sediment drainage patterns during the Late Jurassic. A relative sea-level rise during the Callovian to Early Kimmeridgian caused widespread flooding of the Western Platform, led to reworking of the older Skagerrak palaeovalleys, and resulted in deposition of the Fulmar shallow-marine deposystem. An attractive, subtle strati-

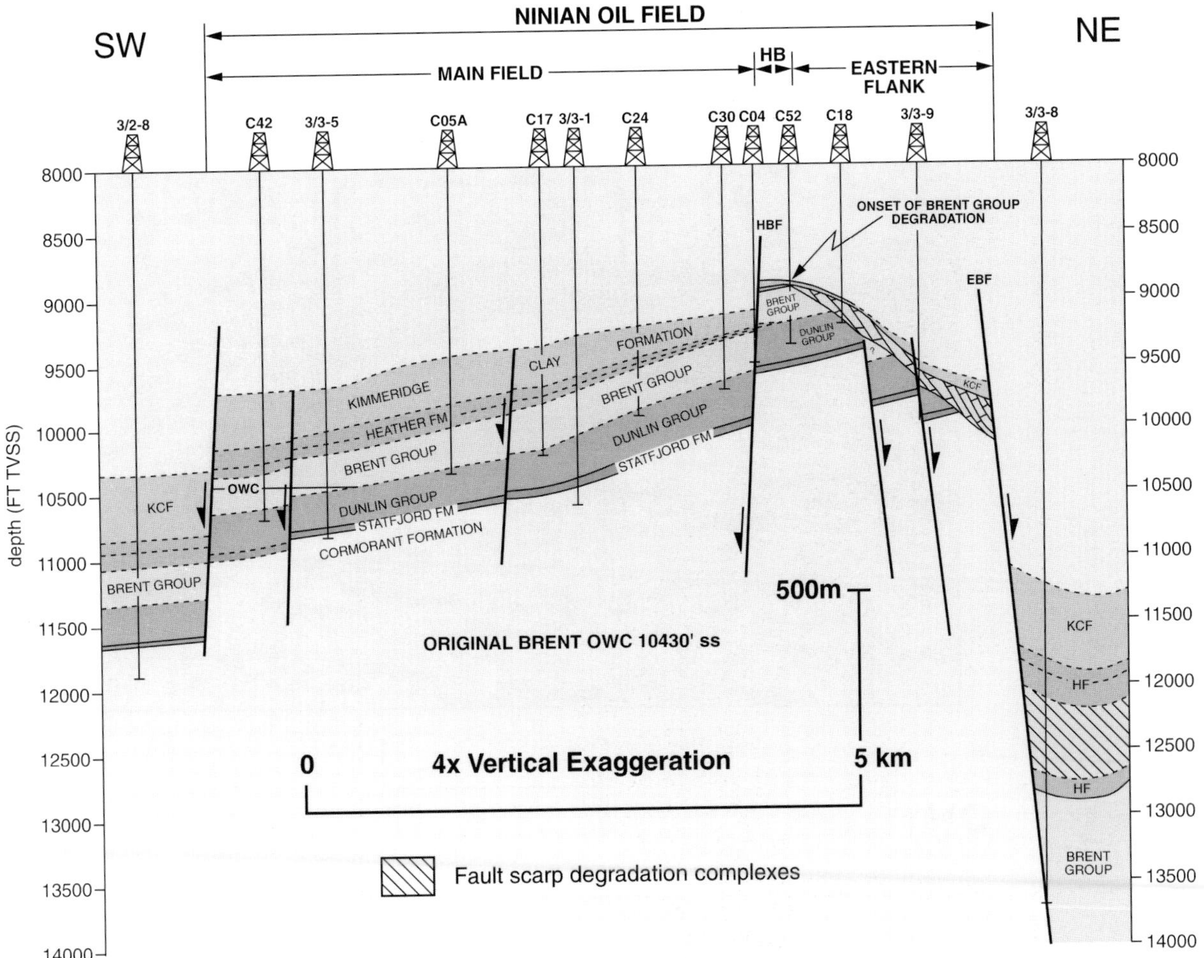

Fig. 8.43 South-west–north-east trending cross-section through the Ninian oilfield, Brent Province, northern North Sea, highlighting the extent and nature of Brent Group truncation and reworking due to fault-scarp degradation in the immediate footwall to its eastern boundary fault (EBF) (after Underhill *et al.*, 1997). HB, Horst Block fault.

graphic play has resulted, with sandstones of the Fulmar Formation forming linear reservoir bodies that are laterally sealed by shales of the Smith Bank Formation and capped by the Kimmeridge Clay Formation, as exemplified by the Gannet and Kittiwake fields (Armstrong *et al.*, 1987; Glennie and Armstrong, 1991).

The Fulmar Formation is an intensely bioturbated, heterolithic, shallow-marine sedimentary unit (Johnson *et al.*, 1986). Study of trace-fossil assemblages has provided a powerful and useful tool with which to attempt to subdivide and correlate the Fulmar Formation. A depth- and substrate-related succession of ichnofacies and ichnofabrics has recently been determined (Gowland, 1996; Martin and Pollard, 1996), which demonstrates that deposition of the Fulmar Formation was consistent with a series of stacked high-energy progradational shorelines within an overall transgressive (retrogradational) depositional system (e.g. Gowland, 1996).

Integration of the ichnological studies with biostratigraphy has provided the framework in which the application of sequence stratigraphy to the Fulmar Formation can be attempted (e.g. Taylor and Gawthorpe, 1993). Although sequence-stratigraphic interpretations are less precise, due to many uncertainties with the data (e.g. Veldkamp *et al.*, 1996), advances have been made in its application, which have led not only to a better understanding of the shallow-marine systems but also, importantly, to an improved prediction of Late Jurassic turbidite plays in the Central Graben (e.g. the Freshney Sandstone Member; Price *et al.*, 1993; Carruthers *et al.*, 1996).

Reservoir properties vary greatly in the Fulmar Formation. Its quality is thought to depend largely upon the original distribution of facies and their influence on diagenesis. Textural controls, such as subtle variations in grain size and detrital clay content, and the degree of bioturbation, with its consequent effect on sorting, have both had a considerable effect on poroperm characteristics of the Fulmar Formation. Rapidly deposited unbioturbated sandstones have permeabilities an order of magnitude higher than bioturbated sandstones (Veldkamp *et al.*, 1996). Secondary effects include permeability reduction by early diagenesis through kaolinite and late illite formation, and porosity reduction by various stages of quartz and carbonate cementation. It is only very locally that secondary leaching of Rhaxella spicules has led to porosity enhancement (e.g. well 22/30a-1; Veldkamp *et al.*, 1996). The best poroperm characteristics occur within the Fulmar field itself, where the average porosity is 23% and permeabilities occur in the range 50–800 mD.

An additional factor in determining the reservoir quality and exploration potential of the Fulmar Formation is its depth of burial and ambient temperature and pressure conditions. The unit is buried to a depth in excess of 4 or

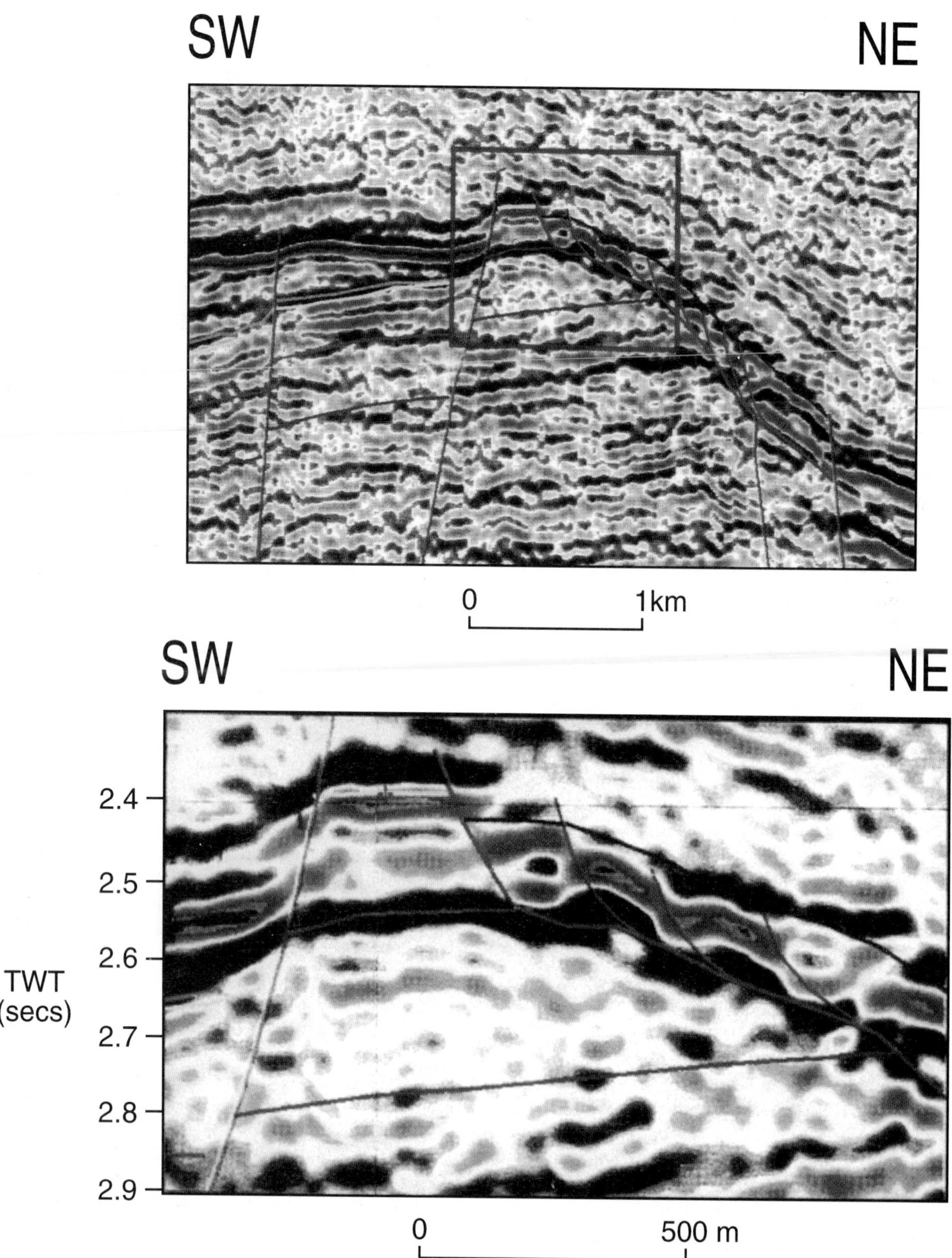

Fig. 8.44 Seismic data illustrating the main internal structure of the degradation complex in the Ninian oilfield (after Underhill *et al.*, 1997). The sections highlight the importance of fault-bounded glide blocks and the corrugated nature of the basal decollement surface.

5 km locally, making it one of the deepest Jurassic exploration targets in the North Sea. Burial to such depths brings with it associated problems, the most notable of which is high pressure (HP) and high temperature (HT), leading to some areas being referred to as the HP/HT Province. Indeed, the first Fulmar well was abandoned (with known oil) because Shell could not control the high pressure. It was only 2 years later that the Fulmar 'discovery' well was officially made, after a high-pressure stack to contain the reservoir pressures had been constructed. Even today,

safety considerations in these areas are such that many operators have been reluctant to drill for such targets.

Piper Formation

The Piper Formation is the term assigned to the marine, sandstone-dominated sedimentary unit that lies between the Kimmeridge Clay Formation and Middle Jurassic continental sediments in the Outer Moray Firth Basin, and is typified by units in the Piper oilfield itself (Maher, 1981). The Piper Formation forms the main reservoir unit for the Piper, Scott, Ivanhoe, Rob Roy, Chanter, Saltire, Hamish and Saltire oilfields (e.g. Maher, 1980; Parker, 1991; Schmitt and Gordon, 1991; Currie, 1996; Harker and Rieuf, 1996).

The formation is not only age-equivalent to lower parts of the Kimmeridge Clay, when it is termed the Chanter Member, but also locally equivalent to upper parts of the Heather Formation, when it is known as the Pibroch Member (Richards *et al.*, 1993). As such, the Piper Formation not only comprises marine sandstones of Deegan and Scull's (1977) original unit but also now includes the Scott Member of Harker *et al.* (1993).

The Piper Formation is interpreted as a fluvial to wave-influenced shallow-marine deltaic to shelfal depositional system derived from basin-margin and intrabasinal highs, like the Fladen Ground Spur and the Halibut Horst (O'Driscoll *et al.*, 1990; Harker *et al.*, 1993; Freer *et al.*, 1996). The dominant shallow-marine sandstone lithofacies passes southwards into more open-marine basinal sediments, including shales of the Heather and Kimmeridge formations, and northwards into non-marine clastic sediments (Davies *et al.*, 1996). The widespread areal distribution of each of the major sandstone bodies within the Piper Formation suggests that they represent individual progradational cycles, which probably represent short-lived regressive pulses that punctuated an overall Late Jurassic transgressive phase (Underhill and Partington, 1993; Davies *et al.*, 1996).

The shallow-marine sandstones of the Piper Formation are characterized by extensive bioturbation, which has often eradicated its original depositional fabric (Boote and Gustav, 1987; Harker *et al.*, 1993). Where this has occurred, the formation commonly consisits of massive, structureless, high-quality sandstones, with porosities in excess of 25% and permeabilities in the range of 1–2 D (Harker and Rieuf, 1996).

Late Jurassic sedimentation in the Danish, German and Dutch sectors of the Central Graben

Stratigraphic relationships have been difficult to determine in the southern part of the Central Graben, not least because they lie across four international borders. Recent sequence-stratigraphic work, however, has aided understanding of the depositional system, and it is now believed that they largely reflect deposition during a punctuated transgression. Sandstones built out from the palaeoshoreline and were supplied to basinal locations during times of relative sea-level fall, while shale deposition reflects times of rapid and widespread marine flooding. Trangression continued in the area through deposition of the Kimmeridge Clay Formation.

As in the UK Central Graben, marine deposition commenced in axial areas of the Norwegian Central Graben (such as the Feda Graben) and subsequently expanded onto the surrounding Cod and Steinbit Terraces (see Fig. 8.36). By the Late Oxfordian, most of the Central Trough area had been transgressed. Sandstone deposition characterized the neighbouring terrace areas with the development of stacked, upward-coarsening, progradational depositional cycles (for example, in the Ula and Gyda fields; see Fig. 8.45).

Johannessen and Andsbjerg (1993) have shown that deposition of widespread Middle Jurassic alluvial-plain sedimentation (Bryne Formation) was followed during the Callovian by a transgressive succession of coastal-plain (which also form part of the Bryne Formation) and shallow-marine deposits, comprising several important, prograding clastic wedges in the Danish Central Graben. As seen elsewhere in the North Sea, transgression continued in the Late Jurassic, causing shorelines to backstep, with coeval deposition of the Heno Formation clastics and Lola Formation claystones. Subsequent deposition was dominated by deep-water clastics of the Poul Formation and fine-grained equivalents of the Farsund Formation.

Kimmeridge Clay Formation

The term Kimmeridge Clay Formation was imported from onshore exposures in South Dorset to describe organic-rich mudstones of Late Jurassic age in the UK North Sea Basin. The lithostratigraphic term is synonymous with the Draupne Formation of Norwegian waters. Importantly, the stratigraphic unit consists not only of claystones but also contains several reservoir sandstones, which were deposited in submarine-apron fans, basin-floor fans or shallow-marine shelves. In general, the claystone facies represent marine hemipelagic deposition in an environment in which bottom waters were commonly anoxic (see Chapter 11), which favoured the accumulation and preservation of organic material. As a result, the Late Oxfordian–Ryazanian stratigraphic interval was an exceptional period of very widespread source-rock deposition in the Boreal Realm. In the North Sea Basin, tectonic setting also appears to have had an influence on deposition of the Kimmeridge Clay Formation by promoting oxygen deficiency and by effecting a marked decrease in sediment accumulation rates in half-graben depocentres. Higher sedimentation rates generally resulted in dilution of the source-rock potential of the Kimmeridge Clay, as evidenced in basin-margin areas. It has been demonstrated that the occurrence and stratigraphic distribution of source-rock potential in its fine distal facies are strongly related to the prevailing palaeo-oxygenation regime (Tyson *et al.*, 1979), especially during deposition of condensed sections (Tyson, 1995). The long-term shallow–deep–shallow relative sea-level fluctuation appears to have influenced the mean palaeo-oxygenation regime, which changed from oxic to dysoxic–anoxic and back to oxic through time, with a corresponding increase in TOC, from around 1 wt% to 4–6 wt% and back again (Tyson, 1995).

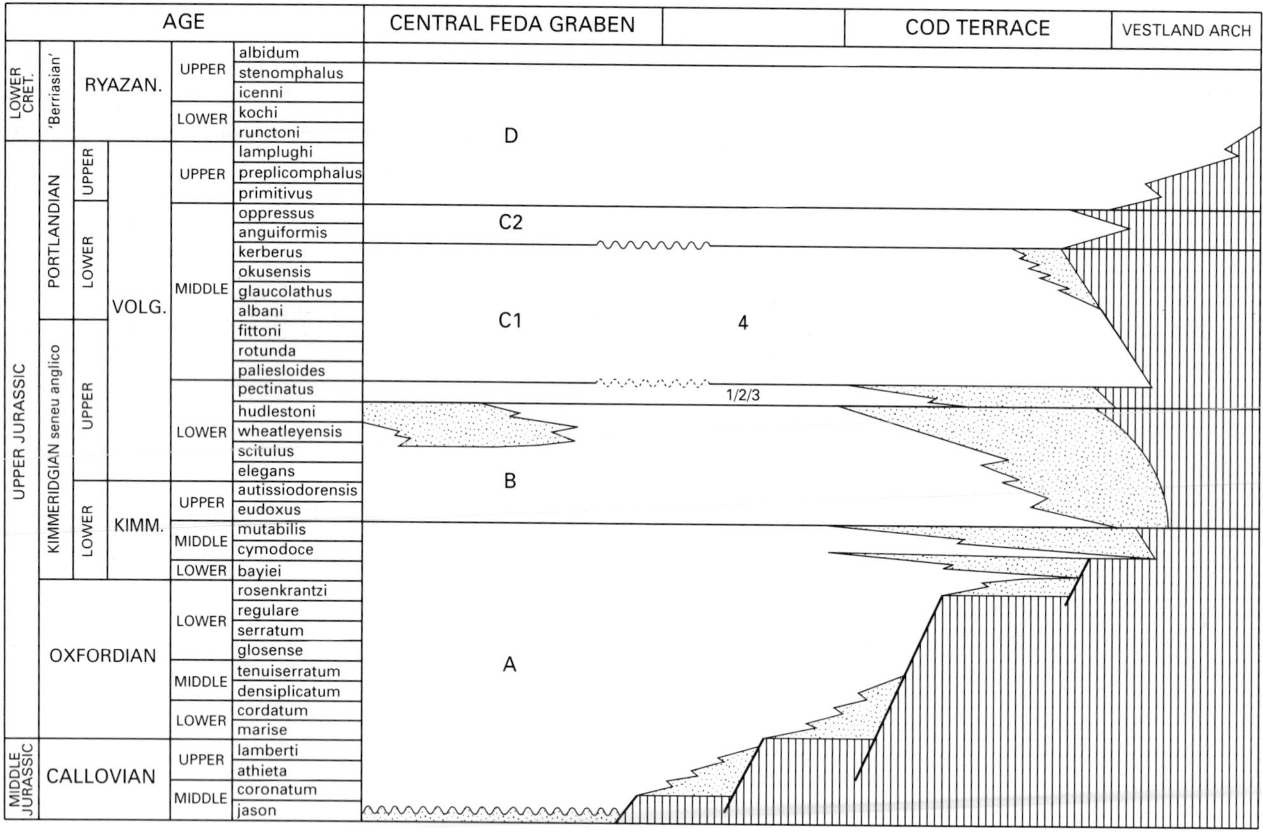

Fig. 8.45 Upper Jurassic sequences recognized in the Central Feda Graben, Cod Terrace and Vestland Arch areas of the Norwegian Central Trough.

The top of the Kimmeridge Clay Formation is typically marked by a strong acoustic-impedance contrast on seismic data. The apparent truncation and onlap of seismic reflectors at this level have led to the Base Cretaceous being interpreted as a regional unconformity (the 'Late Cimmerian Unconformity'). It has become clear, however, that there is no evidence for missing stratigraphy in basin-centre locations (e.g. Rawson and Riley, 1982). Instead, the seismic event marks a time of severe condensation, the apparent truncation representing tuning as seismic events approach and destructively interfere with each other. Furthermore, stratigraphic sampling of the zone of condensation demonstrates that it lies within the Ryazanian, close to the boundary between the *Stenomphalus* and overlying *Albidium* ammonite zones (see Fig. 18.6; Rawson and Riley, 1982; Rattey and Hayward, 1993). In other words, except in basin-margin locations and over local intrabasinal highs, the 'Base Cretaceous Unconformity' is not an unconformity nor does it coincide with the Jurassic–Cretaceous boundary (Rawson and Riley, 1982; Rattey and Hayward, 1993).

Brae Formation

A series of basinward-thinning and fining sedimentary wedges characterize Kimmeridgian–Volgian hangingwall deposition along the margins of the South Viking Graben (Stow *et al.*, 1982) and the Helmsdale Fault in the Inner Moray Firth (Bailey and Weir, 1932; Crowell, 1961; Pickering, 1983, 1984). The coarse clastic wedges form important hydrocarbon reservoirs in the producing field of the Brae trend (Figs 8.46 and 8.47), including East, North, Central and South Brae, Toni, Tiffany and Thelma fields and the Birch, Elm, Larch and Pine discoveries (e.g. Stephenson, 1991; Kerlogue *et al.*, 1995). In all cases, the wedges are interpreted to have formed proximal, deep-water slope-apron complexes immediately adjacent to point-source sediment supply off the neighbouring Fladen Ground Spur and Horda Platform (Fig. 8.48; Stow *et al.*, 1982, Turner *et al.*, 1984, 1987), through conduits possibly controlled by reactivated basement features. The fans show important lateral facies variations and interfinger with the shale-prone Kimmeridge Clay Formation, which forms the all-important top and lateral seals and the source rock for hydrocarbon charge (Figs 8.48 and 8.49). Porosities in reservoirs of the Brae trend average between 11 and 15% and permeabilities lie in the range 1–4000 mD.

Complex structural geometries characterize the area immediately adjacent to the basin boundary faults in the South Viking Graben and Inner Moray Firth areas, both in cross-section and in plan view. Cross-sections demonstrate that structural dips often increase to 60–70° against the main basin-bounding faults in the South Viking Graben and the Inner Moray Firth. The area adjacent to the fault trace takes the form of a dissected monocline, the synclinal core of which often displays contractional features, such as steep reverse faults or low-angle thrusts. Although these have often been taken as evidence for a short-lived phase of

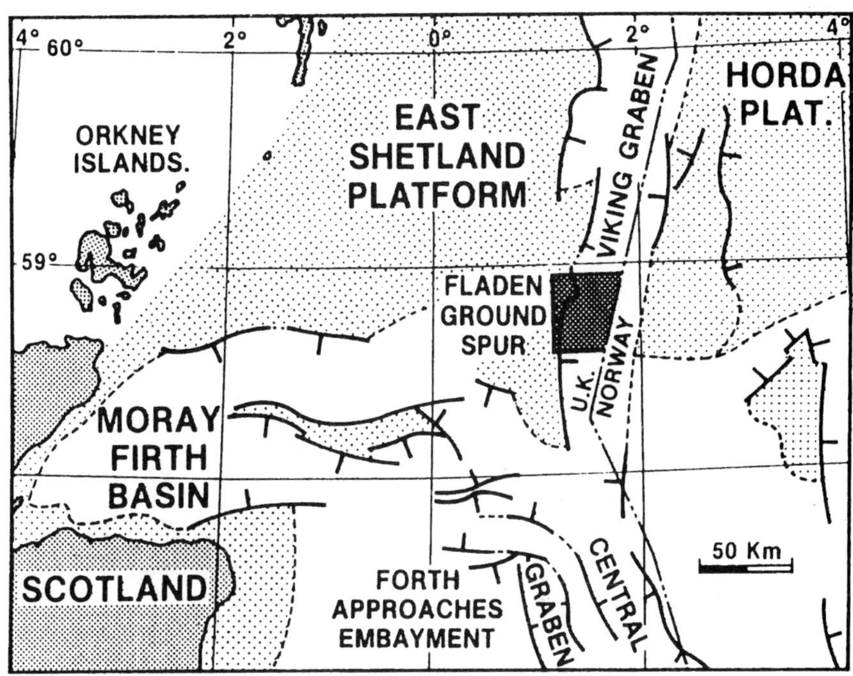

Fig. 8.46 Regional setting for the Brae area (boxed and stippled), South Viking Graben.

compression affecting both areas during the latest Jurassic or Early Cretaceous, field studies in other synrift settings (e.g. Oligo-Miocene of Gulf of Suez and Sardinia) demonstrate that some of the geometries are consistent with extensional provinces affected by synsedimentary fault growth and bed rotation ('forced folds') above upward- and laterally propagating normal faults (e.g. Withjack *et al.*, 1990). Thus the geometries may not necessarily be associated with any Late Cimmerian phase of compressional deformation.

Important lateral variation along strike, as demonstrated by the interpetation of 3D seismic data, has been ascribed to reactivation of basement lineaments, and has led in part to the development of a highly indented and segmented margin interpreted by some workers to represent a series of transfer faults (e.g. Fig. 8.50; Cherry, 1993). Whilst it is not disputed that compartmentalization of the margin probably controlled the location of major sediment entry points more recent evidence suggests that sediment supply took place along relay ramps between two extensional fault segments and not discrete transfer faults. Biostratigraphic data indicate that the cut-off in clastic supply differed along

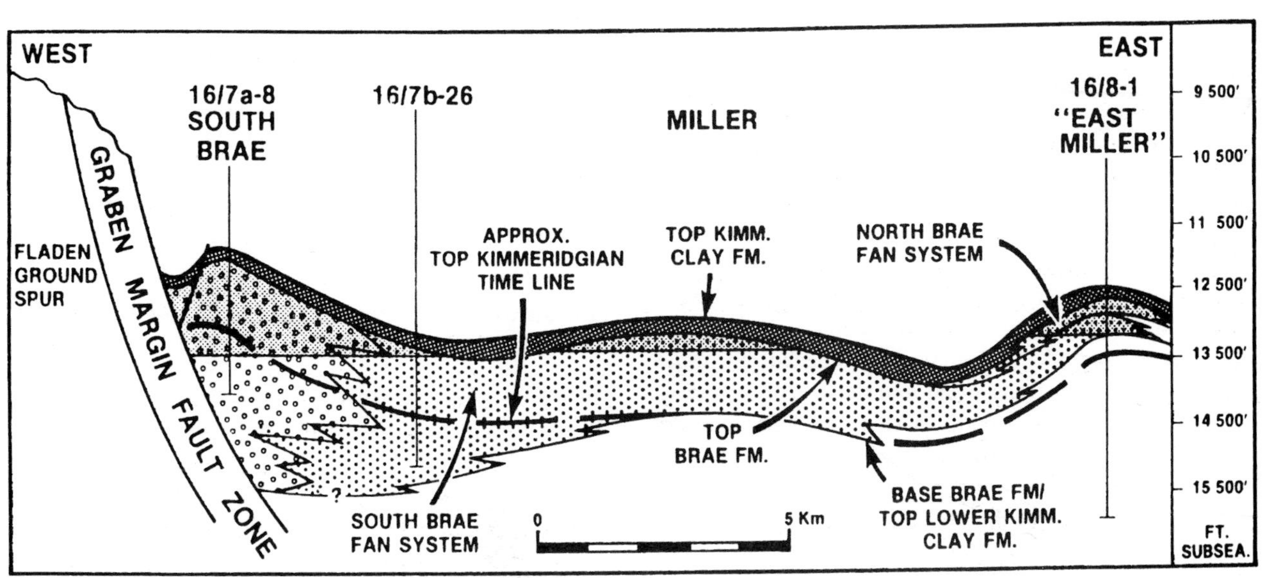

Fig. 8.47 Schematic cross-section through a coarse clastic wedge and submarine-fan sandstone complex characteristic of the Brae Trend, the Miller and East Miller fields, deposited on the western margin of the South Viking Graben (after Turner *et al.*, 1987; Johnson and Stewart, 1985).

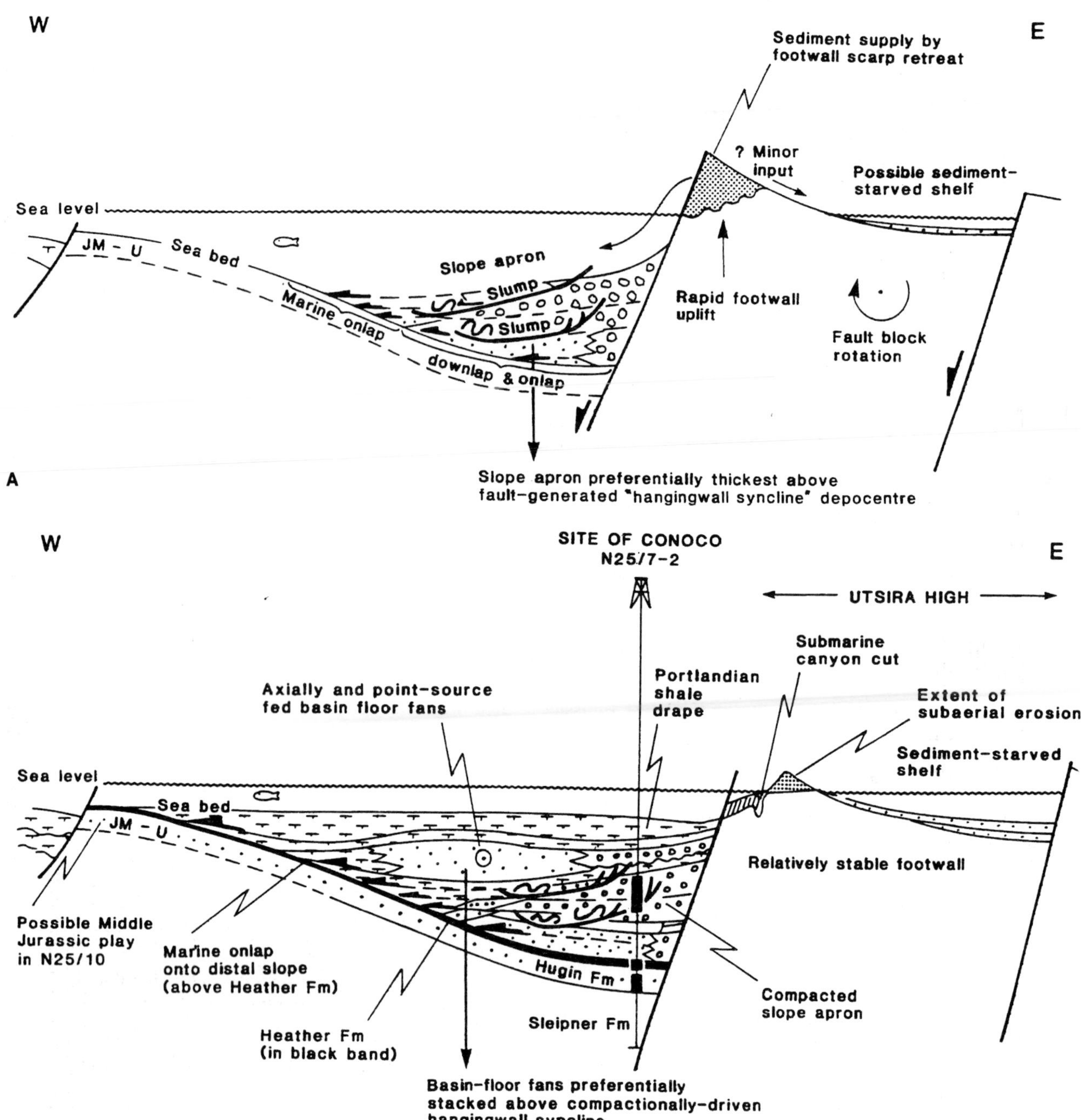

Fig. 8.48 Depositional model for Late Jurassic sedimentation in the Norwegian eastern margin of the South Viking Graben. A. Early Kimmeridgian active syn-rift phase characterized by growth faulting. B. Portlandian early post-rift phase.

the basin margin, perhaps as these sediment supply routes were abandoned, due to fault propagation and/or marine flooding, as Jurassic transgression proceeded. In some cases, sediment was transported into the axis of the basin, to the east of the Brae trend, where it was deposited as a series of submarine fan mounds. Drilling of these features has demonstrated a proved exploration target, as exemplified by the discovery of the Miller and East Miller oilfields (for example, Rooksby, 1991; Garland, 1993), which remains a valid exploration target in the axis of the Viking Graben, both in the UK and in the Norwegian waters. In Miller, porosities average 16% and permeabilities lie in the range 11–200 mD.

Late Jurassic sedimentation in the northern North Sea

One of the effects of synsedimentary footwall uplift of the tilted fault-block structures in the Brent Province during the Late Jurassic was not only to cause incision but also to promote resedimentation of the erosional products down-flank from the crest of the structures. Several sandstone

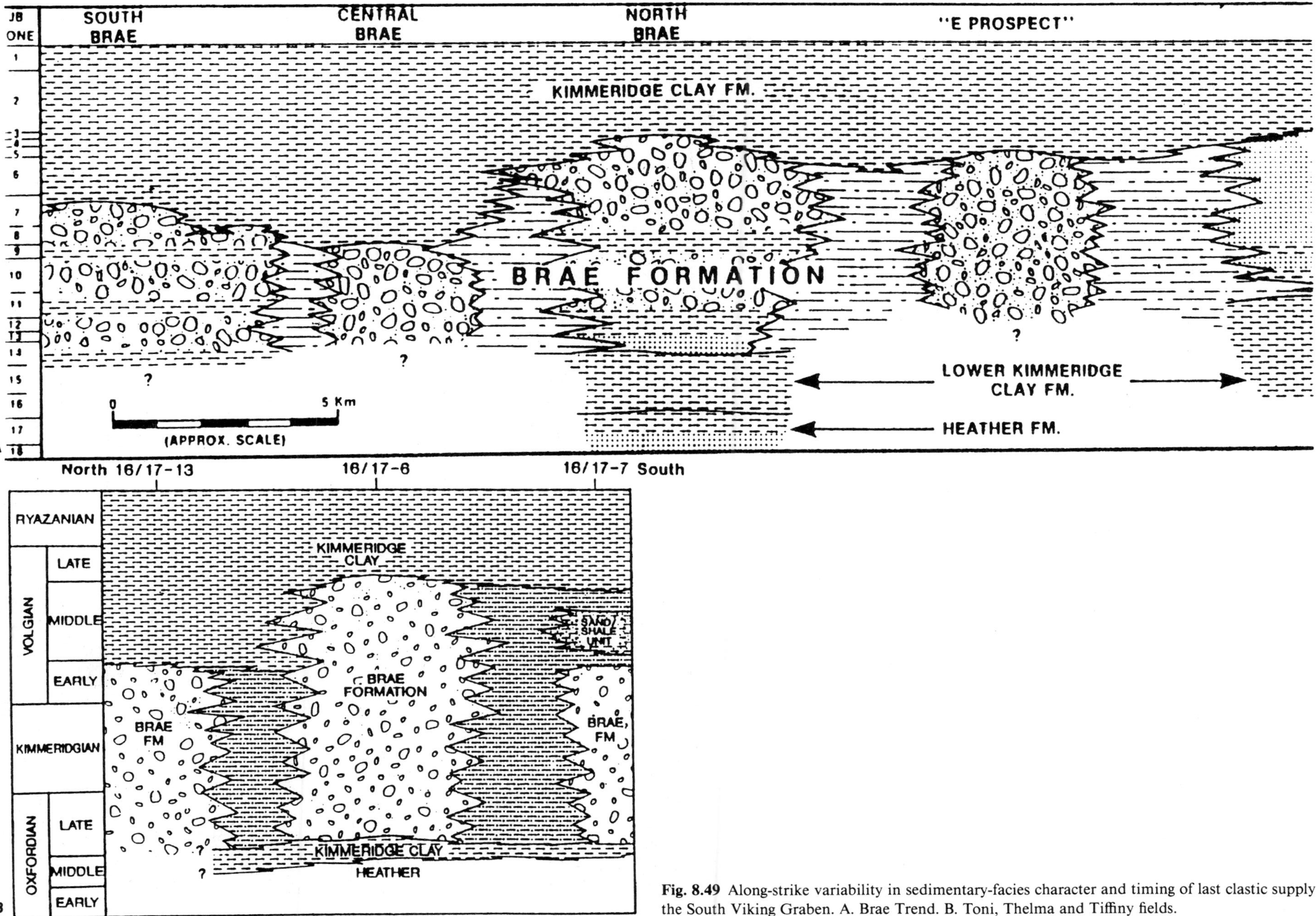

Fig. 8.49 Along-strike variability in sedimentary-facies character and timing of last clastic supply in the South Viking Graben. A. Brae Trend. B. Toni, Thelma and Tiffiny fields.

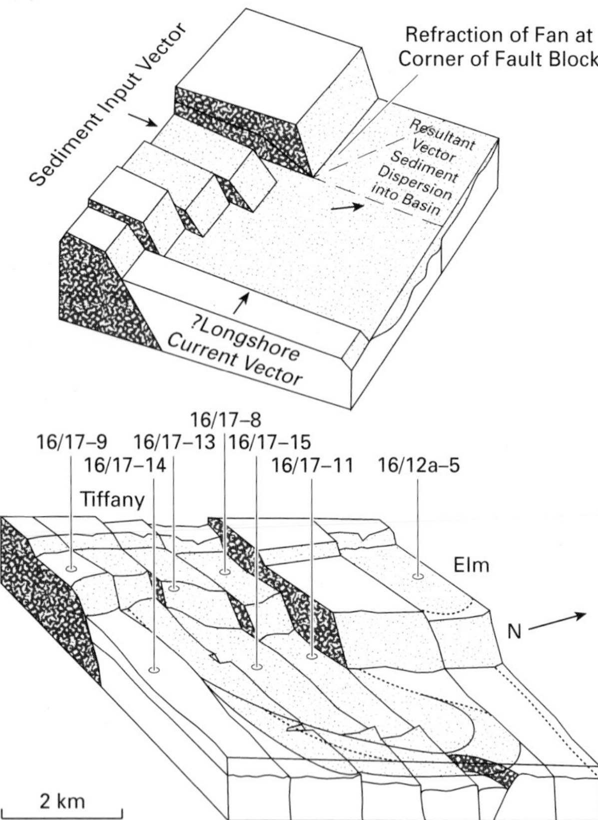

Fig. 8.50 Cherry's (1993) model, which interprets the main sedimentary entry points in Agip's T-field trend of the western South Viking Garben as being controlled by the presence of important reactivated cross- or transfer-fault elements along the East Shetland Platform.

units were deposited in this general setting and they form subtle stratigraphic reservoir targets around many of the main fields of the Brent Province. Examples of the play include high-energy sand belts of the Troll area (Stewart *et al.*, 1995), the informally defined 'Munin Sands' of the North Statfjord field and 'Intra-Draupne Sandstones' or 'Kimmeridge Sand Member' of other areas in the North Viking Graben (Fig. 8.51; and, for example, Dahl and Solli, 1993). It was during the early 1990s that such reservoirs were actively targeted, largely in response to maximizing the opportunity to utilize existing oilfield intrastructure during the decline and eventual abandonment of the main Brent Group fields. Considerable success has been achieved in predicting the distribution of these sands in the Norwegian sector, using a combination of sequence stratigraphy and geophysical modelling (Figs 8.52 and 8.53; Dahl and Soldi, 1993).

Additional prospectivity exists to the north of the Brent Province, as demonstrated by the Magnus field, which remains the most northerly producing field in the UK sector of the North Sea. It produces from Late Jurassic submarine fan sequences, which, unlike the Northern North Sea fields, form a prerift sequence, lying within a tilted fault block formed during the Cretaceous (De'Ath and Schuyleman, 1981), when the locus of active extension had moved to the West Shetlands area (Rattey and Hayward, 1993).

Late Jurassic sedimentation in the Moray Firth

The term Burns Sandstone Member was introduced by Richards *et al.* (1993) to describe thick, deep-marine sandstone units occurring within the Kimmeridge Clay Formation of the Moray Firth Basin, which superseded the shallow-marine Piper depositional system, largely in response to continued synsedimentary extensional faulting. They form reservoir units in the Chanter, Claymore, Ettrick, Galley, Perth and Tartan fields of the Outer Moray Firth (Coward *et al.*, 1991; Partington *et al.*, 1993b) and form important drilling targets, not only in footwall locations but also in hangingwall sites.

The study of samples from the area demonstrates that the change from the shallow-marine deposition of the Piper Formation to deeper-water submarine-fan systems was accompanied by a marked change in sedimentary provenance from previous Devonian sources (Hallsworth *et al.*, 1996) in response to their submergence. The effect that the changes in facies and provenance had on reservoir quality was variable, however, and sandstones of the Burns Formation commonly have porosities up to 19% and permeabilities of around 500 mD (Harker and Rieuf, 1996).

Late Jurassic sandstones have also been an exploration target within the Inner Moray Firth (e.g. UKCS license blocks 12/21 and 12/26). Despite being largely an unsuccessful play to date, additional insight may be gained into the structural control on sediment dispersal of Late Jurassic sandstone units in the North Sea Basin in general, particularly by use of onshore exposures along the Sutherland Coast, including the Allt na Cuile Sandstones, Loth River Shales and Helmsdale Boulder Beds. Use of these onshore analogues demonstrates the important control that a relay ramp between two discontinuous fault segments of the Helmsdale Fault has on supplying sediment into the basin (Fig. 8.54; Underhill, 1994). Comparisons with other well-known extensional fault systems (e.g. Gawthorpe and Hurst, 1993) suggest that similarly important sediment supply routes are likely to occur elsewhere in the North Sea Basin. Their identification may help locate potential sites of clastic deposition and reservoir targets.

Late Jurassic deposition in the Norwegian Atlantic Margin

Exploration extended into areas to the north of the North Viking Graben Brent Province in the early 1980s. Drilling in areas on the Mid Norwegian Shelf, such as the Haltenbanken area and the gas-prone Hammerfest Basin (Troms-I area), has met with success, with several commercial fields being discovered in Middle and Lower Jurassic shallow-marine reservoirs (Gjelberg *et al.*, 1987). The most notable of the discoveries are the Draugen (Ellenor and Mozetic, 1986), Heidrun (Koenig, 1986), Smørbukk (Aasheim *et al.*,

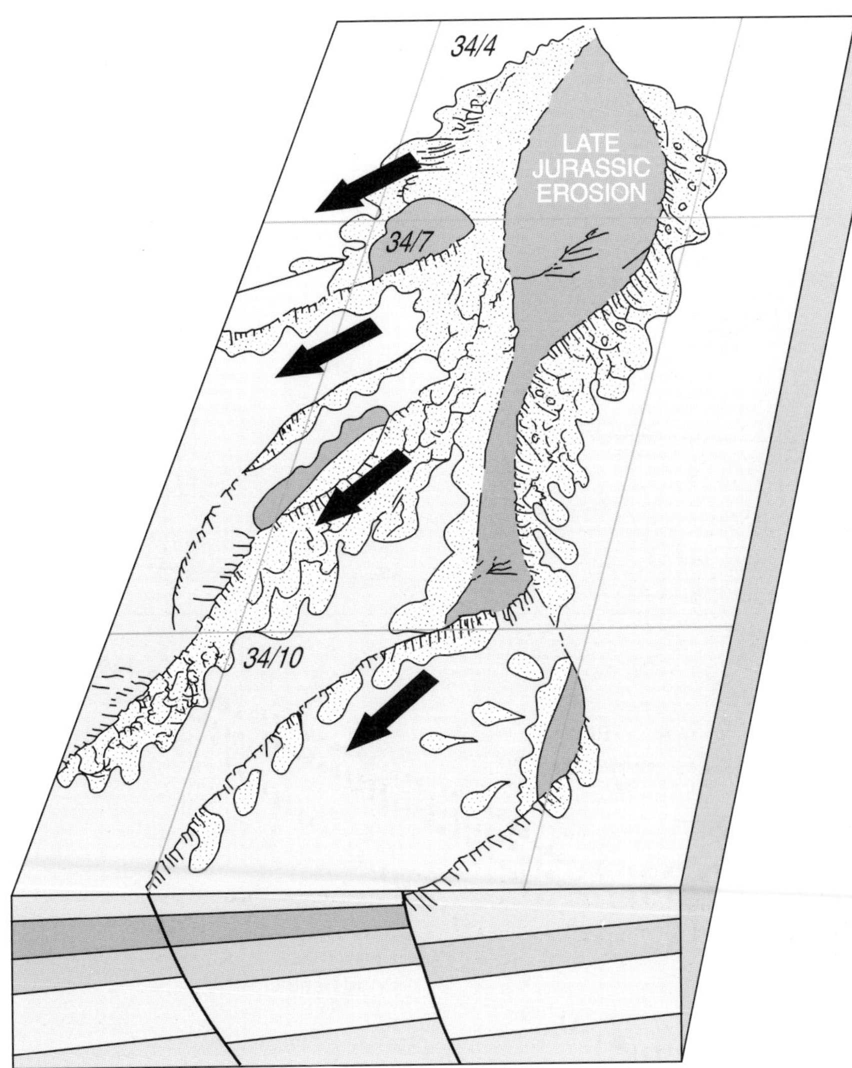

Fig. 8.51 Schematic cartoon demonstrating the importance that structural controls have on dispersal of sediment derived from Late Jurassic erosion of the footwall to the Inner Snorre Fault and shed down the hangingwall dip slope (modified after Dahl and Solli, 1993).

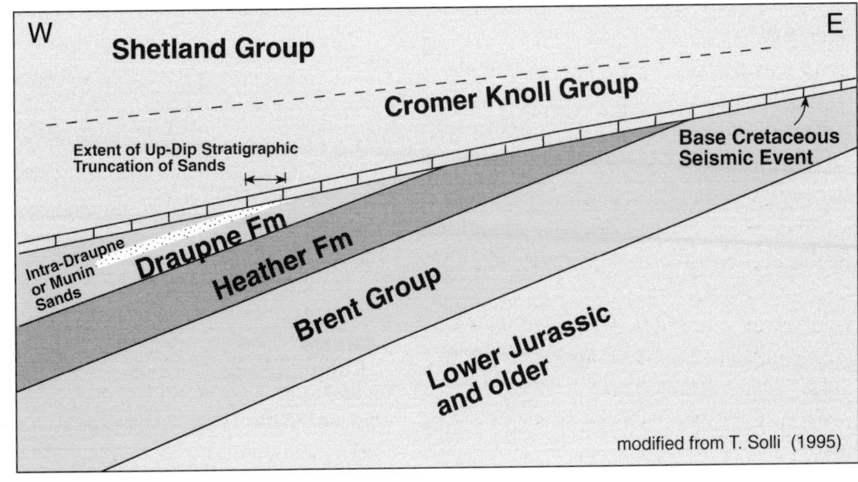

Fig. 8.52 Evidence that the Munin Sandstones, accumulated as a consequence of erosion of the footwall to the Snorre structure, are interbedded with the Draupne Formation (or Kimmeridge Clay Formation) and are erosionally truncated and sealed by sediments belonging to the Cromer Knoll Group has recently led to a programme of exploration activity with the specific aim of a deliberate search for subtle stratigraphic traps in this play fairway (e.g. Norwegian Block 34/7; Solli, 1995; Dawers *et al*., in press).

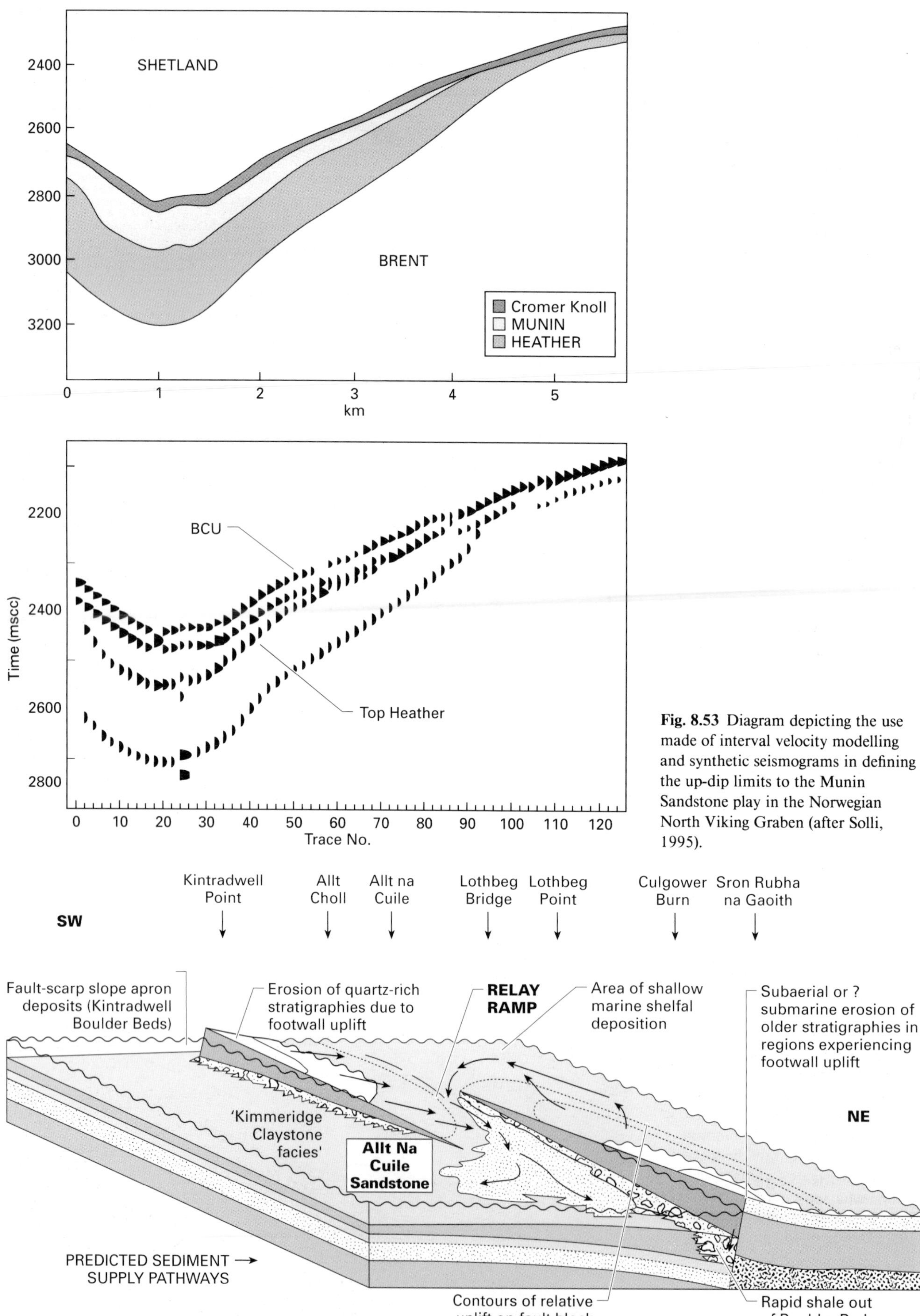

Fig. 8.53 Diagram depicting the use made of interval velocity modelling and synthetic seismograms in defining the up-dip limits to the Munin Sandstone play in the Norwegian North Viking Graben (after Solli, 1995).

Fig. 8.54 Alternative model for sediment supply and dispersal in basin margin locations during the Late Jurassic, in which lateral and vertical growth of extensional fault segments plays the dominant role. The cartoon depicts the control exerted upon deposition of the Allt Na Cuile Sandstone (now exposed at Lothbeg Point on the East Sutherland Coast) by an unfaulted relay ramp formed between two segments along the extensional Helmsdale Fault. (Modified after Underhill, 1994.)

1986), Tyrihans (Larsen *et al.*, 1987), Midgard (Ekern, 1987) and Askaladd fields.

8.5.4 Field analogues in Greenland

Exhumation of East Greenland during the Tertiary, in association with the development of the Iceland hot spot and North Atlantic rifting, has resulted in spectacular exposure of Jurassic outcrops. These exposures have been used effectively as direct structural and sedimentary analogues for the North Viking Graben and Norwegian Shelf areas. For example, outcrops of the Middle Jurassic (Bajocian–Callovian) Vardekløft Formation are considered to represent the best exposed analogue to the deeply buried reservoirs of the northern North Sea and Norwegian Shelf (Engkilde and Surlyk, 1993). It consists of a lower, shallow-marine, sand-rich Pelion Member and an upper, silty, offshore Fossilbjerhet Member (Surlyk *et al.*, 1993).

Large North Sea-scale tilted fault blocks characterize field exposures in Wollaston Forland, East Greenland. These provide excellent analogues for the subsurface exploration targets of the northern North Sea and Moray Firth Basins (e.g. Fig. 8.55). In some cases, exhumation of palaeo-oilfields can even be proved (Price and Whitham, 1997). In other examples, syntectonic sediments of the Wollaston Forland Group can be demonstrated to have been deposited down steep fault scarps to form (Brae-like) thick, wedge-shaped slope aprons adjacent to the extensional-faults. As with the Moray Firth Helmsdale Boulder Beds and the Brae Formation of the South Viking Graben, these sediments pass laterally from resedimented conglomerates and sandstones into siltstones and mudstones (Fig. 8.55). Repeated fault activity led to deepening of the basins, thickening of the slope wedge and the development of a series of fining-up cycles, each of which is several hundreds of metres thick (Surlyk, 1978).

Finally, as well as providing a useful sedimentary and structural analogue for proven reservoirs in the North Sea area, field studies in Greenland have also suggested possible new plays, including prospectivity and source-rock potential, in the Rhaetian–Sinemurian lacustrine interval ascribed to the Kap Stewart Formation (Dam and Christiansen, 1990; Surlyk *et al.*, 1993).

8.6 Postdepositional control on Jurassic prospectivity

That the North Sea is such a prolific hydrocarbon province attests to the fact that all the main criteria for petroleum prospectivity have been met. Not only does the Jurassic contain numerous well-defined structural and stratigraphic traps containing viable reservoir-seal pairs, but it also contains the most prolific source rock. It should be remembered, however, that the overriding factors that make the petroleum system work are its post-Jurassic burial history and the persistence of well-defined migration routes. Post-rift Early Cretaceous and Cenozoic subsidence have enabled the source rock to reach maturity in axial parts of the basins (e.g. Bailey *et al.*, 1987; Cayley, 1987). It is only in areas that have experienced significant Tertiary uplift and exhumation (e.g. western parts of the Inner Moray Firth; Hillis *et al.*, 1994) that migration was arrested and palaeostructures reconfigured or breached.

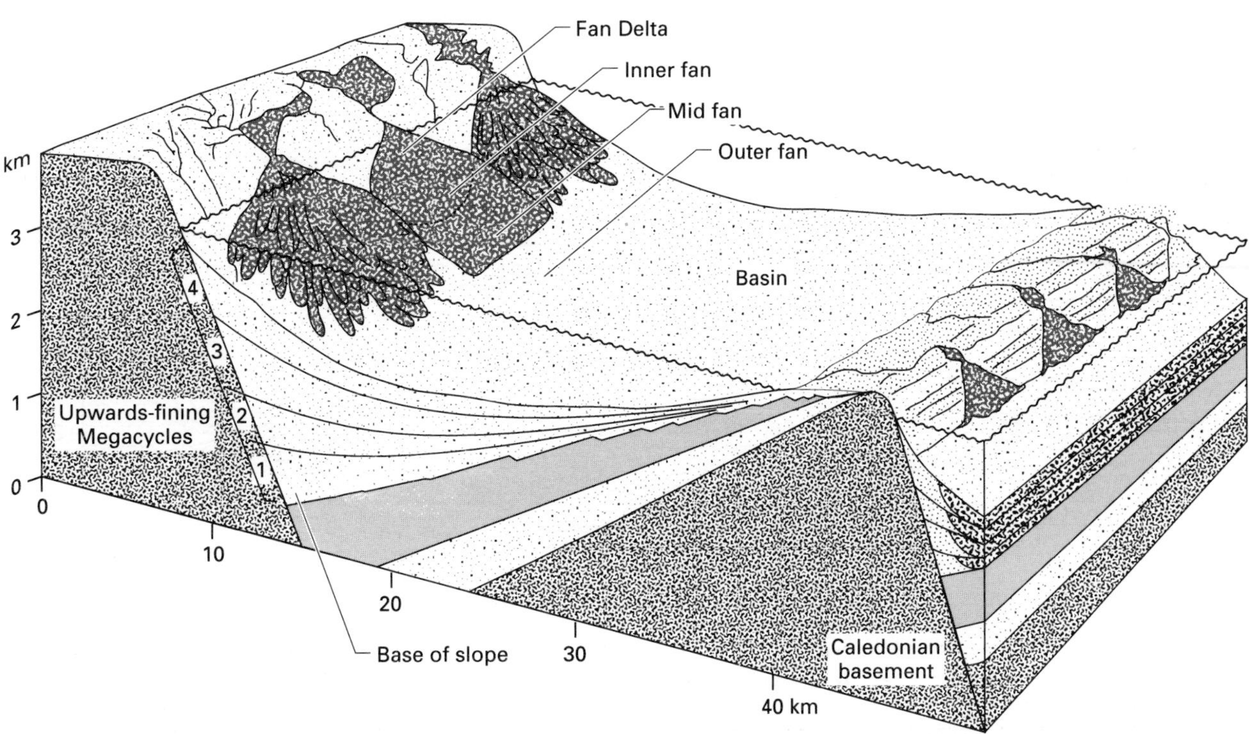

Fig. 8.55 Upper Jurassic faulted slope-apron system from east Greenland (after Surlyk, 1978). The depositional system is consistent with the subsurface examples described from the Brae Trend in the South Viking Graben and with the Helmsdale Boulder Beds of the East Sutherland Coast, Inner Moray Firth.

The tectonic history of the most prolific parts of the basin was ideal for prospectivity, given that the main structures all predate the maturation and migration of hydrocarbons. In the Viking Graben, Central Graben and West Shetlands areas, large-scale vertical and lateral migration occurs along the controlling planar extensional faults and the structural terraces they define. Charge was probably progressive, with more axial structural closures being filled initially and more marginal traps receiving the latest hydrocarbon fill (Fig. 8.56).

8.7 Conclusions

Economically, with almost 50% of the total reservoired oil, the Jurassic forms the most significant stratigraphic interval of the North Sea hydrocarbon province in its own right. Not only does the Jurassic interval contain the main reservoir levels for the basin, in the Kimmeridge Clay Formation and the Liassic Posidoniaschiefer (and its equivalents) it also contains the main source rocks for hydrocarbons in the central and northern North Sea, the West Shetlands, the Wessex, Weald and Paris Basins and other small discoveries in continental north-west Europe. Furthermore, the Jurassic was also the time of major extensional deformation in the North Sea, which created the large structural traps that were charged with hydrocarbons during subsequent post-rift burial.

Recent advances in both the chrono- and sequence-stratigraphic framework of the sedimentary units have improved understanding of the regional development and evolution of the basin and the controls on the local distribution of important reservoir sandstones. As well as improving reservoir field descriptions and recovery, such analysis has also reduced exploration risk, suggested new exploration-play opportunities and provided better reservoir models, all of which have the potential to increase ultimate hydrocarbon recovery and to lengthen the life of the mature North Sea hydrocarbon province.

8.8 Acknowledgements

I would like to acknowledge the help and support of all oil company and research colleagues at Edinburgh who have been instrumental in the development and formulation of ideas presented in this chapter. Particular help and guidance have been received from Richard Davies, Nancye Dawers, John Dixon, Al Fraser, Rob Gawthorpe, Ken Glennie, Dave Latin, Dan McKenzie, Aileen McLeod, Brian Mitchener, Mark Partington, Mark Sawyer, Kevin Stephen, Ken Thomson and Jonathan Turner. Gerry White helped transpose or redraft the diagrams.

8.9 Key references

Abbotts, I.L. (ed.) (1991) *United Kingdom Oil and Gas Fields: 25 Years Commemorative Volume.* Memoir 14, Geological Society, London.

Blundell, D.J. and Gibbs, A.D. (1990) *Tectonic Evolution of the North Sea Rifts.* Clarendon Press.

Brooks, J. and Glennie, K.W. (eds) (1987) *Petroleum Geology of North West Europe.* Graham & Trotman, London.

Hurst, A., Johnson, H.D., Burley, S.D., Canham, A.C. and Mackertich, D.S. (eds) (1996) *Geology of the Humber Group: Central Graben and Moray Firth, UKCS.* Special Publication 114, Geological Society, London.

Illing, L.V. and Hobson, G.D. (eds) (1981) *Petroleum Geology of the Continental Shelf of North-West Europe.* Heyden.

Morton, A.C., Haszeldine, R.S., Giles, M.R. and Brown, S. (1992) *Geology of the Brent Group.* Special Publication 61, Geological Society, London.

Fig. 8.56 Schematic cross-section of the northern North Sea illustrating how present-day deep burial of the stratigraphically younger Kimmeridge Clay following Cretaceous and Cenozoic thermal subsidence has led to the maturation and migration of hydrocarbons out from axial regions of the Viking Graben into structurally higher traps, which were formed during the Late Jurassic rift episode (after Burley, 1993).

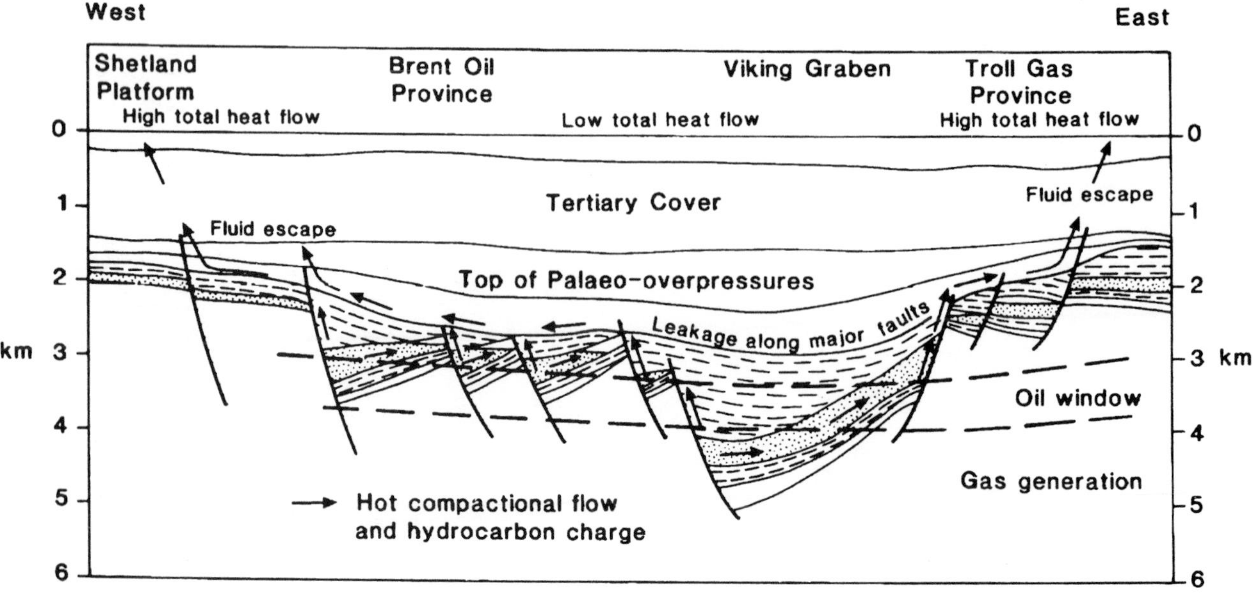

Parker, J.R. (ed) (1993) *Petroleum Geology of Northwest Europe: Proceedings of the 4th Conference*. Geological Society, London.

Spencer, A.M. (ed.) (1986) *Habitat of Hydrocarbons on the Norwegian Continental Shelf.* Graham & Trotman, London.

Spencer, A.M., Campbell, C.J., Hanslien, S.H. *et al.* (eds) (1987) *Geology of Norwegian Oil and Gas fields.* Graham & Trotman, London.

Steel, R.J., Felt, V.L., Johannessen, E.P. and Mathieu, C. (eds) (1995) *Sequence Stratigraphy on the Northwest European Margin.* Elsevier, Amsterdam.

Ziegler, P.A. (1990) *Geological Atlas of Western and Central Europe.* Shell Internationale Petroleum, Maatzchappij BV. The Hague. Distributed by Geological Society, Bath.

9 Cretaceous

C.D. OAKMAN & M.A. PARTINGTON

9.1 Introduction

Figure 9.1 illustrates the distribution of the main Cretaceous hydrocarbon accumulations in the area of the North Sea. Just under half of all Cretaceous fields are within clastic-dominated Lower Cretaceous reservoirs; the remainder have Upper Cretaceous chalk reservoirs. Lower Cretaceous fields are generally confined to two areas—the Moray Firth basins of the UK Central North Sea and the southern part of the Dutch concession area (largely onshore). Most chalk fields occur in the southern Norwegian offshore, with smaller accumulations in the Danish Sector.

Cretaceous sediments were deposited over most of the continental-shelf areas of western Europe, from southern Spain to the Scandinavian Border Zone and from Poland to the margins of the Rockall Trough.

Most of the Cretaceous of the North Sea area can be separated into two major sequences: the Lower Cretaceous Cromer Knoll Group, which is dominantly a siliciclastic succession, ranging in age from Ryazanian at the base to about the Albian–Cenomanian stage boundary; and the Upper Cretaceous Chalk Group. The latter occurs throughout the southern and central areas of the North Sea, but in the north is more argillaceous; it is then assigned to the Shetland Group.

Figure 9.2 illustrates the lithostratigraphy of the Cretaceous for the three sub-areas of the North Sea. Deposition of the major North Sea hydrocarbon source rock, the Kimmeridge Clay, extended from the latest Jurassic Volgian stage into the earliest Cretaceous Ryazanian. Pragmatically, therefore, this description of the Cretaceous begins within the Ryazanian at the point where the source-rock facies is replaced by the organically poorer shales and sandstones of the Cromer Knoll Group.

Over the North Sea Basin, 'base Cretaceous' is commonly taken at a major seismic reflector known variably as:

- (Near) Base Cretaceous;
- Top Humber Group/Kimmeridge Clay Formation (UK);
- Top Mandal or Draupne formations (Norway);
- Top Flekkefjord, Farsund or Fredrikshavn formations (Norway and Denmark);
- Top Bückeberg Formation or German Wealden in Germany;
- Top Delfland Group (Netherlands).

All the above occur within the latest Ryazanian (Fig. 9.2). The seismic boundary represents a dramatic change in the regional tectonic style—from transtension-dominant in the Jurassic to transpression-dominant in the Early Cretaceous (see Fig. 9.6).

The resultant unconformity or multiple condensed sequence is often referred to as the 'Late Cimmerian event' or 'near-Base Cretaceous'. In areas where deposition was continuous and mudrock-dominant, the seismic marker occurs at the change from the low-velocity, low-density and organic-rich ('hot') shales of the Late Volgian/Ryazanian to the higher-velocity and more dense, organically lean ('cool') shales and marls of latest Ryazanian/Valanginian age.

In some areas, the boundary is difficult to pick lithostratigraphically. For example, in the Norwegian–Danish basins and on the Yorkshire coast, mudrock deposition was continuous and exhibits a gradual upwards reduction in organic matter, with a steady increase in sonic velocity (increasing carbonate content). In Lincolnshire, East Anglia, the Netherlands, the South Halibut Basin and some parts of the Viking Graben, rejuvenation of adjacent land masses at about the time of the Late Cimmerian Unconformity resulted in sand deposition that spans the boundary.

The top of the Cretaceous, as considered in this chapter, also lacks a simple stratigraphic definition (Fig. 9.2). Conventionally, the Cretaceous–Tertiary boundary is taken at the top of the chalk comprising the Tor Formation, but lithologically the Chalk Group continues well into the Early Tertiary (Danian stage) as the Ekofisk Formation.

The top Danian chalk is a major seismic marker in the Central North Sea area, but the boundary with the Tertiary clastics of the Montrose Group is diachronous in the Northern North Sea. In the Southern North Sea, the Danian and most of the Maastrichtian are absent, because of inversion and uplift associated with the Alpine orogeny. In the Central North Sea, large areas of previously deposited and semi-lithified Danian (and sometimes Maastrichtian) chalks were redeposited as olistostromes or mélanges and incorporated as geographically extensive masses within Palaeocene sequences (Johnson, 1987). Whether one classifies these as Danian or Palaeocene is highly subjective.

9.2 Tectonic models

The tectonic style of the Cretaceous has been poorly understood throughout most of the exploration history of the North Sea Hydrocarbon Province. Extensional tectonic models for the Jurassic were often extrapolated into the early Cretaceous, and thermal subsidence/halokinetic models for the Tertiary were commonly taken downwards into the chalk sections. Although features normally associated with compressive tectonics were recognized, particularly in the Southern North Sea, it is only recently that plate-wide compressive events and sub-basinal transpressional styles

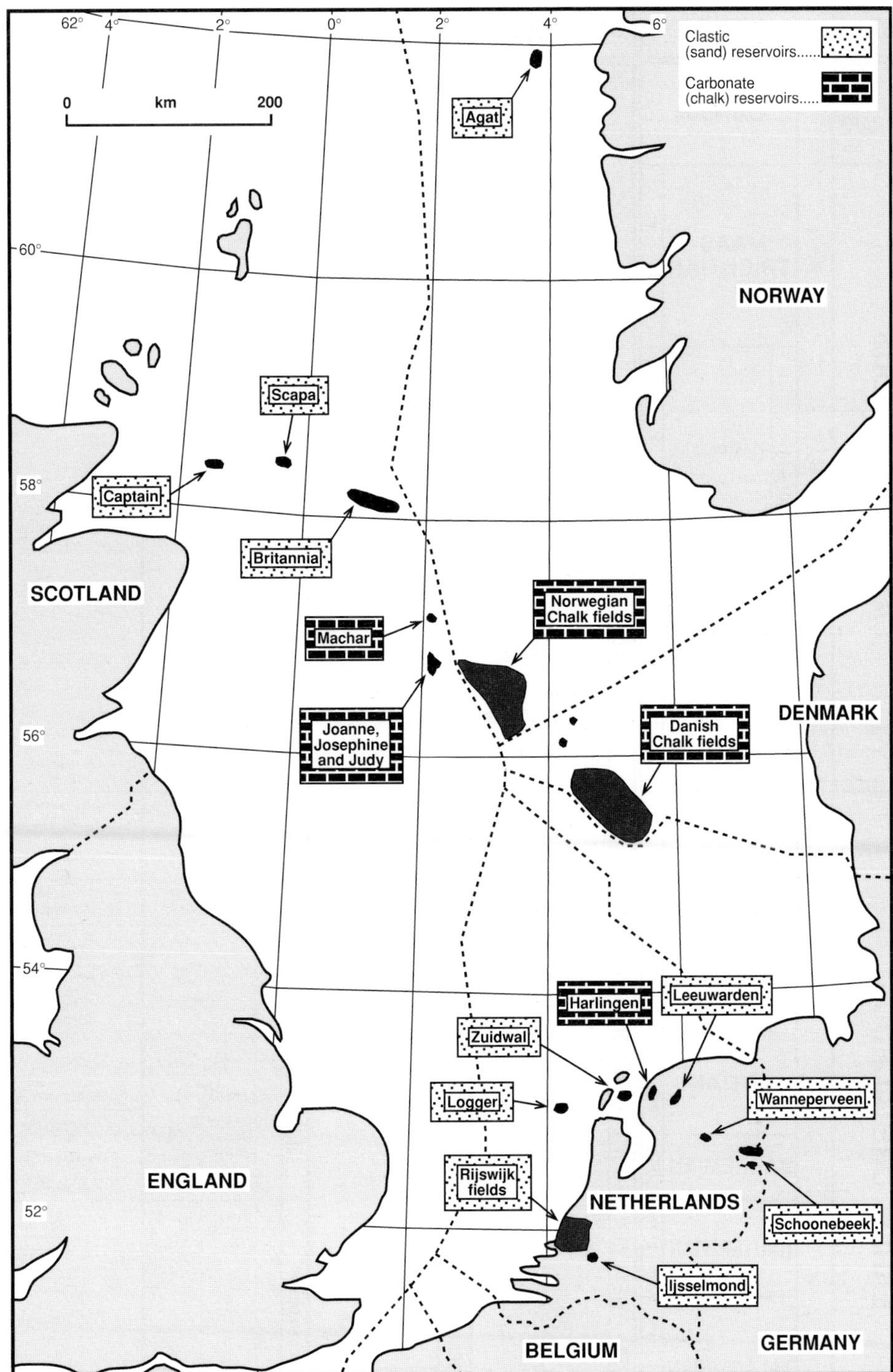

Fig. 9.1 North Sea Cretaceous hydrocarbon accumulations.

have become accepted as the dominant valid Cretaceous tectonic model.

9.2.1 The onset of Cretaceous tectonic styles

The end of the Late Jurassic coincided with the cessation of an east–west tensional phase, with its underfilled rifts and sub-basins. A collapsed thermal dome in the middle of the North Sea area resulted in an overdeepened and greatly underfilled series of transtensional grabens, in which the available accommodation space was not filled with sediment. The axis of extension moved from the North Sea graben systems westwards, where the proto-North Atlantic began to open in the Rockall Trough area. Concurrently, in the south, Tethys was still in an opening (extensional) mode. Most Jurassic rift shoulders were either eroded or submerged during a global rise in sea level.

On a local scale, burial halokinesis was commencing in some of the deeper Jurassic sub-basins and was probably enhanced in the vicinity of transtensional Jurassic shear zones. Local rift shoulders were transgressed and adjacent sub-basins commonly had considerable submarine topography, as a result of underfilling, tilt-block rotation, localized pop-ups and sea-floor rise caused by halokinesis.

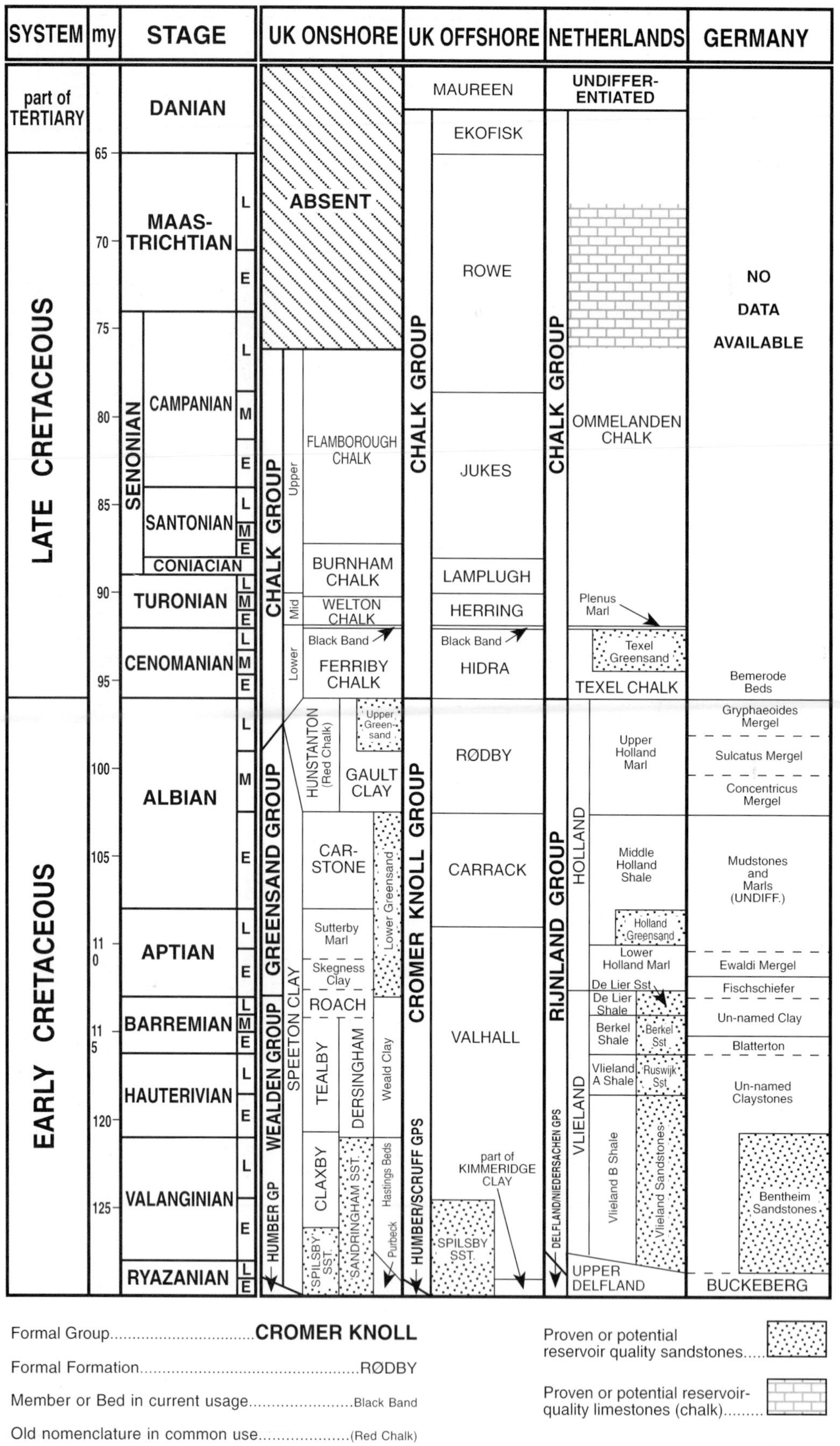

Fig. 9.2A Lithostratigraphic nomenclature–Southern North Sea (from Bodenhausen & Ott, 1981; Cameron *et al.*, 1992; Casey & Gallois, 1973; Chatwin, 1961; Cottençon *et al.*, 1975; Crittenden, 1982, 1987a & 1987b; Deegan & Scull, 1977; Gallois, 1965; Hageman & Hooykaas, 1980; Hamblin *et al.*, 1992; Hancock *et al.*, 1983; Hancock & Rawson, 1992; Herngreen *et al.*, 1992; Kemper, 1974 & 1987; Kent *et al.*, 1980; Lott *et al.*, 1985, 1986 & 1989; Myerscough, 1994; Neal & Catt, 1994; Rawson, 1992b; Rawson & Whitham, 1992a & 1992b; Rhys, 1974; Robinson, 1986; van Adrichem Boogaert & Kouwe, 1993; Whitham, 1993, 1992 & 1994; Wood & Smith, 1978).

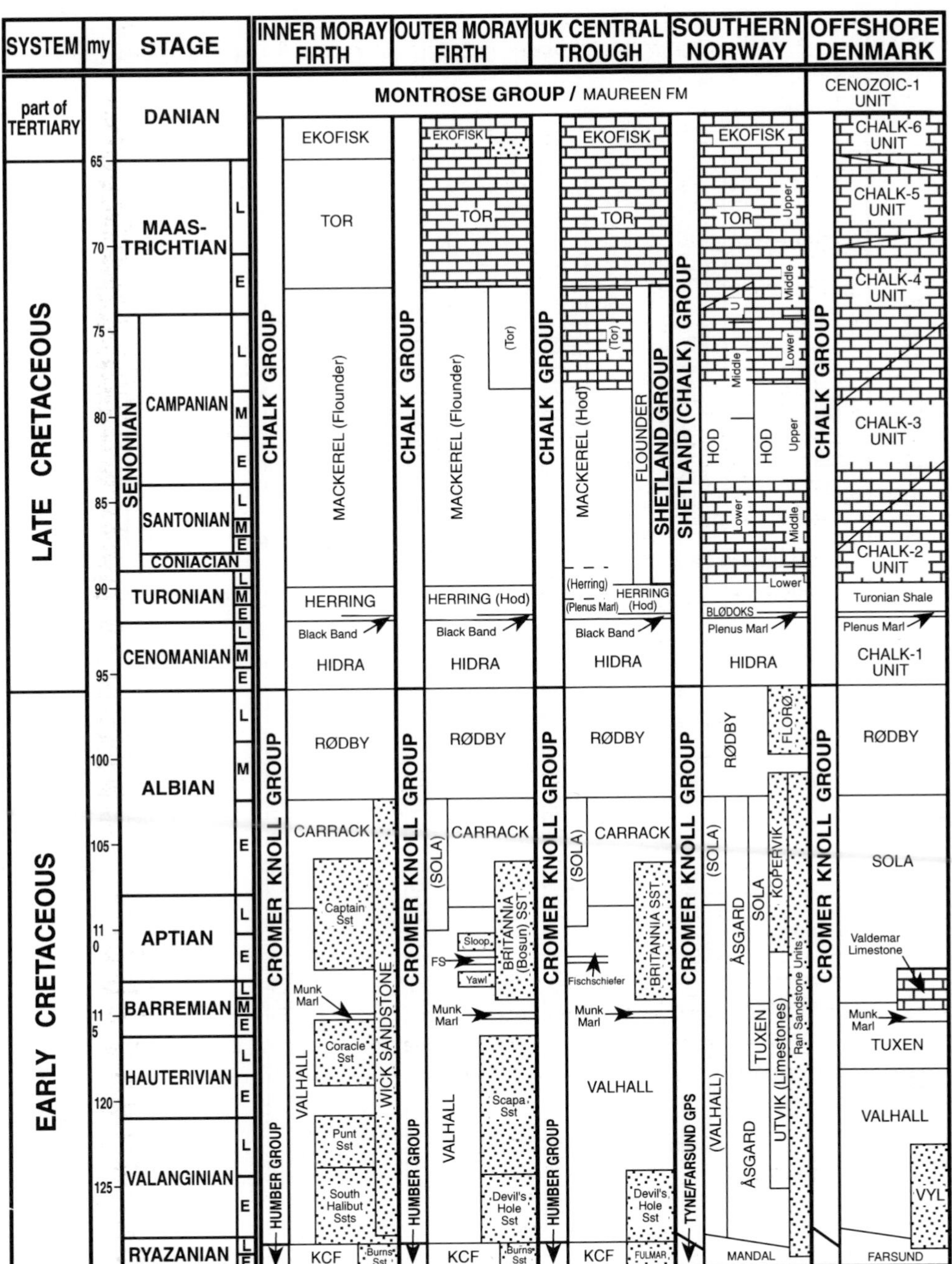

Fig. 9.2B Lithostratigraphic nomenclature—Central North Sea (from Andrews *et al.*, 1990; Birkelund *et al.*, 1983; Bisewski, 1990; Boote & Gustav, 1987; Burnhill & Ramsay, 1981; Deegan & Scull, 1977; Frandsen *et al.*, 1987; Gatliff *et al.*, 1994; Hancock & Kauffmann, 1979; Hancock & Scholle, 1975; Hancock *et al.*, 1983; Hansen & Buch, 1982; Hardman, 1982; Harker *et al.*, 1987; Hesjedal & Hamar, 1983; Isaksen & Tonstad, 1989; Jensen *et al.*, 1986; Johnson & Lott, 1993; Knox & Holloway, 1992; Lieberkind *et al.*, 1982; Megson, 1992; Michelsen *et al.*, 1987; O'Driscoll *et al.*, 1990; Thomsen & Jensen, 1989; Vejbæk, 1986). For legend see Fig. 9.2A.

9.2.2 The Late Cimmerian event

The Late Cimmerian event (see Fig. 9.6) and corresponding 'unconformity', marks a change in regional stress field from east–west extension to northerly-directed compression, imparted by Tethyan sea-floor spreading or rifting in the Bay of Biscay. A northwards-directed compression model partly explains the dominance of Late Jurassic strike–slip faults in the Southern North Sea, which was considerably closer to the compression front and more distal to the pull of Atlantean extension.

In Apto-Albian times, the onset of Tethyan closure and the associated Austrian orogeny accounts for northwards compression throughout the North Sea Basin. In the North Sea Basin, however, the Late Cimmerian event is stronger than the Austrian. This may be partially explained if most Late Jurassic faults had locked before Austrian tectonism. This view seems to be supported by the apparent continuation of Jurassic tectonic trends in many of the earliest Cretaceous sub-basins.

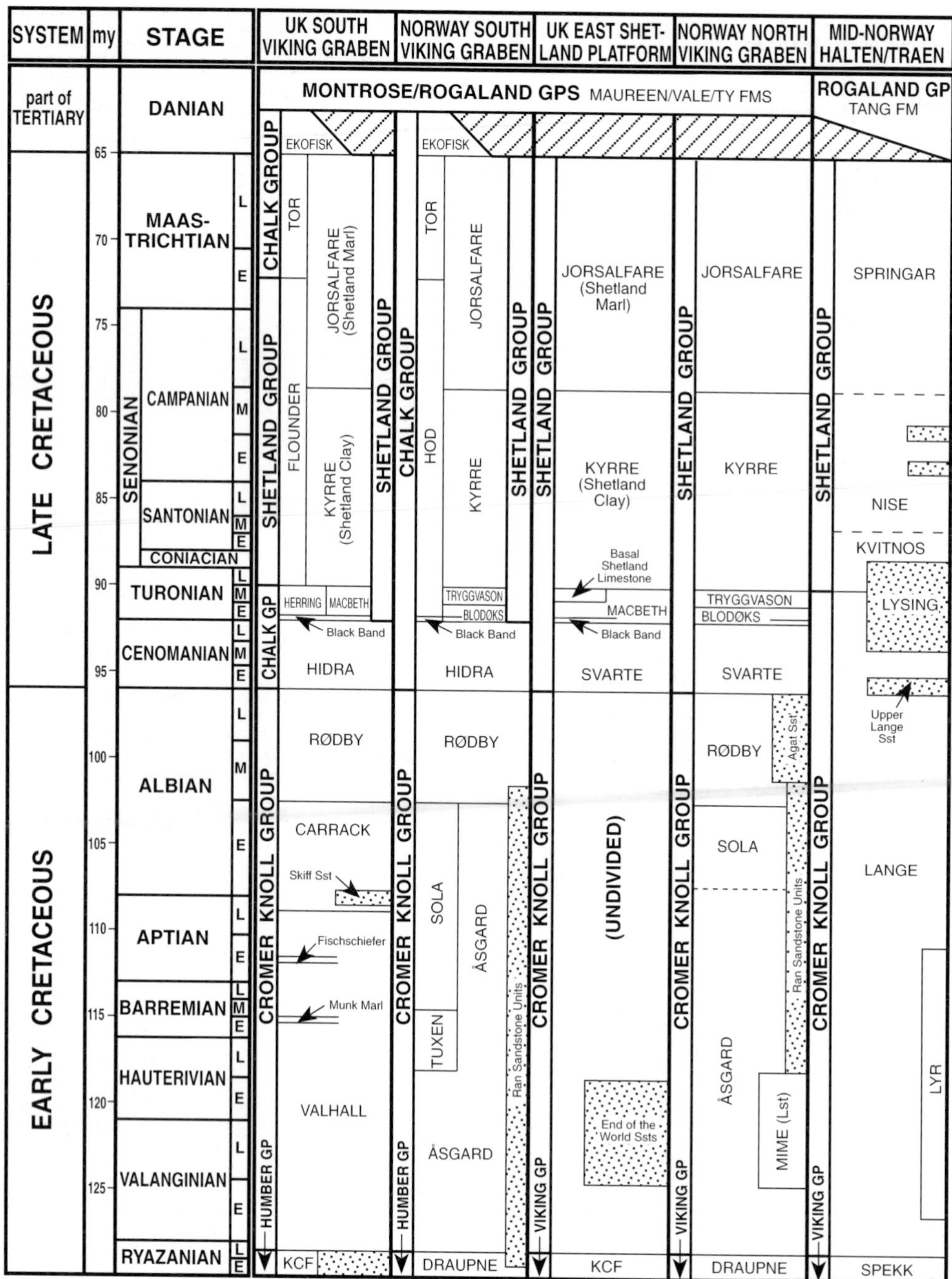

Fig. 9.2C Lithostratigraphic nomenclature—Northern North Sea (from Deegan & Scull, 1977; Isaksen & Tonstad, 1989; Johnson & Lott, 1993; Johnson *et al.*, 1993; Knox & Holloway 1992). For legend see Fig. 9.2A.

9.2.3 New regional compression models

Throughout the Cretaceous, intraplate push–pull stresses created a pattern of structures that were oblique to pre-existing lines of weakness: the pull, or transtension, in an east–west direction, and the push, or transpression, directed from the south. Post-Aptian east–west basinwide extensional components are extremely subtle and weakened progressively during the Cretaceous, whereas northwards-directed compressions, although pulsed, were considerably stronger.

There are also broad regional variations in the response to the variable stress fields. Compressional effects are much greater in the Southern North Sea and more subtle in the Northern North Sea. This has created an apparent 'compressive wave', which decreases in intensity northwards through the basin. Figure 9.3 shows some of the active fault and fold axes resulting from the complex interplay of stress fields and pre-Cretaceous lineaments. Note the weakening of effects in the Late Cretaceous as compared with those of the Early Cretaceous. Figure 9.6 (intraplate stress column) illustrates the stratigraphic distribution of stress fields.

Compressional events occurred at the end of the Jurassic (early Ryazanian), in the Late Valanginian and mid-Hauterivian. These events represent the inversion phase at the end of the Late Cimmerian plate-margin cycle. A weak compressive event occurred in the middle Barremian, with a major pulse in the mid-Aptian. All subsequent compressive events result from the closure of Tethys (alpine orogenesis), which began around the Barremian–Aptian boundary and continued into the Tertiary.

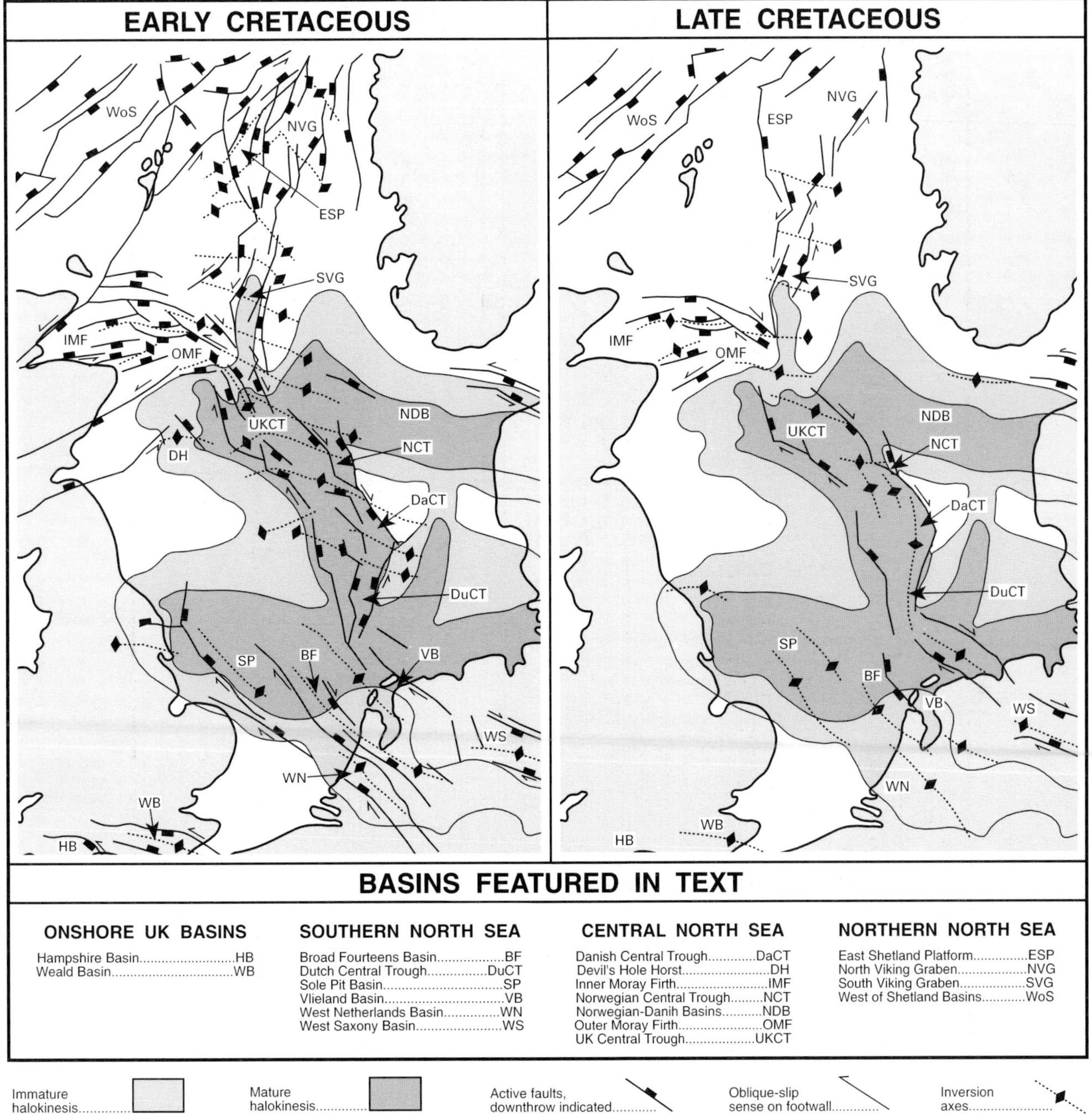

Fig. 9.3 Schematic Cretaceous tectonic controls (modified from Ziegler, 1990a; also from Chapter 1).

The Aptian event is often referred to as the Austrian orogeny or unconformity, although the 'unconformity' is only weakly expressed in the North Sea area. Further compressive pulses occurred in the early to mid-Turonian, early Campanian and around the Campanian–Maastrichtian boundary, the latter two being occasionally referred to as the 'sub-Hercynian event'. A major pulse occurred throughout the North Sea around the middle of the late Maastrichtian, which heralded the onset of continent–continent collision in the Alpine domain (Ziegler, 1990b). Knott *et al.* (1993) suggest that the Late Maastrichtian inversions resulted from Pyrenean rather than Alpine collision. This event directly influences the habitat of hydrocarbons in Late Cretaceous chalk reservoirs.

Cretaceous compressive events created a wide range of structures throughout the North Sea Basin. These can be seen as inversions of variable magnitude and areal extent, complex fault tectonics and a range of north-east–south-west to north-west–south-east folds of variable size. Interaction with older Jurassic lineaments resulted in a continuation of distinct half-graben styles, although these are of very limited areal extent.

Many of the Cretaceous inversion axes are also well documented, particularly in the southern North Sea and onshore UK: for example, the Dutch Central Trough (Clarke-Lowes *et al.*, 1987), the Sole Pit, Broad Fourteens and Saxony basins, in East Anglia, and the Market Weighton inversion. In addition, such axes occur in the Weald and Hampshire basins of southern England. Similar axes can be shown to have existed in the Central North Sea, where they appear to exploit Caledonide grains in the west and Tornquist lineaments in the central and eastern areas

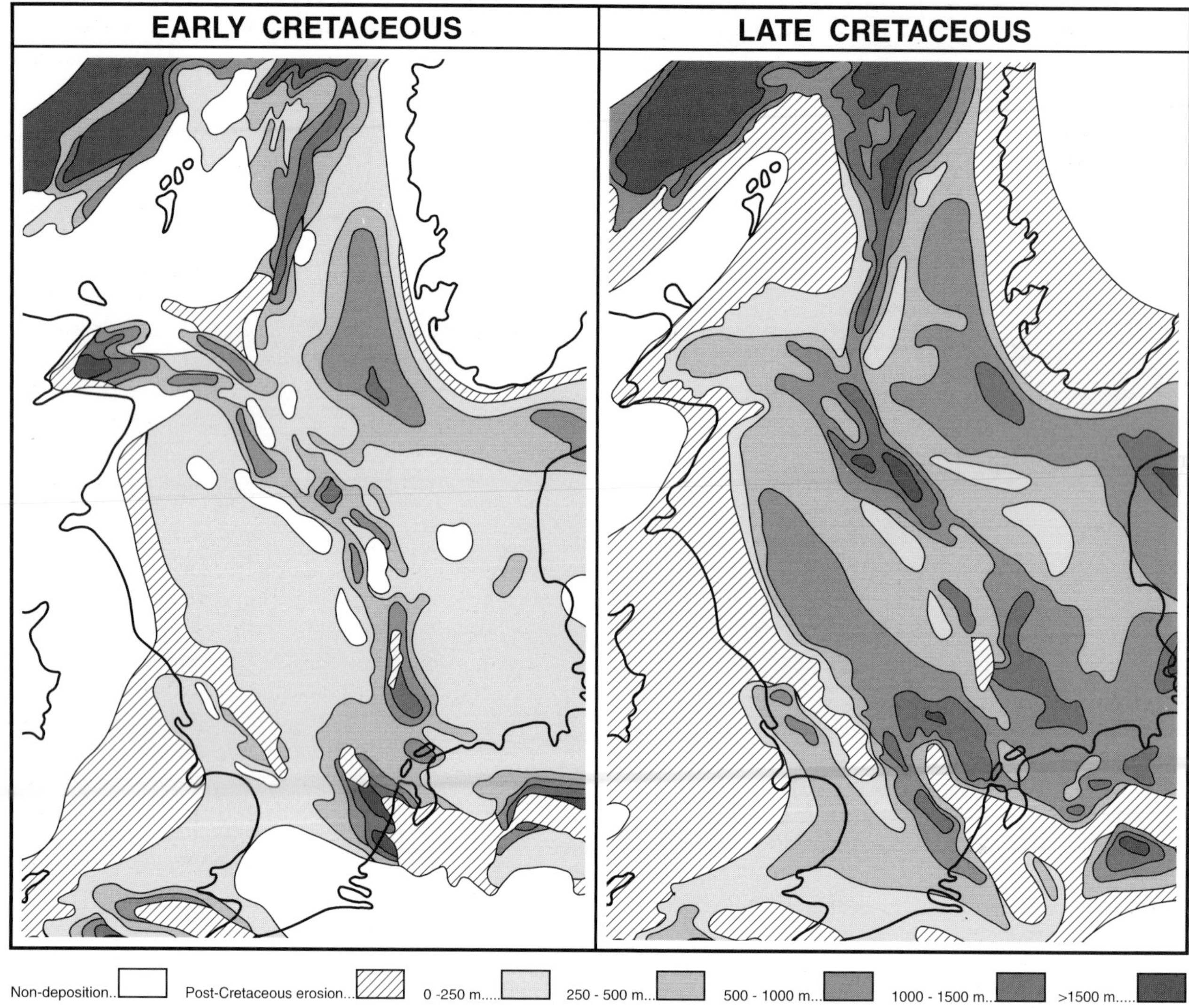

Fig. 9.4 Early and Late Cretaceous thicknesses (modified from Ziegler, 1990a).

(Fig. 9.3). Similar structures, although of lower relief, extend the full length of the Viking Graben, their axes being parallel to the Tornquist lineament.

Compression from the south was oblique to many pre-Cretaceous zones of structural weakness. Strike–slip and oblique-slip movements along reactivated Jurassic faults were extremely common during the Cretaceous. On a local, or block, scale, intersections of Jurassic extensional and transfer faults formed bends of both stress restraints and release, giving rise to positive and negative flower structures, depending on their orientation relative to the regional stress pattern.

Flower structures can be very complex in areas where older, larger, Jurassic grabens intersected (interference zones). Examples include the Fisher Bank Basin at the intersection of the South Viking and the Witch Ground–Central Graben trends; the large area of the Halibut Horst and Halibut Shelf, at the intersection of the Inner Moray Firth/Buchan Graben with the northern Witch Ground Graben; and at the intersection of the Norwegian and Danish central troughs. Other examples can be found in the Dutch Central Graben intersection with the western Vlieland Basin, and the intersection of the Broad Fourteens and West Netherlands basins. During periods of Cretaceous transpression, inversion axes and flower structures were developed on a sub-basin scale.

Low-angle thrust faults are another feature of Cretaceous compression. These are often difficult to attribute to the Cretaceous because of later, and often more intense, Tertiary reactivation (particularly in the Southern North Sea). Many of these structures appear to be reversals of Jurassic extensional listric faults. Classic examples include parts of the Scrubbe Fault in the Norwegian Central Trough (see Figs 9.23 and 9.24) and many faults in the Saxony Basin of Germany.

9.2.4 Residual extensional models

The balance of regional evidence has resulted in a shift away from the more traditional extensional and/or thermal subsidence models for the Cretaceous of the North Sea. Any apparent extensional features, such as local half-grabens and negative flower sub-basins, are now thought of as tectonic accommodation features in a transpressional domain.

There still remain, however, some notable discrepancies

in the adoption of complex transpressional models to the whole of the North Sea Basin. Areas on the northern fringe of the North Sea and Atlantic Margin of Norway contain laterally extensive true syn-rift wedges of variable Early Cretaceous ages. Most thick clastic piles are of Valanginian age. Early Cretaceous tectonic style in these areas is more compatible with dominant transtension associated with North Atlantic rifting.

The Inner Moray Firth also contains a very thick (> 1000 m) and areally extensive (120 × 50 km) 'syn-rift' wedge of Valanginian sand stacked against the Helmsdale and Wick faults (see Fig. 9.22). This basin, however, is much closer to the Atlantean extensional domain than any others in the North Sea, and may be protected from northwards-directed compressive forces by the Precambrian basement of the Scottish Grampian mainland. Protection from the effects of compression by more stable areas of basement can also be seen with the Mid North Sea–Rinkøbing–Fyn High, where compressive features are more extreme to its south (Southern North Sea) than to its north.

Cretaceous tectonic styles were ended by the 'Laramide event' in the late Danian/early Palaeocene. Although this coincides with the onset of the main Alpine orogenic phases, it is also coeval with the onset of sea-floor spreading in the North Atlantic.

9.2.5 Cretaceous halokinesis

Halokinesis and halokinetic structural style in the North Sea Basin have been outlined in Chapters 2 and 6. The presence of thick halites within a basin results in localized modification of plate-wide tectonic features in two ways—passive responsive and active burial halokinesis. Halokinetic models only apply to the Southern North Sea and to the Central Trough area of the Central North Sea (Fig. 9.3). Structural wavelengths are often severely attenuated and amplitudes exaggerated in these salt-prone areas.

Salt acts as an important modifier of otherwise compressional structures, absorbing most underlying stresses and not transmitting them to overburden. This is certainly true of flower structures, which often achieve considerable vertical displacements on their component faults but are of very limited areal extent in sub-basins with a thick salt underlayer when compared with larger, flatter, complexes in non-salt subbasins. Reverse and thrust faults superimposed on older basement-seated Jurassic faults may detach and sole out in salt; such features commonly appear as 'listric' reverse and 'listric' thrust faults on regional and two-dimensional (2D) seismic lines.

The above passive responsive style of salt activity can be contrasted with active burial halokinesis. Progressive burial resulted in the ongoing development of intrusive salt features, from early pillows to diapiric walls, stocks, domes, diapirs and sills. The progression from coherent bedded salt started with the formation of pillows and pods during the Triassic (Hodgson *et al.*, 1992) and continued with intrusive activity in the Late Cretaceous, when diapir chains or salt walls had a significant influence on depositional style and subsequent trap structuration.

It is immensely difficult, however, to favour either passive and/or responsive or active and/or burial styles of halokinesis in the light of proved Cretaceous basin-wide compressions. Most episodes of halokinesis during the Cretaceous are believed to coincide with the discrete compressive events outlined above.

9.3 Palaeogeographic evolution

The palaeogeography of the Cretaceous features in many publications. The work of Ziegler (1981, 1982a, 1987a,b, 1990a,b) provides the best general overview of the entire basin. Tyson and Funnel (1987) provide a more detailed view of Cretaceous shoreline migrations and Sellwood (1979) gives a summary overview of the UK mainland. Jenkyns (1980), Hallam (1984) and Ruffell and Batten (1990) provide descriptions and discussion of Cretaceous climates.

On a more local scale, the UK sector of the Southern North Sea Basin is covered by Cameron *et al.* (1992) and the Weald Basin (on the southern margin of the area) by Hamblin *et al.* (1992). Kemper (e.g. 1982) and Schröder *et al.* (1992) provide details of the German sector. The onshore evolution in the UK is also covered by Hancock and Rawson (1992). In addition to the works of Ziegler (see above), for generalized publications on the Netherlands see Kaaschieter and Reijers (1983) and Rondeel *et al.*, (1996) and references therein.

On the Central North Sea, the UK Offshore Operators Association (UKOOA) and British Geological Survey (BGS) publications provide general information on the UK sector (Knox and Holloway, 1992; Johnson and Lott, 1993). Key UK publications include chapters in Andrews *et al.* (1990) for the Moray Firth and Gatliff *et al.* (1994) on the Central Trough. Boote and Gustav (1987), Harker *et al.* (1987) and Bisewski (1990) provide further details of the Outer Moray Firth basins. The Norwegian sector is featured in Hesjedal and Hamar (1983) and Gowers and Sæbøe (1985). Key references in the Danish sector include Hansen and Buch (1982), Frandsen *et al.* (1987), Michelsen *et al.* (1987) and Nygaard *et al.* (1990).

Data on the Northern North Sea are sparse. Only the UKOOA and BGS publications provide an overview (Knox and Holloway, 1992; Johnson and Lott, 1993; Johnson *et al.*, 1993). Shanmugan *et al.* (1994) provide palaeogeographical modelling of the Early Cretaceous in the northern Norwegian part of the North Viking Graben.

9.3.1 Global controls

There are a series of widely accepted 'global' controls on the palaeogeographical evolution of the Cretaceous.

The progressive shift of climatic belts from northern subtropical at the onset of the Cretaceous to midtemperate by the Danian is well documented, even in school textbooks. This was largely an effect of plate migration. Secondly, the proliferation of the calcareous alga group of coccolithophorids gave rise to temperate carbonates in a wide belt from West Siberia to the American Midwest (Hancock, 1975). Thirdly, there was a global rise in sea levels (Hancock and Kauffmann, 1979; Haq *et al.*, 1988; Christie-Blick, 1990; Ruffell, 1991), from a major low stand during the Late Ryazanian/Early Valanginian, rising to a

Turonian high stand via two smaller high stands (Barremian and latest Albian) and remaining high for the rest of the Cretaceous (see Fig. 9.6).

As a direct consequence of northern-hemisphere high stands, extremely widespread anoxic (basinal) or dysaerobic (shelfal-equivalent) events can be traced throughout almost all Cretaceous sub-basins within the North Sea—and, indeed, throughout almost all of the Atlantic margins (Graciansky *et al.*, 1984; Stein *et al.*, 1986). These include three major events—the early Barremian Munk Marl, or Blatterton, the Early Aptian Fischschiefer and the Cenomanian/Turonian boundary Plenus Marl. These and other less widespread events have been documented and their origin discussed by Farrimond *et al.* (1990), Hart and Leary (1989), Jenkyns (1980, 1985), Kemper (1982, 1987), Lott *et al.* (1989), Schlager and Jenkyns (1976), and Thomsen *et al.* (1983).

9.3.2 Pre-Cretaceous geology

The landmasses that fringed the North Sea Basin at the end of the Jurassic contributed clastics at many different times during the Cretaceous. In the Early Cretaceous, ramp-type basin-margin geometries occurred throughout mainland UK (southern Scotland, Pennine High and what was left of the Welsh–London–Brabant massif). Considerably greater relief existed on the Brabant–Rhenish Massif, parts of the Ringkøbing–Fyn High, the Fenno-Scandian Shield, the Northern Highlands of Scotland and a large area of the East Shetland Platform. All of these areas were in proximity to major Early Cretaceous sub-basins (Fig. 9.4) and contributed large volumes of first-order erosion, pre-Mesozoic, clastics.

In areas where Jurassic facies underlie the Cretaceous, considerable relief occurred in the form of smaller (although volumetrically significant) islands and seamounts. The topography of these local highs was often exaggerated during Cretaceous compressive events, providing access to erosion products from sand-prone areas of the Jurassic, Triassic and, in some local areas, pre-Mesozoic basement. These highs range from the Tampen Spur in the north to the Texel–IJsselmeer High in the south. On a local scale, many inversion axes, transpressive pop-up structures and residual Jurassic footwalls were being continually eroded (either subaerially or submarine) and contributed very localized clastics to Early Cretaceous sub-basins.

The Late Cretaceous presents a very different picture. Almost all available landmasses were flooded during marine transgression and very few islands remained. Almost all landmasses had low relief. Only the Grampian High, a narrow strip of the East Shetland Platform, the core of the Rhenish Massif and the Fenno-Scandian Shield remained above sea level.

The principal Cretaceous sub-basins of the North Sea area are shown on Figs 9.3 and 9.4.

9.3.3 The onset of Cretaceous deposition

The late Ryazanian to Valanginian regional palaeogeographies initially differ little from that established in the Late Jurassic. In the Central and Northern North Sea, some areas suggest transgression, with more widespread marine inundation (Norwegian–Danish Basins, Western Platform of the Central Trough). Other areas show continuations of an apparent rifting style—for example, the Inner Moray Firth, West of Shetlands and the Møre Basin—and some areas show condensation and/or erosion (parts of the East Shetland Platform, Grampian High, Jaeren High and Auk Ridge).

The depositional style over older Late Jurassic platform areas continued, although there is widespread evidence of shallowing. Jurassic basinal areas were often modified by transpressive downwarp, to the extent that some Valanginian sub-basin sections show distinct deepening from the Jurassic template, or they were partially inverted to give rise to sections that illustrate shallowing. Shelfal areas flooded by black shales in the Late Jurassic, if adjacent to reactivated Late Cimmerian landmasses, commonly contain Early Cretaceous shallow-marine sand systems (e.g. Dutch Bank Basin, northern Norwegian–Danish basins and the fringes of the Western Platform). Relic underfilled basinal areas continued the Late Jurassic sedimentation style, with the development of thick slope and basin-floor muds and marls. A new generation of deep-marine sandstones was developed, however, adjacent to relic and new Late Cimmerian highs. Examples of the latter sandstones can be seen in the Inner Moray Firth, around the Halibut Horst, in some South Viking Graben wells, in the Magnus Embayment, in the Northern Viking Graben and in the Western Central Trough.

In the Southern North Sea, the Ryazanian depositional area was considerably less than that of the Late Jurassic. This was due to transpressional rejuvenation of almost all landmasses—for example, the southern Pennine High, the London–Brabant Massif, the Cleaver Bank High, the entire area of the Ringkøbing–Fyn High southwards via the Schill Grund Platform to the Pompeckj Swell, and the Rhenish Massif. This restricted late Ryazanian to Valanginian basin development and confined sedimentation to the Sole Pit, Weald, southern Broad Fourteens, Vlieland and Saxony basins and Dutch Central Trough. All the basins were fringed by progradational clastic wedges, which were deposited in a wide range of environments, from fluvio-deltaic through shoreface to shelfal settings. Deeper marine shelfal, slope and, possibly, basin-floor sands occurred in the northern part of the Dutch Central Trough.

9.3.4 The Hauterivian–Barremian transgression

Late Cimmerian compressional stresses died out during the Late Valanginian. The great global sea-level rise of the Cretaceous started during the Early Hauterivian. Flooding of Valanginian landmasses continued, to the extent that the Cleaver Bank, Ringkøbing–Fyn and Pompeckj Swell almost disappeared from the south of the North Sea area.

In the Central and Northern North Sea, the transgression was less dramatic, with most landmasses being encroached upon but not submerged. Active rifting quietened in the Inner Moray Firth and a short-lived period of apparent thermal subsidence dominated the area. Coarse clastic deposition was much reduced; it appears to have been confined to the Inner Moray Firth area, the Dutch Bank Basin, the northern flanks of the East Shetland Platform and the West of Shetland–Møre Basin–mid-Norway rift. It is often difficult to tell if these sands are shelfal or basin-

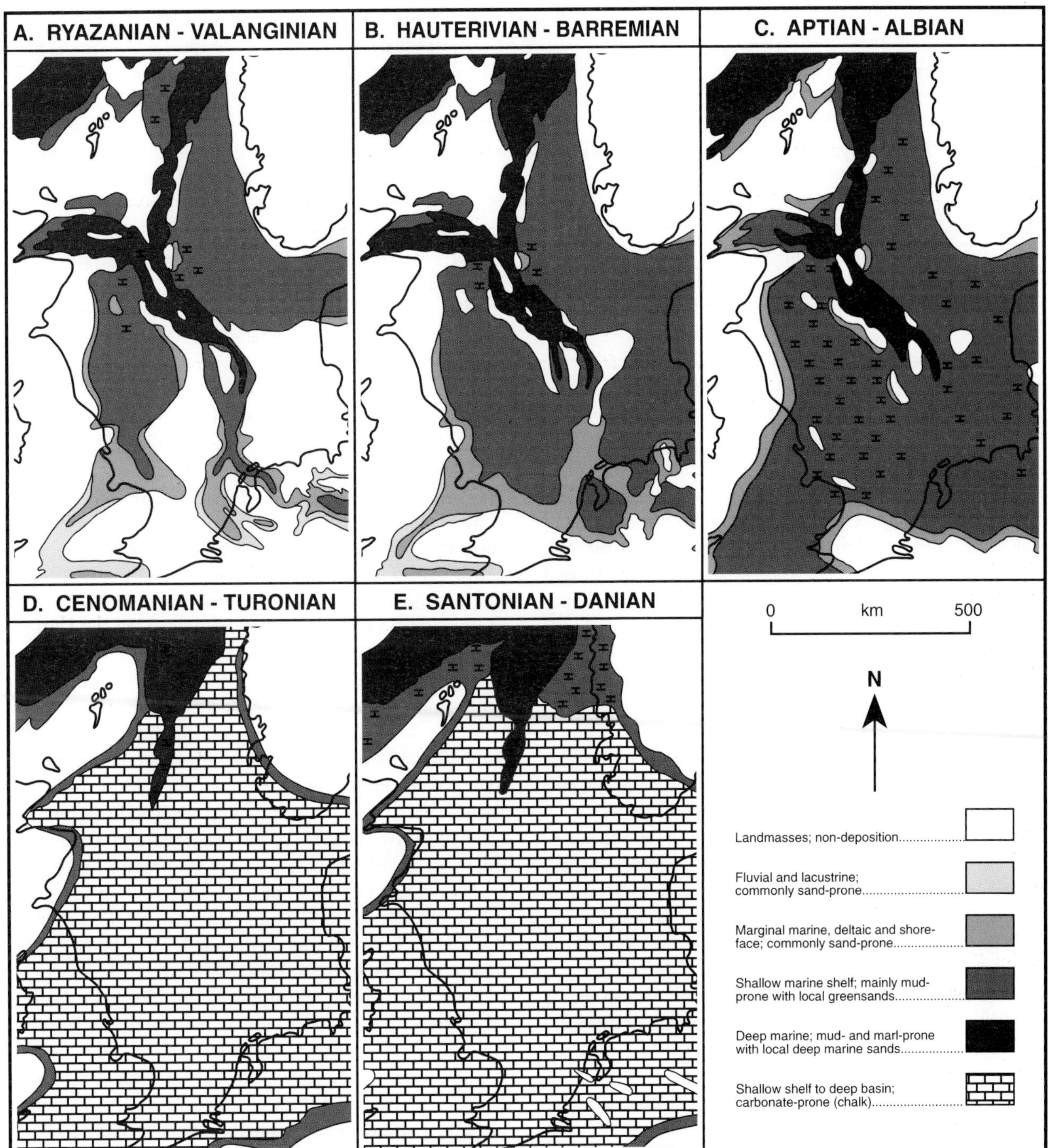

Fig. 9.5 Cretaceous palaeogeographic evolution (modified from Ziegler, 1990a).

floor. Differentiation of basinal and deep shelfal areas in other sub-basins became well established. A lack of coarse clastic input in many of the sub-basins allowed the development of widespread deep-shelf to basinal marls. In some areas of the Central Trough, a distinct deep-marine Barremian limestone (chalk) was developed.

The Southern North Sea was less clearly differentiated into distinct facies belts. Residual landmasses continued to contribute significant volumes of coarse clastics to the southern fringes of the area. In the Weald Basin, lacustrine and paralic facies dominated, reaching as far north as The Wash. Over most of the prospective UK offshore and the Netherlands, shelfal deposition dominated, with some shoreface tracts on the basin fringes. Widespread progradational sheets of shelfal greensands occurred over much of the area, particularly in the Middle and Late Barremian, because of Early Austrian landmass rejuvenation.

In the Barremian, the first of a series of widespread anoxic events, the Munk Marl, is encountered. This provides an isochron for correlation throughout the entire North Sea Basin.

9.3.5 Apto-Albian style change

The passive depositional style of the Barremian changed dramatically at the onset of the Aptian. Major compressive pulses from the south (the onset of the main phase of Austrian tectonics) rejuvenated almost all older landmasses. The Austrian event also formed many new, localized areas of sediment source in the form of pop-up structures.

The Austrian tectonic reconfiguration of the North Sea area occurred against a background of global sea-level rise. By the end of the Albian, the Early Cretaceous seas occupied an area nearly three times that of the Valanginian. The margin of the London–Brabant Massif had retreated to the south, and all south and east England was flooded. Only scattered islands remained in the vicinity of the Mid North Sea and Ringkøbing–Fyn highs. The Aptian contained most of the sandy facies, although, in some areas fringing larger landmasses, shelfal greensand deposition continued into the Cenomanian.

Lacustrine and paralic facies had disappeared from the entire area in the Aptian. In shelfal areas around available landmasses and islands, shoreface and shelfal greensand systems proliferated. In the Southern North Sea, these sands occur in a belt, with its margins limited by the north flank of the West Netherlands Basin and the eastern flank of the Sole Pit Trough. Greensands were also present around the margins of the Dutch Central Trough and in the Auk Shelf/Basin area. Prolific micaceous shelfal sands fringed the south of the Fenno-Scandian Shield and extended into northern Denmark. Well-developed shelfal sands occurred throughout the Inner Moray Firth Basin, and thinner bodies were present over the Halibut Horst, North Halibut Shelf, Dutch Bank Basin and Fladen Ground Spur. Shelfal sand bodies were developed on the flanks of the Central Trough in both the UK and Norwegian sectors. Sand fringes are also documented in wells around the north, east and west margins of the East Shetland Platform.

Slope and basin-floor sand systems are present throughout the deeper subbasins of the Witch Ground Graben, South Halibut Trough, Buchan Graben and Fisher Bank Basin, with rare occurrences penetrated in the Central Trough, South Viking Graben and Sogne Spur area of the North Viking Graben in Norway.

At times of limited coarse-clastic deposition, deep shelfal and basinal marls dominated the North Sea, particularly throughout most of the Albian. Two small late Austrian compressional pulses in the Late Albian and mid-Cenomanian resulted in extremely localized shelfal greensand deposition adjacent to the few remaining landmasses and some of the tiny islands. Over wide areas of the North Sea, marl facies became sufficiently carbonate-enriched to form deep shelf to basinal chalky limestones. These are generally restricted to two horizons—latest early Albian and late Albian. Along with the Fischschiefer anoxic event in the Early Aptian, these provide very useful stratigraphic marker beds.

9.3.6 The early Chalk sea

The onset of Cretaceous limestone deposition, as outlined above, started in deeper sub-basins in the Barremian. Basinwide chalk deposition, however, did not commence until global flooding approached high stand at the end of the Albian. The Albian high stand flooded almost all residual, very low-relief, landmasses and severely curtailed clastic deposition over the whole North Sea Basin.

The early Chalk seas, spanning the Cenomanian, Turonian, Coniacian and Santonian, did not, however, produce the same purity of carbonate deposition as seen in the latest Cretaceous. Shelfal areas became deeper and could accommodate relatively clean carbonates—usually keeping an under-accommodated equilibrium as sea levels rose. The presence of landmasses, although generally considered to have been of low relief (Shetland and Hebrides platforms, Grampian and Welsh highlands, the Cornubian and Rhenish massifs and the Fenno-Scandian Shield) contributed minor shoreface and shelfal sands into the margins of the carbonate basin.

Fringing landmasses also contributed significant volumes of silica dust and/or colloidal silica, variably modified by marine biota and eventually precipitated as chert (flints) within the carbonate succession. Similarly, terrestrially derived muds added to the impurities within the carbonate basin. As the transgression progressed into the Campanian and landmasses became lower and smaller, non-carbonate impurities decreased to produce pure white chalks.

Most of the chalks encountered within the prospective sub-basins of the North Sea were deposited in deep water below storm wavebase. Intrabasinal highs and fringing platform areas commonly show chalks with shallow-marine characteristics. Onshore exposures in the UK and northern France (Melville and Freshney, 1982; Ekdale and Bromley, 1984; Quine and Bosence, 1991) reveal a plethora of sedimentary structures (ichnofabrics, hummocky stratification, bar and channel geometries, cross-lamination, etc.) indicative of shallow-marine to lower shoreface settings, and a biota indicative of shallow wave-dominated clear shelfal seas (corals, bryozoa). The chalk is hydrodynamically deceptive, because of the proliferation of calcareous microorganisms and their breakdown products producing an inbuilt bias to a very fine grain size.

The rapidity of the Cenomanian sea-level rise left many residual Early Cretaceous basinal areas underfilled, particularly in the Central Trough (UK, Norway, Denmark and into Holland), Moray Firth Basins and Viking Graben. In these areas, the sea floor was at or around the limits of carbonate-compensation depths, resulting in the deposition of highly calcareous shales, marls and basinal dolomitic chalks. The deeper the sub-basin, the lower the carbonate content. This can be seen from the influence of the generally northwards-deepening Viking Trough (Fig. 9.5) and major Early and Middle Turonian shales in the deeper parts of the Norwegian and Danish Central Trough.

Both Cenomanian and Turonian clean chalks occur within the above basinal settings. Although very rarely cored, some horizons show slumping, mass- and turbidity-flow structures, suggesting they are allochthonous bodies derived from adjacent deep shelf areas. Mass-flow chalks are not usually recognized until the Campanian. The presence of the Plenus Marl and laterally equivalent Turonian shale (both basin-floor facies in the deep basins) within an otherwise clean carbonate section supports this view.

9.3.7 The mature Chalk sea

Global sea levels continued to rise throughout the Santonian and into the Campanian, reaching a high stand maximum in the latest early Campanian. In the North Sea

Basin, the high stand is masked by a regional compressive event, resulting in reactivation of older inversion structures and, in some areas, the development of new axes. This event is well documented in the Central Trough (e.g. Brewster and Dangerfield, 1984) and in many onshore UK localities (e.g. Gale, 1980). It resulted in the rejuvenation and realignment of shorelines (with thin sands around residual landmasses) and modified the geometry of deeper marine chalk sub-basins. A number of early Campanian biozones are often absent and the Late Campanian is often highly condensed over inversion swells, suggesting considerable local erosion and redeposition in adjacent, deeper sub-basins.

Clear differentiation into large areas of both shelfal and basinal facies belts followed the Campanian event. Most of the more dramatic thickness variations of the Late Cretaceous occurred as the new topography was infilled during the Maastrichtian and Danian. Deeper subbasins were filled with considerable thicknesses of mass-flow and turbiditic chalk relative to thinner, more condensed sections on adjacent shelfal areas. Thick basinal chalks, often of remarkable purity, can be mapped from the Outer Moray Firth basins and throughout the UK, Norwegian and Danish Central Trough, with lesser occurrences in the northern Egersund Basin and North Denmark Basin.

Because of axial inversion, the Dutch Central Trough generally lacks basinal chalks. Such chalks, however, can be found to the east of the inversion axis on the Schill Grund to North Friesland area, to the north of the Broad Fourteens Basin and on a line from the north-eastern flank of the offshore West Netherlands Basin into the eastern part of the Sole Pit area. It should be stressed, however, that basinal areas in the Southern North Sea have less structural definition than those to the north, and many of the mass-flow deposits could be the product of slope changes in deep shelfal (rather than basinal) settings.

The maximum extent of the Chalk sea seems to have occurred in the early part of the late Maastrichtian. Throughout the late Campanian and early Maastrichtian, the increasing severity of tectonic shock from Alpine orogenesis appears to have triggered mass-flow basinal facies with ever-increasing stratigraphic and geographic distribution. A massive basin-wide compression in mid–late Maastrichtian times triggered the onset of extremely thick and regionally widespread chalk submarine-fan complexes in basinal areas. It also appears to have triggered the onset of retreat of the Cretaceous seas and rejuvenation of residual, but low-relief, landmasses adjacent to the basin.

The basinal carbonate depositional style is not found in all areas. In the very north of the Central Trough (Fisher Bank Basin) and throughout the entire Viking Graben area, the basinal facies consisted of mudstones and marls with only thin chalk beds throughout the Coniacian, Santonian and Campanian stages. Carbonate content generally decreases northwards. Shelfal chalks adjacent to these areas are also markedly impure with common shale and marl beds. Maastrichtian chalks are better developed, but also pass northwards to mud-prone facies in the South Viking Graben basins and to dirty shelfal facies adjacent to the North Viking Graben.

9.3.8 The death of the Chalk sea

At the end of the Maastrichtian, in areas of continual deposition, chalk deposition temporarily ceased, with the development of a condensed regional mudrock across the Cretaceous–Tertiary boundary. Coccolithoporid and foraminiferal production were severely curtailed, but became re-established in mid-Early Danian.

Although facies patterns were similar to those of the Maastrichtian, discussed above, the North Sea Basin was becoming progressively smaller. The reduction in depositional area is estimated at 70% of the late Maastrichtian. The rise of all landmasses around the basin resulted in clastic swamping of marginal areas. Most of the clastic wedges were removed during ongoing Tertiary inversion, but pockets within marginal chalk facies can be found around the northern areas of the Witch Ground Graben and parts of the Saxony Basin. Some unusual occurrences of thin greensands within chalk-dominated sections are scattered throughout the Central Trough area. One can only assume that a relative 'global' fall of sea levels, coupled with local inversion tectonics, resulted in older seamounts being subjected to renewed subaerial erosion. Notable examples of early Danian greensands occur near the Fladen Ground Spur in the South Viking Graben, in the area around the Renee Ridge and around the Auk High.

In deeper-marine basinal areas, widespread mass-flow deposition was re-established, particularly in the more rapidly subsiding central part of the North Sea Basin (UK and Norwegian Central Trough area). On adjacent basin margins, Maastrichtian chalks were commonly eroded and reworked into the Danian. Inundation by Tertiary clastics in about the mid-Danian finally extinguished carbonate depositional styles throughout any remaining chalk sub-basins.

9.4 Stratigraphy

Published North Sea Cretaceous stratigraphic data are dominated by lithostratigraphic schemes (see Fig. 9.2), which are often of considerable local detail. Although many attempts have been made to produce robust integrated sequence stratigraphies, these have only focused on proved hydrocarbon plays (Chalk of the Danish Sector and Early Cretaceous of the Outer Moray Firth Basins) or discrete onshore sections, and have never been extrapolated throughout the entire North Sea Basin.

9.4.1 Biostratigraphy

Most biostratigraphic schemes used for the North Sea Cretaceous are hybrids, which reflect the variabilities of both Boreal and Tethyan faunal provincialism. Some (older) schemes regroup the Ryazanian to Barremian into Neocomian and Berriasian. It is also common to see the Coniacian to Campanian (or even Coniacian to Maastrichtian) grouped together as the Senonian, and some literature still refers the Danian to the Lower Palaeocene.

Historically, macrofossil biostratigraphy has utilized ammonite and belemnite schemes in marine onshore sections of the Early Cretaceous, and bivalves and echinoderms in chalk sections. Foraminiferal zonation of the Late Cretaceous became popular in this century and, more recently, the coccolithophorids were integrated into zonation schemes. The latter two microfossil groups dominate Late Cretaceous offshore carbonate biostratigraphy (e.g. Hart, 1983; King *et al.*, 1989), and have been extended to date the marls and thin limestones of the Early Cretaceous (e.g. Birkelund *et al.*, 1983; Thomsen, 1987, 1989).

Duxbury (1977) produced the first valid palynofloral scheme, using dinoflagellate cysts, based on outcrop studies of the Speeton Clay Formation, Filey, East Yorkshire. This scheme was supplemented by further work on Volgian to Barremian dinoflagellate cysts by Davey (1979) and Duxbury (1980) and later on the Lower Greensand dinoflagellate cysts (Aptian to Early Albian) by Duxbury (1983). Together, these complementary schemes provided the basis for most Early Cretaceous palynofloral zonations defined offshore. It is only recently (e.g. Costa and Davey, 1992) that dinocyst palynology has been extended into the Late Cretaceous based mainly on exploration well data from the more argillaceous Shetland Group.

Ostracodes and miospores remain useful in calibrating Early Cretaceous non-marine lacustrine (Wealden) parts of the system (e.g. Anderson, 1985; Neale, 1962).

9.4.2 Lithostratigraphy

There is a tendency for Cretaceous lithostratigraphic terms to be widely accepted chronostratigraphically, which can create considerable confusion. Some lithological units are widespread over large, but not all, parts of the basin and appear to be stratigraphically restricted—for example, the Rødby, Tor and Ekofisk formations. On the other hand, units such as the Munk Marl, Fischschiefer and Plenus Marl are now regarded as isochrons and can be recognized throughout the entire North Sea Basin.

Figure 9.2 presents the lithostratigraphic scheme for the Cretaceous of the North Sea Basin. It is divided into three parts, one for each of the Southern (Fig. 9.2A), Central (Fig. 9.2B) and Northern (Fig. 9.2C) North Sea subareas. Relevant publications used in the compilation are listed in the caption. Key recent publications include the UKOOA/BGS series in the UK (Vol. 1, Knox and Holloway, 1992; Vol. 2, Johnson and Lott, 1993; Vol. 7, Lott and Knox, 1994); Isaksen and Tonstad (1989) for Norway; numerous DGU publications for Denmark (e.g. Frandsen *et al.*, 1987; Thomsen and Jensen, 1989; Nygaard *et al.* 1990); and van Adrichem Boogaert and Kouwe (1993) for the Netherlands.

On these diagrams, potential-reservoir sand and carbonate formations, members, beds or informal units are indicated. Apart from these units, the Early Cretaceous formation nomenclature is based on carbonate content, e.g. Sola shales, Rødby marls, etc. Late Cretaceous nomenclature is based on the content of chalk versus mudrock, e.g. Chalk Group and Shetland Group.

9.4.3 Other stratigraphic tools

Chemostratigraphy, in the early days of North Sea exploration, was an attractive tool in the analysis of apparently monotonous chalk sequences. Most of this work was pioneered in the Danish Sector by Jørgensen (1986a, b), initially on bulk elemental distributions. More recently, this work has been extended by McArthur *et al.* (1993), using strontium isotopes to determine an onshore UK Late Cretaceous chemostratigraphy.

Carbon-isotope fluctuations were investigated for use as a 'global' stratigraphic tool. Scholle and Arthur (1980) proved 'heavy' events near the Aptian/Albian and Cenomanian/Turonian boundaries—the latter being the widely recognized anoxic event of the Plenus Marl. 'Light' events were documented near the Jurassic/Cretaceous, Albian/Cenomanian, Turonian/Coniacian and Cretaceous/Tertiary boundaries. Subsequent work has tended to concentrate on these events and their detailed subdivision (e.g. Jensen and Buchardt, 1987; Gale *et al.*, 1993), using carbon and oxygen isotopes.

Volcanic ashes have also been utilized (Jeans *et al.*, 1977, 1982). These occur within the Late Aptian of the Weald Basin (Hamblin *et al.*, 1992) and Saxony Basin (Zimmerle, 1979) as montmorillonitic Fuller's earths, and can be traced in paralic and shallow shelfal tracts throughout the North Sea Basin.

9.4.4 Sequence stratigraphy

Figure 9.6 illustrates a new prototype sequence stratigraphy for the entire Cretaceous of the North Sea Basin. The scheme illustrated utilizes three integral approaches: the unconformity (sequence boundary) seismic-stratigraphy techniques of Vail *et al.* (1984), the maximum-flood (isochron) techniques, typified in Galloway (1989a), and the synchronous platewide stress models of, for example, Cloetingh (1988).

Two Cretaceous-specific problems have been carefully considered during the compilation of the scheme: firstly, whether it would be possible to apply the traditional techniques of unconformity-driven models and/or the marine-flood approach in compressive—rather than passive-margin or extensional—regimes; and, secondly, how to interpret Late Cretaceous chalks in a sequence-stratigraphic context. These are very different in depositional style when compared with traditional carbonate platforms and basins (e.g. Jacquin *et al.*, 1991; Handford and Loucks, 1993).

The scheme illustrated includes a few assumptions to provide a basic framework within which seismic and well data are integrated: firstly, that global rises in sea level during the Cretaceous are approximately valid in a bulk sense; secondly, that major anoxic events were, or were closely approximated to, isochrons; thirdly, that such anoxic events represented the peaks (maxima) of marine floods; and fourthly, that tectonic events were synchronous (if not always manifest) over the entire North Sea Basin. In practice, the scheme has proved remarkably reliable in linking often disparate stratigraphies and facies belts within and between different sub-basins. The assumptions are therefore considered to be valid, but do require further testing and discussion, which are beyond the scope of this chapter.

9.5 The Early Cretaceous of the Southern North Sea

The Southern North Sea Basin is defined palaeogeographically as the area between the London–Brabant–Rhenish Massif in the south to the Mid North Sea–Ringkøbing–Fyn High in the north. Although somewhat better known as the Permo-Triassic Gas Basin, it encompasses a series of sub-basins in which Early Cretaceous sands are known to be widespread (Figs 9.7 and 9.15).

Published accounts of discoveries are restricted to the onshore areas of the Netherlands Concession (Figs 9.1 and 9.7). They include the Rijswijk complex around The Hague in the onshore West Netherlands Basin (Bodenhausen and Ott, 1981), the Schoonebeek oilfield on the Dutch–German border in the West Saxony Basin (Troost, 1981) and the Zuidwal Gasfield in the Vlieland Basin (Perrot and van der Poel, 1987). Cottençon *et al.* (1975) provide early accounts of other smaller Cretaceous gasfields.

All sub-basins share a common Mesozoic and Tertiary history. Most are underlain by Zechstein halites and potential Carboniferous gas-prone source rocks. All demonstrated a transtensional character during the Jurassic. Most were aligned north-west–south-east to west-north-west–east-south-east. Narrow fault-bounded troughs and broad adjacent highs were formed by extension and associated severe oblique slip (van Wijhe, 1987a).

All basins demonstrate significant Early Cretaceous inversion—the first pulse in the middle of the Ryazanian, with a second regionally significant event at the Valanginian/Hauterivian boundary. Subsequent compressional events occurred at the Barremian/Aptian and the Aptian/Albian boundaries. Each of these events resulted in the development of major packages of reservoir-quality sands, although not yet found in every sub-basin. Compressive pulses continued throughout the Late Cretaceous and into the Tertiary. Tertiary inversions are responsible for most of the structural closure of Early Cretaceous sand reservoirs, although some reservoirs, particularly Valanginian sands, were structurally closed and stratigraphically sealed before Apto-Albian inversions.

Many publications describe the structural evolution and seismic interpretation of the area. For example, van Wijhe (1987) and Hooper *et al.* (1995) discuss the Broad Fourteens Basin, Herngreen *et al.* (1992) the Vlieland Basin and Heybroek (1975) and Clarke-Lowes *et al.* (1987) the Dutch Central Graben and surrounding shelves. The UK sector is featured in Cameron *et al.* (1992) and, to a lesser extent, in Chatwin (1961), Casey and Gallois (1973) and Kent *et al.* (1980). Kemper (1974, 1982) and Betz *et al.* (1987) also provide similar data on the Saxony Basin in Germany.

Although the Cretaceous of the Southern North Sea is presumed to be gas-prone (from Carboniferous source rocks), oil is encountered in many structures. The source rocks for oil are the Posidonia (late Liassic) shales, which are preserved in the West Netherlands and Broad Fourteens basins and the Dutch Central Graben offshore.

Most known reservoirs occur at two horizons—within Vlieland Sandstones (Valanginian) and Rijswijk Sandstones (late Hauterivian)—although not necessarily in every sub-basin. Scattered occurrences of sands predicted by sequence-stratigraphic modelling can be seen in some sub-basins, but are commonly thin and of poor reservoir quality.

9.5.1 West Netherlands Basin

Cretaceous deposition in the onshore part of the West Netherlands Basin did not commence until Late Hauterivian times. Figure 9.8 illustrates a typical well section in the thicker parts of the subbasin. This section shows sand-body development at many different stratigraphic horizons. The Late Hauterivian Rijswijk Sandstone and Late Aptian Holland Greensand are of sufficient thickness, areal extent and reservoir quality to be considered commercially significant.

Sand bodies were deposited in a range of marginal marine settings (Bodenhausen and Ott, 1981; den Hartog Jager, 1996). Basal transgressive sands form the Rijswijk Sandstones, barrier bars and tidal deltas make up the laterally discontinuous Berkel and De Lier sandstones, and shallow-shelf to shoreface progradational high-stand wedges are represented by the Holland Greensand. Intervening mudrocks are usually fully marine in character, with anoxic shelfal marls developed at numerous points. The latter indicate maximum floods within the basin. Incision events into progradational sand packages are apparent in Fig. 9.8.

The Texel Greensand is a progradational high-stand sandstone wedge developed in the Cenomanian. This demonstrates the extension of a typical Early Cretaceous lithofacies into the Late Cretaceous, a common feature in the Southern North Sea. The greensand represents the landwards equivalent of shelfal Texel Chalks (see Fig. 9.2A). The sands are difficult to date precisely, but appear to represent a high-stand event in the Cenomanian or Turonian. In the onshore part of the sub-basin, this sand has produced gas.

Most Early Cretaceous sand bodies appear to have prograded northwards from the adjacent London–Brabant Massif. This was a sediment source area for most of the Early Cretaceous. A second sediment source area, the Ijmuiden High to the north (see Fig. 9.7), also contributed coarse clastics during Hauterivian and Barremian times, but became submerged during the Aptian.

Figure 9.9 illustrates the trapping style of the above sand bodies in the West Netherlands Basin. Onlap of the London–Brabant massif to the south and subsequent inversion created a series of stratigraphic onlap pinch-out traps. Complex inversion tectonics resulted in four-way closure over compressional pop-up structures and up-dip closure against reverse faults. Pure stratigraphic traps are also apparent in barrier-bar sand bodies. Although inversion tectonics facilitated most trapping styles, trap integrity may not have been retained as inversions progressed. The northern flank of the basin commonly lacks a thick seal, because of pre-Chalk inversion and lack of Tertiary cover; most footwall traps in this area have lost previously reservoired hydrocarbons (Bodenhausen and Ott, 1981).

Most of the hydrocarbon accumulations in the West Netherlands Basin are of oil, sourced from the Lower Jurassic Posidonia Shale. Bodenhausen and Ott (1981) report source-rock thickness from 20 m to 50 m and postu-

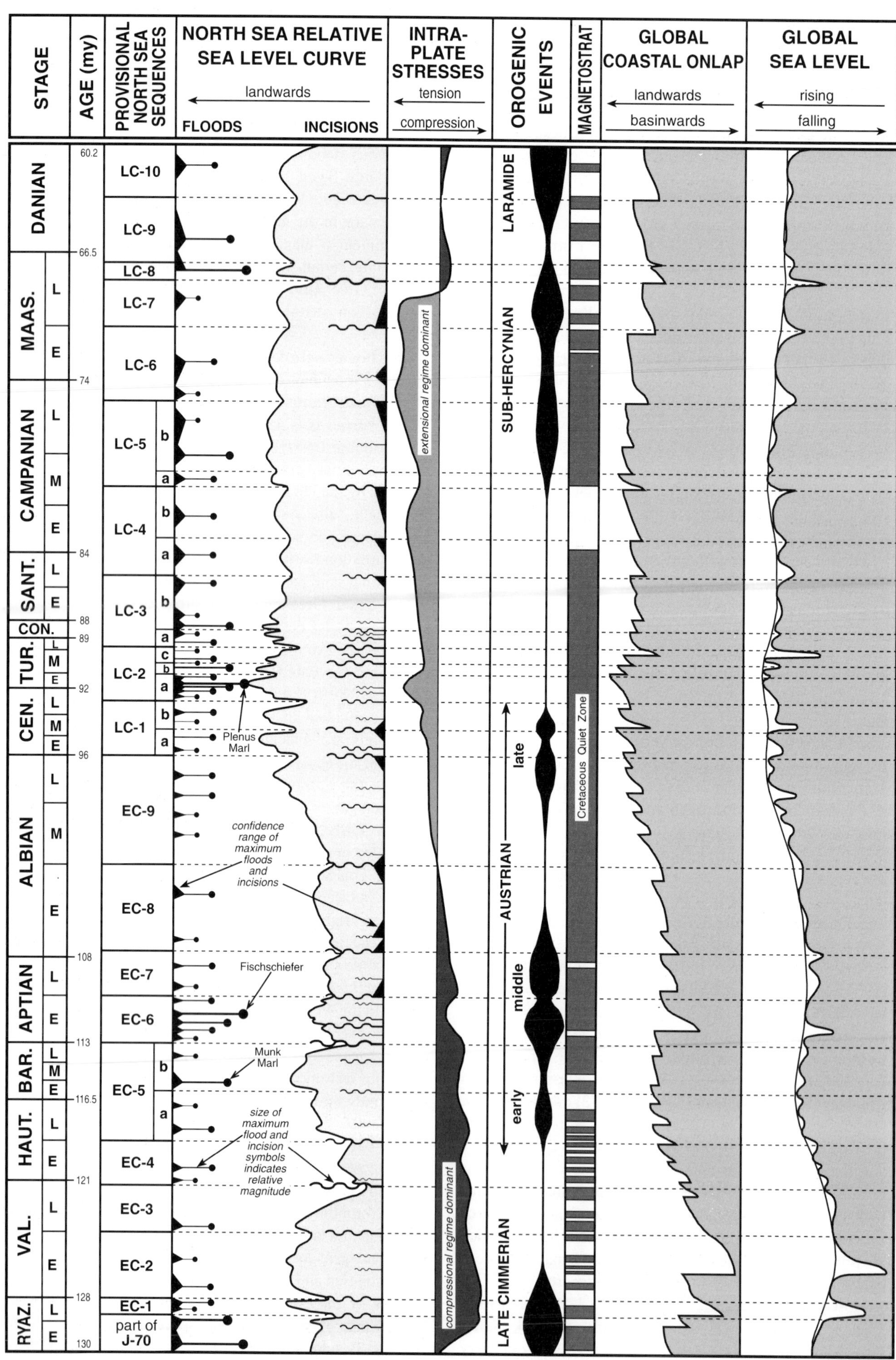
STAGE
AGE (my)
PROVISIONAL NORTH SEA SEQUENCES
NORTH SEA RELATIVE SEA LEVEL CURVE
landwards
FLOODS
INCISIONS
INTRA-PLATE STRESSES
tension
compression
OROGENIC EVENTS
MAGNETOSTRAT
GLOBAL COASTAL ONLAP
landwards
basinwards
GLOBAL SEA LEVEL
rising
falling
DANIAN
MAAS.
CAMPANIAN
SANT.
CON.
TUR.
CEN.
ALBIAN
APTIAN
BAR.
HAUT.
VAL.
RYAZ.
60.2
66.5
74
84
88
89
92
96
108
113
116.5
121
128
130
LC-10
LC-9
LC-8
LC-7
LC-6
LC-5
LC-4
LC-3
LC-2
LC-1
EC-9
EC-8
EC-7
EC-6
EC-5
EC-4
EC-3
EC-2
EC-1
part of J-70
Plenus Marl
confidence range of maximum floods and incisions
Fischschiefer
Munk Marl
size of maximum flood and incision symbols indicates relative magnitude
extensional regime dominant
compressional regime dominant
LARAMIDE
SUB-HERCYNIAN
AUSTRIAN
late
middle
early
LATE CIMMERIAN
Cretaceous Quiet Zone

lated that sufficient maturity was reached at depths of around 2500–3000 m. This occurred in the deeper southern parts of the basin during the latest Early Cretaceous to Late Cretaceous, and again during the Late Tertiary, with loss of hydrocarbons during early Tertiary inversions. The lesser gas accumulations are assumed to have been generated during a second maturity window in the Cenozoic.

The Rijswijk play continues into the offshore parts of the West Netherlands Basin but is separated from UK waters (where the Posidonia Shale is believed to be immature) by the Winterton High.

9.5.2 Vlieland and West (Lower) Saxony basins

The Vlieland and West Saxony basins are located in the northern onshore area of the Netherlands (see Fig. 9.7). Both basins show almost identical Early Cretaceous stratigraphic and structural style (Figs 9.10 and 9.11).

Herngreen *et al.* (1992) proposed that the Vlieland Basin was initially a small transtensional pull-apart basin developed during the Callovian in response to east–west extension in the Dutch Central Trough and Variscide-controlled transtensional slip in the long-established Saxony Basin. Many smaller transtensional features developed during the Late Jurassic, from mini-half-grabens to small eroded pop-up highs and deep compressional troughs.

At the onset of the Early Cretaceous, accommodation space was full. Ryazanian compressions, however, reactivated old Jurassic faults and added transpressional features, to create a new suite of complex mini-basins and

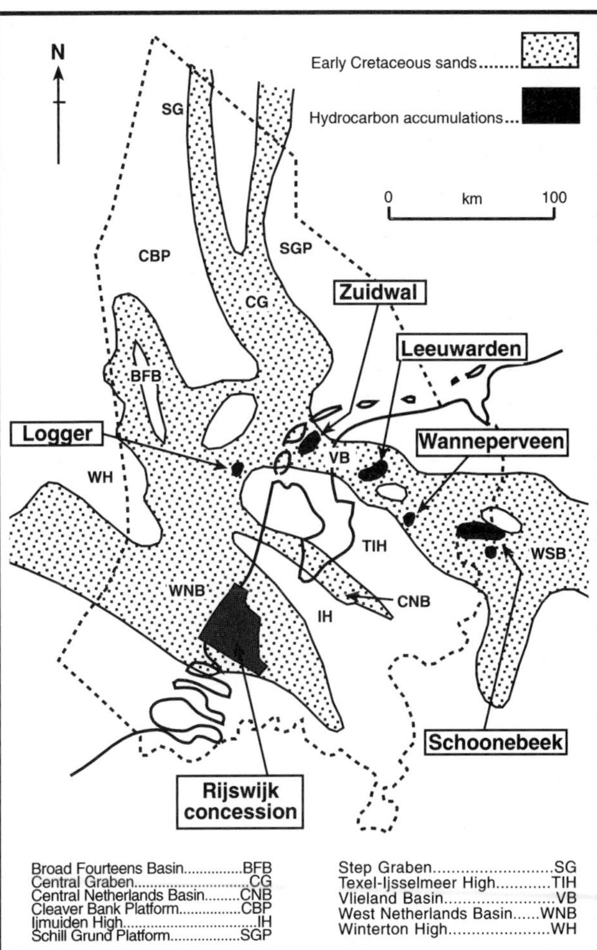

Fig. 9.7 The Netherlands—Early Cretaceous sand distribution and field locations (modified from Perrot and van der Poel, 1987).

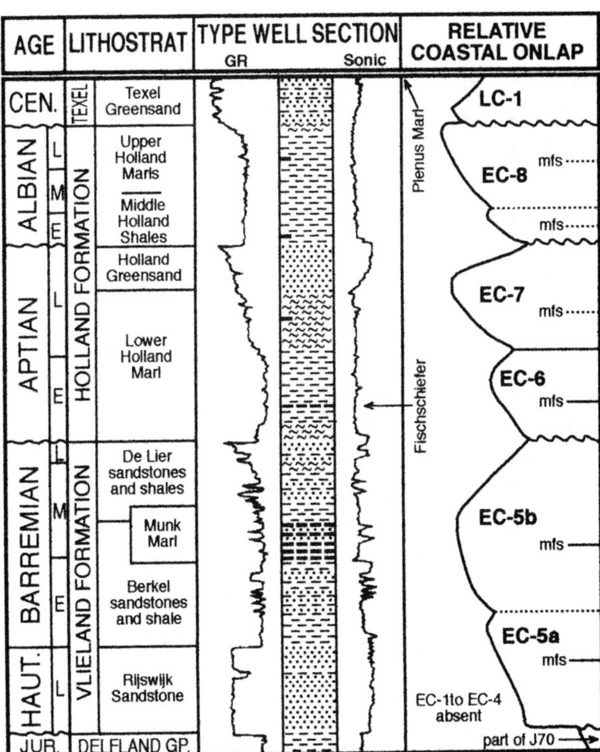

Fig. 9.8 West Netherlands Basin—schematic well section and sequence stratigraphy (modified from data by permission of British Petroleum). Note: arbitrary vertical scale.

Fig. 9.6 (*Opposite*) Generalized Cretaceous sequence stratigraphy (compiled and modified from data by permission of British Petroleum; Cloetingh *et al.*, 1987; Haq *et al.*, 1988; Ziegler, 1990).

Data sources for sequence boundaries (unconformity, incision, condensation events, missing biozones): Andrews *et al.*, 1990; Berridge & Pattinson, 1994; Bisewski *et al.*, 1990; Bodenhausen & Ott, 1981; Brewster & Dangerfield, 1984; Boote & Gustav, 1987; Burnhill & Ramsay, 1981; Cameron *et al.*, 1992; Crittenden *et al.*, 1991; D'Heur, 1986; D'Heur *et al.*, 1985; Eyers, 1991; Frandsen *et al.*, 1987; Gale, 1980; Gallois, 1965; Gowers & Sæbøe, 1985; Guy, 1992; Hamblin *et al.*, 1992; Hardman & Eynon, 1978; Herngreen *et al.* 1992; Hesselbo *et al.*, 1990; Hesselbo & Allen, 1991; Hastings, 1986, 1987; Hatton, 1986; Hooper *et al.*, 1995; Isaksen & Tonstad, 1989; Ineson, 1993; Kent, 1980; Larsen, 1987; Leonard & Munns, 1987; McGann *et al.*, 1988; Munns, 1985; Mortimore & Pomerol, 1991; Norbury, 1987; Nygaard *et al.*, 1990; Ruffell, 1991, 1992; Ruffell & Wach, 1991; Riley *et al.*, 1992; Perrot & van der Poel, 1987; Quine & Bosence, 1991; Svendsen, 1979; Shanmugan *et al.*, 1994 & 1995; van Wijhe, 1987a,b; Vejbæk, 1986; van den Bosch, 1983.

Data sources for maximum flood surfaces: Andrews *et al.*, 1990; Berridge & Pattinson, 1994; Bisewski *et al.*, 1990; Crittenden *et al.*, 1991; D'Heur, 1986; D'Heur *et al.*, 1985; Hesselbo *et al.*, 1990; Hesselbo & Allen, 1991; Hancock & Kauffmann, 1979; Harker & Chermak, 1992; Ineson, 1993; Jenkyns, 1980; Kennedy, 1987; Kent, 1980; Leonard & Munns, 1987; Munns, 1985; Norbury, 1987; Nygaard *et al.*, 1990; Pekot & Gersib, 1987; Ruffell & Wach, 1991; Ruffell, 1992; Skovbro, 1983; van den Bosch, 1983.

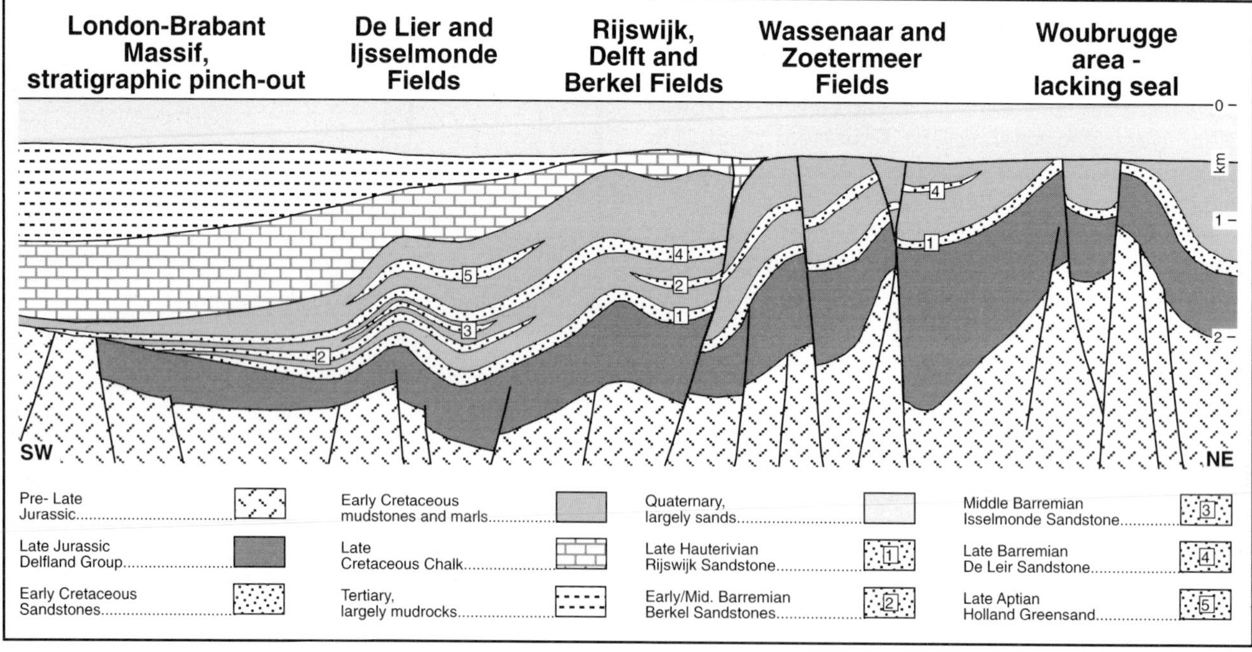

Fig. 9.9 Structural cross-section of the onshore West Netherlands Basin (modified from Bodenhausen and Ott, 1981).

local highs. This became the template for sedimentation at the onset of the Valanginian transgression. From this point on, the structural and stratigraphic style in both the Vlieland and West Saxony basins followed the same route.

In both basins, widespread sands were developed in the latest Ryazanian and throughout the Valanginian. These are variably referred to as Vlieland and Bentheim sandstones (Figs 9.10 and 9.11). The initial sequence, of latest Ryazanian age, contains a basal flood shale followed by thin, laterally discontinuous, shelf and shoreface high-stand sands. After a minor incision, ongoing transgression resulted in widespread deposition of sheet-like shallow-marine and shoreface sands. These overstepped the latest Ryazanian and the Jurassic Delfland Group. A maximum flood surface is seen in offshore sections, although, over most areas, marine shelfal sand became well established as a regionally correlative high-stand tract. A major erosive event at the early/late Valanginian boundary was accompanied by a regional basinwards shift of facies belts, and late Valanginian sands were deposited in the axial areas of evolving mini-basins. Regional inversions then occurred, which resulted in major truncation of pre-Hauterivian sands. The following early Hauterivian trangressive systems tract can be seen in some, but not all, fields.

A series of Early Cretaceous gasfields (principal source rock in the Carboniferous) are currently in production. The largest, Zuidwal, contains gas initially in place (GIIP) of 26.1×10^9 m^3 in Valanginian Vlieland Sandstones. Effective reservoir porosities are around 20% and permeabilities in the range 20–50 millidarcies (mD). Flow rates have been quoted as 716 000 m^3/day (25 mm choke) and 346 050 m^3/day (19 mm choke). Recovery factors are believed to be about 62%. Other fields with production from these sands include Leeuwarden and Wanneperveen (see Fig. 9.7).

In the West Saxony Basin, Valanginian reservoirs form the Schoonebeek Field. Unlike the gasfields of the Vlieland Basin, this contains stock-tank oil initially in place (STOIIP) of 170×10^6 m^3 of 25° API oil (Troost, 1981). The recovery factor of such heavy crude is poor, varying from 6 to 18% in different parts of the structure, from sands with porosities of up to 32%. The source of heavy hydrocarbons has been identified as the hot flood shales and marls within the Early Cretaceous—most notably, the Munk Marl and Fischschiefer. Early-maturity oils migrated from deeper areas of the basin adjacent to the field via up-dip and fault-transfer mechanisms.

9.5.3 Broad Fourteens Basin and Dutch Central Trough

Little attention has been given to the Early Cretaceous of these areas. Drilling histories of both basins reveal mud-prone offshore facies, or their absence over the inversion axes. Inversions in the Broad Fourteens Basin and the Dutch Central Graben occurred during the Late Cimmerian (Ryazanian) and again during the Austrian (Apto-Albian) event. An excellent account of these and subsequent trap-creation inversions, along with detailed truncation maps and discussions of halokinesis, can be found in Clark-Lowes *et al.* (1987).

The Broad Fourteens Basin is exclusively offshore. Largely in Dutch waters, it extends just into the UK sector (see Fig. 9.7). The basin demonstrates a Variscide-aligned transtensional character during the Jurassic. Northern areas are parallel with and merge into north–south-aligned Jurassic basins of the Dutch Central Trough trend, which includes the Dutch Central Graben, Cleaver Bank High, Step Graben and Schill Grund High.

Narrow strips of Early Cretaceous sediments remained in

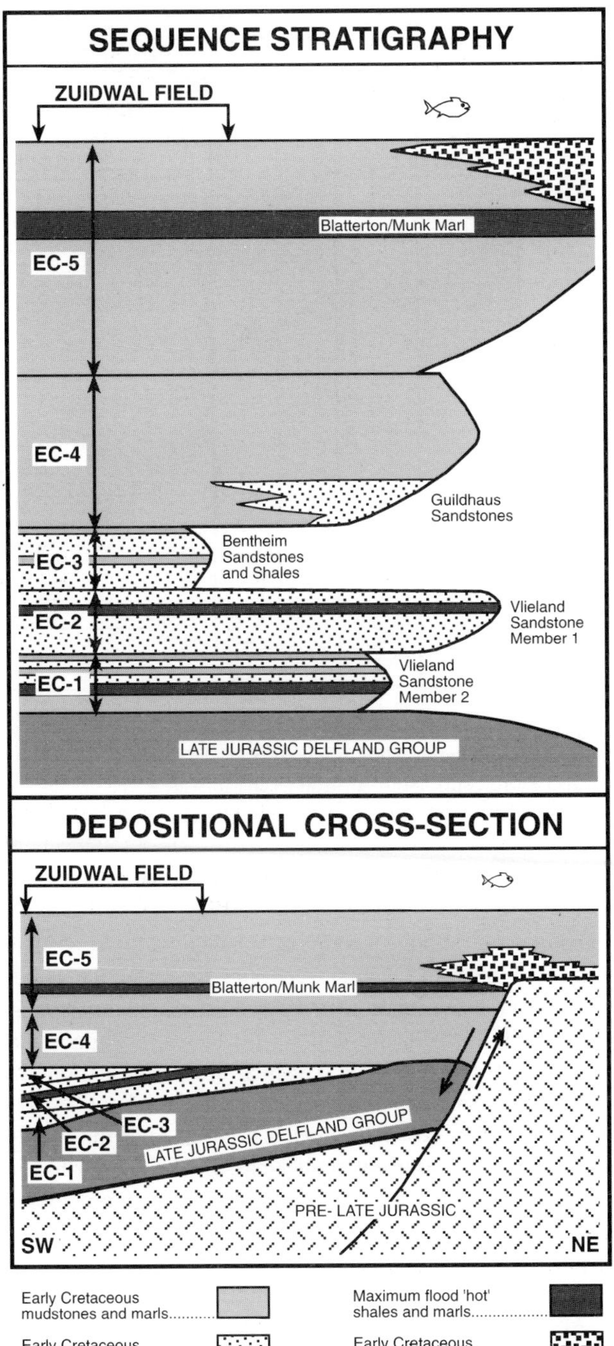

Fig. 9.10 Early Cretaceous sequence stratigraphy and depositional geometry of the Vlieland Basin (modified from data by permission of British Petroleum).

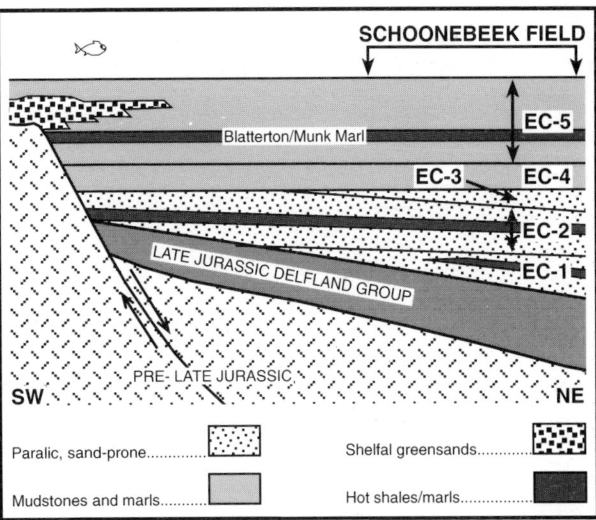

Fig. 9.11 Depositional geometry and sequence stratigraphy of the West Saxony Basin (modified from data by permission of British Petroleum).

structural lows located between axial inverted areas and sites of relic Jurassic basin-bounding faults. In these strips, it is common to see potential diapiric flank traps and small compressional pop-up structures. In addition, stratigraphic pinch-out and associated ripple folding, generated by the more subtle Late Valanginian/Early Hauterivian event (previously unrecognized in this sub-basin), formed potential traps before the Austrian event occurred.

Figure 9.12 illustrates the evolution of the lower part of the Early Cretaceous in an area of the Broad Fourteens Basin that remained unstripped by later inversions. This model can be applied to many parts of the Dutch Central Graben.

Major sand development in the latest Ryazanian and early Valanginian, equivalent to the Vlieland Sandstones of the onshore basins, can be seen in a few wells around the flanks of the sub-basins (Fig. 9.13). Such sections prove the presence of an incised high-stand sand of early Valanginian age. In addition, Fig. 9.13 also illustrates possible slope-facies sand pinch-outs beneath the Munk Marl flood in the early Barremian and the regional incision at the base of the Aptian. The Holland Greensand, of late Aptian age, is also present.

The Logger Field (Goh, 1996), located in Block L16a, is one of the few published examples of an offshore Netherlands Early Cretaceous oil reservoir. It occurs within a complex overthrust anticline, located in the structurally complex junction area of the Central Netherlands, Broad Fourteens and Vlieland basins. The reservoir is within a thin (10–30 m) barrier-bar complex of the Vlieland Sandstone. Oil in place is estimated at 51×10^6 barrels.

9.5.4 Sole Pit area of the United Kingdom offshore

The tectonic history of the Sole Pit area is similar to that of the Dutch offshore basins discussed above, with the exception that both Austrian and pre-Tertiary inversions cut considerably deeper into the axial areas of the older Jurassic and Early Cretaceous trough. The inversion-flank model presented above for the Early Cretaceous of the Broad Fourteens Basin is believed to be directly applicable to the Sole Pit area.

Well drilling has not proved sands to the north-east of the inversion axis, and the Early Cretaceous is represented by only a condensed maximum-flood carbonate facies of late Albian age. Inversions seem to affect not only the axial areas of the old Jurassic Sole Pit Basin, but also the entire eastern flank (Fig. 9.14).

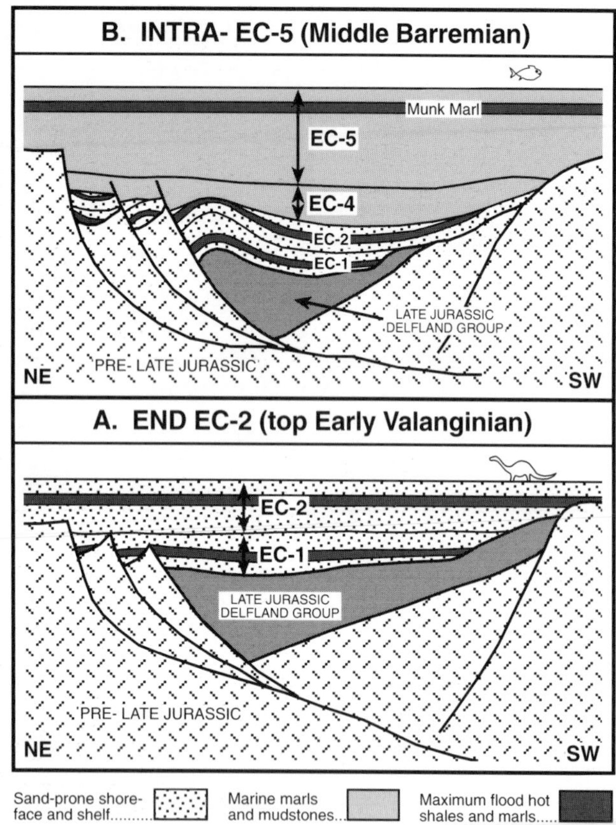

Fig. 9.12 Tectono-stratigraphic evolution of the Early Cretaceous in the Broad Fourteens Basin (modified from data by permission of British Petroleum).

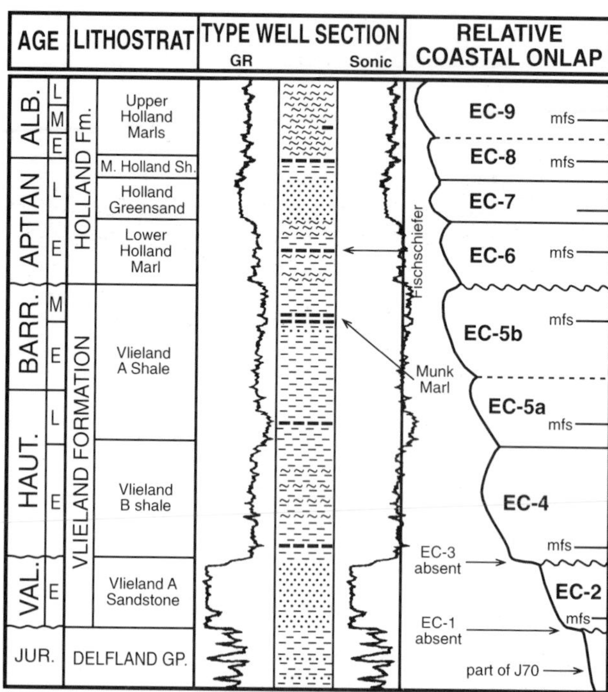

Fig. 9.13 Eastern Broad Fourteens Basin—schematic well section and sequence stratigraphy (modified from data by permission of British Petroleum). Note: arbitrary vertical scale.

The south-western side of the Sole Pit, between the inversion axis and the Dowsing Fault zone, contains Early Cretaceous sands (Fig. 9.15). These dominate the latest Ryazanian and early Valanginian. They are referred to as the Upper Spilsby Sandstone and/or upper Sandringham Sandstones (see Fig. 9.2A). Both sands have 'lower' units, which also occur in the Late Jurassic (late Volgian to early late Ryazanian). They are shelfal 'greensands' and are the UK equivalent of the Vlieland Sandstones. Excellent onshore UK details of these and other Early Cretaceous facies can be found in Gallois (1994). Additional sands have been encountered throughout the Hauterivian and Barremian onshore UK, but their extent offshore is uncertain.

9.6 The Early Cretaceous of the Central and Northern North Sea area

The area of the Central and Northern North Sea lies to the north of the Mid North Sea–Ringkøbing–Fyn High and to the south of the North Atlantic margin at 62°N. All hydrocarbons within the area are sourced from Jurassic shales (with the exception of a minor Devono-Carboniferous source in the Inner Moray Firth), with possible contributions from thin Early Cretaceous source rocks.

The Central North Sea includes a series of depositional basins within the dominant Central Trough (UK, Norway and Denmark). Important marginal areas include the Western Platform and the Norwegian–Danish Basins. In the north-west of the area, the Moray Firth Basin similarly encompasses a large group of smaller sub-basins. The Central Trough in Denmark contains the Valdemar Field, an unusual accumulation in Early Cretaceous limestones. The Moray Firth Basins contain the Scapa, Britannia and Captain fields (see Fig. 9.1), which are accumulations in Early Cretaceous sands.

The Northern North Sea area is dominated by the Viking Graben and its adjacent shelf areas. In the far north, the Tampen Spur Area and the Magnus Embayment are important Early Cretaceous depositional areas. Only one Cretaceous field is known from the area, Agat, on the north-eastern flank of the Viking Graben in Norway.

Apart from descriptions of the above fields and the heavily appraised Danish Sector, publications on the Early Cretaceous are rare. British Geological Survey regional reports provide excellent, though lithostratigraphically-dominated, accounts (Andrews *et al.*, 1990, for the Moray Firth; Gatliff *et al.*, 1994, for the UK Central North Sea; and Johnson *et al.*, 1993, for the UK Northern North Sea). Other key stratigraphic papers include Crittenden *et al.* (1991), on the UK Central North Sea; Deegan and Scull's (1977) regional publication; Isaksen and Tonstad (1989) on the Norwegian Sector; Harker *et al.*, (1987) in the Witch Ground Graben; Lott *et al.* (1985) on the Western Terraces of the UK Central Trough; and UKOOA lithostratigraphic nomenclature and definitions by Johnson and Lott (1993). A few publications are available on regional tectono-stratigraphic evolution, for example, Boote and Gustav (1987) and O'Driscoll *et al.* (1990) in the Witch Ground Graben; Hesjedal and Hamar (1983) and Gowers and Sæbøe (1985) in the southern Norway Sector.

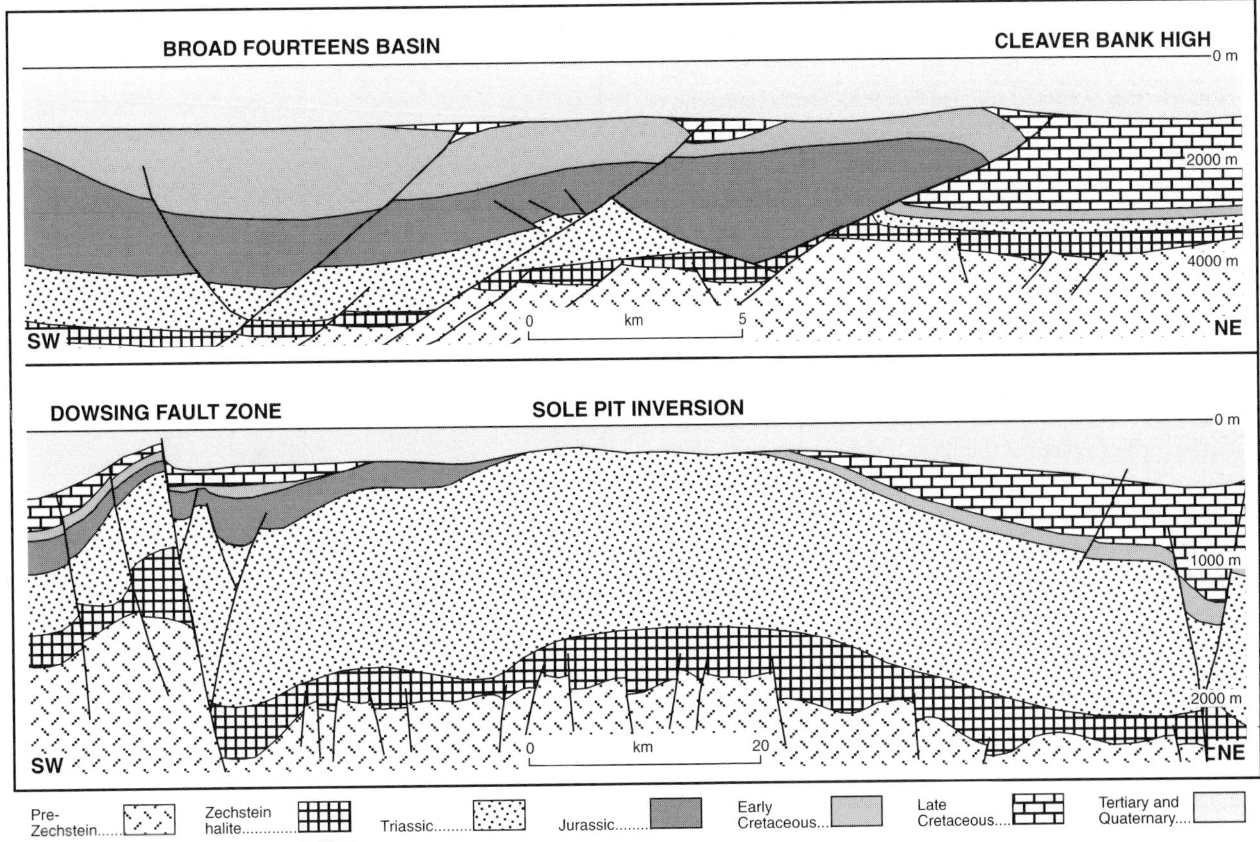

Fig. 9.14 Cartoon geoseismic sections of selected Southern North Sea Early Cretaceous basins (lower section from Hancock, 1990; upper section from Hooper *et al.*, 1995).

9.6.1 Tectono-sedimentological models in the Central North Sea

Figure 9.16 illustrates the depositional sub-basins of the Early Cretaceous in the Outer Moray Firth and Central Trough area. The bottom map shows the 'lower' Early Cretaceous (latest Ryazanian to top Early Hauterivian); the top map is 'upper' Early Cretaceous (Late Hauterivian to Late Albian). The 'global' effect of Early Cretaceous sea-level rise is very clearly demonstrated by these two maps. The former shows relatively tightly constrained sub-basin trends with large adjacent landmass areas; the latter shows the flooding of large amounts of platform as the transgression progressed.

'Lower' Early Cretaceous sub-basins were tightly constrained on a regional scale, apparently by residual Late Cimmerian topography. Regional sub-basin trends occur along sites of older, underfilled Late Jurassic grabens. Most sub-basins accommodated over 600 m of sediment and many contained over 1500 m. The latter are extremely limited geographically. Almost all of the sub-basins are grossly underfilled. Shelfal greensand facies are absent except on some sub-basin margins. The rarity of shelfal tracts and the complete absence of nearshore and paralic facies was probably due to continual stripping of intra- and interbasinal highs. Basin-floor areas remained consistently below storm wavebase.

Many sub-basins are bounded on one side by a major fault, and can appear as apparent half-grabens. Other sub-basins have no distinct half-graben geometry and are often bounded on all sides by apparent monoclines or terraced fault complexes, often too small to resolve without 3D seismic grids. On a local scale, many bounding faults, if traceable into the Jurassic, are reversed. Thus, in some

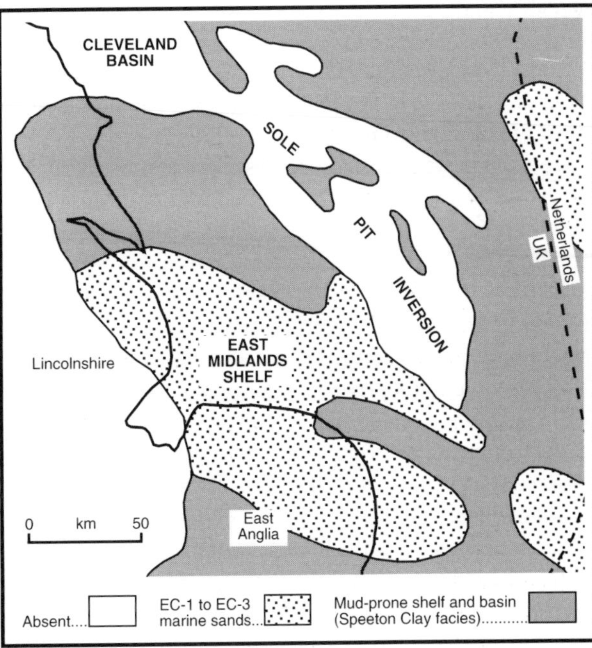

Fig. 9.15 Distribution of Early Cretaceous sands in the Sole Pit Area (modified from Cameron *et al.*, 1992).

cases, thick Jurassic hangingwall sections became heavily incised Early Cretaceous footwalls.

Two compressional fold trends are evident on the isochore maps (Fig. 9.16 and see also Fig. 9.3). One axial trend is orientated approximately west-south-west–east-north-east. This appears to be a latest Ryazanian/earliest Valanginian grain. The second trend appears to be west-north-west–east-south-east, and this became dominant in latest Valanginian/earliest Hauterivian; it is heavily masked by the older north-west–south-east Jurassic trends, particularly in the Central Trough.

Where anticlinal trends cross old Jurassic basins, complex bridges occurred between Early Cretaceous sub-basins to provide rises, seamounts and, at extremes, island highs. Interference of synclinal trends, often superimposed on the residual Late Jurassic template, produced preferential elongation of Early Cretaceous sub-basins.

Complex transpressional faulting was coeval with folding in many sub-basins. In areas where folding is oblique with respect to Jurassic grains, transpressional faults are extremely common and serve to exaggerate one or more of the margins of each Early Cretaceous sub-basin. Small thrust faults are not uncommon. At the many restraining or confining bends formed by the overall transpressional regime, local pop-up structures and overdeepened areas are commonly present. Each sub-basin should be considered as a large-scale extensional duplex (negative, or normal, flower structure) and each high or bridge area as a contractional duplex (positive, or reverse, flower structure).

Additional models are needed to explain the many halokinetic features observed. Salt is not present in the Moray Firth basins. In this area, Early Cretaceous sub-basins are commonly geographically larger and contain lesser sediment thicknesses than those over salt in the Central Trough. Folding is similarly more gentle and transpressional faults all have significantly lower displacements. The presence of salt has clearly resulted in a decrease of syndepositional structural wavelength, with corresponding increase in amplitude. It is likely that this was a response to décollement during compressional pulses. Central Trough Early Cretaceous sub-basins are thus considerably more complex than those in the Moray Firth and most of the Viking Graben.

Tectonism was considerably less active in the latter half of the Early Cretaceous (Fig. 9.16, top map). Most of the activity and therefore thickness variations were generally confined to the Aptian (Austrian tectonism). All basinal areas remained grossly underfilled. Flooded platform and island areas, although often showing tracts of greensand, rarely built out to form distinct shorefaces and therefore also remained underfilled. Transpression-related halokinetic activity was similarly reduced; although Central Trough sub-basins were still considerably smaller than those in the Moray Firth, sediment accumulation rates were virtually identical.

Sequence stratigraphic models for an area with complex tectonics and halokinetics of the Western Central Trough are illustrated in Fig. 9.17. These diagrams clearly show nine Early Cretaceous sequences, each with its own regional maximum flood ties and appropriate seismic sequence boundary. The complexity of sub-basins and intervening

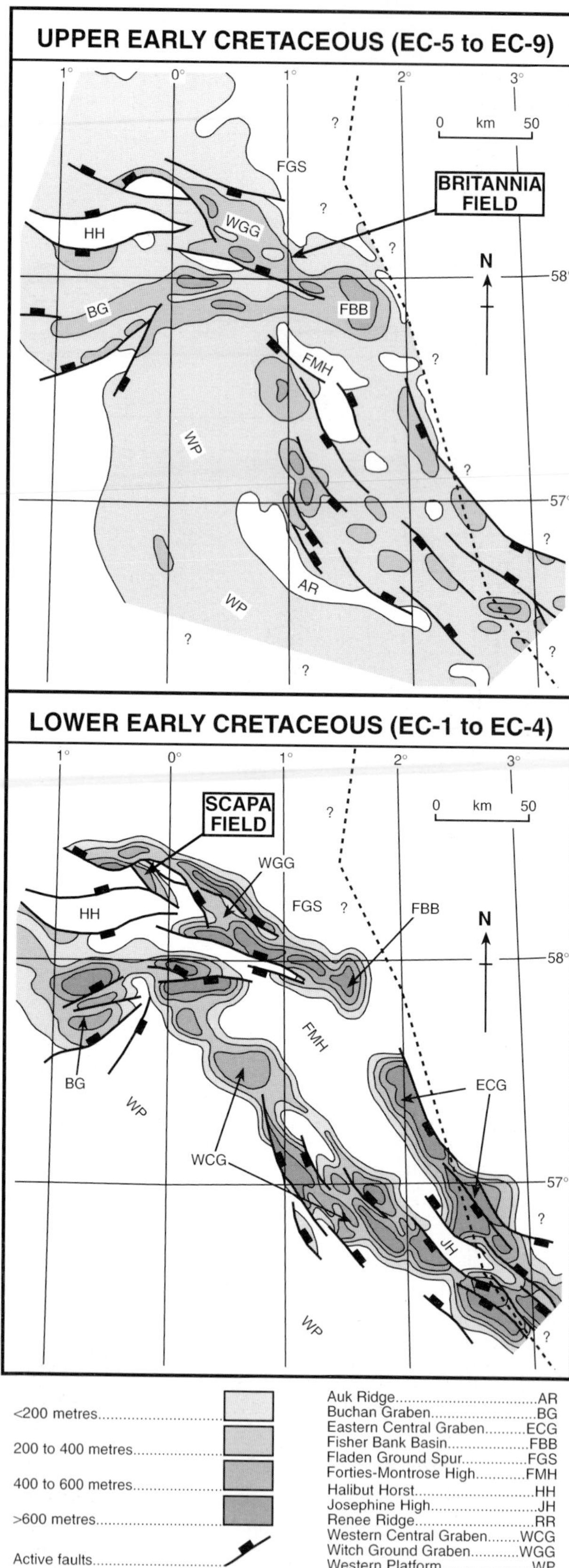

Fig. 9.16 Early Cretaceous sub-basins, isochores and principal active faults: Outer Moray Firth and Central Trough (modified from data by permission of British Petroleum).

highs is clearly illustrated by their variable spatial and stratigraphic distribution. The effect of two pulses of 'global' sea-level rise can also be seen—note the spread of

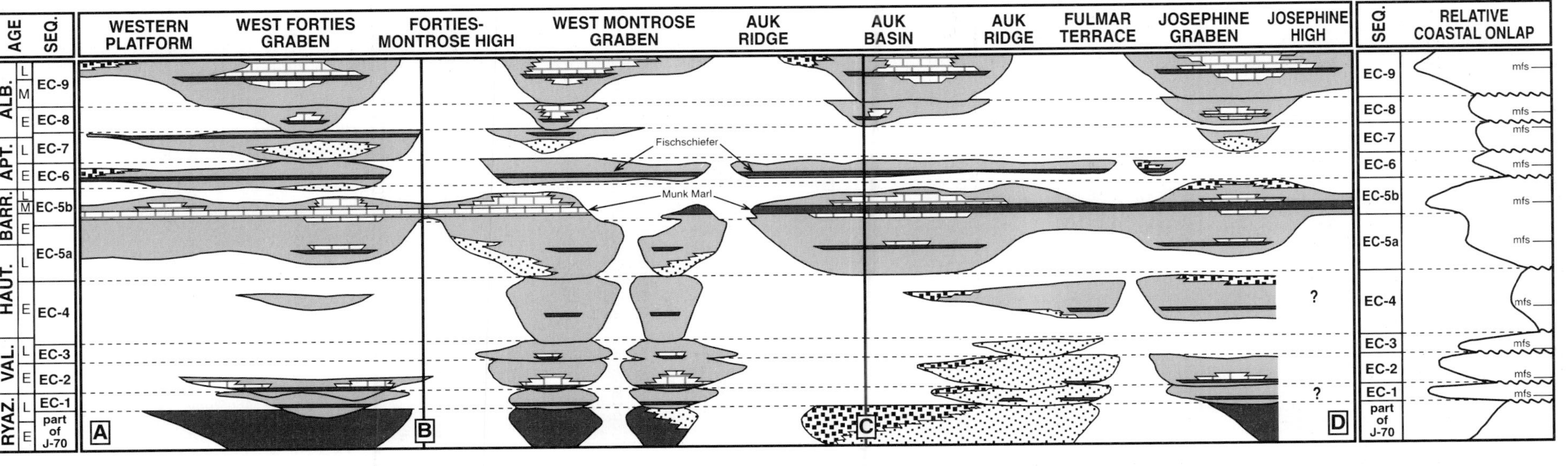

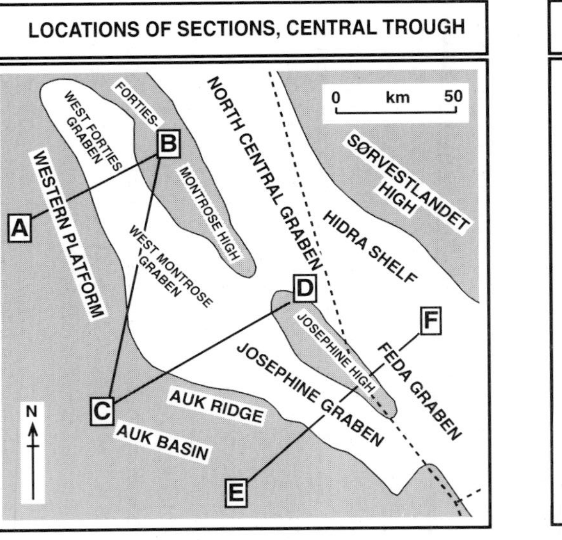

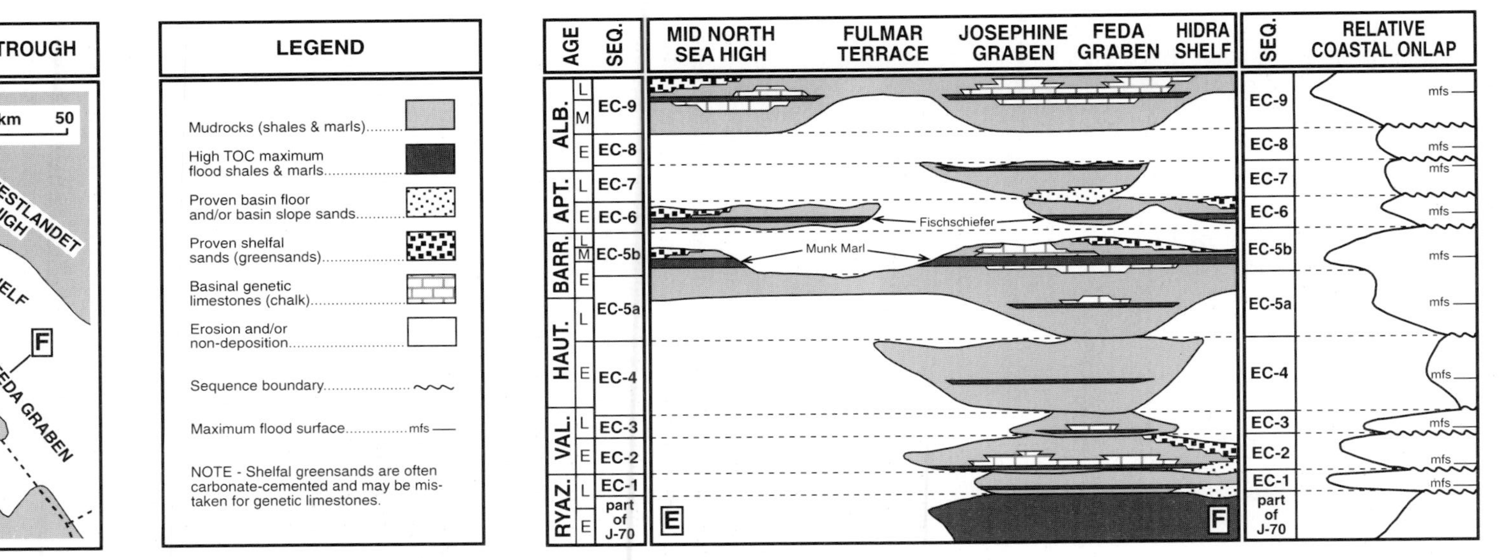

Fig. 9.17 Chronostratigraphic distribution of Early Cretaceous sequences, Central Trough (modified from data by permission of British Petroleum).

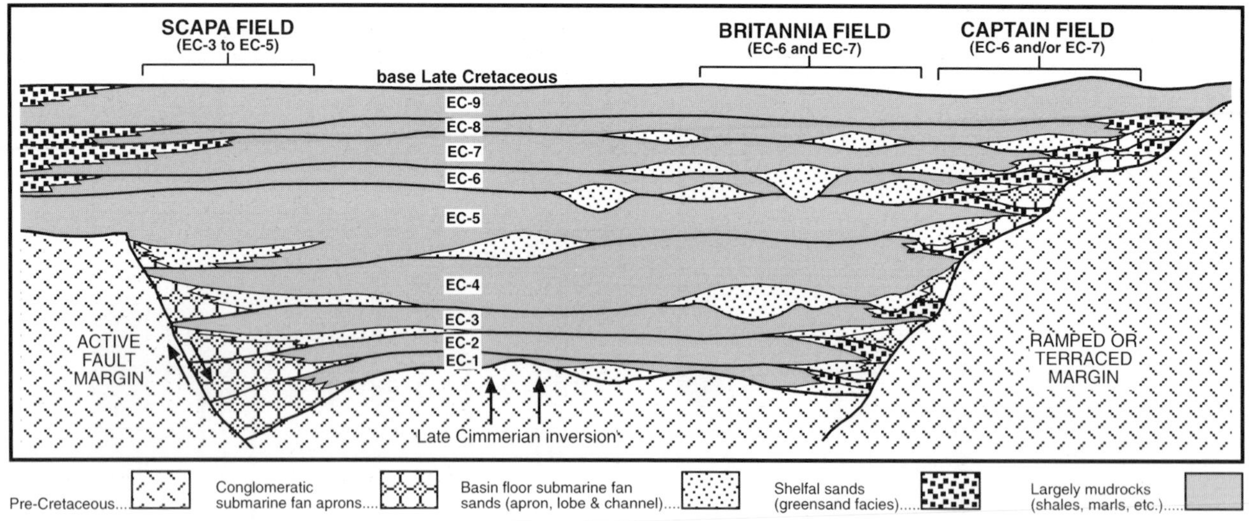

Fig. 9.18 Central North Sea—Early Cretaceous play fairway concepts (modified from data by permission of British Petroleum).

deposition within the Barremian and again in the mid- to late Albian. The restriction of sub-basins by major Late Cimmerian plate-wide compression is well illustrated for the interval from the latest Ryazanian to the Hauterivian, as is the restriction generated by the Austrian compressions during the Aptian to early Albian (see also Fig. 9.6).

Figure 9.17 also illustrates significant sand packages, which have been proved by well penetrations in the area. These fall into three marine depositional tracts—proximal submarine-slope apron fans, usually adjacent to major active faults; basin-floor fan sands, which may be related to fault aprons but also occur as channels and lobes in the axial areas of sub-basins; and shelfal 'greensands'. The latter usually occur around sub-basin margins and up on to adjacent platform areas. The former two associations always occur as low-stand wedges in the basins and can be related to major incision events on adjacent highs. The shelfal greensands commonly occur as progradational high-stand tracts, but can also be present as thin transgressive-system tracts around basins with ramp-style margins, and also on platform areas. In some cases, basin margins show complex interfingering of slope fans with thin shelfal greensands.

Figure 9.18 incorporates data from the Eastern Central Trough, all Moray Firth basins and the Viking Graben; it illustrates the different play-fairway concepts generated by the above tectonic, stratigraphic and sedimentological models. A very large majority of potential reservoirs were deposited in evolving structural lows as stratigraphically sealed basin-slope and basin-floor sands. Significant sands are thus unlikely to be found on any form of structural high—footwall crests, pop-up structures or diapiric features. Most traps will be stratigraphic or, at very least, will exhibit a significant stratigraphic trapping component in their closure.

To date, the Scapa and Britannia fields are the only significant discoveries that illustrate the Early Cretaceous deep-marine play of the Central North Sea.

9.6.2 Scapa Field

The Scapa Field has featured in many regional papers, and a wide range of local literature is available, from general field description (McGann *et al.*, 1991), through stratigraphic palynology, sedimentology and geological evolution (Harker and Chermak, 1992; Riley *et al.*, 1992), to formation evaluation and field-development plans (Chen, 1988; McGann *et al.*, 1988).

The Scapa Basin (or 'syncline') is a relatively small (8 km × 4 km) transpressional feature aligned north-west–south-east, with a major bounding fault on the south-western flank (Fig. 9.19). Basin-slope to basin-floor conglomeratic and sand aprons were sourced from the North Halibut Platform above the bounding fault. The north-eastern flank was a steep ramp formed on the south-eastern flank of the Claymore High.

The formation of the Scapa syncline dates from the Late Jurassic, when transtensional conditions created a large half-graben, with the Claymore High almost entirely within the dip slope. The change to transpression during the latest Ryazanian (Late Cimmerian event) exaggerated the Jurassic half-graben, and inversion elevated most of the older Jurassic basin floor and dip slope above sea level for Early Cretaceous erosion. Hence, Late Jurassic basin-floor fans were elevated to a structural high to form the Claymore Field. Most of the erosional products from this pop-up feature shed northwards into a parallel transpressional basin. Late Jurassic and Early Cretaceous structuration has often been assumed to be a function of tilt-block rotation during a long period of extension. In the light of regional transpression from the Ryazanian onwards, this model has now been modified (S.D. Harker, personal communication). Scapa is not a syn-rift wedge, but a syn-transpressional wedge.

Figure 9.20 illustrates Harker and Chermak's (1992) depositional evolution model, modified to account for transpression. While all nine Early Cretaceous sequences can be recognized in the area, it is often difficult to trace all

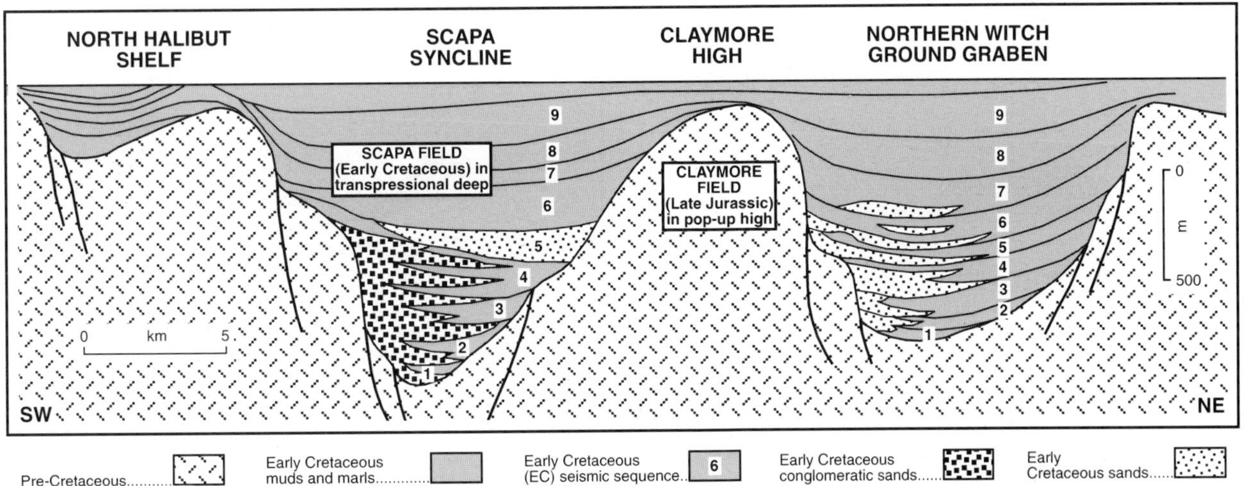

Fig. 9.19 Scapa Field, northern Witch Ground Graben—geoseismic palinspastic restoration to Plenus Marl datum (modified from Andrews *et al.*, 1990; Harker and Chermak, 1992).

the flood markers, or their lateral equivalents, within fan depocentres. This is because of masking by almost continuous coarse-clastic deposition. Individual fan packages, often incised at times of later sequence boundaries, progressively prograded towards the dip slope and became more sand-prone as the source area was worn down. Although tectonic reactivation occurred at all sequence boundaries, the most violent activity occurred during the Late Cimmerian event within the latest Ryazanian.

Figure 9.20 also illustrates depocentre switching between sequences. This is a common phenomenon in submarine fans, and is usually explained by compactional drape creating offlap. This effect, however, can be produced by lateral migration of feeder canyons along the footwall.

Scapa is a true synclinal stratigraphic trap with elements of facies, diagenetic and fault backseal (the latter in older stratigraphic units of the reservoir). Minor structural closure caused by compactional draping is present in the core of the fan complex, although less than 20% of the STOIIP of 206×10^6 barrels (32° API oil) is structurally closed. Recovery factors are moderate, with reserves estimated at 95×10^6 barrels. Most of the production is from sands of Valanginian–Hauterivian age. Conglomerates are of poor reservoir quality. The field has been on production since 1986 from a subsea template tied back to the Claymore Platform, utilizing four producers and four injectors. Average production to mid-1988 is reported as 28 000 barrels of oil/day (McGann *et al.*, 1991). Scapa oils were sourced from mid-mature Late Jurassic source rocks, which occur both under the field and downflank to the south-east. Migration into the trap was up-dip and via the principal south-western bounding fault of the basin.

9.6.3 Britannia Field

Few published details exist about Britannia. Those that do focus on the generalized stratigraphy and particularly on sedimentology (Bisewski, 1990; Guy, 1992; Downie and Stedman, 1993).

The Britannia Field lies within the north flank of the Southern Witch Ground Graben, between the Fladen Ground Spur and Viking Graben to the north and the Renee Ridge and Fisher Bank Basin to the south (Fig. 9.21). Structurally, the area is relatively simple. A large half-graben existed in the Late Jurassic, with the principal basin-bounding fault along the north flank of the Renee Ridge and the dip slope progressing up on to the southern Fladen Ground Spur. Major Late Cimmerian transpression tended to shatter the area into multiple low-relief sub-basins and highs, particularly in the vicinity of the Fisher Bank Basin, where three older Jurassic grabens intersected. Because of very limited oblique slip, older Jurassic features were largely retained. Jurassic basin and ridge lineaments were at right angles to the Early Cretaceous principal compression direction.

Between the Valanginian and the late Barremian, most of the syn-compressional topography of the basin had been smoothed by sediment drape, but it remained a deep, underfilled, elongate trough. Austrian compression, from the onset of the Aptian, rejuvenated adjacent source areas. This resulted in widespread deposition of a complex series of interfingering and overlapping basin-slope and basin-floor fan systems. Maximum spread occurred during the Late Aptian. Thinner, less extensive, sands occurred in the earliest Albian, by which time most of the sediment source areas had been worn down and/or become flooded by the ongoing Early Cretaceous global rise in sea level.

Anoxic shales, which mark major flood events, can be used for a very accurate subdivision and correlation of the complex reservoir sands (see Fig. 9.6). These include the Munk Marl, which underlies the sands, the Fischschiefer, which caps the early Aptian submarine-fan complex, and the 'Hedbergella marker' within the Late Aptian sequence. Apart from the latter, most of these markers are eroded on the north-flank dip slope of the system and over smaller north–south ridges. These ridges subdivide some of the individual fans of the composite system. Smaller anticlinal axes, orientated west–east in the east of the area and generally north–south in interference zones in the west, may also have influenced sedimentation patterns. Some

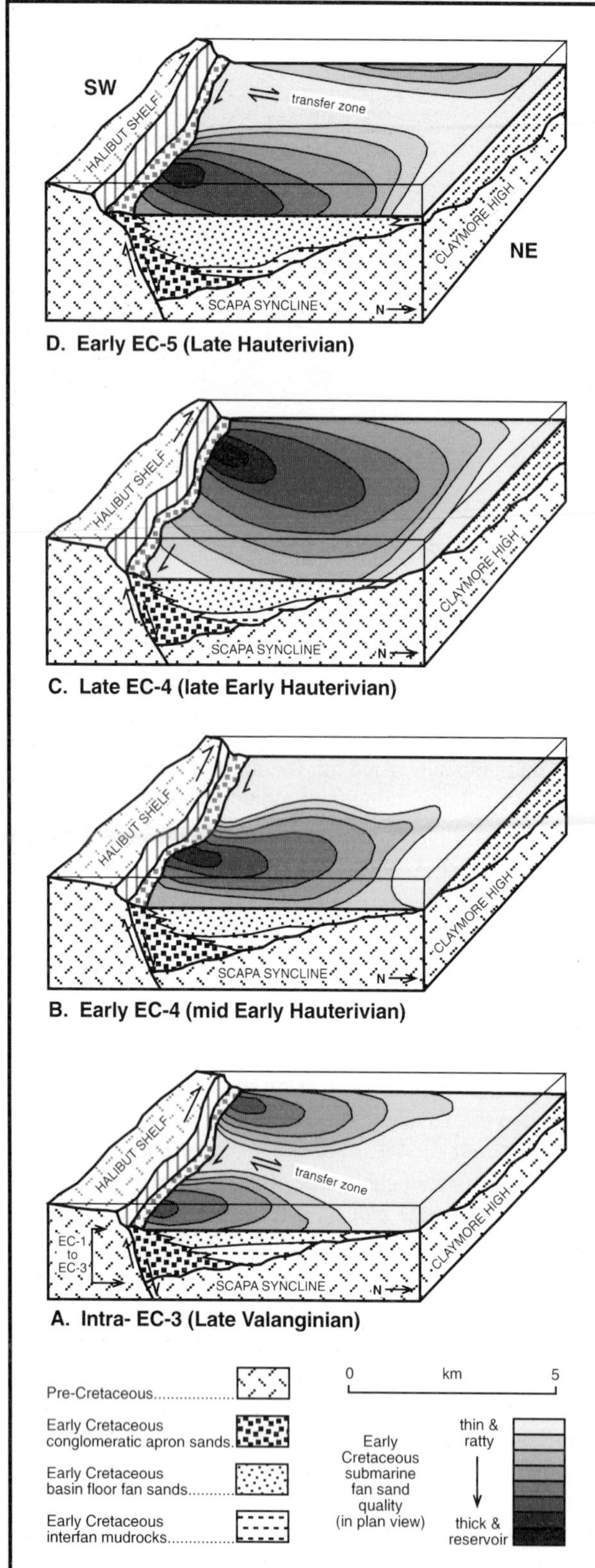

Fig. 9.20 Scapa Field—tectono-sedimentological evolution of a proximal submarine-fan-apron reservoir (modified from Harker and Chermak, 1992).

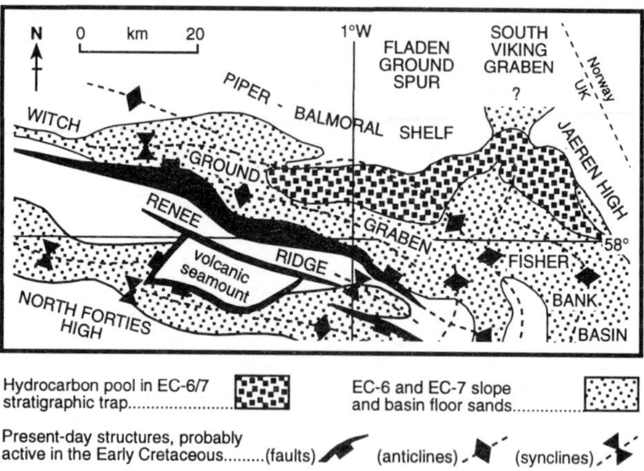

Fig. 9.21 Late Early Cretaceous sand distribution and the Britannia Field submarine-fan stratigraphic trap (modified from data by permission of British Petroleum).

models for the area suggest that the sands are axially sourced from the west via the Witch Ground Graben. Most of the evidence, however, favours many smaller point-sourced fan systems, which interfinger and overlap at different stratigraphic levels. Candidates for local source areas are the Fladen Ground Spur to the north and the northern Jaeren High in Norwegian waters, and probably the Renee Ridge to the south. Fan systems are very extensive in the area; they can be traced south of the Renee Ridge and into the northern sub-basins of the Central Trough and southern Viking Graben.

Although Britannia has some local areas of structural closure from a combination of compactional drape and small anticlinal and domal axes, it is estimated that less than 10% of reserves are trapped this way. Almost all the closure results from northwards up-dip stratigraphic pinch-out. Official reserves are not available, but the field is known to be a condensate accumulation. The field contains approximately $4.3 \times 10^{12}\,\mathrm{ft}^3$ of gas condensate in place (Garrett, 1998). Hydrocarbons are sourced from highly mature Jurassic source rocks, which underlie the area. The presence of asphalt mats in some wells suggests multiple migration and leakage.

9.6.4 Inner Moray Firth Basin

The Inner Moray Firth lies between the western end of the Halibut Horst and the Scottish mainland (Fig. 9.22, top map). During the Early Cretaceous, it had many of the characters of a syn-rift style of tectonism inherited from the Jurassic (see Chapter 8). There is ample evidence, however, to suggest that compression, rather than extension, was dominant in the Early Cretaceous. Transpression on the Great Glen Fault, although apparently inactive during the Jurassic, started again in the Valanginian. Multiple flower structures, often of large size, sprung out of the western areas of the Halibut Horst. Small-scale, but extremely deep, transpressional basins developed to the south of the Halibut Horst and in the vicinity of the Grampian and Peterhead Spurs, superimposed on older Jurassic transtensional half-grabens.

Thick sands are recognized on well logs from the area, with some proximal conglomerates stacked against the active Helmsdale and Wick Faults. These thin over the Smith Bank High, thickening again into the hangingwall of the Smith Bank Fault and then thinning through interdigitation with mud- and marl-prone facies on the southern side of the basin. The net thickness of coarse clastic can exceed 1500 m.

The Inner Moray Firth sands (and conglomerates adjacent to major faults) are enigmatic. Because of a lack of core samples, it is not known whether the depositional environment was shallow- or deep-marine. Shelfal greensands have been encountered on the southern fringes of the basin and continue southwards to the Western Terrace areas of the Central Trough. Most of the coarse clastics are likely to have been deposited in apron-slope and basin-floor environments. They are known to occur in all Early Cretaceous sequences. They seem to become cleaner and more sand-prone and to prograde further across the basin in younger sequences. They have attracted little attention in terms of prospectivity, largely through an obvious lack of seal over most of the western area of the Inner Moray Firth, and the use of geochemical models, which indicate that Jurassic source rocks are immature.

The known early Tertiary uplift led many to think that Jurassic source rocks were never buried deeply enough to become mature. Recent burial models (see Chapter 11) have shown, however, that in the deeper sub-basins maturity was reached before the end of the Cretaceous. The rich Oxfordian/Kimmeridgian (rather than latest Jurassic) source rocks reached depths in excess of 3500 m in the later part of the Early Cretaceous and up to 4500 m by the end of the Maastrichtian.

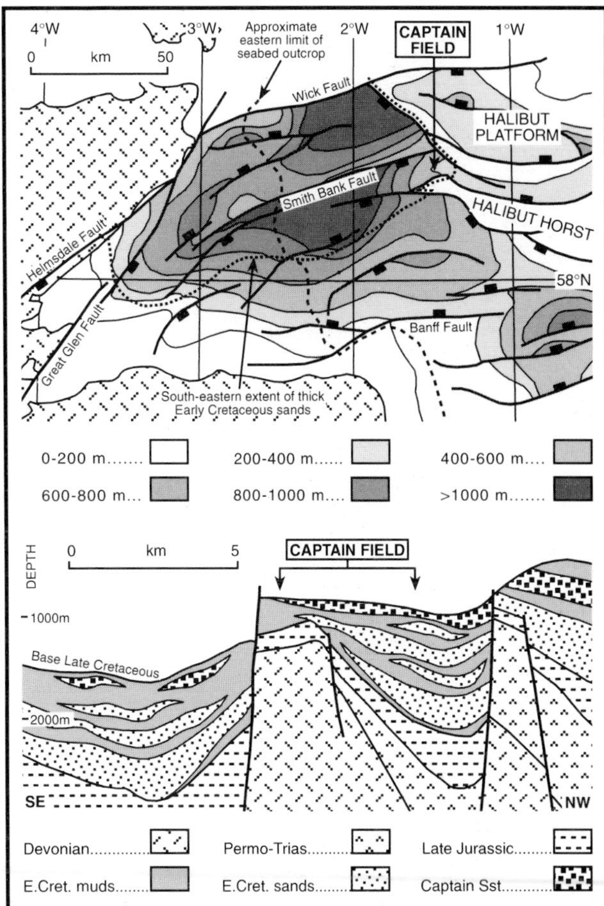

Fig. 9.22 The Inner Moray Firth Basin—Early Cretaceous isochores, sand distribution and cross-section in the vicinity of the Captain Field (modified from Andrews *et al.*, 1990; data by permission of Texaco).

9.6.5 Captain Field

The Captain Field, in UK Block 13/22a, is the only significant proof available at present to illustrate some of the models described above (Fig. 9.22, lower section). The reservoir consists of two thick Apto-Albian sands, separated by thinner flood shales and marls, deposited in basin-slope to basin-floor environments (Rose *et al.*, 1998). The structure of the field suggests that it sits astride a large, but low-relief, pop-up structure. The mid–late Albian mudrock seal is thin and much of the early Late Cretaceous is absent. This suggests that inversion occurred in the Cenomanian or was even a Campanian event.

The field occurs in 100 m of water, with top reservoir at only 900 m (Etebar, 1995). The areal extent of the field is about 10 km × 4 km. Oil in place is 1.5×10^9 barrels, but the crude is viscous and heavy (19.5° API), at low pressure, with a low gas/oil ratio (GOR) and high water cut; recovery is expected to be of the order of $300–350 \times 10^6$ barrels, at 60 000 barrels of oil/day. Because of the large areal extent and thin reservoir (50–100 ft), most development wells are planned to be horizontal. The sands are largely unconsolidated, with average porosities of 30–32% and average permeability of 7 Darcy (Rose *et al.*, 1998).

Additional prospective areas in the Moray Firth basins include the area to the south of the Halibut Horst (South Halibut Graben and South Halibut Terraces) for Scapa-type plays and the complex area of the Buchan Graben for both Scapa and Britannia/Captain-type fields. Many wells have encountered basin-floor sands at pinch-out and (generally sub-commercial) proximal fanglomerates. Few have yet encountered possible commercial finds, as almost all are on structural highs.

9.6.6 The Norwegian Central Trough

Almost all the production and proven reserves in the Norwegian Central Trough are contained within the younger Chalk fields or the two Late Jurassic Ula and Gyda fields. It is an area where the application of compressional tectonic models (see Section 9.2.3) and new sequence-stratigraphic schemes (see Fig. 9.6) could considerably improve the prospectivity of the Early Cretaceous.

Oakman *et al.* (1993) outlined the effects of Early Cretaceous burial history in the area. Figure 9.23 is a summary of these models. The complex base Cretaceous surface (Fig. 9.23, right map) has been visualized by back-stripping of integrated well and seismic data to determine the position and geometry of Late Jurassic facies. From this, Early Cretaceous models have been rebuilt.

Figure 9.23 (left map) shows a Late Jurassic half-graben with a Tornquist trend, progressively dislocated northwards by Caledonide transfer faults. The effects of transtension were evident on both extensional and transfer faults, to the extent that small positive and negative flower structures occurred on restraining and confining bends at their intersections.

The change in the regional stress regime at the end of the Jurassic (Late Cimmerian tectonism) had a dramatic effect. Although the isochores show total Early Cretaceous

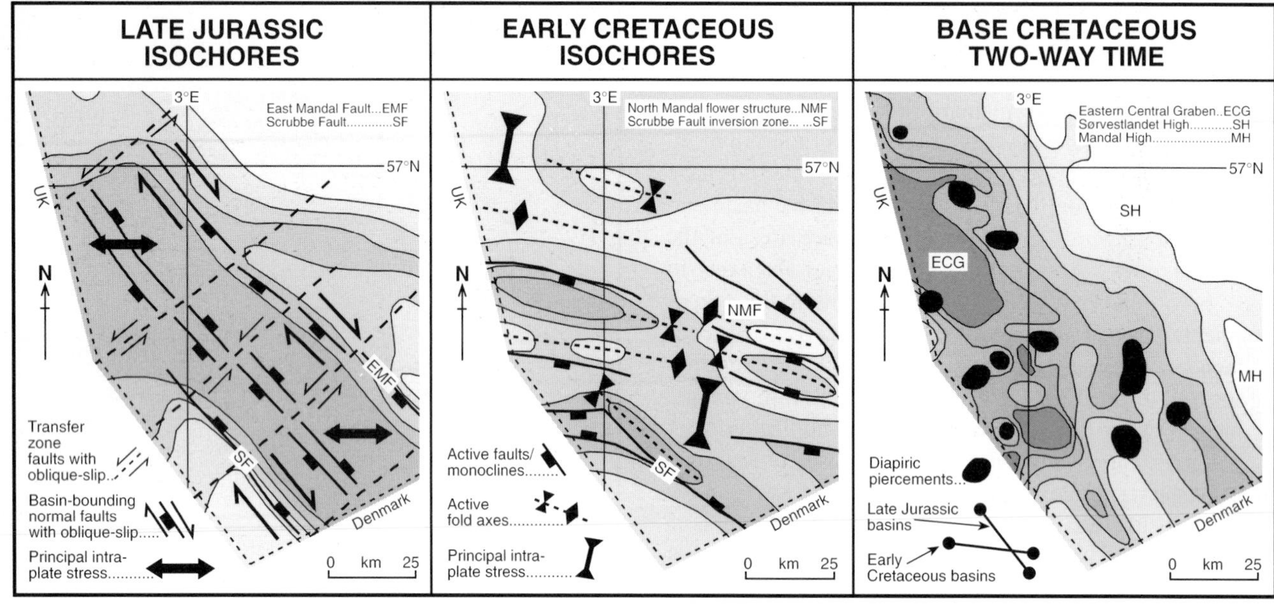

Fig. 9.23 The Norwegian Central Trough—cartoons of the Jurassic tectonic template, Early Cretaceous structuration and present-day basin geometry (modified from data by permission of Amoco Norway Oil Company).

(Fig. 9.23, centre map), most of the thicker areas were well established during the Valanginian. Older Jurassic patterns were almost obliterated by northwards-directed compression. The throw on the Scrubbe Fault was reversed and in places overthrust from a décollement level in salt; fold axes, monoclines and complex reverse faults criss-crossed the area from east to west, and regional flower structures developed from the Mandal High. Very deep but tightly confined en-echelon basinal areas were formed, separated by incised submarine highs and locally emergent footwall areas. Figure 9.24 illustrates a geoseismic section through the southern part of the area, on which most of the above features can be seen.

Proven Valanginian-age sandstones and silts and thin Apto-Albian sands have been encountered on the Grensen Nose (probable shelfal facies), in the Scrubbe Fault complex and adjacent to the Mandal High (probable submarine-fan fringes).

Apart from the Central Trough, thin Apto-Albian sands have also been documented over the eastern flank of the Jaeren High (Hesjedal and Hamar, 1983). These are believed to be shelfal greensands.

9.6.7 The Danish Sector

The structural geology and tectonic evolution of the area are well studied. All the features of Jurassic transtension and Cretaceous transpression and the effects of halokinesis are well documented. For further detail see Frandsen *et al.* (1987), Hansen and Buch (1982), Michelsen *et al.* (1987) and Vejbæk (1986); all of whom also stress the Early Cretaceous palaeogeographic and tectonic evolution. More general descriptions of the geology can be found in Michelsen (1982) and source rocks in Østfeldt (1987). More specific information and discussion on stratigraphy and Early Cretaceous source rocks are found in Jensen *et al.* (1986), Jensen and Buchardt (1987) and Thomsen and Jensen (1989). Damtoft *et al.* (1987, 1992) and Andersen and Doyle (1990) provide excellent overviews of the exploration and development history and give reviews of the main Danish plays. In addition, Ineson (1993) provides a geological account of the Valdemar Field.

The potential of sand-prone Early Cretaceous reservoirs in proximity to prolific mature Jurassic source rocks is, however, rarely mentioned in the literature. The Mid North Sea and Ringkøbing–Fyn highs are potentially prolific coarse-clastic sediment source areas, which are proven to have remained emergent before being finally flooded in early Albian times.

Figure 9.25 illustrates three wells that show some of the newer exploration models for the Early Cretaceous. The central well-log panel is a schematic section typical of the many Early Cretaceous sections drilled on evolving depositional highs (and present-day highs) in the Central Trough area. Thin, poor-quality sands and silts of late Ryazanian and Valanginian age are sometimes present. The Valanginian and Hauterivian, however, are generally represented by muds and marls, and much of the Barremian, Aptian and Albian is highly condensed, mudrock-prone and shows evidence of multiple incision (unconformity) events.

Sequence-stratigraphic modelling suggests that, basinwards of high sites, these unproductive sands could form

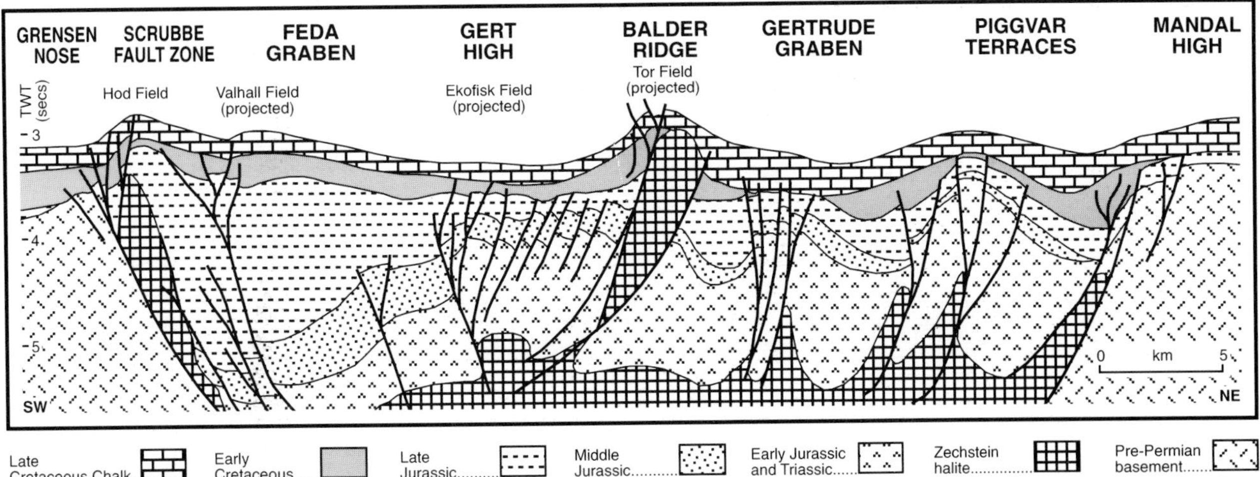

Fig. 9.24 Cartoon geoseismic cross-section through the southern area of the Norwegian Central Trough (modified from data by permission of Amoco Norway Oil Company).

submarine fans, and could be considerably more prospective in larger shelfal areas, such as the Outer Rough Basin and the flanks of the Ringkøbing–Fyn High (see Fig. 9.26). The Vyl-1 well from the latter area is illustrated in Fig. 9. 25 (left panel). Of particular note is the latest Ryazanian and Valanginian Vyl play. The Vyl sands in this area are stacked transgressive shelfal to shoreface sands. They are frequently dismissed as dirty, silty sands, but the figured example shows them to be clean, with a net to gross ratio about 0.5 and average porosity of about 30%.

The Vyl shelfal sands are the same age as the younger Devil's Hole sands in the Western Margin of the UK Central Trough and the Vlieland sands of the Dutch sector to the south. Some extremely thick Early Cretaceous (Valanginian–Hauterivian) sands occur in the eastern parts of the Norwegian–Danish Basin and in the Danish–Polish Trough. These sediments were sourced directly from the Norwegian and Swedish mainland and are not regarded as prospective due to lack of locally mature hydrocarbon source rocks.

9.6.8 The Valdemar play

Although Early Cretaceous sections in the Danish Central Trough are typical of the central log panel in Fig. 9.25, some wells located on inverted basinwards sites, such as North Jens-1 (right panel), illustrate expanded Barremian- and Aptian-age pelagic limestone sections. The pelagic carbonates are developed either side of the two maximum flood markers of the Munk Marl and Fischschiefer (deep-basinal setting) and are capped by condensed anoxic shales and marls. The carbonates are cleanest adjacent to the floods (< 5% non-carbonate) and become dirtier towards sequence boundaries (up to 20% non-carbonate). Porosities in the cleaner limestones average 25%, with permeabilities of < 1 mD. These carbonates are productive and form the Valdemar Field (Fig. 9.26).

Graben-wide Early Tertiary inversions affected many areas of the Danish Central Trough. The deeper Barremian carbonate basins were gently folded against transpressional faults. The principal discovery, Valdemar, has an oil–water contact closure of 200 km^2, with oil in place estimated as 300×10^6 m^3 and gas in place as 100×10^9 m^3. Some wells have tested 100 m^3/day of volatile oil/wet condensate. Figure 9.26 illustrates the location of the Valdemar Field and other structurally closed Barremian limestone plays. The Adda reservoir (at Barremian level) has also been declared commercial, in conjunction with Late Cretaceous reservoirs. Its Barremian reservoir contains gas and lean condensate. Damtoft *et al.* (1992) stress that recovery factors are likely to be extremely low. Water saturations are also high, suggesting inefficient migrational flushing.

It is highly tempting to advocate that the source of the hydrocarbon is from the organic-rich maximum-flood marls and shales contained within the reservoir section of Valdemar. These have a total organic-carbon content of up to 15%. All mudrocks are oil- to gas-prone, with maturity factor (R_o) varying from 0.38 to 0.46 in structurally high sections in the range 2000–2050 m (Jensen and Buchardt, 1987). There is, however, no shortage of mature to overmature Jurassic source rocks and ample migration routeways, via continually active transpressional faults, to account for Barremian hydrocarbon reservoirs. This commercially proven play in the Danish sector is present in both the Norwegian and UK sectors (see Fig. 9.17).

9.6.9 The Viking Graben

Very little information on the Early Cretaceous of the Northern North Sea is published. Available data are dominated by lithostratigraphic nomenclature (Isaksen and Tonstad, 1989; Johnson and Lott, 1993; Johnson *et al.*, 1993) and papers on the Early Cretaceous inversion tectonics of the northern East Shetland Platform (e.g. Thomas and Coward, 1995).

Figure 9.27 illustrates a composite palaeogeography and play-fairway map of the Early Cretaceous of the Viking Graben and surrounding sub-basins. Three tracts can be recognized. The landmasses and islands shown were persistent post-Late Cimmerian highs; any Late Jurassic (mud-prone) cover was stripped during the latest Ryazanian/Early Valanginian to expose a potential source of coarse-clastic sands from Middle Jurassic and older strata. Many

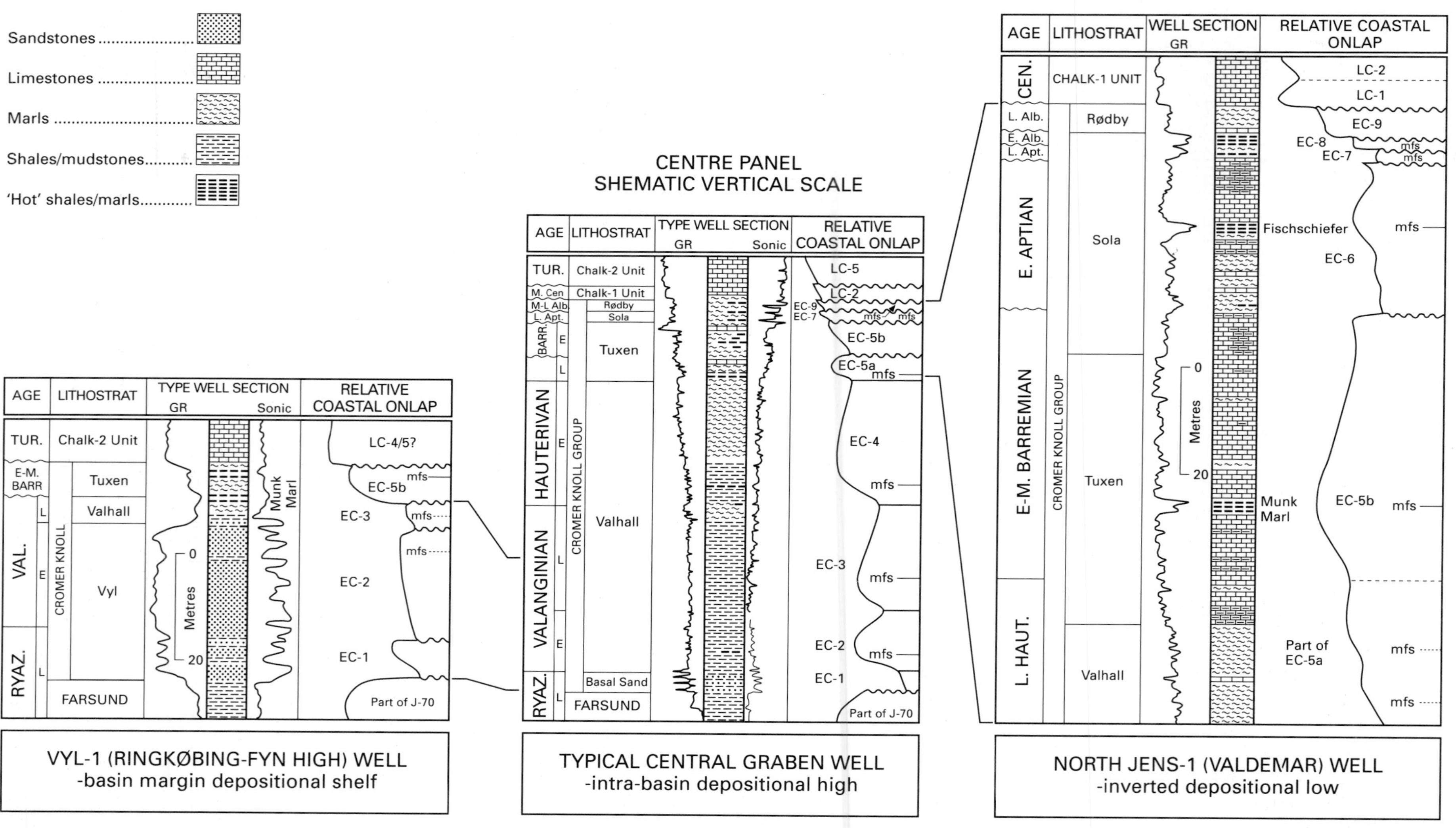

Fig. 9.25 Danish Early Cretaceous well logs—examples of recent new play concepts (modified from data by permission of British Petroleum; Ineson, 1993).

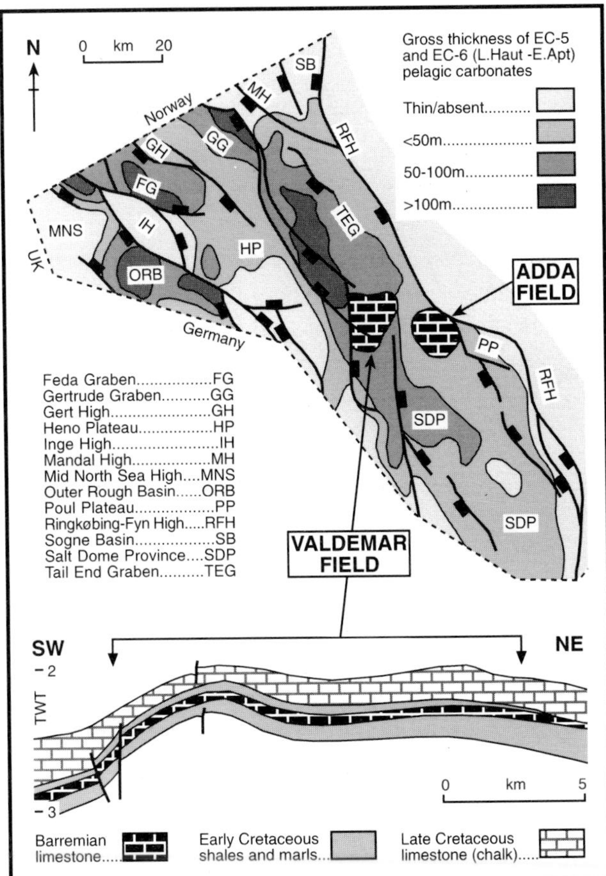

Fig. 9.26 The Barremian carbonate play, Danish Central Trough (modified from Damtoft *et al.*, 1992; Ineson, 1993).

areas, particularly along the western flank of the basins, exposed pre-Mesozoic 'basement'.

Most of the landmasses were eventually flooded in the mid- to late Albian. The areas shown as 'shelfal' tracts are where most Jurassic-targeted wells are located. Over these areas, pre-Cretaceous sections are often eroded (particularly on tilt-block crests), but not deeply enough to cut into Jurassic and older clastic sections. Most of these areas were submarine deep-shelf (characterized by condensed shelfal muds and marls). Some thin greensands (shallow-shelf to shoreface) can be found around emergent footwall crests in either hangingwall or dip-slope sites. The greensands are present in all Early Cretaceous sequences (see Fig. 9.6), but are mostly thin and of poor reservoir quality.

The basin-slope and basin-floor depositional environments are considerably more prospective. Where these occur close to active faults flanking sediment source areas, Scapa-type apron fans can be predicted to occur. Conglomerates have been found in only a few wells (off the Crawford Spur and adjacent to faults within the Magnus Embayment). Apron sands have been encountered in the South Viking Graben and basinwards of the Tampen Spur. Many are thin and water-wet, which tends to suggest migration and/or seal problems. Like Scapa, the main play occurs from the latest Ryazanian to the early Hauterivian. Valanginian sands appear to be the most extensive. Hangingwall mounded facies are commonly seen on seismic data, particularly on the western faulted margins of most sub-basins.

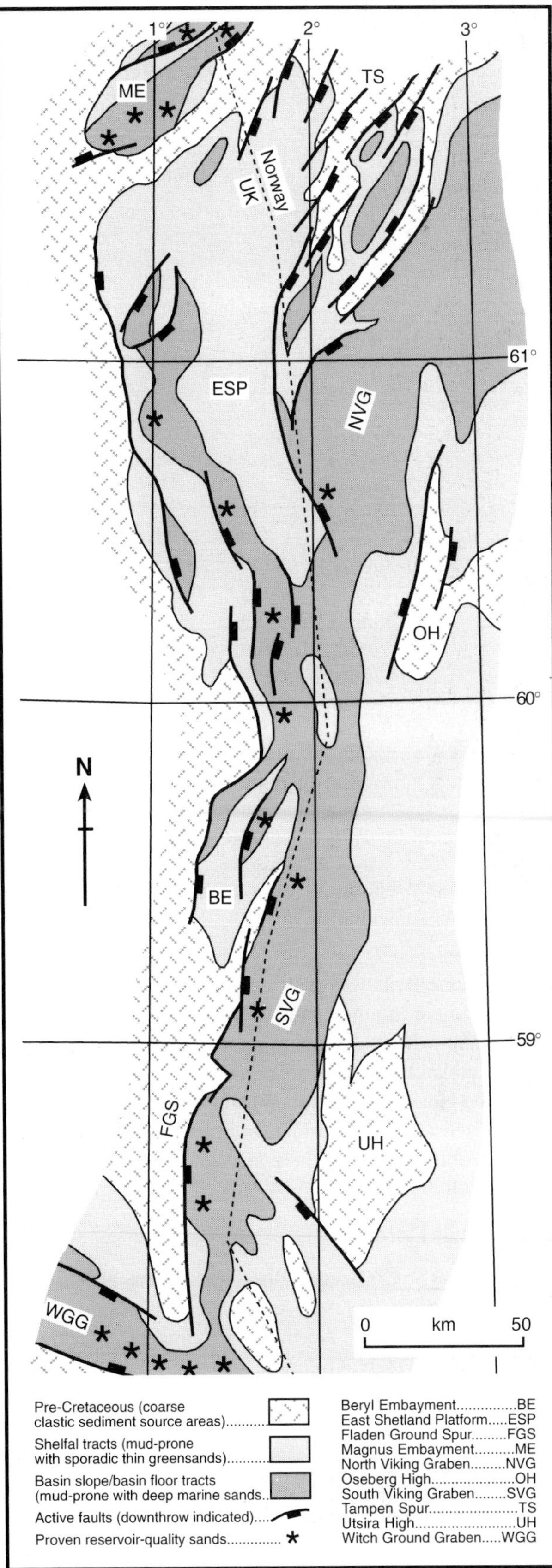

Fig. 9.27 Early Cretaceous palaeogeography of the Viking Graben area and proven reservoir-quality basin-slope and basin-floor sands (modified from data by permission of British Petroleum).

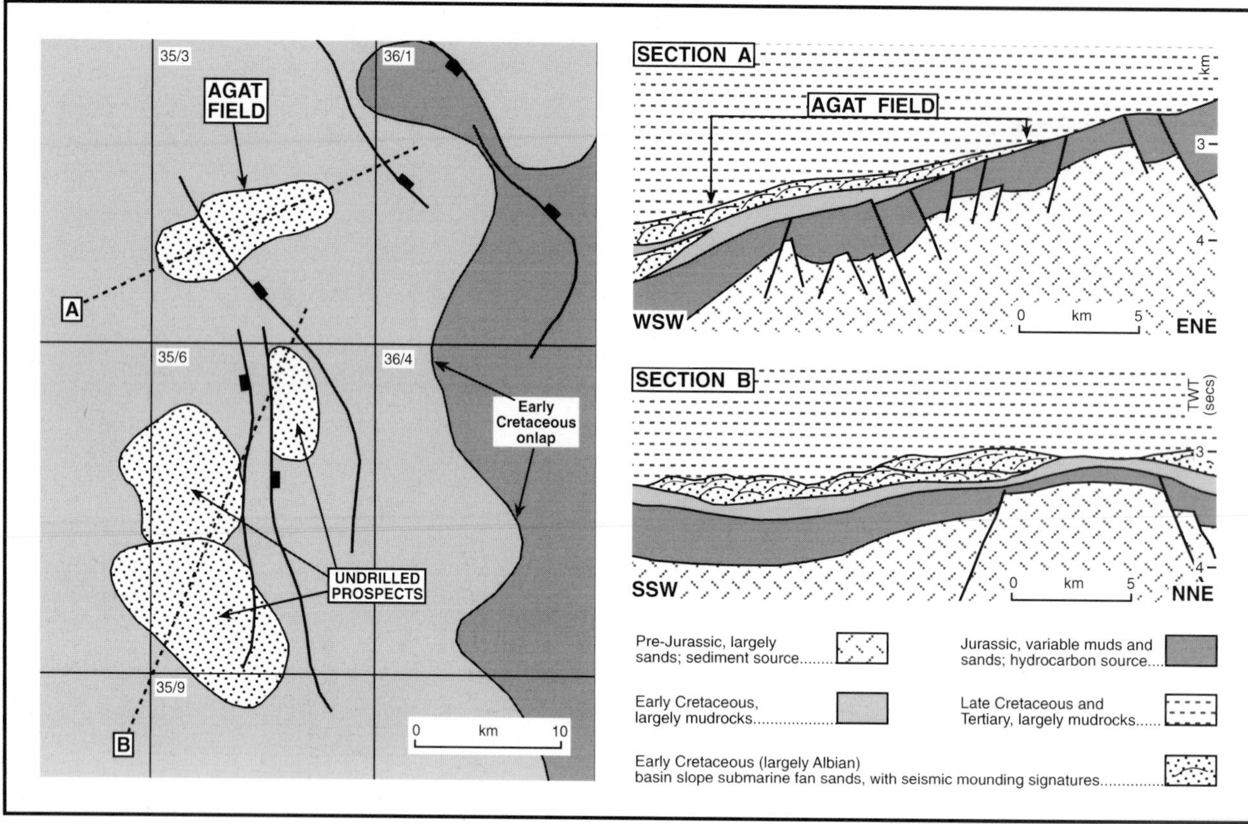

Fig. 9.28 Agat Field, northern Norwegian North Sea—basin-slope stratigraphic traps, proven and prospects (modified from Gulbrandsen, 1987; Shanmugan *et al.*, 1994, 1995).

Sand-prone Britannia-type plays are more elusive. It is common to see mounded and channel-onlap signatures on seismic data within the floors of sub-basins. Very few well ties are available in these areas, which are predominantly within the Norwegian sector. When ties were possible, this seismic facies is of Aptian age, but can occur ranging into the Mid Albian in the northern areas of the Viking Graben.

9.6.10 Agat Field

The only Early Cretaceous discovery in the area of the Viking Graben is the Agat Field. This is located in the Sogne Spur area of the North Viking Graben. A summary of the field is given by Gulbrandsen (1987). Sedimentological and sequence stratigraphic reviews of the Agat sands are featured by Shanmugan *et al.* (1994, 1995) as part of a larger study of submarine-fan systems. The field is illustrated in Fig. 9.28.

The Sogne Spur is a broad west-north-west–east-south-east-aligned nose. It was formed by regional Cretaceous compressions (low-amplitude folding) exploiting older transtensional Jurassic grains. Most of the compressional structuration occurred before the end of the Aptian, although Campanian and some Early Tertiary reactivation slightly modified the gross structures. The Agat discovery lies across the flank of the Sogne Spur, with fan-slope rather than true basin-floor reservoir sands. Although sourced from a major fault complex bordering the Norwegian mainland just to the east, the individual fan sands are thin and of poor reservoir quality (gross fan thickness *c.* 200 m, net-to-gross ratio of 0.1 to 0.3); reservoir quality improves downflank into the basin in the south-west.

The Aptian sediments of the area are shelfal in character, are condensed and onlap to the north-east (Fig. 9.28, top section). Progressive flooding of the Sogne Spur during the Albian allowed continual onlap, with the development of slope fans in low-stand tracts. The Albian sections seem to represent the backstepping and dying parts of a potentially prolific fan system that was deposited deeper in the basin during the Aptian. The deeper, basinward, prospects are also illustrated in the map of Fig. 9.28 and the lower section.

The Agat Field is an example of a stratigraphic pinch-out trap in a slope-apron setting. The crest of the accumulation is at 2800 m subsea, with its deepest hydrocarbon/water contact at 3560 m subsea—a column of around 760 m, the thickest known in the North Sea. The hydrocarbons are lean condensates, but water saturations are high (up to 46%). Source rocks are the highly mature Jurassic shales in the adjacent North Viking Graben. Despite the apparent large size of the accumulation, prediction of sand-body extent and geometry is immensely difficult. Correspondingly, gas in place has been estimated conservatively at 65×10^9 m^3 (2.3×10^{12} ft^3) but may be as high as 175×10^9 m^3 (6.2×10^{12} ft^3).

9.7 The Late Cretaceous chalk

The Late Cretaceous chalks of the Central Graben form the most prolific hydrocarbon fairway in the southern Norwegian sector and all of the Danish Sector. In the Norwegian

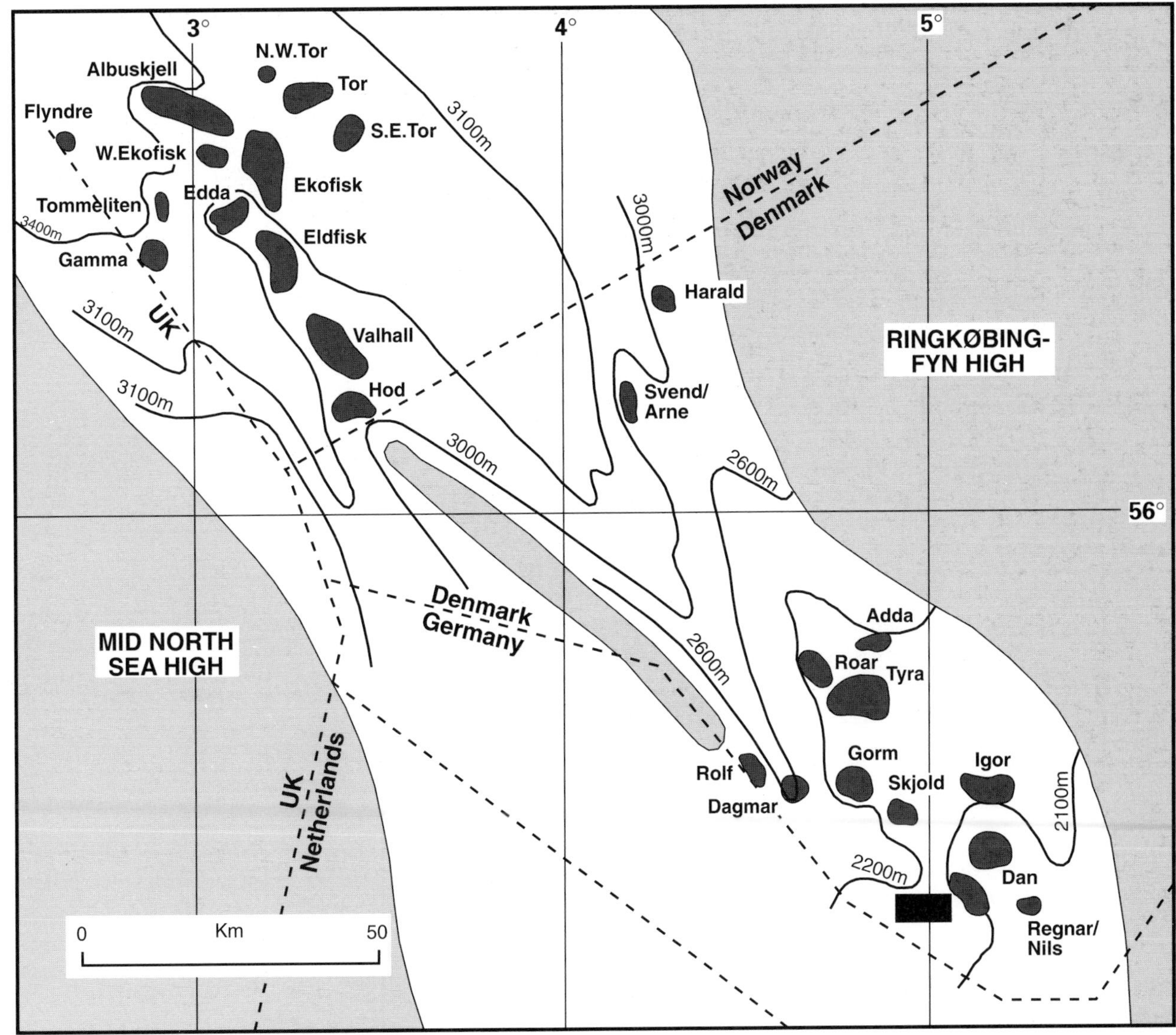

Fig. 9.29 The Norwegian and Danish Chalk fields: location map and regional depth to top reservoir (modified from Megson, 1992).

sector, proven recoverable reserves are in excess of 3.84×10^9 barrels (610×10^6 m^3) of oil and 11.5×10^{12} ft^3 (325×10^9 m^3) of gas (Norsk Hydro, 1994). In the Danish sector, proven recoverable reserves are 786×10^6 barrels (125×10^6 m^3) of oil and 3.8×10^{12} ft^3 (108×10^9 m^3) of gas (Damtoft *et al.*, 1992). In recent years, many individual operators have been revising these numbers upwards in the light of improved geological/geophysical field models and sophisticated water-flood techniques. The Late Cretaceous/Danian reserves account for about 77% of Danish and about 80% of southern Norwegian recoverable reserves. The chalk play is now very well established in the Norwegian and Danish sectors (Fig. 9.29 shows the distribution of fields).

In the early years of discovery and exploitation of the play, few people could believe that such a huge commercial find could form in, and be producible from, chalk sediments. The words of Scholle (1977), 'oil from chalks, a modern miracle?', sum up the sentiments of the decade following the first chalk discoveries (Anne, Valhall, Roar, Tyra and Ekofisk) in the late 1960s.

Many problems occurred in understanding the chalk play. Seismic definition was often very poor because of the presence of gas chimneys over many structures. Some structures were drilled only to encounter very thin chalk, or none at all in the case of diapiric piercement. During development drilling, reservoir-layer continuity and quality appeared to be quite random, hydrocarbon/water contacts were enigmatic and some thick hydrocarbon intervals merely seeped rather than gushed as expected. In other cases, thin poor-quality reservoir produced at prolific rates. Even early on, many wells started producing more reservoir than oil, and platform subsidence appeared as a major problem.

9.7.1 Chalk literature review

A vast amount of literature has been generated on the sedimentology of chalk reservoirs and their diagenesis. As fields were developed, each was described. In more recent years, a new generation of publications on integrated geological and geophysical studies and reservoir engineering has emerged. Papers specific to general field descriptions are given in Table 9.1.

Table 9.1 North Sea Chalk fields—published reserves and further reading.

STATE	FIELD	RECOVERABLE RESERVES	KEY SPECIFIC REFERENCES
NORWAY	ALBUSKJELL	69 mmbbls/564 bcf	D'Heur (1987); Watts (1983a & 1983b); Watts et al. (1980)
	EDDA	19 mmbbls/70 bcf	D'Heur at al. (1985); D'Heur & Michaud (1987)
	EKOFISK	1250 mmbbls/3710 bcf	Bark & Thomas (1981) Brewster et al. (1986); Byrd (1975); Dangerfield & Brown (1987); Fritsen & Corrigan (1990); Pekot & Gersib (1987); Sulak et al. (1990)
	EKOFISK WEST	76 mmbbls/780 bcf	D'Heur (1980 & 1987)
	ELDFISK	302 mmbbls/1094 bcf	Brewster & Dangerfield (1984); Herrington et al. (1991); Maliva & Dickson (1992); Michaud (1987)
	FLYNDRE	NDA	No data published
	HOD	56 mmbbls/247 bcf	Campbell & Gravdal (1995); Hardman & Kennedy (1980); Norbury (1987)
	TOR	120 mmbbls/390 bcf	D'Heur (1987)
	TOR (NW)	NDA	No data published
	TOR (SE)	NDA	No data published
	TOMMELITEN	56 mmbbls/848 bcf	D'Heur & Pekot (1987)
	VALHALL	365 mmbbls/1588 bcf	Hardman & Eynon (1978); Leonard & Munns (1987); Munns (1985)
DENMARK	ADDA	NDA	No data published
	DAGMAR	NDA	No data published
	DAN	100 mmbbls/1000 bcf	Childs & Reed (1975); Svendsen (1979)
	GORM	150 mmbbls/250 bcf	Hurst (1983)
	HARALD	NDA - gas	No data published
	IGOR	NDA - gas	No data published
	KRAKA	NDA - oil & gas	Jørgensen & Andersen (1991)
	ROAR	Condensate & 800 bcf	No data published
	REGNAR/NILS	NDA	No data published
	ROLF	NDA - oil	No data published
	SKJOLD	35 mmbbls & gas	Oen et al. (1986)
	SVEND/ARNE	NDA	No data published
	TYRA	Condensate & 2000 bcf	Doyle & Conlin (1990)
UK	JOANNE	NDA	No data published
	JOSEPHINE	NDA	No data published
	JUDY	NDA	No data published
	MACHAR	approx. 60 mmbbls	Foster & Rattey (1993); Hodgson et al. (1992)
NETHERLANDS	HARLINGEN	up to 80 bcf	Van den Bosch (1983)

mmbbls = millions of barrels of oil bcf = billions of cubic feet of gas NDA = No data available

Papers that give a good overview of chalk sedimentology and diagenesis include Brewster and Dangerfield (1984) on the Lindesnes Ridge area; D'Heur (1984, 1986), Hancock and Scholle (1975), Hancock *et al.* (1983) and Hardman (1983) on the Norwegian sector; Lieberkind *et al.* (1982) in the Danish sector.

Papers specific to sedimentology and depositional processes include Bromley and Ekdale (1987) on debris flows; Ekdale and Bromley (1984) on ichnology; Feazel *et al.* (1985) on the Ekofisk area; Hancock (1976) on petrology; Håkansson *et al.* (1974) on the Maastrichtian; Hancock (1975, 1983) and Hancock and Kauffmann (1979) on European and American chalks; Hatton (1986) on regional Maastrichtian and Danian re-worked chalks; Kennedy (1983, 1987) on the Central Trough; Kennedy and Juignet (1974) on French chalks; Schatzinger *et al.* (1985) on the Central Graben; Skovbro (1983) on the Norwegian sector; Thomsen (1976, 1983, 1989) on Danish bryozoan reefs and plankton production; Voigt (1981) on the Dutch sector; and Watts *et al.* (1980) on the Albuskjell area in Norway.

Publications that cover the many aspects of diagenesis include Bromley and Ekdale (1986) on flints; Feazel and Schatzinger (1985) on cementation inhibition; Herrington *et al.* (1991) on an Eldfisk model; Jørgensen (1986a,b) on the Danish sector; Kennedy and Garrison (1975) on hardgrounds; Mapstone (1975) on older diagenetic models; Neugebauer (1974) on cementation; two classic papers from Scholle (1974, 1977); and Toft (1986) on diagenetic fluorite.

The characteristics of fractures and stylolites are featured in Aguilera and van Poolen (1979) on fracture description; Braithwaite (1989) on stylolites and reservoir engineering models; Burgess and Peter (1985) on stylolites as barriers; Bushinsky (1961) on stylolite description; Fritsen and Corrigan (1990) on Ekofisk reservoir engineering fracture models; Garrison and Kennedy (1977) on early pressure solution diagenesis; Jones *et al.* (1984) on onshore UK fracture and stylolite deformation models; Kirkland (1984),

Nelson (1981, 1982) and Sangree (1969) on integrated core, log and reservoir engineering models and description of fractures and stylolites; Oen *et al.* (1986) on fracture reservoir engineering in the Danish sector; Thomas (1986) on fracture reservoirs; and the classic papers by Watts (1983a,b) on Albuskjell fracture modelling. Some novel approaches to fracture analysis in horizontal wells of the Danish sector are given in Andersen *et al.* (1988, 1990) and Fine *et al.* (1993, 1997).

Papers that are more specific to porosity and permeability prediction include Campbell and Gravdal (1995) on East Hod Field; Jones *et al.* (1990) on the chalk play in general; Maliva and Dickson (1992) on Eldfisk Field; Nygaard *et al.* (1983) in the Danish sector; Sørensen *et al.* (1986) in the Norwegian sector; and Taylor and Lapré (1987) in the Ekofisk area. Following on from predictive studies, there are two quality reservoir engineering applications documented by Jensenius and Munksgaard (1989) in the Danish Sector; and Sulak *et al.* (1990) for the Ekofisk Field—both detailing waterflood models.

Petroleum-geochemical publications on the Chalk do not occur in the literature because of the emphasis on the Jurassic as the only considered source rock (e.g. Hughes *et al.*, 1985). Research, however, has been devoted to intra-chalk source rocks on the Gulf Coast of the USA (e.g. Grabowski, 1984; Hunt and McNichol, 1984), which may have significance for European chalks.

All chalk reservoirs share many common attributes. In the following text, an attempt has been made to explain most of what chalk reservoirs consist of and why they exist.

9.7.2 Primary chalk sediment—biogenic components

Chalk is a biogenic carbonate made up almost entirely of micrite (grains of < 4 μm size) and is therefore classified as a lime mudstone. Most of the carbonate material was contributed by just three groups of plankton, which flourished in the subtropical to temperate seas of the Late Cretaceous—the coccolithophorids, planktonic foraminifera and calcispheres. Coccolithophorids were the dominant group, constructed of two parts—a spherical segment (coccosphere) made of overlapping circular plates (coccoliths), attached to an elongate body made of prismatic calcite crystals. The coccospheres were held together by organic tissue, which easily disintegrated to coccoliths and subsequently to component platelets (Fig. 9.30).

These principal constituents of chalk must have formed an almost constant pelagic rain in Late Cretaceous seas—indeed, seasonal varves can often be seen in some deeper basinal sections. Although whole organisms can be preserved, ingestion within the water column and upper sediment layers commonly resulted in disaggregation to individual component calcitic crystallites. The rain of plankton was reduced to an accumulation of faecal products. Under SEM, some crystallites show etching and slight rounding. This has often been explained by the action of gastric juices during ingestion or, in deep sub-basins, by dissolution within the water column as particles fall below carbonate compensation depths. The sediment formed was therefore dominated by lime mudstone, although occasionally larger foraminifera and calcispheres accumulated in sufficient quantity to produce lime wackestone textures.

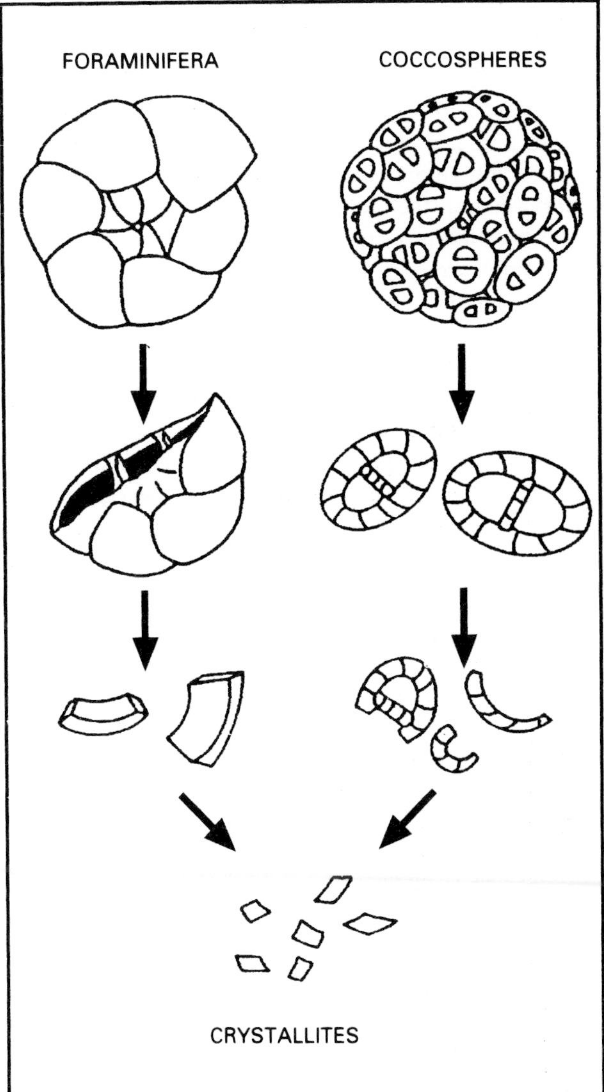

Fig. 9.30 Chalk sediment constituents (modified from Schatzinger *et al.*, 1985).

From a petroleum-geological viewpoint, lime muds are often erroneously grouped with terrestrially derived clays and muds, which are deemed to be non-reservoir right from the point of initial deposition. Detrital muds consist largely of micro-plate minerals, which during settling and early compaction are preferentially aligned to reduce vertical permeability (K_v) quickly to non-reservoir status. Chalk sediment, on the other hand, consists of highly angular crystallites of many different shapes. These will not preferentially pack and therefore resist initial compactive reduction of K_v relative to detrital muds. Experiments on different muds show that, under pressures equivalent to 1000 ft of burial, K_v in detrital muds is commonly < 0.1 mD, whereas in chalk muds they range from 10 to 100 mD.

Early pore pressures in chalk muds are also considerably higher, largely due to the presence of free water, rather than the ionically locked water present in dewatered terrestrially derived muds. Thus, chalk muds can retain three key properties—higher permeability, more free water and thus higher pore pressures—to much greater depths than detrital muds.

Most carbonate muds are aragonitic and are notoriously unstable. Chalk grains, however, are made up largely of calcite and high-magnesium calcite, both of which are considerably more stable than aragonite. Polymorphic transformation therefore cannot occur, and leaching and reprecipitation processes are impaired until pressure solution commences at greater depths.

Although the above planktonic grain types form the dominant constituent of chalk, shelfal areas contributed other carbonate grains, derived from benthos. Most of these were stable calcitic forms and include the echinoids and crinoids, bryozoa, rarer corals and bivalves. Bryozoans can occur in sufficient quantity in some inner shelfal areas to form seagrass-type mats and even poorly developed reefs. Depositional products can range from the entire skeleton down to single-crystal echinoid plates, crinoid ossicles and inoceramid prisms. In some shelfal areas, concentrations of macrofauna can be high enough to produce bimodal lime packstone textures, but these are rare in the vicinity of chalk hydrocarbon fairways.

Although often chaotically textured—from the dominant 'poorly sorted' lime mudstones to rare packstones—the 'clean' nature of these types of carbonate sediment gives the potential to become reservoir rocks. From a reservoir geological viewpoint, these are collectively referred to as 'hemipelagic clean chalks'. Figure 9.31 illustrates the processes that form this type of chalk, and Fig. 9.35 illustrates the depositional locations.

9.7.3 Primary chalk sediment—non-carbonate components

Non-carbonate components are important in the characterization of chalk reservoirs. Most of them, however, are destructive of reservoir quality. Their interaction with depositional processes is shown in Fig. 9.31A.

Silica is present in two forms—firstly, as siliceous sponge spicules and, secondly, in the form of terrestrially derived silt to mud-grade dust. These appear today as very minor constituents (< 1%). Colloidal forms within the water column can settle out and become diagenetically modified in the upper layers of sediment. This process resulted in nodules and discontinuous sheets of chalcedony, which, with ongoing burial diagenesis produce 'beds' or bands of flints. These have a relatively trivial effect on bulk reservoir quality.

Clay minerals (part of the terrestrially derived dust) and organic matter (from terrestrial areas or from marine algal kerogen) are a much greater threat to future reservoir quality. These are often deposited with the pelagic carbonate component as part of seasonal varve sets. Only a few per cent of either non-carbonate matter can destroy reservoir potential. Fortunately, intense bioturbation in the sediment removes most organic matter in shelfal and shallower basinal areas, and disperses what would otherwise result in seasonal clay varves. At times of sea-level high stand, or in deeper subbasins, the production of pelagic carbonate and/or benthic bioturbation may be reduced and a terrestrial pelagic style can predominate. This can result in

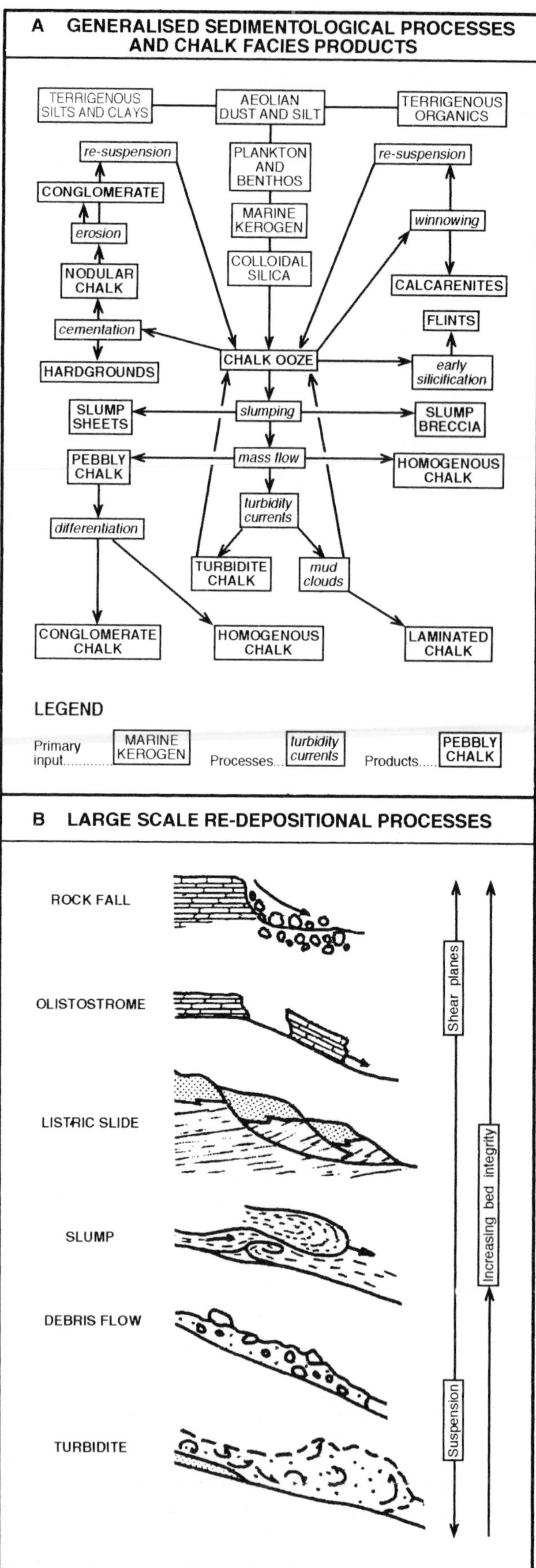

Fig. 9.31 Chalk depositional processes (modified from Schatzinger *et al.*, 1985).

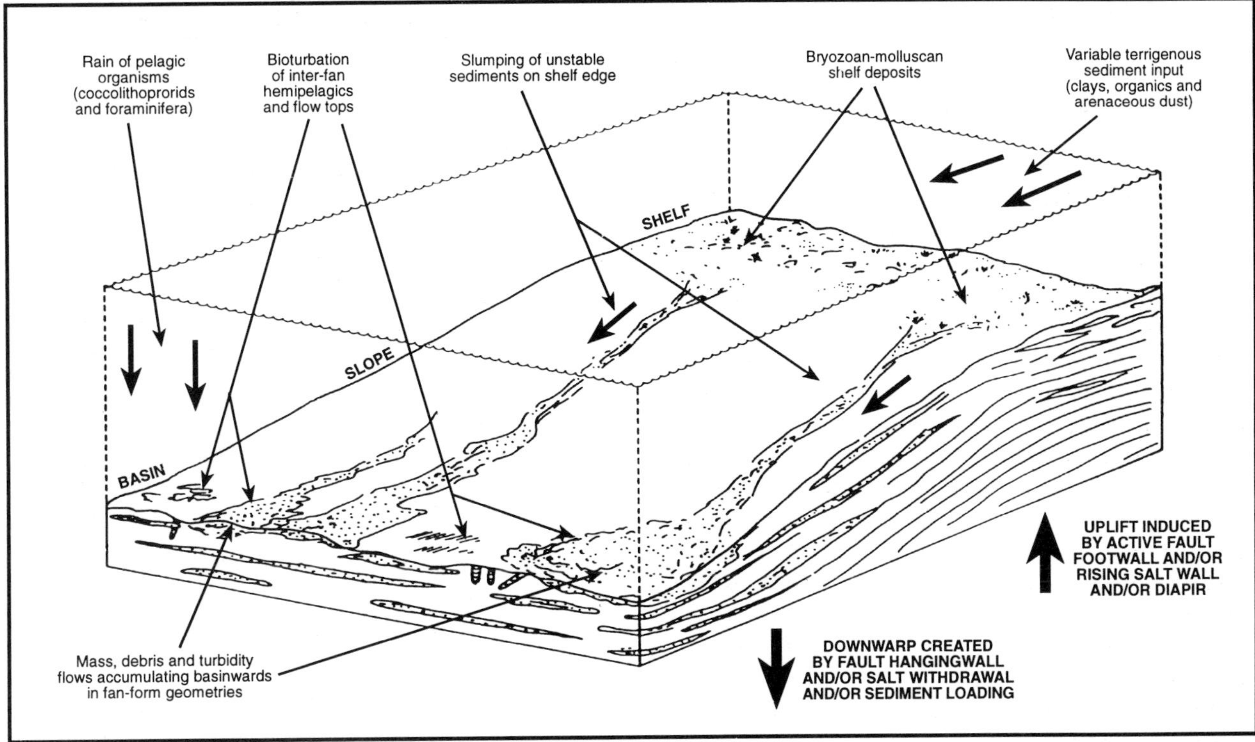

Fig. 9.32 Block diagram of Chalk depositional environments in the Central Trough.

argillaceous chalks (2–10% non-carbonate), marls (10–50% non-carbonate) and calcareous claystones (50–90% non-carbonate), all of which are non-reservoir facies.

Clearly, the terrestrial pelagic influence can occur adjacent to any residual landmass. Palaeogeographic analysis of the North Sea Basin (see Fig. 9.5) can therefore quickly identify areas, both spatially and stratigraphically, where chalks will not be commercially productive on the basis of sedimentology alone. This tends to remove most of the Cenomanian and Turonian across the entire basin and also large areas of the Coniacian and Santonian. It also eliminates almost all of the upper Late Cretaceous in the Northern North Sea, most of the Moray Firth basins and the southern areas of the Southern North Sea. It is this pattern which results in the progressively cleaner (less non-carbonate) character of chalks in stratigraphically younger sections, with a corresponding bulk reduction in the argillaceous facies, at least until major landmass rejuvenation at the end of the Maastrichtian.

9.7.4 Primary chalk sediment—redeposited facies

An initial chalk ooze can be clean or have a varying content of non-carbonate matter. Grain size, although dominated by micrometre-size fines, can vary up to rare but complete bioclasts. Bioturbation commonly resulted in a homogenized mixture of all components (poorly sorted). Numerous synsedimentary processes act on the primary chalk ooze (Fig. 9.31, top diagram A) to produce a wide range of early diagenetic and/or redeposited fabrics. The latter are critical in the formation of chalk plays and reservoirs.

Most redeposition is by hydrodynamic processes, which sort the homogenized chalk ooze. As the flow process of any transport mechanism wanes, coarser grain sizes will settle first, followed by fines later. Larger bioclasts and intraformational intraclasts/faecal pellets usually settle first. Subsequent differentiation then grades the crystallites forming the lime-mud component, to the extent that larger crystallites (of the order of a round 5 μm) are sorted from finer micrite (< 1 μm) and clay minerals. If redeposited masses are thick enough (usually > 2 m) or flow units are rapidly stacked on each other, bioturbation can rehomogenize only the top of the final unit. Thus clean, usually very well-sorted, lime-mud 'grainstones' are developed.

The primary porosity of redeposited chalk is often greater than homogenized ooze; around 80% rather than 70%. They have a larger and more rigid support framework to resist early compactional packing. Pore throats are often larger by almost an order of magnitude. Whereas hemipelagic clean chalks can have permeabilities of 10–100 mD at 1000 ft burial depth, redeposited chalks can have values ranging from 100 mD up to 1 D complete with even higher pore pressures. Redeposited chalks therefore had a much better chance to form a reservoir.

After initial deposition, hemipelagic chalk sediments are often described as a 'hemipelagic ooze', with the pore space occupied by sea water. Such a saturated sediment, if deposited on slopes of less than even 1°, can be easily remobilized. Small, low-velocity, sediment-laden clouds were probably common features on the floor of chalk seas.

Autocyclic processes may include collapse around storm-generated shelfal hummocks, levee collapse in shallow subtidal channels, flank collapse of subtidal bars, and so on. Redeposited chalks within prospective areas of the North Sea more commonly result, however, from allochthonous (external) mechanisms. These may include tectonic flexing over active faults and rising halokinetic features (Fig. 9.32), which introduce and help to maintain unstable palaeoslopes throughout an area. Compressional tectonic events and halokinetic activity are likely to be synchronous throughout the basin, and therefore widespread products of redeposition occur at times of, or closely follow, tectonic activity. The tectono-halokinetic pulses coincide with seismic sequence boundaries.

Redeposition of chalks can form a wide range of features. Figure 9.31 (lower diagram B) illustrates rock falls, olistoliths and listric slides in consolidated sediments (very rare), and slumps, mass flows, debris flows and turbidites from poorly consolidated sediments. The latter four are extremely common, particularly within the Central Trough area. The abundance of chaotic mass-flow and turbiditic structures, seen in conventional cores from thicker bioturbation-free units, has resulted in some workers identifying these deposits as representatives of low-stand basin-floor tracts. Widespread redeposited chalks can form equally well along any regional base of slope: for example, tectonically controlled shoreface/shallow shelf-slope breaks and shallow/deep shelf-slope breaks, as well as the more obvious outer-shelf/basin-slope break. Only an ichnofacies analysis of bioturbated unit tops or intervening hemipelagic chalks can prove a palaeobathymetric location. Almost all Central Trough examples occur in basinal or deep-shelf settings.

9.7.5 Geometry and distribution of redeposited chalks

Figure 9.33 illustrates a typical chalk section from the Central Trough. Note particularly the wireline-log signatures of the redeposited chalk units (very low gamma-ray response, high neutron porosity, low formation density and low velocity). Such signatures are often in dramatic opposition to both hemipelagic clean chalks and argillaceous chalks. It is this physical character which, on high-quality 3D seismic data, allows the redeposited chalk units to be mapped over wide areas. Indeed, thick units can often be clearly seen on older regional seismic lines.

A very wide range of depositional geometries is apparent in single units or stacked complexes of redeposited chalks. Both linear fault- and salt wall-related occurrences have a linear-wedge geometry. Thick units are more commonly distributed as chains of overlapping fans, due to differential degrees of 'submarine incision' into adjacent bathymetrically shallower source areas. Deep-seated active salt domes and diapirs result in downflank ring lenses or ellipses of redeposited chalk. Active, deep-seated flower structures add another local complexity. A pop-up flower structure may shed in any direction, but usually perpendicular to local fault lines. The negative flower will collect very thick piles of redeposited chalk from adjacent shoulders.

The best depositional reservoir quality of redeposited chalks coincides with the greatest grain sizes. These occur where the energy of the flow starts to decrease at the base of the palaeoslope. In most cases, this is in the thickest part of the flow. Therefore, if the proto-structure of a present-day field existed during deposition, the best depositional reservoir quality is downflank into adjacent subbasins. This is illustrated in Fig. 9.34. Crestal areas of evolving structural highs often show 'shelfal' character and polyphase subma-

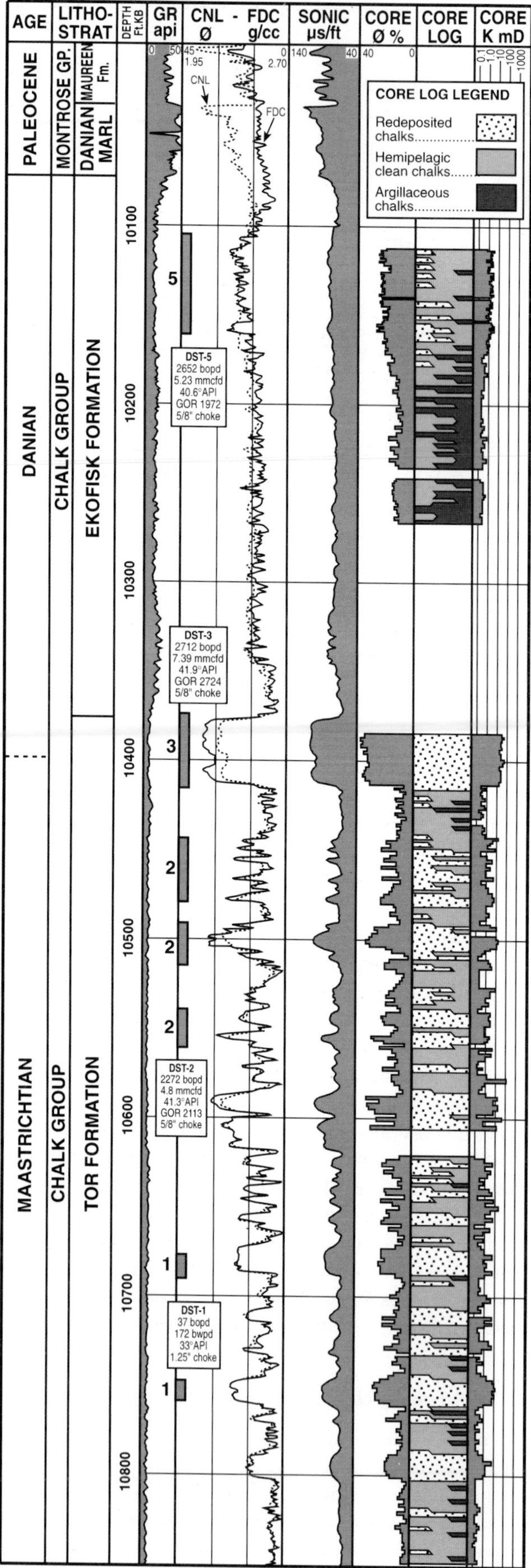

Fig. 9.33 Schematic log of Late Cretaceous and Danian Chalk in a typical Central Trough well.

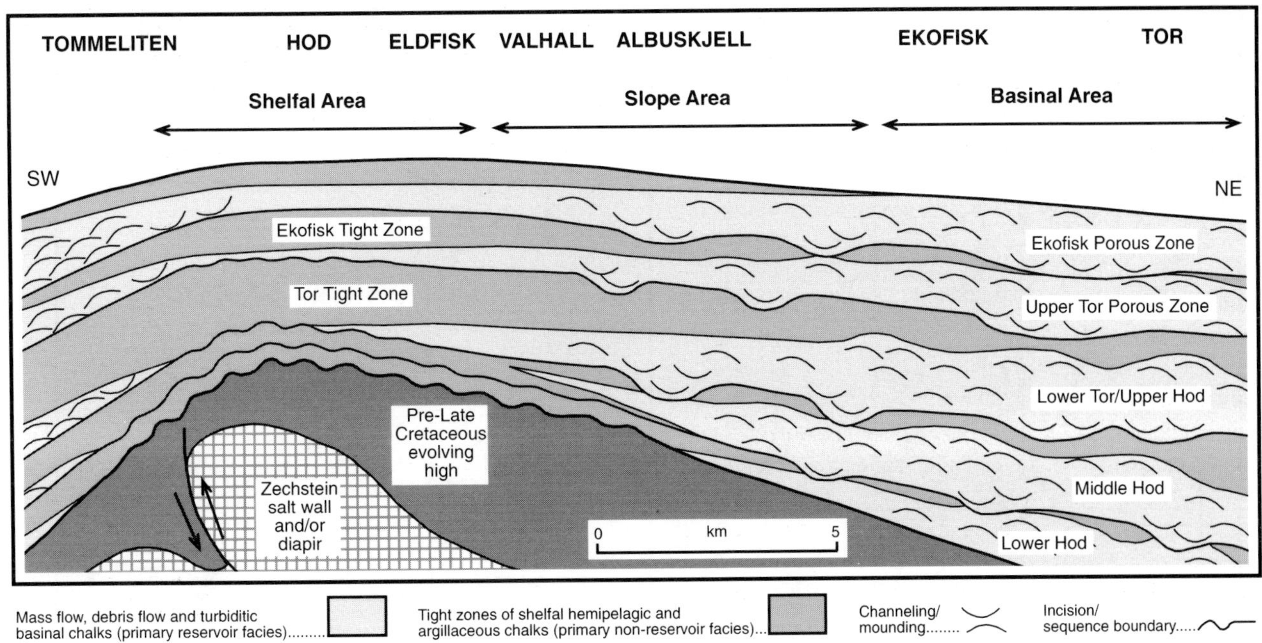

Fig. 9.34 Regional depositional chalk reservoir-quality patterns—an example from the Lindesnes Ridge area of the Norwegian Central Trough (modified from D'Heur, 1987c; D'Heur and Michaud, 1987).

rine incisions, and primary reservoir quality is often poorer. If, however, field structuration is a result only of Tertiary inversions (and/or postdepositional burial halokinesis) acting upon depositional lows, primary reservoir quality is very high throughout the field.

Deeper basin floors adjacent to large shallower shelfal areas have remarkable primary potential. Hatton (1986) mapped the extent of many deep-shelf to basin-floor reworked packages (Fig. 9.35) throughout the Central Trough, using all available wells and a regional seismic grid. This work showed that very few of the optimum primary-quality sites had even been drilled. Hatton's (1986) work also showed that 'redeposition events' formed large fan-form geometries; and that both thickness of individual flow complexes and geographic spread increased with stratigraphic younging—a feature also seen in many chalk fields with complete sections.

The progressive spread of redeposited units (Fig. 9.35, maps A–E), can be ascribed only to ever increasing tectonic shock and regional reactivation of deeper-seated faults. An alternative mechanism of increasing burial halokinesis in response to a progressively thicker overburden would create similar patterns. Halokinesis and transpressional tectonic, however, are inseparable.

9.7.6 Early burial history of chalk sediments

Diagenetically modified reservoir quality is largely controlled by the relative timings of *in situ* water-driven processes and their dynamic interaction with hydrocarbon migration, overpressure development and post-depositional tectonics and halokinetics. Over the prospective areas of the North Sea Basin, all chalk facies followed the same route from shallow-marine phreatic diagenetic environments, through deep-marine phreatic to deep-burial phreatic.

Figure 9.36 illustrates the principal reservoir-forming and reservoir-destroying processes encountered in North Sea chalks. Almost all primary porosity can be classified as matrix, with minor contributions from intragranular pore types (within foraminifera and calcispheres). The type of matrix porosity is highly variable, from the larger micropores within the sorted coarse muds of reworked chalks (primary matrix porosity up to 80%, permeabilities up to 1 D) to a range of micropore sizes in hemipelagic clean chalks (primary matrix porosity up to 70%, permeabilities up to 100 mD).

Initial dewatering and subsequent compaction are by far the most important of the early porosity- and permeability-reducing processes. These continued until pressure solution slowly took over, although physical compaction continues in any weakly cemented horizons to the present day. Even at an early stage, high pore pressures developed within redeposited chalks. This resulted from their higher degree of grain-framework support, and fluid transfer from more rapidly compacting clean and argillaceous chalks. The interbedded and/or interdigitating nature of two or even three chalk textures allowed overpressuring to be dynamically enhanced as depth increased. This process is very similar to overpressure genesis in mudrock-sealed stratigraphic sandstone traps.

Marine phreatic cementation is also an important process. Cementation in the upper layers of sediment is commonly evident in chalk hardgrounds, firmgrounds and burrowed surfaces. Hardground complexes can form almost totally cemented, early diagenetic rocks. In areas with high rates of sedimentation, such as chalk-fan systems in basin-floor and basin-slope settings, this early style of cementation has a negligible effect on reservoir quality. On evolving structural highs, however, sedimentation rates are often low and omission surfaces considerably more frequent, resulting in a higher degree of early marine cementation.

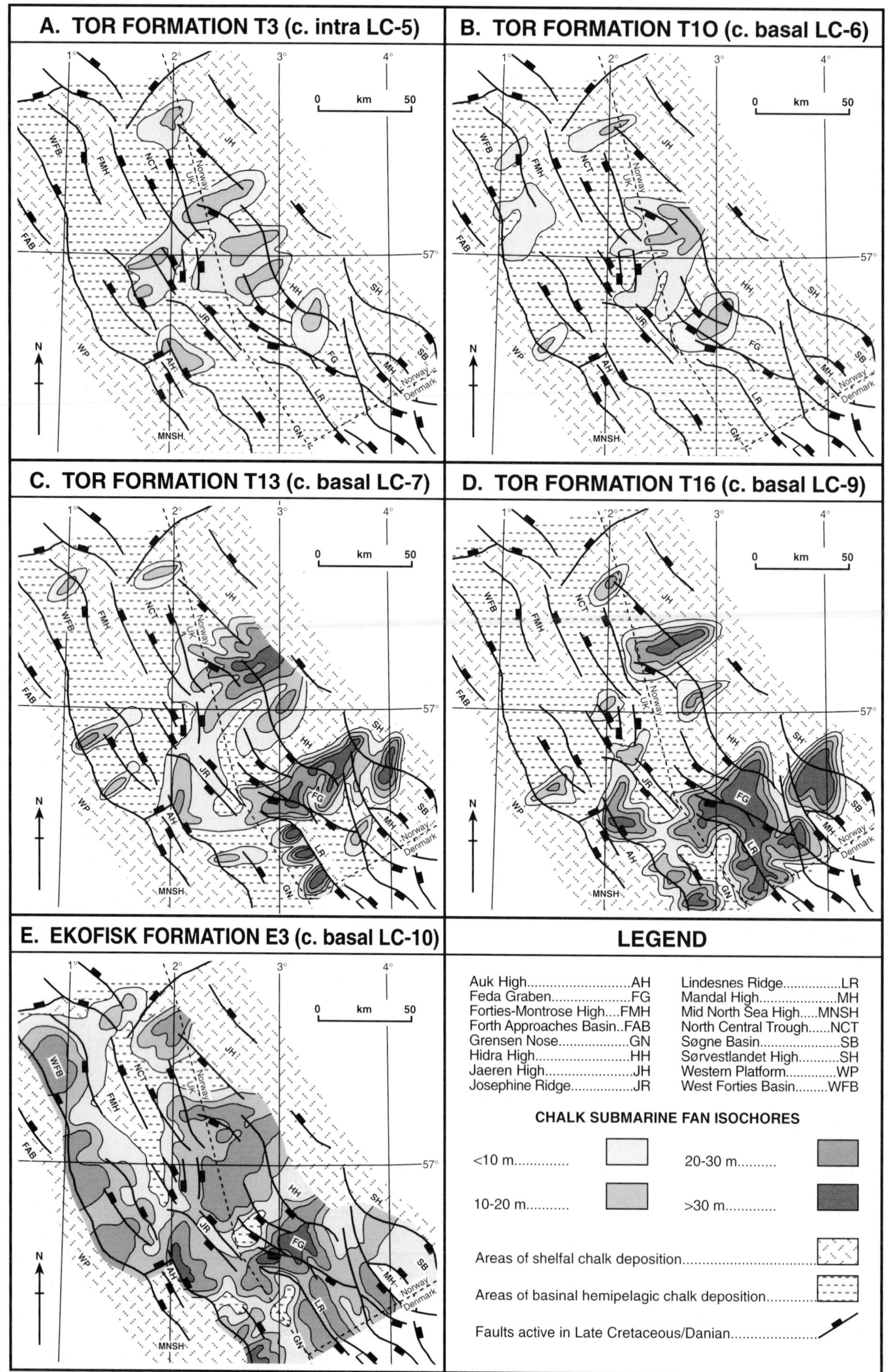

Fig. 9.35 The Central Trough chalk play in the UK sector—Maastrichtian and Danian submarine-fan distribution (modified from Hatton, 1986).

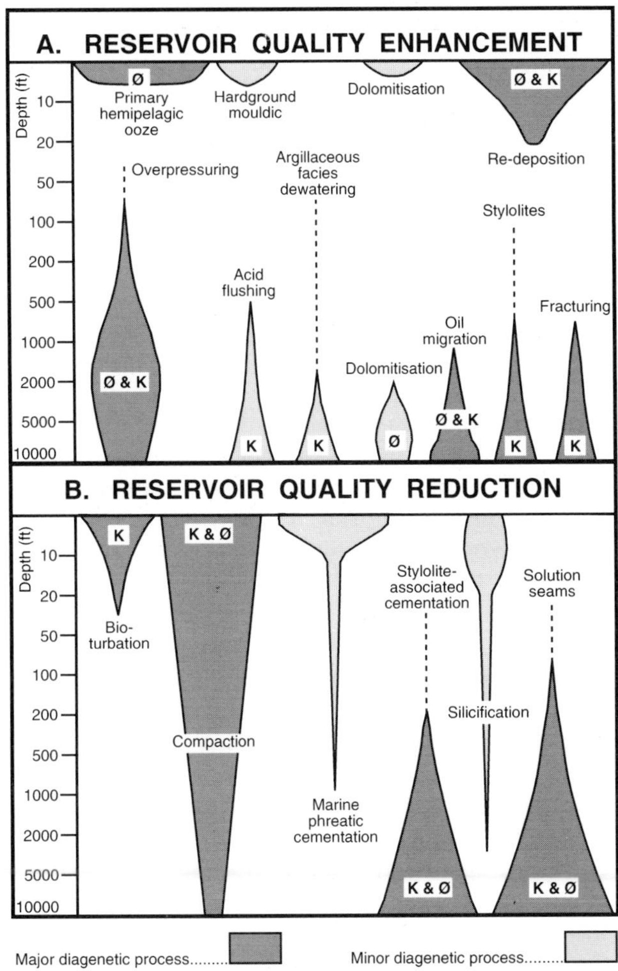

Fig. 9.36 Summary of chalk diagenetic processes and effects on poroperm (modified from Longman, 1980).

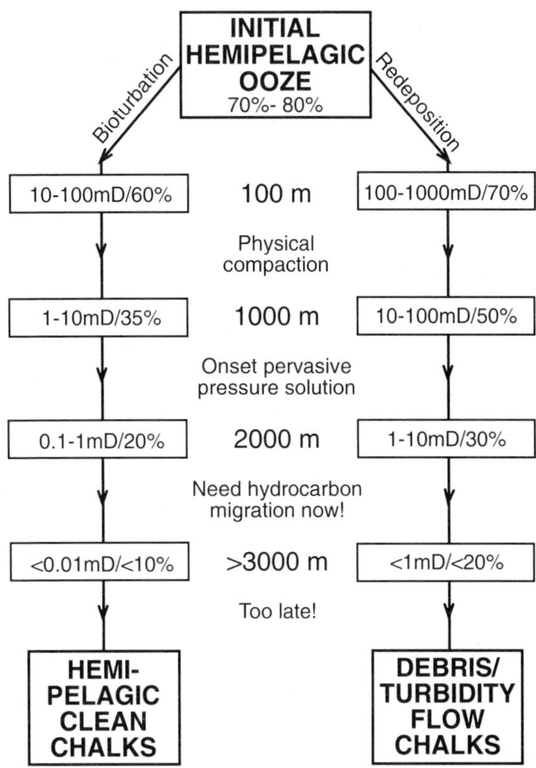

Fig. 9.37 Burial history of redeposited and clean hemipelagic chalks and poroperm destruction.

Minor early-diagenetic effects include silicification, the formation of flints and other chalcedonic replacement/cement microfabrics and deep sea-floor dolomitization. Dolomite occurs as microcrystalline cements and lime-mud replacement fabrics. It is common in association with omission surfaces or in basinal areas at and below carbonate-compensation depths. Both processes can be highly destructive of reservoir quality (poroperm), but have little effect in the prospective chalk fairways. Silica (flint) and dolomite fabrics are more frequently seen in argillaceous chalks. It is rare to find more than trace percentages of dolomite and silica within clean hemipelagic and redeposited chalks.

Figure 9.37 summarizes the effects of early diagenesis (to burial depths of up to 1000 m) on porosity and permeability in clean bioturbated chalks and reworked chalks. Physical compaction and early cementation has reduced the reservoir quality in clean hemipelagic chalks to porosities of 35–40% and permeabilities of less than 10 mD. Because of resedimentation, the debris- and turbidity-flow chalks fare considerably better, with porosities above 50% and permeabilities usually greater than 100 mD, for the same overburden. Slightly argillaceous chalks are no longer viable reservoirs at such depths. Although their porosities can be high (about 25–30%), their permeabilities are commonly less than 0.1 mD.

9.7.7 Late burial history of potential reservoir chalks

Once chalk sediments reach a depth of 2000–3000 ft (600–900 m), overburden pressure becomes high enough for pervasive pressure solution and associated cementation to occur. This style of cementation is highly destructive of reservoir quality. Overpressuring evolved largely because of pressure solution, although compaction and dewatering of the argillaceous facies also contributed. Overpressuring can help to preserve porosity in higher-permeability reservoir facies and restrict cementation to within pressure-solution zones.

During late burial, seals in the form of Tertiary mudrocks were being deposited over most of the Central Trough area. Compressional pulses still interrupted a long period of dominant thermal subsidence. Halokinesis became highly active. Correspondingly, the structuration of traps continued as a dynamic process and burial diagenetic processes started, and continued, to cut across depositional layering. Deeper flank and core areas of structures containing potential reservoir became progressively more indurated than stratigraphically similar crestal areas.

Pressure solution results in a wide range of different phenomena. To form initially, a discontinuity needs to be present in carbonate sediments. This may be as trivial as grain–grain contacts during early compaction (microgranular pressure solution) but is more likely to be lamina or bed boundaries. Once started, the process is self-propagating as part of a dynamic system, with increasing overburden and pore pressures.

In cleaner chalks, pressure solution forms denticular or zigzag structures at a range of amplitudes, from a few millimetres to in excess of a metre (microstylolites or stylolites). In argillaceous chalks, discrete dentae do not develop; instead, wavy seams are commonly formed, with amplitudes from a few millimetres to a few centimetres (solution seams). Dentae are aligned parallel to the principal stress and are therefore often seen perpendicular to bedding features. Some quite remarkable curved stylolites demonstrate evolution in parallel with structural dip development. Others have dentae directions oblique to bedding, indicating development after structural tilting. It is not uncommon to find stylolitized fractures with dentae parallel to overburden stresses, or even perpendicular to the original fracture, indicating lateral compressions in excess of overburden.

Pressure solution is intimately associated with cementation in a dynamic process. It takes over when compaction of the bulk rock or sediment ceases, because of either high pore pressure or the creation of a compact framework. Initial pore-pressure discontinuities are important in allowing the process to commence, with dissolution of calcium carbonate at the developing interface. Pore pressures at the seam become higher than in the adjacent matrix, allowing cementation of the matrix close to the seam (in low-permeability facies) or the movement of fluids out of the system (in higher-permeability facies). Solution seams can, therefore, self-cement in low-permeability facies. Further development cannot proceed until higher overburden pressures increase and trigger the process again. Alternatively, stresses may be accommodated, with the formation of a new seam on the margins of an old stylolite-cemented zone. Hence swarms of solution seams are common in argillaceous chalks. Stylolites adjacent to high-permeability facies often evolve without interruption.

Reworked chalks are rarely involved in cementation associated with pervasive pressure solution, at least until buried deeply. They have few primary discontinuities and a more even distribution of pore pressures. This prevents pressure-solution seams developing within larger flow complexes. It is common, however, to see their lower stylolitized contacts heavily cemented and enhanced cementation from microstylolites in upper laminated parts. Thus, thick flows are more likely to form quality reservoirs than thin ones.

Pressure dissolution can start to dissolve calcium carbonate at very shallow depths. Dolomite and silica dissolution may appear at depths in excess of 2000 m. Correspondingly, 'insolubles' are collected within the evolving solution seam. These commonly include sedimentary clay minerals and silica dust, together with primary organic matter. The insoluble content may vary from a thin veneer in microstylolites to a thickness of 3–4 cm in larger stylolites and solution seams.

Solely from the effects of compaction and later pressure solution, potential reservoir quality in clean hemipelagic chalks is generally destroyed with around 2000 m of overburden, and in debris- and turbidity-flow chalks at around 3000–3500 m (Fig. 9.37).

Potential chalk reservoirs usually had some structure soon after deposition, but were rapidly losing reservoir quality as their depth of burial increased. They needed a mechanism to prevent them from completely losing their reservoir quality. The key was the migration of hydrocarbons into the system. The Jurassic source rock was mature by the latest Cretaceous, although the earliest migrations into the chalk system did not occur until the Late Palaeocene. Hydrocarbon migration effectively inhibited water-zone diagenesis and aided in the preservation of reservoir quality.

9.7.8 Selected matrix reservoir models

The integration models of depositional reservoir quality (see Sections 9.7.2–9.7.5) with dynamic matrix, fluid and pressure histories (see Sections 9.7.6 and 9.7.7) can be used to illustrate and explain a wide range of present-day reservoir phenomena in chalk.

Chalk reservoirs show a range of phenomena related to variations in water saturation and porosity within the hydrocarbon column. This is largely a function of the interaction of matrix porosity and pore-throat size with the differences in relative permeability of fluids (water, oils, condensates and gas). Pore throats vary from about 5 μm in debris- and turbidity-flow chalks to around 1 μm in hemipelagic clean chalks. It is often difficult to directly measure saturation of the different phases of hydrocarbons, so in practice hydrocarbon concentration is expressed by the equation $1 - S_w$ (water saturation).

Using redeposited chalks as an example, migrational flushing by light oils will be less efficient (leaving S_w at, say, 20%) than flushing by clean gas (leaving S_w approaching 0%). In clean bioturbated chalks, light oil flushing may leave a residual S_w of about 40% and gas of, say, 20%. Slightly argillaceous chalks, with their low permeabilities, may have an S_w greater than 60% in an oil environment (technically non-reservoir), and argillaceous chalks may not be flushed at all. Production profiles show similar characteristics, with recovery factors much higher for gas-prone debris-flow chalks (up to 50%) than for light oil in clean hemipelagic chalks (only about 10%). Because of ionic attraction with the carbonate host rock, all reservoirs are water-wet. If they were not, recovery factors would be too low to make even large fields economic, and secondary waterflood recovery would not be possible.

Higher water saturations also created a diagenetic problem. It allowed water-zone diagenesis to continue, even in highly saturated oil reservoirs. The effect is seen to a much lesser degree in gas/condensate fields. This form of intra-hydrocarbon column diagenesis occurred to a much higher degree in hemipelagic clean chalk layers and at extremes in slightly argillaceous chalk. In the latter, pressure solution and associated cementation often proceeded unabated. Sometimes, heavier hydrocarbons became caught in the dynamic diagenetic process and were reduced to thin asphalt mats within and around pressure-solution features. Asphalts can be seen to form mineral cements, subsequently overlain by later generations of calcitic cements. Dynamic pressure solution resulted in a continual reduction in gross reservoir height, and thus contributed further to high reservoir overpressures. The rate of pressure-solution cementation, however, was not as great as in free-water zones beneath a reservoir.

Differences in facies-related changes in water saturation are best seen on the flanks of chalk fields (Fig. 9.38).

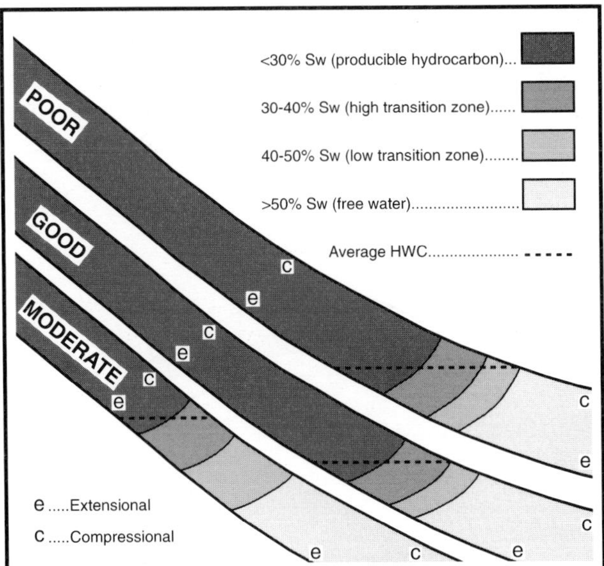

Fig. 9.38 Simplified model of variations in hydrocarbon–water contacts and transition zones in a multilayered reservoir (modified from D'Heur, 1987c).

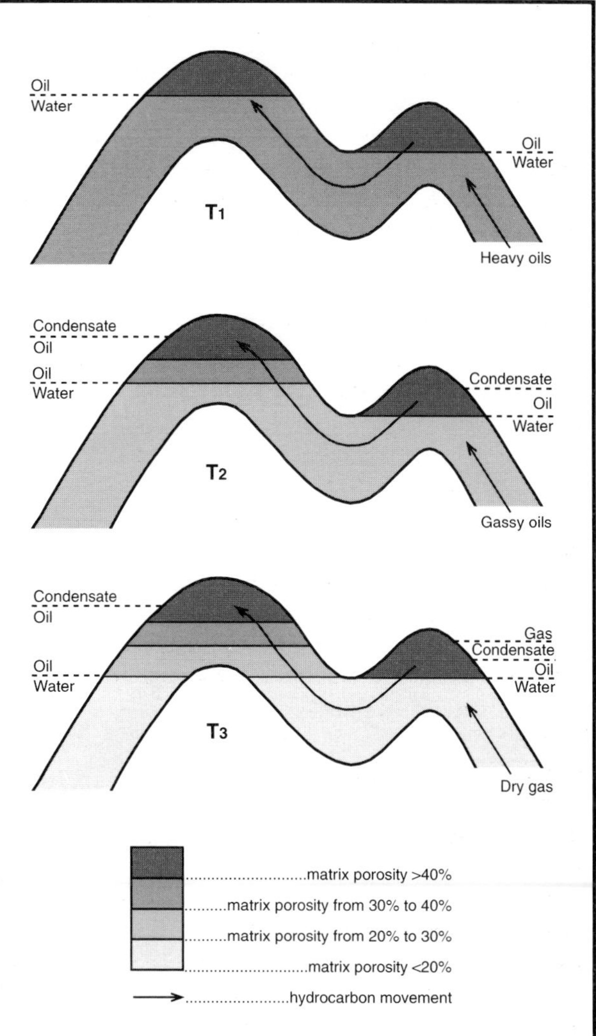

Fig. 9.39 Simplified model of hydrocarbon migration into a twin structure and effects on porosity distribution (modified from D'Heur, 1984).

Hydrocarbon–water contacts, or more rapid water saturation changes, show different elevations in different facies-defined layers. High permeability layers have deeper contacts than low permeability layers. Another effect is also seen in many densely drilled flank areas of fields. Fluid contacts can be shown to curve upwards within any given layer. Structural compression of pores in the upper part of the layer and distension in the lower part have been advocated to explain this phenomenon. On a field or structure scale, tectono-halokinetic compression and distension of pore throats can result in domed or saucer-shaped fluid contacts of significant relief (about 60 m in the case of the oil/water contact of the West Ekofisk Field).

Hydrocarbon migration into actively evolving traps also created a range of phenomena, which are evident from poroperm distribution patterns within present-day fields. If there was just one migration into a structure, a single hydrocarbon–water contact could be expected. Porosity would be well preserved in the hydrocarbon zone and be virtually absent in the water zone, due to enhanced water-zone cementation. What are commonly seen in many fields, however, are sudden jumps in matrix-porosity preservation within the present-day column. Figure 9.39 is a simplified sketch of this phenomenon, which shows how three successive hydrocarbon migrations create this style of porosity-layering, resulting in two palaeo- and one present-day hydrocarbon–water contacts. The figure also illustrates how smaller accumulations, particularly those with a saddle area adjacent to a large structure, need have only one porosity jump (or perhaps two) relative to its neighbour.

The high degree of susceptibility of chalk reservoirs to rapid cementation and the phenomena created by hydrocarbon migration allow sophisticated tectono-migrational basin- and field-history models to be deduced. It is possible to identify four separate regional hydrocarbon migrations in the Central Trough chalk fairway: the earliest in the mid-Palaeocene, a major one in the late Palaeocene/early Eocene, and subsidiary ones in the Late Eocene and mid-Oligocene. These events also correspond to major structural episodes in the creation of chalk traps and tie in with regionally mappable seismic reflectors.

It can be assumed that each migrational pulse was progressively more gas-rich because of increasing Jurassic source-rock maturity in deeper sub-basins. A late gas charge into a trap can displace previously reservoired oil downwards if the structure is not full to spill point. If full, oil can be completely displaced and migrate out of the field, to be lost at the surface or to fill a structurally higher field. Regional fluid-distribution models support this style of pulsed migration. Gasfields, condensate fields and oilfields of varying gravity often occur in close proximity to each other. There are also occurrences where a gas charge has left a residual oil rim to the reservoir.

There will come a point in the burial history when re-migration can no longer take place. This occurs when the matrix in the aquifer becomes too heavily cemented. Earlier migrations can be shown to follow regional facies belts; later ones do not. To complete the models at this stage of the burial history, fracturing needs to be discussed.

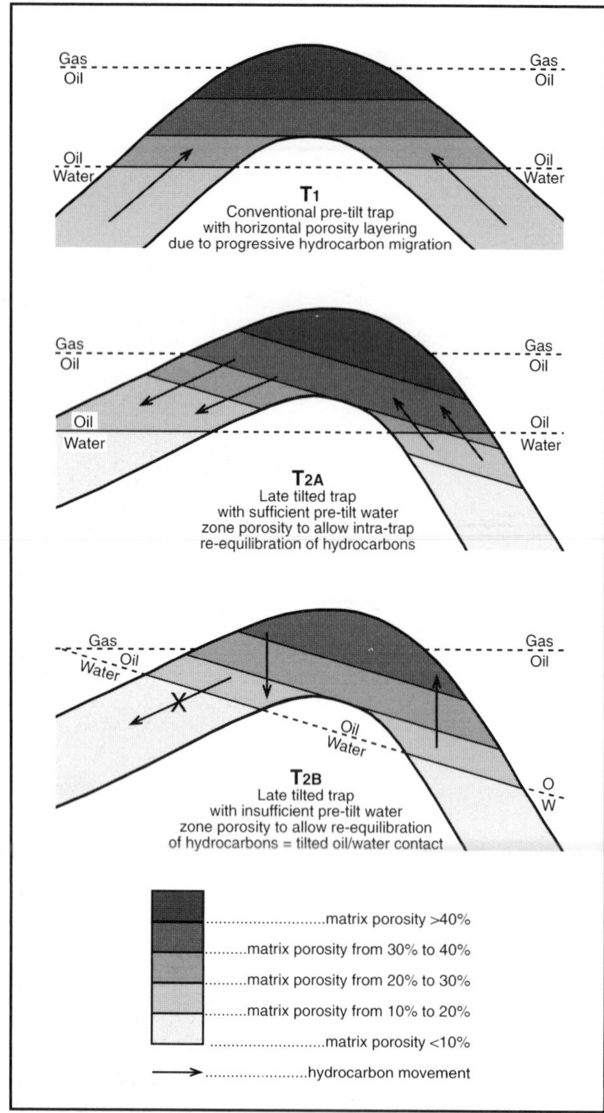

Fig. 9.40 Simplified model of single-layer reservoir showing fluid migration controls on reservoir porosity in conventional and tilted traps (modified from D'Heur, 1984).

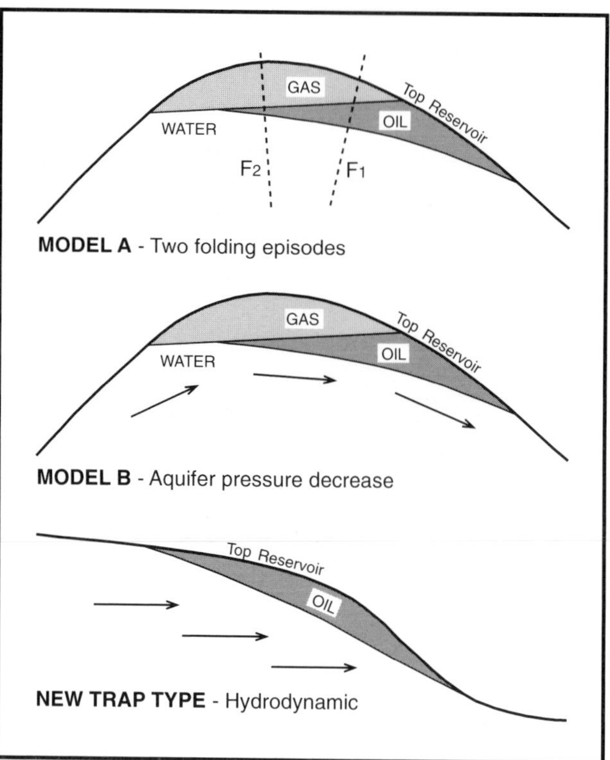

Fig. 9.41 Tilted fluid contacts and hydrodynamic trapping (modified from Megson, 1992).

Today, matrix aquifer response is negligible and almost all waterflood injectors are located above free-hydrocarbon levels. Correspondingly, considerable volumes of hydrocarbons cannot be recovered from the rim area of structures.

Another phenomenon is the tilted trap. After one migration or even a series of successive migrations, some traps become tilted. In a simplified single-layer model (Fig. 9.40), fluids attempt to return to equilibrium (horizontal restratification of gas, oil and water). If relative permeabilities in previous water zones are still high enough, fluid contacts can become horizontal again, but will cross-cut dipping palaeo-porosity layers. If the tilting is late and aquifer permeabilities are too low because of cementation, hydrocarbon–water contacts will have difficulty in reforming horizontally and will remain tilted. Within the reservoir above, however, oils and gases may still be able to re-equilibrate and form near-horizontal fluid layering. An example of this phenomenon, although complicated by facies-selective layering, can be seen in the West Hod Field in the Norwegian sector and the Dan Field in the Danish Central Trough.

Finally, a rather unique model has been proposed by Megson (1992), which is a variation on the dipping-contact model within reservoirs in the Danish sector (Fig. 9.41). Implicit in this model is the assumption that aquifer fluid pressures on one (basinward) side of a structure can push oils preferentially into the opposite flank. This is plausible if regional aquifer response is effective and not impeded by cementation (Danish reservoirs are much shallower) and a strong lateral component to aquifer pressure existed at the time. Regional hydrocarbon migrations up-dip laterally suggest the latter was true. This model can be used only if seismic backstripping can prove the structures did not tilt after migration. Megson (1992) took this concept to its limit and suggested that palaeo-aquifer activity could have wedged a hydrocarbon column in totally unclosed structures, such as a low-relief monocline, and subsequently became fixed by enhanced water-zone diagenesis. On the other hand, if hydrocarbons are frozen into a structural trap that is subsequently tilted by a complete shift of its axis, they can be preserved in a location on the flank of the structure (see the lower diagram of Fig. 9.44).

Figure 9.42 illustrates some of the simplified reservoir-layering models for selected fields in the Norwegian sector. In this figure, layers are expressed in terms of their matrix-porosity distribution evaluated from wells alone. These distributions may be interpreted in the light of many of the models discussed above. They include depositional layering of reworked chalks and diagenetic layering from depth-enhanced burial cementation, multiple hydrocarbon migrations and late tilting of structures.

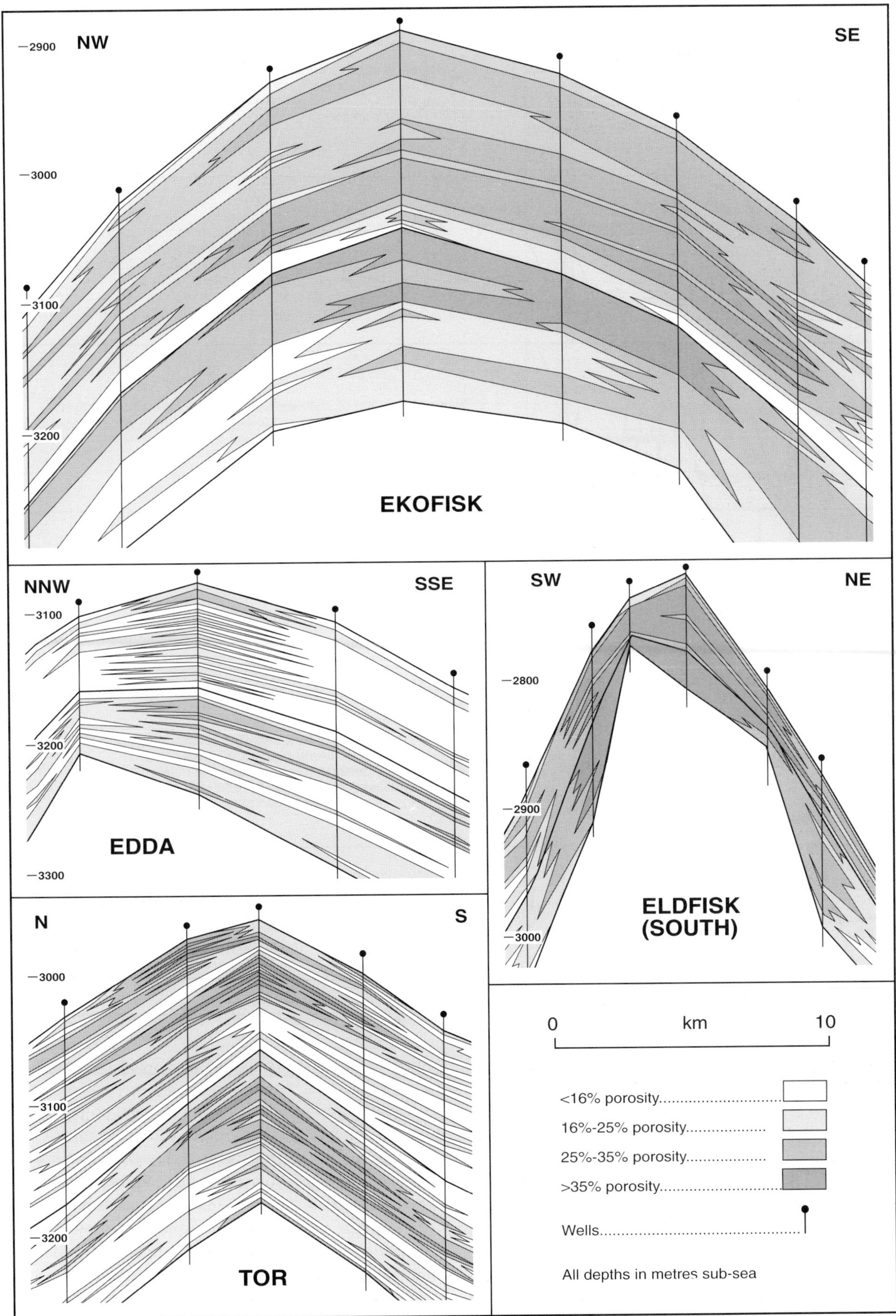

Fig. 9.42 Cross-sectional porosity variations of selected Norwegian chalk fields: Eldfisk, Ekofisk, Tor and Edda (modified from D'Heur, 1987c; D'Heur and Michaud, 1987; Michaud, 1987; Pekot and Gersib, 1987).

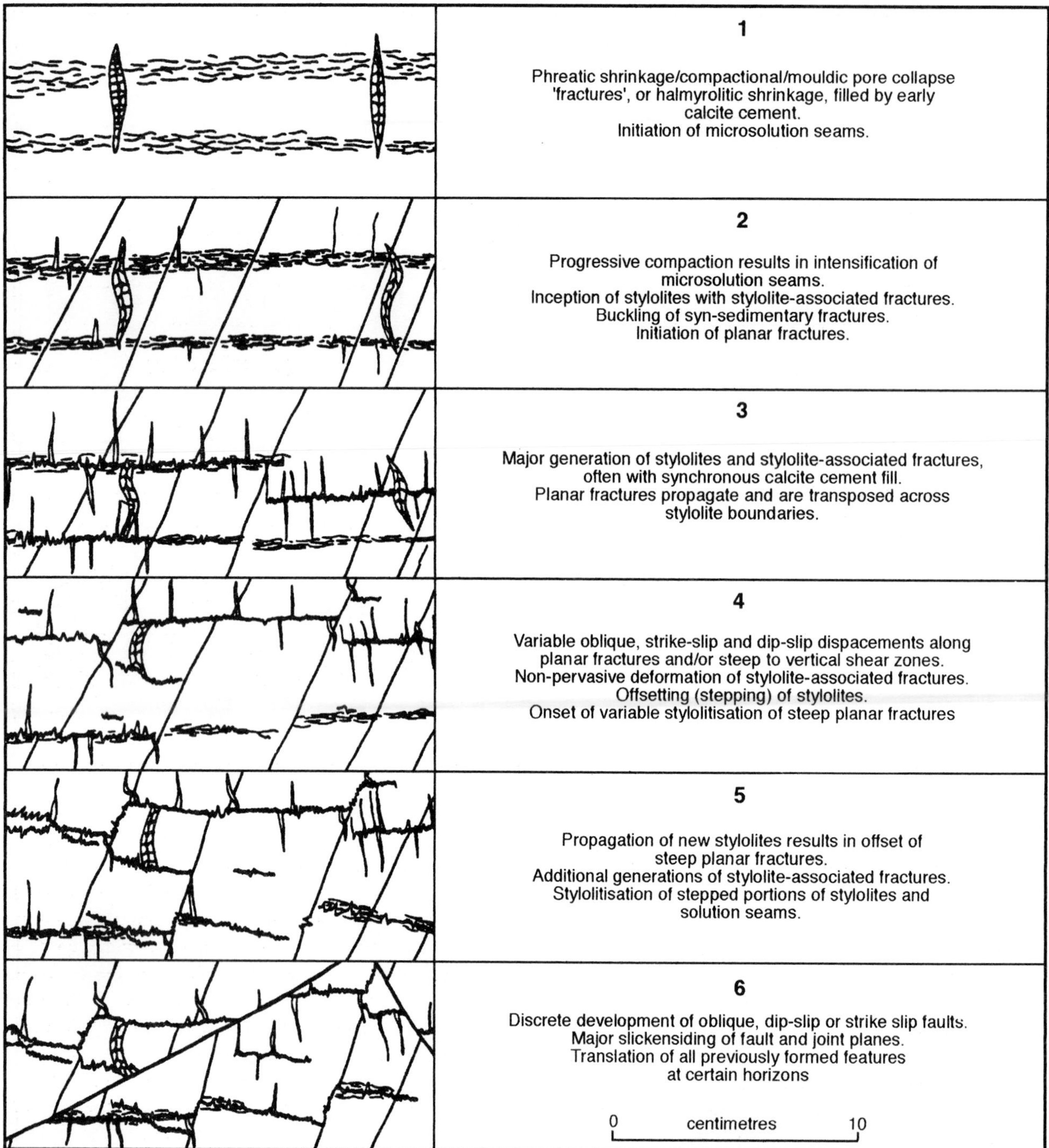

Fig. 9.43 Progressive structural maturity of chalk sediment and the evolution of fractures and stylolites.

9.7.9 Reservoir significance of pressure solution and fracturing

Pressure solution and fracturing are two processes that are dynamically inseparable in chalk reservoirs. Both form in response to stresses generated, principally, from ever-greater overburden and whatever horizontal minimum stress may exist on a local or regional scale. Pressure solution is commonly approached as a bedding-related diagenetic phenomenon. It is, perhaps, just as logical to consider resultant structures to be a class of 'fracture' in carbonate reservoirs. Figure 9.43 illustrates the progressive development of stylolites and fractures, with increasing structural maturity related to burial and regional lateral-stress fields.

Insoluble residues within pressure-solution phenomena, together with adjacent associated calcite cementation, can severely inhibit vertical permeability (K_v). Some thicker swarms of pressure-solution seams can form local permeability barriers in reservoirs. On the other hand, an active stylolite can have a high horizontal permeability (K_h), because of overpressured films of pore fluid at the interface between host rock and insoluble residues. If connections exist into permeable matrix, these can be considered to represent subhorizontal 'fractures', which contribute to enhanced reservoir quality.

Although stylolites can connect directly into matrix porosity, their tortuosity, residue seams and adjacent cemented zones can result in low transmissibility of fluids.

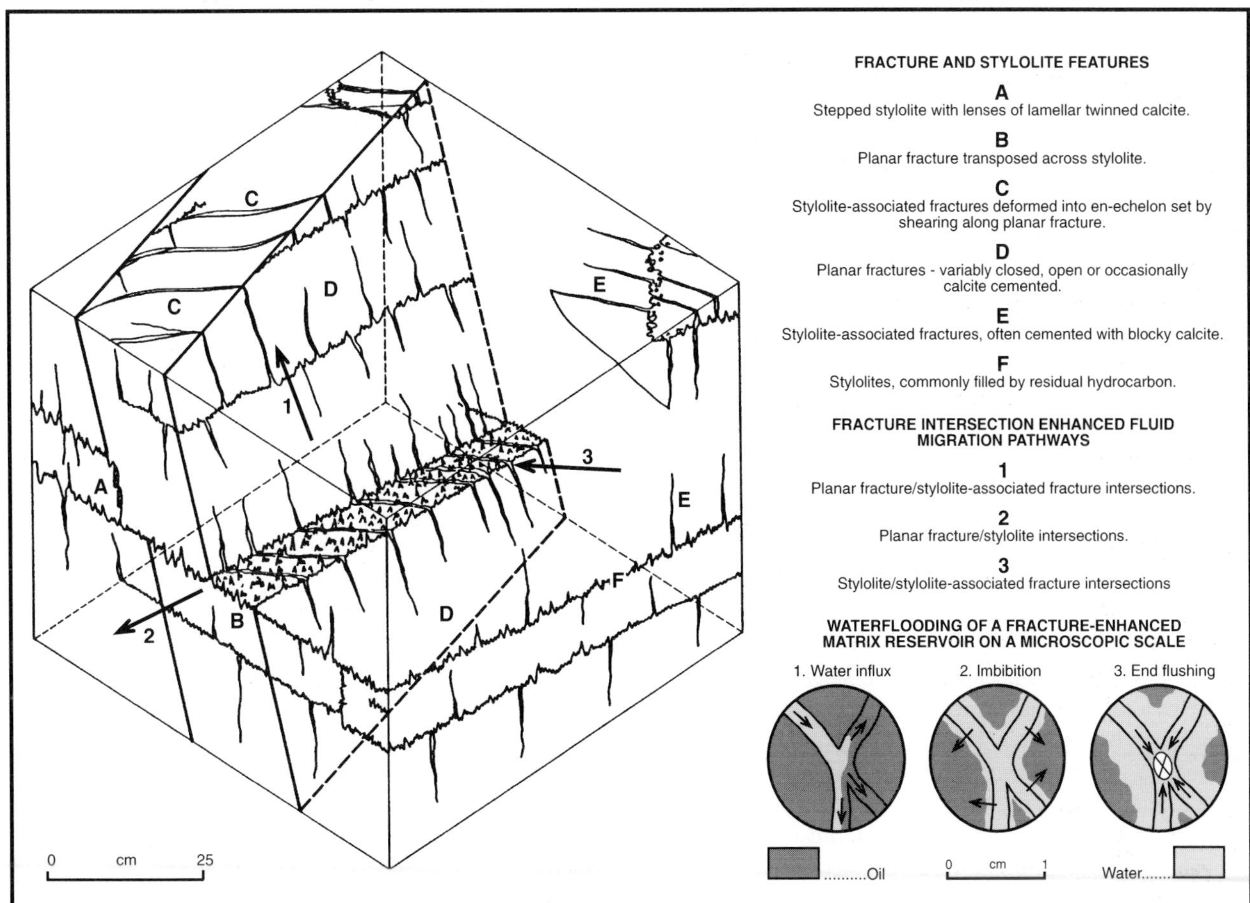

Fig. 9.44 Fracture/matrix reservoirs—block diagram of a typical volume of reservoir and waterflooding profiles (modified from Dangerfield and Brown, 1987).

Higher stresses at the dentae points of stylolites, however, result in propagation of extensional fractures. Early generations of stylolite-associated fractures are often short and merely cross adjacent competent cemented zones to peter out as microfractures within permeable chalks. The orientation of conjugate sets commonly inherits the local or regional influence of minimum horizontal stress. Ongoing structural maturity may result in the total obliteration of the origin of these fractures, as stylolites offset them. Unless an inhibitor (such as hydrocarbon) is present, stylolite-associated cementation can rapidly seal newly formed fractures. Ongoing deeper burial, however, continually reactivates the stylolite–joint partnership until the whole system loses all effective permeability at depths considerably greater than present-day reservoirs. Fracture systems, even if only partially cemented along their walls, can lose their interconnection with matrix fluid systems.

Fractures also form independently in response to regional stresses. These are commonly many orders of magnitude longer than stylolite-associated forms, and are often important in terms of hydrocarbon production. These can take the form of extensional joints, shear fractures and, at optimum extremes, faults. Ample tectonic and/or halokinetic activity occurred during chalk burial to generate these potentially prolific fluid conduits. Pressure-solution-related cementation, however, was continuously active. This provided a local source of cement to partially occlude fracture walls, or even completely seal them, as fast as they could form.

The migration of hydrocarbons into fracture systems would preserve any fracture porosity and permeability available at the time. The timing of hydrocarbon migration and the degree of structural maturity of the trap are therefore important factors. If structuration of the trap ceased before migration, almost all fractures would have been sufficiently occluded to render them ineffective. A large number of reservoir-effective fractures (high density, close spacing) occur in traps where structuration continued for a long time after migration. This is one of the reasons why diapiric traps demonstrate prolific production rates; many are still growing and fracturing today. Hydrocarbons will subsequently be available to fill fracture pores, as they continue to develop, and inhibit fracture cementation.

Hydrocarbon migration alone can serve to enhance fracture porosity and permeability. It is common in many carbonate reservoirs to encounter solution vugs within fractures and stylolites with accompanying patches of saddle dolomite cement and occasionally fluorite. It is generally believed that these vugs originate by flushing with organic acid-rich fluids in the migration front of oils. Crestal areas of fields tend to show this phenomenon more than flank areas. Post-migrational fractures rarely exhibit this feature; if they do, the chances are that there was a second pulse of hydrocarbon migration.

Figure 9.44 illustrates a typical volume of chalk reservoir. In this, the matrix porosity contains in excess of 95% of reservoired hydrocarbon and is criss-crossed by fractures and stylolites, which provide almost all of the effective permeability. Microfractures can tap into and provide a vast surface area for transmissibility of matrix fluids. Microfractures then need to connect, via stylolites and joints, to larger regional fractures to be able to produce matrix fluids in commercial quantities. Most fractures, however, because of the dominance of overburden stresses, are vertical or subvertical, and so are most wells. It is common to find only a few effective microfractures or small joints in conventional cores, despite flow rates that suggest otherwise.

What is more important than the absolute number of fractures is how they interconnect (often as subvertical conjugate sets) to form 'cones' of maximum-flow directions. In almost all fields, these are vertically upwards. Flow can take place within a fracture, but is optimized in 'pipeline tubes' formed at their intersections. Horizontal 'pipelines' also form at fracture and stylolite intersections. Most production in vertical wells is from these intersections. Only with the relatively recent advent of horizontal-well drilling has the true effect of the vertical preferential cones of pipeline tubes been realized. If steered correctly in the top of a reservoir layer, production rates from horizontal wells can be extremely high. In larger fracture systems, intersection tubules can pass through non-reservoir layers to the extent that just one correctly sited horizontal well in the topmost layer of a field can communicate all the way through to the hydrocarbon–water contact.

Fracture density is extremely important in the effective communication between chalk matrix and the wellbore. Figure 9.44 also illustrates the effect of waterflood. Initially, waterflood clears hydrocarbons from fractures, with subsequent imbibition progressively flushing out matrix hydrocarbons. The amount of trapped oil remaining after imbibition depends on three factors: firstly, the differential permeability between matrix and fractures; secondly, the relative viscosity of the hydrocarbon; and, thirdly, the density of effective fracturing.

Chalk reservoirs have highly variable fracture characteristics; these can result in a second group of regional and field models to add to already complex matrix characteristics. Fracture/stylolite/vuggy porosity rarely exceeds 5% bulk-rock volume; it is more commonly of the order of < 1% in most North Sea chalk fields. The effects on reservoir productivity, however, can be from a few hundred barrels/day from unfractured reservoir to over 6000 barrels/day out of a single pair of open joints.

Naturally, considerable effort has gone into predictive models of fracture location and density and the recording of fracture characteristics (width, length, fills, orientation), in both wellbores and conventional cores. Although pressure-solution features are often very obvious in core and on logs (particularly dipmeters), the presence of effective fractures (or lack of them) can often be deduced only from repeat-formation test (RFT) and drill-stem test (DST) results. The search for effective fractures in conventional core is fraught with problems; most 'fractures' identified in core are commonly drilling-induced or result from stress-relief of less effective microfractures. Spinner logs can prove successful in downhole fracture detection. Formation microscanners, available only over the last 5 years, can also be highly successful in detecting larger fractures and their degree of openness. Downhole cameras have also been utilized, albeit experimentally.

9.7.10 Selected fracture-reservoir models

Three key questions are important in fracture-reservoir modelling. Firstly, are fractures likely to be effective throughout the reservoir? Secondly, how big are they and how widely spaced? Thirdly, what is their distribution within the reservoir? A fourth question is often asked. If fractures are so effective in crossing intra-field seals, why do chalk hydrocarbons in traps not leak through the overlying Tertiary?

To be effective, fractures must continue to form after hydrocarbon migration into the reservoir. Continually increasing overburden pressures alone can do this, and variably distributed pore overpressures can create autofracturing. Most fracture orientations reflect regional or local stress fields, generated by compression and/or halokinesis. Invariably, these processes produce large-scale 'fractures', such as joint swarms, faults and shatter (shear) zones. For most chalk fields, these stresses are generating fractures today. Modern 3D seismic can resolve most large fault structures and thus provide data for the siting of development wells in potentially prolific production areas.

Predicting effective fracture density and distribution is more difficult. Firstly, the rate of change of curvature of the structure is evaluated from seismic mapping. This is often conducted at different levels within the reservoir. Areas with higher rates of change of curvature will have greater fracturing—extensional in positive areas and compressional in negative. Rate-of-change maps for the selected horizon are then compared with wellbore fracture characteristics and known production profiles to calibrate the models generated. Crestal areas of fields can have higher fracture density and higher production rates. Unfortunately, not all reservoirs are perfect domes. In some cases, crestal areas can have very low rates of change of curvature when compared with field flanks. Some models, particularly for traps that have been tilted after hydrocarbon migration, need to take account of palaeocurvature through backstripping techniques.

Reservoir-engineering models are also integrated into geophysically derived fracture-density maps. Competent non-reservoir layers are commonly shattered by many closely spaced microfractures, whereas more plastic layers of matrix reservoir have very widely spaced, but large, joints. Fractures are rarely observed in debris-flow units, but are common in tightly cemented stylolitic clean chalks. Thus fracture density can be mapped in all reservoir and non-reservoir layers throughout the field and be integrated into reservoir-engineering models for optimum siting of wells and accurate predictions of waterflood programmes.

Most predictive fracture and structural modelling still relies on the tried, but not always successfully tested, techniques of applying simple blanket models to the field in question to determine fracture orientations. Modern geophysical mapping allows fields to be modelled as compressional pop-up, diapiric or a combination of both. Major faults can be delineated and thus local stress fields predicted, in line with conventional structural modelling. Such analyses produce generalizations of fracture orientations in

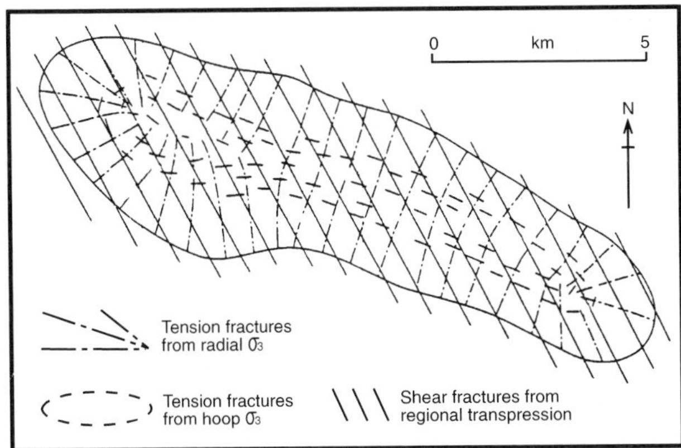

Fig. 9.45 Fracture patterns in chalk fields—an example of salt-wall and regional-transpression effects, Albuskjell Field, Norwegian sector (modified from Watts, 1983b).

different areas of the structure, which can be checked against wellbore and orientated conventional core data.

Many fracture analyses of chalk fields are based on the diapir model, even though many result exclusively from, or are heavily modified by, regional compressive stresses. In the diapir model, both radial and hoop stresses generate ring and radial spoke fractures around the crest of a structure. Fracture density is known to be very high in the steep flanks of diapirs and is often manifest in a series of ring faults, which can result in detached rafts of reservoir on top of the diapir. Both the Skjold and Machar fields show these features. Watts (1983a,b) successfully applied the diapir model to the elongate dome of the Albuskjell Field in the Norwegian sector (Fig. 9.45). In this model, a grain of shear fractures was attributed to the trend of the deeper graben flank. It is now known that these are local shear fractures from a regional north–south Tertiary compression, with conjugate sets coincidental to the previously interpreted 'radial' fractures over the middle areas of the field. The model, however, still stands as the only published work on fracture orientation patterns for a chalk field.

One of the more unusual fracture models is from Kraka Field in the Danish sector (Fig. 9.46). The matrix oil–water contact dips one way and is frozen by enhanced cementation. Recently generated fractures contain producible oil

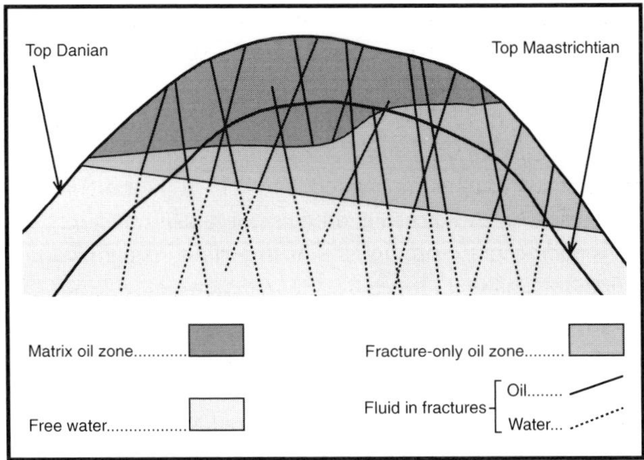

Fig. 9.46 Differential matrix and fracture hydrocarbon–water contacts, Kraka Field, Danish sector (modified from Jørgensen and Andersen, 1991).

well below the contact in the matrix. The free-water level in fractures dips in the opposite direction to that within the matrix. This model proved that migration into traps can take place even after the matrix aquifer is no longer effective. The only question that arises from this model concerns the origin of the opposing contact dips: were there two phases of tilting, like a see-saw, or one phase, with hydrodynamic trapping in fracture aquifers?

In the same way that exploitation today is so dependent on fractures, so also must hydrocarbon migration have been. Simplistic models of connection to the Jurassic source, via large, late-active, Jurassic, basin-bounding faults and then up-dip in the chalk fairways, are currently in need of modification. Local models of direct upwards migration to diapiric traps through shattered flanks are still valid, however. Complex routes through compressional fault and fracture systems and re-migration within and between fields must be taking place today, in the form of a complex pressure-cooker effect, with short sharp pumps interspersed with longer periods of fluid-pressure build-up.

Re-migration taking place today can be seen in the form of gas chimneys above many fields. Gas phases are currently escaping from chalk reservoirs via Tertiary seals that have been fractured by late tectonism and high pore pressures. Pressure-cooker effects within chalk fields allowed pulsed leakage into fractured seal shales, which resealed as pressure depleted. This dynamic system has resulted in dispersed gas above chalk fields, which interferes with seismic resolution.

9.8 Fields in Late Cretaceous and Danian Chalk

Chalk reservoirs are much more complex than many early exploration wells suggested. Although some characteristics can be generalized and specific models illustrated (see Section 9.7 above), the factors that combine to form fields are unique for each structure. It is now known why hydrocarbon-production models differ so much between fields within the Late Cretaceous Chalk play. Apart from basic chalk lithology, differences are much more common than similarities.

This final section provides a brief overview of the major chalk fields and their gross characteristics. Figure 9.1 illustrates the location of Chalk fields in the North Sea Basin; and Fig. 9.29 details the location of fields within the Norwegian and Danish sectors. Table 9.1 summarizes the known reserves within fields and lists specific key publications. Data for fields without specific publications have been derived from general publications on the chalk-play fields. Excellent recommended accounts include those of Damtoft *et al.* (1992) and Megson (1992) in Denmark and Hardman (1982, 1983) and D'Heur (1984, 1987c) in Norway.

9.8.1 The Norwegian fields

The location and depth to top reservoir of the Norwegian chalk fields are illustrated in Fig. 9.47. The fields fall into three tectonic provinces, which were shaped by the original geometry of Late Jurassic transtensional graben systems and subsequent Early Cretaceous transpression (see Figs

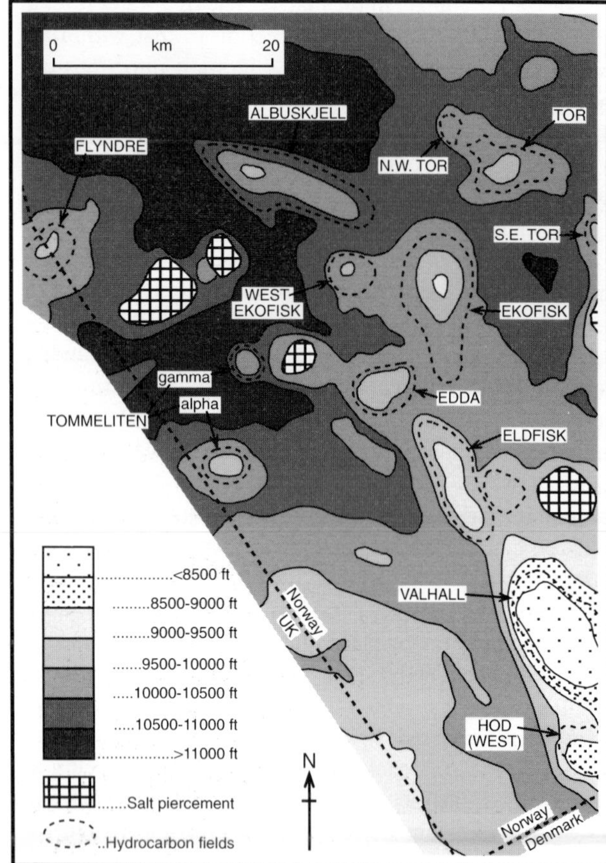

Fig. 9.47 Top chalk-depth map and the location of chalk fields, Norwegian sector (modified from Brewster and Dangerfield, 1984).

9.23 and 9.24). Figure 9.48 illustrates the cross-sectional geometry of selected Norwegian chalk fields.

The first tectonic province is the Lindesnes Ridge, a major Late Cretaceous inversion axis, with an accompanying deep-seated salt wall. A string of fields sits astride this structure, and includes Hod, Valhall, Eldfisk, Edda and Tommeliten (and, arguably, Flyndre on the UK median line). Valhall is the shallowest structure, with depth to top reservoir generally increasing in a north-westerly direction. During deposition, the evolving high of the Lindesnes Ridge resulted in chalk submarine-fan systems being shed north-eastwards. Correspondingly, most of the fields on the Ridge are today within proximal carbonate-fan settings rather than in the higher-reservoir-quality mid-fan lobes.

Most of the fields show Late Cretaceous sequences with a number of units absent, because of syndepositional erosion. For example, the Campanian is absent in Eldfisk, the Danian is absent in East Hod and the entire interval from Campanian to Maastrichtian is absent in West Hod. The principal reservoir in West Hod is of Coniacian and Turonian age, whereas the main reservoirs in all the other fields are within the Maastrichtian and Danian. Campanian reservoirs are present in the cores of Tommeliten Gamma and East Hod. Coniacian and Turonian reservoirs also occur in the core of Eldfisk. The largest field is Valhall, with recoverable reserves of 365×10^6 barrels of oil and 1.5×10^{12} ft^3 of gas; the smallest is the twin-structure field of Tommeliten, with 56×10^6 barrels of oil and 850×10^9 ft^3 of gas.

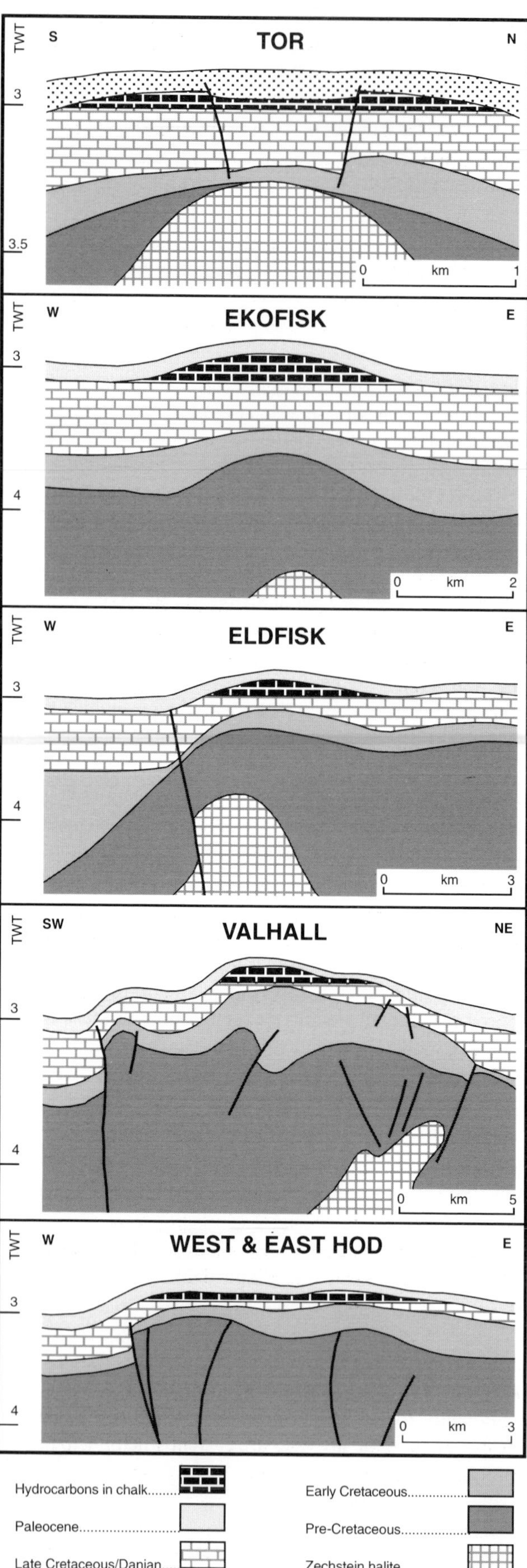

Fig. 9.48 Structural cross-sections of selected Norwegian chalk fields: Ekofisk, Eldfisk, Valhall, Tor and Hod (modified after D'Heur, 1987d; Pekot and Gersib, 1987; Færseth *et al.*, 1986; Leonard and Munns, 1987; Norbury, 1987).

The second chalk structural province is the North Mandal area, which encompasses the three separate accumulations that form the Tor Field complex. This structurally difficult area formed initially from a Late Jurassic transtensional interference zone, which subsequently became a transpressional flower structure during the Early Cretaceous. It was continually reactivated throughout the Late Cretaceous and into the Tertiary. Correspondingly, chalk submarine fans (sourced from the Lindesnes Ridge to the south-west, the Mandal High to the south and the Sørvestlandet High to the north and east) steered tortuous routes through the area. Thus, today, the resultant reservoirs show considerable variability in quality and layer continuity between even closely spaced wells (see Fig. 9.42, Tor cross-section). A large number of satellite prospects occur in the area, which have been somewhat hit-and-miss affairs. Only the north-west and south-east Tor satellites have proved commercial quantities of hydrocarbon.

The final trend is often informally described as the 'Diapir Fields'. These occur in a curved line parallel to the above two trends and include Albuskjell, Ekofisk and West Ekofisk. Ekofisk is one of the few giant fields in the North Sea, with recoverable reserves in excess of 1250×10^6 barrels of oil and nearly 4×10^{12} ft^3 of gas. Albuskjell, on the other hand, as one of the deepest chalk fields, has recoverable reserves of 69×10^6 barrels of oil and 564×10^9 ft^3 of gas. The 'Diapir Fields', as their name suggests, sit astride structurally high halokinetic domes. The salt originated through a deep-seated midgrabenal high initiated in the Late Jurassic, but diapiric emplacement appears not to have taken place until well into the Tertiary. The area was a basin during the Late Cretaceous and received considerable thicknesses of mid-fan chalk sediments (high reservoir quality) from both the Lindesnes Ridge and Sørvestlandet High source areas (see Fig. 9.35). Correspondingly, reservoir layers are of very good quality, thick and areally extensive over the structure (see Fig. 9.42, Ekofisk cross-section).

Although the Norwegian chalk play contains some thirteen separate fields, not all structures have commercially viable reservoirs. Some structures contain little in the way of thick, high-quality, reworked chalks. Flyndre and the Tommeliten fields are barely economic because of this. In other cases, diapirs have pierced through the chalk section and into the Tertiary above, thus destroying the integrity of reservoir layers. Very few diapir-flank wells have been drilled in the Norwegian sector to test the possibility of reserves in these sites. Many fields have proven hydrocarbons below structural closure (for example, West Hod) and the presence of stratigraphic, diagenetic or hydrodynamic components to reservoir seals is now widely accepted. Seismic 'bright-spot' analyses are currently helping to prove up additional preserves trapped in this way, within both established fields and adjacent satellites.

Fields in the north of the area are more gas-prone, because of the proximity to much deeper Jurassic source rock in the northern Feda Graben. Complex migration and re-migration routes between evolving structures, however, have resulted in a wide variation in hydrocarbon gravity, both regionally and even between fields that are close together.

9.8.2 The Danish-sector fields

Chalk fields in the Danish sector are considerably smaller than those in Norway (Table 9.1). From available data on recoverable reserves (Hurst, 1983), the largest oilfield is Gorm, with 150×10^6 barrels of oil, and the largest gas (condensate) field is Tyra Field with 2×10^{12} ft^3. The fields are much shallower than in the Norwegian sector (by almost 1000 m; see Fig. 9.29), which makes smaller fields more economical to exploit.

Because the fields are shallower, they have suffered less burial cementation and pressure solution than Norwegian fields, and higher porosities and permeabilities are preserved. Unlike the deeper Norwegian fields, Danish ones are less dependent on the presence of a redeposited chalk facies to form reservoir. Hemipelagic clean chalks are commonly above the depth at which effective porosity and permeability are destroyed (see Fig. 9.37). Correspondingly, almost all structures in the shallow southern areas of the Central Trough contain reservoir that can be commercially exploited.

This area has been divided into two structural provinces (Fig. 9.49). Structural cross-sections of selected fields are illustrated in Fig. 9.50. The Southern Compression Zone Province, located to the south-east of the Ringkøbing–Fyn High, includes fields such as Adda, Roar, Tyra and lgor. As the name suggests, these structures were initially formed by the inversion of a major Late Jurassic transtensional half-graben (for example, Tyra, Adda and Roar) and continued to be exaggerated by Late Cretaceous and Tertiary transpressive events (for example, Dan and Gorm).

The second area is known as the Salt-Dome Province and, as its name suggests, is dominated by high-level halokinetic features. Fields developed over the diapiric structures can be complex and include salt-raft and diapir-flank components. The Skjold Field is a typical example. Although porosities in salt-dome fields can be highly variable, they are commonly intensely fractured, particularly in off-crest or flank areas. Hydrocarbon production rates can be very high.

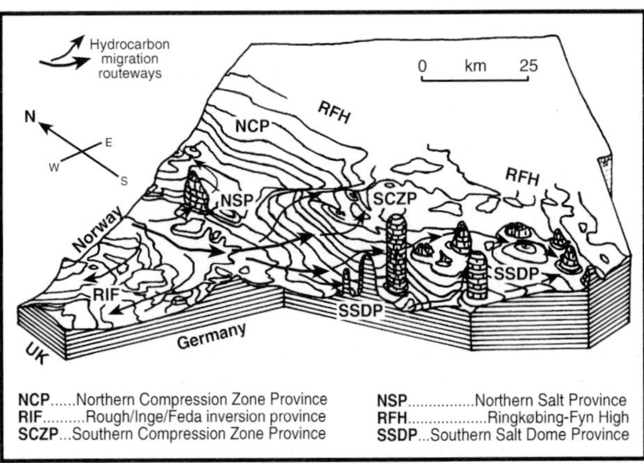

Fig. 9.49 Block diagrams of top chalk depth and the location of chalk-play provinces in the Danish Central Trough (modified from Nygaard *et al.*, 1983; Damtoft *et al.*, 1992; Megson, 1992). For location of fields, see Fig. 9.29.

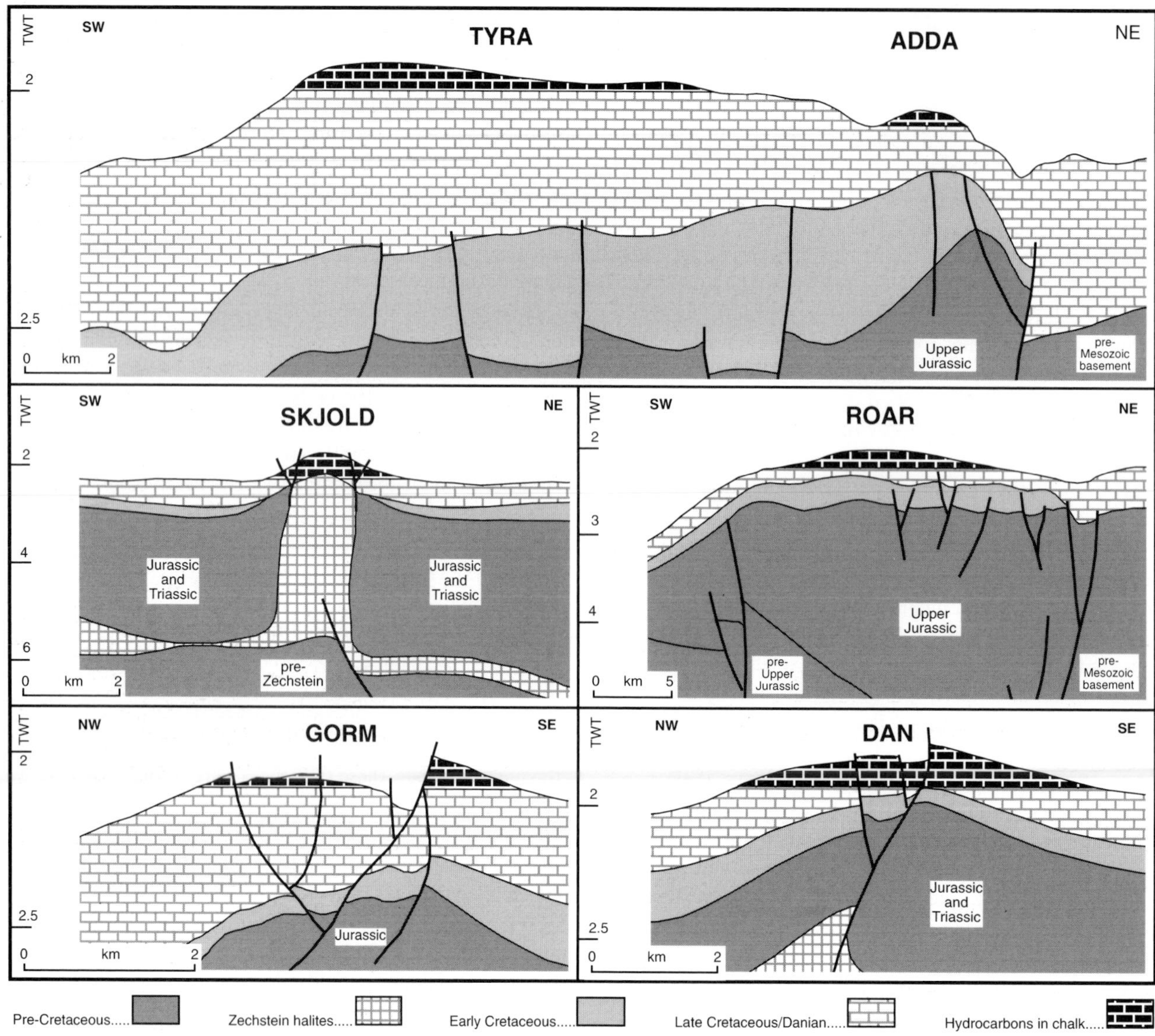

Fig. 9.50 Structural cross-sections of selected Danish chalk fields: Dan, Gorm, Skjold, Tyra, Adda and Roar (modified from Megson, 1992).

The northern, deeper area of the Danish Central Trough has proved more difficult to explore. Structures in this area between the established Danish and Norwegian chalk plays are often very small, because of complex transpression zones that criss-cross the basin. The many pop-up structures throughout the area were well established before deposition. Most wells drilled through the chalk have failed to find significant sequences of redeposited chalk, which would be needed to form reservoirs at deeper structural levels.

The distribution of hydrocarbon type (gravity) is well understood in the Danish sector. There is a direct correlation between gas and light oils in deeper areas, through closer proximity to more mature Jurassic source rocks, and heavier hydrocarbons on the flanks of and in the south of the area. Heavier hydrocarbons from early migrations into the Adda/Roar/Tyra complex were progressively displaced southwards to the Igor, Dan and Regnar areas by a charge of late gas.

9.8.3 Off-trend fields—Machar, the J-Block complex and Harlingen

Machar Field is located in the Eastern Central Graben (Fig. 9.1) and consists of two reservoirs, one in the Chalk (Hod, Tor and Ekofisk formations) and one in Palaeocene sands. Reserves for the two reservoirs are estimated to be 100×10^6 barrels oil equivalent. Foster and Rattey (1993) give an excellent description of the field. Figure 9.51 illustrates a section through the structure. The Cretaceous reservoir is a combination diapiric raft and flank trap (very similar to Skjold in the Danish sector). The origin of the structure is discussed extensively in Hodgson *et al.* (1992) and is illustrated in Fig. 9.53.

Machar's reservoir facies are dominated by hemipelagic chalks. Reworked units are generally absent, suggesting that the diapiric structure was evolving during deposition. Oil migration started early in the Oligocene, when the top of the trap was buried to depths of only 100–300 m. High porosities of up to 35% have been preserved, although matrix permeabilities are usually less than 1 mD. Continual diapiric growth during burial and hydrocarbon emplacement resulted in extensive fracturing of the reservoir.

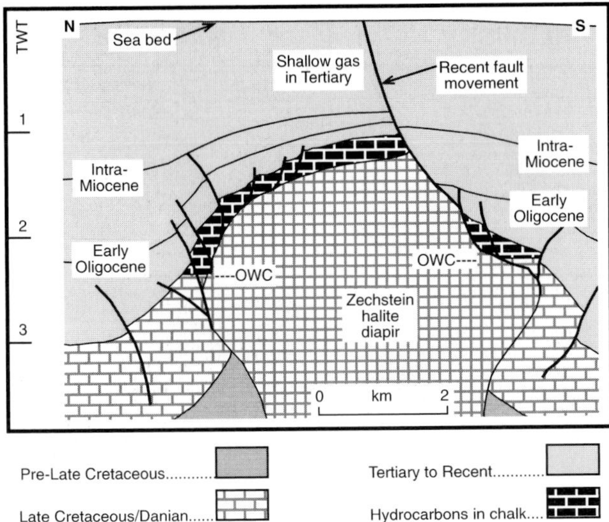

Fig. 9.51 The Machar salt-raft trap (from Hodgson *et al.*, 1992).

Although fracture pores account for less than 1% of the bulk volume, average fracture permeability is of the order of 1 Darcy.

The J-Block complex includes the fields of Josephine, Joanne and Judy. There is no academic publication about the fields, but some information can be extracted from the operators' annual reports. Although these are largely Chalk reservoirs, reserves are also present in Palaeocene, Jurassic and possibly Triassic sands. The three structures appear to be complex transpressional–halokinetic features, which straddle and flank the Josephine High. This was a positive area during chalk deposition—perhaps somewhat similar to the Lindesnes Ridge in the Norwegian chalk fairway. Released wells from the area show the chalk in Josephine to consist of multiple reworked units within a thick Maastrichtian and Danian sequence. Chalk intervals in the Joanne and Judy fields are attenuated and dominated by hemipelagic chalks, in which the reservoir quality is relatively poor. The presence of high-quality light oil in the chalk reservoir and an overlying reservoir in Palaeocene sands indicates that the chalk hydrocarbon fairway need not be restricted to areas where the Tertiary exclusively comprises mudrock, an assumption held by many explorationists.

The Harlingen Field (Fig. 9.52) is discussed by van den Bosch (1983). It is located in the northern onshore area of the Netherlands (see Fig. 9.1). The reservoir contains gas that has been tested at rates of up to 65 000 m^3/day. The reservoir occurs in the Ommelanden Chalk Formation and is of Late Campanian and Maastrichtian age. Reservoir facies consist of hemipelagic chalks; reworked chalks are not present. Unlike many North Sea chalk reservoirs, the chalk has suffered from leaching and extensive fracturing. Matrix porosities can reach 38%, while fractures and small solution vugs account for about 5% of bulk volume. The structure was an inversion high during the late Cretaceous. The structure was further exaggerated by a series of inversions and associated halokinesis during the latest Cretaceous and Tertiary. Recoverable reserves are estimated as up to 80×10^9 ft^3 of gas. There is also a secondary reservoir within the Early Cretaceous Vlieland Sandstones.

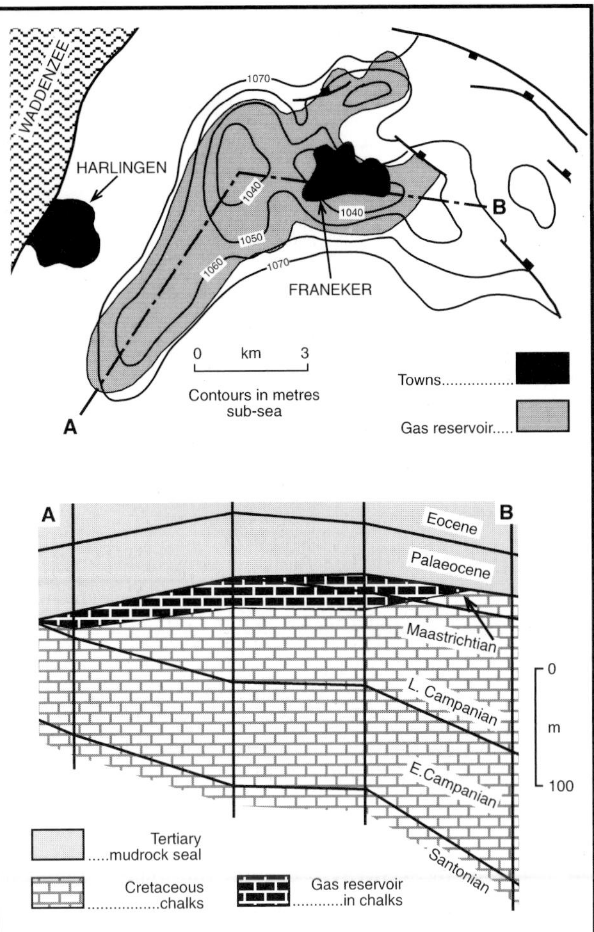

Fig. 9.52 The Harlingen Field, Netherlands sector (modified from van den Bosch, 1983).

9.8.4 Further potential of the Chalk play

Structural models to explain the known chalk fields are well understood. Figure 9.53, although very generalized, illustrates the three different models currently utilized, from the compression of deeper Jurassic transtensional basins, typified by Valhall Field, through the complex transpressive salt-wall model for Eldfisk, to the elegant halokinetic (diapir) model for the Machar Field. Understanding the structure of a chalk field has moved a long way from the older, simple, salt-dome and diapir models of the early 1970s.

Further potential for the Chalk play occurs at three different, although interrelated, scales: firstly, that of exploitation or development, where most work has concentrated on improving recovery factors through a better understanding of the sophisticated burial history models and the recent advances in 3D seismic resolution and processing; secondly, the remote appraisal of the viability of marginal satellite structures and traps, with the aim of supplementing the late-life yield rates of rapidly declining production from current fields; and, thirdly, the application of geological and geophysical models learnt from the fields in the future exploration of the Chalk play away from the principal fairways in Norwegian and Danish areas of the Central Trough.

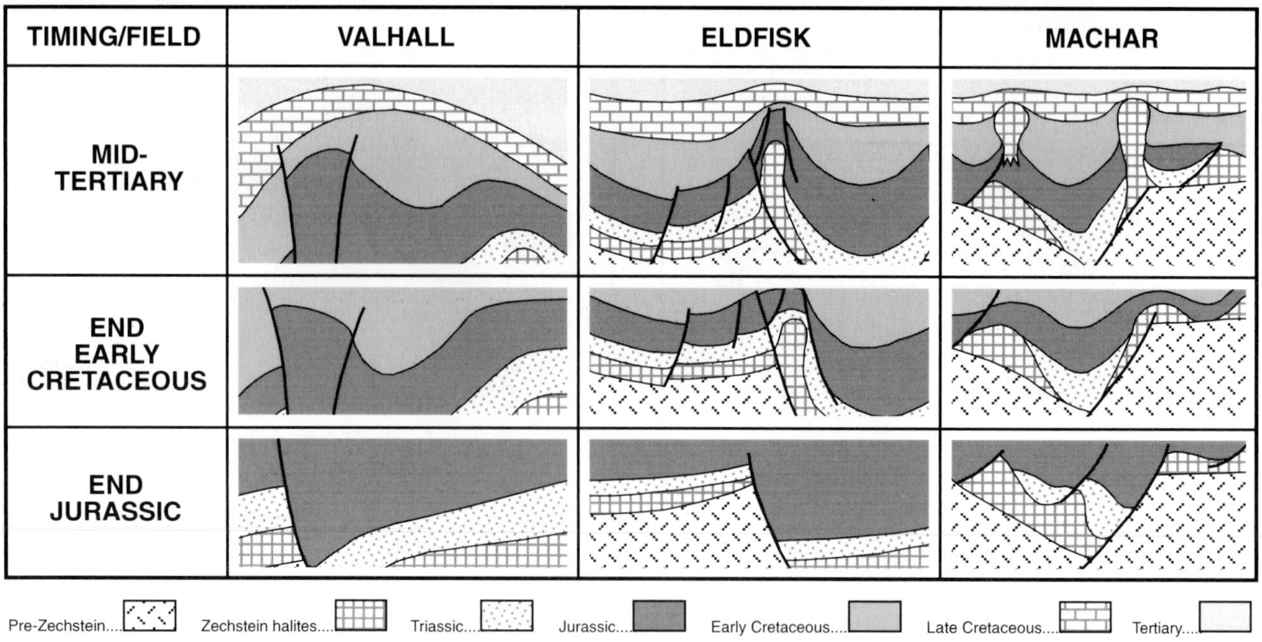

Fig. 9.53 Structural evolution of selected chalk fields: Valhall, Eldfisk and Machar (modified from Leonard and Munns, 1987; Michaud, 1987; Hodgson *et al.*, 1992).

Recent advances in 3D seismic and phase-inversion processing techniques not only allow structures to be defined more precisely, but also allow visualization inside reservoirs and direct mapping of matrix porosity distribution, which may then be quantified at well tie points. This is possible only because of the considerable swing and sharpness of velocity contrasts between debris- or turbidity-flow chalk facies and the largely non-reservoir hemipelagic and argillaceous chalks. Sonic and density logs will show this character in the form of blocky vertical porosity profiles (see Fig. 9.33).

New seismic techniques applied to fieldwide fracture modelling have considerably improved the optimum siting of injector and producer wells, and have provided substantial infill data to dramatically improve reservoir models and production profiles through waterflood. This has resulted in a greater understanding of stratigraphic, diagenetic and hydrodynamic trapping styles within established structures and some of their satellites. The techniques, however, are limited by seismic resolution, although layers as thin as 5 m can be defined in shallower fields

Despite the complex burial histories of individual structures, many large areas within fields can still retain remarkably high porosity (up to and above 40%). In some cases, diagenesis has been so minor (compaction only) and pore pressures so high that reservoirs remain unconsolidated. This problem is present in almost all chalk fields to varying degrees, with the result that many perforated intervals produce chalk along with hydrocarbons. At extremes, over 80% of production consists of chalk matrix. This problem continues to present an engineering challenge to most oil companies; topside structures have to be designed to account for serious seabed subsidence during chalk-field production

All the larger, structurally closed, Late Cretaceous highs have been drilled throughout the Central and Northern North Sea. The (largely field development) models discussed in this chapter, however, demonstrate that hydrocarbon trapping mechanisms in chalk are not exclusively structural. The structural description of many chalk fields now has a rider referring to stratigraphic and diagenetic elements.

Primary stratigraphic trapping resulting from highly variable distributions of reworked chalk units is common within structurally defined fields. Large stratigraphic traps probably exist in some of the major submarine fans (see Fig. 9.35). The best potential stratigraphic traps are likely to drape the flanks of structural highs or slopes and be one-way dip-closed; others may be entirely stratigraphic—as mounded or downlapping 'bright spots' (so-called direct hydrocarbon indicators on a seismic profile) in otherwise flat-bedded basins.

Figure 9.54 illustrates some of the work on regional mapping of gross chalk sequences. This compares two areas: firstly, the salt-free basins of the Moray Firth, where large inverted structures (and/or residual Cimmerian rift shoulders) and basinal sediment drape influence the style of sediment fill. In these settings, basin-slope and basin-floor chalk submarine-fan systems may form stratigraphic traps. The lower two examples are from the Central Trough (in the Norwegian and Danish sectors), basins where a combination of severe inversion and halokinesis has influenced trap styles. Syndepositional tectono-halokinetic activity has resulted in complex fill patterns, often with attenuation of the structural wavelength of the accommodating basins. The potential for stratigraphic traps within both of these structural domains remains considerable.

Stratigraphic trapping within chalk submarine-fan complexes is probably one of the key recent concepts although diagenetic and hydrodynamic trapping styles are widely recognized for their ability to modify even the simplest chalk reservoir. Diagenetic trapping, as a function of differentiated hydrocarbon migrations, relative permeability,

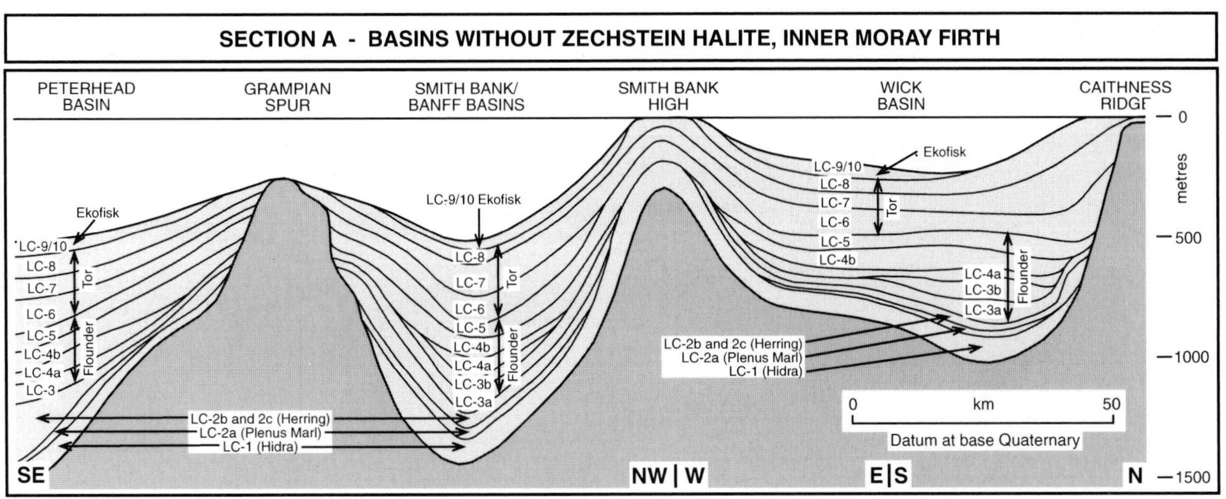

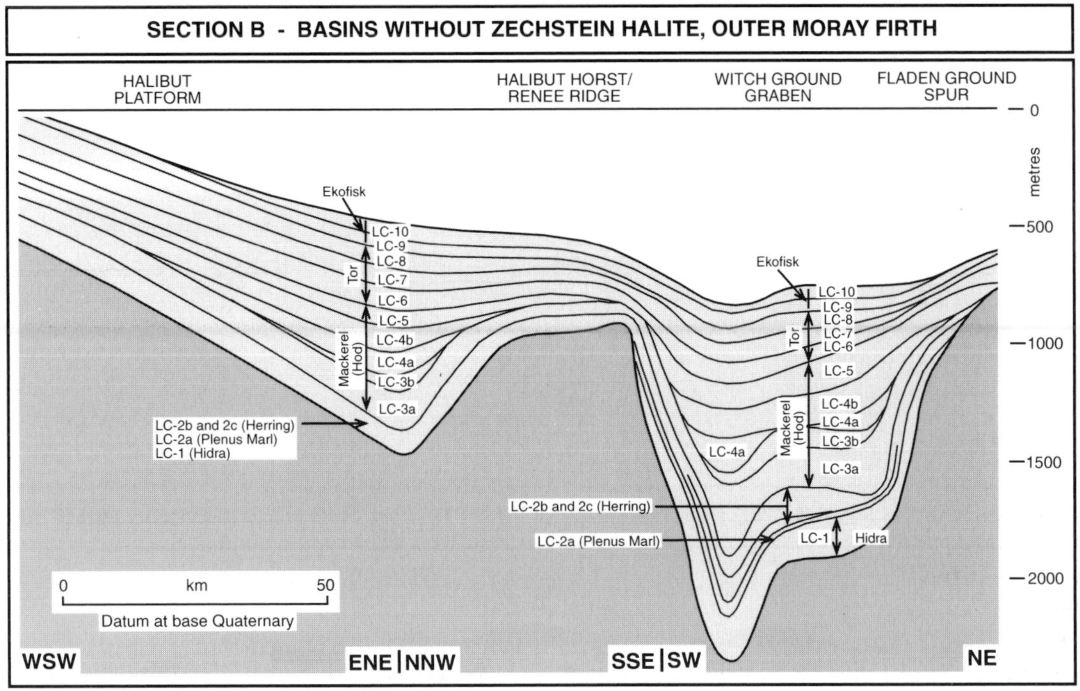

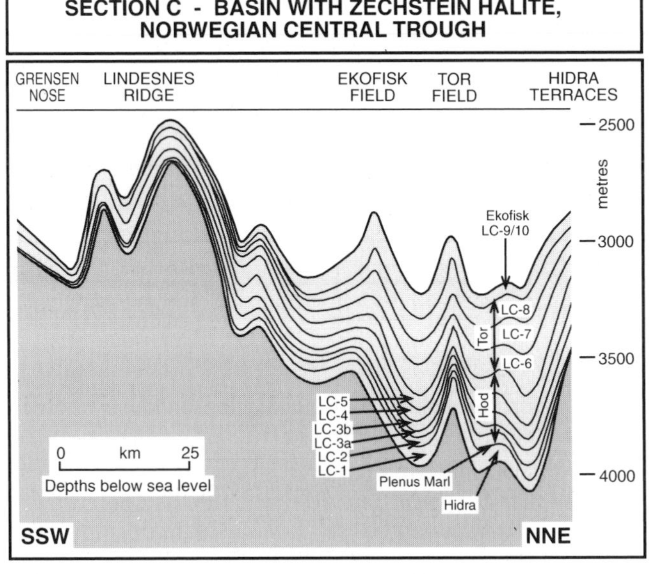

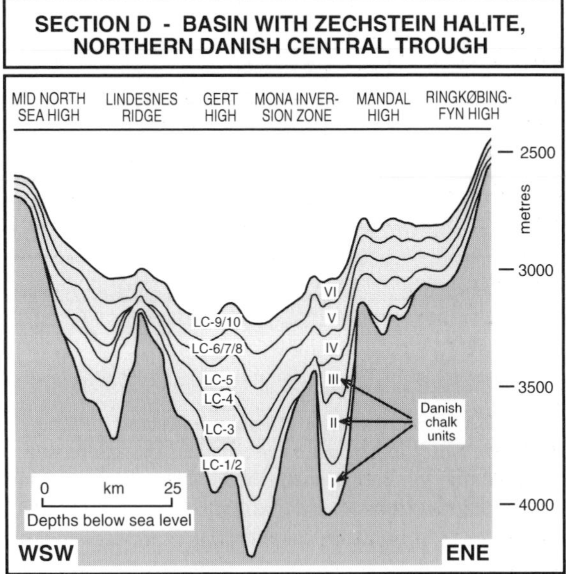

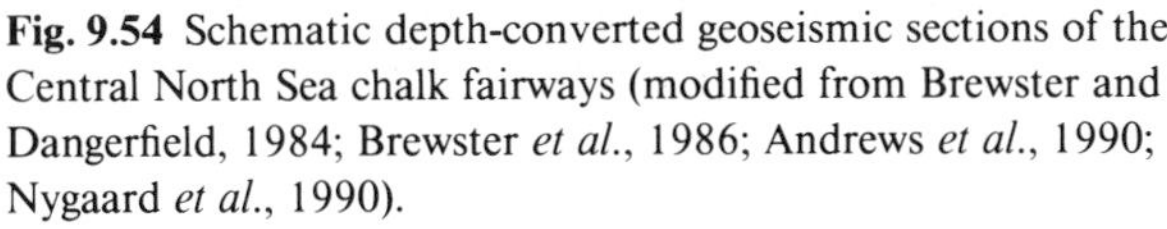
Fig. 9.54 Schematic depth-converted geoseismic sections of the Central North Sea chalk fairways (modified from Brewster and Dangerfield, 1984; Brewster *et al.*, 1986; Andrews *et al.*, 1990; Nygaard *et al.*, 1990).

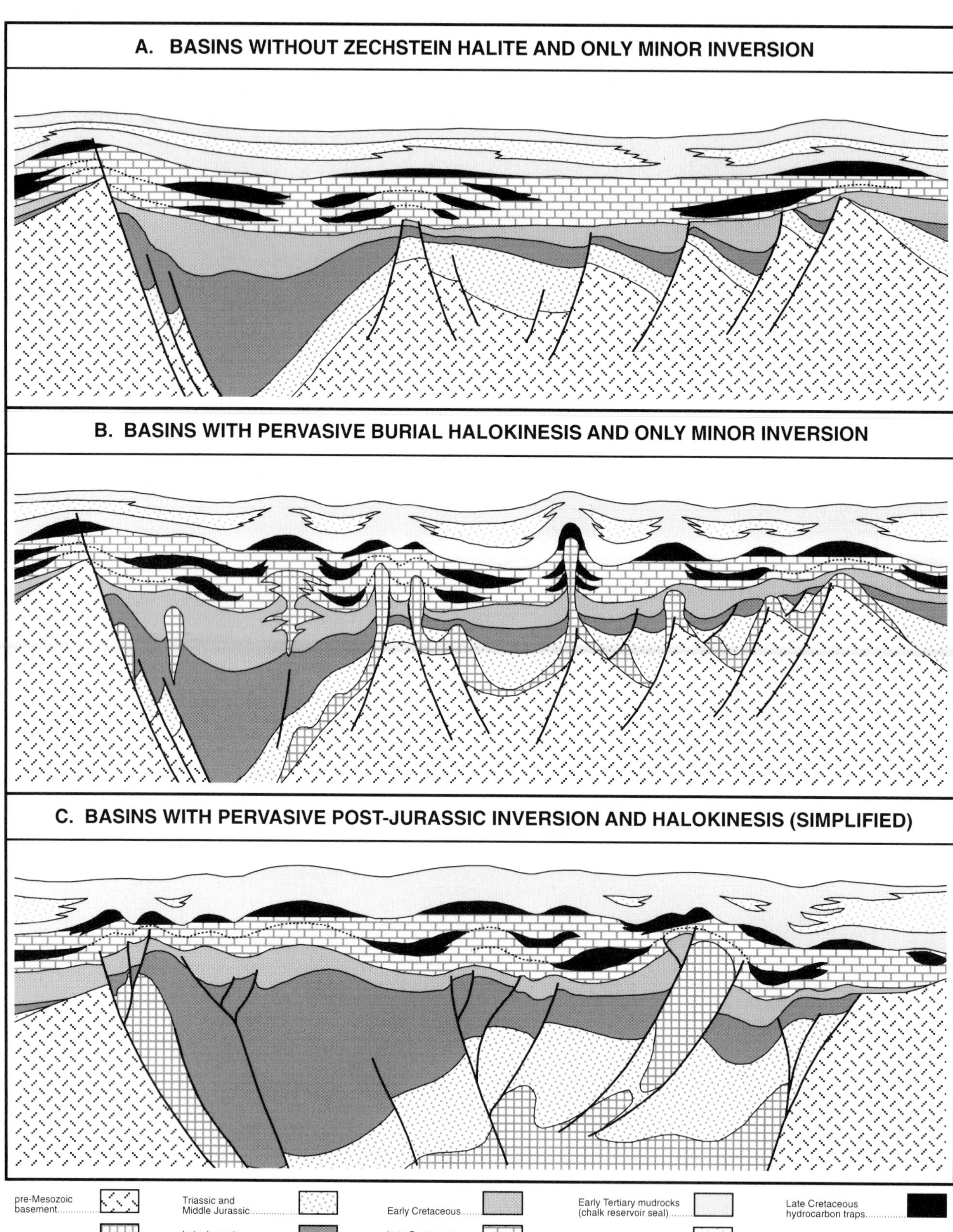

Fig. 9.55 Cartoon summary of Late Cretaceous Chalk play concepts. Note: approximate vertical scale for each section 5000 m from top Early Tertiary; approximate horizontal scale of each panel 100 km.

pore pressure variation and structural evolution, is also highly important. It is common for crestally targeted exploration wells in strongly tilted traps to intersect uncommercial hydrocarbons in poor-quality unproductive reservoir. The bulk of the productive reservoir may remain undrilled downflank. The latter has been suggested as a solo mechanism that can form traps (Megson, 1992). Application of these models away from the traditional and mature Chalk play fairways opens the possibility of new and exciting plays in areas that have been ignored on the grounds of lacking a full Tertiary mudrock seal.

Proven and potential chalk traps are illustrated in Fig. 9.55. These cartoons illustrate three schematic North Sea sub-basin configurations. Almost all exploration wells that passed through the Cretaceous on their way to the Jurassic or older targets were drilled on structural highs. At Chalk level, these highs are either drapes over old Jurassic footwall crests, inversion pop-up structures or halokinetic features. Most of these sites can be proved to have been positive during chalk deposition. Correspondingly, chalk submarine fans with potential high reservoir quality are likely to be basinwards of these structures.

There are no known reservoired hydrocarbons in chalks below present-day depths of about 11 000 ft (3350 m). Combined with the known palaeogeographic extent of favourable chalk-reservoir facies (see Fig. 9.5) and proximity to mature hydrocarbon source rock (oil-prone Jurassic in the Central and Northern North Sea, gas-prone Carboniferous in the Southern North Sea), there are very large areas of Chalk that remain under-explored. Most notable of these are the entire UK Central Trough, large basinal areas of the Outer Moray Firth and selected areas of the Dutch Central Trough.

There may even be areas of the northern North Viking Graben that could contain significant clastic reservoir sands of Late Cretaceous age. This play type occurs to the north of the area in the Halten and Traen provinces of the Atlantic margin in offshore mid-Norway and in the West of Shetlands area.

9.9 Acknowledgements

The authors would like to thank the many individuals and companies who have contributed data for this compilation. We would like to thank British Petroleum (BP) for permission to publish data on the Early Cretaceous in the Southern North Sea and UK Central Trough. We would also like to thank Amoco Norway Oil Company for permission to publish Early Cretaceous data on the Norwegian Central Trough.

CDO would also like to thank Robertson Research International Ltd for a considerable working experience of carbonate reservoirs and, in particular, the opportunity to work in great detail on the Chalk of the North Sea. MAP would also like to thank Pete Turner at BP, without whose assistance the sequence-stratigraphic analysis of the Early Cretaceous could not have been compiled and tested with seismic data. We also thank Ken Glennie for editorial advice and considerable patience in helping to take the rough manuscript through to fruition.

9.10 Key references

D'Heur, M. (1986) The Norwegian chalk fields. In: Spencer, A.M. (ed.) Habitat of hydrocarbons on the Norwegian continental shelf. Graham & Trotman, London, pp. 77–89.

Hatton, I.R. (1986) Geometry of allochthonous Chalk Group members, Central Trough, North Sea. *Marine Petrol. Geol.* **3**, 79–98.

Megson, J.B. (1992) The North Sea Chalk play: examples from the Danish Central Graben. In: Hardman, R.F.P. (ed.) *Exploration Britain: Geological Insights for the Next Decade.* Special Publication No. 67, Geological Society, London, pp. 247–82.

Van Adrichem Boogaert, H.A. (1993) Stratigraphic nomenclature of the Netherlands, revision and update by RGD and NOGEPA. *Mededelingen Rijks Geol Dienst* **50**.

It is anticipated that the following papers, presented at the Barbican Conference in October 1997 and to be published late in 1998, will prove of value as future reading. They will appear in:

Boldy, S.A.R. and Fleet, A.J. (eds) *Petroleum Geology of Northwest Europe: Proceedings of the Fifth Conference.*

Anderson, J.K. The capabilities and challenges of the seismic method in Chalk exploration.

Bramwell, N.P. *et al.* Chalk exploration in the search for a subtle trap.

Jones, L. *et al.* Britannia field, UK Central North Sea: modelling heterogeneities in unusual deep water deposits (UK Central North Sea).

Rose, P.T.S. Reservoir characterisation in the Captain field: integration of horizontal and vertical well data.

Also:

Oakman, C.D., Johnson, H.D. and Martin, J.H. (1998) *Cores from the Northwest European Hydrocarbon Province, Part II.* Geological Society, London.

10 Cenozoic

M.B.J. BOWMAN

10.1 Introduction

Details of Tertiary discoveries and exploration activity since the mid-1980s have only recently become available in the public domain. Much has taken place in exploration and production, together with technological developments, which have changed perceptions of Tertiary play systems for many operators. Recent discoveries by British Petroleum (BP)/Shell and Amerada Hess in the Faeroe–Shetland Province of the Atlantic Margin have led to a high level of renewed exploration activity in the Tertiary. Unfortunately, details of these latest discoveries have yet to be published; they will, in time, doubtless shed further light on the potential of the Tertiary hydrocarbon system. This review builds on the earlier work of Lovell (1990) and focuses on recent discoveries and technology developments, discussing their impact upon prospectivity. It includes a brief update of Cenozoic basin evolution and stratigraphy as they impinge upon the North Sea, together with a description of selected new plays and discoveries. Further reading is also included.

As a successor to the thermal-sag basin of the Late Cretaceous, Cenozoic deposition in the North Sea Basin covered much the same area as the present North Sea but initially extended eastward over Denmark and the North German Plain. With time, its area shrank to nearer its present dimensions. While there was no break in sedimentation in the Central and Northern North Sea at the end of (Danian) chalk deposition, in the south, the effects of Laramide movements resulted in widespread inversion and erosion of the Chalk. Continued compressive pulses induced further inversion movements in the mid-Cenozoic, which were important for trapping both oil and gas in pre-Tertiary reservoirs; this compression also ensured that subsidence in the Southern North Sea, and thus sedimentary fill, remained very limited, thereby reducing its hydrocarbon prospectivity.

The Cenozoic is prospective in the Central and Northern North Sea (north of the Mid North Sea High, *c.* 56°N) (Fig. 10.1), together with parts of the Atlantic Margin (Faeroe–Shetland Basin to South Rockall Trough). It remains largely an oil province, with some wet gas in the Central North Sea and a single, giant dry gasfield (the Frigg Field) in the Northern North Sea. Prospectivity is driven primarily by the presence of reservoir in basinal settings, its probable geometry and internal architecture.

The most favourable combination of reservoir, trap and access to mature Jurassic source lies in the UK sector and the western part of the Norwegian sector from the Central North Sea northwards as far as the Brent Province (East Shetland Basin) and onwards to the Atlantic Margin. Elsewhere, reservoirs are more sporadically developed, limiting prospectivity.

Because of very low relief in the hinterland, sedimentation in the southern part of the North Sea Basin was dominated by shales, which act as seals for the rare reservoir horizon containing hydrocarbons (e.g. basal Eocene Dongen Tuffite in the Dutch De Wijk field (Bruijn, 1996). Because of its very shallow depth (510 m), the gas in the Tertiary part of this field is not being developed at the moment, for fear that associated subsidence might damage recovery from deeper horizons in this multireservoired (also Cretaceous, Triassic, Permian and Carboniferous) structure. Elsewhere in the Southern North Sea area, there seem to be no hydrocarbon prospects and the region will not be discussed any further.

10.2 Exploration and production history

Bain (1993) provides an interesting historical review of exploration in the Tertiary, assessing the various factors that have contributed to success and discussing some of the key play types. Figures 10.2 and 10.3 illustrate the Palaeocene–Eocene discoveries from the onset of exploration to 1991. Early successes were followed by a lean period, with few major discoveries until the early 1980s. Since then, the success rate has improved significantly. This change is in large part due to technological advances, notably improvements in seismic-data acquisition and processing. Combining this with new geological approaches, such as predictive sequence stratigraphy and a more pragmatic awareness of the diversity of submarine-fan plays, has led to the development of a range of new opportunities, where stratigraphy has a significant impact upon trap configuration.

The first successful Tertiary discovery was in 1966, when well UKCS 22/18-1, now the Arbroath Field, encountered *c.* 35 m of oil-bearing sandstone in the Palaeocene. This was quickly followed by the discovery of a number of similar pools, with the discovery of the first giant oilfield, the Forties Field, in 1970 (estimated recoverable reserves: 400×10^6 m^3 of oil). The first Eocene pool was found the following year, when Elf/Total discovered the giant Frigg gasfield, which straddles the UK and Norwegian sectors. Interest in Eocene oil flourished following the discovery of the Alba Field and Harding–Gryphon Fields in the mid-1980s. Recent exploration interest has been rekindled by the discovery of the giant Foinaven ($40–80 \times 10^6$ m^3 of oil) and Schiehallion (68×10^6 m^3; 425×10^6 bbls; Leach *et al.*,

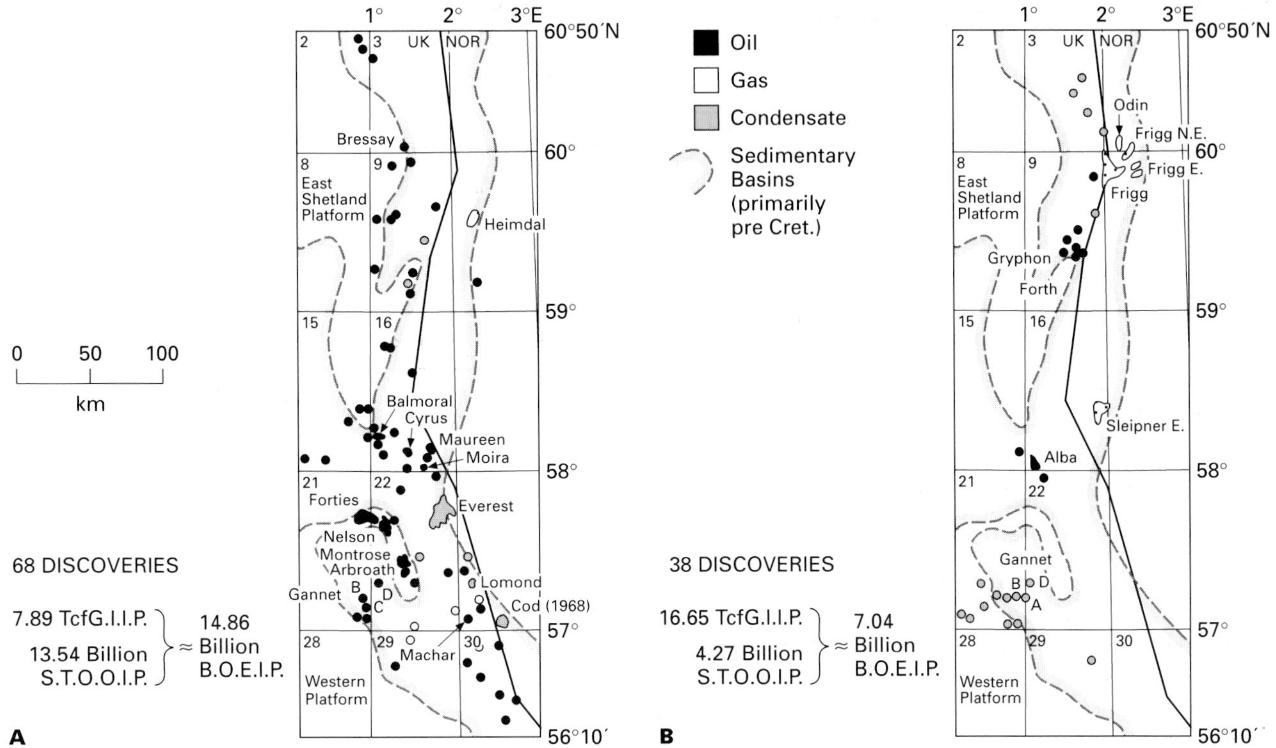

Fig. 10.1 Map of Cenozoic discoveries in the North Sea Basin, 1969–1990. A. Palaeocene. B. Eocene. (From Bain, 1993.)

1997) oilfields in UKCS Quadrant 204 of the Atlantic-Margin Province. These are reservoired in Palaeocene sandstones.

Historically, Palaeocene exploration has been focused on the Central North Sea, the main area of turbidite sand deposition. Here the Eastern Trough and Central Graben salt-diapir and basin-margin pinch-out plays have attracted recent attention (Hodgson *et al.*, 1992; Foster and Rattey, 1993). Examples include BP's 'M' Field complex and Amoco's Everest Field (see Figs 10.1 and 10.22). These are all subtle, partly stratigraphic traps, involving stacked pay zones, within the distal part of the main submarine-fan complexes.

More traditional Palaeocene plays have attracted less attention. However, the successful discovery of the Nelson Field (> 50×10^6 m^3 oil reserves) by Enterprise means that even here opportunities remain. Nelson is a large four-way dip-closed structural trap on the Forties–Montrose ridge along trend from BP's giant Forties Field (see Figs 10.1 and 10.2). Its discovery relied in large part upon application of high-quality, field-development-scale seismic-facies analysis to exploration (Whyatt *et al.*, 1991).

Attention has also been directed towards the Eocene,

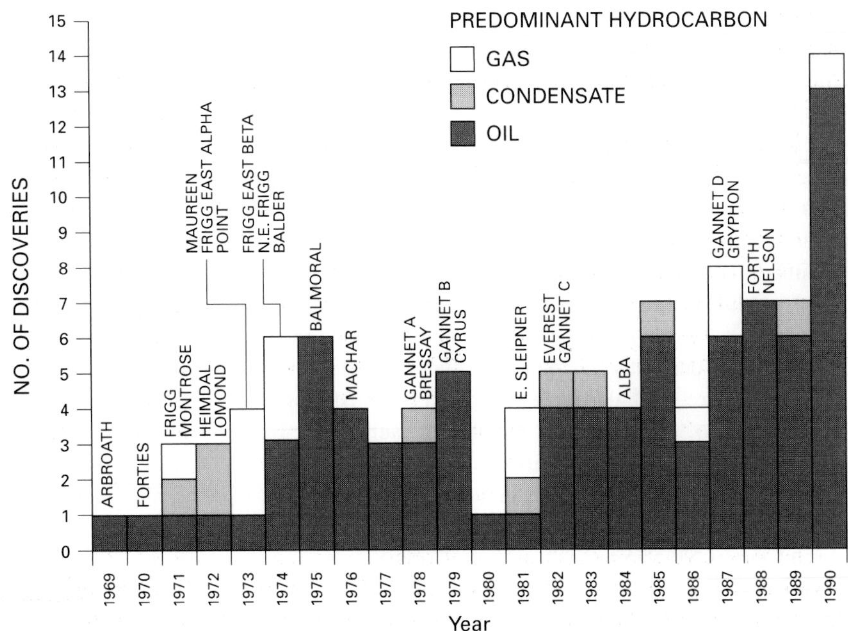

Fig. 10.2 Tertiary play fairway discoveries, 1969–1991. (From Bain, 1993.)

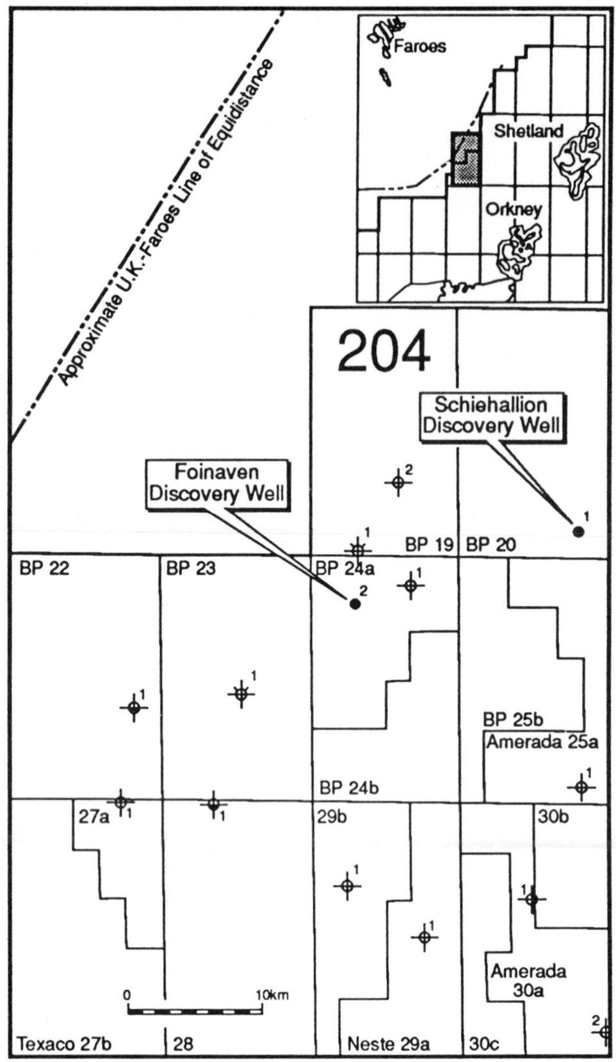

Fig. 10.3 Map of Cenozoic discoveries on the Atlantic Margin (British Petroleum, in-house data).

which, until 1987, had received little interest outside the Frigg and Gannet Field complexes. The Eocene is largely mud-prone, contrasting with the sandy Palaeocene. Exploration has focused on the search for localized submarine-fan plays with stratigraphic trapping potential. Initial successes were achieved in UKCS Quadrant 16, with Chevron's Alba field ($>160 \times 10^6$ m^3 oil-in-place), and UKCS Quadrant 9, with the Gryphon and Harding (formerly Forth) fields (operated by Kerr McGee and BP, respectively). More recent exploration of the Eocene has been less successful, but such subtle plays remain an attractive target for many operators.

Outside the North Sea, there has been success on the Atlantic Margin, with the discovery of the Foinaven and Schiehallion fields (Fig. 10.3). This is a scarcely tested fairway and is currently the focus of a high level of activity. Discoveries are in Palaeocene deepwater sandstones of similar age to many of the North Sea submarine-fan reservoirs. Regional considerations (Knott *et al.*, 1993) suggest that sand development in the Atlantic Margin is probably less extensive than in the Palaeocene correlatives of the Central North Sea.

New-pool exploration is not the focus of activity for the Palaeocene and Eocene. With many North Sea fields approaching decline, small satellite pools, of marginal value as stand-alone developments, become attractive targets for maintaining production rates and pipeline throughput. Also, many of the mature fields have become sites of extensive infill drilling programmes to add reserves by accessing unswept oil. This demands the application of the latest technology, such as seismic-attribute analysis, and non-conventional wells, by integrated teams, to locate and access increasingly small targets for exploitation. The Forties Field infill drilling programme is discussed later in this review as an example of such an approach.

10.3 Regional tectonic framework

Anderton (1993), Galloway *et al.*, (1993) and Knott *et al.* (1993) demonstrate the importance of appreciating the larger-scale (plate) tectonic context of the Cenozoic in understanding the hydrocarbon potential of the early Tertiary in the Central and Northern North Sea and on the Atlantic Margin. During the early Tertiary, both areas were influenced by two separate regions of plate activity. Without these influences, there would be little Cenozoic prospectivity and the early Tertiary would most probably have been a period of largely hemipelagic sedimentation continuing from the late Cretaceous.

Rifting of the Greenland–European plate in the early Palaeocene caused thermal uplift of Scotland and the East Shetland Platform, with rejuvenation of older Mesozoic hinterlands and basin margins (Fig. 10.4). This is the major control on the supply of coarse siliciclastic detritus during the Palaeogene. It changed the relatively deep, sediment-starved basins of the late Cretaceous into major clastic depocentres, dominated by a complex interplay and mosaic of deltaic and submarine-fan systems. Alpine orogenesis had a less dramatic effect on prospectivity, with local inversion mostly along former basin-margin faults.

The publications of Bott (1988), Dewey and Windley (1988), Knox and Morton (1988), White (1988), Milton *et al.* (1990), Morton *et al.* (1993) and Jones and Milton (1994), in conjunction with Anderton (1993) and Knott *et al.* (1993), provide a picture of the plate kinematics and their impact on hydrocarbon prospectivity. Five key events are distinguished; these are summarized below and illustrated in Figs 10.4 and 10.5:

1 Danian/Thanetian: major hinterland rejuvenation related to doming around a mantle hot spot centred under East Greenland.

2 Early Palaeocene: volcanic activity, caused by east–west extension, led to the British and Faeroe–Greenland igneous province, with impact in the North Sea exemplified by the Andrew Tuff of the Witch Ground Graben.

3 Late Palaeocene: volcanic activity, associated with the onset of sea-floor spreading in the Norway–Greenland Sea, led to eruption and deposition of widespread tuff marker beds (the Balder Tuff of the Northern North Sea).

4 Restriction of the Northern North Sea, due to the thermal doming, leading to the development of an anoxic basin in the North Sea during the late Palaeocene and early Eocene.

5 Minor inversion in the early Eocene caused by the final rupture of the North Atlantic. This was followed by passive

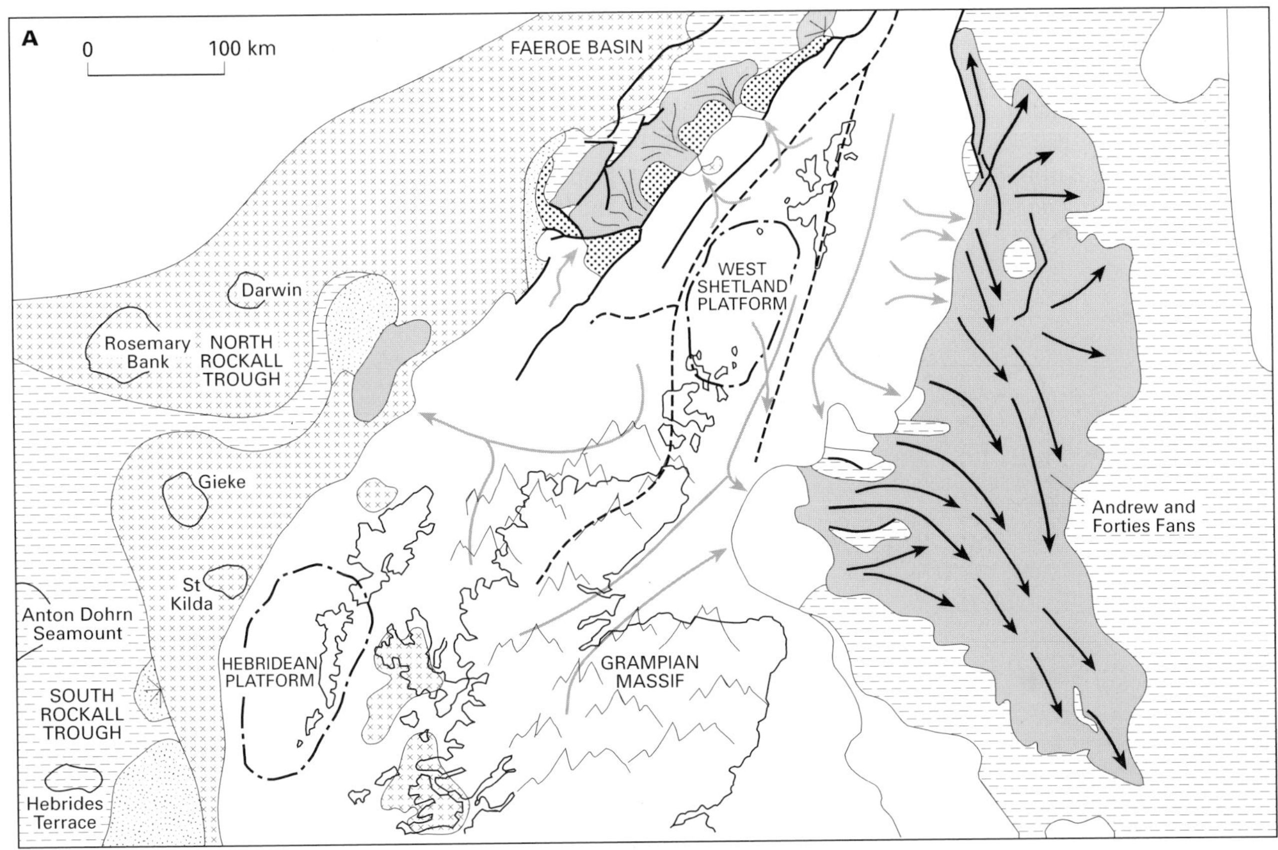

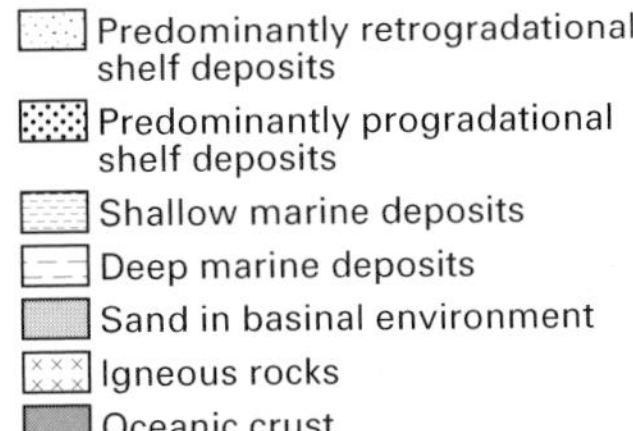

Fig. 10.4 Palaeocene. A. Plate reconstruction. B. Gross depositional environments, North Atlantic region, showing area of thermal upwelling. (From Knott *et al.*, 1993.) EUR, Eurasia; NAM, North America; AFR, Africa; IBA, Iberia; GRN, Greenland. Small half arrows indicate sense of shear across rift zones. See text for elaboration.

subsidence, leading to a clear marine connection of the North Sea with the North Atlantic.

Sedimentation was controlled by a complex interplay between tectonic activity, eustasy and hinterland characteristics. Each operates at both a regional and a local scale, leading to a complex depositional pattern across the basin. The volume and grain size of the clastic detritus increased gradually to a peak in the late Palaeocene (mid-Thanetian). Large volumes of material were fed into the North Sea and Faeroe–Shetland Basins as major submarine-fan depocentres. Provenance data for Palaeocene sands demonstrate that the impact of tectonic activity was not uniform along the basin margins and hinterlands. Differential uplift led to the development of geographically and temporally separate depocentres (Morton *et al.*, 1993).

Following successful rifting of the North Atlantic, the

Eocene and Oligocene are characterized by reduced rates of clastic input along the newly developed passive margin (Fig. 10.5). Sedimentation rates were reduced, the hinterlands gradually denuded, and a more uniform pattern of deposition established across both the North Sea and the Faeroe–Shetland Basins. Relative sea-level changes became the primary control on sedimentation patterns.

10.4 Stratigraphy

Recent years have shown a number of advances in our understanding of Tertiary stratigraphy in the Central and Northern North Sea, building on the original lithostratigraphic framework developed for the UK sector by Deegan and Scull (1977) and later enlarged by Isaksen and Tonstad

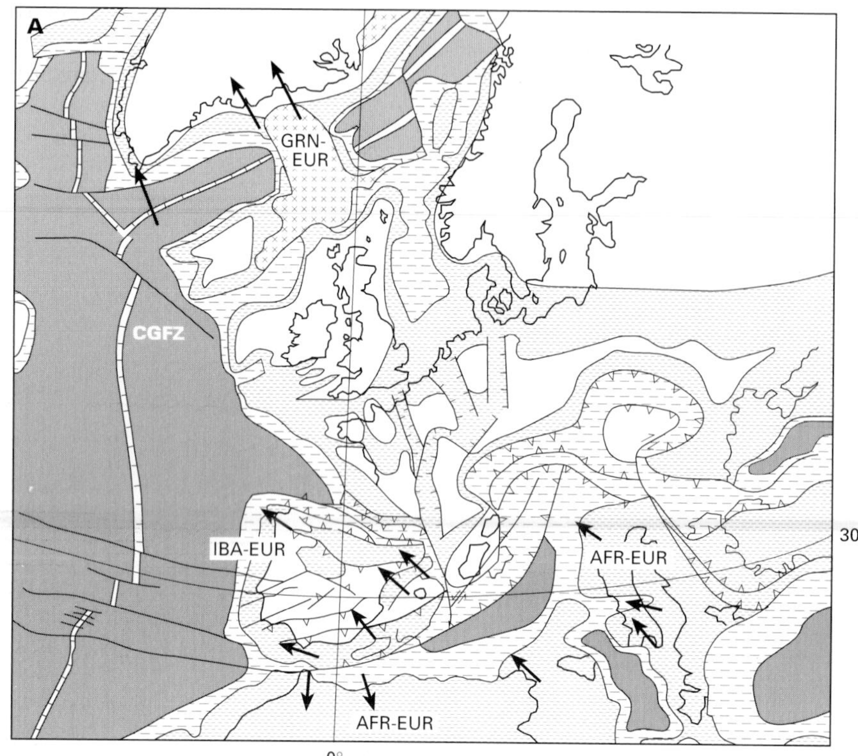

Fig. 10.5 Eocene. A. Plate reconstruction. B. Gross depositional environments, North Atlantic region. (From Knott *et al.*, 1993.) EUR, Eurasia; NAM, North America; AFR, Africa; IBA, Iberia; GRN, Greenland. Small half arrows indicate sense of shear across rift zones. See text for elaboration.

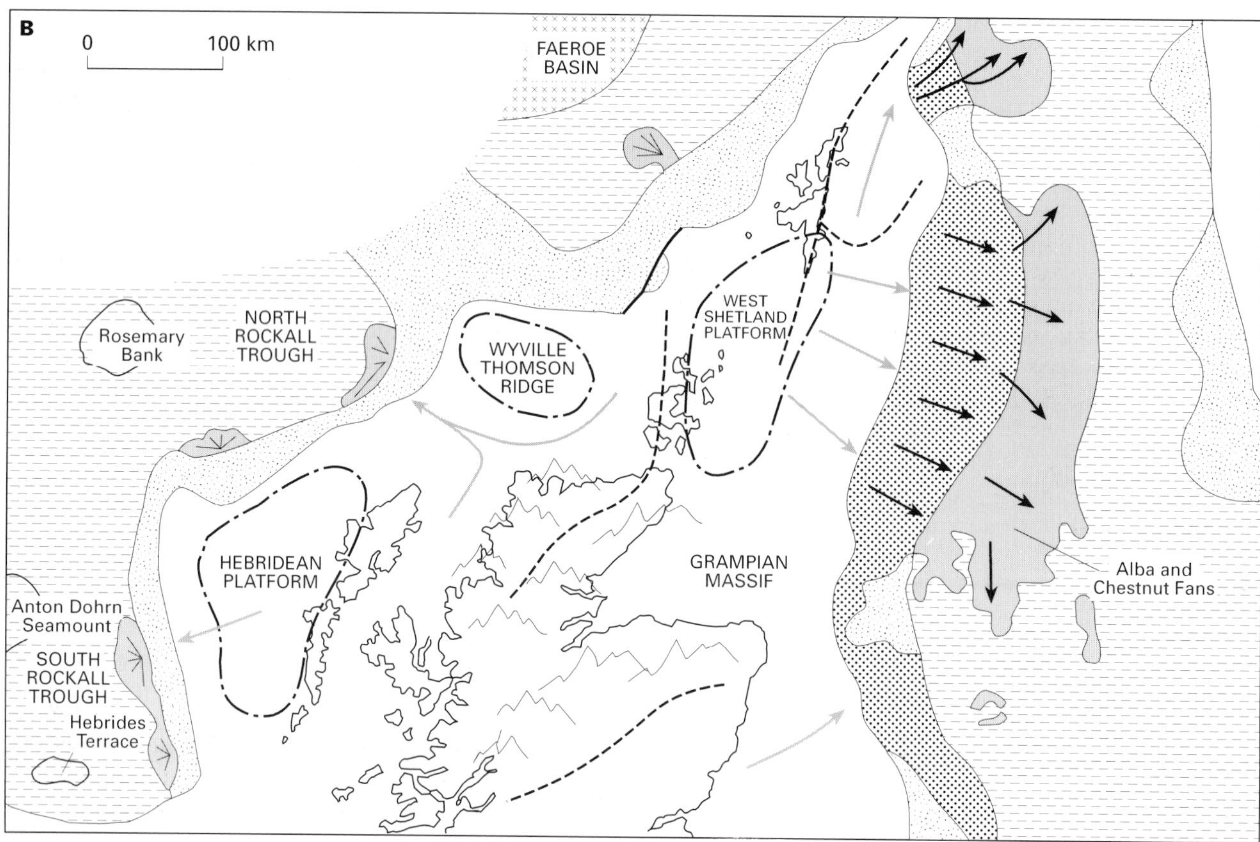

(1989) to include the Norwegian sector. A number of informal and local lithostratigraphic schemes were published during the early period of exploration (e.g. Mudge and Bliss, 1983) for the Northern North Sea. This proliferation caused much confusion and has severely limited the use of lithostratigraphy by many workers as a descriptive tool for the subsurface.

Recent drilling activity has yielded a considerable amount of new stratigraphic information, which has substantially enhanced our awareness of the Palaeogene. This includes the integration of stratigraphic schemes for both the North Sea and the Atlantic Margin. Awareness has been further advanced with the use of seismic data, enabling a better understanding of stratigraphic relationships.

10.4.1 Lithostratigraphy (Fig. 10.6)

A substantial revision of the formal lithostratigraphy has been proposed by Mudge and Copestake (1992a,b), for the Palaeogene and by Mudge and Bujak (1994) for the Eocene. These studies use biostratigraphic and seismic data to supplement the well-based information, leading to more regionally consistent frameworks. Knox and Holloway (1992) draw upon Mudge and Copestake (1992a,b), as well as previous workers, in their UK Offshore Operators Association (UKOOA)-sponsored revision of the Palaeogene lithostratigraphic framework. Collectively, these publications remain the most recent formally defined lithostratigraphic schemes for the Palaeogene. The Knox and Holloway (1992) scheme is that followed here (Fig. 10.6). The differences between this and other schemes (Mudge and Copestake, (1992a,b; Mudge and Bujak, 1994) are relatively minor, concerning mainly the relationships of the late Palaeocene–early Eocene, Dornoch and Beauly sand units in the Northern North Sea.

10.4.2 Biostratigraphy (Fig. 10.7)

There have been substantial advances in biostratigraphic zonation of the Palaeogene during recent years. Most oil companies have developed local modifications to regional biozonation schemes as an aid to exploring different parts of the basin. However, the regional schemes discussed here remain the most widely accepted and consistent available to date. Integration of palynological, foraminiferal and nanofossil data has enabled a five-scale bioevent-based biozonation for both the Palaeogene and the Eocene (Fig. 10.7). This has been a key tool in understanding time-stratigraphic relationships across the basin. Stewart (1987) remains a key reference for the Palaeocene biozonation, later enhanced by Mudge and Copestake (1992a). The most comprehensive Eocene biozonation scheme published to date is that of Mudge and Bujak (1994). Knox and Holloway (1992) also refer to a series of key biomarkers in their lithostratigraphic revision.

10.4.3 Sequence stratigraphy (Fig. 10.8)

The most significant advances in understanding Palaeogene stratigraphy have resulted from the application of sequence-stratigraphic techniques. These provide a powerful predictive and descriptive tool, enabling a clearer understanding of the controls on the development and distribution of different depositional systems. This is now the key to successful exploration in the Tertiary.

Sequence stratigraphy demands an integrated approach, drawing upon a seismic framework, combined with biostratigraphy, lithostratigraphy and a thorough understanding of depositional systems and their temporal relationships. Sequence definition emphasizes time-stratigraphic relationships, controlled by relative sea-level fluctuations, linking the evolution of depositional systems along continuous tracts.

There remains no widely accepted and used standard sequence framework. There are substantial differences in schemes proposed by Galloway *et al.* (1993), BP (Anderton, 1993), Exxon (Vining *et al.*, 1993) and Mobil (Armentrout *et al.*, 1993); a standard frame of reference is needed. Anderton's (1993) scheme, later published by Jones and Milton (1994), is presented here as the benchmark, being based on the most comprehensive, basinwide seismic and well database, incorporating the Atlantic Margin (Fig. 10.8). The scheme builds on Stewart's (1987) original sequence framework, subdividing the Palaeocene–Eocene succession into 13 units, T20–T98. It links with Knott *et al.* (1993), connecting the prospective areas of the North Sea and Atlantic Margin and referring these to onshore outcrops. The sequence framework links shelf and basinal sedimentation and assesses the dynamic evolution of accommodation space. Each sequence is defined by a combination of seismic and wireline logs with biostratigraphic calibration. They are also linked to core-based depositional models and systems-tract analysis. Each is recognizable on seismic as reflector packages bounded by surfaces of marine onlap and downlap (Fig. 10.9; Jones and Milton, 1994). In wells, many of the bounding surfaces can be tied to condensed basinal mudstones or marine-flooding surfaces, which separate major sandstone packages.

Today, development of fine-scale local biozonation schemes is ensuring that biostratigraphy is maximizing its impact at both regional and field scale. Good examples of this 'high-impact biostratigraphy' are provided by Payne *et al.*'s (in press) work on Andrew and other Central North Sea fields.

Future success in exploration and production will rely in large part upon such integrated predictive and descriptive stratigraphy at increasingly higher resolution. As attention switches towards increasingly subtle, often largely stratigraphic traps and small-scale infill drilling targets, success will depend upon rigorous seismic analysis and modelling. This will be linked to a well-based systems-tract framework, enabling division of the succession into regionally correlatable stratigraphic sequences.

10.5 Stratigraphic evolution

10.5.1 Palaeocene–early Eocene (T20–T50)

The stratigraphic evolution summarized in the following text draws mainly upon Milton *et al.* (1990), Anderton (1993), Knott *et al.* (1993), Jones and Milton (1994) and Reynolds (1994; also personal communication).

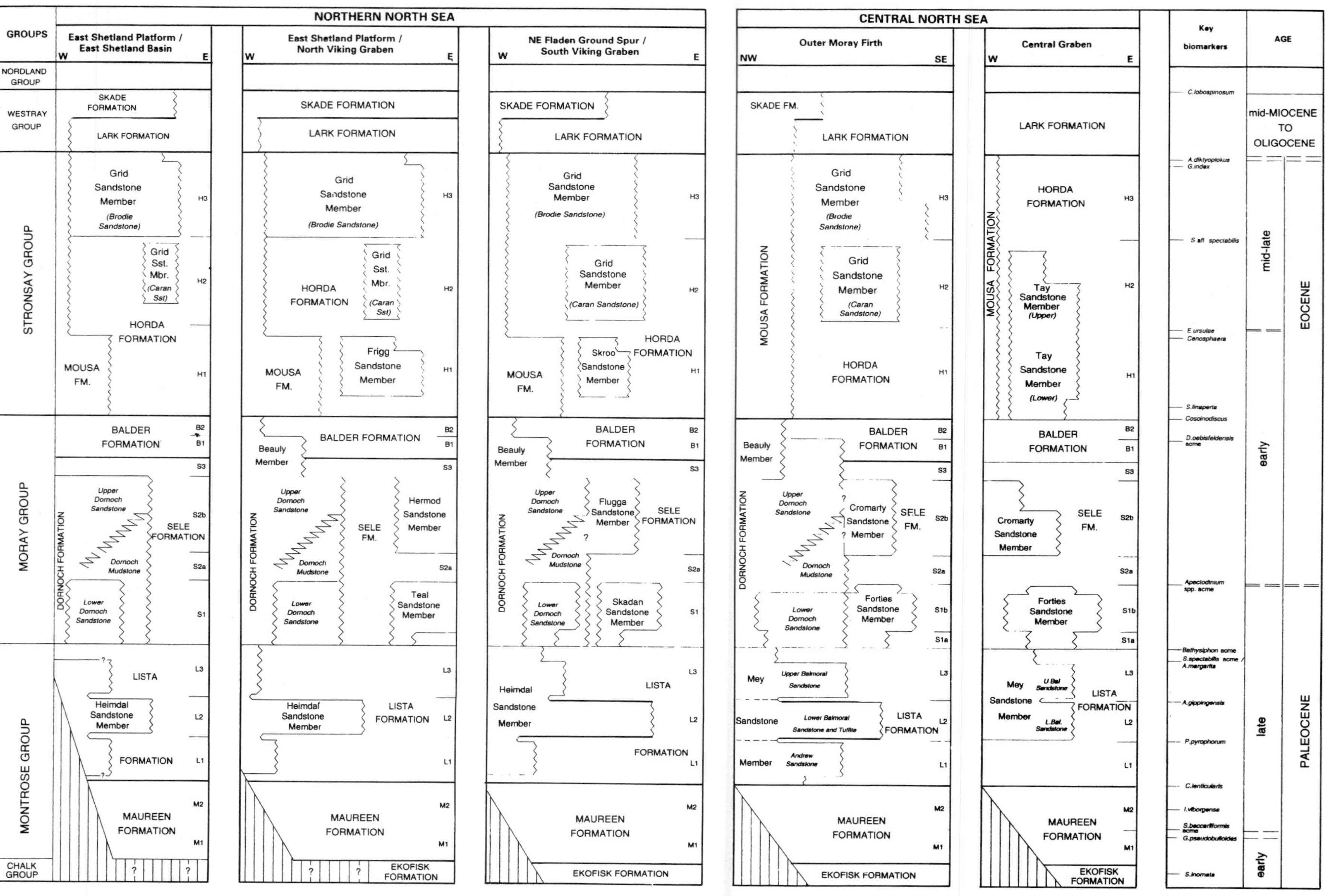

Fig. 10.6 Palaeogene lithostratigraphy (from Knox and Holloway, 1992).

DINOCYST BIOEVENTS

Gs	Top *Glaphyrocysta semitecta*
Ad	Top *Areosphaeridium diktyoplokus*
Ami	Top *Areosphaeridium michoudii*
Hp	Top *Heteraulacacysta porosa*
At	Top *Areoligera tauloma*
Dc	Top consistent *Diphyes colligerum*
Rr	Top *Rhombodinium rhomboideum*
Pd	Top *Phthanoperidinium distinctum*
Spc	Top common *Systematophora placacantha*
Dpf	Top *Diphyes pseudoficusoides*
Spa	Top abundant *Systematophora placacantha*
Pc	Top *Phthanoperidinium clithridium*
Df	Top *Diphyes ficusoides*
Sps	Top acme *Systematophora placacantha*
Cm	Top *Cerebrocysta magna*
Wab	Top *Wetzeliella articulata brevicornuta*
Dp	Top *Dracodinium pachydermum*
Eu	Top *Eatonicysta ursulae*
Eua	Top acme *Eatonicysta ursulae*
Cc	Top *Charlesdowniea columna*
Hot	Top acme *Homotriblium tenuispinosum*
Ec	Top *Eatonicysta compressa*
Am	Top acme *Areoligera medusettiformis*
Dv	Top consistent *Dracodinium varielongitudum*
Drc	Top consistent *Dracodinium condylos*
Dso	Top *Dracodinium solidum*
Ht	Top acme *Hystrichosphaeridium tubiferum*
Do	Top acme *Deflandrea oebisfeldensis*
Cw	Top acme *Cerodinium wardenense*
Aa	Top *Apectodinium augustum*
Alm	Top *Alisocysta margarita*
Ag	Top *Areoligera gippingensis*
Aga	Top abundant *Areoligera gippingensis*
Cs	Top common *Cerodinium speciosum*
Pp	Top consistent *Palaeoperidinium pyrophorum*
Pa	Top *Palaeocystodinium cf.australinum*
Ppa	Top acme *Paleoperidinium pyrophorum*
Iv	Top *Isabelidinium? viborgense*
Td	Top *Thalassiphora cf.delicata*
Sm	Top *Spiniferites 'magnificus'*
Ar	Top *Alisocysta reticulata*
Xd	Top *Xenicodinium delicatum*
Xm	Top *Xenicodinium meandriforme*
Si	Top *Senoniaspharea inornata*
Cac	Top *Carpatella cornuta*

MICROFOSSIL BIOEVENTS

Sas	Top *Spiroplectammina* aff.*spectabilis*
Cda	Top acme *Cenodiscus* spp.
Cs	Top common *Cenosphaera* spp.
Sn	Top *Spiroplectammina navaroanna*
Sl	Top *Subbotina linaperta*
Sla	Top acme *Subbotina linaperta*
C1	Top *Coscinodiscus* sp.1 & sp.2
D	Top acme *Coscinodiscus* spp.2, 4 & 7
IA	Top improverished agglutinate assemblage
DA	Top diverse agglutinate assemblage
Cl	Top *Cenosphaera lenticularis*
Gt	Top *Globigerina trivialis*
Gc	Top *Globigerina* cf.*compressa*
Gs	Top *Globigerina simplicissima*

POLLEN BIOEVENTS

Th	Top acme *Taxodiaceaepollenites hiatus*

Fig. 10.7 Palaeogene biozonation (adapted from Mudge and Bujak, 1996).

The early Palaeogene (Palaeocene to early Eocene) is divisible into two broad sedimentary packages. These are clearly identifiable on seismic and from well data (Figs 10.9 and 10.10). They each provide contrasting opportunities for hydrocarbon prospectivity within the basin. The early stages of deposition (Palaeocene; T20–T40) have an aggradational motif, with submarine fans being the dominant element of the fill. The coeval shelf and deltaic systems were largely zones of bypass, with only thin or erosional remnants preserved. The second depositional package (latest Palaeocene–early Eocene) has a predominantly progradational motif. Shelf/coastal depositional systems predominate, with distinctive clinoform packages on seismic (T45). Basinal sand deposition was smaller-scale and more localized.

T20 'Maureen' sequence (Fig. 10.11)

T20 is broadly comparable to the Maureen Formation of Knox and Holloway (1992), ranging in age from late Danian to early Thanetian. It immediately overlies the Danian Ekofisk Formation, recording the initial influx of coarse clastic material into the Cretaceous Chalk Seas. Pre-existing, rift-related, bathymetry provides the fundamental control on sediment input and distribution patterns across the whole of the North Sea and Atlantic Margin.

T20 is characterized by a major submarine-fan complex in the North Sea Central Graben. This was fed from two sources; the principal feeder was an eastward-flowing drainage system through the Witch Ground Graben. A secondary drainage system came from the Dutch Bank Basin. T20 submarine fans are large, sand-prone systems with an

Age	Series	Stage	Group	Formation	Member	Seismic unit (used here)	Seismic unit (Stewart)
40		Bartonian		Thet.?		T98	
				Belton?		T96	
45		Lutetian	Hordaland			T94	
						T92	
	Eocene					T84	
50						T82	
						T70	
		Ypresian				T60	
55				Be Ba	O	T50	8,9,10
			Moray	D	H	T45	7
	Palaeocene	Thanetian		Sele	Forties	T40	
							6
			Montrose	Lista	Balmoral, Andrew	T30	3-5
60				Maureen		T20	2

Fig. 10.8 Palaeocene and Eocene sequence stratigraphic scheme (from Jones and Milton, 1994).

Fig. 10.9 Line drawing of seismic geometries in the Witch Ground Graben showing stratigraphic sequences (from Jones and Milton 1994).

irregular, asymmetric gross geometry controlled by the basin bathymetry. The depositional signature is predominantly aggradational, with thick 'box-car' sands in axial settings passing rapidly into thinner-bedded units in marginal zones. Coeval shelf and delta complexes acted as bypass zones, transporting large volumes of sediment basinwards. They have low preservation potential, as they were reworked and eroded during the later Palaeocene.

Further north, a series of smaller-scale submarine fans and fan deltas drained eastwards into the Viking Graben. Many of these smaller systems exploited transfer zones in the failed Jurassic rift. The fan systems of the Atlantic Margin also tend to be of a smaller scale and more localized than those of the Central North Sea.

In the Central Graben, the earliest coarse-clastic influxes are intercalated with chalk boulders and turbidites derived from the former rift shoulders, following fault rejuvenation and spalling of carbonate debris at the basin margins. This intercalation of chalk and sandstone often means that the base of T20 is difficult to pick on seismic.

T30 'Andrew' sequence (Fig. 10.11)

This is equivalent to the lower part of the Thanetian Lista Formation of Knox and Holloway (1992). It includes those thick submarine-fan sandstones that historically have been referred to as the Andrew Formation. It also coincides with the first major phase of volcanic activity in the Hebridean Province. This led to the deposition of tuffs and tuffaceous sandstones, which form key marker beds within the basin (the Andrew Tuff).

The T30/T20 boundary is represented by a distinctive high-gamma mudstone on wireline logs, which separates basinal sandstones above and below (the Maureen/Lista

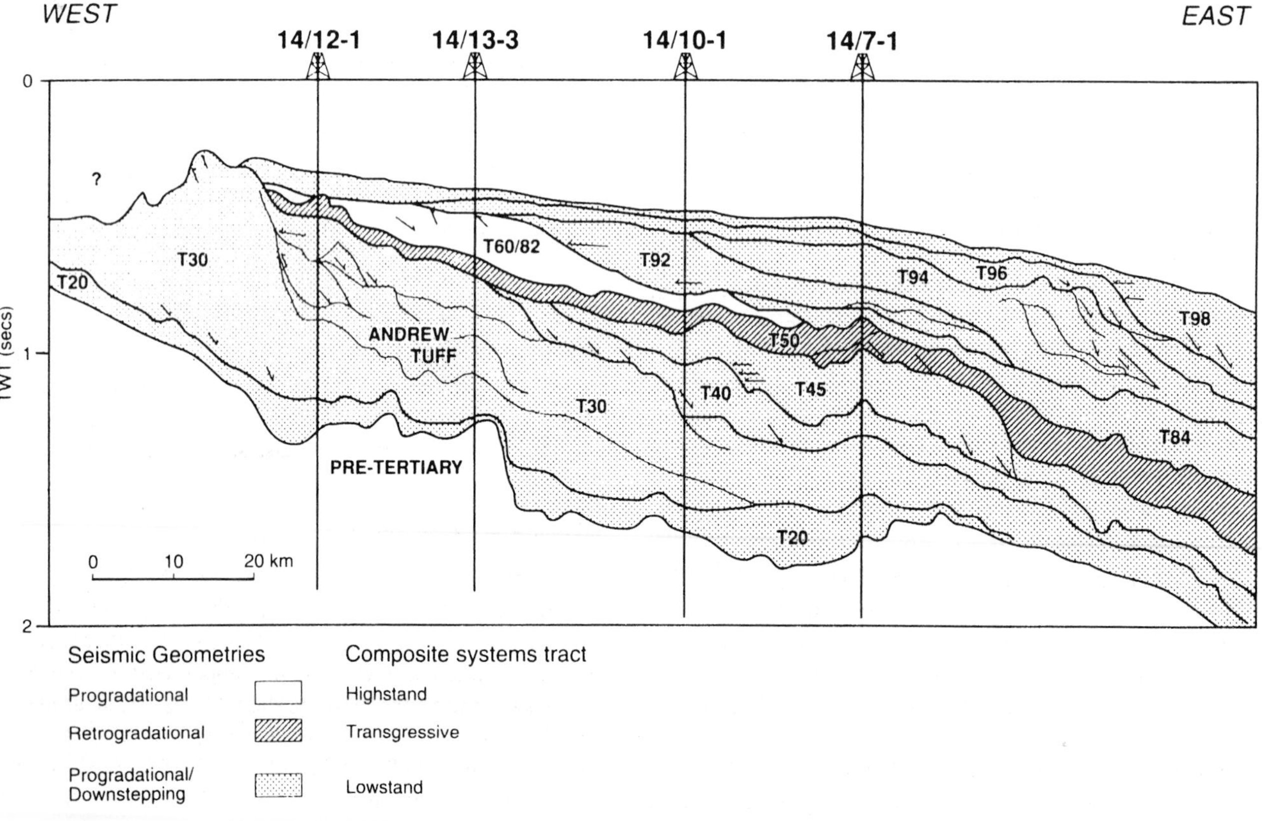

boundary on Fig. 10.10). The mudstone records a temporary halt in coarse-clastic input to the basin, caused by widespread flooding of the adjacent shelf. T30 was the period of most extensive submarine-fan development in the Central Graben. Fans are broadly similar to the Maureen system but have a smoother, less dendritic geometry. This reflects a progressive smoothing and filling of basin bathymetry. The fans are sand-prone over most of the basin. The feeder system for much of the coarse detritus was provided by the Witch Ground Graben, through which major fluvial-dominated deltas drained. Again, an aggradational motif predominates in basinal settings, with only erosional relics of the deltaic feeders remaining.

Further north, in the Viking Graben and along the Atlantic Margin, fan systems were on a smaller scale. Most of the pre-existing fault-scarp topography was subdued. The fans are heterogeneous, mixed sand–mud systems, with the main input points being through former transfer zones. They form the principal reservoirs for the Foinaven and Schiehallion fields

T40 'Forties' sequence (Fig. 10.11)

The T40 sequence is the last major phase of fan aggradation in the Central Graben. It is the upper sandstone unit over large parts of the basin, forming the primary target in many structural closures at top Palaeocene level. It includes the Forties Member, the main reservoir interval for a number of fields, including Forties and Nelson. The sequence is of late Thanetian age, being equivalent to the upper Lista and lower Sele formations of Knox and Holloway (1992).

The late Thanetian was the period when the North Sea Basin was largely cut off from oceanic circulation. This led to a basinwide anoxic phase, with the deposition of organic-rich bituminous muds, which have source potential. By the end of T40, a major relative sea-level rise, associated with the opening of the North Atlantic, re-established a fully marine connection to the north and brought the anoxic phase to an end. This flooding event drowned large areas of the adjacent shelf and temporarily halted deposition of coarse clastics in the basin, leading to a high-gamma mudstone unit on logs.

In the Central Graben, T40 is represented by a more heterogeneous mixed sand and mud fan complex than the T20 and T30 basinal systems, with more extensive delta-feeder systems preserved on the shelf. While the Witch Ground Graben remained as the principal conduit for basinward sediment transport, a new input point was initiated further south, in the area of the Gannet Field. The T40 fan system is more symmetrical and regular in outline than its predecessors. This reflects the continued filling and smoothing of the relic rift topography. The main areas of T40 sand deposition are strongly influenced by constructional topography formed by the underlying T20 and T30 systems. The extent of the fan system is slightly reduced in comparison with T30. Halokinesis along the margin of the Central Graben also strongly influenced sediment thickness and distribution patterns, leading to localized thinning and pinch-out of sands. A similar but less marked halokinetic influence is recognizable at the T20 and T30 levels.

North of the Central Graben and on the Atlantic Margin,

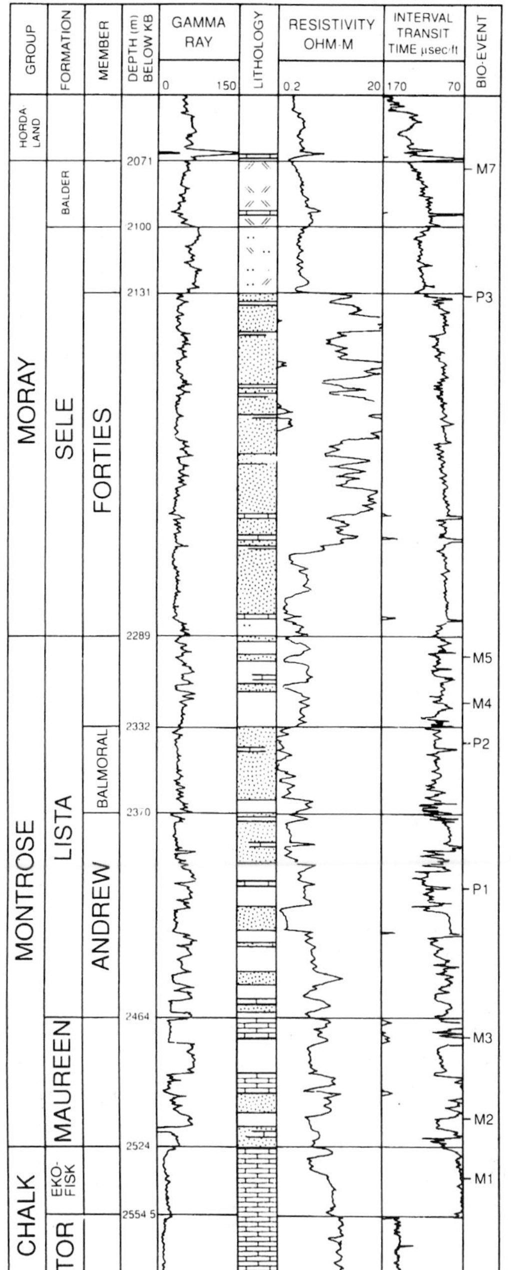

Fig. 10.10 Well 21/10-1: Palaeocene type section (T20–T40 corresponds to the Montrose Group; T45–T50 corresponds to the Moray Group) (from Mudge and Copestake, 1992a).

sand input is much reduced in comparison with the previous sequences. Any fans in the Viking Graben are small, very localized, lobate systems, reflecting an increase in relative subsidence of the East Shetland Platform.

T45 'Sele' Sequence (Fig. 10.11)

T45 records a major reconfigurarion of the basin, with a change to a depositional pattern that is predominantly progradational. The sequence straddles the Palaeocene–Eocene boundary, including the upper part of the Sele Formation mudstones and the shelfal/deltaic sandstones of the Dornoch Formation. Time-equivalent basinal sandstones are on a smaller scale and more localized than in the underlying developments. These often have different

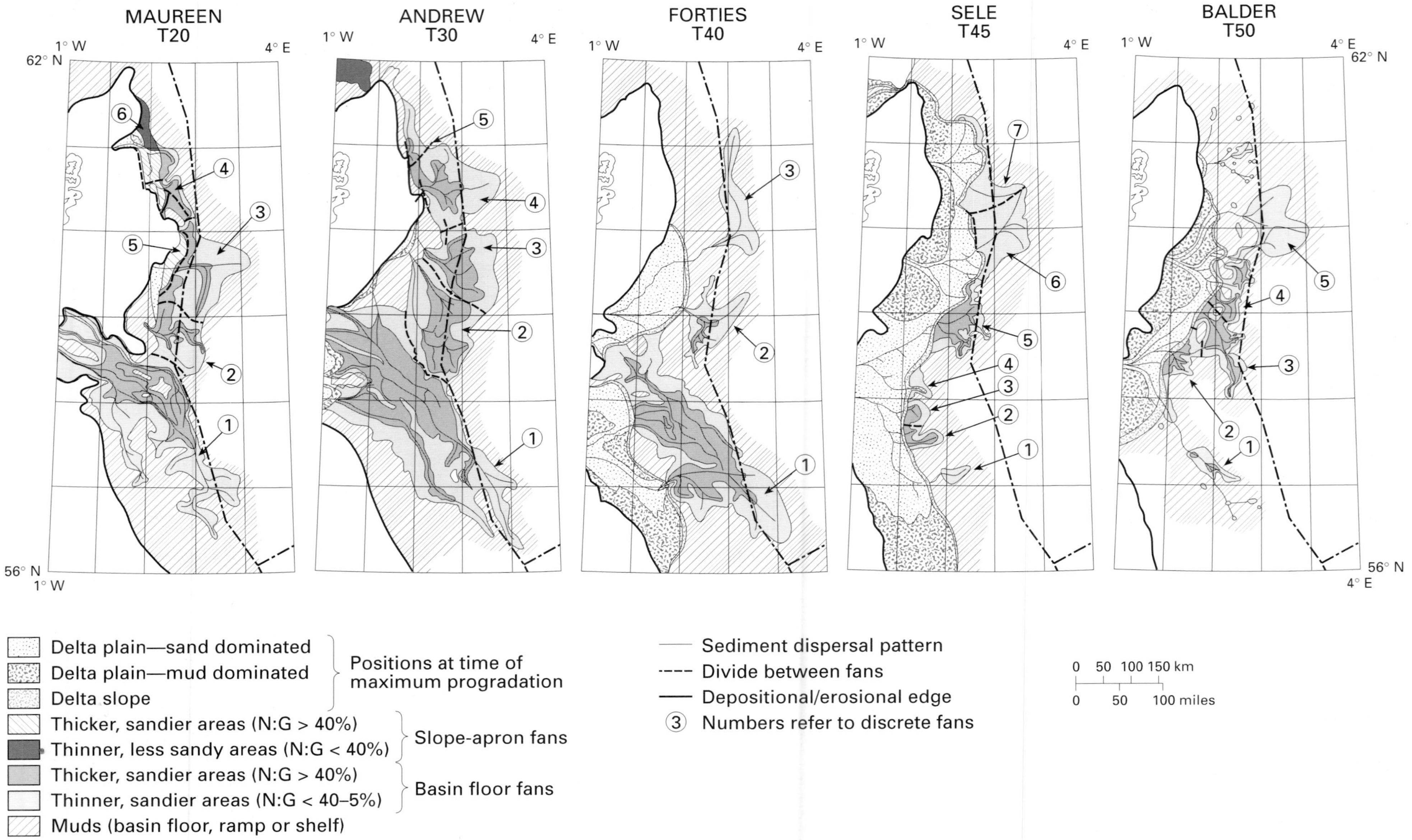

MAUREEN T20

ANDREW T30

FORTIES T40

SELE T45

BALDER T50

1° W 4° E 62° N 56° N

Delta plain—sand dominated
Delta plain—mud dominated
Delta slope
} Positions at time of maximum progradation

Thicker, sandier areas (N:G > 40%)
Thinner, less sandy areas (N:G < 40%)
} Slope-apron fans

Thicker, sandier areas (N:G > 40%)
Thinner, sandier areas (N:G < 40–5%)
} Basin floor fans

Muds (basin floor, ramp or shelf)

Sediment dispersal pattern
Divide between fans
Depositional/erosional edge
③ Numbers refer to discrete fans

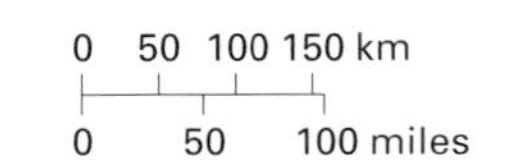

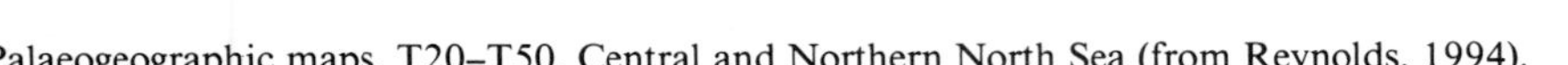

Fig. 10.11 Palaeogeographic maps, T20–T50, Central and Northern North Sea (from Reynolds, 1994).

names, reflecting local terminology (e.g. the Cromarty Member of the Outer Moray Firth (Mudge and Copestake, 1992a) and the Hermod Member of the Northern North Sea (Mudge and Copestake, 1992b).

The change in palaeogeography that characterizes T45 results from the final smoothing of the former rift topography. A major prograding shelf-and-delta complex advanced in a broad belt along the western side of the North Sea Basin. This forms distinctive progradational clinoform packages on seismic. There is little evidence of the Jurassic fault system controlling sediment distribution patterns. T45 submarine fans are smaller-scale and more localized than the older Palaeocene systems, with fan deposition controlled largely by local and regional base-level changes on the delta/shelf, related to both tectonic and eustatic changes.

The only complication to this broad depositional character lies in the North Viking Graben/Bruce–Beryl area (UKCS Quadrants 3 and 9). Here, local differential uplift and subsidence led to partial exhumation and entrenchment of the shelf-and-delta complex. The uplift is attributable to the buoyancy effects of the Bressay granite, which straddles Quadrants 3 and 9. The Bressay Field, in UKCS Quadrant 9, is a biodegraded heavy-oil-bearing stratigraphic trap formed by erosional remnants of the incised T45 shelf complex.

The end of T45 deposition is signalled by a major regional rise in sea level (up to 500 m) and associated coastal onlap (Jones and Milton, 1994). This led to the deposition of a thick and widespread transgressive coal unit, the Dornoch Coal, which forms a key correlation datum on logs (Fig. 10.12).

T50 'Balder' sequence (Fig. 10.11)

T50 is of relatively short duration, being represented by the Balder Formation argillaceous units and the Beauly Formation shelf system. As in T45, basinal sands are localized, small-scale systems. They are largely restricted to the Bruce–Beryl embayment of the Viking Graben and the Gannet Area of the Central Graben. In the Bruce–Beryl area, they include the main reservoir of the Harding and Gryphon Fields. A widespread unit, the Balder Tuff is interbedded with the early T50 and late T45 sandstones. This is probably related to continental rupture and the onset of sea-floor spreading in the Norway–Greenland Sea. The Tuff provides a valuable correlation datum in the basin.

T50 sedimentation is distinguished by a series of small-scale clinoform packages, which overlie the T45 delta top/shelf. These record the advance of shallow-water deltaic complexes over the flooded delta plain. The renewed advance did not reach as far as the T45 shelf edge, reflecting an overall retrogradational or backstepping stratigraphic signature. The end of T50 deposition is marked by another episode of widespread shelf flooding. As with T45, this led to deposition of a distinctive thick transgressive coal unit, the Beauly Coal, which forms an important marker bed (Fig. 10.12).

T50 reservoirs are different from those of previous systems, being localized, thick, massive sandstones with few silt/mud interbeds. They record a change in the style of coarse-clastic deposition within the basin, which persisted until the late Eocene (T94–T98). The sandstones are typically clean and well sorted, very different from the more clay-rich, delta-fed systems of the Palaeocene. The changes are a consequence of reworking sands in shoreline and shelfal settings near to submarine canyon heads (Dixon *et al.*, 1995). The clean sands were carried basinwards, through canyons and incisions in the T45 shell, by high-density, sand-rich turbidity currents. These are low-efficiency fan systems (*sensu* Mutti, 1985), where turbidity currents rapidly dump their sediment load at or near to the base of slope, forming thick piles of sand, with few fine-grained interbeds. Remobilization and liquefaction of the sands often led to extensive injection into the enveloping muds, giving an apparent 'interbedded' character to the upper and lateral boundaries of the fan (Alexander *et al.*, 1992; Dixon *et al.*, 1996). This has led many workers to erroneously describe these systems as channel–levee complexes, using Plio-Pleistocene Gulf of Mexico fan systems as an analogue. The Harding–Gryphon reservoir is of this type, comprising thick clean sands with liquefied and injected envelopes, forming what appear to be discrete 'pods' (see Figs 10.18 and 10.27), deposited at or near to submarine-canyon mouths.

10.5.2 Mid- to late Eocene (T60–T98)

These sequences encompass much of the Ypresian Stage. They are of longer duration than the whole of the Palaeocene succession, immediately overlying the Beauly Coal, and record a resumption of regional shoreline advance. Seismically, T60 is distinguishable from T84 only over the Northern North Sea, where it forms a series of clinoforms separated from the overlying sequences by a major lowstand fan complex (T70 Frigg fan). Elsewhere, it is difficult to differentiate between these overall muddy progradational packages. Mudge and Bujak (1994) include both T60 and T82 in their Frigg Sequence which also contains the T70 Frigg sandstone (their Frigg Sandstone Member).

T60–T82 sequences (Fig. 10.8)

From the early Eocene (T60), the whole of the Central and Northern North Sea, as well as the Atlantic Margin area, were part of the newly subsiding Atlantic passive margin. Rates of sedimentation slowed dramatically, compared with the Palaeocene; hinterlands were reduced and passive subsidence controlled the gross stratigraphic signature. Tectonically controlled sedimentation, which distinguishes the Palaeocene, became suppressed and passive mud deposition dominated for long periods over much of the basin.

The majority of Eocene reservoirs are small, localized submarine fans. T60 and T82 basinal-sand deposition is limited to small, mixed-load fans of limited extent. Thicker, high net : gross sandstones, with blocky gamma profiles, are restricted to parts of UKCS Quadrants 9 and 16 (the Bruce–Beryl embayment), extending eastwards into the

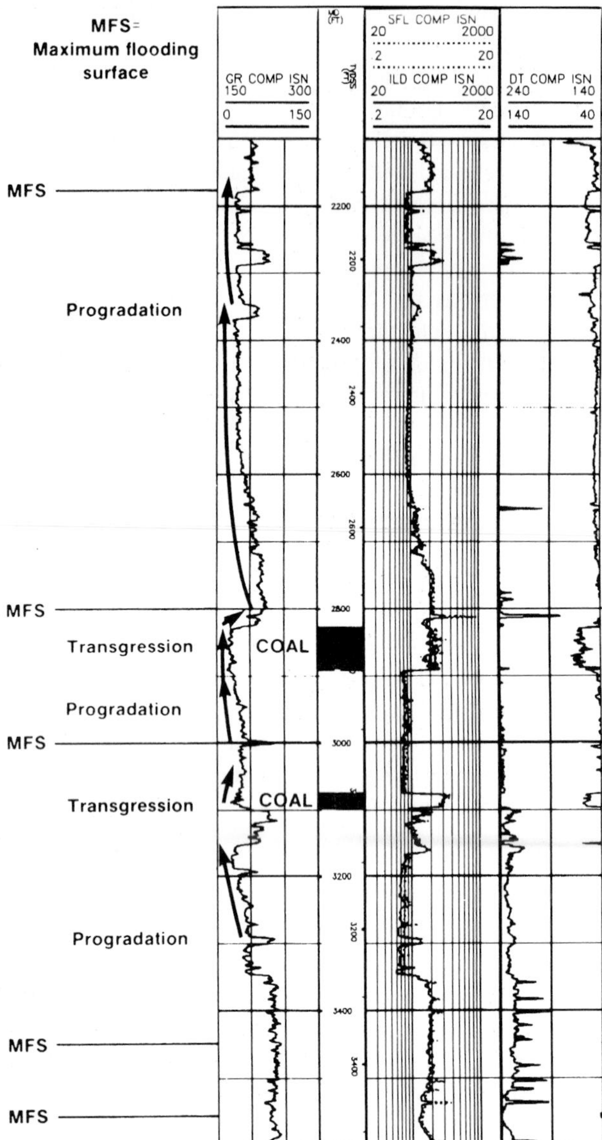

Fig. 10.12 Wireline-log response of late Palaeocene/early Eocene shelfal deposits showing transgressive coal units (from Milton *et al.*, 1990).

Norwegian sector. These have characteristics similar to the underlying T50 Balder Sands.

T70 'Frigg sequence' (Fig. 10.13A)

T70 is restricted to the late Ypresian. It is distinguishable only in the North Viking Graben (the Frigg Sandstone), around the Gannet Field in the Central Graben (the Tay Sandstone) and locally in the Fladen Ground Spur area (the Skroo Sandstone). It resulted from a major fall in relative sea level and in associated hinterland rejuvenation. This led to a resumption of coarse-clastic deposition over selected parts of the basin, in the form of major lowstand fan systems.

The Frigg Fan is the thickest and most widespread fan system, comprising over 200 m of stacked sandstone with an abrupt base and gradational top (Fig. 10.14). It was fed through a major east–west-trending canyon, which can be clearly seen on seismic across parts of UKCS Quadrants 8 and 9. The coeval Alwyn Fan, to the north, is more heterogeneous, with a characteristic lobate geometry. The juxtaposition of these two distinct but coeval fan types is here linked to differentiation of provenance areas.

Further south, the Skroo and Tay sandstones are the products of smaller-scale but commercially important sand-prone fan systems. The Tay fan is clearly identifiable on seismic lines over parts of the Gannet Field Complex (UKCS Quadrants 21 and 22) (Armstrong *et al.*, 1987) (Fig. 10.15).

T92–T98 sequences (Fig. 10.13B)

This encompasses the Middle Eocene, Lutetian Stage, being broadly equivalent to the Alba and Grid Sequences of Mudge and Bujak (1994). Depositionally, it has an overall progradational signature, recording the advance of a mixed-clastic shoreline shelf over much of the Central and Northern North Sea (the middle part of the Moussa Formation). Major sand influxes to the basin are limited to a series of depocentres along the South Viking Graben and the Witch Ground Graben. These appear to be related to basement control and localized reactivation of former rift-margin faults.

The most commercially significant reservoir sandstones are those of the Alba and nearby Chestnut Fields (UKCS Quadrant 16). These are thick, sand-rich fan systems, which can be linked to major submarine-channel/canyon complexes mappable over parts of the Witch Ground Graben and abutting against the Forties–Montrose Ridge. The sandstones resemble the T50 Harding–Gryphon fans in their geometry and architecture. They show evidence of liquefaction, with injection and mixing of sand into surrounding mudstones. It seems likely that supply of sand to the basin was facilitated by a combination of shelf-edge failure and a relative fall in sea level, which occurred at least once during the middle Eocene and again during the late Eocene (Mudge and Bujak, 1994).

T92–T96 deposition was terminated by a regional flooding event, which effectively halted the deposition of major sands in the basin. This boundary defines the shallowest level of hydrocarbon prospectivity in the basin. T98 records a resumption in shelf–shoreline advance, following a relative sea-level fall. The upper boundary of this sequence is also defined by a major flooding event, which corresponds to the early Oligocene global sea-level maximum of Haq *et al.* (1987).

10.6 Hydrocarbon source and migration

This topic is only summarized here, being discussed in detail by Cornford (Chapter 11, this volume). The timing and migration of oil and gas are key issues in Tertiary prospectivity. The source-rock system for the Tertiary is provided by the late Jurassic Kimmeridge Clay Formation. There is also additional source potential within the early Tertiary. Widespread organic-rich muds were deposited over much of the North Sea basin during the late Palaeocene (T40) annexation of the North Sea from the main Atlantic circulation system. These are not mature for hydrocarbon generation at present-day burial depths.

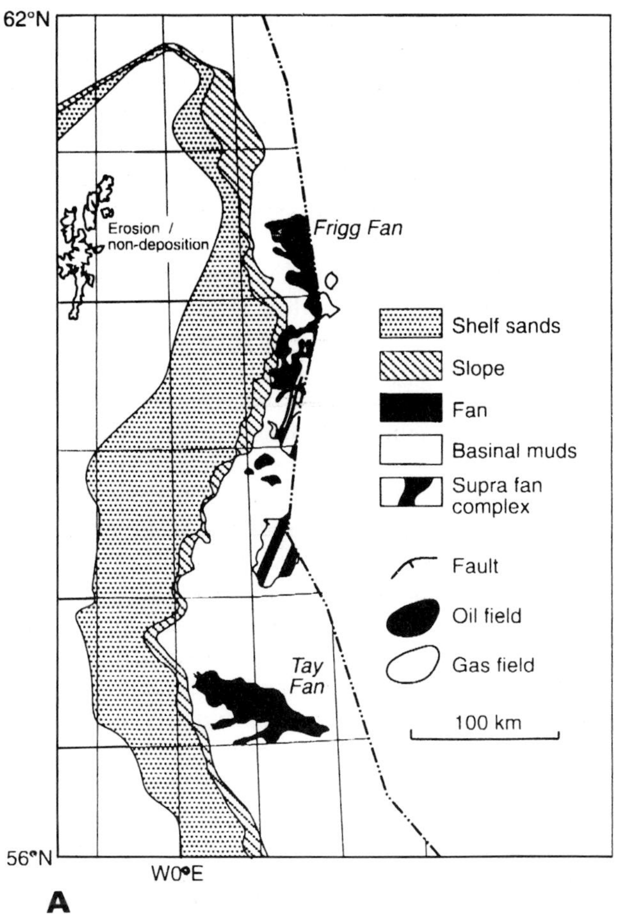

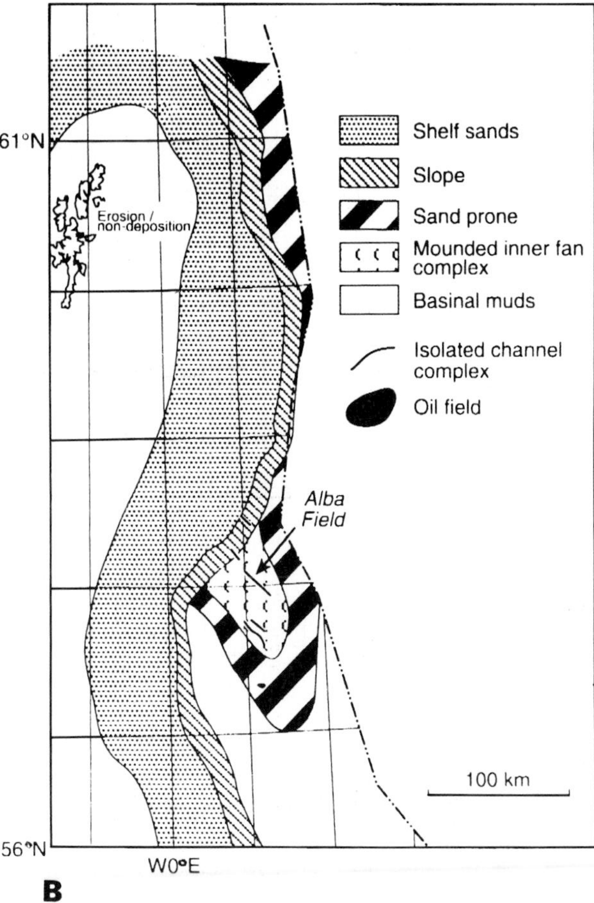

Fig. 10.13 A. T70 'Frigg' sequence. B. T92–T98 gross depositional environment. (Modified from den Hartog Jager *et al.*, 1993.)

Prospectivity is broadly confined to those areas below 1000 m subsea. These are the Palaeocene and Eocene basins. Hydrocarbons in shallow reservoirs are prone to extensive biodegradation (e.g. UKCS Bressay Field). This renders unprospective both Palaeocene and Eocene shelf areas and the post-Eocene rocks. Exceptions to this general rule are localized, stratigraphically isolated pools that received a late charge and hence have not been modified by biodegradation.

The main phase of hydrocarbon migration began during the early Eocene and continues today. Evidence for the early Eocene charge is provided by fossil seeps and reworked oil-impregnated clasts in T50 and T60 sandstones of the Harding Field (UKCS 9/23b). The exact mechanism of migration remains unclear. It is likely that the charge was facilitated by structural discontinuities along the margins of Jurassic tilted fault blocks and associated with halokinesis in the Central North Sea. This interpretation places increased risk on exploration prospects away from the main migration routes, such as the depositional lows, which might be exploited by some of the smaller-scale Eocene fan systems. Alternative interpretations involve overpressure-induced diffusion and hydrodynamic transport, coupled with lateral migration within the Tertiary carrier system. This enables a more widespread hydrocarbon charge and extends Tertiary prospectivity away from areas associated with deeper-seated structural elements. Gas expulsion is considerably more efficient than oil, and poses fewer problems in charging Tertiary reservoirs.

10.7 Reservoirs

10.7.1 Pattern of fill

The uplift of northern Britain and the Shetland Platform exposed large tracts of poorly cemented Palaeozoic and Mesozoic sandstone and also areas of partially granitic basement. These are the source of the coarse-grained siliciclastic material that fed the early Tertiary shelf and basin-floor depositional systems in both the North Sea and the Atlantic Margin (see Figs 10.4 and 10.5). Morton *et al.* (1993) demonstrate that variations in detrital garnet geochemistry within Palaeogene sandstones can be traced to different source-area terrains and periods of unroofing. This aids regional correlation and differentiation of reservoir units across the basin.

The structural grain of the Scottish Highlands, together with the effects of the North Atlantic thermal upwelling, indicate that the watershed for drainage systems from this rejuvenated hinterland probably lay to the west of the Highland region (see Fig. 10.4). Large volumes of sand were transported eastwards into the North Sea Basin to produce the Palaeocene basin-floor fan complexes. Coeval fans in the Atlantic-Margin basins were smaller in both size and volume of coarse-grained sediment. Relatively small

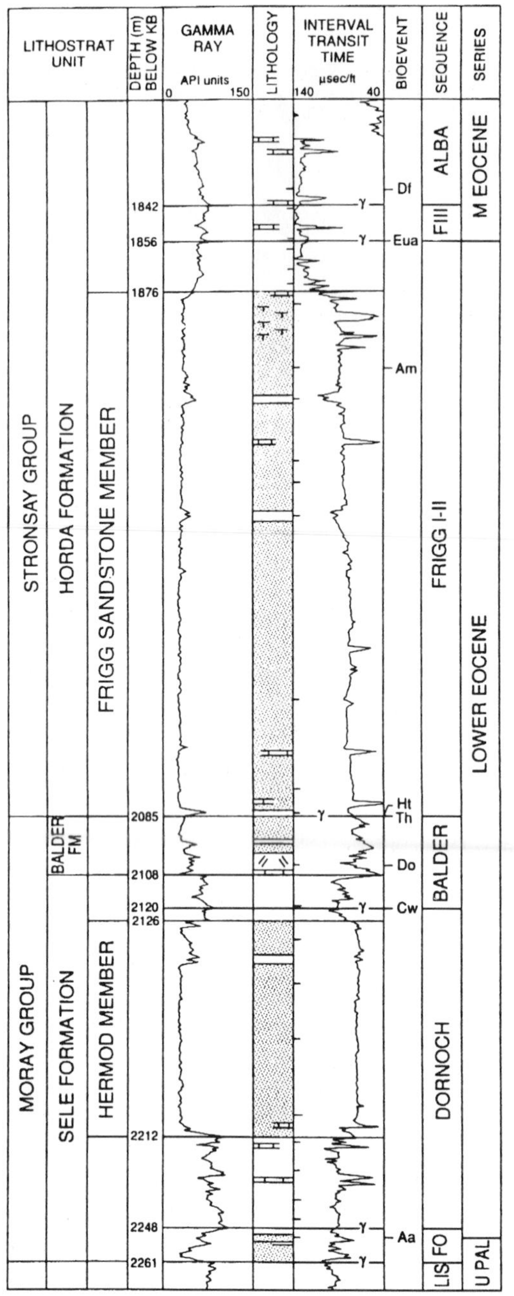

Fig. 10.14 Well 10/1-1A: Frigg Formation type well (T70), Frigg Field (from Mudge and Bujak, 1994).

volumes of sand were derived from western Norway, due to the predominantly argillaceous and metamorphic nature of that hinterland. This explains the relatively poor prospectivity of the Tertiary over large parts of the Norwegian sector. The Balder Field (NOCS 25/10 and 25/11) remains one of the few Tertiary oilfields where the reservoir sands are sourced from local sand-prone areas in the east (Jenssen *et al.*, 1993).

Sediment dispersal patterns and input points were controlled by basin bathymetry, which was strongly influenced by the geometry of the underlying Jurassic failed rift. The overall pattern of fill is complicated by a number of factors related to tectonics and halokinesis over different parts of the basin. Differential uplift of the hinterland affected the overall pattern of basin fill, with temporally and spatially distinct phases of sediment input. Along the margin of the Central Graben, movement of salt diapirs also influenced Palaeocene sediment dispersal and prospectivity.

The combination of seismic, sedimentological and stratigraphic data indicates that early stages of deposition (Palaeocene) are distinguished by great basin depth, steep margins and high sedimentation rates. This led to instability of delta and shelf slopes, enabling direct liberation of large volumes of sediment to the basin. The combination of delta-front failure, relative stability of the margins and rapid basin subsidence led to bypass of the shelf and the direct supply of sand to the basin. In contrast, the late Palaeocene to Eocene progradation results from relatively shallow basin depths. By this time, basin slopes and sedimentation rates were reduced. The relatively stable slopes inhibited direct transfer of sediment to the basin. This led in turn to smaller, more localized, basin-floor systems, which are often more sand-prone than their Palaeocene counterparts.

10.7.2 Depositional geometry

Submarine fans and related deepwater clastic systems are

Fig. 10.15 Strike seismic section through the Tay Fan showing lower mounded and upper sheet-like geometries (from den Hartog Jager *et al.*, 1993).

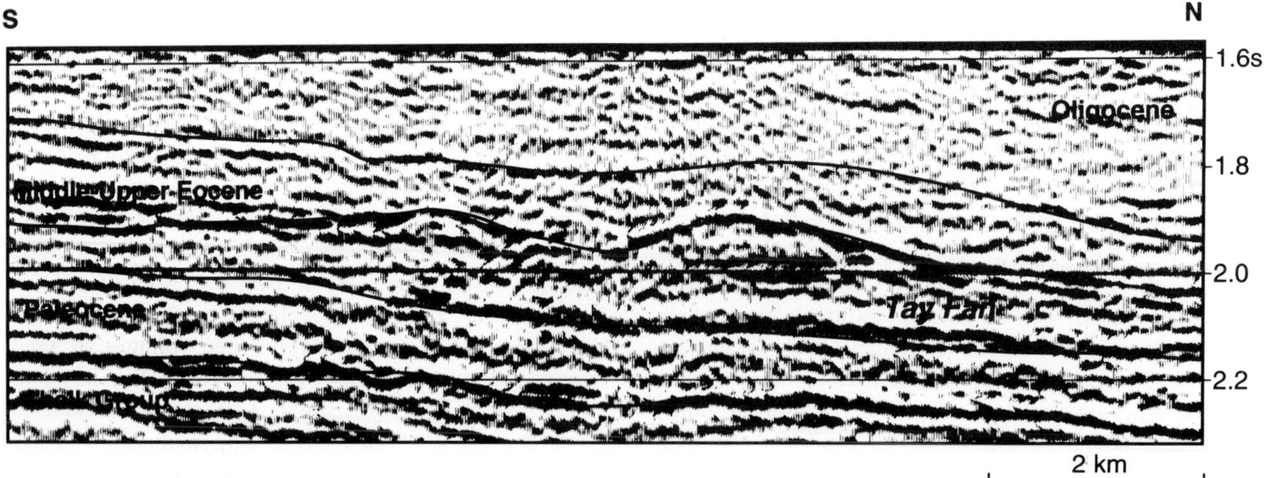

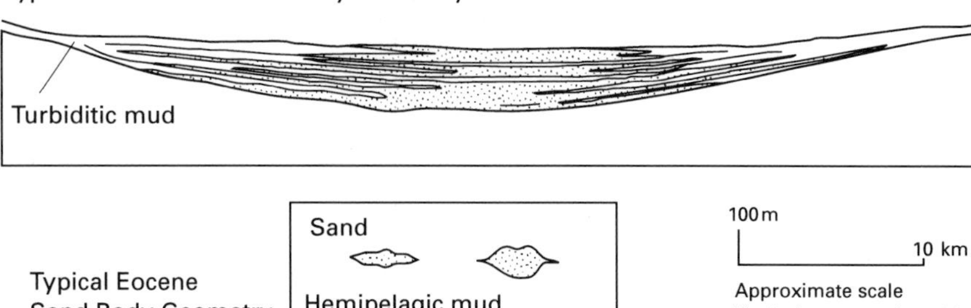

Fig. 10.16 Schematic sand-body geometry of Palaeocene and Eocene submarines fans (from R. Anderton, British Petroleum, unpublished data).

the only truly prospective plays within the Tertiary fairways. Any hydrocarbons reservoired in marginal shelf or coastal systems (e.g. the Bressay Field) typically contain viscous biodegraded oil, due to a combination of shallow burial depth (<1000 m subsea) and access to an active oxygenated aquifer.

The submarine fans are divisible into two broad categories (Fig. 10.16). Each has different characteristics and hence prospectivity. The early Palaeocene T20–T40 fans of the Central Graben are large-scale systems and form major elements of the basin fill. In contrast, the younger T45–T98 fans tend to be localized and on a smaller scale. Each has a distinctive suite of subsurface characteristics in core, log and seismic.

The Palaeocene fans were fed directly via fluvially dominated deltaic shorelines, which liberated large volumes of sediment to the basin, largely by slope failure, such as in the modern Mississippi delta. The fans range from mixed sand–mud to sand-prone types. In the main depocentres, they have large sheet-like geometries (Figs 10.17 and 10.29). They are most sandy along the main axes of flow, becoming muddier distally and laterally over topographic highs. Their gross morphology is strongly influenced by basin geometry. The flows contained large volumes of admixed clay, which enabled the sands to travel large distances. Internal architectural patterns are dominated by sand-prone channel and nested channel fills, together with mixed-load interbedded units between channels (Figs 10.17 and 10.29).

The late Palaeocene–Eocene fans range from small-scale mixed-load systems, closely linked to prograding shelf complexes, to sand-rich pod-like (e.g. Harding–Gryphon Fields) or elongate bodies (e.g. Alba Field). The sand-rich systems typically have steep, abrupt margins and low width-to-thickness ratios (Figs 10.18, 10.27 and 10.28). They comprise homogeneous masses of clean sand, which are highly susceptible to soft-sediment deformation caused by slumping and liquefaction. This modifies the primary depositional geometry. Only the Eocene T70 Frigg and Gannet fans approach the scale of the older Palaeocene systems. They are major lowstand fan complexes.

10.7.3 Reservoir quality

Tertiary sandstones differ from older North Sea reservoirs in being poorly cemented to unconsolidated. There is little evidence that diagenesis and cementation play anything more than a minor role in Palaeogene prospectivity. The main phase of cementation in sandstones of the North Sea and Atlantic Margin began during the early Tertiary, continuing through to the present (Aplin *et al.*, 1993). Most Tertiary sands were hydrocarbon-charged before any cementation had taken effect. Minor clay, quartz and patchy carbonate are the only common cement phases.

Stratigraphically related lithology is the primary control on reservoir quality (porosity and permeability) variations. The gross morphological differences between the Palaeocene and Eocene reservoirs are mirrored by changes in the

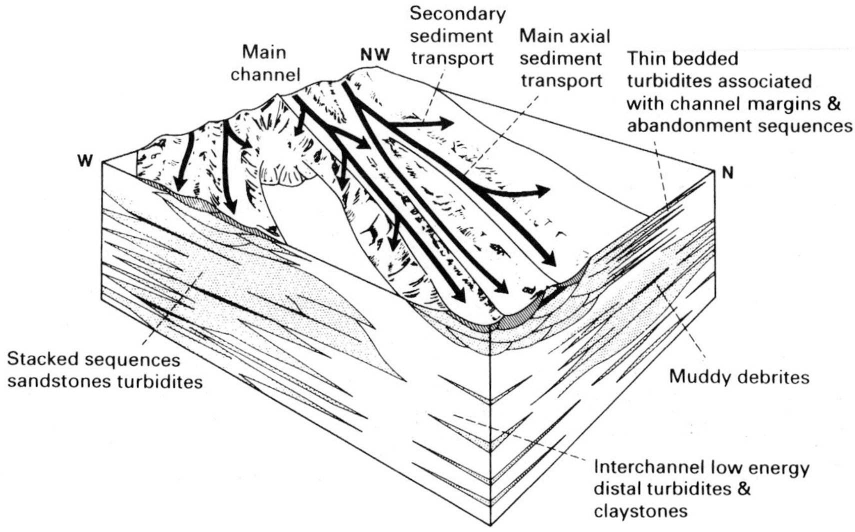

Fig. 10.17 Forties submarine-fan system, depositional model (from Whyatt *et al.*, 1991).

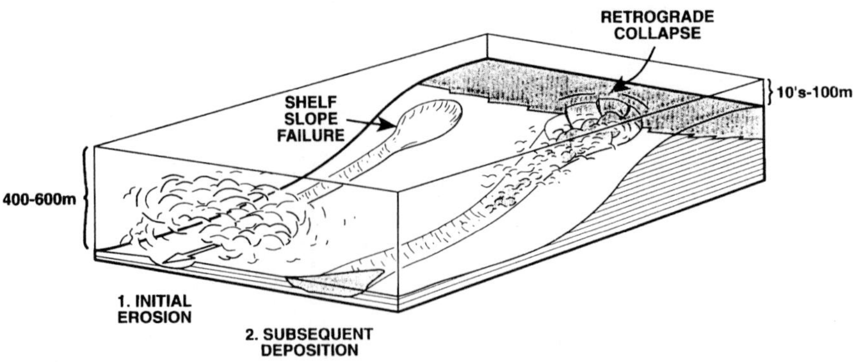

Fig. 10.18 Idealized depositional model, Harding Field (from Dixon *et al.*, 1995).

detrital mineralogy of the sandstones. The large delta-fed Palaeocene fans typically comprise more clay-rich sands. In contrast, the Eocene reservoirs are cleaner sands with little detrital clay, because of extensive shelf reworking prior to resedimentation into the basins. This means that Eocene reservoirs are typically more permeable but have similar porosities to the Palaeocene reservoirs.

10.8 Prospectivity

The contrasting character of the Eocene and Palaeocene fan systems (see Fig. 10.16) has a profound influence on their prospectivity. Palaeocene (T20–T40) fans are typically large-scale basin-floor systems. Associated traps are usually structural, being four-way dip closures developed by differential compaction over deeper buried highs (Jurassic tilted fault blocks). Typical examples are Forties (Fig. 10.19), Nelson, Andrew, Montrose, Arbroath and Donan (UKCS 15/20a). Stratigraphic traps are associated with halokinesis (e.g. the margin of the Central Graben) and depositional pinch-outs (e.g. Everest Field, UKCS Quadrant 22).

The smaller Palaeocene and Eocene fans typically produce stratigraphic traps (e.g. Alba Field, UKCS 16/26) or small four-way dip closures at the tops of mounded sand bodies.

Recent successes in the discovery of new pools and the growth of reserves around existing fields for both the Palaeocene and the Eocene have been achieved by careful depth conversion of seismic data. The Cyrus Field in UKCS 16/28 is a good example of such a pool (Mound *et al.*, 1991).

10.9 Tertiary plays

This section comprises a summary of the more significant Tertiary plays, with emphasis on those discoveries made since 1988. It is by no means comprehensive but aims to illustrate the key plays that have received attention. Detailed descriptions of recent discoveries in the Palaeocene of the Atlantic Margin have yet to be released and so are only briefly summarized here.

Forties Field, UKCS 21/10, 22/6a (Fig. 10.20)

Status. Mature, post-plateau, infill drilling (discovered 1970).
Reserves. 400×10^6 m^3 of oil.
Operator. BP Exploration.
Stratigraphy. Forties Formation (T40).
Trap. Four-way dip-closed compactional drape with two culminations, over buried high (Forties–Montrose ridge).
Reservoir. T40 channel and interchannel turbidite sandstones, divisible into a lower high-net : gross unit and an upper more heterogeneous section. Primary reservoir targets are channelized sandstones, distinguishable from

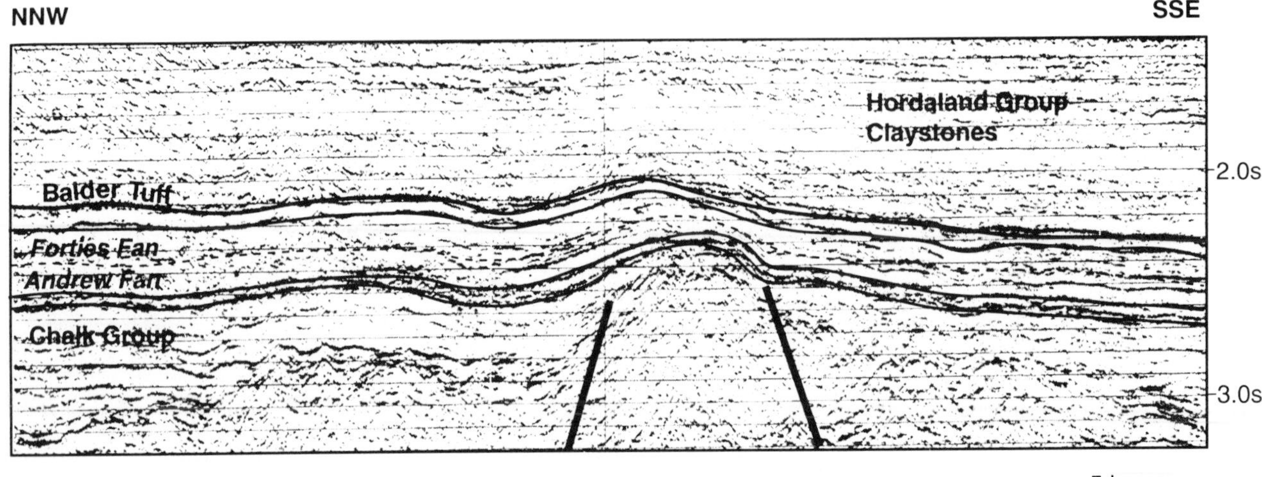

Fig. 10.19 Example of structural closure caused by compactional drape over underlying fault block: the Forties Field (from den Hartog Jager *et al.*, 1993).

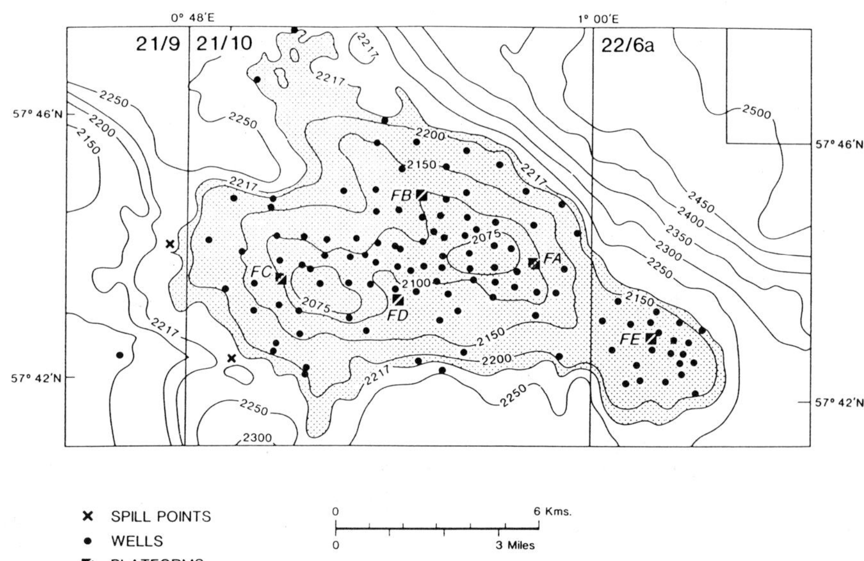

Fig. 10.20 Forties Field, UKCS 21/10, structure map (from Wills, 1991).

Fig. 10.21 Frigg Field, UKCS 10/1, location map and schematic cross-section (from Brewster, 1991).

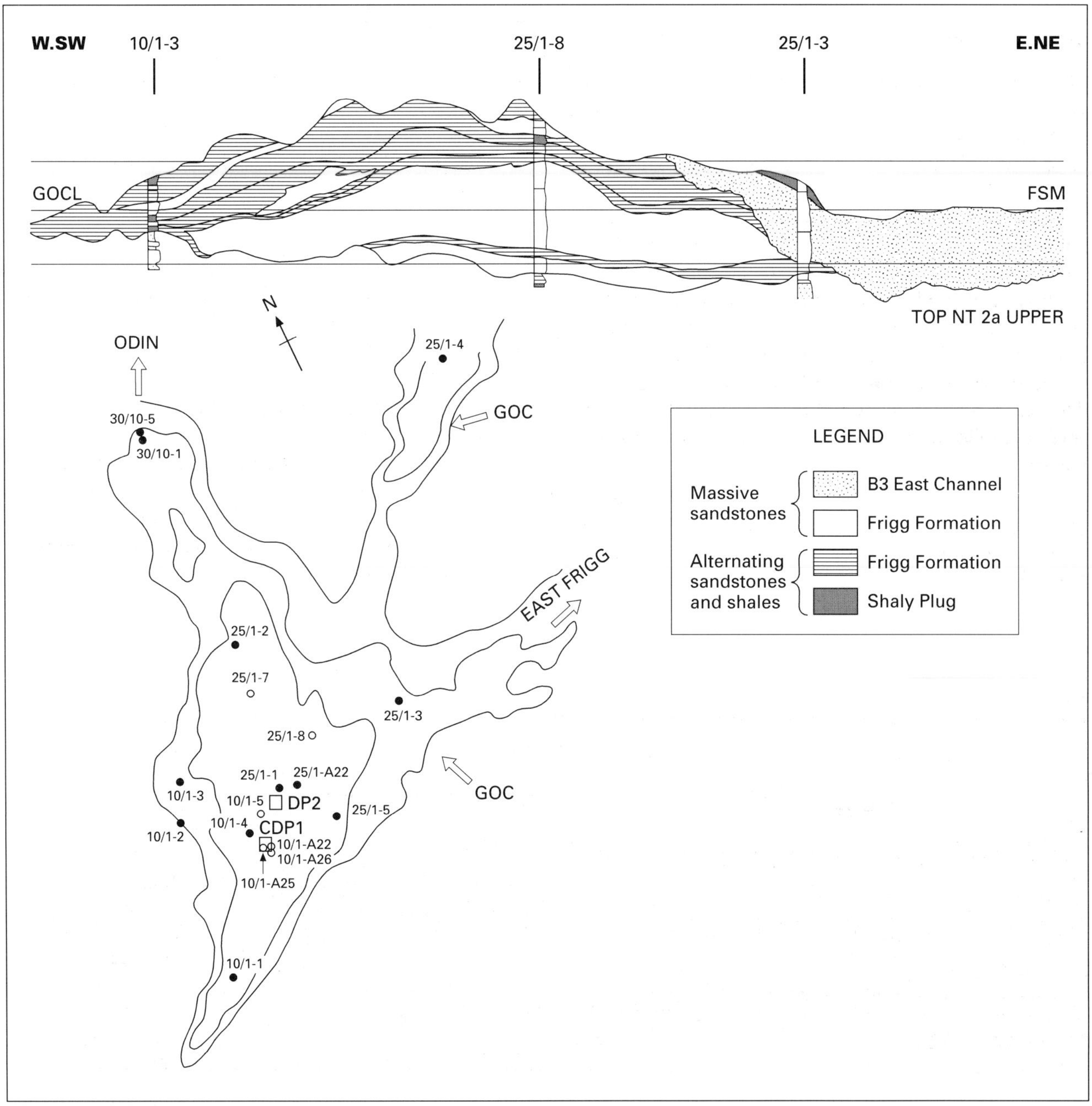

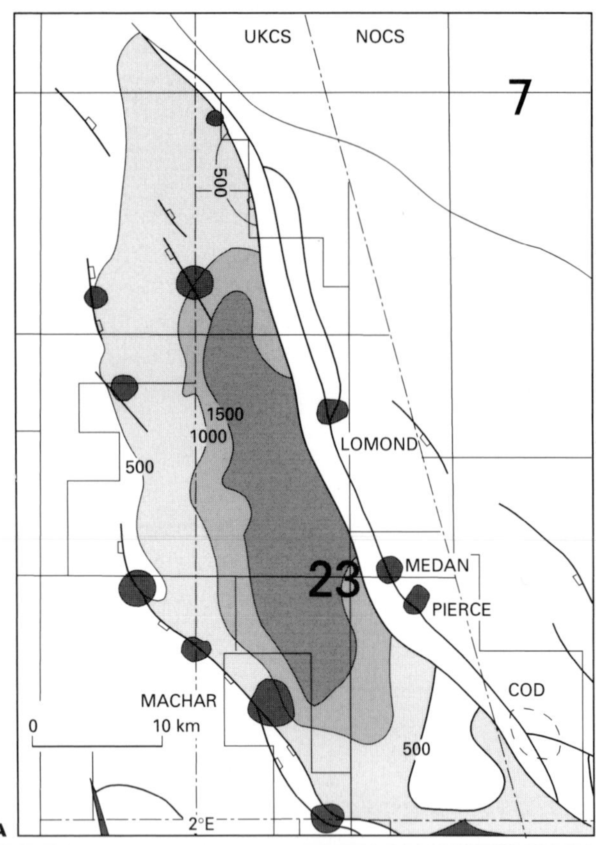

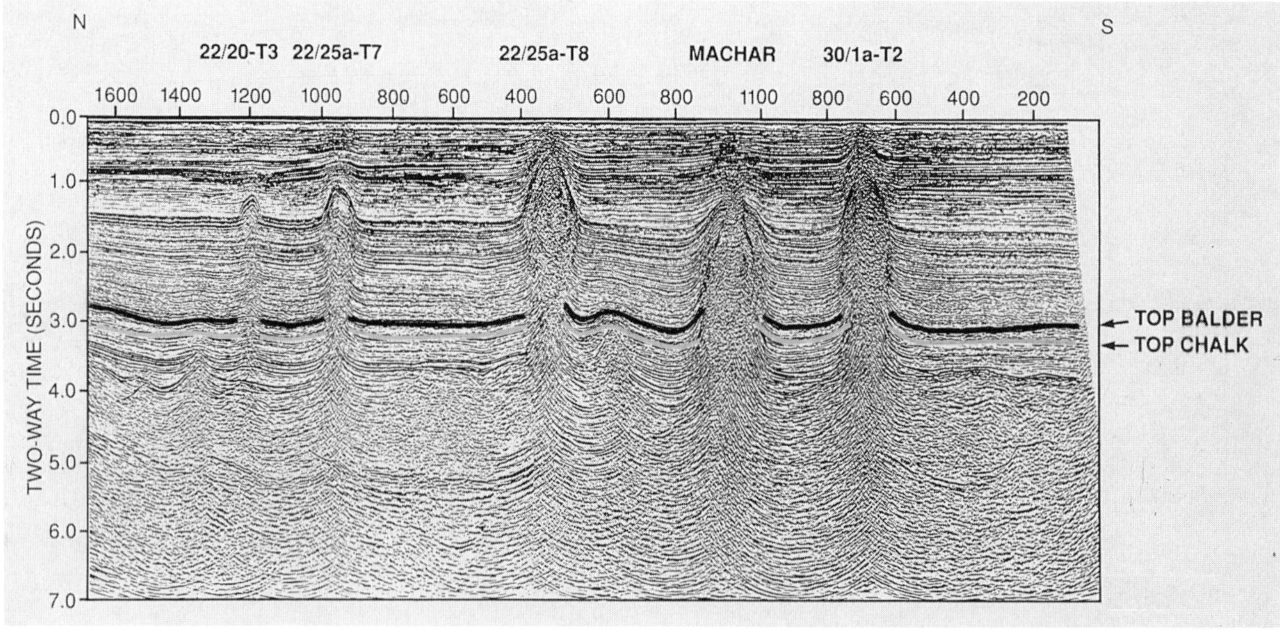

Fig. 10.22 Distribution of salt-diapir fields in the eastern trough, Central Graben. A. Map (from Rattey and Hayward, 1993). B. Seismic section (British Petroleum, unpublished data).

poorer-quality and thinner-bedded interchannel and abandonment facies (see Figs 10.17 and 10.29). Success relies upon identification of channel and interchannel areas. Remaining reserves are largely concentrated in the upper, heterogeneous section (see below).

Related fields. Other smaller-scale producing fields of broadly similar structural and stratigraphic configuration, discovered at around the same time as Forties, include the Montrose and Arbroath Fields (UKCS 22/17 and 22/18) (Crawford *et al.*, 1991).

Frigg Field, NOCS 25/1, UKCS 10/1 (Fig. 10.21)

Status. Post-plateau, infill and satellite drilling (discovered 1970).

Reserves. 184×10^9 m^3 of oil-wet gas.

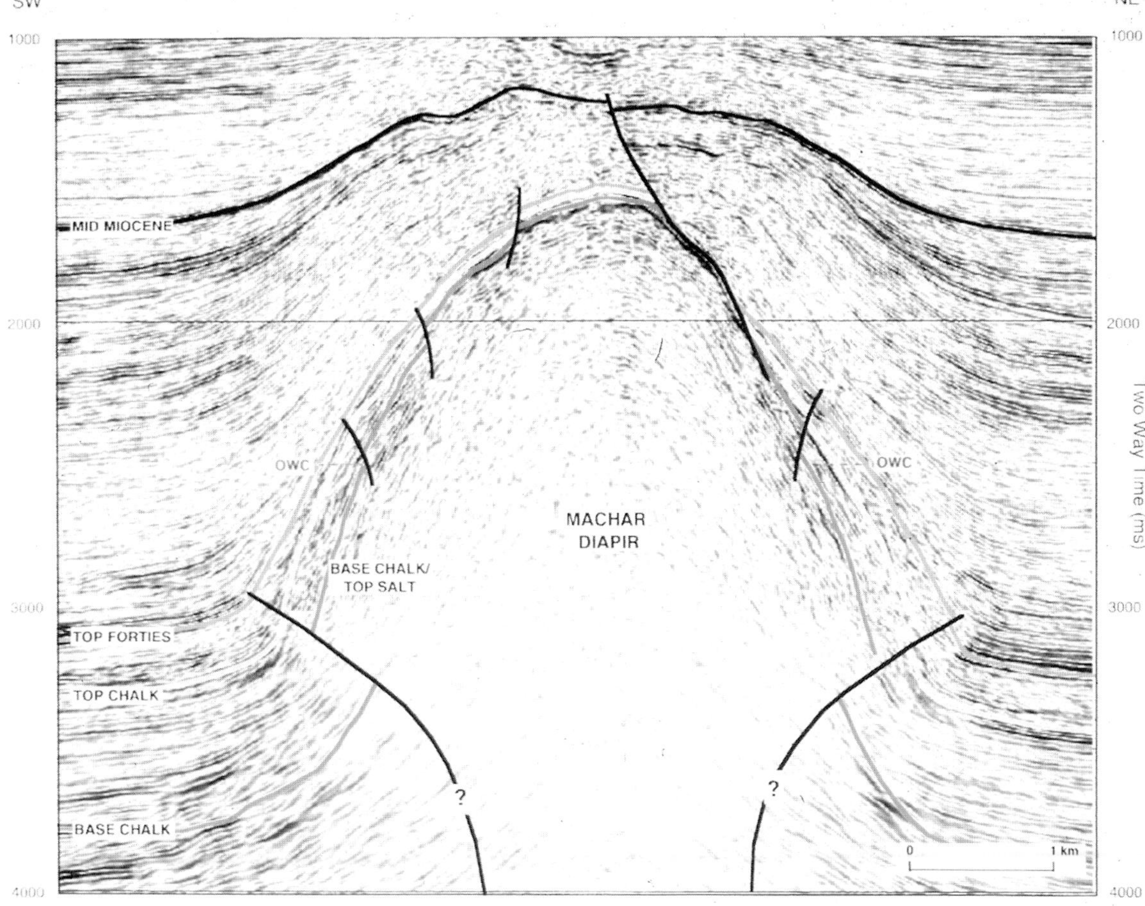

Fig. 10.23 Seismic section of Machar Field, salt-diapir trap (from Foster and Rattey, 1993).

Operator. Elf Norge.

Stratigraphy. Frigg Formation (T70).

Trap. Combined structural and stratigraphic trap related to the depositional topography of the Lower Eocene Frigg sandstone (T70).

Reservoir. The reservoir comprises a series of stacked-channel, lobe and interchannel sandstones, with a gross lobate and mounded morphology. Lithology and reservoir quality change from proximal to distal settings across the field. Reservoir and trap definition rely on a combination of seismic facies and well data.

Diapir fields (Figs 10.22 and 10.23)

Status. Late appraisal.

Reserves. 160×10^6 m^3 barrels of oil equivalent.

Operators. BP Exploration, Shell and others.

Stratigraphy. Late Cretaceous (Chalk) and Palaeocene (T20–T40).

Trap. Combined structural and stratigraphic traps associated with salt diapirs (see Fig. 10.23), occurring in two subparallel chains along the eastern side of the Central Graben, which is adjacent to the Jaeren High. Smaller traps also occur along the western margin of the Graben.

Reservoir. Bedded-turbidite sandstones and fractured chalks. Turbidite sand deposition was controlled by diapir topography and salt movement. This led to rapid lateral changes in reservoir characteristics and quality together with widespread remobilization and slumping of the sands (Fig. 10.24).

Individual structures have hydrocarbon columns up to 1500 m thick. High-relief structures are generally oil-bearing, while shallower structures are, in many cases, gas-flushed.

Nelson Field, UKCS 22/11, 22/6a (Fig. 10.25)

Status. Early production (discovered 1988).

Reserves. $> 71 \times 10^6$ m^3 barrels of oil; *c.* 450×10^6 m^3 of oil.

Operator. Enterprise Oil.

Stratigraphy. Forties Formation (T40).

Trap. Four-way dip-closed compactional drape over buried high (Forties–Montrose ridge).

Reservoir. Upper Forties Formation, channel and interchannel facies. Primary reservoir targets are seismically defined, channelized high-density turbidite sandstones. These are distinguished from poorer-quality thinner-bedded, interchannel and abandonment facies (see Fig. 10.17). Success relies upon identification of channel and interchannel areas, using a combination of seismic facies and isochore mapping.

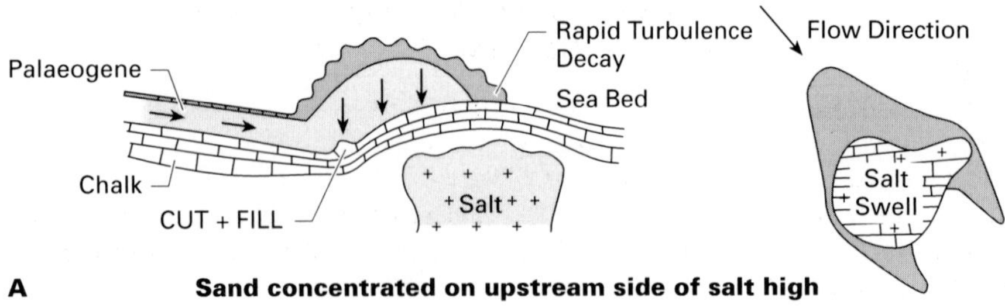

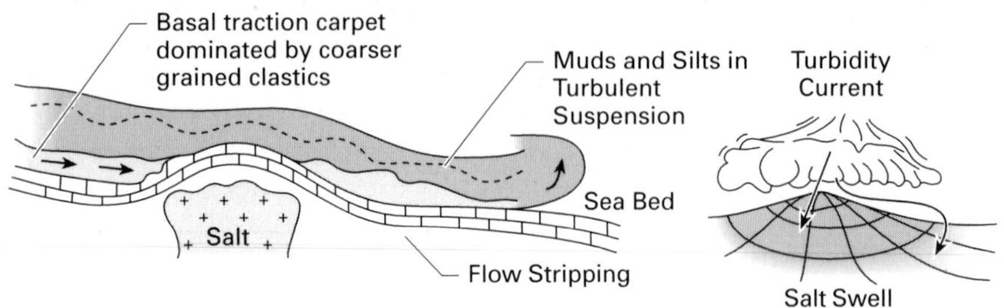

Fig. 10.24 Schematic diagram of turbidite-sand distribution associated with syndepositional halokinesis (from Hodgson *et al.*, 1992).

Everest Field Complex, UKCS 22/5a, 22/9, 22/10a, 22/14a (Fig. 10.26)

Status. Early production (discovered 1982).

Reserves. c. 34×10^9 m^3 of gas (minor oil rim in Everest East).

Operator. Amoco.

Stratigraphy. Andrew Formation (T30), Forties Formation (T40).

Trap. A complex of three separate pools (Everest South, North and East) in composite stratgraphic/structural traps, with eastward pinch-out of sandstones against the

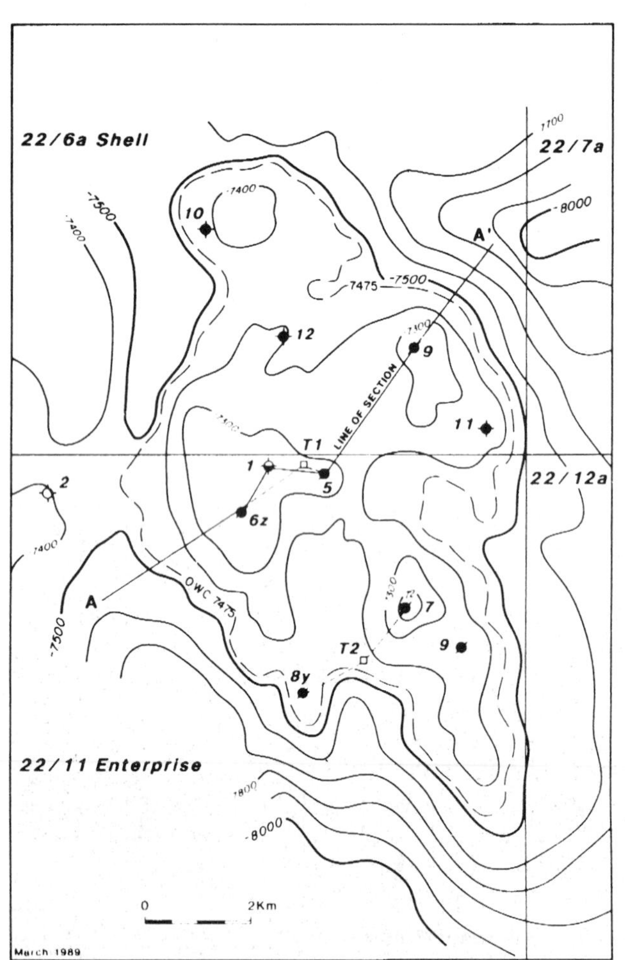

Fig. 10.25 Nelson Field, UKCS 22/11, 22/6a (from Whyatt *et al.*, 1991).

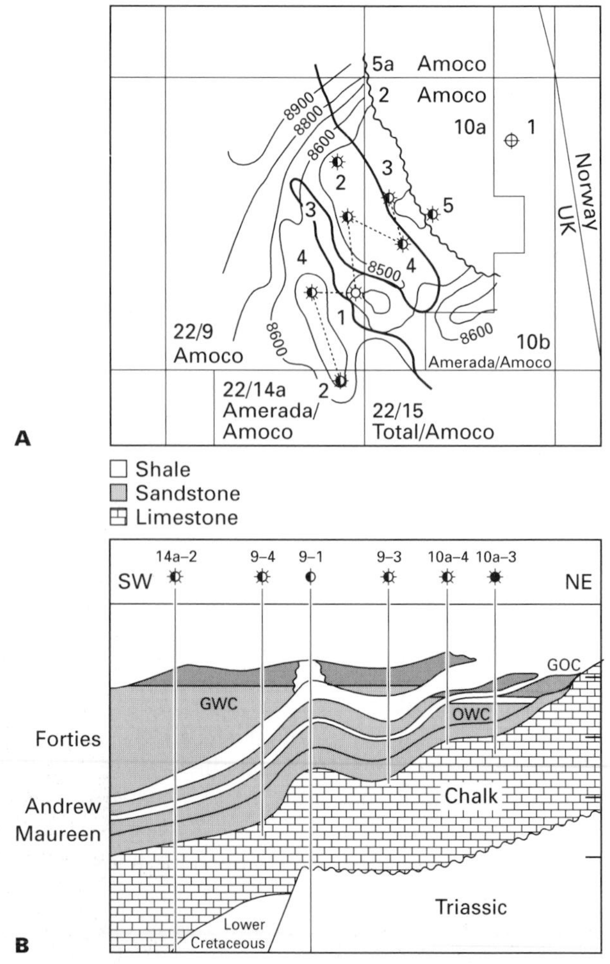

Fig. 10.26 Everest Field, eastern Central Graben, UKCS: map and schematic cross-section (from O'Connor and Walker, 1993).

Jaeren High.

Reservoir. Turbidite sandstones at the edge of the main T40 Forties and T30 Andrew submarine-fan systems. The T30 Andrew reservoir is thinner and comprises a sheet-like unit, with cleaning-upwards sequences interpreted as depositional lobes. The thicker T40 reservoir shows an overall cleaning-upwards character, with two lobes mappable by seismic. Reservoir distribution appears to be controlled by subtle changes in bathymetry. In terms of porosity and permeability, the sandstones are of poorer quality than in the more proximal and axial parts of the submarine-fan complexes. Definition and mapping of reservoir-quality sandstone will be a critical factor in the success of these fields.

Gryphon–Harding (formerly Forth) fields, UKCS 9/23b 9/18b (Fig. 10.27)

Status. Development drilling.

Reserves. 32×10^6 m^3 oil plus gas cap.

Operator. Kerr McGee Oil (9/18b), BP (9/23b).

Stratigraphy. T50 (Balder Formation), with secondary reservoirs in T45 (Sele Formation) and T60 (Frigg Formation).

Trap. The fields comprise a complex, dominantly stratigraphic, trap, with structural elements formed on the northerly plunging nose of the Crawford Ridge.

Reservoir. The principal reservoir units are loosely cemented to unconsolidated sandstones of the T50 Balder Formation. Detailed analysis using tuff and biostratigraphic marker horizons, combined with the results of extensive appraisal drilling, shows the sandstones to be

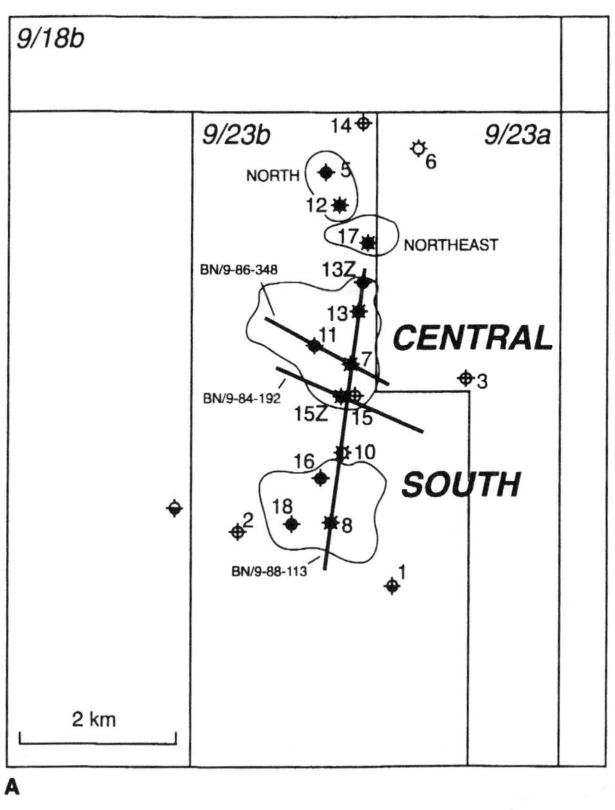

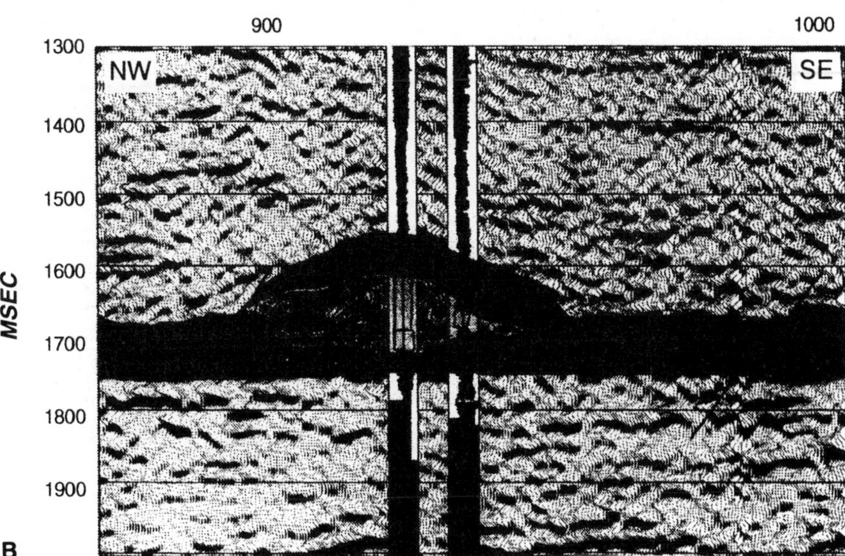

Fig. 10.27 Harding (formerly Forth) Field, UKCS 9/23b. A. Map. B. Seismic section. (From Alexander *et al.*, 1992.)

isolated pod-like bodies, with very abrupt margins. They comprise stacked, high-density, turbidite sandstones with little or no admixed clay. The mechanism of sand emplacement is still not clear, but the general consensus lies with a lowstand fan or slope-failure complex. The masses of homogeneous clean sand are susceptible to soft-sediment deformation by liquefaction; water-escape features, notably large and complex injection of sand into the overlying mud, are common and give a pseudo-interbedded cap to the reservoir. The steep-sided, pod-like nature of the sand bodies is undoubtably due to the deformation, together with differential compaction effects.

Alba Field, UKCS 16/26 (Fig. 10.28)

Status. Development drilling (discovered 1984).
Reserves. > 60×10^6 m^3 of oil.
Operator. Chevron.
Stratigraphy. Middle Eocene (T95).
Trap. A large, apparently stratigraphic, trap, comprising an isolated elongate and lobate sand body encased in basinal sandstone.
Reservoir. The Alba reservoir shows a number of similarities to the Gryphon–Harding complex. It comprises thick, homogeneous, high-density turbidites, which form an elongate lobate body, with abrupt, steep sides. Extensive soft-sediment deformation is also identified. Three main units are recognizable, of which the lower two comprise apparently isolated massive sand bodies, while the upper is an interbedded, possibly injected, unit. As with Harding–Gryphon, reservoir quality is excellent; the sandstones are, at best, poorly cemented. The mode of sand emplacement is unclear. Chevron favour a combination of shelf-edge failure and channel levee systems to produce such isolated elongate bodies of sand (Harding *et al.*, 1990).

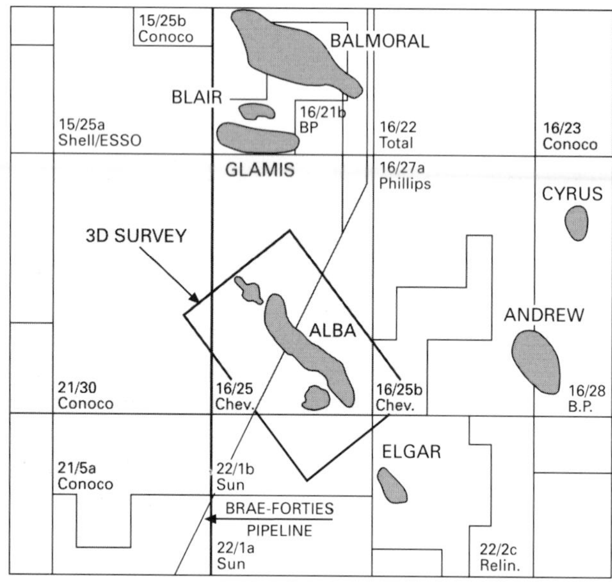

A

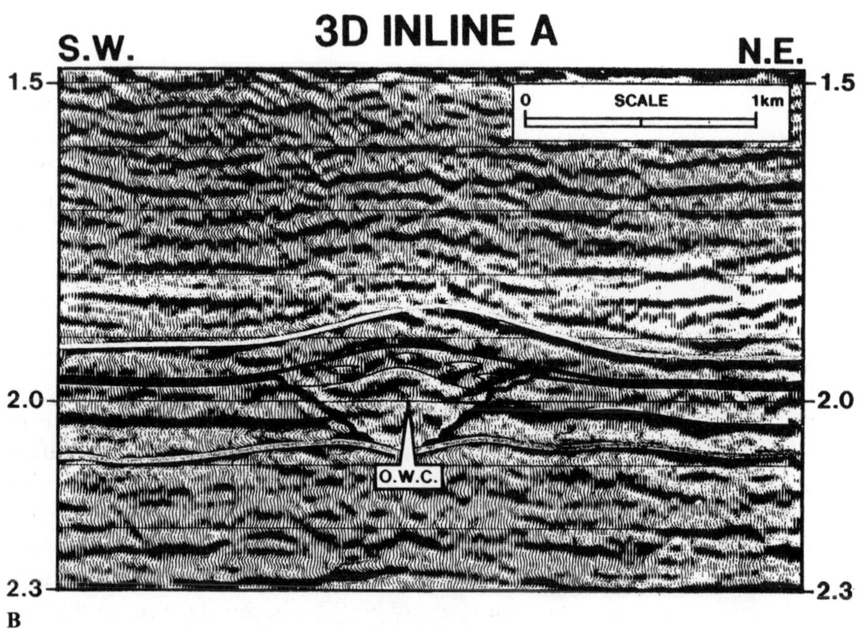

B

Fig. 10.28 Alba Field, UKCS 16/26. A. Structure map. B. Seismic cross-section. (From Newton and Flanagan, 1993.)

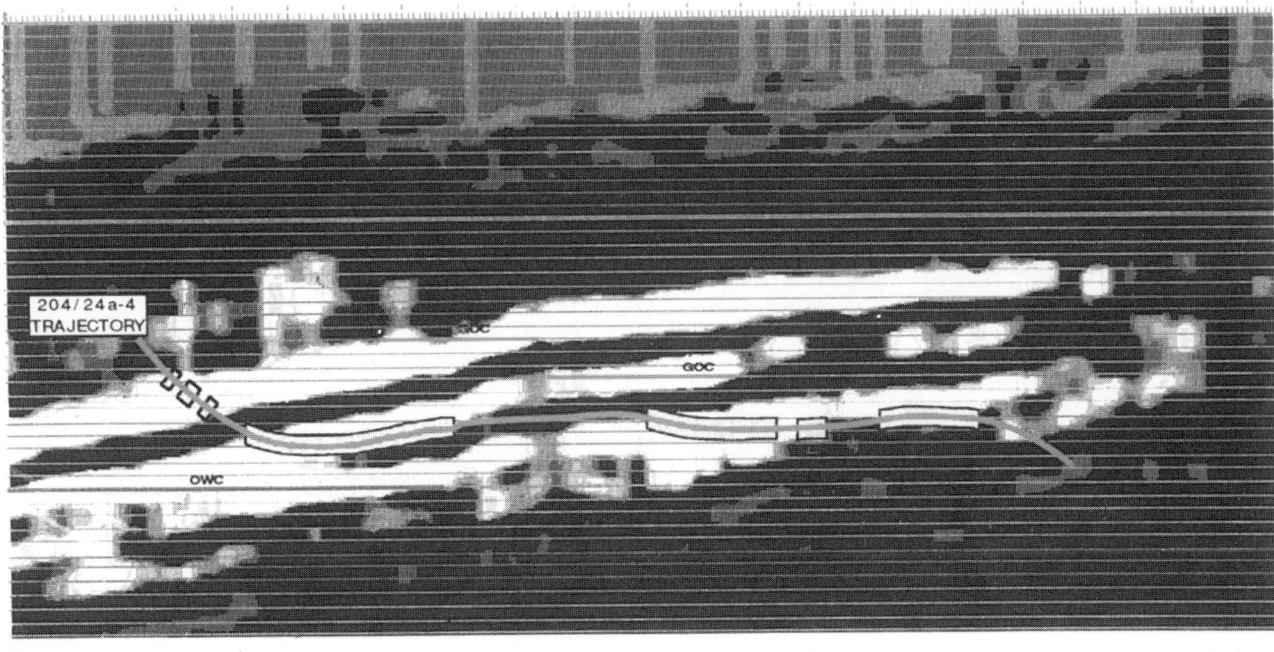

Fig. 10.29 Foinaven Field, UKCS 202/24: seismic section showing the gross form of the main T30 reservoir unit, fluid contacts and the trajectory of horizontal well 204/24a-4 (British Petroleum, unpublished data).

Foinaven and Schiehallion fields, UKCS 204/24 and 204/20 (Figs 10.3 and 10.29)

Status. Appraisal and development drilling (Foinaven discovery well, 204/24a-2: 1992; Schiehallion discovery well, 204/20-1: 1993).

Reserves. 40–80 × 10^6 m^3 of oil in each field.

Operator. BP Exploration.

Stratigraphy. Palaeocene (T30).

Trap. Large Palaeocene traps encased in basinal mudstone.

Reservoir. Regional context (Knott *et al.*, 1993) and the limited information released to date on the discoveries suggest the reservoir to be part of heterogeneous mixed sand–mud fan systems, with clearly defined form and internal architecture on seismic. Figure 10.29 shows the seismic character of the Foinaven reservoir, with bright amplitudes reflecting hydrocarbon-bearing sandstone.

10.10 Mature field operations

Much current Tertiary drilling activity is focused upon accessing additional reserves and maintaining production in mature fields. The growth of reserves from satellite prospects around these fields is also important.

The importance of infill drilling to access additional reserves and maintain production is exemplified by recent successes in the Forties Field (see Fig. 10.20). It is estimated that around 300 × 10^6 barrels of oil reserves remain to be recovered from the field. Most of these reserves lie within the upper heterogeneous part of the T40 reservoir (see Fig. 10.10). Between 1992 and 1994, 18 infill wells were drilled. These collectively contribute around 30% of daily production from the field. This level of infill activity will continue, with the focus on small targets, requiring an increasingly fine-scale reservoir description.

The key to this successful infill programme has been the development of a new detailed reservoir model, linked directly to a fine-grid reservoir simulation (Fig. 10.30). The new model incorporates all available static and dynamic data from the field, together with data from reservoir analogues. A refined deterministic description is calibrated with three-dimensional (3D) seismic data, to discriminate between channel and interchannel facies. Reservoir properties are modelled stochastically. The fine-grid simulation is used to define well locations and to predict performance and reserves distribution. This in turn enables exploitation of the field using non-conventional drilling and completion techniques (Fig. 10.31).

Such approaches are essential to optimize recovery from these mature fields. Current operations in the Frigg Field are similarly focused on producing fine-scale seismic and well-based models to access further reserves. Here the focus is on locating small pools of unswept gas in the main pool and identifying satellite prospects around the field (see Fig. 10.21).

10.11 The future

The recent successes in the Atlantic Margin, together with Eocene stratigraphic plays and Palaeocene diapir and pinch-out plays, mean that the Tertiary will retain exploration interest for the immediate future. Whether the remaining reserves will be in small, localized pools, such as the Cyrus Field, UKCS 16/28, or in further large (> 16 × 10^6 m^3 of oil) fields, such as Foinaven, is a key question. In addition to exploration, Tertiary drilling activity will also continue to focus upon appraisal and development of existing discover-

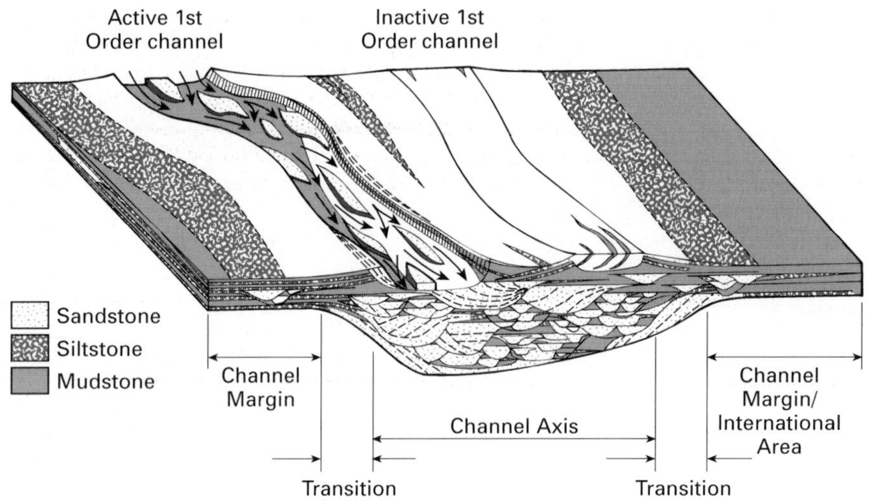

Fig. 10.30 Forties Field reservoir model (from Bennett, British Petroleum, unpublished data).

ies (cf. Payne *et al.*, in press). Reserves growth around existing fields will become increasingly important, with infill drilling and satellite-prospect evaluation (e.g. Frigg Field (Brewster, 1991) and Forties Field).

Success in all of these ventures will rely upon an integrated technical evaluation, involving geoscience and engineering disciplines. Seismic facies and high-technology seismic reprocessing will be essential tools for resolution of the internal characteristics of reservoir bodies. Detailed reservoir description and prediction will be equally important. This will rely upon a combination of high-resolution sequence-stratigraphic techniques and a thorough appreciation of depositional processes and products.

The Tertiary remains an exciting area of opportunity. The key to success will rest upon appropriate use of the

Fig. 10.31 Acoustic-impedance cross-section panel through part of the Forties Field showing proposed track of FA32st horizontal infill well, Forties Field infill drilling programme (from Bennett, British Petroleum, unpublished data).

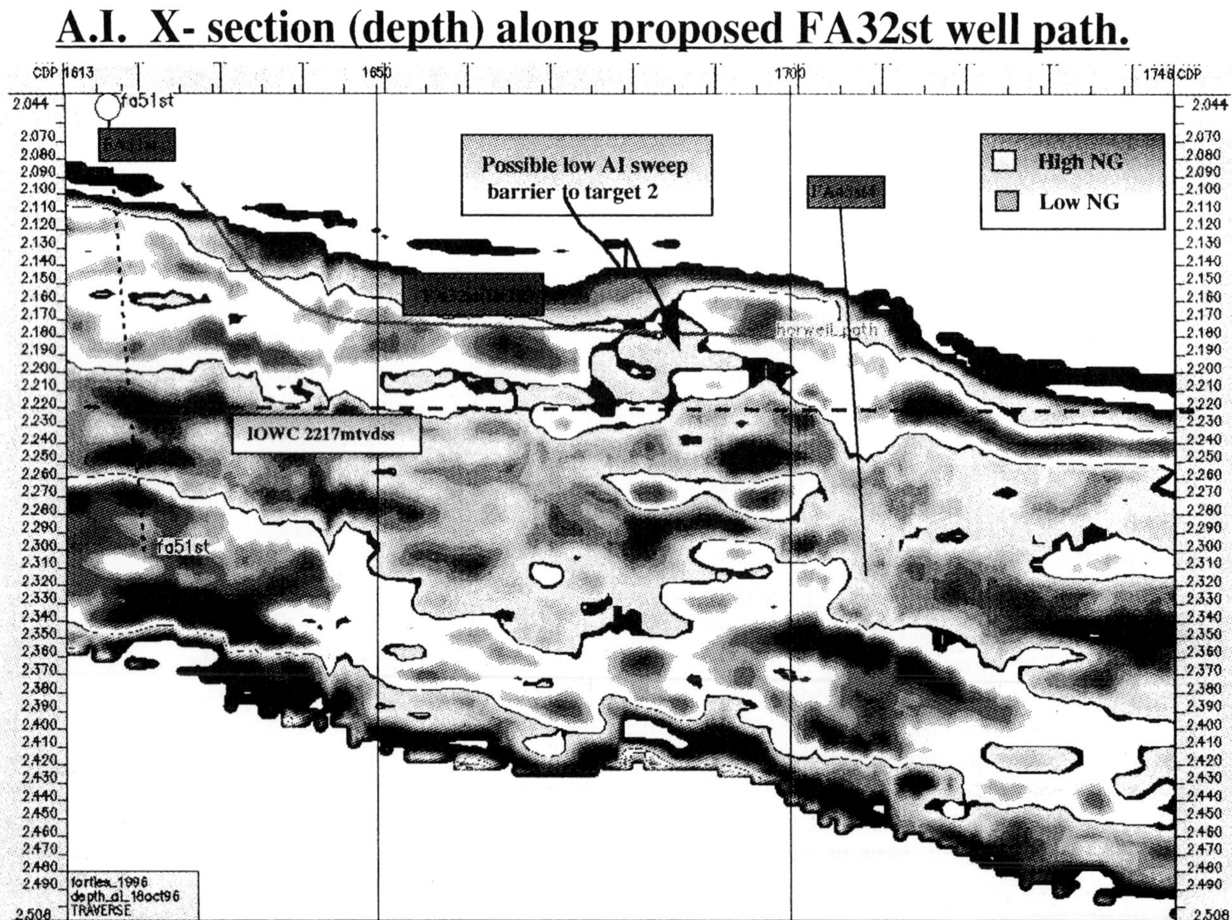

latest technology and the development of new ways for enhancing subsurface reservoir description and prediction.

10.12 Key references

Bain, J.S. (1993) Historical overview of exploration of Tertiary plays in the UK North Sea. In: Parker, J.R. (ed.) *Petroleum Geology of Northwest Europe: Proceedings of the 4th Conference,* Geological Society, London, pp. 5–14.

Milton, N.J., Bertram, C.T. and Vann, I.R. (1990) Early Paleogene tectonics and sedimentation in the central North Sea. In: Hardman, R.F.P. and Brooks, J. (eds) *Tectonic Events Responsible for Britain's Oil and Gas Reserves.* Special Publication 55, Geological Society, London, pp. 339–51.

Reynolds, T. (1994) Quantitative analysis, of submarine-fans in the Tertiary of the North Sea Basin. *Marine and Petrol. Geol.* **11**, 202–7.

11 Source Rocks and Hydrocarbons of the North Sea

C. CORNFORD

11.1 Introduction

This chapter describes the elements of the petroleum systems of the greater North Sea area. To this end, the properties of the known hydrocarbon source rocks of the greater North Sea area are discussed in terms of kerogen type, maturity and hydrocarbon yield. This is followed by a summary of the properties of reservoired oil and gas. Source and reservoir are linked with a discussion of migration paths, established on geological grounds and confirmed by oil/source-rock correlations and petroleum-system efficiency calculations.

As certain aspects of petroleum geochemistry may be unfamiliar to some readers, the initial sections of this chapter review current usage with respect to hydrocarbon genesis, using, where possible, examples from North Sea studies. The North Sea has played a key role in the development of source-rock maturity modelling. This is given particular emphasis in Section 11.4.2 on 'Calculated maturation' where the thermal and stratigraphic aspects of one-dimensional (1D) modelling are discussed in the context of North Sea burial histories.

This text does not attempt an exhaustive treatment of exploration geochemistry. The interested reader is referred, in the first instance, to Tissot and Welte (1984), Hunt (1975) or Bordenave (1993) for a detailed academic treatment, or Waples (1985), Beaumont and Foster (1988), Cooper (1990) or Merrill (1991) for an industry view. Many of these concepts are brought together in the context of basin modelling by Baker (1996) and Welte *et al.* (1997). For definitions and terminology, the *Illustrated Glossary of Petroleum Geochemistry* (Miles, 1989) may form a useful companion for this chapter.

A number of multiauthored volumes now exist covering the oil and gas geology of the Norwegian, Danish, Dutch and UK sectors of the North Sea (Woodland, 1975; Illing and Hobson, 1981; Spencer *et al.*, 1984, 1986b, 1987; Brooks and Glennie, 1987; Abbotts, 1991; Buller *et al.*, 1991; England and Fleet, 1991; Doré *et al.*, 1993; Parker, 1993; Rondeel *et al.*, 1996), volumes in which oils, gases, migration and source rocks are briefly referred to in the context of field or regional studies. Two conference proceedings specifically address source rocks and hydrocarbons (Brooks, 1983; Thomas, 1985), and three papers (in addition to earlier versions of this chapter) have given a brief review of the subject (Barnard and Cooper, 1981; Cooper and Barnard, 1984; Barnard and Bastow, 1991).

The geochemical processes that lead to the accumulation of oil or gas are illustrated in cartoon form in Fig. 11.1 (Cornford *et al.*, 1986). The conditions favouring source-rock accumulation are anoxia and/or high sedimentation rates, which promote the survival of organic matter (Fig. 11.1A), while burial of the source rock to an adequate subsurface temperature is the basis for maturation and generation (Fig. 11.1B and C). The first stage of migration—expulsion from the source rock—is controlled by source-rock richness and permeability (intergranular or fracture), while secondary migration is mainly a consequence of basin geometry and the permeability of long-distance conduits (Fig. 11.1D). Seen in this context, an oil or gas accumulation is merely an interruption of the inexorable escape of these fluids to the surface (sometimes called tertiary or re-migration). The final processes to consider are those such as bacterial degradation (Fig. 11.1E), asphaltene precipitation or thermal cracking, which exert ultimate control over the properties of oils and gases trapped within the reservoir today. All these processes are discussed in more detail below.

While applauding the lateral thought and marketing skills of its proponents (Gold and Soter, 1982; Gold, 1985; Gold and Held, 1987), a significant abiogenic (mantle) origin for commercial hydrocarbons is not, for a number of serious scientific reasons (for example, Vlierboom *et al.*, 1986), considered further. The North Sea provides excellent evidence for the association of oil and gas with thick sedimentary sequences, in contrast to the total absence of significant indigenous hydrocarbons in the surrounding metamorphic shield areas of Scotland and Scandinavia. Overwhelming geological, chemical and isotopic evidence shows that oil and gas derive from organic remains trapped in, and buried with, sedimentary rocks (Tissot and Welte, 1984; Hunt, 1995).

Because of the palaeoclimatic and tectonic setting of the North Sea, this chapter is concerned mainly with clastic source rocks (generally shales or coals); the concepts used and values given for boundary conditions cannot be uncritically applied to areas such as the Middle East, where chemical sediments, such as limestones, dolomites and evaporites, dominate. Considering the whole stratigraphic column of the North Sea area, source rocks are found in lacustrine, coastal-swamp, deltaic, lagoonal, shelf and trench environments (see Table 11.1). Maturation and generation in the Mesozoic and Tertiary North Sea have occurred under a tectonic regime dominated by crustal extension, predisposing a particular temperature and burial history. The maturation and generation processes discussed should thus also be exported with caution to areas subjected to contrasting tectonics.

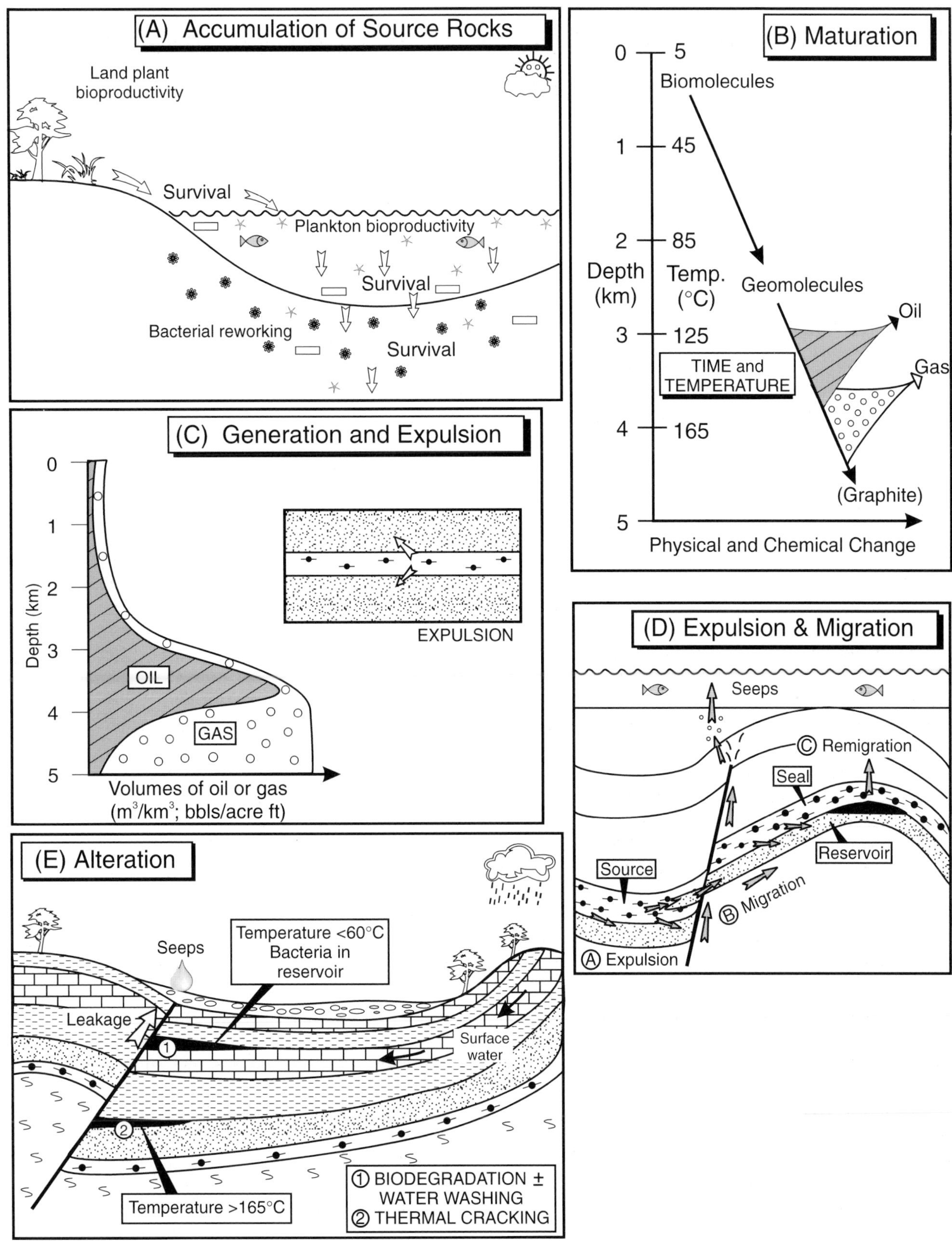

Fig. 11.1 Geochemical processes leading to petroleum accumulation: the modern trend is to attempt to quantify all these processes in order to predict the amounts and composition of oil and gas arriving at the trap and surviving to the present day.

11.2 Elements of hydrocarbon genesis

The genesis of hydrocarbons (oil and gas) can be divided into five processes (Fig. 11.1), and each has geological controls. Each is reviewed below, with emphasis on quantification.

In the broadest sense, sedimentary organic matter comprises kerogen, bitumen, oil and gas. The indigenous (non-migrated) organic matter in source rocks is broadly termed kerogen if solid or solvent-insoluble, bitumen if fluid or solvent-soluble (extractable) and gas (dominantly methane, CH_4) if gaseous. Different techniques (microscopy, pyrolysis, chemical analysis, etc.) may be used to characterize the different organic components. Understanding the contribution that each analytical technique makes to the final characterization of organic matter constitutes one of the basic skills of the petroleum geochemist.

As shown in Fig. 11.1A, indigenous sedimentary organic matter can broadly derive from three sources:

- Higher (land) plants—trees, ferns grass, etc.;
- Lower (mainly aquatic) plants—planktonic algae, etc.;
- Bacteria.

In terms of mass, animal tissue makes a minor contribution to most kerogens (Seifert, 1973). In addition to bioproductivity (Parrish, 1995), the key process operating during transport to the site of deposition and incorporation into the sediment is selective preservation (Schwarzkopf, 1993). It is important to note at this stage that the majority of kerogens are in fact a mixture of organic tissues from all 3 sources. As shown in the subsequent panels of Fig. 11.1, once transformed during burial and generation, and having migrated away from the source rock, organic matter can be divided into bitumens (solids), oil (liquid) and gas (mainly methane). The final fate of all migrated hydrocarbons is seepage to the surface (Fig. 11.1E), with surface seeps being both a readily accessible and highly significant indicator of generation in, and migration through, the subsurface (Clarke and Cleverly, 1991; Selley, 1992; Macgregor, 1993; Schumacher and Abrams, 1996).

11.3 Source rocks—accumulation and identification

Source rocks can be defined as sediments that are (or were) capable of generating significant oil or gas (Miles, 1989). Source rocks have been reviewed in terms of organic facies by Jones (1987) and Huc (1990), of sequence stratigraphy by Katz and Pratt (1993) and of depositional environment by Huc (1995). The classical hydrocarbon source rock is an organic-rich, dark olive-grey to black, laminated mudstone, and from an industrial point of view must be capable of generating and expelling commercial quantities of oil or gas. Whether there is sufficient oil or gas to form a commercial hydrocarbon accumulation depends not only on the volume and richness of the source rock, its maturity history and the geological framework in which it occurs, but also on the current economics and politics of exploitation. Stemming from this definition we can identify:

- Potential (immature) source rocks (those yet to generate hydrocarbons);
- Active (mature) source rocks (those where generation of hydrocarbons has commenced);
- Exhausted (postmature) source rocks (which have already generated, and generally expelled, all hydrocarbons).

In using this convenient tripartite division, it should not be forgotten that even immature source rocks do contain small amounts of hydrocarbon (waxes, resins, aromatics, etc.), inherited more or less directly from living biomass, and that postmature source rocks will have generated oil and gas at some earlier time.

The role of coals as source rocks for gas, gas/condensate and oil has long been actively debated (Durand and Paratte, 1983; Thompson *et al.*, 1985; Khorasani, 1989; Veld and Fermont, 1990; Scott and Fleet, 1994; Thompson *et al.*, 1994). It is now recognized that, while humic coals generate gas, coals with higher concentrations of oil-generating macerals, such as spores, cuticle and resin, can generate gas/condensate and light oils. Exceptionally, algal coals (also termed bog-head coals or torbanites) accumulate and can prove rich oil sources. It is emerging that the limiting factor controlling liquid hydrocarbon sourcing from coals is the escape of the liquid from the organic-rich matrix of coals, rather than the generation of liquids itself (Leythaeuser and Poelchau, 1991). Generating gas from coals proceeds in part via a liquid (viscous bitumen) intermediate. Absorbing the liquid phase within the organic matrix inhibits expulsion from the coal. Hence, with further burial, the retained liquid cracks to gas, this being the final expelled product.

11.3.1 Depositional environments

Organic-rich shales are fairly uncommon in the geological record, since their deposition requires the co-existence of high bioproductivity and high preservation rates for organic tissues (Muller and Suess, 1979). Recent studies suggest that preservation rather than productivity *per se* is the controlling factor (Brooks *et al.*, 1987; Schwarzkopf, 1993; Parrish, 1995; Boussafir and Lallier-Vergés, 1997). Understanding the controls on organic deposition is important for hydrocarbon exploration, in order to predict source-rock properties away from sampled locations (outcrops or wells) particularly in adjacent depocentres where source-rock maturity will be optimum for generation and expulsion. Creation of source-rock facies maps have recently been accomplished via sequence stratigraphy as addressed by Morton (1993) for the Upper Triassic to Middle Jurassic of the North Sea area. Mapping source-rock distribution is thus a key part of modern prospect evaluation.

Uniformitarianism is generally invoked to explain organic-rich ancient sediments. Applying this principle, the depositional environments for organic-rich sediments in the North Sea area are putatively related to modern analogues in Table 11.1 and Fig. 11.2.

High organic preservation is promoted by high sedimentation rates and reduced oxygen concentrations in the water column (Demaison and Moore, 1980; Hay, 1995). As shown in Fig. 11.2, reduced oxygen concentrations are found in stratified lakes (Fleet *et al.*, 1988), in coastal-plain and delta swamps, in shelf and oceanic basins and in oceanic midwater oxygen minima (Demaison *et al.*, 1983). All these situations derive from a combination of oxygen consumption (by respiration) and exceedingly sluggish re-

Table 11.1 Depositional environments for source-rock units in the North Sea region, together with approximate modern analogies.

Number*	North Sea Source Rock	Modern Analogy	Comments
1	Kimmeridge Clay/ Draupne Formation	No good analogy	Baltic Sea has similarities
2	Brent Coals	Mekong Delta	Different vegetation
3	Westphalian Coals	Everglades, USA	Scale problem
4	Orcadian Basin Lacustrine laminites	Lake Junggar	Xinyiang, north-west China
5	Southern Uplands black shales, Scotland	Cariaco Trench	—
6	Alum Shale, Sweden	Miocene offshore California	Arguable analogue

*Numbers relate to Fig. 11.2.

plenishment of oxygen from the atmosphere. However, the whole basis of anoxia favouring source-rock deposition has recently been questioned, on the basis of modern sediment studies, by Pedersen and Calvert (1990) and Calvert *et al.* (1992), although their case is far from proven.

Within the swamp environment, two mechanisms to enrich coals in oil-prone components have been proposed: selective transport and sedimentation of oil-prone macerals (e.g. cuticle) can give rise to more oil-prone coals in a distal deltaic environment (Thompson *et al.*, 1985): on the other hand, preferential subaerial oxidation of the humic component of peats, particularly during episodes of lowered water-tables, may cause enrichment in oil-prone components in back-delta swamps (Powell, 1986).

11.3.2 Quantity of organic matter

An adequate amount of organic matter (normally measured as weight per cent total organic carbon, abbreviated to % TOC) is a necessary but not sufficient prerequisite for a sediment to source commercial quantities of oil or gas. The quantity of organic matter required for a sediment to be considered a source rock is, like most attempts to define a multiparameter system by a single variable, a much disputed point. For a typical, poorly drained, thick, homogeous shale, the TOC ranges detailed in Table 11.2 may be used to rate source-rock potential in terms of quantity (but not quality) of kerogen. Approximate equivalent values for carbonates are given in the right column of the table.

The terms 'fair', 'good', etc. given in Table 11.2 should be used to describe the amount of organic matter and not the hydrocarbon source potential, since this will also depend on kerogen type. In addition to kerogen quantity and type, drainage must be considered in an assessment of the amount of organic matter required for a sediment to be considered a source rock). For example, in a well-drained sequence of interbedded shales and sands, 1% might be considered good and 2% very good.

In using TOC values to rate a source rock, the effects of hydrocarbon maturation and expulsion on the preserved TOC should not be ignored (Cooles *et al.*, 1986; Mackenzie and Quigley, 1988). For example, a 3% TOC immature oil source rock will have its TOC reduced to about 1.9% when it has expelled 80% of its oil by the late mature stage. Measurements made on mature or postmature source rocks may thus be unduly pessimistic and should be corrected (Cornford, 1994). A typical nomogram for correcting measured TOC values to original TOC values at a late immature stage is shown in Fig. 11.3, where:

$$TOC_{original} = TOC_{present} + OC_{lost} \quad (1)$$

This equation assumes no carbon is lost via decarboxylation or other reactions to produce carbon dioxide. Based on a hydrogen mass balance, the lost organic carbon (OC_{lost}) is a function of the ratio of atomic H/C values of the kerogen (H/C_{ker}) and hydrocarbon (H/C_{hc}):

$$OC_{lost} = (TOC_{original}) \times (H/C_{ker} / H/C_{oil}) \quad (2)$$

Thus the original TOC can be approximately calculated as:

$$TOC_{original} = TOC_{present})/(1 - (H/C_{ker} / H/C_{hc}) \quad (3)$$

As shown in Fig. 11.3, the amount of carbon converted to hydrocarbon is a function of the hydrogen content and hence the kerogen type (see next section). Using a similar approach, Skjervøy and Sylta (1993) have corrected Viking Graben TOC values for maturity effects, using pyrolysis data.

It can be argued that the best way of determining the quantity of kerogen with respect to oil and gas generation is to refer to the organic hydrogen (TOH, wt%), since hydrogen is the limiting element in hydrocarbon generation. Stated another way, once the hydrogen-rich phases of oil and gas (general atomic formulae CH_2 and CH_4, respectively) have been generated in and migrated out of a source rock, a carbon-rich graphitic schist remains. The rating of

Table 11.2 Use of total organic carbon (TOC) values to rate hydrocarbon source rock in terms of amount (rather than type) of sedimentary organic matter.

TOC (wt%) clastics	Designation	TOC (wt%) carbonates
< 0.5	Very poor	< 0.3
0.5–1.0	Poor	0.3–0.5
1.0–2.0	Fair	0.5–1.0
2.0–4.0	Good	1.0–2.0
4.0–12.0	Very good	2.0–6.0
> 12.0	Oil shale, carbargillite or bituminous limestone	> 6.0
> 65.0	Coal	—

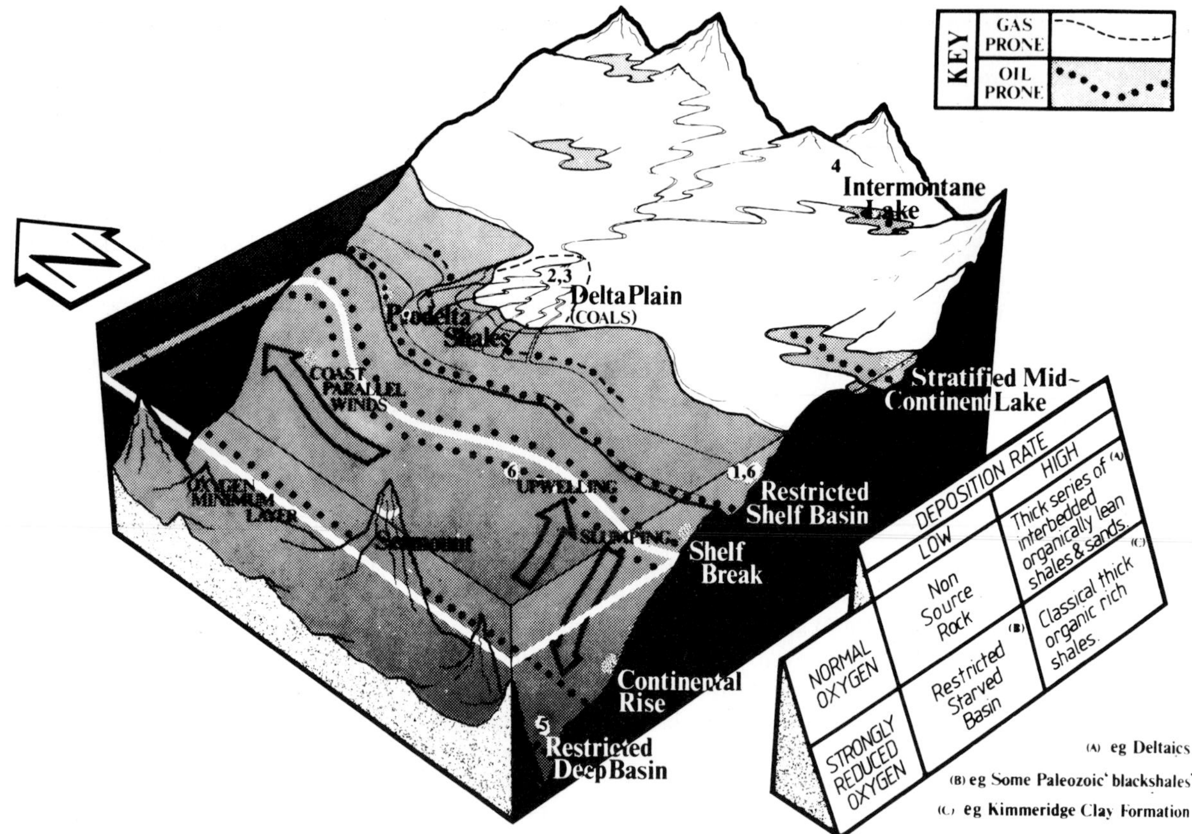

Fig. 11.2 Sedimentary environments favouring the deposition of petroleum source rocks (see Table 11.1 for key to numbers).

source rocks on the basis of organic hydrogen has not been adopted for analytical reasons.

11.3.3 Type of organic matter

For application to hydrocarbon exploration, sedimentary organic matter can conveniently be divided into three types:

- Oil-prone components.
- Gas-prone components.
- Inert components (or dead carbon).

The kerogen of a typical rock will contain a mixture of all three components. When estimating the oil potential of a sediment, only the oil-prone part of the TOC of that sediment should be considered. For example, a 4% TOC sediment with 50% oil-prone kerogen will have 2% oil-prone organic carbon (2% OPOC). If, in addition, it had 25% gas-prone organic matter if would be said to have 1% gas-prone organic carbon (1% GPOC). This would leave 1% dead carbon (DC). Algebraically this can be expressed:

$$\% \text{ TOC} = \% \text{ OPOC} + \% \text{ GPOC} + \% \text{ DC} \qquad (4)$$

In apportioning gas- or oil-generative capacity to kerogen, it should be remembered that oil-prone kerogen (or reservoired oil itself) will crack to yield both wet and dry gas if buried and hence heated sufficiently. In this sense, an oil-prone source rock is also gas-prone; gases from oil-prone kerogen are generally wet gas, while gas-prone kerogen produces only dry gas.

Kerogen type is generally determined by microscopy (organic petrography), by pyrolysis (e.g. the Rock Eval method) or by elemental (C, H, O) analysis. Confirmatory evidence can be provided by stable carbon isotope ($\delta^{13}C$), light hydrocarbons or sediment extract analyses. Because, at present, no single technique is totally reliable or universally applicable, a combination of complementary methods is typically used.

Pyrolysis

Pyrolysis, most commonly undertaken using the Rock Eval apparatus, provides us with the definition of kerogen types I, II, III and IV (Peters, 1986; Bordenave *et al.*, 1993). Rock Eval pyrolysis rapidly yields 'cheap and cheerful' information concerning kerogen type, maturity, potential yield of hydrocarbon and the presence of migrated hydrocarbons. For most applications, TOC data are required for the interpretation of Rock Eval data. The recent version of the Rock Eval machine (termed the Oil Source Analyser (OSA)) does, however, produce an 'organic carbon equivalent' peak, dispensing with the need for separate TOC determinations (Bordenave *et al.*, 1993).

Starting at about 80°C, the Rock Eval machine heats ~200 mg of powdered rock in an inert atmosphere to produce the following sequence of events (Fig. 11.4):

- Release of the 'free' hydrocarbons to give an S_1 peak (to ~300°C).
- Simulation of maturation and generation of pyrolysate (new hydrocarbon) from the kerogen (300–550°C) to yield a second peak, S_2, with maximum at temperature T_{max}°C.
- Release of organically bound carbon dioxide (CO_2) over the temperature range 300–550°C to give an S_3 peak.

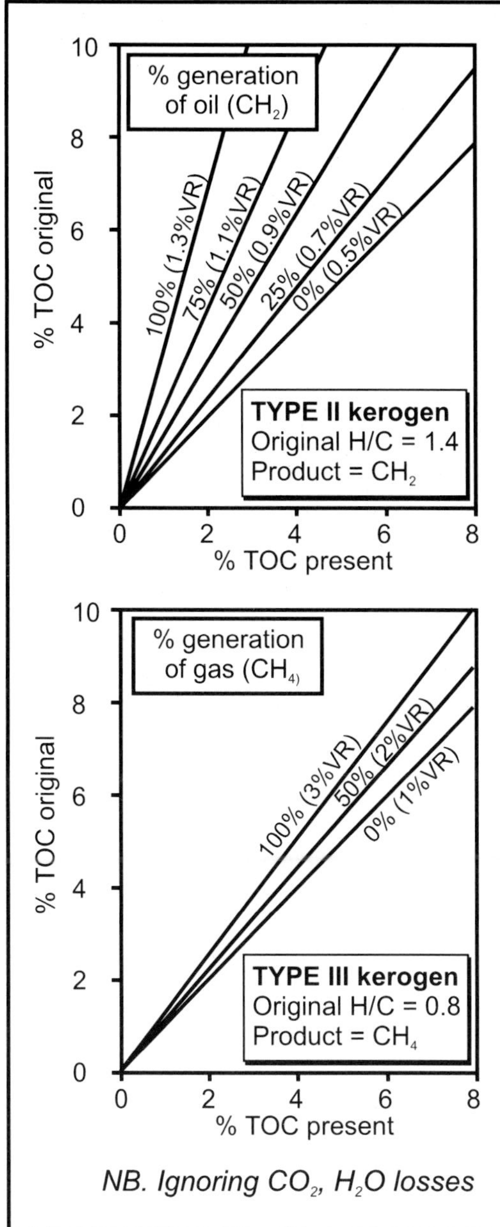

Fig. 11.3 Correcting present-day measured TOC values (horizontal axes) for maturation and expulsion effects in order to recreate the original TOC values at a late-immature stage (vertical axes). The calculation is based on a hydrogen mass balance for Type II (upper) and Type III kerogens (lower), and the diagonals represent progressive conversion (% transformation and equivalent vitrinite reflectance) of solid kerogen to mobile hydrocarbon, which may be expelled from the source rock and in any case does not generally contribute to TOC. (After Cornford, 1994.)

In some literature the French designations S_1, S_2, etc. may be termed P_1, P_2, etc. (P = English 'peak'?).

The hydrocarbon peaks S_1 and S_2 are measured from the response of a flame ionization detector (FID), and the S_3 CO_2 peak is determined using a thermal conductivity detector (TCD). These detectors are mass-sensitive—an important factor for source-rock volumetric calculations. Approximately 200 mg of sample are required for one analysis, this having the advantage of allowing the analysis of a single lithology in drill-cutting samples, but producing problems with respect to representative sampling and sample contamination. Generalizing source-rock properties from a small number of samples may be undertaken using wireline-log responses (Meyer and Nederloff, 1984; Passey *et al.*, 1990; Schwarzkopf, 1992).

Together with a TOC determination, five indices are generated from the values of the three peaks (Fig. 11.4 inset). These indices can be used to determine kerogen type, maturity, potential yield of hydrocarbons and the presence of migrated hydrocarbons. Taken as a whole, Rock Eval interpretation is reliable for non-carbonate rocks with TOC values in the 0.5–12% range. Different calibrations have been established for coals.

Kerogen Types I, II, III and IV are defined in terms of Hydrogen and Oxygen Indices, as shown in Fig. 11.5, with the two parameters defined thus (Py = pyrolysate):

Hydrogen Index = S_2/TOC (mg Py/g TOC)
Oxygen Index = S_3/TOC (mg CO_2/g TOC)

In Fig. 11.5, three main kerogen types are identified: Type I, of algal origin, Type II, derived from land-plant spores, exines, resins, etc., and Type III, of lignocellulosic or humic origin. Type II can equally be a mixture of Types I and III kerogens (Barnard *et al.*, 1981b). In most cases, Type II kerogen is dominated by bacterially degraded algal (originally Type I) organic matter, the Upper Jurassic oil source rocks of the North Sea and onshore UK (for example, Kimmeridge Bay, Dorset) being a world standard for this type of kerogen. Oxidation or sulphidization during deposition can affect the ultimate hydrocarbon potential of this type of kerogen (Khorasani and Michelsen, 1992; Boussafir and Lallier-Vergés, 1997). Though rare in the North Sea area, sulphur-rich Types IS and IIS kerogens have been defined and recorded mainly in carbonate environments (Orr, 1986; Sinninghe-Damsté *et al.*, 1993). A Type IIIb or IV may also be identified: this is altered (oxidized) humic material, termed 'dead carbon' or 'fossil charcoal'.

The use of Oxygen Indices has been questioned, since some carbonates (e.g. iron carbonates) decompose at the temperatures where the 'organic' CO_2 is being collected. High Oxygen Indices are thus suspect if deriving from rocks with high carbonate contents (Peters, 1986).

As pyrolysis is a bulk determination, a Type II kerogen can be (and generally is) a mixture of Type I algal material, degraded by bacteria and mixed with terrigenous Type III material (Barnard *et al.*, 1981b). A pure Type II kerogen typically comprises spores, pollen, cuticle and resin. The hydrocarbon potential of resins (resinite) is for both naphthenic oils and gas/condensate (Mukhopadhyay and Gormly, 1985; Powell, 1986; Powell and Boreham, 1994).

The changes of these indices with maturity are indicated by the heavy black arrows in Fig. 11.5. As a result the general level of maturation may be determined from this diagram, but it is clear that discrimination between different kerogen types is difficult at higher maturities. A more useful plot of kerogen type related to maturity is shown in Fig. 11.6, where the pyrolysis-derived parameters of Hydrogen Index (kerogen type) and T_{max} (maturity) are cross-plotted. Note that, at higher maturities, the relative Hydrogen Indices of the three kerogen types are reversed.

A more detailed approach to kerogen identification is

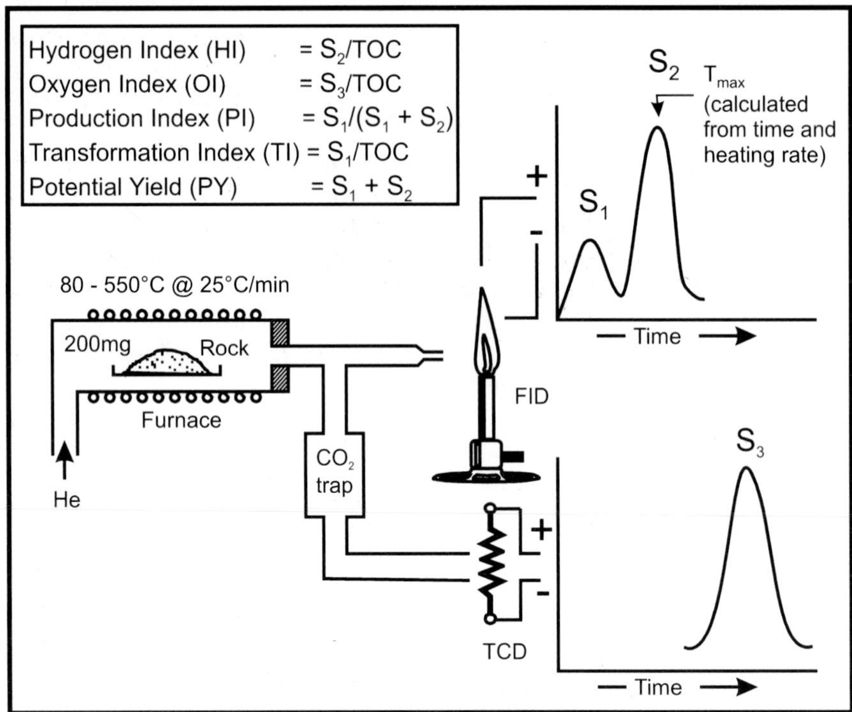

Fig. 11.4 Schematic of the Rock Eval apparatus and its output, together with the major derived parameters. Note that TOC values are required for some derived parameters. FID, flame ionization detector; TCD, thermal conductivity detector.

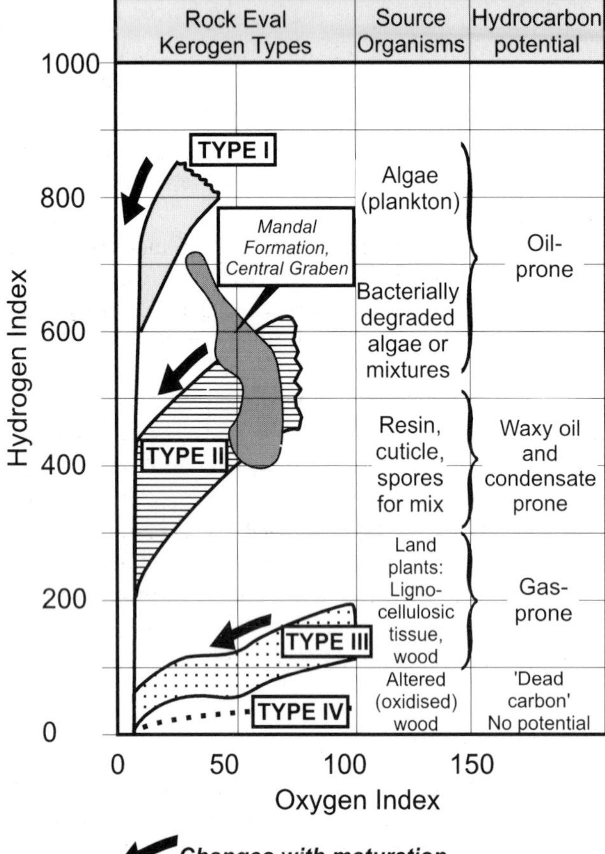

Fig. 11.5 Kerogen type defined using Rock Eval Hydrogen and Oxygen Indices, with typical North Sea (Central Graben) Upper Jurassic oil-prone source-rock data shown. Source organisms are indicated for the three major kerogen types (Types I, II, III), but most kerogens are in fact mixes. A mix of Type I and Type III will plot as Type II kerogen, demanding additional analyses (e.g. microscopy or pyrolysis–gas chromatography) to distinguish the end-members. The black arrows indicate the changes observed with progressive maturation, leading eventually to graphite.

shown in Fig. 11.7A, where a plot of peak S_2 yield versus TOC is used to identify the percentage of dead carbon (%DC) as the intercept on the TOC axis. The TOC is taken as the sum of DC and active carbon (AC):

$$TOC = DC + AC \qquad (5)$$

The slope of the trend is a general Hydrogen Index calculated on a dead-carbon-free basis (HI′ = 585 mg Py/g AC):

$$HI' = S_2/(TOC - DC) = S_2/AC \qquad (6)$$

The triangular plot (Fig. 11.7B) then shows how dead-carbon-free Hydrogen Indices will vary for mixtures of types I, II and III kerogens. The final position on the diagram along the HI′ = 585 mg S_2g AC line) can be established using organic petrographic information. By comparing whole and demineralized samples, the intercept, on the S_2 vs. TOC plot has been attributed to a ‘clay-matrix’ effect (Langford and Blanc-Valeron, 1990), which turns out to be minimal in comparison with the inertinite shift.

Having identified the dead carbon (inertinite) content, an alternative simpler treatment of data can differentiate the HI′ values into a contribution from gas-prone (Type III) kerogen on the one hand and a contribution from oil-prone kerogens on the other. Where F_{oil} and F_{gas} are the fractions of oil- and gas-prone kerogens in the active carbon fraction (%AC), and HI_{oil} and HI_{gas} are the Hydrogen Indices of the pure late immature components, then:

$$HI' = (F_{oil} \times HI_{oil}) + (F_{gas} \times HI_{gas}) \qquad (7)$$

Since $F_{oil} + F_{gas} = 1$, pure HI_{oil} is approximately 750 mg Py/g TOC and pure HI_{gas} = 140 mg Py/g TOC at the late immature stage, the fraction of oil- and gas-prone kerogens can be calculated as:

$$F_{oil} = (HI' - HI_{gas})/(HI_{oil} - HI_{gas}) \qquad (8)$$

$$F_{gas} = 1 - F_{oil} \qquad (9)$$

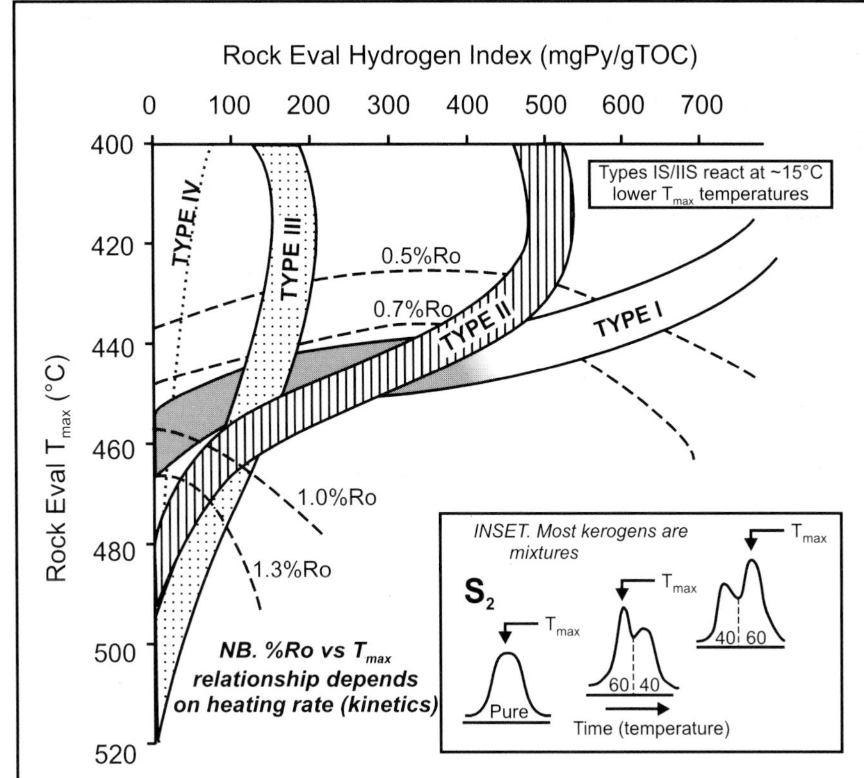

Fig. 11.6 Effects of maturity changes (T_{max}) on Hydrogen Index (HI) values, based mainly on data from the Witch Ground and Viking Grabens, North Sea. Note that the rapid decrease in HI for Type I kerogens indicates a narrower oil window in terms of depth compared with Type II kerogens, and at high maturity levels the gas-prone Type III kerogen has the highest residual Hydrogen Index values (see text).

Thus the measured 585 mg Py/g TOC shown in Fig. 11.7 would comprise 73% of active carbon as oil-prone and 27% of active carbon as gas-prone. For an initial TOC of 6%, this would break down into:

Average dead carbon = 2% (Fig. 11.7A)
Average active carbon = 4%

TOC = 6%

Oil-prone active carbon = 73% of 4% AC = 2.92% OPOC
Gas-prone active carbon = 27% of 4% AC = 1.08% GPOC

The approach is applicable to higher-maturity kerogens if the HI_{oil} and HI_{gas} values are adjusted to those appropriate to the maturity levels encountered (see Fig. 11.6). This approach to quantifying pyrolysis interpretation is important for quantitative prospect-evaluation calculations (see Cooles *et al.*, 1986, for North Sea examples).

Microscopy (organic petrography)

For both academic and industrial convenience, the multitude of possible kerogen types (as diverse as the tissues of the various plant and animal groups contributing to the sediments) are reduced into three broad organic petrographic groups (see Fig. 11.8):

- Liptinite: algal and/or bacterial input with the potential to produce oil together with associated condensate and gas upon further burial. If pure, unaltered algae fall in the Type I kerogen group; if bacterially degraded, in the Type II kerogen group (e.g. Upper Jurassic source rocks of the North Sea). Land plant-derived spores, resin or cuticles can generate oil or condensates of a characteristic type, forming the group collectively termed exinites, and comprising part of the Type II kerogen group.
- Vitrinite: a woody (humic) land-plant input to a sediment generally producing a gas-prone or Type III kerogen (e.g. a typical Carboniferous coal) of NW Europe.
- Inertinite: if altered (e.g. oxidized to charcoal), then land-plant tissues will produce dead carbon or Type IV kerogen.

These kerogen components, identified by microscopy (e.g. Fig. 11.8), are termed macerals, by analogy with the inorganic 'minerals' constituting rocks in general. Most kerogens are in fact mixtures of these three maceral end members.

In contrast to the bulk pyrolysis analysis discussed in the previous section, an organic petrographic description of a kerogen can be highly detailed with subdivisions beyond those shown in Figs 11.8 and 11.9 (Stach *et al.*, 1982; Bustin *et al.*, 1983). Thus detailed microscopy and bulk-pyrolysis analyses together produce a good overview of kerogen properties. Approximate comparisons are made in Fig. 11.9 between the pyrolysis and various organic petrographic nomenclatures for kerogen description.

Elemental analysis

The carbon, hydrogen and oxygen contents can be rapidly and cheaply determined on isolated kerogen (Durand and Monin, 1980), although problems can occur in the cost and effectiveness of the hydrofluoric acid maceration process required for the isolation (Durand and Nicaise, 1980). Interpretation is based on a so-called van Krevelen diagram of H/C_{atomic} versus O/C_{atomic} weight ratios (van Krevelen, 1993). The diagram has the same form as that for Rock Eval interpretation (see Fig. 11.5), where H/C is equivalent to Hydrogen Index and O/C to Oxygen Index. Immature kerogen from the Kimmeridge Clay Formation of the

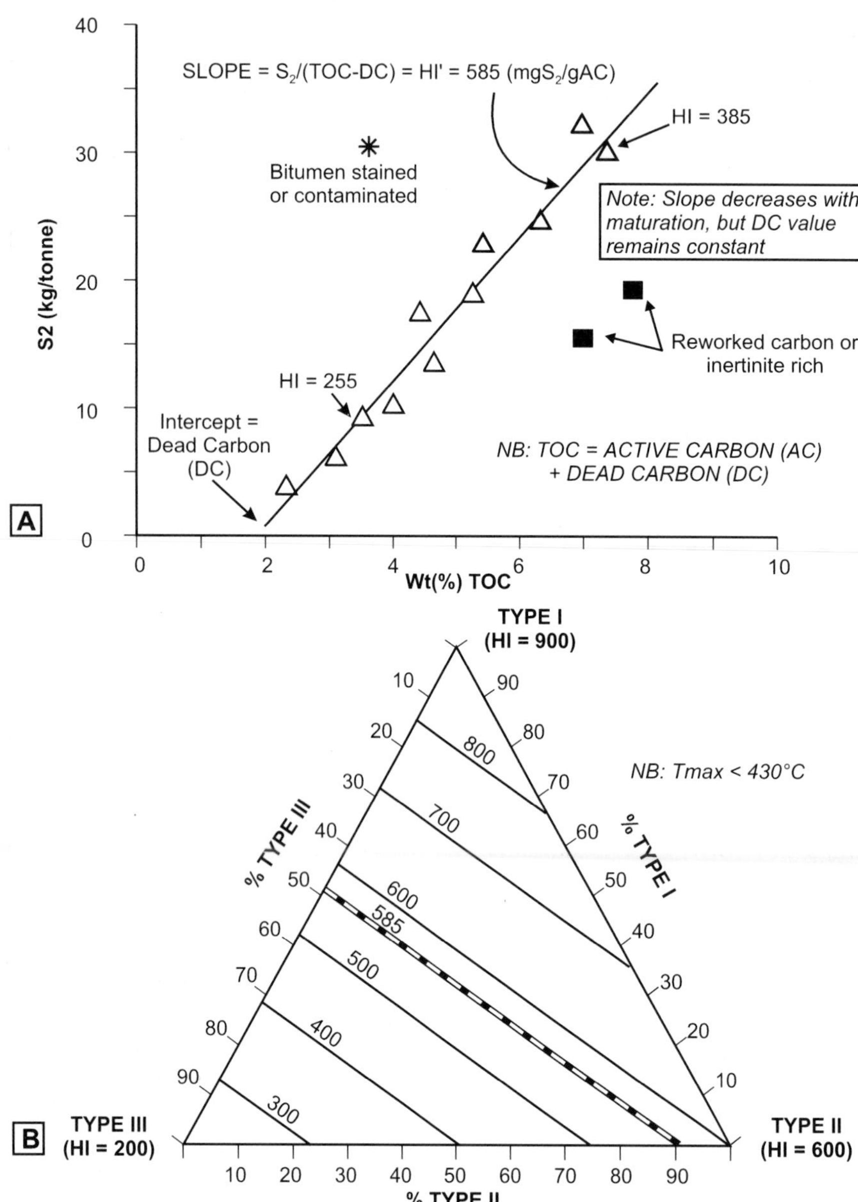

Fig. 11.7 Determination of immature or early-mature kerogen mixes using Rock Eval data. (A) Plot of Rock Eval S_2 versus TOC (Central Graben) to indicate the average dead carbon content (intercept on TOC axis) and dead-carbon-free HI value (slope = HI′). (B) Triangular plot of active kerogen components showing iso-HI lines and kerogen mixes consistent with HI′ = 585 mg Py/g TOC (dashed line).

North Sea is attributed values of $H/C_{atomic} = 1.29$ and $O/C_{atomic} = 0.074$ (Bertrand *et al.*, 1993), while Carboniferous coals of high volatile bituminous rank return H/C_{atomic} values in the ~0.8 and O/C_{atomic} ~0.15 range (Teichmüller and Durand, 1983). Using Upper Jurassic Draupne Formation samples from the North Sea and Middle Triassic Barentsøya Formation samples from Svalbard, Khorasani and Michelsen (1992) have emphasized the role of oxidative alteration of algal kerogen in controlling the elemental analysis and hydrocarbon potential.

Kerogen-type equivalences and applications

A typical problem in applying these concepts regionally, using data from a number of sources, is to understand the approximate equivalence of the many terms used to describe kerogen type. Scientists from different countries, different backgrounds (e.g. coal vs. palynology), different laboratories and, indeed, the same laboratories at different times have used a number of terminologies. Some approximate equivalences are indicated in Fig. 11.9, although no consensus exists over details. In evaluating North Sea organic petrographic reports (particularly older ones), one should be aware of the tendency to describe pyrite as inertinite and the incorrect assumption that all amorphous kerogen is oil-prone.

Finally, applying kerogen-type information to basin or acreage evaluation requires the generalization of analytical information away from the sampled sites (well or outcrop). The recognition of sedimentary environments from the associated kerogen has been termed 'palynofacies' or more generally 'organofacies' (Fig. 11.10). This requires the kerogen type to be attributed to an 'organic facies' or association of macerals, the lateral distribution of which can be predicted using normal sedimentological concepts of transport and deposition (Jones, 1987). This approach has proved useful in reservoir studies and, increasingly, in acreage evaluation, where different kinetic models have been applied to different organic facies (see Pepper and Corvi, 1995, which includes examples from the North Sea).

11.3.4 Hydrocarbon yields from source rocks

The basis for all basin or prospect evaluations is to estimate

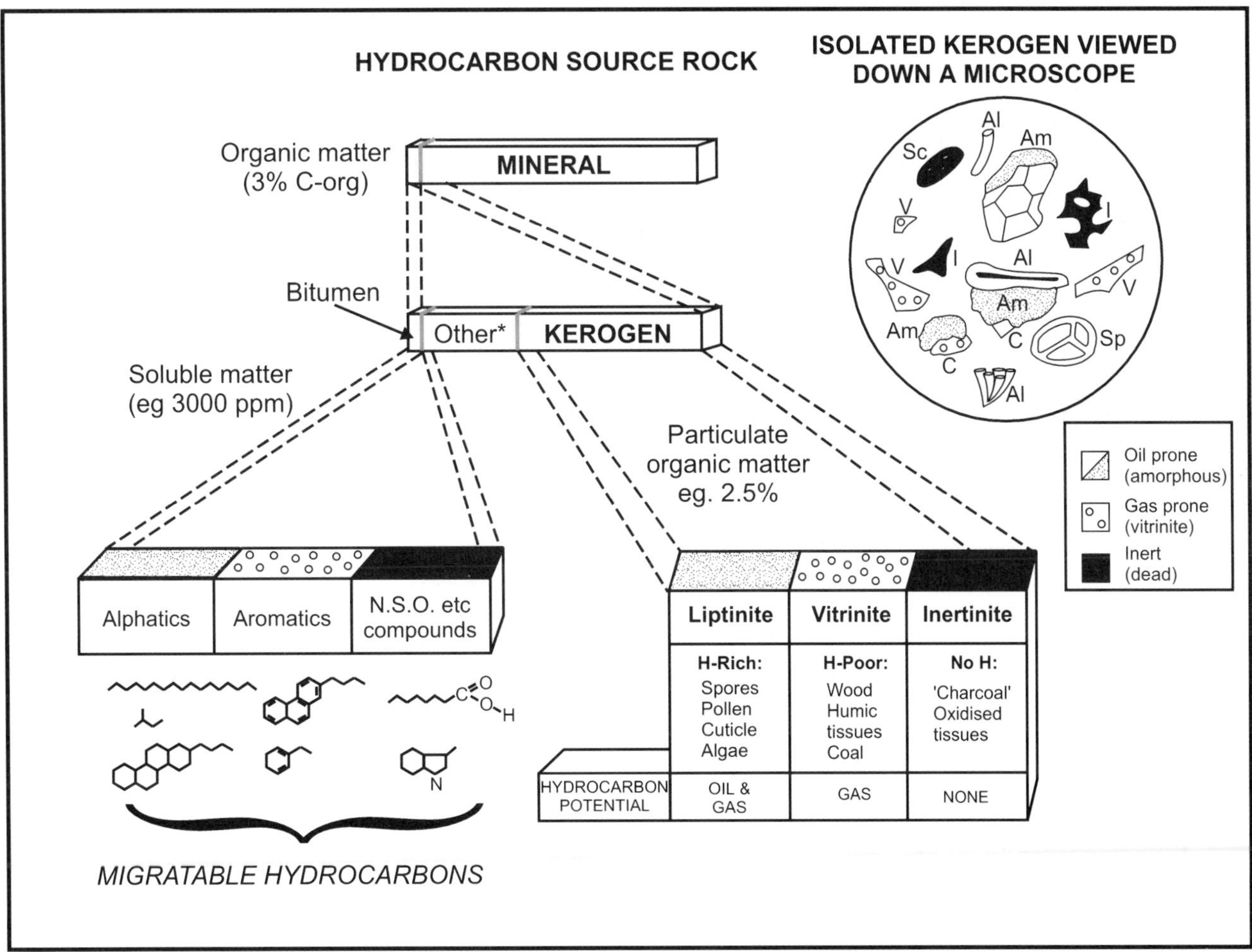

Fig. 11.8 Kerogen characterization based on chemical (left) and microscopic techniques (right). In contrast to the quantitative bulk characterization of Rock Eval, chemical analysis of organic-solvent extracts and organic petrography produces complementary qualitative information on the individual kerogen components. Al, algal; Am, amorphous; C, cutinite; V, vitrinite; I, inertinite; Sc, sclerotinite (fungal spores); Sp, sporinite; Other*, submicroscopic but solvent-insoluble organic matter.

EQUIVALENT KEROGEN NOMENCLATURES

HYDROCARBON POTENTIAL	COAL PETROGRAPHY	PALYNOLOGY	CHEMISTRY	ROCK EVAL	COMMENTS
Oil-prone (cracking to condensate and gas)	Liptinite	Algal/ Amorphous	Sapropel	Type I (Type IS)	Pure algal Includes bacterial biomass
Light oil, condensate-prone	Exinite	Herbaceous		Type II (Type IIS)	Cuticle gives waxy oil, Type II can be a mix of Types I, III and IV
Gas-prone	Vitrinite	"Woody"	Humic	Type III	Amorphous vitrinite and oil-impregnated fluorescent vitrinite exist
Dead carbon	Inertinite	"Coaly"	(not recognised)	Type IV (or IIIB)	Minor gas potential from semi-fusinite

Fig. 11.9 Equivalences of terms used for optical kerogen descriptions and Rock Eval kerogen types (after Cornford, 1984). Equivalences are at best approximate—and no one explicitly recognizes the ubiquitous bacterial contribution.

the hydrocarbon (oil and gas) 'charge' (e.g. millions of barrels of oil) reaching the prospective structure. The starting-point for this calculation is to estimate the amount of oil and gas yield that can be generated in a unit volume (mass) of source rock (Fig. 11.11). Either weight or volume units are used, namely:

- Kilograms of oil or gas per tonne of source rock;
- Cubic metres of oil or gas per cubic kilometre of source rock (or barrels of oil/acre foot of source rock).

Conversions may be made based on density, where general

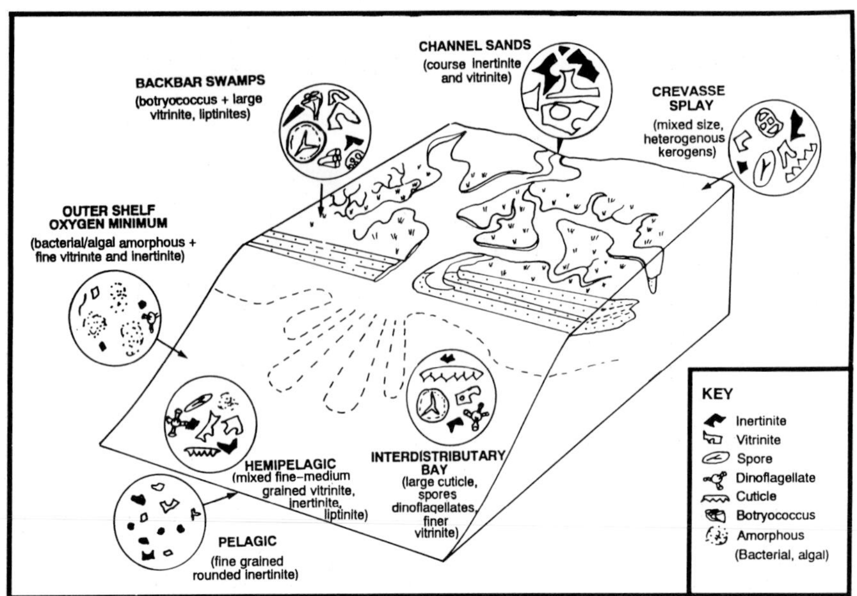

Fig. 11.10 Kerogen type as a palynofacies indicator, illustrating that, like sediments, kerogen particles are carried and sedimented according to their hydrodynamic properties (size, shape and density).

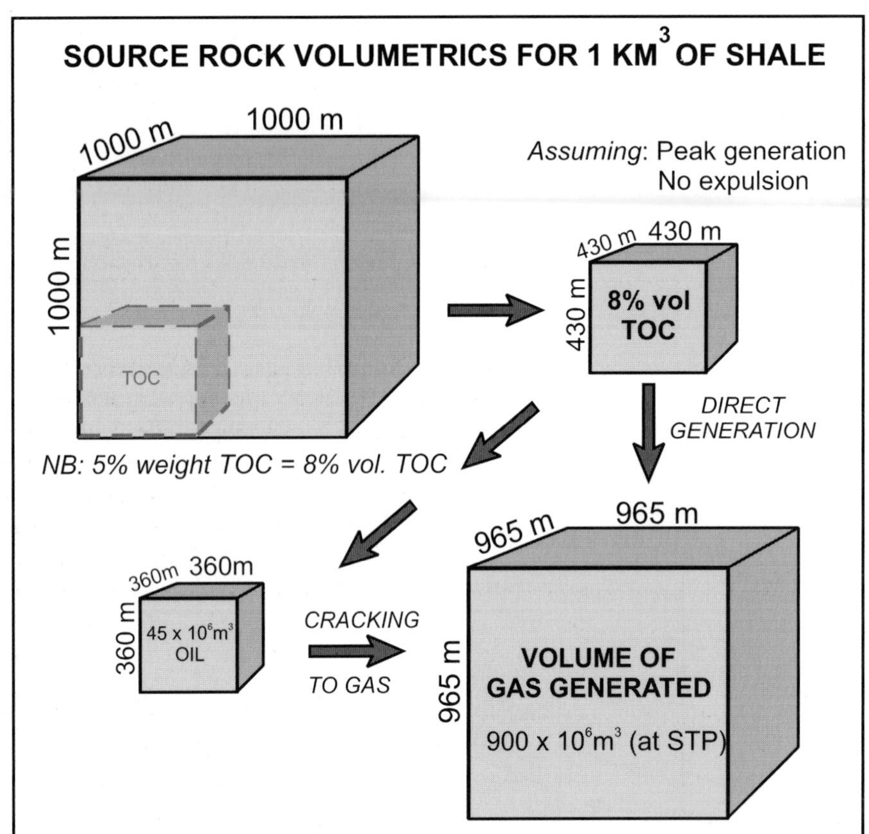

Fig. 11.11 Illustration of a volumetric prediction of generation of oil and gas from 1 km^3 of source rock. Volumes of rock (top left), TOC (top right), oil (bottom left) and gas (bottom right) at standard temperatures and pressures for a typical Upper Jurassic shale from the North Sea, i.e. 5% TOC and Type II kerogen = 30 kg/t yield. Note: Calculation based on extract and Rock Eval data and assumes peak generation and no expulsion.

values of 2.65 g/cm^3 for rock and 0.86 g/cm^3 for oil may be used at standard temperature and pressure (STP) (1 g/cm^3 = 1 t/m^3).

Numbers related to oil and gas generation can be derived from pyrolysis or solvent-extract results (Cooles *et al*., 1986), but the effects of migration in removing hydrocarbon from the source rock must not be ignored. Historically, generation has been demonstrated using carbon-normalized extract (CNE) or carbon-normalized hydrocarbon (CNH) yields expressed in the form of mg extract/g TOC or mg HC/g TOC, respectively (Tissot and Welte, 1984). Today's pyrolysis S_2 peak yields in kg/t are also commonly used.

Depth trends, produced from summing results from a number of techniques, covering the range from methane to waxes, can be converted to more useful industry units, as shown in Fig. 11.12. The horizontal axis is expressed in units of millions of cubic metres per cubic kilometre of rock, based on a rock containing 1% TOC and a Type II bacterially degraded algal kerogen. Proportionally higher yields can be apportioned for higher TOC values. Because of the complexity and expense of extraction, pyrolysis results are much more commonly used to quantify generation.

While pyrolysis is the main analytical approach used to estimate generation, it is often abused. The S_1 peak com-

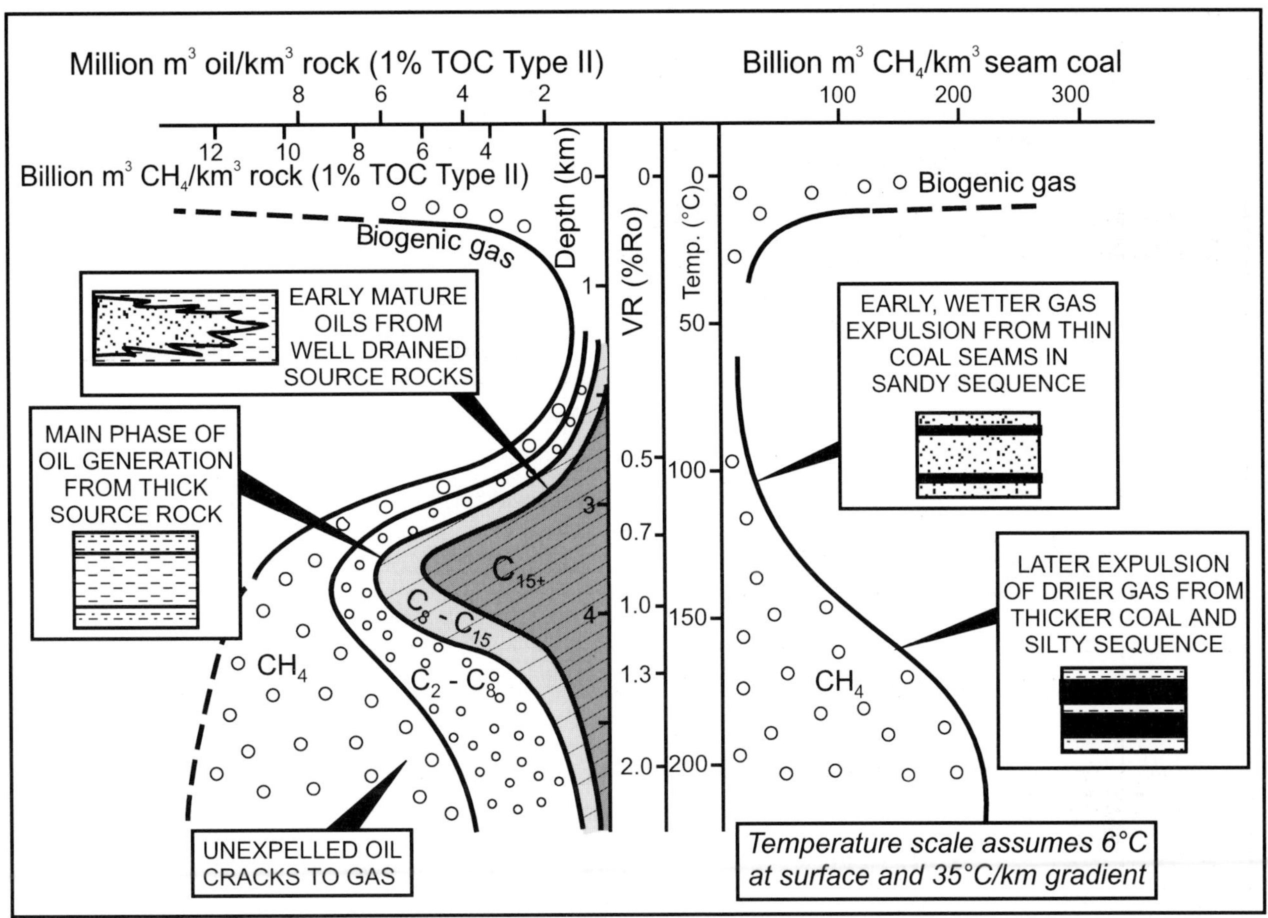

Fig. 11.12 Downhole maturity curves for cumulative generation from 1% TOC of Type II oil-prone kerogen (left) and 1 km^3 of gas-prone kerogen (right), using bulk units. The left curve is based on extract and pyrolysis–gas chromatography results, and the right curve on changes in H/C ratio with rank in Upper Carboniferous seam coal.

prises the free (or migratable) hydrocarbon, while the S_2 peak is the 'chemically bound' hydrocarbon, generated from the kerogen by bond cleavage during the pyrolysis process. The S_2 peak thus represents the future potential of the rock. Both peaks are expressed in units of milligrams of pyrolysate per gram of rock (= kilograms of pyrolysate per tonne of rock).

In the simplest case, one can take the pyrolysis total yield ($= S_1 + S_2$ kg/t) to represent the source-rock yield. This approach assumes:

- that the data come from an immature, uncontaminated rock sample, which has lost no hydrocarbon by expulsion (primary migration);
- that all the pyrolysate (i.e. the material seen by the Rock Eval FID) represents migratable hydrocarbon.

In many cases the first of these assumptions is not met and in all cases the second is unjustified. Gas-chromatographic analysis of the S_2 peak (Py-GC) and comparison with solvent-extract yields (Fig. 11.12) suggests that only some 70% of the S_2 peak comprises hydrocarbon capable of migration.

Two curves resulting from pyrolysis are frequently (though loosely) used to illustrate generation: the trends of Production Index (PI) or Hydrogen Index (HI) with respect to depth. With a uniform organic facies, generation is indicated by the reduction in Hydrogen Index with depth, while the retained hydrocarbons are represented by the Production Index trend. Figure 11.13 shows examples of these trends for sediment samples from nine stratigraphic units from wells in and adjacent to the Moray Firth area of the North Sea. Note that the Hydrogen-Index trend initially increases, as a result of the loss of carbon via decarboxylation reactions, and then falls linearly, while the Production Index rises much more rapidly, reaching a maximum at a value of 0.5–0.6 (off-scale in Fig. 11.13, lower), while the Hydrogen Index continues to decrease. As discussed later, the difference between these two trends can be accounted for by loss of generated hydrocarbon from the system once the PI has reached 'saturation', this being a route to quantifying expulsion efficiencies.

The importance of recognizing these generation trends as a function of depth (or maturity) is that they can be used to determine correctly the original hydrocarbon potential of a sample of mature or even postmature source-rock and to determine the hydrocarbon expulsion efficiency of the system.

11.3.5 Recognizing source rocks using wireline logs

Organic-rich rocks can often be recognized and their richness quantified using a normal suite of wireline logs (Fertl, 1976; Fertl *et al.*, 1986; Carpentier *et al.*, 1989;

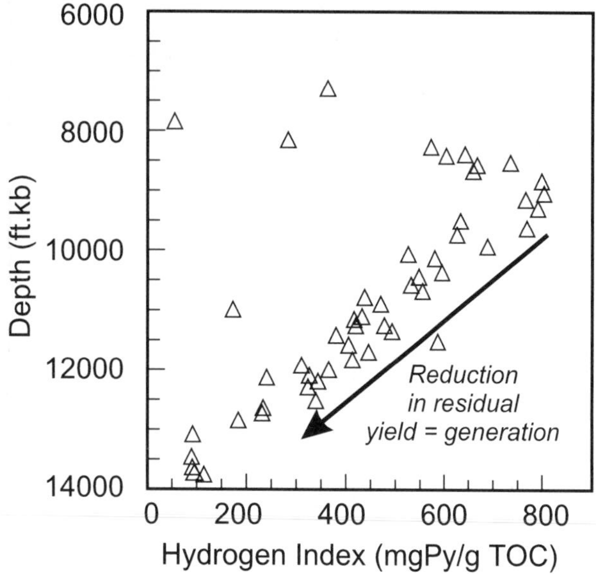

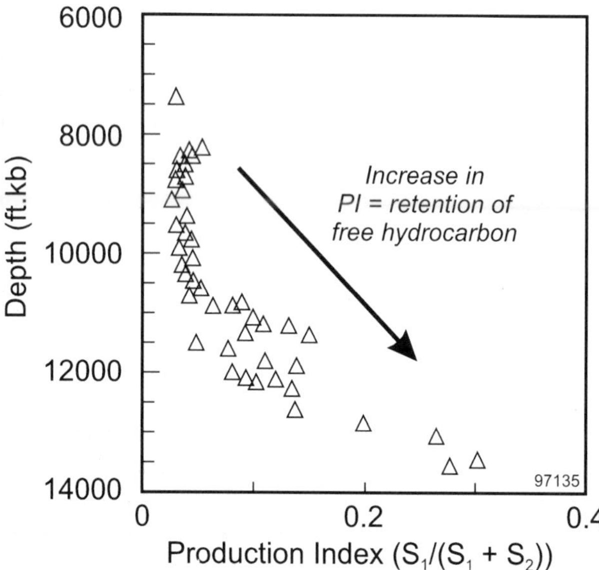

Fig. 11.13 Depth trends for reduction of Rock Eval Hydrogen Index as an indicator of the remaining hydrocarbon potential, and increase in Production Index as an indicator of conversion to, and retention of, free hydrocarbons (nine stratigraphic units intersected by Moray Firth wells).

Herron and Herron, 1990; Passey *et al.*, 1990; Stocks and Lawrence, 1990; Herron, 1991; Bertrand *et al.*, 1993). The log response of a source-rock unit is largely that which would be predicted from its physical properties. Typical log responses through some of the Cretaceous and Jurassic rock intervals of the North Sea are shown in Fig. 11.14, where the characteristic gamma-log kick is seen at the top of the Jurassic–basal Cretaceous organic-rich mudstones of the Kimmeridge Clay/Draupne Formation (Meyer and Nederlof, 1984), and below at the break with the Heather Formation shales (Telnaes *et al.*, 1991). The term 'hot shale' applied to organic-rich sections of the Kimmeridge Clay Formation of the North Sea stems from its high natural radioactivity, deriving from an enhanced uranium, thorium and potassium content (Telnaes *et al.*, 1991).

A typical calibration of natural gamma-ray log response (API units) with %TOC for the Upper Jurassic–basal Cretaceous mudstones of the central North Sea is shown in Fig. 11.15, where an approximately linear relationship is seen. The general correlation stems from the control of anoxia on both the precipitation of uraninite from aqueous uranyl ions and the enhanced preservation of organic matter: little, if any, uranium is actually bonded (chelated) within the organic matter. With gamma-log response in API units, the following linear correlations have been observed for the Upper Jurassic in a limited number of wells in the grabens of the North Sea:

Witch Ground Graben:
$$\%\,TOC = (0.047 \times gamma) + 0.37 \quad (10a)$$
(Correlation coefficient = 0.72)

Northern Viking Graben:
$$\%\,TOC = (0.048 \times gamma) + 0.55 \quad (10b)$$
(Correlation coefficient = 0.64)

As emphasized by Miller (1990), when noting the heterogeneity of gamma-log responses, uncritical application of these equations is not recommended, it being better to derive local correlations for the area investigated. Miller (1990) proposes that the supply of dissolved uranyl ion is from the Boreal ocean to the north, recording a decrease of the gamma-log response of the 'upper hot shale' from 300 API units in mid-Norway, 200 API units in the northern Viking Graben, 190 API units in the Central Graben and 150 API units in the southern North Sea. The more enclosed Moray Firth Basin is reported to yield values of 130 API units, with values dropping to < 100 API units in the Wessex Basin. These trends may indicate a concentration gradient from the north, with uranium supply rather than anoxia and hence amount of organic matter being the controlling factor. When attributing source-rock properties from natural gamma logs, note that sandstones, if enriched in potassium (e.g. in feldspars) and thorium (e.g. in the heavy mineral monzonite), may produce high radioactive responses (Hurst and Milodowski, 1996).

Meyer and Nederlof (1984), Passey *et al.* (1990), Herron (1991) and Schwarzkopf (1992) have discussed the identification of source rocks from an evaluation of gamma-ray, sonic, density and resistivity log responses, with examples given from the North Sea area. The method of Passey *et al.* (1990) subtracts the sonic and resistivity log responses in a 'normal' low-TOC interval from that of the suspected source rock of the same mineralogy and degree of compaction. Thus the Upper Jurassic of the North Sea is an ideal case, with the organically-rich Kimmeridge Clay sandwiched between the organically lean Heather and Valhall shales. Where R_s and R_b (ohm metres) and T_s and T_b (μs/ft) are the resistivity and sonic responses for source and baseline intervals, respectively, TOC_b is the TOC of the baseline interval, the TOC of claystones of a given level of organic maturation (LOM) can be predicted as:

$$TOC = \Delta \log R \times 10^{2.297-(0.1688 \times LOM)} + TOC_b \quad (10c)$$

where

$$\Delta \log R = [\log_{10}(R_s/R_b)] + 0.02(T_s - T_b) \quad (10d)$$

These authors show an excellent correlation between measured and predicted TOC values for a section of relatively

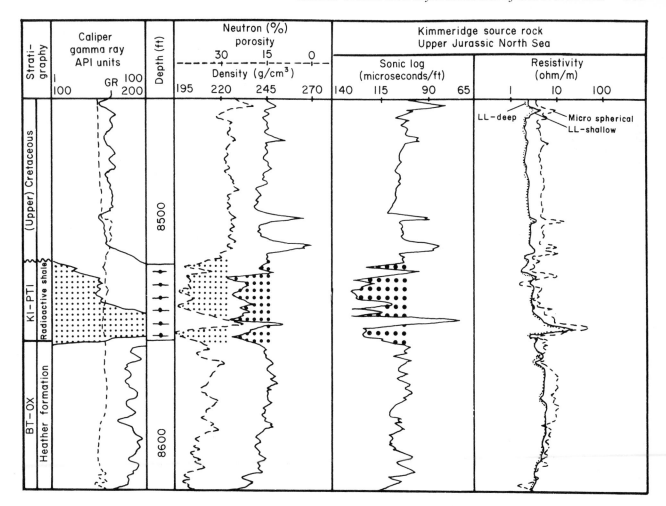

Fig. 11.14 Wireline-log characteristics of immature Upper Jurassic source rocks (Meyer and Nederlof, 1984). The source interval (stippled) shows increased gamma, higher neutron porosity, lower densities, slower sonic velocities but little resistivity contrast at this maturity level.

lean (TOC ~3%), immature (LOM 9–10) Kimmeridge Clay Formation (figured in Herron, 1991), and, in the context of sequence stratigraphy, for a more typical North Sea section from UK well 21/25-2 (Creaney and Passey, 1993).

The reference to LOM (see Section 11.4) in equation 10c acknowledges that the resistivity response, in particular, is maturity-dependent. Thus, if the TOC value is known, maturity can be estimated from wireline-log data. Goff (1983) demonstrates the maturity effect on resistivity of the Kimmeridge Clay Formation of the northern Viking Graben, and interprets it in terms of conversion of solid kerogen to high-resistivity oil, which displaces low-resistivity pore water, followed by expulsion of hydrocarbon from the pore space of the source rock. Upon expulsion, water may re-occupy the available pore space and the partially graphitized kerogen will be increasingly electrically conductive.

The log responses for the top Liassic (Toarcian) Posidonia-schiefer from north-west Germany have been investigated

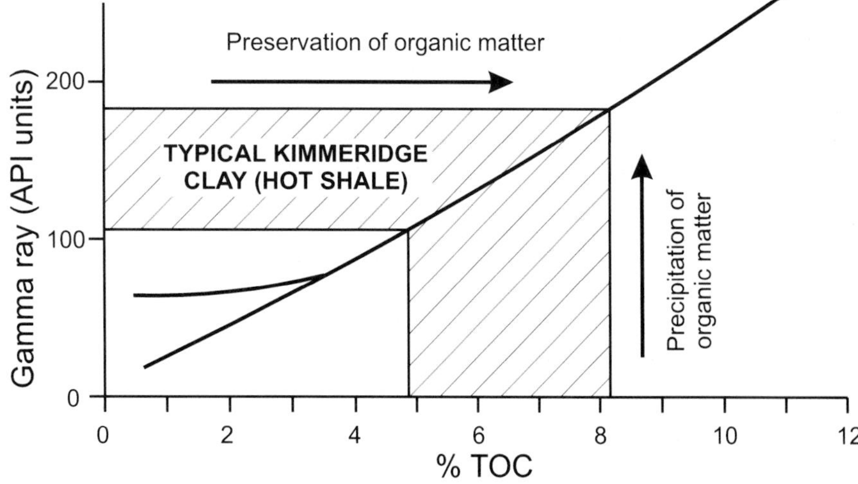

Fig. 11.15 Relationship between gamma-log response and TOC values based on well data from the Central Graben, North Sea. Correlations vary in different basins of the North Sea, this probably being due to the differences in supply and precipitation of uranyl ion in the Upper Jurassic basins. Such locally calibrated relationships may be used to predict source-rock TOC where only wireline-log data are available.

in detail by Mann *et al.* (1986), Mann and Muller (1988) and Schwarzkopf (1992), and the stratigraphic equivalents in the Paris Basin have been discussed in the context of predicting TOC (Herron and Le Tendre, 1990; Bertrand *et al.*, 1993; Creaney and Passey, 1993). The 'Carbolog' technique of Carpentier *et al.* (1991), based on sonic and resistivity log responses, has been applied by Bessereau *et al.* (1995) to predict the TOC values in the whole Liassic section of the Paris Basin, with generalization using sequence stratigraphy. Herron and Le Tendre (1990) compared the effectiveness of laboratory and downhole scans of combinations of gamma, sonic, resistivity and density logs for predicting TOC values in a rich Toarcian core (TOC < 12%). The combination of the first three logs produced the most reliable prediction. The best prediction, however, was obtained from use of a gamma-ray spectrometry tool (GST) to determine the carbon/oxygen ratio and then the TOC value. Using the approach of Passey *et al.* (1990), Murphy *et al.* (1995) noted both good (Irish well 49/9-1) and poor correlations (Irish well 47/29-1) between measured and predicted TOC values in the Toarcian of the North Celtic Sea basin. The poor correlation was attributed to downhole cavings in the cuttings samples and the presence of carbonates and migrated hydrocarbons in the section. Log based prediction of the pyrolysis S_2 peak yields was also made.

The log responses of shales with low TOC and poor kerogen quality from the Sinemurian–Pleinsbachian of Yorkshire and SW England have been described, using separate potassium, uranium and thorium logs, by van Buchem *et al.* (1992) and Bessa and Hesselbo (1997), respectively. Cyclicity was recognized and attributed to Milankovitch cycles of < 5 m thickness, with evidence of eccentricity, obliquity and precession cycles. Using gamma, resistivity and density logs, Herbin *et al.* (1993a,b) provide a continuous monitor of TOC in four boreholes through the Kimmeridge Clay of the UK onshore, noting cycles attributed to a transgressive sequence stratigraphic tract. Cycles of 25 ka and 280 ka are recognized.

The use of logs to define source rocks is particularly appropriate for volumetric studies, since it allows a continuous monitor of source quality over the whole section where only a few discrete analyses may have been made (Herron, 1989, 1991; Murphy *et al.*, 1995). Some conventional analyses are always required in order to calibrate the log responses.

11.4 Maturation of kerogens

Kerogen matures during burial, that is, it undergoes physical and chemical changes that result largely from temperature increases related to depth of burial (see Fig. 11.1B,C). One result of these changes is the generation of hydrocarbons—first liquids (oil) and then gas—as burial proceeds. The results of these processes are shown in Fig. 11.12 in terms of generation.

11.4.1 Measured maturation

Maturation—as opposed to generation—parameters are generally measured in source-rock studies. Parameters are used to measure kerogen maturity, include vitrinite reflectance (Buiskool-Toxopeus, 1982; Senftle and Landis, 1991), the properties of coal in general (Bostick, 1979; Teichmuller and Teichmuller, 1979; Stach *et al.*, 1982; Durand *et al.*, 1986; van Krevelen, 1993), spore or kerogen colour (Smith, 1983), to a lesser extent, fluorescence, and finally the temperature of maximum pyrolysis yield (T_{max}) from the Rock Eval pyrolysis technique (Brosse and Huc, 1986; Espitalié, 1986). Standard correlations between parameters have been established by Heroux *et al.* (1979), Bertrand and Achab (1990) and Miles (1989). Using high-temperature micropyrolysis, Everlien (1996) has extended correlation of vitrinite reflectance with T_{max} to high maturities (7% R_o; T_{max} = 700°C). Velde and Espitalié (1989) demonstrate a correlation between Rock Eval T_{max} and illite/smectite ratios for North Sea Dogger (Middle Jurassic) samples and Smart and Clayton (1985) for the UK Carboniferous.

In the Palaeozoic, graptolite (Goodarzi and Norford, 1985) and chitinozoan (Goodarzi, 1985; Tricker *et al.*, 1992) reflectance has been utilized as indicators of maturity. Using, in part, samples from the UK and France, chitinozoan reflectance (R_{ch}) has been related to vitrinite reflectance by Tricker *et al.* (1992), using the relationship $R_{ch} = 1.152\% R_o + 0.08$. Conodont alteration (colour) index (CAI), measured on a 1–5 scale, has been applied by Bergstrøm (1980) and Aldridge (1986) to determination of the thermal history of the Caledonides of the UK and Scandinavia. Correlations between the reflectance of vitrinite, chitinozoans, graptolites and sclerecodonts are demonstrated by Bertrand and Heroux (1987) and Bertrand (1990), on the basis of samples from the eastern seaboard of Canada.

Molecular ratios, e.g. sterane and triterpane isomer ratios (Mackenzie *et al.*, 1980; Cornford *et al.*, 1983; Pearson *et al.*, 1983; Mackenzie, 1984; van Graas, 1990; Hall and Bjorøy, 1991), have proved most accurate and versatile maturity parameters, as discussed below. More exotic maturity parameters, such as methylphenanthrene isomer ratios (Radke and Welte, 1982; Hall *et al.*, 1985; Schou *et al.*, 1985; Radke, 1988), porphyrins (Huseby *et al.*, 1996), benzothiophenes (Schou and Myhr, 1988), and diffuse infrared reflectance spectroscopy of sediment-extract asphaltenes (Christy *et al.*, 1989), have also been applied to North Sea samples.

Vitrinite reflectance and spore colour are the most widely used microscopy techniques, while the rapidly developing field of biomarker molecular ratios has the important advantage that measurements can be made on both source rock and reservoired oil, and ratios may be readily modelled (e.g. Düppenbecker *et al.*, 1991; Welte *et al.*, 1997). As shown in Fig. 11.16, the vitrinite reflectance technique, though objective in using a stable light source, a high precision standard and a high-sensitivity photomultiplier, is subjective in terms of the operator's decision as to what to measure and the statistical treatment of the resulting measurements (Senftle and Landis, 1991). Opinions differ, as to whether it is better to calculate an arithmetic mean of a few carefully selected telocollinite particles, with selection being made at the microscope (Buiskpool-Toxopeus, 1982, 1983), or to measure large numbers of particles, with selection of the 'true indigenous vitrinite population' being made with reference to the histogram of values. While the former is best undertaken on polished whole-rock samples, the latter

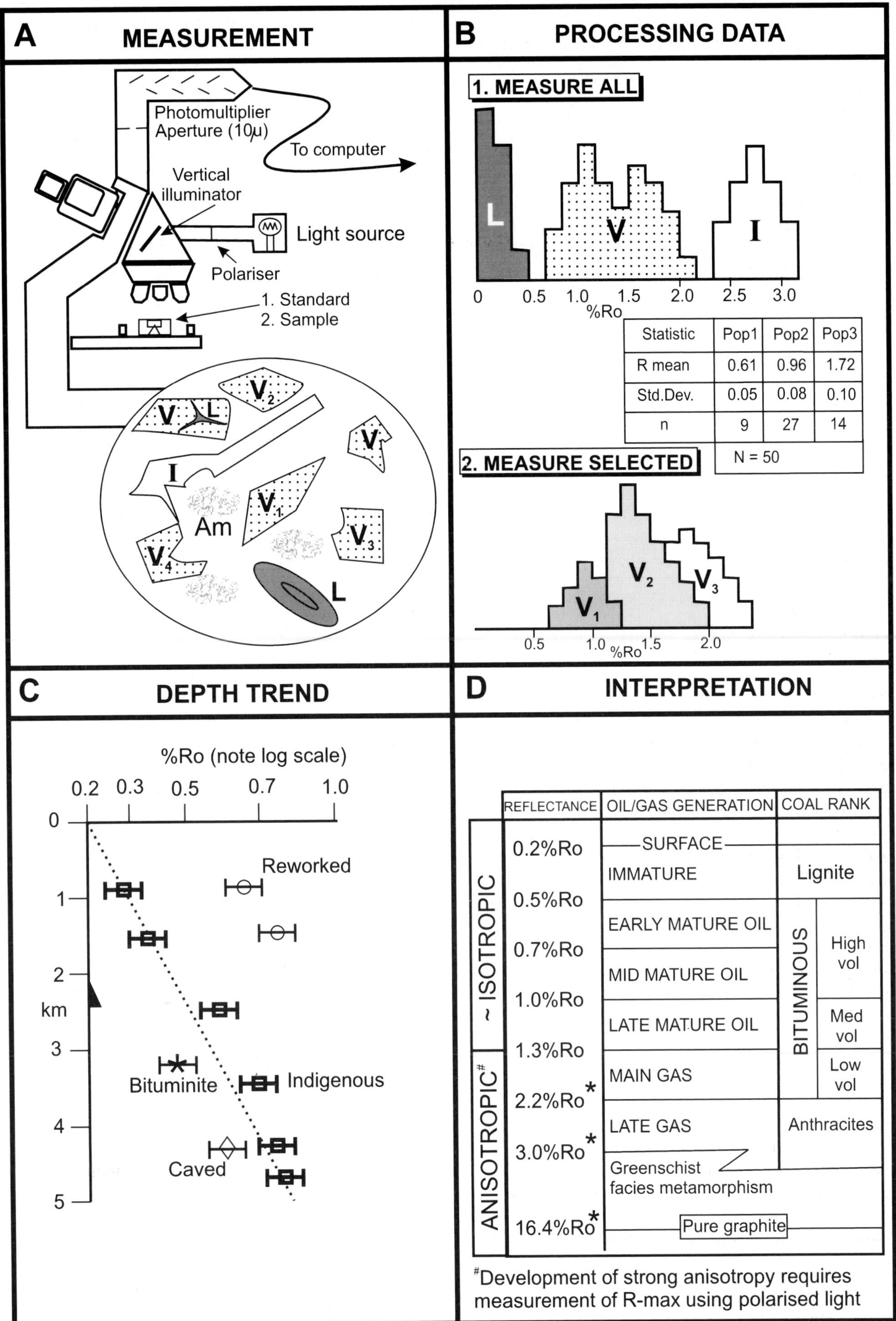

Fig. 11.16 Applying the vitrinite reflectance technique requires (A) measurement on polished vitrinite particles; (B) presentation of a histogram—all particles or selected vitrinites—and calculation of an appropriate arithmetic mean; (C) plotting mean values against sample depth; and (D) interpretation of the trend in terms of oil generation and coal rank. The technique is objective other than in the selection of particles to measure (A), the selection of the vitrinite subpopulation for inclusion in the calculation of the mean (B) and the rejection of data in establishing the depth trend (C).

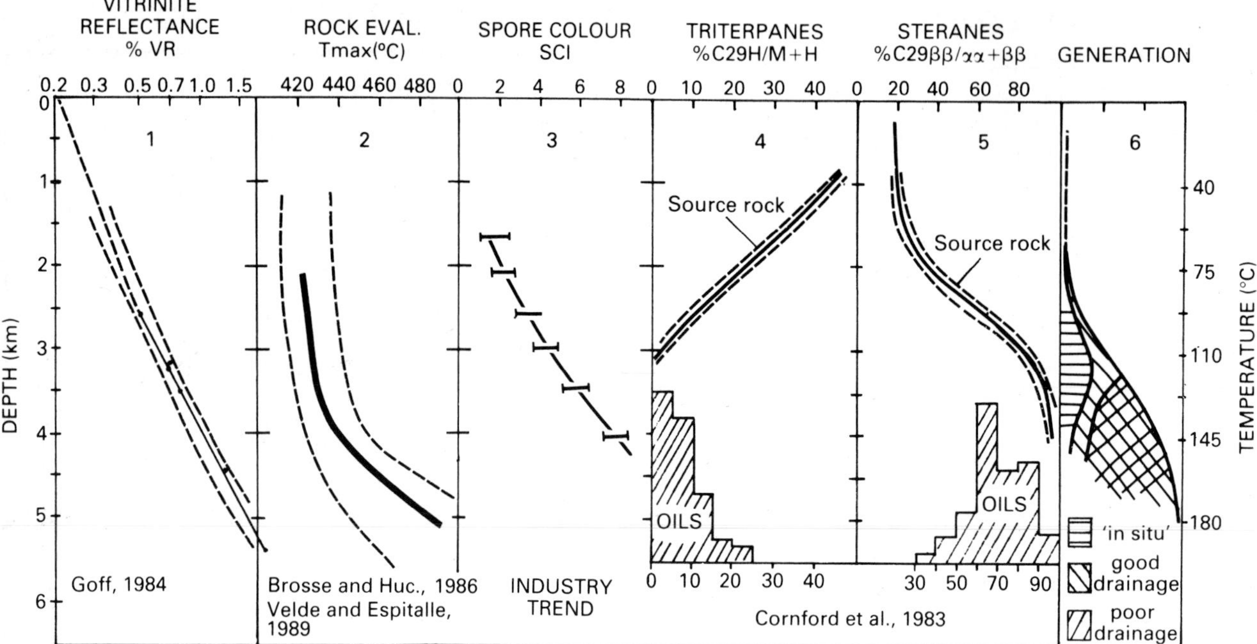

Fig. 11.17 General maturity depth trends for vitrinite reflectance (% R_0), Rock Eval T_{max} (°C), Spore Colour Index (1–10 scale), moretane/hopane and sterane isomerization ratios (a/(a + b)) for North Sea wells. NB: Temperature/depth conversion based on 35°C/km and 5°C at sea floor. H = C_{29} hopane; M = C_{29} moretane; $C_{29}\beta\beta$ = 5α, 14β, 17β 20R and 20S iso-ethylcholestane; αα = 5α, 14α, 17α 20R and 20S regular ethylcholestane.

approach—as adopted by most industry service laboratories—is usually undertaken on mounted and polished kerogen concentrates.

No one maturity technique can be applied universally, and no simple equivalence can be established between the various techniques (cf. Heroux *et al.*, 1979). Figure 11.17 shows a number of maturity trends for the post-Palaeozoic North Sea sediments, the trends being taken from different wells (or groups of wells) in areas with different subsidence and thermal histories. The trends in Fig. 11.17 are thus not strictly comparable, because they ignore both geography and the effects of heating rate in °C per million years (°C/Ma).

A number of equivalences have been proposed between vitrinite reflectance and maximum downhole palaeotemperature (T peak). Ignoring the time spent at maximum temperature, Barker (1996) proposed the following relationship:

$$\text{T Peak (°C)} = (80.6 \times \log_{10}(\%\ R_o)) + 135.5 \quad (11a)$$

Given a generalized heating rate of about 1°C/Ma during continuous steady subsidence, the relationship of measured vitrinite reflectance (%R_o) with corrected downhole temperature in selected North Sea wells (Quigley *et al.*, 1987) follows a steeper gradient but with a cooler intercept (given the log scale, the intercept is the temperature for 1% R_o):

$$\text{T Peak (°C)} = (169 \times \log_{10}(\%\ R_o)) + 139 \quad (11b)$$

Essentially parallel but offset trends are shown for measured reflectance for higher (e.g. Pannonian Basin, 3–5°C/Ma) and lower heating rates (e.g. Scotian Shelf, 0.5°C/Ma). From practical North Sea experience, both equations 11a and 11b predict rather low temperatures for a given % R_o value. Much depends on the nature of the temperature database. Pressure as well as temperature has been suggested as affecting reflectance values (Carr, 1991).

Biomarker maturity-controlled ratios allow the establishment of a direct genetic link via a maturity-based oil/source-rock correlation (Cornford *et al.*, 1983). In the case of the molecular parameters, the range of values found in reservoired oils can also be shown as a histogram on a common horizontal scale for direct comparison with source-rock-extract maturation trends (Fig. 11.17). These molecular measurements also have the advantage of reflecting maturation changes within the oil-prone material directly and do not require empirical correlation with the generation process.

Discrete boundary values can be established, based on published depth/temperature trends. An approximate comparison of maturity parameters is shown in Table 11.3, although details of the equivalences will depend on the origin of the measured data (interlaboratory variation) and the heating rate.

11.4.2 Calculated maturation

Maturity may be calculated from stratigraphic and geothermal information in both drilled and undrilled basins (Welte *et al.*, 1997). In the case of undrilled basins 'pseudo-wells' may be created, based on seismic stratigraphy and geothermal information taken from surface heat-flow measurements, or by analogy with similar tectonic settings elsewhere.

The North Sea and adjacent basins have played an important role in the development of maturity modelling, the copiously drilled and sampled basins being used as natural laboratories to study maturation in an extensional tectonic setting (Thorne and Watts, 1989; Ungerer *et al.*, 1990; Doré *et al.*, 1991; Ungerer, 1993). The progress in maturity modelling has contributed to our understanding

Table 11.3 A comparison of maturity boundaries applicable in the North Sea related to oil and gas generation. Drawing exact equivalences between these parameters ignores the different kinetics of the various changes being monitored. The equivalences shown are true for typical North Sea heating rates of ~1 °C/Ma (average geothermal gradient = 35 °C/km; average sedimentation/burial rate = 29 m/Ma).

Technique	Immature	Early mature	Midmature	Late mature	Postmature
Oil-generation zones					
% R_o	< 0.5	0.5–0.7	0.7–1.0	1.0–1.3	> 1.3
SCI	< 3.5	3.5–4.5	4.5–7.5	7.5–(9?)	> 9?
TAI	< 2.25	2.25–2.3	2.3–2.9	2.9–3.0	> 3.0
LOM	7.8	7.8–8.0	8.0–10.5	10.5–12.0	> 12.0
T_{max}	< 435	435–443	443–458	458–468	> 468
Gas-generation zones					
% R_o	< 0.8	0.8–1.0	1.0–2.2	2.2–3.0	
SCI	< 5.0	5.0–7.0	7.0–10.0	> 10	Greenschist facies
TAI	< 2.5	2.5–2.9	2.9–3.3	3.3–4.0	
LOM	< 9.5	9.5–11.5	11.5–14.5	14.5–18.0	
T_{max}	< 450	450–455	455–525	525–580	

R_o, vitrinite reflectance (%); SCI, Spore Colour Index (1–10); TAI, thermal alteration index (1–5); LOM, level of organic maturation (1–20); T_{max}, Rock Eval T_{max}, °C (see Fig. 11.6 for details).

of the thermal, sedimentological and structural development of basins formed by crustal extension. Specifically, maturation, generation and migration have been addressed, using both 1D and 2D modelling in the Viking Graben (Ungerer *et al.*, 1984, 1990), the Central Graben, (Cornford, 1994) and Haltenbanken in mid-Norway (Ungerer *et al.*, 1990; Forbes *et al.*, 1991; Hermans *et al.*, 1992). Based mainly on defining migration paths through time, 3D modelling has been applied in mid-Norway to explain exisiting (and hopefully predict further) accumulations (Hermans *et al.*, 1992; Grigo *et al.*, 1993).

The inputs for calculating maturation indices are stratigraphy and geothermal information. Stratigraphy (chronostratigraphic ages and thicknesses) can be used to generate a burial-history plot, which is combined with either geothermal gradient or heat-flow modelling, to estimate the time–temperature history of stratigraphic units of interest during burial and uplift (Thorne and Watts, 1989; Welte *et al.*, 1997). Given kerogen kinetic properties, such models are used to show both how much oil and gas have been generated and when generation occurred. These concepts are discussed below.

Depth of burial

In the absence of, or in addition to, measured maturity parameters, a first estimate of the maturity of a source rock can be obtained from its maximum depth of burial. As an example, in the northern North Sea it would be generally true to say that 'all rocks buried below 3200 m are mature for oil generation'. In an area such as the southern North Sea or Inner Moray Firth, where there has been uplift and erosion (i.e. Chalk on the sea floor, as in the Inner Moray Firth), such a statement would be incorrect since the present-day burial is less than the maximum burial for Chalk and pre-Chalk rocks.

The burial (or geo-) history diagram of depth versus geological time has become synonymous with maturity modelling, this providing a method of tracing the burial of all identified horizons through time. The effects on burial of compaction and episodes of uplift are frequently taken into account. Typical burial-history diagrams have been presented for mid- and north Norway (Leadholm *et al.*, 1985; Hermans *et al.*, 1992; Dahl and Augustson, 1993), Viking Graben (Lepercq and Gaulier, 1996), Witch Ground Graben (Mason *et al.*, 1995), Central Graben (Kooi *et al.*, 1989; Cornford, 1994), Danish Central Graben (Jensen *et al.*, 1985), Egersund Basin (Ritter, 1988; Hermanrud *et al.*, 1990b,c), eastern Baltic (Brangulis *et al.*, 1993); Broad Fourteens and West Netherlands Basin (Kooi *et al.*, 1989; Winstanley, 1993), Sole Pit Trough (Glennie and Boegner, 1981; Cooke-Yarborough, 1991; Robinson *et al.*, 1993; Turner *et al.*, 1993), Western Approaches and Cardigan Bay (Menpes and Hillis, 1995), Bristol Channel (Cornford, 1986) and Celtic Sea (Howell and Griffiths, 1995). Largely theoretical discussions in terms of North Sea crustal history (tectonic subsidence) and sediment loading are given by Kooi *et al.* (1989) and Thorne and Watts (1989). Such reconstructions of palaeoburial should be validated by calibration against measured rock properties, such as porosity, vitrinite reflectance, biomarkers or clay-mineral transformations.

Building models based on the preserved stratigraphy is fairly straightforward at well locations, using the stratigraphic information recorded on a typical composite well log, or using regional isopachyte maps, as reproduced in Nielsen *et al.* (1986) or Thorne and Watts (1989) for the North Sea. Where inversion (uplift and erosion) have occurred (Riis and Jensen, 1992; Buchanan and Buchanan, 1995) and hence unconformities are present, the missing section must be estimated. The extent of the missing section is most commonly estimated from offsets in maturity-versus-depth trends (e.g. Armagnac *et al.*, 1989), from anomalies in the density or sonic velocity-versus-depth trends (e.g. Chadwick, 1993; Hillis, 1993; Hillis *et al.*, 1994), from seismic truncation, from apatite fission-track analysis (Lewis *et al.*, 1992; Green *et al.*, 1995a), from offsets in biomarker depth trends (Haszeldine *et al.*, 1984a,b; Scotchman, 1994), from modelling tectonic subsidence (Nadin and Kuznir, 1995; Nadin *et al.*, 1995), from palinspastic reconstruction (Hooper *et al.*, 1995) or from regional topographic inferences (Cope, 1994).

Based on theoretical calculations, Thorne and Watts (1989) have produced a regional map of areas of uplift ('decelerating tectonic subsidence') in the Inner Moray Firth, Sole Pit, Broad Fourteens and Norwegian–Danish

Basin areas, using a deconvolution of burial histories in terms of sediment loading and teconic-subsidence effects. A Palaeocene uplift event has been identified by Nadin and Kuznir (1995) in the Viking Graben, based on forward and reverse modelling of tectonic subsidence where sediment load and water depth are known. Cope (1994) has proposed a regional dome of hot-spot-induced latest Cretaceous uplift centred on the East Irish Sea, although the symmetry of the uplifted area has subsequently been challenged (Thomson, 1995). Kooi *et al.* (1989) argue for extensional tectonics in the Mesozoic North Sea area, in contrast to compressional in the Cenozoic—the North Sea *sensu stricto* being influenced more by Atlantic opening, while inversions in the Broad Fourteens and West Netherlands Basin have responded to Tethyan events. Triassic–Tertiary basement uplift of at least 2.2 km is estimated from 2D palinspastic reconstructions in the Broad Fourteens Basin (Hooper *et al.*, 1995). In the more recent stratigraphy of the North Sea a widespread mid-Miocene break has been identified (Cameron *et al.*, 1993), although the amount of section lost is poorly constrained (Dahl and Augustson, 1993).

Temperatures and geothermal gradients

This initial depth-based opinion on source rock maturity can be refined using a knowledge of the geothermal gradients to determine the maximum temperature a source rock has experienced during burial (Waples, 1984). In continuously subsiding basins, such as the North Sea Graben system, the maximum temperature experienced is related to depth of burial via the local geothermal gradient (°C/km or °F/100 ft). Present-day North Sea well-temperature data have been treated in detail by Harper (1971), Cooper *et al.*, (1975), Cornelius (1975), Carstens and Finstad (1981), Toth *et al.*, (1983), Eggen (1984), Thomas *et al.* (1985), Hermanrud *et al.* (1991) and Brigaud *et al.* (1992), and a composite geothermal-gradient map based on their results is shown in Fig. 11.18. Dependent on the temperature source and correction method, Brigaud *et al.* (1992) report average gradients between 31.8°C/km (Horner-corrected wireline-log temperatures) and 36.3°C/ km drill-stem test (DST) temperatures) for 598 log temperatures and 50 DST temperatures from the North Viking Graben. They also report interval geothermal gradients for the Tertiary, Cretaceous and Jurassic–Triassic intervals, contouring values between 28°C/km and 56°C/km for the interval gradients.

There are a number of ways of correcting raw temperature measurements to estimate the 'virgin rock temperature' prior to disturbance by drilling (Hermanrud and Shen, 1989; Hermanrud *et al.*, 1990a; Brigaud *et al.*, 1992). Recognizing that drilling disturbs the downhole temperature, mathematical correction methods (e.g. the Horner-plot method (Barker, 1996)) all include correction according to a ratio of cooling time (e.g. time of mud circulation to bring last cuttings to surface or time to drill the last 10 m) divided by warming time (e.g. time since mud circulation). The true formation temperature may be determined by extrapolating to infinitely long warming times and infinitely short cooling times. In the Horner-plot method, the extrapolation to infinite times is performed graphically.

The treatment of downhole temperature measurements

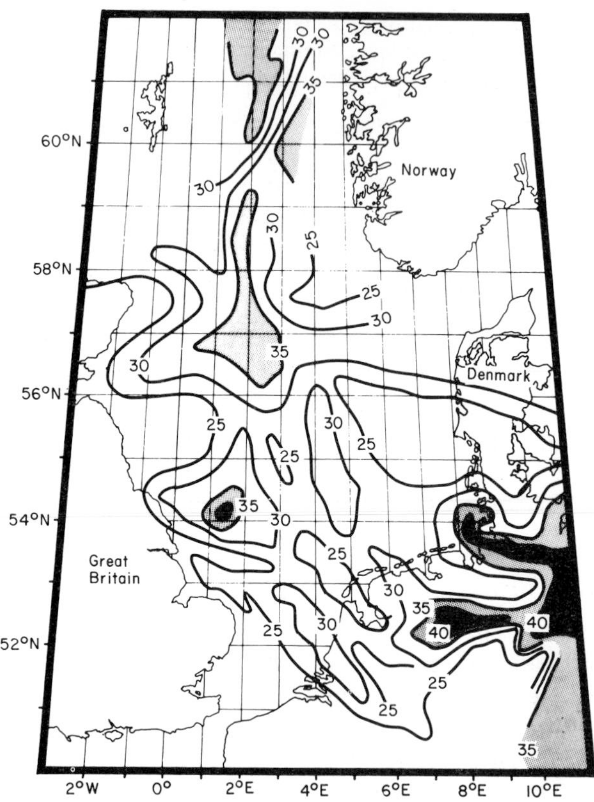

Fig. 11.18 Present-day geothermal-gradient map for the North Sea area. For a constant basement heat flow, first-order controls on geothermal gradients are the depth of the well (compaction effects) and the lithologies penetrated (mineral thermal conductivities), since both affect the bulk thermal conductivity of the section. Hydrodynamic flux also redistributes heat. The map thus only partially reflects the regional variation in heat flow as shown by Burley (1993).

and the effect of errors on calculations of hydrocarbon generation have been discussed for the Horda Platform to the east of the North Viking Graben by Hermanrud *et al.* (1991). The thermal anomaly associated with the Troll Field is attributed by these authors to transient effects resulting from uplift following the late Quaternary deglaciation. This contrasts with the explanation of Eggen (1984) and Hermanrud (1986), who attribute the Troll anomaly to the flux of hot water from deep compacting shales in the main Viking Graben. In addition to the approaches reviewed by Hermanrud *et al.* (1990a), Jensen and Doré (1993) use a simple industry correction method for Haltenbanken wells, where a fixed temperature increment is added (~5°C at 1 km, ~18°C at 3.5 km and ~14°C at 5 km), dependent on the depth of the log-temperature measurement. Barker (1996) reports a similar down-hole temperature correction plot for the Gulf of Mexico.

The approach of estimating present and palaeo-downhole temperatures from geothermal gradients is flawed, for a number of reasons (Hermanrud *et al.*, 1990a):

- Since most models use a single gradient from sediment surface to bottom hole, this ignores the inevitable and predictable non-linearities in the temperature profile (Fig. 11.19A).
- Since shallower wells (generally on the basin margins) penetrate less compacted sediments, higher gradients are inevitably measured due to lower bulk thermal conductivities.

• Using the present-day gradient in the past imposes a gradient on less compacted rocks and, back through time, on different lithologies which may not be capable of supporting this gradient for a given heat flux (see next section).

All the above problems are eliminated if heat-flow modelling is adopted. That temperature and not heat flow is the primary measured parameter favours gradient modeling and, since maturation and generation are a function of temperature rather than heat, the geothermal gradient is in some ways a more direct predictor of reaction progress. On balance, it is generally more reliable to use heat-flow modelling rather than geothermal gradients to establish the temperature history of a sediment section. An exception would be predicting maturity in areas where the lithologies, and hence thermal conductivities, are uncertain.

Heat flow

With the application of 1D and the advent of 2D modelling in the North Sea (Doré *et al.*, 1993), considerable effort has been expended in establishing a heat-flow history for the area. As shown in Fig. 11.19, there is a simple relationship between heat flow (Q; mW/m^2), geothermal gradient (GG; °C/km) and rock thermal conductivity (TC; W/m°C):

$$Q = \mathrm{GG} \times \mathrm{TC} \tag{12}$$

Subsurface temperature (T_z; °C) at depths Z (km) is then calculated, using the geothermal gradient, GG, and surface temperature, T_s (°C):

$$T_z = (\mathrm{GG} \times Z) + T_s \tag{13}$$

The major area of uncertainty in application of these equations lies in the values of the thermal conductivities of the lithologies encountered. Thermal conductivity (the amount of heat transferred through 1 m of rock with a 1°C (1 K) temperature difference between its extremes) can be measured as a bulk property (grains + pore-filling fluids) or, given the porosity, can be calculated from the separate properties of an appropriate mix of mineral grains and the pore-filling fluid(s). The porosity of progressively buried rock decreases with compaction. Since pore fluids (water, oil and gas) have low thermal conductivities, a decrease in porosity produces an increase in thermal conductivity and hence a decrease in geothermal gradient (equation 12). Thermal conductivity of the majority of materials (e.g. rock-forming minerals) decreases with increase in temperature (Sekiguchi, 1984; Ungerer *et al.*, 1990), although the behaviour of water is anomalous (Welte *et al.*, 1997).

For rapid changes in heat flow or sedimentation rates, the heat capacity (thermal mass) of the rock must be considered—so-called time-transient heat-flow modelling. Time-transient heat-flow events occur with igneous intrusions, hydrothermal pulses, rapid burial or uplift, but are not of particular significance in the main North Sea area, where heating rates are relatively low (typically ~1°C/Ma) and intrusions rare. Heating rates may be calculated from the geothermal gradient (°C/m) multiplied by the sedimentation rate (m/Ma).

The detailed variation of geothermal gradient with time and with depth in the sedimentary section is a complex function of temporally fluctuating heat flow, as discussed by Lucazeau and Douaran (1984) for Viking Graben, the continuously changing thermal conductivity of the various lithologies of the progressively compacting strata (e.g. Andrews-Speed *et al.*, 1984; Brigaud *et al.*, 1992), groundwater movement (e.g. Eggen, 1984, for the Troll area), recent uplift (Hermanrud *et al.*, 1990b,c) and a sediment

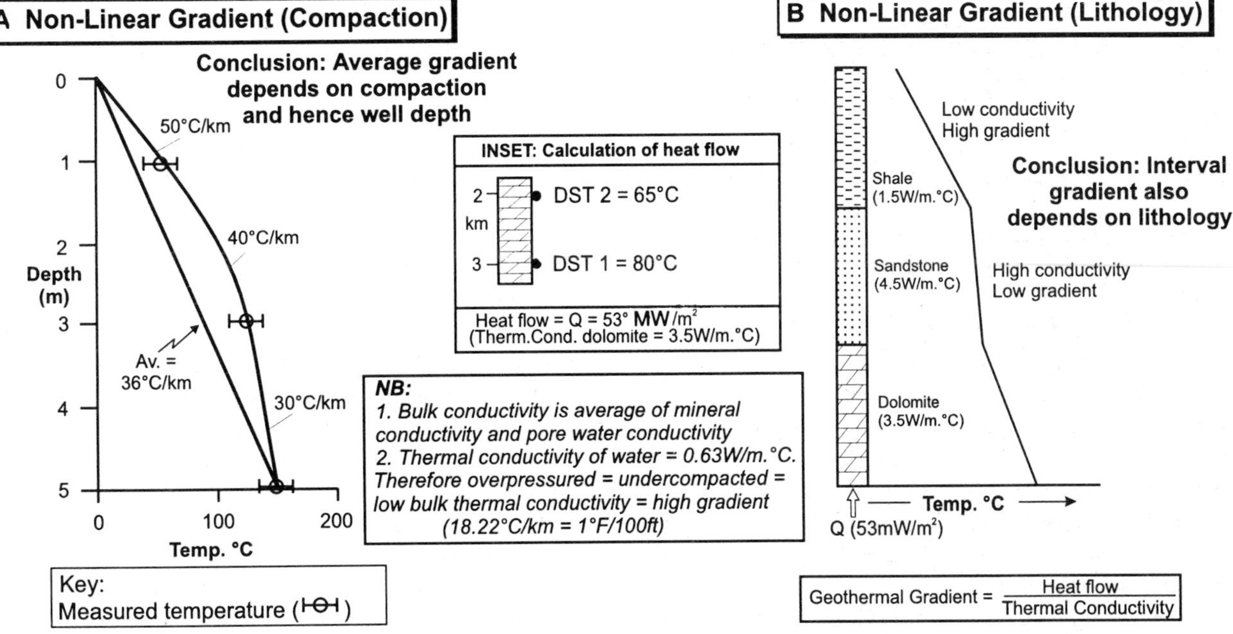

Fig. 11.19 Relationship of heat flow to downhole temperature and geothermal gradient, given compaction effects (A) and lithology/mineralogy controls (B). With compaction (A), interval gradients may be both higher in the shallow section and lower in the deeper section compared with the average gradient. Changes in mineralogy give local changes in gradient for a constant heat flow (B). Note that only temperature can be measured, both gradient and heat flow being derived values.

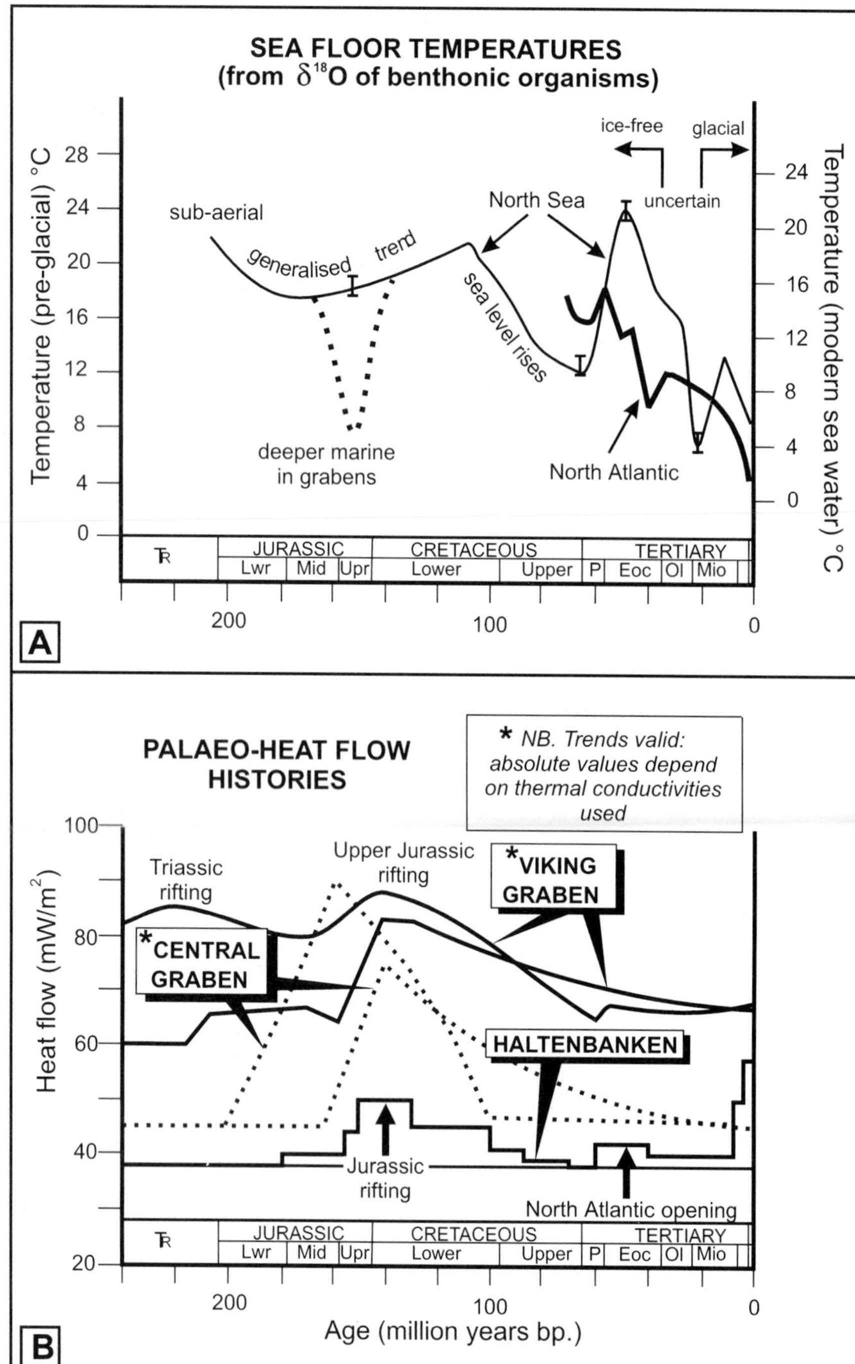

Fig. 11.20 (A) Temporal variation in sea-floor temperatures based on stable oxygen-isotope ratios from benthonic foraminifera and bivalves from the North Sea area (Jensen and Buchardt, 1987; Anderson *et al.*, 1994) and the Atlantic (Miller K.G. *et al.*, 1987), and (B) heat-flow histories for the Viking Graben (Lucazeau and Le Douaran, 1984; Schroeder and Sylta, 1993), Central Graben (Cornford, 1994, and unpublished BasinMod^R results) and Haltenbanken (Jensen and Doré, 1993). The wide range of present-day heat-flow values (45–70 mW/m^2) reflects differences in mineral thermal conductivities and compaction equations used by the various authors, but all are correlated against similar downhole temperatures. A regional present day heat flow map has been compiled by Burley (1993).

surface temperature (Fig. 11.20A) that changes with palaeoclimate (Buchardt, 1978; Hallam, 1985; Miller K.G. *et al.*, 1987). Fjeldskaar *et al.* (1993a) utilized these relationships to predict present and palaeotemperatures in the largely undrilled Barents Sea.

Some published trends of heat flow resulting from these temporal changes are illustrated in Fig. 11.20B. The similarities in the changes in heat flow with time are apparent in this figure, but so are the differences, all of the authors providing internally consistent but not directly comparable models. The absolute values of heat flow depend upon the values of thermal conductivity used. Since the geothermal gradients and lithologies in all three areas are relatively similar, equation 12 shows that the Viking Graben model of Lucazeau and Le Douaran (1984, 1985) must have used thermal conductivities about twice as high as those of Jensen and Doré (1993) for Haltenbanken. In both cases, the models are internally consistent in that they are calibrated against measured downhole temperatures.

Thus rock thermal conductivities become a major issue when comparing models and their predictions. Brigaud *et al.* (1992) have measured and contoured the thermal conductivity for Tertiary, Cretaceous and Jurassic–Triassic intervals of the North Viking Graben, these values reflecting both mineralogy and compaction. The thermal conductivities can be measured in the laboratory for individual lithotypes or mixtures thereof (Brigaud and Vasseur, 1989) derived from well data or calculated from wireline-log data—see Demongodin *et al.* (1991) for the Paris Basin and Brigaud *et al.* (1992) for the North Viking Graben.

Thermal conductivities of rocks and of their component minerals, the effect of rock porosity and the anisotropy of these properties are all the subject of debate (Blackwell and Steele, 1989; Brigaud *et al.*, 1990; Hermanrud *et al.*, 1991; Yalcin *et al.*, 1997). The wide range of heat-flow values quoted in the literature must be treated with caution: local calibration against corrected downhole temperatures is recommended—see Hermanrud and Shen (1989) and Brigaud *et al.* (1992) for a discussion, with examples from North Sea geology.

In the North Sea area, perturbations to a temporally constant heat-flow model derive from the tectonic history, extension and crustal thinning producing higher heat-flow regimes (McKenzie, 1981). The effects of the Upper Jurassic extension (e.g. Lucazeau and Le Douaran, 1984, for the Viking Graben) will be superimposed on top of a possible earlier (Triassic and/or Lower Jurassic) or later (Palaeocene) rifting events (Lepercq and Gaulier, 1996). In addition, the thermal insulating effects of overpressured shales occurring below about 2800 m in the main northern North Sea grabens produce non-linear geothermal gradients, with a rapid rise in temperature below the thermally insulating overpressured zone (Leonard, 1989). Considerable variation in the interval vertical heat flow through the Tertiary (45–70 mW/m^2), Cretaceous (45–85 mW/m^2) Jurassic–Triassic sediment packages (70–140 mW/m^2) of the North Viking Graben has been demonstrated by Brigaud *et al.* (1992), who attribute the observed variations to overpressure-driven fluid flow focusing at discrete locations, equating to clusters of oilfields.

The thermal conductivity of metamorphic basement rocks is high, and Yu *et al.* (1995) have noted a focusing of heat flux by basement horsts in the Danish Central Graben. This gives an increase of some 21 mW/m^2 in basement heat flux at the I-1 well location.

A number of different palaeoheat-flow models for mid-Norway have been compared in terms of calibration against vitrinite reflectance by Jensen and Doré (1993), present-day measured maturity being generally insensitive to elevated palaeoheat-flow values.

Using equations 12 and 13, the variation of subsurface temperatures with time for any given formation can be calculated from the heat-flow and surface-temperature variations, as reproduced in Fig. 11.20 for the North Sea and adjacent areas. In terms of the sediment surface temperature, the North Atlantic benthonic temperature curve may be more appropriate for areas west of the UK and Eire and for Norway north of 62° of latitude.

Palaeohydrothermal events have been noted from diagenetic studies from the Huldra Field in the North Viking Graben (Glasmann *et al.*, 1989) and in an unidentified Brent reservoir (Liewig *et al.*, 1987). Jensen and Doré (1993) have investigated the possibility of a recent heat-flow event in mid-Norway (Haltenbanken). Also in mid-Norway, the significance of transient thermal modelling has been investigated by Vik and Hermanrud (1993) in the context of rapid burial. It was concluded that transient modelling produced better temperature prediction, but that both transient and steady-state models equally fitted the measured maturity trend for well 6407/14-1. Rapid hydrothermal heating has been proposed for the Pennine–Cheshire–East Irish Sea area (Green *et al.*, 1995a,b), and igneous heating is locally common in the areas offshore of western Scotland and Ireland (e.g. Bishop and Abbott, 1993; Farrimond *et al.*, 1996).

Based on the concept that the base of the crust is at 1335°C (Allen and Allen, 1990) and assuming steady state, heat flow at the surface (mW/m^2) is a function of crustal thickness and the amount of radioactive heat introduced within the crust:

$$\text{Surface heat flow} = \text{heat flow from mantle} + \text{crustal radioactive heat} \quad (14a)$$

No data have been published concerning the contribution of radioactive heat to total heat flux in the North Sea, but data from UK onshore, Baltic Shield and Hercynian of France show that the heat flow at the base of the crust comprises between 38% and 61% of the surface heat flow (Ungerer *et al.*, 1990), the difference being introduced by radioactive decay within the crust. Most radioactivity occurs in the upper crust and hence a concept of 'reduced heat flow' as the heat flow at the base of the upper crust (~10 km) has been introduced (Sass *et al.*, 1989). In a worldwide review, the latter authors report that between 13% and 67% of the surface heat flow derives from radioactivity within the upper crust. Since crustal heat flow derives from radioactivity and all radioactive species have a half-life (e.g. ^{238}U of 4.5×10^9a), the basement heat flow (Q_{bas} mW/m^2) will be a function of the age of the basement rocks (T). Using a worldwide heat-flow database available on the Internet, Jessop (1992) has established the relationship:

$$Q_{bas} = [-0.63 \times \sqrt{(T)}] + 73.3 \quad (14b)$$

This predicts that the North Sea graben area, with a basement age of about 1.1×10^9 years, has a basement heat flow of ~55 mW/m^2, not far from measured values (Burley, 1993).

Bücker and Rybach (1996) have pointed out that radioactive heating (μW/m^3) can be derived from the natural gamma wireline-log response (GR) in API units:

$$\text{Radioactive heat} = (0.0158 \times \text{GR}) - 0.8 \quad (15a)$$

The radioactivity of the Tertiary sediments of the central North Sea have been documented by Berstad and Dypvik (1982).

The temporal variation of heat flow may be calculated from the crustal-stretching and thinning history of an area (Steckler, 1985, 1988; Buck *et al.*, 1988). For example, application of the McKenzie model—a model initially validated using North Sea data (McKenzie, 1978, 1981; Barton, 1984; White and Latin, 1993)—produces the following essentially linear relationship between heat flow (mW/m^2) and McKenzie's β-factor for the Central Graben of the North Sea:

$$\text{Heat flow (mW/m}^2) = (38 \times \beta\text{-factor} + 7) \quad (15b)$$

This equation relates the maximum heat flow (in this case reached at 140 Ma) after stretching over the period 156–140 Ma, for β-factors in the range 1–3, with the β-factor defined as initial crustal thickness/final crustal thickness. Based on subsidence analysis of 41 wells penetrating to the Triassic, β-factors up to 2.0 are derived for the 'triple junction' joining the Witch Ground, Central and Viking grabens (White and Latin, 1993). This must be seen in the

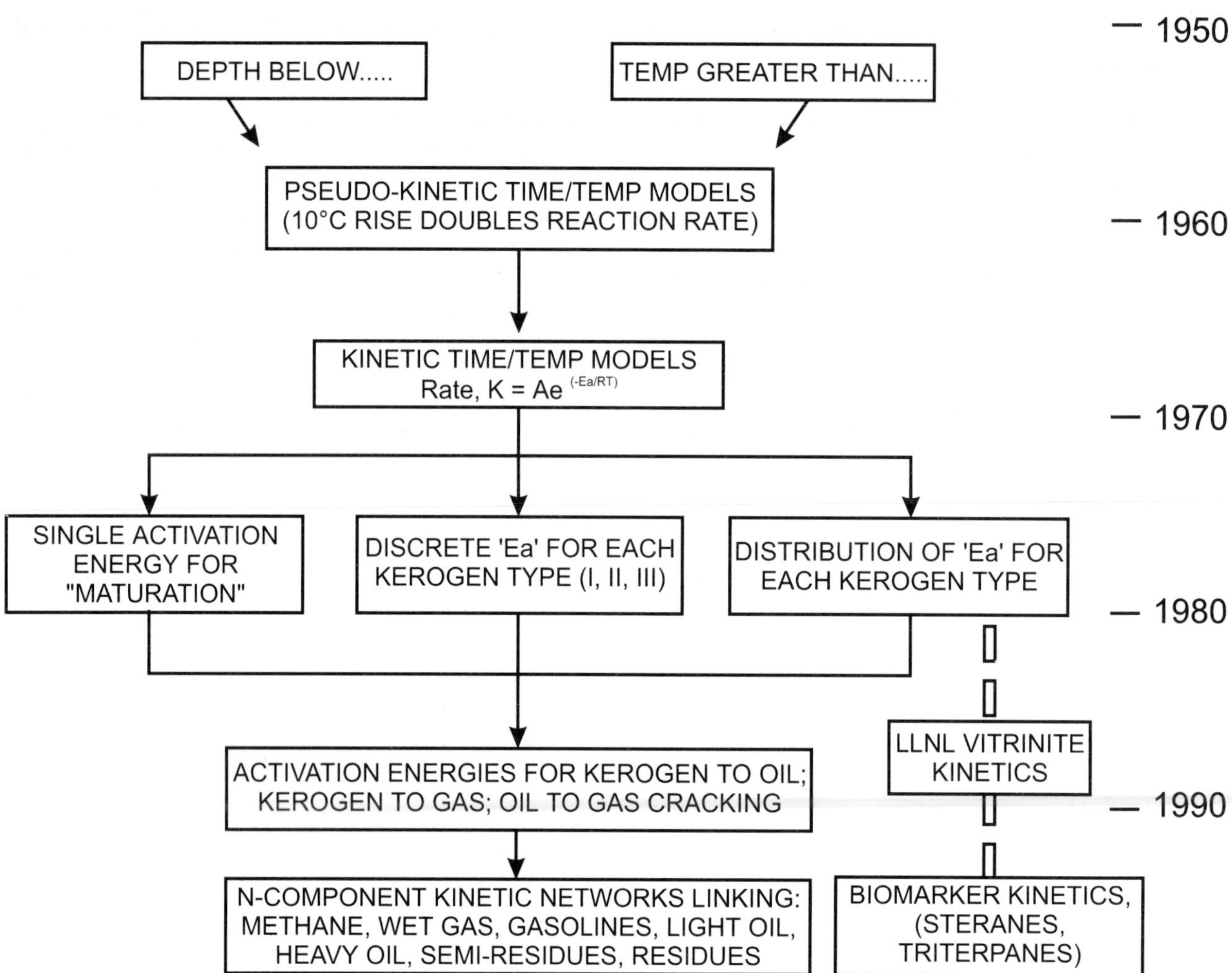

Fig. 11.21 Historical view of the development of maturity modelling concepts, 1950s to 1990s, many of which have been developed and validated using North Sea examples. E_a, activation energy.

context of the earlier thermal updoming event (Underhill and Partington, 1993). In the Central Graben, the effect of Late Jurassic crustal stretching, with β-factor of 1.8 (Barton, 1984), is modelled as having largely dissipated by the Upper Cretaceous, as shown in Fig. 11.20 (lower). Since the apparent β-factor will decrease away from the graben centre (Barton, 1984; White and Latin, 1993), so will the maximum post-stretching heat flow. Eggen (1984) discusses a situation in the present-day northern Viking Graben where the heat flow is higher on the flanks, possibly as a result of fluid (plus associated heat) flux from the basin centre.

An estimate of the location of the model with respect to the stretching centre is thus important. The local β-factor can be calculated by deconvoluting the burial-history curve into contributions from sediment loading and crustal thinning. This approach is discussed by Allen and Allen (1990, p. 275), in the context of the Witch Ground Graben well UK15/30-1, and Lepercq and Gaulier (1996), for the Viking Graben.

In the North Sea graben system in general, post-Cretaceous burial is substantial, which generally renders insignificant the effect of any late Jurassic heat-flow pulse on present-day maturities. The same can be said of the burial and hence thermal effects of syn-rift uplift and erosion (Thomas and Coward, 1995). Thus the North Sea is a poor area to confirm or deny the McKenzie model using standard organic calibrants. In the deeper graben areas, an elevated palaeoheat flow may effect the timing of onset of generation where this occurred during the Cretaceous.

Putting together the concepts of equations 14 and 15, it follows that, with more than 50% of the heat flow deriving from radioactivity within the crust (cf. from the mantle), thinning the crust may decrease the surface heat flow, while tectonic thickening may increase the surface heat flow—the opposite of that predicted by the McKenzie model. Particularly where thin-skin tectonics has occurred, this may explain the absence of evidence for a palaeoheat pulse where evidence for crustal thinning is present.

Time/temperature models

Subsurface temperatures provide the major but not total control of maturity and generation. The time that a source rock has been exposed to a certain temperature has a strong effect on some maturity parameters (Waples, 1984; Guidish *et al.*, 1985; Piggot, 1985; Burnham and Sweeney, 1991: Pepper and Corvi, 1995; but see Landais *et al.*, 1994).

For a given temperature, the reflectance of vitrinite is strongly time-dependent (Hood *et al.*, 1975; Waples, 1980; Gretener and Curtis, 1982; Burnham and Sweeney, 1991), spore colour is less dependent (Barnard *et al.*, 1981a), while some common biomarker isomer ratios appear to be minimally effected (Mackenzie and McKenzie, 1983; Cornford *et al.*, 1983; Radke *et al.*, 1997). The effect of time on the actual volumetric generation of oil itself, modelled as the transformation ratio of kerogen, is significant (Nielsen, 1995; Pepper and Corvi, 1995), although this concept has not gone unchallenged (Price, 1983).

There are a number of ways to calculate maturity based on geological time (the effect of which is essentially linear) and temperature (the effect of which is approximately exponential). Figure 11.21 shows the historical development of the time/temperature models, from a simple concept of source-rock maturation being related to depth of burial to the modern concepts of kinetic networks. The application of these models to the North Sea is discussed below.

The 'effective heating-time' method of Hood *et al.* (1975) requires the calculation of the effective heating time (T_{eff}), which is the time (in millions of years) that the stratum has spent within 15°C (27°F) of its maximum subsurface temperature (Fig. 11.22A). The maturity of the stratum is then obtained from the nomogram reproduced in Fig. 11.22B in terms of the vitrinite reflectance diagonals, given a knowledge of the T_{eff} and maximum temperature values. The plot shown in the Fig. 11.22A is called a thermal geohistory or burial-history plot, as discussed by Thorne and Watts (1989) and Kooi *et al.* (1989) in the context of the North Sea. This type of display has proved of

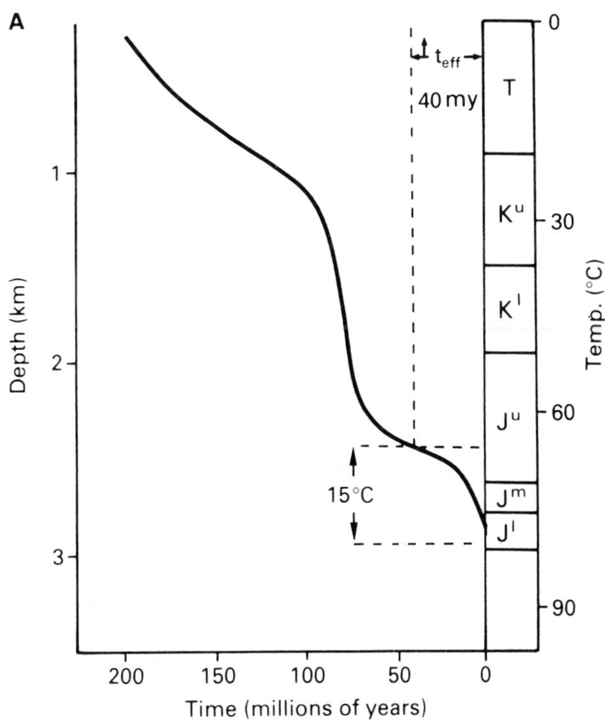

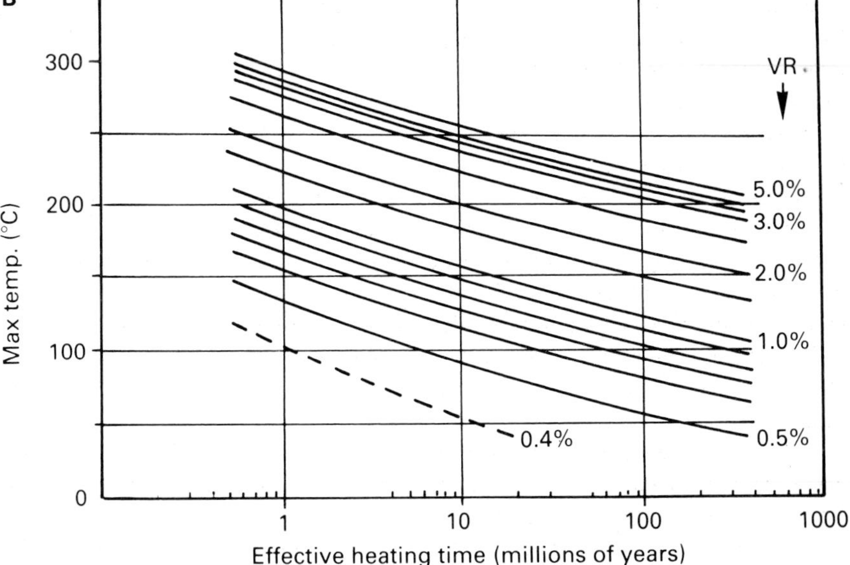

Fig. 11.22 Calculating maturity from the maximum thermal event using Hood *et al.*'s effective heating time (T_{eff}) calculation (Hood *et al.*, 1975). The approach considers the duration (T_{eff}, Ma) of burial leading to the last 15°C rise in temperature (A), and converts it to vitrinite reflectance using a cross-plot against the maximum temperature (B).

Table 11.4 TTI boundaries generally applicable in the North Sea.

TTI	% R_o	Generation stage
6	0.5	Onset early mature (oil)
50	0.7	Onset midmature (oil)
450	1.0	Onset late mature (oil)
2 300	1.3	End late mature (oil)
32 900	2.0	Peak gas generation
404 000	3.0	End gas generation
1 000 000	4.0	Greenschist, etc.

great utility in understanding the thermal history and hence maturity of sediment sections.

A somewhat more rigorous treatment of the complete thermal history of a single stratum can be carried out, using the Lopatin/Waples method to calculate its time–temperature integral (TTI) value (Waples, 1980; Issler, 1984) or using a kinetic model to estimate the rate of change of kerogen to oil and gas (Tissot and Espitalié 1975; Burnham and Sweeney, 1991). A comparison of these two methods has been published by Wood (1988) and Burnham and Sweeney (1991) in terms of worldwide applications, by Ritter (1988) in the context of the Egersund Basin and Throndsen *et al.* (1993) for the northern North Sea. The TTI boundaries in Table 11.4 have been found to be generally applicable in the North Sea.

The TTI-model approach assumes that the rate of the 'maturation/generation' reaction, k, doubles every 10°C rise in temperature. The simplest reaction can be written:

$$\text{Kerogen} \xrightarrow{k} \text{hydrocarbon} + \text{residue} \qquad (16)$$

To implement the model, divide the temperature history of any given stratum into 10°C intervals, and multiply the time spent in each interval by a rate factor that doubles for each 10°C rise in temperature. The interval 100–110°C is conventionally taken as having a rate factor of unity (Table 11.5). Interval TTI values are summed to produce a running total. This can be expressed mathematically as:

$$\text{TTI} = \Sigma\,(t_i)(R_i) \qquad (17)$$

where t_i is the time in millions of years spent in the ith temperature interval, and R_i is the rate factor for that interval. Values of TTI have been calibrated mainly with vitrinite reflectance (Table 11.4). The relationship between TTI and vitrinite reflectance (% R_o) has been expressed as:

Log % R_o = 0.173 log TTI – 0.349 (Royden *et al.*, 1980)
Log % R_o = 0.162 log TTI – 0.429 (Issler, 1984—east coast, Canada)
Log % R_o = 0.208 log TTI – 0.472 (Goff, 1984—North Sea)
Log % R_o = 0.178 log TTI – 0.429 (Maragna *et al.*, 1985—North Sea)

The latter two report data from the North Sea Viking Graben. The Issler (1984) relationship, developed off Canada's east coast, appears to be most generally applicable worldwide, but it must be emphasized that all models should be calibrated with local data as far as possible.

Experience shows that the differences between the equations stem mainly from the use of different temperature databases (e.g. well-corrected, poorly corrected and uncorrected wire line log, drill-stem tests or flowing-reservoir temperatures) and measuring errors in the vitrinite reflectance values. Secondary errors may be introduced by the use of different time-scales (e.g. Harland *et al.*, 1990; Berggren *et al.*, 1995) and uncertainties with stratigraphy in general (e.g. unconformities).

The numbers in Table 11.4 refer to standard stratigraphy, conventionally corrected wireline-log temperatures (Horner-plot correction) and routine service-company vitrinite reflectance data. The calculation shown in Table 11.5 can be interpreted in terms of these boundaries, with the modelled horizon passing the immature/early-mature boundary (TTI = 6) some 55 Ma ago when buried to 1750 m and the onset of the main oil window (TTI = 50) occurring between 15 Ma and 5 Ma ago when the horizon was 2500 m and 2750 m below the palaeosurface.

More sophisticated models of reactions such as shown in equation 16, can be based on a direct application of the Arrhenius equation relating reaction rate (k: Ma^{-1}) to time (t: s) and temperature (T: K), via the activation energy (E_a: kJ/mol or kcal/mol), the pre-exponential factor (A: s^{-1} or Ma^{-1}) and the universal gas constant (R = 0.0083454 kJ/mol.K or 0.00198717 kcal/mol.K):

$$k = A^{(-E_a/RT)} \qquad (18)$$

Table 11.5 Lopatin/Waples TTI calculation for a single horizon.

Depth (m)*	Temperature (°C)†	Time (Ma)‡	Duration (Ma)§	Rate factor‖	Interval TTI¶	Cumulative TTI**
0	10	156	12	1/512	0.02	0.02
250	20	144	24	1/256	0.09	0.12
500	30	120	12	1/128	0.09	0.21
750	40	108	11	1/64	0.17	0.38
1000	50	97	9	1/32	0.28	0.66
1250	60	88	22	1/16	1.44	2.10
1500	70	65	10	1/8	1.25	3.35
1750	80	55	12	1/4	3	6.35
2000	90	43	18	1/2	9	15.35
2250	100	25	10	1	10	25.35
2500	110	15	10	2	20	45.35
2750	120	5	5	4	20	65.35
3000	130	0	—	8	0	65.35

*Depth of horizon subsediment-surface at time t. †Geothermal gradient = 40°C/km + 10°C at surface. ‡Time formation enters each 10°C interval. §Time spent in each 10°C interval. ‖Rate factor set arbitrarily at 1 for the 100–110°C and doubling or halving for each higher and lower 10°C interval. ¶Interval TTI = duration × rate factor. **Cumulative (running) total of TTI interval values.

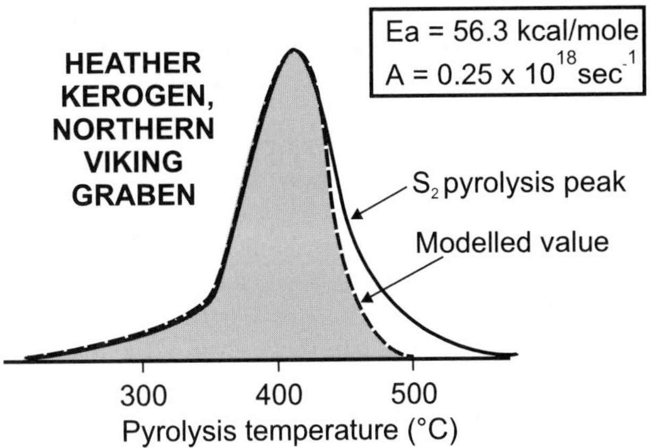

Fig. 11.23 Activation energy and frequency factor derived from an iterative fit of modelled curve to a measured S_2 pyrolysis peak for a Heather Formation sample, northern North Sea (after Maragna *et al.*, 1985). In most cases, the pyrolysis peak is generated at three heating rates (e.g. 1°C/min, 5°C/min and 25°C/min) to increase precision.

Assuming first-order kinetics, the reaction rate, k, is defined in terms of the rate of loss of the reactant (x):

$$dx/dt = -kx \qquad (19)$$

Solving the Arrhenius equation for k yields a number in the range 0–1, indicating the extent of conversion of reactant (kerogen) to product (oil). This is the Transformation Ratio. In terms of kerogen chemistry, the activation energy can be thought of as the strength of the carbon–carbon bonds joining 'oily side-chains' to the kerogen, and the frequency factor as the rate of bond breakage (more correctly the rate of production of moieties in the activated state) per million years or per second.

With modern computer-linked pyrolysis apparatus it is possible to derive a single 'best-fit' pseudo-activation energy and pre-exponential factor for a single sample from the shape of a pyrolysis S_2 peak (Fig. 11.23), greater precision being achieved by pyrolysing at different heating rates (e.g. 0.1, 0.7 and 5.0°C/min; Schenk *et al.*, 1997).

Activation energies are expressed as single values (e.g. Fig. 11.23), as a mean ± standard deviation (e.g. Pepper and Corvi, 1995) or as explicit irregular distributions (e.g. Espitalié *et al.*, 1988; Schaefer *et al.*, 1990; Burnham and Sweeney, 1991; Schenk *et al.*, 1997). Some kinetic parameters have been derived from the more laborious stepwise hydrous pyrolysis methods (Lafargue *et al.*, 1988), where the activation energies for generating multiple reaction products (e.g. methane, wet gas, gasolines, light oil, heavy oil) can be determined independently. Criticism of these methods (Snowdon, 1979; Schenk *et al.*, 1997) stems largely from the differences in temperature and duration (time) between laboratory and natural systems (see Fig. 11.24).

The Arrhenius equation can be most simply applied to predict any maturity relationship where a single precursor is converted to a single product plus residue. As discussed in considerable detail by Nielsen (1995), the imposingly complex Arrhenius equation rather disappointingly predicts a transformation ratio (TR) as a number on the scale zero to one:

$$\text{TR} = \text{mass of product}/\,(\text{mass of residue} + \text{mass of product}) \qquad (20)$$

Kerogen kinetics are generally used to construct a curve of the generation of fluid hydrocarbons from solid kerogen (Barth *et al.*, 1989; Ungerer, 1993; Nielsen, 1995), while, applied to molecular ratios comprising a precursor and product, molecular maturation parameters can be derived (Ungerer and Pelet, 1987; Radke *et al.*, 1997). In the former case, calibration of the model is performed by attempting to simulate a generation trend (e.g. compare Fig. 11.24 with Fig. 11.12). To produce the latter curve, the transformation ratio is multiplied by the hydrocarbon potential of the kerogen (e.g. the pyrolysis S_2 yield expressed in kilogram hydrocarbon/tonne rock). In a discussion of the sensitivities of model input parameters in terms of the predictions of transformation ratios in the area of Norwegian Block 9/2, Hermanrud *et al.* (1990b) identified activation energies, temperature, seismic time–depth conversion and

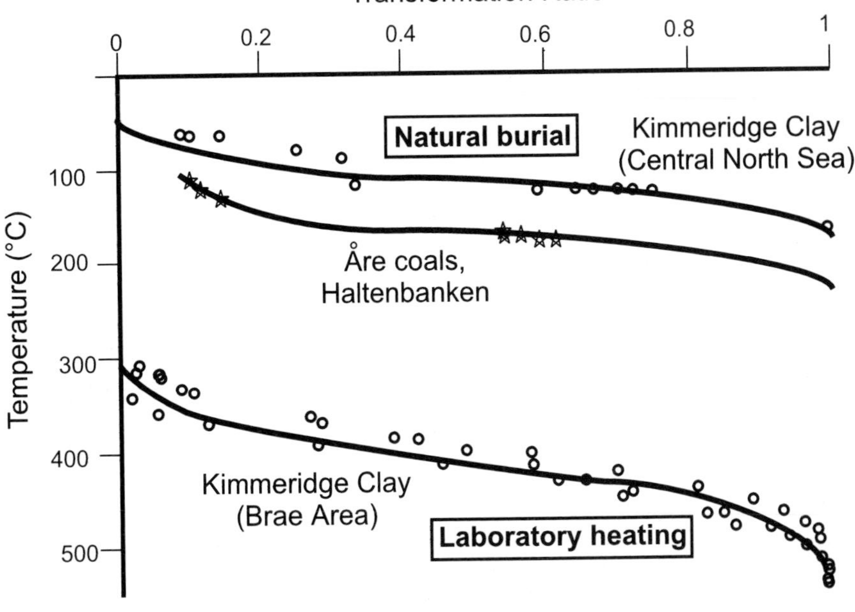

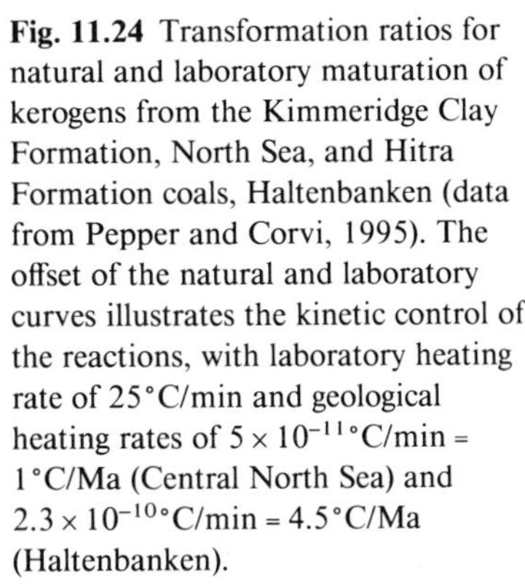

Fig. 11.24 Transformation ratios for natural and laboratory maturation of kerogens from the Kimmeridge Clay Formation, North Sea, and Hitra Formation coals, Haltenbanken (data from Pepper and Corvi, 1995). The offset of the natural and laboratory curves illustrates the kinetic control of the reactions, with laboratory heating rate of 25°C/min and geological heating rates of 5×10^{-11}°C/min = 1°C/Ma (Central North Sea) and 2.3×10^{-10}°C/min = 4.5°C/Ma (Haltenbanken).

palaeoheat flow as the major controls on uncertainty.

It is difficult to calibrate a generation model using natural samples, since, in nature, maturing source rocks cannot be treated as closed systems. Determining the *in situ* hydrocarbon in a sample is a balance between generation and expulsion. For this reason, kinetic generation models are generally calibrated against maturity parameters, such as vitrinite reflectance or Rock Eval T_{max}. As a basis for calibration, a kinetic method of calculating vitrinite reflectance (% R_o) has been developed (Sweeney and Burnham, 1990) and subsequently adopted widely by the oil industry. This method—commonly called the LLNL (Lawrence Livermore National Laboratory) or EasyR_o method—covers the reflectance range 0.2–4.65% R_o and is reviewed by Burnham and Sweeney (1991).

The LLNL method uses the well-constrained and globally applicable reflectance versus H/C_{atomic} plot for vitrinites to convert between the physical (% R_o) and the chemical properties (H/C_{atomic} and O/C_{atomic}) of the maceral:

$$\% R_o = 12 \exp^{[(-3.3(H/C)] - O/C} \quad (21a)$$

Changes in H/C_{atomic} are then modelled by considering four reactions with carbon dioxide (decarboxylation), water (dehydration), gas (methane generation) and oil (oil generation) as the end-products. After these reactions have 'run' for a given time at a given temperature, the new H/C of the vitrinite is calculated and converted back to % R_o. In practice, instead of four parallel reactions, a single activation-energy distribution is used to give a transformation ratio (TR), which is converted to reflectance, using the equation:

$$\% R_o = \exp^{(3.145 \times TR) - 1.6} \quad (21b)$$

A modified polynomial form of this equation is given by Throndsen *et al.* (1993) based on North Sea calibrations.

The LLNL approach has been compared with Lopatin's TTI and other models by Throndsen *et al.* (1993), and extended to higher maturity levels (< 6.23% R_o) by Everlien (1996). In addition, Larter (1989) has proposed a chemical model for vitrinite-reflectance prediction, based on the kinetics of phenol generation. The phenol precursors are monitored and are linearly related to % R_o in the range 0.45% R_o to 1.6% R_o for predictions covering the oil window.

Returning to kerogens in general, each kerogen type contains a variety of chemical bonds (Tissot *et al.*, 1987), and hence is best represented by a range of activation energies. For practical application, two schools of thought exist: to take generalized kerogens derived from a mix of Types I, II and III (Burnham and Sweeney, 1991, and Fig. 11.25A), versus measurements on the specific kerogens from the named source rock (e.g. Tegelaar and Noble, 1994). Global catalogues of activation energies and pre-exponential factors for kerogens from named forrmations—including North Sea examples—have been assembled by Waples (1984) and Tegelaar and Noble (1994). Different activation-energy values may be used for different kerogen components, and the more sophisticated models (e.g. Tissot and Espitalié, 1975; Tissot *et al.*, 1987; Ungerer and Pelet, 1987; Mackenzie and Quigley, 1988; Ritter, 1988; Burnham and Sweeney, 1991; Pepper and Corvi, 1995; Schenk *et al.*, 1997) use a continuous or discrete statistical spread of kinetic values for each kerogen type present in the rock (Fig. 11.25). Given the chemical heterogeneity of kerogen it is not surprising that activation energies and frequency factors are better described by distributions.

The cracking of oil to gas within the reservoir can also be modelled, using kinetic parameters (Ungerer *et al.*, 1988; Andresen *et al.*, 1993; Pepper and Dodd, 1995). Given the relative chemical homogeneity of oil, the cracking of oil to gas may be treated with a single activation energy and frequency factor (Fig. 11.25A). In a detailed discussion of intra-reservoir cracking, Horsfield *et al.* (1992) fitted a distribution of higher activation energies with a mode at 66 kcal/mol, compensated by a higher frequency factor of 3.52×10^{29} Ma for this reaction. A similar range of values was reported for a standard North Sea oil by Andresen *et al.* (1993). In a review of this subject, Schenk *et al.* (1997) found no conclusive evidence for a pressure effect on the kinetics of the cracking reaction.

From a literature review, Kuo and Michael (1994) produced the following relationship between the Arrhenius activation energies (E_a; kcal/mol) and frequency factors (A; /s) for the cracking of oil to gas:

$$\text{Log}(A) = 0.3E_a - 3.048 \quad (22a)$$

This log-linear relationship, with a correlation coefficient of 0.964, illustrates that the same reaction threshold can be produced from many combinations of activation energies and frequency factors. This is only strictly true at a constant heating rate (°C/Ma). Note the similarity with the kerogen relationship shown in Fig. 11.25B.

For no good reason, activation energy is generally singled out in the geological literature as the main controlling factor for kerogen maturation, with typical values in the 50–55 kcal/mol (200–230 kJ/mol) range. There is no theoretical basis for the normal practice of using a distribution of activation energies and a single value for frequency factor; Issler and Snowdon (1990) first figured distributions of both E_a and A for modelling Canadian kerogens, while Skjervøy and Sylta (1993) use distributions of both activation energies and frequency factors to model the generative kerogen components of the Draupne (Kimmeridge Clay), Heather and Dunlin Formations of the Viking Graben. Combinations of constant A and distributed E_a and constant E_a and distributed A, which give equivalent kerogen transformations, are listed by Nielsen (1995).

Using a data set including North Sea samples, Pepper and Corvi (1995) report a linear relationship between activation energy (E_a, kJ/mol) and frequency factor (A, s^{-1}) for five different kerogen types (organofacies), namely:

$$E_a = [15 \times (\log A)] + 10 \quad (22b)$$

Examples of activation energies and frequency factors reported for the North Sea area are listed in Table 11.6. At one extreme, apparent activation energies for the laboratory generation of bitumen from the Kimmeridge Clay have been reported as 17.0 kcal/mol (71.5 kJ/mol) (Barth *et al.*, 1989), but such low values suggest that the rate-limiting process is a physical (e.g. diffusion of the reaction products) rather than a chemical process, such as carbon–carbon bond cleavage.

It can be seen from Table 11.6 that discrepancies be-

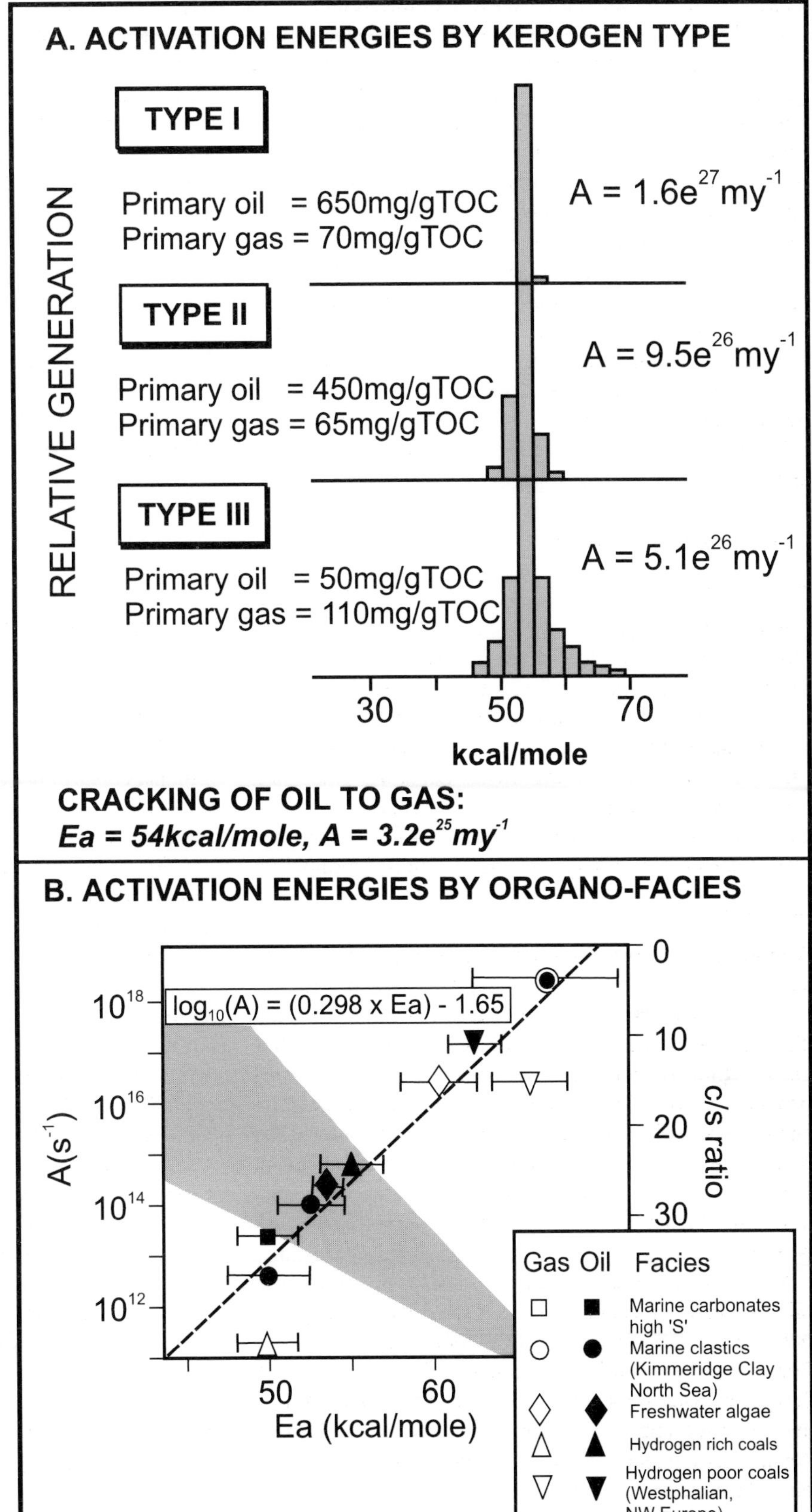

Fig. 11.25 Activation-energy (E_a) distributions for (A) kerogen Types I, II and III and oil-to-gas cracking (Platte River's BasinMod[R] 1D modelling software: Burnham and Sweeney, 1991), and (B) for organofacies associations at least in part controlled by sulphur content (data from Pepper and Corvi, 1995). Sulphur–carbon bonds are weak, so sulphur-rich kerogens (low C/S ratios) correlate with low activation ratios (general shaded trend). See Espitalié *et al.* (1988) for a comparison of E_a and frequency factor (A) values for Brent coals and Kimmeridge Clay kerogens. NB: 1 kcal/mol = 4.2 kJ/mol; oil and gas yields in mg/g TOC; 1 Ma = 31.5×10^{12} secs.

tween the various authors derive in part from compensating for a high activation energy with a high frequency factor (i.e. strong bonds but a high rate of bond breakage). Within a restricted range of heating rate (°C/Ma) a number of combinations of E_a and A can thus produce the same transformation ratio and hence prediction of volumes of hydrocarbon generated.

As can be seen in Fig. 11.25B, the kerogens run from sulphur-rich marine via normal marine (Kimmeridge Clay), lacustrine, hydrogen-rich (Tertiary) coal to humic (Westphalian) coal with increasing values. The authors related this trend in kinetics partially to the abundance of weaker carbon–sulphur bonds in the sulphur-rich Type IIS kerogens. This was discussed in detail by Tegelaar and

Table 11.6 Activation energies and frequency factors for North Sea kerogens.

Kerogen type	Frequency factor (s^{-1})*	Activation energy (kcal/mol)	Reference	Comments
Kimmeridge Clay Formation	8.14×10^{13}	51.2 ± 1.98	Pepper and Corvi, 1995	Liquids
	2.17×10^{18}	66.4 ± 4.38		Gas
	Distributed values		Skjervøy and Sylta, 1993	Draupne, Heather and Dunlin
	1.56×10^{14}	53–> 55	Tegelaar and Noble, 1994	Log distribution
	—	49 ± 1.2	Ritter, 1988†	Tau Formation, Egersund Basin
	5.20×10^{16}	56 (liquid)	Espitalié *et al.*, 1988	Only mode reported here‡
Brent Coals	5.46×10^{16}	62 (gas)	Espitalié *et al.*, 1988	
	Distributed values		Skjervøy and Sylta, 1993	Coals and shales
Toarcian/Posidonia	1.83×10^{14}	54 (mode)	Schaefer *et al.*, 1990	Saxony, Germany
	Distributed values		Burnam, 1990	Germany
Westphalian Coals	1.23×10^{17}	61.7 ± 1.57	Pepper and Corvi, 1995	Liquids
	1.93×10^{16}	65.5 ± 2.36		Gas
Alum Shale	6.37×10^{18}	47.9	Lewan and Buchardt, 1989	Cambrian, Sweden

*1 Ma = 31.5×10^{12} s. †Based on mixes of labile and refractory kerogens (Quigley *et al.*, 1987). ‡Range of activation energies from 48 to 72 kcal/mol.

Noble (1994), using a worldwide data set; they used measured kinetic distributions for high-, medium- and low-sulphur kerogens to show peak conversion to oil at ~0.7% R_o and ~115°C for high-sulphur kerogens to ~1.1% R_o and ~150°C for low-sulphur kerogens.

Kinetics applied to molecular calibrations of aromatization and isomerization reactions are derived from an Arrhenius plot of log(k) versus the reciprocal of absolute temperature (Fig. 11.26). A straight-line relationship demonstrates the first-order behaviour of these reactions, allowing the calibration of thermotectonic models (Mackenzie and McKenzie, 1983; Marzi *et al.*, 1990). The establishment of these molecular relationships had the added utility of allowing the prediction of oil/source-rock correlations on the basis of measured ratios in oils, compared with modelled off-structure maturity within the source-rock kitchen areas. Utilizing the different kinetics of vitrinite reflectance and biomarker reactions, Hermansen (1993) was able to discriminate between high (12.5°C/Ma) and low (0.8°C/Ma) heating rates for a deep North Sea model.

A number of inorganic reactions are now recognized as having geologically significant kinetic (as opposed to thermodynamic) controls. For example, Velde and Vasseur (1992) and Vasseur and Velde (1993) have applied the Arrhenius equation to the transition of smectite + K^+ to illite + Na^+ + quartz in the Paris Basin, with more complicated parallel reactions being proposed (Roaldset and Bjorøy, 1966; Wei *et al.*, 1996).

As discussed above, the use of the Arrhenius equation presumes a first-order reaction (i.e. the rate is proportional to the concentration of a single reactant) and assumes a simple precursor/product relationship. In reality, a number of inter-related reactions must be considered (Cooles *et al.*, 1986; Sweeney *et al.*, 1987; Espitalié *et al.*, 1988; Ungerer, 1993; Pepper and Corvi, 1995), as simulated experimentally for a sample of immature Kimmeridge Clay from Moray Firth well 14/30a-1 (Monin *et al.*, 1990). As shown in Fig. 11.27A–C, three situations must be considered (Braun and Burnham, 1992):

- Generation from a mixed kerogen (modelling separate kinetic calculations and summing the products).
- Generation of a number of products from a single kerogen (modelling parallel reactions and summing the created products).
- A network of generation and destruction with intermediate products (modelling separate reactions, keeping track of destruction as well as creation of products).

The Arrhenius equation has no strict basis in theory (Stannage, 1988), alhough Arrhenius's concept, developed in 1889 from the van't Hoff equation and the law of mass action, and can be derived from the ideal gas equations. Application of the Arrhenius equation should strictly be used over limited temperature and time ranges (Pepper and Corvi, 1995; Schenk *et al.*, 1997), avoiding extrapolation outside the areas of experimental verification (generally restricted to laboratory experiments taking hours or days rather than millions of years—see Figs 11.24 and 11.26). Thus, when well calibrated, kinetic modelling has no greater *a priori* validity than TTI (or other) modelling. Local calibration is of the essence. In short, all approaches are ways of simulating or mimicking, rather than explicitly modelling, the physical chemistry of natural geological processes.

Lopatin-style TTI and Arrhenius-equation kinetic models have been compared (Ritter, 1988; Wood, 1988; Throndsen *et al.*, 1993), and both have been applied to, and calibrated with, data from the North Sea (Table 11.7). Rather disappointingly, the net result of this modelling in

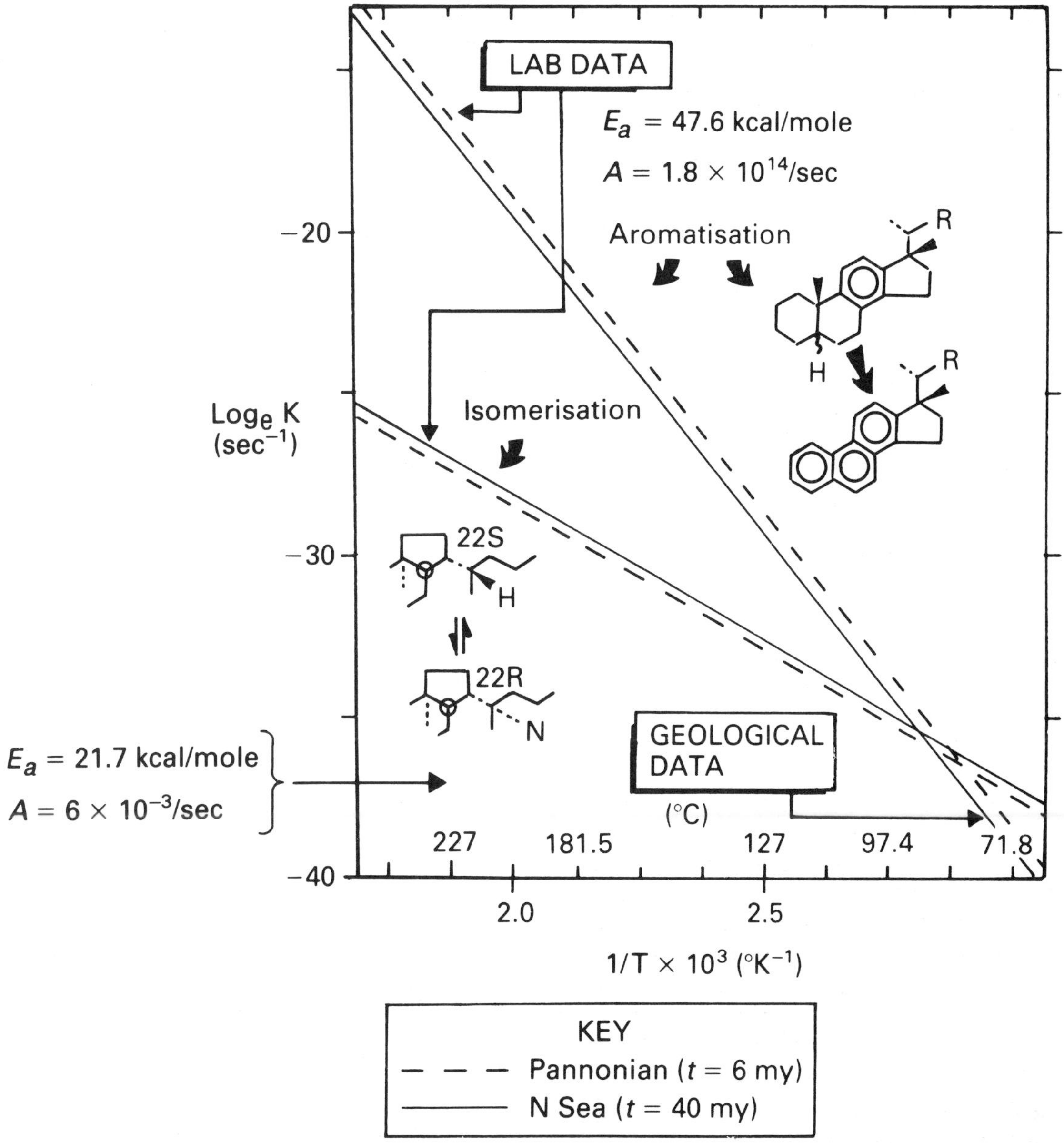

Fig. 11.26 Arrhenius plot applied to molecular kinetics for sterane isomerization (E_a = 21.7 kcal/mol, A = 6 × 10^{-3}/s) and aromatization (E_a = 47.6 kcal/mol, A = 1.8 × 10^{14}/s), showing similar trends for slow heating-rate North Sea (solid line) and high heating-rate Pannonian Basin (dashed line) data (modified after Mackenzie, 1984). A different sterane isomerization trend (E_a = 40.2 kcal/mol, A = 4.7 × 10^8/s) has been proposed by Marzi *et al.* (1990). The kinetic values may be used to calibrate North Sea Basin models and hence constrain thermal histories. E_a, activation energy; A, frequency factor.

the North Sea has been summed by Thorne and Watts (1989) in the following way:

> The apparent ability of simple and sophisticated models alike to match maturation data indicates the need for model calibration on a test data set, followed by rigorous testing of model predictions on data in other areas. Almost any model, simple or complicated, generally can be calibrated (made to fit) the maturation data at one or several wells. The most geologically correct model should best be able to predict the expected variation of maturation away from the well used for calibration.

A different approach to model sensitivities is illustrated by Nielsen in his discussion of the Monte Carlo simulation of generation for the Danish Trough well E-1 (Nielsen, 1993) and a more sophisticated probabilistic treatment of an area of the Norwegian–Danish basin (Nielsen, 1995). In this approach, each of the input variables is attributed a range of values (e.g. mean ± standard deviation), and the outputs are expressed as a series of probabilistic curves for the kerogen Transformation Ratio as a function of time.

Finally, source-rock studies have now moved into the area of 3D modelling of the basic physical and chemical processes (Welte and Yukler, 1981; Ungerer *et al.*, 1984; Yukler and Kokesh, 1984; Welte and Yalcin, 1988; Hermans *et al.*, 1992; Welte *et al.*, 1997). These explicit large-scale computer models are becoming able to predict where and how much oil is generated by thermal modelling, and then where it flows by pressure modelling, coupled with simulating the porosity/permeability characteristics of the

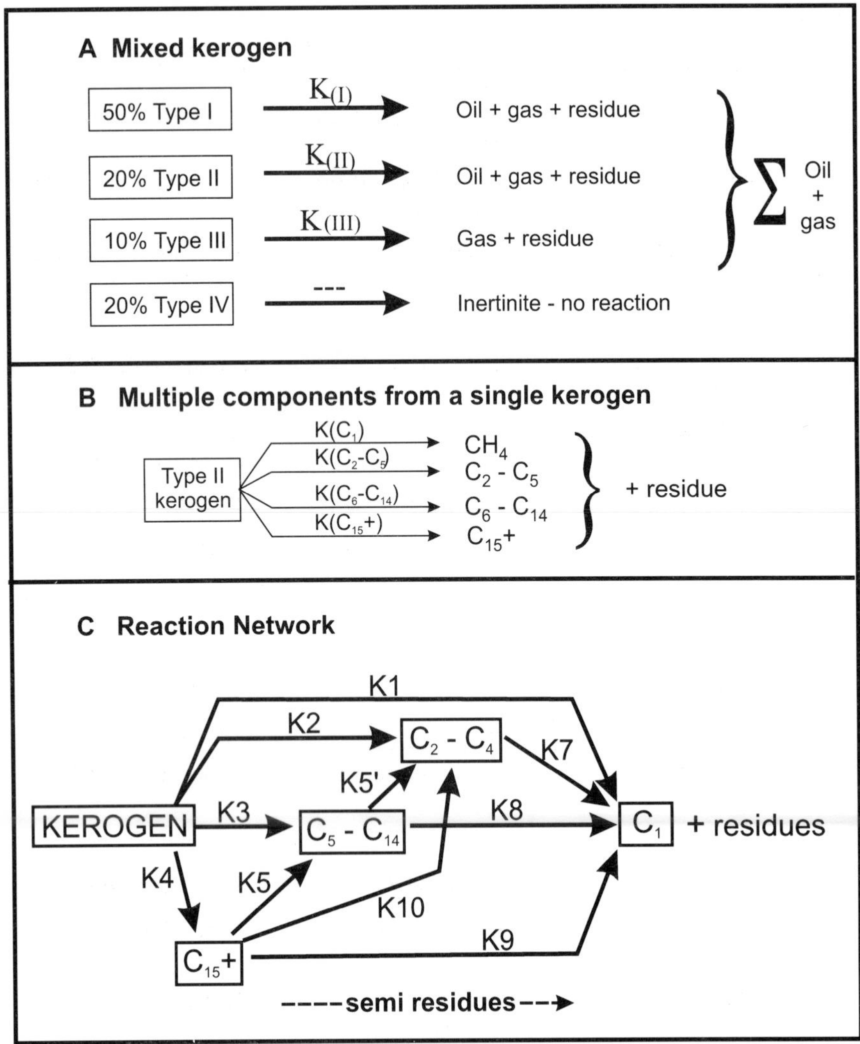

Fig. 11.27 Multicomponent kerogen kinetics: (A) kerogen mixes modelled as separate reactions, the products of which are summed, (B) multiple components modelled from a single kerogen type using discrete parallel reactions (e.g. Espitalié *et al.*, 1988, for North Sea examples) and (C) a reaction network where kerogen produces a number of intermediate liquid phases, which eventually crack to methane (e.g. Lawrence Livermore kerogens in Platte River's BasinMod^R). In all cases, each step has a reaction rate (k), which may be modelled using the Arrhenius equation with single or a distribution of activation energies (E_a) and frequency factors (A).

basin. They do, however, require a large amount of input information and their further development is probably limited as much by the presence of populated databases as by producing appropriate algorithms. Such large quantitative models force the explorationist to view oil generation and accumulation as a unified process, and have greatly benefited our understanding by focusing attention on a number of poorly understood but critical areas of hydrocarbon genesis.

It can be concluded that modelling is most at home with words like 'compatible', 'consistent', 'predicted' and 'calculated', rather than 'truth', 'certainty' and 'correct', and the words 'calibration' and 'sensitivity' should definitely appear. As can be seen from Table 11.7, calibrants range from vitrinite reflectance to molecular ratios and generation curves. For the oil industry, calibration with generation trends, such as derived from the decrease in Rock Eval Hydrogen Index (Fig. 11.13, upper), may be more appropriate for most basinal or prospect evaluations than calibration with indirect measures of generation, such as vitrinite reflectance, which is arguably more appropriate for constraining purely geological models.

The conclusion must be that a simple, well-calibrated 1D maturity model, applied consistently, is the most practical solution for most exploration-related problems. With consistent use, modelling allows the rating of relative rather than absolute results, aiding management decisions on which structure to drill or which block to relinquish.

11.5 Generation of hydrocarbons

The above discussion of source rock characterization and maturity modelling has largely avoided reference to oil and gas generation. In this section, generation (as opposed to maturation) is discussed in the context of North Sea geology.

11.5.1 Oil and gas windows

The 'oil window' is a concept that hides confusion in its apparent simplicity; the 'gas window' is even less well defined. Miles (1989) defines the oil window as 'the maturity zone over which oils are generated from source rocks'. However, it is important for commercial application of this concept to differentiate between the zone of generation and the zone of expulsion, since only upon expulsion from the source rock can commercial accumulations of hydrocarbons develop. Thus the concept of an 'expulsion window' has been advanced.

Within the North Sea, the difference between the generation and expulsion windows is illustrated in Fig. 11.13, where the decrease in Hydrogen Index with depth initiates

Table 11.7 Maturity modelling in North Sea and adacent areas.

Area	Model	Calibrant	Comment	Reference
NOCS 33/6	TTI and kinetic	% R_o and pyrolysis	Tissot and Espitalié model	Maragna *et al.*, 1985
North Sea 56–61°N	TTI	Oil gravity	Used Waples (1980) calibration	Thorn and Watts, 1989
Egersund Basin	TTI and kinetic	T_{max} and biomarkers	Compared kinetics	Ritter, 1988; Hermanrud *et al.*, 1990b
Viking Graben	Modified TTI	% R_o	% R_o	Goff, 1983, 1984
Viking Graben	Kinetics and migration	Pressure and field distribution	2D modelling	Ungerer *et al.*, 1988
North Sea 56–62°N	Kinetic	Biomarker isomerization and aromatization	–	McKenzie and Mackenzie, 1983
Norwegian North Sea	TTI	% R_o	Calibrated to generation	Baird, 1986
Northern North Sea and Haltenbanken	TTI	% R_o	Uses interval velocities	Leadholm *et al.*, 1985
UK onshore and Southern Gas Basin	LLNL kinetics	% R_o and Apatite Fission Tracks	Estimates of uplift	Bray *et al.*, 1992
Danish North Sea	Pseudo-kinetic	% R_o	–	Jensen *et al.*, 1985
Danish sub-basin	TTI	% R_o	–	Sorensen, 1985
Danish Trough		Transformation ratio curves	Risked (probabilistic) modelling	Nielsen, 1993, 1995
North Sea	Kinetic	% R_o	–	Lepoutre, 1984
North Sea	Kinetic	Pyrolysis curves	General review	Tissot *et al.*, 1987
North Sea, Egersund	TTI and kinetics	T_{max}, steranes and triterpanes	Generation as output	Ritter, 1988
Haltenbanken	Kinetics and buoyancy	Oil distribution	Generation and migration model	Hermans *et al.*, 1989

at ~9000 ft, indicating the generation reaction. The Production Index, however, does not start to increase until ~10 500 ft, suggesting that the reduction of the S_2 peak is not balanced by a corresponding increase in the S_1 peak. Logically, there must be matter lost from the S_2 pyrolysate that is not 'seen' by the pyrolyser as mobile hydrocarbon (S_1). This loss reflects expulsion. The expulsion window develops deeper in the section, where generation has built up sufficient hydrocarbon volume and pressure to initiate primary migration. In the North Sea, as in most other basins, the interval from the top of the generation window to the top of the expulsion window approximates to the early mature zone (0.5–0.7% R_o). With well-drained source rocks (e.g. with fractures, or silt and sand interbeds), the top of the generation and expulsion windows can occur at almost the same shallow depth.

A feature of the early mature oil window is the initial decarboxylation of organic matter to generate CO_2 and organic acids. This acidic mix can produce secondary porosity in carbonates and accelerated dissolution of feldspars (Manning *et al.*, 1994). Purvis (1994) has associated albite (but not K-feldspar) dissolution in the Triassic reservoir of the Gannet oilfield, with an early influx of organic acids emanating from the maturing Kimmeridge Clay source rock.

The 'gas window' is more diffuse, since both its onset and base are more gradational. Gas is generated and can escape at almost all stages of burial up to greenschist facies metamorphism, and all oil is generated with some gas. There are two major gas-generating zones—the shallow biogenic zone (Vially, 1989) and the deeper thermogenic zone (see Fig. 11.12). Gas generation from coal seems to be a two-step process, controlled essentially by expulsion: early generation of a heavy asphaltenic–resin phase is not expelled and, upon further burial, is cracked down to gas, which escapes (Monin *et al.*, 1990). Using dry pyrolysis in the laboratory, Andresen *et al.* (1991) have demonstrated this progressive generation by showing that the isotope signature of methane generated from Heather, Brent and Draupne source rocks gets progressively isotopically lighter with oil-proneness.

The base of the thermogenic gas zone will be controlled by the stability of methane. In a reservoir with unreactive mineralogy, methane is stable at temperatures in excess of 200°C, but, in the presence of reactive minerals, destruction may occur at much lower temperatures. For example, in the presence of anhydrite (calcium sulphate), methane can react to produce hydrogen sulphide and carbon dioxide at temperatures in excess of ~130°C via thermochemical sulphate reduction (TSR; Rooney, 1995). This proceeds via two sequential steps to give equation 23 as the net reaction:

$$2SO_4^{2-} + 4H^+ = 2H_2S + 4O_2$$
$$2CH_4 + 4O_2 = 4H_2O + 2CO_2$$

$$2CH_4 + 2SO_4^{2-} + 4H^+ = 2H_2S + 4H_2O + 2CO_2 \quad (23)$$

While this is a common source of sour gas in deeper reservoirs, a similar set of reactions can occur due to bacterial sulphate reduction (BSR) at shallower depths and lower temperatures (< 60°C), as discussed for the UK onshore Wytch Farm Field by Aplin and Coleman (1995).

Late in the gas window, Type III kerogen starts to generate increasing amounts of nitrogen, with falling methane yields (Krooss *et al.*, 1993, 1995). If only late-stage expulsion products are trapped, then high nitrogen contents are to be expected.

11.5.2 Source-rock volumetrics and estimating the 'charge to trap'

For exploration in frontier areas it is probably sufficient to know that you have a source rock and that it is likely to be mature; major structures often being drilled even without the comfort of this information. For smaller or satellite structures, especially in mature exploration areas, a comprehensive prospect evaluation requires evidence that there is sufficient volume of mature source rock of an adequate quality to fill the proposed trap with oil or gas. In addition, it is required to be established that the oil or gas was generated at an appropriate time with respect to trap formation. Indeed, satellite structures adjacent to existing fields may be identified as lying on a migration path established using this approach. The 'charge to trap' approach is of great value as one aspect of rating prospects, that is, deciding which one to drill first or which acreage to bid for or to relinquish.

Attempting a mass balance between the amount of hydrocarbon generated in the source rock and that found 'in place' in the reservoirs (e.g. Fuller, 1975; Goff, 1983; Cousteau *et al.*, 1988; Hermanrud *et al.*, 1989; Cornford, 1994) has produced at least three major benefits: firstly, and probably most importantly, it has generated a dialogue and common purpose between reservoir engineer, explorationist, geophysicist, geochemist and stratigrapher, highlighting the interlinking of these disciplines; secondly, it constitutes an important part of quantitative basin models and prospect evaluation procedures now adopted by many companies (e.g. Welte and Yukler, 1981; Mackenzie and Quigley, 1988; Schmoker, 1994; Barker, 1996; Welte *et al.*, 1997); and, finally, it has placed important constraints on the logistics of the migration process, since it addresses both the supply and the accumulation of hydrocarbons in a quantitative way.

Applying the mass-balance approach to the northern Viking Graben, Schroeder and Sylta (1993) compared the in-place and predicted hydrocarbon volumes and bulk compositions for the Brage, Huldra, Oseberg and West Troll fields. In order to estimate the possible volumes of oil and gas in Norwegian Block 35/9, Sylta (1993) has calculated the charge that may have spilled eastward from the Troll Field. In a comparison of probabilistic, graphical and mass- balance methods of estimating the oil reserves of the East Shetland Basin, Cousteau *et al.* (1988) obtained good agreement between these approaches.

Such a calculation can be undertaken in reverse; for example, knowing the volumes and maturity of the oils of the Dorset Basin, Cornford (1988) identified the only possible source-rock and kitchen area as a defined area of Liassic shales in Bournemouth Bay. Using similar arguments, Hermanrud *et al.* (1989) applied detailed maturity modelling to downgrade the Norwegian 9/2-1 prospect, on the basis of an inadequate volume of mature source rock, a conclusion later proved correct by drilling. Thomsen *et al.* (1990) high-graded the eastern flank of the Danish Tail End Graben using a similar integrated basinal approach.

The simple mass balance (Fig. 11.28) can be stated as:

$$VR = (VG \times ME) - VL \quad (24)$$

where: VR = volume of oil in reservoir (barrels or m^3); VG = volume of oil generated in source rock; ME = overall migration efficiency; VL = volume leaked or lost from reservoir.

The actual calculation is considerably more complex than implied by equation 24, it being possible to break down each of these terms into its component parts.

11.5.3 Estimating volume generated in the source rock

Calculating the volume of oil and gas generated in the source rock (VG) demands a knowledge of a number of source rock properties:

- Quantity of organic matter (TOC, wt% or pyrolysis yield, kg/t).
- Type of organic matter (oil- or gas-prone, or inert).
- Thickness of source-rock unit.
- Areal extent and lateral variation of source facies.
- Regional maturity boundaries on the top/base source-rock surfaces.

The first two items, attributes of kerogen, have been discussed in previous sections. The concepts of source-rock volumetric estimation have been discussed with reference to the North Sea by Fuller (1975), Tissot and Welte (1984), Goff (1983, 1984), Larter (1988), Mackenzie and Quigley (1988) and Cornford (1994). For a simple approach, it consists of three steps:

1 Establishing the areal extent and thickness of source rock(s) draining into the structure.
2 Defining those parts of the drainage area that reached maturity after trap/seal creation.
3 Determining the source-rock yield, i.e. the volume of hydrocarbon (oil or gas) generated per unit volume of source rock (e.g. m^3 oil/km^3 source rock, Fig. 11.11), and hence the volume generated within the mature drainage area.

$$VG = (Area_{SR} \times Thick_{SR}) \times Yield_{SR} \quad (25)$$

The concepts used to establish the source-rock drainage area for a given structure ($Area_{SR}$) are summarized in Fig. 11.29: the area may be bounded by one or all of the following features:

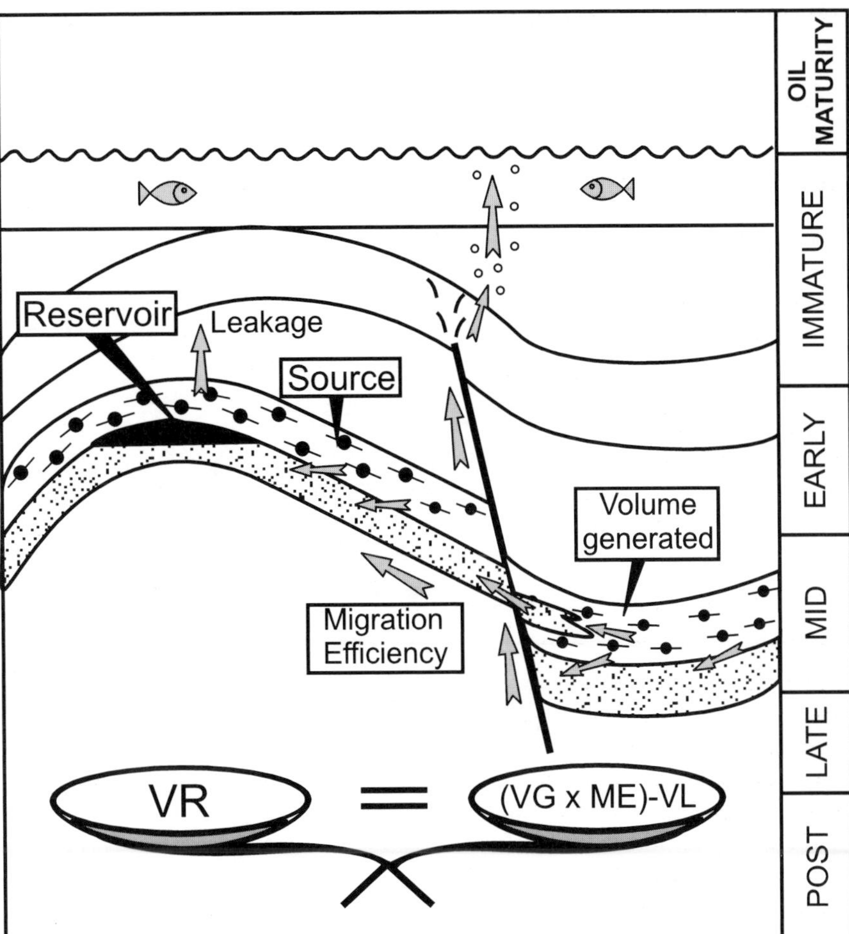

Fig. 11.28 Concept of a mass balance between the volume of oil generated in the source rock (VG), migration efficiency (ME) and reservoir leakage (VL) and the volume found in the reservoir (VR).

- The immature/early-mature or early-mature/mid-mature boundary, shallower than which no hydrocarbon generation and expulsion (see below) will have occurred. It has proved of particular importance in the North Sea to choose this boundary with care for prospects in flank areas of marginal maturity. In uplifted areas (e.g. the Inner Moray Firth to the west or the Stord Basin to the east), maturity boundaries must be generalized using the palaeostructural contours.
- The late-mature/post-mature source-rock boundary as established at the time of trap/seal formation. In a continuously subsiding basin, this will generally be stratigraphically deeper than the present-day late-mature/post-mature boundary. Only that part of the drainage area which reaches maturity after trap formation should be considered. This can be deduced from burial-history curves for representative parts of the drainage area. In Fig. 11.29, part of the source rock in the basin centre (location C) generated its hydrocarbon prior to trap formation at about 110 Ma.
- Sealing faults—the problem is to estimate the probability that a fault is sealing, rather than acting as a migration conduit (Schroeder and Sylta, 1993). While transtensional strike–slip faults tend to be leaky in the North Sea (Brooks *et al.*, 1984) dip–slip faults, with shales juxtaposed in hanging and footwalls, will tend to be sealing. Fault movement synchronous with migration mitigates against a seal.
- Erosional or regressional subcrop of the source-rock unit.
- Facies change of the source rock from, say, oil-prone to gas-prone within a shale, or from shale to sandstone (e.g. Fig. 11.10), as seen on Late Jurassic 'highs' in the Moray Firth Basin of the North Sea (Fisher and Miles, 1983).
- A maximum migration distance may be a limiting factor, but this is contentious. Certainly, the regional extent of the proposed migration pathway (e.g. a sheet sand) could place limits on the effective drainage area (Skjervøy and Sylta, 1993).
- Pressure compartments—either horizontal or vertical (e.g. Cayley, 1987; Buhrig, 1989; Hunt, 1990; Caillet *et al.*, 1997)—as hydrocarbons will not generally migrate from areas of low- to high-pressure potential (Leonard, 1989).

The thickness of the source rock unit ($Thick_{SR}$) can be determined from well data, seismic mapping or regional compilations. If the source-rock unit can be subdivided into a number of horizons with different levels of richness (e.g. 'hot' shale and 'cold' shale units, as well displayed in the Moray Firth Basin), a separate volumetric calculation should be carried out for each unit. Use of calibrated wireline logs, as shown in Figs 11.14 and 11.15, can prove useful to identify the vertical variations in hydrocarbon source potential within the source rock interval.

The volume of hydrocarbon generated per unit volume of source rock ($Yield_{SR}$) can be obtained from pyrolysis data of late-immature (pre-expulsion) source rocks (total yield, $S_1 + S_2$ in kg/t; see Fig. 11.4) or from extract data (see Fig. 11.12), given an average value for the TOC content of the unit (Larter, 1988).

If the quantity or type of kerogen varies laterally or

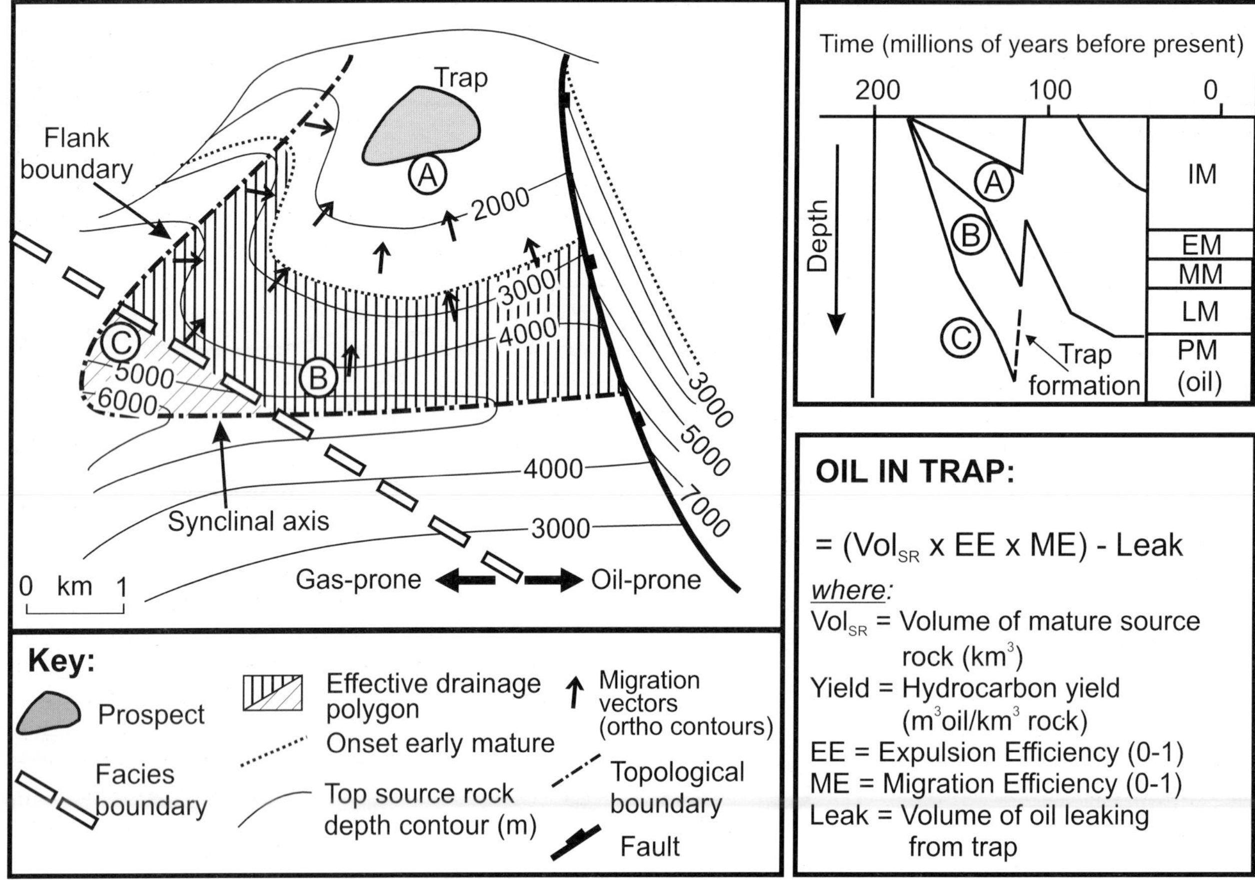

Fig. 11.29 Typical boundary conditions for a volumetric calculation. The drainage polygon charging oil to the trap (A) is defined by a synclinal axis, a facies boundary (oil-prone in north-east, gas-prone in south-west), a fault in the east, a flank boundary (plunging syncline) and an early maturity contour, with simple oil flow shown by arrows at right angles to the structural contours (migrational flux vectors). Burial-history modelling (inset) shows the source rock to be immature on structure at A up to the present day (zero oil charge), maturing synchronous with trap formation at location B (partial oil charge) and mature prior to trap formation at location C (no oil or gas charge).

vertically, the drainage area should be split up and each volume treated separately. A source rock isopachyte map forms the basis for organofacies mapping, allowing prediction of the lateral variation in kerogen type (Cornford, 1994).

In addition to lateral and vertical organofacies variations, it is important to correct analytical data for maturity effects (see Figs 11.3 and 11.6), since mature source rocks have a reduced apparent kerogen quality, determined by Rock Eval Hydrogen Index (S_2/TOC). This problem arises using the pseudo-van Krevelen plot (see Fig. 11.5) to define kerogen type when using the HI values from deeper samples (Fig. 11.13). The reduction in the S_2 peak yield, due to generation, together with a reduced apparent yield ($S_1 + S_2$), due to losses by expulsion, have been discussed in Section 11.3.4.

11.6 Migration

Migration—the movement of hydrocarbons in the subsurface—is conventionally divided into two sequential steps (primary and secondary) although four distinct steps are now recognized (Table 11.8):

1 Discharge: the escape of newly generated oil and gas from the organic phase of the lithostatically loaded kerogen into the generally limited intergranular or fracture porosity of the fine-grained source rock.

2 Drainage: the movement of expelled monophasic oil plus gas out of the fine-grained source rock through a water-wet network of fractures and/or higher-permeability silty streaks into the migration conduit.

3 Migration (formerly secondary migration): longer-distance, largely buoyancy-driven movement of the drained oil and gas through regionally extensive conduits of high permeability, such as sandstones, carbonates and faults, to the reservoir or to the surface.

4 Remigration: the escape of oil and gas from an accumulation, generally through an imperfect seal or via a breaching fault, either to another reservoir or to the surface.

Steps 1 and 2 have been grouped as 'expulsion' or 'primary migration', while step 3 has been termed 'secondary' and step 4 'tertiary' or 'dismigration'. Values for the overall efficiency of all four processes seem to range from about 0.5% to 30% efficiency in natural systems (Magoon and Valin, 1994), and derive from mass-balance calculations equating the volume of generative source rock with the volume of reservoired oils (e.g. Cornford, 1994, for the Central Graben).

The first 3 steps of migration occur under very different physical conditions, and the nature of the processes controlling migration have been the subject of controversy (Roberts and Cordell, 1980; England and Fleet, 1991; England *et al.*, 1991; Palciauskas, 1991). Despite this, it is generally agreed the following simple physical principles are involved:

- That under pressures and temperatures associated with generation the oil and gas will be a monophasic and probably super-critical fluid (Vandenbroucke, 1993).
- Hydrocarbons diffuse from high to low concentrations, with higher rates for the smaller molecules (Leythaeuser *et al.*, 1988).
- They can move by capillary forces through a lipophilic kerogen network in rich source rocks (McAuliffe, 1980) and be squeezed out of kerogen where subjected to lithostatic load (England *et al.*, 1991).
- Generated hydrocarbons will be differentially absorbed on the kerogen network (Sandvik *et al.*, 1992) and mineral matrix of the source rock, and once free to move will be subjected to geochromatography (Krooss *et al.*, 1991).
- They move along a pressure-potential gradient from high to low pressure potentials (Larter, 1988; Mann *et al.*, 1997).
- They rise under buoyant forces if the surrounding medium (formation water) is of higher density (Lehner *et al.*, 1988; Sylta, 1993).
- If dissolved (gas) or entrained (oil) in moving formation water, hydrocarbons can be moved hydrodynamically—although differential solubility effects are seldom seen.

It is paradoxical that we understand little about the mechanism(s) of discharge and drainage of oil and gas from source rocks, while the efficiency of the process as a whole is well quantified (Pepper, 1991). In contrast, the major influence of buoyancy as controlling (secondary) migration is not in dispute, but the efficiency of the process is very poorly constrained.

Uncertainties about migration are a significant source of doubt in many prospect evaluations and a major source of error in volumetric calculations. Mechanisms of expulsion, drainage and migration are currently the topic of animated debate—increasingly in the presence of sound geological observation (Roberts and Cordell, 1980; Durand, 1983; Doligez, 1987; England and Fleet, 1991; Mann *et al.*, 1987; Ungerer *et al.*, 1990). The movement of light (Leythaeuser *et al.*, 1982) and heavy (Makenzie *et al.*, 1987; Leythaeuser *et al.*, 1988a; Wilhelms *et al.*, 1990) hydrocarbons across source-rock/carrier-rock interfaces has emphasized the role of diffusion in expulsion. Efficient expulsion and migration of heavier hydrocarbons is demonstrated by Leythaeuser *et al.* (1988a,b), using molecular parameters, with confirmation from labratory studies (Lafargue *et al.*, 1990, 1994; Rudkiewicz *et al.*, 1994) and computer modelling (Düppenbecker *et al.*, 1991; Ungerer *et al.*, 1990). Larter (1988), with reference to North Sea studies, notes the increasingly accepted view that oil expulsion 'occurs in a single phase, driven by overburden stress, with diffusive effects occurring within a few metres of the source rock–conductor system boundaries'. Oil movement within the Kimmeridge Clay Formation source rock of the North Sea has been discussed in terms of adsorption and molecular sieving through micropores and microfractures (Lindgreen, 1985, 1987a,b).

Secondary migration appears to occur in a discrete oil phase, moving primarily under the influence of buoyant forces. Barnard and Bastow (1991) discuss secondary migration in the context of the North Sea geology. The ability of isotopes (James, 1983), gasoline-range hydrocarbons (Thompson, 1979, 1983; Kurbskiy *et al.*, 1983), aromatics (Radke, 1988) and biomarkers (Cornford *et al.*, 1983; Karlsen *et al.*, 1995) to link reservoired hydrocarbons with their source rocks in terms of maturity has done much to place limits on the vertical and lateral extent of secondary migration. A buoyancy-driven model, visualized using regional seismic mapping to produce 'orthocontours', suggest regional secondary migration pathways in the northern Viking Graben (Skjervøy and Sylta, 1993) and Haltenbanken (Hermans *et al.*, 1992).

11.6.1 Discharge (step 1 of expulsion)

Kerogen decomposes to bitumen and gas as a result of the increase in temperature experienced during burial (see Fig. 11.1B). This process of maturation gives rise to the generation of oil and gas (see Fig. 11.1C), initially within the kerogen. In order to discharge the generated oil (or gas), the hydrocarbon phase must first escape from the oil-wet kerogen particle or network (Barker, 1980).

Very little is known about the discharge of hydrocarbon from the kerogenous phase, but a number of factors must play a role:

- Generation of oil and gas from solid kerogen produces a change in volume (Düppenbecker *et al.*, 1991) and hence generates a pressure gradient promoting movement away from the site of generation (Palciauskas, 1991).
- Diffusion, especially of light hydrocarbons, will initially occur from the high hydrocarbon concentrations in the

Table 11.8 Terminology used to describe migration of hydrocarbons in the subsurface.

	Migration (sensu lato)			
	Explusion (Primary*)			
Nomenclature	Discharge	Drainage	Migration (Secondary*)	Remigration (Tertiary*)†
Processes	Lithostatic load, diffusion	Pressure, diffusion	Buoyancy, hydrodynamics	Pressure, buoyancy
Route	Kerogen to source-rock porosity	Porous source-rock interbeds or fractures to migration conduit	Regionally extensive permeable beds, diagenetic zones or faults	Trap-seal failure via faults, tilting

*Older terminology. †Also termed dismigration.

kerogen to low concentrations in the surrounding rock porosity (Leythaeuser *et al.*, 1987; Krooss, 1987).

- Larger kerogen particles will be supporting the lithostatic load—and hence be subjected to anisotropic stress—'squeezing' out newly generated fluids in the direction of least stress (England *et al.*, 1991; Palciauskas, 1991).

A corollary of the latter process is that discharge will generally occur at right angles to the lithostatic stress (i.e. parallel to bedding), may be promoted by deviatoric stress (e.g. compressional tectonics) and that the process will be most efficient as the pore pressure approaches the lithostatic load. In comparing expulsion from carbonate as opposed to argillaceous source rocks, Vandenbroucke (1993) emphasizes the role of lithostatic load on load-bearing kerogenous filaments and membranes in carbonates.

The escape of oil droplets from kerogen has been observed with the microscope and is referred to as the secondary maceral 'exudatinite' (Teichmüller, 1974). Otherwise, it appears that, once samples are transported to, and subjected to investigation in the laboratory, the transient processes of expulsion are no longer detectable. On the other hand, it is not in doubt that, whatever the process, the hydrocarbon phase discharges from the kerogen and can then drain out of the source rock. Based on studies of the Upper Jurassic of the North Sea (Cooles *et al.*, 1986; Wilhelms *et al.*, 1990), overall expulsion efficiencies up to 85% have been demonstrated.

11.6.2 Drainage (step 2 of expulsion)

Once hydrocarbon is discharged from the kerogen, it must escape through the fine-grained source rock via local intergranular or fracture permeability. This is the process termed 'drainage'. At the temperatures and pressures of a generative source rock (e.g. 140°C and between 5000 psi hydrostatic and 15 000 psi lithostatic pressures) drainage will operate on monophasic mix of oil and gas (Vandenbroucke, 1993), with viscosities dependent on the pressures and temperatures of the source rock and composition of the expelled products (Werner *et al.*, 1996). Rudkiewicz *et al.* (1994), modelling laboratory results from the Toarcian of the Paris Basin, obtained the best agreement maintaining separate water, CO_2 and oil phases.

For drainage to occur, the source rock must be capable of generating enough hydrocarbon to 'saturate' the porosity of the source rock, with any excess being available for expulsion. The most impressive direct evidence for source-rock saturation on expulsion derives from the increase and then decrease in wireline-log resistivity over the oil-expulsion window (Meissner reported in Durand, 1983, and as described by Goff (1984) for the Kimmeridge Clay Formation of the northern Viking Graben).

In an examination of the exhumed clay-marl bitumen/source-rock complex of the Lower Toarcian Hils syncline of north-west Germany, Mann *et al.* (1989) have identified three sequential drainage routes:

- Lithological pathways (more porous interbeds).
- Diagenetic pathways (dissolution features).
- Fracture pathways (formed by volumetric changes during generation).

In terms of bulk parameters, it is difficult to differentiate the discharge from the drainage process, the two processes together being termed here expulsion. Expulsion efficiency (EE) is defined as:

$$EE = \text{mass of oil expelled/mass of oil generated} \quad (26)$$

A trend of increasing expulsion efficiency with source-rock richness has been shown for a data set including North Sea Kimmeridge Clay Formation source rocks. The latter show high values, in the 80% range (Cooles *et al.*, 1986; Pepper, 1991). Using two separate approaches (cemented versus non-cemented shales and silt versus shale bands), Wilhelms *et al.* (1989) have estimated primary migration efficiencies up to 70–75% for Kimmeridge Clay samples from the Norwegian sector. The trend of expulsion efficiency derived from the depth trends given in Fig. 11.13 is shown in Fig. 11.30.

By a different approach, Larter (1988) showed somewhat lower values (up to 50%) for the Kimmeridge Clay Formation shales of the Viking Graben over the depth interval 2.5–4.5 km. The differences may be attributed to the observation of Leythaeuser *et al.* (1988a), who reported relative expulsion efficiencies rising from an arbitrary midshale baseline of zero to 80% over the outer 5 m of a thick mature shale interbedded with sandstone in a Brae area core. Measurements on thin interbedded shales will thus show high efficiencies. Gas (C_1–C_4) expulsion is probably nearly 100% efficient in most cases (Leythaeuser *et al.*, 1982). Models showing the principles of low and high

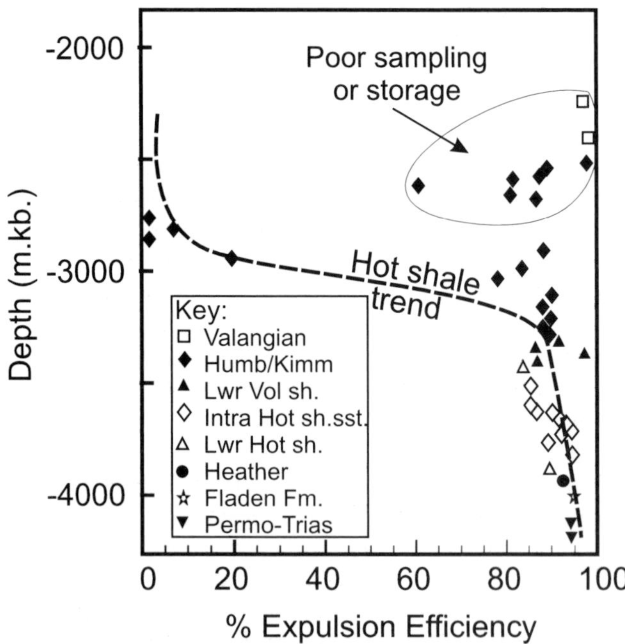

Fig. 11.30 Expulsion efficiencies (EE) for Kimmeridge Clay Formation source rocks in the South Viking graben and Moray Firth basin based substantially on the pyrolysis trends established in Fig. 11.13. The expelled mass of hydrocarbon is determined by comparison of the reduction in Hydrogen Index (indicating generation) with the increase in Production Index (indicating retention) as generation progresses. Lower expulsion efficiencies with an approximate linear depth trend of $\%\ EE = (0.025 \times \text{depth in metres}) - 62.5$, giving EE = 50% at 4500 m, have been calculated by Larter (1988) for a similar suite of North Sea source rocks.

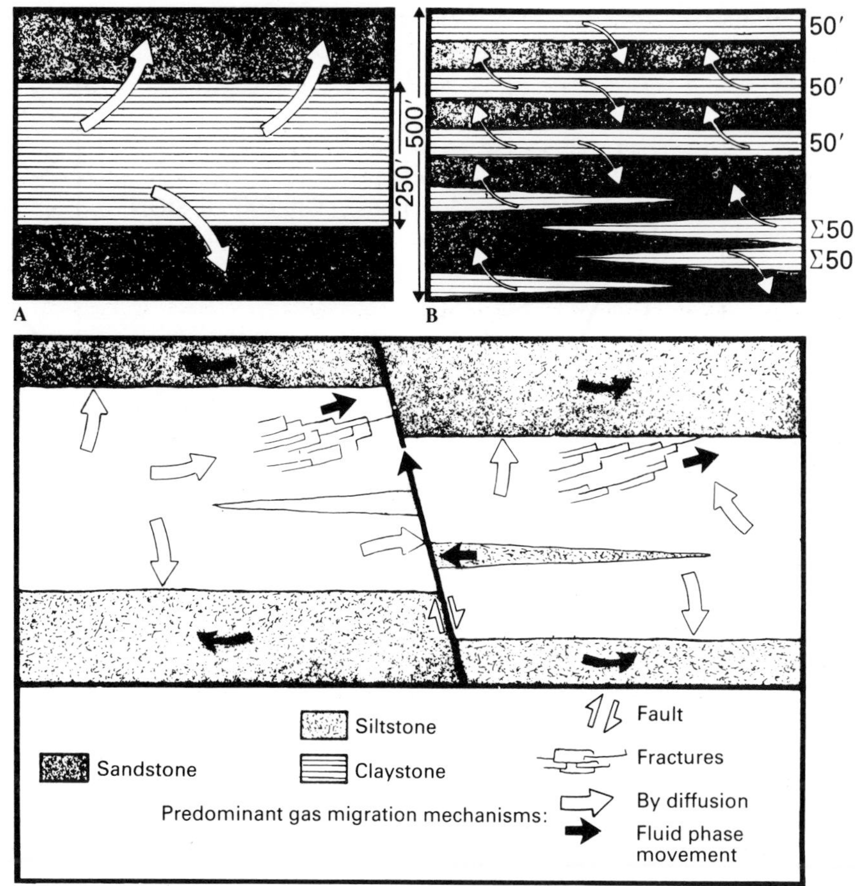

Fig. 11.31 Conduits for source-rock drainage—fractures and interbeds. A poorly drained single massive source rock (A) can be contrasted with the same net thickness of well-drained interbedded source rock (B). Eventually, expulsion efficiencies will be the same in both cases, it being the timing and composition of the expelled product that changes (see text). The lower diagram (C) depicts factors affecting source-rock drainage, such as major faults, silty interbeds and transient microfractures. Diffusional processes may operate at the upper and lower bed margins.

drainage efficiencies are depicted in Fig. 11.31A and B, respectively. A drainage index (DI), relating the volume of the source rock to the total effective surface area available for hydrocarbon escape, can be defined:

DI = number of layers/sum shale thickness in metres (27)

The sum shale thickness can include non-source-rock shales if they lie between source rocks and secondary migration horizons. Values (which are the reciprocal of the average shale-bed thickness) range from about 0.66 in the Brae Field to 0.003 for typical 300-m-thick Kimmeridge Clay sequences at full maturity. Layers thinner than 1 m may be presumed to have DI values of unity. In the case of a fractured source rock a DI may be expressed as fractures per unit area or volume.

The exact mechanism of drainage is not yet established, but microscopy studies of mature source rocks and the relationship of maturation with the development of overpressure point to an overpressure-related fracture mechanism. It is envisaged that the hydrocarbon-generation process develops localized overpressure that produces or reopens transient fractures in the source rock (Lindgreen, 1987a). Conversely, Ungerer *et al.* (1983) have calculated that the conversion of kerogen to hydrocarbon produces no volume change, and hence no 'generation' pressure increase is predicted. The consensus, expressed by Düppenbecker *et al.* (1991, p. 51), is that generation produces 'a considerable overall volume expansion of organic matter [and] represents an excellent potential for pressure build-up in the source rock pore system, which is then the driving force for expulsion'. Though localized at the kerogen particles, generation will not be the only source of overpressure, it being added to compaction- and aquathermal-related pressure increments.

Whether overpressured or not, source-rock drainage is promoted by the presence of sand or silt interbeds (Mackenzie *et al.*, 1983, 1987) and fractures (Fig. 11.31). Interpretation of compositional trends for light hydrocarbons in the Brae field cores shows that diffusive enrichment of the lower-molecular-weight molecules is sometimes seen in the migrated extract (Leythaeuser *et al.*, 1982), but affects only the region of the shale/sand interface. Such fractionation will be observed in extracts or oils only during active migration; once equilibrium is re-established, molecules of higher molecular-weight would be expected to have 'caught up' with the more mobile lighter fractions. Using cemented and uncemented North Sea Upper Jurassic shales, Wilhelms *et al.* (1990) have described the relative expulsion efficiencies (REE) of individual *n*-alkanes. An absence of major fractionation in the $C_{15}+$ fraction confirms expulsion of oil as a single liquid phase (Leythaeuser *et al.*, 1988a).

In practice, expulsion efficiencies for rich source rocks (petroleum potential greater than 5 kg/t) can rise as high as 50–80% at peak maturity, while drainage indices are constant for a given section, being effectively independent of maturity. Combining this approach of source-rock richness and bed thickness (drainage index) can produce a series of iso-expulsion efficiency curves, as shown in Fig. 11.32.

Lindgreen (1985, 1987a,b) has discussed migrational processes in the Kimmeridge Clay/Draupne Formation claystones of the Norwegian Central Graben and the

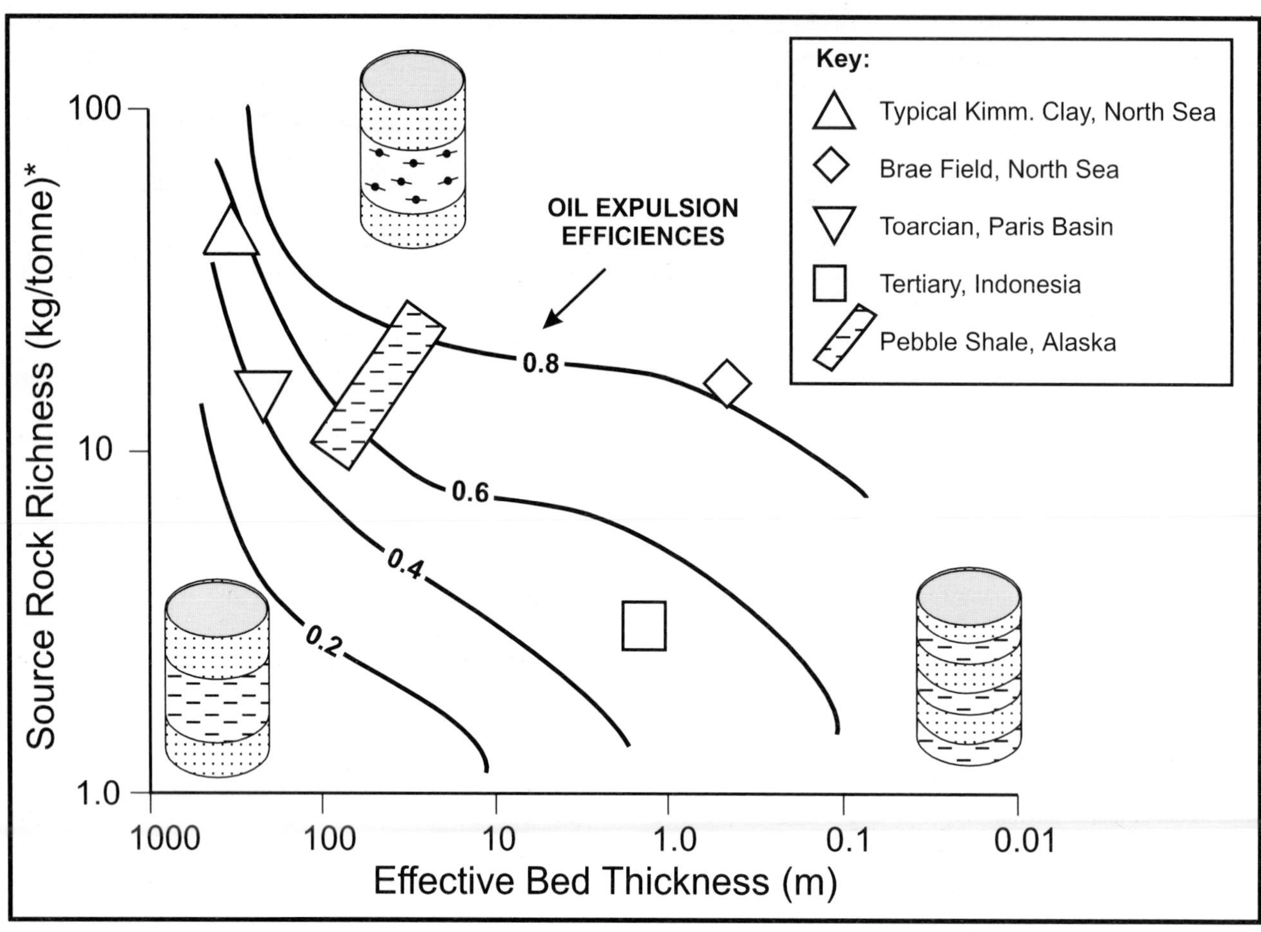

Fig. 11.32 Iso-expulsion-efficiency (EE) curves as a function of source-rock richness (kg hydrocarbon/t rock) and effective bed thickness (m). The EE contours indicate the expelled fraction of the total hydrocarbon generated by the end of the main oil window (% R_o = 1%). Eventually unexpelled oil cracks to gas, the vast majority of which is finally expelled.

Middle Cambrian Alum Shale of Denmark and Sweden, with respect to microporosity, microfracturing and cementation. A long history of overpressuring and fracturing is indicated, with synchronous expulsion of hydrocarbons and more alkaline, saline water through a series of early, compaction-related bed-parallel and bed-normal fractures. While Lindgreen (1987a) favours diffusion as a mechanism for expulsion of smaller molecules, it is suggested that thin shales may not develop overpressures and hence fractures, making them less efficient than thicker shales with respect to expulsion of oil. Expulsion of liquid hydrocarbons from a maturity sequence of coals of the Hitra Formation of mid-Norway has been seen using microscopy and deduced from chemistry to occur through a network of microfractures (Hvoslef *et al.*, 1988). Similar conclusions were drawn concerning expulsion from the Liassic Posidoniaschiefer to form the latest Jurassic–basal Cretaceous Asphaltkalk of the Hils syncline, north-west Germany (Horsfield *et al.*, 1989, 1991; Mann *et al.*, 1989). A mathematical treatment of expulsion from the Posidoniaschiefer of the Lower Saxony Basin is elaborated by Düppenbecker *et al.* (1991), simulating monophasic pressure-driven expulsion through existing pores plus a transient fracture network.

In the North Sea oil area, the highest primary migration efficiencies are to be expected in areas where sand is adjacent to or interbeds with the 'hot shales' of the Kimmeridge Clay Formation, as in Piper of the field of the Moray Firth (Schmitt and Gordon, 1991), Fife of the Central Graben (Mackertich, 1996) and in the Brae (Turner *et al.*, 1987), T-block areas (Kerlogue *et al.*, 1994) and Magnus (Shepherd, 1991a) of the Viking Graben. In the case of low drainage indices from thick, rich, source rocks, the oil remaining *in situ* will be thermally cracked on further burial and expelled as condensate or wet gas (Mackenzie and Quigley, 1988). In this way expulsion controls the gross gas/oil ratio of the total expelled product. Late-stage gas generated from oil-prone source rocks in the basin centre (e.g. sourcing the Frigg, Troll, or Sleipner Fields) is a witness to the inefficiency of oil expulsion in the thick midgraben shale sequences. Leythaeuser and Poelchau (1991) have found evidence for fractionation of heavier hydrocarbon in solution in the gas phase expelled from Type III kerogen, the process being very sensitive to overpressures.

11.6.3 (Secondary) migration

Secondary migration—longer-distance hydrocarbon-phase movement through regionally extensive conduits (permeable beds, faults, etc.)—operates primarily as a result of buoyant forces (England *et al.*, 1987, 1991; Lehner *et al.*, 1988; Dembicki and Anderson, 1989; Thomas and Clouse, 1995), modified by hydrodynamic flow and pressure-potential gradients (Buhrig, 1989; Thomsen *et al.*, 1990; Hvoslef *et al.*, 1995). The efficiency of secondary migration can be quite high (e.g. approx. 80%) in cases of simple upflank migration through a laterally continuous conduit to a structure at the natural focus (Pratsch, 1983) of the migration path.

The general migration of oil and gas out of the deeply buried 'kitchen areas' of the North Sea graben depocentres is shown in Fig. 11.33, a composite map modified after Thomas *et al.* (1985), Field (1985), Cayley (1987) and Cornford (1994). Such maps are the first step towards establishing a source-rock/reservoir mass balance for each kitchen area, as discussed by Cornford (1994) in the context of the Central Graben. A more detailed calculation can be undertaken by dividing each kitchen area into 'drainage polygons', as discussed by Dahl and Yükler (1991) for the greater Oseberg area, Sylta (1993) for the Troll area and Caillet *et al.* (1997) for the greater Ekofisk area and Thomsen *et al.* (1990) for the Danish Central Graben. A drainage polygon is an area of mature source rock that charges a single structure or a group of closely associated structures.

As shown in the lower panels of Fig. 11.34, controls on secondary migration include the distance of migration and the effective permeability of the pathway. Inefficient, tortuous, short-distance migration through low-permeability rock can be contrasted with efficient, long-distance migration through high-permeability, regionally extensive, sandstone beds (Lehner *et al.*, 1988). Distance affects efficiency, due to the amounts of residual oil or gas left in the migration conduit. Experience seems to indicate that oil and gas do not have to saturate the entire rock volume in a secondary migration path (Dembicki and Anderson, 1989; Thomas and Clouse, 1995). Rather, the hydrocarbon moves in the form of a braided stream along specific oil-wet pathways in the roof of the conduit, running from high point to high point and favouring the more permeable strata (England *et al.*, 1991).

The oil column thickness in the migration path is inversely related to the permeability of the carrier rock, with the migrating oil stream restricted to the top few metres of a highly permeable carrier bed. Secondary migration efficiencies will reach zero in cases where low permeability, a tortuous pathway and numerous intermediate highs (microtraps) mean that no oil arrives at the studied structure. An approach for quantifying migration efficiency in terms of the surface area of mineral grains 'seen' during flux from source to reservoir has been proposed by Larter *et al.* (1995) and Li *et al.* (1995) using pyrrolic nitrogen compounds in source-rock extracts and their resulting oils. Migration distances have been predicted to a high precision of ± 327 m (but not necessarily with accuracy) by Christie *et al.* (1993), using multivariate predictive modelling applied to the biomarkers from 12 North Sea oils with migration distances between 2 and 17 km. A number of migration-related molecular fractionations have been recorded by Miles (1990) and Curiale and Bromley (1996).

The role of faults as seals versus migration conduits is contentious. Knott (1993), considering North Sea fault seals on a reservoir scale, concludes that the major factor favouring a seal is displacement greater than reservoir (carrier) thickness, with the sand/shale ratio and sandconnectivity as ancillary controls. The converse should be true for migration conduits. Since Gauthier and Lake (1993) have demonstrated the faults cutting the Brent sandstones in the northern Viking Graben to be fractal in terms of geometry (throw, length, density), secondary migration efficiency will be a function of the relationship between carrier-bed thickness and fault spatial density at seismic and subseismic scales. Moving away from the dominantly dip–slip faulting of the northern Viking Graben, it is a general observation that faults with a major lateral component tend to be poor seals, conversely acting as good vertical-migration conduits.

Figure 11.34 depicts cartoons of some of the routes for secondary migration invoked to explain the majority of North Sea hydrocarbon occurrences (Cornford *et al.*, 1986). The top surface of the migration conduit defines the migration pathway on a structure-by-structure basis (e.g. for the northern Viking Graben, Burrus *et al.*, 1991; Skjervøy and Sylta, 1993). Basin geometries must be re-created for the time at which migration occurred by isopachyte subtraction, plus uplift isopachytes where unconformities exist. Such pathways can then be validated by geochemical means, as discussed later.

The shortest distance of migration, together with the highest oil-expulsion efficiencies (Leythaeuser *et al.*, 1988a) are to be expected in areas where sand interbeds with the 'hot shales' of the Kimmeridge Clay Formation, as in Piper, Brae and Magnus fields (Fig. 11.34A). In some cases the on-structure source rocks are fully mature, further shortening the migration pathways.

Migration in the southern Viking Graben is illustrated in more detail in an east–west cross-section in Fig. 11.35. The asymmetry of the graben is marked at this location, allowing high-efficiency oil migration to the west into the Brae and 'T-Block' fields, while the late-mature associated gas generated from the shales at depth will tend to migrate to the east to charge the Sleipner complex with wet gas. Axial reservoirs (distal fans or turbidites), such as those of the Miller Field, will fill directly from contiguous source rocks.

On a smaller scale (Fig. 11.34B), such simple secondary-migration pathways must exist in the Brent Sands of the East Shetland Basin (Buhrig, 1989; Miles, 1990) and Norwegian northern Viking Graben (Larter and Horstad, 1992; Schroeder and Sylta, 1993; Skjervøy and Sylta, 1993) from highly overpressured source rocks into less overpressured Brent sandstones (Hvoslef *et al.*, 1996). A ray-tracing method has been applied by Schroeder and Sylta (1993) to large areas of the Norwegian North Viking Graben. Migration conduits, flow rates and filling histories are defined for the Oseberg, Huldra and Brage fields. Miles (1990) and Larter and Horstad (1992) have used asphaltenic residues to identify the migration pathways and sterane fractionation ratios to assess the distance (tortuosity) of the proposed migration routes for Ninian, Alwyn, Lyell and other UK Quadrant 3 oils. These secondary-migration routes in the rotated fault blocks of the North Viking Graben have been followed, using 2D fluid-flow models, to estimate hydrocarbon flow rates in the range 5–50 m/Ma (Burrus *et al.*, 1991). In the context of reservoir permeability, Gauthier and Lake (1993) have established fault geometry,

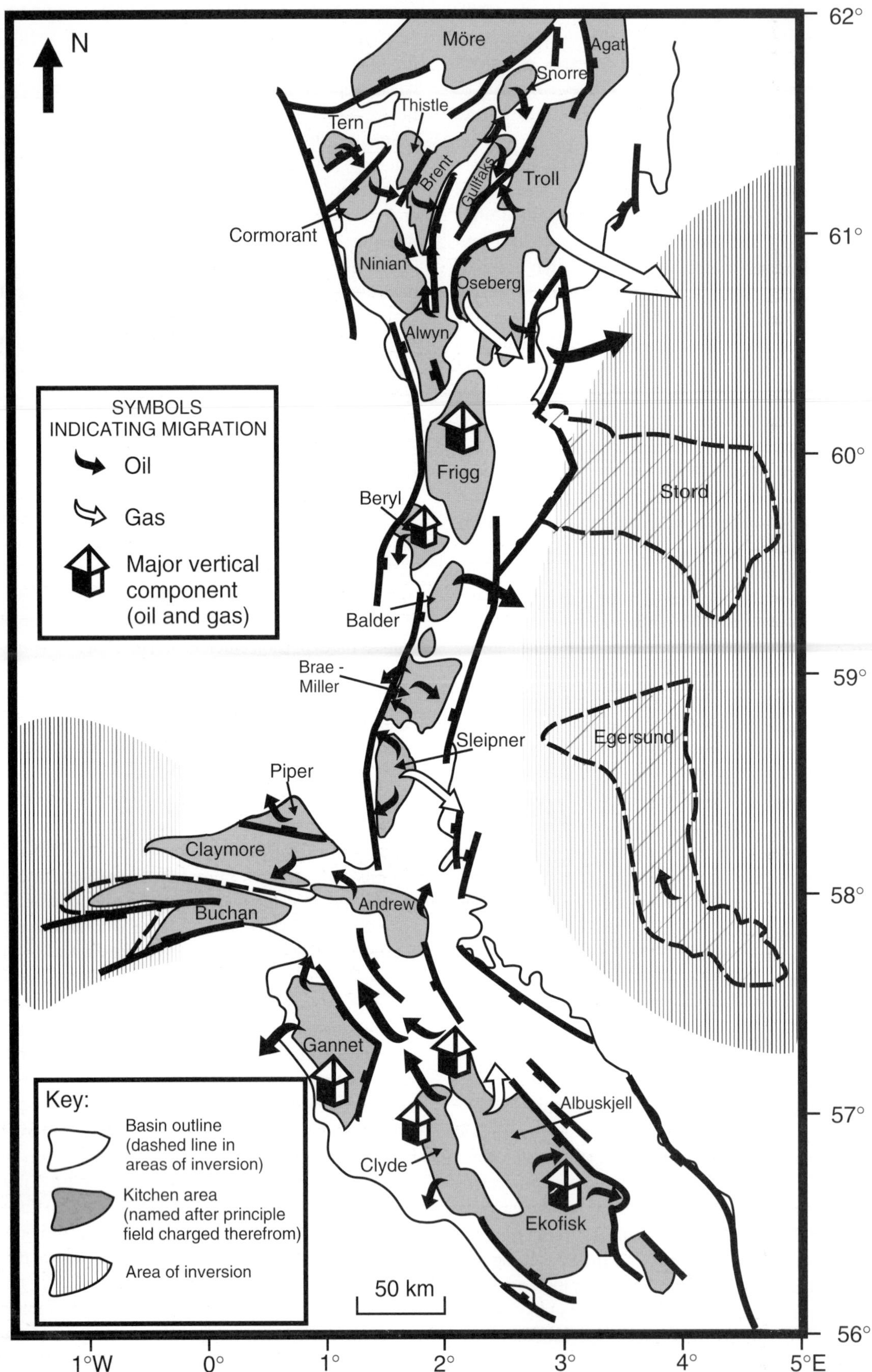

Fig. 11.33 The major hydrocarbon-drainage kitchens (light stipple) and the migration pathways for oil (solid arrows) and gas (open arrows) within the oil province of the North Sea. Areas of inversion, where the original basin outline may have been distorted by uplift, are shown by heavy ruling; establishing basin geometry at maximum burial and generation in order to predict migration requires detailed reconstruction. Areas where vertical migration is indicated (stout arrows) frequently coincide with areas such as graben-margin transfer zones, where basement-controlled wrench faulting occurs.

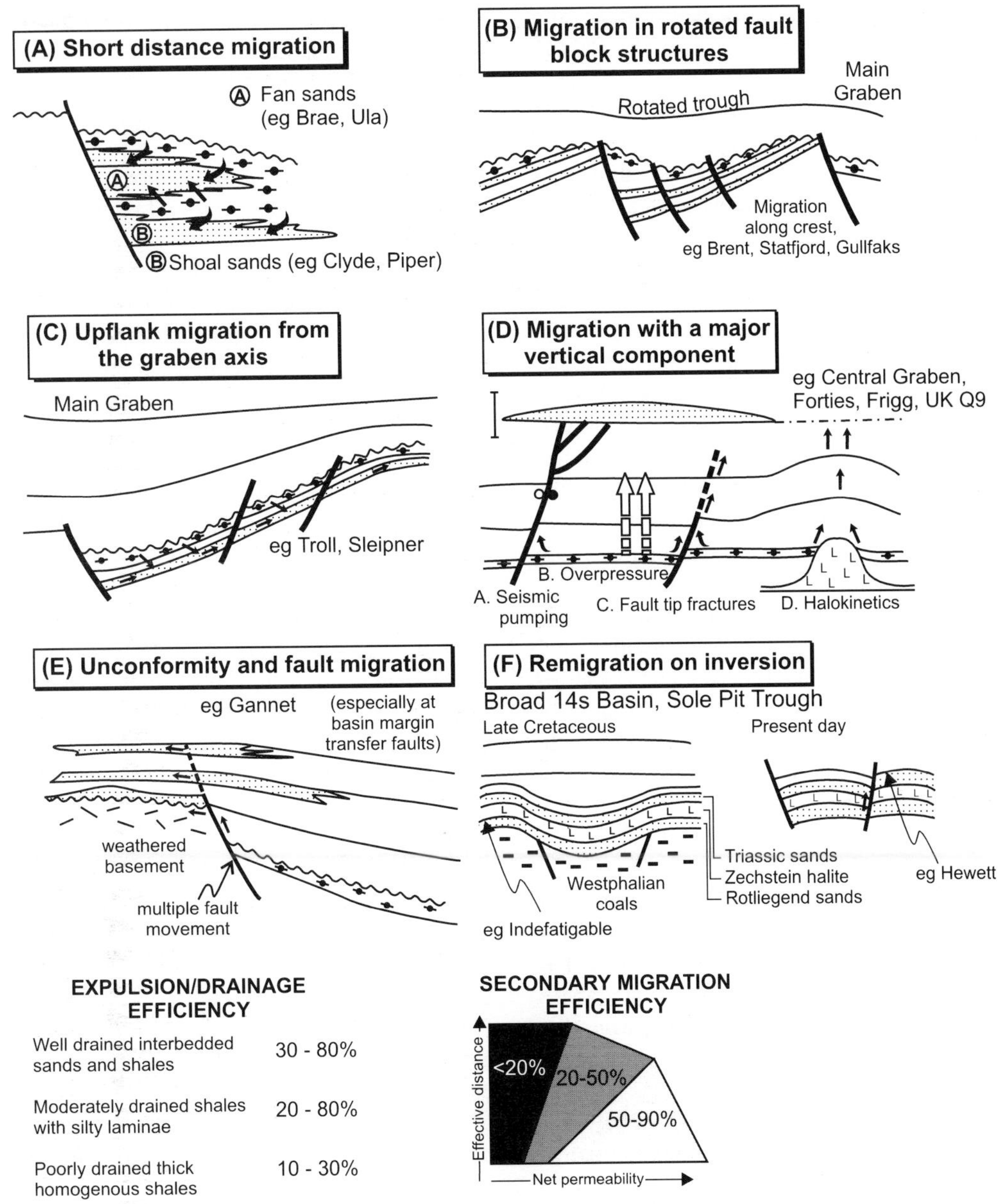

(A) Short distance migration
Ⓐ Fan sands
(eg Brae, Ula)
Ⓑ Shoal sands (eg Clyde, Piper)
(B) Migration in rotated fault block structures
Rotated trough
Main Graben
Migration along crest,
eg Brent, Statfjord, Gullfaks
(C) Upflank migration from the graben axis
Main Graben
eg Troll, Sleipner
(D) Migration with a major vertical component
eg Central Graben,
Forties, Frigg, UK Q9
A. Seismic pumping
B. Overpressure
C. Fault tip fractures
D. Halokinetics
(E) Unconformity and fault migration
eg Gannet
(especially at basin margin transfer faults)
weathered basement
multiple fault movement
(F) Remigration on inversion
Broad 14s Basin, Sole Pit Trough
Late Cretaceous
Present day
Westphalian coals
Triassic sands
Zechstein halite
Rotliegend sands
eg Indefatigable
eg Hewett
EXPULSION/DRAINAGE EFFICIENCY
Well drained interbedded sands and shales 30 - 80%
Moderately drained shales with silty laminae 20 - 80%
Poorly drained thick homogenous shales 10 - 30%
SECONDARY MIGRATION EFFICIENCY
Effective distance
Net permeability
<20%
20-50%
50-90%

Fig. 11.34 Cartoon cross-sections illustrating secondary-migration trends, North Sea, with some controls summarized below. (After Cornford *et al.*, 1986.)

West
UK Norway
East
FAULT SCARP FANS
(e.g. Brae)
AXIAL TURBIDITE
(e.g. Miller)
FLANK
(e.g. Sleipner)
LOWER CRETACEOUS
TRIASSIC
Approx. depth (km)
3
6
Reservoir sand pressure gradient
LOW
HIGH
KEY
Oil, migration
Gas, migration
Mudstones (some SR)
Sands and conglomerates

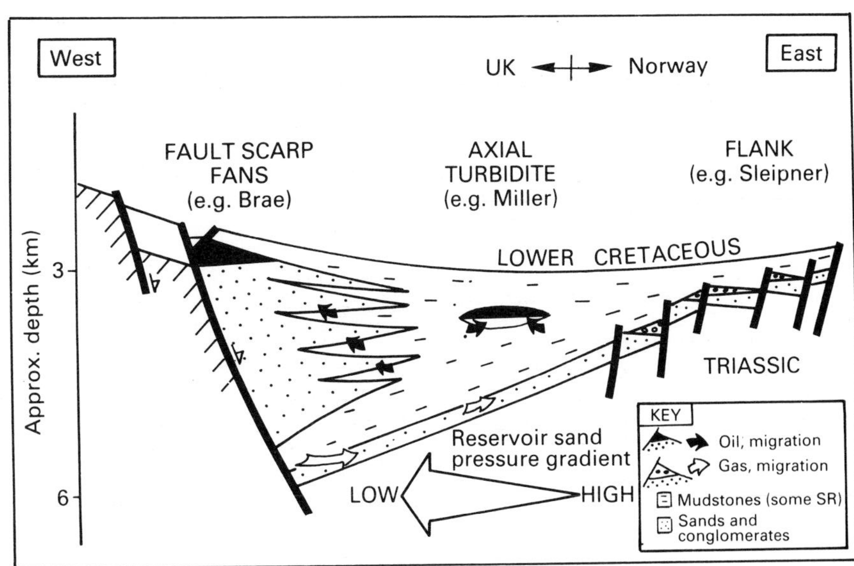

Fig. 11.35 Proposed oil- and gas-migration routes in the southern Viking Graben, North Sea, where basin assymetry favours deep migration of wet gas focusing eastward to the Sleipner structures, mid-depth migration of oil focusing westwards to the Brae and T-block fields, with mid-graben generation being picked up by the mid-basin distal fan/turbidite sands constituting the Miller Field complex. Basin-margin transfer faults trapping deeper source rocks lead to exceptions, such as the wet-gas East Brae Field in the west.

frequency and distribution as fractal (scale-invariant) and identify faults beyond seismic resolution in the Pelican Field of the north Viking Graben.

High secondary-migration efficiencies are also associated with upflank movement through regionally extensive, laterally continuous conduits, to a structure at the natural focus of the migration path (Fig. 11.34C). Super-giant accumulations result, for example as in the Troll and Sleipner fields.

As illustrated in Fig. 11.34D, the presence of oil sourced from the Kimmeridge Clay Formation in the Upper Cretaceous and Tertiary reservoirs of the North Sea can sometimes pose problems in terms of defining a plausible migration route (Barnard and Bastow, 1991). Mainly vertical migration within the Central Graben of the North Sea is addressed by Cayley (1987) and Cornford (1994). In the Ekofisk Field, faults (van den Bark and Thomas, 1981) and overpressure (Gaarenstroom *et al.*, 1993; Leonard, 1993; Caillet *et al.*, 1997) are believed to have played a major role in moving oil from the Upper Jurassic shale source rocks to the Upper Cretaceous chalk reservoirs. Jensenius and Munksgaard (1988) have identified the presence of large-scale hot-water migration systems around the salt diapirs of the Danish Central Graben: oil may travel the same path. An absence of hydrodynamic flow, however, is noted in the halokinetically influenced Ekofisk area (Caillet *et al.*, 1997).

Using high-resolution seismic, Cartwright (1994) identifies a polygonal network of hydrofracture-related faults in the Tertiary mudrocks of the Central Graben, and implies a significance for vertical hydrocarbon migration. Using 1D modelling, Mudford *et al.* (1991) have demonstrated that overpressures in the Central and southern Viking grabens are a transient effect of rapid Neogene burial—suggesting that any pressure-driven migration into Tertiary reservoirs is likely to be a relatively recent process.

Migration into the Forties Field is also the subject of speculation. Figure 11.36 shows two possible migration paths, one from contiguous source rocks, via faults and overlying fracture zones flanking the Forties high (Wills and Peattie, 1990), and the second longer-distance path, first vertically via halokinetic fractures in the Central Graben (Cayley, 1987), and then laterally via the distal facies of the Palaeocene reservoir fan sands.

On the flanks of the Central Graben, migration of light oil and gas–condensate to the Palaeocene reservoirs of the Gannet Field (e.g. Fig. 11.34E) is claimed to be via vertical basin-margin faults into the Tertiary sands, and then laterally into the stacked reservoirs (Armstrong *et al.*, 1987; Cayley, 1987). A degree of fractionation is indicated by a tendency towards lighter oil and condensate in the upper reservoirs.

In the case of the Eocene Frigg Field gas reservoir, overpressure-induced diffusion and hydrodynamic transport of gas in solution have been invoked (Goff, 1984). This mechanism fails to explain the presence of biodegraded, demonstrably Jurassic-sourced oils in apparently closed, lensoid, Tertiary sand bodies, with no faulting apparent on seismic lines. In this case, not only must the oil get into the reservoir, but so also must groundwater to carry the bacteria and their supporting nutrients (Connan, 1984). To explain this, Brooks *et al.* (1984) have proposed a major role for seismic pumping at trans-tensional strike–slip faults (Sibson, 1992a,b). These transfer faults have a role both in the vertical movement of oil and groundwater and in the accumulation of the reservoir sands in the area of UK Quadrant 9. Pegrum and Ljones (1984) have also noted the effects of wrench faulting in the Sleipner Field complex, resulting in gas/condensate migration into the Palaeocene sands of the Norwegian Block 15/9 Gamma structure.

Moving to the south, the gas in some of the gasfields of the southern North Sea must have been trapped since generation in the late Cretacous (Glennie and Boegner, 1981), although in the Broad Fourteens Basin (Oele *et al.*, 1981) and the giant Groningen gasfield (van Wijhe *et al.*, 1980) the gas may still be being ‘topped up’ to the present day. As shown in Fig. 11.34F, tertiary or remigration to new reservoirs may also have occurred as a result of tectonic inversion of former basinal areas (Hillis, 1995a,b). The longevity of the gasfields of the UK southern North Sea, despite inversion, is in large part due to the excellent seal afforded by the Zechstein halite and associated anhydrite deposits.

11.6.4 Tertiary migration—remigration and seal development

Oil and gas not only have to migrate to the trap, but also

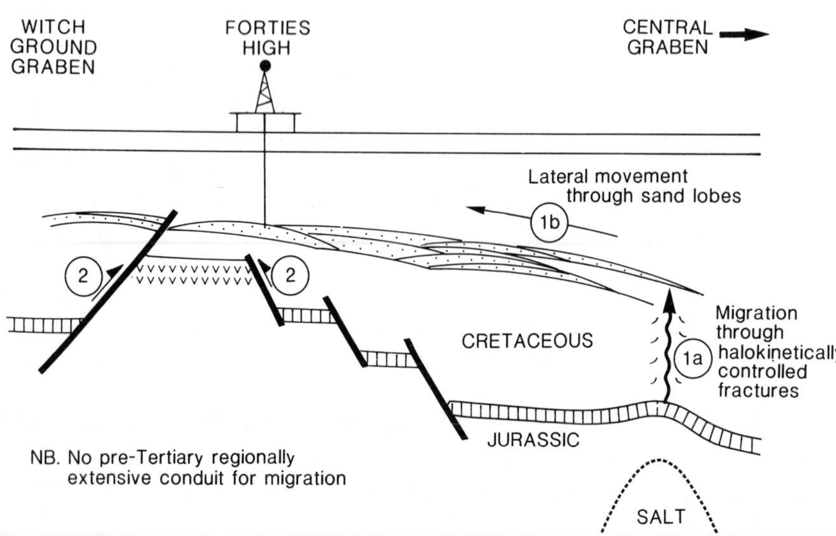

Fig. 11.36 Putative secondary migration routes to Tertiary aged reservoirs in the northern sector of the Central Graben, North Sea. Migration into Palaeocene fan sandstones draped over structural highs (e.g. Forties and Montrose fields) is probably from the main graben to the south, vertically via salt or fault-related fractures (1a) and then laterally up-fan from distal to proximal facies (1b). An alternative pathway via faults bounding the basement high (2) has been proposed (Wills and Peattie, 1990; Wills, 1991).

have to be retained within the structure to the present day. Thus the ability of the cap rock to seal the structure can be the key to commercial success (Jenyon, 1984b; Montel *et al.*, 1993; Nelson and Simmons, 1997). This is of particular significance for the areas of the North Sea where generation occurred some time ago and, because of inversion (uplift and erosion), has not continued to the present day.

In the southern gas province, Cretaceous inversion has cut off the supply of gas from the underlying Coal Measures into the Permian (Rotliegendes) reservoirs. While salt (halite) has been observed to be an almost perfect seal, with the capacity to reseal after fracturing, shale seals for gas accumulations have been calculated, from diffusion rates of light hydrocarbons, to have a half-life of some 60 Ma (Leythaeuser *et al.*, 1982). In this context, the 'half-life' of a gas accumulation is the time required to lose half the accumulation, in the absence of additional influx of gas. Other values for diffusion coefficients have been measured (Welte *et al.*, 1984; Whelan *et al.*, 1984) which give different estimates of half-lives (Krooss and Leythaeuser, 1997). Deming (1994) discusses the ability of shales to retain overpressure in terms of permeability, thickness and geological time, and calculates that typical shales are unable to retain pressure-potential gradients over long geological times (e.g. > 1 Ma), given observed permeabilities. Breaching of the salt seal—either by salt withdrawal or by faulting—is demonstrated by the presence of gas in the Triassic sandstones of the Hewitt Field, with remigration occurring during the late Cretaceous–Tertiary inversion of the Sole Pit Trough (Hillis, 1995a,b). Salt thinning during pillow formation also allowed Carboniferous gas to migrate into the Esmond, Forbes and Gordon fields of the southern North Sea (Ketter, 1991a).

Seals in the oil province of the North Sea are generally mudstone (north Central Graben, Witch Ground and Viking grabens), while in the southern Central Graben the Chalk acts as both reservoir and seal. In the latter area, the presence of numerous 'gas chimneys' attests to the limitations of the Chalk as a seal over time. In the Ekofisk area, van den Bark and Thomas (1981) and Munns (1985) have identified 'gas chimneys' on seismic lines, interpreted as the leakage of gas from the Chalk through the Tertiary (Caillet *et al.*, 1997). Active replenishment is expected in this area due to rapid Neogene and Quaternary burial. Other examples of 'gas chimneys' defined on seismic lines are given by Nordberg (1981) and Brewster and Dangerfield (1984).

The North Sea mudstone seals appear highly effective for oil, as evidenced by a general low level of shows in sand overlying the reservoir and absence of significant multipay reservoirs. Illustrating one extreme, Leith *et al.* (1993) report that oil shows are present in the Lower Cretaceous cap rock up to 400 m above the crest of the Triassic–Lower Jurassic reservoir of the Snorre Field. Early arrival of oil (during or shortly after cap-rock deposition), together with microfracturing, is believed to control seal failure in this case. The case for microfracturing has been made by Caillet (1993), who points out that, with a ~300 m oil column and strong overpressuring, the pore pressure at the top of the Snorre reservoir reaches some 82% of the lithostatic load and equates approximately to the minimum (horizontal) stress.

Sales (1993) has divided North Sea reservoirs into three categories—those with a seal retaining oil and gas (Class 1: Troll East), where Gussow-type gas displacement may be a problem; those with a seal for oil but leaking gas (Class 2: Oseberg), where excess gas will not spill oil; and finally those with a cap rock 'modestly' permeable to both oil and gas (Class 3: Gullfaks and Ekofisk) and unlikely to spill either. Although this is a generalization, seal quality, except where resulting from fault breaching, is rarely a controlling factor in North Sea oil accumulations.

In mid-Norway, tertiary migration has been invoked to explain the presence of shallow gas accumulations and gas chimneys overlying highly overpressured dry-gas accumulations in Jurassic reservoirs (Vik *et al.*, 1991). Here, rapid Pliocene subsidence overpressured the trapped gas, raising pressures to levels approaching the tensile strength of the cap rock. Hydraulic fracturing of the seal ensues, leading to shallow accumulations of thermal gas originating from the coals of the Lower Jurassic Åre Formation.

Tertiary migration has been invoked to explain the presence of gas and residual oil in the Barents Sea area of north Norway (Kjemperud and Fjeldskaar, 1989; Nyland *et al.*, 1992; Sales, 1989; Bakken, 1991). Here, rift-margin uplift of the Norwegian–Greenland Sea in the Palaeogene, coupled with Quaternary postglacial isostacy, has halted generation and uplifted oil- and gas-filled structures by some 1–2 km. Expansion of the gas, together with regional tilting, appears to have spilt much of the oil from the traps, leaving gas-filled structures, together with residual oil and palaeo-oil/water contacts well below the base of the present gas column. The evidence for gas remigration has been reported by Laberg and Andreassen (1996) in terms of seismically imaged free gas in the Palaeocene–Eocene mudstones (gas chimneys) and shallower hydrate zones, often associated with major faults.

Surface seeps form the end-point of secondary or tertiary migration (Schumacher and Abrams, 1996), and form excellent evidence of the existence of mature source rocks and active migration—although the presence of a seal is clearly in question (Horvitz, 1980; Selley, 1991). Some 173 onshore seeps of the UK have been catalogued by Selley (1992), but only the seeps of Dorset (Cornford *et al.*, 1988; Miles *et al.*, 1993), the Lancashire coast (Harriman and Miles, 1995) and Caithness (Parnell, 1983) are associated with oil production.

Offshore seeps have been detected in the North Sea in a number of ways. Hovland and Sommerville (1985) have proved the deep thermal origins of some surface gas seeps in the Central and Witch Ground grabens (Norwegian Block 1/9 and UK Block 15/25), while Buhrig (1989) has attributed gas chimneys to late-stage gas migration from an overpressured reservoir or source. In the North Sea area, surface prospecting for vertically migrated gas, using stable carbon-isotope analyses (Faber and Stahl, 1984), and by analysing seabed sediments for light and heavy hydrocarbons (Emmel *et al.*, 1985; Gervirtz *et al.*, 1985; Hvoslef *et al.*, 1996) have been attempted, together with sea-water 'sniffer' gas analysis (Schiener *et al.*, 1985) and the interpretation of clay diapirism and the presence of 'pock-marks' and carbonate build-ups on the sea floor (Hovland and Judd, 1988; Hovland, 1990, 1993). The major problem with this approach is confidently relating the surface 'anomalies' with generation and migration at depth via intervening lithology and structure.

11.6.5 Timing of migration

The timing of migration of oil or gas into the reservoir is often of great significance to prospectivity, it being possible to determine this from both ends of the migration path. Among 350 giant oilfields of the world, Macgregor (1996) analysed the accumulation history of 15 North Sea examples, and listed six as filling in the Neogene (Brent, Gullfaks, Snorre, Beryl, Cormorant, Claymore), seven in the Eocene–Oligocene (Statfjord, Ekofisk, Forties, Troll, Clare, Ninian, Eldfisk) and two in the Maastrichtian–Palaeocene (Oseberg, Magnus). Using basin modelling in the adjacent depocentres, Dahl and Yükler (1991) also concluded that the first arrival of oil in Oseberg Field occurred during the Maastrichtian, although Miles (1990) argued for at least some Late Cretaceous filling of the Ninian field. In more detail, studies of reservoir diagenesis have defined the timing of oil influx into the Tartan Field of the Witch Ground Graben (Burley *et al.*, 1989) and the Fulmar Sandstone reservoirs of the Central Graben (Saigal *et al.*, 1992). Isotope and diagenetic studies of the Groningen Field in the onshore Netherlands have defined the timing of gas influx into the Rotliegend sandstone (Lee *et al.*, 1985). At the generation end of the migration path, it is possible to estimate the timing of generation and expulsion of oil from the source rock, using thermal-geohistory plots and maturation modelling, as discussed previously (Section 11.4.2).

11.7 Oil and gas accumulation

Accumulation is the final process leading to commercial reservoired hydrocarbons. Hydrocarbon accumulations result from the interruption of the hydrocarbon migration path linking oil and gas kitchens with the earth's surface. An accumulation is thus a time-transient feature of the subsurface. Using a database of 350 worldwide giant oilfields (including a number of North Sea examples), Macgregor (1996) notes a median lifetime (fill to destruction) of 35 Ma. In this sense, an accumulation will occur when hydrocarbon enters a subsurface zone faster than it can escape, this being a special case of migration.

Within an accumulation, there is a certain amount of homogenization both vertical and horizontal (Leythaeuser *et al.*, 1989; Horstad *et al.*, 1990; Wills, 1991; Cubitt and England, 1995). Horizontal barriers are generally low-permeability zones (e.g. shale breaks within a sandstone reservoir), while vertical barriers are generally faults with associated cements and 'shale smears'. Where permeabilities are high, the mixing of hydrocarbons within a single reservoir unit can be both demonstrated and calculated to be a relatively rapid process.

What constitutes a commercial accumulation depends not only on the volume of oil or gas present, but also on how much of it can be produced to the surface, the cost of extraction relative to the size of the deposit, the current price of oil or gas, and the proximity of pipelines and/or an economy capable of utilizing the resource. Additional factors include the tax regime and the political stability of the region. The North Sea is an area of large accumulations, with a nearby market and stable politics, but due to the prospective areas being offshore, with extremes of weather and exacting safety requirements, it is an area with high costs of exploitation.

11.7.1 Hydrocarbon types

Hydrocarbons are roughly divided by density (API gravity) and composition into the groups listed in Table 11.9. Another common subdivision of oils is into paraffinic, naphthenic and aromatic types, based on the most abundant chemical component (Tissot and Welte, 1984). Other parameters, such as sulphur content, are used to separate sweet oils and gases—as are generally the case in the North Sea—from the sour crudes and gases, e.g. as found in the East Irish Sea Basin. The detailed composition of North Sea oils and gases is discussed in a later section. The classification of hydrocarbons in Table 11.9 is readily applicable to most North Sea production.

11.7.2 Intra-reservoir alteration

Once in the reservoir, hydrocarbons can continue to be altered (Palmer, 1991; Blanc and Connan, 1994; Macgregor, 1996), often in a way to decrease their commercial value. One-third of a worldwide database of giant oilfields showed evidence of post-entrapment destruction (Macgregor, 1996). Alteration processes divide into three groups (Fig. 11.37), resulting from:

Table 11.9 Hydrocarbon definitions based on density and composition.

Hydrocarbon type	Density: Specific gravity (g/cm^3)	Density: Gravity* (°API)	Composition (typical carbon number range†)	Comment
Dry gas	0.0007	—	C-1 (methane)	Methane > 95% gas
Wet gas	0.0010	—	C-1 to C-4	Methane < 95% gas
Gas/condensate‡	< 0.76	> 55	C-1 to C-20	Check reservoir P & T
Light oil	0.76–0.82	45–55	C-1 to C25	Most North Sea oil production
Mid-gravity crude	0.82–0.88	30–45	C-1 to C-35	
Intermediate	0.88–0.93	20–30	C-1 to C-35	Often a mixture
Heavy oil	> 0.93	< 20	Unresolved complex mixture, often lacking *n*-alkanes§	Can derive from a range of processes
Bitumen/asphalt	Typically > 1.0	Typically < 10	May reflect original oil	Generally highly aromatic

*API gravity = (141.5/density) – 131.5; density in g/cm^3 = t/m^3. †Range of *n*-alkanes in typical hydrocarbon of this type. ‡'Condensate' has many definitions—see text. §Most 'heavy oils' form from bacterial degradation of normal crude and hence lack *n*-alkanes.

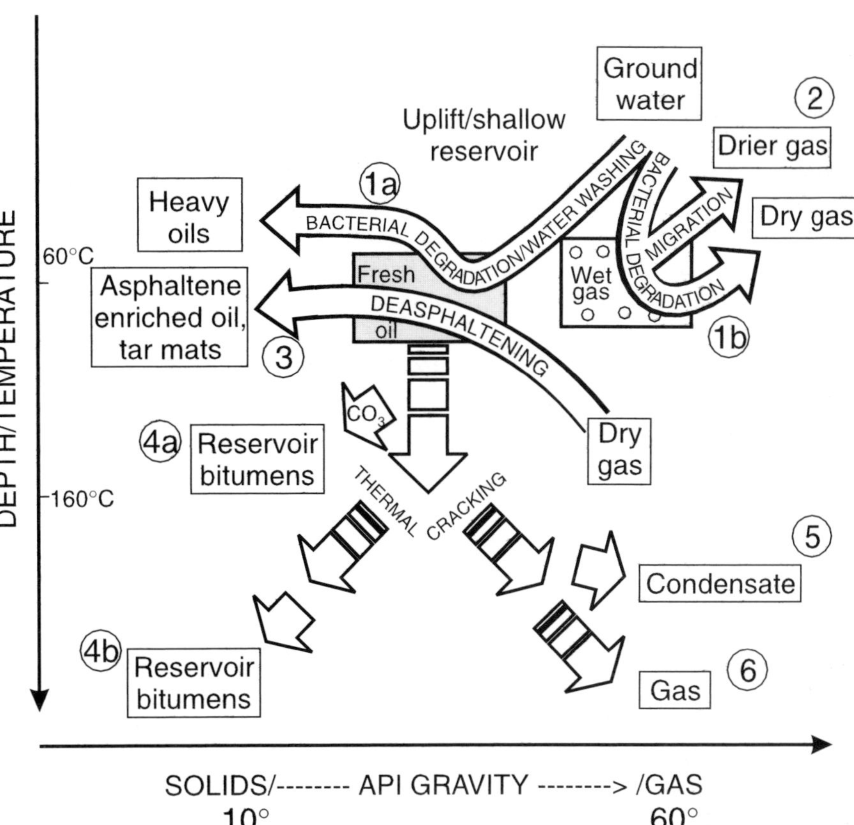

Fig. 11.37 Intra-reservoir alteration processes, where ingress of fresh water into a shallow and hence cool (< 60°C) reservoir produces heavy biodegraded from midgravity oil (1a) and dry from wet gas (1b). Fractionation during migration also tends towards drier gas (2). Influx of gas into fresh oil can precipitate asphaltenes, which settle to form tar mats (3), while reservoir bitumens (4a and 4b) can be formed in carbonates at low temperatures and in clastics at higher temperatures (> 160°C), the cracking reaction producing condensate and gas as the other product. Evaporative fractionation is not shown in this schematic. Together with source-rock maturity, these processes, which affect the API gravity of the hydrocarbon (horizontal axis), help explain the wide variety of oil properties in the North Sea oil province, despite a single uniform source rock.

1 Bacterial degradation, giving heavy oil plus dry gas, at reservoir temperatures less than ~65°C (Connan, 1984).
2 Excess reservoir temperatures, e.g. cracking of oil to gas plus pyrobitumen, at temperatures greater than ~165°C (Ungerer *et al.*, 1988; Pepper and Dodd, 1995).
3 Flux of fluids, e.g. water washing (Lafargue and Barker, 1988; Lafargue and Thiez, 1996), evaporative fractionation (Thompson, 1987a,b, 1991), asphaltene precipitation (Wilhelms and Larter, 1995), sulphate reduction (Rooney, 1995) or simply trap breaching.
Quantitatively, bacterial degradation is the dominating degradation process in the world, it being claimed that, prior to biodegradation, the tar sand of Athabaska (Canada) alone contained three times more oil than the entire conventional reserves of the world (Masters, 1984).

While oil-prone Type II kerogen (as found in the Upper Jurassic–basal Cretaceous source rocks of the North Sea) directly yields mid-gravity crudes, heavy oils result mainly from bacterial degradation of these mid-gravity crudes. Early-generated oils and asphaltene-enriched oils can also be heavy. Examples of asphaltene enrichment have been described from the Oseberg and Ula fields in the Norwegian Viking and Central grabens, respectively (Dahl and Speers, 1985, 1986; Wilhelms and Larter, 1994a,b). Using examples from the Troll, Agat, Ula and Snorre fields from the Norwegian North Sea, Wilhelms *et al.* (1996) have demonstrated an asphaltene precipitation sequence with the first-arriving early-mature oil precipitating early and the later-arriving oils from more mature source rocks precipitating the later asphaltenes.

Bacterial degradation of reservoired oil is now readily recognized, not only from anomalous bulk properties, such as density—which can be misleading—but also from its molecular and isotopic signature (Rullkotter and Wendisch, 1982; Cornford *et al.*, 1983; Volkman *et al.*, 1983; Connan, 1984; Ahsan *et al.*, 1997). In particular, the presence of a homologous series of 25-demethylated hopanes is indicative of intra-reservoir bacterial degradation (e.g. with respect to the Gullfaks field, see Horstad *et al.*, 1990). These latter authors document the mixing of degraded and undegraded oils to produce a so-called 'co-mingled' oil in some compartments of the Gullfaks field, a process repeated during the complex filling history of the Troll Field (Horstad and Larter, 1997). In an attempt to quantify degradation in the Gullfaks Field, Horstad *et al.* (1992) estimated that, of the 14 wt% of the undegraded oil that comprised C_8–C_{35} *n*-alkanes, 80% was removed. They report failure to define a geologically plausible mechanism on the basis that several thousand reservoir volumes of oxygenated water would be required to account for the 90×10^9 m^3 of CO_2 or 60×10^9 m^3 CH_4 produced.

Dry gas can be produced direct from a vitrinitic kerogen or coal (Higgs, 1986), as in the Southern Gas Province of the North Sea. In addition bacterial degradation of wet gas derived at a late-mature stage from oil-prone kerogen also produces dry-gas (James and Burns, 1984), as appears to be the case in the Troll and Frigg fields in the northern Viking Graben. In the latter case, an anomalously heavy, stable carbon-isotope ratio for propane is a key indicator although other factors can affect isotopic composition of hydrocarbons (Clayton, 1991a,b).

The cracking of oil to gas within the reservoir (see Fig. 11.25, and equation 22a) has been modelled, using kinetic parameters (Ungerer *et al.*, 1988; Andresen *et al.*, 1993; Pepper and Dodd, 1995; Schenk *et al.*, 1997). Using oils from Norwegian wells 2/4-14 and 33/9-14, distributed activation energies, with a mode at 66 kcal/mol, compensated by a higher frequency factor of 3.52×10^{29}/Ma, have been given for the general cracking reaction (Horsfield *et al.*, 1992). Laboratory temperatures of 300–500°C equate to geological temperatures of 160–190°C for geological heating rates of 0.53–5.3°C/Ma.

Taking three Norwegian examples from wells 7/11-5 and 2/4-14 and the Block 2/7 Embla Field oil, densities of 0.83 g/cm^3 (39° API), 0.816 g/cm^3 (42° API) and 0.784 g/cm^3 (49° API) are reported for reservoir temperatures of 161°C, 160°C and 165°C, respectively (Strand and Slaatsveen, 1987; Horsfield *et al.*, 1992; Knight *et al.*, 1993). In all these cases, rapid Neogene and Quaternary sedimentation may mean that the observed temperature has only been achieved in the recent past, and hence time (as opposed to temperature alone) may exert a significant kinetic effect.

11.8 Stratigraphic and geographical distribution of source rocks

This section comprises a review of the area (Fig. 11.38). The amount (TOC, yield in kg/t) and type (oil-prone, gas-prone) of the sedimentary organic matter, together with the maturity, are discussed. It should not be forgotten that in the study region the Kimmeridge Clay or Draupne Formation (Upper Jurassic/basal Lower Cretaceous) is overwhelmingly the most important oil source rock, and the Carboniferous Coal Measures (Westphalian) and locally the Middle Jurassic coals are the only well-established source rocks for dry gas.

Barnard and Cooper (1981) have previously reviewed the hydrocarbon source rocks of the North Sea, while Thomsen *et al.* (1983) and Rønnevik *et al.* (1983), have summarized data on Danish and Norwegian acreage, respectively. Many source rocks in the Irish offshore are discussed in Croker and Shannon (1995). To the north of the sixty-second parallel, Mørk and Bjorøy (1984) have reviewed the Mesozoic source rocks of Svalbard, with Spencer *et al.* (1984) providing additional source-rock information in the area of mid- and northern Norway.

11.8.1 Lower Palaeozoic

Cambro-Ordovician black shales are found on the shelves and in the basins of the Iapetus Ocean (Thickpenny and Leggett, 1987; Bharati and Larter, 1991), within and flanking the Caledonides (Fig. 11.39). Excellent source-rock quality has been described, from the Kukersite oil shales of Estonia in the east (Duncan and Swanson, 1965; Leggett, 1980; Thickpenny and Leggett, 1987; Derenne *et al.*, 1989; Brangulis *et al.*, 1993; Zdanavičiūtė and Bojesen-Koefoed, 1997) the Franklin Basin of North Greenland in the north (Christiansen, 1989; Koch and Christiansen, 1993) and the Appalachians in the west (Islam *et al.*, 1982). Within the area of interest, the Cambro-Ordovician black shales outcrop in the Southern Uplands of Scotland and in Scandinavia (Fig. 11.39) and could source both oil and gas (Thomsen *et al.*, 1987). Entrapment to the present day is largely a function of seal quality and timing.

Analogous contiguous deposits are found in south-west Ireland, Wales, the Welsh Borders and the Ardennes (Thickpenny and Leggett, 1987). A TOC value of 5% is reported for an Upper Cambrian black shale west of the Malverns (Smith, 1987), this being the only datum available for the English onshore. As illustrated in the inset in Fig. 11.39, the Alum Shale of southern Sweden, varying between 12 and 80 m thick (Thickpenny and Leggett, 1987), contains as much as 17.5% TOC (Bitterli, 1963; Andersson *et al.*, 1982, 1985; Buchardt *et al.*, 1986). The kerogen is oil-prone, giving 40–80 kg/t oil yield upon pyrolysis as an oil shale and Hydrogen Indices of 668 mg/g TOC for immature samples at Narke (Buchardt *et al.*, 1986). An activation energy of 47.9 kcal/mole and frequency factor of 2×10^{30}/my has been reported for oil generation (Lewan and Buchardt, 1989). In the extreme east, Brangulis *et al.* (1993) report TOC values of 11–23%, Type II kerogen with an H/C ratio of 1.5–1.7 and maturities in the oil window (~0.5–1.15% R_o) for the Alum Shale of the Lithuania–Polish border region. Poorer source quality is reported for the Cambrian to Silurian of wells from onshore and offshore Lithuania (Kanew *et al.*, 1994; Zdanavičiūtė and Bojesen-Koefoed, 1997).

In Sweden, at least part of the outcrop currently falls within the oil-generation window (Bergstrom, 1980; Buchardt *et al.*, 1986; Buchardt and Lewan, 1990; Vigneresse, 1993; Everlien, 1996), with maturities falling towards immature/early-mature levels (0.59–0.77% R_o) on the island of Öland (Buchardt *et al.*, 1986). Thomsen *et al.* (1987) argue on regional grounds that the maturity was attained prior to the Permo/Carboniferous. This is confirmed by a burial-history plot in Brangulis *et al.* (1993), which shows, following Lower Carboniferous uplift, continued burial to the present. Vigneresse (1993) attributes the variable maturity of the Alum Shale to basement (granite) radioactivity and, in the Oslo Graben, to Permian burial and vulcanicity. In excess of 100 ppm of Uranium is reported for the Alum Shale (Lewan and Buchardt, 1989), the resulting 500 mg of radiation reducing the hydrocarbon generative potential by polymerization (cf. Sundaraman and Dahl, 1993). Everlien (1996) has pointed out that gas may be generated even from the late-mature Alum Shale. Vlierboom *et al.* (1986) associate the asphalts and oil seeps of the Siljan meteoric-impact crater in Sweden with localized maturation of the Lower Palaeozoic organic-rich shales of the district during the late Devonian impact event. The presence of astroblemic oil shows in the Siljan area was influential in promoting the unsuccessful drilling of two wells (Siljan-1 and Siljan-2) to test the mantle-methane theory of Thomas Gold (Gold, 1985).

Ordovician organic-rich shales are noted in Ireland, Wales, the Ardennes and the Amorican Massif (Thickpenny and Leggett, 1987). Few TOC or kerogen-type data are available for the Ordovician black shales of the area, although maturity data are available; a TOC value of 2.4% is recorded for a Tremadocian black shale at Nuneaton (Smith, 1987). The Ordovician (Caradocian) black shales of the Southern Uplands of Scotland are postmature for both oil and gas generation at Hartfell and Dobb's Linn and late-mature for gas generation at Mountbenger (average uncorrected graptolite reflectance values of 4.45% R_o, 3.27% R_o and 2.20% R_o, respectively; Watson, 1976). For a discussion of the interpretation of reflectance maturity measurements in Lower Palaeozoic rocks made on particles

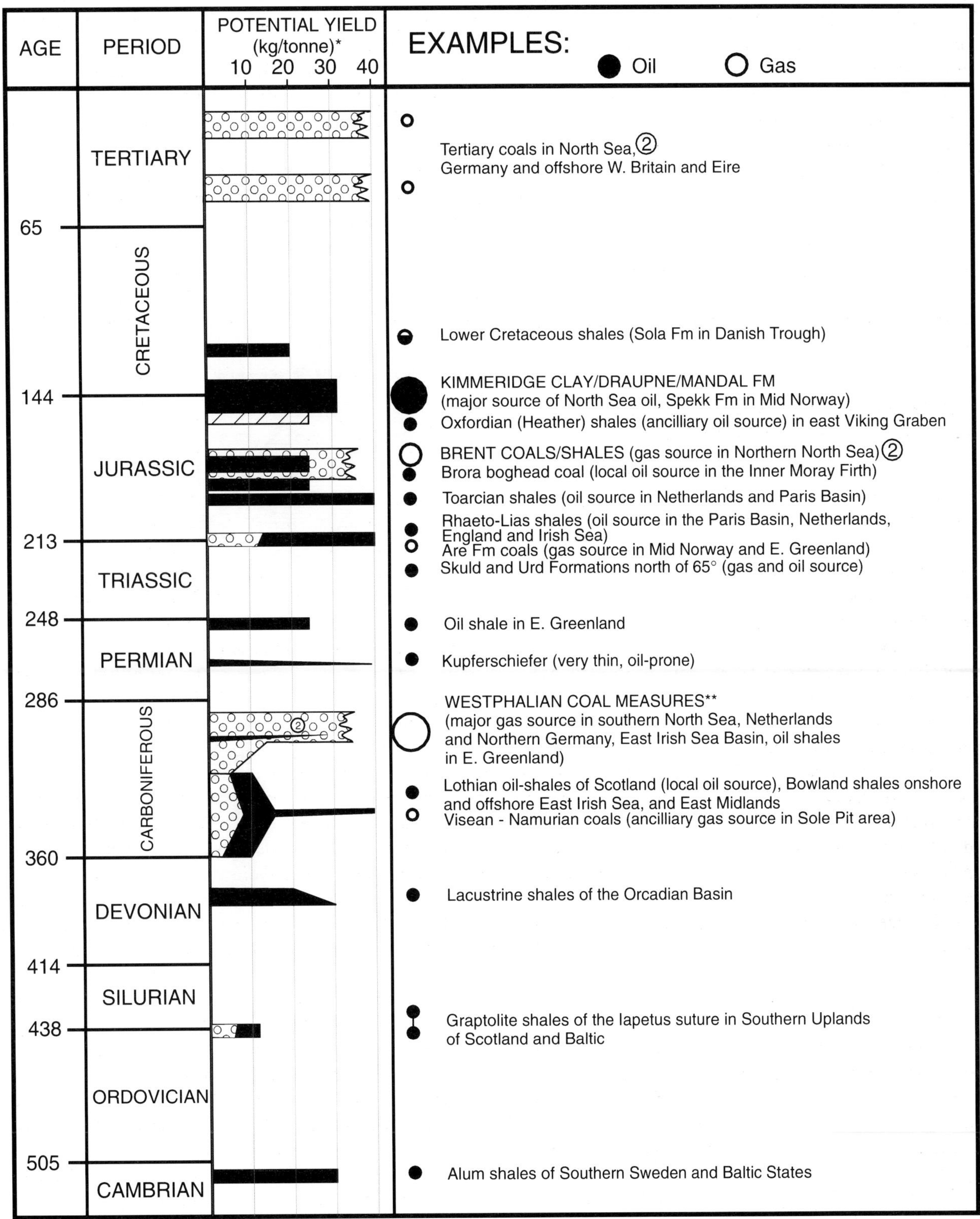

Fig. 11.38 Stratigraphic distribution of source rocks in the North Sea and adjacent areas. Major proved source rocks are in UPPER-CASE and locally developed or putative source rocks are in lower-case text. *Potential yields are corrected to late-immature levels and represent the richest significant units, which are often of restricted area. ** (or ②) Potential yields of coal are difficult to define (see text).

other than vitrinite, see Bertrand and Heroux (1987) and Goodarzi and Higgins (1987). Watson (1976) and Bergstrom (1980) note a decrease in maturity to the west, but Illing and Griffith (1986) report (non-vitrinite) reflectance values between 3 and 8% R_o on the highly folded Ordovician shales of County Down, Northern Ireland. Reflectance values of 1.1–2.0% R_o for two Ordovician inliers in Fermanagh and Tyrone (Illing and Griffith, 1986) confirm lower maturities for the south-west of Northern Ireland.

A range of 2–4% TOC is reported for the Silurian (Llandovery) black shales at Moffat (Thickpenny and Leggett, 1987) with stable carbon isotope data for the nearby Dob's

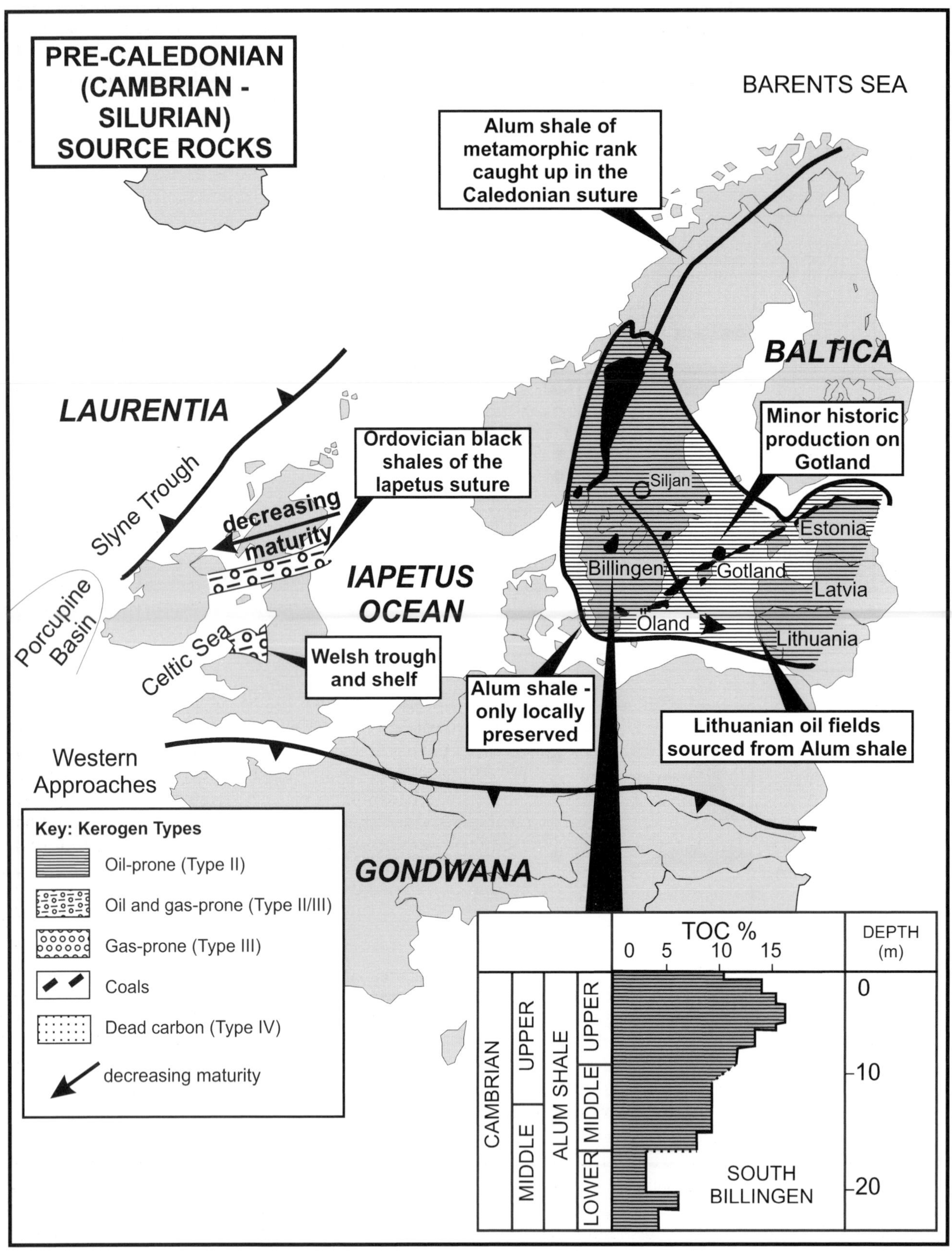

Fig. 11.39 Geographical distribution of pre-Caledonian Palaeozoic source rocks of the North Sea and adjacent areas. The major proven source rock is the Cambrian Alum Shale deposited on the shallow shelf of Baltica. Maturity decreases to the south and east away from the Caledonian suture. The inset shows TOC values through a 23 m section of Alum Shale outcrop in southern Sweden. (From Thickpenny and Leggett, 1987; Kulke, 1994.)

Linn section reported by Underwood *et al.* (1997). Lateral equivalents are found in Northern Ireland, the Lake District and North Wales (Leggett, 1980). The organic and inorganic maturity of the Silurian of the Southern Uplands of Scotland has been defined by Pearce *et al.* (1991). Organically rich (up to 16.46% TOC) Lower Silurian graptolitic black shales are described in the eastern Baltic by Brangulis *et al.* (1993), while well data for undifferentiated Silurian produced TOC values up to 11% TOC and yields up to 55.5 kg/tonne (Zdanavičiūtė and Bojesen-Koefoed, 1997).

Maturity data for Lower Palaeozoic sediments of Wales, the Midland Massif and the Pennines have been derived from vitrinite reflectance (Smith, 1993), the Conodont Alteration Index (CAI) (Aldridge, 1986) and clay mineral crystallinity (Robinson and Bevins, 1986; Oliver, 1988). A strip of lower-maturity strata (diagenetic level of crystallinity, CAI 1–2.5) is identified in the Welsh borderlands. Smith (1987) gives the maturity of the Lower Palaeozoic (Tremadocian) of the English Midland Massif as falling in the range 1.5%–2.8% R_o, with a steep depth gradient, in contrast to the greenschist facies recorded in the lateral equivalents in Wales to the west and East Anglia in the east.

There is no current evidence of a Lower Palaeozoic source for any North Sea hydrocarbon occurrence, but generation from these rocks is known in the palaeocontiguous Appalachians of North America in the west (Longman and Palmer, 1987) and in Estonia to the east (Thomsen *et al.*, 1987; Brangulis *et al.*, 1993; Zdanavičiūtė and Bojesen-Koefoed, 1997) Ordovician-sourced oils, in particular, have a unique molecular signature (Longman and Palmer, 1987; Derenne *et al.*, 1990; Guthrie and Pratt, 1995), from which they could be recognized, if present.

11.8.2 Devonian

The Middle and Upper Devonian lacustrine marls and shales of the Orcadian Basin sequence of north-east Scotland and the Orkneys is the only significant source rock in this interval. The approximate distribution of organic-rich Old Red Sandstone sediments is shown in Fig. 11.40 (Hamilton and Trewin, 1985; Marshall *et al.*, 1985; Parnell, 1985; Duncan and Hamilton, 1988; Hillier and Marshall, 1988; Parnell, 1988a). These lacustrine shales and marls (Donovan, 1988) constitute a minor lithology within the thick prograding sequence (Donovan *et al.*, 1974) and are often associated with oil staining, bitumens and seeps (Parnell, 1983; Robinson *et al.*, 1989). The local thicknesses of Devonian units are discussed in more detail in Chapter 3 (this volume). Hall and Douglas (1983) obtained TOC values of 0.6–5.2% for five Devonian samples, while Duncan and Hamilton (1988) report more data within this range. Histograms of 158 TOC values from the three outcrop locations (Orkneys, Caithness and the Moray Firth area) are shown as an inset in Fig. 11.40 (partly from Parnell, 1985, part new data), together with averages sorted by stratigraphy. Bjorøy *et al.* (1988) contrast rich examples of this facies with a worldwide database of non-marine lacustrine source rocks. Duncan and Hamilton (1988) produced evidence for lacustrine deposition based on kerogen carbon/sulphur ratios (Berner and Raiswell, 1984) while Irwin and Meyer (1990) have put forward an elegant model differentiating lake organofacies and sediment maturity by applying multivariate statistical techniques to a wide range of geochemical parameters.

In the offshore, a Devonian sample from UK well 12/27-1 yields TOC values <2.2% and a present-day Hydrogen Index <600 mg/g TOC (Peters *et al.*, 1989; Marshall, 1998). In a detailed molecular and isotopic study of a Middle Devonian anhydritic lacustrine section penetrated by UK well 9/16-3, Duncan and Buxton (1995) report TOC values from cuttings averaging 0.5%, with a maximum of 0.91%. At the molecular level, any oil generated from this facies would be characterized by abundant gammacerane and β-carotane.

The kerogen type of the Devonian lacustrine source rock is poorly defined and appears variable, from oil-prone to gas-prone: visually, the kerogen appears dominantly amorphous with bitomen (Marshall, 1998). Hydrogen Indices range up to about 700 mg/g/TOC for the least mature samples of the carbonate laminites (Parnell, 1988a). Taking an average of 1.4% TOC gives an oil potential of 11.8 kg/t, the organic-rich lithology does not constitute more than 20% of the lacustrine units.

Onshore maturity is mainly within the oil window based on estimates from Hall and Douglas's (1983) figured sterane and triterpane distributions. Spore colour values from three outcrop locations (Orkneys, Caithness and the Moray Firth area) fall in the late-immature–early-mature range (Duncan and Hamilton, 1988), while vitrinite reflectance values, dominantly in the 0.68–1.09% R_o range, place the Middle and Lower Old Red Sandstone of the eastern Central Valley of Scotland within the oil window (Marshall *et al.*, 1994; Friedman, 1995). Hillier and Marshall (1988) report that the sequences of Walls Sandstone on Fair Isle and south-east Shetland are dominantly overmature. A further area of overmature Middle Devonian sediment (Spore Colour Indices of 9–10) was noted in the Wick–Lybster area on the east coast of Caithness. Rapid lateral changes in maturity in the Rhynie area are attributed to hydrothermal activity (Rice *et al.*, 1995). In Caithness, the Lower Old Red Sandstone is always postmature for oil and gas, there being a puzzling maturity break approximating to the Lower–Mid-Devonian boundary. Within the oil window, a high maturity gradient is reported for the offshore Lower Devonian of the UK well 12/27-1 (Marshall, 1998).

As shown in Fig. 11.40, the Orcadian Basin itself may extend from eastern Scotland and the Orkneys over to the Hornelen Basin exposed on the west coast of Norway (Ziegler, 1982; Duncan and Hamilton, 1988; Hitchen and Ritchie, 1987). The extent to which the lacustrine source rock-facies continues under the North Sea is unknown. On the basis of a somewhat tenuous geochemical oil/source-rock correlation, based mainly on sterane carbon-number distributions, Duncan and Hamilton (1988) suggest that the Beatrice oil in the Inner Moray Firth contains a Middle Devonian-sourced component, a conclusion independently confirmed by Peters *et al.* (1989), on the basis of evidence from stable carbon isotope, β-carotane and desmethyl sterane distributions. Further confirmatory evidence has been supplied by Duncan and Buxton (1995) using offshore Middle Devonian samples. Bailey *et al.* (1990), using a multi-parameter approach, go as far as to specify the Upper Caithness Flags as the specific co-source for the Beatrice oil.

As pointed out in a global review of source rocks of the Upper Devonian (Ormiston and Oglesby, 1995), source rocks are also found in Brittany (Porsquen Formation) and Poland. Upper Devonian freshwater shale, with averages of 10% TOC and initial Hydrogen Indices of 700 mg Py/g TOC has been identified as a putative source rock in East Greenland (Christiansen *et al.*, 1993a).

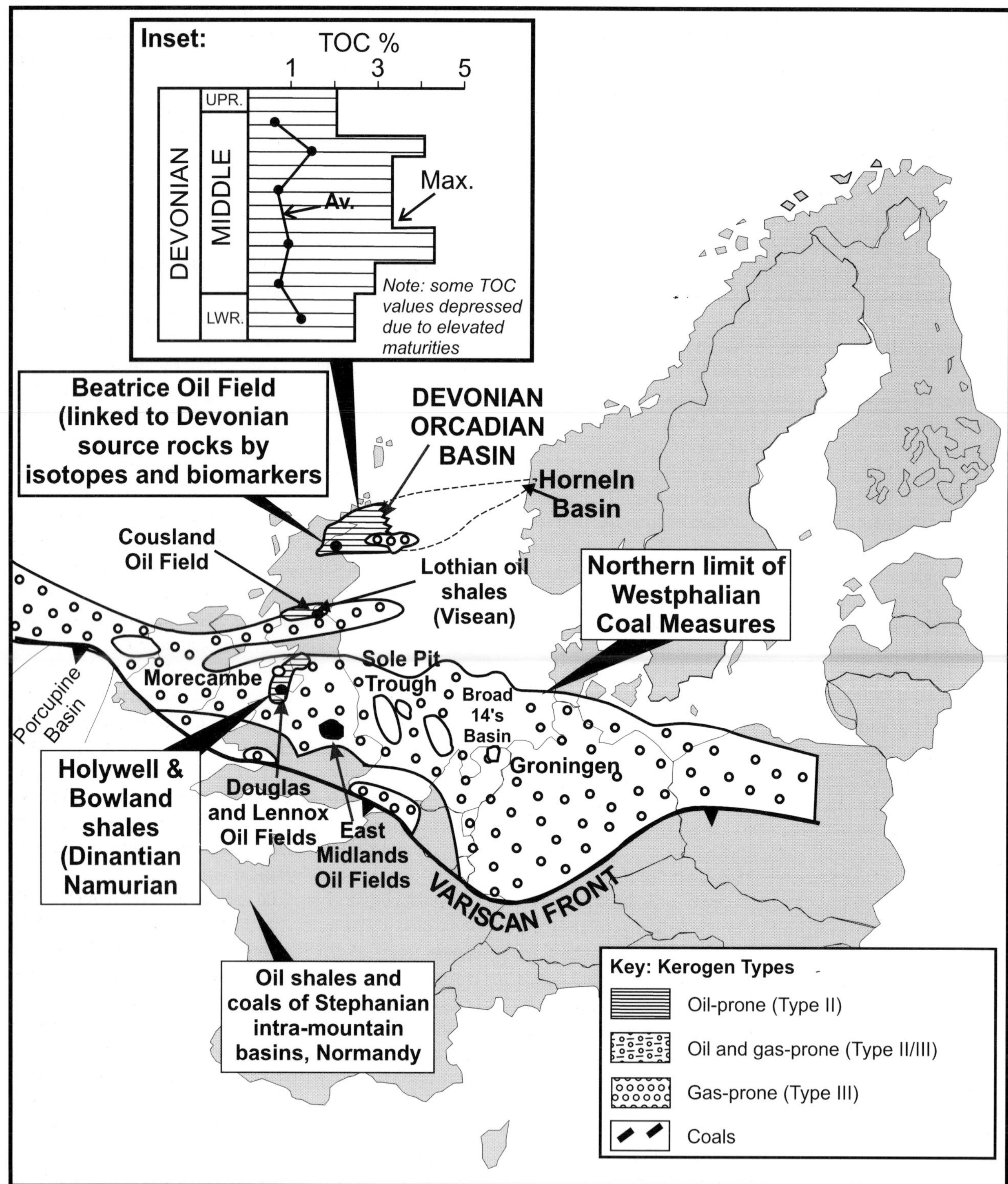

Fig. 11.40 Geographical distribution of post-Caledonian to pre-Variscan source rocks of the North Sea and adjacent areas. Oil seeps in Caithness and Orkneys, together with the Beatrice oilfield, are linked to the Middle Devonian Orcadian lacustrine source rocks. The major source rocks deposited in this time period are the gas-prone Westphalian Coal Measures productive in the inverted basins of the southern North Sea, onshore Netherlands and East Irish Sea. Local development of oil-prone oil shales in the Carboniferous sourced oilfields in the East Irish Sea, West Midlands and Lothians. The inset shows average and maximum TOC values for ~160 outcrop samples of Middle Devonian Orcadian Basin lacustrine source rocks.

11.8.3 Carboniferous

Carboniferous sediments (coals and associated carbonaceous shales) have sourced the major gasfields of the southern North Sea, onshore Netherlands and Germany (Bartenstein, 1979; Tissot and Bessereau, 1982; Barnard and Cooper, 1983; de Jager *et al.*, 1996), while the minor UK onshore oilfields of the East Midlands and Central Valley of Scotland and the East Irish Basin offshore are reputedly sourced from locally developed liptinite-rich sediments

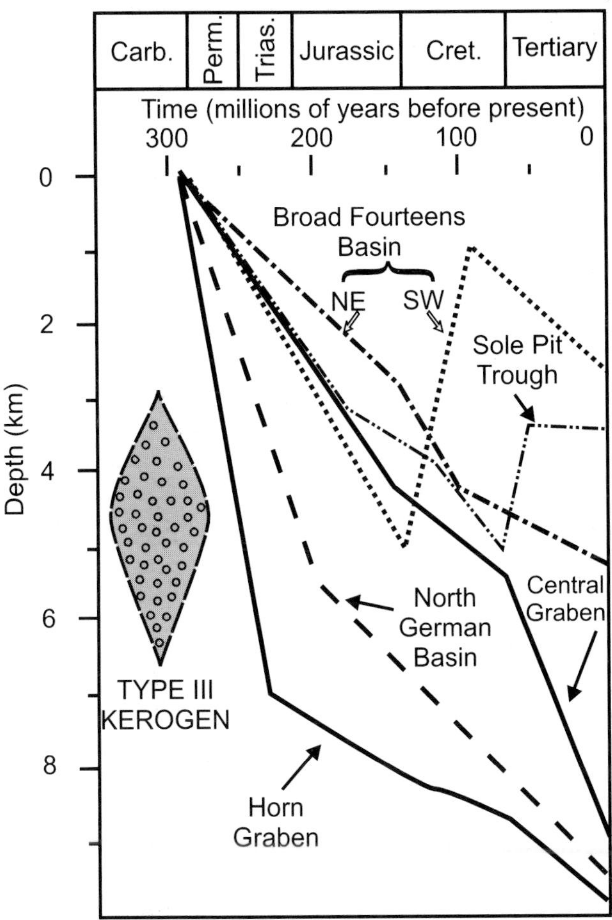

Fig. 11.41 A comparison of published burial-history curves for the top Westphalian source rocks of the southern gas basin of north-west Europe. The curves generally reflect depocentres, with lesser degrees of burial and/or later generation in upflank locations. The progressive shift of the locus of the depocentre with time seen in the Broad Fourteens Basin seems to be characteristic of transpressional Cretaceous–Tertiary inversion.

within the Carboniferous. The distribution of these source rocks and their kerogen type are shown in Fig. 11.40.

Many of the gasfields sourced by the Carboniferous occur in areas subjected either to early Mesozoic subsidence and inversion (Fig. 11.41) as in the Sole Pit Trough (Glennie and Boegner, 1981) and the Broad Fourteens Basin (Oele *et al.*, 1981), or to areas of higher geothermal gradients (Kettel, 1983). Donato (1993) has suggested the presence of granites under the north-eastern flank of the Sole Pit Trough: if radioactive, they may augment the regional heat flow and hence assist maturation. The Carboniferous source-rock story is not so much about kerogen quantity (which is huge) or quality (which is overwhelmingly gas-prone), but about maturity and the timing of gas generation relative to the development of reservoir, structure and seal.

Dinantian (Visean/Tournaisian)

The Visean oil-shale facies of the Central Valley of Scotland—deposited in 'Lake Cadell'—was probably laterally restricted (Loftus and Greensmith, 1988), but outcrops of a similar type are known as far apart as Linwood (south of Glasgow) in the west, to the Lothians and the Firth of Forth in the east. The extension of this facies further east (offshore) is a matter of speculation (Ziegler, 1982a; Leeder and Boldy, 1990), but its presence has not been reported in any published well information. Loftus and Greensmith (1988) show a restricted area for Lake Cadell, even at its maximum transgressive extent.

The Visean oil shales of the Central Valley of Scotland constitute a high-quality oil-prone source rock (Parnell, 1988b). Total organic carbon of 11.2% (Bitterli, 1963) and oil yields of about 30 gal./ton (approx. 100 kg/t) (Duncan and Swanson, 1965) probably represent the richest beds. Bjorøy *et al.* (1988) report TOC values of 34–49% and pyrolysis yields of 277–292 kg/t for Scottish torbanite and boghead coal, and two Lothian shales with TOC values of 11.70 and 16.88% and pyrolysis yields of 71–85 kg/t. More recent data show the richest beds to reach nearly 30% TOC, with Rock Eval pyrolysis yields up to 180 kg/t (Parnell, 1988b). Algal ('boghead') coals or 'torbanites' exist which consist of almost pure compressed bodies of the colonial algae *Botryococcus* (Allan *et al.*, 1980; Parnell, 1988b). At outcrop, these boghead coals and oil shales appear to be early- to mid-mature with respect to oil generation, as estimated from the rank of the surrounding coals (high to medium volatile bituminous), and biomarkers (Bateson and Haszeldine, 1986), although transient contact heating by Stephanian sills has locally raised the maturity (Robinson *et al.*, 1989; George, 1992; Raymond and Murchison, 1992). These oil shales should produce a fairly high-wax oil, judging from the *n*-alkane distribution recorded by Douglas *et al.* (1969) and Allan *et al.* (1980) and given the known oil-generative capacity of freshwater algae. The Lothian oil shales, having generated mineral hydrocarbons (Parnell, 1984; Robinson *et al.*, 1989), may well be the source for the nearby abandoned Cousland oilfeld.

The Dinantian Bowland shales are rich source rocks with TOC values <6% and HI values <300 mg/g TOC (Lawrence *et al.*, 1987; Thompson *et al.*, 1994). As rich, oil-prone source rocks: maturities fall in the peak to late oil-generation range (vitrinite reflectance 0.71–1.46% R_o) for the Dinantian and Namurian. The Bowland shales are believed to be the source of the methane gas that exploded in the Abbeystead water tunnel, killing 16 people in 1984 (Selley, 1992). Maturity levels of the Dinantian of northern England (Burnett, 1987) and Ireland (Jones, 1992) have been defined by conodont colour estimates. In Northern Ireland, Parnell (1991) notes Dinantian and Namurian coals, cannel coals and historic records of oil shales, explaining the bitumens in the overlying Permo-Trias.

In the offshore area, the East Irish Sea Basin has both oil- and gasfields. The oils of the Douglas, Lennox and Formby fields are reported to derive from the Dinantian/Namurian Holywell Shales (Trueblood *et al.*, 1995), although Armstrong *et al.* (1995) report fluctuating properties from onshore samples, ranging from 2 kg/t to > 10 kg/t pyrolysis yields at the single eponymous outcrop. An additional source-rock facies is indicated by the presence of a second oil family based on stable carbon-isotope ratios, as pointed out by Harriman and Miles (1995).

Coals of Visean age have been recorded over the southern part of the Mid North Sea High, and may constitute a source for relatively small gas accumulations. Both oil and gas potential is noted in the early- and mid-oil mature Lower Carboniferous boghead and humic coals of central Svalbard (Abdullah *et al.*, 1988).

The Namurian

Namurian marine shales may have oil potential over a limited area to the north of the London–Brabant High. Here and further north, the Namurian is relatively rich in coal seams (Tubb *et al.*, 1986) and could hence be an additional source for gas in the southern North Sea. Shales within the Upper Namurian of the Silver Pit Trough have TOC values in the 10–20% range, with mixed gas plus oil potential (Bailey *et al.*, 1993). Sparse penetrations of the Lower Namurian of this area indicate up to 2 km of leaner gas-prone shales, with 2–5% TOC. Based on log and geochemical data from UK well 48/3-3, Leeder *et al.* (1990a) have discussed the identification of Namurian radioactive carbon-rich black shales. A Namurian-sourced oil and gas play is also possible on the southern margin of the Mid North Sea High (Collinson *et al.*, 1993). Namurian coals have been attributed gas-source potential in the 'Midland Valley' of Northern Ireland (Illing and Griffiths, 1986). Maddox *et al.* (1995) allude to a possible Namurian oil-source rock in the Central Irish Sea Basin.

As pointed out by Fraser *et al.* (1990), the oil-source rocks of the onshore UK Carboniferous become younger to the south. The oldest Namurian oil-prone source rock is identified as a Pendelian high-gamma black shale in the Gainsborough, Widmerpool and Wellbeck basins of the East Midlands, with 3–4% TOC and hydrocarbon yields in excess of 15 kg/t. The middle unit of Namurian (lower Arnsbergian) age possesses oil potential in the same basins, with hydrocarbon potentials of 5–10 kg/t. An upper leaner unit of late Arnsbergian–Marsdenian is of wider distribution, though of inferior oil potential (5–7 kg/t). Delta-top marine bands and interdistributary shales are the most widespread and, though thin, have typical potentials of 7–10 kg/t (Fraser *et al.*, 1990). Approaching marine bands from a sequence-stratigraphic context, Maynard *et al.* (1991) and Wignall and Maynard (1993) report TOC values up to 14.3% and pyrolysis S_2 yields up to 40.5 kg/t for black shales of the Owd Betts and *Gastrioceras cumbriense* marine bands in the Leeds—Sheffield area of northern England. Using high-temperature micropyrolysis, Everlien (1996) has demonstrated further gas-generation potential of two Namurian black shales (1.92% TOC and 3.71% TOC) from Yorkshire and Wales, even though the latter has reached maturities of 2.89% R_o.

The Namurian shales, such as the Edale Shale, are currently oil-mature in wells Gainsborough-2 and Applehead-1 (Russell and Pearson, 1990), and are modelled (using TTI calculations) to have reached the peak mature levels as early as late Westphalian (Variscan) in some areas (Kirby *et al.*, 1987). Although Fraser *et al.* (1990) put forward evidence for post-Carboniferous maximum burial prior to 900 m of uplift in Bardney-1 well in the East Midlands. Onshore maturities appropriate for the generation of oil from a Lower Namurian source rock are modelled as having been attained in a Palaeogene (~60 Ma) maximum-burial event (Green, 1986; Lewis, 1992; Duddy and Bray, 1993; Green *et al.*, 1993, 1995a,b; Holliday, 1993, 1994; McCulloch, 1994a,b, 1995a,b), with Neogene uplift (Fraser *et al.*, 1990). In addition, Green *et al.* (1993b) have suggested a role for hydrothermal flux in maturing the Carboniferous source rocks, a proposal confirmed with detailed modelling by Cornford and Highton (1995) in the East Irish Sea Basin. The bitumens of Windy Knoll in Derbyshire, though impregnating in the Visean limestones, are believed to derive from the overlying Namurian Edale Shales (Pering, 1973), with the associated mineralization also indicating the influence of hydrothermal activity from adjacent basinal brines (Ewbank *et al.*, 1993, 1995). These latter authors, rejecting any correlation with the East Midlands oils, identify an organic facies of the Lower Namurian plus ?Dinantian of the 'Edale Gulf', with Type II kerogen as the most probable source for the bitumens. Hydrocarbon generation is modelled in the latest Carboniferous, followed by mineralization in the late Carboniferous and early Permian.

Kent (1985) and Storey and Nash (1993) assert that Namurian shales may also have sourced the high-wax mid-gravity oil of the Eakring field of Nottinghamshire, which is reservoired in the associated Millstone Grit (Namurian B to Westphalian A sandstones). This assertion was confirmed by Fraser *et al.* (1990) on the basis of stable carbon-isotope ratios of saturate and aromatic fractions. They identified the isotopically light basin-edge oilfields to derive from the Namurian pro-delta shales, with a major marine-band contribution to the isotopically heavier oils of the shelf fields. Harriman and Miles (1995) point out that this isotopic contrast is mirrored to the west in the Liverpool–East Irish Sea area.

The early oil-mature shales of the Visean/Lower Namurian Limestone and Scremerston Coal Groups of Northumberland are organic-rich (1.08–3.80% TOC), containing approximately equal quantities of exinite and vitrinite (Powell *et al.*, 1976). At appropriate maturities these rocks could generate some oil or condensate, as well as gas. The stratigraphically equivalent Holywell Shale of the East Irish Sea area is discussed with the Visean above. The Visean/Lower Namurian Clare Shale of western Ireland (well Doonbeg-1) is reported to have high TOC contents (2.4–5.4% TOC) and gamma-log responses, although at the drilled location the maturity of ~5% R_o indicates low-grade metamorphism (Croker, 1995b).

The Coal Measures (Westphalian)

The Coal Measures, comprising coal seams and associated shales, were deposited in a large delta system covering much of north-western Europe (Ziegler, 1982a, Glennie, 1986b; Kettel, 1989). In the North Sea region, the area extends from a latitude of about 57°N to the Variscan front in the south, although coals are probably restricted to an area south of 55°30′N (Fig. 11.40). Over much of this area the Coal Measures, where preserved in full, are typically between 1000 and 2500 m thick, of which, in north-west Germany, coal seams comprise about 3% (Lutz *et al.*, 1975). It is still a matter of largely academic debate (e.g.

Rigby and Smith, 1982) as to whether coals or the associated carbonaceous shales constitute the major source of gas: whatever the detailed source, Westphalian strata contain a large thickness of rich, dominantly gas-prone coals and shales. It is of interest to note that Bailey *et al.* (1990) attribute a Westphalian source to the Forcelles field of the Paris Basin, a proposition supported by Kettel (1989) on palaeogeographic grounds.

Carboniferous coal rank in north-western Europe varies from subbituminous to meta-anthracite. The majority of gas generation occurs between vitrinite reflectance values of 1 and 3% R_o (see Fig. 11.12), that is, from medium-volatile bituminous coal to anthracite. The presence of Kupferschiefer at a subbituminous-equivalent rank overlying Westphalian anthracites in north Germany and north-east England (Boigk *et al.*, 1971; Gibbons, 1987; Grice *et al.*, 1997) shows that at least some of the area attained its present maturity by early Zechstein time. The Carboniferous maturity of Irish onshore has been mapped, using conodont colour (CAI), by Jones (1992) and apatite fission tracks by McCulloch (1994b), both of whom demonstrate low maturities in northern and eastern Ireland (McCulloch, 1994b), with meta-anthracite equivalents in the south in the zone of the Variscan front.

In the Variscan tectonic episode, vast amounts of gas and possibly oil must have been generated during the latest Carboniferous and early Permian burial, only to be rapidly lost during the subsequent uplift and erosion. The major Variscan coalification occurred in a series of possibly thrust-soled basins just to the north of the Variscan Front, as evidenced by the anthracites of South Wales (Gayer *et al.*, 1997), Kent and the Ruhr district of Germany. Using high-temperature pyrolysis, Everlien (1996) has pointed out that some methane can continue to be generated from coals, even at ranks in excess of 3% R_o.

To the north of the Variscan Front (Taylor, 1986), many areas survived the Variscan orogeny with their gas-generating potential relatively intact (i.e. having vitrinite reflectance values of less than ~2% R_o), and it is these areas that should be considered as being capable of generating more gas on subsequent burial. South of the Variscan Front, most reported maturities approach the greenschist facies, beyond the gas-preservation deadline (Cornford *et al.*, 1987; Smith, 1993).

In the centre of the Sole Pit Trough, burial of the top Carboniferous up to about 4000 m occurred progressively from the Triassic to the late Cretaceous (Fig. 11.41; Glennie and Boegner, 1981), followed by inversion (van Hoorn, 1987a). Vitrinite reflectance values of 1.0% R_o to > 2.8% R_o are reported by Robert (1980) along the axis of the trough, while values < 1.0% R_o are measured on the flanks (see cartoon in Fig. 11.34F). Values of 1.07–1.39% R_o and 1.6–2.1% R_o are reported in Indefatigable (Pearson *et al.*, 1991) and Leman (Hillier and William, 1991) field wells, respectively. Gas generation and migration will thus have occurred during the Jurassic and Cretaceous in the basin centre, with migration to the basin flanks. Late Cretaceous inversion of up to 1500 m (Cope, 1986; but see also Bulat and Stoker, 1987; Green, 1989; Green *et al.*, 1989; Hillis, 1995a,b) resulted in remigration of gas. Estimates of uplift and erosion depend not only on the technique used (e.g. vitrinite reflectance, sonic velocities, apatite fission tracks), but also on the definition of an 'unexhumed trend' (Hillis, 1995a,b). This latter author, using sonic-velocity analysis identifies some 1.0–1.5 km of regional Tertiary uplift, with an additional ~1.0 km of uplift associated with the defined inversion axes. The presence of 'regional' uplift precludes the establishment of an 'unexhumed trend' from local wells, a conclusion broadly supported by the apatite fission-track interpretation (Bray *et al.*, 1992).

Van Hoorn (1987a) discussed the Sole Pit inversion in a regional context, with Green *et al.* (1989) putting forward evidence for a more complex uplift history. Two separate inversion events are proposed during the mid-Cretaceous and early Tertiary, based on apatite fission-track measurements. In a study of the Hewett Field in the south of the gas basin, Cooke-Yarborough (1991) gives a burial-history plot also showing two phases of uplift—one in the mid-Jurassic and one in the late Cretaceous. This results in two phases of gas generation from Westphalian source rocks, together with two phases of reservoir diagenesis. Gas in former basin-margin structures (e.g. Indefatigable area) is predicted to have remigrated into more axial Rotliegend reservoirs, which had suffered diagenetic damage during maximum burial (e.g. West Sole and Leman Bank fields). The survival of reservoired gas to this day, despite these inversion events, bears witness to the exceptional sealing properties of the Zechstein halites.

Outside the areas of inversion (e.g. UK Quadrant 44), subsidence has continued (Cope, 1986), with gas generation from Westphalian coals being modelled as continuous from the Cretaceous to the present day. Within the Silverpit Basin (UK Quadrants 43 and 44), Bailey *et al.* (1993) report up to 100 m of preserved Westphalian, with coals constituting 5% of Westphalian A and 8% of Westphalian B–C. The top Carboniferous maturity is in the range 1.0–1.2% R_o. On the Mid North Sea High (UK Quadrants 36 to 39), maturity is controlled by a complex interaction of at least three unconformities. The intra-Carboniferous and Variscan uplift and erosions cut down into the basement on the high, while the late Jurassic footwall uplift of the southern Central Graben bounding faults forms an unconformity that intersects with the Variscan surface. Finally, the Tertiary (Neogene) erosion event (e.g. Japsen, 1993) cuts down yet again to place post-Neogene sediments on the Carboniferous. Considering UK Quadrant 38, the Variscan unconformity cuts deeper to the north, the Jurassic deeper to the east and the Neogene deeper to the west.

In contrast to intersecting discrete events, Oele *et al.* (1981) have shown that generation in the tectonically inverted Broad Fourteens Basin continued from late Jurassic to the present, with a progressive north-easterly migration of the depocentre. Present-day reflectance values up to 2.4% R_o are found along the pre-inversion basin axis by van Wijhe (1987a,b), who also suggest that wetter gas found in some areas could be attributed to generation from the more spore-rich coals of Westphalian B and C age. The subcrop and maturity of the Carboniferous of the onshore extension (i.e. the West Netherlands Basin and Roer Graben) are mapped by Winstanley (1993), Veld *et al.* (1996) and de Jager *et al.* (1996), who show Dinantian to Westphalian strata preserved, with the top Carboniferous maturity contours ranging from 0.75% R_o to 2.25% R_o. While de Jager *et al.* (1996) recognize, in addition, Toarcian and Rhaetian source rocks in the Roer Graben, Winstanley (1993) sources both the oil- and the gasfields in the area from the Carboniferous, largely on maturity grounds.

Limits have been placed on the timing of influx of gas into Rotliegend reservoir in the Sole Pit Trough area by Robinson *et al.* (1993), who demonstrated that the major illitization episode, dated as 158 ± 18.6 Ma (39 samples from 11 wells), predated the arrival of the gas flux. Maturity modelling confirmed the timing of gas generation from the Middle Jurassic onwards. Lee *et al.* (1985) have constrained the timing of arrival of gas into the western fault block of the Broad Fourteens Basin, using potassium/argon (K/Ar) dating of diagenetic illite. They estimate that gas displaced water some 140 Ma ago (early Cretaceous) at the investigated location, in confirmation of the model of Oele *et al.* (1981). Two phases of gas generation from Westphalian coals have been proposed for the Zuidwal field (Perrot *et al.*, 1987): a late Jurassic pre-structure phase, related to vulcanicity, and a Tertiary–Recent burial phase. In the UK onshore, the presence of reworked anthracites in the Jurassic of the Weald Basin (Smith, 1993) points to pre-Jurassic (?Variscan) coalification.

Barnard and Cooper (1983) and Bailey *et al.* (1993) have noted that the Westphalian can be deeply eroded in areas of late Carboniferous (Variscan) inversion, as well as on regional highs. It is the same areas affected by Mesozoic inversion that appear to have sourced the major gasfields.

A source-rock/reservoir mass balance can be attempted for the 693×10^9 m^3 (24.5×10^{12} ft^3) of reserves in the gasfields surrounding the Sole Pit Trough (Barnard and Cooper, 1983):

- Area of the Sole Pit Trough, as defined in Glennie and Boegner (1981):

11.12×10^2 km^2 (2.5×10^6 acres)

- Generation of gas (STP) at peak maturity (from Fig. 11.12):

1% TOC shale = 3×10^9 m^3 gas/km^3 (0.14×10^6 ft^3/acre.ft)
Seam coal = 160×10^9 m^3/km^3 (7×10^6 ft^3/acre.ft)

- Volume of gas generated per metre (foot) of:

1% TOC shale = 30.6×10^9 m^3 (0.364×10^{12} ft^3)
Seam coal = 1619×10^9 m^3 (18×10^{12} ft^3).

Thus less than 1.5 m of seam coal or 82 m of 1% TOC shale, if mature over the whole area of the Sole Pit Trough, would be sufficient to yield the $\sim 2500 \times 10^9$ m^3 of in-place gas. The current in-place gas volume is estimated as 1983 reserves (693×10^9 m^3; Barnard and Cooper, 1983) + 20% for subsequent discoveries and assuming a recovery factor of 33%. It is thus argued that there is adequate source potential to account for the reservoired gas, even given a relatively incomplete Westphalian section and a low migration/remigration efficiency. A similar calculation for the West Netherlands Basin indicates only 0.05% of the generated gas has been discovered to date (de Jager, 1996).

A second area of major post-Variscan burial is in north Germany/southern Denmark, which may have sourced the massive Groningen Field in the north-east Netherlands. However Lutz *et al.* (1975) suggest a source to the south-west of the field in the West Netherlands Basin (de Jager *et al.*, 1996). A further source area may have been more precisely defined by Kettel (1983) as overlying the East Gröningen massif. Glennie (1986) notes the possible role of contact metamorphism by early Permian or earliest Cretaceous volcanism or associated dykes as a further agent for gas generation. The first influx of gas into the northern flank of the Groningen Field has been estimated as occurring during the Late Jurassic (some 150 Ma ago), using K/Ar dating of diagenetic illite (Lee *et al.*, 1985, 1989). Younger ages at greater depths within the reservoir section were interpreted as indicating the progressive filling of the reservoir from 150 to 120 Ma before present, at the location of the investigated well.

A third area of gas generation from the Westphalian may exist under the Central Graben and Horn Graben, where Day *et al.* (1981) show local burial of up to 10 km for the base Zechstein. Remnants of early Carboniferous coals—equivalent to the Limestone Coal Group of the Lothian coalfield—are immature with respect to gas generation where penetrated on the flanks of the Central Graben (Cayley, 1987). The Coal Measures, if preserved, are expected, on regional grounds, to be of bituminous-coal rank outside the grabens (Eames, 1975). The timing of gas generation can be estimated as Triassic–Jurassic from Fig. 11.41.

Small areas where the Westphalian overlies post-Variscan intrusives (e.g. Bramsche Massif, the Alston granite; Creary, 1980). Thermal anomalies associated with earlier-intruded but radioactive granites might also prove of exploration interest. A Westphalian source rock has been invoked for the Northumberland–Solway basin, where coals and shales with TOC values in the 1–5% range contain dominantly gas-prone kerogen (Chadwick *et al.*, 1993). Excluding local heating by the Whin Sill (Jones and Creany, 1977; Creany, 1980), maturities in this basin fall within the oil window (Creany *et al.*, 1980; Burnett, 1987).

No Carboniferous of source-rock facies is known from the west of Shetlands (Ridd, 1981). Westphalian coals are mined in Ireland (Griffiths, 1983; Illing and Griffith, 1986; Parnell, 1991), with conodont maturities ranging from high-volatile bituminous in the north-east to meta-anthracite ranks to the south of the Variscan front (Jones, 1992). The Carboniferous is present in gas-prone Coal Measure facies in the Kish Bank Basin (Jenner, 1981, Croker, 1995a) and in the northern (Barr *et al.*, 1981) central (Maddox *et al.*, 1995) and western parts of the Irish Sea, where gas seeps are reported at the Quaternary subcrop (Croker, 1995a). The maturity of the Carboniferous (Westphalian–Namurian) in the Porcupine Basin wells west of Ireland (~0.7–5% R_o) covers the gas window and proceeds into greenschist-grade sediments (Croker, 1995b).

Of the two producing gasfields west of UK, the Morecambe Field (6.75×10^{12} ft^3 in place) is of unknown source, but compositional similarities with the southern North Sea gasfields (Ebbern, 1981) make it almost certain that either the Westphalian coals, which are known to underlie the area (Bushell, 1986; Stuart and Cowan, 1991), or the

Namurian Sabden Shales (Hardman *et al.*, 1993) are the source. Gas generation is modelled by Bushell (1986) as occurring in the Jurassic and again in the Late Cretaceous from this horizon, while Hardman *et al.* (1993) recognize only the latter event at an on-structure modelled location.

In contrast, the Kinsale Head and associated gasfields in Irish waters are believed by Colley *et al.* (1981) to have a complex origin from Mesozoic (Liassic) rather than Carboniferous source rocks. Reviewing the source rocks of the Central Irish Sea Basin, Maddox *et al.* (1995) noted a Westphalian section in two wells (Irish 42/17-1 and 33/22-1) with TOC values ranging from 2% to 74% TOC (coals), with Hydrogen Indices in the 55–400 mg Py/g TOC, indicating oil as well as gas potential. Modelling predicts late Jurassic and early Tertiary generation episodes.

Mature, gas-prone, organic-rich shales and coals of Carboniferous age are present on Bjornoya (Bear Island), offshore northern Norway (Bjorøy *et al.*, 1983). Lower carboniferous coals and coal measures are also present on Svalbard (Spitsbergen). This gas-prone source sequence (cf. Abdullah *et al.*, 1986) is part of a separate North Greenland Basin (Rønnevik, 1981; Whittaker *et al.*, 1997). In east Greenland, Carboniferous lacustrine oil shale, with averages of 5% TOC and 700 mgPy/g TOC, is reported (Piasecki *et al.*, 1990; Stemmerik *et al.*, 1991; Christiansen *et al.*, 1993a,b; Price and Whitham, 1997).

11.8.4 Permian

The Kupferschiefer/Marl Slate horizon—typically 1–3 m (3–10 ft) thick—is the only Permian sediment with a high-grade hydrocarbon-source potential in the North Sea area. It outcrops in Durham, UK and north Germany and extends under much of the central and southern North Sea (Vaughan *et al.*, 1989). Total organic carbon values are typically 5–15%, decreasing upwards, and the kerogen type is oil-prone, with Hydrogen Indices above 600 mg/g TOC from onshore outcrop samples (Gibbons, 1987; Rae and Manning, 1989; Telnaes *et al.*, 1989). Within Germany, Puttmann and Ekhardt (1990) report average TOC values of ~4%, with a range of 0.5–9.1% TOC and 'amorphous' kerogen. Original maximum-pyrolysis yields approach 100 kg/t. Offshore, Cayley (1987) reports TOC values up to 6%, but with an average of 3%. The unit is marginally mature in the area of the Auk Field and in north-west Germany (Puttmann *et al.*, 1989; Schwark and Puttmann, 1990; Grice *et al.*, 1997), although, where affected by intrusives (Krefeld High), elevated maturities are reported (Puttmann and Eckhardt, 1990). Using molecular indicators, Grice *et al.* (1996, 1997) define the kerogen type of a Kupferschiefer core from the Lower Rhine area to be a mix of cyanobacterial and algal input, with specific carotenoids indicative of the Chlorobiacea (green sulphur bacteria) growing under anoxic conditions but within the photic zone.

Any oil generated would be recognizable by its high porphyrin content (Eckhardt *et al.*, 1989; Schwark and Puttmann, 1990), high absolute concentrations of extractable nickel and vanadium (Chicarelli *et al.*, 1990), with a nickel/vanadium (Ni/V) ratio of 1.4–2.8 (Barwise and Park, 1983; Chicarelli *et al.*, 1990). These latter values compare with Ni/V ratios of ~0.2 in the Upper Jurassic (Telnaes *et al.*, 1991; Haseby *et al.*, 1996). The Kupferschiefer/Marl Slate horizon is known to extend at least as far north as the Buchan Field (Taylor, 1981), and the high vanadium and nickel content of Buchan oil may suggest a Kupferschiefer contribution, although the ratio of 0.17 (see Table 11.9) is in poor agreement with Permian source-rock extracts. The absence of demonstrable generation and expulsion from this unit is probably a function of its thinness; conversely, for a structure to be filled with oil from this source rock demands the drainage of a large area into a single structure. For example (see equations 24 and 25), some 741 km^2 of oil-mature Kupferschiefer (2 m thick and with a hydrocarbon yield of 24 kg/t) would be required to fill a field with 100×10^6 barrels in place (gross migration–entrapment efficiency assumed to be 20%).

Oil shows in the Zechstein may derive from the Kupferschiefer, although Cayley (1987) notes the presence of thin dolomitized algal laminites within the Zechstein itself. In the onshore Netherlands, van den Bosch (1983) also reports oil-prone Zechstein shales and carbonates as noted in the Thuringian Basin (Karnin *et al.*, 1996).

To the north, in the Permian Basin on the east coast of Greenland, a probably relatively local facies of source rock is developed (Surlyk *et al.*, 1984: Christiansen *et al.*, 1993a,b; Price and Whittham, 1997). Here, in a lagoonal to shallow-marine setting, thin, dominantly gas-prone claystones were laid down during the Early Permian, while a thicker more persistent oil-prone shale of the Ravnefjeld Formation, with 2–5% TOC in its richest upper part, accumulated in an anoxic facies of the Upper Permian (Piasecki and Stemmerik, 1991) with bitomes in associated carbonate build ups (Stemmerik *et al.*, 1997). Background maturities are in the late-immature to early-mature zone (0.44–0.60% vitrinite reflectance), except where affected by intrusive heating during the Tertiary. To the east of the outcrop, the maturity rises as high as 1.83% R_o. Early mature (T_{max} ~435°C) late Permian shales, with typical TOC values of ~1.5% (plus a single value of 19% TOC) and gas-prone kerogens, are reported by Isaksen (1995) for the Hammerfest Basin well 7120/12-4 (Barents Sea).

11.8.5 Triassic

The largely continental Triassic sediments of north-west Europe have no source potential in the North Sea area. The marginal marine Skuld and Urd Formations of Bjornøya Island south of Svalbard (Spitsbergen) contains shales and silty shales with up to 2.0% TOC of dominantly gas-prone kerogen, which is in the lower part of the oil window (Bjorøy *et al.*, 1983). On Svalbard itself, Middle Triassic shales of the Botneheia Formation contain considerable quantities (2.9–5.5% TOC) of oil- and gas-prone kerogen, which ranges from late immature to peak oil-generating maturities (Forsberg and Bjorøy 1983; Mørk and Bjorøy, 1984; Khorasani and Michelsen, 1992). Small amounts of locally generated hydrocarbon have been described from this formation by Schou *et al.* (1984). This high-quality oil- and gas-prone source may extend as far south as about 65°N, according to facies maps of Rønnevik (1981) and Rønnevik *et al.* (1983).

Oil-source potential has been identified in the lacustrine Upper Triassic–Liassic Kap Stewart Formation of Jamesonland, east Greenland (Dam and Christiansen, 1990; Krabbe *et al.*, 1994; Krabbe, 1996), with TOC values above 8% and S_2 yields in excess of 40 kg/t for the immature–early-mature samples. As discussed in the Liassic section below, this facies is probably a stratigraphic equivalent of the coal-rich Åre Formation of Haltenbanken.

The Rhaetic (latest Triassic) of southern England locally develops in an oil- or gas-prone facies (Macquaker *et al.*, 1985, 1986), while over the Channel in the Paris Basin late Triassic lagoonal carbonates and shales have been proposed as more realistic source rocks for oil than the Toarcian (see section on Lower Jurassic below), on grounds of both maturity and association with the Upper Triassic reservoir play. The Rhaetian Sleen Formation of the Dutch Central Graben has also been described as a fair oil-source rock (Clark-Lowes *et al.*, 1987).

11.8.6 Jurassic

The Jurassic contains the source rocks for almost all North Sea oil, with the Upper Jurassic–basal Cretaceous Kimmeridge Clay and Borglum–Draupne Formations (UK and Norwegian nomenclature respectively) being the dominant source interval. As pointed out by Spencer *et al.* (1996), 71% of the total hydrocarbon (oil or oil equivalent) derives from Jurassic source rocks. In addition, the Jurassic contains a major potential source of gas in the Middle Jurassic coals— although their contribution to the gasfields of the oil province is not proven. Garrigues *et al.* (1989) and Monin *et al.* (1990) have reported a comparison of laboratory-simulated generation from these two source-rock types (Brent and Kimmeridgian kerogens). Ironically, relatively little detailed work has been published on the best oil-source rock, the offshore 'hot shale' of the Kimmeridge Clay Formation (Fuller, 1975; Oudin, 1976; Brooks and Thusu, 1977). The pre-eminence of the Upper Jurassic is clear from the generalized summary of the post-Triassic source rocks given in Fig. 11.42, from Field's (1985) study of the northern North Sea.

Lower Jurassic (Liassic)

Within the North Sea itself, the Lower Jurassic, where it develops source-rock quality at all, is dominantly gas-prone (Fig. 11.43); only the Upper Lias (Toarcian) shows region-wide propensities for the development of an oil-prone facies (Fleet *et al.*, 1987; Farrimond *et al.*, 1989; Bandin, 1990, 1995). The development of Liassic oil- and gas-prone source rocks and coals in the North Sea area has been reviewed, using a suite of stage-by-stage palaeogeographic

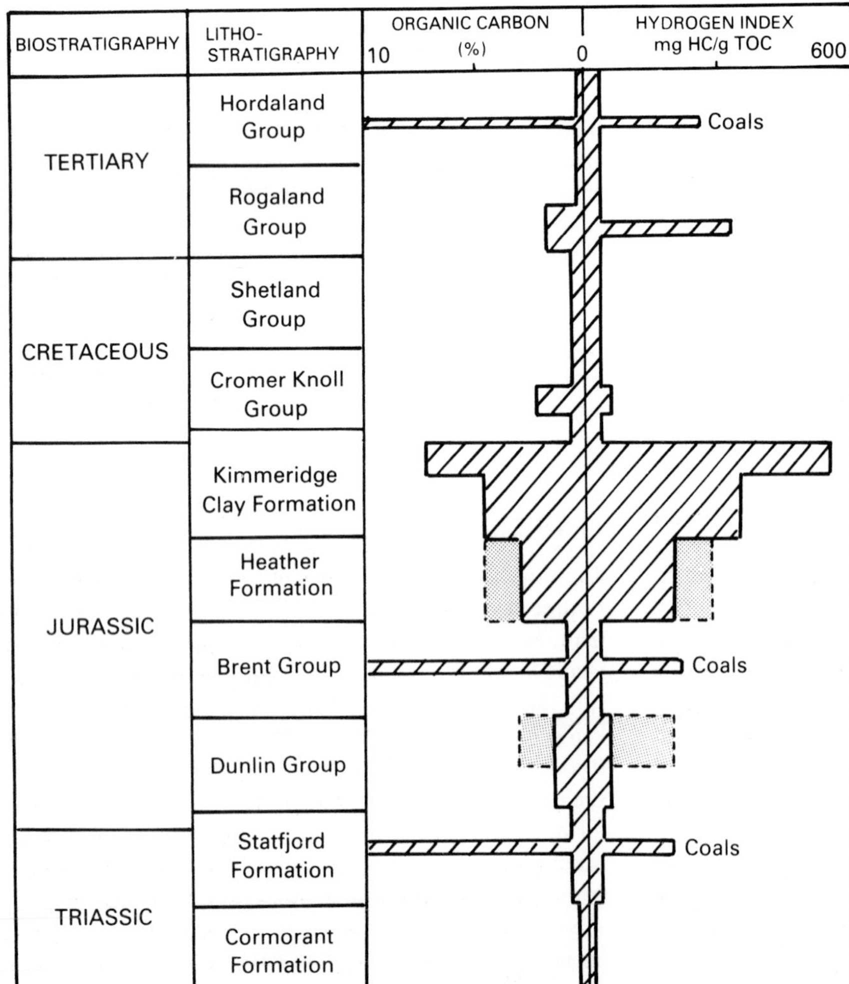

Fig. 11.42 Source-potential summary for the Mesozoic and Tertiary of the North Sea (modified after Field, 1985). This diagram emphasizes the pre-eminence of the basal Cretaceous–Upper Jurassic mudstones as the major oil-source rock in the basin. The recorded Hydrogen Index may be reduced by maturation in the older and hence deeper-buried intervals (see Fig. 11.6).

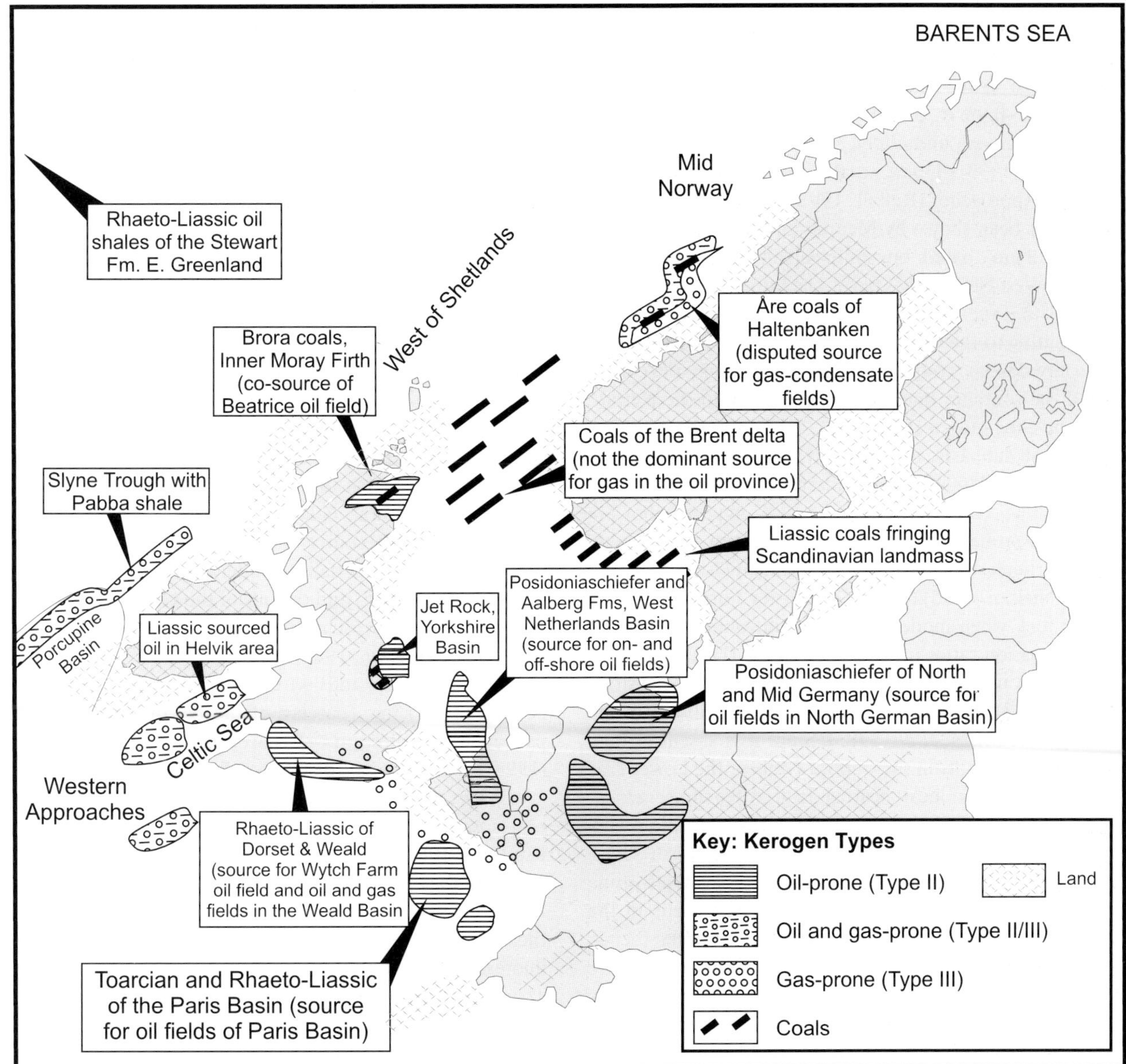

Fig. 11.43 Regional distribution of Lower and Middle Jurassic source rocks in the North Sea and adjacent areas. Within a broad epeiric seaway, oil-prone source rocks accumulated in local basins, developed as a result of the east–west extension associated with Central Atlantic rifting. Gas-prone coal sequences fringe the major land masses. With the exception of the Toarcian Schistes Carton/Posidoniaschiefer/Jet Rock interval, the other oil-source rocks of the Lias comprise relatively thin, rich, oil-prone units interbedded with leaner shales.

maps from Hettangian to Toarcian (Fleet *et al.*, 1987), in the context of sequence-stratigraphic units (Morton, 1993), and as an extension of the western Tethyan shelf (Baudin, 1995).

In the offshore area, the Lower Jurassic is absent, or present only as erosional remnants, in the Central and Witch Ground grabens and the Outer Moray Firth Basin (Ziegler, 1982b; Underhill and Partington, 1993). Using sequence-stratigraphic criteria, Morton (1993) predicts oil-source potential in the Sinemurian (Amundsen Formation of the Dunlin Group) and Toarcian (Drake Formation) of the Viking Graben. In the extreme south of the Central Graben, in the Broad Fourteens Basin and in the area of the Yorkshire–Sole Pit Trough, the Lower Jurassic is preserved and may develop local oil and gas potential.

The Hettangian/Sinemurian Blue Lias of Dorset, north Somerset and south Wales, UK, is present as cyclic sequences of limestone–marl–shale on a decimetre scale with early diagenetic nodules (Wolff *et al.*, 1992). The shales, particularly when laminated, contain good yields (up to 18% TOC) of mixed oil- and gas-prone kerogen, and are late immature to early mature with respect to oil generation (Cornford, 1972, 1986; Ebukanson and Kinghorn, 1985, 1986a,b; Thomas *et al.*, 1993). An average value of 2% TOC would be appropriate for the laminated lithologies. The shales of the Sinemurian Black Venn Marls of Dorset are also rich in mixed oil- and gas-prone organic matter but here are immature. The Lias has been suggested as a source for the Wytch Farm oil (Colter and Havard, 1981) and for the onshore oil seeps of Dorset (Selley and Stoneley, 1987; Cornford *et al.*, 1988; Miles *et al.*, 1993) and may extend as far south as the Brittany Basin (Ruffelt,

1989). A Liassic source has been proposed for the oils of the Weald (Burwood *et al.*, 1991) and specifically Humbly Grove (Hancock and Mithen, 1987). Modelling maturity shows large areas where the Lias is predicted to have reached 0.7–0.9% R_o prior to late Cretaceous/early Tertiary uplift (McLimans and Videtch, 1987; Penn *et al.*, 1987; Chadwick, 1993). Uplift to the south and west in the Western Approaches (Ruffell, 1995) and South Celtic Sea basins has been shown by Menpes and Hillis (1995) to be up to 1.2 km, based on sonic-velocity analysis. In the onshore West Netherlands basin, the Lower Liassic Aalburg shale, with Type II kerogen, de Jager *et al.* (1996) cite as contributing to the oil accumulations of the area.

A fluviodeltaic sequence containing coals developed round the southern margin of Scandinavia in the Hettangian and Sinemurian (Fleet *et al.*, 1987). To the south, in the Danish subbasin, the generally gas-prone open-marine shales of the Upper Liassic Fjerritslev Formation have intermediate TOC contents (1.6% TOC average), but are locally oil-prone, with Hydrogen Indices up to 500 mg/g TOC particularly at the top of the unit (Thomsen *et al.*, 1987). Nielsen (1995) uses the Fjerritslev Formation as the source rock when modelling stochastic values for kerogen transformation ratios in northern Jutland (Denmark).

Apart from some 40 m of Upper Jurassic claystone (Olsen, 1983) the Jurassic is believed to be absent in the Horn Graben (Best *et al.*, 1983). Since it is absent or only present as remnants in the Central Graben, the Lower Lias will not, without local heating (e.g. Altebaumer *et al.*, 1983), have reached gas-generating levels (vitrinite reflectance of 1.0% R_o) in the southern part of the North Sea.

In the Yorkshire–Sole Pit Trough area, lean gas-prone shales (0.8% TOC average) and coals ('jet') occur in the Lower and Middle Liassic sequence. Sections of the Pleinsbachian (*jamesoni* zone) of three UK onshore basins have been geochemically characterized by Weedon and Jenkyns (1990) and van Buchem *et al.* (1992, 1995), showing lean (and dominantly dead carbon) in Yorkshire (0.2–1.5% TOC), richer in Oxfordshire (0.2–2.0% TOC) and richest (dominantly oil prone) in the Belemnite marls of Dorset (0.2–6.0% TOC). This interval can thus be considered an oil-prone source rock only in Dorset.

An Upper Lias (Toarcian) rich oil-prone faces is variously known in north-west Europe as the Jet Rock and Bituminous shales of Yorkshire, the Posidoniaschiefer of Germany and the Netherlands and the Schistes Cartons in France, Belgium and Luxemburg. These rocks are rich in kerogen of 'sapropelic' type, typically with 2–12% TOC (Brand and Hoffman, 1963; Tissot *et al.*, 1971; Huc, 1976; Barnard and Cooper, 1981; Farrimond *et al.*, 1989). More specifically the Lower Toarcian is organic-rich over much of north-west Europe (Huc, 1976; Demaison and Moore, 1980; Farrimond *et al.*, 1989; Murphy *et al.*, 1995). Moldowan *et al.* (1986) and Schoell *et al.* (1989) have attributed fluctuations in TOC contents (up to 14% TOC) to changes in the depositional environments, monitored using molecular parameters. In West Netherlands, a mixing trend of Carboniferous and Toarcian sourced gas is identified by de Jager *et al.* (1996) on the basis of heavy and light carbon isotope signatures respectively. Mixed accumulations occur.

Mackenzie *et al.* (1988) have detailed the maturation of the Posidoniaschiefer from biomarker kinetics, an alternative set of numbers being subsequently put forward by Marzi *et al.* (1990) and Schaefer *et al.* (1990). Efficient expulsion and migration from this unit has been demonstrated by Leythaeuser *et al.* (1988), using molecular parameters, conclusions subsequently confirmed by laboratory investigations (Lafargue *et al.*, 1990, 1994) and 1D computer modelling (Düppenbecker *et al.*, 1991).

The Upper Liassic oil-prone shales between Yorkshire and the Netherlands may, however, have reached early maturity with respect to oil generation in the pre-inversion Sole Pit Trough at the end of the Cretaceous (Glennie and Boegner, 1981), in the Cleveland Basin, Yorkshire (Barnard and Cooper, 1983; Kirby *et al.*, 1987), in the Broad Fourteens Basin from the Cretaceous onwards (Oele *et al.*, 1981), and in the offshore and onshore West Netherlands Basin (de Jager *et al.*, 1996). The Toarcian Posidonia Shale is described as a rich oil-prone source rock in the Dutch Central Graben area (Clark-Lowes *et al.*, 1987), and is believed to be the source for the oils in the Rijswijk oil province of the Netherlands (Bodenhausen and Ott, 1981; de Jager *et al.*, 1996) and the oils of the Gifhorn Trough in northern Germany (Schwarzkopf and Leythaeuser, 1988; Düppenbecker *et al.*, 1989).

Using analyses and wireline-log interpretation (see Section 11.3.5) the Toarcian of the North Celtic Sea—in particular, the lower unit—has been shown by Murphy *et al.* (1995) to contain up to 138 m (450 ft) of oil-prone shale (> 0.5 kg/t yield). The Toarcian (Schistes Cartons) is the notorious source rock in the Paris Basin (Tissot *et al.*, 1971; Espitalié, 1987; Espitalié *et al.*, 1987), although, as discussed in the previous section, the Upper Triassic (Rhaetic) is now believed to be the most probable source for at least the Triassic-reservoired oils. Ultraviolet (UV)-excitation fluorescence colour photographs of the kerogens of the Toarcian of the Paris Basin are reproduced in Bertrand *et al.* (1993, Plate 3).

Jurassic rocks are present in the North Lewis, North and South Minches and Malin basins, which locally develop a black shale facies, particularly in the Hettangian to Sinemurian (Broadfoot Beds), the Sinemurian to Pleinsbachian (Pabba Shale), the Toarcian Portree and Callovian–Kimmeridgian Staffin shales (Fisher and Hudson, 1987; Ambler, 1989; Thrasher, 1992; Fyfe *et al.*, 1993; Morton, 1993; Bishop and Abbott, 1991, 1995). The contiguous offshore sequence is confirmed by shallow borehole data and seismic. The widespread Pabba Shale yields TOCs are values > 1.5%, while the Portree Shales and lateral equivalent produce TOC values up to 4%. The thin Aalenian Dun Caan Shale is locally organic-rich (>7% TOC) and is mature for oil generation adjacent to igneous dykes (Bishop and Abbott, 1995; Farrimond *et al.*, 1996). Appropriate in the heartlands of Celtic Christianity, Bishop and Abbott (1995) detail the organic conversions in the Pabba and Staffin shales and silty shales of the Bearreraig Sandstone. Based on diagenetic studies, Searl (1994) has estimated maximum burial of the Broadfoot beds outcrop to have occurred in the latest Jurassic (~1.6 km) followed by Cretaceous uplift, with renewed burial by extrusive lavas (~1 km) plus associated hydrothermal heating in the early Tertiary.

The Liassic has been demonstrated to be the source rock

for the oil- and gasfields in the Irish waters of the Celtic Sea Basin. The Kinsale Head and associated gasfields are believed by Colley *et al.* (1981) to have been sourced from Liassic shales, as is the oil of the Helvik accumulation (Caston, 1995). The maturity of the Lower Jurassic of the North Celtic Sea Basin is mapped from modelling by Murdoch *et al.* (1995), who show substantial areas to be both oil- and gas-mature.

To the west of Ireland, Trueblood (1992) and Scotchman and Thomas (1995) have characterized the Pabba Shale (Sinemurian) and Portree Shales (Toarcian) as oil-prone source rocks in the Slyne Trough (Irish well 27/13-1). The latter authors report the average TOC values as 4.7% and 3.7% with mean yields (Rock Eval S_2) of 25 kg/t and 11 kg/t, respectively. In the Celtic and Porcupine basins, Morton (1993) uses sequence stratigraphy to identify the same stratigraphic intervals as potential oil-prone source rocks. It has been proposed that the Toarcian contains the source rock for the Connemara Field in the Porcupine Basin to the south, although Macdonald *et al.* (1987) claim that the balance of equivocal carbon-isotopic evidence points to generation from lateral equivalents to the Middle and Upper Jurassic shales interbedded with the reservoir. The presence of Upper Jurassic source rocks of appropriate maturity in the proximity of the Connemara wells favours the latter as the likely source. McCulloch (1993), using apatite fission-track data, points out that the Jurassic section in Irish Block 26 is currently at its maximum temperature, although higher heat flow values are indicated in the Palaeocene (?Thulian igneous activity and associated hydrothermal events) and in the Middle Jurassic (crustal extension and basin formation).

Minor lean, probably gas-prone, shales of the Statfjord and Dunlin groups are preserved in areas of the northern North Sea, such as the East Shetland Basin (e.g. Magnus, Thistle, Heather). In comparison with the overlying Middle Jurassic, this is probably not a significant gas-prone source rock. Rønnevik *et al.* (1983) show Toarcian marine shales of the Drake Formation to be mature (for oil), with an average of 2% TOC, and oil-prone kerogen in the Norwegian Viking Graben and the Horda, Basin and gas-prone kerogen in the Norwegian sector of the Central Graben.

North of 62°N, mature Lower Jurassic coals and, deeper in the basin, marine shales are present in the Åre (Hitra) Formation of Haltenbanken and Traenabanken areas as gas- and oil-prone source rocks, respectivity (Rønnevik *et al.*, 1983; Hollander, 1984; Larsen and Skarpnes, 1984; Elvsborg *et al.*, 1985; Pittion and Gouadain, 1985; Thompson *et al.*, 1985; Cohen and Dunn, 1987; Hvoslef *et al.*, 1988; Khorasani, 1989; Forbes *et al.*, 1991; Isaksen, 1995). It is these coals and associated shales, reported to have unusually high Hydrogen Indices (values in the 200–400 mg/g TOC range), that are believed to have generated the gas–condensate (and even oil: Cohen and Dunn, 1987) in the mid-Norway area. Isaksen (1995) gives a pyrolysis–GC trace showing a mix of oil and gas potential for a 12.8% TOC shale of Upper Åre Formation. In contrast, Khorasani (1989) reports a high abundance of inertinite from Åre coal and shale samples from six Haltenbanken wells.

Thompson *et al.* (1985a,b, 1994), in comparing the Norwegian Åre coals with Tertiary coals of South-East Asia, have attempted to account for the oil-prone nature by considering the palaeolatitudes, delta geometry and hydrodynamics. They claim that, while delta-top-rooted coals are gas-prone, the pro-delta sediments will be preferentially enriched in the more hydrodynamically mobile and hydrogen-rich kerogen components (e.g. cuticle, resin, spores) which produce the observed gas condensate. This distribution is the reverse of that proposed by Powell (1986) in his discussion of the Tertiary deltas of Canada and the Far East. His model would predict the most oil-prone coals on the delta top, where the fluctuating water-table will promote preferential oxidation of the gas-prone humic fraction, as envisaged by Khorasani (1989).

No significant source potential is attributed to the Lower Jurassic in Svalbard (Mørk and Bjorøy, 1984), although Dam and Christiansen (1990) and Krabbe *et al.* (1994) have identified source potential in the Upper Triassic–Liassic of Jamesonland, east Greenland, in the area where Mathiesen *et al.* (1995) have modelled oil and gas generation. The lacustrine Upper Triassic—Lower Jurassic Kap Stewart Formation of eastern Greenland is the conjugate of the coastal plain Åre Formation of Haltenbanken, and shows similarities in biomarker signature (Krabbe, 1996). With TOC values above 8% and S_2 yields in excess of 40 kg/t (Hydrogen Index ~500 mg/g TOC), this unit is a superior oil source compared with its stratigraphic equivalent in mid-Norway.

Middle Jurassic

In the North Sea area, Middle Jurassic deposition is limited (Ziegler, 1982b). With the exception of the oil-prone, algal-rich shales and coals of the Inner Moray Firth Basin (Mackenzie and Quigley, 1988), the Middle Jurassic generally has gas-generating potential in the paralic/deltaic sands, shales and coals of the Yorkshire–Sole Pit Trough area (Hancock and Fisher, 1981), the Moray Firth Basin (e.g. Maher, 1981; Bissada, 1983), the Viking Graben (Eynon, 1981; Pearson *et al.*, 1983) and the Horda, Egersund and North Danish basins (Hamar *et al.*, 1983; Koch, 1983). Coals are shown as minor components in these sequences (Eynon, 1981; Hancock and Fisher, 1981; Parry *et al.*, 1981). Goff (1983) estimates a total thickness of 10 m of Brent coals in the drainage area of the Frigg Field, but suggests that the isotopic composition of Frigg gas indicates generation from the shales, rather than coals (cf. Rigby and Smith, 1982).

The wet gas in the Sleipner complex of fields was claimed to have derived from Middle Jurassic coals and shales (Larsen and Jaarvik, 1981), but these coals frequently contain considerable altered vitrinite and inertinite (Cope, 1980; Wilson *et al.*, 1991), which somewhat downgrades their gas-generating potential. Indeed, the majority of the major gasfields of the North Sea oil province (e.g. Sleipner, Frigg, Troll) have methane stable carbon-isotope compositions incompatible with generation from a fully gas-mature seam coal of the type found in the Brent delta. As illustrated in Fig. 11.35, it is much more likely that the Sleipner gas derives mainly from late-mature oil-prone Kimmeridge Clay/Draupne Formation shales deeply buried in the asymmetric southern Viking Graben. A dual origin is proposed by Østuedt (1987), Ranaweera (1987) and Pegrum and Spencer (1990).

The Middle Jurassic shales commonly have TOC values as high as 5% but, like most Coal-Measure sequences, the TOC values fluctuate considerably. Their gas-prone nature is shown by the kerogen composition of the claystone facies, which is dominated by vitrinitic material (Parry *et al.*, 1981). Interestingly, minor amounts of oil/condensate-prone algae, resin and cuticle are present in these shales within the Viking Graben. Monin *et al.* (1990) have reported the pyrolysis behaviour of early-mature Brent coal from Norwegian well 25/2-7, concluding that its gas-prone nature results from the cracking of heavier hydrocarbons and asphaltic matter within the coal.

The Callovian shales of the Ninian Field are claimed to contain up to 4% TOC and oil-prone kerogen (Albright *et al.*, 1980). The waxy oil (41–45°API) of the nearby gas/condensate Hild Field correlates with Brent coal extracts on the basis of a high C_{29} sterane content (Rønning *et al.*, 1986, 1987). In addition, Bissada (1983) has reported fairly rich (1.4–2.7% TOC) oil-prone rocks, together with coals, in the Middle Jurassic of the Moray Firth in the area of the Piper Field. Some of the reports of oil-prone Middle Jurassic may, if not based on core samples or supported by biostratigraphy, be the result of downhole caving of drill cuttings from the overlying organic-rich Upper Jurassic.

A concentration of the freshwater alga *Botryococcus* is found in the oil-prone Middle Jurassic Parrot coal and associated shales of the Brora section of the Moray Firth coast. Bjorøy *et al.* (1988) report TOC values between 8.4 and 9.7% TOC and pyrolysis S_2 yields up to 60.8 kg/t for onshore samples of the Brora oil shale. The Middle Jurassic of the Inner Moray Firth occurs at about 2000–2100 m (6500–6800 ft) in the Beatrice Field area (Linsley *et al.*, 1980), but additional pre-inversion burial is indicated by the regional seismic truncation mapped by Day *et al.* (1981), offsets in biomarker trends (Pearson and Duncan, 1996) and detailed analysis of sonic velocities in the Chalk and Kimmeridge Clay (Hillis *et al.*, 1994; Smith *et al.*, 1994; Thomson and Hillis, 1995). The various estimates of uplift differ. Barnard and Cooper (1981) have suggested that these algal-rich coals and shales might be the source for the high-wax Beatrice crude, although Duncan and Hamilton (1988), Peters *et al.* (1989), Bailey *et al.* (1990) and Duncan and Buxton (1995) have argued for a Middle Devonian contribution, as discussed in Section 11.8.2. A sample of oil-prone Middle Jurassic sediment was used by Mackenzie and Quigley (1988) for their generation-modelling studies.

In terms of maturity, the Brent coals and shales will be mature for gas generation (> 1% R_o) in the Viking Graben below about 3.9 km (Goff, 1983, 1984). High-quality vitrinite-reflectance trends have been constructed from measurements made on Brent coal samples taken from different depths in a number of wells. Interpolating between the top Trias and base Cretaceous depth maps of Day *et al.* (1981), gas generation will be confined to the central part of the Viking and Witch Ground grabens and excluded from the East Shetlands Basin and from the Horda, Egersund and North Danish basins. Field (1985) shows the top Brent horizon to be at a gas-generating stage (> 1.3% R_o over substantial areas down the axis of the North and South Viking Graben.

Measured UK onshore reflectance data (up to 0.7% R_o) from Barnard and Cooper (1983) and Kirby *et al.* (1987) and offshore subsidence shown by Glennie and Boegner (1981) suggest that the Middle Jurassic of the Yorkshire–Sole Pit Trough area would not have attained gas-generating rank even at pre-inversion maximum burial. The timing of this uplift has been attributed to the early Tertiary, based on regional arguments (Kent, 1980), sonic velocities (Hillis, 1995a,b) and apatite fission-track measurements (Green, 1989; Bray *et al.*, 1992; Green *et al.*, 1993).

To the west of the North Sea, the Middle Jurassic is present as the arenaceous gas-prone coal-bearing Great Estuarine Series in Skye and the Inner Isles, outcrops that may be representative of both of the Minch basins (Ziegler, 1982; Ambler, 1989). Bjorøy *et al.* (1988) report a TOC value of 15% and a pyrolysis yield of 74 kg/t for a single sample of the Sky oil shale. Whilst a similar facies is reported over the Rona Ridge and in the West Shetland Basin (Bailey *et al.*, 1987), thick, dark grey, marine shales are present on the north-west flank of the Rona Ridge (Ridd, 1981). The onshore outcrops of the Callovian Staffin Shale unit contain local developments of bituminous facies, with up to 10.4% TOC, while the generally lean shales of the Bathonian Great Estuarine Group are potential gas-prone source rocks (Fisher and Hudson, 1987; Thrasher, 1992; Fyfe *et al.*, 1993; Morton, 1993). The locally developed Cullaidh Shale oil shale is an excellent oil-prone source rock (< 15% TOC and pyrolysis yield of 12 gal./t). None of these potential source rocks is mature for oil or gas generation at outcrop, except where affected by Thulian (Palaeocene) contact metamorphism. Maximum burial of these rocks may well have been at the end of the Jurassic, while their maximum temperature was probably attained during elevated heat flow associated with the Thulian igneous event.

In the Irish offshore, Middle Jurassic oil-source potential is present in the Porcupine Basin, particularly on the western margin where it correlates with oil shows.

To the north of 62°N, the Middle Jurassic Hestberget member of the Andoya Island outlier contains rich, immature, oil-prone (1.0–11.5% TOC), as well as gas-prone (4.6–12.3% TOC) sediments (Bjorøy *et al.*, 1980). A study of an undifferentiated Jurassic–Cretaceous shale sequence from Spitsbergen (Bjorøy and Vigran, 1980; Mørk and Bjorøy, 1984) indicates good quantities (0.45–16.0% TOC) of oil- and gas-prone kerogen.

The Upper Jurassic and basal Cretaceous (Oxfordian to Ryazanian)

The Kimmeridge Clay / Borglum / Draupne Formation, which falls within this age range, is recognized as the dominant oil (and associated condensate and gas) source rock of the North Sea. The regional distribution of Upper Jurassic source rocks has been reviewed in terms of average TOC values and kerogen type in the context of the palaeogeography of the western Tethys by Baudin (1995). Over much of the area the Kimmeridge Clay overlies the Oxfordian Heather Formation, which, where developed in an argillaceous facies, is typically a fair to good gas-prone source, but locally also has significant oil potential (Gormly *et al.*, 1994).

Oxfordian

The Oxfordian shales of the North Sea (i.e. approximately equating to the Heather Formation of the Viking Graben, the Haugesund Formation of the Central Graben and the Egersund Formation of the Fiskebank subbasin) are generally reported as fair to rich gas-prone source rocks (Barnard and Cooper, 1981). In the Statfjord Field, Kirk (1980) reports an average of 2.24% TOC for the Heather Formation, which appears gas-prone. Larsen and Jaarvik (1981) have suggested that the Heather Formation may be the source for the condensate in the Sleipner Field, but a late-mature Kimmeridge Clay/Draupne Formation source now looks more likely. Locally, Oxfordian shales can have some oil potential (Bissada, 1983) and, where thick (e.g. in the East Shetland Basin), could augment the Kimmeridge Clay Formation (Goff, 1983). The Heather Formation shales are reported oil-prone from wells drilled on the eastern (Norwegian) flanks of the North Viking Graben (Gormly *et al.*, 1994) and locally in the South Viking Graben. Based on carbon-isotope and pristane/phytane ratios, these latter authors claim a pure Heather source rock for most of the Troll and Norwegian Block 35/9 oils, with mixed Draupne/Heather sourcing for some compartments of West Troll, Brage and Block 35/11 oils. To the south, Bissada (1983) found the Oxfordian ('non-Kimmeridge Clay Upper Jurassic') of UK Quadrants 14 and 15 to be richer (1.4–8.8% average TOC values for eight wells) than, and to contain equally oil-prone kerogen as, the overlying Kimmeridge Clay Formation. The possible effects of caving from overlying organic-rich strata must be considered.

In onshore UK, Fuller (1975) reports the Oxford Clay of Yorkshire to be a lean, gas-prone source. While Kenig *et al.* (1994) and Belin and Kenig (1994) show the Lower Oxford clay of Central England to be an immature, rich, oil-prone source rock with TOC values up to 16.6% and Hydrogen Indices generally above 500 mg/g TOC. Macquaker (1994) compares the lithofacies of these organic rich mudstones with those of the Kimmeridge Clay. In southern England, the Oxford Clay is reported to be a rich but immature potential source for oil in Dorset (Colter and Harvard, 1981). Duff (1975) reports mean TOC values of 2.9 and 4.1% for two intervals of the Oxford Clay of Central England, Hudson and Martill (1994) argue for a minimal burial history. Total organic carbon of 1.2–6.4% but with a poor-quality kerogen and low Hydrogen Indices are reported for the flanks of the Faeroe Basin, west of Shetland (Bailey *et al.*, 1987).

In the Troms area of northern Norway, the Upper Jurassic is argillaceous, with the lower part of this unit being as good as, if not better than, the overlying Kimmeridge Clay equivalent (= Nesna) Formation in terms of oil-source potential. It is locally mature on-structure in at least some wells (Bjorøy *et al.*, 1983). A similar situation exists in the Haltenbanken area of mid-Norway (Cohen and Dunn, 1987). On Svalbard, the undifferentiated Jurassic section, presumably containing material representative of the Oxfordian, comprises mature oil-prone (Type II) kerogen (Bjorøy and Vigran, 1980; Mørk and Bjorøy, 1984).

The Kimmeridge Clay, Borglum and Draupne formations

These formations (hereafter termed the Kimmeridge Clay Formation) contain the major oil source rocks of the North Sea (Barnard and Cooper, 1981; Field, 1985). They are developed as organic-rich 'black' shales in most of the graben areas of the North Sea (Ziegler, 1982b; Rønnevik *et al.*, 1983). As previously noted, the high natural gamma-ray log response by which the formation is commonly characterized (see Fig. 11.14) has earned the most organic rich unit the name of 'hot shale', with lower-gamma, lower-TOC units being referred to as the 'cold shale' (Miller, 1990). Typical average TOC values are shown in Table 11.10, which also catalogues studies of the source-rock potential of the Kimmeridge Clay Formation.

There is a natural tendency for source-rock reports to have analysed the richest lithologies. The richest lithologies are often characterized in well descriptions as 'olive-black' or 'dark olive-grey', as opposed to 'dark grey', with sample selection also being based on intervals with the highest natural gamma-log response (Fig. 11.15). In a detailed study in the Outer Moray Firth and southern Viking Graben, Stow and Atkin (1987) differentiated between fissile-laminated, silt-laminated and silty-bioturbated mudrock facies. In terms of chemistry, the fissile-laminated mudrock was characterized by high concentrations of vanadium, uranium and selenium relative to the other two facies, but the source potential of these facies was not discussed.

The laminated olive-black claystones of the Kimmeridge Clay organic-rich facies were deposited in a marine seaway running from the Barents Sea area in the north, down the line of the graben (Fig. 11.44). Prior to and after Kimmeridge Clay Formation deposition, this seaway connected the Boreal Sea in the north and Tethyan Ocean in the south and east. The Norwegian–Danish and Yorkshire–Dorset basins were lateral to, and probably atypical of, sedimentation in the major sea-floor depression extending south from the Boreal Sea (Cornford *et al.*, 1986; Wignall, 1989). The development of the Boreal/Tethyan provincialism in at least some groups of organisms (e.g. ammonites) suggests a partial (?climatic) barrier to the south of the Yorkshire–Dorset Basin (Fig. 11.44). This persisted for the time period between the onset of widespread anoxia in the Oxfordian to basin flushing in the late Ryazanian (Fig. 11.45). Anoxic shale deposition was also well developed right into the Inner Moray Firth Basin (Bissada, 1983; Pearson and Watkins, 1983; Turner *et al.*, 1984; Andrews *et al.*, 1990; Hurst *et al.*, 1996).

Huc *et al.* (1992) review the response of the organic matter in the Kimmeridge Clay of Dorset to its depositional environment. A series of lateral organofacies transitions in the Dorset outcrop are explained by Wignall (1989) in terms of tempestites (storm deposits) with a temperature-stratified water column, while a similar sequence of cycles in Yorkshire have been attributed to orbital climatic cycles by van Buchem *et al.*, (1995). In both cases, relatively shallow water is envisaged in these basins lateral to the main North Sea graben development. According to a sequence-stratigraphic study (Herbin *et al.*, 1995), water

Table 11.10 Some examples of TOC values for the Kimmeridge Clay Formation and equivalents of the North Sea (see also field studies in Spencer *et al.*, 1987; Abbotts, 1991).

Area	TOC (wt%)	Reference	Comments
North Sea (unspec.)	2.7 av.	Fuller, 1980	= 3.25% organic matter
Central Graben	5.5 av.	Cornford, 1994	2–9% TOC range
Unidentified North Sea well	5.6, 4.9 av.	Brooks and Thusu, 1977	Upper and lower intervals
Unst Basin	6–10	Johns and Andrews, 1985	UK 1/4-1
Ekofisk	1.4–2.6 av.	van den Bark and Thomas, 1981	NOCS 2/4-19
Outer Moray Firth	1.0–3.8 av.	Bissada, 1983	UK Q14 and Q15 Piper wells
South Viking Graben	2.5–4.5	Pearson *et al.*, 1983	UK 16/22-2
Inner Moray Firth	3–6	Pearson and Watkins, 1983	UK Quadrant 12 wells
Brae area wells	4.29 av.	Reitsema, 1983	UK 16/7a-19
East Shetland basin	5.4	Goff, 1983	Estimated mean
Ninian	6–9	Albright *et al.*, 1980	Range
Tern	3.4–8.1	Grantham *et al.*, 1980	UK 210/25-3, 6.8% TOC av.
Statfjord	4.58 av.	Kirk, 1980	Licence 037, av.
South Norwegian Sea	7	Hamar *et al.*, 1983	7–17.5% TOC in hot shale
South Norway Shelf	2.1; 5.1 avs	Fuller, 1975; Lindgreen; 1985	NOCS 2/11-1 core
Danish North Sea	3.85 av.	Thomsen *et al.*, 1983	North-west Central Graben, well I-1
	1.59 av.		South-west Central Graben, well M-8
Norwegian North Sea	5 av.	Rønnevik *et al.*, 1983	Range 1–5% TOC
Dorset type section (UK onshore)	3.75 av.	Fuller, 1975	Bituminous shale
	1.6 av.		Grey clay
	15.3–40.0	Cosgrove, 1970	Richest oil-shale bands
	0.9–52.7	Farrimond *et al.*, 1984	Selected oil shales and limestone
Yorkshire outcrop	7.95 av.	Fuller, 1975	Bituminous shale
	30.9		Oil shale
	3.4		Dark grey shale
Sutherland outcrop	5.5 av.		Black shale
Andoya Svalbard	1.4, 4.3	Dypvik *et al.*, 1979	North Norway, Spitzbergen

depth is the dominant control on depositional conditions.

In the Viking Graben, the accumulation was syntectonic, occurring during rapid subsidence along the graben axis (Field, 1985). In the Central Graben, tectonic control of thickness is generally less pronounced, but locally fault control can be demonstrated, as in the case of the 3000 ft of Upper Jurassic mudstone accumulated on the downthrown side of the western boundary fault in the Gannet area (Cayley, 1987).

Within the Kimmeridge Clay Formation, the 'hot-shale' facies can be developed locally, and is diachronous: for example, the Tau 'hot-shale' Member of the Borglum Formation of the Fiske subbasin is of Volgian (uppermost Jurassic) age, while the 'hot shale' of the Norwegian sector of the adjacent Central Graben comprises the Ryazanian (basal Cretaceous) Mandal Formation (Hamar *et al.*, 1983; Doré *et al.*, 1985). Thomsen *et al.* (1983) have shown that, while the most oil-prone shales occur at the top of the Upper Jurassic in the deepest (north-west) part of the Danish Central Graben (Well I-1), the best source quality in a generally poorer section occurs towards the base of the Upper Jurassic on the south-east flank of the graben (well M-8). The organic-rich facies of the UK onshore Yorkshire–Dorset basin (Williams, 1986) was deposited only during the earlier phase of the development of the graben anoxia (Fig. 11.45), the upper Volgian–Ryazanian equivalent being a completely different (shallow-water terrigenous) facies in the UK onshore.

The kerogen of the Kimmeridge Clay Formation of the North Sea is one of the world standards for a marine Type II kerogen. The typical kerogen type of the 'hot shales' (see Table 11.11) is a mixture of bacterially degraded algal debris of marine planktonic origin (amorphous liptinite) and degraded humic matter of terrigenous origin (amorphous vitrinite). This amorphous component is mixed with variable amounts of particulate vitrinite (woody debris) and inertinite—highly altered (oxidized or burnt) material of land-plant origin. Land-plant spores and marine algae,

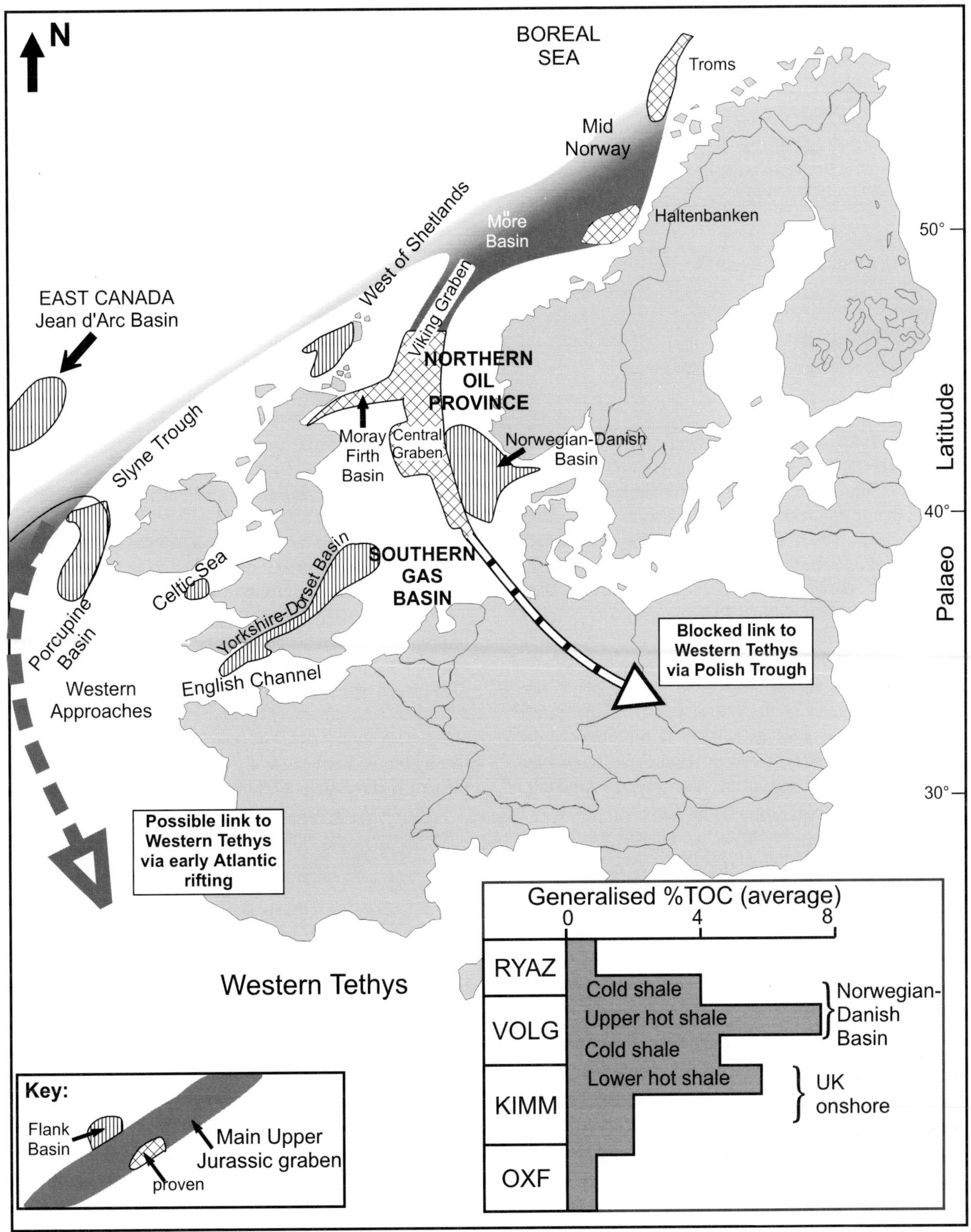

Fig. 11.44 Palaeogeography of the Upper Jurassic of the North Sea and adjacent areas at the time of organic-rich shale deposition (after Cornford, 1986; Miller, 1990). Opening of the Central Atlantic to the west produced a coincidence of high sedimentation rates and anoxia in the developing graben system (see Fig. 11.45). Inset illustrates that organic-rich sedimentation was continuous in the main grabens, in contrast to transient development of anoxicity in the flank basins. Upper and lower 'hot-shale' units are recognized over much of the main graben system.

such as dinoflagellates and hystrichospheres, are present in minor to trace amounts. Framboidal pyrite is common, attesting to the action of sulphate-reducing bacteria.

Under the reflected-light microscope, the kerogen in a polished whole-rock preparation appears as a wispy to diffuse fluorescent background (Combaz, 1980; Gutjahr, 1983), while isolated kerogen in transmitted light appears as a clumpy, amorphous, dull or blotchy fluorescent mass

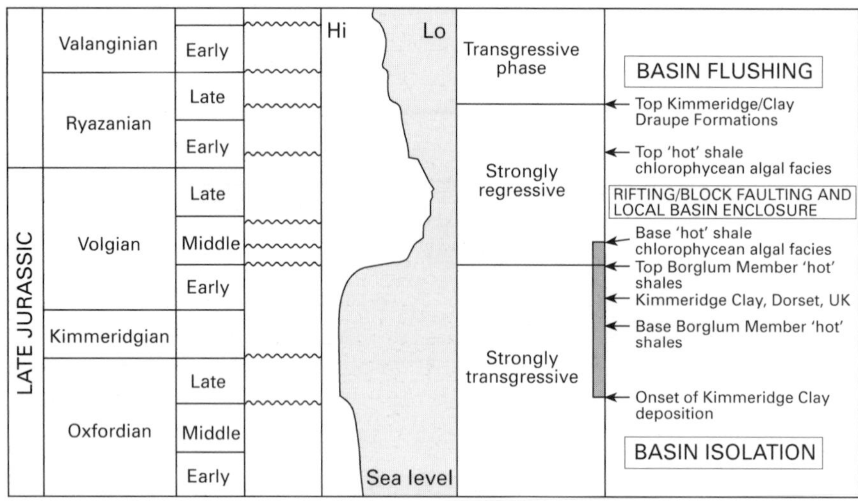

Fig. 11.45 Key events controlling organic deposition in the Upper Jurassic–basal Cretaceous seaway of the North Sea (modified after Rawson and Riley, 1982). The onset of organic-rich deposition was triggered by the development of a barrier cutting the Boreal–Tethyan seaway link. The major regression in the mid–late Volgian produced emergent land (Fossil Forest) in Dorset, but led to increased isolation and anoxia in the main graben system—hence the upper 'hot shale'.

(Combaz, 1980, Plate 3.10; Batten, 1983; Teichmuller, 1986, Plate 2). Under transmission electron microscopy have been shown to comprise 'ultra-laminar' structures attributable to the resistant outer cell walls of microalgae (Largeau *et al.*, 1990). Optical, TOC and pyrolysis characterization of one Kimmeridgian organic cycle in Yorkshire (Marton 87 borehole) has been undertaken in detail (Pradier and Bertrand 1992; Ramanampisoa *et al.*, 1992; Tribovillard *et al.*, 1994; Ramanampisoa and Radke, 1995). The authors found organic cycles on a 30 ka scale, with correlations between TOC and 'orange structureless kerogen'. Subjected to Rock Eval pyrolysis, the kerogen mixture generally plots as a Type II oil-prone kerogen (Barnard *et al.*, 1981), based on Hydrogen Index values (compare Figs 11.5 and 11.6). The average amount of inertinite in a typical Kimmeridge Clay kerogen may be determined using microscopy, or graphically using pyrolysis results as shown in Fig. 11.7A. When deeply buried and hence mature, the simple pyrolysis technique alone is inadequate to define this kerogen type (see Fig. 11.13). In the case of more mature samples, pyrolysis–gas chromatography results can provide diagnostic information (Bjorøy *et al.*, 1985).

The detailed composition of the pyrolysis products of a sample of 'hot shale' from UK well 14/30a-1 has been described in terms of C_1, C_2–C_4, C_6–C_{12}, C_{13} + saturates, C_{13} + aromatics, resins and asphaltenes by Monin *et al.* (1990) and Geibrokk *et al.* (1992).

A map of the Kimmeridgian–Volgian kerogen type in the North Sea Basin, based on 2200 Rock Eval pyrolysis analyses, has been published by Demaison *et al.* (1983). The Chevron group differentiated four essentially arbitrary facies: dead carbon (Type IV); gas-prone (Type III); mixed oil- and gas-prone (Type II/III); and oil-prone (Type II). This map is reproduced as part of Fig. 11.46 and shows the essentially concentric (centripetal: Huc, 1988) nature of the organfacies boundaries. The poor kerogen quality predicted for the Stord, Egersund and Norwegian–Danish basins to the east is currently (1997) being investigated by drilling.

A = Optical microscopy in the south Viking and Witch Ground Grabens shows the Volgian in the depocentres to contain the optimum oil-prone kerogen (Fisher and Miles, 1983).

This concept has been generalized by Barnard *et al.* (1981b), using Rock Eval pyrolysis data to illustrate the variation of TOC and four kerogen types down through a single 180 m (600 ft) section of the Kimmeridge Clay Formation, and spatially in terms of the change of domi-

Table 11.11 A summary of average properties of the kerogen from the 'hot-shale' facies of the Kimmeridge Clay Formation of the North Sea.

Property	Value	Reference	Comments
TOC (%)	5	Table 11.7	Realistic average
H : C ratio (kerogen)	0.9–1.2	—	Immature kerogen
Rock Eval Hydrogen Index	450–600	Barnard *et al.*, 1981b	Immature kerogen
Kerogen Type	II	Barnard *et al.*, 1981b	Rock Eval
$\delta^{13}C$‰ (kerogen) pdb	−27.6 to −28.7	Reitsema, 1983	Brae, Moray–Statfjord areas
	(−25)	Fuller, 1975	Isotopically heavier at the base
*Organic petrography**			
% amorphous liptinite	30–80	—	Bacterially degraded algae
% particulate liptinite	1–10	—	Dinoflagellates, spores, etc.
% vitrinite (am. + partic.)	20–70	—	(Amorphous + particulate)
% inertinite	1–25	—	Fusinite and semifusinite
Oil yield (barrels/acre.ft per 1% TOC)	50	Fig. 11.11	In the source rock at peak maturity

*Percentage values are visual estimates of area per cent of slide occupied by kerogen components.

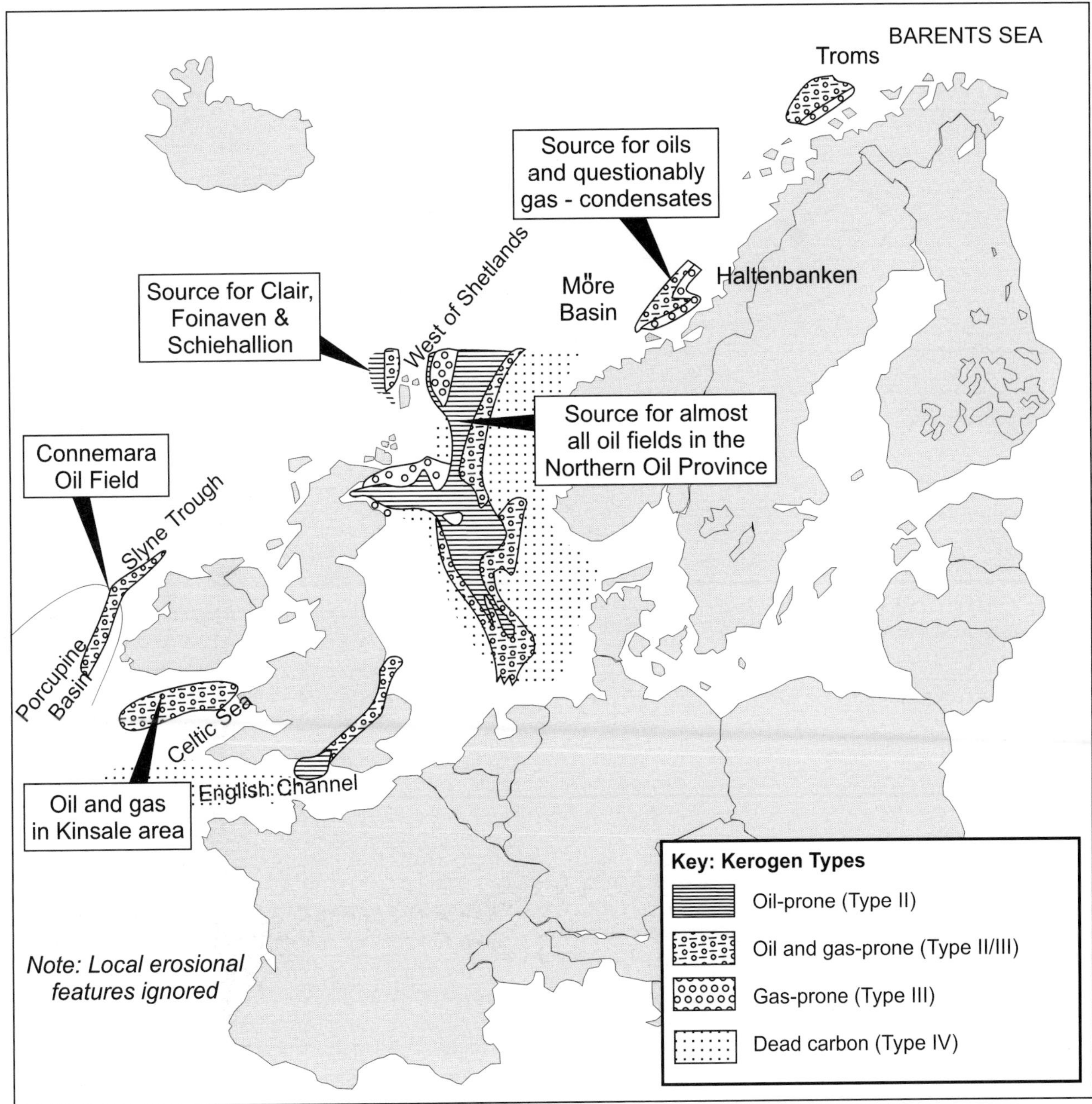

Fig. 11.46 Kerogen types of the Upper Jurassic–basal Cretaceous source rocks of the North Sea derived from Rock Eval pyrolysis characterization (after Demaison *et al.*, 1983). The map identifies the optimum kerogen quality, ignoring the vertical variation at any one location. Pyrolysis results from samples now deeply buried in the main graben require correction, as maturation has reduced the Hydrogen Index; microscopy may be more reliable. See Fig. 11.9 for definitions of kerogen type.

nant kerogen type from platform edge to the graben centre. This model can be rationalized in both space and time as an increase in (gas-prone) terrigenous kerogen and a decrease in (oil-prone) marine planktonic/bacterial kerogen towards the palaeocoastline.

In a review of the different kerogen types in the Upper Jurassic, Cooper and Barnard (1984) distinguish between an algal and a waxy sapropel. Cooper and Barnard's illustrated sterane and triterpane data clearly show that at least one major distinction between the samples is maturity, with the so-called 'algal sapropel' being the least mature, despite the reportedly similar spore colour and vitrinite reflectance values. The extract-gas chromatograms of Stow and Atkin (1987) also show a similar effect, with the fissile-laminated mudrock (marine sapropel-dominant) showing a fully mature distribution. In general, the rocks deposited in the (Upper Jurassic) basin centre will tend to contain the best kerogen quality, however they will also be the most deeply buried (most mature), as subsidence and burial progress to the present day. It is important to compensate for these effects in producing maps of kerogen quality (for example, Fig. 11.46).

It is clear from the above discussion that the characteristics of the 'hot shales' of the Kimmeridge Clay Formation vary considerably. However, for volumetric calculations and broad source-rock/oil correlations, some average properties are summarized in Table 11.11. Such a summary is validated by the high compositional uniformity of North

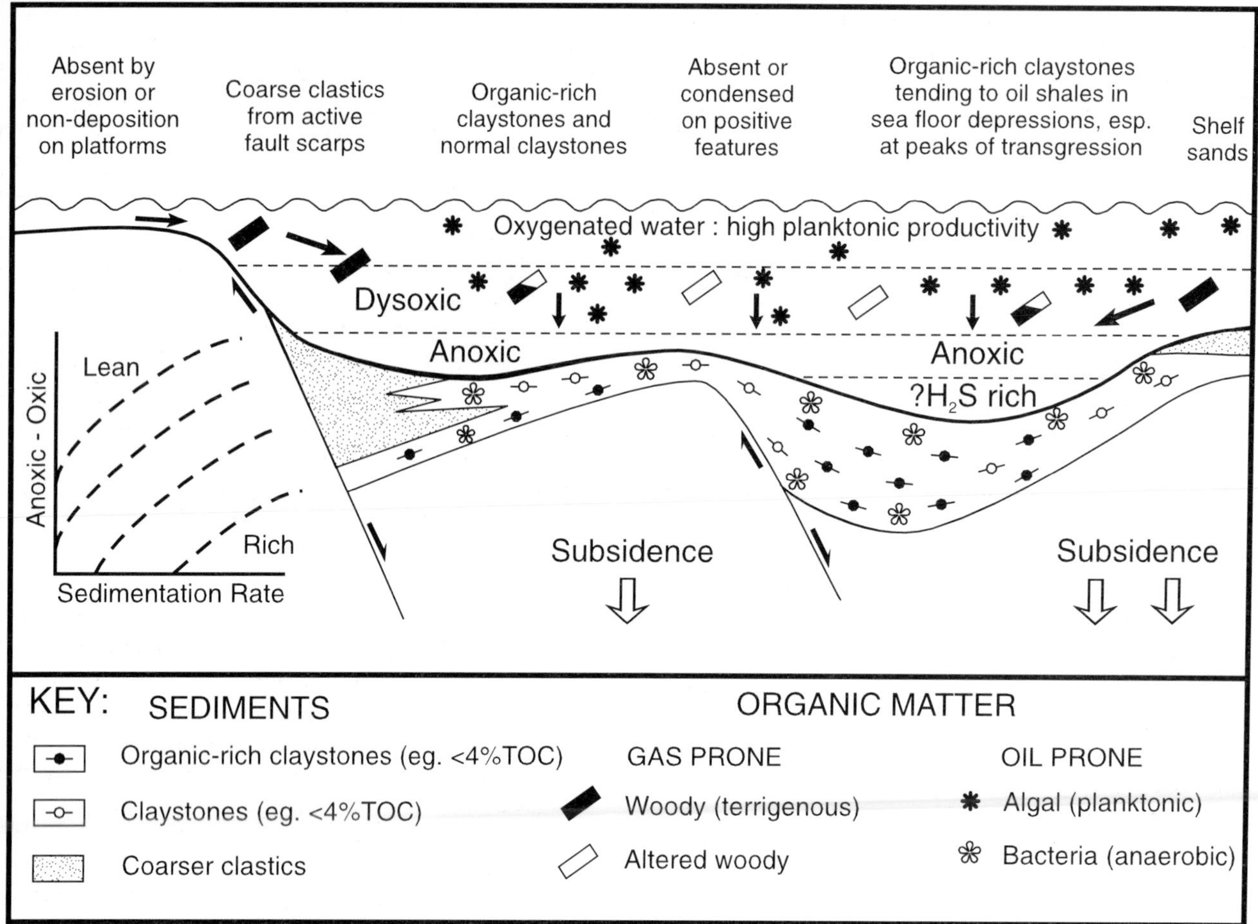

Fig. 11.47 Cartoon of the environments favouring the deposition of organic-rich source rocks in the Upper Jurassic of the North Sea graben system. The coincidence of high sedimentation rates and anoxia in the developing graben system, together with high bioproductivity in the surface waters, produces a world-class source rock.

Sea crude oils from the southern Central Graben to the northern Viking Graben (see Section 11.9).

In onshore UK, the Kimmeridge Clay contains variable quantities of oil-prone kerogen from Yorkshire to Dorset, being most mature in the north (Williams and Douglas, 1980; Douglas and Williams, 1981; Williams, 1986; Herbin *et al.*, 1991, 1993, 1995; Scotchman, 1994). Scotchman (1987, 1989) demonstrates the lateral-facies changes in four laterally persistent horizons from Yorkshire to the Weald and Dorset; surprisingly only three relatively small areas around Marton, Warlingham and Kimmeridge Bay yield optimum Type II kerogen. On a smaller scale, using optical and electron microscopy, Bertrand *et al.* (1990) have related the banded distribution of Kimmeridge Clay kerogen and mineral matrix to the expulsion behaviour of the generated hydrocarbons. To the south-east of the North Sea area, the Kimmeridge Clay Formation is not of optimum quality. In the Rijswijk oil province, the equivalent Delfland Formation may have been a secondary source to the Toarcian (Bodenhausen and Ott, 1981; de Jager, 1996).

To the west of the UK, a generally leaner but locally good-quality organic-rich claystone facies with a high natural gamma-ray response is reported west of Shetlands (Ridd, 1981; Bailey *et al.*, 1987): it thickens to the north-west into the Faeroe Basin, with deposition occurring in a deeper marine setting (Hitchin and Ritchie, 1987). In a detailed study, Bailey *et al.* (1987) found high TOC values (8–11%) and oil-prone kerogen (Hydrogen Indices 300–425 mg/g TOC) in the Volgian of the Solan and Foula basins, but lean, gas-prone kerogens on the rim of the main Faeroe Basin. To the west of Ireland in the Porcupine Trough, Croker and Shannon (1987) and MacDonald *et al.* (1987) described organically rich Upper Jurassic shales, possibly defining an impersistent westerly connection of the Boreal and Tethyan seas. Miller (1990), quoting Pepper, gives values in the range 1.1–2.4% TOC for Porcupine Trough wells. Making a transatlantic comparison, recent analysis has shown that the majority of the more southerly Porcupine wells show a leaner Upper Jurassic facies more reminiscent of the Scotian Shelf of eastern Canada, rather than the oil-prone facies of the North Celtic Sea, Solan Basin and Jeanne d'Arc basins.

The Upper Jurassic is also developed as a rich oil-prone source beyond the North Sea. North of 62°N, Rønnevik *et al.* (1983) predicting marine shales with good source quality. More specifically, Bjorøy *et al.* (1980) report TOC contents of 0.98–6.80% but generally gas/condensate-prone kerogen for the immature Kimmeridgian to Ryazanian of Andoya Island.

In the undifferentiated Jurassic of Svalbard, TOC values are lower, but the kerogen is of better quality and fully mature (Bjorøy and Vigran, 1980; Mørk and Bjorøy, 1984). Hvoslef *et al.* (1986) report rich (up to 11.66% TOC) but gas-prone kerogen in the Upper Jurassic–Lower Cretaceous Janusfjellet Formation of Svalbard. In a more detailed study, Backer-Owe *et al.* (1989) have pointed out that only the paper-shales are potential oil-source rocks on the basis of reaching a minimum richness to allow expulsion to occur.

In the Troms area, the Upper Jurassic contains an organically rich oil- and gas-prone claystone (mixed Types II and III kerogens), which on structure is late immature to early mature with respect to oil generation (Bjorøy *et al.*, 1983; Westre, 1984). Larsen and Skarpnes (1984) reported mixed Type II and III kerogens, with TOC values in the 5% to > 10% range in the Traenabanken area of mid-Norway, while similar values (4–13% TOC and Type II kerogen) were identified by Hollander (1984) and Cohen and Dunn (1987) for the Upper Jurassic Spekk Formation of the Haltenbanken area. Hermans *et al.* (1992) and Grigo *et al.* (1993) have demonstrated a migrational link between the Upper Jurassic Spekk Formation shales and the oil in the Draugen Field, using a form of 3D modelling.

A complete petroleum system of Upper Jurassic source and Middle to Upper Jurassic reservoir has been described in eastern Greenland by Price and Whittham (1997). Here the Oxfordian–Kimmeridgian Bernbjerg Formation shales are locally Type III kerogen and overmature, although lateral equivalents (Hareelv Formation; Christiansen *et al.*, 1992; Stemmerick *et al.*, 1993) are a more promising source for the bitumen-stained breached traps.

Depositional environment. The depositional environment of the organic-rich shales of the Kimmeridge Clay Formation has been much debated. Worldwide, the Upper Jurassic is a time of transgression (Vail and Todd, 1981; Rawson and Riley, 1982), which favours the deposition of organic-rich rocks (Hallam and Bradshaw, 1979). This is due to high nutrient concentrations and hence bioproductivity, high sedimentation rates, and a tendency towards deeper and hence stratified water and low concentrations of dissolved oxygen (Demaison and Moore, 1980; Boussafir and Lallier-Vergès, 1997). Based on the geochemical and radioactive heterogeneity of the unit, Miller (1990) favours the development of halocline-related stratification as promoting anoxia and hence organic-rich sedimentation. A climatic change from humid to semi-arid during the Kimmeridgian (pre- and post-*hudlestoni* zone) has been proposed by Wignall and Ruffell (1990) as controlling organic deposition. They relate climate change to reduced sedimentation rate and change in the dominant soft-sediment biodegradation mechanism from methanogenic fermentation to sulphate reduction with time. The climatic control on Kimmeridgian organic matter has also been discussed by Oschmann (1988, 1990), who invoked a high-latitude monsoonal climate controlling anoxicity, with a broadly similar model proposed by Mann and Myers (1990). Cycles of 25 ka and 280 ka are recognized from TOC trends derived from gamma, resistivity and density logs in four Yorkshire boreholes (Herbin *et al.*, 1993), cycles being attributed to a transgressive sequence-stratigraphic tract. Orbital climatic cycles are put forward by van Buchem *et al.* (1995) as exerting overall control on the organic cyclicity in the Kimmeridge Clay of Yorkshire.

Published work on depositional environments is largely restricted to the UK onshore outcrops, where limestones, marls, dolomic limestones, oil shales, bituminous shale and clay are all recognized (Tyson *et al.*, 1979; Herbin *et al.*, 1991, 1993, 1995; Belin and Brosse, 1992). Gallois (1976) suggested that blooms of phytoplankton—evidenced by the presence of coccolith limestones interbedded with the oil shales—produced an excess of organic matter over available dissolved oxygen, and hence promoted preservation of organic-rich anoxic sediments. It was then suggested (Tyson *et al.*, 1979) that the anoxicity was the cause and not the effect of the coccolith blooms. The model of Tyson *et al.* (1979) envisaged a stratified water body (halocline or thermocline), anoxic at depth, which overturned periodically, liberating nutrients and giving rise to algal blooms.

Based on an isotopic and diagenetic study, a modification of these two models has been suggested by Irwin (1979), who reported bioturbation within the oil-shale facies and noted that biogenic calcite would be rapidly dissolved below any (O_2–H_2S) interface. The level of this interface will be controlled by the availability of sulphate relative to the organic-matter supply: where sulphate is limiting, the less efficient methanogenesis will dominate (Lallier-Vergès *et al.*, 1993; Boussafir and Lallier-Vergès, 1997). These latter authors point out that, where the organic matter produces more H_2S than can be removed as sulphides by the available iron, sulphidization of the organic matter occurs. Noting that TOC tracks their sulphate-reduction index through a Kimmeridge Clay cycle in Yorkshire, Bertrand and Lallier-Vergès (1993) argue for the surface-water productivity as the factor controlling high organic accumulation rates under deep anoxic water. Work on the Dorset onshore section, however, has detailed much more bioturbation, benthonic fossils, currents (storms) and earthquake activity than previously reported (Wignall, 1989). Cornford *et al.* (1980) have suggested that the influx of terrigenous organic debris itself may also have reduced the background oxygen levels in extensive shelf seas, producing inertinite (semifusinite, fusinite) as an end-product. As pointed out by Miller (1990), depositional models developed from samples from the UK onshore flanking basins (e.g. Kimmeridge Bay, Dorset) should not be applied without re-evaluation to the source rocks of the graben depocentres of the North Sea.

It is clear that, in the graben areas, deposition occurred below wave base in an oxygen-deficient environment, with high planktonic bioproductivity and high sedimentation rates (Fig. 11.47). Anoxicity of the Draupne relative to the underlying Heather Formation has been argued on the basis of pristane/phytane and Ni/V ratios (Telnaes *et al.*, 1991). The water depth was probably moderate (e.g. ~ 200 m), since condensed rather than eroded sequences are generally seen on footwall uplift 'highs' in the northern North Sea. Where coarser clastics are associated with the organic-rich facies, such as in the Brae area, there is clear evidence for downslope transport, but water depths still need not be great (Stow *et al.*, 1982; Harris and Fowler, 1987). The association of these coarse clastics with fault scarps is indicative of fault movement immediately before or during organic sedimentation.

The overall picture is one of rapid deposition in a

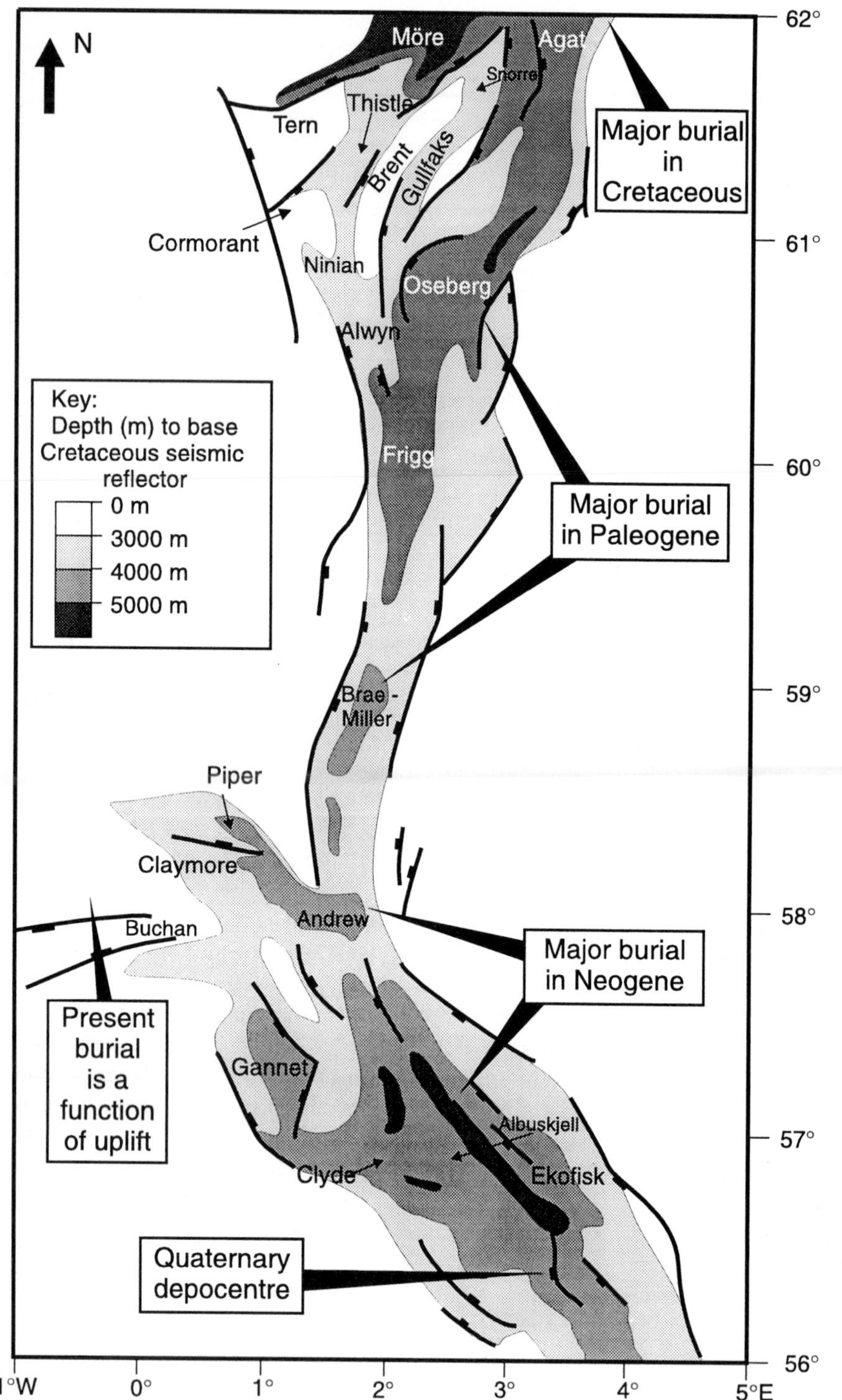

Fig. 11.48 Depth of burial of the near-base Cretaceous seismic reflector (equivalent to the top Kimmeridge Clay/Borglum/Draupne Formation). Dependent on thickness, the base of the source-rock unit may be up to 1200 m deeper than shown on the map. (Simplified from Day *et al.*, 1981, and subsequent published subregional maps—see text.)

relatively shallow sea, with highly productive, oxygenated surface waters and anoxic to hydrogen-sulphide-saturated water at depth, but with occasional overturning and mixing. The anoxic Upper Jurassic trough was effectively cut off from the Tethys to the south and east, and exchange of oxygenated water with the Boreal sea in the north would have been restricted merely by the extreme distance (~2000 km) and the relative shallowness of the seaway. The highly organic 'oil-shale' facies may have accumulated in deeper sea-floor depressions controlled by fault subsidence, where ponding of highly anoxic water and higher sedimentation rates would favour organic preservation. The surrounding land areas must have been rich in higher-plant vegetation (e.g. the Volgian 'Fossil Forest' of Dorset), which contributed the vitrinitic components to the kerogen, particularly on the graben flanks.

Upper Jurassic source-rock maturity. The maturity of the Kimmeridge Clay/Borglum Formation can be estimated from a base Cretaceous, seismic-depth map, such as that produced by Day *et al.* (1981) and reproduced in part in Fig. 11.48. Similar, more detailed maps have been produced by Goff (1983) for the East Shetland Basin and Viking Graben, by Harris and Fowler (1987) for the south Viking Graben, by Michelsen and Andersen (1983) for the Danish Central Graben area and for the whole graben

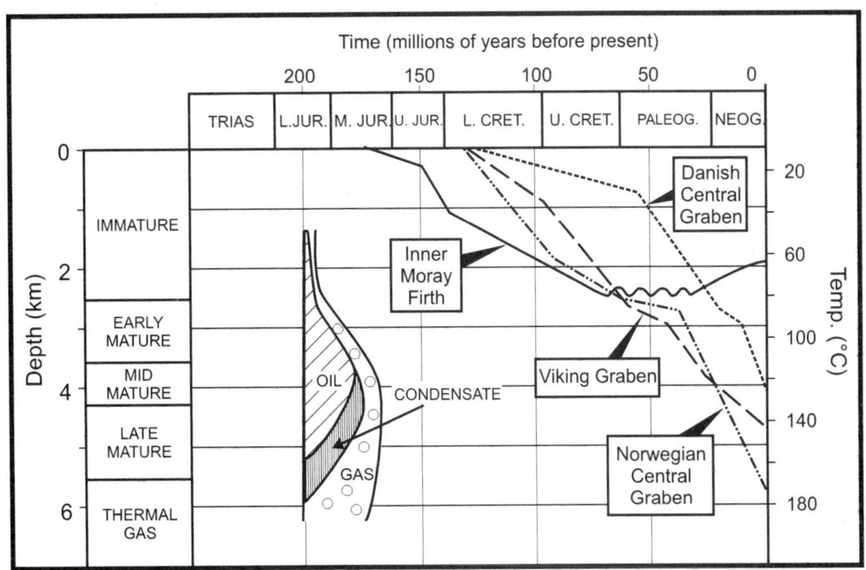

Fig. 11.49 Burial-history plots for Upper Jurassic source rocks in some North Sea mid-graben locations. At mid-basin locations, the burial histories show a rather monotonous linear burial profile (e.g. Pegrum and Spencer, 1990), equating to a constant burial rate of between 30 and 40 m/Ma. Note the absence of uplift and erosion at the midbasin locations, in contrast to the profile of the Inner Moray Firth, where inversion has brought Chalk to the seabed. A schematic generation (left) and temperature profile (right) are based on a regional geothermal gradient of 35°C/km.

system by Pegrum and Spencer (1990). Depth maps can be used to define the area of mature rocks where the temperatures for generation boundaries, taken from Cornford *et al.* (1983) or Mackenzie and Quigley (1988), are converted to depth, using a range of geothermal gradients (i.e. 25, 30, 35°C/km; 1.4, 1.6, 1.9°F/100 ft), plus 4°C (40°F) at sea floor. This spread covers the majority of mapped variation in geothermal gradient in the North Sea, as shown in Fig. 11.18.

Heat-flow mapping, is, however, to be preferred, although only general regionally extensive published maps are available (Burley, 1993). Ritter (1988) using calibrated 1D maturity modelling, with a range of present-day heat-flow values, has calculated the kinetic maturity for three drainage areas in the northern part of Norwegian Quadrant 9 (Egersund Basin).

Equating temperature in the source-rock kitchen with maturity in this way, however, ignores the effect of the heating time. Within the North Sea region, there are areas of both rapid (Central Graben) and slow (Viking Graben) Neogene and Recent subsidence (Fig. 11.49; Nielsen *et al.*, 1986; Thorne and Watts, 1989), as well as of Tertiary uplift and erosion (e.g. Inner Moray Firth; Andrews *et al.*, 1990; Hillis *et al.*, 1993; Pearson and Duncan, 1996). As in all continuously subsiding basins, the sedimentation exerting the maximum maturational effect on deeply buried strata is recent burial (i.e. Neogene and Quaternary), this being the sedimentation that is responsible for burying the source rock to its maximum depth and hence temperature.

Stated simply, the maturity of the Upper Jurassic is largely a function of Neogene to Recent sedimentation rates (Fig. 11.49). Dahl and Augustson (1991) have emphasized this point in producing transformation-ratio maps at the present and at 4 Ma for the Upper Jurassic of the North Viking Graben, Haltenbanken and Hammerfest Basins. It is unfortunate that the oil industry does not collect top-hole samples from exploration wells (e.g. down to ~2000 ft; 600 m.kb), so little detail is acquired concerning the top-hole stratigraphy. High-resolution strontium ($^{87}Sr/^{86}Sr$) stratigraphy has been used to identify a regionally extensive Miocene unconformity within the Cenozoic of the Norwegian sector of the Viking Graben between 60° and 62°N (Rundberg and Smalley, 1989). A major Mid-Miocene unconformity has also been identified at the base of a prograding deltaic sequence, using high-resolution shallow seismic in the Central Graben area (Cameron *et al.*, 1993). The best source of information for shallow stratigraphy in the UK sector is probably the Offshore Regional Reports series of the British Geological Survey (e.g. Johnson *et al.*, 1993 for the northern North Sea; Gatliff *et al.*, 1994 for the Central North Sea; Andrews *et al.*, 1990 for the Moray Firth; Cameron *et al.*, 1992 for the southern North Sea and Jordt *et al.*, 1995 for Norwegian and Danish waters).

Taking a regional view, it can be seen that the locus of maximum sedimentation migrated south with time, moving from the Møre Basin in the north (Upper Cretaceous depocentre) via the Viking Graben in the Palaeogene, and finally to the Central Graben in the Neogene and Quaternary (Nielsen *et al.*, 1986). Using a theoretical approach, Thorne and Watts (1989) have differentiated tectonic subsidence from sediment-loading subsidence, and mapped the former for the entire North Sea. Thus, while present-day burial depths are approximately equivalent (Fig. 11.48), source rocks in the Central Graben have been exposed to oil-generating temperatures for shorter times than in the north (Fig. 11.49).

Another consequence of high Tertiary sedimentation rates is the insensitivity of models to palaeoheat-flow events (see Fig. 11.20). Temperatures felt by an Upper Jurassic source rock under high heat flow associated with Upper Jurassic crustal extension (rifting) will tend to be exceeded by progressively deeper burial under lower heat-flow values. For example, Ritter (1988), using a variety of calibrants, was able to obtain a good maturity correlation for wells in the Egersund Basin, using a constant heat flow (50.3 mW/m^2) with time.

Maps of vitrinite reflectance of the Upper Jurassic have been published by Thomas *et al.* (1985) and Pegrum and Spencer (1990) for the whole Viking–Central Graben system, by Barnard and Bastow (1991) for the area between 58°N and 60°N, by Cayley (1987) for the Central Graben, by Rønnevik *et al.* (1983) for the Norwegian North Sea, by

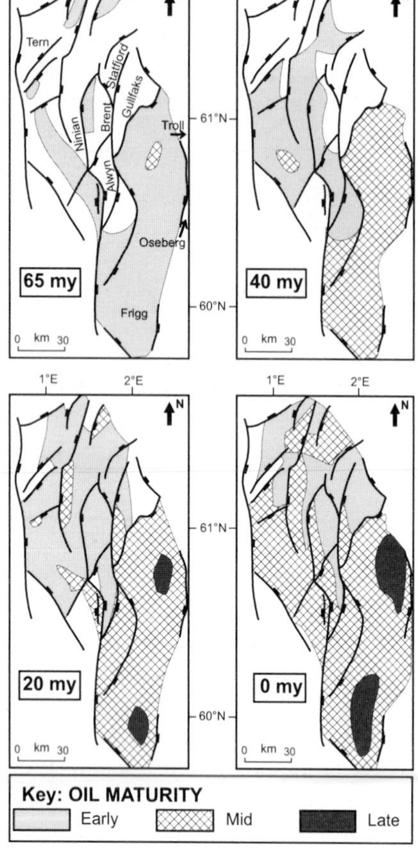

Fig. 11.50 Temporal progression of maturation of the top Kimmeridge Clay Formation surface in the East Shetland Basin, UK North Sea (from Goff, 1983). Early maturation was established in the main graben some 65 Ma ago, with progressive migration of the main oil-generation zone into the East Shetland embayment during the subsequent regional burial. Today, major areas of late (oil) mature source rock occur in the Troll and Frigg gasfield kitchens. Given the thickness of the Viking Group shales, the base of the source-rock unit will exhibit higher levels of maturation than shown.

Field (1985) for the north and south Viking grabens, and for the Moray Firth–South Viking Graben area by Fisher and Miles (1983). A suite of palaeomaturity maps has been prepared by Mason *et al.* (1995) covering the Witch Ground Graben. Goff (1983) has produced a series of maps of the East Shetland Basin to show the progressive expansion of the areas of maturity with time from the basin centre to the basin edge (Fig. 11.50), while a similar set of maturity maps covering the oil area to the south during the end-Eocene and Recent has been prepared by Barnard and Bastow (1991). In this context, note the reservations expressed in Section 11.4 concerning the equivalence of maturation parameters, such as vitrinite reflectance, and the oil/gas-generation reaction itself.

Macgregor (1996) has reported the timing of charge to 15 giant North Sea fields in terms of three time slices from Maastrichtian to Neogene, although the basis for each calculation is not given. Predictions concerning the timing of migration in the Viking Graben have been tested by Glasmann *et al.* (1989), using K/Ar dating of neoformed illites. They established that the Huldra Field was first charged in the latest Palaeocene–early Eocene (~50 Ma), with filling completed by the early Oligocene (~38 Ma). Thermal modelling has been integrated with petrographic and fluid-inclusion studies to demonstrate that the filling history of the Tarbet reservoir of the Oseberg field was initiated 32–27 Ma ago (Walderhaug and Fjeldskaar, 1993). Migration to the Veslefrikk field occurred in the late Oligocene at ~30 Ma. Having failed to separate neoformed from detrital illite as a basis for rubidium/strontium (Rb/Sr) and K/Ar dating, Liewig *et al.* (1987) have constrained the arrival of oil into an unidentified Brent reservoir in the Alwyn area as occurring at about 40–45 Ma, the arrival of oil triggering illitization. This timing was confirmed by Hogg *et al.* (1993), who also estimated the average progression of the illitization front in uniformly permeable reservoirs as 5.5 m/Ma. Attempts to date the arrival of oil into the Ula Field of the Central Graben, using diagenetic criteria, have concluded that diagenesis continues after the arrival of the oil, with a filling history of progressively more mature oil entering the structure (Nedkvittne *et al.*, 1993; Oxtoby *et al.*, 1995). Stoddart *et al.* (1995) and Larter *et al.* (1990) have argued for Neogene–Recent filling of the Eldfisk and Ula fields, respectively, while Leonard and Munns (1987) estimate oil arrival in the Valhaff field as a Pleistocene-Recent event.

As the North Sea Basin moves into a mature phase of exploration, it is of increasing importance to define the early-mature generation phase as precisely as possible, and hence to recognize those geological situations under which the early-mature oil will be effectively drained from the source rocks. This knowledge will optimize the search for oil in the areas of marginal maturity on the flanks of the grabens, areas such as the West Forties Basin, the Egersund Basin (Ritter, 1988), the inner Moray Firth Basin (Hillis *et al.*, 1993) and the western margin of the East Shetland Basin.

Maturity of the Upper Jurassic in the mid-Norway area is controlled by fairly constant Mesozoic and Tertiary burial in basinal areas (Hollander, 1984; Larsen and Skarpnes, 1984; Cohen and Dunn, 1987). Structural highs, as usual, reveal unconformities and non-sequences. Pittion and Gouadain (1985) and Khorasani (1989) have established measured maturity trends in the Haltenbanken area, showing reflectance values of about 0.5%R_o and T_{max} values of 430°C by about 3000 m, with 1.0% R_o and T_{max} of 460°C at 4500 m. Maturation, expressed in terms of modelled generation from the Upper Jurassic Nesna Formation, has been mapped and showed oil expulsion to initiate at a depth of 3080 m (Cohen and Dunn, 1987), while Hermans *et al.* (1992) have shown the progressive development of generation kitchens with geological time in the southern Haltenbanken area. The effect of 1–2 km of Tertiary–Recent uplift on the maturity of the Barents Sea area had been discussed (Nyland *et al.*, 1989; Liu *et al.*, 1992; Theis *et al.*, 1993; Reemst *et al.*, 1994), with evidence for uplift and gas spillage being reported in terms of fault-related gas chimneys and hydrate zones (Laberg and Andreassen, 1996).

Oil and source-rock correlations. Oils can correlate with other oils or with source rocks for a number of reasons, e.g.:

- Source-rock type (e.g. common source-rock lithology,

organic input to kerogen, oxicity of depositional environment, salinity of water).

• Maturation (common range of source-rock maturities or, more specifically, common volumes of source rock at each maturity level contributing to the accumulation).

• Migration (common distance, tortuosity or leakiness of migration paths).

• Alteration in the reservoir (common degree of bacterial degradation, water washing, thermal degradation, etc.).

The modern approach is not just to say that a link between oil and source rock exists, but to say why (e.g. Østfeldt, 1987, for the Danish North Sea; Gormly *et al.*, 1994, for the northern Viking Graben; Horstad *et al.*, 1990, 1992, for the Gullfaks Field; Harstad and Larter, 1997, for the Troll field; Karlsen *et al.*, 1995, for Haltenbanken).

The 'hot shales' of the Kimmeridge Clay Formation have repeatedly been shown to give the best match with North Sea oil properties, using oil/source-rock correlation techniques (e.g. Oudin, 1976; Cornford *et al.*, 1983; Mackenzie *et al.*, 1983, 1984; Northam, 1985; Schou *et al.*, 1985; Gormly *et al.*, 1994). Fisher and Miles (1983) have noted that in the Southern Viking and Witch Ground grabens, a correlation can be made between specific kerogen types, maturity and oil properties. Oil/source-rock correlations have latterly relied heavily on sterane and triterpane fingerprinting, using the computerized gas chromatography–mass spectrometry technique (GCMS), although such correlations are dominantly controlled by maturity rather than kerogen type (Cornford *et al.*, 1983; Gormly *et al.*, 1994; Stoddart *et al.*, 1995). Differential release of biomarkers during hydrous pyrolysis of Kimmeridge Clay kerogen has been described by Abbott *et al.* (1989) and has cast some doubt on their use in establishing correlations. In a study of oils and source rocks, van Graas (1989) has introduced some innovative higher-temperature correlative biomarker ratios (maturity < 1% R_o), based on experiments on Kimmeridge Clay.

Cornford *et al.* (1983), using a suite of some 100 North Sea oils and 'hot-shale' extracts, have demonstrated a detailed maturity-based correlation, emphasizing that the 'hot shale' generates a recognizable oil type (biomarker signature) at each maturity stage (Fig. 11.51). It is concluded that the bulk of reservoirs have received a full spectrum of maturity products from the source rock within the graben, as shown in Fig. 11.51C, a concept proved in Eldfisk and Troll at the field scale by Hall *et al.* (1994) and Harstad and Larter (1997), respectively.

Considering more regional studies, correlation by Gormly *et al.* (1994) in the prolific North Viking Graben has differentiated oils generated from the Draupne Formation, with isotopically lighter carbon (−31‰ to −28‰) and low pristane/phytane ratios (1.0 to 1.5), from those generated from the underlying Heather Formation shales, with isotopically heavier carbon (−28‰ to −25‰) and higher pristane/phytane ratios (2.15 to 4.0). As would be expected, given the complex 'plumbing' of the area, mixes with isotopically intermediate carbon (−28.8‰ to −28.0‰) and pristane/phytane ratios (1.5 to 2.15) are reported. In a detailed study of reservoir heterogeneity in the Gullfaks Field (Norwegian north Viking Graben), Horstad *et al.* (1990) attributed variation to the progressive filling of the reservoir from source rocks of ever-increasing maturity, as in Fig. 11.51A–C. In this case, the primary maturity-related variation was also overprinted by a water-washing and biodegradation event. Using the same approach, Horstad and Larter (1997) were able to identify two fill points in the Troll field, despite variable biodegradation of both oil and gas and remigration within the field during late tilting.

Reitsema (1983) demonstrated that the Brae oils correlate with the adjacent 'hot shales', using *n*- and isoprenoid alkane, sterane and triterpane distributions, as well as the stable carbon-isotope curve-matching technique. He emphasized the need to correlate the properties of the different oils within the Brae complex with specific facies of the hot shale, an approach also emphasized by Scotchman (1981) in the northern Viking Graben. Mackenzie *et al.* (1983) showed a close correlation of the Brae oils with extracts from the interbedded 'hot shales', using GCMS and isotope data. Significantly, the oil appears less mature than the extracts of the interbedded shales, which is what is expected if the reservoir oil is a mix (time integral: Larter, 1988) of early, mid- and late-mature source-rock expulsion products (Fig. 11.51). In the case of on-structure maturity, this assumes that molecular structures are more thermally stable in the 'clean' sandstone pores of the reservoir than in the finer-grained, clay-rich source rock.

Maturity established using whole-oil gas chromatograms, biomarkers and isotopes, has been found to be the most important variable in correlations, between northern North Sea oils and their Upper Jurassic source rocks (Schou *et al.*, 1985; Harstad *et al.*, 1990), while at the southern extreme Østfeldt (1987) came to similar conclusions in the Danish Central Trough area. Here, biodegraded oils were also encountered, but the degradation did not greatly affect the maturity and source-type correlation parameters.

A cautionary note has been injected by Knudsen *et al.* (1988), who have claimed to see the effects of minor allochthonous oil staining in the extracts of an otherwise homogenous series of early-mature Draupne shale samples from the Oseberg area of the northern Viking Graben. Migration of anomalously mature oil into, rather than out of, a fine-grained source rock has been noted at the type locality for the Kimmeridge Clay (Farrimond *et al.*, 1984). Such effects could seriously interfere with any attempt at oil/source-rock correlation.

The utility of establishing a maturity correlation has been confirmed by Radke (1988), who has used a maturity-related methylphenanthrene index (MPI) to attribute an average source-rock maturity for 83 North Sea crude oils in terms of calculated vitrinite reflectance values (Fig. 11.52). Source-rock trends from North Sea wells have been published for comparison by Hall *et al.* (1985) and Schou *et al.* (1985). Relating source-rock and oil data (Fig. 11.52) suggests most generation and expulsion occur at 4–5 km depth. Grantham (1989) has shown, by a comparison of the Kimmeridge Clay source-rock trends with those of carbonates and phosphates, that the use of these aromatic maturity indices should be restricted to rocks with catalytically active, acidic, clay surfaces, such as mudstones. Using artificial laboratory maturation, Garrigues *et al.* (1990) found internal consistency between MPI ratios derived from shales of the Kimmeridge Clay Formation with

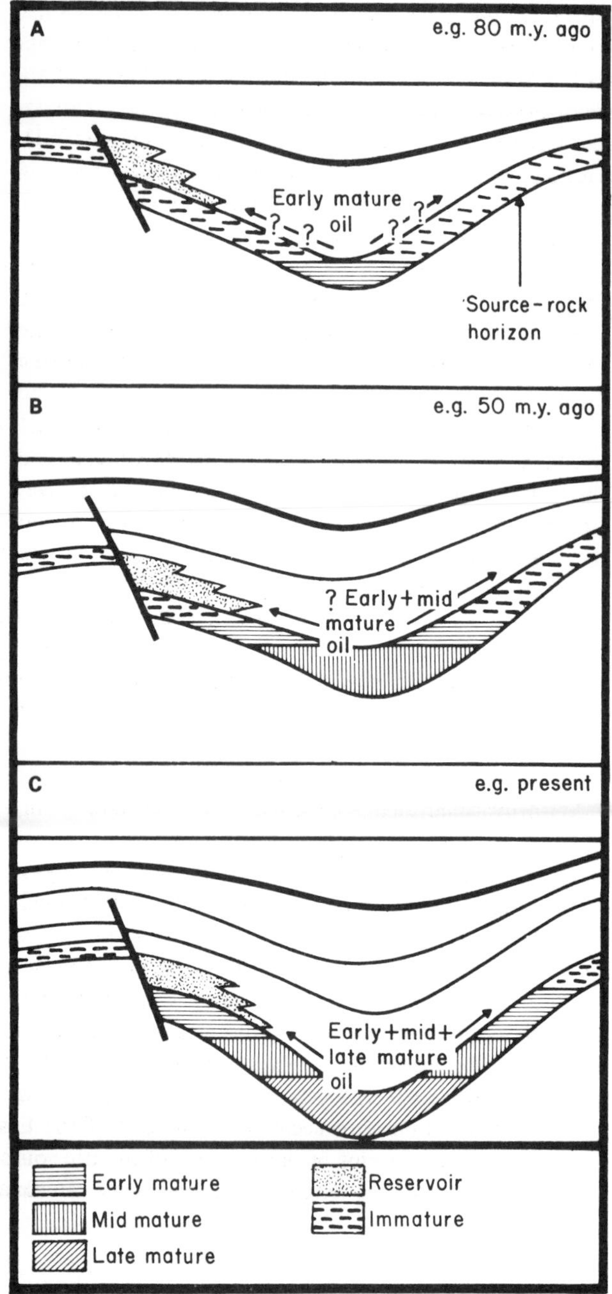

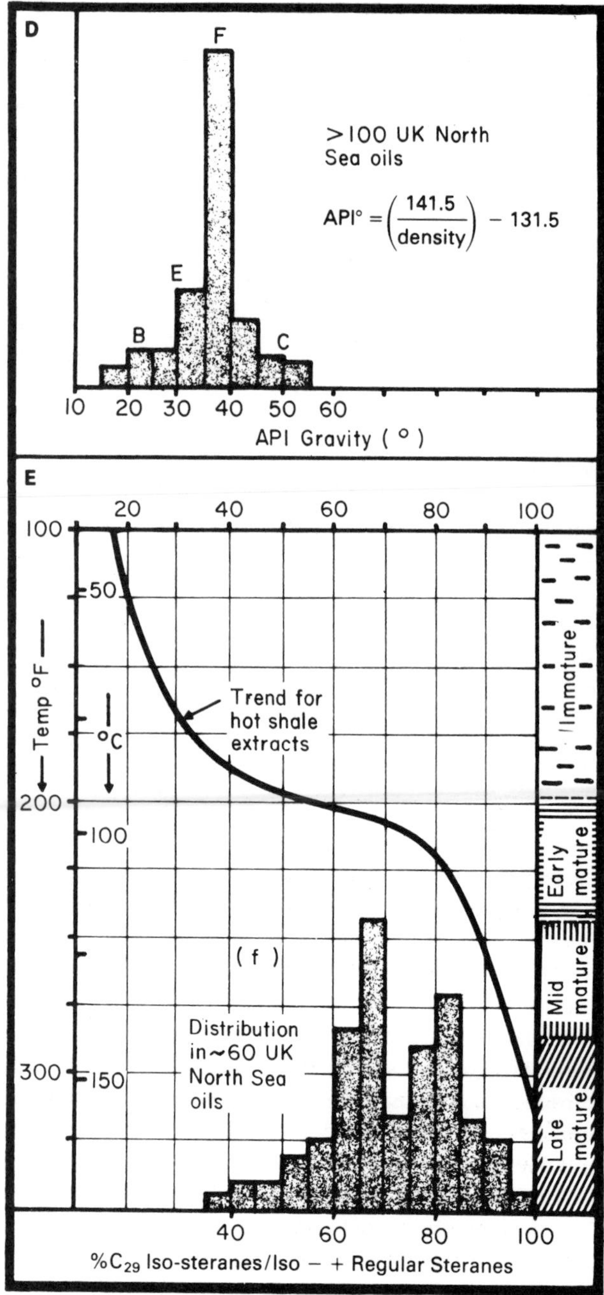

Fig. 11.51 North Sea oil properties are primarily controlled by source-rock maturity (Cornford *et al.*, 1983). Progressive oil generation from a single source-rock horizon being buried in a graben can be shown as a time series of cross-sections (A), (B) and (C). Early-mature oil generated first in the graben centre (A) cannot migrate due to poor drainage. Effective expulsion from the graben centre occurs only when this location reaches the mid- (B) or even late-mature stage (C), by which time early-mature generation is occurring on the basin flanks, where improved drainage may facilitate expulsion. (D) and (E) show the results of this process on oil properties. In the distribution of API gravities of UK North Sea oils (D), where four oil types are identified: B, biodegraded; C, condensate/light oil; E, early-mature; F, full-maturity-spectrum oil. The maturity designation may be derived from comparison of a sterane-ratio trend measured on a series of source rocks (E, trend line), with the same ratio measured on a collection of 60 oils (E, histogram). The histogram shows that early-mature oils are rare (being found on the flanks of the basin generally in Upper Jurassic reservoirs). The bulk of the oils contain molecules characteristic of a full-maturity spectrum (main mode) or just a late-maturity fraction of source-rock products consistent with the poor drainage of a hot-shale source rock. Full-maturity-spectrum oils are found in reservoirs that have received oil from a range of source-rock maturities, as in graben model C.

abundant clays and Brent coals (with little clay), casting doubt on this interpretation.

Using carbon-isotope and pristane/phytane ratios as source-facies indicators, Chung *et al.* (1992) have placed North Sea oils in the context of a global data set of oils derived from marine source rocks. North Sea oils fall in a higher-sulphur (> 0.5% S) category of oils deriving from 'Palaeozoic and Mesozoic marine shales and Palaeozoic carbonates'. Concentrating mainly on stable carbon-isotope ratios, Bailey *et al.* (1990) have attempted to establish facies-related correlations between North Sea oils and specific organic facies of the Upper Jurassic. They noted

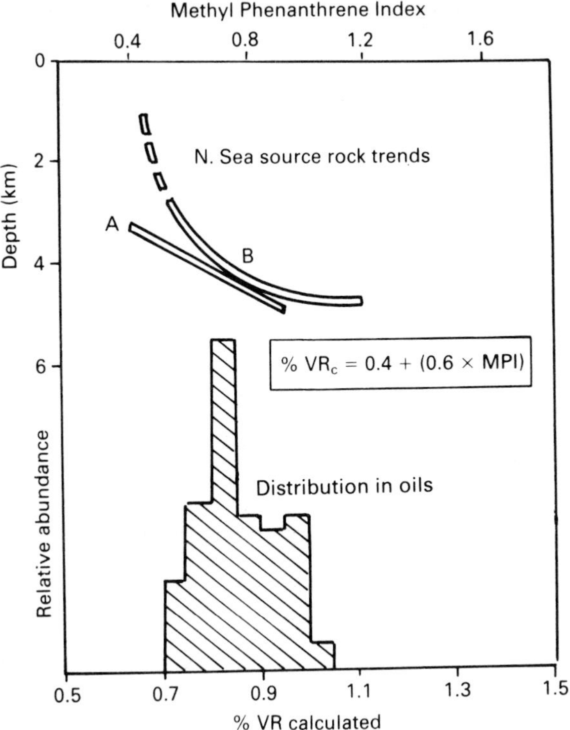

Fig. 11.52 Use of methylphenanthrene indices (MPI) to establish a maturity trend for source-rock extracts (upper, Schou *et al.*, 1985 (A); Hall *et al.*, 1985 (B)) and (lower) a distribution from North Sea oil. Conversion from MPI to % R_o values is according to Radke (1988). A simple interpretation of the two trends and the histogram suggests that the oils were generated in and expelled from source rocks buried between 3 and 5 km, with maturities between 0.7 and 1.0% R_o.

isotopically heavy kerogens in 'cold-shale' intervals, an observation others have correlated with the abundance of vitrinite (Scotchman, 1991) and inertinite, rather than a feature of the oil-prone kerogen component itself. The derived oils would thus not follow this trend. More specifically, an oil/source-rock correlation based on facies-related source-rock parameters was proposed by Grantham *et al.* (1981) in the Tern area (UK Block 210/25), using the C_{27} and C_{28} triterpanes to identify a specific interval of the 'hot shales'. Fuller (1975) demonstrated a good facies-related correlation between North Sea oils and the Upper Jurassic source rocks, using optical rotation and stable carbon isotope data.

Oil/oil correlations based on bulk, isotopic and molecular properties have been reported for the Ekofisk oils (Hughes *et al.*, 1985), where a clear decrease in oil maturity was demonstrated in a north-west to south-east direction away from the Central Graben: variation in source-rock facies in terms of terrigenous input and oxicity was noted as a second-order effect. This conclusion was confirmed by Stoddart *et al.* (1995), who demonstrated a strong correlation with their biomarker-derived first principal component and the maturity-related $T_s/(T_s + T_m)$ trisnor–hopane ratios for 26 greater Ekofisk-area oils. In contrast, using multivariate statistics to extract significant variations in sterane and triterpane peak-area data from a group of 45 oils from the Norwegian North Sea, Telnaes and Dahl (1986) attributed the first and second principal components to source-rock depositional environment and maturity, respectively. It is unusual for more variation to be attributable to facies rather than maturation in North Sea data sets. Principal-component analysis of biomarker (sterane and triterpane) ratios was used by Dahl and Yükler (1991) to differentiate two families of oils in the Oseberg Field, and to compare Oseberg with oils from the adjacent Gullfaks, Snorre, Brage and Veslefrikk fields.

Oil/source-rock correlation in mid-Norway has been used to differentiate between the Upper Jurassic Spekk (Nesna) Formation and the Lower Jurassic Åre (Hitra). Formation as source rocks for the oil and gas/condensate accumulations, respectively (Cohen and Dunn, 1987). In a more recent study, Karlsen *et al.* (1995) claim that the oils of the area fall into two groups, sourced from a deeper and shallower water facies of the Spekk Formation. Based mainly on stable carbon isotopes, they found no direct evidence for oil generation from the coals of the Åre Formation in agreement with Khorasani (1989). In terms of intrareservoir variations, both continuity and barriers have been demonstrated by Stølum *et al.* (1993) within the stacked Jurassic reservoirs of the Smørbukk Field, using strontium-isotope ratios ($^{87}Sr/^{86}Sr$) of the formation waters.

Volumetric modelling. The North Sea oil province, with a well-defined single oil-source rock and a relatively simple history of maturation, is ideal for developing and testing volumetric (mass-balance) models for the generation and accumulation of hydrocarbons (see Section 11.4). One of the world's first published basin-scale calculations was attempted by Fuller (1975) for the North Viking Graben, and Barnard and Bastow (1991) have produced a refined calculation for oil and gas generated and discovered in the Central Graben, Moray Firth Basin and South Viking Graben. Goff (1983, 1984) and Cousteau *et al.* (1988) have published an attempt to account for the oils of the East Shetland Basin in terms of the volumes of mature source rock of known richness. Goff (1993, 1984) uses the equation:

$$\text{Oil generated} = \text{SRV} \times \text{OMV} \times \text{GP} \times \text{FOIH} \times \text{TR} \times \text{VI} \quad (28)$$

where:

SRV = source-rock volume (m^3 or acre × ft)
OMV = organic matter content by volume (from TOC)
GP = genetic potential (fraction of oil-prone kerogen = 0.7 for the Kimmeridge Clay)
FOIH = fraction of oil in hydrocarbon yield (0.8 according to Goff, 1983, 1984)
TR = Transformation ratio (depends on the maturity of the source rock)
VI = volume increase on oil generation (1.2 according to Goff, 1983, 1984)

To this must be added a migration and entrapment factor. From a comparison of the generated versus discovered (in-place) hydrocarbons, Goff (1983, 1984) found the overall efficiency of generation—migration and entrapment—to be 20–30%, having allowed for some additional generation from the Heather Formation. A large number of similar calculations made in the southern Viking Graben produces efficiencies of 25–35% for individual drainage polygons.

Higher values suggest super-efficient migration and entrapment (or error), while lower values indicate that hydrocarbon remains to be discovered (or error).

A similar calculation has been reported for the Central Graben petroleum system as a whole (Cornford, 1993a,b, 1994), with efficiencies of 0.5–~2%. Leonard and Munns (1987) estimate an 8% efficiency for the Valhall field drainage area alone. Barnard and Bastow (1991), using isopachyte and maturity maps for the Upper Jurassic source rocks for the majority of the North Sea oil province south of 60°N, produced oil efficiencies of 14% for the Central Graben, 22% for the Witch Ground Graben, 14% for the greater Maureen (Fisher Bank) area and 21% for the South Viking Graben. Pegrum and Spencer (1990) report more detailed Upper Jurassic isopachyte and maturity maps for the whole North Sea graben system, which may be used as a basis for such calculations.

The geochemical elements of a quantitative prospect evaluation in NOCS Block 33/6 have been described by Maranga *et al.* (1985). A kinetic-generation model was calibrated with laboratory pyrolysis data and used to predict the volumes of oil generated from an unusually lean Upper Jurassic section within each identified fault compartment in the block. Models of generation and migration in the Oseberg area have been reported by Dahl *et al.* (1987), Doligez *et al.* (1987) and Dahl and Yükler (1991). In the former two cases, the quantitative and qualitative results of modelling were validated with biomarker parameters from Oseberg oils. Dahl and Yükler (1991) produced present-day and palaeogeneration maps of the Norwegian northern Viking Graben, and mapped the net generation from the Viking Group in units of tonnes of hydrocarbon/m^2. Values in excess of 10 t/m^2 (30×10^4 barrels/acre) are reported.

Shell have presented an essentially vectorial mass balance for the southern area (Draugen Field and kitchen) of Haltenbanken, where oil generated from the Upper Jurassic Spekk Formation is modelled to migrate rapidly into the flank traps along specific pathways (Hermans *et al.*, 1989). No actual numbers or efficiencies were reported, oil being represented by its equivalent column height at any one location.

Kimmeridge-sourced petroleum systems in the North Sea. A petroleum system (Magoon and Dow, 1994) describes 'the genesis of reservoired hydrocarbons deriving from a single contiguous pod of mature source rock'. Based on a volumetric (or mass) balance, the system efficiency, defined as the discovered hydrocarbon as a percentage of that generated, can be calculated:

$$\text{Efficiency} = \text{volume discovered}/\text{volume generated} \qquad (29)$$

As discussed in Magoon and Valin (1994) in the context of a number of examples worldwide, the efficiencies of petroleum systems range from ~0.5% to ~30%. Demaison and Huizinga (1991), classifying petroleum systems in general, designated the Central Graben of the North Sea as a type example of a vertically impeded petroleum system, recognizing the dominance of vertical migration in the greater Ekofisk and Forties areas. A relatively low efficiency was envisaged. This was confirmed by Cornford (1994), who reported a simplistic system efficiency of 0.56% for the Central Graben, using the method detailed by Schmoker (1994).

A more realistic calculation, using the same data but taking into account the cracking of non-expelled oil to gas, was reported by Cornford (1993a, b), with the results shown in Fig. 11.53. This calculation gives a gross efficiency of 1.43% for hydrocarbons in general, 1.25% for oil and 2.33% for gas. The higher efficiency for gas is taken to result from the high impedance of the largely vertical migration pathways operating in the Central Graben (Demaison and Huizinga, 1991). The tortuosity of this migration path is further implied by the difference between the gross generated and in-place gas/oil ratios (175 m^3/m^3 and 392 m^3/m^3, respectively), where the preferential fractionation of gas over oil is indicated. In contrast, Barnard and Bastow (1991) report figures that indicate oil efficiencies of 14% for the Central Graben, with values up to 22% for adjacent basins.

In the eastern North Viking Graben, Dahl and Yükler (1991) report 9.4×10^9 m^3 of oil and 1300×10^9 m^3 of gas generated in the drainage areas feeding the Oseberg, Brage, Huldra and Veslefrikk fields. The reported in-place hydrocarbons for Oseberg are 0.4×10^9 m^3 of oil and 180×10^9 m^3 of gas (Nipen, 1987) and for Brage are $0.106–0.215 \times 10^9$ m^3 of oil and $13–21 \times 10^9$ m^3 of gas (Hage *et al.*, 1987). Taking averages of the Brage ranges and doubling the resulting sum to account for the other fields in the area gives total in-place volumes of 1.2×10^9 m^3 of oil and 394×10^9 m^3 of gas for all fields in the area. These figures indicate a system efficiency of 13% for oil and 30% for gas for this important and complex area of northern North Sea exploration and production. As with the Central Graben, a higher efficiency for gas indicates a low-permeability migration pathway.

The above estimations of system efficiency are based on single-value calculations. These single-value calculations are of limited use for risk-based exploration decision-making (Hermanrud *et al.*, 1990b). To overcome this problem, Cornford (1993b) assembled nine distributions of the key input variables to produce a probabilistic calculation based on 1000 iterative Latin–Hypercube samplings of the various distributions, three of which are shown in Fig. 11.54.

These distributions derive from analytical data (67 pyrolysis S_2 yields at the late immature level), planimetered areas of the source-rock isopachyte map (distribution of source-rock thickness) and planimetered areas of source-rock depth maps (geobathymetric curve). The distributions relate to the known variability of these input parameters and do not address the question of precision or accuracy in the measurements themselves. Uncertainty in, for example, the seismic picks used to define the source-rock isopachyte is a separate issue. The bi-modal geobathymetric curve for the Central Graben reflects the basin's relatively large trough (with modal depths of 3000–3500 m) plus specific fault-bounded depocentres (with modal depths of 4500–5000 m). It is interesting to note that, as shown in Fig. 11.54C, the risk associated with maturity is largely a function of the shape of the basin, represented by the geobathymetric curve, rather than uncertainty in measured

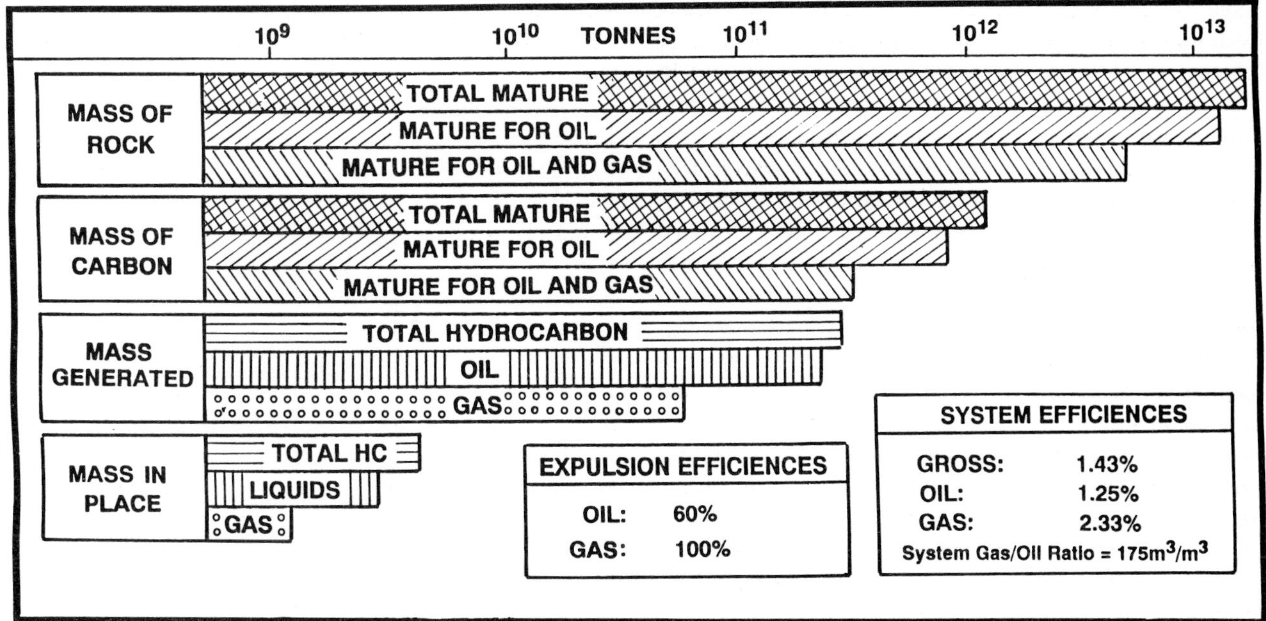

Fig. 11.53 Summary of the elements of a hydrocarbon mass balance for the Central Graben petroleum system (Cornford, 1993a). The overall system efficiency of 1.43% (i.e. 98.57% of generated oil has not been accounted for in currently discovered accumulations) includes the inefficiencies of expulsion, migration, entrapment and seal, not to mention the failure of exploration to discover all the accumulations associated with the source rock. Once fractionation during migration is considered, the predicted system gas/oil ratio of 175 m^3/m^3 compares favourably with the observed mean of 392 m^3/m^3 for 21 reservoired oils. Higher efficiencies for gas and lower for oil confirm a high degree of fractionation during dominantly vertical migration.

or modelled maturity parameters *per se*. On the other hand, source-rock thickness (Fig. 11.54B) is a function of the tectonic conditions at the time of deposition, rather than the basin's subsequent history. The fault-bounded depocentres of the Central Graben provide small areas of very thick source rock, with the large areas of the trough containing a thinner succession; the resulting distribution is log-normal (Fig. 11.55). Thus, the geometry of the basin, both at the

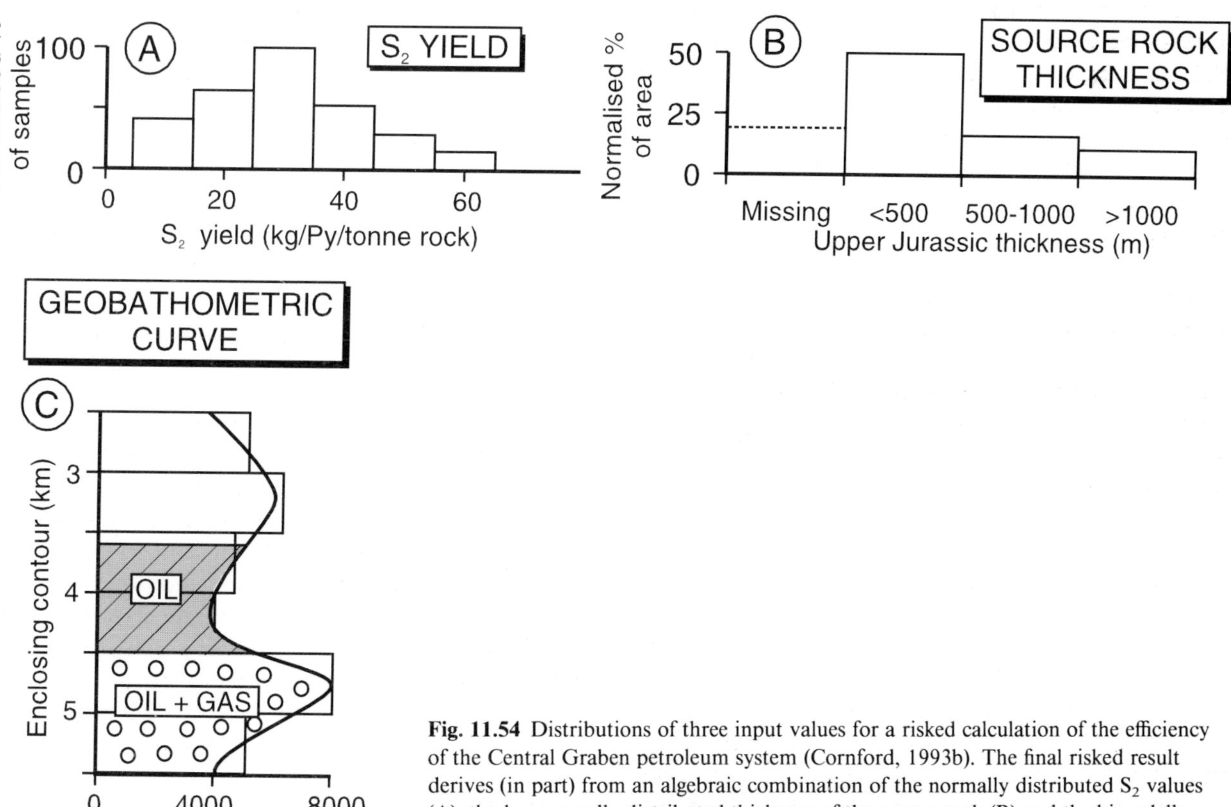

Fig. 11.54 Distributions of three input values for a risked calculation of the efficiency of the Central Graben petroleum system (Cornford, 1993b). The final risked result derives (in part) from an algebraic combination of the normally distributed S_2 values (A), the log-normally distributed thickness of the source rock (B) and the bimodally distributed areas enclosed by the specified depth contours (C).

time of source-rock deposition and during its subsequent tectonic development, contributes significantly to the range of outcomes in terms of hydrocarbon generated and its maturity.

The treatment of uncertainty in such models is referred to as 'sensitivity analysis'. In a discussion of the sensitivities of hydrocarbon-charge calculations in the area of Norwegian Block 9/2, Hermanrud *et al.* (1990b) identified source-rock thickness, transformation ratio, drainage area and initial source-rock potential as the major controls on charge uncertainty. Using triangular distributions (maximum–mean–minimum), their results are presented as cumulative probability curves for generated, expelled and migrated oil and gas (i.e. charge to trap), and recoverable hydrocarbons. These calculations produce mean values for their prospect of 136, 82, 43 and 11×10^6 m^3, respectively. In this example, less than 10% of the initially generated hydrocarbon is producible from the trap. This same area is treated stochastically by Irwin *et al.* (1993) in a pre- and post-drilling study of the Block 9/2 drilling.

Cornford's (1993b) probabilistic treatment of the Central Graben petroleum system (Fig. 11.55) yielded similar mean values to the discrete calculation, but with skewed distributions. The skewedness of the system-efficiency curves for both gas and oil suggest there is a significant probability of higher-than-average values. Higher efficiencies mean that there is more oil and gas to be found, or a downward revision of the volumes generated (equation 29). To address this, creaming curves for each play type were used to predict the ultimate in-place volumes in the Central Graben, assuming no additional play types are discovered (see Chapter 12, this volume). From the exploration point of view, the ultimate distribution (Fig. 11.55C) is strongly log-normal, with a skew to higher efficiencies, suggesting that significant undiscovered Jurassic-sourced reserves remain to be found.

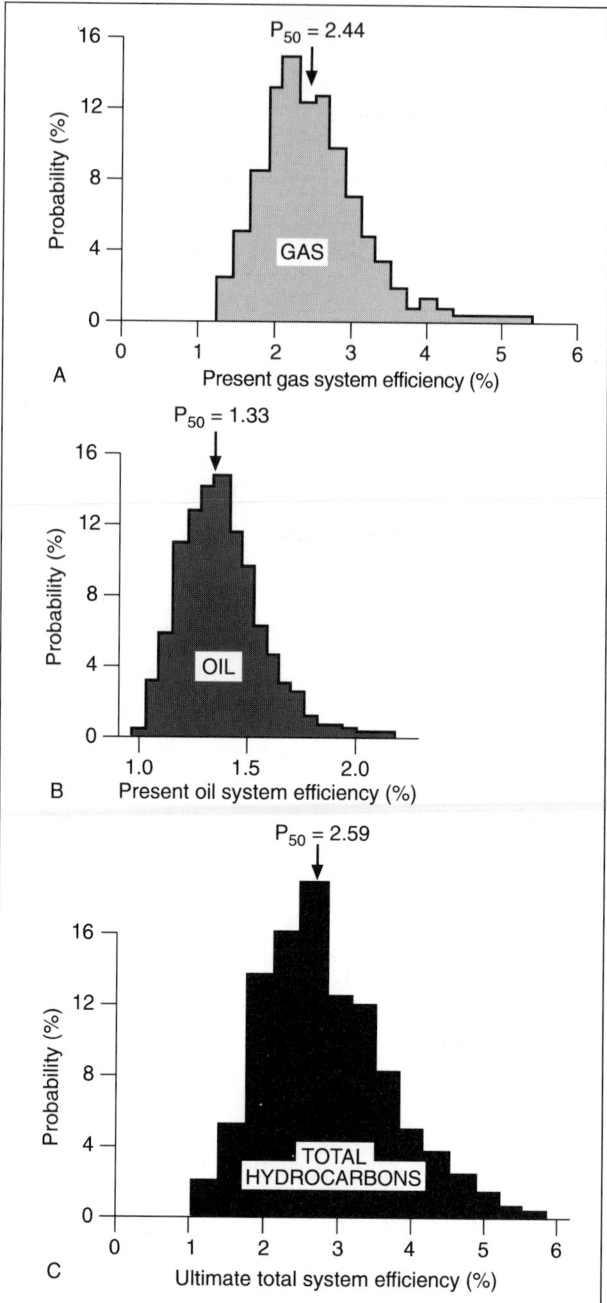

Fig. 11.55 Distributed efficiencies for Central Graben petroleum system: present-day curves for gas (A) and for oil (B) and the ultimate system efficiency (C) based on a projection of the creaming curve for discoveries (Cornford, 1993b). The P_{50} values (50% probability) approximate to the discrete values given in Fig. 11.53.

11.8.7 Cretaceous

The Lower Cretaceous (excluding the Ryazanian; see previous section) is developed as a dominantly shaly facies over much of the North Sea (Ziegler, 1982b; Hesjedal and Hamar, 1983). Major depocentres (> 500 m thickness) occurred in the Moray Firth Basin, in the Viking and Central grabens, in the Horda, Egersund and North Danish basins, and in the Broad Fourteens and Lower Saxony basins (Ziegler, 1982b). In all these basins, the section is believed to contain fair to good amounts of gas-prone kerogen (Barnard and Cooper, 1981).

Total organic carbon values averaging 0.6–1.4% were reported for seven wells in the Moray Firth Basin (Bissada, 1983), with the kerogens of most wells being of gas/condensate-prone land-plant type, except in Block 14/20, where the kerogen is of lower-plant origin. In the Statfjord area, the TOC values of the Lower Cretaceous Cromer Knoll Group average 1.04% (Kirk, 1980) and the kerogens appear to be gas-prone.

In the south, the Barremiann–Albian Sola Formation of the Danish Trough has TOC values in the range 0.36–12.74% TOC, with averages for two wells of 5.71% and 1.50% TOC (Jensen and Buchardt, 1987). The Hydrogen Indices all fall below 400 mg/g TOC, making the unit only a fair oil-prone prospect. In contrast, the thin (< 1.0 m thick) Munk Marl Bed of the underlying Barremian–Hauterivian Tuxen Formation produced two TOC values of 8.55% and 14.55%, giving an average of 11.55% TOC: Hydrogen Indices are about 500 mg/g TOC, making the unit oil-prone but too thin to generate significant volumes of oil. Both units were immature (0.38–0.46% R_o) in the two Danish offshore wells analysed (Adda-1 and E-1).

Locally, particularly during the Albian and Aptian, a thinly developed oil-prone kerogen facies was deposited in the southern half of the North Sea and may be equivalent to the pyritous black Speeton Clay of eastern onshore England.

The Lower Cretaceous is, along with the Jurassic, the source for some of the north-west German gasfields, according to Tissot and Bessereau (1982). The Wealden facies of the Lower Cretaceous, as displayed on the south coast of England, is in a gas-prone facies, with thin lignite seams and stringers and carbonaceous shales. The high sand/shale ratios will favour efficient drainage of any generated hydrocarbons, although the migrated hydrocarbon seeps reported in the Wealden facies at Mupe Bay (Dorset, UK) are of Rhaeto-Liassic origin (Cornford *et al.*, 1987; Miles *et al.*,1993).

North of 62°N the Cretaceous of mid-Norway has low TOC values (mainly in the range 0.5–1.5%), but the low pyrolysis yield indicates no oil-source potential (Cohen and Dunn, 1987). Variable quantities of mature gas- and oil-prone kerogen occur in the uppermost part of the Skarstein Formation (Aptian/Barremian) on Andoya Island (Bjorøy *et al.*, 1980), while rich gas- and oil-prone shales also occur in the Cretaceous of Svalbard (Bjorøy and Vigran, 1980; Leythaeuser *et al.*, 1983).

The regional maturity of the Lower Cretaceous can be deduced from the base Cretaceous depth map (see Fig. 11.48), which indicates that it will be verging on the mature boundary for gas (> 1% R_o, > 4000 m) in the deepest parts of the Central, Viking and Witch Ground grabens only. It also reaches gas-generating maturities in the North German Basin, where, as mentioned above, it is believed to have contributed to commercial gasfields. The Cretaceous would be mature in the Møre Basin, according to seismic mapping (Hamar and Hjelle, 1984; Nelson and Lamy, 1987), and in the Faeroe Basin (Duindam and van Hoorn, 1987); in both locations the base Cretaceous is mapped at or below 10 km.

The Upper Cretaceous is devoid of source potential in the southern part of the North Sea, where it is developed mainly as chalk. The widespread Turonian–Cenomanian anoxic event (Kuhnt *et al.*, 1990; Kuhnt and Wiedmann, 1995) is poorly represented in the North Sea, the Plenus Marl being only occasionally developed in a dark or black lithology (Schlanger *et al.*, 1987; Farrimond *et al.*, 1990; Jeans *et al.*, 1991; Baudin, 1995). These authors report TOC values up to 10.2% and Hydrogen Indices up to 222 mg/g TOC from selected samples from Flixton and South Ferriby (Yorkshire, UK), indicating immature, mainly oxidized, amorphous kerogen, with little hydrocarbon potential remaining. The South Ferriby outcrop is confirmed as immature by sterane isomerization ratios (Farrimond *et al.*, 1990). Nowhere is the Upper Cretaceous known to be fully mature for oil generation in the main North Sea area.

11.8.8 Tertiary

In the North Sea area Tertiary sediments are unlikely to be effective source rocks because of their low degree of maturity. Day *et al.* (1981), Gowers and Sæbøe (1985), and Ziegler (1988) show that the top Cretaceous is buried below 3000 m only in the Central Graben. A similar situation exists in the Møre Basin (Hamar and Hjelle, 1984). Palaeocene shales are in any case generally lean and gas- and condensate-prone. Early work (Pennington, 1975) claimed a Tertiary source for the Argyll Field oil, but this has subsequently been rejected on grounds of both regional maturity and kerogen type. Total organic carbon values average 0.43% in the Palaeocene Rogaland Group in the Statfjord area (Kirk, 1981), while two Palaeocene intervals of Ekofisk well 2/4-12 gave average TOC values of 1.8 and 2.0‰ (van den Bark and Thomas, 1980). Ungerer *et al.* (1984) report TOC values of 0.5–1.0% and Rock Eval Hydrogen Indices of about 210 mg/g TOC for the Palaeocene–Eocene of the Viking Graben. In the Ekofisk area, the shales have reached maturities of 0.59–0.62% R_o at 3030–3070 m (9930–10 060 ft). Since this is in the area of deepest burial of the base Palaeocene, no significant gas or condensate generation is expected from this formation.

North of 62°N, the Tertiary may be a potential source for gas, where burial is sufficient (Cohen and Dunn, 1987). Palaeocene–Eocene coals on Spitsbergen are immature with respect to gas generation (0.4–0.7% vitrinite reflectance), but are interbedded with black marine shales, containing abundant amorphous kerogen (Manum and Throndsen, 1977).

Today's (recent sediment) kerogens consist of reworked and largely modern terrigenous components (Wiesner *et al.*, 1990) as expected in a shelf sea.

11.9 North Sea hydrocarbons

As a generalization, the North Sea produces three broad types of hydrocarbons: the dry gas of the southern gas province, the wet (associated) gas of the northern oil province, and the midgravity crude of the northern oil province. The majority of producing fields in the oil province have been catalogued by Spencer *et al.* (1987) for the Norwegian offshore and by Abbotts (1991) for the UK sector. A typical North Sea oil, sourced by the Kimmeridge Clay Formation, is a low-sulphur, medium-gravity, naphtheno-paraffinic oil (see Table 11.12). Gas sourced from the Westphalian coal Measures of the southern North Sea is sweet and dry in composition (Table 11.13). Where sour gas from a Carboniferous source rock is encountered, the hydrogen sulphide is attributed to thermochemical sulphate reduction (equation 23). Gas in the oil province varies from wet gas–condensate to sweet dry gas, the latter often overlying a thin heavy oil leg) and thought to derive from bacterial degradation of wet gas (James and Burns, 1984).

It is beyond the scope of this chapter to detail the formation-water chemistry of the North Sea area, but reviews of this subject are found in Warren and Smalley (1993, 1994), Smalley and Warren (1994), Warren *et al.* (1994) and Thurlow and Coleman (1997) and references therein.

Using data from the late 1980s, Pegrum and Spencer (1990) gave figures for discovered producible hydrocarbon which break down to some 3.34×10^9 m^3 of oil (almost all light to midgravity crude), 350×10^9 m^3 of wet gas and condensate, and 1443×10^9 m^3 of dry gas in the oil province of the North Sea (volumes at surface temperature and pressure). Cousteau *et al.* (1988) have estimated some 2.14×10^9 m^3 of recoverable Jurassic oil in the UK and Norwegian East Shetland Basin, with between 0.80 and 0.94×10^9 m^3 of reserves yet to be found in Middle and

Lower Jurassic reservoirs. A similar type of calculation by Cornford (1994) for the Central Graben gave 0.99×10^9 m^3 of oil, 0.041×10^9 m^3 of condensate and 301×10^9 m^3 of gas in the discovered reserves, with the majority in the Ekofisk and Forties fields. The most recent estimates of the producible dry gas producible from the southern gas province are 4250×10^9 m^3, of which 2690×10^9 m^3 are attributable to the Groningen structure. This compares with 1250×10^9 m^3 of recoverable gas in the Troll Field in the northern Viking Graben (Horstad and Larter, 1997). On an oil-equivalent basis, the Troll Field contains reserves of 2.2×10^9 m^3 oil. A determination of the total recoverable hydrocarbon (converted to cubic metres of oil equivalent) for the entire North Sea is 16.06×10^9 m^3 (Spencer *et al.*, 1996), of which 71% derive from Jurassic and 29% from Carboniferous source rocks.

11.9.1 Oils

In terms of properties that result from their origin in the uniform kerogen type of the Upper Jurassic–basal Cretaceous source rock, the oils of the North Sea are remarkably homogeneous. The reservoired hydrocarbons, however, are highly variable, due to the influences of source-rock maturity, fractionation during expulsion and migration, and alteration within the reservoirs. Oil and source-rock correlations are discussed in the previous section under Upper Jurassic source rocks.

The range of North Sea oils

Compilations of the composition of North Sea oils have been made by Aalund (1983a,b), Cornford *et al.* (1983), Cooper and Barnard (1984), Northam (1985), Spencer *et al.* (1987) and Abbotts (1991). The range of oil API gravities is reproduced in Fig. 11.56 and gas/oil ratios in Fig. 11.57, show that a typical oil is 37° API (0.84 g/cm^3) with a gas/oil ratio of 244 m^3/m^3. The lighter oils are associated with areas of higher-maturity source rock, the heavier oils being bacterially degraded and hence generally in more shallowly buried reservoirs (for examples in the Witch Ground Graben, see Mason *et al.*, 1995). In addition to the normal, sweet, medium-gravity oils, condensates are known in the North Viking Graben, Beryl Embayment, the Witch Ground Graben and the Central Graben. In all cases associated with the most deeply buried Kimmeridge Clay/Borglum Formation source rocks (see Fig. 11.48), a relationship detailed by Barnard and Cooper (1981).

Heavy oils (API gravity < 20°) are found over much of the North Sea, generally in reservoirs with present-day temperatures less than 60°C, equating to approximately 1800 m at 30°C/km + 6°C at sea floor (Barnard and Cooper, 1981; Cooper and Barnard, 1984). This depth (temperature) limitation means that reservoirs containing heavy oils are mainly restricted to the Tertiary, with a symmetric axial zonation of heavier oils on the flanks and lighter mixed oils along the graben axes (Barnard and Bastow, 1991). There is abundant evidence to show that these heavy oils are the products of intrareservoir bacterial alteration of normal mid-gravity oil sourced by the Kimmeridge Clay Formation (for example, Oudin, 1976; Østfeldt, 1987; Horstad and Larter, 1997). There is currently no commercial production of heavy oil in the North Sea, huge volumes of which occur (e.g. in UK Quadrants 9 and 3). There are, however, plans to produce from a number of heavy-oilfields (e.g. West Brae, Balder), and mixed degraded and fresh oil is produced from the Gullfaks Field (Horstad *et al.*, 1992).

In many cases these intermediate-gravity oils (20–30° API) result from a mixing of a heavy degraded oil (produced from a normal midgravity influx when the reservoir was more shallowly buried) with a more recent influx of fresh oil arriving at the reservoir after it had been buried below the limiting depth (temperature) for active biodegradation (e.g. Norway's Troll Field oil: Horstad and Larter, 1997). The Emerald Field of the 'Transitional Shelf' (East-Shetland Platform) tested 20–25° API gravity oil

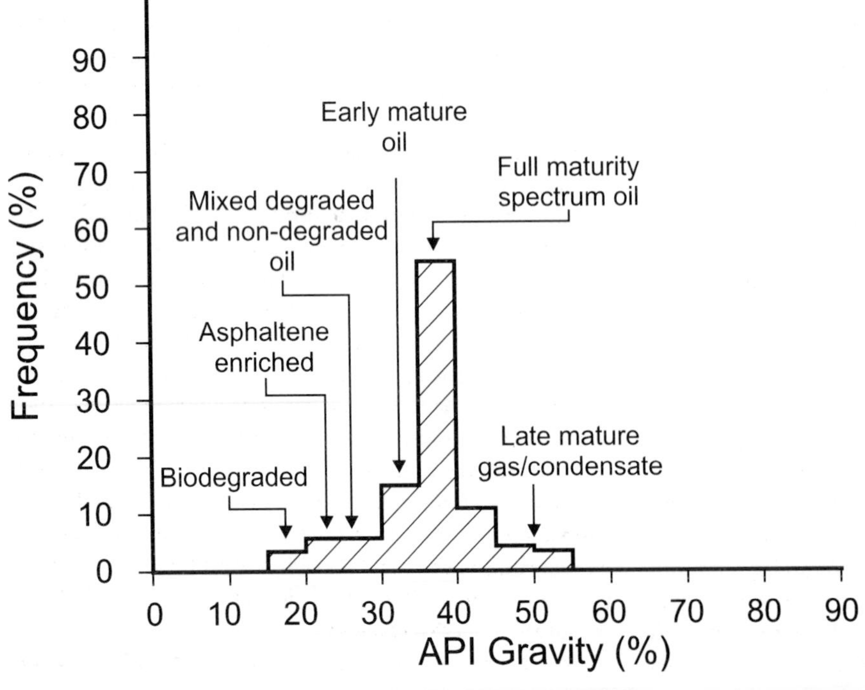

Fig. 11.56 API gravity distribution for a representative set of North Sea oils. Processes leading to the low- and high-gravity oils are indicated, with the main mode at ~38° API resulting from accumulations containing full-maturity-spectrum oils (i.e. a mix of hydrocarbons expelled from early- + mid- + late source rocks) as illustrated in Fig. 11.51.

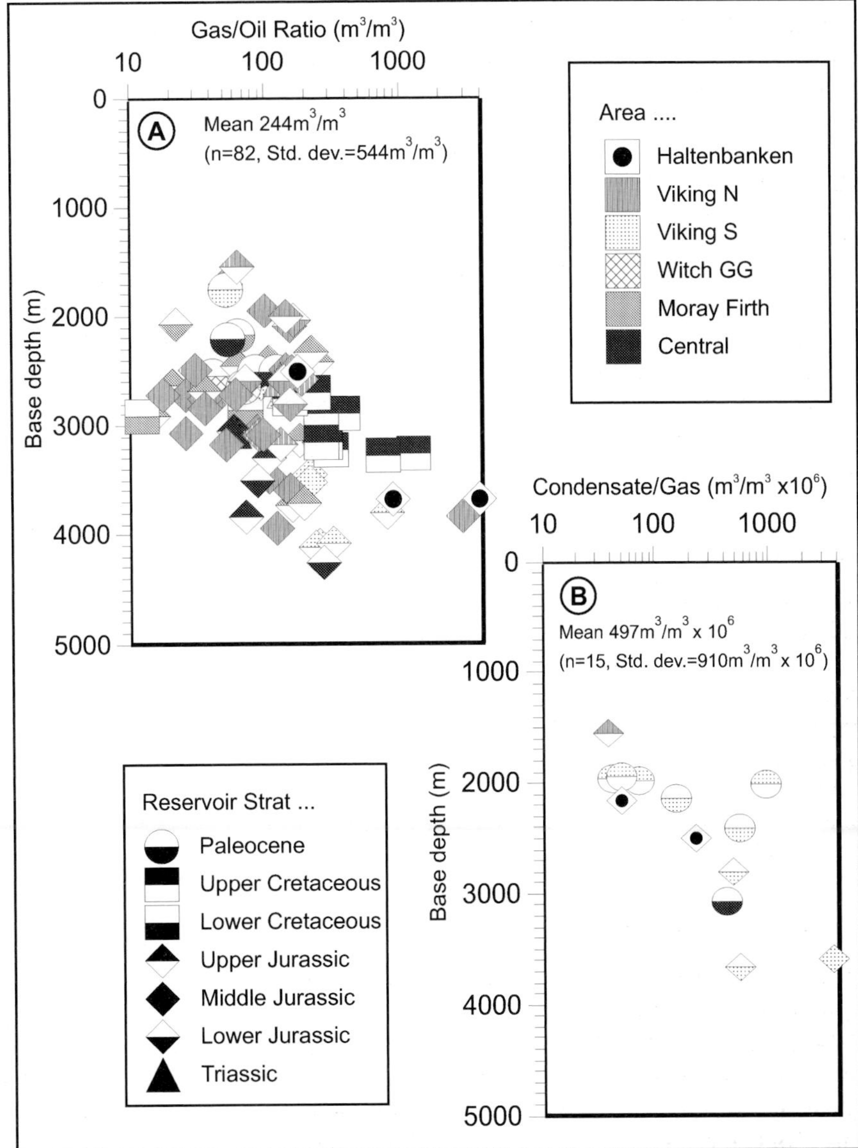

Fig. 11.57 Gas/oil ratios (GOR) and condensate/gas ratios (CGR) for representative North Sea oil accumulations as a function of depth, stratigraphy and area (data from chapters in Spencer *et al.*, 1987, and Abbotts, 1991). The GOR values increase with depth (A), paralleling source-rock maturity, while the CGR values increase with depth, suggesting fractionation during vertical migration. Geographically, there are no high-GOR oils in the North Viking Graben data set, while the majority occur in the Chalk reservoirs of the Central Graben.

(Wheatley *et al.*, 1987), which is probably also a mixture of degraded and non-degraded oil, as the reservoir is at or about the boundary temperature for active bacterial degradation. Variations of oil densities in the intermediate range are noted in North Sea Tertiary reservoirs from 58–60°N, although they are attributed by Barnard and Bastow (1991) to variations in the degree of degradation rather than mixing. Mason *et al.* (1995) report a whole range of API gravities (16–34° API) from the Tertiary-reservoired Alba-trend oils of the Witch Ground Graben, invoking progressive degradation due to meteoric water influx. Intermediate-gravity oils (20–30° API gravity) are known from the Clair Field, West of Shetland (22–25° API gravity), which Barnard and Cooper (1981) and Cooper and Barnard (1984) have classified as a biodegraded oil.

Barnard and Cooper (1981) have also noted that early-mature North Sea oils may fall in the 28–32° API gravity range, confirmed by analytical data presented by Cornford *et al.* (1983). For example, the 26–33° API Claymore oil (Maher and Harker, 1987) would be predicted to contain a major early-mature charge, due to the interbedding of Upper Jurassic reservoir and source, coupled with the limited local maturation at that horizon.

A similar explanation has been proposed to account for the relatively high-sulphur oils of the Bream and Brisling fields, where the 28–34° API oil arguably derives from early-mature Upper Jurassic Tau Formation source rock immediately overlying the reservoir sands (D'Heur and de Walque, 1987; Irwin *et al.*, 1993). A Liassic source is arguably more likely on maturity grounds, since the biomarker ratios of the Bream oil demand at least some generation from an area of late-mature source rock. The degree of uplift in the Egersund Basin is the key to discriminating between these two models (Fjeldskaar *et al.*, 1993b). Oils with higher sulphur levels are found in some of the Witch Ground Graben fields (e.g. Piper, Claymore, Tartan and Buchan), and either represent a more sulphur-rich Upper Jurassic facies or a contribution from the Devonian. Heavy, asphaltene-enriched oil layers are known from a number of fields (e.g. Oseberg and Ula), as discussed by Dahl and Speers (1985, 1986) and Wilhelms and Larter (1994a,b, 1995), although in the North Sea no whole-oil column falls in this category.

The geographical variation in oil properties (e.g. gas/oil

Table 11.12 Some average properties of oils sourced from the Kimmeridge Clay Formation.

Property	Range	Average (± SD)	Comments
API gravity (°)*	17–51	36 (± 6.5)	63 North Sea dead oils
	36–55	42 (± 5.4)	26 Central Graben fields
Sulphur	0.13–0.55 0.56–1.57	0.32 (± 0.12) 1.0	15 oils excluding Piper, Claymore Tartan, Buchan
Gas/oil ratios (scf/bbl)†	216–1547 158–13 219	671 (± 415) 2200 (± 3100)	Various data sources 26 Central Graben fields
Asphaltenes (wt%)	0.1–5.1	1.2 (± 1.2)	Excluding asphaltene-enriched oils (< 35%)
Saturate/aromatics	0.62–8.0	2.02 (± 1.2)	Topped oils
Pristane/*n*C17	0.3–1.0	0.63 (± 0.17)	Excluding biodegraded oils
Phytane/*n*C18	0.2–1.1	0.56 (± 0.18)	—
Pristane/phytane	0.6–1.9	1.24 (± 0.25)	—
δ^{13}C whole oil(‰)	–27.1 to –30.4	–28.9 (± 1.3)	Brae, Statfjord, Moray, Central Graben oils
Vanadium (ppm)	0.53–6.0	3.1 (± 1.8)	Piper area oils (average of nine production crudes), excluding Buchan at 26 ppm V, 4.5 ppm Ni
Nickel (ppm)	0.5–5.0	1.8 (± 1.3)	
Wax (wt%)	4.0–7.7	6.3 (± 1.1)	6 oils only (cf. Beatrice at 17% wax)

Sources of miscellaneous compositional information on North Sea oils		
Benzothiophenes	Bjorøy *et al.*, 1994	Central Graben oils
Polynuclear aromatics	Requejo *et al.*, 1996	Comparison with worldwide marine crudes
Carbazoles (pyrrolic N)	Stoddart *et al.*, 1995	Eldfisk oils (cf. Li *et al.*, 1995, for other oils)

*API° = (141.5/density) – 131.5 (density in g/cc ≡ tonne/m^3). †1 m^3/m^3 = 5.68 scf/barrel (bbl).
SD, standard deviation; ppm, parts per million.

ratios; Fig. 11.57 and Table 11.12) is mainly attributable to maturity and hence differences in source-rock burial (see Fig. 11.48). A mean gas/oil ratio (GOR) value for a typical Middle Jurassic-reservoired North Sea oil from the Viking Graben is ~120 m^3/m^3 with higher average values of ~390 m^3/m^3 for the Central Graben, stemming from deeper source rocks, augmented by fractionation during migration to Chalk and Palaeogene reservoirs (Cornford, 1994).

As evidenced by the oils, variations in source-rock facies are more subtle (cf. Miller, 1990) with consistent variation in conservative non-maturity-related properties such as whole-oil stable carbon isotope ratios (Fig. 11.58) and sterane carbon-number distributions (Fig. 11.59). These variations reflect local differences in the kerogen type, and hence depositional environment from basin to basin, within the Upper Jurassic seaway of the North Sea area (see Figs 11.44–11.47). Among a large number of controls on isotopic composition, the isotopically heavy oils of the Central Graben (Bjorøy *et al.*, 1994) point to derivation from a source rock deposited under highly stratified, strongly anoxic deep water, where bacterial fermentation dominated and the relatively closed nature of the system inhibited fractionation. Fermentation at the Jurassic sea floor produced isotopically light methane, leaving isotopically heavy bacterial biomass, seen now as the exhausted amorphous kerogen and expelled oil. Post-depositional maturity effects on isotopes are minimal (Clayton 1991a).

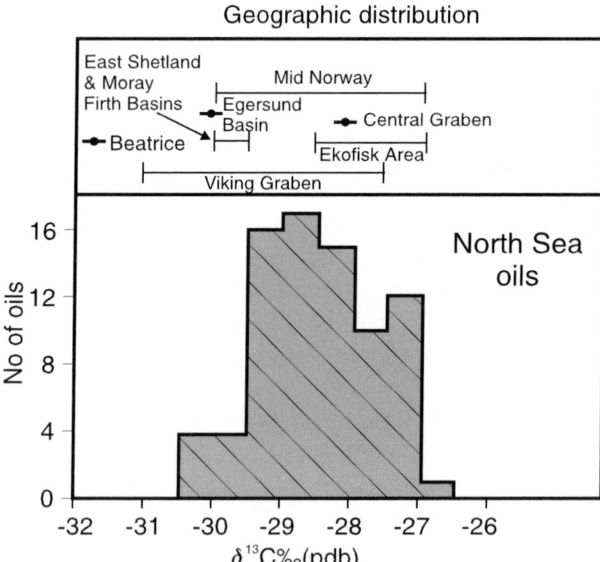

Fig. 11.58 Stable carbon-isotope distributions for North Sea oils reflecting source-rock depositional conditions and kerogen type (data from Mackenzie *et al.*, 1983; Gabrielsen *et al.*, 1985; Hughes *et al.*, 1985; Northam, 1985; Bailey *et al.*, 1990; Wills and Peattie, 1990; Bjorøy *et al.*, 1996). With the exception of the Beatrice oil (Lower–Mid-Jurassic and Devonian co-sourced), detailed correlations are not established between specific groups of oils and specific stratigraphic intervals of the source rock. Some discrimination is possible based on geographical areas.

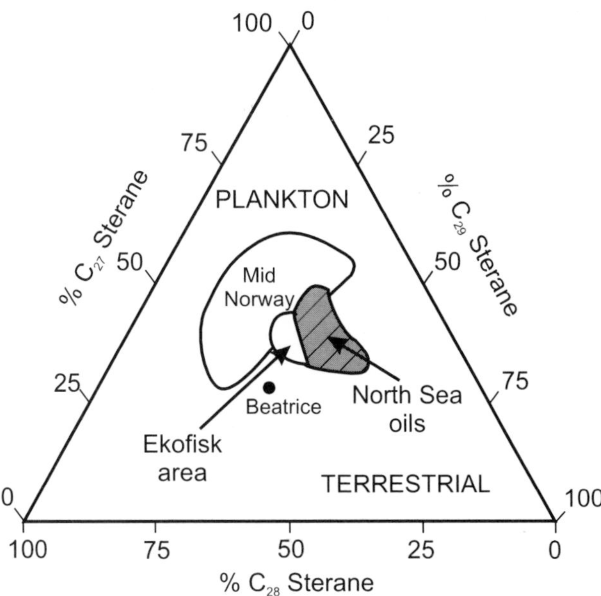

Fig. 11.59 Sterane carbon-number distributions for the North Sea and mid-Norway (data from: Hughes *et al.*, 1985; Northam *et al.*, 1985; Cohen and Dunn, 1987; Peters *et al.*, 1989; Cornford *et al.*, 1993). The grouping of the oils shows the dominance of bacterially degraded algal kerogen in all source rocks.

The sterane carbon-number distribution shown in Fig. 11.59 indicates the balance been planktonic–algal kerogen (C_{27}) and terrigenous matter (C_{29}), with bacterial degradation moving towards the C_{28} sterane apex. Thus, the Ekofisk oils of the Central Graben derive from source rocks with a greater bacterial input than the average North Sea oil, and the Beatrice oil is again distinct.

Reservoir geochemistry

The above comments assume that most accumulations (fields) are chemically homogeneous. More recent studies (e.g. Karlsen and Larter, 1997; Cubitt and England, 1995; Horstad and Larter, 1997) have highlighted not only that most reservoirs are chemically heterogeneous, but also that understanding the causes of the heterogeneity can be a great asset in developing cost-effective production. Understanding the filling history of a field, locating its fill point and defining the compositional variability, both vertically and horizontally, have been popularized under the heading of 'Reservoir' or 'Production Geochemistry' (Larter and Aplin, 1995). Links with fluid inclusion studies have also been reported (Haszeldine *et al.*, 1984; Karlsen *et al.*, 1993). In essence, this boils down to an informed application of oil–oil correlation.

Within an accumulation, heterogeneity can be both vertical and horizontal (Leythaeuser *et al.*, 1989; Horstad *et al.*, 1990, 1995; Hall *et al.*, 1994). Largely vertical variation, associated with asphaltene bands, has been logged in the recently filled Ula Field (Larter *et al.*, 1990, Oxtoby *et al.*, 1995), with both lateral and vertical variation noted for the nearby Eldfisk Field (Hall *et al.*, 1994; Stoddart *et al.*, 1995). The two culminations comprising Eldfisk are shown to contain oils expelled from source rocks of different maturities, and maturity gradients are described within the individual compartments. Compositional variability is attributed to the sporadic arrival of oil and distribution within the field being a function of heterogeneous sedimentology (primary hard grounds and diagenetic tight zones).

Given low and heterogeneous permeabilities (e.g. separating the Forties and South-East Forties fields in the Central Graben), a clear transmissibility barrier exists both over production (e.g. years) and geological time spans (Wills and Peattie, 1990; Wills, 1991; England *et al.*, 1995). Spatially separated fill points and long-term (geological) compartmentalization by faults and stratigraphic boundaries have also been identified in the Norwegian Gullfaks Field (Horstad *et al.*, 1990), a study which when expanded to cover the entire Tampen Spur, recognized five petroleum populations in the Visund, Snorre, Gullfaks, Vigdis and Statfjord fields (Horstad *et al.*, 1995). Most of these fields contain examples of more than one oil type, reflecting charges from discrete source-rock drainage areas. The differentiation of five oil types is thus made on the basis of both source-rock facies (stable carbon-isotope ratios and bisnorhopane abundance) and biomarker-maturity ratios. A complex sequence of gas fill, oil fill, biodegradation and tilting, together with two fill points for the field, is proposed by Horstad and Larter (1997) to account for the substantial vertical and horizontal variations in the dry gas and variable-gravity oil within Norway's giant Troll accumulation.

Oils of the greater North Sea area

The overall variation of oil properties within the greater North Sea area is shown in Fig. 11.60. Ignoring the Beatrice, Bream and Brisling fields (the little 'B's), this illustrates the provincialism that derives from subtle variations in kerogen type, gross variation in source-rock maturity and the generally predictable effects of intrareservoir alteration within the main North Sea Basin. Moving out of the main Jurassic graben system, the diversity of primary oil types increases.

The oil of the Beatrice Field, though light (approx. 38° API) and sweet, has a high wax content (17%) and pour point (65°F; 18°C), together with a low gas/oil (GOR) ratio (Linsley *et al.*, 1980). As suspected from the structural inversion of the area, tar mats are also present in the field. Beatrice oil is consistently different from the bulk of North Sea oils in terms of source-rock kerogen-type indicators (Figs 11.58 and 11.59). As discussed in Section 11.7, a mix of Middle Jurassic and Devonian sourcing has been suggested (Peters *et al.*, 1989). A similar high-wax (16%), low-sulphur (0.3%), 35° API, low-GOR oil is produced from the DeLier Field of the Rijswijk oil province (Bodenhausen and Ott, 1981), which has similarities with the Hild condensate in the northern Viking Graben (Rønning *et al.*, 1986). All three oils are believed to come from non-marine source rocks, though not necessarily of the same stratigraphic age.

To the west of the Shetlands, the complex Clair field, which has tested oil with intermediate gravities (22–34° API) and low gas/oil ratios, comprises a variety of mixes of biodegraded and fresh oil, all derived from an Upper Jurassic source rock. A history of oil charge, uplift, leakage

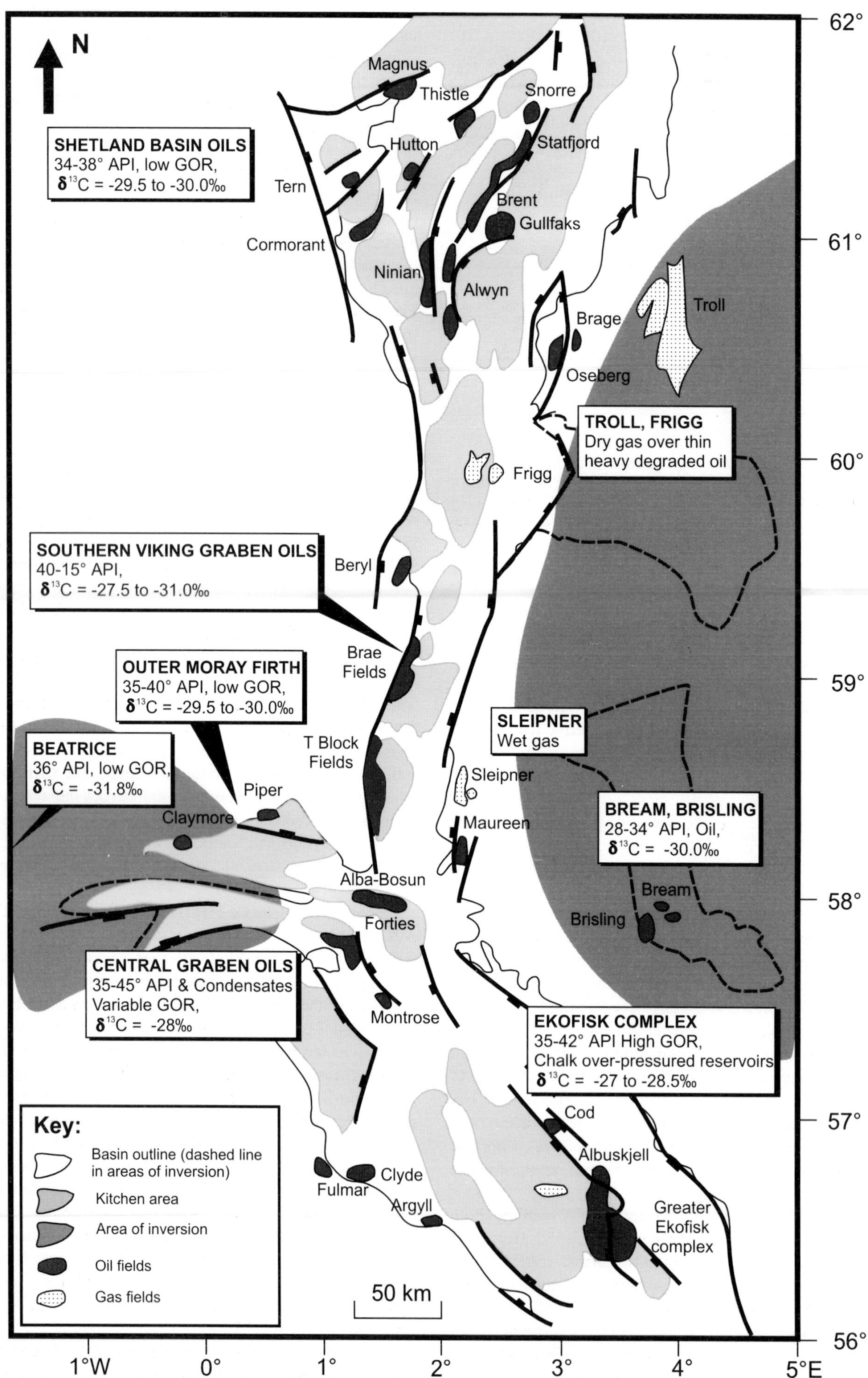

Fig. 11.60 Distribution of hydrocarbon types within the oil province of the North Sea, based on bulk oil properties.

and biodegradation, followed by influx of fresh oil, is indicated. The low-API gravity, low gas/oil-ratio oil in the Solan Field (Block 205/26) may relate to the contiguous area of very rich Upper Jurassic (Volgian) source rock, identified by Bailey *et al.* (1987). The Palaeocene-reservoired Foinaven and Schiehallion fields (UK Blocks 204/24 and 204/25) tested 25–27° API oils with gas/oil ratios in the 246–320 ft^3/barrel range.

The Liassic-sourced onshore oilfield of Wytch Farm (Dorset, UK) produces a sweet 38° API (0.835 g/cm^3) gravity crude from the Sherwood with $\sim 270 \times 10^6$ barrels in place and GOR of 357 ft^3/barrel and Bridport Sandstones and $\sim 30 \times 10^6$ barrels in place and GOR of 150 ft^3/barrel respectively (Cornford *et al.*, 1988; Bowman *et al.*, 1993; Aplin and Coleman, 1995). With additional extended-reach drilling (McClure *et al.*, 1995), higher estimates of in-place volumes have been made. Injection of sea water into the Bridport reservoir has produced hydrogen sulphide via bacterial sulphate reduction (Aplin and Coleman, 1995).

The Namurian-sourced oils of the Eastern Midlands of the UK are high-wax, midgravity crudes and are sometimes sour (e.g. Welton). The Eakring Dukeswood field produces oil of 31.9–37.6° API and wax contents up to 16 wt% from a series of Namurian B to Westphalian A sandstones (Storey and Nash, 1993). Two source-related oil families have been established on the basis of stable carbon-isotope ratios of saturate and aromatic fractions (Fraser *et al.*, 1990; Harriman and Miles, 1995).

Of the two oilfields in the south of the offshore East Irish Sea Basin, the Douglas Field (UK Block 110/13; 225×10^6 barrels in place) tested a 44° API low gas/oil ratio (171 ft^3/barrel) waxy crude with some 88×10^6 barrels recoverable from the Triassic reservoir (Trueblood *et al.*, 1995; Yaliz, 1995). The oil has a high H_2S and mercaptan content, and is believed to be sourced from an offshore equivalent of the Dinantian/Namurian Holywell shale. The Lennox oilfield (UK Block 110/15) contains 40° API oil with a gas/oil ratio of 400 ft^3/barrel, but with up to 2350 ppm of H_2S (Haig *et al.*, 1995). The high hydrogen sulphide probably results from thermochemical sulphate reduction (equation 23) within Triassic evaporites or brines therefrom. Once production commenced, sea-water injection is known to have produced enhanced levels of hydrogen sulphide via bacterial sulphate reduction.

Three oil families are identified in the Celtic Sea Basin, discriminated on the basis of stable carbon isotope and biomarker characteristics:

- Portlandian-sourced oils: isotopically light (about –32‰ pdb), with approximately equal $C_{27} : C_{28} : C_{29}$ steranes and a low biomarker maturity, as found in the Kinsale Head area (Caston, 1995; Howell and Griffiths, 1995).
- Liassic-sourced oils: isotopically average (–28 to –30‰ pdb; Caston, 1995), with steranes dominated by C_{29} ethyl cholestanes, with generally mid- to late-mature biomarkers.
- Third family of oils: very waxy, isotopically average, but deficient in C_{28} methyl sterane.

In many cases these oils have been biodegraded where the reservoir is sufficiently shallow for temperatures to have been below 65°C after arrival of the oil.

The Liassic-sourced family is characterized by oils in the northern part of the basin (Caston, 1995; cf. Craven, 1995). A 44° API oil with a gas/oil ratio of 600 ft^3/barrel was tested from a Middle–Upper Jurassic reservoir in 49/9-1, with heavy oil shows (13–22° API) in the overlying Upper Jurassic and Lower Cretaceous (Craven, 1994). The Helvick Field (49/9-2 and 49/9-3) tested 44° API oil, with a gas/oil ratio of 680 ft^3/barrel, and a total of between 8 and 12×10^6 barrels in place (Caston, 1995). The third family of oils derives from the extreme north of the basin (e.g. well 50/6-1).

In the Porcupine Basin to the west of Ireland the 34° API, 300 ft^3/barrel, low-sulphur oil tested from Upper Jurassic sands in Irish well 26/28-1 (the $\sim 200 \times 10^6$ barrel Connemara Field) is probably from an Upper Jurassic source rock (Croker and Shannon, 1987; MacDonald *et al.*, 1987; Earls, 1995). The adjacent IRL26/28-2 well tested lighter oil (39–41° API). Isotopically, the Connemara oil falls in the –28.5 to –29.2‰ pdb range.

To the north of the North Sea, intermediate to mid-gravity crude is found in the Tyrihans (29° API; Larsen *et al.*, 1987) and the Heidrun (Koenig, 1986) and Draugen (41° API; Ellenor and Mozetic, 1986) fields in the Haltenbanken area (Cohen and Dunn, 1987). Karlsen *et al.* (1995) report API gravities of 39° for Trykstakk (6406/3-2), 47.2° for Tyrihans Sør (6407/1-2), 30.8° for Tyrihans Nord (6407/1-3), 41.3–59.5° for Mikkel (6407/6-3), 38.2–44.7° for Njord (6407/7-1, 2, 3, 4), 40° for Draugen (6407/9-2), 45.4° for Alve (6507/3-1), 14.4-70.6° for Heidrun (6507/7-2, 4, 5), 35.2–52.3° for Midgard (6407/2-2, 6507/11-3) and 38.0–45.4 for Smørbrukk Sør (6506/12-3, 5). Subsurface gas/oil ratios are also listed. A detailed molecular study was also undertaken by Karlsen *et al.* (1995), who identified two Upper Jurassic-sourced families of oil, with gravities from 14.4 to 70.6° API and gas/oil ratios from 18 to 8353 m^3/m^3 (100–47 000 ft^3/barrel). Stable carbon-isotope ratios of –27.5‰ to –30.0‰ pdb. strongly support derivation from the Upper Jurassic Spekk Formation shales.

11.9.2 Gas in the northern oil province

Most of the 'associated-gas' fields in the oil province of the North Sea lie above (Frigg, East Brae) or to the east of the main graben axes (e.g. Cod, Sleipner, Gudrun, Heimdal, Oseberg, Troll). This distribution results from the asymmetry of the graben with respect to migration from its deeper levels (e.g. Fig. 11.35). Gas/condensate is also present in the deeper (Statfjord Formation) reservoir of North Alwyn (Johnson and Eyssautier, 1987), confirming this general model. Some general properties of the gases of the oil province are reported in Table 11.13. Nitrogen quantities are small, while carbon dioxide yields can be strongly related to reservoir depth (James, 1990).

Frigg Field gas is dry and sweet and arguably sourced by Middle Jurassic coal measures (Heritier *et al.*, 1979; Goff, 1983). The balance of evidence, however, suggests that it is associated wet gas from late-mature oil-prone Upper Jurassic shales that has lost its wet- gas components due to bacterial degradation in the reservoir. This mechanism is discussed by James and Burns (1984) and may be identified by a propane stable carbon isotope anomaly. The same process also explains the presence of a thin heavy-oil leg in the Frigg Field, and a similar process may account for the dryish gas underlain by a variable thickness of heavy oil in the giant Troll accumulation (Gray, 1987; Osborne and Evans, 1987; Horstad and Larter, 1997).

The gas of the Sleipner Field, which also contains significant condensate, is wetter, with a high CO_2 content (Larsen and Jaarvik, 1981; Østvedt, 1987; Ranaweera, 1987; James, 1990). This appears to be dominantly an unaltered late-mature Kimmeridge Clay/Draupne Formation source-rock product. James (1990) argues that isotopically heavy CO_2 derives from deep thermal sources. To the south, the Cod gas/condensate field on the east flank of the Central Graben is perhaps a typical example of a late mature oil source-rock product (D'Heur, 1987b), with a gas/liquid ratio in excess of 2000 m^3/m^3 (~11 000 ft^3/barrel). Vertical migration of Jurassic-sourced gas into the Tertiary produces increasingly dry compositions, as in the Oligocene gas play of Norwegian block 2/2 (Ekern, 1986). In this context, many North Sea 'condensates' (i.e. the liquid condensing out of a gas test or production stream) are not, upon analysis, a low-molecular-weight late-mature source-rock product (i.e. a thermal condensate). In many cases, the liquid phase turns out to be fairly normal oil entrained in the gas flow. Differentiating a thermal condensate from entrained oil may have a considerable impact on exploration prospectivity and production plans.

The Oseberg Field of the eastern flank of the North Viking Graben contains a gas cap with a gas/liquid ratio of 3200 m^3/m^3 (18 000 ft^3/barrel). The late-stage influx of gas may have been responsible for precipitating the tar mats found in this field via a de-asphalting mechanism (Dahl and Speers, 1985, 1986; Wilhelms and Larter, 1995). Cayley (1987) notes the presence of pyrobitumen in some gas/condensate reservoirs of the Central Graben, suggesting that the deeply reservoired gas-condensate may be the result of thermal degradation of normal oil buried to greater temperatures once trapped within the reservoir.

Both Heimdal and Odin have thin oil legs, under wet and dry gas respectively (Mure, 1987d; Nordgard-Bolas, 1987). Gas–condensate has been tested from the small Agat field of the North Viking Graben (Gulbrandsen, 1987), which appear to be sourced from late-mature Draupne Formation shales. In contrast, the waxy oil (41–45° API) that condenses from the gas phase of the Hild Field correlates with Brent coal extracts, on the basis of a high C_{29} sterane content (Rønning *et al.*, 1986, 1987).

North of 62°N gas/condensates are common. The Upper Triassic–Lower Jurassic Åre (Hitra) Formation coals are believed to source the gas/condensate of the Midgard Field in Haltenbanken (Elvsborg *et al.*, 1985; Pittion and Gouadain, 1985; Ekern, 1987); although Cohen and Dunn (1987) and Karlsen *et al.* (1995) argue for a contribution from the Upper Jurassic Spekk (Nesna) Formation. The adjacent Tyrihans Field contains oil (29° API) and gas/condensate (Larsen *et al.*, 1987), both source rocks may contribute here. Of the other Haltenbanken fields, Heidrun (Koenig, 1986) and Draugen (Ellenor and Mozetic, 1986) produce oil and Smorbukk gas condensate (Aasheim *et al.*, 1986). The Askeladd Field in the Troms area of northern Norway tested dry gas, with about 5% CO_2 (Olsen and Hanssen, 1987). This area, together with the rest of the Barents Shelf, has suffered uplift, gas expansion and, potentially, oil displacement, with residual oil and tar mats within and below the present-day gas columns (Nyland *et al.*, 1989; Bakken *et al.*, 1991). Seismic evidence for gas spillage is presented in terms of fault-related gas chimneys and hydrate zones in the Børnøya Basin (Laberg and Andreassen, 1996).

11.9.3 Gas in the southern gas province

Limited compositional information is available for the gas of the southern North Sea gas province (Barnard and Cooper, 1983), but the available public data for compositional properties of the dry gases are summarized in Table 11.13. The stable carbon-isotope values of the gases are in the −22 to −28‰ range (pdb), this being characteristic of mid–late-mature gas generated from seam coals (Kettel, 1989; Clayton 1991b). As discussed earlier, the Westphalian Coal Measures (stable carbon-isotope range −22 to −25‰ pdb; Schoell, 1984) are the assumed source, with isotopically lighter methane deriving from transitional and marine-influenced sediments (Kettel, 1989). Since, in a chemical sense, the methane derives from only part of the coal, there is no expectation of a good correlation between coal (or vitrain) and gas.

Boigk and Stahl (1970) and Boigk *et al.* (1971) have noted regional trends in the contents of methane, nitrogen and carbon dioxide within the southern North Sea and North German offshore and land areas, although a number of possible explanations for these trends were offered. Kettel (1982, 1989) and Krooss *et al.* (1993, 1995) present evidence to show that gases with high nitrogen contents derive from the most mature source rocks of terrigenous origin. The nitrogen content is generally higher in the Bunter than in the Rotliegend reservoirs of the UK southern North Sea (Barnard and Cooper, 1983; Cooper and Barnard, 1984), giving rise to the suggestion of enrichment during (re)migration. Both these processes are considered possible in accounting for the differences between the 14% Nitrogen of the Groningen Field and the 4% Nitrogen of the adjacent Annerveen Field in north Netherlands (Veenhof, 1996). Carbon dioxide can reach high concentrations if a deep carbonate source is contributing some inorganic gas, but it tends to dissolve in formation water during long-distance migration, leaving a methane-enriched gas.

11.9.4 Gas in other areas

Gas is produced from the northern part of the East Irish Sea Basin (mainly from the North and South Morecambe fields), sourced from Carboniferous source rocks and trapped in Triassic sandstone reservoirs: hydrogen sulphide is a problem in many accumulations (Hardman *et al.*, 1993). Based on sulphur isotope values, Stuart and Cowan (1991) suggest that the hydrogen sulphide may be derived from Dinantian limestones. Gas wetness in the West Netherlands basin depends on co-mingling of Carboniferous and Jurassic source gas (de Jager *et al.*, 1996), as identified by carbon isotope ratios (cf. Berner and Faber, 1988).

Fairly wet gas (7.3% $C_2{}^+$ is produced from the North and South Morecambe fields (1×10^{12} ft^3 and 4.57×10^{12} ft^3 respectively), with variable H_2S contents (Stuart and

Table 11.13 Some properties of hydrocarbon gases from the greater North Sea area.

Field or test	% Methane	% Ethane	% N_2	% CO_2	$\delta^{13}C_1$	Reference
Gas from the southern gas basin (Carboniferous source rock)						
UK averages	91.2	5.2	3.6	0.27	–	Barnard and Cooper, 1983
UK ranges	83.2–95.0	3.7–8.2	1.0–8.4	0.1–0.5	–	
Groningen (Neth.)	81.6	2.7	14.8	0.9	-36.6	
Waddenzee (Neth.)	77.1–88.7	2.87–6.4	3.1–19.7	< 1.26	-31.1	Cottençon *et al.*, 1975; van den Bosch, 1983
Broad Fourteens (K13)	85.3	6.7	< 7.1	< 1.7	–	Roos and Smits, 1983
Amethyst	91.95	3.63	2.22	0.64	–	Garland, 1991
Barque	94.59	3.65	1.36	0.35	–	Farmer and Hillier, 1991a
Caister (B, C)	84.0, 84.5	1.4, 7.0	14.5, 6.0	0.1, 2.5	–	Ritchie and Pratsides, 1993
Camelot	90.7	6.8	2.4	0.1	–	Holmes, 1991
Cleeton	91.55	6.67	1.33	0.45	–	Heinrich, 1991a
Clipper	95.76	3.05	0.71	0.47	–	Farmer and Hillier, 1991b
Esmond	–	–	> 8	< 1	–	Ketter, 1991a
Forbes	–	–	–	–	–	
Gordon	–	–	14	–	–	
Hewett	–	–	High	–	–	Cooke-Yarborough, 1991
Indefatigable	92	3.4	2.7	0.5	–	Pearson *et al.*, 1991
Leman	94.94	3.74	1.26	0.04	–	Hillier and Williams, 1991
Markham	83	–	–	–	–	Meyres *et al.*, 1995
Ravenspurn	–	–	~2.5	> 1.0	–	Ketter, 1991b
Ravenspurn South	93.15	3.71	1.78	0.96	–	Heinrich, 1991b
Rough	91	6	2.4	0.6	–	Stuart, 1991
Sean	91.06	4.54	3.19	1.21	–	Hobson and Hillier, 1991
Thames, Yare, Bure	92	–	–	0.4	–	Werngren, 1991
Vulcan, Vanguard, Valiant	–	–	–	–	–	Pritchard, 1991
West Sole	94	4.4	1	0.8	–	Winter and King, 1991
North-west Germany						
North-west Germany	65–91	8–30	1	–	-54 to -44	Tissot and Bessereau, 1982
Irish Sea						
North Morecambe	81.02	6.11	6.88	5.89	–	Stuart and Cowan, 1991; Cowan, 1996
South Morecambe	84.84	6.8	7.7	0.56	–	
Kinsale Head*	99.1	0.2	0.4	0.3	-45.5 to -48.3	Colley *et al.*, 1981
Associated gases in the oil province						
Sleipner	75–84	12–19†	0.41–3.17	0.15–16.6	-39 to -44	Larsen and Jaarvik, 1981; James, 1990
Frigg	95	3.6	0.4	0.3	-43.3	Héritier, 1981; Mure, 1987a
East Frigg (alpha, beta)	94.9–95.6	4.00–3.68	0.7–0.6	0.3–0.05	–	Mure, 1987b
NE Frigg	94.2	4.74	0.67	0.3	–	Mure, 1987c
Troll	93				–	Gray, 1987
Block N31/2	92.6	5.4	1.5	0.5	–	Brekke *et al.*, 1981

*Probably from Jurassic to Lower Cretaceous source rocks. †C_2^+ figures include nitrogen.

Cowan 1991). To the south, the Calder Field (UK Block 110/7a) contains gas with up to 4500 ppm H_2S, plus a 20 ft oil rim (Blow and Hardman, 1995). The Millom Field (113/26, 113/27), to the north of Morecambe, produces dry gas from inferior reservoir sands (Lewis and Hardman, 1995). In the southern sector of the basin, the gas of the Hamilton Field is wet (11% C_2^+ with 8% N_2 and variable H_2S (1100 ppm in Hamilton North; 3ppm in Hamilton South).

The gas of the Kinsale Head Field (Irish Block 48/20) is dry (99% methane) with methane stable carbon isotope value of −48.4‰ pdb. Since the Wealden (Albian) sandstone reservoir also contains biodegraded oil stain of the Portlandian Family (Howell and Griffiths, 1995), the dryness of the gas may result from bacterial degradation of wet gas. Its source is uncertain, but is probably Liassic (Howell and Griffiths, 1994). The smaller (0.1×10^{12} ft^3 Ballycotton Field (IRL 48/20-2 and 4), to the north, yields similar dry gas.

11.10 Conclusions

The North Sea is now a classic role model for hydrocarbon genesis under extensional (northern oil province) and inverted (southern gas province) tectonic regimes. As such, it is being used as a natural laboratory for developing and testing new ideas on the origin and occurrence of oil and gas. In the oil provinces, it provides one of the best-documented examples of a simple oil/source-rock association, which, together with a relatively uncomplicated maturational history, produces a wide range of oils and associated gases. Its fundamental simplicity but diversity of hydrocarbon types attests to the detailed controls of source, maturation, migration and alteration within this small area of the North-West European shelf.

11.11 Acknowledgements

I am pleased to thank Ken Glennie for detailed comments on this fourth version of this chapter. For my understanding of the genesis of North Sea hydrocarbons, I would like to thank my many former colleagues in Britoil and diverse contacts and friends made while consulting in Europe and worldwide. In particular, the latter experience has allowed me to put the North Sea situation in its correct global context. Sally Cornford deserves special thanks for drafting the figures.

11.12 Key references

Bordenave, M.L. (ed.) (1993) *Applied Petroleum Geochemistry.* Editions Technip, Paris, 524 pp.

Cooles, G.P., Mackenzie, A.S. and Quigley, T.M. (1986) Calculation of petroleum masses generated and expelled from source rocks. *Organic Geochemistry* **10**(1–3), 235–46.

Cornford, C., (1994) Mandal–Ekofisk petroleum system in the Central Graben of the North Sea. In: Magoon, L.B. and Dow, G.W. (eds) *The Petroleum System from Source to Trap.* Memoir 60, AAPG, Tulsa, pp. 537–72.

Miles, J.A. (1989) *Illustrated Glossary of Petroleum Geochemistry.* Oxford Science Publications, Oxford, 137 pp.

Skervøy, A. and Sylta, O. (1993) Modelling of expulsion and secondary migration along the southwestern margin of the Horda Platform. In: Doré, A.G. *et al.* (eds) *Basin Modelling: Advances and Applications, Proceedings of the Norwegian Petroleum Society Conference.* Special Publication No. 3, Norwegian Petroleum Society (NPF), Elsevier, London, pp. 499–537.

12 North Sea Plays: Geological Controls on Hydrocarbon Distribution

H.D. JOHNSON & M.J. FISHER

12.1 Introduction

This final chapter reviews the geological controls on hydrocarbon occurrences in the North Sea and adjacent areas, including the western margin of the British Isles (Fig. 12.1). Emphasis is placed on the concept of hydrocarbon plays and play fairways, the analysis of which provides the most rigorous means of evaluating remaining exploration potential, particularly in mature basins with abundant subsurface data. The same framework also contributes to a better understanding of existing fields, including the geological controls on fluid flow and hydrocarbon recovery.

Over 30 years of continuous offshore exploration in the North Sea has resulted in the discovery of some 100×10^9 barrels of oil and oil equivalent (Spencer *et al.*, 1996). About 100 commercial oil- and gasfields are in the Northern and Central North Sea, almost entirely within the Norwegian, Danish and UK sectors, and around 30 commercial gasfields are in the Southern North Sea, mainly within the UK and Dutch sectors (see Fig. 1.1). On the UK Continental Shelf (UKCS) alone, some 60 commercial hydrocarbon accumulations contain total recoverable reserves of around 32×10^9 barrels of oil equivalent in the Northern and Central North Sea, while in the Southern North Sea commercial fields contain some 30×10^{12} standard cubic feet (ft^3) of gas.

Although the North Sea Basin is an extremely mature hydrocarbon province, there remains significant potential for increasing the overall reserves through additional, more intensive, exploration and by improving recovery from existing fields. New discoveries are being made in increasingly complex and subtle traps, primarily as a result of the following factors: (i) improved imaging of the subsurface through three-dimensional (3D) seismic; (ii) use of new, computer-based, technologies, particularly those involving seismic modelling (e.g. the Faeroe Basin Palaeocene play), basin modelling (including hydrocarbon maturation/migration), 3D structural modelling/restoration and stratigraphic/basin-fill modelling; (iii) application of high-resolution sequence-stratigraphic techniques (e.g. Rattey and Hayward, 1993); and (iv) creative evaluation ('data mining') of the enormous well database (e.g. between 1965 and 1992, over 3500 exploration and appraisal wells were drilled on the UKCS alone; the total North Sea, including the Norwegian Atlantic Margin, comprises some 3500 exploration wells, plus more than 1300 appraisal wells). These techniques are also contributing to increases in reserves and to improvements in oil recovery, both within producing fields and from undeveloped accumulations, including marginal fields. Developed and undeveloped accumulations are also benefiting from advances in well technology (e.g. extended-reach, horizontal, multilateral high pressure/high temperature (HP/HT) wells, etc.), time-lapse/4D seismic and 3D geological modelling/reservoir characterization.

The analysis of hydrocarbon plays involves the synthesis and mapping of all the key geological parameters controlling the occurrence of oil and gas within the same genetic groups of fields and prospects, most notably (Table 12.1), source, maturation/migration, reservoir, trap, timing, seal, preservation and recovery (White, 1988). The play-fairway (by analogy to the 'fairways' in golf) approach additionally relies on the identification and mapping within each play of regional and local geological parameters (Harbaugh, 1984). This approach enables basin-scale geological factors to be distinguished and evaluated separately from field/prospect/block-scale factors (Fig. 12.2). When considered together, evaluation of these parameters provides the best means of identifying those areas with maximum exploration potential ('high-grading' remaining acreage), estimating hydrocarbon volumes and quantifying the full range of risk factors. This also provides the basis for quantifying the uncertainties associated with drilling new prospects, and the risk of drilling a dry hole, namely the chance of successfully finding the following in the right place and at the right time: (i) a sealed trap; (ii) a porous and permeable reservoir; (iii) hydrocarbon maturation and migration; (iv) trap formation having predated hydrocarbon migration; and (v) an accumulation that has been preserved and contains producible hydrocarbons (Parsley, 1990).

The evaluation of a play fairway is based fundamentally on a rigorous stratigraphic analysis, which provides a consistent, systematic framework for evaluating the key primary regional play parameters. In the context of the North Sea, these comprise reservoir, top seal, source, timing of trap creation and hydrocarbon maturation and migration. Predicting the distribution of the primary depositional parameters (reservoir, top seal and source) relies heavily on understanding the nature of depositional processes and environments within a chronostratigraphic framework. Within the structural framework of the main prospective basins (see Fig. 1.17), this was undertaken initially for the main geological periods and zones (Fig. 12.3), which have been clearly demonstrated in previous reviews of North Sea plays and their hydrocarbon habitat (e.g. Ziegler, 1990a). More recently, the development and application of sequence-stratigraphic concepts to the North Sea offer a higher resolution for evaluating the stratigraphic framework of regional and subregional plays

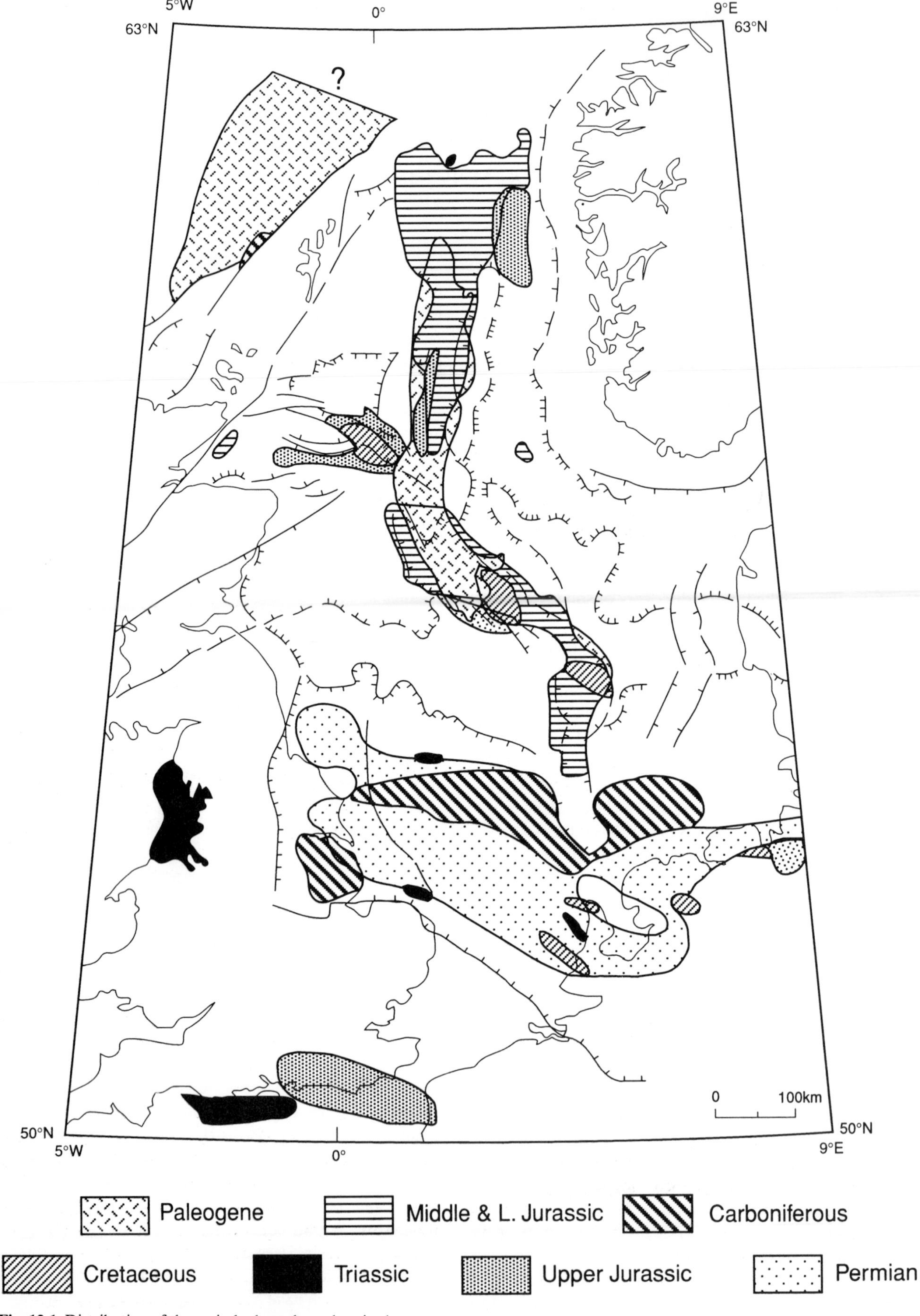

Fig. 12.1 Distribution of the main hydrocarbon plays in the North Sea and adjacent areas (based on Pegrum and Spencer, 1990; Parsley, 1990; Glennie, 1986, 1992).

Table 12.1 Geological hydrocarbon-control factors and related maps (modifed from White, 1988).

Hydrocarbon-control factor	Examples of possible maps
Source	
Bed thickness and area	Effective isopach/facies maps (thickness)
Total organic carbon (TOC)	TOC % to cut-off (source/non-source ratio)
Organic matter/kerogen type	Adequacy edge maps: oil- vs. gas-prone
Maturation/primary migration (within source)	Kitchen areas/oil and gas windows (maturity)
Overmaturation	HC deadline
Combinations	HC volumetric yields
Migration	
Secondary migration	Paths from structure, stratigraphy Limits from source (prospective perimeters)
Reservoir	
Gross thickness	Isopach with limits/effective edge
Net/gross (N/G)	N/G maps (to porosity/permeability cut-off)
Net reservoir/net pay	Isopach (net oil, net gas, etc. to cut-off)
Porosity	Porosity maps
Permeability	Permeability maps
Saturation (water/hydrocarbon)	HC saturation maps
Reservoir quality	Facies/depositional sequences
Trap type and timing	
Closure: area, height/depth/thickness	Structural contours/depth maps
Timing of formation	Stratigraphic maps (thickness/pinch-outs) Trap timing keyed to migration timing
Seal	
Thickness	Isopach with effective edge/limits
Lithology/facies	Facies types
Seal quality/integrity	Capillary pressures/entry pressures
Modifiers	Faults, fractures; hydrodynamics; tar seals; diagenesis; gas hydrates
Preservation	
Flushing	Hydrodynamics; salinity; solubility; depth
Biodegradation	Formation water types; depth; temperature
Diffusion	HC type; seal diffusivity; timing
Viscous oil	Oil viscosity or gravity
Inert-gas dilution (e.g. CO_2)	Inert fraction of gases
Low saturation/concentration	Barrels/acre or barrels/platform, by trap/play type
Combinations	
HC occurrence	Fields, shows, seeps, seismic DHIs Tested and untested closures Cross-sections Play-summary maps (combination maps of key parameters)

HC, Hydrocarbon; CO_2, carbon dioxide; DHI, Direct Hydrocarbon Inhibitor.

in increasing detail, as seen, for example, in the Mesozoic and Palaeogene of the North Sea (e.g. Underhill, Chapter 8, and Bowman, Chapter 10, this volume; see Figs 8.3–8.7). Such an approach enhances the delineation of subtle traps, particularly those depending on a greater stratigraphic component for their closure.

The history of hydrocarbon discoveries in the various national sectors that led to the present play analysis is described in Chapter 1 and the controlling structural development is outlined in Chapter 2. In the area under discussion, hydrocarbons occur in rocks ranging in age from the Precambrian to the Tertiary, but mainly in the Late Palaeozoic, Mesozoic and Palaeogene (Fig. 12.3). This play review considers the distribution of these hydrocarbons in three main areas: the Northern and Central North Sea, Southern North Sea and western British Isles (see Fig. 12.1). The first two areas are mature and data-rich, whereas the latter is less well known, forming part of the much larger and relatively underexplored frontier area of the Atlantic Margin.

12.2 Northern and Central North Sea plays

The key to understanding hydrocarbon distribution in the Northern and Central North Sea area can be simplified in

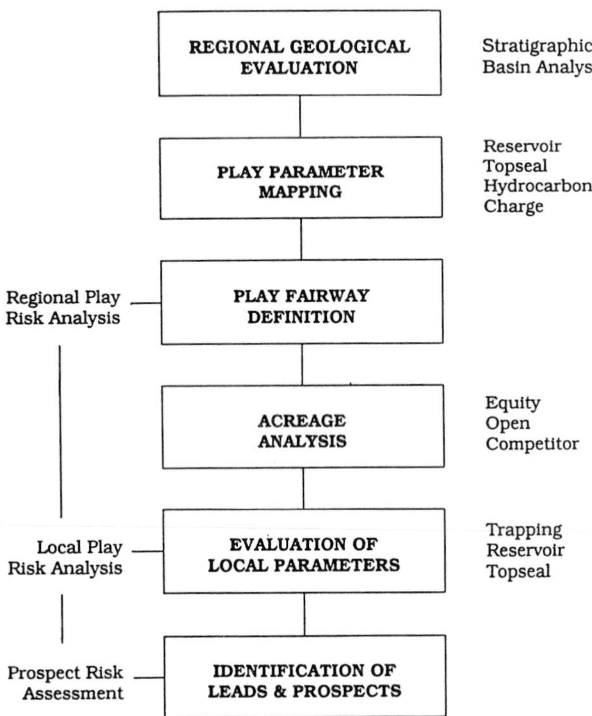

Fig. 12.2 The key stages in the play-fairway evaluation process from regional to local/prospect level.

terms of several tectonostratigraphic events and their relative timing (Fig. 12.4). In this area, plays are commonly grouped according to reservoir age and their relationship to 3 main rift-related tectonic phases relative to the main Late Jurassic rift event (Fig. 12.5): (i) pre-rift play (mainly Lower–Middle Jurassic, Permo-Triassic and older Palaeozoic plays); (ii) syn-rift play (mainly Upper Jurassic plays); and (iii) post-rift play (mainly Lower Cretaceous, Chalk and Palaeogene plays). The close relationship between play type and rifting highlights the great importance exerted by Late Jurassic/Early Cretaceous rifting on the hydrocarbon habitat of the North Sea Basin. Initially, immediately pre- (Middle Jurassic) and syn-rift processes directly influenced the deposition of major Jurassic source, reservoir and seal facies. Subsequently, Late Jurassic and Early Cretaceous rifting was the main process, resulting in the development of structural traps in both Jurassic and pre-Jurassic sediments. Finally, the collapse of this rift system resulted in post-rift thermal subsidence during the rest of the Cretaceous to Recent time span, which was responsible for the maturation and migration of mainly Jurassic-sourced hydrocarbons from the Palaeogene to the Recent. Fundamental to this is the rift-related structural framework (Figs 12.6 and 12.7), because it largely controlled the thickness, facies and distribution of the Upper Jurassic syn-rift succession, including its widespread organic-rich, marine source rock (Fig. 12.8). It also determined the distribution of mature Upper Jurassic source rock following post-rift thermal subsidence, which focused the main oil- and gas-kitchen areas along the old rift axes (Fig. 12.9).

Within each of these rift-related tectonic phases, it is possible to recognize a range of play types, based mainly on the age, nature and depositional environment of the reservoir rock. In this review, 14 main play types are identified, which are distributed over the seven major hydrocarbon-bearing stratigraphic intervals in the Northern and Central North Sea (Table 12.2 (see p. 274)). In detail, the precise relationship between some of these plays and rifting is sometimes complex and uncertain. The main uncertainty with this scheme is the classification of Upper Jurassic and Lower Cretaceous plays, which span the onset and completion of major rifting. Although this rifting process may have begun in the Permian and Triassic, the most important phase of North Sea rifting was a short-lived (*c.* 10 million years (Ma)) and highly diachronous process, which occurred in a 'zipper-like' manner, extending from the North Viking Graben (Middle Oxfordian), through the South Viking, Witch Ground and Central grabens (Kimmeridgian) and terminating in the Inner Moray Firth (Late Volgian/Ryazanian). This diachroneity results in several variations in terms of published interpretations concerning the relationship between depositional and rifting processes.

The play types can be evaluated and risked in terms of three main primary regional play parameters (see Fig. 12.2): reservoir, top seal and hydrocarbon charge (combining the occurrence of mature source rocks and a migration pathway into the reservoir). The most important secondary regional-play parameter is timing, particularly in relation to the Upper Cretaceous and Palaeogene plays. However, timing is generally not a critical parameter for most Mesozoic and older plays, since trap formation considerably predated hydrocarbon charge (main oil generation was during the Eocene, with gas generation during the Neogene to Recent).

The regional play-fairway map defines those areas where all the regional play parameters are favourably developed, where hydrocarbon prospectivity is at a maximum and where the risk of play failure is at a minimum (e.g. Fig. 12.10). Some of the key local play parameters are traps, local presence and/or effectiveness of reservoir and local presence and/or effectiveness of top seal. In the Central and Northern North Sea, trap definition is the most important local parameter. Other local risks include top-seal or reservoir erosion and variations in top-seal or reservoir facies.

Apart from the occasional exception (e.g. Beatrice Field in the Inner Moray Firth and Bream/Brisling in the Stord Basin), the main hydrocarbon accumulations all occur in close proximity to mature Late Jurassic source rocks, which reached maturity only where deeply buried within the Mesozoic grabens. Hydrocarbon migration has been mainly vertical, with lateral migration significant only in Tertiary successions. Hence, the Northern and Central North Sea area is consistent with Demaison's (1984) generative-basin concept, which states that: (i) the highest exploration success occurs in genetically related oil-generative basins; (ii) the largest hydrocarbon accumulations tend to be located close to the centre of generative basins or on structurally high trends immediately adjacent to generative basins; and (iii) migration distances are most commonly around tens of kilometres and are limited by the drainage areas of individual structures. Thus, as in the North Sea, most of the producible hydrocarbons and the largest accumulations occur within the area of mature source rocks, and long-distance migration beyond these limits is rare (see Fig. 11.48).

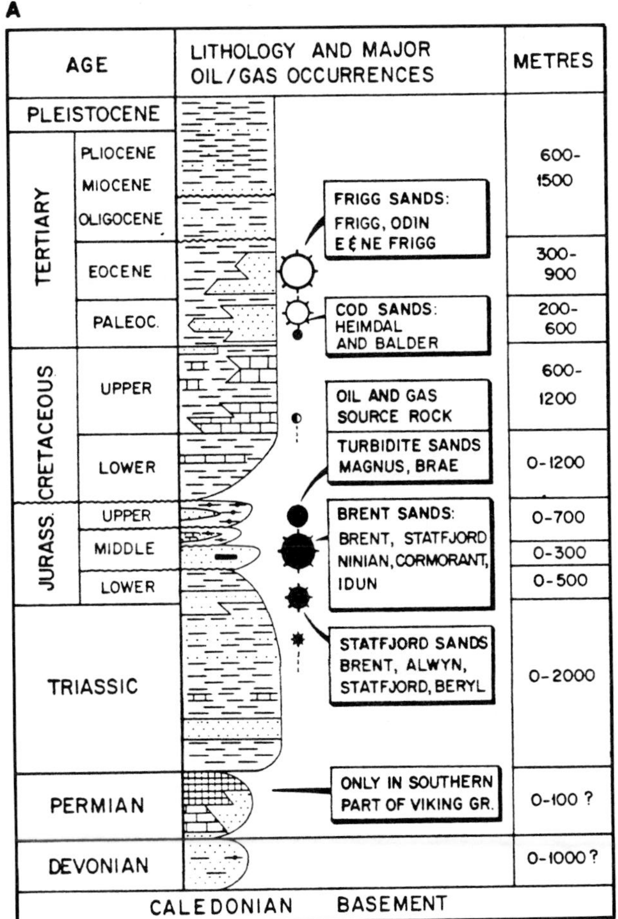

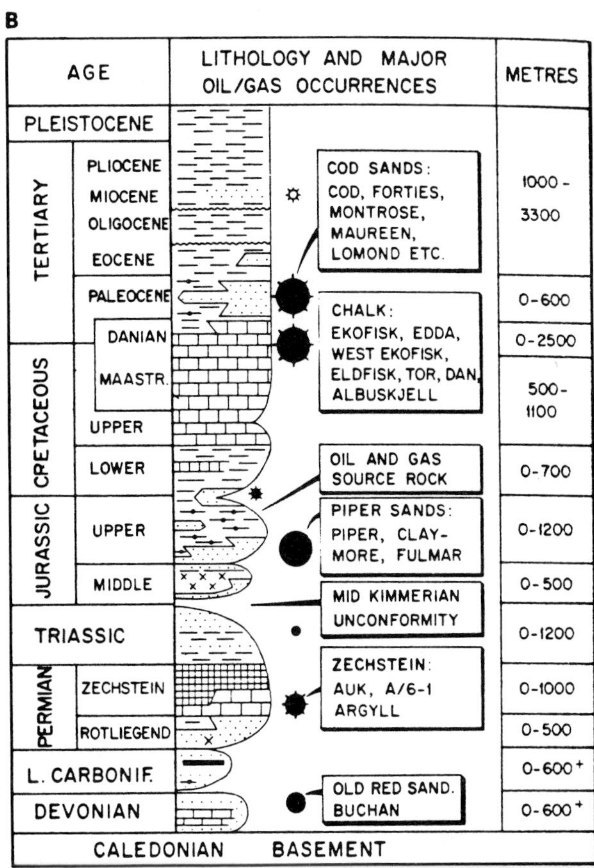

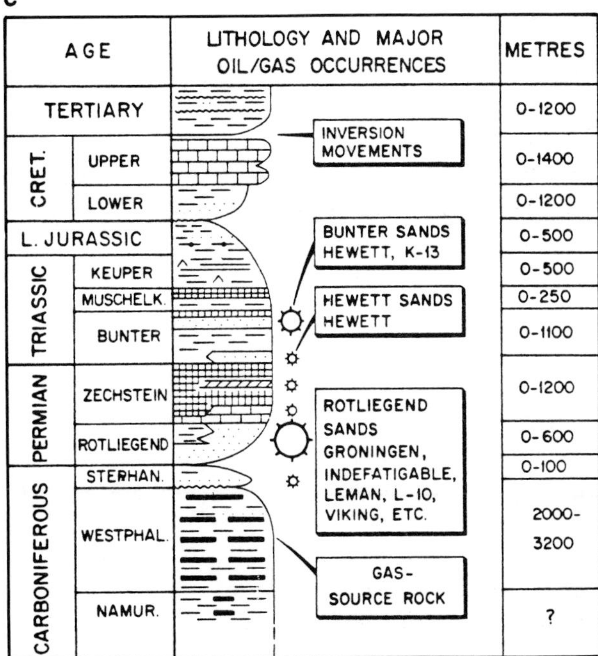

Fig. 12.3 Stratigraphic summary of the main North Sea reservoir rocks and their hydrocarbon occurrences. (A) Northern North Sea. (B) Central North Sea/Moray Firth. (C) Southern North Sea. (After Ziegler, 1977.)

In addition to hydrocarbon charge, the other key geological controls (reservoir, top seal and traps) are specific to each play type and are described below.

12.2.1 Pre-rift plays of the Northern and Central North Sea

The pre-rift plays comprise three main groups: Palaeozoic, Triassic/Lower Jurassic and Middle Jurassic. These plays occur throughout the area but are concentrated in the Northern North Sea, which includes some of the largest hydrocarbon accumulations in the basin (Fig. 12.10). Many of the pre-rift accumulations represent large structural closures, which were visible on early 2D seismic sections and were discovered in the initial phases of exploration (1970s). In an exploration sense, they are relatively mature

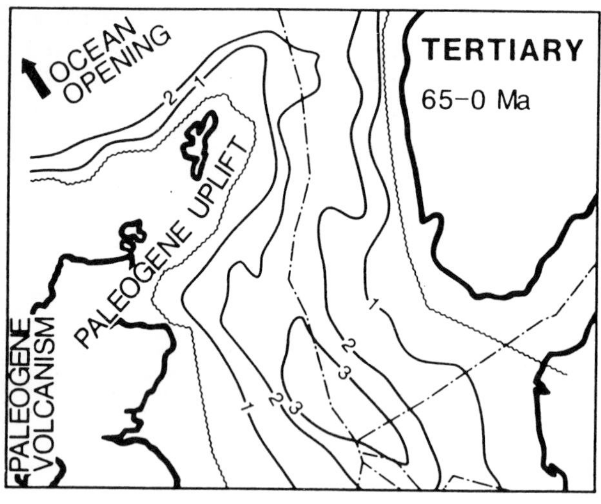

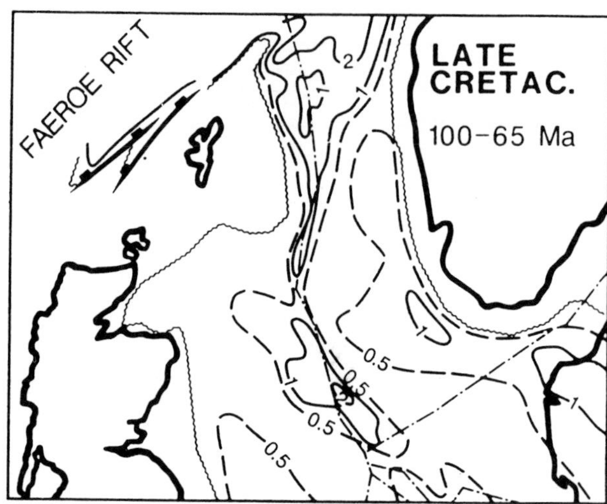

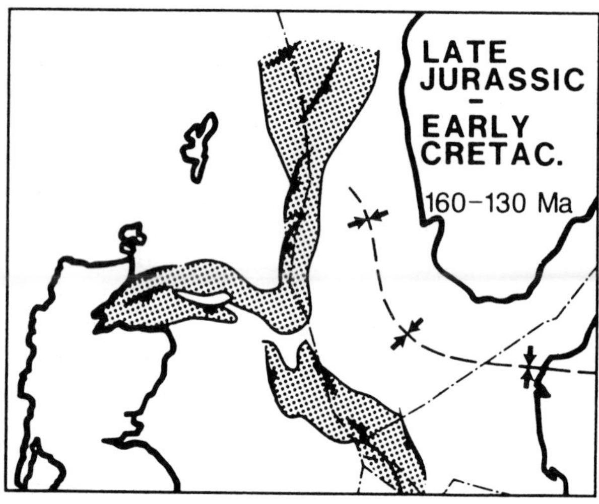

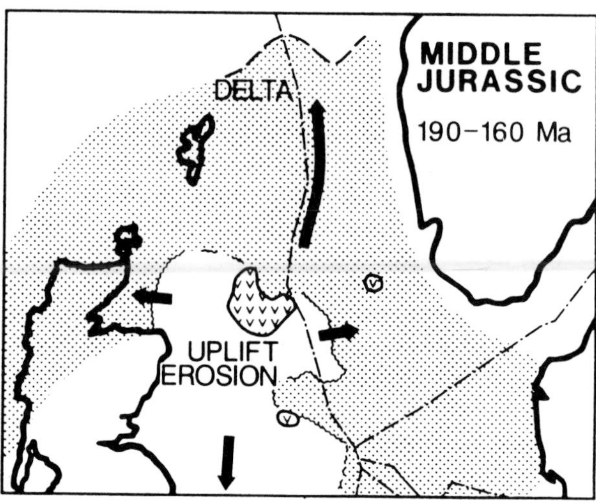

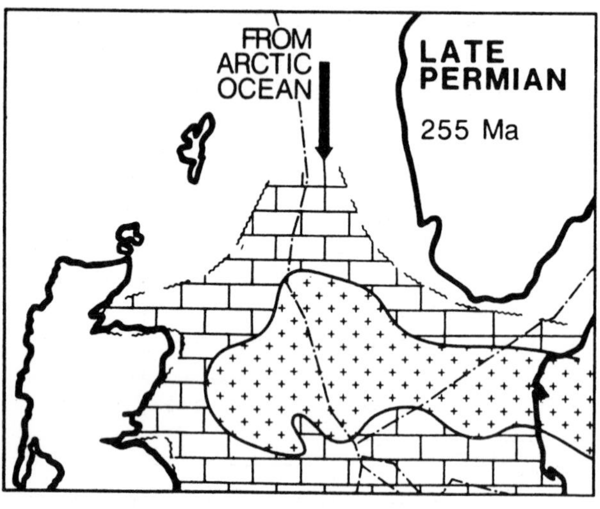

TERTIARY	UPLIFT IN N.W. BASIN-WIDE SUBSIDENCE	COOLING PHASE
LATE CRETAC.	RIFTS INFILLED	COOLING PHASE
EARLY CRETAC. LATE JURASSIC	RIFTING	
MID. JURASSIC	DELTAS IN N.E. VOLCANISM, UPLIFT IN S.	
TRIASSIC	DESERT RED BEDS	
LATE PERMIAN	EVAPORITE BASIN	

Fig. 12.4 Simplified geological history of the Northern and Central North Sea (from Pegrum and Spencer, 1990).

but still contain significant potential for improved hydrocarbon recovery within producing fields.

Palaeozoic play

The Palaeozoic play contains relatively minor but significant hydrocarbon volumes. It is most important in the Central North Sea, where two main types of reservoir occur (Fig. 12.11): (i) Zechstein carbonates (e.g. Auk, Argyll); and (ii) pre-Zechstein sandstones, which comprise several reservoir types, including fractured Devonian sandstones (Buchan, Embla), Rotliegend sandstones (Argyll, Auk) and Carboniferous sandstones. These form part of a more extensive system of Permo-Carboniferous and older oil-bearing plays in northern Europe and adjacent regions

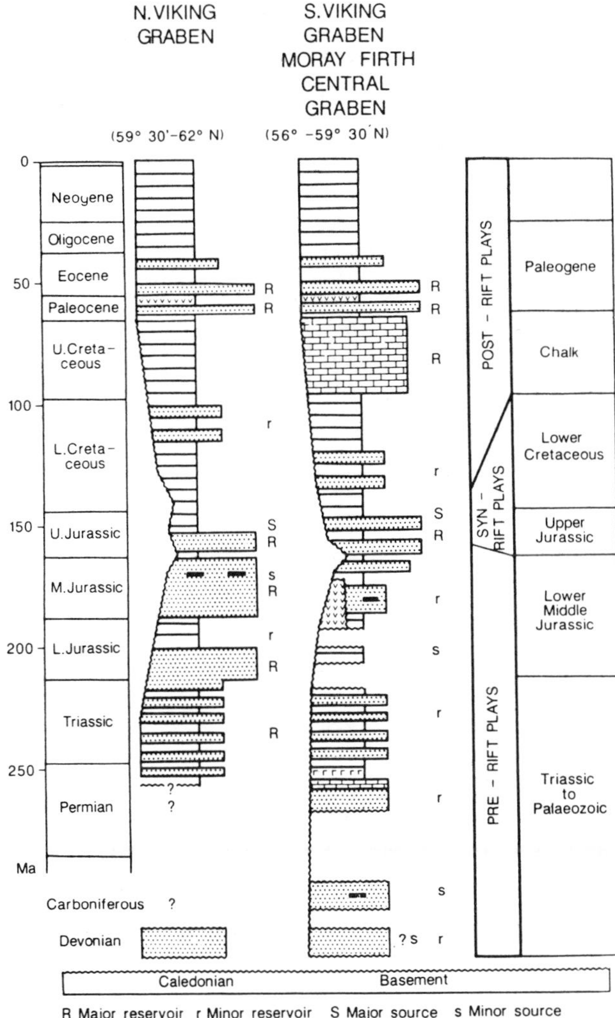

Fig. 12.5 Classification and stratigraphic distribution of the main plays in the Northern and Central North Sea (from Pegrum and Spencer, 1990).

(Leeder and Boldy, 1990; Christiansen *et al.*, 1993; Gérard *et al.*, 1993; Smith, 1993).

Some of the better-known hydrocarbon-bearing Palaeozoic accumulations include the following fields: Auk (Brennand and van Veen, 1975; Heward, 1991; Trewin and Bramwell, 1991), Argyll (Pennington, 1975; Bifani *et al.*, 1987; Robson, 1991) and Buchan (Butler *et al.*, 1976; Richards, 1985; Edwards, 1991). This play also provides secondary reservoirs in the Stirling, West Brae and Ettrick fields. In the Central Graben of the Norwegian sector, the Devonian and other indeterminate sandstones of the Embla Field represent the only development of pre-Jurassic reservoirs in this area (Knight *et al.*, 1993; see also Chapter 3). Reservoirs associated with this play are also oil-bearing west of the Shetland Islands in the Clair Field (Coney *et al.*, 1993) and onshore Britain (Smith, 1993).

Reservoirs. Zechstein carbonates are effective reservoirs in Auk and Argyll, because of extensive leaching, fracturing and the development of vugular porosity, associated in part with weathering below the base Cretaceous unconformity (Brennand and van Veen, 1975; Pennington, 1975; Taylor, 1981; see Section 6.12.5). In both these fields and others (e.g. Ettrick), complex diagenesis of the dolomites results in highly variable reservoir properties (Amiri-Garroussi and Taylor, 1987). In both Auk and Argyll, the relatively thin (*c.* 10–30 m) carbonate reservoir overlies oil-bearing Permian Rotliegend Sandstones of the Northern Permian Basin (Section 5.3.2), whose variable reservoir properties are largely related to primary depositional environment, with the best quality associated with the aeolian dune facies (Buchanan and Hoogteyling, 1979; Heward, 1991; see Section 5.5.2).

Although the sandstone reservoirs are highly variable, they share a common risk in terms of their quality and effectiveness. In the Buchan Field, the main oil-bearing reservoir is the Devonian Old Red Sandstone (Hill and Smith, 1979), which comprises well-developed fluvial sandstones but with largely ineffective primary reservoir properties (matrix permeabilities < 1 millidarcy (mD)). Well productivity is achieved through an open-fracture system, which was developed during rift-related uplift of the Buchan Horst, and has resulted in effective permeabilities of up to 50 mD (Butler *et al.*, 1976).

Traps. All the accumulations in this play are fault-bounded, dip-closed traps, which are located close to mature Upper Jurassic source rocks (Fig. 12.11). Many of these traps are in footwall uplifts, tilted away from major, graben-bounding faults, as in the case of the Auk and Argyll fields. The main structural configuration of these fields was formed during late Jurassic rifting. Internal field complexity is caused mainly by a combination of dense, rift-related faulting and depositional reservoir heterogeneity.

Play-fairway analysis. The most critical regional parameter controlling this play is the hydrocarbon charge from Upper Jurassic Kimmeridge Clay source rocks into the much older reservoirs. Top-seal and reservoir quality may also be critical. These factors are combined most effectively along the graben margin or within intrabasinal highs (Fig. 12.11). Major fault systems provide migration pathways from source to reservoir, but migration away from these major conduits occurred over only short distances, and hence charge can be a significant risk beyond the graben margin. The crestal areas of tilted and eroded fault blocks were preferentially leached and fractured during prolonged Mesozoic exposure, which enhanced reservoir quality particularly in the Zechstein carbonates. This local play parameter is highly variable and difficult to predict. Onlapping by marine mudstones in the Late Jurassic and Early Cretaceous provides an effective top seal, although in places this is provided by the Chalk (e.g. Auk and Argyll).

Deeper structures are difficult to charge by downward migration. Traps located significant distances beyond the main Upper Jurassic kitchen areas (greater than *c.* 10–20 km) also suffer from the risk of charge failure. Risks increase dramatically where Palaeozoic source rocks have to be invoked (see Section 6.13). Devonian source rocks provide a partial source for the Beatrice Field in the Inner Moray Firth (Peters and Moldowan, 1989), but an extensive Palaeozoic-sourced play has not yet been established. Hence, Jurassic sourcing remains the most viable option for any remaining potential in this play.

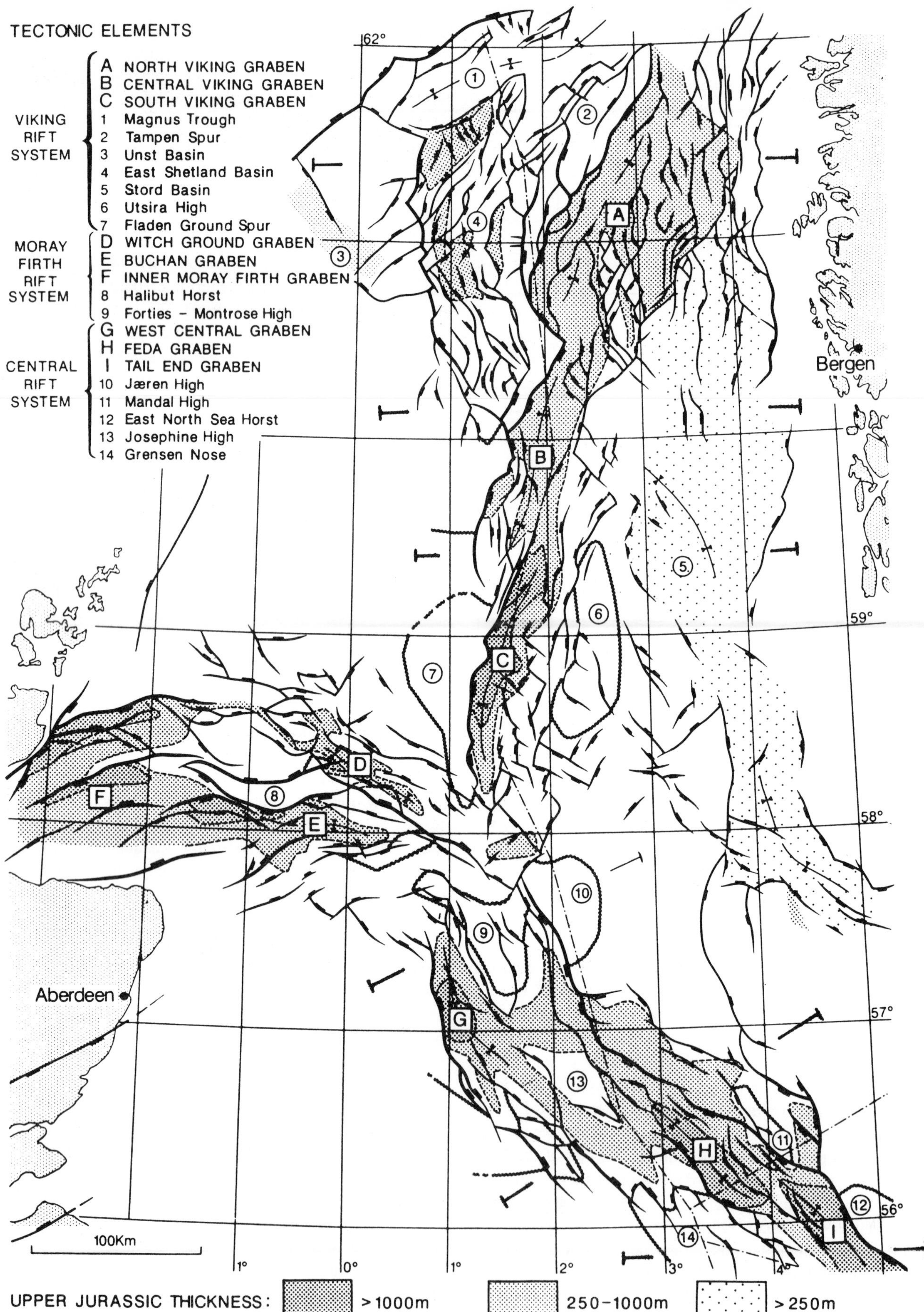

Fig. 12.6 Northern and Central North Sea tectonic framework of the Late Jurassic–Early Cretaceous rift system and associated tectonic elements (from Pegrum and Spencer, 1990). Lines of section refer to the geological cross-sections in Fig. 12.7.

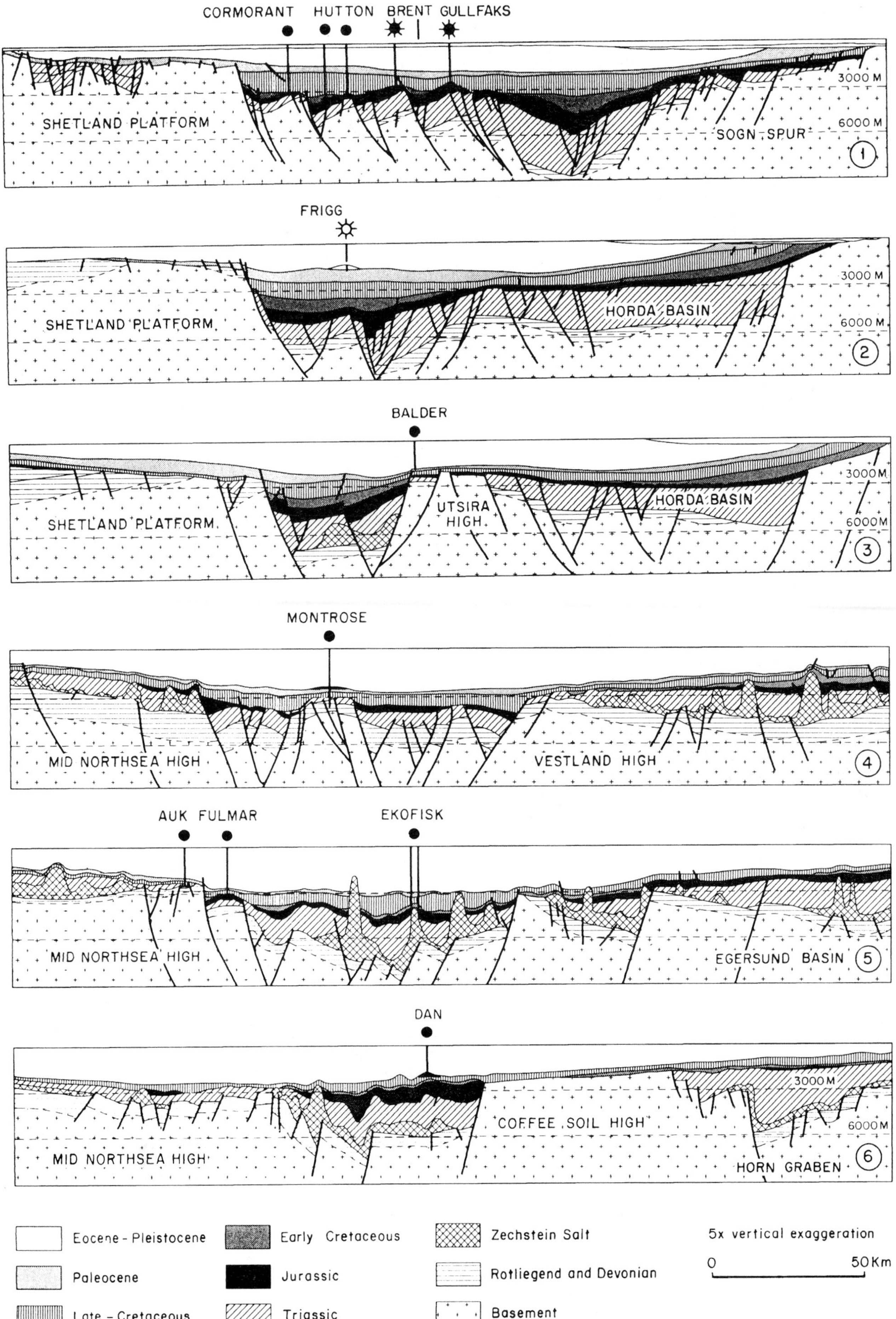

Fig. 12.7 Regional geological cross-sections through the Northern and Central North Sea (after Ziegler, 1982a). See Fig. 12.6 for section locations.

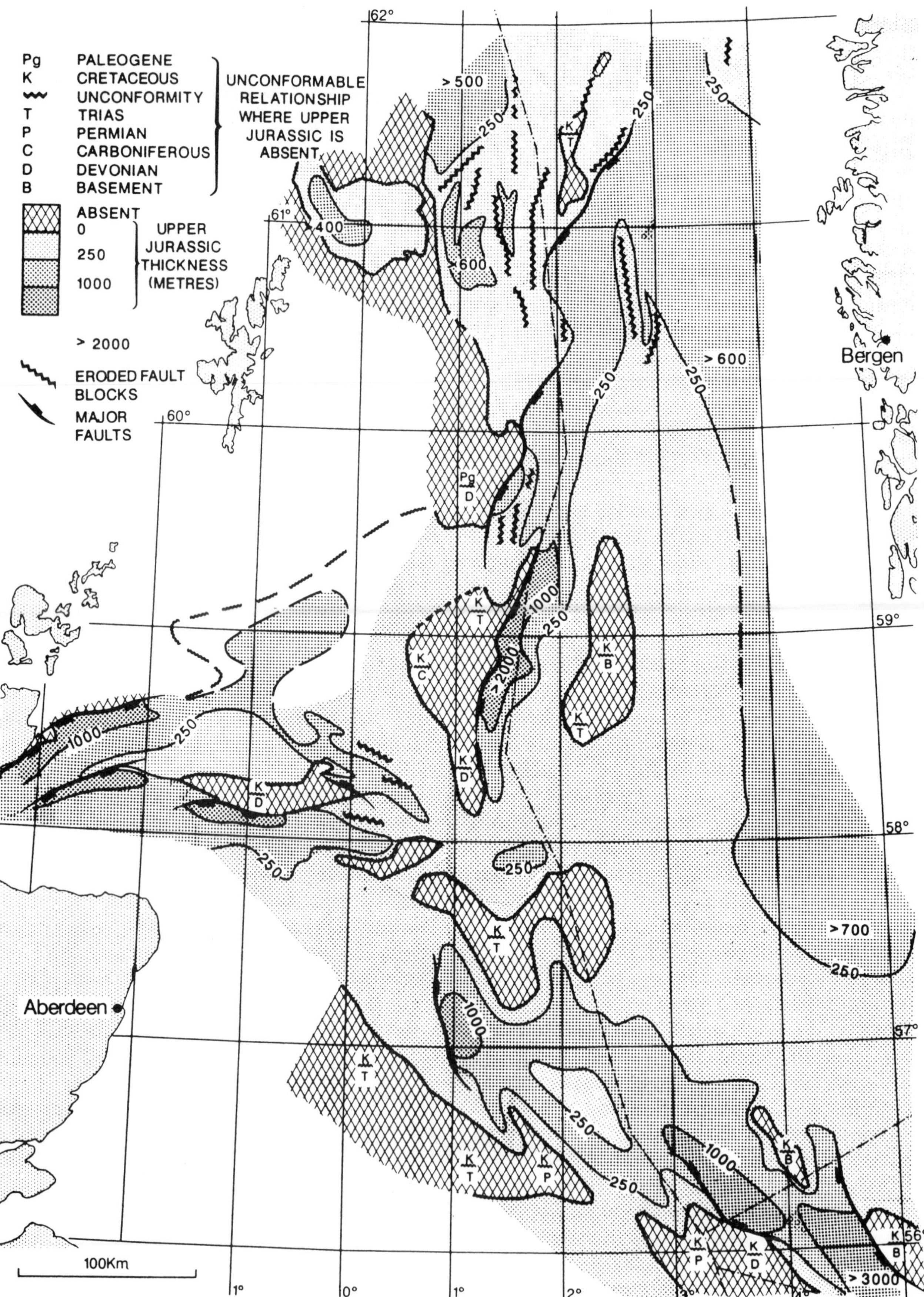

Fig. 12.8 Isopach map of Upper Jurassic strata based on well results (from Pegrum and Spencer, 1990). Note the close relationship between the location of isopach thicks and graben axes and eroded highs where Cretaceous or Palaeogene strata overlie pre-Jurassic strata (cf. Fig. 12.6).

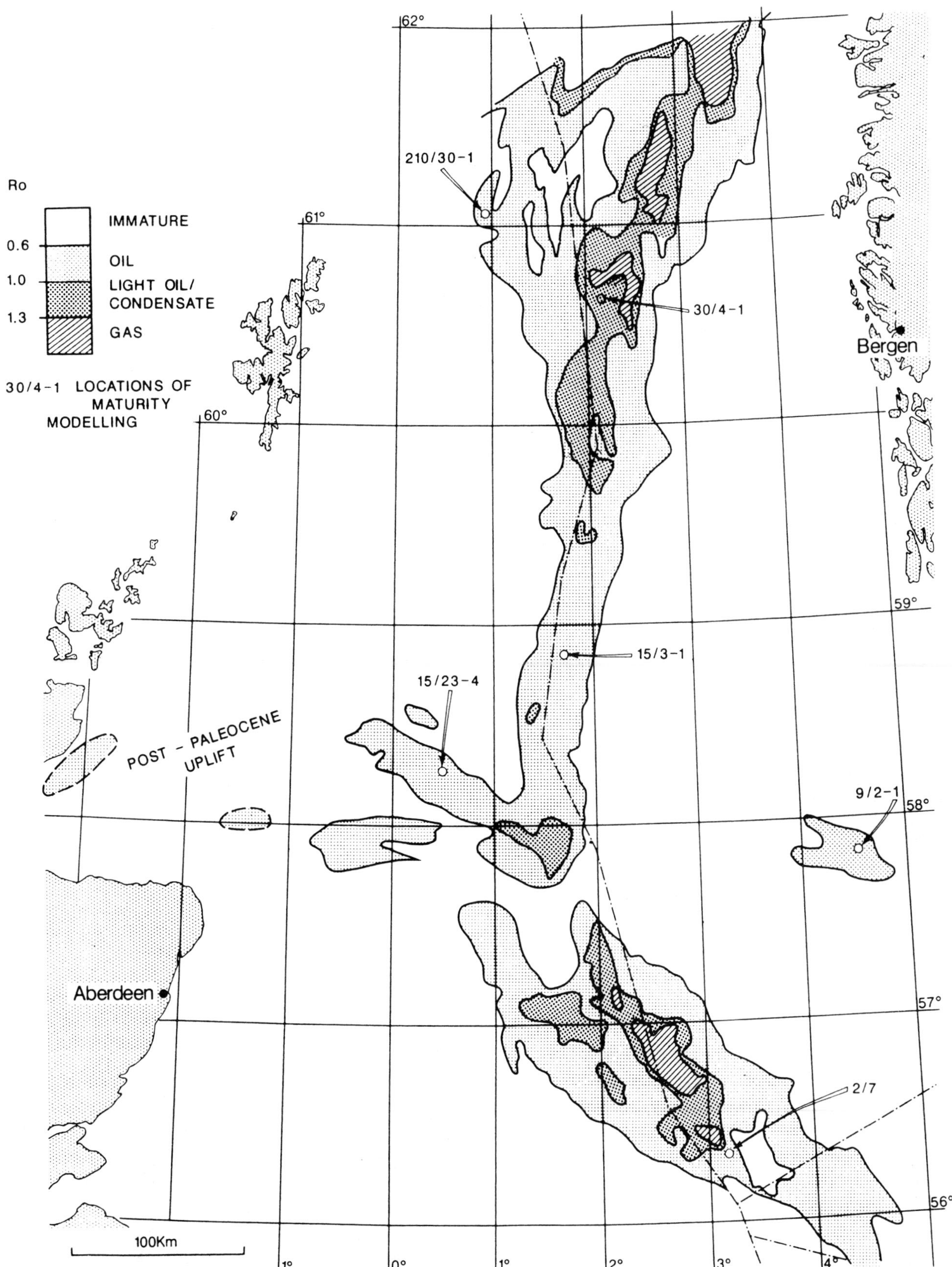

Fig. 12.9 Maturity of the top of the Upper Jurassic–earliest Cretaceous source-rock intervals, based on maturation modelling studies and regional seismic mapping (from Pegrum and Spencer, 1990).

Triassic–Lower Jurassic play

The Triassic–Lower Jurassic play covers a wide range of reservoir and trap types in both the Central and Northern North Sea, often as secondary or marginal reservoirs to more significant Middle or Upper Jurassic accumulations (see Sections 7.3.5 and 7.4.4). The most common hydro-

Table 12.2 Central and Northern North Sea plays and associated reservoirs.

Stratigraphic unit	Play	Reservoir
Palaeogene	Eocene (undifferentiated)	Alba, Tay, Frigg
	Eocene, Balder	Beauly, Odin
	Palaeocene–Eocene	Dornoch, Hermod, Cromarty
	Palaeocene (undifferentiated)	Forties, Balmoral, Andrew, Heimdal, Maureen
Upper Cretaceous	Chalk	Ekofisk, Tor
Lower Cretaceous	Apto–Albian basinal play	Bosun, Kopervik
	Hauterivian–Barremian fault-scarp play	Scapa, Spey Cromer Knoll sandstones
Upper Jurassic	Kimmeridgian–Volgian basinal and fault-scarp plays Deep-marine sandstones	Magnus Brae, Miller Claymore, Ettrick Kimmeridge sandstones
	Callovian–Volgian shelf play Shallow-marine sandstones	Piper, Beatrice, Hugin Fulmar, Ula, Heno Heather sandstones
Middle Jurassic	Bathonian–Callovian Aalenian–Bathonian	Pentland, Sleipner Brent, Bruce, Beryl
Lower Jurassic–Triassic	Rhaetian–Lias Scythian–?Norian	Dunlin sandstone, Cook, Statfjord, Marnock, Gassum, Cormorant, Lunde, Skagerrak
Palaeozoic	Subcrop plays	Zechstein Rotliegend Upper Carboniferous Devonian/Old Red sandstone

carbon occurrences in this play are: (i) subcrops below the Base Upper Jurassic or Base Cretaceous unconformities; (ii) juxtaposition against Upper Jurassic source rocks across major graben faults; and (iii) in the lower part of stacked reservoir sequences, which are overlain by either Middle or Upper Jurassic sandstones.

The largest accumulations occur in major tilted fault blocks in the Northern North Sea, notably in the following fields: Snorre (see Fig. 8.32; Karlsson, 1986; Hollander, 1987; Dahl and Solli, 1993); Gullfaks (Sæland and Simpson, 1982; Fossen, 1989; Petterson *et al.*, 1990); Statfjord (Roberts *et al.*, 1987); Brent (Bowen, 1975; Struijk and Green, 1991); Alwyn North (Johnson and Eyssautier, 1987); and Beryl (Steele and Adams, 1984; O'Donnell, 1993; Robertson, 1993).

Hydrocarbon accumulations in the Central North Sea are generally smaller, they cover a wider range of trap types, and their reservoir properties are more susceptible to significant diagenetic deterioration. Triassic sandstone oil accumulations include the Acorn, Beechnut, Crawford, Heron, Josephine, Joanne, Judy, Skua, Ula (Bailey *et al.*, 1981; Brown *et al.*, 1992) and Kittiwake fields. However, many of these are small (e.g. Crawford) and reservoir quality marginal to non-productive (e.g. Kittiwake), although the latter may be productive through the overlying Jurassic Fulmar sandstones (Glennie and Armstrong, 1991). Gas/condensate-bearing reservoirs occur in Marnock and in the 30/1c fields (Bartholomew *et al.*, 1993; Smith *et al.*, 1993).

Reservoirs. The main reservoir intervals comprise thick, fluvial-dominated sandstones, particularly the Triassic/Lower Jurassic (Rhaetian–Sinemurian) Statfjord, Marnock and Gassum formations and the older Triassic Skaggerak, Lunde and Lomvi formations (Fisher and Mudge, Chapter 7, this volume). The main characteristics of these reservoirs reflect deposition in terrestrial, semiarid environments, although the younger (Rhaetian–Sinemurian) intervals show an increasing marginal marine influence in response to both the Early Jurassic eustatic sea-level rise and a return to more humid climatic conditions (Hodgson *et al.*, 1992). The most common reservoirs are fluvial channel and sheetflood deposits; they are typically highly feldspathic, mainly fine-grained and commonly tightly cemented. Major regional variations in reservoir facies characteristics and thickness patterns reflect the marked differences in structural framework between the Central Graben and the Viking Graben (see Sections 7.3 and 7.4, respectively).

In the Central Graben, Triassic deposition was controlled mainly by halokinetic movements of Zechstein salt during a period of passive thermal subsidence (Fig. 12.12; see Section 7.3). Fine-grained fluvial and sheetflood sandstones are abundant within salt-induced (halokinesis and/or salt dissolution) topographic lows, which form sand-rich pods, intercalated with thick mud-dominated intervals (Fig. 12.13). This results in complex, elongate sand-thickness patterns, abrupt thickness changes and reduced connectivity between adjacent sand-prone systems.

The North Viking Graben, beyond the Beryl Embay-

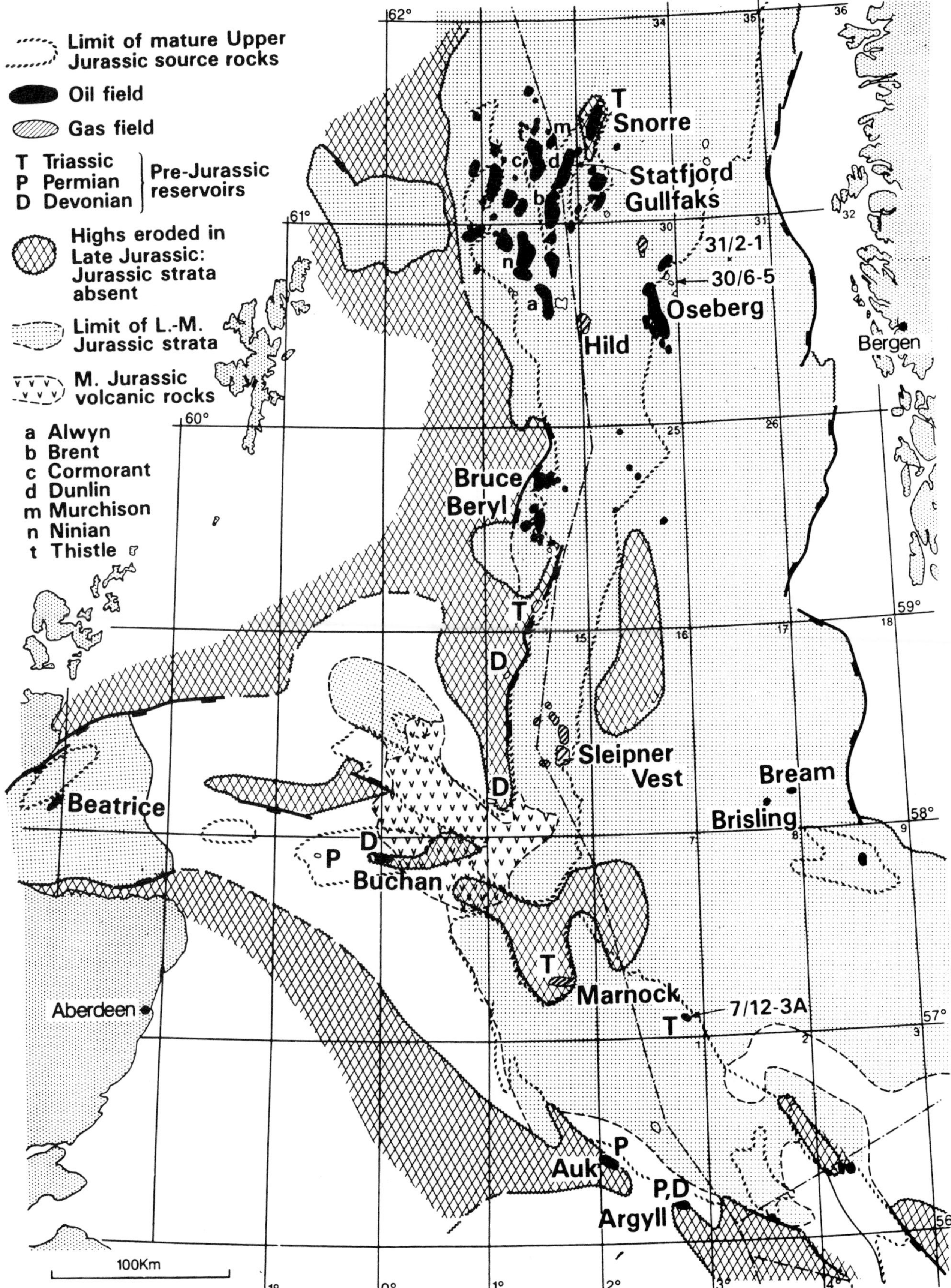

Fig. 12.10 Pre-rift play map of the Northern and Central North Sea (from Fjaeran and Spencer, 1991).

ment, and adjacent areas were devoid of Zechstein salt, and were separated from the Central Graben by a watershed located at the southern end of the South Viking Graben (see Section 8.3.1). As a result, regional continuity of both sandstone- and mudstone-dominated units is much greater

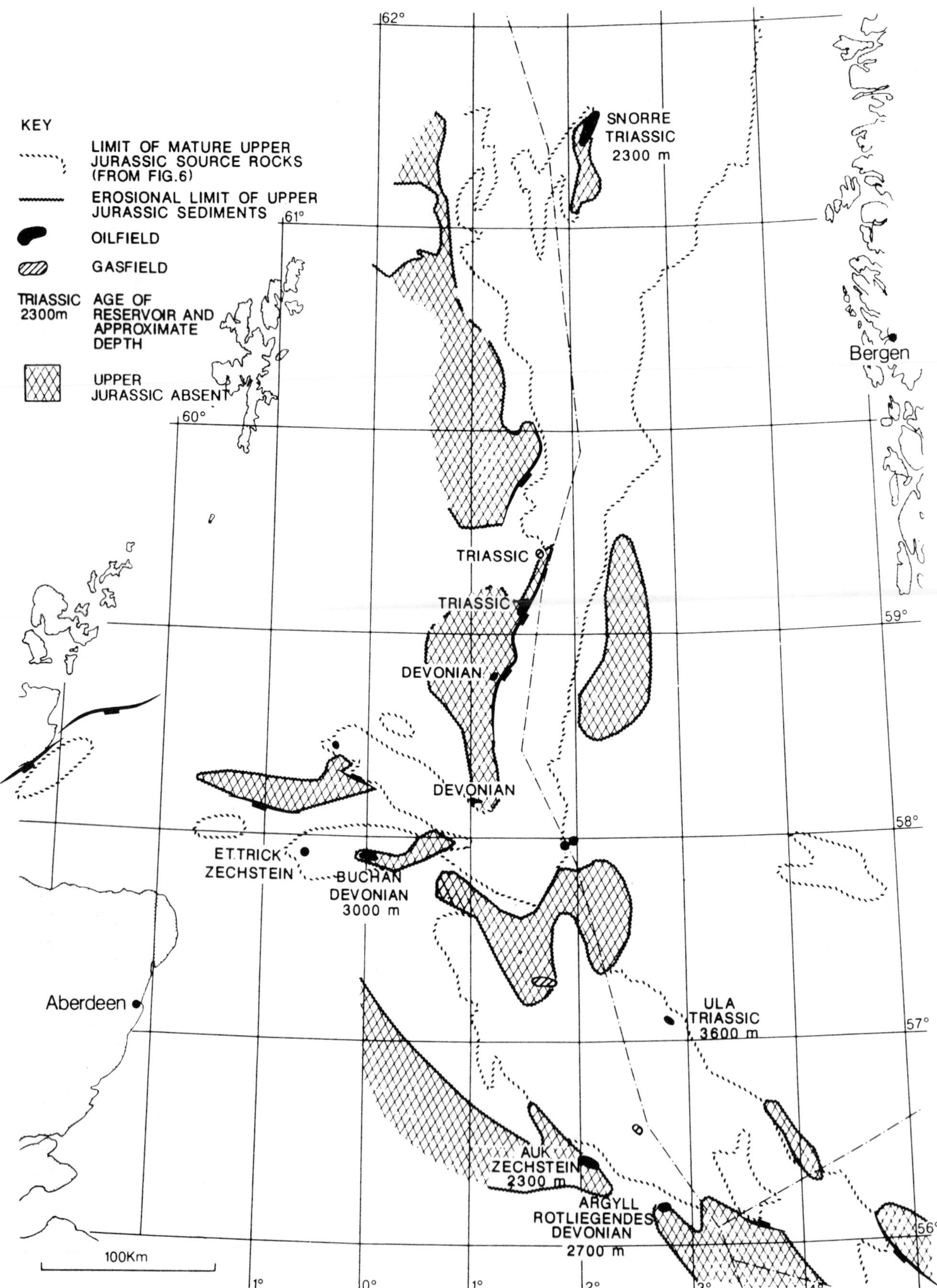

Fig. 12.11 Pre-rift/pre-Jurassic play map (from Pegrum and Spencer, 1990). All hydrocarbon accumulations are located on deeply eroded highs adjacent to the edges of mature Upper Jurassic source rocks (hashed line—see key).

than in the Central Graben area. Fluvial drainage was mainly to the north but with additional sediment supply from the basin margins (including Old Red Sandstone and metamorphic basement). Both syn-rift and post-rift intervals are recognized, with the development of several pro-

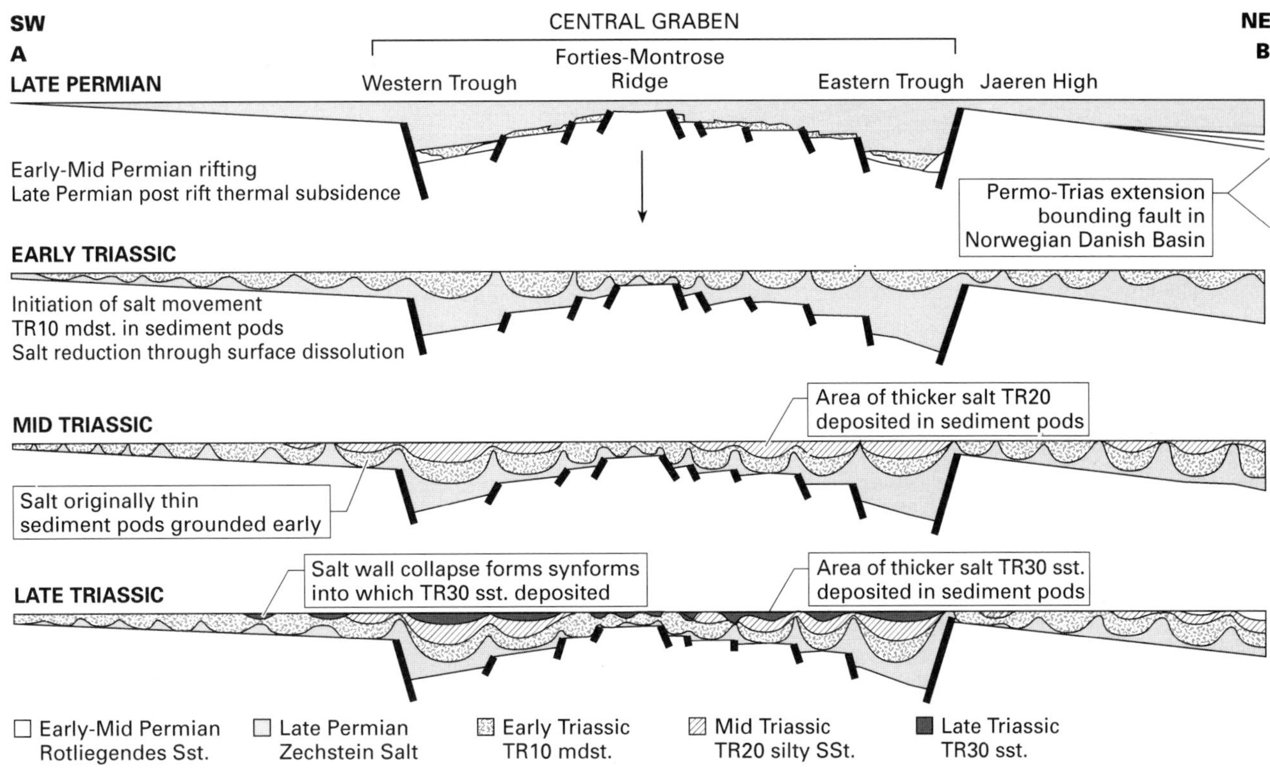

Fig. 12.12 Schematic structural development of the Central Graben during the Late Permian to Late Triassic (from Smith *et al.*, 1993). Note the influence of salt movement on Triassic sediment accommodation.

grading, fluvial-dominated clastic wedges (megasequences), characterized by variations in sand-body stacking patterns (see Fig. 8.14; Steel, 1993). The most important hydrocarbon-bearing interval (Megasequence 3) is represented by the youngest clastic succession in this play, the Statfjord Formation (see Figs 8.12 and 8.13 and Sections 7.4.3 and 8.3.1). Renewed tectonic uplift (part of the Early Cimmerian event of Ziegler, 1990) resulted in the development of low-sinuosity, coarse-grained fluvial systems, which built out over muddy alluvial and coastal plains (Røe and Steel, 1985). This was followed by deposition of a laterally extensive transgressive sandstone (Nansen Formation) during the Sinemurian eustatic rise in sea level. The coarse-grained, sandstone-dominated nature of the Statfjord Formation results in a high-quality reservoir in several major fields (e.g. porosity *c.* 20–24% and permeability *c.* 300–2000 mD in the Brent Field; Johnson and Kroll, 1984). Deposition of the conformably overlying marine mudstone unit (Dunlin Group) provides a regional top seal for the Statfjord Formation.

Traps. Trapping styles in the Central Graben are influenced by both tectonics and halokinesis. Simple dip-closed, anticlinal structures defined at Base Cretaceous level occur above Triassic pods grounded above base-Zechstein surfaces. More complex structural closures at top reservoir level occur where Triassic pods have undergone tilting or lateral displacement, due to Upper Jurassic rift-related faulting, gravity sliding and late-stage salt withdrawal (Erratt, 1993; Penge *et al.*, 1993). The most complex structural traps occur in response to polyphase deformation, where basement lineaments have been reactivated during late Jurassic rifting to produce local transpressional structures with variable and often complicated fault patterns (Bartholomew *et al.*, 1993; Platt and Philip, 1993; Sears *et al.*, 1993).

Traps in the Northern North Sea all occur as fault-bounded dip closures in Late Jurassic rotated fault blocks (Fig. 12.14). These range from large, simple and internally unfaulted structures (e.g. Brent and Statfjord) through to much more complex structures, characterized by intense faulting (e.g. Snorre: see Fig. 8.32; and Gullfaks; Dahl and Solli, 1993) (Fig. 12.14).

Play-fairway analysis. In the Central North Sea, this play is most successful where Triassic reservoir sandstones (mainly Marnock or Gassum formations, locally Skagerrak Formation) subcrop below the Base Upper Jurassic or Base Cretaceous unconformity, which provides a short migration route from Upper or Middle Jurassic source rocks (e.g. the Marnock Field, see Fig. 7.23 and Section 7.3.5). Top seal is provided by either the Kimmeridge Clay source rock or overlying Cretaceous deposits (Cromer Knoll shales or the Chalk). Key uncertainties surround reservoir presence and thickness, mainly due to the variable depth of erosion below the unconformity and to the influence of both syn- and post-depositional salt movements. Reservoir effectiveness is an additional risk and is determined by a combination of facies, compaction and burial-related diagenesis, which can result in extensive authigenic clay mineralization (e.g. chlorite in the Marnock Field) and calcite and dolomite cementation (e.g. the non-productive Skagerrak Formation in Kittiwake; Glennie and Armstrong, 1991). Hence, even subcropping Triassic sandstones may be non-prospective.

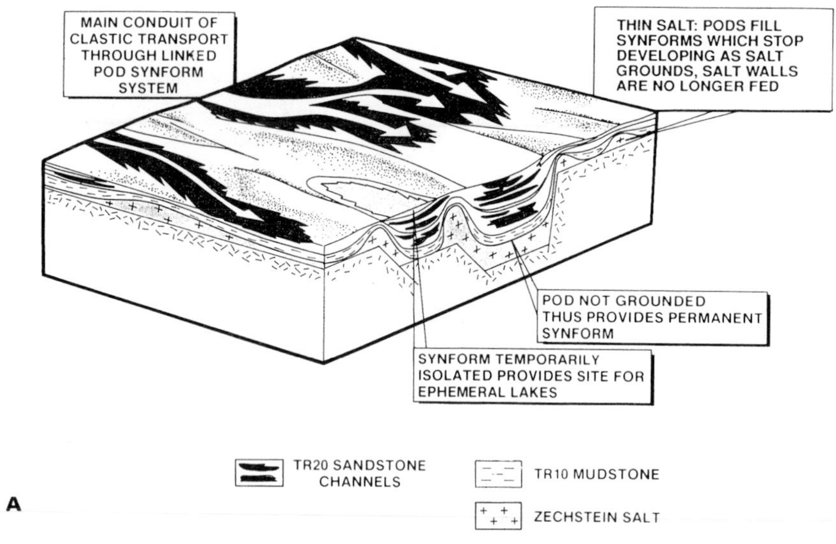
MAIN CONDUIT OF CLASTIC TRANSPORT THROUGH LINKED POD SYNFORM SYSTEM
THIN SALT: PODS FILL SYNFORMS WHICH STOP DEVELOPING AS SALT GROUNDS, SALT WALLS ARE NO LONGER FED
POD NOT GROUNDED THUS PROVIDES PERMANENT SYNFORM
SYNFORM TEMPORARILY ISOLATED PROVIDES SITE FOR EPHEMERAL LAKES
TR20 SANDSTONE CHANNELS
TR10 MUDSTONE
ZECHSTEIN SALT
A

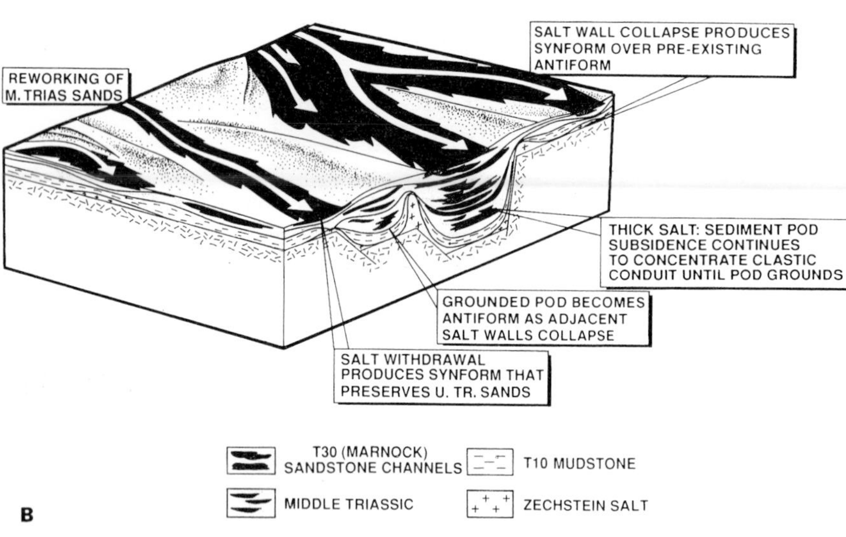
REWORKING OF M. TRIAS SANDS
SALT WALL COLLAPSE PRODUCES SYNFORM OVER PRE-EXISTING ANTIFORM
THICK SALT: SEDIMENT POD SUBSIDENCE CONTINUES TO CONCENTRATE CLASTIC CONDUIT UNTIL POD GROUNDS
GROUNDED POD BECOMES ANTIFORM AS ADJACENT SALT WALLS COLLAPSE
SALT WITHDRAWAL PRODUCES SYNFORM THAT PRESERVES U. TR. SANDS
T30 (MARNOCK) SANDSTONE CHANNELS
T10 MUDSTONE
MIDDLE TRIASSIC
ZECHSTEIN SALT
B

Fig. 12.13 Triassic depositional model for the (A) Mid-Triassic (TR 20) and (B) Late Triassic (TR 30), illustrating the influence of salt-related subsidence on fluvial sedimentation patterns and reservoir geometry (from Smith *et al.*, 1993).

This play also forms a secondary target in the Central North Sea, where Triassic sandstones form the basal part of stacked reservoir sequences, capped by Middle or Upper Jurassic sandstones. Deeper Triassic reservoirs, however, are largely non-prospective because of charge limitations.

In the Northern North Sea, the play relies on either subcrop beneath the Base Cretaceous unconformity or juxtaposition across major faults to provide migration routes from the Kimmeridge Clay. This is most effectively achieved in the large rotated fault blocks with deeply eroded crests (e.g. Snorre) or where fault throw has been sufficient to juxtapose the Kimmeridge Clay against Triassic/Lower Jurassic reservoirs (Fig. 12.14). The latter situation is best developed along the Alwyn–Brent–Statfjord trend where thick (up to *c.* 1000 ft/300 m), high-quality Statfjord Formation sandstones form a major secondary reservoir within large tilted fault blocks. Separation from the shallower Middle Jurassic reservoirs is demonstrated by different fluid contacts, pressure regimes and hydrocarbon composition. The different hydrocarbon fluid properties, including higher gas–oil ratios, also reflect migration from a deeper, more mature Kimmeridge Clay kitchen area. Secondary Triassic reservoirs also occur in the generally much lower-quality Cormorant Formation, such as in the Beryl, Tern and Penguin fields (see Section 7.4.4). The traps are all fault-bounded dip closures, with Lower Jurassic shales (Dunlin Group and its equivalents) forming the cap rock.

Middle Jurassic Play

The Middle Jurassic of the North Viking Graben, East Shetland Basin and Tampen Spur is one of the most productive plays in the North Sea. Although it has now been virtually fully exploited as an exploration target, it remains a major oil-producing play, which is represented by some of the largest fields in the area, including Brent, Statfjord, Cormorant, Ninian and Gullfaks, among many others (Fig. 12.15). The 'creaming-curve' profile for the East Shetland Basin (UK sector of the North Sea) is typical of a highly mature play (see Fig. 12.38A), which suggests that all significant upthrown closures have been tested and that remaining potential is restricted mainly to either small

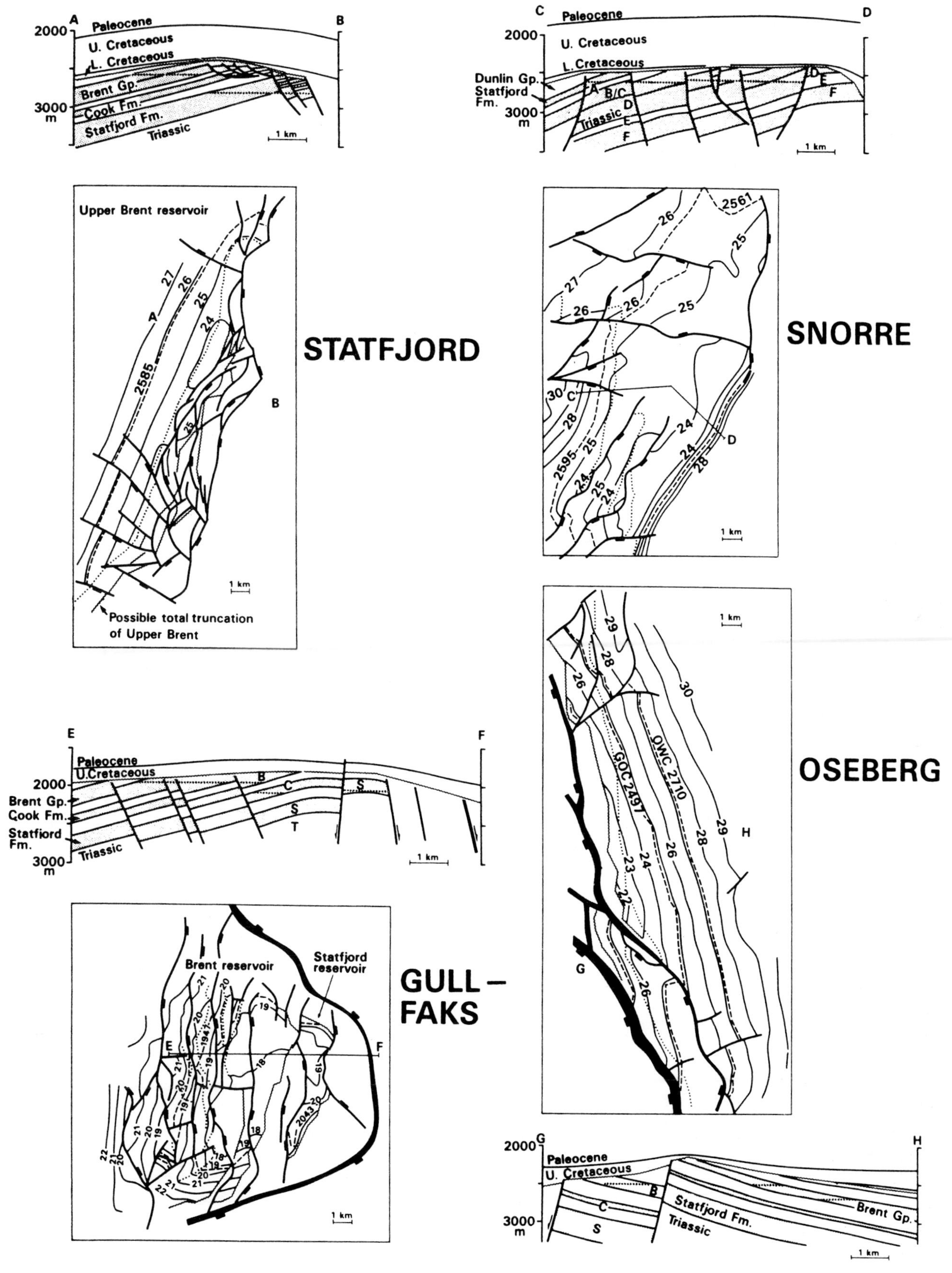

Fig. 12.14 Examples of pre-rift play hydrocarbon accumulations in the Northern North Sea (from Fjaeran and Spencer, 1991).

upthrown traps or to higher-risk downthrown closures (Bowen, 1992; Cordey, 1993). The play is only slightly less mature in the Norwegian sector (Knag *et al.*, in press).

Reservoirs. Middle Jurassic reservoirs form a thick (*c.* 300–1000 ft/90–300 m), laterally extensive (*c.* 400 km × 50–100 km), highly diachronous (Aalenian to Callovian; *c.* 30 Ma duration) clastic wedge comprising laterally interconnected fluvial, deltaic and coastal depositional systems with excellent reservoir properties (net/gross ratios *c.* 0.4–0.8, porosities *c.* 20–30% and permeabilities *c.* 50–

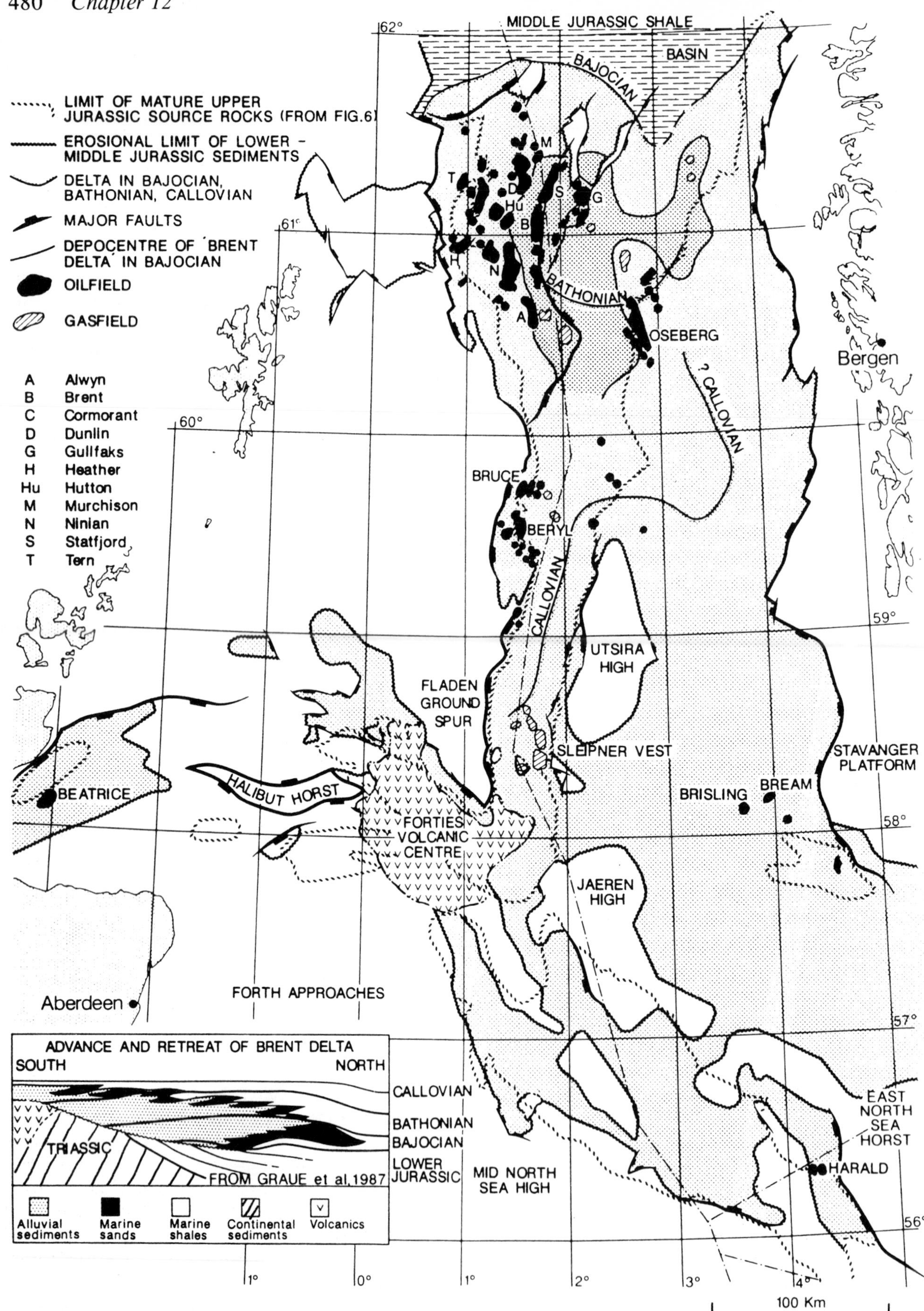

Fig. 12.15 Pre-rift/Lower–Middle Jurassic play map (from Pegrum and Spencer, 1990). Note the widespread distribution of reservoir-bearing strata within the basinal areas and their absence from intrabasinal highs and fault-bounded graben margins. Inset shows the simplified depositional model of the Brent delta (from Graue *et al.*, 1987). The map also shows the northward limit of the Brent delta, and potential sandstone reservoirs in the Bajocian.

5000 mD at depths shallower than 10 000 ft ss/3000 m ss; e.g. Giles *et al.*, 1992). This play is dominated by the Brent Group in the Northern North Sea, although several separate plays can be distinguished on the basis of reservoir age, facies and distribution (Fig. 12.15): Brent play (East Shetland Basin/Tampen Spur/North Viking Graben); Cook play (East Shetland Basin/Tampen Spur/North Viking Graben); Beryl play (Beryl Embayment); Sleipner play (South Viking Graben); Pentland play (Central Graben/Outer Moray Firth); and Brora Coal play (Inner Moray Firth).

Deposition of the Brent clastic wedge (Fig. 12.15 inset) is widely accepted to be related to pre-rift thermal doming centred on the triple junction area of the South Viking, Central and Witch Ground grabens (Ziegler, 1990a; Underhill and Partington, 1993; Chapters 2 and 8 this volume, Section 8.4), with sediment supplied both from the south and from platform areas along the graben flanks to the east and west (see Fig. 8.27; Mearns, 1992; Morton, 1992). This resulted in a radial drainage system, including a major northward-prograding coastal to wave-dominated delta complex (see Fig. 8.19). Most regional palaeogeographic facies maps taken at various time increments reflect coastal and delta-plain deposits to the south, which pass northwards through sand-rich coastal-barrier and wave-dominated delta-front deposits into offshore marine mudstones (see Richards, 1992); for a review of Brent Group depositional models see Cannon *et al.* (1992). Sequence-stratigraphic studies, with their higher-resolution chronostratigraphic frameworks, demonstrate the lateral linkage of the Aalenian to Callovian clastic successions throughout the North and South Viking grabens and adjacent areas, which comprise several depositional sequences (Fält *et al.*, 1989; Galloway, 1989a; Cockings *et al.*, 1992; Mitchener *et al.*, 1992; Milton, 1993; Steel, 1993). Two main tectonostratigraphic units (J20 and J30 of Mitchener *et al.*, 1992; Milton, 1993; Rattey and Hayward, 1993) record the transition from pre-rift thermal uplift (J20/Aalenian–Late Bajocian) through to the onset of rifting (J30/latest Bajocian-Middle Callovian). Seven depositional sequences bounded by maximum flooding surfaces are also recognized, which reflect fluctuations in basin subsidence, sediment supply and eustatic sea level (Partington *et al.*, 1993a). These sequences each record individual progradational events, initially recording the overall northward build-out of the Brent delta (Aalenian to Late Bajocian), followed by punctuated and diachronous southward retreat (Early Bathonian to Callovian) (see Figs 8.26 and 8.28). Early ideas of a single fluviodeltaic supply system along the axis of the Viking Graben have been replaced by a model of multiple sediment-entry points (Brown *et al.*, 1987; Graue *et al.*, 1987; Richards *et al.*, 1988; Rattey and Hayward, 1993).

Although the regressive–transgressive model of the Brent Group is more complex than originally envisaged (see Fig. 8.17; Budding and Inglin, 1981; Johnson and Stewart, 1985), it was important in predicting one of the most critical aspects of reservoir development in the East Shetland Basin and Tampen Spur areas, namely the northward limit of reservoir-sandstone development (Fig. 12.15). This coincides with the point of the maximum advance and subsequent retreat of the Brent delta system, which occurs around 62°N (Brown and Richards, 1989; Helland-Hansen *et al.*, 1992). Reservoirs are thickest along the axial parts of the Viking Graben, where channel sand bodies are best developed, with abrupt thinning across pre-rift syndepositional faults and on to the graben margins. The combination of different vertically stacked depositional environments, mainly coastal barrier and coastal/delta plain (see Fig. 8.19), and syn-depositional faulting has resulted in complex reservoir architecture and thickness patterns, many of which have become apparent only during later stages of field development (e.g. Hallet, 1981; Livera, 1989).

Superimposed on this facies trend is a systematic, depth-related reduction in reservoir quality related to increasing compaction and diagenesis (Giles *et al.*, 1992). In general, porosity decreases linearly with depth (*c.* 25% at 9000 ft/2700 m ss to *c.* 15% at 12 500 ft/3750 m ss), due to compaction and burial cementation (quartz and iron-rich carbonates), apart from localized improvements below the Base Cretaceous Unconformity. Permeability shows a marked linear decrease only below *c.* 10 200 ft/3110 m ss, which corresponds to a sudden onset of increasing quantities of illite with depth. This places important depth constraints on exploration and development in the area, with much reduced reservoir quality (in both oil and water legs) and performance in accumulations below *c.* 10 000 ft/3000 m ss (e.g. Cormorant Block II, Heather, North-west Hutton, Lyell and Pelican; Gray and Barnes, 1981; Stiles and McKee, 1986; Jourdan *et al.*, 1987; Glasmann, 1992; Harris, 1992; Kantorowicz *et al.*, 1992).

The Middle Jurassic play is only sparsely developed outside the Northern North Sea, partly because of limited development of reservoir in the Central North Sea/Outer Moray Firth (Pentland play) and limited preservation following Tertiary erosion in the Inner Moray Firth (Brora Coal play). However, it does occur, albeit as part of a different depositional system, in the several Atlantic Margin fields offshore mid-Norway, such as in Midgard (Ekern, 1987), Tyrihans (Larsen *et al.*, 1987) and other pre-rift rotated fault-block structures, such as Heidrun, Smørbukk and Smørbukk Sør (Fjaeran and Spencer, 1991). Similar-age reservoirs also occur in the Barents Sea region, most notably in the gas-dominated Hammerfest Basin in the far north (e.g. Snøhvit, Askeladden and Albatross fields; Fjaeran and Spencer, 1991).

Traps. The majority of traps in this play are formed by rotated fault blocks, with closure provided by a combination of fault seal (reservoirs juxtaposed against Upper Jurassic and Lower Cretaceous shales), erosional truncation capped by top-seal shales and dip closure down the fault-block flank (see Figs 12.7 and 12.14). Structures vary enormously in both size and internal fault complexity. The Brent (Fig. 12.15) and Statfjord fields are among the largest and simplest, with fault-block trap widths of *c.* 4 km and lengths of *c.* 18 and 24 km, respectively (Kirk, 1980; Roberts *et al.*, 1987; Struijk and Green, 1991). These structures occur where pre-existing structural zones were widely spaced, allowing the structures to develop orthogonally in the direction of Mesozoic extension (Bartholomew *et al.*, 1993).

More complex structures developed in regions where pre-existing structural lineaments were reactivated during Mesozoic extension. For example, Gullfaks is arguably the

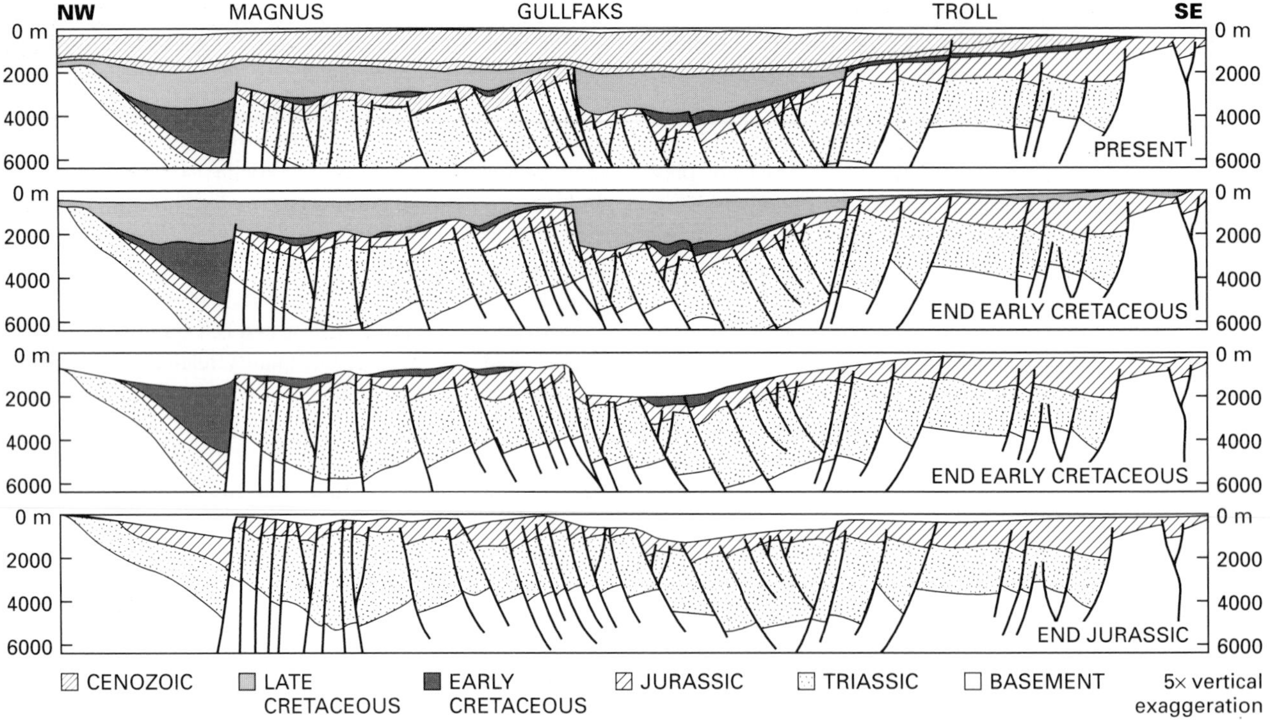

Fig. 12.16 Back-stripped palinspastic reconstruction of a north-west–south-east structural cross-section across the northern Viking Graben, through the Magnus, Gullfaks and Troll fields (from Ziegler, 1990a). The section shows progressive fault-block rotation through the Late Jurassic/Early Cretaceous, and thermal sag subsidence concentrated over the Mesozoic grabens during the late Cretaceous through to the present day.

most complex structure in this area (Figs 12.14 and 12.16), although it is located only 20 km east of the simple fault block of the Statfjord Field. The Gullfaks Field is divided into three distinctive areas, on the basis of structural style and fault patterns (Erichsen *et al.*, 1987; Fossen, 1989; Petterson *et al.*, 1990), and is bounded on its eastern edge by a major fault system, which merges northwards with the leading edge of the Snorre Field (Inner Snorre Fault). Snorre represents another intensely faulted structure but without Middle Jurassic reservoirs (Dahl and Solli, 1993). This structural trend is interpreted as resulting from a combination of extensional faulting, plus dextral strike–slip movement along larger basement faults, with the latter producing compressional structures (Fossen, 1989). A similar interaction between Caledonian basement lineaments and Mesozoic extension is noted in the Cormorant Field (Speksnijder, 1987; Demyttenaere *et al.*, 1993) and along the Tern–Eider Horst Block (Bartholomew *et al.*, 1993).

Play-fairway analysis. The success of the Brent play results from a combination of widespread, high-quality sandstone reservoirs (mainly Brent Group), extensive and well-developed top-seal facies (Upper Jurassic and Lower Cretaceous shales), numerous large tilted fault blocks with effective trapping configurations, and close proximity to mature Upper Jurassic source rocks, providing short and relatively simple migration pathways (see Fig. 12.15). However, the Beryl Field has a much more complex history with at least three separate phases of migration from three slightly different source-rock facies, which has resulted in variable hydrocarbon compositions across the field (Fig. 12.17).

The controlling parameters of this play (reservoir, top seal, source and trap) were all in place by the Late Jurassic, with source-rock maturation and hydrocarbon migration reaching a peak during the Eocene and continuing through to the present day (Goff, 1983; Eggen, 1984; Field, 1985; Larter and Horstad, 1992). Major uncertainties with this play are field- and prospect-scale prediction of reservoir thickness, following erosion accompanying Late Jurassic tilting and fault-block rotation. In some cases, this has been accompanied by slumping and/or reworking of Middle Jurassic deposits from the crest down to lower levels along the leading edge of the fault blocks (e.g. Brent and Statfjord; (see Fig. 12.14, cross-section A–B). In extreme cases, this has formed a series of secondary fault terraces (e.g. Ninian/Columba) or additional fault-blocks (e.g. Cormorant Block IV; Demyttenaere *et al.*, 1993).

The remaining exploration potential is likely to be restricted to those few remaining, but small, upthrown closures. Although more downthrown closures remain untested, they are generally also relatively small and associated with a greater risk due to fault-seal uncertainties. Deeper closures (greater than *c.* 10 200ft/3060 m ss) are severely affected by reduced reservoir effectiveness, while there is a lack of stratigraphic trapping configurations in this play because of the sheet-like geometry and sand-rich nature of the Brent Group. The greatest remaining interest in this play will probably be directed mainly towards maximizing hydrocarbon recovery from fields that display significant stratigraphic and structural complexities.

12.2.2 Syn-rift plays of the Northern and Central North Sea

The syn-rift plays are divided into two main stratigraphic

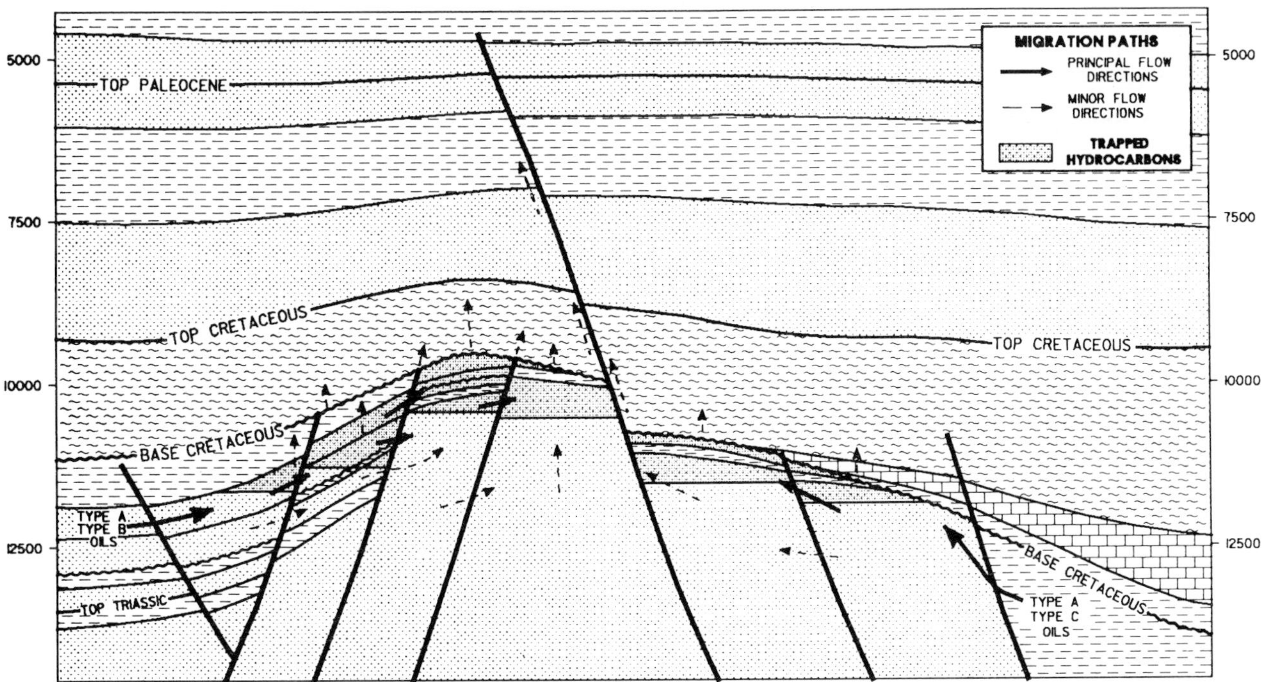

Fig. 12.17 Hydrocarbon migration and entrapment model of the Beryl Field (from Robertson, 1993).

intervals (see Fig. 12.5): Upper Jurassic and Lower Cretaceous. Both plays are dominated by Late Jurassic/Early Cretaceous rifting, which had a major control on reservoir, seal and source-rock distribution, as well as hydrocarbon entrapment. These are among the most complex plays in the North Sea, with significant remaining exploration potential.

Upper Jurassic play

The Upper Jurassic play (Fig. 12.18) owes its importance, firstly, to the occurrence of high-quality sandstone reservoirs, which are distributed throughout the Viking, Central and Moray Firth graben systems, and, secondly, to their intimate stratigraphic relationship with the North Sea's pre-eminent oil-source rock, the Upper Jurassic Kimmeridge Clay Formation and its lateral equivalents (Draupne Formation in the Norwegian sector; Doré *et al.*, 1985). The Kimmeridge Clay Formation commonly acts as both source rock and seal for this play.

Major oil, gas and condensate accumulations occur in a variety of reservoir and trap styles, including approximately 50 potentially commercial fields, of which 14 are already in production. The most important accumulations occur in the Central Graben (e.g. Fulmar, Clyde, Gannet West/Guillemot A, Kittiwake, Ula, Puffin, Duncan and Judy), South Viking Graben (Brae complex, Birch, Miller, Kingfisher and Toni/Tiffany/Thelma) and in the Outer Moray Firth (e.g. Piper, Claymore, Chanter, Tartan, Galley, Ettrick, Glamis, Glenn, Petronella and Scott). Elsewhere this play is represented by more isolated, but extremely important, accumulations, of which the Magnus and Troll fields in the Northern North Sea are most significant (Fig. 12.18).

This play contains significant remaining exploration potential, particularly in the Central Graben and Outer Moray Firth (Cordey, 1993) and also in the hangingwalls of the large fault blocks in the Viking Graben and Marulk Basin (Knag *et al.*, in press), and extends into the offshore areas of mid-Norway, such as in the Draugen Field (Ellenor and Mozetic, 1986; Jackson and Hastings, 1987). Existing accumulations in this play also provide many further development opportunities.

Reservoirs. The nature and distribution of reservoirs in this play are controlled by both depositional environment and the nature and rate of creation of accommodation space. The depositional environment defines two main reservoir types: shallow-marine sandstones (e.g. Fulmar, Hugin and Piper formations) and deep-marine/fan sandstones (e.g. Magnus and Brae formations). The creation of accommodation space varies throughout the basin, both in space and time; it was dominated by syn-depositional rifting and, in places, by salt movement (e.g. Central Graben). Consequently, the distribution and predictability of Upper Jurassic reservoirs are among the most complex in the whole of the North Sea Basin.

The controlling basinal processes at this time were: (i) the onset of rift-related faulting, although the precise timing is variable throughout the basin; and (ii) a rapid rise in relative sea level, which resulted in the drowning of the earlier thermal dome and its associated 'Mid-Cimmerian' unconformity. It remains debatable whether the rapid rise in relative sea level was eustatic, the outcome of regional thermal subsidence or a combination of the two (Rattey and Hayward, 1993). The result was that shorelines moved diachronously along the main graben axes towards the thermal dome ('Triple Junction') and the facies belt orientation changed from being perpendicular to the graben trends (east–west in the Viking Graben) to essentially parallel to the developing graben faults (north–south in the Viking Graben). This provides the framework for consid-

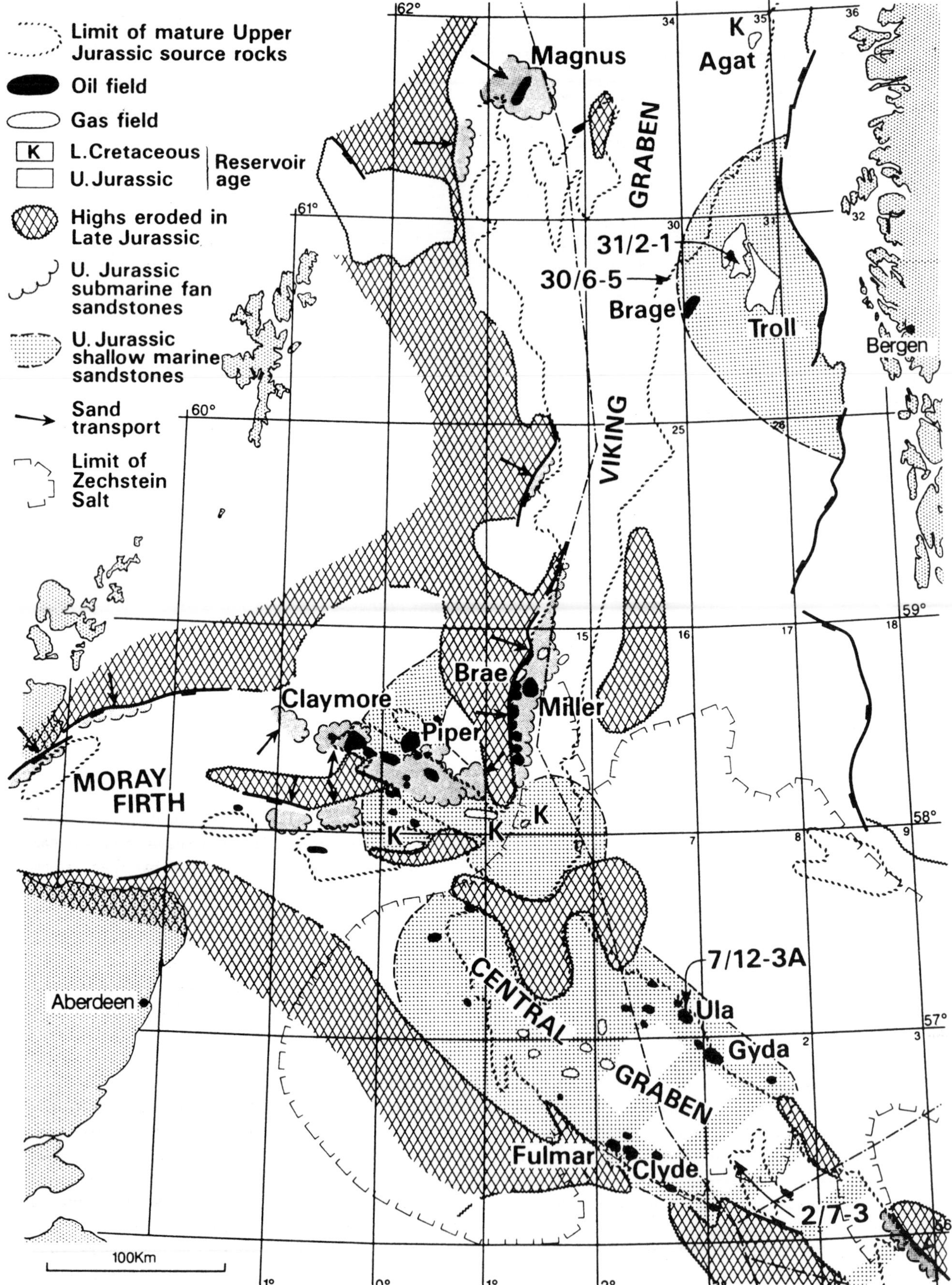

Fig. 12.18 Syn-rift/Upper Jurassic play map (from Pegrum and Spencer, 1990). Note the Late Jurassic eroded highs, which were important sites of sand supply both for the shallow-marine sandstones (e.g. in the Central Graben and Troll Field areas) and for the submarine-fan sandstones (e.g. Moray Firth and southern Viking Graben).

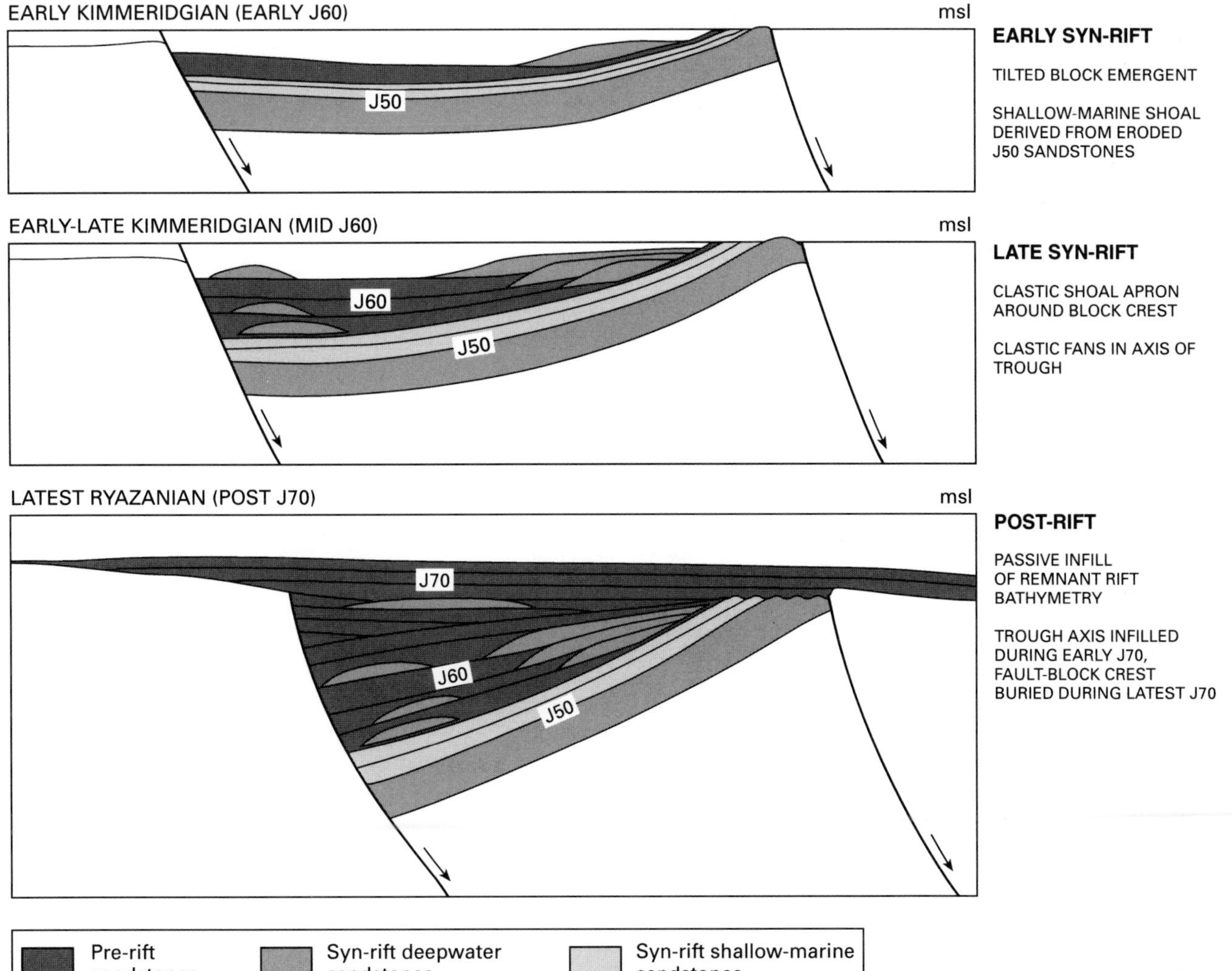

Fig. 12.19 Model of syn-rift reservoirs (early Kimmeridgian to latest Ryazanian) showing the temporal relationships between shallow-marine and submarine-fan sandstones (from Rattey and Hayward, 1993). Note contemporaneous shelf progradation from footwall highs and basin-floor fan deposition in adajcent hangingwalls.

ering in more detail the shallow- and deep-marine sandstone reservoirs.

Shallow-marine sandstones. An important consequence of the rapid rise in relative sea level was the development of an extensive shallow-marine shelf with high-quality sandstone reservoirs deposited around the basin margin, particularly in the Central Graben area (Fig. 12.18), and around prominent fault blocks (Fig. 12.19). Sands were deposited in various shoreline, shoreface and shelf environments, which pass basinwards into a widespread, oxygenated shelf-mudstone facies, with good sealing but poor source-rock potential (e.g. Heather Formation). The latter forms the seal to many Middle Jurassic accumulations. Individual sand bodies are preserved mainly as stacked progradational to aggradational coarsening-upward successions some 200–800 ft (62–240 m) thick. In detail, each succession can often be divided, on the basis of flooding events and, to a lesser extent, sequence boundaries, into a hierarchy of facies successions, which form the basis of most regional- and reservoir-scale subdivisions and correlations. These reflect a combination of eustatic sea-level changes, fluctuations in sediment supply and the history of variations in the creation of local accommodation space. Regionally, these reservoirs form part of a highly diachronous, retrogradational succession, related mainly to the rapid, basin-wide, relative rise in sea level. The onset of this major marine transgression was in the Viking Graben during the Bathonian, which ended deposition of the Brent deltaic system (preserved as the Tarbert Formation in the East Shetland Basin, or J32 of Mitchener *et al.*, 1992). Further south, it coincides with the major early Callovian transgression, which established shallow-marine conditions in the South Viking Graben, Central Graben and Inner Moray Firth areas (see Fig. 12.18) Retrogradational shelf-to-shoreline successions were deposited around the basin margins throughout the Callovian, Oxfordian and Kimmeridgian in the Viking Graben (Rattey and Hayward, 1993), Moray Firth (Stephen *et al.*, 1993), Central Graben (Donovan *et al.*, 1993) and Danish Central Trough (Johannessen and Andsbjerg, 1993). Although these shallow-marine sandstones vary widely in age, their physical characteristics are remarkably similar. The sandstones are commonly structureless, due to extensive bioturbation, but they display a wide range of ichnofacies, which provide the best means of

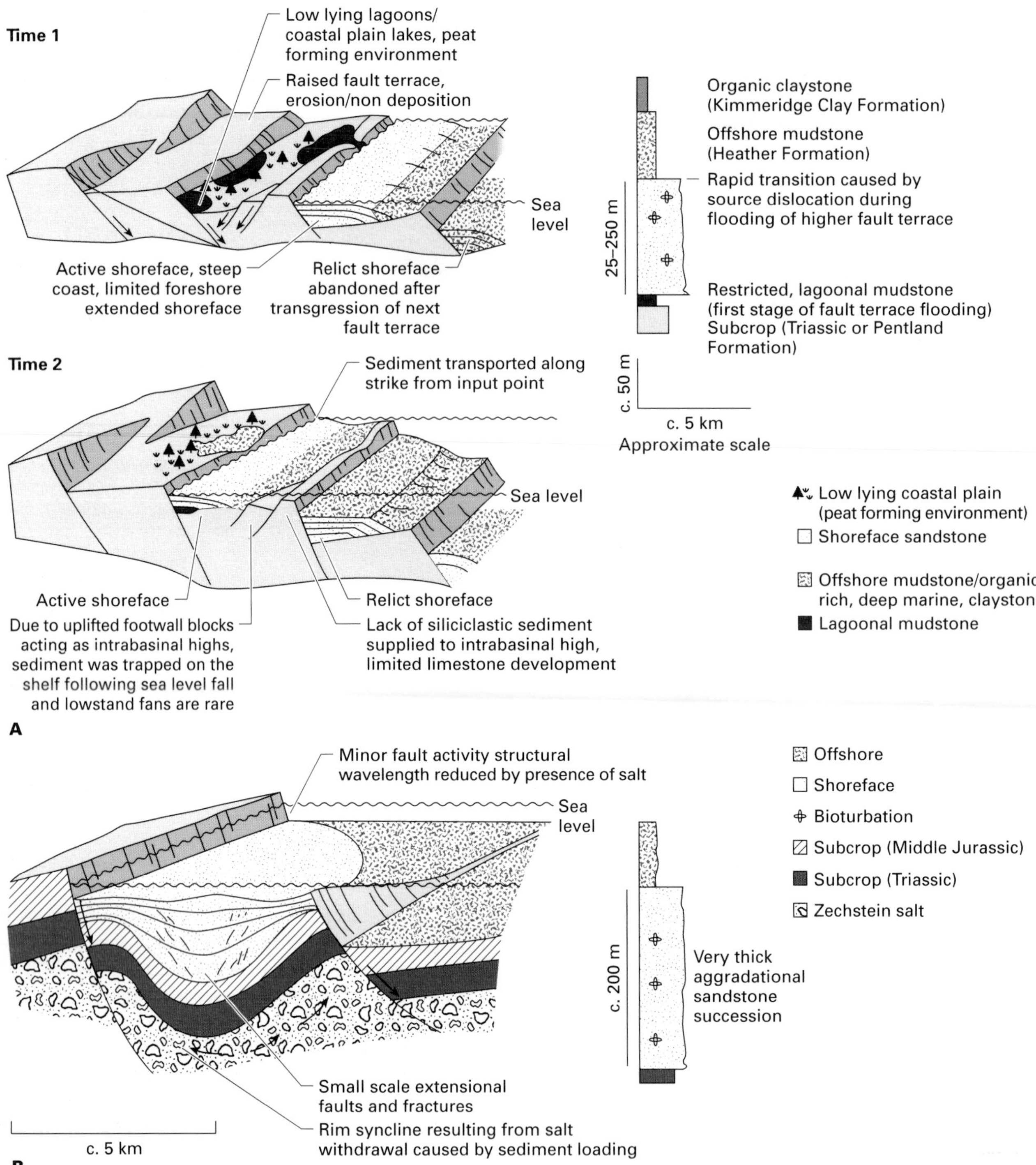

Fig. 12.20 Models of Upper Jurassic shallow-marine sand deposition in the Central Graben–Western Platform area (from Howell *et al*., 1996). (A) retrogradational, back-stepping of shoreface systems across a rifted shelf (cf Times 1 and 2), and (B) development of thick, aggradational shallow-marine sandstone successions, where there is a balance between sand supply and accommodation space, created by a combination of faulting and salt withdrawal (based on the Fulmar Field; Johnson *et al*., 1986).

reconstructing detailed depositional conditions and palaeobathymetry (Martin and Pollard, 1996; Pemberton *et al*., 1992; Donovan *et al*., 1993; Taylor and Gawthorpe, 1993; Gowland, 1996). The physical similarity between these sandstones, combined with their extreme diachroneity, increases the need for an integrated, high-resolution sequence-stratigraphic evaluation to understand their spatial and temporal distribution (Donovan *et al*., 1993; Price *et al*., 1993).

Despite extensive burial, reservoir quality continued to be controlled by primary depositional processes, of which physical reworking (intensity and frequency) and palaeobathymetry were the most important. Hence, the highest-quality sandstones usually occur at the top of the progradational cycles, reflecting deposition mainly in high-energy, shoreface environments. The geometry and preservation of these sands are controlled by (Fig. 12.20): (i) shoreline position and orientation, which are typically fault-related (e.g. Rattey and Hayward, 1993; Howell *et al*., 1996); (ii) accommodation space and relative sea-level history, which reflect a combination of predepositional erosional relief, eustatic sea-level rise, regional thermal subsidence, local fault-block subsidence and, in places, halokinesis and salt withdrawal (e.g. Stewart, 1993; Wakefield *et al.*, 1993) and (iii) post-depositional erosion (e.g. Donovan *et al.*, 1993). It is the latter two factors, rather than regional facies-distribution patterns, which make reservoir prediction particularly difficult in these shallow-marine sandstones. At a reservoir scale, lateral sand-body continuity is normally very good, but there is a decrease in reservoir quality and increase in heterogeneity from the relatively proximal, basin-margin areas to the more distal, lower-energy areas (cf. the proximal to distal characteristics of the Fulmar and Clyde Fields, respectively, in the southwest Central Graben).

Important hydrocarbon accumulations in these shallow-marine sandstones occur mainly in the Outer Moray Firth and Central Graben. In the Outer Moray Firth, these sandstones (e.g. Piper Formation) are of late Oxfordian–Kimmeridgian age and transgressively overlie pre-rift deltaic deposits (Sgiath Formation, Oxfordian). The main fields are: Chanter (Schmitt, 1991), Claymore (Maher and Harker, 1987; Harker *et al.*, 1991), Glamis (Fraser and Tonkin, 1991), Highlander (Whitehead and Pinnock, 1991), Ivanhoe (Parker, 1991), Petronella (Waddams and Clark, 1991), Piper (Williams *et al.*, 1975; Maher, 1980, 1981; Schmitt and Gordon, 1991), Rob Roy (Parker, 1991) and Tartan (Coward *et al.*, 1991). In the Central Graben area, these shallow-marine sandstones transgressively overlie Triassic deposits. The main accumulations occur inside the active graben-boundary faults, such as Angus (Hall, 1992), Clyde (Smith, 1987; Stevens and Wallis, 1991; Turner, 1993), Duncan (Robson, 1991), Fulmar (Johnson *et al.*, 1986; Stockbridge and Gray, 1991; van der Helm *et al.*, 1991); Mjølner (Søderström *et al.*, 1991) and Ula (Bailey *et al.*, 1981; Spencer *et al.*, 1986a; Stewart, 1993). The adjacent Western Platform is represented by a thinner reservoir succession, rapid thickness variations and evidence of some stratigraphic trapping. So far, hydrocarbon accumulations on the Western Platform, such as Gannet West/Guillemot A (Armstrong *et al.*, 1987), Kittiwake (Glennie and Armstrong, 1991), Durward/Dauntless and Mallard have been smaller than those within the Central Graben.

Upper Jurassic shallow-marine reservoirs are rare in the Northern North Sea, although the Emerald Field is a minor exception in the East Shetland Basin (Wheatley *et al.*, 1987; Stewart and Faulkner, 1991). However, a special occurrence of such sands is found in the Troll Field (see Fig. 12.18), a supergiant gas accumulation with a prominent oil rim (1669×10^9 m^3 gas initially in place and 778×10^6 m^3 oil initially in place; Høye *et al.*, 1994). This large, tilted, fault-block structure (Fig. 12.21) is located in the eastern part of the northern North Sea; it was originally identified from its exceptional seismic 'flat spot' (Birtles, 1986). Laterally extensive reservoirs (Sognefjord Formation), up to 400 m thick, formed part of a large, but localized, westward-prograding delta complex during the late Callovian to early Volgian (see Figs 8.38 and 12.18). These poorly consolidated sandstones have been buried to around only 1000 m ss and, consequently, they display exceptionally good, reservoir quality, with porosities up to 34% and permeabilities up to 10 D (Gray, 1987; Osborne and Evans, 1987). In detail, individual reservoir sand bodies display strongly lenticular geometries across the field, as a result of both primary deposition within a tidally influenced shoreline–shelf complex and erosion associated with bounding unconformities and flooding surfaces (Gibbons *et al.*, 1993).

Deep-marine sandstones. Deep-marine sandstones either overlie some of the shallow-marine successions described earlier (e.g. Outer Moray Firth and parts of the Central Graben) or are found as separate occurrences within the Kimmeridge Clay Formation and its equivalents (e.g. South Viking Graben, North Viking Graben and East Shetland Basin). In all cases, the presence of these sand bodies can be related to major rift-related, fault-footwall uplift and erosion, which locally overprinted the regional trend of a rapid rise in relative sea level (see Fig. 8.48). The adjacent deep-water basins were sediment-starved and anoxic, which resulted in the deposition of organic-rich source rocks ('hot shales' of the Kimmeridge Clay Formation; Stow and Atkin, 1987). The timing and location of fan deposition varied throughout the basin, in response to the diachronous rifting process (mainly early Kimmeridgian to Portlandian). The result was thick, high-quality and laterally restricted sand bodies (see Fig. 8.50), which are partially encased in organic-rich muds and juxtaposed updip against major and often sealing fault systems—a near-perfect situation for hydrocarbon migration and entrapment.

The main Upper Jurassic deep-water sandstone play fairways include: (i) the Brae, Tiffany/Toni/Thelma and Miller fan systems in the South Viking Graben; (ii) the Magnus Sands in the East Shetland Basin; (iii) other hangingwall occurrences along the Viking and Central grabens; and (iv) along the edges of horst blocks in the Moray Firth (e.g. Witch Ground Graben and South Halibut Trough). The key play-fairway aspects of these deposits are best illustrated in the South Viking Graben along the Brae Trend, which forms a long (*c.* 100 km) and narrow (*c.* 2–15 km) sand-prone belt along the hangingwall of the western boundary fault. There are two main types of fan: apron-fringe fans and basin-floor fans (Rattey and Hayward, 1993).

The apron-fringe fans (Fig. 12.22; Brae, Toni, Tiffany and Thelma fan systems) are characterized by their small radius (< 10 km) and thick conglomeratic facies, which fine rapidly basinwards through turbiditic sandstones into basinal mudstones (Stow *et al.*, 1982; Stow, 1983; Turner *et al.*, 1987). Some 10 individually point-sourced fans were deposited along this trend during the early Kimmeridgian, with coarse-grained sediment entry points controlled by transfer faults orientated obliquely to the graben faults (Cherry, 1993). Coarse-grained clastic fan sequences range from *c.* 2000 ft to up 5000 ft in thickness (600–1500 m) and form laterally disconnected bodies separated by inter-fan mudstones. This configuration has resulted in several separate oilfields (e.g. North Brae, South Brae, Tiffany, Toni and Thelma). The best reservoir quality occurs in massive sandstones (porosities up to 21% and permeabilities up to 2000 mD), which are located in midfan areas

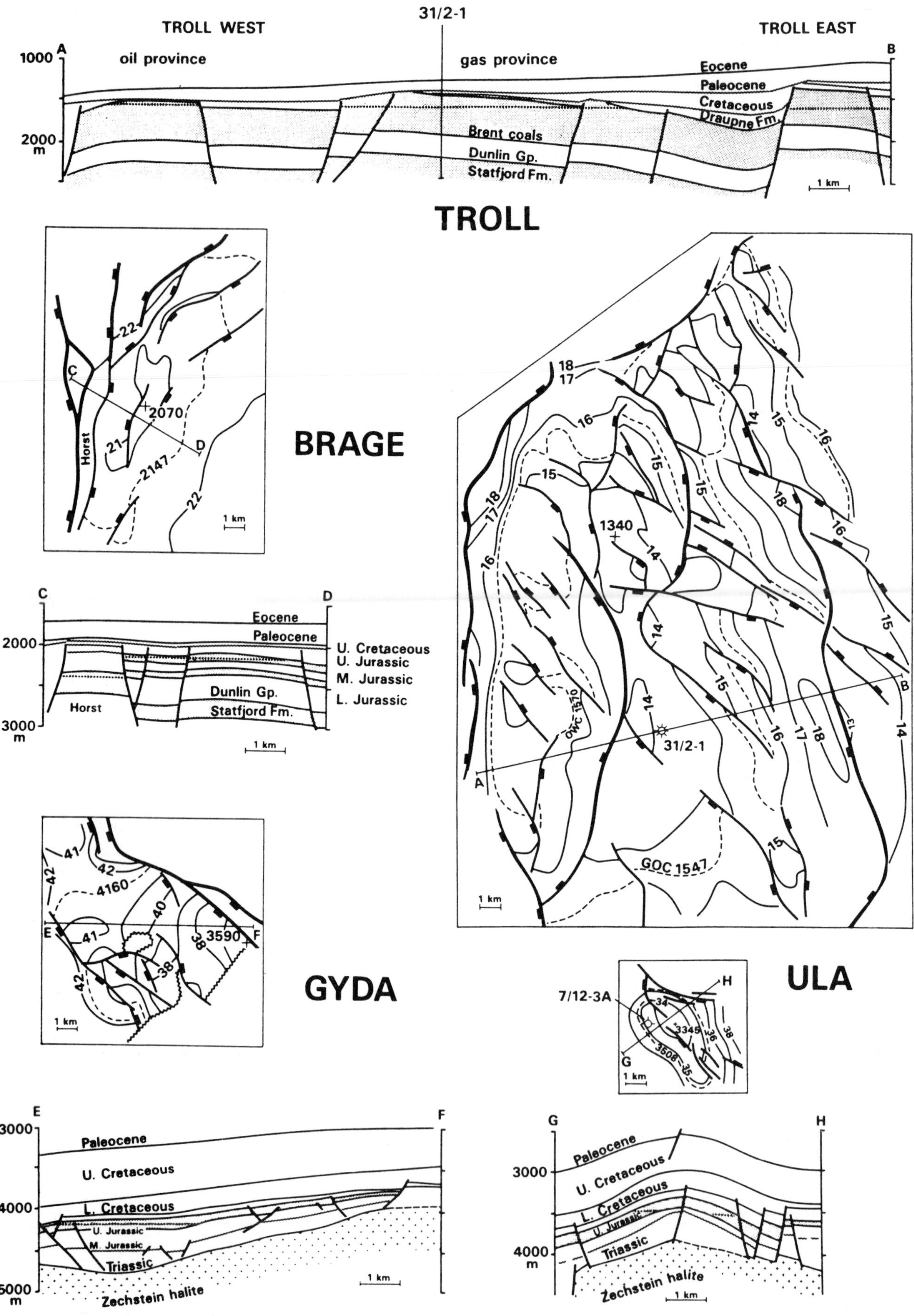

Fig. 12.21 Examples of hydrocarbon accumulations developed in the syn-rift play (from Fjaeran and Spencer, 1991).

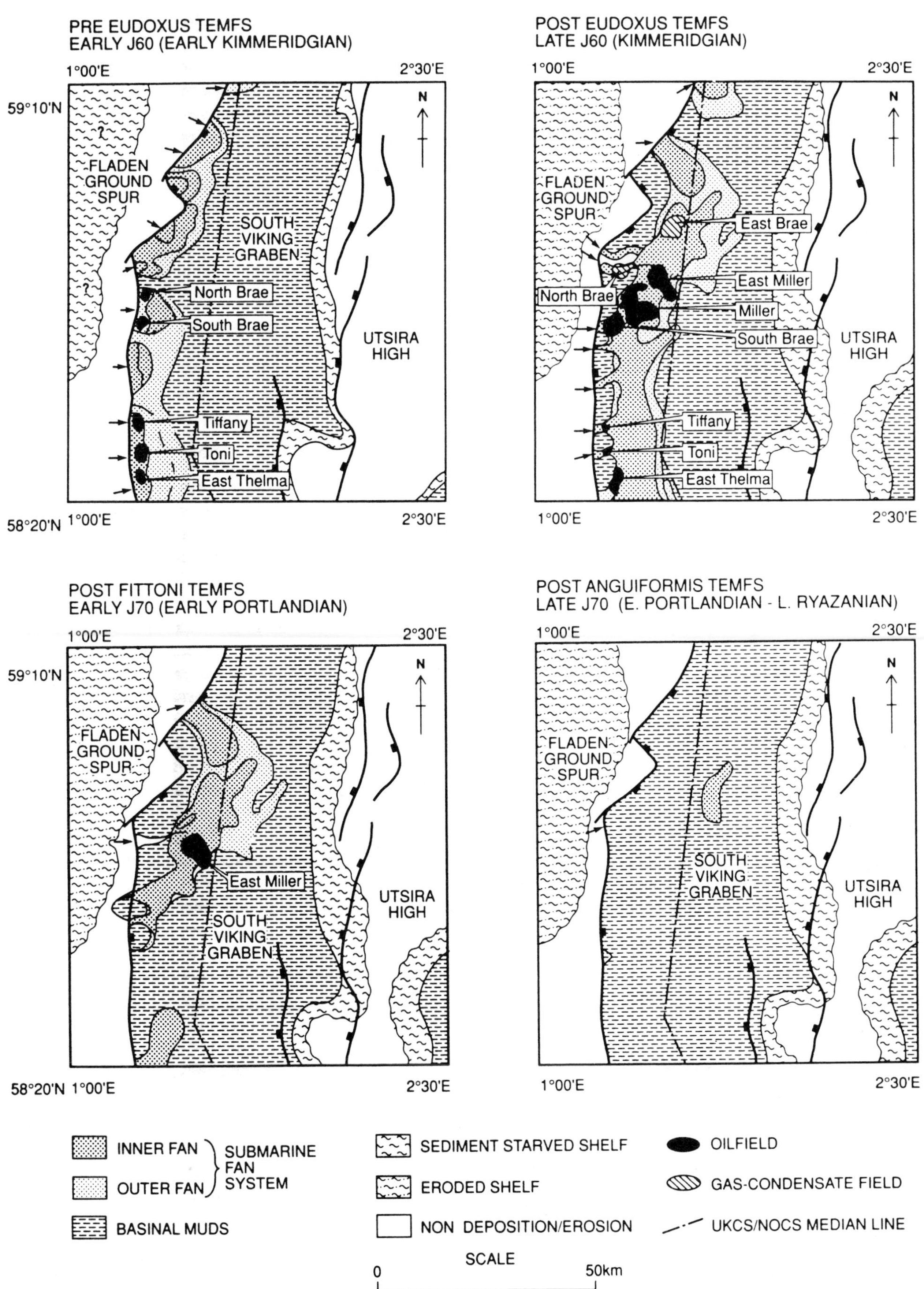

Fig. 12.22 Distribution, geometry and evolution of syn-rift basin-floor fans during the Late Jurassic of the South Viking Graben (from Partington *et al.*, 1993a).

between the proximal conglomerates and the distal fan-fringe areas.

The basin-floor fans (Fig. 12.22; Miller and East Miller (Kingfisher) fan systems) are characterized by the relatively large radius (*c.* 10–15 km), greater basinwards extent, higher proportion of sandstone facies and relatively minor conglomeratic component. Four major basin-floor fans have been identified, which were all deposited after the apron-fringe fans (*c.* late Kimmeridgian). The Miller fan is one of the best-known examples, composed of well-sorted, fine- to medium-grained sandstone, with minor mudstone intercalations. The best reservoirs have porosities of 18–22% and permeabilities up to *c.* 2000 mD and, importantly, display much greater lateral continuity when compared with sandstones within the apron-fringe fans (Garland, 1993). Coarser-grained, proximal equivalents of these sands are found to the west, in the area of North, Central and South Brae. A combination of structural and stratigraphic trapping controls the Miller accumulation (Rooksby, 1991; McClure and Brown, 1992).

The Magnus Sands represent an additional, but somewhat isolated, Upper Jurassic deep-water submarine-fan play in the East Shetland Basin. During the Kimmeridgian, these south-easterly-prograding submarine fans were part of the main rifting phase in this part of the northern North Sea. Within the Magnus Field, the reservoir comprises up to 650 ft (180 m), of good-quality massive sandstone (porosity *c.* 18–24%, permeability *c.* 100–1000 mD), deposited by high-density turbidity currents, with relatively thin interbedded mudstones (De'Ath and Schuyleman, 1981; Shepherd, 1991a). Reservoir quality decreases away from the field, in response both to lateral facies changes along the axis of fan progradation and to depth-related diagenesis. Sands are best developed in basinal areas and are absent from adjacent syndepositional highs. Differential diagenesis, notably illite growth below the oil–water contact, has caused extreme permeability reduction in the aquifer (Heaviside *et al.*, 1983; Pallatt *et al.*, 1984).

Other examples of Upper Jurassic deep-water sands derived from extrabasinal sources occur sporadically along the length of the major graben-bounding faults, such as in the Beryl area (Knutson and Munro, 1991) and adjacent to the Crawford Spur (Yaliz, 1991). Elsewhere, sands found in the hangingwalls of tilted fault blocks were point-sourced from several intrabasinal highs. These locally supplied sands were derived by erosion and slumping of weakly consolidated pre-rift deposits during rift-related fault-block rotation, such as around Statfjord North (Gradijan and Wiik, 1987), around Snorre (Dahl and Solli, 1993) and along the leading edge of the Brent–Statfjord–North Alwyn Trend (Johnson and Eyssautier, 1987; Inglis and Gerard, 1991). This represents a relatively underexplored play, with potential for stratigraphic traps located in the sparsely drilled basinal areas.

Traps. Trap style is highly variable in this play and comprises one or more of the following types of closure: (i) upthrown fault closures, which result mainly from fault-block rotation and are the most successful type of closure (e.g. Magnus, Troll, Piper, Tartan, Claymore, etc.); (ii) downthrown fault closures, which require a lateral fault seal (e.g. Tartan Field, Brae Trend) and, consequently, have a relatively high risk of failure; (iii) anticlinal closures, comprising four-way dip closure, which is common in salt-related areas because of salt withdrawal or halokinesis (e.g. Fulmar and Clyde); (iv) compaction closures, involving a combination of dip closure at top reservoir level and a variable degree of stratigraphic trapping at deeper levels (e.g. Brae Trend); and (v) stratigraphic onlap and subcrop closures occur where the reservoir pinches-out through a combination of erosion and/or depositional limits, together with both lateral and top seals (e.g. Magnus).

Both the Magnus and the Troll fields are located in large, but relatively simple, tilted fault-block structures, analogous to their Middle Jurassic counterparts (see Fig. 12.21). The Magnus trap also involves stratigraphic pinch-out of the submarine-fan sand bodies and reservoir truncation by the Base Cretaceous Unconformity. The top seal is provided both by the Kimmeridge Clay Formation and by Lower Cretaceous marine shales (Cromer Knoll Group).

In contrast, traps within the other occurrences of the deep-water sands play are combined structural–stratigraphic traps, but with a significant stratigraphic-trapping element. The apron-fringe fans along the Brae Trend trap hydrocarbons through a combination of fault seal (against tight Devonian sandstones in the footwall), dip closure at top reservoir level and stratigraphic pinch-out. Hydrocarbon columns considerably greater than structural closure confirm the significant stratigraphic-trapping element. For example, South Brae (see Fig. 8.47) has a maximum hydrocarbon column length of 1670 ft/500 m, of which only 200 ft/60 m is the result of structural closure. Compactional drape over sand-rich basin-floor fans defines the Miller trap, whereas at East Miller (to be developed as the renamed Kingfisher Field) additional structuration is evident (Fig. 8.47). The submarine-fan sandstone traps in the Outer Moray Firth, such as Claymore, also show both stratigraphic and structural elements, with the reservoirs sealed partly by conformably overlying Kimmeridge Clay and partly by truncation and onlap of Lower Cretaceous shales (e.g. Valhall Formation; Harker *et al.*, 1991).

The shallow-marine sandstones of the Central Graben and Western Platform occur in more variable traps, which owe their origin to a combination of halokinesis (Smith, 1987), salt withdrawal (Johnson *et al.*, 1986), gravitational sliding (Gibbs, 1984b) and rift tectonics (e.g. Ula and Gyda fields; see Fig. 12.21, cross-sections E–F and G–H). This highlights a major difference in structural style, and consequently trap size, between the Central and Northern North Sea. This is due to the presence of underlying Zechstein evaporites in the Central North Sea, which forms a major shallow-detachment surface. Although the structural wavelengths of the large tilted fault blocks are similar in the two areas (*c.* 10–40 km), the smaller Central North Sea traps reflect numerous minor shallow detachments and gravitational sliding towards the fault scarps (see Fig. 8.36 and Section 8.5.1). Consequently, traps in the Central North Sea are much smaller (trap width *c.* 2–10 km) compared with the tilted fault-blocks of the Northern North Sea (trap width *c.* 10–40 km) (Rattey and Hayward, 1993). Thus oil-field sizes in the Central North Sea range from 50×10^6 to 500×10^6 stock-tank oil initially in place (STOIIP) and

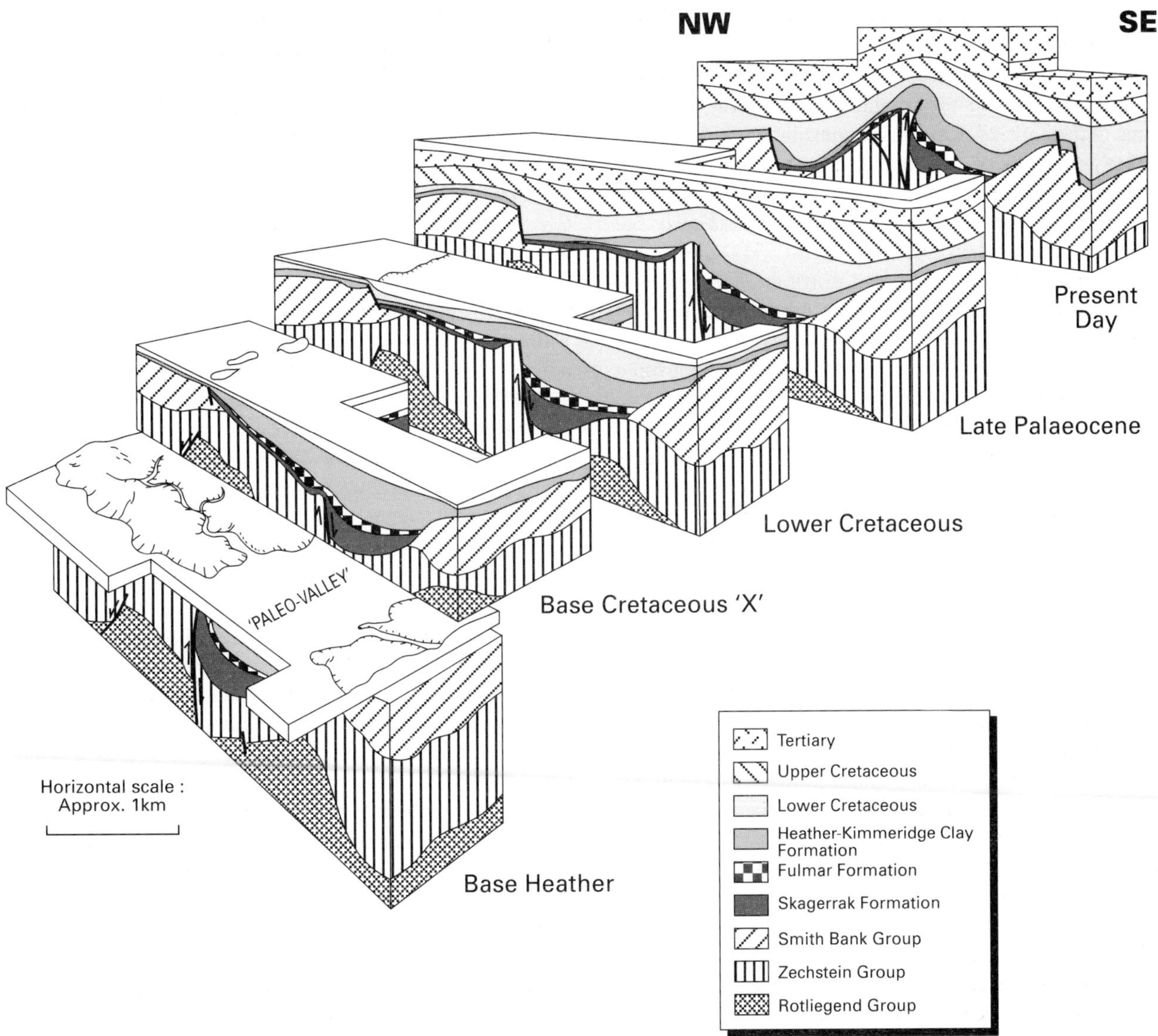

Fig. 12.23 Schematic evolution of the margin of the Central Graben in the Guillemot area (Quadrant 21) showing the initial influence of extensional faulting on Late Jurassic depositional systems (time of Base Heather Formation) and subsequent tectonic inversion during the Late Palaeocene to present day (from Bartholomew *et al.*, 1993).

the billion-barrel (> 1000×10^6 STOIIP) fields of the Northern North Sea (e.g. Brent, Statfjord, Gullfaks and Ninian) are absent.

In the Central North Sea, dip closure at top-reservoir level, partly reflecting the lensoid geometry of the Jurassic–Triassic sediment package, and erosional truncation provide the main trapping mechanism. The top seal is provided mainly by the Kimmeridge Clay Formation and, locally, by the Chalk (e.g. Fulmar Field). Significant, lateral thickness changes, both depositional (erosional topography and salt/fault movements) and postdepositional (erosional truncation), result in some degree of stratigraphic trapping (e.g. Kittiwake; Glennie and Armstrong, 1991). In the Central Graben (e.g. in Fulmar and Clyde), the complex interaction of salt and graben tectonics is difficult to resolve. The lenticular-shaped and heavily faulted Fulmar structure has been interpreted as reflecting mainly withdrawal of Zechstein salt, including the grounding and inversion of the Triassic rim synclines, followed by late structuration due to graben faulting (Johnson *et al.*, 1986). The lensoid sandstone-thickness patterns within the field support syndepositional salt movements (see Fig. 12.20B; 1996; Stockbridge and Gray, 1991; Howell *et al.*, 1996). At Clyde, about 5–10 km further basinwards to the south-east, a combined structural and stratigraphic trap, comprising three-way dip closure and erosional truncation, has been interpreted as resulting from one or more of the following: (i) tectonic inversion caused by halokinetic movements (Smith, 1987); (ii) gravitational sliding (Gibbs, 1984b); and (iii) fault-terrace rotation in response to late Jurassic extension and/or reverse faulting (Stevens and Wallis, 1991). Traps along the Western Platform also show complex interaction between all these processes, including tectonic inversion (Fig. 12.23).

Play-fairway analysis. The controlling parameters of both the shallow-marine and deep-water sandstone plays are reservoir, top seal and hydrocarbon charge. The key success factors of this play comprise: (i) the widespread, although complexly distributed, high-quality sandstone reservoirs; (ii) the overlying and/or interfingering Kimmeridge Clay

Formation, which provides both the top seal and, in graben areas, mature source rock; (iii) short-distance migration routes, generally not more than 8 km up-dip of the mature oil window (Cayley, 1987); and (iv) significant overpressures in the grabens, which have contributed to the preservation of porosity at depths in excess of 13 000 ft/3900 m ss (Gaarenstroom *et al.*, 1993).

The shallow-marine (shelf) sandstone play-fairway has been dominant in the Central North Sea and Outer Moray Firth, because of the widespread distribution and favourable occurrence of reservoir, seal and charge. The reservoir provides a critical risk because of basinwards shale-out of the sand bodies, rapid thickness changes resulting from a variable history of accommodation-space development and complex preservation patterns, including truncation below the Base Cretaceous Unconformity. The play is most successful along graben margins and on adjacent platforms, although long migration routes may downgrade the latter (Cayley, 1987). Major highs (e.g. Forties–Montrose, Jaeren, Auk, Fladen Ground Spur, etc.) suffer from a combination of two main risks, notably a lack of both effective reservoir and top seal.

The deep-water/submarine-fan play occurs throughout the basin, but primarily in the hangingwalls of active fault blocks along the graben margins, which delimit the area of thermally mature Kimmeridge Clay source rock. The sporadic occurrence of this play is linked mainly to reservoir presence, and hence to the entry points of fan systems into the basin. Local play risks include the effectiveness of fault seal and lateral stratigraphic trapping.

Lower Cretaceous play

Within the Central North Sea/Moray Firth area, the Lower Cretaceous play has many similarities to the Upper Jurassic deep-water submarine-fan play but is much more restricted in its extent (Oakman and Partington, Chapter 9, this volume). This reflects the more limited degree of rifting, which by this time was active mainly in the Moray Firth. The main hydrocarbon accumulations of this play within the Moray Firth are as follows: Scapa (McGann *et al.*, 1991; Riley *et al.*, 1991; Harker and Chermak, 1992, see Figs 9.19 and 9.20), Claymore (Harker *et al.*, 1991; see Fig. 9.19), Bosun and Kilda (Guy, 1992), Brittania (see Figs 9.1 and 9.18) and Captain (see Figs 9.1, 9.18 and 9.22). There is also the isolated Agat Field in the Northern North Sea (Gulbrandsen, 1987; see Figs 9.1 and 9.28) and other significant hydrocarbon occurrences in adjacent areas, such as the Southern North Sea and Lower Saxony Basins (see below, 'Lower Cretaceous plays in adjacent areas').

Reservoirs. The abrupt cessation in active rifting and the onset of thermal subsidence during the early Cretaceous resulted in mainly fine-grained suspended sediment forming an onlapping blanket over the inactive syn-rift topography. Hence, over most of the area, Lower Cretaceous sediments form important regional seals (e.g. Cromer Knoll Group, Valhall Formation) for the underlying Jurassic reservoirs (Fig. 12.24). However, slope-apron and basin-floor fans continued to be deposited in basins adjacent to active fault scarps, most notably in the Moray Firth (Fig. 12.24; Boote and Gustav, 1987).

Two main types of deep-water reservoir are recognized: (i) slope-apron and fault-scarp deposits (e.g. Valanginian–Barremian Scapa and Spey sandstones); and (ii) basin-floor fans (e.g. Aptian–Albian Bosun and Kopervik sandstones). The Scapa Sandstone in the Scapa Field comprises a thick (> 1200 ft/360 m), tightly cemented conglomeratic facies, which forms a zone some 200–700 m wide immediately adjacent to the fault scarp and parallel to the edge of the Halibut shelf (see Figs 9.19 and 9.20). This facies passes basinwards through porous turbiditic sandstones, which form a clastic wedge thinning and shaling-out on to the adjacent Claymore tilted fault block (Harker and Chermak, 1992). The pre-existing faulted topography exerted a major influence on fan deposition and sand distribution. The main massive sandstones in this play are up to 600 ft (180 m) thick and display excellent reservoir properties (porosity *c.* 20–30%, permeability *c.* 100–4000 mD), which deteriorate into the more distal facies of the fan fringe.

The Bosun and Kopervik sandstones represent sand-dominated basin-floor fan deposits forming fairways in the Witch Ground Graben/East Forties Basin and North Buchan Graben, respectively. The composition and texture of the Kopervik Sandstones (i.e. fine-grained, well-sorted sandstones with shallow-marine bioclastic debris) indicate derivation from an adjacent high-energy shelf area (Fladen Ground Spur). Facies characteristics support deposition both by high-density turbidity currents and by liquefied flows (debris flows and low-density turbidite flows) on an unstable slope (Guy, 1992). The abundance of dewatering and soft-sediment deformation structures throughout the *c.* 500+ ft (*c.* 180+ m) thick Bosun Sandstone in Blocks 15/28 and 16/28 led Downie and Stedman (1993) to infer deposition mainly by liquefied flows on a gravitationally unstable prograding slope.

Net-to-gross ratios (*c.* 0.7–0.9) and the distribution of reservoir quality within all these sandstone occurrences are controlled by primary depositional processes and by fan geometry. In the Kopervik Sandstones of the Kilda Field (Block 16/26), reservoir quality has been reduced through compaction and depth-related diagenesis, with the best reservoirs displaying porosities of *c.* 15% and permeabilities of *c.* 40 mD, although most sandstones contain some reservoir potential (Guy, 1992).

Traps and play-fairway analysis. This play is controlled primarily by the reservoir distribution within fault-controlled basinal areas (Fig. 12.24). Traps comprise both stratigraphic and combined stratigraphic/structural closures, displaying similar styles and controls to those of the Upper Jurassic deep-water sandstone play (see above).

In the Claymore Field, although trapping is provided by upthrown structural closure, hydrocarbon distribution within the subordinate Lower Cretaceous sandstone reservoir is controlled by reservoir pinch-out within the Valhall marls (Harker *et al.*, 1991; see Fig. 9.14). The Scapa Field is also a combination structural/stratigraphic trap but is located within a syncline (see Fig. 9.20; Green *et al.*, 1991). Trapping is provided by onlap termination of reservoir sands against the Claymore fault block to the north-east, while to the south-west hydrocarbons are trapped by fault

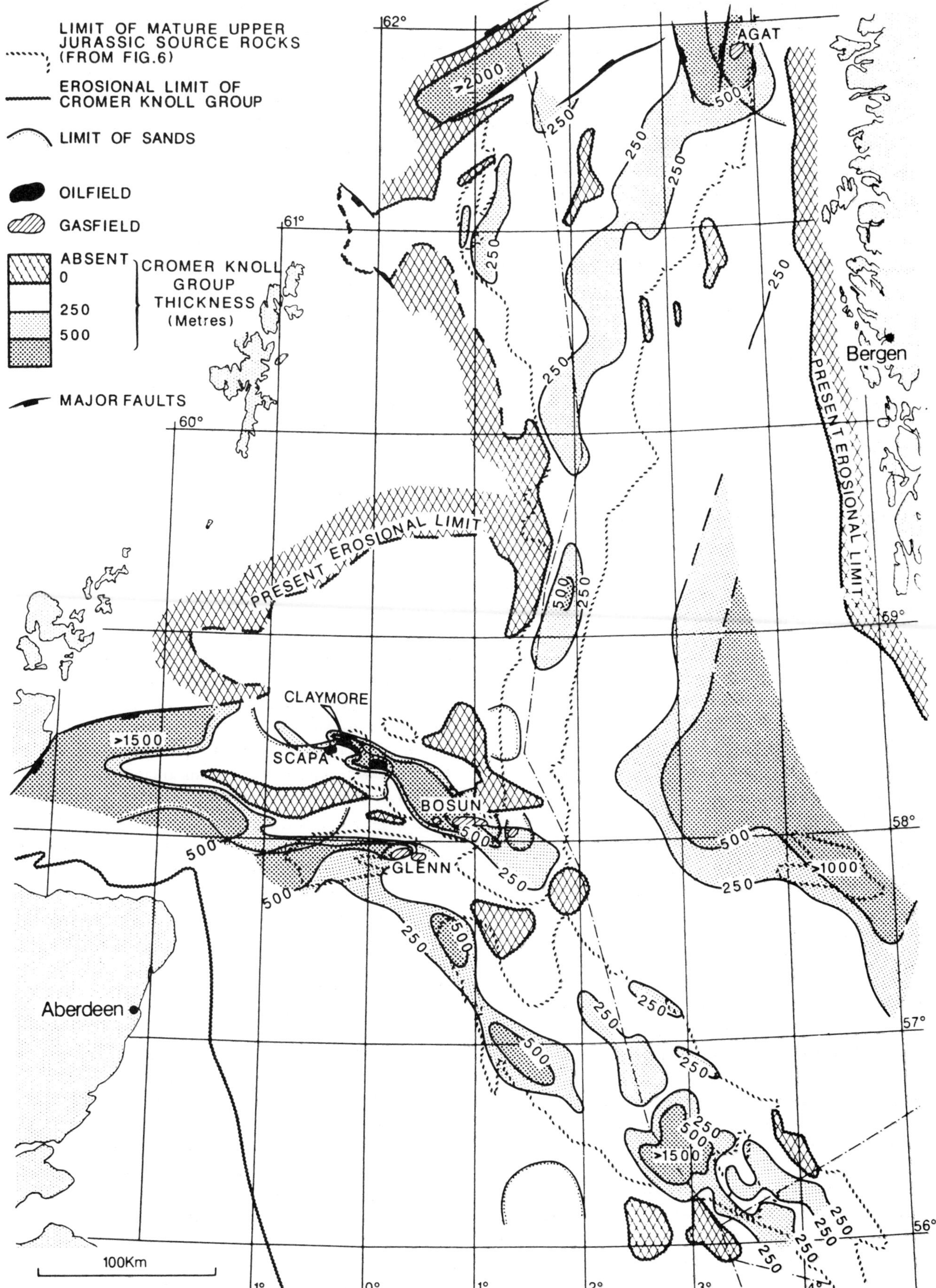

Fig. 12.24 Syn-rift/Lower Cretaceous play map (from Pegrum and Spencer, 1990). Continuing syn-rift submarine-fan deposition is restricted to the Moray Firth area (e.g. along the Claymore–Scapa–Bosun trend).

closure and/or sand pinch-out into tight conglomerates. Hydrocarbon migration was directly updip from the mature Kimmeridge Clay (McGann *et al.*, 1991).

In the Bosun trend, the reservoir overlies mature Kimmeridge Clay source rocks, which are currently generating gas condensate (e.g. Glenn Field tested 53° API condensate at 1915 barrels per day (305 m^3) and gas at 21×10^6 ft^3/day (0.6×10^6 m^3). Complex stratigraphic trapping along this trend is controlled by updip pinch-out of the discontinuous sand bodies.

The Agat Field in the Norwegian sector comprises minor gas accumulations (Fig. 12.24), which are stratigraphically trapped within a westerly-thickening clastic wedge (up to 500 m thick) of Aptian–Albian deep-water sandstones (Gulbrandsen, 1987; see Fig. 9.28).

Lower Cretaceous plays in adjacent areas. A Lower Cretaceous chalk play also exists in the Danish Central Trough, with two commercial oilfields in Valdemar and Adda and relatively common occurrences of hydrocarbon shows (Ineson, 1993; see Section 9.6.7). Lower Cretaceous sandstones are also hydrocarbon-bearing in the Southern North Sea (see Section 12.3.6), Paris Basin, Celtic Sea Basin (see Section 12.4.4) and West Shetland Basin (see Section 12.4.6).

12.2.3 Post-rift plays of the Northern and Central North Sea

Post-rift plays are of two main types (see Fig. 12.5): Upper Cretaceous play and Palaeogene play (Fig. 12.25). Depositional patterns reflect the rapid cessation of rifting and, in response to a widespread reduction in heat flow, the onset of post-rift thermal subsidence, which has dominated the North Sea Basin from the Early Cretaceous to the present day. Subsidence patterns continued to reflect the ongoing influence of the underlying graben systems; maximum subsidence rates and maximum post-rift sediment thicknesses (up to around 4000 m of combined Upper Cretaceous and Tertiary sediments) are located immediately above the axes of the earlier rift basins (e.g. Viking Graben, Moray Firth, Egersund Basin and the Norwegian–Danish Trough). The only departure from this trend is where Late Cretaceous inversion has modified this subsidence pattern (e.g. in the Central Graben).

This period also includes the main phase of Jurassic source-rock maturation and hydrocarbon migration and entrapment (Eocene to present day).

Upper Cretaceous play

The Upper Cretaceous play is represented by the Chalk Group (up to *c.* 1000 m thick), which forms the reservoir in several major oil accumulations, mainly in the Norwegian and Danish sectors of the North Sea (Table 9.1 and Fig. 9.29, Oakman and Partington, Chapter 9, this volume). The play is effective only in a relatively small part of the Central North Sea, where there is a combination of thick, porous and permeable chalk (Late Cretaceous to Danian age), salt-induced structures and an effective early Tertiary top seal (Fig. 12.26).

Major oil accumulations in the Norwegian sector (in Quadrants 1 and 2) include the Ekofisk, Eldfisk, Tor, Albuskjell, Tommeliten, Valhall and Hod fields, which together represent a significant proportion of Norwegian oil reserves (< 20%). In the Danish sector, the Chalk Group is the only current hydrocarbon-producing reservoir, with significant production (*c.* 140 000 barrels of oil per day/ 22 260 m^3/day and 300×10^6 ft^3/day/8.5×10^6 m^3/day) coming from seven main fields: Dan, Gorm, Skjold, Tyra, Kraka, Dagmar and Rolf. Several other Danish Chalk fields are also expected to undergo development, including Valdemar, Adda, Svend, Igor, Harald and Roar (Damtoft *et al.*, 1992; Megson, 1992). The UK Sector contains minor chalk accumulations (e.g. Joanne, Machar and others), but a combination of small volumes, complex geometry and poor reservoir properties has resulted in no commercial hydrocarbon production so far from the Chalk Group.

The dominant aspect of the Upper Cretaceous play is the widespread chalk facies, with its unusual reservoir properties (discussed below). This facies extends throughout the Southern North Sea and the Central Graben, as far north as around 57°N. Further north, the chalk facies gradually gives way to marls and claystones of the widespread Shetland Group (Fig. 12.26), which contains no reservoir rocks. Chalk deposition extended from the Late Cretaceous through to the Palaeocene (Danian).

The key controlling regional parameters for the Chalk play in the Central North Sea are reservoir quality and top-seal effectiveness. Related parameters are overpressures, structures, underlying mature source rocks, early entrapment of hydrocarbons and an absence of overlying Palaeocene sandstones. These aspects restrict the play to a limited part of the Central Graben in the Norwegian and Danish sectors and to a minor extension into Quadrants 22, 29 and 30 in the UK sector (Fig. 12.26).

Reservoirs. Hydrocarbon-bearing Chalk Group reservoirs occur mainly in reworked units within the Tor and Ekofisk formations, but other rare occurrences are also found in the older Hod Formation. Normal pelagic chalk is extremely fine-grained and is usually extensively bioturbated. The very high porosities of up to 70% at the time of deposition are rapidly reduced during burial, initially through compaction-related dewatering (down to *c.* 40–50%) and subsequently during burial-related diagenesis, when porosities of some 5–15% are predicted at depths of *c.* 10 000 ft/3000 m ss (D'Heur, 1986). The accompanying permeabilities of less than 0.1 mD result in these chalks having no reservoir potential, and may even make the Chalk an effective seal (e.g. in Fulmar, Auk and Argyll). Within producing fields, pelagic, or autochthonous, chalk is also tight and commonly forms non-productive intervals, which may compartmentalize the reservoir and even create additional traps (Megson, 1992).

Much better reservoir properties are found within the reworked, or allochthonous, chalk facies, which have higher porosities at equivalent depths of burial than autochthonous chalk (e.g. porosities of 20–30% in allochthonous chalk, compared with porosities of less than 10% in autochthonous chalk at the same depth of burial (Watts *et al.*, 1980; Taylor and Lapré, 1987)). Allochthonous chalks were

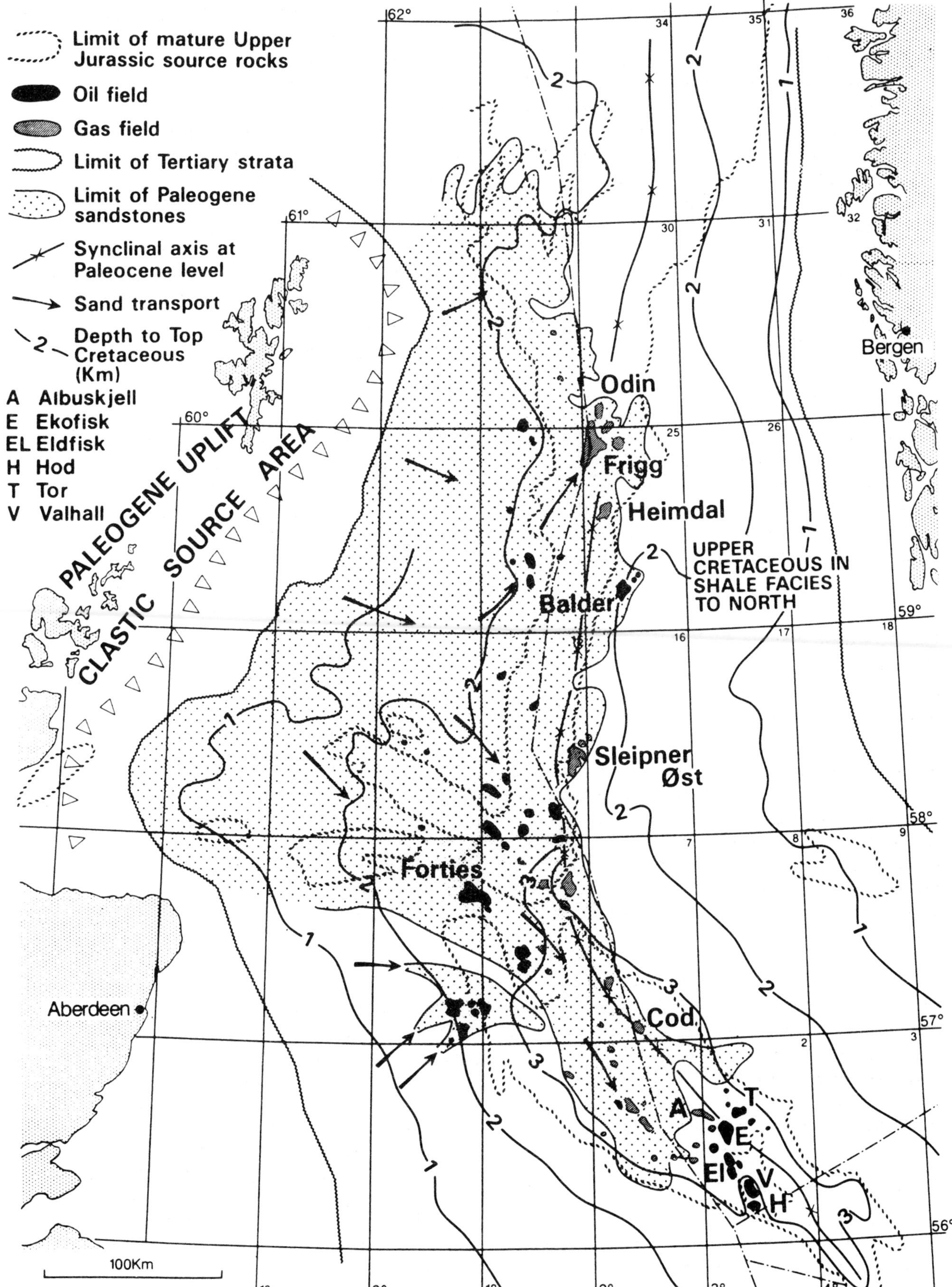

Fig. 12.25 Post-rift play map of the Northern and Central North Sea (from Fjaeran and Spencer, 1991). Note the restriction of Upper Cretaceous–Lower Tertiary chalk reservoir accumulations to the south-eastern part of the Central Graben (denoted by letters). All other fields are in Tertiary sandstone reservoirs.

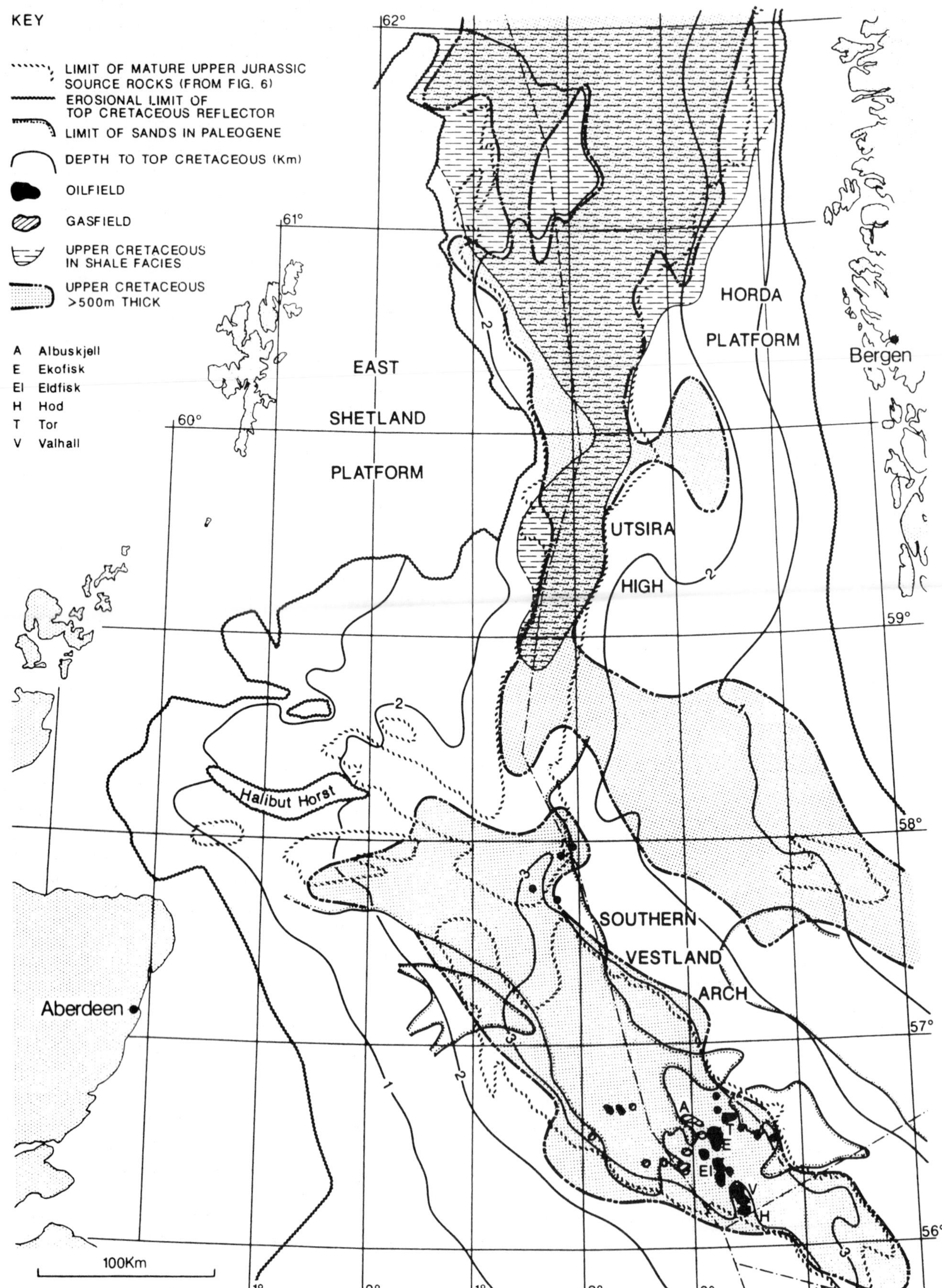

Fig. 12.26 Post-rift/Upper Cretaceous (Chalk) play map (from Pegrum and Spencer, 1990). Note the close relationship between chalk accumulations and the south-eastern limit of Tertiary submarine-fan sandstones and the northward change from chalk facies (Central Graben and adjacent areas) to mudstone facies (Viking Graben).

deposited by a range of gravity-flow processes, including debris flows, slumps and turbidity currents (Schatzinger *et al.*, 1985; Hatton, 1986). These resedimentation processes caused the destruction of early diagenetic cements and deposited a much more poorly packed sediment, which had higher porosities than the original autochthonous pelagic chalk (Taylor and Lapré, 1987; see Fig. 9.37). The redeposited chalks form units up to 250 m thick and may comprise up to 50% of the total Chalk Group thickness. The best-developed areas of thick, redeposited chalk are adjacent to oversteepened late Cretaceous slopes, particularly above rising salt-induced structures, over inversion features and around the graben margins (Kennedy, 1987).

Preservation of the higher depositional porosities at depth is also a function of the subsequent burial, overpressure and hydrocarbon migration history (Scholle, 1977; Hardman, 1982). The retention of high porosity is widely believed to be enhanced by two main processes: (i) the development of overpressures, which minimize physical compaction (e.g. Munns, 1985); and (ii) early hydrocarbon migration into pore space (at *c.* 1000–1500 m depth of burial during the Eocene to early Miocene) to retard chemical compaction and pressure dissolution processes, seen in the form of stylolites and clay flasers (Taylor and Lapré, 1987). In some respects, porosity distribution within these chalks resembles that of clastic reservoirs; heterogeneities developed at the time of deposition have been preserved and amplified during burial history.

The commonly observed decrease in porosity down the flank of chalk structures is also believed to reflect progressively later hydrocarbon emplacement and overpressure support (D'Heur, 1984a). In such accumulations, porosity within the water-bearing zone is negligible and significant reservoir quality exists only on structural crests. However, downflank potential for oil accumulation may occur where palaeostructural crests were bypassed by porous redeposited chalk units (D'Heur, 1986).

Where matrix reservoir properties are too low to achieve effective production from chalk reservoirs, the presence of natural, open fractures is necessary to sustain commercial production. Alternatively, hydraulically induced fractures may help stimulate production. However, many chalk reservoirs display natural fractures, which can result in dual porosity/permeability systems with a significant increase in effective permeability. In the Machar Field (UK Central Graben) the pelagic chalk matrix has a primary porosity of 12–35% and permeability of less than 1 mD (Foster and Rattey, 1993). Secondary fracture porosity may have contributed a modest additional 1% to total pore space, but production tests across open-fractured intervals indicate that effective permeabilities are greater than 1000 mD. In such chalk reservoirs, knowledge of the open-fracture system is the key to a full appreciation of net pay distribution and to effective hydrocarbon drainage, which may be enhanced by using horizontal wells and/or hydraulic fracturing (Damgaard *et al.*, 1989; Conlin *et al.*, 1990; Fjeldgaard, 1990; Fine *et al.*, 1993).

Traps. All current commercial hydrocarbon accumulations within the Chalk are structural traps (e.g. Fig. 12.27). Stratigraphic aspects of the Chalk control mainly reservoir-scale features, such as intra-reservoir heterogeneity, permeability variations and fluid saturation distribution, which reflect differences in the capillary-pressure properties of the chalk matrix (Watts, 1987). However, there is growing evidence from existing chalk fields and recent discoveries to support non-structural trapping mechanisms (Megson, 1992), such as trapping by intrachalk seals (e.g. within the Adda Field), hydrodynamic processes (e.g. tilted oil–water contacts; Jørgensen and Andersen, 1991) and diagenetic variations (Wilson, 1977; D'Heur, 1984a).

The structural traps are mainly four-way dip closures, resulting from several mechanisms, particularly salt diapirism (buoyancy or piercement processes; e.g. Skjold, Nils, Dagmar, Machar), doming over salt swells (e.g. Dan, Kraka, Ekofisk, Albuskjell) and late Cretaceous/Miocene structural inversion (e.g. Tyra, Roar; Doyle and Conlin, 1990). Differentiating between these processes may be difficult, since to some extent they were simultaneous. Three-dimensional seismic data are beginning to highlight the importance of local structural complexities, particularly fault and fracture patterns, and complex variations in the seismic expression of chalk reservoirs (Megson, 1992).

Play-fairway analysis. Despite chalk reservoirs containing the earliest oil discoveries in the North Sea (oil and gas were discovered at Anne in Danish waters in 1966 and oil was discovered at Valhall in Norwegian waters in 1967; Brennand *et al.*, Chapter 1, this volume), this play has proved to be of very limited extent. All the Chalk fields are overpressured by some 2000–2500 pounds per square inch (psi), with porosities of 40–50% preserved in fields such as Ekofisk and Valhall (Hardman and Eynon, 1977; Scholle, 1977; Munns, 1985; D'Heur, 1986). Retention of these overpressures had a major control on the success of this play and emphasizes the importance of an effective top seal.

The two main controlling regional parameters are reservoir and top seal. An important secondary factor is the timing of hydrocarbon migration and its subsequent retention. The main factor limiting the areal extent of the Chalk play is the distribution of Palaeocene deep-water sandstones within the Central Graben; there are no major Chalk fields underlying a sand-rich Lower Tertiary succession. The presence of sands within the background Palaeocene claystone succession rapidly reduces the effectiveness of the top seal, which has three major implications for the Chalk play: (i) hydrocarbons are able to continue along their vertical migration path through the Chalk and into the Tertiary sandstone-bearing succession; (ii) overpressures are unable to develop and hence porosity retention is reduced within the chalk reservoirs; and (iii) absence of an early hydrocarbon pore fill fails to retard burial-related diagenesis. Hence, the south-eastern, distal extent of the Palaeocene deep-water fans forms a major boundary to the Chalk play (Fig. 12.28).

Since early hydrocarbon emplacement is a critical factor in porosity preservation, the relationship between the timing of trap formation and hydrocarbon migration is critical in the analysis of this play at both regional and prospect levels. In the Central North Sea, oil generation probably started during the Late Cretaceous, and significant amounts of oil were being generated during the

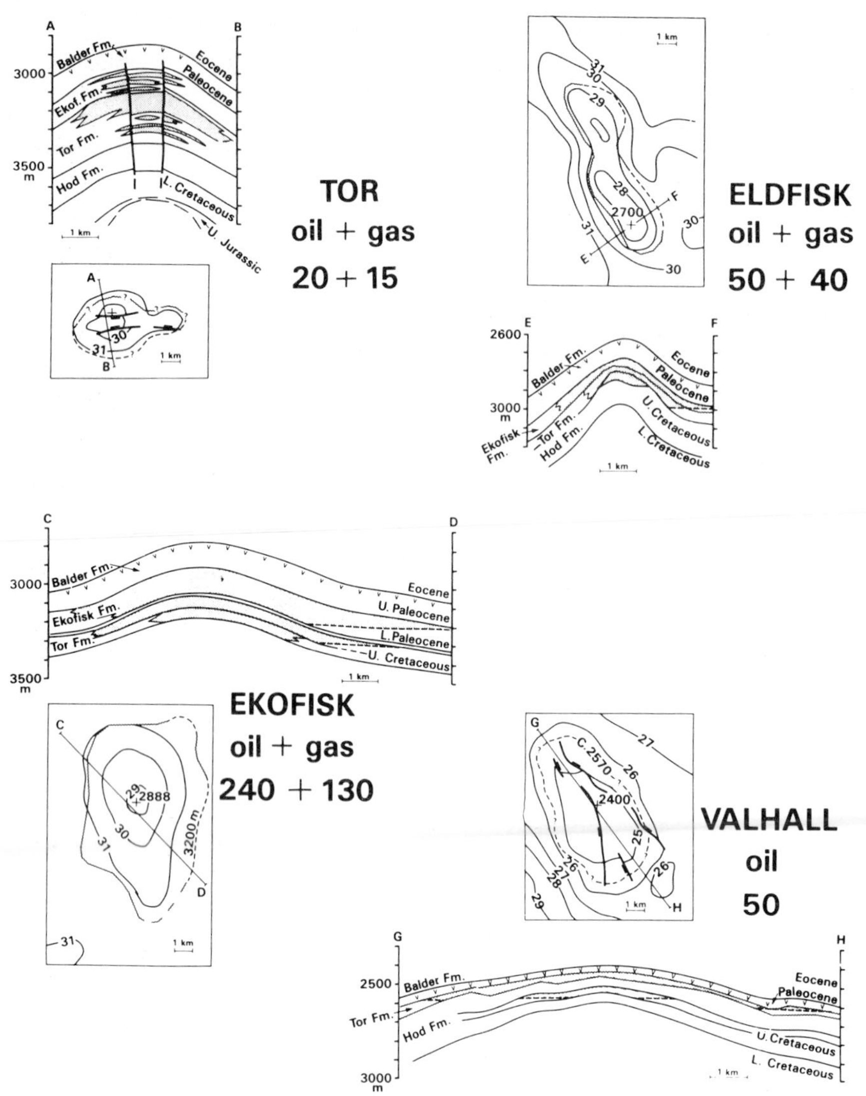

Fig. 12.27 Examples of hydrocarbon accumulations in the post-rift/Upper Cretaceous (Chalk) play in the Central North Sea (from Fjaeran and Spencer, 1991).

Palaeocene from the thick, underlying and thermally mature Upper Jurassic source rocks (see Cornford, Chapter 11, this volume). Hydrocarbon migration was essentially vertical, possibly enhanced by the simultaneous vertical movement of salt and the onset of inversion structures, both of which are known to have created extensive microfracturing within adjacent and overlying successions. Direct evidence of vertical hydrocarbon migration is well displayed in the form of gas chimneys, which extend from the top of several Chalk fields (e.g. Ekofisk) through the overlying Tertiary claystone succession to the seabed. The formation of Chalk-trap configurations must have started immediately after burial in order to have been in place for the time of maximum hydrocarbon generation. Upper Jurassic source rocks are currently at their maximum depth of burial, and vertical hydrocarbon migration is continuing today, albeit very slowly.

Most of the dip closures within the proved area of the Chalk play have now been drilled. The most likely remaining exploration potential remains within the proved play area but alternative trapping configurations need to be considered. For example, there may be hydrocarbon potential within a structural closure but downflank from tight reservoir. This may happen where the structure was already present at the time of deposition, and hence porous redeposited chalk units may be located on the flanks, away from structural culminations. Alternatively, oil accumulations may lie outside present-day structural closure but within palaeostructural closure. The close interaction between gravity-driven sedimentation and structural evolution provides a mechanism for stratigraphic trapping, with redeposited chalk units pinching out and onlapping against evolving structures.

Palaeogene play

The Palaeogene (Palaeocene to Oligocene) play is dominated by both oil and gas accumulations, mainly in Palaeocene and Eocene deep-water sandstones supplied from a westerly source area (Fig. 12.28). Both structural and stratigraphic traps are important. Stratigraphic aspects also have a dominant influence on reservoir characteristics (sandbody thickness, geometry and orientation), hydrocarbon distribution, trap definition and fluid-migration pathways (Fig. 12.29). Of all the North Sea plays, the Palaeogene has remained highly prospective throughout the basin's exploration history and, more recently, has benefited from 3D seismic (e.g. den Hartog Jager *et al.*, 1993; Jenssen *et al.*, 1993; Newton and Flanagan, 1993), improved seismic acquisition and processing, amplitude versus offset analysis

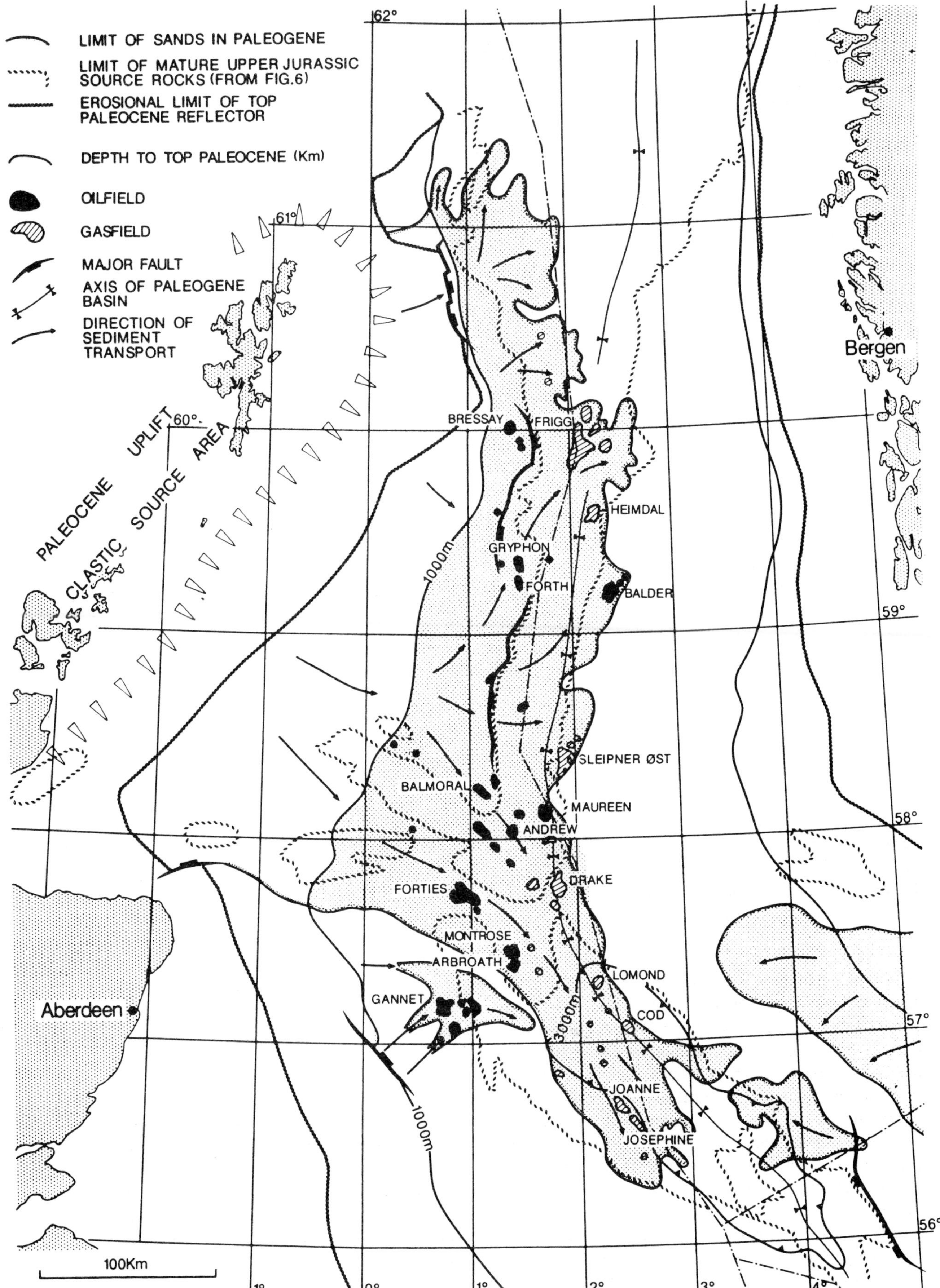

Fig. 12.28 Post-rift/Palaeogene play map (from Pegrum and Spencer, 1990). Note the greater degree of lateral migration compared with the other plays, indicated by accumulations often extending beyond the limits of mature Upper Jurassic source rocks.

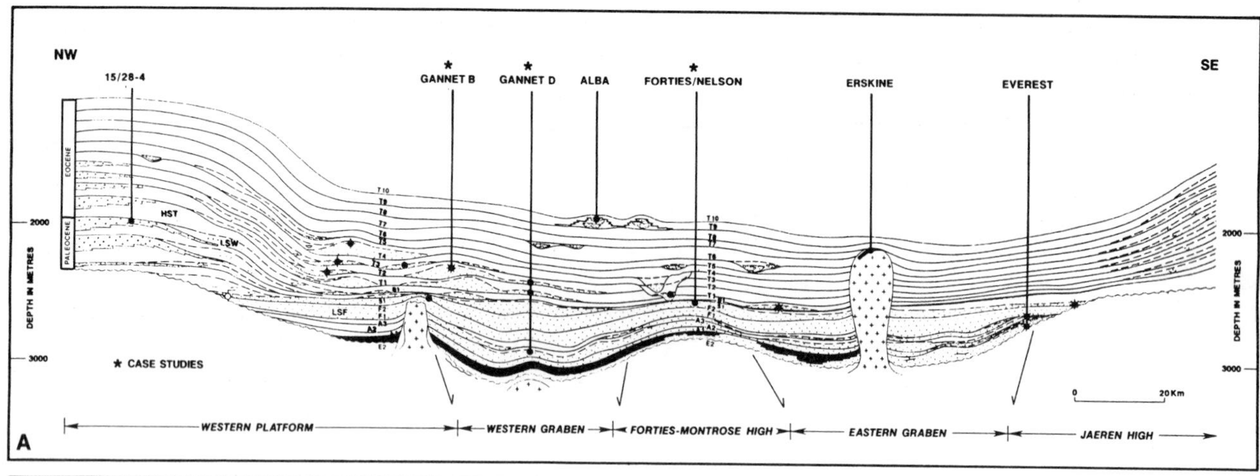

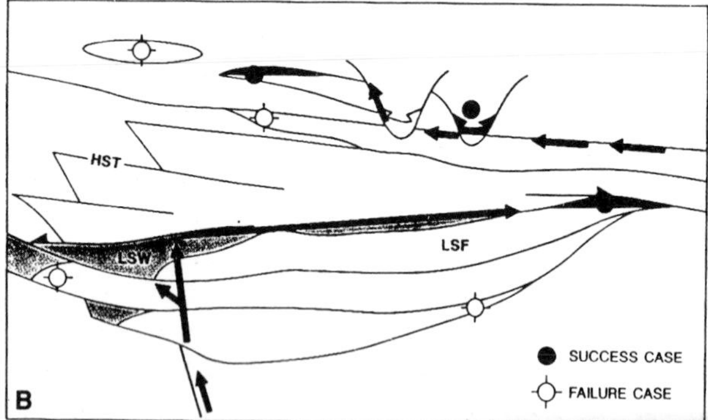

Fig. 12.29 (A) Schematic Central North Sea sequence stratigraphic cross-section through the Palaeocene and Eocene, indicating examples of different trapping types from the proximal base-of-slope area (Western Platform), through the Central Graben area and on to the distal pinch-out area (Jaeren High). (B) Influence of sand-body connectivity/reservoir architecture on hydrocarbon migration and trapping in the Palaeogene play. (From Vining *et al*., 1993.)

(Crook *et al.*, 1993), the application of high-resolution stratigraphic concepts (sequence stratigraphy and very-high-resolution biostratigraphy, which approaches a quarter of a million years for certain parts of the Lower Palaeogene section; e.g. Vining *et al.*, 1993), mineralogical studies and an improved understanding of deep-water submarine-fan sedimentation models. Hence, although the Palaeogene play comprised some of the earliest oil discoveries (Cod, in the Norwegian sector, was discovered in 1968, and Arbroath, in the UK sector, was discovered in 1969; Brennand *et al.*, 1990), this play has provided continued exploration success throughout the 1970s and 1980s both in the North Sea Basin (Bain, 1993) and, more recently, along the Atlantic Margin in the Faeroe Basin (see Bowman, Chapter 10, this volume).

Although linked with the Upper Cretaceous, as one of the two major post-rift plays, the Palaeogene marked a major change in depositional and tectonic conditions. Depositional patterns within the North Sea Basin continued to reflect thermal subsidence over the underlying Mesozoic rift basins, but active rifting and initial opening of the North Atlantic Ocean resulted in thermal uplift of northern Scotland and the East Shetland Platform (see Glennie and Underhill, Chapter 2, this volume; Morton *et al.*, 1993). This was the main event that converted the North Sea and Faeroe basins from deep-water, sediment-starved areas in the Late Cretaceous into sites of coarse siliciclastic sedimentation dominated by deltaic and submarine-fan sedimentation in the Early Palaeogene (Anderton, 1993; den Hartog Jager *et al.*, 1993; Knott *et al.*, 1993). Up to around 3500 m of sediment was deposited during the Cenozoic, with maximum thicknesses and burial depths located over the Central Graben (Ziegler, 1990a).

This play is controlled by three main regional parameters: reservoir, top seal and hydrocarbon charge. The stratigraphic and depositional evolution of the Palaeogene controls both reservoir distribution and top-seal characteristics. The high-resolution sequence-stratigraphic framework of the North Sea Basin enables several (*c.* 10–15), separate, depositional units to be recognized and mapped by means of integrated analysis of well and seismic data (Stewart, 1987; Armentrout *et al.*, 1993; Galloway *et al.*, 1993; Vining *et al.*, 1993). Such an analysis provides the basis for dividing the play into several sequences, which effectively form a series of subplays (Fig. 12.29; see Bowman, Chapter 10, this volume), each requiring separate play-fairway evaluation. This approach is essential in view of the marked spatial and temporal variation in submarine-fan characteristics throughout the Early Palaeogene (e.g. den Hartog Jager *et al.*, 1993). Hydrocarbon charge has been mainly through vertical migration from thermally mature Jurassic source rocks, but the excellent reservoir properties of the Palaeogene sandstones have permitted relatively extensive and complex lateral secondary migration.

Reservoirs. Hydrocarbons are concentrated within Paleocene to Eocene, deep-water, submarine-fan sand bodies, whose distribution closely reflects the gross depositional setting of the different submarine-fan systems (Enjolras *et al.*, 1987). The majority of individual fans were derived from the west and shale-out basinwards towards the east and south-east (Figs 12.28 and 12.29; Morton *et al.*, 1993), although there is evidence of localized supply from the east (e.g. recent Tertiary oil-bearing sands in the Danish sector of the North Sea; see Figs 12.1 and 2.36). In more detail, the succession comprises a series of gently inclined, easterly-dipping, erosionally-bounded depositional sequences of deltaic, slope and basinal sediments, which prograded eastwards across, and to some extent along, the axis of the basin. Sand-rich, blanket-type basin-floor fans dominate the Palaeocene (e.g. Maureen, Andrew and Forties; T20–T40 sequences of Bowman, Chapter 10, this volume). These were gradually replaced in the late Palaeocene-early Eocene by smaller fan systems (e.g. Sele and Balder fans/T45–T50 and Frigg and Tay fans/T70 sequences of Bowman, Chapter 10, this volume), which pass upwards into Middle Eocene mud-rich slope fans, characterized by narrow, elongate, channel-fill sand bodies (e.g. Alba fan system; T92–T98 sequences of Bowman, Chapter 10, this volume). The nature of the individual fan system controls both the thickness and geometry of the reservoir and, in some cases, the type of hydrocarbon-trapping mechanism (Fig. 12.29). Most fan systems become more distal towards the east and south-east, with a gradual reduction in the amount and thickness of sandstones, before passing into basinal mudstones.

In detail, the nature and distribution of North Sea submarine-fan reservoir lithologies are influenced to varying degrees by the following: (i) sub-environments within the fans (cf. channel sand bodies with fan-fringe, levee and mouth-bar deposits; Carman and Young, 1981; Timbrell, 1993); (ii) lateral changes within individual fan systems, such as in the Nelson Field (Whyatt *et al.*, 1992) and Balder Field (Skjold, 1980; Jenssen *et al.*, 1993); (iii) proximal to distal relationships; (iv) channel-stacking patterns (e.g. Quadrant 9, Fig. 12.30; UK sector; Timbrell, 1993); (v) spatial and temporal changes in fan morphology (cf. the sand-rich Palaeocene fans with the mud-rich late Palaeocene–Eocene systems, Fig. 12.29; den Hartog Jager *et al.*, 1993); (vi) pinch-out against intrabasinal highs (e.g. Everest trend, Fig. 12.31; O'Connor and Walker, 1993); (vii) relationship between fan sedimentation and local seabed topography, including that resulting from salt movement (e.g. Gannet/Guillemot complex, Fig. 12.29; Armstrong *et al.*, 1987; Banner *et al.*, 1992; Hodgson *et al.*, 1992); and (viii) postdepositional soft-sediment deformation and dewatering (e.g. Alba, Gryphon, Harding/Forth and Balder fields; Alexander *et al.*, 1993; Jenssen *et al.*, 1993; Newman *et al.*, 1993; Dixon *et al.*, 1995). This variability in depositional style results in a wide range of reservoir characteristics, particularly in terms of sand-body thickness, geometry, orientation and architecture (e.g. Banner *et al.*, 1992; Timbrell, 1993).

Reservoir quality is usually very good, partly because the sands are mineralogically and texturally mature and partly since depth-related diagenesis is relatively minor; burial depths increase from zero around the basin margins to a maximum of 3500 m in the Central North Sea. These sandstones are typically poorly cemented to unconsolidated, with cementation generally occurring after hydrocarbon emplacement (Aplin *et al.*, 1993).

The best and most thickly developed reservoirs were deposited by high-density turbidity currents within submarine-fan channels and within the proximal parts of fan lobes. Sand bodies in these latter environments commonly stack together to form massive reservoir units, up to several hundred feet in thickness (*c.* 200–600 ft/60–180 m). Porosities are commonly between 20 and 35% and permeabilities in the 100s to 1000s of millidarcies range. The main source of reservoir heterogeneity and reduction in reservoir quality results from shale-layer intercalations. In the proximal fan environments described earlier, net-to-gross ratios of *c.* 0.9 are common and shale layers are typically thin (< 2 ft), laterally discontinuous and of minor impact on reservoir quality and production performance (e.g. Tay Sands; Armstrong *et al.*, 1987; Charlie Sand in the Forties Fan/Forties Field: Carman and Young, 1981). The lowest reservoir quality occurs in distal fan environments and within interfan/interchannel areas. Here, net-to-gross ratios are less than *c.* 0.5, individual sand-bed thicknesses are less than 1 ft/0.3 m (thin-bedded turbidites) and these are separated by well-developed, laterally extensive muddy turbidites and hemipelagic claystones, which dramatically reduce both effective horizontal permeability and, especially, vertical permeability (e.g. the interchannel facies in the Forties and Nelson fields; Carman and Young, 1981,

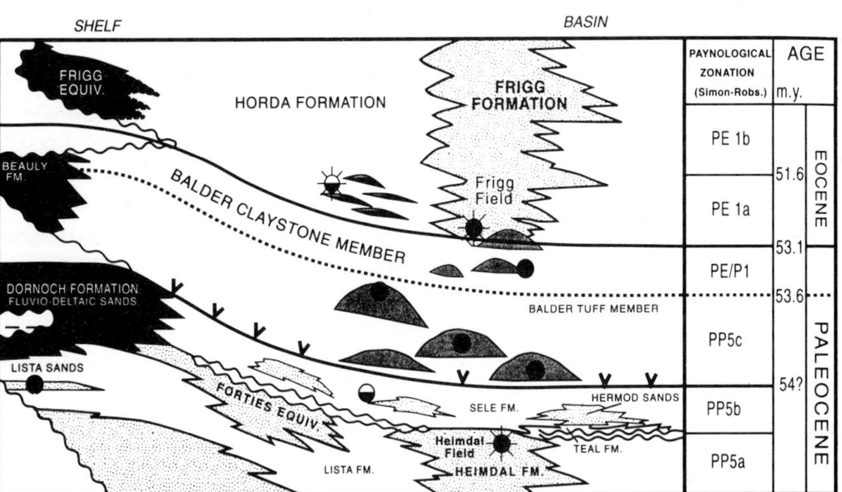

Fig. 12.30 Stratigraphic relationships and reservoir architecture across the South Viking Graben basin margin in the Palaeocene/Eocene (from Timbrell, 1993).

Whyatt *et al.*, 1992). Some of the most extreme reservoir complexity arises from dewatering processes, which can mobilize massive sands at deeper levels and inject them into overlying claystones (e.g. Harding (formerly Forth), Alexander *et al.*, 1992; Balder, Jenssen *et al.*, 1993). Sand distribution is chaotic and largely unpredictable, and may be associated with closely spaced polygonal fault patterns, collapse features and diapiric structures, seen on 3D seismic sections and horizon slices (Jenssen *et al.*, 1993; Lonergan *et al.*, 1998). The latter also highlight the potential importance of subseismic faulting (throws < 30 m) in breaking up otherwise continuous reservoir sands (Cartwright, 1994).

Traps. There are two main successful trapping mechanisms within the Palaeogene sandstone play (Table 12.3; see p. 507): structural and stratigraphic (Bain, 1993). These may occur independently or in various combinations.

Structural traps are most commonly associated with extensive Palaeocene basin-floor fans above pre-Tertiary structural highs, which were reactivated during early Tertiary inversion to form relatively simple, low-relief anticlines, with four-way dip closure. The most prospective interval has proved to be the youngest of these basin-floor fans, the Forties Fan sequence, which is immediately overlain by a major regional seal (Balder/Sele formations). Good examples are fields along the Forties–Montrose High, including Forties (Hill and Wood, 1980; Wills, 1991), Nelson (Whyatt *et al.*, 1992), Montrose (Fowler, 1975) and Arbroath (Crawford *et al.*, 1991) (see Fig. 12.29). Secondary relief over these structures also reflects additional stratigraphically-related, differential compaction (see below). A second group of structural traps is associated with the movement of Zechstein salt within the area of the Central Graben (see Fig. 2.29). This includes: (i) traps against the flanks of piercing salt diapirs (e.g. Gannet North/Gannet C and Gannet Central/Gannet C fields: Armstrong *et al.*, 1987; Banner *et al.*, 1992; Machar Field: Bain, 1993; Foster and Rattey, 1993); and (ii) four-way dip closures above salt diapirs, domes and ridges (e.g. Gannet South/Guillemot C Field: Armstrong *et al.*, 1987; Banner *et al.*, 1992; Maureen Field: Bain, 1993). Top seal is provided by extensive basinal mudstone intervals (e.g. Palaeocene Lista Formation and Eocene Balder/Sele formations and Hordaland Group).

Stratigraphic traps are of two main types: mounded closures (see Fig. 12.30) and sand pinch-outs (e.g. Fig. 12.31). Mounded closures may be considered 'structural' in so far as trap geometry is defined by four-way dip closure. However, the origin of this dip closure is stratigraphic, resulting from the differential compaction of mudstones around sand-dominated parts of submarine fans. Trap variability is mainly the result of differences in original sand-body geometry, which is clearly displayed in accumulations such as Frigg (Fig. 12.32; Héritier *et al.*, 1980, 1981; Brewster and Jeangeot, 1987; de Leebeeck, 1987; Mure, 1987a; Brewster, 1991), Alba (Figs 12.33–12.35; Mattingly and Bretthauer, 1992; Newton and Flanagan, 1993), Balder (see Fig. 12.32; Sarg and Skjold, 1982; Hanslien, 1987, Jenssen *et al.*, 1993), Gannet East/Gannet A (see Fig. 12.29; Armstrong *et al.*, 1987; Banner *et al.*, 1992), Harding (Forth) (Alexander *et al.*, 1992) and Gryphon (Newman *et al.*, 1993).

The most subtle traps are those resulting from sand pinch-out, which occurs in four main depositional settings: base-of-slope/proximal fan, distal fan, against intrabasinal highs (tectonic and/or salt-induced features) and isolated channel fills (see Fig. 12.29). Base-of-slope sand pinch-out contributes to the trap configuration at Gannet West/Guillemot B (Armstrong *et al.*, 1987; Banner *et al.*, 1992), Harding (Forth) and Gryphon (see Figs 12.28 and 12.29). It may also be a factor in the Faeroe Basin (Ebdon *et al.*, 1995; see Bowman, Chapter 10, this volume). Unfortunately, the abundance of sands in these areas, and their often close proximity and, in many cases, lateral connectivity to shallow-water sand bodies has probably caused most westerly-migrating hydrocarbons to escape to the surface. Pinch-out at the distal end of a fan system provides a more favourable trapping configuration and is well illustrated at Balder (see Figs 12.28 and 12.32; Jenssen *et al.*, 1993). However, this distal fan setting is far less attractive than proximal areas in terms of the much lower reservoir thickness and quality. Intrabasinal sand pinch-out against structural highs along the eastern margin of the Central Graben provides the best location for sand pinch-out traps. The best example of this is the Everest complex, in which hydrocarbons are trapped along the edges of the Maureen, Andrew and Forties fan sequences where they onlap against the Jaeren High (see Fig. 12.31). Trapping is provided by both updip shale-out to the east and by dip closure above a structural nose in the underlying Jaeren High basement block (O'Connor and Walker, 1993). This type of trap is extensive along the eastern margin of the Central Graben. The best example of an isolated channel-fill sand body is in the Alba Field, where Eocene sands are located within clearly incised features with elongate geometries (Figs 12.33–12.35).

In general, and particularly in the North Sea Basin, sand pinch-out traps carry high exploration risk, with most potential traps failing due to inadequate seal (bottom, top and/or lateral seal failure). Notwithstanding this risk, these traps also contain significant remaining potential, because they have rarely been deliberately tested in optimal drilling locations.

Play-fairway analysis. Each of the *c.* 10–15 depositional sequences requires separate evaluation in terms of the three controlling regional parameters of reservoir, seal and hydrocarbon charge. Timing is not a controlling parameter, since the generation of hydrocarbons from Upper Jurassic source rocks is continuing at the present day and much of the total generation occurred in post-Palaeocene times.

Reservoir and seal delineation needs to fully utilize sequence-stratigraphic techniques in order to develop a high-resolution stratigraphic framework capable of predicting the location of increasingly subtle traps. Such a stratigraphic framework, combined with regional depositional models, enables areas of optimum reservoir development to be defined, which in practice corresponds to areas of thickly bedded sandstones. Established models of reservoir geometry for such reservoirs include sheet-like, basin-floor-fan sand bodies, elongate, laterally restricted, channel

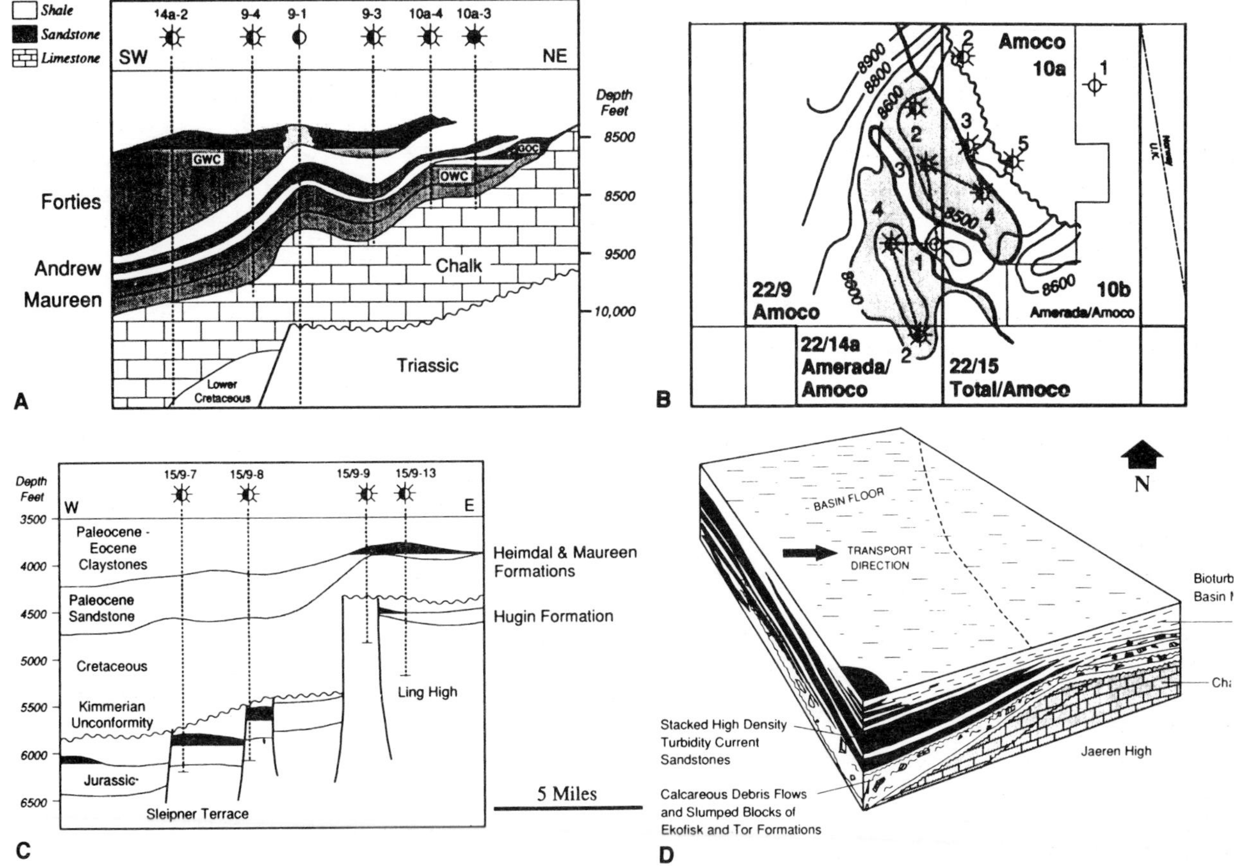

Fig. 12.31 Palaeocene reservoirs of the Everest trend. (A) Everest play concept and hydrocarbon distribution. (B) Top Forties Sandstone depth map showing two main sandstone lobes separated by a non-reservoir interlobe area. (C) Sleipner Field play concept. (D) Depositional model of the Maureen fan sandstones indicating the influence of the Jaeren High on turbidite deposition. (From O'Connor and Walker, 1993.)

sand bodies (in single or multistorey bodies) and ovoid-shaped, sand-prone mounds.

In the case of the sheet-like, basin-floor fans, hydrocarbon accumulations and shows are invariably located in the highest reservoir units in any particular well, below the main regional seals. This implies that intraformational shales in the Palaeogene are not reliable top seals. Hence the top seal is a critical controlling parameter on a regional scale.

Mature Upper Jurassic source rocks are widely distributed beneath most of the Palaeogene fairways, and sufficient hydrocarbons have been generated to fill any traps present. Migration appears to have been along deep faults, associated with the margins of the Mesozoic graben system (Fig. 12.36), or via salt domes and their related fracture systems (Conort, 1986). The latter is significant only in the Central North Sea, where areas of extreme overpressure, combined with defective top seal at the base of the Cretaceous, enabled direct, vertical migration into the Palaeogene. Lateral migration within the Palaeogene has been effective as a result of the widespread, high-quality sandstone reservoirs and their good 3D connectivity.

An important limitation on this play is depth, which affects oil quality. Oils in Palaeogene reservoirs are normally biodegraded at depths of less than 5000 ft (1500 m), with gravities in the range 10–20° API as a result of reservoir temperature and freshwater flushing. This general relationship between depth and API gravity suggests that 5000 ft ss (1500 m) is a critical limit above which biodegradation and oil API become important risk factors in this play. There may also be a relationship between the length of the migration route and API gravity; this is more difficult to quantify but may affect accumulations located at the end of local migration routes (e.g. along the margin of the faulted East Shetland Platform, such as in Bressay). A lengthy migration pathway from the Central Graben across the Ringkøbing High, however, has resulted in little deterioration in the quality of the oil in the Siri discovery.

12.2.4 Summary of the Northern and Central North Sea plays and hydrocarbon resources

This review identifies seven main hydrocarbon plays within the Northern and Central North Sea, whose reservoirs range in age from the Palaeozoic (Devonian–Carboniferous) through to the Palaeogene (Eocene). In line with other play reviews in the North Sea (e.g. Pegrum and Spencer, 1990) and along the Atlantic Margin (Knott *et al.*, 1993), the occurrence of these plays is further simplified into three genetically related, stratigraphic groups, based on their relationship to the main phase of Late Jurassic to Early Cretaceous rifting: (i) pre-rift plays (Palaeozoic, Triassic–Lower Jurassic and Middle Jurassic plays), (ii) syn-rift plays (Upper Jurassic and Lower Cretaceous plays);

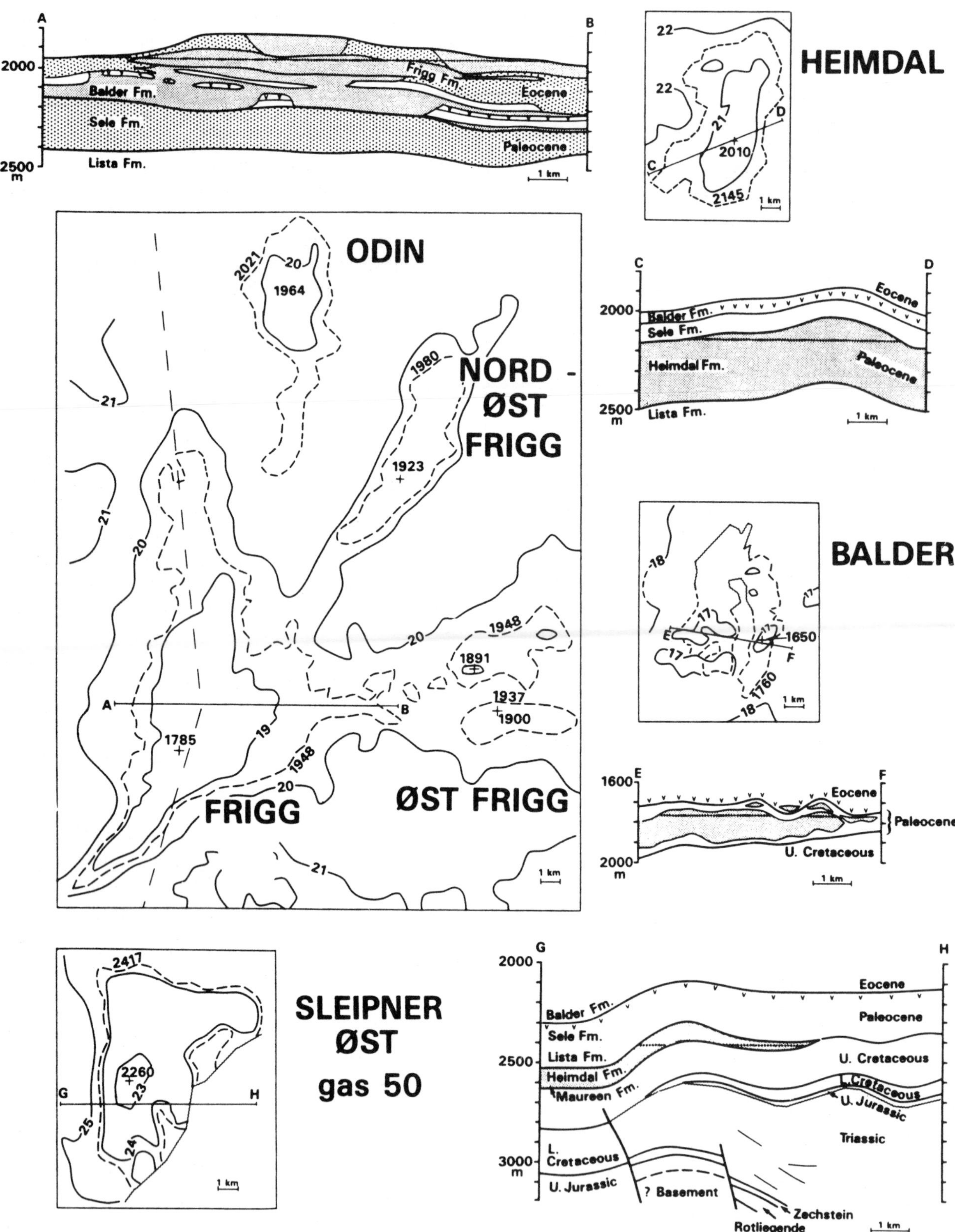

Fig. 12.32 Examples of hydrocarbon accumulations in the post-rift/Palaeogene play in the Northern and Central North Sea (from Fjaeran and Spencer, 1991).

and (iii) post-rift plays (Upper Cretaceous and Palaeogene plays).

This genetic-play classification emphasizes the dominant control exerted by the Mesozoic rifting episode on the hydrocarbon habitat of the Northern and Central North Sea. Hence, the syn-rift phase is by far the most important, because it influenced directly and indirectly key aspects of the North Sea's petroleum geology, including sedimentation patterns (e.g. the nature and distribution of source rocks, reservoirs and seals), structural evolution and trap formation (e.g. structural traps, particularly rotated fault blocks) and the basin's subsidence and burial history (e.g. post-rift subsidence, thermal maturation of the source rocks

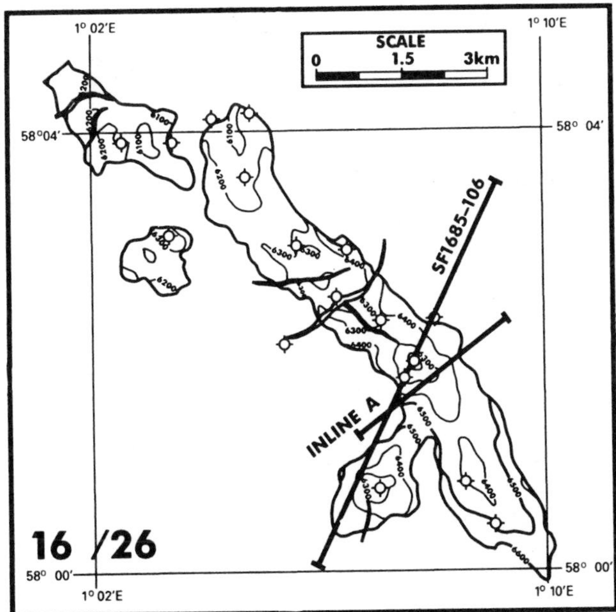

Fig. 12.33 Outline of the Alba Field (Central North Sea) as defined by the Top Nauchlan Sand depth-structure map (from Newton and Flanagan, 1993). Note the pronounced elongate geometry. Inline A is shown in Fig. 12.35.

Fig. 12.34 Schematic south-west–north-east cross-section showing typical off-field and in-field wells across the Alba Field (approximate well spacing 500 m) (modified from Newton and Flanagan, 1993). Note the complete absence of reservoir sands in the off-field well.

and hydrocarbon maturation and migration). Hydrocarbon occurrence in all the plays is closely linked to proximity to the thermally mature Late Jurassic source rocks, which has charged reservoirs mainly through vertical migration and occasionally through short-distance lateral migration. In addition to the stratigraphically adjacent syn-rift reservoirs, the same source rock has also charged pre-rift and post-rift reservoir-bearing horizons. All the main hydrocarbon accumulations occur either within or close to the limits of the Late Jurassic source-rock kitchen areas.

The main play fairways and their hydrocarbon resources are summarized below (Fig. 12.37).

Pre-rift (Palaeozoic to Middle Jurassic)

The pre-rift play is characterized by tilted and rotated fault-block structural traps, which are of variable size but include some giant structures. The most successful aspect of this play is the thick and laterally extensive sandstone reservoirs, of which the most important were those deposited in the Middle Jurassic by the high-energy Brent delta, in response to the pre-rift thermal doming event in the Central North Sea. The absence of this reservoir renders other basins non-prospective at this level (e.g. Rockall Trough, Faeroe Basin, West Shetland Basin, Magnus Embayment, etc.). The pre-rift play fairway is the most prospective play in this region, with around 40–50% of the hydrocarbon reserves. It dominates the Northern North Sea, especially in the East Shetland Basin and Tampen Spur (Brent Province). According to Knott *et al.* (1993) in their review of the Atlantic Margin (including the North Sea,

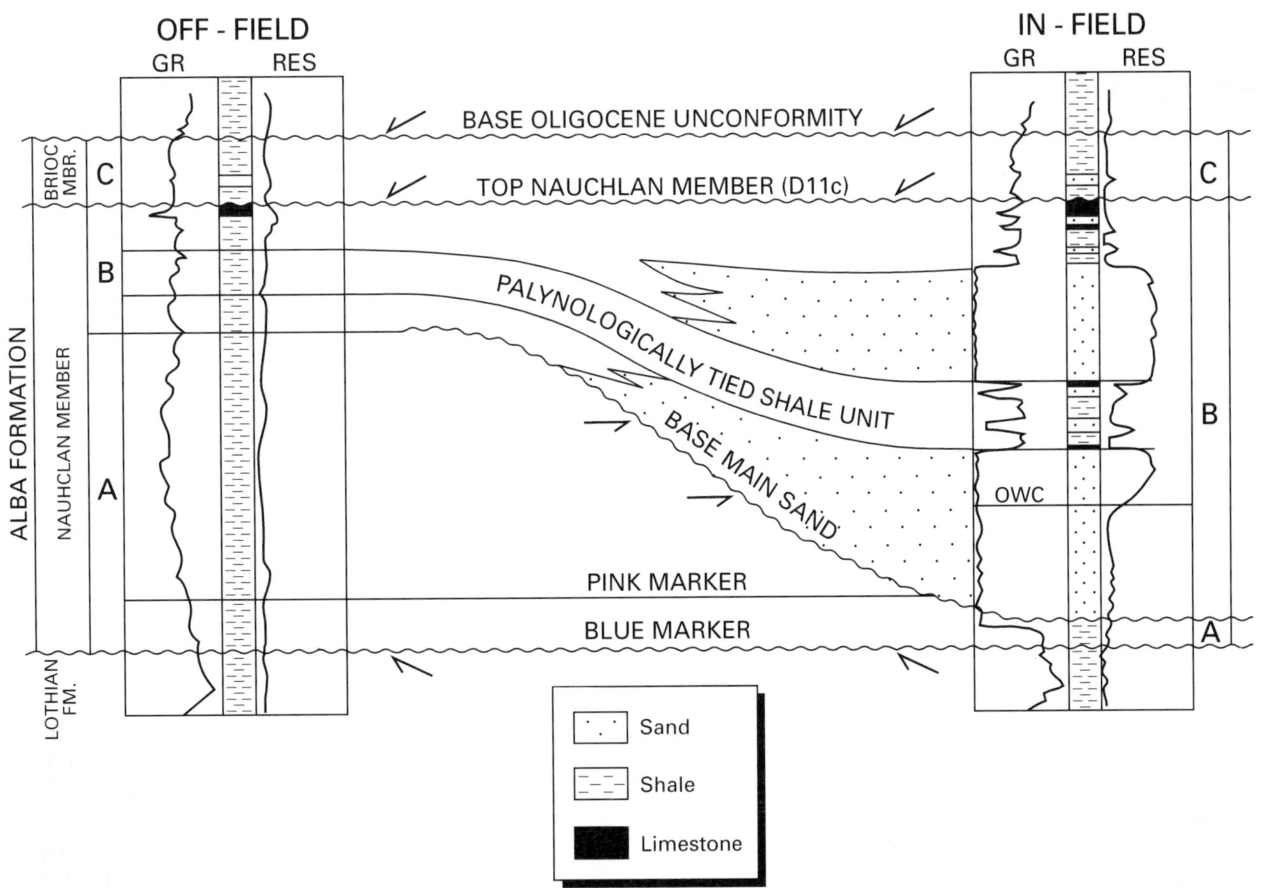

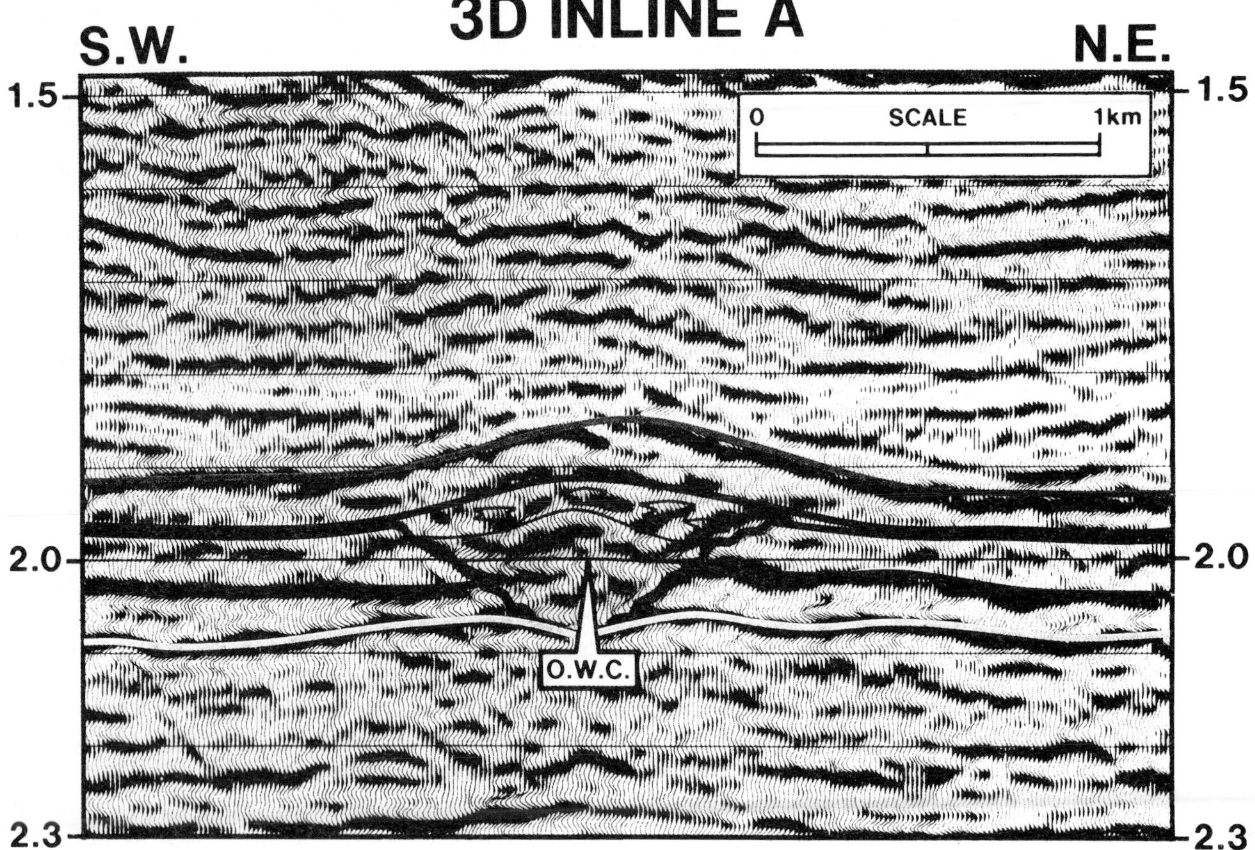

Fig. 12.35 Alba 3D seismic inline 'A' (interpreted; normal polarity) illustrating the incised, channelized nature of the Alba sandstone reservoir (from Newton and Flanagan, 1993). See Fig. 12.33 for location. (With permission of the Geological Society.)

mid-Norway and West of Shetland), this play contains some 32×10^9 barrels of oil equivalent (boe) or 46.5% of the total discovered volume. Both creaming curves (Fig. 12.38A) and field-size distribution (Knott *et al.*, 1993) reflect a mature fairway, which contains several billion-barrel fields and many small fields ($< 80 \times 10^6$ boe; Fig. 12.39). The Brent Province is now extremely mature, with emphasis on evaluating near-field exploration potential and maximizing recovery from fields now experiencing declining production. This play continues to have exploration potential in the Central Graben, Danish Central Trough and mid-Norway.

Syn-rift (late Jurassic to early Cretaceous)

The syn-rift play is dominated by two contrasting sandstone reservoirs, comprising: (i) submarine fans (slope aprons and basin-floor fans); and (ii) retrogradational shelf sandstones. The play is complex, as a result of rapid variations in sandstone presence, thickness and effectiveness, which are partly controlled by the nature of the local source areas (e.g. footwall-rock characteristics from active intrabasinal highs), by accommodation-space history within the basin and by shallow-marine depositional processes. Restricted regional oceanic circulation contributed to widespread anoxia and deposition of the Kimmeridge

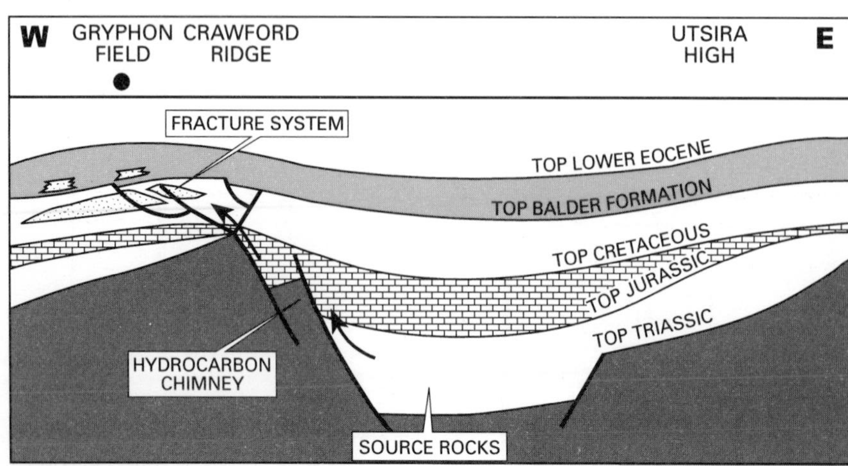

Fig. 12.36 Hydrocarbon play concepts in the Gryphon Field (from Newman *et al.*, 1993).

Clay Formation source-rock facies in basinal settings, which interfinger towards the basin margins with the submarine-fan sandstone facies. This ideal configuration is a major factor in the success of the play, which has widespread occurrence both within the North Sea and elsewhere along the Atlantic Margin (e.g. South Rockall and mid-Norway). Similar play characteristics continue into the Lower Cretaceous, where rifting persisted longer (e.g. Moray Firth) or where rifting began later (e.g. Magnus Embayment, West Shetland Basin, Faeroe Basin and other parts of the Atlantic Margin). The syn-rift play is estimated to contain around 31% of discovered reserves in this region, with some 22×10^9 boe (Knott *et al.*, 1993). Creaming curves suggest that the play is relatively underexplored (see Fig. 12.38B,C; Cordey, 1993). Field sizes are variable and with a higher proportion of small-to-medium-size accumulations ($< 100 \times 10^6$ boe), compared with those associated with the pre-rift plays (Knott *et al.*, 1993). This reflects smaller trap sizes, as a result of both structural and, in particular, stratigraphic complexity, which is an inherent feature of syn-rift plays. Reserves in this play are strongly skewed by a single supergiant field (Troll), which is not typical of the play. It is currently one of the most active exploration plays in the North Sea, with significant remaining potential.

Post-rift (late Cretaceous to Palaeogene)

The post-rift play is dominated by Palaeogene submarine-fan clastic deposits in all the main depocentres and by Chalk reservoirs in one limited part of the Central Graben; in the latter area, these two plays are mutually exclusive. In the North Sea, both plays were controlled by the post-rift thermal subsidence but differ fundamentally in terms of tectonic activity and source-area behaviour; the Chalk play involved shallow- to deep-water conditions starved of clastic sediment, while the Palaeogene records thermal uplift over northern Scotland and the emplacement of thick siliciclastic deposits into adjacent basinal areas, both to the south-east (North Sea Basin) and to the north-west (Faeroe Basin and other adjacent Atlantic Margin basins). The traditional structural traps of the Chalk play have been extensively tested, and remaining exploration potential probably lies in non-conventional traps (stratigraphic, hydrodynamic and diagenetic). Horizontal drilling and well-stimulation techniques are proving to be extremely successful in increasing recovery within proved Chalk accumulations. The Palaeogene play is dominated by stratigraphically variable deep-water sandstones. Drilling results and creaming curves confirm that this play continues to provide significant exploration potential, with increasing emphasis on subtle, stratigraphic traps. The play is estimated to contain between 22% (Knott *et al.*, 1993) and 30% (Pegrum and Spencer, 1990) of reserves in this region, with the latter estimate assigning some 20% to the Palaeogene play and 10% to the Chalk play. Field-size distribution resembles that of the syn-rift play, with a large proportion of fields in the small category ($< 80 \times 10^6$ boe; Fig. 12.39).

12.2.5 Remaining hydrocarbon potential in the Northern and Central North Sea

As the North Sea approaches maturity as an exploration province, comparison with similar provinces worldwide suggests that, once the obvious prospects in the major plays are tested, the majority of new discoveries will be incremental additions to, or extensions of, earlier discoveries.

Remaining potential in the UK sector of the North Sea, derived from an analysis of hydrocarbon reserves and trapping risks for all prospective plays, is summarized in Table 12.3. Risked volumetrics have been calculated from data in the *Energy Report*, Vol. 2 (HMSO, 1995) and the mean trap size and success factors of all available prospects. Reserves are given as risked oil or gas in place.

Table 12.3 Undiscovered hydrocarbons by play, Central North Sea, Moray Firth and Northern North Sea (6×10^9 ft^3 gas = 1×10^6 barrel of oil equivalent (boe)).

Play	Risked oil in place (10^6 barrels)	Risked gas in place (10^9 ft^3)	Risked gas in place (10^6 boe)
Post-rift plays			
Eocene Alba/Tay/Frigg	800	3500	580
Eocene–Palaeocene Balder/Sele/Dornoch	1400	900	150
Palaeocene Forties	1300	1750	290
Palaeocene Balmoral/Andrew/Heimdal/ Maureen	1500	800	130
Post-rift total	5000	6950	1160
Syn-rift plays			
Lower Cretaceous	800	4250	710
Upper Jurassic deep-marine	1750	2350	390
Upper Jurassic shallow-marine	2650	2300	380
Syn-rift total	5200	8900	1480
Pre-rift plays			
Middle Jurassic Pentland	200	4600	766
Middle Jurassic Brent/Beryl/Bruce	300	1100	183
Triassic Marnock	250	700	116
Pre-rift total	750	6400	1065

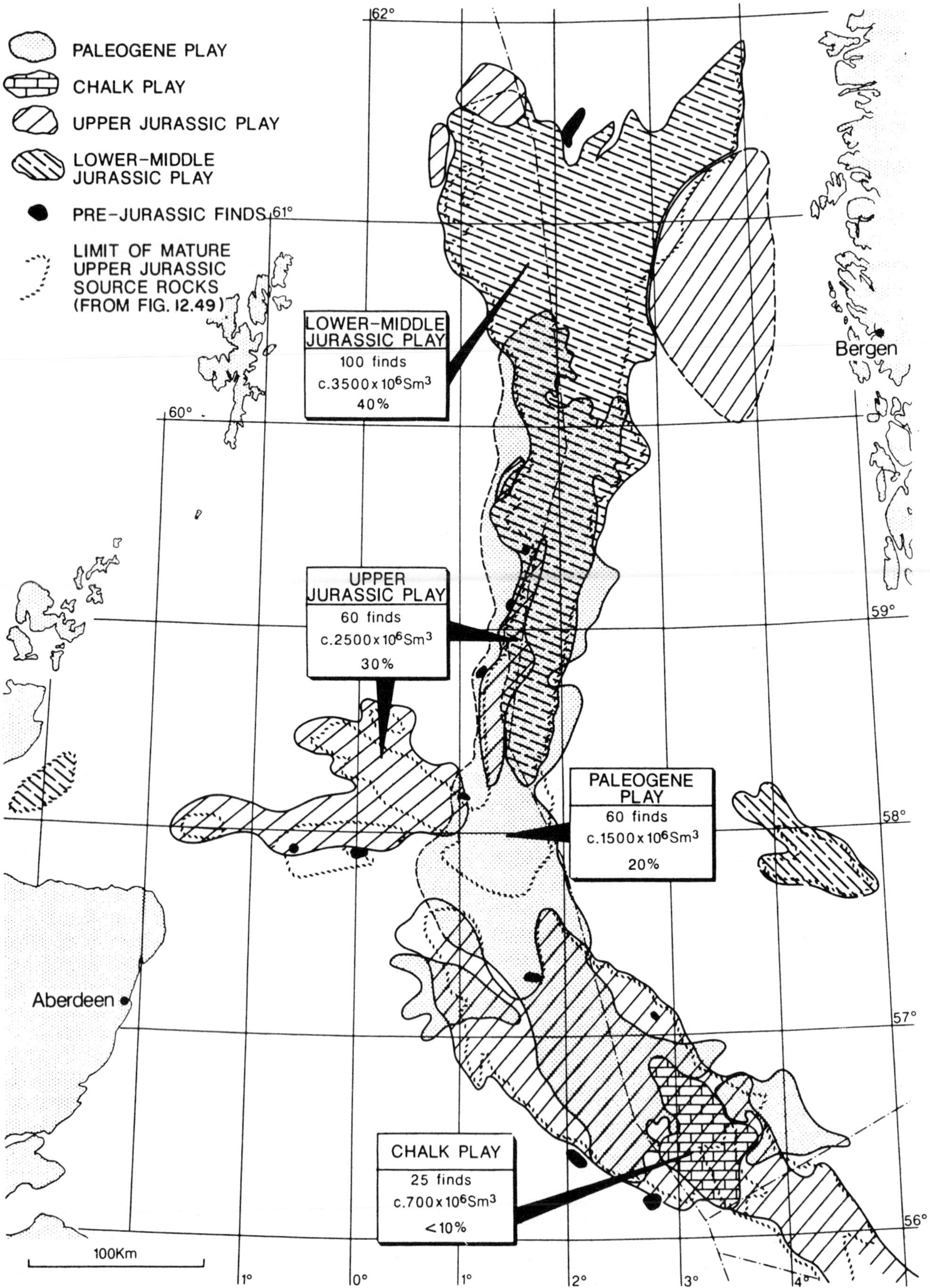

Fig. 12.37 Summary of the distribution of the four main plays in the Northern and Central North Sea (from Pegrum and Spencer, 1990).

An analysis of the risk of drilling exploration wells in the separate plays of the Central and Northern North Sea has been undertaken, based on well results up to 1995 (Table 12.4). It indicates the high success rates associated with anticlinal and upthrown fault-bounded traps (*c.* 60–70%), compared with the much lower rates related to stratigraphic traps and downthrown fault-bounded traps (*c.* 15–25%). The critical risk for each well drilled in the play fairway is attached to the trap.

Pre-rift potential

The Middle Jurassic Bruce, Brent and Beryl plays are very mature and have remaining potential restricted largely to prospects in the range $10–50 \times 10^6$ barrels of oil. More significant potential is seen in Middle Jurassic Pentland structural traps, but these are difficult to define because of poor seismic resolution. They account for > 50% of pre-rift reserves, but with a significant proportion of this being as gas or condensate.

The Triassic plays are relatively unexplored, although the Marnock play has been successful in Quadrant 22 and the play is becoming better understood. Triassic reserves located below Middle Jurassic discoveries in the Viking Graben may also provide a significant incremental addition to existing reserves, with better seismic resolution of reservoir facies (e.g. Tern Field) and multilateral drilling techniques. Undiscovered reserves are estimated at 250×10^6 barrels of oil and 700×10^9 ft^3 gas.

Syn-rift potential

The North Sea oil province is an area where the traditional Mesozoic structural plays are largely mature. Remaining hydrocarbon potential is either in small oil accumulations, with commercial viability in satellite developments or in deep gas/condensate traps. The size distribution of recent Upper Jurassic discoveries indicates that a large proportion of remaining oil and gas accumulations are likely to have oil in place of less than 50×10^6 barrels and gas-in-place of less than 500×10^9 ft^3. Potential may still exist, however, for larger fields over 250×10^6 barrels of oil or 1000×10^9 ft^3 gas.

Upper Jurassic deep- and shallow-marine plays are considered to contain > 75% of undiscovered syn-rift reserves (4400×10^6 barrels of oil and 4650×10^9 ft^3 gas). A significant proportion of these reserves—3500×10^6 barrels of oil and 2500×10^9 ft^3 gas—is predicted in subtle structural and stratigraphic plays, and this unexplored potential results from their poor seismic definition. The Fulmar play, in particular, has seen intense recent exploration activity, with a success rate of *c.* 40%. Notwithstanding difficulty in accurately mapping the limits of the shelf-sand complexes, targeting the coastal onlap of the Fulmar sands onto the basal Jurassic unconformity on the Western Platform has resulted in discoveries in blocks 21/11 and 21/16. With predicted prospect volume ranges of $50–500 \times 10^6$ barrels of oil, this remains an attractive exploration target. Upper Jurassic deep-marine structural and stratigraphic plays have been less actively explored but still contain attractive remaining potential, with predicted prospect volume ranges of $20–300 \times 10^6$ barrels of oil. Upper Jurassic Piper structural traps, with a prospect volume range of $20–250 \times 10^6$ barrels of oil, have also been heavily explored in recent years, with a *c.* 30% success rate. Remaining reserves are estimated at 1000×10^6 boe, approximately 15% of syn-rift reserves.

Additional syn-rift potential is seen in Lower Cretaceous structural and stratigraphic plays, which suffer from poor seismic definition. These account for a further 22.5% of syn-rift reserves, much of it being gas or condensate.

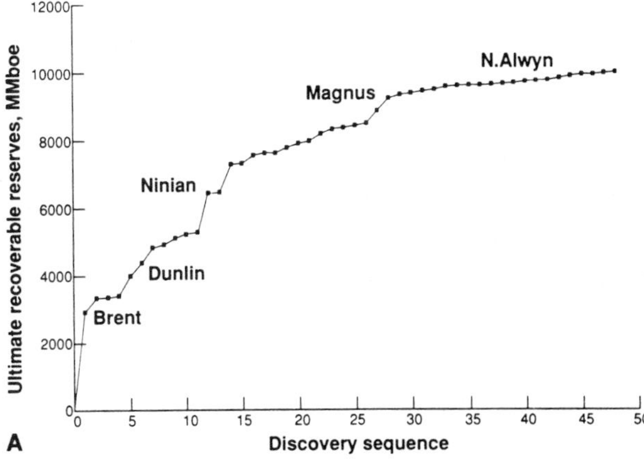

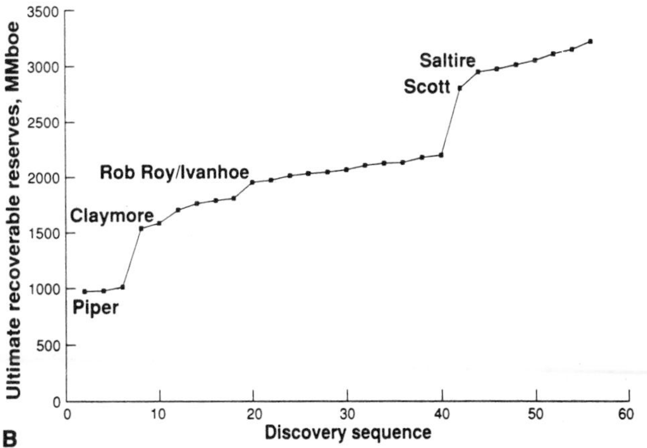

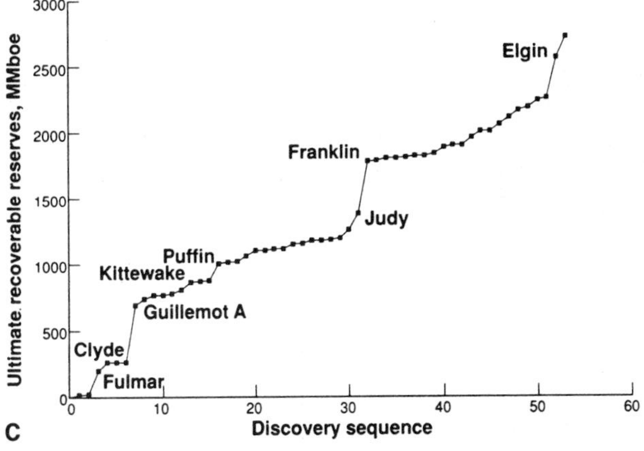

Fig. 12.38 Discovery ('creaming') curves of Jurassic plays: (A) Northern North Sea (UK East Shetland Basin), (B) Outer Moray Firth, and (C) Central North Sea (from Cordey, 1993).

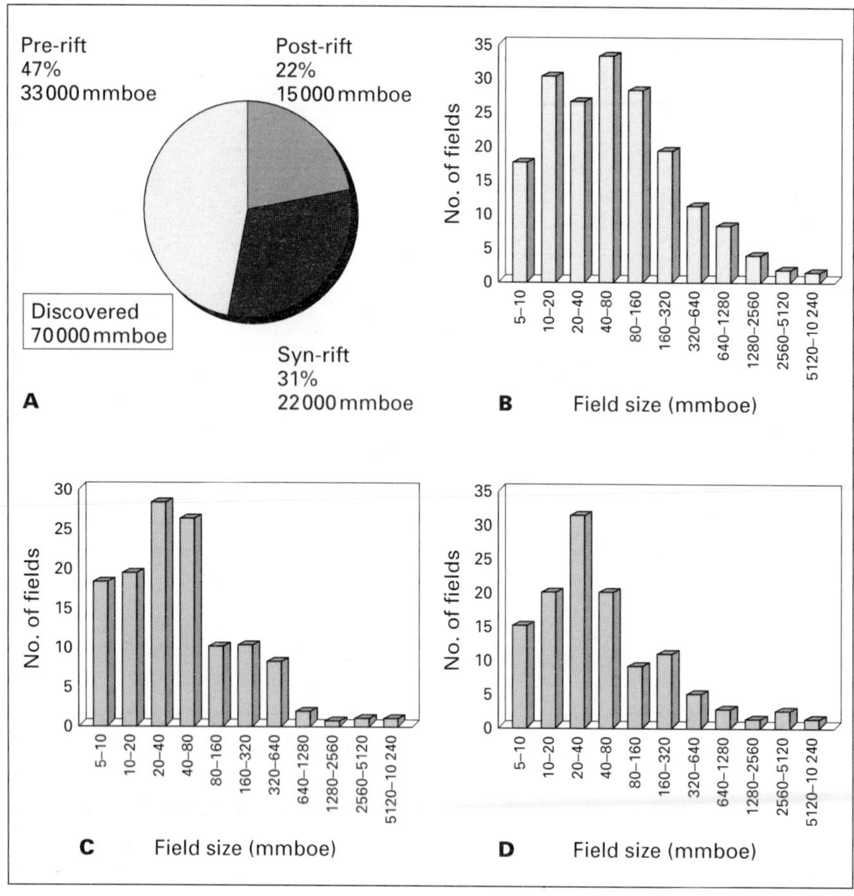

Fig. 12.39 Play fairway reserves statistics for the UKCS and North Atlantic margins: (A) discoveries by play type, (B) pre-rift play field-size distribution, (C) syn-rift play field-size distribution and (D) post-rift play field-size distribution (from Knott *et al.*, 1993).

Post-rift potential

All significant structures at Chalk horizons have been drilled in Norwegian and Danish acreage in the Central Graben. Remaining, albeit marginal, potential may exist in more subtle traps or on the shelf to the east of the graben.

Current exploration has confirmed the hydrocarbon potential of Palaeogene stratigraphic plays, which are still relatively underexplored. In-place risked reserves are estimated at 5000×10^6 barrels of oil and 6950×10^9 ft^3 gas. As these occur at depths generally shallower than 8000 ft (2900 m), a contingent risk is for heavy, biodegraded oils, with subsequent recovery problems. The most attractive Palaeogene reserves are concentrated in stratigraphic traps, which are currently difficult to map because of complex geology and poor seismic resolution. The trend of discoveries in the period 1989–92 confirms a remaining potential for accumulations with oil in place of over 200×10^6 barrels of oil, as well as for smaller accumulations.

The Alba and Tay plays are the subject of continued exploration, although the Alba play is particularly elusive on regional seismic data. Combined reserves are 650×10^6 barrels of oil and 2900×10^9 ft^3 gas, with the potential for large accumulations in the range $50–600 \times 10^6$ barrels of oil for the Alba play and $20–150 \times 10^6$ barrels of oil ($100–1000 \times 10^9$ ft^3 gas) for the Tay play. Remaining Frigg potential is likely to be limited to small accumulations totalling 150×10^6 barrels of oil and 580×10^9 ft^3 gas.

The Balder/Sele oil play has major, partly undiscovered, potential with 1400×10^6 barrels of oil in place of risked

Table 12.4 Relationship between exploration success rate (%) and trap type in the Northern and Central North Sea.

Trapping style	Palaeogene	Lower Cretaceous	Upper and Middle Jurassic	Triassic	Palaeozoic
Structural anticline		70	70	30	30
Salt anticline	60		70	30	
Drape anticline	50				
Upthrown fault		60	60	25	
Downthrown fault		15	15–50*	10–20*	
Stratigraphic shale-out	25	25	15	10	
Stratigraphic onlap/subcrop	25		20	10	
Stratigraphic mound	20–30–60†				
Stratigraphic compaction		40	45		
Fractured basement					10

*The higher success rate reflects the efficiency of salt seal in salt abutment traps.
†20% Alba play; 30% Sele play; 60% Balder play.

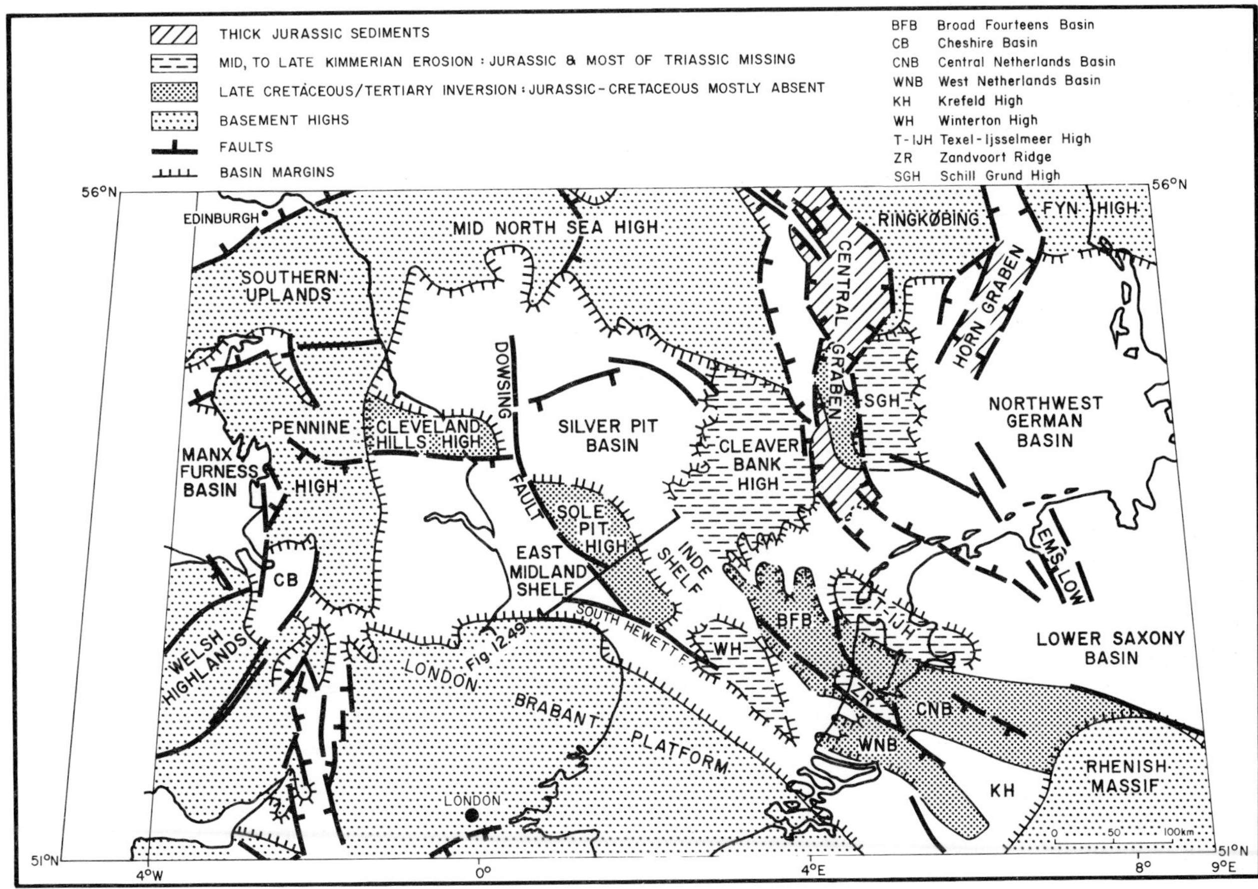

Fig. 12.40 Major tectonic elements and sedimentary basins in the Southern North Sea and adjacent areas (from Glennie and Provan, 1990).

reserves representing 28% of estimated Palaeogene oil reserves. These are predicted to be concentrated in mounded closures of 100–200 × 10^6 barrels of oil in Quadrants 9, 15 and 16.

The Forties play has been successfully explored, although it retains significant potential with undiscovered reserves of 1300 × 10^6 barrels of oil and 1750 × 10^9 ft^3 gas representing 26% of Palaeogene oil and gas reserves. Since 1989, exploration has been particularly successful, with > 60% of wells drilled recovering oil or gas. Additional significant potential is seen in the Balmoral, Andrew, Heimdal and Maureen plays in Quadrants 15, 16, 20, 21, 22 and 30, which contain 30% of remaining Tertiary oil reserves.

In the Norwegian/Danish margin area, the Siri discovery, in an incised channel in the Chalk, has demonstrated the existence of long-range hydrocarbon migration from the Central Graben and the presence of significant Palaeocene reservoir facies.

Some of these Palaeogene plays are difficult to resolve seismically but recent drilling results reveal a combined success rate approaching 50%, reflecting an increasing understanding of the factors controlling the distribution of post-rift prospectivity.

12.3 Southern North Sea plays

The geological setting and resulting hydrocarbon distribution in the now extremely mature Southern North Sea Basin are drastically different from those of the northern basins. The main differences are as follows: (i) the basin is gas-dominated; (ii) there is one dominant play type; (iii) the principal source rock comprises terrestrial organic material; and (iv) the structural setting is generally more complex, due to widespread tectonic inversion (Fig. 12.40).

The four major plays in the UK sector of this basin, which are classified mainly in terms of their stratigraphic age, are (Fig. 12.41) Carboniferous sandstones, Permian Rotliegend sandstones, Permian Zechstein carbonates and Triassic sandstones (Table 12.5). In detail, it is possible to subdivide these plays into various subplays, based on specific reservoir/seal/trap configurations, which will be discussed further. The widespread Permian Rotliegend Sandstone play (Fig. 12.42A) is overwhelmingly dominant, containing some 75% of the UK sector's recoverable gas reserves in this basin (33 × 10^{12} ft^3/0.94 × 10^{12} m^3 of gas), while the basin as a whole is dominated by the giant Groningen gasfield, with reserves of around 97 × 10^{12} ft^3/ 2.75 × $10^{12}$$m^3$ of gas (Breunese and Rispens, 1996). The same play extends further eastwards as a narrow fairway into north-west Germany, where it is the reservoir in more than 40 gasfields (Burri *et al.*, 1993). The other plays in the Southern North Sea contain locally significant but in general much lower gas reserves, although, in the case of the Carboniferous play (gas reserves = 4.75 × 10^{12} ft^3/0.13 × 10^{12} m^3), this is much less mature and underexplored compared with the Rotliegend plays. Of the younger plays, the Triassic is the most significant (gas reserves = 5.3 × 10^{12} ft^3/0.15 × 10^{12} m^3), with the Zechstein carbonate play containing only minor reserves of both oil and gas in the UKCS (gas reserves = 0.17 × 10^{12} ft^3/

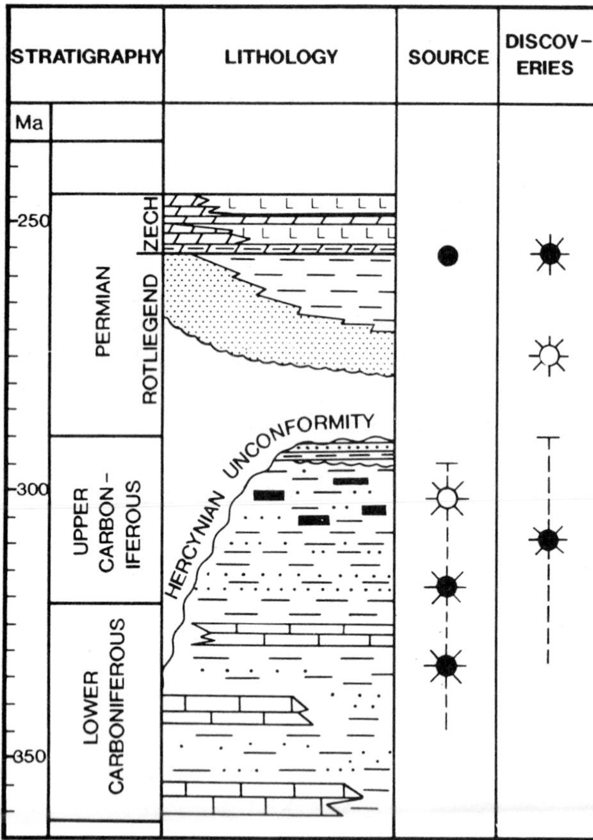

Fig. 12.41 Generalized stratigraphy and source/reservoir/hydrocarbon distribution in the Southern Permian Basin (from Gérard *et al.*, 1993).

0.005×10^{12} m^3), but this play becomes more important further east (in the Netherlands, Germany and Poland).

The single most common factor in all these plays is the Carboniferous Coal Measures, which underlies the whole basin and provides the source rock for all the commercial gasfields, both in the UK sector and in the eastern extension of the basin into the Netherlands and Germany (Fig. 12.42B). This *c.* 2500-m-thick succession, with its abundance of land-plant-derived source rock (coals and carbonaceous shales), is so rich that it could have charged the known accumulations in the basin many times over (Cornford, 1990; see also Chapter 11, this volume); in other words, the basin has a vast oversupply of gas.

In addition to the gas-dominated plays of the South Permian Basin, other mature plays in the Southern North Sea are also found in nearby regions, most notably the oil-bearing Carboniferous sandstones in the English Midlands (Kirby *et al.*, 1987; Fraser and Gawthorpe, 1990; Fraser *et al.*, 1990) and the oil- and gas-bearing Lower Cretaceous sandstones in both onshore and offshore the Netherlands (Cottençon *et al.*, 1975; Bodenhausen and Ott, 1981; Perrot and van der Poel, 1987; Rondeel *et al.*, 1996), which extend eastwards into the Lower Saxony Basin of northern Germany and the eastern Netherlands (Boigk, 1981; Kemper, 1973).

This review mainly considers the four plays of the South Permian Basin: Carboniferous, Rotliegend, Zechstein and Triassic (Table 12.5). In addition, the Mesozoic plays of the southernmost part of the basin and its onshore extension into the Netherlands and north-west Germany are also briefly discussed.

12.3.1 Carboniferous play

The primary exploration significance of the Carboniferous has been its enormous source-rock potential, which has charged all the gas-bearing plays in this region, including the prolific Permian/Rotliegend play, on which the majority of drilling activity has been focused. As a result, the exploration potential of this play has remained relatively underexplored and, consequently, there are only a limited number of commercial gas accumulations within the Carboniferous play: (i) Trent Field and Cavendish area, with reservoirs in the Namurian; (ii) Murdoch and Caister 'C', with reservoirs in the Westphalian Coal Measures (Ritchie and Pratsides, 1993); and (iii) Ketch and Schooner, with reservoirs in the Westphalian red beds (Besly, *et al.*, 1993).

The main play fairway is delimited by the presence of top seal, provided by the Silverpit Claystone Formation, which unconformably overlies reservoirs of Dinantian to Westphalian age, most notably in the Silver Pit Basin (Fig. 12.43; Bailey *et al.*, 1993; see Besly, Chapter 4, this volume). Intra-Carboniferous seals may also have limited capacity to develop secondary plays, including outside the main play fairway. Karstified Dinantian carbonates contain gas shows on the northern flanks of the London–Brabant Massif in Belgium. This play may have minor potential as a secondary objective along the southern margin of the South Hewett Shelf and above other intrabasinal highs (Fig. 12.44). Mature Upper Carboniferous gas-prone source

Stratigraphic unit	Play	Reservoir
Triassic	Triassic (undifferentiated)	Bunter, Hewett
Zechstein	Upper Permian, Zechstein (undifferentiated)	Hauptdolomit, Zechsteinkalk
Rotliegend	Lower Permian, Rotliegend	Rotliegend
Carboniferous	Westphalian A–B and C	Ketch Member, Caister Coal Formation Millstone Grit Formation
	Namurian B–C Dinantian	Scremerston Formation, Fell Sandstone Formation

Table 12.5 Southern North Sea plays and associated reservoirs.

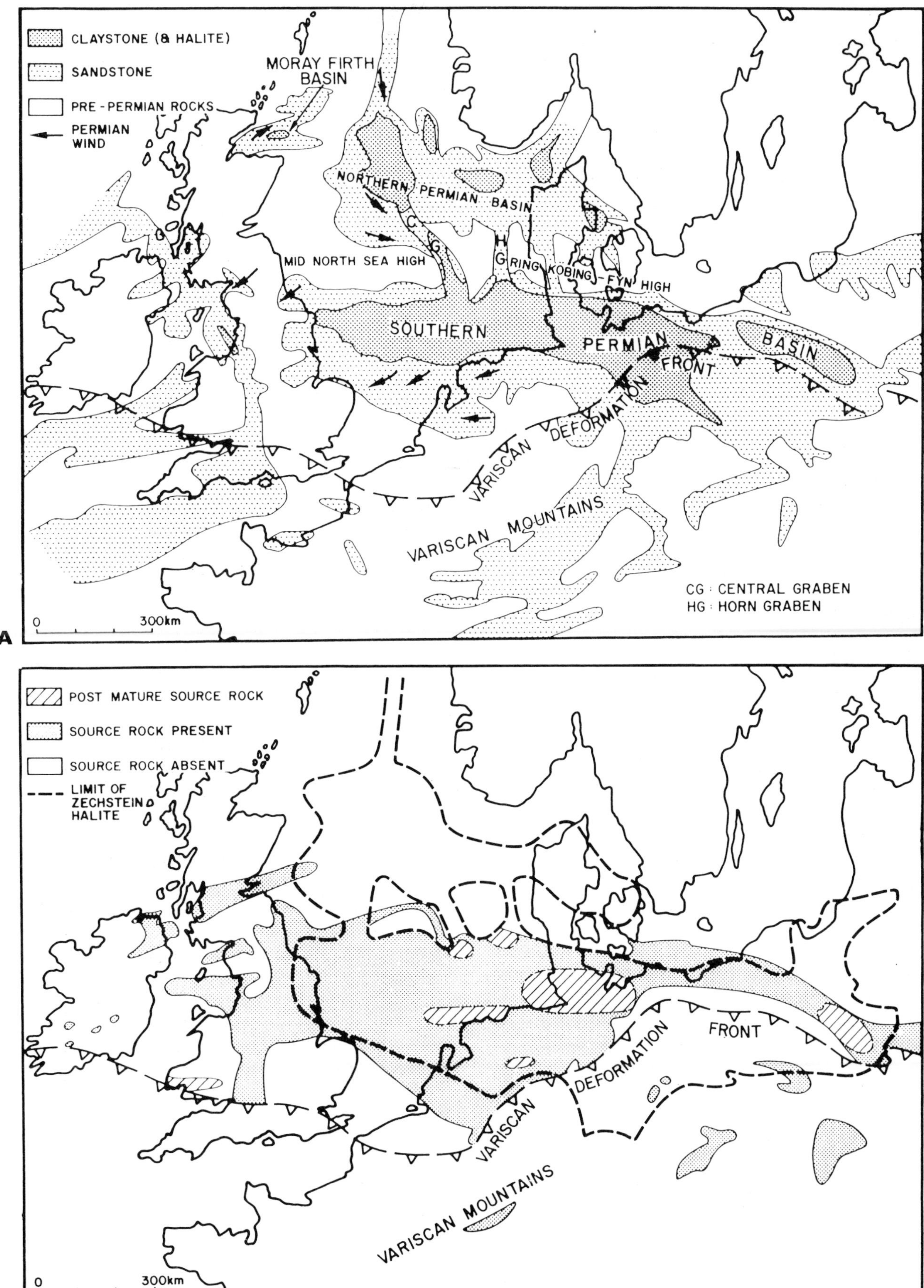

Fig. 12.42 (A) Regional distribution of the major facies in the Permian Rotliegend fluvial/aeolian sandstones and adjacent desert-lake/sabkha claystones in north-west Europe. (B) Distribution of preserved Carboniferous Coal Measures source rocks in north-west Europe, including areas of post-mature source rocks. (From Glennie and Provan, 1990.)

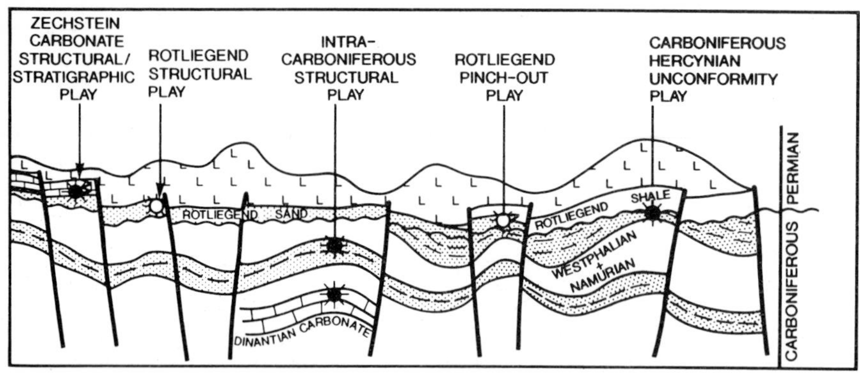

Fig. 12.43 Summary of the major play concepts in the Southern Permian Basin (from Gérard *et al.*, 1993).

rocks have the potential to charge any structures where top seal and reservoir combine.

The main limitations on this play are predicting the presence, thickness and lateral extent of sandstones with sufficient reservoir quality to allow sustained commercial rates of gas-production. Other limitations have been poor seismic imaging of reservoir and traps, particularly within the Carboniferous, the quality and effectiveness of intra-Carboniferous seals and the integrity of fault-bounded structural closures. Improvements in seismic imaging, notably through 3D seismic data, probably hold the key to establishing the true potential of this play.

Reservoirs. Reservoir prediction is faced with all the problems associated with fluvial and deltaic environments, plus the fact that the succession has a cumulative thickness in excess of 9 km, it has been buried to depths of up to 4 km and there are relatively few well penetrations. The depositional environment ensures a complex stratigraphy and a high variability in reservoir distribution. The best reservoirs comprise coarse-grained, multistorey fluvial and distributary-channel sand bodies. The thickest developments occur along the sand-rich fairways of major channel systems, some of which are located within incised valleys (Bailey *et al.*, 1993; Collinson *et al.*, 1993). In other cases, channel sand bodies are thinner and more randomly distributed, which makes even field-scale predictions in sandstone thickness difficult (e.g. Cowan, 1989; Leeder and Hardman, 1990; Ritchie and Pratsides, 1993). Superimposed on this primary depositional variability is a pronounced diagenetic overprint. This includes compaction- and depth-related diagenesis, which significantly reduces reservoir quality, and also two phases of dissolution, which have created important secondary porosity. As a result, reservoir quality is generally relatively low (average porosities *c.* 10% and permeabilities mainly *c.* 0.1–10 mD), but in some places, such as below the Saalian (or Hercynian) Unconformity, anomalously high porosity (> 20%) and permeability (up to 500 mD) can be preserved (Fig. 12.44). Importantly, even the average reservoir properties are still capable of providing commercial rates of gas production. Despite the diagenetic overprint, the best reservoir properties continue to occur within the thick channel sand bodies, but lateral variability can be extreme.

Traps and seals. The only commercial traps established so far are structural closures, of which there are two main types: (i) Variscan folds, which require an intra-Carboniferous seal; and (ii) post-Palaeozoic structures, which comprise Variscan structures reactivated during Cimmerian and Late Cretaceous/Early Tertiary tectonic inversion (see Fig. 12.43). The most successful trapping configuration comprises dip-closed or combined fault/dip-closed structures along the four major north-west–south-east-trending ridges, which are sealed at the Saalian Unconformity by claystones of the Permian Silverpit Claystone Formation (see Fig. 12.43). Structural traps involving intra-Carboniferous and/or fault seals are of considerably higher risk, due to the generally poorer sealing characteristics of intra-Carboniferous mudstones and the close proximity of intercalated sandstones (Fraser *et al.*, 1990).

There is also potential for stratigraphic traps, particularly: (i) Carboniferous sandstones subcropping the Saalian Unconformity against Silverpit Formation claystones or below the Zechstein; (ii) Carboniferous sandstones sealed by intraformational mudstone seals (shale-out/pinch-out traps); and (iii) intraformational onlap of lowstand wedges (Fig. 12.44). These traps carry a high exploration risk of failure, mainly because of the combination of low reservoir quality and variable sealing potential, particularly within the intraformational mudstones. This is further exaggerated by the difficulty of imaging these traps, due to the limited seismic resolution resulting from a complex overburden geology.

Play-fairway analysis. The main established fairway for the Carboniferous play is in the Silver Pit Basin, which coincides with the availability of an effective regional top seal (Silverpit Claystone Formation). Other areas generally carry a much higher exploration risk through a reliance on intra-Carboniferous seals.

Hydrocarbon generation from the Carboniferous probably began before Variscan basin inversion, but it is unlikely that these hydrocarbons have been preserved, due to later tectonic events. The present-day distribution of hydrocarbons is more closely linked to source-rock maturation and migration during Mesozoic burial. Although there are areas of limited maturity, notably around the marginal shelf areas, most parts of the basin contain gas-mature source rocks. The main uncertainties concern the timing of maturation and gas migration relative to the development of structures and seals that were disturbed by tectonic inversion; migration during the Jurassic and Cretaceous occurred from the basin centres to the flanks, but Late Cretaceous/Early Tertiary inversion of the basinal areas (up to 1500 m) resulted in later remigration of gas.

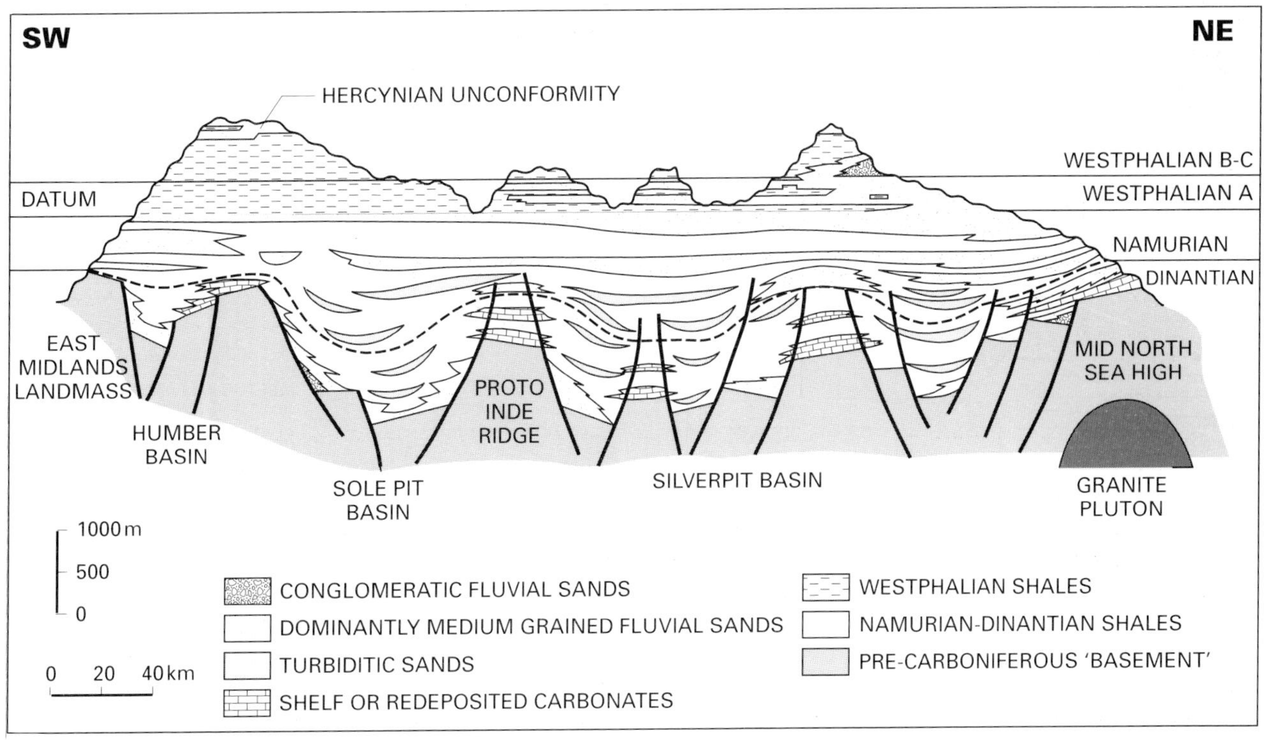

Fig. 12.44 Schematic cross-section through the Carboniferous of the Southern North Sea highlighting reservoir development (from Bailey *et al.*, 1993).

Notwithstanding these charge/retention uncertainties, the main difficulties in this play are associated with seismic imaging and reservoir prediction (see Section 4.8). Problems of seismic imaging relate to: (i) lithological variability (thick succession of sand, shale and coal, with rapid lateral facies changes); (ii) lack of well-defined intra-Carboniferous

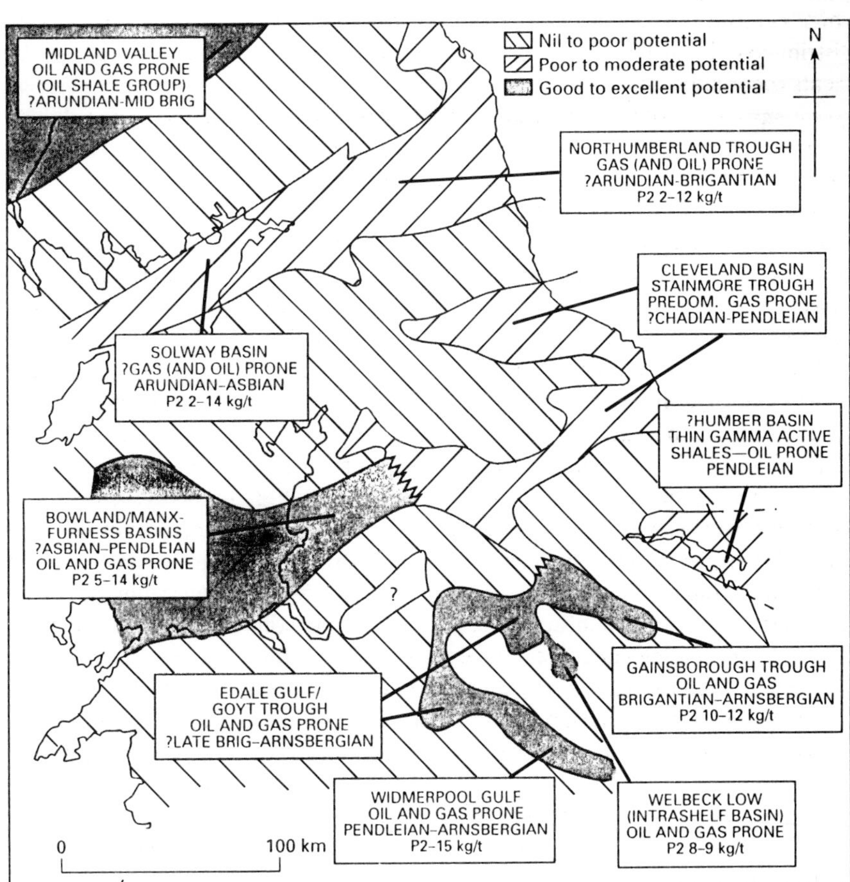

Fig. 12.45 Distribution and quality of source rocks in the Carboniferous of central Britain (from Fraser *et al.*, 1990).

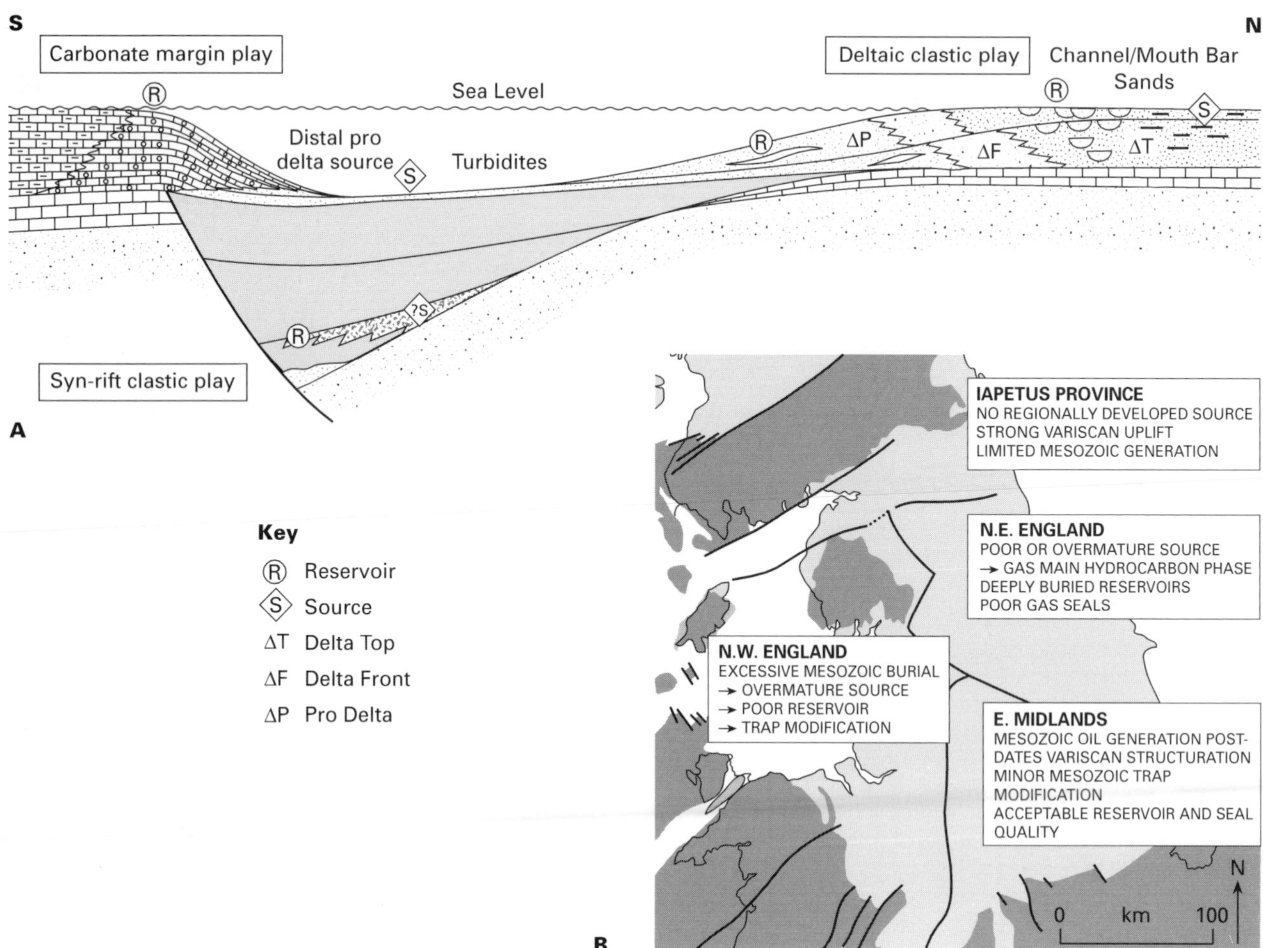

Fig. 12.46 (A) Play types in the Carboniferous of central Britain, and (B) hydrocarbon play summaries (from Fraser *et al.*, 1990).

stratigraphic markers; (iii) lack of acoustic impedance contrast (e.g. at Base Rotliegend/Top Carboniferous level and within the thick Carboniferous succession); and (iv) difficulties in seismic discrimination between hydrocarbon and lithological effects. Uncertainties in reservoir prediction are caused mainly by rapid lateral changes in channel sandstone thickness, variable porosity and permeability, and lack of well-defined, regionally extensive reservoir–seal pairs.

Other Carboniferous plays. A Carboniferous play is present in northern England, although it has only ever yielded small volumes and is now fully mature (Fraser *et al.*, 1990). Oil and gas were derived from Carboniferous source rocks, deposited in prodelta and delta-plain environments, with individual subbasins displaying different source types, quality and maturation histories (Fig. 12.45). Reservoir rocks occur within three main play types (Fig. 12.46): syn-rift clastic play (fault scarp fans), deltaic clastic play (distributary channel/mouth bar and delta-front turbidite sandstones), and carbonate-margin play.

12.3.2 Permian Rotliegend sandstone play

The background and importance of the Permian Rotliegend sandstones to the whole exploration history of the North Sea Basin have been extensively discussed by Brennand *et al.* (Chapter 1, this volume) and by Glennie (Chapter 5, this volume). This is now a very mature play, in which virtually all the footwall closures have been drilled (e.g. Fig. 12.47); it has resulted in some 112 discoveries, mainly in the Sole Pit Basin and Indefatigable Shelf areas (see Fig. 1.1). Remaining potential in this play is now limited, probably mainly to smaller, near-field traps, but also in hangingwall closures against sealing faults. The detailed nature of many of the Rotliegend accumulations on the UKCS are described in Abbotts (1991). In the Dutch sector of the Southern North Sea, which is discussed extensively in Rondeel *et al.* (1996), there are 120 offshore Rotliegend discoveries, with a further 62 onshore (Spencer *et al.*, 1996)

Reservoir, seal, traps and charge. The key element of this play is the combination of: (i) a generally good-quality and widespread Rotliegend sandstone reservoir (see Glennie, Chapter 5, this volume), although with considerable lateral facies variability (Fig. 12.48; George and Berry, 1993; Verdier, 1996); (ii) an overlying and extremely effective seal (see Fig. 12.47); and (iii) an underlying and almost infinitely abundant source rock (see Fig. 12.44), with an extremely high generation capacity and migration efficiency (Cornford, Chapter 11, this volume). Importantly, this configuration (see Fig. 12.43) was also sufficiently effective to withstand several periods of tectonic activity (Late

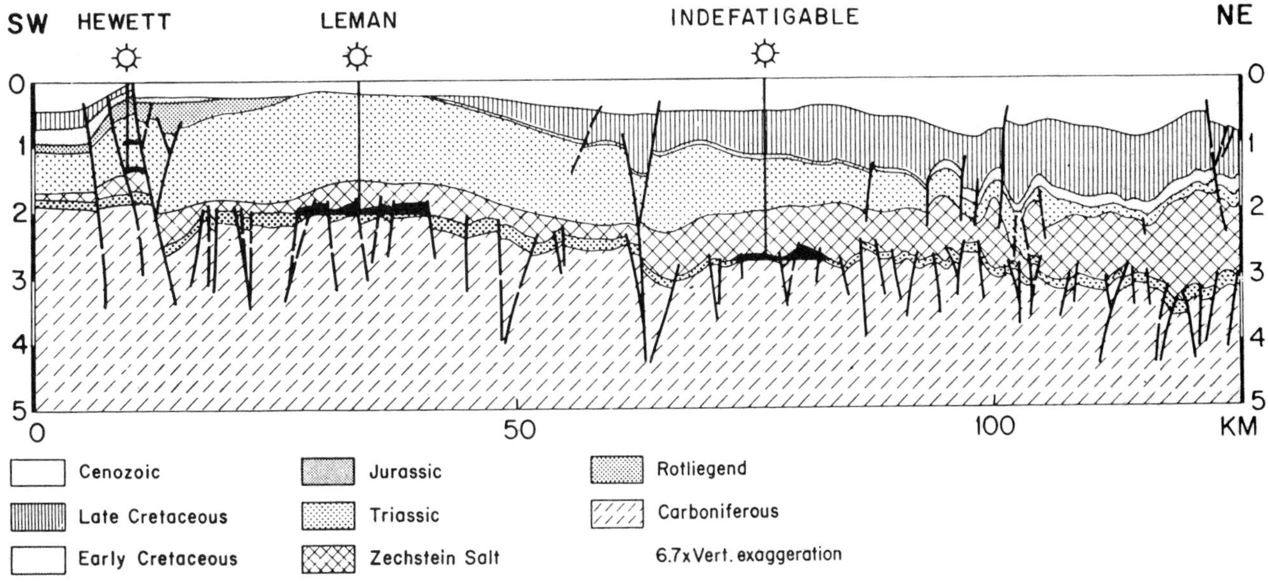

Fig. 12.47 Simplified structural cross-section through the Sole Pit High, illustrating two major Rotliegend accumulations at the Leman and Indefatigable fields (from Ziegler, 1990). Note the occurrence of Triassic gas-bearing sandstones in the Hewett Field, where the Zechstein salt seal has been locally breached.

Jurassic–Early Cretaceous, Late Cretaceous and Tertiary), trap formation and gas maturation and migration. This included the latest phase of transpressive tectonic stresses during the Tertiary, which caused widespread inversion of the basin and the formation of structural fault-bounded closures at Top Rotliegend level (Fig. 12.49). Also crucial was the mobility of the Zechstein salt, which allowed the top seal to remain intact for the most part (minor gas occurrences in the Zechstein and Triassic reflect local breaching of this seal), despite the late phase of tectonic inversion. The latter processes have resulted in a much more complex history of hydrocarbon maturation/ migration and timing of trap formation, compared with the situation in the Central and Northern North Sea. For example, the Westphalian coals reached gas maturation at different times in different parts of the basin, as a result of variations in subsidence history. Generally, most parts of the basin reached maturity in the Late Cretaceous, but in some areas (e.g. Broad Fourteens and Sole Pit basins) this occurred as early as the Late Jurassic (Lutz *et al.*, 1975; Glennie and Boegner, 1981; Lee *et al.*, 1985). This resulted

Fig. 12.48 Silver Pit Basin Rotliegend well correlation showing the relationship between depositional facies and variations in relative lake level (from Bailey *et al.*, 1993).

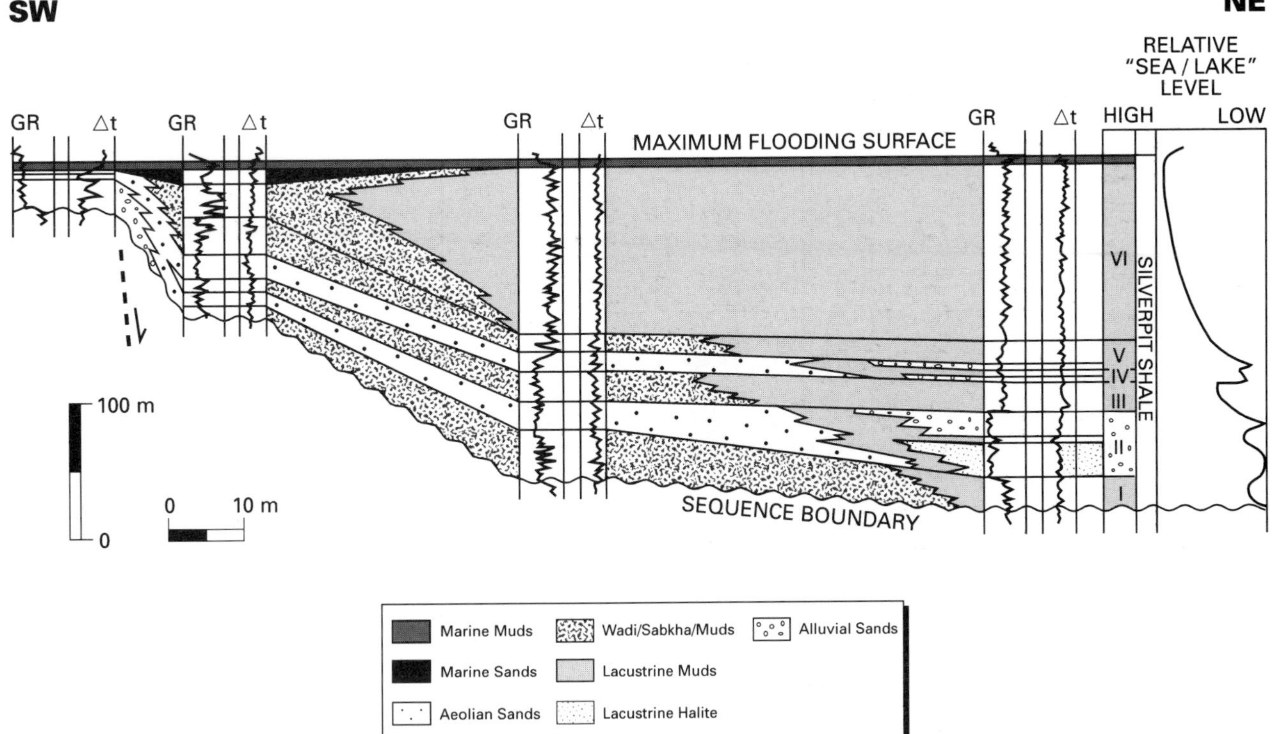

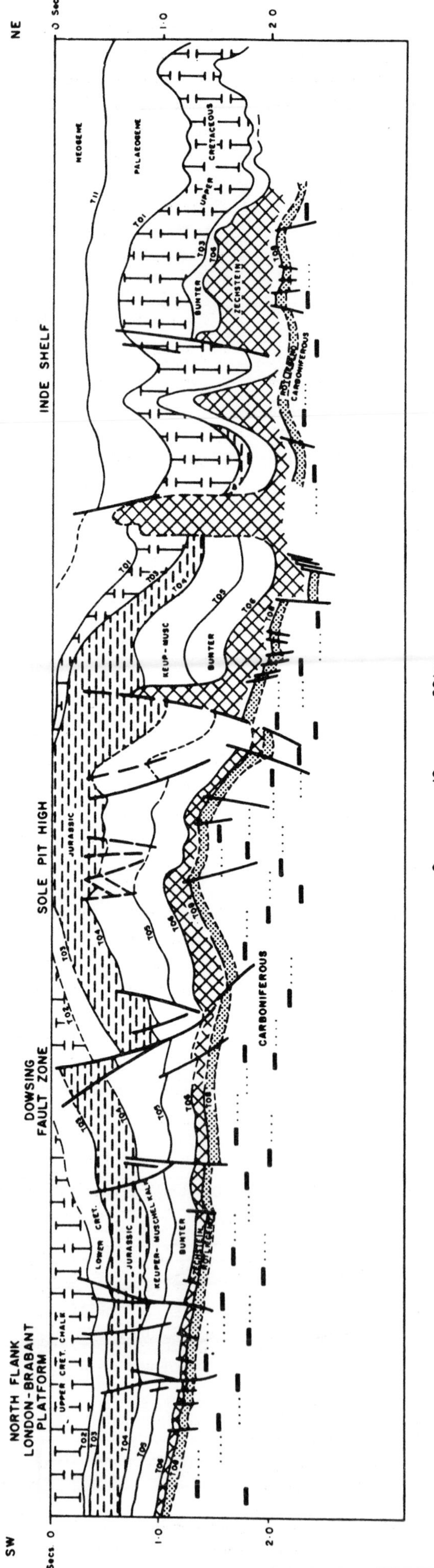

Fig. 12.49 South-west–north-east structural cross-section, Southern Permian Basin (from Glennie and Provan, 1990).

in gas maturation and migration at the same time as the Cimmerian rifting and wrenching was forming faulted anticlinal traps at Top Rotliegend level. Retention of this early phase of gas generation was disturbed by subsequent inversion, initially in the Middle–Late Cretaceous and subsequently during the Tertiary, which affected different parts of the basin to differing degrees and at different times. It is generally believed, for example, that gas trapped in early structures within the shallower, marginal parts of the basins remigrated towards the elevated parts of the inversion axes, where reservoir quality had been reduced due to the more extreme, depth-related diagenesis (e.g. Sole Pit area; but see Glennie, Chapter 5, this volume, Section 5.5.1, for a more complete discussion of this complex process). Elsewhere, such as the offshore Netherlands, the capacity of Westphalian coals to generate gas had not been exhausted during pre-inversion burial, which enabled a later phase of gas generation to occur during later Cretaceous and Tertiary burial (van Wijhe *et al.*, 1980; van Wijhe, 1987a,b). This has resulted in most accumulations being full to spill point (Lutz *et al.*, 1975).

Play-fairway analysis. The key factors determining the distribution of this play fairway are: (i) depositional facies trends (fluvial/wadi and aeolian sandstones located between the sediment source areas and the rapid shale-out into the lacustrine facies of the Silverpit Claystone Formation); (ii) nature and extent of depth-related diagenesis (particularly the degree of illitization); (iii) presence and thickness of Zechstein Salt; and (iv) distribution of the areas of basin inversion. The main risks associated with these factors are: (i) tight reservoirs, resulting from extreme diagenesis (e.g. in the Sole Pit area); (ii) lack of charge and/or loss of early charge, caused by structural inversion (e.g. Sean area and in marginal areas of the basin, such as the South Hewett Shelf, where the Zechstein Salt is thin); (iii) reduced top-seal quality, due to lateral facies changes (e.g. where salt is replaced by carbonate facies, such as in the South Hewett Shelf area); and (iv) seismic depth conversion, giving the depth and height of structural closure, resulting from the complex overburden and generally low relief of the Top Rotliegend closures (many structural closures are visible only after seismic depth conversion).

An additional play also occurs where near-basal Rotliegend sandstones directly overlie the Saalian Unconformity and where they are sealed by overlying claystones of the Silverpit Formation (Bailey *et al.*, 1993). The play fairway in this case is best developed to the north-east of the main Rotliegend play, in the Silver Pit area, where reduced sandstone thickness and quality significantly increase the exploration risk (see Fig. 12.48).

12.3.3 Zechstein carbonate play

The Zechstein is best known because of its sealing potential for the Rotliegend play and for generating a wide range of structures in overlying successions through salt movement. Other parts of this complex lithological succession, however, also contain potential reservoir and source-rock facies (see Taylor, Chapter 6, this volume). The Zechstein potential for oil, condensate and gas is realized in the onshore Netherlands, Germany and Poland, but has never been fully recognized in the UK sector of the Southern North Sea. Limited volumes of gas were produced from fractured Z2 and Z3 carbonates in the onshore wells at Lockton and Eskdale, and the Low Marishes well tested gas at 9.5×10^6 ft^3/day through a 44/64″ choke. In the adjacent offshore, gas and condensate were tested in blocks 41/20, 41/24 and 41/25.

The Zechstein comprises four carbonate/evaporite sedimentary cycles (Z1–Z4), each one corresponding to a rapid marine transgression, followed by a phase of regression, which was generally accompanied by an increase in salinity (Fig. 12.50). An idealized cycle comprises a thin clastic interval, followed by limestones and dolomites (including barrier/shelf-margin reservoir facies), and passing into anhydrite and halite (seal facies) (see Taylor, Chapter 6, this volume, for a full review of the Zechstein cycles, including their source, reservoir and seal characteristics).

The Zechstein sequence contains oil-prone source rocks, but not in sufficient quantity to constitute a major source. The Kupferschiefer, at the base of the succession underlying the Z1–Z4 cycles (Fig. 12.50), has total organic carbon

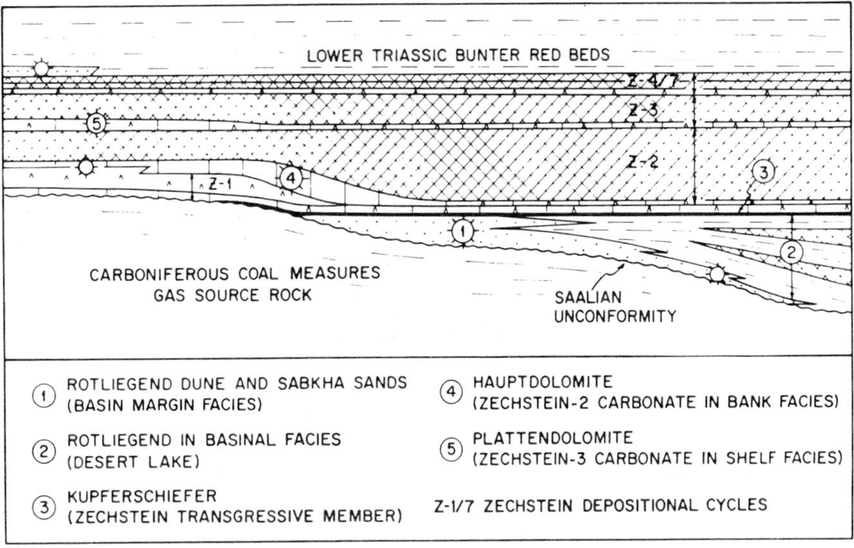

Fig. 12.50 Late Permian Zechstein depositional cycles (Z1 to Z4/7) of the Northern and Southern Permian basins, and the main hydrocarbon occurrences (from Ziegler, 1990a).

(TOC) values of 5–15%, and algal laminates occur within most of the cycles. Although no North Sea oils have definitely been attributed to Zechstein source rocks, circumstantial evidence would suggest that some oil shows do derive from this source. However, the most common established Zechstein carbonate play in the Southern North Sea comprises gas accumulations in structural traps sourced from the Carboniferous (Fig. 12.50). Even so, some of the condensate that occurs in many Rotliegend fields may be derived from the Kupferschiefer or other Zechstein sources.

One of the factors limiting Zechstein potential is prediction of reservoir quality, mainly permeability, which is highly variable and subject to rapid lateral changes (see Taylor, Chapter 6, this volume, for a fuller discussion). For example, well 48/22-1 tested 2000 barrels of oil per day from a fractured Zechstein reservoir (Hauptdolomit), but the appraisal well 48/22-2 encountered a tight reservoir section. Unfortunately, the Zechstein has rarely been a primary target and consequently most drilling activity has not been optimally located for the Zechstein play, with most wells occurring in areas of poor reservoir facies. The best reservoirs occur in oncolitic and oolitic barrier/shelf-margin facies, which are concentrated near the basinward margins of the Zechstein shelf in the Southern North Sea. The Hauptdolomit/Zechsteinkalk is the most common reservoir, which is gas-bearing where the intercalated evaporite seals have been breached, allowing charge from the Carboniferous (Fig. 12.50). This is most clearly illustrated in the Hewett Field, which includes gas production from the fractured carbonates of the Z1 Zechsteinkalk and the Z3 Plattendolomit, within a Late Jurassic–Cretaceous anticline. Most notable about these secondary reservoirs (see Section 12.3.4 for a summary of the main Bunter reservoirs) is the enhanced production (fivefold increase) achieved from optimal completion of the open-fracture systems, which could be applied to other marginal, fractured Zechstein reservoirs (Cooke-Yarborough, 1994). In some parts of the basin, fractured Zechstein reservoirs can produce gas at up to 400 000 m^3/day/well, which is some 10–100 times greater than that of an unfractured reservoir (van der Baan, 1990). Dramatic productivity improvements can also be achieved through artificial fracture stimulation. Elsewhere, reservoir properties may have been improved through leaching and dissolution (e.g. as seen in the Auk and Argyll fields of the Central North Sea (see Section 6.12.4).

12.3.4 Triassic play

A few gas accumulations occur in anticlinal traps within the good-quality Triassic sandstone reservoirs (e.g. Bunter and Hewett sandstones), but only in those limited areas where the Zechstein salt seal has been breached, through a combination of reduced thickness, facies changes and faulting. This has occurred mainly where the Zechstein seal effectiveness decreases towards the basin margins and has allowed vertical migration to charge the Bunter and Hewett sandstones, which form separate, vertically stacked reservoirs in a north-west–south-east-orientated anticline bounded by faults on its north-east and south-west flanks (Hewett Field; Cooke-Yarborough, 1991). In parts of the Dutch sector of the North Sea, gas has also been discovered in the Bunter Sandstone, even where it overlies thick Zechstein salt (Roos and Smits, 1983). Significantly, this occurs in areas of strong Late Cretaceous to Early Tertiary inversion and it is inferred that the Bunter was charged following gas leakage from early-formed Rotliegend accumulations during tectonic inversion (Oele *et al.*, 1981).

A different style of gas-bearing Bunter Sandstone structure occurs on the northern side of the North Sea Basin in the Esmond, Forbes and Gordon fields (UK Quadrant 43). The three separate accumulations occur in small (gas reserves = 535×10^9 ft^3/15.1×10^9 m^3), simple and unfaulted dip-closed anticlines, formed by the pillowing of the underlying Zechstein evaporites and sealed by the Röt Claystone Member (Ketter, 1991a). The high nitrogen content of the gas (8–14%) suggests relatively long-distance remigration from an earlier Rotliegend trap. Areas of salt withdrawal, resulting in thinned or absent Zechstein salt, have acted as migration pathways during initial charging and filling of the structures in the mid- to Late Jurassic. Continued structural growth during the Tertiary increased structural closure, but this was not accompanied by additional gas migration, since the structures are far from full to spill point.

12.3.5 Remaining hydrocarbon potential in the Southern North Sea

The Southern North Sea Basin is widely considered to be at a more mature stage of exploration than the Central and Northern North Sea. To date, with the exception of the Carboniferous plays, all the readily identified structures and potentially prospective horizons appear to have been drilled. Future discoveries either will be of limited volume in existing plays or will test new concepts of trapping or reservoir distribution. Undiscovered gas reserves are estimated at 21×10^{12} ft^3 (600×10^9 m^3), primarily in the Carboniferous play.

Carboniferous play

For such a relatively mature basin, the Carboniferous play is still underexplored (Smith, 1993). Attractive exploration potential remains, although prospective acreage is already licensed. Remaining potential derives largely from the difficulties so far encountered in mapping prospects and determining the occurrence of good-quality reservoir facies. Problems in seismic imaging and depth conversion result primarily from the overlying Zechstein. In addition, it is difficult to differentiate between Carboniferous sequences lacking coals and the overlying Rotliegend, so the position of the unconformity surface is often conjectural.

The most attractive reservoir facies are associated with major channel systems. These may be located in incised palaeovalleys or in less areally constrained fluviodeltaic complexes. The Carboniferous stratigraphic framework in general usage has not solved the problems of the subsurface correlation that would aid in consistently locating these channels. A combination of refined geophysical and stratigraphic resolution and interpretation holds the key to successful future exploitation of this play. The potential for large gas discoveries cannot be discounted.

Permian Rotliegend Sandstone play

The Rotliegend play has all of the regional play parameters particularly favourably combined. As a consequence, cumulative regional exploration success rates indicate that some of the earlier exploration decisions to downgrade prospects when drilling failure was attributed to local parameters, especially reservoir quality, may have been too pessimistic. Some potential may therefore remain in previously rejected prospects. Although it may be assumed that all large Rotliegend structures have been tested, considerable cumulative reserves may remain in small structures and, based on recent experience in the Dutch sector, in hangingwall (lowside) closures against sealing faults. Individually these may not be commercially viable at current prices, but they can be profitable as incremental additions to existing production.

One of the limiting factors in the effectiveness of the Rotliegend reservoir is the degree of diagenetic deterioration. For example, the most significant reservoir facies in the German Rotliegend forms a broad (12–20 km wide) zone of sabkha and aeolian sands adjacent to, and interbedded with, lacustrine shales, which contain 60% of the gas reserves. A similar sequence occurs in the Ravenspurn/Caister area. In both cases, the most important control on reservoir quality is the occurrence of pore-lining radial chlorite cement, which inhibits later diagenetic cementation. Such a sabkha complex has not been specifically targeted by drilling along the margins of the Silverpit Lake, where there would also be the potential for stratigraphic traps where the sands interdigitate with the lacustrine shales (Fig. 12.50).

New play concepts may develop from a more complete understanding of the structural evolution of the basin. Recent interpretation of gas and pore-fluid movements during inversion has focused attention on the potential for deep gas traps. Previously, it was believed that the gas in the Sole Pit Basin was the result of remigration during Late Cretaceous and Tertiary inversion, the gas originally having been expelled during pre-inversion subsidence. The new evidence suggests that on the Leman Bank the gas has been in place since at least the early Middle Jurassic (170–175 Ma before present (BP) and was effectively contained during subsidence by the Zechstein top seal and anhydrite-cemented lateral fault-plane seals. As early emplacement of gas inhibits illite growth and preserves permeability, the implications for deep gas in lateral fault-sealed downdip or synclinal closures with reasonable reservoir are attractive. More comprehensive isotopic dating of diagenetic illite and fault-sealing anhydrite, coupled with a detailed reconstruction of the inversion history of the Sole Pit Basin, could reveal a new generation of Rotliegend objectives.

Zechstein carbonate play

Perhaps the best potential for additional hydrocarbons in Zechstein reservoirs is in the Hauptdolomit shelf-edge barrier carbonates. The principal Rotliegend targets overlie either tight lagoonal facies or tight basinal facies to the north and south of the shelf edge. Coupled with the difficulty of mapping structures at top Hauptdolomit level, the lack of data on the better-quality reservoir facies limits our understanding of this play. As a result, the potential for stratigraphic trapping along the updip margin of the shelf-edge barrier complex should not be overlooked.

The Z1 Zechsteinkalk also has attractive reservoir potential in intertidal and upper submarine-slope carbonates, and this is productive in the Hewett Field (Cooke-Yarborough, 1991). The most productive dolomite intervals are relatively tight, with low matrix permeability (< 1 mD), with a typically high gamma-ray response, but they are highly fractured. The dominant regional fracture trend is from north–south to north-north-east–south-south-west and this play could have further potential, such as along the depositional strike from the Hewett Field (Cooke-Yarborough, 1994). If regional fracture trends can be predicted with confidence, experience gained from the Hewett Field could also encourage a re-examination of some of the Zechstein well sections, where shows were recorded but the formation was dismissed as too tight to warrant testing (e.g. follow-up on fracture porosity of 48/22-1).

Triassic play

When the sealing effectiveness of the underlying Zechstein halites is taken into consideration, the Triassic play has been modestly successful. The obviously attractive prospects have all been tested, however, and the small structures that remain undrilled in areas of breached or absent Zechstein halites would have no stand-alone potential at current gas prices. If there is remaining potential it will be in subtle traps, although the charge problem will still be a major constraint. Faulting and diapiric activity associated with Late Cretaceous inversion created a conduit for gas to escape from Rotliegend reservoirs to form the Triassic K/13 fields. Similarly, fault activity at the periphery of the Zechstein halite basin allowed gas to migrate from Rotliegend reservoirs to form the Hewett Field. The Esmond, Forbes and Gordon (Quadrant 43) and Caister B (block 44/23a) fields are located over salt swells, which have generated areas of thin Zechstein halites. Here, gas has migrated from the Carboniferous, through the Silverpit Claystone facies of the Rotliegend, and into the Triassic structures. It is in these structural settings, which facilitate either direct or indirect connection between reservoir and source, that the search for subtle Triassic traps must start. The prospects need not be of large rock volume because in deep traps, such as the Dutch F/15A Field, the formation-volume factor enables large gas reserves to be trapped in a relatively small closure.

12.3.6 Other Mesozoic plays

Other Mesozoic plays occur in the southernmost part of the Southern North Sea and extend onshore into the Netherlands and north-west Germany (Fig. 12.51). These plays are sourced mainly by Lower Jurassic rocks, which have limited potential as a result of continental to marginal

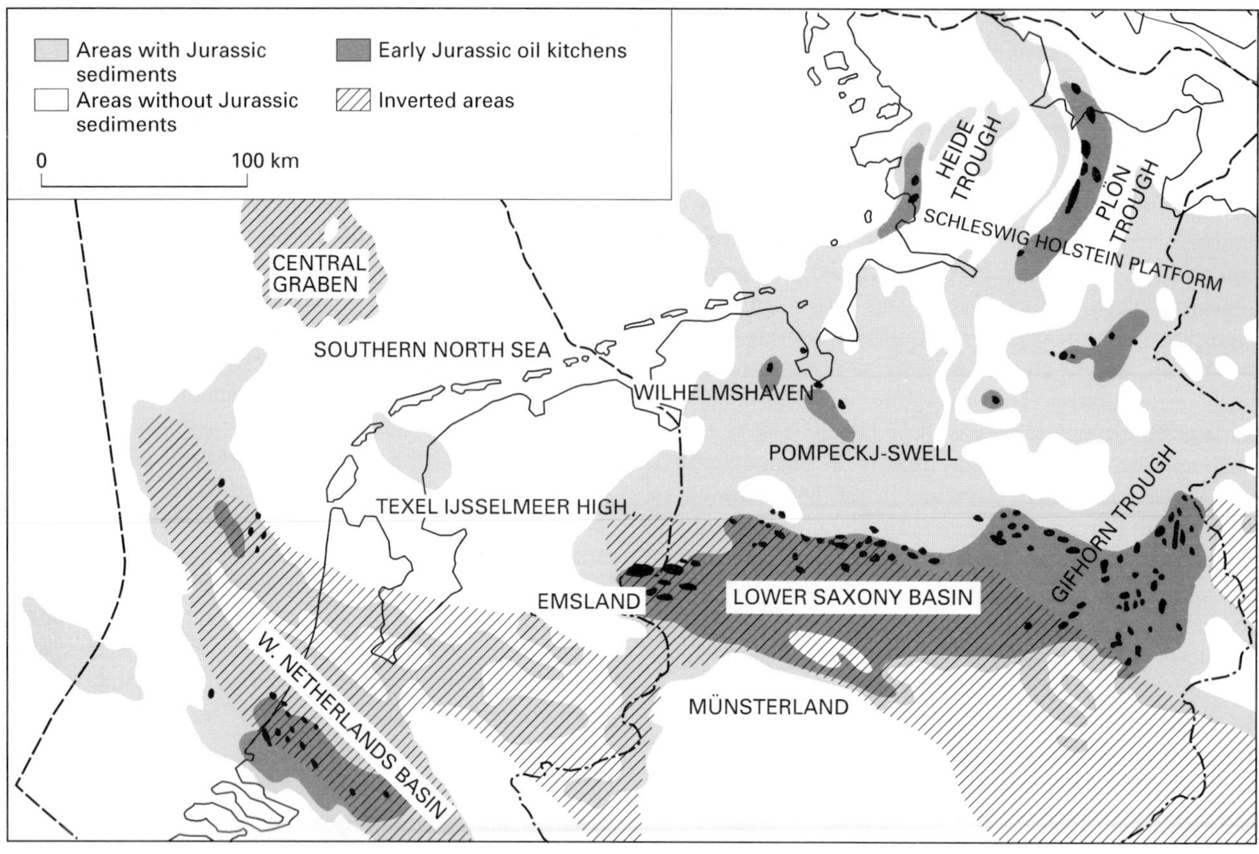

Fig. 12.51 Distribution of the Mesozoic play fairway in the southernmost part of the Southern North Sea and its onshore extension into the Netherlands and Germany (from Ziegler, 1990a).

marine facies, including thin, sporadic coals, and common erosion/non-preservation. Oil sourced from the Liassic Posidonia shales is reservoired in Mesozoic rocks, mainly Upper Jurassic to Lower Cretaceous sands (Fig. 12.52). The Liassic source rock is not a major contributor to prospectivity in the immediately adjacent UK acreage (Southern North Sea), mainly because of its general immaturity and lack of preservation, due to widespread erosion (e.g. Fig. 12.49). However, slightly further afield, most notably in parts of the Wessex Basin, marine Liassic mudstones are the principal source rock for Jurassic and Triassic sandstone reservoirs in the large Wytch Farm oilfield (see Section 12.4.1).

The Lower Cretaceous is a particularly important play in the Dutch sector of the Southern North Sea and in its onshore extension into both the southern and northern parts of the Netherlands (Bodenhausen and Ott, 1981; de Jager *et al.*, 1996; see Oakman and Partington, Chapter 9, this volume, Fig. 9.7) and eastwards into the Lower Saxony Basin of north-west Germany (see Fig. 12.51; Boigk *et al.*, 1963; Boigk, 1981; Binot *et al.*, 1993). The petroleum system in this case is quite different from that of the main North Sea Basin: a succession of mainly progradational coastal sandstone reservoirs are retrogradationally stacked and overlain by transgressive shallow-marine mudstones, which provide local seals (Figs 12.53, 9.2A, 9.10 and 9.11; den Hartog Jager, 1996; Racero-Baena and Drake, 1996). Hydrocarbons were generated from two source rocks; the Lower Jurassic (Liassic to Toarcian) Posidonia Shale is the principal source rock, and the Lower Cretaceous (Ryazanian) 'Wealden' (Delfland Formation; Fig. 12.53) bituminous shales form a secondary source rock (Binot *et al.*, 1993). This play is complicated by Late Cretaceous to Early Tertiary inversion tectonics in a way similar to the Permian gas play in the Southern North Sea Basin (see Section 12.3; Fig. 9.14), which effectively destroyed those early hydrocarbon accumulations resulting from migration in the Late Jurassic–Late Turonian. The majority of the current anticlinal traps (e.g. Fig. 12.54 and Fig. 9.19) formed during the period of Late Cretaceous to Early Tertiary inversion tectonics, with hydrocarbon migration mainly later than 88 Ma, with most having formed in subrecent times or during the Late Tertiary.

The Upper Jurassic to Lower Cretaceous play in the onshore Netherlands is highly mature, and all the main structural traps have probably been discovered. However, the combination of high-resolution sequence stratigraphy, modern hydrocarbon-habitat and modelling studies and, especially, 3D seismic data may revitalize this old play (de Jager *et al.*, 1996; den Hartog Jager, 1996; Racero-Baena and Drake, 1996). This also applies to the other mature Mesozoic plays of the northern and western parts of Netherlands (Bruijn, 1996; Goh, 1996; Rijkers and Geluk, 1996).

In the central and northern part of the Netherlands (Central Netherlands, Texel-Ijsselmeer High, Friesland Platform and Vlieland Basin; Figs 12.40 and 12.55), there is a wide range of play types, ranging from the Carbonife-

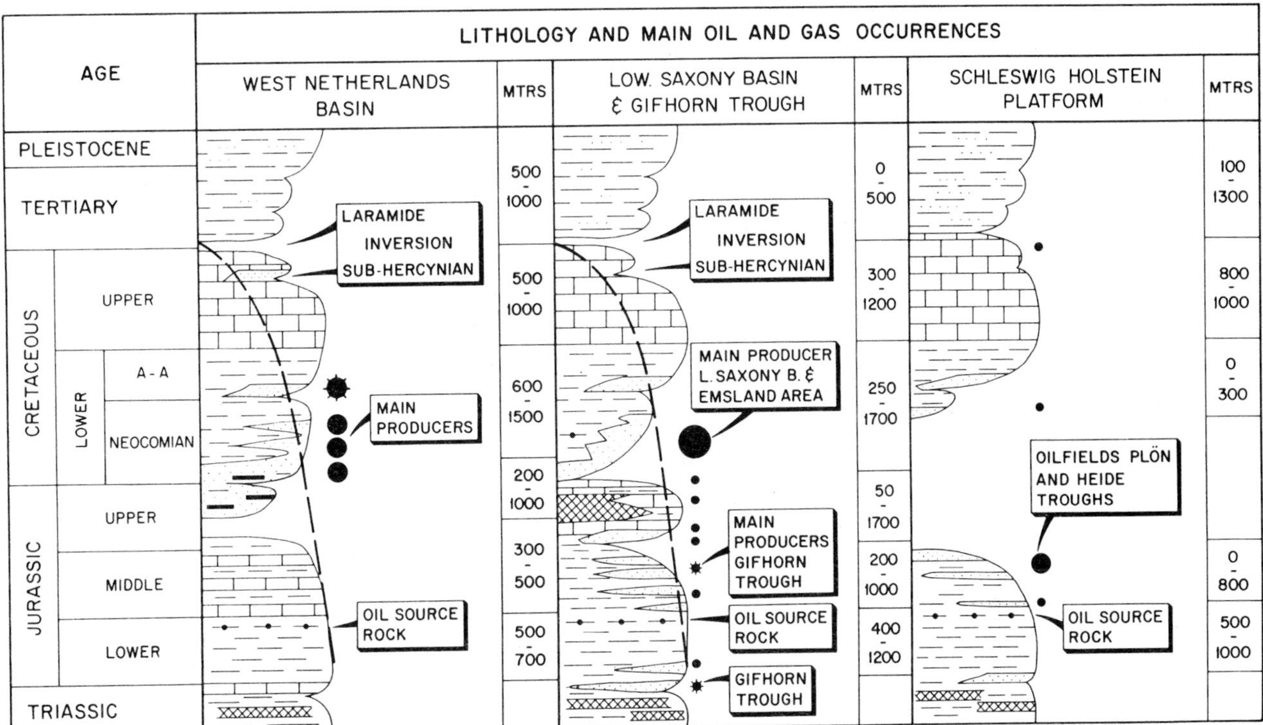

Fig. 12.52 Stratigraphic sections illustrating the main reservoirs and source rocks of the Dutch–German Mesozoic oil province (from Ziegler, 1990a).

rous through to the Tertiary (Fig. 12.56), and including the onshore extension of the Permian Rotliegend Sandstone play, with its giant Groningen gasfield. The province is dominated by gas accumulations sourced from the Carboniferous, with traps including inversion anticlines and subunconformity traps, as in the old de Wijk Field (discovered in 1949), which is unique in the Netherlands for having hydrocarbons in Carboniferous, Permian, Triassic, Cretaceous and Tertiary reservoirs (Fig. 12.57). A further unusual factor is that the main reservoir comprises Triassic claystones, where post-depositional leaching of anhydrite has significantly enhanced reservoir properties (Bruijn, 1996). The stratigraphically widespread occurrence of gas, as also exemplified by the Texel-Ijsselmeer High, is a result of the highly efficient and extensive gas generation from the Carboniferous and the effective vertical and lateral migration through faults and along unconformities (Fig. 12.58; Rijkers and Geluk, 1996).

12.4 Western British Isles plays

The western margin of the British Isles includes a large number of sedimentary basins, which form part of a series of rift-related Mesozoic basins, extending along the whole length of the Atlantic Margin from the South-West Approaches in the south to the Barents Sea in the north (Fig. 12.59). However, the hydrocarbon potential of those sedimentary basins immediately to the west and south-west of the British Isles, including the Irish offshore areas, has failed to achieve many of the early (e.g. 1970s) predictions; certainly there is no second North Sea Basin. Recent activities, however, have highlighted significant remaining potential and, in the case of new discoveries in the Faeroe Channel Basin, have led to a resurgence in exploration activity along the whole Atlantic Margin of the British Isles. In southern Britain, the recognition of a major extension to the Wytch Farm oilfield and the successful use of the latest technological advances in extended-reach drilling have also fuelled renewed exploration interest in the Wessex Basin. The experience gained in extended-reach and horizontal drilling is also being applied to other areas, including the North Sea, and has highlighted the potential for maximizing oil recovery from complex reservoirs and successfully developing marginal fields. All these activities have contributed to the current resurgence in exploration and production throughout the western British Isles. For this reason, the main petroliferous basins and their most significant plays are summarized below.

12.4.1 Wessex Basin plays

The Wessex Basin (synonymous with the Channel Basin of some workers; e.g. Ziegler, 1990a) is the most important onshore hydrocarbon province in the British Isles and, with the Wytch Farm oilfield (with some 300×10^6 barrels of oil reserves; Bowman *et al.*, 1993), contains the largest onshore oil accumulation in north-west Europe. Focused hydrocarbon exploration in this area began in 1934, based on the analysis of numerous oil seeps and one of inflammable gas along the Dorset coast, together with the surface mapping of anticlinal structures (Fig. 12.60A). Continued, if somewhat sporadic, exploration since then has resulted in the basin now estimated to contain over 300×10^6 barrels of oil and some 100×10^9 ft^3 gas, with the Wytch Farm oilfield the overwhelmingly dominant contributor (Bowman *et al.*, 1993; Hawkes *et al.*, 1997). As with most basins in the western British Isles, the history of hydrocarbon generation, migration, entrapment and preservation in the Wessex

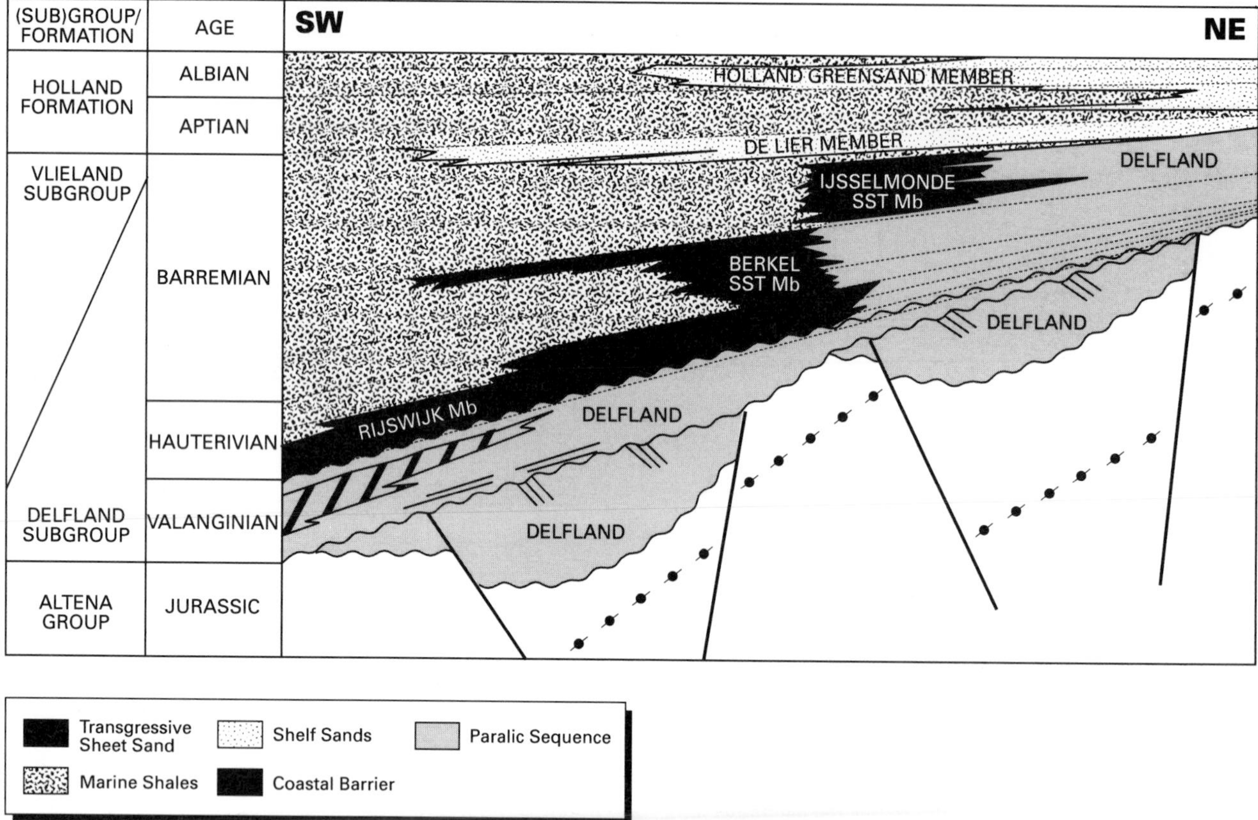

Fig. 12.53 Schematic stratigraphic cross-section through the Lower Cretaceous of the West Netherlands Basin (from Racero-Baena and Drake, 1996). Note the overall retrogradational stacking of the coastal and shelf-sand bodies in response to the widespread relative sea-level rise during the Early Cretaceous.

Basin is complicated by complex stratigraphic and tectonic histories, particularly the effect of Late Tertiary tectonic inversion, which, except in the Southern North Sea and locally in the Central Graben, was largely absent over most of the North Sea Basin.

The stratigraphy, depositional environments and associated tectonic events of the Wessex Basin, and the closely related Weald Basin, have been studied by many workers, but only a few have related surface-outcrop geology to the basin's hydrocarbon habitat (e.g. Colter and Havard, 1981; Selley and Stoneley, 1987). Hawkes *et al.* (1997), who provide one of the most recent reviews of this basin, divide the succession into four major tectonostratigraphic sequences (termed 'plate sequences') and relate these and their associated hydrocarbon play types (Fig. 12.61) to major plate-margin processes in the North Atlantic and Tethyan Provinces. The four sequences are Early Atlantic (Late Permian–Triassic), Atlantic (Early–Late Jurassic), Biscay (Late Jurassic–Late Cretaceous) and Alpine (Late Cretaceous–Tertiary), which control the development of the hydrocarbon system within the Wessex Basin. This provides a framework for predicting reservoir, seal and source-rock distribution and for evaluating the timing of trap formation, source-rock burial/maturation and hydrocarbon migration, which are summarized below. (To aid comparison with other parts of this chapter, stratigraphic nomenclature is placed ahead of the plate-sequence names of Hawkes *et al.*, 1997.)

Late Permian–Triassic (Early Atlantic plate sequence)

This is characterized by two large-scale, fining-upward cycles, which are related to two unsuccessful attempts to open the north Atlantic. These cycles are characterized by continental clastic facies at their base and playa facies at the top: (i) Permian Rotliegend sandstones and overlying marls/evaporites (the Red Marls); and (ii) Trassic Sherwood Sandstone (the major reservoir at Wytch Farm) and the overlying Mercia Mudstone Group. Each cycle provides an excellent reservoir seal pair (Fig. 12.61). There is, however, a marked lack of source-rock potential at this stratigraphic level.

Early–Late Jurassic (Atlantic plate sequence)

This marks the onset of a rapid and widespread marine transgression, which was accompanied by rifting. This tectonic phase is linked to successful rifting in the central Atlantic. Cyclic sedimentation comprises several shallowing-upward shelf to coastal depositional systems, which resulted in the repeated accumulation of several potential reservoir/seal/source intervals. This includes deposition of Liassic marine mudstones, which are a most significant source rock for liquid hydrocarbons, both in the Wessex Basin and in several nearby basins (e.g. Western Approaches, Weald and Paris basins). Other source rocks, most notably the Kimmeridge Clay, have the potential to generate hydrocarbons, but are immature throughout the basin.

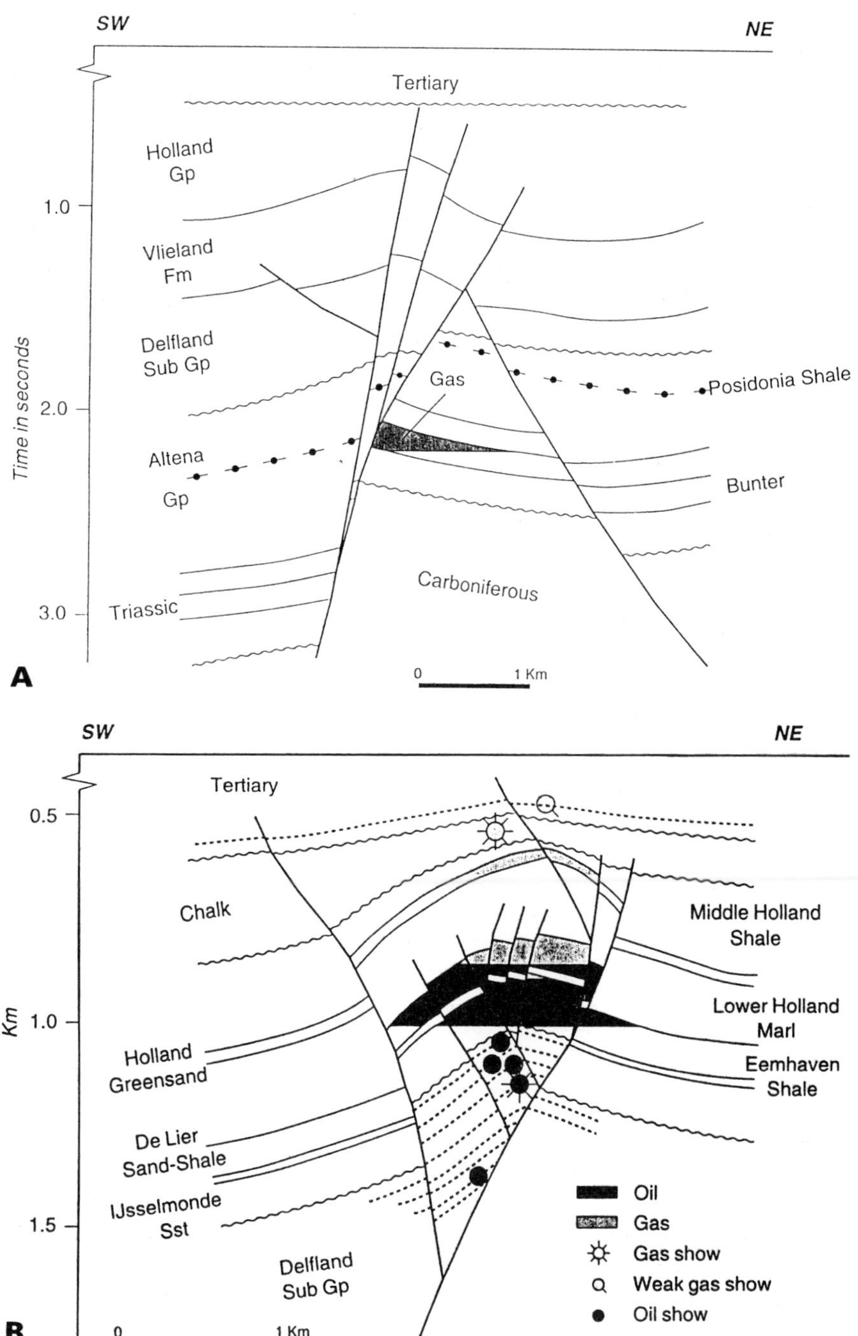

Fig. 12.54 Examples of structural traps in the West Netherlands Basin: (A) structural cross-section through an accumulation in the Triassic Bunter play of the Wassenaar Deep trap; and (B) structural cross-section through an accumulation in the Upper Jurassic–Lower Cretaceous play of the IJsselmonde/Ridderkerk trap (from de Jager *et al.*, 1996).

Thickness patterns within the Jurassic sequences reflect active extensional faulting and fault-block rotation during this syn-rift phase. The most important clastic reservoir is the shallow marine/shoreface Bridport Sands (a secondary reservoir at Wytch Farm), together with several other potential sandstone reservoirs, mainly in the Lower to Middle Jurassic. Carbonate reservoirs are also present, but these commonly rely on fractures to be productive (e.g. Cornbrash and Forest Marble).

Late Jurassic–Late Cretaceous (Biscay plate sequence)

This marks the period of successful rifting and spreading in the Bay of Biscay and Rockall areas. This is represented by the formation of the widespread Austrian breakup unconformity during the Aptian–Albian, presumably related to thermal doming. Marine conditions characteristic of most of the Jurassic gave way to non-marine conditions (e.g. Purbeck facies of the latest Jurassic) and, during the Early Cretaceous, widespread deposition of continental fluvial deposits (Wealden facies) within narrow, fault-bounded extensional basins. Early Cretaceous easterly tilting and erosion were followed by a phase of thermal subsidence, with widespread deposition of the shallow-marine Upper Greensand and the Chalk.

The strong differential subsidence during this phase resulted in burial of Lower Jurassic source rocks to maturation levels in various parts of the basin, with Early Cretaceous oil seepage occurring along active fault planes (e.g. along the Purbeck fault zone) and discharging on to the Wealden floodplains (Selley and Stoneley, 1987; Wimbledon *et al.*, 1996). Rotated fault blocks, which form the trap

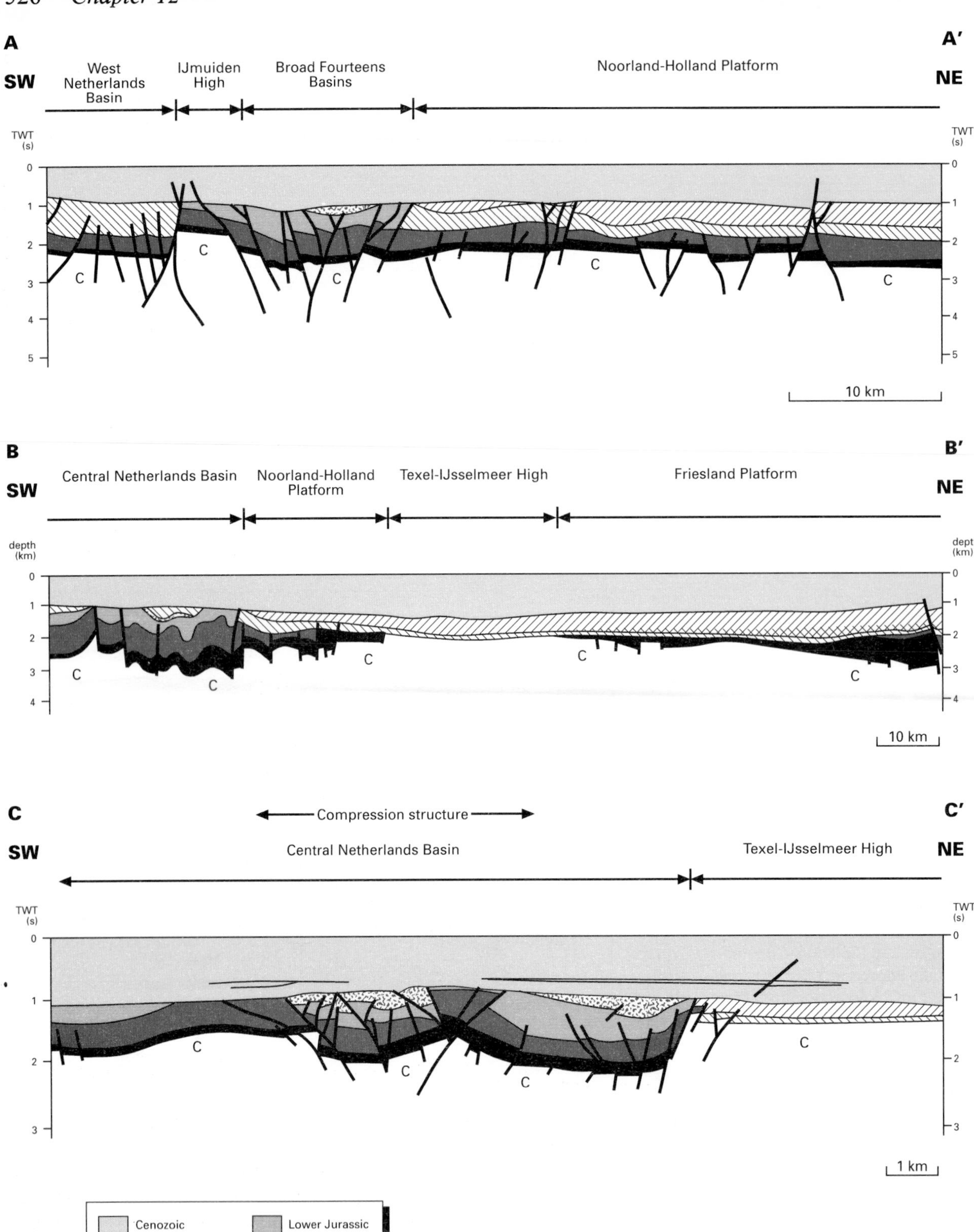

Fig. 12.55 Regional structural cross-sections through the Texel–IJsselmeer High and adjacent basins in the onshore Netherlands (from Rijkers and Geluk, 1996).

at Wytch Farm, were formed at the beginning of this stage and, by the latest Cretaceous, favourable structures with pre-Cretaceous reservoirs were filled with hydrocarbons. Reservoirs deposited during this sequence, however, are not associated with any significant hydrocarbon accumulations.

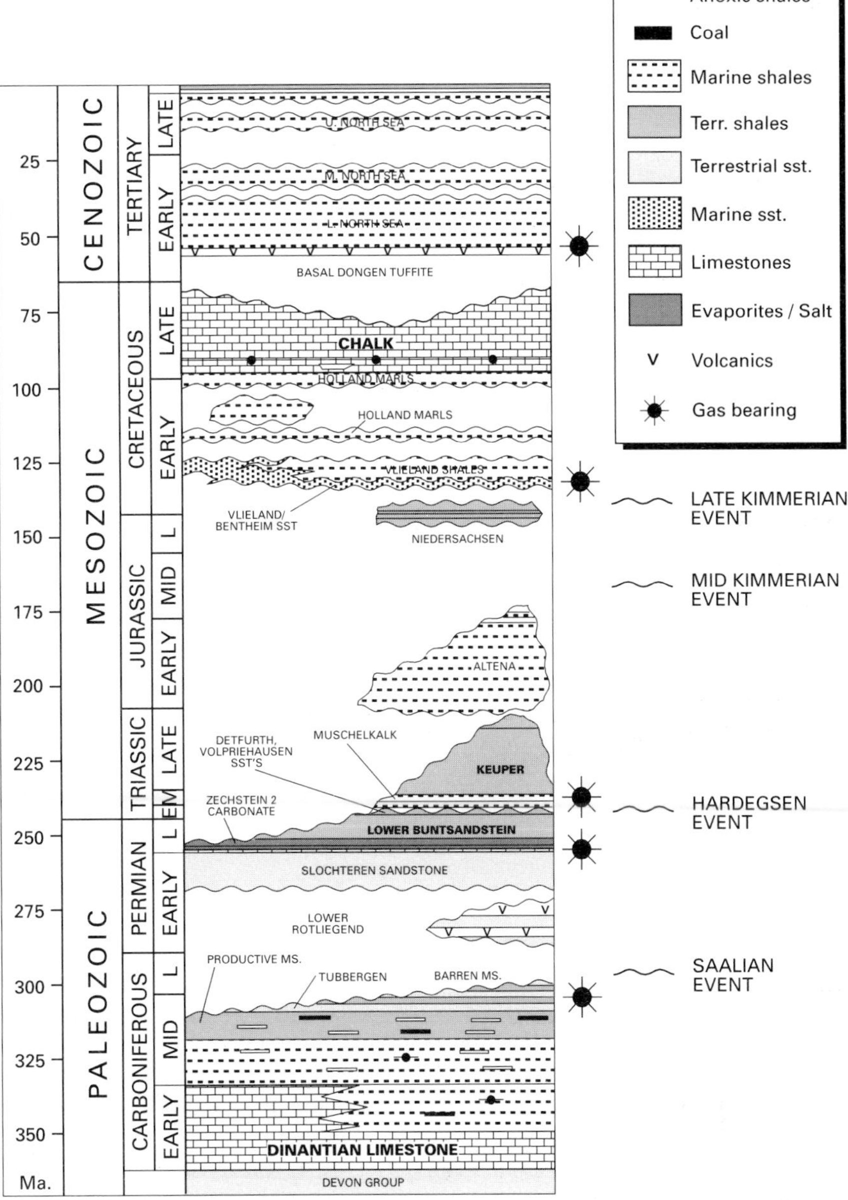

Fig. 12.56 Generalized stratigraphy and source/reservoir/seals of the north-eastern part of the Netherlands (from Bruijn, 1996). Note the association between gas-bearing sandstones and major tectonic events.

Late Cretaceous–Tertiary (Alpine plate sequence)

This is dominated regionally by the north–south closure of Tethys and the subsequent Alpine orogeny, which had a major impact throughout the basins of the southern and south-western British Isles. Strong north–south compression resulted in the inversion of many of the mainly east–west-trending Jurassic–Cretaceous basins, including the Wessex Basin (Stoneley, 1982; Karner *et al.*, 1987), Southern North Sea Basin (see Section 12.3 and Chapter 2) and the numerous basins to the south-west of the British Isles (see Section 12.4.2).

The main result of tectonic inversion in the Wessex Basin was the initiation of reversed movement along earlier extensional faults, most of which were orientated orthogonally to the main compressive forces. This included the formation of prominent inverted anticlines in the hanging-walls of earlier extensional faults (e.g. Purbeck–Isle of Wight Monocline, Portsdown Anticline; see Fig. 12.60B). The main inversion axes became areas of uplift and erosion, while the intervening lows became the sites of new basins infilled with Lower Tertiary fluvial to shallow-marine sediments (e.g. the east–west-trending Hampshire Basin).

The impact of this event on hydrocarbon prospectivity was largely negative. Inverted structures and cap rocks were fractured, faults were reopened and probably large volumes of earlier trapped hydrocarbons were lost to the surface (see Figs 12.60A and 12.63). Source rocks (e.g. the Liassic) which had reached optimum burial depths for maturation by the Late Cretaceous were brought to shallower levels and hydrocarbon generation was reduced and eventually halted. Hence, there is little chance of finding hydrocarbon-bearing Tertiary traps.

Wessex Basin play types

There are two main play types in the Wessex Basin, defined on the basis of structure/trap style and the timing of deformation (Fig. 12.61):

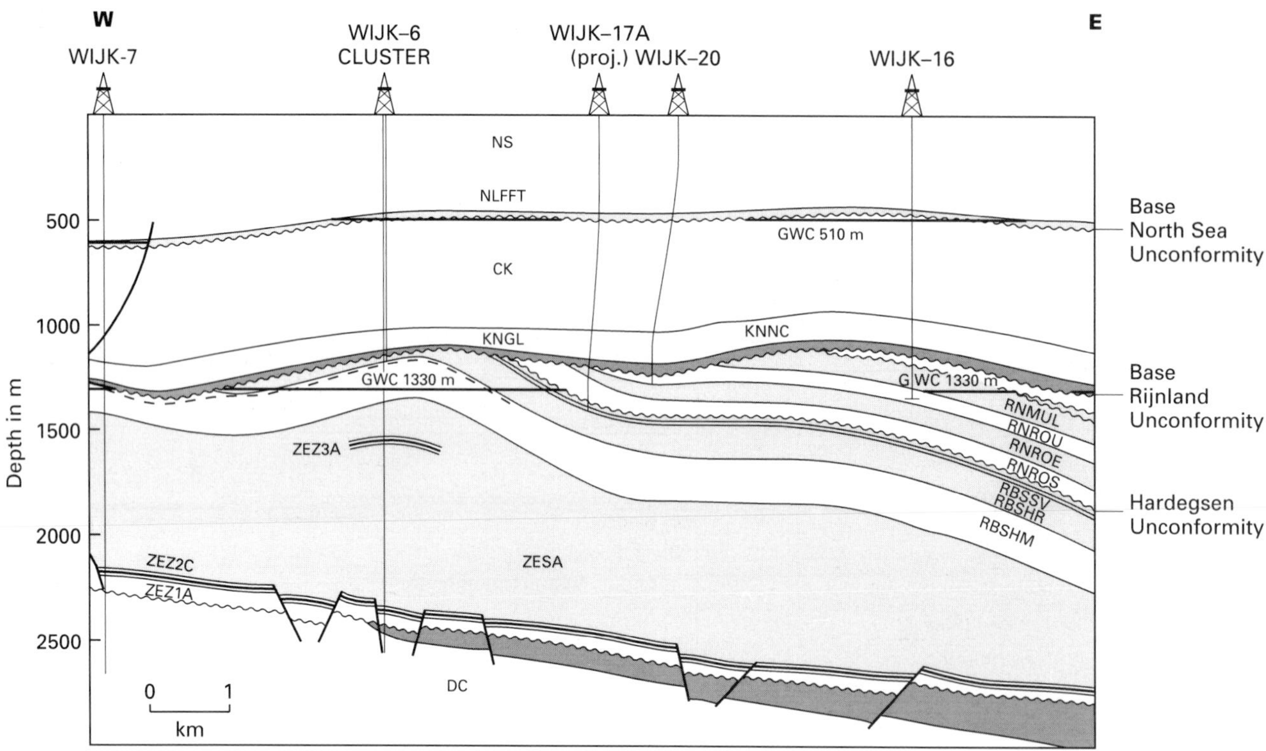

Fig. 12.57 Structural cross-section through the De Wijk Field (from Bruijn, 1996). Major stratigraphic abbreviations are as follows: NS, North Sea Group (Tertiary); CK, Chalk Group; KN, Cretaceous Group; RN, Triassic/Keuper; RB, Triassic/Buntsandstein; ZE, Zechstein; DC, Carboniferous (Westphalian).

1 Jurassic/Early Cretaceous tilted fault blocks/horsts, which are located below the Apto-Albian breakup unconformity and predate hydrocarbon generation. This represents the most successful play type, both in the Wessex and the nearby Weald basins; it includes the Wareham and, most notably, the Wytch Farm tilted-fault-block structures. The latter comprise a primary fluvial sandstone reservoir (Triassic Sherwood Sandstone; reserves of *c.* 260×10^6 barrels of oil) a secondary shallow-marine/shoreface sandstone reservoir (Bridport Sandstone; reserves *c.* 35×10^6 barrels of oil) and a very minor, fractured limestone reservoir ('Frome Clay' limestone; reserves *c.* 5×10^6 barrels of oil). The Wytch Farm accumulation is notable for having survived Tertiary inversion (Fig. 12.62); other similar structures show evidence, in the form of residual oil saturation, of having been once filled but subsequently breached (Fig. 12.63; Selley and Stoneley, 1987).

2 Tertiary inversion anticlines include many structures originally defined by surface mapping and which were the target of early exploration. These structures postdate the main Late Cretaceous phase of hydrocarbon generation and hence many early wells were unsuccessful. One exception is the one-well Kimmeridge oilfield (total cumulative production some 2.5×10^6 barrels of oil since 1961), which is a small inversion anticline located south of the major Purbeck fault zone, with production coming from a fractured-limestone (Cornbrash) reservoir (see Fig. 12.60B). The field is thought to have been charged by late-stage remigration along a fault system that connected with a deeper, Bridport Sandstone trap (Fig. 12.63). Indeed, since the produced volume greatly exceeds trap volume, it is thought that recharging is continuing, albeit at less than the rate of production, at the present time (Hawkes *et al.*, 1997).

Other inversion structures have been found to be dry, with residual oil shows, or to contain minor, non-commercial quantities of gas. This lack of success can be ascribed to: (i) structures/traps located outside kitchen areas and/or beyond any migration pathways; (ii) breaching of traps/cap rocks caused by faulting and fracturing; and (iii) low reservoir quality at various stratigraphic levels, particularly several Jurassic limestone intervals (e.g. Cornbrash/Forest Marble, Inferior Oolite, etc.), but also including lateral facies changes within both the Bridport and Sherwood sandstone reservoirs (Hawkes *et al.*, 1997). In addition, generally poor seismic data quality in the past has hampered pre-Cretaceous trap definition and has contributed to suboptimally located exploration wells.

Compared with many other onshore petroleum provinces (e.g. North America), the Wessex Basin has been only moderately explored.

12.4.2 Basins and plays south-west of the British Isles

Around 20 separate sedimentary basins occur south-west of the British Isles, including the Irish offshore, south-western Britain and north-west France (Fig. 12.64). They range in size from the relatively small basins (some 10s of kilometres width and length) to the huge basins offshore western Ireland (e.g. Porcupine Basin and Rockall Trough, which are some 100s of kilometres long). Exploration activity in this region has been active since around 1970, but results measured in commercial success have been disappointing, with two notable exceptions: the North Celtic Sea Basin

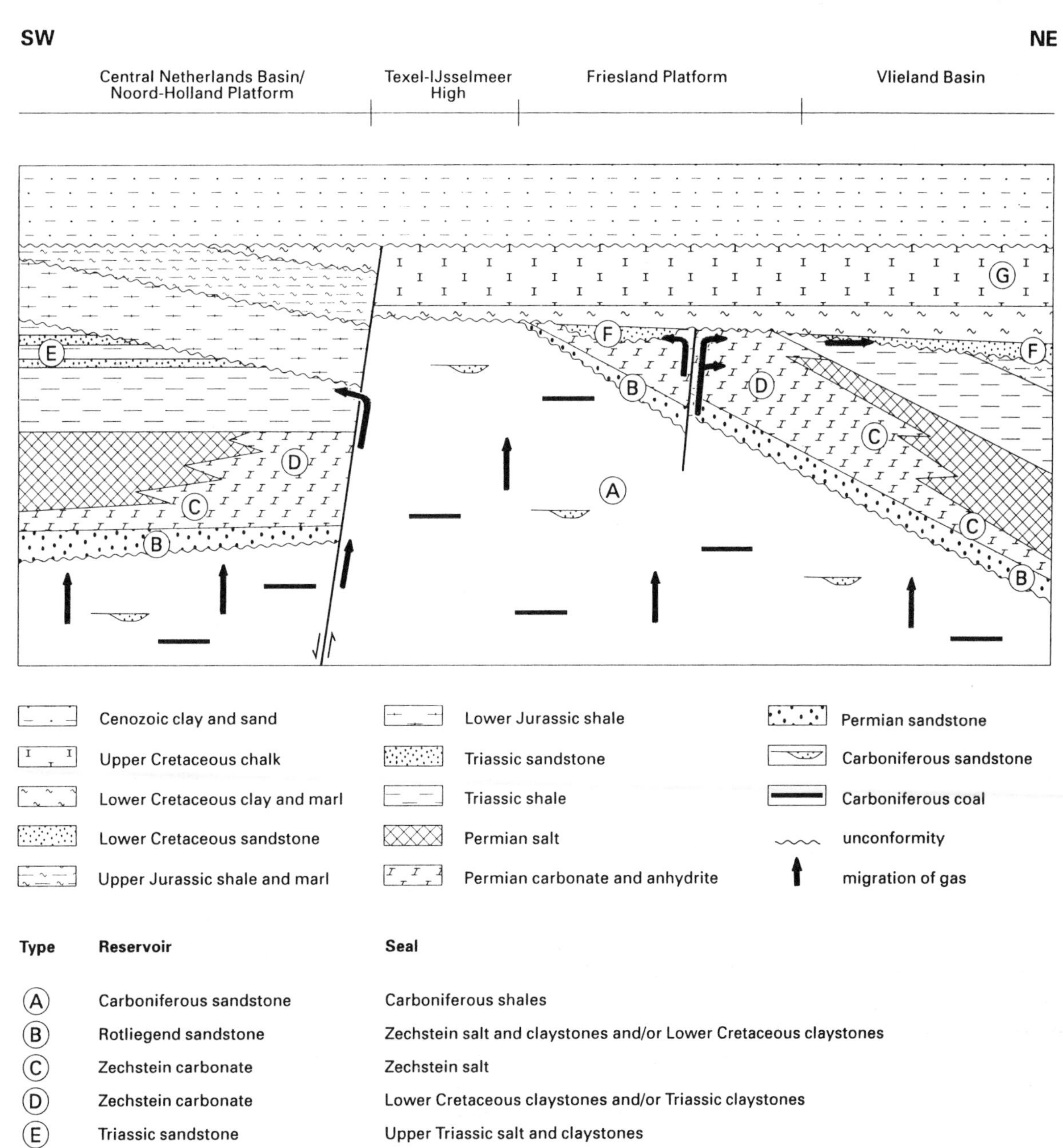

Fig. 12.58 Summary of reservoirs, seals, source rocks and migration pathways around the Texel–IJsselmeer High (from Rijkers and Geluk, 1996).

(this section) and the East Irish Sea Basin (see Section 12.4.3).

The south-western British Isles (excluding western, northern and north-western parts of offshore Ireland) contain several narrow, elongate and partially isolated Mesozoic sedimentary basins, situated in two main provinces (Fig. 12.64): (i) south-west Great Britain to Ireland (the North and South Celtic Sea, Bristol Channel, Haig Fras, St George's Channel, Cardigan Bay, Kish Bank, Central Irish Sea, Cockburn and Fastnet basins); and (ii) south-west Great Britain to north-west France (the Western Approaches, Brittany and South-west Channel basins (Ziegler, 1990a; Croker and Shannon, 1995; Ruffell, 1995).

Despite extensive exploration drilling, the hydrocarbon potential of these basins is limited, mainly due to the largely detrimental effects of widespread tectonic inversion. The timing of inversion varied from basin to basin, but may have started in the earliest Cretaceous, with a later phase during the Late Palaeogene to Neogene in response to Alpine and, especially further west, Pyrenean compressional deformation (Ziegler, 1990a; Knott *et al.*, 1993). The main impact of inversion on hydrocarbon prospectivity was the destruction of pre-existing hydrocarbon accumulations through seal breaching, the remigration of existing oil and gas into new traps, with a consequent large volume loss to the surface, and the interruption, or prevention, of hydrocarbon generation by raising source rocks above their maturation depths. Even so, the hydrocarbon-bearing traps in the North Celtic Sea Basin are all inversion-related anticlines (Figs 12.64 and 12.65).

Fig. 12.59 Late Jurassic geological framework of the Arctic–North Atlantic region (from Ziegler, 1990a).

The geological evolution of this whole region has been summarized by Ziegler (1990a), while more specific aspects of the geological development of the individual basins have been outlined by other workers (e.g. Kamerling, 1979; Gardiner and Sheridan, 1981; Naylor and Shannon, 1982; Millson, 1987; Tucker and Arter, 1987; van Hoorn, 1987b; Petrie *et al.*, 1989; O'Reilly *et al.*, 1991; Shannon, 1991a,b,c; Naylor *et al.*, 1993).

The pre-Cretaceous geology of most of these basins is very similar, each with thick (up to *c.* 4–8 km) Permo-Triassic successions, none of which has ever been fully penetrated by exploration wells. Permian and Early Triassic deposits are dominated by coarse-grained fluvial sandstones and conglomerates, with intercalated aeolian sandstones, providing considerable reservoir potential (cf. Sherwood Sandstone in the Wessex and East Irish Sea basins). The Middle to Late Triassic is dominated by playa-type facies of marls and evaporites, with subordinate limestones. Hence, the Permo-Triassic succession provides a reservoir–seal pair similar to that successfully developed in, for example, the Wytch Farm, Morecambe and Douglas fields. Unfortunately, there is little information on potential Carboniferous source rocks in these basins.

Lower Jurassic marine shales form the most widespread source rock south-west of the British Isles, and have the

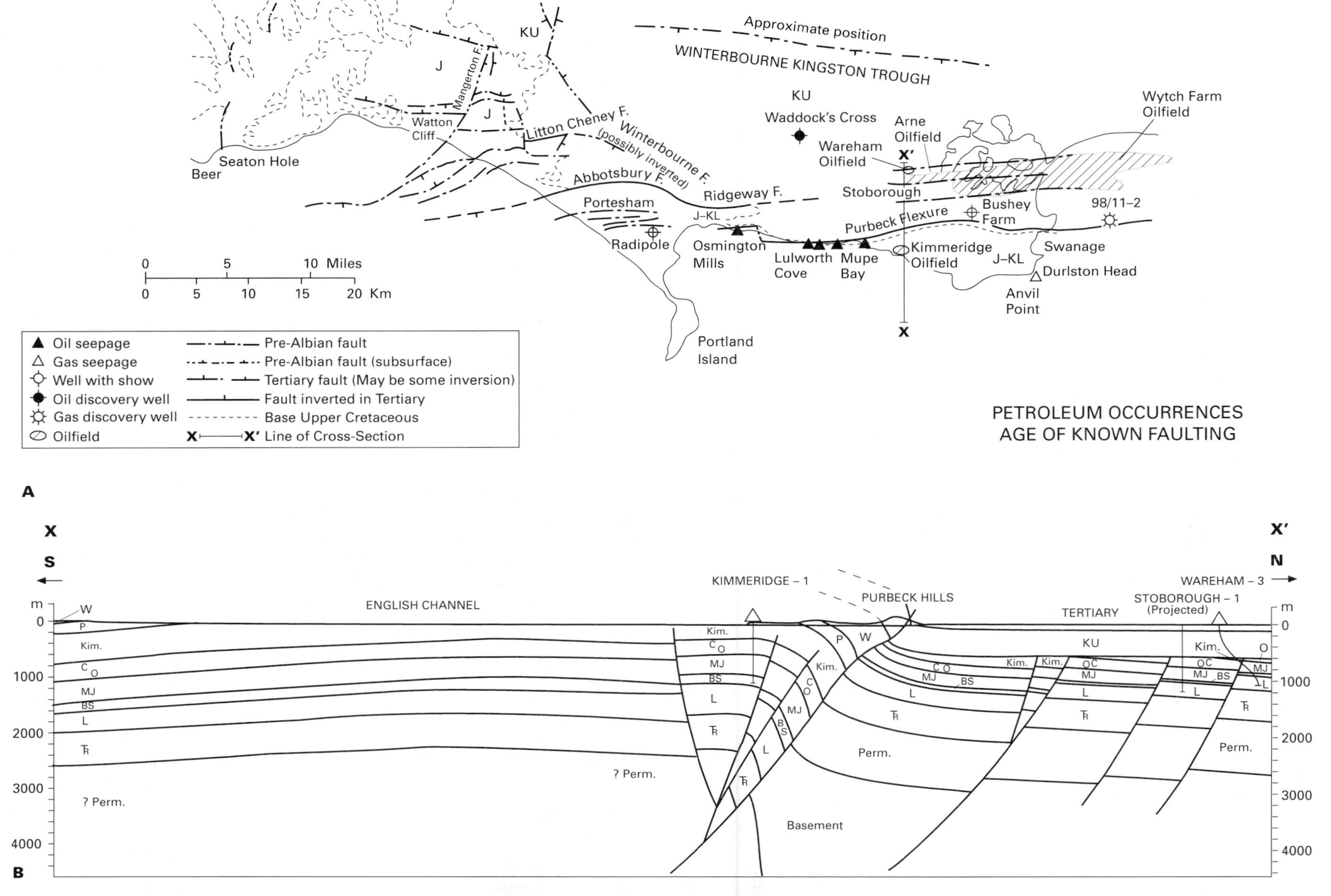

Fig. 12.60 (A) Major structural trends and surface and subsurface hydrocarbon occurrences in the Wessex Basin, including the offshore extension to the large Wytch Farm oilfield. (B) South–north structural cross-section through the small Kimmeridge and Wareham oilfields. (From Selley and Stoneley, 1987.)

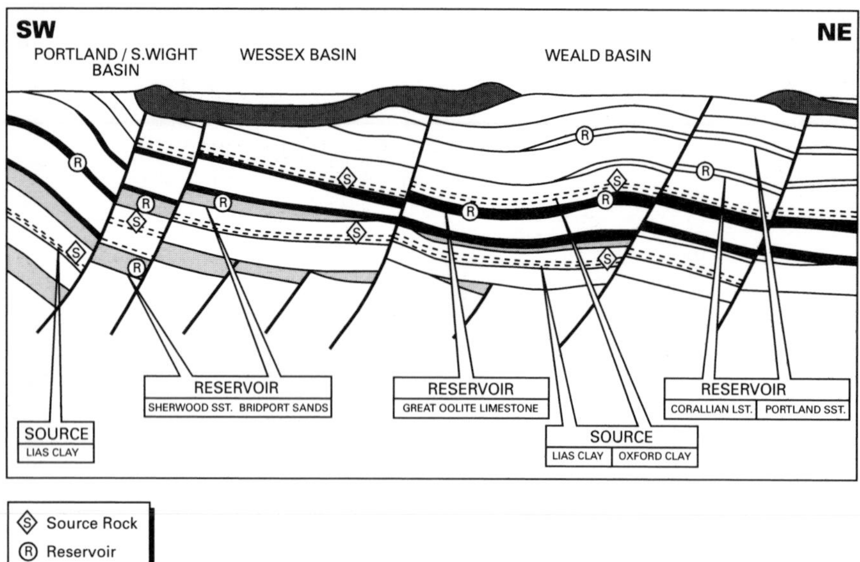

Fig. 12.61 Schematic south-west–north-east cross-section illustrating the major plays in the Wessex and Weald basins (from Hawkes *et al.*, 1997).

greatest potential for generating hydrocarbons. The Upper Jurassic Kimmeridge Clay Formation is also locally organic-rich, but is less extensively preserved. Associated Jurassic sandstone reservoirs are highly variable and their distribution is closely related to syndepositional faulting (e.g. Bajocian–Bathonian successions up to *c.* 1 km thick in hangingwall positions). However, there is no evidence of a regional thermal doming event or of any associated, widespread sandstone-dominated successions, as seen in the North Sea Basin.

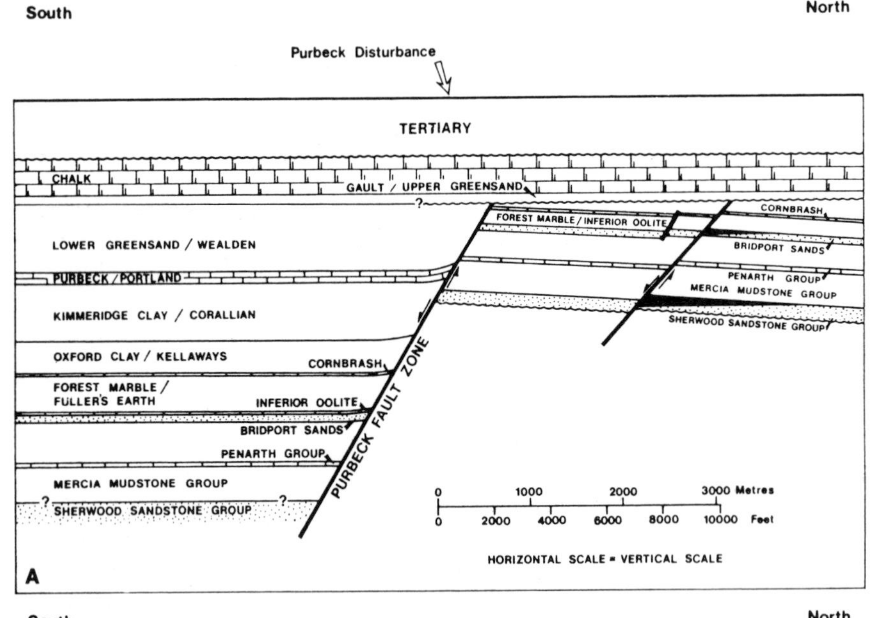

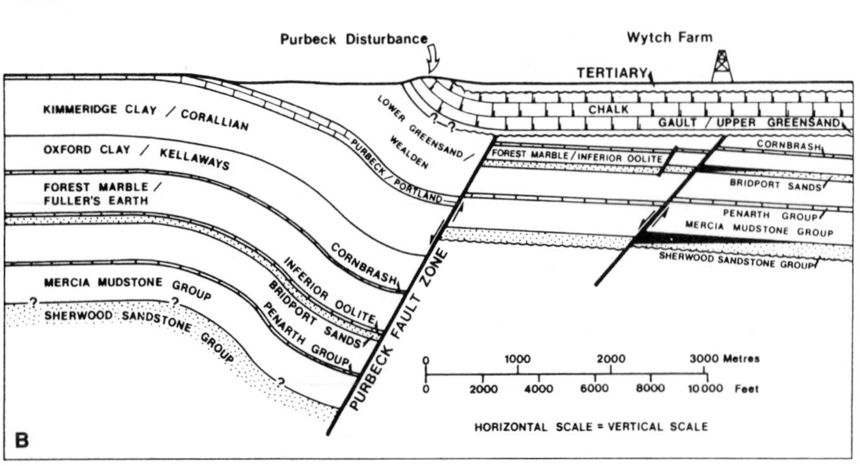

Fig. 12.62 Structural cross-sections illustrating the Wytch Farm oilfield and its relationship to Tertiary tectonic inversion: (A) preinversion section highlighting the active, extensional nature of the Purbeck Fault Zone during the Triassic, Jurassic and Lower Cretaceous (ending by Gault/Upper Greensand times), and (B) present-day section indicating the result of Tertiary inversion with the formation of the Isle of Wight–Purbeck Monocline (from Colter and Harvard, 1981).

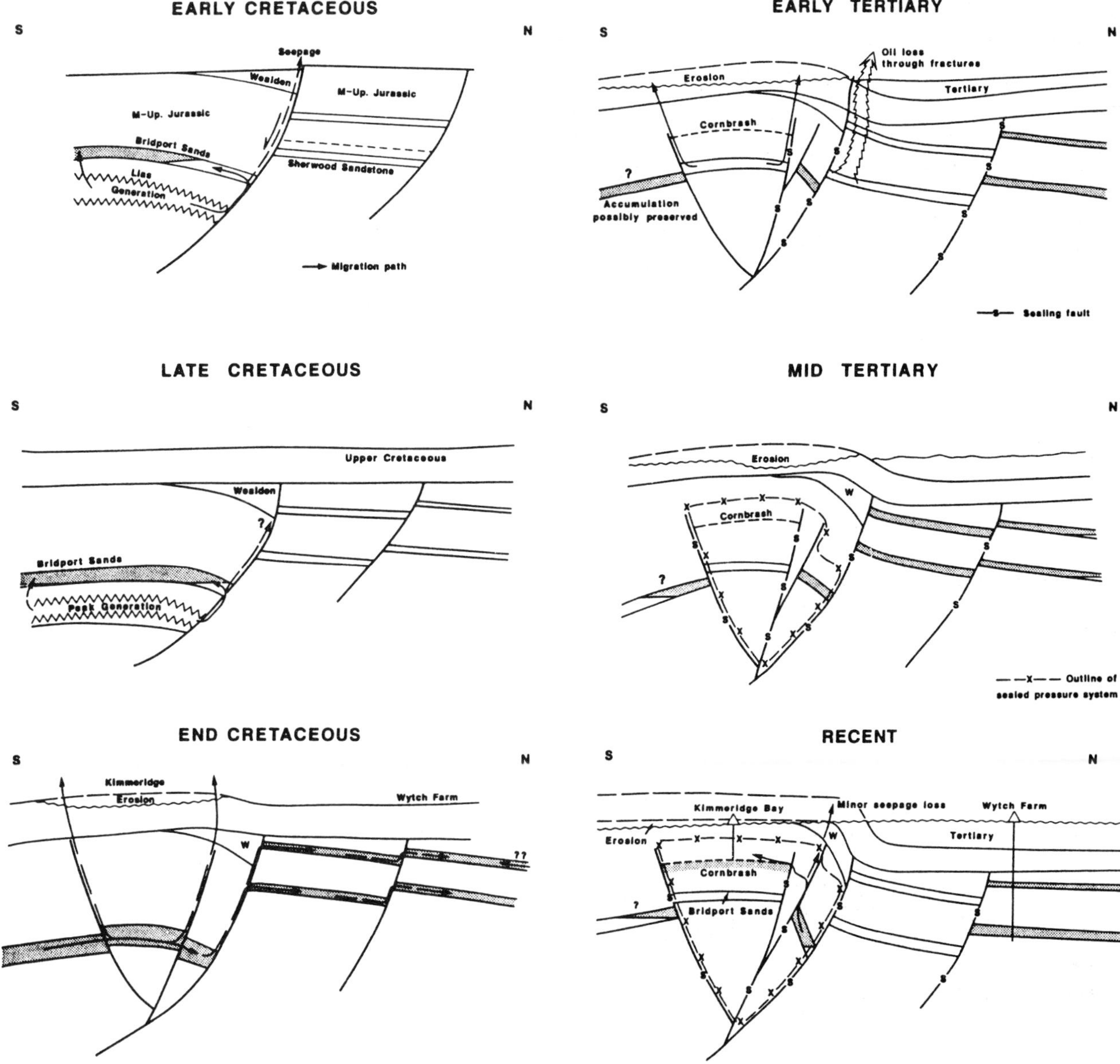

Fig. 12.63 Schematic sections to illustrate the hypothetical history of hydrocarbon generation, migration and entrapment in the area of the Kimmeridge and Wytch Farm oilfields, Wessex Basin (from Selley and Stoneley, 1987).

One of the thickest reservoir successions comprises the Lower Cretaceous Wealden and the overlying Lower Greensand (e.g. in the Celtic Sea and Western Approaches basins), but, in some basins, erosion following Early Cretaceous inversion has removed much of this interval (e.g. Bristol Channel Basin).

The effects of Tertiary inversion are widespread, and these have had a largely negative effect on the hydrocarbon potential in most of these basins, mainly through trap breaching and by raising source rocks above their maturation thresholds. In addition, Chalk and Tertiary successions have been uplifted such that they are generally too thin and too shallow to provide viable reservoirs.

Potential structural traps in these basins range from Permo-Triassic to Jurassic tilted fault blocks through to inversion-related anticlines (see Fig. 12.64 and below).

12.4.3 Basins and plays offshore western and northern Ireland

The basins in this area occur in 2 main regions (Fig. 12.66): (i) west of Ireland, including the Porcupine Basin (25 wells drilled, some with shows but no commercial discoveries; McCann *et al.*, 1995; Moore and Shannon, 1995), Rockall Trough (undrilled in the Irish sector; England, 1995; Shannon *et al.*, 1995), Goban Spur (one well), Slyne/Erris troughs and Donegal Basin (four wells; O'Reilly *et al.*, 1995; Scotchman and Thomas, 1995) and Hatton Basin (no wells); and (ii) north-east of Ireland, including the Rathlin Basin and North Channel Basin (Parnell, 1991; Fitzsimmons and Parnell, 1995; Shelton, 1995).

The main features of the basins and plays west of Ireland are summarized as follows (Shannon *et al.*, 1993; Croker and Shannon, 1995): (i) up to 10 km of Upper Carboniferous to Tertiary sediments; (ii) source rocks occur at various stratigraphic levels (from Carboniferous to Lower Tertiary); (iii) reservoirs tested for hydrocarbons occur in the Middle and Upper Jurassic (fluvial to shallow-marine sandstones),

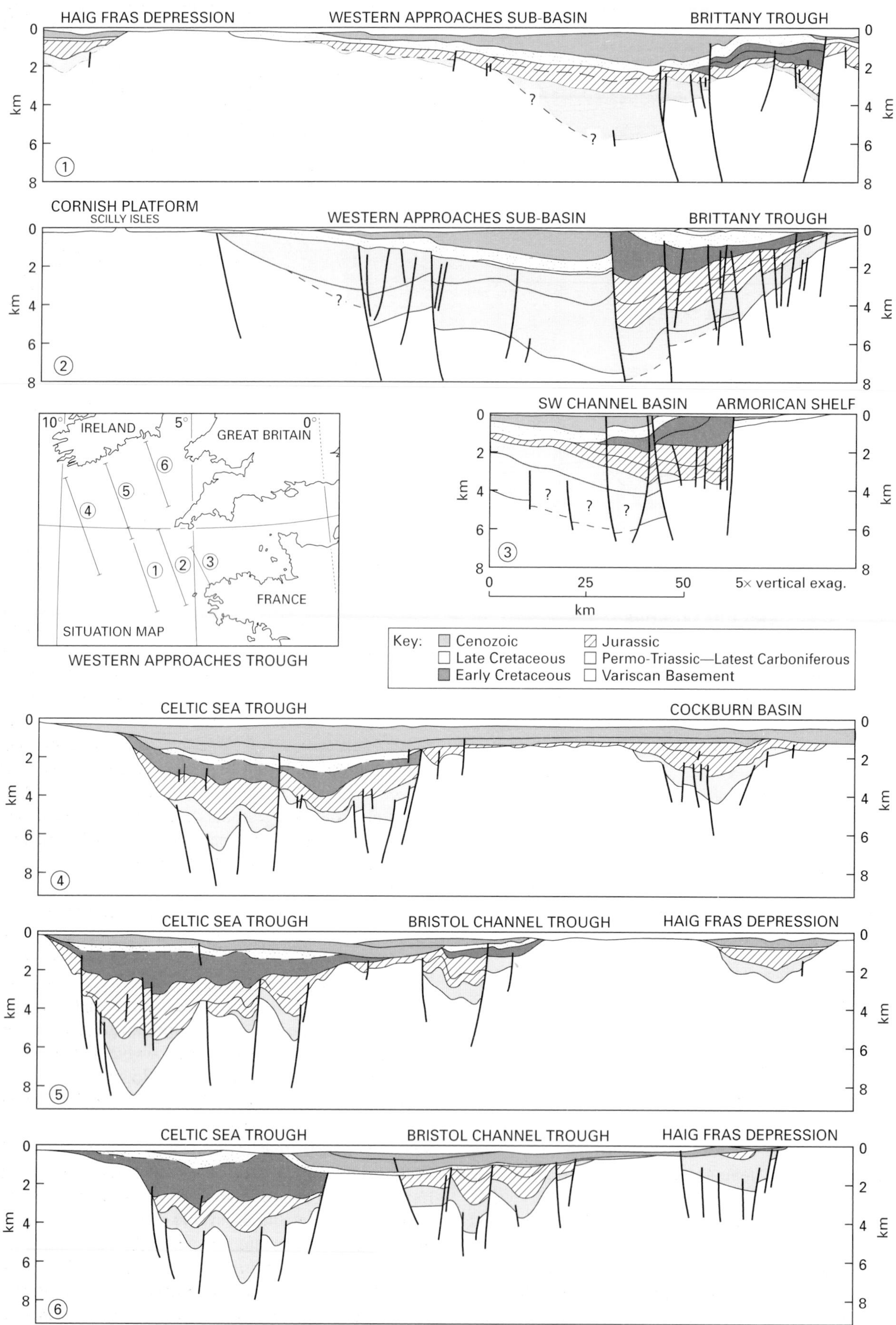

Fig. 12.64 Schematic structural cross-sections through basins of the south-western British Isles and offshore north-western France: 1, 2—Western Approaches subbasin and Brittany Trough; 3—South-west Channel Basin; 5, 6, 7—Celtic Sea Trough, Bristol Channel Trough and Haig Fras Depression (from Ziegler, 1990a).

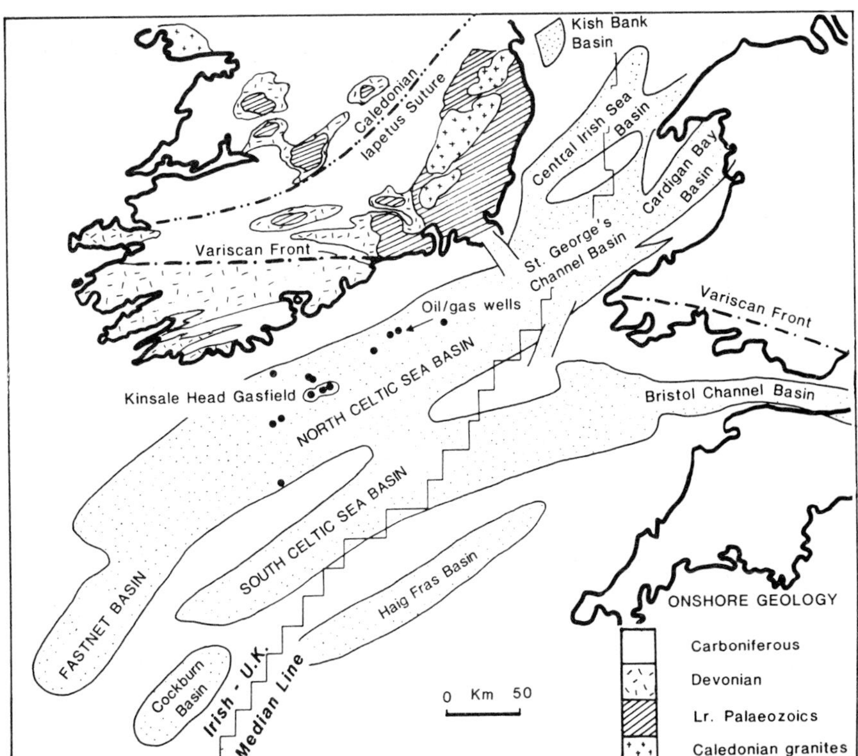

Fig. 12.65 Geological framework of offshore southern Ireland and hydrocarbon occurrences in the North Celtic Sea Basin (from Shannon, 1991a).

Upper Jurassic and Lower Cretaceous (deep-water sandstones); (iv) additional potential reservoirs, but not yet tested with hydrocarbons, include the Upper Carboniferous, Triassic and Lower Tertiary; (v) cap rocks are developed throughout the succession; and (vi) traps range from tilted fault blocks (Carboniferous to Jurassic) to various possible stratigraphic traps, from the Jurassic through to the Tertiary. Hydrocarbon generation and migration are confirmed by four tested wells, which produced good-quality oil (32–41° API), numerous oil shows and the uncommercial Connemara discovery, containing an estimated 195×10^6 barrels of oil in place (Earls, 1995). Hence the basic play ingredients are present in this area and the limited number of exploration wells drilled (*c.* 30) indicates that the area is still underexplored.

12.4.4 North Celtic Sea Basin plays

Lower Cretaceous play in the North Celtic Sea Basin

The Lower Cretaceous coastal to shallow-marine sandstones form the most successful play in the North Celtic Sea Basin (see Fig. 12.65). These reservoirs are productive in two commercial discoveries: the Kinsale Head (1.6×10^{12} ft^3 gas initially in place; *c.* 1×10^{12} ft^3 recoverable gas reserves) and Ballycotton gasfields in the North Celtic Sea Basin, offshore southern Ireland (Colley *et al.*, 1981; Shannon, 1993, Taber *et al.*, 1995). This play also contains one uncommercial gas accumulation in the Seven Heads Field (with reserves of some 100×10^9 ft^3 gas and 2×10^6 barrels of oil; Murray, 1995), together with relatively common occurrences of oil and gas shows in several other exploration wells along the same north-east–south-west trend (Howell and Griffiths, 1995).

The North Celtic Sea Basin contains a *c.* 9 km thickness of Triassic to Cretaceous sediments, which are overlain by a thin veneer of Tertiary deposits. This asymmetric basin has had a long and complex structural history, involving Caledonian, Variscan and Alpine tectonic events, with repeated reactivation of earlier structural features influencing stratigraphic architecture and structural evolution (Petrie *et al.*, 1989; Shannon, 1993; Rowell, 1995). Major phases of rifting occurred during the Permo-Triassic and Late Jurassic–Early Cretaceous, each separated by thermal subsidence events in Early–Middle Jurassic and in the Late Cretaceous to Tertiary.

The successful Lower Cretaceous play is best represented by the Kinsale Head gasfield, where two reservoirs are developed: (i) the primary reservoir is a *c.* 125 ft (46 m) thick interval, comprising a composite coarsening-upward succession of Aptian–Albian coastal to shallow-marine sandstones (Lower Greensand), with good reservoir properties (porosity *c.* 20% and permeability *c.* 420 mD); and (ii) the secondary reservoir comprises a thin (22 ft/8 m), mud-dominated succession of fluvial to deltaic Wealden facies (porosity *c.* 22% and permeability *c.* 280 mD). The depositional environment of the primary reservoir (the 'A' Sand) is widely accepted as being shallow-marine (< 100 ft/30 m water depth), but in detail several models have been proposed: tidal sand ridges/offshore bars (Colley *et al.*, 1981), a storm-dominated sand sheet (Winn, 1994) and a wave-/storm-dominated shoreface (Hartley, 1995). The laterally extensive, sheet-like geometry has been used to support a broadly westerly-prograding shoreface deposited during a punctuated regression, with sands supplied to this narrow north-east–south-west basin from the south, north and east (Hartley, 1995; Taber *et al.*, 1995).

The top seal is provided by the laterally variable Albian to Cenomanian shallow-marine mudstones and siltstones

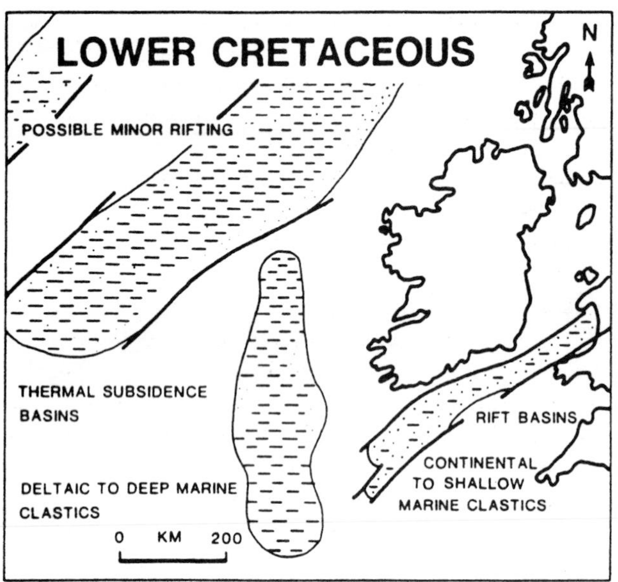

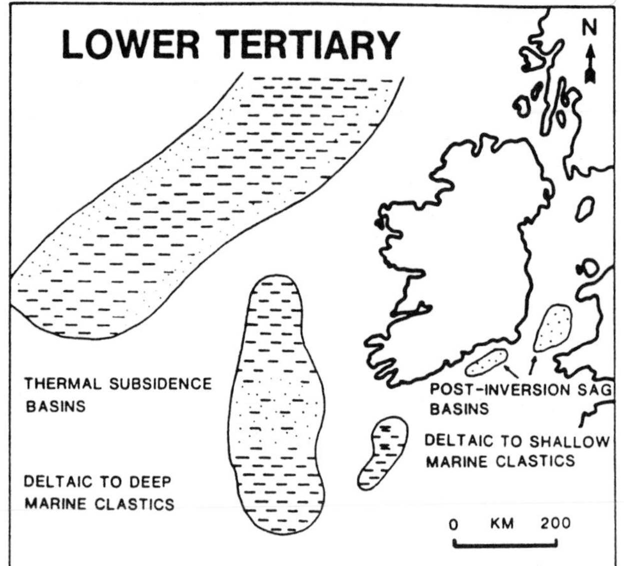

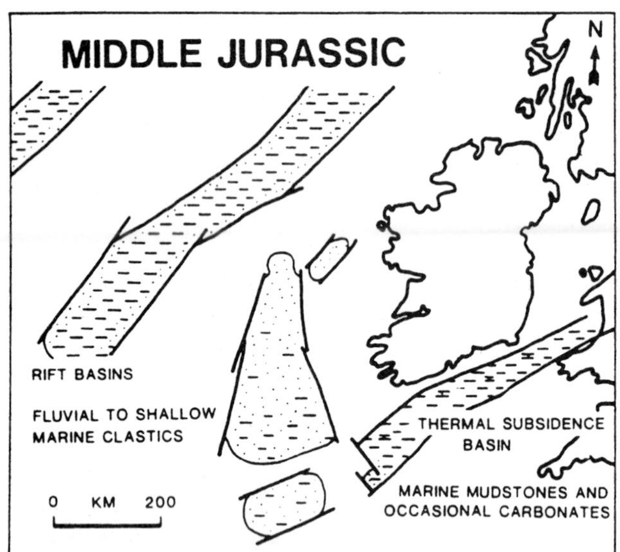

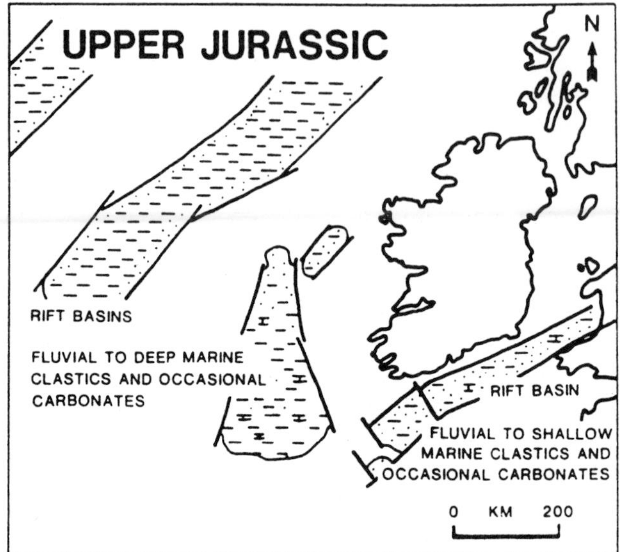

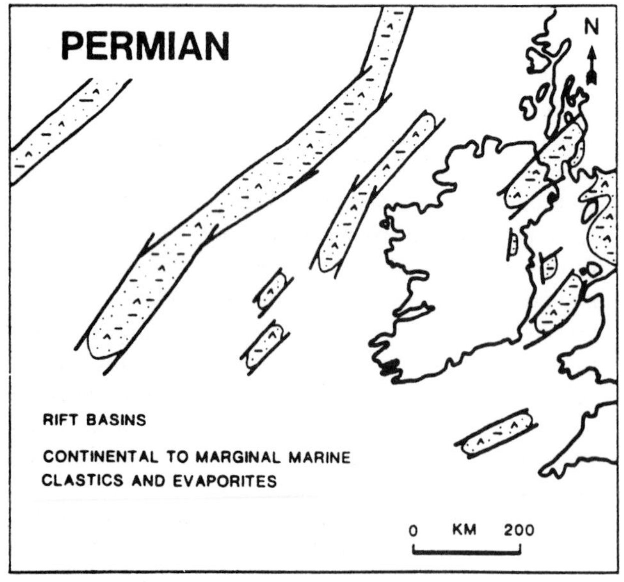

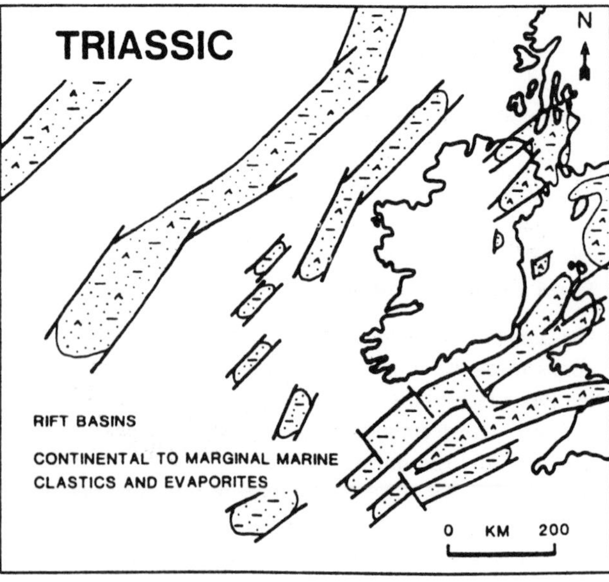

Fig. 12.66 Simplified geological evolution from the Permo-Triassic to Tertiary of the major sedimentary basins of offshore Ireland (from Shannon, 1991b).

(Gault), consisting of optimally sealing shelf-mudstone facies in the Kinsale Head/Ballycotton area, which becomes sandier to the north-east (Taber *et al.*, 1945). The trap is the largest of several simple, elongate east-north-east–west-south-west-trending anticlines, which provide four-way dip closures at the top of the main 'A' Sand. The trap at Kinsale Head has a vertical relief of *c.* 300 ft/91 m, while at Ballycotton this is *c.* 200 ft/61 m. The anticlines are parallel to the underlying Upper Palaeozoic trend and were formed during Palaeogene inversion when the graben axis was uplifted.

Potential source rocks include two main intervals: (i) Upper Jurassic to Lower Cretaceous 'Purbeck' (Kimmeridgian to Berriasian) in lacustrine to lagoonal facies, with several oil-prone layers rich in Type I kerogen; and Lower Jurassic (Liassic) in shelf mudstone facies, with abundant Type II kerogen (Murphy *et al.*, 1995). The gas in the Kinsale Head and Ballycotton fields is of very similar composition: almost pure methane (> 99%). The modelling of source-rock burial points to the Liassic as the main source of gas, with initial maturation having been reached in the Early Cretaceous (Taber *et al.*, 1995). Further migration of dry gas is deduced to have taken place in the Late Cretaceous (Murphy *et al.*, 1995) and during the trap-forming Tertiary inversion, the latter possibly including remigration from much deeper traps. The common occurrence of oil shows suggests that an earlier oil accumulation may have been displaced by the late migration of gas (Taber *et al.*, 1995).

The small, undeveloped Seven Heads oil and gasfield is in a structure similar to the Kinsale Head Field and is located *c.* 60 km along strike to the north-east. Here the Lower Greensand reservoir is tight and hydrocarbons occur at various levels but entirely within the Wealden. The top seal comprises Albian marine mudstones. The gas is thought to be derived from the Liassic, which is mature over much of the basin. Oil is inferred to come from Kimmeridgian to Lower Berriasian source rocks (Kimmeridge Clay Formation), which are mature only locally.

This Lower Cretaceous play extends throughout the North Celtic Sea Basin in similar, low-relief, inversion-related structures. The common occurrence of shows suggests that the majority of remigrated hydrocarbons were lost to the surface, as a result of limited trap/seal integrity.

Other plays in the North Celtic Sea Basin

Initial exploration was directed at the large Tertiary inversion structures, but these traps were breached as a result of penetration by non-sealing faults. Tilted fault blocks provide potential for Triassic (Shannon and Mactiernan, 1993) and Middle Jurassic (Millson, 1987) plays, sourced by mature Jurassic source rocks, mainly from the Lias. However, this has so far been proved only in small uncommercial oilfields in Jurassic fault-block structures around the basin margins, most notable of which is the Helvick oilfield (Caston, 1995).

The Helvick Field was discovered in 1983 and is located *c.* 60 km north-east of the Kinsale Head Field (in Block 49/9). This marginal field contains around $8–12 \times 10^6$ barrels of 44° API oil initially in place (ultimate recovery *c.* $2–4 \times 10^6$ barrels of oil; Caston, 1995). The reservoir is a widespread Middle to Upper Jurassic (Callovian–Oxfordian) braided fluvial sandstone, which is sealed by overlying Upper Oxfordian to Lower Kimmeridgian marine shales and limestones. Hydrocarbons are trapped in a hangingwall closure adjacent to a prominent south-west to east-west structural high, dominated by down-to-basin extensional faults, which forms the northern margin of the North Celtic Sea Basin. This fault trend reflects initial Caledonian and Variscan influences, but the principal displacement in relation to trap formation dates from the latest Jurassic. Subsequent reactivation of this fault during Tertiary inversion has preserved some 1100 ft/335 m of Eocene, Oligocene and Pliocene–Pleistocene deposits in the footwall immediately north of the accumulation, whereas in the inverted hangingwall only a thin veneer of some 200 ft/62 m of Tertiary–Quaternary deposits is preserved. The Helvick oil accumulation is in a fault-bounded dip/closure located in a kink (or 'elbow') within the hangingwall of this reactivated fault system, probably associated with the intersection of earlier fault trends (Caston, 1995).

The field is associated with a variable and complex source and migration history, with at least two phases of oil generation/migration, biodegradation of shallow oil-bearing reservoirs and evidence of mixed biodegraded and non-biodegraded oil. The most likely source rocks are of Liassic age, which here contain a distinctive terrestrially derived herbaceous source. However, *c.* 2 km along strike (well 49/9-4), minor oil occurrences have been calibrated with fresh to brackish water deposits of the Portlandian–Purbeckian sequence.

The traps of all accumulations in the North Celtic Sea Basin have been influenced, or entirely controlled by, inversion tectonics. No undisturbed, pre-inversion traps, such as that responsible for the Wytch Farm accumulation in the Wessex Basin, have been identified, probably since none has been preserved. The likelihood of finding large, undrilled structural traps is probably remote, and subtle stratigraphic traps may not have survived the extensive inversion events.

12.4.5 East Irish Sea Basin plays

The East Irish Sea Basin is located offshore between the coasts of north-west England and North Wales, extending westwards as far as the Isle of Man, the Solway Firth Basin and the Anglesey–Isle of Man Uplift (Fig. 12.67). Exploration in this small basin (*c.* 80 km by 120 km) has been extremely successful, with some 78 exploration and appraisal wells resulting in 19 hydrocarbon discoveries with reserves estimates of approximately 9×10^{12} ft^3 of gas and 150×10^6 barrels of oil. Key discoveries since initial exploration in 1969 have been: (i) the giant Morecambe (or South Morecambe; Stuart and Cowan, 1991) gasfield (5.5×10^{12} ft^3 of gas) in 1974; (ii) the Morecambe North gasfield (1.2×10^{12} ft^3 of gas) in 1976; (iii) the Hamilton gasfield (500×10^9 ft^3 of gas) in 1990 (Fig. 12.68); (iv) the Douglas oilfield (90×10^6 barrels of oil) and Hamilton North gasfield (250×10^9 ft^3 of gas), both in 1990 (Fig. 12.68); and (v) the Lennox oil- and gasfield in 1992 (located only 8 km off the Lancashire coast; Fig. 12.68).

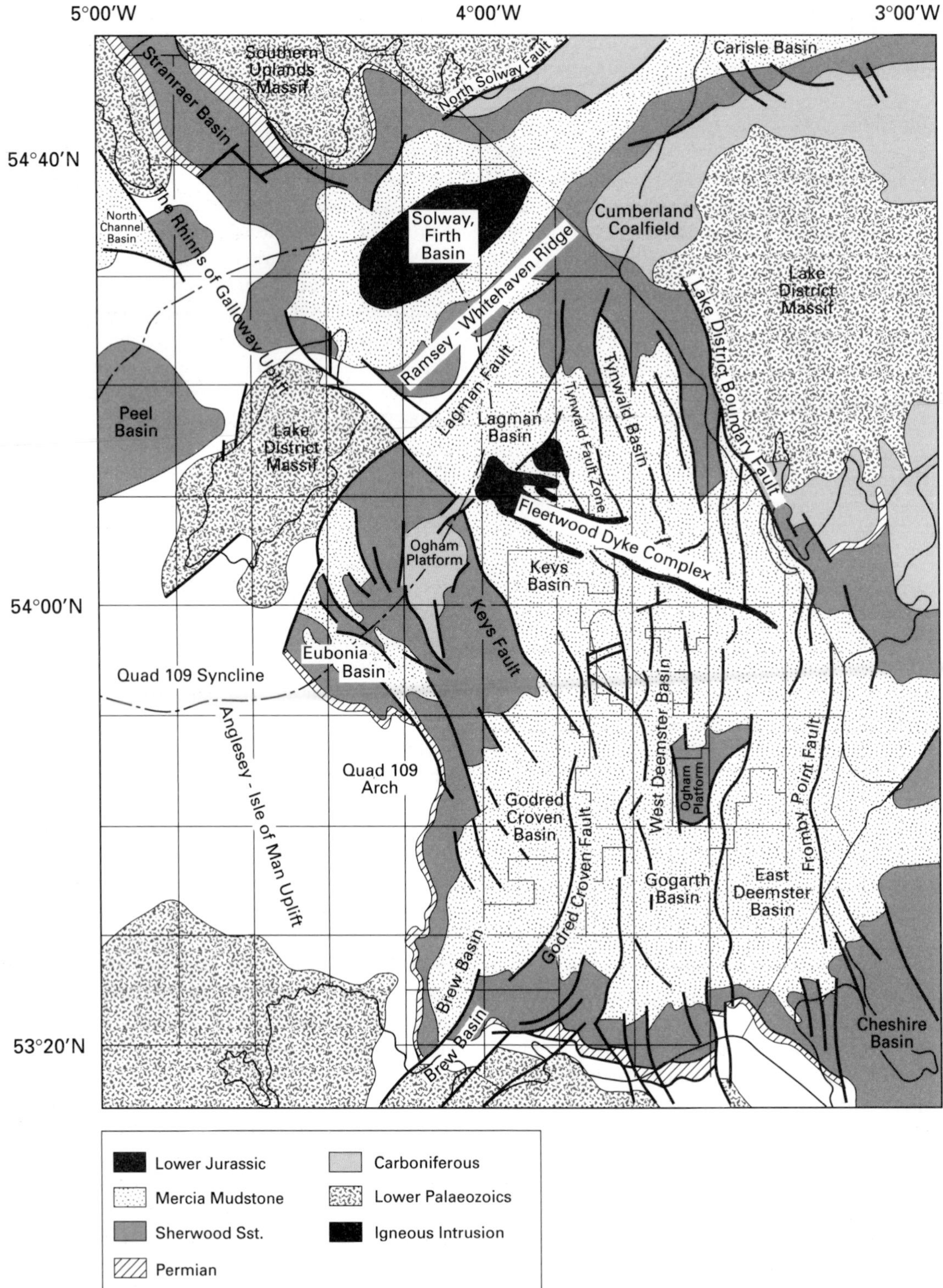

Fig. 12.67 Geological framework and structural elements of the East Irish Sea Basin and adjacent areas (courtesy Steve Pickering, BHP Petroleum).

The basin is one of several Permo-Triassic basins initiated in the early Permian by north-east–south-west rifting, which extend from the Worcester Graben in the English Midlands to the Loch Indall Basin in western Scotland (Jackson and Mulholland, 1993). Rifting was accompanied by deposition of a thick sequence of continental deposits, including aeolian and braided-stream sandstones and an overlying evaporitic succession. The underlying Caledonian structural framework has exerted a strong influence on the distribution of potential structures. The basin is bounded to the south by the Lower Palaeozoic Welsh Massif and to the north by the Ramsey–Whitehaven ridge, which is a remnant Caledonian structure linked to the closure of the Iapetus Ocean. Three main structural trends are present within the basin: north-west–south-east and north–south faulting, both trends being linked to Permo-Triassic rifting, and north-east–south-west fault trends,

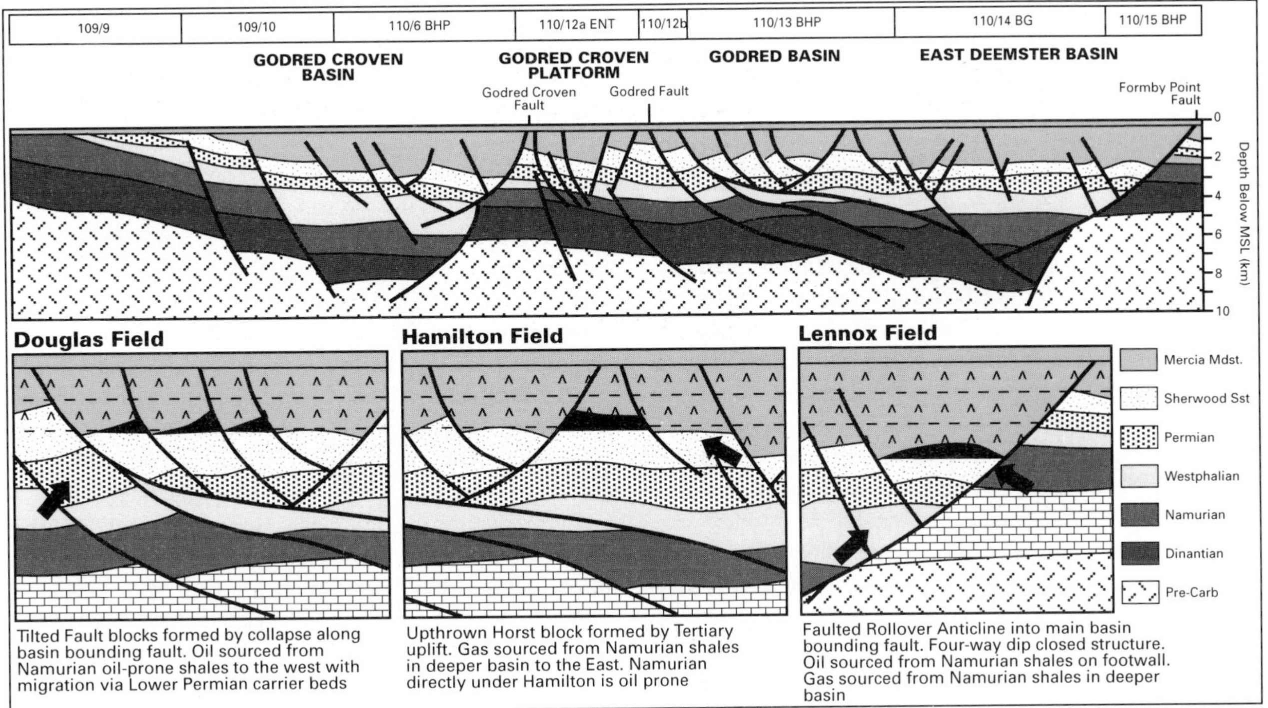

Fig. 12.68 Structural cross-section through the East Irish Sea Basin and a summary of the key play concepts in the Douglas, Hamilton and Lennox fields (courtesy Steve Pickering, BHP Petroleum).

which represent reactivated Caledonian structures (Knipe *et al.*, 1993). The interaction of these complex fault trends subdivides the basin into a number of subbasins, each with a different and often complex structural history.

All the commercial hydrocarbon discoveries have been in the Triassic Sherwood Sandstone Group, most notably the upper part of the Ormskirk Sandstone Formation. The main variations in this Triassic sandstone play are determined by structural style, local subsidence/maturation–migration histories and reservoir/seal-facies differences (Fig. 12.68).

Throughout most of the basin the Sherwood Sandstone Group is between 2000 and 3000 ft/610–915 m thick and at a depth of less than 3000 ft/915 m. The Sherwood Sandstone Group is overlain by a thick sequence of salts and shales of the Mercia Mudstone Group. Within the Ormskirk Sandstone Formation, reservoir quality is excellent in the southern part of the basin, with stacked sequences of aeolian dune facies, aeolian sand sheets and braided fluvial channels, with occasional playa-lake deposits (Cowan, 1993; Meadows and Beach, 1993). Active basin subsidence may have controlled facies distribution by raising and lowering the water-table, causing switching between aeolian-dominated and fluvial sequences. Further north, diagenetic processes, especially pervasive illitization, has caused a deterioration in reservoir quality.

Other potential reservoirs in the basin include the Permian Collyhurst Sandstone, which is overlain by evaporites and shales of the Manchester Marl Formation and, further north, the St Bees Evaporites. The Manchester Marl Formation in the south of the East Irish Sea Basin is too thin to form an effective seal.

The primary source rock for hydrocarbons in the basin is the underlying Carboniferous (Dinantian/Namurian) Holywell Shale Formation, which outcrops in North Wales. Studies of outcrop samples demonstrate the heterogeneous nature of this unit in terms of its source-rock potential (Armstrong *et al.*, 1995). The depositional environment was mainly anoxic, with numerous pyritized goniatites but negligible benthonic fauna. Organic material comprises both marine- and terrestrially-derived material. With the exception of two small Jurassic inliers, there are no post-Trias sediments preserved in the basin, due to regional uplift in the Tertiary.

The structural traps in the basin are in a series of tilted fault blocks, whose style is consistent with hangingwall deformation of a homogeneous sand above a simple listric fault and with basin extension of up to 50% (McClay, 1990). The basin has suffered a complex burial history, including Tertiary inversion, which caused removal of the post-Triassic cover from the area. This is partly reflected in detailed geochemical analyses and thermal modelling, which indicate that the various subbasins of the East Irish Sea Basin have each undergone different burial histories (Hardman *et al.*, 1993). In the southern part of the basin, early oil generation commenced in the Jurassic. Breaching of traps in the late Jurassic resulted in a bitumen residue, which can be seen in cores throughout the region. Oil generation was renewed in the Cretaceous, with gas generation in the deeper subbasins (e.g. in the East Deemster Basin). Tertiary inversion has resulted in uplift, trap breaching, late-stage tilting and gas expansion in preserved traps (e.g. see the Morecambe Field, below).

Some of the key components of the Triassic gas and oil play are illustrated by the Morecambe and Douglas fields, respectively.

The Morecambe gasfields

The Morecambe gasfields (South Morecambe: Stuart and Cowan, 1991; North Morecambe: Stuart, 1993) are located between prominent, approximately north–south-trending regional faults, the Tynwald Fault to the east and the Keys Fault to the west (Cowan *et al.*, 1993; Knipe *et al.*, 1993). The two fields are separated by another complex fault system, which represents a transfer zone trending obliquely to the regional faults. The South Morecambe structure is extremely shallow (crest at 2400 ft ss/732 m ss; gas–water contact at 3750 ft ss/1144 m ss), which has required the use of slant-drilling techniques for its development. Reservoir quality is determined by both depositional environment and diagenesis. The main reservoir sandstones were deposited in rapidly subsiding basins under continental semiarid conditions, including major channel-fill sandstones, secondary channel fills, with associated sheetflood sandstones, and locally developed aeolian sandstones, characterized by their extremely high permeability (> 1000 mD). Complex diagenesis has involved differential compaction and several phases of dolomite and quartz cementation. The main control on reservoir quality is the development of fibrous ('hairy') illite, which has formed extensively at palaeogas/water contacts and resulted in diagenetic layering within the reservoir (e.g. a high-permeability illite-free layer and a low-permeability illite-affected layer; Stuart and Cowan, 1991).

The Morecambe structural trap is located at the intersection of two fundamental structural domains, which have controlled the development of the half-grabens of the East Irish Sea Basin. Marked variations in fault patterns characterize different parts of the fields (e.g. as seen in the northern and southen limbs of South Morecambe). The structural attitudes of the 'illite-affected' and 'illite-free' layers also demonstrate the late-stage tilting of the structure, which was associated with extension and late (Tertiary) fault movements, followed by inversion and uplift (estimated at 1500 ft/458 m by Green, 1986, but possibly as much as 4000–5000 ft/1220–1525 m; Colter and Barr, 1975; Bushell, 1986). The complex burial and diagenetic history of the Morecambe fields can be summarized as follows: (i) early burial during the Triassic (localized grain dissolution and early quartz and feldspar overgrowths); (ii) rapid burial during the Late Mesozoic (compaction, decarboxylation and creation of secondary porosity; (iii) initial hydrocarbon maturation and migration during the Jurassic (filling of low-relief structures and illitization within the aquifers); (iv) uplift and trap breaching during the Cretaceous (Late Cimmerian), followed by reburial and a second phase of hydrocarbon migration (filling of Late Cimmerian structural traps, including earlier illite-affected reservoirs); and (v) Tertiary uplift to present depth (further expansion of gas into illite-affected reservoirs).

The gas at Morecambe is believed to have been sourced from Westphalian coals and carbonaceous shales, possibly with some minor contribution from Dinantian Limestone (i.e. hydrogen sulphide (H_2S)-enriched gas). Hydrocarbon migration occurred mainly along the central fault system.

The Douglas oilfield

The Douglas oilfield reservoir is the Triassic Helsby Sandstone Formation (part of the Sherwood Sandstone Group; Meadows and Beach, 1993), which is made up of intercalated aeolian and fluvial facies. The Mercia Mudstone Group also acts as the top seal, but, unlike in the Hamilton and Morecambe fields, it is developed in a mudstone facies rather than as evaporites. The source is provided by the Holywell Shale Formation, which is mature for oil generation within the area of the field, and has also been responsible for the oils found in the Lennox and Formby fields (Trueblood *et al.*, 1995). The Douglas Field has experienced a complex migration history, involving several phases of filling. The oil is a 44° API waxy crude with a low gas/oil ratio (171 ft^3/barrel) and high H_2S content.

The structure is dominated by north-south-trending faults downthrowing to the east, which divide the field into three separate fault compartments, each with a slightly different oil–water contact (Fig. 12.68). Maximum structural relief is 400 ft/122 m.

12.4.6 Faeroe/Shetland Basin plays and adjacent areas

The Faeroe/Shetland Basin (up to *c.* 125 km wide and 600 km long and comprising the Faeroe Basin, Rona Ridge and West Shetland Basin) forms a small part of the much larger (*c.* 3000 km long) rifted continental margin of north-west Europe, which borders the North Atlantic and extends from the Rockall Trough and Porcupine Basin in the south-west to the Møre, Vøring, Tromsø and Barents Sea basins in the north-east (Fig. 12.69; Ziegler, 1988; Spencer and Eldholm 1993; Knott *et al.*, 1993, Faleide *et al.*, 1993). The area totals some five times that of the prospective parts of the North Sea Basin, but, mainly because of the extreme water depths (up to *c.* 2.5 km), it has been only lightly explored (1992 statistics): Norwegian sector—125 exploration wells/25 discoveries; UK sector—97 wells/seven discoveries; Ireland—29 wells/three discoveries (Spencer and Eldholm, 1993).

The most geologically critical elements of this margin in relation to hydrocarbon prospectivity are similar to those in other parts of the regions discussed elsewhere, but specifically include (Fig. 12.69): (i) Late Jurassie to Early Cretaceous rifting; (ii) the ensuing restricted-marine depositional conditions, which resulted in the widespread accumulation of Late Jurassic oil-prone marine shales (extending continuously along the margin from the Porcupine Basin to the Barents Sea); (iii) Late Jurassic–Early Cretaceous rotated fault-block structures; (iv) major subsidence and the accumulation of thick sedimentary successions (*c.* 5 km thick) during the Cretaceous; (v) major continental breakup during the Late Cretaceous to Early Tertiary, accompanying the northward propagation of sea-floor spreading and the opening of the North Atlantic; (vi) voluminous and extensive volcanic activity, including subaerial volcanism, with thick contemporaneous basin-fill successions during the Tertiary; and (vii) Late Tertiary, intraplate uplift of the Scandinavian landmass and adjacent basins (e.g. Barents Sea).

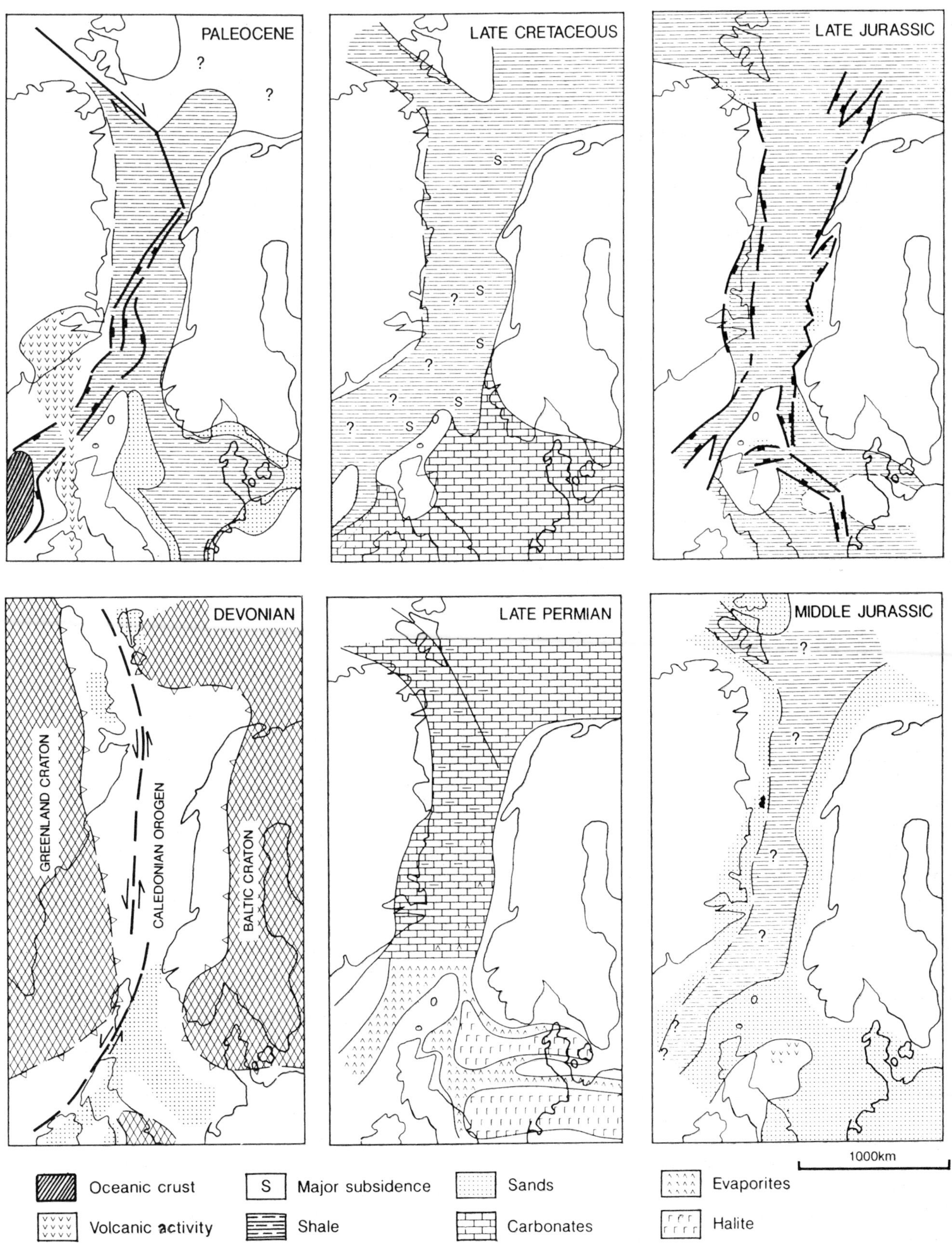

Fig. 12.69 Simplified geological history of the North Atlantic region from the Devonian to the Palaeocene (from Fjaeran and Spencer, 1991).

As with the North Sea Basin, hydrocarbon occurrence and play-fairway distribution can again be classified into pre-, syn- and post-rift plays (Fig. 12.70), which facilitates comparison with the North Sea Basin (Knott *et al.*, 1993). Considering the margin as a whole, the most successful plays are the pre-rift tilted fault blocks with Jurassic and pre-Jurassic reservoirs (Fig. 12.71). Syn-rift accumulations tend to be smaller and more complex, but are locally important. Post-rift plays are associated with major clastic deposition during the Early Tertiary, particularly within deep-water settings, which has been responsible for the

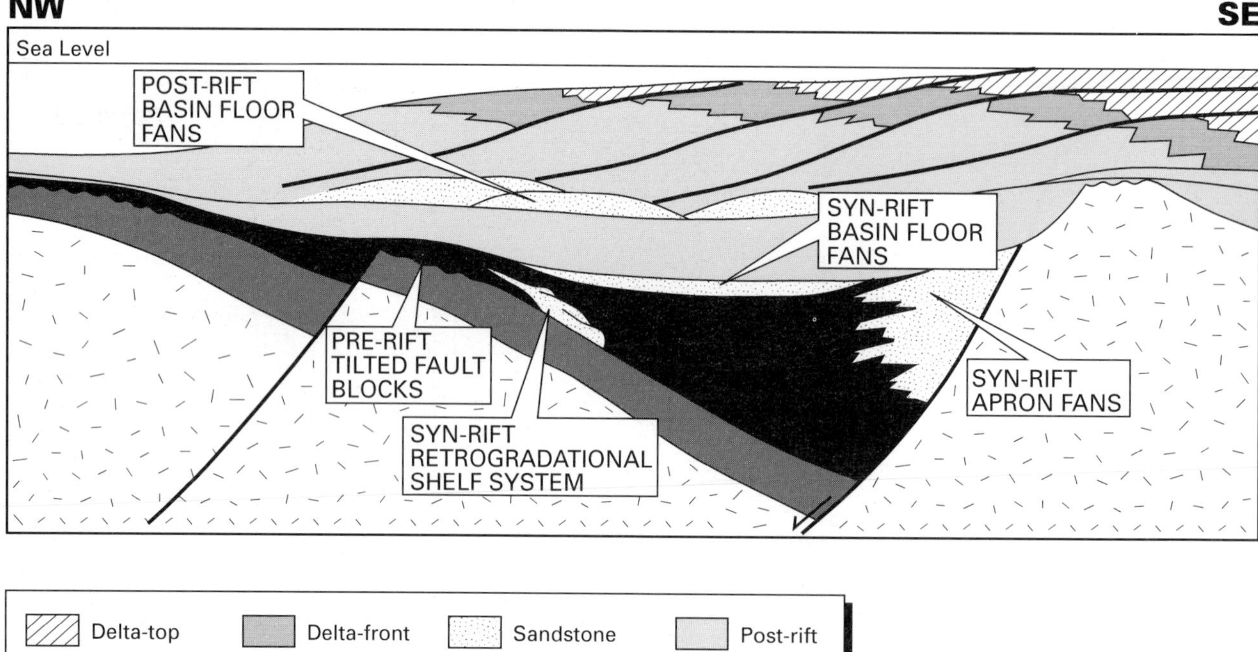

Fig. 12.70 North Atlantic Margin play types (from Knott *et al.*, 1993).

rapidly developing Palaeocene oil and gas play in the Faeroe Basin (Fig. 12.72; see below). The latter has renewed exploration interest throughout the margin, but particularly within the area between the UK and Faeroe continental shelves (see Section 1.4.7). Analogues for some of these plays can be found in the better-understood North Sea Basin (see Section 12.2).

An analysis of hydrocarbon occurrence in relation to gross play type on the UKCS and Norway, including both the North Sea and the Atlantic Margin, indicates the following ranking in terms of discovered volumes: (i) pre-rift (47%; 33×10^9 barrels of oil and oil equivalent); (ii) syn-rift (31%; 22×10^9 barrels of oil and oil equivalent); and (iii) post-rift (22%; 15×10^9 barrels of oil and oil equivalent) (Knott *et al.*, 1993).

Pre-rift plays in the Faeroe/Shetland Basin area

The pre-rift play in this area is dominated by the very large Clair oilfield (Fig. 12.73A). This field has an oil column greater than 800 m thick and contains several billion barrels of oil initially in place (Coney *et al.*, 1993). The reservoir is mainly in Devonian–Carboniferous sandstones (Fig. 12.73B), but it includes oil tested at 960 barrels of oil per day from fractured Lewisian basement (Britain's oldest oil reservoir at *c.* 3300×10^6 Ma!). A combination of relatively heavy oil (*c.* 25° API), low-quality reservoirs, variable and complex fracture distribution and fault com-

Table 12.6 Discovered hydrocarbons by play.

Play area	H/C type*	North Sea	Mid-Norway	Barents Sea	% Total
Post-rift	Oil	3145	No	No	
	Gas	4090	discovery	discovery	
	Total	7235			20
Syn-rift	Oil	1760	440	No	
	Gas	8115	65	discovery	
	Total	9875	505		30
Pre-rift	Oil	7740	1575	125	
	Gas	4090	1700	1760	
	Total	11830	3275	1885	50
Sum	Oil	12645	2015	125	
	Gas	16295	1765	1760	
	Total	28940	3780	1885	100

All figures in 10^6 barrels or barrel of oil equivalent.
*H/C = hydrocarbon.

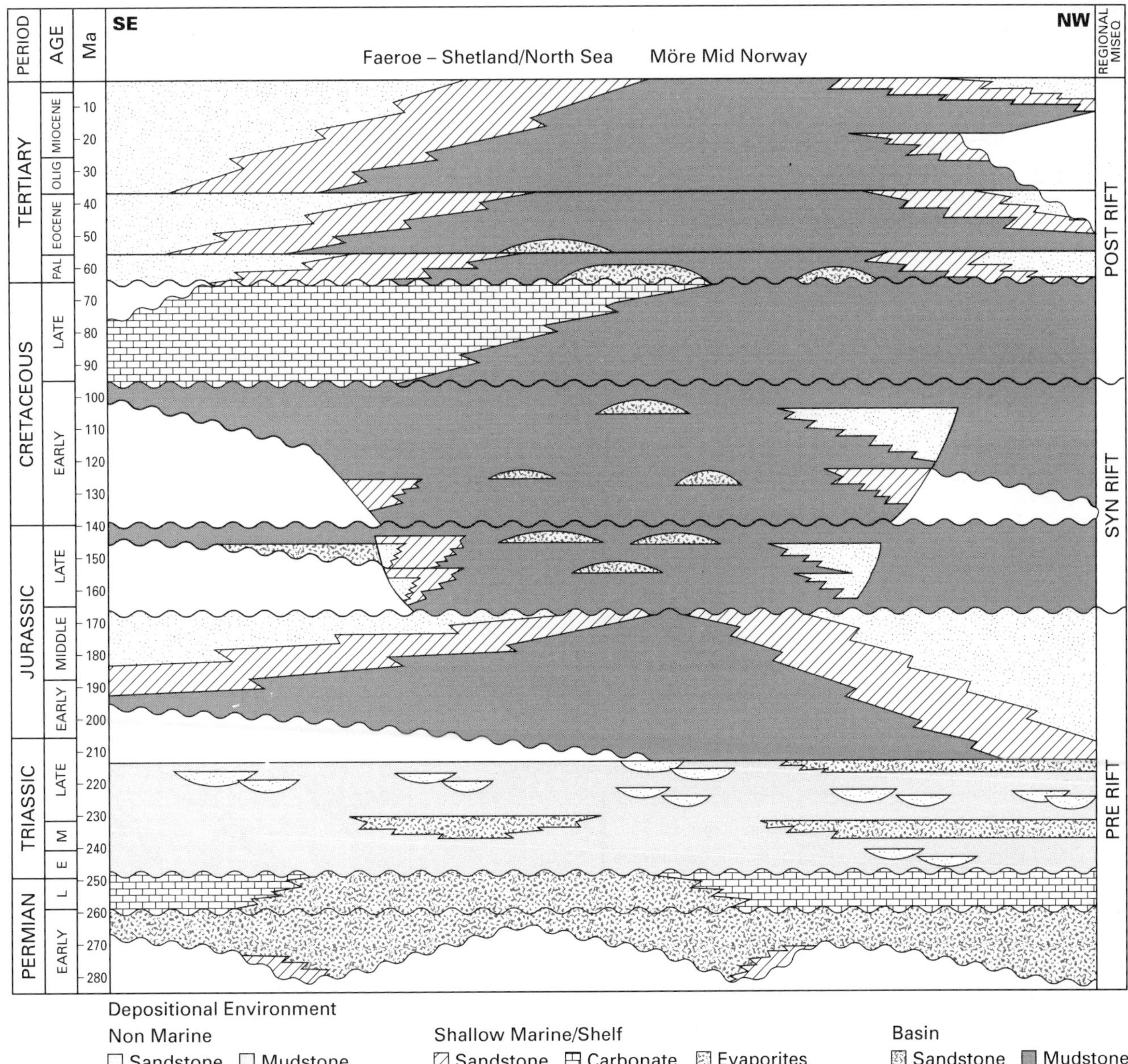

Fig. 12.71 Simplified chronostratigraphic framework of the North Atlantic Margin (from Knott *et al.*, 1993).

partmentalization has prevented commercial development of what is, in terms of hydrocarbons in place, the largest oil accumulation on the UKCS.

Syn-rift plays in the Faeroe/Shetland Basin area

The main syn-rift play in this area involves Lower Cretaceous submarine-fan sandstones (basin-floor and slope-apron fans), which are gas-bearing in the Victory Field and in other minor traps along the north-west side of the Rona Ridge. Conceptually, this play resembles that of the Late Jurassic to Early Cretaceous sandstones and conglomerates of the Brae complex and its lateral equivalents in the South Viking Graben and Outer Moray Firth (see Section 12.2.3). However, these Lower Cretaceous sandstones are of lower reservoir quality and have been found only gas-bearing, generally in subcommercial quantities.

Post-rift plays in the Faeroe/Shetland Basin area

The post-rift play of the Faeroe Basin is centred around Palaeocene deep-water/submarine-fan sandstones. This play has been recognized since initial exploration in the early 1970s, but it was not established as hydrocarbon-bearing until drilling in the northern part of the basin (Quadrants 205/206) in the 1980s discovered gas-bearing sandstones. However, much more significant has been the recent activity in the southern part of the basin (Quadrant 204), which, in the early 1990s, established, for the first time, an oil-bearing Palaeocene sandstone play (the Foinaven and Schiehallion discoveries, with tentative reserve estimates of *c.* 2×10^9 barrels of oil).

A key feature of this play is the distribution of Early Tertiary depositional environments, particularly submarine-fan sandstones along the south-eastern margin of the basin, and their relationship to relative sea-level change, tectonically related subsidence and basin structure and topography. During this time, some 4.5 km of clastic sediments accumulated in the Faeroe Basin. Sequence-stratigraphic studies

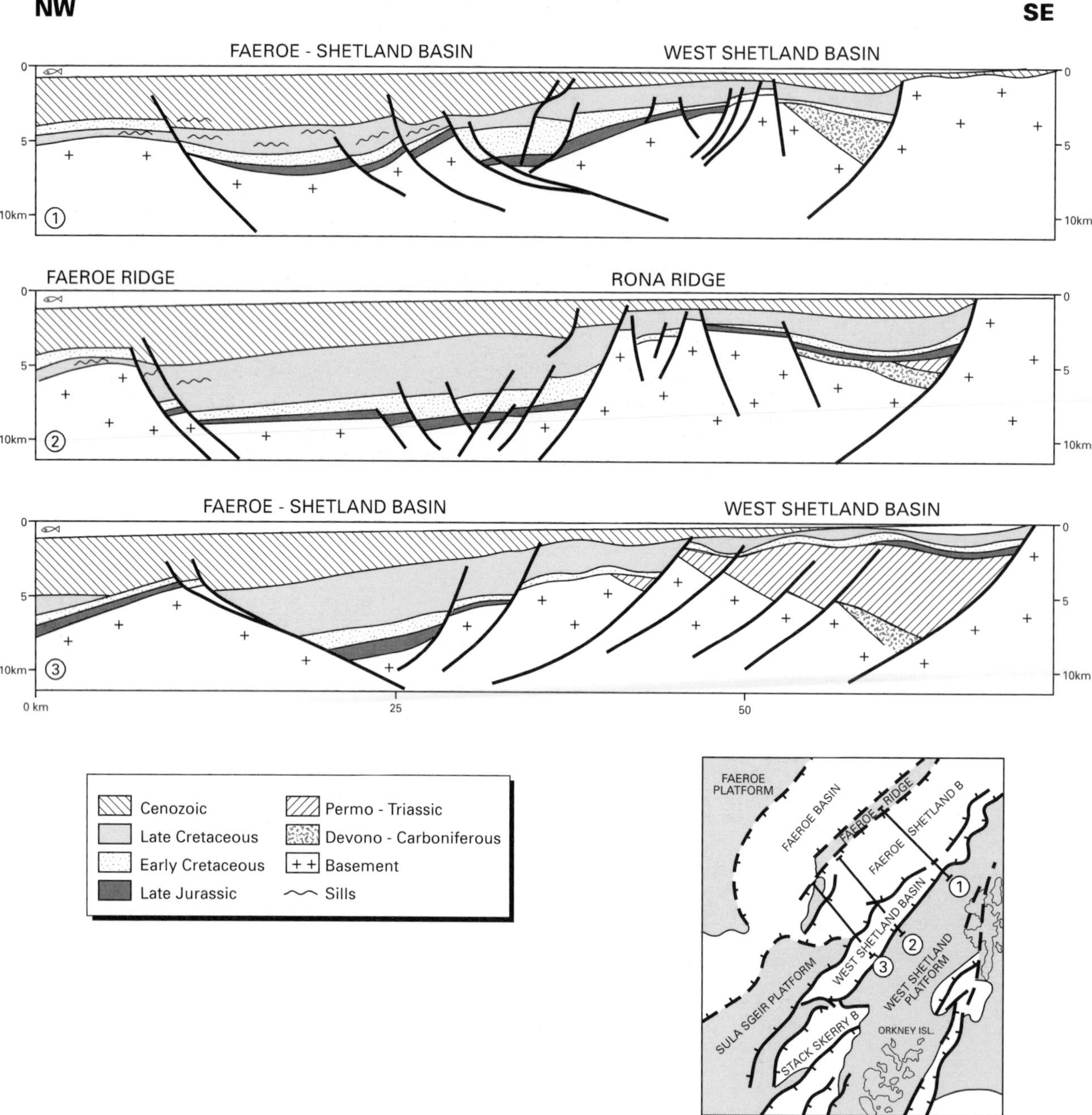

Fig. 12.72 Structural cross-sections across the West Shetland and Faeroe–Shetland basins (from Ziegler, 1990a). Note the position of the Rona Ridge, which marks the location of the Clair oilfield (see Fig. 12.73 for details).

recognize a complex basin margin, with marked basinward and landward shifts in sedimentation determining the location of sand-prone areas (Mitchell *et al.*, 1993). Ebdon *et al.* (1995) recognize a major sequence boundary at the base of the Late Palaeocene and subdivided the succession into several depositional sequences, based mainly on maximum-flooding surfaces, which can be correlated with their equivalents in the North Sea Basin.

The successful play in Quadrant 204 is located within submarine-fan sands (basin-floor lowstand fans). Reservoir effectiveness (thickness and quality) is determined by primary depositional facies, with thick-bedded, massive turbidites displaying the best reservoir characteristics (porosity *c.* 25%; permeability up to 2000 mD). Sands were supplied to the basin from the south and south-east, down a partly tectonically controlled slope, and coalesced in the basin with other contemporaneous fans. Slope orientation and sand entry points varied with time, which has resulted in complex sand-distribution patterns, particularly of those intervals with high reservoir effectiveness (reservoir 'sweet spots'). Factors influencing sand distribution and trap formation are as follows (Ebdon *et al.*, 1995): (i) depositional limits of the Early Tertiary fans were influenced by the underlying syn-rift (Late Jurassic–Early Cretaceous) fault-block topography, which was accentuated by differential compaction of the thick, intervening Late Cretaceous mudstones; (ii) fan deposition was initially aggradational (e.g. during the Palaeocene), followed by periods of progradation (e.g. post-Palaeocene); (iii) intra-Palaeocene tectonic events contributed to slope instability and the emplacement of large gravity slides; (iv) the dominant north-westerly dip in the Late Palaeocene and younger successions reflects regional thermal subsidence (and is

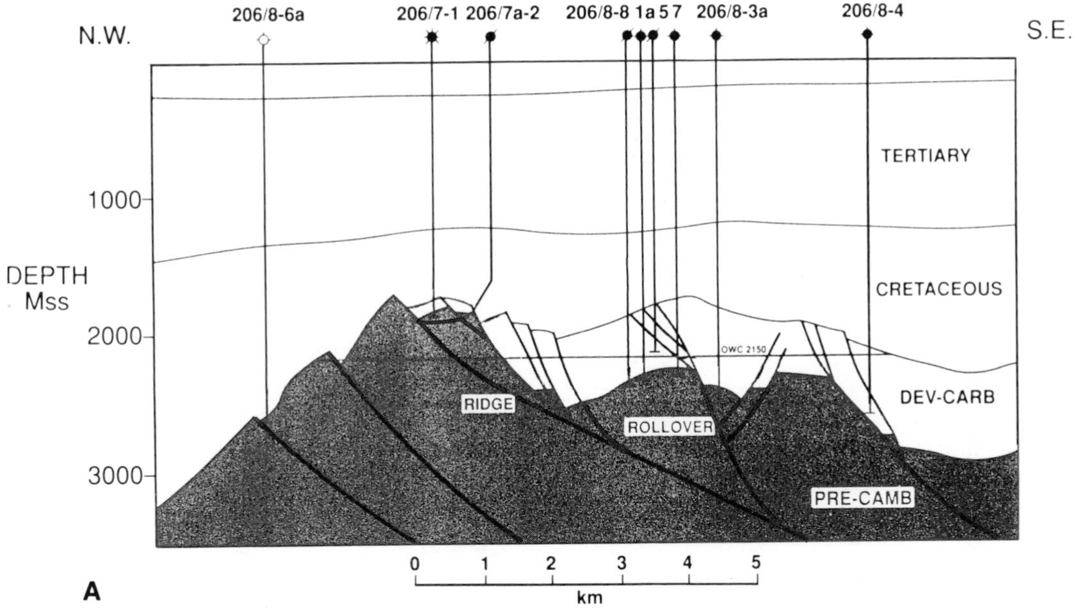

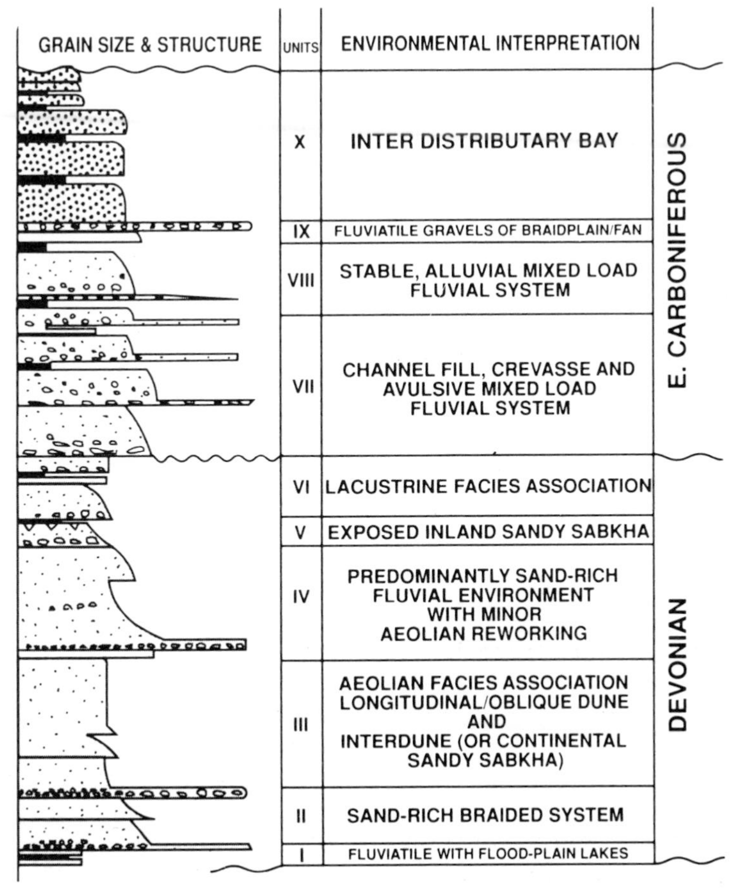

Fig. 12.73 (A) North-west–south-east structural cross-section through the Clair oilfield; (B) summary of the facies types and depositional environments in the Devonian–Carboniferous reservoirs of the Clair Field (from Coney *et al.*, 1993).

mirrored by an oppositely dipping slope on the north-western side of the basin); and (v) Late Tertiary structural inversion is marked by two regionally significant compressional events (Early Oligocene and Late Miocene), which in Quadrant 204 reactivated basin-forming faults and transfer zones and caused some folding of the overlying Palaeocene succession.

In addition to depositional facies, there is also a strong depth dependency effect on absolute reservoir quality. The more deeply buried northern subbasin is characterized by reservoirs of much lower quality, as reflected in disappoint-

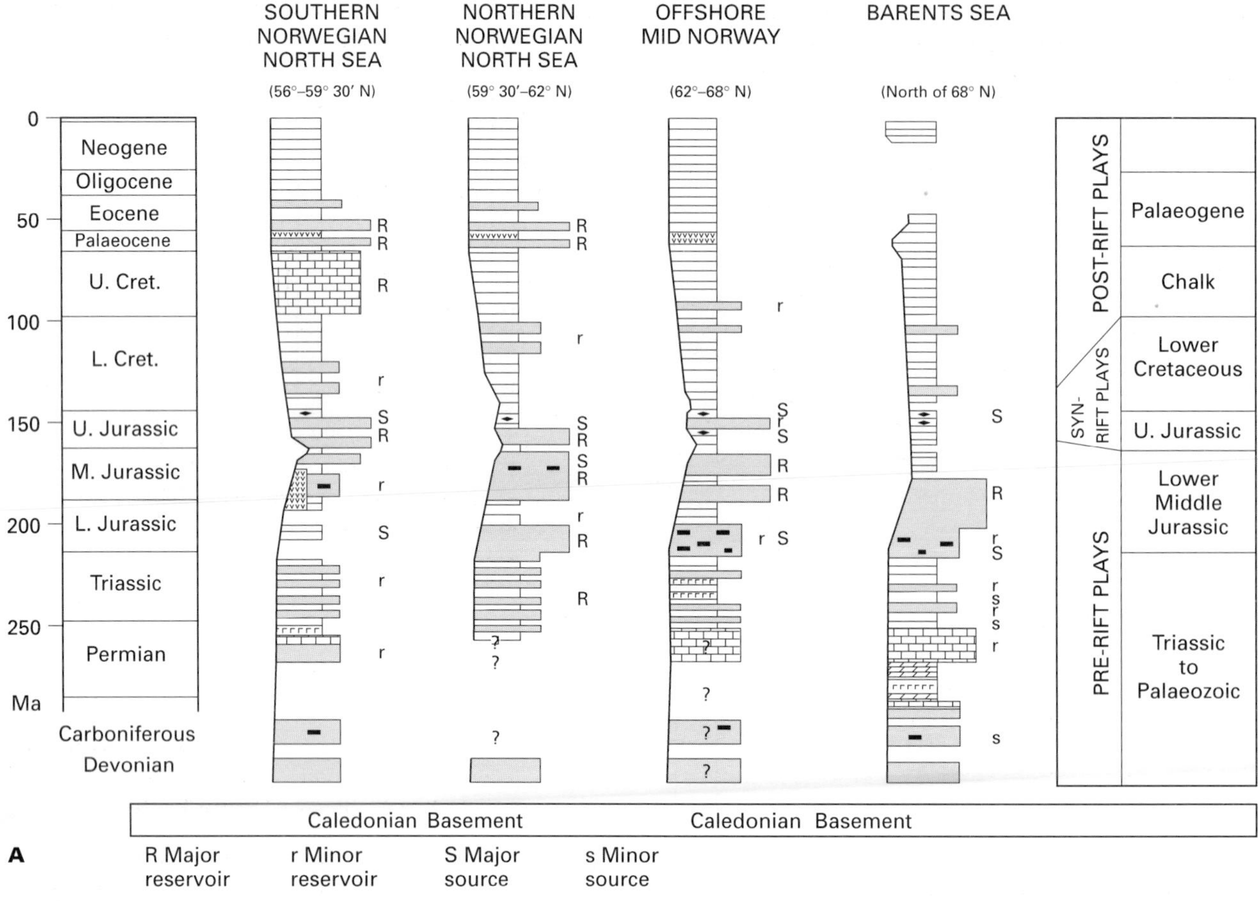

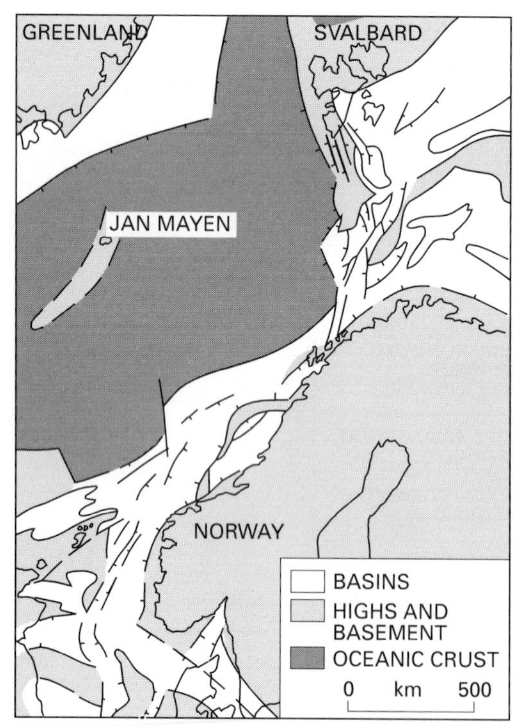

Fig. 12.74 (A) Simplified stratigraphic sections illustrating the distribution of source/reservoir/seal intervals and associated play types in (1) southern Norwegian North Sea, (2) northern Norwegian North Sea, (3) offshore mid-Norway, and (4) Barents Sea; (B) major structural elements of the Norwegian seaboard from the North Sea to the Barents Sea (from Fjaeran and Spencer, 1991).

ing rates of gas flow. In contrast, depth-related reduction in permeability in Quadrant 204 is not significant, because of the much shallower burial depths (mainly < 2500 m ss).

Seals are provided by the basinal mudstones, particularly those associated with regional flooding surfaces, which are capable of sealing hydrocarbons trapped within the basin-floor fans. Lateral seals may be less effective, particularly where traps require seals against prograding slope successions.

Traps within the Palaeocene play-fairway can be purely structural (e.g. drape over Cretaceous structures, Tertiary slides or Oligocene inversion structures), purely stratigraphic (pinch-outs, onlaps, compactional drape, incised channels, etc.) or, most likely, combinations of the two.

The least-known aspects of this play are the source, charge and migration histories. It is generally assumed that the Late Jurassic marine source rocks have generated the oil in Quadrant 204. Essentially vertical oil migration would then be assumed, presumably along some of the deeper, Mesozoic basin-margin faults, in a way similar to that seen in the North Sea Basin. Statistically, there seems to be no reason why other plays, similar to Quadrant 204, should not be discovered in more than one part of the long Atlantic Margin between the Porcupine Basin and the Barents Sea.

The remainder of the Atlantic Margin, and possibly the most prospective part, is located to the north-east of the Faeroe Basin, mainly along the Norwegian seaboard, including the established petroleum province of offshore mid-Norway and extending northwards to the less successful Barents Sea region (Fjaeran and Spencer, 1991). Similar tectonostratigraphic units define the same major play types seen elsewhere along this margin and in the North Sea, which facilitates their comparison (Fig. 12.74). It is uncertain whether or not the same criteria for hydrocarbon generation, entrapment and preservation can be achieved in the unexplored parts of this large area, but the Atlantic Margin remains a most important frontier area for hydrocarbon exploration.

12.5 Acknowledgements

The authors are indebted to Ken Glennie for his encouragement, assistance and (almost) unlimited patience during the preparation of this chapter. Previous contributions to this chapter and the associated course lecture by Alan Parsley and John Parker (Shell International Exploration and Production) are gratefully acknowledged.

12.6 Key references

Croker, P.F. and Shannon, P.M. (eds) (1995) *The Petroleum Geology of Ireland's Offshore Basins.* Special Publication 93, Geological Society, London, 498 pp.

Fjaeran, T. and Spencer, A.M. (1991) Proven hydrocarbon plays, offshore Norway. In: Spencer, A.M. (ed.) *Generation, Accumulation and Production of Europe's Hydrocarbons III.* European Association of Petroleum Geoscientists, Special Publication No. 3, pp. 25–48.

Glennie, K.W. and Provan, D.M.J. (1990) Lower Permian Rotliegend reservoir of the Southern North Sea gas province. In: Brooks, J. (ed.) *Classic Petroleum Provinces.* Special Publication 50, Geological Society, London, pp. 399–416.

Hawkes, P.W., Fraser, A.J. and Einchomb, C.C.G. (1997) The tectono-stratigraphic development and exploration history of the Weald and Wessex Basins, south England. In: Underhill, J.R. (ed.) *Development and Evolution of the Wessex Basin and Adjacent Areas* Special Publication 133, Geological Society, London, pp. 39–66.

Parker, J.R. (ed.) (1993) *Petroleum Geology of Northwest Europe: Proceedings of the 4th Conference.* Geological Society, London, 1542 pp.

Pegrum, R.M. and Spencer, A.M. (1990) Hydrocarbon plays in the northern North Sea. In: Brooks, J. (ed.) *Classic Petroleum Provinces.* Special Publication 50, Geological Society, London, pp. 441–70.

Rondeel, H.E., Batjes, D.A.J. and Nieuwenhuijs, W.H. (eds) (1996) *The Geology of Gas and Oil under the Netherlands.* Royal Geological and Mining Society of the Netherlands. Kluwer Academic Publishers, 284 pp.

Ziegler, P.A. (1990a) *Geological Atlas of Western and Central Europe,* 2nd edn. Shell Internationale Petroleum Maatschappij B.V., Geological Society, Bath (distributors), 239 pp.

References

[References lacking a chapter attribution are of a general nature, but are not specifically cited]

Aalund, L.R. (1983a) Guide to export crudes of the '80s-4: North Sea crudes: Flotta to Thistle. *Oil Gas J.* **81**(23), 99–113. [11]

Aalund, L.R. (1983b) Guide to export crudes of the 80s-3. North Sea now offers 14 export crudes. *Oil Gas J.* **81**(21) 75–9. [11]

Aasheim, S.M., Dalland, A., Netland, A. and Thon, A. (1986) The Smørbukk gas/condensate discovery, Haltenbanken. In: Spencer, A.M., Campbell, C.J., Hanslien, S.H., Nelson, P.H., Nysaether, E. and Ormaasen, E.G. (eds) *Habitat of Hydrocarbons on the Norwegian Continental Shelf.* Graham and Trotman, London, pp. 299–305. [1, 8, 11]

Abbott, G.D., Eglinton, T.I. and Home, A.K. (1989) The kinetics of biological marker release from Kimmeridge kerogen during hydrous pyrolysis. In: *Abstracts of Papers presented at the 14th Int'l Org. Geochem. Conference*, Paper No. 142, Paris, EAOG and IFP. [11]

Abbotts, I.L. (ed.) (1991) *United Kingdom Oil and Gas Fields: 25 Years Commemorative Volume.* Geological Society, London, Memoir No. 14, 573 pp. [1, 5, 7, 9, 11, 12]

Abdullah, W.H., Murchison, D.G., Jones, J.M., Telnaes, N. and Gjelberg, J. (1988) Early Carboniferous coal depositional environments in Spitsbergen, Svalbard. In: Mattavelli, L. and Novelli, L. (eds) *Advances in Organic Geochemistry, 1987 (Proc. 13th Int'l. Mtg on Org. Geochem. Venice,) Part 2.* Pergamon Press, Oxford, pp. 953–64. [11]

Agterberg, F.P., Ogg, J.G., Hardenbol, J., van Veen, P., Thierry, J. and Huang, Z. (1995) A Triassic, Jurassic and Cretaceous time scale. In: Berggren, W.A. *et al.* (eds) *Geochronology: Timescales and Global Stratigraphic Correlation.* Special Publication No. 54, SEPM, Tulsa, OK., pp. 95–126.

Aguilera, R. and van Poolen, H.K. (1979) Porosity and water saturation can be estimated from well logs. *Oil Gas J.* Jan., 101–8. [9]

Ahlbrandt, T.S. and Fryberger, S.G. (1982) Introduction to eolian deposits. In: Scholle, P.A. and Spearing, D. (eds) *Sandstone Depositional Environments.* Memoir No. 31, AAPG, Tulsa, OK., pp. 11–47. [5]

Ahsan, A., Karlsen, D.A., Mitchell, A.W., Dodd, T., Rothwell, N. and Olsen, R. (1997) *Inter- and intra-field hydrocarbon compositional variations in the Ula and Gyda fields (Central Graben -NOCS)—implication for understanding the controls on hydrocarbon distribution within and between these fields.* Paper submitted for the post Maastricht (18th Int'l Meeting) Organic Geochemistry volumes. [11]

Ainsworth, N.R., Burnett, R.D. and Kontrovitz, M. (1990) Ostracod colour change by thermal alteration, offshore Ireland and Western UK. *Marine. Petr. Geol.* 7. [11]

Alberts, M.A. and Underhill, J.R. (1991) The effect of Tertiary structuration on Permian gas prospectivity, Cleaver Bank area, southern North Sea, UK. In: Spencer, A.M. (ed.) *Generation, Accumulation, and Production of Europe's Hydrocarbons.* Special Publication, EAPG, Oxford University Press, Oxford, pp. 161–73. [2]

Albertsen, M. (1992) Oil and gas in Germany—exploration, production and research. *First Break* **10**, 225–31. [6]

Albright, W.A., Turner, W.L. and Williamson, K.R. (1980) Ninian Field, UK Sector, North Sea. In: Halbouty, M.T. (ed.) *Giant Oil and Gas Fields of the Decade 1968–1978.* Memoir No. 30, AAPG, pp. 173–94. [1, 11]

Aldridge, R.J. (1986) Conodont palaeo-biogeography and thermal maturation in the Caledonides. *J. Geol. Soc.* **143**(1), 177–84. [11]

Alexander, R.W.S., Schofield, K. and Williams, M.C. (1992) Understanding the Eocene reservoirs of the Forth Field, UKCS Block 9/23b. In: Spencer, A.M. (ed.) *Generation, Accumulation, and Production of Europe's Hydrocarbons.* Special Publication No. 3, EAPG, Oxford University Press, Oxford, pp. 3–15. [10, 12]

Allan, J., Bjorøy, M. and Douglas, A.G. (1980) A geochemical study of the Exinite Group maceral alginite selected from three Permo-Carboniferous torbanites. In: Douglas, A.G. and Maxwell, J.R. (eds) *Advances in Organic Geochemistry 1979.* Pergamon Press, Oxford, pp. 599–618. [11]

Allan, P., Anderton, R., Davies, M., Marshall, A., Pooler, I., Vaughan, O. and Hossack, J. (1994) Structural development of the ETAP diapirs, central North Sea (abstract). In: Alsop, A. (Convenor) *Salt Tectonics: Programme and Abstracts.* Geological Society, London. [1, 6]

Allen, P.A. and Allen, J.R. (1990) *Basin Analysis.* Blackwell Scientific Publications, Oxford, UK, 451 pp. [11]

Allen, P.A. and Mange-Rajetzky, M.A. (1992) Devonian–Carboniferous sedimentary evolution of the Clair area, offshore north-western UK: impact of changing provenance. *Mar. Petr. Geol.* **9**, 29–52. [3]

Allen, P.A. and Marshall, J.E.A. (1981) Depositional environments and palynology of the Devonian South-east Shetland Basin. *Scott. J. Geol.* **17**, 257–73. [2]

Alsop, A. (1996) Physical modelling of fold and fracture geometries associated with salt diapirism. In: Alsop, G.I., Blundell, D.J. and Davison, I. (eds) *Salt Tectonics.* Special Publication No. 100, Geological Society, London, pp. 277–341. [6]

Altebaumer, F.J., Leythaeuser, D. and Schaefer, R.G. (1983) Effects of geologically rapid heating on maturation and hydrocarbon generation in Lower Jurassic shales from N.W. Germany. In: Bjorøy, M., Albrecht, P. Cornford, C., de Groot, K., Eglinton, G., Galimov, E. *et al.* (eds) *Advances in Organic Geochemistry 1981.* John Wiley, Chichester, pp. 80–6. [11]

Ambler, J. (1989) The organic geochemistry of the Minch basin Jurassic shales. PhD thesis, University of Aberdeen. [11]

Ames, R. and Farfan, P.F. (1996) The environments of deposition of the Triassic Main Buntsandstein Formation in the P and Q quadrants, offshore the Netherlands. In: Rondeel, H.E., Batjes, D.A.J. and Nieuwenhuis, W.H. (eds) *Geology of Oil and Gas under the Netherlands.* Royal Geological and Mining Society, the Netherlands, Kluwer Academic Publishers, Dordrecht, pp. 167–78. [7]

Amiri-Garroussi, K. and Taylor, J.C.M. (1987) Complex diagenesis in Zechstein dolomites of the Ettrick oil field. In: Brooks, J. and Glennie, K.W. (eds) *Petroleum Geology of North West Europe.*

Graham and Trotman, London, pp. 577–89. [6, 12]

Amiri-Garroussi, K. and Taylor, J.C.M. (1992) Displaced carbonates in the Zechstein of the UK North Sea. *Mar. Petr. Geol.* **9**, 186–96. [6]

Andersen, C. and Doyle, C. (1990) Review of hydrocarbon exploration and production in Denmark. *First Break* **8**, 155–65. [9]

Andersen, S.A., Hansen, S.A. and Fjeldgaard, K. (1989) Horizontal drilling and completion, Denmark. Publication No. 18349, SPE. [9]

Andersen, S.A., Conlin, J.M., Fjeldgaard, K. and Hansen, S.A. (1990) Exploiting reservoirs with horizontal wells: the Maersk experience. *Schlumberger Oilfield Rev.* **2**(3). [9]

Anderson, T.F., Popp, B.N., Williams, A.C., Ho, L.-Z. and Hudson, J.D. (1994) The stable isotopic record of fossils from the Peterborough Member, Oxford Clay Formation (Jurassic), UK: palaeoenvironmental implications. *J. Geol. Soc.* **151**(1), 125–38. [11]

Anderton, R (1982) Dalradian deposition and the late Precambrian–Cambrian history of the N. Atlantic region: a review of the early evolution of the lapetus Ocean. *J. Geol. Soc.* **139**, 423–31. [2]

Anderton, R. (1993) Sedimentation and basin evolution in the Paleogene of the northern North Sea and Faeroe–Shetland Basins. Abstract. In: Parker, J.R. (ed.) *Petroleum Geology of Northwest Europe: Proceedings of the 4th Conference.* Geological Society, London, p. 31. [10, 12]

Andersson, A., Dahlman, B. and Gee, D.G. (1982) Kerogen and uranium resources in the Cambrian Alum Shales of Billingen-Falbygden and Narke area. *Sweden. Geol. Stockh. Forh.* **104**(3), 197–209. [11]

Andersson, A., Dahlman, B., Gee, D.G. and Snall, S. (1985) The Scandinavian Alum Shales. *Sveriges Geol. Unders.* **56**, 50. [11]

Andresen, B., Barth, T., Irwin, H. and Throndsen, T. (1991) Yields and isotopic composition of pyrolysis products from different types of source rock from the North Sea area. In: Manning, D. (ed.) *Organic Geochemistry, Advances and Applications in Energy and Natural Environment.* Ext. Abstracts, 15th Meeting EAOG, Manchester University Press, Manchester, 131–4. [11]

Andresen, P., Mills, N., Schenk, H.-J. and Horsfield, B. (1993) The importance of kinetic parameters in modelling generation by cracking of oil to gas—a case study in 1D from well 2/4-14. In: Doré, A.G., Augustson, J.H., Hermanrud, C., Stewart, D.J. and Sylta, O. (eds) *Basin Modelling: Advances and Applications.* Special Publication No. 3, NPF, Elsevier, Amsterdam, pp. 563–71. [11]

Andrews, I.J. and Brown, S. (1987) Stratigraphic evolution of the Jurassic, Moray Firth. In: Brooks, J. and Glennie, K. (eds) *Petroleum Geology of North-West Europe.* Graham and Trotman, London, pp. 785–95. [2, 8]

Andrews, I.J., Long, D., Richards, P.C., Thomson, A.R., Brown, S., Chesher, J.A. and McCormac, M. (eds) (1990) *The Geology of the Moray Firth.* UK Offshore Regional Report, BGS, HMSO, London, 96 pp. [3, 4, 5, 9, 11]

Andrews, J.E., Turner, M.S., Nabi, G., and Spiro, B. (1991) The anatomy of an early Dinantian terraced floodplain: palaeo-environment and early diagenesis. *Sedimentology* **38**, 271–87. [4]

Andrews-Speed, C.P., Oxburgh, E.R. and Cooper, B.A. (1984) Temperature and depth-dependent heat flow in western North Sea. *AAPG Bull.* **68**(11), 1764–81. [11]

Antonowicz, L. and Knieszner, L. (1981) Reef zones of the Main Dolomite, set out on the basis of paleogeomorphic analysis and the results of modern seismic techniques. In: *Proceedings International Symposium on Central European Permian, Jablonna, Poland, 1978.* Geology Institute, Warsaw, pp. 356–68. [6]

Antonowicz, L. and Knieszner, L. (1984) Zechstein reefs of the main dolomite in Poland and their seismic recognition. *Acta Geol. Polonica* **34**, 81–93. [6]

Aplin, A.C. and Coleman, M.L. (1995) Sour gas and water chemistry of the Bridport Sands reservoir, Wytch Farm, UK. In: Cubitt, J. and England, W. (eds) *The Geochemistry of Reservoirs.* Special Publication No. 86, Geological Society, London, pp. 303–14. [11]

Aplin, A.C., Warren, E.A., Grant, S.M. and Robinson, A.G. (1993) Mechanisms of quartz cementation in North Sea reservoir sandstones: constraints from fluid compositions. In: Horbury, A.D. and Robinson, A.G. (eds) *Diagenesis and Basin Development.* Studies in Geology No. 36, AAPG, Tulsa, OK., pp. 5–22. [10, 12]

Armagnac, D., Bucci, J., Kendall, C.G. and Lerche, I. (1989) Estimating the thickness of sediment removed at an unconformity using vitrinite reflectance data. In: Naeser, N.D. and McCulloh, T.H. (eds) *Thermal History of Sedimentary Basins: Methods and Case Histories.* Springer-Verlag, New York, pp. 217–39. [11]

Armentrout, J.M., Malececk, S.J., Fearn, L.B. *et al.* (1993) Log-motif analysis of Paleogene depositional systems tracts, Central and Northern North Sea: defined by sequence stratigraphic analysis. In: Parker, J.R. (ed.) *Petroleum Geology of Northwest Europe: Proceedings of the 4th Conference.* Geological Society, London, pp. 45–58. [10]

Armstrong, J.P., d'Elia, V.A.A. and Loberg, R. (1995) Holywell Shale: a potential source of hydrocarbons in the East Irish Sea. In: Croker, P.F. and Shannon, P.M. (eds) *The Petroleum Geology of Ireland's Offshore Basins.* Special Publication No. 93, Geological Society, London, pp. 37–8. [11, 12]

Armstrong, L.A., Ten Have, A. and Johnson, H.D. (1987) The geology of the Gannet Fields, Central North Sea, UK Sector. In: Brooks, J. and Glennie, K.W. (eds) *Petroleum Geology of North West Europe.* Graham and Trotman, London, pp. 533–48. [1, 6, 8, 10, 12]

Arthur, T.J., Pilling, G., Bush, D. and Macchi, L. (1986) The Leman sandstone formation in U.K. Block 49/28: sedimentation, diagenesis and burial history. In: Brooks, J., Goff, J. and van Hoorn, B. (eds) *Habitat of Palaeozoic Gas in N.W. Europe.* Special Publication No. 23, Geological Society, London, pp. 251–66. [5]

Arthurton, R.S., Gutteridge, P. and Nolan, S.C. (eds) (1989) *The Role of Tectonics in Devonian and Carboniferous Sedimentation in the British Isles.* Occasional Publication No. 6, Yorkshire Geological Society, Ellenbank Press, Maryport, Cumbria. [4]

Arveschoug, N.C., Conn, P.J., Drabble, J., Mead, D. and Bird, A. (1995) Acquisition and processing of an intermediate depth walkaway VSP in a UK North Sea Central Graben well. *First Break* **13**(ii), 435–40. [1]

Ashcroft, W.A. and Ridgway, M.S. (1996) Early discordant diagenesis in the Brent Group, Murchison Field, UK North Sea, detected in high values of seismic-derived acoustic impedance. *Petr. Geosci.* **2**, 75–81. [8]

Ashcroft, W.A., Kneller, B.C., Leslie, A.G. and Munro, M. (1984) Major shear zones and autochthonous Dalradian in the north-east Scottish Caledonides. *Nature* **310**, 760–2. [2]

Astin, T.R. (1982) The Devonian geology of the Walls Peninsula, Shetland. PhD thesis, University of Cambridge.

Astin, T.R (1990) The Devonian lacustrine sediments of Orkney, Scotland: implications for climate cyclicity, basin structure and maturation history. *J. Geol. Soc.* **147**, 141–51. [3]

Austin, R.L. (1973) Modification of the British Avonian conodont zonation and a reappraisal of European Dinantian conodont zonation and correlation. *Ann. Soc. Geol. Belge* **96**, 523–32. [4]

Backer-Owe, K., Dypvik, H. and Larter, S.R. (1989) Organic facies development in the Janusfjellet Subgroup in Nordenskjoeld Land, Svalbard, Norway. In: *Abstracts of Papers presented at the 14th Int'l Org. Geochem. Conference, Paper No. 59, Paris.* EAOG and IFP. [11]

Badley, M.E., Egeberg, T. and Nipen, O. (1984) Development of rift basins illustrated by the structural evolution of the Oseberg

Feature, Block 30/6, offshore Norway. *J. Geol. Soc.* **141**, 639–49. [1]

Badley, M.E., Price, J.D., Rambech Dahl, C. and Agdestein, T. (1988) The structural evolution of the northern Viking Graben and its bearing upon extensional modes of basin formation. *J. Geol. Soc.* **145**, 455–72. [7]

Bailey, C.C., Price, I. and Spencer, A.M. (1981) The Ula Oilfield, Block 7/12. In: *Norwegian Symposium on Exploration.* NPF, Article 18, Norwegian Petroleum Society, 26 pp. [1, 12]

Bailey, E.B. and Weir, J. (1932) Submarine faulting in Kimmeridgian times: east Sutherland. *Trans. Roy. Soc. Edinburgh* **57**, 429–67. [8]

Bailey, J.B., Arbin, P., Daffinoti, O., Gibson, P. and Richie, J.S. (1993) Permo-Carboniferous plays of the Silver Pit Basin. In: Parker, J. (ed.) *Petroleum Geology of Northwest Europe: Proceedings of the 4th Conference.* Geological Society, London, pp. 707–15. [4, 5, 11, 12]

Bailey, N.J.L., Walko, P. and Sauer, M.J. (1987) Geochemistry and source potential of the West of Shetlands. In: Brooks, J. and Glennie, K.W. (eds) *Petroleum Geology of North West Europe.* Graham and Trotman, London, pp. 711–21. [1, 8, 11]

Bailey, N.J.L., Burwood, R. and Harriman, G. (1990) Application of pyrolysate carbon isotope and biomarker technology to organofacies definition and oil correlation problems in North Sea basins. In: Durand, B. and Behar, F. (eds) In: *Advances in Organic Geochemistry. Org. Geochem.* **16**(1–3), 1157–72. [11]

Bain, J.S. (1993) Historical overview of exploration of Tertiary plays in the UK North Sea. In: Parker, J.R. (ed.) *Petroleum Geology of Northwest Europe: Proceedings of the 4th Conference.* Geological Society, London, pp. 5–14. [10, 12]

Baird, A., Kelly, J. and Symonds, R. (1993) The reservoir potential of the Zechstein-3 Carbonate Member Platten, Offshore Netherlands. Abstracts, AAPG International Conference, The Hague, the Netherlands, 17–20 October. *AAPG Bull.* **77**, 1604. [6]

Baird, R.A. (1986) Maturation and source rock evaluation of Kimmeridge Clay, Norwegian North Sea. *AAPG Bull.* **70**(1), 1–11. [11]

Bakken, K.A., Loberg, R. and Theis, N. (1991) Residual oils and other hydrocarbons from the Barents Sea: grouping and sourcing. In: Manning, D. (ed.) *Organic Geochemistry, Advances and Applications in Energy and Natural Environment.* Ext. Abstracts, 15th Meeting EAOG, Manchester University Press, London, pp. 11–13. [11]

Banner, J.A., Chatellier, J.-Y., Feurer, J.R. and Neuhaus, D. (1992) Guillemot D: a successful appraisal through alternative interpretation. In: Hardman, R.F.P. (ed.) *Exploration Britain: Geological Insights for the Next Decade.* Special Publication No. 67, Geological Society, London, pp. 129–49. [1, 12]

Barker, A.J. and Gayer, R.A. (1985) Caledonide–Appalachian tectonic analysis and evolution of related oceans. In: Gayer, R. (ed.) *The Tectonic Evolution of the Caledonide–Appalachian Orogen.* Fried. Vieweg and Sohn, Braunschweig, Wiesbaden, pp. 126–65. [2]

Barker, C. (1996) *Thermal Modelling of Petroleum Generation: Theory and Applications.* Elsevier Science BV., Amsterdam, 512 pp. [11]

Barker, C.E. (1991) Implications for organic maturation studies of evidence for a geologically rapid increase and stabilisation of vitrinite reflectance at peak temperature: Cerro Prieto geothermal system, Mexico. *AAPG Bull.* **75**, 1852–63. [3]

Barker, C.E. and Pawlewicz, M.J. (1994) Calculation of vitrinite reflectance from thermal histories and peak temperatures—a comparison of methods. In: Mukhopadhyay, P.K. and Dow, W.G. (eds) *Vitrinite Reflectance as a Maturity Parameter—Applications and Limitations.* American Chemical Society, pp. 216–29. [11]

Barnard, P.C. and Bastow, M.A. (1991) Petroleum generation, alteration, entrapment and mixing in the Central and Northern North Sea. In: England, W.A. and Fleet, A.J. (eds) *Petroleum Migration.* Special Publication No. 59, Geological Society, London, pp. 167–90. [11]

Barnard, P.C. and Cooper, B.S. (1981) Oils and source rocks of the North Sea area. In: Illing, L.V. and Hobson, G.D. (eds) *Petroleum Geology of the Continental Shelf of North-West Europe.* Heyden, London, pp. 169–75. [11]

Barnard, P.C. and Cooper, B.S. (1983) A review of geochemical data related to the north-west European gas province. In: Brooks, J. (ed.) *Petroleum Geochemistry and Exploration of Europe.* Special Publication No. 12., Geological Society, Blackwell Scientific Publications, Oxford, pp. 19–33. [11]

Barnard, P.C., Collins, A.G. and Cooper, B.S. (1981a) Identification and distribution of kerogen facies in a source rock horizon—examples from the North Sea Basin. In: Brooks, J. (ed.) *Organic Maturation Studies and Fossil Fuel Exploration.* Academic Press, London, pp. 271–82. [11]

Barnard, P.C., Collins, A.G. and Cooper, B.S. (1981b) Generation of hydrocarbons: time, temperature and source rock quality. In: Brooks, J. (ed.) *Organic Maturation Studies and Fossil Fuel Exploration.* Academic Press, London, pp. 337–42. [11]

Barr, D., Strachan, R.A., Holdsworth, R.E. and Roberts, A.M. (1988) Summary of the geology. In: Allison, I., May, F. and Strachan, R.A. (eds) *An Excursion Guide to the Moine Geology of the Scottish Highlands.* Scottish Academic Press, Edinburgh, pp. 11–38. [2]

Barr, K.W., Colter, V.S. and Young, R. (1981) The geology of the Cardigan Bay—St George's Channel Basin. In: Illing, L.V. and Hobson, G.D. (eds) *Petroleum Geology of the Continental Shelf of North-West Europe.* Heyden, London, pp. 432–43. [11]

Barrell, J. (1916) The dominantly fluvial origin under seasonal rainfall of the Old Red Sandstone. *Bull. Geol. Soc. Am.* **27**, 345–86. [2]

Barrett, R.F., Margesson, R.W. and D'Angelo, R.M. (1995) Use of rock properties and AVO in the Everest Field development, UKCS. *Petrol. Geosci.* **1**, 311–7. [1]

Bartenstein, H. (1979) Essay on the coalification and hydrocarbon potential of the Northwest European Palaeozoic. *Geol. Mijnbouw* **58**, 57–64. [5, 11]

Barth, T., Borgund, E. and Hopland, A.L. (1989) Generation of organic compounds by hydrous pyrolysis of Kimmeridge oil shale—bulk results and activation energy calculations. *Org. Geochem.* **14**(1), 69–76. [11]

Bartholomew, I.D., Peters, J.M. and Powell, C.M. (1993) Regional structural evolution of the North Sea: oblique slip and the reactivation of basement lineaments. In: Parker, J.R. (ed.) *Petroleum Geology of Northwest Europe: Proceedings of the 4th Conference.* Geological Society, London, pp. 1109–22. [1, 2, 9, 12]

Barton, D.C. (1933) Mechanics of formation of salt domes with special reference to Gulf Coast salt domes of Texas and Louisiana. *AAPG Bull.* **17**, 1025–83. [6]

Barton, P. (1984) Crustal stretching in the North Sea: implications for thermal history. In: Durand, B. (ed.) *Thermal Phenomena in Sedimentary Basins.* Editions Technip, Paris, pp. 227–34. [11]

Barwise, A.J.G. and Park, P.J.D. (1983) Petroporphyrin fingerprinting as a geochemical marker. In: Bjorøy, M., Albrecht, P., Cornford, C., de Groot, K., Eglinton, G., Galimov, E. *et al.* (eds) *Advances in Organic Geochemistry 1981.* John Wiley, Chichester, pp. 668–75. [11]

Bateson, J.F. and Haszeldine, R.S. (1986) Organic maturation of Scottish Dinantian oil shales: a study using biomarkers. In: Cater, J. (ed.) *Sedimentology and Hydrocarbon Potential of the Dinantian Oil-shales of Northern Britain. Scott. J. Geol.* **22**, 417–29. [11]

Batten, D.J. (1983) Indentification of amorphous sedimentary organic matter by transmitted light microscopy. In: Brooks, J. (ed.) *Petroleum Geochemistry and Exploration of Europe.* Special

Publication No. 12., Geological Society, Blackwell Scientific Publications, Oxford, pp. 275–88. [11]

Batten, D.J., Trewin, N.H. and Tudhope, A.W. (1986) The Triassic–Jurassic junction at Golspie, Inner Moray Firth basin. *Scott. J. Geol.* **22**, 85–98. [8]

Baudin, F., Herbin, J.-P., Bassoullet, J.-P., Dercourt, J., Lachkar, G., Manivit, H. and Renard, M. (1990) Distribution of organic matter during the Toarcian in the Mediterranean Tethys and Middle East. In: Huc, A.Y. (ed.) *Deposition of Organic Facies.* Studies in Geology No. 30, AAPG, pp. 73–92. [1, 11]

Baumann, A. and O'Cathain, B. (1991) The Dunlin Field, Blocks 211/23a, 211/24a. In: Abbotts, I.L. (ed.) *UK North Sea United Kingdom Oil and Gas Fields: 25 Years Commemorative Volume.* Memoir No. 14, Geological Society, London, pp. 95–102. [1]

Beach, A. (1984) Structural evolution of the Witch Ground Graben. *J. Geol. Soc.* **141**, 621–8.

Beaumont, E.A. and Foster, N.H. (compilers) (1988) *Geochemistry. Treatise of Petroleum Geol.* Reprint Series No. 8, AAPG, Tulsa, OK., 660 pp. [11]

Beckly, A., Dodd, C. and Los, A. (1993) The Bruce field. In: Parker, J.R. (ed.) *Petroleum Geology of Northwest Europe: Proceedings of the 4th Conference.* Geological Society, London, pp. 1453–63. [1]

Belin, S. and Brosse, E. (1992) Petrographical and geochemical study of a Kimmeridgian organic sequence Yorkshire area, UK. *Rev. IFP* **47**, 711–25. [11]

Belin, S. and Kenig, F. (1994) Petrographic analysis of organo-mineral relationships: depositional conditions of the Oxford Clay Formation (Jurassic), UK. *J. Geol. Soc.* **151**(1), 153–60. [11]

Bell, J., Holden, J., Pettigrew, T.H. and Sedman, K.W. (1979) The Marl Slate and basal Permian breccia at Middridge, Co. Durham. *Proc. Yorks. Geol. Soc.* **42**(3), 439–60. [5]

Benton, M.J. and Walker, A.D. (1985) Palaeoecology, taphonomy, and dating of Permo-Triassic reptiles from Elgin, North-East Scotland. *Palaeontology* **28**(2), 207–34. [5]

Bentz, A. (1958) Northwest German Sedimentary Basin. In: Weeks, L.G. (ed.) *Habitat of Oil.* Special Publication, AAPG, Tulsa, OK., pp. 1054–66. [1]

Bergan, M., Torudbakken, B. and Wandas, B. (1989) Lithostratigraphic correlation of Upper Jurassic sandstones within the Norwegian Central Graben: sedimentological and tectonic implications. In: Collinson, J.D. (ed.) *Correlation in Hydrocarbon Exploration.* Graham and Trotman, London, pp. 243–51. [8]

Berggren, W.A., Kent, D.V., Aubry, M. and Hardenbol, J. (eds) (1995) *Geochronology, Timescales and Global Stratigraphic Correlation.* Special Publication No. 54, SEPM, Tulsa, OK. [11]

Bergstrøm, J., Bess, M.J.M. and Paproth, E. (1985) The marine Knabberud Limestone in the Oslo Graben: possible implications for the model of Silesian palaeogeography. *Zeitschr. Deutsches Geol. Ges.* **136**, 181–94. [2]

Bergstrom, S.M. (1980) Conodants as paleotemperature tools in Ordovician rocks of the Caledonides and adjacent areas in Scandinavia and the British Isles. *Geol. For. Stockh. Forh.* **102**, 377–92. [11]

Berner, R.A. and Raiswell, R. (1984) C/S method for distinguishing freshwater from marine sedimentary rocks. *Geology,* **12**, 365–8. [4, 11]

Berner, U. and Faber, E. (1988) Maturity related mixing model for methane, ethane and propane, based on carbon isotopes. *Org. Geochem.* **13**, 1–3, 67–72. [11]

Berridge, N.G. and Ivimey-Cook, H.C. (1967) The geology of a Geological Survey borehole at Lossiemouth, Morayshire. *Bull. Geol. Survey GB* **27**, 155–69. [8]

Berridge, N.G. and Pattinson, J. (1994) *Geology of the Country around Grimsby and Patrington.* BGS Memoir, HMSO, London, 96 pp. [9]

Bessereau, G., Guillocheau, F. and Huc, A.-Y. (1995) Source rock occurrence in a sequence stratigraphic framework: the example of the Lias of the Paris Basin. In: A.-Y. Huc (ed.), *Paleogeography, Paleoclimate, and Source Rock,* Studies in Geology No. 40, AAPG, pp. 273–303. [11]

Berstad, S. and Dypvik, H. (1982) Sedimentological evolution and natural radioactivity of Tertiary sediments from the central North Sea. *J. Petr. Geol.* **5**, 77–88. [11]

Bertelsen, F. (1978) The Upper Triassic–Lower Triassic Vinding and Gassum Formations of the Norwegian–Danish Basin. *Danm. Geol. Unders., Ser. B* **3**, 1–26. [7]

Bertelsen, F. (1980) Lithostratigraphy and depositional history of the Danish Triassic. *Danm. Geol. Unders., Ser. B* **4**, 1–59. [7]

Bertrand, P. and Lallier-Verges, E. (1993) Past sedimentary organic matter accumulation and degradation controlled by productivity. *Nature* **364**, 786–88. [11]

Bertrand, P., Lallier-Verges, E., Martinez, L., Pradier, B., Tremblay, P., Huc, A., Jouhannel, R. and Tricart, J.P. (1989) Examples of spatial relationships between organic matter and mineral groundmass in the microstructures of the organic-rich Dorset Formation rocks, Great Britain. *Org. Geochem.* **16** (4–6), 661–75.

Bertrand, P., Bordenave, M.L., Brosse, E., Espitalié, J., Houzay, J.P., Pradier, B. *et al.* (1993) Other methods and tools for source rock appraisal: modelling of petroleum generation and migration. In: Bordenave, M.L. (ed.) *Applied Petroleum Geochemistry.* Editions Technip, Paris, pp. 279–372. [11]

Bertrand, R. (1990) Correlations among the reflectances of vitrinite, chitinozoans, graptolites and scoledonts. *Org. Geochem.* **15** (6), 565–74. [11]

Bertrand, R. and Achab, A. (1990) Equivalences between the reflectances of vitrinite, zooclasts chitinozoans, graptolites and scolecodonts and the colour alteration of palynomorphs spores and acritarchs. *Palynology* **13**, 280. *AAPG Bull.* **71**(8), 951–7. [11]

Besly, B.M. (1987) Sedimentological evidence for Carboniferous and Early Permian palaeoclimates of Europe. *Ann. Soc. Geol. Nord.* **106**, 131–43. [4]

Besly, B.M. (1988) Palaeogeographic implications of late Westphalian to early Permian red-beds, central England. In: Besly, B.M. and Kelling, G. (eds) *Sedimentation in a Synorogenic Basin Complex: the Upper Carboniferous of Northwest Europe.* Blackie, Glasgow, London, pp. 200–21. [4]

Besly, B.M. (1995) Stratigraphy of Late Carboniferous red beds in the Southern North Sea and adjoining land areas. (Abstract) *Stratigraphic advances in the offshore Devonian and Carboniferous rocks, UKCS and adjacent onshore areas.* Geological Society, London. [4]

Besly, B.M. and Cleal, C.J. (1997) Upper Carboniferous stratigraphy of the West Midlands (UK), revised in the light of borehole geophysical logs and detrital compositional suites. *Geol. J.,* **32**, 85–118. [4]

Besly, B.M. and Fielding, C.R. (1989) Palaeosols in Westphalian coal-bearing and red-bed sequences, central and northern England. *Palaeogeogr. Palaeoclimatol. Palaeoecol.* **70**, 303–30. [4]

Besly, B.M. and Kelling, G. (eds) (1988) *Sedimentation in a Synorogenic Basin Complex: the Upper Carboniferous of Northwest Europe.* Blackie, Glasgow, London, 276 pp. [4]

Besly, B.M. and Turner, P. (1983) Origin of red-beds in a moist tropical climate: Etruria Formation, Upper Carboniferous, UK. In: Wilson, R.C.L. (ed.) *Residual Deposits.* Special Publication No. 11, Geological Society, London, pp. 131–47. [4, 5]

Besly, B.M., Burley, S.D. and Turner, P. (1993) The late Carboniferous 'Barren Red Bed' play of the Silver Pit area, Southern North Sea. In: Parker, J.R. (ed.) *Petroleum Geology of Northwest Europe: Proceedings of the 4th Conference.* Geological Society, London, pp. 727–40. [2, 4, 12]

Bessa, J.L. and Hesselbo, S.P. (1997) Gamma-ray character and correlation of the Lower Lias, SW Britain. *Proc. Geol. Assoc.* **108**(2), 113–30. [11]

Best, G., Kockel, F. and Schoeneich, H. (1983) Geological history of the southern Horn Graben. *Geol. Mijnbouw* **62**, 25–33. [5, 7, 11]

Betz, D., Führer, F., Greiner, G. and Plein, E. (1987) Evolution of the Lower Saxony Basin. *Tectonophysics* **137**, 127–70. [9]

Bharati, S. and Larter, S. (1991) Origin, evolution and petroleum potential of a Cambrian source rock: evidence from an organic petrographic study. In: Manning, D. (ed.) *Organic Geochemistry: Advances and Applications in Energy and Natural Environment.* Ext. Abstracts, 15th Meeting EAOG, Manchester University Press, Manchester, pp. 141–3. [11]

Bifani, B., George, G.T. and Lever, A. (1987) Geological and reservoir characteristics of the Rotliegend sandstones in the Argyll Field. In: Brooks, J. and Glennie, K.W. (eds) *Petroleum Geology of North West Europe*. Graham and Trotman, London, pp. 509–22. [3, 5, 12]

Bifani, R. (1985) A Zechstein depositional model for the Argyll field. In: Taylor, J.C.M. (ed.) *The Role of Evaporites in Hyrocarbon Exploration*. Course Notes No. 39, JAPEC, London. [6]

Bifani R. (1986) Esmond gas complex. In: Brooks, J., Goff, J. and van Hoorn, B. (eds) *Habitat of Palaeozoic Gas in N.W. Europe*. Special Publication No. 23, Geological Society, London, pp. 209–21. [5, 6, 7]

Bifani, R. and Smith, C.A. (1985) The Argyll field after a decade of production. In: *SPE of AIME*, Offshore Europe 1985 Conference, Aberdeen, Sept. [6]

Binot, F., Gerling, P. Hiltmann, W., Kockel, F. and Wehner, H. (1993) The petroleum system in the Lower Saxony Basin. In: Spencer, A.M. (ed.) *Generation, Accumulation, and Production of Europe's Hydrocarbons*. Special Publication, EAPG, pp. 121–39. [12]

Birkelund, T., Clausen, C.K., Hansen, H.N. and Holm, L. (1983) The Hectoroceras kochi Zone Ryazanian in the North Sea Central Graben and remarks on the Late Cimmerian Unconformity. In: *Danm. Geol. Unders.*, Arborg, 1982, pp. 53–72. [9]

Birtles, R. (1986) The seismic flatspot and the discovery and delineation of the Troll Field. In: Spencer, A.M., Campbell, C.J., Hanslien, S.H., Nelson, P.H., Nysaether, E. and Ormaasen, E.G. (eds) *Habitat of Hydrocarbons on the Norwegian Continental Shelf.* Norwegian Petroleum Soc. Stavanger, Graham and Trotman, London, pp. 207–15. [12]

Bisewski, H. (1990) Occurrence and depositional environment of the Lower Cretaceous sands in the southern Witch Ground Graben. In: Hardman, R.P.F. and Brooks, J. (eds) *Tectonic Events Responsible for Britain's Oil and Gas Reserves.* Special Publication No. 55, Geological Society, London, pp. 299–323. [9]

Bishop, A.N. and Abbott, G.D. (1991) The effect of minor igneous intrusions on the petroleum potential of Jurassic shales, Isle of Skye, Scotland. In: Manning, D. (ed.) *Organic Geochemistry. Advances and Applications in Energy and the Natural Environment. 15th EAOG Meeting.* Pergamon, Oxford, pp. 278–81. [11]

Bishop, A.N. and Abbott, G.D. (1995) Vitrinite reflectances and molecular geochemistry of Jurassic sediments: the influence of heating by Tertiary dykes northwest Scotland. *Org. Geochem.* **22**, 165–77. [11]

Bishop, D.J., Buchanan, P.G. and Bishop, C.J. (1995) Gravity-driven thick skinned extension above Zechstein Group evaporites in the western Central North Sea: an application of computer-aided section restoration techniques. *Mar. Petr. Geol.* **12** (2), 115–35.

Bissada, K.K. (1983) Petroleum generation in Mesozoic sediments of the Moray Firth Basin, North Sea area. In: Bjorøy, M., Albrecht, P., Cornford, C., de Groot, K., Eglinton, G., Galimov, E. *et al.* (eds) *Advances in Organic Geochemistry 1981.* John Wiley, Chichester, pp. 7–15. [11]

Bitterli, P. (1963) On the classification of bituminous rocks from Western Europe. In: *Proceedings of 6th World Petroleum Congress.* Frankfurt, Section 1, Verein, Welt-Erdol-Kongresses, Hamburg, pp. 155–65. [11]

Bjorøy, M. and Vigran, J.O. (1980) Geochemical study of the organic matter in outcrop samples from Agardhfjellet, Spitzbergen. In: Douglas, A.G. and Maxwell, J.R. (eds) *Advances in Organic Geochemistry 1979.* Pergamon, Oxford, pp. 141–7. [11]

Bjorøy, M., Hall, K. and Vigran, J.O. (1980) An organic geochemical study of Mesozoic shales from Andoya, North Norway. In: Douglas, A.G. and Maxwell, J.R. (eds) *Advances in Organic Geochemistry 1979.* Pergamon, Oxford, pp. 77–91. [11]

Bjorøy. M., Mork, A. and Vigran, J.O. (1983) Organic geochemical studies of the Devonian to Triassic succession on Bjornoya and the implications for the Barent Shelf. In: Bjorøy, M., Albrecht, P., Cornford, C., de Groot, K., Eglinton, G., Galimov, E. *et al.* (eds) *Advances in Organic Geochemistry 1981.* John Wiley, Chichester, pp. 49–59. [11]

Bjorøy, M., Hall, P.B., Loberg, R., McDermott, J.A. and Mills, N. (1988) Hydrocarbons from non-marine source rocks. In: Mattavelli, L. and Novelli, L. (eds) *Advances in Organic Geochemistry 1987 (Proc. 13th Int'l. Mtg. on Org. Geochem. Venice), Part 1.* Pergamon, Oxford, pp. 221–4. [11]

Bjorøy, M., Williams, J.A., Dolcater, D.L. and Winters, J.C. (1988) Variation in hydrocarbon distribution in artificially matured oils. In: Mattavelli, L. and Novelli, L. (eds) *Advances in Organic Geochemistry 1987 (Proc. 13th Int'l. Mtg on Org. Geochem. Venice), Part 2.* Pergamon, Oxford, pp. 901–14. [11]

Bjorøy, M., Hall, P.B. and Moe, R.P. (1994) Stable carbon isotope variation of *n*-alkanes in Central Graben oils. *Org. Geochem.* **22**, (3–5), 355–82. [11]

Bjorøy, M. *et al.* (1996) Maturity assessment and characterization of Big Horn Basin Palaeozoic oils. *Mar. Petr. Geol.* **13**(1) 3–24. [11]

Blackbourn, G.A. (1981) Red bed successions on the western seaboard of Scotland. PhD thesis, University of Strathclyde. [3]

Blackbourn, G.A. (1987) Sedimentary environments and stratigraphy of the Late Devonian–Early Carboniferous Clair Basin, west of Shetland. In: Miller, J., Adams, A.E. and Wright, V.P. (eds) *European Dinantian Environments.* John Wiley, Chichester, pp. 21–32. [3]

Blackwell, D.D. and Steele, J.L. (1989) Thermal conductivity of sedimentary rocks: measurements and significance. In: Naeser, N.D. and McCulloh, T.H. (eds) *Thermal History of Sedimentary Basins: Methods and Case Histories.* Springer-Verlag, New York, pp. 13–37. [11]

Blanc, P. and Connan, J. (1994) Preservation, degradation and destruction of trapped oil. In: Magoon, L.B. and Dow, W.G. (eds) *The Petroleum System–from Source to Trap.* Memoir No. 60, AAPG, Tulsa, OK., pp. 237–51. [11]

Bless, M.J., Bouckaert, J. and Paproth, E. (1984) Migration of climate belts as a response to continental drift during the late Devonian and Carboniferous. *Bull. Soc. Belge Geol.* **93**, 189–95. [4]

Bless, M.J., Bouckaert, J. and Paproth, E. (1987) Fossil assemblages and depositional environments: limits to stratigraphical correlations. In: Miller, J., Adams, A.E. and Wright, V.P. (eds) *European Dinantian Environments.* John Wiley, Chichester, pp. 61–74. [4]

Blow, R.A. and Hardman, M. (1995) The Calder Field: Appraisal Well 110/7A-8, East Irish Sea Basin. Paper presented at Conference on the Petroleum Geology of the Irish Sea and Adjacent Areas, Geological Society, London, Feb., Abstracts, p. 41. [11]

Bluck, B.J. (1978) Sedimentation in a late orogenic basin: the Old Red Sandstone, Scotland. In: Bowes, D.R. and Leake, B.E. (eds) *Crustal Evolution in North Western Europe and Adjacent Regions. Geol. J.* Special Issue No. 10, 249–78. [3]

Blundell, D.J. and Gibbs, A.D. (eds) (1990) *Tectonic Evolution of the North Sea Rifts.* Clarendon Press, Oxford.

Bodenhausen, J.W.A. and Ott, W.F. (1981) Habitat of the Rijswijk oil province, onshore The Netherlands. In: Illing, L.V. and Hobson, G.P. (eds) *Petroleum Geology of the Continental Shelf of North-West Europe.* Heyden, London, pp. 301–9. [1, 9, 11, 12]

Boigk, H. (1981) *Erdöl und Erdgas in der Bundesrepublik Deutschland. Erdölprovinzen, Felder, Forderung, Vorrate, Lagerstattentechnik.* Ferdinand Enke Verlag, Stuttgart, 330 pp. [12]

Boigk, H. and Stahl, W. (1970) Zum Problem der Entstehung Nordwestdeutcher Erdgaslagerstatten. *Erdoel Kohle, Erdgas, Petrochem. Brennst. Chem.* **23**(6), 104–11. [11]

Boigk, H., Hark, H.-U. and Schott, W. (1963) Oil migration and accumulation at the border of the Lower Saxony Basin. In: *Proceedings of the 6th World Petroleum Congress 1963, Frankfurt.* Section 1, pp. 435–55. [12]

Boigk, H., Stahl, W., Teichmuller, M. and Teichmuller, R. (1971) Metamorphism of coal and natural gas. *Forschr. Geol. Rheinld. Westphal.* **19**, 104–11. [11]

Boldy, S.A.R. (ed.) (1995) *Permian and Triassic Rifting in North West Europe.* Special Publication No. 92, Geological Society, London.

Boldy, S.A.R. and Brealey, S. (1990) Timing, nature and sedimentary result of Jurassic tectonism in the Outer Moray Firth. In: Hardman, R.F.P. and Brooks, J. (eds) *Tectonic Events Responsible for Britain's Oil and Gas Reserves.* Special Publication No. 55, Geological Society, London, pp. 259–79. [8]

Boote, D.R.D. and Gustav, S.H. (1987) Evolving depositional systems within an active rift, Witch Ground Graben, North Sea. In: Brooks, J. and Glennie, K.W. (eds) *Petroleum Geology of North West Europe.* Graham and Trotman, London, pp. 819–33. [9, 12]

Booth, J., Swiecicki, T. and Wilcockson, P. (1993) The tectonostratigraphy of the Solan Basin, west of Shetland. In: Parker, J.R. (ed.) *Petroleum Geology of Northwest Europe: Proceedings of the 4th Conference.* Geological Society, London, pp. 987–98. [7]

Bordenave, M.L. (ed.) (1993) *Applied Petroleum Geochemistry.* Editions Technip, Paris 524 pp. [11]

Bordenave, M.L., Espitalié, J., Leplat, P. and Oudin, J.L. (1993) Screening techniques for source rock evaluation: 1993 modelling of petroleum generation and migration. In: Bordenave, M.L. (ed.) *Applied Petroleum Geochemistry.* Editions Technip, Paris, pp. 217–78. [11]

Bostick, N.H. (1979) Microscopic measurement of the level of catagenesis of solid organic matter In: *Sedimentary Rocks to Aid Exploration for Petroleum and to Determine Former Burial Temperatures.* Special Publication No. 26, SEPM, Tulsa, OK., pp. 17–43. [11]

Bostrom, R.C. (1989) Subsurface exploration via satellite: structure visible in seasat images of North Sea, Atlantic continental margin, and Australia. *AAPG Bull.* **73**(9), 1053–64.

Bott, M.H. (1967) Geophysical investigations of the Northern Pennine basement rocks. *Proc. Yorks. Geol. Soc.* **36**, 139–68. [4]

Bott, M.H.P. (1988) The continental margin of central East Greenland in relation to North Atlantic plate tectonic evolution. *J. Geol. Soc.* **144**, 561–8. [10]

Bott, M.P.H. *et al.* (1985) Crustal structure south of the Iapetus suture beneath northern England. *Nature* **314**, 724–7. [2]

Boulton, G.S. (1993) Ice ages and climatic change. In Duff, P. McL. D. (ed.) *Holmes' Principles of Physical Geology*, 4th edn. Chapman and Hall, London, pp. 439–69. [5]

Boussafir, M. and Lallier-Vergès, E. (1997) Accumulation of organic matter in the Kimmeridge Clay Formation KCF: an update fossilisation model for marine petroleum source rocks. *Mar. Petr. Geol.* **14**(1), 75–84. [11]

Bowen, J.M. (1975) The Brent oil field. In: Woodland, A.W. (ed.) *Petroleum and the Continental Shelf of Northwest Europe*, Vol. I, *Geology*. Applied Science Publishers, Barking, pp. 353–60. [1, 7, 8, 12]

Bowen, J.M. (1989) Win some, lose some: minerals industry international. *Bull. Inst. Mining Mineral.* **986**, 6–12. [1]

Bowen, J.M (1991) 25 years of UK North Sea exploration. In: Abbotts, J.L. (ed.) *UK North Sea United Kingdom Oil and Gas Fields: 25 Years Commemorative Volume.* Memoir No. 14, Geological Society, London, pp. 1–7. [1]

Bowen, J.M. (1992) Exploration of the Brent Province. In: Morton, A.C., Haszeldine, R.S., Giles, M.R. and Brown, S. (eds) *Geology of the Brent Group.* Special Publication No. 61, Geological Society, London, pp. 3–14. [12]

Bowler, J.M. (1976) Aridity in Australia: age, origins and expression in aeolian landforms and sediments. *Earth Sci. Rev.* **12**, 279–310. [5]

Bowman, M.B.J., McClure, N.M. and Wilkinson, D.W. (1993). Wytch Farm oilfield: deterministic description of the Triassic Sherwood sandstone. In: Parker, J.R. (ed.) *Petroleum Geology of Northwest Europe: Proceedings of the 4th Conference.* Geological Society, London, pp. 1513–17. [7, 11, 12]

BP Exploration (1995) Buchan's century beats expectations. *BPXPRESS* **52**, 10. [3]

Bradbury, H.J., Smith, R.A. and Harris, A.L. (1976) Older granites as time markers in Dalradian evolution. *J. Geol. Soc.* **132**, 677–84. [2]

Braithwaite, C.J.R. (1989) Stylolites as open fluid conduits. *Mar. Petr. Geol.* **6**, 93–6. [9]

Brangulis, A.P., Kanev, S.V., Margulis, L.S. and Pomerantseva, R.A. (1993) Geology and hydrocarbon prospects of the Paleozoic in the Baltic region. In: Parker, J.R. (ed.) *Petroleum Geology of Northwest Europe: Proceedings of the 4th Conference.* Geological Society, London, pp. 651–6. [11]

Braun, R.L. and Burnham, A.K. (1992) PMOD: a flexible model of oil and gas generation, cracking, and expulsion. *Org. Geochem.* **19**(1–3), 161–72. [11]

Bray, R.J., Green, P.F. and Duddy, I.R. (1992) Thermal history reconstruction using apatite fission track analysis and vitrinite reflectance: a case study from the UK East Midlands and the southern North Sea. In: Hardman, R.F.P. (ed.) *Exploration Britain: Into the Next Decade.* Special Publication No. 67, Geological Society, London, pp. 3–25. [11]

Brekke, H., Furnes, H., Nordos, J. and Hertogen, J. (1984) Lower Palaeozoic convergent plate margin volcanism on Bomlo, SW Norway, and its bearing on the tectonic environments of the Norwegian Caledonides. *J. Geol. Soc.* **141**, 1015–32. [2]

Brekke, T., Pegrum, R.M. and Watts, P.B. (1981) First exploration results in Block 31/2 offshore Norway. In: Norwegian Symposium on Exploration, Bergen, NPF, Article 16, 34 pp. [1, 11]

Brennand, T.P. (1975) The Triassic of the North Sea. In: Woodland, A.W. (ed.) *Petroleum and the Continental Shelf of Northwest Europe.* Vol. I, *Geology.* Applied Science Publishers, Barking, pp. 295–310. [7]

Brennand, T.P. and van Veen, F.R. (1975) The Auk field. In: Woodland, A.W. (ed.) *Petroleum and the Continental Shelf of Northwest Europe.* Vol. I, *Geology.* Applied Science Publishers, Barking, pp. 275–85. [1, 5, 6, 12]

Brennand, T.P., van Hoorn, B. and James K.H. (1990). Historical review of North Sea exploration. In: Glennie, K.W. (ed.) *Introduction to the Petroleum Geology of the North Sea.* Blackwell Scientific Publications, Oxford, pp. 1–33. [12]

Breunese, J.N. and Rispens, F.B. (1996) Natural gas in the Netherlands: exploration and development in historic and future perspective. In: Rondeel, H.E., Batjes, D.A.J. and Nieuwenhuis, W.H. (eds) *Geology of Oil and Gas under the Netherlands.* Royal Geological and Mining Society, the Netherlands, Kluwer Academic Publishers, Dordrecht, pp. 19–30. [1, 12]

Brewer, J.A. and Smythe, D.K. (1984) MOIST and the continuity of crustal reflector geometry along the Caledonian–Appalachian orogen. *J. Geol. Soc.* **141**, 105–20. [2]

Brewster, J. (1991) The Frigg field, Block 10/1 UK North Sea and 25/1 Norwegian North Sea. In: Abbotts, I.L. (ed.) *UK North Sea United Kingdom Oil and Gas Fields: 25 Years Commemorative Volume.* Memoir No. 14, Geological Society, London, pp. 117–26. [1, 10, 12]

Brewster, J. and Dangerfield, J.A. (1984) Chalk fields along the Lindesnes Ridge, Eldfisk. *Mar. Petr. Geol.* **1**, 239–78. [9, 11]

Brewster, J. and Jeangeot, G. (1987) The production geology of the

Frigg field. In: Kleppe, J., Berg, E.W., Buller, A.T., Hjelmelaud, O. and Torsæter, O. (eds) *North Sea Oil and Gas Reservoirs*. Graham and Trotman, London, pp. 75–88. [12]

Brewster, J., Dangerfield, J. and Farrell, H. (1986) The geology and geophysics of the Ekofisk Field waterflood. *Mar. Petr. Geol.* **3**, 139–69. [9]

Brigaud, F. and Vasseur, G. (1989) Minerology, porosity and fluid control on thermal conductivity of sedimentary rocks. *Geophys. J.* **98**, 525–42. [11]

Brigaud, F., Chapman, D.S. and Le Douaran, S. (1990) Estimating thermal conductivity in sedimentary basins using lithologic data and geophysical well logs. *AAPG Bull.* **74**(9), 1459–77. [11]

Brigaud, F., Vasseur, G. and Caillet, G. (1992) Thermal state in the North Viking Graben North Sea determined from oil exploration well data. *Geophysics* **57**(1), 69–88. [11]

Bristow, C.S. (1988) Controls on the sedimentation of the Rough Rock Group Namurian from the Pennine Basin of northern England. In: Besly, B.M. and Kelling, G. (eds) *Sedimentation in a Synorogenic Basin Complex: the Upper Carboniferous of Northwest Europe*. Blackie, Glasgow, London, pp. 113–31. [4]

Brodie, J. and White, N. (1994) Sedimentary basin inversion caused by igneous underplating: Northwest European continental shelf. *Geology* **22**, 147–50. [2]

Bromley, R.G. and Ekdale, A.A. (1986) Flint and fabric in the European Chalk. In: Sieveking, G. de G. and Hart, M.B. (eds) *The Scientific Study of Flint and Chert*. Cambridge University Press, Cambridge, pp. 71–82. [9]

Bromley, R.G. and Ekdale, A.A. (1987) Mass transport in European Cretaceous chalk: fabric criteria for its recognition. *Sedimentology* **34**, 1079–92. [9]

Brookfield, M.E. (1980) Permian intermontane basin sedimentation in southern Scotland. *Sedimentary Geol.* **27**, 167–94. [5]

Brooks, J. (1983) Applications of petroleum geochemistry to basin studies. *J. Geol. Soc.* **140**, 413–14. [11]

Brooks, J. and Glennie, K. (eds) (1987) *Petroleum Geology of North-West Europe*, Vol. 1, 1–598; Vol. 2, 599–1219. Graham and Trotman, London. [1, 9, 11]

Brooks, J. and Thusu, B. (1977) Oil source rock identification and characterisation of the Jurassic sediments in the northern North Sea. *Chem. Geol.* **20**, 283–94. [11]

Brooks, J., Gibbs, A.D. and Cornford, C. (1984) Geological controls on occurrence and composition of Tertiary heavy oils, Northern North Sea. Fossil Fuels of Europe Conference, Geneva, 1984, Abstract. *AAPG Bull.* **68**(6), p. 793. [11]

Brooks, J., Goff, J. and van Hoorn, B. (eds) (1986) *Habitat of Palaeozoic gas in N.W. Europe*. Special Publication No. 23, Geological Society, London, 276 pp. [1]

Brooks, J., Cornford, C. and Archer, R. (1987) The role of hydrocarbon source rocks in petroleum exploration. In: Brooks, J. and Fleet, A.J. (eds) *Marine Petroleum Source Rocks*. Blackwell, Oxford, pp. 17–46. [11]

Brooks, J.R.V. (1977) Exploration status of the Mesozoic of the U.K. Northern North Sea. In: *N.P.F. Mesozoic Northern North Sea Symposium*. MNNSS/2, NPS, Oslo, pp. 1–28. [7]

Brosse, E. and Huc, A.-Y. (1986) Organic parameters as indicators of thermal evolution in the Viking Graben. In: Burrus, J. (ed.) *Thermal Modelling in Sedimentary Basins*. Editions Technip, Paris, pp. 517–30. [11]

Brown, A., Mitchell, A.W., Nilssen, I.R., Stewart, I.J. and Svela, P.T. (1992) Ula Field: relationship between structure and hydrocarbon distribution. In: Larsen, B.T. and Larsen, R.M. (eds) *Structural and Tectonic Modelling and its Application to Petroleum Geology*. NPS, Graham and Trotman, London, pp. 165–78. [12]

Brown, S. (1984) Jurassic. In: Glennie, K.W. (ed.) *Introduction to the Petroleum Geology of the North Sea*. Blackwell Scientific Publications, Oxford, pp. 219–54. [8]

Brown, S. (1991) Stratigraphy of the oil and gas reservoirs: UK Continental Shelf. In: Abbotts, I.L. (ed.) *United Kingdom Oil and Gas Fields, 25 Years Commemorative Volume*. Memoir No. 14, Geological Society, London, pp. 9–18. [5]

Brown, S. and Richards, P.C. (1989) Facies and development of the mid Jurassic Brent delta near the northern limit of its progradation, UK North Sea. In: Whateley, M.K.G. and Pickering, K.T. (eds) *Deltas: Sites and Traps for Fossil Fuels*. Special Publication No. 41, Geological Society, London, pp. 253–67. [12]

Brown, S., Richards, P.C. and Thomson, A.R. (1987) Patterns in the deposition of the Brent Group Middle Jurassic U.K. North Sea. In: Brooks, J. and Glennie, K.W. (eds) *Petroleum Geology of North West Europe*. Graham and Trotman, pp. 899–913. [12]

Browne, M.A.E., Hargreaves, R.L. and Smith, I.F. (1985) The Upper Palaeozoic Basins of the Midland Valley of Scotland. In: *Investigations of the Geothermal Potential of the UK*. Geothermal Energy Research Project, Energy Resources, Report Ref. WJ/GE/85/2, British Geological Survey, Edinburgh, p. 48. [3]

Brueren, J.W.R. (1959). The stratigraphy of the Upper Permian 'Zechstein' formation in the eastern Netherlands. In: *I giacimenti gassiferi dell'Europa Occidentale 1*. Accademia Nazionale dei Lincei, Rome, pp. 243–74. [6]

Bruijn, A.N. (1996) De Wijk gas field Netherlands: reservoir mapping with amplitude anomalies. In: Rondeel, H.E., Batjes, D.A.J. and Nieuwenhuis, W.H. (eds) *Geology of Oil and Gas under the Netherlands*. Royal Geological and Mining Society, the Netherlands, Kluwer Academic Publishers, Dordrecht, pp. 243–53. [7, 10, 12]

Brunstrom R.G.W. and Walmsley, P.J. (1969) Permian evaporites in North Sea basin. *AAPG Bull.* **53**, 870–83. [6]

Bryant, I.D. and Livera, S.E. (1991) Identification of unswept oil volumes by using data analysis: Ness Formation, Brent Field, UK North Sea. In: Spencer, A.M. *Generation, Accumulation and Production of Europe's Hydrocarbon*. Oxford University Press, Oxford, pp. 75–88. [1]

Buchanan, J.G. and Buchanan, P.G. (eds) (1995) *Basin Inversion*. Special Publication No. 88, Geological Society, London, 596 pp. [9, 11]

Buchanan, P.G., Bishop, D.J. and Hood, D.N. (1996) Development of related structures in the Central North Sea: results from section balance. In: Alsop, G.I., Blundell, D.J. and Davison, I. (eds) *Salt Tectonics*. Special Publication 100, Geological Society, London, pp. 111–28. [6]

Buchanan, R. and Hoogteyling, L. (1979) Auk field development: a case history illustrating the need for a flexible plan. *J. Petr. Technol.* **31**, 1305–12. [1, 5, 12]

Buchardt, B. (1978) Oxygen isotope paleo-temperatures from the Tertiary period in the North Sea area. *Nature* **275**, 121–23. [11]

Buchardt, B. and Lewan, M.D. (1990) Reflectance of vitrinite-like macerals as a thermal maturity index in the Cambro-Ordivicean alum shale of Southern Scandinavia. *AAPG Bull.* **74**, 394–406. [11]

Buchardt, B., Clausen, J. and Thomsen, E. (1986) Carbonisotope composition of Lower Palaeozoic kerogen: effects of maturation. In: Leythaeuser, D. and Rullkotter, J. (eds) *Advances in Organic Geochemistry 1985*. Pergamon, Oxford, pp. 127–34. [11]

Buck, W.R., Martinez, F., Steckler, M.S. and Cochran, J.R. (1988) Thermal Consequences of Lithospheric Extension: Pure and Simple. *Tectonics*. **7**(2), 213–34. [11]

Bücker, C. and Rybach, L. (1996) A simple method to determine heat production from gamma-ray logs. In: First Nordic Symposium on Petrophysics, 31st May–1st June 1994, Gothenburg, Sweden. *Mar. Petr. Geol.* **13**(4), 373–6. [11]

Budding, M.C. and Inglin, H.F. (1981) A reservoir geological model of the Brent Sands in Southern Cormorant. In: Illing, L.V. and Hobson, G.P. (eds) *Petroleum Geology of the Continental Shelf of North-West Europe*. Heyden, 326–34. [8, 12]

Budny, M. (1991) Seismic reservoir interpretation of deep Permian carbonates and sandstones, NW German gas province. *First Break* **9**, 55–64. [6]

Buhrig, C. (1989) Geopressured Jurassic reservoirs in the Viking

Graben: modelling and geological significance. *Mar. Petr. Geol.* **6**(1), 31–48. [11]

Buiskool-Toxopeus, J.M.A. (1982) Selection criteria for the use of vitrinite reflectance as a maturity tool in petroleum exploration. *J. Petr. Geol.* **5**, 215. [11]

Buiskool-Toxopeus, J.M.A. (1983) Selection criteria for the use of vitrinite reflectance as a maturity tool. In: Brooks, J. (ed.) *Petroleum Geochemistry and Exploration of Europe.* Special Publication No. 12, Geological Society, Blackwell Scientific Publications, Oxford, pp. 295–308. [11]

Bukovics, C., Cartier, E.G., Shaw, N.D. and Ziegler, P.A. (1984) Structure and development of the Mid-Norway Continental Margin. In: Spencer, A.M. *et al.* (eds) *Petroleum Geology of the North European Margin.* Graham and Trotman, London, pp. 407–23. [2]

Bulat, J. and Stoker, S.J. (1987) Uplift determination from interval velocity studies, UK Southern North Sea. In: Brooks, J. and Glennie, K.W. (eds) *Petroleum Geology of North West Europe.* Graham and Trotman, London, pp. 293–306. [11]

Buller, A.T., Berg, E., Hjelmeland, O., Kleppe, J., Torsæter, O. and Aasen, J.O. (eds) (1990) *North Sea Oil and Gas Reservoirs II.* Norwegian Institute of Technology, Graham and Trotman, London, 453 pp. [9]

Bunch, A.W.H. and Dromgoole, P.W. (1995) Lithology and fluid prediction from seismic and well data. *Petr. Geosci.* **1**, 49–57. [1]

Bungener, M.J.A. (1969) Le Champ de gaz de Groningen. *Rev. Assoc. Techniciens Petr.* **196**, 19–32. [5]

Burgess, C.J. and Peter, C.K. (1985) Formation, distribution, and prediction of stylolites as permeability barriers in the Thamama Group, Abu Dhabi. Publication No. 13698, SPE, Richardson, TX. 165–8. [9]

Burley, S.D. (1987) Diagenetic modelling in the Triassic Sherwood Sandstone Group of England and its offshore eqivalents. United Kingdom Continental Shelf. PhD thesis, University of Hull. [7]

Burley, S.D. (1993) Models of burial diagenesis for deep exploration in Jurassic fault traps of the Central and Northern North Sea. In: Parker, J.R. (ed.) *Petroleum Geology of Northwest Europe: Proceedings of the 4th Conference.* Geological Society, London, pp. 1353–75. [8, 11]

Burley, S.D., Mullis, J. and Matter, A. (1989) Timing diagenesis in the Tartan reservoir (UK North Sea): constraints from combined cathodeluminescence microscopy and fluid inclusion studies. *Mar. Petr. Geol.* **6**(2), 98–120. [11]

Burnett, R.D. (1987) Regional maturation patterns for late Visean Carboniferous, Dinantian rocks of northern England based on mapping on conodont colour. *Irish J. Earth Sci.* **8**, 165–85. [11]

Burnham, A.K. (1990) *Pyrolysis kinetics and composition for Posidonia Shale. Report UCRL-ID-105871.* Lawrence Livenmore National Laboratories USA, 12 pp. [11]

Burnham, A.K. and Sweeney, J.J. (1991) Modelling the maturation and migration of petroleum. In: Merrill, R.K. (ed.) *Source and Migration Processes and Evaluation Techniques.* Treatise Petroleum Geology, AAPG, Tulsa, OK., pp. 55–64. [11]

Burnhill, T.J. and Ramsay, W.V. (1981) Mid-Cretaceous palaeontology and stratigraphy, Central North Sea. In: Illing, L.V. and Hobson, G.P. (eds) *Petroleum Geology of the Continental Shelf of North-West Europe*, Heyden, London, pp. 245–54. [9]

Burri, P., Faupel, J. and Koopman, B. (1993) The Rotliegend in northwest Germany, from frontier to fairway. In: Parker, J.R. (ed.) *Petroleum Geology of Northwest Europe: Proceedings of the 4th Conference.* Geological Society, London, pp. 741–8. [5, 12]

Burrus, J. (ed.) (1986) *Thermal Modeling in Sedimentary Basins. Proc. 1st IFP Expl. Rsch. Conference, Carcans, France, 1985.* Editions Technip, Paris, 600 pp. [11]

Burrus, J., Kuhfuss, A., Doligez, B. and Ungerer, P. (1991) Are numerical models useful in reconstructing the migration of hydrocarbons? A discussion based on the Northern Viking Graben. In: England, W.A. and Fleet, A.J. (eds) *Petroleum Migration.* Special Publication No. 59, Geological Society, London, pp. 89–111. [11]

Burwood, R., Staffurth, J., de Walque, L. and de Witte, S.M. (1991) Petroleum geochemistry of the Weald–Wessex basin of Southern England: a problem in source–oil correlation. In: Manning, D. (ed.) *Organic Geochemistry, Advances and Applications in Energy and Natural Environment.* Ext. Abstracts, 15th Meeting EAOG, Manchester University Press, Manchester, pp. 22–7.

Bushell, T.P. (1986) Reservoir geology of the Morcambe Field. In: Brooks, J., Goff, J. and van Hoorn, B. (eds) *Habitat of Palaeozoic Gas in N.W. Europe.* Special Publication No. 23, Geological Society, London, pp. 189–207. [1, 11, 12]

Bushinsky, G.I. (1961) Stylolites. *Izvestiya Acad. Sci. USSR, Geol. Ser.* (English translation) **8**, 31–45. [9]

Bustin, R.M. and Moffat, I.W. (1983) Groundhog coalfield, central British Columbia: reconnaissance stratigraphy and structure. *Bull. Can. Petr. Geol.* **31**(4), 231–45. [11]

Butler, J.B. (1975) The West Sole gas field. In: Woodland, A.W. (ed.) *Petroleum and Continental Shelf of Northwest Europe.* Vol. I, *Geology.* Applied Science Publishers, Barking, pp. 213–23. [1, 5]

Butler, M. (1998) The geological history of the Wessex basin: a review of new information from oil exploration. In: Underhill, J.R. (ed.) *Development and Evolution of the Wessex Basin.* Special Publication 133, Geological Society, London, pp. 67–86. [2, 8]

Butler, M. and Pullan, C.P. (1990) Tertiary structures and hydrocarbon entrapment in the Weald basin of Southern England. In: Hardman, R.P.F. and Brooks, J. (eds) *Tectonic Events Responsible for Britain's Oil and Gas Reserves.* Special Publication No. 55, Geological Society, London, pp. 371–91. [2]

Butler, M., Phelan, M.J. and Wright, A.W.R. (1976) Buchan Field: evaluation of a fractured sandstone reservoir. *Log Analyst* **18**, 23–31. [1]

Buza, J.W. and Unneberg, A. (1987) Geological and reservoir engineering aspects of the Statfjord field. In: Kleppe, J. *et al.* (eds) *North Sea Oil and Gas Reservoirs.* Graham and Trotman, London, pp. 23–38.

Byrd, W.D. (1975) Geology of the Ekofisk Field, offshore Norway. In: Woodland, A.W. (ed.) *Petroleum and the Continental Shelf of Northwest Europe.* Vol. I, *Geology.* Applied Science Publishers, Barking, pp. 439–45. [1, 9]

Caillet, G. (1993) The caprock of the Snorre Field, Norway: a possible leakage by hydraulic fracturing. *Mar. Petr. Geol.* **10** (1), 42–51. [11]

Caillet, G., Judge, N.C., Bramwell, N.P., Meciani, L., Green, M. and Adam, P. (1997) Overpressure and hydrocarbon trapping in the chalk of the Norwegian Central Graben. *Petr. Geosci.* **3**, 33–42. [11]

Callomon, J.H. (1979) Marine Boreal Bathonian fossils from the northern North Sea and their palaeogeographical significance. *Proc. Geol. Assoc.* **90**, 163–9. [8]

Callomon, J.H., Donovan, D.T. and Trumpy, R. (1972) An annotated map of the Permian and Mesozoic formations of East Greenland. *Meddedeling Groenland* **168**, 1–35. [5]

Calvert, S.E., Bustin, R.M. and Pedersen, T.F. (1992) Lack of evidence for enhanced preservation of sedimentary organic matter in the oxygen minimum of the Gulf of California. *Geology* **20**, 757–60. [11]

Cameron, I.B. and Stephenson, D. (1985) *The Midland Valley of Scotland.* British Regional Geology. BGS, HMSO, London, 172 pp. [2, 4]

Cameron, T.D.J. (1993a) Triassic, Permian and Pre-Permian of the Central and Northern North Sea. In: Knox, R.W.O'B. and Cordey, W.G. (eds) *Lithostratigraphic Nomenclature of the UK North Sea 4.* BGS, HMSO, Nottingham, p. 163 [4, 7, 8]

Cameron, T.D.J. (1993b) Carboniferous and Devonian of the Southern North Sea. In: Knox, R.W.O'B. and Cordey, W.G.

(eds) *Lithostratigraphic Nomenclature of the UK North Sea*. BGS, HMSO, London, 163 pp. [4]

Cameron, T.D.J., Crosby, A., Balson, P.S., Jeffery, D.H., Lott, K.G., Bulat, J. and Harrison, D.J. (1992) *United Kingdom Offshore Regional Report: the Geology of the Southern North Sea*. BGS, HMSO, London. [3, 4, 7, 9, 11]

Cameron, T.D.J., Bulat, J. and Mesdag, C.S. (1993) High resolution seismic profile through a Late Cenozoic delta complex in the Southern North Sea. *Mar. Petr. Geol.* **10**(6), 591–600. [11]

Campbell, C.J. and Ormaasen, E. (1987) The discovery of oil and gas in Norway: an historical synopsis. In: Spencer, A.M., Campbell, C.J., Hanslien, S.H., Nelson, P.H., Nysaether, E., Ormaasen, E.G. (eds) *Geology of the Norwegian Oil and Gas Fields*. Norwegian Petroleum Society, Graham and Trotman, London, pp. 1–37. [1]

Campbell, S.J.D. and Gravdal, N. (1995) The prediction of high porosity chalks in the East Hod Field. *Petr. Geosci.* **1**, 57–69. [9]

Canham, A.C. and Mackertich, D.S. (eds) *Geology of the Humber Group: Central Graben and Moray Firth, UKCS*. Special Publication No. 114, Geological Society, London, pp. 29–45. [1]

Cannon, S.J.C., Giles, M.R., Whitaker, M.F., Please, P.M. and Martin, S.V. (1992) A regional reassessment of the Brent Group, UK sector, North Sea. In: Morton, A.C., Haszeldine, R.S., Giles, M.R. and Brown, S. (eds) *Geology of the Brent Group*. Special Publication No. 61, Geological Society, London, pp. 81–107. [12]

Carman, G.J. and Young, R. (1981) Reservoir geology of the Forties oil field. In: Illing, L.V. and Hobson, G.P. (eds) *Petroleum Geology of the Continental Shelf of North-West Europe*. Heyden, pp. 371–9. [1, 12]

Carpentier, B., Bessereau, G. and Huc, A.Y. (1989) Digraphics et roches mètres. Estimation des teneurs en carbone organique par la méthode Carbolog. *Rev. IFP* **44**, 699–719. [11]

Carpentier, B., Huc, A.Y. and Bessereau, G. (1991) Wire-line logging and source rocks—estimation of organic carbon content by the Carbolog method. *Log Analyst* **32**(3), 279–97. [11]

Carr, A. (1991) A pressure dependent kinetic model for vitrinite reflectance. In: Manning, D. (ed.) *Organic Geochemistry, Advances and Applications in Energy and Natural Environment*. Ext. Abstracts, 15th Meeting EAOG, Manchester University Press, Manchester, pp. 285–7. [11]

Carruthers, A., McKie, T., Price, J., Dyer, R., Williams, G. and Watson, P. (1966) The application of sequence stratigraphy to the understanding of Late Jurassic turbidite plays in the Central North Sea, UKCS. In: Hurst, A., Johnson, H.D., Burley, S.D., Canham, A.C. and Mackertich, D.S. (eds) (1996) *Geology of the Humber Group: Central Graben and Moray Firth, UKCS*. Special Publication No. 114, Geological Society, London, pp. 29–46. [8]

Carstens, H. and Finstad, K.G. (1981) Geothermal gradients of the northern North Sea Basin, 59-62deg N. In: Illing, L.V. and Hobson, G.D. (eds) *Petroleum Geology of the Continental Shelf of North-West Europe*. Heyden, London, pp. 152–61. [11]

Cartwright, J. (1990) The structural evolution of the Ringkøbing–Fyn High. In: Blundell, D.J. and Gibbs, A.D. (eds) *Tectonic Evolution of the North Sea Rifts*. Oxford Science Publishers, Oxford, pp. 200–16. [7]

Cartwright, J. (1991) The kinematic evolution of the Coffee Soil Fault. In: Roberts, A.M., Yielding, G. and Freeman, B. (eds) *The Geometry of Normal Faults*. Special Publication No. 56, Geological Society, London, pp. 29–40. [7]

Cartwright, J. A. (1987) Transverse structural zones in continental rifts—an example from the Danish sector of the North Sea. In: Brooks, J. and Glennie, K.W. (eds) *Petroleum Geology of North West Europe*. Graham and Trotman, London, pp. 441–52. [2]

Cartwright, J.A. (1994) Episodic basin-wide hydrofracturing of overpressured Early Cenozoic mudrock sequences in the North Sea Basin. *Mar. Petr. Geol.* **11**, 587–608. [11, 12]

Casey, B.J., Romani, R.S. and Schmitt, R.H. (1993) Appraisal geology of the Saltire Field, Witch Ground Graben, North Sea. In: Parker, J.R. (ed.) *Petroleum Geology of Northwest Europe: Proceedings of the 4th Conference*. Geological Society, London, pp. 507–19. [1]

Casey, R. and Gallois, R.W. (1973) The Sandringham Sands of Norfolk. *Proc. Yorks. Geol. Soc.* **40**, 1–22. [9]

Casey, R. and Rawson, P.F. (eds) (1974) The boreal Lower Cretaceous. *Geol. J.* Special Issue No. 5. [9]

Casson, N., van Wees, B., Rebel, H. and Reijers, T. (1993) Successful integration of 3-D seismic and multidisciplinary approaches in exploring the Zechstein-2 Carbonates in North-east Netherlands. *AAPG Bull.* **77**, 1612. [6]

Caston, V.N.D. (1995) The Helvick Oil accumulation, Block 49/9, North Celtic Sea Basin. In: Croker, P.F. and Shannon, P.M. (eds) *The Petroleum Geology of Ireland's Offshore Basins*. Special Publication No.93, Geological Society, London, pp. 209–26. [1, 11, 12]

Cayley, G.T. (1987) Hydrocarbon migration in the central North Sea. In: Brooks, J. and Glennie, K.W. (eds) *Petroleum Geology of North West Europe*. Graham and Trotman, pp. 549–55. [8, 11, 12]

Chadwick, R.A. (1993) Aspects of basin inversion in Southern Britian. *J. Geol. Soc.* **150**(2), 311–23. [2, 11]

Chadwick, R.A. and Holliday, D.W. (1991) Deep crustal structure and Carboniferous basin development within the Iapetus convergence zone, northern England. *J. Geol. Soc.* **148**, 41–53. [4]

Chadwick, R.A., Holliday, D.W., Holloway, S. and Hulbert, A.G. (1993) The evolution and hydrocarbon potential of the Northumberland–Solway Basin. In: Parker, J.R. (ed.) *Petroleum Geology of Northwest Europe: Proceedings of the 4th Conference*. Geological Society, London, pp. 717–26. [4, 11]

Chatwin, C.P. (1961) British Regional Geology: East Anglia and Adjoining Areas, 4th edn. Institute of Geological Science, HMSO, London. [9]

Chauvin, A.L. and Valachi, L.Z. (1980) Sedimentology of the Brent and Statfjord Field. In: N.P.F. *The Sedimentation of the North Sea Reservoir Rocks*. Geilo XVI, Norwegian Petroleum Society, Geilo. pp. 1–17. [7]

Chen, H.K. (1988) Field development of the Scapa Field: a marginal North Sea field. In: European Petroleum Conference, London, SPE18347, pp. 553–62. [9]

Chen, Y., Clark, A.H., Farrer, E., Wasteneys, H.A.H.P., Hodgson, M.J. and Bromley, A.V. (1993) Diachronous and independent histories of plutonism and mineralisation in the Cornubian Batholith, southwest England. *J. Geol. Soc.* **1506**, 1183–91. [2]

Cherry, S.J.T. (1993) The interaction of structure and sedimentary process controlling deposition of the Upper Jurassic Brae Formation conglomerate, Block 16/17, North Sea. In: Parker, J.R. (ed.) *Petroleum Geology of Northwest Europe: Proceedings of the 4th Conference*. Geological Society, London, pp. 387–400. [8, 12]

Chicarelli, M.I., Eckardt, C.B., Owen, C.R., Maxwell, J.R., Eglinton, G., Hutton, R.C. *et al.* (1990) Application of inductively coupled plasma-mass spectrometry in the detection of organometallic compounds in chromatographic fractions from organic rich shales. *Org. Geochem.* **15**(3), 267–74. [11]

Childs, F.B. and Reed, P.E.C. (1975) Geology of the Dan Field and the Danish North Sea. In: Woodland, A.W. (ed.) *Petroleum and the Continental Shelf of Northwest Europe*. Vol. I *Geology*. Applied Science Publishers, Barking, pp. 429–38. [1, 9]

Christian, H.E. (1969) Some observations on the initiation of salt structures of the Southern British North Sea. In: Hepple, P. (ed.) *The Exploration for Petroleum in Europe and North Africa*. Institute of Petroleum, London, pp. 231–48. [5, 6]

Christiansen, F.G. (1989) Bitumen occurrence in the Lower Palaeozoic Franklinian Basin in North Greenland. In: *Abstracts of Papers presented at the 14th Int'l Org. Geochem. Conference, Paris, 18-22 Sept. 1989*. EAOG and IFP. [1]

Christiansen, F.G., Dam, G., Piasecki, S. and Stemmerik, L. (1992) A review of Upper Paleozoic and Mesozoic source rocks from onshore East Greenland. In: Spencer, A.M. (ed.) *Generation,*

Accumulation, and Production of Europe's Hydrocarbons. Special Publication, EAPG, Oxford University Press, Oxford, pp. 151–61. [11]

Christiansen, F.G., Larsen, H.C., Marcussen, C., Piasecki, S. and Stemmerik, L. (1993a) Late Paleozoic plays in East Greenland. In: Parker, J.R. (ed.) *Petroleum Geology of Northwest Europe: Proceedings of the 4th Conference*. Geological Society, London, pp. 657–66. [11]

Christiansen, F.G., Piasecki, S., Stemmerick, L. and Telnaes, N. (1993b) Depositional environment and organic geochemistry of the Upper Permian Ravnefjeld Formation source rock in East Greenland. *AAPG Bull.* **77**(9), 1519–38. [11]

Christie, O.H.J., Cornford, C., Endresen, U., Auxietre, J.-L., Pittion, J.-I. and Noyau, A. (1993) Prediction of North Sea source rock burial depth and oil generation depth, and a migration distance modelling technique, based on sterane and triterpane data. In: Øygard, K. (ed.) *Organic Geochemistry, Poster Sessions from 16th Int'l. Meeting*. Falch Hurtigtrykk, Oslo, pp. 212–7. [11]

Christie, P.A. and Sclater, J.G. (1980) An extensional origin of the Buchan and Witchground Graben in the North Sea. *Nature* **283**, 729–32.

Christie-Blick, N. (1990) Sequence stratigraphy and sea-level changes in Cretaceous time. In: Ginsburg, R.N. and Beaudoin, B. (eds) *Createaceous Resources, Events and Rhythms*. Nato ASI Ser. C. **304**, 1–21. [9]

Christy, A.A., Hopland, A.L., Barth, T. and Kvalheim, O.M. (1989) Quantitative determination of thermal maturity in sedimentary organic matter by diffuse reflectance infrared spectroscopy of asphaltenes. *Org. Geochem.* **14**(1), 77–81. [11]

Church, K.D. and Gawthorpe, R.L. (1994) High resolution sequence stratigraphy of the late Namurian in the Widmerpool Gulf East Midlands, UK. *Mar. Petr. Geol.* **11**, 528–44. [4]

Claoue-Long, J.C., Zhang, Z., Ma, G. and Du, S. (1991) The age of the Permo-Triassic boundary. *Earth Plan. Sci. Letts* **105**, 182. [5, 6]

Clark, D.N. (1980a) The sedimentology of the Zechstein 2 carbonate formation of Eastern Drenthe, The Netherlands. In: Füchtbauer, H. and Peryt, T.M. (eds) *The Zechstein Basin with Emphasis on Carbonate Sequences*. Contr. Sedimentology No. 9, Schweitzerbart'sche Verlagsbuchhandlung, Stuttgart, pp. 131–65. [6]

Clark, D.N. (1980b) The diagenesis of Zechstein carbonate sediments. In: Füchtbauer, H. and Peryt, T.M. (eds) *The Zechstein Basin with Emphasis on Carbonate Sequences*. Contr. Sedimentology No. 9, Schweitzerbart'sche Verlagsbuchhandlung, Stuttgart, pp. 167–203. [6]

Clark, D.N. (1985) Diagenesis in evaporite basins in relation to reservoir potential. In: Taylor, J.C.M. (ed.) *The Role of Evaporites in Hydrocarbon Accumulation*. Course No. 39, JAPEC. [6]

Clark, D.N. (1986) The distribution of porosity in Zechstein carbonates. In: Brooks, J., Goff, J. and van Hoorn, B. (eds) *Habitat of Palaeozoic Gas in N.W. Europe*. Special Publication No. 23, Geological Society, London, pp. 120–49. [6]

Clark, D.N. and Tallbacka, L. (1980) The Zechstein deposits of Southern Denmark. In: Füchtbauer, H. and Peryt, T.M. (eds) *The Zechstein Basin with Emphasis on Carbonate Sequences*. Contr. Sedimentology No. 9, Schweitzerbart'sche Verlagsbuchhandlung, Stuttgart, pp. 205–31. [6]

Clark-Lowes, D.D., Kuzemko, N.C. and Scott, D.A. (1987) Structure and petroleum prospectivity of the Dutch Central Graben and neighbouring platform areas. In: Brooks, J. and Glennie, K.W. (eds) *Petroleum Geology of North West Europe*. Graham and Trotman, London, pp. 337–56. [4, 9, 11]

Clarke, R.H. and Cleverly, R.W. (1991) Petroleum seepage and post-accumulation migration. In: England, W.A and Fleet, A.J. (eds) *Petroleum Migration*. P. Publication No. 59, Geological Society, London, pp. 265–72. [11]

Clausen, R.O. and Korstgård, J.A. (1993) Faults and faulting in the Horn Graben Area, Danish North Sea. *First Break* **11**, 127–43. [7]

Clayton, C.J. (1991a) Effect of maturity on carbon isotope ratios of oils and condensates. *Org. Geochem.* **17**(6), 887–99. [3, 11]

Clayton, C.J. (1991b) Carbon isotope fractionation during natural gas generation from kerogen. *Mar. Petr. Geol.* **8**, 232–41. [3, 11]

Clayton, G. *et al.* (1977) Carboniferous miospores of Western Europe: illustration and zonation. *Meded. Rijks. Geol. Dienst* **29**, 1–71. [4]

Cleal, C.J. (1978) Floral biostratigraphy of the Upper Silesian Pennant Measures of South Wales. *Geol. J.* **13**, 165–94. [4]

Clemmensen, L. (1979) Triassic lacustrine red-beds and palaeoclimate: the 'Buntsandstein' of Helgoland and the Malmros Klint Member of East Greenland. *Geol. Rundsch.* **68**, 748–74. [7]

Clemmensen, L., Jacobsen, V. and Steel, R. (1980) Some aspects of Triassic sedimentation and basin development, East Greenland, North Sea. In: N.P.F. *The Sedimentation of the North Sea Reservoir Rocks*. Geilo XVII. Norwegian Petroleum Society, Geilo, pp. 1–21. [7]

Clemmensen, L.B. and Abrahamsen, K. (1983) Aeolian stratification and facies association in desert sediments, Arran basin, Permian Scotland. *Sedimentology* **30**, 311–39. [5]

Cliff, R.A., Drewery, S.E. and Leeder, M.R. (1991) Sourcelands for the Carboniferous Pennine river system: constraints from sedimentary evidence and U–Pb geochronology using monazite and zircon. In: Morton, A.C., Todd, S.P. and Houghton, P.D.W. (eds) *Developments in Sedimentary Provenance Studies*. Special Publication No. 57, Geological Society, London, pp. 137–59. [4]

Clift, P.D., Turner, J. and ODP LEG 152 Scientific Party (1995) Dynamic support by the Iceland Plume and its effect on the subsidence of the northern Atlantic margins. *J. Geol. Soc.* **152**, 935–41. [2]

Cloetingh, S. (1988) Intraplate stress: a tectonic cause for third order cycles in apparent sea level. In: Wilgus, C.K., Hastings, B.S., Kendall, C.G.St.C., Posameutier, H., Ross, C.A. and van Wagoner, J. (eds) *Sea-Level Changes: an Integrated Approach*. Special Publication No. 42, SEPM, Tulsa, OK., pp. 19–30. [9]

Cloetingh, S., Lambeck, K. and McQueen, H. (1987) Apparent sea-level fluctuations and a palaeostress field for the North Sea region. In: Brooks, J. and Glennie, K.W. (eds) *Petroleum Geology of North West Europe*. Graham and Trotman, London, pp. 49–57. [9]

Cockings, J.H., Kessler, L.G., Mazza, T.A. and Riley, L.A. (1992) Bathonian to mid-Oxfordian sequence stratigraphy of the south Viking Graben, North Sea. In: Hardman, R.P.F. (ed.) *Exploration Britain: Geological Insights for the Next Decade*. Special Publication No. 67, Geological Society, London, pp. 65–105. [8, 12]

Cocks, L.R.M. and Fortey, R.A. (1982) Faunal evidence for oceanic separations in the Palaeozoic of Britain. *J. Geol. Soc.* **1394**, 465–78. [2]

Coelewij, P.A.G., Haug, G.M.W. and van Kuijk, H. (1978) Magnesium-salt exploration in the northeast Netherlands. *Geol. Mijnbouw* **57**, 487–502. [6]

Cohen, M.J. and Dunn, M.E. (1987) The hydrocarbon habitat of the Haltenbanken—Traenabank area, offshore Norway. In: Brooks, J. and Glennie, K.W. (eds) *Petroleum Geology of North West Europe*. Graham and Trotman, London, pp. 1091–104. [11]

Colley, M.G., Mcwilliams, A.S.F. and Myers, R.C. (1981) Geology of the Kinsale Head gas field, Celtic Sea, Ireland. In: Illing, L.V. and Hobson, G.P. (eds) *Petroleum Geology of the Continental Shelf of North-West Europe*. Heyden, London, pp. 504–10. [1, 11, 12]

Collier, R.E.L. (1991) The Lower Carboniferous Stainmore Basin, Northern England: extensional basin tectonics and sedimentation. *J. Geol. Soc.* **148**, 379–90. [4]

Collier, R.E.L., Leeder, M.R. and Maynard, J.R. (1990) Trangressions and regressions: a model for the influence of tectonic subsidence, deposition and eustasy, with application to Quaternary and Carboniferous examples. *Geol. Mag.* **127**, 117–28. [4]

Collins, A. (1990) The 1–10 Spore Colour Index (SCI) Scale: a universally applicable colour maturation scale, based on graded, picked palynomorphs. *Meded. Rijks Geol. Dienst* **45**, 39–47. [3]

Collinson, J.D. (1988) Controls on Namurian sedimentation in the Central Province basins of northern England. In: Besley, B.M. and Kelling, G. (eds) *Sedimentation in a Synorogenic Basin Complex: the Upper Carboniferous of Northwest Europe.* Blackie, Glasgow, London, pp. 85–101. [4]

Collinson, J.D. (ed.) (1990) *Correlation in Hydrocarbon Exploration.* NPS, Graham and Trotman, London.

Collinson, J.D., Jones, C.M., Blackbourne, G.A., Besly, B.M., Archard, G.M. and McMahon, A.H. (1993) Carboniferous depositional systems of the Southern North Sea. In: Parker, J.R. (ed.) *Petroleum Geology of Northwest Europe:* Geological Society, London, pp. 677–87. [4, 11, 12]

Colter, V.S. and Barr, K.W. (1975) Recent developments in the geology of the Irish Sea and Cheshire Basins. In: Woodland, A.W. (ed.) *Petroleum and the Continental Shelf of Northwest Europe.* Vol. I, *Geology.* Applied Science Publishers, Barking, pp. 61–73. [6, 12]

Colter, V.S. and Harvard, D.J. (1981) The Wytch Farm Oilfield. In: Illing, L.V. and Hobson, G.P. (eds) *Petroleum Geology of the Continental Shelf of North-West Europe.* Heyden, pp. 494–503. [2, 8, 11, 12]

Colter, V.S. and Reed, G.E. (1980) Zechstein 2 Fordon Evaporites of the Atwick No. 1 borehole, surrounding areas of N.E. England and the adjacent Southern North Sea. In: Füchtbauer, H. and Peryt, T.M. (eds) *The Zechstein Basin with Emphasis on Carbonate Sequences.* Contr. Sedimentology No. 9, Schweitzerbart'sche Verlagsbuchhandlung, Stuttgart, pp. 115–29. [6]

Combaz, A. (1980) Les kerogens vus au microscope. In: Durand, B. (ed.) *Kerogen.* Editions Technip, Paris, pp. 55–112. [11]

Coney, D., Fyfe, T.B., Retail, P. and Smith, P.J. (1993) Clair appraisal: the benefits of a co-operative approach. In: Parker, J.R. (ed.) *Petroleum Geology of Northwest Europe: Proceedings of the 4th Conference.* Geological Society, London, pp. 1409–20. [1, 3, 12]

Conlin, J.M. Hale, J.L., Sabathier, J.C., Faure, F. and Mas, D. (1990) *Multiple-fracture Horizontal Wells: Performance and Numerical Simulation.* Europec 90, SPE, Tulsa, OK. [12]

Connan, J. (1984) Biodegradation of crude oils in reservoirs. In: Brooks, J. and Welte, D. *Advances in Petroleum Geochemistry,* Vol. 1. Academic Press, London, pp. 299–335. [11]

Conort, A. (1986) Habitat of Tertiary hydrocarbons, South Viking Graben: In: Spencer, A.M., Campbell, C.J., Hanslien, S.H., Nelson, P.H., Nysaether, E. and Ormaasen, E.G. (eds) *Habitat of Hydrocarbons on the Norwegian Continental Shelf.* Graham and Trotman, London, pp. 159–70. [12]

Conway, A. (1986) Geology and petrophysics of the Victor gas field. In: Brooks, J., Goff, J. and van Hoorn, B. (eds) *Habitat of Palaeozoic Gas in N.W. Europe.* P. Publication No. 23, Geological Society, London, pp. 237–49. [1, 5]

Cook, A.C., Murchison, D.H. and Scott, E. (1972) Optical biaxial anthracitic vitrinites. *Fuel* **51**, 180–4.

Cooke-Yarborough, P. (1991) The Hewett Field, Blocks 48/28–29–30, 52/4a–5a, UK North Sea. In: Abbots, I.L. (ed.) *UK North Sea United Kingdom Oil and Gas Fields: 25 Years Commemorative Volume.* Memoir No. 14, Geological Society, London, pp. 433–42. [1, 5, 6, 7, 11, 12]

Cooke-Yarborough, P. (1994) Analysis of fractures yields improved gas production from Zechstein carbonates, Hewett Field, UKCS. *First Break* **12**(5), 243–52. [6, 12]

Cooles, G.P., Mackenzie, A.S. and Quigley, T.M. (1986) Calculation of petroleum masses generated and expelled from source rocks. In: Leythaeuser, D. and Rullkotter, J. (eds) *Advances in Organic Geochemistry 1985.* Pergamon, Oxford, pp. 235–46. [11]

Cooper, A.H. (1986) Subsidence and foundering of strata caused by dissolution of Permian gypsum in the Ripon and Bedale areas, North Yorkshire. In: Harwood, G.M. and Smith, D.B. (eds) *The English Zechstein and Related Topics.* Special Publication No. 22, Geological Society, London, pp. 127–39. [6]

Cooper, B.S. (1990) *Practical Petroleum Geochemistry.* Robertson Scientific Publications, London. [11]

Cooper, B.S. and Barnard, P.C. (1984) Source rocks and oils of the Central and Northern North Sea. In: Demaison, G. and Murris, R.J. (eds) *Petroleum Geochemistry and Basin Evaluation.* Memoir No. 35, AAPG, Tulsa, OK., pp. 303–14. [11]

Cooper, B.S., Coleman, S.H., Barnard, P.C. and Butterworth, J.S. (1975) Paleotemperatures in the Northern North Sea basin. In: Woodland, A.V. (ed.) *Petroleum and the Continental Shelf of North West Europe.* Applied Science Publishers, Barking, pp. 487–92. [11]

Cope, J.C.W. (1994) A latest Cretaceous hotspot and the southeasterly tilt of Britain. *J. Geol. Soc.* **151**, 905–8. [11]

Cope, J.C.W., Getty, T.A., Howarth, M.K. and Torrens, H.S. (1980a) *A Correlation of Jurassic Rocks in the British Isles.* Part One: *Introduction and Lower Jurassic.* Special Report No. 14, Geological Society, London.

Cope, J.C.W., Duff, K.L., Parsons, C.F., Torrens, H.S., Wimbledon, W.A. and Wright, J.K. (1980b) *A Correlation of Jurassic Rocks in the British Isles.* Part Two: *Middle and Upper Jurassic.* P. Report No. 15, Geological Society, London.

Cope, M.J. (1980) Physical and chemical properties of coalified and charcoalified phytoclasts from some British Mesozoic sediments: an organic geochemical approach to paleobotany. In: Douglas, A.G. and Maxwell, J.R. (eds) *Advances in Organic Geochemistry 1979.* Pergamon, Oxford, pp. 663–77. [11]

Cope, M.J. (1986) An interpretation of vitrinite reflectance data in the southern North Sea Basin. In: Brooks, J., Goff, J. and van Hoorne, B. (eds) *Habitat of Palaeozoic Gas in N.W. Europe.* Special Publication No. 23, Geological Society, Scottish Academic Press, Edinburgh, pp. 85–100. [11]

Cordey, W.G. (1993) Jurassic exploration history: a look at the past and future. In: Parker, J.R. (ed.) *Petroleum Geology of Northwest Europe: Proceedings of the 4th Conference.* Geological Society, London, pp. 195–8. [12]

Corfield, S.M., Gawthorpe, R.L., Gage, M., Fraser, A.J. and Besly, B.M. (1996) Inversion tectonics of the Variscan Foreland of the British Isles. *J. Geol. Soc.* **153**, 17–32. [2, 4]

Cornelius, C.D. (1975) Geothermal aspects of hydrocarbon exploration in the North Sea Area. *Norges Geol. Unders.* **316**, 29–67. [11]

Cornford, C. (1972) *An organic geochemical approach to the palaeogeography, environment of deposition and maturity of the Lower Lias Limestone-Shale, S. England.* Unpublished M.Sc. Dissertation, University of Newcastle. [11]

Cornford, C. (1986) The Bristol Channel Graben: organic geochemical limits on subsidence and speculation on the origin of inversion. *Proc. Ussher Soc.* **6**, 360–7. [11]

Cornford, C. (1990) Source rocks and hydrocarbons of the North Sea. In: Glennie, K.W. (ed.) *Introduction to the Petroleum Geology of the North Sea.* JAPEC, Blackwell Scientific Publications, Oxford, pp. 294–361. [12]

Cornford, C. (1993a) Hydrocarbon flux efficiencies in the Central Graben of the North Sea. In: Parnell, J., Ruffell, A.H. and Moles, N.R. (eds) *Geofluids '93: Extended Abstracts of the International Conference on Fluid Evolution, Migration and Interaction in Rocks.* Geological Society Publishing House, Bath, pp. 76–81. [11]

Cornford, C. (1993b) Risked basin efficiency calculations for the Central Graben, North Sea. In: Øygard, K. (ed.) *Organic Geochemistry, Poster Sessions from 16th Int'l. Meeting.* Falch Hurtigtrykk, Oslo, pp. 80–6. [11]

Cornford, C. (1994) Mandal–Ekofisk (1) petroleum system in the Central Graben of the North Sea. In: Magoon, L.B. and Dow,

W.G. (eds) *The Petroleum System from Source to Trap.* Memoir No. 60, AAPG, Tulsa, OK., pp. 537–72. [11]

Cornford, C. and Brooks, J. (1990) *Tectonic Controls on Oil and Gas Occurrences in the North Sea.* Memoir No. 46, AAPG, Tulsa, OK., pp. 523–39.

Cornford, C. and Highton, P. (1995) A comparison of uplift versus focussed fluid flow models for the thermal history of the East Irish Sea Basin. Paper presented at the Conference on the Petroleum Geology of the Irish Sea and Adjacent Areas, Geological Society, London, Feb. [11]

Cornford, C., Rullkotter, J. and Welte, D. (1980) A synthesis of organic petrographic and geochemical results from DSDP sites in the eastern central North Atlantic. In: Douglas, A.G. and Maxwell, J.R. (eds) *Advances in Organic Geochemistry 1979.* Pergamon, Oxford, pp. 445–53. [11]

Cornford, C., Morrow, J.A., Turrington, A., Miles, J.A. and Brooks, J. (1983) Some geological controls on oil composition in the U.K. North Sea. In:Brooks, J. (ed.) *Petroleum Geochemistry and Exploration of Europe.* Special Publication No. 12, Geological Society, Blackwell Scientific Publications, Oxford, pp. 175–94. [11]

Cornford, C., Needham, C.E.J. and de Walque, L. (1986) Geochemical habitat of North Sea oils and gases. In: N.P.S. *Habitat of Hydrocarbons on Norwegian Cont. Shelf.* Graham and Trotman, London, pp. 39–54. [11]

Cornford, C., Christie, O., Endresen, U., Jensen, P. and Myhr, M.B. (1988) Source rock and seep oil maturity in Dorset, Southern England. *Org. Geochem.* **13**, 399–409. [8, 11]

Cornford, C., Leonard, J.E. and Cornford, D. (1993) *Modelling short-term time-transient heat flow effects.* Ancillary extended abstract for Geofluids '93, Torquay, May. [11]

Cornford, C., Christie, O.H.J., Endresen, U., Jensen, P. and Myhr, M-B. (1997) Source rock and seep oil maturity in Dorset, Southern England. *Org. Geochem.* **13**(1–3), 399–409. [11]

Cosgrove, M.E. (1970) Iodine in the bituminous Kimmeridge shales of the Dorset Coast, England. *Geochim. Cosmochim. Acta* **34**, 830–6. [11]

Costa, L.I. and Davey, R.J. (1992) Dinoflagellate cysts of the Cretaceous system. In: Powell, A.J. (ed.) *A Stratigraphic Index of Dinoflagellate Cysts.* British Micropalaeontological Society Series, Chapman and Hall, London, pp. 99–153. [9]

Cottençon, A., Parant, B. and Flacelière, G. (1975) Lower Cretaceous gas fields in Holland. In: Woodland, A.W. (ed.) *Petroleum and the Continental Shelf of Northwest Europe.* Vol. I, *Geology.* Applied Science Publishers, Barking, pp. 403–12. [9, 11]

Courtney, R.C. and White, R.S. (1986) Anomalous heat flow and geoid across the Cape Verde Rise: evidence for dynamic support from a thermal plume in the mantle. *Geophys. J. Roy. Astron. Soc.* **87**, 815–67. [2]

Cousteau, H., Lee, P.J., Dupuy, J. and Junca, J. (1988) The Jurassic oil resources of the East Shetland basin, North Sea. *Bull. Can. Petr. Geol.* **36**(2), 177–85. [11]

Coutts, S.D., Larsson, S.Y. and Rosman, R. (1996) Development of the slumped crestal area of the Brent Reservoir, Brent Field: an integrated approach. *Petr. Geosci.* **2**, 219–29. [1, 8]

Cowan, G. (1989) Diagenesis of Upper Carboniferous sandstones: southern North Sea basin. In: Whateley, M.K. and Pickering, K.T. (eds) *Deltas: Sites and Traps for Fossil Fuels.* Special Publication No. 41, Geological Society, London, pp. 57–73. [4, 12]

Cowan, G. (1993) Identification and significance of aeolian deposits within the predominantly fluvial Sherwood Sandstone Group of the East Irish Sea Basin UK. In: North, C.P. and Prosser, D.J. (eds) *Characterization of Fluvial and Aeolian Reservoirs.* Special Publication No. 73, Geological Society, London, pp. 231–45. [12]

Cowan, G. (1996) The development of the North Morecambe Gas Field, East Irish Sea Basin, UK. *Petr. Geosci.* **2**(1), 43–52. [1, 11]

Cowan, G., Ottesen, C. and Stuart, I.A. (1993) The use of dipmeter logs in the structural interpretation and palaeocurrent analysis of the Morecambe Fields, East Irish Sea Basin. In: Parker, J.R. (ed.) *Petroleum Geology of Northwest Europe: Proceedings of the 4th Conference.* Geological Society, London, pp. 867–82. [12]

Coward, M.P. (1990) Caledonian framework. In: Brooks, J. and Hardman, R.F.P. (eds) *Tectonic Events Responsible for Britain's Oil and Gas Reserves.* Special Publication, Geological Society, London. [1, 2]

Coward, M.P. (1993) The effect of Late Caledonian and Variscan continental escape tectonics on basement structure, Palaeozoic basin kinematics and subsequent Mesozoic basin development in NW Europe. In: Parker, J.R. (ed.) *Petroleum Geology of Northwest Europe: Proceedings of the 4th Conference.* Geological Society, London, pp. 1095–108. [1, 4]

Coward, M.P. (1995) Structural and tectonic setting of the Permo-Triassic basins of northwest Europe. In: Boldy, S.A.R. (ed.) *Permian and Triassic Rifting in North West Europe.* Special Publication No. 92, Geological Society, London, pp. 7–39. [7]

Coward, R.N., Clark, N.M. and Pinnock, S.J. (1991) The Tartan Field, Block 15/16, UK North Sea. In: Abbotts, I.L. (ed.) *UK North Sea United Kingdom Oil and Gas Fields: 25 Years Commemorative Volume.* London, Memoir No. 14, Geological Society, pp. 377–84. [1, 8, 12]

Cowper, D.R., Lynch, D.J. and Neville, G. (1995) *Foinaven: Successful Reservoir Appraisal and Characterisation through the use of Horizontal Well and EWT Technology.* SPE 30565, Tulsa, OK., 7 pp. [1]

Craven, J.E. (1995) Exploration for Middle and Upper Jurassic reservoirs in Quadrants 49 and 50 in the North Celtic Graben: an overview. In: Croker, P.F. and Shannon, P.M. (eds) *The Petroleum Geology of Ireland's Offshore Basins.* Special Publication No. 93, Geological Society, London, pp. 277–8. [11]

Crawford, R., Littlefair, R.W. and Affleck, L.G. (1991) The Arbroath and Montrose Fields, Blocks 22/17,18, UK North Sea. In: Abbotts, I.L. (ed.) *UK North Sea United Kingdom Oil and Gas Fields: 25 Years Commemorative Volume.* Memoir No. 14, Geological Society, London, pp. 211–17. [1, 10, 12]

Creaney, S. (1980) Petrographic texture and vitrinite reflectance variation on the Alston Block, north-east England. *Proc. Yorks, Geol. Soc.* **42**, 553–80. [11]

Creaney, S and Passey, Q.R. (1993) Recurring patterns of total organic carbon and source rock quality within a sequence stratigraphic framework. *AAPG Bull* **77**(3), 386–401. [11]

Creaney, S., Jones, J.M., Holliday, D.W., and Robson, P. (1980) The occurrence of bitumen in the Great Limestone around Matfen, Northumberland—its characterization and possible genesis. *Proc. Yorks. Geol. Soc.* **43**(1), No. 5, 69–79. [11]

Crittenden, S. (1982) Lower Cretaceous lithostratigraphy NE of the Sole Pit area in the UK southern North Sea. *J. Petr. Geol.* **5**, 191–202. [9]

Crittenden, S. (1987a) Aptian lithostratigraphy and biostratigraphy (foraminifera) of Block 49 in the southern North Sea (UK sector). *J. Micropalaeontol.* **6**, 11–20. [9]

Crittenden, S. (1987b) The 'Albian transgression' in the southern North Sea Basin. *J. Petr. Geol.* **10**, 395–414. [9]

Crittenden, S., Cole, J. and Harlow, C. (1991) The Early to 'middle' Cretaceous lithostratigraphy of the Central North Sea UK Sector. *J. Petr. Geol.* **14**, 387–416. [9]

Crittenden, S., Cole, J.M. and Kirk, M.J. (1997) The distribution of Aptian sandstones in the Central and Northern North Sea (UK sectors): a lowstand systems tract 'play'. *J. Petrol. Geol.* **20**(1), 3–25. [9]

Croker, P.F. (1995a) Shallow gas accumulation and migration in the western Irish Sea. In: Croker, P.F. and Shannon, P.M. (eds) *The Petroleum Geology of Ireland's Offshore Basins.* Special Publication No. 93, Geological Society, London, pp. 41–58. [11]

Croker, P.F. (1995b) The Clare Basin: a geological and geophysical outline. In: Croker, P.F. and Shannon, P.M. (eds) *The Petroleum*

Geology of Ireland's Offshore Basins. Special Publication No. 93, Geological Society, London, pp. 327–40. [11]

Croker, P.F. and Shannon, P.M. (1987) The evolution in hydrocarbon prospectivity of the Porcupine Basin, Offshore Ireland. In: Brooks, J. and Glennie, K.W. (eds) *Petroleum Geology of North West Europe*. Graham and Trotman, London, pp. 633–42. [11]

Croker, P.F. and Shannon, P.M., (eds) (1995) *The Petroleum Geology of Ireland's Offshore Basins*. Special Publication No. 93, Geological Society, London, 498 pp. [11, 12]

Croker, P.F. and Shannon, P.M. (1995) The petroleum geology of Ireland's offshore basins: introduction: In: Croker, P.F. and Shannon, P.M. (eds) *The Petroleum Geology of Ireland's Offshore Basins*. Special Publication No. 93. Geological Society. London, pp. 1–8. [1]

Crook, H., Palmer, J.E., Painter, D.J. and Uden, R.C. (1993) Acquisition and processing techniques to improve seismic resolution of Tertiary targets. In: Parker, J.R. (ed.) *Petroleum Geology of Northwest Europe: Proceedings of the 4th Conference*. Geological Society, London, pp. 85–96. [12]

Crookall, R. (1955–76) *Fossil Plants of the Carboniferous Rocks of Great Britain*. Memoir, Geological Survey of Great Britain, Palaeontology, 1004 pp. (publ. in 7 parts) HMSO, London. [4]

Crowell, J.C. (1961) Depositional structures from the Jurassic Boulder Beds, east Sutherland. *Trans. Roy. Soc. Edinburgh* **18**, 202–19. [8]

Cubitt, J.M. and England, W.A. (eds) (1995) *The Geochemistry of Reservoirs*. Special Publication No. 86, Geological Society, London. [11]

Cumming, A.D. and Wyndham, C.L. (1975) The geology and development of the Hewett Gas Field. In: Woodland, A.W. (ed.) *Petroleum and the Continental Shelf of Northwest Europe*. Vol. I, *Geology*, Applied Science Publishers, Barking, pp. 313–25. [1, 5, 7]

Curiale, J.A. and Bromley, B.W. (1996) Migration induced compositional changes in oils and condensates of a single field. *Org. Geochem.* **24**(12), 1097–114. [11]

Currie, S. (1996) The development of the Ivanhoe, Rob Roy and Hamish Fields, Block 15/21A, UK North Sea. In: Hurst, A., Johnson, H.D., Burley, S.D., Canham, A.C. and Mackertich, D.S. (eds) *Geology of the Humber Group: Central Graben and Moray Firth, UKCS*. Special Publication No. 114, Geological Society, London, pp. 329–41. [8]

Cutts, P.L. (1991) The Maureen Field, Block 16/29a, UK North Sea. In: Abbotts, I.L. (ed.) *UK North Sea United Kingdom Oil and Gas Fields: 25 Years Commemorative Volume*. Memoir No. 14, Geological Society, London, pp. 347–52. [1]

Dahl, B. and Augustson, J.H. (1993) The influence of Tertiary and Quarternary erosion on hydrocarbon generation in Norwegian offshore basins. In: Doré, A.G., Auguston, J.H., Hermanrud, C., Stewart, D.J. and Sylta, O. (eds) *Basin Modelling: Advances and Applications*. Special Publication No. 3, NPF, Elsevier, Amsterdam, pp. 419–32. [11]

Dahl, B. and Speers, G.C. (1985) Organic geochemistry of the Oseberg Field (I). In: Thomas, B.M., Doré, A.G., Eggen, S.S., Home, P.C. and Larsen, R.M. (eds) *Petroleum Geochemistry in Exploration of the Norwegian Shelf.* Graham and Trotman, London, pp. 185–96. [11]

Dahl, B. and Speers, G.C. (1986) Geochemical characterization of a tar mat in the Oseberg Field, Norwegian Sector, North Sea. In: Leythaeuser, D. and Rullkotter, J. (eds) *Advances in Organic Geochemistry 1985*. Pergamon, Oxford, pp. 547–58. [11]

Dahl, B. and Yükler, A. (1991) The role of petroleum geochemistry in basin modelling of the Oseberg area, North Sea. In: Merrill, R.K. (ed.) *Source and Migration Processes and Evaluation Techniques*. Treatise Petroleum Geology, AAPG, Tulsa, OK., pp. 65–86. [11]

Dahl, B., Nysaether, E., Speers, G.C. and Yukler, A. (1987) Oseberg area—integrated basin modelling. In: Brooks, J. and Glennie, K.W. (eds) *Petroleum Geology of North West Europe*. Graham and Trotman, London, pp. 1029–38. [11]

Dahl, N. and Solli, T. (1993) The structural evolution of the Snorre Field and surrounding areas. In: Parker, J.R. (ed.) *Petroleum Geology of Northwest Europe: Proceedings of the 4th Conference*. Geological Society, London, pp. 1157–66. [8, 12]

Daly, M.C., Bell, M.S. and Smith, P.J. (1996) The remaining resources of the UK North Sea and its future development. In: Glennie, K.W. and Hurst, A. *AD 1995: NW Europe's Hydrocarbon Industry*. Geological Society, London, pp. 187–93 [1]

Dam, G. and Christiansen, F.G. (1990) Organic geochemistry and source potential of the lacustrine shales of the Upper Triassic/Lower Jurassic/Lower Jurassic Kap Stewart Formation, East Greenland. *Mar. Petr. Geol.* **7**(4), 428–43. [8, 11]

Damgaard, A. *et al.* (1989) *A Unique Method for Perforating, Fracturing and Completing Horizontal Wells*. SPE, Offshore Europe. [12]

Damtoft, K., Andersen, C. and Thomsen, E. (1987) Prospectivity and hydrocarbon plays of the Danish Central Trough. In: Brooks, J. and Glennie, K.W. (eds) *Petroleum Geology of North West Europe*. Graham and Trotman, London, pp. 403–17. [6, 9]

Damtoft, K., Nielsen, L.H., Johannessen, P.N., Thomsen, E. and Andersen, P.R. (1992) Hydrocarbon plays of the Danish Central Trough. In: Spencer, A.M. (ed.) *Generation, Accumulation, and Production of Europe's Hydrocarbons*. EAPG, Oxford University Press, Oxford, pp. 35–58. [9, 12]

Dangerfield, J.A. and Brown, D.A. (1987) The Ekofisk Field. In: Kleppe, J., Berg, E.W. Buller, A.T., Hjelmeland, O. and Torsæter, O. (eds) *North Sea Oil and Gas Reservoirs*. Graham and Trotman, London, pp. 3–22. [9]

Davey, R.J. (1979) The stratigraphic distribution of dinocysts in the Portlandian (latest Jurassic) to Barremain (Early Cretaceous) of northwest Europe. *Am. Assoc. Stratigr. Palynol. Contrib. Ser.* SB, pp. 49–81. [9]

David, F. (1987) Sandkoerper in fluviatilen Sandsteinen des Unteren Westphal D Oberkarbon am Piesberg bei Osnabrück. *Facies* **17**, 51–8. [4]

David, F. (1990) *Sedimentologie und Beckenanalyse im Westphal C und D des nordwestdeutschen Oberkarbons*. DGMK-Dericht 384–3. Deutsche wissenschaftlichte Gesellschaft für Erdol, Erdgas und Kohle, Hamburg, 271 pp. [4]

David, F., Gast, R. and Kraft, T. (1993) Relation between facies, diagenesis, and reservoir quality of Rotliegende Reservoirs in North Germany. *AAPG. Bull.* **77** (9), 1617. [5]

Davies, E.J. and Watts, T.R. (1977) The Murchison oil field. In: Finstad, R.C. and Selley, R.C. (eds) *Mesozoic Northern North Sea Symposium*, Paper MNNSS/15, Norwegian Petroleum Society, Oslo. [11]

Davies, J.R. (1991) Karstification and pedogenesis on a late Dinantian carbonate platform, Anglesey, North Wales. *Proc. Yorks, Geol. Soc.* **48**, 297–321. [4]

Davies, R.J., Stephen, K.J. and Underhill, J.R. (1996) The use of sequence stratigraphy in determining the tectono-stratigraphic evolution of the Bathonian to Kimmeridgian of the Moray Firth basin, Central North Sea. In: Hurst, A., Johnson, H.D., Burley, S.D. Canham, A.C. and Mackertich, D.S. (eds) (1996) *Geology of the Humber Group: Central Graben and Moray Firth, UKCS*. Special Publication No. 114, Geological Society, London, pp. 81–108. [8]

Davies, S.R. and McClean D. (1996) Spectral gamma-ray and palynological characterisation of Kinderscoutian marine bands in the Namurian of the Pennine Basin. *Proc. Yorks. Geol. Soc.*, **51**, 103–14. [4]

Davis, B.K. (1987) Velocity changes and burial diagenesis in the chalk of the southern North Sea Basin. In: Brooks, J. and Glennie, K.W. (eds) *Petroleum Geology of North West Europe*. Graham and Trotman, London, pp. 307–13. [4]

Davison, I., Weston, P.J. and Insley, M.W. (1993) Physical modelling of North Sea Salt diapirism. In: Parker, J.R. (ed.) *Petroleum*

Geology of Northwest Europe: Proceedings of the 4th Conference. Geological Society, London, pp. 559–68.

Day, G.A., Cooper, B.A., Anderson, C., Burgers, W.F.J., Rønnevik, H.C. and Schöneich, H. (1981) Regional seismic structure maps of the North Sea. In: Illing, L.V. and Hobson, G.P. (eds) *Petroleum Geology of the Continental Shelf of North-West Europe.* Heyden, London, pp. 76–84. [5, 6, 7, 11]

Dean, G. (1996) 'Undiscovery' wells of the UK continental shelf. In: Glennie, K.W. and Hurst, A. (eds), *AD 1995: NW Europe's Hydrocarbon Industry.* Geological Society, London, pp. 69–80. [1]

Dean, K.P. (1993) Sedimentation of Upper Triassic reservoirs in the Beryl Embayment: lacustrine sedimentation in a semi-arid environment. In: Parker, J.R. (ed.) *Petroleum Geology of Northwest Europe: Proceedings of the 4th Conference.* Geological Society, London, p. 581. [7]

De'Ath, N.G. and Schuyleman, S.F. (1981) The geology of the Magnus Oilfield. In: Illing, L.V. and Hobson, G.P. (eds) *Petroleum Geology of the Continental Shelf of North-West Europe.* Heyden, London, pp. 342–51. [1, 8, 12]

Deegan, C.E. (1973) Tectonic control of sedimentation at the margin of a Carboniferous depositional basin in Kirkcudbrightshire. *Scott. J. Geol.* **9**, 1–28. [4]

Deegan, C.E. and Scull, B.J. (1977) *A Standard Lithostratigraphic Nomenclature for the Central and Northern North Sea.* Institute of Geological Science Report No. 77/25, HMSO, 36 pp. [7, 8, 9, 10]

de Jager, J., Doyle, M.A., Grantham, P.J. and Mabillard, J.E. (1996) Hydrocarbon habitat in the West Netherlands Basin. In: Rondeel, H.E., Batjes, D.A.J. and Nieuwenhuis, W.H. (eds) *Geology of Oil and Gas under the Netherlands.* Royal Geological and Mining Society, the Netherlands, Kluwer Academic Publishers, Dordrecht, pp. 191–209. [8, 11, 12]

de Leebeeck, A. (1987) The Frigg Field reservoir: characteristics and performance. In: Kleppe, J., Berg, E.W., Buller, A.T., Hjelmelaud, O. and Torsæter, O. (eds) *North Sea Oil and Gas Reservoirs.* Graham and Trotman, London, pp. 89–100. [12]

Demaison, G. (1984) The generative basin concept. In: Demaison, G. and Murris, R.J. (eds) *Petroleum Geochemistry and Basin Evaluation.* AAPG Memoir No. 35, Tulsa, OK., pp. 1–14. [12]

Demaison, G. and Huizinga, B.J. (1991) Genetic classification of petroleum systems. *AAPG Bull.* **75**(10), 1626. [11]

Demaison, G.J. and Moore, G.T. (1980) Anoxic environments and oil source bed genesis. *Org. Geochem.* **2**, 9–31. [11]

Demaison, G., Holck, A.J.J., Jones, R.W. and Moore, G.T. (1983) Predictive source bed stratigraphy: a guide to regional petroleum occurrence. *Proc. Basin and Eastern North American Continental Margin.* World Petroleum Congress, London. [11]

Dembicki, H., Jr and Anderson, M.J. (1989) Secondary migration of oil: experiments supporting efficient movement of separate, buoyant opil phase along limited conduits. *AAPG Bull.* **73**, 1018–21. [11]

Deming, D. (1994) Factors necessary to define a pressure seal. *AAPG Bull.* **78**(6), 1005–10. [11]

Demongodin, L., Pinoteau, B., Vasseur, G. and Gable, R. (1991) Thermal conductivity and well logs: a case study in the Paris Basin. *Geophys. J. Int.* **105**, 675–91. [11]

Demyttenaere, R.R.A., Sluijk, A.H. and Bentley, M.R. (1993) A fundamental reappraisal of the structure of the Cormorant Field and its impact on field development strategy. In: Parker, J.R. (ed.) *Petroleum Geology of Northwest Europe: Proceedings of the 4th Conference.* Geological Society, London, pp. 1151–7. [12]

den Hartog Jager, D., Giles, M.R., and Griffiths, G.R. (1993) Evolution of Paleogene submarine fans of the North Sea in space and time. In: Parker, J.R. (ed.) *Petroleum Geology of Northwest Europe: Proceedings of the 4th Conference.* Geological Society, London, pp. 59–72. [10, 12]

den Hartog Jager, D.G. (1996) Fluviomarine sequences in the Lower Cretaceous of the West Netherlands Basin: correlation and seismic expression. In: Rondeel, H.E., Batjes, D.A.J. and Nieuwenhuis, W.H. (eds) *Geology of Oil and Gas under the Netherlands.* Royal Geological and Mining Society, the Netherlands, Kluwer Academic Publishers, Dordrecht, pp. 19–30. [9, 12]

Department of Energy (1996) *Development of the Oil and Gas Resources of the United Kingdom 1996.* Department of Energy Brown Book, HMSO, 96 pp. [1]

Depowski, S. (1978) *Lithofacies–Palaeogeographical Atlas of the Permian Platform Areas of Poland.* Geological Institute, Warsaw, 30 pp. [5, 6]

Depowski, S. (1981) The geological factors of hydrocarbon accumulations in the Permian in the Polish Lowland. In: *Proceedings of an International Symposium on Central Europe Permian, Jablonna, Poland, 1978.* Geological Institute, Warsaw, pp. 547–67. [6]

Depowski, S. and Peryt, T.M. (1985) Carbonate petroleum reservoirs in the Permian dolomites of the Zechstein, Fore-Sudetic Area. Western Poland. In: Roehl, P.O. and Choquette, P.W. (eds) *Carbonate Petroleum Reservoirs.* Springer-Verlag, Berlin, pp. 253–64. [6]

Depowski, S., Peryt, T.M., Piatkowski, S. and Wagner, R. (1981) Palaeogeography versus oil and gas potential of the Zechstein Main Dolomite in the Polish Lowlands. In: *Proceedings of an International Symposium on Central Europe Permian, Jablonna, Poland, 1978.* Geological Institute, Warsaw, pp. 587–95. [6]

Derenne, S., Largeau, C., Casadevall, E., Leeuw, J.E.de and Tegelaar, E.W. (1989) Study of kerogen-rich Ordovician deposits: chemical structure, mechanism of formation and affinities with modern organisms. In: *Abstracts of Papers presented at the 14th Int'l Org. Geochem. Conference, Paris, Sept. 1989.* EAOG and IFP France. [11]

Derenne, S., Largeau, C., Casadevall, E., Sinninghe-Damste, J.S., Tegelaar, E.W. and Leeuw, J.E.de. (1990) Characterization of Estonian Kukersite by spectroscopy and pyrolysis: evidence for abundant alkyl phenolic moieties in Ordivician, marine, Type II/I kerogen. In: Durand, B. and Behan, F. (eds) *Advances in Organic Geochemistry 1989*, Pergamon Press, Oxford, pp. 873–88. [11]

Dewey, J.F. (1982) Plate tectonics and the evolution of the British Isles. *J. Geol. Soc.* **139**(4), 371–412. [2]

Dewey, J.F. and Shackleton, R.M. (1984) A model for the evolution of the Grampian tract in the early Caledonides and Appalachians. *Nature* **312**, 115–21. [2]

Dewey, J.F. and Windley, B.F. (1988) Paleocene–Oligocene tectonics of NW Europe. In: Morton, A.C. and Parson, L.M. (eds) *Early Tertiary Volcanism and the Opening of the NE Atlantic.* Special Publication No. 39, Geological Society, London, pp. 25–31. [10]

D'Heur, M. (1980) Chalk reservoir of the West Ekofisk Field. In: NPF (ed.) *The Sedimentation of the North Sea Reservoir Rocks.* NPF, Geilo, 20 pp. [9]

D'Heur, M. (1984) Porosity and hydrocarbon distribution in North Sea chalk reservoirs. *Mar. Petr. Geol.* **9**, 211–38. [9, 12]

D'Heur, M. (1986) The Norwegian chalk fields. In: Spencer, A.M., Campbell, C.J., Hanslien, S.H., Nelson, P.H., Nysaether, E. and Ormaasen, E.G. (eds) *Habitat of hydrocarbons on the Norwegian Continental Shelf.* Graham and Trotman, London, pp. 77–89. [9]

D'Heur, M. (1987a) Albuskjell. In: Spencer, A.M., Campbell, C.J., Hanslien, S.H., Nelson, P.H., Nysaether, E. and Ormaasen, E.G. (eds) *Geology of the Norwegian Oil and Gas Fields.* Graham and Trotman, London, pp. 39–50. [1, 9]

D'Heur, M. (1987b) Cod. In: Spencer, A.M., Campbell, C.J., Hanslien, S.H., Nelson, P.H., Nysaether, E. and Ormaasen, E.G. (eds) *Geology of the Norwegian Oil and Gas Fields.* Graham and Trotman, London, pp. 51–62. [1, 11]

D'Heur, M. (1987c) The Norwegian chalk fields. In: Spencer, A.M., Campbell, C.J., Hanslien, S.H., Nelson, P.H., Nysaether, E. and Ormaasen, E.G. (eds) *Geology of the Norwegian Oil and Gas Fields.* Graham and Trotman, London, pp. 77–89.

D'Heur, M. (1987d) Tor. In: Spencer, A.M., Campbell, C.J., Hanslien, S.H., Nelson, P.H., Nysaether, E. and Ormaasen, E.G. (eds) *Geology of the Norwegian Oil and Gas Fields*. Graham and Trotman, London, pp. 129–42. [1, 9]

D'Heur, M. (1987e) West Ekofisk. In: Spencer, A.M., Campbell, C.J., Hanslien, S.H., Nelson, P.H., Nysaether, E. and Ormaasen, E.G. (eds) *Geology of the Norwegian Oil and Gas Fields*. Graham and Trotman, London, pp. 165–76. [1, 9]

D'Heur, M. and de Walque, L. (1987) Bream and Brisling. In: Spencer, A.M., Campbell, C.J., Hanslien, S.H., Nelson, P.H., Nysaether, E. and Ormaasen, E.G. (eds) *Geology of the Norwegian Oil and Gas Fields*. Graham and Trotman, London, pp. 185–92. [1, 11]

D'Heur, M. and Michaud, F. (1987) Edda. In: Spencer, A.M., Campbell, C.J., Hanslien, S.H., Nelson, P.H., Nysaether, E. and Ormaasen, E.G. (eds) *Geology of the Norwegian Oil and Gas Fields*. Graham and Trotman, London, pp. 63–72. [1, 9]

D'Heur, M. and Pekot, L.J. (1987) Tommeliten. In: Spencer, A.M., Campbell, C.J., Hanslien, S.H., Nelson, P.H., Nysaether, E. and Ormaasen, E.G. (eds) *Geology of the Norwegian Oil and Gas Fields*. Graham and Trotman, London, pp. 117–28. [1, 9]

D'Heur, M., de Walque, L. and Michaud, F. (1985) Geology of the Edda Field reservoir. *Mar. Petr. Geol.* **2**, 327–40. [9]

di Primio, R. and Horsfield, B. (1995) Predicting the generation of heavy oils in carbonate/evaporitic environments using pyrolysis methods. Submitted for *Adv. Org. Geochem*

Dixon, J.E., Fitton, J.G. and Frost, R.T.C. (1981) The tectonic significance of post-Carboniferous igneous activity in the North Sea Basin. In: Illing, L.V. and Hobson, G.P. (eds) *Petroleum Geology of the Continental Shelf of North-West Europe*. Heyden, London, pp. 121–37.

Dixon, R.J., Schofield, K., Anderton, R., Reynolds, A.D., Alexander, R.W.S., Williams, M.C. and Davies, K.G. (1995) Sandstone diapirism and clastic intrusion in the Tertiary fans of the Bruce–Beryl embayment, Quadrant 9, UKCS. In: Hartley, A.J. and Prosser, D.J. (eds) *Characterisation of Deep Marine Clastic Systems*. Special Publication No. 94, Geological Society, London, pp. 75–92. [12]

Doligez, B. (ed.) (1987) *Migration of Hydrocarbons in Sedimentary Basins*. Editions Technip, Paris, 682 pp. [11]

Donato, J.A. (1993) A buried granite batholith and the origin of the Sole Pit Basin, UK Southern North Sea. *J. Geol. Soc.* **150**(2) 255–9. [11]

Donato, J.A., Martindale, W. and Tully, M.C. (1983) Buried granites within the Mid North Sea High. *J. Geol. Soc.* **140**, 825–37. [2, 4, 6]

Donovan, A.D., Djakic, A.W., Ioannides, N.S., Garfield, T.R. and Jones, C.R. (1993) Sequence stratigraphic control on Middle and Upper Jurassic reservoir distribution within the UK Central North Sea. In: Parker, J.R. (ed.) *Petroleum Geology of Northwest Europe: Proceedings of the 4th Conference*. Geological Society, London, pp. 251–69. [8, 12]

Donovan, R.N., and Meyerhoff, A.A. (1982) Comment on 'Paleomagnetic evidence for a large ~2000 km sinistral offset along the Great Glen Fault during Carboniferous time'. *Geology* **10**, 604–5. [2]

Donovan, R.N., Foster, R.J. and Westoll, T.S. (1974) A stratigraphic revision of the Old Red Sandstone of north-eastern Caithness. *Trans. Roy. Soc. Edinburgh* **69**, 167–201. [3, 11]

Doré, A.G. and Gage, M.S. (1987) Crustal alignments and sedimentary domains in the evolution of the North Sea, North-east Atlantic Margin and Barents Shelf. In: Brooks, J. and Glennie, K.W. (eds) *Petroleum Geology of North West Europe*. Graham and Trotman, London, pp. 1131–48. [6]

Doré, A.G., Vollset, J. and Hamar, G.P. (1985) Correlation of the offshore sequences referred to the Kimmeridge Clay Formation: relevance to the Norwegian sector. In: Thomas, B.M., Doré, A.G., Eggen, S.S., Home, P.C. and Magne Larsen, R. (eds) *Petroleum Geochemistry and Exploration of the Norwegian Shelf*. NPS, Graham and Trotman, London, pp. 27–37. [11, 12]

Doré, A.G. and Lundin, E.R. (1996) Cenozoic compressional structures on the NE Atlantic margin: nature, orgin and potential significance for hydrocarbon exploration. *Petroleum Geoscience* **2**(4), 299–311. [1]

Doré, A.G., Augustson, J.H., Hermanrud, C., Stewart, D.J. and Sylta, O. (eds) (1993) *Basin Modelling: Advances and Applications*. Proceedings of NPF Conferences, Special Publication No. 3, Elsevier, Amsterdam, 675 pp. [11]

Douglas, A.G. and Williams, P.F.V. (1981) Kimmeridge oil shale: a study of organic maturation. In: Brooks, J. (ed.) *Organic Maturation Studies and Fossil Fuel Exploration*. Academic Press, London, pp. 256–69. [11]

Douglas, A.G., Eglinton, G. and Maxwell, J.R. (1969) The organic geochemistry of certain samples from the Scottish Carboniferous formation. *Geochim. Cosmochim. Acta.* **33**, 579–90. [11]

Downie, R.A. (1989) Controls on the reservoir quality of the Old Red Sandstone of the Orcadian Basin. MSc thesis, University of Aberdeen. [3]

Downie, R.A. and Stedman, C.I. (1993) Complex deformation and fluidisation structures in Aptian sediment gravity flow deposits of the Outer Moray Firth. In: Parker, J.R. (ed.) *Petroleum Geology of Northwest Europe: Proceedings of the 4th Conference*. Geological Society, London, pp. 185–8. [9, 12]

Doyle, M.C. and Conlin, J.M. (1990) The Tyra Field. In: Buller, A.T., Berg, E., Hjelmelaud, O., Kleppe, J., Torsæter, O. and Aasen, J.O. (eds) *North Sea Oil and Gas Reservoirs II*. Norwegian Institute of Technology, Graham and Trotman, London, pp. 47–65. [9, 12]

Drinkwater, N.J., Pickering, K.T. and Siedlecka, A. (1996) Deep-water fault-controlled sedimentation, Arctic Norway and Russia: response to Late Proterozoic rifting and the opening of the Iapetus Ocean. *J. Geol. Soc.* **153**, 427–36. [2]

Drong, H.J., Plein, E., Sannemann, D., Schuepbach, M.A. and Zimdars, J. (1982) Der Schneverdingen Sandstein des Rotliegenden—eine aolische Sedimentfullung alter Graben Strukturen. *Zeitschr Deutsches Geol Gesampt* **133**, 699–725. [5]

Dronkers, A.J. and Mrozek, F.J. (1991) Inverted basins of The Netherlands. *First Break* **9**, 409–25. [2]

Dronkert, H. and Remmelts, G. (1996) Influence of salt structures on reservoir rocks in Block L-2, Dutch Continental Shelf. In: Rondeel, H.E., Batjes, D.A. G. and Nieuwenhius, W.H. (eds) *Geology of Oil and Gas under the Netherlands*. Royal Geological and Mining Society, the Netherlands, Kluwer Academic Publishers, Dordrecht, pp. 159–66. [7]

Drumgoole, P. and Speers, T. (1997) Geoscore: a method for quantifying uncertainty in field reserve estimates. *Petroleum Geoscience* **3**(1), 1–12.

Duddy, I.R., and Bray, R.J. (1993) Early Tertiary heating in Northwest England: fluids or burial or both? In: Parnell, J., Raffell, A.H. and Moles, N.R. (eds) *Geofluids '93, Extended Abstracts to International Conference on Fluid Evolution, Migration and Interaction in Rocks*. Geological Society, Publishing House, Bath, pp. 119–23. [11]

Duindam, P. and van Hoorn, B. (1987) Structural evolution of the West Shetland continental margin. In: Brooks, J. and Glennie, K.W. (eds) *Petroleum Geology of North West Europe*. Graham and Trotman, London, pp. 765–73. [1, 2, 3, 11]

Dunay, R.E. and Hailwood, E.A. (eds) (1995) *Non-biostratigraphical Methods of Dating and Correlation*. Special Publication No. 89, Geological Society, London.

Duncan, A.D. and Hamilton, R.M.F. (1988) Palaeolimnology and organic geochemistry of the Middle Devonian in the Orcadian Basin. In: Fleet, A.J., Kelts, K. and Talbot, M.R. (eds) *Lacustrine Petroleum Source Rocks*. Special Publication No. 40, Graham and Trotman, London, pp. 173–202. [3, 11]

Duncan, D.C. and Swanson, V.E. (1965) Organic-rich shale of the United States and world land areas. *US Geol. Survey Circular 523*, Washington. [11]

Duncan, W.I. and Buxton, N.W.K. (1995) New evidence for evaporitic Middle Devonian lacustrine sediments with hydrocarbon source potential on the East Shetland Platform, North Sea. *J. Geol. Soc.* **152**(2), 229–50. [3, 11]

Dunham, K.C. and Wilson, A.A. (1985) *Geology of the North Pennine Orefield*, Vol. 2: *Stainmore to Craven*. Economic Memoir, BGS, HMSO, London, 247 pp. [4]

Düppenbecker, S., Horsfield, B. and Welte, D.H. (1989) Integration of field studies, basin modelling and simulation experiments to evaluate petroleum generation in two parts of the Lower Saxony Basin, In: *Abstracts of Papers presented at the 14th Int'l Org. Geochem. Conference, Paris, 18–22 Sept. 1989.* EAOG and IFP (France) Abstract No. 93. [11]

Düppenbecker, S.J., Dohmen, L. and Welte, D.H. (1991) Numerical modelling of petroleum expulsion in two areas of the Lower Saxony Basin, Northern Germany. In: England, W.A. and Fleet, A.J. (eds) *Petroleum Migration.* Special Publication No. 59, Geological Society, London, pp. 47–65. [11]

Durand, B. (1983) Present trends in organic geochemistry in research in migration of hydrocarbons. In: Bjorøy, M., Albrecht, P., Cornford, C., de Groot, K., Eglinton, G., Galimov, E., *et al. Advances in Organic Geochemistry, 1981.* John Wiley, Chichester, pp. 117–28. [11]

Durand, B. and Monin, J.C. (1980) Elemental analysis of kerogens: C, H, O, N, S, Fe. In: Durand, B. (ed.) *Kerogen—Insoluble Organic Matter from Sedimentary Rocks.* Editions Technip, Paris, pp. 113–42. [11]

Durand, B. and Nicaise, G. (1980) Procedures for kerogen isolation. In: Durand, B. (ed.) *Kerogen—Insoluble Organic Matter from Sedimentary Rocks.* Editions Technip, Paris, pp. 35–53. [11]

Durand, B. and Paratte, M. (1983) Oil potential of coals: a geochemical approach. In: Brooks, J. (ed.) *Petroleum Geochemistry and Exploration of Europe.* Special Publication 12, Geological Society, Blackwell Scientific Publications, Oxford, pp. 255–65. [11]

Durand, B., Alpern, B., Pittion, J.L. and Pradier, B. (1986) Reflectance of vitrinite as a control of thermal history of sediments. In: Burrus, J. (ed.) *Thermal Modelling in Sedimentary Basins.* Editions Technip, Paris, pp. 441–74. [11]

Dypvik, H., Rueslatten, H.G. and Throndsen, T. (1979) Composition of organic matter from North Atlantic Kimmeridgian shales. *AAPG Bull.* **63**(12), 2222–6. [11]

Duxbury, S. (1977) A palynostratigraphy of the Berriasian to Barrema in of the Speeton Clay of Speeton, England. *Palaeontographica, Abt. B. Bd.* **160**, 17–67, pl. 1–15. [9]

Duxbury, S. (1980) Barremian phytoplankton from the Speeton, East Yorkshire. *Palaeontographica, Abt. B. Bd.* **173**, 107–46, pl. 1–3. [9]

Duxbury, S. (1983) A study of dinoflagellate cysts and acritarchs from the Lower Greensand (Aptian to Lower Albian) of the Isle of Wight, southern England. *Palaeontographica, Abt, B. Bd.* **186**, 18–80, 1–10. [9]

Eames, T.D. (1975) Coal rank and gas source relationships—Rotliegendes reservoirs. In: Woodland, A.W. (ed.) *Petroleum and the Continental Shelf of Northwest Europe*, Vol. I, *Geology.* Applied Science Publishers, Barking, pp. 191–201. [11]

Earle, M.M., Jankowski, E.J. and Vann, I.R. (1989) Structural and stratigraphic evolution of the Faeroe–Shetland Channel and Northern Rockall Trough. In: Tankard, A.J. and Balkwill, H.R. (eds) *Extensional Tectonics and Stratigraphy of the North Atlantic Margins.* Memoir No. 46, AAPG, pp. 461–9. [2]

Earls, T.C. (1995) Potential for development of the Connemara Field—Block 26/28. In: Croker, P.F. and Shannon, P.M. (eds) *The Petroleum Geology of Ireland's Offshore Basins.* Special Publication No. 93, Geological Society, London, pp. 343–4. [11, 12]

Ebbern, J. (1981) The geology of the Morecambe Gas Field. In: Illing, L.V. and Hobson, G.D. (eds) *Petroleum Geology of the Continental Shelf of North-West Europe.* Heyden, London, pp. 485–93. [11]

Ebdon, C.C., Fraser, A.J., Higgins, A.C., Mitchener, B.C., and Strank, A.R.E. (1990) The Dinantian stratigraphy of the East Midlands: a seismo-stratigraphic approach. *J. Geol. Soc.* **147** 519–36. [4]

Ebdon, C.C., Granger, P.J., Johnson, H.D. and Evans, A.M. (1995) Early Tertiary evolution and sequence stratigraphy of the Faeroe—Shetland Basin: implications for hydrocarbon prospectivity. In: Scrutton, R.A., Stoker, M.S., Shimmield, G.B. and Tudhope, A.W. (eds) *The Tectonics Sedimentation and Palaeoceanography of the North Atlantic Region.* Special Publication No. 90, Geological Society, London, pp. 51–69. [2, 12]

Ebukanson, E.J. and Kinghorn, R.R.F. (1985) Kerogen facies in the major Jurassic mudrock formations of S. England and the implications on the depostional environments of their precursors. *J. Petr. Geol.* **8**(4), 435–62. [11]

Ebukanson, E.J. and Kinghorn, R.R.F. (1986a) Maturity of organic matter in the Jurassic of Southern England and its relation to the burial history of the sediments. *J. Petr. Geol.* **9**(3), 259–80a. [11]

Ebukanson, E.J. and Kinghorn, R.R.F. (1986b) Oil and gas accumulations and their possible source rocks in Southern England, *J. Petr. Geol.* **9**(4), 413–28. [11]

Eckardt, C.B., Wolf, M. and Maxwell, J.R. (1989) Porphyrins in Permian Kupferschiefer from NW Germany: preliminary assessment of facies and thermal history. In: *Abstracts of Papers presented at the 14th Int'l Org. Geochem. Conference, Paris, 18–22 Sept. 1989.* EAOG and IFP (France) Abstract No. 93. [11]

Edwards, C.W. (1991) The Buchan Field, Blocks 20/5a and 21/1a, UK North Sea. In: Abbotts, I.L. (ed.) *UK North Sea United Kingdom Oil and Gas Fields. 25 Years Commemorative Volume.* Memoir No. 14, Geological Society, London, pp. 253–9. [1, 3, 12]

Edwards, D.A., Warrington, G., Scrivener, R.C., Jones, N.S., Haslam, H.W. and Ault, L. (1997) The Exeter Group, south Devon, England: a contribution to the early post-Variscan stratigraphy of northwest Europe. *Geological Magazine* **34**(2), 177–97.

Energistyrelsen (1997) København, pp. 64. [1]

Eggen, S. (1984) Modelling of subsidence, hydrocarbon generation, and heat transport in the Norwegian North Sea. In: Durand, B. (ed.) *Thermal Phenomena in Sedimentary Basins.* Technip, Paris, pp. 271–86. [11, 12]

Eggink, J.W., Riegstra, D.E. and Suzanne, P. (1996) Using 3D seismic to understand the structural evolution of the UK Central North Sea. *Petr. Geosci.* **2**, 83–96. [1, 2]

Ekdale, A.A. and Bromley, R.G. (1984) Comparative ichnology of shelf-sea and deep-sea chalk. *J. Paleontol.* **58**, 322–32. [9]

Ekern, O.F. (1986) Late Oligocene gas accumulations, block 2/2, Norway. In: Spencer, A.M., Campbell, C.J., Hanslien, S.H., Nelson, P.H., Nysaether, E. and Ormaasen, E.G. (eds) *Habitat of Hydrocarbons on the Norwegian Continental Shelf.* NPF, Graham and Trotman, London, pp. 143–9. [11]

Ekern, O.F. (1987) Midgard. In: Spencer, A.M., Campbell, C.J., Hanslien, S.H., Nelson, P.H., Nysaether, E. and Ormaasen, E.G. (eds) *Geology of the Norwegian Oil and Gas Fields.* Graham and Trotman, London, pp. 403–10. [1, 8, 11, 12]

Ellenor, D.W. and Mozetic, A. (1986) The Draugen oil discovery. In: Spencer, A.M., Campbell, C.J., Hanslien, S.H., Nelson, P.H., Nysaether, E. and Ormaasen, E.G. (eds) *Habitat of Hydrocarbons on the Norwegian Continental Shelf.* NPF, Graham and Trotman, London, pp. 313–16. [1, 8, 11, 12]

Ellis, D. (1993) The Rough gas field: distribution of Permian aeolian and non-aeolian reservoir facies and their impact on field development. In: North, C.P. and Prosser, D.J. (eds) *Characterization of Fluvial and Aeolian Reservoirs.* Special Publication No. 73, Geological Society, London, pp. 265–77. [5]

Elvsborg, A., Ekern, O.F., Gabrielsen, R.H. and Ulvoen, S. (1984) A geochemical evaluation of Block 2/2 offshore Norway. In:

Doré, A.G., Eggen, S.S., Home, P.C. and Larsen, R.M. (eds) *Petroleum Geochemistry in Exploration of the Norwegian Shelf.* Graham and Trotman, London, pp. 213–22.

Elvsborg, A., Hagevang, T. and Throndsen, T. (1985) Origin of the gas-condensate of the Midgard Field at Haltenbanken. In: Thomas, B.M., Doré, A.G., Eggen, S.S., Home, P.C. and Larsen, R.M. (eds) *Petroleum Geochemistry in Exploration of the Norwegian Shelf.* Graham and Trotman, London, pp. 213–22. [11]

Emmel, R.H., Bjorøy, M. and van Grass, G. (1985) Geochemical exploration on the Norwegian Continental Shelf by analysis of shallow cores. In: Thomas, B.M., Doré, A.G., Eggen, S.S., Home, P.C. and Larsen, R.M. (eds) *Petroleum Geochemistry in Exploration of the Norwegian Shelf.* Graham and Trotman, London, pp. 239–46. [11]

Enay, R. and Mangold, C. (1982) Dynamique biogéographie d'évolution des faunes d'ammonites au Jurassique. *Bull. Soc. Geol. France* **24**, 1025–46. [8]

Enfield, M.A. and Coward M.P. (1987) The structure of the West Orkney Basin, northern Scotland. *J. Geol. Soc.* **144**, 871–84. [3]

Engelstad, N. (1987) Murchison. In: Spencer, A.M., Campbell, C.J., Hanslien, S.H., Nelson, P.H., Nysaether, E. and Ormaasen, E.G. (eds) *Geology of the Norwegian Oil and Gas Fields*, Graham and Trotman, London, pp. 295–305. [1]

Engkilde, M. and Surlyk, F. (1993) The Middle Jurassic Vardekloft Formation of East Greenland—analogue for reservoir units of the Norwegian shelf and the Northern North Sea. In: Parker, J.R. (ed.) *Petroleum Geology of Northwest Europe: Proceedings of the 4th Conference.* Geological Society, London, pp. 533–42. [8]

England, R.W. (1995) Westline: a deep near-normal incidence reflection profile across the Rockall Trough. In: Croker, P.F. and Shannon, P.M. (eds) *The Petroleum Geology of Ireland's Offshore Basins.* Special Publication No. 93, Geological Society, London, pp. 423–7. [12]

England, W.A. and Fleet, A.J. (1991) *Petroleum Migration.* Special Publication No. 59, Geological Society, London, 280 pp. [11]

England, W.A., Mackenzie, A.S., Mann, D.M. and Quigley, T.M. (1987) The movement and entrapment of petroleum fluids in the subsurface. *J. Geol. Soc.* **144**, 327–47. [11]

England, W.A., Mann, A.L. and Mann, D.M. (1991) Migration from source to trap. In: Merrill, R.K. (ed.) *Source and Migration Processes and Evaluation Techniques.* Treatise Petroleum Geology, AAPG, Tulsa, OK. pp. 23–46. [11]

England, W.A. Muggenridge, A.H., Clifford, P.J. and Tang, Z. (1995) Modelling density driven mixing rates in petroleum reservoirs on geological time scales, with application to detection of barriers in the Forties field UKCS. In: Cubitt, J.M. and England, W.A. (eds) *The Geochemistry of Reservoirs.* Special Publication No. 86, Geological Society, London, pp. 185–202. [11]

Enjolras, J.M., Gouadain, J., Mutti, E. and Pizon, J. (1987) New turbiditic model for the Lower Tertiary sands in the South Viking Graben. In: Spencer, A.M., Campbell, C.J., Hanslien, S.H., Nelson, P.H., Nysaether, E. and Ormaasen, E.G. (eds) *Geology of the Norwegian Oil and Gas Fields.* Graham and Trotman, London, pp. 171–8. [12]

Epting, M. (1996) The Rotliegend play in the northeast Netherlands: an exploration evergreen. In: Glennie, K.W. and Hurst, A. (eds) *AD 1995: NW Europe's Hydrocarbon Industry.* Geological Society, London, pp. 65–7.

Epting, M., Walzebuck, J., Reijers, T., Kosters, M., Huis Int't Veld, R., Pipping, K. and Ormerod, M. *et al.* (1993) Regional trends in reservoir quality of the Rotliegende in the Dutch on- and offshore. *AAPG Bull.* **77**(9), 1621.

Erichsen, T., Helle, M., Henden, J. and Rognebakke, A. (1987) Gullfaks. In: Spencer, A.M., Campbell, C.J., Hanslien, S.H., Nelson, P.H., Nysaether, E. and Ormaasen, E.G. (eds) *Geology of the Norwegian Oil and Gas Fields.* Graham and Trotman, London, pp. 273–86. [1, 8, 12]

Erickson, J.W. and van Panhuys, C.D. (1991) The Osprey Field, Blocks 211/18a and 211/23a, UK North Sea. In: Abbotts, I.L. (ed.) *UK North Sea United Kingdom Oil and Gas Fields: 25 Years Commemorative Volume.* Memoir No. 14, Geological Society, London, p. 83. [1]

Erratt, D. (1993) Relationship between basement faulting, salt withdrawal and Late Jurassic rifting, UK Central North Sea. In: Parker, J.R. (ed.) *Petroleum Geology of Northwest Europe: Proceedings of the 4th Conference.* Geological Society, London, pp. 1211–20. [2, 6, 12]

Eschner, T.B. and Kocurek, G. (1988) Origins of relief along contacts between eolian sandstone and overlying marine strata. *AAPG Bull.* **72**(8), 932–43. [5]

Espitalié, J. (1986) Use of Tmax as a maturation index for different types of organic matter: comparison with vitrinite reflectance. In: Burrus, J. (ed.) *Thermal Modelling in Sedimentary Basins.* Editions Technip, Paris, pp. 475–96. [11]

Espitalié, J. (1987) Exploration of the Paris Basin. In: Brooks, J. and Glennie, K.W. (eds) *Petroleum Geology of Northwest Europe, Volume 1.* Graham and Trotman, London, pp. 71–86. [11]

Espitalié, J., Marquis, F., Sage, L. and Barsony, I. (1987) Geochemie organique du Bassin de Paris. *Revue de l'IFP.* **42**(3), 271–302. [11]

Espitalié, J., Ungerer, P., Irwin, H. and Marquis, F. (1988) Primary cracking of kerogens. Experimenting and modeling C1, C2–C5, C6–C15 + classes of hydrocarbons formed. In: Mattavelli, L. and Novelli, L. (eds) *Advances in Organic Geochemistry 1987 (Proc. 13th Int'l. Mtg on Org. Geochem. Venice), Part 2.* Pergamon, Oxford, pp. 893–900. [11]

Etebar, S. (1995) Captain Field: a case study. In: Petroleum Economist/Lloyds List (eds) *The North Sea: Proceedings of the 3rd Annual Conference.* [9]

Evans, D.J., Meneilly, A. and Brown, G. (1992) Seismic facies analysis of Westphalian sequences of the Southern North Sea. *Mar. Petr. Geol.* **9**, 578–89. [4]

Evans, J., Jenkins, D.G. and Gluyas, J. (1998) The Kimmeridge Oilfield. In: Underhill, J.R. (ed.) *Development and Evolution of the Wessex Basin.* Special Publication, Geological Society, London. [8]

Everlien, G. (1996) High temperature programmed pyrolysis of Paleozoic source rocks from Northern Germany and adjacent areas and its thermodynamic constraints. *Org. Geochem.* **24** (10/11), 985–98. [11]

Ewbank, G., Manning, D.A.C. and Abbott, G.D. (1993) An organic geochemical study of bitumens and their potential source rocks from the South Pennine Orefield, Central England. *Org. Geochem.* **20**(5), 579–98.

Eyers, J. (1991) The influence of tectonics on early Cretaceous sedimentation in Bedfordshire, England. *J. Geol. Soc.* **148**, 405–14. [2, 9]

Eynon, G. (1981) Basin development and sedimentation in the Middle Jurassic of the Northern North Sea. In: Illing, L.V. and Hobson, G.P. (eds) *Petroleum Geology of the Continental Shelf of North-West Europe.* Heyden, London, pp. 196–204. [8, 11]

Ewbank, G., Manning, D.A.C. and Abbott, G.D. (1995) The relationship between bitumens and mineralization in the South Pennine Orefield, Central England. *J. Geol. Soc.* **152**, 751–66. [11]

Faber, E. and Stahl, W. (1984) Geochemical surface exploration for hydrocarbons in the North Sea. *AAPG Bull.* **68**(3), 363–86.

Færseth, R.B., Oppebøen, K.A. and Saebøe, A. (1987) Trapping styles and associated hydrocarbon potential in the Norwegian North Sea. In: Spencer, A.M., Campbell, C.J., Hanslien, S.H., Nelson, P.H., Nysaether, E. and Ormaasen, E.G. (eds) *Geology of the Norwegian Oil and Gas Fields.* Graham and Trotman, London, pp. 585–97.

Fagerland, N. (1983) Tectonic analysis of a Viking Graben Border Fault. *Am. Assoc. Petr. Geol. Bull.* **6711**, 2125–36. [2]

Faleide, J.I., Gudlaugsson, S.T. and Jacquart, G. (1984) Evolution of the western Barents Sea. *Mar. Petr. Geol.* **1**(2), 123–50. [5]

Faleide, J.I., Vagnes, E. and Gudlaugsson, S.T. (1993) Late

Mesozoic–Cenozoic evolution of the southwestern Barents Sea. In: Parker, J.R. (ed.) *Petroleum Geology of Northwest Europe: Proceedings of the 4th Conference*. Geological Society, London, pp. 933–50. [12]

Falke, H. (1971) Zur Palaogeographie des kontinentalen Perms in Suddeutschland. *Abh. Hess. L-Amt Bodenforsch* **60**, 223–34. [5]

Fall, H.G., Gibb, F.G.F. and Kanaris-Sotiriou, R. (1982) Jurassic volcanic rocks of the northern North Sea. *J. Geol. Soc.* **139**, 277–92. [8]

Fält, L.M., Helland, R., Wiik Jacobsen, V. and Renshaw, D. (1989) Correlation of transgressive–regressive depositional sequences in the Middle Jurassic Brent/Vestland Group Megacycle, Viking Graben, Norwegian North Sea. In: Collinson, J.D. (ed.) *Correlation in Hydrocarbon Exploration*. Graham and Trotman, London, pp. 191–200. [12]

Farmer, R.T. and Hillier, A.P. (1991a) The Barque Field, Blocks 48/13a, 48/14, UK North Sea. In: Abbotts, I.L. (ed.) *UK North Sea United Kingdom Oil and Gas Fields: 25 Years Commemorative Volume*. Memoir No. 14, Geological Society, London, p. 395. [1, 5, 11]

Farmer, R.T. and Hillier, A.P. (1991b) The Clipper Field, Block 48/19a, 48/19c, UK North Sea. In: Abbotts, I.L. (ed.) *UK North Sea United Kingdom Oil and Gas Fields: 25 Years Commemorative Volume*. Memoir No. 14, Geological Society, London, p. 395. [11]

Farrimond, P., Comet, P., Eglinton, G., Evershed, R.P., Hall, M.A., Park, D.W., *et al.* (1984) Organic geochemical study of the Upper Kimmeridge Clay of the Dorset type area. *Mar. Petr. Geol.* **1**(4), 340–54. [11]

Farrimond, P., Eglinton, G., Brassell, S.C. and Jenkyns, H.C. (1989) Toarcian anoxic event in Europe: an organic geochemical study. *Mar. Petr. Geol.* **6**(2), 136–47. [11]

Farrimond, P., Eglinton, G., Brassell, S.C. and Jenkyns, H.C. (1990) The Cenomanian/Turonian anoxic event in Europe: an organic geochemical study. *Mar. Petr. Geol.* **7**, 75–89. [9, 11]

Farrimond, P., Bevan, J.C. and Bishop, A.N. (1996) Hopanoid hydrocarbon maturation by an igneous intrusion. *Org. Geochem.* **25**(3/4), 149–64. [11]

Feazel, C.T. and Schatzinger, R.A. (1985) Prevention of carbonate cementation in petroleum reservoirs. In: Schneidermann, N. and Harris, P.M. (eds) *Carbonate Cements*. Special Publication No. 36, Society of Economic Palaeontologists and Mineralogists, Tulsa, OK., pp. 97–106. [9]

Feazel, C.T., Keany, J. and Peterson, R.M. (1985) Cretaceous and Tertiary chalks of the Ekofisk area, central North Sea. In: Roehl, P.O. and Choquette, P.W. (eds) *Carbonate Petroleum Reservoirs*. Springer-Verlag, Berlin, pp. 495–507. [9]

Fertl, W.H. (1976) Elucidation of oil shales using geophysical well logging techniques. In: Yen, T.F. and Chilingarian, G.V. (eds) *Oil Shale*. Elsevier, Barking, pp. 199–213.

Fertl, W.H., Chillingarian, G.V. and Yen, T.F. (1986) Organic carbon content and source rock identification based on geophysical well logs. *Energy Sources* **8**, 381–437. [11]

Fewtrell, M.D., Ramsbottom, W.H.C. and Strank, A.R.E. (1981) Carboniferous. In: Murray, J.W. and Jenkins, G. (eds) *A Stratigraphic Atlas of Fossil Foraminifera*. Ellis Horwood, London. [4]

Field, J.D. (1985) Organic geochemistry in exploration of the northern North Sea. In: Thomas, B.M., Doré, A.G., Eggen, S.S., Home, P.C. and Larsen, R.M. (eds) *Petroleum Geochemistry and Exploration of the Norwegian Shelf*. Graham and Trotman, London, pp. 39–57. [11, 12]

Fielding, C.R. (1984a) Upper delta plain lacustrine and fluviolacustrine facies from the Westphalian of the Durham coalfield. *Sedimentology* **31**, 547–67. [4]

Fielding, C.R. (1984b) A coal depositional model for the Durham Coal Measures of NE England. *J. Geol. Soc.* **141**, 919–31. [4]

Fielding, C.R. (1986) Fluvial channel and overbank deposits from the Westphalian of the Durham coalfield, NE England. *Sedimentology* **33**, 119–40. [4]

Fine, S.E., Yusas, M.R. and Jorgensen, L.N. (1993) Geological aspects of horizontal drilling in chalks from the Danish Sector of the North Sea. In: Parker, J.R. (ed.) *Petroleum Geology of Northwest Europe: Proceedings of the 4th Conference*. Geological Society, London, pp. 1483–90. [9, 12]

Fine, S.E., Yusas, M.R. and Jorgensen, L.N. (1997) Geological aspects of horizontal drilling: chalk cores from the Danish Central Trough. In: Oakman, C.D., Martin, J.H. and Corbett, P. (eds) *Cores from the Northwest European Hydrocarbon Province: an Illustration of Geological Applications from Exploration to Development*. Geological Society, London pp. 225–32. [9]

Fisher, I.St.J. and Hudson, J.D. (1987) Pyrite formation in Jurassic shales of contrasting biofacies. In: Brooks, J. and Fleet, A.J. (eds) *Marine Petroleum Source Rocks*. Special Publication No. 26, Geological Society, Blackwell Scientific Publications, Oxford, pp. 69–78. [11]

Fisher, M.J. (1979) *The Triassic Palynofloral Succession in the Canadian Arctic Archipelago*. Contribution Series 5B, AASP, pp. 83–100. [7]

Fisher, M.J. (1985) Palynology of sedimentary cycles in the Mercia Mudstone and Penarth groups (Triassic) of southwest and central England. *Pollen et Spores* **27**, 95–111. [7]

Fisher, M.J. and Dunay, R.E. (1981) Palynology and the Triassic–Jurassic boundary. *Rev. Palaeobot. Palynot.* **34**, 129–35. [7]

Fisher, M.J. and Hancock, N.J. (1985) The Scalby Formation M. Jurassic, Ravenscar Group of Yorkshire: reassessment of age and depositional environment. *Proc. Yorks. Geol. Soc.* **45**, 293–8. [8]

Fisher, M.J. and Jeans, C.V. (1982) Clay mineral stratigraphy in the Permo-Triassic red-bed sequences of BNOC 72/10–1A, Western Approaches, and the south Devon coast. *Clay Minerals* **17**, 79–89. [7]

Fisher, M.J. and Mudge, D.C. (1990) Triassic. In: Glennie, K.W. (ed.) *Introduction to the Petroleum Geology of the North Sea*. Blackwell Scientific Publications, Oxford, pp. 191–218. [7]

Fitzsimmons, S. and Parnell, J. (1995) Diagenetic history and reservoir potential of Permo Triassic sandstones in the Rathlin Basin. In: Croker, P.F. and Shannon, P.M. (eds) *The Petroleum Geology of Ireland's Offshore Basins*. Special Publication No. 93, Geological Society, London, pp. 21–35. [12]

Fjaeran, T. and Spencer, A.M. (1991) Proven hydrocarbon plays, offshore Norway. In: Spencer, A.M. (ed.) *Generation, Accumulation, and Production of Europe's Hydrocarbons*. Special Publication, EAPG, Oxford University Press, Oxford. pp. 25–48. [12]

Fjeldgaard, K. (1990) Steep learning curve for horizontal wells. *Expl. Prod. Technol. Int.* 29–35. [12]

Fjeldskaar, W., Prestholm, E., Guargena, C. and Stephenson, M. (1993a) Mineralogical and diagenetic control on the thermal conductivity of the sedimentary sequences in the Bjornoya Basin, Barents Sea. In: Doré, A.G., Augustson, J.H., Hermanrud, C., Stewart, D.J. and Sylta, O. (eds) *Basin Modelling: Advances and Applications*. Special Publication No. 3, NPF, Elsevier, Amsterdam, 445–68. [11]

Fjeldskaar, W., Prestholm, E., Guargena, C. and Gravdal, N. (1993b) Isostatic and tectonic subsidence of the Egersund Basin. In: Doré, A.G., Augustson, J.H., Hermanrud, C., Stewart, D.J. and Sylta, O. (eds) *Basin Modelling: Advances and Applications*. Special Publication No. 3, NPF, Elsevier, Amsterdam, pp. 549–62. [11]

Fleet, A.J., Kelts, K. and Talbot, M.R. (eds) (1988) *Lacustrine Petroleum Source Rocks*. Special Publication 26, Geological Society, Blackwell Scientific Publications, Oxford, pp. 391. [11]

Fleet, A.J., Clayton, C.J., Jenkyns, H.C. and Parkinson, D.N. (1987) Liassic source rock deposition in western Europe. In: Brooks, J and Glennie, K.W. (eds) *Petroleum Geology of North West Europe*. Graham and Trotman, London, pp. 59–70. [8, 11]

Flinn, D. (1969) A geological interpretation of the aeromagnetic maps of the continental shelf around Orkney and Shetland. *Geol. J.* **6**, 279–92.

Flinn, D. (1993) New evidence that the high temperature

hornblende-schists below the Shetland ophiolite include basic igneous rocks intruded during obduction of the 'cold' ophiolite. *Scott. J. Geol.* **292**, 159–65. [2]

Flinn, D., Miller, J.A. and Roddom, D. (1991) The age of the Norwick hornblende schists of Unst and Fetlar and the obduction of the Shetland Ophiolite. *Scott. J. Geol.* **271**, 11–19. [2]

Follows, E. (1997) Integration of inclined pilot hole core with horizontal image logs to appraise an aeolian reservoir, Auk Field, Central North Sea Petroleum. *Geoscience* **3**(1), 43–55. [1]

Fontaine, J.M., Guastella, G., Jonalt, P. and de la Vega, P. (1993) F15-A: a Triassic gas field on the eastern limit of the Dutch Central Graben. In: Parker, J.R. (ed.) *Petroleum Geology of Northwest Europe: Proceedings of the 4th Conference*. Geological Society, London, pp. 583–94. [7]

Forbes, P.L., Ungerer, P.M., Kuhfuss, A.B., Riis, F. and Eggens, S. (1991) Compositional modelling of petroleum generation and expulsion: trial application to a local mass balance in the Smorbukk Sor Field, Haltenbanken area, Norway. *AAPG Bull.* **75**(5), 873–93. [11]

Ford, M.E., Bains, A.J. and Tarron, R.D. (1974) Log analysis by linear programming—an application to the exploration for salt cavity storage locations. In: *Transactions 3rd Europe Formation Evaluation Symposium*. Society of Professional Well Log Analysts, Paper B. [6]

Forney, G.G. (1975) Permo-Triassic sea level changes. *J. Geol.* **83**, 773–9. [7]

Forsberg, A. and Bjorøy, M. (1983) A sedimentological and organic geochemical study of the Botneheia formation, Svalbard with special emphasis on the effects of weathering on the organic matter in the shales. In: Bjorøy, M., Albrecht, P., Cornford, C., de Groot, K., Eglinton, G., Galimov, E., *et al.* (eds) *Advances in Organic Geochemistry, 1981*. John Wiley, Chichester, pp. 60–68. [11]

Fossen, H. (1989) Indication of transpressional tectonics in the Gullfaks oilfield, northern North Sea. *Mar. Petr. Geol.* **6**, 22–30. [12]

Foster, P.T. and Rattey, P.R. (1993) The evolution of a fractured chalk reservoir: Machar Oilfield, UK North Sea. In: Parker, J.R. (ed.) *Petroleum Geology of Northwest Europe: Proceedings of the 4th Conference*. Geological Society, London, pp. 1445–52. [1, 2, 9, 10, 12]

Fowler, C. (1975) The geology of the Montrose field. In: Woodland, A.W. (ed.) *Petroleum and the Continental Shelf of Northwest Europe*, Vol. I, *Geology*. Applied Science Publishers, Barking, pp. 467–77. [1, 12]

France, D.S. (1975) The geology of the Indefatigable gas field. In: Woodland, A.W. (ed.) *Petroleum and the Continental Shelf of Northwest Europe*. Vol. I, *Geology*, Applied Science Publishers, Barking, pp. 233–9. [1]

Frandsen, N., Vejboek, O.V., Møeller, J.J. and Michelsen, O. (1987) A dynamic geological model of the Danish Central Trough during the Jurassic–Early Cretaceous. In: Brooks, J. and Glennie, K.W. (eds) *Petroleum Geology of North West Europe*. Graham and Trotman, London, pp. 453–68. [9]

Franssen, R.C.M.W., Brint, J.F., Sleeswijk Visser, T.J. and Beecham, A. (1993) Fracture characterization and diagenesis in the Clipper Field, Sole Pit Basin, Southern North Sea. *AAPG Bull.* **77**(9), 16–23. [5]

Fraser, A.J. and Gawthorpe, R.L. (1990) Tectono-stratigraphic development and hydrocarbon habitat of the Carboniferous in Northern England. In: Hardman, R.P.F. and Brooks, J. (eds) *Tectonic Events Responsible for Britain's Oil and Gas Reserves*. Special Publication No. 55, Geological Society, London, pp. 49–86. [2, 4, 12]

Fraser, A.J., Nash, D.F., Steele, R.P. and Ebdon, C.C. (1990) A regional assessment of the intra-carboniferous play of Northern England. In: Brooks, J. (ed.) *Classic Petroleum Provinces*. Special Publication No. 50, Geological Society, London, pp. 417–40. [4, 11, 12]

Fraser, A.R. and Tonkin, P.C. (1991) The Glamis Field, Block 16/21a, UK North Sea. In: Abbotts, I.L. (ed.) *UK North Sea United Kingdom Oil and Gas Fields: 25 Years Commemorative Volume*. Memoir No. 14, Geological Society, London, pp. 317–22. [1, 12]

Freer, G., Hurst, A. and Middleton, P. (1996) Upper Jurassic sandstone reservoir quality and distribution on the Fladen Ground Spur. In: Hurst, A., Johnson, H.D., Burley, S.D., Canham, A.C. and Mackertich, D.S. (eds) *Geology of the Humber Group: Central Graben and Moray Firth, UKCS*. Special Publication No. 114, Geological Society, London, pp. 235–49. [8]

Friedman, G.M. (1995) Discussion of vitrinite reflectivity and the structure and burial history of the Old Red Sandstone of the Midland Valley of Scotland. *J. Geol. Soc.* **152**(1), 196–7. [11]

Frikken, H. (1996) Subhorizontal drilling: remedy for underperforming Rotliegend gasfields, central offshore Netherlands. In: Rondeel, H.E., Batjes, D.A.J. and Nieuwenhuis, W.H. (1996) *Geology of Gas and Oil under the Netherlands*. Kluwer Academic Publishers, Dordrecht. [5]

Fritsen, A. and Corrigan, T. (1990) Establishment of a geological fracture model for dual porosity simulation of the Ekofisk Field. In: Buller, A.T., Berg, E., Hjelmelaud, O., Kleppe, J., Torsæter, O. and Aasen, J.O. (eds) *North Sea Oil and Gas Reservoirs II*. Norwegian Institute of Technology, Graham and Trotman, London, pp. 173–84. [9]

Frodesen, S. (1979) *Lithology: Wells 9/4-1, 9/4-2 and 9/4-3*. Paper No. 24, NPD, Stavanger, Norway. [7]

Frodesen, S., Moe, A., Ofstad, K., Ormaasen, E., Sjulsen, S.-E. and Ulleberg, K. (1981) *The Balder Area*. Lithology Paper No. 28, NPD, Stavanger, Norway. [1]

Frostick, L.E., Reid, I., Jarvis, J. and Eardley, H., (1988) Triassic sediments of the Inner Moray Firth, Scotland: early rift deposits. *J. Geol. Soc.* **145**, 235–48. [5, 7]

Frostick, L.E., Linsey, T.K. and Reid, I. (1992) Tectonic and climatic control of Triassic sedimentation in the Beryl Basin, northern North Sea. *J. Geol. Soc.* **149**, 13–26. [7]

Füchtbauer, H. (1968) Carbonate sedimentation and subsidence in the Zechstein Basin northern Germany. In: Muller, G. and Friedman, G.M. (eds) *Recent Developments in Carbonate Sedimentology in Central Europe*. Springer-Verlag, Berlin, pp. 196–204. [6]

Füchtbauer, H. and Peryt, T.M. (eds) (1980) *The Zechstein Basin with Emphasis on Carbonate Sequences*. Contr. Sedimentology No. 9, Schweitzerbart'sche Verlagsbuchhandlung, Stuttgart, 328 pp.

Fuglewicz, J. (1987) Upper Permian assemblages of Gondwana miospores in sediments of Lower Buntsandstein. *Przgl. Geol.* **11**, 415. [7]

Fuller, J.G.C.M. (1975) Jurassic source rock potential—and hydrocarbon correlation, North Sea. *Proceedings of the Symposium on Jurassic—Northern North Sea, Norwegian Petroleum Society meeting, 1975*. MPS Norway, pp. 11–38. [11]

Fuller, J.G.C.M. (1980) Progress report on fossil fuels—exploration and exploitation. *Proc. Yorks Geol. Soc.* **42**, 581–93. [11]

Fyfe, J.A., Long, D. and Evans, D. (1993) *The Geology of the Malin–Hebrides Sea Area*. BGS UK Offshore Regional Report, HMSO, 91 pp. [11]

Gaarenstroom, L., Tromp. R.A.J., de Jong, M.C. and Brandenburg, A.M. (1993) Overpressures in the Central North Sea: implications for trap integrity and drilling safety. In: Parker. J.R. (ed.) *Petroleum Geology of Northwest Europe: Proceedings of the 4th Conference*. Geological Society, London, pp. 1305–13. [11, 12]

Gabrielsen, R.H., Ekern, O.F. and Edvardsen, A. (1986) Structural development of hydrocarbon traps, block 2/2, Norway. In: Spencer, A.M., Campbell, C.J., Hanslien, S.H., Nelson, P.H., Nysaether, E. and Ormaasen, E.G. (eds) *Habitat of Hydrocarbons on the Norwegian Continental Shelf*. Graham and Trotman, London, pp. 129–41.

Gabrielsen, R.H., Ulvoen, S., Elvsborg, A. and Fredrik, O. (1985) The geological history and geochemical evaluation of Block 2/2, Offshore Norway. In: Thomas, B.M., Doré, A.G., Eggen, S.S., Home, P.C. and Larsen, R.M. (eds) *Petroleum Geochemistry in Exploration of the Norwegian Shelf.* Graham and Trotman, London, pp. 165–78. [11]

Gabrielsen, R.H., Faerseth, R.B., Steel, R.J., Idil, S. and Klovjan, O.S. (1990) Architectural styles of basin fill in the northern Viking Graben. In: Blundell, D.J. and Gibbs, A.D. (eds) *Tectonic Evolution of the North Sea Rifts.* Oxford Scientific Publications, Oxford, pp. 158–79. [7]

Gage, M. (1980) A review of the Viking Gas Field. In: Halbouty, M.T. (ed.) *Giant Oil and Gas Fields of the Decade, 1968–1978.* Memoir No. 30, AAPG, Tulsa, OK., pp. 39–55. [1, 5]

Gale, A.S. (1980) Penecontemporaneous folding, sedimentation and erosion in Campanian chalk near Portsmouth, England. *Sedimentology* **27**, 137–51. [9]

Gale, A.S., Jenkyns, H.C., Kennedy, W.J. and Cornfield, R.M. (1993) Chemostratigraphy versus biostratigraphy: data from around the Cenomanian–Turonian boundary. *J. Geol. Soc.* **150**, 29–32. [9]

Gallois, R.W. (1965) *British Regional Geology: the Wealden District*, 4th edn. BGS, HMSO, London. [9]

Gallois, R.W. (1976) Coccolith blooms in the Kimmeridge Clay and origin of North Sea oil. *Nature* **259**, 474–5. [11]

Gallois, R.W. (1994) *Geology of the Country around King's Lynn and the Wash.* Geological Survey Memoir, HMSO, London, 210 pp. [9]

Galloway, R.W. (1965) Late Quaternary climates in Australia. *J. Geol.* **73**, 603–18. [5]

Galloway, W.E. (1989a) Genetic stratigraphic sequences in basin analysis I: architecture and genesis of flooding surface bounded depositional units. *AAPG Bull.* **73**, 125–42. [8, 9]

Galloway, W.E. (1989b) Genetic stratigraphic sequences in basin analysis II: application to the northwest Gulf of Mexico. *AAPG Bull.* **73**, 143–54. [8]

Galloway, W., Garber, J.L., Xijin Lu and Sloan, B.J. (1993) Sequence stratigraphic framework of the Cenozoic fill, Central and Northern North Sea Basin. In: Parker, J.R. (ed.) *Petroleum Geology of Northwest Europe: Proceedings of the 4th Conference.* Geological Society, London, pp. 33–44. [10, 12]

Gardiner, P.R.R. and Sheridan, D.J.R. (1981) Tectonic framework of the Celtic Sea and adjacent areas with special reference to the location of the Variscan Front. *J. Struct. Geol.* **3**, 317–31. [12]

Garland, C.R. (1991) The Amethyst Field, Blocks 47/8a, 47/13a, 47/14a, 47/15a, UK North Sea. In: Abbotts, I.L. (ed.) *UK North Sea United Kingdom Oil and Gas Fields: 25 Years Commemorative Volume.* Memoir No. 14, Geological Society, London, p. 387. [1, 5, 11]

Garland, C.R. (1993) Miller Field: reservoir stratigraphy and its impact on development. In: Parker, J.R. (ed.) *Petroleum Geology of Northwest Europe: Proceedings of the 4th Conference.* Geological Society, London, pp. 401–4. [1, 8, 12]

Garrett, S., Guy, M. and Jones, L. (1998) Britannia Field, UK Central North Sea: Reservoir Description and Development. In: Boldy, S.A.R. *et al.* (eds) *Abstract of the 5th Conference of the Petroleum Geology of NW Europe*, London, p. 197. [9]

Garrigues, P., Oudin, J.L., Parlanti, E., Monin, J.C., Robcis, S. and Bellocq, J. (1989) Alkylated phenanthrene distribution in artificially matured kerogens from Kimmeridge Clay and the Brent Formation North Sea. In: Durand, B. and Behar, F. (eds) *Advances in Organic Geochemistry. Org. Geochem.* **16**(1–3), 167–73. [11]

Garrison, R.E. and Kennedy, W.J. (1977) Origin of solution seams and flaser structures in Upper Cretaceous chalks of southern England. *Sedimentary Geol.* **19**, 107–37. [9]

Gast, R.E. (1988) Rifting in Rotliegenden Niedersachsens. *Geowissenschaften* **6**, 115–22. [2, 5]

Gast, R.E. (1991) The perennial Rotliegend saline lake in NW Germany. *Geol. Jahrbuch A* **119**, 25–59. [5]

Gast, R. (1993a) Sequence stratigraphy of the North German Rotliegende. *AAPG Bull.* **77**(9), 1624. [5]

Gast, R. (1993b) Sequenzanalyse von ao elischen Abfolgen im Rotliegenden und deren Verzahnung mit Küstensedimenten. *Geol. Jahrbuch A* **131**, 117–39.

Gatliff, R.W., Richards, P.C., Smith, K., Graham, C.C., McCormack, M., Smith, N.J.P. *et al.* (1994) *The Geology of the Central North Sea.* BGS UK Offshore Regional Report, HMSO, London. [9, 11]

Gaupp, R., Matter, A., Platt, J., Ramseyer K. and Walzebuck, J. (1993) Diagenesis and fluid evolution of deeply buried Permian Rotliegende gas reservoirs, Northwest Germany. *AAPG Bull.* **77** (7), 1111–28. [5]

Gauthier, B.D.M. and Lake, S.D. (1993) Probabilistic modelling of faults below the limit of seismic revolution in Pelican Field, North Sea, offshore United Kingdom. *AAPG Bull.* **77**(5), 761–78. [11]

Gawthorpe, R.L. and Hurst, J.M. (1993) Transfer zones in extensional basins: their structural style and influence on drainage basin and stratigraphy. *J. Geol. Soc.* **150**, 1137–52. [8]

Gayer, R., Fowler, R. and Davies, G. (1997) Coal rank variations with depth related to major thrust detatchments in the South Wales coalfield; implications for fluid flow and mineralization. In: Gayer, R. and Pesel, J. (eds) *European Coal Geology and Technology.* Geological Society, Special Publication 125, London, pp. 161–78. [11]

Gdula, J.E. (1983) Reservoir geology, structural framework and petrophysical aspects of the De Wijk gas field. *Geol. Mijnbouw* **62**, 191–202. [7]

Gebhardt, U. (1994) Zur Genese der Rotliegend-Salinare in der Norddeutschen Senke Oberrotliegend II. *Perm Freiberger Forschungsheft* **C452**, 3–22. [5]

Gebhardt, U., Schneider, J. and Hoffmann, N. (1991) Modelle zur Stratigraphie und Beckenentwicklung im Rotliegenden der Norddeutschen Senke. *Geol. Jahrbuch* **127**, 405–27. [5]

Geiger, M.E. and Hopping, C.A. (1968) Triassic stratigraphy of the Southern North Sea Basin. *Phil. Trans. Roy. Soc. London Ser. B* **154**, 1–36. [1, 7]

Geikie, Sir A. (1879) The Old Red Sandstone of Western Europe. *Trans. Roy. Soc. Edinburgh* **28**, 345. [2]

Geil, K. (1991) The development of salt structures in Denmark and adjacent areas: the role of basin floor dip and differential pressure. *First Break* **9**, 467–83. [6]

Geil, K. (1992) Reply by the author to Comment by Madirazza 1992. *First Break* **10**, 134–5. [6]

Geluk, M.C., Plomp, A. and van Doorn, T.H.M. (1996) Development of the Permo-Triassic succession in the basin fringe area, southern Netherlands. In: Rondeel, H.E., Batjes, D.A.J. and Nieuwenhius, W.H. (eds) *Geology of Oil and Gas under the Netherlands.* Royal Geological and Mining Society, the Netherlands, Kluwer Academic Publishers, Dordrecht, pp. 57–78. [6, 7]

Geochem Group, Ltd. (1992) *A Geochemical Atlas of North Sea Oils.* Report, Geochem Group Ltd. Chester, UK.

George, G.T. and Berry, J.K. (1993) A new lithostratigraphy and depositional model for the Upper Rotliegend of the UK Sector of the Southern North Sea. In: North, C.P. and Prosser, D.J. (eds) *Characterization of Fluvial and Aeolian Reservoirs.* Special Publication No. 73, Geological Society, London, pp. 291–319. [5, 12]

George, G.T. and Berry, J.K. (1994) A new palaeogeographic and depositional model for the Upper Rotliegend, offshore The Netherlands. *First Break* **12**(3) 147–58. [5]

George, G.T. and Berry, J.K. (1997) Permian Upper Rotliegend synsedimentary tectonics, basin development and palaeogeography of the southern North Sea. In: Ziegler, P., Turner, P. and Daines, S.R. (eds) *Petroleum Geology of the Southern North Sea.* Special Publication No. 123, Geological Society, London pp. 31–61. [5]

George, S.C. (1992) Effect of igneous intrusion on the organic geochemistry of a siltstone and an oil shale horizon in the Midland Valley of Scotland. *Org. Geochem.* **18**(5), 705–23. [11]

George, T.N., Johnson, G.A.L., Mitchell, M., Prentice, J.E., Ramsbottom, W.H.C., Sevastopulo, G.D. *et al.* (1976) *A Correlation of Dinantian Rocks in the British Isles.* Special Report No. 7, Geological Society, London, 87 pp. [4]

Gérard, J. and Buhrig, C. (1990) Seismic facies of the Permian section of the Barents Shelf: analysis and interpretation. *Mar. Petr. Geol.* **7**, 234–52.

Gérard, J. Wheatley, T.J., Ritchie, J.S., Sullivan, M. and Bassett, M.G. (1993) Permo-Carboniferous and older plays, their historical development and future potential. In: Parker, J.R. (ed.) *Petroleum Geology of Northwest Europe: Proceedings of the 4th Conference.* Geological Society, London, pp. 641–50. [12]

Gervirtz, J.L., Carey, B.D. and Blanco, S.R. (1985) Regional geochemical analysis of the southern portion of the Nowegian sector of the North Sea. In: Thomas, B.M., Doré, A.G., Eggen, S.S., Home, P.C. and Larsen, R.M. (eds) *Petroleum Geochemistry in Exploration of the Norwegian Shelf.* Graham and Trotman, London, pp. 247–62. [11]

Gibbons, K., Hellem, T., Kjemperud, A., Nio, S.D. and Vebenstad, K. (1993) Sequence architecture, facies development and carbonate-cemented horizons in the Troll Field reservoir, offshore Norway. In: Ashton, M. (ed.) *Advances in Reservoir Geology.* Special Publication No. 69, Geological Society, London, pp. 1–31. [12]

Gibbons, M.J. (1987) The depositional environment and petroleum geochemistry of the marl slate/Kupferschiefer. In: Brooks, J. and Fleet, A.J. (eds) *Marine Petroleum Source Rocks.* Blackwell, Oxford, pp. 249–50. [11]

Gibbons, W. and Harris, A.L. (1994) Precambrian rocks in Britain and Ireland: an overview. In: Gibbons, W. and Harris, A.L. (eds) *A Revised Correlation to Precambrian Rocks in British Isles.* Special Publication No. 22, Geological Society, London, pp. 1–5. [2]

Gibbs, A.D. (1984a) Structural evolution of extensional basin margins. *J. Geol. Soc.* **141**(4), 609–20. [2]

Gibbs, A.D. (1984b) Clyde Field growth fault secondary detachment above basement faults in North Sea. *AAPG Bull.* **68**, 1029–39. [1, 2, 12]

Gibbs, A.D. (1987) Deep seismic profiles in the northern North Sea. In: Brooks, J. and Glennie, K. (eds) *Petroleum Geology of North West Europe*, Vol. 2. Graham and Trotman, London, pp. 1025–8.

Giffard, H.P. (1923) The recent search for oil in Great Britain. *Trans. Inst. Min. Eng.* **65**(1922–3), 221–50. [4]

Giles, M. R. *et al.* (1992) The reservoir properties and diagenesis of the Brent Group: a regional perspective. In: Morton, A.C., Haszeldine, R.S., Giles, M.R. and Brown, S. (eds) *Geology of the Brent Group.* Special Publication No. 61, Geological Society, London, pp. 289–327. [12]

Gjelberg, J., Dreyer, T., Hoie, A., Tjelland, T. and Lilleng, T. (1987) Late Triassic to Mid Jurassic sandbody development on the Barents and Mid-Norwegian shelf. In: Brooks, J. and Glennie, K.W. (eds) *Petroleum Geology of North West Europe.* Graham and Trotman, London, pp. 1105–29. [8]

Glasmann, J.R. (1992) The fate of feldspar in Brent Group reservoirs, North Sea: a regional synthesis of diagenesis in shallow, intermediate and deep burial environments. In: Morton, A.C., Haszeldine, RS., Giles, M.R. and Brown, S. (eds) *Geology of the Brent Group.* Special Publication No. 61, Geological Society, London, pp. 329–50. [12]

Glasmann, J.R. and Wilkinson, G.C. (1993) Clay mineral stratigraphy of Mesozoic and Palaeozoic red beds, Northern North Sea. In: Parker, J.R. (ed). *Petroleum Geology of Northwest Europe: Proceedings of the 4th Conference.* Geological Society, London, pp. 625–36 [3]

Glasmann, J.R., Clark, R.A., Larter, S., Briedis, N.A. and Lundegard, P.D. (1989) Diagenesis and hydrocarbon accumulation, Brent Sandstone Jurassic, Bergen High area, North Sea. *AAPG Bull.* **73**(11), 1341–60. [11]

Glennie, K.W. (1970) Desert sedimentary environments. Developments in Sedimentology 14, Elsevier, Amsterdam, 222 pp. [5]

Glennie, K.W. (1972) Permian Rotliegendes of North-West Europe interpreted in light of modern desert sedimentation studies. *AAPG Bull.* **56**(6), 1048–71. [1, 5, 7]

Glennie, K.W. (1983a) Early Permian Rotliegendes palaeowinds of the North Sea. *Sedimentary Geol.* **34**, 245–65. [5]

Glennie, K.W. (1983b) Lower Permian Rotliegend sedimentation in the North Sea area. In: Brookfield, M.E. and Ahlbrandt, T.S. (eds) *Eolian Sediments and Processes.* Elsevier, Amsterdam, pp. 521–41. [5]

Glennie, K.W. (1986) Development of NW Europe's Southern Permian gas basin. In: Brooks, J., Goff, J. and van Hoorn, B. (eds) *Habitat of Palaeozoic Gas in N.W. Europe.* Special Publication No. 23, Geological Society, London, pp. 3–22. [5, 11]

Glennie, K.W. (1987) Desert sedimentary environments, present and past: a summary. *Sedimentary Geol.* **50**, 135–65. [5]

Glennie, K.W. (1989) Some effects of the Late Permian Zechstein transgression in northwestern Europe. In: Boyle, R.W., Brown, A.C., Jefferson, C.W., Jowett, E.C. and Kirkham, R.V. (eds) *Sediment-hosted Copper Deposits.* Special Paper No. 36, Geological Association of Canada, Ottawa, pp. 557–65. [5]

Glennie, K.W. (1990) Lower Permian—Rotliegend. In: Glennie, K.W. (ed.) *Introduction to the Petroleum Geology of the North Sea.* Blackwell Scientific Publications, Oxford, pp. 120–49. [6]

Glennie, K.W. (1992) Some geological advances resulting from North Sea exploration. *First Break* **10**, 161–73. [12]

Glennie, K.W. (1994) Quaternary dunes of SE Arabia and Permian (Rotliegend) dunes of NW Europe: some comparisions. *Zbl. Geol. Paläont.* Stuttgart, pp. 1199–215. [5]

Glennie, K.W. (1997a) History of exploration in the Southern North Sea. In: Ziegler, K., Turner, P. and Daines, S.R. (eds) *Petroleum Geology of the Southern North Sea.* Special Publication No. 123, Geological Society, London, pp. 5–16. [1, 2]

Glennie, K.W. (1997b) Recent advances in understanding the southern North Sea Basin: a summary. In: *Petroleum Geology of the Southern North Sea.* Special Publication No. 123, Geological Society, London, pp. 17–29. [2, 5]

Glennie, K.W. and Armstrong, L.A. (1991) The Kittiwake Field, Block 21/18, UK North Sea. In: Abbotts, I.L. (ed.) *UK North Sea United Kingdom Oil and Gas Fields: 25 Years Commemorative Volume.* Memoir No. 14, Geological Society, London, pp. 339–45. [1, 6, 7, 8, 12]

Glennie, K.W. and Boegner, P. (1981) Sole Pit inversion tectonics. In: Illing, L.V. and Hobson, G.P. (eds) *Petroleum Geology of the Continental Shelf of North-West Europe.* Institute of Petroleum, Heyden, London, pp. 110–20. [2, 5, 11, 12]

Glennie, K.W. and Buller, A.T. (1983) The Permian Weissliegend of NW Europe: the partial deformation of aeolian dune sands caused by the Zechstein transgression. *Sedimentary Geol.* **35**, 43–81. [2, 5, 6, 7]

Glennie, K.W. and Hurst, A. (ed.) (1996a) *AD 1995: NW Europe's Hydrocarbon Industry.* Geological Society, London. [1]

Glennie, K.W. and Hurst, A. (1996b) Hydrocarbon exploration and production in NW Europe: an overview of some key factors. In: Glennie, K.W. and Hurst, A. (eds) *AD 1995: NW Europe's Hydrocarbon Industry.* Geological Society, London, pp. 5–14. [1]

Glennie, K.W. and Provan, D.M.J. (1990) Lower Permian Rotliegend reservoir of the Southern North Sea Gas Province. In: Brooks, J. (ed.) *Classic Petroleum Provinces.* Special Publication No. 50, Geological Society, London, pp. 399–416. [5, 12]

Glennie, K.W., Mudd, G.C. and Nagtegaal, P.J.C. (1978) Depositional environment and diagenesis of Permian Rotliegendes sandstones in Leman Bank and Sole Pit areas of the UK southern North Sea. *J. Geol. Soc.* **135**, 25–34. [5]

Glennie, K.W., Brooks, J. and Brooks, J.R.V. (1987) Hydrocarbon exploration and geological history of North-West Europe. In: Brooks, J. and Glennie, K. (eds) *Petroleum Geology of North West Europe.* Graham and Trotman, London, pp. 1–10. [2]

Glover, B.W., Leng, M.J. and Chisholm, J.I. (1996) A second major fluvial sourceland for the Silesian Pennine Basin of Northern England. *J. Geol. Soc. London.* **153**, 901–906. [4]

Goff, J.C. (1983) Hydrocarbon generation and migration from the E. Shetland Basin and Viking Graben of the Northern North Sea. *J. Geol. Soc.* **140**, 445–74. [7, 11, 12]

Goff, J.C. (1984) Hydrocarbon habitat of East Shetland Basin and North Viking Graben of the Northern North Sea. Fossil Fuels of Europe Conference, Geneva, 1984. *Abstract in AAPG Bull.* **68**(6), 794. [11]

Goggin, D.J., Chandler, M.A., Kocurek, G.A. and Lake, L.W. (1986) Patterns of permeability in eolian deposits. In: *5th Symposium on EOR, Apr.*, SPE/DOE, 14893, Tulsa, OK., pp. 181–8. [5]

Goh, L.S. (1996) The Logger oil field Netherlands: reservoir mapping with amplitude anomalies. In: Rondeel, H.E., Batjes, D.A.J. and Nieuwenhius, W.H. (eds) *Geology of Oil and Gas under the Netherlands.* Royal Geological and Mining Society, the Netherlands, Kluwer Academic Publishers, Dordrecht, pp. 255–63. [1, 9, 12]

Gold, T. (1985) The origin of natural gas and petroleum and the prognosis for future supplies. *Ann. Rev. Energy.* **10**, 53–77.]11]

Gold, T. and Held, M. (1987) Ne-N-CH4 systematics in natural gases of Texas and Kansas. *J. Petr. Geol.* **10**(4), 415–24. [11]

Gold, T. and Soter, S. (1982) Abiogenic methane and the origin of petroleum. *Energy Exploration and Exploitation.* **1**(2), 89–104. [11]

Goldsmith, P.J., Rich, B. and Standring, J. (1995) Triassic correlation and stratigraphy in the south Central Graben, UK North Sea. In: Boldy, S.A.R. (ed.) *Permian and Triassic Rifting in North West Europe.* Special Publication No. 92, Geological Society, London, pp. 123–43. [7]

Goodall, I.G., Harwood, G.M., Kendall, A.C. and McKie, T. (1992) Discussion on sequence stratigraphy of carbonate–evaporite basins: models and application to the Upper Permian Zechstein of northeast England and adjoining North Sea. *J. Geol. Soc.* **149**, 1050–4. [6]

Goodarzi, F. (1985) Reflected light microscopy of chitinozoan fragments. *Mar. Petr. Geol.* **2**(1), 72–8. [11]

Goodarzi, F. and Higgins, A.C. (1987) Optical properties of scolecodonts and their use as indicators of thermal maturity. *Marine and Petroleum Geology* **4**(4), 353–9. [11]

Goodarzi, F. and Norford, B.S. (1985) Graptolites as indicators of the temperature history of rocks. *J. Geol. Soc.* **142**, 1089–99. [11]

Goodchild, M. and Bryant, P. (1986) The geology of the Rough Field. In: Brooks, J., Goff, J. and van Hoorn, B. *Habitat of Palaeozoic Gas in N.W. Europe.* Special Publication No. 23, Geological Society, London, pp. 223–35. [1, 5]

Gormly, J.R., Buck, S.P. and Chung, H.M. (1994) Oil–source rock correlation in the North Viking Graben. *Adv. Org. Geochem.* **22** (3–5), 403–15. [11]

Gowers, M.B. and Sæbøe, A. (1985) On the structural evolution of the Central Trough in the Norwegian and Danish sectors of the North Sea. *Mar. Petr. Geol.* **2**(4), 298–318. [9, 11]

Gowers, M.B., Holtar, E. and Swensson, E. (1993) The structure of the Norwegian Central Trough Central Graben Area. In: Parker, J.R. (ed.) *Petroleum Geology of Northwest Europe: Proceedings of the 4th Conference.* Geological Society, London, pp. 1245–54. [2, 6, 7]

Gowland, S. (1996) Facies characteristics and depositional models of highly bioturbated shallow marine siliciclastic strata: an example from the Fulmar Formation Late Jurassic, UK Central Graben. In: Hurst, A., Johnson, H.D., Burley, S.D., Canham, A.C. and Mackertich, D.S. (eds) (1996) *Geology of the Humber Group: Central Graben and Moray Firth, UKCS.* Special Publication No. 114, Geological Society, London, pp. 185–214. [8]

Grabowski, G.J., Jr (1984) Generation and migration of hydrocarbons in Upper Createaceous Austin Chalk, south-central Texas. In: Palacas, J.G. (ed.) *Petroleum Geochemistry and Source Rock Potential of Carbonate Rocks.* Studies in Geology No. 18, AAPG, Tulsa, OK., p. 116. [9]

Graciansky, P.C. de *et al.* (1984) Ocean-wide stagnation episode in the late Cretaceous. *Nature* **308**, 346–9. [9]

Gradijan, S.J. and Wiik, M. (1987) Statfjord Nord. In: Spencer, A.M., Campbell, C.J., Hanslien, S.H., Nelson, P.H., Nysaether, E. and Ormaasen, E.G. (eds) *Geology of the Norwegian Oil and Gas Fields.* Graham and Trotman, London, pp. 341–50. [1, 12]

Gradstein, F.M. and Ogg, J. (1996) A Phanerozoic time scale. *Episodes* **19**, 3–5. [5]

Gradstein, F.M., Agterberg, F.P., Ogg, J.G., Hardenbol, J., van Veen, P., Thierry, J., *et al.* (1995) A Triassic, Jurassic and Createaceous Time Scale. In: Berggren, W.A., Kent, D.V., Aubry, M. and Hardenbol, J. (eds) *Geochronology, Timescales and Global Stratigraphic Correlation.* Special Publication No. 54. SEPM, Tulsa, OK., pp. 95–126. [7]

Gralla, P. (1988) Das Oberrotliegende in NW Deutschland Lithostratigraphie und Faziesanalyse. *Geol. Jahrbuch A* **106**, 3–59. [5]

Gralla, P. (1993) Structure and facies development of the Dutch/North German Rotliegende Basin. *AAPG Bull.* **77**(9), 1627. [5]

Grantham, P.J., Posthuma, J. and Groot, K. de, (1980) Variation and significance of the C27 and C28 triterpane content of a North Sea core and various North Sea crude oils. In: Douglas, A.G. and Maxwell, J.R. (eds) *Advances in Organic Geochemistry, 1979.* Pergamon, Oxford, pp. 29–38. [11]

Graue, E., Helland-Hausen, W., Johnson, J., Lomo, L., Nottvedt, A., Rønning, K. *et al.* (1987) Advance and retreat of the Brent delta system, Norwegian North Sea. In: Brooks, J. and Glennie, K.W. (eds) *Petroleum Geology of North West Europe.* Graham and Trotman, London, pp. 915–37. [8, 12]

Graverson, O. (1994) Interrelationship between basement structure and salt tectonics in the salt dome province, Danish Central Graben, North Sea (abstract). In: Alsop, A. (convenor) *Salt Tectonics: Programme and Abstracts.* Geological Society, London, pp. 13. [6]

Gray, D.I. (1987) Troll. In: Spencer, A.M., Campbell, C.J., Hanslien, S.H., Nelson, P.H., Nysaether, E. and Ormaasen, E.G. (eds) *Geology of the Norwegian Oil and Gas Fields.* Graham and Trotman, London, pp. 389–401. [1, 8, 11, 12]

Gray, I. (1975) Viking gas field. In: Woodland, A.W. (ed.) *Petroleum and the Continental Shelf of Northwest Europe.* Vol. I, *Geology.* Applied Science Publishers, Barking, pp. 241–9. [1]

Gray, W.D.T. and Barnes, G. (1981) The Heather oil field. In: Illing, L.V. and Hobson, G.P. (eds) *Petroleum Geology of the Continental Shelf of North-West Europe.* Heyden, London, pp. 335–41. [1, 12]

Grayson, R.F. and Oldham, L. (1987) A new structural framework for the northern British Dinantian as a basis for oil, gas and mineral exploration. In: Miller, J., Adams, A.E. and Wright, V.P. (eds) *European Dinantian Environments.* John Wiley, Chichester, pp. 33–60. [4]

Green, P.F. (1986) On the thermotectonic evolution of northern England: evidence from fission-track analysis. *Geol. Mag.* **123**, 493–506. [11, 12]

Green, P.F. (1989) Thermal and tectonic history of the East Midlands Shelf onshore UK and surrounding regions assessed by apatite fission track analysis. *J. Geol. Soc.* **146**(5), 755–73. [11]

Green, P.F., Duddy, I.R., Bray, R.J. and Catlin, T.J. (1989) Apatite fission track analysis and 3-D basin modelling in the study of complex basin history. *Proc. NPF Conference on Structural and Tectonic Modelling and its Application to Petroleum Geology, Stavanger, Norway, 18–20 October, 1989.* [11]

Green, P.F., Duddy, I.R., Bray, R.J. and Lewis, C.L.E. (1993) Elevated palaeotemperatures prior to Early Tertiary cooling throughout the UK region: implications for hydrocarbon generation. In: Parker, J.R. (ed.) *Petroleum Geology of Northwest Europe: Proceedings of the 4th Conference*. Geological Society, London, pp. 1067–74. [11]

Green, P.F., Duddy, I.R. and Bray, R.J. (1995a) Further discussion on Mesozoic cover over northern England: interpretation of apatite fission track data. *J. Geol. Soc.* **152**(2), 416–17. [11]

Green, P.F., Duddy, L.R. and Bray, R.J. (1995b) Applications of thermal history reconstructions in inverted basins. In: Buchanan, J.G. and Buchanan, P.G. (eds) *Basin Inversion*. Special Publication No. 88, Geological Society, London, pp. 149–66.

Green, S.C.H., Harker, S.D. and Romani, R.S. (1991) The Scapa Field, Block 14/19, UK North Sea. In: Abbotts, I.L. (ed.) *UK North Sea United Kingdom Oil and Gas Fields: 25 Years Commemorative Volume*. Memoir No. 14, Geological Society, London, p. 369. [12]

Greibrokk, T., Radke, M., Skurdal, M. and Willsch, J. (1992) Multistage supercritical fluid extraction of petroleum source rocks: application to samples from Kimmeridge Clay and Posidonia Shale formation. *Org. Geochem.* **18**(4), 447–55.

Gretener, P.E. and Curtis, C.D. (1982) Role of temperature and time on organic metamorphism. *AAPG Bull.* **66**(8), 1124–9.

Grice, K., Schaeffer, P., Schwark, L. and Maxwell, J.R. (1996) Molecular indicators of palaeoenvironmental conditions in an immature Permian shale Kupferschiefer, Lower Rhine Basin, north-west Germany from free and S-bound lipids. *Org. Geochem.* **25**(3/4), 131–48. [11]

Grice, K., Schaeffer, P., Schwark, L. and Maxwell, J.R. (1997) Changes in palaeoenvironmental conditions during deposition of the Permian Kupferschiefer, (Lower Rhine Basin, north-west Germany) inferred from molecular and isotopic compositions of biomarker components. *Org. Geochem.* **26**(11/12), 677–90. [11]

Griffith, A.E, (1983) The search for petroleum in Northern Ireland. In: Brooks, J. (ed.) *Petroleum Geochemistry and the Exploration of Europe*. Blackwell Scientific Publications, Oxford, pp. 213–22. [11]

Griffiths, R.W. and Campbell, I.H. (1991) Interaction of mantle plume heads with the earth's surface and onset of small-scale convection. *J. Geophys. Res.* **96**, 18295–310. [2]

Grigo, D., Maragna, B., Arienti, M.T., Fiorani, M., Parisi, A., Marrone, M. *et al.* (1993) Issues in 3D sedimentary basin modelling and application to Haltenbanken, offshore Norway. In: Doré, A.G.,Maragna, B., Arienti, M.T., Fiorani, M., Parisi, A., Marrone, M. *et al.* (eds) *Basin Modelling: Advances and Applications*. Special Publication No. 3, NPF, Elsevier, Amsterdam, pp. 455–68. [11]

Grung Olsen, R. and Hanssen, O.K. (1987) Askeladd. In: Spencer, A.M., Campbell, C.J., Hanslien, S.H., Nelson, P.H., Nysaether, E. and Ormaasen, E.G. (eds) *Geology of the Norwegian Oil and Gas Fields*. Graham and Trotman, London, pp. 419–28. [1]

Guidish, T.M., Kendall, C.G.St.C., Lerche, I., Toth, D.J. and Yarzab, R.F. (1985) Basin evaluation using burial history calculations: an overview. *AAPG Bull.* **69**(1), 92–105. [11]

Guion, P.D. and Fielding, C.R. (1988) Westphalian A and B sedimentation in the Pennine Basin, UK. In: Besly, B.M. and Kelling, G. (eds) *Sedimentation in a Synorogenic Basin Complex: the Upper Carboniferous of Northwest Europe*. Blackie, Glasgow, London, pp. 153–77. [4]

Gulbrandsen, A. (1987) Agat. In: Spencer, A.M., Campbell, C.J., Hanslien, S.H., Nelson, P.H., Nysaether, E. and Ormaasen, E.G. (eds) *Geology of the Norwegian Oil and Gas Fields*. Graham and Trotman, London, pp.363–70. [1, 9, 11, 12]

Gunn, R., Hwang, N.J., Murray, I. and Rezigh, A. (1993) The Murdoch Field development. *PESGB Newsl. Abstr.* March, 4–7. [4]

Gussow, W.C. (1968) Salt diapirism: importance of temperature and energy source of emplacement. Memoir No. 8, *AAPG*, Tulsa, OK., pp. 16–52. [6]

Guthrie, J.M. and Pratt, L.M. (1994) Gochemical indicators of depositional environment and source rock potential for the Upper Ordovician Maquoketa Group, Illinois Basin. *AAPG Bull.* **78**(5), 744–58. [11]

Guthrie, J.M. and Pratt, L.M. (1995) Geochemical character and origin of oils in Ordovician reservoir rock, Illinois and Indiana, USA. *AAPG Bull.* **79**(11), 1631–49.

Gutjahr, C.C.M. (1983) Incident-light microscopy of oil and gas source rocks. *Geol. Mijinbouw* **62**(3), 417–25. [11]

Guy, M. (1992) Facies analysis of the Kopervik sand interval, Kilda Field, Block 16/26, UK North Sea. In: Hardman, R.F.P. (ed.) *Exploration Britain: Geological Insights for the Next Decade*. Special Publication No. 67, Geological Society, London, pp. 187–220. [1, 9, 12]

Habicht, J.K.A. (1979) *Palaeoclimate, Palaeomagnetism and Continental Drift*. Studies in Geology, AAPG, 31 pp. [2]

Hage, A., Bromstad, K. and Strand, J.E. (1987) Brage. In: Spencer, A.M., Campbell, C.J., Hanslien, S.H., Nelson, P.H., Nysaether, E. and Ormaasen, E.G. (eds) *Geology of the Norwegian Oil and Gas Fields*. Graham and Trotman, London, pp. 371–8. [1, 11]

Hageman, B.P. and Hooykaas, H. (1980) *Stratigraphic Nomenclature of The Netherlands*. Nederlandse Aardolie Maatschappij bv (NAM) and Rijks Geologische Dienst (RGD) Verhandelingen van het Koninklijk Nederlands Geologisch Mijnbouwkunding Genootschap 32, The Netherlands, 77 pp. [9]

Haig, D.B. (1991) The Hutton Field, Blocks 211/28, 211/27, UK North Sea. In: Abbotts, J.L. (ed.) *UK North Sea United Kingdom Oil and Gas Fields: 25 Years Commemorative Volume*. Memoir No. 14, Geological Society, London, p. 135. [1]

Haig, D.B., Pickering, S.C. and Probert, R. (1995) The Lennox oil and gas field. Paper presented at Petroleum Geology of the Irish Sea and Adjacent Areas. In: *Geological Society Abstracts*, Geological Society, London, p. 41. [11]

Håkansson, E., Bromley, R. and Perch-Nielsen, K. (1974) Maastrichtian chalk of north-west Europe—a pelagic-shelf sediment. In: Hsü, K.J. and Jenkyns, H.C. (eds) *Pelagic Sediments: On Land and Under the Sea*. Special Publication No. 1, International Association of Sedimentologists, Blackwell Scientific Publications, Oxford, pp. 211–33. [9]

Hall, P.B. and Bjorøy, M. (1991) Biomarkers and organic facies of source rocks and oils of the North Sea Central Graben. In: Manning, D. (ed.) *Organic Geochemistry, Advances and Applications in Energy and Natural Environment*. Ext. Abstracts, 15th meeting EAOG, Manchester University Press, Manchester, pp. 192–4. [11]

Hall, P.B. and Douglas, A.G. (1983) The distribution of cyclic alkanes in two lacustrine deposits. In: Bjorøy, M., Albrecht, P., Cornford, C., de Groot, K., Eglinton, G., Galimov, E., *et al.* *Advances in Organic Geochemistry, 1981*. John Wiley, Chichester, pp. 576–87. [11]

Hall, P.B., Schou, L. and Bjorøy, M. (1985) Aromatic hydrocarbon variations in North Sea Wells. In: Thomas, B.M., Doré, A.G., Eggen, S.S., Home, P.C. and Larsen, R.M. (eds) *Petroleum Geochemistry in Exploration of the Norwegian Shelf.* Graham and Trotman, London, pp. 293–302. [11]

Hall, P.B., Stoddart, D., Bjorøy, M., Larter, S.R. and Brasher, J.E. (1994) Detection of petroleum heterogeneity in Eldfisk and satellite fields using thermal extraction, pyrolysis-GC, GC-MS and isotope techniques. *Adv. Org. Geochem.* **22**(3–5), 383–483. [11]

Hall, S.A. (1992) The Angus Field, a subtle trap. In: Hardman, R.F.P. (ed.) *Exploration Britain: Geological Insights for the Next Decade*. Special Publication No. 67, Geological Society, London, pp. 151–81. [1, 12]

Hallam, A. (1977) *Biogeographic Evidence Bearing on the Creation of Atlantic Seaways in the Atlantic*. Special Publication in Biology/Geology 2, Milwaukee Public Museum, Milwaukee, pp. 23–34. [2]

Hallam, A. (1984) Continental humid and arid zones during the Jurassic and Cretaceous. *Palaeogeog. Palaeoclim. Palaeoecol.* **47**, 195–223. [9]

Hallam, A. (1985) A review of Mesozoic climates. *J. Geol. Soc.* **142**(3), 433–46. [11]

Hallam, A. and Bradshaw, M.J. (1979) Bituminous shales and oolitic ironstones as indicators of transgressions and regressions. *J. Geol. Soc.* **136**, pp. 157–64. [11]

Hallam, A. and El Shaarawy, Z. (1982) Salinity reduction of the end-Triassic sea from the Alpine region into north-western Europe. *Lethaia* **15**, 169–78. [7]

Hallam, A. and Sellwood, B.W. (1976) Middle Mesozoic sedimentation in relation to tectonics in the British area. *J. Geol.* **84**, 301–21. [2, 8]

Hallet, D. (1981) Refinement of the geological model of the Thistle field. In: Illing, L.V. and Hobson, G.P. (eds) *Petroleum Geology of the Continental Shelf of North-West Europe*. Heyden, London, pp. 315–25. [1, 12]

Hallett, D., Durant, G.P. and Farrow, G.E. (1985) Oil exploration and production in Scotland. *Scott. J. Geol.* **21**, 547–70. [4]

Hallsworth, C.R., Morton, A.C. and Doré, G. (1996) Contrasting mineralogy of Upper Jurassic sandstones in the Outer Moray Firth, North Sea: implications for the evolution of sediment dispersal patterns. In: Hurst, A., Johnson, H.D., Burley, S.D., Canham, A.C. and Mackertich, D.S. (eds) *Geology of the Humber Group: Central Graben and Moray Firth, UKCS*. Special Publication No. 114, Geological Society, London, pp. 131–44. [8]

Hamar, G.P. and Hjelle, K. (1984) Tectonic framework of the More Basin and the Northern North Sea. In: Spencer, A.M., Johnsen, S.O., Mork, A., Nysaether, E., Songstad, P. and Spinnangr, A. (eds) *Petroleum Geology of the North European Margin*. Graham and Trotman, London, pp. 349–58. [11]

Hamar, G.P., Jakobsson, J.H., Ormaasen, D.E. and Skarpnes, O. (1980) Tectonic development of the North Sea north of the Central Highs. In: NPF *The Sedimentation of the North Sea Reservoir Rocks*. NPF, Norway, pp. 1–11. [7]

Hamar, G.P., Fjaeran, T. and Hesjedal, A. (1983) Jurassic stratigraphy and tectonics of the south-southeastern Norwegian offshore. *Geol. Mijnbouw* **62**(1), 103–114. [11]

Hamblin, R.J.O., Crosby, A., Balson, P.S., Jones, S.M., Chadwick, R.A., Penn, I.E. *et al.* (1992) *United Kingdom Offshore Regional Report: the Geology of the English Channel*. BGS, HMSO, London. [9]

Hamilton, R.M.F. and Trewin, N.H. (1985) Excursion guide to the Devonian of Caithness. PESGB Field Guide (unpubl.). [3, 11]

Hancock, F.R.P. and Mithen, D.P. (1987) The geology of Humbly Grove oil field, Hampshire, UK. In: Brooks, J. and Glennie, K.W. (eds) *Petroleum Geology of North West Europe*. Graham and Trotman, London, pp. 161–70. [11]

Hancock, J.M. (1975) The sequence of facies in the Upper Cretaceous of northern Europe compared with that in the western interior. In: Caldwell, W.G.F. (ed.) *Cretaceous System in the Western Interior of North America*. Special Paper No. 13, Geological Association of Canada, Ottowa, pp. 83–118. [9]

Hancock, J.M. (1976) The petrology of the chalk. *Proc. Geol. Assoc.* **86**, 499–535. [9]

Hancock, J.M. (1983) The setting of the Chalk and its initial accumulation. In: Hancock, J.M., Hamblin, R.J.O., Crosby, A., Balson, P.S., Jones, S.M., Chadwick, R.A., *et al.* (eds) *Chalk in the North Sea*. UK Course Notes No. 20, JAPEC, London, pp. A1–A32. [9]

Hancock, J.M. (1990) Cretaceous. In: Glennie, K.W. (ed.) *Introduction to the Petroleum Geology of the North Sea*. Blackwell Scientific Publications, Oxford, pp. 255–72. [9]

Hancock, J.M. and Kauffmann, E.G. (1979) The great transgressions of the Late Cretaceous. *J. Geol. Soc.* **136**, 175–86. [9]

Hancock, J.M. and Rawson, P.F. (1992) Cretaceous. In: Cope, J.C.W., Ingham, J.K. and Rawson, P.F. (eds) *Atlas of Palaeogeography and Lithofacies*. Memoir No. 13, Geological Society, London, pp. 131–9. [9]

Hancock, J.M. and Scholle. P.A. (1975) Chalk of the North Sea. In: Woodland, A.W. (ed.) *Petroleum and the Continental Shelf of Northwest Europe*. Vol. I, *Geology*. Applied Science Publishers, Barking, pp. 413–27. [9]

Hancock, J.M., Hamblin, R.J.O., Crosby, A., Balson, P.S., Jones, S.M., Chadwick, R.A., *et al.* (eds) (1983) *Chalk in the North Sea*. UK Course Notes No. 20, JAPEC, London. [9]

Hancock, N.J. (1978) Possible causes of Rotliegend sandstone diagenesis in northern West Germany. *J. Geol. Soc.* **135**, 35–40. [5]

Hancock, N.J. and Fisher, M.J. (1981) Middle Jurassic North Sea deltas with particular reference to Yorkshire. In: Illing, L.V. and Hobson, G.P. (eds) *Petroleum Geology of the Continental Shelf of North-West Europe*. Heyden, London, pp. 186–95. [8, 11]

Handford, C.R. and Loucks, R.G. (1993) Carbonate depositional sequences and systems tracts—responses of carbonate platforms to relative sea-level change. In: Loucks, R.G. and Sarg, J.F. (eds) *Carbonate Sequence Stratigraphy—Recent Developments and Applications*. Memoir No. 57, AAPG, pp. 3–41. [9]

Hansen, J.M. and Buch, A. (1982) Early Cretaceous. In: Michelson, O. (ed.) *Geology of the Danish Central Graben*. Series B, No. 8, G.S. Denmark, pp. 45–9. [9]

Hanslien, S. (1987) Balder. In: Spencer, A.M., Campbell, C.J., Hanslien, S.H., Nelson, P.H., Nysaether, E. and Ormaasen, E.G. (eds) *Geology of the Norwegian Oil and Gas Fields*. Graham and Trotman, London, pp. 193–201. [1, 12]

Haq, B.U. (1991) Sequence stratigraphy, sea-level change, and significance of the deep sea. In: McDonald, D.I.M. (ed.) *Sedimentation, Tectonics and Eustacy*. Special Publication No. 12, International Association of Sedimentologists, Blackwell Scientific Publications, Oxford, pp. 3–39. [9]

Haq, B.U. and van Eysinga, W.B. (1987) *Geological Time Table*. Elsevier, Amsterdam. [5]

Haq, B.U., Hardenbol, J. and Vail, P.R. (1987) The chronology of fluctuating sea level since the Triassic. *Science* **235**, 1156–67. [7, 8, 10]

Haq, B.U., Hardenbol, J. and Vail, P.R. (1988) Mesozoic and Cenozoic chronostratigraphy and cycles of sea level change. In: Wilgus, C.K., Hastings, B.S., Posamentier, H., van Wagoner, J., Ross, C.A. and Kendall, C.G.St.C. (eds) *Sea-Level Changes: An Integrated Approach*. Special Publication No. 42, SEPM, Tulsa, pp. 71–108. [8, 9]

Harbaugh, J.W. (1984) Quantitative estimation of petroleum prospect outcome probabilities: an overview of procedures. *Mar. Petr. Geol.* **1**, 298–312. [12]

Hardie, L.A., Smoot, J.P. and Eugster, H.P. (1978) Saline lakes and their deposits: a sediment approach. Special Publication No. 2, International Association of Sedimentologists, Utrecht, pp. 7–41. [7]

Harding, A.W., Humphrey, T.J., Latham, A., Lunsford, M.K. and Strider, M.H. (1990) Controls on Eocene submarine-fan deposition in the Witch Ground Graben. In: Hardman, R.P.F. and Brooks, J. (eds) *Tectonics Events Responsible for Britain's Oil and Gas Reserves*. Special Publication No. 55, Geological Society, London, pp. 353–67. [10]

Hardman, M., Buchanan, J., Herrington, P. and Carr, A. (1993) Geochemical modelling of the East Irish Sea Basin: its influence on predicting hydrocarbon type and quality. In: Parker, J.R. (ed.) *Petroleum Geology of Northwest Europe: Proceedings of the 4th Conference*. Geological Society, London, pp. 809–21. [11, 12]

Hardman, R.P.F. (1982) Chalk reservoirs of the North Sea. *Bull. Geol. Soc. Denmark* **30**, 119–37. [9, 12]

Hardman, R.P.F. (1983) Chalk hydrocarbon reservoirs of the North Sea—an introduction. In: Hancock, J.M., Hamblin, R.J.O., Crosby, A., Balson, P.S., Jones, S.M., Chadwick, R.A. *et al.* (eds) *Chalk in the North Sea*. UK Course Notes No. 20, JAPEC, London, pp. D1–D33. [9]

Hardman, R.P.F. (ed.) (1992) *Exploration Britain: Geological Insights for the Next Decade*. Special Publication No. 67, Geological Society, London, 312 pp. [1]

Hardman, R.P.F. and Brooks, J. (1990) *Tectonic Events Responsible for Britain's Oil and Gas Reserves.* Special Publication No. 55, Geological Society, London. [1]

Hardman, R.P.F. and Eynon, G. (1977) Valhall Field—a structural/stratigraphic trap. In: NPF *Mesozoic Northern North Sea Symposium, Oslo, 1977.* MNNSS/14, Geilo, 33 pp. [9, 12]

Hardman, R.F.P. and Kennedy, W.J. (1980) Chalk reservoirs in the Hod fields. In: NPF *Sedimentation of the North Sea Reservoir Rocks.* NPF, Norway, Article 11, 31 pp. [1, 9]

Harker, S.D. and Chermak, A. (1992) Detection and prediction of Lower Cretaceous sandstone distribution in the Scapa Field, North Sea. In: Hardman, R.F.P. (ed.) *Exploration Britain: Geological Insights for the Next Decade.* Special Publication No. 67, Geological Society, London, pp. 221–46. [1, 9, 12]

Harker, S.D. and Rieuf, M. (1996) Genetic stratigraphy and sandstone distribution of the Moray Firth Humber Group Upper Jurassic. In: Hurst, A., Johnson, H.D., Burley, S.D., Canham, A.C. and MacKertich, D.S. (eds) (1996) (eds) *Geology of the Humber Group: Central Graben and Moray Firth, UKCS.* Special Publication No. 114, Geological Society, London, pp. 109–30. [8]

Harker, S.D., Gustav, S.H. and Riley, L.A. (1987) Triassic to Cenomanian stratigraphy of the Witch Ground Graben. In: Brooks, J. and Glennie, K.W. (eds) *Petroleum Geology of North West Europe.* Graham and Trotman, London, pp. 809–18. [7, 9]

Harker, S.D., Green, S.C.H. and Romani, R.S. (1991) The Claymore Field, Block 14/19, UK North Sea. In: Abbotts, I.L. (ed.) *UK North Sea United Kingdom Oil and Gas Fields: 25 Years Commemorative Volume.* Memoir No. 14, Geological Society, London, pp. 269–78. [1, 4, 12]

Harker, S.D., Mantell, K.A., Morton, D.J. and Riley, L.A. (1993) The stratigraphy of Oxfordian–Kimmeridgian late Jurassic reservoir sandstones in the Witch Ground Graben, United Kingdom North Sea. *AAPG Bull.* **77**(10), 1693–709. [8]

Harland, W.D. and Gayer, R.A. (1972) The Arctic Caledonides and earlier oceans. *Geol. Mag.* **109**, 289–314. [1]

Harms, J.C., Tackenberg, P., Pickle, E. and Pollock, R.E. (1981) The Brae oilfield area. In: Illing, L.V. and Hobson, G.P. (eds) *Petroleum Geology of the Continental Shelf of North-West Europe.* Heyden, London, pp. 352–7. [1]

Harper, M.L. (1971) Approximate geothermal gradients in the North Sea Basin. *Nature* **230**, 235–6. [11]

Harriman, G. and Miles, J. (1995) A new oil seep in the heart of Liverpool. *PESGB Newsl.* July, 88–9. [11]

Harris, A.L. (1983) The growth and structure of Scotland. In: Craig, G.Y. (ed.) *Geology of Scotland.* Scottish Academic Press, Edinburg, pp. 1–22. [2]

Harris, A.L. (ed.) (1985) *The Nature and Timing of Orogenic Activity in the Caledonian Rocks of the British Isles.* Memoir No. 9, Geological Society, London, 53 pp. [2]

Harris, A.L., Haselock, P.J., Kennedy, M.J. and Mandum, J.R. (1994) The Dalradian Supergroup in Scotland, Shetland and Ireland. In: Gibbons, W. and Harris, A.L. (eds) *A Revised Correlation of Precambrian Rocks in the British Isles.* Special Report 22, Geological Society, London, 110 pp. [2]

Harris, J.P. and Fowler, R.M. (1987) Enhanced prospectivity of the mid-Late Jurrassic sediments of the South Viking Graben, Northern North Sea. In: Brooks, J. and Glennie, K.W. (eds) *Petroleum Geology of North West Europe.* Graham and Trotman, London, pp. 879–98. [11]

Harris, N.B. (1992) Burial diagenesis of Brent sandstones: a study of Statfjord, Hutton and Lyell fields. In: Morton, A.C., Haszeldine, R.S., Giles, M.R. and Brown, S. (eds) *Geology of the Brent Group.* Special Publication No. 61, Geological Society, London, pp. 351–75. [12]

Hart, M.B. (1983) The Chalk of the North Sea—biostratigraphy. In: Hancock, J.M., Kennedy, W.J., Hart, M.B., Hardman, R.F.P., Eynon, G. and Watts, N.L. (eds) *Chalk in the North Sea.* UK Course Notes No. 20, JAPEC, London, pp. C1–C67. [9]

Hart, M.B. and Leary, P.N. (1989) The stratigraphic and palaeogeographic setting of the late Cenomanian 'anoxic' event. *J. Geol. Soc.* **146**, 305–10. [9]

Hartley, A. (1995) Sedimentology of the Cretaceous Greensand, Quadrants 48 and 49, North Celtic Sea Basin: a progradational shoreface deposit. In: Croker, P.F. and Shannon, P.M. (eds) *The Petroleum Geology of Ireland's Offshore Basins.* Special Publication No. 93, Geological Society, London, pp. 245–57. [12]

Hartung, M., Heinke, J. and Reichel, B. (1993) Evaluating gas in-place from 3D deismic data—Rotliegend formation, North Germany. *First Break* **11**(6), 241–5. [1, 5]

Harvey, M.A. and Stewart, S.A. (1998) Influence of salt on the structural evolution of the channel basin. In: Underhill, J.R. (ed.) *Development, Evolution and Petroleum Geology of the Wessex Basin.* Special Publication 133, Geological Society, London, pp. 243–66. [2]

Harwood, G.M. (1986) The diagenetic history of the Cadeby Formation carbonate EZ1Ca, Upper Permian, eastern England. In: Harwood, G.M. and Smith, D.B. (eds) *The English Zechstein and Related Topics.* Special Publication No. 22, Geological Society, London, pp. 75–86. [6]

Harwood, G.M. and Smith, D.B. (eds) (1986) *The English Zechstein and Related Topics.* Special Publication No. 22, Geological Society, London.

Hastings, A., Murphy, P. and Stewart, L. (1991) A multidisciplinary approach to reservoir characterization: Helm field, Dutch North Sea. In: Spencer, A.M. (ed.) *Generation, Accumulation, and Production of Europe's Hydrocarbons.* Special Publication, EAPG, pp. 193–202. [1]

Hastings, D.S. (1986) Cretaceous stratigraphy and reservoir potential, mid Norway continental shelf. In: Spencer, A.M., Campbell, C.J., Hanslien, S.H., Nelson, P.H., Nysaether, E. and Ormaasen, E.G. (eds) *Habitat of Hydrocarbons on the Norwegian Continental Shelf.* Graham and Trotman, London, pp. 287–98. [9]

Hastings, D.S. (1987) Sand-prone facies in the Cretaceous of Mid Norway. In: Brooks, J. and Glennie, K.W. (eds) *Petroleum Geology of North West Europe.* Graham and Trotman, London, pp. 1065–78. [9]

Haszeldine, R.S. (1984) Carboniferous North Atlantic palaeogeography: stratigraphic evidence for rifting, not megashear or subduction. *Geol. Mag.* **121**, 442–63. [2, 4]

Haszeldine, R.S. and Russell, M.J. (1987) The Late Carboniferous northern North Atlantic Ocean: implications for hydrocarbon exploration from Britain to the Arctic. In: Brooks, J. and Glennie, K.W. (eds) *Petroleum Geology of North West Europe.* Graham and Trotman, London, pp. 1163–75. [6]

Haszeldine, R.S., Samson, I.M. and Cornford, C. (1984a) Quartz diagenesis and convective fluid movement. Beatrice oilfield. UK North Sea. *Clay Minerals* **19**, 391–402. [11]

Haszeldine, R.S., Samson, I.M. and Cornford, C. (1984b) Dating diagenesis in a petroleum basin, a new fluid inclusion method. *Nature* **307**(5949), 354–7. [11]

Haszeldine, R.S., Ritchie, J.D. and Hitchin, K. (1987) Seismic and well evidence for the early development of the Faeroe–Shetland Basin. *Scott. J. Geol.* **23**, 283–300. [2]

Hatton, I.R. (1986) Geometry of allochthonous Chalk Group members, Central Trough, North Sea. *Mar. Petr. Geol.* **3**, 79–98. [9, 12]

Haughton, P.D.W., Rogers, G. and Halliday, A.N. (1990) Provenance of Lower Old Red Sandstone conglomerates, SE Kincardineshire: evidence for the timing of Caledonian terrane accretion in central Scotland. *J. Geol. Soc.* **147**, 105–20. [3]

Hauk, V.M., Petersen, H.H.F., Spoerker, H.F. and Moritz, J. (1979) Deep European H_2S is handled with special muds, cement and tubulars. *Oil Gas J.* **77**, 62–9. [6]

Hawkes, P.W., Fraser, A.J. and Einchomb, C.C.G. (1998) The tectono-stratigraphic development and exploration history of the Weald and Wessex Basins, southern England. In: Underhill, J.R.

(ed.) *Development, Evolution and Petroleum Geology of the Wessex Basin.* Special Publication 113, Geological Society, London, pp. 317–27. [12]

Hawkins, P.J. (1978) Relationship between diagenesis, porosity reduction and oil emplacement in late Carboniferous sandstone reservoirs, Bothamsall Oilfield, East Midlands. *J. Geol. Soc.* **135**, 7–24. [4]

Hay, J.T.C. (1977) The Thistle oilfield. In: *Mesozoic Northern North Sea Symposium.* NPS, Oslo, Section 11. [1]

Hay, J.T.C. (1978) Structural development in the Northern North Sea. *J. Petr. Geol.* **1**(1), 65–77.

Hay, W.W. (1995) Paleoceanography of marine organic-carbon-rich sediments. In: Huc, -A.Y. (ed.) *Paleogeography, Paleoclimate, and Source Rock.* AAPG Studies in Geology No. 40, pp. 21–60. [11]

Hazeu, G.J.A. (1981) 34/10 Delta structure geological evaluation and appraisal. In: *Norwegian Symposium on Exploration.* NPF, Norway, Article 13, 36 pp.

Heaviside, J., Langley, G.O. and Pallatt, N. (1983) Permeability characteristics of Magnus reservoir rock. In: *8th SPWLA London Chapter European Evaluation Symposium Transactions.* [12]

Hedberg, H.D. (1980) Methane generation and petroleum migration. In: Roberts, W.H. and Cordell, R.J. (eds) *Problems of Petroleum Migration.* Studies in Geology No. 10, AAPG, Tulsa, OK., pp. 179–206. [11]

Hedemann, H.-A. (1980) Die Bedeutung des Oberkarbons für die Kohlenwasserstoffvorkommen im Nordseebecken. *Erdöl-Kohle-Erdgas* **33**, 255–66. [4]

Hedemann, H.-A., Mascheck, W., Paulus, B. and Plein, E. (1984a) Mitteilung zur lithostratigraphischen Gleiderung des Oberrotliegenden in nordwestdeutschen Becken. *Nachschr. Deutsches Geol. Gesampt* **30**, 100–7.

Hedemann, H.-A., Schuster, A., Stancu-Kristoff, G. and Loesch, J. (1984b) Die Verbreitung der Kohlenfloze des Oberkarbons in Nordwestdeutschland und ihre stratigraphische Einstufung. *Fortschr. Geol. Rheinld. Westf.* **32**, 39–88. [4]

Heinrich, R.D. (1991a) The Cleeton Field, Block 42/29, UK North Sea. In: Abbotts, I.L. (ed.) *UK North Sea United Kingdom Oil and Gas Fields: 25 Years Commemorative Volume.* Memoir No. 14, Geological Society, London, pp. 409–15. [1, 5, 11]

Heinrich, R.D. (1991b) The Ravenspurn South Field, Blocks 42/29, 42/30, 43/26, UK North Sea. In: Abbotts, I.L. (ed.) *UK North Sea United Kingdom Oil and Gas Fields: 25 Years Commemorative Volume.* Memoir No. 14, Geological Society, London, pp. 469–75. [1, 5, 11]

Helland-Hansen, W., Ashton, M., Lomo, L. and Steel, R. (1992) Advance and retreat of the Brent delta: recent contributions to the depositional model. In: Morton, A.C., Haszeldine, R.S., Giles, M.R. and Brown, S. (eds) *Geology of the Brent Group.* Special Publication No. 61, Geological Society, London, pp. 109–27. [8, 12]

Hellinger, S.J., Sclater, J.G. and Giltner, J. (1988) Mid-Jurassic through mid-Cretaceous extension in the Central Graben of the North Sea—part 1: estimates from subsidence. *Basin Res.* **1**, 191–200.

Hemingway, J.E. and Riddler, G.P. (1982) Basin inversion in North Yorkshire. *Trans. Instn. Min. Metall.* **91**, 175–86. [2]

Herber, R., Elders, C., Lamens, J., Bachman, M. and Sanchez Ferrer, F. (1993) Blanket 3-D coverage—its successful application in exploring a mature area, Central Dutch offshore. *AAPG Bull.* **77**(9), 1630.

Herbin, J.P. and Geyssant, J.R. (1993) Ceintures organiques au Kimmeridgian/Tithonian en Angleterre Yorkshire, Dorset, et en France, Boulonnais. *CR Acad. Sci. Paris* **317**, 1308–16. [11]

Herbin, J.P., Geyssant, J.R., Mélières, F., Müller, C., Penn, I.E. *et al.* (1991) Hétérogénéité quantitative et qualitative de la matière organique dans les argiles du Kimmeridgian du val de Pickering Yorkshire, UK. *Rev. IFP* **46**(6), 1–39. [11]

Herbin, J.P., Müller, C., Geyssant, J.R., Mélières, F., Penn, I.E. *et al.* (1993) Variation of the distribution of organic matter within a transgressive systems tract: Kimmeridge Clay (Jurassic) England. In: Katz, B. and Pratt, L. (eds) *Petroleum Source Rocks in a Sequence Stratigraphic Framework.* Studies in Geology No. 37, AAPG, pp. 67–100. [11]

Herbin, J.P., Fernandez-Martinez, J.L., Geyssant, J.R., EL Albani, A., Deconick, J.F., Proust, J.N. *et al.* (1995) Sequence stratigraphy of source rocks applied to the study of the Kimmeridgian/Tithonian in the north-west European shelf, Dorset/UK, Yorkshire UK and Boulonnais, France. *Mar. Petr. Geol.* **12**(2), 177–94. [11]

Héritier, F.E., Lossel, P. and Wathne, E. (1980) Frigg Field—large submarine-fan trap in Lower Eocene rocks of the Viking Graben, North Sea. *AAPG Bull.* **63**(11), 1999–2020. [1, 11, 12]

Héritier, F.E., Lossel, P. and Wathne, E. (1981) The Frigg gas field. In: Illing, L.V. and Hobson, G.P. (eds) *Petroleum Geology of the Continental Shelf of North-West Europe.* Heyden, 380–91. [1, 12]

Hermanrud, C. (1986) On the importance to the petroleum generation of heating effects from compaction derived water: an example from the Northern North Sea. In: Burrus, J. (ed.) *Thermal Modeling in Sedimentary Basins.* Proc. 1st IFP Expl. Rsch. Conference, Carcans, France, 1985, Editions Technip, Paris, pp. 247–70. [11]

Hermanrud, C. and Shen, P.Y. (1989) Virgin rock temperatures from well logs—accuracy analysis for some advanced inversion models. *Mar. Petr. Geol.* **6**(4), 360–3. [11]

Hermanrud, C., Cao, S. and Lerche, I. (1990a) Estimates of virgin rock temperature derived from BHT measurements: bias and error. *Geophysics* **55**(7), 924–31. [11]

Hermanrud, C., Eggen, S., Jacobsen, T., Carlsen, E.M. and Pallesen, S. (1990b) On the accuracy of modelling hydrocarbon generation and migration: the Egersund Basin oil find, Norway. In: Durand, B. and Behar, F. (eds) *Advances in Organic Geochemistry. Org. Geochem.* **16**(1–3), 389–99. [11]

Hermanrud, C., Eggen, S. and Larsen, R.M. (1991) Investigation of the thermal regime of the Horda Platform by basin modelling: implications for the hydrocarbon potential of the Stord Basin, northern North Sea. In: *Generation, Accumulation and Production of Europe's Hydrocarbons.* Spencer, A.M. (ed.) Special Publication No. 1, EAPG, Oxford University Press, Oxford, pp. 65–73. [11]

Hermanrud, C., Lerche, I. and Meisingset, K.K. (1991) Determination of virgin rock temperature from drillstem tests. *J. Petroleum Tech.* September 1126–31. [11]

Hermanrud, C., Eggen, S. and Larsen, R.M. (1991) Investigation of the thermal regime of the Horda Platform by basin modelling: implications for the hydrocarbon potential of the Stord Basin, northern North Sea. In: Spencer, A.M. (ed.) *Generation, Accumulation, and Production of Europe's Hydrocarbons.* Special Publication No. 1, EAPG, Oxford University Press, Oxford, pp. 65–73. [11]

Hermans, L., van Kuyk, A.D., Lehner, F.K. and Featherstone, P.S. (1992) Modelling secondary hydrocarbon migration in Haltenbanken, Norway. In: Larsen, R.M., Brekke, H., Larsen, B.T. and Talleraas, E. (eds) *Structural and Tectonic Modelling and its Applications to Petroleum Geology.* Special Publication No. 1, NPS, Proc. of NPF Workshop, 18–20 Oct. 1989, Stavanger, Norway, pp. 305–23. [11]

Hermansen, D. (1993) Optimization of temperature history aspects of vitrinite reflectance and sterane isomerisation. In: Doré, A.G., Augustson, J.H., Hermanrud, C., Stewart, D.J. and Sylta, O. (eds) *Basin Modelling: Advances and Applications.* Special Publication No. 3, NPF, Elsevier, Amsterdam, pp. 119–26. [11]

Herngreen, G.F.W., Smit, R. and Wong, T.E. (1992) Stratigraphy and tectonics of the Vlieland Basin, The Netherlands. In: Spencer, A.M. (ed.) *Generation, Accumulation, and Production of Europe's Hydrocarbons.* Special Publication, EAPG, pp. 175–92. [9]

Heroux, Yvon, Chagnon, A. and Bertrand, R. (1979) Compilation and correlation of major thermal maturation indicators. *AAPG Bulletin* **63**(12), 2128–44. [11]

Herrington, P.M., Pederstad, K. and Dickson, J.A.D. (1991) Sedimentology and diagenesis of resedimented and rhythmically bedded chalks from the Eldfisk Field, North Sea Central Graben. *AAPG Bull.* **75**, 1661–74. [9]

Herron, M.M. and Herron, S.L. (1990) Geological applications of geochemical well logging. In: Hurst, A., Lovell, M.A. and Morton, A.C. (eds) *Geological Applications of Wireline Logs*. Special Publication No. 48, Geological Society, London, pp. 165–75. [11]

Herron, S.L. (1989) Potential application of wireline logs to basin analysis. In: *Abstracts of Papers presented at the 14th Int'l Org. Geochem. Conference, Paris, 18–22 Sept. 1989*. Abstract No. 32 EAOG and IFP, Paris. [11]

Herron, S.L. (1991) *In situ* evaluation of potential source rocks by wireline logs. In: Merrill, R.K. (ed.) *Source and Migration Processes and Evaluation Techniques*. Treatise on Petroleum Geology, AAPG, Tulsa, OK., pp. 127–34. [11]

Herron, S.L. and Le Tendre, L. (1990) Wireline source-rock evaluation in the Paris basin. In: Hue, A.Y. (ed.) *Deposition of Organic Facies*. Studies in Geology No. 30, AAPG, Tulsa, OK., pp. 57–72. [11]

Hesjedal, A. and Hamar, G.P. (1983) Lower Cretaceous stratigraphy and tectonics of the south-southeastern Norwegian offshore. *Geol. Mijnbouw* **62**, 135–44. [9, 11]

Hesselbo, S.P. and Allen, P.A. (1991) Major erosion surfaces in the basal Wealden Beds, Lower Cretaceous, south Dorset. *J. Geol. Soc.* **148**, 103–13. [9]

Hesselbo, S.P., Coe, A.L. and Jenkyns, H.C. (1990) Recognition and documentation of depositional sequences at outcrop: an example from the Aptian and Albian on the eastern margin of the Wessex Basin. *J. Geol. Soc.* **147**, 549–59. [9]

Heward, A.P. (1991) Inside Auk—the anatomy of an eolian oil reservoir. In: Miall, D. and Tyler, N. (eds) *The Three Dimensional Facies Architecture of Clastic Sediments and its Implications for Hydrocarbon Discovery and Recovery*. Concepts and Models in Sedimentology and Paleontology 3, SEPM, Tulsa, OK., pp. 44–56. [1, 5, 12]

Heybroek, P. (1975) On the structure of the Dutch part of the Central North Sea Graben. In: Woodland, A.W. (ed.) *Petroleum and the Continental Shelf of Northwest Europe*. Vol. I, *Geology*, Applied Science Publishers, Barking, pp. 339–51. [9]

Heybroek, P., Haanstra, U. and Erdman, D.A. (1967) Observations on the geology of the North Sea area. In: *Proceedings of the 7th World Petroleum Congress*, No. 2, pp. 905–16. [1, 6]

Higgins, A.C. (1976) Conodont zonation of the late Visean early Westphalian strata of the south and central Pennines of northern England. *Bull. Geol Surv. GB*, **53**, 1–90. [4]

Higgs, M.D. (1986) Laboratory studies in the generation of natural gas from coals. In: Brooks, J., Goff, J. and van Hoorn, B. (eds) *Habitat of Palaeozoic Gas in N.W. Europe*. Special Publication No. 23, Geological Society, London, pp. 113–20. [11]

Highton, A.J. (1992) The tectonostratigraphical significance of pre-750 Ma metagobbros within the northern Central Highlands, Inverness-shire. *Scott. J. Geol.* **28**, 71–6. [2]

Hill, P.E. and Smith, G. (1979) Geological aspects of the drilling of the Buchan Field. In: *Offshore Europe '79 Conference, Aberdeen*. Paper 8153.1, SPE, p. 7. [3, 4, 12]

Hill, P.J. and Wood, G.V. (1980) Geology of the Forties Field, UK Continental Shelf, North Sea. In: Halbouty, M.T. (ed.) *Giant Oil and Gas Fields of the Decade 1968–1978*. Memoir No. 30, AAPG, Tulsa, OK., pp. 81–94. [1, 12]

Hillebrand, T. and Leythaeuser, D. (1992) Reservoir geochemistry of Stockstadt oilfield: compositional heterogeneities reflecting accumulation history and multiple source input. *Org. Geochem.* **19**(1–13), 119–31.

Hillier, A.P. and Williams, B.P.J. (1991) The Leman Field, Blocks 49/26, 49/27, 49/28 53/1, UK North Sea. In: Abbotts, I.L. (ed.) *UK North Sea United Kingdom Oil and Gas Fields: 25 Years Commemorative Volume*. Memoir No. 14, Geological Society, London, pp. 451–8. [1, 5, 11]

Hillier, S. and Marshall, J.E.A. (1992) Organic maturation, thermal history and hydrocarbon generation in the Orcadian Basin, Scotland. *J. Geol. Soc.* **149**, 491–502. [3]

Hillier, S.J. (1989) Clay mineral diagenesis and organic maturity indicators in Devonian lacustrine mudrocks from the Orcadian Basin, Scotland. PhD thesis, University of Southampton. [3]

Hillier, S.J. and Marshall, J.E.A. (1988) Hydrocarbon source rocks, thermal maturity and burial history of the Orcadian Basin, Scotland. In: Fleet, A.J., Kelts, K. and Talbot, M.R. (eds) *Lacustrine Petroleum Source Rocks*. Special Publication, Geological Society, Blackwell Scientific Publications, Oxford, pp. 203. [11]

Hillis, R.R. (1991) Chalk porosity and Tertiary uplift, Western Approaches Trough, SW UK and NW French continental shelves. *J. Geol. Soc.* **148**, 669–79.

Hillis, R.R. (1992) A two layer lithospheric compressional model for the Tertiary uplift of the southern United Kingdom. *Geophys. Res. Lett.* **19**, 573–6.

Hillis, R.R. (1993) Quantifying erosion in sedimentary basins from sonic velocities in shales and sandstones. *Expl. Geophys.* **24**, 561–6. [11]

Hillis, R.R. (1995a) Regional Tertiary exhumation in and around the United Kingdom. In: Buchanan, J.G. and Buchanan, P.G. (eds) *Basin Inversion*. Special Publication No. 88, Geological Society, London, pp. 167–90. [11]

Hillis, R.R. (1995b) Quantification of Tertiary exhumation in the United Kingdom Southern North Sea using sonic velocity data. *AAPG Bull.* **79**, 130–52. [11]

Hillis, R.R., Thomson, K. and Underhill, J.R. (1994) Quantification of Tertiary erosion in the Inner Moray Firth by sonic velocity data from the Chalk and Kimmeridge Clay. *Mar. Petr. Geol.* **11**, 283–93. [1, 11]

Hinch, H.H. (1993) The nature of shales and the dynamics of hydrocarbon expulsion in the Gulf Coast Tertiary section. In: Roberts, W.H. and Cordell, R.J. (eds) *Problems of Petroleum Migration*. Studies in Geology No. 10, AAPG, Tulsa, OK., pp. 1–18.

Hitchen, K. and Ritchie, J.D. (1987) Geological review of the West Shetland area. In: Brooks, J. and Glennie, K.W. (eds) *Petroleum Geology of North West Europe*. Graham and Trotman, London, pp.737–49. [1, 2, 11]

Hitchon, B. (1980) Some economic aspects of water-rock interaction. In: Roberts, W.H. and Cordell, R.J. (eds) *Problems of Petroleum Migration*. Studies in Geology No. 10, AAPG, Tulsa, OK., pp. 109–20.

Hobson, G.D. and Hillier, A.P. (1991) The Sean North and Sean South Fields, Block 49/25a, UK North Sea. In: Abbotts, I.L. (ed.) *UK North Sea United Kingdom Oil and Gas Fields: 25 Years Commemorative Volume*. Memoir No. 14, Geological Society, London, p. 485. [1, 5, 11]

Hodgson, G.W. (1980) Origin of petroleum: in-transit conversion of organic compounds in water. In: Roberts, W.H. and Cordell, R.J. (eds) *Problems of Petroleum Migration*. Studies in Geology No. 10, AAPG, Tulsa, OK., pp. 169–78.

Hodgson, N.A., Farnsworth, J. and Fraser, A.J. (1992) Salt-related tectonics, sedimentation and hydrocarbon plays in the Central Graben, North Sea, UKCS3. In: Hardman, R.F.P. (ed.) *Exploration Britain: Geological Insights for the Next Decade*. Special Publication No. 67, Geological Society, London, pp. 31–63. [2, 7, 8, 9, 10, 12]

Hoffman, J. and Hower, J. (1979) Clay mineral assemblages as low grade metamorphic geothermometers: applications to the thrust faulted disturbed belt of Montana, U.S.A. In: Scholle, P.A. and Schluger, R.P. (eds) *Aspects of Diagenesis*. Special Publication No. 26, SEPM, pp. 55–79.

Hoffmann, N., Kamps, H.-J. and Schneider, J. (1988) Neuerkenntnisse zur Biostratigraphie und Palaodynamik des Perms in der Nordostdeutschen Senks-ein Diskussionsbeitrag. *Zeitschr. angewandte Geol.* **35**(7), 189–207. [5]

Hogg, A.J.C., Hamilton, P.G. and Macintyre, R.M. (1993) Mapping diagenetic fluid flow within a reservoir. K–Ar dating in the

Alwyn area UK North Sea. *Mar. Petr. Geol.* **10**(3), 279–95. [11]

Høiland, O., Kristensen, J. and Monson, T. (1993) Mesozoic evolution of the Jaeren High area, Norwegian Central North Sea. In: Parker, J.R. (ed.) *Petroleum Geology of Northwest Europe: Proceedings of the 4th Conference.* Geological Society, London, pp. 1189–95. [6, 7]

Holdsworth, B.K. and Collinson, J.D. (1988) Millstone Grit cyclicity revisited. In: Besly, B.M. and Kelling, G. (eds) *Sedimentation in a Synorogenic Basin Complex: the Upper Carboniferous of Northwest Europe.* Blackie, Glasgow, London, pp. 132–52. [4]

Holdsworth, R.E., Strachan, R.A. and Harris, A.L. (1994) Precambrian rocks in northern Scotland east of the Moine Thrust: the Moine Supergroup. In: Gibbons, W. and Harris, A.L. (eds) *A Revised Correlation of Precambrian Rocks in the British Isles.* Special Report No. 22, Geological Society, London, pp. 23–32. [2]

Hollander, N.B. (1984) Geohistory and hydrocarbon evaluation of the Haltenbanken area. In: Spencer, A.M., Johnsen, S.O., Mork, A., Nysaether, E., Songstad, P. and Spinnangr. A. (eds) *Petroleum Geology of the North European Margin.* Graham and Trotman, London, pp. 383–8. [11]

Hollander, N.B. (1987) Snorre. In: Spencer, A.M., Campbell, C.J., Hanslien, S.H., Nelson, P.H., Nysaether, E. and Ormaasen, E.G. (eds) *Geology of the Norwegian Oil and Gas Fields.* Graham and Trotman, London, pp. 307–18. [1, 7, 12]

Holliday, D.W. (1993a) Geophysical log signatures in the Eden-shales (Permo-Trassic) of Cumbria and their regional significance. *Proc. Yorks. Geol Soc.* **49**, 345–54. [6]

Holliday, D.W. (1993b) Mesozoic cover over northern England: interpretation of apatite fission-track data. *J. Geol. Soc.* **150**, 657–60. [6, 11]

Holliday, D.W. (1994a) Reply to discussion of Holliday, D.W. (1993) Geophysical log signatures in the Eden Shales (Permo-Trassic) of Cumbria and their regional significance. *Proc. Yorks. Geol. Soc.* **50**, 183–4. [6]

Holliday, D.W. (1994b) Discussion on Mesozoic cover over northern England: interpretation of apatite fission-track data reply. *J. Geol. Soc.* **151**, 735–6. [6, 11]

Hollingworth, N.T.J. and Tucker, M.E. (1987) The Upper Permian Zechstein Tunstall Reef of North East England: palaecoecology and early diagenesis. In: Peryt, T.M. (ed.) *The Zechstein Facies in Europe.* Springer-Verlag, Berlin, New York, pp. 23–50. [6]

Holloway, S., Reay, D.M., Donato, J.A. and Beddoe-Stephens, B. (1991) Distribution of granite and possible Devonian sediments in part of the East Shetland Platform, North Sea. *J. Geol. Sci.* **148**, 635–8. [3]

Hollywood, J.M. and Whorlow, C.V. (1993) Structural development and hydrocarbon occurrence of the Carboniferous in the UK Southern North Sea Basin. In: Parker, J.R. (ed.) *Petroleum Geology of Northwest Europe: Proceedings of the 4th Conference.* Geological Society, London, pp. 689–96. [4]

Holm, G.M. (1996) The Central Graben: a dynamic overpressure system. In: Glennie, K.W. and Hurst, A. (eds) *AD 1995: NW Europe's Hydrocarbon Industry.* Geological Society, London, pp. 107–22. [1]

Holmes, A.J. (1991) The Camelot Fields, Blocks 53/1a, 53/2, UK North Sea, UK North Sea. In: Abbotts, I.L. (ed.) *UK North Sea United Kingdom Oil and Gas Fields: 25 Years Commemorative Volume.* Memoir No. 14, Geological Society, London, pp. 401–8. [1, 11]

Holser, W.T. and Wilgus, C.K. (1981) Bromide profiles of the Rowet Salt. Triassic of Northern Europe, as evidence of its marine origin. *Neues Jahr. Min. Monats.* **6**, 267–76. [7]

Home, P.C. (1987) Ula. In: Spencer, A.M., Campbell, C.J., Hanslien, S.H., Nelson, P.H., Nysaether, E. and Ormaasen, E.G. (eds) *Geology of the Norwegian Oil and Gas Fields.* Graham and Trotman, London, pp. 143–51. [1]

Hood, A., Gutjahr, C.C.M. and Heacock, R.L. (1975) Organic metamorphism and the generation of petroleum. *AAPG Bull.* **59**, 986–96. [11]

Hooper, R.J., Goh, L.S. and Dewey, F. (1995) The inversion history of the northeastern region of the Broad Fourteens Basin. In: Buchanan, J.G. and Buchanan, P.G. (eds) *Basin Inversion.* Special Publication No. 88, Geological Society, London, pp. 307–18. [9,11]

Horsfield, B., Heckers, J., Leythaeuser, D., Littke, R. and Mann, U. (1991) A study of the Holzener Asphaltkalk, N. Germany: observations regarding the distribution, composition and origin of organic matter in an exhumed petroleum reservoir. *Mar. Petr. Geol.* **8**(2), 198–212. [11]

Horsfield, B., Littke, R., Mann, U., Leythaeuser, D. and Heckers. J. (1989) The Holzner asphaltkalk of the Hils syncline, northern Germany: an exhumed petroleum reservoir? In: *Abstracts of Papers presented at the 14th Int'l Org. Geochem. Conference, Paris, 18-22 Sept. 1989.* EAOG and IFP, Paris. Paper 137. [11]

Horsfield, B., Schenk, H.J., Mills, N. and Welte, D.H. (1992) An investigation of the in-reservoir conversion of oil to gas: compositional and kinetic findings from closed-system programmed-temperature pyrolysis. *Org. Geochem.* **119**(1–3), 191–204. [11]

Horsfield, B., Curry, D.J., Bohacs, K., Little, R., Rullkotter, J., Schenk, H.J. *et al.* (1994) Organic geochemistry of freshwater and alkaline lacustrine sediments in the Green River Formation of the Washakie Basin, Wyoming. *Org. Geochem.* **22**(3–5), 415–40.

Horstad, I. and Larter, S.R. (1997) Petroleum migration, alteration and re-migration within the Troll Field, Norwegian North Sea. *AAPG. Bull.* **81**(2), 222–48. [1, 11]

Horstad, I., Larter, S.R., Dypvik, H., Aagaard, P., Bjørnvik, A.M., Johansen, P.E. *et al.* (1990) Degradation and maturity controls on oil field petroleum column heterogeneity in the Gullfaks field, Norwegian North Sea. In: Durand, B. and Behar, F. (eds) *Advances in Organic Geochemistry. Org. Geochem.* **16**(1–3), 497–510. [11]

Horstad, I., Larter, S.R. and Mills, N. (1992) A quantitative model of biological petroleum degradation within the Brent Group reservoir in the Gullfaks Field, the North Sea. In: Eckardt, C.B., Maxwell, J.R., Larter, S.R. and Manning, D.A.C. (eds) Advances in Organic Geochemistry, Part 11: Advances and Applications in Energy and the Natural Environment. *Organic Geochemistry,* **19**(4–6) pp. 107–33.

Horstad, I., Larter, S.R.and Mills, N. (1995) Migration of hydrocarbons in the Tampen Spur area, Norwegian North Sea: a reservoir geochemical evaluation. In: Cubitt, J.M. and England, W.A. (eds) *The Geochemistry of Reservoirs.* Special Publication No. 86, Geological Society, London, pp. 159–84. [11]

Horvitz, L. (1980) Near-surface evidence of hydrocarbon movement from depth. In: Roberts, W.H. and Cordell, R.J. (eds) *Problems of Petroleum Migration.* Studies in Geology No. 10, AAPG, Tulsa, OK., pp. 241–70. [11]

Hospers, J., Rathore, J.S., Jianhua, F., Finnstrom, E.G. and Holthe, J. (1988) Salt tectonics in the Norwegian–Danish Basin. *Tectonophysics* **149**, 35–60. [6]

House, M.R., Richardson, J.B., Chaloner, W.G., Allen, J.R.L., Holland, C.H. and Westoll, T.S. (1977) *A Correlation of Devonian Rocks of the British Isles.* Special Report No. 8, Geological Society, London, p. 110. [3]

Houseman, G. and England, P. (1986) A dynamical model of lithosphere extension and sedimentary basin formation. *J. Geophys. Res.* **91**, 719–29. [2, 8]

Hovland, M. (1990) Suspected gas-associated clay diapirism on the seabed off Mid Norway, *Mar. Petr. Geol.* **7**(3), 267–76. [11]

Hovland, M. (1993) Submarine gas seepage in the North Sea and adjacent areas. In: Parker, J.R. (ed.) *Petroleum Geology of Northwest Europe: Proceedings of the 4th Conference.* Geological Society, London, pp. 1333–8 [11]

Hovland, M. and Judd, A.G. (1988) *Seabed Pockmarks and Seepages: Impact on Geology, Biology and Marine Environment.* Graham and Trotman, London, 293 pp. [11]

Hovland, M. and Sommerville, J.H. (1985) Characteristics of two natural gas seepages in the North Sea. *Marine and Petroleum Geol.* **2**, 319–26.

Howell, J.A., Flint, S.S. and Hunt, C. (1996) Sedimentological aspects of the Humber Group Upper Jurassic of the South Central Graben, UK North Sea. *Sedimentology* **43**, 89–114. [12]

Howell, T.J. and Griffiths, P.S. (1994) A study of the Blocks 48/18, 48/19 and 48/15 Lower Cretaceous Greensand prospectivity, North Celtic Sea Basin. In: Croker, P.F. and Shannon, P.M. (eds) (1995) *The Petroleum Geology of Ireland's Offshore Basins*, Special Publication No. 93, Geological Society, London, pp. 261–76. [11]

Howell, T.J. and Griffiths, P.S. (1995) A study of the hydrocarbon distribution and Lower Cretaceous Greensand prospectivity in Blocks 48/15, 48/17, 48/18 and 48/19, North Celtic Sea Basin. In: Croker, P.F. and Shannon, P.M. (eds) *The Petroleum Geology of Ireland's Offshore Basins*. Special Publication No. 93, Geological Society, London, pp. 261–75. [11, 12]

Høye, T., Damsleth, E. and Hollund, K. (1994) Stochastic structural modelling of Troll West, with special emphasis on the thin oil zone. In: Yarus, J.M. and Chambers, R.L. (eds) *Stochastic Modelling and Geostatistics: Principles, Methods and Case Studies*. Computer Applied Geology No. 3, AAPG, Tulsa, OK., pp. 217–39. [12]

Hsü, K.J. (1972) Origin of the saline giants: a critical review after discovery of the Mediterranean evaporite. *Earth Sci. Rev.* **8**, 371–96. [6]

Hsü, K.J. and Jenkyns, H.C. (eds) (1974) *Pelagic Sediments: On Land and Under the Sea*. Special Publication No. 1, International Association of Sedimentologists, Blackwell Scientific Publications, Oxford, 447 pp.

Huc, A.-Y. (1976) Mise en evidence de provinces geochemeques dans les schistes bitumineux de Toarcien de l'est du Bassin de Paris. *Rev. IFP.* **XXX1**(6), 933–53. [11]

Huc, A.-Y. (1990) Understanding organic facies: a key to improved quantitative petroleum evaluation of sedimentary basins. In: Huc, A.-Y. (ed.) *Deposition of Organic Facies*. Studies in Geology No. 30, AAPG, Tulsa, OK., pp. 1–12. [11]

Huc, A.-Y. (ed.) (1995) *Paleogeography, Paleoclimate, and Source Rocks*. Studies in Geology No. 40, AAPG, Tulsa, OK., 347 pp. [11]

Huc, A.-Y., Lallier-Verges, E., Bertrand, P., Carpentier, B. and Hollander, D.J. (1992) Organic matter response to change of depositional environment in Kimmeridgian shales, Dorset, UK. In: Whelan, J. and Farrington, J. (eds) *Organic Matter, Productivity, Accumulation and Preservation in Recent Sediments*. pp. 469–86. [11]

Hudson, J.D. and Martill, D.M. (1994) The Peterborough Member (Callovian, Middle Jurassic) of the Oxford Clay formation at Peterborough, UK. *J. Geol. Soc.* **151**(1), 113–25. [11]

Hudson, N.F.C. (1985) Conditions of Dalradian metamorphism in the Buchan area, NE Scotland. *J. Geol. Soc.* **142**(1), 63–76. [2]

Hughes, W.B., Holba, A.G., Miller, D.E. and Richardson, J.S. (1985) Geochemistry of greater Ekofisk crude oils. In: Thomas, B.M., Muller-Pedersen, P., Whitaker, M.F. and Shaw, N.D. (eds), *Petroleum Geochemistry in Exploration of the Norwegian Shelf.* NPS, Graham and Trotman, London, pp. 75–92. [9, 11]

Humphreys, B., Smith, S.A. and Strang, G.E. (1989) Authigenic chlorite in late Triassic sandstones from the Central Graben, North Sea. *Clay Minerals* **24**, 427–44. [7]

Hunt, D. and Tucker, M.E. (1992) Stranded parasequences and the forced regressive wedge systems tract: deposition during base-level fall. *Sed. Geol.* **81**, 1–9. [8]

Hunt, J.M. (1990) AAPG Generation and migration of petroleum from abnormally pressured fluid compartments. *AAPG Bull.* **74**(1), 1–12. [11]

Hunt, J.M. (1995) *Petroleum Geochemistry and Geology.* (2 edn) W.H. Freeman and Co., New York, NY. [11]

Hunt, J.M. and McNichol, A.P. (1984) The Cretaceous Austin Chalk of south Texas—a petroleum source rock. In: Palacas, J.G. (ed.) *Petroleum Geochemistry and Source Rock Potential of Carbonate Rocks*. Studies in Geology No. 18, AAPG, Tulsa, OK., pp. 97–116. [9]

Hunter, R.E. (1977) Basic types of stratification in small eolian dunes. *Sedimentology* **24**, 361–87. [5]

Hurst, A. (1981) Mid Jurassic stratigraphy and facies at Brora, Sutherland. *Scott. J. Geol.* **17**, 169–77. [8]

Hurst, A. and Milodowski, A. (1966) Thorium distribution in some North Sea sandstones: implications for petrophysical evaluation. *Petr. Geosci.* **2**, 59–68. [11]

Hurst, A., Johnson, H.D., Burley, S.D., Canham, A.C. and Mackertich, D.S. (eds) (1996) *Geology of the Humber Group, Central Graben and Moray Firth, UKCS.* Special Publication No. 114, Geological Society, London. [8, 11]

Hurst, C. (1983) Petroleum geology of the Gorm Field, Danish North Sea. *Geol. Minjbouw* **62**, 157–68. [1, 9]

Huttel, P. (1989) Das Stassfurt-Karbonat Ca2 in Süd-Oldenburg Fazies und Diagenese eines Sediments am Nordhang der Hunter Schwelle. *Gottinger Arb. Geol. Paldont.* **39**, 1–94. [6]

Hutton, D.H.W. and Alsop, G.I. (1996) The Caledonian strike-swing and associated lineaments in MW Ireland and adjacent areas: sedimentation, deformation and igneous intrusion patterns. *J. Geol. Soc.* **153**(3), 345–60. [2]

Huseby, B., Barth, T. and Ocampo, R. (1996) Porphyrins in Upper Jurassic source rocks and correlations with other source rock descriptors. *Organic Geochem.* **25**(5/7), 273–94. [11]

Hvoslef, S., Dypvik, H. and Solli, H. (1986) A combined sedimentological and organic geochemical study of the Jurassic/ Cretaceous Janusfjellet Formation (Svalbard), Norway. In: Leythaeuser, D. and Rullkotter, J. (eds) *Advances in Organic Geochemistry, 1985.* Pergamon, Oxford, pp. 101–12. [11]

Hvoslef, S., Larter, S.R. and Leythaeuser, D. (1988) Aspects of generation and migration of hydrocarbons from coal bearing strata of the Hitra formation, Haltenbanken area, offshore Norway. In: Mattavelli, L. and Novelli, L. (eds) *Advances in Org. Geochem., 1987 (Proc. 13th Int'l. Mtg on Org. Geochem. Venice), Part 1.* Pergamon, Oxford, pp. 525–36. [11]

Hvoslef, S., Christie, O.H.J., Sassen, R., Kennicutt, M.C.II, Requeto, A.G. and Brooks, J.M. (1996) Test of a new surface geochemistry tool for resource prediction in frontier areas. *Mar. Petr. Geol.* **13**(1), 107–24. [11]

Illing, L.V. and Griffith, A.E. (1986) Gas prospects in the 'Midland Valley' of Northern Ireland. In: Brooks, J., Goff, J and van Hoorn, B. (eds) *Habitat of Palaeozoic Gas in N.W. Europe.* Special Publication No. 23, Geological Society, London, pp. 73–84. [2, 6, 11]

Illing, L.V. and Hobson, G.D. (eds) (1981) *Petroleum Geology of the Continental Shelf of North-West Europe.* Institute of Petroleum, Heyden, London, 521 pp. [1]

IMNES/Netherlands Geological Survey (1984) *Geologische dwardsdoorsneden van Nederland en het bijbehorend continental plat.* IMNES, The Hague. [2]

Ineson, J.R. (1993) The Lower Cretaceous chalk play in the Danish Central Trough. In: Parker, J.R. (ed.) *Petroleum Geology of Northwest Europe: Proceedings of the 4th Conference.* Geological Society, London, pp. 175–83. [1, 2, 5, 7, 9, 12]

Inglis, I. and Gerard, J. (1991) The Alwyn North Field, Blocks 3/9a, 3/4a, UK North Sea. In: Abbotts, I.L. (ed.) *UK North Sea United Kingdom Oil and Gas Fields: 25 Years Commemorative Volume.* Memoir No. 14, Geological Society, London, pp. 21–32. [1, 7, 12]

Irwin, H. (1979) An environmental model for the Type Kimmeridge Clay (comment and reply). *Nature* **279**, 819–20. [11]

Irwin, H. and Meyer, T. (1990) Lacustrine organic facies. A biomarker study using multivariate statistical analysis. In: Durand, B. and Behar, F. (eds) *Advances in Organic Geochemistry.* Pergamon Press, Oxford. pp. 197–210. [11]

Irwin, H., Hermanrud, C., Carlsen, E.M., Vollset, J. and Nordvall,

I. (1993) Basin modelling of hydrocarbon charge in the Egersund Basin, Norwegian North Sea: pre- and post drilling assessments. In: Doré, A.G., Augustson, J.H., Hermanrud, C., Stewart, D.J. and Sylta, O. (eds) *Basin Modelling: Advances and Applications*. Special Publication No. 3, NPF, Elsevier, Amsterdam, pp. 539–48. [11]

Isaksen, D. and Tonstad, K. (1989) A revised Cretaceous and Tertiary lithostratigraphic nomenclature for the Norwegian North Sea. *NPD Bull.* **5**, 1–59. [9, 10]

Isaksen, G.H. (1995) Organic geochemistry of paleodepositional environments with a predominance of terrigenous higher-plant organic matter. In: Huc, A.-Y. (ed.) *Paleogeography, Paleoclimate, and Source Rock.* Studies in Geology No. 40, AAPG, Tulsa, OK., pp. 81–105. [11]

Islam, S., Hesse, R. and Chagnon, A. (1982) Zonation of diagenesis and low grade metamorphism in Cambro-Ordovician flysch of Gaspe Peninsula, Quebec Appalachians. *Canadian Mineralogist* **20**, 155–67. [11]

Issler, D.R. (1984) Calculation of organic maturation levels for offshore Eastern Canada—implications for general application of Lopatin's method. *Canadian Journal of Earth Sciences* **21**, 477–88. [11]

Issler, D.R. and Snowdon, L.R. (1990) Hydrocarbon generation kinetics and thermal modelling, Beaufort–Mackenzie Basin. *Bull. Canadian Petr. Geol.* **38**(1), 1–16. [11]

Jackson, D.I. (1994) Discussion of Holliday, D.W. (1993) Geophysical log signatures in the Eden Shales (Permo-Triassic) of Cumbria and their regional significance. *Procs. Yorks. Geol. Soc.* **50**, 175–83. [6]

Jackson, D.I. and Mulholland, P. (1993) Tectonic and stratigraphic aspects of the East Irish Sea Basin and adjacent areas: contrasts in their post-Carboniferous structural styles. In: Parker, J.R. (ed.) *Petroleum Geology of Northwest Europe: Proceedings of the 4th Conference.* Geological Society, London, pp. 791–808. [2, 12]

Jackson, D.I., Mulholland, P., Jones, S.M. and Warrington, G. (1987) The geological framework of the East Irish Sea Basin. In: Brooks, J. and Glennie, K.W. (eds) *Petroleum Geology of North West Europe.* Graham and Trotman, London, pp. 191–203. [6]

Jackson, J.A. and McKenzie, D.P. (1983) The geometric evolution of the normal fault system. *J. Struct. Geol.* **5**, 471–82. [2, 8]

Jackson, J.S. and Hastings, D.S. (1987) Hydrocarbon trapping styles in mid-Norway. In: Brooks, J. and Glennie, K.W. (eds) *Petroleum Geology of North West Europe.* Graham and Trotman, London, pp. 1079–90. [12]

Jackson, M.P.A. and Talbot, C.J. (1986) External shapes, strain rates, and dynamics of salt structures. *AAPG Bull.* **97**, 305–23. [6]

Jacob, H. (1993) Nomenclature, classification, characterization and genesis of natural solid bitumen (migrabitumen). In: Parnell, J., Kucha, H. and Landais, P. (eds) *Bitumens in Ore Deposits.* Special Publication No. 9, Soc. Geol. Appl. Minl. Dps., Springer-Verlag, London. pp. 11–27. [11]

Jacobsen, F. (1982) Triassic. In: Michelsen, O. (ed.) *Geology of the Danish Central Graben.* Denmark Geol. Unders. Series B, No. 8, pp. 32–7. [7]

Jacobsen, V.W. and van Veen P. (1984) The Triassic offshore Norway north of 62°N. In: Spencer, A.M. (ed.) *Petroleum Geology of the North European Margin.* Proceedings of the Northern Europe Margin Symposium, NEMS, Trondheim, 1983, Graham and Trotman, London. [7]

Jacquin, T., Arnaud-Vanneau, A., Arnaud, H., Ravenne, C. and Vail, P.R. (1991) Systems tracts and depositional sequences in a carbonate setting: a study of continuous outcrops from platform to basin at the scale of seismic lines. *Mar. Petr. Geol.* **8**, 122–39. [9]

Jager, J. de, Doylke, M.A., Grantham, P.J. and Mabillard, J.E. (1996) Hydrocarbon habitat of the West Netherlands Basin. In: Rondeel, H.E., Batjes, D.A.J. and Nieuwenhuijs, W.H. (eds) *Geology of Gas in and under the Netherlands.* Kluwer Academic Publishers, Dordrecht, 284 pp. [11]

Jakobsson, K.H., Hamar, G.P., Ormaasen, O.E. and Skarpnes, O. (1980) Triassic facies in the North Sea north of the Central Highs. In: NPF (ed.) *The Sedimentation of the North Sea Reservoir Rocks.* Geilo XVIII, NPS, pp. 1–10. [7]

James, A.T. (1983) Correlation of natural gas by use of isotope distribution between components. *AAPG Bull.* **67**, 1176–91. [11]

James A.T. (1990) Correlation of reservoired gases using the carbon isotopic compositions of wet gas components. *AAPG Bull.* **74**(9), 1441–58. [11]

James, A.T. and Burns, B.J. (1984) Microbial alteration of subsurface natural gas accumulations. *AAPG Bulletin* **68**(8), 957–60. [11]

James, W.C. (1985) Early diagenesis, Atherton Formation Quaternary: a guide for understanding early cement distribution and grain modifications in non-marine deposits. *J. Sed. Petrol.* **55**(1), 135–46. [5]

Jankowski, B. (1991) *Sedimentologie des Oberkarbons in Nordwest-Deutschland.* DGMK-Bericht 384–5, Deutsche wissenschaftlichte Gesellschaft für Erdol, Erdgas und Kohle, Hamburg, 317 pp. [4]

Japsen, P. (1993) Influence of lithology and Neogene uplift on seismic velocities in Denmark: implications for depth conversion of maps. *AAPG Bull.* **77**(2), 194–212. [11]

Jeans, C.V. (1978) The origin of the Triassic clay assemblages of Europe with special reference to the Keuper Marl and Rhaetic of parts of England. *Phil. Trans. Roy. Soc. London* **289**(1365), 549–639. [7]

Jeans, C.V., Merriman, R.J. and Mitchell, J.G. (1977) Origin of Middle Jurassic and Lower Cretaceous Fuller's earths in England. *Clay Mineralogy* **12**, 11–44. [2, 9]

Jeans, C.V., Merriman, R.J. Mitchell, J.G. and Bland, D.J. (1982) Volcanic clays in the Cretaceous of southern England and Northern Ireland. *Clay Mineralogy* **17**, 105–56. [9]

Jeans, C.V., Long, D., Hall, M.A., Bland, D.J. and Cornford, C. (1991) The geochemistry of the Plenus Marls at Dover, England: evidence of fluctuating oceanographic conditions and of glacial control during the development of the Cenomanian–Turonian $\delta^{13}C$ anomaly. *Geol. Mag.* **28**(6), 603–32. [11]

Jeans, C.V., Reed, S.J.B. and Xing, M. (1993) Heavy mineral stratigraphy in the UK Trias: Western Approaches, onshore England and the Central North Sea. In: Parker, J.R. (ed.) *Petroleum Geology of Northwest Europe: Proceedings of the 4th Conference.* Geological Society, London, pp. 609–24. [1, 7]

Jeans, C.V., Mitchell, J.G., Scherer, M. and Fisher, M.J. (1994) Origin of the Permo-Triassic clay mica assemblage. *Clay Minerals* **29**, 575–89. [7]

Jenkyns, H.C. (1980) Cretaceous anoxic events: from continents to oceans. *J. Geol. Soc.* **137**, 171–88. [9]

Jenkyns, H.C. (1985) The Early Toarcian and Cenomanian/Turonian anoxic events in Europe: comparisons and contrasts. *Geol. Rundschau* **74**, 505–18. [9]

Jenner, J.K. (1981) The structure and stratigraphy of the Kish Bank Basin. In: Illing, L.V. and Hobson, G.D. (eds) *Petroleum Geology of the Continental Shelf of North-West Europe.* Heyden, London, pp. 426–31. [11]

Jensen, P.K., Holm, L. and Thomsen, E. (1985) Modelling burial history, temperature and maturation: In: Thomas, B.M., Doré, A.G., Eggen, S.S., Home, P.C. and Larsen, R. (eds) *Petroleum Geochemistry in Exploration of the Norwegian Shelf.* Graham and Trotman, London, pp. 145–52. [11]

Jensen, R.P. and Doré, A.G. (1993) A recent Norwegian Shelf heating event—fact or fantasy? In: Doré, A.G., Augustson, J.H., Hermanrud, C., Stewart, D.J. and Sylta, O. (eds) *Basin Modelling: Advances and Applications.* Special Publication No. 3, NPF, Elsevier, Amsterdam, pp. 85–106. [11]

Jensen, T.F. and Buchardt, B. (1987) Sedimentology and geochemistry of the organic carbon-rich Lower Cretaceous Sola Formation (Barremian–Albian), Danish North Sea. In: Brooks, J. and Glennie, K.W. *Petroleum Geology of North West Europe*. Graham and Trotman, London, pp. 431–40. [9, 11]

Jensen, T.F., Holm, L., Frandsen, N. and Michelsen, O. (1986) Jurassic–Lower Cretaceous lithostratigraphic nomenclature for the Danish Central Trough. *Danmarks Geol. Underssogelse Ser.* **12**, 7–65. [9]

Jensenius, J. and Munksgaard, N.C. (1989) Large scale hot water migration systems around salt diapirs in the Danish Central Trough and their impact on diagenesis in chalk reservoirs. *Geochim. Cosmochim. Acta* **53**, 79–88. [9, 11]

Jenssen, A.I., Bergslien, D., Rye-Larsen, M. and Lindholm, R.M. (1993) Origin of complex mound geometry of Paleocene submarine-fan sandstone reservoirs, Balder Field, Norway. In: Parker, J.R. (ed.) *Petroleum Geology of Northwest Europe; Proceedings of the 4th Conference*. Geological Society, London, pp. 135–43. [10, 12]

Jenyon, M.K. (1984a) Seismic response to collapse structures in the Southern North Sea. *Mar. Petr. Geol.* **1**, 27–36. [6]

Jenyon, M.K. (1984b) Upper Carboniferous gas indications and Zechstein features in Southern North Sea. *Oil Gas J.* **82**(20), 135–44. [11]

Jenyon, M.K. (1985a) Fault-associated salt flow and mass movement. *J. Geol. Soc.* **142**, 547–53. [6]

Jenyon, M.K. (1985b) Differential movement in salt rock. *Oil Gas J.* **83**, 73–5. [6]

Jenyon, M.K. (1985c) Basin edge diapirism and updip salt flow in Zechstein of Southern North Sea. *AAPG Bull.* **69**, 53–64. [6]

Jenyon, M.K. (1986a) *Salt Tectonics*. Elsevier Applied Science Publishers, London, New York. [6]

Jenyon, M.K. (1986b) Some consequences of faulting in the presence of a salt rock interval. *J. Petr. Geol.* **9**, 29–52. [6]

Jenyon, M.K. (1987) Regional salt movement effects in the English Southern Zechstein Basin. In: Peryt, T.M. (ed.) *The Zechstein Facies in Europe*. Springer-Verlag, Berlin, New York, pp. 77–92. [6]

Jenyon, M.K. (1988a) Seismic expression of salt dissolution-related features in the North Sea. *Bull. Canadian Petr. Geol.* **36**, 274–83. [6]

Jenyon, M.K. (1988b) Fault-salt wall relationships, Southern North Sea. *Oil Gas J.* Sept., 76–81. [6]

Jenyon, M.K. (1988c) Overburden deformation related to pre-piercement development of salt structures in the North Sea. *J. Geol. Soc.* **145**, 445–54. [6]

Jenyon, M.K. (1988d) Re-entrants of Zechstein 2 salt in the Mid North Sea High. *Mar. Petr. Geol.* **5**, 352–8. [6]

Jenyon, M.K. (1990) *Oil and Gas Traps*. John Wiley and Sons, London. [6]

Jenyon, M.K. and Cresswell, P.M. (1987) The Southern Zechstein Salt Basin of the British North Sea, as observed in regional seismic traverses. In: Brooks, J. and Glennie, K.W. (eds) *Petroleum Geology of North West Europe*. Graham and Trotman, London, pp. 277–92. [6]

Jenyon, M.K. and Taylor, J.C.M. (1983) Hydrocarbon indications associated with North Sea Zechstein shelf features. *Oil Gas J.* **81**, 155–60. [6]

Jenyon, M.K. and Taylor, J.C.M. (1987) Dissolution effects and reef-like features in the Zechstein across the Mid North Sea High. In: Peryt, T.M. (ed.) *The Zechstein Facies in Europe*. Springer-Verlag, Berlin, New York, pp. 51–75. [6]

Jenyon, M.K., Cresswell, P.M. and Taylor. J.C.M. (1984) Nature of the connection between the Northern and Southern Zechstein basins across the Mid North Sea High. *Mar. Petr. Geol.* **1**, 355–63. [5, 6]

Jessop, A.M. (1992) Thermal input from the basement of the Western Canada Sedimentary Basin. *Bull. Canadian Petr. Geol.* **40**(3), 198–206. [11]

Johannessen, P.N. and Andsbjerg, J. (1993) Middle to Late Jurassic basin evolution and sandstone reservoir distribution in the Danish Central Trough. In: Parker, J.R. (ed.) *Petroleum Geology of Northwest Europe: Proceedings of the 4th Conference*. Geological Society, London, pp. 271–83. [8, 12]

Johnes, L.H. and Gauer, M.B. (1991) The Northwest Hutton Field, Block 211/27, UK North Sea. In: Abbots, I.L. (ed.) *UK North Sea United Kingdom Oil and Gas Fields: 25 Years Commemorative Volume*. Memoir No. 14, Geological Society, London, p. 145. [1]

Johns, C.R. and Andrews, I.J. (1985) The petroleum geology of the Unst Basin, North Sea. *Marine and Petroleum Geology* **2**, 361–72. [1, 11]

Johnson, A. and Eyssautier, M. (1987) Alwyn North field and its regional geological context. In: Brooks, J. and Glennie, K.W. (eds) *Petroleum Geology of North West Europe*. Graham and Trotman, London, pp. 963–77. [1, 11, 12]

Johnson, G.A.L. (1984) Subsidence and sedimentation in the Northumberland Trough. *Proc. Yorks. Geol. Soc.* **45**, 71–83. [2]

Johnson, H. (1987) Seismic expression of major chalk reworking events in the Palaeocene of the central North Sea. In: Brooks, J. and Glennie, K.W. (eds) *Petroleum Geology of North West Europe*. Graham and Trotman, London, pp. 591–8. [9]

Johnson, H. and Lott, G.K. (1993) 2. Cretaceous of the Central and Northern North Sea. In: Knox, R.W.O'B. and Cordey, W.G. (eds) *Lithostratigraphic Nomenclature of the UK North Sea*. BGS, UKOOA, Nottingham. [9]

Johnson, H., Richards, P.C., Long, D. and Graham, C.C. (1993) *United Kingdom Offshore Regional Report: the Geology of the Northern North Sea*. BGS, HMSO, London, 110 pp. [9, 11]

Johnson, H., Warrington, G. and Stokes, S.J. (1994) Permian and Triassic of the Southern North Sea. In: Knox, R.W.O'B. and Cordey, W.G. (eds) *Lithostratigraphic Nomenclature of the UK North Sea*. BGS, UKOOA, Nottingham. [6]

Johnson, H.D. and Kroll, D.E. (1984) Geological modelling of a heterogeneous sandstone reservoir: Lower Jurassic Statfjord Formation, Brent Field. In: 59th Annual Technical Conference and Exhibition, Houston, SPE Paper 13050. [12]

Johnson, H.D. and Stewart, D.J. (1985) Role of clastic sedimentology in the exploration and production of oil and gas in the North Sea. In: Brenchley, P.J. and Williams, B.P.J. (eds) *Sedimentology: Recent Developments and Applied Aspects*. Special Publication No. 17, Geological Society, London, pp. 249–310. [1, 8, 11, 12]

Johnson, H.D., Mackay, T.A. and Stewart, D. (1986) The Fulmar Oilfield Central North Sea: geological aspects of its discovery, appraisal and development. *Mar. Petr. Geol.* **3**(2), 99–125. [1, 6, 7, 8, 12]

Johnson, M.R.W. (1983a) Torridonian–Moine. In: Craig, G.Y. (ed.) *Geology of Scotland*. Scottish Academic Press, Edinburgh, pp. 49–75. [2]

Johnson, M.R.W. (1983b) Dalradian. In: Craig, G.Y. (ed.) *Geology of Scotland*. Scottish Academic Press, Edinburgh, pp. 77–104. [2]

Johnson, R.J. and Dingwall, R.G. (1981) The Caledonides: their influence on the stratigraphy of the Northwest European Shelf. In: Illing, L.V. and Hobson, G.P. (eds) *Petroleum Geology of the Continental Shelf of North-West Europe*. Heyden, London, pp. 88–97. [2]

Johnstad, S.E., Uden, R.C. and Dunlop, K.N.B. (1993) Seismic reservoir modelling over the Oseberg field. *First Break* **11**(5), 177–85. [1]

Johnston, S.C., Smith, R.I. and Underhill, J.R. (1995) The Clair Discovery, west of the Shetland Isles. *Scott. J. Geol.* **31**, 187–90. [3]

Johnstone, G.S. (1966) *The Grampian Highlands*. British Regional Geology, Institute of Geological Science, HMSO, Edinburgh, 107, pp. [2]

Jones, E.L., Raveling, H.P. and Taylor, H.R. (1975) Stratfjord Field. In: Jurassic Northern North Sea Symposium, NPF, Article 20. [1]

Jones, G.L. (1992) Irish Carboniferous conodonts record maturation levels and the influence of tectonism, igneous activity and mineralisation. *Terra Nova* **4**, 238–44. [11]

Jones, J.M. and Creaney, S. (1977) Optical character of thermally metamorphosed coals of northern England. *J. Microsc.* **109**, 105–18. [11]

Jones, M.E., Bedford, J. and Clayton, C. (1984) On natural deformation mechanisms in the Chalk. *J. Geol. Soc.* **141**, 675–83. [9]

Jones, M.E., Leddra, M.J., Goldsmith, A., Berget, O.P. and Tappel, I. (1990) The geotechnical characteristics of weak North Sea reservoir rocks. In: Buller, A.T., Berg, E., Hjelmeland, O., Kleppe, J., Torsæter, O. and Aasen, J.O. (eds) *North Sea Oil and Gas Reservoirs II*. Norwegian Institute of Technology, Graham and Trotman, London, pp. 201–11. [9]

Jones, P.H. (1980) The role of geopressure in the hydrocarbon and water system. In: Roberts, W.H. and Cordell, R.J. (eds) *Problems of Petroleum Migration*. Studies in Geology No. 10, AAPG, Tulsa, OK., pp. 207–16.

Jones, R.W. (1980) Some mass balance and geological constraints on migration mechanisms. In: Roberts, W.H. and Cordell, R.J. (eds) *Problems of Petroleum Migration*. Studies in Geology No. 10, AAPG, Tulsa, OK., pp. 47–68.

Jones, R.W. (1987) Organic facies. In: Brooks, J. and Welte, D. (eds) *Advances in Petroleum Geochemistry, Vol. 2*. Academic Press, London, pp. 1–90. [11]

Jones, R.W. and Milton, N.J. (1994) Sequence development during uplift: Paleogene stratigraphy and relative sea-level history of the outer Moray Firth, UK North Sea. *Mar. Petr. Geol.* **11**, 157–63. [10]

Jordt, H., Faleide, J.J. Bjorlykke, K. and Ibrahim, M.T. (1995) Cenozoic sequence stratigraphy of the central and northern North Sea Basin: tectonic development, sediment distribution and provenance areas. *Mar. Petr. Geol.* **12**(8), 845–81. [11]

Jørgensen, L.N. and Andersen, P.M. (1991) *Integrated Study of the Kraka Field*. Publication No. 23082, SPE, pp. 461–74. [9, 12]

Jørgensen, N.O. (1986a) Geochemistry, diagenesis and nannofacies of chalk in the North Sea Central Graben. *Sedimentary Geol.* **48**, 267–94. [9]

Jørgensen, N.O. (1986b) Chemostratigraphy of Upper Cretaceous Chalk in the Danish sub-basin. *AAPG Bull.* **70**, 309–17. [9]

Jourdan, A., Thomas, M., Brevart, O., Robson, P., Sommer, F. and Sullivan, M. (1987) Diagenesis as the control of the Brent sandstone reservoir properties in the Greater Alwyn area (East Shetland Basin). In: Brooks, J. and Glennie, K.W. (eds) *Petroleum Geology of North West Europe*. Graham and Trotman, London, pp. 951–61. [12]

Kaaschieter, J.P.H. and Reijers, T.J.A. (eds) (1983) *Petroleum Geology of the Southeastern North Sea and Adjacent Onshore Areas*. Royal Geological Mining Society, the Hague, 239 pp. [1]

Kaldi, J. (1986a) Sedimentology of sandwaves in an oolite shoal complex in the Cadeby Magnesian Limestone Formation Upper Permian of eastern England. In: Harwood, G.M. and Smith, D.B. (eds) *The English Zechstein and Related Topics*. Special Publication No. 22, Geological Society, London, pp. 63–74. [6]

Kaldi, J. (1986b) Diagenesis of nearshore carbonate rocks in the Sprotbrough Member of the Cadeby Magnesian Limestone Formation Upper Permian of eastern England. In: Harwood, G.M. and Smith, D.B. (eds) *The English Zechstein and Related Topics*. Special Publication No. 22, Geological Society, London, pp. 87–102. [6]

Kamerling, P. (1979) Geology and hydrocarbon habitat of the Bristol Channel Basin. *J. Petr. Geol.* **2**, 75–93. [12]

Kanev, S.V., Margulis, L.S., Bojesen-Koefoed, J.A., Weil, W.A., Merta, H. and Zdanaviciute, O. (1994) Oils and hydrocarbon source rocks of the Baltic Syneclise. *Oil and Gas J.* **92**, 69–73. [11]

Kantorowicz, J.D., Eigner, M.R.P., Livera, S.E., van Schijndel-Goester, F.S. and Hamilton, P.J. (1992) Integration of petroleum engineering studies of producing Brent Group fields to predict reservoir properties in the Pelican Field, UK North Sea. In: Morton, A.C., Haszeldine, R.S., Giles, M.R. and Brown, S. (eds) *Geology of the Brent Group*. Special Publication No. 61, Geological Society, London, pp. 453–69. [12]

Karlsen, D.A. and Larter, S.R. (1991) Analysis of petroleum fractions by TLC-FID: applications to petroleum reservoir description. *Org. Geochem.* **17**(5), 603–17. [11]

Karlsen, D.A., Nedkvitne, T., Larter, S. and Bjorlykke, K. (1993) Hydrocarbon composition of authigenic inclusions: application to elucidation of petroleum reservoir filling history. *Geochim. Cosmochim. Acta* **57**, 3641–59. [11]

Karlsen, D.A., Nyland, B., Flood, B., Ohm, S.E., Brekke, T., Ohlsen, S. *et al.* (1995) Petroleum geochemistry of the Haltenbanken, Norwegian continental shelf. In: Cubitt, J.M. and England, W.A. (eds) *The Geochemistry of Reservoirs*. Special Publication No. 86, Geological Society, London, pp. 203–56. [11]

Karlsson, W. (1986) The Snorre, Statfjord and Gullfaks oil-fields and the habitat of hydrocarbons on the Tampen Spur, offshore Norway. In: Spencer, A.M., Campbell, C.J., Hanslien, S.H., Nelson, P.H., Nysaether, E. and Ormaasen, E.G. (eds) *Habitat of Hydrocarbons on the Norwegian Continental Shelf*. NPF, Graham and Trotman, London, pp. 181–97. [7, 12]

Karner, G.D., Lake, S.D. and Dewey, J.F. (1987) The thermal and mechanical development of the Wessex Basin, southern England. In: Coward, M.P., Dewey, J.F. and Hancock, P.L. (eds) *Continental Extensional Tectonics*. Special Publication No. 28, Geological Society, London, pp. 517–36.

Karnin, W.-D., Rockenbauch, K. and Ruijtenberg, P.A. (1991a) Zechstein exploration and appraisal in NW Germany with 3D seismic: an example. In: Abstracts 3rd EAPG Conference 26–30 May 1991, Florence, Italy. [6]

Karnin, W.-D., Rockenbauch, K. and Ruijtenberg, P.A. (1991b) The effect of the success of 3D seismic data on the exploration and appraisal of Zechstein targets in NW Germany. *First Break* **10**(6), 233–40. [1]

Karnin, W.-D., Idiz, E., Merkel, D. and Ruprecht, E. (1996) The Zechstein Stassfurt Carbonate hydrocarbon system of the Thuringian Basin, Germany. *Petr. Geosci.* **2**(1), 53–8. [11]

Kassler, P. (1996) Global energy challenges over the next fifty years. In: Glennie, K.W. and Hurst, A. (eds) *AD 1995: NW Europe's Hydrocarbon Industry*. Geological Society, London, pp. 195–201. [1]

Katz, B.J. and Pratt, L.M. (eds) (1993) *Source Rocks in a Sequence Stratigraphic Framework*. Studies in Geology No. 37, AAPG, Tulsa, OK., 247 pp. [11]

Kelling, G. (1988) Silesian sedimentation and tectonics in the South Wales Basin: a brief review. In: Besly, B.M. and Kelling, G. (eds) *Sedimentation in a Synorogenic Basin Complex: the Upper Carboniferous of Northwest Europe*. Blackie, Glasgow, London, pp. 38–42. [4]

Kelling, S.B. (1992) Milankovitch cyclicity recorded from Devonian non-marine sediments. *Terra Nova* **4**, 578–84. [3]

Kelly, P.G. (1996) Strike–slip faulting and block rotations at Kilve, Somerset. *Geoscientist* **6**(6), 14–16.

Kemp, A.G. and Stephen, L. (1996) Prospective developments, production and revenues from the UKCS 1996–2000: a financial and regional simulation. In: Glennie, K.W. and Hurst, A. (eds) *AD 1995: NW Europe's Hydrocarbon Industry*. Geological Society, London, pp. 215–25. [1]

Kemper, E. (1973) Das Berrias tiefe Unterkreide in NW-Deutschland. *Geol. Jahrbuch* **A9**, 47–67. [12]

Kemper, E. (1974) The Valanginian and Hauterivian stages in north-west Germany. In: Casey, R. and Rawson, P.F. (eds) *The Boreal Lower Cretaceous, Geol. J.* P. Issue No. 5, 327–44. [9]

Kemper, E. (1982) Palaeogeographie und Umveltfaktoren zur Zeit des spoten Apt und Frühen Alb. *Geol. Jahrbuch Reihe A* **45**, 651–3. [9]

Kemper, E. (1987) Das Lkima der Kreide-Zeit. *Geologische Jahrbuch Reihe A*. **96**, 5–185. [9]

Kenig, F., Hayes, B.N., Popp, B.N. and Summons, R.E. (1994) Isotopic biogeochemistry of the Oxford Clay Formation (Jurassic), UK. *J. Geol. Soc.* **151**(1), 139–53. [11]

Kennedy, W.J. (1983) Depositional mechanisms in North Sea Chalks. In: Hancock, J.M., Hamlin, R.J.O., Crosby, A., Balson, P.S., Jones, S.M., Chadwick, R.A. *et al.* (eds) *Chalk in the North Sea*. UK Course Notes No. 20, JAPEC, London, pp. B1–B16 and 4 figs. [9]

Kennedy, W.J. (1987) Sedimentology of Late Cretaceous–Palaeocene Chalk reservoirs, North Sea Central Graben. In: Brooks, J. and Glennie, K.W. (eds) *Petroleum Geology of North-West Europe*. Graham and Trotman, London, pp.469–81. [9]

Kennedy, W.J. and Garrison, R.E. (1975) Morphology and genesis of nodular chalks and hardgrounds in the Upper Cretaceous of southern England. *Sedimentology* **22**, pp. 311–86. [9]

Kennedy, W.J. and Juignet, P. (1974) Carbonate banks and slump beds in the Upper Cretaceous (Upper Turonian–Santonian) of Haute Normandie, France. *Sedimentology* **21**, pp. 1–42. [9]

Kennedy, W.Q. (1946) The Great Glen Fault. *Quart. J. Geol. Soc.* **102**, 41–76. [2]

Kent, D.V. and Gradstein, F.M. (1985) A Cretaceous and Jurassic geochronology. *Geological Society of America Bulletin* **96**, 1419–27.

Kent, P.E. (1967a) Outline geology of the southern North Sea Basin. *Proc. Yorks. Geol. Soc.* **36**, 1–22. [6]

Kent, P.E. (1967b) Progress of exploration in North Sea. *Am. Assoc. Petrol. Geol. Bull.* **51**, 731–41. [6]

Kent, P.E. (1980) Subsidence and uplift in East Yorkshire and Lincolnshire: a double inversion. *Proc. Yorks. Geol. Soc.* **42**, 505–24. [5, 9, 11]

Kent, P.E. (1985) Onshore oil exploration, 1930–1964. *Marine and Petroleum Geology*, **2**(1), 56–64. [1, 4, 11]

Kent, P.E., Gaunt, G.D. and Wood, C.J. (1980) *British Regional Geology: Eastern England from the Tees to The Wash*. HMSO for Institute of Geological Sciences, London, 2nd Edn. [9]

Kent, P.K. (1981) The history of the Northeast Atlantic Margin in a world setting. In: Kerr, J.W. and Fergusson A. *Geology of the North Atlantic Borderlands*. Memoir 7, Canadian Society of Petroleum Geology, Canada.

Kerlogue, A., Cherry, S., Davies, H., Quine, M. and Spotti, G. (1995) The Tiffany and Toni oil fields, Upper Jurassic submarine fan reservoirs, South Viking Grahen, UK North Sea. *Petroleum Geoscience* **1**, 279–85. [11]

Kerth, M. and Hailwood, E.A. (1988) Magnetostratigraphy of the Lower Cretaceous Vectis Formation (Wealden Group) on the Isle of Wight, southern England. *Journal of the Geological Society* **145**, 351–60. [9]

Kessler, L.G., Zang, R.D., Englehorn, J.A. and Eger, J.D. (1980) Stratigraphy and sedimentology of a Palaeocene submarine fan complex, Cod Field, Norwegian North Sea. In: *Sedimentation of the North Sea Reservoir Rocks*. Geilo, Elsevier, Amsterdam, Article 8. [1]

Kettel, D. (1982) Norddeutsche Erdgase: Stickstoff Gehalt und Isotopenvariationen als Reife-und Faziesindikatoren. *Erdol und Kohle, Erkgas, Petrochemie mit Brenstoff-Chemie* **35**(12), 557–9. [11]

Kettel, D. (1983) The East Groningen Massif: detection of an intrusive body by means of coalification. *Geol. Mijnbouw* **60**(1), 203–10. [11]

Kettel, D. (1989) Upper Carboniferous source rocks north and south of the Variscan Front NW and Central Europe. *Mar. Petr. Geol.* **6**, 170–81. [2, 4, 11]

Ketter, F.J. (1991a) The Esmond, Forbes and Gordon Fields, Blocks 43/8a, 43/13a, 43/15a, 43/20a, UK North Sea. In: Abbotts, I.L. (ed.) *UK North Sea United Kingdom Oil and Gas Fields: 25 Years Commemorative Volume*. Memoir No. 14, Geological Society, London, pp. 425–32. [1, 5, 7, 11, 12]

Ketter, F.J. (1991b) The Ravenspurn North Field, Blocks 42/30, 43/26a, UK North Sea. In: Abbotts, I.L. (ed.) *UK North Sea United Kingdom Oil and Gas Fields: 25 Years Commemorative Volume*. Memoir No. 14, Geological Society, London, pp. 459–67. [1, 5, 11]

Khorasani, G.K. (1989) Factors controlling source rock potential of the Mesozoic coal-bearing strata from offshore Central Norway: application to petroleum exploration. *Bull. Canadian Petr. Geol.* **37**(4), 417–27. [11]

Khorasani, G.K. and Michelsen, J.K. (1992) Primary alteration–oxidation of marine algal organic matter from oil source rocks of the North Sea and Norwegian Arctic: new findings. *Org. Geochem.* **19**(4–6), 327–45. [11]

Kiersnowski, H., Paul, J., Peryt, T. and Smith, D.B. (1995) Facies, paleogeography, and sedimentary history of the Southern Permian Basin in Europe. In: Scholle, P.A., Peryt, T.M. and Ulmer-Scholle, D.S. (eds) *The Permian of Northern Pangaea*: Vol. 1. *Paleogeography, Paleoclimates, Stratigraphy*. Springer-Verlag, Berlin, pp. 119–36. [7]

Kimbell, G.S., Chadwick, R.A., Holliday, D.W. and Werngren, O.C. (1989) The structure and evolution of the Northumberland Trough from new seismic reflection data, and its bearing on modes of continental extension. *J. Geol. Soc.* **146**, 775–87. [4]

King, C., Bailey, H.W., Burton, C.A. and King, A.D. (1989) Cretaceous of the North Sea. In: Jenkins, D.G. and Murray, J.W. (eds) *Stratigraphical Atlas of Fossil Foraminifera*. Ellis Horwood, Chichester, for British Micropal. Society, pp. 372–417. [9]

King, R.E. (1977) North Sea joins ranks of World's major oil regions. *World Oil* **185**, 35–45. [6]

Kirby, G.A., Smith, K., Smith, N.J. and Swallow, P.W. (1987) Oil and gas generation in eastern England. In: Brooks, J. and Glennie, K.W. (eds) *Petroleum Geology of North West Europe*. Graham and Trotman, London, pp. 171–80. [4, 11, 12]

Kirk, R.H. (1980) Statfjord Field—a North Sea giant. In: Halbouty, M.T. (ed.) *Giant Oil and Gas Fields of the Decade 1968–1978*. Memoir No. 30, AAPG, Tulsa, OK., pp. 95–116. [1, 11, 12]

Kirkland, J.T. (1984) How to recover, label, and evaluate fractured core. *Oil Gas J.* Dec. 118–20. [9]

Kirton, S.R. and Hitchen, K. (1987) Timing and style of crustal extension north of the Scottish Mainland. In: Coward, M.P., Dewey, J.F. and Hancock, P.L. (eds) *Continental Extensional Tectonics*. Special Publication No. 39, Geological Society, London, pp. 407–19. [2]

Kjemperud, A. and Fjeldskaar, W. (1989) Glacial isostacy—a tectonic effect neglected by petroleum geologists. In: *Proc. NPF Conference on Structural and Tectonic Modelling and its application to Petroleum Geology, Stavanger, Norway, Oct. '89*. NPS, Geilo. [11]

Kleppe, J., Berg, E.W., Buller, A.T., Hjelmelaud, O. and Torsæter, O. (eds) (1987) *North Sea Oil and Gas Reservoirs*. Graham and Trotman, London, 252 pp. [1]

Knag, G.O., South, D. and Spencer, A.M. (in press) Exploration trends in the Northern North Sea. In: Hanslien, S. (ed.) *Petroleum Exploration and Exploitation in Norway*. Special Publication No. 4, NPS. [12]

Kneller, B.C. (1987) A geological history of North-East Scotland. In: Trewin, N.H., Kneller, B.C. and Gillen, C. (eds) *Excursion Guide to the Geology of the Aberdeen Area*. Scottish Academic Press, Edinburgh, pp. 1–50. [2]

Knight, I.A. *et al.* (1993) The Embla Field. In: Parker, J. (ed.) *Petroleum Geology of Northwest Europe: Proceedings of the 4th Conference*. Geological Society, London, pp. 1433–44. [3, 4, 11, 12]

Knipe, R., Cowan, G. and Baledran, V.S. (1993) The tectonic history of the East Irish Sea Basin with reference to the Morecambe Fields. In: Parker, J.R. (ed.) *Petroleum Geology of Northwest Europe: Proceedings of the 4th Conference*. Geological Society, London, pp. 857–66. [12]

Knott, S.D. (1993) Fault seal analysis in the North Sea. *AAPG Bull.* **77**(5), 778–93. [11]

Knott, S.D., Burchell, M.T., Jolley, E.J. and Fraser, A.J. (1993) Mesozoic to Cenozoic plate reconstructions of the North Atlantic and hydrocarbon plays of the Atlantic margins. In: Parker, J.R. (ed.) *Petroleum Geology of Northwest Europe: Proceedings of the 4th Conference*. Geological Society, London, pp. 953–74. [2, 9, 10, 12]

Knox, R.W.O'B. and Holloway, S. (1992) Paleogene of the Central and Northern North Sea. In: Knox, R.W.O'B. and Cordey, W.G. (eds) *Lithostratigraphic Nomenclature of the UK North Sea*. BGS, Nottingham, 133 pp. [9, 10]

Knox, R.W.O'B. and Morton, A.C. (1988) The record of early Tertiary N. Atlantic volcanism in sediments of the North Sea Basin. In: Morton, A.C. and Parson, L.M. (eds) *Early Tertiary Volcanism and the Opening of the NE Atlantic*. Special Publication No. 39, Geological Society, London, pp. 407–19. [10]

Knudsen, K. and Meisingset, K. (1991) Evaporative fractionation effects for the oils in the Gullfaks South area. In: Manning, D. (ed.) *Organic Geochemistry, Advances and Applications in Energy and Natural Environments: Extended Abstracts, 15th Meeting EAOG*. Manchester University Press, Manchester, pp. 163–5. [11]

Knudsen, K., Leythaeuser, D., Dale, B., Larter, S. and Dahl, B. (1988) Variation in organic matter quality and maturity in Draupne formation source rocks from the Oseberg field region, offshore Norway. In: Mattavelli, L. and Novelli, L. (eds) *Org. Geochem.* **13**(4–6), 1051–60. [11]

Knutson, C.A. and Munro, I.C. (1991) The Beryl Field, Block 9/13, UK North Sea. In: Abbotts, I.L. (ed.) *UK North Sea United Kingdom Oil and Gas Fields: 25 Years Commemorative Volume*. Memoir No. 14, Geological Society, London, pp. 33–42. [1, 7, 12

Koch, C.B. and Christiansen, F.G. (1993) Maturation of Lower Palaeozoic kerogens from North Greenland. *Org. Geochem.* **20**(3), 405–15. [11]

Koch, J.O. (1983) Sedimentology of the Middle and Upper Jurassic reservoirs of Denmark. *Geol. Mijnbouw* **62**(1), 115–29. [11]

Kockel, F. (1995) *Structural and Palaeogeographical Development of the German North Sea Sector*. Beitraege zur Regionalen Geologie der Erde. Geb. Borntraeger, Berlin, 96 pp. [5, 7]

Koenig, R.H. (1986) Oil discovery in 6507/7, an initial look at the Heidrun Field. In: Spencer, A.M., Campbell, C.J., Hanslien, S.H., Nelson, P.H., Nysaether, E. and Ormaasen, E.G. (eds) *Habitat of Hydrocarbons on the Norwegian Continental Shelf*. Graham and Trotman, London, pp. 307–11. [8, 11]

Kooi, H., Cloetingh, S. and Remmelts, G. (1989) Intraplate stresses and the stratigraphic evolution of the North Sea Central Graben. *Geol. Mijnbouw* **68**, 49–72. [7, 11]

Kozur, H. and Reinhardt, P. (1969) Charophyten aus dem Muschelkalk und dem Unteren Keuper Mecklenburgs und Thuringens. *Monatsber. deutsch. Akad. Wiss. Berlin* **11**, 369–86. [7]

Krabbe, H. (1996) Biomarker distribution in the lacustrine shales of the Upper Triassic–Lower Jurassic Kap Stewart Formation, Jameson Land, Greenland. *Mar. Petr. Geol.* **13**(7), 741–55. [11]

Krabbe, H.F.G., Christiansen, G., Dam, S., Piasecki, S. and Stemmerick, L. (1994) Organic geochemistry of the Lower Jurassic Sortehat formation, Jameson Land, East Greenland. *Gronlands Geol. Und. Rapport* **164**, 5–18. [11]

Krinsley, D.H. and Smith, D.B. (1981) A selective study of grains from the Permian Yellow Sands of North East England. *Proc. Geol. Assoc.* **93**(3), 189–96. [5]

Krooss, B., Littke, R., Müller, B., Frielingsdorf, J., Schwochau, K. and Idiz, R.F. (1995) Generation of nitrogen and methane from sedimentary organic matter: implications on the dynamics of natural gas accumulations. *Chem. Geol.* **126**, 291–318. [11]

Krooss, B.M. (1987) Experimental investigation of the diffusion of light hydrocarbons in sedimentary rocks. In: Doligez, B. (ed.) *Migration of Hydrocarbons in Sedimentary Basins*. Editions Technip, Paris, pp. 329–52. [11]

Krooss, B.M. and Leythaeuser, D. (1997) Diffusion of methane and ethane through the reservoir cap rock: implications for the timing and duration of catagenesis, Discussion. *AAPG Bull.* **81**(1), 155–61. [11]

Krooss, B.M., Brothers, L. and Engel, M.H. (1991) Geochromatography in petroleum migration: a review. In: England, W.A. and Fleet, A.J. (eds) *Petroleum Migration*. Special Publication No. 59, Geological Society, London, pp. 149–66. [11]

Krooss, B.M., Leythauser, D. and Lillack, H. (1993) Nitrogen-rich natural gases: qualitative and quantitative aspects of natural gas accumulation in reservoirs. *Erdol Kohle-Erdgas-Petrochem* **46**, 271–6. [11]

Kruis, E. and Donzae, A. (1993) Revival of the Northeast Netherlands onshore Rotliegende play. *AAPG Bull.* **77**(9), 1638–9. [5]

Krystofiak, S., Koopmann, B., Degro, T., Beusekom, G.V. and Pool, V.D. (1993) The application of amplitude anomalies to Rotliegende reservoir delineation in Northern Germany. *AAPG Bull.* **77**(9), 1639. [5]

Kuhnt, W. and Wiedmann, J. (1995) Cenomanian–Turonian source rocks paleobiogeographic and paleoenvironmental aspects. In: Huc, A.-Y. (ed.) *Paleogeography, Paleoclimate, and Source Rock*. Studies in Geology No. 40, AAPG, Tulsa, OK., pp. 213–32. [11]

Kuhnt, W., Herbin, J.P., Thurow, J. and Wiedmann, J. (1990) Distribution of Cenomanian–Turonian organic facies in the Western Mediterranean and along the adjacent Atlantic margin. In: Huc, A.-Y. (ed.) *Deposition of Organic Facies*. Studies in Geology No. 30, AAPG, Tulsa, OK., pp. 133–60. [11]

Kulke, H. (ed.) (1994) *Regional Petroleum Geology of the World, Part 1*, Borntraeger, Berlin, 931 pp. [11]

Kuo, Lung Chuan and Michael, G.E. (1994) A multi-component oil-cracking kinetics model for modeling preservation and composition of reservoired oils. *Org. Geochem.* **21**(8/9), 911–27. [11]

Kurnsky, G.P., Bogdanchikov, P.I. and Abushayeva, V.V. (1983) Changes in hydrocarbon composition of benzine straight-run gasoline fractions during biodegradation. *Geochemistry International* **20**(1), 158–64. [11]

Laberg, J.S. and Andreassen, K. (1996) Gas hydrate and free gas indications within the Cenozoic succession of the Bjornoya Basin, western Barents Sea. *Marine and Petr. Geol.* **13**(8), 921–40. [11]

Lafargue, E. and Barker, C. (1988) Effect of water washing on crude oil compositions. *AAPG Bull.* **72**(3), 263–76. [11]

Lafargue, E. and Le Thiez, P. (1996) Effect of waterwashing on light ends compositional heterogeneity. *Org. Geochem.* **24**(12), 1141–50. [11]

Lafargue, E., Espitalié, J., Jacobsen, T. and Eggen, S. (1990) Experimental simulation of hydrocarbon expulsion. In: Durand, B. and Behar, F. (eds) *Advances in Organic Geochemistry. Org. Geochem.* **16**(1–3), 121–31. [11]

Lafargue, E., Espitalié, J., Brooks, T.M. and Nyland, B. (1994) Experimental simulation of primary migration. *Adv. Org. Geochem.* **22**(3–5), 575–86. [11]

Lagios, E. (1983) A gravity study of the E. Berwickshire Devonian basins. *Scott. J. Geol.* **19**, 189–203. [4]

Lallier-Vergès, E., Bertrand, P., Huc, A.Y., Buckel, D. and Tremblay, P. (1993) Control of the preservation of organic matter by productivity and sulphate reduction in Kimmeridgian shales from Dorset. *Mar. Petr. Geol.* **10**(6), 600–6. [11]

Lam, K. and Porter, R. (1977) The distribution of palynormphs in the Jurassic rocks of the Brora outlier, NE Scotland. *J. Geol. Soc.* **134**, 45–55. [8]

Lambert, R.A. (1991) The Victor Field, Blocks 49/17, 49/22, UK North Sea. In: Abbotts, I.L. (ed.) *UK North Sea United Kingdom Oil and Gas Fields: 25 Years Commemorative Volume*. Memoir No. 14, Geological Society, London, pp. 503–8. [1, 5]

Lambert, R. St. J. and McKerrow, W.S. (1976) The Grampian Orogeny. *Scott. J. Geol.* **12**, 271–92. [2]

Lambiase, J.J. (ed.) (1995) *Hydrocarbon Habitat in Rift Basins.* Special Publication No. 80, Geological Society, London.

Laming, D.J.C. (1966) Imbrication, palaeocurrents and other sedimentary features in the Lower Permian New Red Sandstone, Devonshire, England. *J. Sed. Petrol.* **36**(4), 940–59. [5]

Landais, P., Michels, R. and Elie, M. (1994) Are time and temperature the only constraints to the simulation of organic matter maturation? *Org. Geochem.* **22**(3–5), 617–30. [11]

Langbein, R. (1987) The Zechstein sulphates: the state of the art. In: Peryt, T.M. (ed.) *The Zechstein Facies in Europe.* Springer-Verlag, Berlin, New York, pp. 143–88. [6]

Langford, F.F. and Blanc-Valleron, M.-M. (1990) Interpreting Rock-Eval pyrolysis data using graphs of pyrolizable hydrocarbons vs. total organic carbon. *AAPG Bull.* **74**(6), 799–804. [11]

Largeau, C., Derenne, S., Casadevall, E., Berkaloff, C., Corolleur, M., Lugardon, B. *et al.* (1990) Occurrence and origin of 'ultralaminar' structures in 'amorphous' kerogens of various source rocks and oil shales. In: *Org. Geochem.* **16**(4–6), 889–95. [11]

Larsen, R.M. and Jaarvik, L.J. (1981) The geology of the Sleipner Field Complex. In: *Norwegian Symposium on Exploration.* NPF, Article 15, Norway, 31 pp. [11]

Larsen, R.M. and Skarpnes, O. (1984) Regional interpretation and hydrocarbon potential of the Traenabanken area. In: Spencer, A.M., Johnsen, S.O., Mork, A., Nysaether, E., Songstad, P. and Spinnangr. A. (eds) *Petroleum Geology of the North European Margin.* Graham and Trotman, London, pp. 217–36. [11]

Larsen, V., Aaasheim, S.M. and Masset, J.M. (1981) 30/6-Alpha structure: a field case study in the silver block. In: *Norwegian Symposium on Exploration, Bergen.* NPF, Article 14, pp. 1–34.

Larsen, V., Morkeseth, P.O. and Aasheim, S.M. (1987) Tyrihans. In: Spencer, A.M., Campbell, C.J., Hanslien, S.H., Nelson, P.H., Nysaether, E. and Ormaasen, E.G. (eds) *Geology of the Norwegian Oil and Gas Fields.* Graham and Trotman, London, pp. 411–18. [1, 8, 11, 12]

Larsen, V.B. (1987) A synthesis of tectonically-related stratigraphy in the North Atlantic–Arctic region from Aalenian to Cenomanian time. *Norsk Geol.* **67**, 281–93. [9]

Larter, S. (1988) Some pragmatic perspectives in source rock geochemistry. *Mar. Petr. Geol.* **5**(3), 194–204. [11]

Larter, S. (1989) Chemical models of vitrinite reflectance evolution. *Geol. Rundsch.* **78**, 349–59. [11]

Larter, S., Bjorlykke, K.O., Karlsen, D.A., Nedkvitne, T., Eglinton, T., Johansen, P.E., Mitchell, A.W. *et al.* (1990) Determination of petroleum accumulation histories: examples from the Ula Field, Central Graben, Norwegian North Sea. In: Buller, A.T., Berg, E., Hjelmeland, O., Kleppe, J., Torsæter, O. and Aasen, J.O. (eds) *North Sea Oil and Gas Reservoirs II.* Norwegian Institute of Technology, Graham and Trotman, London, pp. 319–30. [11]

Larter, S.R. and Aplin, A.C. (1995) Reservoir geochemistry, methods, applications and opportunities. In: Cubitt, J.M. and England, W.A. (eds) *The Geochemistry of Reservoirs.* Special Publication No. 86, Geological Society, London, pp. 5–32. [11]

Larter, S.R. and Horstad, I. (1992) Migration of petroleum into Brent Group reservoirs: some observations from the Gullfaks field, Tampen Spur area, North Sea. In: Morton, A.C. *et al.* (eds) *Geology of the Brent Group.* Special Publication No. 61, Geological Society, London, pp. 441–52. [11, 12]

Larter, S.R., Stoddart, D. and Bjorøy, M. (1995) Fractionation of pyrrolic nitrogen compounds in petroleum during migration: derivation of migration-related geochemical parameters. In: Cubitt, J.M. and England, W.A. (eds) *The Geochemistry of Reservoirs.* Special Publication No. 86, Geological Society, London, pp. 103–24. [11]

Latin, D.M., Dixon, J.E. and Fitton, J.G. (1990a) Rift-related magmatism in the North Sea basin. In: Blundell, D.J. and Gibbs, A.D. (eds) *Tectonic Evolution of the North Sea Rifts.* Oxford Scientific Publications, Oxford, pp. 101–44.

Latin, D.M., Dixon, J.E., Fitton, J.G. and White, N. (1990b) Mesozoic magmatic activity in the North Sea Basin: implications for stretching history. In: Hardman, R.P.F. and Brooks, J. (eds) *Tectonic Events Responsible for Britain's Oil and Gas Reserves.* Special Publication No. 55, Geological Society, London, pp. 207–27.

Lawrence, S.R., Coster, P.W. and Ireland, R.J. (1987) Structural development and petroleum potential of the northern flanks of the Craven Basin (Carboniferous), North West England. In: Brooks, J. and Glennie, K.W. (eds) *Petroleum Geology of North West Europe.* Graham and Trotman, London, pp. 225–34. [11]

Leach, H.M., Herbert, N., Los, A. and Smith, R.L. (1997) The Schiehallion Field. Abstract, *5th Conference on Petroleum Geology of NW Europe.* Barbican Centre, London. [1]

Leadholm, R.H., Ho, T.T.Y. and Sahai, S.K. (1985) Heat flow, geothermal gradients and maturation modelling on the Norweigian Continental Shelf using computer methods. In: Thomas, B.M., Doré, A.G., Eggen, S.S., Home, P.C. and Larsen, R.M. (eds) *Petroleum Geochemistry in Exploration of the Norwegian Shelf.* Graham and Trotman, London, pp. 131–44.

Lee, M., Aronson, J.L. and Savin, S.M. (1985) K/Ar dating of time of gas emplacement in Rotliegendes Sandstone, Netherlands, *AAPG Bull.* **69**(9), 1381–5. [5, 11, 12]

Lee, M., Aronson, J.L. and Savin, S.M. (1989) Timing and conditions of Permian Rotliegende sandstone diagenesis, southern North Sea: K/Ar and oxygen isotope data. *AAPG Bull.* **73**(2), 195–215. [11]

Lee, M.J. and Y.J. Hwang (1993) Tectonics and structural styles of the East Shetland Basin. In: Parker, J.R. (ed.) *Petroleum Geology of Northwest Europe: Proceedings of the 4th Conference.* Geological Society, London, pp. 1137. [2, 8]

Lee, M.R. (1993) Formation and diagenesis of slope limestones within the Upper Permian Zechstein Raisby Formation, northeast England. *Proc. Yorks. Geol. Soc.* **49**, 215–27. [6]

Leeder, M.R. (1982) Upper Palaeozoic basins of the British Isles: Caledonide inheritance versus Hercynian plate margin processes. *J. Geol. Soc.* **139**, 479–91. [2, 4]

Leeder, M.R. (1983) Lithospheric stretching and the North Sea Jurassic sourcelands. *Nature* **303**, 510–14. [2, 8]

Leeder, M.R. (1987) Tectonic and palaeogeographic models for Lower Carboniferous Europe. In: Miller, J.M., Adams, A.E. and Wright, V.P. (eds) *European Dinantian Environments.* John Wiley, Chichester, pp. 1–20. [4]

Leeder, M.R. (1988) Recent developments in Carboniferous geology: a critical review with implications for the British Isles and N.W. Europe. *Proc. Geol. Assoc.* **99**(2), 74–100. [2, 4, 5]

Leeder, M.R. and Boldy, S.R. (1990) The Carboniferous of the Outer Moray Firth Basin, quadrants 14 and 15, Central North Sea. *Mar. Petr. Geol.* **7**, 29–37. [3, 4, 11, 12]

Leeder, M.R. and Hardman, M. (1990) Carboniferous geology of the southern North Sea basin and controls on hydrocarbon propsectivity. In: Hardman, R.P.F. and Brooks, J. (eds) *Tectonic Events Responsible for Britain's Oil and Gas Reserves.* Special Publication No. 55, Geological Society, London, pp. 87–105. [2, 4, 12]

Leeder, M.R. *et al.* (1989) Sedimentary and tectonic evolution of the Northumberland Basin. In: Arthurton, R.S., Gutteridge, P. and Norlan, S.C. (eds) *The Role of Tectonics in Devonian and Carboniferous Sedimentation in the British Isles.* Occasional Publication No. 6, Yorkshire Geological Society, Ellembank Press, Maryport, Cumbria, pp. 207–24. [4]

Leeder, M.R., Raiswell, R., Al-Batty, H., McMahon, A. and Hardman, M. (1990a) Carboniferous stratigraphy, sedimentation, and correlation of well 48/3-3 in southern North Sea Basin: integrated use of palynology, natural gamma/sonic logs and carbon/sulphur geochemistry. *J. Geol. Soc.* **147**, 287–300. [4, 11]

Leeder, M.R., Boldy, S.R., Raiswell, R. and Cameron, R. (1990b) The Carboniferous of the Outer Moray Firth Basin, quadrants 14 and 15, Central North Sea. *Mar. Petr. Geol.* **7**(1), 29–37.

Leggett, J.K. (1980) British Palaeozoic black shales and their paleo-oceanographic significance. *J. Geol. Soc.* **137**, 139–56. [11]

Leggett, J.K., McKerrow, W.S. and Soper, N.J. (1983) A model for the crustal evolution of Southern Scotland. *Tectonics* **2**, 187–210. [2]

Lehner, F.K., Marsal, D., Hermans, L. and Kuyk, A. van (1988) A model of secondary hydrocarbon migration as a buoyancy-driven separate phase flow. *Revue de l'Institut Fracais du Petrole* **43**(2), 155–64. [11]

Leith, T.L., Kaarstad, I., Connan, J., Pierron, J. and Caillet, G. (1993) Recognition of caprock leakage in the Snorre Field, Norwegian North Sea. *Mar. Petr. Geol.* **10**(1), 29–42. [11]

Leonard, R.C. (1989) Distribution of subsurface pressure in the Norwegian Central Graben, and its effects on hydrocarbon migration and accumulation. Proc. *NPF Conference on Structural and Tectonic Modelling and its Application to Petroleum Geology, Stavanger, Norway, 18–20 October, 1989.* NPF, Geilo. [11]

Leonard, R.C. (1993) Distribution of sub-surface pressure in Norwegian Central Graben and applications for exploration. In: Parker, J. (ed.) *Petroleum Geology of Northwest Europe: Proc. 4th Conf.* Geological Society, London, pp. 1295–304. [11]

Leonard, R.C. and Munns, J.W. (1987) Valhall. In: Spencer, A.M., Campbell, C.J., Hanslien, S.H., Nelson, P.H., Nysaether, E. and Ormaasen, E.G. (eds) *Geology of the Norwegian Oil and Gas Fields.* Graham and Trotman, London, pp. 153–64. [1, 9]

Lepercq, Y.-Y. and Gaulier, J.-M. (1996) Two-stage rifting in the North Viking Graben area North sea: inferences from a new three-dimensional subsidence analysis. *Mar. Petr. Geol.* **13**(2) 129–48. [11]

Lepoutre, M. (1984) Diagen: a numerical model for appreciation of chemical evolution of organic matter during time. In: Durand, B. (ed.) *Thermal Phenomena in Sedimentary Basins.* Editions Technip, Paris, pp. 247–56.

Lervik, K.S., Spencer, A.M. and Warrington, G. (1990) Outline of Triassic stratigraphy and structure in the central and northern North Sea. In: Collinson, J.D. (ed.) *Correlation in Hydrocarbon Exploration.* Graham and Trotman, London, pp. 173–89. [7]

Lewan, M.D. and Buchardt, B. (1989) Irradiation of organic matter by uranium decay in the Alum Shale, Sweden. *Geochim. Cosmochim. Acta.* **53**, 1307–22. [11]

Lewis, A. and Hardman, M. (1995) Realising the upside: multidisciplinary field appraisal. The Millom Field, East Irish Sea Well 113/26a-2. Paper presented at Conference on the Petroleum Geology of the Irish Sea and Adjacent Areas, Geological Society, London, Feb., Abstracts, p. 42. [11]

Lewis, C.L.E., Green, P.F., Carter, A and Hurford, A.J. (1992) Elevated Late Cretaceous to Early Tertiary palaeotemperatures throughout northwest England: three kilometres of Tertiary erosion? *Earth Plan. Sci. Lett.* **112**, 131–45. [11]

Leythaeuser, D. and Poelchau, H.S. (1991) Expulsion of petroleum from Type III kerogen source rocks in gaseous solution: modelling of solubility fractionation. In: England, W.A. and Fleet, A.J. (eds) *Petroleum Migration.* Special Publication No. 59, Geological Society, London, pp. 33–46. [11]

Leythaeuser, D. and Ruckheim, J. (1989) Heterogeneity of oil composition within a reservoir as a reflection of accumulation history. *Geochim. Cosmochim. Acta* **53**(8), 2119–23. [11]

Leythaeuser, D., Schaefer, R.G. and Yukler, A. (1982) The role of diffusion in primary migration of hydrocarbons. *AAPG Bull.* **66**(4), 408–29. [11]

Leythaeuser, D., Mackenzie, A.S., Schaefer, R.G., Altebaumer, F.J. and Bjorøy, M. (1983) Recognition of migration and its effects within two coreholes in shale/sandstone sequences from Svalbard, Norway. In: Bjorøy, M., Albrecht, P., Cornford, C., Groot, K. de, Eglinton, G., Galimov, E. *et al.* (eds) *Advances in Organic Geochemistry, 1981.* John Wiley, Chichester, pp. 136–46. [11]

Leythaeuser, D., Littke, R., Muller, P.J., Radke, M. and Schaefer, R.G. (1987) Effects of primary migration and petroleum expulsion recognized by geochemical analysis. In: *Abstracts from the 13th International Meeting on Organic Geochemistry, Venice, Sept. 21–25th 1987.* EAOG, pp. 275–6. [11]

Leythaeuser, D., Radke, M. and Willsch, H. (1988) Geochemical effects of primary migration of petroleum in Kimmeridge source rocks from Brae Field area, North Sea. Part II. *Geochim. Cosmochim. Acta* **52**, 2879–91. [11]

Leythaeuser, D., Schaefer, R.G. and Radke, M. (1988) Geochemical effects of primary migration of petroleum in Kimmeridge source rocks from Brae Field area, North Sea. Part I. *Geochim. Cosmochim. Acta* **52**, 1–13. [11]

Leythaeuser, D., Littke, R., Radke, M. and Schaefer, R.G. (1988a) Geochemical effects of petroleum migration and expulsion from Toarcian rocks in the Hils syncline Area, NW Germany. In: Mattavelli, L. and Novelli, L. (eds) *Advances in Org. Geochem., 1987 (Proc. 13th Int'l. Mtg on Org. Geochem. Venice,), Part 1.* Pergamon, Oxford, pp. 489–502. [11]

Leythaeuser, D., Mackenzie, A., Schaefer, R.G. and Bjorøy, M. (1988b) A novel approach for recognition and quantification of hydrocarbon migration effects in shale–sandstone sequences. In: Beaumont, E.A. and Foster, N.H. (compilers) *Geochemistry.* Treatise of Petroleum Geology reprint series No. 8. AAPG, Tulsa, OK., pp. 540–64. [11]

Li, M., Larter, S.R., Taylor, P., Jones, M., Bowler, B. and Bjorøy, M. (1995) Biomarkers or not biomarkers? A new hypothesis for the origin of pristane involving derivation from methyltrimethyltridecylchromans (MTTCs) formed during diagenesis from chlorophyll and alkylphenols. *Org. Geochem.* **23**(2), 139–59. [11]

Liewig, N., Clauer, N. and Sommer, F. (1987) Rb–Sr and K–Ar dating of clay diagenesis in Jurassic sandstone oil reservoir, North Sea. *AAPG Bull.* **71**, 1467–74. [11]

Lindgreen, H. (1985) Diagenesis and primary migration in Upper Jurassic claystone source rocks in the North Sea. *AAPG Bull.* **69**(4), 525–36. [11]

Lindgreen, H. (1987a) Experiments on adsorption and molecular sieving and interferences on primary migration in Upper Jurassic claystone source rocks, North Sea. *AAPG Bull.* **71**(3), 308–21. [11]

Lindgreen, N. (1987b) Molecular sieving and primary migration in Upper Jurassic and Cambrian claystone source rocks. In: Brooks, J. and Glennie, K.W. (eds) *Petroleum Geology of North West Europe.* Graham and Trotman, London, pp. 357–64. [11]

Lindquist, S.J. (1988) Practical characterization of eolian reservoirs for development: Nugget Sandstone, Utah–Wyoming thrust belt. *Sed. Geol.* **56**(1/4), 315–39. [5]

Lindsay, N.G., Haselock, P.J. and Harris, A.L. (1989) The extent of Grampian orogenic activity in the Scottish Highlands. *J. Geol. Soc.* **146**, 733–5. [2]

Linsley, P.N., Potter, H.C., McNab, G. and Racher, D. (1980) The Beatrice Field, Inner Moray Firth, UK North Sea. In: Halbouty, M.T. (ed.) *Giant Oil and Gas Fields of the Decade 1968–1978.* Memoir No. 30, AAPG, Tulsa, OK., pp. 117–30. [1, 8, 11]

Lippolt, H.J., Hess, J.C. and Burger, K. (1984) Isotopische alter von pyroclastischen Sanidinen aus kaolin Kohlensteinene als Korrelationsmarken für das Mitteleuropaische Oberkarbon. *Fortschr. Geol. Westf.* **32**, 119–50. [2, 4, 5]

Liu, G. *et al.* (1992) Quantitative geodynamic modelling of Barents Sea Cenozoic uplift and erosion. *Norsk Geol. Tidsskrift* **72**, 313–16. [11]

Livberg, F. and Mjøs, R. (1989) The Cook Formation, an offshore sand ridge in the Oseberg area, northern North Sea. In: Collinson, J.D. (ed.) *Correlation in Hydrocarbon Exploration.* Graham and Trotman, London, pp. 299–312. [8]

Livera, S.E. (1989) Facies associations and sand body geometries in the Ness Formation of the Brent Group, Brent Field. In: Whateley, M.K.G. and Pickering, K.T. (eds) *Deltas: Sites and*

Traps for Fossil Fuels. Special Publication No. 41, Geological Society, London, pp. 269–86. [12]

Llewellyn, P.G. and Stabbins, R. (1970) The Hathern anhydrite series, Lower Carboniferous, Leicestershire, England. *Trans. Inst. Min. Metall.* **79B**, B1–B15. [4]

Loftus, G.W. and Greensmith, J.T. (1988) The lacustrine Burdiehouse Limestone Formation: key to the deposition of the Dinantian Oil Shales of Scotland. In: Fleet, A.J., Kelts, K. and Talbot, M.R. (eds) *Lacustrine Petroleum Source Rocks*. Special Publication No. 40, Geological Society, London, pp. 219–34. [4, 11]

Logan, B.W. (1987) *The MacLeod Evaporite Basin, Western Australia*. Memoir No. 44, AAPG, Tulsa, 140 pp. [6]

Lohmann, H.H. (1972) Salt dissolution in subsurface of British North Sea as interpreted from seismograms. *AAPG Bull.* **56**, 472–9. [6]

Lonergan, L., Cartwright, J.A., Laver, R. and Staffurth, J. (1998) Polygonal faulting in the Tertiary of the Central North Sea—implications for reservoir geology. In: Coward, M.P., Daltaban, T.S. and Johnson, H.D. (eds) *Structural Geology and Reservoir Characterisation*. Special Publication, Geological Society, London. [12]

Longman, C.D., Bluck, B.J. and Van Breeman, O. (1979) Ordovician conglomerates and the evolution of the Midland Valley. *Nature* **280**, 578–80. [2]

Longman, M.W. (1980) Carbonate diagenetic textures from nearshore diagenetic environments. *AAPG Bull.* **64**, 461–87. [9]

Longman, M.W. and Palmer, S.E. (1987) Organic geochemistry of mid-continent Middle and Late Ordovician oils. *AAPG Bull.* **71**(8), 938–50. [11]

Lorenz, V. and Nicholls, I.A. (1984) Plate and intra-plate processes of Hercynian Europe during the Late Paleozoic. *Tectonophysics* **107**, 25–56. [2, 6]

Lott, G.K. and Knox, R.W.O'B. (1994) Post Triassic of the Southern North Sea. In: Knox, R.W.O'B. and Cordey, W.G. (eds) *Lithostratigraphic Nomenclature of the UK North Sea*. BGS, Nottingham. [9]

Lott, G.K., Ball, K.C. and Wilkinson, I.P. (1985) Mid-Cretaceous stratigraphy of a cored borehole in the western part of the Central North Sea Basin. *Proc. Yorks. Geol. Soc.* **45**, 235–48. [9]

Lott, G.K., Fletcher, B.N. and Wilkinson, I.P. (1986) The stratigraphy of the Lower Cretaceous Speeton Clay Formation in a cored borehole off the coast of north-east England. *Proc. Yorks. Geol. Soc.* **46**, 39–56. [9]

Lott, G.K., Thomas, J.E., Riding, J.B., Davey, R.J. and Butler, N. (1989) Late Ryazanian black shales in the Southern North Sea Basin and their lithostratigraphical significance. *Proc. Yorks. Geol. Soc.* **47**, 321–4. [9]

Loutit, T.S., Hardenbol, J., Vail, P.R. and Baum, G.R. (1988) Condensed sections: the key to age dating and correlation of continental margin sequences. In: Wilgus, C.K., Hastings, B.S., Posamentier, H., van Wagoner, J., Ross, C.A. and Kendall, C.G.St.C. (eds) *Sea-Level Changes: an Integrated Approach*. Special Publication No. 42, SEPM, Tulsa, OK., pp. 183–213. [8]

Lovell, J.P.B. (1983) Permian and Triassic. In: Craig, G.Y. (ed.) *Geology of Scotland*, 2nd edn. Scottish Academic Press, pp. 325–42. [5]

Lovell, J.P.B. (1990) The Cenozoic. In: Glennie, K.W. (ed.) *Introduction to the Petroleum Geology of the North Sea*. Blackwell Scientific Publications, Oxford, pp. 273–93. [10]

Lucazeau, S. and Douaran, Le S. (1984) Numerical model of sediment thermal history: comparison between the Gulf of Lion and the Viking Graben. In: Durand, B. (ed.) *Thermal Phenomena in Sedimentary Basins*. Editions Technip, Paris, pp. 211–18. [11]

Lucazeau, S. and Douaran, Le S. (1985) The blanketing effect of sediments in basins formed by extension: a numerical model. Application to the Gulf of Lion and the Viking Graben. *Earth Plan. Sci. Lett.* **74**, 92–102. [11]

Lupe, R. and Ahlbrandt, T.S. (1979) Sediments of ancient eolian environments—reservoir inhomogeneity. In: McKee, E.D. (ed.) *A Study of Global Sand Seas*. Professional Paper No. 1052, USGS, Washington, DC., pp. 241–51. [5]

Luthi, S.M. and Banavar, J.R. (1988) Application of borehole images to three dimensional modelling of eolian sandstone reservoirs, Permian Rotliegende, North Sea. *AAPG Bull.* **72**(9), 1074–89. [5]

Lutz, M., Kaaschieter, J.P.H. and van Wijhe, D.H. (1975) Geological factors controlling Rotliegend gas accumulation in the Mid European Basin. *Proc. 9th World Petrol. Congr.* **22**, 93–7. [5, 11, 12]

Lützner, H. (1969) Uber die verbreitung der Manebacher Schichten im Rotliegenden des Thuringer Waldes. *Geologie* **18**(7), 815–27. [5]

Lützner, H. (1988) Sedimentology and basin development of intramontane Rotliegend basins in Central Europe. *Zeit. Geol. Wiss.* **16**(9), 845–63. [5]

McArthur, J.M., Thirlwall, M.F., Gale, A.S., Kennedy, W.J., Burnett, J.A., Mattey, D. *et al.* (1993) Strontium isotope stratigraphy for the Late Cretaceous: a new curve, based on the English chalk. In: Hailwood, E.A. and Kidd, R.B. (eds) *High Resolution Stratigraphy*. Special Publication No. 70, Geological Society, London, pp. 195–209. [9]

McAuliffe, C.D. (1980) Oil and gas migration: chemical and physical constraints. In: Roberts, W.H., III and Cordell, R.J. (eds) *Problems of Petroleum Migration*. Studies in Geology 10, AAPG, Tulsa, OK., pp. 89–107. [11]

McBridge, J.J. (1992) The diagenesis of Middle Jurassic reservoir sandstones of Bruce Field, U.K. North Sea. PhD thesis, University of Aberdeen. [3]

McCann, T. (1998) The Rotliegend of the NE German Basin: background and prospectivity. *Petr. Geosci.* **4**(i), 17–27. [5]

McCann, T., Shannon, P.M. and Moore, J.G. (1995) Fault styles in the Porcupine Basin, offshore Ireland: tectonic and sedimentary control. In: Croker, P.F. and Shannon, P.M. (eds) *The Petroleum Geology of Ireland's Offshore Basins*. Special Publication No. 93, Geological Society, London, pp. 371–83. [12]

McClay, K.R. (1990) Extensional fault systems in sedimentary basins: a review of analogue studies. *Mar. Petr. Geol.* **7**, 206–33. [12]

McClay, K.R., Norton, M.G., Coney, P. and Davis, G.H. (1986) Collapse of the Caledonian orogen and the Old Red Sandstone. *Nature* **323**, 147–9. [3]

McClean, D. (1995) A palynostratigraphic classification of the Westphalian of the Southern North Sea Carboniferous Basin. (Abstract) In: *Stratigraphic advances in the offshore Devonian and Carboniferous rocks, UKCS and adjacent onshore areas*. Geological Society, London. [7]

McClure, N.M. and Brown, A.A. (1992) Miller Field. In: Halbouty, M.T. (ed.) *Giant Oil and Gas Fields of the Decade 1978–1988*. Memoir No. 54, AAPG, Tulsa, OK. [12]

McCulloch, A.A. (1993) Apatite fission track results from Ireland and the Porcupine Basin and their significance for the evolution of the North Atlantic. *Mar. Petr. Geol.* **10**(6), 572–91. [11]

McCulloch, A.A. (1994a) Discussion on Mesozoic cover over northern England: interpretation of apatite fission-track data. *J. Geol. Soc.* **151**, 735–6. [11]

McCulloch, A.A. (1994b) Low temperature thermal history of eastern Ireland: effects of fluid flow. *Mar. Petr. Geol.* **11**, 389–99. [11]

MacDonald, A.C. and Halland, E.K. (1993) Sedimentology and shale modelling of a sandstone-rich fluvial reservoir: Upper Statfjord Formation, Statfjord Field, Northern North Sea. *AAPG Bull.* **77**, 1016–40. [8]

MacDonald, H., Allan, P.M. and Lovell, J.P.B. (1987) The geology of an oil accumulation in Block 26/28, Porcupine Basin, offshore Ireland. In: Brooks, J. and Glennie, K.W. (eds) *Petroleum Geology of North West Europe*. Graham and Trotman, London, pp. 643–52. [1, 11]

McGann, G.J., Riches, H.A. and Renoult, D.C. (1988) Formation evaluation in a thinly bedded reservoir. A case history: Scapa Field, North Sea. In: Society of Professional Well Logging Engineers 29th Annual Logging Symposium, San Antonio, Texas. [9]

McGann, G.J., Green, S.C.H., Harker, S.D. and Romani, R.S. (1991) The Scapa Field, Block 14/19, UK North Sea. In: Abbotts, I.L. (ed.) *UK North Sea United Kingdom Oil and Gas Fields: 25 Years Commemorative Volume.* Memoir No. 14, Geological Society, London, pp. 369–76. [1, 9, 12]

McGovney, J.E. and Radovich, B.J. (1985) Seismic stratigraphy and facies of Frigg fan complex. In: Berg, O.R. and Woolverton, D.G. (eds) *Seismic Stratigraphy II: an Integrated Approach.* Memoir No. 39, AAPG, Tulsa, OK., pp. 139–54. [1]

Macgregor, D.S. (1993) Relationships between seepages, tectonics, and subsurface petroleum reserves. *Mar. Petr. Geol.* **10**(6), 606–20. [11]

Macgregor, D.S. (1996) Factors controlling the destruction or preservation of giant light oilfields. *Petr. Geosci.* **2**(3), 197–218. [11]

McKee, E.D. (ed.) (1979) *A Study of Global Sand Seas.* Professional Paper No. 1052, USGS, Washington, DC., 429 pp. [5]

Mackenzie, A.S. (1984) Applications of biological markers in petroleum geochemistry. In: Brooks, J. and Welte, D. (eds) *Advances in Petroleum Geochemistry, Volume 1.* Academic Press, London, pp. 115–214. [11]

Mackenzie, A.S. and McKenzie, D.P. (1983) Isomerization and aromatization of hydrocarbons in sedimentary basins formed by extension. *Geological Magazine* **120**, 417–70. [11]

Mackenzie, A.S. and Quigley, T.M. (1988) Principles of geochemical prospect appraisal. In: Beaumont, E.A. and Foster, N.H. (compilers) *Geochemistry.* Treatise of Petroleum Geology reprint series No. 8, AAPG, Tulsa, OK., pp. 231–48. [11]

Mackenzie, A.S., Patience, R.L., Maxwell, J.R., Vandenbrouke, M. and Durand, B. (1980) Molecular parameters of maturation in the Toarcian Shales, Paris Basin, France. 1. Changes in the configuration of the acyclic isoprenoid alkanes, steranes and triterpanes. *Geochim. Cosmochim. Acta.* **44**, 1709–21. [11]

Mackenzie, A.S., Leythaeuser, D., Schaefer, R.G. and Bjorøy, M. (1983) Expulsion of petroleum hydrocarbons from shale source rocks. *Nature* **301**, 506–9. [11]

Mackenzie, A.S., Maxwell, J.R., Coleman, M.L. and Deegan, C.E. (1984) Biological marker and isotope studies of North Sea crude oils and sediments. *Proc. 11th World Petroleum Congress, London 1983.* [11]

Mackenzie, A.S. *et al.* (1987) The expulsion of petroleum from Kimmeridge Clay source rocks in the area of the Brae Oilfield, UK Continental Shelf. In: Brooks, J. and Glennie, K.W. (eds) *Petroleum Geology of North West Euurope.* Graham and Trotman, London, 865–77. [11]

McKenzie, D.P. (1978) Some remarks on the development of sedimentary basins. *Earth Plan. Sci. Lett.* **40**, 25–32. [1, 2, 8, 11]

McKenzie, D.P. (1981) The variation of temperature with time and hydrocarbon maturation in sedimentary basins formed by extension. *Earth Plan. Sci. Lett.* **55**, 87–98. [11]

McKenzie, D.P. (1983) The Earth's mantle. In: *The Dynamic Earth. Sci. Am.* **249**, 26–38. [2]

McKenzie, D.P., Watts, A., Parsons, B. and Roufosse, B. (1980) Platform of mantle convection beneath the Pacific Ocean. *Nature* **288**, 442–6. [2]

McKerrow, W.S. (1988) Wenlock to Givetian deformation in the British Isles and the Canadian Appalachians. In: Harris, A.L. and Fettes, D.J. (eds) *The Caledonian–Appalachian Orogen.* Special Publication No. 38, Geological Society, London, pp. 437–48. [4]

McKerrow, W.S. and Elders, C.F. (1989) Movements on the Southern Upland Fault. *J. Geol. Soc.* **146**, 393–5. [2]

McKerrow, W.S., Leggatt, J.K. and Eales, M.H. (1977) An imbricate thrust model for the Southern Uplands of Scotland. *Nature* **267**, 237–9. [2]

Mackertich, D. (1996) The Fife Field, UK. Central North Sea. *Petr. Geosc.* **2**(4), 373–80. [11]

McKie, T., Aggett, J. and Hogg, A.J.C. (1998) Reservoir architecture of the upper Sherwood Sandstone. Wytch Farm Field, southern England. In: Underhill, J.R. (ed.) *Development and Evolution of the Wessex Basin.* Special Publication, Geological Society, London. [8]

Maclean, I., Eglinton, G., Douraghi-Zadeh, K., Ackman, R.G. and Hooper, S.N. (1968) Correlation of stereoisomerism in present day and geologically ancient isoprenoid fatty acids. *Nature* **218**, 1019–24.

MacLennan, A.M. and Trewin, N.H. (1989) Palaeoenvironments of the late Bathonian to mid-Callovian in the Inner Moray Firth. In: Batten, D.J. and Keen, M.C. (eds) *Northwest European Micropalaeontology and Palynology.* Ellis Horwood, Chichester, pp. 92–117. [8]

McLimans, R.K. and Videtich, P.E. (1987) Reservoir diagenesis and oil migration: Middle Jurassic great oolite limestone, Wealden Basin, Southern England. In: Brooks, J. and Glennie, K.W. (eds) *Petroleum Geology of North West Europe.* Graham and Trotman, London, 119–28. [11]

McNeil, B., Shaw, H.F. and Rankin, A.H. (1995) Diagenesis of the Rotliegend sandstones in the V-Fields, southern North Sea: a fluid inclusion study. In: Cubitt, J.M. and England, W.A. (eds) *The Geochemistry of Reservoirs.* Special Publication No. 86, Geological Society, London, pp. 125–41.

Macquaker, J.H.S. (1994) A lithofacies study of the Peterborough Member, Oxford Clay Formation (Jurassic), UK: an example of sediment bypass in a mudstone succession. *J. Geol. Soc.* **151**(1), 161–72. [11]

Macquaker, J.H.S., Farrimond, P. and Brassel, S.C. (1985) Rhaetian black shales; potential oil source rocks of SW. Britain. *Paper presented to 12th Intl. Meeting on Org. Geochem., Julich, Germany.* [11]

Macquaker, J.H.S., Farrimond, P. and Brassell, S.C. (1986) Biological markers in the Rhaetian black shales of South West Britain. In: Leythaeuser, D. and Rullkotter, J. (eds) *Advances in Organic Geochemistry, 1985.* Pergamon, Oxford, pp. 93–100. [11]

Maddox, S.J., Blow, R. and Hardman, M. (1995) Hydrocarbon propsectivity of the Central Irish Sea Basin with reference to Block 42/12, offshore Ireland. In: Croker, P.F. and Shannon, P.M. (eds) *The Petroleum Geology of Ireland's Offshore Basins.* Special Publication No. 93, Geological Society, London, pp. 59–77. [11]

Mader, D. (1982) Aeolian Sands in continental red-beds of the Middle Buntsandstein Lower Triassic at the western margin of the German Basin. *Sediment. Geol.* **31**, 191–230. [7]

Madirazza, I. (1992) Comment on 'The development of salt structures in Denmark and adjacent areas: the role of basin floor dip and differential pressure' by K. Geil. *First Break* **10**, 134. [6]

Magara, K. (1980) Agents for primary hydrocarbon migration. In: Roberts, W.H. and Cordell, R.J. (eds) *Problems of Petroleum Migration.* Studies in Geology No. 10, AAPG, Tulsa, OK., pp. 33–46.

Magoon, L.B. and Dow, W.G. (eds) (1994) *The Petroleum System —from Source to Trap.* Memoir No. 60, AAPG, 655 pp. [11]

Magoon, L.B. and Valin, Z.C. (1994) Overview of petroleum system case studies. In: Magoon, L.B. and Dow, W.G. (eds) *The Petroleum System—from Source to Trap.* Memoir No. 60, AAPG, Tulsa, OK., pp. 329–39. [11]

Maher, C.E. (1980) The Piper Oilfield. In: *Giant Oil and Gas Fields of the Decade: 1968–1978.* Memoir No. 30, AAPG, 131–72. [8, 12]

Maher, C.E. (1981) The Piper oil field. In: Illing, L.V. and Hobson, G.P. (eds) *Petroleum Geology of the Continental Shelf of North-West Europe.* Heyden, London, pp. 358–70. [1, 8, 11, 12]

Maher, C.E. and Harker, S.D. (1987) Claymore oil field. In:

Brooks, J. and Glennie, K.W. (eds) *Petroleum Geology of North West Europe*. Graham and Trotman, London, pp. 835–45. [1, 4, 6, 11, 12]

Maliva, R.G. and Dickson, J.A.D. (1992) Microfacies and diagenetic controls of porosity in Cretaceous/Tertiary chalks, Eldfisk Field, Norwegian North Sea. *AAPG Bull.* **76**, 1825–38. [9]

Mange-Rajetsky, M.A. (1995) Subdivision and correlation of monotonous sandstone sequences using high resolution heavy mineral analysis, a case study: the Traissic of the Central Graben. In: Dunay, R.E. and Hailwood, E.A. (eds) *Non-biostratigraphical Methods of Dating and Correlation*. Special Publication No. 89, Geological Society, London, pp. 23–30. [7]

Mann, A.L. and Myers, K.J. (1990) The effect of climate on the geochemistry of the Kimmeridge Clay formation. In: *Biomarkers in Petroleum: Memorial Symposium for W. Seifert*. Division of Petroleum Chemistry, 10, ACS, pp. 139–42. [11]

Mann, U. and Muller, P.J. (1988) Source rock evaluation by well log analysis (Lower Toarcian, Hils Syncline). In: Mattavelli, L. and Novelli, L. (eds) *Advances in Org. Geochem. 1987 (Proc. 13th Int'l Mtg on Org. Geochem. Venice,), Part 1*. Pergamon, Oxford, pp. 109–20. [11]

Mann, U., Leythaeuser, D. and Muller, P.J. (1986) Relation between source rock properties and wireline log parameters: an example from Lower Jurassic Posidonia Shale, NW Germany. In: Leythaeuser, D. and Rullkotter, J. (eds) *Advances in Organic Geochemistry, 1985*. Pergamon, Oxford, pp. 1105–12. [11]

Mann, U., Dueppenbecker, S., Langen, A., Ropertz, B. and Welte, D.H. (1989) Petroleum pathways during primary migration: evidence and implications (Lower Toarcian, Hils Syncline, NW Germany). *Paper presented at the 14th Int'l Org. Geochem. Conference, Paris, 18–22 Sept. 1989*. [11]

Mann, U., Hantschel, T., Schaefer, R.G., Krooss, B., Leythaeuser, D., Littke, R. *et al.* (1997) Petroleum migration: mechanisms, pathways, efficiencies. In: Welte, D.H., Horsfield, B. and Baker, D.R. (eds) *Petroleum and Basin Evolution*. Springer, Berlin, pp. 403–89. [11]

Manning, D. (ed.) (1991) *Organic Geochemistry, Advances and Applications in Energy and Natural Environment*. Extended Abstracts, 15th Meeting EAOG, Manchester University Press, Manchester.

Manning, D.A.C., Rae, E.I.C. and Gestsdottir, K. (1994) Appraisal of the use of experimental and analogue studies in the assessment of the role of organic acid anions in diagenesis. *Marine and Petr. Geol.* **11**(1), 10–20. [11]

Manum, S.B. and Throndsen, T. (1977) Rank of coal and dispersed organic matter and its geological bearing in the Spitzbergen Tertiary. In: *Norsk Polarinst. Arbok 1977*. pp. 159–77, 179–87. [11]

Mapstone, N.B. (1975) Diagenetic history of North Sea chalk. *Sedimentology* **22**, 601–14. [9]

Maragna, B., Zaro, G. and Pessina, P. (1985) Thomas, B.M., Doré, A.G., Eggen, S.S., Home, P.C. and Larsen, R.M. (eds) *Block 33/6 geochemical evaluation*. Graham and Trotman, London, pp. 197–204. [11]

Marcussen, C. *et al.* (1987) Studies of the onshore hydrocarbon potential in East Greenland 1986–7: field work from 72° to 74°N. *Gronlands Geol. Und. Rapport* **135**, 72–81.

Margaritz, M. and Peryt, T.M. (1994) Mixed evaporative and meteoric water dolomitization: isotope study of the Zechstein Limestone (Upper Permian), southwestern Poland. *Sediment. Geol.* **92**, 257–72. [6]

Marie, J.P.P. (1975) Rotliegendes stratigraphy and diagenesis. In: Woodland, A.W. (ed.) *Petroleum and the Continental Shelf of Northwest Europe*, Vol. I, *Geology*. Applied Science Publishers, Barking, pp. 205–10. [5]

Marjanac, T. (1995) Architecture and sequence stratigraphic perspectives of the Dunlin Group formations and proposal for new type- and reference-wells. In: Steel, R.J., Felt, V.L., Johannessen, E.P. and Mathieu, C. (eds) *Sequence Stratigraphy on the Northwest European Margin*. Special Publication No. 5, NPS, pp. 143–65. [8]

Marjanac, T. and Steel, R.J. (1997) Dunlin Group sequence stratigraphy in the Northern North Sea: a model for Cook Sandstone deposition. *AAPG Bull.* **81**, 276–92. [8]

Marsden, G., Yielding, G., Roberts, A.M. and Kuznir, N.J. (1990) Application of a flexural cantilever simple-shear/pure shear model of continental lithosphere extension to the formation of the northern North Sea basin. In: Blundell, D.J. and Gibbs, A.D. (eds) *Tectonic Evolution of the North Sea Rifts*. Oxford Scientific Publishers, Oxford, pp. 241–72. [7]

Marshall, J.E.A. (1995) *Newlett. Micropalaentol.* **53**, 19–22. [3]

Marshall, J.E.A. (1998) The recognition of multiple hydrocarbon generation episodes: an example from Devonian lacustrine source rocks in the Inner Moray Firth. *J. Geol. Soc.* **155**(2), 335–52. [11]

Marshall, J.E.A., Brown, J.F. and Hindmarsh, S. (1985) Hydrocarbon source rock potential of the Devonian rocks of the Orcadian Basin. *Scott J. Geol.* **21**, 301–20. [3, 11]

Marshall, J.E.A., Haughton, P.D.W. and Hiller, S.J. (1994) Vitrinite reflectivity and the structure and burial history of the Old Red Sandstone of the Midland Valley of Scotland. *J. Geol. Soc.* **151**(3), 425–39. [11]

Marshall, J.E.A., Rogers, D.A. and Whitely, M.J. (1996) Devonian marine incursions into the Orcadian Basin, Scotland, *J. Geol. Soc.* **153**(3), 451–66. [2]

Martin, J.H. and Evans, P.F. (1988) Reservoir modeling of marginal aeolian/sabkha sequences, southern North Sea, UK Sector. SPE, 18155, pp. 473–86. [5]

Martin, M.A. and Pollard, J.E. (1996) The role of trace fossil ichnofabric analysis in the development of depositional models for the Upper Jurassic Fulmar Formation of the Kittiwake Field Quadrant 21 UKCS. In: Hurst, A., Johnson, H.D., Burley, S.D., Canham, A.C. and Mackertich, D.S. (eds) *Geology of the Humber Group: Central Graben and Moray Firth, UKCS*. Special Publication No. 114, pp. 163–83. [8]

Martinsen, O.J. (1993) Namurian late Carboniferous depositional systems of the Craven–Askrigg area, northern England: implications for sequence stratigraphic models. In: Posamentier, H.W., Summerhayes, C.P., Haq, B.U. and Allen, G.P. (eds) *Stratigraphy and Facies Association in a Sequence Stratigraphic Framework*. Special Publication No. 18, International Association of Sedimentologists, Blackwell Scientific Publications, Oxford, pp. 247–81. [4]

Marzi, R., Rullkotter, J. and Perriman, W.S. (1990) Application of the change of sterane isomer ratio to the reconstruction of geothermal histories: implications of results of hydrous pyrolysis experiments. In: Durand, B. and Behar, F. (eds) *Advances in Organic Geochemistry. Org. Geochem.* **16**(1–3), 91–102. [11]

Mason, P.C., Burwood, R. and Mycke, B. (1995) The reservoir geochemistry and petroleum charging histories of Paleogene reservoired fields in the Outer Witch Ground Graben. In: Cubitt, J.M. and England, W.A. (eds) *The Geochemistry of Reservoirs*. Special Publication No. 86, Geological Society, London, pp. 281–302. [11]

Masters, J.A. (1984) Lower Cretaceous oil and gas in Western Canada. In: Masters, J.A. (ed.) *Elmworth, Case Study of a Deep Basin Gas Field*. Memoir No. 38, AAPG, Tulsa, OK., pp. 1–34. [11]

Mathisen, M.E. and Budny, M. (1990) Seismic lithostratigraphy of deep subsalt Permo-Carboniferous gas reservoirs. *Geophysics* **55**, 1357–65. [6]

Mattingly, G.A. and Bretthauer, H.B. (1992) The Alba field—a Middle Eocene deepwater channel system in the U.K. North Sea. In: Halbouty, M.T. (ed.) *Giant Oil and Gas Fields of the Decade 1978–1988*. Special Publication No. 54, AAPG, Tulsa, OK., pp. 297–305. [12]

Maureau, G.T.F.R. and van Wijhe, D.H. (1979) The prediction of porosity in the Permian Zechstein carbonate of eastern Netherlands using seismic data. *Geophysics* **44**, 1502–17. [6]

May, B.T. and Covey, J.D. (1983) Structural inversion of salt dome flanks. *Geophysics* **48**, 1039–50. [6]

Maynard, J.R. (1992) Sequence stratigraphy of the Upper Yeadonian of northern England. *Mar. Petr. Geol.* **9**, 197–207. [4]

Maynard, J.R. and Leeder, M.R. (1992) On the periodicity and magnitude of Late Carboniferous glacio-eustatic sea-level changes. *J. Geol. Soc.* **149**, 303–11. [4]

Maynard, J.R., Wignall, P.B. and Varker, W.J. (1991) A 'hot' new shale facies from the Upper Carboniferous of Northern Ireland. *J. Geol. Soc.* **148**, 805–8. [4, 11]

Maync, W. (1961) The Permian of Greenland. In: *Geology of the Arctic 1.* University of Toronto Press, Toronto, pp. 214–23. [6]

Meadows, N.S. and Beach, A. (1993) Controls on reservoir quality in the Triassic Sherwood sandstone of the Irish Sea. In: Parker, J.R. (ed.) *Petroleum Geology of Northwest Europe: Proceedings of the 4th Conference.* Geological Society, London, pp. 823–33. [12]

Mearns, E.W. (1992) Samarium–neodymium isotopic constraints on the provenance of the Brent Group. In: Morton, A.C., Haszeldine, R.S., Giles, M.R. and Brown, S. (eds) *Geology of the Brent Group.* Special Publication No. 61, Geological Society, London, pp. 213–25. [1, 12]

Mearns, E.W., Knarud, R., Raestad, N., Stanley, K.O. and Stockbridge, S.P. (1989) Samarium–neodymium isotope stratigraphy of the Lunde and Statfjord Formation of Snorre Oil Field, Northern North Sea. *J. Geol. Soc.* **146**, 217–28. [7]

Megson, J.B. (1992) The North Sea Chalk play: examples from the Danish Central Graben. In: Hardman, R.F.P. (ed.) *Exploration Britain: Geological Insights for the Next Decade.* Special Publication No. 67, Geological Society, London, pp. 247–82. [1, 9, 12]

Meier, R. (1981) Clastic resedimentation phenomena of the Werra Sulphate Zechstein 1 at the eastern slope of the Eichsfeld Swell Middle European Basin—an information. In: Proceedings of an International Symposium on Central Europe Permian, Jablonna, Poland, 1978. Geological Institute, Warsaw, pp. 369–73. [6]

Meissner, F.F., Woodward, J. and Clayton, J.L. (1984) *Stratigraphic Relationships and Distribution of Source Rocks in the Greater Rocky Mountain Region.* Rocky Mountain Association Geologists, Denver, CO., 34 pp.

Melville, R.V. and Freshney, E.C. (1982) *British Regional Geology: the Hampshire Basin and Adjoining Areas*, 4th edn. Institute of Geological Science, HMSO, London. [9]

Menning, M. (1991) Rapid subsidence in the Central European Basin during the initial development Permian–Triassic boundary sequences, 258–240 Ma. *Zbl. Geol. Palaontol.* **1**(4), 809–24. [5]

Menning, M. (1992) A numerical time scale for the Permian and Triassic lithostratigraphic units of Central Europe. In: *IAS 13th Regional Meeting on Sedimentology, Jena.* Abstracts, pp. 175–7. [5]

Menning, M. (1995) A numerical time scale for the Permian and Triassic periods: an integrated time analysis. In: Scholle, P.A., Peryt, T.M. and Ulmer-Scholle, D.S. (eds) *The Permian of Northern Pangaea*: Vol. 1. *Paleogeography, Paleoclimates, Stratigraphy.* Springer-Verlag, Berlin, pp. 77–97. [2, 5]

Menning, M., Katzung, G. and Lützner, H. (1988) Magnetostratigraphic investigations in the Rotliegendes 300–252 Ma of central Europe. *Z. Geol. Wiss.* **16**, 1045–63. [5]

Menpes, R.J. and Hillis, R.R. (1995) Quantification of Tertiary exhumation from sonic velocity data, Celtic Sea/South Western Approaches. In: Buchanan, J.G. and Buchanan, P.G. (eds) *Basin Inversion.* Special Publication No. 88, Geological Society, London, pp. 191–210. [11]

Menpes, R.J. and Hillis, R.R. (in press) *Quantification of Tertiary Erosion in the Celtic Sea/South-Western Approaches, Basin Inversion.* Special Publication, Geological Society, London.

Meyer, B.L. and Nederlof, M.H. (1984) Identification of source rocks on wireline logs by density/resistivity and sonic transit time-resistivity cross plots. *AAPG Bull.* **68**(2) 121–9. [11]

Michaud, F. (1987) Eldfisk. In: Spencer, A.M., Campbell, C.J., Hanslien, S.H., Nelson, P.H., Nysaether, E. and Ormaasen, E.G. (eds) *Geology of the Norwegian Oil and Gas Fields.* Graham and Trotman, London, pp. 89–105. [1, 9]

Michelsen, O. (1982) *Geology of the Danish Central Graben.* Series B, No. 8, G.S., København, Denmark. [9]

Michelsen, O. and Andersen, C. (1983) Mesozoic structural sedimentary development of the Danish Central Graben. *Geol. Mijnbouw* **62**, 93–102. [7, 11]

Michelsen, O. and Nielsen, L.H. (1993) Structural development of the Fennoscandian Border Zone, offshore Denmark. *Mar. Petr. Geol.* **10**, 124–34. [5]

Michelsen, O., Frandsen, N., Holm, L., Møller, J.J. and Vejbdæk, O.V. (1987) *Jurassic–Lower Cretaceous of the Danish Central Trough: Depositional Environments, Tectonism, and Reservoirs.* Series A, No. 16, G.S., København, Denmark, 45 pp. [9]

Mijnssen, F.C.J. (1997) Modelling of sandbody connectivity in the Schooner Field. In: Ziegler, K., Turner, P. and Daines, S. (eds) *Petroleum Geology of the Southern North Sea: Future Potential.* Special Publication No. 123, Geological Society, London, pp. 169–80. [4]

Miles, J.A. (1989) *Illustrated Glossary of Petroleum Geochemistry.* Oxford Scientific Publications, Oxford, 137 pp. [11]

Miles, J.A. (1990) Secondary migration routes in the Brent sandstones of the Viking Graben and East Shetland Basin: evidence from oil residues and subsurface pressure data. *AAPG Bull.* **74**(11), 1718–35. [11]

Miles, J.A., Downes, C.J. and Cook, S.E. (1993) The fossil oil seep in Mupe Bay, Dorset: a myth investigated. *Mar. Petr. Geol.* **10**(1), 58–71. [11]

Miller, J., Adams, A.E. and Wright, V.P. (1987) *European Dinantian Environments.* John Wiley, Chichester, 402 pp. [4]

Miller, K.G., Fairbanks, R.G. and Mountain, G.S. (1987) Tertiary oxygen isotope synthesis, sea level history, and continental margin erosion. *Paleoceanography* **2**(1), pp. 1–19.

Miller, R.G. (1990) A paleoceanographic approach to the Kimmeridge clay formation. In: Huc, A.Y. (ed.) *Deposition of Organic Facies.* Studies in Geology No. 30, AAPG, Tulsa, OK., pp. 13–26. [11]

Millson, J.A. (1987) The Jurassic evolution of the Celtic Sea basins. In: Brooks, J. and Glennie, K.W. (eds) *Petroleum Geology of North West Europe.* Graham and Trotman, London, pp. 599–610. [12]

Milton, N.J. (1993) Evolving depositional geometries in the North Sea Jurassic rift. In: Parker, J.R. (ed.) *Petroleum Geology of N.W. Europe, Proceedings of the 4th Conference.* Geological Society, London, pp. 425–42. [12]

Milton, N.J., Bertram, G.T. and Vann, I.R. (1990) Early Paleogene tectonics and sedimentation in the Central North Sea. In: Hardman, R.P.F. and Brooks, J. (eds) *Tectonic Events Responsible for Britain's Oil and Gas Reserves.* Special Publication No. 55, Geological Society, London, pp. 339–51. [10]

Mitchell, S.M., Beamish, G.W.J., Wood, M.V., Malacek, S.J., Armentrout, J.A., Damuth, J.E. *et al.* (1993) Paleogene sequence stratigraphic framework of the Faeroe Basin. In: Parker, J.R. (ed.) *Petroleum Geology of Northwest Europe: Proceedings of the 4th Conference*, Geological Society, London, pp. 1011–23. [2, 12]

Mitchener, B.C., Lawrence, D.A., Partington, M.A., Bowman, M.B.J. and Gluyas, J. (1992) Brent Group: sequence stratigraphy and regional implications. In: Morton, A.C., Haszeldine, R.S., Giles, M.R. and Brown, S. (eds) *Geology of the Brent Group.* Special Publication No. 61, Geological Society, London, pp. 45–80. [8, 12]

Moldowan, J.M., Sundararaman, P. and Schoell, M. (1986) Sensitivity of biomarker properties to depositional environment and/or source input in the Lower Toarcian of SW Germany. In: Leythaeuser, D. and Rullkotter, J. (eds) *Advances in Organic Geochemistry, 1985*. Pergamon, Oxford, pp. 15–926. [11]

Monin, J.C., Connan, J., Oudin, J.L. and Durand, B. (1990) Quantitative and qualitative experimental approach to oil and gas generation: application to North Sea source rocks. In: Durand, B. and Behar, F. (eds) *Advances in Organic Geochemistry. Org. Geochem.* **16**(1–3), 133–42. [11]

Montel, F. and Caillet, G. (1993) Diffusion model for predicting reservoir gas losses. *Marine and Petroleum Geology* **10**(1), 51–8. [11]

Moore, J.G. and Shannon, P.M. (1995) The Cretaceous succession in the Porcupine Basin, offshore Ireland: facies distribution and hydrocarbon potential. In: Croker, P.F. and Shannon, P.M. (eds) *The Petroleum Geology of Ireland's Offshore Basins*. Special Publication No. 93, Geological Society, London, pp. 345–70. [12]

Morgan, C.P. (1991) The Viking Complex Field, Blocks 49/12a, 49/16, 49/17, UK North Sea. In: Abbotts, I.L. (ed.) *UK North Sea United Kingdom Oil and Gas Fields: 25 Years Commemorative Volume*. Memoir No. 14, Geological Society, London, p. 509. [1]

Mørk, A. and Bjorøy, M. (1984) Mesozoic source rocks on Svalbard. In: Spencer, A.M., Johnsen, S.O., Mørk, A., Nysaether, E., Songstad, P. and Spinnangr, A. (eds) *Petroleum Geology of the North European Margin*. Graham and Trotman, London, pp. 371–82. [11]

Mørk, A. and Worsley, D. (1979) The Triassic and Lower Jurassic succession of Svalbard: a review. In: *Norwegian Sea Symposium Tromso*, NPF, NSS/29, 1–22. [7]

Morrison, D., Bennet, G.G. and Bayat, M.G. (1991) The Don Field, Block 211/13a, 211/18a, 211/14, 211/19a, UK North Sea. In: Abbotts, I.L. (ed.) *UK North Sea United Kingdom Oil and Gas Fields: 25 Years Commemorative Volume*. Memoir No. 14, Geological Society, London, p. 89. [1]

Mortimore, R.N. and Pomerol, B. (1991) Upper Cretaceous tectonic disruptions in a placid chalk sequence in the Anglo-Paris Basin. *J. Geol. Soc.* **148**, 391–404. [9]

Morton, A.C. (1992) Provenance of Brent Group sandstones: heavy mineral constraints. In: Morton, A.C., Haszeldine, R.S., Giles, M.R. and Brown, S. (eds) *Geology of the Brent Group*. Special Publication No. 61, Geological Society, London, pp. 227–44. [12]

Morton, A.C. and Berge, C. (1995) Heavy mineral suites in the Statfjord and Nansen Formations of the Brent Field, North Sea: a new tool for reservoir subdivision and correlation. *Petr. Geosci.* **1**, 355–64. [1, 8]

Morton, A.C. and Parson, L.M. (eds) (1988) *Early Tertiary Volcanism and the Opening of the NE Atlantic*. Special Publication No. 39, Geological Society, London.

Morton, A.C., Hazeldine, R.S., Giles, M.R. and Brown, S. (1992) *Geology of the Brent Group*. Special Publication No. 61, Geological Society, London. [1]

Morton, A.C., Hallsworth, C.R. and Wilkinson, G.C. (1993) Stratigraphic evolution of sand provenance during Paleocene deposition in the Northern North Sea area. In: Parker, J.R. (ed.) *Petroleum Geology of Northwest Europe: Proceedings of the 4th Conference*. Geological Society, London, pp. 73–84. [10, 12]

Morton, N. (1993) Potential reservoir and source rocks in relation to Upper Triassic to Middle Jurassic sequence stratigraphy, Atlantic basins of the British Isles. In: Parker, J.R. (ed.) *Petroleum Geology of Northwest Europe: Proceedings of the 4th Conference*. Geological Society, London, pp. 285–97. [11]

Mound, D.G., Robertson, I.D. and Wallis, R.J. (1991) The Cyrus Field, Block 16/28, UK North Sea. In: Abbotts, I.L. (ed.) *UK North Sea United Kingdom Oil and Gas Fields: 25 Years Commemorative Volume*. Memoir No. 14, Geological Society, London, pp. 295–300. [1, 10]

Mudford, B.S., Gradstein, F.M., Katsube, T.J. and Best, M.E. (1991) Modelling ID compaction-driven flow in sedimentary basins: a comparison of the Scotian Shelf, North Sea and Gulf Coast petroleum migration. In: England, W.A. and Fleet, A.J. (eds) *Petroleum Migration*. Special Publication No. 59, Geological Society, London, pp. 65–89. [11]

Mudge, D.C. and Bliss, G.M. (1993) Stratigraphy and sedimentation of the Paleocene sands in the northern North Sea. In: Parker, J.R. (ed.) *Petroleum Geology of Northwest Europe: Proceedings of the 4th Conference*. Geological Society, London, pp. 95–111. [10]

Mudge, D.C. and Bujak, J.P. (1994) Eocene stratigraphy of the North Sea basin. *Mar. Petr. Geol.* **11**, 166–81. [10]

Mudge, D.C. and Bujak, J.P. (1996) An integrated stratigraphy for the Palaeocene and Eocene of the North Sea. In: Knox, R.W.O'B., Corfield, R.M. and Dunay, R.E. (eds) *Correlation of the Early Paleogene in Northwest Europe*. Special Publication 101, Geological Society, London, pp. 91–113.

Mudge, D.C. and Copestake, P. (1992a) Revised Lower Paleogene lithostratigraphy for the Outer Moray Firth, North Sea. *Mar. Petr. Geol.* **9**, 52–69. [2, 10]

Mudge, D.C. and Copestake, P. (1992b) Lower Paleogene lithostratigraphy of the northern North Sea. *Mar. Petr. Geol.* **9**, 287–301. [2, 10]

Mudge, D.C. and Rashid, B. (1987) The geology of the Faeroe Basin area. In: Brooks, J. and Glennie, K.W. (eds) *Petroleum Geology of North West Europe*. Graham and Trotman, London, pp. 751–63. [1, 2]

Mukhopadhyay, P.K. and Gormly, J.R. (1984) Hydrocarbon potential of two types of resinite. *Org. Geochem.* **6**, 439–54. [11]

Mukhopadhyay, P.K., Sumanta, U. and Jassal, J. (1985) Origin of oil in a lagoon environment: desmocollinite/bituminite source-rock concept. Ninth Carboniferous Congress, Washington, Champaign-Urbana, May 1979. In: Cross, A.T. (ed.) *Compte Rendu, Volume 4*, pp. 753–64. [11]

Müller, E.P., Dubslaff, H., Eiserbeck, W. and Sallum, R. (1993) Zur Entwicklung der Erdöl-und gasexploration zwischen Ostsee und Thüringer Wald. In: Müller, E.P. (ed.) *Zur Geologie und Kohlenwasserstoff-Fuhrung des Permim Ostteil der Norddeutschen Senke. Geol. Jahrbuch Reihe A* **131**, 5–30. [5]

Muller, P.J. and Suess, E. (1979) Productivity, sedimentation rate and sedimentary organic matter in the oceans—I. organic carbon preservation. *Deep Sea Research* **26**A, 1347–62. [11]

Munns, J.W. (1985) The Valhall Field: a geological overview. *Mar. Petr. Geol.* **2**, 23–43. [1, 9, 11, 12]

Murdoch, L.M., Musgrove, F.W. and Perry, J.S. (1995) Tertiary uplift and inversion history in the North Celtic Sea Basin and its influence on source rock maturity. In: Croker, P.F. and Shannon, P.M. (eds) *The Petroleum Geology of Ireland's Offshore Basins*. Special Publication No. 93, Geological Society, London, pp. 297–319. [11]

Mure, E. (1987a) Frigg. In: Spencer, A.M., Campbell, C.J., Hanslien, S.H., Nelson, P.H., Nysaether, E. and Ormaasen, E.G. (eds) *Geology of the Norwegian Oil and Gas Fields*. Graham and Trotman, London, pp. 203–13. [1, 11, 12]

Mure, E. (1987b) Ost Frigg—Alpha and Beta. In: Spencer, A.M., Campbell, C.J., Hanslien, S.H., Nelson, P.H., Nysaether, E. and Ormaasen, E.G. (eds) *Geology of the Norwegian Oil and Gas Fields*. Graham and Trotman, London, pp. 215–22. [1, 11]

Mure, E. (1987c) Nord-Ost Frigg. In: Spencer, A.M., Campbell, C.J., Hanslien, S.H., Nelson, P.H., Nysaether, E. and Ormaasen, E.G. (eds) *Geology of the Norwegian Oil and Gas Fields*. Graham and Trotman, London, pp. 223–8. [1, 11]

Mure, E. (1987d) Heimdal. In: Spencer, A.M., Campbell, C.J., Hanslien, S.H., Nelson, P.H., Nysaether, E. and Ormaasen, E.G. (eds) *Geology of the Norwegian Oil and Gas Fields*. Graham and Trotman, London, pp. 229–34. [1, 11]

Murphy, N.J., Sauer, M.J. and Armstrong, J.P. (1995) Toarcian source rock potential in the North Celtic Sea Basin, offshore

Ireland. In: Croker, P.F. and Shannon, P.M. (eds) *The Petroleum Geology of Ireland's Offshore Basins*. Special Publication No. 93, Geological Society, London, pp. 193–207. [11, 12]

Murray, M.V. (1995) Development of small gas fields in the Kinsale Head area. In: Croker, P.F. and Shannon, P.M. (eds) *The Petroleum Geology of Ireland's Offshore Basins*. Special Publication No. 93, Geological Society, London, pp. 259–60. [1, 12]

Mussett, A.E., Dagley, P. and Skelthorn, R.R. (1988) Time and duration of activity in the British Tertiary Igneous Province. In: Morton, A.C. and Parson, L.M. (eds) *Early Tertiary Volcanism and the Opening of the NE Atlantic*. Special Publication No. 39, Geological Society, London, pp. 337–48. [2]

Mutti, E. (1985) Turbidite systems and their relations to depositional sequences. In: Zuffa, G.G. (ed.) *Provenance of Arenites*. ASI Series 148, NATO, pp. 65–93. [10]

Myerscough, R. (1994) The Chalk of Flamborough Head. In: Scrutton, C. (ed.) *Yorkshire Rocks and Landscape: A Field Guide*. Yorkshire Geological Society, Ellenbank Press, Maryport, Cumbria, pp. 192–9. [9]

Myhre, L. (1975) *Lithology: Well 8/3-1*. Paper No. 1, NPD, Stavanger. [7]

Myhre, L. (1978) *Lithology: Well 9/8-1*. Paper No. 5, NPD, Stavanger. [7]

Mykura, W. (1960) The replacement of coal by limestone and the reddening of Coal Measures in the Ayrshire Coalfield. *Bull. Geol. Surv. GB* **16**, 69–109. [4]

Mykura, W. (1976) *Orkney and Shetland*. British Regional Geology, Institute of Geological Science, HMSO, Edinburgh, 149 pp. [2]

Mykura, W. (1983a) Old Red Sandstone. In: Craig, G.Y. (ed.) *Geology of Scotland*. Scottish Academic Press, Edinburgh, pp. 205–51.

Mykura, W. (1983b) *The Old Red Sandstone East of Loch Ness, Inverness-shire*. Report No. 83/13, Institute of Geological Science, Edinburgh. [3]

Mykura, W. (1991) Old Red Sandstone. In: Craig, G.Y. (ed.) *Geology of Scotland*, 2nd edn. Geological Society, London, pp. 297–344. [3]

Mykura, W. and Owens, B. (1983) *The Old Red Sandstone of the Mealfluarvonie Outlier, West of Loch Ness, Inverness-shire*. Report No. 83/7, Institute of Geological Science, Edinburgh. [3]

Myreland, R., Messell, K. and Raestad, N. (1981) Gas discovery in 35/3—an exploration history. In: *Norwegian Symposium on Exploration, Bergen*. NPF, Article 17, 17 pp. [1]

Myres, J.C., Jones, A.F. and Towart (1995) The Markham Field: UK Blocks 49/5a and 49/10b, Netherlands Blocks J3b and J6. *Petr. Geosci.* **1**(4), 303–9. [1, 5]

Nachsel, G. and Franz, E. (1983) Zur Ausbildung der Wippertal-Sturungszone in Bereich der Grubenfelder des Kaliwerks 'Gluckauf' Sondershausen. On the construction of the Wippertal fault zone in the region of the Carnallite salt mine, 'Gluckauf' Sonderhausen. *Z. Geol. Wiss.* **11**, 1005–21.

Nadin, P.A. and Kusznir, N.J. (1995) Palaeocene uplift and Eocene subsidence in the northern North Sea Basin from 2D forward and reverse stratigraphic modelling. *J. Geol. Soc.* **152**(5), 833–48. [11]

Nadin, P.A., Kusznir, N.J. and Toth, J. (1995) Transient regional uplift in the Early Tertiary of the northern North Sea and the development of the Iceland Plume. *J. Geol. Soc.* **152**(6), 953–8. [11]

Nagtegaal, P.J.C. (1979) Relationship of facies and reservoir quality in Rotliegendes desert sandstones, Southern North Sea Region. *J. Petr. Geol.* **2**, 145–58. [5]

Naylor, D. (1996) History of oil and gas exploration in Ireland. In: Glennie, K. and Hurst, A. (eds) *Northwest Europe's Hydrocarbon Industry*, Geological Society, London, pp. 43–52. [1]

Naylor, D. and Shannon, P.M. (1982) *The Geology of Offshore Ireland and West Britain*. Graham and Trotman, London, 161 pp. [1, 12]

Naylor, D., Haughey, N., Clayton, G. and Graham, J.R. (1993) The Kish Bank Basin, offshore Ireland. In: Parker, J.R. (ed.) *Petroleum Geology of Northwest Europe: Proceedings of the 4th Conference*. Geological Society, London, pp. 845–55. [12]

Naylor, H., Turner, P., Vaughan, D.J. and Fallick, A.E. (1989) The Stotfield Cherty Rock, Elgin: a petrographic and isotopic study of a Permo-Triassic calcrete. *Geol. J.* **24**, 205–21. [8]

Neal, J.E. (1996) A summary of Paleogene sequence stratigraphy in northwest Europe and the North Sea. In: Knox, R.W.O'B., Corfield, R.M. and Duney, R.E. (eds) *Correlation of the Early Paleogene in Northwest Europe*. Special Publication No. 101, Geological Society, London, pp. 15–42.

Neale, J. and Catt, J. (1994) Jurassic, Cretaceous and Quaternary rocks of Filey Bay and Speeton. In: Scrutton, C. (ed.) *Yorkshire Rocks and Landscape: a Field Guide*. Yorkshire Geological Society, Ellenbank Press, Maryport, Cumbria, pp. 183–91. [9]

Neale, J.W. (1962) Ostracoda from the type Speeton Clay (Lower Cretaceous) of Yorkshire. *Micropaleontology*. **8**(4) 425–84. [9]

Nederlandse Aardolie Maatschappij B.V., (NAM) and Rijks Geologische Dienst (RGD) (1980) *Stratigraphic Nomenclature of the Netherlands*. Transactions of the Roy. Dutch Geological and Mining Society, Delft, 77 pp. [4, 5, 8]

Nedkvittne, T., Karlson, D.A., Bjorlykke, K. and Larter, S.R. (1993) Relationship between reservoir diagenetic evolution and petroleum emplacement in the Ula Field, North Sea. *Mar. Petr. Geol.* **10**(3), 255–71. [11]

Nelson, J.S. and Simmons, E.C. (1995) Diffusion of methane and ethane through the reservoir cap rock: implications for the timing and duration of catagenesis. *AAPG Bull.* **79**(7), 1064–75. [11]

Nelson, J.S. and Simmons, E.C. (1997) Diffusion of methane and ethane through the reservoir cap rock: implications for the timing and duration of catagenesis: Reply. *AAPG Bull.* **81**(1), 162–7. [11]

Nelson, P.H.H. and Lamy, J.M. (1987) The Møre/West Shetland area: a review. In: Brooks, J. and Glennie, K.W. (eds) *Petroleum Geology of North West Europe*. Graham and Trotman, London, pp. 775–84. [1, 9, 11]

Nelson, R.A. (1981) Significance of fracture sets associated with stylolite zones. *AAPG Bull.* **65**, 2417–25. [9]

Nelson, R.A. (1982) An approach to evaluating fractured reservoirs. *J. Petr. Technol.* **34** (Sept.), 2167–70. [9]

Nettleton, L.L. (1934) Fluid mechanics of salt domes. *AAPG Bull.* **18**, 1175–204. [6]

Neugebauer, J. (1974) Some aspects of cementation in chalk. In: Hsü, K.J. and Jenkyns, H.C. (eds) *Pelagic Sediments: on Land and Under the Sea*. Special Publication No. 1, International Association of Sedimentologists, Blackwell Scientific Publications, Oxford, pp. 149–76. [9]

Neves, R. and Selley, R.C. (1975) A review of the Jurassic rocks of North-East Scotland. In: Finstad, K.G. and Selley, R.C. (eds) *Proceedings of the Jurassic Northern North Sea Symposium*. NPF, Stavanger, JNNSS/5, pp. 1–29.

Neves, R., Gueinn, K.J., Clayton, G., Joannides, N. and Neville, R.S. (1972) A scheme of miospore zones for the British Dinantian. In: *Comptes rendues 7-ième Cong. int. Strat. Geol. Carb., Krefeld, 1971*, Vol. 1, pp. 347–53. [4]

Neves, R., Gueinn, K.J., Clayton, G., Joannides, N., Neville, R.S. and Kruszewska, K. (1973) Palynological correlations within the Lower Carboniferous of Scotland and northern England. *Trans. Roy. Soc. Edinburgh* **69**, 23–70. [4]

Newman, M.St.J., Reeer, M.L., Woodruff, A.H.W. and Hatton, I.R. (1993) The geology of the Gryphon oil field. In: Parker, J.R. (ed.) *Petroleum Geology of Northwest Europe: Proceedings of the 4th Conference*. Geological Society, London, pp. 123–34. [1, 10, 12]

Newton, S.K. and Flanagan, K.P. (1993) The Alba Field: evolution of the depositional model. In: Parker, J.R. (ed.) *Petroleum Geology of Northwest Europe. Proceedings of the 4th Conference*. Geological Society, London, pp. 161–71. [1, 10, 12]

Nicholson, R. (1979) Caledonian correlations: Britain and Scandinavia. In: Harris, A.L., Holland, C.H. and Leake, B.E. (eds) *The Caledonides of the British Isles Reviewed.* Special Publication No. 8, Geological Society, London, pp. 3–18. [2]

Nielsen, L.H. and Japsen, P. (1991) *Deep Wells in Denmark 1935–1990: Lithostratigraphic Subdivision.* Series A, No. 31, Danmarks Geol. Unders., 177 pp. [2, 4, 5, 6]

Nielsen, S.B. (1995) Reliable modelling of the source rock transformation ratio. *Mar. Petr. Geol.* **12**(1) 82–101. [11]

Nielsen, S.B. (1995) Fast Monte Carlo simulation of hydrocarbon generation. In: Doré, A.G., Augustson, J.H., Hermanrud, C., Stewart, D.J. and Sylta, O. (eds) *Basin Modelling: Advances and Applications.* Special Publication No. 3, NPF, Elsevier, Amsterdam, pp. 265–76. [11]

Nielsen, O.B., Sorensen, S., Thiede, J. and Skarbo, O. (1986) Cenozoic differential subsidence of the North Sea. *AAPG Bull.* **70**(3), 276–98. [11]

Nilsen, K.T., Hendriksen, E. and Larssen, G.B. (1992) Exploration of the Late Palaeozoic carbonates in the southern Barents Sea: a seismic stratigraphic study. In: Vorren, T.O. *et al.* (eds) *Arctic Geology and Petroleum Potential.* Special Publication No. 2, NPF, Elsevier, Amsterdam, pp. 393–403. [6]

Nipen, O. (1987) Oseberg. In: Spencer, A.M., Campbell, C.J., Hanslien, S.H., Nelson, P.H., Nysaether, E. and Ormaasen, E.G. (eds) In: *Geology of the Norwegian Oil and Gas Fields.* Graham and Trotman, London, pp. 378–87. [1, 11]

Nordberg, H.E. (1981) Seismic hydrocarbon indicators in the North Sea. *Norweigian Symposium on Exploration, Bergen.* NPS, Geilo, pp. 8.1–8.40. [11]

Norbury, I. (1987) Hod. In: Spencer, A.M., Campbell, C.J., Hanslien, S.H., Nelson, P.H., Nysaether, E. and Ormaasen, E.G. (eds) *Geology of the Norwegian Oil and Gas Fields.* Graham and Trotman, London, pp. 107–16. [1, 9]

Nordgård Bolås, H.M. (1987) Odin. In: Spencer, A.M., Campbell, C.J., Hanslien, S.H., Nelson, P.H., Nysaether, E. and Ormaasen, E.G. (eds) *Geology of the Norwegian Oil and Gas Fields.* Graham and Trotman, London, pp. 235–42. [1]

Norsk Hydro, A.S. (1994) *Facts on Norwegian Oil and Gas Fields.* Norske Hydro Public Affairs, Stabekk, Norway, 49 pp. [9]

Northam, M.A. (1985) Correlation of northern North Sea oils: The different facies of their Jurassic source. In: Thomas, B.M., Doré, A.G., Eggen, S.S., Home, P.C. and Larsen, R.M. (eds) *Petroleum Geochemistry in Exploration of the Norwegian Shelf.* Graham and Trotman, London, pp. 93–100. [11]

Norton, M.G., McClay, K.R. and Way, N.R. (1987) Tectonic evolution of Devonian basins in northern Scotland and southern Norway. *Norsk Geol.* **67**, 323–38. [3]

Nummedal, D. and Swift, D.J.P. (1987) Transgressive stratigraphy at sequence-bounding unconformities: some principles derived from Holocene and Cretaceous examples. In: Nummedal, D., Pilkey, O.H. and Howard, J.D. (eds) In: *Sea Level Fluctuation and Coastal Evolution.* Special Publication No. 41, Society of Economic Palaeontologists and Mineralogists, Tulsa, OK. pp. 241–60. [8]

Nyberg, I.T. (1987) Statfjord Ost. In: Spencer, A.M., Campbell, C.J., Hanslien, S.H., Nelson, P.H., Nysaether, E. and Ormaasen, E.G. (eds) *Geology of the Norwegian Oil and Gas Fields.* Graham and Trotman, London, pp. 351–62. [1]

Nygaard, E., Lieberkind, K. and Frykman, P. (1983) Sedimentology and reservoir parameters of the Chalk Group in the Danish Central Graben. *Geol. Mijnbouw* **62**, 177–90. [9]

Nygaard, E., Andersen, C., Moller, C., Clausen, C.K. and Stouge, S. (1990) Integrated multidisciplinary stratigraphy of the Chalk Group: an example from the Danish Central Trough. In: *Proceedings of an International Chalk Symposium, Brighton, UK, 1989.* Thomas Telford, London, pp. 195–201. [9]

Nyland, B., Jensen, L.N., Skagen, J.I., Skarpnes, O. and Vorren, T. (1989) Tertiary uplift and erosion in the Barents Sea area: magnitude, timing and consequences. *Proc. NPF Conference on Structural and Tectonic Modelling and its Application to Petroleum Geology, Stavanger, Norway, 18–20 October, 1989.* [11]

Nyland, B., Jensen, L.N., Skagen, J., Skarpnes, O. and Vorren, T. (1992) Tertiary uplift and erosion in the Barents Sea; magnitude, timing and consequences. In: Larsen, R.M. *et al.* (eds) *Structural and Tectonic Modelling and its Application to Petroleum Geology,* NPF Special Publication No. 1, pp. 153–62.

Nystuen, J.P., Knarud, R., Jorde, K. and Stanley, K.O. (1990) Correlation of Triassic to Lower Jurassic sequences. Snorre Field and adjacent areas, northern North Sea. In: Collinson, J.D. (ed.) *Correlation in Hydrocarbon Exploration.* Graham and Trotman, London, pp.273–89. [7, 8, 9]

Oakman, C.D., Rudser, R.J., Macintyre, D.M., Humphreville, R.G., Haigh, J.A. and Rasmussen, S. (1993) Jurassic submarine fans in the Norwegian Central Trough—an exciting new play concept. In: *Extended abstracts of papers, European Association of Petroleum Geoscientists and Engineers 5th conference.* Stavanger, Norway. [9]

O'Connor, S.J. and Walker, D. (1993) Paleocene reservoirs of the Everest Trend. In: Parker, J.R. (ed.) *Petroleum Geology of Northwest Europe: Proceedings of the 4th Conference.* Geological Society, London, pp. 145–60. [10, 12]

O'Donnell, D. (1993) Enhancing the oil potential of secondary Triassic reservoirs in the Beryl A Field, UK North Sea. In: Spencer, A.M. *et al.* (ed.) *Generation, Accumulation, and Production of Europe's Hydrocarbons.* Special Publication, EAPG., Springer-Verlag, Berlin. [12]

O'Driscoll, D., Hindle, A. D. and Long, D.C. (1990) The structural controls on Upper Jurassic and Lower Cretaceous reservoir sandstones in the Witch Ground Graben, UK North Sea. In: Hardman, R.P.F. and Brooks, J. (eds) *Tectonic Events Responsible for Britain's Oil and Gas Reserves.* Special Publication No. 55, Geological Society, London, pp. 299–323. [9]

Oele, J.A., Hol, A.C.P.J. and Tiemans, J. (1981) Some Rotliegend gas fields of the K and L blocks, Netherlands offshore 1968–1978—a case history. In: Illing, L.V. and Hobson, G.P. (eds) *Petroleum Geology of the Continental Shelf of North-West Europe.* Heyden, London, pp. 289–300. [5, 11, 12]

Oen, P.M., Engell-Jensen, M. and Barendregt, A.A. (1986) *Skjold Field, Danish North Sea: Early Evaluation of Oil Recovery Through Water inhibition in a Fractured Reservoir.* 15569, SPE, 11 pp. [9]

Ofstad, K. (1981) *The Eldfisk Area.* Paper No. 30, NPD, Stavanger, Norway.

Ofstad, K. (1983) *The Southernmost Part of the Norwegian Section of the Central Trough.* Paper No. 32, NPD, 40 pp. [4]

Oftedahl, C. (1976) Northern end of European Continental Permian—the Oslo region. In: Falke, H. (ed.) *The Continental Permian in Central, West and South Europe.* Advanced Study Institute Series C, NATO, Reidel, Dordrecht, pp. 2–13. [2]

Olaussen, S., Larsen, B.T., Midtkandal, P.A. and Steel, R. (1982) Sedimentation in Upper Palaeozoic Oslo Graben. Abstract for KNGMG Conference on Petroleum Geology in the SE North Sea and Adjacent Onshore Areas, The Hague, Nov. [2]

Oliver, G.J.H. (1988) Arenig to Wenlock regional metamorphism in the Paratectonic Caledonides of the British Isles—a review. In: Harris, A.L. and Fattes, D.J. *The Caledonian–Appalachian Orogen.* Special Publication No. 38, Geological Society, London, pp. 347–63. [11]

Olsen, J.C. (1983) The structural outline of the Horn Graben. *Geol. Mijnbouw* **62**, 47–50. [7, 11]

Olsen, J.C. (1987) Tectonic evolution of the North Sea region. In: Brooks, J. and Glennie, K.W. (eds) *Petroleum Geology of North West Europe.* Graham and Trotman, London, pp. 389–401. [7]

Olsen, R.C. (1979) *Lithology: Well 17/11-1.* Paper No. 12, NPD, Stavanger. [7]

O'Reilly, B.M., Shannon, P.M. and Vogt, U. (1991) Seismic studies in the North Celtic Sea Basin: implications for basin development. *J. Geol. Soc.* **148**, 191–5. [12]

O'Reilly, B.M. *et al.* (1995) The Erris and eastern Rockall Troughs: structural and sedimentological development. In: Croker, P.F. and Shannon, P.M. (eds) *The Petroleum Geology of Ireland's Offshore basins*. Special Publication No. 93, Geological Society, London, pp. 413–21. [12]

Ormaasen, O.E., Hamar, G.P., Jakobsson, K.H. and Skarpnes, O. (1980) Permo-Triassic correlations in the North Sea area north of the Central Highs. In: NPF (ed.) *The Sedimentation of the North Sea Reservoir Rocks*. Geilo XIX, NPF, Elsevier, Amsterdam, pp. 1–15. [7]

Ormiston, A.R. and Oglesby, R.J. (1995) Effect of Late Devonian paleoclimate on source rock quality and location. In: Huc, A.-Y. (ed.) *Paleogeography, Paleoclimate, and Source Rock*. Studies in Geology No. 40. AAPG, Tulsa, OK., pp. 105–33. [11]

Orr, W.L. (1986) Kerogen/asphaltene/sulfur relationships in sulfur-rich Moneterey oils. In: Leythaeuser, D. and Rullkotter, J. (eds) *Advances in Organic Geochemistry*. Pergamon, Oxford, pp. 499–516. [11]

Osborne, P. and Evans, S. (1987) The Troll field: reservoir geology and field development planning. In: Kleppe, J., Berg, E.W., Buller, A.T., Hjelmelaud, O. and Torsæter, O. (eds) *North Sea Oil and Gas Reservoirs*. Graham and Trotman, London, pp. 39–60. [11]

Oschmann, W. (1988) Kimmeridge Clay sedimentation—a new cyclic model. *Palaeogeog. Palaeoclim. Palaeoecol.* **65**, 217–51. [11]

Oschmann, W. (1990) Environmental cycles in the Late Jurassic northwest European epeiric basin: interaction with atmospheric and hydrospheric circulations. *Sed. Geol.* **69**, 313–32. [11]

Østfeldt, P. (1987) Oil source rock correlation in the Danish North Sea. In: Brooks, J. and Glennie, K.W. (eds) *Petroleum Geology of North West Europe*. Graham and Trotman, London, pp. 419–30. [9, 11]

Østvedt, O.J. (1987) Sleipner Ost. In: Spencer, A.M., Campbell, C.J., Hanslien, S.H., Nelson, P.H., Nysaether, E. and Ormaasen, E.G. (eds) *Geology of the Norwegian Oil and Gas Fields*. Graham and Trotman, London, pp. 243–52. [1, 11]

Oudin, J.L. (1976) Etude geochimique du Bassin de la Mere du Nord. *Bull. Centre Rech. Pau-SNPA* **10**(1), 339–58. [11]

Oudmayer, B.C. and de Jager, J. (1993) Fault reactivation and oblique-slip in the Southern North Sea. In: Parker, J.R. (ed.) *Petroleum Geology of Northwest Europe: Proceedings of the 4th Conference*. Geological Society, London, pp. 1281–90. [2, 4, 6]

Owens, B., Neves, R., Gueinn, K.J., Mishell, D.R., Sabry, H.S. and Williams, J.E. (1977) Palynological division of the Namurian of northern England and Scotland. *Proc. Yorks. Geol. Soc.* **41**, 381–98. [4]

Owens, B., Hewett, A.J. and Gueinn, N.J. (1995) Discovery of Carboniferous deposits in Quadrant 12, Inner Moray Firth Basin, Scotland. (Abstract) *Stratigraphic advances in the offshore Devonian and Carboniferous rocks, UKCS and adjacent onshore areas*. Geological Society, London. [4]

Oxtoby, N.H., Mitchell, A.W. and Gluyas, J.G. (1995) The filling and emptying of the Ula oilfield: fluid inclusion constraints. In: Cubitt, J.M. and England, W.A. (eds) *The Geochemistry of Reservoirs*. Special Publication No. 86, Geological Society, London, pp. 141–59. [11]

Paine, J.G. (1993) Subsidence of the Texas coast: inferences from historical and late Pleistocene sea levels. *Tectonophysics* **222**, 445–58. [5]

Palciauskas, V.V. (1991) Primary migration of petroleum. In: Merill, R.K. (ed.) *Source and Migration Processes and Evaluation Techniques*. Treatise Petroleum Geology, AAPG, Tulsa, OK., pp. 13–22. [11]

Pallatt, N., Wilson, J. and McHardy, W. (1984) *The Relationship Between Permeability and the Morphology of Diagenetic Illite in Reservoir Rocks*. No. SPE-12798, SPE of AIME. [12]

Palmer, S.E. (1991) Effect of biodegradation and water washing on crude oil composition. In: Merrill, R.K. (ed.) *Source and Migration Processes and Evaluation Techniques*. Treatise Petroleum Geology, AAPG, Tulsa, OK., pp. 47–54. [11]

Park, R.G., Cliff, R.A., Fettes, D.J. and Stewart, A.D. (1994) Precambrian rocks in northwest Scotland west of the Moine Thrust: the Lewisian Complex and the Torridonian. In: Gibbons, W. and Harris, A.L. (eds) *A Revised Correlation of Precambrian Rocks in the British Isles*. Special Report No. 22, Geological Society, London, pp. 6–22. [2]

Parker, J.R. (1975) Lower Tertiary sand development in the Central North Sea. In: Woodland, A.W. (ed.) *Petroleum and the Continental Shelf of Northwest Europe*, Vol. I, *Geology*. Applied Science Publishers, Barking, pp. 447–53. [2]

Parker, J.R. (ed.) (1993) *Petroleum Geology of Northwest Europe: Proceedings of the 4th Conference*. 2 vols. Geological Society, London, 1542 pp. [1, 6, 11]

Parker, R.H. (1991) The Ivanhoe and Rob Roy Fields, Block 15/21a, b, UK North Sea. In: Abbotts, I.L. (ed.) *UK North Sea United Kingdom Oil and Gas Fields: 25 Years Commemorative Volume*. Memoir No. 14, Geological Society, London, pp. 331–8. [1, 8, 12]

Parkinson, D.N. and Hines, F.M. (1995) The Lower Jurassic of the North Viking Graben in the context of western European Lower Jurassic stratigraphy. In: Steel, R.J., Felt, V.L., Johannessen, E.P. and Mathieu, C. (eds) *Sequence Stratigraphy on the Northwest European Margin*. Special Publication No. 5, NPS, Norway, pp. 97–107. [8]

Parnell, J. (1982a) Genesis of the graphite deposit at Seathwaite, Cumbria. *Geol. Mag.* **119**, 511–12. [2]

Parnell, J. (1982b) Comment on 'Paleomagnetic evidence for a large ~2000 km offset along the Great Glen Fault during Carboniferous time'. *Geology* **10** 605. [2]

Parnell, J. (1983) The distribution of hydrocarbon minerals in the Orcadian Basin. *Scott. J. Geol.* **19**, 205–13. [3, 11]

Parnell, J. (1984) Hydrocarbon minerals in the Midland Valley of Scotland with particular reference to the oil-shale group. *Proc. Geol. Assoc.* **95**(3), 275–86. [11]

Parnell, J. (1985) Hydrocarbon source rocks, reservoir rocks and migration in the Orcadian Basin. *Scottish Journal of Geology* **21**, 321–35.

Parnell, J. (1988a) Significance of lacustrine cherts for the environment of source rock deposition in the Orcadian Basin, Scotland. In: Fleet, A.J., Kelts, K. and Talbot, M.R. (eds) *Lacustrine Petroleum Source Rocks*. Blackwell Scientific Publications, Oxford, pp. 205–18. [11]

Parnell, J. (1988b) Lacustrine petroleum source rocks in the Dinantian Oil Shale Group, Scotland: A Review. In: Fleet, A.J., Kelts, K. and Talbot, M.R. (eds) *Lacustrine Petroleum Source Rocks*. Blackwell Scientific Publications, Oxford, pp. 235–46. [11]

Parnell, J. (1991) Hydrocarbon potential of Northern Ireland. Part 1. Burial histories and source-rock potential. *J. Petr. Geol.* **14**(1), 65–78.

Parrish, J.T. (1982) Upwelling and petroleum source beds with reference to the Palaeozoic. *AAPG Bull.* **66**, 750–74. [4]

Parrish, J.T. (1995) Paleogeography of C org-rich rocks and the preservation versus production controversy. In: Huc, A.Y. (ed.) *Paleogeography, Paleoclimate, and Source Rock*. Studies in Geology No. 40, AAPG, Tulsa, OK., pp. 1–20. [11]

Parrish, J.T., Ziegler, A.M. and Scotese, C.R. (1982) Rainfall patterns and the distribution of coals and evaporites in the Mesozoic and Cenozoic. *Palaeogeog. Palaeoclimatol. Palaeoecol.* **40**, 67–101. [7]

Parry, C.C., Whitley, P.K.J. and Simpson, R.D.H. (1981) Integration of palynological and sedimentological methods in facies analysis of the Brent Formation. In: Illing, L.V. and Hobson, G.D. (eds) *Petroleum Geology of the Continental Shelf of North-West Europe*. Heyden, London, pp. 205–15. [11]

Parsley, A.J. (1990) North Sea hydrocarbon plays. In: Glennie, K.W. (ed.) *Introduction to the Petroleum Geology of the North*

Sea. Blackwell Scientific Publications, Oxford, pp. 362–88. [12]

Parsons, B. and McKenzie, D. (1978) Mantle convection and thermal structure of the plates. *J. Geophys. Res.* **83**, 4485–96.

Partington, M.A., Mitchener, B.C., Milton, N.J. and Fraser, A.J. (1993a) Genetic sequence stratigraphy for the North Sea Late Jurassic and early Cretaceous: distribution and prediction of Kimmeridgian–Late Ryazanian reservoirs in the North Sea and adjacent areas. In: Parker, J.R. (ed.) *Petroleum Geology of Northwest Europe: Proceedings of the 4th Conference*. Geological Society, London, pp. 347–70. [8]

Partington, M.A., Copestake, P., Mitchener, B.C. and Underhill, J.R. (1993b) Biostratigraphic calibration of genetic stratigraphic sequences in the Jurassic–lowermost Cretaceous Hettangian to Ryazanian of the North Sea and adjacent areas. In: Parker, J.R. (ed.) *Petroleum Geology of Northwest Europe: Proceedings of the 4th Conference*. Geological Society, London, pp. 371–86. [8]

Passey, Q.R., Creaney, S., Kulla, J.B., Moretti, F.J. and Stroud, J.D. (1990) A practical model for organic richness from porosity and resistivity logs. *AAPG Bull.* **74**(12), 1777–94. [11]

Pattison, J., Smith, D.B. and Warrington, G. (1973) A review of the Late Permian and Early Triassic biostratigraphy in the British Isles. In: Logan, A.V. and Mills, L.V. (eds) *The Permian and Triassic Systems and their Mutual Boundary*. Memoir 19, Canadian Society, *Petrol. Geol.* **2**, 220–60. [6]

Paul, J. (1980) Upper Permian algal stromatolite reefs, Harz Mountains F.R. Germany. In: Füchtbauer, H. and Peryt, T.M. (eds) *The Zechstein Basin with Emphasis on Carbonate Sequences*. Contr. Sedimentology No. 9, Schweitzerbart'sche Verlagsbuchhandlung, Stuttgart, pp. 253–68. [6]

Peacock, J.D., Berridge, N.G., Harris, A.L. and May, F. (1968) *The Geology of the Elgin District*. Memoir, Geological Survey, HMSO, Edinburgh. [8]

Peace, G.R. and Besly, B.M. (1997) End-Carboniferous fold-thrust structures, Oxfordshire, UK: implications for the structural evolution of the late Variscan foreland of south-central England. *J. Geol. Soc. Lond.* **154**, 225–37. [4]

Pearce, R.B., Clayton, T. and Kemp, A.E.S. (1991) Illitization and organic maturity in Silurian sediments from the southern Uplands of Scotland. *Clay Min.* **26**, 199–210. [11]

Pearson, J.F.S., Youngs, R.A. and Smith, A. (1991) The Indefatigable Field, Blocks 48/18, 48/19, 48/23, 48/24, UK North Sea. In: Abbotts, I.L. (ed.) *UK North Sea United Kingdom Oil and Gas Fields: 25 Years Commemorative Volume*. Memoir No. 14, Geological Society, London, pp. 443–50. [1, 11]

Pearson, M.J. and Duncan, A.D. (1996) Biomarker maturity profiles in the Inner Moray Firth basin and implications for inversion estimates. In: Hurst, A., Johnson, H.D., Burley, S.D., Canham, A.C. and Mackertich, D.S. (eds) *Geology of the Humber Group: Central Graben and Moray Firth*, UKCS, Special Publication No. 114. Geological Society, London, pp. 287–98.

Pearson, M.J. and Watkins, D. (1983) Organofacies and early maturation effects in Upper Jurassic sediments from the Inner Moray Firth Basin, North Sea. In: Brooks, J. (ed.) *Petroleum Geochemistry and Exploration of Europe*. Special Publication No. 12. Geological Society, Blackwell Scientific Publications, Oxford, pp. 147–60. [11]

Pearson, M.J., Watkins, D., Pittion, J.-L., Caston, D. and Small, J.S. (1983) Aspects of burial diagenesis, organic maturation and palaeothermal history of an area in the South Viking Graben, North Sea. In: Brooks, J. (ed.) *Petroleum Geochemistry and Exploration of Europe*. Special Publication No. 12. Geological Society, Blackwell Scientific Publications, Oxford, pp. 161–74. [11]

Pedersen, T.F. and Calvert, S.E. (1990) Anoxia vs. productivity: what controls the formation of organic-carbon rich sediments rocks? *AAPG Bull.* **74**(4), 454–66. [11]

Pegrum, R.M. and Ljones, T.E. (1984) 15/9 Gamma gas field offshore Norway: a new trap type for North Sea basin with regional implications. *AAPG Bull.* **68**(7), 874–902. [1, 2, 11]

Pegrum, R.M. and Spencer, A.M. (eds) (1990) Hydrocarbon plays in the North Sea. In: Brooks, J. (ed.) *Classic Petroleum Provinces*. Special Publication No. 50, Geological Society, London, pp. 441–70. [11, 12]

Pekot, L.J. and Gersib, G.A. (1987) Ekofisk. In: Spencer, A.M., Campbell, C.J., Hanslien, S.H., Nelson, P.H., Nysaether, E. and Ormaasen, E.G. (eds) *Geology of the Norwegian Oil and Gas Fields*. Graham and Trotman, London, pp. 73–88. [1, 9]

Pemberton, S.G., Maceachern. J.A. and Frey, R.W. (1992) Trace fossil models: environmental and allostratigraphic significance. In: Walker, R.G. and James, N.P. (eds) *Facies Models—Response to Sea-Level Change*. Geological Association of Canada, Canada, pp. 47–73. [12]

Penge, J., Taylor, B., Huckerby, J.A. and Munns, J.W. (1993) Extension and salt tectonics in the East Central Graben. In: Parker, J.R. (ed.) *Petroleum Geology of Northwest Europe: Proceedings of the 4th Conference*. Geological Society, London, pp. 119–210. [2, 6, 7, 12]

Penn, I.E., Chadwick, R.A., Holloway, S., Roberts, G., Pharaoh, T.C. and Alsop, J.M. (1987) Principal features of the hydrocarbon prospectivity of the Wessex Channel Basin, UK. In: Brooks, J. and Glennie, K.W. (eds) *Petroleum Geology of North West Europe*. Graham and Trotman, London, pp. 109–18. [11]

Pennington, J.J. (1975) The geology of the Argyll Field. In: Woodland, A.W. (ed.) *Petroleum and the Continental Shelf of Northwest Europe*. Vol. I, *Geology*. Applied Science Publishers, Barking, pp. 285–91. [1, 3, 5, 6, 11, 12]

Penny, B. (1991) The Heather Field, Block 2/5, UK North Sea. In: Abbotts, I.L. (ed.) *UK North Sea United Kingdom Oil and Gas Fields: 25 Years Commemorative Volume*. Memoir No. 14, Geological Society, London, p. 127. [1]

Pepper, A.S. (1991) Estimating the petroleum expulsion behaviour of source rocks: a novel quantitative approach. In: England, W.A. and Fleet, A.J. (eds) *Petroleum Migration*. Special Publication No. 59, Geological Society, London, pp. 9–32. [11]

Pepper, A.S. and Corvi, P.J. (1995) Simple kinetic models of petroleum formation. Part I: Oil and gas generation from kerogen. *Mar. Petr. Geol.* **12**(3), 291–321. [11]

Pepper, A.S. and Dodd, T.A. (1995) Simple kinetic models of petroleum formation. Part II: Oil–gas cracking. *Mar. Petr. Geol.* **12**, 291–341. [11]

Pering, K.L. (1973) Bitumens associated with lead, zinc and fluorite ore minerals in North Derbyshire, England. *Geochim. Cosmochim. Acta* **37**, 401–17. [11]

Perrot, J. and van der Poel, A.B. (1987) Zuidwal—a Neocomian gas field. In: Brooks, J. and Glennie, K.W. (eds) *Petroleum Geology of North West Europe*. Graham and Trotman, London, pp. 325–35. [9, 12]

Peryt, T.M. (1985) A Permian beach in the Zechstein dolomites of Western Poland: influence on reservoirs. *J. Petr. Geol.* **8**, 463–74. [6]

Peryt, T.M. (ed.) (1987) *The Zechstein Facies in Europe*. Springer-Verlag, Berlin, New York.

Peryt, T.M. (1992) Debris-flow deposits in the Zechstein (Upper Permian) Main Dolomite of Poland: significance for the evolution of the basin. *N. Jb. Geol. Paläont. Abh.* **185**(1), 1–19. [6]

Peryt, T.M. (1994) The anatomy of a sulphate platform and adjacent basin system in the Leba sub-basin of the Lower Werra Anhydrite (Zechstein, Upper Permian), northern Poland. *Sedimentology* **41**, 83–113. [6]

Peryt, T.M. and Dyjaczn'ski, K. (1991) An isolated carbonate bank in the Zechstein Main Dolomite basin, Western Poland. *J. Petr. Geol.* **14**, 445–58. [6]

Peryt, T.M., Orti, F. and Rosell, L. (1993) Sulphate platform–basin transition of the Lower Werra Anhydrite (Zechstein, Upper Permian) SW Poland: facies and petrography. *J. Sediment. Petrol.* **63**, 646–58. [6]

Peters, K.E. (1986) Guidelines for evaluating petroleum source rock using programmed pyrolysis. *AAPG Bull.* **70**(3), 318–29. [11]

Peters, K.E., Moldowan, J.M., Driscole, A.R. and Demaison, G.J. (1989) Origin of Beatrice oil by co-sourcing from Devonian and Middle Jurassic source rocks, Inner Moray Firth, United Kingdom. *AAPG Bull.* **73**, 454–71. [2, 3, 11, 12]

Petersen, K. and Lerche, I. (1996) Temperature dependence of thermal anomalies near evolving salt structures: importance for reducing exploration risk. In: Alsop, G.I., Blundell, D.J. and Davison, I. (eds) *Salt Tectonics.* Special Publication 100, Geological Society, London, pp. 275–90. [6]

Petit-Maire, N. (1994) Natural variability of the Asian, Indian and African monsoons over the last 130 ka. In: Desbois, M. and Daesalmand, F. (eds) *Global Precipitation and Climate Change.* ASI Series 126, NATO, Springer-Verlag, Berlin, pp. 3–26. [5]

Petrie, S.H., Brown, J.R., Granger, P.J. and Lovell, J.P.B. (1989) Mesozoic history of the Celtric Sea basins. In: Tankard, A.J. and Balkwill, H.R. (eds) *Extensional Tectonics and Stratigraphy of the North Atlantic Margins.* Memoir No. 46, AAPG, Tulsa, OK., pp. 433–44. [12]

Petroleum Geological Circle (1995) *Synopsis: Petroleum Geology of the Netherlands.* Royal Geological and Mining Society, Dordrecht. [1, 5] Also in: Rondeel, H.E., Batjes, D.A.J. and Niewenhuis, W.H. (eds) (1996) *Geology of Gas and Oil under the Netherlands-KNGMG.* Kluwer Academic Publishers, The Netherlands, pp. S1–S20. [1, 5]

Petterson, O., Storli, A., Ljosland, E. and Massie, I. (1990) The Gullfaks Field: geology and reservoir development. In: Buller, T., Berg, E., Hjelmeland, O., Kleppe, J., Torsæter, O. and Aasen, J.O. (eds) *Norwegian Oil and Gas Reserviors–II.* Graham and Trotman, London, pp. 67–90. [12]

Pharaoh, T.C., Merriman, R.J, Webb, P.C., and Beckinsale, R.D. (1987) The concealed Caledonides of eastern England: preliminary results of a multidisciplinary study. *Proc. Yorks. Geol. Soc.* **46**, 355–69. [3]

Phemister, J. (1960) *Scotland: the Northern Highlands*, 3rd edn. British Regional Geology, Institute of Geological Science, HMSO, Edinburgh, 104 pp. [2]

Phillips, W.E.A., Stillman, C.J. and Murphy, T. (1976) A Caledonian plate tectonics model. *J. Geol. Soc.* **132**, 579–609. [2]

Piasecki, S. and Stemmerik, L. (1991) Late Permian anoxia in Central East Greenland. In: Tyson, R.V. and Pearson, T.H. (eds) *Modern and Ancient Continental Shelf Anoxia.* Special Publication No. 58, Geological Society, London, pp. 275–90. [11]

Piasecki, S., Christiansen, F.G. and Stemmerik, L. (1990) Depositional history of an Upper Carboniferous organic-rich lacustrine shale from East Greenland. *Bull. Canadian Petr. Geol.* **38**(3), 273–87. [11]

Pickering, K.T. (1983) Small-scale syn-sedimentary faults in the Upper Jurassic 'Boulder Beds'. *Scott. J. Geol.* **19**, 169–81. [8]

Pickering, K.T. (1984) The Upper Jurassic Boulder Beds and related deposits: a fault controlled submarine slope. *J. Geol. Soc.* **141**, 357–74. [8]

Pigott, J.D. (1985) Assessing source rock maturity in frontier basins: importance of time, temperature and tectonics. *AAPG Bull.* **69**(8), 1269–74. [11]

Pittion, J.L. and Gouadain, J. (1985) Maturity studies of the Jurassic 'coal unit' in three wells from the Haltenbanken area. In: Thomas, B.M., Doré, A.G., Eggen, S.S., Home, P.C. and Larsen, R.M. (eds) *Petroleum Geochemistry in Exploration of the Norwegian Shelf.* Graham and Trotman, London, pp. 205–212. [11]

Platt, J.D. (1994) Geochemical evolution of pore waters in the Rotliegend Early Permian of northern Germany. *Mar. Petr. Geol.* **7**(1), 66–78. [5]

Platt, N.H. (1995) Structure and tectonics of the northern North Sea: new insights from deep penetration regional seismic data. In: Lambiase, J.J. (ed.) *Hydrocarbon Habitat in Rift Basins.* Special Publication No. 80, Geological Society, London, pp. 103–13. [7]

Platt, N.H. and Philip, P. (1993) Comparison of Permo-Triassic and deep structure of the Forties–Montrose and Jaeren highs, Central Graben, UK and Norwegian North Sea. In: Parker, J.R. (ed.) *Petroleum Geology of Northwest Europe: Proceedings of the 4th Conference.* Geological Society, London, pp. 1221–30. [12]

Plein, E. (1978) Rotliegend Ablagerungen im Norddeutschen Becken. *Zeitschr. deutsches geol. Gesampt* **129**, 71–97. [5]

Plein, E. (1993) Bemerkungen zum Ablauf der palèogeographisches Entwicklung in Stefan und Rotliegend des Norddeutschen Beckens. *Geol. Jahrubuch* **A131**, 99–116. [2, 5]

Plein, E. (compiler) (1995) *Stratigraphie von Deutschland I. Norddeutsches Rotligendbecken.* Rotliegend-Monographie Teil II, Courier Forschungsinstitut Senckenberg 183, Frankfurt, 193 pp. [5]

Plumhoff, M. (1966) Marines Ober Rotliegendes Perm in Zentrum des nordwestdeutschen Rotliegend Beckens: Neue Beweise und Folgerungen. *Erdol Kohle* **19**(10), 713–20. [5]

Pokorski, J. (1989) Evolution of the Rotliegendes basin in Poland. *Bull. Polish Acad. Sci., Earth Sci.* **37**(1–2), 49–55. [5]

Pokorski, J. and Wagner, R. (1993) Geology of the oil and gas bearing Permian formation in the Polish Lowlands. *AAPG Bull.* **77**(9), 1654. [6]

Pollard, J.E. (1981) A comparison between the Triassic trace fossils of Cheshire and South Germany. *Palaeontology* **24**, 555–88. [7]

Posamentier, H.W. and Vail, P.R. (1988) Eustatic controls on clastic deposition II: sequence and systems tract models. In: Wilgus, C.K., Hastings, B.S., Posamentier, H., van Wagoner, J., Ross, C.A. and Kendall, C.G.St.C. (eds) *Sea-Level Changes: an Integrated Approach.* Special Publication No. 42, SEPM, Tulsa, OK., pp. 125–54. [8]

Posaementier, H.W., Jervey, M.T. and Vail, P.R. (1988) Eustatic controls on clastic deposition I—conceptual framework. In: Wilgus, C.K., Hastings, B.S., Posamentier, H., van Wagoner, J., Ross, C.A. and Kendall, C.G.St.C. (eds) *Sea-Level Changes: An Integrated Approach.* Special Publication No. 42, SEPM, Tulsa, OK., pp. 108–24. [8]

Posamentier, H.W., Allen, G.P., James, D.P. and Tesson, M. (1992) Forced regressions in a sequence stratigraphic framework: concepts, examples and exploration signifiance. *AAPG Bull.* **76**, 1687–709. [8]

Powell, D. and Phillips, W.E.A. (1985) Time of deformation in the Caledonide orogen of Britian and Ireland. In: Harris, A.L. (ed.) *The Nature and Timing of Orogenic Activity in the Caledonian Rocks of the British Isles.* Memoir No. 9, Geological Society, London, pp. 17–39. [2]

Powell, T.G. (1986) Developments in concepts of hydrocarbon generation from terrestrial organic matter. In: *Proc. Beijing Petroleum Symposium.* AAPG, Tulsa, OK. [11]

Powell, T.G. and Boreham, C.J. (1994) Terrestrial source oils: where do they exist and what are our limits of knowledge? In: Scott, A.C. and Fleet, A.J. (eds) *Coal and Coal-bearing Strata as Oil-prone Source Rocks?* Special Publication No. 77. Geological Society, pp. 11–30. [11]

Powell, T.G., Douglas, A.G. and Allan, J. (1976) Variations in the type and distribution of organic matter in some Carboniferous sediments from Northern England. *Chemical Geology* **18**, 137–48. [11]

Pradier, B. and Bertrand, P. (1992) Etude à haute résolution d'un cycle du carbone organique de rochemère du Kimmeridgian du Yorkshire GB: relation entre composition pétrographique du contenu organique observé *in situ* teneur en carbone organique et qualité pétroligène. *CR Acad. Sci. Paris* **315**, 187–92.

Pratsch, J.-C. (1983) Gasfields, NW German basin: secondary gas migration as a major geologic parameter. *J. Petroleum Geology* **5**(3), 229–44. [11]

Price, J., Dyer, R., Goodall, I., McKie, T., Watson, P. and

Williams, G. (1993) Effective stratigraphical subdivision of the Humber Group and the Late Jurassic evolution of the UK Central Graben. In: Parker, J.R. (ed.) *Petroleum Geology of Northwest Europe: Proceedings of the 4th Conference*. Geological Society, London, pp. 443–58. [8, 12]

Price, L.C. (1983) Geological time as a parameter in organic metamorphism and vitrinite reflectance as an absolute paleogeothermometer. *Journal of Petroleum Geology* **6**(1), 5–38. [11]

Price, L.C. and Barker, C.E. (1985) Suppression of vitrinite reflectance in amorphous rich kerogen—a major unrecognized problem. *Journal of Petroleum Geology* **8**(1), 59–84. [11]

Price, S.P. and Whitham, A.G. (1997) Exhumed hydrocarbon traps in East Greenland: analogs for the Lower–Middle Jurassic plat of northwest Europe. *AAPG Bull.* **81**(2), 196–221. [8, 11]

Primio, R. di and Horsfield, B. (1995) Predicting the generation of heavy oils in carbonate/evaporitic environments using pyrolysis methods. *Org. Geochem.* **24**(10/11), 999–1095. [3]

Pritchard, M.J. (1991) The V Fields, Blocks 49/16, 49/21, 48/20a, 48/25b, UK North Sea. In: Abbotts, I.L. (ed.) *UK North Sea United Kingdom Oil and Gas Fields: 25 Years Commemorative Volume*. Memoir No. 14, Geological Society, London, pp. 497–502. [1, 5, 11]

Prosser, D.J. and Maskall, R. (1993) Permeability variation within aeolian sandstones: a case study using core cut sub-parallel to slip-face bedding, the Auk Field, central North Sea, UK. In: North, C.P. and Prosser, D.J. (eds) *Characterization of Fluvial and Aeolian Reservoirs*. Special Publication 73, Geological Society, London, pp. 377–97. [5]

Purvis, K. (1990) The clay mineralogy of the Upper Triassic Skagerrak Formation, Central North Sea. In: Sci. Geol. Memoir No. 88, pp. 125–34. [7]

Purvis, K. (1994) Extensive albite dissolution in Triassic reservoir sandstones from the Gannet field, UK North Sea. *Mar. Petr. Geol.* **11**(5), 624. [11]

Purvis, K. and Okkerman, J.A. (1996) Inversion of reservoir quality by early diagenesis: an example from the Triassic Buntsandstein, offshore the Netherlands. In: Rondeel, H.E., Batjes, D.A.J. and Niewenhuis, W.H. (eds) *Geology of Gas and Oil under the Netherlands*. Kluwer Academic Publishers, the Netherlands, pp. 179–89. [7]

Puttmann, W. and Eckardt, C.B. (1990) Influence of an intrusion on the extent of isomerism in acyclic isoprenoids in the Permian Kupferschiefer of the Lower Rhine basin, NW Germany. In: Durand, B. and Behar, F. (eds) *Advances in Organic Geochemistry*. Pergamon, pp. 651–8. [11]

Puttmann, W., Eckardt, C.B. and Schwark, L. (1989) Use of biological marker distributions to study thermal history of the Permian Kupferschiefer of the Lower Rhine basin. *Geol. Rd* **78**, 411–26. [11]

Quigley, M., Mackenzie, A.S. and Gray, J.R. (1987) Kinetic theory of petroleum generation. In: Doligez, B. (ed.) *Migration of Hydrocarbons in Sedimentary Basins*. Technip, Paris, pp. 649–66. [11]

Quine, M. and Bosence, D. (1991) Stratal geometries, facies and sea-floor erosion in Upper Cretaceous Chalk, Normandy, France. *Sedimentology* **38**, 1113–52. [9]

Quirk, D.G. (1993) Interpreting the Upper Carboniferous of the Dutch Cleaver Bank High. In: Parker, J.R. (ed.) *Petroleum Geology of Northwest Europe: Proceedings of the 4th Conference*. Geological Society, London, pp. 697–706. [4]

Quirk, D.G. (1997) Sequence stratigraphy of the Westphalian in the northern part of the Southern North Sea. In: Ziegler, K., Turner, P. and Daines, S. (eds) *Petroleum Geology of the Southern North Sea: Future Potential*. Special Publication No. 123, Geological Society, London, pp. 153–68. [4]

Racero-Baena, A. and Drake, S.J. (1996) Structural style and reservoir development in the West Netherlands oil province. In: Rondeel, H.E., Batjes, D.A.J. and Nieuwenhuis, W.H. (eds) *Geology of Oil and Gas under the Netherlands*. Royal Geological and Mining Society, the Netherlands, Kluwer Academic Publishers, Dordrecht, pp. 211–27. [12]

Radke, M. (1988) Application of aromatic compounds as maturity indicators in source rocks and crude oils. *Marine and Petroleum Geology* **5**(3), 224–36. [11]

Radke, M. and Welte, D.H. (1983) The Methylphenanthrene Index (MPI): a maturity parameter based on aromatic hydrocarbons. In: Bjorøy, M., Albrecht, P., Cornford, C., de Groot, K., Eglinton, G., Galimov, E., *et al.* (eds) *Advances in Organic Geochemistry, 1981*. John Wiley, Chichester, pp. 504–12. [11]

Radke, M., Horsfield, B., Littke, R. and Rulkotter, J. (1997) Maturation and petroleum generation. In: Welte, D.H., Horsfield, B. and Baker, D.R. (eds) *Petroleum and Basin Evolution*. Springer-Verlag, Berlin, pp. 169–230. [11]

Rae, E.I.C. and Manning, D.A.C. (1989) Experimental studies of fluid–rock reaction during petroleum genesis. In: *Book of Abstracts from 14th Int'l Org. Geochem. Conference, Paris, 18–22 Sept. 1989*. Abstract No. 144, EAOG. [11]

Ramanampisoa, L. and Disnar, J.R. (1994) Primary control of paleo-production on organic matter preservation and accumulation in the Kimmeridge rocks of Yorkshire, UK. *Org. Geochem.* **21**(12), 1153–69.

Ramanampisoa, L. and Radke, M. (1995) Extractable aromatic hydrocarbons in a short-term organic cycle of the Kimmeridge Clay formation, Yorkshire, UK: relationship to primary production and thermal maturity. *Org. Geochem.*, **23**(9), 803–18. [11]

Ramanampisoa, L., Bertrand, P., Disnar, J.-R., Lallier-Verges, E., Pradier, B. and Tribovillard, N.J. (1992) High resolution study of an organic-carbon cycle in the Kimmeridge rocks of Yorkshire GB: preliminary results of organic geochemistry and petrography. *CR Acad. Sci. Paris* **314**(11), 1493–8. [11]

Ramsbottom, W.H.C. (1979) Rates of transgression and regression in the Carboniferous of NW Europe. *J. Geol. Soc.* **136**, 147–54. [4]

Ramsbottom, W.H.C., Calver, M.A., Eagar, R.M.C., Hodson, F., Hollidays, D.W., Stubblefield, C.J. *et al.* (1978) *A Correlation of Silesian Rocks in the British Isles*. Special Report 10, Geological Society, London, p. 81. [4]

Ranaweera, H.K.A. (1987) Sleipner Vest. In: Spencer, A.M., Campbell, C.J., Hanslien, S.H., Nelson, P.H., Nysaether, E. and Ormaasen, E.G. (eds) *Geology of the Norwegian Oil and Gas Fields*. Graham and Trotman, London, pp. 253–64. [1, 11]

Randall, B.A.O. (1980) The Great Whin Sill and its associated dyke suite. In: Robson, D.A. (ed.) *Geology of North East England*. Special Publication, Natural History Society of Northumbria, pp. 67–75. [5]

Rattey, R.P. and Hayward, A.B. (1993) Sequence stratigraphy of a failed rift system: the Middle Jurassic to Early Cretaceous basin evolution of the Central and Northern North Sea. In: Parker, J.K. (ed.) *Petroleum Geology of Northwest Europe: Proceedings of the 4th Conference*. Geological Society, London, pp. 215–49. [2, 8, 10, 12]

Rawson, P.F. (1992a) The Cretaceous. In: Duff, P.McL.D. and Smith, A.J. (eds) *Geology of England and Wales*. Geological Society, London.

Rawson, P.F. (1992b) Itinerary X: Reighton Gap to Speeton Cliffs. In: Rawson, P.F. and Wright, J.K. (eds) *Guide No. 34, Yorkshire Coast*. Geologists' Association, pp. 88–94. [9]

Rawson, P.F. and Riley, L. (1982) Latest Jurassic–Early Cretaceous Events and the Late Cimmerian Unconformity in the North Sea Area. *AAPG Bull.* **66**, 2628–48. [2, 8, 11]

Rawson, P.F. and Whitham, F. (1992a) Itinerary XI: Thornwick Bay and North Landing, Flamborough. In: Rawson, P.F. and Wright, J.K. (eds) *Guide No. 34, Yorkshire Coast*. Geologists' Association, pp. 94–9. [9]

Rawson, P.F. and Whitham, F. (1992b) Itinerary XII: Flamborough Head. In: Rawson, P.F. and Wright J.K. (eds) *Guide No. 34, Yorkshire Coast*. Geologists' Association, pp. 99–103. [9]

Rawson, P.F. *et al.* (1976) *A Correlation of Cretaceous Rocks in the*

British Isles. Geological Society, London, Special Report No. 9, 70 pp.

Raymond, A. (1985) Floral diversity, phytogeography and climatic amelioration during the early Carboniferous Dinantian. *Paleobiology* **11**, 293–309. [4]

Raymond, A.C. and Murchison, D.G. (1992a) Effect of igneous activity on molecular-maturation indices in different types of organic matter. *Org. Geochem.* **18**(5), 725–35.

Raymond, A.C. and Murchison, D.G. (1992b) Organic maturation and its timing in a Carboniferous sequence in the central Midland Valley of Scotland: comparisons with northern England. *FUEL* **68**, 328–34. [11]

Rea, D.K. and Janacek, T.R. (1982) Late Cenozoic changes in atmospheric circulation deduced from North Pacific eolian sediments. *Mar. Petr. Geol.* **49**, 149–67. [5]

Read, W.A. (1988) Controls on Silesian sedimentation in the Midland Valley of Scotland. In: Besly, B.M. and Kelling, G. (eds) *Sedimentation in a Synorogenic Basin Complex: the Upper Carboniferous of Northwest Europe*. Blackie, Glasgow, London, pp. 222–41. [4]

Read, W.A. (1991) The Millstone Grit Namurian of the southern Pennines viewed in the light of eustatically controlled sequence stratigraphy. *J. Geol.* **26**, 157–65. [4]

Read, W.A. and Forsyth, I.H. (1989). Allocycles and autocycles in the upper part of the Limestone Coal Group Pendleian E1 in the Glasgow–Stirling region of the Midland Valley of Scotland. *J. Geol.* **24**, 121–37. [4]

Reemst, P., Cloetingh, S. and Fanavoll, S. (1994) Tectonostratigraphic modelling of Cenozoic uplift and erosion in the southwestern Barents Sea. *Mar. Petr. Geol.* **11**, 478–90. [11]

Reid, H.H. (1961) Aspects of the Caledonian magmatism in Britain. *Proc. Liverpool Manchester Geol. Soc.* **2**, 653–83. [3]

Reilly, J.M. (1992) Seismic imaging adjacent to and beneath salt diapirs. UK North Sea. *First Break* **10**, 383–97. [6]

Reitsema, R.H. (1983) Geochemistry of North and South Brae areas, North Sea. In: Brooks, J. (ed.) *Petroleum Geochemistry and Exploration of Europe*. Special Publication No. 12. Blackwell Scientific Publications, Oxford, pp. 203–12. [11]

Remmelts, G. (1996) Salt tectonics in the southern North Sea, the Netherlands. In Rondeel, H.E., Batjes, D.A.J. and Niewenhuis, W.H. (eds) *Geology of Gas and Oil under the Netherlands*. Kluwer Academic Publishers, Dordrecht, pp. 143–58. [8]

Requejo, A.G., Wielchowsky, C.C., Klosterman, M.J. and Sassen, R. (1994) Geochemical characterization of lithofacies and organic facies in Cretaceous organic-rich rocks from Trinidad, East Venezuela Basin. *Org. Geochem.* **22**(3–5), 441–60.

Requejo, A.G., Sassen, R., McDonald, T., Denoux, G., Kennicutt II, M.C. and Brooks, J.M. (1996) Polynuclear aromatic hydrocarbons PAH as indicators of the source and maturity of marine crude oils. *Org. Geochem.* **24**(10/11), 1017–33. [11]

Reynolds, T. (1994) Quantitative analysis of submarine-fans in the Tertiary of the North Sea Basin. *Mar. Petr. Geol.* **11**, 202–7. [10]

Rhys, G.H. (compiler) (1974) *A Proposed Stratigraphic Nomenclature for the Southern North Sea and an Outline Structural Nomenclature for the Whole of the UK North Sea*. Report No. 74/8, Institute of Geological Science, HMSO, 14 pp. [5, 6, 7, 8, 9]

Rice, C.M., Ashcroft, W.A., Batten, D.J., Boyce, A.J., Caulfield, J.B.D., Fallick, A.E. *et al.* (1995) A Devonian auriferous hot spring system, Rhynie, Scotland. *J. Geol. Soc.* **152**(2), 229–50. [11]

Richards, P.C. (1985) Upper Old Red Sandstone sedimentation in the Buchan oilfield, North Sea, Scott. *J. Geol.* **21**, 227–37. [3, 12]

Richards, P.C. (1990) Devonian. In: Glennie, K.W. (ed.) *Introduction to Petroleum Geology of the North Sea*. Blackwell Scientific Publications, Oxford, 1990. pp. 78–89. [3]

Richards, P.C. (1992) An introduction to the Brent Group: a literature review. In: Morton, A.C., Haszeldine, R.S., Giles, M.R. and Brown, S. (eds) *Geology of the Brent Group*. Special Publication No. 61, Geological Society, London, pp. 15–26. [12]

Richards, P.C., Brown, S., Dean, J.M. and Anderton, R. (1988) A new palaeogeographic reconstruction for the Middle Jurassic of the northern North Sea. *J. Geol. Soc.* **145**, 883–6. [12]

Richards, P.C., Lott, G.K., Johnson, H., Knox, R.W.O'B. and Riding, J.B. (1993) Jurassic of the Central and Northern North Sea. In: Knox, R.W.O'B. and Cordey, W.G. (eds) *Lithostratigraphic Nomenclature of the UK North Sea*. BGS, Nottingham. [8]

Richter-Bernburg, G. (1955) Stratigraphische Gliederung des deutschen Zechstein. *Zeitschr. deutsche geol. Ges.* **105**, 843–54. [6]

Richter-Bernburg, G. (1959) Zur Palaegeographie der Zechsteins. In: *I giacimenti gassiferi dell'Europa Occidentale 1*, Accademia Nazionale dei Lincei, Rome, pp. 88–9. [6]

Richter-Bernburg, G. (1980) Salt tectonics: interior structures of salt bodies. *Bull. Centre Rech. Explor. Prod. Elf-Aquitaine* **4**, 373–93. [6]

Richter-Bernburg, G. (1985) Zechstein Anhydrite—Fazies und Genese. *Geol. Jahrbuch Reihe A* **85**, 3–82. [6]

Richter-Bernburg, G. (1986a) Zechstein 1 and 2 Anhydrites: facts and problems of sedimentation. In: Harwood, G.M. and Smith, D.B. (eds) *The English Zechstein and Related Topics*. Special Publication No. 22, pp. 157–63. [6]

Richter-Bernburg, G. (1986b) Zechstein salt correlation: England–Denmark–Germany. In: Harwood, G.M. and Smith, D.B. (eds) *The English Zechstein and Related Topics*. Special Publication No. 22, Geological Society, London, pp. 165–8. [5]

Ridd, M.F. (1981) Petroleum geology west of the Shetlands. In: Illing, L.V. and Hobson, G.P. (eds) *Petroleum Geology of the Continental Shelf of North-West Europe*. Heyden, London, pp. 414–25. [7, 11]

Ridd, M.F. (1983) Aspects of the Tertiary geology of the Faeroe–Shetland Channel. In: Bott, M.H.P., Saxov, S., Talwani, M. and Thiede, J. (eds) *Structure and Development of the Scotland–Greenland Ridge: New Methods and Concepts*. Plenum Press, New York, London, pp. 91–108. [2]

Rigby, D. and Smith, J.W. (1982) A reassessment of stable carbon isotopes in hydrocarbon exploration. Erdol Kohle, Erdgas. *Petrochemie* **35**(9), 415–7. [11]

Riis, F. and Jensen, L.N. (1992) Introduction: measuring uplift and erosion—proposal for a terminology. *Norsk Geol. Tidsskrift* **72**, 223–8. [11]

Riise, R. (1978) *Lithology: Wells 2/3-1, 2/3-2 and 2/3-3*. Paper No. 17, NPD, Stavanger. [7]

Rijkers, R.H.B. and Geluk, M.C. (1996) Sedimentary and structural history of the Texel–IJsselmeer High, the Netherlands. In: Rondeel, H.E., Batjes, D.A.J. and Nieuwenhuis, W.H. (eds) *Geology of Oil and Gas under the Netherlands*. Royal Geological and Mining Society, the Netherlands, Kluwer Academic Publishers, Dordrecht, pp. 265–84. [12]

Riley, L.A., Roberts, M.J. and Connell, E.R. (1989) The application of palynology in the interpretation of Brae Formation stratigraphy and reservoir geology in the South Brae Field area, British North Sea. In: Collinson, J.D. (ed.) *Correlation in Hydrocarbon Exploration*. Graham and Trotman, London, pp. 339–56. [8]

Riley, L.A., Harker, S.D. and Green, S.C.H. (1991) Valhall Sandstones Lower Cretaceous distribution through time: the application of palynology in the Scapa field, U.K. North Sea. *J. Petr. Geol.* **15**, 97–110. [9, 12]

Riley, L.A., Harker, S.D. and Green, S.C.H. (1992) Lower Cretaceous palynology and sandstone distribution in the Scapa Field, UK North Sea. *J. Petrol. Geol.* **15**, 97–110. [9]

Rippon, J.H. (1996) Sand body orientation, palaeoslope analysis and basin fill implications in the Westphalian A-C of Great Britain. *J. Geol. Soc. Lond.* **153**, 881–900. [4]

Ritchie, J.D. and Hitchen, K. (1996) Early Paleogene offshore igneous activity to the northwest of the UK and its relationship to the North Atlantic Igneous Province. In: Knox, R.W.O'B., Corfield, R.M and Dunay, R.E. (eds) *Correlation of the Early Paleogene in Northwest Europe*. Special Publication 101, Geological Society, London, pp. 63–78. [2]

Ritchie, J.D., Swallow, J.L., Mitchell, J.G. and Morton A.C. (1988) Jurassic ages for intrusives and extrusives within the Forties Igneous Province. *Scott. J. Geol.* **24**, 81–8. [8]

Ritchie, J.S. and Pratsides, P. (1993) The Caister Fields, Block 44/23a, UK. North Sea. In: Parker, J.R. (ed.) *Petroleum Geology of Northwest Europe: Proceedings of the 4th Conference*. Geological Society, London, pp. 759–69. [1, 4, 11, 12]

Ritter, U. (1988) Modelling of hydrocarbon generation patterns in the Egersund sub-basin, North Sea. In: Mattavelli, L. and Novelli, L. (eds) *Advances in Org. Geochem., 1987 (Proc. 13th Int'l. Mtg on Org. Geochem. Venice) Part 1.* Pergamon, Oxford, pp. 165–74. [11]

Roaldset, H.W.B. and Bjorøy, M. (1996) Parallel reaction kinetics of smectite to illite conversion. *Clay Minerals* **31**, 365–76.

Robert, P. (1980) The optical evolution of kerogen and geothermal histories applied to oil and gas exploration. In: Durand, B. (ed) *Kerogen, Insoluble Organic Matter from Sedimentary Rocks.* Editions Technip, Paris, p. 385. [11]

Roberts, A.M. Price, J.D. and Badley, M.E. (1989) Discussion on Triassic sediments of the Inner Moray Firth, Scotland: early rift deposits. *J. Geol. Soc.* **146**, 361–3. [7]

Roberts, A.M., Badley, M.E., Price, J.D. and Huck, I.W. (1990a) The structural evolution of a transtensional basin: Inner Moray Firth, NE Scotland. *J. Geol. Soc.* **147**, 87–103. [2]

Roberts, A.M., Price, J.D. and Olsen, T.S. (1990b) Late Jurassic half-graben control on the siting and structure of hydrocarbon accumulations. UK/Norwegian Central Graben. In: Hardman, R.P.F. and Brooks, J. (eds) *Tectonic Events Responsible for Britain's Oil and Gas Reserves*. Special Publication No. 55, Geological Society, London, pp. 229–57. [8]

Roberts, A.M., Yielding, G. and Badley, M.E. (1990c) A kinematic model for the orthogonal opening of the late Jurassic North Sea rift system, Denmark–mid Norway. In: Blundell, D.J. and Gibbs, A.D. (eds) *Tectonic Evolution of the North Sea Rifts*. Oxford Scientific Publications, Oxford, pp. 180–99.

Roberts, A.M., Yielding, G. and Freeman, B. (eds) (1991) *The Geometry of Normal Faults*. Special Publication No. 56, Geological Society, London. [7]

Roberts, A.M., Yielding, G. Kusznir, N.J., Walker, I. and Dorn-Lopez, D. (1993) Mesozoic extension in the North Sea: constraints from flexural backstripping, forward modelling and fault populations. In: Parker, J.R. (ed.) *Petroleum Geology of Northwest Europe: Proceedings of the 4th Conference*. Geological Society, London, pp. 1123–36. [2, 7, 8]

Roberts, A.M., Yielding, G., Kuznir, N.J., Walker, I. and Dorn-Lopez, D. (1995) Quantitative analysis of Triassic extension in the northern Viking Graben. *J. Geol. Soc.* **152**, 15–26. [7]

Roberts, J.D., Mathieson, A.S. and Hampson, J.M. (1987) Statfjord. In: Spencer, A.M. *et al.* (eds) *Geology of the Norwegian Oil and Gas Fields*. Graham and Trotman, London, pp. 319–40. [1, 7, 8, 12]

Roberts, M.J. (1991) The South Brae Field, Block 16/7a, UK North Sea. In: Abbotts, I.L. (ed.) *UK North Sea United Kingdom Oil and Gas Fields: 25 Years Commemorative Volume*. Memoir No. 14, Geological Society, London, p. 55. [1]

Roberts, W.H. (1980) Design and function of oil and gas traps. In: Roberts, W.H. and Cordell, R.J. (eds) *Problems of Petroleum Migration*. Studies in Geology No. 10, AAPG, pp. 217–40. [11]

Roberts, W.H. and Cordell, R.J. (eds) (1980). Problems of Petroleum Migration. Studies in Geology No. 10, AAPG, Tulsa, OK. [11]

Robertson, G. (1993) Beryl Field: geological evolution and reservoir behaviour. In: Parker, J.R. (ed.) *Petroleum Geology of Northwest Europe: Proceedings of the 4th Conference*. Geological Society, London, pp. 1491–502. [1, 12]

Robinson, A.E. (1981) Facies types and reservoir quality of the Rotliegendes Sandstone, North Sea. In: *56th Annual SPE AIME Technological Conference, San Antonio, Texas*. SPE 10303, 10 pp. [5]

Robinson, A.G. and Gluyas, J. (1992) Duration of quartz cementation in sandstones, North Sea and Haltenbanken Basins. *Mar. Petr. Geol.* **9**(3), 324–27. [11]

Robinson, A.G., Coleman, M.L. and Gluyas, J.G. (1993) The age of illite cement growth, Village Fields area, Southern North Sea: evidence from K–Ar ages and ^{18}O/^{16}O ratios. *AAPG Bull.* **77**(1), 68–81. [11]

Robinson, D. and Bevins, R.E. (1986) Incipient metamorphism in the Lower Palaeozoic marginal basin of Wales. *J. Metamorphic Geol.* **4**, 101–13. [11]

Robinson, M.A. (1985) Palaeomagnetism of volcanics and sediments of the Eday Group, Southern Orkney. *Scott. J. Geol.* **21**, 285–300.

Robinson, N., Parnell, J. and Brassell, S. (1989) Hydrocarbon compositions of bitumens from mineralized Devonian lavas and Carboniferous sedimentary rocks, central Scotland. *Mar. Petr. Geol.* **6**(4), 316–23. [11]

Robinson, N.D. (1986) Lithostratigraphy of the Chalk Group of the North Downs, southeast England. *Proc. Geol. Assoc.* **97**, 141–70. [9]

Robson, D. (1991) The Argyll, Duncan and Innes Fields, Blocks 30/24 and 30/25a, UK North Sea. In: Abbotts, I.L. (ed.) *UK North Sea United Kingdom Oil and Gas Fields: 25 Years Commemorative Volume*. Geological Society, London, pp. 14, 219–25. [1, 3, 4, 5, 6, 12]

Robson, D.A. (ed.) (1980) *The Geology of North East England.* Natural History Society of Northumbria, Newcastle upon Tyne, 113 pp. [2]

Rochow, K.A. (1981) Seismic stratigraphy of the North Sea 'Palaeocene' deposits. In: Illing, L.V. and Hobson, G.P. (eds) *Petroleum Geology of the Continental Shelf of North-West Europe*. Heyden, London, pp. 255–66. [2]

Røe, S.L. and Steel, R. (1985) Sedimentation, sea-level rise and tectonics at the Triassic–Jurassic boundary: Statfjord Formation, Tampen Spur, northern North Sea. *J. Petr. Geol.* **8**, 163–86. [7, 8, 12]

Roelofsen, J.W. and de Boer, W.D. (1991) Geology of the Lower Cretaceous Q/1 oil fields, Broad Fourteens Basin, The Netherlands. In: Spencer, A.M. (ed.) *Generation, Accumulation, and Production of Europe's Hydrocarbons*. Special Publication, EAPG, Oxford University Press, Oxford, pp. 203–16. [1]

Rogers, D.A. (1987) Devonian correlations, environments and tectonics across the Great Glen Fault. PhD thesis, University of Cambridge. [3]

Rogers, D.A. and Astin, T.R. (1991) Ephemeral lakes, mud pellet dunes and wind-blown sand and silt: reinterpretations of Devonian lacustrine cycles in north Scotland. *Sp. Publ. Int. Assoc. Sedimentol.* **13**, 199–221. [3]

Rogers, D.A., Marshall, J.E.A. and Astin, T.R. (1989) Devonian and later movements on the Great Glen fault system, Scotland. *J. Geol. Soc.* **146**, 369–72. [3]

Rogers, G. and Pankhurst, R.J. (1993) Unravelling dates through the ages: geochronology of the Scottish metamorphic complexes. *J. Geol. Soc.* **150**, 447–64. [2]

Rogers, G., Dempster, T.J., Bluck, B.J. and Tanner, P.W.G. (1989) A high precision U–Pb age for the Ben Vuirich granite: implications for evolution of the Scottish Dalradian Supergroup. *J. Geol. Soc.* **146**, 789–98. [3]

Rondeel, H.E., Batjes, D.A.J. and Nieuwenhuis, W.H. (1996) *Geology of Gas and Oil under the Netherlands*. Kluwer Academic Publishers, the Netherlands. [11, 12]

Rønnevik, H. and Johnsen, S. (1984) Geology of the greater Troll area. *Oil Gas J.* **82**, 100–6.

Rønnevik, H., Eggen, S. and Vollset, J. (1983) Exploration of the Norwegian Shelf. In: Brooks, J. (ed.) *Petroleum Geochemistry and Exploration of Europe.* Special Publication No. 12. Geological Society, Blackwell Scientific Publications, Oxford, pp. 71–93. [11]

Rønnevik, H.C. (1981) Geology of the Barents Sea. In: Illing, L.V. and Hobson, G.D. (eds) *Petroleum Geology of the Continental Shelf of North-West Europe.* Heyden, London, pp. 395–406. [11]

Rønning, K., Johnston, C.D., Johnstad, S.E. and Songstad, P. (1986) Geology of the Hild Field. In: Spencer, A.M., Campbell, C.J., Hanslien, S.H., Nelson, P.H., Nysaether, E. and Ormaasen, E. (eds) *Habitat of Hydrocarbons on the Norwegian Continental Shelf.* Norwegian Petroleum Society, Graham and Trotman, London, pp. 199–206. [11]

Rønning, K., Johnston, C.D., Johnstad, S.E. and Songstad, P. (1987) Hild. In: Spencer, A.M., Campbell, C.J., Hanslien, S.H., Nelson, P.H., Nysaether, E. and Ormaasen, E. (eds) *Geology of the Norwegian oil and Gas Fields.* Graham and Trotman, London, pp. 287–94. [11]

Rooksby, S.K. (1991) The Miller Field, Blocks 16/7b, 16/8b, UK North Sea. In: Abbotts, I.L. (ed.) *UK North Sea United Kngdom Oil and Gas Fields: 25 Years Commemorative Volume.* Memoir No. 14, Geological Society, London, pp. 159–64. [1, 8, 12]

Rooney, M.A. (1995) Carbon isotope ratios of light hydrocarbons as indicators of thermochemical sulfate reduction. Extended Abstracts 17th International Meeting of the Organization of Geochemistry, San Sebastian, pp. 523–5. [11]

Roos, B.M. and Smits, B.J. (1983) Rotliegend and Main Buntsandstein gas fields in block K/13: a case history. *Geol. Mijnbouw* **62**, 75–83. [1, 5, 7, 11, 12]

Rose, P.T.S., Masco, C., Lofts, J. and Malcay, M. (in press) Reservoir characterisation in the Captain field: integration of horizontal and vertical well data. In: Boldy, S.A.R. and Fleet, A.J. (eds) *Petroleum Geology of Northwest Europe: Proceedings of the Fifth Conference*, London. [9]

Ross, C.A. and Ross, J.R. (1985) Late Palaeozoic depositional sequences are synchronous and worldwide. *Geology* **13**, 194–7. [4]

Rothwell, N.R. and Quinn, P. (1987) The Welton Oilfield. In: Brooks, J. and Glennie, K.W. (eds) *Petroleum Geology of North West Europe.* Graham and Trotman, London, pp. 181–9. [4]

Rowell, P. (1995) Tectono-stratigraphy of the North Celtic Sea Basin. In: Croker, P.F. and Shannon, P.M. (eds) *The Petroleum Geology of Ireland's Offshore Basins.* Special Publication No. 93, Geological Society, London, pp. 101–37. [12]

Rowley, D.B., Raymond, A., Parrish, J.T., Lottes, A.L., Scotese, C.R. and Ziegler, A.M. (1985) Carboniferous palaeogeographic, phytogeographic and palaeoclimatic reconstruction. *Int. J. Coal Geol.* **5**, 7–42. [4]

Royden, L., Sclater, J.G. and Herzen, R.P., von (1980) Continental margin subsidence and heat flow: important parameters in formation of petroleum hydrocarbons. *AAPG Bull.* **64**, 173–82. [11]

Rudkiewicz, J.L., Brevart, O., Connan, J. and Montel, F. (1994) Primary migration behaviour of hydrocarbons: from laboratory experiments to geological situations through fluid flow models. *Adv. Org. Geochem.* **22**(3–5), 587–616. [11]

Ruffell, A.H. (1991) Sea-level events during the Early Cretaceous in Western Europe. *Cretaceous Res.* **12**, 527–51. [9]

Ruffell, A.H. (1992) Early to mid-Cretaceous tectonics and unconformities of the Wessex Basin (southern England). *J. Geol. Soc.* **149**, 443–54. [9]

Ruffell, A.H. (1995) Evolution and hydrocarbon prospectivity of the Brittany Basin Western Approaches Trough, offshore northwest France. *Mar. Petr. Geol.* **12**, 387–407.

Ruffell, A.H. and Batten, D.J. (1990) The Barremian–Aptian arid phase in western Europe. *Palaeogeog. Palaeoclim. Palaeoecol.* **80**, 197–212. [9]

Ruffell, A.H. and Wach, G.D. (1991) Sequence stratigraphic analysis of the Aptian–Albian Lower Greensand in southern England. *Mar. Petr. Geol.* **8**, 341–53. [9]

Rullkotter, J. and Wendisch, D. (1982) Microbialalteration of 17alpha(H)-hopanes in Madagascar asphalts: removal of C-10 methyl group and ring opening. *Geochim. Cosmochim. Acta* **46**, 1545–53. [11]

Rumph, B., Reaves, C.M., Orange, V.G. and Robinson, D.L. (1993) Structuring and transfer zones in the Faeroe Basin in a regional tectonic context. In: Parker, J.R. (ed.) *Petroleum Geology of Northwest Europe: Proceedings of the 4th Conference.* Geological Society, London, pp. 999–1009. [2]

Rundberg, Y. and Smalley, P.C. (1989) High resolution dating of Cenozoic sediments from Northern North Sea using 87Sr/86Sr stratigraphy. *AAPG Bull.* **73**(3), 298–308. [11]

Russell, M. and Pearson, M.J. (1990) Correlation of vitrinite reflectance with aromatic maturity parameters in the Pennine Carboniferous basin, England. In: Fermont, W.J.J. and Weegink, J.W. (eds) *International Symposium on Organic Petrology–Zeist, NL. Proceedings.* Mededelingen rijks geologische, 115 pp. [11]

Russell, M.J. (1976) A possible Lower Permian age for the onset of ocean floor spreading in the North Atlantic. *Scott. J. Geol.* **12**(4), 315–23. [5]

Russell, M.J. and Smythe, D.K. (1983) Origin of the Oslo Graben in relation to the Hercynian–Alleghanian orogeny and lithospheric rifting in the North Atlantic. *Tectonophysics* **94**, 457–72. [2]

Ryseth, A. and Ramm, M. (1996) Alluvial architecture and differential subsidence in the Statfjord Formation, North Sea: prediction of reservoir potential. *Petr. Geosci.* **2**, 271–87. [8]

Sæland, G.T. and Simpson, G.S. (1982) Interpretation of 3-D data in delineating a subunconformity trap in block 34/10, Norwegian North Sea. In: Halbouty, M.T. (ed.) *The Deliberate Search for the Subtle Trap.* Memoir No. 32, AAPG, Tulsa, OK., pp. 235–71. [1, 12]

Saigal, G.C., Bjorlykke, K. and Larter, S. (1992) The effect of oil emplacement on diagenetic processes—examples from the Fulmar Reservoir Sandstones, Central North Sea. *AAPG Bull.* **76**(7), 1024–34. [11]

Sales J. (1989) Late Tertiary uplift—west Barents Shelf: character, mechanism and timing. NPF: *Proc. NPF Conference on Structural and Tectonic Modelling and its Application to Petroleum Geology, October 1989*, Stavanger, Norway. [11]

Sales, J.K. (1993) Closure vs. seal capacity—a fundamental control on the distribution of oil and gas. In: Doré, A.G., Augustson, J.H., Hermanrud, C., Stewart, D.J. and Sylta, O. (eds) *Basin Modelling: Advances and Applications.* Special Publication No. 3, NPF, Elsevier, Amsterdam, pp. 399–414. [11]

Sandvik, E.I., Young, W.A. and Curry, D.J. (1992) Expulsion from hydrocarbon sources: the role of organic absorption. In: Eckardt, C.B., Maxwell, J.R., Larter, S.R. and Manning, D.A.C. (eds) *Advances in Organic Geochemistry, Part II: Advances and Applications in Energy and the Natural Enviroment.* Pergamon, Oxford, pp. 77–89.

Sangree, J.B. (1969) What you should know to analyze core fractures. *World Oil Sp. Expl. Rep.* April, 69–72. [9]

Sanneman, D. (1968) Salt-stock families northwestern Germany. In: Braunstein, J. and O'Brien, G.D. (eds) *Diaprisim and Diapirs.* Memoir No. 8, AAPG, pp. 261–70. [6]

Sarg, J.F. and Skjold, L.J. (1982) Stratigraphic traps in Palaeocene sands in the Balder area, North Sea. In: Halbouty, M.T. (ed.) *The Deliberate Search for the Subtle Trap.* Memoir No. 32, AAPG, Tulsa, OK., pp. 197–206. [12]

Sass, J.H., Blackwell, D.D., Chapman, D.S., Costain, J.K., Decker, E.R., Lawyer, L.A. *et al.* (1989) Heat flow from the crust of the United States. In: Touloukian, Y.S., Judd, W.R. and Roy, R.F. (eds) *Physical Properties of Rocks and Minerals, Volume II-2.* Hemisphere, New York, NY., pp. 503–48. [11]

Sawyer, M.J. and J.B. Keegan (1996) Use of palynofacies characterisation in sand-dominated sequences, Brent Group, Ninian Field, UK North Sea. *Petr. Geosci.* **2**, 289–97. [8]

Schäfer, A. and Sneh, A. (1983) Lower Rotliegend fluvio-lacustrine sequences in the Saar–Nahe Basin. *Geol. Rundschau* **72**(3), 1135–46. [5]

Schaefer, R.G., Schenk, H.J., Hardelauf, H. and Harms, R. (1990) Determination of gross kinetic parameters for petroleum formation from Jurassic source rocks of different maturity levels by laboratory experiments. In: Durand, B. and Behar, F. (eds) *Adv. Org. Geochem. Org. Geochem.* **16**(1–3), 115–20. [11]

Schatzinger, R.A. Feazel, C.T. and Henry, W.E. (1985) Evidence of resedimentation in Chalk from the Central Graben, North Sea. In: Crevello, P.D. and Harris, P.M. (eds) *SEPM Core Workshop No. 6*, Society of Economic Paleonological Mineralogists, New Orleans, pp. 342–85. [9, 12]

Schenk, H.J., Horsfield, B., Krooss, B., Schaefer, R.G. and Schwochau, K. (1997) Kinetics of petroleum formation and cracking. In: Welte, D.H., Horsfield, B. and Baker, D.R. (eds) *Petroleum and Basin Evolution*. Springer Verlag, Berlin. [11]

Schiener, E.J., Stober, G. and Faber, E. (1985) Surface geochemical exploration of hydrocarbons in offshore areas—principles, methods and results. In: Thomas, B.M., Doré, A.G., Eggen, S.S., Home, P.C. and Larsen, R.M. (eds) *Petroleum Geochemistry in Exploration of the Norwegian Shelf.* Graham and Trotman, London, pp. 223–38. [11]

Schlager, S.O. and Jenkyns, H.C. (1976) Cretaceous oceanic anoxic events: causes and consequences. *Geol. Mijnbouw* **55**, 179–84. [9]

Schlager, W. and Bolz, H. (1977) Clastic accumulations of sulphate evaporites in deep water. *J. Sediment. Petrol.* **42**, 600–9. [6]

Schlanger, S.O., Arthur, M.A., Jenkyns, H.C. and Scholle, P.A. (1987) The Cnomanian-Turonian oceanic event, I. Stratigraphy and distribution of organic carbon-rich beds and the marine $\delta^{13}C$ excursion. In: Brooks, J. and Fleet, A.J. (eds) *Marine Petroleum Source Rocks.* Blackwell Scientific Publications, Oxford, pp. 371–400. [11]

Schmitt, H.R.H. (1991) The Chanter Field, Block 15/17, UK North Sea. In: Abbotts, I.L. (ed.) *UK North Sea United Kingdom Oil and Gas Fields: 25 Years Commemorative Volume.* Memoir No. 14, Geological Society, London, pp. 261–8. [1, 12]

Schmitt, H.R.H. and Gordon, A.F. (1991) The Piper Field, Block 15/17, UK North Sea. In: Abbotts, I.L. (ed.) *UK North Sea United Kingdom Oil and Gas Fields: 25 Years Commemorative Volume.* Memoir No. 14, Geological Society, London, pp. 361–8. [1, 8, 11, 12]

Schmoker, J.W. (1994) Volumetric calculation of hydrocarbons generated. In: Magoon, L.B. and Dow, W.G. (eds) *The Petroleum System—from Source to Trap.* Memoir No. 60, AAPG, Tulsa, OK., pp. 323–9. [11]

Schneider, J. and Gebhardt, U. (1994) Litho- und Biofaziesmuster in intra- und extramontanen Senken des Rotliegend Perm, Nord und Ostdeutschland. *Geol. Jahrbuch* **A131**, 57–98. [5]

Schneider, J.W., Roessler, R. and Gaitsch, B. (1995) Time lines of the Late Variscan volcanism—a holostratigraphic synthesis. *Zentralbl. Geol. Palaeontol.* **1**(5/6), 477–90. [5]

Schoell, M. (1984) Stable isotopes in petroleum research. In: Brooks, J. and Welte, D.H. (eds) *Advances in Organic Geochemistry, Volume 1.* Academic Press, London, pp. 215–46. [11]

Schoell, M., Moldowan, J.M., Sundararaman, P. and Teerman, S.C. (1989) Toarcian shale, SW Germany: anatomy of an euxinic event. In: *14th Int'l Org. Geochem. Conference, Paris, Sept. 1989.* Abstract No. 244. [11]

Scholle, P.A. (1974) Diagenesis of Upper Cretaceous chalks from England, Northern Ireland and the North Sea. In: Hsü, K.J. and Jenkyns, H.C. (eds) *Pelagic Sediments: On Land and Under the Sea.* Special Publication No. 1, International Association of Sedimentologists, Blackwell Scientific Publications, Oxford, pp. 177–210. [9]

Scholle, P.A. (1977) Chalk diagenesis and its relation to petroleum exploration: oil from chalks, a modern miracle? *AAPG Bull.* **61**, 982–1009. [9, 12]

Scholle, P.A. and Arthur, M.A. (1980) Carbon isotope fluctuations in Cretaceous pelagic limestones: potential stratigraphic and petroleum exploration tool. *AAPG Bull.* **64**, 67–87. [9]

Scholle, P.A., Stemmerik, L. and Ulmar, D.S. (1991) Diagenetic history and hydrocarbon potential of Upper Permian carbonate buildups, Wegener Halvo Area, Jameson Land Basin, East Greenland. *AAPG Bull.* **75**, 701–25. [6]

Scholle, P.A., Peryt, T.M. and Ulmer-Scholle, D.S. (eds) (1995) *The Permian of Northern Pangaea*: Vol. 1. *Paleogeography, Paleoclimates, Stratigraphy*, 261 pp; Vol. 2. *Sedimentary Basins and Economic Resources*, 312 pp. Springer-Verlag, Berlin. See also Menning, 1995. [2, 5]

Schopf, T.J.M. (1974) Permo-Triassic extinctions: relations to 'seafloor spreading'. *J. Geol.* **82**, 129–43. [7]

Schou, L. and Myhr, M.B. (1988) Sulfur aromatic compounds as maturity parameters. *Org. Geochem.* **13**(1–3), 61–6. [11]

Schou, L., Mørk, A. and Bjorøy, M. (1984) Correlation of source rocks and migrated hydrocarbons by GC-MS in the Middle Triassic of Svalbard. *Organic Geochemistry* **6**, 513–20. [11]

Schou, L., Eggen, S. and Schoell, M. (1985) Oil–oil and oil–source rock correlation, northern North Sea. In: Thomas, B.M., Doré, A.G., Eggen, S.S., Home, P.C. and Larsen, R.M. (eds) *Petroleum Geochemistry in Exploration of the Norwegian Shelf.* Graham and Trotman, London, pp. 101–20. [11]

Schreiber, B.C. (1987) Arid shorelines and evaporites. In: Reading, H.G. (ed.) *Sedimentary Environments and Facies.* Blackwell Scientific Publications, Oxford, pp. 189–228. [6]

Schroeder, F.W. and Sylta, O. (1993) Modelling the hydrocarbon system of the North Viking Graben: a case study. In: Doré, A.G., Augustson, J.H., Hermanrud, C., Stewart, D.J. and Sylta, O. (eds) *Basin Modelling: Advances and Applications.* Special Publication No. 3, NPF, pp. 469–84. [11]

Schröder, L., Lösch, J., Schöneich, H, Stancu-Kristoff, G. and Tafel, W.-D. (1991) Oil and gas in the north-west German basin. In: Spencer, A.M. (ed.) *Generation, Accumulation, and Production of Europe's Hydrocarbons.* Special Publication, EAPG, Oxford University Press, Oxford, pp. 139–48. [1]

Schröder, L., Plein, E., Bachmann, G.H., Gast, R.E., Gebhardt, U., Graf, R. *et al.* (1995) Stratigraphische Neugliederung des Rotliegend im Norddeutschen Becken. *Geol. Jahrbuch* **148A**, 21 pp. [5]

Schulte, W.M., Van Rossem, P.A.H. and van de Vijver, W. (1994) Current challenges in the Brent Field. *J. Petr. Technol.* **46**, 1073–9. [8]

Schumacher, D. and Abrams, M.A. (eds) (1996) *Hydrocarbon migration and its near surface expression.* AAPG Memoir 66, AAPG Tulsa OK., 446 pp. [11]

Schuster, A. (1966) Karbonstratigraphie nach Borlochmessungen. *Erdöl-Erdgas-Zeitschr.* **84**, 439–57. [4]

Schuster, A. (1971) Die Westphal-Profile der Bohrungen Bockraden 1 bis 5 bei Ibbenbüren und ihre Parallelisierung mit dem Bohrprofil Norddeutschland 8 und dem jüngsten Ruhrkarbon nach Bohrlochmessungen. *Fortschr. Geol. Rheinland Westf.* **18**, 233–56. [4]

Schwark, L. and Puttmann, W. (1990) Aromatic hydrocarbon composition of the Permian Kupferschiefer in the Lower Rhine Basin, NW Germany. In: Durand, B. and Behar, F. (eds) *Advances in Organic Geochemistry.* Pergamon, Oxford, pp. 749–61. [11]

Schwarzkopf, T.A. (1992) Source rock potential TOC + Hydrogen Index evaluation by integrating well log and geochemical data. *Org. Geochem.* **19**(4–6), 545–53. [11]

Schwarzkopf, T.A. (1993) Model for prediction of organic carbon content in possible source rocks. *Mar. Petr. Geol.* **10**, 478–93. [11]

Schwartzkopf, Th. and Leythaeuser, D. (1988) Oil generation and migration in the Gifhorh Trough, NW Germany. In: Mattavelli, L. and Novelli, L. (eds) *Advances in Org. Geochem., 1987 (Proc. 13th Int'l. Mtg on Org. Geochem. Venice), Part 1.* Pergamon, Oxford, pp. 245–54. [11]

Scotchman, I.C., (1987) Relationship between clay and organic maturation in the Kimmeridge Clay Formation, onshore UK. In: Brooks, J. and Glennie, K.W. (eds) *Petroleum Geology of North West Europe.* Graham and Trotman, London, pp. 251–62. [11]

Scotchman, I.C. (1989) Diagenesis of Kimmeridge Clay Formation, onshore UK. *J. Geol. Soc. Lond.* **146**(2), 283–304. [11]

Scotchman, I.C. (1991) Organic facies variations in the Upper Jurassic Kimmeridge Clay Formation of the East Shetland Basin, UK Northern North Sea. In: Manning, D. (ed.) *Organic Geochemistry: Advances and Applications in Energy and Natural Environment.* Extended Abstracts, 15th Meeting EAOG, Manchester University Press, Manchester, pp. 104–8. [11]

Scotchman, I.C. (1994) Maturity and burial history of the Kimmeridge Clay Formation, onshore UK: a biomarker study. *First Break* **12**(4), pp. 193–202. [11]

Scotchman, I.C. and Thomas, J.R.W. (1995) Maturity and hydrocarbon generation in the Slyne Trough, northwest Ireland. In: Croker, P.F. and Shannon, P.M. (eds) *The Petroleum Geology of Ireland's Offshore Basins.* Special Publication No. 93, Geological Society, London, pp. 385–411. [11, 12]

Scott, A.C. and Fleet, A.J. (eds) (1994) *Coal and Coal-bearing Strata as Oil-prone Source Rocks*? Special Publication No. 77, Geological Society, London, 213 pp. [11]

Scott, D.L. and Rosendahl, B.R. (1989) North Viking Graben: an East African perspective. *AAPG Bull.* **73**(2), 155–65. [7]

Scott, E.S. (1992) The palaeoenvironments and dynamics of the Rannoch–Etive nearshore and coastal succession, Brent Group, Northern North Sea. In: Morton, A.C, Haszeldine, R.S., Giles, M.R. and Brown, S. (eds) *Geology of the Brent Group.* Special Publication No. 61, Geological Society, London, pp. 129–47. [8]

Scott, J. and Colter, V.S. (1987) Geological aspects of current Great Britain onshore exploration plays. In: Brooks, J. and Glennie, K.W. (eds) *Petroleum Geology of North West Europe.* Graham and Trotman, London, pp. 95–107. [4, 6]

Scott, W.B. (1986) Nodular carbonates in the Lower Carboniferous Cementstone Group of the Tweed Embayment, Berwickshire: evidence for a former sulphate evaporitic facies. *Scott. J. Geol.* **22**, 325–45. [4]

Searl, A. (1994) Diagenetic destruction of reservoir potential in shallow marine sandstones of the Broadford Beds Lower Jurassic, NW Scotland: depositional versus burial and thermal history controls on porosity destruction. *Mar. Petr. Geol.* **11**(2), 131–48. [11]

Sears, R.A., Harbury, A.R., Protoy, A.J.G. and Stewart, D.J. (1993) Structural styles from the Central Graben in the UK and Norway. In: Parker, J.R. (ed.) *Petroleum Geology of Northwest Europe: Proceedings of the 4th Conference.* Geological Society, London, pp. 1231–44. [6, 12]

Seifert, W.K. (1973) Steroid acids in petroleum. Animal contribution to the origin of petroleum. *Pure Applied Chemistry* **34**, 633–41. [11]

Sekiguchi, K. (1984) A method for determining terrestrial heat flow in oil basinal areas. *Tectonophysics* **103**, 67–79. [11]

Selley, R. (1991) How to find oil without really trying. *N. Sci.* Jan. 44–7 [11]

Selley, R.C. (1992) Petroleum seepages and impregnations in Great Britain. *Mar. Petr. Geol.* **9**, 226–45. [11]

Selley, R.C. and Stoneley, R. (1987) Petroleum habitat in South Dorset. In: Brooks, J. and Glennie, K.W. (eds) *Petroleum Geology of North West Europe.* Graham and Trotman, London, pp. 139–48. [11, 12]

Sellwood, B.W. (1979) The Wealden rivers and chalky seas of Cretaceous Britain. In: Anderton, R., Bridges, P.H., Leeder, M.R. and Sellwood, B.W. (eds) *A Dynamic Stratigraphy of the British Isles.* Allen and Unwin, London, pp. 227–44. [9]

Selter, V. (1989) Palaöböden im obersten Westphal bis Stephan des nordwestdeutschen Oberkarbon-Beckens. *Z. deutsch Geol. Ges.* **140**, 249–58. [4]

Selter, V. (1990) *Sedimentologie und Klimaentwicklung im Westphal C/D und Stephan des nordwestdeutschen Oberkarbon-Beckens.* DGMK-Bericht 384-4, Deutsche Wiss. Ges. Erdöl, Erdgas und Kohle, Hamburg, 311 pp. [4]

Senftle, J.T. and Landis, C.R. (1991) Vitrinite reflectance as a tool to assess thermal maturity. In: Merrill, R.K. (ed.) *Source and Migration Processess and Evaluation Techniques.* Treatise Petr. Geol. AAPG, Tulsa, OK., pp. 119–26. [11]

Seni, S.J. and Jackson, M.P.A. (1983) Evolution of salt structures, East Texas diapir province, Pt. 2: patterns and rates of halokinesis. *AAPG Bull.* **67**, 1245–74. [6]

Shanley, K.W. and McCabe, P.J. (1994) Perspectives on the sequence stratigraphy of continental strata. *AAPG Bull.* **78**, 544–68. [5, 7]

Shanmugan, G., Lehtonen, L.R., Straume, T., Syvertsen, S.E., Hodgkinson, R.J. and Skibeli, M. (1994) Slump and debris-flow dominated upper slope facies in the Cretaceous of the Norwegian and Northern North Seas 61–67°N: implications for sand distribution. *AAPG Bull.* **78**, 910–37. [9]

Shanmugan, G., Bloch, R.B., Mitchell, S.M., Beamish, G.W.J., Hodgkinson, R.J., Damuth, J.E., *et al.* (1995) Basin-floor fans in the North Sea: sequence stratigraphic models vs. sedimentary facies. *AAPG Bull.* **79**, 477–512. [9]

Shannon, P.M. (1991a) Tectonic framework and petroleum potential of the Celtic Sea, Ireland. *First Break* **9**, 107–21. [12]

Shannon, P.M. (1991b) The development of Irish offshore sedimentary basins. *J. Geol. Soc.* **148**, 181–9. [12]

Shannon, P.M. (1991c) The Irish offshore basins: geological development and petroleum plays. In: Spencer, A.M. (ed.) *Generation, Accumulation, and Production of Europe's Hydrocarbons.* Special Publication, EAPG, Oxford University Press, Oxford, pp. 99–109. [12]

Shannon, P.M. (1993) Oil and gas in Ireland—exploration, production and research. *First Break* **11**, 429–33. [12]

Shannon, P.M. (1996) Current and future potential of oil and gas exploration in Ireland. In: Glennie, K.W. and Hurst, A. (eds) *AD 1995: NW Europe's Hydrocarbon Industry.* Geological Society, London, pp. 53–64. [1]

Shannon, P.M. and Mactiernan, B. (1993) Triassic prospectivity in the Celtic Sea, Ireland—a case history. *First Break* **11**, 47–57. [12]

Shannon, P.M., Moore, J.G., Jacob, A.W.B. and Makris, J. (1993) Cretaceous and Tertiary basin development west of Ireland. In: Parker, J.R. (ed.) *Petroleum Geology of Northwest Europe: Proceedings of the 4th Conference.* Geological Society, London, pp. 1057–66. [12]

Shannon, P.M., Jacob, A.W.B., Makris, J., O'Reilly, B., Hauser, F. and Vogt, U. (1995) Basin development and petroleum prospectivity in the Rockall and Hatton region. In: Croker, P.F. and Shannon, P.M. (eds) *The Petroleum Geology of Ireland's Offshore Basins.* Special Publication No. 93, Geological Society, London, pp. 435–58. [3, 12]

Sharpnes, O., Briseid, E. and Milton, D.I. (1982) The 34/10 Delta Prospect of the Norwegian North Sea: exploration study of an unconformity trap. In: Halbouty, M.T. (ed.) *The Deliberate Search for the Subtle Trap.* Memoir No. 32, AAPG, Tulsa, OK., pp. 207–16.

Shelton, R. (1995) Mesozoic basin evolution of the North Channel: preliminary results. In: Croker, P.F. and Shannon, P.M. (eds) *The Petroleum Geology of Ireland's Offshore basins.* Special Publication No. 93, Geological Society, London, pp. 17–20. [12]

Shepherd, M. (1991a) The Magnus Field, Blocks 211/12a, UK North Sea. In: Abbotts, I.L. (ed.) *UK North Sea United Kingdom Oil and Gas Fields: 25 Years Commemorative Volume.* Memoir No. 14, Geological Society, London, pp. 153–7. [1, 11, 12]

Shepherd, M. (1991b) The Arbroath and Montrose Field, Blocks 211/7a, 211/12a, UK North Sea. In: Abbotts, I.L. (ed.) *UK North Sea United Kingdom Oil and Gas Fields: 25 Years Commemorative Volume.* Memoir No. 14, Geological Society, London, pp. 211–17. [1]

Sibson, R.H. (1990) Rupture nucleation on unfavourably oriented faults. *Bull. Seismol. Soc. Am.* **80**(6), 1580–604.

Sibson, R.H. (1991) Loading of faults to failure. *Bull. Seismol. Soc. Am.* **81**(6), 2493–7.

Sibson, R.H. (1992a) Implications of fault-valve behaviour for rupture nucleation and recurrence. *Tectonophysics* **211**, 283–93. [11]

Sibson, R.H. (1992b) Earthquake faulting, induced fluid flow, and fault-hosted gold-quartz mineralisation. In: Batholomew, M.J. *et al.* (eds) *Basement Tectonics 8, Proceedings of 8th International Conference on Basement Tectonics, Bute, Montana, 1988.* pp. 603–14. [11]

Simon, J.B. and Bluck, B.J. (1982) Palaeodrainage of the southern margin of the Caledonian mountain chain in the Northern British Isles. *Trans. Roy. Soc. Edinburgh Earth Sci.* **73**, 11–15. [2]

Sinninghe-Damsté, J.S., de las Heras, F.X., van Bergen, P.F. and de Leeuw, J.W. (1993) Characterization of Tertiary Catalan lacustrine oil shales: discovery of extremely organic sulphur-rich type 1 kerogens. *Geochim. Cosmochim. Acta* **57**, 389–415. [11]

Skervøy, A. and Sylta, O. (1993) Modelling of expulsion and secondary migration along the southwestern margin of the Horda Platform. In: Doré, A.G., Augustson, J.H., Hermanrud, C., Stewart, D.J. and Sylta, O. (eds) *Basin Modelling: Advances and Applications.* Special Publication No. 3, NPF, Elsevier, Amsterdam, pp. 499–537. [11]

Skjerven, J., Rijs, F. and Kalheim, J.E. (1983) Late Palaeozoic to early Cenozoic structural development of the south-southeastern Norwegian North Sea. *Geol. Mijnbouw* **62**, 35–45. [7]

Skjold, L.J. (1980) Palaeocene sands for the Balder field. In: *The Sedimentation of the North Sea Reservoir Rocks. Proceedings of the 12th Geilo Conference.* Norwegian Petroleum Society, 7 pp. [12]

Skovbro, B. (1983) Depositional conditions during chalk sedimentation in the Ekofisk area Norwegian North Sea. *Geol. Mijnbouw* **62**, 169–75. [9]

Smalley, P.C. and Warren, E.A. (1994) North Sea formation waters: implications for diagenesis and production chemistry. *Marine and Petr. Geol.* **11**(1), 2–5. [11]

Smart, G. and Clayton, T. (1985) The progressive illitization of interstratified illite–smectite from Carboniferous sediments of northern England and its relationship to organic maturity indicators. *Clay Minerals* **20**, 455–66. [11]

Smith, A.G. Hurley, A.M. and Briden, J.C. (1981) *Phanerozoic Palaeocontinental World Maps.* Cambridge University Press, Cambridge, 101 pp. [2]

Smith, A.H.V. and Butterworth, M. (1967) *Miospores in the Coal Seams of the Carboniferous of Great Britain.* Special Paper on Palaeontology 1. [4]

Smith, D.B. (1970a) The palaeogeography of the British Zechstein. In: Rau, J.L. and Dellwig, L.F. (eds) *Third Symposium on Salt, Vol. 1. North Ohio Geol. Soc.* 20–3. [6]

Smith, D.B. (1970b) Submarine slumping and sliding in the Lower Magnesian Limestone of Northumberland and Durham. *Proc. Yorks. Geol. Soc.* **38**, 1–36. [6]

Smith, D.B. (1972a) The Lower Permian in the British Isels. In: Falke, H. (ed.) *Rotliegend: Essays on European Lower Permian.* Brill, Leiden, pp. 1–32. [5]

Smith, D.B. (1972b) Foundered strata, collapse-breccias and subsidence features of the English Zechstein. In: *Geology of Saline Deposits, Proceedings of Hannover Symposium, 1968.* Earth Science 7, UNESCO.

Smith, D.B. (1975) Gravitational movements in Zechstein carbonate rocks in North East England. In: Taylor, J.C.M. *et al.* (eds) *The Role of Evaporites in Hydrocarbon Exploration.* Course Notes No. 39, JAPEC, London, pp. F1–F12. [6]

Smith, D.B. (1979) Rapid marine transgressions and regressions of the Upper Permian Zechstein Sea. *J. Geol. Soc.* **136**, 155–6. [5, 6]

Smith, D.B. (1980a) Permian and Triassic rocks. In: Robson, D.A. (ed.) *The Geology of North East England.* Special Publication Natural History Society of Northumbria, Newcastle upon Tyne, pp. 36–48. [5]

Smith, D.B. (1980b) The evolution of the English Zechstein Basin. In: Füchtbauer, H. and Peryt, T.M. (eds) *The Zechstein Basin with Emphasis on Carbonate Sequences.* Contributions to Sedimentology No. 9, Schweitzerbart'sche Verlagsbuchhandlung, Stuttgart, pp. 7–34. [6]

Smith, D.B. (1981a) The Magnesian Limestone Upper Permian reef complex of Northeastern England. In: Special Publication No. 30, Soc. Econ. Pal. Min., pp. 161–86. [6]

Smith, D.B. (1981b) Bryozoan–algal Patch Reefs in the Upper Permian Lower Magnesian Limestone of Yorkshire, Northeast England. Special Publication No. 30, Soc. Econ. Pal. Min., pp. 187–202. [6]

Smith, D.B. (1986) The Trow Point Bed—a deposition of Upper Permian marine oncoids, peloids and columnar stromatolites in the Zechstein of NE England. In: Harwood, G.M. and Smith, D.B. (eds) *The English Zechstein and Related Topics.* Special Publication No. 22, Geological Society, London, pp. 113–25. [6]

Smith, D.B. (1989) The Late Permian palaeogeography of North East England. *Proc. Yorks. Geol. Soc.* **47**, 285–312. [6]

Smith, D.B. (1994) Discussion of Holliday, D.W. (1993) Geophysical log signatures in the Eden Shales (Permo-Triassic) of Cumbria and their regional significance. *Proc. Yorks. Geol. Soc.* **50**, 173–85. [6]

Smith, D.B. (1996) Deformation in the late Permian Boulby Halite (EZ3Na) in Teeside, NE England. In: Alsop, G.I., Blundell, D.J. and Davison, I. (1996) *Salt Tectonics.* Special Publication 100, Geological Society, London, 77–87. [6]

Smith, D.B. and Crosby, A. (1979) The regional and stratigraphical context of Zechstein 3 and 4 potash deposits in the British sector of the Southern North Sea and adjoining land area. *Econ. Geol.* **74**, 397–408. [6]

Smith, D.B. and Francis, E.A. (1967) *Geology of the Country between Durham and West Hartlepool.* Memoir, Geological Survey G.B., HMSO, London, 354 pp. [5, 6]

Smith, D.B. and Taylor, J.C.M. (1989) A north-west passage to the Southern Zechstein Basin of the UK North Sea. *Proc. Yorks. Geol. Soc.* **47**, 313–20. [5, 6]

Smith, D.B. and Taylor, J.C.M. (1992) Permian. In: Cope, J.C.W., Ingham, J.K. and Rawson, P.F. (eds) *Atlas of British Palaeogeography and Lithofacies.* Geological Society, London, pp. 87–96. [5, 6]

Smith, D.B., Brunstrom, R.G.W., Manning, P.I., Simpson, S. and Shotton, F.W. (1974) *Correlation of the Permian Rocks of the British Isles.* Special Report No, 5, Geological Society, Blackwell Scientific Publications, Oxford, 45 pp. [2, 5]

Smith, D.B., Harwood, G.M., Pattison, J. and Pettigrew, T.H. (1986) A revised nomenclature for Upper Permian strata in eastern England. In: Harwood, G.M. and Smith, D.B. (eds) *The English Zechstein and Related Topics.* Special Publication No. 22, Geological Society, London, pp. 9–17. [6]

Smith, D.I. and Watson, J. (1983) Scale and timing on the Great Glen Fault, Scotland. *Geology* **11**, 523–6. [2]

Smith, K. and Ritchie, J.D. (1993) Jurassic volcanic centres in the Central North Sea. In: Parker, J.R. (ed.) *Petroleum Geology of Northwest Eruope: Proceedings of the 4th Conference.* Geological Society, London, pp. 519–31. [2, 8]

Smith, K., Smith, N.J. and Holliday, D.W. (1985). The deep structure of Derbyshire. *Geol. J.* **20**, 215–25. [4]

Smith, K., Gatliff, R.W. and Smith, N.J.P. (1994) Discussion on the amount of Tertiary erosion in the UK estimated using sonic analysis: reply by Hillis, R.R. *J. Geol. Soc.* **151**, 1041–5. [11]

Smith, N.J.P. (1993) The case for exploration of deep plays in the Variscan fold belt and its foreland. In: Parker, J.R. (ed.) *Petroleum Geology of Northwest Europe: Proceedings of the 4th*

Conference. Geological Society, London, pp. 667–75. [2, 11, 12]

Smith, P.M.R. (1983) Spectral correlation of spore colour standards. In: Brooks, J. (ed.) *Petroleum Geochemistry and Exploration of Europe.* Blackwell Scientific Publications, Oxford. pp. 289–94. [11]

Smith, R.I., Hodgson, N. and Fulton, M. (1993) Salt controls on Triassic reservoir distribution, UKCS Central North Sea. In: Parker, J.R. (ed.) *Petroleum Geology of Northwest Europe: Proceedings of the 4th Conference.* Geological Society, London, pp. 547–58. [6, 7, 12]

Smith, R.L. (1987) The structural development of the Clyde field. In: Brooks, J. and Glennie, K.W. (eds) *Petroleum Geology of North West Europe.* Graham and Trotman, London, pp. 523–31. [1, 6, 11, 12]

Smythe, D.K., Dobinson, A., McQuillan, R., Brewer, J.A., Matthews, D.H., Blundell, D.J. *et al.* (1982) Deep structures of the Scottish Caledonides revealed by the MOIST reflection profile. *Nature* **229**, 338–40.

Sneh, A. (1988) Permian dune patterns in northwestern Europe challenged. *J. Sediment. Petrol.* **58**(1), 44–51. [5]

Snowdon, L.R. (1979) Errors in extrapolation of experimental kinetic parameters of organic geochemical systems. In: Welte, D.H., Horsfield, B. and Baker, D.R. (eds) *Petroleum and Basin Evolution.* Springer Verlag, Berlin. [11]

Snyder, D.B., England, R.W. and McBride, J.H. (1997) Linkage between mantle and crustal structures and its bearing on inherited structures in northwestern Scotland. *J. Geol. Soc.* **154**, 79–82. [2]

Söderström, B., Forsberg, A., Holtar, E. and Rasmussen, B.A. (1991) The Mjolner Field, a deep Upper Jurassic oil field in the Central North Sea. *First Break* **9**, 156–71. [12]

Solli, T. (1995) Upper Jurassic play concept—an integrated study in Block 34/7, Norway. *First Break* **13**, 21–30. [8]

Sønderholm, M. (1987) Facies and geochemical aspects of the dolomite–anhydrite transition zone Zechstein 1–2 in the Batum 13-well, northern Jutland, Denmark: a key to the evolution of the Norwegian–Danish. In: Peryt, T.M. (ed.) *The Zechstein Facies in Europe.* Springer-Verlag, Berlin, New York, pp. 93–122. [6]

Soper, N.J. and Anderton, R. (1984) Did the Dalradian slides originate as extensional faults? *Nature* **307**, 357–60. [2]

Soper, N.J., Webb, B.C. and Woodcock, N.H. (1987) Late Caledonian Acadian transpression in north-west England: timing, geometry and geotectonic significance. *Proc. Yorks. Geol. Soc.* **46**, 175–92. [3, 4]

Sørensen, K. (1986) Danish Basin subsidence by Triassic rifting on a lithosphere cooling background. *Nature* **319**, 660–3. [2,7]

Sørensen, S. (1985) Basin analysis and maturation modelling onshore Demmark, a case study of the Danish First Round. In: Thomas, D.M., Doré, A.G., Eggen, S.S., Home, P.C. and Larsen, R.M. (eds) *Petroleum Geochemistry in Exploration of the Norwegian Shelf.* Grahmam and Trotman, London, pp. 153–60. [11]

Sørensen, S. and Martinsen, B.B. (1987) A palaeogeographic reconstruction of the Rotliegendes deposits of the Northeastern Permian Basin. In: Brooks, J. and Glennie, K.W. (eds) *Petroleum Geology of North West Europe.* Graham and Trotman, London, pp. 497–508. [5]

Sørensen, S., Jones, M., Hardman, R.P.F., Leutz, W.K. and Schwarz, P.H. (1986) Reservoir characteristics of high- and low-productivity chalks from the Central North Sea. In: Spencer, A.M., Campbell, C.J., Hanslien, S.H., Nelson, P.H., Nysaether, E. and Ormaasen, E.G. (eds) *Habitat of Hydrocarbons on the Norwegian Continental Shelf.* Graham and Trotman, London, pp. 91–110. [9]

Sorgenfrei, T. (1969) A review of petroleum development in Scandinavia. In: Hepple, P.W. (ed.) *The Exploration for Petroleum in Europe and North Africa.* Institute of Petrology, London, pp. 191–203. [6]

Sorgenfrei, T. and Buch, A. (1964) *Deep Tests in Denmark 1935–1959.* 3rd Series No. 36, Geol. Survey of Denmark, København. [6]

Southward, D.A., Morgan, R.K. and Hill, R. (1993) The Zechsteinkalk reservoir of the Hewett Field, southern North Sea (abstract). *AAPG Bull.* **77**, 1666. [6]

Southwood, D.A., Morgan, R.K. and Hill, R. (1993) The Zechsteinkalk reservoir of the Hewett Field, Southern North Sea. In: *Abstracts*, AAPG International Conference, The Hague, the Netherlands. *AAPG Bull.* **77**, 1666. [6]

Spark, I.S.C. and Trewin, N.H. (1986) Facies-related diagenesis in the main Claymore Oilfield sandstones. *Clay Minerals* **21**, 479–96. [7]

Speksnijder, A. (1987) The structural configuration of Cormorant Block IV in the context of the northern Viking Graben structural framework. *Geol. Mijnbouw* **65**, 357–79. [12]

Spencer, A.M. (ed.) (1984) *Petroleum Geology of the North European Margin: Proceedings of North European Margin Symposium (NEMS), Trondheim, 1983.* Graham and Trotman, London.

Spencer, A.M. (ed.) (1991) *Generation, accumulation, and production of Europe's hydrocarbons.* Special Publication, EAPG, Oxford University Press, Oxford, 459 pp.

Spencer, A.M. and Eldholm, O. (1993) Atlantic margin exploration: Cretaceous–Tertiary evolution, basin development and petroleum geology: introduction and review. In: Parker, J.R. (ed.) *Petroleum Geology of Northwest Europe: Proceedings of the 4th Conference.* Geological Society, London, p. 899. [12]

Spencer, A.M. and Pegrum, R.M. (1990) Hydrocarbon plays in the northern North Sea. In: Brooks, J. (ed) *Classic Petroleum Provinces.* Special Publication No. 50, Geological Society, London, pp. 441–70. [12]

Spencer, A.M., Johnsen, S.O., Mørk, A., Nysaether, E., Songstad, P. and Spinnangr, A. (eds) (1984) Petroleum geology of the north European margin. In: *Proceedings of the North European Margin Symposium, Norwegian Petroleum Society, 1983.* Graham and Trotman, London. [11]

Spencer, A.M., Home, P.C. and Wiik, V. (1986a) Habitat of hydrocarbons in the Jurassic Ula Trend, Central Graben, Norway. In: Spencer, A.M., Campbell, C.J., Hanslien, S.H., Nelson, P.H., Nysaether, E. and Ormaasen, E.G. (eds) *Habitat of Hydrocarbons on the Norwegian Continental Shelf.* NPF, Graham and Trotman, London, pp. 111–27. [12]

Spencer, A.M., Campbell, C.J., Hanslien, S.H., Nelson, P.H., Nysaether, E. and Ormaasen, E.G. (1986b) *Habitat of Hydrocarbons on the Norwegian Continental Shelf.* NPF, Graham and Trotman, London, 354 pp. [11]

Spencer, A.M., Campbell, C.J., Hanslien, S.H., Nelson, P.H., Nysaether, E. and Ormaasen, E.G. (eds) (1987) *Geology of the Norwegian Oil and Gas Fields.* Graham and Trotman, London, 493 pp. [11]

Spencer, A.M., Leckie, G.G. and Chew, K.J. (1996) North Sea hydrocarbon plays and their resources. In: Glennie, K. and Hurst, A. (eds) *Northwest Europe's Hydrocarbon Industry.* Geological Society, London, pp. 25–41. [1, 5, 8, 11, 12]

Spiers, C.J. (1994) Microphysical aspects of the flow of rocksalt (abstract). In: Alsop, A. (convenor) *Salt Tectonics; Programme and Abstracts.* Geological Society, London, 32. [6]

Spiers, C.J., Urai, J.L., Lister, G.S. and Zwart, H.J. (1984) Water weakening and dynamic recrystallization in salt. In: Abstracts with Programmes No. 16, Geological Society of America, Abstract No. 52601, p. 665. [6]

Spiers, C.J., Urai, J.L., Lister, G.S., Boland, J.N. and Zwart, H.J. (1986) *The Influence of Fluid–Rock Interaction on the Rheology of Salt Rock and on Ionic Transport in the Salt.* Nuclear Science Technology 1, EUR 10399 EN, Commission of European Communities. [6]

Stach, E., Machowsky, M.T., Teichmüller, M., Taylor, G.H., Chandra, D. and Teichmuller, R. (1982) *Stach's textbook of coal petrography.* 3rd edn. Gebruder Borntraeger, Berlin. [11]

Stannage, W. (1988) Some problems in petroleum geochemistry. *J. Pet. Geol.* **11**(4), 415–28. [11]

Stäuble, A.J. and Milius, G. (1970) Geology of Groningen gas field. In: Memoir No. 14, AAPG, pp. 359–69. [5]

Stearns, D.W. and Friedman, M. (1972) Reservoirs in fractured rock. In: King, R.E. (ed.) *Stratigraphic Oil and Gas Fields*. Memoir No. 16, AAPG, Tulsa, OK., pp. 82–106. [3]

Steckler, M.S. (1985) Uplift and extension at the Gulf of Suez: indications of induced mantle conversion. *Nature* **317** (6033), 135–9. [11]

Steckler, M.S., Berthelot, F., Lyberis, N. and le Pichon, X. (1988) Subsidence in the Gulf of Suez: implications for rifting and plate kinematics. *Tectonophysics*. **153**, 249–70. [11]

Stedman, C. (1988) Namurian E1 tectonics and sedimentation in the Midland Valley of Scotland: rifting versus strike–slip influence. In: Besly, B.M. and Kelling, G. (eds) *Sedimentation in a Synorogenic Basin Complex: the Upper Carboniferous of Northwest Europe*. Blackie, Glasgow, London, pp. 242–54. [4]

Steel, R.J. (1993) Triassic–Jurassic megasequence stratigraphy in the northern North Sea: rift to post-rift evolution. In: Parker, J.R. (ed.) *Petroleum Geology of Northwest Europe: Proceedings of the 4th Conference*. Geological Society, London, pp. 299–315. [1, 7, 8, 12]

Steel, R.J. and Ryseth, A. (1990) The Triassic–early Jurassic succession in the northern North Sea: megasequence stratigraphy and intra-Jurassic tectonics. In: *Tectonic Events Responsible for Britain's Oil and Gas Reserves*. Special Publication No. 55, Geological Society, London, pp. [7]

Steel, R.J. and Wilson, A.C. (1975) Sedimentation and tectonism? Permo-Triassic on the margin of the North Minch Basin, Lewis. *J. Geol. Soc.* **131**, 183–202.

Steel, R.J. and Worsley, D. (1984) Svalbard's post-Caledonian strata: an atlas of sedimentational patterns and palaeogeographic evolution. In: Spencer, A.M. (ed.) *Petroleum Geology of the North European Margin: Proceedings of the North European Margin Symposium* (NEMS) Trondheim, 1983. Graham and Trotman, London, pp. 109–35. [6]

Steele, L.E. and Adams, G.E. (1984) A review of the northern North Sea's Beryl Field after seven years' production. In: *Proceedings of European Petroleum Conference, London*. Paper 12960, SPE. [12]

Steele, R.P. (1983) Longitudinal draa in the Permian Yellow Sands of north east England. In: Brookfield, M.E. and Ahlbrandt, T.S. (eds) Eolian Sediments and Processes. *Developments in Sedimentology* **38**, pp. 543–50. [5]

Steele, R.P. (1988) The Namurian sedimentary history of the Gainsborough Trough. In: Besly, B.M. and Kelling, G. (eds) *Sedimentation in a Synorogenic Basin Complex: the Upper Carboniferous of Northwest Europe*. Glasgow, London, pp. 102–13. [4]

Steele, R.P., Allan, R.M., Allinson, G.J. and Booth, A.J. (1993) Hyde: a proposed field development on the Southern North Sea using horizontal wells. In: Parker, J.R. (ed.) *Petroleum Geology of Northwest Europe: Proceedings of the 4th Conference*. Geological Society, London, pp. 1465–72. [1]

Stein, R., Rullkotter, J. and Welte, D.H. (1986) Accumulation of organic-carbon-rich sediments in the Late Jurassic and Cretaceous Atlantic Ocean—a synthesis. *Chem. Geol.* **56**, pp. 1–32. [9]

Stemmerik, L. (1987) Cyclic carbonate and sulphate from the Upper Permian Karstryggen Formation, East Greenland. In: Peryt, T.M. (ed.) *The Zechstein Facies in Europe*. Springer-Verlag, Berlin, New York, pp. 5–22. [6]

Stemmerik, L. (1995) Permian history of the Norwegian–Greenland Sea area. In: Scholle, P.A., Peryt, T.M. and Ulmer-Scholle, D.S. (eds) *The Permian of Northern Pangaea*: Vol. 1. *Paleogeography, Paleoclimates, Stratigraphy*. Springer-Verlag, Berlin, pp. 98–118. [5]

Stemmerik, L., Frykman, P., Christensen, O.W. and Stentoft, N. (1987) The Zechstein carbonates of Southern Jylland, Denmark. In: Brooks, J. and Glennie, K.W. (eds) *Petroleum Geology of North West Europe*. Graham and Trotman, London, pp. 365–74. [6]

Stemmerik, L., Christiansen, F.G. and Piasecki, S. (1991) Carboniferous lacustrine shale in East Greenland—additional source rock in northern North Atlantic. In: *Lacustrine Basin Exploration—Case Studies and Modern Analogs*. Memoir No. 50, AAPG, Tulsa, OK., pp. 277–86. [11]

Stemmerik, L., Jensen, S.M. and Pedersen, M. (1997) Hydrocarbon-associated mineralisation of Upper Permian carbonate buildups, Wegener Halvo, East Greenland. In: Hendry, J.P., Carey, P.F., Parnell, J., Ruffell, A.H. and Worden, R.H. (eds) *Geofluids 11 Extended Abstracts*. Queen's University, Belfast. [11]

Stephen, K., Underhill, J.R., Partington, M. and Hedley, R. (1993) Hettangian–Oxfordian sequence stratigraphy of the Inner Moray Firth basin. In: Parker, J.R. (ed.) *Petroleum Geology of Northwest Europe: Proceedings of the 4th Conference*. Geological Society, London, pp. 485–505. [8, 12]

Stephenson, D. and Gould, D. (1995) *The Grampian Highlands*, 4th edn., British Regional Geology, HMSO, London, 261 pp. [2]

Stephenson, M.A. (1991) The North Brae Field, Block 16/17a, UK North Sea. In: Abbotts, I.L. (ed.) *UK North Sea United Kingdom Oil and Gas Fields: 25 Years Commemorative Volume*. Memoir No. 14, Geological Society, London, pp. 43–8. [1, 8]

Stevens, D.A. and Wallis, R.J. (1991) The Clyde Field, Block 301/17b, UK North Sea. In: Abbotts, I.L. (ed.) *UK North Sea United Kingdom Oil and Gas Fields: 25 Years Commemorative Volume*. Memoir No. 14, Geological Society, London, pp. 279–86. [1, 12]

Stevens, V. (1991) The Beatrice Field, Block 11/30a, UK North Sea. In: Abbotts, I.L. (ed.) *UK North Sea United Kingdom Oil and Gas Fields: 25 Years Commemorative Volume*. Memoir No. 14, Geological Society, London, pp. 245–52. [1, 8]

Stewart, A.D. (1982) Late Proterozoic rifting in NW Scotland: the genesis of the Torridonian. *J. Geol. Soc. Lond.* **139**(4), 413–20. [2]

Stewart, D.J., Schwander, M. and Bolle, L. (1995) Jurassic depositional systems of the Horda Platform, Norwegian North Sea: practical consequences of applying sequence stratigraphic techniques. In: Steel, R.J., Felt, V.L., Johanesson, E.P. and Mathieu, C. (eds) *Sequence Stratigraphy on the Northwest European Margin*. Special Publication No. 5, NPS, pp. 291–323. [8]

Stewart, D.M. and Faulkner, A.J.G. (1991) The Emerald Field, Blocks 2/10a, 2/15a, 3/11b, UK North Sea. In: Abbotts, I.L. (ed.) *UK North Sea United Kingdom Oil and Gas Fields: 25 Years Commemorative Volume*. Memoir No. 14, Geological Society, London, p. 111. [1, 12]

Stewart, I.J. (1987) A revised stratigraphic interpretation of the Early Paleogene of the Central North Sea. In: Brooks, J. and Glennie, K.W. (eds) *Petroleum Geology of North West Europe*. Graham and Trotman, London, pp. 557–76. [1, 10, 12]

Stewart, I.J. (1993) Structural controls on the Late Jurassic age shelf system, Ula Trend, Norwegian North Sea. In: Parker, J.R. (ed.) *Petroleum Geology of Northwest Europe: Proceedings of the 4th Conference*. Geological Society, London, pp. 469–84. [12]

Stheeman, H.A. and Thiadens, A.A. (1969) A history of the exploration for hydrocarbons within the territorial boundaries of the Netherlands. In: Hepple, P. (ed.) *The Exploration for Petroleum in Europe and North Africa*. Institute of Petroleum, London, pp. 259–69. [1]

Stiles, J.H. and Mckee, J.W. (1986) Cormorant: development of a complex field. In: *61st Annual Technology Conference and Exhibition*. SPE, New Orleans, Paper No. 15504. [12]

Stockbridge, C.P. and Gray, D.I. (1991) The Fulmar Fields, Blocks 30/16, 30/11b, UK North Sea. In: Abbotts, I.L. (ed.) *UK North Sea United Kingdom Oil and Gas Fields: 25 Years Commemorative Volume*. Memoir No. 14, Geological Society, London, p. 309. [1, 12]

Stocks, A.E. and Lawrence, S.R. (1990) Identification of source rocks from wire line logs. In: Hurst, A., Lovell, M.A. and Morton, A.C. (eds) *Geological Applications of Wireline Logs*. Special Publication No. 48, Geological Society, London, pp. 241–52. [11]

Stoddart, D.P., Hall, P.B., Larter, S.R., Brasher, J. and Bjorøy, M. (1995) The reservoir geochemistry of the Eldfisk field, Norwegian North Sea. In: Cubitt, J.M. and England, W.A. (eds) *The Geochemistry of Reservoirs*. Special Publication No. 86, Geological Society, London, pp. 203–56. [11]

Stoker, M.S., Hitchen, K. and Graham, C.C. (1993) *United Kingdom Offshore Regional Report: the Geology of the Hebrides and West Shetland Shelves, and Adjacent Deep-water Areas*. HMSO for BGS, London. [11]

Stollhofen, H. and Stanistreet, I.G. (1994) Interaction between bimodal volcanism, fluvial sedimentation and basin development in the Permo-Carboniferous Saar–Nahe Basin southwest Germany. *Basin Res.* **6**(4), 245–67. [5]

Stølum, H.-H., Smalley, P.C. and Hanken, N.-M. (1993) Prediction of large-scale communication in the Smorbukk fields from strontium fingerprinting. In: Parker, J. (ed.) *Petroleum Geology of Northwest Europe: Proceedings of the 4th Conference*. Geological Society, London, pp. 1421–32. [11]

Stoneley, R. (1982) The structural development of the Wessex Basin. *J. Geol. Soc.* **139**, 543–54. [2, 8, 12]

Storey, M.W. and Nash, D.F. (1993) The Eakring Dukeswood oil field: an unconventional technique to describe a field's geology. In: Parker, J.R. (ed.) *Petroleum Geology of Northwest Europe: Proceedings of the 4th Conference*. Geological Society, London, pp. 1527–37. [4, 11]

Stow, D.A.V. (1983) Sedimentology of the Brae oilfield, North Sea: a reply. *J. Petr. Geol.* **6**, 103–4. [12]

Stow, D.A.V. and Atkin, B.P. (1987) Sediment facies and geochemistry of Upper Jurassic mudrocks in the Central North Sea area. In: Brooks, J. and Glennie, K.W. (eds) *Petroleum Geology of North West Europe*. Graham and Trotman, London, pp. 797–808. [11, 12]

Stow, D.A.V., Bishop, C.D. and Mills, S.J. (1982) Sedimentology of the Brae oil field, North Sea: fan models and controls. *J. Petr. Geol.* **5**, 129–48. [8, 11, 12]

Strakhov, N.M. (1962) *Principles of lithogenesis III*. Fitzsimmons, J.P., Tomkieff, S.I. and Hemingway, J.E. (eds) Oliver and Boyd, Edinburgh, 577 p. English translation by J.P. Fitzsimmons. [5]

Strand, J.E. and Slaatsveen, P.R. (1987) 7/11-5 discovery. In: Spencer, A.M., Campbell, C.J., Hanslien, S.H., Nelson, P.H., Nysaether, E. and Ormaasen, E.G. (eds) *Geology of the Norwegian Oil and Gas Fields*. NPS, Graham and Trotman, London, pp. 177–83. [11]

Strank, A.R. (1987) The stratigraphy and structure of Dinantian strata in the East Midlands. In: Miller, J., Adams, A.E. and Wright, V.P. (eds) *European Dinantian Environments*. John Wiley, Chichester, pp. 157–76. [4]

Strehlau, K., and David, F. (1989) Sedimentologie und Flözfacies im Westfal C des nordlichen Ruhrkarbons. *Z. dt. Geol. Ges.* **140**, 231–47. [4]

Strohmenger, C., Jäger, G., Mitchell, J.C., Love, K.M., Antonini, M., Gast, R., *et al.* (1993a) An integrated approach to Zechstein Ca2 carbonate reservoir facies prediction in the South Oldenburg area, Upper Permian, Northwest Germany. *AAPG Bull.* **77**, 1668. [6]

Strohmenger, C., Love, K.M. and Mitchell, J.C. (1993b) Sedimentology and diagenesis of the Zechstein Ca 2 Carbonate, Late Permian, Northwest Germany (abstract). *Am. Assoc. Petrol. Geol. Bull.* **77**, p. 1668. [6]

Strohmenger, C., Antonini, M., Jäger, G., Rockenbauch, K. and Strauss, C. (1996) Zechstein 2 Carbonate Reservoir facies distribution in relation to Zechstein sequence stratigraphy (Upper Permian, Northwest Germany): an integrated approach. *Bull. Centres Rech. Explor-Prod. Elf Aquitaine* **20**(1), 1–35. [6]

Struijk, A.P. and Green, R.T. (1991) The Brent Field, Block 211/29, UK North Sea. In: Abbotts, I.L. (ed.) *UK North Sea United Kingdom Oil and Gas Fields: 25 Years Commemorative Volume*. Memoir No. 14, Geological Society, London, pp. 63–72. [1, 7, 12]

Stuart, I.A. (1991) The Rough Gas Storage Field, Blocks 47/3d, 47/8b, UK North Sea. In: Abbotts, I.L. (ed.) *UK North Sea United Kingdom Oil and Gas Fields: 25 Years Commemorative Volume*. Memoir No. 14, Geological Society, London, pp. 477–84. [1, 5, 11]

Stuart, I.A. (1993) The geology of the North Morcambe Gas Field, East Irish Sea Basin. In: Parker, J.R. (ed.) *Petroleum Geology of Northwest Europe: Proceedings of the 4th Conference*. Geological Society, London, pp. 883–95. [1, 12]

Stuart, I.A. and Cowan, G. (1991) The South Morecambe Field, Blocks 110/2a, 110/8a, UK East Irish Sea. In: Abbotts, I.L. (ed.) *UK North Sea United Kingdom Oil and Gas Fields: 25 Years Commemorative Volume*. Memoir No. 14, Geological Society, London, pp. 527–45. [1, 7, 11, 12]

Stute, M., Forster, M., Frischkorn, H., Serejo, A., Clark, J.F. and Schlosser, P. (1995) Cooling of tropical Brazil 5°C during the last glacial maximum. *Science* **269**, 379–83.

Sulak, R.M., Nossa, G.R. and Thompson, D.A. (1990) Ekofisk Field enhanced recovery. In: Buller, A.T., Berg, E., Hjelmelaud, O., Kleppe, I., Torsæter, O. and Aasen, J.O. (eds) *North Sea Oil and Gas Reservoirs II*. Norwegian Institute of Technology, Graham and Trotman, London, pp. 281–95. [9]

Sullivan, M.D., Haszeldine, R.S., Boyce, A.J., Rogers, G. and Fallick, A.E. (1994) Late anhydrite cements mark basin inversion: isotopic and formation water evidence, Rotliegend Sandstone, North Sea. *Mar. Petr. Geol.* **11**(1), 46–54. [5]

Sundararaman, P. and Dahl, J.E. (1993) Depositional environment, thermal maturity and irradiation effects on porphyrin distribution: Alum Shale, Sweden. *Org. Geochem.* **20**(3), 333–7. [11]

Sundsbø, G.O. and Megson, J.B. (1993) Structural styles in the Danish Central Graben. In: Parker, J.R. (ed.) *Petroleum Geology of Northwest Europe: Proceedings of the 4th Conference*. Geological Society, London, pp. 1255–67. [6, 7]

Surlyk, F. (1978) Submarine fan sedimentation along fault scraps on tilted fault blocks Jurassic–Cretaceous boundary, East Greenland. *Gronlands Geol. Undersogelse Bull.* **28**, 1–103. [8]

Surlyk, F., Piasechi, S., Rolle, F., Stemmerik, K., Thomsen, E. and Wrang, P. (1984) The Permian Basin of East Greenland. In: Spencer, A.M. (ed.) *Petroleum Geology of the North European Margin: Proceedings of the North European Margin Symposium (NEMS) Trondheim, 1983*. Graham and Trotman, London, pp. 303–15. [2, 5, 6, 11]

Surlyk, F., Noe-Nygaard, N. and Dam, G. (1993) High and low resolution sequence stratigraphy in lithological prediction examples from the Mesozoic around the northern North Atlantic. In: Parker, J.R. (ed.) *Petroleum Geology of Northwest Europe: Proceedings of the 4th Conference*. Geological Society, London, pp. 199–214. [8]

Svendsen, N. (1979) The Tertiary/Cretaceous chalk in the Dan Field of the Danish North Sea. In: Birkelund, T. and Bromley, R.G. (eds) *Cretaceous/Tertiary Boundary Events Symposium*. Vol. 2. University of Copenhagen, Copenhagen, pp. 112–19. [9]

Sweeney, J.J. and Burnham, A.K. (1990) Evaluation of a simple model of vitrinite reflectance based on chemical kinetics. *AAPG Bull.* **74**(10), 1559. [11]

Sweeney, J.J., Burnham, A.K. and Braun, R.L. (1987) A model of hydrocarbon generation from Type I kerogen: application to Uinta Basin, Utah. *AAPG Bull.* **71**(8), 967–985. [11]

Swift, D.J.P. (1968) Coastal erosion and transgressive stratigraphy. *J. Geol.* **76**, 444–56. [8]

Sylta, O. (1993) New techniques and their applications in the analysis of secondary migration. In: Doré, A.G., Augustson, J.H., Hermanrud, C., Stewart, D.J. and Sylta, O. (eds) *Basin Model-*

ling: Advances and Applications. Special Publication 3, NPF, Elsevier, Amsterdam, pp. 385–98. [11]

Taber, D.R., Vickers, M.K. and Winn, R.D. (1995) The definition of the Albian 'A' sand reservoir fairway and aspects of associated gas accumulations in the North Celtic Sea Basin. In: Croker, P.F. and Shannon, P.M. (eds) *The Petroleum Geology of Ireland's Offshore Basins*. Special Publication No. 93, Geological Society, London, pp. 227–44. [12]

Talbot, C.J., Tully, C.P. and Woods, P.J.E. (1982) The structural geology of the Boulby potash mine, Cleveland, United Kingdom. *Tectonophysics* **85**, 167–204. [6]

Tarling, D.H. (1985) Palaeomagnetic studies in the Orcadian Basin. *Scott. J. Geol.* **21**, 261–73. [3]

Tate, M.P. and Dobson, M.R. (1989) Late Permian to early Mesozoic rifting and sedimentation offshore NW Ireland. *Mar. Petr. Geol.* **6**, 49–59. [6]

Taylor, A.M. and Gawthorpe, R.L. (1993) Application of sequence stratigraphy and trace fossil analysis to reservoir description: examples from the Jurassic of the North Sea. In: Parker, J.R. (ed.) *Petroleum Geology of Northwest Europe: Proceedings of the 4th Conference*. Geological Society, London, pp. 317–35. [8, 12]

Taylor, B.J., Burgess, I.C., Land, D.H., Mills, D.A.C., Smith, D.B. and Warren, P.T. (1971) *Northern England. British Regional Geology*, Institute of Geological Sciences, HMSO, London, 121 pp. [2]

Taylor, D.J. and Dietvorst, J.P.A. (1991) The Cormorant Field, Block 211/21a, 211/26a, UK North Sea. In: Abbotts, I.L. (ed.) *UK North Sea United Kingdom Oil and Gas Fields: 25 Years Commemorative Volume*. Memoir No. 14, Geological Society, London, p. 73. [1, 8]

Taylor, J.C.M. (1980) Origin of the Werraanhydrit in the Southern North Sea—a reappraisal. In: Füchtbauer, H. and Peryt, T.M. (eds) *The Zechstein Basin with Emphasis on Carbonate Sequences*. Contributions to Sedimentology No. 9 Schweitzerbart'sche Verlagsbuchhandlung, Stuttgart, pp. 91–113. [6]

Taylor, J.C.M. (1981) Zechstein facies and petroleum prospects in the central and northern North Sea. In: Illing, L.V. and Hobson, G.P. (eds) *Petroleum Geology of the Continental Shelf of North-West Europe*. Heyden, London, pp. 176–85. [5, 6, 11, 12]

Taylor, J.C.M. (1986) Gas prospects in the Variscan thrust province of southern England. In: Brooks, J., Goff, J. and van Hoorn, B. (eds) *Habitat of Palaeozoic gas in NW Europe*. Special Publication No. 23, Geological Society, London, pp. 37–53. [3, 5, 11]

Taylor, J.C.M. (1990) Upper Permian–Zechstein. In: Glennie, K.W. (ed.) *Introduction to the Petroleum Geology of the North Sea*. Blackwell Scientific Publications, Oxford, pp. 153–90. [5]

Taylor, J.C.M. (1993) Pseudo-reefs beneath Zechstein salt on the northern flank of the Mid North Sea High. In: Parker, J.R. (ed.) *Petroleum Geology of Northwest Europe: Proceedings of the 4th Conference*. Geological Society, London, pp. 749–57. [6]

Taylor, J.C.M. and Colter, V.S. (1975) Zechstein of the English sector of the Southern North Sea Basin. In: Woodland, A.W. (ed.) *Petroleum and the Continental Shelf of Northwest Europe*, Vol. I, *Geology*. Applied Science Publishers, Barking, pp. 249–63. [1, 6]

Taylor, S.R. and Lapré, J.F. (1987) North Sea chalk diagenesis: its effect on reservoir location and properties. In: Brooks, J. and Glennie, K.W. (eds) *Petroleum Geology of North West Europe*. Graham and Trotman, London, pp. 483–95. [6, 9, 12]

Tegelaar, E.W. and Noble, R.A. (1994) Kinetics of hydrocarbon generation as a function of the molecular structure of kerogen as revealed by pyrolysis–gas chromatography. *Adv. Org. Geochem.* 22(3–5), 543–74. [11]

te Groen, D.M.W. and Steenken, W.F. (1968) Exploration and delineation of the Groningen gas field. *Verhaudelingen Koninklyk Nederlands Geol. minjnbouwkun dig Genootschap* **25**, 9–20. [5]

Teichmüller, M. (1974) Generation of petroleum-like substances in coal seams as seen under the microscope. In: Tissot, B. and Bienner, F. (eds) *Advances in Organic Geochemistry, 1973*. Technip, Paris, pp. 379–407. [11]

Teichmüller, M. (1986) Organic petrology of source rocks, history and state of the art. In: Leythaeuser, D. and Rullkotter, J. (eds) *Advances in Organic Geochemistry, 1985*. Pergamon, Oxford, pp. 581–600. [11]

Teichmüller, M. and Durand, B. (1983) Fluorescence microscopical rank studies on liptinites and vitrinites in peat and coals, and comparison with results of the Rock-Eval pyrolysis. *International Journal of Coal Geology*. **2**, 197–230. [11]

Teichmüller, M. and Teichmüller, R. (1979) Diagenesis of coal (coalification). In: Larsen, G. and Chilingar, G.V. (eds) *Diagenesis in Sediments and Rocks*. Applied Science Publishers, Barking, pp. 207–46. [11]

Teichmüller, M., Teichmüller, R. and Bartenstein, H. (1979) Inkohlung und Erdgas in Nordwestdeutchland. Eine Inkohlungkante der Oberflacher des Oberkarbons. *Fortschr. Geol. Rheinld. u. Westf.* **27**, 137–70. [11]

Telnæs, N. and Dahl, B. (1986) Oil–oil correlation using multivariate techniques *Org. Geochem.* **10**, 425–32. [11]

Telnaes, N., Isaksen, G., Douglas, A. and Van Veen, P. (1989) Geochemistry of marl slate/Kupferschiefer. *List of Abstracts from the 14th Int'l Org. Geochem. Conference (EAOG)*, Paris, Sept. 1989. Abstract no. 291. [11]

Telnæs, N., Cooper, B.S. and Jones, B. (1991) Kerogen facies, biomarkers, trace metal contents and spectral logs as indicators of oxicity and salinity, Upper Jurassic, North Sea. In: Manning, D. (ed.) *Organic Geochemistry, Advances and Applications in Energy and Natural Environment*. Extended Abstracts, 15th Meeting EAOG, Manchester University Press, Manchester, pp. 391–4. [11]

Ten Have, A. and Hillier, A.P. (1986) Reservoir geology of the Sean North and South gas fields, UK Southern North Sea. In: Brooks, J., Goff, J and van Hoorn, B. (eds) *Habitat of Palaeozoic Gas in NW Europe*. Special Publication No. 23, Geological Society, London, pp. 267–73. [1, 5]

Theis, N.J., Mielsen, H.H., Sales, J.K. and Gail, G.J. (1993) Impact of data integration on basin modelling in the Barents Sea. In: Doré, A.G., Augustson, J.H., Hermanrud, C., Stewart, D.J. and Sylta, O. (eds) *Basin Modelling: Advances and Applications*. Special Publication No. 3, NPF, Elsevier, Amsterdam, pp. 433–44. [11]

Thickpenny, A. and Leggett, J.K. (1987) Stratigraphic distribution and palaeo-oceanographic significance of European early Palaeozoic organic-rich sediments. In: Brooks, J. and Fleet, A.J. (eds) *Marine Petroleum Source Rocks*. Blackwell, Oxford, pp. 231–48. [11]

Thomas, B.M., Doré, A.G., Eggen, S.S., Home, P.C. and Larsen, R.M. (eds) (1985) Petroleum geochemistry in exploration of the Norwegian Shelf. *Proceedings of the NPF conference, Stavanger, 22–24 October 1984*. Graham and Trotman, London. [11]

Thomas, B.M., Mueller-Pedersen, P., Whittaker, M.F. and Shaw, N.D. (1985) Organic facies and hydrocarbon distributions in the Norwegian North Sea. In: Thomas, B.M. (ed.) *Petroleum Geochemistry in Exploration of the Norwegian Shelf*. Proceedings of NPF Conference, Graham and Trotman, London, pp. 3–26. [11]

Thomas, D.W. and Coward, M.P. (1995) Late Jurassic–Early Cretaceous inversion of the northern East Shetland Basin, northern North Sea. In: Buchanan, J.G. and Buchanan, P.G. (eds) *Basin Inversion*. Special Publication No. 88, Geological Society, London, pp. 275–306. [9, 11]

Thomas, E.P. (1986) Understanding fractured oil reservoirs. *Oil Gas J.* July, 75–9. [9]

Thomas, J.B., Marshall, J., Mann, A.L., Summons, R.E. and Maxwell, J.R. (1993) Dinosteranes 4, 23, 24-trimethylsteranes and other biological markers in dinoflagellate-rich marine sediments of Rhaetian age. *Org. Geochem.* **20**(1), 91–104. [11]

Thomas, M.M. and Clouse, J.A. (1995) Scaled physical model of secondary oil migration. *AAPG Bull.* **79**(1), 19–29. [11]

Thomsen, E. (1976) Depositional environments and development of Danian bryozoan biomicrite mounds, Karlby Klint, Denmark. *Sedimentology* **23**, 485–509. [9]

Thomsen, E. (1983) Relationships between currents and growth of Palaeocene reef mounds. *Lethaia* **16**, 165–84. [9]

Thomsen, E. (1987) *Lower Cretaceous Calcareous Nanofossil Biostratigraphy in the Danish Central Trough.* Series A, Vol. 20. Danmarks Geol. Underse., Copenhagen, Denmark. [9]

Thomsen, E. (1989) Seasonal variability in the production of Lower Cretaceous calcareous nanoplankton. *Geology* **17**, 715–17. [9]

Thomsen, E. and Jensen, T.F. (1989) Aptian to Cenomanian stratigraphy in the Central Trough of the Danish North Sea sector. *Geol. Jahrbuch Reihe A* **113**, 337–58. [9]

Thomsen, E., Lindgreen, H. and Wrang, P. (1983) Investigation of the source rock potential of Denmark. *Geol. Mijnbouw* **62**, 221–39. [2, 9, 11]

Thomsen, E., Damtoft, K. and Andersen, C. (1987) Hydrocarbon plays in Denmark outside the Central Trough. In: Brooks, J. and Glennie, K.W. (eds) *Petroleum Geology of North West Europe.* Graham and Trotman, London, pp. 375–88. [6, 11]

Thomsen, R.O. and Lerche, I. (1997) Relative contributions to uncertainties in reserve. *Marine and Petroleum Geology* **14**(1), 65–74. [11]

Thomsen, R.O., Lerche, I. and Kostgard, J.A. (1990) Dynamic hydrocarbon predictions for the northern part of the Danish Central Graben: an integrated basin analysis assessment. *Marine and Petroleum Geology* **7**, 123. [11]

Thomson, K. (1995) Discussion on a latest Cretaceous hotspot and the southeasterly tilt of Britain. *J. Geol. Soc.* **152**(4), 729–31. [11]

Thomson, K. and Hillis, R.R. (1995) Tertiary structuration and erosion in the Inner Moray Firth. In: Scrutton, R.A., Stoker, M.S., Shimmield, G.B. and Tudhope, A.W. (eds) *The Tectonics, Sedimentation and Palaeooceanography of the North Atlantic Region.* Special Publication No. 90, Geological Society, London, pp. 249–69. [11]

Thomson, K. and Underhill, J.R. (1993) Controls on the development and evolution of structural styles in the Inner Moray Firth Basin. In: Parker, J.R. (ed.) *Petroleum Geology of Northwest Europe: Proceedings of the 4th Conference.* Geological Society, London, pp. 1167–78. [2, 4, 7, 8]

Thompson, K.F.M. (1979) Light hydrocarbons in subsurface sediments. *Geochim. Cosmochim. Acta* **43**, 657–72. [11]

Thompson, K.F.M. (1983) Classification and thermal history of petroleum based on light hydrocarbons. *Geochim. Cosmochim. Acta* **47**, 303–16.

Thompson, K.F.M. (1987a) Gas–condensate migration and oil-fractionation in deltaic systems. *Mar. Petr. Geol.* **5**, 237–46. [11]

Thompson, K.F.M. (1987b) Fractionated aromatic petroleums and the generation of gas–condensates. *Org. Geochem.* **11**(6), 573–90. [11]

Thompson, K.F.M. (1991) Contrasting characteristics attributed to migration observed in petroleums reservoired in clastic and carbonate sequences in the Gulf of Mexico. In: England, W.A. and Fleet, A.J. (eds) *Petroleum Migration.* Special Publication No. 59. Geological Society, London, pp. 191–207. [11]

Thompson, P.J. and Butcher, P.D. (1991) The geology and geophysics of the Everest Complex. In: Spencer, A.M., Campbell, C.J., Hanslien, S.H., Nelson, P.H., Nysaether, E. and Ormaasen, E.G. (eds) *Generation, Accumulation, and Production of Europe's Hydrocarbons.* Special Publication, EAPG, Oxford University Press, Oxford, pp. 89–98. [1]

Thompson, S., Cooper, B.S., Morley, R.J. and Barnard, P.C. (1985) Oil generating coals. In: Thomas, B.M., Doré, A.G., Eggen, S.S., Home, P.C. and Larsen, R.M. (eds) *Petroleum Geochemistry in Exploration of the Norwegian Shelf.* NPF, Graham and Trotman, London, pp. 59–73. [11]

Thompson, S., Cooper, B.S. and Barnard, P.C. (1994) Some examples and possible explanations for oil generation from coals and coaly sequences. In: Scott, A. and Fleet, A. (eds) *Coal and Coal-bearing Strata as Oil-prone Source Rocks?* Special Publication no. 77. Geological Society, pp. 119–37. [11]

Thorne, J.A. and Watts, A.B. (1989) Quantitative analysis of North Sea subsidence. *AAPG Bull.* **73**(1), 88–116. [7, 11]

Thrasher, J. (1992) Thermal effect of the Tertiary Cullins intrusive complex in the Jurassic of the Hebrides: an organic geochemical study. In: Parnell, J. (ed.) *Basins on the Atlantic Seaboard: Petroleum Geology, Sedimentology and Basin Evolution.* Special Publication No. 62. Geological Society, London, pp. 35–52. [11]

Throndsen, T., Andresen, B. and Unander, A. (1993) Comparison of different models for vitrinite reflectance evolution using laboratory calibration and modelling of well data. In: Doré, A.G., Augustson, J.H., Hermanrud, C., Stewart, D.J. and Sylta, O. (eds) *Basin Modelling: Advances and Applications.* Special Publication No. 3, NPF, pp. 127–35. [11]

Thurlow, J.E. and Coleman, M.L. (1997) Heterogeneity and evolution of formation waters in the oilfields of the Central North Sea. In: *Geofluids 11 Extended Abstracts.* [11]

Timbrell, G. (1993) Sandstone architecture of the Balder Formation depositional system, UK Quadrant 9 and adjacent areas. In: Parker, J.R. (ed.) *Petroleum Geology of Northwest Europe: Proceedings of the 4th Conference.* Geological Society, London, pp. 107–21. [12]

Tissot, B. and Espitalié, J. (1975) L'evolution thermique de la matiere organique des sediments: applicationd'une simulation mathematique. *Rev. Inst. Francais du Petrol.* **30**, 743–77. [11]

Tissot, B.P. and Bessereau, G. (1982) Geochimie des gaz naturels et origine des gisements de gaz en Europe occidentale. *Rev. Inst. Francais du Petrole.* **37**(1), 63–77. [11]

Tissot, B.P. and Welte, D.H. (1984) *Petroleum Formation and Occurrence.* Springer-Verlag, Berlin, 538 pp. [11]

Tissot, B., Califet-Debyser, Y., Deroo, G. and Oudin, J.L. (1971) Origin and evolution of hydrocarbons in Early Toarcian shales, Paris Basin, France. *AAPG Bull.* **55**(12), 2177–93. [11]

Tissot, B.P., Pelet, R. and Ungerer, Ph. (1987) Thermal history of sedimentary basins, maturation indices and kinetics of oil and gas generation. *AAPG Bull.* **71**(12), 1445–66. [11]

Toft, P.C. (1986) Diagenetic fluorite in chalks from Stevns Klint and Mons Klint, Denmark. *Sed. Geol.* **46**, 311–23. [9]

Tonkin, P.C. and Fraser, A.R. (1991) The Balmoral Field, Blocks 16/21, UK North Sea. In: Abbotts, I.L. (ed.) *UK North Sea United Kingdom Oil and Gas Fields: 25 Years Commemorative Volume.* Memoir No. 14, Geological Society, London, pp. 237–43. [1]

Torrens, H.S. (1994) 300 years of oil: mirrored by developments in the West Midlands. In: *The British Association Lectures, 1993.* Geological Society, London, pp. 4–8. [4]

Torsvik, T.H., Smethurst, M.A., Meert, J.G., van der Voo, R., McKerrow, W.S., Brasier, M.D., *et al.* (1996) Continental breakup and collision in the Neoproterozoic and Palaeozoic—a tale of Baltica and Laurentia. *Earth Sci. Rev.* **40**, 229–59. [2]

Torvund, T. and Nipen, O. (1987) The Oseberg field: potential for increased oil recovery by gas injection in a high permeability environment. In: Kleppe, J., Berg, E.W., Buller, A.T., Hjelmeland, O. and Torsæter, O. (eds) *North Sea Oil and Gas Reservoirs.* Graham and Trotman, London, pp. 61–74. [1]

Toth, D.J., Lerche, I., Petroy, D.E., Meyer, R.J. and Kendall, C.G.St.C. (1983) Vitrinite reflectance and the derivation of heat flow changes with time. In: Bjorøy, M., Albrecht, P., Cornford, C., de Groot, K., Eglinton, G., Galimov, E. *et al.* (eds) *Advances in Organic Geochemistry, 1981.* John Wiley, Chichester, pp. 588–96. [11]

Toth, J. (1980) Cross-formational gravity flow of groundwater: a mechanism of the transport and accumulation of petroleum. In: Roberts, W.H. and Cordell, R.J. (eds) *Problems of Petroleum*

Migration. Studies in Geology No. 10, AAPG, Tulsa, OK., pp. 121–68.

Trewin, N.H. (1986) Palaeoecology and sedimentology of the Achanarras fish bed of the Middle Old Red Sandstone, Scotland. *Trans. Roy. Soc. Edinburgh* **77**, 21–46. [3]

Trewin, N.H. (1989) The petroleum potential of the Old Red Sandstone of northern Scotland. *Scott. J. Geol.* **25**, 201–25. [3]

Trewin, N.H. and Bramwell, M.G. (1991) The Auk Field, Block 30/16, UK North Sea. In: Abbotts, I.L. (ed.) *UK North Sea United Kingdom Oil and Gas Fields: 25 Years Commemorative Volume*. Memoir No. 14, Geological Society, London, pp. 227–36. [1, 3, 5, 6, 12]

Tribovillard, N.-P. *et al.* (1994) Geochemical study of organic matter rich cycles from the Kimmeridge clay formation in Yorkshire UK: productivity versus anoxia. *Palaeogeog. Palaeoclimatol. Palaeoecol.* **108**, 165–81. [11]

Tricker, P.M., Marshall, J.A.E.A. and Badman, T.D. (1992) Chitinozoan reflectance: a Lower Palaeozic thermal maturity indicator. *Mar. Petr. Geol.* **9**, 302–8. [11]

Tromp, D. (1996) The offshore industry in the North Sea: environmental aspects. In: Glennie, K.W. and Hurst, A. (eds) *AD 1995: NW Europe's Hydrocarbon Industry*. Geological Society, London, pp. 227–35. [1]

Troost, P.J.P.M. (1981) Schoonebeek oil field: the RW-2E steam injection project. *Geol. Mijnbouw* **60**, 531–9. [9]

Trotter, F.M. (1953) Reddened beds of Carboniferous age in north-west England and their origin. *Proc. Yorks. Geol. Soc.* **29**, 1–20. [4]

Trotter, F.M. (1954) Reddened beds in the Coal Measures of South Lancashire. *Bull. Geol. Sutr. GB* **5**, 61–80. [4]

Trueblood, S. (1992) Petroleum geology of the Slyne Trough and adajcent basins. In: Parnell, J. (ed.) *Basins on the Atlantic Seaboard: Petroleum Geology, Sedimentology and Basin Evolution*. Special Publication No. 62, Geological Society, London, pp. 315–26. [11]

Trueblood, S., Bryan, C. and Pickering, S. (1995) The Douglas oil field and its implications for exploration on the Irish Continental Shelf. In: Croker, P.F. and Shannon, P.M. (eds) *The Petroleum Geology of Ireland's Offshore Basins*. Special Publication No. 93, Geological Society, London, pp. 39–40. [1, 7, 11, 12]

Trueman, A.E. and Weir, J. (1946–58) Weir, J. (1960–68) *A Monograph on British Carboniferous Non-marine Lamellibranchia.* Monograph, Palaeontography Society, 449 pp. [4]

Trusheim, F. (1960) Mechanism of salt migration in northern Germany. *AAPG Bull.* **44**, 1519–40. [6]

Tsoar, H. (1983) Dynamic processes acting on a longitudinal seif dune. *Sedimentology* **30**, 567–78. [5]

Tubb, S.R., Soulsby, A. and Lawrence, S.R. (1986) Palaeozoic prospects on the north flank of the London–Brabant Massif. In: Brooks, J., Goff, J. and van Hoorn, B. (eds) *Habitat of Palaeozoic gas in NW Europe*. Special Publication No. 23, Geological Society, London, pp. 55–72. [4, 11]

Tucker, M.E. (1991) Sequence stratigraphy of carbonate evaporite basins: models and application to the Upper Permian Zechstein of northeast England and adjoining North Sea. *J. Geol. Soc.* **148**, 1019–36. [6]

Tucker, R.M. and Arter, G. (1987) The tectonic evolution of the North Celtic Sea and Cardigan Bay Basins with special reference to basin inversion. *Tectonophysics* **137**, 291–307. [12]

Turner, B.R., Younger, P.L. and Fordham, C.E. (1993) Fell sandstone lithostratigraphy south-west of Berwick-upon-Tweed: implications for the regional development of the Fell Sandstone. *Proc. Yorks. Geol. Soc.* **49**, 269–81. [4]

Turner, C.C. and Allen, P.J. (1991) The Central Brae Field, Block 16/7a. In: Abbotts, I.L. (ed.) *UK North Sea United Kingdom Oil and Gas Fields: 25 Years Commemorative Volume*. Memoir No. 14, Geological Society, London, p. 49. [1]

Turner, C.C., Richards, P.C., Swallow, J.L. and Grimshaw, S.P. (1984) Upper Jurassic stratigraphy and sedimentary facies in the Central Outer Moray Firth Basin, North Sea. *Mar. Petr. Geol.* **1**, 105–17. [8, 11]

Turner, C.C., Cohen, J.M., Connell, E.R. and Cooper, D.M. (1987) A depositional model for the South Brae oilfield. In: Brooks, J. and Glennie, K.W. (eds) *Petroleum Geology of North West Europe*. Graham and Trotman, London, pp. 853–64. [1, 8, 11, 12]

Turner, J.D. and Scrutton, R.A. (1993) Subsidence patterns in western margin basins: evidence from the Faeroe–Shetland Basin. In: Parker, J.R. (ed.) *Petroleum Geology of Northwest Europe: Proceedings of the 4th Conference*. Geological Society, London, pp. 975–83. [2]

Turner, J.S. (1949) The deeper structure of central and northern England. *Proc. Yorks. Geol. Soc.* **27**, 280–97. [3]

Turner. P. (1980) *Continental Red Beds*. Developments in Sedimentology No. 29, Elsevier, Amsterdam, 562, pp. [5, 7]

Turner, P., Jones, M., Prosser, D.J., Williams, G.D. and Searl, A. (1993) Structural and sedimentological controls on diagenesis in the Ravenspurn North Gas reservoir, UK southern North Sea. In: Parker, J.R. (ed.) *Petroleum Geology of Northwest Europe: Proceedings of the 4th Conference*. Geological Society, London, pp. 771–85. [1, 5]

Turner, P.J. (1993) Clyde: reapprasial of a producing field. In: Parker, J.R. (ed.) *Petroleum Geology of Northwest Europe: Proceedings of the 4th Conference*. Geological Society, London, pp. 1503–12. [1, 12]

Tyson, R.V. (1995) Sequence stratigraphical interpretation of organic facies variations in marine siliciclastic systems: general principles and application to the onshore Kimmeridge Clay Formation, UK. In: Hesselbo, S.P. and Parkinson, D.N. (eds) *Sequence Stratigraphy in British Geology*. Special Publication No. 103, Geological Society, London, pp. 75–96. [8]

Tyson, R.V. and Funnel, B.M. (1987) European Cretaceous shorelines, stage by stage. *Palaeogeog. Palaeoclim. Palaeoecol.* **59**, 69–91. [9]

Tyson, R.V., Wilson, R.C.L. and Downie, C. (1979) A stratified water column environment model for the Type Kimmeridge Clay. *Nature* **277**, 377–80. [8, 11]

Underhill, J.R. (1991a) Implications of Mesozoic–Recent basin development in the western Inner Moray Firth, UK. *Mar. Petr. Geol.* **8**, 359–69. [2, 8]

Underhill, J.R. (1991b) Controls on Late Jurassic seismic sequences, Inner Moray Firth, UK North Sea: a critical test of a key segment of Exxon's original global cycle chart. *Basin Res.* **3**, 79–98. [2, 8]

Underhill, J.R. (1994) Discussion on the palaeoecology and sedimentology across a Jurassic fault scrap, NE Scotland. *J. Geol. Soc.* **151**, 729–31. [8]

Underhill, J.R. and Brodie, J.A. (1993) Structural geology of the Easter Ross Peninsula, Scotland: implications for on the Great Glen Fault Zone. *J. Geol. Soc.* **150**, 515–27. [2]

Underhill, J.R. and Partington, M.A. (1993) Jurassic thermal doming and deflation in the North Sea: implications of the sequence stratigraphic evidence. In: Parker, J.R. (ed.) *Petroleum Geology of Northwest Europe: Proceedings of the 4th Conference*. Geological Society, London, pp. 337–46. [2, 8, 9, 11, 12]

Underhill, J.R. and Partington, M.A. (1994) Use of maximum flooding surfaces in determining a regional control on the Intra-Aalenian Mid Cimmerian sequence boundary: implications of North Sea basin development and Exxon's Sea-Level Chart. In: Posamentier, H.W. and Wiemer, P.J. (eds) *Recent Advances in Siliciclastic Sequence Stratigraphy*. Memoir No. 58, AAPG, Tulsa, OK., pp. 449–84. [2, 8]

Underhill, J.R. and Paterson, S. (in press) Genesis of tectonic inversion structures: seismic evidence for the development of key structures along the Purbeck–Isle of Wight monocline. *J. Geol. Soc.* [2]

Underhill, J.R. and Stoneley, R. (1998) Introduction to the development, evolution and petroleum geology of the Wessex Basin. In: Underhill, J.R. (ed.) *Development, Evolution and Petroleum*

Geology of the Wessex Basin. Special Publication 133. Geological Society, London, pp. 1–18. [2]

Underhill, J.R., Gayer, R.A., Woodock, N.H., Donnelly, R., Jolley, E.J. and Stimpson, I.G. (1988) The Dent Fault System, northern England—reinterpreted as a major oblique-slip fault zone. *Journal Geol. Soc. London.* **145**(2), 303–16. [2]

Underhill, J.R. Sawyer, M.J., Hodgson, P., Shallcross, M.D. and Gawthorpe, R.L. (1997) Implications of fault scarp degradation for Brent Group prospectivity, Ninian Field, Northern North Sea. *AAPG Bull.* **81**, 295–311. [2, 8]

Underwood, C.J., Crowley, S.F., Marshall, J.D. and Brenchley, P.J. (1997) High-resolution carbon isotope stratigraphy of the basal Silurian Stratotype (Dob's Linn, Scotland) and its global correlation. *J. Geol. Soc.* **154**(4), 709–18. [11]

Ungerer, P. (1993) Modelling of petroleum generation and migration. In: Bordenave, M.L. (ed.) *Applied Petroleum Geochemistry.* Editions Technip, Paris, pp. 395–442. [11]

Ungerer, P. and Pelet, R. (1987) Extrapolation of the kinetics of oil and gas formation from laboratory experiments to sedimentary basins. *Nature* **327**, 52–4. [11]

Ungerer, P., Behar, E. and Discamps, D. (1983) Tentative calculation of the overall volume expansion of organic matter during hydrocarbon genesis from geochemistry data. Implications for primary migration. In: Bjorøy, M., Albrecht, P., Cornford, C., de Groot, K., Eglinton, G., Galimov, E., *et al.* (eds) *Advances in Organic Geochemistry, 1981.* John Wiley, Chichester, pp. 129–35. [11]

Ungerer, P., Bessis, F., Chenet, P.Y., Durand, B., Nogaret, E., Chiarelli, A. *et al.* (1984) Geological and geochemical models in oil exploration: principles and practical examples. In: Demaison, G. and Murris, R.J. (eds) *Petroleum Geochemistry and Basin Evaluation.* Memoir No. 35. AAPG, Tulsa, OK., 53–77. [11]

Ungerer, P., Behar, F., Villalba, M., Heum, O.R. and Audibert, A. (1988) Kinetic modelling of oil cracking. *Org. Geochem.* **13**(4–6), 857–68. [11]

Ungerer, P., Burrus, J., Doligez, B., Chenet, P.Y. and Bessis, F. (1990) Basin evaluation by integrated two-dimensional modeling of heat transfer, fluid flow, hydrocarbon generation, and migration. *AAPG Bull.* **74**(3), 309–35. [11]

Vail, P.R. and Todd, R.G. (1981) Northern North Sea Jurassic unconformities, chronostratigraphy and sea-level changes from seismic stratigraphy. In: Illing, L.V. and Hobson, G.P. (eds) *Petroleum Geology of the Continental Shelf of North-West Europe.* Heyden, London, pp. 216–35. [8, 11]

Vail, P.R., Hardenbol, J. and Todd, R.G. (1984) Jurassic unconformities, chronostratigraphy, and sea-level changes from seismic stratigraphy and biostratigraphy. In: Schlee, J.S. (ed.) *International Unconformities and Hydrocarbon Accumulation.* Memoir No. 36, AAPG, pp. 129–44. [8, 9]

van Adrichem Boogaert, H.A. and Burgers, W.K.J. (1983) The development of the Zechstein in the Netherlands. *Geol. Mijnbouw* **62**, 83–92. [6]

van Adrichem Boogaert, H.A. and Kouwe, W.F.P. (1993) *Stratigraphic Nomenclature of the Netherlands: Revision/Updated by RGD and NOGEPA.* Mededelingen Rijks Geol. Dienst Vol. 50. [6, 9]

van Alstine, D.R. and Butterworth, J.E. (1993) Palaeomagnetic orientation of fractues and bedding in Rotliegende and Zechstein cores from the Southern Permian Basin, North Sea (abstract). *AAPG Bull.* **77**(9) 1676. [6]

van Buchem, F.S.P., Melnyk, D.H. and McCave, I.N. (1992) Chemical cyclicity and correlation of Lower Lias mudstones using gamma ray logs. Yorkshire, UK. *J. Geol. Soc.* **149**, 991–1002. [11]

van Buchem, F.S.P., de Boer, P.L., McCave, I.N. and Herbin, J.-P (1995) The organic carbon distribution in Mesozoic marine sediments and the influence of orbital climatic cycles: England and western North Alantic. In: *Paleogeography, Paleoclimate and Source Rock.* Studies in Geology No. 40, AAPG, Tulsa, OK., pp. 303–36. [11]

van den Bark, E. and Thomas, O.D. (1981) Ekofisk: first of the giant oil fields in western Europe. *AAPG Bull.* **65**, 2341–63. [1, 9, 11]

van den Bosch, W.J. (1983) The Harlingen Field, the only gas field in the Upper Cretaceous Chalk of The Netherlands. *Geol. Mijnbouw* **62**, 145–56. [9, 11]

Vandenbroucke, M. (1993) Migration of hydrocarbons: 1993 modelling of petroleum generation and migration. In: Bordenave, M.L. (ed.) *Applied Petroleum Geochemistry.* Editions Technip, Paris, pp. 123–48. [11]

van der Baan, D. (1990) Zechstein reservoirs in The Netherlands. In: Brooks, J. (ed.) *Classic Petroleum Provinces.* Special Publication No. 50, Geological Society, London, pp. 379–98. [6, 12]

van der Helm, A.A., Gray, D.I. Cook, M.A. and Schulte, W.M. (1991) Fulmar: the development of a large North Sea field. In: Buller, A.T. *et al.* (eds) *North Sea Oil and Gas Reservoirs.* Norwegian Institute of Technology, Graham and Trotman, London, pp. 25–45. [12]

van der Laan, G. (1968) Physical properties of the reservoir and volume of gas initially in place. *Verhandelingen Koningklijk NL geol. mijnboukundig Genootschap* **25**, 25–33. [5]

van der Pal, R., Bacon, M. and Pronk, D. (1996) 3D walkaway VSP, enhancing seismic resolution for development optimization of the Brent Field. *First Break* **14**, 463–9. [1, 8]

van der Sande, J.M.M., Reijers, T.J.A. and Casson, N. (1996) Multidisciplinary exploration strategy in the northeast Netherlands Zechstein 2 Carbonate play, guided by 3D seismic. In: Rondeel, H.E., Batjes, D.A.J. and Nieuwenhuis, W.H. (eds) *Geology of Gas and Oil under the Netherlands.* Kluwer Academic Publishers, Dordrecht, pp. 125–42. [6]

van der Voo, R. and Scotese, C. (1981) Paleomagnetic evidence for a large ~2000 km sinistral offset along the Great Glen fault during Carboniferous time. *Geology* **9**, 583–9. [2]

van der Zwan, C.J. (1981) Palynology, phytogeography and climate of the Lower Carboniferous. *Palaeogeog. Palaeoclimatol. Palaeoecol.* **33**, 279–310. [4]

van der Zwan, C.J. and Speak, P. (1992) Lower to Middle Triassic sequence stratigraphy and climatology of the Netherlands, a model. *Palaeogeog., Palaeoclim., Palaeoecol.* **91**, 277–90.

van der Zwan, C.J., Boulter, M.C. and Hubbard, R.N. (1985) Climatic change in the Lower Carboniferous in Euramerica, based on multivariate statistical analysis of palynological data. *Palaeogeog. Palaeoclimatol. Palaeoecol.* **52**, 1–20. [4]

van der Zwan, C.J, van der Laar, J.G.M., Pagnier, H.J.M. and van Ameron, H.W.J. (1993) Palynological, ecological and climatological synthesis of the Upper Carboniferous of the well De Lutte-6 East Netherlands. In: *Comptes Rendues 12ème Congrès International Strat. Geol. Carb., Buenos Aires*, Vol. 1, pp. 167–86. [4]

van Graas, G.W. (1990) Biomarker maturity parameters for high maturities: calibration of the working range up to the oil/condensate threshold. *Org. Geochem.* **16**(1–3), 1025–32. [11]

van Hoorn, B. (1987a) Structural evolution, timing and tectonic style of the Sole Pit inversion. *Tectonopysics* **137**, pp. 239–84. [5, 11]

van Hoorn, B. (1987b) The South Celtic Sea/Bristol Channel Basin: origin, deformation and inversion history. *Tectonophysics* **137**, 309–34.

Van Lith, J.G.J. (1983) Gas fields of the Bergen Concession, The Netherlands. *Geol. Mijnbouw* **62**, 63–74. [1, 5, 6]

van Panhuys-Sigler, M., Baumann, A. and Holland, T.C. (1991) The Tern Field, Block 210/25a, UK North Sea. In: Abbotts, I.L. (ed.) *UK North Sea United Kingdom Oil and Gas Fields: 25 Years Commemorative Volume.* Memoir No. 14, Geological Society, London, pp. 191–7. [1, 7]

van Veen, F.R. (1975) Geology of the Leman gas field. In: Woodland, A.W. (ed.) *Petroleum and the Continental Shelf of*

Northwest Europe, Vol. I, *Geology*. Elsevier Applied Science Publishers, pp. 477–87. [1, 5]

van Wagoner, J.C., Posamentier, H.W., Mitchum, R.M., Vail, P.R., Sarg, J.F., Loutit, T.S., *et al.* (1988) An overview of the fundamentals of sequence stratigraphy, and key definitions. In: Wilgus, C.K., Hastings, B.S., Posamentier, H.W., van Wagoner, J.C., Ross, C.A. and Kendall, C.G.St.C. (eds) *Sea-Level Changes: an Integrated Approach*. Special Publication No. 42, SEPM, Tulsa, OK., pp. 39–45. [4]

van Wagoner, J.C., Mitchum, R.M., Campion, K.M. and Rahmanian, V.D. (1990) *Siliciclastic Sequence Stratigraphy in Well Logs, Cores and Outcrops: Concepts for High Resolution Correlation of Time and Facies*. Methods in Exploration Series No. 7, AAPG, Tulsa, OK. [8]

van Wessem, E.J. and Gan, T.L. (1991) The Ninian Field, Blocks 3/3 and 3/8, UK North Sea. In: Abbotts, I.L. (ed.) *UK North Sea United Kingdom Oil and Gas Fields: 25 Years Commemorative Volume*. Memoir No. 14, Geological Society, London, pp. 175–82. [1]

van Wijhe, D.H. (1981) The Zechstein 2 carbonate exploration in the eastern Netherlands. In: *Proceedings of an International Symposium on Central European Permian, Jablonna, Poland, 1978*. Geological Institute, Warsaw, pp. 574–86. [6]

van Wijhe, D.H. (1987a) Structural evolution of inverted basins in the Dutch offshore. *Tectonophysics* **137**, 171–219. [9]

van Wijhe, D.H. (1987b) The structural evolution of the Broad Fourteens Basin. In: Brooks, J. and Glennie, K.W. (eds) *Petroleum Geology of North West Europe*. Graham and Trotman, London, pp. 315–23. [2, 9]

van Wijhe, D.H. and Bless, M.J. (1974) The Westphalian of the Netherlands with special reference to miospore assemblages. *Geol. Mijnbouw* **53**, 295–327. [4]

van Wijhe, D.H., Lutz, M. and Kaasschieter, J.P.H. (1980) The Rotliegend in the Netherlands and its gas accumulations. *Geol. Mijnbouw* **59**, 3–24. [5, 7, 11, 12]

Vasseur, G. and Velde, B. (1993) A kinetic interpretation of the smectite-to-illite transformation. In: Doré, A.G., Augustson, J.H., Hermanrud, C., Stewart, D.J. and Sylta, O. (eds) *Basin Modelling: Advances and Applications*. Special Publication No. 3, NPF, Elsevier, Amsterdam, pp. 173–84. [11]

Vaughan, D.J., Sweeney, M., Diedel, G.F.R. and Haranczyk, C. (1989) The Kupferschiefer: an overview with an appraisal of the different types of mineralisation. *Econ. Geol.* **84**, 1003–27.

Veenhof, E. (1996) Geological aspects of the Annerveen Field, the Netherlands. In: Rondeel, H.E., Batjes, D.A.J. and Nieuwenhius, W.H. (eds) *Geology of Oil and Gas under the Netherlands*. Royal Geological and Mining Society, the Netherlands, Kluwer Academic Publishers, Dordrecht, pp. 79–92. [1, 5, 11]

Vejbæk, O.V. (1986) *Seismic Stratigraphy and Tectonic Evolution of the Lower Cretaceous in the Danish Central Trough*. Series A, No. 11, Danmarks Geol. Unders., Denmark. 45 pp. [9]

Vejbæk, O.V. and Andersen, C. (1987) Cretaceous–early Tertiary inversion tectonism in the Danish Central Trough. *Tectonophysics* **137**, 221–38.

Veld, H. and Fermont, W.J.J. (1990) The effect of marine transgression on vitrinite reflectance values. *Mededelingen rijks geologische dienst*. **45**, 151–70. [11]

Veld, H., Fermont, W.J.J., Kerp, H. and Visscher, H. (1996) Geothermal history of the Carboniferous in South Limberg, the Netherlands. In: Rondeel, H.E., Batjes, D.A.J. and Nieuwenhuijs, W.H. (eds) *Geology of Gas and Oil under the Netherlands*. KNMG, Kluwer, Dordrecht. pp. 31–44. [11]

Velde, B. and Espitalié, J. (1989) Comparison of kerogen maturation and illite/smectite composition in diagenesis. *J. Petroleum Geology* **12**(1), 103–10. [11]

Velde, B. and Vasseur, G. (1992) Estimation of the diagenetic smectite-to-illite transformation in time–temperature space. *Am. Mineralogist* **77**, 967–76. [11]

Veldkamp, J.J., Gaillard, M.G., Jonkers, H.A. and Levell, B.K. (1996) A Kimmeridgian time-slice through the Humber Group of the central North Sea: a test of sequence stratigraphic models. In: Hurst, A., Johnson, H.D., Burley, S.D., Canham, A.C. and Mackertich, D.S. (eds) *Geology of the Humber Group: Central Graben and Moray Firth, UKCS*. Special Publication No. 114, Geological Society, London, pp. 1–28. [8]

Vendeville, B.C. and Jackson, M.P.A. (1992a) The rise of diapirs during thin-skinned extension. *Mar. Petr. Geol.* **9**, 331–53. [6]

Vendeville, B.C. and Jackson, M.P.A. (1992b) The fall of diapirs during thin-skinned extension. *Mar. Petr. Geol.* **9**, 354–71. [6]

Verdier, J.P. (1996) The Rotliegend sedimentation history of the southern North Sea and adjacent countries. In: Rondeel, H.E., Batjes, D.A.J. and Niewenhuis, W.H. (eds) *Geology of Oil and Gas under the Netherlands*. Royal Geological and Mining Society, the Netherlands, Kluwer Academic Publishers, Dordrecht, pp. 45–56. [5, 12]

Vially, R. (ed) (1992) Bacterial Gas. In: *Proc. Milan Conference, Sept. 1989*. IFP, [11]

Vigneresse, J.L. (1993) Reflectance of vitrinite-like macerals as a thermal maturity index for Cambrian–Ordovician Alum Shale, Southern Scandinavia. *AAPG Bull.* **77**(3), 509–12. [11]

Vik, E. and Hermanrud, C. (1993) Transient thermal effects of rapid subsidence in the Haltenbanken area: a recent Norwegian Shelf heating event—fact or fantasy? In: Doré, A.G., Augustson, J.H., Hermanrud, C., Stewart, D.J. and Sylta, O. (eds) *Basin Modelling: Advances and Applications*. Special Publication No. 3, NPF, Elsevier, Amsterdam, pp. 107–17. [11]

Vik, E., Heum, O.R. and Amaliksen, K.G. (1991) Leakage from deep reservoirs: possible mechanisms and relationship to shallow gas in the Haltenbanken area, mid Norwegian Shelf. In: England, W.A. and Fleet, A.J. (eds) *Petroleum Migration*. Special Publication No. 59, Geological Society, London, p. 273. [11]

Vining, B.A., Ioannides, N.S. and Pickering, K.T. (1993) Stratigraphic relationships of some Tertiary lowstand depositional systems in the Central North Sea. In: Parker, J.R. (ed.) *Petroleum Geology of Northwest Europe: Proccedings of the 4th Conference*. Geological Society, London, pp. 17–30. [1, 10, 12]

Visser, W.A. and Sung, G.C.L. (1958) Northeastern Netherlands basin. In: Weeks, L.G. (ed.) *Habitat of Oil*. Special Publication, AAPG, Tulsa, OK., pp. 1067–90. [1]

Vlierboom, F.W., Collini, B. and Zumberge, J.E. (1986) The occurrence of petroleum in sedimentary rocks of the meteor Impact Crater at Lake Siljan, Sweden. In: Leythaeuser, D. and Rullkotter, J. (eds) *Advances in Organic Geochemistry, 1985*. Pergamon, Oxford, pp. 153–62. [11]

Voigt, E. (1981) Upper Cretaceous bryozoan seagrass association in the Maastrichtian of the Netherlands. In: Larwood, G.P. and Nielsen, C.C. (eds) *Recent and Fossil Bryozoa*. Olsen and Olsen, pp. 281–98. [9]

Volkman, J.K., Alexander, R., Kagi, R.I. and Woodhouse, G.W. (1983) Demethylated hopanes in crude oils and their application in petroleum geochemistry. *Geochim. Cosmochim. Acta* **47**, 785–94. [11]

Vollsett, J. and Doré, A.G. (1984) *A revised Triassic and Jurassic Lithostratigraphic Nomenclature of the Norwegian North Sea*. Bulletin No. 59, NPD, Stavanger, Norway, 53 pp. [7, 8]

Waddams, P. and Clark, N.M. (1991) The Petronella Field, Block 14/20b, UK North Sea. In: Abbotts, I.L. (ed.) *UK North Sea United Kingdom Oil and Gas Fields: 25 Years Commemorative Volume*. Memoir No. 14, Geological Society, London, p. 353. [1, 12]

Wagner, R.H. (1983) A lower Rotliegend flora from Ayrshire. *Scott. J. Geol.* **19**, 135–55. [4]

Wagner, T., Peryt, T.M. and Piatkowski, T.S. (1981) The evolution of the Zechstein sedimentary basin in Poland. In: *Procedings of an International Symposium on Central European Permian, Jablonna, Poland, 1978*. Geological Institute Warsaw, pp. 69–83. [6]

Wakefield, L.L., Droste, H., Giles, M.R. and Janssen, R. (1993)

Late Jurassic plays along the western margin of the Central Graben. In: Parker, J.R. (eds) *Petroleum Geology of Northwest Europe: Proceedings of the 4th Conference*. Geological Society, London, pp. 459–68. [8, 12]

Walderhaug, O. and Fjeldskaar, W. (1993) History of hydrocarbon emplacement in the Oseberg Field determined by fluid inclusion microthermometry and temperature modelling. In: Doré, A.G., Augustson, J.H., Hermanrud, C., Stewart, D.J. and Sylta, O. (eds) *Basin Modelling: Advances and Applications*. Special Publication No. 3, NPF, Elsevier, Amsterdam, pp. 485–98. [11]

Walker, I.M. and Cooper, W.G. (1987) The structural and stratigraphic evolution of the north-east margin of the Sole Pit Basin. In: Brooks, J. and Glennie, K.W. (eds) *Petroleum Geology of North West Europe*. Graham and Trotman, London, pp. 263–75. [6]

Walker, T.R. (1967) Formation of redbeds in modern and ancient deserts. *Bull. Geol. Soc. Am.* **78**, 353–68. [5]

Walker, T.R. (1976) Diagenetic origin of redbeds. In: Falke, H. (ed.) *The Continental Permian in Central, Western and Southern Europe*. Reidel, Dordrecht, pp. 240–82. [5]

Walker, T.R., Waugh, B. and Crone, A.J. (1978) Diagenesis in first-cycle alluvium of Cenozoic age, south-western United States and north-western Mexico. *AAPG Bull.* **89**, 19–32. [7]

Walmsley, P.J. (1975) The Forties field. In: Woodland, A.W. (ed.) *Petroleum and the Continental Shelf of Northwest Europe*, Vol. I, *Geology*. Applied Science Publishers, Barking, pp. 477–87. [1]

Walmsley, P.J. (1983) The role of the Department of Energy in petroleum exploration of the United Kingdom. In: Brooks, J. (ed.) *Petroleum Geochemistry and Exploration of Europe*. Special Publication No. 10, Geological Society, London, pp. 3–10. [1]

Walzebuck, J. (1993) Ranking of geological factors controlling the well productivity in Rotliegende Gas Fields of the Dutch Offshore Area. *AAPG Bull.* **77**(9), 1675. [5]

Waples, D.W. (1980) Time and temperature in petroleum formation: application of Lopatin's method to petroleum exploration. *AAPG Bull.* **64**, 916–26. [11]

Waples, D.W. (1984) Thermal models for oil generation. In: Brooks, J. and Welte, D. (eds) *Advances in Petroleum Geochemistry, Vol. 1*. Academic Press, London, pp. 7–68.]11]

Warren, E.A. and Smalley, P.C. (1993) The chemical composition of North Sea formation waters: a review of their heterogeneity and potential applications. In: Parker, J. (ed.) *Petroleum Geology of Northwest Europe: Proc. 4th Conf.* Geological Society, London, pp. 1347–52. [11]

Warren, E.A. and Smalley, P.C. (eds) (1994) *North Sea formation Waters Atlas*. Memoir No. 15. Geological Society, London, 208 pp. [11]

Warren, E.A., Smalley, P.C. and Howarth, R.J. (1994) Compositional variations in North Sea formation waters. In: Warren, E.A. and Smalley, P.C. (eds) *North Sea Formation Waters Atlas*. Memoir No. 15. Geological Society, London, pp. 119–206. [11]

Warrender, J. (1991) The Murchison Field, Blocks 211/19a, UK North Sea. In: Abbotts, I.L. (ed.) *UK North Sea United Kingdom Oil and Gas Fields: 25 Years Commemorative Volume*. Memoir No. 14, Geological Society, London, pp. 165–73. [1]

Watson, S. (1976) Sedimentary geochemistry of the Moffat shales; a carbonaceous sequence in the southern Uplands of Scotland. Unpublished PhD, University of St. Andrews, Scotland. [11]

Watts, A.B. and Thorne, J. (1984) Tectonics, global changes in sea level and their relationship to stratigraphical sequences at the US Atlantic Continental Margin. *Marine and Petroleum Geology* **1**(4), 319–39. [11]

Watts, G.F.T., Jizba, D., Gawith, D.E. and Gutteridge, P. (1996) Reservoir monitoring of the Magnus Field through 4D time-lapse seismic analysis. *Petroleum Geoscience* **2**, 361–72. [1]

Watts, N.L. (1983a) Microfractures in chalks of Albuskjell Field, Norwegian Sector, North Sea: possible origin and distribution. *AAPG Bull.* **67**(9), 201–34. [9]

Watts, N.L. (1983b) Fractures in North Sea Chalks: geological modelling and a case example from the Albuskjell Field. In: Hancock, J.M., Hamlin, R.J.O., Crosby, A., Balson, P.S., Jones, S.M., Chadwick, R.A. *et al.* (eds) *Chalk in the North Sea*. Course Notes No. 20, JAPEC UK, London, pp. E1–E17. [9]

Watts, N.L. (1987) Theoretical aspects of cap-rock and fault seals for single- and two-phase hydrocarbon columns. *Mar. Petr. Geol.* **4**, 274–307. [12]

Watts, N.L., Lapre, J.F., van Schijndel-Goester, F.S. and Ford, A. (1980) Upper Cretaceous and Lower Tertiary chalks of the Albuskjell area, North Sea: deposition in a slope and base-of-slope environment. *Geology* **8**, 217–21. [1, 9, 12]

Webb, B. (1983) Imbricate structure of the Ettrick area, Southern Uplands. *Scott. J. Geol.* **193**, 387–400. [2]

Weber, K.J. (1987) Computation of initial well productivities in aeolian sandstone on the basis of a geological model, Leman gas field, UK. In: Tillman, R.W. and Weber, K.J. (eds) *Reservoir Sedimentology*, Special Publication No. 40, Society of Economic Paleontologists and Mineralogists, Tulsa, OK., pp. 333–54. [5]

Weedon, G.P. and Jenkyns, H. (1990) Regular and irregular climatic cycles and the Belemnite marls Pliensbachian, Lower Jurassic, Wessex Basin. *J. Geol. Soc.* **147**, 915–18. [11]

Wehner, H., Gerling, P., Muller, P. and Bleschert, K.-H. (1993) Facies-controlled compositions of petroleum and natural gas in Zechstein carbonates in Eastern Germany: abstracts, AAPG International Conference, The Hague, the Netherlands. *AAPG Bull.* **77**, 1676. [6]

Welte, D.H. and Yalcin, M.N. (1988) Basin modelling--a new comprehensive method in petroleum geology. In: Mattavelli, L. and Novelli, L. (eds) *Advances in Org. Geochem., 1987 (Proc. 13th Int'l Mtg on Org. Geochem. Venice,), Part 1*. Pergamon, Oxford, pp. 141–52. [11]

Welte, D.H. and Yukler, M.A. (1981) Petroleum origin and accumulation in basin evolution—a quantitative model. AAPG Bull. **65**(8), 1387–96. [11]

Welte, D.H., Schaefer, R.G., Stoessingeer, W. and Radke, M. (1984) Gas generation and migration in the Deep Basin of western Canada. In: Masters, J.A. (ed.) *Elmworth, Case Study of a Deep Basin Gas Field*. Memoir, No. 38, AAPG, Tulsa, OK. pp. 35–48. [11]

Welte, D.H., Horsfield, B. and Baker, D.R. (eds) (1997) *Petroleum and Basin Evolution*. Springer, Berlin. 524 pp. [11]

Wensrich, M.D., Eastwood, K.M., van Panhuys, C.D. and Smart, J.M. (1991) The Eider Field, Blocks 211/16a and 211/21a, UK North Sea. In: Abbotts, I.L. (ed.) *UK North Sea United Kingdom Oil and Gas Fields: 25 Years Commemorative Volume*. Memoir No. 14, Geological Society, London, pp. 103–9. [1]

Werner, A., Behar, F., de Hemptinne, J.C. and Behar, E. (1996) Thermodynamic properties of petroleum fluids during expulsion and migration from source rocks. *Org. Geochem* **24**(10/11), 1079–95. [11]

Werngren, O.C. (1991) The Thames, Yare and Bure Fields, Blocks 47/3d, 47/8b, UK North Sea. In: Abbotts, I.L. (ed.) *UK North Sea United Kingdom Oil and Gas Fields: 25 Years Commemorative Volume*. Memoir No. 14, Geological Society, London, pp. 491–6. [1, 11]

Weston, P.J., Davison, I. and Insley, M.W. (1993) Physical modelling of North Sea salt diapirism. In: Parker, J.R. (ed.) *Petroleum Geology of Northwest Europe: Proceedings of the 4th Conference*. Geological Society, London, pp. 559–68. [6]

Westre, S. (1984) The Askeladden gas find—Troms 1. In: Spencer, A.M., Johnsen, S.O., Mørk, A., Nysaether, E., Songstad, P. and Spinnangr, A. (eds) *Petroleum Geology of the North European Margin*. Graham and Trotman, London, pp. 33–40. [11]

Wheatley, T.J., Biggins, D., Buckingham, J. and Holloway, N.H. (1987) The geology and exploration of the Transitioinal Shelf, an area to the west of the Viking Graben. In: Brooks, J. and Glennie, K.W. (eds) *Petroleum Geology of North West Europe*. Graham and Trotman, London, pp. 979–89. [11, 12]

Whelan, J.K., Hunt, J.M., Jaspar, J. and Huc, A. (1984) Migration of C-1–C-8 hydrocarbons in marine sediments. In: Schenck, P.A., de Leeuw, J.W. and Lÿnbach, G.W.M. (eds) *Advances in Organic Geochemistry, 1983. Org. Geochem.* **6**, 683–94. [11]

White, D.A. (1988) Oil and gas play maps in exploration and assessment. *AAPG Bull.* **72**, 944–9. [12]

White, N. (1987) Constraints on the measurement of extension in the brittle upper crust. *Norsk Geol. Tidsskr.* **67**, 269–79.

White, N. (1990) Does the uniform stretching model work in the North Sea? In: Blundell, D.J. and Gibbs, A.D. (eds) *Tectonic Evolution of the North Sea Rifts.* Oxford Science Publications, Oxford, pp. 217–40. [7]

White, N. and Latin, D. (1993) Subsidence analyses from the North Sea 'Triple-junction'. *J. Geol. Soc.* **150**(3), 473–89. [7, 11]

White, R.S. (1988) A hot-spot model for early Tertiary volcanism in the N Atlantic. In: Morton, A.C. and Parson, L.M. (eds) *Early Tertiary Volcanism and the Opening of the NE Atlantic.* Special Publication No. 39, Geological Society, London, pp. 3–13. [10]

White, R.S. (1989) Initiation of the Iceland Plume and the opening of the North Atlantic Margins. In: Tankard, A.J. and Balkwill, H.R. (eds) *Extensional Tectonics and Stratigraphy of the North Atlantic Margins.* Memoir No. 46, AAPG, pp. 149–54. [2]

Whitehead, M. and Pinnock, S.J. (1991) The Highlander Field, Block 14/20b, UK North Sea. In: Abbotts, I.L. (ed.) *UK North Sea United Kingdom Oil and Gas Fields: 25 Years Commemorative Volume.* Memoir No. 14, Geological Society, London, p. 323. [1, 12]

Whiteman, A., Naylor, D., Pegrum, R. and Rees, G. (1975a) North Sea troughs and plate tectonics. *Tectonophysics* **26**, 39–54.

Whiteman, A.J., Rees, G., Naylor, D. and Pegrum, R.M. (1975b) North Sea troughs and plate tectonics. In: Whiteman, A.J., Roberts, D. and Jellevolle, M.A. (eds) *Petroleum Geology and Geology of the North Sea and NE Atlantic Continental Margin.* Norwegian Geol. Unders. 316, pp. 317–62.

Whithan, F. (1992) Itinerary XIII: South Landing to Sewerby. In: Rawson, P.F. and Wright, J.K. (eds) *Guide No. 34. Yorkshire Coast.* Geologists' Association, pp. 103–9. [9]

Whitham, F. (1993) The stratigraphy of the Upper Cretaceous Flamborough Chalk Formation north of the Humber, north-east England. *Proc. Yorks. Geol. Soc.* **49**, 235–59. [9]

Whitham, F. (1994) Jurassic and Cretaceous rocks of the Market Weighton area. In: Scrutton, C. (ed.) *Yorkshire Rocks and Landscape: a Field Guide.* Yorkshire Geological Society, Ellenbank Press, Maryport, Cumbria, pp. 142–9. [9]

Whittaker, A., Holliday, D.W. and Penn, I.E. (1985) *Geophysical Logs in British Stratigraphy.* Special Report No. 18, Geological Society, London, 74 pp. [4]

Whittaker, R.C., Hamann, N.E. and Pulvertaft, T.C.R. (1997) A new frontier province offshore northwest Greenland: structure, basin development, and petroleum potential of the Melville Bay area. *AAPG Bull.* **81**(6), 978–98. [11]

Whyatt, M., Bowen, J.M. and Rhodes, D.N. (1991) Nelson successful application of a development model in North Sea exploration. *First Break* **9**, 265–80. [10]

Whyatt, M., Bowen, J.M. and Rhodes, D.N. (1992) The Nelson Field: a successful application of a development geoseismic model in North Sea exploration. In: Hardman, R.F.P. (ed.) *Exploration Britain: Geological Insights for the Next Decade.* Special Publication No. 67, Geological Society, London, pp. 283–305. [1, 12]

Wiesner, M.G., Haake, B. and Wirth, H. (1990) Organic facies of surface sediments in the North Sea. *Org. Geochem.* **15**(4), 419–32. [11]

Wignall, P.B. (1989) Sedimentary dynamics of Kimmeridge Clay: tempests and earthquakes. *J. Geol. Soc. Lond.* **146**(2), 273–82. [11]

Wignall, P.B. and Maynard, J.R. (1993) The sequence stratigraphy of transgressive black shales. In: Katz, B.J. and Pratt, L.M. (eds) *Source Rocks in a Sequence Stratigraphic Framework.* Studies in Geology No. 37, AAPG, Tulsa, OK., pp. 35–48. [11]

Wignall, P.B. and Maynard, J.R. (1996) High resolution sequence stratigraphy in the early Marsdenian (Namurian, Carboniferous) of the central Pennines and adjacent areas. *Proc. Yorks. Geol. Soc.*, **51**, 127–40. [4]

Wignall, P.B. and Ruffell, A.H. (1990) The influence of a sudden climatic change on marine deposition in the Kimmeridgian of northwest Europe. *J. Geol. Soc.* **147**, 365–71. [11]

Wilgus, C.H., Hastings, B.S., Posamentier, H.W., van Wagoner, J.C., Ross, C.A. and Kendall, C.G.St.C. (eds) *Sea Level Changes: an Integrated Approach.* Special Publication No. 42, Society of Economic Palaeontologists and Mineralogists, Tulsa, OK.

Wilhelms, A. and Larter, S.R. (1994a) Origin of tar mats in petroleum reservoirs. Part I: introduction and case studies. *Mar. Petr. Geol.* **11**(4), 418–42. [11]

Wilhelms, A. and Larter, S.R. (1994b) Origin of tar mats in petroleum resevoirs. Part II: formation mechanisms for tar mats. *Mar. Petr. Geol.* **11**(4), 442–57. [11]

Wilhelms, A. and Larter, S.R. (1995) Overview of the geochemistry of some tar mats from the North Sea and USA: implications for tar mat origin. In: Cubitt, J.M. and England, W.A. (eds) *The Geochemistry of Reservoirs.* Special Publication No. 86, Geological Society, London, pp. 87–101. [11]

Wilhelms, A., Larter, S., Leythaeuser, D. and Dypvik, H. (1990) Recognition and quantification of the effects of primary migration in a Jurassic clastic source-rock from the Norwegian continental shelf. In: Durand, B. and Beher, F. (eds) *Advances in Organic Geochemistry. Org. Geochem.* **16**(1–3), 103–13. [11]

Williams, G.D., Powell, C.M. and Cooper, M.A. (1989) Geometry and kinematics of inversion tectonics. In: Cooper, M.A. and Williams, G.D. (eds) *Inversion Tectonics.* Special Publication 44, Geological Society, London, pp. 3–15. [1]

Williams, J.J., Conner, D.C. and Peterson, K.E. (1975) The Piper oil field, UK North Sea: a fault-block structure with Upper Jurassic beach-bar reservoir sands. In: Woodland, A.W. (ed.) *Petroleum and the Continental Shelf of Northwest Europe.* Vol. I, *Geology.* Applied Science Publishers, Barking, pp. 363–79. [1, 12]

Williams, P.F.V. (1986) Petroleum geochemistry of the Kimmeridge Clay of Southern and Eastern England. *Mar. Pet. Geol.* **3**, 258–81. [11]

Williams, P.F.V. and Douglas, A.G. (1980) A preliminary organic geochemical investigation of the Kimmeridge oil shales. In: Douglas, A.G. and Maxwell, J.R. (eds) *Advances in Organic Geochemistry, 1979.* Pergamon, Oxford, pp. 531–45. [11]

Williams, R.R. (1991) The Deveron Field, Block 211/18a, UK North Sea. In: Abbotts, I.L. (ed.) *UK North Sea United Kingdom Oil and Gas Fields: 25 Years Commemorative Volume.* Memoir No. 14, Geological Society, London, pp. 83–7. [1]

Williams, R.R. and Milne, A.D. (1991) The Thistle Field, Blocks 211/18a and 211/19, UK North Sea. In: Abbotts, I.L. (ed.) *UK North Sea United Kingdom Oil and Gas Fields: 25 Years Commemorative Volume.* Memoir No. 14, Geological Society, London, pp. 199–207. [1]

Wills, J.M. (1991) The Forties Field, Block 21/10, 22/6a, UK North Sea. In: Abbotts, I.L. (ed.) *UK North Sea United Kingdom Oil and Gas Fields: 25 Years Commemorative Volume.* Memoir No. 14, Geological Society, London, pp. 301–8. [1, 10, 11, 12]

Wills, J.M. and Peattie, D.K. (1990) The Forties Field and the evolution of a reservoir management strategy. In: Buller, A.T., Berg, E., Hjelmeland, O., Kleppe, J., Torsaeter, O. and Aasen, J.O. (eds) *North Sea Oil and Gas Reservoirs.* Norwegian Inst. Tech., Graham and Trotman, London. p. 1. [11]

Wills, L.J. (1951) *A Palaeogeographic Atlas of the British Isles and adjacent parts of Europe.* Blackie and Sons, London. [1]

Wilson, H.H. (1977) 'Frozen-in' hydrocarbon accumulations or diagenetic traps—exploration targets. *AAPG Bull.* **61**, 463–91. [12]

Wilson, M.A., Vassallo, A.M., Gizachew, D. and Lafargue. E. (1991) A high resolution solid state nuclear magnetic resonance study of some coaly source rocks from the Brent group North Sea. *Org. Geochem* **17**(1), 107–11. [11]

Wimbledon, W.A., Allen, P. and Fleet, A.J. (1996) Penecontemporaneous oil-seep in the wealden early Cretaceous at Mupe Bay, Dorset, UK. *Sediment. Geol.* **102**, 213–20. [12]

Winchester, J.A. (1988) Later Proterozoic environments and tectonic evolution in the northern Atlantic lands. In: Winchester, J.A. (ed.) *Later Proterozoic Stratigraphy of the Northern Atlantic Regions*. Blackie, Glasgow, London, pp. 253–70. [2]

Winn, R.D., Jr (1994) Shelf sand-sheet reservoir of the Lower Cretaceous Greensand, North Celtic Sea Basin, offshore Ireland. *AAPG Bull.* **78**, 1775–89. [12]

Winstanley, A.M. (1993) A review of the Triassic play in the Roer Valley Graben, southeast onshore Netherlands. In: Parker, J.R. (ed.) *Petroleum Geology of Northwest Europe: Proceedings of the 4th Conference*. Geological Society, London, pp. 595–608. [11]

Winter, D.A. and King, B. (1991) The West Sole Field, Block 48/6, UK North Sea. In: Abbotts, I.L. (ed.) *UK North Sea United Kingdom Oil and Gas Fields: 25 Years Commemorative Volume*. Memoir No. 14, Geological Society, London, p. 517. [5, 11]

Withjack, M.O., Olson, J. and Petersen, E. (1990) Experimental models of extensional forced folds. *AAPG Bull.* **74**, 1038–54. [2, 8]

Wolff, G.A. and Rukin, N. and Marshall, J.D. (1992) Geochemistry of an early diagenetic concretion from the Birchi Bed. (L. Lias, W. Dorset, U.K.). *Org. Geochem.* **19**(4–6), 431–45. [11]

Wood, C.J. and Smith, E.G. (1978) Lithostratigraphical classification of the Chalk in North Yorkshire, Humberside and Lincolnshire. *Proc. Yorks. Geol. Soc.* **42**, 263–88. [9]

Wood, D.A. (1988) Relationships between thermal maturity indices calculated using Arrhenius Equation and Lopatin method: implications for petroleum exploration. *AAPG Bull.* **72**(2), 115–35. [11]

Woodland, A.W. (ed.) (1975) *Petroleum and the Continental Shelf of Northwest Europe*, Vol. I. *Geology*. Applied Science Publishers, Barking, 501 pp. [1, 5, 11]

Woollam, R. and Riding, J.B. (1983) *Dinoflagellate Cyst Zonation of the English Jurassic*. Report No. 83/2, Institute of Geological Science. [8]

Wright, V.P (1990) Equatorial aridity and climatic oscillations during the early Carboniferous, southern Britain. *J. Geol. Soc.* **147**, 359–63. [4]

Wrigley, R., Philling, D. and Melvin, A. (1993) Rotliegende Group Event Stratigraphy, Quadrants 42 and 47, Southern North Sea. *AAPG Bull.* **77**(9), 1676–7. [5]

Yalçin, M.N., Littke, R. and Sachsenhofer, R.F. (1997) Thermal history of sedimentary basins. In: Welte, D.H., Horsfield, B. and Baker, D.R. (eds) *Petroleum and Basin Evolution*. Springer Verlag, Berlin. [11]

Yaliz, A. (1991) The Crawford Field, Block 9/28a, UK North Sea. In: Abbotts, I.L. (ed.) *UK North Sea United Kingdom Oil and Gas Fields: 25 Years Commemorative Volume*. Memoir No. 14, Geological Society, London, pp. 287–93. [1, 5, 7, 12]

Yang, C.S. and Baumfalk, Y.A. (1994) Milankovitch cyclicity in the Upper Rotliegend Group of The Netherlands offshore. In: de Boer, P.I. and Smith, D.G. (eds) *Orbital Forcing and Cyclic Sequences*. Special Publication 19, International Association of Sedimentologists, Blackwell Scientific Publications, Oxford. pp. 47–61. [5]

Yielding, G. (1990) Footwall uplift associated with Late Jurassic normal faulting in the northern North Sea. *J. Geol. Soc.* **147**, 219–22. [2, 8]

Yielding, G., Badley, M.E. and Roberts, A.M. (1992) The structural evolution of the Brent Province. In: Morton, A.C., Haszeldine, R.S., Giles, M.R. and Brown, S. (eds) *Geology of the Brent Group*. Special Publication 61, Geological Society, London, pp. 27–43. [8]

Yu, Z., Thomsen, R.O. and Lerche, I. (1995) Crystalline basement focussing of heat versus fluid flow/compaction effects: a case study of the 1-1 well in the Danish North Sea. *Petr. Geosci.* **1**, 31–6. [11]

Yukler, M.A. and Kokesh, F. (1984) A review of models used in petroleum resource estimation and organic geochemistry. In: Brooks, J. and Welte, D. (eds) *Advances in Petroleum Geochemistry Vol. 1*. Academic Press, London, pp. 69–114. [11]

Zdanaviciute, O. and Bojesen-Koefoed, J.A. (1997) Geochemistry of Lithuanian oils and source rocks: a preliminary assessment. *J. Petr. Geol.* **20**(4), 381–402. [1]

Zeck, H.P., Andriessen, P.A.M., Hansen, K., Jensen, P.K. and Rasmussen, B.L. (1988) Paleozoic paleo-cover of the southern part of the Fennoscandian Shield—fission track constraints. *Tectonophysics* **149**, 61–6. [7]

Ziegler, K., Turner, P. and Doines, S.R. (eds) (1997) *Petroleum Geology of the Southern North Sea: Future Potential*. Special Publication 123, Geological Society, London. [5]

Ziegler, M.A. (1989) North German facies patterns in relation to their substrate. *Geol. Rundschau* **78**, 105–27. [6]

Zeigler, P.A. (1975) Geologic evolution of North Sea and its tectonic framework. *AAPG Bull.* **59**, 1073–97. [7]

Ziegler, P.A. (1977) Geology and hydrocarbon provinces of the North Sea. *Geojournal* **1**, 7–32. [2, 12]

Ziegler, P.A. (1978) North Western Europe: tectonics and basin development. *Geol. Mijnbouw* **57**, 487–502. [5, 7]

Ziegler, P.A. (1980a) Northwestern Europe: Geology and Hydrocarbon Provinces. In: Miall, A.D. (ed.) *Facts and Principles of World Petroleum Occurrence*. Memoir No. 6, Canadian Society of Petroleum Geology, Calgary, pp. 653–706. [5]

Ziegler, P.A. (1980b) North Western Europe: subsidence patterns of Post Variscan basins. In: *Proceedings International Geological Congress, Paris*. pp. C3–C5. [5]

Ziegler, P.A. (1981) Evolution of sedimentary basins in north-west Europe. In: Illing, L.V. and Hobson, G.P. (eds) *Petroleum Geology of the Continental Shelf of North-West Europe*. Heyden, London, pp. 3–39. [6, 9]

Ziegler, P.A. (1982a) *Geological Atlas of Western and Central Europe*. Shell Internationale Petroleum Mij, Elsevier, Amsterdam, 130 pp. [1, 2, 9]

Ziegler, P.A. (1982b) Faulting and graben formation in Western and Central Europe. *Phil. Trans. Roy. Soc. London* **A305**, 113–43. [2]

Ziegler, P.A. (1987) Evolution of the Arctic–North Atlantic borderlands. In: Brooks, J. and Glennie, K.W. (eds) *Petroleum Geology of North West Europe*. Graham and Trotman, London, pp. 1201–4. [11]

Ziegler, P.A. (1988) *Evolution of the Arctic–North Atlantic and the Western Tethys*. Memoir No. 43, AAPG, Tulsa, OK. [1, 2, 3, 12]

Ziegler, P.A. (1990a) *Geological Atlas of Western and Central Europe*, 2nd edn. Shell Internationale Petroleum Maatschappij B.V., Geological Society, Bath (distributors), 239 pp. [1, 2, 3, 4, 5, 8, 9, 12]

Ziegler, P.A. (1990b) Tectonic and palaeogeographic development of the North Sea rift system. In: Blundell, D.J. and Gibbs, A.D. (eds) *Tectonic Evolution of the North Sea Rifts*. Oxford Science Publications, Oxford, pp. 1–36. [2, 8]

Ziegler, W.H. (1975) Outline of the geological history of the North Sea. In: Woodland, A.W. (ed.) *Petroleum and the Continental Shelf of Northwest Europe*. Vol. I, *Geology*. Applied Science Publishers, Barking, pp. 165–87. [7]

Zimmerle, W. (1979) Lower Cretaceous tuffs in northwest Germany and their geotectonic significance. In: *Aspekte der kriede Europas*. Series A, No. 4, IUGS, pp. 385–402. [9]

Ziolkowski, A.M., Underhill, J.R. and Johnston, R.G.K. (1998) Wavelets, well-ties, and the search for stratigraphic traps. *Geophysics* **63**, 297–313. [8]

Index

Note: Page references in *italic* indicate figures, those in **bold** indicate tables.

Many terms are grouped under broad headings without cross-reference: Basins, Continents, and depositional environments, Faults, Formations, Gas fields, Geographical locations, Graben(s), Groups, Inversion, Members and Beds, Mountains, Highs (Horsts), Oceans and Seas, Oil fields, Orongenies Platforms (shelves), Reservoirs, Source rocks, Stages and Epochs, Transgressions, and Troughs. North Sea Blocks, Quadrants and Wells are listed numerically by country.